MW00845961

FORAGES

THE SCIENCE
OF GRASSLAND
AGRICULTURE

VOLUME II

7TH EDITION

Contents

Preface

Forages in Transition

It is daunting to consider how to increase the food supply while conserving natural resources to feed the expected 10 billion people worldwide by 2050. This must occur with less land, less water, less fossil fuel, higher costs of labor, and will require more efficient use of inputs. And, it must be done while protecting the environment in the face of global climate change and greater public demand for sustainability. Forages and pastures will play a critical role by effectively using lower quality land resources, while simultaneously, supplying an adequate quantity of high-quality and safe products, especially animal products. Emphasis will increase for forages and pastures to contribute specific ecosystem services.

Much of the land resource of North America is occupied by grasslands and forages managed by ranchers and farmers for yield, quality and persistence. However, the social climate surrounding agriculture is rapidly changing as the public becomes more concerned and even distrustful about motives and priorities of land management for income over sustainability. How will research and technical advancement of forages and pastures address the non-production factors while moving the discipline forward?

Volume I of the 7th edition of Forages, an Introduction to Grassland Agriculture (2018), serves primarily as an undergraduate textbook. It emphasizes basic roles of the diverse array of forage plants, their adaptation, and principles of management practices used for efficient animal production that is sustainable. Volume II of the 7th edition of Forages, the Science of Grassland Agriculture (2020) gives more detail on how biological and physical processes in cells and tissues affect growth, forage quality and persistence of individual plants. We then integrate the basic knowledge about individual plants to their interaction in plant communities, whether harvested mechanically or grazed by animals, and how they contribute ecosystem services.

Forage yield in research plots has increased very little over the past half century. Relative focus is changing from increasing yield to reducing input costs and improving and retaining forage quality. New cultivars and strategies for disease and insect control help protect yield and improve both animal performance and stand persistence. New harvesting equipment improves leaf retention and shortens drying time to reduce weathering losses. Improved bale wrapping and silage preservation technologies further help retain digestible components. Global positioning and drones will find important uses for precision farming to increase management efficiency.

At the same time, the public desires increased emphasis on ecology, climate change, ecosystem services, animal welfare, and sustainable forage and pasture management. These concerns have led to stronger links among forage scientists, animal scientists, ecologists, climatologists and social scientists to form transdisciplinary foundations for managing forages and pastures. The broader role of forages and pastures will lead to new policies to provide quality animal products as well as valuable ecosystem services. New science will establish the best policies and practices.

Forages Need Innovation

Forages and pastures can effectively use land resources that do not compete directly with grain and oilseed crop production. Ruminants are critical since they have natural advantages in converting fibrous plant material into high nutritional value meat and milk products. Hundreds of plants could become significant forages in specific environments, and biotechnology will help improve species already used. Direct use of perennials for renewable energy sources can reduce dependence on fossil fuels. Forages will be more integral components of crop rotations, cover crops and vegetative waterways for feed sources and erosion control. Perennial legumes in crop rotations will protect the soil and support wildlife while providing fixed nitrogen for subsequent crops.

Fortunately, there are many new technologies in the pipeline such as global positioning systems, precision agriculture, drones, improved harvesting and packaging machinery, safer pesticides, improved efficiency of fertilizer use and many findings from biotechnology that are leading to major changes in plant and animal agriculture. Scientists are learning about managing marginal soils, how ecosystems work, how new technologies might be transferrable to other areas and how the benefits of plant diversity assist in maintaining ecosystem services. The private sector will continue to help by developing new cultivars, improved farm machinery, new research methodologies and instruments for monitoring hayfields, pastures and animal behavior.

Forages and the Role of Volume II

For Volume II of Forages, The Science of Grassland Agriculture, authors assembled a thorough review of relevant literature to glean, evaluate and integrate the most important factors for current and potential use. Unfortunately, the number of forage researchers in the US and Canada is decreasing, similar to trends in Europe, Australia, New Zealand and South America. This requires more use of international literature when the information is transferable or is validated or modified in the new environment. In addition, especially at basic levels, there is a need to use data and evaluations from non-forage species to provide insight to important features of forage and pasture plants. More transdisciplinary research with social and environmental scientists has aided evaluation of applications for economic viability and social acceptance.

As a first priority, authors considered how research improves adaptation, quality and persistence of forage and pasture plants. Second, authors evaluated technologies and management systems for sustainability within a field or pasture. In systems chapters, they considered forages and pastures as components when scaled to cropping or livestock systems within a larger area. Third, authors considered potential effects of resource limitations and pending climate change to support production and provide ecosystem services. Collectively, Volume II presents a comprehensive assessment of forages and their roles in agricultural systems that are changing in character and function.

Thanks to Contributors

The editors are very appreciative of the contributions of the 93 authors who delivered this work through their vision, commitment and knowledge. Their generosity, good will and talent made this 7th edition of Volume II possible. The completed edition also continues the tradition of providing the most comprehensive reference book available on forages and grasslands that is written by national leaders in their areas of education, extension, and research expertise.

In some chapters, concepts and descriptions include material from chapters on similar topics in earlier editions, especially the 5th and 6th editions. The current authors and editors are indebted to those authors who helped form the foundation and format for chapters in the 7th edition. With great respect, we thank those earlier authors for their efforts to advance the science of grassland agriculture and the roles of forages and pastures in dynamic ecosystems.

Ken Moore provided administrative leadership for the project and also edited and co-authored chapters. Michael Collins, Jerry Nelson, and Daren Redfearn shared in the editorial work and also co-authored chapters. We hope you can learn from and be reassured and stimulated by the publication. We welcome your responses about our collective effort, both negative and positive.

Kenneth J. Moore
Michael Collins
C. Jerry Nelson
Daren D. Redfearn

List of Contributors

Paul Adler
Research Agronomist,
Pasture Systems and Watershed Management Research Unit,
USDA-Agricultural Research Service,
University Park, PA, USA

Montgomery W. Alison
Extension Forage Specialist,
Louisiana State University Agricultural Center,
Winnsboro, LA, USA

L. Niel Allen
Associate Professor and Irrigation Specialist,
Utah State University,
Logan, UT, USA

David J. Barker
Professor of Horticulture and Crop Science,
The Ohio State University,
Columbus, OH, USA

Vern S. Baron
Research Scientist,
Agriculture and Agri-Food Canada,
Lacombe, AB, Canada

Gilles Bélanger
Research Scientist,
Agriculture and Agri-Food Canada,
Sainte-Foy, PQ, Canada

David P. Belesky
Clinical Associate Professor & Director Davis College Farm System,
West Virginia University,
Morgantown, WV, USA

Geoffrey E. Brink
Research Agronomist, USDA- Agricultural Research Service,
US Dairy Forage Research Center,
Madison, WI, USA

Sylvie M. Brouder
Wickersham Chair of Excellence in Agricultural Research and Professor of Agronomy,
Purdue University,
West Lafayette, IN, USA

E. Charles Brummer
Professor,
University of California,
Davis, CA, USA

G. David Buntin
Professor of Entomology,
University of Georgia,
Griffin, GA, USA

Michael D. Casler
USDA-Agricultural Research Service, US Dairy Forage Research Center
Madison, WI, USA

Kimberly A. Cassida
Forage Extension Specialist, Michigan State University
East Lansing, MI, USA

Debbie J. Cherney
Professor of Animal Science,
Cornell University,
Ithaca, NY, USA

Jerome H. Cherney
Professor of Soil and Crop Sciences,
Cornell University,
Ithaca, NY, USA

Wayne K. Coblentz
Institute for Environmentally Integrated Dairy Management,
US Dairy Forage Research Center,
Marshfield, WI, USA

Robert C. Cochran
Professor, Kansas State University,
Manhattan, KS, USA

Michael Collins
Professor and Director Emeritus, Division of Plant Sciences,
University of Missouri,
Manchester, KY, USA

Steven W. Culman
Professor, School of Environment and Natural Resources,
The Ohio State University,
Columbus, OH, USA

Tim DelCurto
Professor and Nancy Cameron Chair,
Montana State University,
Bozeman, MT, USA

Matthew Digman
Assistant Professor of Agricultural Engineering,
University of Wisconsin,
River Falls, WI, USA

Gerald W. Evers
Professor Emeritus of Soil and Crop Sciences,
Texas A&M University,
Overton, TX, USA

Steven L. Fales
Emeritus Professor of Agronomy,
Iowa State University,
Ames, IA, USA

E. Scott Flynn
Agronomist, Corteva Agriscience,
Lees Summit, MO, USA

Jamie Foster
Associate Professor of Forage Agronomy, Texas A&M AgriLife Research,
Beeville, TX, USA

John A. Guretzky
Grassland Systems Ecologist,
University of Nebraska,
Lincoln, NE, USA

Richard J. Grant
President and Research Scientist,
The William H. Miner Agricultural Research Institute,
Chazy, NY, USA

Marvin H. Hall
Professor of Crop and Soil Sciences,
Pennsylvania State University
University Park, PA, USA

Mary Beth Hall
Research Animal Scientist, USDA-Agricultural Research Service,
US Dairy Forage Research Center
Madison, WI, USA

Dennis W. Hancock
Center Director, USDA-Agricultural Research Service,
US Dairy Forage Research Center,
Madison, WI, USA

Keith R. Harmoney
Range Scientist, Kansas State University,
Hays, KS, USA

Ronald D. Hatfield
Research Plant Physiologist, USDA-Agricultural Research Service,
US Dairy Forage Research Center,
Madison, WI, USA

John R. Hendrickson
Research Rangeland Management Specialist, USDA-Agricultural Research Service
Mandan, ND, USA

Nicholas S. Hill
Professor of Crop and Soil Sciences,
The University of Georgia,
Athens, GA, USA

John Jennings
Professor of Animal Science-Forages,
University of Arkansas,
Little Rock, AR, USA

Jacob M. Jungers
Assistant Professor of Agronomy and Plant Genetics,
University of Minnesota,
St. Paul, MN, USA

Robert L. Kallenbach
Associate Dean, Agriculture and Environment Extension,
University of Missouri,
Columbia, MO, USA

Kenneth F. Kalscheur
Research Dairy Scientist, USDA- Agricultural Research Service,
US Dairy Forage Research Center,
Madison, WI, USA

Douglas L. Karlen
Soil Scientist (Retired),
USDA-Agricultural Research Service,
Ames, IA, USA

Limin Kung, Jr.
Professor of Animal and Food Sciences,
University of Delaware,
Newark, DE, USA

William O. Lamp
Professor of Entomology,
University of Maryland,
College Park, MD, USA

Gregory Lardy
Department Head, Animal Sciences,
North Dakota State University,
Fargo, ND, USA

Karen L. Launchbaugh
Heady Professor of Rangeland Ecology,
University of Idaho,
Moscow, ID, USA

Andrew W. Lenssen
Professor of Agronomy,
Iowa State University,
Ames, IA, USA

Mark A. Liebig
Soil Scientist, USDA- Agricultural Research Service,
Northern Great Plains Research Laboratory,
Mandan, ND, USA

Jennifer W. MacAdam
Professor of Plants, Soils and Climate,
Utah State University,
Logan, UT, USA

Bisoondat Macoon
Research Professor,
Mississippi State University,
Raymond, MS, USA

Neal P. Martin
Director (Retired), USDA- Agricultural Research Service,
US Dairy Forage Research Center,
Madison, WI, USA

Robert A. Masters
Rangeland Scientist (Retired),
Corteva Agriscience,
Indianapolis, IN, USA

Tim A. McAllister
Principal Research Scientist, Agricultural and Agri-Food Canada,
Lethbridge, AB, Canada

David R. Mertens
President, Mertens Innovation & Research LLC,
Belleville, WI, USA and
Research Dairy Scientist (Retired), USDA- Agricultural Research Service,
US Dairy Forage Research Center,
Madison, WI, USA

Robert B. Mitchell
Research Agronomist, USDA-Agricultural Research Service,
Lincoln, NE, USA

Corey Moffet
Research Rangeland Management Specialist,
USDA-Agricultural Research Service,
Woodward, OK, USA

Kenneth J. Moore
Charles F. Curtiss Distinguished Professor in Agriculture and Life Sciences and Pioneer Hi-Bred Professor of Agronomy,
Iowa State University,
Ames, IA, USA

Richard E. Muck
Agricultural Engineer, USDA-Agricultural Research Service (Retired),
US Dairy Forage Research Center,
Madison, WI, USA

James P. Muir
Professor Grassland Ecology,
Texas A&M AgriLife Research & Extension Center,
Stephenville, TX, USA

C. Jerry Nelson
Professor Emeritus of Plant Sciences,
University of Missouri,
Columbia, MO, USA

Yoana C. Newman
Associate Professor of Plant and Earth Science,
University of Wisconsin,
River Falls, WI, USA

Renata N. Oakes
Assistant Professor of Forage Systems and Management,
University of Tennessee,
Spring Hill, TN, USA

John F. Obrycki
ORISE Fellow, USDA National Laboratory for Agriculture and the Environment,
Ames, IA, USA

David Parsons
Professor of Crop Science,
Swedish University of Agricultural Sciences (SLU)
Umeå, Sweden

Carlos G.S. Pedreira
Associate Professor of Animal Science,
University of São Paulo,
São Paulo, Brazil

Valentín D. Picasso Risso
Assistant Professor of Agronomy,
University of Wisconsin,
Madison, WI, USA

William D. Pitman
Professor,
Louisiana State University Agricultural Center,
Homer, LA, USA

Daniel H. Putnam
Forage Extension Specialist,
University of California,
Davis, CA, USA

Daren D. Redfearn
Associate Professor of Agronomy,
University of Nebraska,
Lincoln, NE, USA

Gabriel Ribeiro
Assistant Professor
Animal and Poultry Science,
University of Saskatchewan,
Saskatoon, SK, Canada

Esteban F. Rios
Assistant Professor of Agronomy,
University of Florida,
Gainesville, FL, USA

Craig A. Roberts
Professor of Agronomy,
University of Missouri,
Columbia, MO, USA

C. Alan Rotz
Agricultural Engineer, USDA-Agricultural Research Service,
US Dairy Forage Research Center,
Madison, WI, USA

Michael P. Russelle
Soil Scientist (Retired), USDA-Agricultural Research Service,
St. Paul, MN, USA

Matt A. Sanderson
Research Agronomist and Research Leader (Retired),
USDA-Agricultural Research Service,
State College, PA, USA

Craig C. Sheaffer
Professor of Agronomy and Plant Genetics,
University of Minnesota,
St. Paul, MN, USA

Kevin J. Shinners
Professor of Agricultural Engineering,
University of Wisconsin,
Madison, WI, USA

Byron B. Sleugh
Agronomist, Corteva Agriscience,
Indianapolis, IN, USA

Alexander J. Smart
Professor and Rangeland Management Specialist,
South Dakota State University,
Brookings, SD, USA

Lynn E. Sollenberger
Distinguished Professor of Agronomy,
University of Florida,
Gainesville, FL, USA

Tim L. Springer
Research Agronomist, USDA-Agricultural Research Service,
Woodward, OK, USA

Kim Stanford
Research Scientist, Alberta Agriculture and Forestry,
Lethbridge, AB, Canada

Jeffrey J. Steiner
Associate Director, Global Hemp Innovation Center,
Oregon State University,
Corvalis, OR, USA

R. Mark Sulc
Professor of Horticulture and Crop Science,
The Ohio State University,
Columbus, OH, USA

Eric S. Vanzant
Associate Professor, University of Kentucky,
Lexington, KY, USA

João M.B. Vendramini
Associate Professor of Agronomy, Range Cattle Research and Education Center,
University of Florida,
Ona, FL, USA

Kenneth P. Vogel
USDA-Agricultural Research Service (retired),
Lincoln, NE, USA

Jeffrey J. Volenec
Professor of Agronomy, Purdue University,
West Lafayette, IN, USA

John W. Walker
Professor and Resident Director of Research, Texas A&M AgriLife Research and Extension Center,
San Angelo, TX, USA

Yuxi Wang
Research Scientist, Agriculture and Agri-Food Canada,
Lethbridge, AB, Canada

Zeng-Yu Wang
Professor,
Qingdao Agricultural University,
Yantai, China

Marcelo O. Wallau
Assistant Professor of Agronomy,
University of Florida,
Gainesville, FL, USA

Richard Waterman
Research Animal Scientist, USDA- Agricultural Research Service, Fort Keogh Livestock and Range Research Laboratory,
Miles City, MT, USA

David A. Wedin
Professor,
School of Natural Resources,
University of Nebraska,
Lincoln, NE, USA

William P. Weiss
Professor of Animal Sciences,
Ohio Agricultural Research and Development Center,
The Ohio State University,
Wooster, OH, USA

Jessica A. Williamson
Assistant Professor of Crop and Soil Sciences,
Pennsylvania State University,
University Park, PA, USA

Dedication

This volume is dedicated to the memory of Drs. Steven Louis Fales, Lowell E. Moser and Walter F. Wedin. Devoted and passionate grasslanders all, they were also highly productive researchers and enthusiastic educators. They inspired and trained many of the authors contributing to this volume. Their lives and careers crossed paths many times over the years and all three were contributors to earlier editions of Forages.

Steven L. Fales

Lowel E. Moser

Walter F. Wedin

Lowell edited several important books and monographs related to forages including ***Cool-Season Forage Grasses*** and ***Warm-Season (C4) Grasses***, both published by the Tri-Societies (ASA-CSSA-SSSA). Walt and Steve co-edited ***Grassland: Quietness and Strength for a New American Agriculture*** their homage to ***Grass, the 1948 Yearbook of Agriculture***.

This volume is also respectfully dedicated[1]:

To the Memory of Those gone on before, who, envisioning the needs of the future and the possibility of better things, lived purposively, giving of themselves.

In Recognition of Those of our own day, who, endowed with leadership ability in research and education, continue to stimulate us to more productive effort.

For the Inspiration of Those who today follow on, but who tomorrow, building upon established foundations, will be charged with the responsibility of solving problems with which those of their day will be confronted.

[1]From , Hughes, H.D., Heath, M.E., and Metcalfe, D.S. (eds.) (1951). *Forages: The Science of Grassland Agriculture*, 1e. Ames, IA: The Iowa State College Press.

PART I

FORAGE PLANTS

A mixed stand of alfalfa and timothy. Timothy mixtures with alfalfa in Kentucky provide mixed forage on the first cutting or grazing but nearly pure alfalfa through the remainder of the growing season. *Source:* Photo courtesy of Mike Collins.

Part I covers basic physiologic and physical properties of forage species at the cellular and whole-plant levels that guide genetic improvement and underscore management practices. The goals are to improve yield and quality of the biomass and resistance to biotic and abiotic stresses. These processes often have negative correlations that are species dependent, and responses of spaced-plants may not reflect their properties when grown in dense stands or mixtures. Critical topics such as photosynthesis, root growth, canopy architecture, lignification of cell walls and presence of antiquality factors such as alkaloids in leaves need to continue to be evaluated. Most perennial forage plants are polyploids and cross-pollinated, making it difficult to identify and transfer genes using biotechnology, but

Forages: The Science of Grassland Agriculture, Volume II, Seventh Edition.
Edited by Kenneth J. Moore, Michael Collins, C. Jerry Nelson and Daren D. Redfearn.

CRISPR-Cas9 and other new technologies are opening new ways to supplement traditional breeding methods.

Genetic potential for growth and persistence set the upper limits for yield. Management strategies utilize resources efficiently to achieve the actual annual yield, but it rarely nears the genetic potential. Reducing the yield gap by more intense management may not be economically feasible or environmentally friendly. Thus, increasing efficiency of energy, radiation, nutrient, water and other natural resources are objectives. These processes are integrated to understand and optimize plant growth, flowering and seed development. The integrated system is what the manager must understand to achieve the desired objective in a way that is sustainable for now and the future.

CHAPTER

1

Perspectives, Terminology, and Classification

C. Jerry Nelson, Professor Emeritus, *Plant Sciences, University of Missouri, Columbia, MO, USA*
Kenneth J. Moore, Distinguished Professor, *Agronomy, Iowa State University, Ames, IA, USA*
Michael Collins, Professor Emeritus, *Plant Sciences, University of Missouri, Columbia, MO, USA*
Daren D. Redfearn, Associate Professor, *Agronomy, University of Nebraska, Lincoln, NE, USA*

As it has for millennia, the earth is changing physically, especially during the past few decades, while human population is growing very rapidly. Forage management has advanced to help meet the expanding needs for ruminant animal products, nitrogen acquisition, fuel resources and environmental stewardship. However, changes in climate, conflicts and shortages of water supplies, increased public emphasis on ecosystem management, and the challenges of world hunger and energy remain in the news almost daily. Other concerns include food safety, food quality and animal welfare. Each raises questions about how to deal with hunger, the environment and quality of human life; especially how management of pastures, forage fields and the products they support can help provide solutions.

Need for Consistent Terminology

Clear communication depends on terminology that is common among the individuals involved. Many terms are common to production of all crops and animals. In this book, however, emphasis is on those terms unique to forage crops, pastures, range, and livestock that describe their underlying science and practical use. Terms in bold face are defined in a comprehensive glossary in the appendix.

While many terms have a history of usage, they can be confusing when moved from one culture or location to another. New terms appear regularly along with new technologies, and need to be clear and used correctly. For example, a few years ago, a drone would have referred to a male bee, which it still does, but with the advent of precision agriculture, a drone is also now an unmanned aircraft guided by remote control or onboard computers using global positioning systems (GPSs). Drones can carry instruments that measure plant health, forage quality, forage production and monitor animal behavior in a pasture. Many other applications will soon follow.

Most definitions are written for the practitioner and may not be fully understood by the general public or policy makers. Practitioners are more aware than the public or legislators about the intrinsic values of forages

Forages: The Science of Grassland Agriculture, Volume II, Seventh Edition.
Edited by Kenneth J. Moore, Michael Collins, C. Jerry Nelson and Daren D. Redfearn.

and grasslands. They have a vested interest in technical and economic aspects that help them be better managers or marketers. Some specific or technical terms for communication among researchers are in the glossary for use by practitioners involved with technical communications. Scientists, extension specialists, consultants, and journalists need to be aware of differences in knowledge levels between practitioners and the public, especially in urban areas.

Terms in Grassland Management

Professionals in forages and grasslands have responsibility to develop consistency of definitions so communication is clear. Endophyte-free or E^- tall fescue, glandular-haired alfalfa, and no-till seeding are terms that are becoming common. Conversely, there is debate as to what constitutes animal rights, labor laws, use of water, and others, including how to measure these factors and assign or estimate economic values.

When allowed to develop unabated, local, and generic terms take on a local meaning. For example, the public may observe a pasture that is "rundown" or "overgrazed." The practitioner might suggest the pasture was "grazed heavily," whereas the scientist might say an inappropriate **stocking rate**, **stocking density**, or **grazing pressure**, respectively, was the cause. Each scientific term has some features in common with the more general descriptor, but focuses on a more specific factor to add clarity using biological reasons for the pasture condition. For example, overgrazing could be due to poor plant growth, having too many animals, or retaining them on the pasture too long, all with the result of leaving too little residual **forage mass**.

"Grazing heavily" suggests too many animals for the forage available such that too much forage was removed. The scientist would use terms such as **stocking rate** (number of animals per unit land area for a period of time) and **grazing pressure** (mass of forage available per animal at a given time) to understand the situation in quantifiable terms.

Sometimes a term used routinely needs to be modified to lead to change. "Intensive grazing management" was commonly used for decades and generally connoted the use of management practices involving "rotational grazing," now called **rotational stocking**, but it also implied that the "grazing intensity," now called **stocking rate**, was managed. Earlier interpretations could involve rotating periods of grazing and rest, or encouraging faster bite rate or larger bite size of animals, i.e. "grazing with intensity."

Research on the technologies introduced new terms that helped increase producer interest in pasture management. The need to be biologically accurate, and consistent with other terminology regarding grazing methods, led professionals to shift the term from intensive grazing management to **management-intensive grazing** (Nation 2004). This focuses properly on the grazing method that is managed intensively based on knowledge about plants, animals, fencing, water supplies, and other technologies used as inputs (Gerrish 2004).

In another case, professionals early on used accumulated forage for **deferred grazing**; meaning the forage that accumulated during active growth was allowed to stand until needed for grazing. Yet, practitioners and technology transfer specialists also coined the terms "stockpiling" and "grazing on-the-stump," neither of which was functionally descriptive of forage accumulated during active growth, usually during fall, and subsequently grazed in winter when growth was slow or had stopped. Even so, the term **stockpiling** was gradually accepted, clearly defined and is now widely adopted (Figure 1.1).

Terms for Soil and Its Functions

Soil has long been defined as "unconsolidated mineral or organic material on the immediate surface of the earth that serves as the natural medium for the growth of land plants." However, this definition raised concerns among soil scientists that soils are not limited to earth, some parts of soil may be rocks or other consolidated material, soils contain liquids, gases and biological organisms, including plants, and that soils are dynamic due to soil-forming factors that differ depending on their use and management (van Es 2017).

Under leadership by the Soil Science Society of America, ideas and concepts were coalesced to a new definition: **Soil** is now "the layer(s) of generally loose mineral and/or organic material that are affected by physical, chemical, and/or biological processes at or near the planetary surface and usually holds liquids, gases and biota and support plants" (van Es 2017). The new definition clearly places more emphasis on the physical makeup of the soil and broadens the definition and uses beyond agriculture, i.e. more than just supporting plants.

Soil Quality

A number of years ago, the term soil quality was introduced and considered as "the capacity (of soil) to function" (Karlen et al. 2003). Soil quality depends on physical, chemical, and biological features of upper layers, and how they interact to provide a given function, be it for road construction, crop production or a home lawn. A change in one feature results in a different soil. Scientists are developing methods to assess the indicators, and then use mathematical equations to combine several physical, chemical, and biological features, including organic matter that changes with human activity, into a numeric index (Friedman et al. 2001). The desired numeric index based on physical, chemical, and biological properties would vary depending on the purpose, e.g. agriculture or civil engineering.

Fig. 1.1. Beef cattle in Saskatchewan extending the grazing season by using accumulated forage. *Source:* Photo courtesy of Vern Baron.

Understanding how the components of the soil quality index interact, while being a noble goal, has been difficult to measure and interpret over a range of soil types and topographies (Laishram et al. 2012). This led to interest in soil health, a simpler concept for evaluating "soil value" that is related more directly to content of organic matter (Doran and Zeiss 2000). This seemed more practical for agricultural uses, especially in the short term, since organic matter is responsive to management and affects the structure of the soil and its capacity for holding water and nutrients.

Soil Health

Soil health is the continued capacity of soil to function as a vital living ecosystem that sustains plants, animals, and humans. The term may be more useful than soil quality to describe the "health state" of a soil in terms of its productivity and roles in environmental conservation and the many ecosystem services of pastures and forages since it is based mainly on organic matter content that is relatively easy to measure and quantify. Many practitioners and the USDA Natural Resources Conservation Services have adopted the term soil health using organic matter as the main component for evaluating soil conditions associated with crop and forage management. Soil health and its emphasis on organic matter is the term championed by organic agriculturalists.

Sustainability of Grassland Agriculture

Sustainability of agriculture has been a critical issue for farmers and ranchers for generations, but in the 1960's public concern grew about the increased emphasis on primary production of food and fiber based on use of chemical fertilizers and pesticides. Agriculture was perceived as mining natural resources for economic benefit with little concern for short- and long-term sustainability based on health and well-being of consumers and the environment. The public raised real concerns about government regulations and management practices for use of chemicals on farms.

The Delaney amendment in 1958 prohibited any compound in feeds or foods that caused cancer in animals or humans. To help meet these concerns, the government developed stricter regulations on use of chemical fertilizers and especially pesticides based on better diagnostic procedures. This was coupled with increased public interest in organic agriculture that prohibits use of chemical inputs. To address the growing concerns, the concept of sustainable agriculture emerged and, early on, was somewhat linked to the use of organic practices.

Roles of the Public

The public was concerned about sustainability even though it was clear organic agriculture alone could not meet the total food needs then (Pesek et al. 1993) or later (Reganold and Wachter 2016). Soon a more holistic perspective of sustainable agriculture emerged that involved more than food production and was defined based on three major components: (i) the economic return to the producer, (ii) the conservation of the environment, and (iii) use of practices that are accepted socially (American Society of Agronomy 1989).

Federal and state governments began cost-share programs to encourage and reward producers who adopted management practices to reduce soil erosion, maintain water quality, increase plant diversity, enhance wildlife and reduce negative effects of chemical nutrients and pesticides. Industry also accepted the challenges by working on the broader issues before submitting chemicals for registration.

Today, there is growing concern about social aspects like animal rights, worker safety, food safety and labeling of contents in food as components of sustainability. In many cases today, the consumer can get some reassurance by purchasing food directly from Farmer's Markets or track products back to the farm or ranch from which it was produced.

Consideration of Ecosystem Services

After a detailed international analysis (Millennium Ecosystem Assessment 2005), sustainability of agriculture today also includes providing a wide range of **ecosystem services**, a more inclusive and more comprehensive set of ecosystem components and interactions affecting human well-being. This four-part framework, led mainly by ecologists and social scientists, consisted of four outputs or services from the land (Figure 1.2).

The desired outputs, all expected from agriculture, include (i) Supporting services like primary production, nutrient cycling and soil formation; (ii) Provisioning services like food, fresh water, wood, and fuel; (iii) Regulating services like influences on climate, quantity and quality of water, and diseases of plants and animals; and (iv) Cultural services like spiritual issues, education, and esthetics. Currently, a major goal for scientists is to learn the breadth and determine values of individual ecosystem services and their interrelationships.

The millennium report on sustainable agriculture is gradually being accepted internationally as a goal, while more sub-components are added on a regular basis. Costanza et al. (2017) reported on the explosion of research by ecologists and economists wanting to assign values to ecosystems, encourage policies and document applications of the ideas. Sustainability now extends beyond the farm gate to the entire food chain and includes a myriad of environmental, social, and cultural issues rarely considered a few decades ago. Agricultural scientists need to continue to be involved in all aspects.

Unfortunately, agricultural science has often not kept up to provide a scientific basis for leadership to make good policy decisions. The public is now the major player, often without scientific evidence, in decisions and regulations for the entire food system and the preservation of natural resources.

Assessing and understanding the complexity involved with agricultural sustainability will likely require mathematical modeling and transdisciplinary approaches in research. Forage and pasture management and animal welfare issues need science-based cooperation with social scientists and practitioners to understand relationships, provide education and satisfy public demands for sustainability.

Industrialization of agriculture via new technologies from both the public and private sectors has raised concerns about ethical and economic motivation among the players. The question arises; are commercial motives parallel with those of the public, and based on science? Scientists are in the early stages of establishing an index that includes measurable variables associated with the Millennium Assessment to achieve sustainability in ways that are socially acceptable. As incomes increase in developed countries, demands for fresh and safe foods with good taste will continue to rise. Many will believe, with little or no scientific evidence, that organically produced foods are safer, healthier, and taste better. The balance between organic and other production systems will evolve (Tillman et al. 2002).

The Role of Organic Agriculture

Organic foods and beverages are a small, but rapidly growing market segment in the global food industry including meat and milk products, primarily due to health and nutrition concerns. A recent study analyzed 40 years of science comparing organic and conventional agriculture across the four goals of sustainability, productivity, environmental impact, economic viability, and social well-being (Crowder and Reganold 2015). In summary, organic systems produced lower yields compared with conventional agriculture, yet it was more profitable because consumers pay 12–50% more for the products.

Overall, organic farms tend to store more soil carbon, have better soil quality, and reduced soil erosion. Initial evidence indicates that organic agricultural systems deliver greater ecosystem services and social benefits.

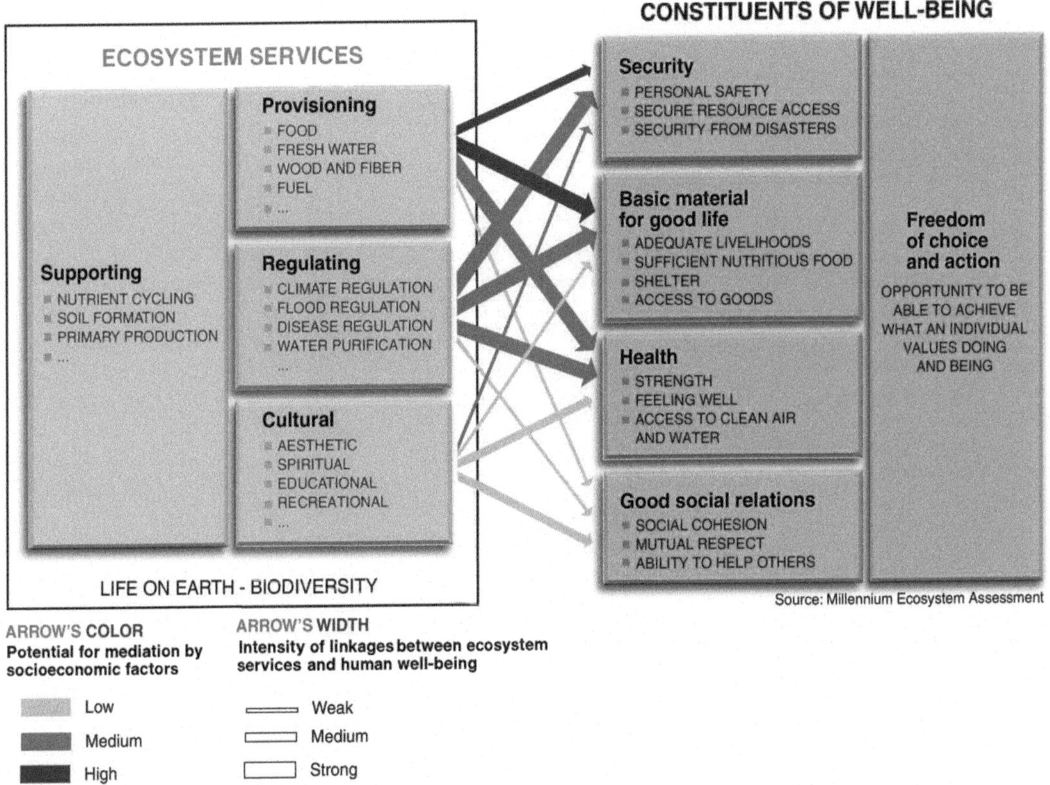

Fig. 1.2. Left, ecosystem services are divided into four boxes related to major services within each. Right, list of components of well-being. Line darkness indicates the potential effects of socioeconomic factors whereas line thickness shows intensity of effects of ecosystem services on human well-being. *Source:* From Millennium Ecosystem Assessment (2005), presentation; Credited to Millennium Ecosystem Assessment (2005).

Although organic agriculture has an untapped role to play when it comes to the establishment of sustainable farming systems, no single approach will safely feed the planet. Rather, a blend of organic and other innovative farming systems is needed. Significant barriers exist to understanding and adopting these systems, and a diversity of policies will be required to facilitate their development and implementation.

A major social issue is that organic agriculture is conceptually associated with small farms with a mix of crops and livestock that are owned or operated by a family. Also, organic products are often marketed in nearby farmer's markets where freshness and relationships with the producer are valued. Due to increasing public demand for organic products, they are offered in most supermarkets at prices from 15% to 50% or more than products produced traditionally.

In contrast with organic plant agriculture, more than 90% of livestock products are produced on large farms using traditional practices for plant and animal management that tend to be focused on only one or two commodities. Most are owned and managed by a family, but have several employees who do much of the work. Due to size, these operations are more economic than smaller farms because inputs are often purchased directly from suppliers and products are marketed through pre-arranged contracts to obtain a higher price. This leads to greater net income for the family or owner.

Many consumers have negative perceptions of production agriculture that is highly mechanized with large fields that deter wildlife or has a high density of farm animals. They criticize use of genetically modified plants offered by private industry, safety-approved pesticides, economic rates of chemical fertilizers, and confinement housing for animals, even if all practices comply with federal regulations. Operators of very large farms are accused of exploiting government assistance programs and disregarding animal rights, worker welfare and environmental regulations. Consumer demand in the long term will determine the proportion of foods produced by organic or traditional means on either small or large farms.

Classification Systems Based on Crop Use

Some systems of plant classification arose out of convenience while others arose out of necessity, such as the binomial system of plant nomenclature based on morphologic features. New classification categories continue to emerge as technologies and uses change. For example, a **medicinal crop** is grown for its natural products from the leaves, flowers or roots that are used for medical purposes. A **pharmaceutical crop,** sometimes genetically engineered, is grown primarily as a biologic synthesizer of a specific compound used for medical purposes.

Terms for Agronomic Uses

Agronomy, derived from the Greek term for "field," deals with field crops including wheat, corn, soybean, cotton, and forages. These are grown on a large scale using relatively extensive management compared with horticultural crops. **Forage** includes edible parts of plants, other than separated grain, that can provide feed for animals, or that can be harvested for "feeding." Thus, it includes leaves, twigs, stems, roots, nuts, and other parts of a wide range of plant species.

Primary uses of forages associated with feed for animals are **pasture**, **hay**, **silage**, and **soilage**. Pasture is a **grazing management unit** that is enclosed and separated from other areas by fencing or other barriers and is managed to produce forage that is harvested primarily by grazing. Range is land supporting native vegetation that is grazed or has the potential to be grazed and, in contrast to pasture, is usually managed extensively as a natural ecosystem. In addition to grasses, legumes, and other **forbs** (Smith and Collins 2003), range includes shrubs and trees that provide **browse** for animals.

Hay is forage preserved by field drying to moisture levels low enough to prevent microbial activity that leads to spoilage. In contrast, silage is forage preserved in a succulent condition at low pH due to microbial production of organic acids by anaerobic fermentation of **sugars** in the forage. Soilage or green chop is forage that is cut and fed fresh within a few hours. Browse is leaf and twig growth of shrubs, woody vines, trees, cacti, and other vegetation that is available for direct animal consumption by "browsing." A catch crop is a forage crop, usually an annual like sudangrass that is used short-term in a rotation with one or more row crops. Catch crops are used when severe winter injury or other situation arises and more forage is needed in the short-term.

In some cases, forage species are grown for primary purposes other than animal feed. A **green manure crop** is allowed to produce vegetation to be tilled under for soil improvement; a **grassed waterway** is planted in surface drainage areas of large fields to accept surface water and channel it off the field to reduce erosion and gully formation. A **smother crop** is a strongly competitive crop that is grown in monoculture to control weeds until it is harvested or grazed. A **companion crop** (the preferred term over *nurse crop*), such as oat or spring wheat, can be sown at a reduced rate along with a forage crop that emerges and develops slowly. The companion crop establishes quickly to reduce erosion and compete with weeds. In all these cases, the forage or grain can be harvested if removal does not interfere with the primary objective.

Cover crops are used to stabilize the topsoil and reduce water runoff and erosion between successive annual crops, often over winter (Finney et al. 2017). Usually, a winter grain, winter legume or root crop like radish or turnip is planted in autumn after the previous crop was harvested. Growth of roots help hold the soil particles together while the tops intercept rainfall and reduce impact of water droplets that can dislodge soil particles. The forage can be harvested in spring or killed to leave mulch for direct planting of the next crop.

Detailed studies in Pennsylvania indicated positive and negative effects from different cover crops. Legumes are usually preferred because they fix some N, whereas N applied to grass remains sequestered in the killed tissue due to slow mineralization, can lead to low N supply and low yields in the subsequent crop (White et al. 2017). But yields and other ecosystem services like weed and insect control were better with non-legume species, so tradeoffs need to be considered. Also, changes in planting dates and seeding rates altered the ecosystem values of cover crops (Murrell et al. 2017).

Terms for Economic Land Uses

Cropland forage is cultivated in some way, usually as part of a rotation with a grain, fiber, or oilseed crop that includes forages and short-term pastures. Forages help control erosion, increase soil organic matter, improve aeration, and legumes leave residual N in the soil for subsequent crops. Cropland forages, including cornstalks or other crop residues that are part of a crop rotation, can be harvested for hay or silage, or can be grazed as pastures. These areas also serve as sites for application of manures.

Grazingland includes both pastureland and rangeland, the former more common in the humid areas of North America east of the 98° meridian, using introduced forage species in systems that are more intensively managed. **Rangeland** consists largely of native species in the semi-arid western parts of North America that are managed more extensively. Native grassland species are more drought tolerant, usually lower in herbage yield, and more sensitive to grazing management than are the **introduced species** that predominate in the East. Availability of water and competition with other crop species for land use in the East often relegate forages and pastures to land classes that are less productive or too erosive for crop production.

Forestland consists of somewhat open, tree-covered areas that support forage and grassland species that can be grazed (Garrett et al. 2000) or browsed (Figure 1.3). Grazing offers some animal production and helps control understory vegetation. These systems in the West can expand use of forestland, or in the East can add a few years of crop or forage use before the tree canopy closes in a planned approach to forest management. These systems, of particular importance in the pine forests of the southern and southeastern US, are relatively complex to design and manage, but can be very productive (Child and Pearson 1995).

Agroforestry is a designed management system in which trees are purposely spaced to allow planting of crops or forages among them or in alleys between tree rows. The combined objectives are short-term animal or crop production from the alleys for a few years followed by intermediate-term income from nuts for food or needle production for mulch until the timber is harvested. These are often used with production of high-value trees such as walnut or pecan that produce nut crops each year. This agronomic use for the first years, provides erosion control, N for the ecosystem if legumes are used, increased biodiversity of the area and habitat options for wildlife.

Silvopasture systems are agroforestry systems using pasture plants that occupy open areas or alleys between tree rows that can be grazed (Clason and Sharrow 2000). The young trees need protection from damage, be cared for and pruned regularly (Figure 1.4). The ultimate goal is to provide income from livestock products in the short term while the canopy gradually closes and provides too much shade. The longer-term goal of the system is to produce nuts or other products until the harvest of well-shaped trees of high value.

Terms for Ecological Land Uses

Forages and grasslands play major roles in environmental stability by reducing erosion (see Chapter 12 and Sharp et al. 1995), improving water quality (Chapter 12), increasing biodiversity (Chapter 13), and providing food and habitat for wildlife (Clubine 1995; Sollenberger et al. 2012). The Conservation Reserve Program, a federal program to pay US landowners to remove highly erosive lands from crop production, is based on these principles. The 10-year contract requires land managers to plant adapted perennial forage species to maintain year-round ground cover. The result is reduced water runoff, enhanced water quality and improved wildlife habitat on the conserved land. In addition, the program helps reduce overproduction of crops and the need for subsidy payments by the government.

Grassed waterways provide drainage channels for crop fields whereas planting forages in **riparian buffers** protecting streams helps control soil erosion and capture nutrients and other materials carried in runoff water (Chapter 12). Desired widths of waterways and riparian strips depend on scientific estimates of expected rates and volumes of runoff. Waterways and riparian areas can be harvested for hay or silage at appropriate times during the

Fig. 1.3. A forestland pasture system in which trees shade the pasture and cattle. The large trees resist animal damage and are spaced to produce quality timber. *Source:* Photo courtesy of National Agroforestry Center.

Fig. 1.4. A silvopastoral system with winter rye planted among trees and in the alley. Other crops can be grown in the alleys until the trees get larger and alleys gradually narrow. *Source:* Photo courtesy of Rob Kallenbach.

growing season, if adequate stubble is left for subsequent regrowth to still provide the needed protection.

Relationship to Precipitation

The ecologic basis for land types depends on climatic factors such as precipitation and soil moisture. **Desert** is obviously an arid land classification. Moving toward areas of higher and higher precipitation, or to areas where **evapotranspiration** decreases, the natural vegetation progressively phases to **shrubland**, **steppe,** and **prairie** (Chapter 8). Desert plants often have crassulacean acid metabolism in which stomata open for CO_2 uptake only at night to conserve water and tolerate drought (see Chapter 4). In addition, several shrubs avoid herbivory due to spines or taste factors (Chapter 46) to survive and dominate in a dry area. Steppes usually occupy drier areas than prairie and consist mainly of deep-rooted short grasses. Eastern portions of the prairies in North America consist naturally of tall grasses. **Range** is more encompassing and includes areas such as desert, shrubland, steppes, and prairie.

In high precipitation areas, unless burned or managed correctly, the natural grassland vegetation will gradually be overcome by wooded vegetation or forest. **Marshland** and **wetlands** are areas of high precipitation or poor soil drainage such that a high-water table is maintained for much of or all year. These areas can be grazed when sufficiently dry, but serve primarily to reduce flooding, provide wildlife habitat and maintain biodiversity. **Meadows** are grassland sites, often with native or naturalized species that exist as long-term stands, but productivity is affected strongly by the landscape topography and water-holding capacity. As such, they are often naturally sub-irrigated in the West or depend on natural rainfall in the East. They are usually grazed or harvested for hay during dry periods. Often an adjective such as hay, mountain, native, or wet is used to help describe the meadow, i.e. mountain meadow.

Relationship to Temperature

In North America, high air temperatures in July and low air temperatures in January are primary factors affecting adaptation of grassland species (Chapter 8). **Tundra** is treeless grassland that occupies large areas of arctic regions of North America, Asia, and Europe. In warmer areas, plants with good winter hardiness and active growth at low temperatures, i.e. cool-season species with C_3 photosynthesis (Chapter 4), dominate eastern temperate grasslands. Due to high temperatures and dry conditions, however, warm-season species with C_4 photosynthesis and good winter hardiness grow actively in summer and dominate many temperate regions of the North American prairie.

Further south, the **transition zone** consists of some areas of cool-season species with C_3 photosynthesis, and others with warm-season species with C_4 photosynthesis. High summer temperatures restrict many cool-season grasses in this area especially when grazed to a short stubble height when soil temperatures are high. Conversely, several native warm-season grasses are highly productive and some subtropical grasses can survive the milder winters.

Depending on the mildness of winter temperatures, subtropical perennial grasses and some **herbaceous** legumes occupy the gulf region. Farther south, most forages are tropical species that tolerate heat, but are very sensitive to cold (Chapter 18). Woody plants, especially tree legumes such as leucaena, can be a valuable forage component in subtropical and tropical areas (Chapter 15). **Savannas** describe grasslands with scattered trees, often legume trees in the tropics and subtropics, or hardwood trees in temperate areas.

Terms Describing Vegetation Types

Forages, rangelands, and pastures consist of vegetation that coexists in different stages or conditions. Plant functional types in a diverse mixture consist of grasses, forbs, brush or shrubs, and trees. Each provides a food source for harvest or direct grazing by animals. In the broad sense, **forbs** are **herbaceous** (non-woody) broad-leafed plants that include the legumes (Smith and Collins 2003). However, when describing forages, legumes are usually considered separately from forbs due to their higher economic value. Thus, in general, forb refers to non-legume, herbaceous, broad-leafed plants such as dandelion and several *Brassicas* including rape, turnip, and kale.

Forbs include some naturally occurring poisonous plants and others commonly considered as weeds, but forage quality of several "weeds" is as good and, in some cases, even better than seeded species (Marten et al. 1987). Some forbs such as dandelion and goldenrod are invasive and need to be kept in check with good management (Chapter 28). Many forbs known to be weeds with good forage quality are strong competitors with crop plants. Therefore, they are considered undesirable in pastures and forage fields nearby or in a rotation with crop species. The positive roles and potential negative consequences of these forbs need further evaluation.

Forage can be seeded and harvested as a **monoculture**, i.e. a single species, or as a mixture of two or more species. **Herbage**, the aboveground material that consists of leaves and stems, usually refers to **forage mass** harvested mechanically, whereas **forage available** refers to the mass that can be grazed to a defined height. **Aftermath** describes the regrowth after harvest, which can be high quality, or the residue left in the field after seed harvest that is usually low quality.

Stockpiled forage results from a special management strategy to graze the aftermath or deferred growth of cool-season or warm-season grasses during a part of the year when plants are no longer growing rapidly. Stockpiling is used regularly to accumulate vegetative (leafy) growth of cool-season species during late summer and fall to extend the grazing season into the winter-dormant period (see Figure 1.1). This can be a cost-saving alternative to reduce feeding conserved forage as hay or silage that requires harvest and some form of storage (Chapter 20).

Terms Describing Life Cycles and Stand Persistence

Some forage and grassland plants are **annuals** that complete their life cycle in one year. **Summer annuals** germinate in spring, grow actively, produce seed, and then die. Some die as a direct result of **flowering** and seed production that triggers a coordinated and programmed death. This involves reallocation of organic and mineral resources from the stem, root, and leaves to the seed. Other summer annuals, such as crabgrass and annual lespedezas, continue to grow, in an indeterminate manner after flowering and producing seed, until killed by cold temperature.

Sudangrass is a summer annual that flowers in summer but differs from corn in that, like crabgrass, it tillers actively and regrows after flowering or cutting. Sudangrass eventually dies because this subtropical grass is sensitive to frost and does not develop winter hardiness. Stand persistence of summer annuals like korean lespedeza or crabgrass depend on their ability to produce seed that must overwinter to germinate the following spring (Beuselinck et al. 1994).

Winter annuals like crimson clover germinate in fall, grow vegetatively overwinter, and then die after flowering the following spring. They generally have programmed **senescence** processes that begin shortly after flowering and seed production. They depend on seed survival over summer to germinate the following fall to provide stand persistence.

Some forbs, including sweetclover, are true **biennials**. They germinate in spring and remain vegetative by producing only leaves and stems and form a large taproot during the first year. After a cold-induction period during winter, they flower and produce seed in spring. They have little, if any, storage of organic resources in the root or crown in year two, produce little or no regrowth and do not survive the second winter. Since the plants survive only two seasons, long-term stand persistence depends on seed production and seedling development. Sweetclover has adapted by having a high percentage of **hard seed**, some of which does not germinate for several years. No known grass is a true biennial.

Stand persistence, the longevity of a planting, can depend on innate longevity of the seedlings that become established plants that survive (i.e. plant persistence) or the ability of short-lived plants to spread vegetatively or by seed (Beuselinck et al. 1994). Plants like alfalfa and sericea lespedeza are long-lived **crown** formers with a seedling root that survives for several years and maintains a crown of buds at the soil surface for overwintering and regrowth (Chapter 3). Birdsfoot trefoil, also a crown former, is intermediate in that the seedlings survive for up to four years in cool environments but less than two years in hot environments, so stand persistence depends on both plant survival and reseeding.

In contrast, **clone** formers like white clover survive and spread by **stolons** or **rhizomes** that both form new roots at nodes and produce new shoots from axillary buds at rooted nodes. Thus, when the seedling root dies, the newly rooted stolons or rhizomes of the clonal plant continue to grow and produce vegetation to perennate.

Terms Involving Biodiversity

Mixed swards allow species with different growth patterns and maturity to minimize disease and pest problems of monocultures and help balance production rates throughout the year. Several studies have demonstrated the positive value of species diversity on production, especially in natural ecosystems (Loreau et al. 2002; Figure 1.5). **Biodiversity** refers to the number of species or functional groups in a habitat, which has an effect on several key ecologic processes including biomass productivity, rates of mineralization of soil nutrients, and stability or longevity of the system.

Biodiversity of several species is usually characterized as a function of species richness, i.e. the total number of species present, and the proportional abundance of each species within the community. Evenness refers to the distribution of species; high evenness indicates the proportions of species are similar or homogeneous within the canopy. In many diverse plant communities, a relatively few species predominate (Figure 1.6). These communities have a heterogeneous or uneven distribution of species and, therefore, low evenness. The inverse of evenness is dominance, so plant communities with low evenness have high dominance and vice versa (Peet 1974).

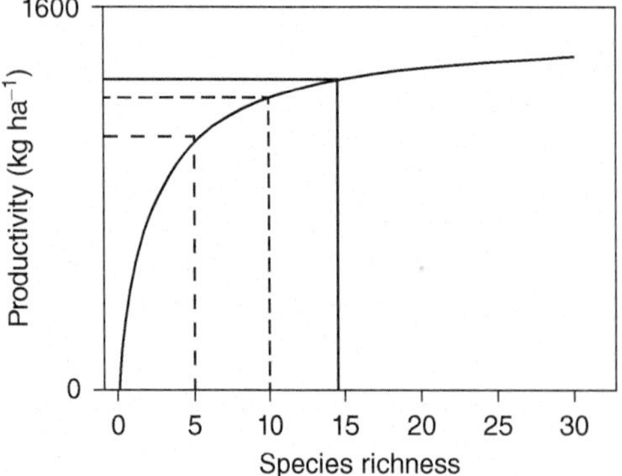

Fig. 1.5. The effect of species richness (number of species present) on productivity of a natural grassland. *Source:* Compiled from several data sets from the Great Plains.

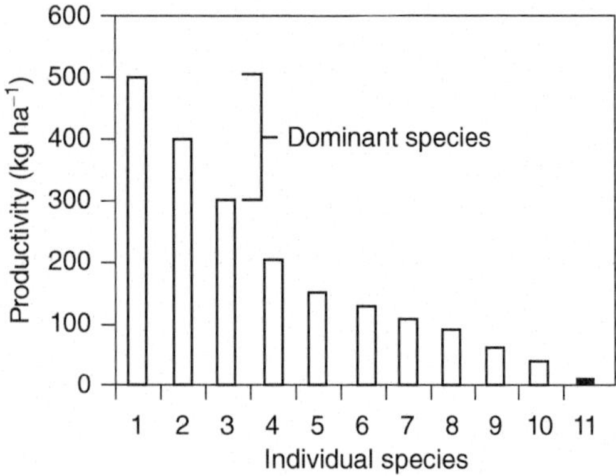

Fig. 1.6. Productivity of individual species in a mixed-species pasture or grassland. Most production comes from a few species that dominate the stand. Though not highly productive due to low density or low yield per plant, the remaining species contribute to stability and resilience of the ecosystem. *Source:* Compiled from several data sets from the Great Plains.

Plants can be classified into functional types, based either on their responses or effects. Response-functional types consist of plant species that respond in a similar manner to abiotic and biotic conditions. In general, since the species have similar functional traits, they are theoretically interchangeable in the plant community. For example, they may have similar reproductive strategies, growth habits, or carbon metabolism that enable them to persist in the population in response to herbage removal by cutting or grazing.

Effect-functional types are plant species that alter their processes such as their productivity or nutrient cycling in a similar way. Types that alter productivity in a similar way could include C_3 grasses, C_4 grasses, legumes, non-legume forbs, and woody species. For example, in North American grasslands and shrub lands, temperature is usually the main factor affecting abundance of C_3 versus C_4 grasses (Paruelo and Lauenroth 2005). C_4 grasses are favored by warmer climates in the southeastern US. Abundance of forbs is less affected by geographic and climatic variables, but the species of forb in the population can change.

High productivity of natural grassland ecosystems in western areas where soil fertility and/or soil moisture are limiting (Huston 1994) is often associated with increased diversity of species. Data in Figure 1.5 are for natural grassland systems with an annual production of about $1.4\,\mathrm{Mg\,ha^{-1}}$; the situation common in the Great Plains. Generally, production increases with plant diversity up to about 15–20 species, after which it remains near constant. This is expected in areas where plant density is low and plants of different growth forms and life cycles share resources. Open niches are occupied by annuals and small perennials that exist as subordinate species among the dominant species, usually tall perennials. Subordinate species are often transient (i.e. come and go), such as annual or perennial forbs (Tracy and Sanderson 2000).

A few dominant species are generally the most abundant in the mixture (Figure 1.6). If biomass is the major goal of the ecosystem function, then these dominant species are also most important functionally. This is usually the case in pasture and range situations. However, subordinate species can be important as they may occupy niches and, under even small changes in environmental or management conditions, may increase in biomass production, whereas the dominant species may produce less, thereby providing continuous ecosystem function (Walker et al. 1999).

Diverse canopies may be more resistant than monocultures to invasive or exotic (non-native) species. Not all exotic species are invasive, and invasive species vary in aggressiveness. Invasive species are classified as weak or strong invaders; they can be similar in the beginning as seedlings but differ in capability to gradually become a dominant component. In native Montana grassland containing 24 exotic grass and forb species, 13 were found to be weak and 11 were found to be strong invaders (Ortega and Pearson 2005). Weak invaders were primarily annual forbs, whereas strong invaders tended to be perennial forbs and grasses.

Invasiveness may result from differences in plant vigor, competition for resources, ability to store carbohydrate or nitrogen resources, **allelopathy**, or responses to **herbivory** and other environmental stresses (Maron and Vila 2001). The mechanisms governing strong invasiveness and the resultant long-term impact on the original or resident native species are not well known and usually are approached using mathematical models (e.g. Finnhoff and Tschirhart 2005).

Due to differences in soil types and **microclimates**, species diversity is usually not uniform across a mixed-species hayfield or pasture (Guretzky et al. 2005), and especially across a range site (Benedetti-Cecchi 2005). This reduces the value of knowing species composition over a wide area because the sward or population is operating as a series of small units or patches, each with its own diversity. Localized climates and soil properties alter the relative response of each species or functional group. Spatial variation in species richness or frequency needs consideration in the assessment.

Since forages and grasslands are managed primarily to produce quality feed for livestock, management systems are devised to take advantage of species diversity, even though each is not contributing equally. Thus, inputs like harvest frequency or stocking rate in more intensively managed systems in humid areas usually have the goal of maintaining density while favoring productivity of the dominant, preferred species.

Terms Describing Pasture and Grazing Management

An international committee (Allen et al. 2017) defined technical terms that are specific to pasture and grazing management. These terms are accepted internationally by professionals and form the basis for scientist-to-scientist communication. Other less specific terms are in common use by practitioners and the public. The glossary at the end of the book lists these technical terms along with many others, and defines them in the context of forage and grassland use.

Terms for Plant Nomenclature

Common names of plants can differ among countries and often vary regionally within a country. This is a critical issue when communicating about a specific plant species, so an explanatory background in plant classification and use of scientific names is presented. The appendix includes a listing of approved scientific names and synonyms for most forage and grassland species used on farms and ranches in North America.

Variation in Common Names

Due to geographic separation and differences in languages, most species have several common names in use by practitioners (Table 1.1). To illustrate, there are several common names for three forage species. Alfalfa originated in the area of what is now Iran, Afghanistan, and the Caucasus area of southern Russia. Due to its vast genetic variability, it is used as a forage crop around the world. Tall fescue originated in the Atlas mountain area of Morocco and Algeria and is used as a pasture and forage crop in transition zones. Dandelion, a ubiquitous forb in most pastures and meadows in humid climates, originated in Europe and Asia. Each has a unique growth form.

Terms for Taxonomic Relationships

Phylogeny describes genetic relationships between closely related species and pertains to evolution, ancestry, and descent. A general phylogenetic classification based on visual appearance of the seed would be grasses have one seed leaf or cotyledon, i.e. monocots, and forbs have two seed leaves or cotyledons, i.e. dicots. Forbs can be further separated into legumes and other plant groups according to biologic features such as leaf shape, presence of rhizomes or stolons and, especially, morphologic characteristics of flowers and other reproductive structures.

Table 1.1 Representative common names (and language) for the three species discussed in taxonomic detail in this chapter

alfalfa	Also known as blaue Luzerne (German), lucerne (English), luzerna (Portuguese), luzerne (French), mielga (Spanish), murasaki-umagoyashi (Japanese)
common dandelion	Also known as achicoria amarga (Spanish), amargón (Spanish), blowballs (English), dandelion (English), dent de lion (French), dente-de-leão (Portuguese), diente de león (Spanish), lion's tooth (English), Löwenzahn (German), pissenlit vulgare (French)
tall fescue	Also known as cañuela alta (Spanish), coarse fescue (English), erva-corneira (Portuguese), fétuque élevée (French), fétuque roseau (French), reed fescue (English), Rohrschwingel (German)

Source: Griffiths (1992); USDA-NRCS (2005); and Wiersema and León (1999).

The two-name system begun by Linnaeus, a Swedish botanist (1707–1778), is based on flower parts, especially **stamens**, to distinguish genera. Linnaeus evaluated and grouped species based on morphologic features of the flower parts, leaves or stems. Flower and seed characters were preferred since their morphology remains relatively consistent across environments and management conditions (Bailey 1963).

The Binomial (Linnean) System

With a wide variation in common names and little hope of standardizing them among geographic locations and languages, professionals use the approved scientific name that is consistent worldwide. The scientific name consists of (i) genus, (ii) specific epithet (species) and (iii) initial or name of the authority who classified the plant. The genus name is capitalized; the specific epithet is lowercase. A capital L as the authority indicates that Linnaeus, the father of plant taxonomy, classified the species. It is remarkable that so many of his original classifications and assigned scientific names have withstood the test of time.

In addition to the genus and species, several other abbreviations and symbols can be used in the scientific names of plants (Table 1.2). **Cultivar** is currently used instead of **variety** to designate a "cultivated variety"; i.e. one that has been genetically improved by sexual crosses or selection from the general population to have specific features or unique characteristics. This leaves the term variety (var. in Table 1.2) to designate a naturally occurring subgroup of a species that is distinguishable.

Criteria used for Classification

Early taxonomists recognized that many morphologic features, such as height or leaf shape, often changed with environmental factors, but the reproductive structures

Table 1.2 Abbreviations and symbols used in some scientific names of plants to give clarity or greater distinction

Abbreviation	Meaning
cv.	cultivar. The 1995 ICNCP specifies that this abbreviation is no longer being used.
ex	Latin for "from." In the authority for a name, it indicates that the publishing author (following ex) attributed the Latin name to someone else (preceding ex). The ICBN requires only the publishing author; the other is a courtesy.
p.p.	*pro parte*, Latin for "in part." It indicates that only part of the taxon circumscribed by a synonymous name is included in the taxon.
s.l.	*sensu lato*, Latin for "in the broad sense." It indicates that the present name circumscribes two or more taxa recognized by others; it is used when "lumping" taxa.
subsp./ssp.	subspecies. A less inclusive taxon than species, it is often applied to taxa that are incipient species, possibly due to geographic isolation.
syn.	synonym. It is an alternative scientific name that was either published later than the correct name or that could be the correct name in a different taxonomy.
var.	variety. A less inclusive taxon than species, it is often applied to taxa that are physiologically unique, corresponding roughly to ecotypes. According to the ICNCP, it applies to natural or wild taxa and should not be used for cultivars.
( )	Parentheses are used in the authority when a name has been changed. The original author is within the parentheses; the author name making the change follows outside the parentheses.

Source: ICBN is Greuter et al. (2000); ICNCP is Brickell et al. (2016).

like floral parts remained remarkably similar across environments. Today, flowers and reproductive structures remain the main criteria, but as knowledge about plants increased, plant taxonomists have supplemented morphology with the use of anatomy, enzymology, genetics, and now they have added **genome** analysis. Each technology made successive contributions to defining and refining the taxonomic relationships that support modern classifications for plants.

In most cases, the added technologies have supported the original classification based on morphology but, in some, there has been reclassification. This underscores the need for multiple methods to fully understand the phylogenetic relationships and classify the plants accordingly. As with most disciplines, there is overlap among representative species in each classification. These overlaps are accepted by some taxonomists, yet others strive for separation. For example, tall fescue is closely related to perennial ryegrass and interspecific crosses can be made. There is other evidence such that some taxonomists suggest they should be in the same genus, while others want them kept separate. Thus, the system is dynamic and needs to be updated constantly. The list of scientific and common names at the end of the book serves as a reference for technical names of forage crops that are correct to use in most situations.

Higher-level Taxonomic Groups

Table 1.3 shows the classification system for alfalfa beginning with the Plant Kingdom through to cultivar groups that differ in fall dormancy. Fall **dormancy** is directly associated with a capability to develop winter hardiness. Dormancy of alfalfa is designated for cultivars that have little fall growth (Groups 1 and 2), moderate fall growth (Groups 3–5), and those that are non-dormant and grow rapidly during fall (Groups 8–10).

Note that alfalfa belongs to the pea family of the dicotyledonous class, indicating the pod-like fruit structure and two cotyledons. Other plants in the same class would be similar for the cotyledon number, and those in the family would have similar pods. Thus, plants within each group are similar in key features. Alfalfa also has many subspecies, which represent small but consistent variants within the species. Subspecies can usually

Table 1.3 The full taxonomic classification of alfalfa with the preferred name in bold type

Kingdom	Plantae – Plants
Subkingdom	Tracheobionta – Vascular plants
Superdivision	Spermatophyta – Seed plants
Division	Magnoliophyta – Flowering plants
Class	Magnoliopsida – Dicotyledonous plants
Subclass	Rosidae
Order	Fabales
Family	Fabaceae – the pea family [also named Leguminosae]
Genus	*Medicago* L. – alfalfa, Linneus is authority
Species	***Medicago sativa* L.** – alfalfa
Subspecies	*Medicago sativa L. subsp. sativa* – alfalfa
Synonyms	*Medicago mesopotamica* Vassilcz.
	Plus 6 others
Subspecies	*Medicago sativa* L. subsp. *falcata* (L.) Arcang. – yellow alfalfa
Synonyms	*Medicago falcata* L.
	Plus 6 others
Subspecies	*Medicago sativa* L. nothosubsp. *Varia* (Martyn) Arcang. – variegated alfalfa
Synonyms	*Medicago* x*varia* Martyn
	Medicago sativa L. var. *varia* (Martyn) Urb.
	Plus 10 others
Subspecies	Plus 5 other subspecies
Cultivar group	Fall Dormancy Group 2 (very dormant),
Cultivars	"Alfagraze," "Mariner II," "Vernal," etc.
Cultivar group	Fall Dormancy Group 4
Cultivars	"Alliant," "Pioneer 54Q53," "ProLeaf," "Select," etc.
Cultivar group	Fall Dormancy Group 10 (non-dormant)
Cultivars	"Sedona," "WL711WF"

Source: Alfalfa Council (2018); USDA-ARS (2005); USDA-NRCS (2005).

cross-pollinate and can serve as additional sources of genetic variation for breeding programs. Subspecies variation may explain why alfalfa can be adapted successfully to such a wide range of locations and environments.

Dandelion, a common perennial forb in pastures and hayfields of eastern North America, is invasive due to its means of seed dispersal and competitiveness due to its large, flat leaf blades. Being a dicot, the classification scheme for dandelion is similar to alfalfa through the class level (Table 1.4), after which alfalfa diverges as a legume forb from the non-legume forbs. Alfalfa is a rosid (Figure 1.7) and belongs to the subclass Rosidae (Table 1.3), whereas dandelion is an asterid and belongs to the subclass Asteridae (Table 1.4). The dendrogram continues to show the divergence between these species as the hierarchy continues to develop.

Tall fescue, a monocotyledonous plant, is classified similarly to alfalfa and dandelion above the class level (compare Table 1.5 with Tables 1.3 and 1.4), but then diverges as sharp distinction occurs between dicots and monocots (Figure 1.7). Grasses follow the same dendrogram pattern through the family Poaceae, which includes both cool-season (C_3) and warm-season (C_4) types. Subfamilies of Poaceae include Pooideae (Table 1.5; also known as Festucoideae) and Panicoideae; the first has C_3 photosynthetic metabolism, the second has C_4. Chapman and Peat (1992) present a thorough discussion of grass taxonomy including the characteristics that distinguish subfamilies.

Changing the Scientific Name

To assign a scientific name of a new species or make changes in an approved scientific name, taxonomists conduct appropriate research, publish the findings, verify the uniqueness and then propose a name. As science progresses, the evidence may indicate either or both the genus or species needs to change. The International Code of Nomenclature for Cultivated Plants, also known as the Cultivated Plant Code (Brickell et al. 2016) gives guidelines for naming new species or renaming species.

Naming and renaming genera or species is especially a challenge with species that are polyploid, particularly if they are **allopolyploid** like tall fescue (2n = 6x = 42 chromosomes). Gradually, the original diploid progenitors that merged to form the 6x polyploid plant become

Table 1.4 The full taxonomic classification of common dandelion with the preferred name in bold type

Kingdom	Plantae – Plants
Subkingdom	Tracheobionta – Vascular plants
Superdivision	Spermatophyta – Seed plants
Division	Magnoliophyta – Flowering plants
Class	Magnoliopsida – Dicotyledonous plants
Subclass	Asteridae
Order	Asterales
Family	Asteraceae – the aster family [also named Compositae]
Genus	*Taraxacum* G.H. Weber ex Wiggers – dandelion
Species	***Taraxacum officinale*** **G.H. Weber ex Wiggers s.l.** – common dandelion
Subspecies	*Taraxacum officinale* G.H. Weber ex Wiggers subsp. *officinale* – common dandelion subspecies
Synonyms	*Taraxacum atroglaucum* M.P. Christens
	Plus 18 others
Subspecies	*Taraxacum officinale* G.H. Weber ex Wiggers subsp. *vulgare* (Lam.) Schinz & R. Keller – common dandelion subspecies
Synonyms	*Leontodon latiloba* (DC.) Britt.
	Leontodon taraxacum L. p.p.
	Taraxacum latiloba DC.
	Taraxacum palustre (Lyons) Symons var. *vulgare* (Lam.) Fern.
	Taraxacum vulgare Lam.
Subspecies	*Taraxacum officinale* G.H. Weber ex Wiggers subsp. *ceratophorum* (Ledeb.) Schinz ex Thellung – common dandelion subspecies
Synonyms	*Taraxacum ambigens* Fern.
	Plus 28 others
Cultivar group	None for primary forage use
Cultivar group	Culinary
Cultivars	"Thick Leaf," "Improved Giant," etc.

Source: Parmenter (2002); USDA-ARS (2005); USDA-NRCS (2005).

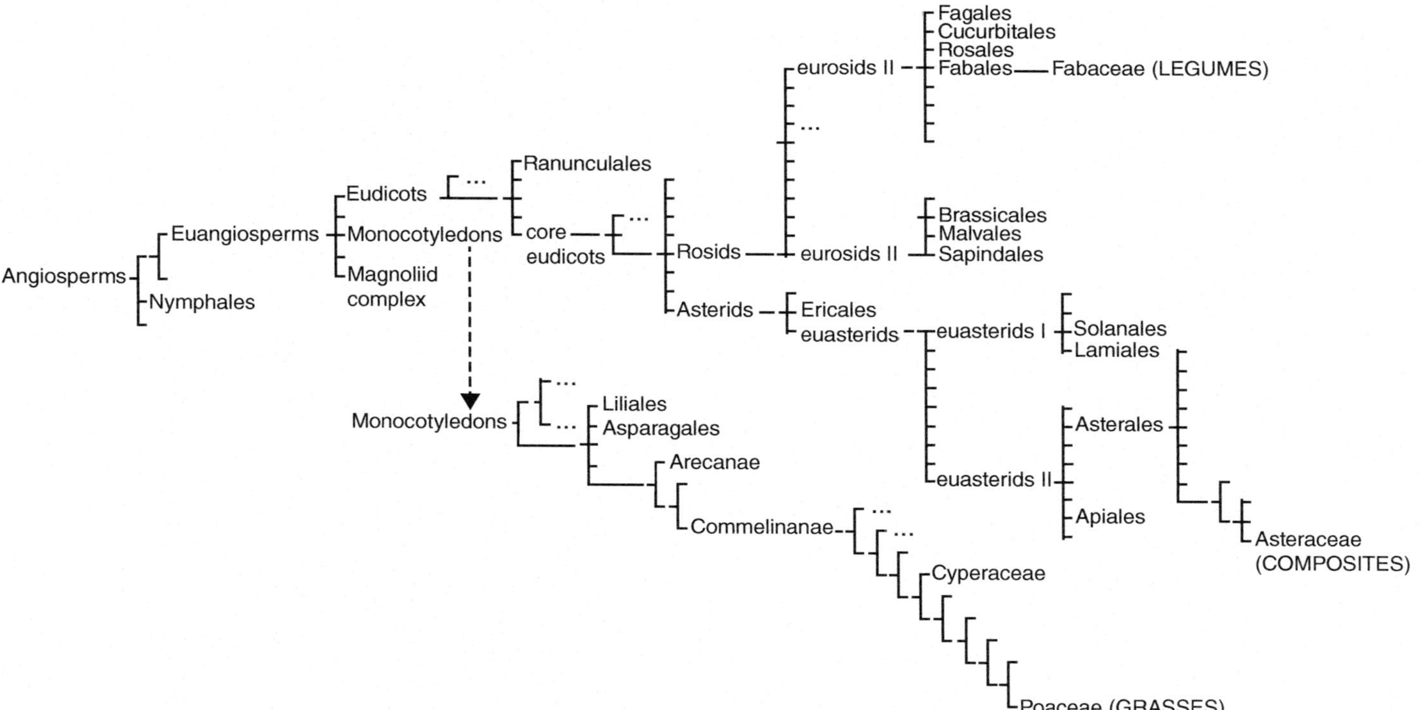

Fig. 1.7. Modified dendrogram showing the relationship of legumes, composites, and grasses based on recent cladistic analyses. Note that cladistic terminology can be associated with the categories of traditional nomenclature used in Tables 1.3–1.5. *Source:* Adapted from Tree of Life Web Project http://tolweb.org/tree/phylogeny.html confirmed 16 April 2018.

Table 1.5 The full classification of tall fescue with the preferred name in bold type

Kingdom	Plantae – Plants
Subkingdom	Tracheobionta – Vascular plants
Superdivision	Spermatophyta – Seed plants
Division (Phylum)	Magnoliophyta – Flowering plants
Class	Liliopsida – Monocotyledonous plants
Subclass	Commelinidae
Order	Cyperales
Family	Poaceae – the grass family [also named Gramineae]
Subfamily	Pooideae
Tribe	Poeae
Genus	*Festuca* L. – fescue
Species	***Festuca arundinacea* Schreb.** – tall fescue
Synonyms	*Lolium arundinaceum* (Schreb.) S.J. Darbyshire
Synonyms to be rejected	
	Festuca elatior L. p.p.
	Festuca elatior L. subsp. *arundinacea* (Schreb.) Hack.
	Festuca elatior L. var. *arundinacea* (Schreb.) C.F.H. Wimmer
Subspecies	7 subspecies are listed by USDA-ARS
Cultivar group	Forage Group
	Cultivars "Alta," "Fawn," "Kentucky 31," "Stargrazer," etc.
Cultivar group	Turf Group
Cultivars	"Carefree," "Finelawn 88," "Pixie," "Tribute," etc.

Source: Cowan (1956); USDA-ARS (2005); USDA-NRCS (2005).

known. Alfalfa is also a polyploid, but is an **autopolyploid** formed by doubling the original chromosome set of 16 (2n = 4x = 32). The doubling did not introduce new chromosomes from another source, but increased the dosage of **gene** expressions by doubling the number of gene copies involved. Linneus named this legume *Medicago sativa* L. and it still remains.

Linneus assigned tall fescue the name *Festuca elatior* L. Later, according to Bailey (1949), William Hudson (1730–1793) a British botanist renamed it *Festuca pratensis* Huds. which is now the official name for meadow fescue. In 1824, the Belgian botanist Barthélemy Charles Joseph Dumortier proposed tall fescue be moved to the genus *Schedonorus.* In the 1948 USDA Yearbook of Agriculture, it was named *F. elatior var. arundinacea,* with no authority designation (Stefferud 1948). In the first edition of "Forages", Bailey (1951) used the same scientific name, adding that arundinacea indicates it is a sub-species and not a variety of meadow fescue (*F. elatior*). It was officially named by Cowan (1956) as *Festuca arundinacea* Schreb., presumably based on the discovery of the name assigned in 1771 by Johann Christian Daniel von Schreber, a German naturalist. In the third edition of "Forages", Buckner and Cowan (1973) again used *F. arundinacea* Schreb. But the saga was not finished.

Cytogeneticists in Great Britain and other locations began detailed microscopic comparisons of the 42 chromosomes of tall fescue with those of related species to learn the progenitors. They proposed that the diploid (2x, n = 7) donor was *F. pratensis* while the tetraploid donor (4x, n = 7) was *F. arundinacea* var. *glaucescens* (Chandrasekharan and Thomas 1971). Later, others discovered the two pairs (2X) of chromosomes in the hexaploid species (6X) appeared to be from a *Lolium* species. Then, based on **chloroplast** DNA analysis, S.J. Darbyshire (1993), a Canadian taxonomist, renamed the species *Lolium arundinaceum* (Schreb.) Darbysh.

That name was adopted in 2002 by the USDA-NRCS along with alternate names (synonyms) of *Schedonorus phoenix* (Scop.) Holub and *F. arundinacea* Schreb. In a treatise of names for species in the Great Plains, Nowick (2015) used *L. arundinaceum* (Schreb.) S.J. Darbyshire as the official name for tall fescue. Based on recent phylogenetic and DNA analysis, tall fescue has now been renamed Schedonorus arundiaceus (Schreb.) Dumort. However, it is still contentious among taxonomists whether *Schedonorus* is a true genus or is a subgenus of either Lolium or Festuca (Soreng et al. 2001). Further analysis (Cheng et al. 2016) of *Festuca-Lolium* hybrids crosses suggested *Schedonorus* is closely related with *Lolium.* It is well-known that hybrids of *Festuca-Lolium* have significant forage value and seed is available.

The USDA Germplasm Resources Information Network (GRIN) system maintains a current listing of important agricultural plants, including the approved scientific name (USDA-ARS 2005). In that database, the one this book uses as the authority, tall fescue is identified as *F. arundinacea* Schreb. with synonyms of both *L. arundinaceum* (Schreb.) S.J. Darbyshire and *Schedonorus arundinaceus* (Schreb.) Dumort. The GRIN database lists 12 subfamilies and 45 tribes in the grass family (USDA-ARS 2005). Genera are grouped into subfamilies to associate general morphologic and physiologic characteristics (e.g. C_3 vs. C_4), while tribes separate more subtle properties.

Summary and Conclusions

Terminology for describing forage plants and management systems will continue to evolve and become more detailed and descriptive. It will take effort and communication among a range of professionals and practitioners to decide on the best terms. The scientific community must also continue to develop terms for specific processes or management practices that have international meaning and acceptance.

The classification of plants into various groups and subgroups will also be an ongoing effort, especially as new techniques develop in genetics and molecular biology that allow taxonomists to determine phylogeny more accurately. This will likely lead to more splitting of groupings based on DNA and molecular features within current species. In addition, there will be more "novel plant types" and "uniqueness" developed, including genetically modified organisms (GMOs), by combining diverse traits from totally unrelated species, even from microbes and animals. These new "hybrid organisms" may or may not fit the current classification system or, for economic, cultural, or political reasons, will be classified out of the normal systems.

References

Alfalfa Council (2018). *Fall Dormancy and Pest Resistance Ratings for Alfalfa Varieties*. Kansas City, MO: Alfalfa Council http://alfalfa.org/pdf/2018_Variety_Leaflet.pdf.

Allen, V.G., Batello, C., Berretta, E.J. et al. (2017). An international terminology for grazing lands and grazing animals. *Grass Forage Sci*. https://doi.org/10.1111/j.1365-2494.2010.00780.x.

American Society of Agronomy. (1989). Decision reached on sustainable ag. *Agronomy News* (January), p. 15.

Bailey, L.H. (1949). *Manual of Cultivated Plants*. New York: MacMillan Co.

Bailey, L.H. (1963). *How Plants Get Their Names*. New Tork: Dover Publications, Inc.

Bailey, R.Y. (1951). The fescues. In: *Forages, the Science of Grassland Agriculture* (eds. H.D. Hughes, M.E. Heath and D.S. Metcalfe), 327–335. Ames, IA: Iowa State College Press.

Benedetti-Cecchi, L. (2005). Unanticipated impacts of spatial variance of biodiversity on plant productivity. *Ecol. Lett*. 8: 791–799.

Beuselinck, P.R., Bouton, J.H., Lamp, W.O. et al. (1994). Improving legume persistence in forage crop systems. *J. Prod. Agric*. 7: 311–322.

Brickell, C.D., Alexander, C., Cubey, J.A. et al. (2016). *International Code of Nomenclature for Cultivated Plants*. Leuven, Belgium: International Society for Horticultural Science.

Buckner, R.C. and Cowan, J.R. (1973). Tall fescue. In: *Forages, The Science of Grassland Agriculture* (eds. M.E. Heath, D.S. Metcalfe and R.F Barnes), 297–306. Ames, IA: Iowa State University Press.

Chandrasekharan, P. and Thomas, H. (1971). Studies in *Festuca*. 5. Cytogenetic relationships between species of *Bovinae* and *Scariosae*. *Z. Planzenzuchtg*. 65: 345–354.

Chapman, G.P. and Peat, W.E. (1992). *An Introduction to the Grasses (Including Bamboos and Cereals)*. London, UK: CAB International.

Cheng, Y., Zhou, K., Humphreys, M.W. et al. (2016). Phylogenetic relationships in the *Festuca-Lolium* complex (Loliinae;Poaceae): New insights from chloroplast sequences. *Front. Ecol. Evol*. 4: 89.

Child, R.D. and Pearson, H.A. (1995). Rangeland and agroforestry. In: *Forages, Vol. II: The Science of Grassland Agriculture* (eds. R.F Barnes, D.A. Miller and C.J. Nelson), 225–242. Ames, IA: Iowa State University Press.

Clason, T.R. and Sharrow, S.H. (2000). Silvopastoral pastures. In: *North American Agroforestry: An Integrated Science and Practice* (ed. H.E. Garrett), 119–147. Madison, WI: American Society of Agronomy.

Clubine, S.E. (1995). Managing forages to benefit wildlife. In: *Forages, Vol. II: The Science of Grassland Agriculture* (eds. R.F Barnes, D.A. Miller and C.J. Nelson), 263–275. Ames, IA: Iowa State University Press.

Costanza, R., de Groot, R., Braat, L. et al. (2017). Twenty years of ecosystem services: how far have we come and how far do we still need to go? *Ecosyst. Serv*. 28: 1–16.

Cowan, J.R. (1956). Tall fescue. *Adv. Agron*. 8: 283–320.

Crowder, D. and Reganold, J. (2015). Financial competitiveness of organic agriculture on a global scale. *Proc. Natl. Acad. Sci. U.S.A*. https://doi.org/10.1073/pnas.1423674112.

Darbyshire, S.J. (1993). Realignment of *Festuca* subgenus *Schedonorus* with the genus *Lolium* (Poaceae). *Novon* 3: 239–243.

Doran, J.W. and Zeiss, M.R. (2000). Soil health and sustainability: managing the biotic component of soil quality. *Appl. Soil Ecol*. 15: 3–11.

van Es, H. (2017). A new definition of soil. *CSA News Magazine* 62 (10): 20–21.

Finney, D.M., Murrell, E.G., White, C.M. et al. (2017). Ecosystem services and disservices are bundled in simple and diverse cover cropping systems. *Agric. Environ. Lett*. 2: 170033. https://doi.org/10.2134/ael2017.09.0033.

Finnhoff, D. and Tschirhart, J. (2005). Identifying, preventing, and controlling invasive plant species using their physiological traits. *Ecolog. Econ.* 52: 397–413.

Friedman, D., Hubbs, M., Tugel, A. et al. (2001). *Guidelines for Soil Quality Assessment in Conservation Planning*. Washington, DC: USDA-NRCS, Soil Quality Institute.

Garrett, H.E., Rietvald, W.J., and Fisher, R.F. (2000). *North American Agroforestry: An Integrated Science and Practice*. Madison, WI: American Society of Agronomy.

Gerrish, J. (2004). *Management-Intensive Grazing, the Grassroots of Grass Farming*. Ridgeland, MS: Green Park Press.

Greuter, W., McNeill, J., Barrie, F.R. et al. (2000). *International Code of Botanical Nomenclature (St. Louis Code) Regnum Vegatabile 138*. Königstein, Ger.: Koeltz Scientific Books.

Griffiths, M. (ed.) (1992). *The New Royal Horticultural Society Dictionary of Gardening*. London: MacMillan.

Guretzky, J.A., Moore, K.J., Brummer, E.C., and Burras, C.L. (2005). Species diversity and functional composition of pastures that vary in landscape position and grazing management. *Crop Sci.* 45: 282–289.

Huston, M. (1994). *Biological Diversity: The Coexistence of Species on Changing Landscapes*. Cambridge, UK: Cambridge University Press.

Karlen, D.L., Ditzler, C.A., and Andrews, S.S. (2003). Soil quality: why and how? *Geoderma* 114: 45–156.

Laishram, J., Saxena, K.G., Miakhuri, R.K., and Rao, K.S. (2012). Soil quality and soil health: a review. *Int. J. Ecol. Environ. Sci.* 38: 19–37.

Loreau, M., Naeem, S., and Inchausti, P. (2002). *Biodiversity and Ecosystem Functioning: Synthesis and Perspectives*. New York: Oxford University Press.

Maron, J.L. and Vila, M. (2001). When do herbivores affect plant invasion? Evidence for the natural enemies and biotic resistance hypotheses. *Oikos* 95: 361–373.

Marten, G.C., Sheaffer, C.C., and Wyse, D.L. (1987). Forage nutritive value and palatability of perennial weeds. *Agron. J.* 79: 980–986.

Millennium Ecosystem Assessment (2005). *Ecosystems and Human Well-Being: Synthesis*. Washington, DC: Island Press.

Murrell, E.G., Schipanski, M.E., Finney, D.M. et al. (2017). Achieving diverse cover crop mixtures: effects of planting date and seeding rate. *Agron. J.* 109: 259–271. https://doi.org/10.2134/agronj2016.03.0174.

Nation, A. (2004). Foreword. In: *Management-Intensive Grazing, the Grassroots of Grass Farming* (ed. J. Gerrish), 9–10. Ridgeland, MS: Green Park Press.

Nowick, E. (2015). *Historical Common Names of Great Plains Plants, with Scientific Names Index. Vol. I: Common Names*. Lincoln NE: Zea Books, University of Nebraska–Lincoln.

Ortega, Y.K. and Pearson, D.E. (2005). Weak vs. strong invaders of natural plant communities: assessing invasibility and impact. *Ecol. Applic.* 15: 651–661.

Parmenter, G. (2002). *Taraxacum officinale: Common Dandelion, Lion's Tooth*. Christchurch, NZ: Crop and Food Research Ltd. https://www.scribd.com/document/19139849/Dandelion Verified on 15 April, 2018.

Paruelo, J.M. and Lauenroth, W.K. (2005). Relative abundance of plant functional types in grasslands and shrublands of North America. *Ecol. Applic.* 6: 1212–1224.

Peet, R.K. (1974). The measurement of species diversity. *Annu. Rev. Ecol. Systemat.* 5: 285–307.

Pesek, J., Brown, S., Clancy, K. et al. (1993). *Alternative Agriculture/Committee on the Role of Alternative Farming Methods in Modern Production Agriculture*, Board on Agriculture, National Research Council. Washington, D.C.L National Academy of Science.

Reganold, J.P. and Wachter, J.M. (2016). Organic agriculture in the twenty-first century. *Nature Plants* 2 (15221). doi:10.1038/nplants.2015.221.

Sharp, W.C., Schertz, D.L., and Carlson, J.R. (1995). Forages for conservation and soil stabilization. In: *Forages, Vol. II: The Science of Grassland Agriculture* (eds. R.F Barnes, D.A. Miller and C.J. Nelson), 243–262. Ames, IA: Iowa State University Press.

Smith, D.H. and Collins, M. (2003). Forbs. In: *Forages, Vol. I: An Introduction to Grassland Agriculture* (eds. R.F Barnes, C.J. Nelson, M. Collins and K.J. Moore), 215–236. Ames, IA: Iowa State Press.

Sollenberger, L.S., Agouridis, C.T., Vansant, F.S. et al. (2012). Prescribed grazing on pasturelands. In: *Conservation Outcomes from Pastureland and Hayland Practices: Assessment, Recommendations, and Knowledge Gaps* (ed. C.J. Nelson), 113–204. Lawrence, KS: Allen Press.

Soreng, R.J., Terrell, E.E., Wiersema, J., and Darbyshire, S.J. (2001). Proposal to conserve the name *Schedonorus arundinaceus* (Schreb.) Dumort. against *Schedonorus arundinaceus* Roem. & Schult. (Poaceae: Poeae). *Taxon* 50: 915–917.

Stefferud, A. (1948). *Grass, The Yearbook of Agriculture 1948*. Washington, DC: United States Government Printing Office.

Tillman, D., Cassman, K.G., Matson, P.A. et al. (2002). Agricultural sustainability and intensive production practices. *Nature* 418: 671–677.

Tracy, B.F. and Sanderson, M.A. (2000). Patterns of plant species richness in pasture lands of the northeast United States. *Plant Ecol.* 149: 169–180.

USDA-ARS. (2005). National genetic resources program. Germplasm resources information network (GRIN). National Germplasm Resources Laboratory, Beltsville, MD. https://www.ars.usda.gov/northeast-area/beltsville-md-barc/beltsville-agricultural-research-center/national-germplasm-resources-laboratory/ (accessed 22 November 2019).

USDA-NRCS. (2005). The PLANTS database, Version 3.5. National Plant Data Center, Baton Rouge, LA. http://plants.usda.gov (accessed 22 November 2019).

Walker, B., Kinzig, A., and Langridge, J. (1999). Plant attribute diversity, resilience, and ecosystem function: the nature and significance of dominant and minor species. *Ecosyst.* 2: 95–113.

White, C.M., DuPont, S.T., Hautau, M. et al. (2017). Managing the trade off between nitrogen supply and retention with cover crop mixtures. *Agric. Ecosyst. Environ.* 237: 121–133. https://doi.org/10.1016/j.agee.2016.12.016.

Wiersema, J.H. and León, B. (1999). *World Economic Plants*. Boca Raton, FL: CRC Press.

CHAPTER

2

Grass Morphology

C. Jerry Nelson, Professor Emeritus, *Plant Sciences, University of Missouri, Columbia, MO, USA*
Kenneth J. Moore, Distinguished Professor, *Agronomy, Iowa State University, Ames, IA, USA*

Introduction

Morphology refers to the structure and arrangement of plant parts that characterize plant shapes and are primary factors for species identification and management. Structures of flowers, seed, leaves, and stems form the base of the Linnaean system of plant classification. Morphologic features result from the way these above-ground structures are initiated, enlarged and displayed and how roots, **tubers** and rhizomes grow below ground. The relative shapes and sizes of these structures and their growth in natural and managed conditions help determine adaptation, productivity, quality, and persistence of forage grasses.

In Volume I, Moore and Nelson (2018) introduced the structure and morphology of grasses including the seed, seedling development, vegetative growth, and reproductive growth. Applications covered use of general morphology for hayland and pasture management. This chapter describes in more detail how grass plants achieve their shape and how these processes respond to environment and management. Much of the fundamental research has been done with perennial cool-season grasses, usually perennial ryegrass or tall fescue, but has applications and provides insight for a range of cool- and warm-season grass species. In addition to agricultural issues, morphologic features associated with yield potential, plant-to-plant competition and plant–herbivore interactions also affect environmental conservation (Chapter 13).

Plant growth and shape depend on a series of specific meristems that produce cells followed by expansion and finally specialization of those cells within a specific plant part. All grass plants have a **terminal meristem** located at the tip of every stem and root. Growth of grasses is unique and heavily dependent on **intercalary meristems** that produce and enlarge cells between non-growing areas and serve to elongate stem internodes, leaf blades, and leaf **sheaths**. Understanding the mechanisms, functions and management of these meristems is key to optimization of species selection and their practical uses.

Abiotic factors such as soil properties, rainfall patterns and temperature variation (Chapters 4–8) affect adaptation and growth rates involving morphology. Biotic variables of grasses affect **herbivory**, plant shape, growth processes, and competitive interactions with other plants. These include **allelopathy**, associations with **mycorrhiza**, species signaling and recognition, plant parasites and pathogens (Chapter 9). Negative biotic interactions occur more frequently than positive ones.

Overview of Forage Grass Morphology

The combination of morphology (growth habit and structure) and **physiology** (growth rates and metabolic processes) is critical for understanding the multiplicity of uses and diversity for adaptation of forage grasses to the environment and management (Matthew 2017). About ten grasses have been domesticated and genetically improved

Forages: The Science of Grassland Agriculture, Volume II, Seventh Edition.
Edited by Kenneth J. Moore, Michael Collins, C. Jerry Nelson and Daren D. Redfearn.

for cereal crops (Hartley and Williams 1956). In contrast, scores of cool-season (Moser et al. 1996) and warm-season (Moser et al. 2004) grasses have been evaluated as forage crops and many have been genetically improved (Vogel and Burson 2004). Countless other unimproved grasses are managed as components of natural ecosystems.

Due to effects of morphologic and physiologic adaptation, some members of the grass family, Poaceae, with over 800 genera and 10 000 known species (Watson and Dallwitz 1992), exist in nearly every land habitat in the world. Economically and ecologically, Poaceae is the most important plant family on earth (Bouchenak-Khelladi et al. 2010). Variation among grasses is complex due to the high level of genetic **polymorphism** within the family in terms of geographic distribution, metabolic pathways, and morphologic structures. This includes the transition, thousands of years ago, of some grass families with C_3 photosynthesis to have C_4 photosynthesis that further added to their adaptation to drought and high temperature (Chapter 4).

The life cycle of most cool-season grass plants used as forage involves two distinct seasonal growth forms; early growth in spring consists mainly of leaves from nodes on short stems. Later in spring or early summer, in response to warmer temperatures and length of day, growth changes to stem elongation, flowering and seed production. Stem elongation and flowering lead to higher yield than leaves alone, but lower quality due to older leaves and stronger cell walls of stems, which have higher lignin content. Compared with cool-season grasses, development of adapted warm-season grasses used for forage is generally shifted to later in the growing season. After flowering, most perennial grasses resume growth from tillers that continue vegetative growth during the rest of the year.

While many concepts associated with cool-season grasses are transferable to warm-season or tropical grasses, there are exceptions (da Silva et al. 2015). For example, forage quality of young leaves with C_4 photosynthesis is usually lower than for C_3 grasses due to its lower protein and higher fiber contents, ages more rapidly and reduces its value for plant growth. Some warm-season grasses can be tall, grow rapidly and have a long lifespan for leaves. Nevertheless, long durations between grazing periods have a high cost due to reduced quality (Lemaire et al. 2009).

Current emphasis is on relationships of leaf area index (leaf area per ground area), grazing more frequently, and even grazing continuously, to match leaf production rate if a minimal amount of basal leaf area is retained to support growth of new tillers. Grazing the higher quality leaves frequently results in a better balance among rate of leaf growth, reduced leaf senescence and amount of leaf removal by the animals (da Silva et al. 2013).

The Concept of a Phytomer

Grass plants, whether annual or perennial, have a common basic morphologic unit called a **phytomer** (Figure 2.1). Each phytomer is composed of (i) a leaf consisting of a **blade**, a ligule (collar) and a sheath, (ii) an internode, (iii) a node and (iv) an axillary bud. There is some disagreement among botanists regarding whether the axillary bud of a specific phytomer originates at the node associated with the base of the sheath (Sharman 1945) or one node below (Clark and Fisher 1987). Many

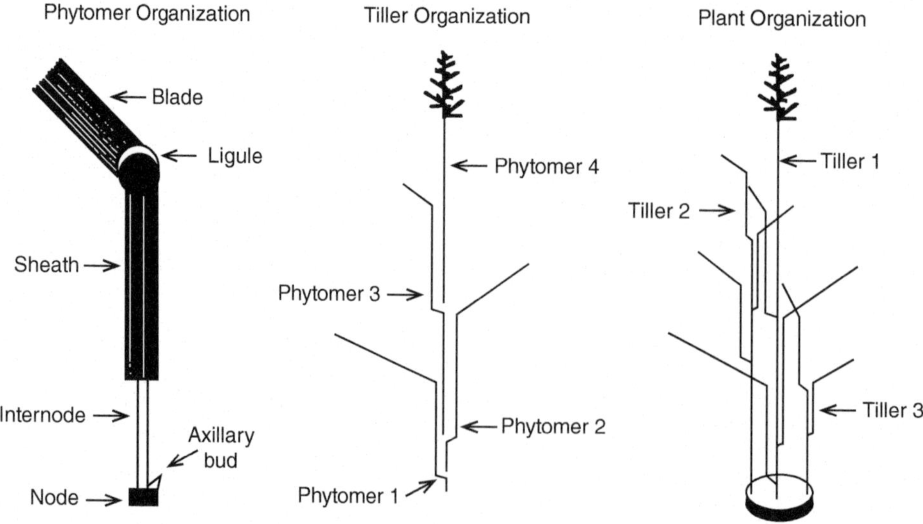

Fig. 2.1. Hierarchical levels of phytomers, the vegetation organization of grass plants that characterize pasture, forage, and rangeland ecosystems. Each phytomer is initiated sequentially by the shoot apex and develops its structure. New tillers develop from axillary buds and grow by sequential phytomers to form higher organizational levels of the plant. *Source:* Adapted from Briske (2007).

scientists follow the Clark and Fisher model since the axillary bud is delayed and forms last in the phytomer. However, the distinction is not critical at the scale of presentation here.

Each individual phytomer is initiated by the **shoot apex** (terminal meristem), then develops by active intercalary meristems located at the base of each blade, **sheath**, and stem internode. Lastly, an axillary bud develops at each node in the axil of the leaf sheath and internode. An accumulation of successive phytomers, each differentiated by the same shoot apex, comprises the main tiller or stem. New juvenile or daughter tillers are developed by axillary buds that have their own shoot apex to produce additional phytomers. The tillers form root apices from intercalary meristems at the base of the non-elongated internodes. Cell production occurs in the root apex after which cell elongation pushes the root tip and its apex through the soil. Apices for root branches form in the **pericycle** of the root and grow outwardly through the cortex to reach the soil.

Shapes and sizes of each component of the phytomer differ among grass species, but the organization is the same and each phytomer has the same components. For example, even though underground and growing laterally by internode elongation, **rhizomes** have phytomers with extended internodes and a small, modified leaf at each node that covers and helps protect the axillary bud. **Stolons** also consist of phytomers.

Structure of the Shoot Apex

The shoot apex at the tip of the stem produces cells for its own enlargement and production of sequential phytomers (Figure 2.2). In a stepwise manner, it initiates each new leaf **primordium** and forms the nodes and basal cells for the associated internode and axillary bud. After several phytomers have formed on the apex, each with its leaf, the shoot apex responds to environmental signals, stops initiating leaves and differentiates to form an **inflorescence** that is pushed upward and above the leaves by intercalary meristems that elongate the stem internodes. If the shoot apex is elevated enough to be removed by cutting or grazing, that shoot dies and needs to be replaced by a tiller. If the stem produces flowers, the shoot dies naturally since there is no shoot apex.

Cellular Organization and Function

The surface of the shoot apex includes cell layers or cell types that provide various functions. Grasses have either one or two **tunica** layers; for example, tall fescue has one layer (Figure 2.2a) whereas for unknown reasons, quackgrass and wheat have two (Williams 1975), and some grasses have three layers (Cleland 2001). The tunica axial cells (Figure 2.2b) show anticlinal divisions (90° from direction of growth), followed by minimal cell enlargement, mainly to slowly extend the tip of the shoot apex.

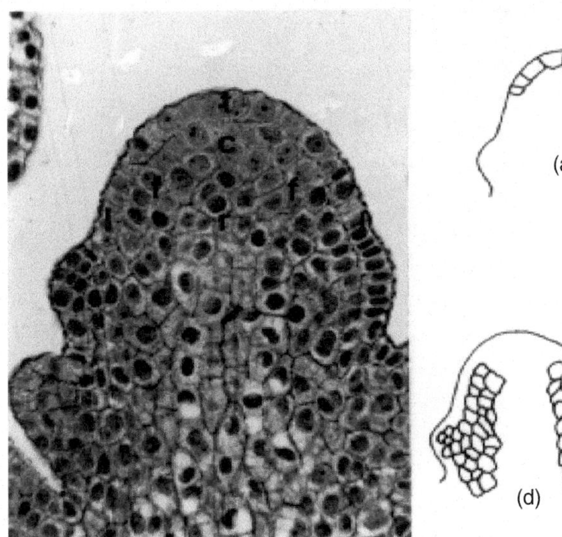

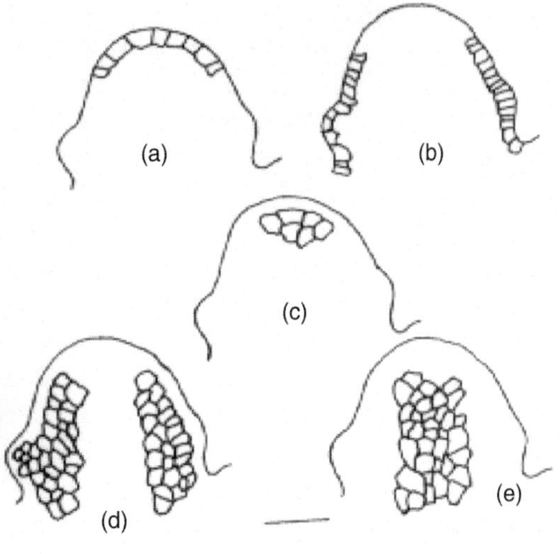

Fig. 2.2. A median longitudinal section of the vegetative shoot apex of tall fescue (left) and delineations showing locations of the five major types of apical cells. Tunica cells (a) extend the apical dome, tunica axil cells (b) initiate leaf primordia, large corpus initials (c) support growth of the dome, flank corpus initials (d) support growth in diameter while the leaf primordia begins to differentiate, and rib corpus initials (e) increase growth in height and girth. The node and vascular structures develop later. *Source:* From Vassey (1986).

Similarly, the tunica lateral cells (Figure 2.2b) extend the shoulders of the dome and initiate the leaf **primordia.** As the apex grows upward, by adding new cells below the tunica layer, a cell in the outer tunica layer divides along with adjacent cells to form a protrusion (note left side of Figure 2.2d) that develops into a leaf primordium. Gradually, cells in the tunica layer adjacent to the developing primordium, divide along the perimeter of the apex causing lateral spread of the cell division zone in both directions to circumscribe the apex. Edges meet on the opposite side (note the "bump" on the upper right), which is the last part of the primordium to be initiated. At that position, the circumscribed cells within the apex differentiate to develop a new node. The circumference of the apex helps determine the number of cells and subsequent width of the **blade** and sheath (Rademacher and Nelson 2001).

Apex cells just above the new node divide and differentiate with little elongation to form the **meristematic** base of the next internode. Later, cells at the base of the leaf primordium, that now surrounds the apex, begin to divide with little elongation to form the intercalary meristem that provides cells for elongation of the leaf. The large corpus initials (Figure 2.2c) produce cells to support the dome as it enlarges while the flank corpus initials (Figure 2.2d) support growth in diameter. Both are supported by rib corpus initials (Figure 2.2e) of larger cells that add further growth in height and girth of the shoot apex. A few days later, a new leaf primordium is initiated above and opposite to the previous one to continue the production of new phytomers. Regardless of the mechanism, the leaf arrangement on alternate sides, 180° apart, is very fixed and characterizes all grasses (Cleland 2001).

Plastochrons and Phyllochrons

The plastochron is the time interval between development of sequential primordia on the **shoot apex**, whereas the phyllochron is the time interval between sequential appearances of leaves, usually determined as the time between consecutive **ligule** (collar) appearances above the whorl of older leaf sheaths. The plastochron at the shoot apex is often slightly shorter by a few hours, than the phyllochron. This allows a gradual accumulation of primordia, nodes and potential sites for axillary buds associated with the apex. This is especially important during stress periods such as drought since the accumulated primordia can elongate into leaves to assist with plant recovery.

Apex Size and Leaf Growth Rates

Vassey (1986) conducted a morphometric study of the shoot apices of tall fescue genotypes selected for high (HYT) and low (LYT) yield per tiller (Figure 2.2). The HYT genotype had 30–40% faster leaf growth rates (mm d^{-1}) and 40–50% longer leaf blades. Apex structure and organization were similar for both genotypes, but cell sizes in each zone of the apex were consistently larger in the HYT genotype. Similarly, volume of the dome above the leaf primordia was 75% larger, due mainly to greater apex diameter and more cells at the base of each primordium, which directly increased leaf width. Conversely, the smaller shoot apex of the LYT genotype had, on average, 5.0 leaf primordia and 2.1 visible axillary buds compared with the HYT genotype with 3.8 leaf primordia and 1.4 visible axillary buds. This reflected a faster rate of leaf appearance, i.e. a shorter phyllochron, and production of more tillers in the LYT genotype.

The shoot apex of quackgrass and perennial ryegrass can accumulate as many as seven leaf primordia (Langer 1972). These species have slower leaf growth rates, shorter leaves, and more tillers, similar to the LYT genotype of tall fescue above. A shortened plastochron leading to more leaf primordia is most common during periods of drought, whereas warm temperatures lengthen the plastochron (Wilhelm and McMaster 1995). Unless the apex is damaged, dies or shifts to reproductive growth, every primordium in sequence has potential to develop a leaf, oldest first, and each leaf has an associated axillary bud.

The practical significance for accumulation of primordia is unknown except the apex has a larger "reserve" of primordia and tiller sites. **Organic reserves** also increase in the stem bases of tall fescue (Horst and Nelson 1979) and orchardgrass (Volaire and Lelievre 1997) during drought to help support regrowth of surviving **primordia** and tillers (Volaire et al. 1998).

Growth of Leaves

Leaf elongation rate (Horst et al. 1978) and mature leaf length (Barre et al. 2015) are major morphologic components of forage grasses that are breeding and management objectives (Edwards and Cooper 1963; Tan et al. 1973; Reeder et al. 1984; Humphreys 2005). Long, erect leaves that form a taller vegetative plant canopy of high-quality forage also facilitates large bite sizes and improves grazing efficiency, especially with rotational stocking (Gastal and Lemaire 2015). Leaf growth rates of grasses are affected by genetics, water stress, N nutrition, shading and harvest or grazing management.

Formation of Leaf Growth Zones

Elongation of the leaf blade begins with activation of basal cells in the oldest primordium formed on the shoot apex (Node 3 in Figure 2.3). Basal cells of each leaf primordium of tall fescue and perennial ryegrass develop a cell division zone (Schnyder et al. 1987). The recently divided, non-elongated cells on the apex are about 20–30 μm long and barely large enough to enclose the nucleus (Figure 2.2). In leaf primordia of tall fescue, the epidermal and mesophyll cells elongate to about 50–60 μm, enough to divide forming two daughter cells. Both daughter cells then divide again to have four cells in a linear column

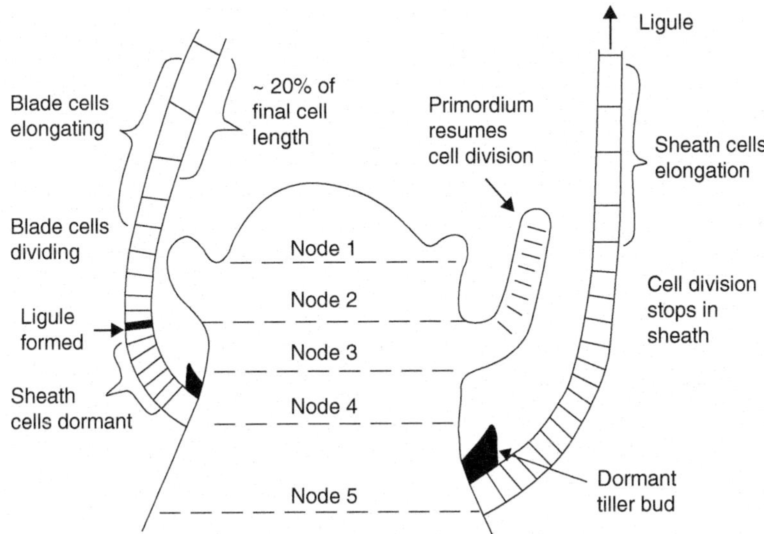

Fig. 2.3. Drawing of the shoot apex showing developing primordia and developing leaves for each phytomer. Axillary buds form last and remain dormant until environmental cues cause tiller growth and emergence. See text for details. Drawing based on several evaluations of tall fescue. *Source:* Adapted from Nelson (2000).

and, these four divide again, to form eight, etc. to form the number of cells needed (MacAdam et al. 1989).

The signal is unknown, but when the cell division zone in tall fescue is about 3.0 mm long, depending on the genotype and growth conditions (Volenec and Nelson 1981, 1983; MacAdam et al. 1989; Rademacher and Nelson 2001), a few epidermal and inner cells differentiate to form a ligule. This precursor to the collar appears as a narrow band of specialized cells across the middle of the cell division zone (node 4 in Figure 2.3). The ligule cells are nearly colorless; they stop dividing or elongating and separate the dividing cells into those above the ligule that will form the blade and those below the ligule that will later form the **sheath**. The process is similar in perennial ryegrass (Schnyder et al. 1990) and annual cereals.

Development of Leaf Length

Genotypes of many grass species differ in leaf growth rates, which serve as a valuable selection criterion (Jones et al. 1979; Barre et al. 2015). Plants with slow leaf growth (~18 mm d^{-1}) had fewer cells in the cell division zone than genotypes with rapid leaf growth (~32 mm d^{-1}) (Figure 2.4). Final lengths of epidermal cells of the **blade** were about 20% shorter (~100 μm) for slow leaf growth genotype compared with fast leaf growth genotype (120 μm). This is consistent with many studies indicating cell production rate (cells d^{-1}) is more important than final cell size in determining leaf growth rates. It is unknown what determines final length of epidermal cells, but some evidence suggests it is regulated by a peroxidase reaction (MacAdam et al. 1992).

The blade angle with the sheath also depends on light; if the exposed ligule is shaded the **blade** will retain a more vertical orientation. If exposed to high light and little competition, some cells of the ligule will elongate to subtend the blade more horizontally which serves to capture more light and increase competition by shading smaller neighboring plants. This response is associated with **phenotypic plasticity**, i.e. the ability of an organ to change its shape in response to environmental cues to be more competitive.

Cell elongation of the blade takes place in the region from about 5 to about 30 mm above the apex for tall fescue and many other cool-season perennial grasses (Figure 2.4), but the elongation zone is shortened under N, drought or heat stress due to fewer cells. Cells located just above the ligule continue to divide and elongate to push the blade upward until the tip emerges into light above the sheath of the previous leaf (Begg and Wright 1962). At that time, cells below the ligule resume cell division and elongation to extend the leaf sheath and push its blade though the whorl of older leaves until the collar reaches light just above the sheath of the previous leaf.

Sheath growth does not stop immediately since cell elongation continues until they reach their final length leaving the collar slightly above the previous one. Thus, length of sequential leaf blades increases in seedlings and during regrowth since the effective whorl length increases

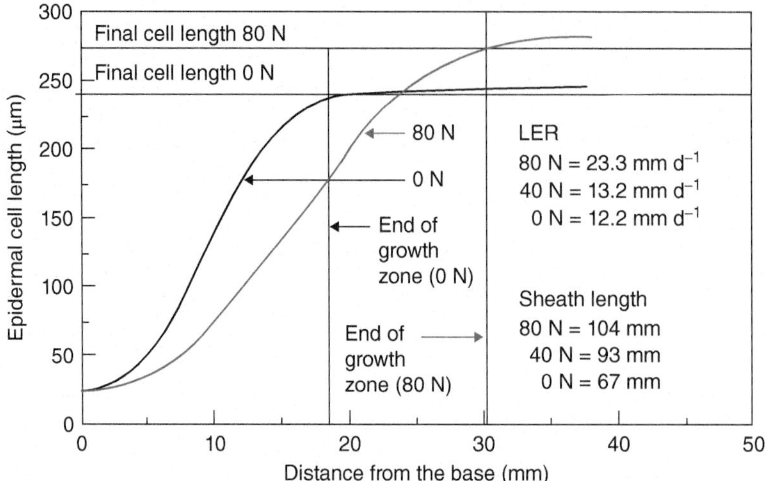

Fig. 2.4. Epidermal cell lengths in leaf tissue above the connection to the shoot apex when N rates equivalent to 0 and 80 kg ha^{-1} are applied. N increased cell production in the lower 2–4 mm and lengthened the cell elongation zone by about 50% due to more cells elongating to their final cell length that was about 12% greater with high N. The major factor was due to greater cell production (Nelson 2000). *Source:* Adapted from Volenec and Nelson (1983), MacAdam et al. (1989), and Gastal and Nelson (1994).

for each subsequent leaf. Leaves develop in a hierarchical sequence according to age. Sheath length is an important regulator of final size of the leaf blade (Casey et al. 1999). Due to stiffness and low quality, sheath height also deters grazing by most ruminants (Gastal and Lemaire 2015).

Development of Leaf Width

Due to the larger shoot apex there are more cells in the circumference of the leaf primordium in plants with higher leaf growth rates (Rademacher and Nelson 2001). In addition, the cell division zone shows a further increase in leaf width (Figure 2.5) due to lateral cell divisions at the base of the blade to form more cell columns plus some lateral expansion of the elongating cells. The widening and thickening of leaf blades are also caused by secondary cell wall deposition in expanding fiber cells around the parallel veins (MacAdam and Nelson 2002). This growth process expands cross-sectional dimensions of the veins during their growth in the whorl and contributes to forming the ridges on the upper surface of the leaf blade.

The increase in leaf width above the shoot apex also causes the blade edges to overlap in the sheath to later display rolled leaf blades such as kentucky bluegrass, tall fescue and smooth bromegrass, or the whole tiller becomes somewhat flattened with elongation of folded leaves like orchardgrass. Regardless, when cells of the leaf blade are actively elongating, the parallel nature of the edges and veins is clearly evident throughout most of the leaf blade.

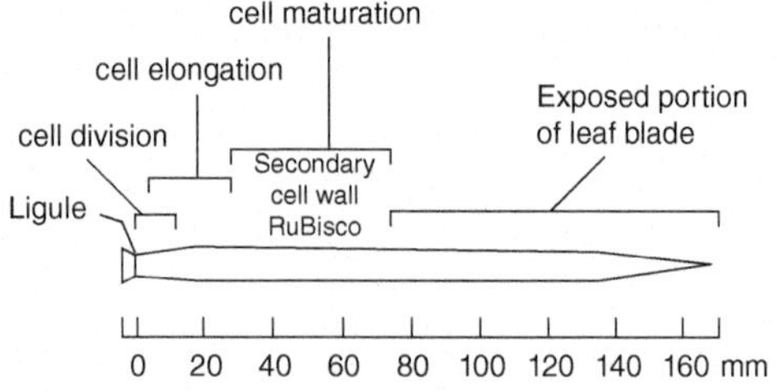

Fig. 2.5. Schematic drawing of an elongating grass leaf blade showing the ligule and sequential zones of cell division, elongation, and maturation before the cells are exposed above the whorl of older leaves. Some cells divide laterally in the division and early elongations zones to add more cell rows and widen the leaf blade. *Source:* Adapted from Volenec and Nelson (1983), Schnyder et al. (1987), MacAdam et al. (1989), and Gastal and Nelson (1994).

Rates of leaf area expansion (mm^2 of leaf area per day) over many experiments are correlated with both leaf width (r = 0.45) and leaf elongation rate (r = 0.75). Based on several experiments, leaf elongation rate has 1.6–1.8 times more effect than leaf width on area expansion rate of tall fescue leaves (Nelson 2000). Very similar relationships of width and elongation rate were reported for wheat (Gauthier et al. 2018) and a wide range of forage grasses (Barre et al. 2015).

Blade Volume and Specific Leaf Weight

During active cell division at the leaf base, the mass per unit area (**specific leaf weight**, SLW) is high (Figure 2.6) due to cells consisting mainly of a cell wall, a dense nucleus, no vacuole and no air spaces (MacAdam and Nelson 1987, 2002). The SLW on a dry weight basis decreases rapidly during cell elongation since elongating cells add solutes and mainly water to greatly expand the length and volume of the cells. When cell elongation stops and air spaces have formed, SLW of the leaf blade increases as veins enlarge and secondary cell wall and **chloroplast** material are deposited before the leaf blade is visible above the previous sheath (Figure 2.6).

Total leaf volume in tall fescue and other grass leaves is determined by the surface area and thickness of the leaf, and depends mainly on the rate and duration of the longitudinal expansion and associated thickening of the leaf. Final epidermal cell lengths are similar for genotypes of tall fescue that differ in leaf elongation rate. While epidermal and vein cells are elongating, adjacent mesophyll cells alongside continue to divide about every 12–13 hours until the elongating epidermal cells are about 50% of their final length. Then, the closely packed mesophyll cells stop dividing and gradually separate to form air spaces during the latter half of epidermal cell elongation (MacAdam et al. 1989; Rademacher and Nelson 2001). Air spaces reduce the SLW of the blade.

Ridges on the upper surface, associated with major and minor veins, are developed prior to exposure, probably by secondary thickening of the fiber cells associated with the **vascular tissues** in the veins (MacAdam and Nelson 2002). The prominent parallel ridges on the upper surface of tall fescue leaves allow the leaf blade to be rolled in the whorl with the lower epidermis on the outside. The veins are constructed like girders to give the blade strength to display the leaf blade.

In a mature leaf of C_3 grasses, mesophyll cells contribute most to the cross-sectional area (40–60%), followed by epidermal cells (20–30%), intercellular air space (10–30%), **vascular tissue** (5–15%), and fiber tissue (1–5%) (Cohen et al. 1982; Allard et al. 1991a; Garnier and Laurent 1994). However, the proportion of these tissues varies greatly with environment, grass species and developmental stage of the plant. Protein is mainly concentrated in the mesophyll cells where photosynthesis occurs and the thin-walled cells are easily digestible in the rumen after microorganisms work past the less digestible structural tissues.

Early Effects on Forage Quality

Components of forage quality that are major contributors to mass of the leaf are deposited during leaf growth within the previous sheath (Figure 2.7). **Neutral detergent fiber (NDF)**, the residue after extracting soluble **sugars**, nitrogen compounds and lipids (Van Soest et al. 1991), is high in the cell division (0–4 mm) zone and increased in the cell elongation zone (3–25 mm). **Cellulose** and **hemicellulose** (the difference between NDF and **acid detergent fiber (ADF)**, are major carbohydrates of cell walls as they thicken. **Lignin** that binds the cell wall components and is hard to digest increases gradually through the growth process. In general, the two types of fiber, ADF and NDF, are general estimates of energy and intake, respectively.

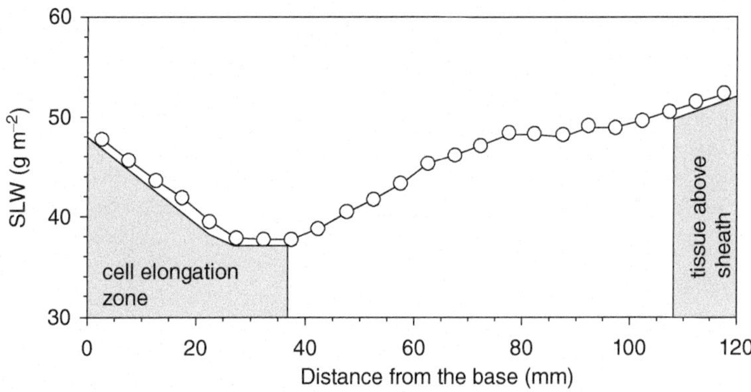

Fig. 2.6. The specific leaf weight (SLW) of tall fescue is high in the cell division zone at the base and decreases as cells with only primary cell walls absorb water and elongate. Deposition of secondary cell wall material increases the specific leaf weight within and beyond the whorl of sheath. Data based on several experiments. *Source:* Adapted from MacAdam and Nelson (1987), MacAdam and Nelson (2002), and C.J. Nelson (unpublished data).

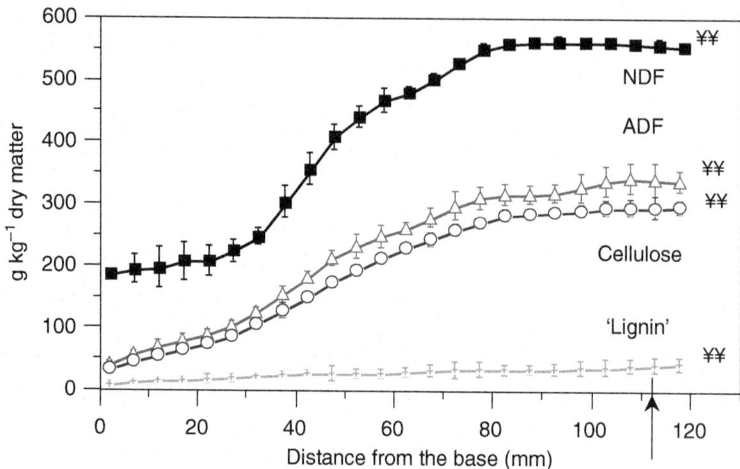

Fig. 2.7. Pattern of deposition of cell wall components associated with forage quality during leaf development of tall fescue within the whorl of sheaths (C.J. Nelson unpublished). Arrow indicates the height of the leaf sheaths; the YY symbol indicates the value of the component in mid-length of the mature leaf that is exposed. Note, except for lignin which doubles, most of the quality characters of the mature leaf are developed while within the previous sheath.

Deposition of secondary cell wall material, including some lignin, increases during drought, which reduces forage quality (MacAdam and Nelson 1987). Thus, many cell features that affect forage quality are deposited in the growing leaf before it emerges (MacAdam and Nelson 2002). Once emerged, continued changes in these constituents will further decrease forage quality as the leaf blade and sheath age. To date, there are few studies on the effects of leaf growth rates and leaf size on forage quality, but there is a vast array of research related to advantages of a high leaf:stem ratio (Chapter 39).

Compared with cool-season grasses (C_3 photosynthesis), there are few data on leaf growth characteristics of warm-season grasses (C_4 photosynthesis). Currently, there are research groups actively working on this topic in many countries that are evaluating different management approaches for warm-season species based on plant morphology (da Silva et al. 2015). Already, there are some data on anatomic structures of developed leaf blades. Comparisons are often based on the proportions of tissues that are rapidly and nearly 100% degraded (e.g., mesophyll cells) relative to those with thicker or lignified walls that are slowly degraded (e.g., vascular bundles and epidermal cells). In general, compared to C_3 species, leaves of C_4 species have a higher proportion of slowly degradable cell walls and a lower proportion of rapidly degradable mesophyll cells and cell solubles (Akin et al. 1973; Van Soest et al. 1991).

In addition to content, rate of digestion of cell walls of C_3 grasses like tall fescue, smooth bromegrass, orchardgrass, and kentucky bluegrass is faster, perhaps due to less lignification (Buxton 1990), than that of C_4 grasses like bahiagrass, bermudagrass, dallisgrass, and pangolagrass (Akin and Burdick 1975). Based on cross-sectional analyses of leaf blades for the above C_3 grasses, 60% of the tissue was rapidly degradable, 26% was slowly degradable, and 14% was non-degradable. Averages for the C_4 grasses above were 39%, 50%, and 11%, respectively. Slower degradation increases the time the material spends in the rumen, reducing the rate of passage and subsequent forage intake, which further reduces animal performance.

Of importance is that breeding goals to reduce lignin in leaves, and especially in stems, can be successful in improving forage quality (see Chapters 30 and 31), but may be correlated with negative effects. Casler et al. (2002) compared plant survival of smooth bromegrass, orchardgrass, and switchgrass after selection for low lignin in herbage (mix of stems and leaves). As expected, lignin concentration decreased with selection and ***in vitro* dry matter digestibility** increased. Survival of smooth bromegrass plants in the field for four years was affected very little, perhaps due to its inherent stress tolerance. In contrast, orchardgrass survival declined during the first year and then stabilized, and that of switchgrass gradually declined over the four-year period.

The causative relationship between herbage lignin and plant survival, mainly overwinter, was not determined. But if the low-lignin plants in the heterogeneous population do not survive, the plant population will gradually revert to the high-lignin survivors.

Other Leaf Blade Features

Serrations along the leaf blade are formed after emergence of cellular outgrowths from edges of epidermal cells. Later they harden and become sharp. Some grass leaves form **trichomes**, hardened cellular appendages on the upper surface, that can be sharp. Some trichomes produce exudates that cause pain or have odors, but little is known about these on grass leaves. The functions of these structures are not fully known, but are thought to reduce forage preference by ruminants or act as deterrants to herbivory by small animals and insects.

An interesting observation and question is "why do grass leaves of some species like smooth bromegrass and quackgrass have a noticeable constriction near the middle of the blade"? It may look like an W (for Wisconsin?) or an M (for Missouri?) depending on how the blade is viewed. The constriction is formed by the rather rigid cells of the ligule (collar) of the preceding leaf since its sheath surrounds and clasps the new leaf blade while they elongate through the whorl at the same time and rate (see Figure 2.3). The collar of the sheath restricts lateral expansion of the new blade at that position.

Light Effects on Leaf Growth

Light or radiation affects both morphology and anatomy of grass plants by either the amount of light or quality of the light. Shading occurs from dense canopies of both reproductive and vegetative canopies due to less light passing through the canopy to the base of the tillers where leaf growth occurs. Shading in vegetative canopies of perennial ryegrass led to reduced photosynthetic rate of emerging leaves (Woledge 1977). Shaded leaves have lower stomatal density (Woledge 1971), lower protein content and fewer **chloroplasts** (Dean and Leech 1982).

In field experiments, leaf blades of tall fescue under 30% full sun were about 50% longer and had 66% higher area than those grown in full sun (Allard et al. 1991a). This demonstrated phenotypic plasticity by altering leaf dimensions such that leaf blades grown at 30% full sun were 12% thinner and had about 22% lower SLW than those at full sun. There was more air space in the leaves and less root growth in the shaded plants, indicating allocation of resources was altered to maximize development of leaf area.

The light reactions of photosynthesis, CO_2 diffusion and overall rate of photosynthesis were about 20% lower in the shaded plants (Allard et al. 1991b). More detailed analyses showed CO_2 diffusion, light and dark reactions at the **chloroplast** level were all lower in the shade-grown leaves indicating the tight coordination among processes associated with photosynthesis. A similar study with *Panicum maximum*, a warm-season grass, gave similar conclusions that shade increased leaf growth and reduced photosynthesis (Paciullo et al. 2016). In addition, they found N fertilizer further increased leaf growth and biomass under shade, perhaps offsetting the lowered photosynthesis rate.

Sanderson and Nelson (1995) evaluated the effects of gradual changes in shading over time on leaves, i.e., to mimic effects of canopy development, and in reverse, the acclimation after harvest. Relative responses were similar for genotypes of tall fescue with rapid leaf growth (40% faster) and slow leaf growth. As shading increased over time (became more shaded), the length of the cell division and elongation zone for blade growth increased from about 19 mm to about 24 mm. The main response was in the cell division zone, i.e., more cells to increase elongation rate, which is consistent with other shading studies.

The relative Red/Far-red ratio of the spectrum of light decreases due to photosynthesis, mostly the Red component, as it passes through the canopy. The Far-red component reaches the base of the plant where the leaf and tiller meristems are located. Using fiber optics to add supplemental Far-red light to leaf blades or to tiller bases, Skinner and Simmons (1993) found that like shading, adding Far-red light to the exposed tips of the elongating leaf (EL) blade or the tiller base caused the leaf to elongate faster. If a fully elongated leaf was exposed, the elongating leaf did not respond. This verified the elongating blade and its base were the primary reactors to the Far-red stimulus and shading.

Water Stress Effects on Leaf Growth

The primary walls of the dividing and early elongating cells of the growing leaf are thin and flexible. Therefore, the cell is dependent on **turgor** pressure from water influx into the cells for strength and cell expansion until this process is complete and secondary wall formation begins. As discussed in Chapter 6, the **osmotica** in the elongating cells are primarily sugars, **fructan**, K^+ and NO_3^- (Spollen and Nelson 1994; Gastal and Nelson 1994). When cell elongation stops, there is active synthesis of **enzymes** and other proteins needed for photosynthesis and metabolism. Similarly, sugars and fructan, first used passively as osmotica during cell elongation, are used again when elongation stops to synthesize secondary cell wall material, primarily cellulose, and some lignin that hardens the walls (Allard and Nelson 1991). This allows emerging cells to be somewhat rigid when exposed above the previous sheath.

Drought stress reduces leaf elongation rate since there is less uptake of water in the cell elongation zone. When shifted from dark to light the stomata open, **transpiration** begins and the leaf loses water, creating a deficit. Leaf elongation rate that depends on turgor, quickly decreases by about 30% of the dark rate and then remains steady (Volenec and Nelson 1982). When the dark period began 12 hours later, stomata closed, leaf elongation rapidly recovered and was even enhanced for a few minutes until the steady state was resumed. During the light, the water potential and turgor pressure in the leaf growth zone decreased (Durand et al. 1995). The reduced rate of leaf elongation was due about equally to reductions in cell division and cell elongation.

Transpiration of the leaf blade in light reduces **turgor** in the cell elongation zone, and more **osmotica**, especially **sugars** and fructans, are required to maintain turgor pressure to drive cell elongation (Spollen and Nelson 1988, 1994). Drought stress caused the leaf growth zone of tall fescue to gradually shorten by about 50% when cell elongation stopped (Durand et al. 1995). Soluble sugars remained high in the leaf growth zone, and when watered two days later, epidermal cells gradually resumed elongation to near full recovery after two more days. This mechanism allows plants to endure short-term drought. Tall fescue infected by an endophyte fungus (Chapter 35) is more tolerant of long-term drought than endophyte-free tall fescue (West et al. 1993), due partly to its ability to accumulate more carbohydrate in the stem-base tissue (Nagabhyru et al. 2013).

Among warm-season grasses, salt stress reduced leaf elongation of rhodesgrass, with near equal effects on cell division and cell elongation (Taleisnik et al. 2009). Growth forms of grasses can alter the response to drought, seemingly with conservation of leaf growth. For example, drought effects on kleingrass, an upright growing warm-season species, reduced stem elongation and **tillering** more than leaf area (Bade et al. 1985). In the same study, bermudagrass, a low-growing species had greater reductions in stolon growth than leaf area.

Nitrogen Effects on Leaf Growth

After cell division pushes cells to about 3 mm from the base of the grass leaf, the epidermal cells of the blade begin to elongate until they reach a somewhat-fixed final length (Figure 2.4). Final lengths of epidermal cells of genotypes with high N were about 10% longer than with low N, whereas leaf elongation rate was 90% higher. Therefore, the main effect of N on leaf elongation is due to increased cell division to provide more cells in the cell elongation zone. Actual elongation rates of individual cells are similar, but more cells are elongating causing the cell elongation zone to be longer (Volenec and Nelson 1981).

Based on several experiments, high N increased rate of leaf elongation by about 80% in tall fescue genotypes selected for either slow or rapid leaf growth rate. Again, this was mainly due to a similar increase in cell production (Volenec and Nelson 1983; MacAdam et al. 1989; Schnyder et al. 2000; Rademacher and Nelson 2001). Fertilizing with N also increased the number of mesophyll cells per unit leaf area (MacAdam et al. 1989) which would be a partial explanation for higher leaf photosynthesis when N is applied.

Treatment with N increased the N content of the dividing cells at the leaf base to as high as 7% of dry weight (Gastal and Nelson 1994), which equates to about 45% protein and nuclear material in the dividing cells (Figure 2.8). Cell elongation by accumulation of water dilutes the overall N content while accumulating NO_3^-, K^+, and soluble carbohydrates. The latter can make up to 40% of the dry weight of the cell elongation zone (Spollen and Nelson 1988). These **osmotica** increase

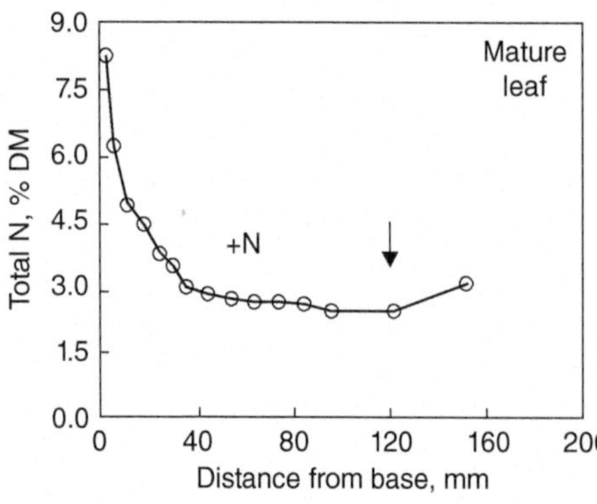

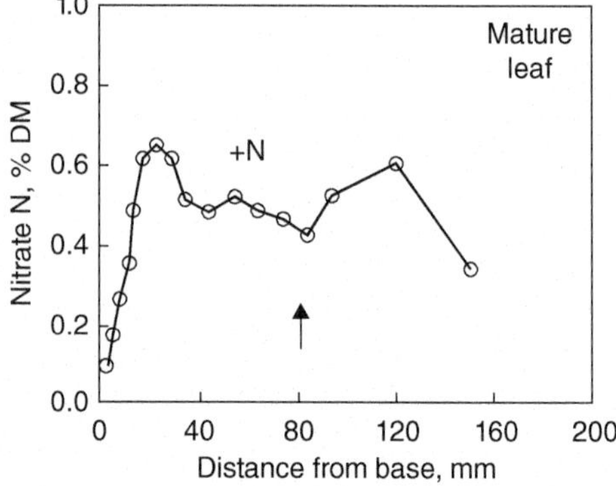

Fig. 2.8. Concentration of total N is very high in the base of the leaf growth zone where cells are actively dividing and then decreases as the cells elongate. There is rapid accumulation of NO_3^- in the cell elongation zone where it serves as osmoticum and is finally reduced in the exposed leaf. Vertical arrows indicate the sheath height. *Source:* Adapted from Gastal and Nelson (1994).

water uptake and generate the turgor pressure needed to further drive elongation of the thin–walled cells. Later, the proteins used during cell division are recycled for synthesis of ribulose bis-phosphate carboxylase and other **enzymes** needed for photosynthesis (Gastal and Nelson 1994; Xu et al. 1996). Likewise, the carbohydrates, used temporally for turgor, are used for synthesis of secondary cell walls and lignification of the elongated cells (Allard and Nelson 1991).

Number of Expanding Leaves per Tiller

Most early information on leaf growth of perennial grasses came from experiments with perennial ryegrass and tall fescue. It is now known that many other cool-season forage and pasture grasses follow similar processes (Matthew 2017) as does vegetative growth of small grains like wheat (Malinowski et al. 2018; Touati et al. 2018). For example, nearly all cool-season grasses have two leaves developing at the same time, but they are at different stages and two or three fully developed leaves (Figure 2.9). Older plants of many cool-season grasses have three green leaves; as the fourth leaf is developing, the lower leaf is senescing (Fulkerson and Donaghy 2001).

The shoot apex and leaf primordia on vegetative tillers are near soil level (Figure 2.9). The first elongating leaf is partially emerged above the whorl, the second is partly formed and still in the whorl, the two previous leaves are fully developed and functional, while the oldest leaf is partially senesced. This gives a pattern of three live-leaf equivalents per tiller, which repeats as more leaves develop. In some conditions, e.g., cool temperatures in autumn, perennial ryegrass can accumulate four green leaves (Poff et al. 2011). Observations with tall fescue indicate the rate of die back from the tip of the senescing leaf is nearly the same as elongation rate of a new leaf (Davidson and Nelson 1980).

The mechanism regulating senescence of the older leaf is unknown. Several studies have used N fertilizer or other strategies to delay senescence with little or no success (e.g. Wilhelm and Nelson 1978; Davidson and Nelson 1980). N-balance studies suggest the third leaf, which is oldest and lowest in the canopy, senesces to release amino acids and other nitrogenous compounds to support growth of the emerging leaf to optimize N use. The emerging leaf will be less shaded and much younger with higher photosynthetic potential (Wilhelm and Nelson 1978).

Many studies have shown fertilization of grasses with N increases yield through enhanced leaf growth rate by higher cell production. Another response is a longer phyllochron, which allows production of longer lengths of leaf blades while the older leaf still senesces (Lemaire et al. 2009). The result is the three, or possibly four leaves per tiller are much larger and there would be more total-N required from senescing leaves circulating within the tiller to support continued growth of the new larger blade. Thus, yield is higher because each of the live leaves is larger.

Overall, the longer phyllochron reduces tiller bud formation and the taller canopy will increase shading of the lower plant to reduce tiller development which further alters the **sinks** for recycled N. Thus, perennial grasses seem programed to reuse forms of reduced-N

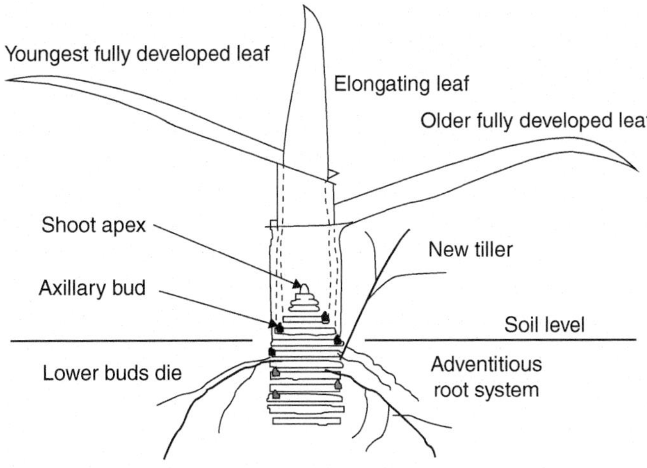

Fig. 2.9. Drawing of a grass tiller showing the elongating leaf and two developed leaves. The next oldest leaf is beginning to senesce and not shown. The tiller shows the shoot apex, live axillary buds, a young tiller from an axillary bud, and adventitious roots that developed from the bases of non-elongated internodes. Older axillary buds at the base are dying after missing their window to form a tiller. *Source:* Figure developed from Skinner and Nelson (1992, 1994) and Matthew et al. (2016).

from the older, aging leaf in preference to costly uptake and reduction of soil N to make more N available in the plant. This strategy is common in a range of **herbaceous** plants (Field 1983).

Warm-season C_4 grasses, like cool-season grasses, always have at least two leaves elongating, but when growth conditions include favorable temperatures, adequate nutrition, and water, some C_4 grasses like maize may have three, four or more leaves elongating at one time in the vegetative canopy that greatly increase rate of leaf area development. The C_4 species have a lower N requirement for leaf growth and have more energy from efficient C_4 photosynthesis to reduce NO_3^- in the leaves. Some C_4 grasses also have a longer leaf lifespan (Lemaire et al. 2009). The synchrony for development and growth of leaf primordia and how they are regulated for leaf growth in C_4 perennial forage grasses is not fully understood.

Formation and Types of Tillers

An accumulation of successive phytomers, each differentiated from a single-shoot apex, defines a tiller (Briske and Derner 1998). Juvenile or daughter tillers initiated from axillary buds of previous tiller generations have the potential to become a free-living structure. Regulation of tiller production by the shoot apex in cool-season grasses has been described for tall fescue (Skinner and Nelson 1995) and perennial ryegrass (Matthew et al. 1998).

Each axillary bud has a shoot apex that, similar to the main shoot, forms its phytomers in sequence. The shoot apex of the tiller, although smaller, adds cells at the terminal end to maintain its structure, while it activates new meristematic cells to initiate leaf primordia, nodes, internodes and more axillary buds. Roots are initiated later from the basal part of the internodes of each tiller and appear to have turnover similar to the shoots above ground (Matthew et al. 2016).

Intravaginal tillers grow upright within the sheath at the same node to form a bunch-type growth habit with open spaces between plants like orchardgrass or big bluestem (Figure 2.10). Conversely, many grass species like kentucky bluegrass, smooth bromegrass and tall fescue produce **extravaginal** tillers that grow laterally from the apex to penetrate the surrounding sheaths to reach light and then usually angle upward. These species tend to gradually spread laterally to fill in open areas and have a sod-type growth habit. The lateral spread of leaf area covers a higher percentage of ground area increasing overall competitiveness for light and nutrients.

Some tall grasses with short rhizomes and extravaginal tillers (e.g., tall fescue and big bluestem) form loose bunches. Grasses that are relatively short, but with long rhizomes like kentucky bluegrass or stolons like bermudagrass tolerate frequent and close cutting or grazing better than do bunchgrasses. Bunchgrasses may be more compatible with legumes and provide better wildlife habitat.

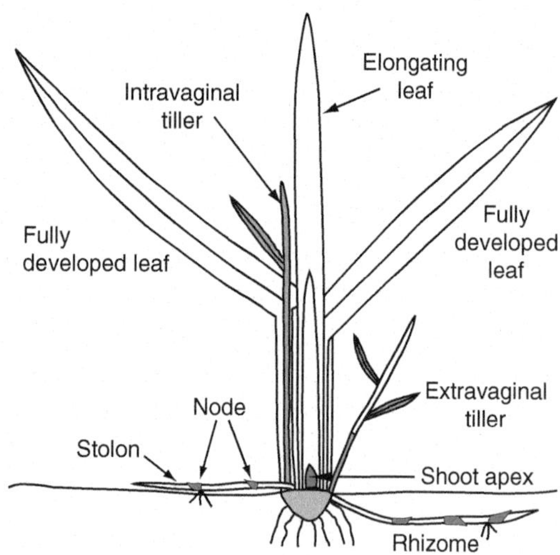

Fig. 2.10. Drawing to illustrate positions on an idealized grass plant where axillary buds at the shoot apex can produce new phytomers for intravaginal tillers or extravaginal tillers, stolons and rhizomes. Roots are produced adventitiously from intercalary meristems for non-elongated internodes of the main plant, stolons, and rhizomes. Later, in a similar way, each new tiller will produce its own roots from lower nodes.

Axillary Bud Development to a New Tiller

Most information on axillary bud development comes from experiments with tall fescue, perennial ryegrass, some range species and several annual cereals. However, the mechanisms tend to be somewhat similar for most grasses including upright warm-season grasses. Signals associated with formation of the tiller are related directly to development of the phytomer (Figure 2.1). It is unclear when the axillary bud is initiated in the axil of the leaf primordia, but it is visible in tissue sections at each node by the time the associated leaf primordium is actively elongating (Figure 2.3). If, and when the axillary bud at the shoot apex is activated, and grows into a new tiller is determined later (Skinner and Nelson 1994).

Activation of tillers was first associated with **apical dominance** and auxin (Jewiss 1972; Yeh et al. 1976) based largely on noting the reduced tillering when the stem was elongating. Later, Murphy and Briske (1992) showed the auxin reaction likely was controlled by a cytokinin that is activated by light, especially if the relative proportions of Red to Far-red radiation are high. The young tiller itself, within the whorl of leaf sheaths, is the direct receiver of the light signal (Skinner and Simmons 1993; Casal 2013).

Red light in the solar spectrum is actively absorbed in the leaf canopy by chlorophyll and used in photosynthesis, whereas the Far-red light is transmitted with little loss as it passes through the canopy and lower sheaths to reach the axillary bud near soil level. Under a full canopy, the ratio of Red:Far-red is low and the bud remains dormant (Casal et al. 1986). Conversely, if plants were recently grazed or cut, the ratio will be high (large amounts of red light) at ground level. Therefore, the bud will sense little shade, break dormancy, reactivate cell division and allow growth of the tiller leaf blades and sheaths. The new tiller will continue growth if light, water and nutrients are adequate.

Axillary Bud Release

Axillary buds in many cool-season (Skinner and Nelson 1994) and warm-season grasses (Hendrickson and Briske 1997) remain dormant until released by the Red:Far-red signal. The release to initiate tiller growth from an axillary bud occurs in a narrow time-window; if the window is missed, only rarely will the axillary bud at that site develop into a tiller. Based on these data and observations, Skinner and Nelson (1994) proposed "apical coordination" rather than "apical dominance" best describes the relationship between growth of the leaf and development of its axillary bud.

In general, leaf growth of successive phytomers is coordinated in an overlapping and integrated way among tissues of successive phytomers (Figure 2.3). The axillary bud on Node 2 begins elongation very near the same time that cell division of the sheath ends on Node 2. This event coincides with time of ligule development on the young leaf attached at Node 3 and beginning of cell elongation of the blade at Node 4. The window for the axillary bud to begin growth is thus "open" at the time when the preceding leaf is forming its ligule. Interestingly, the leaf and axillary bud at that node are initiated 180° from the axillary bud that has the window to grow, but the reason for that association is unknown.

Current evidence suggests the signal is based on high red light reaching the bud, but if the bud area receives low red light, indicating shading during its window, that bud is skipped. Very rarely, is a skipped bud reactivated. Instead, when the sequence of phytomer events occur again a few days later and 180° away, the next bud on the apex will get its window of opportunity.

Filling of potential tiller sites ranges from near 0.0 (no tiller formed) to 1.0 if every axillary bud forms a tiller, but that is almost impossible (Skinner and Nelson 1992). As plants develop, the site usage gradually decreases as the canopy closes, especially with grasses with high leaf growth rates that close the canopy faster (Zarrough et al. 1984).

For example, micro-swards of vegetative tillers of tall fescue genotypes, established in a greenhouse, were moved to the field in Missouri in mid-April. Nearly all new tillers survived during the first 10–12 weeks (Figure 2.11). Density of live tillers for all genotypes was maximum during late spring, but this was gradually offset by tiller death that was highest during summer, probably due to temperature and water stress, especially for the slow leaf growth genotype, which has a smaller root system. Live tiller density increased again in fall.

Overall, tiller density was highest for the genotype with slow leaf growth (LYT) and lowest for HYT genotype with rapid leaf growth (Zarrough et al. 1983). Regardless, the appearance of new tillers for each genotype tended to balance the death of older ones. Equilibrium density of live tillers was about 30, 35, and 44 tillers dm^{-2} for the HYT, medium yield tiller (MYT), and LYT genotypes, respectively. Tiller death over the season was 19%, 43%, and 49% of the total tillers produced by the HYT, MYT, and LYT genotypes, respectively. Several studies indicate decreased tillering is due, in order, to longer phyllochrons (fewer potential sites), reduced survival and reduced site usage. Tiller death is important since the investment in their development is not realized.

Tiller Production and Survival

Perennial grass plants are considered perennial because they are able to produce short-lived tillers in a sequential manner that allows the plant to persist for several years. Thus, perennation of individual grass plants and sustainable productivity of grasslands depend on regular tiller replacement from axillary buds, as described above, and survival of some of the tillers.

Size and Density of Vegetative Tillers

In nearly all studies, there is a negative relationship between plant size and tiller density, for which Harper (1977) proposed a negative slope of 3/2 based on several species. This has been supported with most forage species with a few exceptions (Sackville-Hamilton et al. 1995). For example, genotypes of tall fescue selected from a broad-based breeding population (Figure 2.12, closed squares) show the typical negative trade-off during vegetative growth in the field in Missouri. Further, all six genotypes responded in a similar way to the split applications of N fertilizer. Combined data for the genotypes showed weight per tiller increased by 43% and tiller density by 59% in response to the first increment of N, 90 kg ha^{-1}. As N rates increased, the maximum tiller density was approached and relative response to higher N rates shifted nearly exclusively to longer leaves and increased tiller weight.

In another experiment, four generations of **recurrent selection** in tall fescue for high- and low-weight per tiller gave different genetic populations to further evaluate the relationship with tiller density. Vegetative plants in seeded plots in Missouri received split applications

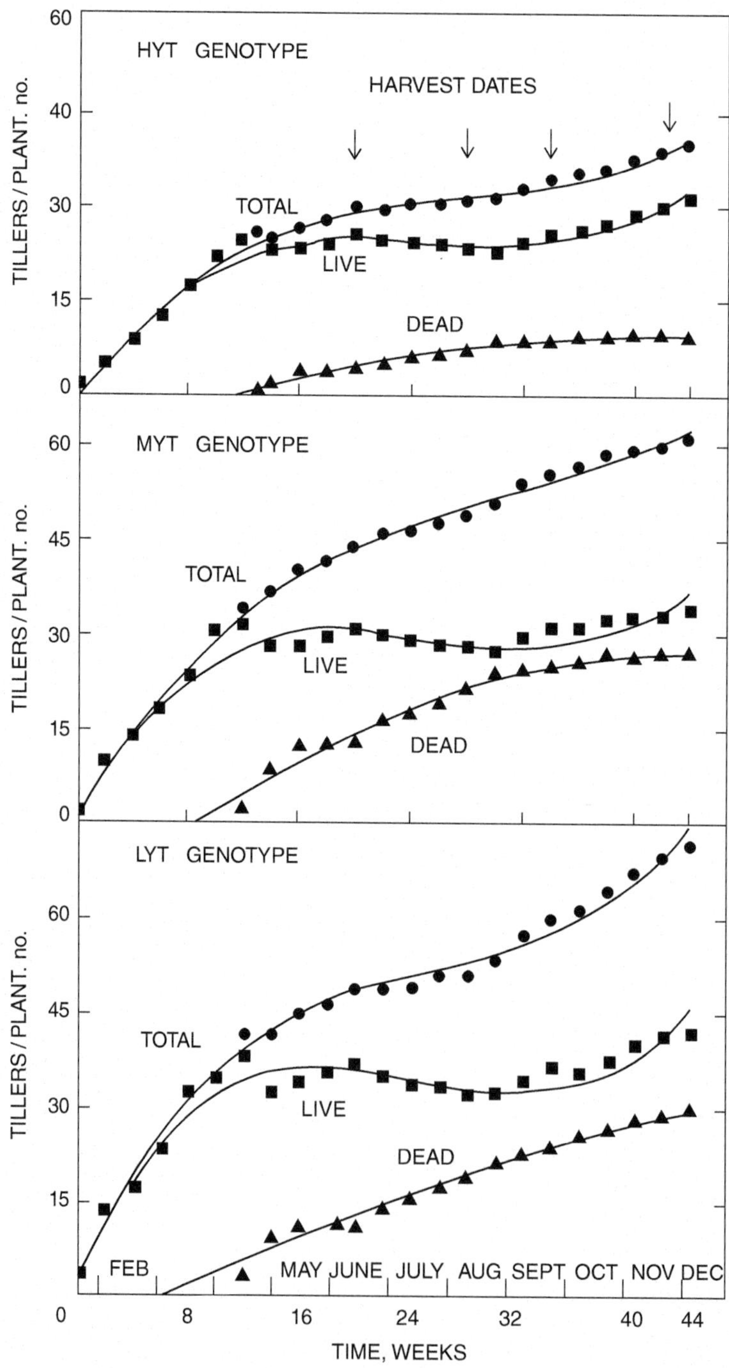

Fig. 2.11. Tiller populations for vegetative growth of tall fescue plants in the field in Missouri. Genotypes had low (LYT), medium (MYT), and high (HYT) leaf elongation rates. Note the overproduction of tillers that died. Live tiller density for all genotypes was high in early spring, reduced in summer and then increased again in fall. *Source:* From Zarrough et al. (1983).

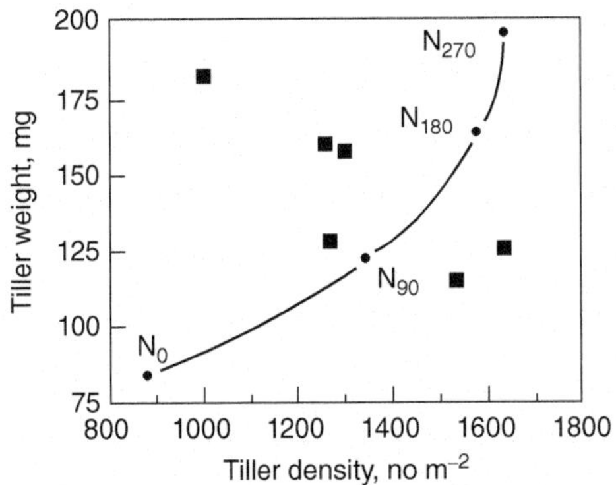

Fig. 2.12. Relationship between weight and density of vegetative tillers in a field study of tall fescue. Squares indicate the negative relationship for six genotypes selected for a range of tiller weights. The line is the mean response of the six genotypes to rates of N in kg ha^{-1}. *Source:* Adapted from Nelson and Zarrough (1981).

of N. Genetic change in weight per tiller in the low direction was more rapid than in the high direction, but the indirect effect of selection on tiller density was clearly evident (Figure 2.13). Multiplying tiller weight by tiller density indicates the vegetative yield of the H_2 population (650 mg m^{-2}) was greater than the C_0 (522 mg m^{-2}) and L_4 (418 mg m^{-2}) populations indicating the need for emphasis on tiller weight and leaf growth in breeding and management.

There is a similar negative genetic relationship in perennial ryegrass and most other cool-season grasses between tillering and leaf length that is related to leaf elongation rate and weight per tiller (Barre et al. 2015). Conversely, there are data that show a slow rate of leaf growth combined with high rate of tillering, while being lower yielding, may improve persistence with continuous stocking. However, most grass-breeding programs are focused on yield in early generations, often taken at early heading stages, that is based on spaced plants that can freely tiller.

Longevity of Tillers

An individual tiller of most perennial grasses possesses a maximum longevity of two years, but many survive only for a few days or through the growing season in which they are produced (Briske and Derner 1998). Tiller longevity is related to when it appears, if it survives and when it dies, due to competition or flowering. Tillers that are older going into fall are most likely to be vernalized and then flower and die the following spring. Tillers developing later in the year may not be vernalized. If not, they will likely be vernalized the following year, produce an inflorescence and die.

A newly emerged tiller depends on water, nutrient and carbohydrate resources from the mother tiller until it produces its own leaf area and root system. In high radiation, the young tiller will use photosynthesis for food and initiate root growth from intercalary meristems of the lower internodes to gain water and nutrients (Matthew et al. 2001). Roots of most species will not grow into dry soil, in which case, the tiller gradually dies.

In some C_4 bunchgrasses like little bluestem that are adapted to dry conditions, plants can accumulate as many as three linked tiller generations that maintain vascular connections to the mother tiller for food and water before the oldest tiller dies and decomposes (Briske and Derner 1998). These patterns of physiologic integration with the mother tiller greatly enhance establishment success and survival of juvenile tillers.

Production and Growth of Stolons and Rhizomes

In addition to tillers, some grass plants produce stolons and rhizomes to increase lateral expansion of plant size and enhance its competitiveness. In general, stolon formation is a primary mechanism of foraging for light whereas rhizome formation is a primary mechanism for storage of organic reserves, protection during winter stress and foraging for a more favorable habitat for water and nutrients (Briske 2007).

Rhizomes and stolons originate from axillary buds similar to extravaginal tillers, except they grow laterally instead of mainly growing upward (Figure 2.11). Both have identifiable nodes and internodes. Stolons can initiate roots and produce an axillary bud at each node from which new shoots or stolon branches can arise. As with rhizomes, grasses with stolons can propagate vegetatively and form a sod. Bermudagrass is a major grass species that produces both rhizomes and stolons. Other warm-season grasses with stolons are rhodesgrass, st. augustinegrass, zoysia, buffalograss, and centipedegrass. Stoloniferous grasses are often considered invasive.

Rhizomatous grasses include johnsongrass, red top, creeping red fescue, kentucky bluegrass, quackgrass, sideoats grama, smooth bromegrass, reed canarygrass, and western wheatgrass. These species differ in their rhizome vigor, which can lead to various degrees of invasiveness. Other grasses such as big and little bluestem, indiangrass, switchgrass and tall fescue have short rhizomes, which give individual plants a loose, bunchlike appearance. Grasses with intervaginal tillers and without rhizomes (e.g., orchardgrass and timothy) have little lateral spread and form tight bunches, which allow legumes and other species to occupy the open spaces.

As with upright tillers, stolons of grasses such as bermudagrass are sensitive to ratios of Red:Far-red light for initiating growth of the axillary bud, but they are

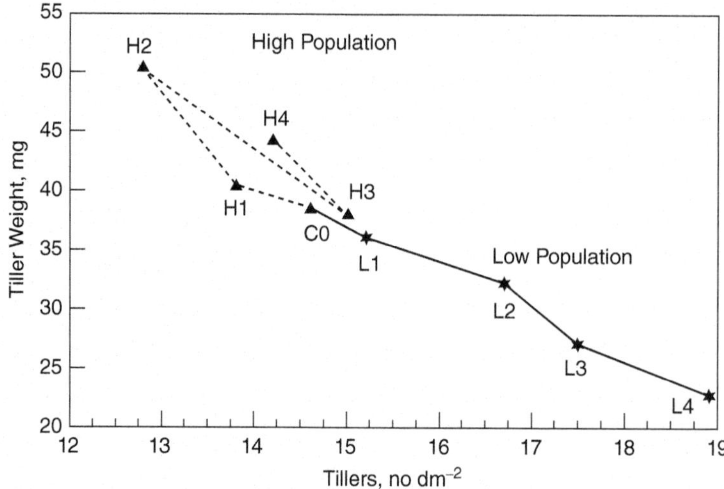

Fig. 2.13. Effect of four cycles of recurrent selection for high and low leaf area expansion rate in a base population of tall fescue (C0). Data were taken during vegetative growth stages in the field. Parental plants from each generation were crossed and planted in the field for the next generation, then all generations were tested in the field. For unknown reasons the third generation of selection in the high direction reverted, but the overall inverse relationship between tiller weight and tiller density was retained. *Source:* Adapted from Reeder et al. (1984) and Nelson (2000).

not sensitive to **phototropism** (grow upward toward light) or **gravitropism** (grow toward gravity) as they grow laterally. The lateral spread is controlled somewhat by red light since the stolon tip will rarely grow into a shaded area where the Red:Far-red ratio is low. Instead, it will turn toward the area with greatest light or stop its growth and initiate branches from axillary buds to grow toward higher light. Unlike young tillers, stolons have extended internodes that separate the nodes of the phytomers, with an axillary bud behind the modified leaf located at each node. Each node can initiate roots to anchor the stolon and support water and nutrient needs as leaf area on the stolon increases. Axillary buds on the stolon can be activated by red light to form a stolon branch or an upright tiller to expand leaf area.

Like stolons, rhizomes on grasses originate from axillary buds located on the older, lower stem base and grow downward at an angle to below ground level. The rhizome has internodes and nodes that have an incomplete leaf (very short blade) that covers and protects the axillary bud at each node. The bud can break dormancy and grow upright to develop leaves above ground, or grow laterally as a rhizome to continue to spread the plant. It is not clear how rhizome growth is regulated, especially how it remains at a given depth in the soil.

Rhizomes can be several centimeters below soil level making it unlikely that depth and development of new shoots is a direct effect of Red and Far-red light. Light can penetrate sandy soils up to 2 mm, but to lesser depths in clay or loam soil (Woolley and Stoller 1978). Ciani et al. (2005) used sophisticated instruments to evaluate light attenuation of 19 soils and learned most particulate minerals and soil depths of less than 1 mm reduced light by 99%.

Thus, it is more likely the response is to a chemical like ethylene (Briggs 2016). Ethylene, a gaseous growth regulator accumulates in the soil and gradually diffuses out. High ethylene concentrations in the soil increase cell-wall strength to support rhizome growth through dense layers of compacted soil. Whether ethylene concentration is involved indirectly with light responses or other environmental controls of rhizome growth is unknown.

Reproductive Tillers

Forage grasses have two distinct forms of vertical stems; vegetative and reproductive. Both have a shoot apex at the tip, but the reproductive stems also have active stem intercalary meristems. In seedlings and non-reproductive (vegetative) tillers, stems are very short, consisting of nodes and basal, non-elongated internodes (Figure 2.9). This adaptive mechanism keeps the shoot apex near ground level and enclosed within the whorl of older leaf sheaths, where the apex usually escapes removal by grazing or cutting. The vegetative shoot apex continues to initiate leaves and axillary buds until there is a stimulus to flower.

Transition from vegetative tillers to flower tillers is usually a response to changes in daylength and/or temperature. The signal causes the shoot apex to stop producing leaf primordia and transitions to develop the reproductive structure. As the apex changes, dormant internodes below the apex begin stem elongation by dividing and elongating of cells in their intercalary meristems. The growth from below pushes the shoot apex upward from near the soil surface while it differentiates into the **inflorescence**.

Mowing or grazing before the transition allows the shoot apex to remain close to soil level and avoid removal or physical damage. In contrast, if the apex is elevated it can be removed by grazing or cutting, causing that tiller to die. In that case, plant growth must be maintained by new tillers.

Leaves on Reproductive Tillers

In contrast with vegetative tillers, more than three leaves on reproductive tillers are frequent and the blades are progressively shorter. After several leaves of increasing size have formed on the vegetative tiller, the shoot apex can respond to environmental cues to differentiate into the inflorescence.

Intercalary meristems on lower internodes of the reproductive tiller are activated and the developing inflorescence is pushed upward within the whorl. This shortens the effective distance and time during which the final leaves are enclosed. Therefore, the longest leaf is the last to develop prior to the apex change. This is easily noted in mature corn plants in which the ear leaf is the largest and subsequent leaves are gradually smaller. That leaf arrangement allows light to penetrate deeper into the canopy to increase photosynthesis to support growth of the grain, or seed in the case of forage grasses.

Stems on Reproductive Tillers

Elongated stems (**culms**) of a flowering grass plant are divided into distinct nodes and internodes (Figure 2.1). The last leaf produced (directly below the inflorescence) is the **flag leaf**; the topmost internode that supports the inflorescence is the **peduncle**. Nodes or joints of grass stems are always solid and very high in fiber that discourages grazing and is low-quality feed. The internodes of most C_3 forage grasses are hollow, but stems of C_4 grasses are pithy, usually solid and rather low in nutritional value. All of the C_4 grasses occur in the following tribes: Panicoideae, Arundinoideae sensu stricto, Chlorideae, Centothecoideae, Aristidoideae, and Danthonioideae (Barkworth et al. 2003). Solid stems are very common in species in each of these tribes. However, some species in these tribes have hollow stems.

Cells of the stem intercalary meristem, a zone of cell division and elongation at the base of each internode, retains **meristematic** potential the longest of any stem tissue. The activated intercalary meristems between nodes elongate the stem after a flowering stimulus. Lodged stems can bend upward again because cell growth of the intercalary meristem can be more rapid on the lower side. Adventitious roots arise from these intercalary meristem areas on young vegetative tillers as a step for a tiller to become independent (Figure 2.9).

Mature grass stems are generally highly lignified, causing swards with a high proportion of stems to be low in quality. Even though leaves of elongated stems maintain relatively high quality as they age, they are separated by the internodes and grazing animals often cannot maintain sufficient intake. Bite size and forage intake by ruminants are highest when the grass canopy is composed of long leaves on vegetative tillers that are densely packed. With very short canopies, the bite size is restricted by the height of the older sheaths, and grazing animals cannot maximize forage intake even though the bite rate is increased (Laca et al. 1992; Schacht et al. 2001). Though low in forage quality, the elongated stems produce an inflorescence and seed that has economic value.

The vertical stems of many perennial grasses have thickened lower internodes and tillers forming a crown (Figure 2.9) where reserve carbohydrates and proteins accumulate. Axillary buds on lower nodes in this crown area develop into new tillers, rhizomes, or stolons, depending on the species. The combination of an energy source (storage in the lower stem) and active meristems (axillary buds) near ground level, allow perennial and winter annual grasses to persist through the winter, a dormant season or grow late into the autumn under suitable environmental conditions. Summer annuals must develop seed to pass from one season to the next (Chapter 1). With most grasses, axillary buds near or below soil level develop into tillers, though in some grasses, like big bluestem and reed canarygrass, axillary buds located in leaf axils well above soil level can give rise to aerial tillers (Begg and Wright 1962).

Roles of Vernalization and Photoperiod in Flowering

Stem and flower development from tillers in most upright cool-season grasses occur only in spring. This restricts or prevents flowering in the late summer and fall when the remaining time is insufficient to produce seed. If all tillers flowered and died in fall, there would be no perennation. To both perennate and optimize seed production, plants have evolved a mechanism of using decreasing daylength during fall and duration of cold exposure during winter to induce (prepare) the shoot apex of the plant to flower. The floral **induction** process called **vernalization** serves only as an enabling precondition prior to the second set of environmental signals, namely lengthening photoperiod in spring to initiate differentiation and flowering of the apices that have been vernalized.

Early evidence of the vernalization mechanism came from noting whether early spring seedings of perennial grasses remained vegetative through the year. If so, the species had a vernalization requirement and could not flower until the next year. Conversely, if the plant elongated its stems and flowered in summer after planting, it had a low or no vernalization requirement (Gardner and Loomis 1953). When planted in early spring in Iowa, orchardgrass, reed canarygrass, redtop, kentucky bluegrass, red fescue, and crested wheatgrass did not flower that summer. Tall fescue and perennial ryegrass had sparse flowering, while northern bromegrass, southern bromegrass, and mountain bromegrass had medium flowering. Canada wildrye, big bluestem, side-oats grama, weeping lovegrass, and switchgrass had heavy flowering. This gave insight that some species had a fixed vernalization requirement while others had a lower requirement and flowering responded mainly to spring daylength.

Gardner and Loomis, (1953) evaluated established orchardgrass plants grown in the field into fall and winter in Iowa while experiencing decreasing daylength and air temperatures. They transferred some plants every 15 days to a warm greenhouse with long daylength. Plants transferred prior to November 1 did not flower, whereas flowering was maximized in those transferred on December 1. In Missouri, where the days in late fall are longer and air temperatures are higher than Iowa, a similar experiment was conducted on tall fescue. Those removed December 1, had some flowers and those remaining in the field until December 15 and beyond had abundant flowers (Nelson unpublished). This indicates tillers of species vernalize at different rates.

Current evidence indicates average daily air temperatures needed for vernalization are optimum from 0 to 10 C and act directly on the shoot apex. Conversely, the separate photoperiod signal of short days in fall is detected by the leaves and transmitted through the phloem to the apex. Further, the relative influences of the two signals for vernalization differ among grass species. For example, smooth bromegrass and orchardgrass respond primarily to short days, while perennial ryegrass responds primarily to low temperature, and meadow fescue responds almost exclusively to low temperature (Havstad et al. 2004). They separated the environmental signals and found the light factor in a vernalized tiller could be transferred, presumably in the phloem, to an attached non-vernalized tiller, but not the low temperature component.

It is now known that grass plants have a gene complex that produces an inhibitor that blocks flowering during summer and fall until vernalization occurs (Yan et al. 2003; Fjellheim et al. 2014). Vernalization does not stimulate flowering, but unblocks the inhibitor in the shoot apex of older tillers with several leaves by exposure of the plant to shorter days and decreasing temperature during fall and winter. In this case, the unknown product, referred to as **vernalin**, unblocks the shoot apices of tillers that are beyond a minimal size or age. Thus, the smaller and younger juvenile tillers are not unblocked and survive over winter to produce only vegetative growth during the next year. This appears to be nature's way to maintain plant persistence of the genotype while the unblocked tillers flower and die.

The original question remains, "How do plants know that winter is over and a better life plus seed production can begin"? Or, "How is tiller size or age a mitigating factor"? i.e., "what proportion of tillers should be unblocked"? There is clearly a tradeoff between continued perennation of the mother plant genotype through tillers, or using seed production to alter the genetic base at that location of spread by seed to other areas. So far, our understanding of mechanisms for sensing duration and level of coldness for vernalization has been investigated mainly in winter-annual cereal grasses (Evans 1987; Heide 1994). However, this mechanism is currently being studied using molecular biology to regulate production and activities of the responsible inhibitor and the unblocker (Yan et al. 2003; Fjellheim et al. 2014).

In spring, the lengthening photoperiod changes the relative amounts of Red and Far-red radiation inducing accumulation of **florigen** in leaves and its transport to or synthesis of florigen in the shoot apices. However, only those apices that are vernalized (unblocked) will respond to differentiate the shoot apex, elongate the stem internodes and produce an inflorescence. The overall combination of these seasonal responses ensures that winter annual and perennial forage grasses remain vegetative before winter, avoiding floral development that would be subject to frost damage in fall, and then flower at the appropriate time in spring. Inflorescence development and seed production can then occur before the onset of heat and water limitations in summer.

Due to the requirements for environmental stimuli, most cool- and warm-season grasses, especially those that grow upright, have only one main time-period of stem elongation and reproductive growth. However, some vegetative tillers of smooth bromegrass, reed canarygrass and perhaps other species regrow in summer by internode elongation that elevates the vegetative shoot apex up to 15 cm or more above soil level. The elongated tillers have shorter leaf blades, remain vegetative and can support more than three live leaves. These tillers are subject to removal if grazed or cut for forage. Possibly related, but not tested, these two species have active rhizome production.

Timothy is also unique in that from mid-latitudes of the US northward, where it is well-adapted biologically, it can flower twice a year. The plants flower when daylength gets longer in spring and, then again in summer, when new tillers regrow after cutting or grazing. These timothy types do not require cold treatment and develop flowering stems based only on daylength. This regrowth character allows timothy to be harvested as hay rather than as pasture and the second growth can be used for seed production. Summer regrowth of switchgrass also flowers. Though having a strong cold temperature vernalization requirement, tall fescue at mid-latitudes will flower sparsely a second time in late summer if there has been a prolonged period of slow growth, especially due to drought. Drought stress severe enough to stop growth may mimic the cold period needed for vernalization.

Species differ in time of maturity based on daylength needed for flowering and inflorescence emergence. For example, in mid-latitudes of the US, orchardgrass flowers earlier than tall fescue, followed in order by smooth bromegrass, timothy, and reed canarygrass. Similarly, among common warm-season grasses, switchgrass flowers earlier than big bluestem followed by indiangrass. Managing stem elongation is a major factor in pasture and hay management. However, species differences in flowering time allow managers to sequence pastures based on these developmental differences to optimize forage and animal production. Halevy (2017) published a handbook on flowering including the requirements for several grass species.

Root Initiation and Growth

Established grasses have adventitious, fibrous root systems. As each new tiller emerges into light, it develops a few phytomers and small leaves while remaining dependent on the root system of the mother tiller for water and minerals. Adventitious roots must develop from lower internodes of the tiller for it to become independent. If rooting is successful, the tiller will be able to grow, form leaf area and be self-supporting. The tiller dies if it does not become independent. As mentioned above, there is usually a continuous pattern of tiller development and death.

When a reproductive tiller dies after flowering or if cutting removes the shoot apex, its root system also dies. The decaying roots contribute organic matter and open channels in the soil resulting in the high soil organic matter and porous structure characteristic of grassland soils.

The root system of grasses is heavily branched, especially in the upper soil horizons, making it well adapted for using intermittent rainfall, holding soil particles together to aid in soil conservation, and taking up top-dressed fertilizers. Most grass roots are in the upper meter of soil. Greater depth of rooting is favored in soils that have low physical strength and good aeration. Deeper rooting allows the plants to extract water from a larger soil volume, which aids during drought.

Species also differ in the depth and distribution of roots in the same soils, which can affect their drought tolerance. Root systems of some native warm-season grasses, such as big bluestem and indiangrass, have roots of larger diameter that reach depths of 2–3 m, contributing to their ability to tolerate dry conditions (Weaver 1926, 1954). Under drought conditions, roots of grasses become thinner to facilitate ease of penetration and extend further through the dry soil to reach deeper water.

Stoloniferous plants, such as bermudagrass and bahiagrass, form adventitious roots at nodes of the stolons (Figure 2.10). These roots anchor the stolon and take up water and nutrients for that part of the plant. Likewise, roots can form at rhizome nodes to help absorb water and nutrients (Figure 2.10).

Symbiotic Relationships

Most grass roots form a symbiotic relationship with arbuscular mycorrhizal fungi (Wilson and Hartnett 1998). The mycorrhizal mycelia become an effective extension of the plant root system to increase plant uptake of mineral nutrients (especially P) (Brejda et al. 1993), increase drought tolerance (Sylvia and Williams 1992), and may reduce root diseases and damage from root nematodes (Linderman 1992). Many grasses, especially dryland species, are highly dependent on mycorrhizae. Without infection, they would likely not establish or compete effectively with plants that are not mycorrhizal dependent (Wilson and Hartnett 1998).

Effects of Carbohydrate Supply

On a whole plant basis, the plant balances growth of shoots and roots with photosynthesis and respiration to optimize use of carbohydrate. If the balance is very positive, the above and below ground growth will be optimized for the conditions. Nearly always, seed production has the highest priority. For vegetative plants in a good environment, leaf growth has a higher priority followed by growth of vegetative tillers with root growth being lowest. If light intensity is low due to cloudy weather or plants are shaded, photosynthesis is reduced while leaves continue to grow, perhaps even faster at the expense of roots and carbohydrate storage. If there is drought stress, the leaves will slow growth while maintaining or even increasing transport of **sugars** for storage or root growth. Removing leaf area by harvesting or grazing reduces carbohydrate storage and root growth while top growth is being restored. Since yield is a high priority, pastures and hay fields are often managed in ways that are detrimental to storage and especially to root growth.

Close and frequent defoliation reduces both shoot and root development because there is less leaf area to produce the carbohydrate necessary for growth, mineral uptake and respiration of the roots. Shallow roots explore less soil volume so there is less access to nutrients and especially to soil water. This results in less shoot production, which reduces photosynthesis, further increasing the problem of a smaller root system, and leads the grass into a downward spiral. If defoliation is not relaxed, plants become weak and may eventually die (Chapter 4). This affects both cool- and warm-season grasses (Dawson et al. 2000).

Root Response to Nutrient Distribution

Nutrient distribution within soils has substantial horizontal and vertical heterogeneity. Concentrations of NO_3^- and NH_4^+ can vary threefold within 3 cm and tenfold within 50 cm of a plant root (Jackson and Caldwell 1993). Where soil resources are abundant, roots of many species tend to proliferate, mainly by increased formation and growth of lateral roots (Caldwell 1994; Fitter 1994; Johnson and Biondini 2001). Downward growth of the main root axis may be unaffected by proliferation of lateral roots, suggesting the soil is continuously explored by the main axis, while lateral roots contribute to exploitation of nutrient rich patches.

The mechanism for detection of nutrient-rich areas is not understood, but species differ in their expression of root morphology, and some species appear unable to grow selectively into nutrient-rich patches (Fitter 1994; Hutchings and de Kroon 1994). Species grown in fertile habitats often possess higher levels of root **plasticity** than species from infertile habitats (Johnson and Biondini 2001). Habitats characterized by small and short-lived nutrient patches favor species with a fine root system and a high degree of morphologic plasticity (Fitter 1994; De Kroon and Hutchings 1995). In contrast, long-lived nutrient patches are best exploited by species with long-lived coarse root systems that are physiologically more costly to produce. Therefore, species diversity and soil heterogeneity can produce an array of root proliferation strategies even within similar habitats.

In addition to values in forage or pasture production and use, plant interactions based on morphology are an integral component of ecosystem functions that encompass interrelationships among associated organisms. In productive habitats with good soils and moisture supplies, natural selection favors plant traits that increase resource acquisition and productivity that increases above-ground competition. In less-productive habitats, below-ground competition is dominant causing selection for traits that increase tolerance to resource limitations like water or nutrients (Casper and Jackson 1997; Burke et al. 1998).

The occurrence of alternative plant traits to optimize resource acquisition at various locations along resource gradients imposes a tradeoff that strongly influences plant interactions and life-history strategies. In the future, when forages and pastures may be moved to less productive sites, there will be greater need for understanding how genetics and management affect rooting strategies.

Morphology of the Inflorescence

The grass inflorescence (Figure 2.14) consists of a group or cluster of spikelets, the basic reproductive unit. **Spikelet** characteristics and the organization of the inflorescence offer convenient traits for identifying grasses (Figure 2.14a–c). The **spike** inflorescence has a strong central rachis and is characteristic of wheat, western wheatgrass, and perennial ryegrass. Spikelets are **sessile** because they are attached directly to the rachis (Figure 2.14a). The **raceme** differs from the spike in that **spikelets** are connected to the rachis by short stalks called **pedicels** (Figure 2.14b). The simple raceme is the least common inflorescence type for grasses, yet there are numerous modifications. For example, the raceme of big bluestem has a sessile fertile spikelet positioned with a sterile spikelet that is on a pedicel. Crabgrass has a digitate cluster of racemes.

The **panicle**, the most common grass inflorescence, has branches with pedicelled spikelets (Figure 2.14c). Panicles may be open and diffuse-like in smooth bromegrass, kentucky bluegrass, and switchgrass. Alternatively, panicles with very short branches and pedicels can be compact as with timothy and pearl millet, almost looking like spikes.

Morphology of Grass Seed

The seed unit of all grasses is the **caryopsis.** In contrast with wheat or corn, in which the seed unit is the bare caryopsis, the caryopsis of most forage grasses remains enclosed by the **lemma** and **palea**, even during seed harvest and processing (Moore and Nelson 2018). The lemma, palea and seed coat (ovary wall) protect the caryopsis against mechanical damage, moisture loss in storage, and attack by biologic pests. The caryopsis consists largely of **endosperm**, the starchy tissue that maintains and protects the **embryo** during storage and provides energy for the seedling during **germination** and emergence. Compared with the endosperm, the embryo is relatively small in forage grasses and consists of two major parts. The **cotyledon** (or scutellum) is a modified leaf (first leaf) that surrounds the **embryo axis**

Table 2.1 Seed weight and storage time for maintaining good germination and estimated years to time of 50% germination (T_{50})

Species	Seed wt.[a] (no. g^{-1})	Storage[b] (yr)	T_{50}[b] (yr)
Meadow foxtail	895	2–3	6.2
Smooth bromegrass	315	—	3.4
Orchardgrass	945	2–3	6.6
Perennial ryegrass	530	3–4	7.2
Reed canarygrass	1185	3–5	11.0
Timothy	2565	2–4	5.7
Kentucky bluegrass	3065	1–3	6.6

[a] *Source:* From Assoc. Off. Seed Anal. (1983).
[b] *Source:* From Priestly (1986).

and stores energy-rich compounds. It secretes enzymes during germination to digest starch in the endosperm, and transports energy-rich compounds to the root and shoot (McDonald et al. 1996).

Although grass seeds are generally small and may look uniform, weights of individual caryopses differ markedly within a single inflorescence. Viable seed of some **florets** may weigh less than one-tenth that of other florets, the difference being largely due to the endosperm component that develops later than the embryo (Walton 1983). Germination and seedling vigor are generally lower for smaller seed because endosperm substrates needed to support cell division and growth may be limited. For example, reed canarygrass with a relatively large seed unit (Table 2.1) exhausted the endosperm in 10–14 days of germination. However, there may be threshold levels as relatively small differences in seed weight (0.20 vs. 0.145 g per 100 seed) had only minor short-term effects on seedling establishment of switchgrass, and both sizes of seed established equally acceptable stands (Smart and Moser 1999). While reserve substrates are important, they are only one factor involved in seedling vigor (Aamlid et al. 1997).

Diaspores is a term for describing seed dispersal units that include the lemma and palea and other appendages that normally remain after harvest and processing. These retained appendages add considerable weight to the small-sized caryopsis that they protect. Several cool-season grasses adapted to more arid areas have diaspore weights that are similar to those from humid areas (Table 2.1). In contrast, many warm-season grasses for more arid areas have much larger diaspore weights, about 350 seed g^{-1} for big bluestem, 385 for indiangrass, and 860 for switchgrass (Masters et al. 2004). Some diaspores like sideoats grama and buffalograss contain multiple florets

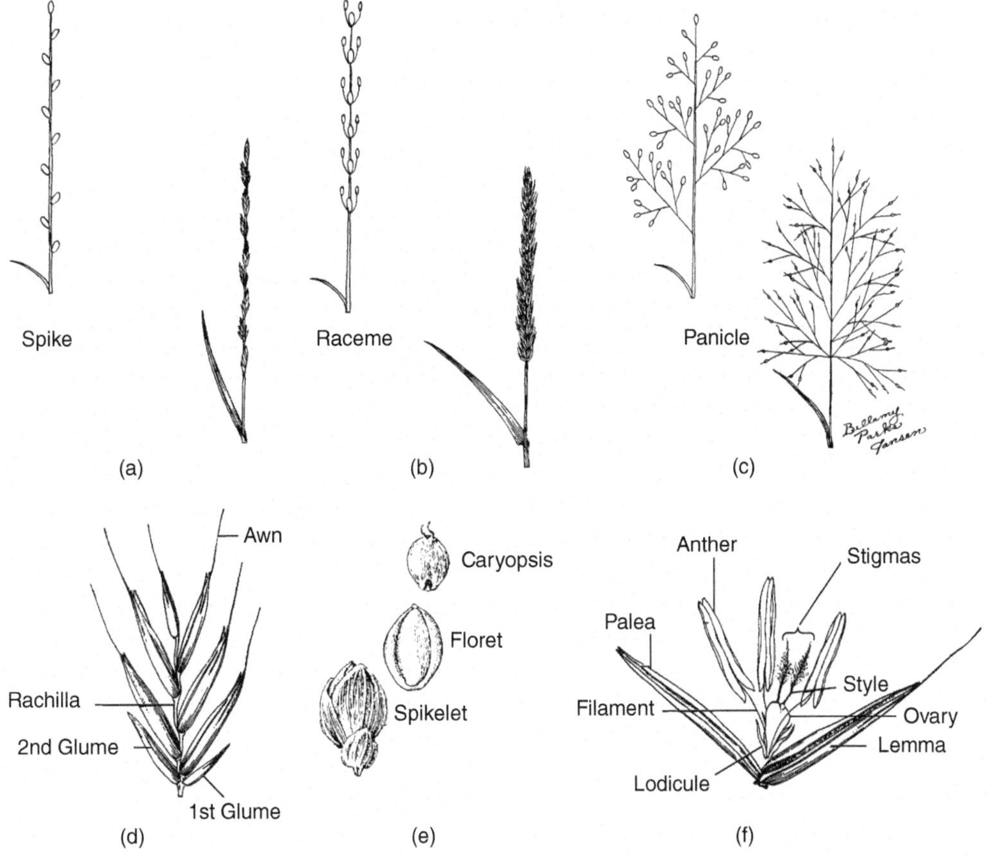

Fig. 2.14. Reproductive structures of grasses. (a, b, c) Diagrammatic and drawn inflorescences and flag leaves. (d) Spikelet subtended by glumes that contains six florets arranged sequentially on the rachilla (central stalk). (e) Spikelet with only one floret with glumes removed to show the floret, and with the lemma and palea removed to show the caryopsis. (f) Floret at anthesis showing floral parts enclosed by the lemma and palea. *Source: a, b, c* drawn by Bellamy Parks Jansen, adapted from Stubbendieck et al. (2003); *d, e, f* adapted from Dayton (1948).

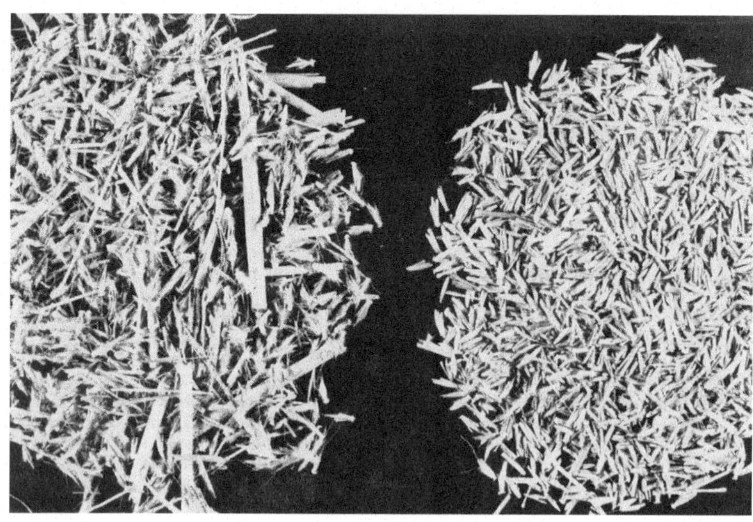

Fig. 2.15. Diaspores (seed units) of a commercial seed lot of big bluestem before (*left*) and after being debearded and cleaned (*right*). Debearding removes appendages from the seed and cleaning removes debris and small florets to increase the percentage of pure seed. *Source:* Photo courtesy of USDA-NRCS.

with fertile caryopses giving potential to produce more than one seedling per seed dispersal unit.

During caryopsis development, the green lemmas and paleas may provide photosynthate while protecting the caryopsis from dehydration, pathogens, and insects. Several forage species, especially warm-season species in dry areas, have awns attached to the lemma. Awns have been considered an advantage in deterring herbivory of seed. In addition, they contribute photosynthate and transpire to move water and minerals to the caryopsis.

Many grasses, such as big bluestem and indiangrass, are difficult to seed using mechanical equipment because the lemma and palea are pubescent (hairy) and have appendages such as awns, **rachis** sections, and pedicels that make the seed mass light and fluffy (Figure 2.15). Mechanical removal of the hair and appendages facilitates seeding, but the small caryopsis may be affected if the protective lemma and palea are damaged. Seed processing also breaks up many multiple diaspores (Figure 2.15).

Awns assist seed dispersal by attaching to animals and increasing buoyancy for wind transport. Chapman (1996) suggests awns may help wind movement of diaspores along the soil surface in sparsely vegetated areas and then help anchor the seed in a soil crack or other place where the **germination** environment is more favorable. While not true for most awned species (Peart and Clifford 1987), some (e.g. needle-and-thread) have hygroscopic awns that move upon wetting and drying and can push the seed into the ground. Due to protective seed coats, many diaspores of grasses are ingested by animals, transported, and viable seed excreted. Survival of seed in the digestive tract of animals depends on seed-coat characteristics and rate of passage through the animal's gastrointestinal tract (Ortmann et al. 1998).

In some species, the lemma and palea interfere with absorption of water and exchange of gases needed for germination. Also, they may contain inhibitors that delay or slow germination until the chemical is leached or degraded by soil microorganisms. Dormancy also can be within the embryo, such as with immature embryos or dormant embryos. Dormancy characters are usually reduced or eliminated by genetic selection in cultivated forage grasses, but many are still operative in grasses found in native ecosystems. Generally, cool-season grasses have low amounts of seed dormancy and differ in seed longevity (Table 2.1). Smooth bromegrass seed with large appendages and a relatively small embryo is short-lived, whereas reed canarygrass with heavier seed and small appendages has a long storage life.

Summary

Knowledge about vegetation organization and structure, and how they are achieved, is important for understanding the scale of plant morphology and its significance on plant interactions that are critical components of pasture, forage, and rangeland ecosystems. Herbage production, its quality and effects on plant persistence are the most recognized interactions, but plants also interact with biotic and abiotic variables to influence ecologic processes and vegetation responses to management practices.

Most management practices on forage plants and their functions focus on how to improve the environment for plant growth and ultimately animal performance. However, these practices may not be mutually beneficial and compromises are needed to achieve the hoped for economic and environmental outcomes. Understanding variations in morphology offers options that may be beneficial to the overall system.

Plants also interact with the abiotic environment by showing plastic growth responses of leaves, tillers and roots to enhance resource capture in habitats characterized by heterogeneous resource distribution. Plant species have evolved significant adaptations to major abiotic variables that are distributed along environmental gradients. When understood, some of these features can be genetically enhanced. Even now, these adaptations can be managed to optimize plant growth and functions in specific habitats, and contribute to sustainable production of animal products and conservation of the environment.

References

Aamlid, T.S., Heide, O.M., Christie, B.R., and McGraw, R.L. (1997). Reproductive development and the establishment of potential seed yield in grasses and legumes. In: *Forage Seed Production*, Temperate Species, vol. I (eds. D.T. Fairey and J.G. Hampton), 9–44. Wallingford, Oxon, UK: CAB International.

Akin, D.E. and Burdick, D. (1975). Percentage of tissue types in tropical and temperate grass leaf blades and degradation of tissue by microorganisms. *Crop Sci.* 15: 661–668.

Akin, D.E., Amos, H.E., Barton, F.E. II, and Burdick, D. (1973). Rumen microbial degradation of grass tissue revealed by scanning electron microscopy. *Agron. J.* 65: 825–828.

Allard, G. and Nelson, C.J. (1991). Photosynthate partitioning in basal zones of tall fescue leaves. *Plant Physiol.* 95: 663–668.

Allard, G., Nelson, C.J., and Pallardy, S.G. (1991a). Shade effects on growth of tall fescue. I. Leaf anatomy and dry matter partitioning. *Crop Sci.* 31: 163–167.

Allard, G., Nelson, C.J., and Pallardy, S.G. (1991b). Shade effects on growth of tall fescue. II. Leaf gas exchange characteristics. *Crop Sci.* 31: 167–172.

Association of Official Seed Analysts (1983). Seed vigor testing handbook. In: *The Handbook of Seed Testing. Assoc. Off* (eds. B.E. Clark et al.) 1–89. Lincoln, NE: Seed Anal.

Bade, D.H., Conrad, B.E., and Holt, E.C. (1985). Temperature and water stress effects on growth of tropical grasses. *J. Range Manage.* 38: 321–324.

Barkworth, M.E., Capels, K.M., Long, S., and Piep, M.B. (eds.) (2003). *Flora of North America: North of Mexico*, vol. 25. New York: Oxford University Press.

Barre, P., Turner, L.B., and Escobar-Gutierrez, A.J. (2015). Leaf length variation in perennial forage grasses. *Agriculture* 5: 682–696.

Begg, J.E. and Wright, M.J. (1962). Growth and development of leaves from intercalary meristems of *Phalaris arundinacea*. *Nature* 194: 1097–1098.

Bouchenak-Khelladi, Y., Verboom, G.A., Savolainen, V., and Hodkinson, T.R. (2010). Biogeography of the grasses (Poaceae): a phylogenetic approach to reveal evolutionary history in geographical space and geological time. *Bot. J. Linn. Soc.* 162: 543–557. https://doi.org/10.1111/j.1095-8339.2010.01041.x.

Brejda, J.J., Yocum, D.H., Moser, L.E., and Waller, S.S. (1993). Dependence of 3 Nebraska Sandhills warm-season grasses on vesicular-arbuscular mycorrhizae. *J. Range Manage.* 46: 14–20.

Briggs, W.R. (2016). Plant biology: seedling emergence through soil. *Curr. Biol.* 26: R68–R69.

Briske, D.D. (2007). Plant interactions. In: *Forages, the Science of Grassland Agriculture* (eds. R.F Barnes, C.J. Nelson, K.J. Moore and M. Collins), 105–122. Oxford, U.K.: Blackwell.

Briske, D.D. and Derner, J.D. (1998). Clonal biology of caespitose grasses. In: *Population Biology of Grasses* (ed. G.P. Cheplick), 106–135. Cambridge, UK: Cambridge University Press.

Burke, I.C., Lauenroth, W.K., Vinton, M.A. et al. (1998). Plant-soil interactions in temperate grassland. *Biogeochemistry* 42: 121–143.

Buxton, D.R. (1990). Cell-wall components in divergent germplasm of four perennial forage grass species. *Crop Sci.* 30: 402–408.

Caldwell, M.M. (1994). v. In: *Exploitation of Environmental Heterogeneity by Plants: Ecophysiological Processes above – and Belowground* (eds. M.M. Caldwell and R.W. Pearcy), 325–347. San Diego: Academic.

Casal, J.J. (2013). Canopy light signals and crop yield in sickness and in health. *ISRN Agron.* 2013 650439.

Casal, J.J., Sanchez, R.A., and Deregibus, V.A. (1986). The effect of plant density on tillering: the involvement of R/FR ratio on the proportion of radiation intercepted per plant. *Env. Exp. Bot.* 26: 365–371.

Casey, I.A., Brereton, A.J., Laidlaw, A.S., and McGilloway, D.A. (1999). Effects of sheath tube length on leaf development in perennial ryegrass (*Lolium perenne* L.). *Ann. Appl. Biol.* 134: 251–257.

Casler, M.D., Buxton, D.R., and Vogel, K.P. (2002). Genetic modification of lignin concentration affects fitness of perennial herbaceous plants. *Theor. Appl. Genet.* 104: 127–131.

Casper, B.B. and Jackson, R.B. (1997). Plant competition underground. *Annu. Rev. Ecol. Syst.* 28: 545–570.

Chapman, G.P. (1996). *The Biology of Grasses*. Wallingford, Oxon, UK: CAB International.

Ciani, A., Goss, K.-U., and Schwarzenbach, R.P. (2005). Light penetration in soil and particulate minerals. *Eur. J. Soil Sci.* 56: 561–574.

Clark, L.G. and Fisher, J.B. (1987). Vegetative morphology of grasses: shoots and roots. In: *Grass Systematics and Evolution* (eds. T.R. Soderstrom, K.W. Hilu, C.S. Campbell and M.E. Barkworth), 37–45. Washington, D.C.: Smithsonian Institute Press.

Cleland, R.E. (2001). Unlocking the mysteries of leaf primordia formation. *Proc. Natl. Acad. Sci. U.S.A.* 98: 10981–10982.

Cohen, C.J., Chilcote, D.O., and Frakes, R.V. (1982). Leaf anatomy and stomatal characteristics of four tall fescue selections differing in forage yield. *Crop Sci.* 22: 704–708.

Davidson, D.J. and Nelson, C.J. (1980). Nitrogen effects on photosynthesis and leaf senescence of tall fescue. *Agron. Abstr.*: 78.

Dawson, L.A., Grayston, S.J., and Peterson, E. (2000). Effects of grazing on the roots and rhizosphere of grasses. In: *Grassland Ecophysiology and Grazing Ecology* (eds. G. Lamaire, J. Hodgson, A. de Moraes, et al.), 61–84. Wallingford, Oxon, UK: CAB International.

Dayton, W.A. (1948). The family tree of Gramineae. In: *Grass: The Yearbook of Agriculture 1948*, 637–639. USDA (ed.). Washington, D.C., US Government Printing Office.

De Kroon, H. and Hutchings, M.J. (1995). Morphological plasticity in clonal plants: the foraging concept reconsidered. *J. Ecol.* 83: 143–152.

Dean, C. and Leech, R.M. (1982). Genome expression during normal leaf development. *Plant Physiol.* 69: 904–910.

Durand, J.-L., Onillon, B., Schnyder, H., and Rademacher, I. (1995). Drought effects on cellular and spatial parameters of leaf growth in tall fescue. *J. Exp. Bot.* 46: 1147–1155.

Edwards, D. and Cooper, J.P. (1963). The genetic control of leaf development in Lolium. II. Response to selection. *Heredity* 18: 307–317.

Evans, L.T. (1987). Short day induction of inflorescence initiation in some winter wheat varieties. *Australian Journal of Plant Physiology* 14 (3): 277–285.

Field, C. (1983). Allocating leaf nitrogen for the maximisation of carbon gain: leaf age as a control on the allocation program. *Oecologia* 56: 341–347.

Fitter, A.H. (1994). Architecture and biomass allocation as components of the plastic response of root systems to soil heterogeneity. In: *Exploitation of Environmental Heterogeneity by Plants: Ecophysiological Processes Above- and Belowground* (eds. M.M. Caldwell and R.W. Pearcy), 305–323. San Diego: Academic Press.

Fjellheim, S., Boden, S., and Trevaskis, B. (2014). The role of seasonal flowering responses in adaptation of grasses to temperate climates. *Front. Plant Sci.* 5: 431. https://doi.org/10.3389/fpls.2014.00431.

Fulkerson, W.J. and Donaghy, D.J. (2001). Plant-soluble carbohydrate reserves and senescence – key criteria for developing an effective grazing management system for ryegrass-based pastures: a review. *Aust. J. Exp. Agric.* 41: 261–273.

Gardner, F.P. and Loomis, W.E. (1953). Floral induction and development in orchard grass. *Plant Physiol.* 28: 201–217.

Garnier, F. and Laurent, G. (1994). Leaf anatomy, specific mass and water content in congeneric annual and perennial grass species. *New Phytol.* 128: 725–736.

Gastal, F. and Lemaire, G. (2015). Defoliation, shoot plasticity, sward structure and herbage utilization in pasture: review of the underlying ecophysiological processes. *Agriculture* 5: 1146–1171. https://doi.org/10.3390/agriculture5041146.

Gastal, F. and Nelson, C.J. (1994). Nitrogen use within the growing leaf blade of tall fescue. *Plant Physiol.* 105: 191–197.

Gauthier, M., Barillot, R., Schneider, A. et al. (2018). Towards a model of wheat leaf morphogenesis at plant scale driven by organ-level metabolites. 6th International Symposium on Plant Growth Modeling, Simulation, Visualization and Applications (PMA), Hefei.

Halevy, A.H. (2017). *Handbook of Flowering: Vol I.* Boca Raton, FL: CRC Press.

Harper, J.L. (1977). *Population Biology of Plants.* London: Academic Press.

Hartley, W. and Williams, R.J. (1956). Centers of distribution of cultivated pasture grasses and their significance for plant introductions. In: Proc. 7th Int. Grassl. Congr., Palmerston North, NZ, 6–17 November (ed. G.J. Neale), 190–201. Wellington, NZ: Wright and Carmen Ltd.

Havstad, L.T., Aamlid, T.S., Heide, T.S., and Junttila, O. (2004). Transfer of flower induction stimuli to non-exposed tillers in a selection of temperate grasses. *Acta Agric. Scand. Sect. B* 54: 23–30.

Heide, O.M. (1994). Control of flowering and reproduction in temperate grasses. *New Phytol.* 128: 347–362.

Hendrickson, J.R. and Briske, D.D. (1997). Axillary bud banks of two semiarid perennial grasses: occurrence, longevity, and contribution to population persistence. *Oecologia* 110: 584–591.

Horst, G.L. and Nelson, C.J. (1979). Compensatory growth of tall fescue following drought. *Agron. J.* 71: 559–563.

Horst, G.L., Nelson, C.J., and Asay, K.H. (1978). Relationship of leaf elongation to forage yield of tall fescue genotypes. *Crop Sci.* 18: 715–719.

Humphreys, M.O. (2005). Genetic improvement of forage crops – past, present and future. *Can. J. Agr. Sci.* 143: 441–446.

Hutchings, M.J. and de Kroon, H. (1994). Foraging in plants: the role of morphological plasticity in resource acquisition. *Adv. Ecol. Res.* 25: 159–238.

Jackson, R.B. and Caldwell, M.M. (1993). The scale of nutrient heterogeneity around individual plants and its quantification with geostatistics. *Ecology* 74: 612–614.

Jewiss, O.R. (1972). Tillering in grasses – its significance and control. *J. Br. Grassland Soc.* 27: 65–82.

Johnson, H.A. and Biondini, M.E. (2001). Root morphological plasticity and nitrogen uptake of 59 plant species from the Great Plains grasslands, U.S.A. *Basic Appl. Ecol.* 2: 127–143.

Jones, R.J., Nelson, C.J., and Sleper, D.A. (1979). Seedling selection for morphological characters associated with yield of tall fescue. *Crop Sci.* 19: 631–634.

Laca, E.A., Ungar, E.D., Seligman, N., and Demment, M.W. (1992). Effects of sward height and bulk density on bite dimensions of cattle grazing homogeneous swards. *Grass Forage Sci.* 47: 91–102.

Langer, R.H.M. (1972). *How Grasses Grow*. London: Edward Arnold Ltd.

Lemaire, B., Da Silva, S.C., Agnusdei, M. et al. (2009). Interactions between leaf lifespan and defoliation frequency in temperate and tropical pastures: a review. *Grass Forage Sci.* 64: 341–353.

Linderman, R.G. (1992). Vesicular-arbuscular mycorrhizae and soil microbial interactions. In: *Mycorrhizae in Sustainable Agriculture*, American Society of Agronomy Special Publication Number 54 (eds. G.J. Bethlenfalvay and R.G. Linderman), 45–70. Madison, WI: American Society of Agronomy.

MacAdam, J.W. and Nelson, C.J. (1987). Specific leaf weight in zones of cell division, elongation and maturation in tall fescue leaf blades. *Ann. Bot.* 59: 369–376.

MacAdam, J.M. and Nelson, C.J. (2002). Secondary cell wall deposition causes radial growth of fibre cells in the maturation zone of elongating tall fescue leaf blades. *Ann. Bot.* 89: 89–96.

MacAdam, J.M., Volenec, J.J., and Nelson, C.J. (1989). Effects of nitrogen on mesophyll cell division and epidermal cell elongation in tall fescue leaf blades. *Plant Physiol.* 89: 549–556.

MacAdam, J.M., Nelson, C.J., and Sharp, R.E. (1992). Peroxidase activity in the leaf elongation zone of tall fescue. II. Spatial distribution of apoplastic peroxidase activity in genotypes differing in length of the elongation zone. *Plant Physiol.* 99: 879–885.

Malinowski, D.P., Rudd, J.C., Pinchak, W.E., and Baker, J. (2018). Determining morphological traits for selecting wheat (*Triticum aestivum* L.) with improved early-season forage production. *J. Adv. Agric.* 9: 1508–1530.

Masters, R.A., Mislevy, P., Moser, L.E., and Rivas-Pantoja, F. (2004). Stand establishment. In: *Warm-Season Grasses*, American Society of Agronomy Monograph 45 (eds. L.E. Moser, B.L. Burson and L.E. Sollenberger), 145–177. Madison, WI: American Society of Agronomy.

Matthew, C. (ed.) (2017). *Forage Plant Ecophysiology*. Basal, Switzerland: MDPI AG https://www.mdpi.com/journal/agriculture/special_issues/forage_plant_ecophysiology.

Matthew, C., Yang, J.Z., and Potter, J.F. (1998). Determination of tiller and root appearance in perennial ryegrass (*Lolium perenne*) swards by observation of the tiller axis and potential application in mechanistic modelling. *N. Z. J. Agric. Res.* 41: 1–10.

Matthew, C., van Loo, E.N., Thom, E.R. et al. (2001). Understanding shoot and root development. In: Proc. XIX Int. Grassl, Cong. (eds. J.A. Gromide, W.R.S. Mattos and S.C. de Silva), 19–27. Sao Pedro, Brazil.

Matthew, C., MacKay, A.D., and Robin, A.H.K. (2016). Do phytomer turnover models of plant morphology describe perennial ryegrass root data from field swards? *Agriculture* 6: 28. https://doi.org/10.3390/agriculture6030028.

McDonald, M.B., Copeland, L.O., Knapp, A.D., and Grabe, D.F. (1996). Seed development, germination, and quality. In: *Cool-Season Forage Grasses*, American Society of Agronomy Monography 34 (eds. L.E. Moser, D.R. Buxton and M.D. Casler), 15–70. Madison, WI: American Society of Agronomy.

Moore, K.J. and Nelson, C.J. (2018). Structure and morphology of grasses. In: *Forages, An Introduction to Grassland Agriculture*, 7e, vol. 1 (eds. M. Collins, K.J. Moore, C.J. Nelson and R.F Barnes), 19–34. Hoboken, NJ: Wiley.

Moser, L.E., Buxton, D.R., and Casler, M.D. (eds.) (1996). *Cool-Season Forage Grasses*, American Society of Agronomy Monography 34. Madison, WI: American Society of Agronomy.

Moser, L.E., Burson, B.L., and Sollenberger, L.E. (eds.) (2004). *Warm-Season (C_4) Grasses*, American Society of Agronomy Monography 45. Madison, WI: American Society of Agronomy.

Murphy, J.S. and Briske, D.D. (1992). Regulation of tillering by apical dominance: chronology, interpretive value, and current perspectives. *J. Range Manage.* 45(5): 419–430.

Nagabhyru, P., Dinkins, R.D., Wood, C.L. et al. (2013). Tall fescue endophyte effects on tolerance to water-deficit stress. *BMC Plant Biol.* 13: 127–144.

Nelson, C.J. (2000). Shoot morphological plasticity of grasses: leaf growth vs. tillering. In: *Grassland Ecophysiology and Grazing Ecology* (eds. G. Lemaire, J. Hodgson, A. de Moraes, et al.), 101–126. Wallingford, Oxon, UK: CAB International.

Nelson, C.J. and Zarrough, K.M. (1981). Tiller density and tiller weight as yield determinants of vegetative swards. In: Plant Physiol. and Herb. Prod., Brit.

Grassld. Soc. Occ. Symp. 13. (ed. C.E. Wright), 25–29. Hurley, U.K.

Ortmann, J., Schacht, W.H., and Stubbendieck, J. (1998). The “foliage is the fruit” hypothesis: complex adaptations in buffalograss (*Buchloe dactyloides*). *Am. Midl. Nat.* 140: 252–263.

Paciullo, D.S.C., Gomide, C.A.M., Castro, C.R.T. et al. (2016). Morphogenesis, biomass and nutritive value of *Panicum maximum* under different shade levels and fertilizer nitrogen rates. *Grass Forage Sci.* https://doi.org/10.1111/gfs.12264.

Peart, M.H. and Clifford, H.T. (1987). The influence of diaspore morphology and soil surface properties on the distribution of grasses. *J. Ecol.* 75: 569–576.

Poff, J.A., Balocchi, O.A., and López, I. (2011). Sward and tiller growth dynamics of *Lolium perenne* L. as affected by defoliation frequency during autumn. *Crop & Pasture Science* 62: 346–354.

Priestly, D.A. (1986). *Seed Aging*. Ithaca, NY: Cornell University Press.

Rademacher, I.F. and Nelson, C.J. (2001). Nitrogen effects on leaf anatomy within the intercalary meristems of tall fescue leaf blades. *Ann. Bot.* 88: 893–903.

Reeder, L.R., Sleper, D.A., and Nelson, C.J. (1984). Response to selection for leaf area expansion rate of tall fescue. *Crop Sci.* 24: 97–100.

Sackville-Hamilton, N.R., Matthew, C., and Lemaire, G. (1995). In defence of the 3/2 boundary rule. A reevaluation of self-thinning concepts and status. *Ann. Bot.* 76: 569–577.

Sanderson, M.A. and Nelson, C.J. (1995). Growth of tall fescue leaf blades in various irradiances. *Eur. J. Agron.* 4: 197–203.

Schacht, W.H., Smart, A.J., and Mousel, E.M. (2001). Using artificial swards to demonstrate plant-grazing animal interactions. *J. Nat. Resour. Life Sci. Educ.* 30: 89–92.

Schnyder, H., Nelson, C.J., and Coutts, J.H. (1987). Assessment of spatial distribution of growth in the elongation zone of grass leaf blades. *Plant Physiol.* 85: 290–293.

Schnyder, H., Seo, S., Rademacher, I.F., and Kuhbauch, W. (1990). Spatial distribution of growth rates and of epidermal cell length in the elongation zone during leaf development in *Lolium perenne* L. *Planta* 181: 423–431.

Schnyder, H., Schaufele, R., de Visser, R., and Nelson, C.J. (2000). An integrated view of C and N uses in leaf growth zones of defoliated grasses. In: *Grassland Ecophysiology and Grazing Ecology* (eds. G. Lemaiire, J. Hodgson, A. de Moraes, et al.), 41–59. Wallingford, Oxon, UK: CAB International.

Sharman, B.C. (1945). Leaf and bud initiation in the Gramineae. *Bot. Gaz.* 106: 269–289.

da Silva, S.C., Gimenes, F.M.A., Sarmento, D.O.L. et al. (2013). Grazing behaviour, herbage intake and animal performance of beef cattle heifers on marandu palisade grass subjected to intensities of continuous stocking management. *J. Agric. Sci.* 151: 727–739.

da Silva, S.C., Sbrissia, A.F., and Pereira, L.E.T. (2015). Ecophysiology of C4 grasses – understanding plant growth for optimising their use and management. *Agriculture* 5: 598–625.

Skinner, R.H. and Nelson, C.J. (1992). Estimation of potential tiller production and site usage during tall fescue canopy development. *Ann. Bot.* 70: 493–499.

Skinner, R.H. and Nelson, C.J. (1994). Epidermal cell division and the co-ordination of leaf and tiller development. *Ann. Bot.* 74: 9–15.

Skinner, R.H. and Nelson, C.J. (1995). Elongation of the grass leaf and its relationship to the phyllochron. *Crop Sci.* 35: 4–10.

Skinner, R.H. and Simmons, S.R. (1993). Modulation of leaf elongation, tiller appearance, and tiller senescence in spring barley by far-red light. *Plant Cell Environ.* 16: 555–562.

Smart, A.J. and Moser, L.E. (1999). Switchgrass seedling development as affected by seed size. *Agron. J.* 91: 335–338.

Spollen, W.G. and Nelson, C.J. (1988). Characterization of fructan from mature leaf blades and elongation zones of developing leaf blades of wheat, tall fescue and timothy. *Plant Physiol.* 88: 1349–1353.

Spollen, W.G. and Nelson, C.J. (1994). Response of fructan to water deficit in growing leaves of tall fescue. *Plant Physiol.* 106: 329–336.

Stubbendieck, J., Hatch, S.L., and Landholt, L.M. (2003). *North American Wildland Plants: A Field Guide*. Lincoln, NE: University of Nebraska Press.

Sylvia, D.M. and Williams, S.E. (1992). Vesicular-arbuscular mycorrhizae and environmental stress. In: *Mycorrhizae in Sustainable Agriculture*, American Society of Agronomy Special Publication Number 54 (eds. G.J. Bethlenfalvay and R.G. Linderman), 101–124. Madison, WI: American Society of Agronomy.

Taleisnik, E., Rodriguez, A.A., Bustos, D. et al. (2009). Leaf expansion in grasses under salt stress. *J. Plant Physiol.* 166: 1123–1140.

Tan, W., Tan, G., and Walton, P.D. (1973). Genotype x environment interactions in smooth bromegrass. II. Morphological characters and their associations with forage yield. *Can. J. Genet. Cytol.* 21: 73–80.

Touati, M., Kameli, A., Yabir, B. et al. (2018). Drought effects on elongation kinetics and sugar deposition in the elongation zone of durum wheat (*Triticum durum* Desf.) leaves. *Iran. J. Plant Physiol.* 9: 2619–2628.

Van Soest, P.J., Robertson, J.B., and Lewis, B.A. (1991). Methods for dietary fiber, neutral detergent fiber, and nonstarch polysaccharides in relation to animal nutrition. *J. Dairy Sci.* 74: 3583–3597.

Vassey, T.L. (1986). Morphological, anatomical, and cytohistological evaluations of terminal and axillary meristems of tall fescue. Ph.D. Thesis. University of Missouri.

Vogel, K.P. and Burson, B.L. (2004). Breeding and genetics. In: *Warm-Season Grasses*, American Society of Agronomy Monography 45 (eds. L.E. Moser, B.L. Burson and L.E. Sollenberger), 51–94. Madison, WI: American Society of Agronomy.

Volaire, F. and Lelievre, F. (1997). Production, persistence, and water-soluble carbohydrate accumulation in 21 contrasting populations of *Dactylis glomerata* L. subjected to severe drought in the south of France. *Aust. J. Agic. Res.* 48: 733–744.

Volaire, F., Thomas, H., and Lelievre, F. (1998). Survival and recovery of perennial forage grasses under prolonged Mediterranean drought: growth, death, water relations and solute content in herbage and stubble. *New Phytol.* 140: 439–449.

Volenec, J.J. and Nelson, C.J. (1981). Cell dynamics in leaf meristems of contrasting tall fescue genotypes. *Crop Sci.* 21: 381–385.

Volenec, J.J. and Nelson, C.J. (1982). Diurnal leaf elongation of contrasting tall fescue genotypes. *Crop Sci.* 22: 531–535.

Volenec, J.J. and Nelson, C.J. (1983). Responses of tall fescue leaf meristems to nitrogen fertilization and harvest frequency. *Crop Sci.* 23: 720–724.

Walton, P.D. (1983). *Production and Management of Cultivated Forages*. Reston, VA: Reston Pub. Co.

Watson, L. and Dallwitz, M.J. (1992). *The Grass Genera of the World*. Wallingford, Oxon, UK: CAB International.

Weaver, J.E. (1926). *Root Development of Field Crops*. New York: McGraw-Hill.

Weaver, J.E. (1954). *North American Prairie*. Lincoln, NE: Johnson Publ. Co.

West, C.P., Izekor, E., Turner, K.E., and Elmi, A.A. (1993). Endophyte effects on growth and persistence of tall fescue along a water-supply gradient. *Agron. J.* 85: 264–270.

Wilhelm, W.W. and McMaster, G.S. (1995). Importance of the phyllochron in studying development and growth in grasses. *Crop Sci.* 35: 1–3.

Wilhelm, W.W. and Nelson, C.J. (1978). Leaf growth, leaf aging, and photosynthetic rates of tall fescue genotypes. *Crop Sci.* 18: 769–772.

Williams, R.F. (1975). *The Shoot Apex and Leaf Growth*. Cambridge, UK: Cambridge University Press.

Wilson, G.W.T. and Hartnett, D.C. (1998). Interspecific variation in plant response to mycorrhizal colonization in tallgrass prairie. *Am. J. Bot.* 85: 1732–1738.

Woledge, J. (1971). The effect of light intensity during growth on the subsequent rate of photosynthesis of leaves of tall fescue (*Festuca arundincea* Schreb.). *Ann. Bot.* 35: 311–322.

Woledge, J. (1977). The effects of shading and cutting treatments on the photosynthetic rate of ryegrass leaves. *Ann. Bot.* 41: 1279–1286.

Woolley, J.T. and Stoller, E.W. (1978). Light penetration and light-induced seed germination in soil. *Plant Physiol.* 61: 597–600.

Xu, Q., Nelson, C.J., and Coutts, J.H. (1996). Chloroplast development in tall fescue leaves. *Plant Physiol.* 111 (2): 140.

Yan, L., Loukoianov, A., Tranquilli, G. et al. (2003). Positional cloning of the wheat vernalization gene *VRN1*. *Proc. Natl. Acad. Sci. U.S.A.* 100: 6263–6268. https://doi.org/10.1073/pnas.0937399100.

Yeh, R.Y., Matches, A.G., and Larson, R.L. (1976). Endogenous growth regulators and summer tillering of tall fescue. *Crop Sci.* 16: 409–413.

Zarrough, K.M., Nelson, C.J., and Coutts, J.H. (1983). Relationships between tillering and forage yield of tall fescue. II. Patterns of tillering. *Crop Sci.* 23: 338–342.

Zarrough, K.M., Nelson, C.J., and Sleper, D.A. (1984). Interrelationships between rates of leaf appearance and tillering in tall fescue populations. *Crop Sci.* 24: 565–569.

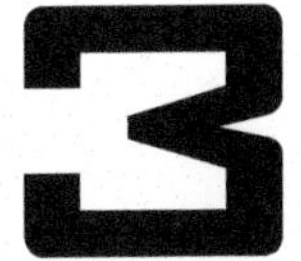

Legume Structure and Morphology

John Jennings, Professor, *Animal Science-Forages, University of Arkansas, Little Rock, AR, USA*
Jamie Foster, Professor, *Forage Agronomy, Texas A&M AgriLife Research, Beeville, TX, USA*

Introduction

Morphology refers to the structure and arrangement of plant parts. Familiarity with morphology and growth processes is essential for identifying plants and understanding management and environmental effects on forage yield, quality, and persistence. Legumes are important contributors to forage and pasture production but require management practices based on morphology and physiology. Morphology of legumes is distinctly different from grasses beginning with the seed, followed by the seedling and growth forms at vegetative and reproductive stages. In this chapter, we focus on legumes grown as monocultures like alfalfa or in mixtures of other legumes, forbs, and grasses.

The purpose of this chapter is to describe the variations among major legumes used in forage and pasture systems. Special emphasis is on comparative analysis of morphologic features associated with seed, seedling growth, vegetative growth, reproductive growth and perennation mechanisms. Then, these features are explained in terms of management for yield, forage quality, and stand persistence.

The Legume Family

There are nearly 700 genera and 18 000 species of known legumes (Polhill and Raven 1981). This group is second only to the grasses in providing food crops for agriculture. Legumes are a subgroup of forbs, the large group of plants with seed containing two cotyledons, i.e., dicots. Forbs contain mainly plants with broad leaves, of which, legumes are unique due to their high forage quality, ability to fix atmospheric nitrogen for plant use and wide adaptation to climate and management practices.

Legumes are important food sources for man and livestock and include several oilseed crops such as soybean and peanut. In addition, legumes usually enhance animal performance, contribute to soil improvement, provide wildlife habitat and beautify landscapes. Many species of major agricultural interest in humid areas of North America were known by early settlers who introduced them from Europe and Asia. Currently, there are efforts to document legumes that are native to North America and evaluate their economic and cultural use.

The term *legume* is defined as "a pod such as that of a pea or bean that splits into two valves with a single row of seeds attached to the lower edge of one of the valves." Legumes are dicots and include herbaceous annuals, biennials, perennials, and several woody shrubs, vines, and trees. Most, but not all, legumes grow symbiotically with nitrogen-fixing bacteria (*Rhizobium* or *Bradyrhizobium* species) that live in nodules attached to the roots. This symbiotic relationship of some legume forage species can fix more than 100 kg ha^{-1} N per year. Nitrogen fixation makes legumes valuable components in forage mixtures

Forages: The Science of Grassland Agriculture, Volume II, Seventh Edition.
Edited by Kenneth J. Moore, Michael Collins, C. Jerry Nelson and Daren D. Redfearn.

with grasses and in rotation with cereal grain crops to decrease dependence on fertilizer N.

In general, legumes have a higher N (crude protein) and mineral content and higher forage quality compared with grasses. Most forage legumes are best adapted to soils with near neutral or slightly acidic pH (Bordeleau and Prevost 1994). However, rhizoma peanut produces higher forage yield in acidic soils (Venuto et al. 1999).

Some common annual forage legumes include cool-season species such as arrowleaf clover, crimson clover, rose clover, and vetch and warm-season species such as lespedeza, soybean, cowpea, and lablab. All legumes have the C_3 system of photosynthesis so the cool-season and warm-season temperature classification is based on typical production periods. Common perennial forage legumes include white clover, red clover, alfalfa, birdsfoot trefoil, and rhizoma peanut. Some other native and naturalized legume species are grown such as Illinois bundleflower.

Seed Characteristics

Seed size varies widely among common forage legume species from 1.9 million kg^{-1} for low hop clover to 36 000 kg^{-1} for hairy vetch, whereas seed of soybean and cowpea are even larger (Martin et al. 1976). The typical legume seed consists of the seed coat (testa), hilum, lens, micropyle, and **embryo**. Coloration of the seed coat can be black, brown, reddish, purple, green, yellow, or mottled. The seed coat generally has a hard, smooth surface and is characterized by the hilum, or scar from where the seed was attached to the pod by a short funiculus. The micropyle opening, adjacent to the hilum, is where the pollen tube entered the embryo to transfer the male gametes. The lens is the weakest point of the palisade layer of the seed coat and is located at the apex of the cotyledons. This serves as a point of entry for water during the imbibition phase of **germination** (Teuber and Brick 1988). Seeding rates and seed placement in the soil are both affected by seed size, which must be considered in establishment.

The embryo consists of two cotyledons that enclose the embryo axis made up by the **radicle** or primary root, hypocotyl, and an **epicotyl** or **plumule**. Legume seed contains little or no endosperm; therefore, the cotyledons serve as food storage organs to support the developing embryo during germination and emergence. In small-seeded species such as red or white clover, energy reserves in the cotyledons are not sufficient to support emergence from deep planting so seed should be planted shallow (<6 mm) or broadcast and pressed into the soil surface to help ensure seed-soil contact and good stand establishment. Larger seeded species such as vetch or cowpea may be planted 25 mm deep without reducing emergence.

Legumes produce a percentage of **hard seed** that do not germinate immediately after planting. The seed coat of hard seed is impervious to water, thus limiting water imbibition and germination. Hard seed may survive passage through the digestive tracts of animals and can remain viable in a soil seed bank for up to 30 years (Gibson and Cope 1985). Delayed germination of hard seed of many annual and perennial legumes grown in harsh environments (Gibson and Holowell 1966) helps maintain stands via volunteer seedlings following years not favorable for seed production.

Delayed germination of hard seed is a mechanism which allows for survival and persistence in environments which are not optimal, such as the cool-season burr medic naturalized in the hot, humid Southern US. Armadillo burr medic has 93% and black medic has up to 96% hard seed that may not germinate until 2 years after planting (Clark 2007). This trait helps maintain stands as cover crops and as a reseeding forage. However, hard seed is not advantageous for alfalfa due to **autotoxicity** of established alfalfa plants on the seedlings emerging later. Alfalfa seed germinating in an established stand are usually weak and seldom survive or become productive (Bass et al. 1988; Jennings and Nelson 2002a).

Production of hard seed tends to be greater in environments that favor long seed-fill periods. Hard seed coats become softened or permeable over time in soil by wetting and drying, freezing and thawing, and the action of microorganisms allowing germination. In commercial seed production, hard-seededness is most commonly overcome by mechanical **scarification** of the seed coat, increasing time in storage, or blending with seed lots having low hard-seed content (Bass et al. 1988). Storage of dry seed at temperatures of 35 ° C for up to 84 months also decreases hard seededness and increases germination rates of alfalfa (Acharya et al. 1999).

Seedling Development

Legume seedlings emerge either by the epigeal or hypogeal mechanism. Most forage legumes have epigeal emergence by which the cotyledons are pulled above the soil surface (*epi* = above, *geal* = earth) by hypocotyl growth. In contrast, the hypocotyl of hypogeal-emerging legumes like the garden pea does not elongate leaving the cotyledons below ground (*hypo* = below) at planting depth.

Epigeal Emergence

In *epigeal-emerging* legumes, the seed swells by water absorption until it reaches about 50% water content when the primary root emerges and forms an anchor with the soil. At that time, the hypocotyl consists of columns of 100 or more very short epidermal cells that were formed during seed development. Elongation of the epidermal cells, mainly on the upper side, cause the arch to form while epidermal cells around the cylinder and directly below the arch elongate to push the arch upward until it reaches light at the soil surface. Length of epidermal cells below the arch can exceed 0.9 mm, those in the center of the hypocotyl reach 0.5 mm, and those closest to the root elongate to about 0.2 mm.

In light, the hypocotyl arch straightens, the cotyledons continue to expand and photosynthesis begins. The initial growth is supported entirely by seed reserves that are used for both energy and osmotica to drive cell elongation of the hypocotyl. The tender plumule, the first true leaves enclosed and protected during emergence by the cotyledons, is exposed above ground when the cotyledons open. The first true leaves also expand and normal growth begins from the shoot apex at the base of the first true leaf or leaves.

Photosynthesis of the expanded cotyledons is very beneficial to early seedling growth (Shibles and MacDonald 1962). Cotyledons shrivel as their nutrients are expended to support seedling growth. Early loss of both cotyledons can severely reduce seedling development and was noted to be fatal to *Medicago falcata*. Overall, cotyledons provide nutrition to the seedling from reserves and through photosynthesis up to 28 days from imbibition and may also stimulate first leaf expansion through a hormonal-type signal (Hongxiang et al. 2008).

Downward **radicle** penetration is improved when upward seed movement is restrained by soil during germination (Caradus 1990). Therefore, epigeal emergence can be a disadvantage for legume seed planted by overseeding if it is not incorporated in the soil by harrowing, frost action, or rain.

Hypogeal Emergence

In *hypogeal-emerging* species the seed (cotyledons) remains at planting depth in the soil as an anchor while the **epicotyl** pushes the plumule above ground. Hypogeal emergence is also an advantage if the emerged seedling tip is damaged by cutting, insect pests, or frost. Regrowth can come from axillary buds of the cotyledons (food supply) located below ground. But hypogeal seedlings do not benefit from photosynthesis of the cotyledons and this is a disadvantage when nutrient reserves are not sufficient, such as in pigeonpea which has small seed size and hypogeal emergence (Brakke and Gardner 1987).

The first true leaf from the **plumule** can be unifoliolate, as in alfalfa, cowpea, clovers, and soybean, or trifoliolate, as in birdsfoot trefoil, bundleflower, and crownvetch (Figure 3.1). In annual lespedeza, the first two true leaves are unifoliolate. Subsequent leaves of legumes are consistent with characteristics of the species. The epicotyl continues to grow by initiating a new leaf at each node along the developing stem. Secondary stems arising from the axillary buds of the cotyledons begin formation of the crown (Figure 3.1) (Teuber and Brick 1988).

Root Systems

The **radicle**, or primary root, penetrates the seed coat during **germination** and develops as a simple tapering taproot. In cross section, the primary taproot consists of an epidermis, cortex, and stele, which is the inner solid portion of vascular tissue. The taproot penetrates soil to approximately 1 m for most species; however, alfalfa taproots can penetrate to depths of 8 m in unrestricted soil (Martin et al. 1976). Most forage legumes form numerous lateral roots arising from the pericycle that surrounds the stele. Taproots of spring-seeded alfalfa may become branched 3–4 months after planting, depending on genetics.

There is great interest in root traits for plant improvement (Paez-Garcia et al. 2015). While long taproots can be beneficial in dry conditions, taprooted plants with many lateral roots or branches may be more effective in nutrient uptake. Branch-rooted plants are less susceptible to frost heaving than those with taproots and few laterals (Garner 1922). The degree of lateral root development is dependent on environment. Root branching is more pronounced in unfavorable soil conditions such as dense upland soils (Garner 1922; Weaver 1926), acidic soils (Joost and Hoveland 1986), or those with poor drainage (Chapter 25). Alfalfa seedlings affected by autotoxicity have shortened taproots, increased root branching, and other effects, leading to lower productivity (Jennings and Nelson 2002b).

Species differ in their degree of natural root branching. For example, roots of red clover and rhizoma peanut are extensively branched, but among alfalfa species, *Medicago sativa* generally produces deep taproots with few branches, whereas *M. falcata* produces many root branches and is more winter hardy. Root branching is also under genetic control. In Minnesota, Lamb et al. (2000) found alfalfa strains selected for branching did not differ in disease **resistance** or fall dormancy, but had 9–16% higher yield than the tap-rooted counterparts. The ecologic functions of root morphology of legumes need more research.

Root Hairs

Root hairs are short tubular extensions of epidermal cells that develop behind the zone of elongation of the growing seedling root tip and on subsequent branches. Root hairs increase absorbing surface of the root and are the site of infection by **rhizobia** bacteria for N_2 fixation (Teuber and Brick 1988). Greater root surface area through greater branching and root hairs enhances both nutrient absorption and nodulation. Symbiotic rhizobia bacteria that infect root hairs cause development of nodules on roots. The bacteria in the nodules use carbohydrate from the plant and N from the soil air to fix N into organic forms that are available to the plant.

Most nodules develop in the upper 20 cm of soil (Bowley et al. 1984), but nodule development is poor

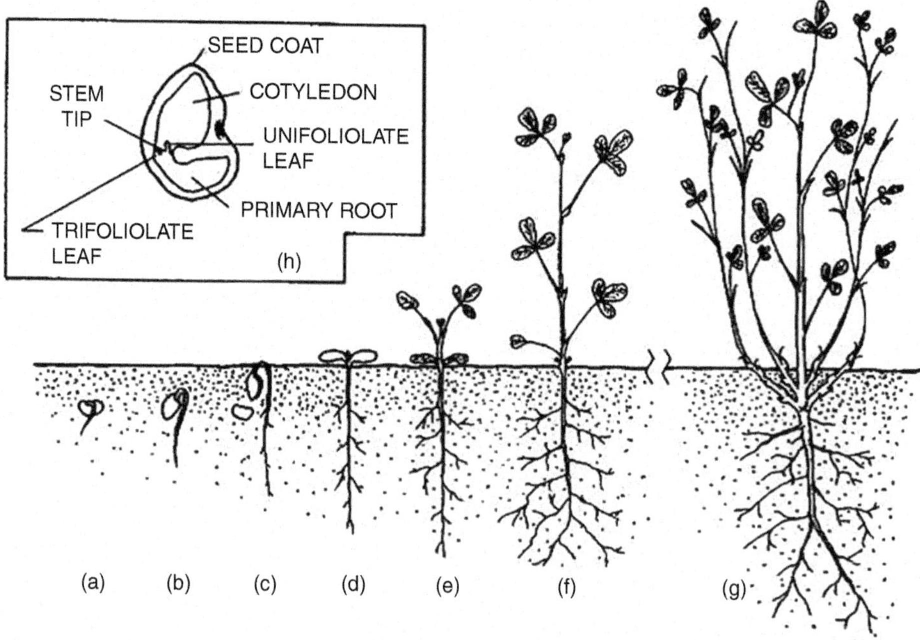

Fig. 3.1. Germination and development of a legume seedling, such as alfalfa, with epigeal emergence. (a) Seed imbibes water and primary root emerges. (b) Hypocotyl becomes active and forms an arch to penetrate the soil. (c) Elongation of hypocotyl stops when the arch reaches light. (d) Arch straightens and cotyledons open for photosynthesis, exposing the epicotyl (plumule) that was protected by the cotyledons during emergence. (e) Primary root continues to elongate, enlarge, and develop secondary (branch) roots while the unifoliolate leaf develops, followed by the first trifoliolate leaf. (f) Cotyledons fall off while stem continues to elongate, forming a leaf at each node. Axillary buds at the cotyledonary node begin to expand to form new shoots. (g) Contractile growth has occurred, and taproot morphology is developing. Crown is forming, with branches at some of the nodes. (h) Alfalfa seed dissected longitudinally to show one cotyledon, the primary root, the small shoot apex, and embryonic leaves.

in acid soils because acidity reduces both the population and N-fixing capacity of many rhizobia strains. Nodule shape varies from basically elongated to spherical. Those containing active N-fixing bacteria have a pink to red interior due to the presence of leghemoglobin that protects the key enzymes from O_2. Ineffective nodules (those that do not fix N) are generally more numerous, smaller, more round-shaped, and have a greenish or white interior (Burton 1985).

Associated Root Functions

Roots function to absorb water and take up soil nutrients but are also involved in nitrogen fixation and storage of organic compounds to support growth when leaf area is low. These activities are described in more detail in Chapters 4, 5, and 6. Here we focus on how morphologic components are related to these functions.

Nitrogen fixation is higher for vigorous, actively growing legume plants and contributes to improved yields of companion grasses through recycling of fixed N into the soil. Defoliation of legumes causes both root growth and N_2 fixation to stop for a few days before resuming. Thus, frequent repeated defoliation can result in reduced plant vigor and subsequent sloughing of roots and nodules (Kendall and Stringer 1985). Alfalfa releases more N through decomposing roots than decomposing nodules, whereas birdsfoot trefoil contributes more N to soil through decomposing nodules than roots (Duback and Russelle 1994).

In perennial legumes, roots also serve as storage organs for proteins and carbohydrates needed during regrowth after defoliation and for winter survival (Nelson and Smith 1968b; Blaser et al. 1986; Chapter 4). Carbohydrate storage in alfalfa roots reaches a maximum at or just after full bloom (Bowley et al. 1984). It is utilized to provide food for the regrowing plant after harvest and is then gradually restored. Frequent removal of plant topgrowth before adequate recharge of carbohydrate reserves weakens plants and reduces plant regrowth and persistence.

In annuals such as common lespedeza, little carbohydrate is stored in roots, thus regrowth is reduced if leaf area is completely removed (Davis et al. 1995). Timing of fall harvest and grazing management are important for perennial legumes to develop and maintain adequate carbohydrate and N reserves for winter survival, rapid spring growth and stand persistence. Defoliation of alfalfa during mid-October greatly increased winter injury in the north central US compared to defoliation during mid-September (Haagenson et al. 2003).

Crown Development and Contractile Growth

The transitional structure between the shoots and roots is referred to as the crown. During early crown development, shortly after seedling emergence (Figure 3.2), seedling roots of many epigeal legumes undergo contractile growth. Hypogeal legumes retain axillary buds at the cotyledons and on short internodes below soil that help the **epicotyl** emerge and usually do not have well-developed crowns.

Contractile growth of the hypocotyl and upper root helps anchor the plant (Esau 1977) and causes the cotyledonary node of epigeal seedlings to be pulled near or below the soil surface where the crown develops (Figure 3.1). During contractile growth, the elongated cells of the hypocotyl remain partially attached to each other but are forced apart laterally by secondary growth of the stele and outer phloem parenchyma cells. When the long cells of the upper root and hypocotyl are stretched

Fig. 3.2. Typical development of red clover seedlings exhibiting cotyledons, followed by the unifoliolate leaf, followed next by the first trifoliolate leaf. *Source*: Photo courtesy of John Jennings.

laterally by radial growth, they become larger in diameter and shorter in length, thus shortening the hypocotyl and upper root to pull the cotyledonary node downward (Cresswell et al. 1999).

Most perennial legumes like white clover end up with the cotyledonary node slightly below ground level (~0.6 cm) whereas red clover may have the node lowered 1.2 cm or more. Alfalfa cultivars with greater fall dormancy exhibit more pronounced contractile growth (2.0 cm) than do less dormant cultivars. Sweetclover is very winter hardy and can lower the cotyledonary node by more than 3 cm. Deeper crown placement due to contractile growth improves plant persistence, tolerance to grazing or wheel traffic, and **resistance** to heat and freezing (Teuber and Brick 1988). There is little evidence that seeding depth affects the amount of contractile growth.

Crown Development

The first shoots on the crown arise from axils of the two cotyledons and the first true leaf. But when each of the three shoots develop, the lower few internodes remain very short and their axillary buds are covered and protected by an incomplete leaf (no blade). If sufficient light reaches the growing crown, new shoots can arise from these buds increasing the number of shoots per crown. Each time a new shoot develops, it produces short internodes and protected buds that can further expand the crown and the number of shoots per plant.

Crowns may be simple or may develop many bud-laden branches. In perennial legumes, most regrowth comes from buds formed on the base of the cut shoot. This causes the buds for each cutting or grazing to be slightly further from the root as the crown spreads. Spring growth originates from the overwintering buds formed lower on the crown that developed during the previous fall (Nelson and Smith 1968a). Grazing-tolerant legumes like white clover have more crown buds and stems or stolons and have low-set, broad crowns. Grazing-tolerant alfalfa cultivars tend to have thicker stems and many more crown buds than grazing-intolerant cultivars (Brummer and Bouton 1991).

Stolons, Rhizomes, and Adventitious Stems

Plants may spread vegetatively by means of specialized stems such as stolons, adventitious stems or rootstalks, and rhizomes. Like in grasses, stolons of legumes (Figure 3.3) arise from axillary buds of the primary stem near the crown. These elongated, horizontal, aboveground stems grow along the soil surface by elongated internodes and root at the nodes. Axillary buds on these rooted stolons can give rise to new plants. White clover is a good example.

Plants may "creep" by means of adventitious stems, or rootstalks. Adventitious stems are different from stolons

Fig. 3.3. Some legumes, such as white clover, spread by stolons. Adventitious roots develop at each node. Leaflets are supported by long petioles, and inflorescences are supported by long peduncles, which are attached to nodes on the stolon. Axillary buds behind scale leaves at nodes have the potential to produce new branches. *Source:* Adapted from Isely (1951).

Fig. 3.4. Kura clover spreads by rhizomes. Stems and adventitious roots develop at each node. The parent plant produced from seed is on the left. *Source:* Photo courtesy of Michael Collins, University of Missouri.

or rhizomes since they do not develop from an axillary bud. Instead, they arise from dividing cells that organize and develop a shoot apex from either the taproot or from lateral roots to produce aboveground shoots. Crownvetch is an example of the latter, but the stimulus to originate the shoot on the lateral root is unknown.

Rhizomes are underground stems that are usually initiated from axillary buds of the lower crown or occasionally from an adventitious bud of the taproot. Rhizomes assist in vegetative propagation and are capable of storing carbohydrate reserves for plant growth. Roots and rhizomes are the primary carbohydrate storage location for rhizoma peanut (Saldivar et al. 1992). The rhizome, consisting of nodes, internodes, buds, and scale leaves, can produce adventitious roots or shoots at each node similar to aboveground stolons (Figure 3.4). Again, it is not understood how underground buds of legume rhizomes are stimulated to break dormancy and to form above ground shoots. See Chapter 2 for theory on grasses.

Rhizomes are differentiated from aerial shoots by presence of starch in plastids of parenchyma cells, in contrast to chloroplasts found in parenchyma cells of aerial shoots (Li and Beuselinck 1996). Light is required for plastids to develop into chloroplasts to gain their photosynthetic potential. Rhizomatous forages rely on rhizome storage, in addition to root storage, of protein and carbohydrate for regrowth after defoliation or winter survival. Rhizoma peanut spreads primarily by a dense mat of rhizomes that radiate from a woody taproot. When rhizoma peanut is cut for hay every 2 weeks, rhizome carbohydrate reserves are not decreased, but forage yield is, because morphology shifts from erect to decumbent and leaf area is closer to the ground than when harvests occur at 6 or 8 week intervals (Saldivar et al. 1992).

In contrast to frequent harvest for hay, frequent grazing causes a decline in rhizome carbohydrate reserve in rhizoma peanut because the leaves from decumbent growth are harvested by livestock causing reduced photosynthetic capacity (Ortega-S et al. 1992). Rhizoma peanut rarely produces seed, so field establishment is accomplished by sprigging rhizome fragments much the same as for planting hybrid bermudagrass.

For crown-forming legumes like alfalfa, birdsfoot trefoil, or bundleflower, the primary taproot generally survives until the plant dies. However, if part of the primary taproot is injured by diseases or frost heaving, plants may survive from intact lateral roots. In the case of clone-forming legumes like white clover, the taproot usually dies from diseases within two years, but plant survival is attained through production of rooted stolons (Figure 3.3) (Gibson and Holowell 1966). Each stolon node bears two adventitious root primordia, which can produce shallow, fibrous, adventitious roots (Thomas 1987). White clover cultivars with short stolon internodes have more nodes and a higher capacity to form nodal roots to compensate for the death of the taproot (Caradus 1990).

Stolon growth is regulated by light so they proliferate in open patches in grass stands. However, plants originating from shallow-rooted stolons are more prone to drought stress than are taprooted plants. Contractile growth by the nodal roots pulls the stolon partially into the soil, improving anchoring of the stolon and increasing grazing tolerance (Cresswell et al. 1999). As white clover stands age, they become a mixture of fibrous-rooted ramets from stolons and taprooted plants from volunteer seedlings. White clover **ecotypes** adapted to **continuous grazing** have more branched and rooted stolons (Brink et al. 1998). Some companies have developed this trait into commercial cultivars.

Shoots and Shoot Growth

A typical legume shoot consists of stems (internodes and nodes), branches, leaves (leaflets and petioles), trichomes or hairs in some species, inflorescences and fruits (seedpods). Stems may be hollow or solid, erect or decumbent, and climbing or trailing. Shoot length is generally 1 m or less for most herbaceous species used for hay or pasture production but can reach lengths over 2 m in some species. Stems in some species, such as bundleflower and shrub lespedeza, can be 1–2 m tall and become semi-woody providing excellent wildlife cover. In both red clover and alfalfa, the number of shoots per plant increases annually with age as the crown develops, due partly to death of nearby plants, but shows a general decline over successive regrowths within a given year (Bowley et al. 1984).

Shoot Effects on Management

The development of the shoot begins as vegetative (only stems and leaves), followed by bud, bloom, and mature (seed production) stages. In the vegetative stage, no floral buds, flowers, or seedpods are present. The primary growing points of legume shoots are the **terminal** bud (shoot apex) located at the tip of the stem and at the ends of stem branches, stolons, and rhizomes. On upright **determinate** plants like red clover, growth from the terminal bud continues until it differentiates into the inflorescence. On **indeterminate** plants like alfalfa, the terminal bud continues growing, but at a slow rate, when lateral apices produce flowers. After stems flower, new growth must come from crown buds. The developmental cycle from new shoot after cutting or grazing to flowering is typically about 30 days for species such as alfalfa and red clover. Within species, there may be genetic variation in vegetative growth during reproduction; for example, there are both determinate and indeterminate cowpea and soybean cultivars. Forage cowpea and soybean cultivars are indeterminate because they continue to produce leaves from the shoot apex after flowers develop from buds at nodes of the stems.

Upward growth of the main shoot is halted when the terminal bud is removed by cutting or grazing. The terminal bud is at or near the top of the plant for upright species like alfalfa, cowpea, red clover, soybean, and sweetclover and is nearly always removed by grazing or hay harvest. In contrast, the terminal bud in stoloniferous species like white or strawberry clover is located at the end of the stolon on or near the soil surface. The most visible growth is the leaf blades at the ends of upright petioles which are grazed or harvested reducing the probability of stolons being grazed.

In the case of red clover, the stem and shoot apex of plants in the vegetative stage remain low in the canopy and somewhat protected while the leaf blades displayed higher in the canopy on long petioles are grazed. Just prior to flowering, the shoot apex develops the flower head and internode growth of the stem elevates the inflorescence above the canopy where it is cut or grazed off and the plant must regrow from lower axillary buds. Similarly, in rhizomatous species like rhizoma peanut, the terminal bud is more protected from grazing. When the terminal bud is removed, successive growth must come from axillary buds in leaf axils or at nodes of cut stems or from crown buds. Regrowth from axillary buds may be upright such as in alfalfa or can be lateral or nearly prostrate such as in annual lespedeza or birdsfoot trefoil.

Plant hairs (trichomes) or glandular hairs may be present on leaves and stems; however, some species are completely glabrous (smooth) while others are weakly or densely pubescent (Figure 3.5). Leaves and stems of soybean are pubescent and the color may be gray or beige.

Fig. 3.5. Pubescence, dense white hairs from epidermal cells, occurs on some legumes, such as on this red clover stem.

Stems of common lespedeza are uniformly covered with hairs that point downward. In contrast, stems of Korean lespedeza have fewer hairs, only on the side opposite the leaf axis, that point upward (McKee 1944; Henson and Hanson 1961). Glandular hairs can provide defense against pests such as potato leafhopper (*Empoasca fabae* Harris) on alfalfa and soybean mosaic virus (Retallack and Willison 1988; Ren et al. 2000). Pubescence may reduce plant water stress in arid environments (Lenssen et al. 1989) or can slow the drying rate of harvested forages like red clover, thus increasing time needed for curing hay.

Shoot Effects on Quality

Plant age and population influence forage quality. Upper internodes and stem tips of clovers and alfalfa are lower in cell wall concentration than lower portions of the stem (Akin and Robinson, 1982; Buxton and Hornstein 1986). Shoots of alfalfa at low plant populations are higher yielding, larger in diameter, and have more nodes, but they are more lignified and less digestible than at high populations (Volenec et al. 1987). The harvest interval often recommended for optimum quality and yield of alfalfa is approximately 30 days between cuttings, but may be shorter during warmer temperatures or water stress during summer because shoots mature more quickly than during spring (Sanderson and Wedin 1988).

Plants vary widely in shoot maturity within a sward due to localized differences in light, moisture, nutrients, grazing pressure, or cutting height. Because crown and axillary buds can develop at different times, a single plant may have stems in several growth stages or maturity classes. This uneven maturity can make it difficult to accurately describe the growth stage for application of certain management practices. For alfalfa, the Mean Stage Weight (MSW) and Mean Stage Count (MSC) classification systems (Kalu and Fick 1981; Fick and Mueller 1989) improve precision of stage description. The MSC is easier to use in the field but may underestimate the true maturity due to new vegetative shoots that develop from axillary crown buds. Their presence may even cause the mean growth stage to decline in mature canopies. The MSW is not influenced as strongly by presence of new crown shoots and is more precise and useful for research applications to document the growth stage of canopies (Kalu and Fick 1981).

Leaf Effects on Quality

Most leaves of forage legumes are relatively young at 30 days since the average lifespan of a leaf from emergence to senescence is about 40 days. Leaves of mature legume plants are compound, i.e., more than one blade per petiole, and arise from near alternate positions on the stem. The leaf includes the petiole (stalk) that connects to the node and the paired, leaflike appendages called stipules that are fused at the base of the petiole. Stipules at the nodes help protect the axillary buds. They may have acute (pointed) tips or long slender points and may have margins ranging from entire (smooth) to lacerate (irregular indentations).

Petioles can be short as with alfalfa, in which the leaflets are displayed near the stem, or they can be several centimeters long, like with red clover or rhizoma peanut, to display the leaflets at a higher position in the canopy. Stoloniferous and rhizomatous legumes like white clover and rhizoma peanut depend on the petiole to elevate the leaflets in response to light to achieve a favorable position in the canopy for photosynthesis. Bundleflower has doubly compound leaves in which each leaf has multiple pairs of even-pinnate leaflets and each even-pinnate leaflet has multiple pairs of subleaflets giving the plant a fern-like appearance (Wynia 2007; Hilty 2017). Certain legumes with this fern-like leaf configuration such as bundleflower, sensitive briar, and partridge pea are touch-sensitive where leaves fold when touched.

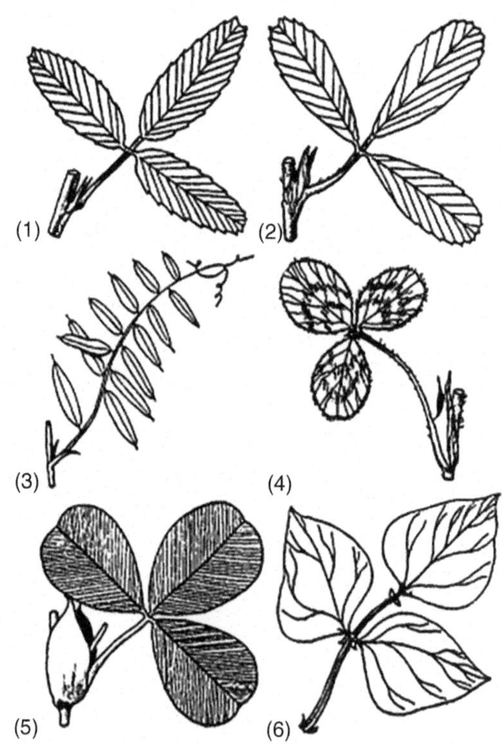

Fig. 3.6. Legume leaves differ in arrangement and shape of leaf blades, length of petioles and petiolules, presence or absence of pubescence, amount of serration on leaf blades, and size and shape of the stipules. Leaves of (1) sweetclover, (2) alfalfa, (3) hairy vetch, (4) red clover, (5) korean lespedeza, (6) cowpea. *Source:* Drawings 1 and 2 adapted from Isely (1951).

Individual leaflets are attached directly to the distal end of the petiole or are connected to the petiole by short-stalked petiolules. Leaf arrangement can be pinnately trifoliolate with unequal length petiolules (e.g., alfalfa, cowpea, and soybean), palmately trifoliolate with equal length petiolules (e.g., red clover), tetrafoliolate (e.g., rhizoma peanut), odd-pinnate (e.g., birdsfoot trefoil), or even-pinnate with terminal tendrils (e.g., hairy vetch) (Figure 3.6). Species that are normally trifoliolate can occasionally be multifoliolate, giving rise to the occasional "four-leaf clover," or in the case of alfalfa, some genetic lines may bear as many as nine leaflets per leaf. This character is one way of improving leafiness and has been used to market multi-leaf cultivars for superior quality alfalfa.

Unlike most monocots, legume leaflets do not have parallel venation (Figure 3.7). There may be a single midrib vein with secondary branches (pinnately netted venation) or several midribs originating at the base and spreading out toward the leaflet edges (palmately netted venation). This difference from monocots is relevant to the increased forage quality of legumes versus grasses. The portion of the leaf between veins that ruminal microbes can degrade easily, and thus utilized for animal gain or milk production, is more accessible to rumen microbes and more rapidly degraded compared with parallel venation of grasses.

Leaf shape can be cuneate (wedge-shaped with pointed end at point of attachment), elliptic (oval and narrowed to rounded ends), ovate (like a cross-section of a hen's egg with the broader portion near the base), obovate (like ovate, but inverted), obcordate (somewhat heart-shaped with attachment at the point), or rhomboid-ovate (generally oval but with four flattened sides) (Figure 3.8). Leaf margins may be entire (smooth) or finely serrate along all or portions of the margins. For example, sweetclover leaflets are serrate all around the margin, but alfalfa leaflets are finely serrate along only about one-third of the margin near the tip.

A whitish, inverted *V*-shaped "watermark" often occurs on the leaf surface of some clover species. Leaves may be pubescent with a watermark, as in red clover, or pubescent without the watermark as in crimson clover. Some species also exhibit occasional specks of reddish pigment on leaves. Leaf shape may change with plant age as with arrowleaf clover where young leaflets are short obovate, similar to white clover, but mature plants have later leaves exhibiting the characteristic elongated arrowhead shape (Figure 3.8) .

Leaf orientation of most legumes tends to be nearly horizontal resulting in a relatively low **critical leaf area index** of 3–4, the amount of leaf area per ground area required to intercept 95% of the solar radiation. In contrast, most grasses present leaves in a more vertical orientation, allow more light penetration into the canopy and have a critical leaf area index of 6–8. This makes legumes susceptible to shading from the companion grasses if allowed to develop a canopy above the legume plants.

Fig. 3.7. Leaf venation of a tall fescue (top) leaf blade, a grass, with parallel venation compared to the netted venation of a cowpea leaf blade, a legume (bottom). Cowpea has pinnately trifoliate leaves with palmately netted venation. The leaf tissue between the veins, or ribs, is more rapidly degraded by ruminal microorganisms than is the vein. Leaves with netted morphology have increased surface area between veins for microbial attachment and rapid degradation in the rumen.

Inflorescences and Flowering

As growth advances, flower buds of determinate plants are differentiated by the altered terminal meristem, usually in response to photoperiod. Flowers of indeterminate plants are formed from axillary buds at nodes along the main stem, especially from those near the apical meristems of the main stem and branches.

Fig. 3.8. Variation in leaf shape from a young trifoliate (top); to an older trifoliate (bottom) of arrowleaf clover. *Source:* Photos courtesy of John Jennings.

Onset of flowering, signals a continuing decline in forage quality because leaf growth slows, developed leaves continue to age, stems comprise a higher proportion of the aboveground weight, and stems become more fibrous and lignified (Buxton and Hornstein 1986). The combination of these factors causes a significant decline in voluntary intake by livestock, protein content and digestibility in legume forage with advancing maturity. Digestibility of leaves is higher than stems due to lower cell wall concentration and it declines less rapidly than stems with advanced maturity (Albrecht et al. 1987; Buxton and Redfearn 1997). Due to higher lignification with stem age and loss of leaves, the lower sections of the canopy have the lowest forage quality (Buxton and Hornstein 1986). Legumes with greater leaf: stem due to species or maturity have greater nutritive value and forage quality. New alfalfa cultivars have been developed using biotechnology to have lower lignin content in stems (Sheaffer and Undersander 2014).

Fig. 3.9. Inflorescences of legumes are generally one of three types: The inflorescence of red clover (*top*) is a compact raceme with short pedicels that connect to the peduncle; that of sweetclover (*middle*) is an elongated raceme with short pedicels; and birdsfoot trefoil (*bottom*) is an umbel with flowers attached radially to the end of the peduncle by short pedicels. *Source:* Photos courtesy of John Jennings.

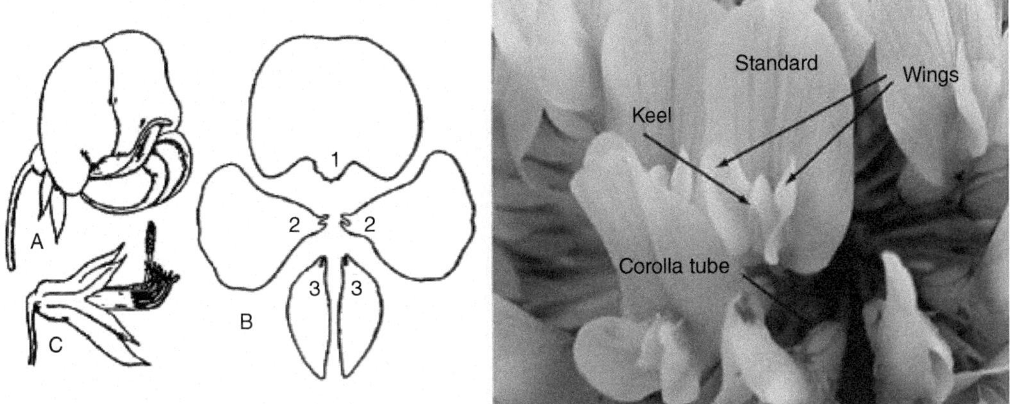

Fig. 3.10. Drawing of (a) a typical legume flower, (b) its five petals (1 = standard, 2 = wings, 3 = keel), and (c) the corolla tube that is inside the fused petals of the keel. Picture on right shows floral parts and exposed corolla tube of a white clover flower. *Source:* Photo courtesy of John Jennings.

The physical *inflorescence* (mode of flower-bearing) arises from a bud either at the stem apex (determinate plants) or in a leaf axil (mainly indeterminate plants). Inflorescences of legumes are usually a raceme or an umbel (Figure 3.9). Raceme inflorescences may be spikelike, as in alfalfa or sweetclover, or very compact, as in the true clovers. In red and white clovers, the raceme is not evident due to the compact structure of the inflorescence (Gillett 1985) and is commonly referred to as a **head**. Flowers mature from the bottom of the inflorescence toward the top of a head (Gibson and Holowell 1966). Number of flowers per inflorescence may range from fewer than 20 for alfalfa to more than 150 for some clovers. Flowering is influenced mainly by photoperiod and temperature.

Flowers and Pollination

Individual legume flowers are *papilionaceous*, meaning the corolla is butterfly- or pea-like and consists of five petals, namely a large standard or banner, two lateral wings, and two that are fused to form the keel (Figure 3.10). The five petals are partially joined at the base to form the corolla tube. The greenish calyx that protects the flower in the bud is normally four or five toothed and encircles the petals at the base of the corolla tube. Lespedeza flowers can be petaliferous (visible petals) or apetalous (without noticeable petals). Petaliferous flowers predominate with high temperatures, whereas apetalous flowers are more common later in summer or early fall (Martin et al. 1976).

The **keel** encloses the sexual column (**pistil** and **stamens**) of the flower. Each flower has ten stamens, nine of which are fused into a tube that encircles the style and **ovary** of the pistil (Figure 3.10). The tenth stamen is free. The keel petals are joined together by fingerlike projections (Viands et al. 1988) that fit together, much like a zipper, to protect the pistil and stamens.

Most forage legumes are cross-pollinated but may be self-fertile or self-incompatible. Subterranean clover, like soybean, is an exception since it is naturally self-pollinated (Rincker and Rampton 1985). Cross-pollination is normally by insect pollinators that are attracted to nectar secreted from cells in the corolla tube at the base of the **ovary**. The tube length influences cross-pollination, depending on the ability of the insect to reach the nectar. Honeybees are the most commonly used pollinator in commercial seed production, but leafcutter bees, alkali bees, and bumblebees, when nearby, are more effective on species having long corolla tubes like red clover (see Chapter 32).

Flowers of many species must be tripped (keel opened) by pollinating insects because the **stamens** are too short to allow the anthers and pollen to reach the **stigma**. Tripping occurs when a pollinator lands on the keel petals. The weight of the insect ruptures the junction of the keel petals, thus releasing the sexual column and exposing the pollinator to pollen (Figure 3.11). In an alfalfa flower, the sexual column is under pressure and, during tripping, springs out of the **keel** petals, dusting the insect with pollen. The tripping process for alfalfa is irreversible. In red clover, the sexual column returns to the original position when the weight of the pollinator is released from the keel petals (Bowley et al. 1984).

Fruits and Seed

After fertilization, the corolla withers and drops off, leaving the calyx at the base of the developing pod. Growing seed enlarges and the **ovary** wall grows into the pod that is characteristic of the species. Legume pods

Fig. 3.11. A tripped alfalfa flower showing the sexual column, which was under tension, has been tripped out of the fused keel petals to expose the hair-like stigma and yellow anthers. In nature, the column is released from the keel by pressure from an insect pollinator. *Source:* Photo courtesy of John Jennings.

are monocarpellary (one-chamber) fruits that contain a single seed or multiple seed in a single row; pods vary considerably among species (Figure 3.12). Seeds mature 22–30 days after fertilization. Pods may be dehiscent or indehiscent. At maturity, the dehiscent pod splits along both sutures or ribs. **Dehiscence** allows natural seed dispersal but is a problem for seed producers because mature pods can open, causing **seed shatter** before harvest. Dehiscence is especially difficult to manage for seed production in indeterminate species.

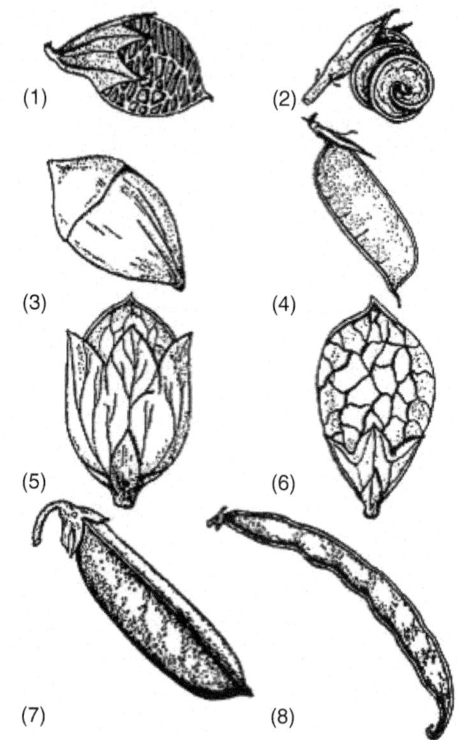

Fig. 3.12. Different types of legume seedpods (ovary wall) showing the residual of the stigma at the tip and dried sepals that remain attached at the base. 1, 3, 5, and 6 have one seed per pod. Seedpods are from (1) sweetclover, (2) alfalfa, (3) red clover, (4) hairy vetch, (5) common lespedeza, (6) korean lespedeza, (7) fieldpea, and (8) cowpea. The pod wall of the lespedezas remains intact during harvest and cleaning.

Summary and Conclusions

Knowledge about morphology of forage plants is important for developing superior cultivars and developing management practices that utilize the properties. As noted, there is considerable variability among species and within each species there is considerable plant-to-plant variation.

Plant breeders have narrowed the species variation within a cultivar, making the plants more consistent and predictable in their response to management. But nearly all pastures and range sites, and many harvested forages, are managed as mixtures of one or more legumes and one or more grasses. Thus, compatibility of morphology needs to be considered in selecting species for the mixture and then managing to favor desired but weaker competitors.

The goal of this chapter is to introduce the important morphologic features of legumes. In subsequent chapters this information will be integrated and applied to management practices and breeding strategies to improve yield, quality, and persistence of forage systems. Morphology will also be a useful foundation for improving management to best utilize the economic and environmental values of forages and grasslands.

References

Acharya, S.N., Stout, D.G., Brooke, B., and Thompson, D. (1999). Cultivar and storage effects on germination and hard seed content of alfalfa. *Can. J. Plant. Sci.* 79: 201–208.

Akin, D.E. and Robinson, E.L. (1982). Structure of leaves and stems of arrowleaf and crimson clovers as related to in vitro digestibility. *Crop Sci.* 22: 24–29.

Albrecht, K.A., Wedin, W.F., and Buxton, D.R. (1987). Cell-wall composition and digestibility of alfalfa stems and leaves. *Crop Sci.* 27: 735–741.

Bass, L.N., Gunn, C.R., Hesterman, O.B., and Roos, E.E. (1988). Seed physiology, seedling performance, and seed sprouting. In: *Alfalfa and Alfalfa Improvement* (eds. A.A. Hanson, D.K. Barnes and R.R. Hill), 961–983. Madison, WI: American Society of Agronomy.

Blaser, R.E., Hammes Jr., R.C., Fontenot, J.P. et al. (1986). Forage-animal management systems. Virginia Agricultural Experiment Station Bulletin No. 86–7.

Bordeleau, L.M. and Prevost, D. (1994). Nodulation and nitrogen fixation in extreme environments. *Plant Soil* 161: 115–124.

Bowley, S.R., Taylor, N.L., and Dougherty, C.T. (1984). Physiology and morphology of red clover. *Adv. Agron.* 37: 317–341.

Brakke, M.P. and Gardner, F.P. (1987). Juvenile growth in pigeonpea, soybean, and cowpea in relation to seed and seedling characteristics. *Crop Sci.* 27: 311–316.

Brink, G.E., Fairbrother, T.E., and Rowe, D.E. (1998). Seasonal variation in morphology of continuously stocked white clover. *Crop Sci.* 38: 1224–1228.

Brummer, E.C. and Bouton, J.H. (1991). Plant traits associated with grazing-tolerant alfalfa. *Agron. J.* 83: 996–1000.

Burton, J.C. (1985). *Rhizobium* relationships. In: *Clover Science and Technology*, American Society Agronomy Monograph, vol. 25 (ed. N.L. Taylor), 161–184. Madison, WI: American Society of Agronomy.

Buxton, D.R. and Hornstein, J.S. (1986). Cell-wall concentration and components in stratified canopies of alfalfa, birdsfoot trefoil and red clover. *Crop Sci.* 26: 180–184.

Buxton, D.R. and Redfearn, D.D. (1997). Plant limitations to fiber digestion and utilization. *J. Nutr.* 127: 814–818.

Caradus, J.R. (1990). The structure and function of white clover root systems. *Adv. Agron.* 43: 1–45.

Clark, A. (2007). Medics. In: *Managing Cover Crops Profitably*, Sustainable Agriculture Research and Education (SARE) program handbook series, 3e (ed. A. Clark), 152–159. College Park, MD: University of Maryland.

Cresswell, A., Sackville-Hamilton, N.R., Thomas, H. et al. (1999). Evidence for root contraction in white clover (*Trifolium repens* L.). *Ann. Bot.* 84: 359–369.

Davis, D.K., McGraw, R.L., Beuselinck, P.R., and Roberts, C.A. (1995). Total nonstructural carbohydrate accumulation in roots of annual lespedeza. *Agron. J.* 87: 89–92.

Duback, M. and Russelle, M.P. (1994). Forage legume roots and nodules and their role in nitrogen transfer. *Agron. J.* 86: 259–266.

Esau, K. (1977). The root: Primary state of growth. In: *Anatomy of Seed Plants*, 2e, 215–242. New York: Wiley.

Fick, G.W. and Mueller, S.C. (1989). Alfalfa: Quality, maturity, and mean stage of development. Cornell University. Cooperative Extension Information Bulletin No. 217.

Garver, S. (1922). *Alfalfa root studies*. Washington, DC: USDA Bulletin 1087.

Gibson, P.B. and Cope, W.A. (1985). White clover. In: *Clover Science and Technology*, American Society Agronomy Monograph, vol. 25 (ed. N.L. Taylor), 471–490. Madison, WI: American Society of Agronomy.

Gibson, P.B. and Holowell, E.A. (1966). *White Clover*. Agric. Handbook No. 314. Washington D.C.: USDA-ARS.

Gillett, J.M. (1985). Taxonomy and morphology. In: *Clover Science and Technology*, American Society Agronomy Monograph, vol. 25 (ed. N.L. Taylor), 7–47. Madison, WI: American Society of Agronomy.

Haagenson, D.M., Cunningham, S.M., Joern, B.C., and Volenec, J.J. (2003). Autumn defoliation effects on alfalfa winter survival, root physiology, and gene expression. *Crop Sci.* 43: 1340–1348.

Henson, P.R. and Hanson, C.H. (1961). *Annual Lespedeza Culture and Use*. Washington DC: USDA Farmer's Bulletin 2113.

Hilty, J. (2017). Illinois bundleflower. http://www.illinoiswildflowers.info/prairie/plantx/il_bundleflowerx.htm (accessed 8 October 2019).

Hongxiang, Z., Wu, Y., Matthew, C. et al. (2008). Contribution of cotyledons to seedling dry weight and development in *Medicago falcata* L. *N. Z. J. Agric. Res.* 51 (2): 107–114. https://doi.org/10.1080/00288230809510440.

Isely, D. (1951). The Leguminosae of the north central United States: Loteae and Trifolieae. *Iowa State Coll. J. Sci.* 25: 439–482.

Jennings, J.A. and Nelson, C.J. (2002a). Zone of autotoxic influence around established alfalfa plants. *Agron. J.* 94: 1104–1111.

Jennings, J.A. and Nelson, C.J. (2002b). Rotation interval and pesticide effects on establishment of alfalfa after alfalfa. *Agron. J.* 94: 786–791.

Joost, R.E. and Hoveland, C.S. (1986). Root development of sericea lespedeza and alfalfa in acid soils. *Agron. J.* 78: 711–714.

Kalu, B.A. and Fick, G.W. (1981). Quantifying morphological development of alfalfa for studies of herbage quality. *Crop Sci.* 21: 267–271.

Kendall, W.A. and Stringer, W.C. (1985). Physiological aspects of clover. In: *Clover Science and Technology*, American Society Agronomy Monograph vol. 25 (ed. N.L. Taylor), 111–159. Madison, WI: American Society of Agronomy.

Lamb, J.F.S., Samac, D.A., Barnes, D.K., and Henjum, K.I. (2000). Increased herbage yield in alfalfa associated with selection for fibrous and lateral roots. *Crop Sci.* 40: 693–699.

Lenssen, A.W., Sorenson, E.L., Posler, G.L., and Harbers, L.A. (1989). Sheep preference for perennial glandular-haired and glandular *Medicago* populations. *Crop Sci.* 29: 65–68.

Li, B. and Beuselinck, P.R. (1996). Rhizomatous *Lotus corniculatus* L.: II. Morphology and anatomy of rhizomes. *Crop Sci.* 36: 407–411.

Martin, J.H., Leonard, W.H., and Stamp, D.L. (1976). Alfalfa. In: *Principles of Field Crop Production*, 3e, 621–644. New York: Macmillan.

McKee, R. (1944). *Kobe: A Superior Lespedeza.* Washington DC: USDA Leaflet 240.

Nelson, C.J. and Smith, D. (1968a). Growth of birdsfoot trefoil and alfalfa. II: Morphological development and dry matter distribution. *Crop Sci.* 8: 21–24.

Nelson, C.J. and Smith, D. (1968b). Growth of birdsfoot trefoil and alfalfa. III: Changes in carbohydrate reserves and growth analysis under field conditions. *Crop Sci.* 8: 25–28.

Ortega-S, J.A., Sollenberger, L.E., Bennett, J.M., and Cornell, J.A. (1992). Rhizome characteristics and canopy light interception of grazed rhizoma peanut pastures. *Agron. J.* 84: 804–809.

Paez-Garcia, A., Motes, C.M., Scheible, W.R. et al. (2015). Root traits and phenotyping strategies for plant improvement. *Plants (Basel)* 15: 334–355.

Polhill, R.M. and Raven, P.H. (1981). *Advances in Legume Systematics.* Kew, UK: Royal Botanic Gardens.

Ren, Q., Pfeiffer, T.W., and Ghabrial, S.A. (2000). Relationship between soybean pubescence density and soybean mosaic virus field spread. *Euphytica* 111: 191–198.

Retallack, B. and Willison, J.H.M. (1988). Morphology, anatomy, and distribution of capitate glandular trichomes on selected *Trifolium* species. *Crop Sci.* 28: 677–680.

Rincker, C.M. and Rampton, H.H. (1985). Seed production. In: *Clover Science and Technology*, American Society Agronomy Monograph 25 (ed. N.L. Taylor), 417–443. Madison, WI: American Society of Agronomy.

Saldivar, A.J., Ocumpaugh, W.R., Gildersleeve, R.R., and Prine, G.M. (1992). Total nonstructural carbohydrates and nitrogen of 'Florigraze' rhizoma peanut. *Agron. J.* 84: 439–444.

Sanderson, M.A. and Wedin, W.F. (1988). Cell wall composition of alfalfa stems at similar morphological stages and chronological age during spring growth and summer regrowth. *Crop Sci.* 28: 342–347.

Sheaffer, C. and Undersander, D. (2014). *New Reduced Lignin Alfalfa Varieties: A Potential Forage Quality Breakthrough.* St. Paul, MN: University of Minnesota Extension Services.

Shibles, R.M. and MacDonald, H.M. (1962). Photosynthetic area and rate in relation to seedling vigor in birdsfoot trefoil. *Crop Sci.* 2: 299–302.

Teuber, L.R. and Brick, M.A. (1988). Morphology and anatomy. In: *Alfalfa and Alfalfa Improvement* (eds. A.A. Hanson, D.K. Barnes and R.R. Hill), 125–162. Madison, WI: American Society of Agronomy.

Thomas, R.G. (1987). The structure of the mature plant. In: *White Clover* (eds. M.J. Baker and W.M. Williams). Wallingford, Oxon, UK: CAB International.

Venuto, B.C., Pitman, W.D., Redfearn, D.D. et al. (1999). *Rhizoma Peanut - A New Forage Option for Louisiana.* Baton Rouge, LA: Louisiana State University Agricultural Center.

Viands, D.R., Sun, P., and Barnes, D.K. (1988). Pollination control: Mechanical and sterility. In: *Alfalfa and Alfalfa Improvement*, American Society Agronomy Monograph 29 (eds. A.A. Hanson, D.K. Barnes and R.R. Hill), 931–960. Madison, WI: American Society of Agronomy.

Volenec, J.J., Cherney, J.H., and Johnson, K.D. (1987). Yield components, plant morphology, and forage quality of alfalfa as influenced by plant population. *Crop Sci.* 27: 321–326.

Weaver, J.E. (1926). *Root Development of Field Crops.* New York: McGraw-Hill.

Wynia, R. (2007). *Illinois Bundleflower.* Manhattan, KS: USDA NRCS Manhattan Plant Materials Center https://plants.sc.egov.usda.gov/factsheet/pdf/fs_deil.pdf.

CHAPTER

4

Carbon Metabolism in Forage Plants

Jeffrey J. Volenec, Professor, *Agronomy, Purdue University, West Lafayette, IN, USA*
C. Jerry Nelson, Professor Emeritus, *Plant Sciences, University of Missouri, Columbia, MO, USA*

Introduction

Carbon (C) metabolism of forage plants is a physiologic process that begins with photosynthetic activity, mainly in leaves, to convert atmospheric CO_2 to **sugars** that are translocated to other sites for respiration, growth or storage. In addition, sugars are respired to provide energy to drive physiologic processes associated with mineral uptake and developing **resistance** to abiotic and biotic stresses. Carbon metabolism is critical since cellulose and hemicellulose consist of about 40% C, and C is a major component of proteins, lignin and lipids. Collectively, these carbon compounds make up about 95% of forage yield, with the remainder being soil minerals.

This chapter focuses on capturing carbon dioxide (CO_2) from the atmosphere by green leaves, and passing it through cellular- and tissue-level processes of photosynthesis to make useful compounds. Some of these are metabolized within the leaf for respiration and wasteful photorespiration, while most are transported to other areas of the plant for synthesis of new tissues and storage. We also cover dark respiration involved in synthesis of new tissue in meristems and maintenance of cellular integrity in mature tissues, how it is regulated in various tissues and how it can be effectively managed.

Morphologic features of plants were introduced in Chapters 2 and 3. These aspects are considered here relative to canopy photosynthesis and use of carbohydrates for growth, stress resistance and storage for future use. Forage and pasture plants differ markedly in leaf size, orientation and photosynthetic capacity. We cover key factors in management to increase the amount of solar energy captured and how the products of photosynthesis are allocated and used in various plant parts. Other chapters address effects of mineral metabolism (Chapter 5), water relations (Chapter 6), and growth and development (Chapter 7), which all depend on C metabolism.

Photosynthesis and Photorespiration

Photosynthesis is critical since it is the beginning of the process of capturing CO_2 from the air and converting it to sugars and other products. The capture process begins in the chloroplast of green cells, which are green due to chlorophyll, the primary pigment for absorbing radiation. Chloroplasts also contain xanthophylls, carotenoids, and other pigments that absorb radiation and activate electrons that are initially captured as high-energy compounds capable of transforming CO_2 into organic materials. Collectively, these pigments absorb about 70% of the incoming visible radiation (wavelengths between 400 and 700 nm) to activate electrons for photosynthesis, about 10% is absorbed by non-photosynthetic tissue such as cell walls, about 10% is reflected back to the atmosphere, and about 10% is transmitted through the leaf (Woolley 1971).

The CO_2 concentration of air in the field and outside the leaf is currently about 410–420 ppm and is gradually

Forages: The Science of Grassland Agriculture, Volume II, Seventh Edition.
Edited by Kenneth J. Moore, Michael Collins, C. Jerry Nelson and Daren D. Redfearn.

increasing leading to increases in global temperature (Chapter 8). In light, stomata open to allow diffusion of CO_2 into the leaf when active chloroplasts in the mesophyll cells reduce its internal concentration to 60 ppm or less. The gradient causes CO_2 diffusion through the stomata and intercellular spaces to the mesophyll cells. It then moves though the liquid cytoplasm to the chloroplast where CO_2 is captured and incorporated into a 3-carbon acid that is metabolized into sugars and other molecules for use in the cell. Sugars in excess of the cellular need are transported to other areas of the plant and stored as **starch** or **fructan.**

Cool- and Warm-Season Species

Both grass and broadleaf forage crops are commonly described as **cool-season** or **warm-season** species because they grow better under cool (optimum of 20–25 °C) or warm temperatures (optimum of 28–35 °C), respectively. The growth rate is primarily dependent on rates of enzymatic reactions and turgor pressure that are involved with cell division, cell elongation, and secondary cell wall formation. Growth rates differ at low and high temperatures due to physiologic factors regulating growth processes in various parts of the plant and storage organs. How this allocation is regulated at the cellular levels is not fully understood, but is affected by genetics and crop management.

Regardless, growth processes of both cool- and warm-season species depend on a supply of carbon compounds, mainly carbohydrates from **photosynthesis**, that serve as substrates for **respiration** and synthesis of carbon compounds for new growth.

Cool-season grasses, legumes, and most **forbs** have the same photosynthetic pathway (C_3), that includes **photorespiration**; the light-driven loss of CO_2. The rate of photorespiration is relatively low at cool temperatures (<10 °C), but increases markedly with higher temperatures. Though warm- and cool-season legumes and forbs all have photorespiration, warm-season legumes like soybean and perennial peanut gradually adapted to function at high temperature by altering allocation of sugars and regulation of rates of various growth processes. Alfalfa appears unique among cool-season legumes since it has photorespiration, but retains a very wide range of temperature adaptation (Bula and Massengale 1972).

C_4 Photosynthesis Overcomes Photorespiration

Warm-season grasses overcame problems associated with photorespiration millions of years ago by evolving a different pathway for photosynthesis that does not support wasteful photorespiration and allows these species to thrive in warm, dry climates (Sage et al. 2012). In general, both cool- and warm-season species have similar capacity for absorbing radiation using chlorophyll and converting it to the initial high-energy compounds, but the downstream fixation of CO_2 into **sugars** differs markedly (Table 4.1).

Cool-season species of grasses, legumes and some forbs have the C_3 photosynthetic system, so named because the first product of CO_2 fixation is a 3-carbon acid, namely 3-phosphoglyceric acid (3-PGA) (Figure 4.1). The key **enzyme** in C_3 photosynthesis, ribulose bisphosphate carboxylase/oxygenase (RuBP carboxylase), fixes CO_2 directly using CO_2 in solution near the enzyme active site (Table 4.1). Rate of CO_2 movement to the active site depends on a diffusion gradient to move it from the atmosphere to the chloroplast and ultimately to the active site of RuBP carboxylase. For photosynthesis, the CO_2 reacts with ribulose bisphosphate, a 5-carbon sugar, in the enzyme active site to form two molecules of 3-PGA, which are used in subsequent enzymatic steps to form 6-carbon sugars.

Unfortunately, O_2 from the atmosphere also reacts in a competitive manner with CO_2 at the RuBP carboxylase active site such that the 5-carbon sugar splits to form only one 3-PGA molecule and a 2-carbon compound, glycolic acid. Glycolic acid is not useful for carbon gain, so it is reoxidized to release the CO_2 so energy is lost in the wasteful process of photorespiration. This process usually consumes 20–30% of a plant's potential photosynthesis, but under some conditions, especially high temperatures, it can reduce photosynthetic potential by as much as 50% (Peterhansel and Maurino 2011).

Research shows photorespiration is reduced in high CO_2 environments because the CO_2 competes better with O_2 to enter the active site of RuBP carboxylase (Bazzaz 1990). Early in earth history, CO_2 concentrations in the atmosphere were high and photorespiration was not a problem. But gradual decreases in atmospheric CO_2 beginning about 40 million years ago accompanied by simultaneous increases in concentrations of plant-generated O_2 in the atmosphere stimulated photorespiration causing C_3 plants to became less suited for hot environments especially when combined with high **evapotranspiration.** Gradually, the several biochemical steps and anatomical changes in leaves, including more and larger veins and specialized bundle sheath cells, led to C_4 photosynthesis.

Biochemical and Anatomical Changes

In contrast to C_3 species, warm-season grasses with C_4 photosynthesis also have a specialized CO_2 concentrating process in mesophyll cells that uses the **enzyme** carbonic anhydrase (CA) to first capture CO_2 in the cytoplasm to form carbonic acid (HCO_3^-) (Figure 4.1). Phosphoenolpyruvate carboxylase (PEPc), the key enzyme in C_4 plants, has a different active site than RuBP carboxylase and does not react with O_2. Instead, PEPc transfers CO_2 from carbonic acid to a 3-carbon acid to form a 4-carbon

Table 4.1 A comparison of leaf characteristics of forage legumes and grasses with C_3 or C_4 photosynthetic metabolism

Characteristic	Tropical C_4 grasses	Tropical C_3 legumes	Temperate C_3 grasses	Temperate C_3 legumes
Representative genera	*Panicum, Pennisetum, Sorghum, Seteria*	*Stylosanthes, Vigna, Desmodium, Macroptilium*	*Festuca, Lolium, Dactylis, Agropyron, Bromus*	*Trifolium, Medicago, Onobrychis, Lotus*
CO_2 fixing enzyme and location	PEP carboxylase in cytoplasm of mesophyll cells	RuBP carboxylase in chloroplasts of mesophyll cells	RuBP carboxylase in chloroplasts of mesophyll cells	RuBP carboxylase in chloroplasts of mesophyll cells
Initial product of CO_2 fixation in mesophyll	4 carbon organic (oxaloacetate) acid	3-phosphoglyceric acid	3-phosphoglyceric acid	3-phosphoglyceric acid
Bundle sheath cells	Yes, with chloroplasts	Possible, no chloroplasts	Possible, no chloroplasts	Possible, no chloroplasts
CO_2 concentration in mesophyll cells	0–10 ppm at high solar radiation	40–100 ppm at high solar radiation	40–100 ppm at high solar radiation	40–100 ppm at high solar radiation
Light saturation of leaf photosynthesis	No, photosynthesis increases with increasing radiation	Yes, photosynthesis does not increase at radiation levels greater than 1/3 to ½ full sun	Yes, photosynthesis does not increase at radiation levels greater than 1/3 to ½ full sun	Yes, photosynthesis does not increase at radiation levels greater than 1/3 to ½ full sun
Light utilization efficiency (μg CO_2 J^{-1} PAR)	15.7	9.3	10	6.3
Conversion of solar radiation to dry matter, %	4.5	na[a]	3.1	1.9
Extinction coefficient for light passing through a canopy	0.66	1.04	0.52	0.92
Leaf photosynthetic rate (μmol CO_2 m^{-2} s^{-1})	61	24	17	21
Photorespiration	Not detected	15–30% of fixed C lost at moderate temperatures	15–30% of fixed C lost at moderate temperatures	15–30% of fixed C lost at moderate temperatures
Stomatal resistance (m^2 s mol^{-1})	2.3	2.3	5.8	na
Intracellular resistance (m^2 s mol^{-1})	1.1	7.2	10.3	na
Canopy dark respiration rate (mmol CO_2 m^{-2} s^{-1})	3.0	na	6.0	9.0
Maximum crop growth rates (g m^{-2} d^{-1})	50	13	33	23
Water use efficiency, kg H_2O kg^{-1} DM	220–840[b,c]	350–1100[d]	600–1800[b,c,e]	400–1250[c,f]

Source: Adapted from Ludlow (1985) and other sources.
[a] na, not available.
[b] Erickson et al. (2012); Zhang et al. (2018);
[c] Fairbourn (1982)
[d] Bell et al. (2013)
[e] Power (1985)
[f] Saeed and El-Nadi (1997); Bauder et al. (1978); Bolger and Matches (1990); Grimes et al. (1992)

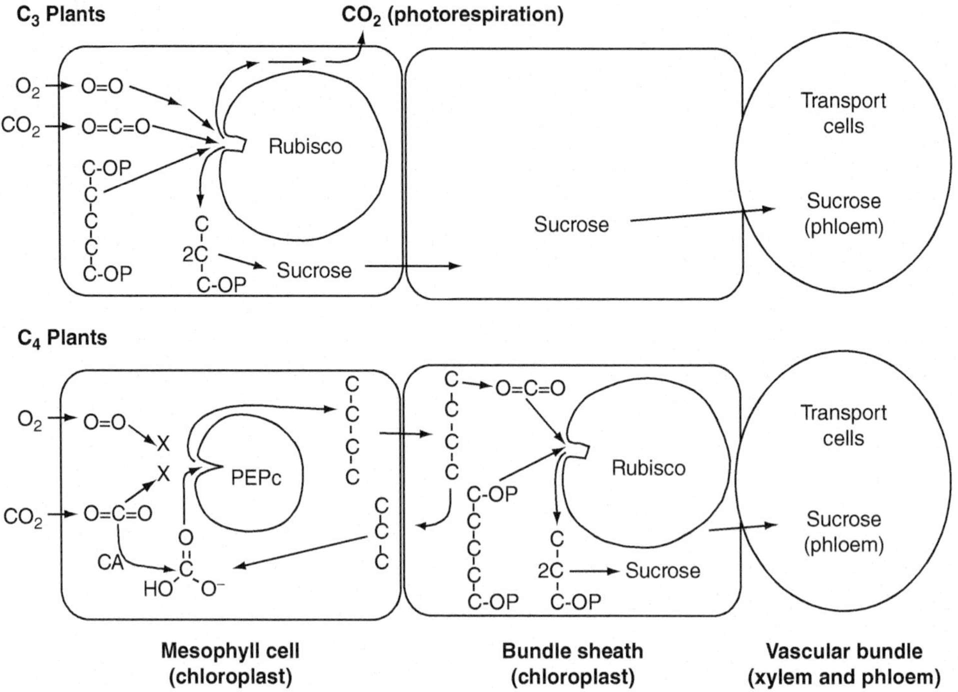

Fig. 4.1. In mesophyll cells of C_3 plants, RuBP carboxylase (Rubisco) catalyzes the reaction of CO_2 with ribulose bisphosphate, a 5-carbon sugar, to form two molecules of 3-phosphoglyceric acid (3-PGA). Glucose and then sucrose are formed from the 3-PGA and used in the cell or transported. If O_2 reacts with ribulose bisphosphate, one 3-PGA and one 2-carbon acid are formed, the latter is quickly oxidized to CO_2 and lost. In contrast, the PEP carboxylase in mesophyll cells of C_4 plants does not react with O_2 to initiate the photorespiration process. The CO_2 fixed by carbonic anhydrase (CA) and transferred to PEP carboxylase forms 4-carbon acids that are moved to the bundle sheath cells, where the CO_2 is released. The high CO_2 concentration at the active site of Rubisco in the bundle sheath cells does not allow O_2 to react and cause photorespiration. *Source:* Adapted from several sources.

acid (oxaloacetate, OAA) as the first product of C_4 photosynthesis (and thus the name). The OAA can easily be converted to malate or aspartate (species dependent) and transported to the bundle sheath cells.

These 4-carbon acids are transported from the mesophyll cells through pores in cells called **plasmodesmata** to the **bundle sheath** cells, a row of cells surrounding each vein, where chloroplasts containing RuBP carboxylase are located. In the bundle sheath cells, an enzyme cleaves the captured CO_2 from the 4-carbon acid, which also reforms the 3-carbon acid that returns to the mesophyll cells to pick up another CO_2, repeating the "pumping" process. Due to this pumping or concentrating mechanism, the CO_2 concentration in the bundle sheath cells of C_4 species is much higher than in the mesophyll cells of C_3 species. Thus, RuBP carboxylase in the bundle sheath cells of C_4 species reacts nearly exclusively with CO_2, excluding O_2 from its active site, which effectively eliminates photorespiration.

In addition to being spatially separated, the two main photosynthetic **enzymes** in C_3 and C_4 species differ in size and other properties (Table 4.1). RuBP carboxylase is a very large molecule and the most abundant protein in chloroplasts, often making up about 50% of the total protein in leaves of C_3 plants. It is likely the single most abundant protein in nature. It has a relatively slow reaction time, however, so a large amount of enzyme is needed. In contrast, mesophyll cells of C_4 species have PEPc, a protein that is about 70% as large, but is much more active than RuBP carboxylase. Lower amounts of RuBP carboxylase are needed in bundle sheath cells of C_4 leaves due to the absence of photorespiration. Thus, the leaf protein concentration in C_4 plants is usually lower than in C_3 plants (von Caemmerer 2000; Barbehenn et al. 2004).

The C_4 process of adaptation probably coincided with warmer and dryer conditions during an earlier period in Earth's history. Very early the CO_2 concentration of the air was several-fold higher than today, so the C_3 system worked well. But as the climate shifted and large amounts of carbon from plant and animal residue were gradually converted into oil and coal, the CO_2 concentration of the air decreased. This allowed O_2 to enter the active site of RuBP carboxylase with greater frequency leading to CO_2 loss via photorespiration. At high temperatures, O_2 diffuses through aqueous solutions, like the cytoplasm, more readily than CO_2.

Currently, photorespiration is not a serious problem in C_3 plants at low temperatures (generally less than 10 °C), but as leaf temperature increases, solubility of O_2 increases such that at 30 °C photorespiration can account for up to 40% of the photosynthetic potential. At higher temperatures, photorespiration can exceed 50% which, along with dark respiration, is a serious detriment to productivity and survival of C_3 plants (Volenec et al. 1984).

Physical separation of the enzymatic processes like in C_4 plants overcame photorespiration. In addition, the high affinity for HCO_3^- and the superior reaction rate of PEP carboxylase of C_4 plants allowed for lower operating CO_2 concentrations in the leaf and faster diffusion, which placed less demand on having fully open stomata. Most C_4 plants have a lower density and smaller stomata than do C_3 plants. This allows C_4 plants to be better adapted to both warm and dry environments by losing less water through stomata as CO_2 diffuses into leaves during photosynthesis (Table 4.1). This adaptation improved carbon and water economies in dry areas with warm temperatures resulting in C_4 plants generally having higher water-use-efficiencies than C_3 plants (Chapter 6).

While photosynthesis of C_4 plants is generally not enhanced with elevated atmospheric CO_2 (Figure 4.2), photosynthetic rates of C_3 plants increase in high CO_2 environments, and especially at elevated temperatures, because of reduced photorespiration. The continued increase in Earth's atmospheric CO_2 levels may improve yield and competitiveness of C_3 plants in regions where they currently do not compare well with C_4 plants.

As mentioned above, the radiation-harvesting process for C_3 and C_4 species is similar at low radiation and both respond in a near linear manner up to about 1000 μmol **photon**s $m^{-2} s^{-1}$ of photosynthetic **photon** flux density (PPFD), about half full-sun (Figure 4.3). Electron activation in the light reactions of C_4 species is linear as radiation flux increases except at very high fluxes where some energy is lost due to high chlorophyll **fluorescence** and heat production. The dark respiration load is similar for all radiation fluxes if leaf temperature remains similar.

With less than about 30% of full sun (600 μmol photons $m^{-2} s^{-1}$), radiation is the primary limit for C_3 plants because the demand for CO_2 is low, the diffusion rate is adequate and the CO_2 concentration at the active site of RuBP carboxylase is sufficient to limit photorespiration. Net photosynthesis of C_3 leaves becomes radiation saturated at about 50% full sun with no further increases (Table 4.1). The diffusion rate of CO_2 is not adequate to saturate the active site of RuBP carboxylase, allowing O_2 to react and increase photorespiration. In contrast, C_4 leaves remain radiation limited beyond full sun (2000 μmol photons $m^{-2} s^{-1}$).

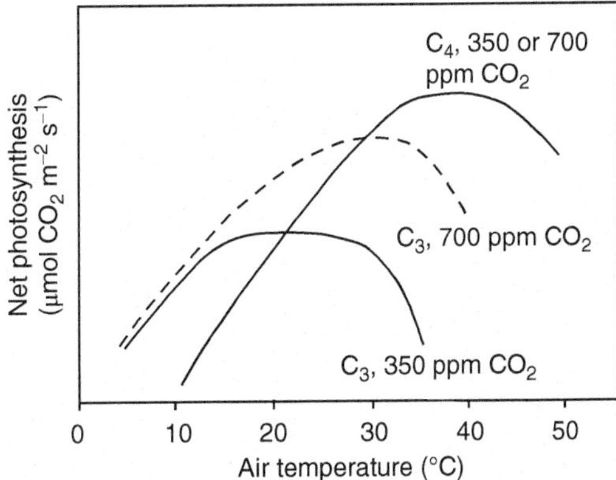

Fig. 4.2. Effect of air temperature on net photosynthesis of C_3 and C_4 species. Increasing the CO_2 concentration in the air increases both the rate and temperature optimum for net photosynthesis of C_3 species but has little effect on C_4 species. *Source:* Adapted from several sources.

Thus, canopy structure that permits even distribution of radiation among many leaves is especially critical to having high per-plant photosynthesis of C_3 plants. Differences in canopy structure and light interception between legumes and grasses also impact whole-plant photosynthesis. In contrast with grasses that usually have vertically oriented leaves in the upper canopy, legumes generally have horizontal leaf angles and higher extinction coefficients than grasses with each leaf layer absorbing progressively more light (Table 4.1). As a result, legume canopies tend to have a lower light utilization efficiency than grasses measured as μg CO_2 fixed per joule of **photosynthetically active radiation** (PAR).

The optimum temperature for photosynthesis of C_3 species is lower than for C_4 species (Figure 4.2, Table 4.1), mainly due to photorespiration. As described above, this can be largely overcome by increasing the concentration of CO_2 around the leaf. Therefore, the optimum temperature for C_3 leaves also increases at high CO_2 concentrations. But both the photosynthesis rate and optimum temperature remain lower than those for a C_4 species because the ribulose bisphosphate

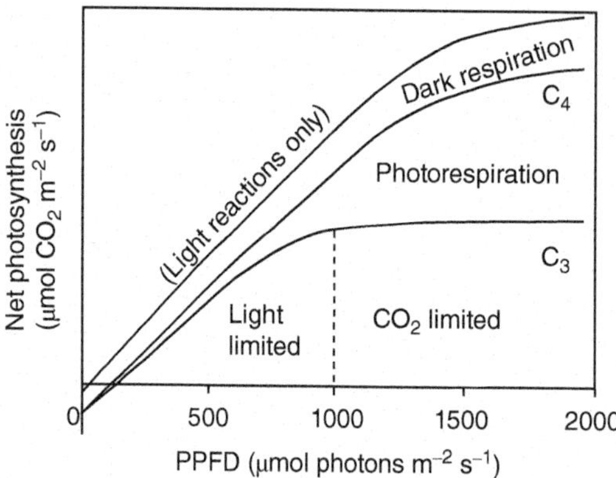

Fig. 4.3. Effect of visible radiation (photosynthetic photon flux density, PPFD) on rates of net photosynthesis of leaves. The rate of the light reactions in chloroplasts is closely related to PPFD, whereas dark respiration in **mitochondria** is little affected. C_4 photosynthesis remains mainly light-limited, whereas C_3 photosynthesis is light-limited to about 50% of full sun (full sun = 2000 µmol photons m^{-2} s^{-1}), after which it becomes CO_2-limited and photorespiration is a major problem. *Source:* Adapted from several sources.

molecules needed to react with RuBP carboxylase cannot be regenerated quickly enough. It is clear, however, that a doubling in CO_2 and subsequent effects on global warming will increase photosynthesis and yields of C_3 forages up to 30% (Bazzaz 1990; Ryle et al. 1992; Hebeisen et al. 1997; Kimball 2016). However, that increase will be partially offset if there is a relative increase in air temperature. Today, scientists are trying to repeat the complex steps of transition in the hope of changing C_3 crops to have C_4 photosynthesis.

CAM Photosynthesis

A third photosynthetic system, called crassulacean acid metabolism (CAM), occurs mainly in **cladodes**, the succulent flattened stem of plants such as cacti, and gives them an advantage in very dry areas. These species keep stomata closed during the day to conserve water, but stomata open for a short time at night to take in CO_2, which is captured rapidly by PEPc as in C_4 plants. The 4-carbon acid formed is stored overnight as malate in vacuoles. The next day, with stomata closed, the CO_2 is released from malate, energy from sunlight drives the light reactions and RuBP carboxylase captures the CO_2 in the same cells to form 3-carbon acids and sugars. The high CO_2 concentration limits the reaction with O_2 and photorespiration. Some cacti **cladodes** are used as forage in very dry areas after removing the needles mechanically or with fire.

Dark Respiration

Dark or oxidative respiration occurs in **mitochondria** of all living cells in leaves, stems and roots to use sugars, primarily sucrose, and other organic substrates to gain energy in the form of reduced pyridine nucleotides (**nicotinamide adenine dinucleotide + hydrogen [NADH]** and **nicotinamide adenine dinucleotide + phosphate [NADPH]**) and adenosine triphosphate (ATP) that support the actual work in the cell. Despite the popular use of the term, dark respiration, it is misleading because oxidative respiration occurs 24 hours each day in all tissues, but in leaves it could be lower in light than in darkness. Plant biologists have looked at the broad functions of respiration and have divided total respiration into growth and maintenance components (Amthor 1984).

Like animals, plants require respiration to maintain tissues that are not growing. For example, dormant seeds that are not growing yet need **maintenance respiration** to retain life. In contrast, **growth respiration** is associated with construction of new tissue and occurs mainly in meristems that are active for new growth and in storage organs that are synthesizing storage compounds and adding dry weight. Uptake of mineral nutrients requires respiration and is generally assigned to the growth component since there is weight gain. All non-growing tissues like older roots or fully expanded leaves have mainly maintenance respiration to repair DNA, proteins, and membranes, and to maintain ion balance in the cell. In most plants, maintenance and growth respiration occur simultaneously but activities of each are concentrated in specific tissues.

Maintenance Respiration

Cells need to respire to produce ATP and **NADH** for work in the cell, to repair components such as DNA, and to resynthesize proteins. Proteins gradually degrade, especially at higher temperatures, and need to be synthesized again in order to maintain basic cell functions. There is no net weight gain (e.g., growth) associated with this process that depends on respiration. Similarly, membranes may degrade and need repair. And, the mineral concentration inside the cell needs to be maintained at a higher concentration than outside, which requires respiration to pump mineral ions back into the cell when they leak out.

Maintenance respiration increases dramatically with temperature, having a Q_{10} of about 2.0, indicating the rate doubles with each 10 °C increase in temperature. Also, the proportion of fully-grown plant parts and associated maintenance respiration increases rapidly as the canopy develops relative to that in **meristematic** tissues of growing plant parts (Figure 4.4). Rate of maintenance respiration is also a direct function of protein content

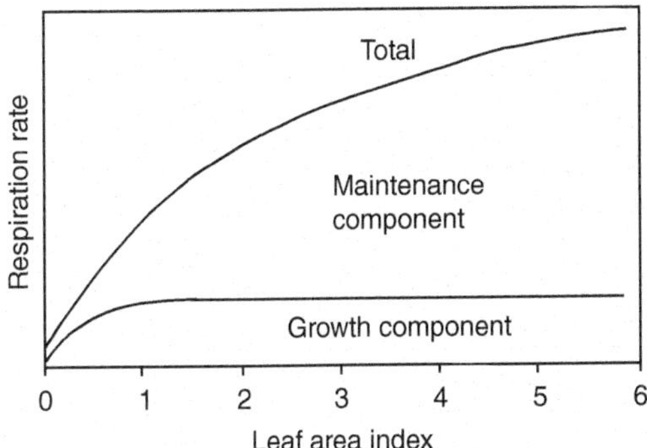

Fig. 4.4. During regrowth the number and size of meristems and amount of growth respiration required are established early and remain relatively constant as the plant grows unless there is a decrease of increase in meristems like tillers. Maintenance respiration occurs in all tissue, so it increases per land area as the canopy gets larger and the mass of developed leaf and stem tissue increases. The total respiration is not linear since older leaves that are no longer growing gradually reduce their maintenance respiration as they age and the protein content decreases. Both growth and maintenance respiration are related to temperature and to protein content of the tissue. *Source:* Adapted from Amthor (1989, 1994) and Nelson (1994).

as there is greater demand for repair and replacement of proteins, especially at high temperatures.

It is generally concluded that maintenance respiration has priority over growth respiration (Amthor 1994), such that if carbohydrate supplies decrease, growth respiration slows and stops while maintenance respiration continues. Based on this relationship, there have been attempts to reduce maintenance respiration in forage grasses (Wilson and Jones 1982), but the increase in yield has not been overly encouraging (Robson 1982; Pilbeam and Robson 1992). Decreasing maintenance respiration may have deleterious effects on plants such as reduced disease resistance or reduced ability to adapt to changes in the environment (Amthor 1989).

Growth Respiration

Rate of growth respiration is tightly linked to synthesis rate and composition of the products formed. Experiments have shown that biochemical reactions are very efficient and the summation of reactions involved is a good estimate of the total cost of tissue construction. Costs to synthesize a given product, however, depend on its composition. For example, it takes about 1.2 g of glucose to synthesize 1.0 g of a carbohydrate compound like starch, cellulose, or fructan. In comparison, it requires about 1.8 g of glucose to synthesize 1.0 g of protein, about 3.0 g to synthesize 1.0 g of lipid or fat, and about 2.1 g to synthesize 1.0 g of lignin (Amthor 1994). Organic acids are more oxidized than glucose, so it takes only 0.9 g of glucose to synthesize 1.0 g of organic acids.

The cost of construction of a plant part can be calculated from its composition. In general, leaves are higher in protein and generally have higher construction costs than stems that are lower in protein, but higher in lignin. Roots have low concentrations of both protein and lignin and, generally, the lowest construction costs (Poorter 1994). In comparison, the grain in cereals like corn or wheat is mainly starch and synthesis costs are low, whereas soybean seed is mainly oil and protein so cost of synthesis is much higher.

Rates of growth respiration are nearly always a direct function of growth rate in the meristematic tissues unless there is a major change in tissue composition. Thus, environmental factors affecting growth rate alter growth respiration independent of the effect of the same factor on maintenance respiration. Therefore, the rate of growth respiration decreases when growth rate is slowed by drought, heat, or biotic stress, while maintenance respiration increases. Because growth respiration is mainly in meristems, on a canopy basis it depends on the number and activity of meristems per unit land area whereas maintenance respiration depends on the mass of the plant per unit land area (Figure 4.4).

During early regrowth after cutting, there is active growth respiration in the meristems, yet maintenance respiration is low because there is little older tissue. As the canopy regrows, the number of active meristems tends to remain similar, but as herbage accumulates, it requires an increasingly higher maintenance respiration per unit land area. During the regrowth period, maintenance respiration would be further increased if temperatures increased. In contrast, growth respiration would decrease if the growth rate slowed due to higher temperatures or water stress. However, high temperatures and water stress also increase the lignin component of growth, and because it has a higher construction cost than cellulose, the amount of growth respiration needed to construct a given amount of leaf tissue would increase proportionally. Most physiologists agree that growth respiration is consistent with actual synthesis costs of biochemical reactions and is a good plant investment.

High-Temperature Stress

Most forage crops are mesophyllic, preferring intermediate growing temperatures of 10–30 °C. The threshold temperature for onset of high-temperature stress varies with the species, but is in the range of 35–45 °C

(Table 4.2). In addition, exposure time interacts with temperature to cause injury. Long exposures to moderately high temperatures (i.e. 40 °C for 60 minutes) can cause more injury than short-term exposure to high temperatures (i.e. 50 °C for 5 minutes).

High-temperature stress is also influenced by the extent of water deficit stress and reduced evaporative cooling with closed stomata. Temperatures of aboveground plant tissue often differ markedly from the surrounding air temperatures and need to be measured directly to determine the degree of stress. If ample water is available, transpirational cooling can reduce leaf temperatures by 5 °C or more compared with the actual air temperature. Latent heat of vaporization consumes 540 cal mol^{-1} of water evaporated with the source of energy being the leaf. When water deficit stress occurs, stomata close partially to limit **transpiration** and its associated cooling, and leaf temperatures increase due to continued absorption of solar radiation.

Under more severe stress, **buliform cells** on the upper surface of grass leaves decrease in volume and the blade rolls, reducing the surface area and limiting tissue exposure to radiation. Broad-leaf plants usually lose turgor and partially wilted leaves droop downward to intercept less solar radiation. Infra-red telethermometry can be used to detect increased tissue temperatures, which can be lowered by irrigation or surface wetting ("syringing") to enhance evaporative cooling.

Heat stress has several negative consequences for metabolism of forages, including reduced photosynthetic rates and elevated rates of maintenance respiration and higher photorespiration. In C_3 forage grasses like tall fescue, photosynthesis declines at 27 °C or higher, while dark respiration increases markedly at 30 °C (Volenec et al. 1984). Less water-soluble carbohydrate accumulates in stem bases of heat-stressed tall fescue, and herbage regrowth after defoliation is less than that of plants grown at 20 °C. These changes in photosynthesis, respiration, and storage water-soluble carbohydrates are often secondary responses to high-temperature stress.

One primary effect of high temperature is increased membrane fluidity. This alters membrane characteristics and functions, including changes in permeability and spatial arrangements of integral membrane proteins like those functioning in the light reactions of photosynthesis and oxidative phosphorylation of respiration (Niu and Xiang 2018). Loss of compartmentalization of cell solutes and reduced ATP synthesis associated with changes in membranes and their integral proteins can have rapid, devastating effects on plants.

Another direct consequence of high temperature on forage is protein denaturation causing lower functionality, and reduced synthesis or enhanced degradation of proteins (Bita and Gerats 2013), both of which increase maintenance respiration. Loss or reduction of even a single **enzyme** in a vital metabolic pathway can slow the process and cause reduced growth or even death if adaptation is slow or not possible. Accumulation of some pathway intermediates can lead to toxicity and cell death.

Resisting high-temperature stress falls into two general categories: heat avoidance and heat tolerance. Avoiding high-temperature stress occurs when plants do not experience stress, even though elevated air temperatures exist. One avoidance mechanism is to increase transpiration/evaporation from plant surfaces, something accomplished by using irrigation or application of water to plant surfaces. A plant mechanism for avoiding heat stress is to decrease radiation interception by leaf rolling or altering leaf orientation relative to the sun.

Heat tolerance occurs when plants have an enhanced ability to survive when experiencing high temperatures. Examples of heat-tolerance mechanisms include maintenance of proper membrane fluidity by an increase in saturation of membrane fatty acids, synthesis of heat-stable isozymes (alternate forms) of key **enzymes**, accumulation of carbohydrate reserves, and synthesis of heat-shock proteins that stabilize existing proteins or assist with cellular targeting of newly synthesized proteins (Wang et al. 2017). Although an understanding of the role of heat-shock proteins in tolerance to environmental stress continues to evolve, an increase in synthesis of heat-shock proteins has been correlated with heat tolerance across several plant species (Jacob et al. 2017).

Low-Temperature Stress

Low-temperature stress can be divided into chilling and freezing stress. Chilling stress occurs at temperatures cool enough to produce injury but without ice formation in plant tissues. Freezing injury occurs when ice crystals form.

Chilling Injury

For most forage plants **chilling injury** occurs at air temperatures between 0 and 10 °C, mainly at night. However, a few species (e.g., rice and sugarcane) are exceptionally sensitive and incur **chilling injury** at temperatures below 15 °C. Like high-temperature stress, both the intensity of low temperature and duration of exposure influence the extent of injury.

Direct injury such as a decrease in membrane fluidity and impaired function of integral membrane proteins can occur within hours of exposure to low, nonfreezing temperatures. Indirect injuries are manifested in many ways, including reduced membrane transport; inhibition of photosynthesis and respiration; accumulation to toxic levels of metabolites like ethanol and **ammonia**; and changes in rates of protein synthesis and degradation leading to metabolic imbalance. For example, forage plants possessing the C_4 photosynthetic pathway tend to be sensitive to chilling temperatures (Table 4.2), in

Table 4.2 Optimum and range in temperature tolerance of forage groups

Species	Low-temperature tolerance	Optimum temperature	High-temperature tolerance
Legumes	LT_{50} = −6.3 to −7.4 °C[a] Seedling survival at −10 °C for 8 h = 4–60%[b] Seedling survival for 8 h at −4.7 to −7.4 °C = 32–96%[c]	Whole-plant growth highest at 16 °C/10 °C[d] or 20 °C/15 °C (day/night)[e] Shoot growth highest at 27 °C[f] Shoot growth highest at a crown temp. of 32 °C[g]	Herbage growth at 35 °C/29 °C is 19% that at 16 °C/10 °C[d] Herbage growth at 32 °C is 44% that at 16 °C; root growth at 32 °C is 23% that at 16 °C[h]
Cool-season (C_3) perennial grasses	Seedling survival at −10 °C for 8 h = 31–98%[b]	Tiller growth highest at 25 °C[i]	Death at 34 °C; growth at 30 °C is 51% that at 25 °C[i]
Cool-season (C_3) annual grasses	Leaf growth ceases at 0 °C[j]	Rate of leaf appearance highest at 22 °C; rate of development greatest at 21 °C[k]	Rate of leaf appearance ceases at 45 °C; growth ceases at 33 °C[k]
Warm-season (C_4) perennial grasses	Survive −3 °C but death or severe injury at −6 °C[l] LT_{50} of −9 to −22 °C[m] Growth at 5 °C less than 10% that at 20 °C[n] Seedling base temperatures range from 2.6 to 7.3 °C[o]		
Warm-season (C_4) annual grasses	Growth ceases at 5 °C[k] Sorghum growth ceases at 8.5 °C[p]	Growth highest at 31 °C in maize; rate of development greatest at 27 °C in sorghum[k]	Growth ceases at 41 °C in maize; 34 °C in sorghum[k]

[a]Meyer and Badaruddin (2001); [b]Arakeri and Schmid (1949); [c]Tysdal and Pieters (1934); [d]Lee and Smith (1972); [e]Pearson and Hunt (1972); [f]Leach (1971); [g]Evenson (1979); [h]Gist and Mott (1957); [i]Volenec et al. (1984); [j]Kirby (1995); [k]Yan and Hunt (1999); [l]Chamblee et al. (1989); [m]Qian et al. (2001); [n]Clifton-Brown and Jones (1997); [o]Madakadze et al. (2003); [p]Craufurd et al. (1998).

part, because activity of PEP carboxylase, a key photosynthetic enzyme in these species, has reduced activity at temperatures below 10 °C (Table 3.1).

Forage plants can acquire resistance to **chilling injury** if they are acclimated by cool temperatures for several consecutive days prior to exposure to chilling temperatures that would normally injure plants. Chilling acclimation involves several physiologic adjustments, including increasing membrane fluidity by reducing the frequency of saturated fatty acids in membrane lipids and synthesis of isozymes of key enzymes that are stable at chilling temperatures (Zheng et al. 2016; Lee and Lee 2000).

Freezing Injury

As temperature declines below 0 °C, ice formation in living tissues is possible; thus, tolerance of freezing stress becomes a critical factor in survival of biennial and perennial forages. Seeds, pollen, and other tissues that have low water content are able to survive exposure to very low temperatures, some near −273 °C. However, hydrated seeds exposed to temperatures of −3 or −4 °C usually form ice crystals resulting in death. This indicates that ice formation in tissue and not low temperature *per se* is the cause of death.

Cell death is almost certain if ice formation occurs within cells (intracellular ice). Extracellular ice formation is the norm and can be tolerated if cell walls are flexible and tolerate the deforming effects of ice crystal formation between cells. In that case, cells can resume their initial shape after temperatures rise above freezing and the ice melts. Water in the extracellular spaces usually has less solute content and freezes before that in the cell interior. As the extracellular water freezes, water moves from the intracellular location to the cell exterior, which further concentrates cell solutes and depresses the freezing point of the cell.

Freezing injury occurs via several mechanisms. Mechanical damage can occur when the force exerted by enlarging ice crystals crushes cells. If excessive liquid water is drawn from within the cell as extracellular ice crystals grow, cells can dehydrate to the point that proteins and other constituents change shape and precipitate leading to cell death. Intracellular ice formation tears cell membranes, compromises compartmentalization of metabolites, and damages integral proteins and cell death results.

Resistance Mechanisms

Forage plants may resist freezing using an array of mechanisms that either avoid exposure to freezing temperatures or prevent ice formation when temperatures decline below 0 °C (Pearce 2001). Because soil can act as a large source of thermal energy during winter, crowns of legumes and stem bases of grasses that are placed at or below the soil surface can avoid exposure of crowns to freezing temperatures. Sweetclover is extremely winter hardy, in part, because it possesses a deep-set crown that places overwintering buds as deep as 4 cm below the soil surface.

Many forage species avoid ice formation when exposed to low temperatures by accumulating solutes in cells during cold acclimation. Nonreducing sugars (e.g. sucrose, raffinose, and stachyose), mineral ions, amino acids, and other solutes accumulate in autumn, and their concentrations are often higher in overwintering organs of winter-hardy cultivars when compared to cultivars that are winter-injured (Cunningham et al. 2003). Being a colligative property, freezing point depression is generally determined by the total concentration of solute present in cells and not by the nature of the solute per se. For aqueous solutions like those in plant cells, this relationship is described mathematically as follows:

$$\Delta T_f = -1.86 \times m$$

where ΔT_f is the freezing point and m is molality of the solution (moles kg^{-1} solvent). Both dehydration of overwintering tissues and loss of water to extracellular ice formation increase the molality of the cytoplasm and decrease the freezing point, but can effectively depress the freezing point only to about −4 °C.

In the absence of nucleators, liquid water may supercool to temperatures well below 0 °C without freezing. Water found in xylem tissues of trees in winter may lack nucleators, in which case these tissues may supercool until they reach −38 °C, which is the homogenous ice nucleation temperature for pure water. Practical application of supercooling may be limited because overwintering crowns, stem bases, and roots often have a great abundance of soil particles, bacteria, and other nucleating agents on them.

Managing for improved low-temperature tolerance and winter survival begins with planting species and cultivars that can effectively cold acclimate when exposed to shortening photoperiods and cool temperatures. Differences in low-temperature tolerance between species and cultivars within species can be large (Table 4.2).

Timely planting in late summer is necessary because small seedlings (<3-leaf stage for cool-season grasses such as timothy and bromegrass; <6-leaf stage for alfalfa and red clover) are more prone to freezing injury (Arakeri and Schmid 1949). Adequate potassium (K) fertility is important to winter survival of many forage species, whereas high nitrogen (N) fertility may reduce winter survival because of its growth-stimulating effects that interfere with cold acclimation (Adams and Twersky 1960).

Untimely harvesting in autumn can reduce leaf area or promote regrowth to reduce storage of nonstructural carbohydrates and may limit accumulation of sugars, amino acids, and proteins that serve as solutes to depress the freezing point (Haagenson et al. 2003). In addition, herbage remaining aboveground over winter provides a direct mulch effect and traps snow, which insulates soil from temperature change and enhances winter survival (Leep et al. 2001; Chapter 8).

Water Stress

More than any other ecological factor, water determines where and what type of vegetation grows on Earth. In addition, water stress often causes reduced transpiration rates resulting in high-temperature stress that further intensifies the overall stress level for plants. The importance of water is due almost exclusively to its key functions in plants:

1. Is the solvent for most reactions in plant cells and the transport stream among plant parts.
2. Provides turgor pressure that supports growth and determines shape and form of most plant organs.
3. Is a reactant in hydrolytic reactions.
4. Its **transpiration** cools plant tissues.

Measuring Water Stress

Water potential (Ψ) is a measure of the free energy status of water. In plant tissues, Ψ has three components that influence its value: solute potential (Ψ_s), pressure potential (Ψ_p), and matric potential (Ψ_m). These components are related to Ψ as follows:

$$\psi = \psi_s + \psi_p + \psi_m$$

The Ψ_m results from interactions of water and cell constituents (e.g. cell walls, proteins, polysaccharides). Its value is often ignored because its overall effect on Ψ is generally small relative to the other terms. Increased Ψ_s results from the solute concentrations of the cell sap. Plants hydrolyze polymers like fructan, starch, and proteins to their respective sugar or amino acid constituents in order to elevate cell solute concentrations to retain water and prevent loss of turgor; a process referred to as "**osmotic adjustment**." During moderate water stress, plants will often prevent wilting by closing stomata to reduce water loss and to maintain $\Psi_p \geq 0$. However, stomatal closure stops transpiration and usually results in increased leaf and canopy temperature and possible heat injury as described above (Jones et al. 2009).

Water movement in plants is from high to low (more negative) water potentials. At field capacity, Ψ_{soil} may be −0.2 MPa (i.e. −2 bars). At the same time, the Ψ_{root} (−0.3 MPa), Ψ_{stem} (−0.5 MPa), and Ψ_{leaf} (−0.7 MPa) each become increasingly more negative, causing water movement from roots to leaves. Finally, water vapor exits the leaf surface primarily through stomata to the atmosphere, which has a very negative Ψ (−150 MPa or lower).

Water stress affects plants in a process-specific manner (Figure 4.5). Cell expansion and cell wall synthesis are among the processes most sensitive to mild water stress. Accumulation of proline, abscisic acid (ABA), and sugars intensifies with increasing water stress, as plants adjust osmotically. Respiration rates can be reduced by initial water stress, but increases in respiration are often observed under moderate to severe water stress as ATP synthesis becomes uncoupled from O_2 consumption. Root growth is not as sensitive as leaf and stem growth, and usually mild to moderate stress first reduces shoot growth and increases allocation of sugars to increase root growth and storage.

Adaptation to Drought

Drought avoidance and drought tolerance are the two categories used to describe plant adaptation to water stress. Drought avoidance occurs when plants are never exposed to water stress and is generally brought about by increased water uptake or decreased water loss during a seemingly stressful period. Specific drought avoidance mechanisms include extensive root systems, controlling transpiration with stomatal control, cuticle structure, leaf shedding, and adjustment of the growing season to periods when water is more abundant. Drought tolerance occurs when plant tissues are able to function and/or survive drought stress. This is not common among forages, with the exception of sorghum and millet, which can stop growing during extended periods of drought and resume growth when water stress is relieved.

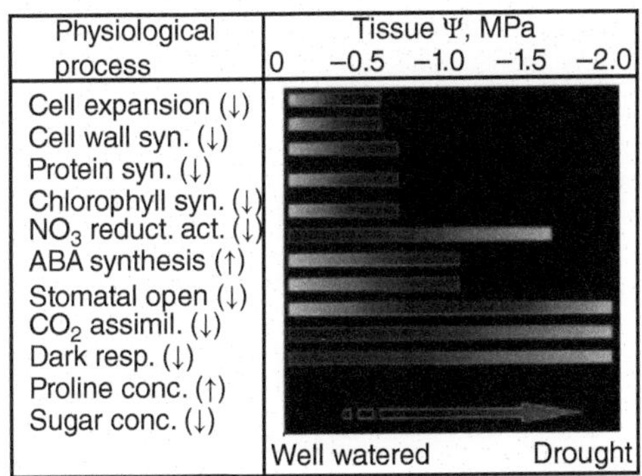

Fig. 4.5. Influence of tissue water potential on rates of important physiological processes in forage plants. Arrows indicate whether the effect is an increase or decrease. The darker the bar coloration, the greater the response of each process. *Source:* Compiled from several sources.

There are several research reports indicating that onset of drought stress affects leaf growth first and allows root growth to continue, and in some cases actually increase. Exposure of plants to moderate drought can increase storage of carbohydrate and proteins in the taproot of legumes or in the stem base of grasses. Roots that continue to grow are thinner and grow faster through the dry soil to reach more water. The ability to down-regulate allocation to shoot growth in favor of root growth and storage is beneficial in the short-term to reach more water and in the long-term to accumulate carbohydrate storage to support regrowth when the drought is relieved.

Water-use efficiency (WUE) is the amount of water used by a plant to produce an amount of dry matter during a defined time period (see Chapter 6 for details). Forages vary widely in WUE but generally separate into two categories based on the leaf photosynthetic mechanism. Due to higher photosynthetic rates, C_4 plants use less water than C_3 plants for each unit of dry matter produced. For example, legume and grass species with the C_3 photosynthetic pathway have WUE values that range from 350 to 1800 kg H_2O kg^{-1} dry matter produced. In contrast, grasses possessing the C_4 pathway generally have WUE values that range from 220 to 840 kg H_2O kg^{-1} dry matter produced; on average, about twice as productive per unit of water as C_3 plants. The wide variation within each group reflects environmental effects (e.g. relative humidity, wind, plant density) on water use and photosynthesis (Fairbourn 1982). Genotypic variation in WUE also exists within forage species. However, improved WUE is usually associated with higher forage yield rather than lower water use per se.

Management strategies for increasing WUE often involve manipulating the growing environment to stimulate dry matter production rather than reducing water use. For example, increasing soil fertility stimulates growth markedly with only a slight increase in total water use, thus improving WUE values (Smika et al. 1965; Power 1985). Use of shelterbelts that reduce wind speeds and transpiration also tend to improve WUE of crops (Davis and Norman 1988).

Adaptation to Excess Water

Stress can occur due to flooding when shoots are at least partially inundated or to waterlogging when the soil air spaces remain filled with water. In both cases, the excess water results in stress due to low O_2 levels in the soil and reduced maintenance respiration of the roots leading to deterioration. Large differences in tolerance of excess water exist among forage species, with alfalfa being very sensitive (Barta 1988) and reed canarygrass being tolerant of flooded conditions for weeks at a time (McKenzie 1951).

The direct effects of waterlogging are due to low O_2 availability to support **aerobic respiration** of the roots and other submerged organs, leading to a shift of the tissues to anaerobic dark respiration. The products of anaerobic dark respiration, including lactic acid and ethanol, can accumulate to toxic concentrations if tissues remain waterlogged for extended time periods (Chapter 27).

Several secondary injuries occur as a result of waterlogging, including pathogen invasion, mineral toxicities (e.g. Mn), reduced water uptake and sugar translocation to roots, changes in phytohormone production (cytokinin, ethylene, ABA) and subsequent effects on growth and **gene** expression. Because high temperatures usually accelerate respiration rates, flooding accompanied by high water temperatures tends to exacerbate the negative consequences of flooding.

Tolerance to waterlogging and flooding is related to several physiological and morphological adaptations, including high levels of alcohol dehydrogenase that metabolizes ethanol that accumulates in cells; presence of aerenchyma cells formed by coordinated breakdown of cell columns in stems and roots. This is common in rice, and permits O_2 transport through the open cell columns in the plant from shoots to inundated roots. There is also genetic **resistance** to pathogens that flourish under anaerobic conditions. Plants adapted to inundation usually produce secondary roots near or above the soil surface where some O_2 can diffuse (Chapter 27).

Organic Reserves in Forages

An attribute of forages that increases mowing and grazing tolerance is the accumulation of **organic reserves** before leaf area is removed. Vast quantities of carbon (C) and N reserves accumulate to 40% or more of the dry weight of specialized storage organs located near or below the soil surface that are remobilized to support regrowth. Three types of C reserves that generally accumulate in forages can be grouped according to C_3 versus C_4 photosynthetic mechanism and whether the forage is a grass or legume (Table 4.3).

Perennial and annual cool-season (C_3) grasses and chicory accumulate **fructan** in storage organs as the principal reserve carbohydrate. Fructan is a water-soluble polysaccharide composed of a sucrose molecule with one to several hundred fructose moieties attached. Legume taproots and storage organs of warm-season (C_4) perennial grasses store starch as large molecules (up to 500 000 glucose units each) that are insoluble in water at 25 °C. **Starch** exists as two forms: **amylopectin**, a highly branched polymer that is approximately 80% of most starches, and **amylose**, an essentially linear polymer that comprises the other 20% of most starches. Sucrose accumulates in the lower stem of C_4 annual grasses like maize, sorghum, sugarcane, and millet.

Understanding the role of N reserves in forage regrowth and stress tolerance has grown over the last two decades

Table 4.3 Form and primary site of accumulation of nonstructural carbohydrates for several forage groups

Group	Example species	Carbohydrate	Site(s)	References
Legumes	Alfalfa, red clover, white clover, birdsfoot trefoil, sweetclover, kura clover	Starch	Taproots, stolons, rhizomes	Smith and Graber (1948), Smith (1962), Peterson et al. (1994), Li et al. (1996), Gallagher et al. (1997)
Cool-season (C_3) perennial grasses	Tall fescue, crested wheatgrass, reed canarygrass, timothy, orchardgrass, kentucky bluegrass, smooth bromegrass	Fructan	Stem bases, rhizomes, stolons	Sullivan and Sprague (1943), Okajima and Smith (1964), Smith and Grotelueschen (1966), Smith (1968)
Cool-season (C_3) annual grasses	Wheat, oats, barley, rye	Fructan	Stem bases, lower stem	Olien and Clark (1995)
Warm-season (C_4) perennial grasses	Bermudagrass, big bluestem, indiangrass, switchgrass, bahiagrass	Starch	Stem bases, stolons, rhizomes	Smith (1968), Smith and Greenfield (1979)
Warm-season (C_4) annual grasses	Maize, sorghum, sudangrass, millet	Sucrose	Lower stem	Setter and Meller (1984)

Table 4.4 Principal storage organ, molecular mass (size), and putative function of vegetative storage proteins identified in forage species

Species	Organ	Mass, kDa	Function	References
Alfalfa	Taproot	15	Unknown	Hendershot and Volenec (1993)
		19	Chitinase	Hendershot and Volenec (1993), Volenec unpub.
		32	Chitinase	Volenec et al. (2002), Meuriot et al. (2003)
		57	β-amylase	Gana et al. (1998)
White clover	Stolon	17.3	Pathogenesis-related	Corre et al. (1996), Goulas et al.(2001), Corbel et al. (1999)
Birdsfoot trefoil	Taproot	25	Unknown	Li et al. (1996)
Chicory	Root	18	Unknown	Cyr and Bewley (1990)
Wheat	Peduncle and flag leaf	Amino acids	NA	MacKown et al. (1992)

(Volenec et al. 1996; Aranjuelo et al. 2015; Dierking et al. 2017; Lu et al. 2018; Meuriot et al. 2018). With the exception of wheat, N accumulates in vegetative tissues as a part of relatively small proteins (Table 4.4). These polypeptides are depleted from taproots of legumes during spring growth and regrowth after defoliation, and then re-accumulate as reproductive development begins (Li et al. 1996). These polypeptides fulfill the criteria used to define **vegetative storage proteins** (VSPs) (Volenec et al. 1996):

1. Proteins exhibit preferential synthesis and accumulation during development of storage organs.
2. Proteins are depleted from storage organs during the reactivation of meristems.
3. Proteins whose abundance greatly exceeds that of other proteins in perenniating organs.

Three of four alfalfa taproot VSPs have been identified as chitinases and β-amylase based on *in vitro* assays and sequence homology (Table 4.4). The 17-kDa VSP from white clover stolons possesses sequence homology to a pathogenesis-related protein. It is unclear if these VSPs serve a dual role in forage legumes; an N source for regrowth and their respective catalytic role within the plant.

Similar to seeds of many plant species, several forage legumes accumulate phytic acid in taproots as a phosphorus (P) reserve (Campbell et al. 1991). Like C and N reserves, taproot phytate concentrations in perennial legumes increase extensively between September and December and decline markedly in spring when shoot growth resumes (Li et al. 1996). In December, phytate-P is over 10% of the total root-P pool in taproots of birdsfoot trefoil, whereas it is between 4% and 7% of total root-P in roots of alfalfa, red clover, and sweetclover. By comparison, despite its positive role in winter-hardiness, taproot-K concentrations do not increase during cold acclimation.

Whether it is a VSP or a nonstructural carbohydrate, their seemingly coordinated pattern of organic reserve depletion and restoration is similar in storage organs of many forage species. After mowing or grazing, the concentration of **organic reserves** declines rapidly for 10–14 days, then remains low for another week before re-accumulating during weeks 4 and 5 of regrowth (Figure 4.6). It is important to not impose additional mowing or close grazing between weeks 2 and 4 when reserves are low because plants may be weakened or

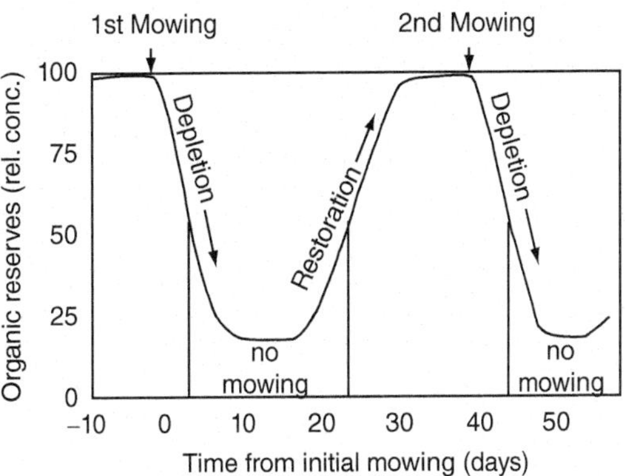

Fig. 4.6. Effect of mowing or grazing on use and restoration of organic reserves in forage plants. Rapid depletion occurs during the two weeks following defoliation, and extensive re-accumulation begins after Week 3. Reserves are low between weeks 1 and 3, when plant survival may be compromised if additional mowing or grazing occurs. *Source:* Adapted from several sources.

die when reserves are exhausted. Certain forage species, including birdsfoot trefoil, do not accumulate high concentrations of root reserves during the growing season (Nelson and Smith 1968; Li et al. 1996). The mowing and grazing height of such forages must be raised to 10 cm or more in order to maintain leaf area so plants have a readily available source of C from photosynthesis for regrowth (Figure 4.6).

Light, Phytohormones, and Shoot Development

Little is known regarding regulation of bud dormancy and bud development of forages despite the obvious agricultural importance of this information. The extensive crown bud formation observed in autumn in sweetclover is due primarily to shortened photoperiods irrespective of temperature (Kasperbauer et al. 1963). Later work showed that phytochrome and the ratio of red: far-red light plays a significant role in shoot development of sweetclover (Kasperbauer and Borthwick 1964). Recently, Du et al. (2018) isolated nearly 4000 proteins from apical meristems of alfalfa sampled in the field in early autumn. Of these, 90 proteins were differentially accumulated between apical meristems of fall dormant versus non-fall dormant cultivars, and were involved in several biochemical and physiological processes including metabolism of pyruvate and amino acids, indole acetic acid (IAA) metabolism and transport, flavonoid biosynthesis, regulation of circadian rhythm and phenylpropanoid biosynthesis.

Hodgkinson (1973) suggested that rapidly developing buds that begin growth soon after defoliation exert **apical dominance** over other buds, thereby reducing their growth rates. Growth substances like auxins may control the number of early shoots that become "dominant" by altering the partitioning of carbohydrates, amino acids, and minerals among buds. The continued supply of these nutrients to dominant shoots facilitates their rapid growth. By restricting the number of shoots from eight to two developing per alfalfa crown, Leach (1971) demonstrated intershoot competition for growth substrates from roots.

Dormancy of axillary buds can be defined as a state in which visible growth is temporarily suspended in a plant structure that is capable of growth (Wareing and Phillips 1970; Lang 1987). Three major forms of dormancy have been described. In **ecodormancy** growth is suspended because of environmental constraints such as temperature, water stress, or light. Growth resumes when these constraints are alleviated. This form of dormancy has also been referred to as *imposed* or *forced dormancy*. In **paradormancy**, the cause of bud dormancy resides in the plant itself, but outside of the dormant tissue per se. For example, paradormancy occurs in axillary buds because of apical dominance from the shoot apex, also known as *correlative inhibition*. **Endodormancy** (also known as *innate dormancy*) is caused by factors found within the dormant tissues themselves.

Seeds, **corms**, tubers, rhizomes, as well as axillary buds, are all capable of possessing endodormancy. In these underground storage organs and seeds, a period of after-ripening (storage at room temperature for several months) will often break endodormancy, making shoot development possible. It is thought that axillary buds of forages exhibit ecodormancy and paradormancy, but they do not appear to possess endodormancy.

Apical dominance of axillary bud growth is a unique form of dormancy. During apical dominance, apical meristem growth is much more vigorous than is that of subtending axillary buds. When the apical meristem is removed during forage harvest, one or more of the subtending axillary buds begins to develop and eventually replaces the functions of the **shoot apex**, including its apical dominance over other subtending axillary meristems.

IAA and other auxins have been a recurring theme in concepts describing apical dominance. Axillary buds can be released from apical dominance within 30 minutes of shoot removal, but apical dominance continues if IAA is applied to the cut surface of the shoot (Hall and Hillman 1975). Although cytokinins transported from the roots are involved in cell division of leaves and stems, and are powerful initiators of axillary bud growth, their effects appear to be secondary to the primary effects of inhibitors (like IAA) originating in shoots (Cline 1991). Beveridge et al. (1994) speculated that the ratio of IAA to cytokinin was more influential than IAA or cytokinin levels alone in controlling apical dominance. Thus, it may be the interaction of growth regulators that determines the extent of axillary bud dormancy.

Considerable effort has been expended to identify compound(s) that could serve as an inhibitor of IAA action in dormant axillary buds. Since the discovery of ABA and its known effects on several IAA-mediated responses, it has been a popular focus of research on bud dormancy. Application of ABA to axillary meristems after shoot removal can inhibit bud development as effectively as application of IAA (White and Mansfield 1977). Despite the widespread belief that growth regulators are involved in apical dominance and axillary bud dormancy, many issues and inconsistencies remain unresolved (see reviews by Cline 1991; Murphy and Briske 1992).

Managing Leaf Area and Photosynthesis

Most forage and pasture plants are perennial and must survive a range of stresses like low temperature, high temperature, drought and salinity, all of which also reduce photosynthesis. Removal of herbage, the economic product, by cutting or grazing, reduces the amount of young leaves at the top of the canopy leaving the older residual leaf area and stored substrates, mainly carbohydrate and storage proteins, to support regrowth. After cutting or grazing, leaf growth gradually restores the canopy with young leaves that have active photosynthesis. But grasses and legumes differ in growth form and leaf area needed to intercept and efficiently convert solar energy to usable forms and to allocate the photosynthate among areas involved with optimal growth or storage.

Managing Forages Based on Leaf Area

Since leaf area is the primary source of photosynthesis, in theory forages and pastures should be managed to intercept as much radiation as efficiently as possible during the growing season. Several experiments with many crops show a curvilinear relationship between leaf area index (LAI) and radiation interception because internal shading and average leaf age increase as LAI increases. Rice has rather upright leaf orientation so a LAI of 1, 3, 6, and 9 intercepts, respectively, 39%, 78%, 95%, and 99% of the visible light that drives photosynthesis (Figure 4.7).

The maximum crop growth rate for most plants occurs when 95% of incident light is intercepted. In addition, because of the curvilinear shape of the light response curve of photosynthesis (Figure 4.3), intercepting as much light as possible also is advantageous. Most forage grasses have upright leaf angles that allow for good light distribution

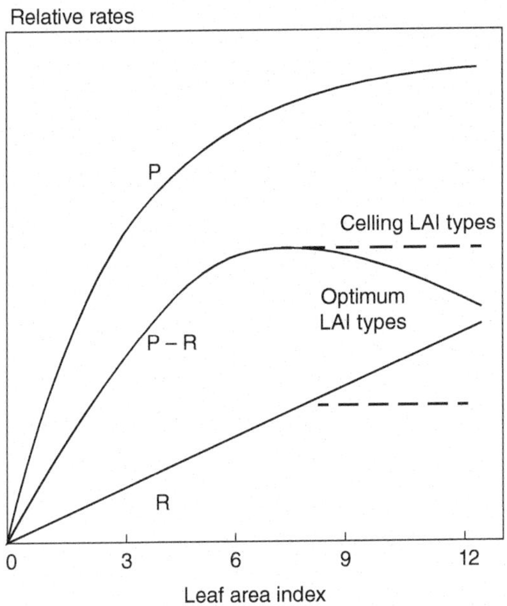

Fig. 4.7. Relative rates of CO_2 exchange for photosynthesis (P, curvilinear), Respiration (R, linear), and the net difference (P-R) for a grass canopy. In this generalized illustration based on rice, the critical LAI for maximum light interception is 9 whereas the optimum LAI, best LAI when adjusted for R, is 6.

within a canopy and an LAI of 6-7 to intercept 95% of the light. By comparison, the greater horizontal orientation of leaves of alfalfa, and especially clovers, results in less leaf area needed to intercept 95% of the light (LAI of 5-6 for alfalfa and 3-4 for clovers), and a lower potential crop growth rate.

In theory, the best management would be to graze enough forage to maintain a season-long average of 90% or higher light interception through the season. Grazing lightly to 85% interception each time the canopy reaches 95% interception would achieve that goal, but grazing or harvest removes mainly the upper, younger leaves which have the highest photosynthesis. The older leaves with low photosynthesis that remain are less active to support regrowth.

As the removal process is repeated the residual leaf area becomes less and less productive and there will be little storage in the basal parts. In addition, canopy respiration is a function of leaf mass and needs to be considered as aging of the leaf area reduces the net amount of photosynthesis available for transport to meristems and plant parts. In contrast with photosynthesis that responds to radiation in a curvilinear manner, respiration is a linear function of leaf area since it is related to the mass of tissue. In that case the canopy should be allowed to grow to reach the ceiling LAI that maximizes net production of photosynthate and then harvested regularly to maintain that level. But the ceiling LAI will gradually decrease.

Unfortunately, the theory does not work well for hay harvest or for grazing pasture legumes or grasses due mainly to their variation in growth habits and needs for multiple harvest. Researchers soon found that it was superior to use some active photosynthesis to support storage in roots of legumes or lower stem bases of grasses. Even though storage required synthesis respiration and reduced photosynthesis for immediate yield, the use of the storage for regrowth after the topgrowth was removed provided young leaves for active regrowth to reach 95% light interception and restoration of C and N reserves. This system of growth and removal usually gave higher forage yield and increased stand persistence.

There have been many science-based arguments among researchers both ways as to whether rotational stocking of pastures, i.e., grazing most of the forage sequentially, and then allowing a regrowth period, is superior to continuous stocking, i.e. grazing frequently but lightly to retain leaf area to support regrowth, for animal production. In a detailed meta-analysis of data over a wide range of species and environments, Sollenberger et al. (2014) concluded average daily gains were similar for rotational and continuous stocking in humid areas of the U.S., but gain per ha was about 30% higher for rotational stocking since yield of forage was also about 30% higher.

This corroborated conclusions by several researchers including Parsons et al. (1988) who found the percentage of young leaves increased more rapidly in pastures grazed rotationally than continuously. This raised the importance of how to optimize the balance of the residual leaf area left after the grazing period and especially with the duration of the regrowth interval. In addition, this underscored the need to incorporate how and when photosynthate is allocated to meristems for growth of leaves, stems, tillers and roots relative to that allocated to storage organs to support regrowth after grazing of cutting. This further emphasized the need to incorporate the knowledge of **canopy architecture**, i.e. leaf angles and tiller or shoot development, to optimize use of photosynthesis for forage adaptation and its management for yield, quality, animal performance, and plant persistence.

Integrating Physiology and Morphology

Due to morphology, in early spring leaf meristems of grasses produce long blades that are oriented vertically. When the proper photoperiod is reached, growth shifts to include intercalary meristems for stem elongation and development of the inflorescence (Chapter 2). This elevates the leaf area and slows new tiller development. After cutting or grazing the regrowth is short-leafed vegetative growth from tillers that were not vernalized to provide leaf growth and further **tillering** in summer and longer leaves in fall. Leaf growth is from the blade and sheath meristems that like the shoot apex remain near soil level and are not removed. Thus, the architecture of the canopy and location of meristems involved need to be considered in terms of managing leaf area and photosynthesis at different times of the season (Lemaire et al. 2009). In addition, young leaves are most efficient in photosynthesis, but gradually age and become shaded by the newer leaves that are developing. The lower leaves intercept some radiation, but from a photosynthetic point of view, are not as effective as upper leaves.

In contrast, legume morphologies like alfalfa always have the shoot apex at the top of the canopy and when removed by grazing or cutting the plant regrows from axillary buds on the crown near soil level that are supported by C and N reserves stored in the root. Many clovers like red clover have long petioles so the **shoot apex** is in mid canopy and it avoids light grazing to continue producing leaves. If more leaf area and the shoot apex are removed, the plant uses C and N reserves stored in the root to supplement regrowth. In contrast, white clover has stolons with long petioles to display the leaf blades well above the shoot apices at the stolon tips that remain near soil level and are seldom removed, some leaf area remains and the plants regrow.

Thus, the leaf area needed to achieve the critical LAI is higher for grasses than legumes. But critical LAI includes all the leaf area ages in the canopy and the lower leaves are senescing and less functional. Some attempts to use leaf area as a management guide were based on "green" leaf area, but that was not a great improvement. In addition, when the plants were cut or grazed, the leaves at the base for regrowth were old and not active photosynthetically. Regrowth had to be supported by storage forms of C and N.

Carbohydrates also serve as substrates for respiration to provide cellular energy and carbon chains for synthesis of new growth in **meristematic** areas (growth respiration) and maintenance of enzymes and structural materials that gradually turn over (maintenance respiration). In many instances, the carbon compounds synthesized can be stored for short or long periods of time before they are transported and utilized in another growth process.

Balancing the dynamic and diverse needs for photosynthate that change with time and stage of growth requires compromises in management that are based on integration of photosynthetic capacity, morphology of the plant and efficient allocation of the photosynthate to growing regions and storage sites within the plant (Poorter 1994). More research is needed, especially with changes in climate, to optimize these relationships to improve radiation capture and forage production through the growing season that includes consideration of stored reserves and plant morphology.

Summary and Conclusions

Despite good progress, more understanding is needed about the physiology of forage plants and its role in management of photosynthesis and carbon metabolism and potential for genetic improvement. New molecular biology techniques (Chapter 31) are adding powerful tools that physiologists can use in their studies, especially those related to genetic improvement of a specific species. Even so, there are hundreds of important forage species to study, each with its unique **physiology** and morphology, plus inherent genetic variation among individuals.

Contributions related to temperature stress, water stress and inundation will be important in the future as the climate changes, in addition to management technologies that will improve the greenhouse gas environment and especially water quality. These will include C sequestration in soil, reductions in soil erosion, and phytoremediation of contaminated areas. In addition, influences of pests and pathogens and the plant mechanisms needed for resistance will continue to be major problems as the climate changes.

Physiological issues related to forage quality and mechanisms of plant adaptation to challenging soil and landscape positions need to be addressed. Further, the positive impacts of biotechnology will need to be understood by the public and considered as tools for forage and grassland improvement. Advances in management will depend on understanding the plant, its genetics, and how it interacts with the environment; the fundamental goals of physiology research.

Much of the above will likely require scaling from individual plant research to its application at the field or pasture and eventually at community and ecosystem levels. The challenge is exacerbated by having populations of mixed species in most hayfields, pastures, and rangelands. Gathering basic information on each species and then evaluating those responses as components of a diverse biological community and the spatial variation in soil properties in a small area add to the complexity. There will likely be needs for modeling approaches to solve many of the problems associated with capture of solar radiation, its effective use in supporting growth and storage of reserve C and N, and how to harvest and use it in ways that enhance the overall value of forages and grasslands for food production and environmental conservation.

References

Adams, W.E. and Twersky, M. (1960). Effect of soil fertility on winter killing of coastal bermudagrass. *Agron. J.* 52: 325–326.

Amthor, J.S. (1984). The role of maintenance respiration in plant growth. *Plant Cell Environ.* 7: 561–569.

Amthor, J.S. (1989). *Respiration and Crop Productivity*. New York, NY: Springer-Verlag.

Amthor, J.S. (1994). Respiration and carbon assimilate use. In: *Physiology and Determination of Crop Yield* (eds. K.J. Boote, J.M. Bennett, T.R. Sinclair and G.M. Paulsen), 221–250. Madison, WI: American Society of Agronomy.

Arakeri, H.R. and Schmid, A.R. (1949). Cold resistance of various legumes and grasses in early stages of growth. *Agron. J.* 41: 182–185.

Aranjuelo, I., Molero, G., Erice, G. et al. (2015). Effect of shoot removal on remobilization of carbon and nitrogen during regrowth of nitrogen-fixing alfalfa. *Physiol. Plant.* 153: 91–104.

Barbehenn, R.V., Chen, Z., Karowe, D.N., and Spickard, A. (2004). C_3 grasses have higher nutritional quality that C_4 grasses under ambient and elevated atmospheric CO_2. *Global Change Biol.* 10: 1565–1575.

Barta, A.L. (1988). Response of field-grown alfalfa to root waterlogging and shoot removal. I. Plant injury and mineral content of roots. *Agron. J.* 80: 889–892.

Bauder, J.W., Bauer, A., Ramirez, J.M., and Cassel, D.K. (1978). Alfalfa water use and production on dryland and irrigated sandy loam. *Agron. J.* 70: 95–99.

Bazzaz, F.A. (1990). The response of natural ecosystems to the rising global CO_2 levels. *Annu. Rev. Ecol. Syst.* 21: 167–196.

Bell, L.W., Lawrence, J., Johnson, B., and Whitbread, A. (2013). Exploring short-term ley legumes in subtropical grain systems: production, water-use, water-use efficiency and economics of tropical and temperate options. *Crop Pasture Sci.* 63: 819–832.

Beveridge, C.A., Ross, J.J., and Murfet, I.C. (1994). Branching mutant rms-2 in *Pisum sativum*. Grafting studies and endogenous indole-3-acetic acid levels. *Plant Physiol.* 104: 953–959.

Bita, C. and Gerats, T. (2013). Plant tolerance to high temperature in a changing environment: scientific fundamentals and production of heat stress-tolerant crops. *Front. Plant Sci.* 4: 273. https://doi.org/10.3389/fpls.2013.00273.

Bolger, T.P. and Matches, A.G. (1990). Water-use efficiency and yield of sainfoin and alfalfa. *Crop Sci.* 30: 143–148.

Bula, R.J. and Massengale, M.A. (1972). Environmental physiology. In: *Alfalfa Science and Technology. Agronomy Monograph 15* (ed. C.H. Hansen), 167–184. Madison WI: American Society of Agronomy. ISBN: 978-0-89118-210-8.

Campbell, M., Dunn, R., Ditterline, R. et al. (1991). Phytic acid represents 10–15% of total phosphorus in alfalfa root and crown. *J. Plant Nutr.* 14: 925–937.

Chamblee, D.S., Mueller, J.P., and Timothy, D.H. (1989). Vegetative establishment of three warm-season perennial grasses in late fall and late winter. *Agron. J.* 81: 687–691.

Clifton-Brown, J.C. and Jones, M.B. (1997). The thermal response of leaf extension rate in genotypes of the C_4-grass *Miscanthus*: an important factor in determining the potential productivity of different genotypes. *J. Exp. Bot.* 48: 1573–1581.

Cline, M.G. (1991). Apical dominance. *Bot. Rev.* 57: 318–358.

Corbel, G., Robin, C., Frankow-Lindberg, B.E. et al. (1999). Regrowth of white clover after chilling: assimilate partitioning and vegetative storage proteins. *Crop Sci.* 39: 1756–1761.

Corre, N., Bouchart, V., Ourry, A., and Boucaud, J. (1996). Mobilization of nitrogen reserves during regrowth of defoliated *Trifolium repens* L. and identification of potential vegetative storage proteins. *J. Exp. Bot.* 47: 1111–1118.

Craufurd, P.Q., Qi, A., Ellis, R.H. et al. (1998). Effect of temperature on time to panicle initiation and leaf appearance in sorghum. *Crop Sci.* 38: 942–947.

Cunningham, S.M., Nadeau, P., Castonguay, Y. et al. (2003). Raffinose and stachyose accumulation, galactinol synthase expression, and winter injury of contrasting alfalfa germ-plasms. *Crop Sci.* 43: 562–570.

Cyr, D.R. and Bewley, J.D. (1990). Proteins in the roots of the perennial weeds chicory (*Cichorium intybus* L.) and dandelion (*Taraxacum officinale* Weber) are associated with overwintering. *Planta* 182: 370–374.

Davis, J.E. and Norman, J.M. (1988). Effects of shelter on plant water use. *Agric. Ecosyst. Environ.* 22: 393–402.

Dierking, R.M., Allen, D.J., Cunningham, S.M. et al. (2017). Nitrogen reserve pools in two *Miscanthus* x *giganteus* genotypes under contrasting N managements. *Front. Plant Sci.* 8: 1618. https://doi.org/10.3389/fpls.2017.01618.

Du, H., Shi, Y., Li, D. et al. (2018). Proteomics reveals key proteins participating in growth difference between fall dormant and non-dormant alfalfa in terminal buds. *J. Proteomics* 173: 126–138. https://doi.org/10.1016/j.jprot.2017.11.029.

Evenson, P.D. (1979). Optimum crown temperatures for maximum alfalfa growth. *Agron. J.* 71: 798–800.

Erickson, J.E., Soikaew, A., Sollenberger, L.E., and Bennett, J.M. (2012). Water use and water-use efficiency of three perennial bioenergy grass crops in Florida. *Agriculture* 2: 325–338.

Fairbourn, M.L. (1982). Water use by forage species. *Agron. J.* 74: 62–66.

Gallagher, J.A., Volenec, J.J., Turner, L.B., and Pollock, C.J. (1997). Starch hydrolytic enzyme activities following defoliation of white clover. *Crop Sci.* 37: 1812–1818.

Gana, J.A., Kalengamaliro, N.E., Cunningham, S.M., and Volenec, J.J. (1998). Expression of beta-amylase from alfalfa taproots. *Plant Physiol.* 118: 1495–1505.

Gist, G.R. and Mott, G.O. (1957). Some effects of light intensity, temperature, and soil moisture on the growth of alfalfa, red clover and birdsfoot trefoil seedlings. *Agron. J.* 49: 33–36.

Goulas, E., Le Dily, F., Teissedre, L. et al. (2001). Vegetative storage proteins in white clover (*Trifolium repens* L.): quantitative and qualitative features. *Ann. Bot.* 88: 789–795.

Grimes, D.W., Wiley, P.L., and Sheesley, W.R. (1992). Alfalfa yield and plant water relations with variable irrigation. *Crop Sci.* 32: 1381–1387.

Haagenson, D.M., Cunningham, S.M., Joern, B.C., and Volenec, J.J. (2003). Autumn defoliation effects on alfalfa winter survival, root physiology, and gene expression. *Crop Sci.* 43: 1340–1348.

Hall, S.M. and Hillman, J.R. (1975). Correlative inhibition of lateral bud growth in *Phaseolus vulgaris* L. timing of bud growth following decapitation. *Planta* 123: 137–143.

Hebeisen, T., Lüscher, A., Zanetti, S. et al. (1997). Growth response of *Trifolium repens* L. and *Lolium perenne* L. as monocultures and bi-species mixture to free air CO_2 enrichment and management. *Global Change Biol.* 3: 149–160.

Hendershot, K.L. and Volenec, J.J. (1993). Nitrogen pools in taproots of *Medicago sativa* L. after defoliation. *J. Plant Physiol.* 141: 129–135.

Hodgkinson, K.C. (1973). Establishment and growth of shoots following low and high cutting of lucerne in relation to the pattern of nutrient uptake. *Aust. J. Agric. Resour.* 24: 497–510.

Jacob, P., Hirt, H., and Bendahmane, A. (2017). The heat-shock protein/chaperone network and multiple stress resistance. *Plant Biotechnol J.* 15: 405–414.

Jones, H.G., Serraj, R., Loveys, B.R. et al. (2009). Thermal infrared imaging of crop canopies for the remote diagnosis and quantification of plant responses to water stress in the field. *Funct. Plant Biol.* 36: 978–989.

Kasperbauer, M.J. and Borthwick, H.A. (1964). Photo-reversibility of stem elongation in *Melilotus alba* Desr. *Crop Sci.* 4: 42–44.

Kasperbauer, M.J., Gardner, F.P., and Johnson, I.J. (1963). Taproot growth and crown bud development in biennial sweetclover as related to photoperiod and temperature. *Crop Sci.* 3: 4–7.

Kimball, B.A. (2016). Crop responses to elevated CO_2 and interactions with H_2O, N, and temperature. *Curr. Opin. Plant Biol.* 31: 36–43.

Kirby, E.J.M. (1995). Factors affecting rate of leaf emergence in barley and wheat. *Crop Sci.* 35: 11–19.

Lang, G.A. (1987). Dormancy: a universal terminology. *HortSci.* 22: 817–820.

Leach, G.J. (1971). The relation between shoot growth and temperature. *Aust. J. Agric. Resour.* 22: 49–59.

Lee, C.T. and Smith, D. (1972). Influence of soil nitrogen and potassium levels on the growth and composition of lucerne grown to first flower in four temperature regimes. *J. Sci. Food Agric.* 23: 1169–1181.

Lee, D.H. and Lee, C.B. (2000). Chilling stress-induced changes of antioxidant enzymes in the leaves of cucumber: in gel enzyme activity assays. *Plant Sci.* 159: 75–85.

Leep, R.H., Andresen, J.A., and Jeranyama, P. (2001). Fall dormancy and snow depth effects on winterkill of alfalfa. *Agron. J.* 93: 1142–1148.

Lemaire, G., Da Silva, S.C., Agnusdei, M. et al. (2009). Interactions between leaf lifespan and defoliation frequency in temperate and tropical pastures: a review. *Grass Forage Sci.* 64: 341–353.

Li, R., Volenec, J.J., Joern, B.C., and Cunningham, S.M. (1996). Seasonal changes in nonstructural carbohydrates, protein, and macronutrients in roots of alfalfa, red clover, sweetclover, and birdsfoot trefoil. *Crop Sci.* 36: 617–623.

Lu, X., Ji, S., Hou, C. et al. (2018). Impact of root C and N reserves on shoot regrowth of defoliated alfalfa cultivars differing in fall dormancy. *Grassland Sci.* 64: 83–90.

Ludlow, M.M. (1985). Photosynthesis and dry matter production in C_3 and C_4 pasture plants, with special emphasis on tropical C_3 legumes and C_4 grasses. *Funct. Plant Biol.* 12: 557–572.

MacKown, C.T., van Sanford, D.A., and Zhang, N. (1992). Wheat vegetative nitrogen compositional changes in response to reduced reproductive sink strength. *Plant Physiol.* 99: 1469–1474.

Madakadze, I.C., Stewart, K.A., Madakadze, R.M., and Smith, D.L. (2003). Base temperatures for seedling growth and their correlation with chilling sensitivity for warm-season grasses. *Crop Sci.* 43: 874–878.

McKenzie, R.E. (1951). The ability of forage plants to survive early spring flooding. *Sci. Agric.* 31: 358–367.

Meuriot, F., Avice, J.C., Decau, M.L. et al. (2003). Accumulation of N reserves and vegetative storage proteins (VSP) in taproots of non-nodulated alfalfa (*Medicago sativa* L.) are affected by mineral N availability. *Plant Sci.* 165: 709–718.

Meuriot, F., Morvan-Bertrand, A., Noiraud-Romy, N. et al. (2018). Short-term effects of defoliation intensity on sugar remobilization and N fluxes in ryegrass. *J. Exp. Bot.* 69: 3975–3986.

Meyer, D.W. and Badaruddin, M. (2001). Frost tolerance of ten seedling legume species at four growth stages. *Crop Sci.* 41: 1838–1842.

Murphy, J.S. and Briske, D.D. (1992). Regulation of tillering by apical dominance: chronology, interpretive value, and current perspectives. *J. Range Manage.* 45: 419–429.

Nelson, C.J. (1994). Apparent respiration and plant productivity. In: *Physiology and Determination of Crop Yield* (eds. K.J. Boote, J.M. Bennett, T.R. Sinclair and G.M. Paulsen), 251–258. Madison, WI: American Society of Agronomy.

Nelson, C.J. and Smith, D. (1968). Growth of birdsfoot trefoil and alfalfa. III. Changes in carbohydrate reserves and growth analysis under field conditions. *Crop Sci.* 8: 25–28.

Niu, Y. and Xiang, Y. (2018). An overview of biomembrane functions in plant responses to high-temperature stress. *Front. Plant Sci.* 9: 915. https://doi.org/10.3389/fpls.2018.00915.

Okajima, H. and Smith, D. (1964). Available carbohydrate fraction in the stem bases and seed of timothy, smooth bromegrass, and several other northern grasses. *Crop Sci.* 4: 317–320.

Olien, C.R. and Clark, J.L. (1995). Freeze-induced changes in carbohydrates associated with hardiness of barley and rye. *Crop Sci.* 35: 496–502.

Parsons, A.J., Johnson, I.R., and Williams, J.H.H. (1988). Leaf age structure and canopy photosynthesis in rotationally and continuously grazed pastures. *Grass Forage Sci.* 43: 1–14.

Pearce, R.S. (2001). Plant freezing and damage. *Ann. Bot.* 87: 417–424.

Pearson, C.J. and Hunt, L.A. (1972). Effects of temperature on primary growth and regrowths of alfalfa. *Can. J. Plant. Sci.* 52: 1017–1027.

Peterhansel, C. and Maurino, V.G. (2011). Photorespiration redesigned. *Plant Physiol.* 155: 49–55.

Peterson, P.R., Sheaffer, C.C., Jordan, R.M., and Christians, C.J. (1994). Responses of kura clover to sheep grazing and clipping. II. Belowground morphology, persistence, and total nonstructural carbohydrates. *Agron. J.* 86: 660–667.

Pilbeam, C.J. and Robson, M.J. (1992). Response of populations of *Lolium perenne* cv. S23 with contrasting rates of dark respiration to nitrogen supply and defoliation regime. 2. Grown as mixtures. *Ann. Bot.* 69: 79–86.

Poorter, H. (1994). Construction costs and payback time of biomass: a whole plant perspective. In: *A Whole Plant Perspective on Carbon–Nitrogen Interactions* (eds. E. Royand and E. Garnier), 111–127. The Hague, The Netherlands: SPB Academic Publishing.

Power, J.F. (1985). Nitrogen- and water-use efficiency of several cool-season grasses receiving ammonium nitrate for 9 years. *Agron. J.* 77: 189–192.

Qian, Y.L., Ball, S., Tan, Z. et al. (2001). Freezing tolerance of six cultivars of buffalo-grass. *Crop Sci.* 41: 1174–1178.

Robson, M.J. (1982). The growth and carbon economy of selection lines of *Lolium perenne* cv. S23 with differing rates of dark respiration. 2. Grown as simulated swards during a regrowth period. *Ann. Bot.* 49: 331–339.

Ryle, G.J.A., Powell, C.E., and Tewson, W. (1992). Effect of elevated CO_2 on the photosynthesis, respiration and growth of perennial ryegrass. *J. Exp. Bot.* 43: 811–813.

Saeed, I.A.M. and El-Nadi, A.H. (1997). Irrigation effects on the growth, yield, and water use efficiency of alfalfa. *Irr. Sci.* 17: 63–68.

Sage, R.F., Sage, T.L., and Kocacinar, F. (2012). Photorespiration and the evolution of C_4 photosynthesis. *Annu. Rev. Plant Biol.* 63: 19–47.

Setter, T.L. and Meller, V.H. (1984). Reserve carbohydrate in maize stem: [^{14}C]glucose and [^{14}C]sucrose uptake characteristics. *Plant Physiol.* 75: 617–622.

Smika, D.E., Haas, H.J., and Power, J.F. (1965). Effects of moisture and nitrogen fertilizer on growth and water use by native grass. *Agron. J.* 57: 483–486.

Smith, D. (1962). Carbohydrate root reserves in alfalfa, red clover, and birdsfoot trefoil under several management schedules. *Crop Sci.* 2: 75–78.

Smith, D. (1968). Classification of several native north American grasses as starch or fructosan accumulators in relation to taxonomy. *J. Brit. Grassl. Soc.* 23: 306–309.

Smith, D. and Graber, L.F. (1948). Influence of top growth removal on the root and vegetative development of biennial sweetclover. *Agron. J.* 40: 818–831.

Smith, D. and Greenfield, S.B. (1979). Distribution of chemical constituents among shoot parts of timothy and switchgrass at anthesis. *J. Plant Nutr.* 1: 81–99.

Smith, D. and Grotelueschen, R.D. (1966). Carbohydrates in grasses. I. Sugar and fructosan composition of the stem bases of several northern-adapted grasses at seed maturity. *Crop Sci.* 6: 263–266.

Sollenberger, L.E., Agouridis, C.T., Vanzant, E.S. et al. (2014). Prescribed grazing on pasturelands. In: *Conservation Outcomes from Pastureland and Hayland Practices; Assessment, Recommendations, and Knowledge Gaps* (ed. C.J. Nelson), 111–204. Lawrence, KS: Allen Press.

Sullivan, J.T. and Sprague, V.G. (1943). Composition of the roots and stubble of perennial ryegrass following partial defoliation. *Plant Physiol.* 18: 656–670.

Tysdal, H.M. and Pieters, A.J. (1934). Cold resistance of three species of lespedeza compared to that of alfalfa, red clover, and crown vetch. *Agron. J.* 26: 923–928.

Volenec, J.J., Nelson, C.J., and Sleper, D.A. (1984). Influence of temperature on leaf dark respiration of diverse tall fescue genotypes. *Crop Sci.* 24: 907–912.

Volenec, J.J., Ourry, A., and Joern, B.C. (1996). A role for nitrogen reserves in forage regrowth and stress tolerance. *Physiol. Plant.* 97: 185–193.

Volenec, J.J., Cunningham, S.M., Haagenson, D.M. et al. (2002). Physiological genetics of alfalfa improvement: past failures, future prospects. *Field Crops Res.* 75: 97–110.

von Caemmerer, S. (2000). *Biochemical Models of Leaf Photosynthesis*. Melbourne, Australia: CSIRO.

Wang, J., Juliani, H.R., Jespersen, D., and Huang, B. (2017). Differential profiles of membrane proteins, fatty acids, and sterols associated with genetic variations in heat tolerance for a perennial grass species, hard fescue (*Festuca trachyphylla*). *Environ. Exp. Bot.* 140: 65–75.

Wareing, P.E. and Phillips, I.D.J. (1970). *The Control of Growth and Differentiation in Plants*, 122–126. New York, NY: Pergamon Press.

White, J.C. and Mansfield, T.A. (1977). Correlative inhibition of lateral bud growth in *Pisum sativum* L. and *Phaseolus vulgaris* L.: studies on the role of abscisic acid. *Ann. Bot.* 41: 1163–1170.

Wilson, D. and Jones, J.G. (1982). Effects of selection for dark respiration rate of mature leaves on crop yields of *Lolium perenne* cv. S23. *Ann. Bot.* 49: 313–320.

Woolley, J.T. (1971). Reflectance and transmittance of light by leaves. *Plant Physiol.* 47: 656–662.

Yan, W. and Hunt, L.A. (1999). An equation for modeling the temperature response of plants using only the cardinal temperatures. *Ann. Bot.* 84: 607–614.

Zheng, G., Li, L., and Li, W. (2016). Glycerolipidome responses to freezing-and chilling-induced injuries: examples in Arabidopsis and rice. *BMC Plant Biol.* 16: 70. https://doi.org/10.1186/s12870-016-0758-8.

Zhang, Q., Bell, L.W., Shen, Y., and Whish, J.P.M. (2018). Indices of forage nutritional yield and water use efficiency amongst spring-sown annual forage crops in north-West China. *Eur. J. Agron.* 93: 1–10.

CHAPTER

Mineral Nutrient Acquisition and Metabolism

Sylvie M. Brouder, Wickersham Chair and Professor, *Agronomy, Purdue University, West Lafayette, IN, USA*
Jeffrey J. Volenec, Professor, *Agronomy, Purdue University, West Lafayette, IN, USA*

Nutrients Required by Forage Crops

Plant mineral nutrition has been studied intensively for centuries and remains conceptually linked to Liebig's Law of the Minimum. Inadequate supply of any required nutrient can severely restrict normal plant growth and development. In addition, when consumed, plants used for forage often supply ruminant livestock and horses with mineral nutrients and energy necessary for profitable production of meat, milk, and fiber, and for companionship. However, when supplied in excess, nutrients can become toxic to livestock and/or have negative environmental consequences. In this chapter, we review the mineral nutrients in forage plants, including nutrient uptake and assimilation processes. We also discuss the process of dinitrogen fixation because many forages are leguminous plants that fix their own nitrogen. We close by briefly discussing emerging concepts of stored mineral nutrients in forages and nutrient-use efficiency.

Nutrients Required by Forage Crops

Like all plants, forages require 17 chemical elements for growth and development (Table 5.1). Macronutrients are required in relatively large quantities; micronutrients or **trace elements** are required in lower amounts. A few other elements called "beneficial" are required by some plants, but are not yet proven essential for all plants. The nine required macronutrients are carbon (C), hydrogen (H), oxygen (O), nitrogen (N), potassium (K), phosphorus (P), calcium (Ca), magnesium (Mg), and sulfur (S). Carbon enters plants through the stomata as CO_2 and is fixed by photosynthesis. Hydrogen generally enters crop plants via the roots as water. Oxygen enters plants primarily as a component of air and as part of the water molecule, and is released in the plant as a co-product of photosynthesis.

Other macronutrients generally enter the plant through the root system, but some nutrients can also enter through leaves after foliar application. In hydroponic solutions, they are generally provided at concentrations above 0.1 μM, whereas the eight essential micronutrients (iron [Fe], manganese [Mn], zinc [Zn], copper [Cu], boron [B], molybdenum [Mo], chlorine [Cl], and nickel [Ni]) are required at concentrations of <0.1 μM. Beneficial elements include sodium (Na), silicon (Si), vanadium (V), cobalt (Co), and aluminum (Al) (Kirkby 2012). Of these, Co is required by legumes because it is required for synthesis of **leghemoglobin** in nodules and is a component of nitrogenase; an essential protein and **enzyme**, respectively, in N_2 fixation.

Forages: The Science of Grassland Agriculture, Volume II, Seventh Edition.
Edited by Kenneth J. Moore, Michael Collins, C. Jerry Nelson and Daren D. Redfearn.

Table 5.1 Typical functions and concentrations of mineral elements in shoot tissues of plants

Element	Symbol	Concentration dry wt.	dry wt.	Number of atoms with Mo = 1	Typical plant function(s)
Macronutrients		**μmol g^{-1}**	**mg kg^{-1}**		
Molybdenum	Mo	0.001	0.1	1	Constituent of nitrate reductase and nitrogenase in N_2 fixation.
Nickel	Ni	0.001	0.1	1	Required for ureide metabolism in ureide-transporting legumes
Copper	Cu	0.1	6	100	Constituent of several enzymes and plastocyanin in the light reactions of photosynthesis
Zinc	Zn	0.3	20	300	Required for activity of many dehydrogenases, anhydrases, others
Manganese	Mn	1.0	50	1 000	Required for enzyme activities; photosynthetic evolution of O_2
Iron	Fe	2.0	100	2 000	Required by many enzymes; constituent of chlorophyll
Boron	B	2.0	20	2 000	Involved in carbohydrate transport and meristem development
Chlorine	Cl	3.0	100	3 000	Required for the photosynthetic reactions involved in O_2 evolution
Macronutrients			**g kg^{-1}**		
Sulfur	S	30	1	30 000	Component of cysteine, cystine, methionine, and thus proteins.
Phosphorus	P	60	2	60 000	Component of sugar phosphates, nucleic acids, ATP, membranes
Magnesium	Mg	80	2	80 000	Required by P-transferring enzymes; constituent of chlorophyll
Calcium	Ca	125	5	125 000	Constituent of cell wall middle lamella; calmodulin messaging system
Potassium	K	250	10	250 000	Cofactor for >40 enzymes; stomatal movement; cell electro-neutrality
Nitrogen	N	1 000	15	1×10^6	Constituent of chlorophyll, amino acids, proteins, nucleic acids (DNA, RNA)
Oxygen	O	30 000	450	30×10^6	Constituent of organic matter
Carbon	C	40 000	450	40×10^6	Constituent of organic matter
Hydrogen	H	60 000	60	60×10^6	Constituent of organic matter

Source: Adapted from Kirkby (2012).
Concentrations found in forages will vary with species, stage of growth, management, and environment.

Macronutrients Required by Forage Crops

Nitrogen (N)

Nitrogen concentrations in forages vary from <10 g kg^{-1} dry wt in grasses to over 50 g kg^{-1} dry wt in protein-rich legumes. Nitrogen has multiple roles in plants including as a constituent of amino acids, proteins, and enzymes, it is abundant in nucleic acids like DNA and RNA, and is a structural component of chlorophyll. For these reasons, N can limit growth and yield of forage crops. Major sources of N for plants in forage-livestock systems include fertilizers, animal manures, and biologic N_2 fixation (Figure 5.1). In addition, atmospheric deposition of N occurs providing up to 35 kg N ha^{-1} yr^{-1} (Stevens et al. 2004; Storkey et al. 2015). System N losses include removal of N in forage, NO_3^- leaching to ground and surface waters, and denitrification of N including loss as **greenhouse gases (GHGs)** like N_2O.

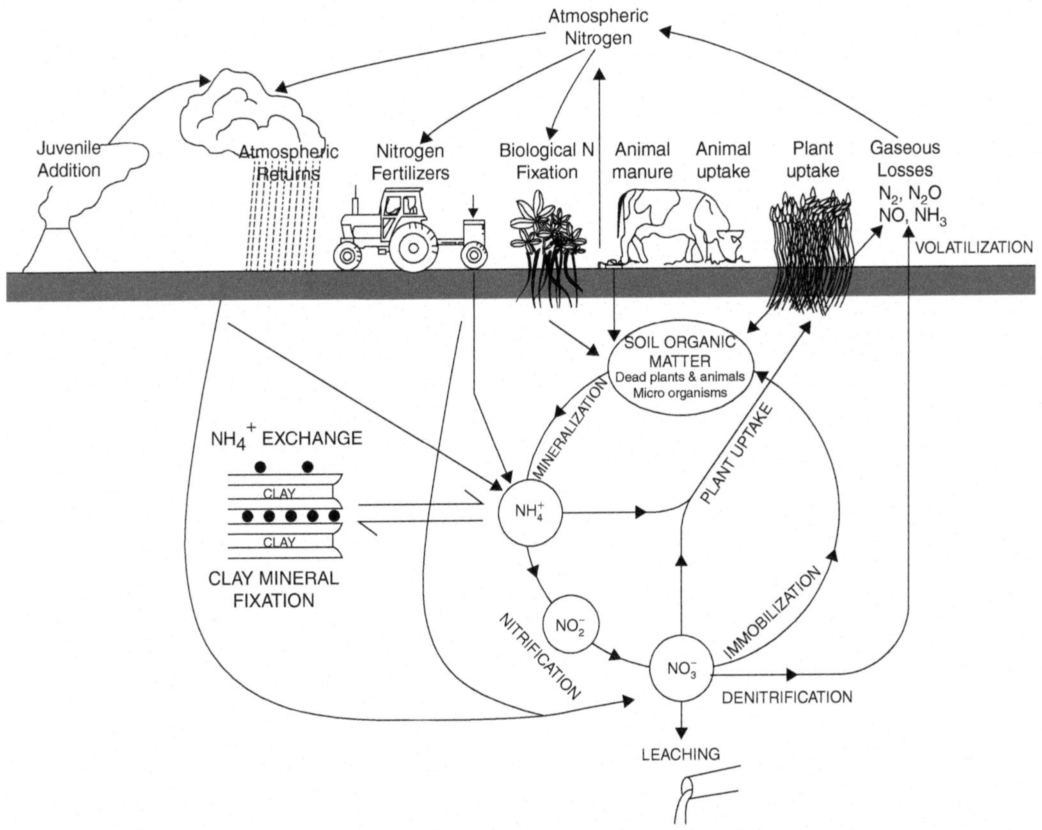

Fig. 5.1. The nitrogen cycle in forage-livestock systems. *Source:* From Andrews et al. (2011), used with permission.

Nitrogen is generally taken up as nitrate (NO_3^-) or ammonium (NH_4^+) in non-leguminous plants, or can be fixed as atmospheric N_2 by legumes like alfalfa and clovers (discussed in detail below). Assimilation of NH_4^+ is rapid, involving two enzymes present in root and shoot tissues, including chloroplasts. Glutamine synthetase (GS) catalyzes the synthesis of glutamine from ammonium and glutamate, a step that also requires **adenosine triphosphate (ATP)**. The newly formed glutamine reacts with 2-oxoglutarate to form two glutamate molecules through a reaction catalyzed by glutamate synthase (GOGAT) (Figure 5.2).

From a bioenergetics standpoint, NH_4^+ assimilation is relatively efficient requiring an ATP and reducing power. However, because of rapid **nitrification** (Figure 5.1), soil NH_4^+ concentrations tend to be low relative to soil NO_3^-. Thus, NO_3^- is often the dominant form of soil N acquired by forages. Nitrate must be reduced in the plant to NH_4^+ prior to assimilation into amino acids by GS-GOGAT. Nitrate reduction is a two-step process (Figure 5.3). Step 1 typically occurs in the cytosol of leaf cells and is catalyzed by nitrate reductase, a molybdenum (Mo)-containing enzyme. The nitrite (NO_2^-) produced by nitrate reductase enters the chloroplast where it is converted to NH_4^+ by nitrite reductase. Conversion of NO_2^- to NH_4^+ requires substantial energy and reducing power, and these are provided by the light reactions of photosynthesis. Therefore, Step 2 only occurs in light. Nitrite is also toxic to plants in small quantities so nitrate reductase is inactivated in darkness to prevent accumulation of its product, NO_2^-, when light is not available for photosynthesis.

High rates of N fertilization can elevate forage NO_3^- concentrations to levels that endanger livestock health (Wedin 1974). In addition, high N fertilization rates reduce N-use efficiency (NUE, kg forage dry wt kg^{-1} N fertilizer applied), can contaminate ground and surface waters, and may even reduce forage yields in some environments. For example, forage yield of reed canarygrass doubled with the first increment of N fertilizer application (56 kg N ha^{-1} prior to each of two harvests for a total of 112 kg N ha^{-1}) (Figure 5.4). Increasing the

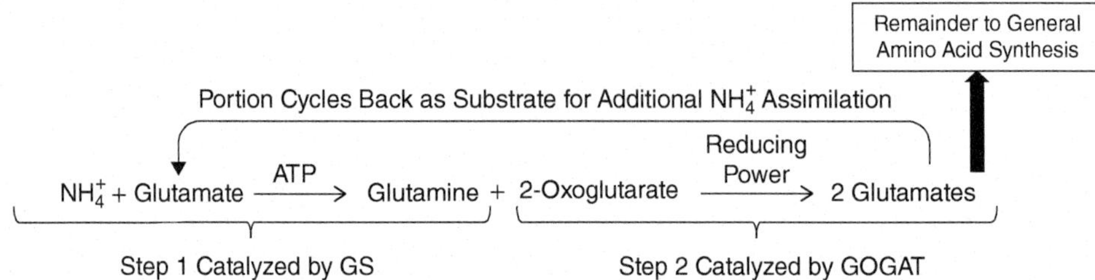

Fig. 5.2. Assimilation of ammonium (NH_4^+) occurs in chloroplasts and is a two-step process catalyzed by glutamine synthetase (GS) and glutamate synthase (GOGAT). A portion of the glutamate produced in Step 2 is used as substrate in Step 1 for additional NH_4^+ assimilation. The remainder of the glutamate produced in Step 2 is used as substrate for synthesis of the other amino acids required by plants.

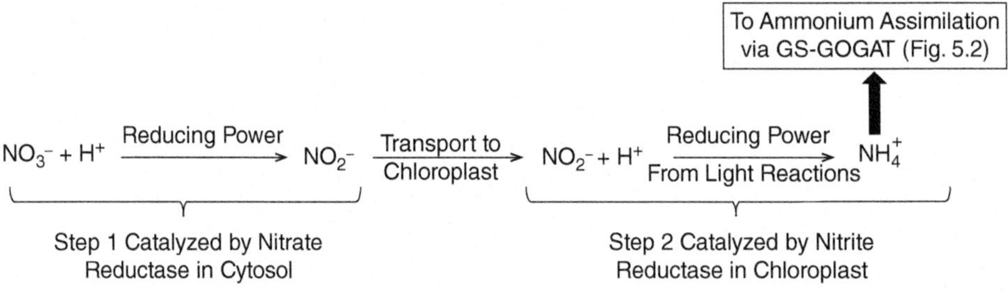

Fig. 5.3. Nitrate (NO_3^-) reduction is a two-step process catalyzed by nitrate reductase (Step 1) and nitrite reductase (Step 2). Nitrate reductase is located in the cytosol, while nitrite reductase in located in the chloroplast. The ammonium (NH_4^+) produced by nitrite reductase is assimilated into amino acids by glutamine synthetase (GS) and glutamate synthase (GOGAT) as described in Figure 5.2.

N to 112 kg N ha^{-1} prior to each harvest (224 kg N ha^{-1} total) increased yield, but also elevated forage NO_3^- concentrations to 3300 mg kg^{-1}, a level considered potentially toxic to livestock (Vetsch et al. 1999). Additional N fertilizer failed to enhance forage yield, but increased both forage NO_3^- concentrations and residual soil-NO_3^- levels increasing risks to both animal health and surface water contamination.

Potassium (K)

Unlike N, K assimilation does not require biochemical modification following uptake. Shoot-K concentrations can vary widely from <10 to >30 g kg^{-1} dry wt. The critical K concentration varies with species, tissue, maturity, and environment but, for many, forage species is in the range of 15–20 g kg^{-1} of forage dry wt (Brown 1957; Kresge and Younts 1962; Lissbrant et al. 2010). Concentrations in excess of those needed for high forage yield and stand persistence (the "critical concentration") are considered "luxury consumption" (Bartholomew and Janssen 1929). High K fertilizer application rates can cause problems in forage-livestock production systems. High rates of KCl fertilizer application can ultimately reduce forage yield, a response thought to be associated with Cl toxicity (Rominger et al. 1976). In addition, luxury consumption of K can increase the risk of hypomagnesemia (grass tetany) in livestock (Pelletier et al. 2008); an issue discussed in detail below in the magnesium nutrition section.

As an abundant univalent cation, K plays a critical role in ionic balance in cells and the development of **turgor**. For example, cell growth in meristems, stomatal opening, and plant movements all require K influx to increase turgor and cellular expansion, and collectively enhance water-use efficiency (Martineau et al. 2017). Potassium also enhances long-distance transport of minerals and organic nutrients in xylem and phloem. Potassium activates >75 enzymes including starch synthase (Murata and Akazawa 1968) and is involved in cell protein synthesis (Blevins 1985).

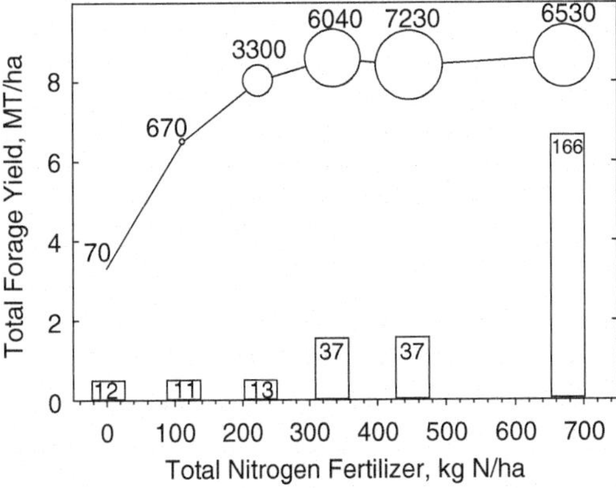

FIG. 5.4. The influence of N fertilizer on reed canarygrass yield (line), NO_3^- concentrations of forage (bubbles in mg kg^{-1}), and residual NO_3^- accumulation in soil (bars in kg NO_3^- ha^{-1}). Nitrogen was applied as ammonium nitrate fertilizer in equal amounts prior to the first (June) and second (July) forage harvests and the total amount of N received is provided on the x-axis (e.g. the 112 kg N rate is two applications of 56 kg N ha^{-1} prior to both Harvest 1 and 2). Total forage yield is the sum of Harvest 1 and 2. Bubble diameters are proportional to the NO_3^- concentration of forage at Harvest 2; concentrations are shown above the bubbles. A NO_3^- concentration of 3000 mg kg^{-1} or greater is considered potentially toxic to ruminant livestock. Bars represent residual soil NO_3^- summed over the soil profile to a depth of 0.9 m from cores taken in November. Numerals in bars represent actual amounts. *Source:* Adapted from Vetsch et al. (1999).

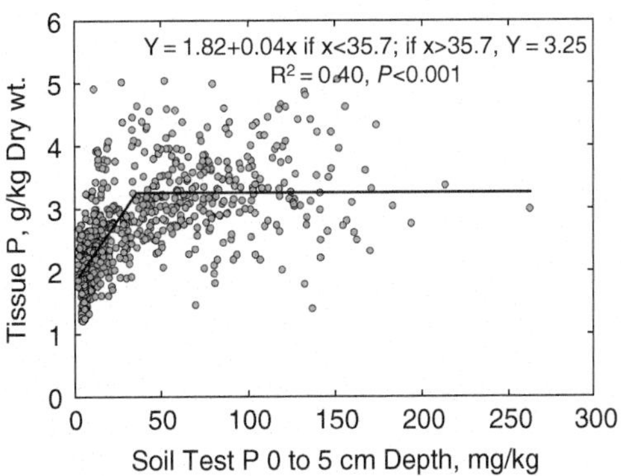

FIG. 5.5. Relationship between soil-test P and tissue-P concentrations of alfalfa. Data (640 observations) are from eight forage harvests obtained over two years (2003–2004) following application of P fertilizer (0, 25, 50, or 75 kg P ha^{-1} yr^{-1}) from 1997 to 2004. Linear-plateau regression was used to model the relationship with the inflection point at 35.7 mg kg^{-1} of soil-P. *Source:* Adapted from Berg et al. (2012).

Phosphorus (P)

Forage-P concentrations generally range from 1.5 to 3 g kg^{-1} dry wt and depend on several factors including stage of plant development, species, management, and environment (Mays et al. 1980). Unlike N and K, a high rate of P fertilizer application does not generally result in luxury consumption. For example, long-term application of high rates of P fertilizer that elevate soil-test P concentrations to levels in excess of 36 mg P kg^{-1} soil in the 0–5-cm depth result in little change in tissue-P concentrations above 3.25 g P kg^{-1} dry wt (Figure 5.5). At soil-test P concentrations between 1 and 35 mg P kg^{-1} soil, however, tissue-P concentrations increased linearly at a rate of 0.04 g P kg^{-1} dry wt for each 1 mg kg^{-1} increase in soil-test P levels.

Important structural and functional roles are played by P. Phospholipids are the predominant component of membranes, and P forms the backbone of DNA and RNA molecules. One important and unique property of P is the formation of high-energy phosphate (PO_4) bonds in molecules like ATP. The high-energy PO_4 bonds provide short-term storage of energy from radiation captured during photosynthesis and for energy recovered in mitochondrial respiration. The rapid turnover of high energy, P-containing molecules indicates their importance in plant metabolism and growth (Table 5.2).

Phosphorus also has an important role in post-translational regulation of other enzymes (Figure 5.6). For example, the dark-period inactivation of nitrate reductase described above involves attaching a phosphate group to a serine in the nitrate-reductase polypeptide. This attachment is catalyzed by a protein kinase and consumes ATP. The PO_4^{2-} is removed in light by a protein phosphatase, and nitrate reductase function returns. Numerous other key plant processes are regulated, in part, by protein kinase/phosphatase systems (Rodriguez et al. 2010).

Sugar phosphates play a central role in plant metabolism, and as such, a P deficiency can severely restrict growth. In leaves, P plays a critical role in photosynthesis because all initial precursors and products of CO_2 fixation and reduction are sugar phosphates. Often overlooked is the need for P to enter the chloroplast to exchange with triose phosphates that are exiting the

Table 5.2 Concentration, turnover rate, and rate of synthesis of organic P compounds in *Spirodela* spp. (duckweed)

Phosphorus fraction	Concentration (nmol g^{-1} fresh wt)	Turnover time (min)	Synthesis rate (mol P g^{-1} fresh wt min^{-1})
ATP	170	0.5	340
Glu-6-phosphate	670	7.0	95
Phospholipids	2700	130	20
RNA	4900	2800	2
DNA	560	2800	0.2

Source: Adapted from Bieleski and Ferguson (1983).

Turnover time estimates the minutes needed for complete breakdown and resynthesis of the P fraction amount listed in the table.

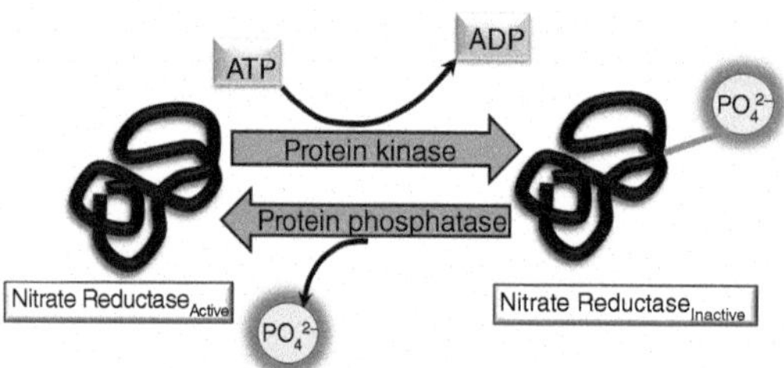

Fig. 5.6. Post-translational regulation of enzyme activity using P. In this example, nitrate reductase activity is lessened in darkness when a protein kinase uses ATP to attach phosphorus (PO_4^{2-}) to the nitrate reductase polypeptide. Activity returns in light when the PO_4^{2-} is removed by a protein phosphatase.

chloroplast to be used for sugar synthesis for transport to the rest of the plant. Post-harvest mobilization of starch in taproots of alfalfa is also dependent on adequate P supply and causes a C-limited reduction in shoot regrowth after harvest (Li et al. 1998). An important P-containing molecule, **phytic acid**, accumulates in seeds of many plants, including forages, to provide P for the developing seedling when P uptake by roots is limited. Phytic acid also accumulates in taproots of forage legumes in autumn and declines coincident with the onset of spring shoot growth suggesting that, as in seeds, taproot phytate serves as a stored source of P for shoot development when P uptake from soil is limited (discussed in detail below) (Li et al. 1996).

Root morphology and function can also be altered by P deficiency. Haynes and Ludecke (1981) reported reductions in root length, root hair size and frequency, and branching associated with P deficiency in birdsfoot trefoil and white clover. Changes in root architecture leading to more extensive root development in the upper soil profile also also been observed for plants grown under P-limiting conditions (Bonser et al. 1996). This response might be an adaptive response that enhances root exploration in parts of the soil profile where P is often more abundant.

Another morphologic adaptation to P deficiency is exemplified by proteoid root formation of lupine (Neumann et al. 2000). **Proteoid roots** can release H+ ions to acidify the **rhizosphere**, excrete P chelators, and induce high-affinity P transporters to aid P uptake under extremely low soil-P conditions. Several P-deficient plants develop a symbiotic relationship with arbuscular mycorrhizal fungi (Figure 5.7); a relationship that alters root morphology, and often enhances P, and to a lesser extent N, uptake by the plant (George et al. 1995). Other symbioses can also impact forage adaptation to P deficiency. For example, endophyte-infected tall fescue plants have longer root hairs than do endophyte-free plants, and under P deficiency, narrower root diameters (Malinowski et al. 1999). These changes in root morphology may enhance P uptake and adaptation of endophyte-infected tall fescue to low fertility soils.

Calcium (Ca)

Calcium sufficiency levels vary widely with species, tissue, and development and range from 0.5 g kg^{-1} dry wt for sugarcane internodes to 35 g kg^{-1} for cotton leaves (Clark 1984). For alfalfa shoot tissues (uppermost 15 cm), Ca concentrations ranging from 5 to 30 g kg^{-1} dry wt are considered sufficient. In general, Ca concentrations in grasses are lower than legumes. Spears (1994) summarized Ca concentrations in forage submitted for analysis by producers and found values ranging from 5 g kg^{-1} dry wt in grasses to 12 g kg^{-1} dry wt in legumes.

Calcium is generally supplied as **lime** ($CaCO_3$), so the impact of Ca on forage performance can be confounded with the neutralizing effect of carbonate on soil acidity. For example, Jackson et al. (1948) reported enhanced alfalfa persistence associated with K fertilization only if soil pH was raised from 5.5 to 7.0 with lime. Tissue-Ca concentrations varied inversely with K as lime application rates of up to 16 Mg ha^{-1} enhanced plant growth and

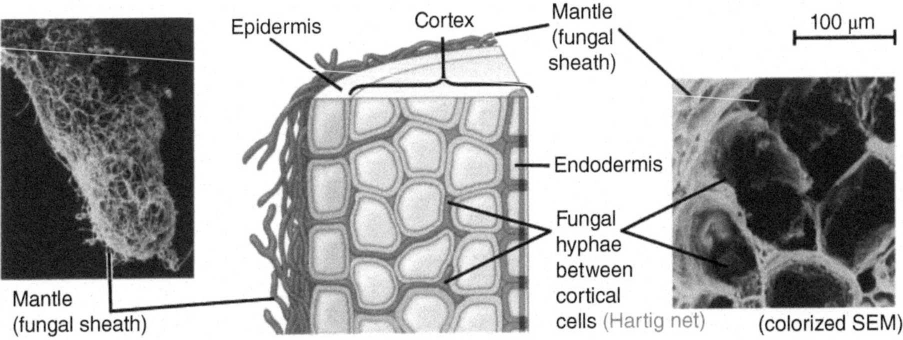

Fig. 5.7. Roots colonized by ectomycorrhizal fungi. The fungal hyphae increase root length and surface area permitting the plant-fungi system to explore a greater soil volume and acquire nutrients that have poor mobility in soil. The Hartig net is a hyphal network that extends into the root between the epidermal and cortical cells of ectomycorrhizal plants. This network is a site of nutrient exchange between the fungus and the host plant. *Source:* From: https://en.wikipedia.org/wiki/Hartig_net#/media/File:Ectomycorrhiza_illustration.jpg; Creative Commons BY-SA 3.0.

responsiveness to K application. In other studies (Berg et al. 2007; Lissbrant et al. 2010), when soil pH was adequate for alfalfa growth (~pH = 6.6), herbage Ca concentrations declined slightly with K fertilization (Figure 5.8, regression) whereas application of P fertilizer within each K rate had little impact on herbage Ca levels. The slight reduction in Ca with K fertilization may reflect a dilution effect because of the several-fold increase in forage yield that resulted from the K application.

Calcium has multiple and diverse roles in plants (Clark 1984). Calcium plays a structural role in cell walls of many forage crops by bridging cell wall components, especially **pectins** in the middle lamellae between cells. Calcium also stabilizes membranes and improves their integrity and selectivity. Activity of an array of enzymes, including amylases, protein kinases, and ATPase, is enhanced by Ca by altering enzyme conformation. Calcium also interacts with phytohormones changing both their synthesis and effectiveness. The role of Ca in signal transduction in plants is very important (Poovaiah et al. 1993). Membranes have Ca-efflux pumps that keep the cytoplasmic-Ca concentrations below 10^{-7} M. This low internal-Ca concentration allows a "pulse" of Ca to be an effective "signal" that often involves Ca-binding proteins such as calmodulin.

Magnesium (Mg)

Plants typically contain approximately 2 g kg^{-1} dry wt of Mg (Table 5.1). Forage grasses contain less Mg than forage legumes, especially when sampled in spring (Todd 1961). Potassium fertilizer application also reduces herbage-Mg concentrations. For example, increasing annual K fertilization from 0 to 400 kg K ha^{-1} reduced alfalfa herbage Mg concentrations (Figure 5.9). Within a K-fertilizer rate, there was no consistent effect of P fertilization on herbage Mg concentrations. Low forage-Mg can reduce serum Mg in livestock and result in a serious disease called grass tetany (hypomagnesemia, see Chapters 35 and 47). Symptoms of grass tetany in cattle include irritability, muscle twitching, staring, poor coordination, and staggering followed by collapse, coma, and death, if left untreated. Incidence of grass tetany is increased in K-fertilized grasses including annual forages like winter wheat that are grazed in spring when soils are cold and Mg uptake is reduced.

Significant variation in herbage-Mg concentrations has been reported among grass and legume species as well as consistent cultivar-specific variation within species (Gross and Jung 1978). These authors also observed large differences among genotypes in how tissue-Mg concentrations respond to rates of Mg fertilizer application. This suggests that breeding for improved forage-Mg concentrations may be a possible strategy for reducing the incidence of grass tetany. Subsequent plant breeding has successfully increased herbage-Mg concentrations in tall fescue forage (Crawford et al. 1998).

As a central component of chlorophyll, Mg is a vital component of the light-harvesting complexes in photosynthesis (Clark 1984). Chloroplast organization, specifically **grana** stacking and formation of light-harvesting complexes of chlorophyll a and b, requires Mg. The increase in stromal-Mg in chloroplasts during the day activates Rubisco (ribulose bisphosphate carboxylase/oxygenase), the key enzyme converting CO_2 to sugars, leading to high rates of photosynthesis. Like the monovalent cation, K, the divalent cation, Mg, is involved

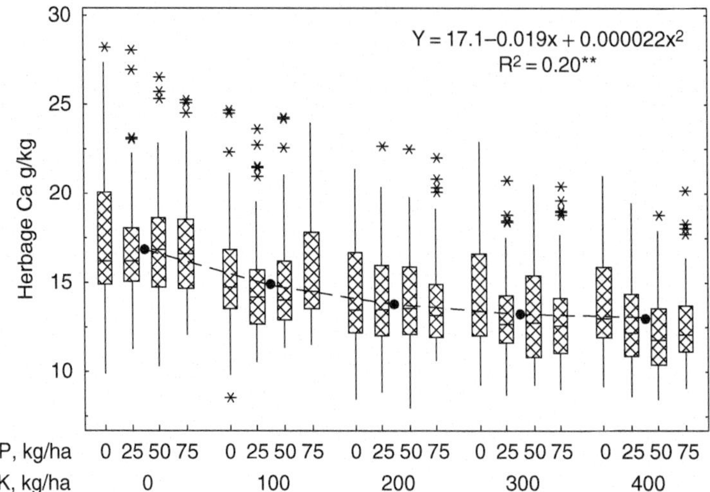

Fig. 5.8. Boxplots illustrating the influence of phosphorus (P) and potassium (K) fertilizer application on calcium (Ca) concentrations of alfalfa shoots averaged over four harvests per year for four years. Fertilizer applications shown are total annual amounts with one-half applied after the first harvest in May and the remaining half applied after the fourth harvest in September. The median is shown as a line within each box. Each box length represents the first quartile around the mean and the vertical lines extend to the second quartiles. Asterisks represent outlier observations. The quadratic regression of herbage Ca concentration on K fertilizer application was significant and plotted at the midpoint of the K treatments. *Source:* Adapted from Berg et al. (2012).

in activation of many enzymes (Marschner 2012). Interestingly, utilization and metabolism of Mg in plants are closely linked with that of P. Almost all metabolic reactions involving ATP (and similar high-energy compounds) also require Mg. Magnesium is also an important factor in ribosome stability and, therefore, protein synthesis.

Sulfur (S)

Sulfur concentrations in plants vary widely, ranging from 1 to 5 g kg^{-1} dry wt (Table 5.1). Highest concentrations are generally observed in crucifers, lowest in grasses, with legumes having intermediate concentrations (Marschner 2012). Martin and Walker (1966) reviewed the S requirements of forage and pasture plants and concluded that critical levels for herbage of several forages was in the range of 2–3 g S kg^{-1} dry wt (Table 5.3).

Sulfur assimilation requires reduction of sulfate SO_4^{2-} to plant-useable S and this occurs in a multi-step fashion similar to NO_3^- reduction described above. Sulfate is activated by first reacting with ATP to form APS, an S-analog to ATP. The APS is reduced in a two-step process, first to sulfite then to sulfide (Figure 5.10). The sulfide finally reacts with *O*-acetylserine to form cysteine that can itself serve as substrate for an array of S-containing compounds including methionine, glutathione, sulfonolipids, etc.

Sulfur is found in the S-containing amino acids (cysteine and methionine) and is, therefore, a critical component of virtually all polypeptides, proteins, and enzymes. In addition S, along with iron (Fe), is frequently found as prosthetic groups in active sites of enzymes involved in electron transport reactions such as light reactions of photosynthesis and oxidative phosphorylation in dark respiration. Lipids containing S (sulfolipids) are found in photosynthetic membranes of plastids where their negative charge stabilizes photosynthetic complexes. Sulfur is also a component of glutathione that can minimize damage due to oxidative stresses associated with superoxides and peroxides formed as a result of environmental stress.

Glucosinolates are S-containing compounds abundant in *Brassica* species used as cover crops and winter forage. These compounds are generally viewed as undesirable in livestock feed, including forages, because they can reduce palatability, feed intake, and animal performance (Gustine and Jung 1985).

Efforts to improve air quality by trapping sulfur emissions from power plants have been successful; annual SO_2 emissions in the US have declined from 28 MT yr^{-1} in 1970 to 7 MT yr^{-1} in 2010 (Hand et al. 2012). While this is a positive outcome for the environment, concerns have arisen recently that present-day SO_2 deposition rates

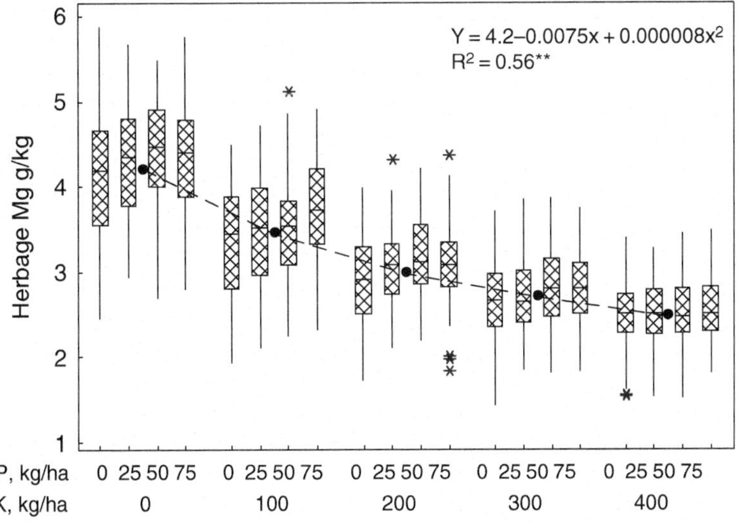

Fig. 5.9. Boxplots illustrating the influence of phosphorus (P) and potassium (K) fertilizer application on magnesium (Mg) concentrations of alfalfa shoots. See Figure 5.8 for details on presentation. The quadratic regression of herbage Mg concentration on K fertilizer application was significant and is plotted at the midpoint of the K treatments. *Source:* Adapted from Berg et al. (2012).

Table 5.3 Deficient, critical, and adequate herbage sulfur (S) concentrations (g S kg^{-1} dry wt) for several commonly grown forages

Species	Stage	Deficient	Critical level	Adequate
			S (g kg^{-1})	
Alfalfa	Early bloom	1.1–1.9	2.2	1.7–3.0
	1/3 bloom	1.0–1.5	2.2	3.0–4.0
	Full bloom	1.0–3.3	—	1.6–2.2
Crimson clover		1.6	2.0	2.1
Rose clover	Full bloom	1.0	—	1.4
Subterranean clover	Flowering	1.6	2.1	3.0
White clover		1.3	2.6	
Coastal bermudagrass		1.4	—	1.9
Mixed grasses			2.3–3.0	
Range grasses	Heading	1.0		2.2

Source: Adapted from Martin and Walker (1966).
Where available developmental stage of the forage is provided.

may lead to S deficiencies in plants, including forages. This has prompted a resurgence of S fertility research and a re-evaluation of S fertilizer recommendations in the Midwest US (Sawyer et al. 2015). Recent mass balance analysis suggests that more S is being removed from grasslands in the UK than is being supplied through fertilizers, manure, and atmospheric deposition (Webb et al. 2016).

Micronutrients Required by Forage Crops

Iron (Fe)

Iron is present in plants at concentrations of approximately 100 mg kg^{-1} (Table 5.1). Spears (1994) summarized Fe concentrations of forages analyzed at a commercial testing laboratory and reported average values of 222 (n = 992) and 184 (n = 352) mg kg^{-1} for legume and grass forage, respectively. These levels

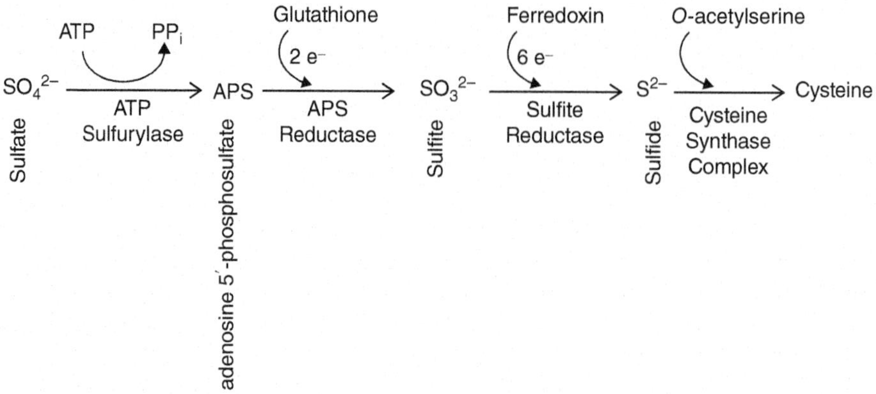

ATP PP$_i$
SO_4^{2-} → APS → SO_3^{2-} → S^{2-} → Cysteine
ATP Sulfurylase
Glutathione 2 e$^-$ APS Reductase
Ferredoxin 6 e$^-$ Sulfite Reductase
O-acetylserine Cysteine Synthase Complex
Sulfate
adenosine 5′-phosphosulfate
Sulfite
Sulfide

Fig. 5.10. Reduction of sulfate in plants and assimilation into cysteine is a four-step process. Sulfate is initially activated by converting it to 5′-adenylylsulfate (APS) an analog to ATP. The APS is reduced to sulfite using glutathione as an electron donor. The sulfite is reduced to sulfide by sulfite reductase using ferredoxin as the electron donor. Finally, the sulfide reacts with *O*-acetylserine to form cysteine. *Source:* Adapted from Ravilious et al. (2012).

are considered in excess of the nutritional requirements of ruminant animals. While Fe is abundant in soils, it is present largely as ferric Fe (Fe^{3+}) and its solubility in **soil solution** is very low. Many plants exude chemicals that reduce Fe^{3+} to Fe^{2+} (ferrous iron) before uptake (Marschner 2012). Plants differ in their iron uptake mechanism (Figure 5.11). Most dicots use Strategy I, in which proton pumps on the root membranes simultaneously acidify the **rhizosphere** while also chelating the external Fe^{3+} with citrate or similar organic acid. Ferric chelate reductase (FCR) reduces the chelate-Fe^{3+} complex to Fe^{2+} that is subsequently taken up by iron-regulated transporters (IRTs) on the root epidermis. Graminaceous plants use Strategy II for Fe uptake. This involves secretion of phytosiderophores like mugineic acid that chelate (attach to) Fe^{3+}. The Fe^{3+}-mugineic acid complex is soluble and transported through the plasmalemma via transport channels before the Fe^{3+} is subsequently released into the cytoplasm (Krohling et al. 2016).

Iron has numerous key roles in plants including forages. Deficiency of Fe often causes leaf **chlorosis** because chlorophyll synthesis requires Fe. It is also required in the heme group of many enzymes (e.g. nitrate reductase, catalase, and peroxidases). Non-heme Fe proteins are important components of nitrogenase, a key enzyme for N_2 fixation in root nodules of forage legumes. In proteins, Fe is often bound to the S of the amino acid cysteine, forming Fe-S centers that are important in metal binding and electron transport in cells. Although present in plants at relatively high concentrations for a micronutrient, large amounts of soluble-Fe in plant cells can lead to production of superoxide (O_2^-) radicals that can damage lipids, proteins, and other cell constituents. Production of phytoferritin, a protein-Fe complex, minimizes Fe-stimulated superoxide production while still releasing Fe for key plant processes.

Manganese (Mn)

Concentrations of Mn in plants are typically 50 mg kg^{-1} dry wt (Table 5.1). Manganese concentrations in grass forage samples are generally higher (76 mg kg^{-1}) than those found in samples of legume forage (44 mg kg^{-1}) (Spears 1994). Like Fe, Mn exists in several oxidation states; Mn^{2+} is more plant-available than Mn^{3+} and Mn^{4+}. Soil pH, aeration, organic matter, and microbial activity can influence the oxidation state and therefore plant availability of Mn in soils (Mulder and Gerretsen 1952). For example, flooding alfalfa plants with >5 cm of sugar-rich deproteinized alfalfa juice elevated tissue-Mn concentrations from 58 to 275 mg Mn kg^{-1} dry wt; levels that contributed to reduced plant growth (Walgenbach et al. 1977).

Manganese is an important component of some enzymes and activates others. For example, Mn is involved in water splitting during the light reactions in photosynthesis. During seed germination, arginase, a Mn enzyme, breaks down arginine in seed storage proteins to release N needed for seedling growth. Root nodules of warm-season legumes may synthesize and transport N in the form of the ureide allantoic acid. In leaves, allantoic acid is degraded by allantoate amidohydrolase, a Mn-activated enzyme (Lukaszewska et al. 1992). The C_4 photosynthetic process is very dependent on malic

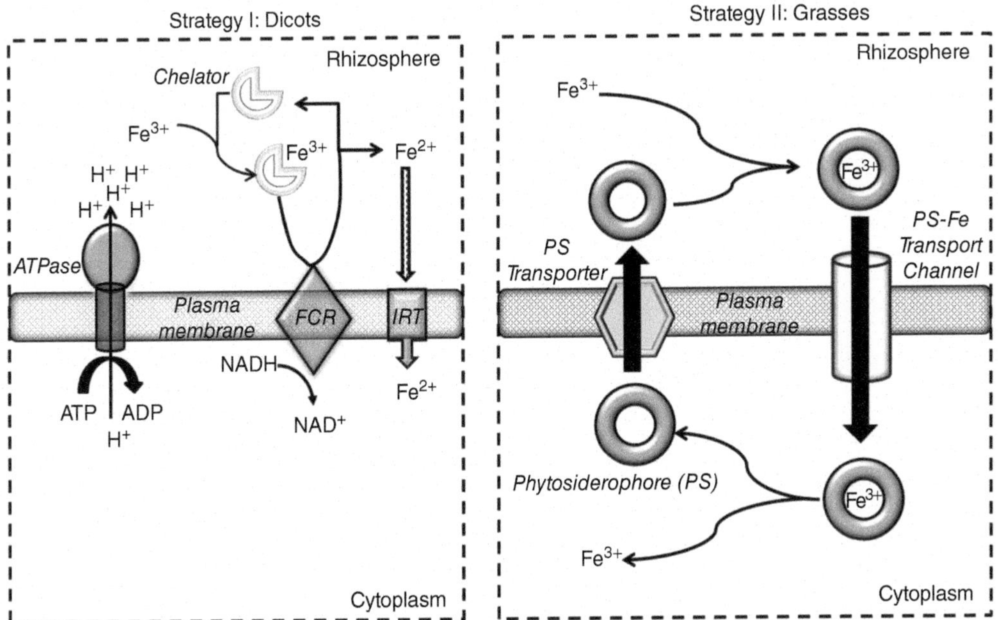

Fig. 5.11. Strategies for iron (Fe) uptake by plants. Strategy I represents Fe uptake in most dicot plants. The rhizosphere is acidified by active pumping of protons across the plasma membrane using ATPase. This increases the solubility of Fe^{3+} that is subsequently chelated with an organic acid like citrate. Ferric chelate reductase (FCR) reduces the chelated Fe^{3+} to Fe^{2+}. The Fe^{2+} enters the root cytoplasm through an iron-regulated transporter (IRT). Strategy II represents Fe uptake in most grasses. A phytosiderophore (PS) transporter excretes the PS into the rhizosphere where it binds with Fe^{3+}. The PS-Fe^{3+} complex is transported into the root cytoplasm through a PS-Fe transport channel where the PS and Fe^{3+}subsequently disassociate. *Source:* Adapted from Krohling et al. (2016).

enzyme and phosphoenolpyruvate (PEP) carboxykinase, both Mn-activated enzymes (Graham et al. 1988).

Boron (B)

Plants typically contain 20 mg B kg^{-1} dry wt. Plants may be grouped according to their B requirement: (i) lactifers require the most (70–100 ppm in leaf tissue); (ii) cole crops (*Brassica*) require 50–70 ppm; (iii) legumes, other forbs and monocots of the lily family require 20–50 ppm; and (iv) grasses (until flowering) require only 2–5 ppm (Blevins and Lukaszewski 1998). These general concentrations agree with species-specific estimates of B deficiency and sufficiency for major forage species (Gupta 2007; Table 5.4). Sufficiency values for alfalfa and clovers are 13–52 mg B kg^{-1} dry wt, whereas grasses, including small grains, have sufficiency ranges that are lower (3–10 mg B kg^{-1} dry wt) than those for legumes. In addition, B toxicity has been reported in many species with small grains being more sensitive to high B than perennial forages.

Root growth is inhibited by B deficiency and is associated with low rates of cell division and elongation (Gupta 2007). Boron effects on growth are mediated, in part, through its impact on auxin metabolism. Seed set and seed yield are often reduced with B deficiency due to poor pollen tube growth and low rates of pollination (Gauch and Dugger 1954). For example, B may increase alfalfa hay yield by 3%, while increasing seed yield by 600% (Piland and Ireland 1944). Membrane stability and transport processes across membranes also depend on B (Gupta 2007). Boron deficiency also affects sugar and starch metabolism, but the direction of the response (e.g. increase vs decrease) varies with the sugar pool (**reducing sugars**, nonreducing sugars, starch), tissue, and plant species. Nitrogen assimilation associated with nitrate reductase and nitrogenase is low when B is limiting (Gupta 2007).

Zinc (Zn)

Plant tissues with 20 mg Zn kg^{-1} dry wt are considered Zn-sufficient (Table 5.1). This generally agrees

Table 5.4 Deficient, critical, and toxicity levels for boron (B) in herbage of several commonly grown forages

Species	Stage	Deficiency	Sufficiency	Toxicity
		mg B kg^{-1} dry wt		
Alfalfa	Early bloom	<20	17–52	200
Red clover	Bud	12–20	21–45	> 59
White clover	Vegetative		13–16	53
Pasture grass	Anthesis		10–50	> 800
Ryegrass	Vegetative		9–38	> 39
Timothy	Heading		3–93	> 102
Small grains				
Barley	Boot	1.9–3.5	10	> 20
Oats	Boot	<1–3.5	8–50	> 30
Wheat	Boot	2.1–5.0	8	> 16

Source: Adapted from Gupta (2007).

Results from several studies are included and their individual ranges for deficiency-sufficiency-toxicity overlapped for some species (e.g. oats).

with the forage-specific values reported for legumes (18 mg Zn kg^{-1} dry wt) and grasses (28 mg Zn kg^{-1} dry wt) (Spears 1994). Many enzymes are activated by Zn, and Zn is a component of over 200 metalloenzymes, including carbonic anhydrase (important in C_4 plants), alcohol dehydrogenase, CuZn-superoxide dismutase, alkaline phosphatase, phospholipase, carboxypeptidase, and RNA polymerase. It also is an important component of Zn-binding proteins that are critical to DNA transcription and, therefore, protein synthesis. Healthy membranes depend on Zn, perhaps owing to its role in enzymes that control membrane damage from superoxides and peroxides (Marschner 2012).

Copper (Cu)

Plant tissues with 6 mg Cu kg^{-1} dry wt are generally considered Cu-sufficient (Table 5.1). Spears (1994) reported that forage grasses and legumes had similar Cu concentrations that average about 13 mg Cu kg^{-1} dry wt. Recent results from a meta-analysis suggest that **arbuscular mycorrhizae** may enhance Cu uptake of many plant species (Lehmann and Rillig 2015). Often the functions of Cu in plants involve Cu-metalloproteins that catalyze oxidation-reduction reactions. For example, Cu is a constituent of plastocyanin, a protein that shuttles electrons between Photosystems II and I in photosynthesis. Other examples include superoxide dismutase, cytochrome-c oxidase, ascorbate oxidase, diamine oxidase, and polyphenol oxidase (Marschner 2012). Low lignin concentrations have been observed in leaves of Cu-deficient plants (Robson et al. 1981), presumably because of the low polyphenol oxidase activity. Extremely high concentrations of Cu (~30 mg Cu kg^{-1} dry wt) in forage can result in liver failure and hemolytic crisis where red blood cells are destroyed faster than they are synthesized (Todd 1969). These Cu-related health issues are generally more common in sheep than cattle.

Chlorine (Cl)

Plant tissues with 100 mg Cl kg^{-1} dry wt are generally considered Cl-sufficient (Table 5.1). As with K and N, luxury consumption of Cl can occur in many forage species. Rominger et al. (1976) reported very high tissue-Cl concentrations (>20 000 mg Cl kg^{-1} dry wt) and yield reductions of alfalfa fertilized with KCl at rates exceeding 448 kg K ha^{-1}. As with Mn, Cl plays a role in the water splitting in the light reactions of photosynthesis. In addition, Cl-stimulated tonoplast H^+-ATPase is important in stomatal opening in many species (Marschner 2012).

Molybdenum (Mo)

Plant tissues with 0.1 mg Mo kg^{-1} dry wt are generally considered Mo-sufficient (Table 5.1). Roots take up and transport Mo to the leaves as molybdate, MoO_4^{2-}. Molybdenum is important in many electron transport reactions. Plants and associated microorganisms contain enzymes that share a common FeMo Component (FeMoCo). These FeMoCo enzymes include nitrate reductase, dinitrogenase, xanthine dehydrogenase, xanthine oxidase, sulfite oxidase, and aldehyde oxidase. These Mo-containing enzymes play important roles in N_2 fixation and metabolism, S metabolism, and synthesis of the plant hormones, abscisic acid and auxin (Marschner 2012).

Nickel (Ni)

Nickel is a component of urease, an abundant enzyme in seed of several species (Fabiano et al. 2015). Urea accumulates in Ni-deficient plants and may arise from several sources, including the ornithine cycle, arginase action, and polyamine and ureide degradation. Nickel also may have a role in synthesis of glutathione, a cell constituent that helps minimize oxidative stress in plants. In general, Ni requirements are low and deficiencies are rarely seen.

Beneficial Elements

Cobalt (Co) is required for dinitrogen fixation in legumes. Cobalt is needed for **leghemoglobin** synthesis in nitrogen-fixing nodules. Like human hemoglobin, leghemoglobin binds O_2 preventing denaturation of dinitrogenase, the key enzyme responsible for conversion of N_2 to NH_3. Cobalt deficiency, though rare, depresses leghemoglobin synthesis and with it, dinitrogen fixation rates in nodules (Figure 5.12). However, legumes acquiring N from fertilizers do not have a Co requirement.

Sodium (Na) is required at micro-nutrient levels in many, but not all C_4 plants, and is not beneficial to growth of C_3 plants (Marschner 2012). Maize and sugarcane are two C_4 plants whose growth is similar with and without Na. In responsive species, Na enhances photosynthesis by stimulating synthesis of PEP, one of the substrates involved in the initial fixation of CO_2 in C_4 plants. In non-halophyte plants Na can substitute for K for osmotic adjustment. Silicon (Si) can benefit growth and stress tolerance of some plants. Interaction of Si among lignin molecules is thought to enhance stem strength and promote the erect orientation of leaf blades.

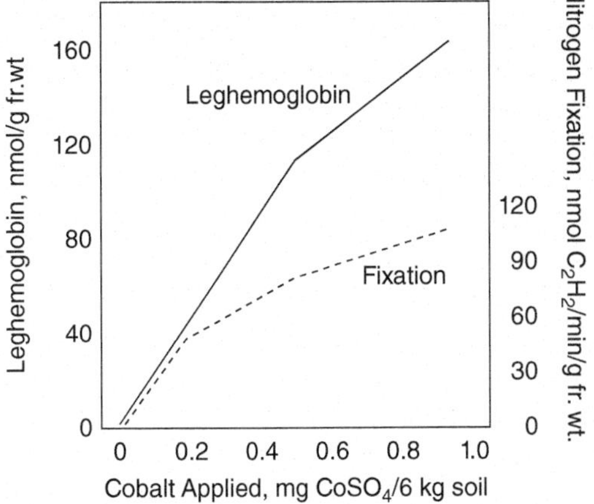

Fig. 5.12. Impact of cobalt application on dinitrogen fixation rate and leghemoglobin concentration of nodules of six-week-old lupin plants. Dinitrogen fixation was measured using acetylene (C_2H_2) reduction and is expressed on a nodule fresh weight basis. Leghemoglobin concentration is based on nodule fresh weight basis. *Source:* Adapted from Riley and Dilworth (1985).

It also enhances resistance to plant diseases like powdery mildew (Marschner 2012). Selenium (Se) is required by animals in low concentrations, but essentiality in plants has not been established. Because of its chemical similarity to S, it may substitute for S in active sites of some enzymes that have Fe-S centers (Marschner 2012).

Nutrient Uptake

Nutrient uptake is the result of four processes: transport in the free space of soil, root-nutrient contact; active uptake into root cells; and transport from roots to shoots.

Root-Nutrient Contact

Three processes are potentially involved in root-nutrient contact: root interception, solute diffusion, and mass flow. Root interception includes nutrients that come into physical contact with root surfaces as roots grow through the soil. This process potentially provides Ca and a portion of the needed Mg because these nutrients are very abundant in many soils, but root interception alone cannot provide adequate quantities of other nutrients (Table 5.5).

Mass flow includes nutrients brought to the root surface in the convective flow of water to roots caused by plant transpiration. The transpiration rate of the plant and the average concentration of nutrients in the soil solution can approximate the amount of each nutrient brought to the root surface by this mechanism. In general, anions suspended in the **soil solution** like NO_3^- and SO_4^{2-} move to roots via mass flow as do abundant cations like Ca^{2+} and Mg^{2+}. Micronutrients are generally provided via mass flow in amounts that meet plant requirements.

Diffusion requires a concentration gradient between the root surface and the bulk soil. This gradient develops when plant demand exceeds the supply at the root surface via mass flow and root interception. Diffusion works primarily for nutrients like P and K that are required by plants in large quantities, but are poorly mobile in soils.

Plant species can differ in the relative contribution of these processes to root-nutrient contact. Baligar (1985) compared K uptake by maize, onion, and wheat. While root interception consistently accounted for less than 1% of K uptake for all species, mass flow accounted for 61% of K uptake in onion with 39% via diffusion, while maize and wheat acquired >96% of their K via diffusion.

Nutrient Transport in the Root Free Space

Absorbed nutrients need to reach the root xylem in order to be transported to shoots. Along with the sugar-transporting phloem, the nutrient- and water-transporting xylem is located inside the stele, the interior of the root surrounded by the endodermis (Figure 5.13). The cell walls of the endodermis are impregnated with a suberin-like material and this forms what is referred to as the Casparian Band that prevents free movement of nutrients from outer cells to the xylem.

Table 5.5 Summary of contributions of root interception, mass flow, and solute diffusion to root-nutrient contact for maize

		Amount supplied by process (kg ha^{-1})		
Nutrient	Crop uptake (kg ha^{-1})	Root interception	Mass flow	Diffusion
Nitrogen	190	2	150	38
Phosphorus	40	1	2	37
Potassium	195	4	35	156
Calcium[a]	40	60	165	0
Magnesium[a]	45	15	110	0
Sulfur	22	1	21	0
Copper[a]	0.1	—	0.4	—
Zinc	0.3	—	0.1	—
Boron[a]	0.2	—	0.7	—
Iron	1.9	—	1.0	—
Manganese[a]	0.3	—	0.4	—
Molybdenum[a]	0.01	—	0.02	—

Source: Adapted from Lambers et al. (1998).

Mass flow provides nutrients with an (*a*) at levels in excess of crop uptake. Under these conditions, the nutrient accumulates at the root surface or diffuses back into the soil.

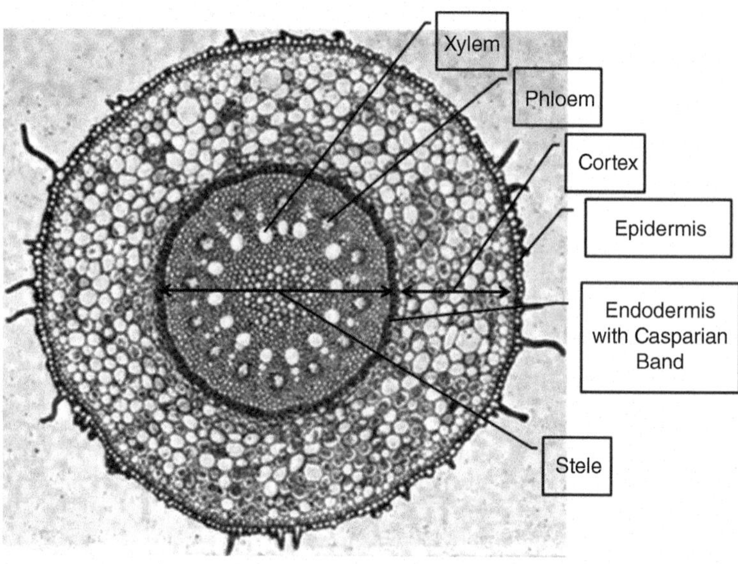

Fig. 5.13. Cross-section of a monocot root showing the epidermis, cortex, stele, endodermis, phloem and xylem. *Source:* Adapted from: http://nanelson.weebly.com/ch-23-flowering-plants-lab.html.

This anatomic design results in most nutrient uptake occurring in the cortex. The root cortex represents the cells between the epidermis and the endodermis that surround the stele. Approximately 15% of the volume of the cortex is intercellular gaps between adjacent cells called the free space. Nutrients can move freely between cells in this liquid-filled region.

Pectins and other cell wall polymers with a net negative charge impart a cation exchange capacity (CEC) to roots (Haynes 1980). High CEC and exchange absorption have generally been shown to enhance nutrient uptake in many plants (Marschner 2012). However, other factors like root-length density also impact CEC-nutrient uptake relationships among species. For example, roots of legumes have greater pectin and higher internal CEC than do roots of grasses, but when grown in association, legumes often cannot effectively compete with grasses for soil K (Haynes 1980 and references cited therein). Differences in root CEC are one of the underlying mechanisms contributing to differential aluminum tolerance among trefoil species (Blamey et al. 1990). Transport in the free space is generally not a rate-limiting step in nutrient uptake.

Active Nutrient Uptake

Nutrient uptake into root cells is an active process requiring expenditure of energy usually in the form of ATP. It also involves carrier proteins in cell membranes, such as the **plasmalemma** located on the inner side of the cell wall, that are often nutrient-specific (Figure 5.14). Energy is required because the concentration of nutrients inside a root cell is higher than in the free space outside the cell. Thus, simple diffusion from the free space into the root will not occur.

The energy used for nutrient uptake is in the form of an electrochemical gradient, with differences in both charge and pH, between the cell interior and the free space outside the cell. This electrochemical gradient is created by ATPase, also known as Coupling Factor that uses ATP to pump protons (H^+) outside the cell. The result is a higher concentration of H^+ (lower pH) and more abundant positive charges outside the root cell when compared to the cell interior. The energy in this gradient is used to move nutrients across the plasmalemma through nutrient-specific carrier proteins called symports (Figure 5.14). Alternatively, nutrients and other ions can move from the cell interior to the apoplast via nutrient-specific antiport proteins (in Figure 5.14, Na^+ is being transported out) using the energy in the H^+ gradient.

Nutrient-specific uniports also exist in cell membranes that permit movement into and out of the cell. Unlike symports and antiports, movement through uniports is not linked to the H^+ gradient. Instead, movement through uniports is driven by the electrochemical gradient across the plasmalemma of the nutrient itself. Finally, channels exist in membranes of root cells that permit ion movement into and out of cells, and like uniports, the flow through channels is based on the electrochemical gradient of the specific nutrient.

Channels are distinguished from uniports by their very high rate of ion transport that can exceed 10 million ions per second (Marschner 2012). Carrier/channel specificity is determined by the charge and hydrated radius of the nutrients, and how these interact with the molecular

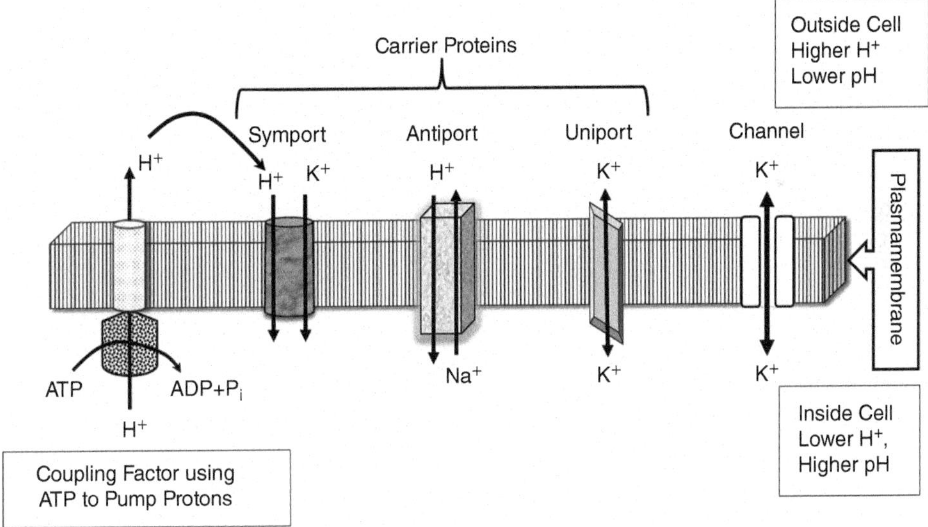

Fig. 5.14. Diagram depicting nutrient uptake processes at the plasmalemma of a root cell. Coupling Factor uses ATP as an energy source to pump protons (H^+) into the intercellular space, lowering the pH and creating an electro-chemical gradient. The energy in this gradient is used to cotransport nutrients (K^+ is used here as an example nutrient) through nutrient-specific carrier proteins (symports). Alternatively, nutrients can move counter to the influx of H^+ through nutrient-specific antiport proteins. Uniport carrier proteins also permit movement of nutrients between the **apoplast** and cell interior without the use of energy/H^+ gradient. Finally, channels can exist in the plasmalemma that allow nutrients to enter without involving nutrient-specific carriers or requiring an expenditure of energy. *Source:* Adapted from Figure 2.7 of Marschner (2012).

structure of the carrier/channel. Experimentally, this relationship has been exploited in ion uptake studies by using non-essential elements of similar charge and radius (e.g. Li^+ or Rb^+ for K^+) to predict uptake rates of required plant nutrients (Ravenek et al. 2016).

The metabolic cost of nutrient uptake for roots of **sedge** (*Carex* spp.) has been estimated to be 29% of all the root ATP in 40-day-old plants, declining to 9% of root ATP in 95-day-old plants as root growth slows and maintenance respiration consumes most metabolic energy (Werf et al. 1988). For most nutrients both high- and low-affinity carriers have been identified adding greater flexibility in nutrient acquisition rates under variable soil-nutrient conditions. These carriers generally follow Michaelis-Menten kinetics in how carrier affinity for a nutrient and the maximum nutrient uptake rate vary with soil nutrient concentration (Barber 1995). While carrier function remains interesting, sensitivity analysis using models (Barber 1995) and experimental evidence (Ravenek et al. 2016) both indicate that root-length density (root growth rates) is a critically important factor influencing nutrient uptake.

Nutrient Transport from Roots to Shoots

Nutrient movement from the site of active uptake in the cortex to the xylem located in the stele (Figure 5.13) involves active processes when membranes must be crossed and passive movement via cytoplasmic streaming and transpiration-driven water flow. Once in the xylem, nutrients move quickly with the water moving to shoot tissues, and most often to leaves where water loss via stomata is high. Nutrient needs of fully expanded, functioning leaves are modest compared to the active meristematic regions so many nutrients arriving in leaf blades are loaded into the sugar-conducting phloem and transported, along with carbohydrates, to **sink** tissues (growing leaves, stems, roots, seeds, ...) where both the sugars and mineral nutrients are incorporated into new plant tissues (Simpson et al. 1982).

Dinitrogen Fixation by Leguminous Forages

In N fertilizer manufacturing, the incredible stability of the covalent triple bond in N≡N molecules require high temperatures (between 400 and 500 °C) and pressures (15–25 MPa; 2200–3600 lb in.$^{-2}$) to chemically convert N_2 to NH_3 via the Haber-Bosch process. Erisman et al. (2008) estimated that 1% of global energy resources are required annually to synthesize the fertilizer N needed for global agriculture via this process. Thus, the conversion of dinitrogen gas (N_2) to plant useable forms via dinitrogen fixation is a remarkable feat accomplished by a subset of plants and microbes.

Rhizobia and Nodule Formation

Non-anthropogenic N_2 fixation is classified as free-living, associative and symbiotic systems; all three types of fixation can be found in agricultural systems (Table 5.6). Free-living organisms are not associated with plants, instead they are free-living in the upper soil profile. The majority of these N_2 fixing organisms are heterotrophic, use organic residues as substrates, and thus, their fixation capacity is limited by substrate availability (1–2 kg N ha^{-1} yr^{-1}; Table 5.6). Associative systems generally fix more N than free-living systems (10–200 kg N ha^{-1} yr^{-1}). These organisms grow in close association with the root system and use root exudates (sugars, organic acids, amino acids, ...) as an energy source. Most forage legumes form a symbiotic system with N_2 fixing bacteria in the root nodules by providing C in the form of sugars to the bacteria, which in turn, reduce gaseous N_2 to useable forms that are available to the plant. Rates of N_2 fixation of symbiotic systems are generally high relative to other systems (50–400 kg N ha^{-1} yr^{-1}).

Table 5.6 The three categories of biological N_2 fixation based on the nature of plant-microbe interaction

Characteristic	Nature of plant-microbe interaction		
	Free-living	Associative	Symbiotic
Micro-organisms involved	*Azotobacter, Klebsiella, Rhodospirillium*	*Azospirillium, Azotobacter*	*Rhizobium* spp., *Actinomycetes*
Location of microbes	Distributed in soil; not associated with plants	Rhizosphere surrounding roots	Nodules attached to roots
Energy source	Heterotrophs: plant residue; autotrophs, photosynthesis; chemolithotrophs, inorganic compounds	Root exudates	Sucrose and other sugars from attached plant
N fixation rate (kg ha^{-1} yr^{-1})	Heterotrophs: 1–2; autotrophs: 10–80	10–200	50–400

Table 5.7 Cross-inoculation groups for *Rhizobium* species used to inoculate common forage legumes and soybean

Cross-inoculation group	*Rhizobium* species	Host genera	Leguminous plant
Clover group	*Rhizobium leguminosarum* biovar. *trifolii*	*Trifolium*	True clovers including red, white, alsike
Alfalfa group	*Sinorhizobium meliloti*	*Medicago, Melilotus, Trigonella*	Alfalfa, sweetclover, fenugreek
Trefoil group	*Mesorhizobium loti*	*Lotus, Lupinus*	Birdsfoot trefoil, lupin
Soybean group	*Bradyrhizobium japonicum*	*Glycine, Vigna*	Soybean, cowpea

Symbiotic relationships are specific between *Rhizobium* bacteria and their plant host. Therefore, it is critical to apply the appropriate commercial inoculant to the seed prior to planting if the legume of interest has not been grown in the field in the last few years. Cross-inoculation groups are shown in Table 5.7. Some like the Clover Group inoculate plant species consistent with the name; the true clovers. Most cross-inoculation groups, however, inoculate several legumes species. For example, the Alfalfa Group infected by *Sinorhizobium meliloti* forms effective nodules on species in the *Medicago*, *Melilotus*, and *Trigonella* genera. Commercial **inoculation** products will often include two or more *Rhizobium* species in order to create a product that will work on commonly grown forage legumes in the region (e.g. *Rhizobium leguminosarum* biovar. *trifolii* and *S. meliloti* are mixed 50 : 50 for the US where both clover and alfalfa are common forage legumes). It is important not to expose inoculum to high temperatures prior to use because heat may reduce microbe viability.

Nodule formation begins with *Rhizobium* infection of root cells. This multi-step process starts with recognition of host-*Rhizobium* compatibility at the root surface. This includes "nod" factors produced by roots that induce genes in *Rhizobium* that initiate infection. In excess of 30 nod factors are known to be involved in the infection process. Once compatibility is confirmed, bacteria enter root hairs using infection threads. These threads grow through epidermal and cortical cell walls until bacteria are deposited in cortical cells (Figure 5.13) where they proliferate and form bacteriods, double-membrane structures that contain from 1 to >20 bacteria.

Bacteriods are filled with leghemoglobin, similar to human hemoglobin, that carefully reduces the concentration of oxygen (O_2) surrounding the *Rhizobium*. This prevents denaturation of the O_2-labile nitrogenase, the key enzyme involved in conversion of N_2 to NH_3, while simultaneously providing sufficient O_2 to meet the respiratory needs of surrounding bacteroid/nodule cells. Cobalt is needed for leghemoglobin synthesis and is one reason legumes have a requirement for this beneficial nutrient (see preceding section on Co nutrition).

Dinitrogen Fixation by Nitrogenase

The reduction of N_2 to NH_3 in the nodule requires electrons, protons, and a substantial amount of energy. The balanced reaction is:

$$N_2 + 8\,H^+ + 8\,e^- + 16\,ATP \rightarrow 2\,NH_3 + H_2 + 16\,ADP + 16\,P_i$$

Dark respiration provides the ATP necessary for this process. Molero et al. (2014) used stable isotope labeling to determine the partitioning of photosynthate between respiration and tissue synthesis in nodulated alfalfa. They reported nearly all (88%) of the carbohydrate entering the nodules was used in dark respiration to form ATP. They also found that, despite their small mass, relative to other plant organs, this carbohydrate pool used for nodule respiration represented 9% of gross CO_2 fixation via photosynthesis. This clearly documents that N fixation is not free, but the fixed N is all readily available.

Nitrogenase Is a Multimeric Protein

All nitrogenases are two-component systems made up of Component I (also known as dinitrogenase) and Component II (also known as dinitrogenase reductase) (Figure 5.15).

Nitrogenase is located in bacteriods in host-root cells and are surrounded by leghemoglobin to regulate O_2 levels and prevent denaturation of nitrogenase into its constituent subunits. Nitrogenase reductase receives electrons from ferredoxin and using ATP passes these electrons to dinitrogenase, the subunit that ultimately coverts N_2 to NH_3. Like nitrate reductase, dinitrogenase contains Mo illustrating the interdependency of this specific micronutrient on N assimilation in crops.

Factors Influencing N_2 Fixation.

High soil-N levels and/or application of fertilizer N can greatly reduce N_2 fixation rates. Using ^{15}N-labeled fertilizer, McAuliffe et al. (1958) quantified N derived from applied fertilizer vs N_2 fixation for white clover plants grown in two low-N soils (Figure 5.16). Three weeks post-N application the percent N from N_2 fixation had

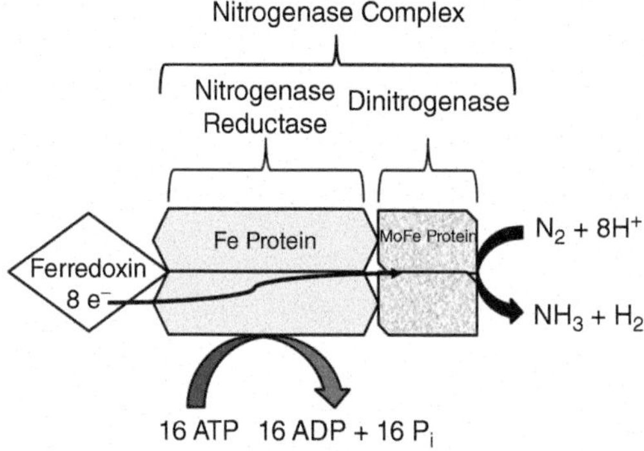

Fig. 5.15. Diagrammatic representation of the nitrogenase complex that catalyzes the reduction of dinitrogen gas to **ammonia**. Ferredoxin supplies electrons to nitrogenase reductase that are subsequently passed to dinitrogenase and, ultimately, to dinitrogen gas to form ammonia. Dinitrogenase contains molybdenum (Mo) and iron (Fe) at its active site. The subunits of the nitrogenase complex are O_2 labile and will disassociate if exposed to O_2.

declined from 66% of total plant N for plants provided $28\,kg\,N\,ha^{-1}$ to 14% for plants receiving $224\,kg\,N\,ha^{-1}$. Forage yield and harvested N mass totaled for the three harvests after N application were not affected by fertilizer N. Subsequent work showed both nitrogenase activity and nodule formation of white clover were reduced with nitrate-N application (Carroll and Gresshoff 1983). The underlying physiologic basis for these responses includes the more energetically favorable uptake and assimilation of fertilizer N (ammonium, nitrate; Figures 5.2 and 5.3) when compared to N_2 fixation (Figure 5.16).

Other nutrients that may impact nodule formation and N_2 fixation include P, S, Ca, B, and Fe. In addition to P, needed as ATP to drive the nitrogenase reaction, nodules themselves have a high P demand and outcompete other vegetative **sinks** for P. Meeting P demand of nodules can be satisfied by mycorrhizae illustrating a three-way symbiosis that may be prevalent among perennial systems like forages that grow with minimal soil disturbance. Likewise, Ca, S, B, and Fe are all required for key processes in development and/or effective functioning of nodules; deficiencies of these nutrients must be corrected for high rates of N_2 fixation to occur.

The environment impacts rates of N_2 fixation. Drought reduces both nodulation and rate of N_2 fixation by nodules (Serraj et al. 1999). Bacteria are generally more drought tolerant than are host plants so *Rhizobium* survival is not likely to be a factor limiting legume nodulation. However, growth and movement of ***Rhizobia*** may be reduced in dry soils and may slow nodule formation. Reduced transport of organic nutrients in phloem to nodules appears to be a major constraint under drought. This could slow new nodule development and limit energy available for N_2 fixation within existing nodules. Flooding reduces N_2 fixation and growth of several forage legumes in a species-specific manner: *Medicago* > *Trifolium* > *Lotus* (Striker and Colmer 2017). Flooding-tolerant legumes maintain O_2 diffusion through roots to nodules by developing **aerenchyma** in the root cortex in which contiguous cells break down to form channels that allow gas diffusion downward to the nodule. These morphologic adjustments enable O_2 supply to the bacteriods for respiration to supply energy necessary for N_2 fixation.

Increased soil temperature can directly inhibit N_2 fixation by slowing nodule development, decreasing nodule function and accelerating nodule senescence (Aranjuelo et al. 2014). Indirect factors, including reduced root hair formation and modified adherence of *Rhizobia* to root hairs at high temperatures, can also impair symbiotic N_2 fixation. Finally, these abiotic stresses in many cases, interact with photosynthetic rates and water flow to further reduce N_2 fixation. Few studies have evaluated the direction and magnitude of these interactions, but such knowledge is becoming increasingly relevant with a changing climate.

Defoliation by cutting or grazing forage legumes reduces N_2 fixation. Shoot removal reduces leaf area and photosynthesis per plant and a substantial reduction in N_2 fixation follows within hours (Figure 5.17). Rate of N_2 fixation remains very low for 7–10 days post-harvest, then increases as leaf regrowth occurs and photosynthesis per plant has increased. The reduction in N_2 fixation is attributed to dependency on current photosynthate to provide the sugars necessary for dark respiration and ATP production in nodules. N_2 fixation is also low in spring when post-winter growth resumes; another time when leaf area and photosynthesis per plant are low.

Nutrient Reserves

Forages are unique among plants used in agriculture because they are exposed to periodic, near complete defoliation numerous times during the growing season. In temperate regions, biennial, or perennial forages also endure winter, and grow the following spring. Reserves accumulated in storage organs (e.g. taproots, stolons, rhizomes, stem bases) are critical to survival, and it has been widely recognized that sugars, starches, fructans, and other nonstructural carbohydrate pools are an important part of forage regrowth and plant survival (see Chapter 4).

Recently, it has become evident that forages also accumulate sources of inorganic nutrients like N and P in storage organs that, like carbohydrate reserves, are mobilized to shoots during regrowth after harvest or when growth resumes in spring when soils are cold and nutrient

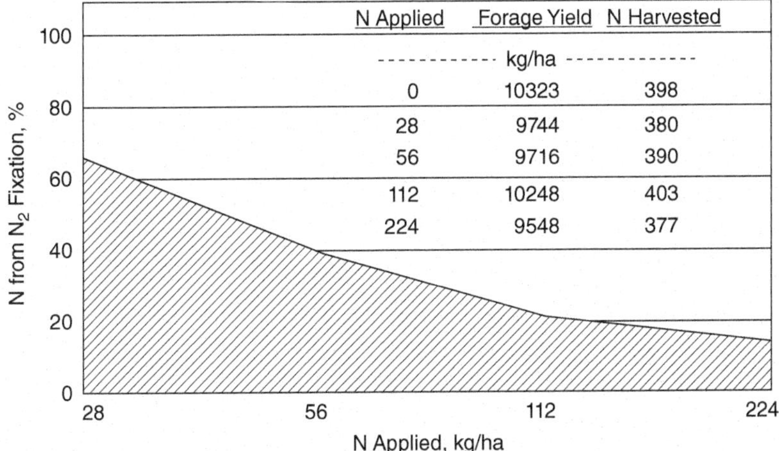

FIG. 5.16. Influence of nitrogen (N) fertilizer application (as ^{15}N-labeled ammonium sulfate) on percent of tissue N derived from dinitrogen (N_2) fixation in white clover. Harvest occurred three weeks post-N application. Forage yield and harvested N mass are the sum of three harvests following N application. Application of N reduced the percent N derived from N_2 fixation and did not enhance forage or N yield when compared to plants not receiving N fertilizer. *Source:* Adapted from McAuliffe et al. (1958).

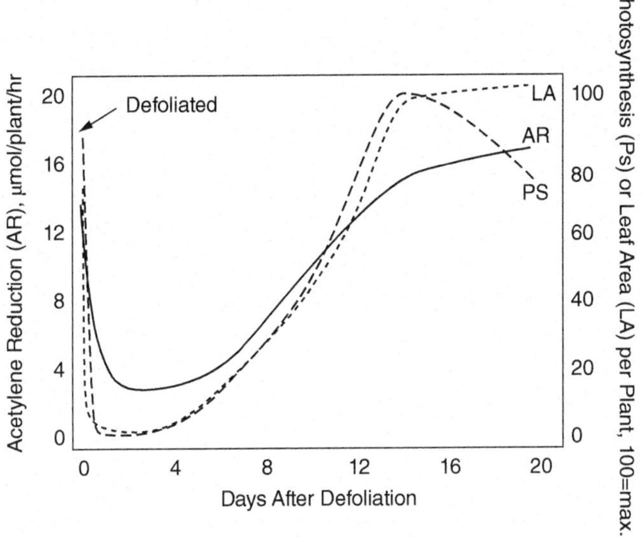

FIG. 5.17. Time course of defoliation on Day 0 on leaf area (LA), photosynthesis (Ps), and acetylene reduction (AR) per plant of alfalfa. Rate of acetylene reduction is an estimate of nitrogenase activity. Defoliation reduces both Ps and LA immediately, followed by a marked decline in AR. Rates of AR increase as both LA and Ps per plant increase. *Source:* Adapted from Fishbeck and Phillips (1982).

movement from soil to roots is reduced. Nitrogen reserves often accumulate as amino acids or specific proteins called vegetative storage proteins (VSPs) (Volenec et al. 1996). Reserve-P can accumulate in storage organs as phytic acid (Li et al. 1996), the same form of P reserve in seed.

Reserves of key minerals are important because nutrient uptake and assimilation require energy as described above and this may be limiting when plants have little/no leaf area for photosynthesis. For example, concentration of protein in legume taproots increases markedly after the final forage harvest in autumn (Figure 5.18). These high concentrations remain relatively unchanged over winter (December to March), but decline rapidly once shoot growth resumes in spring. Except for birdsfoot trefoil, taproot protein concentrations increase again in May. When defoliated in early June, protein concentrations decline for two to three weeks and then increase during late stages of shoot development; a pattern mimicking what occurs with taproot carbohydrate reserves.

Similarly, ^{15}N-labeling studies have confirmed transfer of N from taproots to regrowing shoots (Volenec et al. 1996). Because N_2 fixation is drastically reduced following harvest (Figure 5.17), these stored N pools, called VSPs, can contribute significantly to shoot N nutrition during regrowth providing up to 50% of total mass of shoot N. In addition, the VSPs in alfalfa taproots possess sequence homology with chitinase and β-amylase and may also play important adaptive roles in plant protection against abiotic (low temperature) and biotic (pathogen attack) stresses (Avice et al. 2003).

Less is known about P reserves in forages. Phytic acid is inositol hexakisphosphate; essentially a hexagon-shaped sugar alcohol with six phosphates attached. It is a common form of P accumulated in seeds making up as much as 90% of seed P. It also can constitute 10–15% of root- and crown-P in alfalfa (Campbell et al. 1991). As expected, taproot phytic acid concentrations increase with P fertilization (Li et al. 1998). Also, like carbohydrate and protein reserves, phytate accumulates in taproots of forage legumes in autumn, and is depleted when shoot growth resumes in spring (Figure 5.19). However, unlike C and N reserves (compare with Figure 5.18), cutting the shoots does not result in depletion of taproot phytic acid concentrations expressed on a dry weight basis (Li et al. 1996, 1998).

Departing from patterns exhibited by nonstructural carbohydrates, N, and P, K do not exhibit net accumulation in storage organs in autumn, nor are K concentrations depleted from taproots and crowns of alfalfa in spring and after harvest when shoot growth resumes (Li et al. 1996). Instead, K concentrations in storage organs generally decline as other reserves accumulate, and increase as other reserves are depleted in spring and after harvest. This suggests that K mass per taproot is relatively constant and concentration changes observed on a dry weight basis are a result of losses of other taproot constituents. This seems unique since fertilization with K is known to improve stress tolerance, especially winter hardiness. Additional work expressing K concentrations on a structural dry weight basis (Moser et al. 1982) would

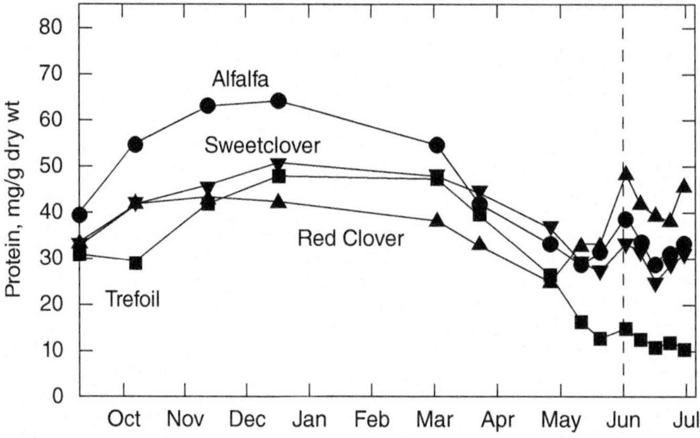

Fig. 5.18. Concentrations of soluble protein in taproots of alfalfa, sweetclover, red clover, and birdsfoot trefoil in Indiana US increase as plants harden for winter, then decrease as plants resume shoot growth in March and after defoliation in June (dotted line). Taproot protein concentrations increase markedly between September and December, and undergo extensive depletion when shoot growth resumes in spring. Protein concentrations increase in taproots of all species except trefoil during May prior to harvest in June. After harvest, protein concentrations in taproots of most species decline for two to three weeks, then increase during late regrowth; again, trefoil is the exception. *Source:* From Li et al. (1996).

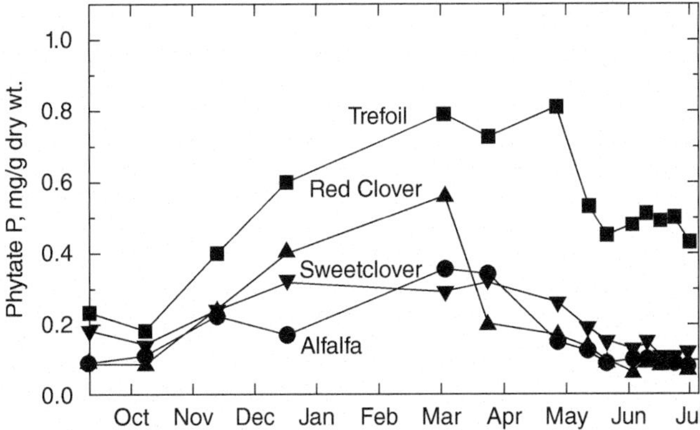

Fig. 5.19. Concentration of phytic acid in taproots of alfalfa, sweetclover, red clover, and birdsfoot trefoil as plants harden for winter in Indiana US and resume shoot growth in March. Plants were harvested in early June. Phytic acid concentrations increase markedly between September and December, and undergo extensive depletion when shoot growth resumes in spring. *Source:* From Li et al. (1996).

further inform patterns of K accumulation and use in storage organs of forages.

Nutrient Use Efficiency (NUE)

Improving efficient use of nutrients in cropping systems remains a critical issue facing agriculture, including the forage-livestock sector. There are many ways to characterize NUE in crops, and these fall into two broad categories; agronomic aspects and physiologic aspects (Brouder and Volenec 2017). A subset of these include Agronomic Efficiency (AE), Physiologic Efficiency (PE), and Uptake Efficiency (UE). The AE is an indicator of increased yield per unit of applied fertilizer (Table 5.8). It is a product of PE and UE. The PE is an indicator of the responsiveness of yield to fertilizer uptake and depends on crop species or genotype, the environment, and how nutrients and the crop are managed. The UE is an indicator of the ability of plants to acquire the nutrients from fertilizer and is influenced by fertilizer management (placement, form, timing, and rate) and crop nutrient need (growth rate, weather, genotype, stress levels). These indices are calculated from differences in yield and nutrient content of fertilized vs unfertilized plots ("difference method," Table 5.8) or by using stable isotopes (e.g. ^{15}N) or surrogate nutrients (e.g. Rb for K) to estimate nutrient uptake.

In general, the first increment of fertilizer is most effective at increasing forage yield, with a diminishing response to additional fertilizer until a yield plateau is attained or, in some cases, yield declines. For example, in 1993, forage yield of reed canarygrass in Minnesota increased up to approximately 300 kg N ha^{-1} (Figure 5.20). The AE in 1993 was 37 kg forage dry matter per kg N fertilizer with application of 56 kg N ha^{-1}, and declined to 0 at 448 kg N ha^{-1}, when additional N did not increase yield. Yields in 1994 were generally lower and yield plateaued with 224 kg N ha^{-1}. The AE also was lower with 26 kg forage dry matter per kg N produced with the first 56 kg N ha^{-1} fertilizer increment. Above this N rate, AE generally declined and at high N fertilizer applications, AE values were at or below 0. Excessive N fertilizer rates and low AE result in excessive nitrate in forage that poses a risk to livestock and high soil nitrate levels that can contribute to eutrophication of surface waters as described previously (Figure 5.4).

A suite of management practices known collectively as 4R Nutrient Stewardship provide opportunities to improve NUE in cropping systems (Christianson and Harmel 2015). These include: **R**ight source; **R**ight placement; **R**ight timing, and **R**ight rate. Most of these 4R principles generally apply to forages and row crops alike. Though nutrient placement studies in perennials are uncommon, Peterson and Smith (1973) reported that surface application of K_2SO_4 to alfalfa was as or more effective than placing this fertilizer at soil depths ranging from 15 to 90 cm. Other work determined that K_2SO_4 was the preferred source over KCl when high (>448 kg K ha^{-1} yr^{-1}) rates of K were applied (Rominger et al. 1976). Timing of K application has generally shown that multiple K applications vs a single high-K fertilizer application results in higher yield and limits luxury consumption (Kresge and Younts 1962).

Timing and rate of N application to forages also is important to maximize AE in forage grasses. Multiple application of moderate N rates during the growing season can enhance yield and alter the seasonal distribution

Table 5.8 Definitions and calculations for estimating agronomic efficiency (AE), physiologic efficiency (PE), and uptake efficiency (UE) of nutrient use by crop plants

Performance index	Definition	Calculation	Interpretation
Agronomic efficiency (AE)	Yield increase (kg) per kg nutrient applied	$AE = (Y - Y_0)/F$ or $AE = UE \times PE$	Product of nutrient uptake from the soil and the efficiency of nutrient use to produce new plant dry matter. Depends on management practices that impact PE and UE.
Physiologic efficiency (PE)	Yield increase (kg) per kg increase in nutrient uptake from fertilizer	$PE = (Y - Y_0)/(U - U_0)$	Ability of plant to transform nutrients from fertilizer into yield. Depends on nutrient management, crop genotype, and environment.
Uptake efficiency (UE)	Increase in nutrient uptake (kg) per kg nutrient applied	$UE = (U - U_0)/F$	Ability of plant to acquire nutrients from fertilizer. Influenced by fertilizer, application management and nutrient needs of crop.

Source: Adapted from Dobermann (2007).
Y = crop yield (kg ha^{-1}) with applied nutrients.
Y_0 = crop yield (kg ha^{-1}) of unfertilized control plot.
F = amount (kg ha^{-1}) of fertilizer applied.
U = nutrient (kg ha^{-1}) in biomass from fertilized plot.
U_0 = nutrient (kg ha^{-1}) in biomass of unfertilized plot.

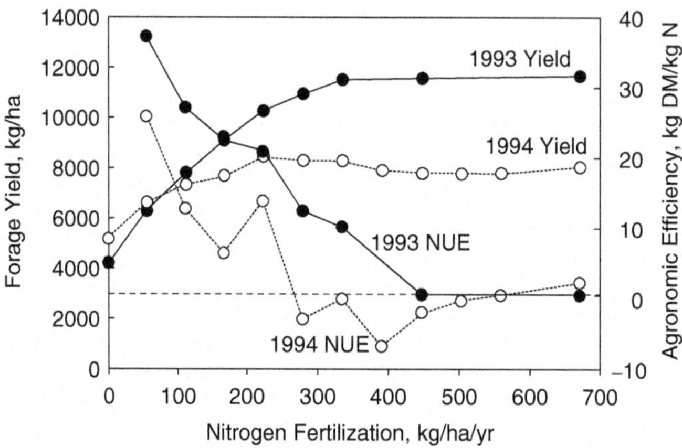

Fig. 5.20. Influence of nitrogen fertilization on forage dry matter (DM) yield and agronomic nitrogen-use-efficiency (NUE) of reed canarygrass in 1993 and 1994. Agronomic efficiency (AE) was calculated as the increment of yield increase divided by the additional N applied. The dashed horizontal line identified the 0 NUE below which yield is reduced by addition of N fertilizer. *Source:* Adapted from Vetsch et al. (1999).

of forage production (Feyter et al. 1985). Source of N can also affect yield responses and unintended environmental impacts. For example, forage yields were occasionally greater with urea than an equivalent rate of calcium ammonium nitrate, but this response is weather dependent (Herlihy and O'Keeffe 1987; Stevens et al. 1989). Forage yields and AE with urea N sources can be lower unless this fertilizer is treated with a urease inhibitor to restrict **volatilization** losses, especially when surface-applied in summer (Dawar et al. 2010; Antille et al. 2015).

Improvement in AE and its components remains a trait of interest to many agronomists and is now a selection criterion in some breeding programs. While species can differ in AE, there is a tight linkage between N uptake/ha and biomass/ha within species (Lemaire et al. 2007). This suggests that nutrient uptake will scale with increases in biomass yield irrespective of whether yield is enhanced via breeding or management. This is due, in large part, to production of N-rich leaves that create the photosynthate needed to drive crop growth rates, increase biomass yields, and for forages, have a critical effect on forage nutritive value. For example, across a range of environments including N fertilization, Lemaire et al. (2007) reported a linear relationship ($R^2 \geq 0.97$) between leaf area index (LAI) of alfalfa and shoot-N accumulation/ha with a slope of 26.6 kg N ha^{-1} per unit of LAI.

Surprisingly, other crops (sunflower, canola, and sorghum) had N accumulation-LAI slopes nearly identical to alfalfa (range from 25.6 to 27.0), whereas the slope of this relationship was higher for maize (30.3) indicating that more N uptake is necessary to produce leaf area in this species. Because of the strong relationship between nutrient uptake and yield across species, management/breeding strategies aimed at reducing tissue-nutrient accumulation in order to increase PE will likely reduce forage yield and alter nutritive value.

The above indicates that nutrient-use-efficiency may hinge more on reducing losses to the environment through management practices, especially for nutrients that are water-soluble or environmentally labile. The suite of options focused on improved nutrient management have been assembled into what is referred to as the 4R Nutrient Stewardship Principles. These include using the Right source, Right rate, Right timing, and Right placement of fertilizers (http://www.nutrientstewardship.com/4rs). These principles are intuitive, have been studied extensively for decades, and are key features of most soil fertility recommendations aimed at improving NUE and environmental stewardship. For most non-leguminous species including maize, the rate of N fertilizer application is viewed as the most important and accessible 4R strategy, however, it alone is unlikely to meet the co-objectives of high yield and reduced environmental N losses (Christianson and Harmel 2015).

Recent large-scale data syntheses indicate that, while N placement method (incorporated, injected, surface applied) could impact maize grain yield, these placement strategies did not reduce N loads to surface waters (Christianson and Harmel 2015). Similarly, although timing of N application to maize (pre-season, pre-plant, at planting, side-dress) had a significant impact on maize yield, it had no significant impact on N losses to surface waters. Similar synthesis studies focused on N management of forages have not been reported.

Multiple loss pathways exist for N (e.g. surface waters vs GHG losses as N_2O, NH_3), and it is rare to find studies where all possible pathways are monitored simultaneously. Where this has been done, complex interactions can occur that underscore the challenges associated with nutrient management. For example, Hernandez-Ramirez et al. (2009, 2011) showed that, while fall- vs spring-applied manure both resulted in high maize grain yields, the fall-injected manure resulted in substantial N losses to surface waters and low GHG emissions whereas spring–injected manure had higher GHG losses and less N lost to surface waters. Only by monitoring both pathways was it evident that there was no single manure management option available that achieved ideal agronomic and environmental outcomes.

A strategy for improving NUE and nutrient stewardship, while meeting future food/feed needs, may emerge from meta-analysis using disparate datasets from past nutrient management research that report both crop and environmental impacts as described by Eagle et al. (2017). These data could also be used to enhance calibration and validation of models focused on improving nutrient management in forage-livestock systems (Holly et al. 2018). This latter point is critical since pastures also include the animal component with harvest efficiency, grazing behavior, urine and manure deposition and the daily or seasonal gain of the animal. In the long-term, another expression of nutrient use efficiency may be weight of animal product/unit of nutrient applied.

Conclusions

Mineral nutrients remain a critical input into forage-livestock agriculture. The needed minerals and concentrations in the tissues are now fairly well-established. Yet, meeting the nutritional needs of a growing human population predicted to reach nearly 10 billion people by 2050, whose diet is more meat intensive, will require substantial inputs of mineral nutrients because forage nutrient accumulation is proportional with forage biomass production. The challenge is how to accomplish these goals with minimal loss to the environment.

Simultaneously, there is recognition that loss of mineral nutrients from agro-ecosystems to surface waters and GHGs must be reduced. Both of these challenges will be exacerbated as row crop agriculture continues to occupy and expand on the best agricultural soils, while forage production is relegated to ever more marginal soil resources that are inherently low in fertility and erodible. Informed nutrient management, including the 4R Nutrient Management framework and similar concepts, will be important as forage-livestock producers strive to meet the dual goals of increasing production of forage and animal product with minimal inputs, while protecting the environment in a changing climate.

References

Andrews, M., Edwards, G.R., Ridgway, H.J. et al. (2011). Positive plant microbial interactions in perennial ryegrass dairy pasture systems. *Ann. Appl. Biol.* 159: 79–92.

Antille, D.L., Hoekstra, N.J., and Lalor, S.T. (2015). Field-scale evaluation of calcium ammonium nitrate, urea, and urea treated with N-(n-butyl) thiophosphoric triamide applied to grassland in Ireland. *Commun. Soil Sci. Plant Anal.* 46: 1345–1361.

Aranjuelo, I., Arrese-Igor, C., and Molero, G. (2014). Nodule performance within a changing environmental context. *J. Plant Physiol.* 171: 1076–1090.

Avice, J.C., Dily, F.L., Goulas, E. et al. (2003). Vegetative storage proteins in overwintering storage organs of forage legumes: roles and regulation. *Can. J. Bot.* 81: 1198–1212.

Baligar, V.C. (1985). Potassium uptake by plants, as characterized by root density, species and K/Rb ratio. *Plant Soil* 85: 43–53.

Barber, S.A. (1995). *Soil Nutrient Bioavailability: A Mechanistic Approach*. New York: Wiley.

Bartholomew, R.P. and Janssen, G. (1929). Luxury consumption of potassium by plants and its significance. *Agron. J.* 21: 751–765.

Berg, W.K., Cunningham, S.M., Brouder, S.M. et al. (2007). The long-term impact of phosphorus and potassium fertilization on alfalfa yield and yield components. *Crop Sci.* 47: 2198–2209.

Berg, W.K., Lissbrant, S., Volenec, J.J. et al. (2012). Phosphorus and potassium influence on alfalfa nutrition. Dataset. Purdue University.

Bieleski, R.L. and Ferguson, I.B. (1983). Physiology and metabolism of phosphate and its compounds. In: *Inorganic Plant Nutrition*, Encyclopedia of Plant Physiology New Series, vol. 15A (eds. A. Lauchli and R.L. Bieleski), 422–449. New York: Springer-Verlag.

Blamey, F.P.C., Edmeades, D.C., and Wheeler, D.M. (1990). Role of root cation-exchange capacity in differential aluminum tolerance of *Lotus* species. *J. Plant Nutr.* 13: 729–744.

Blevins, D.G. (1985). Role of potassium in protein metabolism in plants. In: *Potassium in Agriculture*, (ed. R.D. Munson), 413–424. Madison, WI, ASA, CSSA, SSSA.

Blevins, D.G. and Lukaszewski, K.M. (1998). Boron in plant structure and function. *Annu. Rev. Plant Biol.* 49: 481–500.

Bonser, A.M., Lynch, J.P., and Snapp, S. (1996). Effect of phosphorus deficiency on growth angle of basal roots in *Phaseolus vulgaris*. *New Phytol.* 132: 281–288.

Brouder, S.M. and Volenec, J.J. (2017). Future climate change and plant macro-nutrient use efficiency. In: *Plant Macro-Nutrient Use Efficiency: Molecular and Genomic Perspectives* (eds. M.A. Hossain, T. Kamiya, D.J. Burrit, et al.), 357–379. Elsevier. ISBN: 9780128113080.

Brown, B.A. (1957). Potassium fertilization of ladino clover. *Agron. J.* 49: 477–480.

Campbell, M., Dunn, R., Ditterline, R. et al. (1991). Phytic acid represents 10 to 15% of total phosphorus in alfalfa root and crown. *J. Plant Nutr.* 14: 925–937.

Carroll, B.J. and Gresshoff, P.M. (1983). Nitrate inhibition of nodulation and nitrogen fixation in white clover. *Z. Pflanzenphysiol.* 110: 77–88.

Christianson, L.E. and Harmel, R.D. (2015). 4R water quality impacts: an assessment and synthesis of forty years of drainage nitrogen losses. *J. Environ. Qual.* 44: 1852–1860.

Clark, R.B. (1984). Physiological aspects of calcium, magnesium, and molybdenum deficiencies in plants. In: *Soil Acidity and Liming*, 2e (ed. F. Adams), 99–170. Madison, WI, ASA, CSSA, and SSSA.

Crawford, R.J., Massie, M.D., Sleper, D.A., and Mayland, H.F. (1998). Use of an experimental high-magnesium tall fescue to reduce grass tetany in cattle. *J. Prod. Agric.* 11: 491–496.

Dawar, K., Zaman, M., Rowarth, J.S. et al. (2010). The impact of urease inhibitor on the bioavailability of nitrogen in urea and in comparison with other nitrogen sources in ryegrass (*Lolium perenne* L.). *Crop Pasture Sci.* 61: 214–221.

Dobermann, A. (2007). Nutrient use efficiency–measurement and management. In: *Fertilizer Best Management Practices*, 1e, 1–28. Paris, France: IFA. ISBN: ISBN 2-9523139-2-X.

Eagle, A.J., Christianson, L.E., Cook, R.L. et al. (2017). Meta-analysis constrained by data: recommendations to improve relevance of nutrient management research. *Agron. J.*: 2441–2449.

Erisman, J., Sutton, M., Galloway, J. et al. (2008). How a century of ammonia synthesis changed the world. *Nat. Geosci.* 1: 636–639.

Fabiano, C.C., Tezotto, T., Favarin, J.L. et al. (2015). Essentiality of nickel in plants: a role in plant stresses.

Front. Plant Sci. 6: 754. https://doi.org/10.3389/fpls.2015.00754.

Feyter, C., O'Connor, M.B., and Addison, B. (1985). Effects of rates and times of nitrogen application on the production and composition of dairy pastures in Waikato district, New Zealand. *N. Z. J. Exp. Agric.* 13: 247–252.

Fishbeck, K.A. and Phillips, D.A. (1982). Host plant and *Rhizobium* effects on acetylene reduction in alfalfa during regrowth. *Crop Sci.* 22: 251–254.

Gauch, H.G. and Dugger, W.M. Jr. (1954). *The Physiological Action of Boron in Higher Plants: A Review and Interpretation*, Bulletin A (Maryland Agricultural Experiment Station), vol. 80, 44. College Park, MD: University of Maryland, Agricultural Experiment Station.

George, E., Marschner, H., and Jakobsen, I. (1995). Role of arbuscular mycorrhizal fungi in uptake of phosphorus and nitrogen from soil. *Crit. Rev. Biotechnol.* 15: 257–270.

Graham, R.D., Hannam, R.J., and Uren, N.C. (1988). *Manganese in Soils and Plants*. Boston: Kluwer.

Gross, C.F. and Jung, G.A. (1978). Magnesium, Ca, and K concentration in temperate-origin forage species as affected by temperature and Mg fertilization. *Agron. J.* 70: 397–403.

Gupta, U.C. (2007). Boron. In: *Handbook of Plant Nutrition* (eds. A.V. Barker and D.J. Pilbeam), 241–277. New York, NY: Taylor & Francis. ISBN: 0-8247-5904-4.

Gustine, D.L. and Jung, G.A. (1985). Influence of some management parameters on glucosinolate levels in *Brassica* forage. *Agron. J.* 77: 593–597.

Hand, J.L., Schichtel, B.A., Malm, W.C., and Pitchford, M.L. (2012). Particulate sulfate ion concentration and O_2 emission trends in the United States from the early 1990s through 2010. *Atmos. Chem. Phys.* 12 (8): 19311–19347.

Haynes, R.J. (1980). Ion exchange properties of roots and ionic interactions within the root apoplasm: their role in ion accumulation by plants. *Bot. Rev.* 46: 75–99.

Haynes, R.J. and Ludecke, T.E. (1981). Root morphology and chemical composition of two pasture legumes as influenced by lime and phosphorus applications to an acid soil. *Plant Soil* 62: 241–254.

Herlihy, M. and O'keeffe, W.F. (1987). Evaluation and model of temperature and rainfall effects on response to N sources applied to grassland in spring. *Fert. Res.* 13 (3): 255–267.

Hernandez-Ramirez, G., Brouder, S.M., Smith, D.R., and Van Scoyoc, G.E. (2009). Greenhouse gas fluxes in an eastern corn belt soil: weather, nitrogen source, and rotation. *J. Environ. Qual.* 38: 841–854.

Hernandez-Ramirez, G., Brouder, S.M., Ruark, M.D., and Turco, R.F. (2011). Nitrate, phosphate, and ammonium loads at subsurface drains: agroecosystems and nitrogen management. *J. Environ. Qual.* 40: 1229–1240.

Holly, M.A., Kleinman, P.J., Bryant, R.B. et al. (2018). Identifying challenges and opportunities for improved nutrient management through the USDA's Dairy Agroecosystem Working Group. *J. Dairy Sci.* 101: 6632–6641.

Jackson, M.L., Evans, C.E., Attoe, O.J. et al. (1948). Soil fertility level in relation to mineral and botanical composition of forage. *Soil Sci. Soc. Am. J.* 12: 282–288.

Kirkby, E. (2012). Introduction, definition and classification of nutrients. In: *Mineral Nutrition of Higher Plants*, 3e (ed. P. Marschner), 3–5. New York: Academic Press.

Kresge, C.B. and Younts, S.E. (1962). Effect of various rates and frequencies of potassium application on yield and chemical composition of alfalfa and alfalfa-orchardgrass. *Agron. J.* 54: 313–316.

Krohling, C.A., Eutrópio, F.J., Bertolazi, A.A. et al. (2016). Ecophysiology of iron homeostasis in plants. *Soil Sci. Plant Nutr.* 62: 39–47.

Lambers, H., Chapin, F.S. III, and Pons, T.L. (1998). Mineral nutrition. In: *Plant Physiological Ecology*, 239–298. New York, NY: Springer.

Lehmann, A. and Rillig, M.C. (2015). Arbuscular mycorrhizal contribution to copper, manganese and iron nutrient concentrations in crops–a meta-analysis. *Soil Biol. Biochem.* 81: 147–158.

Lemaire, G., van Oosterom, E., Sheehy, J. et al. (2007). Is crop N demand more closely related to dry matter accumulation or leaf area expansion during vegetative growth? *Field Crops Res.* 100: 91–106.

Li, R., Volenec, J.J., Joern, B.C., and Cunningham, S.M. (1996). Seasonal changes in nonstructural carbohydrates, protein, and macronutrients in roots of alfalfa, red clover, sweetclover, and birdsfoot trefoil. *Crop Sci.* 36: 617–623.

Li, R., Volenec, J.J., Joern, B.C., and Cunningham, S.M. (1998). Effect of phosphate nutrition on carbohydrate and protein metabolism in alfalfa roots. *J. Plant Nutr.* 21: 459–474.

Lissbrant, S., Brouder, S.M., Cunningham, S.M., and Volenec, J.J. (2010). Identification of fertility regimes that enhance long-term productivity of alfalfa using cluster analysis. *Agron. J.* 102: 580–591.

Lukaszewski, K.M., Blevins, D.G., and Randall, D.D. (1992). Asparagine and boric acid cause allantoate accumulation in soybean leaves by inhibiting manganese dependent allantoate amidohydrolase. *Plant Physiol.* 99: 1670–1676.

Malinowski, D.P., Brauer, D.K., and Belesky, D.P. (1999). The endophyte *Neotyphodium coenophialum* affects root morphology of tall fescue grown under phosphorus deficiency. *J. Agron. Crop Sci.* 183: 53–60.

Marschner, P. (2012). *Marschner's Mineral Nutrition of Higher Plants*, 3e. NY: Academic Press.

Martin, W.E. and Walker, T.W. (1966). Sulfur requirements and fertilization of pasture and forage crops. *Soil Sci.* 101: 248–257.

Martineau, E., Domec, J.C., Bosc, A. et al. (2017). The effects of potassium nutrition on water use in field-grown maize (*Zea mays* L.). *Environ. Exp. Bot.* 134: 62–71.

Mays, D.A., Wilkinson, S.R., and Cole, C.V. (1980). Phosphorus nutrition of forages. In: *Role of Phosphorus in Agriculture* (eds. F.E. Khasawneh, E.C. Sample and E.J. Kamprath), 805–846. Madison, WI: ASA-CSSA-SSSA. doi: 10.2134/1980.roleofphosphorus.

McAuliffe, C., Chamblee, D.S., Uribe-Arango, H., and Woodhouse, W.W. (1958). Influence of inorganic nitrogen on nitrogen fixation by legumes as revealed by N^{15}. *Agron. J.* 50: 334–337.

Molero, G., Tcherkez, G., Araus, J.L. et al. (2014). On the relationship between C and N fixation and amino acid synthesis in nodulated alfalfa (*Medicago sativa*). *Funct. Plant Biol.* 41: 331–341.

Moser, L.E., Volenec, J.J., and Nelson, C.J. (1982). Respiration, carbohydrate content, and leaf growth of tall fescue. *Crop Sci.* 22: 781–786.

Mulder, E.G. and Gerretsen, F.C. (1952). Soil manganese in relation to plant growth. *Adv. Agron.* 4: 221–277.

Murata, T. and Akazawa, T. (1968). Enzymic mechanism of starch synthesis in sweet potato roots: I. Requirement of potassium ions for starch synthetase. *Arch. Biochem. Biophys.* 126: 873–879.

Neumann, G., Massonneau, A., Langlade, N. et al. (2000). Physiological aspects of cluster root function and development in phosphorus-deficient white lupin (*Lupinus albus* L.). *Ann. Bot.* 85: 909–919.

Pelletier, S., Belanger, G., Tremblay, G.F. et al. (2008). Timothy mineral concentration and derived indices related to cattle metabolic disorders: a review. *Can. J. Plant. Sci.* 88: 1043–1055.

Peterson, L.A. and Smith, D. (1973). Recovery of K_2SO_4 by alfalfa after placement at different depths in a low fertility soil. *Agron. J.* 65: 769–772.

Piland, J.R. and Ireland, C.F. (1944). The importance of borax in legume seed production in the south. *Soil Sci.* 57: 75–84.

Poovaiah, B.W., Reddy, A.S.N., and Feldman, L. (1993). Calcium and signal transduction in plants. *Crit. Rev. Plant Sci.* 12: 185–211.

Ravenek, J.M., Mommer, L., Visser, E.J. et al. (2016). Linking root traits and competitive success in grassland species. *Plant Soil* 407: 39–53.

Ravilious, G.E., Nguyen, A., Francois, J.A., and Jez, J.M. (2012). Structural basis and evolution of redox regulation in plant adenosine-5′-phosphosulfate kinase. *Proc. Natl. Acad. Sci. U.S.A.* 109: 309–314.

Riley, I.T. and Dilworth, M.J. (1985). Cobalt requirement for nodule development and function in *Lupinus angustifolius* L. *New Phytol.* 100: 347–359.

Robson, A.D., Hartley, R.D., and Jarvis, S.C. (1981). Effect of copper deficiency on phenolic and other constituents of wheat cell walls. *New Phytol.* 89: 361–371.

Rodriguez, M.S.C., Petersen, M., and Mundy, J. (2010). Mitogen-activated protein kinase signaling in plants. *Annu. Rev. Plant Biol.* 61: 621–649.

Rominger, R.S., Smith, D., and Peterson, L.A. (1976). Yield and chemical composition of alfalfa as influenced by high rates of K topdressed as KC1 and K_2SO_4. *Agron. J.* 68: 573–577.

Sawyer, J.E., Lang, B.J., and Barker, D.W. (2015). *Sulfur Management for Iowa Crop Production.* CROP 3072, 12. Ames, IA, USA: Iowa State University https://store.extension.iastate.edu/Product/CROP3072-pdf.

Serraj, R., Sinclair, T.R., and Purcell, L.C. (1999). Symbiotic N_2 fixation response to drought. *J. Exp. Bot.* 50: 143–155.

Simpson, R.J., Lambers, H., and Dalling, M.J. (1982). Translocation of nitrogen in a vegetative wheat plant (*Triticum aestivum*). *Physiol. Plant.* 56: 11–17.

Spears, J.W. (1994). Minerals in forages. In: *Forage Quality, Evaluation, and Utilization* (ed. G.C. Fahey et al.), 281–317. Madison, WI: ASA, CSSA, SSSA https://doi.org/10.2134/1994.foragequality.

Stevens, R.J., Gracey, H.I., Kilpatrick, D.J. et al. (1989). Effect of date of application and form of nitrogen on herbage production in spring. *J. Agric. Sci.* 112: 329–337.

Stevens, C.J., Dise, N.B., Mountford, J.O., and Gowing, D.J. (2004). Impact of nitrogen deposition on the species richness of grasslands. *Science* 303: 1876–1879.

Storkey, J., Macdonald, A.J., Poulton, P.R. et al. (2015). Grassland biodiversity bounces back from long-term nitrogen addition. *Nature* 528: 401–404.

Striker, G.G. and Colmer, T.D. (2017). Flooding tolerance of forage legumes. *J. Exp. Bot.* 68: 1851–1872.

Todd, J.R. (1961). Magnesium in forage plants I. Magnesium contents of different species and strains as affected by season and soil treatment. *J. Agric. Sci.* 56: 411–415.

Todd, J.R. (1969). Chronic copper toxicity of ruminants. *Proc. Nutr. Soc.* 28: 189–198.

Vetsch, J.A., Randall, G.W., and Russelle, M.P. (1999). Reed canarygrass yield, crude protein, and nitrate N response to fertilizer N. *J. Prod. Agric.* 12: 465–471.

Volenec, J.J., Ourry, A., and Joern, B.C. (1996). A role for nitrogen reserves in forage regrowth and stress tolerance. *Physiol. Plant.* 97: 185–193.

Walgenbach, R.P., Smith, D., and Ream, H.W. (1977). Growth and chemical composition of alfalfa fertilized in greenhouse trials with deproteinized alfalfa juice. *Agron. J.* 69: 690–694.

Webb, J., Jephcote, C., Fraser, A. et al. (2016). Do UK crops and grassland require greater inputs of sulphur fertilizer in response to recent and forecast reductions in sulphur emissions and deposition? *Soil Use Manage.* 32: 3–16.

Wedin, W.F. (1974). Fertilization of cool-season grasses. In: *Forage Fertilization* (ed. D.L. Mays), 95–118. Madison, WI: American Society of Agronomy. ISBN: ISBN: 978-0-89118-239-9.

Werf, A., Kooijman, A., Welschen, R., and Lambers, H. (1988). Respiratory energy costs for the maintenance of biomass, for growth and for ion uptake in roots of *Carex diandra* and *Carex acutiformis*. *Physiol. Plant.* 72: 483–491.

Plant-Water Relations in Forage Crops

Jennifer W. MacAdam, Professor, *Plants, Soils and Climate, Utah State University, Logan, UT, USA*

C. Jerry Nelson, Professor Emeritus, *Plant Sciences, University of Missouri, Columbia, MO, USA*

Grasslands are the most extensive **biome** on Earth, and persistence of natural grasslands actually depends on periodic disturbances such as drought, fire, and grazing (Blair et al. 2014). For this reason, forage-animal production ecosystems are inseparable from the dynamics of their plant-water relations. Plants used as forages represent several functional groups including warm- and cool-season grasses, temperate and tropical legumes, and non-leguminous forbs. Each functional group varies in morphologic traits important to plant-water relations, such as rooting depth, plant height and leaf area. They also differ in physiologic traits associated with regulating transpiration, photosynthesis, and growth rates of leaves, stems, seed, and roots.

Forage Plants and the Hydrologic Cycle

The hydrologic cycle describes the movement of water through the biosphere (Figure 6.1). Water evaporates from leaves, the soil, bodies of water, or from snow and ice to become water vapor in the surrounding air. As water vapor rises into cooler air, it condenses into clouds that can travel around the globe and eventually returns to Earth as rain or snow that replenishes the surface and groundwater that supports growth of plants.

Evaporation from the soil surface depends mainly on the proportion of exposed soil surface relative to the leaf canopy, while transpiration from a crop (stomata open and water drawn from inside leaves) increases with rooting depth, crop height and leaf surface area. Rate of transpiration is driven by the difference between the water vapor concentration of the air surrounding the leaf and the water vapor concentration within the leaf. **Evapotranspiration (ET)**, expressed as mm water d^{-1} unit land area^{-1}, is the combined loss of water vapor from leaves and soil. It is higher at high soil water contents, increases with wind speed, solar radiation and temperature, and decreases with increasing relative humidity (Kirkham 2005).

Evapotranspiration and precipitation are influenced by climate and topography. A "humid" climate is one where precipitation exceeds potential evapotranspiration (PET), the total amount of water loss under well-watered conditions. Based on the ratio of annual precipitation to annual PET, the US can be divided into the humid East (ratio greater than one) and the subhumid-to-arid West (ratio less than one) at about 98 °W longitude (Figure 6.2). Movement of this line to the east is being driven by climate change.

Sustainability of Water Use for Irrigation

In environments where precipitation does not meet the needs of cultivated plants, irrigation may support the economic production of crops or pastures. Transient snowpack reservoirs in the Sierra Nevada Mountains of northern California have been sufficiently large to provide

Forages: The Science of Grassland Agriculture, Volume II, Seventh Edition.
Edited by Kenneth J. Moore, Michael Collins, C. Jerry Nelson and Daren D. Redfearn.

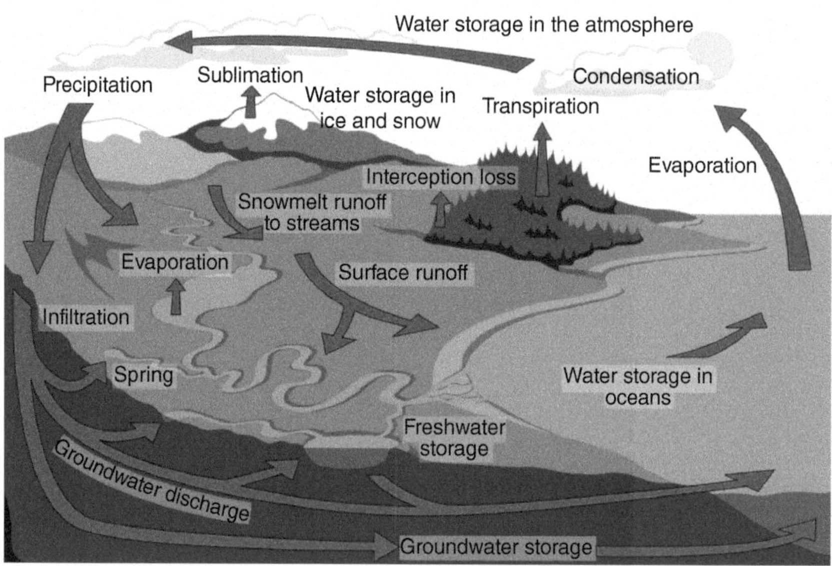

Fig. 6.1. The hydrologic cycle illustrates the driving factors and forms in which water moves through the biosphere. *Source:* Used with permission according to Copyright: University of Waikato. All Rights Reserved.

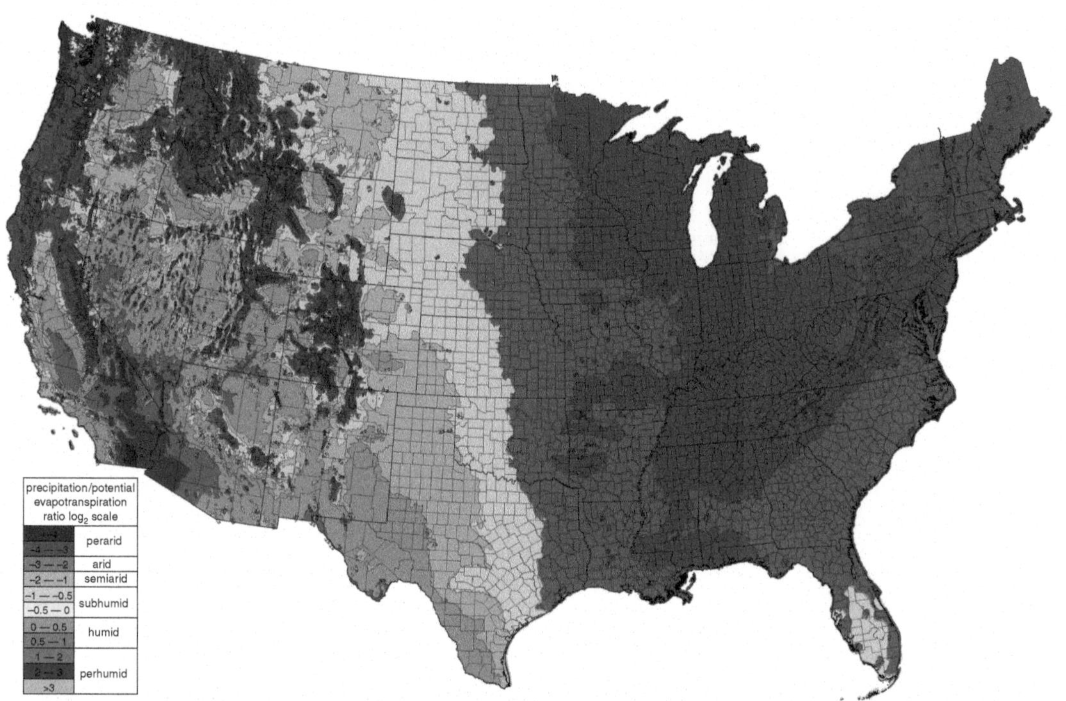

Fig. 6.2. The ratio of precipitation to potential evapotranspiration (PET) in the contiguous 48 states. Precipitation and PET result primarily from climate, independent of plant water use. Precipitation exceeds PET in humid regions and PET exceeds precipitation in semi-arid and arid regions. *Source:* Used with permission. Retrieved from http://www.bonap.org/Climate%20Maps/ClimateMaps.html.

irrigation for the Central Valley as far south as Bakersfield. Likewise, the Rocky Mountain snowpack has supplied Colorado River irrigation to Wyoming, Colorado, Utah, New Mexico, Arizona, Nevada, and southern California, as well as northern Mexico. Water flowing eastward from Rocky Mountain snowfall helped form the Ogallala aquifer under the Great Plains.

There is growing competition from urban areas for water supplies and enforced regulations on irrigation in sensitive watersheds. There is also concern about the sustainability of agricultural production systems where the supply of irrigation water is diminishing, such as the High Plains Ogallala and other aquifers (ground water stored in permeable rock) (Figure 6.3). The massive Ogallala aquifer stretches from South Dakota to the Texas panhandle and supplies 30% of the groundwater used for irrigation in the US (USDA NRCS 2012). It was formed following glaciation and for centuries supported the vast short- and tall-grass native prairie that was grazed by native and then domestic ruminants. But the water is now being consumed to irrigate large areas of wheat, corn, soybean, and cotton production.

This underground "lake" consists of a gravelly mix of clay, silt, and sand with water filling the spaces between the grains. Water is usually found 15–90 m below the land surface. The water-saturated thickness of the aquifer exceeds 300 m in west-central Nebraska and represents about 65% of the total ground water. The thickness around the shrinking perimeter is rapidly decreasing, especially in the southern part. Natural precipitation in the southern Great Plains does not support intensive cropping, and aquifer recharge cannot keep up with demand. On average, the saturated thickness of the Ogallala is decreasing about 1 m per year and the retreating water level increases the cost of pumping (Brauer et al. 2017).

A multi-disciplinary group is currently developing and sharing practical, science-supported information to guide management practices for optimizing water use across the Ogallala region (Brauer et al. 2017). Improved water-use efficiency (WUE) through more efficient irrigation systems may reduce the volume of water use for cropping. The inclusion of higher-quality forages such as alfalfa in mixtures with warm-season grasses can increase liveweight gain per unit land area and thereby increase the WUE of cattle production (Baxter et al. 2017). It is likely that cropping will decrease in this region and cow-calf and yearling beef production will re-occupy much of the southern Great Plains.

Water Potential and the Movement of Water Through Plants

Water potential (Ψ_w), or the potential energy of water, can be used to understand the movement of water in plants. A Ψ_w of zero is defined as pure water at atmospheric pressure and ambient temperature. Water potential is reported in units of pressure, such as atmospheres, bars, or inches of mercury, but the preferred unit is megapascal (MPa).

Water containing dissolved salts, such as nutrient anions and cations, sugars or other solutes, has a lower potential energy than pure water. Water moves from a compartment with higher Ψ_w to compartments with lower Ψ_w; this attraction is termed **solute** or **osmotic potential**. Diffusion is the tendency for a solute to move from a more concentrated to a less concentrated area. Osmosis is a special case of diffusion where relatively pure water is drawn through a differentially permeable membrane toward water where solutes are concentrated. The membrane of a plant cell is a differentially permeable membrane. Plant cells use solute potential to draw water into root cells from the soil, and to draw water into newly divided cells to produce the turgor (**pressure potential**) that is needed to drive cell growth (Figure 6.4).

Water Movement in the Soil

Soil is an ecosystem composed of mineral particles, organic matter, water, air, and living creatures such as microbes and earthworms (Chapter 1). Water applied to the soil surface infiltrates in response to both gravity and the composition and structure, e.g. size of peds of a given soil. If drainage is not impeded, water that replaces air in soil macropores after rain or irrigation will only be transiently available to plant roots before the soil drains. However, smaller soil pores continue to hold water in a soil at field capacity, and most plant water uptake comes from soil meso- and micropores. The living cover of perennial forages, the channels in the soil left by root turnover, and the resulting soil organic matter derived from dead and dying plant material all aid in the infiltration and retention of water.

Soil mineral and organic components (the soil matrix) have a tightly bound surface film of water that is not available to plants and can be determined by weight loss of the "dried out" soil by oven-drying soil at 105 °C. The strength with which this water is held is termed the **soil matric potential**, and the volume of this **bound water** in a soil varies with soil particle surface area; finely divided particles of a clay soil have a greater total surface area per volume of soil than the large particles of a sandy soil. Thus, a clay soil can hold or retain more matric water than a sandy soil.

Soil Ψ_w is the sum of matric, solute and pressure/tension potentials and is nearly always a negative number compared with the Ψ_w of pure water. Rain has a Ψ_w near zero, and will diffuse through the soil toward the rhizosphere, where water uptake by plant roots has left concentrated salts. The water will be used by the plant until the soil Ψ_w is near −1.5 MPa, defined as the permanent wilting point.

Root cells take up water when their Ψ_w is lower than the Ψ_w of the **soil solution**. The membranes enclosing

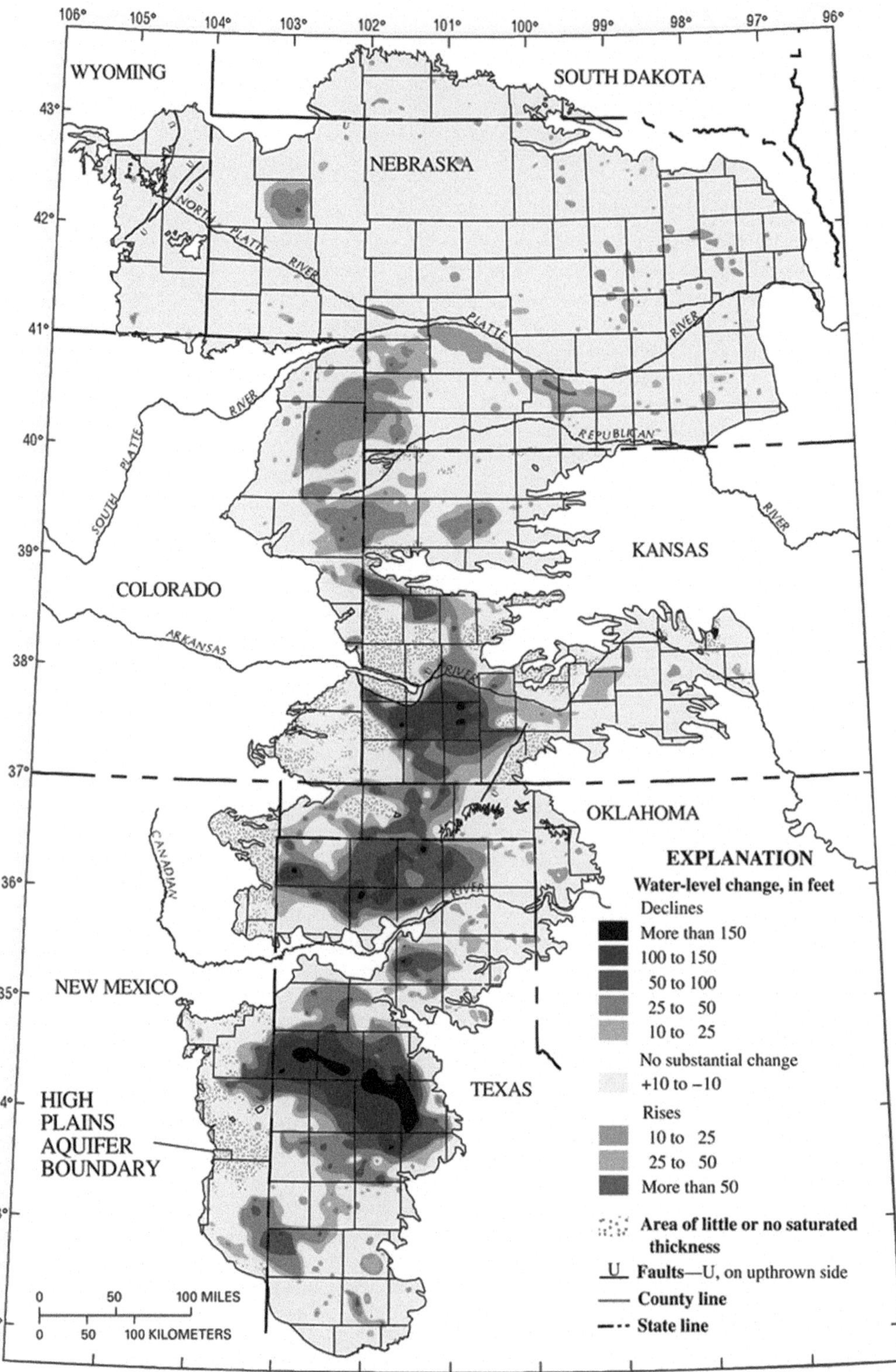

Fig. 6.3. Map of the Great Plains Ogallala Aquifer. Recharge is occurring along the Platte River and in the eastern half of Nebraska; depletion is most severe in northwest Texas.

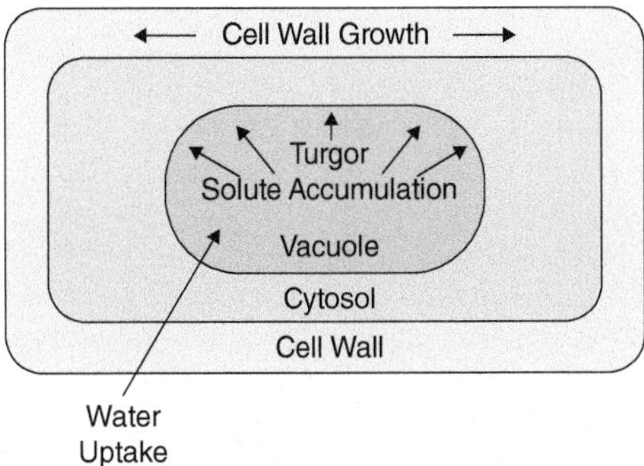

Fig. 6.4. Solute accumulation in the vacuole and cytosol of a plant cell is used to attract water. The resulting uptake of water by osmosis retains water in a drying soil, or produces turgor pressure that can be used to expand the cell wall during growth (MacAdam 2009).

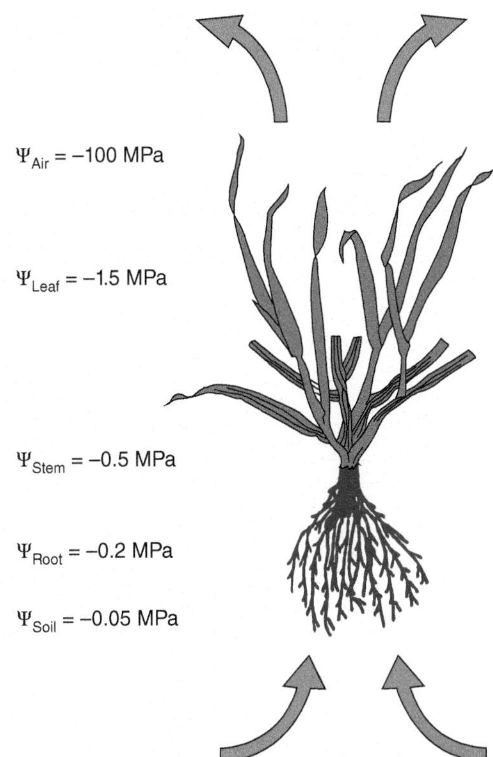

Fig. 6.5. Water transpires from the inner surfaces of leaves to surrounding air when stomata open for photosynthesis. The transpired leaf water is replaced by water from xylem in the stem, which is replaced by water absorbed by the root system from the soil. Water movement into and through plants is driven by plant water potential, with water moving in the direction of more negative values of water potential (MacAdam 2009).

both the cell and the vacuole can regulate the uptake of solutes; ions that are not needed for cellular metabolism can be sequestered in the vacuole to lower the osmotic potential. Water can move across the cell membrane quickly through water channels called **aquaporins** or more slowly through the cell membrane itself. Mineral ions such as potassium (K^+) can be actively absorbed from the soil solution through selective ion channels to serve as cell solutes. In addition, when stressed, the root cells can synthesize and accumulate compatible organic solutes such as proline or fructans to increase the solute concentration of the cytosol, reducing their internal osmotic potential and forming the gradient needed for water uptake (Bray 1997). In these cases, energy is used for synthesis of compounds giving a cost for water uptake.

In a well-watered, non-saline soil, a reasonable Ψ_w for the soil solution surrounding plant roots can be a very small negative number such as −0.05 MPa (Figure 6.5). A reasonable Ψ_w for roots in this soil would be −0.2 MPa. If Ψ_w if the **soil solution** is decreased by fertilizer, salinity, or uptake of water that is more rapid than restoration from precipitation or irrigation, the root Ψ_w will need to decrease (i.e. increase solute concentration in root cells), for water uptake to continue.

Water uptake occurs close to the root tip, where cells are newly elongated and may develop root hairs, i.e. thin-walled extensions of root epidermal cells that increase the root surface area for water uptake. Older roots, especially roots that have experienced drying soil, are protected by a layer of lignified cells formed under the epidermis, the exodermis, to prevent roots from losing water to dry soil as it's transported from deeper roots to the shoot. In some species, however, the upper root parts may leak water to the upper dry soil to support associated plants, a drought avoidance mechanism associated with **hydraulic lift** (see Chapter 8).

Water Movement in Plants

In the shoot of the plant, water vapor in the air spaces of leaves has a relative humidity near 100% and a Ψ_w of 0.0, whereas the air surrounding a leaf can have a Ψ_w as low as −10 to −100 MPa depending on the outside relative humidity. When stomata open to allow CO_2 uptake via photosynthesis, the gradient in Ψ_w from the leaf to the air causes leaves to transpire. Leaf cells need to have solute potentials low enough to remain turgid while stomata are open and losing water to transpiration, so a reasonable leaf Ψ_w is about −1.5 MPa (Figure 6.5).

The water transpired from leaves needs to be replaced with water from the xylem of the stem, so tension (negative values of pressure) develops in the open-ended xylem cells that are carrying water to the leaf. The tube-like vessels of the xylem form a continuous pathway for water flow from roots through stems to leaves. These xylem vessels are non-living and have thick walls permeated with lignin to make them rigid and non-leaky. A reasonable stem Ψ_w is −0.5 MPa in a well-watered soil (Figure 6.5). The tension in the xylem increases if a plant loses water faster by transpiration than it can be replaced by root uptake from the soil.

During a drought, the uptake of relatively pure water from the soil solution decreases the solute potential of the soil, causing leaf and root cells to require and create an even more negative solute potential so they can continue to absorb water and maintain cell turgidity (Figure 6.6; Slatyer 1967). At the same time, the residual solutes near the root increase the gradient causing more soil water to flow to the root. If water uptake cannot keep up with water loss, the stomata close during the day, usually in the afternoon when solar radiation and air temperature are highest and relative humidity is lowest. The closed stomata reduce photosynthesis while leaf water potential recovers and stomata reopen.

At night, when stomata close naturally due to low light intensity, plant Ψ_w gradually equilibrates with soil Ψ_w and xylem flow nearly stops. This daily cycle continues in a drying soil until the root system is no longer capable of accumulating enough solutes to compete for water with the decreasing Ψ_w of the soil and the permanent wilting point is reached. Measuring the water status of soil, especially if being measured using plant measures, should be done pre-dawn when stomata are closed and the plant water has re-achieved equilibrium with the soil water.

One notable exception to the general rule of increasingly negative Ψ_w as water from the soil solution passes through roots, stems, and leaves into the atmosphere can be observed in well-watered and well-fertilized soils after a sunny day. Plants that have transpired, and therefore depleted internal water during the day, close their stomata as night falls, but continue to replace transpired water until its internal solute potential is satisfied. If xylem vessels contain abundant nutrient ions, the resulting negative osmotic potential may result in excessive water uptake and the formation of "root pressure." Many leaves contain a pore, called a hydathode at the tip of the midrib or other large veins, and the visual evidence of root pressure is water droplets (guttation) seen in the early morning (Figure 6.7).

Water-Use Efficiency of Forage Plants

WUE describes the production of plant biomass per unit of water lost by ET. It is commonly used to compare

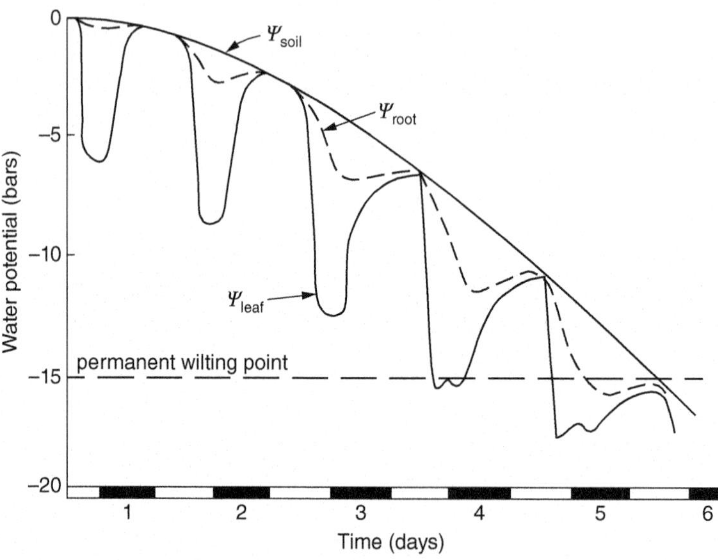

Fig. 6.6. As soil dries, soil Ψ_w (upper line, units in bars = 10× MPa) decreases over time. In response, root Ψ_w (dashed line) decreases to attract move water from the soil solution while leaf Ψ_w (lower line) decreases more than root Ψ_w to move water from the xylem in the root and stem. Light and dark sections of the X-axis indicate day and night periods. At night, when stomata close, plant root and leaf Ψ_w equilibrate with soil Ψ_w. When soil water reaches the permanent wilting point (horizontal dashed line), root Ψ_w can no longer decrease enough to attract water *Source:* Figure 9.1, p. 276 in Slatyer (1967), used with permission.

Fig. 6.7. Guttation at the tip of orchardgrass leaves occurs when plants in well-fertilized, well-watered soil absorb sufficient mineral nutrients or other osmotica to create a pressure potential within the xylem vessels, which extends from roots to leaves. At night, when stomata close, these plants had accumulated enough solutes that excess soil water was absorbed, so water pressure was relieved by excreting water droplets through one-way valves at leaf tips. This is different from dew that originates from water vapor in the atmosphere and is deposited over the leaf surface.

Table 6.1 Water use efficiency (WUE) of a range of warm- and cool-season cereals and forage plants

	Crop	g dry matter (kg water)$^{-1}$	g dry matter (kg water)$^{-1}$ (mm d^{-1})
de Wit (1958)			
	Sorghum	3.43	20.7
	Wheat	2.23	11.5
	Alfalfa	1.24	5.5
Neal et al. (2011a,b)			
Warm-season annual cereals	Maize	3.95	
	Sorghum	2.94	
Cool-season annual cereals	Wheat	3.57	
	Oats	2.15	
Warm-season perennial grasses	Kikuyu	2.50	
	Paspalum	1.97	
Cool-season perennial grasses	Tall fescue	1.92	
	Orchardgrass	1.84	
	Perennial ryegrass	1.84	
Perennial legumes	Alfalfa	1.86	
	Birdsfoot trefoil	1.48	
	Red clover	1.38	
	White clover	1.13	
Annual forbs	Forage rape	2.92	
	Forage radish	1.86	
Perennial forbs	Chicory	1.70	
	Plantain	1.38	

de Wit (1958) reported data of Briggs and Shantz (1913a,b, 1914); Shantz and Piemiesel (1927) and Dillman (1931) and recalculated assimilation (net photosynthesis) of sorghum, wheat, and alfalfa data from four locations as a function of transpiration weighted for mean seasonal **free water** evaporation. Both the original mean WUE and the weighted data is reported, along with data collected during a three-year study at three irrigation levels at Camden, New South Wales, Australia (Neal et al. 2011a,b).

management systems and practices in the western US, where PET exceeds precipitation (Hatfield et al. 2001). For cereal crops and forage seed, the biomass of economic interest is grain or seed, but in forage production, the economic crop is the harvestable aboveground biomass, and the inherent WUE of forages (kg DM kg^{-1} water) is less than that of grain crops. de Wit (1958), working with data from Briggs and Shantz (1913a,b, 1914); Shantz and Piemiesel (1927) and Dillman (1931), reported that the WUE of wheat, a cool-season species, was twice that of alfalfa, and the WUE of grain sorghum, a warm-season species, was nearly twice that of wheat. In these container studies, the soil surface was sealed so transpiration could be accurately measured, and in his calculations, de Wit weighted transpiration for **free water** evaporation in each of the four locations and ten years. Perennial forages are less water-use efficient than grain crops because of the lower density of the crop (i.e. vegetation with little or no grain) and because of their extensive investment of photosynthate in perennial rooting systems.

In Table 6.1, de Wit's calculations are contrasted with the original WUE data, and presented alongside WUE data for a range of annual and perennial crops from a multiple-year Australian study (Neal et al. 2011a,b). C_4 grains and grasses produce more dry matter per unit of water used, but modern cultivars of wheat are similarly water-use efficient. Alfalfa is as water-use efficient as perennial cool-season grasses but of much higher nutritive value. Other temperate legumes are less productive per unit of water, and annual non-leguminous forbs are more productive per unit of water used than perennial forbs.

Transpiration is closely related to photosynthesis since stomata open during the light period to facilitate the inward flow of carbon dioxide (CO_2) but this also results in water vapor inside the leaf escaping to the atmosphere. The gradient of water vapor from the interior of the plant leaf to the atmosphere drives diffusion and the transpiration rate; therefore, WUE is greater in more humid environments where this gradient is smaller and transpiration rate is lower (Tanner and Sinclair 1983). Altering the stomatal density on the leaf surface or altering the response of stomatal conductance to drought have been identified as targets for improvement of forage WUE (Sinclair et al. 1984), but there has been limited progress toward this goal. Current focus is on altering root traits to simultaneously allow more effective soil exploration and acquisition of water and nutrients using non-destructive root phenotyping and high-throughput genotype evaluation (Paez-Garcia et al. 2015).

Viets (1962) found that phosphate and nitrogen fertilization linearly improved the WUE of irrigated smooth bromegrass (and orchardgrass in North Dakota, mountain meadow hay in Colorado, and alfalfa in Arizona). If nutrient status is adequate and there is unused soil water-holding capacity (i.e. more air spaces) irrigation can also improve the WUE of forages by increasing the rate of leaf canopy development more than the associated increase in transpiration (Tanner and Sinclair 1983; White and Snow 2012). Net photosynthesis, the difference between carbon fixation and carbon expended via respiration, can be increased by optimizing light, temperature, and CO_2 concentration, thereby increasing WUE. However, these factors are controlled primarily by the environment (cloud cover, proportion of diffuse radiation, nighttime temperature, and atmospheric CO_2 concentration) rather than management (Passioura and Angus 2010).

When soil moisture is insufficient to supply a plant with water for transpiration, the plant will experience water stress. This difference between actual ET and PET results in a high level of plant adaptation in native ecosystems, or the use of irrigation in agroecosystems. The grasses and forbs native to the semi-arid Great Basin and western Great Plains are better adapted to tolerate environmental stresses than to tolerate competition from other plants (Chapters 2 and 8). Thus, opportunistic plants of low forage value that make inefficient but effective use of water can quickly degrade these grasslands.

Cheatgrass or downy brome is a shallow-rooted winter annual native to Eurasia that out-competes western native rangeland forages and cultivated winter grains for nitrogen and water. Cheatgrass seed matures and its leaves die early in the growing season, creating a wildland fire hazard (Skinner et al. 2008). Root and shoot production of native species in the Great Plains increase with increasing precipitation from west to east, i.e. from the shortgrass to the tallgrass prairie of the Great Plains, and there is an accompanying shift in strategy from survival to competition (Sims and Singh 1978; Chapter 8).

Many warm-season (C_4) forage species are native to the short- and tall-grass prairie. These C_4 grasses have 25–35% greater WUE than cool-season (C_3) forages because use of available water is similar, but photosynthesis rates per unit leaf area are higher; more CO_2 is captured to support growth, respiration, and storage. Enzymes in the chloroplasts of C_4 plants have a higher affinity for capturing CO_2 so the gradient from the atmosphere through stomata to a C_4 mesophyll cell is steeper than for C_3 species, so CO_2 diffuses into a C_4 leaf more rapidly than into a C_3 leaf. The capture of CO_2 in C_4 plants is mediated by the enzyme phosphoenolpyruvate (PEP) carboxylase which does not react with O_2 and can reduce the CO_2 concentration in the stomatal cavity to near zero. In contrast, C_3 plants have photorespiration and reduced net photosynthesis (Chapter 4). Growth of plants with C_4 photosynthesis transpire less water per unit of CO_2 incorporated in sugars to support growth. However, compared with C_3 forages, C_4 forages have a lower proportion of protein and fewer highly digestible mesophyll cells, and thus have lower forage nutritive value (see Chapter 39).

The Challenge of Improving Forage Water-Use Efficiency

Passioura (1996) lists three factors in the optimal crop response to limited environmental water as (i) effective competition by a plant for water from the environment, (ii) minimal water loss when stomata are open for CO_2 uptake, and (iii) optimal use of the carbohydrate produced by photosynthesis to form a harvestable crop. Relevant to the first point, if genetic variation exists for WUE, it should be detected by superior performance in very low-rainfall environments. The relative growth rate (RGR) of aboveground biomass is the increase in dry mass per unit of existing dry mass d^{-1}. Eight species of perennial warm-season grasses native to the Chihuahuan Desert had RGRs ranging from 0.029 to 0.158 $g\,g^{-1}\,d^{-1}$ but when subjected to severe drought, all species decreased their biomass production and increased their root-to-shoot ratios, with no differential effects of species on WUE (Fernández and Reynolds 2000). The conservative (low RGR) grasses were no more tolerant of drought than their less-conservative co-habitants, suggesting that faster-growing plants simply use the available soil water supply more readily than slower-growing plants.

To the second point, the WUE of grain crops, whether C_3 or C_4, can be improved by reducing the proportion of stem and leaves while maintaining grain production, i.e. by investing less dry matter in stems and reducing transpiration. However, except for forage seed production, stems and leaves are the economic product in forages. C_4 grasses produce more shoot biomass for a given amount of water than C_3 grasses, but only in climates to which they're well-adapted.

Since the value of forages varies according to feeding quality, Passioura's third point suggests it would be more relevant to pro-rate forage WUE by the nutritive value or ruminant liveweight gain produced, rather than evaluate it simply on the basis of herbage dry matter production (e.g. Baxter et al. 2017).

Forage root systems are sufficiently plastic to direct their growth toward available water in preference to increasing WUE (Kirkham 2005). Alfalfa is a temperate legume with geographic origin in the Middle East. Compared with many other forages, alfalfa is both tall and deep-rooted, but it allocates a smaller proportion of biomass to roots than most cultivated forages. In a field study, Bray (1963) reported the proportion of biomass found above ground was 70% for alfalfa, 52% for timothy and only 30% for perennial ryegrass. The depth of water extraction by alfalfa root systems was studied in New Zealand under dryland and irrigated conditions (Brown et al. 2009). As cumulative transpiration demand increased over the course of the growing season, the depth of water extraction progressed to deeper soil layers as shallow soil moisture was depleted. Water extraction of dryland alfalfa reached a depth of 2.7 m by the end of the growing season, while under irrigation, alfalfa water requirements were met by withdrawal of water to a depth of only 1.8 m.

The greater availability of both oxygen and nutrients near the soil surface causes roots to concentrate in shallow soil layers. When soil water becomes limiting, however, the shoot biomass production of shallow-rooted forages is reduced by water stress to a greater degree than that of deep-rooted forages. In a Saskatchewan dryland study, monocultures of crested wheatgrass rooted to 2.3 m while alfalfa rooted to at least 3 m. Alfalfa avoided mid-day water stress by maintaining higher shoot water potentials via osmotic adjustment compared with the more stress-tolerant crested wheatgrass. Nitrogen fixation allowed alfalfa to produce more dry matter per ha^{-1} per mm^{-1} of soil water used. Over the course of the six-year study, alfalfa had 30% greater WUE than crested wheatgrass (Jefferson and Cutforth 2005). As in other field studies of forage WUE, rooting below the practical depth of soil water measurement could not be controlled.

In an Australian study, both forage production and WUE of alfalfa was maintained significantly better across a range of irrigation levels than that of 14 warm- and cool-season grasses, other legumes and non-legume forbs (Neal et al. 2009; 2011a,b). Maximum extractable water differed little among temperate legumes and cool-season grasses to a depth of 1.5 m, although alfalfa rooted more deeply than other cool-season perennial forages (Neal et al. 2012). The perennial warm-season grasses kikuyu and dallisgrass did root more deeply, but they also extracted more water than alfalfa. The annuals forage radish grain sorghum and winter wheat also extracted more total water than alfalfa even though their growing seasons were shorter.

Where drought limits growth but does not threaten plant survival, deeper-rooting grasses such as tall fescue, which have access to water in deep, low-nutrient soil layers, have better WUE than orchardgrass, which has an extensive shallow root system with access to nutrient-rich upper soil layers (Lemaire 2001). Growth of deep-rooted legumes, however, is not limited by the nitrogen available in shallow soil layers because they not only fix atmospheric N_2 they can capture N compounds that might otherwise leach from heavily fertilized or grazed systems.

There are few studies of the WUE of grass-legume mixtures. Hendrickson et al. (2013) in North Dakota compared the WUE of monocultures of a C_3, western wheatgrass, and a C_4, switchgrass, to a mixture of western wheatgrass and alfalfa. While WUE was numerically much greater for the C_4 grass under well-watered conditions, there was no significant difference in WUE among treatments. When soil moisture was reduced in May and June, the WUE of both the C_4 grass and the mixture was greater than that of the C_3 grass, but when the deficit occurred in July and August, only the C_4 grass had greater WUE. Over the two-year study, the C_4 grass had far greater WUE than the C_3 grass, but in the second year, the WUE of the C_3 grass–alfalfa mixture was similar to that of the C_4 grass. Overall, the C_4 grass was both more efficient and more productive in July and August, and therefore used significantly more soil water than either the C_3 grass or the C_3 grass–alfalfa mixture.

Water Use by Complex Forage Mixtures

A study comparing the water use of seeded grasslands with low (4) or high (16) plant species richness found no differences in shoot biomass production with increased diversity of species (Milcu et al. 2016), but ET was greater for the more-diverse mixture (Guderle et al. 2017). This was largely due to the poor productivity of many of the non-leguminous broadleaf species that were contributing to biodiversity (i.e. weeds). As was seen for alfalfa monocultures (Brown et al. 2009), water uptake by roots in the mixture shifted from upper to lower soil layers during the course of the growing season. In both the high- and low-diversity grasslands, grasses and legumes represented only about 40% of the leaf canopy, while short and tall non-leguminous forbs accounted for the remaining 60% of leaf area.

Skinner et al. (2004) evaluated the role of species richness (2 and 5 species) and the contribution of drought-resistant pasture species on root mass and rooting depth in Pennsylvania. They used rain-out shelters and drip irrigation to produce deficient, normal, and excessive root moisture. Plots were harvested by clipping to simulate intensive grazing, and data for all irrigation levels were averaged for each mixture (Table 6.2). After three years, the more shallow-rooted binary mixture A (kentucky bluegrass/white clover) had greater total root mass, primarily in the upper 15 cm, than deeper-rooted binary mixture B (orchardgrass/red clover) where more rooting occurred between 15 and 30 cm. Likewise, complex mixture D (kentucky bluegrass/ perennial ryegrass/tall fescue/red clover/narrowleaf plantain), without a dominant deep-rooted species, had greater total root mass than complex mixture C (kentucky bluegrass/perennial ryegrass/orchardgrass/white clover/chicory). Mixture C had only one-third the total root mass of mixture D but greater yield and much more rooting at a depth of 30–60 cm, likely due to the presence of chicory.

This study suggests that chicory obtains water primarily from greater depths and competes less with shallow-rooted species at upper soil levels. The mixture may benefit from this "niche separation" in depth of water that results in "**hydraulic lift**," where deep chicory roots obtain deep water and then leak water from the root to dry soil at upper levels to support shallow-rooted species. Skinner et al. (2004) concluded, in concordance with Tilman (1999), that complex forage mixtures can result in niche separation and utilize water to a greater depth of the soil profile. In a deep-pot study of single plants, Skinner and Comas (2010) found that after 13 weeks of growth, chicory roots were shallow and prolific, while at the end of a two-year study of mixtures in the field (Skinner et al. 2004), chicory produced proportionally less root mass but contributed greatly to the productivity of the mixture by shifting its water uptake to deeper soil layers.

Forage Survival Under Drought

There is interest in genetically increasing the drought tolerance of grasses and legumes, which might be most feasible in the humid parts of the US and Canada. In these locations, drought can be severe but more seasonal than in the West where droughts can persist for multiple years. There is strong evidence that deeper-rooted grasses such a tall fescue or smooth bromegrass are better adapted to severe drought than shallow-rooted grasses such as kentucky bluegrass or timothy (Burns and Chamblee 1979; Fales et al. 1996; Thomas 1986). Survival methods in prolonged water-stress include the induction of dormancy, found in kentucky bluegrass (Suplick-Ploense and Qian 2005) and tall fescue (Volaire and Norton 2006).

In tall fescue, plants that were drought stressed during summer yielded more herbage during fall than did plants irrigated during summer, perhaps due to increased carbohydrate storage in stem bases during drought stress (Horst and Nelson 1979). Others have attributed the response to reduced self-shading and access to carbohydrates stored in rhizomes (van Staalduinen and Anten 2005), or simply to carbohydrates conserved by the cessation of root and shoot growth with onset of stress (Newton et al. 1996). Some suggested such compensatory growth may be due to differential protection of the shoot meristems compared to other organs (Volaire et al. 1998). The most important traits vary among these contrasting strategies due to tradeoffs between resistance to moderate moisture stress and survival under intense drought.

Drought survival can be fundamental to the persistence of perennial forages, so the ability to recover from profound drought can be critical to quality and productivity of the stand in the long term. Perennial forages may be required to perform as strong competitors when soil water is abundant, and shift into survival mode during

Table 6.2 Root distribution under two-species (Mixtures A and B) and five-species (Mixtures C and D) mixtures

Soil depth	Mixture A	Mixture B	Mixture C	Mixture D
		kg ha^{-1}		
0–15 cm	1380 (85%)	609 (71%)	602 (52%)	1790 (73%)
15–30 cm	125 (8%)	168 (20%)	73 (6%)	323 (13%)
30–60 cm	54 (3%)	70 (8%)	390 (34%)	224 (9%)
60–90 cm	4 (<1%)	10 (1%)	86 (7%)	128 (5%)
Total[a]	1624b	857c	1150bc	2465a

[a]Total root biomass data followed by the same letter are not significantly different at $P = 0.05$.

Data were collected on 21 May 2002 following the final harvest for species composition determinations. Mixture (A) kentucky bluegrass/white clover, (B) orchardgrass/red clover, (C) kentucky bluegrass/perennial ryegrass/orchardgrass/white clover/chicory, and (D) kentucky bluegrass/perennial ryegrass/tall fescue/red clover/narrow-leaf plantain. Numbers in parentheses represent the contribution of roots at a given depth to total root biomass. (Skinner et al. 2004) periods of low water availability and high temperatures (Chapter 8). This temporary dormancy may be critical to the extent that growth requires photosynthesis, stomata must open to supply CO_2 for photosynthesis, and plant growth cannot continue without coinciding with water loss.

If soil water decreases below the permanent wilting point of about −1.5 MPa, plants will either perish or enter a state of dormancy, where the protection of plant meristems from dehydration is key to recovery following summer dormancy (Volaire et al. 2014). In the tillers that form after reproductive growth, the stem and leaf meristems of most grasses are located near or below the soil surface (Chapter 2), as are the crowns, rhizomes, and stolons of perennial legumes (Chapter 3). These meristems are co-located with the site of plant carbohydrate storage, which can be accessed for energy and **osmotica** to maintain cell turgor.

Spollen and Nelson (1994) demonstrated that as drought stress of tall fescue increased, fructan, the storage carbohydrate of cool-season grasses, was hydrolyzed to increase concentrations of sucrose and hexoses as solutes to serve as osmotica in the **meristematic** tissues of leaves, concentrating especially in elongating cells. Volaire et al. (2014) further described the progressive response to drought that begins with growth cessation and evolves into mobilization of protective compounds such as dehydrin proteins and hydrolysis of sugars from storage carbohydrates in meristematic tissues. These changes occur in concert with leaf and tiller death, progressing to full summer dormancy and the ability to survive a tolerable level of dehydration (Figure 6.8).

Alternative Forages for Extremely Dry Environments

About 5% of the alfalfa hay produced in the US is exported, but a greater portion (~15%) of the alfalfa hay produced in western states (Putnam et al. 2017) is shipped to China, Saudi Arabia, the United Arab Emirates, Japan, and South Korea. As the climate in western states trends drier, there is concern that the irrigation water and nutrients used for hay production are being "virtually" exported along with the alfalfa (Culp and Glennon 2012). Saudi Arabia has the two largest dairies in the world and is a regional exporter of dairy products, but the level of alfalfa production needed to support these dairies is not sustainable in Saudi Arabia; therefore, forage production there will be phased out by 2020 (USDA FAS 2017), further increasing the export market for western US alfalfa.

Depending on the value and scale of ruminant production, the use of germinated cereal grains such as barley "fodder" can serve as a hybrid forage-concentrate feed. Al-Karaki and Al-Momami (2011) report that a kg of **fodder** barley dry matter can be produced using about 3% of the water required to produce a kg of alfalfa dry matter. Barley grain is sprouted on trays that are sprayed periodically with the water needed for seedling growth. When the fodder is harvested after approximately seven days, it has a moisture content of about 810 g kg^{-1} (Fazaeli et al. 2012).

Since most of the water used to sprout barley is subsequently fed to ruminants in the fodder, it remains within the landscape. Compared with ruminant production systems that include the field production of forages, where transpired water escapes to the atmosphere, a fodder-based ruminant production system has far greater WUE.

Production of the barley grain used to produce **fodder** requires water, of course, but the WUE of barley grain production is greater than the WUE of alfalfa production, due to the short, winter annual lifecycle of barley and the density of the grain yield. In extremely water-limited environments, ruminant feed rations can be balanced around freshly sprouted barley, which has a protein content of about 137 g kg^{-1} DM, NDF content of 313 g kg^{-1} DM, and a non-fibrous carbohydrate concentration of 490 g kg^{-1} DM (Fazaeli et al. 2012). Fodder barley contributes both fiber and energy to a high-quality ration with minimal water use in excessively dry environments.

Conclusion

Forages have evolved using a range of strategies to survive the disturbances that are characteristic of grasslands, such as drought, fire, and grazing. Forages that are adapted to arid and semi-arid environments survive by investing heavily in root biomass, strategic reproduction, and drought survival mechanisms that protect the regrowth meristems. The C_4 grasses adapted to humid environments with high growth-season temperatures and long droughts are often the deepest-rooted forages, but have sacrificed a portion of their nutritive value by use of the C_4 photosynthesis system to increase dry matter production.

Alfalfa invests in a leaner, deeper root system than forage grasses, avoiding drought more successfully than C_3 grasses that root in shallow, nutrient-rich upper soil layers. Mixtures of grasses and broadleaf forages can make more complete use of soil water than monocultures, through intense competition for the water in upper soil layers, motivating tap-rooted species to use water deeper in the soil.

While transpiration is directly related to leaf surface area, the efficiency of water use is greater when plants are not challenged by drought or nutrient deficiency. When drought cannot be avoided, some forage plants survive by ceasing growth and allocating all available carbohydrate resources to infuse growing points with solutes, protecting them from dehydration as soils continue to dry and to provide a ready source of energy to support recovery.

As highly productive but input-intensive cropping systems including hay and pasture production are increasingly challenged by climate extremes, it should become apparent that perennial forages and ruminant production can provide resilience of crop agriculture through greater

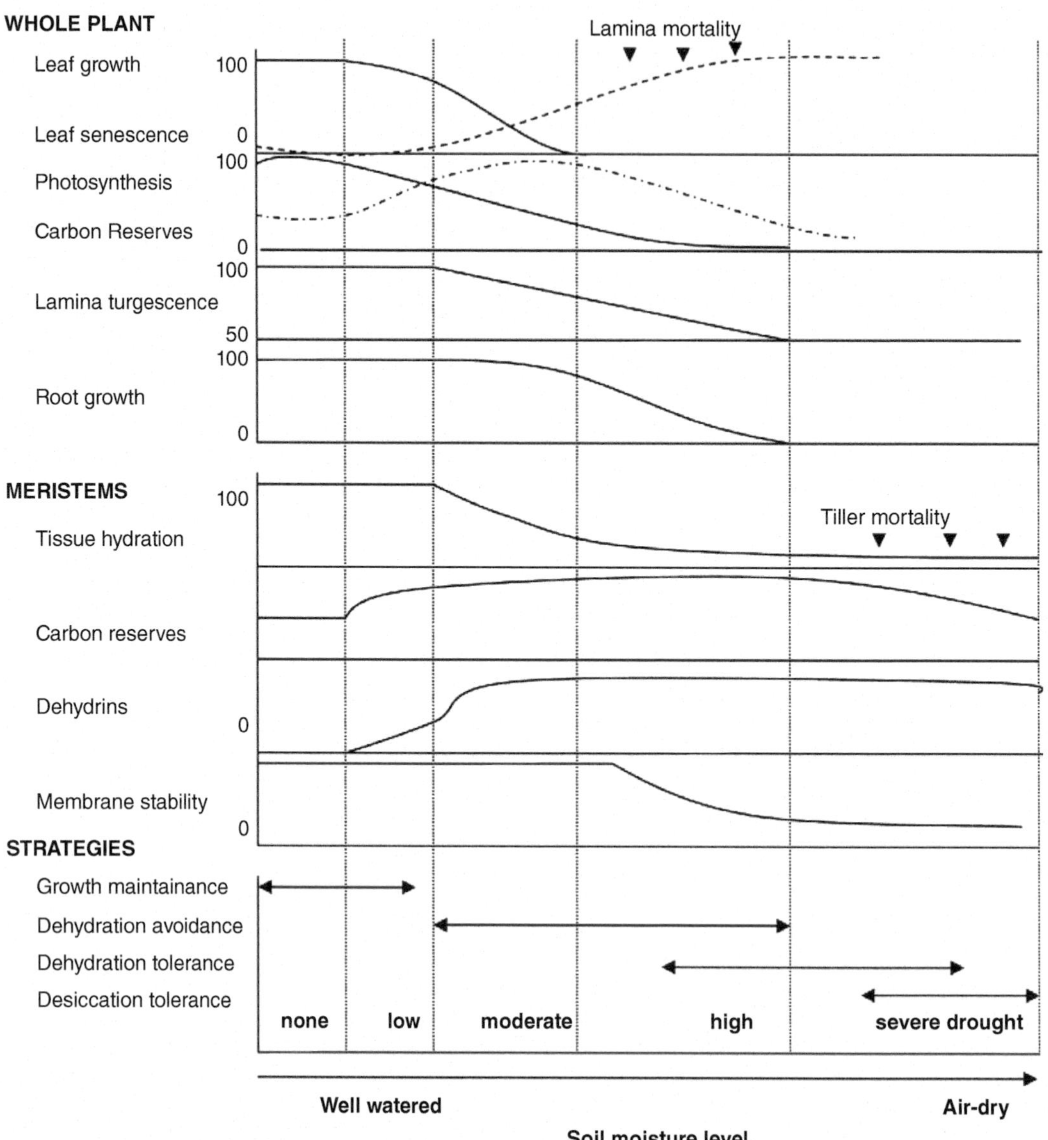

Fig. 6.8. Schematic responses to intensifying drought, from left to right, of a perennial grass at the levels of the whole plant and aerial meristems, and the resulting ecophysiologic strategies to initially avoid, and eventually survive, dehydration *Source:* from Volaire et al. (2009). Scales are arbitrary.

soil protection, and wider range of adaption to variation in soil water availability and WUE. Lastly, the wide variation in inherent forage quality, as well as the variation in quality resulting from management, should be factored into any calculation of WUE if the ultimate use is ruminant production.

References

Al-Karaki, G.N. and Al-Momani, N. (2011). Evaluation of some barley cultivars for green fodder production and water use efficiency under hydroponic conditions. *Jordan J. Agric. Sci.* 7: 448–456.

Baxter, L.L., West, C.P., Sarturi, J.O. et al. (2017). Stocker beef production on low-water-input systems in response to legume inclusion: II. Water footprint. *Crop Sci.* 57: 2303–2312.

Blair, J., Nippert, J., and Briggs, J. (2014). Grassland ecology. In: *The Plant Sciences: Ecology and the Environment*, vol. 8 (ed. R.K. Monson), 389–423. New York: Springer-Verlag.

Brauer, D., Devlin, D., Wagner, K. et al. (2017). Ogallala aquifer program: a catalyst for research and education to sustain the Ogallala Aquifer on the Southern High Plains (2003-2017). *J. Contemp. Water Res. Educ.* 162: 4–17.

Bray, J.R. (1963). Root production and estimation of net productivity. *Can. J. Bot.* 41: 65–72.

Bray, E.A. (1997). Plant responses to water deficit. *Trends Plant Sci.* 2: 48–54.

Briggs, L.J. and Shantz, H.L. (1913a). The water requirements of plants. I. Investigations in the great plains in 1910 and 1911. *USDA Bur. Plant Ind. Bull.* 284.

Briggs, L.J. and Shantz, H.L. (1913b). The water requirements of plants. II. A review of literature. *USDA Bur. Plant Ind. Bull.* 285.

Briggs, L.J. and Shantz, H.L. (1914). Relative water requirements of plants. *J. Agric. Res.* 3: 1–65.

Brown, H.E., Moot, D.J., Fletcher, A.L., and Jamieson, P.D. (2009). A framework for quantifying water extraction and water stress responses of perennial lucerne. *Crop Pasture Sci.* 60: 785–794.

Burns, J.C. and Chamblee, D.S. (1979). Adaptation. In: *Tall Fescue*, Agronomy Monograph 20 (eds. R.C. Buckner and L.P. Bush), 9–30. Madison WI: ASA, CSSA, and SSSA.

Culp, P. and Glennon, R. (2012). Parched in the West but Shipping Water to China, Bale by Bale. *The Wall Street Journal* (5 October). https://www.wsj.com/articles/SB10000872396390444517304577653432417208116 (accessed 08 October 2019).

Dillman, A.C. (1931). Water requirement of certain crop plants and weeds in the Northern Great Plains. *J. Agric. Res.* 42: 187–238.

Fales, S.L., Laidlaw, A.S., and Lambert, M.G. (1996). Cool-season grass ecosystems. In: *Cool-Season Forage Grasses*, Agronomy Monograph 34 (eds. L.E. Moser et al.), 267–296. Madison WI: ASA, CSSA, SSSA.

Fazaeli, H., Golmohammadi, H.A., Tabatabayee, S.N., and Asghari-Tabrizi, M. (2012). Productivity and nutritive value of barley green fodder yield in hydroponic system. *World Appl. Sci. J.* 16: 531–539.

Fernández, R.J. and Reynolds, J.F. (2000). Potential growth and drought tolerance of eight desert grasses: lack of a trade-off? *Oecologia* 123: 90–98.

Guderle, M., Bachmann, D., Milcu, A. et al. (2017). Dynamic niche partitioning in root water uptake facilitates efficient water use in more diverse grassland plant communities. *Funct. Ecol.* 32: 214–227.

Hatfield, J.L., Sauer, T.J., and Prueger, J.H. (2001). Managing soils to achieve greater water use efficiency: a review. *Agron. J.* 93: 271–280.

Hendrickson, J.R., Schmer, M.R., and Sanderson, M.A. (2013). Water use efficiency by switchgrass compared to a native grass or a native grass alfalfa mixture. *Bioenergy Res.* 6: 746–754.

Horst, G.L. and Nelson, C.J. (1979). Compensatory growth of tall fescue following drought. *Agron. J.* 71: 559–563.

Jefferson, P.G. and Cutforth, H.W. (2005). Comparative forage yield, water use, and water use efficiency of alfalfa, crested wheatgrass and spring wheat in a semi-arid climate in southern Saskatchewan. *Can. J. Plant Sci.* 85: 877–888.

Kirkham, M.B. (2005). *Principles of Soil and Plant Water Relations*. Burlington, MA: Elsevier Academic Press.

Lemaire, G. (2001). Ecophysiology of grasslands: dynamic aspects of forage plant populations in grazed swards. In: Proc. 19th Intern. Grassld. Congr., Piracicaba, Brazil, 10–21 Feb. 2001. Brazilian Soc. Anim. Husb., 29–37. Piracicaba, Brazil.

MacAdam, J.W. (2009). *Structure and Function of Plants*. Ames, IA: Wiley-Blackwell.

Milcu, A., Eugster, W., Bachmann, D. et al. (2016). Plant functional diversity increases grassland productivity-related water vapor fluxes: an Ecotron and modeling approach. *Ecology* 97: 2044–2054.

Neal, J.S., Fulkerson, W.J., Lawrie, R., and Barchia, I.M. (2009). Difference in yield and persistence among perennial forages used by the dairy industry under optimum and deficit irrigation. *Crop Pasture Sci.* 60: 1071–1087.

Neal, J.S., Fulkerson, W.J., and Hacker, R.B. (2011a). Differences in water use efficiency among annual forages used by the dairy industry under optimum and deficit irrigation. *Agric. Water Manag.* 98: 759–774.

Neal, J.S., Fulkerson, W.J., and Sutton, B.G. (2011b). Differences in water-use efficiency among perennial forages used by the dairy industry under optimum and deficit irrigation. *Irrig. Sci.* 29: 213–232.

Neal, J.S., Murphy, S.R., Harden, S., and Fulkerson, W.J. (2012). Differences in soil water content between perennial and annual forages and crops grown under deficit irrigation and used by the dairy industry. *Field Crop Res.* 137: 148–162.

Newton, P.C.D., Clark, H., Bell, C.C., and Glasgow, E.M. (1996). Interaction of soil moisture and elevated CO_2 on the above-ground growth rate, root length density and gas exchange of turves from temperate pasture. *J. Exp. Bot.* 47: 771–779.

Paez-Garcia, A., Motes, C., Scheible, W.-R. et al. (2015). Root traits and phenotyping strategies for plant improvement. *Plants* 4: 334–355.

Passioura, J.B. (1996). Drought and drought tolerance. *Plant Growth Regul.* 20: 79–83.

Passioura, J.B. and Angus, J.F. (2010). Improving productivity of crops in water-limited environments. *Adv. Agron.* 10: 37–75.

Putnam, D. H., Matthews, B., and Summer, D.A. (2017). Growth in Saudi hay demand likely to have world-wide impacts – and illustrates the global nature of water limitations. *Alfalfa & Forage News* (7 March). https://ucanr.edu/blogs/blogcore/postdetail.cfm?postnum=23472 (accessed 8 October 2019).

Shantz, H.L. and Piemiesel, L.N. (1927). The water requirements of plants at Akron. *Colorado J. Agric. Res.* 34: 1093–1190.

Sims, P.L. and Singh, J.S. (1978). The structure and function of ten western North American grasslands: III. Net primary production, turnover and efficiencies of energy capture and water use. *J. Ecol.* 66: 573–597.

Sinclair, T.R., Tanner, C.B., and Bennett, J.M. (1984). Water-use efficiency in crop production. *BioScience* 34: 36–40.

Skinner, R.H. and Comas, L.H. (2010). Root distribution of temperate forage species subjected to water and nitrogen stress. *Crop Sci.* 50: 2178–2185.

Skinner, R.H., Gustine, D.I., and Sanderson, M.A. (2004). Growth, water relations, and nutritive value of pasture species mixtures under moisture stress. *Crop Sci.* 44: 1361–1369.

Skinner, M., Ogle, D.G., John, L.S. et al. (2008). *Cheatgrass Plant Guide*. Berkeley, CA: USDA Natural Resources Conservation Service, Boise, ID and University of California.

Slatyer, R.O. (1967). *Plant Water Relationships*. New York: Academic Press (Elsevier).

Spollen, W.C. and Nelson, C.J. (1994). Response of fructan to water deficit in growing leaves of tall fescue. *Plant Physiol.* 106: 329–336.

van Staalduinen, M.A. and Anten, N.P.R. (2005). Differences in the compensatory growth of two co-occurring grass species in relation to water availability. *Oecologia* 146: 190–199.

Suplick-Ploense, M.R. and Qian, Y. (2005). Evapotranspiration, rooting characteristics, and dehydration avoidance: comparisons between hybrid bluegrass and Kentucky bluegrass. *Int. Turfgrass Soc.* 10: 891–898.

Tanner, C.B. and Sinclair, T.R. (1983). Efficient water use in crop production: research or re-search? In: *Limitations to Efficient Water Use in Crop Production* (eds. H.M. Taylor et al.), 1–27. Madison, WI: ASA.

Thomas, H. (1986). Water use characteristics of *Dactylis glomerata* L., *Lolium perenne* L. and *L. multiforum* Lam. *Ann. Bot.* 57: 211–223.

Tilman, D. (1999). The ecological consequences of changes in biodiversity: a search for general principles. *Ecology* 80: 1455–1474.

USDA FAS (2017). *Saudi Arabian Alfalfa Hay Market*. Riyadh, Kingdom of Saudi Arabia: Foreign Agriculture Service Global Agricultural Information Network Retrieved from https://www.fas.usda.gov/data/saudi-arabia-saudi-arabian-alfalfa-hay-market.

USDA NRCS (2012). Ogallala aquifer initiative 2011 report. Natural Resources Conservation Service. https://www.nrcs.usda.gov/wps/portal/nrcs/detailfull/national/programs/initiatives/?cid=stelprdb1048809 (accessed 8 October 2019).

Viets, F.G. (1962). Fertilizers and the efficient use of water. *Adv. Agron.* 14: 223–264.

Volaire, F. and Norton, M. (2006). Summer dormancy in perennial temperate grasses. *Ann. Bot.* 98: 927–933.

Volaire, F., Thomas, H., and Lelievre, F. (1998). Survival and recovery of perennial forage grasses under prolonged Mediterranean drought: growth, death, water relations and solute content in herbage and stubble. *New Phytol.* 140: 439–449.

Volaire, F., Seddaiu, G., Luigi, L., and Lelievre, F. (2009). Water deficit and induction of summer dormancy in perennial Mediterranean grasses. *Ann. Bot.* 103: 1337–1349.

Volaire, F., Barkaoui, K., and Norton, M. (2014). Designing resilient and sustainable grasslands for a drier future: adaptive strategies, functional traits and biotic interactions. *Eur. J. Agron.* 52: 81–89.

White, T.A. and Snow, V.O. (2012). A modelling analysis to identify plant traits for enhanced water-use efficiency of pasture. *Crop Pasture Sci.* 63: 63–76.

de Wit, C.T. (1958). *Transpiration and Crop Yields*. Wageningen, The Netherlands: Institute of Biological and Chemical Research on Field Crops and Herbage.

CHAPTER

7

Growth and Development

Robert B. Mitchell, Research Agronomist, *Agricultural Research Service, USDA, Lincoln, NE, USA*
Daren D. Redfearn, Associate Professor, *Agronomy and Horticulture, University of Nebraska, Lincoln, NE, USA*
Kenneth J. Moore, Distinguished Professor, *Agronomy, Iowa State University, Ames, IA, USA*

Introduction

The growth and development of forage plants is an amazing process. In some annual grasses such as cereal rye, plants can go from the late vegetative stage to fully-flowered in less than two weeks. Conversely, some perennial grasses like indiangrass can go from the vegetative stage to the elongation stage, then enter a quiescent phase for several weeks until adequate moisture is available which then moves plants into the flowering stages to complete the seed production process.

Understanding the developmental morphology of forage plants is important for making good management decisions. Many such decisions involve timing the initiation or termination of a management practice at a specific stage of development in the plant's life cycle. Physiologic responses to defoliation and subsequent growth potential are affected by growth stage and strongly influence subsequent developmental morphology (Parsons 1988; Brueland et al. 2003).

Leaf appearance rate during seedling development has been used to evaluate stand establishment and is strongly related to seedling root development (Moser 2000). Leaf development on established tillers of perennial grasses can be used to time management practices such as defoliation, burning, fertilization, and growth regulator and pesticide application (Moore et al. 1991). Decisions regarding grazing and harvest management are often based on plant development (Frank et al. 1993; Brueland et al. 2003).

This chapter addresses the initiation, expansion, and maturation of leaves, stems, and roots and how they regulate the transition from vegetative to reproductive growth and subsequent production of reproductive tissues. The interaction of these processes has profound effects on forage yield, quality, and stand longevity. Emphasis is given to the interactions of developmental morphology on these processes. The authors thank Dr. Howard Skinner for his work on the previous edition of this chapter.

Growth and Development of Plant Organs

The growth processes of each organ depend on cell division and elongation for plant tissue development and biomass accumulation. The elongated cells then differentiate in different ways to form specific organs and accommodate associated physiologic functions. Interactions among leaf, tiller, and root meristems are coordinated to assure the orderly development of the plant, providing opportunities to predict plant development.

Forages: The Science of Grassland Agriculture, Volume II, Seventh Edition.
Edited by Kenneth J. Moore, Michael Collins, C. Jerry Nelson and Daren D. Redfearn.

Development of Leaf Structures

Production of leaf tissue requires the initiation, elongation, and maturation of new cells. Leaf development has been extensively described for grasses because growth is mostly linear, resulting in large increases in leaf length accompanied by relatively small increases in width and thickness. In a grass leaf, cell division, elongation, and maturation zones occur sequentially along the base of the developing leaf. Subsequently, the youngest leaf tissues are located at the leaf base and the oldest at the leaf tip (Figure 7.1).

The cell division zone is at the very base of the leaf, where modest elongation and repeated divisions of **meristematic** cells produce a region with average cell length of about 20 μm. Epidermal cell division is restricted to the basal 2–3 mm of the elongating leaf (Skinner and Nelson 1995), whereas mesophyll cell division continues throughout the basal 10–15 mm of the leaf (MacAdam et al. 1989). Epidermal cells that have ceased dividing continue to elongate until they reach a mature cell length of 100–1000 μm depending on their position on the leaf and a host of environmental, management, and genetic factors (MacAdam et al. 1989; Erwin et al. 1994; Palmer and Davies 1996; Schaufele and Schnyder 2000). The length of the epidermal cell elongation zone is usually related to leaf elongation rate.

Both cell division and elongation of grasses are affected by the environmental and management factors that alter leaf elongation. Thus, defoliation (Schaufele and Schnyder 2000), **hypoxia** (Smit et al. 1989), water deficits (Lecoeur et al. 1995; Granier and Tardieu 1999), and N stress (MacAdam et al. 1989; Palmer et al. 1996) reduce cell division, cell elongation, or both. Nitrogen stress mainly reduces cell division. Water and other stresses have the greatest effect on cell division when leaves are small, whereas cell elongation can be affected by stress at any time during the leaf growth process.

Unlike grass leaves, which essentially grow in one direction, leaves of forbs, which include all legumes, have large increases in both length and width, which makes growth analysis more difficult. Also, cell division and elongation processes co-occur over a larger portion of the forb leaf and for a longer duration than in grass leaves.

Forb leaf growth is a three-phase process (Granier and Tardieu 1999). During the first phase, leaf area and cell number increase in tandem like the cell division zone of grasses. However, cell division in forbs, which occurs mainly along the leaf perimeter, can continue until the leaf is as much as 95% of its final size (Dale 1988). The second phase of leaf expansion begins as the cell division zone advances outward, leaving the existing cells on the inward side to expand rapidly. In general, cell division ceases first at the leaf tip and continues longest at the leaf base. During the third phase, cell elongation rate declines and eventually ceases as all cells reach their final mature length.

The cell growth zone of grasses is generally located within a whorl of older leaf sheaths, providing some protection against removal by grazing as well as buffering against adverse environmental conditions. In contrast, elongating forb leaves are exposed to environmental stress (Radin 1983). Thus, forb defoliation by grazers is

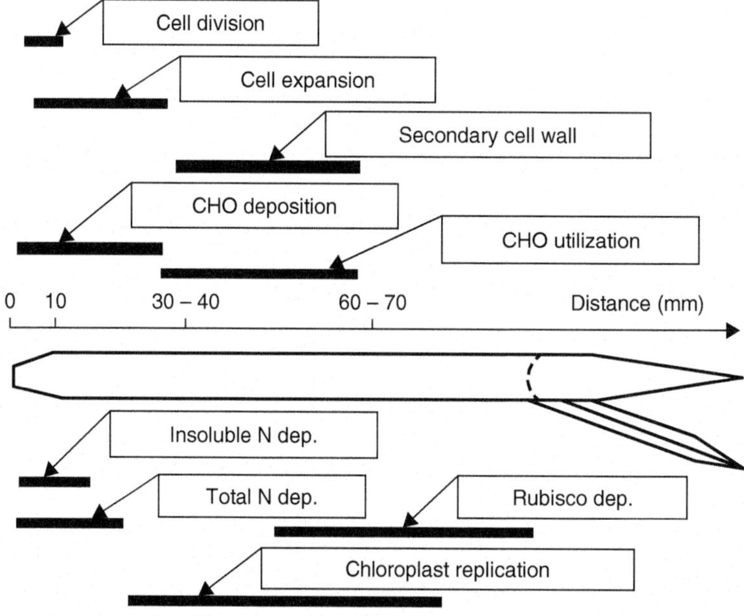

Fig. 7.1. Growth zones and zones of carbon and nitrogen deposition and utilization of elongating tall fescue leaf blades. *Source:* From Skinner and Nelson (1995).

more likely to remove all rapidly expanding leaf material, requiring regrowth to be initiated from new buds or undeveloped leaves. Grazing or mechanical harvest of grass leaves tends to remove mature leaf blades, leaving intact, the fully developed and functional growth zones that can rapidly elongate the remaining leaf and reestablish photosynthetic area.

The biophysical processes associated with cell expansion have been summarized through a framework proposed by Lockhart (1965) that relates cell expansion to the driving force generated by water uptake and to the ability of cell walls to yield to that force. Water uptake is a function of cell membrane hydraulic conductivity, the osmotic pressure difference between a cell and its surrounding tissues, and cellular hydrostatic pressure (Cosgrove 1986). Wall yielding, in turn, depends on the ability of the cell to generate sufficient **turgor** pressure to overcome the initial resistance to expansion (the yield threshold) and subsequent extensibility of cell wall components. Elongating cells have only a primary cell wall, so the yield threshold is low. Cells do not elongate after secondary cell wall material is deposited. While short-term cell elongation that increases plant size is controlled by cell wall yielding and water uptake, long-term growth in weight depends on carbon assimilation, nutrient absorption, and the synthesis of the structural cell wall components and other cellular constituents (Cosgrove 1986).

Biomass Accumulation

The cell division and elongation zones are sites of high metabolic activity and dry matter (DM) accumulation (Figure 7.2). The high biomass deposition in growth zones is mainly due to accumulation of water-soluble carbohydrates (Allard and Nelson 1991) which can reach concentrations of 300–400 $mg\,g^{-1}$ dry weight, or as much as five times the concentration of mature leaf tissue in field-grown plants (MacAdam and Nelson 1987). Similarly, N content, which in the cell division zone can be very high, ranging from 30 to 75 $mg\,g^{-1}$, depending on N-fertility regime (Gastal and Nelson 1994), occurs mainly as proteins and nucleic acids. Given that N content by weight is nearly 16% for both compounds, proteins and nucleic acids can account for nearly half the DM in the cell division zone.

As with C and N accumulation, the growing region is also the strongest **sink** for the mineral nutrients K, Mg, Cl, Ca, and P (Meiri et al. 1992) and for water deposition (Schnyder and Nelson 1987). The rapid influx of water associated with cell elongation means that fresh weight of the leaf elongation zone can be as much as 97% water (Meiri et al. 1992). The high-water content, combined with the high percentage of nonstructural carbohydrate and N compounds and relatively low proportion of cell wall material, makes the grass growth zone extremely delicate and susceptible to damage if not protected by the enclosing sheaths of older leaves.

Nonstructural carbohydrate and N concentrations are much higher in the growth zone compared to mature tissues. As cells cease elongating and enter the cell maturation zone, the nonstructural carbohydrates can be recycled to provide energy and C skeletons for secondary cell wall formation (Allard and Nelson 1991), whereas recycled N can be used for synthesis of photosynthetic proteins (Gastal and Nelson 1994). Even though the rate of DM accumulation is greatly reduced compared to elongating cells, non-elongating cells continue to differentiate and accumulate additional biomass, mostly as secondary cell wall material and in sclerenchyma tissue.

As cells mature and their photosynthetic apparatus develops, they undergo a transition from a C sink to a C source for the rest of the leaf. Similarly, as leaf development continues, the leaf ceases to be a **sink** and becomes a source for younger leaves. This change, which marks a fundamental transition in leaf **physiology**, tends to occur in forb leaves when they reach about 30–60% of their final length and is concurrent with the maturation of minor veins in the leaf (Turgeon 1989). This transition is marked by the cessation of carbohydrate import from mature leaves and is usually, but not necessarily, associated with the achievement of positive C balance in the leaf, i.e. when photosynthesis first exceeds the growth and respiratory needs of the leaf (Turgeon 1984). This can occur simultaneously for several leaves (Gagnon and Beebe 1996) or for only one leaf at a time (Turgeon and Webb 1973).

The sink-to-source transition occurs later in the development of grass leaves than in forbs. For example, tall fescue leaves remain a **sink** until they have reached about 80% of their final length (Bregard and Allard 1999). The delayed transition in grasses occurs because early blade development happens in relative darkness within the whorl of mature sheaths, whereas all stages of forb leaf development occur in full light exposure.

Following defoliation, leaf elongation of grasses often continues at rates equal to or greater than elongation rates prior to defoliation (Morvan-Bertrand et al. 2001). Increased elongation occurs at the same time DM and carbohydrate concentrations in the growth zone decrease (De Visser et al. 1997). Increased elongation is driven by continued high rates of water deposition in the growth zone accompanied by the hydrolysis of fructan, a polymer of fructose that serves as a storage carbohydrate, to support construction of structural materials (Volenec 1986).

The increase in leaf length is accompanied by reduced growth in leaf width and thickness. This shift in growth to produce thinner leaves allows for more rapid establishment of functional leaf area per unit of substrate to quickly capture sunlight and reestablish a positive C balance for the plant. Similarly, narrow and thin leaves occur at low irradiance, allowing increased leaf elongation to occur, despite reduced DM import into the elongation zone (Schnyder and Nelson 1989; Sanderson

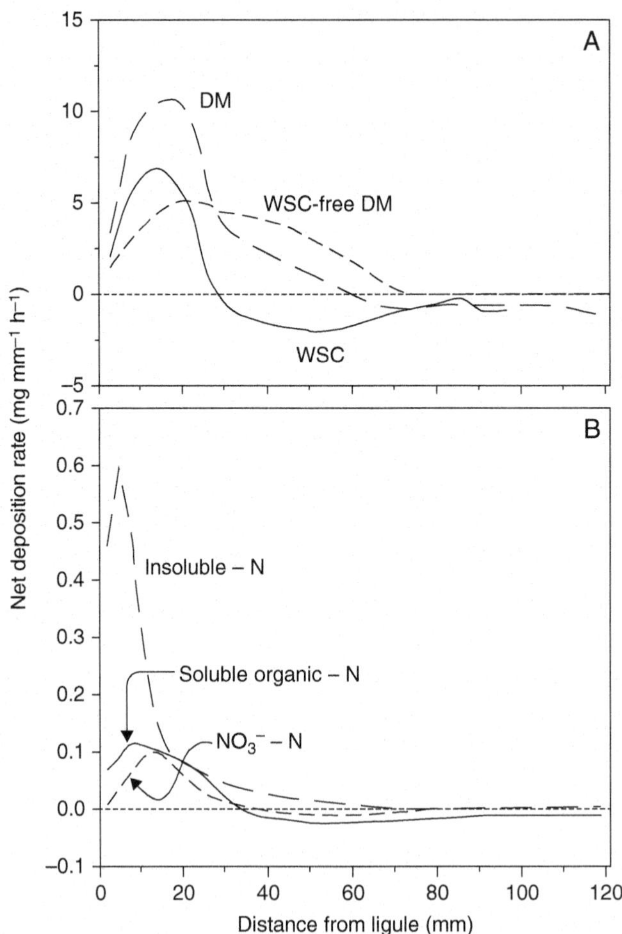

FIG. 7.2. Zones of net deposition and depletion for dry matter (DM), water-soluble carbohydrates (WSC), water-soluble carbohydrate-free dry matter (WSC-free DM), insoluble-N (polypeptides and nucleic acids), soluble organic-N (free amino acids, nucleic acids, and small peptides), and NO_3^--N in elongating tall fescue leaves. The sheath of the previous leaf would enclose the basal 100 mm. *Source:* Adapted from Allard and Nelson (1991) and Gastal and Nelson (1994).

and Nelson 1995). Leaves of forbs also increase in specific leaf area (area wt^{-1}) under shade, resulting in larger but thinner leaves (Dale 1988). Frequent cutting increased white clover leaf elongation rates (Li 2000), but the effect on leaf thickness was not reported.

Location, Activity, and Synchronization of Meristems

In addition to growth of individual leaves, forage production and stand longevity rely on initiation of new leaves and stems (tillers). The basic unit of grass development is the **phytomer,** which consists of a leaf, internode, axillary bud, and one or more root **primordia**. Within each phytomer, the leaf primordium produces both a blade and sheath, separated by a membranous layer of cells called the ligule, while a branch or tiller arises from the axillary bud (Skinner and Nelson 1994b). The internodes remain relatively short during vegetative growth but elongate to elevate the inflorescence during reproductive development. Elongation of the internode tends to inhibit axillary bud elongation as evidenced by a strong negative correlation between axillary bud and internode length (Williams and Langer 1975). Root elongation follows tiller initiation within a given phytomer, generally about three phyllochrons after initiation of the leaf blade (Carman and Briske 1982; Klepper et al. 1984).

As with leaf growth, developmental relationships among leaves and tillers have been more extensively studied in grasses than in forbs. Generally, grass tillers begin to elongate after the leaf that originates from the same node as the tiller has reached full size, giving rate of leaf appearance ultimate control over the rate of tiller appearance (Davies and Thomas 1983; Skinner and Nelson 1994a).

Major transitions in leaf and tiller development in tall fescue appear to be synchronized among at least three adjacent nodes (Table 7.1). Cessation of cell division in the leaf sheath at a given node, e.g. node 4, is accompanied by the initiation of cell division and elongation of the tiller bud at the same node. Simultaneously, the transition between blade and sheath formation begins at the next youngest node (node 5), while elongation of the new blade begins at node 6 (Table 7.1).

The transition between blade and sheath elongation for a given leaf occurs gradually as the ligule, which is visible early in development and marks the boundary between blade and sheath tissue, moves through the leaf elongation zone (Schnyder et al. 1990). The sheath first forms near the base of the cell division zone when the blade of the same leaf is 20% or less of its final length (Skinner and Nelson 1994b). Elongation of the sheath is initially slow compared to the blade, but as sheath elongation rate increases, the ligule above it is displaced through the elongation zone, causing blade elongation to decrease as cell supply is depleted.

The close relationship between leaf and tiller initiation makes it possible to mathematically describe tiller production as a function of leaf appearance rate and of site filling, which provides a measure of the ability of axillary buds to develop into new tillers (Davies 1974). Assuming that buds are produced in each leaf axil and that each bud has the potential to develop into a new tiller, i.e. fill the site, Davies (1974) determined that tiller number can potentially increase by a factor of 1.618 during each leaf appearance interval on the main stem. However, Neuteboom and Lantinga (1989) reported tiller buds can develop in the axil of the prophyll, a small scaly leaf at the base of each tiller.

When prophyll tillers are accounted for, tiller number has the potential to increase by a factor of 2.0 for each leaf appearance interval. In other words, the number of tillers per plant can double with the appearance of each new leaf on the main stem. This potential tiller appearance rate assumes a new tiller appears in the axil of the second-youngest fully emerged leaf on the parent tiller. An analogous concept to site filling called nodal probability, with values ranging from 0 to 1, has been developed to describe the probability of a tiller developing at any individual site (Matthew et al. 1998).

During periods of rapid tiller development, tillers appear in highly synchronized cohorts with the potential size of each cohort doubling with each successive leaf appearance interval (Figure 7.3). Tiller buds that lose synchronization with the remainder of the cohort become progressively less likely to appear (Skinner and Nelson 1992). Growth of tiller buds appears to be constrained by surrounding tissues such that tillers that emerge must escape from the cavities in which they develop before becoming trapped by the maturation and hardening of surrounding tissues (Williams and Langer 1975). This suggests that a window of opportunity exists for each tiller to emerge and that delayed development results in a missed opportunity for rapid growth and eventual emergence.

Adventitious root development is also closely tied to leaf and tiller development since these roots originate from nodes associated with leaves and developing tillers. Adventitious roots usually begin to appear when the main stem or individual tiller has about three developed leaves, and then appear sequentially at each successive node about three plastocrons after the leaf at that node first appears (Carman and Briske 1982; Rickman et al. 1985). Appearance of roots on a tiller is generally an indication that the tiller has become independent of the main stem and is a necessary step for long-term survival of the tiller. Severe defoliation during initial tiller development may decrease tiller root establishment, causing newly initiated tillers to die (Carman and Briske 1982).

As with grasses, growth and development of legumes and forbs also occur through the sequential production of phytomers consisting of a leaf, internode, axillary bud, and one or more root primordia (Gautier et al. 2001).

Table 7.1 Synchronization of major developmental transitions involving epidermal cell division and elongation during initiation and appearance of tall fescue leaves and tillers

Haun index	Node	Event	Haun index	Node	Event
1.9	4	Division ends in sheath of leaf 2	2.8	5	Division ends in sheath of leaf 3
2.0	4	Elongation of tiller 1 begins	2.7	5	Elongation of tiller 2 begins
1.9–2.1	5	Ligule is initiated on leaf 3	2.8–3.0	6	Ligule is initiated on leaf 4
2.0	6	Elongation begins for blade 4	2.8	7	Elongation begins for blade 5

Source: From Skinner and Nelson (1994b).

The cotyledon is located at node 1 and the **coleoptile** at node 2. Thus, leaf 2 develops from node 4, leaf 3 from node 5, and so on.

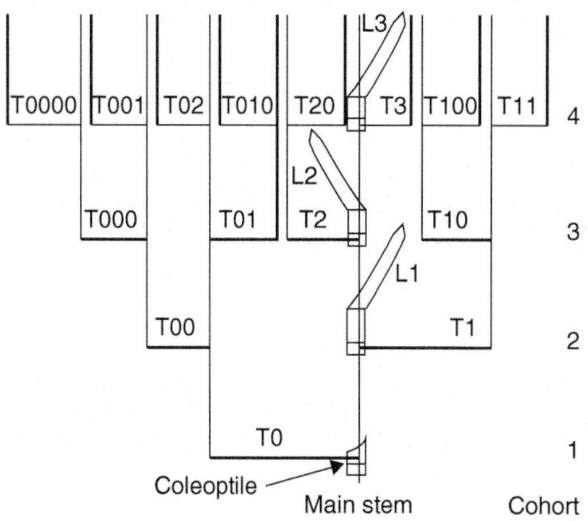

Fig. 7.3. Appearances of tiller cohorts are synchronized with leaf appearance on the main stem (MS). Tillers (T) are named for the leaf axil in which they appear; T0 appears in the axil of the **coleoptile** and tillers; T × 0 appears in the prophyll axil of each tiller. Tillers within a cohort that emerge do so at almost the same time, usually within 0.6–1.0 phyllochron after appearance of the main-stem leaf that is two phytomers younger than the tiller, i.e. T1 appears after appearance of L3. *Source:* Adapted from Skinner and Nelson (1992).

As with grasses, the leaf from a given phytomer for alfalfa expands to nearly full size before the associated internode begins rapid extension (Brown and Tanner 1983). In crown-forming species such as alfalfa, axillary bud development from the cotyledonary node and other basal nodes on developed stems results in the formation of a well-defined crown containing multiple stems (Barnes and Sheaffer 1995). Regrowth following defoliation can occur from basal axillary buds located on the crown or from upper axillary buds along the stem. In contrast, clonal species such as white clover have two distinctive morphologic stages. First, a seminal taproot develops after establishment and is followed by stolon growth to form a dispersed clonal stage one to two years later (Brock et al. 2000). Death of the taproot and primary stolon initiates the fragmentation of the parent plant into numerous independent clones that are rooted at nodes of the surviving stolons. These clonal plants can have a lateral spread of 1 m or more (Brock et al. 2000; Gustine and Sanderson 2001).

Describing Developmental Morphology

Developmental morphology refers to the series of changes in structure and arrangement of plant components associated with plant maturity (Esau 1960). Developmental morphology is similar among grass species with only minor variations separating growth forms (Briske 1991). Temperature and photoperiod are important in controlling the rate of plant morphologic development (Briske 1991; Gillen and Ewing 1992). Developmental morphology within a species has been reported to have strong linear relationships to accumulated growing degree days (GDD) and day of the year (Kalu and Fick 1981; Buxton and Marten 1989; Hendrickson 1992). The relationship between developmental morphology and day of the year can be partially attributed to the process of floral **induction** which occurs in response to photoperiodic stimulus (Briske 1991). Many plant species have photoperiod requirements for floral induction (Salisbury and Ross 1985). Smooth bromegrass is photoperiod sensitive and requires a primary floral induction period of short days and a secondary period of long days and is, therefore, classified as a short-**long-day plant** (Heide 1984). Other perennial, cool-season grasses, such as intermediate wheatgrass are photoperiod sensitive and have a dual induction requirement for flowering (Heide 1994). Switchgrass and big bluestem are photoperiod sensitive and require short days for floral induction (Benedict 1941). Vegetative growth of smooth bromegrass, intermediate wheatgrass, switchgrass, and big bluestem terminates with inflorescence development and is, therefore, determinate in growth habit (Dahl and Hyder 1977). Following floral **induction**, the grass tillers advance to the seed-ripening stages, growth stops, and tiller senescence occurs.

The architectural organization, palatability and accessibility to herbivores, and regrowth potential following defoliation are determined by the developmental morphology of plants (Briske 1991). Production practices such as grazing management, cutting, and seed production should be based on an accurate assessment of developmental stage (Moore et al. 1991). Several systems have been developed to describe developmental stages of forage species and have been used as aids to help schedule management practices.

Developmental Stages

The life cycles of forage plants are characterized by distinct changes in plant morphology. The ontogeny of most forage plants involves seedling, vegetative, and reproductive stages of development. These occur in a predictable manner and are useful for describing the maturity of individual plants as well as populations or stands.

The ***vegetative stage*** encompasses the period during which major activity is in leaf growth and development, which can be characterized by the successive appearance of leaves. In grasses, stem internodes are laid down and differentiated during this period but do not elongate. In many forbs, including most forage legumes, stem growth occurs throughout the vegetative stage. However, in others, such as chicory or plantain, internodes remain short and a leafy **rosette** is formed.

The interval of time between appearances of successive leaves is called the **phyllochron** and is sometimes used as an index for describing vegetative growth (Wilhelm and McMaster 1995). During the time a tiller or stem remains vegetative, the apical meristem is indeterminate and, theoretically, can produce an infinite number of new nodes and leaves. In grasses, stem elongation, a process commonly referred to as **jointing**, is considered a transition state between vegetative and reproductive development (Waller et al. 1985). Elongation of stem internodes is accompanied by differentiation of the **shoot apex** meristem into the inflorescence.

The reproductive stage begins with the initiation of inflorescence development and continues through seed ripening and shatter. Seed ripening is sometimes considered a distinct developmental period, in which case, the reproductive period terminates with fertilization (Moore and Moser 1995).

Metcalfe and Nelson (1985) described several growth stages that are commonly used to indicate the maturity of grass and legume forages (Table 7.2). These useful descriptors are easily understood and applied, but they do not provide a way to quantify maturity, which is essential for mathematical modeling of developmental morphology and describing maturity of populations of forages.

Table 7.2 Morphologic descriptors for growth stages of forage grasses and legumes

Terminology	Definition
Grasses	
First growth	
Vegetative	Leaves only; stems not elongated
Stem elongation	Stems elongated
Boot	Inflorescence enclosed in flag leaf sheath and not showing
Heading	Inflorescence emerging or emerged from flag leaf sheath, but not shedding pollen
Anthesis	Flowering stage; anthers shedding pollen
Milk stage	Seed immature, endosperm milk
Dough stage	Well-developed seed; endosperm doughy
Ripe seed	Seed ripe; leaves green to yellow brown
Post-ripe seed	Seed post-ripe; some dead leaves; some heads shattered
Stem-cured	Leaves cured on stem; seed mostly cast
Regrowth	
Vegetative	Leaves only; stems not elongated
Jointing	Green leaves and elongated stems
Late growth	Leaves and stems weathered
Legumes	
Spring and summer growth	
Vegetative (or pre-bud)	No buds visible
Bud	Buds visible, but no flowers
First flower	First flowers appear on plants
Bloom (flower)	Plants flowering
Pod (or green seed)	Green seedpods developing
Ripe seed	Mostly mature brown seedpods with lower leaves dead and some leaf loss
Fall recovery growth	Vegetative or with floral development

Source: From Metcalfe and Nelson (1985).

Quantifying Developmental Morphology

Numerous systems have been developed to accurately quantify the growth and development of plants (Vanderlip 1972; Haun 1973; Zadoks et al. 1974; Fehr and Caviness 1977; Kalu and Fick 1981; Simon and Park 1983; Moore et al. 1991; Sanderson 1992). Many of these systems were intraspecific or difficult to apply in the field. For example, the soybean staging system described by Fehr and Caviness (1977) described at least five vegetative stages and eight reproductive stages and the reproductive stages varied between determinate and indeterminate cultivars. Kalu and Fick (1981) presented a staging system for alfalfa which included ten growth stages ranging from early vegetative to ripe seed pod. This system included two methods to quantify morphologic stage of alfalfa shoot populations based on mean stage by count (MSC) and mean stage by weight (MSW).

In grasses, Phillips et al. (1954) used six general stages ranging from vegetative stage to seeds at **dough stage**. The vegetative stage included a broad spectrum of tillers which ranged from early elongation to the **boot stage** and lacked identification of non-elongated tillers. Haun (1973) developed a system for quantifying wheat development which integrated the number of leaves developed and the rate of development of the next older plant part into plant development. Simon and Park (1983) modified an earlier system developed by Zadoks et al. (1974) which included eight primary growth stages subdivided into secondary stages. They noted the variability of growth stages in cross-pollinated forage grasses was much larger than in cultivars of self-pollinated cereals. This system was applicable to most grass species but is complex and difficult to apply under field conditions (Moore et al. 1991).

The comprehensive growth-staging system developed by Moore et al. (1991) is applicable to most annual and perennial grasses, is easily applied in the field, and produced repeatable results (Hendrickson 1992). This comprehensive growth-staging system contains four primary-growth stages for quantifying the developmental morphology of established perennial grasses: vegetative, elongation, reproductive, and seed ripening (Moore et al. 1991). Secondary stages within each primary stage describe specific events and are given numerical indices to quantify tiller population development. A representative sample of tillers is collected from the **sward** to determine the mean growth index for the tiller population based on MSC or MSW (Kalu and Fick 1981). The system can be used to quantify the relationship between developmental morphology and forage quality of the population (Mitchell et al. 2001).

The systems developed to describe and quantify morphologic development of forage species share some common characteristics, including a defined series of morphologic descriptors that have an associated numerical index. The numerical index can be used to develop mathematical relationships between forage maturity and variables such as forage quality and yield (Kalu and Fick 1981; Hendrickson et al. 1997). Conversely, maturity indices can be used as dependent variables to predict forage maturity based on chronology or accumulated heat units (Mitchell et al. 1997; Sanderson and Moore 1999). These phenologic relationships are useful for timing management practices that depend on maturity.

Attempts to develop a universal system for describing and quantifying morphologic development of forage crops has not been successful (Sanderson et al. 1997). A committee appointed by the Crop Science Society of America to identify and recommend a growth-staging system that was generally applicable to crops and weeds was unsuccessful in identifying any that could be used with acceptable precision (Frank et al. 1997). Instead, the committee recommended growth-staging systems specific to individual crops, including forages (Table 7.3).

Table 7.3 Staging systems recommended for use with forage crops

Forage crop	Reference
Alfalfa	Kalu and Fick (1981); Fick and Mueller (1989)
Cool-season grasses	Haun (1973); Moore et al. (1991)
Red clover	Ohlsson and Wedin (1989)
Stoloniferous grasses	West (1990)
Warm-season grasses	Moore et al. (1991); Sanderson (1992)

Source: Adapted from Frank et al. (1997).

Alfalfa

The recommended system for staging alfalfa was originally developed by Kalu and Fick (1981) and was later modified by Fick and Mueller (1989). It recognizes ten stages of development that occur within four growth phases (Table 7.4). Vegetative stages consist of leaf and stem development and are defined in terms of stem length. Stages during flower-bud development are defined by the appearance and number of flower buds on the stems. Flowering stages correspond to the number of open flowers present on a stem. Seed-production stages are defined by the number and color of seedpods. Many of these morphologic descriptors are specific to alfalfa but can be modified for other species. However, they are not generally directly applicable to most other legumes.

Red Clover

The staging system developed by Ohlsson and Wedin (1989) for red clover is an adaptation of the alfalfa system (Table 7.4) with descriptors for vegetative and flower-bud development stages being nearly identical for the two systems. The main differences are in the flowering and seed-production stages, reflecting differences in inflorescence morphology between the species. Ohlsson and Wedin (1989) also evaluated another system for red clover that includes 18 stages and has the advantage of having more logical morphologic descriptors. It performed well. The ten-stage systems for both alfalfa and red clover include length descriptors that are not strictly morphologic (Fick and Mueller 1989). Stem length varies among cultivars of both species, so vegetative stages may be inconsistent with regard to the number of nodes and length of internodes of the plant. Thus, Ohlsson and Wedin (1989) recommended using the 18-stage system for research studies on red clover, especially those focused on early stages of development.

Cool-Season Grasses

The Haun system was developed to quantify wheat development (Haun 1973) but has been used to quantify development of cool-season perennial grasses (Frank et al. 1993). Numerical indices correspond to the number of developed leaves on the primary tiller; that is, tillers with one, two, and three fully expanded leaves are assigned index values of 1, 2, and 3, respectively. Partially expanded leaves are assigned a fractional value relative to the most recent fully expanded leaf. For example, a tiller with three fully expanded leaves and a developing fourth leaf that is one-half the length of the third is assigned an index of 3.5. The Haun system applies only to leaf development through stem elongation, so its use is primarily limited to vegetative growth. It has been used to predict grazing readiness of native and introduced pastures (Frank et al. 1993).

Table 7.4 Developmental stages, numerical indices, and morphologic descriptors for alfalfa and red clover

Index	Stage	Alfalfa descriptors[a]	Red clover descriptors[b]
Vegetative phase			
0	Early vegetative	Stem length ≤ 15 cm, no buds, flowers, or seedpods	Stem length ≤ 15 cm, no buds, flowers, or seedpods
1	Mid-vegetative	Stem length 16–30 cm, no buds, flowers, or seedpods	Stem length > 15 to <30 cm, no buds, flowers, or seedpods
2	Late vegetative	Stem length ≥ 31 cm, no buds, flowers, or seedpods	Stem length ≥ 31 cm, no buds, flowers, or seedpods
Flower bud development			
3	Early bud	1–2 nodes with buds, no flowers or seedpods	1–2 nodes with buds, no flowers or seedpods
4	Late bud	≥3 nodes with buds, no flowers or seedpods	≥3 nodes with buds, no flowers or seedpods
Flowering phase			
5	Early flower	1 node with 1 open flower, no seedpods	Open flower (standard open) on main stem, no seed in flower head
6	Late flower	≥2 nodes with open flowers, no seedpods	Open flowers (standard open) on main and axillary stems, no seed in flower heads
Seed production			
7	Early seedpod	1–3 nodes with green seedpods	Seeds developing in the flower of the main stem
8	Late seedpod	≥4 nodes with green seedpods	Seeds developing in the flowers of the main and axillary stems
9	Ripe seedpod	Nodes with mostly brown mature seedpods	Sepals of flowers brown

[a]From Fick and Mueller (1989).
[b]From Ohlsson and Wedin (1989).

Moore et al. (1991) developed a system for quantifying the developmental morphology of grasses for use in forage and range-management studies. Their system, called the Nebraska system, is based on the ontogeny of individual tillers, which is divided into four primary growth stages: (i) vegetative, (ii) elongation, (iii) reproductive, and (iv) seed ripening (Table 7.5). Within each primary stage, substages are defined that correspond to specific morphologic events. Thus, each growth stage consists of a primary and secondary stage and has a numerical index associated with it that can be used for quantitative purposes. The vegetative and elongation substages are open ended, with the number of substages being equivalent to the number of morphologic events (*N*) that occur for that species or environment. The reproductive and seed-ripening primary stages each have six secondary or substages, numbered 0 through 5, which pertain to particular events in the ontogeny of the primary **shoot** or tiller. The substages for these primary stages describe specific events that occur similarly in most grasses.

In addition to the numerical index, the Nebraska system associates a mnemonic code with each growth stage. The codes can be easily memorized and are useful for applying the system in the field. Each code consists of two characters: a capital letter denoting the primary growth stage, followed by a number denoting the substage within that primary stage. Growth stages as denoted by the mnemonic codes are consistent across species.

Warm-Season Grasses

The Nebraska system (Moore et al. 1991) described above was developed for both warm- and cool-season grasses and works well for both (Mitchell et al. 1998). Another system recommended for warm-season grasses is the TAES system, which was developed specifically to describe and quantify development of determinate and indeterminate flowering warm-season bunchgrasses (Sanderson 1992). It uses a numerical index similar to the Haun (1973) scale during vegetative development.

The numerical index of the TAES system is discontinuous between the vegetative and stem elongation stages, and between the elongation and reproductive stages of development. These discontinuities result from inclusion of enough indices within a major growth stage to allow for variation in development that occurs among species and growth environments. The Nebraska system avoids this problem by linearizing indices within the vegetative and elongation growth stages according to the number of

Table 7.5 Growth stages of perennial grasses, their numerical indices, and descriptions

Stage	Numerical index	Description
Vegetative stage – Leaf development		
VE or V0	1.0	Emergence of first leaf
V1	$(1/N^a) + 0.9$	First leaf collared
V2	$(2/N) + 0.9$	Second leaf collared
Vn	$(n/N) + 0.9$	*N*th leaf collared
Elongation stage – Stem elongation		
E0	2.0	Onset of stem elongation
E1	$(1/N) + 1.9$	First node palpable/visible
E2	$(2/N) + 1.9$	Second node palpable/visible
En	$(n/N) + 1.9$	*N*th node palpable/visible
Reproductive stage – Floral development		
R0	3.0	Boot stage
R1	3.1	Inflorescence emergence/1st spikelet visible
R2	3.3	Spikelets fully emerged/ peduncle not emerged
R3	3.5	Inflorescence and peduncle fully elongated
R4	3.7	Anther emergence/anthesis
R5	3.9	Post-anthesis/fertilization
Seed development and ripening stage		
S0	4.0	Caryopsis visible
S1	4.1	Milk
S2	4.3	Soft dough
S3	4.5	Hard dough
S4	4.7	Endosperm hard/ physiological maturity
S5	4.9	Endosperm dry/seed ripe

Source: From Moore et al. (1991).

[a]Where n equals the event number (number of leaves or nodes) and N equals the number of events within the primary stage (total number of leaves or nodes developed). General formula is $P + (n/N) - 0.1$, where P equals primary stage number (1 or 2 for vegetative and elongation, respectively) and n equals the event number. When $N > 9$, the formula $P + 0.9(n/N)$ should be used.

morphologic events that occur within them (Moore et al. 1991).

Discontinuous scales can result in significant numerical shifts in transitions between stages, resulting in nonlinear responses (Sanderson et al. 1997). Another problem occurs when demographic statistics are calculated for a population of tillers that include discontinuous growth stages. Under these circumstances, it is possible to calculate a mean index associated with a morphologic descriptor that does not occur for the species. For example, the mean stage might indicate a stem with seven nodes for a species that elevates only four (Moore and Moser 1995).

Discontinuous scales can be useful, but caution should be exercised when interpolating across discontinuous growth stages. Indeed, the TAES system may be more useful than the Nebraska system for detailed studies on vegetative development because it uses a greater number of indices to describe growth during this period.

Stoloniferous Grasses

Grasses that produce predominantly horizontal stems cannot be described well using systems recommended for staging upright grasses. West (1990) developed a system for staging the development of bermudagrass that is applicable to other stoloniferous grasses. The primary difference from other systems is that vegetative stages are defined in terms of development of nodal zones rather than leaves. Descriptors for other stages of development are analogous to other grass-staging systems, though the coding of the numerical index to descriptors varies among systems.

Predicting Developmental Morphology

Continuous numerical indices can be used to develop mathematical relationships between developmental stages and temporal and climatic variables. These relationships can be descriptive or predictive in nature, depending on the intended use of the resulting equations. In many cases, staging systems are used to accurately describe the development of forages within the context of a specified period of time with no intention of making predictions about the development of the forage at another time (Sanderson 1992; Brueland et al. 2003). The goal is simply to provide a clear account of the maturity of the forage in relation to other factors of interest.

A potentially more powerful use of numeric indices is developing phenologic models for predicting forage development. Such models relate developmental morphology to climatic variables, such as photoperiod and accumulated heat units. Development of robust phenologic models would enable forage producers to predict the occurrence of important morphologic events using climate data. This is significant because many important management decisions are based on maturity of the forage. Unfortunately, few such models have been developed and validated for general use.

Empirical models for predicting morphologic development of switchgrass and big bluestem have been developed and validated for use in the central US (Mitchell et al. 1997; Sanderson and Moore 1999). Equations were developed for predicting MSC using the Nebraska system as a function of day of year (DOY) and GDDs. Under Nebraska conditions, switchgrass development was best predicted ($r^2 = 0.96$) using a linear equation based on day of the year. This relationship indicates that photoperiod is the main determinant of switchgrass morphologic development (Mitchell and Moser 2000). In contrast, big bluestem development was more accurately predicted ($r^2 = 0.83$) using a nonlinear equation based on GDDs, suggesting that its development is less determinate than that of switchgrass.

Prediction equations were developed in Nebraska based on data collected over two growing seasons for 'Trailblazer' switchgrass and 'Pawnee' big bluestem (Mitchell et al. 1997). Prediction equations for MSC and MSW were developed based on DOY and GDD. The equations were subsequently validated over two additional growing seasons in Nebraska and Kansas (Figure 7.4). Switchgrass and big bluestem MSC and MSW were related linearly in all environments. Linear DOY calibration equations accounted for 96% of the variation in switchgrass MSC across four environments, which indicates that switchgrass development was related to photoperiod and that general management recommendations could be based on DOY in the central Great Plains. Quadratic GDD calibration equations accounted for 83% of the variation in big bluestem MSC across four environments, which indicates that big bluestem development is more difficult to predict and management recommendations in the central Great Plains should be based on morphologic development which is best predicted by GDD. The switchgrass equation was further evaluated for use with 'Cave-in-Rock' and 'Kanlow' switchgrass in Iowa, and Cave-in-Rock and 'Alamo' switchgrass in Texas (Sanderson and Moore 1999). The Nebraska equation performed well for predicting development of the two cultivars in Iowa but did not do as well

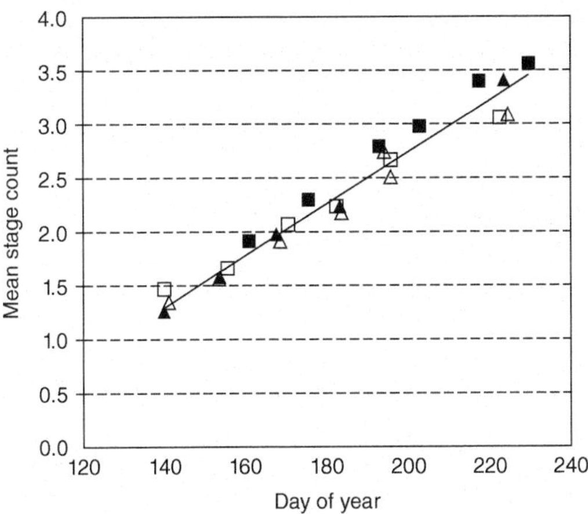

Fig. 7.4. Actual and predicted mean stage count of 'Trailblazer' switchgrass grown in Kansas (■) and Nebraska (Δ) during 1992 (*open symbols*) and 1993 (*closed symbols*). Predicted MSC = 0.024(Day) − 2.063. *Source:* Adapted from Mitchell et al. (1997).

in Texas due to large differences in daylength and climate. These studies suggest that there is good potential for developing reliable and robust equations for predicting grass development on a regional basis. Developing similar equations for important forage species, within different regions, could be of great benefit to producers.

Plant Maturity and Relationships to Forage Quality

Quantifying maturity of perennial grass tiller populations is essential to characterize nutrient content throughout the developmental cycle. As plant maturity increases, the quality for ruminant animals decreases because of an increase in cell wall concentration and decrease in crude protein (CP) concentration. Quantifying the growth and development of forage grasses and determining relationships with forage quality is essential for making forage management decisions.

Forage quality is affected by genetic, physiologic, environmental, and plant developmental factors (Van Soest 1982). The influences of these factors on forage quality are highly integrated and often difficult to isolate. Plant maturity is the major factor affecting developmental morphology and forage quality (Nelson and Moser 1994). The existence of relationships between plant maturity and forage quality of perennial grasses has long been recognized (Phillips et al. 1954). However, the environment can modify the impact of plant maturity on forage quality (Buxton and Fales 1994). Factors such as high temperature, high-solar irradiation, and abundant water may accelerate the maturation process, while factors such as clipping, grazing, and disease may retard the maturation process (Van Soest 1982).

Environmental factors that affect plant growth have a profound effect on forage quality (Van Soest 1985). High temperatures reduce forage quality at similar physiologic ages (Wilson 1983), possibly through decreases in leaf: stem ratios with high temperatures promoting stem growth over leaf growth (Buxton and Fales 1994). Metabolic activity increases as temperature increases which results in higher accumulations of cellulose, hemicellulose, and lignin, while forages grown in cooler climates have higher carbohydrate reserves and protein concentrations associated with needs to develop winter-hardiness (Van Soest 1985). Cell-wall materials deposited at lower temperatures are less lignified and more digestible (Nelson and Moser 1994).

Increasing irradiation stimulates photosynthetic activity which promotes synthesis of soluble sugars and starches which dilute cell-wall material (Buxton and Casler 1993). High irradiance or extended photoperiods for short timeframes generally increases forage quality (Buxton and Casler 1993). Prolonged periods of shade may reduce photosynthate availability which reduces secondary cell wall deposition, resulting in lower lignin concentrations (Buxton and Casler 1993).

The effect of water on forage quality is variable. If water is limiting, plant advancement toward maturity may be hindered, resulting in higher forage quality. Mild water stress increases forage quality by increasing leaf: stem ratios and increasing the digestibility of leaves and stems (Nelson and Moser 1994). However, if water stress is too severe, perennial plants may go dormant and translocate reserves into the roots and crown, reducing the forage quality of the plant (Van Soest 1985).

Factors influencing plant maturity may be site specific (weather, water, and management) or vary on a geographic basis (light quality, light quantity, soil, and climate) (Van Soest 1982). Therefore, quantifying relationships between developmental morphology and forage quality in different environments is necessary to provide information for developing strategies for improving utilization and seasonal distribution of perennial forage grasses.

Factors that influence forage quality are complex and interactive (Van Soest 1982; Akin 1989). The anatomic organization of C_4 grasses causes forage quality to be lower than for C_3 grasses (Akin 1989). The loose arrangement of mesophyll cells in C_3 grasses increases intercellular air space allowing more rapid penetration by rumen bacteria into the leaf, increasing digestion (Hanna et al. 1973). Cool-season species typically accumulate more **total nonstructural carbohydrates (TNC)** in the form of fructan than warm-season species, particularly at low temperatures, which greatly improves the quality of cool-season species (Nelson and Moser 1994). However, plant maturity is the major factor determining forage quality within a species (Nelson and Moser 1994). As plant maturity increases, forage quality for ruminant animals decreases through an increase in cell-wall concentration and a decrease in crude protein concentration (CP). The decline in forage quality associated with increased maturity may be partially explained by a decrease in leaf material and an increase in stem material. Stem elongation and inflorescence development form lower quality stem material that dilutes the higher quality leaf material (Nelson and Moser 1994).

Griffin and Jung (1983) reported the percentage of total DM production of leaf tissue of switchgrass declined from 71% to 31% and big bluestem declined from 64% to 21% as maturity progressed. With advancing maturity, stems decreased in quality faster than leaves (Griffin and Jung 1983; Nelson and Moser 1994). The lower forage quality of grass stems compared to grass leaves can be attributed to differences in anatomic characteristics of grass leaves and stems. Grass stems are composed of an epidermis covered with a thick waxy cuticle which is nearly impervious to microbial penetration (Monson et al. 1972), plus more sclerenchyma and parenchyma tissue, resulting in a more rigid, less digestible tissue than leaves (Akin 1989).

Dry Matter Digestibility

The ***in vitro*** procedure for estimating DM digestibility (Tilley and Terry 1963), later modified with direct acidification by Marten and Barnes (1980), has allowed researchers to rapidly quantify the digestibility of large numbers of forage species (Vogel et al. 1981). *In vitro* DM disappearance (IVDMD) of perennial forage grasses declined as the growing season progressed and maturity advanced (Anderson and Matches 1983; Jung and Vogel 1986; Sanderson and Wedin 1989; Mitchell et al. 1994). Cool-season grass species tended to be higher in IVDMD than warm-season grass species at similar maturities (Akin 1989). Vascular bundles of leaves of C_4 grasses are closely spaced and surrounded by a thick-walled parenchyma bundle sheath, whereas vascular bundles of C_3 grasses are widely spaced with less distinct parenchyma bundle sheaths and loosely arranged mesophyll cells which are rapidly digested (Buxton and Casler 1993).

The IVDMD concentrations of smooth bromegrass and timothy leaf blades, stems, and herbage declined linearly with increasing maturity (Sanderson and Wedin 1989). Maturity accounted for more IVDMD variation in timothy leaf blades and herbage than in bromegrass. The IVDMD of the stems of both smooth bromegrass and timothy declined more rapidly than did the IVDMD of the leaf blades of each species. The IVDMD of smooth bromegrass herbage ranged from approximately 500–750 g kg^{-1}.

There was a linear decline in IVDMD of two cultivars each of orchardgrass, smooth bromegrass, reed canarygrass, and tall fescue as maturity advanced (Buxton and Marten 1989). Total herbage IVDMD of the two smooth bromegrass cultivars harvested between 10 May and 5 July ranged from approximately 440–780 g kg^{-1}. With four grass species, the top leaf blades were most digestible, the inflorescences were intermediate, and the stems were least digestible (Buxton and Marten 1989). Day of the year, GDD, and morphologic stage accounted for at least 95%, 92%, and 89%, respectively, of the variation associated with IVDMD for all species during the two-years study.

Similar decreases in warm-season grass IVDMD with advancing maturity have been observed. The nutritional value of perennial warm-season grasses is primarily limited by digestible energy (Moore et al. 1993). The IVDMD of switchgrass leaves declined linearly throughout the growing season (Anderson 1985). The whole-plant IVDMD declined approximately 20 g kg^{-1} per week as switchgrass and caucasian bluestem matured from the vegetative to the **heading** stages (Anderson and Matches 1983). They also noted that switchgrass whole-plant IVDMD was higher than caucasian bluestem IVDMD at similar growth stages, but the IVDMD of the two species was nearly equal on a given date (Anderson and Matches 1983). Balasko et al. (1984) reported switchgrass

IVDMD declined with maturity with IVDMD at the **boot stage** ranging between 504 and 576 g kg^{-1}.

Switchgrass and big bluestem leaf and stem IVDMD declined throughout the growing season, and IVDMD was higher during a dry year than during a year with above normal precipitation (Perry and Baltensperger 1979). Switchgrass IVDMD was usually higher than big bluestem IVDMD when harvested on a common day of the year, and stage of maturity had more influence on IVDMD than did unfavorable precipitation (George and Hall 1983). The IVDMD of big bluestem harvested from tallgrass prairies declined as the growing season progressed and ranged from 710 to 508 g kg^{-1} in mid-June and mid-August, respectively (Mitchell et al. 1994). The highest IVDMD of 20 elite switchgrass populations ranged from 650 g kg^{-1} in vegetative growth stages to 492 g kg^{-1} at heading (Hopkins et al. 1995). Switchgrass IVDMD was best predicted by GDD which accounted for 86% of the variation, whereas big bluestem IVDMD was best predicted by MSW which accounted for 90% of the variation (Mitchell et al. 2001).

Fiber Concentration

Warm-season grasses tend to have higher fiber concentrations than cool-season grasses at similar maturities (Griffin et al. 1980; Jung and Vogel 1986). Increased fiber concentrations in perennial warm-season grasses would result in lower digestibility and reduced intake (Kilcher 1981).

Eight species of cool-season grasses increased in lignin and **crude fiber** up to the flowering stage and, in some species, to the seed-dough stage (Phillips et al. 1954). They concluded on the basis of the changes in lignin and crude fiber concentration that lignin was preferred over crude fiber as a criterion for **feeding value** (Phillips et al. 1954).

Neutral detergent fiber concentrations (NDF) of switchgrass and big bluestem leaves changed little with advanced maturity (Griffin and Jung 1983). However, NDF accumulation in the stem tissue of switchgrass and big bluestem increased rapidly with maturation. Switchgrass leaves and stems averaged 23 and 49 g kg^{-1} higher NDF than big bluestem leaves and stems, respectively, at early head emergence. Lignin concentrations in switchgrass and big bluestem leaves and stems increased with maturity. However, lignin concentrations in the stems increased at a much faster rate than lignin concentrations in the leaves. At early head emergence, lignin concentrations for switchgrass leaves and stems was 47 and 83 g kg^{-1}, respectively, whereas lignin concentrations for big bluestem leaves and stems were 46 and 61 g kg^{-1}, respectively. However, lignin continued to accumulate in big bluestem after seedheads emerged, indicating the importance of harvesting prior to heading.

The NDF, acid detergent fiber (ADF), and lignin concentrations increased more than three times faster in switchgrass stems than in the leaves during the first 25 days of stem collection (Anderson 1985). At similar growth stages, leaves that developed early in the growing season contained less NDF and ADF than leaves that developed late in the growing season. Switchgrass leaves never contained less than 600 g kg^{-1} NDF, and average NDF increased 0.13 g kg^{-1} d^{-1} from the two-leaf stage in May until late July, whereas ADF concentration increased less consistently. Lignin concentrations ranged from 21 to 128 g kg^{-1} in leaves and from 58 to 152 g kg^{-1} in stems and was consistently low in leaves in the whorl (Anderson 1985).

Hendrickson (1992) reported the NDF and ADF concentrations of prairie sandreed and sand bluestem leaves did not vary in response to morphologic advancement as measured by MSC or MSW. Prairie sandreed leaf NDF was higher than sand bluestem leaf NDF, but leaf ADF of the two species was similar throughout the growing season. Neither MSC nor MSW had consistently high correlation coefficients with NDF and ADF concentrations. Leaf lignin was highly variable and neither MSC nor MSW had a consistently good relationship with leaf lignin. He concluded the stable leaf NDF and ADF concentrations of both species indicated a decline in cell-wall digestibility rather than a decrease in cell contents was responsible for declines in digestibility.

Switchgrass NDF was best predicted by MSC and MSW (Mitchell et al. 2001). Mean stage weight accounted for 74% of the variability in big bluestem NDF. The model adequately predicted forage quality due primarily to the determinate growth habit of these species. Morphologic development accurately predicted forage quality in many instances.

Crude Protein Concentration

CP concentration of perennial forage grasses typically decreased as maturity progressed (Kamstra 1973; Perry and Baltensperger 1979; Griffin and Jung 1983; Mitchell et al. 1994), and was higher for cool-season grasses than for warm-season grasses at similar growth stages (Kamstra 1973; Griffin et al. 1980). Kamstra (1973) reported that the CP of two cool-season and two warm-season grasses decreased with maturity. The CP of western wheatgrass declined linearly as maturity progressed and ranged from approximately 120–69 g kg^{-1}. Kilcher and Troelsen (1973) reported smooth bromegrass CP ranged from 250 g kg^{-1} in the very immature stage to 80 g kg^{-1} in the mature stage. Perry and Baltensperger (1979) concluded leaf maturation was primarily responsible for declining CP rather than plant development. Griffin and Jung (1983) concluded quality of leaf tissue was responsible for the declining whole-plant forage quality of switchgrass and big bluestem.

Rehm et al. (1971) evaluated the influence of nine fertility levels on smooth bromegrass CP. Smooth bromegrass whole-plant CP ranged from 87 to 240 g

kg^{-1} when harvested at the early inflorescence growth stages. They concluded that CP generally increased with increasing rates of N. Residual effects of yearly N applications had no effect on smooth bromegrass CP.

Newell and Moline (1978) evaluated the CP trends of intermediate wheatgrass throughout the growing season. The CP of intermediate wheatgrass was 297 g kg^{-1} in the very early vegetative growth and continued through the summer with averages well above 100 g kg^{-1}. The extended day-length and high temperatures of the summer were responsible for the low summer CP. Intermediate wheatgrass CP increased with shorter days and cooler night temperatures to 170 g kg^{-1} in mid-August and reached 220 g kg^{-1} in early October from samples taken above 20 cm.

The CP declined in two cultivars each of orchardgrass, smooth bromegrass, reed canarygrass, and tall fescue as maturity advanced (Buxton and Marten 1989). The CP was consistently greatest in reed canarygrass and least in tall fescue. Total herbage CP of the two smooth bromegrass cultivars harvested between 10 May and 5 July ranged from 77 to 314 g kg^{-1}, respectively. The CP in all four species was greatest in the top leaves, intermediate in the inflorescences, and least in the bottom leaves (Buxton and Marten 1989). They concluded that CP was closely related to day of the year, GDD, and morphologic stage. Day of the year, GDD, and growth stage accounted for at least 88%, 77%, and 74%, respectively, of the variation associated with CP during the two-year study.

Switchgrass and big bluestem leaf CP decreased with plant maturation an average of 7 and 11 g kg^{-1} between harvests conducted at 14-day intervals (Perry and Baltensperger 1979). Switchgrass leaf CP was higher than big bluestem on common days of the year, except on the first harvest date when big bluestem leaf CP was highest. Big bluestem leaf CP declined more than switchgrass throughout all harvests. They concluded the decline in CP of forage topgrowth was apparently associated with both leaf maturation and increased stem growth.

Switchgrass and big bluestem leaf CP decreased with plant maturation on average of 15 g kg^{-1} between weekly harvests (Griffin and Jung 1983). Switchgrass averaged 17 g kg^{-1} lower in CP than big bluestem on common days of the year, but big bluestem stem CP declined more rapidly than switchgrass with increased maturity. At early head emergence, switchgrass leaf and stem CP averaged 85 and 38 g kg^{-1}, respectively, whereas big bluestem leaf and stem CP averaged 108 and 48 g kg^{-1}, respectively. The CP of switchgrass leaves declined as maturity progressed and the decline was most rapid between a leaf's emergence in the whorl until collaring of the following leaf (Anderson 1985). The decline in CP of the stems was more rapid than in most leaves. Switchgrass and big bluestem CP were best predicted by GDD which accounted for 91% and 90% of the variation in CP, respectively (Mitchell et al. 2001). Although no universal parameter adequately predicted concentrations of IVDDM, CP, and NDF, it was possible to accurately predict quality with readily available environmental data and measures of plant maturity (Mitchell et al. 2001).

Rumen Undegradable Protein

CP concentration alone may not be adequate to identify dietary protein for nutritional purposes (Mangan 1982). Dietary protein consumed by ruminant animals is degraded by microbial fermentation in the rumen or "escapes" to the small intestine. Protein protected from ruminal degradation allows more amino acids to reach the small intestine, increasing animal performance (Chalupa 1975). The rumen degradability of forage protein is highly variable among forage species (Petit and Tremblay 1992) and varies with maturity (Mullahey et al. 1992).

Rumen degradable protein (RDP) is highly variable between species harvested at similar stages of developmental morphology. Warm-season grasses tend to degrade more slowly in the rumen than cool-season grasses (Akin 1989). Anatomic differences between C_3 and C_4 grasses may explain some of the variability in ruminal protein degradation (Mullahey et al. 1992). Whole-plant rumen undegradable protein (RUP) was greater in switchgrass than smooth bromegrass, except at the last harvest when RUP was similar in both species (Mullahey et al. 1992). The RUP for switchgrass ranged from 52 to 18 g kg^{-1} DM and declined with maturity. The RUP for smooth bromegrass ranged from 28 to 18 g kg^{-1} DM and was lowest for the most immature growth stage. They attributed the differences in ruminal protein degradation between switchgrass (C_4) and smooth bromegrass (C_3) to anatomic differences. RUP for switchgrass leaves was greater than stems at each harvest date, and both leaf and stem **escape protein** decreased linearly with advancing maturity (Mullahey et al. 1992). **Escape protein** of smooth bromegrass was consistently greater in leaves than stems. Greater RUP occurred at later harvests for smooth bromegrass leaves but occurred early in the growing season for stems. Changes in the leaf-stem ratio had a significant impact on whole-plant RUP (Mullahey et al. 1992).

Hoffman et al. (1993) evaluated the influence of maturity on ruminal DM and CP degradation of three legume species and five cool-season grass species. They reported that legumes exhibited more extensive ruminal DM degradation than did grasses, and mature grasses were lowest in RDP. Smooth bromegrass ruminal DM degradation was 620 g kg^{-1} at emergence of the second node, 555 g kg^{-1} at the **boot stage**, and 410 g kg^{-1} at full heading (Hoffman et al. 1993). Smooth bromegrass ruminal CP degradation was 760 g kg^{-1} at emergence of the second node, 720 g kg^{-1} at the boot stage, and 644 g kg^{-1} at full **heading** (Hoffman et al. 1993). They concluded the relative relationship and range among forage species and maturities should be of primary interest.

Mitchell et al. (1997) quantified the relationships between the morphologic development and RDP, RUP, and microbial protein of intermediate wheatgrass, smooth bromegrass, switchgrass, and big bluestem. The mean stage of cool-season grasses was higher than that of warm-season grasses throughout the growing season. The RDP decreased as plant maturity increased for all species. The RUP expressed as a percentage of CP for the cool-season grasses was lower than that for warm-season grasses. The RUP for intermediate wheatgrass, smooth bromegrass, and switchgrass remained constant across maturities, but RUP for big bluestem decreased as maturity increased. Microbial augmentation of RUP decreased as CP decreased in all species. The RUP corrected for acid detergent insoluble N and microbial protein was relatively constant across plant maturities. Quantifying RUP across a range of plant maturities provides a starting point for incorporating RUP of forage grasses into animal diets.

Canopy Architecture and Tiller Demographics

Canopy architecture influences many plant canopy processes and must be considered when describing the interaction between plants and the environment (Welles and Norman 1991; Redfearn et al. 1997). Canopy architecture affects forage plant physiology, quality of forage offered to grazing animals, and animal grazing patterns (Nelson and Moser 1994). Canopy architectural measurements such as leaf area index (LAI) and mean leaf inclination angle (Welles and Norman 1991) can be related to relative light interception, forage productivity, forage availability, and forage accessibility to grazing livestock (Redfearn et al. 1997).

The phytomer is the basic modular unit of growth in grass plants and consists of a leaf blade, leaf sheath, node, internode, and axillary bud (Hyder 1972; Briske 1991). A series of phytomers forms the grass tiller, which consists of a single growing point, a stem, leaves, roots, nodes, dormant buds, and if reproductive, a potential inflorescence (Hyder 1972; Vallentine 1990). Grass tillers are further organized into anatomically attached groups which form the grass plant (Vallentine 1990; Walton 1983). Grass plants collectively form a sward.

A grass leaf is composed of a sheath and blade. New leaves are generated by cell division and pushed upward by expansion at the basal meristem which results in the linear aspect of the entire leaf (Mauseth 1988). Leaf blades emerge through the whorl and extend to the top of the canopy in vegetative grass canopies (Allard et al. 1991). The oldest leaves of a grass tiller have the lowest level of insertion from the plant base, while new leaves have a higher insertion level on the plant (Wilson 1976; Walton 1983). Leaf length in grass species is controlled by the transport limitations of the vascular bundles (Mauseth 1988). In green panic, leaf length and area increased progressively up to leaf 10, then decreased to the **flag leaf** (Wilson 1976). Leaves of high-insertion levels developed more slowly, stayed green longer, and senesced more slowly than those of a low-insertion level (Wilson 1976). When corn leaves reach a predetermined length, the basal meristem disorganizes, and leaf growth stops (Mauseth 1988).

Grasses are efficient forage producers because of the location of the meristematic tissue, growth habits of the plant, and the ability of the plant to tiller (Rechenthin 1956). The number of live tillers within a plant or per unit area is determined by the seasonality of tiller recruitment in relation to tiller longevity (Briske 1991). Tiller density is controlled by the recruitment rate of new tillers, the mortality of existing tillers, and the interaction of recruitment and mortality (Langer et al. 1964; Briske 1991). In a smooth bromegrass sward, tiller density was highest in early spring and decreased as spring growth progressed (Krause and Moser 1980). The reduction in tiller density resulted from the lack of light penetration through the canopy to the depth of the small tillers which caused many of the small tillers to cease functioning and the number of functional tillers to decline (Krause and Moser 1980). However, tiller recruitment in perennial cool-season grasses like smooth bromegrass typically involves at least two tiller generations annually, with tillering episodes occurring in the early spring and a more active tillering episode immediately following **anthesis** (Lamp 1952; Krause and Moser 1980).

Numerical indices are useful for describing the demography of forage populations (Mitchell et al. 1998). This is important because there is often significant variation in morphology among plants comprising a population of a given species. Many important forage species are cross-pollinated and are propagated as synthetic cultivars that represent an assemblage of related genotypes. Hence, there is more variation in developmental morphology within a population of perennial forages than would be observed with most annual grain crops (Moore and Moser 1995).

Most staging systems applied to perennial forage crops are not applied at the whole plant or population level. Rather, they are applied to modular subunits, which are usually tillers in grasses and stems in legumes. This approach arises from the difficulty in distinguishing among plants in dense swards and the fact that, in many species, significant variation in maturity exists among subunits arising from a single plant. Thus, a forage plant can be considered a metapopulation of tillers to which demographic principles can be applied (Harper 1980; White 1979).

A notable exception to the above approach would be in studies of seedling development where the whole plant is the subject of interest. For example, Moser et al. (1993) developed a system for describing the development of grass seedlings that includes morphologic descriptors for the whole plant, including roots.

The developmental morphology of a population of established forage plants can be characterized using numerical indices and descriptive statistics. A random sample of plants (or tillers) is selected and the growth stage of each tiller in the sample is determined. The mean developmental stage can be calculated using the following equation:

$$MSC = \sum_{i=1} \frac{S_i \times N_i}{C}$$

Where MSC = mean stage count, S_i = growth stage index, N_i = number of plants in stage S_i, and C = total number of plants in the sample population (Moore et al. 1991). A weighted mean stage, referred to as MSW, can be calculated using this formula by replacing N with the dry weight of the plants in each stage and C with the total dry weight of the sample (Kalu and Fick 1981). The MSW gives more influence to later growth stages since plants accumulate more dry weight as they mature. Therefore, MSW accounts for the contribution of each growth stage to the total biomass of the population. In some studies, MSW is more useful than MSC for quantifying the relationship between maturity and forage quality (Ohlsson and Wedin 1989).

The standard deviation of the MSC (S_{MSC}) is useful for interpreting the variability in maturity existing within a population of one or many forage species (Moore et al. 1991). Higher values of S_{MSC} indicate greater variation in maturity within the population. Small values of S_{MSC} indicate that most plants in the population are of similar maturity and have a value near the MSC. The S_{MSC} can be calculated from the formula

$$S_{MSC} = \sqrt{\sum_{i=1} \frac{(S_i - MSC)^2 \times N_i}{C}}$$

using parameters from the equation for MSC. Calculating a similar statistic for MSW is not as easy because it is the product of two variables (stage and weight), which are not independent (Moore et al. 1991).

The MSC and S_{MSC} were used to describe tiller population maturity for intermediate wheatgrass and big bluestem in mid-June near Mead, NE, and staged using the Nebraska system (Table 7.5). The four vegetative stages, V1, V2, V3, and V4, for big bluestem coded numerically as 1.15, 1.40, 1.65, and 1.9 (Figure 7.5). The MSC was 1.51, indicating the average tiller in this population had between two and three fully collared leaves. Intermediate wheatgrass, a cool-season grass, had a higher MSC, indicating it was more mature on the sampling date. The higher S_{MSC} indicated it also had a wider range of stages present than did big bluestem, a warm-season grass.

Systems for staging developmental morphology can be used to quantify and describe the seasonal demography of forage populations. A demographic analysis of a population of intermediate wheatgrass tillers (Figure 7.6) shows the change in number of tillers in each primary growth stage with respect to time. At the first four sampling dates,

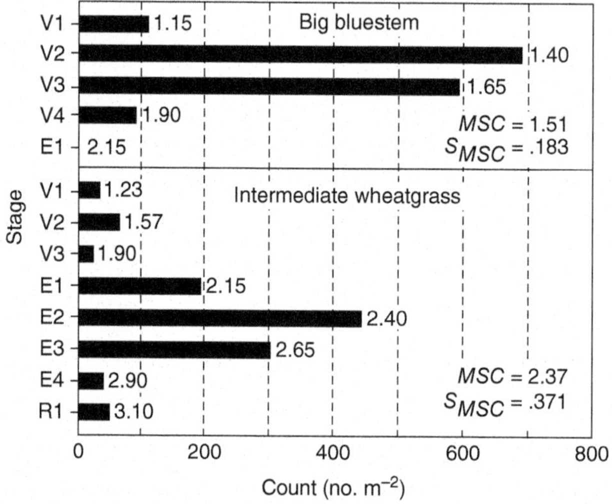

FIG. 7.5. Frequency distribution of tiller growth stages for big bluestem and intermediate wheatgrass populations sampled in mid-June near Mead, NE. *Source:* From Moore and Moser (1995).

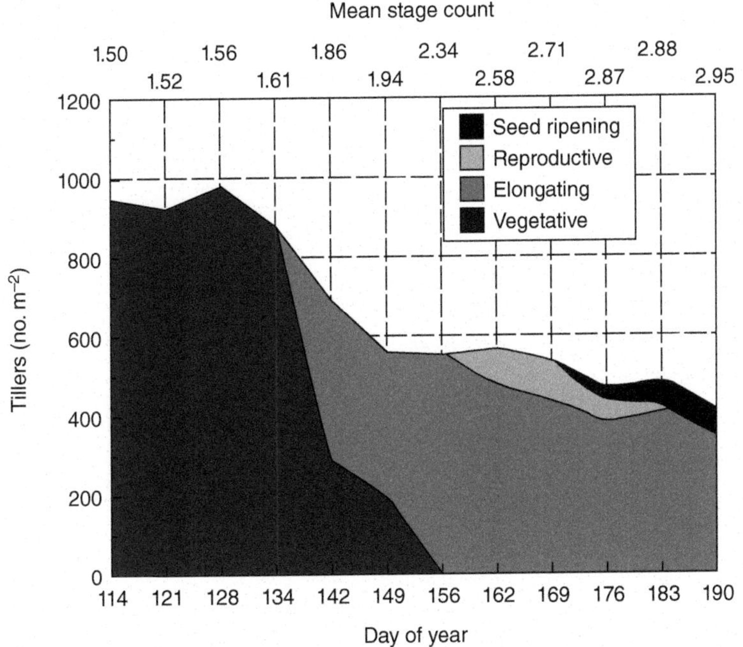

Fig. 7.6. Developmental morphology and demography of an intermediate wheatgrass tiller population during the 1991 growing season near Mead, NE. *Source:* From Moore and Moser (1995).

all tillers were vegetative. In a period of one week, however, over half the tillers began to elongate and in another three to four weeks, some tillers were advancing into reproductive stages. Coincident with the onset of elongation was an increase in tiller mortality that resulted in an almost 40% decrease in tiller density by day 149.

Interestingly, only a relatively small proportion of tillers advanced through the reproductive to seed-ripening stages (Figure 7.6). This population would have been described as fully headed based on visual observation during the reproductive and seed-ripening phases when, in reality, less than 20% of the culms produced inflorescences. It is evident from this example that MSC should not be interpreted as the actual growth stage of the population but rather as the mean representing all the growth stages present in a population.

Quantifying tiller population morphology on a unit area basis allows changes in tiller demography to be monitored over time. Tiller density and demographics is highly variable across species, but tiller density in perennial grasses typically declines as MSC advances and the growing season progresses (Moore and Moser 1995; Mitchell et al. 1998). Intermediate wheatgrass tiller density generally declined as MSC increased, but smooth bromegrass tiller density followed no clear patterns with increased MSC. Tiller demographics was highly variable by year for intermediate wheatgrass and smooth bromegrass which indicates grazing management should be based on current tiller populations. Tiller populations with a large proportion of vegetative tillers provide grazing livestock the opportunity to select less mature and higher quality tillers. Vegetative tillers declined most rapidly for smooth bromegrass, followed by intermediate wheatgrass, switchgrass, and big bluestem. Switchgrass and big bluestem tiller density generally declined as MSC increased and demographics were more uniform and predictable across years. Big bluestem tiller mortality averaged as many as 47 tillers m^{-2} d^{-1} for the first four weeks. LAI of intermediate wheatgrass, smooth bromegrass, switchgrass, and big bluestem tiller populations increased as morphology advanced (Mitchell et al. 1998). The LAI for all species increased as MSC increased. Maximum LAI for intermediate wheatgrass, smooth bromegrass, switchgrass, and big bluestem was 4.7, 5.1, 4.9, and 5.8, respectively. Integrating tiller demographics and LAI indicates initial grazing order for a four-species complementary grazing system should be smooth bromegrass in early spring followed by intermediate wheatgrass in about two-weeks, switchgrass in late spring, and big bluestem in early summer.

References

Akin, D.E. (1989). Histological and physical factors affecting digestibility of forages. *Agron. J.* 81: 17–25.

Allard, G. and Nelson, C.J. (1991). Photosynthate partitioning in basal zones of tall fescue leaf blades. *Plant Physiol.* 95: 663–668.

Allard, G., Nelson, C.J., and Pallardy, S.G. (1991). Shade effects on growth of tall fescue: I. Leaf anatomy and dry matter partitioning. *Crop Sci.* 31: 163–167.

Anderson, B. (1985). The influence of aging on forage quality of individual switchgrass leaves and stems. In: *Proceedings of 15th International Grassland Congress, Kyoto, Japan. 24–31 August, 1985*, 947–949. Nishi-nasuno, Tochigi-ken, Japan: Japanese Society of Grassland Science.

Anderson, B. and Matches, A.G. (1983). Forage yield, quality, and persistence of switchgrass and caucasian bluestem. *Agron. J.* 75: 119–124.

Balasko, J.A., Burner, D.M., and Thayne, W.V. (1984). Yield and quality of switchgrass grown without soil amendments. *Agron. J.* 76: 204–208.

Barnes, D.K. and Sheaffer, C.C. (1995). Alfalfa. In: *Forages: An Introduction to Grassland Agriculture* (eds. R.F Barnes, D.A. Miller and C.J. Nelson), 205–216. Ames, Iowa: Iowa State University Press.

Benedict, H.M. (1941). Effect of day length and temperature on the flowering and growth of four species of grasses. *J. Agric. Res.* 61: 661–672.

Bregard, A. and Allard, G. (1999). Sink to source transition in developing leaf blades of tall fescue. *New Phytol.* 141: 45–50.

Briske, D.D. (1991). Developmental morphology and physiology of grasses. In: *Grazing Management: An Ecological Perspective* (eds. R.K. Heitschmidt and J.W. Stuth), 85–108. Portland, OR: Timber Press.

Brock, J.L., Albrecht, K.A., Tilbecht, J.C., and Hay, M.J.M. (2000). Morphology of white clover during development from seed to clonal populations in grazed pastures. *J. Agric. Sci.* 135: 103–111.

Brown, P.W. and Tanner, C.B. (1983). Alfalfa stem and leaf growth during water stress. *Agron. J.* 75: 799–805.

Brueland, B.A., Harmoney, K.R., Moore, K.J. et al. (2003). Developmental morphology of smooth bromegrass growth following spring grazing. *Crop Sci.* 43: 1789–1796.

Buxton, D.R. and Casler, M.D. (1993). Environmental and genetic effects on cell wall composition and digestibility. In: *Forage Cell Wall Structure and Digestibility* (eds. H.G. Jung, D.R. Buxton, R.D. Hatfield and J. Ralph), 685–714. Madison, WI: ASA/CSSA/SSSA.

Buxton, D.R. and Fales, S.L. (1994). Plant environment and quality. In: *Forage Quality, Evaluation, and Utilization* (eds. G.C. Fahey Jr., M. Collins, D.R. Mertens and L.E. Moser), 155–199. Madison, WI: ASA/CSSA/SSSA.

Buxton, D.R. and Marten, G.C. (1989). Forage quality of plant parts of perennial grasses and relationship to phenology. *Crop Sci.* 29: 429–435.

Carman, J.G. and Briske, D.D. (1982). Root initiation and root and leaf elongation of dependent little bluestem tillers following defoliation. *Agron. J.* 74: 432–435.

Chalupa, W. (1975). Rumen bypass and protection of proteins and amino acids. *J. Dairy Sci.* 58: 1198–1218.

Cosgrove, D. (1986). Biophysical control of plant cell growth. *Annu. Rev. Plant Physiol.* 37: 377–405.

Dahl, B.E. and Hyder, D.N. (1977). Developmental morphology and management implications. In: *Rangeland Plant Physiology* (ed. R.E. Sosebee), 257–290. Denver, CO: Society for Range Management.

Dale, J.E. (1988). The control of leaf expansion. *Annu. Rev. Plant Physiol.* 39: 267–295.

Davies, A. (1974). Leaf tissue remaining after cutting and regrowth in perennial ryegrass. *J. Agric. Sci.* 82: 165–172.

Davies, A. and Thomas, H. (1983). Rates of leaf and tiller production in young spaced perennial ryegrass plants in relation to soil temperature and solar radiation. *Ann. Bot.* 57: 591–597.

De Visser, R., Vianden, H., and Schnyder, H. (1997). Kinetics and relative significance of remobilized and current C and N incorporation in leaf and root growth zone of *Lolium perenne* after defoliation: assessment by ^{13}C and ^{15}N steady-state labeling. Plant. *Cell Environ.* 20: 37–46.

Erwin, J., Velguth, P., and Heins, R. (1994). Day/night temperature environment affects cell elongation but not division in *Lilium longiflorum* Thunb. *J. Exp. Bot.* 45: 1019–1025.

Esau, K. (1960). *Anatomy of Seed Plants*. New York: Wiley.

Fehr, W.R., and Caviness, C.E. (1977). Stages of soybean development. Cooperative Extension Service. Special Report. Iowa State University.

Fick, G.W., and Mueller, S.C. (1989). Alfalfa quality, maturity, and mean stage of development. Cornell University Cooperative Extension. Information Bulletin No. 217.

Frank, A.B., Sedivic, K.K., and Hofmann, L. (1993). Determining grazing readiness for native and tame pastures. North Dakota State University External Bulletin No. R-106.

Frank, A.B., Cardwell, V.B., Ciha, A.J., and Wilhelm, W.W. (1997). Growth staging in research and crop management (letter to the editor). *Crop Sci.* 37: 1039–1040.

Gagnon, M.-J. and Beebe, D.U. (1996). Establishment of a plastochron index and analysis of the sink-to-source transition in leaves of *Moricandia arvensis* (L.) DC (Brassicaceae). *Int. J. Plant Sci.* 157: 262–268.

Gastal, F. and Nelson, C.J. (1994). Nitrogen use within the growing leaf blade of tall fescue. *Plant Physiol.* 105: 191–197.

Gautier, H., Varlet-Grancher, C., and Membre, J.M. (2001). Plasticity of petioles of white clover (*Trifolium repens*) to blue light. *Physiol. Plant.* 112: 293–300.

George, J.R. and Hall, K.E. (1983). Herbage quality of three warm-season grasses with nitrogen fertilization. *Iowa State J. Res.* 58: 247–259.

Gillen, R.L. and Ewing, A.L. (1992). Leaf development of native bluestem grasses in relation to degree-day accumulation. *J. Range Manage.* 45: 200–204.

Granier, C. and Tardieu, F. (1999). Water deficit and spatial pattern of leaf development. Variability in responses can be simulated using a simple model of leaf development. *Plant Physiol.* 119: 609–619.

Griffin, J.L. and Jung, G.A. (1983). Leaf and stem quality of big bluestem and switchgrass. *Agron. J.* 75: 723–726.

Griffin, J.L., Wangsness, P.J., and Jung, G.A. (1980). Forage quality evaluation of two warm-season range grasses using laboratory and animal measurements. *Agron. J.* 72: 951–956.

Gustine, D.L. and Sanderson, M.A. (2001). Quantifying spatial and temporal genotypic changes in white clover populations by RAPD technology. *Crop Sci.* 41: 143–148.

Hanna, W.W., Monson, W.G., and Burton, G.W. (1973). Histological examination of fresh forage leaves after in vitro digestion. *Crop Sci.* 13: 98–102.

Harper, J.L. (1980). Plant demography and ecological theory. *Oikos* 35: 244–253.

Haun, J.R. (1973). Visual quantification of wheat development. *Agron. J.* 65: 116–119.

Heide, O.M. (1984). Flowering requirements in *Bromus inermis*, a short-long-day plant. *Physiol. Plant.* 62: 59–64.

Heide, O.M. (1994). Control of flowering and reproduction in temperate grasses. *New Phytol.* 128: 347–362.

Hendrickson, J.R. 1992. Developmental morphology of two Nebraska Sandhills grasses and its relationship to forage quality. Master thesis. University of Nebraska.

Hendrickson, J.R., Moser, L.E., Moore, K.J., and Waller, S.S. (1997). Leaf nutritive value related to tiller development in warm-season grasses. *J. Range Manage.* 50: 116–122.

Hoffman, P.C., Sievert, S.J., Shaver, R.D. et al. (1993). In situ dry matter, protein, and fiber degradation of perennial forages. *J. Dairy Sci.* 76: 2632–2643.

Hopkins, A.A., Vogel, K.P., Moore, K.J. et al. (1995). Genotype effects and genotype by environment interactions for traits of elite switchgrass populations. *Crop Sci.* 35: 125–132.

Hyder, D.N. (1972). Defoliation in relation to vegetative growth. In: *The Biology and Utilization of Grasses* (eds. V.B. Youngner and C.M. McKell), 302–317. New York: Academic Press.

Jung, H.G. and Vogel, K.P. (1986). Influence of lignin on digestibility of forage cell wall material. *J. Anim. Sci.* 62: 1703–1712.

Kalu, B.A. and Fick, G.W. (1981). Quantifying morphological stage of maturity as a predictor of alfalfa herbage quality. *Crop Sci.* 23: 267–271.

Kamstra, L.D. (1973). Seasonal changes in quality of some important range grasses. *J. Range Manage.* 26: 289–291.

Kilcher, M.R. (1981). Plant development, stage of maturity and nutrient composition. *J. Range Manage.* 34: 363–364.

Kilcher, M.R. and Troelsen, J.E. (1973). Contribution of stems and leaves to the composition and nutrient content of irrigated bromegrass at different stages of development. *Can. J. Plant. Sci.* 53: 767–771.

Klepper, B., Belford, R.K., and Rickman, R.W. (1984). Root and shoot development in winter wheat. *Agron. J.* 76: 117–122.

Krause, J.W. and Moser, L.E. (1980). Tillering in irrigated smooth bromegrass (*Bromus inermis* Leyss.) as affected by elongated tiller removal. In: *Proceedings of 13th International Grassland Congress, Leipzig, German Democratic Republic. 18–27 May 1977* (eds. E. Wojahn and H. Thons), 189–191. Berlin: Akademie-Verlag.

Lamp, H.F. (1952). Reproductive activity in *Bromus inermis* in relation to phases of tiller development. *Bot. Gaz.* 113: 413–438.

Langer, R.H.M., Ryle, S.M., and Jewiss, O.R. (1964). The changing plant and tiller populations of timothy and meadow fescue swards: I. Plant survival and the pattern of tillering. *J. Appl. Ecol.* 1: 197–208.

Lecoeur, J., Wery, J., Turc, O., and Tardieu, F. (1995). Expansion of pea leaves subjected to short water deficit: cell number and cell size are sensitive to stress at different periods of leaf development. *J. Exp. Bot.* 46: 1093–1101.

Li, F. (2000). The effects of frequency of cutting and cultivar on the period of leaf expansion in white clover grown in mixed swards. *Grass Forage Sci.* 55: 280–284.

Lockhart, J.A. (1965). An analysis of irreversible plant cell elongation. *J. Theor. Biol.* 8: 264–275.

MacAdam, J.W. and Nelson, C.J. (1987). Specific leaf weight in zones of cell division, elongation and maturation in tall fescue leaf blades. *Ann. Bot.* 59: 369–376.

MacAdam, J.W., Volenec, J.J., and Nelson, C.J. (1989). Effects of nitrogen on mesophyll cell division and epidermal cell elongation in tall fescue leaf blades. *Plant Physiol.* 89: 549–556.

Mangan, J.L. (1982). The nitrogenous constituents of fresh forage. In: *Forage Protein in Ruminant Animal*

Production. Occ. Publ. no. 6 (eds. D.J. Thomson, D.E. Beever and R.G. Gunn), 25–40. Edinburgh, UK: BSAP.

Marten, G.C. and Barnes, R.F (1980). Prediction of energy digestibility of forages with in vitro rumen fermentation and fungal enzymes systems. In: *Proceedings of International Workshop on Standardization of Analytical Methodology of Feeds, Ottawa, Canada* (ed. W.J. Pigden), 61–71. New York: International Journal of Development Research (IRDC) Rep. 134e. Unipub.

Matthew, C., Yang, J.Z., and Potter, J.F. (1998). Determination of tiller and root appearance in perennial ryegrass (*Lolium perenne*) swards by observation of the tiller axis, and potential application in mechanistic modeling. *N. Z. J. Agric. Res.* 41: 1–10.

Mauseth, J.D. (1988). *Plant Anatomy*. Menlo Park, CA: Benjamin/Cummings Publishing Co., Inc.

Meiri, A., Silk, W.K., and Lauchli, A. (1992). Growth and deposition in inorganic nutrient elements in developing leaves of *Zea mays* L. *Plant Physiol.* 99: 972–978.

Metcalfe, D.S. and Nelson, C.J. (1985). The botany of grasses and legumes. In: *Forages: The Science of Grassland Agriculture* (eds. M.E. Heath, D.S. Metcalfe and R.F Barnes), 52–63. Ames, IA: Iowa State University Press.

Mitchell, R.B. and Moser, L.E. (2000). Developmental morphology and tiller dynamics of warm-season grass swards. In: *Native Warm-Season Grasses: Research Trends and Issues*. CSSA Spec. Pub. 30. (eds. K.J. Moore and B.E. Anderson), 49–66. Madison, WI: Crop Science Society of America.

Mitchell, R.B., Masters, R.A., Waller, S.S. et al. (1994). Big bluestem production and forage quality response to burning date and fertilizer in tallgrass prairie. *J. Prod. Agric.* 7: 355–359.

Mitchell, R.B., Moore, K.J., Moser, L.E. et al. (1997). Predicting developmental morphology in switchgrass and big bluestem. *Agron. J.* 89: 827–832.

Mitchell, R.B., Moser, L.E., Moore, K.J., and Redfearn, D.D. (1998). Tiller demographics and leaf area index of four perennial pasture grasses. *Agron. J.* 90: 47–53.

Mitchell, R., Fritz, J., Moore, K. et al. (2001). Predicting forage quality in switchgrass and big bluestem. *Agron. J.* 93: 118–124.

Monson, W.G., Powell, J.G., and Burton, G.W. (1972). Digestion of fresh forage in rumen fluid. *Agron. J.* 64: 231–233.

Moore, K.J. and Moser, L.E. (1995). Quantifying developmental morphology of perennial grasses. *Crop Sci.* 35: 37–43.

Moore, K.J., Moser, L.E., Vogel, K.P. et al. (1991). Describing and quantifying growth stages of perennial forage grasses. *Agron. J.* 83: 1073–1077.

Moore, K.J., Vogel, K.P., Hopkins, A.A. et al. (1993). Improving the digestibility of warm-season perennial grasses. In: *Proceedings of 17th International Grassland Congress, Palmerston North, New Zealand. 8–21 February 1993* (eds. M.J. Baker et al.), 447–448. The New Zealand Grassland Association.

Morvan-Bertrand, A., Boucaud, J., Le Saos, J., and Prud'homme, M.-P. (2001). Roles of the fructans from leaf sheaths and from the elongating leaf bases in the regrowth following defoliation of *Lolium perenne* L. *Planta* 213: 109–120.

Moser, L.E. (2000). Morphology of germinating and emerging warm-season grass seedlings. In: *Native Warm-Season Grasses: Research Trends and Issues*. CSSA Spec. Pub. 28. (eds. K.J. Moore and B.E. Anderson), 35–47. Madison, WI: Crop Sci. Soc. Am.

Moser, L.E., Moore, K.J., Miller, M.S. et al. (1993). A quantitative system for describing the developmental morphology of grass seedling populations. In: *Proceedings of 16th International Grassland Congress, New Zealand*, 317–318. The New Zealand Grassland Association.

Mullahey, J.J., Waller, S.S., Moore, K.J. et al. (1992). In situ ruminal protein degradation of switchgrass and smooth bromegrass. *Agron. J.* 84: 183–188.

Nelson, C.J. and Moser, L.E. (1994). Plant factors affecting forage quality. In: *Forage Quality, Evaluation, and Utilization* (eds. G.C. Fahey Jr., M. Collins, D.R. Mertens and L.E. Moser), 115–154. Madison, WI: ASA/CSSA/SSSA.

Neuteboom, J.H. and Lantinga, E.A. (1989). Tillering potential and relationship between leaf and tiller production in perennial ryegrass. *Ann. Bot.* 63: 265–270.

Newell, L.C., and Moline, W.J. (1978). Forage quality evaluations of twelve grasses in relation to season for grazing. University of Nebraska. Agricultural Experiment Station Research Bulletin No. 283.

Ohlsson, C. and Wedin, W.F. (1989). Phenological staging schemes for predicting red clover quality. *Crop Sci.* 29: 416–420.

Palmer, S.J. and Davies, W.J. (1996). An analysis of relative elemental growth rate, epidermal cell size and xyloglucan endotransglycosylase activity through the growing zone of ageing maize leaves. *J. Exp. Bot.* 47: 339–347.

Palmer, S.J., Berridge, D.M., McDonald, A.J.S., and Davies, W.J. (1996). Control of leaf expansion in sunflower (*Helianthus annuus* L.) by nitrogen nutrition. *J. Exp. Bot.* 47: 359–368.

Parsons, A.J. (1988). The effects of season and management on the growth of grass swards. In: *The Grass Crop: The Physiological Basis of Production* (eds. M.B. Jones and A. Lazenby), 129–177. New York: Chapman and Hall.

Perry, L.J. Jr. and Baltensperger, D.D. (1979). Leaf and stem yields and forage quality of three N-fertilized warm-season grasses. *Agron. J.* 71: 355–358.

Petit, H.V. and Tremblay, G.F. (1992). In situ degradability of fresh grass and grass conserved under different harvesting methods. *J. Dairy Sci.* 75: 774–781.

Phillips, T.G., Sullivan, J.T., Loughlin, M.E., and Sprague, V.G. (1954). Chemical composition of some forage grasses: I. changes with plant maturity. *Agron. J.* 46: 361–369.

Radin, J.W. (1983). Control of plant growth by nitrogen: differences between cereals and broadleaf species. *Plant Cell Environ.* 6: 65–68.

Rechenthin, C.A. (1956). Elementary morphology of grass growth and how it affects utilization. *J. Range Manage.* 9: 167–170.

Redfearn, D.D., Moore, K.J., Vogel, K.P. et al. (1997). Canopy architecture and morphology of switchgrass populations differing in forage yield. *Agron. J.* 89: 262–269.

Rehm, G.W., Moline, W.J., Schwartz, E.J. et al. (1971). Effect of fertilization and management on the production of bromegrass in northeast Nebraska. University of Nebraska. Agricultural Experiment Station Research Bulletin No. 247.

Rickman, R.W., Klepper, B., and Belford, R.K. (1985). Developmental relationships among roots, leaves and tillers in winter wheat. In: *Wheat Growth Modeling* (eds. W. Day and R.K. Atkin), 83–98. New York: Plenum Press.

Salisbury, F.B. and Ross, C.W. (1985). *Plant Physiology*. Belmont, CA: Wadsworth Publ. Co.

Sanderson, M.A. (1992). Morphological development of switchgrass and kleingrass. *Agron. J.* 84: 415–419.

Sanderson, M.A. and Moore, K.J. (1999). Switchgrass morphological development predicted from day of the year or degree day models. *Agron. J.* 91: 732–734.

Sanderson, M.A. and Nelson, C.J. (1995). Growth of tall fescue leaf blades in various irradiances. *Eur. J. Agron.* 4: 197–203.

Sanderson, M.A. and Wedin, W.F. (1989). Phenological stage and herbage quality relationships in temperate grasses and legumes. *Agron. J.* 81: 864–869.

Sanderson, M.A., West, C.P., Moore, K.J. et al. (1997). Comparison of morphological development indexes for switchgrass and bermudagrass. *Crop Sci.* 37: 871–878.

Schaufele, R. and Schnyder, H. (2000). Cell growth analysis during steady and non-steady growth in leaves of perennial ryegrass (*Lolium perenne* L.) subject to defoliation. *Plant Cell Environ.* 23: 185–194.

Schnyder, H. and Nelson, C.J. (1987). Growth rates and carbohydrate fluxes within the elongation zone of tall fescue leaf blades. *Plant Physiol.* 85: 548–553.

Schnyder, H. and Nelson, C.J. (1989). Growth rates and assimilate partitioning in the elongation zone of tall fescue leaf blades at high and low irradiance. *Plant Physiol.* 90: 1201–1206.

Schnyder, H., Seo, S., Rademacher, I.F., and Kuhbauch, W. (1990). Spatial distribution of growth rates and of epidermal cell lengths in the elongation zone during leaf development in *Lolium perenne* L. *Planta* 181: 423–431.

Simon, U. and Park, B.H. (1983). A descriptive scheme for stages of development in perennial forage grasses. In: *Proceedings of 14th International Grassland Congress, Lexington, KY. 15–24 June 1981* (eds. J.A. Smith and V.W. Hays), 416–418. Boulder, CO: Westview Press.

Skinner, R.H. and Nelson, C.J. (1992). Estimation of potential tiller production and site usage during tall fescue canopy development. *Ann. Bot.* 70: 493–499.

Skinner, R.H. and Nelson, C.J. (1994a). Role of leaf appearance rate and the coleoptile tiller in regulating tiller production. *Crop Sci.* 34: 71–75.

Skinner, R.H. and Nelson, C.J. (1994b). Epidermal cell division and the coordination of leaf and tiller development. *Ann. Bot.* 74: 9–15.

Skinner, R.H. and Nelson, C.J. (1995). Elongation of the grass leaf and its relationship to the phyllochron. *Crop Sci.* 35: 4–10.

Smit, B., Stachowiak, M., and van Volkenburgh, E. (1989). Cellular processes limiting leaf growth in plants under hypoxic root stress. *J. Exp. Bot.* 40: 89–94.

Tilley, J.M.A. and Terry, R.A. (1963). A two-stage technique for the in vitro digestion of forage crops. *J. Br. Grassland Soc.* 18: 104–111.

Turgeon, R. (1984). Termination of nutrient import and development of vein loading capacity in albino tobacco leaves. *Plant Physiol.* 76: 45–48.

Turgeon, R. (1989). The sink-source transition in leaves. *Annu. Rev. Plant Physiol. Plant Mol. Biol.* 40: 119–138.

Turgeon, R. and Webb, J.A. (1973). Leaf development and phloem transport in *Cucurbita pepo*: transition from import to export. *Planta* 113: 179–191.

Vallentine, J.F. (1990). *Grazing Management*. San Diego, CA: Academic Press, Inc.

Van Soest, P.J. (1982). *Nutritional Ecology of the Ruminant*. Corvallis, OR: O & B Books.

Van Soest, P.J. (1985). Composition, fiber quality, and nutritive value of forages. In: *Forages: The Science of Grassland Agriculture* (eds. M.E. Heath, D.S. Metcalfe and R.F Barnes), 412–421. Ames, IA: Iowa State University Press.

Vanderlip, R.L. (1972). How a sorghum plant develops. Cooperative Extension Service, Kansas State University. Manhattan Research Bulletin No. C-447.

Vogel, K.P., Haskins, F.A., and Gorz, H.J. (1981). Divergent selection for in vitro dry matter digestibility in switchgrass. *Crop Sci.* 21: 39–41.

Volenec, J.J. (1986). Nonstructural carbohydrates in stem base components of tall fescue during regrowth. *Crop Sci.* 26: 122–127.

Waller, S.S., Moser, L.E., and Reece, P.E. (1985). *Understanding Grass Growth: The Key to Profitable Livestock Production*. Kansas City, MO: Trabon Printing Co.

Walton, P.D. (1983). *Production and Management of Cultivated Forages*. Reston, VA: Reston Publishing Co.

Welles, J.M. and Norman, J.M. (1991). Instrument for indirect measurement of canopy architecture. *Agron. J.* 83: 818–825.

West, C.P. (1990). A proposed growth stage system for bermudagrass. In: *Proceedings of American Forage Grassland Conference Blacksburg, VA. 6–9 June 1990*, 38–42. Georgetown, TX: AFGC.

White, J. (1979). The plant as a metapopulation. *Annu. Rev. Ecol. Syst.* 10: 109–145.

Wilhelm, W.W. and McMaster, G.S. (1995). Importance of the phyllochron in studying development and growth in grasses. *Crop Sci.* 35: 1–3.

Williams, R.F. and Langer, R.H.M. (1975). Growth and development of the wheat tiller. II. The dynamics of tiller growth. *Aust. J. Bot.* 23: 745–759.

Wilson, J.R. (1976). Variation of leaf characteristics with level of insertion on a grass tiller: I. Development rate, chemical composition and dry matter digestibility. *Aust. J. Agric. Res.* 27: 343–354.

Wilson, J.R. (1983). Effects of water stress on herbage quality. In: *Proceedings of 14th International Grassland Congress, Lexington, KY. 15–24 June 1981* (eds. J.A. Smith and V.W. Hays), 470–472. Boulder, CO: Westview Press.

Zadoks, J.T., Chang, T.T., and Konzak, C.F. (1974). A decimal code for the growth stages of cereals. *Weed Res.* 14: 415–421.

PART II

FORAGE ECOLOGY

Tall grass prairie in the Konza Prairie of eastern Kansas. Tall grass species such as big bluestem, switchgrass, indiangrass and several other warm and cool season grass species originally covered large areas in the US. *Source*: Photo courtesy of Mike Collins.

Forage plants interact with their environment in ways that influence their productivity and survival. They are one component of many comprising the ecosystem of which they are a part. Climate has a large impact on how well a forage species will grow within a given region. The adaptation of forages is in large part related to the extremes of climatic conditions they can tolerate. Forage plants are often grown in mixtures and are subject to interspecific competition from neighboring plants. How such interactions play out within the plant community is dynamic and influenced by abiotic and biotic factors affecting their growth and development.

Herbivory by grazing livestock can have a large impact on plant performance. Some forage species are better adapted to being grazed than others and, thus, the dynamics of the plant community are often altered

Forages: The Science of Grassland Agriculture, Volume II, Seventh Edition.
Edited by Kenneth J. Moore, Michael Collins, C. Jerry Nelson and Daren D. Redfearn.

significantly under grazing. Besides providing nutrients for livestock production, forages confer many other positive benefits to the environments where they are grown.

Due to their perennial growth habit and often extensive root systems, forage plants are regularly used for soil conservation. They contribute to improved water quality by reducing runoff and erosion, and because they develop extensive root systems and are tilled infrequently, help to sequester carbon in the soil where they are grown. They also provide habitat diversity for wildlife that enhances the populations of desirable species and generally improve the aesthetics of the environment. All of these topics and more are covered in this part on Forage Ecology.

Climate, Climate-Change and Forage Adaptation

Vern S. Baron, Research Scientist, *Agriculture and Agri-Food Canada, Lacombe, Canada*
Gilles Bélanger, Research Scientist, *Agriculture and Agri-Food Canada, Sainte-Foy, Canada*

Climate is the long-term (decades) pattern of several weather events or parameters that may be described on a regional basis. **Weather** is the current (daily, hourly or instantaneous) components of climate, including temperature, all forms of precipitation, relative humidity, wind and solar radiation. **Climate change** refers to a change in the state of the climate that can be identified by changes in the mean and variability that persist for decades or longer (Hartmann et al. 2013). Climatologists summarize weather records statistically to determine long-term climatic averages (trends) as well as standard deviations (variability) of the trends. Climatologists and ecologists recognize that climate has a direct relationship with the ecologic regions of the world. Thus, climatic trends are more useful when classified or mapped on a regional basis.

Forage species and plant populations can adjust or adapt physiologically and morphologically over a certain range of climatic variability (e.g. temperature and rainfall) to produce and persist. The range of adaptability is genetically controlled so most species are confined to a region or adaptation zone where short- and long-term production are sustainable. The adaptation zones of many species overlap, providing species choices. In addition, management techniques are developed and used to mitigate climatic constraints or stresses to further extend areas where productive genotypes may be grown economically.

Earths Energy Balance Affects Climate

Solar radiation drives the climatic system. The earth's net energy balance (Figure 8.1) determines the global environment to which all living things must adapt (Le Treut et al. 2007). The atmosphere, held in place by gravity, reflects, filters, absorbs, stores, and radiates incoming short-wave solar energy and outgoing long-wave energy from the earth-ocean surface. Satellite measurements at the top of the atmosphere of net incoming short-wave and outgoing long-wave radiation verify the energy balance.

The terrestrial energy budget (summarized from Kiehl and Trenberth 1997; Mavi and Tupper 2004; Le Treut et al. 2007) is a balance between incoming solar short-wave radiation and outgoing long-wave radiation (Figure 8.1). About 1370 W m^{-2} of solar energy reaches the atmosphere. Averaged over the whole day for a year, 342 W m^{-2} of short-wave energy reaches the atmosphere. Of this, 77 and 30 W m^{-2} of solar energy are reflected back to space by clouds and the earth's surface (albedo), respectively. About half (168 W m^{-2}) is absorbed by the

Forages: The Science of Grassland Agriculture, Volume II, Seventh Edition.
Edited by Kenneth J. Moore, Michael Collins, C. Jerry Nelson and Daren D. Redfearn.

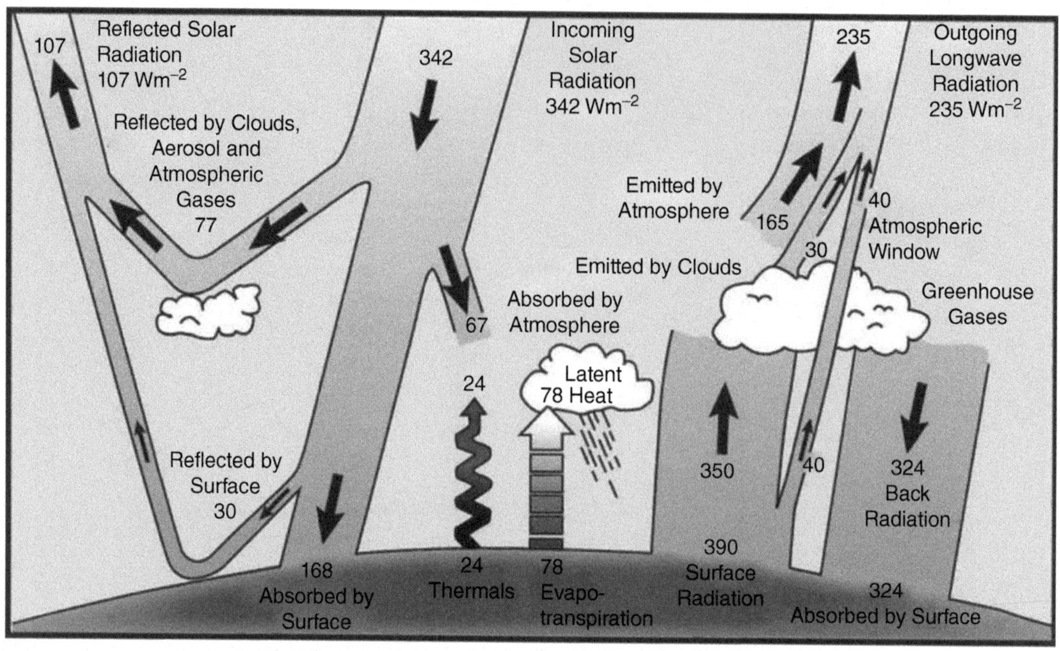

Fig. 8.1. The terrestrial energy budget. Many factors influence the balance of incoming and outgoing radiation including greenhouse gasses that are associated with climate change. See explanation in text. *Source:* Adapted from Kiehl and Trenberth (1997) and Le Treut et al. (2007).

earth's surface and 67 W m^{-2} by the atmosphere, leaving 235 W m^{-2} in the earth-atmosphere system.

Long-wave energy moving toward space balances the absorption of short-wave energy, but the pathway is not simple and direct. The earth, which acts as a black body, radiates 350 W m^{-2} to the atmosphere and 40 W m^{-2} directly to space as long wavelength radiation (total of 390 W m^{-2}). The atmosphere now contains 24 W m^{-2} gained from thermal transfer and 78 W m^{-2} gained from latent heat of vaporization from the earth's surface plus 77 W m^{-2} of short-wave energy gained on entry and the 350 W m^{-2} of long-wave energy radiated from the earth's surface (total of 529 W m^{-2}). However, atmospheric greenhouse gases (GHG) re-radiate 324 W m^{-2} long wave energy back to earth. The remaining 205 W m^{-2} (529–324 W m^{-2}) is transferred indirectly to space through the atmosphere and clouds. Adding the 40 W m^{-2} transferred directly to space from earth, balances the energy budget with 245 W m^{-2} (205 + 40 W m^{-2}) of outgoing energy. As the concentrations of water vapor and CO_2, the two most prominent GHGs (Le Treut et al. 2007) increase, more back-transfer of energy occurs, gradually warming the earth year after year. The current balance slightly favors storage in the earth-atmosphere by 1–3 W m^{-2} annually (Kiehl and Trenberth 1997; Mavi and Tupper 2004), which may be due to an increase in GHG or changes in heat storage associated with El Nino events (Kiehl and Trenberth 1997).

The greenhouse effect is the re-radiation of energy stored in the atmosphere back to the earth. If no energy was captured by the atmosphere the average global temperature would be approximately −19 °C. The additional long wave or back-radiated thermal energy due to normal CO_2 and water vapor increases the average temperature by 33 °C to a relatively stable 14 °C. Water vapor is the most important GHG; CO_2 is second. Concentration of CO_2 has increased more than 35% from 1890 until present (Le Treut et al. 2007) and is continuing to increase. **Methane** (CH_4), nitrous oxide (N_2O), ozone and other gases also contribute to the greenhouse effect.

The net energy balance is not uniform over the earth's surface due to daily, seasonal and annual differences in solar radiation, properties of surface reflection (e.g. water, vegetation cover, bare soil, and snow) and capacities of the surface to absorb and store energy. More solar energy is received and stored annually at the equator than the poles, but the daily and seasonal range and variability in temperature increases with **latitude** from the equator due to daylength and seasonal effects (Trewartha 1968; Oliver and Hidore 1984).

A phenomenon known as solar dimming and brightening may occur in decades-long cycles. Less incident solar radiation reached the earth's surface from the 1960s to the mid-1980s than from the mid-1980s until recently, perhaps due to reduction of atmospheric aerosols associated with industrialization. Tollenaar et al. (2017) estimated solar brightening between 1984 and 2013 was responsible for increasing solar irradiance by 8 W m^{-2} per decade which may have accounted for 27% of the maize crop yield improvement over that time. Over this same time period, solar dimming was reported over regions of India and China, possibly due to pollution.

Energy moves from the equator to the poles via atmospheric and oceanic circulation. Water vapor concentration is greater near the equator than in northern latitudes. Increases in average global temperature enhances the capacity for the atmosphere to contain more water vapor along with the capacity to absorb more long-wave energy in the atmosphere that contributes to the greenhouse effect. This is a feedback mechanism in that the relative increase in water vapor at the equator is not as large as near the poles, so warming is greater at high latitudes (Le Treut et al. 2007).

Atmospheric circulation is generally more east-west than north-south due partly to the earth's rotation which also transports heat toward the poles. The dynamics are influenced by position of continents and mountains. Water has a greater thermal capacity than land so land areas near water bodies have more moderate annual and diurnal temperature ranges than inland regions at the same latitude.

Altitude affects the latitude-based assessments since a thin atmosphere allows more direct solar radiation to reach the land surface in mountainous regions than at sea level. However, night-time heat loss is more rapid due to lower resistance to heat loss causing a larger diurnal temperature range, but a lower mean daily temperature than at lower elevations at the same latitude (Trewartha 1968; Oliver and Hidore 1984; Bailey 1996).

Feedbacks to the dynamics of climate and climate change due to changes in the global energy balance may accelerate or slow down climate change. Increased cloud cover results in cooling due to reflection (albedo) of incoming radiation. Clouds store and re-radiate heat energy due to absorption of long-wave energy in water vapor. Melting of ice and snow reveals darker land surfaces that absorb energy increasing heating, rather than reflecting the energy. Industrialization increases reflection and cooling through aerosols (dimming) and may increase atmospheric CO_2 through combustion of fossil fuels.

The worldwide patterns of temperature and precipitation determine the global distribution of vegetation. Land use change such as tillage and land management of crops for agriculture are parts of the system that feed back to influence global climate, needed adaptation of vegetation and its distribution.

Status of Climate Classification Systems

Ecologic Maps

Modern systems are generally modifications of the Köppen Climate Classification system (Köppen 1931). Bailey (1996) related climatic classification to continental distributions of climax vegetation called ecoregions (Table 8.1 and Figure 8.2). Ecoregions are subdivided into domains, divisions and provinces that describe areas of defined vegetation within the same climatic group. Bailey (1996) added Prairie, designated 250, to the Köppen climate (Table 8.1 and Figure 8.2).

Later, Bailey (2005) described the basis for constructing boundaries based on differences in vegetation responses to temperature and moisture. For example, the northern limit for the boreal forest is associated with the daily temperature for the warmest month that is too cold for tree growth. The boundary between the boreal forest and the northern grasslands is controlled by dryness. A hierarchy consisting of four levels fall within each of the macro-climatic regions (Omernik and Griffith 2014) expanding the total into 967 ecologic subunits in the conterminous US.

Ecoregions differ from **biomes** in that they are continuous and contain climax communities, all successional stages and many ecosystems. Similarly, **watersheds** may pass through more than one ecoregion. Biomes are associated with climax vegetation and the same or similar biomes can be found in more than one ecoregion (Bailey 2005).

Boundaries of ecoregions are stable relative to climate and are delineated based on natural vegetation or what the vegetation would be without disturbance. However, climax vegetation was changed dramatically by settlement, which brought agriculture and urbanization. This had a large impact on vegetation in continental areas east of 98 W long (e.g. humid temperate domain, Div. 220 and 230 in Table 8.1 and Figure 8.2) and coastal areas (Div. 240 and 260) of North America, which were originally forested. Further, land use changes due to cropping, grazing or mining may impact climate, vegetation and climate change at finer levels within the macro climatic region (Sleeter et al. 2012).

Climate change alters the geographic area covered by ecozones, but not uniformly across all ecozones due to factors such as continental position, altitude and latitude. Actual changes in temperature and precipitation from 1950 to 2000, especially from 1986 to 2000 have been associated with a decreasing ice cap and tundra (Div. 110 and 120) and a northern shift of the boreal forest (Div. 130). Simultaneously, an increasing dry domain resulted in a net decrease in the polar domain, which includes the boreal forest (Beck et al. 2005). Estimates of

Table 8.1 Regional climates based on the Köppen system (1931), as modified by Trewartha et al. (1967) and Trewartha (1968); and approximate ecoregion equivalents modified from Bailey (1996)

Koppen classification group and types[a]		Ecoregion equivilents[b]	
A	**Tropical and humid climates**	**400**	**Humid tropical domain**
Ar	Tropical wet	420	Rainforest
Aw	Tropical wet-dry	410	Savanna
B	**Dry climates**	**300**	**Dry domain**
BSh	Tropical/subtropical semiarid	310	Steppe
BWh	Tropical/subtropical arid	320	Desert
BSK	Temperate semiarid	330	Steppe
BWk	Temperate arid	340	Desert
C	**Subtropical climates**	**200**	**Humid temperate domain**
Cs	Subtropical dry summer	260	Mediterranean
Cf	Humid subtropical	230	Subtropical
		250	Prairie
D	**Temperate climates**		
Do	Temperate oceanic	240	Marine
Dca	Temperate continental, warm summer	220	Hot continental
Dcb	Temperate continental, cool summer	210	Warm continental
		250	Prairie
E	**Boreal climates**	**100**	**Polar domain**
E	Subarctic	130	Subarctic
F	**Polar climates**	120	Tundra
Ft	Tundra		
Fi	Ice cap	110	

[a]Köppen did not recognize the prairie as a distinct climatic type.
[b]Bailey's (1996) ecoregion classification system places prairie at the arid sides of the Cf, Dca and Dcb types.

climate change predict that dry areas of North America will become drier spreading the deserts in Div. 320 (Arizona – New Mexico) east (Div. 310) and north into west Texas and (Division 330) southern Colorado.

Agroecologic Maps

Ecosystems display similarities among components such as soil, climate and landscape. However, where the ecologically-based map delineates **climax vegetation**, the agroecologic map delineates crop production environments that reflect similar responses to management for both native and **introduced species**. For example, Padbury et al. (2002) described 14 agroecosystems for the Northern Great Plains, an area which contains dry, temperate semiarid steppe, prairie and boreal climatic types and ecoregion divisions (Bailey 1996). Agriculture and urbanization have blurred boundaries of climax vegetation because local microclimates have been altered, extending adaptation areas to include both native and crop species.

Adaptation Zones

Today, native and domesticated forage species occur within and among the agroecosystems and ecoregions. Delineations of adaptation zones for introduced species have been made through consensus among experts. Figure 8.3 is an example of adaptation zones for four prominent species that occupy different, although overlapping, geographic regions. The species are adapted to these regions because of their variation in physiologic and morphologic responses to climate and soil conditions.

The original definitions of adaptation zones were based on species and cultivar trials carried out for a limited number of years and management requirements (e.g. cutting regimes) by state and provincial agronomists. Data were later organized by agroclimatic regions, often including soil types, and consensus-based recommendations were made for management.

Uniform construction of climatically-based adaptation maps for introduced forage species of North America is incomplete. Nearly all estimates of adaptation zones for introduced forage species have been made through consensus among experts as indicated in Figure 8.3 and in Chapters 14–19. Some general correlations between regional climatic and species abundance data have been used to delineate zones for functional groups, e.g. warm- and cool-season species or short- and tallgrass species (Looman 1983; Sims 1988).

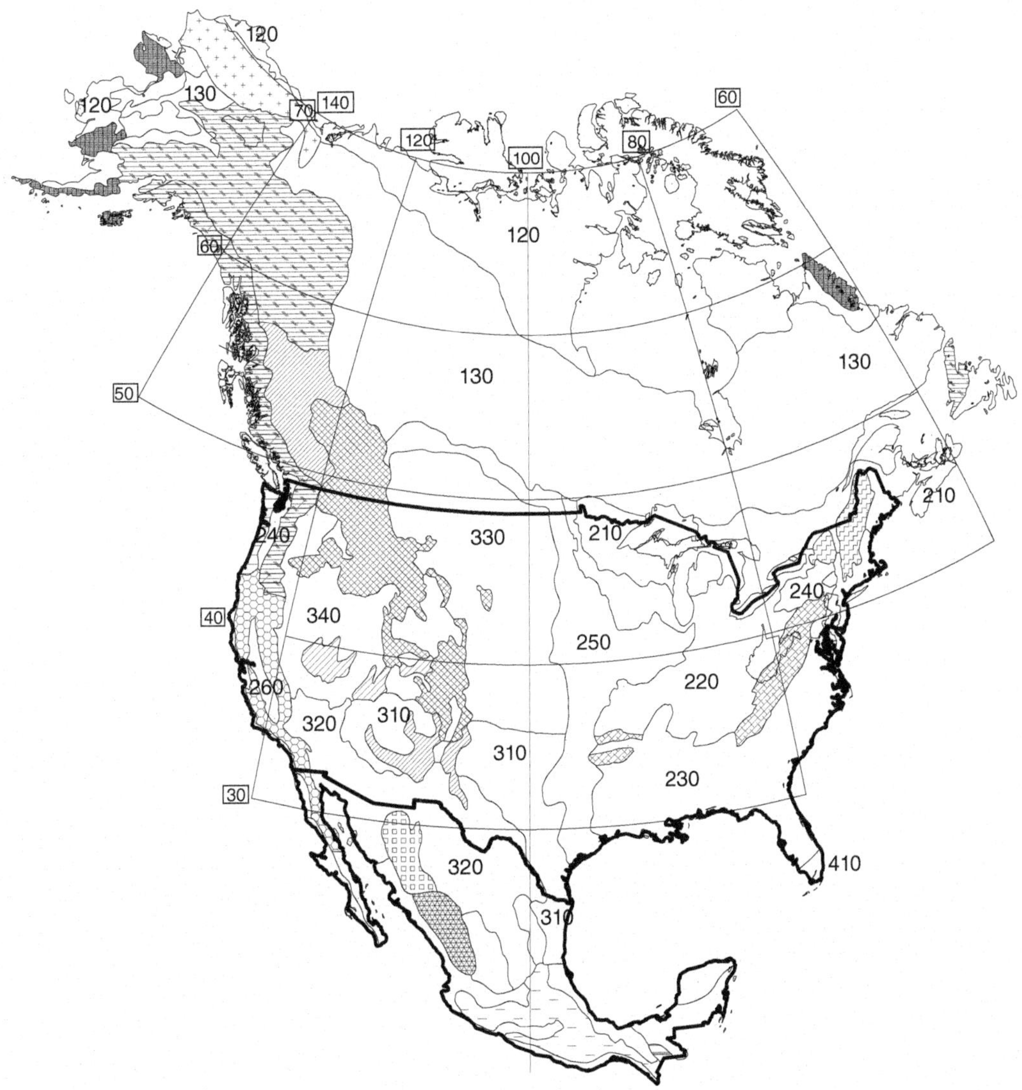

Fig. 8.2. Climatic ecoregions of North America. Cross-hatched areas represent altitude effects within adjacent ecoregions. *Source:* Adapted from Bailey (1996). Ecoregion climates are described in Table 8.1 with boundary definitions in Table 8.2.

Agroecologists combine geographic information systems (GIS) technology and crop simulation models to determine adaptation zones for forage and native species (Hill 1996; Hill et al. 2000; Hannaway 2009). Boundaries are predicted using species abundance data from surveys and preexisting collections. Where species composition data are scarce, but climatic data exist, potential adaptation is modeled spatially using GIS techniques and mathematical relationships between species productivity or survival based on critical temperatures and precipitation: evapotranspiration ratios. In this manner, Thompson et al. (1999) delineated adaptation zones for tree and shrub species throughout North America, using map correlations (latitude, longitude, and altitude) with relatively simple climatic data and indices (e.g. temperature of coldest and warmest months). Hannaway et al. (2009) used a similar approach to map adaptation of tall fescue in the US and China.

In the future, a combination of consensus, correlation and GIS modeling technologies will likely be used to

Table 8.2 Criteria for climatic regions and boundary definitions of the Köppen-Trewartha system

Criteria for climatic regions	
Ar	All months above 18 °C and no dry season
Aw	Same as Ar, but with 2 mo dry in winter
BSh	Potential evaporation exceeds precipitation, all months above 0 °C
BWh	One-half precipitation of BSh, all months above 0 °C
BSk	Same as BSh, but with at least 1 mo below 0 °C
BWk	Same as BWh, but with at least 1 mo below 0 °C
Cs	8 months 10 °C, coldest month below 18 °C and summer dry
Cf	Same as Cs, but no dry season
Do	4–7 mo above 10 °C
Dca	4–7 mo above 10 °C, coldest month below 0 °C and warmest month above 22 °C
Dcb	Same as Dca, but warmest month below 22 °C
E	Up to 3 mo above 10 °C
Ft	All months below 10 °C
Fi	All months below 0 °C
Boundary Definitions	
A/C	Tropical/Subtropical = Equatorial limits of frost; in marine locations, 18 °C for coolest month
C/D	Subtropical/Temperate = 8 months 10 °C
D/E	Temperate/Boreal = 4 months 10 °C
E/F	Boreal/Polar = 10 °C for warmest month

Demarcation of boundaries between dry B and tropical A or B and temperate C or B and boreal E is where potential evapotranspiration = precipitation. Precipitation is successively lower for the division between B/A than B/C and B/E; note prairie, Division 250, is between B/C and B/E in North America (Figure 8.2).

Source: After Bailey (1996).

delineate adaptation zones for individual crop or forage species. Agreement on delineation protocols and terminology used to define map boundaries is as essential in this process as it has been in climatic classification.

Climate and Forage Species Distribution

High Temperature

High temperature extremes that occur in July throughout North America affect plant species distribution and, therefore, are critical in mapping adaptation zones of plant taxa (Thompson et al. 1999). Some modifications may be more functional since maximum August temperature best facilitated the delineation of forage and range species in western Canada (Hill et al. 2000).

Mean July temperature across North America increases from Northwest to Southeast and from the subarctic to tropical domains (Bailey 1996). The trend is disrupted by altitude and proximity to large bodies of water (Figure 8.4), but four July-temperature zones, at 5 °C intervals (e.g. 10–15 or 25–30 °C mean July temperature), exist along this track. Mean temperature is the average of the daily maximum and minimum, so daily extremes are often 5–10 °C below and above the mean.

Change in altitude at the same latitude causes **macroclimate** scale changes over short lateral distances (mesoclimate) in mountainous areas (Trewartha 1968; Bailey 1996). In addition, temperature change varies among regions, occurring rapidly with altitude at mid-latitudes compared to the equator. This affects the altitude at which ecoclimatic zones appear. For example, the upper limit to the tree line at the equator may be 4000 m, but only 2000 m at 55–60 N lat (Bailey 1996). Global warming will increase the tree-line altitude more moving north and south from the equator. In general, the current average annual temperature decreases 6.4 °C for every 1000 m in altitude above sea level. Elevation of the tree line due to climate change will be highly site and species specific, but estimates are 640 m for a mean temperature rise of 4–5 °C in the northern mid latitudes (Grace et al. 2002). Changes in vegetation are not immediate, lagging behind climatic indicators, because old growth suppresses new growth until the vegetation favored by the transition dominates.

Due to inclination of the sun during the growing season, in the northern hemisphere south facing slopes receive more total solar insolation than north facing slopes. Thus, on south facing slopes, snow melt occurs

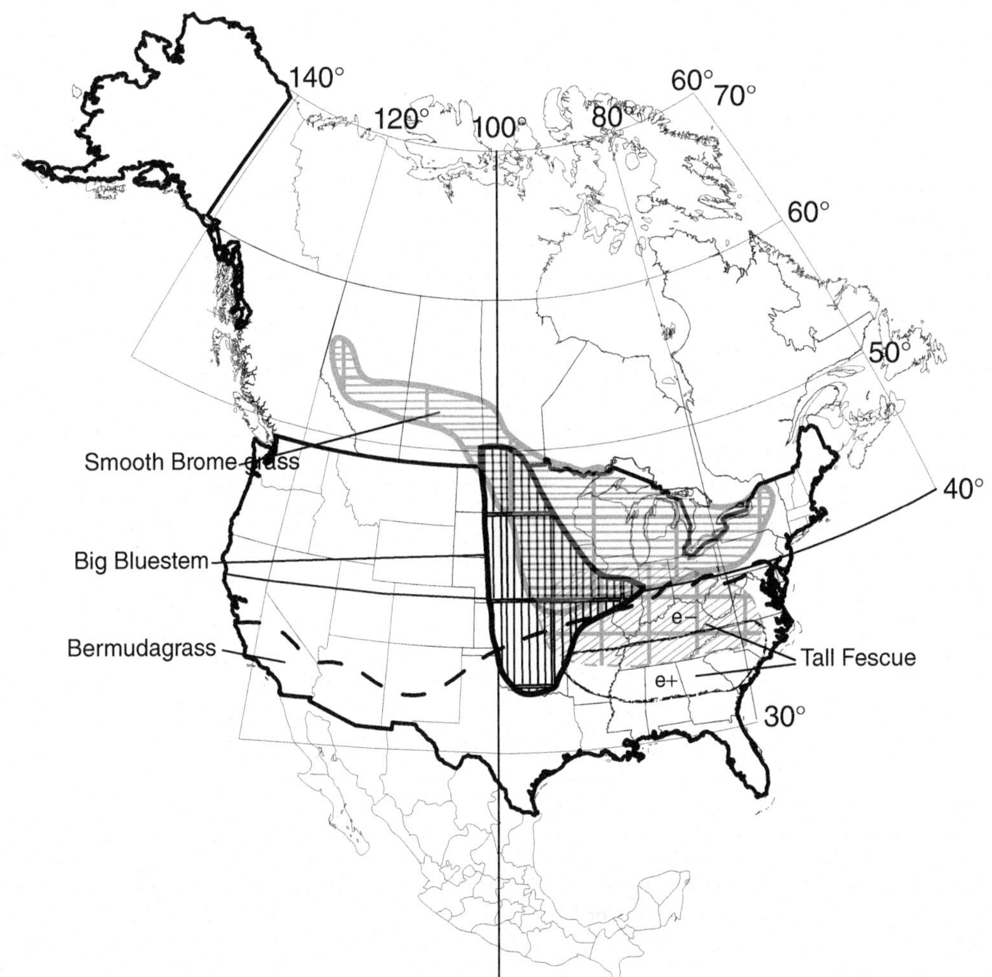

Fig. 8.3. Adaptation zones for smooth bromegrass, big bluestem, bermudagrass, endophytic (e+) and non-endophytic tall fescue (e-) as estimated by the consensus of authors and editors of this edition.

earlier in the spring and growth resumes sooner after cutting, but soils may dry out faster than on north facing slopes (Barbour et al. 1987).

Adaptation to temperature is important as it affects species diversity by latitude (Nelson and Moser 1994). North of approximately 44 °N lat cool-season grasses predominate (southern borders of Wyoming and Minnesota). Few C_4 species occur at locations with mean minimum July temperatures below 8 °C (Barbour et al. 1987). These boundaries are not absolute as some **ecotypes** of C_4 species (genetic variation) are found north and some of C_3 species south of 44 °N lat. For example, blue gramagrass, a C_4 species, is found as far north as 53 °N lat near the Saskatchewan-Alberta border (Hill et al. 2000).

The primary adaptation zone (mostly Div. 220 and 230, Figure 8.2) of tall fescue extends as far south as 32 °N lat (Figure 8.3) and is grown as far south as northern Florida (Sleper and Buckner 1995). This tall fescue area (Figure 8.3) predominates a region known as the transition zone, an area where both cool- and warm-season species are found. In the southern part of the transition zone, tall fescue survives the temperature extremes through a symbiotic relationship with an endophytic (e^+) fungus that infects the plant tissues (see Chapter 33). Here, it overlaps with bermudagrass, a dominant C_4 species (Figure 8.3).

In the northern part of the transition zone, tall fescue cultivars which are endophyte-free (e^-) are preferred and

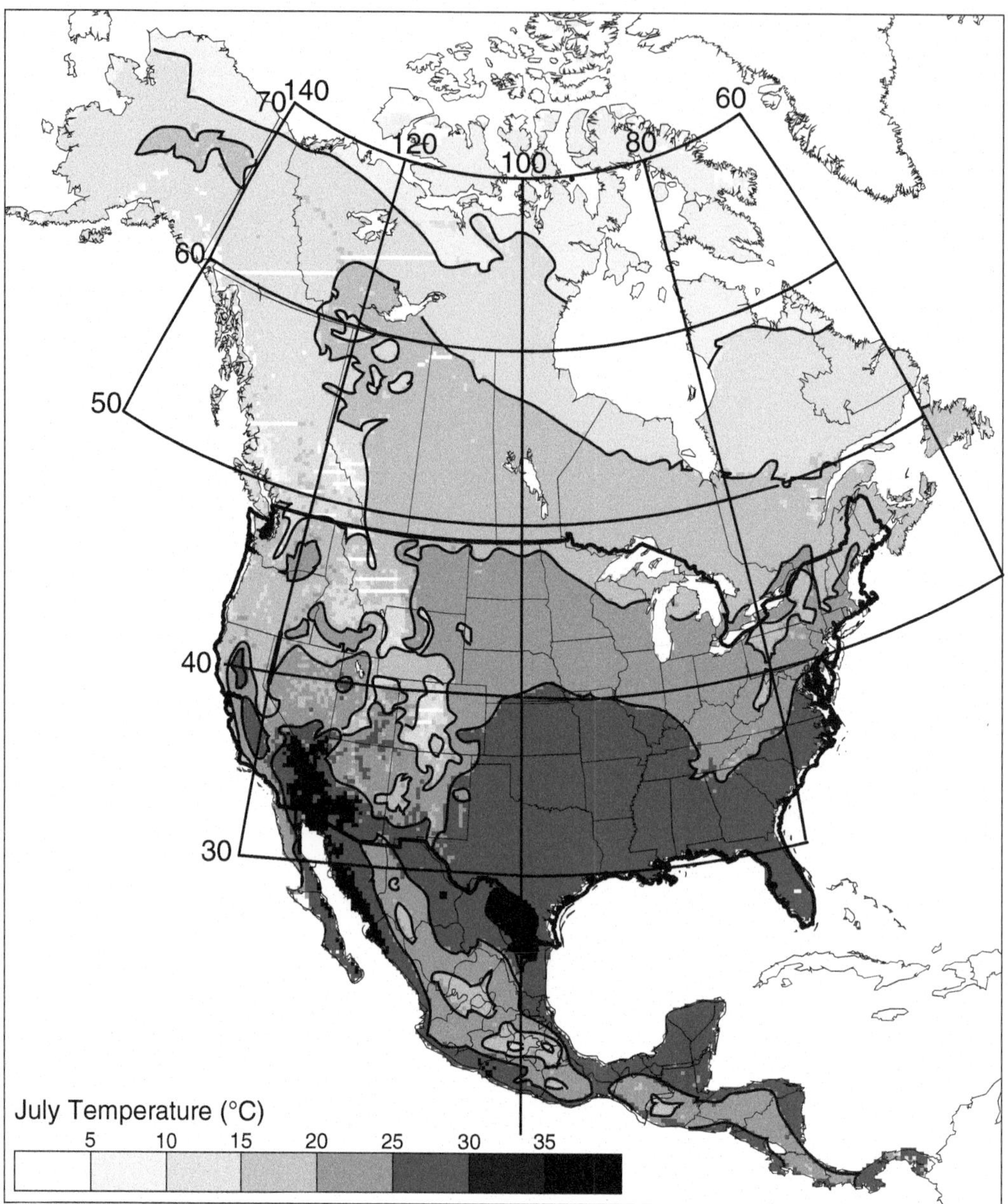

Fig. 8.4. Mean July temperature (°C) for North America on the 25-km grid. *Source:* Adapted from Thompson et al. (1999).

overlap with the adaptation zone for smooth bromegrass, which is not as heat tolerant, but more drought tolerant.

To the west, other species such as big bluestem that are adapted to more arid environments overlap the adaptation zones of bermudagrass, smooth bromegrass, and tall fescue to offer management alternatives (Figure 8.3) near common boundaries of ecoregions (Figure 8.2). Cultivars differ in distribution due to plasticity for climatic and environmental stress.

Climate change may move ecoregion borders slowly due to gradual changes in average temperature and precipitation, but temperature change may be the major factor affecting species abundance and dynamics within regions (Fig. 8.4). Individual species within a larger population

Table 8.3 Optimum and range in temperature tolerance of forage groups after Volenec and Nelson (2007)

Species	Low temp. tolerance	Optimum temp.	High temp. tolerance
Legumes	LT_{50} = −6.3 to −7.4 °C (Meyer and Badaruddin 2001) Seedling survival at −10 °C for 8 h = 4 to 60% (Arakeri and Schmid 1949) Seedling survival for 8 h at −4.7 to −7.4 °C = 32 to 96% (Tysdal and Pieters 1934)	Whole-plant growth highest at 16/10 (4) or 20/15 °C (day/night) (Pearson and Hunt 1972) Shoot growth highest at 27 °C (Leach 1971) Shoot growth highest at a crown temp. of 32 °C (Evenson 1979)	Herbage growth at 35/29 °C is 19% that at 16/10 °C (Lee and Smith 1972) Herbage growth at 32 °C is 44% that at 16 °C; root growth at 32 °C is 23% that at 16 °C (Gist and Mott 1957)
Cool-season (C_3) perennial grasses	Seedling survival at −10 °C for 8 h = 31 to 98% (Arakeri and Schmid 1949)	Tiller growth highest at 25 °C (Volenec et al. 1984)	Death at 34 °C, growth at 30 °C is 51% that at 25 °C (Volenec et al. 1984)
Cool-season (C_3) annual grasses	Leaf growth ceases at 0 °C (Kirby 1995)	Rate of leaf appearance highest at 22 °C; rate of development greatest at 21 °C (Yan and Hunt 1999)	Rate of leaf appearance ceases at 45 °C; growth ceases at 33 °C (Yan and Hunt 1999)
Warm-season (C_4) perennial grasses	Survive −3 °C but death or severe injury at −6 °C (Chamblee et al. 1989) LT_{50} of −9 to −22 °C (Qian et al. 2001); Growth at 5 °C less than 10% that at 20 °C (Clifton-Brown and Jones 1997); Seedling base temperatures range from 2.6 to 7.3 °C (Madakadze et al. 2003)		
Warm-season (C_4) annual grasses	Growth ceases at 5 °C (Yan and Hunt 1999) Sorghum growth ceases at 8.5 °C (Craufurd et al. 1998)	Growth highest at 31 °C in maize; rate of development greatest at 27 °C in sorghum (Yan and Hunt 1999)	Growth ceases at 41 °C in maize; 34 °C in sorghum (Yan and Hunt 1999)

have specific temperature thresholds (Table 8.3). Those with higher thresholds, called cardinal temperatures, may eventually dominate. However, the entire population may not adapt in unison, resulting in a population in a continual transition in species composition as conditions change (Izaurralde et al. 2011).

Species that thrive and develop a relatively large leaf area due to efficient use of light and CO_2 within a geographic area should reside for longer durations at optimum temperatures for their vegetative and reproductive growth and development (Hatfield et al. 2008). These adapted species may dominate mixtures and populations or yield more than alternatives which are less adapted. The extreme temperatures, both above or slightly below the optimal range, usually determine the persistence of the species. Thus, the limits of the temperature range should define species adaptation zone boundaries.

Optimal growth temperatures for cool-season species are usually from 20 to 25 °C and for warm-season species are usually from 30 to 35 °C (Table 8.3). In addition, warm-season types grow slowly below 15 °C, while cool-season types can grow at temperatures as low as 0 °C, but grow slowly below 5–7 °C. Both can coexist in the same environment, but growth for either predominates at different times of the year or under different management regimes (Nelson and Moser 1994; Nelson and Volenec 1995).

Climate change may cause species composition to change due to an interaction of many factors including species responses to increasing CO_2 and reduced soil moisture, plus intolerance or adaptation to high temperatures (Table 8.3) may be positive or negative to species abundance or economic viability. Regional species presence may result from frequency of years when temperatures exceed thresholds for rapid growth (Hatfield et al. 2011) resulting in periods of dormancy, reduced growth rates or death. Features of temperature that may cause a change in a regional species complex are prevalence of high temperatures at night, frequency of very hot days with maximum >30 °C, hot days coinciding with sensitive stages of development, e.g., pollination, and a longer growing season (Hatfield et al. 2011).

Alfalfa is a C_3 species that tolerates relatively high temperatures (over 30 °C) and appears intermediate in temperature response between warm- and cool-season legumes (Buxton 1989; Nelson and Moser 1994). Conversely, red clover and white clover are tolerant to moderate and low summer temperatures, respectively (Nelson and Moser 1994; Chapter 14). Among cool-season grasses, tall fescue, smooth bromegrass and perennial ryegrass are tolerant to relatively high, medium and low temperatures, respectively (Chapter 16), although temperature optima are below those of C_4 species. Similarly, warm-season tropical and subtropical species have temperature optima greater than 30 °C, but are less likely to tolerate cool temperatures and are not very productive at or below 15 °C.

In general, average air temperatures have risen slowly from the beginning of the twentieth century and continuing, but average temperature increases have been largest in the southwest, north and central parts of the continent and least in the southeast (Walsh et al. 2014). Temperature increases in the north may not be very consequential in the short term (2020–2050) as the current average and maximum temperatures are not high. In some cases, the normal (1901–1960) expectation for extreme temperatures within an ecozone may simply move from the optimum of one species (e.g. red clover) to the optimum of another (e.g. alfalfa). This may be the case for the north-western Canadian Prairies (Div. 250; 53 °N) as prairie parkland encroaches into the subpolar or boreal forest (Div. 130) over the next 30 years. However, south and east in the Northern Great Plains where average annual temperature increase for the Red River Valley and Southern South Dakota (1981–2010) has been 0.6 and 0.8 °C, respectively (Ojima et al. 2015), the mean temperature may increase by 5–6 °C by the end of the twenty-first century.

Compared to current conditions (Fig. 8.4), more hot days will be experienced in western Canada and at higher altitudes in the western US, but many more hot days will be observed south of the South Dakota-Nebraska border (Ojima et al. 2015; Derner et al. 2018). In central Alberta, the average number of hot days (> 28 °C) during a year may increase from 7 to 16 by mid-century (Thivierge et al. 2017). However, in the southern half of the Northern Great Plains the number of days >35 °C may increase to 30 by 2050 and continue up to 70 by 2085 (Derner et al. 2018) necessitating use of C_4 over C_3 species.

Low Temperature

Cold temperatures during winter have a greater effect on species distribution than high temperatures during summer (Nelson and Moser 1994). Coldest temperatures invariably occur in January in North America and were used to delineate five adaptation zones for species ranging from a mean of −30 °C in the subarctic to 30 °C in Central America (Thompson et al. 1999). **Winterhardiness**, the ability to survive a winter, includes more than just cold temperatures, **freezing tolerance**, the ability to survive a cold temperature, and **cold tolerance**, the ability of top growth to tolerate low temperatures (usually night) and resume growth per se, are not necessarily synonymous, but all impact plant distribution.

North of the 40th parallel, adaptation to winter temperatures which involves frozen soil, affects forage distribution. Where snow depth is minimal, winter minimum temperature drives winter kill. The semi-arid region (Div. 330, Figure 8.2) of the Northern Great Plains requires plants adapted to highest freezing tolerance as snow depth, which insulates crowns from extreme air temperatures, may be sparse and intermittent (Ouellet 1976; Sims 1988). By contrast, east of 98 °W long, low winter temperature is important, but is one of several factors which cumulatively cause winterkill and loss of plant stands. Species and cultivars (largely cool-season) grown in Div. 210 (warm continental, Figure 8.3) and the maritime provinces of Canada require more freezing tolerance than those grown in Div. 220 (hot continental, Figure 8.3), but are not as cold and drought tolerant as those found in Div. 330.

Tolerance to cold, but not necessarily freezing temperatures, also affects distribution of warm-season species. Stargrass cannot persist where temperatures go below −4 °C and is not recommended for northern Florida (See Fig. 8.5; Chapter 18). Tropical species are adapted further north along coastal areas where low temperatures are moderated by water (See Chapter 15).

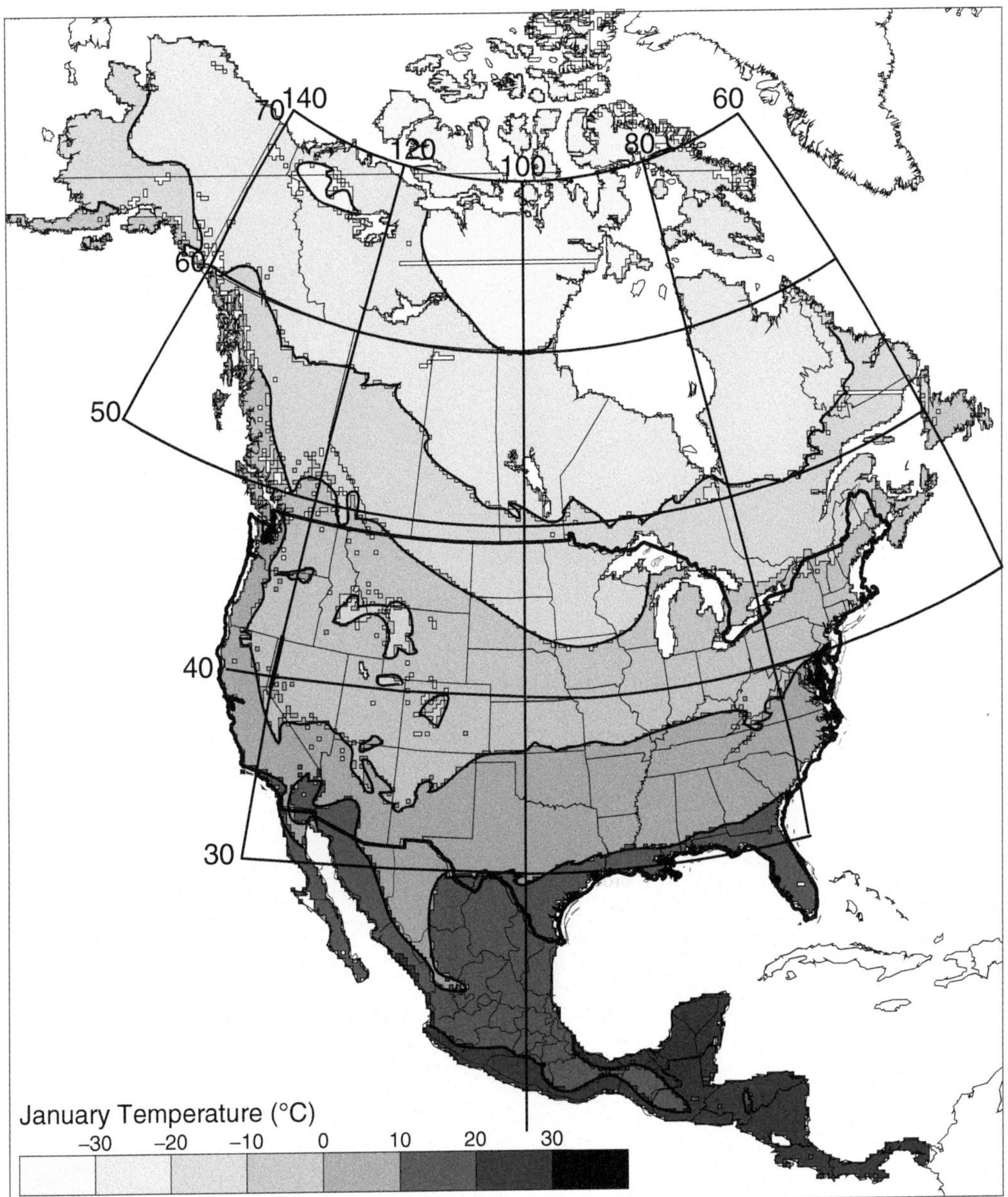

Fig. 8.5. Mean January temperature (°C) for North America on the 25-km grid. *Source:* Adapted from Thompson et al. (1999).

Climatic indices are used to describe and quantify the most probable climatic causes of plant loss during fall and winter (Durling et al. 1995; Bélanger et al. 2002). The USDA Plant Hardiness Map (Figure 8.6) is based on the minimal annual temperature (Cathey 1990) and is similar to Figure 8.4. It provides some information on the coldness of different regions of North America, which can be related to the distribution of forage species.

Due to effects of global warming, winter temperatures have and will continue to increase more in Northern than in Southern States. Over the last 30 years, winter temperatures have increased by 3.9 °C in the Midwest and Northern Great Plains, but by only 0.5 °C in West Texas.

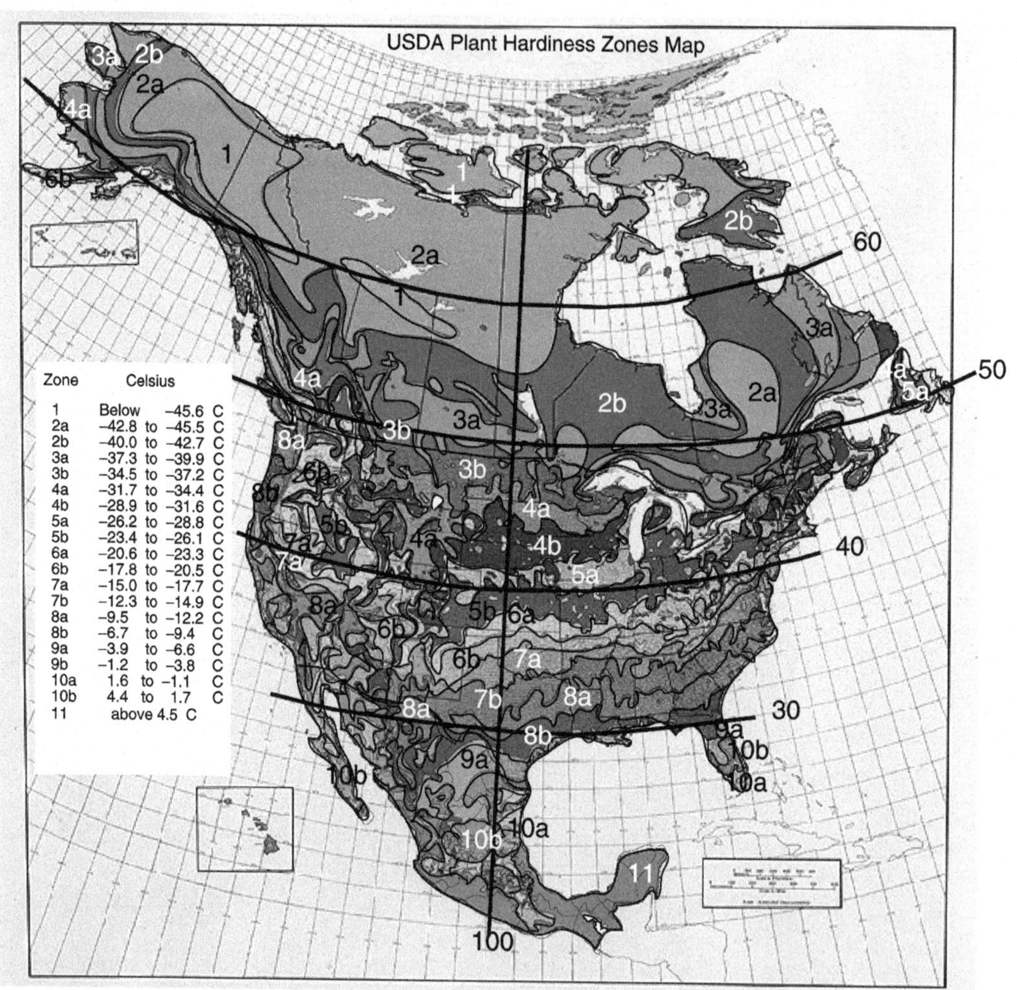

Fig. 8.6. USDA Plant Hardiness Zone Map. *Source:* Adapted from Cathey (1990). USDA Miscellaneous Pub. # 1475. Updated to 2002. https://planthardiness.ars.usda.gov/PHZMWeb/

An increased number of frost-free days in spring and fall due to climate change increases length of the growing season and may allow frost-sensitive crops to be grown further north. Compared to the first 60 years of the twentieth century the growing season (days >0.0 °C) has increased by 19 and 16 days in the Southwest and Northwest US and by 6 and 10 days in the Southeast and Northeast (Walsh et al. 2014). The number of days above freezing will likely increase by 20–40 days by 2050 and by 50–75 days before 2085 on the Northern Great Plains (Derner et al. 2018). By 2050, the growing season (for perennial crops) is expected to increase about 21 days over the average for 1970–2000 in both central Alberta (53 °N) and eastern Quebec 48 °N (Thivierge et al. 2017).

Moisture and Species Distribution

At the macroclimatic scale, dry climates have annual evapotranspiration greater than precipitation (Bailey 1996). Currently, dry climates are separated from subhumid climates roughly along 98 °W long. The division reflects annual precipitation (Figure 8.6) and the potential for crop and rangeland soils to dry out based on the ratio of actual (AE) to potential evapotranspiration (PET) (Figure 8.7).

More precipitation occurs near the equator than the poles, and the colder air of the polar regions does not hold as much water in the form of water vapor. In North America, the drying out of ecoregions from east to west occurs due to the predominant air flow from the Pacific coast. As moist Pacific air rises to cross the Sierra Nevada – Cascade

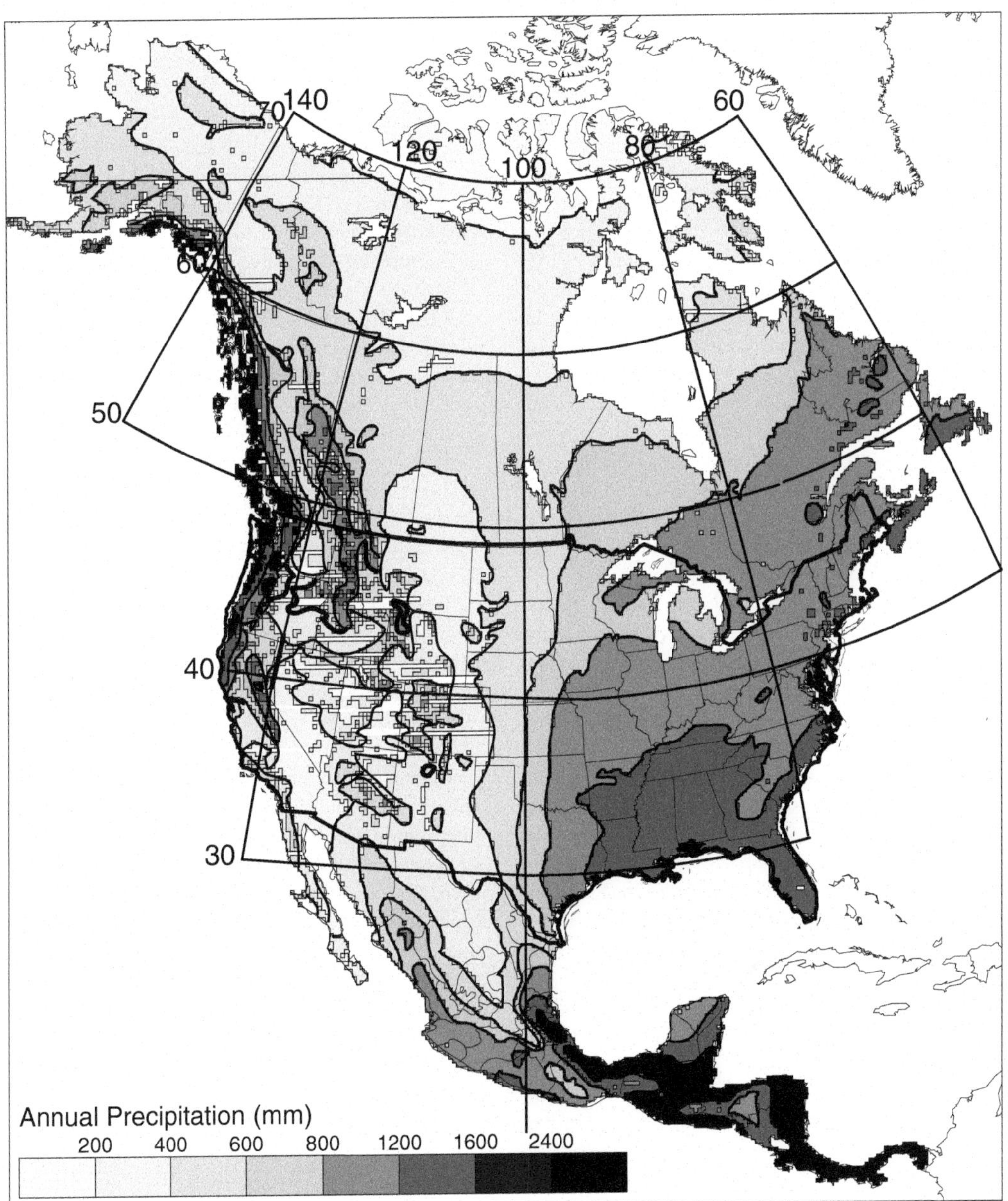

Fig. 8.7. Mean annual precipitation (mm) for North America on the 25-km grid. *Source:* Adapted from Thompson et al. (1999).

and Rocky Mountain ranges, precipitation occurs on the coastal side creating rain forests. As the drier air passes over successive ranges to the east, it warms and takes up water, which leads to drying of the eastern slopes, creating deserts and semi-arid regions in the rainshadow (Oliver and Hidore 1984). The overall effect alters climax vegetation and agricultural management across the continent.

East of 98°W long annual precipitation increases from 510 to 1020 mm across the prairie and exceeds 1300 mm close to the Atlantic Ocean (Bailey 1996). Going west from 98°W long, precipitation decreases to approximately 330 mm in the semi-arid regions (e.g. Colorado Springs, CO and Swift Current, SK). Weaker north-south trends exist with precipitation increasing from 600 mm close to the Manitoba-Ontario border to over 1200 mm in

Alabama on the east side of 98° W long. However, further west, the precipitation decreases from approximately 450 mm in Northern Alberta to less than 300 mm in the South Western Deserts of the US and Mexico (Figure 8.6).

Climate change affects the quantity and seasonal distribution of precipitation regionally, but the uncertainty of future estimates is large. In the future, wetter portions of the US will become more wet (Northeast) and the drier locations (Southwest) more dry (Walsh et al. 2014). Annual precipitation, averaged over the US, has increased about 5% over the last 50 years. Recent increases in precipitation in the Northeast, Midwest and Southern Great Plains range from 8% to 9%, the Southeast and Southwest have been variable and the Northern Great Plains has seen small increases (Walsh et al. 2014). In the future, heavy, intense rainfall is expected to contribute to greater proportions of the total annual precipitation, particularly east of the 98th long. Heavy rainfall events may comprise 71% and 37% of annual totals in the Northeast and Midwest, but only 5% in the Southwest and 16% over the Great Plains from Texas to North Dakota (Walsh et al. 2014). The periods of intense precipitation may increase runoff but, in general, may be offset by periods (weeks) of no precipitation or drought.

By 2090, a higher proportion of precipitation will occur during winter and spring (Walsh et al. 2014). Most current agricultural areas of the US and Canada will experience hotter, drier summer growing seasons, with less precipitation. Decreases in northern precipitation patterns will be due to atmospheric warming allowing air to hold more moisture and to changes in seasonal weather patterns, which affect where precipitation occurs. In the Southwest, drying occurs due to northern expansion of subtropical high pressure, which suppresses rainfall.

Areas with the largest increases (>20% over current average) in precipitation will occur in the Northwest Territories of Canada and Alaska. A seasonally-dynamic north to south transition band of reduced precipitation is expected across the middle latitudes of the US. During winter, the southern edge of the moist zone (>10% over current average) will reach the lower Midwest and Northern Great Plains, while the northern edge of the dry zone runs from Southern California to Louisiana. However, in summer, the southern perimeter of the moist zone moves north to the border between the boreal forest (Division 130) and Prairie Parkland (Div. 250). During summer, the central portion of the Great Plains is expected to be dried out to the southern Prairie provinces, northern Ontario and Quebec; the most severe drying out in summer will occur in the Southern Plains and Northwest US (Walsh et al. 2014).

Moisture Efficiency

Thornthwaite (1948) recognized that precipitation alone did not explain the geographic distribution of plant species or their management as crops, and introduced the concept of moisture "efficiency" based on the annual balance of actual precipitation and PET. The relationship is represented by the ratio of AE:PET (Figure 8.8).

PET is water evaporated under high soil-moisture conditions (field capacity) which have no physical limitation to uptake by vegetation (Branson et al. 1981). PET can be measured using "pan evaporation," but is almost always calculated and is dependent on net radiation, air temperature, water vapor density and wind speed (Branson et al. 1981; Oliver and Hidore 1984). Actual evapotranspiration (AE) is the fraction of PET that is evaporated from the soil and crop, given any stage of crop and available soil moisture content. It can be measured directly using "water balance" techniques such as a **lysimeter**; or estimated by the difference between precipitation and available soil moisture or by using energy balance methods (Branson et al. 1981; Oliver and Hidore 1984). Both AE and PET can be estimated using modeling approaches (Thornthwaite and Mather 1957; Kolka and Wolf 1998) which facilitates mapping (e.g. Figure 8.8).

The ratio, AE: PET, is highest or approaches 1.0 in areas where AE, PET and rainfall are high and vegetation is rarely drought stressed, such as coastal areas of south and eastern US and parts of Central America and Mexico. West of 98°W long AE is reduced to near 0.5 because of low precipitation and available soil moisture such that plants are usually water stressed while PET remains relatively high due to high summer solar radiation, high temperature and low water vapor density. Above 50°N lat, (especially moving north-west from Manitoba) the decreasing net daily radiation (with increasing lat) and cooler summer temperatures (with lat and altitude) result in a "less-drying" AE:PET ratio (Figure 8.7). The resultant soil moisture conditions resemble the northern edges of Div. 210 (Minnesota), but under a cooler (Figure 8.3) and drier (Figure 8.6) climatic regime. This allows the adaptation zone of smooth bromegrass (Figure 8.3) to move west and north into the prairie and subartic ecozones (Div. 250 and 130).

Climax grassland in North America generally falls between 95°W and 105°W long (Barbour et al. 1987). The grasslands are delineated into areas of tall grass (prairie), mixed grass, and short grass (steppe) reflecting natural decreases in annual precipitation and productivity. Big bluestem, a tall warm-season species, is currently adapted to a zone bracketing the 98°W long north of Texas and is found as far north as 50°N lat in southern Manitoba and Saskatchewan (Figure 8.3). Warm-season grasses (e.g. big bluestem) with tolerance to both drought and high temperatures are found in the Great Plains (Moser and Vogel 1995) and overlap with cool-season species to the north. North of 40°N lat, but west of the big bluestem zone (Div. 330), dryland C_3 grasses such as crested wheatgrass, russian wildrye, native species (C_3 and C_4 mixed grasses) and alfalfa are found. Under more severe conditions of heat and drought (Div. 310),

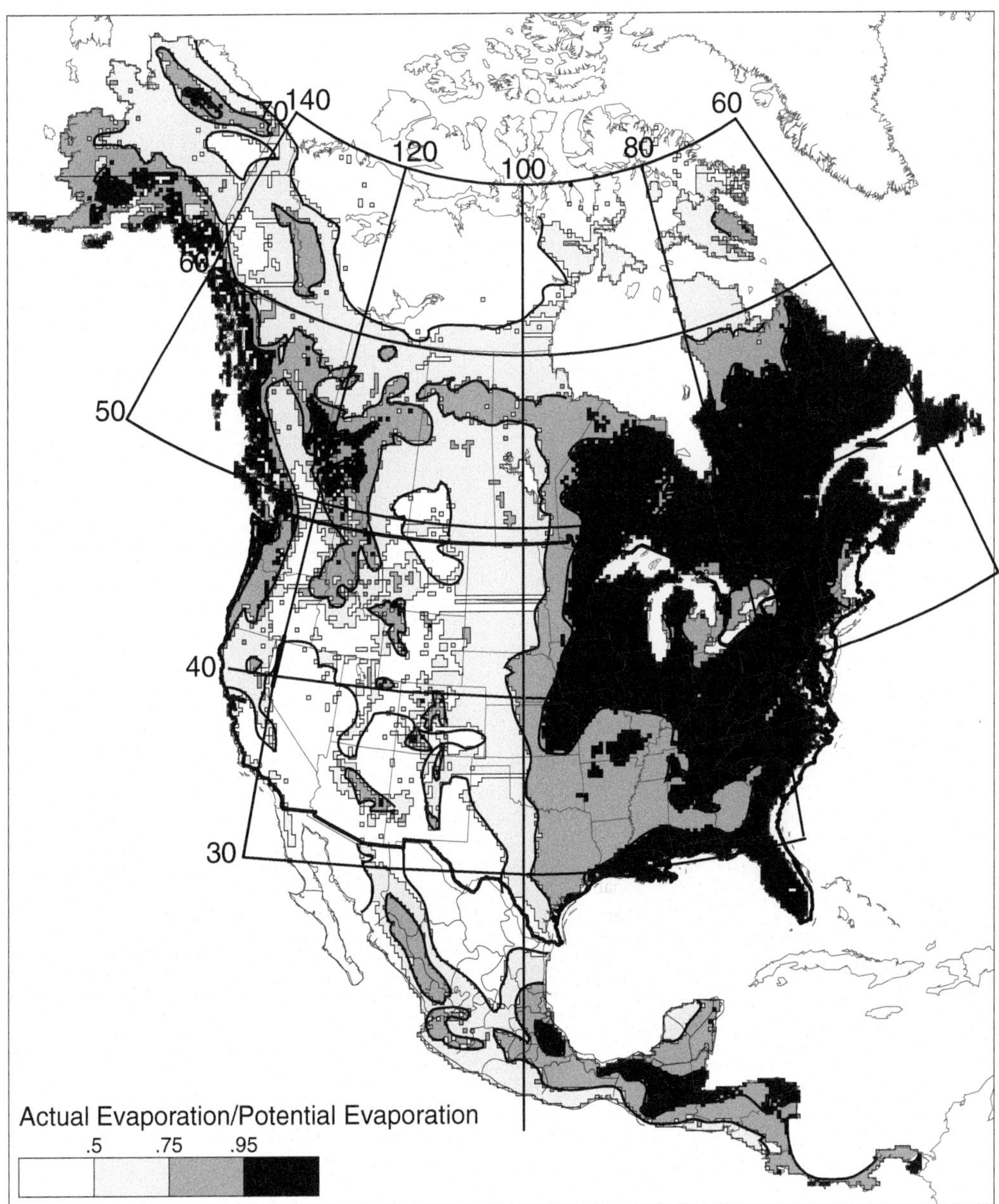

Fig. 8.8. Actual evaporation as proportion of potential evaporation on the 25-km grid with four divisions shown. *Source:* Adapted from Thompson et al. (1999).

warm-season C_4 species such as blue grama grass are found in the southwest (Voigt and Sharp 1995).

North of the smooth bromegrass zone (Div. 130), but west of 98°W long, soil moisture during the growing season is sufficient for timothy, red fescue, clovers and occasionally orchardgrass. This is more typical of northern Minnesota, Wisconsin, Ontario and Quebec (Div. 210), but cultivars found in the northern zone are more winter hardy.

Using Modeling to Predict Adaptation

Based on the Variable Infiltration Capacity Model to forecast, future trends indicate decreased soil moisture, especially south of South Dakota and west of 98th long by mid-century. Slight increases in soil moisture are predicted in eastern Washington, Northern Idaho and western Montana (Walsh et al. 2014). With drier conditions expected, the dry B ecozones of the southwest move north and east so that the demarcation of dry vs moist zones (currently 95° long) moves east and into the tall fescue grassland and north such that the Prairie (250) encroaches the boreal forest (130) in Canada.

The future distribution and performance of tame and native species is expected to depend on moisture more than any other variable (Izaurralde et al. 2011), except where low winter temperatures limit survival of perennials. Prospective changes in species distribution among ecoregions are highly speculative, because fundamental variables (moisture, CO_2 concentration and temperature) interact with plant adaptive features, soil physical characteristics, inherent nutrient supply, continental position and altitude (i.e. temperature and photoperiod) to name a few. Each species and genotype has specific responses that facilitate **competition** with counterparts in a population resulting in survival or equilibrium.

Because of the speed with which climate change could occur, native or naturalized populations could be in a perpetual flux. The seasonal change in distribution of precipitation is pivotal to the distribution of species and will change crop management practices (Izaurrulde et al. 2011). Species which emerge or grow first in the cooler, earlier spring in the new, longer growing season will dominate. These are likely to be C_3 species that can complete annual or life cycles prior to the intense summer heat that is more suited to C_4 species.

Modeling research with alfalfa (Izaurralde et al. 2011) indicated that, in general, alfalfa yield will decrease in the western US vs the east, since yields will likely decrease 1% for every 4 mm reduction in annual precipitation. Timothy is suited to regions with low temperature and long photoperiods, but is very sensitive to moisture deficits (Bertrand et al. 2008; Jing et al. 2013).

Climate change simulation across Canada involving timothy projected an average annual increase of 53 mm of precipitation across ten locations. However, precipitation during summer regrowth by 2040–2069 was predicted to be less than under present conditions. The predicted soil moisture stress reduced timothy regrowth in northern areas (48–58°N lat), though first-growth yields were generally increased. This suggests timothy will need to be replaced by species with greater tolerance of high temperature and low moisture conditions. Similarly, tall fescue on the western side of the tall fescue zone might suffer more from moisture stress. Whereas tall fescue is heat resistant especially in association with endophytes, it is not particularly drought tolerant. However, due to heat tolerance, tall fescue may move as far north as Quebec and eastern Ontario, but endophyte-infected tall fescue cultivars may be required further north of the current transition zone.

Plant Adaptation

Adaptation is the process by which individual plants, populations or species change morphologically or physiologically in a way to better survive growing conditions (Allard 1966). In a natural system, genotypes can survive only if their descendants also fit the microenvironment in to which they were introduced (Harper 1977). Many species may be adapted to a site or region, the difference among them being their ability to reproduce and compete to survive the effects of the range of climatic and management interactions (Chapter 9). Generally, but not necessarily, this involves completing a reproductive cycle and spreading seed (Harper 1977).

Introduced forage and grassland species have management objectives involving economic returns and must have the ability to regrow and persist after cutting or grazing. Generally, management for high-forage quality dictates that they don't spread seed (Nelson and Moser 1994). At the center of an adaptation zone, species can survive the extremes of temperature and moisture as well as the stress of lax management (Nelson and Moser 1994). However, at the edge of the adaptation zone, the stresses of climate and management may be too great to overcome by the array of adaptive features set by the genetic makeup of the species. Near the edge, the species will be weaker and will require excellent management or it will be eliminated from the forage stand by a more aggressive and better-adapted species (Donald 1963). The result is that the intended stand may be terminated in favor of a more economic cropping option.

Nelson (2000) summarized the genotypic and phenotypic components of forage crop adaptation, i.e. plasticity (Figure 8.9). The genetic component of adaptation of a forage cultivar is limited by the range of heterogeneity. For example, most cool-season forage grasses are cross-pollinated, descendants of several parental clones, and often are polyploids. This genetic heterogeneity confers and transfers a large range in basic adaptive mechanisms to the species. By comparison, hybrid corn is a diploid formed from two inbreds, and is genetically uniform with a narrow adaptive range. In either case, **genotypic plasticity** is set by selection criteria of the breeder. Genotypic plasticity is irreversible. When exposed to conditions outside of the range of adaptation, loss of weaker plants reduces the genotypic plasticity of the population for that environmental stress (e.g. winterhardiness).

Phenotypic plasticity reflects the ability of an individual plant to change morphologically or physiologically in response to climate or management stress. A morphologic response to drought stress is the ability for roots to narrow in diameter, but continue elongation growth to access a greater volume of soil. Osmoregulation by

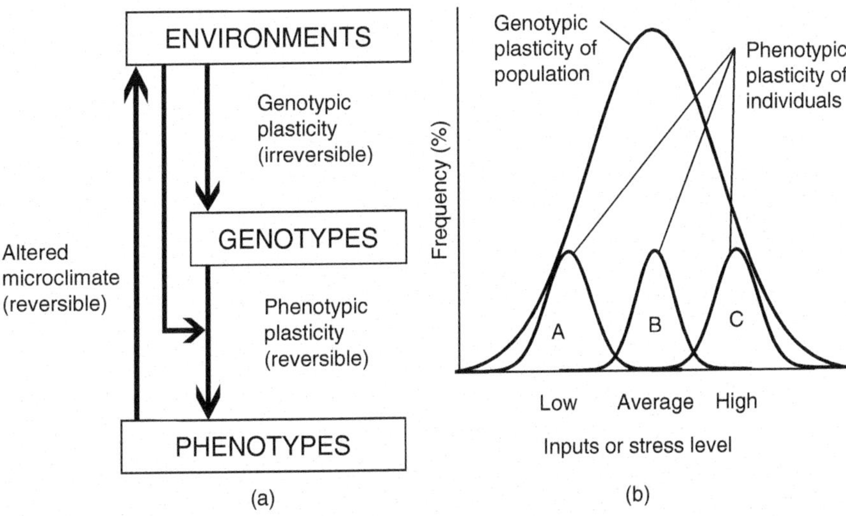

Fig. 8.9. Genotypic and phenotypic plasticity influence adaptation of crops used for forage. (a) Environments affect the survival of genotypes in a population and interact with each genotype to alter its morphology to improve adaptation. (b) Genotypic plasticity depends on preferential survival of genotypes within the population. Phenotypic plasticity is expressed as morphologic and physiologic changes of a genotype that improves its survival and broadens its adaptation. *Source:* After Nelson (2000).

roots is an example of a physiologic response to facilitate water uptake by roots from dry soil. In either case, the phenotypic **plasticity** exhibited is reversible as either morphologic or physiologic trait can revert to the original status when conditions change back to normal.

Alfalfa is an example of a species with great genotypic and phenotypic plasticity allowing it to be adapted throughout North America and in many ecoregions of other continents. But any one cultivar does not have this adaptation range so the breeder develops the genotypic plasticity for the average climate of a region with the potential for phenotypic plasticity to accommodate short-term needs for adaptation within that region.

Role of Management to Improve Plant Adaptation

Climate and climate change (Karl et al. 2009) also affect animal performance and management (Chapter 44, 45). Within the constraints of the climate and management system, choice of both animal and plant species is critical. Plant management techniques such as fertilizing or altering **defoliation** height broaden plant adaptation zones by mitigating the combination of climatic and management-induced stress (e.g. grazing method or frequent defoliation).

Seasonal patterns of dry matter production are signatures of regional climate (Figure 8.10). In turn, these patterns dictate animal and pasture management such as stocking rate and grazing method. Snaydon (1981) summarized seasonal pasture distribution on a world climatic scale, indicating broad climatic limitations to productivity. Largely, his illustration for "humid temperate" (Figure 8.10) approximates patterns for east of 98°W long in the US and eastern Canada, and north of 52°N lat in western Canada (Div. 210 to 250, Figure 8.2). This distribution is typical of a cool-season grass growth pattern. The area is characterized (Bailey 1996) by distinct winter and summer seasons; more precipitation occurs during summer than winter, and winters and summers have a range in cold and warm temperatures, respectively, from south to north (Figures 8.4 and 8.5).

Production in "Continental" (Figure 8.10) semiarid climates (mostly Div. 330 in Figure 8.2) occurs during early to mid-summer and is dictated by rainfall. This ecoregion (Texas to southern Saskatchewan) is typified by dry, cool or cold winters and hot summers with most precipitation occurring in summer; yet droughts are common. Thus, peak production is during early summer. By contrast, pasture production for "Mediterranean" climates (Div. 260, Figure 8.2) is limited by consistent summer drought, although winter precipitation is sufficient to support forest vegetation (Bailey 1996). Therefore, this grassland is dormant during summer and productive mainly during winter and spring. Pasture yield during fall and winter is limited by low solar radiation (Snaydon 1981). Pastures in wet tropical (Florida) climatic regions (Div. 410 and 420, Figure 8.2) are productive all year.

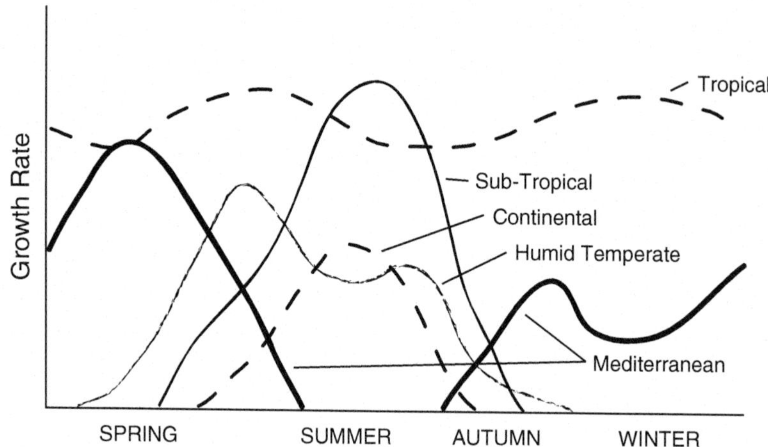

Fig. 8.10. Generalized patterns of seasonal dry matter production rates of pastures in several climatic zones. *Source:* Adapted from Snaydon (1981).

Limitations to production are mostly due to microclimate and landscape effects.

Differences in annual precipitation, PET and summer temperatures dictate that management philosophies differ east and west of 98°W long in the US and north and south of 52°N lat in western Canada. In the drier regions, there are fewer forage species used (Padbury et al. 2002) and forages occupy longer durations in the rotation sequence than in humid regions. This is mainly due to stand establishment problems (Entz et al. 1995).

Within the humid domain, there are opportunities to maximize seasonal production through filling in production gaps by using complementary species, although this potential differs according to latitude. From north to south within the humid and subhumid zones, length of the growing season, July temperatures, time of precipitation events and winter conditions influence the choice of species and the management strategy.

Length of the growing season increases from north to south ranging from 90 to 120 days in western Canada (Padbury et al. 2002) to 300 days or more in the southern transition zone of the US. Growth is limited, in order, by average dates of spring and fall frosts, mean air temperature below 0 °C and then by frozen soil, all of which occur at above 45° N lat, while only frosts occur in the southern transition zone lower than 33°N lat.

Cool-season grass production begins and peaks (Figure 8.11) between April and June over the Midwestern US and southern Ontario (Div. 210), but only from May to early July in northern Canada (Div. 250 and 130). High temperatures and low rainfall during summer (Figure 8.4) in the Midwest US cause cool-season grasses to grow slowly or be dormant. However, peak production of warm-season grasses occurs from June through September and complements the low production of cool-season grasses. Cool-season species can resume active growth from September to November due to cooler temperatures and higher rainfall (Nelson and Moser 1994).

In the northern Canadian Parklands and Prairies, alfalfa is grown to fill the production gap in mid to late summer (Figure 8.11) when cool-season grass species grow slowly due to above-optimal temperature and low soil moisture conditions. However, July temperatures are cooler than those in the Midwest and southern US (Figure 8.4) such that reduction of mid-season production by cool-season grasses in the northern areas is less pronounced and of shorter duration.

In the southern transition zone (Div. 230), warm-season species predominate from May through October. Tall fescue, one of the most heat-tolerant cool-season grasses (Nelson and Moser 1994) can survive the summer, but production is limited to winter and shorter spring and fall periods (Figure 8.11). Winter cereals such as winter wheat, triticale, winter rye and annual or italian ryegrass fill gaps from September to December and from January to May from Alabama to Texas and north to Alberta. The duration of annual forage use decreases from South to North.

Climate change will affect forage distribution causing management practices to adjust and choice of forage species may change regionally. As the drier ecozones (Div. 310 and 320) become hotter and drier in summer they may extend into the permanent grassland areas of the current Northern Great Plains (Div. 330) and eastern Prairie (Div. 250). There, they will have lower productivity and lower stocking rates will become necessary.

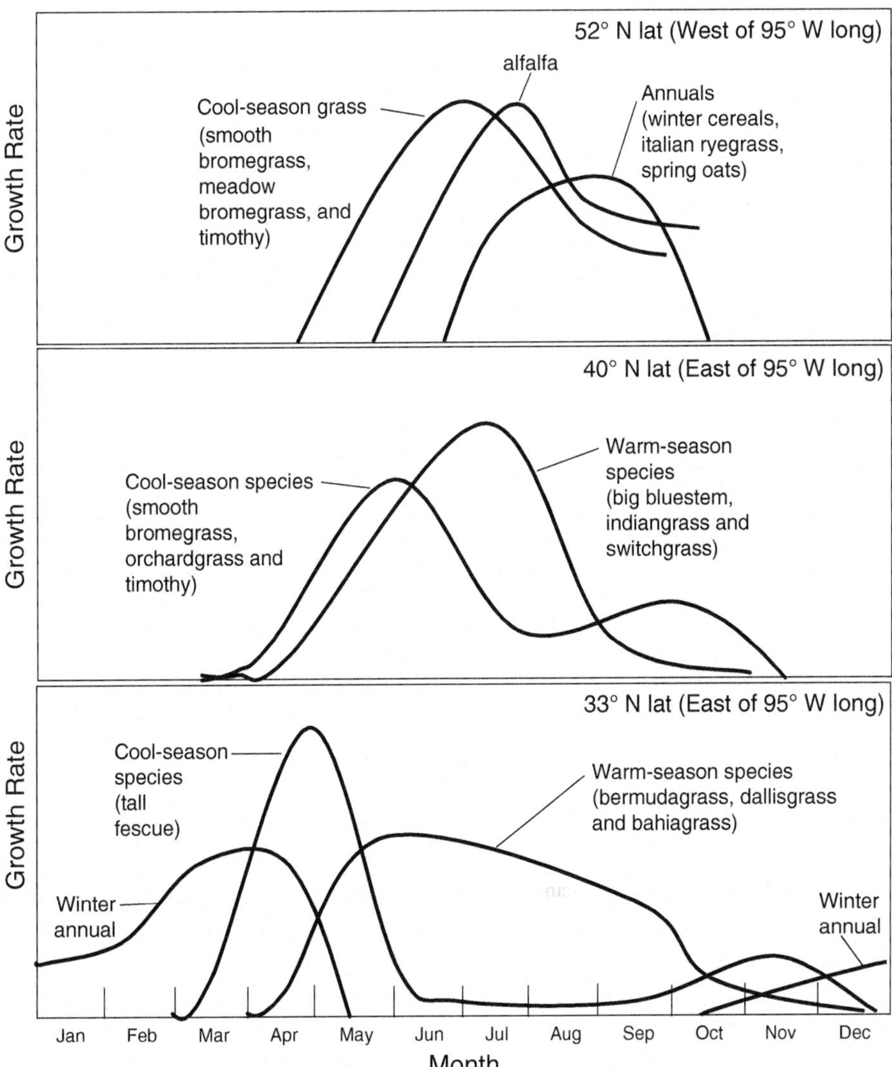

Fig. 8.11. Generalized patterns of seasonal distribution of forage dry matter production as affected by latitude and typical management within humid zones. Top is typical of western Canada; middle is typical of Midwest US, and bottom is typical of southeastern US. *Source:* Modified from Nelson and Moser (1994).

Establishing new and identical populations of native and naturalized species may be impossible (Izaurralde et al. 2011) because grassland populations may be in rapid transition without time for a typical equilibrium among species to be achieved until a tipping point occurs. The tipping point may be caused by frequency of drought or the reverse, prevalence of heavy rainfall and flooding that simply eliminates some species and creates a new ecosystem (Karl et al. 2009).

Growing seasons will be longer in the North and extend later into the fall, probably resulting in higher seasonal production in areas east of 100° lat. However, in the northwest, late summer precipitation may be lower and temperatures higher than optimum to support high regrowth yields in cool-season grasses like timothy (Jing et al. 2013).

In more humid ecoregions like 210, 220, and 130 north of 40°lat, yields of alfalfa may increase regionally due to increased rainfall, moving genotypes with reduced dormancy north with the possibility of additional cuttings per season (Thivierge et al. 2016). Simulated production of timothy under three cuttings compared to a normal two-cut system (Jing et al. 2014) resulted in greater annual yield, yield increases >1.75 Mg ha^{-1}, for five eastern locations, but much smaller yield increases, <1.15 Mg ha^{-1}, for six locations in western Canada, even though the growing season was extended in all locations. It was speculated that carbohydrate reserves of timothy might be sacrificed by taking the extra cut and that initial yield in subsequent years might be reduced.

Winter Hardiness

Winter hardiness of perennial forage species is determined by their ability to tolerate a wide range of environmental stresses, including subfreezing temperatures, excess soil moisture, ice encasement, heaving, and low-temperature pathogens (Andrews 1987). Entire stand loss can impact yield suddenly over a single winter if a combination of unfortunate climatic and management conditions occurs, or gradual loss of component species may occur over a period of years. The Midwest US, western and eastern Canada, and northern Europe are regions most affected by poor winter hardiness of winter-sensitive species like alfalfa (Castonguay et al. 2006).

Role of Cold Tolerance

Cold tolerance is the single most important factor involved in winter hardiness. The increase in cold tolerance occurs progressively during fall in response to shorter photoperiods and lower air temperatures that trigger developmental, biochemical, and molecular changes (McKenzie et al. 1988; Stout and Hall 1989). Cryoprotective sugars and cold-regulated proteins accumulate during this cold hardening (Chapter 4). The acquisition of cold tolerance accelerates upon exposure to air temperatures below 5 °C and continues under snow cover (Paquin and Pelletier 1980; Castonguay et al. 1993). A net accumulation of 100 days below 5 °C is deemed necessary to maximize the cold tolerance of alfalfa (Bélanger et al. 2006). Maximum cold tolerance is achieved in early winter after the soil has frozen.

In several areas of North America where perennial forage species are grown, air temperature frequently drops below the maximum cold tolerance. For example, the potential LT_{50} (temperature that causes 50% mortality) of field-grown alfalfa exposed to optimal hardening conditions is around −15 °C (Paquin and Mehuys 1980). Snow cover is therefore essential to provide insulation and protection against minimum air temperatures that can be as low as −44 °C (Figure 8.12). A snow cover of 0.1 m is considered sufficient to maintain air and soil temperature at the crown level near the freezing point (Leep et al. 2001).

Several environmental stresses cause winter damage indirectly through their effect on the capacity of the forage plants to develop sufficient tolerance to survive sub-freezing temperatures over winter. Wet soils will prevent forage plants from reaching their full hardening potential in the fall (Paquin and Mehuys 1980; McKenzie and McLean 1980). The amount of rain during fall, which directly affects soil moisture when evapotranspiration is low, is negatively correlated with winter survival (Ouellet and Desjardins 1981). The acquisition of cold tolerance can also be affected by dry conditions in the fall through a decrease in photosynthesis and accumulation of **organic reserves** important for cold tolerance.

Exposure of plant roots and crowns to temperatures above 0 °C in winter causes a loss of cold tolerance (Sakai and Larcher 1987). This dehardening occurs at a much faster rate than hardening, and winter damage can occur by exposure to even a few days of warm temperatures during winter if plants are not protected by stubble or snow (Eagles et al. 1997).

Roles of Ice, Frozen Soils and Pathogens

Rainfall or snow melt during winter may induce ice-sheet formation at the soil surface and ice encasement of plants (Gudleifsson 1993). The ice restricts gas exchange between the atmosphere and respiring root and crown tissues, plus oxygen consumption by soil microbes and decaying organic matter can result in toxic anoxic conditions. Ice encasement from alternate freezing and thawing may also physically damage plants through soil heaving.

Perennial forage legumes and grasses are prone to several diseases. Winter survival is likely affected by the severity of crown and root diseases because of a reduction in **organic reserves**. Conversely, sublethal injuries by freezing weaken plants and increase their sensitivity to other stresses such as infection by soil pathogens. Collectively, subfreezing temperatures, rapid loss of cold hardiness in spring, ice encasement, and soil heaving result in loss of stands and yield reduction of perennial forage crops in many northern forage-growing areas.

Indices were developed to estimate the relative risks associated with the most probable climatic causes of damage to perennial forage species (Bélanger et al. 2002). Fall indices express the influence of air temperature and soil moisture on the acquisition of cold tolerance. Winter indices assess the impact of subfreezing air temperatures, the loss of cold hardiness due to temperatures above 0 °C, and the potential damage to the root system by soil heaving and ice encasement. Those fall and winter indices can be used to assess the risks of winter injury due to the current and future climatic conditions (Bélanger et al. 2002; Thorsen and Höglind 2010).

Forage Species Differ in Cold Tolerance

The choice of species, cultivars, and crop management practices based on the entire growing season, but primarily in the fall, affect the risk of winter damage to

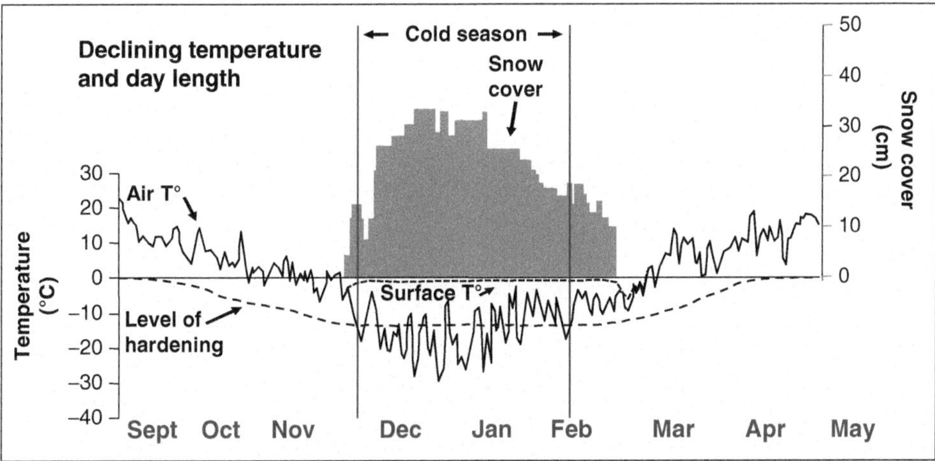

Fig. 8.12. Cold tolerance of plants increases as air temperatures decrease and daylength shortens during fall. Note soil temperature remains near zero during winter due to insulation by snow. Cold tolerance decreases rapidly in spring when air temperatures increase. *Source:* Modified from Belanger et al. (2006).

perennial crops. Generally cool-season grasses are more cold-tolerant than legumes (Ouellet 1976). Some alfalfa cultivars can tolerate temperatures at the crown level of −20 to −26 °C for a few hours but will be damaged when crowns are exposed to temperature of −8 to −10 °C for a few days (Paquin 1984). Timothy, however, can tolerate crown temperature as low as −30 °C (Paquin 1984).

In decreasing order, grasses with high cold tolerance are crested wheatgrass, russian wildrye, smooth bromegrass, and kentucky bluegrass; medium tolerance includes timothy and intermediate wheatgrass; while orchardgrass and reed canarygrass have low tolerance (Ouellet 1976; Gudleifsson et al. 1986; Limin and Fowler 1987), and tall fescue and perennial ryegrass have very low tolerance. Cold tolerance for legumes, in decreasing order, is sweetclover (very high), alfalfa (high), alsike clover, red clover and birdsfoot trefoil (intermediate) and "ladino" white clover (low) (Ouellet 1976).

Forage species also differ in their sensitivity to **anoxia** caused by the presence of ice on fields. Red clover is most sensitive followed in order by alfalfa, orchardgrass and timothy (Bertrand et al. 2001). Timothy delays development of anoxic conditions by maintaining a relatively slow metabolic rate during winter (Bertrand et al. 2003).

The winter hardiness, ability to survive stresses in winter, of species and cultivars within species depends on cold tolerance, fall dormancy, withstanding water-logging and resistance to root and crown diseases. As the result of specific adaptation to winter conditions, breeding and then selection, includes a range of resistance traits that mitigate the cumulative components of winterkill; i.e. genotypic plasticity, at the cultivar and clonal level within most species. Modern cultivars have genetically improved winter hardiness and disease resistance giving them a full range of genotypic and phenotypic plasticity which is important in dealing with winter hardiness.

Management to Improve Winter Survival

Special fall harvest management of winter-sensitive species like alfalfa is necessary to allow hardening and accumulation of adequate energy reserves to survive winter (Bélanger et al. 1999). Fall cutting management of alfalfa has long been centered around a critical fall rest period of 4–6 weeks for non-harvested growth preceding the expected date of the first **killing frost**. The regrowth interval between the fall harvest and the preceding one, however, appears to be a better determinant of winter survival and spring regrowth than calendar dates (Sheaffer et al. 1986; Bélanger et al. 1999; Dhont et al. 2004). An interval between the fall harvest and the previous one of at least 500 growing degree-days (5 °C basis) decreases the risk of winter damage. Not taking a fall harvest, however, remains the best strategy to avoid winterkill and ensure the long-term productivity. This allows plants to harden without interruption and retains the top growth overwinter to catch snow and provide a mulching effect on temperature of the soil and crown area of the plant (Figure 8.12).

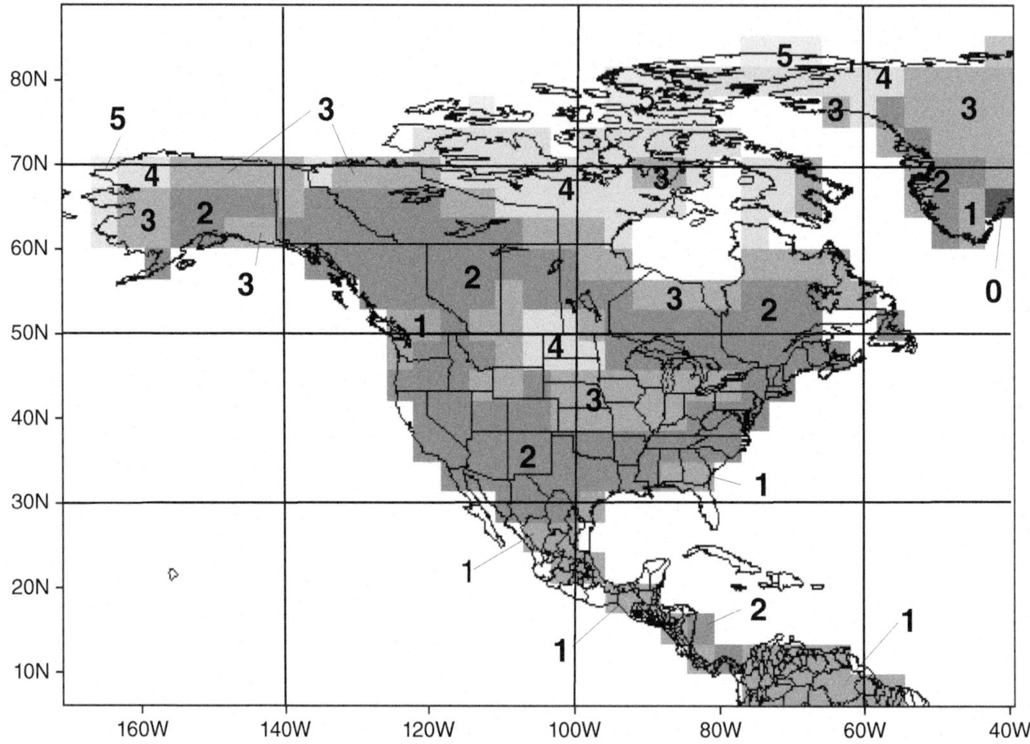

Fig. 8.13. Average annual temperature changes due to greenhouse gases as determined by Canadian Climate Impact Scenarios (2015) considering a modest increase in greenhouse gas emission rate. Numbers indicate temperature difference (°C) between 2000 and 2050. http://www.cics.uvic.ca/scenarios/index.cgi

High levels of soil-K are essential for development of maximum cold tolerance, but high levels of soil-N can promote growth in fall and prevent maximum development of cold tolerance, particularly in grasses. Drainage of heavy soils will reduce injury from **ice sheets** and soil heaving. Snow fencing, leaving residues on fields and windrowing at different stubble heights helps keep snow from blowing off fields and reduces the chance of low temperature injury (McKenzie et al. 1988).

Pending Effects Due to Climate Change

Use of **winterhardiness** zones and dormancy ratings as a basis to choose cultivars and manage perennial forages depends on the certainty of hardening and de-hardening patterns as impacted by environment within and among regions. Climate change factors disrupt and complicate the expected patterns and the increasing rate of climate change provides little time for natural selection processes to provide stable genetic populations needed for agricultural production.

In traditionally cold regions (northern Great Plains, northern Corn Belt; Ontario and Quebec), warmer fall temperatures will delay hardening (Fig. 8.13). In the northeastern US, increased soil moisture during fall due to intensive rainfall events delays hardening and may increase soil heaving. During winter, a reduction of snow depth, lack of snow cover (days of snow depth > 0.1 m.) relative to cold period (days air temperature < – 15 °C), more rainfall and ice encasement exposes plants to lower than expected soil temperatures. De-hardening may occur in late winter as air temperatures rise intermittently above 0.0° C. Daylength increases in late winter may reduce the chance of plants to re-harden. A general loss in average snow depth exposes de-hardened plants to low temperatures and reduces chance of survival.

In eastern Canada, climatic indices were used to assess climate change risks of winter damage to perennial crops such as alfalfa. Risks of winter injury to perennial forage crops in eastern Canada and the neighboring regions will increase because of less cold hardening during fall and reduced protective snow cover during the coldest period leading to increased exposure of plants to **killing frosts**, soil heaving, and ice encasement. Compared to current conditions, the hardening period in 2040–2069 would be four days shorter with less accumulation of hardiness-inducing cool temperatures (Bélanger et al. 2002). The cold period during which a temperature of less than −15 °C can occur would be reduced by 24 days, but with 39 fewer days having at least 10 cm of snow cover. Consequently, the duration of a protective snow cover during the cold period would be reduced by 16 days.

As climate changes, winter injury due to repeated freezing and thawing of the soil will move north of 40°N lat (Colorado to Kentucky). The northward advancement and retreat transition line of frozen soil in winter will be erratic over years and the latitude uncertain. Adapting to temperature increase by simply moving southern cultivars and populations north to accommodate a longer growing season, while increasing productivity, will meet with limited success without extensive research to reveal genetic response to environment, develop genomic-assisted selection, and have time for breeding and development of optimum agronomic practices.

Daylengths Do Not Change

The phenologic response to daylength and its effects on survival is an issue for both native and tame forage species. Tall grass species such as switchgrass, big bluestem and indiangrass are photoperiod sensitive with adaptation zones based on latitude (Moser and Vogel 1995). Latitude of origin impacts productivity and survival of switchgrass **ecotypes** (Casler et al. 2004), most likely based on adaptive genetic characters contributing to fitness (Milano et al. 2016). Flowering is induced by decreasing daylength (Lowry et al. 2014). Moving northern types south advances flowering date and decreases biomass yield. Moving southern types north delays flowering and increases biomass yield, but the late flowering date interrupts the fall hardening process causing winter injury and plant death (Moser and Vogel 1995).

Casler et al. (2004) observed that survival of switchgrass populations originating from divergent latitudes (range 36–46°N lat) were most sensitive to movement in opposite directions. Thus, while some ecotypes may be more tolerant to northern relocation than others, Casler et al. (2004) concluded that movement north or south of a single hardiness zone of origin could result in losses in biomass yield and survival. A successful move may be the equivalent of 500 km (Moser and Vogel 1995).

Movement of non-dormant alfalfa genotypes north in the mid-latitude US may increase seasonal yield by taking advantage of greater growing season length. However, in regions where soils freeze, sufficient hardening will be required. There is uncertainty about the needed alfalfa hardiness and dormancy requirement while moving northward due to a changing temperature, combined with an existing higher rate of change in daylength in the fall (Castonguay et al. 2006). The positive relationships among dormancy, winterhardiness and reduced fall growth are well known (Smith 1961). Dormancy and hardening are triggered by decreasing photoperiod and temperature, but the rate of change in photoperiod appears more important than photoperiod per se (Castonguay et al. 2006). Rapid changes in photoperiod as observed during fall in Alaska were not perceived by mid-lat US alfalfa cultivars to induce hardening (Bula et al. 1956; Klebesadel 1971). As latitude increases, the rate at which daylength declines in the fall increases such that growth slows at increasingly earlier dates as plants are moved north, resulting in almost no fall regrowth in existing winter hardy *Medicago falcata* genotypes.

The presence of dormancy in the most winterhardy alfalfa genotypes, e.g. falcata in the northern US and Canada reduced fall growth to the extent that only one cut per season is recommended (Hendrickson et al. 2008; McLeod et al. 2009). Selection of low dormancy genotypes (taller plants) from dormant populations under artificial 12-hour daylengths at 15.5° C (Castonguay et al. 2006) may be a step toward developing populations with lower dormancy, while maintaining winterhardiness for northern alfalfa production. Sources of low dormancy in *M. falcata* germplasm have been identified (Brummer et al. 2000). Likewise, genetic sources of winterhardiness have been found in non-dormant alfalfa populations (Weishaar et al. 2005). But, none of these avenues has been used in cultivar development.

High Temperatures

The C_4 photosynthetic system has a distinct advantage over the C_3 system in regions with high day and night temperatures. The absence of photorespiration and higher stomatal resistance to water loss of C_4 species confers advantages over C_3 species in warm, dry summers (Nelson and Moser 1994), mainly south of 40°N lat (Figures 8.3 and 8.4).

Most forage crops are mesophyllic, preferring intermediate temperatures between 10 and 30 °C. The threshold temperature for onset of high-temperature stress varies with the species, but is in the range of 35–45 °C (Table 8.3). In addition, duration of exposure interacts with temperature to cause injury. Note that long exposures to moderately high temperatures (i.e. 40 °C for 60 min) can cause more injury than short-term exposure to a higher temperature (e.g. 50 °C for 5 min).

High temperatures increase tissue respiration rates and can reduce carbohydrate storage, growth rate and plant survival (Table 8.3). This is particularly limiting to C_3 species as they also express photorespiration (Nelson and Volenec 1995). High-temperature stress frequently occurs concurrently with moisture stress (e.g. West Texas, Oklahoma and Kansas, Figures 8.4 and 8.8). Limited water availability decreases **transpiration** and evaporative cooling, so high-temperature stress is exacerbated (Nelson and Volenec 1995).

High-temperature stress also is influenced by extent of water deficit stress. Temperatures of above-ground plant tissue often differ markedly, both lower and higher than the surrounding air temperatures and need to be measured directly by infra-red telethermometry to determine the degree of stress. If ample water is available, transpirational cooling can reduce leaf temperatures by 5 °C or more. Latent heat of vaporization consumes 540 cal mole^{-1} of water evaporated, the heat coming from the leaf mass. When water deficit stress occurs, stomata close partially to limit transpiration and its associated cooling, and leaf temperatures increase due to continued absorption of solar radiation. Under more severe stress, buliform cells between the veins of grass leaves decrease in volume and the blade rolls inward reducing the surface area and heat load.

Resisting high-temperature stress falls into two general categories: heat avoidance and heat tolerance. Avoiding high-temperature stress occurs when plants do not experience stress while exposed to high-air temperature. One avoidance mechanism is to lower temperature of plant surfaces by irrigation or surface wetting ("syringing") to enhance evaporation. Plant temperatures of an irrigated crop under arid conditions can be 10 °C cooler than ambient air temperature (Hatfield et al. 2008).

Heat tolerance occurs when plants have an enhanced ability to survive when experiencing high temperatures. Examples of heat-tolerance mechanisms include: maintenance of proper membrane fluidity by increase in saturation of membrane fatty acids, synthesis of heat-stable isozymes (alternate forms) of key enzymes, accumulation of carbohydrate reserves, and synthesis of heat-shock proteins that stabilize existing proteins or assist with cellular targeting of newly synthesized proteins. Though the underlying mechanism is not known, an increase in synthesis of heat-shock proteins has been correlated with heat tolerance across several plant species.

In warm environments, plant development is accelerated; plants tend to be shorter and bloom earlier than in cool environments. This contributes to declines in yield during the hot summer period for cool-season grasses and legume species such as alfalfa and red clover. Conversely, tropical species with C_4 photosynthesis are best adapted and most productive under hot, humid conditions (Nelson and Volenec 1995).

Cardinal temperatures, which appear to place limits on plant process-efficiency to the extent that growth and reproduction are impaired, are given in air temperature not actual plant tissue temperature. Plant species may be able to adjust or compete to advantage by altering flowering time (phenology) to take advantage of a cooler vs warmer time of the year to facilitate seed growth and dispersion.

Temperature changes due to climate change will affect processes such as photosynthesis, respiration and phenology directly and the plant environment indirectly. Indirect effects are extending the growing season, increasing soil mineralization, altering soil water content and ultimately shifting species composition (Hatfield et al. 2008).

Gradually, C_4 species will be predominant in warmer and C_3 species in cooler locations (Teeri and Stowe 1976) or C_4 species will grow primarily in warmer (summer) and C_3 will grow in cooler times (spring and fall) of the year. Direct effects of high-plant temperature generally reduce yield as respiration and photorespiration are affected more than photosynthesis, plants flower and mature earlier, leaf area is reduced and less absorption of radiation leads to lower **net primary productivity** (NPP).

The indirect effects of temperature or other environmental drivers may be gradual and only be revealed in grassland systems after decades of observation during which time natural genetic selection will occur (Thornley and Cannell 1997). For example, modeling results for temperate pastures in a humid zone by Thornley and Cannell (1977) indicated that rising temperatures, alone, resulted in enhanced CO_2 generated mostly by increased soil respiration and rapid leaf senescence, which reduced leaf area. Conversely, rising CO_2 concentration increased NPP, resulting in a carbon sink. Combining the effects also resulted in a carbon sink. However, even though soil organic matter degradation lags behind the C-inputs from NPP, a soil–carbon equilibrium should eventually occur, minimizing additional soil-C sequestration.

Drought

Drought occurs when soil moisture supplies are reduced below expected values that affect plant growth. A drought may range from a few days in a moist subtropical region to months without precipitation in semiarid regions. For example, droughts persisted for a decade in the Great Plains (Basara et al. 2013) from Texas to Saskatchewan in the 1930s and 1950s. These long droughts were associated with periods of higher than normal temperature, but were not considered a by-product of climate change at the time. The causes of regional drought trends are highly complex, not identical, and their timing not highly predictable. The recorded droughts of the past century in the Great Plains are legendary, but more severe and longer droughts likely occurred in prior decades and centuries (Lettenmaier et al. 2008) as indicated by records of fossilized tree rings.

Many factors combine to cause droughts. Studies by NASA indicate that one-third of the historic variation in precipitation on the Great Plains has been due to increased sea-surface temperatures (SST) and two-thirds due to soil-atmospheric interactions (Basara et al. 2013). Agriculture on the Great Plains depends on moisture moving north from the Gulf of Mexico as far as Canada.

The common summer drought phenomenon is known as subsidence or sinking of the atmosphere, which causes air pressure aloft to hold the moist gulf air closer to the earth's surface. The moist air cannot rise, cool, condense, and fall as rain as it normally would. Instead, the air gains heat from the earth's surface, resulting in a warm, dry lower atmosphere, which draws additional water from the moister surface vegetation.

Highly impactful are teleconnections which are correlated weather events that are geographically far apart, but affect precipitation and drought in a distant location. Warm pacific sea-surface temperature causes increasing precipitation (El Niño), while cooler sea-surface temperatures cause drought (La Niña) on the Great Plains. Due to subsidence or La Niña, the dry soil becomes warmer than normal, exposed to direct solar radiation by reduced green leaf area due to drought stress and lower latent heat of vaporization that would have cooled a moister soil. The warm, dry soil initiates soil-atmospheric feedbacks by passing a higher sensitive heat flux to the air. The factors build on one another in sustained drought.

The effects of long-term drought may linger. Reduced vegetation cover due to plowing, aridity and thin wheat stands in the 1930s contributed to the length of that drought. Farmers recognize it takes years to recover from a drought; rate of recovery from drought is inversely related to its duration. In fact, the duration of drought is affected by the quantity of precipitation necessary to restore soil moisture and watersheds to original levels. Estimates are the 1930s or 1950s drought in Nebraska would require seven times the average two-month precipitation in winter (lower quantity than summer) and three times the average two-month precipitation in spring and summer to restore normalcy (Basara et al. 2013).

Plant Responses to Drought

Species grown near centers of their adaptation zones should best survive droughts. Crested wheatgrass plants survived 15–20 years of fluctuating precipitation in the northern Great Plains (Looman and Heinrichs 1973). However, stand loss of some recommended species can occur in dry areas (Currie and White 1982). Recommendations are based on normal variance in climatic conditions and they do not consider the rare extreme events.

Drought avoidance and drought tolerance are the two categories used to describe plant adaptation to water stress (Volenec and Nelson 2007). Drought avoidance occurs when plants are exposed to water stress, and is generally expressed by increased water uptake or decreased water loss during a seemingly stressful period. Specific drought avoidance mechanisms of plants include extensive root systems, modifying **transpiration** with stomatal control, cuticle structure or leaf shedding. Stress can also be managed by using parts of the growing season when water is more abundant; for example, using winter-annual or summer-annual species.

Drought tolerance occurs when plant tissues are able to function and/or adjust to survive drought stress. Continued functioning is not common among forages, with the exception of sorghum and millet, which can stop growing during extended periods of drought and quickly resume growth when water stress is relieved. Instead, forage grasses often become dormant and gradually resume growth once water is again available.

Plant physiologic responses to drought as described by Frank et al. (1996) are more closely associated with gradual soil-water depletion than with rapid short-term changes in plant-water status. Examples of morphologic adaptation to drought include changes in root morphology (Nelson 2000) and reduced crown size and development of smaller daughter crowns from rhizomes of western wheatgrass (Currie and White 1982). In general, the deep taproot of alfalfa allows it to extract water from deeper soil zones than grasses (Shaeffer et al. 1988). Slow growing, creeping-rooted, falcata-type alfalfa cultivars are adapted to the semiarid northern Great Plains (Melton et al. 1988). A combination of small plant size, low leaf area and deep roots provides moisture procurement and conservation for dryland grasses such as western wheatgrass (Frank et al. 1996).

Severe drought and close grazing can diminish ground cover and shift species dominance in native grasslands. During long-term drought, western wheatgrass replaced needle and thread as the dominant species under grazing in eastern Montana (Rogler and Haas 1946). However, changes among grassland species and populations must be considered over a long-term as population dynamics routinely ebb and flow due to climatic cycles. For example, vegetation studies over 84 years at Mandan, ND showed that species composition for blue grama under moderate grazing fluctuated from 19% of the stand in 1916 to 64% in 1964 to only 3% in 1998 (Frank et al. 1999).

Drought affects seasonal yield through reduced growth, and affects long-term yield through lack of persistence in cultivated species. In the north central US, drought reduced regrowth yield of reed canarygrass, orchardgrass, smooth bromegrass and timothy by 33%, 37%, 24%, and 34%, respectively, but reduced total yield by 54%, 60%, 81%, and 62% of the irrigated controls, respectively (Sheaffer et al. 1992). In another Minnesota study under drought, annual yields of alfalfa were 120% of birdsfoot trefoil and cicer milkvetch and 165% of red clover (Peterson et al. 1992). Drought reduced persistence of reed canarygrass and timothy, while smooth bromegrass persistence was unaffected by drought (Sheaffer et al. 1992). Enhanced growth rates of grasses following drought may be compensatory and are related to increased carbohydrate storage during the dry period (Horst and Nelson 1979).

Matching Species with Climatic Regions

Growth patterns of adapted species in response to drought are consistent with drought patterns of their adaptation zones. In the southwest, central plains and prairie states of the US (below 40° N lat in Div. 330, 250, and 310), native C_4 grasses have a larger root system and are most drought hardy, with growth patterns matching up with peak solar radiation during summer (Nelson and Moser 1994; Figure 8.11). The C_4 species have greater CO_2 uptake per unit stomatal conductance than C_3 species (Buxton and Fales 1994). This makes C_4 species ideal for regions with periodic high temperature and drought conditions.

In cooler semiarid states bordering Canada and the Palliser Triangle of the southern Prairie Provinces (Div. 330 and 250), crested wheatgrass (C_3 species) survives drought by water procurement through deep roots (Frank and Bauer 1991) and has rapid leaf senescence as a means to reduce transpirational area.

In cool, humid areas such as Minnesota, Wisconsin, Southern Ontario, and Quebec (Div. 210), droughts are intermittent and shorter than in semiarid areas, so adaptive strategies are not focused on survival as much as on maintaining plant size until the drought ends. Thus, mesic species such as timothy and orchardgrass extract water from near the soil surface, but orchardgrass tends to maintain green leaf area during drought to support rapid recovery after the drought ends (Frank et al. 1996).

In the prairies of Canada and the US located north of 40°N lat (Div. 250), drought is intermediate between that in humid and semiarid regions. Smooth bromegrass occupies these transitional prairie zones because it is less susceptible to leaf senescence than crested wheatgrass (Frank et al. 1996), but is more likely to reduce growth during drought to a greater degree than orchardgrass and timothy (Shaeffer et al. 1992). Alfalfa is more tolerant than birdsfoot trefoil, cicer milkvetch and red clover of droughts experienced in the north central US (Div. 210) (Peterson et al. 1992) and is found in ecoregions from semiarid to subarctic (Div. 330, 250, and 130) in western Canada (Hill et al. 2000).

Variability in Climates

Among the effects of climate change, droughts are difficult to attribute to GHG emissions per se. However, the associated effects of climate change support the contention that droughts will be more severe and common than in recent decades (Walsh et al. 2014). The summer drought in Texas of 2011 was primarily driven by lack of rainfall, but the human contribution of high GHG concentrations increased the duration of intense heat (96 and 70 days > 38° C in San Angelo and Dallas, TX, respectively) and doubled the probability of drought severity (Walsh et al. 2014).

As in the past, future drought will not be identical geographically. Forages and other crops will be stressed to their adaptation limits due to drier conditions relative to time of year and phenology. In the north and west, the cumulative factors are the annual, repetitive drying out in the fall, longer growing seasons with less precipitation and lower soil moisture as a result of low summer rainfall, warmer day and night temperatures, and less snowfall.

Intense weather events exacerbate and overlay the general climatic trends in all regions. Whereas precipitation increases from New England and the Great Lakes to Florida, most of it will come in intense storms with runoff, soil puddling and flooding of low areas. These large storms may be matched by record numbers of dry and hot days in the form of heatwaves. The longer dry periods may cause drought and heat stress for currently adapted species.

In the southwest, southern Great Plains and the western side of the Prairie region, the intense weather events may be in the form of heatwaves, less rainfall, consecutive dry days and the generally drying out of soil. The lack of a cold winter, less prevalent snow and presence of a dry atmosphere may predispose grass and shrub land to more wildfires in these regions.

Climate Change

Averaged globally, land and ocean surface temperatures have increased by 0.85 °C from 1880 to 2012, and each of the past three decades has been successively warmer than the last (Hartmann et al. 2013). The International Panel on Climate Change (IPCC) in its Fifth Assessment Report indicated that North America could warm by more than 2 °C over the next century (Romero-Lankao et al. 2014). The US average temperature has increased by 0.73 °C to 1.1 °C since 1895 with most of this increase occurring since about 1970 (Walsh et al. 2014). In Canada, the annual air temperature has warmed by 1.5 °C from 1950 to 2010 (Bush et al. 2014) which is consistent with predictions that effects will increase more in northern latitudes.

Warming is projected for all parts of the US and Canada during this century. Under lower emissions, scenarios involving substantial reductions in emissions in the US, this warming will be 1.1 °C to 2.2 °C in the next few decades and 1.7 °C to 2.8 °C by the end of the century. However, if current emission trends continue, average annual temperature is projected to increase 2.8 °C to 5.6 °C in the longer term (Walsh et al. 2014). The largest temperature increases in the US are projected for the upper Midwest and Alaska.

How Do We Know that Climate Change Is Real?

Future climate change can only be estimated using models for which there is no control, however model verification with past climate and weather data is possible. The entire body of information, including estimated changes in the global energy balance, support the contention that climate change is real and that human activities are playing a role (Hartmann et al. 2013; Walsh et al. 2014).

Mean temperature change of the earth's surface over the last 50 years cannot be explained by natural causes such as solar forcing or dimming, or the reflection of solar radiation back to space by particles from volcanoes or pollution (Walsh et al. 2014). Accumulation of human-derived GHG are correlated with the increasing temperature. There is a cause and effect relationship between the increasing concentration of GHG (Table 8.4) and capacity to absorb and back-radiate long-wave energy to the earth. Reconstruction of past climates based on tree ring, ice core and coral data indicate that temperatures have risen at an otherwise unaccountable rate over the last several decades.

Research shows that when human factors are removed as model-inputs the earth would have cooled instead of warmed over the last several decades. Other observed changes supporting the idea that human-induced climate change is active are: temperature increases in the lower stratosphere show the entire ocean-earth-atmosphere system is warming; areas occupied by glaciers, snow cover and sea ice are shrinking or melting; sea levels have increased because water expands as temperature rises and sea ice melts; changing precipitation patterns and increasing humidity (Hartmann et al. 2013; Walsh et al. 2014).

Greenhouse Effect

The greenhouse effect is a natural feature of the climate system and refers to the tendency of the atmosphere to create a warmer climate than would otherwise be the case (Mearns 2000). The greenhouse effect is due to the re-radiation of long-wave **infrared** radiation (Figure 8.1) absorbed by water vapor, CO_2 and other GHG. The downward emission serves to heat the earth. Without it, the average temperature on our planet would be about 33 °C colder, and life as we know it could not exist.

The amount of long-wave radiation that is absorbed and then reradiated downward is a function of the constituents of the atmosphere (Mearns 2000). The GHG, water vapor, CO_2, CH_4, some chlorofluorocarbons (CFCS) and N_2O, are particularly good at absorbing long-wave radiation. An increase in the concentrations of those GHG results in more of the long-wave infrared radiation from earth being absorbed by the atmosphere and then reradiated back to the earth. Because of human activities, atmospheric concentrations of GHG (except water vapor) have increased considerably since the beginning of the industrial revolution (Table 8.4). It has been estimated that between 10 to 20% of the greenhouse effect is related to agricultural activities.

Greenhouse Gas Emission from Agriculture

Agriculture in the US and Canada is responsible for 8–9% of total national emissions, respectively (Environment and Climate Change Canada (ECCC) 2018; Environmental Protection Agency (EPA) (2018). The emission profiles are proportionately similar and reflective of industrialized nations. The three main GHG (CO_2, CH_4, and N_2O) generated by agriculture differ in their global warming potential and are represented on a CO_2 equivalent (CO_2e) basis. One kg of N_2O has a global warming potential that is 221 times that of CO_2, whereas one kg of CH_4 has the warming potential of 27 kg CO_2 (Table 8.4). Of all GHG emissions in the US, 81% are from CO_2, 6% from N_2O and 10% CH_4 (EPA 2018) on a CO_2 equivelent basis.

In the US and Canada, the majority of total N_2O emissions comes from agriculture (ECCC 2018; EPA 2018). The N_2O is produced primarily when excess nitrates in soil undergo denitrification. Sources are fertilizer nitrogen (N) application, N additions to soil from legumes and N inputs from decay of above and below ground residues including roots from all crops (EPA 2018). Cropland in the US, which includes hay (alfalfa and grass-legume mixtures) and grazed cropland, was responsible for 73% while long-term grassland and pasture was responsible for 27% of N_2O emissions from the soil (EPA 2018). In Canada, N_2O derived from inorganic N-fertilizers accounted for 22% of total agricultural emissions (ECCC 2018).

About 80% of national and global GHG emission is due to fossil fuel combustion from all sources (ECCC 2018; EPA 2018). Energy-based emissions, as CO_2e in agriculture, originate from farm mechanization (diesel fuel and electricity), manufacture of farm equipment and inputs such as herbicides and fertilizer, including their packaging and transport to the farm. In Canada, use of the fossil fuel equivalent for energy in agriculture represented less than 1% of all GHG emissions compared to 77% for N_2O (ECCC 2018).

Methane (CH_4) from enteric fermentation, as a result of ruminant digestion, is responsible for 30% and CH_4 from manure is 12% of the total US agricultural emissions (EPA 2018). In Canada, CH_4 from enteric fermentation accounted for 41% of total agricultural emissions (ECCC 2018).

Enteric CH_4 emission from ruminants is an important GHG issue that is closely related to forage and pasture utilization. Methane (CH_4) production (eructation by ruminants) is a natural by-product of **anaerobic respiration** and its production serves as the principal electron sink within the rumen. **Methane** (CH_4) represents a

Table 8.4 Atmospheric composition of greenhouse gases produced by agriculture activities and their effect on global climate

Gas[a]	CO_2	CH_4	N_2O
Pre-industrial concentration	278 ppm	722 ppb	270 ppb
Atmospheric lifetime[b]	>200 yr	9.1 yr	131 yr
Global warming potential[c]	1	27	221

Source: Hartmann et al. (2013).
[a] CO_2 is carbon dioxide. CH_4 is methane. N_2O is nitrous oxide.
[b] Time duration for turnover.
[c] Data are relative with CO_2 having a base of 1.

significant loss of dietary energy (Beauchemin et al. 2008). US beef and dairy cattle populations contribute 71% and 25%, respectively, of the total CH_4 from enteric fermentation (EPA 2018). In the Canadian beef production cycle, from birth to harvest, the cow herd is responsible for 80%, largely CH_4, and the feedlot sector for 20% of the on-farm GHG emissions (Beauchemin et al. 2010; Basarab et al. 2012).

Ruminants consuming roughage or forage emit more CH_4 per unit intake than those consuming concentrate or grain diets (Beauchemin et al. 2008). Improved forage quality reduces CH_4 emission. Some legumes containing condensed tannins, such as big trefoil, birdsfoot trefoil and sainfoin, have the potential to reduce CH_4 emission per unit forage fed, though all have some agronomic or adaptive deficiencies which prevent them from being grown widely (Beauchemin et al. 2008). Also, C_3 grass species yielded less CH_4 per kg fed than C_4 species, and feeding corn silage may result in less CH_4 emission than grass silage since starch reduces methanogenesis (Beauchemin et al. 2008). However, reducing national herd size through improved efficiency of beef and dairy production systems, reduces the national CH_4 emission in both countries much more than improving dietary efficiencies per se (EPA 2018).

Carbon Sequestration

Carbon sequestration in soils, originating from crop and grassland photosynthesis, may offset some of the GHG emission from agriculture and society. Carbon sequestration in soil organic matter (SOC) is the equivalent of removing CO_2 from the atmosphere and, therefore, is considered a sink, and a GHG mitigation option. Each kg of C sequestered in SOC offsets its equivalent (CO_2e mass = 3.66 kg × 1 kg C) in GHG emitted.

Grasslands and forages are important for storing and increasing SOC pools or stores (West and Six 2007; Council for Agricultural Science and Technology [CAST] 2011; Conant et al. 2017). The pools are organic, therefore dynamic, and subject to growth (by sequestration) and loss (microbial respiration and erosion) over years until SOC reaches a saturation or equilibrium point (West and Six 2007). Organic carbon inputs to soil consisting of residues and roots from crops and grasslands undergo several iterations of microbial degradation or metabolism, all giving off CO_2 (respiratory processes) before a stable quantity of SOC is attained (West and Six 2007).

Most long-term grasslands are at a steady-state equilibrium having a SOC growth rate of zero or being subject to small sequestration rates and small carbon losses depending on year-to-year changes in climate and management (West and Six 2007; CAST 2011; EPA 2018). Grasslands in the US have had SOC gains tempered by losses due to drought and wild fires (EPA 2018). The 265 million ha of permanent grasslands in the US (Bigelow and Borchers 2017) make their management significant to the maintenance and dynamics of SOC storage (Derner and Schuman 2007). Generally, Canadian permanent grasslands have lost small amounts of SOC annually (< 1% of annual total agricultural emissions) over the last 10 years and are small sources (ECCC 2018).

Land-use change or conversion from cropland to grassland systems and the reverse (breaking) tend to have large impacts on rates of C-sequestration maintenance and loss of the agriculturally-based SOC pools. After the original breaking of native grassland and subsequent years of cultivation, 40% or more of original SOC mass has been lost (Voroney et al. 1981; Schimel et al. 1985).

Potentially high rates of SOC accumulation may occur in newly established pasture or hay land on crop fields and in restoration of previously degraded grasslands (Franzluebbers 2007 2010; CAST 2011; Conant et al. 2017). Establishment of perennial forage on previously eroded cropland in southeast US averaged an accumulation rate of 1.0 $Mg\,ha^{-1}\,yr^{-1}$ SOC over 15 year (Franzluebbers 2007). Grassland management change (e.g. rotational grazing, fertilization) may have positive effects on SOC stores, but the magnitude of annual sequestration is inversely proportional to the state of original soil degradation (Franzluebbers 2007, 2010; CAST 2011) and is smaller than that due to land use change (i.e. cropland to grassland).

Cropland per se in Canada and in the US (159 M ha, Bigelow and Borchers 2017), consistently acts as a C-sink. In 2016, C-sequestration by Canadian cropland offset agriculture emissions by 18%. However, in both Canada and the US, this C-sink is shrinking due to the way land is managed and used (CAST 2011; EPA 2018; ECCC 2018). Soil-C stocks increased annually due to sequestration when US cropland was converted to the Conservation Reserve Program (CRP) or hay lands, especially if conservation tillage was adopted or there was a reduction in fallow. However, the annual contribution of sequestered SOC by US cropland in 2016 was only 44% of that from 1990, due to reduction in area occupied by **cropland pasture**, hay land and less area being enrolled in the CRP (CAST 2011; EPA 2018).

Reduction in SOC sequestration in Canadian cropland is due to reductions in areas of alfalfa and alfalfa-grass mixtures for hay, and in tame pasture areas in favor of more lucrative grain and oilseed production. US cropland pasture area peaked in 1968 at 36 million ha, decreased to 25 million ha by 2002 and was 5.3 million ha in 2012 (Bigelow and Borchers 2017). Hay area in the US has declined 33% since 1944, and area of alfalfa hay has decreased 37% since 1979 (Zulauf 2018). Perennial forage is a valued GHG offset but the drastic loss of cropland pasture and perennial hay land constitutes a significant net loss of SOC accumulation. Also, in the long-term CRP program (9.5 million ha), soil-C sequestration rates are slowing down as SOC stocks reach equilibrium about 30 year after establishment (CAST 2011).

Effects of Climate Change on Forage Production

Most global climate change scenarios indicate that higher latitudes in North America would undergo warming that would lengthen the frost-free season under climate change, ranging from a minimum of one week to a maximum of nine weeks (Brklacich et al. 1997, cited in IPCC 2001). A limited northward shift in production areas in both the US and Canada is therefore likely to occur as a consequence of climate warming. Substantial changes to the distribution of ecosystems and to disturbance regimes will likely increase probabilities of fire and drought (IPCC 2001). Subtropical conditions are expected to extend further North into the lower portions of the US, with changes in the distribution of C_3 and C_4 species.

Carbon dioxide enrichment and climate warming are predicted to increase canopy photosynthesis and NPP on most range and forage land (Nösberger et al. 2000; Polley et al. 2000; Izaurralde et al. 2011; Soussana and Lüscher 2017; McGranahan and Yurkonis 2018). The doubling of ambient CO_2 concentration could increase annual grassland production by 17% (Campbell and Smith 2000; Ainsworth and Long 2005). The size of this increase will vary with climatic and crop management factors as well as location. Combined increases of temperature and CO_2 concentration may enhance production in temperate regions (Soussana and Lüscher 2017; Izaurralde et al. 2011), but could vary with forage species. Reduced water loss and enhanced WUE have been observed at the canopy level which would improve plant and soil water relations and, increase plant production under water limitation. In principle, this lengthens the effective growing season (Soussana and Lüscher 2017).

Forage yield response to climate change, however, is expected to vary across North America (Izaurralde et al. 2003; Izaurralde et al. 2011). The positive effect of warmer temperatures and higher CO_2 concentration on production may be lessened by an accompanying increase in evapotranspiration rate in drier areas, as well as by interacting factors that feed back to limit net productivity (Izaurralde et al. 2011; Polley et al. 2011). For example, higher respiratory responses to temperature relative to photosynthesis (Volenec and Nelson 2007) will likely partially limit the yield expected from the higher CO_2 benefits. Impacts of temperature and photoperiod on phenology may reduce leaf area, resulting in lack of seasonal synchronization with advantageous soil moisture, radiation interception, soil mineralization, acclimation for winter, or use of the extended growing season (Izaurralde et al. 2011; Hatfield et al. 2011).

Izaurralde et al. (2011) concluded precipitation contributed most to seasonal yield of alfalfa, followed by CO_2 concentration and then temperature. Alfalfa yields in the US are not expected to change at the national level (Hatfield et al. 2014); the increase in eastern regions will offset the decline in the central and western US. Predicted future conditions of temperature and CO_2 concentration had little effect on of timothy (Piva et al. 2013).

Due to climate change, annual yields of alfalfa-timothy mixtures are expected to increase from 5 to 35% in eastern Canada by 2050–2079 (Thivierge et al. 2016). In another study, annual yields of timothy will increase by 0.46 to 2.47 Mg DM ha^{-1} by 2040–2069, being greater in eastern than in western Canada (Jing et al. 2014). The increase in Canada is due mostly to the longer growing season and more cuttings. Forage yields of first cuttings may increase, but the summer re-growths of timothy and alfalfa-timothy mixtures are expected to decrease because of an increased water stress (Jing et al. 2013; Thivierge et al. 2016).

Many studies on temperature and CO_2 concentration effects on perennial crop productivity are relatively short term and therefore may underestimate long-term climatic feedbacks on productivity (Thornley and Cannel 1997). Variability to time sensitive soil mineralization rates and SOC accumulation, reinforce the need for field and pasture verification of models. Low N availability could limit the response of grasslands to elevated atmospheric

CO_2 concentrations (Soussana and Lüscher 2017) as soil C : N ratios adjust to management (Thornley and Cannel 1997). In the longer term, those species that can adapt using genotypic plasticity for flowering, seed production and seedling recruitment may survive better.

In western US rangelands, climate change is expected to increase production in northern regions, but decrease production in southern regions while increasing interannual variability of production levels (Reeves et al. 2017). Water limits productivity of most rangelands; therefore, changes in frequency or amount of precipitation will have a significant effect (Polley et al. 2000, 2011). Arid and semi-arid lands will be most sensitive to changes in precipitation, while wet mountain meadows will be less affected.

Elevated UV-B radiation associated with climate change is expected to have little effect on primary production of grasslands (Norton et al. 1999; Papadopoulos et al. 1999). Increased ozone (O_3) concentration, however, has been shown to decrease forage yield and the contribution of legumes in grass-legume mixtures (Nösberger et al. 2000). Fortunately, the O_3 situation has been reduced by international bans on use of many aerosol sprays and refrigeration gasses.

It is hard to predict if positive or negative changes in forage quality will occur because of interactions among changes in lifeform distributions, species distributions, and/or plant biochemical properties (Polley et al. 2000). The relative abundance of woody species and grasses, and of C_3 and C_4 species, will be affected. For example, global warming may decrease forage quality by favoring C_4 over C_3 grasses in an area whereas increased CO_2 concentration may improve forage quality by favoring legumes over grasses (Dumont et al. 2015). Increased CO_2 will likely decrease crude protein concentration and increase non-structural carbohydrates of forages (Polley et al. 2000; Körner 2002). A meta-analysis revealed that elevated atmospheric CO_2 increases nonstructural carbohydrates of forages in mountainous and Mediterranean areas by an average of 25%, decreases N concentration by 8%, but has little effect on forage digestibility and fiber concentration (Dumont et al. 2015). In the semiarid, mixed-grass prairie, elevated atmospheric CO_2 decreased both N concentration and forage digestibility, except in wetter years (Augustine et al. 2018).

Increased CO_2 concentration or higher temperatures are also expected to increase fiber concentration (Owensby et al. 1996). Increased temperatures were shown to reduce the ***in vitro*** digestibility of the neutral detergent fiber in timothy (Bertrand et al. 2008; Jing et al. 2013) and the *in vitro* dry matter digestibility of alfalfa (Sanz-Sáez et al. 2012). However, management can adjust to offset increased temperature and precipitation changes on forage digestibility of timothy and alfalfa-timothy mixtures by changing the seasonal harvest schedule through additional forage cuttings (Jing et al. 2014; Thivierge et al. 2016).

The plant species composition of a region is largely determined by climate and soils, with fire regime, grazing and other land uses being more important at local levels (Polley et al. 2000). Water balance is the primary climatic control on the distribution and abundance of plants, especially on rangelands, where species composition is highly correlated with both plant water use and its availability in time and space (Polley et al. 2000). Increases in WUE should favor progressively taller and less drought-tolerant plants due to greater leaf area and **competition** for light. Conversely, slower **evapotranspiration (ET)** and wetter soils enhance reproduction and survival of drought-sensitive species; and increased deep percolation could favor deep-rooted species. In western US rangelands, a shift from woody dominance toward grassier vegetation types is expected (Reeves et al. 2017). In the Scandinavian countries, better overwintering conditions will make it possible to grow perennial ryegrass in areas where it is not grown currently (Thorsen and Höglind 2010).

Data are unclear on climate change and cold tolerance. In one study, cold tolerance of alfalfa was increased by elevated CO_2 (Bertrand et al. 2003), though in another study, elevated CO_2 concentration may have reduced the acquisition of cold tolerance (Bertrand et al. 2007). Piva et al. (2013) found no change in cold tolerance in timothy at elevated CO_2 and temperature. In climate change simulations for intensive systems, particularly in temperate areas, CO_2 enrichment is predicted to increase legume content of mixed grass-legume swards (Campbell and Smith 2000; Thivierge et al. 2016; Soussana and Lüscher 2017).

Summary and Conclusion

Global and regional climatic patterns such as seasonal temperature and precipitation can be explained on the basis of the global energy balance whereas climate change due to the "greenhouse effect" may be attributed to changes in the earth's net energy balance. The Köppen Climate Classification System is the basis for modern ecologic classification, and on a broader geographic scale, the principles can be used to delineate current and future agroecologic and species adaptation zones.

Boundaries for ecoregion divisions, however, are not identical to agroecologic and species adaptation zones (Table 8.1). Adaptation zones for several species may overlap but, on a broad scale, all are based on climatic characteristics and the associated responses of the plants. Adaptation zones for forage species can be explained and delineated on the basis of regional temperature, precipitation and "water balance" data. In addition, forage species have a range of genotypic and phenotypic plasticity that allows survival within an adaptation zone.

While science is determining the cause and effect relationships with greater precision, it is quite clear that observed and estimated future changes in global climate are heavily influenced by human activities. The effects of global climate change, due mainly to the increase in GHG in the atmosphere, will likely move the adaptation regions of most forage crops. This will require improved methodologies for associating adaptive traits of crop species and their collective responses to climatic parameters with regional and global climatic characteristics. Since forage quality and animal adjustment are also involved, the nature and management of forage-livestock systems will require integrative research and approaches for decision-making on management.

References

Ainsworth, E.A. and Long, S.P. (2005). What have we learned from 15 years of free-air CO_2 enrichment (FACE)? A meta-analytic review of the responses of photosynthesis. *New Phytol.* 165: 351–371.

Allard, R.W. (1960). *Principles of Plant Breeding*, 1e. New York: Wiley.

Andrews, C.J. (1987). Low-temperature stress in field and forage crop production. *Can. J. Plant. Sci.* 67: 1121–1133.

Arakeri, H.R. and Schmid, A.R. (1949). Cold resistance of various legumes and grasses in early stages of growth. *Agron. J.* 41: 182–185.

Augustine, D.J., Blumenthal, D.M., Springer, T.L. et al. (2018). Elevated CO_2 induces substantial and persistent declines in forage quality irrespective of warming in mixed grass prairie. *Ecol. Appl.* 28: 721–735.

Bailey, R.G. (1996). *Ecosystem Geography*. New York: Springer-Verlag.

Bailey, R.G. (2005). Identifying ecoregion boundaries. *Environ. Manage.* 34 (Suppl. 1): S14–S26. Springer Science+Business Media, Inc. DOI: https://doi.org/10.1007/s00267-003-0163-6.

Barbour, M.G., Burk, J.H., and Pitts, W.D. (1987). *Terrestrial Plant Ecology*, 2e. Melano Park, CA: Benjamin-Cummings Publications Co.

Basara, J.B., Maybourn, J.N., Peirano, C.M. et al. (2013). Drought and associated impacts in the great plains of the United States – a review. *Int. J. Geosci.* 4: 72–81.

Basarab, J.A., Baron, V.S., Lopez-Campos, O. et al. (2012). Greenhouse gas emissions from calf-and yearling-fed beef production systems with and without the use of growth promotants. *Animals* 2: 195–220. https://doi.org/10.3390/ani2020195.

Beauchemin, K.A., Kreuzer, M., O'Mara, F., and McAllister, T.A. (2008). Nutritional management for enteric methane abatement: a review. *Aust. J. Exp. Agric.* 48: 21–27.

Beauchemin, K.A., Janzen, H.H., Little, S.M. et al. (2010). Life cycle assessment of greenhouse gas emissions from beef production in western Canada: a case study. *Agric. Syst.* 103: 371–379.

Beck, C., Grieser, J., Kottek, M. et al. (2005). Characterizing global climate change by means of Koppen climate classification. *Technical Report.*

Bélanger, G., Kunelius, T., McKenzie, D. et al. (1999). Fall cutting management affects yield and persistence of alfalfa in Atlantic Canada. *Can. J. Plant. Sci.* 79: 57–63.

Bélanger, G., Rochette, P., Castonguay, Y. et al. (2002). Climate change and winter survival of perennial forage crops in Eastern Canada. *Agron. J.* 94: 1120–1130.

Bélanger, G., Castonguay, Y., Bertrand, A. et al. (2006). Winter damage to perennial forage crops in eastern Canada: causes, mitigation, and prediction. *Can. J. Plant. Sci.* 86: 33–47.

Bertrand, A., Castonguay, Y., Nadeau, P. et al. (2001). Molecular and biochemical responses of perennial forage crops to oxygen deprivation at low temperature. *Plant Cell Environ.* 24: 1085–1093.

Bertrand, A., Castonguay, Y., Nadeau, P. et al. (2003). Oxygen deficiency affects carbohydrate reserves in overwintering forage crops. *J. Exp. Bot.* 54: 1721–1730.

Bertrand, A., Prevost, D., Bigras, F.J., and Castonguay, Y. (2007). Elevated atmospheric CO_2 and strain of Rhizobium alter freezing tolerance and cold-induced molecular changes in alfalfa (*Medicago sativa*). *Ann. Bot.* 99: 275–284. https://doi.org/10.1093/aob/mcl254.

Bertrand, A., Tremblay, G.F., Pelletier, S. et al. (2008). Yield and nutritive value of timothy as affected by temperature, photoperiod and time of harvest. *Grass Forage Sci.* 63: 421–432. https://doi.org/10.1111/j.1365-2494.2008.00649.x.

Bigelow, P., and Borchers, A. (2017). Major uses of land in the United States, 2012. United States Department of Agriculture. Economic Information Bulletin No.178.

Branson, F.A., Gifford, G.F., Renard, K.G., and Hadley, R.F. (1981). Evaporation and transpiration. In: *Rangeland Hydrology*, 2e (eds. F.A. Branson et al.), 179–200. Denver, CO: Society for Range Manage.

Brklacich, M., Bryant, C., Veenhof, B., and Beauchesne, A. (1997). Implications of global climatic change for Canadian agriculture: a review and appraisal of research from 1984-1997. In: *Chapter 4, Canada Country Study: Climate Impacts and Adaptation*, 220–256. Toronto, ON: Environment Canada.

Brummer, E.C., Shah, M.M., and Luth, D. (2000). Reexamining the relationship between fall dormancy and winter hardiness in alfalfa. *Crop Sci.* 35: 509–516.

Bula, R.J., Smith, D., and Hodgson, H.J. (1956). Cold resistance and chemical composition in overwintering alfalfa, red clover and sweetclover. *Agron. J.* 46: 397–401.

Bush, E.J., Loder, J.W., James, T.S. et al. (2014). *An Overview of Canada's Changing Climate; in Canada in a Changing Climate: Sector Perspectives on Impacts and Adaptation* (eds. F.J. Warren and D.S. Lemmen), 23–64. Ottawa, ON: Government of Canada.

Buxton, D.R. (1989). Major edaphic and climatic stresses in the United States. In: *Persistence in Forage Legumes* (eds. G.C. Marten, A.G. Matches and R.F Barnes), 217–232. Madison, WI: ASA.

Buxton, D.R. and Fales, S.L. (1994). Plant environment and quality. In: *Forage Quality, Evaluation, and Utilization* (eds. G.C. Fahey et al.), 155–199. Madison, WI: ASA, CSSA and SSSA.

Campbell, B.D. and Smith, D.M.S. (2000). A synthesis of recent global change research on pasture and rangeland production: reduced uncertainties and their management implications. *Agric. Ecosyst. Environ.* 82: 39–55.

Canadian Climate Impact Scenarios (2015). Annual mean temperature change – 2050s. https://www.nrcan.gc.ca/environment/resources/publications/impacts-adaptation/reports/assessments/2008/ch2/10321 (accessed 9 October 2019).

Casler, M.D., Vogel, K.P., Taliaferro, C.M., and Wynia, R.L. (2004). Latitudinal adaptation of switchgrass populations. *Crop Sci.* 44: 293–303.

Castonguay, Y.S., Laberge, S., Brummer, E.C., and Volenec, J.J. (2006). Alfalfa winter hardiness. A research retrospective and integrated perspective. *Adv. Agron.* 90: 203–265.

Castonguay, Y., Nadeau, P., and Laberge, S. (1993). Freezing tolerance and alteration of translatable mRNAs in alfalfa (*Medicago sativa* L.) hardened at subzero temperatures. *Plant Cell. Physiol.* 34: 31–38.

Cathey, H.M. (1990). USDA Plant hardiness zone map. Miscellaneous Publication No. 1475. https://planthardiness.ars.usda.gov/PHZMWeb/ (accessed 22 November 2019).

Chamblee, D.S., Mueller, J.P., and Timothy, D.H. (1989). Vegetative establishment of three warm-season perennial grasses in late fall and late winter. *Agron. J.* 81: 687–691.

Clifton-Brown, J.C. and Jones, M.B. (1997). The thermal response of leaf extension rate in genotypes of the C_4-grass *Miscanthus*: an important factor in determining potential productivity of different genotypes. *J. Exp. Bot.* 48: 1573–1581.

Conant, R.T., Cerri, C.E.P., Osborne, B.B., and Paustian, K. (2017). Grassland management impacts on soil carbon stocks: a new synthesis. *Ecol. Appl.* 27: 662–668.

Council for Agricultural Science and Technology (2011). Carbon sequestration and greenhouse gas fluxes in agriculture: Challenges and opportunities. CAST Task Force Report No. 142.

Craufurd, P.Q., Qi, A., Ellis, R.H. et al. (1998). Effect of temperature on time to panicle initiation and leaf appearance in sorghum. *Crop Sci.* 38: 942–947.

Currie, P.O. and White, R.S. (1982). Drought survival of selected forage grasses commonly seeded in the northern Great Plains. *Can. J. Plant. Sci.* 62: 949–955.

Derner, J.D. and Schuman, G.E. (2007). Carbon sequestration and rangelands: a synthesis of land management and precipitation effects. *J. Soil Water Conserv.* 62: 77–84.

Derner, J., Briske, D., Reeves, M. et al. (2018). Vulnerability of grazing and confined livestock in the northern Great Plains to projected mid- and late-twenty-first century climate. *Clim. Change* 146: 19–32. https://doi.org/10.1007/s10584-017-2029-6.

Dhont, C., Castonguay, Y., Nadeau, P. et al. (2004). Untimely fall harvest affects dry matter yield and root organic reserves in field-grown alfalfa. *Crop Sci.* 44: 144–157.

Donald, C.M. (1963). Competition among crop and pasture plants. *Adv. Agron.* 15: 1–118.

Dumont, B., Andueza, D., Niderkorn, V. et al. (2015). A meta-analysis of climate change effects on forage quality in grasslands: specificities of mountain and Mediterranean areas. *Grass Forage Sci.* 70: 239–254.

Durling, J.C., Hesterman, O.B., and Rotz, C.A. (1995). Predicting first-cut alfalfa yields from preceding winter weather. *J. Prod. Agric.* 8: 254–259.

Eagles, C. F., Thomas, H., Volaire, F. et al. (1997). Stress physiology and crop improvement. Proceedings of the 18th International Grassland Congress in Winnipeg, MB (18–19 June 1997) (CD-ROM).

Entz, M.H., Bullied, W.J., and Katepa-Mupondwa, F. (1995). Rotational benefits of forage crops in Canadian prairie cropping systems. *J. Prod. Agric.* 8: 521–529.

Environment and Climate Change Canada (2018). National Inventory Report 1990–2016. Greenhouse Gas Sources and Sinks in Canada. Cat. No.: En81-4/1E-PDF https://www.canada.ca/en/environment-climate-change/services/climate-change/greenhouse-gas-emissions.html (accessed 9 October 2019).

Environmental Protection Agency (2018). Inventory of US greenhouse gas emissions and sinks 1990–2016. https://www.epa.gov/ghgemissions/sources-greenhouse-gas-emissions#agriculture (accessed 9 October 2019).

Evenson, P.D. (1979). Optimum crown temperatures for maximum alfalfa growth. *Agron. J.* 71: 798–800.

Frank, A.B. and Bauer, A. (1991). Rooting activity and water use during vegetative development of crested and western wheatgrass. *Agron. J.* 83: 906–910.

Frank, A.B., Bittman, S., and Johnson, D.A. (1996). Water relations of cool-season grasses. In: *Cool-Season Forage Grasses*, Agronomy Monograph 34 (eds. L.E.

Moser et al.), 127–164. Madison, WI: ASA, CSSA and SSSA.

Frank, A.B., Karn, J.F., and Berdahl, J.D. (1999). *Vegetative Changes After 82 years of Grazing at Mandan, ND.* Denver, CO: Soc. Range Manage.

Franzluebbers, A.J. (2007). Integrated crop-livestock systems in the Southeastern USA. *Agron. J.* 99: 361–372.

Franzluebbers, A.J. (2010). Achieving soil organic carbon sequestration with conservation agricultural systems in the Southeastern United States. *Soil Sci. Soc. Am. J.* 74: 347–357.

Gist, G.R. and Mott, G.O. (1957). Some effects of light intensity, temperature and soil moisture on the growth of alfalfa, red clover and birdsfoot trefoil seedlings. *Agron. J.* 49: 33–36.

Grace, J., Berninger, F., and Nagy, L. (2002). Impacts of climate change on the tree line. *Ann. Bot.* 90: 537–544.

Gudleifsson, B.E. (1993). Metabolic and cellular impact of ice encasement on herbage plants. In: *Interacting Stresses on Plants in a Changing Climate*, NATO ASI Series, vol. 16 (eds. M.B. Jackson and C.R. Black), 407–421. Berlin: Springer-Verlag.

Gudleifsson, B.E., Andrews, C.J., and Bjornsson, H. (1986). Cold hardiness and ice tolerance of pasture grasses grown and tested in cold environments. *Can. J. Plant. Sci.* 66: 601–608.

Hannaway, D.B., Daly, C., Halbleib, M.B. et al. (2009). Development of suitability maps with examples for the United States and China. In: *Tall Fescue for the Twenty-First Century*, Agronomy Monograph 53 (eds. H.A. Fribourg, D.B. Hannaway and C.P. West), 33–47. Madison, WI: ASA, CSSA, SSSA https://doi.org/10.2134/agronmonogr53.

Harper, J.L. (1977). *Population Biology of Plants.* New York: Academic Press Ch. 2.

Hartmann, D.L., Klein Tank, A.M.G., Rusticucci, M. et al. (2013). Observations: atmosphere and surface. In: *Climate Change 2013: The Physical Science Basis. Contribution of Working Group I to the Fifth Assessment Report of the Intergovernmental Panel on Climate Change* (eds. T.F. Stocker et al.). Cambridge, UK: Cambridge University Press.

Hatfield, J.L., Boote, K.J., Fay, P. et al. (2008). Agriculture. In: *The Effects of Climate Change on Agriculture, Land Resources, Water Resources, and Biodiversity in the United States.* Washington, DC: US Climate Change Science Program and the Subcommittee on Global Change Research.

Hatfield, J.L., Boote, K.J., Kimball, B.A. et al. (2011). Climate impacts on agriculture: implications for crop production. *Agron. J.* 130: 351–370.

Hatfield, J., Takle, G., Grotjahn, R. et al. (2014). Agriculture. In: *Climate Change Impacts in the United States: The Third National Climate Assessment* (eds. J.M. Melillo, T.C. Richmond and G.W. Yohe), 150–174. Washington, DC: U.S. Government Printing Office.

Hendrickson, J.R., Liebig, M.A., and Berdahl, J. (2008). Response of *Medicago sativa* and *M. falcata* type alfalfas to different defoliation times and grass competition. *Can. J. Plant. Sci.* 89: 61–69.

Hill, M.J. (1996). Defining the white clover zone in eastern mainland Australia using a model and a geographic information system. *Ecol. Modell.* 86: 245–252.

Hill, M.J., Willms, W.D., and Aspinall, R.J. (2000). Distribution of range and cultivated grassland plants in southern Alberta. *Plant Ecol.* 147: 59–76.

Horst, G.L. and Nelson, C.J. (1979). Compensatory growth of tall fescue following drought. *Agron. J.* 71: 559–563.

Intergovernmental Panel on Climate Change (2001). Climate change 2001: impacts, adaptation, and vulnerability: contribution of working group II to the third assessment report of the intergovernmental panel on climate change. In: *Published for the Intergovernmental Panel on Climate Change* (eds. J.J. McCarthy et al.), 735–800. Ottawa, Canada: Intergovernmental Panel on Climate Change.

Izaurralde, R.C., Rosenberg, N.J., Brown, R.A., and Thomson, A.M. (2003). Integrated assessment of Hadley Center (HadCM2) climate-change impacts on agricultural productivity and irrigation water supply in the conterminous United States. Part II. Regional agricultural production in 2030 and 2095. *Agric. For. Meteorol.* 117: 97–122.

Izaurralde, R.C., Thompson, A.M., Morgan, J.A. et al. (2011). Climate impacts on agriculture: implications for forage and rangeland production. *Agron. J.* 103: 371–381.

Jing, Q., Bélanger, G., Qian, B., and Baron, V. (2013). Timothy yield and nutritive value under climate change in Canada. *Agron. J.* 105: 1683–1694. https://doi.org/10.2134/agronj2013.0195.

Jing, Q., Bélanger, G., Qian, B., and Baron, V. (2014). Timothy yield and nutritive value with a three-harvest system under the projected future climate in Canada. *Can. J. Plant. Sci.* 94: 213–222. https://doi.org/10.4141/cjps2013-279.

Karl, T.R., Melillo, J.M., and Peterson, T.C. (eds.) (2009). *Global Climate Change Impacts in the United States.* New York: Cambridge Univetisty Press.

Kiehl, J.T. and Trenberth, K.E. (1997). Bulletin of the Amer. *Meteorol. Soc.* 78: 197–208.

Kirby, E.J.M. (1995). Factors affecting rate of leaf emergence in barley and wheat. *Crop Sci.* 35: 11–19.

Klebesadel, L.J. (1971). Selective modification of alfalfa towards acclimation in a subartic area of severe winter stress. *Crop Sci.* 11: 609–614.

Kolka, R.K., and Wolf, A.T. (1998). Estimating actual evapotranspiration for forested sites: Modifications to the Thornthwaite model. United States Department of Agriculture. Forest Service Research Note SRS6.

Köppen, W. (1931). *Grundriss der Klimakunde*. Berlin: Walter de Gruyter.

Körner, C. (2002). Grassland in a CO_2-enriched world. In: *Multi-Function Grasslands: Quality Forages, Animal Products and Landscapes*. Proc. 19th General Meeting of the European Grassland Federation (eds. J.L. Durand, J.C. Emile, C. Huyghe and G. Lemaire), 611, 27–624, 30. La Rochelle, France: May 2002.

Le Treut, H., Somerville, R., Cubasch, U. et al. (2007). Historical overview of climate change. In: *Climate Change: 2007: The Physical Science Basis. Contribution of Working Group I to the Fourth Assessment Report of the Intergovernmental Panel on Climate Change* (eds. S. Solomon et al.). Cambridge, UK: Cambridge Univeristy Press.

Leach, G.J. (1971). The relation between shoot growth and temperature. *Aust. J. Agric. Res.* 22: 49–59.

Lee, C.T. and Smith, D. (1972). Influence of soil nitrogen and potassium levels on growth and composition of lucerne grown to first flower in four temperature regimes. *J. Sci. Food Agric.* 23: 1169–1181.

Leep, R.H., Andresen, J.A., and Jeranyama, P. (2001). Fall dormancy and snow depth effects on winterkill of alfalfa. *Agron. J.* 93: 1142–1148.

Lettenmaier, D.P., Major, D., Poff, L., and Running, S. (2008). Water resources. In: *The Effects of Climate Change on Agriculture, Land Resources, Water Resources, and Biodiversity in the United States*, 362. Washington, DC: U.S. Climate Change Science Program and the Subcommittee on Global Change Res.

Limin, A.E. and Fowler, D.B. (1987). Cold hardiness of forage grasses grown on the Canadian Prairies. *Can. J. Plant. Sci.* 67: 1111–1115.

Looman, J. (1983). Distribution of plant species and vegetation types in relation to climate. *Vegetation* 54: 17–25.

Looman, J. and Heinrichs, D.H. (1973). Stability of crested wheatgrass pastures under long-term pasture use. *Can. J. Plant. Sci.* 53: 501–506.

Lowry, D.B., Behrman, K.D., Grabowski, P. et al. (2014). Adaptations between ecotypes and along environmental gradients in *Panicum virgatum*. *Am. Nat.* 183: 682–692. https://doi.org/10.1086/675760.

Madakadze, I.C., Stewart, K.A., Madakadze, R.M., and Smith, D.L. (2003). Base temperatures for seedling growth and their correlation with chilling sensitivity for warm-season grasses. *Crop Sci.* 43: 874–878.

Mavi, S. and Tupper, G.J. (2004). Solar radiation and its role in plant growth. In: *Agrometeorology Principles and Applications of Climate Studies in Agriculture* (eds. S. Mavi and G.J. Tupper), 13–41. Binghampton, NY: Haworth Press.

McGranahan, D.A. and Yurkonis, K.A. (2018). Variability in grass forage quality and quantity in response to elevated CO_2 and water limitation. *Grass Forage Sci.* 73: 517–521.

McKenzie, J.S. and McLean, G.E. (1980). Some factors associated with injury to alfalfa during the 1977-78 winter at Beaverlodge, Alberta. *Can. J. Plant. Sci.* 60: 103–112.

McKenzie, J.S., Paquin, R., and Duke, S.H. (1988). Cold and heat tolerance. In: *Alfalfa and Alfalfa Improvement*, Agronomy Monograph 29 (eds. A.A. Hanson et al.), 259–302. Madison, WI: ASA, CSSA and SSSA.

McLeod, J.G., Muri, R., Jefferson, P.G. et al. (2009). Yellowhead alfalfa. *Can. J. Plant. Sci.* 89: 653–655.

Mearns, L.O. (2000). Climate change and variability. In: *Climate Change and Global Crop Productivity* (eds. K.R. Reddy and H.F. Hodges), 7–35. Wallingford, UK: CABI Publishing.

Melton, B., Moutray, J.B., and Bouton, J.H. (1988). Geographic adaptation and cultivar selection. In: *Alfalfa and Alfalfa Improvement*, Agronomy Monograph 29 (eds. A.A. Hanson et al.), 595–620. Madison, WI: ASA, CSSA and SSSA.

Meyer, D.W. and Badaruddin, M. (2001). Frost tolerance of ten seedling legume species at four growth stages. *Crop Sci.* 41: 1838–1842.

Milano, E.R., Lowry, D.B., and Juenger, T.E. (2016). The genetic basis of upland/lowland ecotype divergence in switchgrass (*Panicum virgatum*). *Genes/Genomes/Genetics* 6: 3561–3570.

Moser, L.E. and Vogel, K.P. (1995). Switchgrass, big bluestem, and indiangrass. In: *Forages. An Introduction to Grassland Agriculture*, 5e, vol. I (eds. R.F Barnes et al.), 409–420. Ames, IA: Iowa State University Press.

Nelson, C.J. (2000). Shoot morphological plasticity of grasses: leaf growth vs. tillering. In: *Grassland Ecophysiology and Grazing Ecology* (eds. G. Lemaire, J. Hodgson and A. de Moraes), 101–125. Wallingford, UK: CABI Publishing.

Nelson, C.J. and Moser, L.E. (1994). Plant factors affecting forage quality. In: *Forage Quality, Evaluation, and Utilization* (eds. G.C. Fahey et al.), 115–154. Madison, WI: ASA, CSSA and SSSA.

Nelson, C.J. and Volenec, J.J. (1995). Environmental and physiological aspects of forage management. In: *Forages. An Introduction to Grassland Agriculture*, 5e, vol. I (eds. R.F Barnes et al.), 55–69. Ames, IA: Iowa State University Press.

Norton, L.R., McLeod, A.R., Greenslade, P.D. et al. (1999). Elevated UV-B effects on experimental grassland communities. *Global Change Biol.* 5: 601–608.

Nösberger, J., Blum, H., and Fuhrer, J. (2000). Crop ecosystem responses to climatic change: productive grasslands. In: *Climate Change and Global Crop Productivity* (eds. K.R. Reddy and H.F. Hodges), 271–291. Wallingford, Oxon, UK: CABI Publications.

Ojima, D.S., Steiner, J., McNeeley S. et al. (2015). Great Plains Regional Technical Input Report. Washington, DC: Island Press.

Oliver, J.E. and Hidore, J.J. (1984). *Climatology: An Introduction*. Columbus, OH: Merrill.

Omernik, J.M. and Griffith, G.E. (2014). Ecoregions of the conterminous United States: evolution of a hierarchical spatial framework. *Environ. Manage.* 54: 1249–1266. https://doi.org/10.1007/s00267-014-0364-1.

Ouellet, C.E. (1976). Winter hardiness and survival of forage crops in Canada. *Can. J. Plant. Sci.* 56: 679–689.

Ouellet, C.E. and Desjardins, R.L. (1981). Interprétation des relations entre le climat et la survie à l'hiver de la luzerne par l'analyse des corrélations. *Can. J. Plant. Sci.* 61: 945–954.

Owensby, C.E., Cochran, R.C., and Auen, L.M. (1996). Effects of elevated carbon dioxide on forage quality for ruminants. In: *Carbon Dioxide, Populations, and Communities* (eds. C. Körner and F.A. Bazzaz), 363–371. San Diego, CA: Academic Press.

Padbury, G., Waltman, S., Caprio, J. et al. (2002). Agroecosystems and land resources of the Northern Great Plains. *Agron. J.* 94: 251–261.

Papadopoulos, Y.A., Gordon, R.J., McRae, K.B. et al. (1999). Current and elevated levels of UV-B radiation have few impacts on yields of perennial forage crops. *Global Change Biol.* 5: 847–856.

Paquin, R. (1984). Influence of the environment on cold hardening and winter survival of forage plants and cereals; proline as a metabolic marker of hardening. In: *Being Alive on Land* (eds. N.S. Margaris, M. Arianoustou-Faraggitali and W.C. Oechel), 137–154. Dordrecht: Springer.

Paquin, R. and Mehuys, G.R. (1980). Influence of soil moisture on cold tolerance of alfalfa. *Can. J. Plant. Sci.* 60: 139–147.

Paquin, R. and Pelletier, H. (1980). Influence de l'environnement sur l'acclimatation au froid de la luzerne (*Medicago media* Pers.) et sa re' sistance au gel. *Can. J. Plant. Sci.* 60: 1351–1366.

Pearson, C.J. and Hunt, L.A. (1972). Effects of temperature on primary growth and regrowth of alfalfa. *Can. J. Plant. Sci.* 52: 1017–1027.

Peterson, P.R., Sheaffer, C.C., and Hall, M.H. (1992). Drought effects on perennial forage legume yield and quality. *Agron. J.* 84: 774–779.

Piva, A., Bertrand, A., Bélanger, G. et al. (2013). Growth and physiological response of timothy to elevated carbon dioxide and temperature under contrasted nitrogen fertilization. *Crop Sci.* 53: 704–715.

Polley, H.W., Morgan, J.A., Campbell, B.D., and Stafford Smith, M. (2000). Crop ecosystem responses to climatic change: rangelands. In: *Climate Change and Global Crop Productivity* (eds. K.R. Reddy and H.F. Hodges), 293–314. Wallingford, UK: CAB International.

Polley, H.W., Morgan, J.A., and Fay, P.A. (2011). Application of a conceptual framework to interpret variability in rangeland responses to atmospheric CO_2 enrichment. *J. Agric. Sci.* 149: 1–14.

Qian, Y.L., Ball, S., Tan, Z. et al. (2001). Freezing tolerance of six cultivars of buffalo-grass. *Crop Sci.* 41: 1174–1178.

Reeves, M.C., Bagne, K.E., and Tanaka, J. (2017). Potential climate change impacts on four biophysical indicators of cattle production from Western US Rangelands. *Rangeland Ecol. Manage.* 70: 529–539.

Rogler, G.A. and Haas, H.J. (1946). Range production as related to soil moisture and precipitation on the Northern Great Plains. *J. Am. Soc. Agron.* 39: 378–379.

Romero-Lankao, P. et al. (2014). North America. In: *Climate Change 2014: Impacts, Adaptation, and Vulnerability. Part B: Regional Aspects. Contribution of Working Group II to the Fifth Assessment Report of the Intergovernmental Panel on Climate Change* (eds. Field, C.B. et al.), 1439–1498. Cambridge, UK: Cambridge University Press.

Sakai, A. and Larcher, W. (1987). *Frost Survival of Plants: Responses and Adaptation to Freezing Stress*. New York: Springer-Verlag.

Sanz-Sáez, Á., Erice, G., Aguirreolea, J. et al. (2012). Alfalfa forage digestibility, quality and yield under future climate change scenarios vary with *Sinorhizobium meliloti* strain. *J. Plant Physiol.* 169: 782–788.

Schimel, D.S., Coleman, D.C., and Horton, K.A. (1985). Soil organic matter dynamics in paired rangeland and cropland toposequences in North Dakota. *Geoderma* 36: 201–214.

Sheaffer, C.C., Wiersma, J.V., Warnes, D.D. et al. (1986). Fall harvesting and alfalfa yield, persistence and quality. *Can. J. Plant. Sci.* 66: 329–338. https://doi.org/10.4141/cjps86-047.

Sheaffer, C.C., Tanner, C.B., and Kirkam, M.B. (1988). Alfalfa water relations and irrigation. In: *Alfalfa and Alfalfa Improvement*, Agronomy Monograph 29 (eds. A.A. Hanson et al.), 373–409. Madison, WI: ASA, CSSA and SSSA.

Sheaffer, C.C., Peterson, P.R., Hall, M.H., and Stordahl, J.B. (1992). Drought effects on yield and quality of perennial grasses in the North Central United States. *J. Prod. Agric.* 5: 556–561.

Sims, P.L. (1988). Grasslands. In: *North American Terrestrial Vegetation* (eds. M.G. Barbour and W.D. Billings), 265–286. New York: Cambridge University Press.

Sleeter, B.M., Sohl, T.L., Bouchard, M.A. et al. (2012). Scenarios of land use and land cover change in the conterminous United States: utilizing the special report on emission scenarios at ecoregional scales. *Global Environ. Change* 22: 896–914.

Sleper, D.A. and Buckner, R.C. (1995). The fescues. In: *Forages. An Introduction to Grassland Agriculture*, 5e, vol. I (eds. R.F Barnes et al.), 345–356. Ames, IA: Iowa State University Press.

Smith, D. (1961). Association of fall growth habit and winter survival in alfalfa. *Can. J. Plant. Sci.* 41: 244–251.

Snaydon, R.W. (1981). The ecology of grazed pastures. In: *Grazing Animals. World Animal Science. B. Disciplinary Approach* (ed. F.H.W. Morley), 13–31. New York, NY: Elsevier Scientific Publications Co.

Soussana, J.F. and Lüscher, A. (2017). Temperate grasslands and global atmospheric change. A review. *Grass Forage Sci.* 62: 127–134.

Stout, D.G. and Hall, W.J. (1989). Fall growth and winter survival of alfalfa in interior British Columbia. *Can. J. Plant. Sci.* 69: 491–499.

Teeri, J.A. and Stowe, L.G. (1976). Climatic patterns and the distribution of C_4 grasses in North America. *Oecologia (Berl.)* 23: 1–12.

Thivierge, M., Jego, G., Bélanger, G. et al. (2016). Predicted yield and nutritive value of an alfalfa-timothy mixture under climate change and elevated atmospheric carbon dioxide. *Agron. J.* 108: 1–19. https://doi.org/10.2134/agronj2015.0484.

Thivierge, M., Jego, G., Bélanger, G. et al. (2017). Projected impact of future climate conditions on the agronomic and environmental performance of Canadian dairy farms. *Agric. Syst.* 157: 241–257. https://doi.org/10.1016/j.agsy.2017.07.003.

Thompson, R.S., Anderson, K.H. and Bartlein, P.J. (1999). Climate-vegetation atlas of North America – introduction. USGS Professional Paper 1650-A. http://pubs.usgs.gov/pp/1999/p1650-a/atlas_intro.html (accessed 9 October 2019).

Thornley, J.H.M. and Cannell, M.G.R. (1997). Temperate grassland responses to climate change: An analysis using the Hurley Pasture Model. *Annals of Botany* 80: 205–221.

Thornthwaite, C.W. (1948). An approach toward a rational classification of climate. *Geogr. Rev.* 38: 55–94.

Thornthwaite, C.W. and Mather, J.R. (1957). Instructions and tables for computing potential evapotranspiration and the water balance. In: *Publications in Climatology*, vol. X, No. 3. Centerto, NJ: Drexel Institute of Technology.

Thorsen, S.M. and Höglind, M. (2010). Assessing winter survival of forage grasses in Norway under future climate scenarios by simulating potential frost tolerance in combination with simple agroclimatic indices. *Agric. Forest Meteorol.* 150: 1272–1282.

Tollenaar, M., Fridgen, J., Priyanka, T. et al. (2017). The contribution of solar brightening to the US maize yield trend. *Nat. Clim. Change* 7: 275–278. https://doi.org/10.1038/NCLIMATE3234.

Trewartha, G.T. (1968). *An Introduction to Climate*, 4e. New York: McGraw-Hill.

Trewartha, G.T., Robinson, A.H., and Hammond, E.H. (1967). *Physical Elements of Geography*, 5e. New York: McGraw-Hill.

Tysdal, H.M. and Pieters, A.J. (1934). Cold resistance of three species of lespedeza compared to that of alfalfa, red clover, and crown vetch. *Agron. J.* 26: 923–928.

Voigt, P.W. and Sharp, W.C. (1995). Grasses of the plains and southwest. In: *Forages. An Introduction to Grassland Agriculture*, 5e, vol. I (eds. R.F Barnes et al.), 395–408. Ames, IA: Iowa State University Press.

Volenec, J.J. and Nelson, C.J. (2007). Physiology of forage plants. In: *Forages. The Science of Grassland Agriculture*, 6e, (eds. R.F Barnes et al.), 37–52. Ames, IA: Blackwell.

Volenec, J.J., Nelson, C.J., and Sleper, D.A. (1984). Influence of temperature on leaf dark respiration of diverse tall fescue genotypes. *Crop Sci.* 24: 907–912.

Voroney, R.P., Van Veen, J.A., and Paul, E.A. (1981). Organic C dynamics in grassland soils. 2. Model validation and simulation of long-term effects of cultivation and rainfall erosion. *Can. J. Soil Sci.* 61: 211–224.

Walsh, J. and Wuebbles, D. et al. (2014). Our Changing Climate. In: *Climate Change Impacts in the United States* (eds. J.M. Melillo et al.), 19–67. The Third National Climate Assessment, US Global Change research program http://www.nrc.gov/docs/ML1412/ML14129A233.pdf.

Weishaar, M.A., Brummer, E.C., Volenec, J.J. et al. (2005). Improving winter hardiness in nondormant alfalfa germplasm. *Crop Sci.* 45: 60–65.

West, T.O. and Six, J. (2007). Considering the influence of sequestration duration and carbon saturation on estimates of soil carbon capacity. *Clim. Change* 80: 25–41. https://doi.org/10.1007/s10584-006-9173-8.

Yan, W. and Hunt, L.A. (1999). An equation for modeling the temperature response of plants using only the cardinal temperatures. *Ann. Bot.* 84: 607–614.

Zulauf, C. (2018). US hay market over the last 100 years. https://farmdocdaily.illinois.edu/2018/09/us-hay-market-over-the-last-100-years.html (accessed 22 November 2019).

Plant Interactions

John A. Guretzky, Grassland Systems Ecologist, *Department of Agronomy and Horticulture, University of Nebraska-Lincoln, NE, USA*

Overview

Interactions with neighboring plants and other organisms influence the ability of forages to grow, live healthy, and reproduce. Common interactions include competition, commensalism, mutualism, and parasitism depending on nature of the outcomes (Figure 9.1). Competition describes interactions that negatively affect the forage plant as well as one or more of its neighbors; while commensalism describes interactions where effects on the focal plant remain neutral but the partner benefits.

Mutualism describes interactions where both individuals benefit, and it usually involves exchange of resources that are relatively cheap to acquire or produce for resources that would be difficult or impossible to acquire (Bronstein 2009). Although mutualisms might appear to have a positive impact, the focal plant incurs costs as well as benefits, and the difference between mutualism and that of parasitism, where an organism takes advantage of its host, may be blurred (Bronstein 2009; Cheplick and Faeth 2009; Moenne-Loccoz et al. 2015).

A symbiosis describes a mutualism where two species physiologically dependent on each other exist in an intimate physical association for most of their lifetimes (Bronstein 2009; Moenne-Loccoz et al. 2015). If one species penetrates the other, the interaction may be described as an endosymbiosis, but if the two species live outside each other, the interaction would be described as an ectosymbiosis (Moenne-Loccoz et al. 2015). An obligatory symbiosis, meanwhile, describes the interaction when partners cannot live without each other, while a facultative symbiosis describes the interaction when the partners can live without the other (Moenne-Loccoz et al. 2015). Whether symbiosis produces positive or negative effects often depends on environmental conditions and genetics of the host plant and symbiont (Cheplick and Faeth 2009).

Predation, as well as parasitism, describes interactions that are negative for the host plant but beneficial to the partner (Figure 9.1). Predator-prey interactions typically have a short duration that leads to destruction of the prey and its genetic information after a short duration. Parasites, on the other hand, normally cause harm but not immediate death of their hosts (Begon 2009). They have sustained, physical associations with their hosts, at least for part of their life cycle, and the genetic information of the partners remains in intimate contact as molecular signals between the two can persist for months or years. Parasitism differs from predation because predators kill and consume many prey in their lifetime, and it differs from herbivory because herbivores often take many small parts from many different prey. A pathogen would be a parasite that gives rise to clearly harmful symptoms or a disease (Begon 2009). An insect or rodent would be an herbivore.

Plant-Plant Interactions

Competition

Plants compete with each other for nutrients, water, and light (Grace and Tilman 1990; Dybzinski and Tilman

Forages: The Science of Grassland Agriculture, Volume II, Seventh Edition.
Edited by Kenneth J. Moore, Michael Collins, C. Jerry Nelson and Daren D. Redfearn.

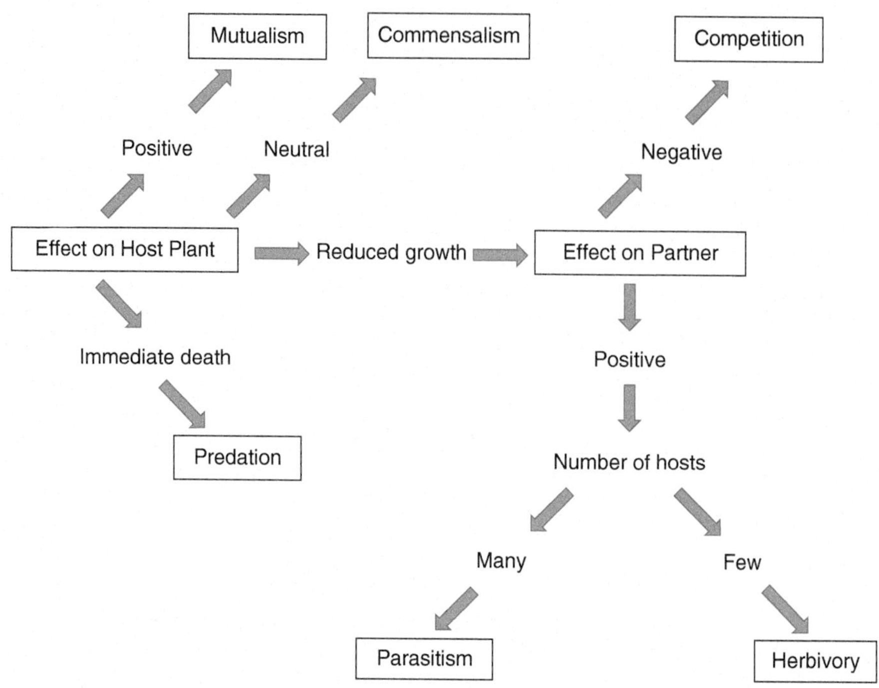

Fig. 9.1. Variation of plant interactions and their relative dependence on effects to the host plant and its partner.

2009; Craine and Dybinski 2013). As resources dwindle, rates of supply can no longer meet demands, and resource consumption by one individual will reduce its availability to another, thereby decreasing its fitness (Dybzinski and Tilman 2009). Outcomes of resource competition may be exclusion of the inferior competitor or coexistence depending on the dynamics of resource supply and resource requirements of interacting species (Dybzinski and Tilman 2009). In competition, though, negative effects on the interacting species must occur (Craine 2009). When one plant acquires limiting resources that could have been acquired by the other, and vice versa, competition occurs, stress increases, and growth of one, and usually both individuals slows (Craine 2009).

In the past 50 years, ecologists have developed different models to explain plant competition. The Competitors-Stress Tolerators-Ruderals (CSR) model described by Grime (2001) sought to recognize evolution of different plant strategies. The model recognized that some evolved to compete in low-stress and low-disturbance environments; some evolved to tolerate stressful environments; and some evolved to rapidly invade disturbed environments. Evolution of these strategies required tradeoffs, the dilemma whereby genetic change conferring increased fitness in one circumstance was associated with a reduction in fitness in another (Grime 2001). The CSR model indicates strong competitors grow best in the absence of stress and disturbance, have large capacities for above- and belowground resource capture, and have large maximum **relative growth rates**.

Mathematical models described by Tilman (1990) predicted that species displaced others due to their ability to reduce resource concentrations in the environment (Grace and Tilman 1990; Craine 2009). Resource utilization would draw concentrations down to an equilibrium resource concentration, upon which, part of the population would be unable to sustain itself. The models defined the minimum resource requirement a population needed to sustain itself as R^*, and the species with the smallest R^* would be the superior competitor at equilibrium (Grace 1990; Tilman 1990; Craine 2009). Determining which among a suite of species is competitively superior is based on measured growth of individual species in monocultures at different rates of nutrient supply. Growth rates of a species will be positive when the resource concentration exceeds R^* and negative when the resource concentration falls below R^*. When two species compete, they will both reduce the resource concentration. The stronger competitor will be the species with the smaller R^* for that particular limiting resource.

Table 9.1 Summary of developmental and physiologic response to reduced R:FR ratios in various plant species

Physiologic process	Response to reduced R:FR[a]
Germination	Retarded
Extension growth	Accelerated
Internode extension	Rapidly increased (lag c. 5 min)
Petiole extension	Rapidly increased
Leaf extension	Increased in cereals
Leaf development	Retarded
Leaf area growth	Marginally reduced
Leaf thickness	Reduced
Chloroplast development	Retarded
Chlorophyll synthesis	Reduced
Chlorophyll a:b ratio	Balance changed
Apical dominance	Strengthened
Branching	Inhibited
Tillering (grasses)	Inhibited
Flowering	Accelerated
Rate of flowering	Markedly increased
Seed set	Severe reduction
Fruit development	Truncated
Assimilate distribution	Marked change
Storage organ deposition	Severe reduction

Source: Adapted from Smith and Whitelam (1997).
[a]These responses to light quality are collectively referred to as the shade avoidance syndrome.

Grace (1990) presented an analysis of the theories of Grime and Tilman. He explained that the operational definition of competition by Grime assumed that neighboring plants utilized the same resources and that successful competitors had large capacities for resource capture and large maximum relative growth rates. He further assumed that good competitors rapidly develop leaves and roots, capture sunlight, and remove nutrients and water from soil, while minimizing investment in sexual reproduction. Grime had also assumed that to exploit resources, plants evolved tradeoffs between ability to tolerate low resource supplies and ability to grow rapidly, creating a divide between stress-tolerant and competitive species, respectively.

Traits of strong competitors defined by Grime, therefore, differed fundamentally from the mechanistic model developed by Tilman, which predicted that the species with the smallest resource requirement, R*, will be the superior competitor at equilibrium. Grime defined competition in terms of resource capture while Tilman defined competition in terms of tolerance to limited resources (Grace 1990). Tilman assumed that plants compete for soil resources in unproductive habitats and light in productive habitats and therefore have developed evolutionary tradeoffs in abilities to compete for different resources. The CSR model, though, predicts a positive correlation, while the resource concentration reduction model predicts a negative correlation, between traits that enhance competition for different resources (Grace 1990).

Craine (2009) further reviewed the debate about plant competition mechanisms and offered support for a supply preemption theory of competition. He critiqued the concentration reduction hypothesis as it applied to nutrients on grounds that it theorized that competitive superiority resulted from the ability of plants to reduce nutrient concentrations in well-mixed solutions. But he pointed out that soils have heterogeneous nutrient supplies and nutrient uptake depends on concentrations at the root surface.

The supply preemption theory of competition posits that plants do not outcompete each other by reducing the concentrations of nutrients but rather by preempting supplies from coming in contact with roots of other species. Superior competitors do so by maximizing root length. The greater the fraction of root length of an individual plant, the greater the fraction of nutrients it will preempt and acquire. Craine (2009) further identified an analog to R*, that is, nutrient supply per unit root length (S_L*), that would allow a comparison of the competitive ability of two species for nutrients. The species with the smaller S_L* would be the superior competitor.

Craine and Dybinski (2013) discussed roles of supply pre-emption and availability reduction in competition

Fig. 9.2. Superior competitors for light grow taller and elevate their leaves above others in plant canopies.

for nutrients, water, and light when supplied evenly in space and time. Plants compete for nutrients by pre-empting nutrient supplies from coming into contact with neighbors, which requires maximizing root length. While competition for water generally proceeds through reducing availability, with superior competitors being those that withstand the lowest **water potential**. Competition for light, on the other hand, depends on plants growing taller and placing leaves above those of their neighbors (Figure 9.2). Individuals that develop leaves placed above those of neighbors benefit directly from increased photosynthetic rates and indirectly by shading neighbors and reducing their growth.

Briske (2007) and Pierek and de Wit (2014) discussed how plants use shade avoidance strategies and summarized developmental and physiologic responses to competition for light (Table 9.1). Responses include accelerated **hypocotyl**, internode, and petiole elongation; having leaves higher in the canopy and displayed horizontally; reduced shoot branching; earlier flowering; and less investment in belowground organs.

Cues that plants use to detect the presence and proximity of competitors and avoid shade include light quality and quantity signals, mechanical stimuli, and volatile organic compounds. Plants detect the presence of neighbors through capture of horizontal reflection of far-red light in the 700–800 nm waveband from the neighbor that is greater proportionally from direct light. As vegetation density increases, upper leaves deplete red light for photosynthesis and reduce the red to Far-Red ratio of light, a signal captured by the **phytochrome** family of photoreceptors. Capture of Far-Red light signals the presence of neighbors even before they become a competitive threat; nearby plants perceiving this signal use it to initiate shade avoidance responses.

Other light signals include decreased availability of photosynthetically active radiation in the lower canopy, which results in increased internode length, **hyponastic** leaf growth, and elongated hypocotyls. The light spectrum under a dense canopy shows less reduction of green light (500–580 nm) while blue (400–500 nm) and **ultraviolet**-blue light (280–315 nm) are absorbed and therefore depleted by the canopy.

Other shade avoidance strategies include responses to release of volatile organic compounds such as ethylene from plants or the soil. Other non-light cues for shade avoidance include touch-induced leaf movements in stands where vertical structure is lacking in early stages of canopy development.

Niche Complementarity and Coexistence

Niche complementarity reduces competition for resources and allows species to coexist (Ashton et al. 2010; Faget et al. 2013). Plants partition resources by having characteristic differences in **phenology** and rooting depths, capacities for N fixation, and **plasticity** in chemical forms of N used (Ashton et al. 2010). Plants first produce roots in unoccupied soil higher in nutrients and free of other roots (Gersani et al. 2001). Then, roots occupy soil of competitors and lastly soil already occupied by their own roots (Gersani et al. 2001).

O'Brien et al. (2007) developed a spatially-explicit model of belowground competition for nutrients. The model predicts that rooting areas of individual plants overlap. In response to nutrient competition, individual plants concede some but not all space to roots of their neighbors. With increasing soil fertility, the model predicts root proliferation and overlap should increase.

Cahill et al. (2010) discovered plants integrate signals about nutrients and neighbors. When grown without competitors, plants adopt a broad-rooting strategy regardless of resource distribution (Cahill et al. 2010). When grown with competitors, plants produce roots with a narrower distribution modified by nutrient distribution. In soil with uniform nutrient distribution and presence of competitors, plants produce narrowly distributed, spatially segregated roots. In soil with variable nutrient distribution, plants produce broadly distributed roots. In summary, when growing alone, plants adopt a broad-foraging strategy. If neighbors are present, the plant adopts a restricted-foraging strategy modified by resource distribution (Cahill et al. 2010).

Root responses to nutrient distribution also may depend on the competitive strength of neighboring species (Mommer et al. 2012). Root growth of inferior competitors shifts toward nutrient-poor soil when roots of a superior competitor occupy the nutrient-rich soil (Mommer et al. 2012). Plasticity in uptake of different chemical forms of resources, for example soil N, may also be a manner by which coexisting plant species exhibit niche complementarity (Ashton et al. 2010).

Niche complementarity with respect to water use may also allow species to coexist. Nippert and Knapp (2007) hypothesized that C_3 forbs and shrubs persist in C_4-dominated mesic grassland by soil water partitioning. Although the upper soil layer (0–25 cm) contains the majority of root biomass in grassland, they observed that C_4 grasses relied on shallow soil water (upper 5 cm) across the growing season regardless of soil water availability at greater depths. In contrast, C_3 forbs and shrubs only used shallow soil water only when plentiful, increasing their reliance on soil water from greater depths as upper soil layers dried. Essentially, C_3 forbs and shrubs show niche complementarity in water use strategies to avoid competition with C_4 grasses when water is limiting (Nippert and Knapp 2007).

Silvertown et al. (2015) proposed three types of constraints that underlie development of hydrologic niches in plant communities. First, a soil or topographic constraint creates a tradeoff that forces species to specialize in acquisition of O_2 by roots compared with water and nutrients. Second, a biophysical constraint to gas exchange at the stomatal level leads to a tradeoff between CO_2 acquisition and water loss. Third, a structural constraint is related to high transpiration and rapid water conduction in the xylem to grow faster leads to another tradeoff. Maximum transpiration allows plants to grow faster and compete with neighbors but increases risk of **embolism**.

In a review of field studies across vegetation types ranging from arid to wet, Silvertown et al. (2015) found hydrologic niche complementarity to be widespread but mechanisms underlying it were unclear. Temporal partitioning of water use promoted species coexistence in arid communities but was not shown elsewhere. Several studies observed species partitioned soil water by having different predominant rooting depths (e.g. Nippert and Knapp 2007).

Facilitation

Facilitation describes interactions where one species modifies some component of the abiotic or biotic environment that then enhances the colonization, recruitment, and establishment of another (Bronstein 2009). In these commensal interactions, the facilitated species benefits while effects on the facilitator remain neutral. Facilitation has commonly been found during studies of **succession** in plant communities where some species improve soil conditions for future plants or act as nurses sheltering seedlings of other plants from weather extremes.

The stress-gradient hypothesis predicts that physically and biologically stressful environments such as deserts, intertidal habitats, saltmarshes, and seagrass beds would have several examples where facilitation expands the range of habitats for organisms to live (Bertness and Callaway 1994). These environments typically contain lethal conditions, reducing the fundamental niche of solitary organisms. However, when organized into groups, the individual organisms show better survival, greater distribution, and increased size of their fundamental niche. In these conditions, the positive benefits of growing in groups must outweigh negative effects of growing in close proximity. Benefits of growing as solitary organisms tends to occur more in environments with moderate physical conditions (Bertness and Callaway 1994).

In addition to stressful environments, those with intense herbivory may also promote facilitation. In this case, palatable species derive associational benefits from living with less palatable neighbors (Bertness and Callaway 1994).

Interference Competition

Interference competition describes how plants negatively affect the ability of other plants to grow or reproduce, not by preempting or exploiting limiting resources, but rather by physically or chemically altering the environment of the plant (Craine 2009). **Allelopathy** is the most commonly studied form of interference competition (Craine 2009; Aschehoug et al. 2016). It occurs when specific plants release chemicals into the soil or aerial environment that decrease the ability of other plants to function (Craine 2009). Allelopathic chemicals are released directly as volatiles or exudates from living plants or indirectly from decomposing leaf and root tissue. The chemicals released vary widely in structure, mode of action, effects on different plants, and longevity in the environment (Bais et al. 2006).

Study of sorghum, rye, and black walnut trees, have provided some of the best insights into how allelopathy negatively affects other plants (Aschehoug et al. 2016). Sorgoleone, a major component of the oily, hydrophobic root exudate produced by root hairs of sorghum, as well as sudangrass and johnsongrass, suppresses the growth of a large number of small-seeded plants (Dayan et al. 2010). Modes of action include inhibition of several molecular target sites including root meristems and photosynthesis in germinating seedlings (Dayan et al. 2010).

Allelopathic potential of rye results from production of benzoazinones, which inhibit germination and seedling growth of a wide-range of weeds and crops (Schulz et al. 2013). Meanwhile, walnut trees produce small concentrations of juglone, a relatively stable compound in soil that has phytotoxic effects (Bais et al. 2006). **Autotoxicity**, such as with alfalfa, is a special type in which chemicals are produced by established plants that interfere with germination and seedling growth of alfalfa seedlings. In effect, the chemical reduces the potential of more competition from alfalfa. The chemical has less effect on other species and generally dissipates in the soil within one year in humid areas and longer in drier areas (see Chapter 25).

Plant-Microbial Interactions

Plants host diverse microbial communities that span across above- and belowground organs, as well as inside the plant (Berg et al. 2016). In some instances, interactions with microorganisms can promote germination, stimulate growth, and enhance plant health through suppression of pathogens and pests, and increase tolerance to stress (Berendsen et al. 2012), while in others they may have little if any effects (Berg et al. 2016). Biologic nitrification inhibition (BNI), the phenomenon by which certain plants, such as koroniviagrass, release inhibitors from roots that block enzymatic pathways of nitrifying-bacteria represents a plant-microbial interaction that if exploited could become a powerful strategy to reduce N losses from agricultural systems (Subbarao et al. 2009). The following sections describe a few plant-microbial interactions that influence growth, health, and function of **herbaceous** plants in agroecosystems.

Associative and Endophytic Bacteria

Interactions with associative and endophytic bacteria, a taxonomically diverse group sometimes referred to as plant-growth promoting rhizobacteria (PGPR), can enhance nutritional status (Carvalho et al. 2014), prevent pathogen colonization (Berendsen et al. 2012), boost defenses (Berendsen et al. 2012), and improve abiotic stress tolerance of plants (Vacheron et al. 2013; Venturi and Keel 2016). Associative bacteria generally reside in the **rhizosphere** (Figure 9.3), a narrow zone of soil that surrounds roots (Philippot et al. 2013), and the rhizoplane, where they adhere to the outer surface of roots. Endophytic bacteria, meanwhile, colonize intercellular spaces, xylem vessels, and xylem parenchyma of roots, a region defined as the endosphere. Endophytic bacteria invade plant tissues but differ from endosymbiont bacteria (described below) in that they do not reside intracellularly in living plant cells, and their colonization does not induce the formation of differentiated structures such as nodules (Carvalho et al. 2014).

Interactions with associative and endophytic bacteria start in the rhizosphere where exudation of carbohydrates and other compounds by roots attract the bacteria (Compant et al. 2010). After migration to the root, release of **rhizodeposits** including nutrients, exudates, border cells, and mucilage from the roots continue to feed and mediate the interactions (Philippot et al. 2013) while polysaccharides present in the bacterial wall adhere to the root surface (Reinhold-Hurek and Hurek 2011). The rhizodeposits can attract deleterious as well as beneficial and neutral bacteria, among other soil organisms (Compant et al. 2010). Genetic and biochemical mechanisms involving bacterial signals, plant receptors, and developmental processes regulate the perception and recognition of beneficial and pathogenic interactions (Carvalho et al. 2016).

Endophytic bacteria enter the roots through root hairs, emergence points of lateral roots and, to some extent, the zone of cell differentiation and elongation near the root tip (Reinhold-Hurek and Hurek 2011; Carvalho et al. 2014). They then colonize intercellular spaces in the **epidermis** and **cortex** of roots and in disintegrated plant cells before spreading through the xylem or nutrient-rich intercellular spaces to colonize stems and leaves and to a lesser extent, flowers, fruits, and seeds (Compant et al. 2010).

Abiotic factors governing assembly of the microbial community include soil physical and chemical characteristics, climate, and weather conditions (Berg and Smalla 2009). Soil type commonly influences microbial composition but biotic factors including plant species, cultivar, age, developmental stage, and health also can be strong determinants (Berg et al. 2016). Each plant species recruits its own specific microbial community from the surrounding bulk soil outside the rhizosphere (Berg and Smalla 2009) by producing roots that vary in morphology and amounts and types of **rhizodeposits** (Philippot et al. 2013).

Associative and endophytic bacteria that live near the root surface or within intercellular spaces and **vascular tissues** of monocots can fix atmospheric N. The process involves similar genomic and biochemical mechanisms to those which occur in *Rhizobium* species (Carvalho et al. 2014). Associative and endophytic bacteria also improve N uptake from soil and produce phytohormones (auxin, cytokinin, and gibberellin) that help regulate plant growth. Through signals, associative and endophytic bacteria can regulate root system architecture, root structure, and growth of roots and shoots (Vacheron et al. 2013; Carvalho et al. 2014). Reduction in growth rate of the primary root and increase in number and length of lateral roots and root hairs is also commonly observed. It has been suggested that modification of root cell wall and root tissue structural properties contribute to pathogen and disease suppression (Vacheron et al. 2013).

Rhizobial Endosymbiont Bacteria

Symbiosis of legumes with rhizobial endosymbiont bacteria represents a well-studied plant-microbial interaction

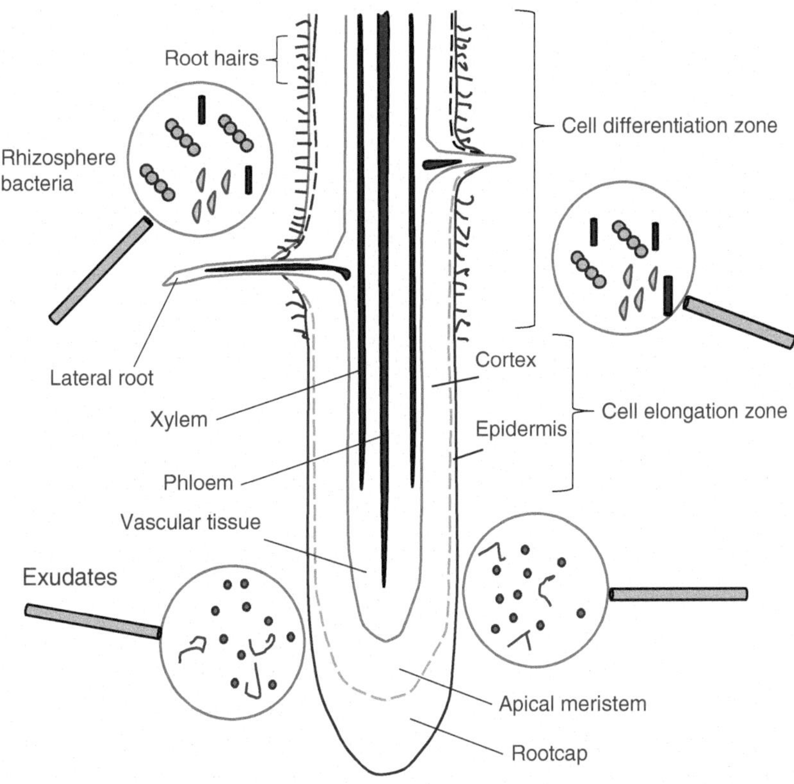

Fig. 9.3. Exudation of phytochemicals attracts beneficial and deleterious rhizosphere bacteria that inhabit the surface of the root, the rhizoplane and invade inside of roots, the endosphere, at points of emergence of lateral roots and root hairs.

that often benefits both partners. Nitrogen fixed by **rhizobia** serves the plant which, in return, provides a source of energy. Oldroyd et al. (2011) reviewed mechanisms by which legumes allow rhizobial infection, promote nodule formation, and force the bacteria into a N-fixing organelle-like state within the plant (Figure 9.4). Essentially, the symbiosis relies on two coordinated developmental processes: bacterial infection and nodule organogenesis. Rhizobia attach to the root surface and commonly gain entry through root hairs that turn inward and entrap rhizobia in an infection pocket. Release of flavonoids by the plant and nodulation factors by **rhizobia** regulates host specificity.

The entrapped rhizobia divide, form colonies, degrade plant cell walls, and invade through an infection thread that grows as a hollow tube through the root hair and multiple, underlying cortical cells. Proliferation of cells from the cortex and the **pericycle**, regulated by cytokinin and auxin levels, forms the nodule and provides an environment suitable for N fixation. Once enclosed by plant cells, the rhizobia, which began as free-living bacteria, become surrounded by a plant-derived membrane, differentiate into **bacteroids**, and become host dependent for metabolism.

Nitrogen fixation requires considerable energy input, regulation of N_2-fixation genes, and protection of nitrogenases, the O_2-sensitive **enzymes** that catalyze the biologic reduction of N_2 to NH_3. The nodule cortex provides an O_2 diffusion barrier, and leghemoglobin, a plant protein in nodules that reversibly binds with O_2 to facilitate its diffusion and support respiration of the bacteroids, without interfering with nitrogenases. These specialized N-fixing organelles exchange photosynthate to fix N for the plant (Dixon and Kahn 2004). Ferguson and Mathesius (2014) provide more detail about the roles and interactions of phytohormones and signaling peptides in the regulation of nodule infection, initiation, positioning, and development. Similarly, Oldroyd (2013) compared signaling processes used by plants during mutualistic interactions with mycorrhizal fungi and rhizobial bacteria.

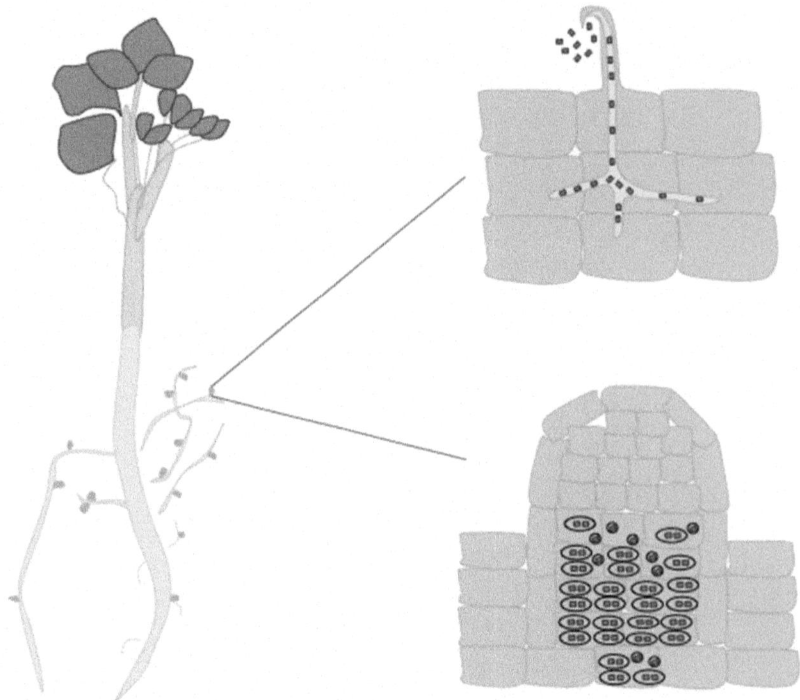

Fig. 9.4. Interaction of rhizobium bacteria and legume roots facilitate symbiotic nitrogen fixation. Bacteria form an infection tube through root hairs that allow their movement into root cortical cells. Cell division and formation of symbiosomes lead to formation of nodules and the exchange of resources between plant cells and bacteria.

Arbuscular Mycorrhizal Fungi

Mycorrhizal fungi form symbiotic interactions with almost all terrestrial plant species (Figure 9.5) where they play a key role in nutrient exchange (van der Heijden et al. 2015). Four major types based on their structure and interactions with plants include ectomycorrhiza (EM), arbuscular mycorrhiza (AM), orchid mycorrhiza, and ericoid mycorrhiza (van der Heijden et al. 2015). Ectomycorrhizal fungi develop extensively at root tips of temperate woody species, covering the surface with a loose mantle of fungal mycelium that extends into the roots between epidermal and cortical cells. Arbuscular mycorrhiza colonize intercellular spaces in the cortex but do not form a mantle around the roots. Distinctive features include formation of arbuscules, tree-shaped structures that serve as intracellular sites of exchange with the plant partner, and vesicles, intracellular or intercellular storage structures. Interactions with AM fungi mainly occur in **herbaceous** and woody species, while orchid and ericoid mycorrhiza colonize orchids and members of the Ericaceae (heath or heather family) and some liverworts, respectively (van der Heijden et al. 2015).

Arbuscular mycorrhizal fungi provide plants with minerals from soil in exchange for C substrates derived from photosynthesis from the host plant (Walder and van der Heijden 2015). They form long, branching, tubular structures known as hyphae (Smith and Smith 2011) which grow out from roots into soil where they forage for nutrients (van der Heijden et al. 2015). They can obtain C from roots on the same or different plants (Smith and Smith 2011), and linking of common mycorrhizal networks facilitate movement of resources among neighboring plants (Walder et al. 2012). Collins Johnson et al. (2015) found that AM symbiosis can completely eliminate P but not N limitations in grasslands. Carbon limitations, such as what might occur for plants growing in shade, shift AM interactions from mutualistic to parasitic (Collins Johnson et al. 2015).

Walder and van der Heijden (2015) questioned whether benefits of reciprocal exchange best characterize AM interactions, where both species trade resources or services and benefits of doing so are weighed against costs. They noted that occurrence of parasitic interactions, lack of specific partnerships, simultaneous interactions with

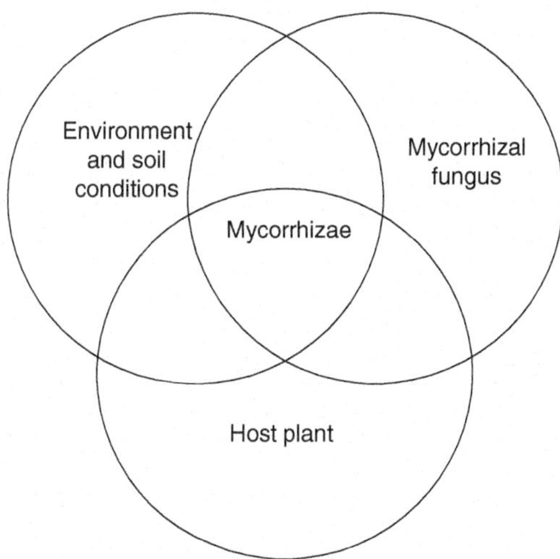

Fig. 9.5. The occurrence and ecologic significance of mycorrhizal symbioses are a result of interactions among mycorrhizal fungi, host plants, and associated biotic and abiotic variables within the environment. Variation in mycorrhizal dependencies among plant species has been identified as a unifying concept to explain the ecologic effects of mycorrhizae on species composition and diversity. *Source:* Adapted from Brundett (1991); Briske (2007); Figure 7.4 Forages Vol. II.

several partners, functional diversity among partners, and lack of partner discrimination in AM interactions challenges the concept of reciprocal exchange. Despite this, others have noted that non-nutritional benefits associated with AM fungi such as increased soil aggregation, water flow, and disease resistance can be as beneficial as the provisioning of nutrients in exchange for C (Delavaux et al. 2017).

Fungal Endophytes

Fungal endophytes infect and profoundly impact fitness of nearly all plants in natural ecosystems (Rodriguez et al. 2009). They often confer abiotic and biotic stress tolerance, increase biomass, decrease water consumption and alter resource allocation, sometimes to the detriment of the host (Rodriguez et al. 2009). They also can enhance host resistance to herbivory from insects and **nematodes**, while being toxic to livestock. Effects on host reproduction vary depending on the particular grass-endophyte symbiosis (Cheplick and Faeth 2009), but fungal endophytes can improve sexual and asexual reproduction in tall fescue. The fungal associate does not exist independently of its host and, therefore, it benefits from the symbiosis regardless of negative (parasitic or pathogenic), positive (mutualistic), or neutral (commensal) effects on the host plant (Cheplick and Faeth 2009).

Fungal endophytes reside entirely within plant tissues, growing throughout roots, stems, and leaves. Two major groups of fungal endophytes, clavicipitaceous, and nonclavicipitaceous differ in evolutionary relatedness, taxonomy, plant hosts, and ecologic functions (Rodriguez et al. 2009). Clavicipitaceous endophytes, which form systemic intercellular infections in grasses, rushes, and **sedges**, generally have a narrow host range with interactions ranging from mutualistic to pathogenic depending on host species, host genotype, and environmental conditions (Clay and Schardl 2002). Fungal transmission to offspring from maternal plants occurs vertically through seed infection and horizontally through vegetative propagation (Rodriguez et al. 2009). Nonclavicipitaceous endophytes, on the other hand, have a broad host range but largely unknown and poorly defined ecologic roles (Rodriguez et al. 2009).

Three functional classes of nonclavicipitaceous endophytes have been differentiated based on host colonization patterns, tissues colonized, extent of host colonization, mechanisms of transmission between host generations, biodiversity in plants, and ecologic function (Rodriguez et al. 2009). The following discussion focuses on clavicipitaceous endophytes which include sexually-reproducing species of *Epichloë* and their asexual descendants, *Neotyphodium*, the principal symbionts of grasses (Leuchtmann et al. 2014).

Epichloë species produce yellow-orange stromata on the leaf sheath (Leuchtmann et al. 2014) and grow over and arrest development of the inflorescence, a phenomenon known as choke disease, when the host is flowering (Rodriguez et al. 2009; Leuchtmann et al. 2014). Once mature, fungal fruiting bodies transmit spores horizontally much like pathogens with infection of reproductive structures resulting in sterilization of the host (Clay and Schardl 2002). *Epichloë* species that produce stromata on all or most of the tillers have been defined as Type I endophytes and generally occur in C_4 grasses (Clay and Schardl 2002). Those that exhibit stromata only in a fraction of the tillers, allowing partial seed production and thus vertical transmission within seeds, have been defined as Type II endophytes. They occur only in C_3 grasses (Clay and Schardl 2002).

Vertically **transmitted**, **asexual** endophytes in the genus *Neotyphodium* have lost the capacity for development of the sexual stage and have no regular mechanism for genetic recombination. They infect leaves, culms, and rhizomes but display no obvious symptoms at any stage of plant development (Clay and Schardl 2002). Vertical transmission involves fungal growth into **ovules** of developing inflorescences and ultimately into seeds. There it colonizes the **scuttelum** and **embryo** axis of the seed

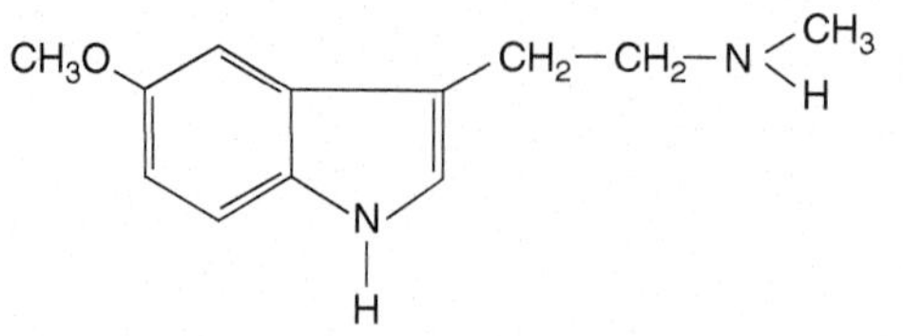

5-methoxy-N-methyltryptamine

N-acetylloline

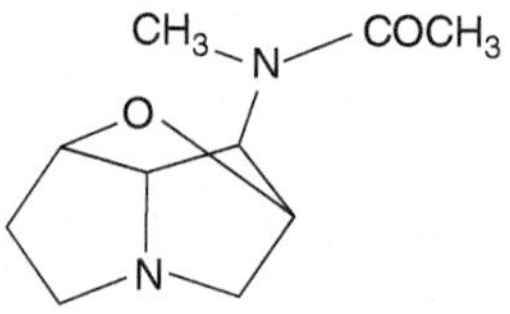

ergovaline

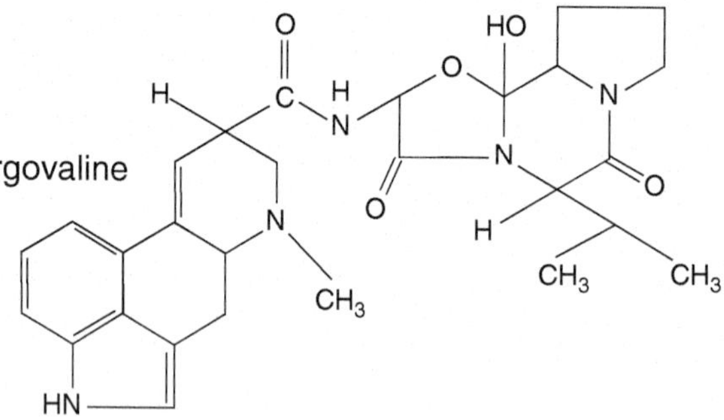

peramine

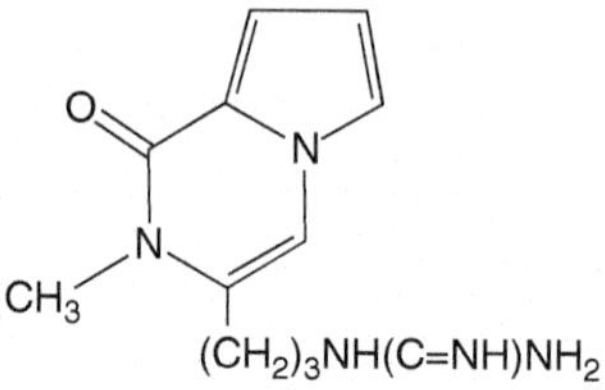

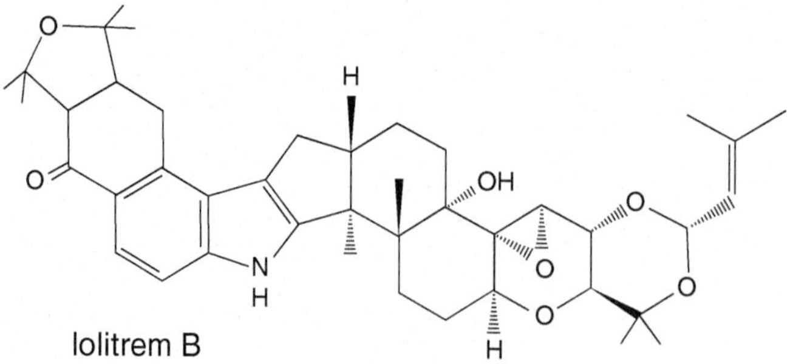

lolitrem B

Fig. 9.6. Some cool-season grasses accumulate toxic alkaloids that reduce animal performance. Those illustrated are important representatives of each alkaloid group. Others in the group have slight changes in the molecule and are usually lower in concentration or have less effect on the animal. (Balasko and Nelson 2003. Figure 6.2 Forages Vol. I).

and reemerges in the germinating seedling to provide the mechanism for infection of the new plant. Though they exhibit less genetic variation than sexually-recombining *Epichloë* endophytes, they do exhibit greater frequencies of infection within host populations, which principally includes C_3 grasses.

Clay and Schardl (2002) defined these as Type III endophytes, which includes *Neotyphodium coenophialum* and *Neotyphodium lolii*, the ubiquitous symbionts of tall fescue and perennial ryegrass, respectively (Clay and Schardl 2002, Cheplick and Faeth 2009). Leuchtmann et al. (2014) proposed that most previously described *Neotyphodium* species including *N. coenophialum* and *N. lolii* be considered synonymously within the genus *Epichloë* and be renamed as *E. coenophialum* and *E. festucae,* respectively. They explained that the dual naming system provides more of an impediment than benefit to understanding the diversity of evolutionary histories, life histories, and host interactions of each species and their various strains. In addition, the species share similar morphologies and behaviors such as intercellular, systemic colonization of aerial plant tissues and seed transmissibility.

Many *Epichloë* species produce toxic alkaloids (Figure 9.6) (Leuchtmann et al. 2014) which frequently results in portrayal of endophytes as defensive mutualists (Clay and Schardl 2002; Schardl et al. 2004, 2009; Cheplick and Faeth 2009). Four classes of alkaloids known to deter herbivory include saturated aminopyrrolizidine (loline) alkaloids, indolediterpenoid (lolitrem) alkaloids, ergot alkaloids, and pyrrolopyrazine (peramine) alkaloids (Clay and Schardl 2002; Bush et al. 2007, and Schardl et al. 2012). Though resistance to herbivory would be a benefit from serving as an endophyte host, most grass-endophyte symbioses come with costs (Rodriguez et al. 2009), and mutualism based on defense from herbivores tends to be less applicable to indigenous grasses in natural settings where the frequency of infection shows wide variation (Cheplick and Faeth 2009; Rodriguez et al. 2009).

Thus, grass-endophyte symbioses span the entire range from mutualism to parasitism depending in part on their associations with flowering tillers, particular host genotype-endophyte combinations, and environmental conditions (Schardl et al. 2004; Cheplick and Faeth 2009). When abnormally high-endophyte contents coincide with stressful conditions and greater nutrient demands for plants, the costs of hosting endophytes may exceed their benefits (Rodriguez et al. 2009). Hosting the endophyte may also be costly in environments absent of herbivores (Rodriguez et al. 2009).

Fungal endophytes can increase tolerance to abiotic stresses, as well as competitive ability of hosts, but inconsistencies exist among studies and few have evaluated the interactions in long-term studies (Cheplick and Faeth 2009). Many observed increases in growth of mutualistic grass-endophyte symbioses, have occurred in controlled environments under conditions highly favorable to plant growth (Cheplick and Faeth 2009). Few studies of grass-endophyte symbioses have been conducted in natural environments, especially with native grasses, and relationships between physiologic variables and those of survival, growth, reproductive fitness, competitive ability, or abiotic stress tolerance of hosts remain unknown. Some evidence that endophyte-infected tall fescue and perennial ryegrass may be more competitive than their uninfected counterparts exists, but few have investigated host competitive ability in other grass-endophyte symbioses in communities they inhabit (Cheplick and Faeth 2009).

Summary

Plants interact with multiple organisms throughout their life cycle. Many of these interactions have negative consequences but some serve to enhance plant growth and function. Traditional studies have focused on those occurring aboveground, including competition between two or more plant species, mutualistic and parasitic grass-endophyte interactions, and endosymbiotic N_2-fixation. Increasingly, though, exciting areas of new research have moved belowground to examine plant-microbial interactions in the **rhizosphere**. This research has found that plants and microbes strongly influence each other through secretion and detection of signaling compounds but the microbial community is complex (Venturi and Keel 2016). Emerging tools, such as metabolomics, may help us better understand chemical communication between plants and their microbiome (van Dam and Bouwmeester 2016).

References

Aschehoug, E.T., Brooker, R., Atwater, D.Z. et al. (2016). The mechanisms and consequences of interspecific competition among plants. *Annu. Rev. Ecol. Evol. Syst.* 47: 263–281.

Ashton, I.W., Miller, A.E., Bowman, W.D., and Suding, K.N. (2010). Niche complementarity due to plasticity in resource use: plant partitioning of chemical N forms. *Ecology* 91: 3252–3260.

Bais, H.P., Weir, T.L., Perry, L.G. et al. (2006). The role of root exudates in rhizosphere interactions with plants and other organisms. *Annu. Rev. Plant Biol.* 57: 233–266.

Balasko, J.A. and Nelson, C.J. (2003). Grasses for northern areas. In: *Forages Vol. I: An Introduction to Grassland Agriculture*, 6e (eds. R.F Barnes, C.J. Nelson, M. Collins and K.J. Moore), 125–148. Ames, IA: Iowa State Press.

Begon, M. (2009). Ecological epidemiology. In: *The Princeton Guide to Ecology* (ed. S.A. Levin), 220–226. Princeton, NJ: Princeton University Press.

Berendsen, R.L., Pieterse, C.M.J., and Bakker, P.A.H.M. (2012). The rhizosphere microbiome and plant health. *Trends Plant Sci.* 17: 478–486.

Berg, G. and Smalla, K. (2009). Plant species and soil type cooperatively shape the structure and function of microbial communities in the rhizosphere. *FEMS Microbiol. Ecol.* 68: 1–13.

Berg, G., Rybakova, D., Grube, M., and Köberl, M. (2016). The plant microbiome explored: implications for experimental botany. *J. Exp. Bot.* 67: 995–1002.

Bertness, M.D. and Callaway, R. (1994). Positive interactions in communities. *Trends Ecol. Evol.* 5: 191–193.

Briske, D.D. (2007). Plant interactions. In: *Forages, Vol. II: The Science of Grassland Agriculture*, 6e (eds. R.F Barnes, C.J. Nelson, K.J. Moore and M. Collins), 105–122. Ames, IA: Blackwell Publishing.

Bronstein, J.L. (2009). Mutualism and symbiosis. In: *The Princeton Guide to Ecology* (ed. S.A. Levin), 233–238. Princeton, NJ: Princeton University Press.

Brundrett, M.C. (1991). Mycorrhizas in natural ecosystems. *Adv. Ecol. Res.* 21: 171–313.

Bush, L., Roberts, C.A., and Schultz, C. (2007). Plant chemistry and antiquality components in forage. In: *Forages Vol. II: The Science of Grassland Agriculture*, 6e (eds. R.F Barnes, C.J. Nelson, K.J. Moore and M. Collins), 509–528. Ames, IA: Blackwell Publishing.

Cahill, J.F. Jr., McNickle, G.G., Haag, J.J. et al. (2010). Plants integrated information about nutrients and neighbors. *Science* 328: 1657.

Carvalho, T.L.G., Balsemã-Pires, E., Saraiva, R.M. et al. (2014). Nitrogen signaling in plant interactions with associative and endophytic diazotrophic bacteria. *J. Exp. Bot.* 65: 5631–5642.

Carvalho, T.L.G., Ballesteros, H.G.F., Thiebault, F. et al. (2016). Nice to meet you: genetic, epigenetic and metabolic controls of plant perception of beneficial associative and endophytic diazotrophic bacteria in non-leguminous plants. *Plant Mol. Bio.* 90: 561–574.

Cheplick, G.P. and Faeth, S.H. (2009). *Ecology and Evolution of the Grass-Endophyte Symbiosis*. New York, NY: Oxford University Press.

Clay, K. and Schardl, C. (2002). Evolutionary origins and ecological consequences of endophyte symbiosis with grasses. *Am. Nat.* 160: S199–S127.

Collins Johnson, N., Wilson, G.W.T., Wilson, J.A. et al. (2015). Mycorrhizal phenotypes and the law of the minimum. *New Phytol.* 205: 1473–1484.

Compant, S., Clément, C., and Sessitch, A. (2010). Plant growth-promoting bacteria in the rhizo- and endosphere of plants: their role, colonization, mechanisms involved and prospects for utilization. *Soil Biol. Biochem.* 42: 669–678.

Craine, J. (2009). *Resource Strategies of Wild Plants*. Princeton, NJ: Princeton University Press.

Craine, J.M. and Dybinski, R. (2013). Mechanisms of plant competition for nutrients, water, and light. *Funct. Ecol.* 27: 833–840.

van Dam, N.M. and Bouwmeester, H.J. (2016). Metabolomics in the rhizosphere: tapping into belowground chemical communication. *Trends Plant Sci.* 21: 256–265.

Dayan, F.E., Rimando, A.M., Pan, Z. et al. (2010). Sorgoleone. *Phytochemistry* 71: 1032–1039.

Delavaux, C.S., Smith-Ramesh, L.M., and Kuebbing, S.E. (2017). Beyond nutrients: a meta-analysis of the diverse effects of arbuscular mycorrhizal fungi on plants and soils. *Ecology* 98: 2111–2119.

Dixon, R. and Kahn, D. (2004). Genetic regulation of biological nitrogen fixation. *Nat. Rev. Microbiol.* 2: 621–631.

Dybzinski, R. and Tilman, D. (2009). *The Princeton Guide to Ecology* (ed. S.A. Levin). Princeton, NJ: Princeton University Press.

Faget, M., Nagel, K.A., Walter, A. et al. (2013). Root-root interactions: extending our perspective to be more inclusive of the range of theories in ecology and agriculture using *in-vivo* analyses. *Ann. Bot.* 112: 253–266.

Ferguson, B.J. and Mathesius, U. (2014). Phytohormone regulation of legume-rhizobia interactions. *J. Chem. Ecol.* 40: 770–790.

Gersani, M., Brown, J.S., O'Brien, E.E. et al. (2001). Tragedy of the commons as a result of root competition. *J. Ecol.* 89: 660–669.

Grace, J.B. (1990). On the relationship between plant traits and competitive ability. In: *Perspectives on Plant Competition* (eds. J.B. Grace and D. Tilman), 51–65. San Diego, CA: Academic Press.

Grace, J.B. and Tilman, D. (1990). Perspectives on plant competition: some introductory remarks. In: *Perspectives on Plant Competition* (eds. J.B. Grace and D. Tilman), 3–7. San Diego, CA: Academic Press.

Grime, J.P. (2001). *Plant Strategies, Vegetation Processes, and Ecosystem Properties*, 2e. Chichester, West Sussex, England: Wiley.

van der Heijden, M.G.A., Martin, F.M., Selosse, M.-A., and Sanders, I.R. (2015). Mycorrhizal ecology and evolution: the past, present, and the future. *New Phytol.* 205: 1406–1423.

Leuchtmann, A., Bacon, C.W., Schardl, C.L. et al. (2014). Nomenclatural realignment of *Neotyphodium* species with genus *Epichloë*. *Mycologia* 106: 202–215.

Moenne-Loccoz, Y., Mavingui, P., Combes, C. et al. (2015). Microorganisms and biotic interactions. In: *Environmental Microbiology: Fundamentals and Applications* (eds. J.-C. Bertrand, P. Caumette, P. Lebaron, et al.), 395–444. Dordrecht: Springer Science.

Mommer, L., van Ruijven, J., and Jansen, C. (2012). Interactive effects of nutrient heterogeneity and competition: implications for root foraging theory? *Funct. Ecol.* 26: 66–73.

Nippert, J.B. and Knapp, A.K. (2007). Soil water partitioning contributes to species coexistence in tallgrass prairie. *Oikos* 116: 1017–1029.

O'Brien, E.E., Brown, J.S., and Moll, J.D. (2007). Roots in space: a spatially explicit model for belowground competition in plants. *Proc. Royal. Soc. Brit.* 274: 929–934.

Oldroyd, G.E.D. (2013). Speak, friend, and enter: signalling systems that promote beneficial symbiotic associations in plants. *Nat. Rev. Microbiol.* 11: 252–263.

Oldroyd, G.E.D., Murray, J.D., Poole, P.S., and Downie, J.A. (2011). The rules of engagement in the legume-rhizobial symbiosis. *Annu. Rev. Genet.* 45: 119–144.

Philippot, L., Raaijmakers, J.M., Lemanceau, P., and van der Putten, W.H. (2013). Going back to the roots: the microbial ecology of the rhizosphere. *Nat. Rev. Microbiol.* 11: 789–799.

Pierek, R. and de Wit, M. (2014). Shade avoidance: phytochrome signalling and other aboveground neighbour detection cues. *J. Exp. Bot.* 65: 2815–2824.

Reinhold-Hurek, B. and Hurek, T. (2011). Living inside plants: bacterial endophytes. *Curr. Opin. Plant Biol.* 14: 435–443.

Rodriguez, R.J., White Jr, J.F., Arnold, A.E., and Redman, R.S. (2009). Fungal endophytes: diversity and functional roles. *New Phytol.* 182: 314–330.

Schardl, C.L., Leuchtmann, A., and Spiering, M.J. (2004). Symbioses of grasses with seedborne fungal endophytes. *Annu. Rev. Plant Biol.* 55: 315–340.

Schardl, C.L., Balestrini, R., Florea, S. et al. (2009). Epichloë endophytes: clavicipitaceous symbionts of grasses. In: *Plant Relationships. The Mycota (A Comprehensive Treatise on Fungi as Experimental Systems for Basic and Applied Research)*, vol. 5 (ed. H.B. Deising), 275–306. Heidelberg: Springer.

Schardl, C.L., Young, C.A., Faulkner, J.R. et al. (2012). Chemotypic diversity of epichloae, fungal symbionts of grasses. *Fungal Ecol.* 5: 331–344.

Schulz, M., Marocco, A., Tabaglio, V. et al. (2013). Benzoxazinoids in rye allelopathy – from discovery to application in sustainable weed control and organic farming. *J. Chem. Ecol.* 39: 154–174.

Silvertown, J., Araya, Y., and Gowing, D. (2015). Hydrological niches in terrestrial plant communities: a review. *J. Ecol.* 103: 93–108.

Smith, S.E. and Smith, F.A. (2011). Roles of arbuscular mycorrhizas in plant nutrition and growth: new paradigms from cellular to ecosystem scales. *Annu. Rev. Plant Biol.* 62: 227–250.

Smith, H. and Whitelam, G.C. (1997). The shade avoidance syndrome: multiple responses mediated by multiple phytochromes. *Plant Cell Environ.* 20: 840–844.

Subbarao, G.V., Nakahara, K., Hurtado, M.P. et al. (2009). Evidence for biological nitrification inhibition in *Brachiaria* pastures. *Proc. Natl. Acad. Sci. U.S.A.* 106: 17302–17307.

Tilman, D. (1990). Mechanisms of plant competition for nutrients: the elements of a predictive theory of competition. In: *Perspectives on Plant Competition* (eds. J.B. Grace and D. Tilman), 117–141. San Diego, CA: Academic Press.

Vacheron, J., Desbrosses, G., Bouffaud, M.-L. et al. (2013). Plant growth-promoting rhizobacteria and root system functioning. *Front. Plant Sci.* 4: 1–19.

Venturi, V. and Keel, C. (2016). Signaling in the rhizosphere. *Trends Plant Sci.* 21: 187–198.

Walder, F. and van der Heijden, M.G.A. (2015). Regulation of resource exchange in the arbuscular mycorrhizal symbiosis. *Nat. Plant.* 1: 1–7.

Walder, F., Niemann, H., Natarajan, M. et al. (2012). Mycorrhizal networks: common goods of plants shared under unequal terms of trade. *Plant Physiol.* 159: 789–797.

Chapter

10

Plant-Herbivore Interactions

Lynn E. Sollenberger, Distinguished Professor, *Department of Agronomy, University of Florida, Gainesville, FL, USA*
Marcelo O. Wallau, Associate Professor, *Department of Agronomy, University of Florida, Gainesville, FL, USA*

Nature and Complexity of Grassland-Herbivore Interactions

Herbivores create and respond to grassland vegetation patterns in dynamic, interactive ways that can be beneficial or detrimental (Laca 2008). Interactions among herbivores and grasslands occur in several scales of space and time (Frank 2006; Bailey and Provenza 2008). At each scale of **herbivore** decision-making, different grassland characteristics affect their response, and these responses affect feeding efficiency and animal performance.

Due to the diversity and complexity of the grazed landscape, experimentation to determine causes and effects has evolved slowly. The bulk of experimentation dates from the mid-1970s (e.g. Stobbs 1973a,b, 1974) to early 1990s (e.g. Laca et al. 1992, 1994), with most of the focus on animal response to vegetation patterns (Frank 2006). Components of intake were measured in single-species or simple-mixture planted pastures or **microswards**, while ecologists focused on plant-species drivers of animal movement and on general effects of grazing on species-rich rangelands. Mechanistic models have been used to explore the relationship between plants and herbivores. These models have enhanced knowledge of **ecosystem** processes, but the link with grassland management is still weak and application to on-farm management is limited (Weisberg et al. 2006).

Given the breadth of possible topics, our focus will be livestock production systems, generally using cattle as an example herbivore. The objective is to characterize the dynamics that occur in the grazed landscape and the reaction of plants and herbivores to those dynamics. The use of models to explore this complexity will be discussed. Intake regulation and other more animal behavior-oriented topics are beyond the scope of this chapter. For wildlife and **rangeland** perspectives on this topic, we suggest reading the reviews by Frank (2006), Weisberg et al. (2006), and Fuhlendorf et al. (2017).

Scales of Grassland-Herbivore Interactions

To facilitate the understanding of grassland-herbivore interactions, the spatial and temporal dimensions are divided into hierarchical levels (Bailey and Provenza 2008). At each hierarchical level, a different behavioral process takes place in response to canopy characteristics, defining a response variable within a specific timeframe (Figure 10.1; Bailey et al. 1996; Bailey and Provenza 2008; Carvalho 2013).

The finest scale of interaction is the *bite*, occurring every one to two seconds. The major response variable is bite mass, estimated mainly from bite volume which is influenced by canopy architecture and herbivore prehensile anatomy (Shipley et al. 1994; Carvalho et al. 2008).

Forages: The Science of Grassland Agriculture, Volume II, Seventh Edition.
Edited by Kenneth J. Moore, Michael Collins, C. Jerry Nelson and Daren D. Redfearn.

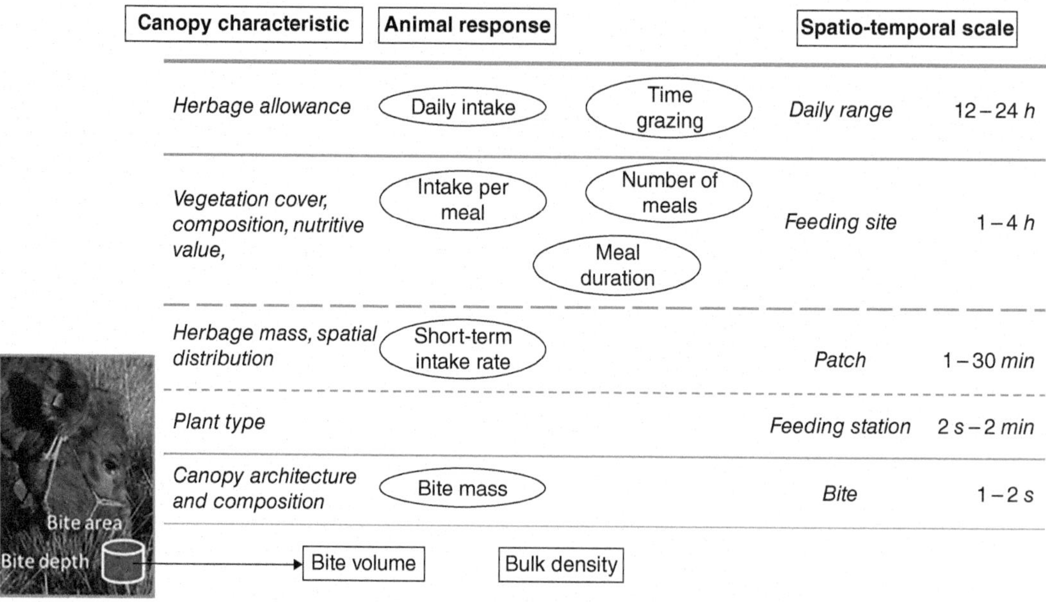

Fig. 10.1. Relationships between canopy characteristics, components of animal grazing behavior, and intake at different spatio-temporal scales in grazed swards. *Source:* Adapted from Bailey et al. (1996) and Carvalho et al. (2013). Animal photo courtesy of Olivier Bonnet.

At the bite level, the animal determines which plant parts to consume, basing its choice on plant size and physical and chemical characteristics. The second level is the *feeding station*, defined as the area to which an animal has access without moving the front legs, generally 0.1 to 1 m^2. Movement to a new feeding station generally occurs every 5–100 seconds, where bite rate is the main factor driving the change in feeding station (Laca and Demment 1990; Bailey et al. 1996; Carvalho et al. 2013). High residence time in each feeding station results in less time searching and is indicative of good pasture structure and high intake rate.

The next scale is the *patch*, defined as a continuum (cluster) of feeding stations of similar characteristics, varying from 1 m^2 to 1 ha, with temporal choices ranging from every 1 to 45 minutes (Bailey et al. 1996). The factor determining when to switch patches is the decline in short-term intake rate (Gordon and Benvenutti 2006; Carvalho et al. 2013). The more heterogeneous the pasture, the smaller the patches will be. A *feeding site* consists of a collection of patches in a contiguous area (Bailey et al. 1996), generally ranges from 1 to 10 ha, and is defined by grazing bouts or periods of concentrated grazing. There is often one to four hours between changes in feeding site. They are influenced especially by **forage mass** (Spalinger and Hobbs 1992), but also by topography, distance to water, and forage nutritive value. *Daily range* is the area where animals drink and rest between grazing bouts within a landscape unit. It is generally the broadest scale of interaction for livestock, represented by areas between 10 and 100 ha and timescales of 12–24 hours. Larger scales, especially for wildlife, are the *seasonal range* (100–1000 ha, 3–12 mo) and *lifetime range* (>1000 ha in a time scale of several years). Seasonal range is a response to seasonality of forage production, and dictates migration behavior within a landscape type. Lifetime range is a function of long-term weather and vegetation patterns within the geographic region.

Importance of Intake and Components of Intake in Grazed Grasslands

Intake is the primary determinant of animal performance, and bite mass of grazing livestock is the short-term measure of grazing behavior that is most positively related to intake (Figure 10.2; Hodgson 1982a; Sollenberger and Burns 2001). Relationships between intake and components of animal grazing behavior can be summarized (Stobbs 1973b; Hodgson 1982b) as:

$$\text{Intake} = \text{intake rate} \times \text{grazing time}$$

Where

$$\text{Intake rate} = \text{intake per bite} \times \text{rate of biting}$$

Thus

$$\text{Daily intake} = \text{intake per bite} \times \text{rate of biting} \times \text{grazing time}$$

It is important, therefore, to identify grazing management strategies and canopy characteristics that maximize intake per bite (bite mass) and **forage intake per unit** of grazing time. This is challenging because the canopy is changing continually due to growth, harvest, treading, fouling, and **senescence**.

Evaluation of components of intake shows that **sward** characteristics and canopy structure directly affect intake per bite and rate of biting (Sollenberger and Burns 2001). Intake per bite is a function of bite area, bite depth, and herbage bulk density. Bites day^{-1} (grazing time × rate of biting) is considered a compensatory response, increasing up to some maximum as bite mass decreases. In general, as sward height and subsequently bite mass decrease, animals may compensate by increasing grazing time and, to a lesser extent, biting rate (Forbes 1988). Increased grazing time, as a response to decreased bite mass, is constrained by other drives such as the need to socialize, ruminate, and rest, whereas the ability to increase biting rate may be constrained by sward structure because the animal spends more time searching for desired plant parts.

This view of ingestive behavior and its impact on intake is more useful for elucidating how sward characteristics influence short-term grazing behavior and relative herbage intake, rather than as a means of estimating daily intake (Coleman and Sollenberger 2007). Scaling from short-term intake to daily intake through daily grazing time may not give rational estimates (Macoon et al. 2003), though it has been suggested that such scaling may be useful in homogeneous grazing environments (Gregorini et al. 2006). This type of scaling is constrained because short-term intake rate is usually controlled by bite mass, which is more related to canopy structure and not necessarily to forage mass or allowance (Gross et al. 1993; Gonçalves et al. 2009). In contrast, daily intake is more likely to be limited by digestive constraints when vegetation is abundant, or diet selection when vegetation is scarce (Wilmshurst et al. 1999; Bailey and Provenza 2008; Laca 2008).

Others have concluded that there is a need for incorporation of ingestive behavior measurements into long-term animal response studies to assess more definitively which measurements have greatest impact on efficiency of the animal enterprise (Burns and Sollenberger 2002). A thoughtful assessment of the study of ingestive behavior was provided by Ungar (1996) who indicated that "while scientifically fascinating,...it remains to be seen to what extent the descent down spatial and temporal scales in the study of ingestive behavior will enable a more agriculturally useful synthesis." There are examples, however, of application of ingestive behavior research for identifying canopy-based triggers for initiation and cessation of grazing in both temperate (Amaral et al. 2013; Mezzalira et al. 2017) and subtropical grasslands (Carvalho 2013).

Grassland Characteristics Affect Animal Response

Herbage Mass

When considered across a relatively wide range of herbage mass, bite mass and herbage intake generally increase linearly with canopy height and herbage mass in both temperate and tropical pastures (Coleman and Sollenberger 2007). Sixty to ninety percent of the variation in individual animal performance is explained by herbage mass when evaluated across a wide range in herbage mass (Hernández Garay et al. 2004; Sollenberger and Vanzant 2011). This can be explained by feeding theory because when herbage mass is limited, large herbivores begin to eat less preferred foods and switch from preferred to less preferred feeding sites (Bailey and Provenza 2008). Slope of the increase in animal response to increasing mass is different among forage species (Forbes 1988) probably because forage **nutritive value**, not mass, determines the slope of the regression of average daily gain on **grazing intensity** (Sollenberger and Vanzant 2011).

Conversely, there may be no detectable relationship between herbage mass or canopy height and individual animal performance if mass and height are evaluated only when in surplus. Herbage mass increased linearly with increasing canopy height from 20 to 60 cm for continuously stocked limpograss pastures, but cattle (*Bos* sp.) daily gain increased with height only up to 40 cm (Newman et al. 2002). Lower gains at 60 cm were due to canopy factors other than mass, including lesser accessibility of leaf due to trampling. For heifers, grazing monocultures of *Cynodon* sp. and black oat, intake rate was greater for intermediate than short or tall sward heights (Mezzalira et al. 2017). The decrease of intake rate was due to lesser bite mass in tall than in intermediate canopies due to a reduction of bite volume, possibly caused by the greater proportion of stem and sheath acting as a physical barrier to bite formation. Bite mass and, consequently, short-term intake rate showed a negative quadratic relationship with canopy height for heifers and ewes (*Ovis aries*) grazing native grasslands (height ranging from 4 to 16 cm and mass from 1360 to 2820 kg DM ha^{-1}; Gonçalves et al. 2009). Although bite depth was greater for taller canopies, it was not sufficient to compensate for the lower herbage bulk density.

Intake rates of sheep grazing hand-constructed wimmera ryegrass microswards (Black and Kenney 1984) and of cattle grazing old world bluestem pastures (Forbes and Coleman 1993) were closely related to herbage mass when mass was low, but the relationship reached an asymptote at a herbage mass of about 1000–1100 kg ha^{-1}. Penning

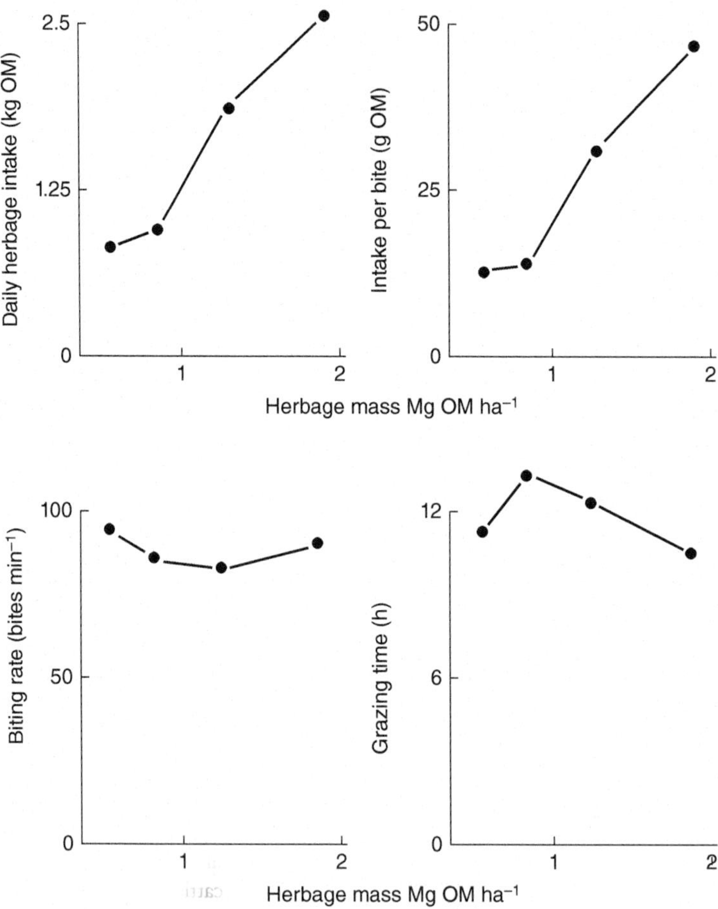

Fig. 10.2. The responses of herbage intake (*top left*), bite weight (*top right*), biting rate (*bottom left*), and grazing time per day (*bottom right*) to increased herbage mass of perennial ryegrass. *Source:* From Hodgson (1982b); used by permission from the British Grassland Society.

(1985) presented evidence that intake rate of sheep grazing perennial ryegrass reached an upper limit when sward height exceeded 70 mm. In highly heterogeneous rangelands, no simple relationships between intake or performance and herbage mass or sward height were found (Laca and Demment 1990; Gordon and Lascano 1993; Bonnet et al. 2015). In those cases, density of preferred bites can be low, affecting feeding station behavior and bite rate through the degree of diet selection (Bonnet et al. 2015). Of possible relevance, is the forage maturation hypothesis that is based on the temporal dynamics of forage quantity and quality of grasslands (Drescher et al. 2006). It explains foraging behavior of large herbivores as an optimal solution between forage ingestion and digestion, leading to maximum daily intake rates in patches of intermediate herbage mass.

Canopy Structure

The sward canopy is defined as the aboveground parts of a sward and includes consideration of distribution and arrangement of plant parts. In addition to herbage mass, sward canopy attributes that affect animal grazing behavior include herbage bulk density, species, and plant-part proportion, spatial arrangement of species and plant parts within the canopy, and chemical composition of selected parts (Gordon and Lascano 1993; Newman et al. 2003).

Leaf bulk density of C_4 grass canopies is often lower than that of temperate grasses (Sollenberger and Burns 2001). This was suggested as one reason for the lesser performance of livestock grazing C_4 compared with C_3 grass pastures (Stobbs 1973a, 1973b). The relationship between leaf bulk density and animal response can be highly negative or positive, however, depending on the spatial arrangement of leaf relative to stem (Sollenberger and Burns 2001). For example, switchgrass pastures had a lower percentage of leaf (29% vs 37%) and higher percentage of stem (54% vs 47%) than did bermudagrass (Burns et al. 1991; Fisher et al. 1991). However, switchgrass leaves were positioned higher in the canopy than stems, making it possible for cattle to select leaf and produce greater daily gain (0.59 kg) compared with those grazing bermudagrass (0.22 kg). Others have suggested that instantaneous intake rate depends upon the proportion and density of both leaves and stems (Van Langevelde et al. 2008). Presence of reproductive stems can act as a barrier to selection (Benvenutti et al. 2006), and the effort required to gather a fixed quantity of leaves increases with increasing canopy stem density (Drescher et al. 2006). Gonçalves et al. (2009) reported a linear decline in total and leaf bulk density as canopy height increased from 4 to 16 cm (3.40–1.75, and 1.58–0.57 mg cm^{-3}, respectively). At the shorter canopy height, bite mass was limited by bite depth, but in taller canopies, bulk density was the main constraint.

Vertical heterogeneity in the distribution of nutrients within the sward canopy occurs as a consequence of the spatial arrangement of different species and plant parts in grazed swards. Fisher et al. (1991) compared *in vitro* dry matter **digestibility** by 5-cm vertical strata for three C_4 (bermudagrass, flaccidgrass, and switchgrass) and one C_3 grass species (tall fescue). All grasses were continuously stocked and canopies were of comparable height when sampled. They found that from the bottom to the top layer, *in vitro* digestibility increased by 21 g kg^{-1} for tall fescue, 31 g kg^{-1} for bermudagrass, 58 g kg^{-1} for flaccidgrass, and 68 g kg^{-1} for switchgrass. Leaves predominated in the tall fescue canopy, while stem and dead material were more prominent throughout the C_4 grass canopies. Limpograss canopies were 33% leaf in the top half by height compared with 10% leaf in the bottom half (Holderbaum et al. 1992), resulting in herbage **crude protein** in the top half being twice as great as in the bottom half. In general, C_4 grass canopies possess greater vertical heterogeneity in terms of plant-part proportion and nutritive value than C_3 grass canopies.

Animal Grazing Behavior Affects Grassland Response

The behavioral components of grazing describe how various species of herbivores search for, gather, and process plant tissue in a range of spatio-temporal scales (Carvalho 2013). Impacts of grazing behavior on grasslands vary among herbivore species and can be generalized as accruing from **defoliation**, selection, treading, and excretion.

Differences Among Species of Livestock Herbivores

Preferential browsers and preferential grazers vary in modes of foraging and degree of selectivity and, therefore, they may encounter very different food resources (Skarpe and Hester 2008). Grasses and herbaceous legumes consumed in planted pastures by grazers constitute a food resource that is relatively homogeneous with generally consistent bite mass. Food selection in browsers is more complex and less well understood, with widely varying bite mass, nutrient concentration, and associated plant defense mechanisms (Skarpe and Hester 2008). Grazers have longer rumen retention time to maximize cell wall digestion by microbes, while browsers derive more nutrients from cell contents and pass forage more rapidly through the rumen (Duncan and Poppi 2008). Browsing ruminants' inability to subsist on predominantly grass diets is attributed to weaker rumen musculature which cannot cope with the rumen load associated with grass diets (Clauss et al. 2002).

Cattle and sheep are preferential grazers with short lips, broad muzzles, and a cornified tongue for protection during tearing of abrasive plant tissue (Van Soest 1994). Goats (*Capra aegagrus hircus*) are preferential browsers with a narrow but deep mouth opening and mobile lips and tongue allowing selective ingestion of plants and plant parts, including leaves and twigs of woody plants (Van Soest 1994). Sheep have narrower mouths and a highly curved incisor arcade making them better suited anatomically than cattle for diet selection and close grazing (Walker 1994). Sheep generally prefer grazing herbaceous material if quantity is not limiting (Benavides et al. 2009). Horses (*Equus caballus*) have mobile lips and a large mouth, ingesting forage by severing it between their upper and lower incisors. Horses prefer shorter pasture than cattle and are notorious spot grazers.

Bite mass varies with the volume of the canopy the animal can enclose in each bite and with the bulk density of the **grazed horizon** (Illius 1997). Illius and Gordon (1987) predicted that the allometric exponent relating bite mass to the animal's body mass (Wt) changed from 0.72 to 0.36 as canopy height was reduced progressively. Comparing bite mass with the animal's metabolic requirements, which scale with $Wt^{0.75}$, large animals are at a disadvantage compared with smaller ones when grazing short swards because each bite represents a smaller proportion of daily requirements.

Defoliation

Herbivory generally reduces the competitive strength of a plant, although responses vary (Skarpe and Hester 2008). In species-diverse landscapes, tissue removal changes the competitive dynamics among plants (Hobbs 2006). As herbage diversity increases, the animal has more choices and may select a diet of greater nutritive value. However, diversity often occurs in less-dense herbage canopies and, thus, bite mass and daily intake may be limited. Under these circumstances, the trade-off between maximizing selectivity and maximizing bite mass is likely influenced by animal species and physiologic state (growing, mature, lactating, pregnant), as well as the diversity and density of the sward.

Rate of herbage growth following **defoliation** generally fits a sigmoidal curve (Bryan et al. 2000), with an asymptotic maximum during reproductive growth stages (Figure 10.3). Leaf defoliation of immature plants interrupts this growth curve, and at the instant of defoliation, new processes and dynamics take place (Richards 1993). Upon defoliation of photosynthetically active tissue, C gain is reduced and translocation of previously fixed C ceases. Other adaptive processes include compensatory photosynthesis and phenotypic plasticity (this volume; Chapter 44). Compensatory photosynthesis is the ability of mature leaves to rejuvenate photosynthetic capacity to that of younger leaves, or of younger leaves to slow the normal decline that occurs with aging. Phenotypic plasticity is a longer-term response to defoliation stress and refers to changes in size, structure, and spatial positioning of plant organs that result in increased grazing tolerance or avoidance (Gastal and Lemaire 2015).

Defoliation directly reduces whole-plant photosynthesis or daily C gain, but not necessarily in direct proportion to leaf-area loss, because canopy microclimate changes after each defoliation bite. When a portion of the canopy is removed, light penetration and interception by the new canopy affects the photosynthetic contribution of different ages and classes of leaves, some of which may be more efficient than those in the old canopy. If mature, previously shaded, and less photosynthetically efficient tissue predominates on the defoliated plant, then subsequent canopy photosynthesis is likely to be reduced greatly. However, if young tissue remains, then the decline in photosynthesis is more proportional to leaf area removed.

Selection

Species of herbivore and the seasonal pattern of grazing are important factors influencing the differential utilization of plant species and hence the floristic composition of mixed swards (Grant et al. 1996). Patches grazed by cattle were taller than those grazed by sheep, as sheep were able to maintain their live weight at a lower canopy height (Benavides et al. 2009). Sheep prefer broadleaf plants; both legumes and forbs, and patches grazed by sheep had five to seven percentage units less white clover and three to six units less forbs than those grazed by cattle (Abaye et al. 1997). White-clover contribution increased in perennial ryegrass–white-clover swards grazed by goats but not by sheep (del Pozo et al. 1997). Under similar sward conditions, cattle and goats utilized more of a tussock-forming grass than did sheep (Grant et al. 1996). Differences between sheep and cattle diets were explained by (i) the height at which the animals consumed herbage in relation to the distribution of plant species within the canopy, (ii) the greater ability of sheep to select from fine-scale mixtures; and (iii) the greater extent to which cattle graze tall, more fibrous plant components (Grant et al. 1985). For each 1% increase in tussock cover, heifers reduced grazing time in the inter-tussock stratum by 0.6%, while grazing time of ewes was reduced only 0.35% (Bremm et al. 2012). Ewes were able to adjust foraging strategy to sustain high-intake rate. Goats often exhibit a preference for woody forbs or browse over other types of forage, including plants with thorns (Gordon 2003). In a grass-legume mixture where sand blackberry was present, the proportion of total blackberry biomass increased 10 percentage units in pastures grazed by cattle only, but blackberry proportion decreased 11 percentage units where goats grazed alone and 13 units where both species grazed concurrently at a high stocking rate (Krueger et al. 2014). Proportion of total bites that were blackberry ranged from nearly 0% for cattle to 62% for goats.

Senft et al. (1987) suggested that selection by large herbivores is based on solving two opposing problems: obtaining maximal quality and adequate quantity. The strategy chosen is influenced by herbivore characteristics that affect energy needs and by plant diversity within the landscape. Upon initiation of a grazing period, the first decision is where to graze. In sown pastures, the choice may be limited because of uniformity in the plant community. However, patch grazing, dung fouling, location of water and feeding stations, and shade may affect the patch choice. In extensive rangelands and savannas, spatial selection of plant communities and patches by the animal is influenced by the features of the landscape (Senft et al. 1987). These include landscape boundaries, distribution of plant communities, and accessibility and distribution of water, shade, and bedding sites. In familiar environments, preferences for feed resources result from food imprinting, social learning, and individual learning (Bailey and Provenza 2008).

Treading

Hoof action may affect pasture plants directly by damaging, severing, or partially burying plant tissue. Indirect effects of treading are mediated through changes in soil characteristics that influence plant growth and persistence (Pott et al. 1983). Plants' ability to tolerate the direct effects of treading are related to growth habit and morphology. However, grazing management (e.g. rotational stocking) may help overcome the vulnerabilities. Vine-forming species are very susceptible to treading damage. Plants with protected bud sites and greater tensile strength are often more tolerant. For example, digitgrass, a stoloniferous perennial, was more tolerant of treading by sheep across a range of stocking rates than the legume lotononis (Pott et al. 1983). The annual legume aeschynomene showed fewer adverse effects of treading if grazed when it was 20–40 cm tall than when stems were taller and more susceptible to breakage when stepped on (Sollenberger et al. 1987).

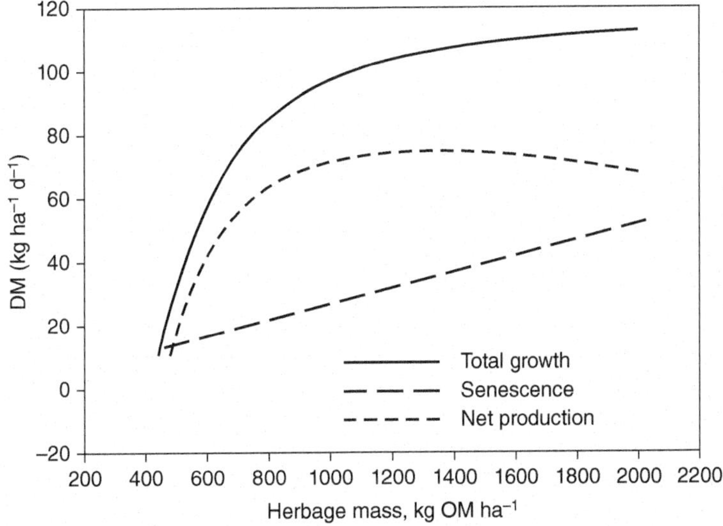

Fig. 10.3. Components of a ryegrass sward. The relationship between rate of total growth, rate of senescence, and net production rate of green herbage to herbage mass. *Source:* After Bircham and Hodgson (1983); used by permission from the British Grassland Society.

There are mediated, or indirect effects of treading. Soil bulk density is increased by animal traffic, but soil texture determines the degree to which compaction occurs (Krenzer et al. 1989). For example, bulk density increased for fine-textured soils as **grazing intensity** increased, but there was no effect of grazing intensity on bulk density of coarse-textured soils. In a white clover–perennial ryegrass pasture growing on a silt-loam soil, surface-soil bulk density and penetrometer resistance were increased by the second year of grazing because of treading (Kelly 1985). By the third and fourth years, these soil changes were associated with 1.5 and 2.3 $Mg\,ha^{-1}$ reduction in herbage accumulation. Similarly, treading compacted the soil under a white clover-ryegrass mixture, especially during periods of high-water table in spring (Phelan et al. 2012).

Grazed winter wheat pastures in Oklahoma had greater soil bulk density and strength than did ungrazed swards (Krenzer et al. 1989). Soil water concentration was less in grazed areas due to a reduction in the number of larger soil pores and total pore space. Reduced water infiltration can also occur with greater grazing pressure because of a reduction in vegetation and litter, both of which decrease the impact of raindrops and increase infiltration (Naeth et al. 1990).

Excretion

Herbivory interferes with nutrient cycling via changes in above- and below-ground litter quality and quantity and by deposition of dung and urine (Skarpe and Hester 2008). Excretion affects nutrient cycling, pasture growth, and animal grazing patterns. In meat- and fiber-producing animals, the percentage of ingested nutrients retained and exported in body tissue is quite low, and most mineral nutrients consumed are excreted in feces and urine. A single urination from mature cattle may provide the equivalent liquid of 5 mm of rain and 400–500 $kg\,N\,ha^{-1}$ on the 0.4 m^2 of ground that it covers, while dung usually covers about 0.1 m^2 and supplies the equivalent of 110 kg P and 220 $kg\,K\,ha^{-1}$ (Haynes and Williams 1993). Nutrients excreted in urine are either volatilized (NH_3), plant available, or mineralized in a few days, while the nutrients in dung generally become plant available more slowly (Mathews et al. 1996). The pattern of dung and urine distribution to the pasture is non-uniform, and the nutrients contained are subject to loss from the system in a variety of ways (Dubeux et al. 2007).

Those nutrients retained in the system stimulate plant growth. Urine patches in a little bluestem–kentucky bluegrass mixture contained 112 $g\,m^{-2}$ more aboveground biomass and 2.5 $g\,m^{-2}$ more plant N than unaffected areas (Day and Detling 1990). Where urine patches covered only 2% of the surface area, they contributed 7–14% of herbage mass (Day and Detling 1990). A urine deposit increased herbage accumulation for ≥84 days and increased crude protein for ≥28 days (White-Leech et al. 2013).

Modeling Grassland-Herbivore Interactions

Mechanistic models are valuable tools for exploring grassland-herbivore relationships that are otherwise impractical or impossible to address with field research. They can aid in interpreting field data and developing new hypotheses for testing. Modeling the plant–animal interface, however, is very challenging because of the number and diversity of interacting components.

Models have been developed to study many aspects of the grassland-herbivore interaction, from diet selection (Belovsky 1984; Parsons et al. 1994) and grassland stability (Noy-Meir 1975; Johnson and Parsons 1985), to impacts of the environment (Coughenour 1993; Hahn et al. 2005; Richardson et al. 2010). Most available models were developed to assess effects of herbivores on vegetation (plant-focused) or vegetation on animal behavior and population dynamics (animal-focused; Weisberg et al. 2006). Traditional ecologic models for range and pasture management represent the plant-focused approach, where herbivory is simply vegetation removal (Schwinning and Parsons 1999; and Van Langevelde et al. 2008). In animal-focused models, vegetation is portrayed as a single input variable which will drive animal behavior, performance, and population dynamics. Examples include the optimal foraging, linear programing, and functional response models (Westoby 1974; Owen-Smith and Novellie 1982; Stephens and Krebs 1986; Spalinger and Hobbs 1992; and Illius and O'Connor 2000).

Approaches that integrate both plant and animal are more complex and are highly demanding of input data and processing capacity. These models struggle to transfer data across scales without losing information or magnifying errors. Large herbivores influence vegetation over very fine spatio-temporal scales, but the effects on vegetation dynamics are amplified over large areas and long periods (Weisberg et al. 2006). Coping with such complexity is a significant challenge for models of grassland-herbivore systems. In the sections that follow, vegetation growth, functional response, animal intake, **diet selection**, and vegetation pattern models will be described.

Vegetation Growth

Following the elucidation of ecophysiologic processes of photosynthesis, resource allocation, and herbage production (Parsons et al. 1983), vegetation growth was simulated using simple logistic curves based on maximum relative growth rate and biomass (Noy-Meir 1975; Woodward et al. 1993) or **leaf area index** (Johnson and Thornley 1983). The former was limited by representation of the vegetation as a single component, but laid the groundwork for models that provided more detailed simulations. These were based on factors such as leaf area expansion, leaf-age structure, and senescence processes (Johnson and Thornley 1983). Leaf-area index was considered an independent variable because different canopy structures may have the same leaf-area index. Later, multiple-species models (Richardson et al. 2010) and models expressing growth as a function of environmental and/or management variables (Herrero et al. 1998) allowed more in-depth exploration of plant–animal interactions and responses to levels of disturbance.

Functional Response

The functional response is "the cornerstone principle of all foraging models" (Fryxell 2008). It refers to the link between consumer and resource, and how the abundance of resources will influence the rate of intake by the consumer (Holling 1959, 1965; Figure 10.4). An asymptotic increase in intake rate as bite mass or vegetation condition increases, is generally the response most closely associated with short-term intake rate in herbivore studies (Mezzalira et al. 2017). Short-term intake rate may be limited by density of potential bites or bite mass and competition between severing and processing bites (Fryxell 2008). In the latter, intake rate is regulated by handling rate, i.e. the capacity of the animal to harvest food, which is determined by bite rate and bite mass.

A Type IV function response (Wilmshurst and Fryxell 1995) represents a decline in intake rate as herbage mass increases, due to declining forage nutritive value (Van Langevelde et al. 2008) or bulk density (Mezzalira et al. 2017). This functional response can promote competitive coexistence and facilitation between herbivore species of different body size (Van Langevelde et al. 2008; Mezzalira et al. 2017). Smaller herbivores are more effective in selecting food of higher nutritive value, thus, intake is less when herbage allowance is high and nutritive value is low. In contrast, larger herbivores benefit from patches with greater herbage mass, simultaneously creating grazing lawns for the smaller herbivores (Wilmshurst et al. 2000). Models based on short-term intake rate in homogenous vegetation (Spalinger and Hobbs 1992) are more difficult to scale up to total intake from a landscape perspective (Weisberg et al. 2006), especially because they do not account for forage nutritive value and species composition as parameters affecting functional response (Van Langevelde et al. 2008) (Figure 10.4).

Animal Intake

The way intake is simulated has a major impact on the outputs of the model (Herrero et al. 1998). It can be represented from systems of energy requirements, by establishing relationships between herbage mass and intake, or using grazing behavior measurements. Systems of energy requirements use animal physiologic characteristics to determine amount of forage consumed and are generally applied to animal-oriented models (Conrad et al. 1964; Newman et al. 1995; Parsons et al. 1994; Gregorini et al. 2015). Early grassland models (Johnson and Parsons 1985; Parsons et al. 1988; Blackburn and Kothmann 1989) simulated intake using simple empirical relationships between herbage mass and potential intake based on animal body weight or energy requirements.

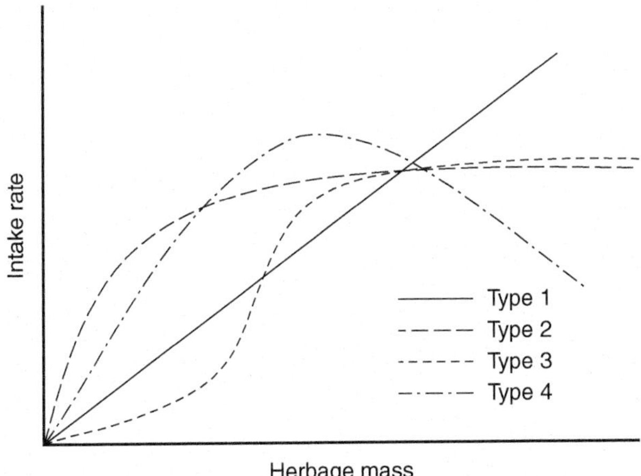

Fig. 10.4. Functional response of intake to herbage mass abundance. Type 1 represents a linear increment of intake as herbage mass increases, while Types 2 and 3 have a saturation phase above certain levels of herbage mass. Type 4 represents a decline in intake when herbage mass is overly-abundant, likely because of a decrease in nutritive value (and consequently an increase in selectivity by the herbivore) or in bulk density. *Source:* Adapted from Spalinger and Hobbs (1992), Wilmshurst and Fryxell (1995), Mezzalira et al. (2017).

The third approach uses grazing behavior measurements (Allden and Whittaker 1970; Stobbs 1973a, 1974; Arnold 1987) to simulate components of intake (i.e. bite mass, bite rate, grazing time). Intake is still a function of herbage mass, but now the different canopy characteristics (e.g. bulk density, leaf: stem ratio, proportion of dead material, digestibility, botanical composition) affect the components of intake separately. This approach gives a more detailed representation of both the intake process and the effects on vegetation, but it is generally species- or grassland type-specific and requires significant effort and care when scaling up to daily intake rate.

Diet Selection

One of the most-used approaches that incorporates diet selection is based on optimal foraging theory (Charnov 1976; Pyke 1984; Stephens and Krebs 1986). It offers a relatively simple, consistent mathematical approach with a solid theoretic base, clear assumptions, and well-defined objectives for simulating foraging strategy (Laca and Demment 1996). The two main questions considered by the foraging models are which items will the animal consume and when will it exit a patch. The goal is to use linear programming to maximize an objective function, generally, the intake rate of a resource, or minimize the time needed to acquire the resource (Owen-Smith and Novellie 1982; Stephens and Krebs 1986).

The decision of which food items should be consumed is based on the marginal value theorem (Charnov 1976). Food items are ranked by decreasing order based on a criterion such as amount of digestible energy per unit of time processing or digesting. The herbivore will add to the diet only items that result in a marginal increase in net intake rate of the selected criterion. The patch model determines how long the herbivore should remain in a patch given diminishing returns as the patch is exploited in relation to the whole-food environment available.

Still, the optimal foraging approach focuses on one function, while foragers respond based on more than one (Prins and van Langevelde 2008; Hengeveld et al. 2009). Linear programing uses two currency constraints (e.g. minimum intake of protein and energy, and maximum intake of toxins), and the optimal solution is at the intersection of the constraint equations (Belovsky 1984). Within this context, analytic approaches based on post-ingestive feedback, dietary experiences, and sensory stimuli are viable based on accepted grazing theories (Prache et al. 1998; Provenza and Cincotta 1993), but the large numbers of variables and empiric assumptions make modeling more challenging (Gregorini et al. 2015).

Vegetation Patterns

Creation and maintenance of vegetation patterns depends mainly on intensity, frequency, and distribution of defoliation. Both intensity and frequency are closely related to animal density and selectivity, while the distribution of defoliation (in spatially-explicit models) is also determined by the initial state of heterogeneity. If stocking rate is at or below grassland carrying capacity, an initially heterogeneous sward will generally maintain or increase heterogeneity (Mouissie et al. 2008), but this depends upon diet selection rules in the model. For example, if rules include avoidance of tall patches or preference for high-nutritive value patches, a bimodal canopy structure will be created with avoided tall and lower quality patches and more preferred grazing lawns (Parsons and Dumont 2003).

In spatially-explicit models, movement is generally represented using a determined or random path (Laca and Demment 1990; Schwinning and Parsons 1999), a decision rule (Oom et al. 2004; Mouissie et al. 2008), or using correlated random walk models (Vincenot et al. 2015). Using random selection, each cell has the same probability of being selected, so representation of grazing behavior is limited. When adding decision rules, at each time step, the grazer moves to the next cell that will result in the best compromise between the chosen constraints (generally digestive or ingestion) and traveling energy costs (Wilmshurst et al. 2000). Those models assume that the grazer has complete knowledge of the entire pasture (Mouissie et al. 2008) or neighborhood (Oom et al. 2004). Correlated random walks use measured parameters for angle and length of the next stride, to determine the direction and distance to the next cell (Vincenot et al. 2015).

Optimizing Grassland-Herbivore Interactions

Grassland-herbivore interactions are remarkably complex, and there are limitations to the ability of field experiments to describe these interactions effectively. Models have significantly enhanced our understanding of ecosystem dynamics, processes, and functioning, but their value for use in grassland management is limited due to complexity and computation requirements, and many times because of the communication gaps between researchers and land managers (Derner et al. 2012). Additionally, the modeler's choices and goals may lead to different results. This dictates that care be taken when extrapolating fine-scale processes to the landscape level (Weisberg et al. 2006) and that some degree of skepticism be a part of interpreting the results.

References

Abaye, A.O., Allen, V.G., and Fontenot, J.P. (1997). Grazing sheep and cattle together or separately: effect on soils and plants. *Agron. J.* 89: 380–386.

Allden, W.G. and Whittaker, I.A.M. (1970). The determinants of herbage intake by grazing sheep: the interrelationship of factors influencing herbage intake and availability. *Aust. J. Agric. Res.* 21: 755–766.

Amaral, M.F., Mezzalira, J.C., Bremm, C. et al. (2013). Sward structure management for a maximum short-term intake rate in annual ryegrass. *Grass Forage Sci.* 68: 271–277.

Arnold, G.W. (1987). Influence of the biomass, botanical composition and sward height of annual pastures on foraging behaviour by sheep. *J. Appl. Ecol.* 24: 759–772.

Bailey, D.W. and Provenza, F.D. (2008). Mechanisms determining large-herbivore distribution. In: *Resource Ecology: Spatial and Temporal Dynamics of Foraging* (eds. H.H.T. Prins and F. van Langevelde), 7–28. New York: Springer.

Bailey, D.W., Gross, J.E., Laca, E.A. et al. (1996). Mechanisms that result in large herbivore grazing distribution patterns. *J. Range Manage.* 49: 386–400.

Belovsky, G.E. (1984). Herbivore optimal foraging: a comparative test of three models. *Am. Nat.* 124: 97–115.

Benavides, R., Celaya, R., Ferreira, L.M.M. et al. (2009). Grazing behavior of domestic ruminants according to flock type and subsequent vegetation changes on partially improved heathlands. *Span. J. Agric. Res.* 7: 417–430.

Benvenutti, M.A., Gordon, I.J., and Poppi, D.P. (2006). The effect of the density and physical properties of grass stems on the foraging behaviour and instantaneous intake rate by cattle grazing an artificial reproductive tropical sward. *Grass Forage Sci.* 61: 272–281.

Bircham, J.S. and Hodgson, J. (1983). The influence of sward condition and rates of herbage growth and senescence in mixed swards under continuous stocking management. *Grass Forage Sci.* 38: 323–331.

Black, J.L. and Kenney, P.A. (1984). Factors affecting diet selection by sheep. II. Height and density of pasture. *Aust. J. Agric. Res.* 35: 551–563.

Blackburn, H.D. and Kothmann, M.M. (1989). A forage dynamics model for use in range or pasture environments. *Grass Forage Sci.* 44: 283–294.

Bonnet, O.J.F., Meuret, M., Tischler, M.R. et al. (2015). Continuous bite monitoring: a method to assess the foraging dynamics of herbivores in natural grazing conditions. *Anim. Prod. Sci.* 55: 339–349.

Bremm, C., Laca, E.A., Fonseca, L. et al. (2012). Foraging behaviour of beef heifers and ewes in natural grasslands with distinct proportions of tussocks. *Appl. Anim. Behav. Sci.* 141: 108–116.

Bryan, W.B., Prigge, E.C., Lasat, M. et al. (2000). Productivity of Kentucky bluegrass pasture grazed at three heights and two intensities. *Agron. J.* 92: 30–35.

Burns, J.C. and Sollenberger, L.E. (2002). Grazing behavior of ruminants and daily performance from warm-season grasses. *Crop Sci.* 42: 873–881.

Burns, J.C., Pond, K.R., and Fisher, D.S. (1991). Effects of grass species on grazing steers. II. Dry matter intake and digesta kinetics. *J. Anim. Sci.* 69: 1199–1204.

Carvalho, P.C.F. (2013). Can grazing behavior support innovations in grassland management. *Trop. Grassl. Forrajes Tropicales* 1: 137–155.

Carvalho, P.C.F., Gonda, H.L., and Wade, M.H. (2008). Características estruturais do pasto e o consumo de forragem: o quê pastar, quanto pastar e como se mover para encontrar o pasto. In: *O manejo estratégico da pastagem. UFV* (eds. O. Pereira, J.A. Obeid, D.M. da Fonseca, et al.), 101–130. Brazil: Viçosa.

Carvalho, P.C.F., da Trindade, J.K., Bremm, C. et al. (2013). Comportamento ingestivo de animais em pastejo. In: *Forragicultura: Ciência, Tecnologia e Gestão dos Recursos Forrageiros. UNESP* (ed. R.A. Reis), 525–545. Brazil: Jaboticabal.

Charnov, E.L. (1976). Optimal foraging, the marginal value theorem. *Theor. Popul. Biol.* 9: 129–136.

Clauss, M., Lechner-Doll, M., and Streich, W.J. (2002). Faecal particle size distribution in captive wild ruminants: an approach to the browser/grazer dichotomy from the other end. *Oecologia* 131: 343–349.

Coleman, S.W. and Sollenberger, L.E. (2007). Plant-herbivore interactions. In: *Forages – the Science of Grassland Agriculture* 6e (eds. R.F Barnes, C.J. Nelson, K.J. Moore and M. Collins), 123–136. Ames, IA: Blackwell Publishing.

Conrad, H.R., Pratt, A.D., and Hibbs, J.W. (1964). Regulation of feed intake in dairy cows. I. Change in importance of physical and physiological factors with increasing digestibility. *J. Dairy Sci.* 47: 54–62.

Coughenour, M.B. (1993). Savanna - Landscape and Regional Ecosystem Model. http://www.nrel.colostate.edu/assets/nrel_files/labs/coughenour-lab/pubs/Manual_1993/modeldescription_plusfigs.pdf (accessed 9 October 2019).

Day, T.A. and Detling, J.K. (1990). Grassland patch dynamics and herbivore grazing preference following urine deposition. *Ecology* 71: 180–188.

Derner, J.D., Augustine, D.J., Ii, J.C.A., and Ahuja, L.R. (2012). Opportunities for increasing utility of models for rangeland management. *Rangeland Ecol. Manage.* 65: 623–631.

Drescher, M., Heitkonig, I.M.A., van den Brink, P.J., and Prins, H.H.T. (2006). Effects of sward structure on herbivore foraging behavior in a South African savanna: an investigation of the forage maturation hypothesis. *Aust. Ecol.* 31: 76–87.

Dubeux, J.C.B. Jr., Sollenberger, L.E., Mathews, B.W. et al. (2007). Nutrient cycling in warm-climate grasslands. *Crop Sci.* 47: 915–928.

Duncan, A.J. and Poppi, D.P. (2008). Nutritional ecology of grazing and browsing ruminants. In: *The Ecology of Browsing and Grazing* (eds. I.J. Gordon and H.H.T. Prins), 89–116. New York: Spring Publishing Co.

Fisher, D.S., Burns, J.C., Pond, K.R. et al. (1991). Effects of grass species on grazing steers: 1. Diet composition and ingestive mastication. *J. Anim. Sci.* 69: 1188–1198.

Forbes, T.D.A. (1988). Researching the plant–animal interface: the investigation of ingestive behavior in grazing animals. *J. Anim. Sci.* 66: 2369–2379.

Forbes, T.D.A. and Coleman, S.W. (1993). Forage intake and ingestive behavior of cattle grazing old world bluestems. *Agron. J.* 85: 808–816.

Frank, D.A. (2006). Large herbivores in heterogeneous grassland ecosystems. In: *Large Herbivore Ecology, Ecosystem Dynamics and Conservation* (eds. K. Danell et al.), 326–343. Cambridge, UK: Cambridge University Press.

Fryxell, J.M. (2008). Predictive modelling of patch use by terrestrial herbivores. In: *Resource Ecology: Spatial and Temporal Dynamics of Foraging* (eds. H.H.T. Prins and F. van Langevelde), 105–124. Springer.

Fuhlendorf, S.D., Fynn, R.W.S., McGranahan, D.A., and Twidwell, D. (2017). Heterogeneity as the basis for rangeland management. In: *Rangeland Systems: Processes, Management and Challenges* (ed. D.D. Briske), 169–196. New York: Springer.

Gastal, F. and Lemaire, G. (2015). Defoliation, shoot plasticity, sward structure and herbage utilization in pasture: review of underlying ecophysiological processes. *Agriculture* 5: 1146–1171.

Gonçalves, E.N., César, P., Carvalho, D.F. et al. (2009). Relações planta-animal em ambiente pastoril heterogêneo: processo de ingestão de forragem. *Rev. Bras. Zootec.* 38: 1655–1662.

Gordon, I.J. (2003). Browsing and grazing ruminants: are they different beasts? *For. Ecol. Manage.* 181: 13–21.

Gordon, I.J. and Benvenutti, M. (2006). Food in 3D: how ruminant livestock interact with sown sward architecture at the bite scale. In: *Feeding in Domestic Vertebrates: From Structures to Behaviour* (ed. V. Bells), 263–277. Wallingford, UK: CABI Publishing.

Gordon, I.J. and Lascano, C. (1993). Foraging strategies of ruminant livestock on intensively managed grasslands: potentials and constraints. In: *Proceedings of 17th International Grassland Congress, Palmerston North, New Zealand and Rockhampton, Australia. 8–21 Feb. 1993* (ed. M.J. Baker), 681–689. Palmerston North, NZ: Keeling and Mundy, Ltd.

Grant, S.A., Suckling, D.E., Smith, H.K. et al. (1985). Comparative studies of diet selection by sheep and cattle: the hill grasslands. *J. Ecol.* 73: 987–1004.

Grant, S.A., Torvell, L., Sim, E.M. et al. (1996). Controlled grazing studies on Nardus grassland: effects of between-tussock sward height and species of grazer on Nardus utilization and floristic composition in two fields in Scotland. *J. Appl. Ecol.* 33: 1053–1064.

Gregorini, P., Tamminga, S., and Gunter, S.A. (2006). Behavior and daily grazing patterns of cattle. *Prof. Anim. Sci.* 22: 201–209.

Gregorini, P., Villalba, J.J., Provenza, F.D. et al. (2015). Modelling preference and diet selection patterns by grazing ruminants: a development in a mechanistic model of a grazing dairy cow, MINDY. *Anim. Prod. Sci.* 55: 360–375.

Gross, J.E., Shipley, L.A., Hobbs, N.T. et al. (1993). Functional response of herbivores in food-concentrated patches: tests of a mechanistic model. *Ecology* 74: 778–791.

Hahn, B.D., Richardson, F.D., Hoffman, M.T. et al. (2005). A simulation model of long-term climate, livestock and vegetation interactions on communal rangelands in the semi-arid Succulent Karoo, Namaqualand, South Africa. *Ecol. Modell.* 183: 211–230.

Haynes, R.J. and Williams, P.H. (1993). Nutrient cycling and soil fertility in the grazed pasture ecosystem. *Adv. Agron.* 49: 119–199.

Hengeveld, G.M., van Langevelde, F., a Groen, T., and de Knegt, H.J. (2009). Optimal foraging for multiple resources in several food species. *Am. Nat.* 174: 102–110.

Hernández Garay, A., Sollenberger, L.E., McDonald, D.C. et al. (2004). Nitrogen fertilization and stocking rate affect stargrass pasture and cattle performance. *Crop Sci.* 44: 1348–1354.

Herrero, M., Dent, J., and Fawcett, R.H. (1998). The plant–animal interface in models of grazing systems. In: *Agricultural Systems Modelling and Simulation* (eds. B. Currie and R. Peart), 495–542. New York, NY: Marcel Dekker.

Hobbs, N.T. (2006). Large herbivores as source of disturbance in ecosystems. In: *Large Herbivore Ecology, Ecosystem Dynamics and Conservation* (eds. K. Danell, R. Bergström, P. Duncan and J. Pastor), 261–288. Cambridge, UK: Cambridge University Press.

Hodgson, J. (1982a). Influence of sward characteristics on diet selection and herbage intake by the grazing animal. In: *Nutritional Limits to Animal Production from Pastures* (ed. J.B. Hacker), 153–166. Farnham Royal, UK: Commonwealth Agricultural Bureaux.

Hodgson, J. (1982b). Ingestive behavior. In: *Herbage Intake Handbook* (ed. J.D. Leaver), 113–138. Hurley, Berkshire, UK: The British Grassland Society.

Holderbaum, J.F., Sollenberger, L.E., Quesenberry, K.H. et al. (1992). Canopy structure and nutritive value of rotationally-grazed limpograss pastures during mid-summer to early autumn. *Agron. J.* 84: 11–16.

Holling, C.S. (1959). The components of predation as revealed by a study of small mammal predation of the European pine sawfly. *Can. Entomol.* 91: 293–320.

Holling, C.S. (1965). The functional response of predators to prey density and its role in mimicry and population regulation. *Mem. Entomol. Soc. Can.* 97: 5–60.

Illius, A.W. (1997). Advances and retreats in specifying the constraints on intake in grazing ruminants. In: *Proceedings of 18th International Grassland Congress, Winnipeg, Man., Saskatoon, Sask., Canada, 8–19 June 1997* (eds. J. Buchanan-Smith, L.D. Bailey and P. McCaughey), 39–44. Calgary, AB, Canada: Association Management Centre.

Illius, A.W. and Gordon, I.J. (1987). The allometry of food intake in grazing ruminants. *J. Anim. Ecol.* 56: 989–999.

Illius, A.W. and O'Connor, T.G. (2000). Resource heterogeneity and ungulate population dynamics. *Oikos* 89: 283–294.

Johnson, I.R. and Parsons, A.J. (1985). Use of a model to analyse the effects of continuous grazing managements on seasonal patterns of grass production. *Grass Forage Sci.* 40: 449–458.

Johnson, I.R. and Thornley, J.H. (1983). Vegetative crop growth model incorporating leaf area expansion and senescence, and applied to grass. *Plant Cell Environ.* 6(9): 721–729.

Kelly, K.B. (1985). Effects of soil modification and treading on pasture growth and physical properties of an irrigated red-brown earth. *Aust. J. Agric. Res.* 36: 799–807.

Krenzer, E.G. Jr., Chee, C.F., and Stone, J.F. (1989). Effects of animal traffic on soil compaction in wheat pastures. *J. Prod. Agric.* 2: 246–249.

Krueger, N.C., Sollenberger, L.E., Blount, A.R. et al. (2014). Mixed grazing by cattle and goats for blackberry control in rhizoma peanut-grass pastures. *Crop Sci.* 54: 2864–2871.

Laca, E.A. (2008). Foraging in heterogeneous environment: intake and diet choice. In: *Resource Ecology: Spatial and Temporal Dynamics of Foraging* (eds. H. Prins and F. van Langevelde), 81–100. New York: Springer.

Laca, E.A. and Demment, M.W. (1990). *Modelling Intake of a Grazing Ruminant in a Heterogeneous Environment*, 57–76. Yokohama, Japan: Intecol, V International Congress of Ecology.

Laca, E.A. and Demment, M.W. (1996). Foraging strategies of grazing animals. In: *The Ecology and Management of Grazing Systems* (eds. J. Hodgson and A.W. Illius), 137–158. Wallingford, UK: CABI Publishing.

Laca, E.A., Ungar, E.D., Seligman, N., and Demment, M.W. (1992). Effects of sward height and bulk density on bite dimensions of cattle grazing homogeneous swards. *Grass Forage Sci.* 47: 91–102.

Laca, E.A., Distel, R.A., Griggs, T.C., and Demment, M.W. (1994). Effects of canopy structure on patch depression by grazers. *Ecology* 75: 706–716.

Macoon, B., Sollenberger, L.E., Moore, J.E. et al. (2003). Comparison of three techniques for estimating forage intake of lactating dairy cows on pasture. *J. Anim. Sci.* 81: 2357–2366.

Mathews, B.W., Sollenberger, L.E., and Tritschler, J.P. II (1996). Grazing systems and spatial distribution of nutrients in pastures: soil considerations. In: *Nutrient Cycling in Forage Systems* (eds. R.E. Joost and C.A. Roberts), 213–229. Manhattan, Kansas: Potash and Phosphate Institute/The Foundation for Agronomic Research.

Mezzalira, J.C., Bonnet, O.J.F., Carvalho, P.C.d.F. et al. (2017). Mechanisms and implications of a type IV functional response for short-term intake rate of dry matter in large mammalian herbivores. *J. Anim. Ecol.* 86: 1159–1168.

Mouissie, A.M., Apol, M.E.F., Heil, G.W., and van Diggelen, R. (2008). Creation and preservation of vegetation patterns by grazing. *Ecol. Modell.* 218: 60–72.

Naeth, M.A., Rothwell, R.L., Chanasyk, D.S., and Bailey, A.W. (1990). Grazing impacts on infiltration in mixed prairie and fescue grassland ecosystems of Alberta. *Can. J. Soil Sci.* 70: 593–605.

Newman, J.A., Parsons, A.J., Thornley, J.H.M. et al. (1995). Optimal diet selection by a generalist grazing herbivore. *Funct. Ecol.* 9: 255–268.

Newman, Y.C., Sollenberger, L.E., Kunkle, W.E., and Chambliss, C.G. (2002). Canopy height and nitrogen supplementation effects on performance of heifers grazing limpograss. *Agron. J.* 94: 1375–1380.

Newman, Y.C., Sollenberger, L.E., and Chambliss, C.G. (2003). Canopy characteristics of continuously stocked limpograss swards grazed to different heights. *Agron. J.* 95: 1246–1252.

Noy-Meir, I. (1975). Stability of grazing systems : an application of predator-prey graphs. *J. Ecol.* 63: 459–481.

Oom, S.P., Beecham, J.A., Legg, C.J., and Hester, A.J. (2004). Foraging in a complex environment: from foraging strategies to emergent spatial properties. *Ecol. Complexity* 1: 299–327.

Owen-Smith, N. and Novellie, P. (1982). What should a clever ungulate eat. *Am. Nat.* 119: 151–178.

Parsons, A.J. and Dumont, B. (2003). Spatial heterogeneity and grazing processes. *Anim. Res.* 52: 161–179.

Parsons, A.J., Laefe, E.L., Collett, B. et al. (1983). The physiology of grass production under grazing. II. Photosynthesis, crop growth and animal intake of continuously-grazed swards. *J. Appl. Ecol.* 20: 127–139.

Parsons, A.J., Johnson, I.R., and Harvey, A. (1988). Use of a model to optimize the interaction between frequency and severity of intermittent defoliation and to provide a fundamental comparison of the continuous and intermittent defoliation of grass. *Grass Forage Sci.* 43: 49–59.

Parsons, A.J., Thornley, J.H.M., Newman, J., and Penning, P.D. (1994). A mechanistic model of some physical determinants of intake rate and diet selection in a 2-species temperate grassland sward. *Funct. Ecol.* 8: 187–204.

Penning, P.D. (1985). Some effects of sward conditions on grazing behaviour and intake in sheep. In: *Grazing Research at Northern Latitudes* (ed. O. Gudmundsson), 219–226. New York: Plenum Press.

Phelan, P., Keogh, B., Casey, I.A. et al. (2012). The effects of treading by dairy cows on soil properties and herbage production for three white clover-based grazing systems on a clay loam soil. *Grass Forage Sci.* 68: 548–563.

Pott, A., Humphreys, L.R., and Hales, J.W. (1983). Persistence and growth of *Lotononis bainesii-Digitaria decumbens* pastures. *J. Agric. Sci.* 101: 9–15.

del Pozo, M., Wright, I.A., and Whyte, T.K. (1997). Diet selection by sheep and goats and sward composition changes in a ryegrass/white clover sward previously grazed by cattle, sheep or goats. *Grass Forage Sci.* 52: 278–290.

Prache, S., Gordon, J., Rook, A.J., and Theix, C.D.C. (1998). Foraging behaviour and diet selection in domestic herbivores. *Ann. Zootech.* 47: 335–345.

Prins, H.H.T. and van Langevelde, F. (2008). Assembling a diet from different places. In: *Resource Ecology: Spatial and Temporal Dynamics of Foraging* (eds. H.H.T. Prins and F. van Langevelde), 129–156. New York: Springer.

Provenza, F. and Cincotta, R.P. (1993). Foraging as a self-organized learning process: accepting adaptability at the expense of predictability. In: *Diet Selection: An Interdisciplinary Approach to Foraging Behavior* (ed. R.N. Hughes), 78–101. Hoboken, NJ: Blackwell.

Pyke, G. (1984). Optimal foraging theory. *Annu. Rev. Ecol. Syst.* 15: 532–575.

Richards, J.H. (1993). Physiology of plants recovering from defoliation. In: *Proceedings of 17th International Grassland Congress, Palmerston North, New Zealand and Rockhampton, Australia. 8–21 Feb. 1993* (ed. M.J. Baker), 85–94. Palmerston North, NZ: Keeling and Mundy, Ltd.

Richardson, F.D., Hoffman, M.T., and Gillson, L. (2010). Modelling the complex dynamics of vegetation, livestock and rainfall in a semiarid rangeland in South Africa. *Afr. J. Range Forage Sci.* 27: 125–142.

Schwinning, S. and Parsons, A.J. (1999). The stability of grazing systems revisited: spatial models and the role of heterogeneity. *Funct. Ecol.* 13: 737–747.

Senft, R.L., Coughenour, M.B., Bailey, D.W. et al. (1987). Large herbivore forage and ecological hierarchies. *Bioscience* 37: 789–799.

Shipley, L.A., Gross, J.E., Spalinger, D.E. et al. (1994). The scaling of intake rate in mammalian herbivores. *Am. Nat.* 143: 1055–1082.

Skarpe, C. and Hester, A. (2008). Plant traits, browsing and grazing herbivores, and vegetation dynamics. In: *The Ecology of Browsing and Grazing* (eds. I.J. Gordon and H.H.T. Prins), 217–261. New York: Springer Publishing Co.

Sollenberger, L.E. and Burns, J.C. (2001). Canopy characteristics, ingestive behaviour, and herbage intake in cultivated tropical grasslands. In: *Proceedings of 19th International Grassland Congress, São Pedro, Brazil, 10–21 Feb. 2001* (eds. J.A. Gomide, W.R.S. Mattos and S.C. DaSilva), 321–327. Piracicaba, Brazil: Brazilian Society of Animal Husbandry.

Sollenberger, L.E. and Vanzant, E.S. (2011). Interrelationships among forage nutritive value and quantity and individual animal performance. *Crop Sci.* 51: 420–432.

Sollenberger, L.E., Moore, J.E., Quesenberry, K.H., and Beede, P.T. (1987). Relationships between canopy botanical composition and diet selection in aeschynomene-limpograss pastures. *Agron. J.* 79: 1049–1054.

Spalinger, D.E. and Hobbs, N.T. (1992). Mechanisms of foraging in mammalian herbivores: new models of functional response. *Am. Nat.* 140: 325–348.

Stephens, D.W. and Krebs, J.R. (1986). *Foraging Theory*. Princeton, NJ: Princeton University Press.

Stobbs, T.H. (1973a). The effect of plant structure on the intake of tropical pastures. I. Variation in the bite size of grazing cattle. *Aust. J. Agric. Res.* 24: 809–819.

Stobbs, T.H. (1973b). The effect of plant structure on the voluntary intake of tropical pastures. II. Differences in sward structure, nutritive value and bite size of animals grazing *Setaria anceps* and *Chloris gayana* at various stages of growth. *Aust. J. Agric. Res.* 24: 821–829.

Stobbs, T.H. (1974). Rate of biting by Jersey cows as influenced by the yield and maturity of pasture swards. *Trop. Grassland* 8: 81–86.

Ungar, E.D. (1996). Ingestive behavior. In: *Ecology and Management of Grazing Systems* (eds. J. Hodgson and A.W. Illius), 185–218. Oxfordshire, UK: CABI.

Van Langevelde, F., Drescher, M., Heitkonig, I.M.A., and Prins, H.H.T. (2008). Instantaneous intake rate of herbivores as function of forage quality and mass: effects on facilitative and competitive interactions. *Ecol. Modell.* 213: 273–284.

Van Soest, P.J. (1994). *Nutritional Ecology of the Ruminant*. Ithaca, NY, USA: Cornell University Press.

Vincenot, C.E., Mazzoleni, S., Moriya, K. et al. (2015). How spatial resource distribution and memory impact foraging success: a hybrid model and mechanistic index. *Ecol. Complexity* 22: 139–151.

Walker, J.W. (1994). Multispecies grazing: the ecological advantage. *J. Sheep Res.* (Special Issue): 52–64.

Weisberg, P.J., Coughenour, M.B., and Bugmann, H. (2006). Modelling of large herbivore-vegetation interactions in a landscape context. In: *Large Herbivore Ecology, Ecosystem Dynamics and Conservation* (eds. K. Danell, R. Bergström, P. Duncan and J. Pastor), 348–382. Cambridge, UK: Cambridge University Press.

Westoby, M. (1974). An analysis of diet selection by large herbivores. *Am. Nat.* 108: 290–304.

White-Leech, R., Liu, K., Sollenberger, L.E. et al. (2013). Excreta deposition on grassland. II. Spatial pattern and duration of forage responses. *Crop Sci.* 53: 696–703.

Wilmshurst, J.F. and Fryxell, J.M. (1995). Patch selection by red deer in relation to energy and protein-intake – a reevaluation of Langvant and Hanley (1993) results. *Oecologia* 104: 297–300.

Wilmshurst, J.F., Fryxell, J.M., and Colucci, P.E. (1999). What constrains daily intake in Thomson's gazelles? *Ecology* 80: 2338–2347.

Wilmshurst, J.F., Fryxell, J.M., and Bergman, C.M. (2000). The allometry of patch selection in ruminants. *Proc. Biol. Sci.* 267: 345–349.

Woodward, S.J.R., Wake, G.C., Pleasants, A.B., and McCall, D.G. (1993). A simple model for optimizing rotational grazing. *Agric. Syst.* 41: 123–155.

CHAPTER

11

Nutrient Cycling in Forage Production Systems

David A. Wedin, Professor, *School of Natural Resources, University of Nebraska, Lincoln, NE, USA*
Michael P. Russelle, Soil Scientist (Retired), *USDA-Agricultural Research Service, St. Paul, MN, USA*

Introduction – The Systems Approach to Nutrient Cycles

In most forage production systems, the nutrients needed for plant growth are provided by the microbially-mediated breakdown and release of plant-available mineral nutrients from dead plant tissues, livestock excreta, soil organic matter, and geochemically-bound mineral forms. Even in fertilized forage systems, determining appropriate fertilizer or manure application rates requires a "systems" approach on the part of the manager (Rotz et al. 2005; Wood et al. 2012). Fertilizer additions are simply one input in the system of inputs, outputs, pools, and fluxes that characterize nutrient cycling in a particular ecosystem.

In a systems approach, the size of the system is determined by the observer, and it is often management driven. It could be a particular field (Stout et al. 2000; Simpson et al. 2015), an entire farm (Rotz et al. 2005; Powell and Rotz 2015), a watershed (Howarth et al. 1996; Loecke et al. 2017) or, as is the case for global biogeochemical cycles, the entire earth (Smil 2000; Galloway et al. 2008). Whereas harvestable forage and livestock have traditionally been the outputs driving management decisions in forage systems, outputs of nutrients such as NO_3^- leaching, N_2O gaseous emissions, and P run off are becoming increasingly important (Vitousek et al. 2009).

Central to nutrient cycling in any ecosystem is the concept of mass balance. Nutrient inputs must balance nutrient outputs and/or nutrient storage. Societal concerns over nutrient pollution in the environment and economic pressures on the profitability of forage systems are forcing scientists and managers to document nutrient budgets more completely and precisely (Nord and Lanyon 2003; Wood et al. 2012). The C dynamics of forage systems can be analyzed with the same "systems" approach outlined here, but are beyond the scope of this chapter (see Conant et al. 2017 for a review of grassland carbon budgets).

A nutrient cycle or budget is a network of pools (amounts) of a particular element, joined by fluxes (transfers) connecting those pools (Chapin et al. 2011). Though most elements have either a large atmospheric (e.g. C and N) or geologic (e.g. P and K) pool, the fluxes or transfer rates of elements from those pools into organic forms are usually low. The microbially-mediated fixation of atmospheric N into organic forms by legumes is an obvious and important exception to that generalization.

Most discussions of nutrient cycling in forage systems emphasize the following pools: (i) soil organic matter, which, in more complex analyses, may be considered as multiple pools or fractions; (ii) living plant biomass,

Forages: The Science of Grassland Agriculture, Volume II, Seventh Edition.
Edited by Kenneth J. Moore, Michael Collins, C. Jerry Nelson and Daren D. Redfearn.

including above- and belowground tissues; (iii) plant residues (dead, relatively undecomposed plant tissues); (iv) living animal biomass, the most obvious being the grazing animal, but the most abundant being above- and belowground invertebrates and microbial populations; and (v) a small but critical pool of plant-available mineral forms of elements necessary for plant growth.

This last pool, the concentration of soil NO_3^- and NH_4^+ in the case of N, deserves special attention. This pool is often measured as an index of site fertility or nutrient availability but, technically speaking, a pool or concentration is not a measure of nutrient availability, which is a flux or rate. Though the concentration of mineral soil N in a grassland may be very low on average, this tells us little about the rate at which N is being made available for plant uptake, which could be high in a fertile soil and low in an infertile soil (Robertson et al. 1999).

Simply put, pools have units of mass (kg ha^{-1}, g m^{-2}, mg kg^{-1}, etc.) whereas fluxes have units of mass transferred per unit time (kg ha^{-1} yr^{-1}, g m^{-2} d^{-1}, etc.). In a systems approach, residence times are the ratios of pools to fluxes and have units of time, because the units of mass cancel. Pools with short residence times are dynamic and are expected to change as management or environmental fluctuations affect the system. For example, consider a hypothetical grassland in which the only source of mineral N for plant uptake is net N mineralization, the flux from soil organic N to soil mineral N, and in which the soil organic matter pool of N contains 5000 kg N ha^{-1}, the soil mineral N pool contains 5 kg N ha^{-1}, and the annual net N mineralization rate is 50 kg ha^{-1} yr^{-1}. In this case, the residence time of N in soil organic matter would be 100 years, whereas the residence time of mineral soil N would be 0.1 year or 36.5 days. The turnover rate of a nutrient pool is simply the inverse of the residence time. In this example, the mineral soil N pool "turns over" 10 times, whereas only 1% of the soil organic N pool turns over per year.

Calculations of residence times assume a steady state or equilibrium. Although never completely valid, it is often a useful starting point in analyzing system behavior (Chapin et al. 2011). In a steady state, pool sizes and flux rates are constant, and fluxes into and out of each pool must balance. This includes net fluxes into and out of the total system.

A system dominated by internal recycling of nutrients with relatively small inputs (e.g. fertilizer or **N fixation**) and outputs (e.g. leaching or animal and forage offtake) is considered relatively closed. As management intensity increases in forage systems, nutrient cycles inevitably become more open. Because nutrients such as N and P behave differently, one element in a system may have a relatively open nutrient cycle, whereas another element's cycle is relatively closed. For example, grasslands receiving animal manures may be managed to minimize N losses, yet still have significant P losses (Wood et al. 2012).

Why Does Nitrogen Frequently Limit Forage Production?

Nitrogen is the dominant nutrient constraint on primary production in most forage systems, though a study replicated across several continents suggests that N and P collectively constrain productivity in many grasslands (Vitousek 2015). All terrestrial ecosystems have access to a near infinite pool of N in the atmosphere, which contains 78% N_2 gas. Many genera of bacteria are able to break the triple bonds of N_2 and reduce ("fix") it to NH_4^+. These bacteria include both symbiotic N fixers such as *Rhizobium* (associated with legumes) and *Frankia* (associated with woody species including *Alnus* and *Ceanothus*), and free-living N fixers such as *Azotobacter* and *Nostoc* (Paul 2015). Despite the abundant source of N, and a pathway for its incorporation into the ecologic cycle, most natural and managed ecosystems are N limited (Houlton et al. 2008).

Hypotheses for widespread N limitation involve the mass balance of inputs and outputs of N from terrestrial ecosystems. Until the advent of fossil fuel combustion, atmospheric inputs of N to ecosystems were generally small to negligible (1–5 kg N ha^{-1} yr^{-1}). Sources of NO_3^- and NH_4^+ deposition included fixation in the atmosphere by lightning, and **volatilization** from oceanic sources in coastal regions (Galloway et al. 2008). Biologic N fixation, in contrast, can potentially add >200 kg N ha^{-1} yr^{-1} to ecosystem N cycles (Figure 11.1).

Biologic N fixation has three general constraints. First, N fixation is energetically expensive. Thus, legumes fixing N divert energy from growth, giving them a disadvantage in competition for light with non-N fixers. N fixation is generally restricted to open, high-light environments such as deserts, grasslands, and savannas (Houlton et al. 2008). Leguminous trees in dense forests are rarely nodulated and probably contribute little to forest N cycles.

Second, biologic N fixation may frequently be limited by the availability of other elements. The biochemistry of N-fixation requires significant P, iron, sulfur, and molybdenum. In highly weathered and low-pH soils, these elements, though present, may be immobilized in a variety of geochemical forms. Increased grassland productivity in many tropical and subtropical regions may ultimately be limited by non-N nutrient constraints on legumes, especially P. Moore (1970) concluded that N is almost universally deficient in humid tropical and subtropical grasslands. However, "for the successful establishment of tropical grass and legume mixtures, every encouragement must be given to the legumes" (Moore 1970). In tropical grasslands, which are often affected by low P and micronutrient availability, P and micronutrient fertilizer additions are

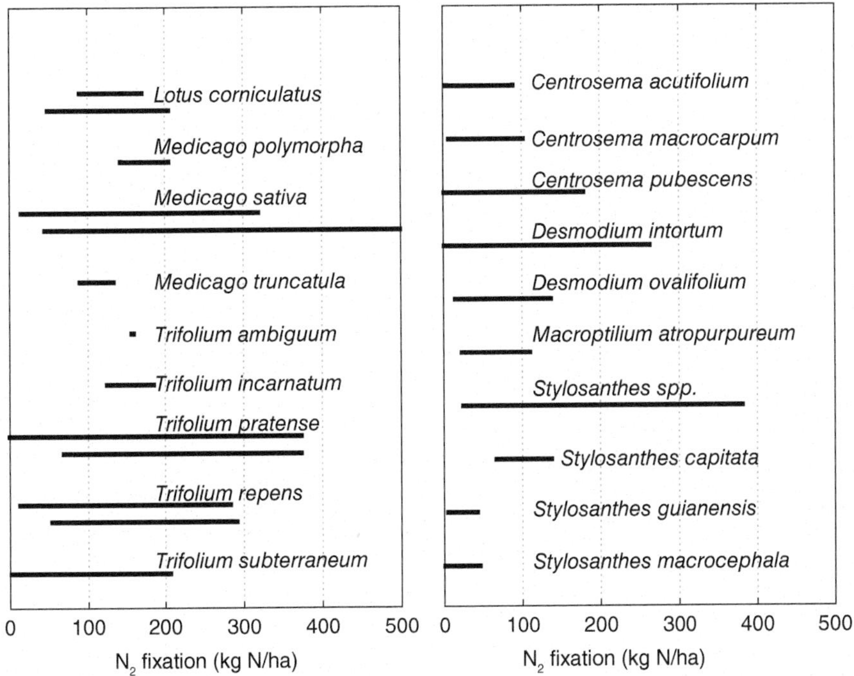

Fig. 11.1. Ranges of reported symbiotic N_2 fixation by temperate (left panel) and tropical (right panel) forage legumes (Russelle 2008). Dinitrogen fixation by temperate legumes in mixtures with nonlegumes is shown by the upper line of a pair, whereas N_2 fixation in pure stands is shown by the lower line.

critical to the establishment of legumes and subsequent improvements in the N budget.

The third general constraint on the abundance of N-fixing plants is herbivory. Plant productivity in most temperate terrestrial ecosystems is N-limited, and, as a consequence, the protein concentration of **available forage** is low. Legumes, which generally have high leaf N concentrations, are often targeted by both generalist herbivores, such as large ruminants, and specialist herbivores, such as many invertebrates. Reducing herbivory has led to increased legume abundance and greater N fixation in a variety of ecosystems. In areas with a long evolutionary history of grazing, such as Africa, legumes have often countered the threat of herbivory with physical (e.g. thorns) or chemical (e.g. alkaloid) defenses (see Chapters 46 and 47).

Nitrogen loss from ecosystems may be as important as constraints on N inputs in explaining the chronic N limitation found in many temperate, terrestrial ecosystems. Because the N cycle is prone to both gaseous losses (NH_3 volatilization, **denitrification**, combustion losses during fire) and leaching losses (NO_3^- and, to a lesser degree, dissolved organic N [DON]), it is inherently leakier than the cycles of P, K, Ca, and various micronutrients (Chapin et al. 2011). The availability of P or Fe may decrease over time in a particular ecosystem, as those elements are chemically immobilized by reactions with soil and subsoil minerals, but, unless erosion or surface runoff occurs, those elements are rarely exported from the local system. In contrast, N losses inevitably increase when ecosystems are disturbed (e.g. tillage, grazing, or fire) and plant uptake from the soil mineral N (NO_3^- and NH_4^+) pool is disrupted (Houlton et al. 2008; Vitousek et al. 2009).

Nitrogen in the Plant-Soil System

In the long-term (centuries to millennia), net inputs and outputs of N play a large role in determining a particular ecosystem's fertility. In the short-term, however, the supply rate of plant-available mineral soil N in an unfertilized ecosystem is regulated by soil biologic activities. A diverse community of soil invertebrates, bacteria, and fungi is responsible for physically and chemically breaking down large organic molecules into smaller organic molecules, CO_2, and various mineral nutrients (Robertson and Groffman 2015). The list of new techniques for assessing the functional, taxonomic, and genetic diversity of soil communities is growing rapidly, but will not be discussed here (Fierer and Jackson 2006).

By far the largest pool of N (excluding the atmosphere) in grassland and forage systems is soil organic matter. The key flux from that pool is net N **mineralization**, defined as the microbially mediated release of NH_4^+ and NO_3^- from soil organic matter and plant residues. Various net N mineralization assays provide key insights into soil fertility and the degree to which N may limit plant productivity (Robertson et al. 1999). Mineralization together with biologic N fixation, N returned by grazing animals, and fertilizer or atmospheric N inputs, make up an ecosystem's N supply rate.

Treating net N mineralization as a single process or flux ignores much of the complexity of soil N dynamics. In the transition from organic matter to mineral N, organic substrates must first be broken down into soluble compounds, the DON pool. The DON pool is the focus of recent attention for several reasons (Jones et al. 2004; Chapin et al. 2011). First, organic compounds must be broken down to DON before they can be absorbed and mineralized by microbes. Second, leaching losses of DON, although rarely measured, may be a significant component of the N budget in some ecosystems (Jones et al. 2004). Third, direct uptake of DON by plant roots or associated mycorrhizae has been documented in numerous ecosystems. Most of the reported cases involve uptake of neutrally charged amino acids such as glycine in cold, wet, and/or acidic environments such as tundra and conifer forests, where up to 65% of plant N uptake has been attributed to DON (Chapin et al. 2011). Because the direct uptake of DON short-circuits the role of N mineralization and the importance of NH_4^+ and NO_3^- availability, researchers are reconceptualizing N cycling where DON uptake has been documented. Plant DON uptake in temperate grasslands is documented (e.g. Wilkinson et al. 2015), but its role in the N cycle of managed forage systems is still unsettled.

Microbial decomposers break down DON as an energy source, respiring CO_2, and releasing NH_4^+ as a by-product. In **aerobic** soils, much of this NH_4^+ is subsequently nitrified to NO_3^- by bacteria that oxidize NH_4^+ as an energy source (Robertson and Groffman 2015). This is the key step in N mineralization and the total amount of mineral N released is called gross mineralization. Much of this NH_4^+ and NO_3^- may be reabsorbed or immobilized by the microbial community, however, in order to meet nutritional needs. If the C:N ratio of decomposing organic matter is high, N is limiting for microbes relative to labile organic C (their energy source) and little if any net release, or net mineralization, of NH_4^+ into the soil occurs.

A C:N ratio of 25–35 is generally accepted as a critical ratio for net N mineralization from decomposing plant residues. This is somewhat higher than the C:N ratio of microbial biomass (generally about 10), but also reflects microbial growth efficiency (the proportion of consumed C incorporated into growth vs respired) (Robertson and Groffman 2015). At C:N ratios less than the critical level, the **sink** for NO_3^- and NH_4^+ provided by microbial **immobilization** disappears and net mineralization increases sharply. The presence of this critical ratio or breakpoint in N cycling (the shift from immobilization to net mineralization) means that soil N availability and ecosystem N losses may respond non-linearly to gradual changes in fertilization, herbivory, or other processes in forage systems (Wedin and Tilman 1996).

Because of the strong role of plant tissue chemistry in regulating the N cycle, it is not valid to consider soil N availability as an abiotic or soil property in isolation from the characteristics of past and present vegetation (Wedin 1995). The C:N ratios of plant residues affect both the rate of decomposition and the balance between N immobilization by microbes and net N mineralization (Chapin et al. 2011). In addition, the C chemistry of plant tissue strongly affects how it decomposes and contributes to formation of soil organic matter.

Lignin in aboveground tissues and **suberin** in roots are energetically expensive to break down for microbes and slow to decompose. Much of the polyphenolic ring structure of **lignin** is not broken-down during decomposition, but is instead transformed and incorporated into large-molecular-weight amorphous compounds known as soil humus. During this transformation, considerable N is tied up in the transformed C rings. Thus, though the C:N ratio of humus is quite low (10–20), the energetic costs for microbes utilizing humus-bound N are high and its contribution to net mineralization is often low. As humus binds with clay or is protected in soil aggregates, its availability for decomposition and mineralization decreases further.

A number of decomposition studies suggest that approximately 20% of decomposing plant residues become stabilized as soil organic matter (Chapin et al. 2011). Using a simple model of N immobilization and soil organic matter formation, Knops et al. (2002) suggested that no net mineralization occurs in decomposing plant residues if they initially contain less than 0.75% N. All of the N becomes incorporated into soil organic matter. Although 0.75% N is low for aboveground plant tissues in managed cool-season pastures, it is typical for aboveground senesced tissues of unfertilized C_4 grasses. It is also a typical N concentration for roots in unfertilized stands of both cool season and warm season grasses. The low rates of net N mineralization observed in many grasslands, and their ability to build soil organic matter rich in N are related, especially considering that roots make up over one-half of **net primary production** in most grasslands.

The N in soil organic matter in grasslands generally ranges from 5000 to over 20 000 kg N ha^{-1}. Net N mineralization rates generally range from 20 to 80 kg N ha^{-1} yr^{-1}, so the residence time of N in soil organic matter would be centuries in most grasslands. Thus, soil organic matter does not appear to be a dynamic pool. However, numerous studies have shown that net N mineralization in grassland soils is dynamic, responding within months to fire, grazing, or changes in plant species composition. This conflict illustrates the point that soil organic matter does not behave as a single pool when considering N, C, or other elemental cycles.

Numerous methods have been published for partitioning soil organic matter into chemical, physical, or functional fractions or pools. Many grassland studies follow the CENTURY model (Parton et al. 1987), which partitions soil organic matter into three fractions. The "active" fraction contains low-molecular-weight fractions of recently added plant residues and live microbial biomass. It makes up 2–8% of total soil organic matter and has a residence time of 1–5 years. The "slow" pool makes up 40–60% of soil organic matter and has a residence time of 20–50 years. The "passive" pool makes up 30–50% of soil organic matter and has a residence time of over 1000 years. The slow and passive pools are strongly affected by soil texture and climate. These two pools comprise the vast majority of soil organic matter, yet they contribute less than 30% of the net N mineralization from grassland soils (Schimel et al. 1994). Various methods of soil organic matter fractionation all indicate that a small, highly active soil organic matter fraction (e.g. CENTURY's "active" fraction) dominates soil biologic activity, including N cycling (McLauchlan and Hobbie 2004).

Referring to tropical grasslands and savannas, Huntley and Walker (1982) said "N has been shown to be of great significance ... but despite many thousands of N measurements, in all its forms, an understanding of the N cycle still eludes us." Subsequent N cycling research in grassland/forage systems has emphasized the strong linkages between vegetation and the small active fraction of soil organic matter. In unfertilized humid and subhumid grasslands, this plant-soil interaction reinforces low soil N availability (Wedin 1995; Dubeux et al. 2007). The low tissue N concentrations of senesced grass leaves and roots lead to microbial N **immobilization**, reducing net N mineralization, which, in turn, reduces both forage production and forage quality. Low soil moisture in semi-arid and arid grasslands constrains both soil microbes and plants, and the role of plant-soil interactions in regulating N cycling is less clear (Burke et al. 1998; McCulley et al. 2009). To address the natural tendency toward N limitation in grasslands, forage production in humid regions has relied on increasing N inputs (N fixation by legumes, animal wastes, inorganic N fertilizer) and managing the plant-soil-grazer (livestock) system to enhance N cycling (Dubeux et al. 2007).

Legumes and N_2 Fixation

Dinitrogen fixation by legumes depends on many factors, including host species and genotype, rhizobial strain and population size, developmental stage of the host, inorganic N (mainly NO_3^-) supply, yield of the host, nutrient supply, and toxic element level, and abiotic growing conditions (Russelle 2008; Suter et al. 2015).

There is considerable uncertainty about how much N_2 a particular legume will fix. In general terms, N_2 fixation by forage legumes usually ranges from 50 to 200 kg N ha^{-1} yr^{-1} (Figure 11.1). Estimates of N_2 fixation in white clover-perennial ryegrass mixtures range from 0 to more than 300 kg N ha^{-1} yr^{-1} (Russelle 2008), and N_2 fixation in alfalfa-bermudagrass pastures range from 80 to 222 kg N ha^{-1} yr^{-1} (Haby et al. 2006). Dinitrogen fixation in pastures tends to be less than in mown forages (Figure 11.1) because of feedback through excreta.

Constraints to N_2 Fixation

Three conditions are necessary for large amounts of symbiotic N_2 fixation in mixed forage stands (Soussana and Tallec 2010): (i) high forage yield; (ii) high proportion of legume in the mixture; and (iii) high reliance of the legume on N_2 fixation. Legume production may vary from one year to the next, in part because of oscillations in soil N availability (Loiseau et al. 2001). Maintenance of sufficient legume populations has been difficult in many pastures, due to selective grazing, inadequate soil fertility, stand declines due to pest pressures, and the availability of inexpensive N fertilizers. However, as the economic and environmental costs of producing N fertilizer increase, interest in using mixed stands of legumes and grass is once again increasing (Wood et al. 2012; Lüscher et al. 2014).

Pathways of N Transfer

The transfer of N from legumes to non-legumes is due to: (i) exudation and leakage of N from roots and nodules; (ii) senescence and degradation of nodules or roots; (iii) direct transfer from legume roots to nonlegume roots through connections made by arbuscular mycorrhizal fungal hyphae; (iv) NH_3 loss from legume herbage and reabsorption by grass herbage; (v) movement of N from legume herbage to the soil by leaching or decomposition of surface litter; and (vi) redeposition of consumed N by livestock (Russelle 2008). Oscillations in legume population also contribute to N transfer to nonlegumes.

Of these, the two most important appear to be decomposition of plant residues, both below and above ground, and the return of N through deposition of livestock excreta. Ledgard (1991), for instance, found N transfer below ground from white clover to perennial ryegrass in a pasture (70 kg N ha^{-1} yr^{-1}) was similar to that transferred through excreta (60 kg N ha^{-1} yr^{-1}). Nearly half of the annual N_2 fixed by clover (270 kg N ha^{-1} yr^{-1}) was transferred to the grass under these conditions.

What proportion of a mixed stand must be comprised of legumes to provide sufficient N to the nonlegume? In grazed white clover/perennial ryegrass, (Sheehy 1989) estimated 41 kg N ha^{-1} yr^{-1} was needed to sustain the system, and this may be achieved with clover contents of about 10% on an area basis. In Brazil, calopo should make up 13–23% of the forage dry mass for the sustainability of a mixture with *Brachiaria* (Cadisch et al. 1994). The required proportion of legume in a stand varies with how the forage is used, which depends on livestock species, stocking rate, management, and forage palatability (Lüscher et al. 2014).

Palatable legumes are grazed selectively and need to comprise 20–30% of the pasture herbage dry matter when pasture utilization (consumption by livestock) is between 10% and 40%. However, with higher utilization rates (40–70%), legumes must comprise up to 45% of total dry matter (Thomas 1992). Decreasing the palatability of legumes by planting species or genotypes with higher tannin concentrations, for example, may provide a partial solution to the problem of maintaining legume populations at desirable levels. Factors affecting palatability are discussed in Chapters 46 and 47.

Transfer of Fixed N in Mixtures

It is unclear how much fixed N is transferred from legumes to nonlegumes growing in mixtures because a wide range of estimates has been reported. This is likely due to the large number of interacting conditions that affect N_2 fixation. Transfer of fixed N is positively related to the proportion of legume N derived from the atmosphere; therefore, more fixed N is transferred under low-N fertility conditions. More N transfer occurs with a higher proportion of legumes in the stand (Russelle 2008). This is due both to greater competition for soil N by the nonlegume and a larger "pool" of fixed N being added to the system.

Transfer of N increases with stand age in perennial forage mixtures, presumably because of increased reliance of the legume on N_2 fixation and cumulative decomposition of above and belowground tissue (Jorgensen et al. 1999). Maximum N transfer from alfalfa to meadow bromegrass was 55 kg N ha^{-1} yr^{-1} (Walley et al. 1996) and from white clover to perennial ryegrass was 43 kg N ha^{-1} yr^{-1} (McNeill and Wood 1990), though a lower value (18 kg N ha^{-1} yr^{-1}) was reported for an alfalfa-bermudagrass mixture (Haby et al. 2006).

Nitrogen in the Plant-Soil-Grazer System

Cattle, sheep, and other large herbivores affect plant growth rates, plant species abundance, and plant elemental composition by removing herbage, trampling vegetation, compacting soil, and excreting waste. All these effects alter the rate of N transformations, the fate of N, and, ultimately, the N balance of pastures.

Growing ruminants utilize 5–10% of the feed N they consume, and lactating dairy cows utilize 15–30% for milk production (Haynes and Williams 1993); the remainder is excreted. Fecal N is mostly insoluble in water and comprises microbial cells (50–65%), undigested plant residues (15–25%), and products of livestock metabolism (Haynes and Williams 1993). Urinary N is largely soluble and in the form of urea (60–90%) and other metabolic products, such as hippuric acid, creatine/creatinine, and allantoin. Consequently, fecal N contributes mainly to medium- to long-term N cycling processes, whereas urinary N is subject to rapid cycling and loss.

Nitrogen use efficiency (NUE) by the animal is low, and more N is excreted in urine when the diet is rich in degradable protein and low in available energy. Conversely, proper supplementation of pastures with digestible energy improves NUE and reduces N excretion. On the other hand, diet composition causes little change in fecal N output. Urinary N output by sheep was lower on perennial ryegrass/white clover swards (54 g N d^{-1}) than on perennial ryegrass fertilized with 420 kg N ha^{-1} yr^{-1} (82 g N d^{-1}), but there was no change in fecal N output (Parsons et al. 1991).

Patchiness of Nitrogen Distribution in Pastures

Concentrated excreta patches generally affect from 14% to 30% of the land area of a pasture annually, assuming the patches do not overlap (Whitehead 2000; Moir et al. 2011). Soil sampling must be more intensive than in mechanically harvested forages to produce accurate maps of nutrient distribution. Optimum fertilization of grazed pastures with N requires site-specific application, but most farmers in North America have not adopted this practice with forages. More research is needed on this topic, because benefits of site-specific N applications in pastures have not been consistent (Cuttle et al. 2001).

More excreta are "deposited" in areas where livestock spend time, such as shelter from sun and wind, near field gates, or near watering tanks (Bogaert et al. 2000; Augustine et al. 2013). Moving the water supply, or using moveable shade structures, improves nutrient distribution in the pasture, as does short-term, high-stocking rate grazing systems (Peterson and Gerrish 1996).

Nitrogen Losses in Pastures

In urine spots, the combination of high soil pH from urea hydrolysis, high NH_4^+ concentration, and high osmotic strength increases NH_3 volatilization and slows **nitrification**. Gaseous NH_3 losses increase with soil temperature and lower soil moisture, making it the primary pathway of N loss in grazed semiarid grasslands. Under subhumid and humid pasture conditions, NH_3 losses account for between 2 and 25% of urinary N (Mulvaney et al. 2008). Higher NH_3 loss rates from urine and manure occur for concentrated or confined animals (Powell and

Rotz 2015). Gaseous N loss by denitrification can be significant when soils become waterlogged and anoxic (Robertson and Groffman 2015), but generally accounts for only a few percent of urinary-N loss (Luo et al. 1999).

Nitrate leaching loss may be larger under grazing than mechanical harvesting, but this depends on the amount and timing of excess soil water, soil texture, the general level of N fertility, and crop growth. The amount of available N in a urine spot (up to 250 kg N ha^{-1} for sheep and 1000 kg N ha^{-1} for dairy cattle; Steele 1987) greatly exceeds the N needs of neighboring plants. High NO_3^- leaching losses occur when precipitation or irrigation occurs during periods of high NO_3^- concentrations (Wood et al. 2012).

Intensive grassland management in humid climates has been implicated in NO_3^- contamination of ground water and surface water (Galloway et al. 2008; NRC 2009; Vitousek et al. 2009). Because N is redeposited by livestock, the probability of NO_3^- leaching losses is higher with higher N fertilizer or manure rates under grazing than under mowing. In New Zealand, for example, critical N application rates were 200–300 kg N ha^{-1} yr^{-1} lower for grazed than mown forages to maintain leachate NO_3^- concentrations below the drinking water standard (Di and Cameron 2000). Leaching losses may also be large for pastures on shallow soils in the humid eastern US, especially with high addition rates of N fertilizer, manure, or biosolids (Stout et al. 2000). This problem has increased in the southeastern US in recent decades with the growth of the poultry industry (Wood et al. 2012). In the Midwest, however, where deeper soils and lower rainfall are typical, NO_3^- leaching losses from forage systems are small if N addition rates are low to moderate (Russelle 1996; Powell and Rotz 2015).

Excellent management of legume/grass mixtures can yield moderate to high animal production levels with modest N losses (Russelle 2008). As indicated above, it is often difficult to maintain sufficient legume populations in mixed stands under grazing. The solution to this site-specific problem requires integrated knowledge of plant characteristics, soil conditions, weather, livestock management, pest pressure, and fertilizer and **lime** management.

Phosphorus Cycling in Forage System

After N, P is the nutrient receiving most attention in forage systems. Though plant tissue concentrations of P are much lower than N, P can nevertheless limit plant productivity under some circumstances. Like N, concern over runoff and leaching of P from agricultural landscapes has also increased dramatically in recent decades (Wood et al. 2012). However, the P cycle has important differences from the N cycle that must be considered whether the goal is optimizing P supply for plant and animal production, minimizing P losses to the environment, or, as is increasingly the case, both.

The various transformations that regulate soil N availability (i.e. N mineralization) are almost entirely microbially driven (Robertson and Groffman 2015). Abiotic soil factors, such as low pH, impact N availability through their effects on microbes and plants. In contrast, phosphate ions (PO_4^{3-}, the main form of available P in soils) easily form chemical bonds with various minerals (Smil 2000). The resulting precipitates are generally unavailable to plants and are known as occluded P. The chemical reactions that PO_4^{3-} undergoes depend on the concentrations of other minerals and pH. At low pH, PO_4^{3-} binds with oxides of Fe, Al, and Mn to form insoluble precipitates. As rock weathers (a process that occurs over millennia), the abundance of Fe, Al, and Mn oxides increases. Thus, highly weathered, ancient soils such as those found throughout the tropics, have a high potential to chemically immobilize available P (Chapin et al. 2011). At high pH, PO_4^{3-} binds with Ca to form various calcium phosphates that also precipitate and are relatively unavailable for plant uptake. Thus, P availability is highest at soil pH values around 6.5 and is less available at both higher and lower values.

The rapid geochemical immobilization of PO_4^{3-} in most soils also explains why leaching of PO_4^{3-} into groundwater is rare (Smil 2000). When P inputs to the soil are high, for instance, with repeated additions of animal wastes to forage systems, the geochemical potential of upper soil horizons to immobilize or precipitate P may be reduced. Soil solution concentrations of PO_4^{3-} may increase near the surface under these circumstances. In regions of high precipitation, PO_4^{3-} and P associated with dissolved organic matter may leach into lower soil horizons, but P is usually immobilized at that point. This contrasts sharply with NO_3^-, which readily moves with percolating water to great depths and frequently enters groundwater. Like N, high concentrations of soluble and particulate P near the soil surface are vulnerable to loss through runoff and associated soil erosion (Wood et al. 2012).

In contrast to N_2 gas for N, there is no atmospheric or gaseous pool of P to replenish terrestrial and aquatic ecosystems. Rather, the ultimate source of P cycling in natural ecosystems is rock **weathering**, a process that is very slow compared with N_2 fixation by legumes and other N-fixing organisms (Chapin et al. 2011). P is abundant in many of the minerals, such as apatite, that form rock, but the solubility of these minerals is low. Because P has no atmospheric pool and the solubility and transport of PO_4^{3-} in soil solutions is low, the linkages between terrestrial P and aquatic P cycles are weak. Simply put, natural terrestrial ecosystems do not leak P to nearby freshwater ecosystems the way they leak N. In addition, in aquatic ecosystems P is limited by the lack of a biotic mechanism for P inputs equivalent to N-fixing cyanobacteria in the plankton. Thus, freshwater ecosystems are often highly responsive and vulnerable to human-caused P loading (Chapin et al. 2011).

Because of concerns over eutrophication of aquatic systems (NRC 2009), P management is becoming increasingly important in forage and livestock management (Rotz et al. 2002; Jarvie et al. 2015). With the development of Total Maximum Daily Loads (TMDLs) for P pollution in surface-water bodies, the Natural Resources Conservation Service (NRCS) and state agencies have begun to institute limitations on P application to agricultural and residential land (reviewed by Wood et al. 2012). In particular, long-term additions of animal manures or biosolids to pasture and hayland near concentrated animal feeding operations have created chronic P pollution problems. Because animal manures, particularly poultry, have relatively high P concentrations, manure application rates calculated to meet plant N needs of forage systems result in the application of 2–4 times greater P than plants are able to use (Wood et al. 2012). Some states limit P addition to fields based on high soil test P levels, whereas others use a P risk index that assesses the likelihood of P loss from a field. A P risk index typically includes many factors known to affect runoff, including slope, soil cover, and distance to surface water (Butler et al. 2010). Regardless of the approach, producers who manage manure are being affected by concerns about P losses.

Behind concerns about P runoff, are the widespread increases in soil test P levels that have been observed (Wood et al. 2012; Jarvie et al. 2015). Such buildup can be attributed to repeated applications of livestock and poultry waste, overapplication of fertilizer P, and large amounts of imported P in livestock rations that end up in waste. Because of the relatively high P content of some animal wastes (e.g. poultry litter), soil test P levels will continue to increase even when manure application rates are matched to crop N requirements (Wood et al. 2012). Where soil test P levels are high, it may take many years to "crop down" fields high in P by harvesting forages. The P removal in animal products is only 10–35% of that for harvested forages. Thus, **hay** sales will send more P off farm than meat or milk. The best long-term solution to P accumulation is to reduce the net import of P to the farm. This can generally be achieved only by reducing input of off-farm P sources (feed, fertilizer, manure, etc.) and increasing export of P in animal and plant products (Rotz et al. 2002).

Though well-managed perennial pastures provide better soil protection than most annual cropping systems, P losses from damaged vegetation, thatch, and dung are environmentally important. Loss rates for P of several kg ha^{-1} yr^{-1} have been measured in snowmelt runoff from hay fields and pastures in cold regions. Surface applications of manure, either as non-incorporated broadcast manure from storage or as dung from grazing stock, are a rich reservoir of water-soluble or biologically available P. As with N, P distribution on a farm is generally heterogeneous because of long-term management decisions (e.g. fields nearest the manure source receive the most manure) and animal behavior (more dung is deposited in areas where livestock rest than in other areas). Decision support tools, such as soil test P levels or P risk indices, need to be used at both the field and the landscape scale to make appropriate decisions about where and when to apply nutrient-rich animal waste (Wood et al. 2012).

The Challenge of "Balancing" Nutrient Budgets

Lanyon (1995) published a provocative paper entitled "Does nitrogen cycle?: Changes in the spatial dynamics of nitrogen with industrial nitrogen fixation." The simple nutrient cycle diagram found in many ecology or agronomy texts (e.g. N flowing from soil to plant to animal and back to soil within an idealized field) rarely exists in modern agricultural landscapes. Many, if not most, forage systems have relatively small losses of N to the atmosphere, groundwater or surface water when compared to arable land at the field level. In contrast, P losses from intensively managed forage systems may approach or exceed values for arable land (NRC 2009). Forage systems are an integral component of modern agriculture, which has dramatically changed local, regional, and global nutrient cycles over the last century (Vitousek et al. 2009). Nutrient outputs (forage, grain, livestock, milk) from one field become intentional or unintentional nutrient inputs to landscapes dozens or hundreds of kilometers away.

This spatial uncoupling of nutrient cycles is combined with unprecedented increases in the magnitude of global nutrient cycles. Human activities (industrial N fertilizer production, inadvertent N fixation during fossil fuel combustion, and agricultural management of legumes) have more than doubled the pre-industrial global rate at which atmospheric N_2 was transferred (i.e. fixed) to biologically active pools (Galloway et al. 2008). Though the sources of P inputs differ (e.g. mining), changes in the global P cycle are of similar magnitude (Smil 2000).

The potential risk of environmental damage from farming systems may be estimated from nutrient budgets. Assuming conservation of mass, the difference between inputs and outputs indicates the mass that is unaccounted for (Chapin et al. 2011). If one assumes steady-state conditions, mass that is not accounted for is presumed to be a net nutrient loss from the system. The simplest approach at the whole-farm level is to measure the difference between purchased inputs and marketed outputs of a given nutrient and to assume steady-state conditions, (e.g. no change in the size of nutrient pools in the soil).

This approach, however, is unlikely to be valid for most situations because management systems (tillage, residue removal, crop rotations, fertilizer management, etc.) vary and interact at timescales shorter than those required for equilibrium of the soil pools. In addition, there can be transfers within the farm, such as occur with sediment runoff and deposition that disrupt equilibrium within the farm. The simple balance approach also fails to partition net nutrient losses into specific fluxes, which is critical in determining the broader environmental impacts of local management decisions. For example, while both NH_3 **volatilization** and N_2O emissions are N losses to the atmosphere, the former has a short residence time in the atmosphere and relatively local negative impacts, whereas the latter is long-lived in the atmosphere and is a potent greenhouse gas (Robertson and Groffman 2015).

Given the large spatial and temporal heterogeneity in nutrient fluxes, many have used simulation models to estimate flows. For example, Rotz et al. (2002) projected that long-term whole-farm P balance could be achieved for northeastern US dairy farms by feeding the minimum dietary P and by maximizing the production and use of forages. Reducing animal N intake or supplementing a grazing herd with **metabolizable energy** also reduces environmental risk (Powell and Rotz 2015). Models have been used to estimate watershed or regional results (e.g. Rotz et al. 2005) and these can lead to crucial insights. For example, Nord and Lanyon (2003) found that changing the production strategy (e.g. heavy reliance on purchased feeds) on one farm can have larger effects on watershed nutrient balances than changing farm operations (e.g. field-specific manure application rates) on a number of farms.

As more parameters are used in a model (i.e. symbiotic N_2 fixation, net N mineralization, NO_3^- leaching, or gaseous losses), more can be inferred about likely nutrient transfers and other pathways of loss, but the number of estimated and uncertain parameters also increases. The nature and magnitude of these uncertainties are important, especially when nutrient budgets are used as policy instruments (Oenema et al. 2003). As farm-scale budgets are aggregated, it is possible to derive general conclusions relevant to watershed and regional spatial scales.

It is difficult to measure nonpoint nutrient losses at large scales, though some pathways are more amenable than others to measurement. P loss (Butler et al. 2010), N_2O emission (Uchida et al. 2008), NH_3 volatilization (Marshall et al. 1998), and NO_3^- loss through tile drains (Watson et al. 2000) can be measured on field scales. Nutrient losses to streams or groundwater are measurable at the watershed scale (Loecke et al. 2017). Many of these approaches, however, are expensive, difficult to replicate, or restricted to a limited suite of sites. Nevertheless, significant advances in the remote sensing of land cover and land use, the computational power of geographic information systems, and the instrumentation available for environmental monitoring offer potential. Perhaps most of all, the conceptual integration of traditionally separate disciplines such as soil science, hydrology, agronomy, atmospheric science, and ecology provide hope that our ability to understand, predict, and manage nutrient cycles will continue to progress rapidly.

References

Augustine, D.J., Milchunas, D.B., and Derner, J.D. (2013). Spatial redistribution of nitrogen by cattle in semiarid rangeland. *Rangeland Ecol. Manage.* 66: 56–62.

Bogaert, N., Salomez, J., Vermoesen, A. et al. (2000). Within-field variability of mineral nitrogen in grassland. *Biol. Fertil. Soils* 32: 186–193.

Burke, I.C., Lauenroth, W.K., Vinton, M.A. et al. (1998). Plant-soil interactions in temperate grasslands. *Biogeochemistry* 42: 121–143.

Butler, D.M., Franklin, D.H., Cabrera, M.L. et al. (2010). Assessment of the Georgia phosphorus index on farm at the field scale for grassland management. *J. Soil Water Conserv.* 65: 200–210.

Cadisch, G., Schunke, R.M., and Giller, K.E. (1994). Nitrogen cycling in a pure grass pasture and a grass-legume mixture on a red Latosol in Brazil. *Trop. Grasslands* 28: 43–52.

Chapin, F.S. III, Matson, P.A., and Vitousek, P.M. (2011). *Principles of Terrestrial Ecosystem Ecology*, 2e. New York: Springer.

Conant, R.T., Cerri, C.E., Osborne, B.B., and Paustian, K. (2017). Grassland management impacts on soil carbon stocks: a new synthesis. *Ecol. Appl.* 11: 343–355.

Cuttle, S.P., Scurlock, R.V., and Davies, B.M.S. (2001). Comparison of fertilizer strategies for reducing nitrate leaching from grazed grassland, with particular reference to the contribution from urine patches. *J. Agric. Sci. (Cambridge)* 136: 221–230.

Di, H.J. and Cameron, K.C. (2000). Calculating nitrogen leaching losses and critical nitrogen application rates in dairy pasture systems using a semi-empirical model. *N.Z. J. Agric. Res.* 43: 139–147.

Dubeux, J.C.B., Sollenberger, L.E., Mathews, B.W. et al. (2007). Nutrient cycling in warm-climate grasslands. *Crop Sci.* 47: 915–928.

Fierer, N. and Jackson, R.B. (2006). The diversity and biogeography of soil bacterial communities. *Proc. Natl. Acad. Sci. U.S.A.* 103: 626–631.

Galloway, J.N., Townsend, A.R., Erisman, J.W. et al. (2008). Transformations of the nitrogen cycle: recent trends, questions, and potential solutions. *Science* 320: 889–892.

Haby, V.A., Stout, S.A., Hons, F.M., and Leonard, A.T. (2006). Nitrogen fixation and transfer in a mixed stand of alfalfa and bermudagrass. *Agron. J.* 98: 890–898.

Haynes, R.J. and Williams, P.H. (1993). Nutrient cycling and soil fertility in the grazed pasture ecosystem. *Adv. Agron.* 49: 119–199.

Houlton, B.Z., Wang, Y.-P., Vitousek, P.M., and Field, C.B. (2008). A unifying framework for dinitrogen fixation in the terrestrial biosphere. *Nature* 454: 327–330.

Howarth, R.W., Billen, G., Swaney, D. et al. (1996). Regional nitrogen budgets and riverine N and P fluxes for the drainages to the North Atlantic Ocean: natural and human influences. *Biogeochemistry* 35: 75–139.

Huntley, B.J. and Walker, B.H. (eds.) (1982). *Ecology of Tropical Savannas*. Ecological Studies no. 42. Berlin: Springer-Verlag.

Jarvie, H.P., Sharpley, A.N., Flaten, D. et al. (2015). The pivotal role of phosphorus in a resilient water-energy-food security nexus. *J. Environ. Qual.* 44: 1049–1062.

Jones, D.L., Shannon, D., Murphy, D.V., and Farrar, J. (2004). Role of dissolved organic nitrogen (DON) in soil N cycling in grasslands soils. *Soil Biol. Biochem.* 36: 749–756.

Jorgensen, F.V., Jensen, E.S., and Schjoerring, J.K. (1999). Dinitrogen fixation in white clover grown in pure stand and mixture with ryegrass estimated by the immobilized N-15 isotope dilution method. *Plant Soil* 208: 293–305.

Knops, J.M.H., Bradley, K.L., and Wedin, D.A. (2002). Mechanisms of plant species impacts on ecosystem nitrogen cycling. *Ecol. Lett.* 5: 454–466.

Lanyon, L.E. (1995). Does nitrogen cycle?: Changes in the spatial dynamics of nitrogen with industrial nitrogen fixation. *J. Prod. Agric.* 8: 70–78.

Ledgard, S.F. (1991). Transfer of fixed nitrogen from white clover to associated grasses in swards grazed by dairy cows, estimated using ^{15}N methods. *Plant Soil* 131: 215–223.

Loecke, T.D., Burgin, A.J., Riveros-Iregui, D.A. et al. (2017). Weather whiplash in agricultural regions drives deterioration of water quality. *Biogeochemistry* 133: 7–15.

Loiseau, P., Soussana, J.F., Louault, F., and Delpy, R. (2001). Soil N contributes to the oscillations of the white clover content in mixed swards of perennial ryegrass under conditions that simulate grazing over five years. *Grass Forage Sci.* 56: 205–217.

Luo, J., Tillman, R.W., and Ball, P.R. (1999). Grazing effects on denitrification in a soil under pasture during two contrasting seasons. *Soil Biol. Biochem.* 31: 903–912.

Lüscher, A., Mueller-Harvey, I., Soussana, J.F. et al. (2014). Potential of legume-based grassland-livestock systems in Europe: a review. *Grass Forage Sci.* 69: 206–228.

Marshall, S.B., Wood, C.W., Braun, L.C. et al. (1998). Ammonia volatilization from tall fescue pastures fertilized with poultry litter. *J. Environ. Qual.* 27: 1125–1129.

McCulley, R.L., Burke, I.C., and Lauenroth, W.K. (2009). Conservation of nitrogen increases with precipitation across a major grassland gradient in the Central Great Plains of North America. *Oecologia* 159: 571–581.

McLauchlan, K.K. and Hobbie, S.E. (2004). Comparison of labile soil organic matter fractionation techniques. *Soil Sci. Soc. Am. J.* 68: 1616–1625.

McNeill, A.M. and Wood, M. (1990). ^{15}N estimates of nitrogen fixation by white clover (*Trifolium repens* L.) growing in a mixture with ryegrass (*Lolium perenne* L.). *Plant Soil* 128: 265–273.

Moir, J.L., Cameron, K.C., Di, H.J., and Fertsak, U. (2011). The spatial coverage of dairy cattle urine patches in an intensively grazed pasture system. *J. Agric. Sci.* 149: 473–485.

Moore, R.M. (1970). *Australian Grasslands*. Canberra: Australian National University Press.

Mulvaney, M.J., Cummins, K.A., Wood, C.W. et al. (2008). Ammonia emissions from field-simulated cattle defecation and urination. *J. Environ. Qual.* 37: 2022–2027.

Nord, E.A. and Lanyon, L.E. (2003). Managing material transfer and nutrient flow in an agricultural watershed. *J. Environ. Qual.* 32: 562–570.

NRC (2009). *Nutrient Control Actions for Improving Water Quality in the Mississippi River Basin and the Northern Gulf of Mexico*. Washington, DC: National Academies Press.

Oenema, O., Kros, H., and de Vries, W. (2003). Approaches and uncertainties in nutrient budgets: implications for nutrient management and environmental policies. *Eur. J. Agron.* 20: 3–16.

Parsons, A.J., Orr, R.J., Penning, P.D., and Lockyer, D.R. (1991). Uptake, cycling and fate of nitrogen in grass-clover swards continuously grazed by sheep. *J. Agric. Sci. (Cambridge)* 116: 47–61.

Parton, W.J., Schimel, D.S., Cole, C.V., and Ojima, D.S. (1987). Analysis of factors controlling soil organic matter levels in Great Plains grasslands. *Soil Sci. Soc. Am. J.* 51: 1173–1179.

Paul, E.A. (2015). *Soil Microbiology, Ecology and Biochemistry*, 4e. Burlington, MA: Academic Press.

Peterson, P.R. and Gerrish, J.R. (1996). Grazing systems and spatial distribution of nutrients in pastures: livestock management considerations. In: *Nutrient Cycling in Forage Systems Symposium*, 7–8 March 1996. Columbia, MO, vol. 1 (eds. R.E. Joost and C.A. Roberts), 203–212. Manhattan, KS: Potash and Phosphate Institute and Foundation for Agronomic Research.

Powell, J.M. and Rotz, C.A. (2015). Measures of nitrogen use efficiency and nitrogen loss from dairy production systems. *J. Environ. Qual.* 44: 336–344.

Robertson, G.P. and Groffman, P.M. (2015). Nitrogen transformations. In: *Soil Microbiology, Ecology and Biochemistry*, 4e (ed. E.A. Paul), 421–446. Burlington, MA: Academic Press.

Robertson, G.P., Wedin, D., Groffman, P.M. et al. (1999). Soil carbon and nitrogen availability: nitrogen mineralization, nitrification, and soil respiration potentials. In: *Standard Soil Methods for Long-Term Ecological Research* (eds. G.P. Robertson, D.C. Coleman, C.S. Bledsoe and P. Sollins), 258–271. New York: Oxford University Press.

Rotz, C.A., Sharpley, A.N., Satter, L.D. et al. (2002). Production and feeding strategies for phosphorus management on dairy farms. *J. Dairy Sci.* 85: 3142–3153.

Rotz, C.A., Taube, F., Russelle, M.P. et al. (2005). Whole-farm perspectives on nutrient flows in grassland agriculture. *Crop Sci.* 45: 2139–2159.

Russelle, M.P. (1996). Nitrogen cycling in pasture systems. In: *Nutrient Cycling in Forage Systems Symposium*, 7–8 March 1996. Columbia, MO, vol. 1 (eds. R.E. Joost and C.A. Roberts), 125–166. Manhattan, KS: Potash and Phosphate Institute and Foundation for Agronomic Research.

Russelle, M.P. (2008). Biological dinitrogen fixation in agriculture. In: *Nitrogen in Agricultural Soils*, 2e (eds. J.S. Schepers and W.R. Raun), 281–359. Madison, WI: ASA, CSSA, SSSA.

Schimel, D.S., Braswell, B.H., Holland, E.A. et al. (1994). Climatic, edaphic, and biotic controls over storage and turnover of carbon in soils. *Global Biogeochem. Cycles* 8: 279–293.

Sheehy, J.E. (1989). How much dinitrogen fixation is required in grazed grassland? *Ann. Bot. (London)* 64: 159–161.

Simpson, R.J., Stefanski, A., Marshall, D.J. et al. (2015). Management of soil phosphorus fertility determines the phosphorus budget of a temperate grazing system and is the key to improving phosphorus efficiency. *Agric. Ecosyst. Environ.* 212: 263–277.

Smil, V. (2000). Phosphorus in the environment: natural flows and human interferences. *Annu. Rev. Energy Env.* 25: 53–88.

Soussana, J.F. and Tallec, T. (2010). Can we understand and predict the regulation of biological N_2 fixation in grassland ecosystems? *Nutr. Cycling Agroecosyst.* 88: 197–213.

Steele, K.W. (1987). Nitrogen losses from managed grassland. In: *Managed Grasslands: Analytical Studies* (ed. R.W. Snaydon), 197–204. Amsterdam: Elsevier.

Stout, W.L., Fales, S.L., Muller, L.D. et al. (2000). Water quality implications of nitrate leaching from intensively grazed pasture swards in the northeast US. *Agric. Ecosyst. Environ.* 77: 203–210.

Suter, M., Connolly, J., Finn, J.A. et al. (2015). Nitrogen yield advantage from grass-legume mixtures is robust over a wide range of legume proportions and environmental conditions. *Global Change Biol.* 21: 2424–2438.

Thomas, R.J. (1992). The role of the legume in the nitrogen cycle of productive and sustainable pastures. *Grass Forage Sci.* 47: 133–142.

Uchida, Y., Clough, T.J., Kelliher, F.M., and Sherlock, R.R. (2008). Effects of aggregate size, soil compaction, and bovine urine on N_2O emissions from a pasture soil. *Soil Biol. Biochem.* 40: 924–931.

Vitousek, P.M. (2015). Complexity of nutrient constraints. *Nat. Plants* 1: 1–2.

Vitousek, P.M., Naylor, R., Crews, T. et al. (2009). Nutrient imbalances in agricultural development. *Science* 324: 1519–1520.

Walley, F.L., Tomm, G.O., Matus, A. et al. (1996). Allocation and cycling of nitrogen in an alfalfa-bromegrass sward. *Agron. J.* 88: 834–843.

Watson, C.J., Jordan, C., Lennox, S.D. et al. (2000). Inorganic nitrogen in drainage water from grazed grassland in Northern Ireland. *J. Environ. Qual.* 29: 225–232.

Wedin, D.A. (1995). Species, nitrogen and grassland dynamics: the constraints of stuff. In: *Linking Species and Ecosystems* (eds. C. Jones and J.H. Lawton), 253–262. New York: Chapman and Hall.

Wedin, D.A. and Tilman, D. (1996). Influence of nitrogen loading and species composition on the carbon balance of grasslands. *Science* 274: 1720–1723.

Whitehead, D.C. (2000). *Nutrient Elements in Grasslands: Soil-Plant-Animal Relationships*. Wallingford, Oxon: CABI Publishing.

Wilkinson, A., Hill, P.W., Vaieretti, M.V. et al. (2015). Challenging the paradigm of nitrogen cycling: no evidence of in situ resource portioning by coexisting plant species in grasslands of contrasting fertility. *Ecol. Evol.* 5: 275–287.

Wood, C.W., Moore, P.A., Joern, B.C. et al. (2012). Nutrient management on pastures and haylands. In: *Conservation Outcomes from Pastureland and Hayland Practices: Assessment, Recommendations, and Knowledge Gaps* (ed. C.J. Nelson), 257–314. Lawrence, KS: Allen Press.

CHAPTER

12

Forages for Conservation and Improved Soil Quality

John F. Obrycki, ORISE Fellow, *USDA-National Laboratory for Agriculture and the Environment, Ames, IA, USA*
Douglas L. Karlen, Soil Scientist (Retired), *USDA-National Laboratory for Agriculture and the Environment, Ames, IA, USA*

Overview

Forages provide several soil benefits, including reduced **soil erosion**, reduced water runoff, improved soil physical properties, increased soil carbon, increased soil biologic activity, reduced soil salinity, and improved land stabilization and restoration when grown continuously or as part of a crop rotation. Ongoing research and synthesis of knowledge have improved our understanding of how forages alter and protect soil resources, thus providing producers, policymakers, and the general public information regarding which **forage crops** are best suited for a specific area or use (e.g. hay, grazing or bioenergy **feedstock**).

Forages can be produced in **forestland**, **range**, **pasture**, and **cropland** settings. These land use types comprise 86% of non-Federal United States rural lands (Table 12.1). In the United States, active forage production occurs on 22.6 million ha and is used for **hay**, **haylage**, grass **silage**, and **greenchop** (Table 12.2). Forages are used as cover crops in several production systems, and approximately 4.2 million ha were recently planted in cover crops (Table 12.3). Currently, the highest cover crop use rates, as a percentage of total cropland within a given state, occur in the northeastern United States.

Globally, permanent meadows and pastures account for over 3.3 billion ha, greater than arable land and permanent crops combined (Table 12.4). Within all regions of the world, except Europe, permanent meadows and pastures are a greater proportion of land cover than permanent crops. Pasture management information and resources are available for countries around the world (FAO 2017a,b). As seen in Tables 12.1–12.4, forages are used globally and can provide soil benefits across varied soil and climate types.

Forages Reduce Water and Wind Erosion

Forages as part of a comprehensive soil management plan can reduce erosion, particularly, if living plants are maintained on the landscape during most of the year. Compared to agricultural fields with limited residue cover, permanent ground cover reduces soil loss (Figure 12.1). Research during the 1930s and early 1940s across several research sites in the United States showed that row crops

USDA is an equal opportunity provider and employer. This research was supported, in part, by an appointment to the Agricultural Research Service (ARS) Research Participation Program administered by the Oak Ridge Institute for Science and Education (ORISE).

Forages: The Science of Grassland Agriculture, Volume II, Seventh Edition.
Edited by Kenneth J. Moore, Michael Collins, C. Jerry Nelson and Daren D. Redfearn.

Table 12.1 Land cover in non-federal rural land (2012)

Cover type	Hectares (000s)
Forest	167 300
Range	164 300
Crop	146 900
Other rural land	49 040
Conservation reserve	18 400
Total rural land	555 700

Source: Adapted from USDA National Resources Inventory Summary Report, August 2015, Table 2.

Table 12.2 Number of farms, hectares, and megagrams used for hay, haylage, grass silage, and greenchop in the United States and sorted by states with greatest number of hectares (2012)

Location	Number of farms	Hectares	Dry Mg
United States	813 583	22 581 037	115 501 930
Texas	86 456	2 052 461	8 656 202
Missouri	50 279	1 356 011	4 781 446
Oklahoma	32 781	1 095 202	3 411 413
South Dakota	14 695	1 058 781	3 305 505
Nebraska	20 034	1 007 009	4 289 189
Kansas	25 710	999 594	3 932 886
Wisconsin	37 020	970 300	6 547 600
Montana	11 728	917 894	3 609 240
North Carolina	10 141	879 651	2 847 363
Kentucky	43 757	826 784	3 771 345
New York	19 182	749 385	4 007 071

Source: Adapted from USDA (2012) Census of Agriculture, Table 26.

had between 98- and 1277-times as much soil loss as permanent cover (Figure 12.2). Please note that the data highlighted in Figure 12.2 were originally summarized by Browning (1951).

With increased adoption of conservation practices reducing tillage frequency and intensity, combined with increasing surface residue cover, the average erosion from agricultural land decreased (Figure 12.3). Current estimates for sheet and rill erosion from cultivated cropland are approximately 6.7 Mg ha^{-1} yr^{-1}, 1.6 Mg ha^{-1} yr^{-1} from pastureland, and 0.9 Mg ha^{-1} yr^{-1}from **conservation reserve** land. Erosion is seven-fold higher in cultivated cropland compared to conservation reserve land. Wind erosion removes 4.9 Mg soil ha^{-1} yr^{-1}from cultivated cropland, 0.4 Mg ha^{-1} yr^{-1} from pastureland, and 2.0 Mg ha^{-1} yr^{-1}from conservation reserve land (USDA 2015). The long-term trends in wind erosion are similar to the sheet and rill erosion data (Figure 12.3).

This reduction in erosion documents a significant improvement compared to the erosion rates seen in Figure 12.2. Figure 12.1 shows that erosion potential is a constant issue that must be addressed. Furthermore, even though average erosion rates are useful for general comparisons, generalized interpretation of these data has been long-recognized to mask variable landscape-level erosion rates (Bennett and Chapline 1928).

Similar trends in sediment reductions occur when forages are incorporated into row crops as conservation buffers (Figure 12.4) (Helmers et al. 2012) or when forages, rotational grazing, and conservation buffer strips are combined (Pilon et al. 2017). Within a no-till corn – soybean rotation, various prairie filter strip configurations reduced sediment loss by over 90%. These reductions were achieved by taking either 10% or 20% of the watershed area out of crop production and placing it in conservation buffers (Helmers et al. 2012).

The prairie filter strips reduced visible ephemeral gully formation. Data from 2008–2010 are presented in Figure 12.4. All treatments in 2007 (the first year of the study), had sediment loss below 0.1 Mg ha^{-1}. The study

Table 12.3 Cover crop use in the United States by farm type, state, and proportion of cropland hectares (2012)

Cropland planted to a cover crop (excluding CRP)	Hectares
United States	4 162 264
By North American Industry Classification System	
Oilseed and grain farming	1 800 024
Sugarcane farming, hay farming, and all other crop farming	543 201
Dairy cattle and milk production	408 593
Beef cattle and ranching	376 017
Vegetable and melon farming	264 363
Cotton farming	312 372
Fruit and tree nut farming	161 615
By State	
Texas	368 851
Indiana	241 321
Wisconsin	223 889
Pennsylvania	180 686
Michigan	177 004
By Proportion of Cropland Hectares	Percent
United States	3
Maryland	23
Delaware	16
Connecticut	14
Rhode Island	11
New Jersey	11

Source: Adapted from USDA (2012) Census of Agriculture, Tables 8, 50 (national and individual states), and 68.

Table 12.4 Worldwide land cover data for arable land, permanent crops, and permanent meadows and pastures

	Arable land	Permanent crops	Permanent meadows and pastures
Location	Hectares (000s)		
Worldwide	1 399 212	152 098	3 362 738
Africa	219 624	31 317	885 058
Americas	366 435	28 287	817 561
Asia	486 772	75 166	1 090 456
Europe	278 516	15 834	178 996
Oceania	47 865	1 494	390 668

Source: From FAOSTAT, composition of global agricultural area, 2000–2014 average, item codes 6621, 6650, 6655.

area was previously planted to smooth bromegrass for at least ten years. Planted species in the buffer included more than 20 species. Indiangrass, little bluestem, and big bluestem were the predominate species (Helmers et al. 2012). Incorporating forages into a cropland landscape can provide large sediment reductions relative to the total area used by the forages.

Forages Improve Soil Properties

At a regional scale, soil C and soil aggregate stability are generally higher in non-cultivated soils than in cultivated soils, as documented for western (Kemper and Koch 1966), central plains (Haas et al. 1957), and southern US (McCracken 1959) areas. These comparisons included important data regarding fields that had been cultivated

Fig. 12.1. Forages can be effective as surface mulches to reduce runoff and erosion. Top photo: *Forages: The Science of Grassland Agriculture* (Browning 1951). Bottom photo: Recent soil erosion in Iowa (Iowa NRCS, undated). *Source:* Figure name adapted from caption used by Browning (1951).

and fields that had never been in production. More recent data, indicates how actively managed forages can help improve numerous soil properties as discussed below. Forages can have positive effects on soil properties even on fields that remain in cultivation. Forages may help reduce site-specific variation in soil properties and processes caused by soil mismanagement, such as can occur from ephemeral gullies.

Soil Carbon

Increased soil C in fields planted to forages relative to other agricultural land uses was documented in a survey of the southeastern United States (Causarano et al. 2008). Within the 0 to 20-cm layer, soil organic C was highest in pasture (39 Mg ha^{-1}), followed by conservation tillage (28 Mg ha^{-1}), and conventional tillage (22 Mg ha^{-1}) fields (Causarano et al. 2008). The differences among

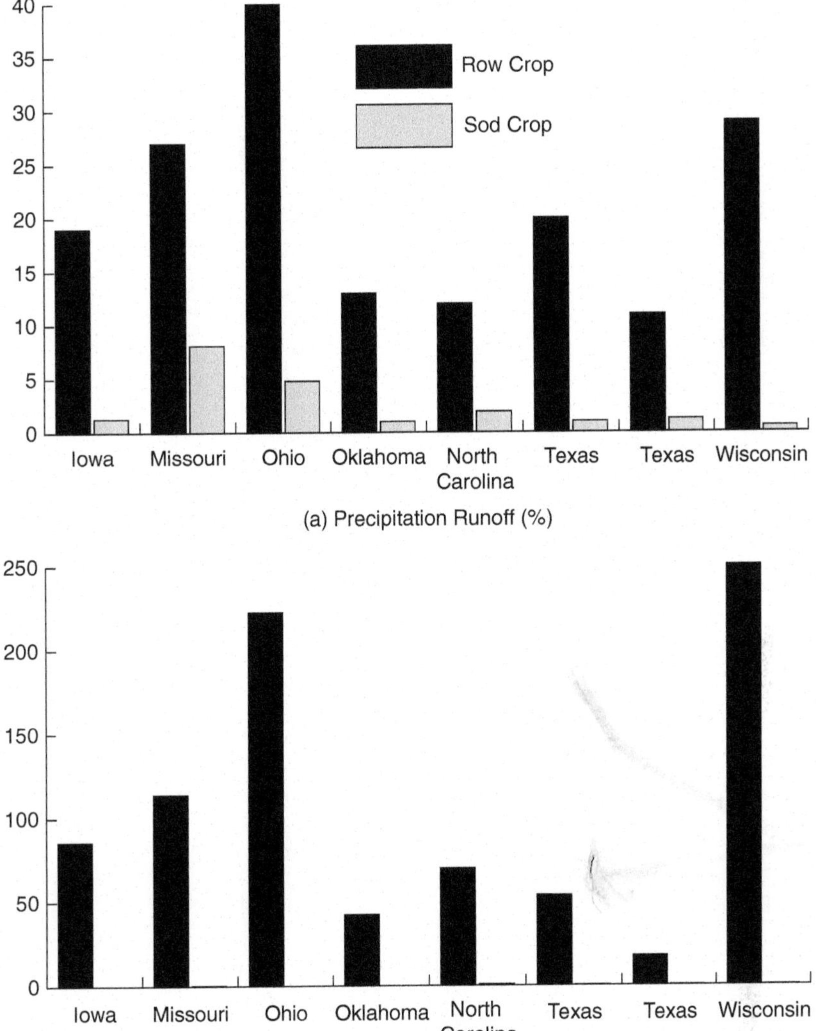

Fig. 12.2. Precipitation runoff (a) and soil loss (b) data between row crop and sod across sites collected during the 1930s and 1940s in the United States. *Source:* Browning (1951).

these soil management systems was largest within the 0 to 5-cm layer which had average soil organic C concentrations of 25, 15, and 7.5 g kg^{-1} for pasture, conservation tillage, and conventional tillage, respectively (Causarano et al. 2008). The study did not find an interaction in soil C effects with **major land resource areas.**

Similar results were reported from long-term field plots in Missouri (Veum et al. 2014). Long-term timothy pasture without manure fertilizer had approximately two-fold more soil organic C than soil in a moldboard-plowed continuous corn system (Table 12.5) (Veum et al. 2014). Other treatments such as reducing tillage and adding manure also increased soil C (Table 12.5). Several studies have confirmed that incorporating forages into a rotation is an effective long-term method for increasing soil C (Chan et al. 2011).

If fields in long-term forage are then cultivated, some C loss will occur (Grandy and Robertson 2007; Reicosky et al. 1995). For example, one tillage activity using inversion plowing to a 20-cm depth in a dairy-based perennial grassland farming system resulted in a 32 Mg C ha^{-1} loss (−22%) in soils collected from

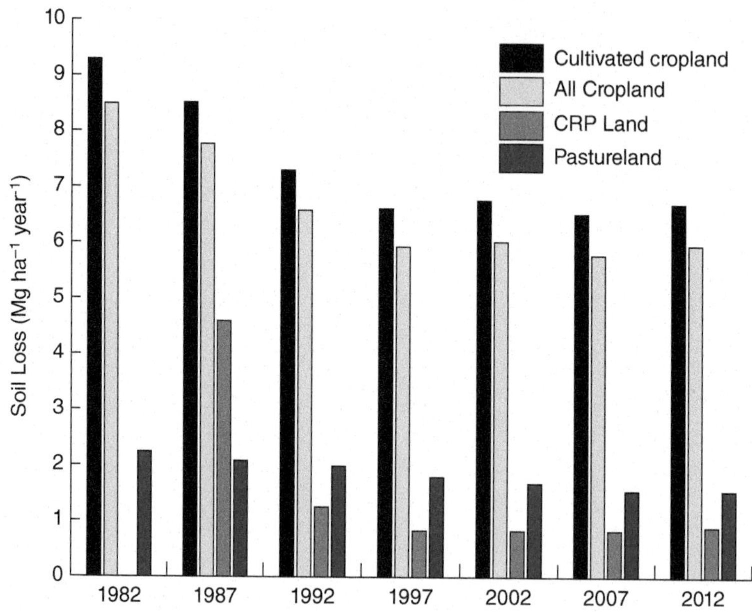

Fig. 12.3. Sheet and rill erosion from cropland, **conservation reserve program** (CRP) land, and pasture land in the United States. *Source:* USDA (2015), National Resources Inventory Report.

Table 12.5 Soil organic C analyzed from long-term Sanborn Field (Boone County, Missouri, US) plots evaluated in 2008

Crop	Tillage	Fertilizer	Year established	Soil organic C (g kg^{-1})
Timothy	None	Manure	1888	27.4
Corn	No-till	Annual N,P,K	1950	22.3
Timothy	None	None	1888	22.2
Wheat	Moldboard plow	Manure	1888	19.9
Corn	Moldboard plow	Manure	1888	17.0
Wheat	Moldboard plow	Annual N,P,K	1888	15.1
Corn	Moldboard plow	Annual N,P,K	1950	13.3
Wheat	Moldboard plow	None	1888	10.3
Corn	Moldboard plow	None	1888	9.4

Source: From Veum et al. (2014).

the 0 to 30-cm depth increment (Necpálová et al. 2013). This difference remained consistent for 2.5 years after the single tillage event and reseeding to forage. These C losses can be reduced if different tillage practices are used. Soils converted from long-term grass alfalfa to malt barley using a 10-cm rototill treatment had a 1.6-fold higher soil surface CO_2 flux average compared to soils remaining in grass-alfalfa. No difference in soil CO_2 flux occurred between malt barley converted using no-till and long-term grass-alfalfa (Jabro et al. 2008).

Variations in soil C under pasture can occur by soil type, as was documented in New Zealand (Schipper et al. 2014) and Canada (Wang et al. 2014). Pasture soil profiles that had been previously surveyed were resampled throughout New Zealand (Figure 12.5). Soil C decreases in 0 to 30-cm soil samples were found in Allophanic and Gley soils, with reductions of approximately 1.4 kg C m^{-2} and 0.8 kg C m^{-2}, respectively (Schipper et al. 2014). These C decreases could have occurred due to increased artificial drainage on the inherently poorly-drained Gley

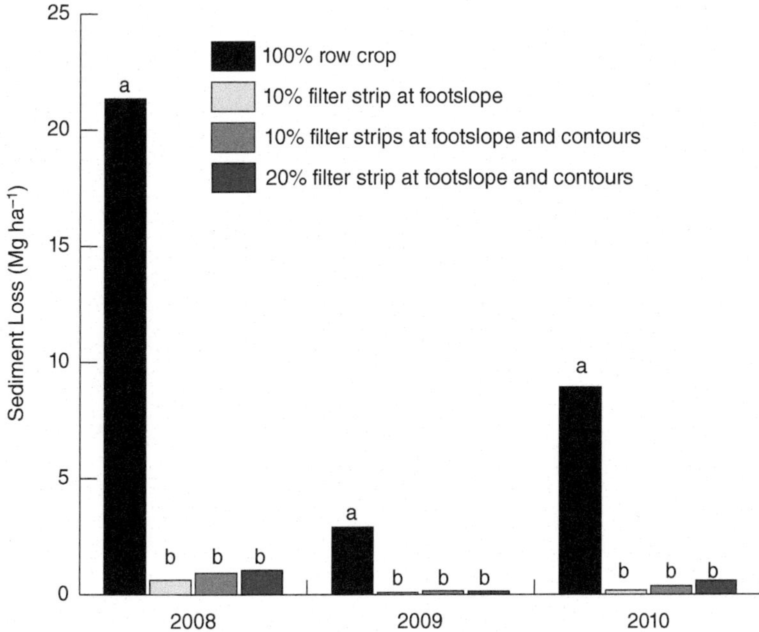

Fig. 12.4. Prairie filter strips effect on reducing sediment loss from no-till corn-soybean fields in Iowa (Helmers et al. 2012).

soils and higher than anticipated soil C degradation in Allophanic soils (Schipper et al. 2014). Soil C differences were not seen in soils under forages used for either dairy or drystock production. Increases in soil C occurred more frequently on sloped fields, compared to flat fields, due to reduced erosion on the sloped fields (Schipper et al. 2014).

Grazing intensity can also affect soil C concentrations. In the southeastern United States, surface soils (0 to 15 cm) frequently trafficked by **grazers** contained 1.3-fold more soil C than soil under non-harvested forages. The frequently trafficked areas also had 1.6-fold more soil C than hayed forages (Franzluebbers and Stuedemann 2009). This trend was reversed in southern Brazil. Compared to an area grazed to a sward height of 10 cm, a nongrazed area had 1.1-fold higher soil carbon stocks at a depth of 30 cm (Carvalho et al. 2010b). At other research sites in the southeastern US, lower grazing intensity increased particulate organic carbon fractions 1.2-fold compared with higher grazing intensities as a greater amount of plant matter remained in the field (Silveira et al. 2013). The root mass available to potentially contribute to soil C can be increased by grazing (Russell and Bisinger 2015) or by maintaining different forages with a variety of rooting depths (McNally et al. 2015).

Surface soil C, such as collected from 0 to 15 cm, tends to increase under permanent forage compared to other cropping systems. Holding other landscape variables constant, these soil C differences between forage and continuous field crop systems is largest when comparing two treatments with a larger range in soil disruption, such as was seen in Table 12.5 for soils collected from 0 to 10 cm. Site-specific factors can change the magnitude of the soil C benefit from forages. For example, soil C changes in integrated crop-livestock systems in Brazil depended on several factors including crops grown, climatic conditions, condition of land transitioned, and amount of time in the management system (Carvalho et al. 2010a; Salton et al. 2014).

Soil Aggregation

Soil aggregation increases under permanent forage when compared to forage in a cropping rotation or continuous crop production system (Figure 12.6) (Harris et al. 1966). The **soil aggregates** formed under forages tend to be larger in size, such as greater than 1 or 2 mm in diameter (Jokela et al. 2011, Angers 1992, Wilson et al. 1948). These larger aggregates are also important sites for increasing soil C (Grandy and Robertson 2007). Seasonal variability in soil aggregation occurs (Bach and Hofmockel 2016; Rasiah and Kay 1994; Perfect et al. 1990). Soil aggregate sizes are reduced when a soil under permanent forage is converted to continuous crop production (van Bavel and Schaller 1951). The difference in

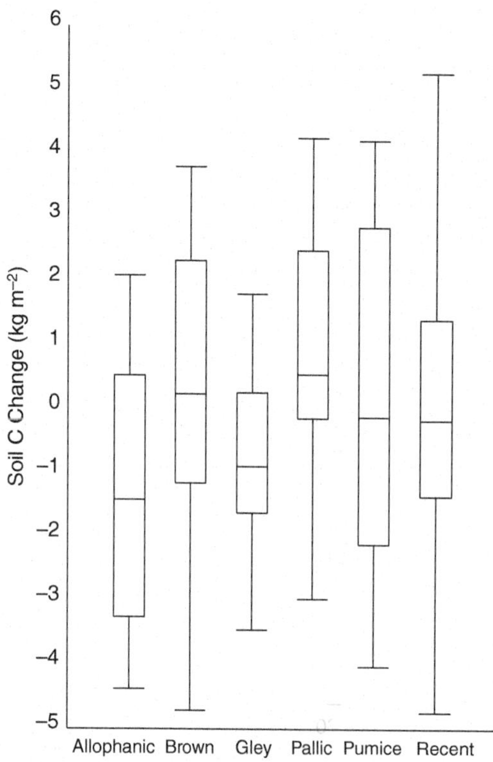

Fig. 12.5. Changes in soil C by soil order for long-term forage systems in New Zealand for 0 to 30-cm samples (Schipper et al. 2014). Figure generated using supplemental data provided in article. Sample size from left to right: 32, 27, 25, 32, 15, 19.

soil aggregate sizes between cultivated and non-cultivated soils can depend on the method used to evaluate soil aggregates. Differences are most pronounced in a test simulating slaking (Elliott 1986).

Across the United States, a 2014 survey of rangeland soils documented a range of aggregate stabilities when evaluated using an in-field soil aggregate stability slake test (Figure 12.7) (USDA NRCS 2014). Soils with a rating of 4 or less were unstable in water. Figure 12.7 shows aggregate stability for rangeland soils plotted by the number of rangeland acres reported in the 2012 National Resources Inventory (USDA 2015). Soils in Arizona, Nevada, and New Mexico were the least stable. These three states also had the highest percentages of bare ground, with approximately 38%, 29%, and 26%, respectively (USDA NRCS 2014). Rangeland practices that promote forage cover and reduce bare ground, such as avoiding **overgrazing**, may help reduce the erosion potential of these soils.

Forage species can impact soil aggregation. Changes in soil wetting and drying conditions caused by smooth bromegrass growth reduced soil aggregate stability compared with a control soil without bromegrass (Caron et al. 1992). Forage species with greater root mass can increase soil aggregation (>0.5 mm, <2.0 mm) (Stone and Buttery 1989). In an 80-day laboratory experiment using Brookston clay loam soil, soil wet aggregate stability ranged from 34 to 45% across nine forage species. Higher wet aggregate stability occurred in reed canarygrass (45%) and produced 17.8 g of root dry weight (Stone and Buttery 1988). Water stable aggregates (>0.25 mm) in an 84-day experiment using coastal plain Georgia soils tended to be similar to control soils under field pH conditions for **warm-** and **cool-season plants** (Karki and Goodman 2011).

Water Infiltration

Forages increase water infiltration into the soil by increasing the amount of roots in the soil and providing canopy interception for rainfall (Angers and Caron 1998; Meek et al. 1989 1992, Miller et al. 1963). The magnitude of these changes can be affected by several factors, including soil type, number of years following management changes, and yearly management activities.

Following four growing seasons of alfalfa, water infiltration rate was higher in no traffic and minimal traffic areas compared to alfalfa areas that received more frequent traffic (Meek et al. 1989). Compared to baseline conditions, increases in infiltration rates were 2.6-fold for no traffic, 2.2-fold for preplant traffic, 1.6-fold for traffic reflecting current grower practices, and 1.2-fold for repeated traffic activities (Meek et al. 1989). In orchard soils, cumulative infiltration was reduced by approximately one-half between traveled and non-traveled areas, and larger infiltration rates occurred in soils with cover crops than in those without (Miller et al. 1963). Site-specific preferential flow can also affect infiltration under forages (Harman et al. 2011).

Conversely, cover crops grown as part of a corn-soybean rotation did not have consistent water infiltration after three years (Kaspar et al. 2001). Instead, wheel traffic affected infiltration to a greater extent than the presence of oat or rye. Averaged across all three years, areas without traffic allowed 1.85-fold higher infiltration than tracked areas ($20.3\,g\,m^{-2}\,s^{-1}$ vs. $10.9\,g\,m^{-2}\,s^{-1}$) (Kaspar et al. 2001). The potential soil structural benefits from growing cover crops for one year may be masked by the machinery traffic required to manage the cover crops (Rücknagel et al. 2016).

Animal hoof action is another form of site traffic that can compact soils and reduce water infiltration rates (Russell and Bisinger 2015; Cuttle 2008). Bare soils are prone to being compacted compared to soils with at least 10 cm sward height, and these compacted soils will have

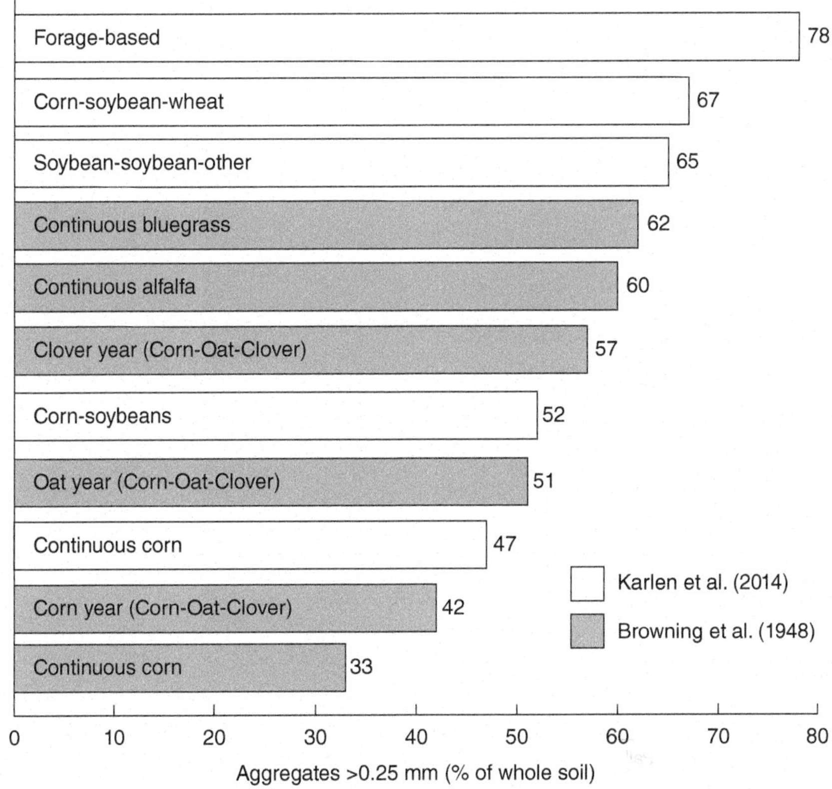

Fig. 12.6. Aggregate stability data for several cropping systems collected from Iowa (Browning et al. 1948, Karlen et al. 2014), Indiana, Missouri, and Ohio (Karlen et al. 2014). *Source:* Data from Karlen et al. (2014) for 0 to 5-cm samples.

lower water infiltration rates (Russell and Bisinger 2015). Grazed soils with sward heights of 10 cm or greater did not have greater phosphorus (P) loss or runoff volume compared to nongrazed areas (Russell and Bisinger 2015). Walking cattle can exert pressures on the soil potentially three times higher than an unloaded tractor. The greatest amounts of soil compaction occur close to the soil surface due to the smaller surface area of hooves compared to tractor tires. Depending on pasture size, length of grazing, and pasture location, each unit of soil could be walked on 10 times during a 140-day grazing season (Russell and Bisinger 2015).

Soil Microbial and Biological Activity

Forages increase soil microbial activity and abundance. Long-term pasture soils had higher values for several soil microbial indicators, particularly for soils collected from 0 to 2.5-cm depth (Haynes 1999). Compared to long-term row-cropped land, long-term pasture had an estimated two-fold higher organic C, 1.5-fold higher microbial quotient, 1.7-fold higher metabolic quotient, 3.5-fold higher fluorescein diacetate (FDA) hydrolytic activity, and five-fold higher acid phosphate activity (Haynes 1999).

Similar microbial differences occurred between long-term corn and pasture plots (Veum et al. 2014) and between vegetable crops and pasture (Bandick and Dick 1999). Timothy plots had 2.5-fold and 7.9-fold higher dehydrogenase activity than wheat and corn plots, respectively. Phenol oxidase activity under timothy was 2.7-fold higher than wheat and two-fold higher than corn (Veum et al. 2014). Pasture soils (0 to 20-cm) had between 1.7- and 3.6-fold higher FDA hydrolysis compared to vegetable crop rotations in western Oregon (Bandick and Dick 1999). Permanent fescue had higher enzyme activities than winter fallow plots in western Oregon for β-galactosidase (1.3-fold higher), amidase (1.4-fold), deaminase (1.04-fold), invertase (1.3-fold), and urease (1.4-fold). At the same research location, cover crops included in the vegetable rotation had similar enzyme activities as the fescue plots (Bandick and Dick 1999).

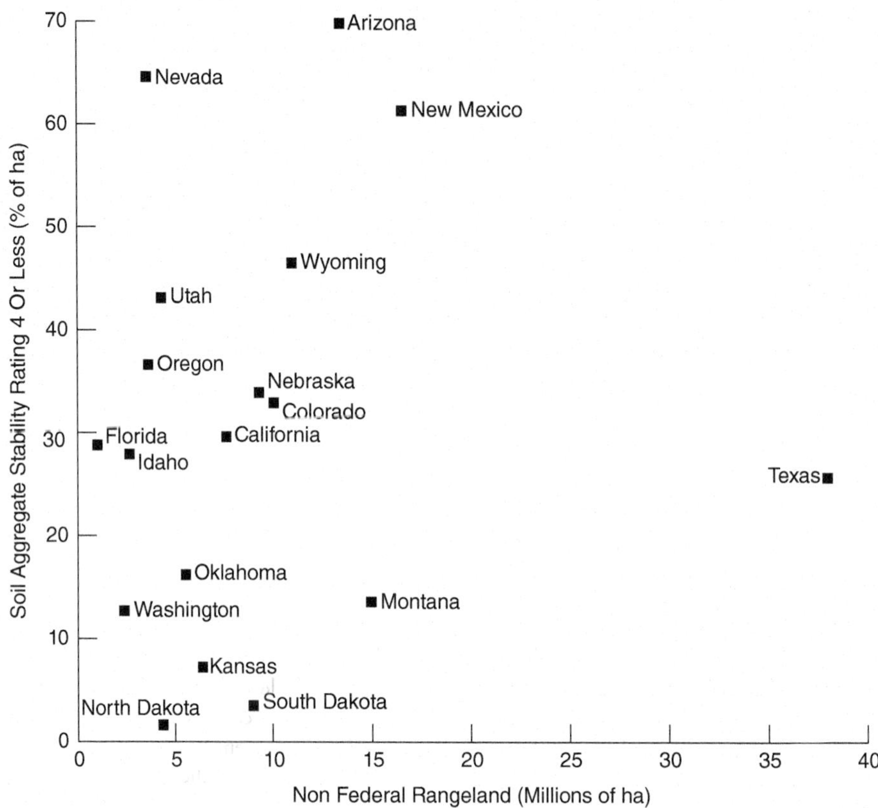

Fig. 12.7. Soil aggregate stability ratings for rangeland soils in the United States. Note: A rating of four or less indicates at most 25% of soil remains on the sieve following aggregate stability test. *Source:* USDA National Resources Inventory (2014) Rangeland Resource Assessment.

Forage species influence soil fauna differences. Perennial ryegrass tended to support more herbivorous invertebrates compared to chicory, red clover, and white clover. These invertebrates included greater abundances of herbivorous nematodes and certain Collembola superfamilies. Higher abundances occurred in perennial ryegrass fertilized with 200 kg N ha^{-1} yr^{-1} compared to ryegrass plots receiving 80 kg N ha^{-1} yr^{-1}. The other three forages supported higher populations of decomposer invertebrates compared to perennial ryegrass (Crotty et al. 2015). White clover had between 1.7- and 2.2-fold higher earthworm abundance compared to the other forages. Total nematode populations were consistent across the forages (Crotty et al. 2015).

Quantifying interactions among soil, forage roots, soil enzymes, and soil organisms requires substantial research to properly evaluate direct and indirect effects (Crotty et al. 2015). For example, soil pH can influence soluble P cycling, as lower pH forage soils tend to have lower activities of phosphodiesterase and higher labile organic P concentrations (Turner and Haygarth 2005). In lower pH soils, fungi tend to occur in greater abundance than bacteria. Subtle linkages between soil microbial activity and nutrient availability could affect forage soil management. Additional opportunities exist for increased understanding of and linking soil biology and plant species to achieve disease-suppressive pasture systems (Dignam et al. 2016).

Three years of forage growth followed by spring wheat and winter barley growth indicated soil fungal community effects could be detected during the cereal crop growth (Detheridge et al. 2016). The anti-fungal isothiocyanates released by forage radish during decomposition did not reduce arbuscular mycorrhizal fungi colonization in maize roots (White and Weil 2010). Compared to no cover crop, cereal rye increased AMF colonization 12–16% in three of six site-years when measured at V4 corn growth stage, with no cover crop effect found at V8 (White and Weil 2010).

Nutrient Cycling

Long-term crop rotations that include forages, such as alfalfa, are well known as a method for maintaining and increasing crop yields while reducing external N inputs (Osterholz et al. 2017; Ross et al. 2008; Page and Willard 1947). Cover crops, such as forage radish and winter pea may help cycle nitrogen (N) and P for subsequent crops (White and Weil 2011; Jahanzad et al. 2016). Cereal rye, a widely planted cover crop, might not cycle agronomically important amounts of N for future crops, but it often does reduce soil N loss (Jahanzad et al. 2016; Pantoja et al. 2016).

Nitrification rates of forages vary by species and **cultivar** (Bowatte et al. 2015). Forage species and cultivars with a low rate of soil nitrification may help reduce soil N losses, and in combination with high **biomass** production could be useful management tools (Bowatte et al. 2015). An analysis of 126 cultivars and 26 species used in temperate grasslands indicated variation between and within cultivars. These findings highlighted the potential use of additional breeding selection for specific characteristics, in addition to improving nitrification test methodology (Bowatte et al. 2015).

Forage grazing introduces N, P, and K to the soil through animal waste, as approximately 70–90% of nutrients will be recycled by animals and not removed from the field (Haynes and Williams 1993). Nutrient loss from the soil may occur due to an imbalance in plant uptake and an excess of animal supplied fertilizer. Across sixteen grasses common to New Zealand, Moir et al. (2012) reported a range of N leaching losses from applied dairy cow urine. Perennial ryegrass, tall fescue, and kentucky bluegrass had lower plant N uptake and increased soil water N leaching (Moir et al. 2012). Pairing perennial ryegrass with white clover in pasture systems, along with nitrification inhibitors and fungal **inoculation** may help reduce N loss from perennial ryegrass systems and reduce the need for N fertilizer applications (Andrews et al. 2011).

Combined Soil Health Effects of Forages

While soil physical, chemical, and biologic effects from forages can be measured and interpreted separately, forage systems can systematically change all of these properties (Jokela et al. 2011) (Table 12.6). **Soil quality** or **soil health** assessments measure multiple soil physical, chemical, and biologic properties and aggregate these values to create a system-level assessment of soil function. Several different methods can be used for this process, including the soil management assessment framework (SMAF) (Andrews et al. 2004), the soil conditioning index (SCI) (Zobeck et al. 2007), the comprehensive assessment of soil health (CASH) (Moebius-Clune et al. 2016), and

Table 12.6 Soil physical, chemical, and biologic properties between pasture and other cropping systems after 18 years from Arlington, WI

		Pasture		Other cropping systems	
Sampling depth (cm)	Units	0–5	5–20	0–5	5–20
pH		6.20	6.50	6.66	6.63
P	$mg\,kg^{-1}$	49.8	36.0	54.8	42.1
K	$mg\,kg^{-1}$	163	75	173	90
Total organic C	$g\,kg^{-1}$	33.6[a]	22.4	24.5	21.7
Total N	$g\,kg^{-1}$	3.40[a]	2.10	2.36	2.14
Active C	$mg\,kg^{-1}$	2350[a]	1380	1850	1640[a]
Potentially mineralizable N	$mg\,kg^{-1}$	61.2[a]	24.4	34.5	24.4
Total microbial biomass	$nmol\,kg^{-1}$	523[a]	193	194	131
Bulk density	$kg\,cm^{-3}$	1.24	1.42	1.21	1.41
Water content	$kg\,kg^{-1}$	0.29[a]	0.23	0.27	0.26
Water stable aggregates (2–8 mm)	$g\,kg^{-1}$	649[a]	639[a]	332	443
Water stable aggregates (0.25–2 mm)	$g\,kg^{-1}$	240	234	439	413[a]

Source: Jokela et al. (2011).

[a]Noted statistical significance for difference between pasture and other cropping systems at a given depth. Symbol next to higher value. See Jokela et al. (2011) for details.

several other tests including those offered by commercial laboratories.

Generally, these assessment methods yield similar results. High-functioning soils are indicated as such across the methods used, though there can be variation in the magnitude of differences found among calculation methods (Congreves et al. 2015; Karlen et al. 2017; Zobeck et al. 2008). Soil quality index calculation methods for cropping systems in Brazil indicated soils under pasture tended to have an intermediate level of soil functioning below the highest functioning native vegetation and the lowest functioning sugarcane production. Several calculation methods also indicated pasture and sugarcane production had similar functioning soils (Cherubin et al. 2016). The difference in results occurred depending on which soil properties were included in a given index calculation and how values were added together (Cherubin et al. 2016).

Forages in rotations with reduced tillage or in permanent pasture tend to have higher soil quality index values compared to other systems (Karlen et al. 2014; Veum et al. 2014, Zobeck et al. 2007). Perennial timothy plots receiving manure applications had approximately 1.5-fold higher SMAF scores than moldboard plow corn receiving no fertilizers. Timothy plots without manure had 1.4-fold higher SMAF scores than the same corn treatment (Veum et al. 2014). Forage-based rotations throughout sites in the Midwest US had 1.1-fold higher soil quality index scores compared to continuous corn sites (Karlen et al. 2014).

Increased soil quality index [illegible] do not always occur, however, especially if all comparison sites are inherently highly productive (Jokela et al. 2011). For example, a comprehensive evaluation of several cropping systems in Wisconsin identified differences in soil physical, chemical, and biologic properties. Pasture soils had larger >2 mm water stable aggregates, higher C and N measurements, and similar P and K soil concentrations. Other cropping systems had higher water stable aggregates <2 mm, potentially due to more frequent soil disruption that reduced larger aggregate formation. The other cropping systems tested included continuous corn, grain rotations, forage systems, and a grass-legume pasture (Jokela et al. 2011). Most of the differences found between pasture soils and the other cropping systems occurred in the 0 to 5-cm depth range, though the aggregate stability difference was also seen at samples collected from 5 to 20-cm.

Saline Seeps

Forages are effective for controlling the soil chemical management problem of saline seeps. Saline seeps occur in areas where water flows through the soil accumulating salts in the **soil solution** until it comes into contact with some form of impermeable layer and gradually moves back to the soil surface. This series of steps often occurs in downslope soil gradients. Then, as water evaporates from the soil surface, excess salts are left behind in the soil rooting zone (Halvorson and Richardson 2011; Miller et al. 1981). These soils can be classified as saline soils if the electrical conductivity is greater than 4 dS m^{-1}, but soil electrical conductivity levels do not need to reach 4 dS m^{-1} for yields to be reduced (Steppuhn et al. 2005; Maas and Grattan 1999).

Forages can help manage saline seeps by lowering the water table through increased plant transpiration. One well-known forage for managing saline seeps is alfalfa because the species has a deep rooting system and can use over 600 mm of soil water after several years of growth (Halvorson and Richardson 2011). Other potential forages include tall wheatgrass, slender wheatgrass, altai wildrye, and tall fescue (Franzen 2013; McCauley and Jones 2005). This water use lowers soil water content throughout the soil surface profile. Forages grown at soil water recharge and discharge areas would be most beneficial to reduce the water table across the entire area of the seep's formation. Rainfall will further help flush surface salts away from the surface in the location of the saline seep.

Plant breeding continues to develop more salt-tolerant crops that are well suited to managing saline seeps. Some alfalfa varieties were suitable for growing in 15.6 dS m^{-1} soils, the highest tested rate (Steppuhn et al. 2012). Beyond an individual saline seep, maintaining permanent vegetative cover on agricultural soils can help reduce saline soil risks at a landscape scale (Wiebe et al. 2007).

Selecting Forages for a Management Goal

Forages can serve many purposes, each of which modifies soils to a different extent. For example, forages can be grazed by livestock or harvested and fed to livestock. The plant material has the same end use, but the type, frequency, and other harvest factors affecting the soil resource are very different. Forages can be used as cover crops or incorporated into a crop rotation. For restoring saline soils, forages can be used to modify water dynamics and reduce soil salinity over time. Forages can also be used to restore cover to landscapes as part of ecologic restoration. When placed within a cropping system as a conservation buffer, forages help reduce sediment loss. Furthermore, the specific forage species that will be best suited to address a specific management goal will differ. Two examples illustrating the selection of forages to address specific management goals and the resultant forage-soil interactions are described below.

Cover Crops

Farmers report several motivations for using cover crops to improve soil physical, chemical, and biologic properties (Table 12.7). This includes their ability to improve soil health through changes in various soil carbon and

Table 12.7 Farmer reported motivations for using cover crops

Motivating factors	% of respondents
Increase overall soil health	87
Increase soil organic matter	83
Reduce soil compaction	76
Reduce soil erosion	76
Scavenge nitrogen	59
Provide another nitrogen source	57
Choose diverse rooting systems	51

Source: From SARE/CTIC 2015–2016 Cover Crop Survey, p. 15, (SARE 2017a).

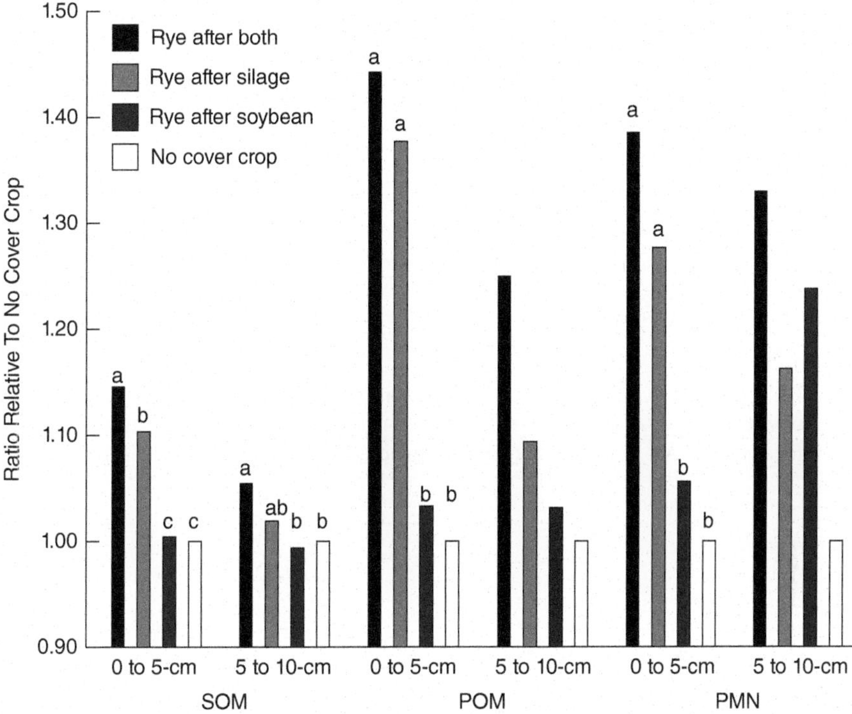

Fig. 12.8. Cover crop effects on soil organic matter (SOM), particulate organic matter (POM), and potentially mineralizable nitrogen (PMN) in a corn silage system measured after 10 years (Moore et al. 2014). Note: Letters indicate statistically significant differences within a given depth and soil property.

nitrogen fractions (Figure 12.8). Forages can also be incorporated into prairie strips which have been shown to reduce runoff and soil erosion (Schulte-Moore et al. 2020). As part of a crop rotation, long-term data evaluating winter rye as a cover crop indicated no reduction in subsequent crop yields (Basche et al. 2016). However, there were significant soil changes that included a 10 to 11% increase in field capacity soil water content and a 21–22% increase in plant available water (Basche et al. 2016). As reported by cover crop survey participants, the most widely planted cover crop is cereal rye, followed by forage radish (Table 12.8).

In recent years, several resources have been developed to help producers select appropriate forage species for a variety of management goals including their use as cover crops. Both state- and regionally-specific decision tools

Table 12.8 Farmer reported cover crops used in the United States

Crop	2015 Planted	2016 Projected
	Hectares	
cereal rye	75 726	88 082
radish	39 589	49 060
winter wheat	33 408	30 588
rapeseed	23 235	24 842
annual ryegrass	20 256	26 471
oat	20 227	25 670
crimson clover	19 123	21 005

Source: From SARE/CTIC 2015-2016 Cover Crop Survey, p. 21-27 (SARE 2017a).

have been developed for the Midwest (MCCC 2017), Pacific Northwest (USDA NRCS 2016), Pennsylvania (Penn State n.d.), and New York (Cornell University 2010). Information resources such as factsheets and other online Extension materials are also available from individual states or nationwide (SARE 2017b). An example of the selection process for choosing cover crops to meet specific soil and forage goals is presented in Table 12.9. Using the Midwest Cover Crops Council decision tool, different cover crops were suggested for goals such as soil building, forage harvest, or grazing. The specific forages recommended demonstrate the range of plant species available for providing forage and improving soil conservation.

Cover crops can also be incorporated as permanent ground cover between crop rows where they function as a living mulch. This practice is more common for fruit and tree crops, though corn was successfully intercropped

Table 12.9 Midwest Cover Crops Council decision-tool selections for cover crops in row crop production combining goals of soil builder, forage harvest value, and good grazing[a]

Cover crop[b]	Locations[c]
Alfalfa	IL, MI, MN, OH, ON
Annual ryegrass	IL, MO, OH
Barley	KS
Berseem clover	IN, OH
Crimson clover	IL, IN, MI, OH
Oat	IL, ON
Pearl millet	IL, IN, KS, MO, OH, WI
Red clover	IL, IN, MI, OH, ON, WI
Rye	IA, KS, MI, MN, MO, ON, WI
Sorghum-sudangrass	IA, IN, KS, MN, OH, ON, WI
Sudangrass	IA, IL, IN, MI, MN, OH, WI
Triticale	IA, KS, MO
Winter wheat	IA, KS, MO

Source: Midwest Cover Crops Council Row Crop Decision Tool (MCCC 2017).

[a]Generated for each state using all counties average.

[b]Listed cover crops are those that received "excellent" ratings in all three goals or "excellent" ratings in at least two of three goals and "very good" ratings in one of three goals. These are not the only cover crops that will meet producers' needs in these locations.

[c]Iowa (IA), Illinois (IL), Indiana (IN), Kansas (KS), Michigan (MI), Minnesota (MN), Missouri (MO), Ohio (OH), Ontario (ON), Wisconsin (WI).

with kura clover in Wisconsin (Ziyomo et al. 2013). Having sufficient plant available soil water, however, can be a major management challenge and result in significant crop yield loss if the cover crop and cash crop are both growing simultaneously (Qi et al. 2011; Ziyomo et al. 2013).

Water-induced yield losses might not occur every year (Bartel et al. 2017) and can also be reduced if some portion of the living mulch is terminated (Kumwenda et al. 1993). Planting drought-resistant crops into living mulches can overcome these potential yield losses. Drought-resistant corn hybrids had approximately three-fold higher yields when grown in living or killed kura clover mulch treatments compared to drought-susceptible corn varieties (Ziyomo et al. 2013).

From a soil-health perspective, fields with living mulches may support higher soil fungal populations and often have higher litter decomposition rates compared to fields without living mulches (Nakamoto and Tsukamoto 2006). Living mulches may also allow higher rates of corn **stover** removal for bioenergy and bio-product **feedstock** due to increased soil cover from the permanent groundcover (Bartel et al. 2017).

Site Restoration

When restoring an entire site, such as occurs in mineland reclamation, forages must survive in a wide range of soil conditions. First, the soil itself must be deep enough to establish plants. Plant production can depend on soil depth of the restoration site and the type of underlying material being remediated. For example, greater soil replacement, such as up to 1.5 m of new material, can increase plant growth over acid spoil materials and sodic materials, but may have less of an impact on other spoil materials (Barth and Martin 1984). Fast-establishing plant species can also help reduce competition by other non-planted species (Wick et al. 2011). Numerous forage species are suitable for mineland reclamation because they can survive a range of soil conditions. The examples, presented in Table 12.10 include species that have the potential for mineland reclamation and a forage source.

Forages provide multiple benefits in site restoration. Potential contaminant migration from the site can be reduced with forages that limit surface erosion and water runoff (Pilon-Smits 2005). Forages can also stimulate soil microbial degradation of contaminants, including fescue and ryegrass (Pilon-Smits 2005). Grazing forages grown in site restoration locations, such as from saline soils, requires evaluating plant uptake that could harm animal health (Masters et al. 2007). The forages should be tested and compared to recommended tolerance ranges to ensure animal health (National Research Council 2005).

Landscape Effects of Forages on Soil

The magnitude of forage effects on larger landscape processes is site specific. The cover management factor (C-factor) used in soil erosion calculations across Europe could be reduced using forages as cover crops (Panagos et al. 2015). This reduced factor would lead to reduced soil erosion estimates from a given location. Cover crops could reduce this C-factor more than 8% in some locations, and less than 0.5% in others (Panagos et al. 2015). For comparison, changing tillage practices could reduce the C-factor up to 30% in some locations and less than 5% in others (Panagos et al. 2015). These modeled data are useful for demonstrating how forages can be helpful in reducing landscape level soil erosion, and how forages are one piece of larger management changes that can help reduce erosion.

Forages in conservation buffers were shown to be effective in reducing ephemeral gully formation (Helmers et al. 2012). Ephemeral gully erosion is implicated in increasing the sedimentation of reservoirs (Fox et al. 2016). However, larger watershed level impacts of conservation buffers can vary by watershed characteristics. Stream sediments can come from several sources including channel and bank erosion, as well as previously eroded sediments that become re-suspended (legacy sources) (Fox et al. 2016; Tomer and Locke 2011).

Increased nutrient and sediment losses in the Lincoln Lake (Arkansas and Oklahoma) watershed evaluated between 1992 and 2004, resulted from a combination of increased urbanization and increased N applications to pasture areas (Chiang et al. 2010). Variations occurred among the three sub-watersheds evaluated. In two of the three sub-watersheds, land use change increased total N losses to a greater extent than pasture management (3.8–5.1 kg N ha^{-1} vs 1–1.7 kg N ha^{-1}). In the other sub-watershed, pasture management had a higher impact on N losses compared to urbanization (4.3 kg N ha^{-1} vs 0.8 kg N ha^{-1}). In this sub-watershed, N applications to pasture increased from 147 kg N ha^{-1} to 332 kg N ha^{-1} between 1992 and 2004 (Chiang et al. 2010).

Nutrient models for this watershed estimated that between 3% and 39% increases in N losses would occur if excess nutrients were available in combination with minimal plant growth. This situation occurred if **overgrazing** and summer or fall applications of poultry litter were used. The range in estimated N losses was influenced by weather variability (Chaubey et al. 2010). These landscape level effects highlight the importance of combining forages with comprehensive soil conservation management.

Sound Management Maximizes Soil Conservation Benefits

Soil conservation benefits from forages are generally determined by site-specific management practices. The need

Table 12.10 Examples of plants suitable for both mineland reclamation and forage potential

Common name	Type	United States region	Note	Suitable **soil textures**	Forage
Alkali sacaton	Perennial, **warm-season**, bunchgrass	Arid, semi-arid Western	Tolerates saline soils	Light to heavy	Good source
Great basin wildrye	Perennial, **cool-season**, bunchgrass	Western	Tolerates acidic and heavy metals	Loamy to sandy	Provides winter forage, avoid heavy grazing during active growth
Black grama	Warm-season, stoloniferous	Southwest		Dry-gravelly, sandy, sandy-loam	Sustained forage production
Coastal panicgrass	Perennial, Warm-season, bunchgrass	Mid-Atlantic coastal plain, and piedmont regions	Salt-tolerant	Coarse to fine, including coastal dunes	Fair palatable graze
Florida paspalum	Perennial, warm-season, bunchgrass	Southeast United States	Facultative wetland plant	Coarse to moderately fine	Palatable, forage value highest for earlier growth stages, not recommended as grazing monoculture
Orchardgrass	Perennial, cool-season, bunchgrass	Nationwide	Not adapted for saline or high-water table soils, invasive potential	Clay to gravelly loams	Cool season forage
Sideoats grama	Perennial, warm-season, bunchgrass	Nationwide	Moderately tolerable of saline soils	Coarse to fine	Plants need 12 months development before grazing

Source: Adapted from USDA NRCS Plant Materials Program Conservation Plant Releases (USDA NRCS 2017).

for additional quantitative information on how soil management practices affect a range of soil types remains a long-recognized issue (Pierre 1946). For example, forages tend to increase aggregation, particularly for larger soil aggregates (>1 or >2 mm). Forage effects on soil properties within grazing management systems are affected by stocking rates, soil moisture content, crop cover, and landscape position (Russell and Bisinger 2015). Soil conservation benefits may be seen at the soil surface, such as from 0 to 5-cm, and may extend down the soil profile, as occurred in alfalfa water use in saline seep areas.

Theoretically, the forage effect on soil properties depends on which forages are grown, site management (planting, harvesting, etc.), and how forages are utilized (cover crop, crop rotation, animal forage, etc). This chapter has highlighted several ways forages affect soil properties. Well-informed site-specific management can ensure forages help maintain and build a stable soil resource. For example, as noted by Pierre in 1946, there was a "relatively new forage crop combination" available in the Midwest US that included smooth bromegrass and alfalfa. This combination produced quality forage, and had "high value in soil improvement and conservation." However, Pierre noted, "whether or not (the forage crop combination) makes the contribution to soil improvement in the Corn Belt that it can, however, will probably depend largely on its management" (Pierre 1946, p. 5). This management included incorporating the forage into a crop rotation rather than planting it permanently, separate from other crops. Seventy years later, this recommendation remains the same.

References

Andrews, S.S., Karlen, D.L., and Cambardella, C.A. (2004). The soil management assessment framework: a quantitative soil quality evaluation method. *Soil Sci. Soc. Am. J.* 68: 1945–1962.

Andrews, M., Edwards, G.R., Ridgway, H.J. et al. (2011). Positive plant microbial interactions in perennial ryegrass dairy pasture systems. *Ann. Appl. Bio.* 159: 79–92.

Angers, D.A. (1992). Changes in soil aggregation and organic carbon under corn and alfalfa. *Soil Sci. Soc. Am. J.* 56: 1244–1249.

Angers, D.A. and Caron, J. (1998). Plant-induced changes in soil structure: processes and feedbacks. *Biogeochemistry* 42: 55–72.

Bach, E.M. and Hofmockel, K.S. (2016). A time for every season: soil aggregate turnover stimulates decomposition and reduces carbon loss in grasslands managed for bioenergy. *GCB Bioenergy* 8: 588–599.

Bandick, A.K. and Dick, R.P. (1999). Field management effects on soil enzyme activities. *Soil Biol. Biochem.* 31: 1471–1479.

Bartel, C.A., Banik, C., Lenssen, A.W. et al. (2017). Establishment of perennial groundcovers for maize-based bioenergy production systems. *Agron. J.* 109: 822–835.

Barth, R.C. and Martin, B.K. (1984). Soil depth requirements for revegetation of surface-mined areas in Wyoming, Montana, and North Dakota. *J. Environ. Qual.* 13 (3): 399–404.

Basche, A.D., Kaspar, T.C., Archontoulis, S.V. et al. (2016). Soil water improvements with the long-term use of a winter rye cover crop. *Agr. Water Manage* 172: 40–50.

van Bavel, C.H.H. and Schaller, F.W. (1951). Soil aggregation, organic matter, and yields in a long-time experiment as affected by crop management. *Soil Sci. Soc. Am. J.* 15: 399–404.

Bennett, H.H. and Chapline, W.R. (1928). *Soil Erosion a National Menace*. USDA Circular No. 33. Washington: United States Government Printing Office.

Bowatte, S., Newton, P.C.D., Hoogendoorn, C.J. et al. (2015). Wide variation in nitrification activity in soil associated with different forage plant cultivars and genotypes. *Grass Forage Sci.* 71: 160–171.

Browning, G.M. (1951). Forages and soil conservation. In: *Forages: The Science of Grassland Agriculture* (eds. H.D. Hughes, M.E. Heath and D.S. Metcalfe), 38–53. Ames, Iowa: The Iowa State College Press.

Browning, G.M., Norton, R.A., McCall, A.G., and Bell, F.G. (1948). *Investigation in Erosion Control and the Reclamation of Eroded Land at the Missouri Valley Loess Conservation Experiment Station, Clarinda, Iowa, 1931-1942*. Washington, D.C.: USDA Soil Conservation Service USDA Technical Bulletin No. 959. Available online: https://naldc.nal.usda.gov/naldc/download.xhtml?id=CAT86200951&content=PDF.

Caron, J., Kay, B.D., and Perfect, E. (1992). Short-term decrease in soil structural stability following bromegrass establishment on a clay loam soil. *Plant Soil* 145: 121–130.

Carvalho, J.L.N., Raucci, G.S., Cerri, C.E.P. et al. (2010a). Impact of pasture, agriculture and crop-livestock systems in soil C stocks in Brazil. *Soil Till. Res.* 110: 175–186.

Carvalho, P.C.F., Anghinoni, I., Moraes, A. et al. (2010b). Managing grazing animals to achieve nutrient cycling and soil improvement in no-till integrated systems. *Nutr. Cycl. Agroecosyst.* 88: 259–273.

Causarano, H.J., Franzluebbers, A.J., Shaw, J.N. et al. (2008). Soil organic carbon fractions and aggregation in the Southern Piedmont and Coastal Plain. *Soil Sci. Soc. Am. J.* 72 (1): 221–230.

Chan, K.Y., Conyers, M.K., Li, G.D. et al. (2011). Soil carbon dynamics under different cropping and pasture management in temperature Australia: results of three long-term experiments. *Soil Res.* 49: 320–328.

Chaubey, I., Chiang, L., Gitau, M.W., and Mohamed, S. (2010). Effectiveness of best management practices in improving water quality in a pasture-dominated watershed. *J. Soil Water Conserv.* 65 (6): 424–437.

Cherubin, M.R., Karlen, D.L., Cerri, C.E.P. et al. (2016). Soil quality indexing strategies for evaluating sugarcane expansion in Brazil. *PLoS One* 11 (3): e0150860. https://doi.org/10.1371/journal.pone.015086.

Chiang, L., Chaubey, I., Gitau, M.W., and Arnold, J.G. (2010). Differentiating impacts of land use changes from pasture management in a CEAP watershed using the SWAT model. *Trans. ASABE* 53 (5): 1569–1584.

Congreves, K.A., Hayes, A., Verhallen, E.A., and Van Eerd, L.L. (2015). Long-term impact of tillage and crop rotation on soil health at four temperate agroecosystems. *Soil Tillage Res.* 152: 17–28.

Cornell University (2010). Cover crops for vegetable growers. http://covercrops.cals.cornell.edu/decision-tool.php (accessed 9 October 2019).

Crotty, F.V., Fychan, R., Scullion, J. et al. (2015). Assessing the impact of agricultural forage crops on soil biodiversity and abundance. *Soil Biol. Biochem.* 91: 119–126.

Cuttle, S.P. (2008). Impacts of pastoral grazing on soil quality. In: *Environmental Impacts of Pasture-based Farming* (ed. R.W. McDowell), 33–74. Wallingford: CABI.

Detheridge, A.P., Brand, G., Fychan, R. et al. (2016). The legacy effect of cover crops on soil fungal populations in a cereal rotation. *Agr. Ecosyst. Environ.* 228: 49–61.

Dignam, B.E.A., O'Callaghan, M., Condron, L.M. et al. (2016). Challenges and opportunities in harnessing soil disease suppressiveness for sustainable pasture production. *Soil Biol. Biochem.* 95: 100–111.

Elliott, E.T. (1986). Aggregate structure and carbon, nitrogen, and phosphorus in native and cultivated soils. *Soil Sci. Soc. Am. J.* 50: 627–633.

FAO (2017a). Grasslands, rangelands, and forage crops. http://www.fao.org/agriculture/crops/thematic-sitemap/theme/spi/grasslands-rangelands-and-forage-crops/en/ (accessed 9 October 2019).

FAO (2017b). FAOSTAT. http://www.fao.org/faostat (accessed 9 October 2019).

Fox, G.A., Sheshukov, A., Cruse, R. et al. (2016). Reservoir sedimentation and upstream sediment sources: perspectives and future research needs on streambank and gully erosion. *Environ. Manage.* 57: 945–955.

Franzen, D. (2013). Managing saline soils in North Dakota. North Dakota State University Extension Publication SF1087. https://www.ag.ndsu.edu/publications/landing-pages/crops/managing-saline-soils-in-north-dakota-sf-1087 (accessed 9 October 2019).

Franzluebbers, A.J. and Stuedemann, J.A. (2009). Soil-profile organic carbon and total nitrogen during 12 years of pasture management in the Southern Piedmont USA. *Agr. Ecosyst. Environ.* 129: 28–36.

Grandy, A.S. and Robertson, G.P. (2007). Land-use intensity effects on soil organic carbon accumulation rates and mechanisms. *Ecosystems* 10: 58–73.

Haas, H.J., Evans, C.E., and Miles, E.F. (1957). *Nitrogen and Carbon Changes in Great Plains Soils as Influenced by Cropping and Soil Treatments*. USDA ARS Technical Bulletin 1164. Washington: U.S. Government Printing Office.

Halvorson, A.D. and Richardson, J.L. (2011). Management of dryland saline seeps. In: *Agricultural Salinity Assessment and Management*, 2e (eds. W.W. Wallender and K.K. Tanji), 561–589. Reston, VA: US: ASCE.

Harman, M.B., Thompson, J.A., Pena-Yewtukhiw, E.M. et al. (2011). Preferential flow in pastures on benchmark soils in West Virginia. *Soil Sci.* 176 (10): 509–519.

Harris, R.F., Chesters, G., and Allen, O.N. (1966). Dynamics of soil aggregation. *Adv. Agron.* 18: 107–169.

Haynes, R.J. (1999). Size and activity of the soil microbial biomass under grass and arable management. *Bio. Fertil. Soils* 30: 210–216.

Haynes, R.J. and Williams, P.H. (1993). Nutrient cycling and soil fertility in the grazed pasture ecosystem. *Adv. Agron.* 49: 119–199.

Helmers, M.J., Zhou, X., Asbjornsen, H. et al. (2012). Sediment removal by prairie filter strips in row-cropped ephemeral watersheds. *J. Environ. Qual.* 41: 1531–1539.

Jabro, J.D., Sainju, U., Stevens, W.B., and Evans, R.G. (2008). Carbon dioxide flux as affected by tillage and irrigation in soil converted from perennial forages to annual crops. *J. Environ. Manage.* 88: 1478–1484.

Jahanzad, E., Barker, A.V., Hashemi, M. et al. (2016). Nitrogen release dynamics and decomposition of buried and surface cover crop residues. *Agron. J.* 108: 1735–1741.

Jokela, W., Posner, J., Hedtcke, J. et al. (2011). Midwest cropping system effects on soil properties and on a soil quality index. *Agron. J.* 103: 1552–1562.

Karki, U. and Goodman, M. (2011). Short-term soil quality response to forage species and pH. *Grass Forage Sci.* 66: 290–299.

Karlen, D.L., Stott, D.E., Cambardella, C.A. et al. (2014). Surface soil quality in five midwestern cropland conservation effects assessment project watersheds. *J. Soil Water Conserv.* 69 (5): 393–401.

Karlen, D.L., Goeser, N.J., Veum, K.S., and Yost, M.A. (2017). On-farm soil health evaluations: challenges and opportunities. *J. Soil Water Conserv.* 72 (2): 26A–31A.

Kaspar, T.C., Radke, J.K., and Laflen, J.M. (2001). Small grain cover crops and wheel traffic effects on infiltration, runoff, and erosion. *J. Soil Water Conserv.* 56 (2): 160–164.

Kemper, W.D. and Koch, E.J. (1966). *Aggregate Stability of Soils from Western United States and Canada.* USDA ARS Technical Bulletin No. 1355. Washington: U.S. Government Printing Office.

Kumwenda, J.D.T., Radcliffe, D.E., Hargrove, W.L., and Bridges, D.C. (1993). Reseeding of crimson clover and corn grain yield in a living mulch system. *Soil Sci. Soc. Am. J.* 57: 517–523.

Maas, E.V. and Grattan, S.R. (1999). Crop yields as affected by salinity. In: *Agricultural Drainage*, Agronomy Monograph 38 (eds. R.W. Skaggs and J. van Schilfgaarde), 55–108. Madison, WI: ASA, CSSA, SSSA.

Masters, D.G., Benes, S.E., and Norman, H.C. (2007). Biosaline agriculture for forage and livestock production. *Agr. Ecosyst. Environ.* 119: 234–248.

McCauley, A. and Jones, C. (2005). Salinity & sodicity management. Land Resources and Environmental Studies. http://landresources.montana.edu/swm/documents/SW 2 updated.pdf (accessed 9 October 2019).

McCracken, R.J. (1959). *Certain Properties of Selected Southeastern United States Soils and Mineralogical Procedures for Their Study.* Southern Regional Bull. 61. Blacksburg: Virginia Agricultural Experiment Station, Virginia Polytechnic Institute.

McNally, S.R., Laughlin, D.C., Rutledge, S. et al. (2015). Root carbon inputs under moderately diverse sward and conventional ryegrass-clover pasture: implications for soil carbon sequestration. *Plant Soil* 392: 289–299.

Meek, B.D., Rechel, E.A., Carter, L.M., and DeTar, W.R. (1989). Changes in infiltration under alfalfa as influenced by time and wheel traffic. *Soil Sci. Soc. Am. J.* 53: 238–241.

Meek, B.D., Rechel, E.R., Carter, L.M. et al. (1992). Infiltration rate of a sandy loam soil: effects of traffic, tillage, and plant roots. *Soil Sci. Soc. Am. J.* 56: 908–913.

Midwest Cover Crops Council (MCCC) (2017). Midwest cover crops council – cover crop decision tool. http://mccc.msu.edu/covercroptool/covercroptool.php (accessed 9 October 2019).

Miller, D.E., Bunger, W.C., and Proebsting, E.L. Jr. (1963). Properties of soil in orchard as influenced by travel and cover crop management systems. *Agron. J.* 55 (2): 188–191.

Miller, M.R., Brown, P.L., Donovan, J.J. et al. (1981). Saline seep development and control in the North American Great Plains – hydrogeological aspects. *Agr. Water Manage.* 4: 115–141.

Moebius-Clune, B.N., Moebius-Clune, D.J., Gugino, B.K. et al. (2016). *Comprehensive Assessment of Soil Health – The Cornell Framework Manual*, 3e, 1. Geneva, NY: Cornell University http://www.css.cornell.edu/extension/soil-health/manual.pdf.

Moir, J.L., Edwards, G.R., and Berry, L.N. (2012). Nitrogen uptake and leaching loss of thirteen temperate grass species under high N loading. *Grass Forage Sci.* 68: 313–325.

Moore, E.B., Wiedenhoeft, M.H., Kaspar, T.C., and Cambardella, C.A. (2014). Rye cover crop effects on soil quality in no-till corn silage-soybean cropping systems. *Soil Sci. Soc. Am. J.* 78: 968–976.

Nakamoto, T. and Tsukamoto, M. (2006). Abundance and activity of soil organisms in fields of maize grown with a white clover living mulch. *Agr. Ecosyst. Environ.* 115: 34–42.

National Research Council (2005). *Mineral Tolerance of Animals*, 2e. Washington: National Academies Press.

Necpálová, M., Li, D., Lanigan, G. et al. (2013). Changes in soil organic carbon in a clay loam soil following ploughing and reseeding of permanent grassland under temperate moist climatic conditions. *Grass Forage Sci.* 69: 611–624.

Osterholz, W.R., Rinot, O., Liebman, M., and Castellano, M.J. (2017). Can mineralization of soil organic nitrogen meet maize nitrogen demand? *Plant Soil* https://doi.org/10.1007/s11104-016-3137-1.

Page, J.B. and Willard, C.J. (1947). Cropping systems and soil properties. *Soil Sci. Soc. Am. J.* 11: 81–88.

Panagos, P., Borrelli, P., Meusburger, K. et al. (2015). Estimating the soil erosion cover-management factor at the European scale. *Land Use Policy* 48: 38–50.

Pantoja, J.L., Woli, K.P., Sawyer, J.E., and Barker, D.W. (2016). Winter rye cover crop biomass production, degradation, and nitrogen recycling. *Agron. J.* 108: 841–853.

Penn State (n.d.). Forage selection tool. http://www.forages.psu.edu/selection_tool/index.html (accessed 9 October 2019).

Perfect, E., Kay, B.D., van Loon, W.K.P. et al. (1990). Rates of change in soil structural stability under forages and corn. *Soil Sci. Soc. Am. J.* 54: 179–186.

Pierre, W.H. (1946). A look forward in the management of corn belt soils. *Soil Sci. Soc. Am. J.* 10: 3–8.

Pilon, C., Moore, P.A. Jr., Pote, D.H. et al. (2017). Long-term effects of grazing management and buffer strips on soil erosion from pastures. *J. Environ. Qual.* 46: 364–372.

Pilon-Smits, E. (2005). Phytoremediation. *Annu. Rev. Plant Biol.* 56: 15–39.

Qi, Z., Helmers, M.J., and Kaleita, A.L. (2011). Soil water dynamics under various agricultural land covers on a subsurface drained field in north-central Iowa, USA. *Agr. Water Manage.* 98: 665–674.

Rasiah, V. and Kay, B.D. (1994). Characterizing changes in aggregate stability subsequent to introduction of forages. *Soil Sci. Soc. Am. J.* 58: 935–942.

Reicosky, D.C., Kemper, W.D., Langdale, G.W. et al. (1995). Soil organic matter changes resulting from tillage and biomass production. *J. Soil Water Conserv.* 50 (3): 253–261.

Ross, S.M., Izaurralde, R.C., Janzen, H.H. et al. (2008). The nitrogen balance of three long-term agroecosystems on a boreal soil in western Canada. *Agr. Ecosyst. Environ.* 127: 241–250.

Rücknagel, J., Götze, P., Koblenz, B. et al. (2016). Impact on soil physical properties of using large-grain legumes for catch crop cultivation under different tillage conditions. *Eur. J. Agron.* 77: 28–37.

Russell, J.R. and Bisinger, J.J. (2015). Improving soil health and productivity on grasslands using managed grazing of livestock. *J. Anim. Sci.* 93: 2626–2640.

Salton, J.C., Mercante, F.M., Tomazi, M. et al. (2014). Integrated crop-livestock in tropical Brazil: toward a sustainable production system. *Agr. Ecosyst. Environ.* 190: 70–79.

SARE (2017a). Cover crop surveys. http://www.sare.org/Learning-Center/Topic-Rooms/Cover-Crops/Cover-Crop-Surveys (accessed 9 October 2019).

SARE (2017b). Cover crops: selection and management. http://www.sare.org/Learning-Center/Topic-Rooms/Cover-Crops/Cover-Crops-Selection-and-Management (accessed 9 October 2019).

Schipper, L.A., Parfitt, R.L., Fraser, S. et al. (2014). Soil order and grazing management effects on changes in soil C and N in New Zealand pastures. *Agr. Ecosyst. Environ.* 184: 67–75.

Lisa Schulte-Moore, Lisa, de Kok-Mercado, Omar, and Love, Fred. (2020). ISU researchers pave the way to make prairie strips eligible option for federal conservation program. https://www.news.iastate.edu/news/2020/01/09/stripscrp (accessed 25 February 2020).

Silveira, M.L., Liu, K., Sollenberger, L.E. et al. (2013). Short-term effects of grazing intensity and nitrogen fertilization on soil organic carbon pools under perennial grass pastures in the southeastern USA. *Soil Biol. Biochem.* 58: 42–49.

Steppuhn, H., van Genuchten, M.T., and Grieve, C.M. (2005). Root-zone salinity: II. Indices for tolerance in agricultural crops. *Crop Sci.* 45: 221–232.

Steppuhn, H., Acharya, S.N., Iwaasa, A.D. et al. (2012). Inherent responses to root-zone salinity in nine alfalfa populations. *Can. J. Plant Sci.* 92: 235–248.

Stone, J.A. and Buttery, B.R. (1989). Nine forages and the aggregation of a clay loam soil. *Can. J. Soil Sci.* 69: 165–169.

Tomer, M.D. and Locke, M.A. (2011). The challenge of documenting water quality benefits of conservation practices: a review of USDA-ARS's conservation effects assessment project watershed studies. *Water Sci. Technol.* 64 (1): 300–310.

Turner, B.L. and Haygarth, P.M. (2005). Phosphatase activity in temperate pasture soils: potential regulation of labile organic phosphorus turnover by phosphodiesterase activity. *Sci. Total Environ.* 344: 27–36.

USDA (2015). *Summary Report: 2012 National Resources Inventory*. Washington, DC and Ames, Iowa: Natural Resources Conservation Service and Center for Survey Statistics and Methodology, Iowa State University https://www.nrcs.usda.gov/Internet/FSE_DOCUMENTS/nrcseprd396218.pdf.

USDA NRCS (2014). Bare ground, inter-canopy gaps, and soil aggregate stability. National resources inventory rangeland resource assessment. https://www.nrcs.usda.gov/wps/portal/nrcs/detail/national/technical/nra/nri/results/?cid=stelprdb1254158 (accessed 9 October 2019).

USDA NRCS (2016). Pacific Northwest cover crop selection tool. https://www.nrcs.usda.gov/wps/portal/nrcs/detail/plantmaterials/technical/toolsdata/plant/?cid=nrcseprd894840 (accessed 9 October 2019).

USDA NRCS (2017). Conservation plant releases. https://www.nrcs.usda.gov/wps/portal/nrcs/releases/plantmaterials/technical/cp/release/ (accessed 9 October 2019).

Veum, K.S., Goyne, K.W., Kremer, R.J. et al. (2014). Biological indicators of soil quality and soil organic matter characteristics in an agricultural management continuum. *Biogeochemistry* 117: 81–99.

Wang, X., VandenBygaart, A.J., and McConkey, B.C. (2014). Land management history of Canadian grasslands and the impact on soil carbon storage. *Rangeland Ecol. Manage.* 67: 333–343.

White, C.M. and Weil, R.R. (2010). Forage radish and cereal rye cover crop effects on mycorrhizal fungus colonization of maize roots. *Plant Soil* 328: 507–521.

White, C.M. and Weil, R.R. (2011). Forage radish cover crops increase soil test phosphorus surrounding radish taproot holes. *Soil Sci. Soc. Am. J.* 75: 121–130.

Wick, A.F., Merrill, S.D., Toy, T.J., and Liebig, M.A. (2011). Effect of soil depth and topographic position on plant productivity and community development on 28-year-old reclaimed mine lands. *J. Soil Water Conserv.* 66 (3): 201–211.

Wiebe, B.H., Eilers, R.G., Eilers, W.D., and Brierley, J.A. (2007). Application of a risk indicator for assessing trends in dryland salinization risk on the Canadian Prairies. *Can. J. Soil Sci.* 87: 212–224.

Wilson, H.A., Gish, R., and Browning, G.M. (1948). Cropping systems and season as factors affecting aggregate stability. *Soil Sci. Soc. Am. J.* 12: 36–38.

Ziyomo, C., Albrecht, K.A., Baker, J.M., and Bernardo, R. (2013). Corn performance under managed drought stress and in a kura clover living mulch intercropping system. *Agron. J.* 105 (3): 579–586.

Zobeck, T.M., Crownover, J., Dollar, M. et al. (2007). Investigation of soil conditioning index values for southern high plains agroecosystems. *J. Soil Water Conserv.* 62 (6): 433–442.

Zobeck, T.M., Halvorson, A.D., Wienhold, B. et al. (2008). Comparison of two soil quality indexes to evaluate cropping systems in northern Colorado. *J. Soil Water Conserv.* 63 (5): 329–338.

Chapter 13

Forages and the Environment

Matt A. Sanderson, Research Agronomist and Research Leader (Retired), *USDA-Agricultural Research Service, State College, PA, USA*
Mark A. Liebig, Soil Scientist, *Northern Great Plains Research Laboratory, Agricultural Research Service, USDA, Mandan, ND, USA*

Forage crops are usually regarded as environmentally friendly or benign. Historically, forage and grasslands have been valued for multiple benefits such as soil conservation, water quality protection, and an esthetically pleasing landscape (Sanderson et al. 2012). These benefits, among many others, are recognized in the concept of ecosystem services, which are "benefits that human populations derive from ecosystem functions" (Costanza et al. 2014).

Farmers, ranchers, and other land managers must often account for environmental outcomes in their management decision making. Governmental programs frequently require an assessment of environmental impact, and society-at-large expects farmers and ranchers to protect the environment (Nelson et al. 2012). Thus, considerations of how forage and grasslands provide ecosystem services and affect environmental quality are very relevant.

In this chapter, we examine how forage and grassland systems contribute ecosystem services, discuss managing for multiple ecosystem services, and consider tradeoffs and potential synergies involved in realizing these services.

Ecosystem Services from Forage and Grasslands

Four broad categories of ecosystem services are recognized: (i) provisioning services, which include many of the economic outputs of forage systems such as food (and food quality), feed, fiber, and fuel; (ii) regulating services, such as climate regulation, maintenance of soil fertility, and water purification; (iii) habitat or supporting services (which also enable other ecosystem services) such as habitat for wildlife and pollinators along with nutrient cycling, landscape stability, and biodiversity; and (iv) cultural services encompassing intangibles such as esthetic or recreational experiences (Millennium Ecosystem Assessment 2005; Figure 13.1).

Provisioning Services

The standard provisioning services of food, feed, fiber, and fuel (including bioenergy) production are covered in other chapters. Globally, forage and grasslands are central to human food production from ruminant livestock systems (Herrero et al. 2013). An important aspect of food production is food quality.

Milk and meat products from livestock consuming forage diets (mainly pasture) have shown increased levels of certain fatty acids such as omega-3 fatty acids, polyunsaturated fatty acids (PUFAs), and conjugated linoleic acid (CLA), which may have human health benefits including anti-cancer properties. Feeding fresh or green forage can increase antioxidant (e.g. alpha-tocopherol, beta-carotene) levels in meat, which can increase shelf life. Forage diets also have large effects on flavors and appearance of foods to provide traditional specialty products

Forages: The Science of Grassland Agriculture, Volume II, Seventh Edition.
Edited by Kenneth J. Moore, Michael Collins, C. Jerry Nelson and Daren D. Redfearn.

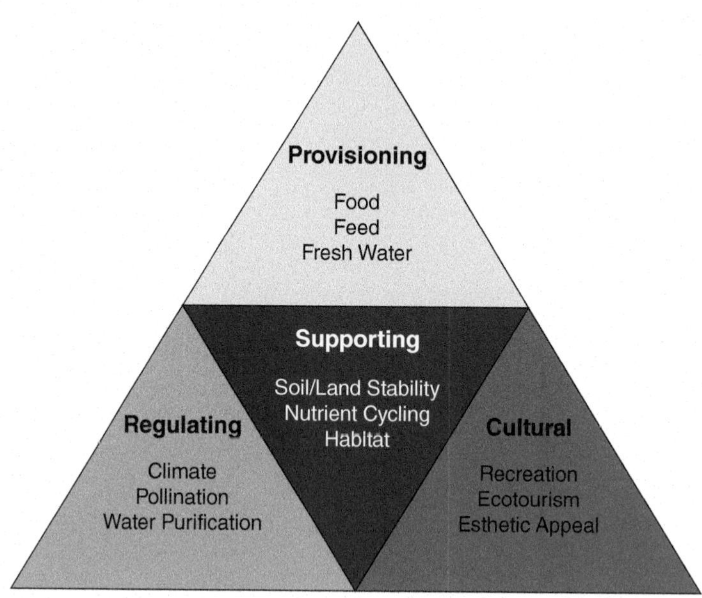

Fig. 13.1. Common ecosystem services derived from forage production systems. *Source:* Adapted from Millennium Ecosystem Assessment (2005).

that are unique to particular geographic environments in certain countries (Moloney et al. 2008).

Regulating Services

Climate

Human population growth through the mid-twenty-first century is projected to increase rates of greenhouse gas (GHG) emissions and exacerbate consequences from climate change into the twenty-second century (IPCC 2014). This reality has underscored the importance of managing land resources to mitigate causes of climate change while adapting to its consequences (Lal et al. 2011).

Globally, agriculture, and associated land-use change is a source of all three major biogenic GHGs, namely carbon dioxide (CO_2), methane (CH_4), and nitrous oxide (N_2O) (Paustian et al. 2016). Forage production systems, however, have significant potential to mitigate GHG emissions from agriculture due to limited soil disturbance and increased C inputs as organic matter from decaying roots and **rhizodeposits** in comparison to annual crops (Franzluebbers 2012). Use of forage production systems to provide climate regulation services is particularly attractive given their long-term ground cover, multi-functional capacity and potential for adoption across broad geographic regions (Sanderson and Adler 2008; Singer et al. 2009). Effective GHG mitigation, however, requires robust estimates of soil organic carbon (SOC) and GHG flux dynamics for representative forage production systems within ecoregions (Eagle et al. 2012).

Conversion of native grassland vegetation to annual cropping has resulted in widespread opportunities to restore depleted carbon stocks in agricultural soils. Within the US Great Plains, a historic evaluation along

Table 13.1 Soil organic C (SOC) at 0–15.2 cm under native rangeland and annual cropping for sites in the northern, central, and southern US Great Plains

		Native vegetation	Cropping	SOC loss	
Location	Cropping period (yr)	g C kg^{-1} soil			(%)
Dickinson, ND	41	36.4	15.1	21.3	59
Archer, WY	35	13.3	7.8	5.5	41
Dalhart, TX	40	7.2	4.4	2.8	39

Source: Adapted from Haas et al. (1957).

a longitudinal transect found relative SOC losses of 39–59% (Table 13.1). Losses of SOC were due to intensive tillage and fallow for the production of corn and small grains (Haas et al. 1957). When scaled to cropland area in the US Great Plains, the absolute SOC change from conversion of native vegetation to cropping reflects a loss of approximately 1100 Teragram (1Tg = 10^9 kg) of C, or one-fifth of the total SOC lost in the US (Lal et al. 1999).

Among land management practices available to agricultural producers, forage production systems are an effective option to restore lost SOC (Franzluebbers et al. 2012). Compared to annual crops, perennial forages, particularly grasses, allocate a higher proportion of C to underground parts and actively grow for a greater number of days over the course of a year, resulting in potential for high biomass production and SOC accrual (Eagle et al. 2012). Ranges of SOC accrual under perennial forages are broad (0.0–10.1 Mg C ha^{-1} yr^{-1}), owing to differences in climatic conditions, forage type, management, and land-use history (Franzluebbers 2012; Schmer et al. 2011).

Soil organic C accrual under forage production systems can occur in both near-surface (0–10 cm) and sub-surface (30–90 cm) soil depths (Anderson-Teixeira et al. 2009; Culman et al. 2010), with the latter an important outcome in the context of permanence of climate regulation, as C stored below 30 cm is less susceptible to mineralization and loss (Schmer et al. 2015). Greater soil-C accrual with increased pasture species diversity has been observed in the northeast US (Skinner and Dell 2016), though other reports have found weak or no relationship between SOC and species composition (De Deyn et al. 2011; Bonin et al. 2014).

Inclusion of perennial forages in production systems maintains or increases SOC (Bremer et al. 2002; Syswerda and Robertson 2014), even with the use of short-duration perennial phases (<4 years). Persson et al. (2008) found replacing annual crops with two-year perennial grass and mixed grass/legume leys in the crop rotation reduced losses of SOC without decreasing yield of the annual crop. Intermediately labile pools of soil-C were enhanced by perennial forage systems or in restored grasslands (Potter and Derner 2006; Cates et al. 2016). Bremer et al. (2002) found much of the increase in SOC under perennial grasses occurred in easily mineralizable and C pools; however, these pools are readily lost upon conversion to annual cropping (DuPont et al. 2010). Accordingly, conversion of perennial phases to annual cropping can substantially alter climate regulation services derived from soil, underscoring the importance of identifying appropriate management practices to mitigate SOC loss during transition to annual cropping.

Forage production systems are minor CH_4 sinks and moderate sources of N_2O (Rochette and Janzen 2005; Oertel et al. 2016). Mechanical application of manures or fertilizers and natural application from grazing animals can increase N_2O emissions (Liebig et al. 2012), whereas grass interseeding to thicken the stand can decrease N_2O emissions from pastures receiving manure (Sauer et al. 2009). Incorporation of forage legumes enhances mineral N in soil with corresponding increases in N_2O emissions compared to non-legume residues (Miller et al. 2008).

Inclusive GHG assessments are rare, but Soussana et al. (2007) found net uptake of GHGs (based on net exchange of CO_2, CH_4, and N_2O) was near zero (-0.85 ± 0.77 Mg CO_2 ha^{-1} yr^{-1}) for seeded pastures, intensively-managed permanent pastures and semi-natural grasslands in Europe. Similar assessments are needed elsewhere to accurately determine the efficacy of climate regulation services from forage production systems, while also taking into consideration potential GHG emissions associated with ruminant livestock and previous production practices.

Soil Fertility

Well-managed forage production systems can improve soil fertility, with synergistic benefits to other ecosystem services (Singer et al. 2009). Increases in quantity of soil organic matter in forage production systems are frequently linked to improvements in soil quality, as reflected by increased labile nutrient fractions, microbial biomass, and cation exchange capacity (Gregorich et al. 1994). Such changes in soil imply an improved capacity to retain and cycle nutrients, which can contribute to forage production and decreased reliance on external inputs.

Forage legumes improve soil fertility via their inherent capacity to fix atmospheric N_2. Amounts of N_2 fixation by forage legumes vary greatly (e.g. 20–200 kg N ha^{-1} yr^{-1} for alfalfa; Russelle and Birr 2004), but general soil fertility enhancement occurs broadly across different production systems and agricultural landscapes. Nitrogen contributions by forage legumes can reduce fertilizer and energy requirements (Hoeppner et al. 2006), while providing a positive feedback to provisioning services through increased crop yields (Entz et al. 2002).

Improvements in soil fertility from forage production systems may also be manifested in changes to soil physical conditions. Perennial forages, with greater root biomass and **rhizodeposits** compared to annual crops, coupled with limited soil disturbance during the perennial cropping phase, foster a biophysical environment favoring larger and more stable soil aggregates, increased soil pore space and continuity, and improved water entry and flow through the soil profile (Blanco-Canqui 2010; Franzluebbers et al. 2014; Figure 13.2).

Water Purification

Water purification services from forage production systems accrue from inherent morphologic attributes and growth habits of grasses and legumes. Rooting depth and density of perennial forages typically exceed those of annual crops (Mensah et al. 2003). When coupled with a longer period of active root growth (Blank & Morgan 2012), these forages become effective management tools

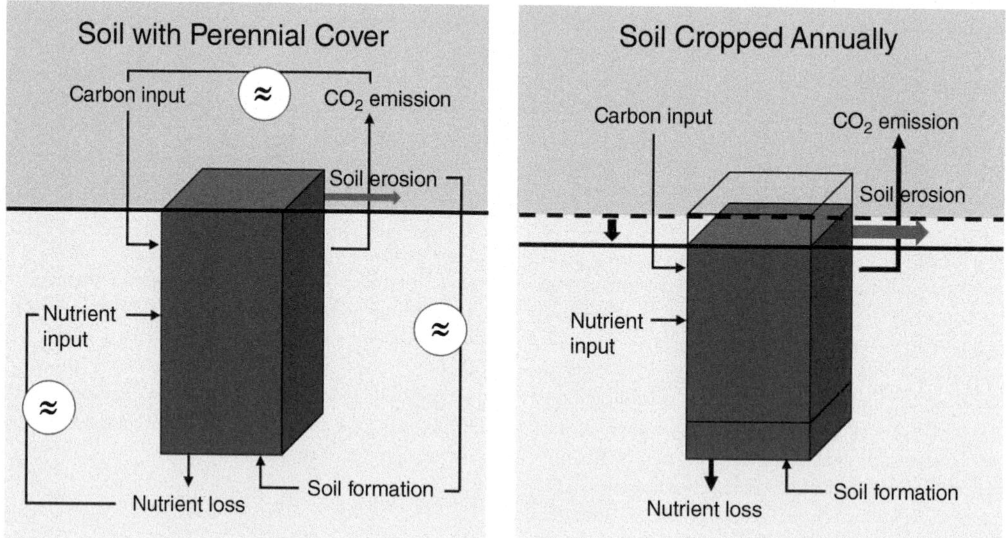

Fig. 13.2. Contrast of basic soil processes under perennial cover and annual cropping. *Source:* Adapted from Amundson et al. (2015).

for improving water quality by decreasing available nutrients in the soil profile. Tap-rooted perennial legumes (e.g. alfalfa), in particular, are capable of scavenging water-soluble nutrients from soil depths beyond the maximum rooting depth of most annual crops. Rooting depth of alfalfa has been observed to exceed 5 m (Karlen et al. 1994), thereby allowing for the extraction of mobile nutrients that may otherwise be leached to the environment. In doing so, alfalfa can serve to mitigate degradation of groundwater resources while improving nutrient-use efficiency (Grant et al. 2002).

Because of their belowground morphology and extended period of growth, perennial forages can modulate hydrologic attributes to an extent that salinity can be reduced within agricultural landscapes. Due to greater maximum rooting depth, perennial pastures can reduce soil water content and decrease the amount of deep drainage compared to annual pastures (Ward et al. 2002). Greater water uptake by perennial pastures also limits salt accumulation in near-surface soil depths by capillary rise. Salinity mitigation by perennial forages serves to increase the likelihood of maintaining ground cover by actively growing plants.

Aboveground characteristics of forages can also serve to improve water quality. Depending on harvest management and plant architecture, forages can trap snow in the winter months (Leep et al. 2001), thereby increasing the likelihood of water transfer to the soil. More continuous ground cover by forages through the seasons, in comparison to annual crops, serves to slow overland flow, increase the potential for temporary ponding, and enhance infiltration (Sollenberger et al. 2012). Even when planted in contour strips, perennial forages can reduce water runoff from cropland, decreasing the delivery of nutrients and sediment to nearby water bodies (Liebman et al. 2013; Eghball et al. 2000).

Habitat or Supporting Services

Nutrient Cycling

Nutrient cycling is central to the functioning of other ecosystem services. Efficient assimilation, retention, and delivery of nutrients allows for efficient production of biomass while mitigating negative environmental outcomes. While biologically available nutrients are generally limited in natural ecosystems, this limitation is overcome and simplified in agricultural production systems through the addition of exogenous nutrients. Though beneficial for provisioning services, system simplification can result in significant nutrient losses resulting in negative impacts on environmental quality and human health (Follett et al. 2010). Reactive-N not accumulated in harvested biomass can be lost by **volatilization**, leaching, erosion, denitrification, and emission as NO_x and N_2O, whereas soluble- and soil-bound P can be lost by runoff and erosion. Accordingly, optimal functioning of this ecosystem service is particularly critical for agricultural systems.

Many forages, due to their morphology and growth habit, foster improved nutrient-use efficiency as dedicated perennials or as part of annual crop production systems (Singer et al. 2009). However, efficient nutrient use and cycling by forages is predicated on good management, particularly when nutrients are derived from animal wastes (Wood et al. 2012). Excessive manure application in fields and/or uneven excreta distribution in pastures can contribute to the accumulation of reactive-N and mobile forms of P (Cartwright et al. 1991).

Frozen plant biomass can release significant amounts of inorganic P when thawed, particularly from forage biomass with low C : N ratios (Liu et al. 2014). Though challenges associated with nutrient cycling in agroecosystems are significant, they are not insurmountable. Using appropriate forage species and management can serve to efficiently deliver nutrients to growing plants, while retaining nutrients within agroecosystems in order to minimize loss to the environment.

Landscape Stability

Once established, perennial forages can enhance ecosystem services by minimizing soil erosion, thereby stabilizing agricultural landscapes. Continuous ground cover, coupled with improvements in soil physical condition, increases the likelihood of efficient transfer of precipitation to the soil profile. Consequently, forage production systems typically have much lower surface runoff and erosion losses compared to other production systems (Sollenberger et al. 2012). As most forms of phosphorus (P) adhere to soil particles, P loss from agricultural lands, associated with erosion, can be decreased considerably when using perennial forages (Wood et al. 2012). In addition to decreasing sediment load and nutrient delivery to nearby waterways, reduced soil erosion from forage production systems serves to retain soil functions supporting other ecosystem services.

Impacts of erosion on key soil processes (and vice versa) are highly complex (Sollenberger et al. 2012). If left unchecked, erosion can decrease **solum** depth such that critical thresholds of soil attributes may limit plant production, water purification, and nutrient retention. Accordingly, it is important to deploy agricultural practices favoring continuous ground cover, particularly given anticipated climate-driven shifts favoring a more vigorous hydrologic cycle (Shafer et al. 2014).

Biodiversity

Agriculture and all of society depend on the biodiversity of the earth. Forages are critical to maintaining biodiversity (a key attribute of properly functioning ecosystems) in intensive agriculture ecosystems (Asbjornsen et al. 2014). Managed grazing lands often consist of scores of different plant species that support diversity of other biologic organisms (Goslee and Sanderson 2010). Greater diversity in grassland plant communities has been linked to increased primary production, greater stability in response to disturbance, and better nutrient retention (Wrage et al. 2011). The quality of rural landscapes can be enhanced with management practices that increase biodiversity, such as planting a diversity of crops, encouraging several plant species in pastures, and using diverse native grasslands (Pilgrim et al. 2010).

Certain useful forage plants have been classified as **invasive species** as they may compromise biodiversity in certain situations. Reed canarygrass, with its vigorous rhizomes is considered a major invader of wetlands in the temperate US (Galatowitsch et al. 1999). Kentucky bluegrass, a short sod-forming grass and major component of pastures in temperate regions, is considered invasive in native rangelands where it can disrupt nutrient and water cycling (Toledo et al. 2014b) Thus, depending on the circumstance, highly productive introduced forage species can be unwanted in other regions or systems.

Wildlife

Wildlife often benefit from increased forage crops and grassland on the landscape. Pasture and hayland managed appropriately can serve as nesting habitat for several species of birds, vegetative cover for reptiles and mammals, and as a food supply for several classes of wildlife (Sollenberger et al. 2012). Farmers and ranchers can actively manage hay and pastureland to attract wildlife by planting food crops or altering harvest and grazing management to provide habitat and food sources for certain types of wildlife. Goals may be for meat production, to provide hunting opportunities and income or for observation and enjoyment (Sayre et al. 2012).

Pollinators

Forage and grasslands support diverse and abundant populations of native pollinators (e.g. bumblebees [Bombus spp.], hoverflies [*Syrphidae* spp.], and butterflies [*Lepidoptera* spp.]) along with introduced honeybees (*Apis mellifera*; Kimoto et al. 2012). The diversity of flowering plants present on native and managed grasslands supplies pollen and nectar for many pollinators and provides an undisturbed habitat for ground-nesting bees (Spivak et al. 2017).

Commercial beekeeping operations (apiaries) rely on perennial forage and grazing lands for bee pasture. A survey of 320 apiaries across North Dakota revealed that about 80% were located in or adjacent to hayland, pastureland, or rangeland (Sanderson 2016). Honeybee health and productivity of commercial apiaries were improved when surrounded by areas of forage and grassland (Smart et al. 2016). Adding forage legumes and forbs to plant mixtures is recommended for pollinator plantings that are used to supplement natural habitat (Orford et al. 2016).

Perennial grasslands supported more ecosystem services in the form of pollinator habitat, grassland bird habitat, and pest suppression than did corn plantings for ethanol (Werling et al. 2014). Native plants in grasslands rely on their corresponding native pollinators and also enhance **pollination** services for nearby agricultural crops (Black et al. 2011).

Cultural Services

Esthetics

A key attribute of forages and grasslands is their visual appeal to the public. The public expects that agricultural enterprises be diversified to maintain the economic and social viability of rural landscapes. Green space and scenic landscapes can bring visitors to rural areas and provide other tourist-related agricultural enterprises such as birding, hiking, and farm stays. As noted above, the increase in plant biodiversity also adds to wildlife diversity and enhanced interest by hunters.

In addition to their visual appeal, forages, and grasslands offer positive esthetic attributes associated with their acoustic characteristics. Agroecosystems possess unique "soundscapes" composed of acoustic properties representing periodicities and frequencies of emitted sounds from the vegetation and the animals it hosts (Gage et al. 2015). In contrast to sounds derived from **anthropogenic** sources, biologic contributions to soundscapes have been documented to promote relaxation and wellbeing (Gould van Praag et al. 2017).

Managing for Multiple Ecosystem Services

Management tools are needed to enable land managers to realize multiple ecosystem services from forage and grassland systems. These tools may include the use of forages in crop rotations, integrating crops and livestock, aids to landscape assessment and design, and support from governmental programs.

Forages in Crop Rotations

Crop rotation management that includes forages improves soil structure, water infiltration, and retention, and reduces soil erodibility, among other benefits, compared with continuous cropping. Rotation of perennial crops with annual crops may provide an opportunity for soil tillage to mix and disperse nutrients in a greater soil volume, thus reducing surface concentrations of nutrients prone to runoff and erosion (e.g. soil P; Karlen et al. 1994). The use of crop rotations, however, has declined drastically as cropping systems have become simplified and based on nearly continuous monocultures grown in large fields and maintained with commercial fertilizers and pesticides.

Integrated Systems

Integrated agricultural systems are recognized as an innovative tool to sustainably intensify agricultural production (Lemaire et al. 2014). These systems are defined as a form of agriculture whereby multiple agricultural enterprises interact in space and/or time, and the interactions result in synergistic resource transfer among enterprises (Hendrickson et al. 2008a). Integrated agricultural systems, with their inclusion of multiple enterprises, broaden production options for a defined area of agricultural land, thereby better matching land use with land potential (Liebig et al. 2017). Additionally, integrating multiple enterprises within systems enhances adaptability to increasingly variable weather and market conditions (Hanson et al. 2007), thereby enhancing resilience of agricultural production.

Forage production occupies a central role in integrated agricultural systems. Agricultural land unsuitable for annual cultivation can provide provisioning services through the production of perennial forages as hay, silage or grazed pasture. Use of perennial forages on fragile agricultural land can also contribute to numerous conservation benefits and enhance ecosystem services across variable landscapes.

Despite purported benefits of integrated agricultural systems, their adoption lags behind specialized production systems with low labor and management inputs (Hendrickson et al. 2008b). Addressing the lack of production and environmental outcome data associated with integrated agricultural systems would serve to objectively discern their effectiveness to improve multiple ecosystem services in comparison to specialized production systems. Moreover, metrics of system performance need to evolve beyond traditional crop, animal, and soil variables to include characterizations of ecosystem services, as well as socio-economic aspects related to producer adoption of integrated practices.

Landscape Design Tools for Management

Incorporating multiple forage and crop species into multifunctional landscapes requires unique analytical tools to aid the selection of appropriate species and to guide the placement, arrangement, and sequence of species, crops, and mixtures across farms. New developments in the analysis of grassland landscapes (Goslee et al. 2014), grazingland assessment (Toledo et al. 2014a), and cropping system performance (Liebig et al. 2001) will contribute to innovative system designs that enhance the delivery of ecosystem services.

Government Programs

Enabling management for multiple ecosystems services from forage and grasslands often requires the investment of public resources via government conservation programs. The most enduring program to date in the US is the **Conservation Reserve Program** (CRP), designed to convert fragile, highly erodible cultivated lands into grasslands for ten years to reduce soil erosion by water and wind and to provide wildlife habitat (USDA-FSA 2017). First implemented in 1989, the program has evolved to include provision of other ecosystem services such as soil accumulation and potentially for biofuels feedstock. Recent years have seen steep declines in renewal of CRP acreage contracts as farmers take land out of the program for crop production because of increased returns from commodity crops (Wright and Wimberly 2013).

Other governmental programs in the US that encourage use of forages to enhance ecosystem services include conservation guidance promulgated by the USDA-NRCS which provides assistance to farmers and ranchers to implement conservation practices such as permanent wildlife habitat and vegetative filter strips (USDA-NRCS 2017). The USDA organic agriculture standards require use of pastures in ruminant animal production systems for environmental and animal health reasons (USDA National Organic Program 2017). Forage crops are also included in programs to enhance pollinator habitat (USDA-NRCS 2015). A recent assessment of USDA-NRCS conservation practices (the Conservation Effects Assessment Project [CEAP]; Nelson et al. 2012) detailed the environmental outcomes of forage and grassland conservation practices in the US.

Europe, through the European Union, has also implemented many programs (referred to as "agri-environmental schemes") to pay farmers for providing ecosystem services from forage and grasslands (European Commission 2017). These schemes include implementing low-intensity cutting or grazing management on semi-natural grasslands, reducing inputs to grasslands, and maintaining landscape diversity (e.g. a mixture of grasslands, wetlands, and woodlands) that is part of the cultural and natural heritage.

Tradeoffs and Synergies in Realizing Multiple Ecosystem Services

Farmers want to profit from provisioning services (i.e. production of agricultural products) and, ideally, simultaneously satisfy society's preference for environmental ecosystem services. This is difficult. Frequently, compromises in forage and grassland management must be made resulting in economic or environmental tradeoffs (Sanderson and Watzold 2010). In some instances, however, there are win-win solutions (i.e. synergies among ecosystem services) that attain the ideal. Good research information helps reach these diverse goals.

Forage and grasslands generally accumulate soil carbon, which can offset some CO_2 emissions, until an equilibrium level of soil-C is reached. Maximizing C accumulation of grasslands may require changes in cutting or grazing management, which could involve an economic tradeoff in livestock performance (Skinner 2008; Liebig et al. 2010). Intensifying management to increase plant productivity (and soil C) via greater inputs often increases emissions of other GHGs (e.g. NO_2, CH_4), an unwelcome tradeoff.

Establishing and maintaining pollinator habitat in perennial systems may require a change in management to favor certain plant species, such as flowering forbs, but may involve an economic tradeoff in productivity and quality of forage for livestock. Commercial beekeepers or wildland managers may value the perennial landscape and its management for pollinator habitat, whereas the farmer or rancher may value the same landscape for its ability to support livestock.

Synergies also occur in forage and grassland systems. Including more perennial crops in the landscape provides more forage for livestock but also is associated with greater groundwater recharge and flood regulation (Qiu and Turner 2013). Healthy pollinator populations resulting from greater diversity of flowering plants could benefit reproduction and persistence of some grassland plants and provide spillover **pollination** services to nearby crops (Black et al. 2011).

Summary

Forage and grassland systems provide many ecosystem services including improved soil quality, protection of surface and subsurface water supplies, sinks for GHGs, habitat for wildlife and pollinators, and esthetic and recreational benefits for the land operator and the public. But there usually are tradeoffs between achieving economic sustainability and environmental sustainability. Intensifying production often results in reduced environmental sustainability, although some synergies (win-win solutions) can be realized. Some tradeoffs may be mitigated via governmental programs to offset economic costs. In the long-term, however, enhancing and sustaining multiple ecosystem services requires more focused research data on efficient use and conservation of the natural resources that support productive forage and grasslands.

References

Amundson, R., Berhe, A.A., Hopmans, J.W. et al. (2015). Soil and human society in the 21st century. *Science* 348: 1261071-1–6.

Anderson-Teixeira, K.J., Davis, S.C., Masters, M.D., and Delucia, E.H. (2009). Changes in soil organic carbon under biofuel crops. *GCB Bioenergy* 1: 75–96.

Asbjornsen, H., Hernandez-Santana, V., Liebman, M. et al. (2014). Targeting perennial vegetation in agricultural landscapes for enhancing ecosystem services. *Renewable Agric. Food Syst.* 29: 101–125.

Black, S.H., Shepherd, M., and Vaughn, M. (2011). Rangeland management for pollinators. *Rangelands* 33: 9–13.

Blanco-Canqui, H. (2010). Energy crops and their implications on soil and environment. *Agron. J.* 102: 403–419.

Blank, R.R. and Morgan, T. (2012). Suppression of *Bromus tectorum* L. by established perennial grasses: potential mechanisms—part one. *Appl. Environ. Soil Sci.* 2012: 1–9.

Bonin, C.L., Lal, R., and Tracy, B.F. (2014). Evaluation of perennial warm-season grass mixtures managed for grazing or biomass production. *Crop Sci.* 54: 2373–2385.

Bremer, E., Janzen, H.H., and McKenzie, R.H. (2002). Short-term impact of fallow frequency and perennial grass on soil organic carbon in a Brown Chernozem in southern Alberta. *Can. J. Soil Sci.* 82: 481–488.

Cartwright, N., Clark, L., and Bird, P. (1991). The impact of agriculture on water quality. *Outlook Agric.* 20: 145–152.

Cates, A.M., Ruark, M.D., Hedtcke, J.L., and Posnera, J.L. (2016). Long-term tillage, rotation and perennialization effects on particulate and aggregate soil organic matter. *Soil Tillage Res.* 155: 371–380.

Costanza, R., de Groot, R., Sutton, P. et al. (2014). Changes in the global value of ecosystem services. *Global Environ. Change* 26: 152–158.

Culman, S.W., DuPont, S.T., Glover, J.D. et al. (2010). Long-term impacts of high-input annual cropping and unfertilized perennial grass production on soil properties and belowground food webs in Kansas, USA. *Agric. Ecosyst. Environ.* 137: 13–24.

De Deyn, G.B., Shiel, R.S., Ostle, N.J. et al. (2011). Additional carbon sequestration benefits of grassland diversity restoration. *J. Appl. Ecol.* 48: 600–608.

DuPont, S.T., Culman, S.W., Ferris, H. et al. (2010). No-tillage conversion of harvested perennial grassland to annual cropland reduces root biomass, decreases active carbon stocks, and impacts soil biota. *Agric. Ecosyst. Environ.* 137: 25–32.

Eagle, A.J., Henry, L.R., Olander, L.P. et al. (2012). *Greenhouse Gas Mitigation Potential of Agricultural Land Management in the United States: A Synthesis of the Literature*, 3e. Durham, NC: Nicholas Institute for Environmental Policy Solutions. Duke University http://nicholasinstitute.duke.edu/ecosystem/land/TAGGDLitRev (Accessed 31 Aug 2018).

Eghball, B., Gilley, J., Kramer, L.A., and Moorman, T.B. (2000). Narrow grass hedge effects on phosphorus and nitrogen in runoff following manure and fertilizer application. *J. Soil Water Conserv.* 55: 172–176.

Entz, M.H., Baron, V.S., Carr, P.M. et al. (2002). Potential of forages to diversify cropping systems in the Northern Great Plains. *Agron. J.* 94: 240–250.

European Commission (2017). Agri-environment measures. https://ec.europa.eu/agriculture/envir/measures_en (accessed 9 October 2019).

Follett, J.R., Follett, R.F., and Herz, W.C. (2010). Environmental and human impacts of reactive nitrogen. In: *Advances in Nitrogen Management for Water Quality* (eds. J.A. Delgado and R.F. Follett), 1–35. Ankeny, IA: Soil and Water Conservation Society.

Franzluebbers, A.J. (2012). Grass roots of soil carbon sequestration. *Carbon Manage.* 3: 9–11.

Franzluebbers, A.J., Owens, L.B., Sigua, G.C. et al. (2012). Soil organic carbon under pasture management. In: *Managing Agricultural Greenhouse Gases: Coordinated Agricultural Research Through GRACEnet to Address Our Changing Climate* (ed. M.A. Liebig, A.J. Franzluebbers, and R. Follet), 93–110. San Diego, CA: Academic Press.

Franzluebbers, A.J., Sawchik, J., and Taboada, M.A. (2014). Agronomic and environmental impacts of pasture-crop rotations in temperate North and South America. *Agric. Ecosyst. Environ.* 190: 18–26.

Gage, S.H., Joo, W., Kasten, E.P. et al. (2015). Acoustic observations in agricultural landscapes. In: *The Ecology of Agricultural Landscapes: Long-Term Research on the Path to Sustainability* (ed. S.K. Hamilton, J.E. Doll and G.P. Robertson), 360–377. New York: Oxford University Press.

Galatowitsch, S.M., Anderson, N.O., and Ascher, P.D. (1999). Invasiveness in wetland plants in temperate North America. *Wetlands* 19: 733–755.

Goslee, S.C. and Sanderson, M.A. (2010). Landscape context and plant community composition in grazed agricultural systems of the Northeastern United States. *Landscape Ecol.* 25: 1029–1039.

Goslee, S.C., Sanderson, M.A., Herrick, J.E., and Ogles, K. (2014). A new landscape classification system for monitoring and assessment of pastures. *J. Soil Water Conserv.* 69: 17A–21A.

Gould van Praag, C.D., Garfinkel, S.N., Sparasci, O. et al. (2017). Mind-wandering and alterations to default mode network connectivity when listening to naturalistic versus artificial sounds. *Sci. Rep.* 7: 45273.

Grant, C.A., Peterson, G.A., and Campbell, C.A. (2002). Nutrient considerations for diversified cropping systems in the northern Great Plains. *Agron. J.* 94: 186–198.

Gregorich, E.G., Carter, M.R., Angers, D.A. et al. (1994). Towards a minimum data set to assess soil organic matter quality in agricultural soils. *Can. J. Soil Sci.* 74: 367–385.

Haas, H.J., Evans, C.E., and Miles, E.F. (1957). Nitrogen and carbon changes in Great Plains soils as influenced by cropping and soil treatments. United States Government Printing Office. Technical Bulletin No. 1164.

Hanson, J.D., Liebig, M.A., Merrill, S.D. et al. (2007). Dynamic cropping systems: increasing adaptability amid an uncertain future. *Agron. J.* 99: 939–943.

Hendrickson, J.R., Hanson, J.D., Tanaka, D.L., and Sassenrath, G. (2008a). Principles of integrated agricultural systems: introduction to processes and definition. *Renewable Agric. Food Syst.* 23: 265–271.

Hendrickson, J.R., Liebig, M.A., and Sassenrath, G.F. (2008b). Environment and integrated agricultural systems. *Renewable Agric. Food Syst.* 23: 304–313.

Herrero, M., Havlik, P., Valin, H. et al. (2013). Biomass use, production, feed efficiencies, and greenhouse gas emissions from global livestock systems. *Proc. Natl. Acad. Sci. U.S.A.* 110: 20888–20893.

Hoeppner, J.W., Entz, M.H., McConkey, B.G. et al. (2006). Energy use and efficiency in two Canadian organic and conventional crop production systems. *Renewable Agric. Food Syst.* 21: 60–67.

IPCC (Intergovernmental Panel on Climate Change) (2014). Climate change 2014: synthesis report. In: *Contribution of Working Groups I, II and III to the Fifth Assessment Report of the Intergovernmental Panel on Climate Change* (eds. Core Writing Team, R.K. Pachauri and L.A. Meyer), 151. Geneva, Switzerland: IPCC.

Karlen, D.L., Varvel, G.E., Bullock, D.G., and Cruse, R.M. (1994). Crop rotations for the 21st century. *Adv. Agron.* 53: 1–45.

Kimoto, C., DeBano, S.J., Thorp, R.W. et al. (2012). Short-term responses of native bees to livestock and implications for managing ecosystem services in grasslands. *Ecosphere* 3: 1–19.

Lal, R., Momka, D., and Lowery, B. (1999). Relation between soil quality and erosion. In: *Soil Quality and Soil Erosion* (ed. R. Lal), 237–258. Ankeny, IA: Soil and Water Conservation Society and CRC Press, Boca Raton, FL.

Lal, R., Delgado, J.A., Groffman, P.M. et al. (2011). Management to mitigate and adapt to climate change. *J. Soil Water Conserv.* 66: 276–285.

Leep, R.H., Andresen, J.A., and Jeranyama, P. (2001). Fall dormancy and snow depth effects on winterkill of alfalfa. *Agron. J.* 93: 1142–1148.

Lemaire, G., Franzluebbers, A., Carvalho, P.C.d.F., and Dedieu, B. (2014). Integrated crop-livestock systems: strategies to achieve synergy between agricultural production and environmental quality. *Agric. Ecosyst. Environ.* 190: 4–8.

Liebig, M.A., Varvel, G., and Doran, J. (2001). A simple performance-based index for assessing multiple agroecosystem functions. *Agron. J.* 93: 313–318.

Liebig, M.A., Gross, J.R., Kronberg, S.L., and Phillips, R.L. (2010). Grazing management contributions to net global warming potential: a long-term evaluation in the Northern Great Plains. *J. Environ. Qual.* 39: 799–809.

Liebig, M.A., Dong, X., McLain, J.E.T., and Dell, C.J. (2012). Greenhouse gas flux from managed grasslands in the U.S. In: *Managing Agricultural Greenhouse Gases: Coordinated Agricultural Research Through GRACEnet to Address Our Changing Climate* (ed. M.A. Liebig, A.J. Franzluebbers, and R. Follet), 183–202. San Diego, CA: Academic Press.

Liebig, M.A., Herrick, J.E., Archer, D.W. et al. (2017). Aligning land use with land potential: the role of integrated agriculture. *Agric. Environ. Lett.* 2: 170007. https://doi.org/10.2134/ael2017.03.0007.

Liebman, M., Helmers, M.J., Schulte, L.A., and Chase, C.A. (2013). Using biodiversity to link agricultural productivity with environmental quality: results from three field experiments in Iowa. *Renewable Agric. Food Syst.* 28: 115–128.

Liu, J., Ulen, B., and Bergkvist, G. (2014). Freezing-thawing effects on phosphorus leaching from catch crops. *Nutr. Cycling Agroecosyst.* 99: 17–30.

Mensah, F., Schoenau, J.J., and Malhi, S.S. (2003). Soil carbon changes in cultivated and excavated land converted to grasses in east-central Saskatchewan. *Biogeochemistry* 63: 85–92.

Millennium Ecosystem Assessment (2005). *Ecosystems and Human Well-Being: Synthesis*. Washington, DC: Island Press.

Miller, M.N., Zebarth, B.J., Dandie, C.E. et al. (2008). Crop residue influence on denitrification, N_2O emissions and denitrifier community abundance in soil. *Soil Biol. Biochem.* 40: 2553–2562.

Moloney, A.P., Fievez, V., Martin, B. et al. (2008). Botanically diverse forage-based rations for cattle: implications for product composition, product quality and consumer health. *Grassl. Sci. Eur.* 13: 361–374.

Nelson, C.J., Sanderson, M.A., and Jolley, L. (2012). Introduction to pastureland CEAP. In: *Environmental Outcomes of Conservation Practices Applied to Pasture and Hayland in the U.S. The Pastureland Conservation Effects Assessment Project (CEAP)* (ed. C.J. Nelson), 1–4. Lawrence, KS: Allen Press.

Oertel, C., Matschullat, J., Zurba, K. et al. (2016). Greenhouse gas emissions from soils–a review. *Chem. Erde-Geochem.* 76: 327–352.

Orford, K.A., Murray, P.J., Vaughan, I.P., and Memmott, J. (2016). Modest enhancements to conventional grassland diversity improve the provision of pollinator services. *J. Appl. Ecol.* 53: 906–915.

Paustian, K., Lehmann, J., Ogle, S. et al. (2016). Climate-smart soils. *Nature* 532: 49–57.

Persson, T., Bergkvist, G., and Katterer, T. (2008). Long-term effects of crop rotations with and without perennial leys on soil carbon stocks and grain yields of winter wheat. *Nutr. Cycling Agroecosyst.* 81: 193–202.

Pilgrim, E.A., Macleod, C.J.A., Blackwell, M.S.A. et al. (2010). Interactions among agricultural production and other ecosystem services derived from European temperate grassland systems. *Adv. Agron.* 109: 117–154.

Potter, K.N. and Derner, J.D. (2006). Soil carbon pools in central Texas: prairies, restored grasslands, and croplands. *J. Soil Water Conserv.* 61: 124–128.

Qiu, J. and Turner, M.G. (2013). Spatial interactions among ecosystem services in an urbanizing agricultural watershed. *Proc. Natl. Acad. Sci. U.S.A.* 110: 12149–12154.

Rochette, P. and Janzen, H.H. (2005). Towards a revised coefficient for estimating N_2O emissions from legumes. *Nutr. Cycling Agroecosyst.* 73: 171–179.

Russelle, M.P. and Birr, A.S. (2004). Large-scale assessment of symbiotic dinitrogen fixation by crops: soybean and alfalfa in the Mississippi River Basin. *Agron. J.* 96: 1754–1760.

Sanderson, M.A. (2016). Forage and grasslands as pollinator habitat in North Dakota. *Crop, Forage Turfgrass Manage.* [online] 2 https://doi.org/10.2134/cftm2016.08.0061.

Sanderson, M.A. and Adler, P.R. (2008). Perennial forages as second generation bioenergy crops. *Int. J. Mol. Sci.* 9: 768–788.

Sanderson, M.A. and Watzold, F. (2010). Balancing tradeoffs in ecosystem functions and services in grassland management. *Grassl. Sci. Eur.* 15: 639–648.

Sanderson, M.A., Jolley, L.M., and Dobrowolski, J.P. (2012). Pastureland and hayland in the USA. In: *Environmental Outcomes of Conservation Practices Applied to Pasture and Hayland in the U.S: The Pastureland Conservation Effects Assessment Project (CEAP)* (ed. C.J. Nelson), 25–40. Lawrence, KS: Allen Press.

Sauer, T.J., Compston, S.R., West, C.P. et al. (2009). Nitrous oxide emissions from bermudagrass pasture: interseeded winter rye and poultry litter. *Soil Biol. Biochem.* 41: 1417–1424.

Sayre, N.F., Carlisle, L., Huntsinger, L. et al. (2012). The role of rangelands in diversified farming systems: innovations, obstacles, and opportunities in the USA. *Ecol. Soc.* 17 (4): 43. https://doi.org/10.5751/ES-04790-170443.

Schmer, M.R., Liebig, M.A., Vogel, K.P., and Mitchell, R. (2011). Field-scale soil property changes under switchgrass managed for bioenergy. *GCB Bioenergy* 3: 439–448.

Schmer, M.R., Jin, V.L., and Wienhold, B.J. (2015). Sub-surface soil carbon changes affects biofuel greenhouse gas emissions. *Biomass Bioenergy* 81: 31–34.

Shafer, M., Ojima, D., Antle, J.M. et al. (2014). Great plains, Chapter 19. In: *Climate Change Impacts in the United States: The Third National Climate Assessment* (ed. J.M. Melillo, T.T.C. Richmond and G.W. Yohe), 441–461. U.S. Global Change Research Program https://doi.org/10.7930/JOD798BC.

Singer, J.W., Franzluebbers, A.J., and Karlen, D.L. (2009). Grass-based farming systems: soil conservation and environmental quality. In: *Grassland: Quietness and Strength for a New American Agriculture* (eds. W.F. Wedin and S.L. Fales), 121–136. Madison, WI: ASA-CSSA-SSSA.

Skinner, R.H. (2008). High biomass removal limits carbon sequestration of mature temperate pastures. *J. Environ. Qual.* 37: 1319–1326.

Skinner, R.H. and Dell, C.J. (2016). Yield and soil carbon sequestration in grazed pastures sown with two and five forage species. *Crop Sci.* 56: 2035–2044.

Smart, M.D., Pettis, J.S., Euliss, N., and Spivak, M.S. (2016). Land use in the northern great plains region of the U.S. influences the survival and productivity of honey bee colonies. *Agric. Ecosyst. Environ.* 230: 139–149.

Sollenberger, L.E., Agouridis, C.T., Vanzant, E.S. et al. (2012). Prescribed grazing in pasturelands. In: *Environmental Outcomes of Conservation Practices Applied to Pasture and Hayland in the U.S: The Pastureland Conservation Effects Assessment Project (CEAP)* (ed. C.J. Nelson), 111–204. Lawrence, KS: Allen Press.

Soussana, J.F., Allard, V., Pilegaard, K. et al. (2007). Full accounting of the greenhouse gas (CO_2, N_2O, CH_4) budget of nine European grassland sites. *Agric. Ecosyst. Environ.* 121: 121–134.

Spivak, M., Browning, Z., Goblirsch, M. et al. 2017. Why does bee health matter? The science surrounding honey bee health concerns and what we can do about it. Council for Agricultural Science and Technology. Commentary QA2017-1.

Syswerda, S.P. and Robertson, G.P. (2014). Ecosystem services along a management gradient in Michigan (USA) cropping systems. *Agric. Ecosyst. Environ.* 189: 28–35.

Toledo, D.T., Sanderson, M.A., Herrick, J.E., and Goslee, J.C. (2014a). An integrated approach to grazingland ecological assessments and management interpretations. *J. Soil Water Conserv.* 69: 110A–114A.

Toledo, D.T., Sanderson, M.A., Spaeth, K.E. et al. (2014b). Extent of Kentucky bluegrass and its effect on native plant species diversity and ecosystem services in the Northern Great Plains of the USA. *Invasive Plant Sci. Manage.* 7: 543–552.

USDA Farm Service Agency (2017). Conservation Reserve Program Statistics (May 2017). https://www.fsa.usda.gov/programs-and-services/conservation-programs/reports-and-statistics/conservation-reserve-program-statistics/index (accessed 9 October 2019).

USDA National Organic Program 2017. Organic regulations. https://www.ams.usda.gov/rules-regulations/organic (accessed 9 October 2019).

USDA Natural Resources Conservation Service (NRCS) (2015). Using 2014 Farm Bill programs for pollinator conservation. Biology Technical Note 78. https://directives.sc.egov.usda.gov/OpenNonWebContent.aspx?content=37370.wba (accessed 9 October 2019).

USDA Natural Resources Conservation Service (NRCS) (2017). National Handbook of Conservation Practices.

https://www.nrcs.usda.gov/wps/portal/nrcs/main/national/technical/cp/ncps (accessed 9 October 2019).

Ward, P.R., Dunin, F.X., and Micin, S.F. (2002). Water use and root growth by annual and perennial pastures and subsequent crops in a phase rotation. *Agric. Water Manage.* 53: 83–97.

Werling, B.P., Dickson, T.L., Isaacs, R. et al. (2014). Perennial grasslands enhance biodiversity and multiple ecosystem services in bioenergy landscapes. *Proc. Natl. Acad. Sci. U.S.A.* 111: 1652–1657.

Wood, C.W., Moore, P.A., Joern, B.C. et al. (2012). Nutrient management on pastures and hayland. In: *Environmental Outcomes of Conservation Practices Applied to Pasture and Hayland in the U.S: The Pastureland Conservation Effects Assessment Project (CEAP)* (ed. C.J. Nelson), 257–324. Lawrence, KS: Allen Press.

Wrage, N., Strodthoff, J., Cuchillo, H.M. et al. (2011). Phytodiversity of temperate perennial grasslands:ecosystem services for agriculture and livestock management for diversity conservation. *Biodivers. Conserv.* 20: 3317–3339.

Wright, C.K. and Wimberly, M.C. (2013). Recent land use change in the Western Corn Belt threatens grasslands and wetlands. *Proc. Natl. Acad. Sci. U.S.A.* 110: 4134–4139.

Part III

FORAGE SPECIES

Crimson clover is a winter annual clover used for winter and early spring forage production and as a cover crop for soil conservation and improvement. *Source*: Photo courtesy of Mike Collins.

There are different species used as forages in different areas and regions to meet specific forage and pasture production objectives. Many of the differences in forage types across a broad region are based on their growth response and production under different climate and environmental constraints. We refer to this as "adaptation." Many of these growth limitations to defined boundaries of adaptation are based on physio-ecologic processes that have been described in detail in the previous chapters. Generally, the legume species have greater production limitations

Forages: The Science of Grassland Agriculture, Volume II, Seventh Edition.
Edited by Kenneth J. Moore, Michael Collins, C. Jerry Nelson and Daren D. Redfearn.

than many of the grasses. This is especially true regarding soil characteristics, such as soil pH, depth, and drainage. The most common use for all of the species described is forage production. However, there are additional ecosystem services provided that should be considered, including erosion control, soil C-sequestration, nutrient cycling, pollinator habitat, and food and cover for wildlife. This part provides the most up-to-date information on forage species and cultivar use across North America.

CHAPTER

14

Cool-Season Legumes for Humid Areas

Craig C. Sheaffer, Professor, *Agronomy and Plant Genetics, University of Minnesota, St. Paul, MN, USA*
Gerald W. Evers, Professor Emeritus, *Soil and Crop Sciences, Texas A&M University, Overton, TX, USA*
Jacob M. Jungers, Associate Professor, *Agronomy and Plant Genetics, University of Minnesota, St. Paul, MN, USA*

Introduction

Forage legumes are key components of livestock rations whether grazed or harvested as hay or silage. In addition, legumes provide soil improvement, soil conservation, wildlife food plots, nitrogen (N) for grasses in mixture or rotation, and they also can be used to produce honey (Perfect et al. 1990; Decker et al. 1994; Zemenchik et al. 1996; Evers 2011a).

The predominant forage legumes grown in humid regions of the US, Canada, and northern Mexico belong to the *Medicago*, *Trifolium*, and *Lotus* genera. Several other temperate legume genera are adapted to and used in unique environments. Cool-season species are most productive at average temperatures of about 20 °C and where moisture is adequate. For most, herbage productivity is greatest in the temperature range of 5–30 °C. Therefore, productivity of these legumes is limited by hot summers and cold winters. Where adapted, winter-hardy perennials are most frequently grown and harvested in spring and summer. In regions with mild winters and hot and dry summers, winter annuals are used and harvested in late winter and spring.

Common Traits

Most forage legumes important to agriculture are **perennials** or annuals. A **biennial legume**, sweetclover, was important at one time, but is not widely cultivated today. Alfalfa, red clover, birdsfoot trefoil, and sainfoin are crown-forming perennials whose stands do not persist indefinitely, and periodic reseeding is required. Kura clover, crownvetch, cicer milkvetch and alfalfa (*falcata* types) have rhizomes and regenerate vegetatively. White clover regenerates via stolons. All legumes have the potential to produce seed, and some perennials like white clover and birdsfoot trefoil and annuals like ball clover and persian clover can be managed so that stands can be maintained using this seed production. Perennial legumes are grown in monocultures or often in mixture with perennial grasses (Sleugh et al. 2000). Grass mixtures provide advantages such as protection against winter injury, dilution of **antiquality** factors in legumes, insurance against stand failure and quicker hay drying (Sheaffer et al. 2009).

Cool-season **annual** legumes are primarily used in the southeastern US where cool-season perennial legumes do

Forages: The Science of Grassland Agriculture, Volume II, Seventh Edition.
Edited by Kenneth J. Moore, Michael Collins, C. Jerry Nelson and Daren D. Redfearn.

not persist well because of hot and dry summers and/or low fertility, acid soils. Small and intermediate perennial white-clover types are the exception because of their ability to persist as **reseeding annuals**. Cool-season annuals are sown in fall and mature from mid-spring to early summer, depending on species and latitude.

In the southeastern US, annual legumes are primarily used for overseeding into tropical perennial grasses which form the basis of pasture systems (Hoveland and Evers 1995). Cool-season annual legumes provide earlier forage production that extends the grazing season, are a N source for the tropical perennial grasses that follow the cool-season legume, and produce a higher quality forage than the tropical grasses resulting in improved animal performance. Cool-season annual legumes like hairy vetch are used as winter cover crops in the northern US.

Cool-season legumes conduct symbiotic **nitrogen fixation** through a relationship with specific soil bacteria, ***Rhizobium***, that allows plants to convert atmospheric N to protein for their growth (Sheaffer et al. 2018). Therefore, legumes have seed and forage high in N content, do not require N fertilization, and can contribute N to subsequent crops in rotation. Estimated N_2-fixation ranges widely because of effects of soil and climate and differences in the effectiveness of the symbiosis. Maximum levels observed for alfalfa and annual clovers are near 300 kg N ha^{-1} and are affected by soil fertility and management (Evers and Parsons 2011; Sheaffer and Moncada 2012). Herbage-N concentrations are highest in leaves and vegetative herbage and decrease as herbage matures (Evers and Parsons 2010; Sheaffer et al. 2018). Root-N concentrations are about one-half that of the **shoot**-N concentration and decline slowly during the growing season.

Adaptation

Species vary in response to specific environmental factors such as temperature, soil moisture, drainage, and soil pH (Tables 14.1 and 14.2). Alfalfa, kura clover, and cicer milkvetch have extensive belowground crowns and roots and are very winter hardy, whereas, other perennials, such as red, white, and alsike clover, are less winter hardy. Winter annuals like crimson clover and hairy vetch have moderate tolerance of freezing temperatures, but cannot tolerate the lower temperature extremes in northern climates. Deep-rooted legumes such as alfalfa, sweetclover, and kura clover tolerate moisture-deficit stress better than shallow-rooted legumes such as white clover. Except for birdsfoot trefoil, alsike clover, white clover, ball clover, and persian clover, legumes do not tolerate excess soil moisture. Legume productivity is generally greatest at neutral soil pH and at high levels of soil K and P. However, birdsfoot trefoil, alsike clover, hairy vetch, and crimson clover have tolerance to lower soil pH.

Yield and nutritive value of forage, stand persistence, and ultimately distribution of cool-season legumes are affected by pathogens and insects (Manglitz 1985; Leath et al. 1988; Manglitz and Ratcliffe 1988; Miller and Hoveland 1995; Samac et al. 2014). Breeding for pest **resistance** has increased the range of adaptation of many legumes.

Medicago

The genus *Medicago* comprises more than 60 species, of which, about two-thirds are annuals and one-third perennials (Quiros and Buachan 1988). The center of origin appears to be the broad region adjoining present-day Iran, Iraq, Turkey, and southern Russia. But diverse annual and perennial species also developed in the broad region encompassing the eastern Mediterranean on the west, Siberia on the north and central Asia on the east. In depth discussion of important *Medicago* spp. is provided in Hanson et al. (1988), and Barnes and Sheaffer (1995).

Alfalfa

Alfalfa is the oldest crop grown solely for forage, having been cultivated for about 9000 years (Russelle 2001). It was used as livestock feed by Middle Eastern civilizations and later spread through the Old World by traders and armies. It was introduced to the Americas by the Spanish in the 1500s and into the eastern US by colonists in the early 1700s. Over time, nine sources of germplasm varying in fall dormancy, winter hardiness and disease resistance were introduced into the US (Barnes et al. 1977). These have been used in development of modern alfalfa cultivars. It is the leading perennial forage legume in the US, and is grown in pure stands or in mixture with grasses on about 8 million ha (USDA-FSA 2019). Cultivars varying in fall dormancy and winter hardiness have been developed for adaptation to regional climates (Barnes et al. 1977; Ventroni et al. 2010). Alfalfa is grown in rotation with cereals and can reduce pest populations and supply N (Coulter et al. 2011; Yost et al. 2015; Goplen et al. 2018).

Leaves are normally pinnately trifoliolate, but some cultivars marketed as multileaf have more than three leaflets per leaf. Shoots are normally indeterminate, and the shoot apex continues to produce leaves and flowers until the stem dies or is removed. Flowers are borne in a **raceme** and are most frequently purple. However, cultivars containing both *Medicago sativa* and *Medicago falcata* germplasm can have a range of flower colors including purple, white, cream, yellow, and variegated (shades of blue and green).

Alfalfa has a distinct crown containing axillary buds from which herbage regrowth originates. Alfalfa roots are mostly found in the top 0.5 m of soil but taproots can penetrate from 7 to 9 m deep (Sheaffer et al. 1988b). Some cultivars produce **adventitious** stems from lateral roots.

Table 14.1 Preferred soil characteristics, plant traits, and seeding rates of cool-season annual legumes

	Preferred soil characteristics			Plant characteristics				Pure stand seeding rate[a] (kg ha $^{-1}$)
Legume species	pH	Texture	Drainage	Maturity	Bloat potential	Reseeding potential	Cold tolerance	
Annual medics	6.5–8.0	Loam, clay	Fair	Early–medium	High	Good	Fair	6–11
Arrowleaf clover	6.0–7.0	Sandy loam	Good	Late	Low	Good	Good	9–11
Ball clover	6.5–8.5	Loam, clay	Fair	Medium	High	Good	Good	3
Berseem clover	6.5–8.5	Loam, clay	Fair	Medium	Low	Poor	Poor	13–18
Crimson clover	6.0–7.0	Sandy loam, clay	Good	Early	Low	Poor	Good	17–22
Persian clover	6.5–8.0	Loam, clay	Poor	Medium	High	Good	Fair	6–8
Rose clover	6.0–8.0	Sand, loam, clay	Good	Early–medium	Low	Good	Good	13–18
Trifolium								
subterraneum *cv.*(Karridale, Denmark)	6.0–7.3	Loam, clay	Fair	Early–medium	Low	Poor–medium	Fair	17–22
brachycalycinum cv. (Clare, Nuba)	6.5–7.5	Loam, clay	Fair	Medium	Low	Poor–medium	Poor	17–22
Vetch, hairy	6.0–8.0	Sand-clay	Fair	Early–medium	Low	Medium	Good	22–28
Vetch, common	6.0–8.0	Sand-clay	Fair	Early–medium	Low	Medium	Poor	33–44

[a]Use two-thirds of recommended seeding rate when planting in mixtures.
Source: Adapted from Evers and Smith (1998).

Table 14.2 Perennial forage legume tolerance to soil factors, climatic conditions, defoliation, bloat potential and seeding rates

	Tolerance to soil limitations				Tolerance to:				Pure stand seeding rate[a] (kg ha $^{-1}$)
Legume species	Salinity	Acidity	Alkalinity	Wet	Drought	Cold	Frequent defoliation	Bloat potential	
Alfalfa	Fair	Poor	Fair	Poor	Excellent	Good	Fair	High	11–17
Alsike clover	Fair	Good	Good	Excellent	Poor	Poor	Poor	?	4–7
Birdsfoot trefoil	Fair	Good	Good	Excellent	Fair	Fair	Good	Low	6–9
Cicer milkvetch	Fair	Fair	Excellent	Fair	Good	Excellent	Fair	Low	22–28
Crownvetch	Fair	Good	Poor	Poor	Good	Fair	Poor	Low	9–17
Kura clover	Fair	Fair	Fair	Good	Fair	Excellent	Excellent	High	9–11
Red clover	Fair	Good	Poor	Fair	Fair	Fair	Fair	Medium	9–11
Sweetclover	Good	Poor	Excellent	Poor	Excellent	Excellent	Poor	Medium	9–17
White clover									
Large type	Fair	Good	Poor	Good	Poor	Fair	Excellent	High	4
Medium type	Fair	Good	Poor	Excellent	Poor	Good	Excellent	High	4

[a]Use two-thirds of recommended seeding rate when planting mixtures.

Alfalfa grows best on soils that are well drained, neutral in pH and have high fertility (Table 14.2). It is poorly adapted to wet or saline soils and will not tolerate flooding. Cold tolerance is a major factor influencing cultivar adaptation in the North Central region. Very winter hardy cultivars can survive air temperatures as low as −25 °C. Water use by alfalfa exceeds that of many annual row crops that have a shorter period of vegetative growth. From 5.6 to 8.3 cm ha^{-1} of water are required to produce a metric ton of dry forage, but water use is increased by high air temperatures and low humidity (Sheaffer et al. 1988b).

During regrowth following dormancy or harvesting, herbage-DM yield increases until flowering and then declines due to leaf loss (Sheaffer et al. 1988a). Depending on length of the growing season, climate, and maturity at harvest, alfalfa will have from 2 to 10 regrowth cycles each year. The forage yield and quality of alfalfa are greatly influenced by the maturity at harvest for hay and silage. Producers who value high-quality forage harvest more frequently, often at bud stage, and sacrifice some yield and persistence. Harvesting later at early flowering provides greater yield, acceptable forage nutrient content for most feeding situations, and reasonable stand persistence. Alfalfa provides high-quality pasture, but should be mixed with grasses to reduce incidence of **bloat**. Grazing-tolerant cultivars have been developed for use under grazing (Smith et al. 2000; Catalano et al. 2019). Domestic and international markets exist for hay and dehydrated pellets, cubes, or meal products.

Incorporation of resistance to multiple diseases has been important for improved stand persistence and yield of modern cultivars (Lamb et al. 2006). Cultivars have also been genetically modified for **resistance** to the herbicide glyphosate, which allows the post-emergence application of glyphosate to control both grass and broadleaf weeds during establishment and in mature stands (Sheaffer et al. 2007). Though attempts have been made to improve cultivar forage quality through conventional breeding without much success (Sheaffer et al. 1998), genetically modified cultivars with reduced lignin concentration have consistently greater forage digestibility and allow growers more flexibility to delay harvest to avoid rain and to achieve greater forage yields (Grev et al. 2017).

Annual Medics

Annual medics (*Medicago* spp.) grow as winter annuals in their native range of North Africa, the Near East, and Mediterranean region, all having hot, dry summers and mild, wet winters. In winter rainfall regions of the US, medics are used as a source of winter forage for livestock and wildlife. Burr medic was introduced into California during the 1500s by the Spanish and now volunteers as a winter annual in pastures and rangeland. Burr medic, button medic, and little burr medic were introduced into Texas in the 1800s and have become naturalized in Texas and the Southern Plains. Spotted burr medic and button medic volunteer in pastures throughout the South (Ball et al. 1991). Medics have been studied for use in pastures in south Texas and "Amarillo" burr medic was released (Occumpaugh 2001). Burr medic has also been used for revegetation of inactive landfills in California.

In the midwestern US, burr and barrel medics from Australia provide high-quality summer or fall forage when spring and summer seeded (Zhu et al. 1998). Annual medics have also been evaluated as intercropped smother crops in corn and soybeans (Jeranyama et al. 1998). Black medic is widely distributed in waste areas, permanent pastures and lawns, in the northern US and southern Canada (Wilson et al. 2017). "George," a selected black medic **ecotype**, has been grown as a replacement for summer fallow in Montana and other states in the central Great Plains (Sims and Slinkard 1991).

Most annual medics are adapted to neutral and alkaline soils (Table 14.1). Medic plants are yellow flowered and produce seed in distinctive pods, some having spines. Leaf structure is similar to alfalfa; however, the stems tend to be less erect and are sometimes prostrate and hairy.

Trifolium

The genus *Trifolium* includes about 250 species of true clovers with about one-third of them perennial and two-thirds annual (Gillett and Taylor 2001). Leaves are borne in spiral phyllotaxy around the apex of the solid stems, are typically palmately compound with three leaflets, but can have four to eight leaflets in some species. Many clovers have white variegation or watermarking on the leaflets. The inflorescence (commonly called a blossom or head) is a globular or conical-shaped raceme with 200 or more flowers that are pedicellate or sessile.

The true clovers originated in a broad region from southern Europe to Asia and were introduced into the US (Taylor 1985; Ball et al. 1991; Hoveland and Evers 1995).

Red Clover

Red clover's name is derived from its red or purple flower color. It has numerous stems arising from a narrow crown and a determinate growth habit (Taylor and Smith 1995). The first flower is produced terminally on each stem with later flowers produced on lateral branches that arise from axillary buds. Leaflets are hairy and often variegated. Red clover has a taproot with secondary branches, but its roots do not penetrate the soil to the extent of alfalfa.

There are three types of red clover (Taylor and Smith 1995). Most red clover grown in the northern regions of the US is a "medium type," with early flowering that can produce two or three hay crops per year. It has a biennial or short-lived growth habit. In contrast, the late flowering, one-cut, "mammoth type" usually produces a single harvest plus an aftermath. Mammoth red clover typically does not flower in the seedling year. The third type is wild red clover found in England.

Medium red clover is grown on about 4 million ha in the US, with greatest production in the north central and northeastern US (Taylor and Smith 1995). The short life span of about 2 yr is associated with lack of disease resistance and lack of winter hardiness. In the southern US, red clover is sometimes grown as a winter annual.

Red clover is best adapted where summer temperatures are moderately cool to warm and moisture is sufficient. It has relatively less drought and heat tolerance than alfalfa but is adapted to a wide range of soil types, except those in areas prone to drought or excess moisture (Table 14.2). It tolerates a pH as low as 5.5 and is used as an alternative to alfalfa on low-pH soils.

Red clover is used for pasture and can produce high-quality hay and silage (Buxton et al. 1985). Protein degradation in the rumen is less than that of alfalfa, allowing more to bypass the rumen and be digested in the lower digestive tract, which is advantageous (Owens et al. 1999).

Animal performance on red-clover pasture is similar to that on alfalfa (Marten et al. 1990). However, long-term carrying capacity and production per hectare are less than for alfalfa due to lower yields and shorter stand life. Persistence is greatly increased by rotational grazing.

Harvesting at about 20% bloom for hay or silage offers the best compromise between forage quality and yield. From two to four harvests can be taken each year, depending on latitude and moisture. Red clover is high in moisture and sometimes slow to dry to moisture levels safe for storage as hay. Red clover, white, and alsike clover can be infected by the black patch fungus (*Rhizoctonia leguminicola*) that produces the mycotoxin slaframine that irritates the mouths of horses. Horses eating infected pasture and hay drool excessively, a condition known as slobbers.

Red-clover seedlings are very competitive, making it easy to establish using minimum tillage equipment or frost seeding (Blaser et al. 2012). It is especially useful for pasture renovation. When interseeded into late growing corn, red clover can provide winter cover and reduce soil nitrate N leaching (Noland et al. 2018).

White Clover

White clover spreads by **stolons**. Leaves and white blossoms are suspended by long **petioles** and **peduncles** that are attached to nodes on the stolons. The original taproot and crown are short-lived, but stolons root from nodes on the stolons to maintain the plant. Thereafter, white clover persists in pastures, lawns, and wastelands either by reseeding or new growth from stolons after death of the original crown (Gibson and Cope 1985).

Three types of white clover are distinguished based on morphologic characteristics (Pederson 1995). Small types have short petioles, small leaflets, short peduncles, and small flowers. They produce very little forage yield because of a prostrate growth habit but are very persistent under close defoliation in pastures, lawns, and recreational areas. Intermediate types have petiole and peduncle length, leaflet, and flower size, and stolon diameter intermediate between small and large types. High seed production with a high percentage of hard seed, allows it to persist in hot summer climates. Ecotypes described as common or white dutch are usually intermediate types. Large types have the largest petioles, peduncles, leaflets, flowers, and stolons and, therefore, are the highest yielding. However, they do not persist as well as the small and intermediate types under **continuous stocking** because of their taller growth habit and limited seed production. Large types have historically been referred to as ladino clovers because the first large-type introduction was an **ecotype** from Italy called Ladino *Gigante Lodigiano*. Large-leafed white clovers originating in New Zealand have been developed for use in northern grazing systems (Woodfield et al. 1998)

It is estimated that white clovers are found in over half of the 45 million ha of humid or irrigated pastures in the US (Pederson 1995). Because of the shallow root system, white clover grows best in humid, moist regions and on soils where moisture is available throughout the growing season. Plants have poor drought and heat tolerance and do not grow well on sandy soils in low rainfall regions (Table 14.2). White clovers are moderately winter hardy and will not consistently overwinter in northern regions without snow cover (Catalano et al. 2019). In the lower South, white clover usually does not survive the hot summers and persists as a **reseeding annual**. For this reason, intermediate types with good seed production are used in that area. White clover summer survival is best where soils have greater moisture retention than dryer upland soils.

White clover is primarily used for pasture. The growth habit allows persistence under continuous stocking (Brink et al. 1999), but rotational stocking with **rest periods** provides for greatest forage yield and persistence. Forage is very high in nutritive value because it is mostly leaves (Brink and Fairbrother 1992).

Alsike Clover

Alsike clover is named after a location in Sweden where it was cultivated as early as 1750 (Townsend 1985). It is a short-lived perennial legume with an upright growth habit like red clover, but with finer stems that lodge easily. Plants have white to pinkish **florets** borne in heads that originate in leaf axils up and down the stem and have leaflets that are finely serrated along the entire edge and never variegated. The indeterminate growth habit can result in uncut stems up to 2-m long that bear flowers as well as ripe seed along their length. The result is a large supply of hard seed that can contribute to stand maintenance through natural reseeding.

Alsike clover has greater adaptation to wet and acidic soils than other perennial clovers or alfalfa, but it does not tolerate drought or high temperatures (Table 14.2). Consequently, it is grown in northern regions of the US and high mountain regions of the western US. It is included in mixtures with other legumes because of its adaptation to unique environmental conditions. It can be grazed or cut for hay but it will not persist under frequent defoliation. In northern regions where it is best adapted, it usually is harvested once or twice per season after flowering so that it produces seed for regeneration.

Kura Clover

This long-lived perennial is also called Caucasian, pellet's, or honey clover (Sheaffer et al. 2017). It is native to the Caucasus region of Europe and is named for the Kura River. Plants have a deep, branching taproot and rhizomes that enable it to spread vigorously. Seedling vigor is poor, and when established from seed productive stands can take several years to develop. Kura clover has been successfully established at the field scale by incorporating rhizome fragments (Baker 2012).

Initial growth in spring is upright stems supporting one or two large pink-white heads that bear individual flowers. Regrowth after harvest consists of petioles and leaves arising from the crown and rhizomes. Leaflets are oblong, often variegated, and acutely serrated at the margin.

Kura clover has excellent winter hardiness, drought tolerance, and has no major disease problems (Table 14.2). In the northern US, kura clover has greater persistence under rotational and **continuous stocking** than any commonly grown legume (Sheaffer et al. 1992). It is most productive in the northern US in areas where summer moisture supply is abundant and uniform. Under moisture stress and high heat, it becomes dormant (Sheaffer and Seguin 2009).

Kura clover is best suited for grazing because of its prostrate growth habit and very leafy, high-moisture forage. Forage quality is very high because the forage is predominantly leaves, but it can cause ruminant bloat and should be planted with perennial grasses (Peterson et al. 1994; Mourino et al. 2003). It has good potential for soil stabilization projects because the dense rhizome mat holds soil and prevents erosion. With adequate suppression, and no-tillage equipment, kura clover can be managed as a living mulch for corn production to reduce nitrate leaching (Zemenchik et al. 2000; Ochsner et al. 2010).

Arrowleaf Clover

This winter annual is grown extensively in the southern US (Hoveland and Evers 1995). Its use in the southeastern US declined because of virus and root rot diseases (Pemberton et al. 1998). The greatest use is in Oklahoma and Texas. Stems can grow over 2 m tall if not grazed or cut. The leaflets are shaped like arrowheads, thus the name. Leaflets are hairless and often variegated. The high percentage of hard seed, up to 90%, gives excellent reseeding ability if herbage is not cut or grazed to a height shorter than 15–20 cm. Arrowleaf clover is best adapted to well-drained loam and sandy soils with a pH near 6.5 but does not grow well on poorly drained soils (Table 14.1). Seedlings emerge slowly on soils with a pH below 6 and can be affected by iron chlorosis on soils with a pH above 7 (Evers 2003). Seedling vigor and growth are poor, with the young plants staying in a **rosette** stage until late February with little forage production until early March (Evers 1999).

Arrowleaf clover is the latest-maturing and the highest-yielding annual clover, with growth continuing through June and early July if moisture is adequate (Evers and Newman 2008). Total **shoot** yield usually decreases with defoliation. Root growth, however, is similar to crimson, rose, and subterranean clovers. It can be used for pasture or hay. For hay production, arrowleaf clover should be grazed until early to mid-April because of poor drying conditions for hay production in early spring, and then harvested at early- to mid-bloom in May. Annual ryegrass is frequently planted in mixture with arrowleaf clover to provide earlier forage production (Evers 2011b).

Crimson Clover

This annual, crown-forming species is grown extensively for grazing and sometimes hay throughout the southeastern US (Knight 1985a). It is the earliest maturing annual clover and, therefore, an ideal annual clover species for winter cover crop mixtures and as a N source. Leaves and stems are hairy, but size, shape, and hairiness of stems and leaves varies greatly among genotypes. The elongated, conical head inflorescence consisting of up to 125 flowers is typically scarlet or crimson, but white and yellow-flowered variants exist. It is an excellent seed producer, but most seeds are soft; therefore, crimson clover does not reseed well. New cultivars such as "Sabine" have been selected for greater hard seed production to improve reseeding.

Seeding vigor of crimson clover is superior to most other annual clovers because of larger seed resulting in rapid establishment (Evers 1999). In pure stands, crimson clover reached 100% light interception six to eight weeks earlier than arrowleaf and rose clovers (Evers and Newman 2008). It grows on soil types ranging from sands to well-drained clays with pHs of 5–7 (Evers 2003). It does not tolerate poor drainage (Table 14.1). Crimson clover is one of the most cold-tolerant cool-season annual legumes which allows it to be used in winter cover crop mixtures in the Midwest US (Brandsaeter et al. 2002). The combination of good seedling vigor and early maturity makes it ideal for fall overseeding into warm-season perennial grasses such as bermudagrass (Knight 1970). Crimson clover is used to a limited extent for roadside beautification in the South.

Subterranean Clover

Subclover is the common name for three *Trifolium* species that produce mature seed in burrs near or below the soil surface (McGuire 1985). *Trifolium subterraneum* L. has black seed and some pubescence on leaves and stems. It originated in the Mediterranean basin and southern England. *Trifolium yanninicum* Katzn and Morley has cream-colored seed and glabrous plant parts except for the upper leaf surface. It is adapted to areas of waterlogged soils. All plant parts of *Trifolium brachycalycinum* Katzn and Morley are glabrous, but the calyx only covers the base to lower third of the pod rather than most of the pod as in other species. It is adapted to alkaline soils, but is not as cold-hardy as the other two species (Table 14.1). All three species are self-fertilized and **cleistogamous**, with 2n = 16 chromosomes.

Globally, subclover is the most widely used annual clover for livestock grazing. It is grown, to a limited extent, in Oregon and the southeastern US. To date, all cultivars were developed in Australia and vary in maturity to fit the length of the rainfall season (McGuire 1985). Subclover's greatest potential in the US is for over-seeding into warm-season perennial grasses.

Subclover produces dense, prostrate growth from non-rooting stems and stolons that allows it to tolerate close grazing by sheep, horses, and wildlife. Leaves and petioles are the primary plant parts grazed, and the forage consumed is very high quality. When not cut, plants can grow 20–25 cm tall. Because of its prostrate growth habit, annual production increases with defoliation in contrast to the upright growth habits of arrowleaf, crimson, and rose clovers (Evers and Newman 2008). The inflorescence is inconspicuous and usually contains four white to pinkish flowers. After fertilization, a seed-bearing burr is formed and is buried in the soil surface as the peduncle elongates. The stiff forked bristles on the seed burr serve as mechanisms to bury a portion of the burr. In Mediterranean climates with pronounced summer dry periods during seed formation, subclover produces a significant amount of hard seed and is an effective reseeder. Hard seed production is limited in the southeastern US where rain occurs throughout the year (Evers et al. 1988). Grazing to a 2–5 cm height during flowering enhances seed production. Because of its shallow root system, subclover does not tolerate drought and grows best on sandy loam to clay soils.

Because of relatively large seed, subclover has excellent seedling vigor and early nodulation, enhancing establishment and making it competitive in mixtures with cool-season annual grasses and when over-seeded on tropical perennial grasses (Evers 1999). Pure stands of subclover reached 100% light interception six to eight weeks after planting and recovered more rapidly after defoliation compared to clovers with upright growth (Evers and Newman 2008). Reviews on subclover are available from McGuire (1985) and Evers (1988).

Berseem Clover

Also called Egyptian clover, berseem clover likely originated in Syria and was first introduced into Egypt about the sixth century (Knight 1985b). It is widely grown as a winter annual in Egypt and India. Berseem clover is grown on limited hectares during the winter in the southeast US and under irrigation in the southwestern US. It is also seeded in the spring and used as a summer annual in the Midwest. Plants have an erect growth habit, oblong leaflets, and white flowers. Cultivars vary in branching behavior (Knight 1985b). Those with basal branching can be cut multiple times during the growing season. Those with basal and apical branching can be cut twice, and those with just apical branching can be cut just once.

Berseem clover lacks drought tolerance because of a short taproot. It is the least cold tolerant among annual clovers and should not be planted where air temperatures repeatedly reach −6 °C or lower (Table 14.1). It is best adapted to loam and clay soils with good internal drainage with maximum growth at pH of 7–8 (Evers 2003). It can tolerate temporary flooding. If inundated with water for three to four days, the top growth may senesce, but new growth will be initiated from the crown.

Berseem clover can be harvested for hay or be grazed, and has potential as a green manure crop (Westcott et al. 1995; Sheaffer et al. 2001). Cutting or grazing should be initiated when the clover reaches a height of 20–25 cm. During spring, berseem clover forage has a high proportion of stems relative to leaves that results in lower protein and digestibility than other clovers (Brink and Fairbrother 1992).

Ball Clover

This species is grown as a winter annual in the southeastern US (Hoveland and Evers 1995). Ball clover plays a unique role on loam or clay soils that may be wet during the winter but are droughty in the summer, though it does not tolerate wet soils as well as white clover (Table 14.1) (Hoveland and Evers 1995). It also does well on sandy soils with adequate rainfall during the growing season. Ball clover has smooth, egg-shaped leaflets and small, white to yellowish-white flowers. Maturity is about one to three weeks later than crimson clover. The seeds are very small (2 200 000 kg^{-1}) resulting in small seedlings that are not very competitive. It is best suited for overseeding into tropical perennial grasses. Ball clover is an excellent reseeder because of its ability to flower and produce seed when grazed to a 5-cm height and has a high hard-seed percentage. Reseeding potential of ball clover is superior to other annual clover species and annual medics (Muir et al. 2005). This is the main reason for its expanded use in recent years. Stems are prostrate to partially erect because they are succulent and hollow. Without defoliation, stems can exceed 75 cm in length but are short under grazing.

Persian Clover

This winter annual grows well on loam and clay soils in low-lying areas in the southeastern US (Table 14.1). Preferred soil pH is 7–8, with seedling growth and nodulation decreasing as soil pH decreases (Evers 2003). Plants are glabrous, have a decumbent growth habit when grazed, but do not root at the nodes. Small, light purple flowers are produced on short axillary peduncles. Stems are hollow and reach 60 cm if not defoliated. About 90% of the seed is hard, which contributes to natural reseeding. Persian clover is used for grazing or harvested as hay or silage with greatest productivity in March and April. Seed is available from Australia; there is no commercial seed production of persian clover in the US.

Rose Clover

Rose clover is a winter annual grown for grazing in north central Texas, Oklahoma, and rangelands in California (Love 1985; Hoveland and Evers 1995). Seedling growth and early forage production are less than most other annual clovers (Evers 1999); however, it can grow and persist on droughty, infertile soils, and in climates that are too dry for other clovers (Table 14.1). It grows best on soils with a pH of 6–7 but will grow on soils as low as pH 5.5 and as high as 8 (Evers 2003). Stems and leaves are pubescent and pale green. Leaflets are wedge-shaped with rounded tops and may have a pale crescent with a dark line above it. Growth habit is semi prostrate to erect, reaching a height of 40–50 cm. Rooting depth reaches 2 m, making rose clover more drought tolerant than shallow-rooted species. The head inflorescence is pale pink, about 2 cm in diameter, and contains 20–40 flowers. About 90% of the seed have hard coats, contributing to excellent reseeding.

When defoliated once or twice, shoot and root growth of rose clover are not as productive as arrowleaf, crimson, and subterranean clovers (Evers and Newman 2008). Shoot-N concentrations of rose clover are similar to the other clovers but root-N concentration is usually less, especially when cut once or twice (Evers and Parsons 2010).

Strawberry Clover

This stoloniferous perennial is similar to intermediate white clover in growth habit and shape of leaves (Townsend 1985). Seed heads are round and mostly pink and white, resembling a strawberry. Strawberry clover is unique in its adaptation to wet, saline, or alkaline soils (Townsend 1985). Because of its growth habit, it is well adapted to close defoliation and continuous stocking in the western US.

Lotus

Broadleaf birdsfoot trefoil is the primary *Lotus* sp. grown in the US. Narrowleaf trefoil and big trefoil are also grown but are less winter hardy and less widely adapted (Beuselinck 1999). Narrowleaf trefoil closely resembles birdsfoot trefoil, except the flowers are smaller and fewer and the leaflets are narrower. Big trefoil has relatively larger leaves and rhizomes. These legumes probably originated in southern Europe or the Mediterranean basin. Birdsfoot trefoil was initially brought to the US in the 1800s but did not become important until the 1930s.

Two types of birdsfoot trefoil, European and Empire, are identified based on growth habit. Empire types are related to a naturalized ecotype discovered in New York (Beuselinck 1999). Empire-type cultivars are generally finer stemmed, more prostrate in growth habit, later in flowering, and recover more slowly after harvest than the European types.

Birdsfoot trefoil has many fine stems that lodge. In the spring, regrowth comes from crown buds, but following defoliation, it regrows from axillary buds on lodged stems. Leaves have five leaflets: three grouped together at the end of the petiole and two at the petiole base near the stem. It is taprooted with lateral branching. Roots can produce adventitious buds and shoots when the crown is removed.

Flowers, borne in an umbel inflorescence, are various shades of yellow and may be tinged with red. Birdsfoot trefoil derives its name from the claw-like arrangement of four to eight 2.5- to 4.0-cm long cylindrical seedpods borne at right angles to the peduncle. The pods readily shatter when mature. Plants have an indeterminate growth habit with flowering occurring over an extended period of time. Birdsfoot trefoil will grow on a diversity of soils, is tolerant of wet soils, and can endure several weeks of flooding. It tolerates acid soil (pH 5.0) and is moderately tolerant of high alkalinity and salinity (Table 14.2). Root and crown diseases reduce individual plant longevity to four years or less in cool environments. If stressed by frequent defoliation, winter injury, or high temperatures, plants may live only two years It is less winter hardy than alfalfa, and stand persistence is aided by natural reseeding resulting from **seed shattering**.

Birdsfoot trefoil can be harvested for hay and silage, but because of its weak stems, it is best suited for pasture. Though its yield and, therefore, carrying capacity are less than for alfalfa, it has high forage quality to provide high daily gains for grazing animals. It does not cause bloat in grazing ruminants because plant tannins precipitate the soluble proteins that produce stable foams in the rumen. Tannins also protect the forage protein from degradation in the rumen (**bypass protein**) thereby increasing protein utilization.

Rotational stocking enhances yield and persistence of birdsfoot trefoil (Van Keuren and Davis 1968). If stocked continuously, plants can persist if leafy material is left to support rapid regrowth. In both rotational and continuous stocking, reseeding can occur if seed forms on prostrate stems that escape grazing. In northern regions, with less incidence of foliar disease, forage can be accumulated in place (stockpiling), with grazing delayed until mid-summer when forage supplies are short. Birdsfoot trefoil has also been widely used for conservation and beautification plantings along highways. However, it may be invasive into native plantings because of high seed production.

Other Temperate Legumes

Sweetclover

Biennial and annual sweetclovers originated in Eurasia and were grown in the US as early as 1739 (Miller and Hoveland 1995). There are white and yellow flower types of both annuals and biennials. Biennial types produce a single, well-branched stem of moderate height the first year but that rarely flowers (Smith and Gorz 1965). In late summer and autumn, after contractile growth places the basal node below soil level, there is a rapid increase in root size and enlargement of crown buds below the soil surface. The following spring, the buds produce numerous rapidly growing stems that can reach a height of 2 m. The name sweetclover is derived from the sweet vanilla-like odor that arises when leaves are crushed.

Annual types produce more top growth and less root growth than biennials do the first year. Annuals reach a height of 1–2 m in the seedling year. Sweetclover has a deep penetrating taproot that is probably responsible for its soil improvement benefits and its ability to survive under unfavorable climatic and edaphic conditions.

Sweetclover flowers are borne in loose racemes. The species names reflect their flower color. Yellow sweetclover (*Melilotus officinalis*) is later maturing, finer stemmed, and has better forage quality than white sweetclover (*Melilotus alba*). Leaves are pinnately trifoliolate with the center leaflet on an elongated **petiolule**. Leaflets are broadly obovate, rounded, and serrated on the margins. Sweetclover produces a high percentage of hard seed, with seed surviving in the soil for over 20 years (Smith and Gorz 1965). It is best adapted to soils with a pH greater than 6.5 and is adapted to low-rainfall regions (Table 14.2).

Sweetclover was used as a green manure and for weed suppression (Blackshaw et al. 2001), but that use has declined due to its invasion potential (Lesica and DeLuca 2000) and the availability of synthetic N fertilizers. Sweetclover volunteers in pastures, wastelands, and roadsides and is used in pasture in dryland regions. It is a source of nectar for honey production. Its use for hay is very limited because the course forage is hard to evenly dry and often contains molds. Consumption of moldy sweetclover hay causes "bleeding disease." Sweetclover herbage contains coumarin, an aromatic compound that reduces herbage palatability and that is converted to dicoumarol by mold. Dicoumarol reduces the blood clotting of animals, which can result in death from internal or external bleeding (Miller and Hoveland 1995). Low-coumarin sweetclover has been developed (Meyer 2005). "Silver River" annual sweetclover has improved rust (*Uromyces striatus* Schroet.) resistance and has been developed for use on the southern plains (Smith et al. 2017). An extensive review of sweetclover is available (Smith and Gorz 1965).

Crownvetch

This native of central Europe was available in the US in the 1890s (Miller and Hoveland 1995). Its name is derived from the crownlike arrangement of white-purple flowers in an **umbel** inflorescence. Leaves are pinnately compound, each containing 8–15 pairs of leaflets plus one **terminal** leaflet. Flowering is indeterminate; thus, flowering and seed production continue on the trailing stems throughout the growing season. Seeds are borne in nonshattering pods that break into sections when dry. It has a deep and creeping root system.

Crownvetch is a long-lived perennial but winter hardiness and persistence are reduced by frequent harvest. It is tolerant of low fertility and low pH (Table 14.2). It is used in the northern US for beautification and erosion control on highway embankments and mine spoil sites. It has a creeping, prostrate grown habit that makes it best suited for pasture, but poor palatability due to antiquality components and poor persistence have limited its use as a forage crop. If harvested for forage, two harvests per season at flowering produce the greatest yield.

Cicer Milkvetch

This perennial legume is found from the Caucasus Mountains across southern Europe to Spain (Miller and Hoveland 1995). It is very winter hardy and well adapted to low rainfall regions in the Great Plains and western US and to soils that are slightly acidic to moderately alkaline (Table 14.2). Cicer milkvetch is a vigorous, persistent, high-yielding legume that spreads by rhizomes and has a deep root system.

Cicer milkvetch can be grazed or harvested as hay. Herbage consists of long trailing stems that can reach 1.5 m if not cut. Leaves are pinnately compound with 8–17 pairs of leaflets plus one **terminal** leaflet. Yellow-white flowers are borne on racemes. Seedpods are bladderlike and contain up to 12 seeds. Protein and fiber levels of cicer milkvetch are similar to alfalfa, but herbage can be less palatable and cause **photosensitization** in grazing ruminants (Sheaffer et al. 2018).

Sainfoin

This native to Europe and western Asia is best adapted to dry calcareous soils in the western US and will tolerate soils low in P, but not waterlogging or acid soils. Sainfoin has a deep, branched taproot, a narrow crown, and erect stems that can grow in height to 1 m. Leaves are pinnately compound with 10–29 leaflets. Elongated racemes may contain up to 80 pink flowers (Miller and Hoveland 1995). Sainfoin is harvested as hay or by grazing. Pastures of sainfoin, a non-bloating legume, can provide excellent daily gains for cattle or sheep. Low to medium stocking rates at the bud or flower stages or rotational stocking are necessary for good stand persistence. An extensive symposium on sainfoin is reported by Cooper and Carleton (1968).

Annual Vetches

The genus *Vicia* contains about 150 species, including 15 that are native to the US (Hoveland and Townsend 1985). Most vetches are native to the Mediterranean region. Jannink et al. (1997) reported that commercially available *Vicia villosa* in the US is of two types, subsp. *villosa* and subsp. *varia* with the hairy subsp. *villosa* the more winterhardy. Vetches have stems up to 2 m long that trail on the soil or cling on companion plants. Leaves have from 8 to 10 leaflets and terminate in tendrils that aid climbing of vetch. The flowers on elongated racemes vary in color from white, pink, and purple. Vetches are among the more acid-tolerant forage legumes (Table 14.1). Hairy vetch was found to be the most cold tolerant winter annual legume (Brandsaeter et al. 2002); however, it lacks **winterhardiness** to reliably overwinter in northern regions (Harbur et al. 2009). Vetches are mostly cultivated as winter annuals for winter cover, green manure and forage (Frye et al. 1988; Noland et al. 2018). Hairy vetch is used as a winter cover crop following annual grain crops for soil erosion control and weed suppression (Clark 2007; Leavitt et al. 2011; Noland et al. 2018) and efforts are underway to identify and evaluate new germplasm that will reliably overwinter (Maul et al. 2011; Wiering et al. 2018). Vetches are often grown with small grains to facilitate forage harvest. Grazing should not begin until plants are 15 cm tall, and plants should not be grazed closer than the lowest leaf axil. Clipping to a 2 to 4-cm height three times reduced yield but increased tillering (Lynd et al. 1980). Vetch is typically harvested for hay at early flowering. Vetches also volunteer in wastelands and along roadsides.

Miscellaneous Legumes

There are many legumes that are grown as forages on low hectarage or that have become naturalized in pastures and wastelands throughout the US. Field pea, lupine, roughpea, and cowpea are large-seeded annual legumes (Miller and Hoveland 1995). Miscellaneous small-seeded annual clovers described by Hoveland and Evers (1995) include hop clover that acts as a winter annual on infertile soils in the South and a summer annual in the North. Lappa clover is adapted to low-lying, calcareous soils in the lower South, cluster clover to the Gulf Coast, and striate clover to clay soils in the lower Southeast.

References

Baker, J.M. (2012). Vegetative propagation of kura clover: a field scale test. *Can. J. Plant. Sci.* 92: 1245–1251.

Ball, D.M., Hoveland, C.S., and Lacefield, G.D. (1991). *Southern Forages*. Atlanta, GA: Potash and Phosphate Institute and the Foundation for Agronomic Research.

Barnes, D.K., Bingham, E.T., Murphy, R.P. et al. (1977). Alfalfa germplasm in the United States: Genetic vulnerability, use, improvement and maintenance. United States Department of Agriculture. Technical Bulletin No. 1571.

Barnes, D.K. and Sheaffer, C.C. (1995). Alfalfa. In: *Forages: An Introduction to Grassland Agriculture*, 5e (eds. R.F Barnes, D.A. Miller and C.J. Nelson), 205–216. Ames, IA: Iowa State University Press.

Beuselinck, P.R. (ed.) (1999). *Trefoil: The Science and Technology of the Lotus*. CSSA Special Pub. 28. Madison, WI: ASA, CSSA, and SSSA.

Blackshaw, R.E., Moyer, J.R., Doram, R.C., and Boswell, A.L. (2001). Yellow sweetclover, green manure, and its residues effectively suppress weeds during fallow. *Weed Sci.* 49: 406–413.

Blaser, B.C., Singer, J.W., and Gibson, L.R. (2012). Winter wheat and red clover intercrop response to tillage and compost amendment. *Crop Sci.* 52: 320–326.

Brandsaeter, L.O., Olsmo, A., Tronsmo, A.M., and Fykse, H. (2002). Freeze resistance of winter annual and biennial legumes at different development stages. *Crop Sci.* 42: 437–443.

Brink, G.E. and Fairbrother, T.E. (1992). Forage quality and morphological components of diverse clovers during primary spring growth. *Crop Sci.* 32: 1043–1048.

Brink, G.E., Pederson, G.A., Alison, M.W. et al. (1999). Growth of white clover ecotypes, cultivars, and germplasms in the southeastern USA. *Crop Sci.* 39: 1809–1814.

Buxton, D., Hornstein, J.S., Wedin, W.F., and Marten, G.C. (1985). Forage quality in stratified canopies of alfalfa, birdsfoot trefoil, and red clover. *Crop Sci.* 25: 273–279.

Catalano, D., Sheaffer, C., Grev, A. et al. (2019). Yield, forage nutritive value, and preference of legumes under horse grazing. *Agron. J.* 111: 1312–1322. https://doi.org/10.2134/agronj2018.07.0442.

Clark, A. (ed.) (2007). Managing cover crops profitably. In: *National SARE Outreach Handbook Series Book 9*, 3e, 1–244. Beltsville, MD: National Agricultural Laboratory. (Available online at: http://www.sare.org/publications/covercrops.htm) (verified 24 March 2010).

Cooper, C.S. and Carleton, A.E. (eds.). (1968). Sainfoin Symposium. Montana Agricultural Experiment Station. Bulletin No. 627.

Coulter, J.A., Sheaffer, C.C., Wyse, D.L. et al. (2011). Agronomic performance of cropping systems with contrasting crop rotations and external inputs. *Agron. J.* 103: 189–192.

Decker, A.M., Clark, A.J., Meisinger, J.J. et al. (1994). Legume cover crop contributions to no-tillage corn production. *Agron. J.* 86: 126–135.

Evers, G.W. (ed.) (1988). *Subterranean Clover: Establishment, Management, and Utilization in Texas.* College Station, TX: Texas Agricultural Experiment Station Bulletin. MP-1640.

Evers, G.W. (1999). Seedling growth comparison of arrowleaf, crimson, rose, and subterranean clovers. *Crop Sci.* 39: 433–440.

Evers, G.W. (2003). Emergence and seedling growth of seven cool-season annual clovers as influenced by soil pH. *J. Sustainable Agric.* 23: 89–107.

Evers, G.W. (2011a). Forage legumes: forage quality, fixed nitrogen, or both. *Crop Sci.* 51: 403–407.

Evers, G.W. (2011b). The interaction of annual ryegrass and nitrogen on arrowleaf clover in the southeastern United States. *Crop Sci.* 51: 1353–1360.

Evers, G.W. and Newman, Y.C. (2008). Arrowleaf, crimson, rose, and subterranean clover growth with and without defoliation in the southeastern United States. *Agron. J.* 100: 221–230.

Evers, G.W. and Parsons, M.J. (2010). Nitrogen partitioning in arrowleaf, crimson, rose, and subterranean clovers without and with defoliation. *Crop Sci.* 50: 1562–1575.

Evers, G.W. and Parsons, M.J. (2011). Estimated N_2-fixation of cool-season annual clovers by the difference method. *Crop Sci.* 51: 2276–2283.

Evers, G.W. and Smith, G.R. (1998). Research Center Technical Report No. 98-2, Texas Agricultural Experiment Station.

Evers, G.W., Smith, G.R., and Beal, P.E. (1988). Subterranean clover reseeding. *Agron. J.* 80: 855–859.

Frye, W.W., Blevins, R.L., Smith, M.S. et al. (1988). Role of annual legume cover crops in efficient use of water and nitrogen. In: *Cropping Strategies for Efficient Use of Water and Nitrogen* (ed. W.L. Hargrove), 129–154. Madison, WI: ASA.

Gibson, P.R. and Cope, W.A. (1985). White clover. In: *Clover Science and Technology*. Agron. Monogr. 25 (ed. N.L. Taylor), 471–490. Madison, WI: ASA, CSSA, SSA.

Gillett, J.M. and Taylor, N.L. (2001). *The World of Clovers* (ed. M. Collins). Ames, IA: Iowa State University Press.

Goplen, J.J., Coulter, J.A., Sheaffer, C.C. et al. (2018). Economic performance of crop rotations in the presence of herbicide-resistant giant ragweed. *Agron. J.* 110: 260–268.

Grev, A.M., Wells, M.S., Samac, D.A. et al. (2017). Forage accumulation and nutritive value of reduced lignin and reference alfalfa cultivars. *Agron. J.* 109: 1–13.

Hanson, A.A., Barnes, D.K., and Hill, R.R. Jr. (eds.) (1988). *Alfalfa and Alfalfa Improvement*. Madison, WI: ASA, CSSA, SSSA.

Harbur, M.M., Sheaffer, C.C., Moncada, K.M., and Wyse, D.L. (2009). Selecting hairy vetch ecotypes for winter hardiness in Minnesota. *Crop Manage.* https://doi.org/10.1094/CM-2009-0831-01-RS.

Hoveland, C.S. and Evers, G.W. (1995). Arrowleaf, crimson, and other annual clovers. In: *Forages: An Introduction to Grassland Agriculture*, 5e, vol. 1 (eds. R.F Barnes, D.A. Miller and C.J. Nelson), 249–260. Ames, IA: Iowa State University Press.

Hoveland, C.S. and Townsend, C.E. (1985). Other legumes. In: *Forages: The Science of Grassland Agriculture*, 4e (eds. M.E. Heath et al.), 146–153. Ames, IA: Iowa State University Press.

Jannink, J.L., Merrick, L.C., Liebman, M. et al. (1997). Management and winter hardiness of hairy vetch in Maine. In: *Maine Agricultural and Forest Experiment Station* Technical Bulletin 167, 1–35.

Jeranyama, P., Hesterman, O.B., and Sheaffer, C.C. (1998). Medic planting date effect on dry matter and nitrogen accumulation when clear-seeded or intercropped with corn. *Agron. J.* 90: 616–622.

Knight, W.E. (1970). Productivity of crimson and arrowleaf clovers grown in a coastal bermudagrass sod. *Agron. J.* 62: 773–775.

Knight, W.E. (1985a). Crimson clover. In: *Clover Science and Technology*. Agron. Monogr. 25 (ed. N.L. Taylor), 491–502. Madison, WI: ASA, CSSA, SSSA.

Knight, W.E. (1985b). Miscellaneous annual clovers. In: *Clover Science and Technology*. Agron. Monogr. 25 (ed. N.L. Taylor), 547–562. Madison, WI: ASA, CSSA, and SSSA.

Lamb, J.F.S., Sheaffer, C.C., Rhodes, L.H. et al. (2006). Five decades of alfalfa cultivar improvement: impact on forage yield, persistence and nutritive value. *Crop Sci.* 46: 902–909.

Leath, K.T., Erwin, D.C., and Griffin, G.D. (1988). Diseases and nematodes. In: *Alfalfa and Alfalfa Improvement*. Agron. Monogr. 29 (eds. A.A. Hanson, D.K. Barnes and R.R. Hill), 621–670. Madison, WI: ASA, CSSA, and SSSA.

Leavitt, M.J., Sheaffer, C.C., Wyse, D.L., and Allan, D.L. (2011). Rolled winter rye and hairy vetch cover crops lower weed density but reduce vegetable yields in no-tillage organic production. *Hortscience* 46: 387–395.

Lesica, P.L. and DeLuca, T.H. (2000). Sweetclover, a potential problem for the northern Great Plains. *J. Soil Water Conserv.* 55: 259–261.

Love, R.M. (1985). Rose clover. In: *Clover Science and Technology*. Agron. Monogr. 25 (ed. N.L. Taylor), 535–546. Madison, WI: ASA, CSSA, and SSSA.

Lynd, J.Q., McNew, R.W., and Odell, G.V. Jr. (1980). Defoliation effects on regrowth, nodulation, and nitrogenase activity at anthesis with hairy vetch. *Agron. J.* 72: 991–994.

Manglitz, G.R. (1985). Insects and related pests. In: *Clover Science and Technology*. Agron. Monogr. 25 (ed. N.L. Taylor), 269–294. Madison, WI: ASA, CSSA, and SSSA.

Manglitz, G.R. and Ratcliffe, R.H. (1988). Insects and mites. In: *Alfalfa and Alfalfa Improvement*. Agron. Monogr. 29 (eds. A.A. Hanson, D.K. Barnes and R.R. Hill), 671–704. Madison, WI: ASA, CSSA, and SSSA.

Marten, G.C., Jordan, R.M., and Ristau, E.A. (1990). Performance and adverse response of four legumes. *Crop Sci.* 30: 860–866.

Maul, J., Mirsky, S., Emche, S., and Devine, T. (2011). Evaluating a germplasm collection of the cover crop hairy vetch for use in sustainable farming systems. *Crop Sci.* 51: 2615–2625.

McGuire, W.S. (1985). Subterranean clover. In: *Clover Science and Technology*. Agron. Monogr. 25 (ed. N.L. Taylor), 515–534. Madison, WI: ASA, CSSA, and SSSA.

Meyer, D. (2005). Sweetclover production and management. North Dakota State University R-862. https://library.ndsu.edu/ir/bitstream/handle/10365/9132/R862_2005.pdf?sequence=1&isAllowed=y (accessed 22 November 2019).

Miller, D.A. and Hoveland, C.S. (1995). Other temperate legumes. In: *Forages: An Introduction to Grassland Agriculture*, 5e, vol. 1 (eds. R.F Barnes, D.A. Miller and C.J. Nelson), 273–281. Ames, IA: Iowa State Univ. Press.

Mourino, F., Albrecht, K.A., Schaefer, D.M., and Berzaghi, P. (2003). Steer performance on kura clover-grass and red clover-grass mixed pastures. *Agron. J.* 95: 652–659.

Muir, J.P., Ocumpaugh, W.R., and Butler, T.J. (2005). Trade-offs in forage and seed parameters of annual *Medicago* and *Trifolium* spedies in north-central Texas as affected by harvest intensity. *Agron. J.* 97: 118–124.

Noland, R.L., Wells, M.S., Sheaffer, C.C. et al. (2018). Establishment and function of cover crops interseeded into corn. *Crop Sci.* 58: 1–11.

Ocumpaugh, W.R. (2001). Developing annual medics for Texas. Proceedings of the 56th Southern Pasture Forage Crop Improvement Conference, Springdale, AR, USA (21–22 April 2001).

Ochsner, T.E., Albrecht, K.A., Schumacher, T.W. et al. (2010). Water balance and nitrate leaching under corn in kura clover living mulch. *Agron. J.* 102: 1169–1178.

Owens, V.N., Albrecht, K.A., Muck, R.E., and Duke, S.H. (1999). Protein degradation and fermentation characteristics of red clover and alfalfa silage harvested with varying levels of total nonstructural carbohydrates. *Crop Sci.* 39: 1873–1880.

Pederson, G.A. (1995). White clover and other perennial clovers. In: *Forages: An Introduction to Grassland Agriculture*, 5e, vol. 1 (eds. R.F Barnes, D.A. Miller and C.J. Nelson), 227–236. Ames, IA: Iowa State University Press.

Pemberton, I.J., Smith, G.R., Philley, G.L. et al. (1998). First report of *Pythium ultimum, P. irregular, Rhizoctonia solani* AG4 *and Fusarium proliferatum* from arrowleaf clover: a disease complex. *Plant Dis.* 82: 128.

Perfect, E., Kay, B.D., van Loom, W.K.P. et al. (1990). Rates of change in soil structural stability under forages and corn. *Soil Sci. Soc. Am. J.* 54: 179–186.

Peterson, P.R., Sheaffer, C.C., and Jordan, R.M. (1994). Responses of kura clover to sheep grazing and clipping. 1. Yield and forage quality. *Agron. J.* 86: 660–667.

Quiros, C.F. and Buachan, G.R. (1988). The genus *Medicago* and the origin of the *Medicago sativa* complex. In: *Alfalfa and Alfalfa Improvement* (eds. A.A. Hanson, D.K. Barnes and R.R. Hill), 93–124. Madison, WI: ASA, CSSA, SSSA Monogr. 29. ASA.

Russelle, M.P. (2001). Alfalfa. *Am. Sci.* 89: 252–261.

Samac, D.A., Rhodes, L.H., and Lamp, W.O. (2014). *Compendium of Alfalfa Diseases and Pests*. St. Paul, MN: Am. Phytopath. Soc.

Shaeffer, C.C., Wells, M.S., and Nelson, C.J. (2017). Legumes for northern areas. In: *Forages: An Introduction to Grassland Agriculture*, 7e, vol. 1 (eds. M. Collins, C.J. Nelson, K.J. Moore and R.F Barnes), 117–131. Wiley Blackwell.

Sheaffer, C.C. and Moncada, K.M. (2012). *Introduction to Agronomy*, 2e. Clifton, Park, NY: Delmar Cengage Learning.

Sheaffer, C.C. and Seguin, P. (2009). Kura clover response to drought. *Forage and Grazinglands* 7 (1). doi:10.1094/FG-2009-1231-01-RS.

Sheaffer, C.C., Lacefield, G.D., and Marble, V.L. (1988a). Cutting schedules and stands. In: *Alfalfa and Alfalfa Improvement*. Agron. Monogr. 29 (eds. A.A. Hanson, D.K. Barnes and R.R. Hill), 411–437. Madison, WI: ASA, CSSA, SSSA.

Sheaffer, C.C., Tanner, C.B., and Kirkham, M.B. (1988b). Alfalfa water relations and irrigation. In: *Alfalfa and Alfalfa Improvement*. Agron. Monogr. 29 (eds. A.A. Hanson, D.K. Barnes and R.R. Hill), 373–409. Madison, WI: ASA, CSSA, SSSA.

Sheaffer, C.C., Marten, G.C., Jordan, R.M., and Ristau, E.A. (1992). Forage potential of kura clover and birdsfoot trefoil when grazed by sheep. *Agron. J.* 84: 176–180.

Sheaffer, C.C., Cash, D., Ehlke, N.J. et al. (1998). Entry X environment interactions for alfalfa forage quality. *Agron. J.* 90: 774–780.

Sheaffer, C.C., Simons, S.R., and Schmitt, M.A. (2001). Annual medic and berseem clover dry matter and nitrogen production in rotation with corn. *Agron. J.* 93: 1080–1086.

Sheaffer, C.C., Undersander, D.J., and Becker, R.L. (2007). Comparing Roundup Ready and conventional systems of alfalfa establishment. *Forage and Grazinglands* 5 (1). doi:10.1094/FG-2007-0724-01-RS.

Sheaffer, C.C., Sollenberger, L.E., Hall, M.H., West, C.P., and Hannaway, D.B. (2009). Grazinglands, forages and livestock in humid regions. In: *Grassland* (eds. W.F. Wedin and S.L. Fales), 95–118. Madison, WI: ASA, CSSA, SSSA.

Sheaffer, C.C., Ehlke, N.J., Albrecht, K.A. et al. (2018). *Forage Legumes*. Bulletin 608-2018. St. Paul, MN: Minnesota Agricultural Experiment Station.

Sims, J.R. and Slinkard, A.E. (1991). Development and evaluation of germplasm and cultivars of cover crops. In: *Cover Crops for Clear Water* (ed. W.L. Hargrove), 121–129. Jackson, TN: Proceedings of Soil Conservation Society of America. 9–11 Apr. Soil Water Conservation Society, Ankeny, IA.

Sleugh, B., Moore, K.J., George, J.R., and Brummer, E.C. (2000). Binary legume–grass mixtures improve forage yield, quality, and seasonal distribution. *Agron. J.* 2000 (92): 24–29.

Smith, W.K. and Gorz, H.J. (1965). Sweetclover improvement. In: *Advances in Agronomy*, vol. 17 (ed. A.F. Norman), 163–231. New York, NY: Academic Press.

Smith, S.R., Bouton, J.H., Singh, A., and McCaughey, W.P. (2000). Development and evaluation of grazing tolerant alfalfa cultivars: a review. *Can. J. Plant Sci.* 80: 503–512.

Smith, G.R., Evers, G.W., Ocumpaugh, W.R. et al. (2017). Registration of 'Silver River' sweetclover. *J. Plant Regist.* 11: 112–115.

Taylor, N.L. (ed.) (1985). *Clover Science and Technology*. Agron. Monogr. No. 25. Madison, WI: ASA, CSSA and SSSA.

Taylor, N.L. and Smith, R.R. (1995). Red clover. In: *Forages Vol. 1. An Introduction to Grassland Agriculture* (eds. R.F Barnes, D.A. Miller and C.J. Nelson), 217–225. Ames, IA: Iowa State University Press.

Townsend, C.E. (1985). Miscellaneous perennial clovers. In: *Clover Science and Technology*. Agron. Monogr. No. 25 (ed. N.L. Taylor), 563–575. Madison, WI: ASA, CSSA, SSSA.

United States Department of Agriculture, FSA (2019). Crop acreages. https://www.fsa.usda.gov/news-room/efoia/electronic-reading-room/frequently-requested-information/crop-acreage-data/index (accessed 9 October 2019).

Van Keuren, R.W. and Davis, R.R. (1968). Persistence of birdsfoot trefoil, *Lotus corniculatus* L. as influenced by plant growth habit and grazing management. *Agron. J.* 60: 92–95.

Ventroni, L.M., Volenec, J.J., and Cangiano, C.A. (2010). Fall dormancy and cutting frequency impact on alfalfa yield and yield components. *Field Crop Res.* 119: 252–259.

Westcott, M.P., Welty, L.E., Knox, M.L., and Prestbye, L.S. (1995). Managing alfalfa and berseem clover for forage and plowdown nitrogen in barley rotations. *Agron. J.* 87: 1176–1181.

Wiering, N.P., Flavin, C., Sheaffer, C.C. et al. (2018). Winter hardiness and freezing tolerance in a hairy vetch collection. *Crop Sci.* 58: 1594–1604.

Wilson, L.C., Braul, A., and Entz, M.H. (2017). Characteristics of black medic seed dormancy loss in Western Canada. *Agron. J.* 109: 1404–1413.

Woodfield, D., Albrecht, K.A., Bures, E.J. et al. (1998). Characterization and performance of Wisconsin white clover ecotypes. Proceedings of the 15th Trifolium Conference, Madison, WI, USA (10–12 June 1998).

Yost M.A., Coulter, J.A., and Russelle, M.P. (2015). Managing the Rotation from Alfalfa to Corn. University of Minnesota Extension. https://extension.umn.edu/corn-cropping-systems/managing-rotation-alfalfa-corn (accessed 22 November 2019).

Zemenchik, R.A., Wollenhaupt, N.C., Albrecht, K.S., and Bosworth, A.H. (1996). Runoff, erosion, and forage production from established alfalfa and smooth bromegrass. *Agron. J.* 88: 461–466.

Zemenchik, R.A., Albrecht, K.A., Boerboom, C.M., and Lauer, J.G. (2000). Corn production with kura clover as a living mulch. *Agron. J.* 92: 698–705.

Zhu, Y., Sheaffer, C.C., Russelle, M.P., and Vance, C.P. (1998). Dry matter accumulation and dinitrogen fixation of annual *Medicago* species. *Agron. J.* 90: 103–108.

Chapter 15

Legumes for Tropical and Subtropical Areas

William D. Pitman, Professor, *Louisiana State University Agricultural Center, Homer, LA, USA*
João M. B. Vendramini, Associate Professor, *Agronomy, Range Cattle Research and Education Center, University of Florida, Ona, FL, USA*

Introduction

A distinct separation of the highly diverse **legume** family into two contrasting categories of tropical and temperate species is not generally recognized, though some species and genera are readily distinguishable as either primarily **tropical** or temperate. In general, the temperate forage legumes include only a few species and are leafy, **herbaceous** plants with optimum growth temperatures below 30 °C. Tropical forage legumes include numerous genera and species that range widely in growth form and have optimum growth temperatures slightly above 30 °C. Many tropical forage legume genera include species that are primarily **subtropical** or even warm-season species of temperate climates. Thus, a few species included in this chapter are primarily useful as warm-season species in temperate regions, though emphasis is on forage legumes of the tropics and subtropics.

The greatest diversity of tropical forage legume germplasm is in the American tropics. While native legume populations have undoubtedly contributed to forage for livestock, natural legume populations in tropical America tend to be more prevalent outside pastures than within them. Beginning with commercial use of the **genus** *Stylosanthes* Sw. in the 1930s (Humphries 1967), Australia led in the adoption of tropical forage legumes. Early successes led to subsequent collection, assessment, and evaluation of diverse tropical legume germplasm.

During the last half of the twentieth century, numerous tropical forage legume cultivars were developed. Early cultivars, which were subsequently distributed throughout tropical and subtropical regions of the world, were developed in Australia (Jones 2001). Cultivar development consisted primarily of the selection and commercialization of naturally occurring genotypes with very limited contribution from plant breeding (Clements 1989).

Early cultivars were often selected from plot evaluations, where primary emphasis was placed on forage yield. Outcomes of insufficient tolerance to commercial grazing management, limited competitive ability with aggressive tropical grasses, and narrow areas of adaptation led to many failures for initial pasture evaluations and commercial plantings. **Persistence** in commercial plantings was, at times, also limited by the high incidences of diseases and insects not encountered in smaller-scale plantings. Nonetheless, vigorous early growth of the widely distributed cultivars repeatedly demonstrated the tremendous potential of tropical forage legumes. This potential continues to propel the search for tropical forage legumes with increased productivity and persistence in various environments across tropical and subtropical locations (Bell et al. 2015). Narrower ranges of adaptation and poorer persistence than the available tropical grasses have limited the use of tropical legumes in general (Marten et al. 1989; Muir et al. 2014).

Forages: The Science of Grassland Agriculture, Volume II, Seventh Edition.
Edited by Kenneth J. Moore, Michael Collins, C. Jerry Nelson and Daren D. Redfearn.

Tropical Legume Forage Characteristics

In contrast to the low-quality, unpalatable roughage typical of mature tropical grasses, many tropical legumes can enhance dietary protein, intake, rate of passage, and subsequent animal performance on tropical grass pastures. Thus, a primary objective of legumes in tropical grass pastures is provision of increased crude protein (CP). As little as 10% legume in a tropical grass diet has overcome CP deficiency and increased grass intake (Minson and Milford 1966; Minson 1980). Another major objective for these legumes is N_2 fixation to support increased forage production. Tropical forage legumes include more genetic diversity than either tropical grasses or temperature forage legumes. Tropical legumes range from low-growing, prostrate types to trees and vines that can grow to several meters in height. Limitations of some tropical legumes for use as forage include presence of thorns, accumulation of toxic secondary metabolic products, or production of **foliage** at a height beyond reach of livestock. A few tropical legumes thrive in waterlogged soils, and others require well-drained soils for survival. There are a few grazing-tolerant species and numerous species susceptible to stand loss from excessive defoliation, especially in pastures with competitive tropical grasses.

In general, collection, storage, and initial evaluation of tropical legume germplasm is similar to that for grasses. The need for associated N_2-fixing bacteria, however, requires consideration of legume–bacteria strain associations. These legumes are more specific in edaphic and climatic requirements than are the associated grasses. Additional considerations for collection, storage, and assessment of tropical legume germplasm have been presented by Stace and Edye (1984), Bray (1985b), Schultze-Kraft and Benavides (1988), Kretschmer (1989), and Sonoda et al. (1991).

Tropical Legume Germplasm

There are about 18 000 species in the family Fabaceae (Leguminosae) with subfamilies Caesalpinioideae (2800–2900 spp.), Mimosoideae (2800–2900 spp.), and Papilionoideae (11 000–12 000 spp.). They are separated into about 47 tribes and 686 genera. Though evolving primarily in Africa prior to continental shifts, the majority of tropical legume genera and species are currently found in tropical America (Summerfield and Bunting 1980; Polhill and Raven 1981; Schultze-Kraft and Keller-Grein 1999; Loch and Ferguson 1999). The tropical forage legumes that have been released for commercial use are included mainly in six tribes. Almost all commercial tropical legumes are found in the subfamily Papilionoideae. Exceptions include leucaena and desmanthus in Mimosoideae, and roundleaf cassia in Caesalpinioideae.

The diversity of tropical legume species with potential as forage is much greater than that of temperate legumes. This added diversity of plant growth form provides both additional forage uses and management requirements. Despite successful use of some individual legume species in suitable environments, particularly in locations in Australia and Asia, many promising legumes and tropical environments have proven challenging for the germplasm evaluated and approaches used. The tremendous diversity of tropical legume germplasm, with much of this diversity held in existing but inadequately evaluated germplasm collections, provides continuing promise for further developments with tropical and subtropical pastures. Following a period of several years, with decreasing emphasis on tropical and subtropical legumes from some previously leading organizations, renewed enthusiasm has been introduced with proposal of a new germplasm strategy (Pengelly 2015). Proposed coordination of germplasm conservation and evaluation efforts could greatly enhance productivity from limited resources. Along with successful cultivars of a number of species, particular potential has recently been identified for some genera as illustrated by Bell et al. (2015).

Though common names are recognized for some species in some locations, these names differ among locations and contribute to confusion in species identification. For some widely used legume genera, particularly *Stylosanthes* (Calles and Schultze-Kraft 2016), ongoing taxonomic contributions on portions of the genus have provided uncertainty in appropriate classification of some germplasm in field evaluations. Use of genus and species is the commonly used nomenclature for identification of tropical legume germplasm with appropriate taxonomic treatment essential for communication of results from widely dispersed locations.

Tropical Legume Genera and Species

Aeschynomene L.

Aeschynomene is primarily a tropical and subtropical genus with about 160 species (Rudd 1955; Bishop et al. 1988). It is native to American tropical areas between about 30 °N and 30 °S latitudes and up to about 2500 m; however, a few of the nine American species are found on the Atlantic Coast of the US to 40 °N latitude and in South America to about 35 °S latitude. The range on the Pacific Coast is from 28 °N to 17 °S latitudes. Other natural distributions are principally in Africa, with a few species native to Asia and the Pacific islands.

About half of the species are **xeric** and the rest are hydrophytes. General morphology ranges from herbs several centimeters in height to small trees up to 8 m high, with intermediate forms being erect or prostrate subshrubs and shrubs. No species is known to be toxic to cattle, though at least one [(*A. elaphroxylon*) Taub.] has sharp thorns. Diseases of many aeschynomene species have been reported by Sonoda and Lenne (1986). Species, cultivars, and zones of adaptation have been described for Queensland, Australia (Jones 2001), with additional potential for the genus recently recognized (Bell et al. 2015).

American Jointvetch (A. americana L.)

This species is a self-regenerating **annual** herbaceous legume similar in appearance to some genotypes of *Aeschynomene villosa*. It is erect to branching with height up to 1–2 m. Stems become woody with maturity. Large quantities of easily harvested seed are produced, however, flowering is only in response to short days. A common type has been used successfully in Florida. In Australia 'Glenn' and 'Lee' have been released (Loch and Ferguson 1999). In Florida, Hodges et al. (1982) suggested aeschynomene is best suited on moist, seasonally waterlogged soils that temporarily flood. Lack of persistence is partially due to erratic early season rainfall, which is often sufficient for germination. Subsequent dry periods can be fatal to the slow-developing seedlings. While growing, CP and digestibility of leaves range from 150 to 250 and 600 to 700 $g\,kg^{-1}$, respectively. With onset of seed formation and plant maturation, however, quality and intake by grazing livestock begin to decrease until plant death. Since animals only graze the pinnately compound leaves and tender stems, there is little relationship between animal performance and yield or quality of the total plant.

Evenia Jointvetch (A. evenia C. Wright)

Evenia jointvetch is native of southern Texas, Cuba, and northern South America into Brazil. It is a short-lived perennial that is somewhat similar to american jointvetch. However, evenia jointvetch survives better under waterlogging on sandy soils, persists through mild winters in subtropical climates and, being day-length neutral, may produce seed during much of the growing season. With proper management, two seed harvests can be made when soil moisture and temperature are adequate.

Seed weight and seedling vigor of evenia jointvetch are greater than for american jointvetch providing establishment advantages. Seed germination and moderate seedling growth occur throughout the mild winter in south Florida, while american jointvetch usually germinates only in the spring with rapid seedling growth not until late May or June. As with most species of the genus, mature woody growth is not **palatable** to most grazing livestock.

Jointvetch (A. falcata Poir.)

Jointvetch is a perennial native to elevations of 1800 m in northwestern and east-central South America. Prostrate and herbaceous, with stems to about 1 m long, jointvetch is drought tolerant and adapted to the subtropics, but not to the humid tropics. 'Bargoo,' an Australian cultivar, has been used since 1973 and persists well where adapted (Wilson et al. 1982).

Villosa Jointvetch (A. villosa Poir.)

Villosa jointvetch has both annual and perennial cultivars. Growth can be prostrate to erect. The species occurs from Mexico south through most of tropical America. Several genotypes have been selected in Australia from which 'Kretschmer' and 'Reid' were released in 1995 (Loch and Ferguson 1999). This species has persisted in Florida evaluations under a variety of grazing management approaches, but it does not tolerate waterlogging.

Alysicarpus Neck. Ex Desv.

The genus *Alysicarpus* is native to tropical Asia and contains about 30 species. The genus includes annual and perennial types, both of which have potential forage value (Gramshaw et al. 1987). To date, commercial planting has been limited primarily to a single species, though recent assessments of evaluations across Australian tropical locations suggest potential for several species of the genus on productive sites (Bell et al. 2015).

Alyceclover, (A. vaginalis [L.] DC)

Alyceclover, also commonly called buffalo clover in Australia, is naturalized in most countries of tropical America. There are two distinct forms, one being primarily erect and annual (cultivated form), the other being prostrate and **stoloniferous**, with some genotypes being perennial. Both forms are palatable to cattle, with most collections from tropical America being the prostrate type. Several prostrate introductions overwinter in peninsular Florida, but they do not tolerate as much waterlogging as does american jointvetch. The annual cultivated form has been used in tropical America with some success, though regeneration from seed is not always consistent. Alyceclover requires about 1000 mm of annual rainfall and a soil pH above about 5.0 for best growth. Stems can reach a length of 1.5 m and more. It can be an excellent, though stemmy, hay crop. Commercial types are damaged by root-knot nematodes (*Meloidogyne* spp.), and a cultivar with resistance to this pest, 'FL-3', was released in Florida in 1989.

Arachis L.

There are more than 70 species of *Arachis*, of which, at least two produce edible seed. Most have prostrate growth habits and four leaflets per leaf (tetrafoliolate). Primarily, they are native to Brazil, with some occurring in Argentina and Paraguay (Krapovickas 1990; IBPGR and ICRISAT 1990, 1992; Krapovickas and Gregory 1994). Except for leucaena, this genus has higher CP and **digestibility** than other tropical forages. Several species are being used commercially for grazing, hay, or cover crops.

Data from Florida indicate that *A. stenosperma* Krapov. & W. Gregory produces the largest quantities of nuts and is adapted as a cover crop in establishing citrus groves (Kretschmer et al. 1994). This species and *A. kretschmeri* are excellent wildlife feed (Kretschmer et al. 2001). Only a few of the estimated 420 accessions (in the US) have been thoroughly evaluated (Kerridge and Hardy 1994).

Rhizoma Peanut (A. glabrata Benth.)

Rhizoma peanut is highly persistent due largely to dense rhizomes. The cultivars 'Florigraze' (Prine et al. 1981), 'Arb', 'Arblick', and 'Arbrook' have been released by the Florida Agricultural Experiment Station. 'UF Tito' and 'UF-Peace' were subsequently released to provide improved tolerance to peanut stunt virus (Quesenberry et al. 2010). Ecoturf is a germplasm release originally evaluated for turf but now used in north and central Florida as forage. 'Latitude 34' was released by Texas A&M AgriLife Research for cooler and drier climates (Muir et al. 2010). Rhizoma peanut must be planted vegetatively by using rhizomes that are typically harvested during short day-length periods (dry season or winter in Florida), and plants may require two to three years to become well established. It is used as a pure stand for hay, in association with commonly used warm-season grasses for grazing, and as a cover crop for citrus groves. Its growth is reduced on waterlogged sites, with best growth occurring on well-drained sites receiving 1000 mm or more of annual rainfall. It survives winters in north Florida and southern Georgia.

Pinto Peanut (A. pintoi Krapov. & W.C. Greg.)

Pinto peanut is native to central Brazil. It was first collected in 1954 and since then has been distributed throughout the tropics. 'Amarillo', first released in Australia in 1987, has been released in several countries of tropical America as 'Mani Forrajero Perenne' (Colombia in 1992); 'Mani Mejorador' (Costa Rica in 1994); 'Pico Bonito' (Honduras in 1993); 'Belmonte' (Brazil in 1999); 'Amarillo MG-100' (Brazil in 1994); and 'BRS Mandobi' (Brazil in 2011). An additional accession was released as 'Porvenir' in Costa Rica in 1998, and a multiline cultivar was released with the previously used name 'Mani Forrajero Perenne' in Panama in 1997 (CIAT 2003). Pinto peanut is a vigorous stoloniferous species that has prostrate stems when growing in dense swards. Flowers are yellow. It can produce 3–5 $Mg\,ha^{-1}$ of seed, though until recently most plantings were established vegetatively (Grof 1985; Cook and Franklin 1988; Oram 1990a; Kerridge and Hardy 1994; Bowman and Wilson 1996; Smith 1996; Kerridge and Loch 1999). Establishment is more rapid by seeding than by vegetative planting (Zdravko and Fisher 1996; Lavers and Silver 2000). Seedling vigor and acceptability to animals are excellent (Jones 1993; Hernandez et al. 1995). The CP and digestibility range from about 150 to 230 and 500 to 750 $g\,kg^{-1}$, respectively, but decrease when large amounts of stem are included. When grazed on the western slopes of the Colombian Llanos (soil pH about 4.3), the legume component percentage was the most important factor in diet selection of a pinto peanut – *Brachiaria humidicola* [(Rendle) Schweick] pasture (Mosquera and Lascano 1992). Because of its shade tolerance, pinto peanut may have potential for grazing or as a cover crop in plantations and orchards (Firth and Wilson 1995; Firth et al. 2002). In Florida, Amarillo pinto peanut was slower to establish than other nut-producing wild peanut species, but was competitive after establishment. Foliage of some pinto peanut accessions may become light green to almost yellow in color, while foliage of other *Arachis pintoi* accessions remains dark green under the same environmental conditions.

Calopogonium Desv.

Calopogonium, a genus of the American tropics and subtropics, consists of eight species with little variation among most of them. Unique among these species, *C. caeruleum* has vigorous twining shoots that are not eaten by cattle and allow them to be highly persistent and dominant in mixtures with tropical grasses.

Calopo (C. mucunoides Desv.)

Calopo is an annual or short-lived perennial indigenous in tropical America. It is grazed inadvertently in many tropical pastures, where it has become naturalized and widespread in Africa and Asia. In the tropics, it persists well with annual, well-distributed or seasonal rainfall above 1000 mm. The short-day flowering response limits seed production and use in the subtropics. Though not highly palatable, acceptability of calopo to grazing livestock has not been a limitation in natural stands, particularly where dry-season forage is limiting. Similar to kudzu, calopo has been used in Brazil for more than 70 years as a cover crop in plantation agriculture, in rotation as a green manure and grazing crop, and has been used in Australia to a limited extent for grazing.

Centrosema (DC.) Benth.

Centrosema is primarily a tropical genus of 34 perennial and annual named species. Collection and evaluation of germplasm was intense for about three decades (Schultze-Kraft and Clements 1990). About 3700 accessions make up the world's germplasm collection. Most species are twining perennials with trifoliolate leaves and large flowers with color ranging from white to purple or red. In spite of the extensive research effort, including breeding programs, the use of *Centrosema* species has remained limited to small environmental niches. It has been used very little for grazing in tropical America, but some use as cover in plantation crops has occurred with some grazing in Australia. An 18-site evaluation of *Centrosema* accessions on acid soils (down to pH 4.2) in Colombia showed several were highly tolerant of soil acidity (Keller-Grein et al. 2000). Only two of the species have been widely identified by common names, whereas others are commonly identified by the genus and species.

C. *acutifolium* Benth.

Centrosema acutifolium is a perennial, twining species which has, only recently, been developed for pasture use in Colombia. 'Vichada', released for the hot tropics with annual rainfall of 1000–2500 mm (Schultze-Kraft et al. 1987), has been evaluated for use as a protein bank for dairy operations (Mosquera and Lascano 1992). It is similar to centro in appearance and nutritional quality, but it can tolerate soils that are more acid (pH down to about 4.3) and high in Al and Mn.

C. brasilianum (L.) Benth.

Centrosema brasilianum is an extensively researched species, similar in growth to centro, which has not been commercialized. Evaluations indicate promise for semi-arid to dry sub-humid regions. Though tolerant of acid soil and drought, lack of persistence under grazing has been a limitation.

C. *macrocarpon* Benth.

Centrosema macrocarpon is native from Mexico through the northern latitudes of Brazil. This species is similar in appearance and adaptation to *C. acutifolium* (Schultze-Kraft 1986). It is the most disease resistant of the species evaluated to date. It is, however, similar to the other species in susceptibility to damage from cutting ants.

Centurion (*C. pascuorum* Mart. ex Benth.)

Centurion is a herbaceous twining annual found mainly in semi-arid areas of Central and South America that have a long dry season. Probably the most successful forage species of the genus, it is used for grazing in the Northern Territory of Australia. Australian cultivars 'Cavalcade' (Oram 1990b) and 'Bundey' (Loch and Ferguson 1999) have been released. Centurion requires a reliable four to six-month wet season, can tolerate temporary flooding or waterlogging, and is moderately drought tolerant once established. The species is very palatable and is a heavy seed producer under good moisture conditions. If spring rains are delayed or are sporadic, seedlings can suffer and plant populations may be reduced for the year. It does not compete well in mixtures with vigorous, perennial grasses in Florida.

Centro (*C. pubescens* Benth.)

Centro is very widely distributed in sub-humid areas of the American tropics. It will not grow well on soils with high Al and Mn contents or when soil pH is below about 5.0. It can persist with as little as 750 mm of annual rainfall but is more productive with 1000 mm or more. In Australia, the common commercial genotype has been successfully used for several decades in a small coastal area of Queensland (Teitzel and Burt 1976). It does not persist well in south Florida because of late flowering and low seed production.

C. schiedeanum ([Schltdl.] R.J. Williams & R.J. Clem)

Centrosema schiedeanum includes the cultivar 'Belalto', initially classified as *Centrosema pubescens*, released in Australia in 1971. This cultivar provided improved cool-season forage production compared with common centro. Stoloniferous growth and later flowering also distinguish Belalto from common centro.

Spurred Butterfly Pea (*C.*) *virginianum* (L.) Benth.

This species has many characteristics similar to centro, though it is more subtropical. Because it is day' length insensitive, vegetative growth and reproductive development continue throughout the warm season. It has been collected from Maryland (possibly New Jersey) to Oklahoma and south into Argentina. It is widely distributed in Florida and the Gulf Coast states. In spite of long-term evaluations of many genotypes and cross-breeding resulting in a few hybrids, there is no commercial use (Jones and Clements 1987). Natural outcrossing was estimated to be about 18% (Maass and Torres 1998).

Chamaecrista (L.) Moench

The *Chamaecrista* genus consists of herbs and subshrubs possessing sensitive leaves and pods that dehisce suddenly. Most of the approximately 250 species are neotropical. Plants in this genus have been classified within the genus *Cassia* L. in some taxonomic treatments. The only released forage cultivar in this genus was initially identified as a species of *Cassia*.

Roundleaf Cassia (*C. rotundifolia* [Pers.] Greene)

Roundleaf cassia (formerly *Cassia rotundifolia*) is an annual or short-lived perennial, semi-erect to prostrate herb with bifoliolate leaves. It is native to Mexico, the Caribbean, and into Brazil and Uruguay. The species is self-pollinated, but natural outcrossing can be as high as 13% (Maass and Torres 1998). It has been very successful and persistent in Queensland, Australia, where 'Wynn' was developed (Strickland et al. 1985). When availability of the associated grass is low, roundleaf cassia can contribute to pasture quality, production, and animal gains (Cook 1988; Tarawali 1995; Clements et al. 1996; Jones et al. 1998; Larbi et al. 1999). Persistence depends primarily on sufficient soil seed reserves, which primarily reflect rainfall patterns and grazing management (Partridge and Wright 1992; Jones and Bunch 1995a; McDonald and Jones 2000). Wynn and other types may be susceptible to leaf diseases (Chakraborty et al. 1994).

Clitoria L.

The genus *Clitoria* contains about 70 species of shrubs and vines, mostly from the neotropics, with pinnately compound leaves and large axillary flowers. Only one single species is important as a forage plant.

Asian Pigeonwings (C. ternatea)

This perennial twining legume with a woody crown has become widely naturalized in the tropics, mainly between latitudes 20 and 24 °N. It does not tolerate flooding. It can survive with about 500 mm of annual rainfall but grows best with about 1500 mm. 'Tehuana' was released in Mexico in 1988 and again as 'Clitoria' in Honduras in 1990. It has been used primarily in cultivated soils by small landholders. In Australia, 'Milgarra' was released in 1990. Despite apparent adaptation, it has not been widely used there or in tropical Asia (Reid and Sinclair 1980; Hall 1985; 't Mannetje and Jones 1992). Lack of persistence, associated with high palatability and consequent **selective grazing**, is a limitation. Recent assessments across tropical locations in Australia indicate potential usefulness on sites with low production potential (Bell et al. 2015).

Desmanthus Willd. nom. cons.

The genus *Desmanthus* has been classified to include as many as 40 to as few as 24 species (Luckow 1993). All species are from the Western Hemisphere with the possible exception of one widely distributed weedy species. Species range from distinctly tropical to warm-season plants of temperate zones. Most are perennial herbs, though annual herbs and subshrub to shrub growth forms are included. Plants are non-toxic and range from erect to prostrate in growth. Leaves are bipinnately compound. Only some have accepted common names.

Early forage work with this genus rather loosely followed taxonomic classifications where the *D. virgatus* complex included a large number of botanic varieties. Prior to detailed taxonomic assessment, most tropical *Desmanthus* genotypes were initially identified as *D. virgatus,* including the initial three cultivars released in Australia in 1991. Recently, Ocumpaugh et al. (2003) indicated that each of these cultivars belongs to a different species according to the classification of Luckow (1993). 'Marc' was retained in *D. virgatus* due to upward nyctinastic leaf movements of both the petiole and pinnae, leaf petiole length of only 1–3 mm, filiform peduncles of inflorescences, minimum elongation of seedling hypocotyl and epicotyl and blue-green glaucous foliage. These taxonomic distinctions have no recognized agronomic importance. The three Australian cultivars have been marketed as a physical mixture under the trade name Jaribu. A recent blend of *Desmanthus* species commercialized in Australia for semi-arid clay soils, PROGARDES™, has been sown on several thousand hectares (Gardiner et al. 2013). Recent assessment in Australia indicates that several species of *Desmanthus* have potential for development as pasture plants across a range of environments, particularly on some difficult sites (Bell et al. 2015).

Bundleflower (D. bicornutus S. Watson)

Bundleflower is a tropical species represented by the four cultivars, 'BeeTAM-06', 'BeeTAM-08', 'BeeTAM-37', and 'BeeTAM-57', which were released in 2003 for subtropical South Texas (Ocumpaugh et al. 2003). They have been marketed as a blend under the trademarked name BeeWild. These genotypes were initially referred to as *D. virgatus* due to general similarities, but were later distinguished from other *Desmanthus* cultivars by an elongated hypocotyl with minimal epicotyl elongation of seedlings along with downward nyctinastic movement of pinnae and petiole at night. The name BeeWild reflects the anticipated value of the mixture for wildlife in addition to use as a forage plant.

Illinois Bundleflower (D. illinoensis [Michx.] MacMill. ex B.L. Rob. & Fernald)

Illinois bundleflower is a warm-season legume native to the temperate US and widely distributed from Texas to Minnesota. A 1984 release in Texas named 'Sabine' has been planted for forage, wildlife food and cover, and reclamation, primarily within its native region (Muir et al. 2018). Overlap of range of adaptation with the tropical *Desmanthus* species is minimal, but in such situations, Illinois bundleflower can be distinguished by minimal elongation of seedling hypocotyl and epicotyl and downward nyctinastic movement of the pinnae at night with no movement of the petiole. Illinois bundleflower is best adapted to soils of sandy-loam to clay texture, with pH 6.0–7.5 and an annual rainfall exceeding 500 mm.

D. leptophyllus Kunth

Desmanthus leptophyllus occurs over a native range overlapping and extending beyond that of *Desmanthus pubescens,* though the range is not as extensive as that of *D. virgatus.* 'Bayamo', from Australia, is now classified in this species rather than *D. virgatus,* according to the taxonomic descriptions of Luckow (1993). In *D. leptophyllus,* each pinna has many narrow leaflets, and nyctinastic leaf movements result in an arched rachis with pinnae folded downward at night. This contrasts with *D. virgatus,* which has pinnae folding upward at night. The pubescence is similar to *D. pubescens,* but *D. leptophyllus* has fewer flowers per head, less conspicuous **auricles**, and no sterile flowers.

D. pubescens B.L. Turner

Desmanthus pubescens is a tropical species from Mexico, Belize, and Guatemala. This species includes 'Uman', which was originally released in Australia as *D. virgatus.* Agronomically, there is little, if any, difference between *D. pubescens* and *D. virgatus,* but taxonomically, *D. pubescens* is distinguished from the other agronomically important tropical species by pubescent ovaries and hairs along the valves of pods. Other species have glabrous ovaries and pods (Luckow 1993).

D. virgatus (L.) Willd.

Desmanthus virgatus includes a range in morphologic variation that can vary from prostrate to upright, with growth up to 2 m high depending on extent of defoliation. This is similar to variation represented by the other tropical species of the genus. This species has been collected from areas with 250–2500 mm of annual rainfall. Evaluations of the tropical *Desmanthus* species, with all initially designated as *D. virgatus,* have indicated adaptation to sand or clay soils (Jones et al. 1997), though adaptation of the Australian cultivars is better on soils above pH 5.5. The CP of succulent stems can be about 65 $g\,kg^{-1}$, whereas that of leaves is often more than 200 $g\,kg^{-1}$. Plant crowns can survive frost in some subtropical climates, though foliage is susceptible to damage.

Desmodium Desv.

This diverse genus comprises more than 300 species of tropical to temperate species. Several have been developed into successful cultivars (Ohashi 1973). Evaluation of many temperate species is much more recent than that of the tropical and subtropical germplasm (Muir et al. 2018).

Carpon Desmodium (D. heterocarpon [L.] DC.)

Carpon desmodium is native to subtropical and tropical India, southeastern and eastern Asia, Australia, and many Pacific islands. It is a long-lived perennial, erect to prostrate (depending on grazing pressure) herb, with a woody crown upon maturation. 'Florida' carpon desmodium has persisted under favorable conditions when grown in association with bahiagrass and other commonly used grasses in peninsular Florida (Kretschmer et al. 1979). A few thousand hectares of this legume in mixture with warm-season grasses have been planted in central and south Florida, though only sparse populations of the legume persist over much of this area. Other genotypes that are more erect also become prostrate when heavily grazed. Most accessions are **short-day plants** that flower in Florida from September to October, with seed ready for harvest in late October or November. Plants are self-pollinating, with an estimated natural outcrossing of about 4% within the species (Maass and Torres 1998). Florida carpon desmodium is better adapted to the humid subtropics than to more tropical areas.

Carpon desmodium is the only seeded, long-lived, perennial, tropical forage legume used in Florida. It can survive moderate droughts and temporary flooding, but grows best in moderately drained, moist sandy soils. Initial seedling growth is slow, but long-term persistence of individual plants and seedling recruitment make it a more dependable species in grass pastures than the adapted annual legumes. Nutrient requirements for carpon desmodium, as for most other tropical legumes, generally are lower than those for temperate legumes. It is not adapted to soils with pH below 5.0 or those high in Al and Mn. Palatability in mixtures is similar to most associated grasses despite tannin concentrations of 20–30 $g\,kg^{-1}$.

Ovalifolium (D. heterocarpon subsp. ovalifolium [Prain] H. Ohashi)

Ovalifolium is a perennial, creeping, stoloniferous legume that is well adapted to the humid and subhumid tropics (Schultze-Kraft and Benavides 1988; Ohashi 1991). It has been used for decades in tropical America and Southeast Asia as a cover crop in plantation agriculture and for grazing. 'Itabela', released in Brazil in 1989, is less palatable and later flowering than Florida carpon desmodium. It produces few seeds in southern Florida prior to defoliation by frost. Ramirez (2002) compiled the agronomic attributes and research on CIAT 13651 that was released as 'Maquenque'.

Hetero Desmodium (D. heterophyllum [Willd.] DC.)

Hetero desmodium is a perennial species which has been used to a limited extent. 'Johnstone' was released in Australia in 1971. This prostrate or slightly ascending herb has a woody root-stock (Partridge 1986). It is native to Southeast Asia and surrounding islands and has spread to many other humid tropical areas where annual rainfall exceeds about 1500 mm. Leaves are trifoliolate and can form a thick mat. Stems are freely branching to 1 m in length, and flowers are small and pink, developing smooth three- to six-jointed pods. Indeterminate flowering and production of seed pods buried in **foliage** make it difficult to harvest seed. Generally, it is planted vegetatively. In Florida, initial evaluation indicated potential adaptation.

Creeping Beggarweed (D. incanum DC.)

Creeping beggarweed, also known as Kaimi clover or Pega-pega, is native from the southern US to Uruguay and Argentina. Creeping beggarweed is now ubiquitous in the wet tropics with 1000–3000 mm of annual rainfall and in many Pacific islands. Like many other *Desmodium* species, its fruit stick to clothing and animals, contributing to seed dispersal. Plants form woody stems and root-stocks with scattered shoots growing to about 60 cm. It is an invader and can survive air temperatures as low as −5 °C in addition to several foliage-killing frosts annually. It is very persistent under most stocking rates. Acceptability to grazing livestock is not as high as that of carpon desmodium, probably due to the higher tannin content and lower percentage of leaf.

Greenleaf Desmodium (*D. intortum* [Mill.] Urb.)

Greenleaf desmodium is indigenous to highlands, from 800 to 2500 m, from Mexico to southern Brazil. Trailing but not twining, this large perennial can grow in lower elevations in the subtropics, but it is not adapted to the humid or sub-humid tropics. It requires about 1000–1500 mm of annual rainfall and a soil pH above about 5.0. 'Green-leaf', released in Australia in 1963, is used in the Atherton Tablelands of subtropical Australia. Its range in the subtropics is limited because it is late flowering and provides inadequate seed for needed seedling recruitment to sustain stands. This also limits its use in Florida.

Silverleaf Desmodium (*D. uncinatum* [Jacq.] DC.)

Silverleaf desmodium is native to Central and South America. It has a similar growth habit to that of greenleaf desmodium. 'Silverleaf', released in Australia in 1962, is earlier flowering, less leafy, and less widely used in Australia than greenleaf. However, it has been used in Kenya and probably elsewhere in small climate niches.

Lablab Adans.

This genus contains only one species, lablab, which was previously classified in the genus *Dolochis*. Despite the single species, considerable variation exists with probably 20–50 botanic varieties (Duke 1981).

Lablab (*L. purpureus* L.)

This species originated in the Old-World tropics where an extended history of use has occurred in plantation agriculture across India, Africa, and Southeast Asia. This climbing annual or short-lived perennial is drought tolerant and grows well in slightly acid to alkaline soils, but it is not tolerant of waterlogging. Growth forms vary from non-vigorous to a climbing mass of leaves and stems when not grazed. It can produce large quantities of seed when growing aggressively and not defoliated. Lablab has been marketed in Texas under the trade name Tecomate primarily as wildlife feed (Loch and Ferguson 1999). 'Rio Verde', selected for defoliation tolerance and forage and seed production, has recently been released in Texas (Smith et al. 2008).

Lablab has been used for grazing in Australia and for human consumption in the Caribbean and elsewhere. It is usually strip-grazed, with cattle eating only the leaves. Cattle are removed after leaf defoliation to permit regrowth. Consumption by dairy cattle may pass an undesirable flavor into non-pasteurized milk. Australian cultivars, which are used primarily for forage, are 'Rongai', released in 1962; 'Highworth', released in 1973; and 'Koala', released in 1996. Because it can grow as a warm-season annual in temperate areas, additional germplasm evaluation is warranted (Fribourg et al. 1984).

Lespedeza Michx.

The approximately 40 species in the genus *Lespedeza* include species native to North America, Asia, and Australia. The most widely known species are herbaceous and become woody with maturity. These species are primarily warm-season plants adapted to mild temperate climates.

Sericea Lespedeza (*L. cuneata* [Dum. Cours.]G. Don)

Sericea lespedeza is native to Asia and has been rather extensively planted in the southeastern US for forage, soil conservation, and reclamation. This warm-season plant of temperate grasslands and open woodlands is adapted north into Iowa. From previous plantings in the southern Great Plains, sericea lespedeza has recently been recognized as an invasive weed of native grasslands. Though potential invasiveness has also been recognized in the forested region of the southeastern US (Pitman 2009), sericea lespedeza remains a useful forage and conservation plant for southeastern pastures and hay fields.

In contrast to the recent designation as invasive, early agronomic efforts revealed establishment difficulties and sensitivity of young stands to loss from excessive defoliation. Low palatability has been a limitation under grazing and is associated with tannin concentrations in leaves and increased stem content with maturity. Breeding efforts in Alabama for increased grazing tolerance resulted in 'AU Grazer' (Mosjidis 2001). Earlier releases from Alabama included 'Serala' in 1962, 'Interstate' in 1969, 'Serala 76' and 'Interstate 76' in 1978, 'AU Lotan' in 1980, and 'AU Donnelly' in 1989. These cultivars had increased leafiness and forage value; the latter two also had reductions in tannin. The role of tannins in sericea lespedeza, and in some other legumes, is complex with benefits such as reducing protein loss from digestion in the rumen and anthelmintic effects on small ruminants contrast with reduced palatability (Muir 2011). 'Appalow' was released by the USDA-NRCS in 1978 primarily for conservation and reclamation uses.

Leucaena Benth.

The *Leucaena* genus may consist of as few as 10 (National Academy of Science 1984) or as many as 50 species (Whiteman 1980). The natural range is from Texas to Ecuador. Where occurring naturally, most species have been used as browse, fuel wood, shade, and fence posts. Several species have potential for forage or browse production.

Leucaena (*L. leucocephala* [Lam.] De Wit)

Leucaena has been widely evaluated and planted for browse production (Jones 1979; Anonymous 1983; Bray 1985a; Brewbaker 1987; Jones and Bunch 1995b; Jones and Palmer 2002). Among the many browse shrubs and trees evaluated, beef gains from leucaena have been the highest. Where not native, leucaena has become naturalized in most tropical areas that have moderately drained soils and a pH greater than 5.5. This very long-lived perennial can reach a height of about 16 m. It has smaller stems than other species, and the small stems are readily consumed by cattle. Cultivation of leucaena for grazing began in Australia with Australian releases including 'El Salvador' and 'Peru' in 1962, 'Cunningham' in 1977, and 'Tarramba' in 1994. 'Romelia' was released in Colombia in 1991.

The major drawback of leucaena has been the adverse effect of mimosine, an amino acid found in small quantities in the leaf tissue. Excess consumption of mimosine can affect the thyroid and lead to loss of hair, goiter, and sub-clinical symptoms including reduced rates of gain. A ruminal bacteria that can detoxify the effect of mimosine in ruminants was discovered leading to increased evaluation and use of leucaena in Australia, Brazil, Asia, and most other tropical areas (Jones 1985; Jones and Megarrity 1986). Recent assessments have revealed more complicated interactions among leucaena, livestock consuming the legume, and the detoxifying bacteria with both increased management of feeding systems and additional research recommended (Halliday et al. 2013).

The leucaena psyllid (*Heteropsylla cubana*), a leaf-feeding insect, devastated foliage several years ago in Australia and elsewhere (Austin et al. 1996; Bray and Woodroffe 1997; Castillo et al. 1997). Establishment difficulties and psyllid damage limited commercial use of the species for a few years. Grazing management and parasitic insects were evaluated as potential approaches to relieve the problem. In 2015, 'Redlands', a psyllid-resistant variety was released in Australia. This cultivar was described as vigorous and high-yielding with tolerance of grazing. In Australia, Asia, Africa, and Brazil, evaluations and commercial use have increased in recent years (Jones 2001).

Lotononis (*DC.*) Eckl. & Zeyh (Proposed *Listia* E. Mey.)

Species of the genus classified as *Lotononis* have recently been proposed to belong in the genus *Listia*. These species primarily occur in southern Africa with their range extending into central Africa. Species are stoloniferous with lupinoid (sleeve-like) root nodules (Boatwright et al. 2011).

Lotononis (L. bainesii Baker)

Lotononis is the major species of the genus recognized for its forage value. It has become naturalized in Namibia, South Africa, and Zimbabwe from 2 to 30 °S latitude, and naturalized forms can be found along highways in Queensland, Australia, and elsewhere as escapes from research.

Lotononis is a non-toxic perennial with a main taproot and adventitious roots. The species is similar in growth and quality to temperate *Trifolium* species. Preliminary testing suggested no substantial variation within the available germplasm collection.

Lotononis is best sown on top of cultivated soil at a rate of about 1 kg ha^{-1} (the hard seed is very small, with 3000–4000 seed g^{-1}) and rolled to prevent covering too deeply. Sowing in the hottest months should be avoided. Seedling emergence is often slow and erratic. It requires special *Bradyrhizobium* bacteria for N_2 fixation. Lotononis is most susceptible to diseases during very wet periods if growth is lush. It is susceptible to *Cercospora* leafspot, *Botrytis* flower blight, *Sclerotium rolfsil*, *Pithium* rots, and legume little leaf virus.

Being drought tolerant, stands have persisted well in areas of Australia, but extreme yearly fluctuations have been noted in stands, indicating the need to replenish populations of this short-lived perennial plant each year (Bryan et al. 1971). In southern Florida, after oversowing 3- to 4-m-wide strips into an establishing bahiagrass pasture, stands persisted for at least 10 years and spread laterally to 15 m wide under high rates of **intermittent stocking**. Consistent with these observations, grazing management, to limit defoliation during early season seed production, followed by heavy grazing can open the grass stand for seedling development during late summer and autumn (Fujita and Humphreys 1992).

In addition to potential in Australia, *Lotononis* may be adaptable to the moderately drained, subtropical soils of Florida and the southern Gulf Coast. Developing better and consistent methods of seed harvest would contribute to this potential. 'Miles' was released in Australia in 1962 (Bryan 1961) and 'Best' was released in Zimbabwe in 1972.

Macroptilium (Benth.) Urb.

This genus of about 12 species is native to the tropics and subtropics of the Western Hemisphere. It is closely related to *Phaseolus* (edible bean) and includes species with twining, climbing, erect, and prostrate growth forms. Research has included extensive germplasm collection, field-plot evaluation, breeding, grazing evaluation, and commercial use of two species. Results have been less positive than expected, so other *Macroptilium* species have been evaluated. Recent assessment suggests potential for several species of the genus (Bell et al. 2015).

M. gracile ([Poepp. & Benth.] Urb.)

Macroptilium gracile is a somewhat variable species with prostrate types that may withstand greater grazing pressure than early cultivars of the species. It is a short-lived herb native in low altitude sites of tropical America. 'Maldonado' was released in Australia in 1990 with potential as a pioneer species in plant mixtures.

Siratro (M. atropurpureum [DC.] Urb.)

Siratro occurs naturally from extreme southern Texas to, but not including, Argentina (Perez et al. 1999) and northern Brazil. Mexico is probably the center of diversity. It has been evaluated in almost all tropical and subtropical areas of the world. Its popularity increased from about 1967 in Australia with release of the cultivar 'Siratro'. Demand for seed declined slowly because of lack of persistence under high stocking rates. Poor seedling recruitment has been identified as a contributing factor in the lack of persistence (Jones 2014). Siratro leaves and stems are susceptible to *Uromyces appendiculatus* (rust), which can be very damaging to forage and seed productivity. Germplasm collections in Mexico and subsequent selection and breeding for rust resistance (Bray et al. 1991) led to the 1993 Australian release of 'Aztec'. A rust-resistant genotype from the University of Florida has been registered (Kretschmer et al. 1992). Siratro is susceptible to foliar leaf blight caused by *Rhizoctonia solani* (Sonoda et al. 1991). Under grazing, success of this **herbaceous**, twining legume has not been very good, especially in the humid tropics. In Australia, siratro has been used primarily in coastal Queensland (Jones and Jones 1976; English 1999a). It persists well in south Florida when cut every 45–60 days, but it does not survive under commonly used grazing practices (Kretschmer et al. 1985). In addition to lack of persistence in Florida, Brazil, and elsewhere, siratro can become a costly weed in citrus groves because of its viney growth habit and difficulty of eradication.

Phasey Bean (M. lathyroides [L.] Urb.)

Phasey bean is a short-lived perennial legume widely distributed in tropical and subtropical regions. It has been used as a pioneer species in planting mixtures in Australia, Florida, and elsewhere (Pitman et al. 1986). Though possessing many desirable forage characteristics, limited stand life greatly reduces its use. It can become a difficult weed to control in citrus groves because of its twining growth habit during plant maturation. Recent assessments of evaluations across tropical locations in Australia indicate potential for increased use on sites with relatively low-yield potential (Bell et al. 2015).

Pueraria DC.

This genus is native to Asia and islands of the Pacific Ocean. Adaptation of the approximately 20 species extends from the tropics to warm temperate regions where kudzu can be a particularly aggressive summer-growing forage plant or weed. Plants in the genus are characterized by vigorous climbing, viney foliage that can dominate a landscape.

Tropical Kudzu (P. phaseoloides [Roxb.] Benth.)

Tropical kudzu or puero is a robust, trailing, perennial species native to East and Southeast Asia. It has been used throughout the tropical world beginning about 100 years ago as a cover crop, which was sometimes grazed, in rubber and oil palm plantations. It is adapted and widely distributed in areas of the tropics receiving more than 1200 mm of annual rainfall, where it is tolerant of flooding and waterlogging. It can also survive droughts. It is not adapted to the subtropics. One of the commercially used genotypes is not very palatable, allowing this genotype to persist well since cattle consume it only as a last resort. Other genotypes are reported to be moderately palatable. Though tropical kudzu produces seed that has been harvested only by hand, almost all plantings in tropical America have been made vegetatively.

Kudzu (P. montana [Lour.] Merr.)

Kudzu is native to Japan, China, and Korea. It is a well-adapted warm-season plant in the southeastern US, where its range extends up to near 38 °N latitude on the Atlantic Coast west to just below 35 °N latitude in southeastern Oklahoma. Though it has been successfully used for soil conservation and forage, kudzu has become an invasive weed of non-grazed areas, especially in forested landscapes where the viney growth can spread over tall trees (Pitman 2009). Late flowering prevents seed production in most temperate areas, where it has been propagated vegetatively. Palatability and nutritive value are high, resulting in good gains of grazing livestock when grazing pressure is not excessive. Both animal gains and plant stand density quickly decrease with excessive grazing pressure.

Stylosanthes Sw.

This genus with trifoliolate leaves has about 30 annual and perennial species. The genus is characterized by small yellowish flowers. *Stylosanthes* can extract P from soils when it is not available to other species. Most species of *Stylosanthes* are not tolerant of prolonged waterlogging. Susceptibility to anthracnose, caused by *Colletotrichum gloeosporioides*, has been a problem with many of the original cultivars and most species, which have varying degrees of resistance (Irwin et al. 1984; Lenne and Trutman 1994). More species of this genus have been successful under grazing management systems than species from any other tropical legume genus. Available cultivars, with some marketed as cultivar mixtures, continue to be useful in specific environments particularly in Australia and Asia. In Brazil, a mixture of 20% *Stylosanthes macrocephala* and 80% *Stylosanthes capitata* named Campo Grande has been developed for well-drained acidic soils with extended dry seasons. Campo Grande is highly resistant to anthracnose. Potential for additional use of the genus in some specific locations was noted by Bell et al. (2015).

Need for a comprehensive taxonomic update for the genus has been recommended (Calles and Schultze-Kraft 2016).

Pencilflower (S. biflora [L.] Britton, Sterns & Poggenb.)

This species ranges from 27 to 41 °N latitude in the eastern US. It occurs in central and north Florida and the Gulf Coast states on moderate to well-drained soils. Despite wide distribution across the region, low-production potential limits forage value of this species.

Capitata (S. capitata Vogel)

Capitata is a perennial subshrub native to Brazil and Venezuela. Accessions include prostrate and erect forms. Production of both vegetation and seed of some accessions has indicated potential forage value, especially with observed resistance to anthracnose. The five-accession blend 'Capica' released in Colombia failed to persist in pasture plantings (Miles and Lascano 1997). Potential for use in breeding programs as a source of anthracnose resistance has been indicated (Santos-Garcia et al. 2012).

Stylo (S. guianensis [Aubl.] Sw.)

Stylo is native to Brazil and comprises seven botanic varieties. The botanic variety *guianensis* is adapted from 23 °N to 27 °S latitudes and has been the source of several cultivars. These include 'Schofield', released in the 1930s; 'Cook', released in 1971; 'Endeavour', released in 1971; and 'Graham', released in 1979, all from Australia; 'Deodora', 'Deodora II', and 'IR-1022', released in Colombia in the early 1970s; 'Mineirao', released in 1993 from Brazil (Anonymous 1993c); 'Pucallpa', released in Peru in 1985; 'Reyan', released in China in 1991; and 'Savanna', released in Florida in 1992. As a group, they are moderately drought tolerant, seldom withstand continuous waterlogging, survive light frosts, and require a soil pH of greater than 5.0. This species was becoming an important pasture plant in Australia but is now seldom used there, or elsewhere, because of devastation by anthracnose. In small plantings in southern Florida, the disease has reduced seed production to non-harvestable quantities and also killed plants. Stylo is competitive with associated grasses, has good palatability, and provides adequate quality to improve performance of livestock grazing most tropical grass pastures.

Fine Stem Stylo (S. guianensis [Aubl.] Sw. var. intermedia [Vogel] Hassl.)

Fine stem stylo has also been classified as the species *S. hippocampoides* (Vander Stappen et al. 2002). It is represented in Australia by a commercial type referred to as common fine stem stylo, which was initially distributed in 1965. 'Oxley' was released in 1969. Fine stem stylo has been successfully used on low-fertility soils in Australia where limited frost tolerance and grazing tolerance are important.

Tardio Stylo (S. guianensis [Aubl.] Sw. var. pauciflora M.B. Ferreira & Sousa Costa)

Tardio stylo is the third botanic variety of *Stylosanthes guianensis* developed as a tropical forage legume. 'Bandeirante' was released in Brazil in 1983 as a cultivar of the variety *pauciflora*. This cultivar is prostrate, more subtropically adapted (from 18 to 36 °S latitude), persistent, and later flowering than variety *guianensis* (Gardener and Ash 1994).

Caribbean Stylo (S. hamata [L.] Taub.)

Caribbean stylo is native to south Florida and the Caribbean into Colombia, Venezuela, and Brazil. This annual and short-lived perennial is adapted from 28 °N to 15 °S latitudes. The Australian tetraploid cultivars 'Verano', released in 1973, and 'Amiga', released in 1997, are adapted to Queensland, Australia. Being more tolerant to anthracnose in Australia, Caribbean stylo replaced large areas of townsville stylo, which had been killed by the disease. Caribbean stylo has also been successfully used in Thailand and several West African countries. Verano is moderately drought tolerant but has not persisted in south Florida, possibly a result of periodic high-water tables, which native genotypes can tolerate.

Townsville Stylo (S. humilis Kunth)

Townsville stylo is an annual, similar in appearance to Caribbean stylo. It became naturalized in northern Australia at the beginning of the twentieth century (Humphries 1967). Later, it was seeded to large areas where it became the major legume used. It is adapted from 20 °N to 21 °S latitudes. Recently, other species replaced townsville stylo after large areas were devastated by anthracnose. Though anthracnose did not damage commercial plantings of townsville stylo in the 1970s in south Florida, the populations did not have long-term persistence. Several cultivars from Australia, Thailand, and Brazil have been developed.

Shrubby Stylo (S. scabra Vogel)

Shrubby stylo is highly variable and strongly perennial, growing up to 2 m high. It is the most drought tolerant of the commonly used stylos. Because of its low leaf-to-stem ratio, as high as 96% stem by weight under extreme drought, shrubby stylo is the least nutritious species of the genus. It is adapted to areas from 10 °N to 30 °S latitudes with annual rainfall from 500–2000 mm. In northern Australia, 'Seca', released in 1976, 'Fitzroy', released in 1979, and 'Siran', released in 1990, replaced Caribbean stylo in the drier climates because they are variably resistant to anthracnose. Seca has been the most commonly used (Edye et al. 1998; English 1999b). 'Q10042' was released in India for use in a cut-and-carry system. In the past, limited cross-pollination, **genotypic** variation in collections, and species integration have occurred. This has led to hybrids in the general population that are difficult to distinguish. An overview and list of the world collections of shrubby stylo was published in 1989 (Maass 1989).

S. seabrana B.L. Maass & 't Mannetje

Stylosanthes seabrana includes selections made from the variable shrubby stylo (*Stylosanthes scabra*) collections. These include 'Primar' and 'Unica' released in Australia in 1996 (Anonymous 1996). These cultivars were superior to Seca in several traits (Edye et al. 1998).

Vigna savi, nom. cons.

There are approximately 150 trifoliolate-leafed species in the genus. Both annual and perennial species exist. Some are edible, and many have a growth form similar to that of *Phaseolus*, that is, have trailing or climbing shoots. Some are robust, resembling siratro or glycine, while others are less robust with smaller leaves and stems. Foliage of most accessions evaluated is acceptable and nutritious to cattle. Palatability is generally high and may prevent long-term persistence of some species because of selective grazing.

V. adenantha (G. Mey.) Marechal, Mascherpa & Stanier

Vigna adenantha is native to Mexico, Central America, and through tropical areas of Argentina. It is found in tropical Africa, Taiwan, and other tropical islands. It is a perennial, climbing legume with stems to about 4 m long. Its growth in pastures is similar to that of siratro and glycine, but it is much more tolerant of waterlogging than these species and is not attacked by rust. *V. adenantha* has potential in the seasonally dry to humid subtropics, but seed production is limited. Indeterminate flowering occurs only during shorter days at the end of the growing season shortly before frost. It is persistent in Florida when not stocked at high rates. It should not be planted near citrus.

Creeping Vigna (V. parkeri Baker)

Creeping vigna is represented by three subspecies. Subspecies *maranguensis* originated in Africa and became naturalized in the highlands of Papua, New Guinea. This legume is short, climbing, stoloniferous, moisture loving, and is tolerant of infertile soils. It can grow in association with white clover. 'Shaw', released in Australia in 1984, has leaflets that often have crescent-shaped white watermarks. Plants flower in response to short days as the growing season ends in the subtropics but indeterminately, thus mechanically harvested seed yields are low. Shaw creeping vigna has been successful in some areas in Australia (Cook and Jones 1987) and has persisted for a few years in a subtropical humid environment with seasonally waterlogged soils in peninsular Florida. Establishment in Florida has been rapid on clean cultivated sites and in establishing grass pastures, with most evaluations conducted in bahiagrass mixtures. Shaw creeping vigna persisted for a few years in Florida under heavy grazing pressure forming low, dense growth. It also survived in tall non-grazed conditions with viney growth climbing over associated species. It can be eliminated by burning (Pitman and Adjei 1994). Lack of seed maturity before frost apparently contributed to eventual loss of stands in Florida plantings by precluding seedling recruitment. Because little is known of the **genotypic** variation of the species, further collections may provide an opportunity to develop genotypes with a wider range of adaptation. Earlier and more concentrated flowering would contribute to its success in the subtropics.

Additional Species Information

Along with the selected list and brief descriptions of some rather widely evaluated and/or commercially used tropical legumes presented in the text above, Table 15.1 provides an introduction to some additional species. Sources of additional information include Whyte et al. 1953; Blumenstock 1957; Bogdan 1977; Anonymous 1979; Minson 1980; Reid 1980; Summerfield and Bunting 1980; Polhill et al. 1981; Hegarty 1982; Williams 1983; Anonymous 1984; Bray 1985a; Hague et al. 1986; Gramshaw et al. 1989; Loch 1991; 't Mannetje and Jones 1992; Gutteridge and Shelton 1993; McDonald and Clements 1999; Jones et al. 2000; Pengelly and Conway 2000; Kretschmer and Pitman 2001; Lapointe and Sonoda 2001; and Adjei and Muir 2001; Jones 2001.

Commercial Use

Commercial plantings of tropical forage legumes have been made in a wide range of environments and production enterprises. As should be expected from the long history of research and development with tropical legumes in Australia, commercially proven legume cultivars are widely recognized for some specific tropical and subtropical environments across the country (Bell et al. 2015). For other environments, even with this history of tropical legume development, the need for further evaluation and development has recently been acknowledged (Bell et al. 2015). Poor legume survival from commercial plantings in various environments across the tropics has resulted in decreased interest in legume planting, lack of demand for seed, and inability to sustain a seed industry. In contrast, the success with a number of species in Australia contributed to development of tropical forage legumes across Asia. Development of leucaena pastures in Brazil has stimulated renewed interest in forage legumes in tropical America in recent years.

Lack of long-term persistence has been a limitation of tropical legumes in grazed pastures in many environments. Reasons for poor survival include lack of tolerance of defoliation, drought, flooding, infertile soils, grass

Table 15.1 Additional legume species of warm climates with potential forage value in specific environments or production systems in tropical, subtropical, or warm-temperate regions

Species	Distinguishing characteristics
Mimosa, silk tree (*Albizzia julibrissin*, often misspelled *Albizia*)	Bipinnately leaved small tree; adapted and invasive across southeastern United States (Pitman 2009); palatable and of high nutritive value; requires appropriate management (Bransby et al. 2003); forage potential of other species of the genus suggested by Larbi et al. (1996)
Gliricidia (*Gliricidia sepium*)	Small deciduous tree, rarely to 12 m height; neotropic origin, now widely distributed across the tropics; primarily used in cut-and-carry forage systems; has been widely evaluated for forage (Withington et al. 1987; Adejumo 1992; Cobbina and Atta-Krah 1992; Bray et al. 1993; 1997)
Hairy indigo (*Indigofera hirsuta*)	Native to tropical Africa and Asia; herbaceous erect-growing annual with woody stems developing at maturity; developing stems and short bristly hairs on leaves limit palatability; adapted to well-drained, sandy soils; can develop long-term dormant seed banks (Baltensperger et al. 1985); 'Flamingo' developed in Florida as soft-seeded, non-persistent type
Korean lespedeza (*Kummerowia stipulacea*, formerly *Lespedeza stipulacea*)	Annual, upright-growing warm-season herb; origin in eastern Asia; adapted in temperate, lower Midwest and eastern United States; useful on acid, low-fertility soils; short-day flowering can limit reseeding for sustained stands; 'Summit' was released from Missouri and Arkansas evaluations in 1963 for productivity and disease resistance
Striate lespedeza (*Kummerowia striata*, formerly *Lespedeza striata*)	Annual, upright-growing warm-season herb; origin in eastern Asia; adapted in temperate, lower Midwest and eastern United States; useful on acid, low-fertility soils; short-day flowering can limit reseeding for sustained stands even more than with the earlier flowering Korean lespedeza; 'Kobe' and the more recent 1989 release 'Marion' from Missouri and Arkansas evaluations are improved cultivars; Marion flowers earlier and provides resistance to leaf disease
Axillaris, perennial horsegram (*Macrotyloma axillare*)	Perennial adapted to frost-free subtropics and tropics with at least 1000 mm annual rainfall (Blumenthal and Staples 1993); heat and drought tolerant; limited palatability contributes to persistence in grazed pastures
Glycine, perennial soybean (*Neonotonia wightii*, formerly *Glycine wightii*)	African and perhaps Asian origin; twining growth habit; less palatable and more persistent than Siratro; rust tolerant; 'Tinaroo', 'Cooper', and 'Clarence' released in Australia in 1962 and 'Malawi' released in 1975 with mixed success (Cameron 1984); tolerates drought and temporary waterlogging; seed production is concentrated with little shattering
Sesban (*Sesbania sesban*)	Annual or short-lived perennial woody plant; can tolerate grazing but often used in cut-and-carry systems; establishes rapidly, is flood-tolerant, and is palatable and of high-nutritive value (Jones 2001); 'Mount Cotton' was released in Australia in 1993 for use in the wet subtropics (Evans and Rotar 1987)
Kenya white clover (*Trifolium semipilosum*)	Adapted to high elevations in tropical Africa and naturalized in some subtropical areas; cultivated in humid areas of northern New South Wales and southern Queensland, Australia where no killing frosts occur
Zornia (*Zornia latifolia*)	Widely adapted to a range of soils and climate; leaves primarily bifoliolate; erect to prostrate growth; can tolerate heavy grazing pressure; does not survive long-term flooding; seed harvest is limited by sticky, single-seeded pods

competition, diseases, and other stresses particularly in grass–legume pastures. Predictability of legume suitability for pasture plantings from typical initial clipping evaluations has not been good. Thus, evaluation of response to grazing has been identified as an essential early step in the evaluation scheme of tropical legume germplasm for pasture use (Kretschmer 1989).

Despite the recognized limitations of many legume species, simulation modeling and systems analysis approaches, based on field experimental data in northern (tropical) Australia, indicate that introducing tropical legumes into native C_4 grass pastures constitutes a major component of improved technologies with potential to provide future viability to beef enterprises under stress from increasing costs of production and declining beef prices (Ash et al. 2015). A strategy to use tropical legumes in mixed cropping and agroforestry systems throughout the tropics to enhance resilience of existing systems and restore degraded lands has been proposed to provide sustainable intensification of forage-based systems which can reduce the ecologic footprint of livestock production and provide multiple ecosystem services (Rao et al. 2015). Use of adapted tropical legumes has been proposed as a viable approach to overcome the widespread degradation of overstocked and/or poorly managed grass pastures across tropical America (Montagnini 2008). On appropriate sites, *A. pintoi* in Costa Rica and *Pueraria phaseoloides* in Colombia were noted as appropriate for such use. *Gliricidia sepium*, in a cut-and-carry system, has effectively provided ecosystem benefits including erosion control, nitrogen fixation, soil improvement, firewood, fodder production, and function as a living fence (Montagnini 2008). The Brazilian government recently instituted a program to integrate crop, livestock, and forest production to reduce carbon emissions by agriculture (Almeida et al. 2013). *P. phaseoloides* in the Rain Forest biome of Central Brazil and *G. sepium* and *Leucaena leucocephala* in hot, semi-arid, northeastern Brazil are the primary species recognized as available legume options to contribute to these systems (Almeida et al. 2013). In the Gulf Coast states of the southeastern US, selected annual tropical legume species have potential to contribute to pasture systems in specialized uses such as creep grazing (Pitman et al. 2015).

Potential value of legumes in tropical grasslands has been illustrated by White et al. (2013) with estimated economic impacts of $2 million per year from 5000 ha planted in *Centrosema pascuorum* and $22 million per year from 1 500 000 ha of *Stylosanthes* species in Australia. In Brazil, distinctly different species of *Stylosanthes* provided an estimated $1.7 million per year from 200 000 ha, while economic impact of *P. phaseoloides* was reported at $33 million per year from 480 000 ha. An on-line interactive tool to assist in selection of the most suited tropical forages, including legumes, for particular environments (Cook et al. 2005) can contribute to increased success of plantings. This tool has been available in recent years with revision currently in progress.

Summary and Conclusions

As with the available germplasm overall, successfully used tropical and subtropical forage legume cultivars include a range of plant growth forms from low-growing herbaceous species to shrubs and trees. Successful cultivars include some with broad distribution and adaptation across tropical environments and others narrowly adapted to very specific niches. Development of superior genotypes of tropical forage legumes continues to extend the range of environments where tropical legumes contribute to productivity of native and improved pastures. Early evaluations of *Centrosema, Pueraria*, and the initial *Stylosanthes* and *Desmodium* genotypes primarily for productivity across different environments gradually shifted to assessment of new species and other genera. Specifically identified characteristics or adaptations were often assessed. Especially sought have been genotypes with attributes necessary for persistence in pastures.

As tropical forage legumes have been increasingly investigated, for use by small landholders and for use in integrated agricultural systems, a broader range of ecosystem services and management options have been addressed. Widespread emphasis on the evaluation of leucaena and targeted evaluations of other shrub and tree legumes, including the genera *Gliricidia* and *Desmanthus*, addressed needs for specific production approaches. Use of the woody legumes has required increased management and labor in both subsistence farming and commercial production.

Since the late 1980s, support for tropical and subtropical legume research and development has decreased. Fortunately, the germplasm banks hold a treasure of untested germplasm. Several Australian conferences have provided information on the future tropical pasture and forage trends in that country and elsewhere (Anonymous 1993a, 1993b; Leslie 1996). Tropical pasture research worldwide has followed the pattern set by Australia, but this source of leadership and research continuity has declined drastically during recent decades. Sufficient tropical forage legume germplasm evaluations have been conducted over the past few decades to provide a considerable body of information and this information base continues to be made available and used.

Broad classifications of adaptation and potential uses can now be assigned to many species. In the past, entries in germplasm evaluations were, to some extent, dependent upon timely availability. Future evaluations must be more carefully planned to make more efficient use of limited resources and to increase the probability of successful results. Many previous evaluations have failed to produce even potentially useful genotypes. The narrow range of adaptation of many tropical pasture legume species has become apparent. Thorough preliminary study can, in many cases, narrow the pool of realistically probable species to a rather small group. Any predetermined management constraints or existing utilization systems may further limit the likely prospects. Identification of the agronomic and social limitations and thorough assessment of the edaphic and climatic factors will offer insight and should result in a rather small set of species to consider.

Assessment of core collections of the identified species can further increase efficiency. The available information base, more sophisticated computer capabilities to analyze data, tremendous germplasm resources assembled, and potential benefits provide a compelling opportunity for continued evaluation and development of tropical forage legume germplasm.

Perceptions of the environmental effects of tropical and subtropical forage legumes, in general, have influenced resource availability for development and deployment of available germplasm. Recognition that these legumes can contribute stability, resilience, and sustainability to many disturbed and deteriorated ecosystems across the tropics is rather recent and increasing. Awareness of the tremendous impact of these legumes on livelihoods, ecosystem services, and regional economies in some tropical locations with limited economic alternatives is increasing. Opportunities to build on regional successes with currently available tropical and subtropical legume germplasm provide promise of improvements for both individual subsistence farmers and commercial livestock enterprises across a range of environments.

References

Adejumo, J.O. (1992). Effect of plant age and harvest date in the dry season on yield and quality of *Gliricidia sepium* in southern Nigeria. *Trop. Grassl.* 26: 21–24.

Adjei, M.B. and Muir, J.P. (2001). Current developments from tropical forage research in Africa. In: *Tropical Forage Plants: Development and Use* (eds. A. Sotomayor-Rios and W.D. Pitman), 331–355. Boca Raton: CRC Press.

de Almeida, R.G., de Andrade, C.M.S., Paciullo, D.S.C. et al. (2013). Brazilian agroforestry systems for cattle and sheep. *Trop. Grassl.–Forrajes Trop.* 1: 175–183.

Anonymous (1979). *Tropical Legumes: Resources for the Future*. Washington, DC: National Academy Press.

Anonymous (1983). *Leucaena Research in the Asian-Pacific Region*. Ottawa, Canada: International Development Research Center.

Anonymous (1984). *Forage and Browse Plants for Arid and Semi-Arid Africa*. UK: Royal Botanic Gardens, Kew and IBPGR, Rome.

Anonymous (1993a). Tropical pasture establishment. *Trop. Grassl.* 27: 257–259.

Anonymous (1993b). Northern dairy feed base 2001. *Trop. Grassl.* 27: 129–130.

Anonymous (1993c). *Stylosanthes guianensis* cv. Mineirao, nova leguminosa para agropecuaria dos Cerrados. *Pasturas Trop.* 15: 32. (in Portuguese).

Anonymous (1996). Caatinga stylo *Stylosanthes*-'Primar' and 'Unica'. *Plant Var. J.* 9: 19.

Ash, A., Hunt, L., McDonald, C. et al. (2015). Boosting the productivity and profitability of northern Australian beef enterprises: exploring innovation options using simulation modelling and systems analysis. *Agric. Syst.* 139: 50–65.

Austin, M.T., Williams, M.J., Hammond, A.C. et al. (1996). Psyllid population dynamics and plant resistance of *Leucaena* selections in Florida. *Trop. Grassl.* 30: 223–228.

Baltensperger, D.D., E.C. French, G.M. Prine, O.C. Ruelke, and K.H. Quesenberry. (1985). Hairy indigo, a summer legume for Florida. Florida Agric. Exp. Sta. Circ. S-318, Gainesville, FL.

Bell, L., Fainges, J., Darnell, R. et al. (2015). *Stocktake and Analysis of Legume Evaluation for Tropical Pastures in Australia*. North Sydney, NSW: Meat and Livestock Australia Ltd.

Bishop, H.G., Pengelly, B.C., and Ludke, D.H. (1988). Classification and description of a collection of the legume genus *Aeschynomene*. *Trop. Grassl.* 22: 160–175.

Blumenstock, D.I. (1957). Distribution and characteristics of tropical climates. In: *Proceedings of the 9th Pacific Science Congress*, vol. 20, 3–23.

Blumenthal, M.J. and Staples, L.B. (1993). Origin, evaluation and use of *Macrotyloma* as forage-a review. *Trop. Grassl.* 27: 16–29.

Boatwright, J.S., Wink, M., and van Wyk, B.-E. (2011). The generic concept of *Lotononis* (Crotalarieae, Fabaceae): reinstatement of the genera *Euchlora, Leobordea and Listia* and the new genus *Exoloba*. *Taxon* 60: 161–177.

Bogdan, A.J. (1977). *Tropical Pasture and Fodder Plants*. London: Longmans.

Bowman, B.A.M. and Wilson, G.P. (1996). Persistence and yield of forage peanuts (*Arachis spp.*) on the New South Wales north coast. *Trop. Grassl.* 30: 402–413.

Bransby, D.I., Solaiman, S.G., Kerth, C.R. et al. (2003). Defoliation patterns of goats browsing mimosa. In: *Proceedings of the American Forage and Grassland Council*, Lafayette, LA. 26–30 Apr. 2003 (ed. K. Cassida), 177–181. Georgetown, TX: American Forage and Grassland Council.

Bray, R.A. (1985a). Selecting and breeding better legumes. In: *Nutritional Limits to Animal Production from Pastures. Proceedings of an International Symposium*, St. Lucia, Queensland, Australia. 24–28 Aug. 1981 (ed. J.B. Hacker), 287–303. Slough, UK: CABI.

Bray, R.A. (1985b). Germplasm sources for genetic improvement of forage legumes in Australia and New Zealand. In: *Forage Legumes for Energy Efficient Animal Production. Proceedings of a Trilateral Workshop*, Palmerston North, New Zealand. 30 Apr.–4 May 1984 (ed. R.F Barnes), 278–284. Washington, DC: USDA, ARS.

Bray, R.A. and Woodroffe, T.D. (1997). Effect of leucaena psyllid on yield of *Leucaena leucocephala* cv. Cunningham in southeast Queensland. *Trop. Grassl.* 25: 356–357.

Bray, R.A., Sonoda, R.M., and Kretschmer, A.E. Jr. (1991). Pathotype variability of rust caused by

Uromyces appendiculatus on *Macroptilium atropurpureum*. *Plant Dis.* 75: 430.

Bray, R.A., Ibrahim, T., Palmer, B., and Schlink, A.C. (1993). Yield and quality of *Gliricidia sepium* accessions at two sites in the tropics. *Trop. Grassl.* 27: 30–36.

Bray, R.A., Palmer, B., and Ibrahim, T.M. (1997). Performance of shrub legumes at four sites in Indonesia and Australia. *Trop. Grassl.* 31: 31–39.

Brewbaker, J.L. (1987). Species in the genus *Leucaena*. *Leucaena Res. Rep.* 7: 6–20.

Bryan, W.W. (1961). *Lotononis bainesii* Baker. A legume for the subtropics. *Aust. J. Exp. Agric. Anim. Husb.* 1: 4–10.

Bryan, W.W., Sharpe, J.P., and Haydock, K.P. (1971). Some factors affecting the growth of lotononis (*Lotononis bainesii*). *Aust. J. Exp. Agric. Anim. Husb.* 11: 29–34.

Calles, T. and Schultze-Kraft, R. (2016). New species, nomenclatural changes and recent taxonomic studies in the genus *Stylosanthes* (Leguminosae): an update. *Trop. Grassl.–Forrajes Trop.* 4: 122–128.

Cameron, D.G. (1984). Tropical and subtropical pasture legumes. 4. Glycine (*Neonotonia wightii*): an outstanding but soil specific legume. *Queensland Agric. J.* 10: 311–316.

Castillo, A.C., Cuyugan, O.C., Fogarty, S., and Shelton, H.M. (1997). Growth, psyllid resistance and forage quality of *Leucaena leucocephala x L. pallida*. *Trop. Grassl.* 31: 188–200.

Chakraborty, S., Charudattan, R., and DeValerio, J.T. (1994). Reaction of selected accessions of forage *Cassia* spp. to some fungal pathogens. *Trop. Grassl.* 28: 32–37.

CIAT (2003). Forages. CIAT. https://ciat.cgiar.org/what-we-do/forages-livestock/ (accessed 22 November 2019).

Clements, R.J. (1989). Developing persistent pasture legume cultivars for Australia. In: *Persistence of Forage Legumes. Proceedings of a Trilateral Workshop*, Honolulu, HI. 18–22 July 1988 (eds. G.C. Marten, R.F Barnes, R.W. Brougham and R.J. Clements), 505–519. Madison, WI: American Society of Agronomy.

Clements, R.J., Jones, R.M., Valdes, L.R., and Bunch, G.A. (1996). Selection of *Chamaecrista rotundifolia* by cattle. *Trop. Grassl.* 30: 389–394.

Cobbina, J. and Atta-Krah, A.N. (1992). Forage productivity of *Gliricidia* accessions on a tropical Alfisol in Nigeria. *Trop. Grassl.* 26: 248–254.

Cook, B.G. (1988). Persistent new legumes for heavy grazing. 2. Wynn round-leafed cassia. *Queensland Agric. J.* 114: 119–121.

Cook, B.G. and Franklin, T.G. (1988). Crop management and seed harvesting of *Arachis pintoi* Krap. et Greg. nom. nud. *J. Appl. Seed Prod.* 6: 26–30.

Cook, B.G. and Jones, R.M. (1987). Persistent new legumes for intensive grazing. 1. Shaw creeping vigna. *Queensland Agric. J.* 113: 89–91.

Cook, B.G., Pengelly, B.C., Brown, S.D. et al. (2005). *Tropical Forages: An Interactive Selection Tool*. Brisbane, Australia: CSIRO, DPI&F (Qld.), CIAT, and ILRI. Available at www.tropicalforages.info.

Duke, J.A. (ed.) (1981). *Handbook of Legumes of World Economic Importance*. New York, NY: Plenum Press.

Edye, L.A., Hall, T.J., Clem, R.L. et al. (1998). Sward evaluation of eleven *Stylosanthes seabrana* and *S. scabra* cv. Seca at five subtropical sites. *Trop. Grassl.* 32: 243–251.

English, B.H. (1999a). *Macroptilium atropurpureum* in Australia. In: *Forage Seed Production. 2. Tropical and Subtropical Species* (eds. D.S. Loch and J.E. Ferguson), 407–412. Wallingford, Oxon, UK: CABI.

English, B.H. (1999b). *Stylosanthes scabra* in Australia. In: *Forage Seed Production. 2. Tropical and Subtropical Species* (eds. D.S. Loch and J.E. Ferguson), 413–419. Wallingford, Oxon, UK: CABI.

Evans, D.O. and Rotar, P.P. (eds.) (1987). *Sesbania in Agriculture*. London: Westview Press.

Firth, D.J. and Wilson, G.P.M. (1995). Preliminary evaluation of species for use as permanent ground cover in orchards on the north coast of New South Wales. *Trop. Grassl.* 29: 18–27.

Firth, D.J., Jones, R.M., McFadyen, L.M. et al. (2002). Selection of pasture species for groundcover suited to shade in mature macadamia orchards in subtropical Australia. *Trop. Grassl.* 36: 1–12.

Fribourg, H.A., Overton, J.R., McNeill, W.W. et al. (1984). Evaluations of the potential of hyacinth bean as an annual warm-season forage in the mid-South. *Agron. J.* 76: 905–910.

Fujita, H. and Humphreys, L.R. (1992). Variation in seasonal stocking rate and the dynamics of *Lotononis bainesii* in *Digitaria decumbens* pastures. *J. Agric. Sci.* (Cambridge) 118: 47–53.

Gardener, C.J. and Ash, A.M. (1994). Diet selection in six *Stylosanthes*-grass pastures and its implications for pasture stability. *Trop. Grassl.* 28: 109–119.

Gardiner, C., Kempe, N., Hannah, I., and McDonald, J. (2013). PROGARDES™: a legume for tropical/subtropical semi-arid clay soils. *Trop. Grassl.–Forrajes Trop.* 1: 78–80.

Gramshaw, D., Pengelly, B.C., Muller, F.W. et al. (1987). Classification of a collection of the legume *Alysicarpus* using morphological and preliminary agronomic attributes. *Aust. J. Agric. Res.* 38: 355–366.

Gramshaw, D., Reed, J.W., Collins, W.J., and Carter, E.D. (1989). Sown pastures and legume persistence: an Australian overview. In: *Persistence of Forage Legumes. Proceedings of a Trilateral Workshop*, Honolulu, HI. 18–22 July 1988 (eds. G.C. Marten, R.F Barnes, R.W. Brougham and R.J. Clements), 1–22. Madison, WI: American Society of Agronomy.

Grof, B. (1985). Forage attributes of the perennial groundnut, *Arachis pintoi*, in a tropical savanna environment in Colombia. In: *Proceedings of the 15th International Grassland Congress*, 168–170. Kyoto, Japan.

Gutteridge, R.C. and Shelton, H.M. (eds.) (1993). *Forage Tree Legumes in Tropical Agriculture*. Tucson, Arizona: CABI.

Hague, I., Jutzi, S., and Neate, P.J.H. (eds.) (1986). *Potentials of Forage Legumes in Farming Systems of Sub-Saharan Africa. Proceedings of a Workshop*, Addis Ababa, Ethiopia. 16–19 Sept. 1985. Addis Ababa, Ethiopa: ILCA.

Hall, T.J. (1985). Adaptation and agronomy of *Clitoria ternatea* L. in northern Australia. *Trop. Grassl.* 19: 156–163.

Halliday, M.J., Padmanabha, J., McSweeney, C.S. et al. (2013). Leucaena toxicity: a new perspective on the most widely used forage tree legume. *Trop. Grassl. – Forrajes Trop.* 1: 1–11.

Hegarty, M.P. (1982). Deletereous factors in forages affecting animal production. In: *Nutritional Limits to Animal Production from Pastures. Proceedings of an International Symposium*, St. Lucia, Queensland, Australia. 24–28 Aug. 1981 (ed. J.B. Hacker), 134–150. Slough, UK: CABI.

Hernandez, M., Argel, P.J., Ibrahim, M.A., and Mannetje, L.'t. (1995). Pasture production, diet selection and liveweight gains of cattle grazing *Brachiaria brizantha* with or without *Arachis pintoi* at two stocking rates in the Atlantic Zone of Costa Rica. *Trop. Grassl.* 29: 134–141.

Hodges, E.M., A.E. Kretschmer, Jr., P. Mislevy, R.D. Roush, O.C. Ruelke, and G.H. Snyder. (1982). Production and utilization of the tropical legume *Aeschynomene*. Florida Agric. Exp. Stn. Circ. S-290. Gainesville.

Humphries, L.R. (1967). Townsville lucerne: history and prospect. *J. Aust. Inst. Agric. Sci.* 33: 3.

IBPGR and ICRISAT (ed.) (1990). *Report of a Workshop on the Genetic Resources of Wild Arachis Species*, Cali, Colombia. 28 Feb.–2 Mar. 1989. Cali, Colombia: CIAT.

IBPGR and ICRISAT (ed.) (1992). *Descriptors for Groundnuts*. Rome: IBPGR.

Irwin, J.A., Cameron, D.G., and Lenne, J.M. (1984). Responses of *Stylosanthes* to anthracnose. In: *Biology and Agronomy of Stylosanthes* (eds. H.M. Stace and L.A. Edye), 295–310. Sydney: Academic Press.

Jones, R.J. (1979). The value of *Leucaena leucocephala* as a feed for ruminants in the tropics. *World Anim. Rev.* 31: 1–11.

Jones, R.J. (1985). *Leucaena* toxicity and the ruminant degradation of mimosine. In: *Plant Toxicology. Proceedings of the Australia/U.S.A. Poisonous Plants Symposium*, Brisbane, Australia. 14–18 May 1984 (ed. A.A. Seawright), 111–118. Brisbane: CSIRO.

Jones, R.J. (2001). Current developments from tropical forage research in Australia. In: *Tropical Forage Plants: Development and Use* (eds. A. Sotomayor-Rios and W.D. Pitman), 295–329. Boca Raton, FL: CRC Press.

Jones, R.J. and Jones, R.M. (1976). The ecology of siratro based pastures. In: *Plant Relations in Pastures* (ed. J.R. Wilson), 353–367. Melbourne, Australia: CSIRO.

Jones, R.J. and Megarrity, R.G. (1986). Successful transfer of DHP-degrading bacteria from Hawaiian goats to Australian ruminants to overcome the toxicity of leucaena. *Aust. Vet. J.* 63: 69–78.

Jones, R.J. and Palmer, B. (2002). Assessment of the condensed tannin concentration in a collection of *Leucaena* species using 14C-labled polyethylene glycol (PEG 4000). *Trop. Grassl.* 36: 47–58.

Jones, R.J., Bishop, H.G., Clem, R.L. et al. (2000). Measurements of nutritive value of a range of tropical legumes and their use in legume evaluation. *Trop. Grassl.* 34: 78–90.

Jones, R.M. (1993). Persistence of *Arachis pintoi* cv. Amarillo on three soil types at Samford, south-eastern Queensland. *Trop. Grassl.* 27: 11–15.

Jones, R.M. (2014). The rise and fall of Siratro (*Macroptilium atropurpureum*) – what went wrong and some implications for legume breeding, evaluation and management. *Trop. Grassl.–Forrajes Trop.* 2: 154–164.

Jones, R.M. and Bunch, G.A. (1995a). Yield and population dynamics of *Chamaecrista rotundifolia* Wynn in coastal south-eastern Queensland as affected by stocking rate and rainfall. *Trop. Grassl.* 29: 65–73.

Jones, R.M. and Bunch, G.A. (1995b). Long-term records of legume persistence and animal production from pastures based on Safari Kenya clover and leucaena in subtropical coastal Queensland. *Trop. Grassl.* 29: 74–80.

Jones, R.M. and Clements, R.J. (1987). Persistence and productivity of *Centrosema virginianum* and *Vigna parkeri* cv. Shaw under grazing on the coastal lowlands of south-east Queensland. *Trop. Grassl.* 22: 55–64.

Jones, R.M., Jones, R.J., and Rees, M.C. (1997). Evaluation of tropical legumes on clay soils at four sites in southern inland Queensland. *Trop. Grassl.* 31: 95–106.

Jones, R.M., Bunch, G.A., and McDonald, C.K. (1998). Ecological and agronomic studies on *Chamaecrista rotundifolia* cv. Wynn related to modeling of persistence. *Trop. Grassl.* 32: 153–165.

Keller-Grein, K., Schultze-Kraft, R., Franco, L.H., and Ramirez, G. (2000). Multilocational agronomic evaluation of *Centrosema pubescens* germplasm on acid soils. *Trop. Grassl.* 34: 65–77.

Kerridge, P.C. and Hardy, B. (eds.) (1994). *Biology and Agronomy of Forage Arachis*. Cali, Colombia: CIAT.

Kerridge, P.C. and Loch, D.S. (1999). *Arachis pintoi* in Australia and Latin America. In: *Forage Seed Production*,

Vol. 2. Tropical and Subtropical Species (eds. D.S. Loch and J.E. Ferguson), 427–434. Wallingford, Oxon, UK: CABI.

Krapovickas, A. (1990). A proposed taxonomic summary of the genus Arachis. Report of a workshop on the genetic resources of wild Arachis species in Cali, Colombia (28 February–2 March 1989).

Krapovickas, A. and Gregory, W.C. (eds.) (1994). Taxonomy of the genus Arachis (Leguminosae). *Bonplandia* 16: 1–205.

Kretschmer, A.E. Jr. (1989). Tropical forage legume development, diversity, and methodology for determining persistence. In: *Persistence of Forage Legumes. Proceedings of a Trilateral Workshop*, Honolulu, HI. 18–22 July 1988 (eds. G.C. Marten, R.F Barnes, R.W. Brougham and R.J. Clements), 117–138. Madison, WI: American Society of Agronomy.

Kretschmer, A.E. Jr. and Pitman, W.D. (2001). Germplasm resources of tropical forage legumes. In: *Tropical Forage Plants: Development and Use* (eds. A. Sotomayor-Rios and W.D. Pitman), 41–57. Boca Raton, FL: CRC Press.

Kretschmer, A.E., Jr., J.B. Brolmann, G.H. Snyder, and S.W. Coleman. (1979). 'Florida' carpon desmodium [*Desmodium heterocarpon* (L.) DC], a perennial tropical forage legume for use in south Florida, Florida Agric. Exp. Stn. Circ. S-260. Gainesville, FL.

Kretschmer, A.E. Jr., Sonoda, R.M., Bullock, R.C. et al. (1985). Diversity in *Macroptilium atropurpureum* (DC) Urb. In: *Proceedings of the 15th International Grassland Congress*, 155–157. Kyoto, Japan.

Kretschmer, A.E. Jr., Sonoda, R.M., Bullock, R.C., and Wilson, T.C. (1992). Registration of IRFL 4655 *Macroptilium atropurpureum* germplasm. *Crop Sci.* 32: 836.

Kretschmer, A.E. Jr., Simpson, C.E., Wilson, T.C., and Pitman, W.D. (1994). Evaluation of wild nut-producing *Arachis* species for forage. In: *Proceedings of the 17th International Grassland Congress*, 2122–2123. Rockhampton, Australia.

Kretschmer, A.E. Jr., Wilson, T.C., Kalmbacher, R.S. et al. (2001). Evaluation of growth, yield, flowering, and seed production of nut-producing peanuts in south Florida. *Soil Crop Sci. Soc. Florida Proc.* 60: 105–113.

Lapointe, S.L. and Sonoda, R.M. (2001). The effects of arthropods, diseases, and nematodes on tropical pastures. In: *Tropical Forage Plants: Development and Use* (eds. A. Sotomayor-Rios and W.D. Pitman), 201–218. Boca Raton, FL: CRC Press.

Larbi, A., Smith, J.W., Adekunle, I.O., and Kurdi, I.O. (1996). Studies on multipurpose fodder trees and shrubs in West Africa: Variation in determinants of forage quality in *Albizia* and *Paraserianthes* species. *Agrofor. Syst.* 33: 29–39.

Larbi, A., Adekunle, I.O., Awojide, A., and Akinlade, J. (1999). Identifying *Chamaecrista rotundifolium* and *Centrosema* species for bridging seasonal feed gaps in smallholder mixed farms in the West African derived savanna. *Trop. Grassl.* 33: 91–97.

Lavers, M.H. and Silver, B.A. (2000). Establishment of pinto peanut direct-drilled into a tropical grass pasture. *Trop. Grassl.* 34: 303.

Lenne, J.M. and Trutman, P. (1994). *Diseases of Tropical Pasture Plants*. Wallingford, Oxon, UK: CABI.

Leslie, J.K. (1996). Pastures for prosperity. Proc. Fifth Trop. Pasture Conf. *Trop. Grassl.* 30: 1.

Loch, D.S. (1991). Tropical herbage and seed production – origins, progress, and opportunities. *J. Appl. Seed Prod.* 9: 14–20.

Loch, D.S. and Ferguson, J.E. (eds.) (1999). *Forage Seed Production. Vol. 2. Tropical and Subtropical Species*. Wallingford, Oxon, UK: CABI.

Luckow, M. (1993). Monograph of *Desmanthus* (Leguminosae-Mimosoideae). *Syst. Bot. Monogr.* 38: 1–166.

Maass, B.L. (1989). Die troposche weideleguminose *Stylosanthes scabra* Vog. variabilitat, leistungstand und moglichkeiten zuchterischer verbesserung. *Landbau-forschung Volkenrode, Sondereft.* 97.

Maass, B.L. and Torres, A.M. (1998). Off-types indicate natural out crossing in five tropical legumes in Colombia. *Trop. Grassl.* 32: 124–130.

Mannetje, L.'.T. and Jones, R.M. (eds.) (1992). *Plant resources of South-East Asia. 4. Forages*. Wageningen, The Netherlands: PUDOC Science Publishing.

Marten, G.C., Matches, A.G., Barnes, R.F et al. (eds.) (1989). *Persistence of Forage Legumes*. Madison, WI: ASA, CSSA, and SSSA.

McDonald, C.K. and Clements, R.J. (1999). Occupational and regional differences in perceived threats and limitations to the future use of sown tropical pasture plants. *Trop. Grassl.* 33: 129–137.

McDonald, C.M. and Jones, R.M. (2000). Managing seed set of Wynn cassia in grazed cassia-grass pastures. *Trop. Grassl.* 34: 304.

Miles, J.W. and Lascano, C.E. (1997). Status of *Stylosanthes* development in other countries. I. *Stylosanthes* development and utilization in South America. *Trop. Grassl.* 31: 454–459.

Minson, D.J. (1980). Nutritional differences between tropical and temperate pastures. In: *Grazing Animals* (ed. F.H.W. Morley), 143–157. Amsterdam: Elsevier Science Publishing.

Minson, D.J. and Milford, R. (1966). The energy values and nutritive value indices of *Digitaria decumbens, Sorghum almum*, and *Phaseolus atropurpureum*. *Aust. J. Agric. Res.* 17: 411–423.

Montagnini, F. (2008). Management for sustainability and restoration of degraded pastures in the neotropics. In: *Post-Agricultural Succession in the Neotropics* (ed. R.W. Myster), 265–295. New York: Springer.

Mosjidis, J.A. (2001). Registration of 'AU Grazer' sericea lespedeza. *Crop Sci.* 41: 262.

Mosquera, P. and Lascano, C. (1992). Produccion de leche de vaca en pasturas de *Brachiaria decumbens* y con acceso controlado a bancos de proteins. *Pastures Trop.* 14: 2. CIAT, Cali, Colombia (in Spanish).

Muir, J.P. (2011). The multi-faceted role of condensed tannins in the goat ecosystem. *Small Ruminant Res.* 98: 115–120.

Muir, J.P., Butler, T.J., Ocumpaugh, W.R., and Simpson, C.E. (2010). 'Latitude 34', a perennial peanut for cool, dry climates. *J. Plant Regist.* 4: 106–108.

Muir, J.P., Pitman, W.D., Dubeux, J.C. Jr., and Foster, J.L. (2014). The future of warm-season, tropical, and subtropical forage legumes in sustainable pastures and rangelands. *Afr. J. Range Forage Sci.* 31: 187–198.

Muir, J.P., Pitman, W.D., Smith, F.S. et al. (2018). Challenges to developing native legume seed supplies: the Texas experience as a case study. *Native Plants J.* 19: 224–238.

National Academy of Science (1984). *Leucaena: Promising Forage and Tree Crop for the Tropics*, 2e. Washington, DC: National Academy Press.

Ocumpaugh, W.R., Kunz, D., Rahmes, J. et al. (2003). *Desmanthus bicornutus*: a new summer-growing perennial shrubby legume for South Texas and Mexico. In: *Proceedings of American Forage and Grassland Council*, Lafayette, LA. 26–30 Apr. 2003 (ed. K. Cassida), 151–155. Georgetown, TX: American Forage and Grassland Council.

Ohashi, H. (ed.) (1973). *Ginkgoana – Contributions to the Flora of Asia and the Pacific Region, No. I*. Tokyo: Tokyo Academia Scientific Books.

Ohashi, H. (1991). Taxonomic studies in *Desmodium heterocarpon* (L.) DC. (Leguminosae). *J. Jpn. Bot.* 66: 14–25.

Oram, R.J. (1990a). Legumes 21. *Arachis* (a), *Arachis pintoi* Krap. et. Greg nom. nud. (Pinto peanut) cv. Amarillo. *Aust. J. Exp. Agric.* 30: 445–446.

Oram, R.J. (1990b). B. Legumes. *Centrosema pascuorum* Mart. ex Benth. (centurion). In: *Register of Australian Herbage Plant Cultivars* (ed. C.J. Barnard), 272. Queensland, Australia: CSIRO.

Partridge, I.N. (1986). Effect of stocking rate and superphosphate level on oversown fire climax grassland *Pennisetum polystachion* in Fiji. *Trop. Grassl.* 20: 166–173.

Partridge, I.N. and Wright, J.W. (1992). The value of round-leafed cassia (*Cassia rotundifolia*) cv. Wynn in a native pasture grazed with steers in south-east Queensland. *Trop. Grassl.* 26: 263–269.

Pengelly, B. (2015). *A Global Strategy for the Conservation and Utilization of Tropical and Sub-Tropical Forage Genetic Resources*. Pengelly Consultancy Pty. Ltd. Available at www.croptrust.org.wp-content/uploads/2014/12/Forages-Strategy.pdf.

Pengelly, B.C. and Conway, M.J. (2000). Pastures on cropping soils: which tropical pasture legume to use. *Trop. Grassl.* 34: 162–168.

Perez, S.M., Camardelli, M.C., Juarez, F. et al. (1999). Geographic distribution of *Macroptilium* species in Argentina. *Trop. Grassl.* 33: 22–33.

Pitman, W.D. (2009). Invasiveness of species useful as warm-season pasture legumes in the southeastern United States. In: *Invasive Species: Detection, Impact and Control* (eds. C.P. Wilcox and R.B. Turpin), 145–159. Hauppauge, NY: Nova Science Publishers.

Pitman, W.D. and Adjei, M.B. (1994). Response of Shaw creeping vigna and a perennial alyceclover accession to burning. *Trop. Grassl.* 28: 53–55.

Pitman, W.D., A.E. Kretschmer, Jr., and C.G. Chambliss. (1986). Phaseybean, a summer legume with forage potential for Florida flatwoods, Florida Agric. Exp. Stn. Circ. S-330. Gainesville, FL.

Pitman, W.D., Walker, R.S., Scaglia, G. et al. (2015). Summer legumes for creep grazing in cow-calf production on bermudagrass pastures. *J. Agric. Sci.* 7 (8): 8–17.

Polhill, R.M. and Raven, P.H. (eds.) (1981). *Advances in Legume Systematics, Pt. 1* & *2*, vol. 2. Surrey, England: Royal Botanic Gardens, Kew.

Polhill, R.M., Raven, P.H., and Sirton, C.H. (1981). Evolution and systematics of the *Leguminosae*. In: *Advances in Legume Systematics (Pt. 1)* (eds. R.M. Polhill and P.H. Raven), 1–26. Surrey, England: Royal Botanic Gardens, Kew.

Prine, G.M., L.S. Dunavin, J.E. Moore, and R.D. Roush. (1981). 'Florigraze' rhizoma peanut, a perennial forage legume. Florida Agric. Exp. Stn. Circ. S-275, Gainesville, FL.

Quesenberry, K.H., Blount, A.R., Mislevy, P. et al. (2010). Registration of 'UF Tito' and 'UF Peace' rhizoma peanut cultivars with high dry matter yields, persistence, and disease tolerance. *J. Plant Regist.* 4: 17–21.

Ramirez, P.A. (ed.) (2002). *Maquenque Desmodium heterocarpon (L.) DC. Subsp. ovalifolium (Prain.) Ohashi, (CIAT 13651)*. Cali, Colombia: CIAT/CIA, Imagenes Graficas, S.A.

Rao, I., Peters, M., Castro, A. et al. (2015). Livestock-Plus – The sustainable intensification of forage-based agricultural systems to improve livelihoods and ecosystem services in the tropics. *Trop. Grassl.–Forrajes Trop.* 3: 59–82.

Reid, R. (1980). Collection and use of climatic data in pasture plant introduction. In: *Collecting and Testing Tropical Forage Plants* (eds. R.J. Clements and D.G. Cameron), 1–10. Melbourne, Australia: CSIRO.

Reid, R. and Sinclair, D.F. (1980). *An Evaluation of a Collection of Clitoria ternatea for Forage and Grain Production*, Genetic Resources Communication No. 1. Brisbane, Australia: CSIRO.

Rudd, V.E. (ed.) (1955). *American Species of Aeschynomene*, U.S. Natl. Herb., vol. 32, 1–72. Washington, D.C.: Smithsonian Institute.

Santos-Garcia, M.O., de Toledo-Silva, G., Sassaki, R.P. et al. (2012). Using genetic diversity information to establish core collections of *Stylosanthes capitata* and *Stylosanthes macrocephala*. *Genet. Mol. Biol.* 35: 847–861.

Schultze-Kraft, R. (1986). Natural distribution and germplasm collection of the pasture legume *Centrosema macrocarpa* Benth. *Angew. Botanik* 60: 407–419.

Schultze-Kraft, R. and Benavides, G. (1988). *Germplasm Collection and Preliminary Evaluation of Desmodium ovalifolium Wall*, Genetic Resources Communication 12. Brisbane, Australia: CSIRO.

Schultze-Kraft, R. and Clements, R.J. (eds.) (1990). *Centrosema: Biology, Agronomy, and Utilization*. Cali, Colombia: CIAT.

Schultze-Kraft, R. and Keller-Grein, G. (1999). Crop growth and development: legumes. In: *Forage Seed Production. 2. Tropical and Subtropical Species* (eds. D. Loch and J.E. Ferguson), 57–80. Wallingford, Oxon, UK: CABI.

Schultze-Kraft, R., Benavides, R.G., and Arias, A. (1987). Collection of germplasm and preliminary evaluation of *Centrosema acutifolium*. *Pasturas Trop.* 9: 12–20. CIAT, Cali, Colombia.

Smith, P. (1996). Pastures for prosperity – Seeds forum, 6. What we want from the seed industry in the future – merchant's viewpoint. *Trop. Grassl.* 30: 88–89.

Smith, G.R., Rouquette, F.M. Jr., and Pemberton, I.J. (2008). Registration of 'Rio Verde' lablab. *J. Plant Regist.* 2: 15.

Sonoda, R.M. and Lenne, J.M. (1986). Diseases of *Aeschynomene* species. *Trop. Grassl.* 20: 30–34.

Sonoda, R.M., Kretschmer, A.E. Jr., and Wilson, T.C. (1991). Evaluation of *Macroptilium atropurpureum* (DC) Urb. germplasm for reaction to foliar diseases. *Soil Crop Sci. Soc. Florida Proc.* 51: 25–27.

Stace, H.M. and Edye, L.A. (eds.) (1984). *The Biology and Agronomy of Stylosanthes*. Sydney: Academic Press.

Strickland, R.W., Greenfield, R.G., Wilson, G.P.M., and Harvey, G.L. (1985). Morphological and agronomic attributes of *Cassia rotundifolia* Pers., *C. pilosa* L., and *C. trichopoda* Benth., potential forage legumes for northern Australia. *Aust. J. Exp. Agric.* 25: 100–108.

Summerfield, R.J. and Bunting, A.H. (eds.). (1980). Advances in legume science, Vol. 1. Proceedings of the International Legume Conference, Kew, Surrey, England (31 July–4 August 1978). Royal Botanic Gardens, Kew, Surrey, England.

Tarawali, S.A. (1995). Evaluation of *Chamaecrista rotundifolia* accessions as a fodder resource in subhumid Nigeria. *Trop. Grassl.* 29: 129–133.

Teitzel, J.K. and Burt, R.L. (1976). *Centrosema pubescens* in Australia. *Trop. Grassl.* 10: 5–14.

Vander Stappen, J., De Laet, J., Gama-Lopez, S. et al. (2002). Phylogenetic analysis of *Stylosanthes* (Fabaceae) based on the internal transcribed spacer region (ITS) of nuclear ribosomal DNA. *Plant Syst. Evol.* 234: 27–51.

White, D.S., Peters, M., and Horne, P. (2013). Global impacts from improved tropical forages: a meta-analysis revealing overlooked benefits and costs, evolving values and new priorities. *Trop. Grassl.–Forrajes Trop.* 1: 12–24.

Whiteman, P.C. (1980). *Tropical Pasture Science*. New York: Oxford University Press.

Whyte, R.O., Nilsson-Leissner, G., and Trumble, H.C. (eds.) (1953). *Legumes in Agriculture*, FAO Agricultural Studies No. 21. Rome: FAO.

Williams, R.J. (1983). Tropical legumes. In: *Genetic Resources of Forage Plants* (eds. J.G. McIver and R.A. Bray), 17–37. Melbourne, Australia: CSIRO.

Wilson, G.P.M., Jones, R.M., and Cook, B.G. (1982). Persistence of jointvetch (*Aeschynomene falcata*) in experimental sowings in the Australian subtropics. *Trop. Grassl.* 16: 155–156.

Withington, D., Glover, N., and Brewbaker, J. (1987). *Gliricidia sepium (Jacq.) Walp: Management and Improvement*, 87–101. Waimanalo, Hawaii: NFTA Special Publication.

Zdravko, B. and Fisher, M.J. (1996). Effect of planting method and soil texture on the growth and development of *Arachis pintoi*. *Trop. Grassl.* 29: 395–401.

CHAPTER

16

Cool-Season Grasses for Humid Areas

Michael D. Casler, Research Geneticist, *U.S. Dairy Forage Research Center, USDA-ARS, Madison, WI, USA*
Robert L. Kallenbach, Associate Dean, *University of Missouri, Columbia, MO, USA*
Geoffrey E. Brink, Research Agronomist, *USDA-Agricultural Research Service, US Dairy Forage Research Center, Madison, WI, USA*

Introduction

Cool-season grasses used for livestock production in humid areas of North America are mainly introduced species from Europe, Western Asia, and North Africa. Most were introduced into eastern North America in the mid- to late-eighteenth century and spread westward by settlers during the nineteenth century and by agricultural research networks and marketing channels during the twentieth century. Many of these grasses are now ubiquitous throughout eastern North America. Despite their non-native status, genetic diversity within these species allows natural selection for adaptation and life-history traits, leading to highly naturalized and adapted populations. Indeed, several cool-season grasses, both native and introduced, are classified as "invasive"and colonize natural areas where they are unwanted and often difficult to control or eradicate.

The humid areas of North America in which cool-season grasses are important livestock feeds include nearly the entire area east of the Rocky Mountains to the Atlantic Seaboard, from the Canadian taiga to the southern edge of the Transition Zone. The Transition Zone is generally considered to extend from the Carolinas to Missouri, extending as far south as central Georgia and Alabama and north to the Ohio River. Due to mild maritime climates, cool-season grasses are also highly adapted and utilized between the Pacific Ocean and the Great Basin, including plains and coastal mountain ranges from Northern California to South-central Alaska.

Cool-season grasses are primarily used for livestock production and conservation in the humid areas of North America. Their uses include hay and haylage, silage, pasture, soil conservation, wildlife food and cover, land reclamation and restoration, buffer zones, revegetation of **riparian** zones, nutrient recycling, pollution abatement, and disposal zones for manure, **effluent**, and wastewater. Cool-season grasses are utilized as monocultures in specialized production systems and as components of pasture, hay, or conservation mixtures.

The Fescues

Tall Fescue

Origin and Description

Tall fescue is native to Europe, Northern Africa, and parts of Asia (Buckner et al. 1979). In addition to its importance for pasture and hay production, it is renowned for its ability to reduce erosion, tolerate drought stress and provide high quality fall growth that can be pastured in late fall and winter to reduce need for harvested forage. Its main area of use in the US is in the humid transition zone between cool-temperate and the subtropical zones where it is grown on about 14 million hectares (Buckner et al. 1979; Hoveland 1993).

Forages: The Science of Grassland Agriculture, Volume II, Seventh Edition.
Edited by Kenneth J. Moore, Michael Collins, C. Jerry Nelson and Daren D. Redfearn.

Tall fescue is **perennial**, upright, coarse, and tufted (**bunch-type growth habit**). While strictly classified as a bunchgrass, if mowed or grazed, it produces an even sod (Sleper and Buckner 1995). Leaf blades are 3 to 12 mm wide, scabrous on the margin, have short rudimentary **auricles**, distinct longitudinal veins or ribs, and can reach lengths of 1 m. **Sheaths** are smooth and have membranous ligules approximately 2 mm long. Most types of tall fescue have short rhizomes, but there is a wide range of genetic variation for this trait (Sleper and West 1996). Tall fescue is typically propagated from seed.

Tall fescue is an allohexaploid (2n = 6x = 42 chromosomes) (Sleper and West 1996). Reproductive **tillers** can be up to 2.0 m tall, up to 5 mm in diameter, and bear seed on panicles that are 10–50 cm long. **Panicle** shapes range from loosely branched to narrow with short branches. Tall fescue flowers later than orchardgrass, somewhat earlier than smooth bromegrass (Olson et al. 2015), and earlier than timothy.

Adaptation, Use and Management

Tall fescue is used for pasture, hay, silage, and soil conservation, primarily in the transition zone (Chapter 23). In this region, it persists longer than most other cool-season grasses, due largely to its adaptability to a wide range of environmental conditions, including soils that are acid, infertile, drought-prone, and somewhat poorly drained. It also tolerates heat, over-grazing and improper management, but is not as cold tolerant as smooth bromegrass or timothy. In the humid-temperate region, tall fescue produces about two-thirds of its annual growth in spring and about one-third in autumn. While tall fescue withstands **continuous stocking** better than most cool-season grasses, rotational grazing with short-rest periods in spring and longer **rest periods** in summer provides better utilization and productivity.

In the past, tall fescue was considered to be of lower quality than other cool-season grasses despite having comparable measurements of forage quality. Research beginning in the late 1970s showed this to be a result of the presence of the endophytic fungus *Epichloë coenophiala* ([Morgan-Jones and W. Gams] Glenn, Bacon, and Hanlin). This endophytic fungus present in infected tall fescue produces ergopeptine alkaloids that cause a number of livestock disorders. Collectively, these disorders are called tall fescue toxicosis. More than 90% of the tall fescue pastures in the humid region are infected with an ergopeptine-**alkaloid**-producing or "toxic" **endophyte** (Sleper and West 1996).

Symptoms from livestock consuming toxic tall fescue include reduced feed intake, reduced rate of gain or milk production, rapid breathing, increased core body temperature, depressed blood serum prolactin, low-conception rate, rough hair coat, and an overall unthrifty appearance (Aiken and Strickland 2013). Symptoms are most noticeable when livestock are under heat or cold stress because capillary blood flow is reduced by ergopeptide compounds (Schmidt and Osborn 1993; Tor-Agbidye et al. 2001).

Two general approaches have been used to deal with tall fescue toxicosis. The first approach uses forage/animal management techniques that minimize toxin intake. Adding other forages to tall fescue pastures (often clover spp.), preventing seedhead formation, feeding supplements, and preventing over-grazing are most widely practiced (Roberts and Andrae 2010; Kallenbach 2015). Further, concentrations of toxins in tall fescue decline by 40–50% during haymaking as forage dries in the windrow (Roberts et al. 2009).Once the forage is baled, however, the remaining toxin concentrations do not decline appreciably during storage.

The second approach is to develop new **cultivars** of tall fescue that have low or nil ergot alkaloid concentrations. In the 1980s, scientists developed several endophyte-free cultivars of tall fescue. Stuedemann and Hoveland (1988) reviewed 13 studies in which stocker cattle grazed either endophyte-infected or endophyte-free cultivars. Gains were reduced about 0.35 kg d^{-1} when stockers grazed endophyte-infected compared to endophyte-free cultivars. Beef and dairy cows produced substantially less milk when fed endophyte-infected tall fescue (Sleper and West 1996).

While endophyte-free cultivars of tall fescue alleviate the symptoms of fescue toxicosis, at locations south of 39° latitude, they generally do not persist as well as do endophyte-infected cultivars (Sleper and West 1996). Endophyte-infected tall fescue is more tolerant of water stress (West et al. 1988), nematodes (Kimmons et al. 1990), insects (Funk et al. 1993), and diseases or poor management (West and Gwinn 1993). Several examples of growth enhancements and stress tolerances are shown in Tables 16.1 and 16.2.

Table 16.1 Differences in growth and stress tolerance between endophyte-infected and endophyte-free tall fescue

Characteristic	Endophyte infected	Endophyte free	Reference
Stand loss during drought	4%	54%	Read and Camp (1986)
Herbage mass after drought, relative	10 g	6 g	Arachevaleta et al. (1989)
Root mass relative	1.3 g	1.0 g	de Battista et al. (1990)
4-yr stands under grazing and heat stress	64%	23%	Hopkins and Alison (2006)

Table 16.2 Typical season-long livestock production under rotational stocking in endophyte-free and endophyte-infected tall fescue-based systems

Scenario	Endophyte-free	Endophyte-infected
	kg ha^{-1}	kg ha^{-1}
Poorly managed	270	190
Well managed	395	280

Pasture management to improve nutritive value increases livestock production by as much as 30 percentage units. Endophyte infection generally reduces season-long livestock production by 30 percentage units regardless of forage management (Bailey and Kallenbach 2010; Aiken et al. 2012).

More recently, researchers infected tall fescue cultivars with endophytes that do not produce ergot-like alkaloids (Latch and Christensen 1985; Bouton et al. 2002). Often, these are referred to as "novel" endophtyes (Roberts and Andrae 2010). The balance of research shows that these novel-infected cultivars show improved plant **persistence** over endophyte-free cultivars but livestock performance is similar to that of endophyte-free cultivars (Bouton et al. 2002; Nihsen et al. 2004).

A valuable attribute of tall fescue is that it grows longer into the autumn than most other cool-season grasses. In addition, tall fescue leaves have a relatively thick cuticle layer, which helps them remain green through early winter. Because of its ability to remain green into early winter, growth in late summer and autumn can be accumulated *in situ* to extend the grazing season into winter (Matches 1979; Fribourg and Bell 1984; Kallenbach et al. 2003b). Livestock producers call this "stockpiled" or "autumn-saved" pasture. Removal of reproductive stems by cutting or grazing leads to increased forage quality throughout the remainder of the grazing season (Figure 16.1).

Several diseases and insects attack tall fescue, but except in rare cases, none poses serious problems for forage-livestock producers in the humid region.

Meadow Fescue

Origin and Description

Meadow fescue originates mainly from northern Europe and mountainous regions of southern Europe. Meadow fescue was introduced to the US and Canada in the early 1800s. During the early 1900s, it was deemed sufficiently important by the USDA to maintain annual records for US meadow fescue seed production. Seed production of meadow fescue averaged 45 400 kg per year from 1929 to 1951. The forage seed trade was not very sophisticated in the late 1800s or early 1900s, so there are no records of particular cultivars or strains that were popular at the time. Meadow fescue seed was most likely propagated, produced and sold by species name only. One cultivar from eastern Canada, "Ensign," was named in the early twentieth century.

Meadow fescue was "rediscovered" as part of the grass-roots grazing movement in the central US during the late twentieth century (Casler et al. 2008). An early on-farm grazing study identified several European cultivars as having superior **palatability** and acceptability to grazing livestock, despite forage yields lower than tall fescue (Casler et al. 1998). It was then discovered that meadow fescue had been an integral component of pasture-based livestock systems in the Driftless Area of the Upper Mississippi River Valley since the time of European settlement (Duncan et al. 2013). Numerous introductions of meadow fescue were made to the region, originating from different regions of Europe, most likely with the influx of different ethnic groups to the region, as they brought seed from their respective homelands.

Meadow fescue is very similar in morphology to tall fescue, differing largely by having softer leaves, a shiny abaxial leaf surface, and by a more diminutive plant structure (shorter leaves, shorter stems, narrower stems, larger panicles, and larger seeds; Hitchcock 1971). Meadow fescue is a diploid ($2n = 2x = 14$) with a tetraploid race ($2n = 4x = 28$) that is generally referred to as Appenine fescue, originating largely in the Appenine Mtns. of central Italy.

Adaptation, Use, and Management

Meadow fescue is a long-lived perennial that tends to be better adapted to colder climates than tall fescue, originating from European environments that are further north and at higher altitudes than tall fescue. It is recommended for management-intensive rotational grazing systems in humid temperate regions of North America, generally hardiness zones 3 through 5, with some adaptation to zone 6. Meadow fescue is about 5–8% lower in forage yield than tall fescue, but about 5% higher in *in vitro* **neutral detergent fiber (NDF) digestibility** (Brink and Casler 2009; Brink et al. 2010; Table 16.3). Two cultivars have been released (Casler et al. 2015, 2017) with several other candidate cultivars nearly ready for release. Most meadow fescue cultivars have an endophytic fungus, *Epichloë uncinata* (W. Gams, Petrini and D. Schmidt) Leuchtm. and Schardl. This fungus differs from the tall fescue endophyte in that it does not produce ergopeptine compounds, so it is non-toxic to livestock. No studies have been conducted to date on the benefits of the endophyte to the host.

The Ryegrasses

Worldwide, perennial ryegrass and annual ryegrass are two of the most important grasses. Perennial ryegrass is native to Europe, Asia and North Africa (Jung et al. 1996). Annual ryegrass originates from northwestern Italy, resulting in the more common name, italian ryegrass (Jung et al. 1996). There are at least six other *Lolium* species found throughout the world, but most are classified as weeds (Terrell 1968).

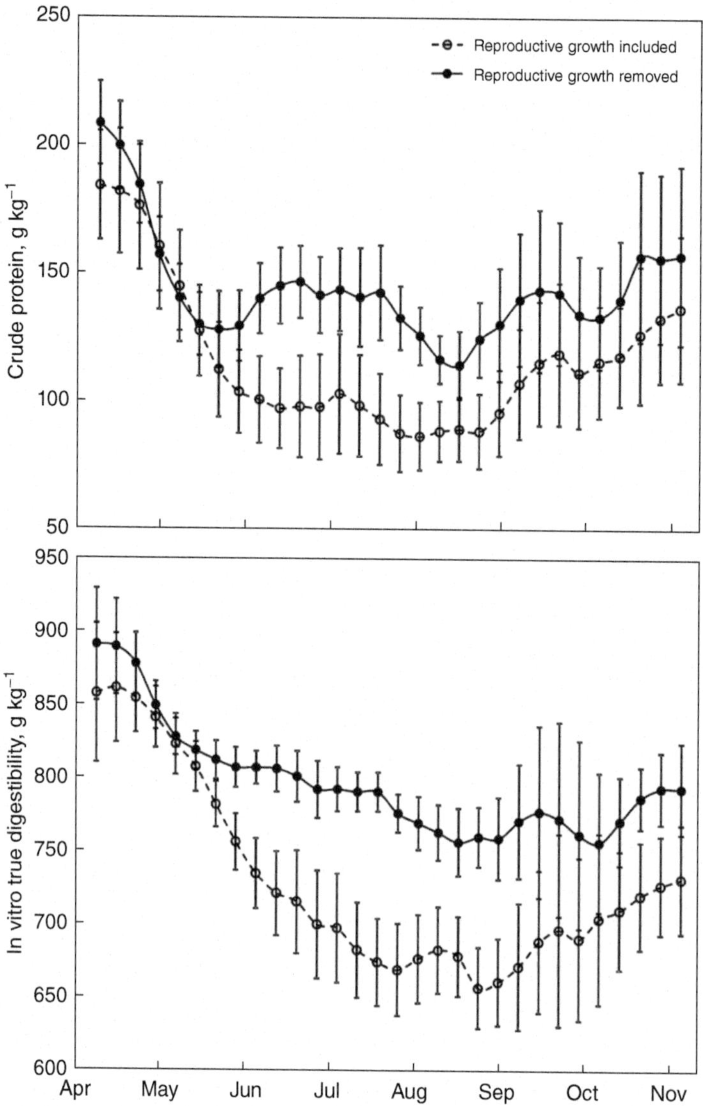

Fig. 16.1. Removing reproductive growth improves nutritive value of tall fescue pastures in late spring and most of the summer. Stocking strategies that match forage nutritive value to **animal nutrient** needs often maximize system productivity (Kallenbach et al. 2003b).

Table 16.3 Annual forage yield and average neutral detergent fiber digestibility (NDFD) of meadow fescue, tall fescue, and orchardgrass harvested according to hay (three harvests) or grazing (six harvests) management (Brink et al. 2010)

	Hay management		Grazing management	
Grass species	Forage yield	NDFD	Forage yield	NDFD
	$Mg\,ha^{-1}$	$g\,kg^{-1}$	$Mg\,ha^{-1}$	$g\,kg^{-1}$
Meadow fescue	6.04	668	4.85	762
Tall fescue	6.65	615	5.28	708
Orchardgrass	7.00	605	5.50	714

Perennial Ryegrass

Origin and Description

While perennial ryegrass is the predominant grass species in the wet, temperate climates of northern Europe and New Zealand, it is grown on a relatively small scale in humid North America. Perennial ryegrass is a bunchgrass with a highly branched root system that produces adventitious roots at the basal nodes of tillers. Leaf blades are approximately 4 mm wide, 180 mm long, shiny, and folded in the bud. The **sheath** is usually compressed and red or purple at the base. The membranous ligule is notched or toothed at the apex and **auricles** are small and claw-like (Balasko et al. 1995). The **spike inflorescence** has 5–40 alternating, awnless, and **sessile spikelets**.

Perennial ryegrass is cross-pollinated and is typically propagated from seed. Both diploid and tetraploid cultivars are available. Like tall fescue, perennial ryegrass can be infected with a fungal endophyte (*Neotyphodium lolii* [Latch, Christensen and Samuels] Glenn, Bacon and Hanlin) that reduces livestock performance.

Adaptation, Use, and Management

Perennial ryegrass is used primarily for pasture but also for conserved forage in humid regions. Perennial ryegrass has low tolerance for hot summers and cold winters of the continental US where it does not persist as long as adapted grasses.

Under ideal growing conditions, forage quality of perennial ryegrass is often superior to other grasses (Collins 1991; Pysher and Fales 1992), providing sufficient quality to support dairy operations. Perennial ryegrass is typically grown in a mixture with a legume, usually white clover. Perennial ryegrass also works well with other legumes, and, in West Virginia, was more compatible with alfalfa than was orchardgrass (Jung et al. 1982). In the northern US, many perennial ryegrass cultivars do not have enough cold tolerance to survive for many years, but hybridization with meadow fescue, combined with selection and breeding, have produced more winter hardy types (Table 16.4).

Table 16.4 Improvement of festulolium (ryegrass × fescue hybrids) by selection for winterhardiness to produce better adapted cultivars (Casler et al. 2002)

Festulolium cultivars	Survival at −11 °C freezing test	Survival in the field (hardiness zones 3 and 4)	Forage yield
	%	%	Mg ha^{-1}
Unselected	18	39	3.74
Improved	51	51	3.91

Perennial ryegrass is typically harvested two or three times annually for hay or silage. Long-term yields were higher when cut to a 7-cm vs a 4-cm height (Motazedian and Sharrow 1986), indicating residual leaf area was important for regrowth. Perennial ryegrass does not generally persist well under continuous grazing but lasts longer under rotational grazing. Because of its high quality and palatability, perennial ryegrass is easily overgrazed which is particularly detrimental to persistence when there is moisture or temperature stress.

In the humid region of the US, there are no diseases or insects specific to perennial ryegrass. However, it does host a number of pests common to perennial cool-season grasses. Lack of tolerance to drought, heat, and cold temperatures limit the distribution and use of perennial ryegrass in the humid US.

Annual Ryegrass

Origin and Description

Annual ryegrass in the US is a specialized form of italian ryegrass. This form has been developed by many years of breeding for the annual growth habit for winter production in the southern US. Because there is still latent variability for perenniality within this form of ryegrass, "annual" ryegrass cultivars sometimes persist through the summer, if they have sufficient heat tolerance. As such, these cultivars are not true annuals in the botanic sense, but annuals largely based on how they are commonly used.

Annual ryegrass grows on about 1 million ha in the humid, southern US, being used primarily for winter pasture in clear seedings and in dormant bermudagrass sods (Evers 1995). Annual ryegrass is a bunchgrass that usually behaves as an annual or winter annual, but under favorable conditions can act as a short-lived perennial. It has a highly branched root system with multiple, fibrous, adventitious roots. Leaf blades are 4–8 mm wide and rolled in the bud. The upper surface of the leaf is ribbed and the lower surface is shiny and has a distinct midrib. The membranous ligule is notched or toothed at the apex. **Auricles** are usually longer and more developed than those of perennial ryegrass.

Annual ryegrass is a prolific seed producer with a spike inflorescence that contains about 40 alternating and **sessile spikelets**. It is cross-pollinated and typically propagated from seed. It has slender awns that are up to 15 mm long.

Adaptation, Use, and Management

Annual ryegrass is used primarily for fall, winter and early spring pasture in southern latitudes (Evers 1995), being preferred over other annual grasses because it grows rapidly during autumn. Growth rates in the southern US have reached 49 kg ha^{-1} d^{-1} (Mooso et al. 1990; Cuomo et al. 1999).

Table 16.5 Forage yield, neutral detergent fiber (NDF), and crude protein (CP) of annual ryegrass stockpiled during winter at two locations in Missouri (Kallenbach et al. 2003a)

Harvest date	Forage yield	NDF	CP
	$Mg\,ha^{-1}$	$g\,kg^{-1}$	$g\,kg^{-1}$
December	1.6	273	195
January	1.4	353	203
February	1.5	370	200
March	1.9	373	198

Forage quality of the **vegetative** growth is outstanding. Concentrations of **crude protein** (CP) often exceed $200\,g\,kg^{-1}$, while those of acid detergent fiber (ADF) remain below $220\,g\,kg^{-1}$ and those of NDF are below $400\,g\,kg^{-1}$ (Mooso et al. 1990). The highly digestible forage, due to low fiber, supports milk yields from dairy cows of $34\,kg\,ha^{-1}\,d^{-1}$ or more (McCormick et al. 2001), and gains of stocker calves of $0.5–1.5\,kg\,ha^{-1}\,d^{-1}$ (Riewe et al. 1963; Hoveland et al. 1991). Annual ryegrass can be stockpiled in the central US, with favorable yields of high-quality forage for winter grazing (Table 16.5).

Annual ryegrass responds favorably to applications of N, with annual rates of $280\,kg\,ha^{-1}$ being economic in the southern part of the humid US (Dunavin 1975). The species is typically grown in monoculture, but can be mixed with annual legumes to decrease N fertilizer needs (Morris et al. 1986).

Annual ryegrass can tolerate continuous stocking, but rotational stocking or flexible stocking options give greatest productivity (Rouquette 1995). For conserved-feed, it is typically harvested once in spring after the stand has been grazed in fall or winter. Silage is the preferred storage method because drying conditions during spring discourage hay making in most of the humid US. Whether grazed or harvested mechanically, a 7- to 10-cm residual gives the greatest productivity over time.

In the humid region of the US, crown rust (*Puccinia coronata* Pers. Cda.) is the major disease affecting annual ryegrass, however, many resistant cultivars are available. Insects generally pose no major threat to production.

Orchardgrass

Origin and Description

Orchardgrass, also known as cocksfoot, is native to Europe, Northern Africa, and parts of Asia (Christie and McElroy 1995). Its main area of use in North America is in the humid transition zone between cool-temperate and the **subtropical** zones.

Orchardgrass is a perennial, upright bunchgrass. Leaf blades are 3–10 mm wide, typically 10–30 cm long, folded in the bud, and when cross-sectioned appear to be "V" shaped (Christie and McElroy 1995). The membranous ligule is 2–8 mm long. There are no **auricles.** Tillers are dark-green to blue and flattened from the base to the glabrous **collar.** Reproductive tillers are usually 3–5 mm in diameter, have two to four nodes, and are 1–1.5 m tall. The panicle is 8–15 cm long with spikelets that contain two to five florets. In the humid zone of the US, orchardgrass flowers 10–14 days before tall fescue, and 14–21 days before smooth bromegrass (Van Santen and Sleper 1996).

Adaptation, Use, and Management

Orchardgrass is popular for pasture, hay, and silage in the central part of the humid region. It is moderately tolerant of infertile, acid, drought-prone and/or poorly-drained soils. It is best adapted to areas of moderate to high rainfall and moderate winters, being less winter hardy than smooth bromegrass or timothy. Orchardgrass tolerates shade better than most cool-season grasses (Blake et al. 1966), making it an ideal companion grass in mixtures.

Because orchardgrass reaches anthesis earlier in spring, it often has lower CP and energy contents than do tall fescue, smooth bromegrass, perennial ryegrass or timothy if all are harvested on the same date. Thus, orchardgrass should be grazed or harvested early in the season to remove reproductive growth when it is still high quality and to encourage vegetative tillers. It can withstand short periods of frequent grazing or clipping (Davies 1988), but is gradually weakened and replaced by kentucky bluegrass or other low-growing species if continually overgrazed. Autumn growth can be stockpiled, but it is generally inferior to tall fescue (Archer and Decker 1977). Orchardgrass does not contain alkaloid compounds or toxins that warrant management consideration.

As a conserved-feed, orchardgrass is typically harvested two or three times annually, though it is often harvested more frequently when grown in mixture with alfalfa (Table 16.6). Orchardgrass typically regrows more rapidly than smooth bromegrass or timothy making it a preferred species for alfalfa/grass mixtures.

Rusts (*Puccinia* spp.), various leaf spots, and brown stripe (*Cercosporidium graminis* [Fuckel] Deighton) occur frequently (Henning and Risner 1990) and typically damage leaves and reduce forage quality. The only practical means of control is to use a resistant cultivar. Several insects feed on orchardgrass but none is specific.

The Bromegrasses

Origin and Description

There are about 100 species of annual and perennial bromegrasses, some of which are native to North America (Hitchcock 1971). Some species are highly aggressive, reproduce by prolific seed production or **rhizome** development, and provide conservation value, in many circumstances causing them to be classified

Table 16.6 Stand percentages and contributions to total mixture yield of three grasses in mixture with alfalfa at two locations in Minnesota under two harvest managements (Sheaffer et al. 1990)

	Percentage stand		Percentage contribution to mixture yield	
Grass species	Three cuts	Four cuts	Three cuts	Four cuts
	%	%	%	%
Smooth bromegrass	8	15	26	26
Reed canarygrass	29	17	25	25
Orchardgrass	60	65	45	52

as weeds or invasive species. Smooth bromegrass and meadow bromegrass, both introduced, are the only species extensively cultivated in North America (Vogel et al. 1996).

Smooth bromegrass, introduced from Hungary and Russia in the late nineteenth century, became very popular in the central US during the 1930s, due to its survival of the droughts and dust storms. In the late 1930s, it quickly spread through eastern North America, spawning research on management and breeding. It is one of the most drought tolerant of the cool-season grasses (Table 16.7).

Plants are tall, leafy perennials that reproduce by seed or by rhizomes to form a sod. Smooth bromegrass is a short-day, long-day plant, i.e. it does not require **vernalization** per se, but requires short days of autumn for floral **induction** followed by long days of spring for floral initiation and development (Vogel et al. 1996). Flowering is highly controlled by **photoperiod**, varying slightly from year-to-year based on temperature, but with little genetic variability among ecotypes or cultivars.

Most cultivated smooth bromegrass is octoploid with a chromosome number of $2n = 8x = 56$, though there are tetraploid forms with $2n = 28$ (Armstrong 1973). There are two ecologic forms (Vogel et al. 1996). The meadow or northern climatype originated in northeastern Europe and is used largely in Canada and the northern edge of the US. The steppe or southern climatype originated in the Ukraine and Trans-Caucasus region west to the dry steppes of Khazakstan and the southern Altai Mountains. It dominates the market in the US and is used less frequently in Canada. Southern climatypes have lower forage yield when moved too far north, while northern climatypes have lower forage yield when moved too far south (Casler et al. 2000, 2001).

Table 16.7 Forage yield under drought as a percentage of non-drought conditions for four forage grasses grown in Minnesota (Sheaffer et al. 1992)

	Forage yield under drought as a percentage of non-drought conditions	
Grass species	Harvest in which drought occurred	Season total
Reed canarygrass	39	53
Orchardgrass	49	62
Smooth bromegrass	50	80
Timothy	41	62

Adaptation, Use, and Management

Smooth bromegrass is adapted to a wide range of soils, but prefers deep, well-drained silt loam or clay loam soils. It is adapted and productive from North Texas to Central Alaska. Plants can survive prolonged periods of drought and tolerate intense cold, desiccating winter winds, deep snow, and short-term ice-sheet formation. Growth rates are optimal at daytime temperatures of 18–25 °C, with reduced growth and dormancy above 35 °C (Baker and Jung 1968). It is highly susceptible to the "summer slump" that characterizes most cool-season grasses, typically producing more than half of its dry matter in the first few weeks of spring (Casler and Carlson 1995). Hot and dry summers can result in prolonged periods of dormancy for pastures or hayfields.

Timing of cutting or harvesting during spring has long-term effects on productivity and persistence of smooth bromegrass. Forage removal at jointing and early heading stages often reduces regrowth because plants are not producing new tillers and carbohydrate reserves are relatively low (Reynolds and Smith 1962; Eastin et al. 1964). Cutting during this time reduces rate of recovery and regrowth yields, and can have long-term effects on stand persistence and forage production. Thus, smooth bromegrass is best adapted to a later harvest and fewer hay cuttings than is orchardgrass or tall fescue.

Due to this critical growth stage, smooth bromegrass is often a poor choice for multi-species mixtures. For example, it is typically in the critical growth stage when the first cutting of an alfalfa-grass mixture is made, causing regrowth of smooth bromegrass to be suppressed by the rapid regrowth and subsequent shading of the alfalfa canopy. When cut three or four times annually, alfalfa has more rapid recovery and faster regrowth than smooth bromegrass. Some progress has been made in developing new cultivars with better regrowth and persistence in mixture with alfalfa (Casler 1988; Casler and Drolsom 1995).

Smooth bromegrass can be utilized in grazing systems, but is not well adapted to most grazing systems. First, reproductive growth of smooth bromegrass is difficult or nearly impossible to suppress simply by defoliation. Early grazing defoliates leaves from flowering stems that are produced later in the spring, resulting in a high density of short, leafless, unpalatable stems. Delaying grazing to the jointing stage may result in defoliation below meristems of the primary **shoot**, but only if livestock are sufficiently hungry or trained to consume stem tissue. Even if it is grazed at this stage, regrowth will be suppressed due to the critical nature of the jointing growth stage. Following spring growth and flowering, grazing management becomes more flexible, limited largely by N fertility, temperature and precipitation.

Smooth bromegrass requires adequate N fertility to maintain a healthy and productive stand. Older stands can become sod-bound, a condition that results directly from inadequate N fertilization (Anderson et al. 1946). Smooth bromegrass responds well to N, ranging from 6 to 35 kg DM kg^{-1} N applied, depending on soil type, temperature, and moisture conditions (Casler and Carlson 1995). The seasonal distribution of smooth bromegrass dry matter can be improved with timely N applications, particularly those in late spring and mid-summer that increase summer production if rainfall is adequate (Krueger and Scholl 1970; Narasimhalu et al. 1981).

Rates of 90–110 kg N ha^{-1} annually are generally adequate to maintain a healthy and productive stand. Rates above 110 kg N ha^{-1} may result in acidification of upper soil horizons and leaching of NO_3-N into ground water (Harapiak et al. 1992; Malhi et al. 1991; Nuttall et al. 1980). Rates above 200 kg N ha^{-1} increase the risk of grass tetany or NO_3^- toxicity to livestock (Chapter 47).

Smooth bromegrass is susceptible to several diseases, two of which are serious potential threats. Brown leafspot (*Drechslera bromi*) is favored by cool, wet conditions in spring. Heavy infections can persist through mid-summer reducing both yield and quality of forage. Resistant cultivars are now available (Casler et al. 2000). Crown rust (*P. coronata*) is favored by warm, humid conditions and typically affects early to mid-summer regrowth, seriously reducing intake and **digestibility** of leaves.

Seed is produced in several regions of North America, especially the Central Great Plains, the Canadian Prairie, and the Palouse of Eastern Washington (Vogel et al. 1996). Seed fields are typically established in rows at a 60- to 90-cm spacing. Yields typically range from 300 to 500 kg ha^{-1} with adequate N. Quackgrass is a very troublesome weed as it is impossible to remove quackgrass seed from the seed crop, severely limiting marketability due to its noxious status. In vegetative stages, quackgrass is very compatible and morphologicially similar to smooth bromegrass, making it difficult to detect and control in seed fields.

Meadow bromegrass is a long-lived perennial with better regrowth potential and a more favorable seasonal dry matter distribution than smooth bromegrass. It is better adapted to northern environments and used largely in the northwestern US and the Canadian Prairie. Its superior regrowth makes it better suited to grazing than smooth bromegrass. Several native bromegrasses have demonstrated potential as forage or conservation crops, including nodding brome, California brome, fringed brome, mountain brome, pumpelly brome, and sitka brome.

Timothy

Origin and Description

Timothy is the only species of economic importance within the *Phleum* genus. It is native to northern Europe, though there are numerous species that originated in other parts of Europe, Southwestern Asia, and North Africa. Timothy appears to be one of the first grasses introduced into North America during the colonial period (Berg et al. 1996). It is most popular in the Northeastern US and the Atlantic Provinces of Canada.

Timothy is a perennial bunchgrass that is characterized by **corms**, bulblike swellings of stem bases that serve as storage organs for carbohydrate reserves (Berg et al. 1996). New tillers form from axillary buds at nodes on the base of corms, resulting in a relatively non-aggressive spreading habit. The root system is the shallowest of the economically important cool-season grasses.

The **corm** structure of timothy makes it relatively easy to identify in the vegetative state. The panicle inflorescence is dense and has the appearance of small cat-tail rushes, making identification relatively simple at the reproductive growth stage. Seedlings are most readily identified by their twisted leaf blades, which can have up to two or three complete twists over their full length.

Timothy does not require vernalization for floral **induction**, but requires long days for floral initiation and development (Hay and Pedersen 1986; Heide 1982). Floral development is also affected by temperature during spring (Bootsma 1984). More northern types require longer days for floral initiation and flower considerably later when moved to lower latitudes. Timothy typically produces inflorescences on the first two growth cycles of spring, regardless of whether grazed or managed for hay.

Cultivated timothy, a hexaploid with 2n = 6x = 42 chromosomes, is the most genetically complex species within the genus. Diploid species are common and one, *P. bertolonii* DC, is receiving increasing interest as a turfgrass. It appears to be one of the natural parents of cultivated timothy (Cai and Bullen 1991).

Adaptation, Use, and Management

Timothy is one of the most cold-tolerant cool-season grasses, commonly occurring north of the Arctic Circle in its native Scandinavia. It exists in natural or agricultural areas between 35 and 55 °N lat in North America, Europe, Asia, and Japan. Plants can withstand extended ice encasement, and the LT_{50} (temperature of **shoot** apex that kills 50% of the plants) can be as low as −20 °C (Andrews and Gundleifsson 1983; Suzuki 1989).

Timothy is often grown in monoculture for harvest as hay or silage. Reproductive maturity is late compared to other cool-season grasses, though there is a three-week range in heading date among cultivars (Berg et al. 1996; McElroy and Kunelius 1995). It is compatible with several forage legumes in hay or silage systems, but timothy cultivars should be chosen based partly on heading date to match the maturity and harvest dates of the companion legume.

As a conserved feed, timothy is typically harvested two or three times annually. Forage quality of timothy hay, measured as IVDMD, can be higher than most other cool-season grasses due to its later maturity (Collins and Casler 1990). It makes high-quality silage at early growth stages when concentrations of water-soluble carbohydrate (WSC) are high enough to support good fermentation (Burgess and Grant 1974; Kunelius and Halliday 1989). Nitrogen fertilization decreases WSC, increases buffering capacity, and increases nitrates, all of which act to decrease silage quality (Table 16.8).

Timothy can tolerate frequent defoliation if temperatures are cool and there is adequate precipitation (Berg et al. 1996). However, it generally does not persist well under rotational grazing (Berg et al. 1996; McElroy and Kunelius 1995), suggesting that the shallow root system, combined with the fragile corm reproduction system, limits its ability to survive the stresses associated with grazing livestock (Edmond 1964). The turf-type *P. bertolonii* is more persistent under grazing than cultivated timothy (Cenci 1980).

Timothy can host several diseases and insects, but none is known to cause significant or economic reductions in forage yield, forage quality, or persistence. Drought and heat appear to be the most important factors limiting its distribution and use.

Table 16.8 Forage yield and characteristics of timothy herbage under four rates of nitrogen fertilizer and four growth stages at harvest in Québec (Tremblay et al. 2005)

Nitrogen rate (kg ha^{-1})	Stem elongation	Early heading	Late heading	Early flowering
	Forage yield (Mg ha^{-1})			
0	2.01	3.15	4.46	5.95
60	2.81	5.10	5.71	6.88
120	3.29	4.82	6.93	6.81
180	4.45	5.57	6.81	6.44
	Water soluble carbohydrates (g kg^{-1} DM)			
0	110	92	72	69
60	62	56	57	64
120	60	57	54	71
180	50	58	57	66
	Buffering capacity (g lactic acid kg^{-1} DM)			
0	45	42	39	38
60	51	51	46	43
120	54	53	48	42
180	50	51	48	45
	Nitrates (g NO_3-N kg^{-1} DM)			
0	0.10	0.10	0.13	0.09
60	0.23	0.14	0.26	0.16
120	0.87	0.71	0.42	0.29
180	1.35	1.71	0.73	0.66

Reed Canarygrass

Origin and Description

Reed canarygrass is native to temperate regions of North America, Europe, and Asia (Anderson 1961). It occurs throughout the temperate regions of North America, often as the dominant vegetation in **riparian** areas, river and stream banks, marshes, pond margins, drainage ditch banks, and roadside ditches. The **seed shatters** at time of ripening and can be readily dispersed by moving water or wildlife to colonize additional areas. Once established, the aggressive and persistent rhizomes help the stand to persist indefinitely.

The perennial plants are tall, coarse, and sod-forming. Plants are up to 2 m tall, leaf blades are up to 2 cm wide, and stems are up to 6 mm in diameter. Reed canarygrass can be propagated by seed, rhizomes, or stem nodes. In contrast with most other cool-season grasses, the axillary bud at a stem node of reed canarygrass can break dormancy when apical dominance is removed by cutting the panicle or by seed ripening. Axillary buds will produce adventitious roots and new shoots. Reed canarygrass can be propagated for conservation purposes by disking freshly harvested and chopped forage into soil with adequate moisture.

Seed of reed canarygrass are borne on a cylindric panicle that bleaches in color as it ripens. The gray, waxy seed shatters quickly after ripening, which begins at the top of the panicle. **Seed shattering** is the most important problem for seed production. A gene for shattering resistance discovered in phalaris or hardinggrass, a related species, has been incorporated into several cultivars of that species (Oram and Lodge 2003). Hardinggrass is cultivated in the Southeastern US.

Reed canarygrass, like many perennial grasses has a natural anti-herbivory mechanism to deter feeding by mammalian herbivores and insects. Its herbage contains a wide array of alkaloids, naturally occurring compounds with a relatively complex cyclic structure. All plants of reed canarygrass contain some form of alkaloids, though there is a large amount of genetic variation for concentration and type. Alkaloid type is under relatively simple genetic control (Marum et al. 1979).

Total alkaloid concentration is negatively related to palatability and intake by sheep (Figure 16.2). Concentrations above 6 g kg^{-1} DM lead to consistent rejection by sheep (Sheaffer and Marten 1995). All alkaloids appear to contribute to the palatability and intake limitations, and selection and breeding for low concentrations can result in improved average daily gain of livestock (Marten 1989). Tryptamine and ß-carboline alkaloids can have severe effects on livestock, causing diarrhea and severe reductions in average daily gain (Marten 1989).

Adaptation, Use, and Management

Reed canarygrass is well adapted to a very wide range of environmental conditions, including wetlands and drought-prone soils. It tolerates flooding and waterlogging better than other cool-season grasses (McKenzie 1951) and has a wide pH range of 4.9–8.2 (Vose 1959). It is one of the most cold tolerant of the cool-season grasses, rarely showing winter injury.

Reed canarygrass is used for hay, silage, pasture, soil conservation, and wastewater disposal or pollution

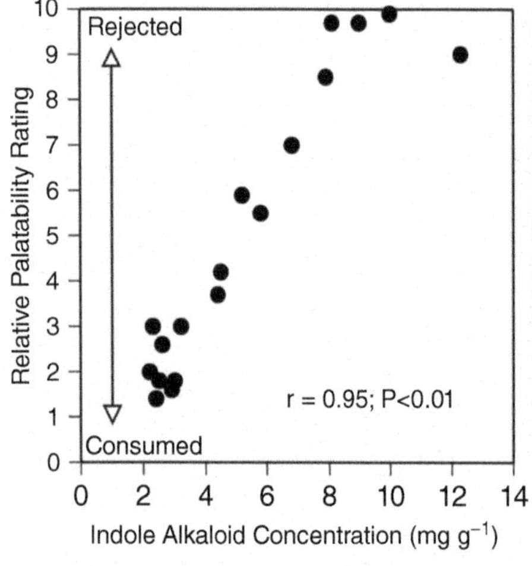

Fig. 16.2. Lambs grazing genotypes of reed canarygrass that differ in alkaloid concentration *Source:* right, photo courtesy of A.W. Hovin, University of Minnesota; relationship of palatability, measured by grazing lambs, to indole alkaloid concentration of 18 reed canarygrass genotypes (left, Simons and Marten 1971).

abatement. As a hay or silage crop, it can be grown in monoculture or in mixture with perennial legumes (Carlson et al. 1996; Sheaffer and Marten 1995). It persists well in mixture with perennial legumes in upland areas without becoming the dominant species as often occurs in wet or low-lying areas. Native or naturalized stands in wetlands are often difficult to manage as a hay or silage crop due to the persistent wet soils.

Reed canarygrass is well suited as a pasture grass. Frequent and early grazing in the spring can be highly effective for eliminating or reducing production of stems and panicles. Rotational stocking with short rest periods in spring and longer **rest periods** in mid-summer provide the best utilization when pastured. It is the highest yielding cool-season grass under rotational stocking and good stands are highly resistant to trampling damage, even on wetland sites (Sheaffer and Marten 1995).

Reed canarygrass responds well to application of N fertilizer, making an excellent grass for N removal from municipal and food-processing wastewater **effluent** (Marten et al. 1979; Quin 1979). It removed more N from wastewater than other grasses, including corn (Marten 1985). Furthermore, NO_3-N concentrations were lower in ground water under reed canarygrass than other grass-based systems (Quin 1979).

Reed canarygrass has several disease and insect pests, some of which can become serious on some cultivars or ecotypes. However, pests rarely seriously threaten its production, forage quality, or persistence (Carlson et al. 1996).

Stand establishment is one of the more difficult problems for acceptance and use of reed canarygrass. **Seedbed** preparation, time and depth of seeding, and weed control are critical factors in successful establishment (Carlson et al. 1996). Seed is slow to germinate and seedlings grow very slowly, often lagging far behind annual weed infestations. Late summer seedings can be successful if made sufficiently early to allow seedlings to cold-harden in autumn. However, late summer seedings carry a greater risk of inadequate moisture and the stand will require more time to reach maximum productivity.

The Bluegrasses

Origin and Description

Kentucky bluegrass is the most important species in the genus *Poa*, which comprises approximately 300 species. It is adapted and distributed throughout most temperate regions of the world (Wedin and Huff 1996). **Germplasm** has been introduced into North America numerous times beginning with the earliest European immigrants (Bashaw and Funk 1987). There are reports that the species is native to North America, having been loosely described by early Appalachian settlers, but no clear determinations have been made to document its putative North American origin (Duell 1985; Gray 1908; Wedin and Huff 1996).

This perennial, rhizomatous, sod-former is a relatively short-growing grass with a large proportion of its above-ground dry matter close to the soil surface. It also has an extremely shallow root system. Tillers that are vernalized during short days and low temperatures of autumn produce an open panicle when days lengthen in spring. Seed are produced by apomixis, an abnormality to meiosis that results in all seed being genetically identical to the mother plant. Chromosome number of kentucky bluegrass ranges from 24 to 124 in a complex polyploid and aneuploid series.

Adaptation, Use, and Management

Kentucky bluegrass is ubiquitous in agricultural systems and disturbed lands throughout the temperate regions of the world. It colonizes disturbed lands including hay fields and pastures from long-lived seed banks or long-dormant rhizomes. It is regarded as a superb pasture and turfgrass. It is broadly adapted to a wide range of environmental conditions, but the 24 °C isotherm for mean July temperature is an upper thermal limit to its adaptation (Duell 1985). Recent hybrids show potential to combine favorable traits of kentucky bluegrass with the heat tolerance of texas bluegrass, a native of central Texas (Read and Anderson 2003).

Kentucky bluegrass is compatible in mixtures with a wide array of species including other perennial grasses (Balasko et al. 1995) and legumes (Wedin and Huff 1996). It is favored by frequent harvesting or grazing, particularly when rest periods for pastures are insufficient for tall-growing grasses. However, bluegrass-dominant pastures are highly sensitive to drought, due to their extremely shallow root system.

Summary

Cool-season grasses are an important foundation for livestock agriculture and soil conservation in humid areas of North America. With an array of morphologies, growth habits, and stress tolerances, they are, as a group, adapted to nearly every soil-based environmental habitat in this broad region. They provide a long-term and reliable feed source for many types of livestock, helping to sustain and preserve rural families, economies, and communities.

Acknowledgment

We thank Ryan Lock, University of Missouri, for assistance in collecting data and reviewing the contents of this chapter.

References

Aiken, G.E. and Strickland, J.R. (2013). Forages and pastures symposium: managing the tall fescue-fungal endophyte symbiosis for optimum forage-animal production. *J. Anim. Sci.* 91: 2369–2378.

Aiken, G.E., Goff, B.M., Witt, W.W. et al. (2012). Steer and plant responses to chemical suppression of seed-head emergence in toxic endophyte-infected tall fescue. *Crop Sci.* 52: 960–969.

Anderson, D.E. (1961). Taxonomy and distribution of the genus *Phalaris. Iowa State J. Sci.* 36: 1–96.

Anderson, K.L., Krenzin, R.E., and Heide, J.C. (1946). The effect of nitrogen fertilizer on bromegrass in Kansas. *J. Am. Soc. Agron.* 38: 1058–1067.

Andrews, C.J. and Gundleifsson, B.E. (1983). A comparison of cold hardiness and ice encasement tolerance of timothy grass and winter wheat. *Can. J. Plant. Sci.* 63: 429–435.

Arachevaleta, M., Bacon, C.W., Hoveland, C.S., and Radcliffe, D.E. (1989). Effect of the tall fescue endophyte on plant response to environmental stress. *Agron. J.* 81: 83–90.

Archer, K.A. and Decker, A.M. (1977). Autumn-accumulated tall fescue and orchardgrass. I. Growth and quality as influenced by nitrogen and soil temperature. *Agron. J.* 69: 601–605.

Armstrong, K.C. (1973). Chromosome pairing in hexaploid hybrids from *Bromus erectus* (2n = 28) x *Bromus inermis* (2n = 56). *Can. J. Genet. Cytol.* 15: 427–436.

Bailey, N.J. and Kallenbach, R.L. (2010). Comparison of 3 tall fescue-based stocker systems. *J. Anim. Sci.* 88: 1880–1890.

Baker, B.S. and Jung, G.A. (1968). Effect of environmental conditions on the growth of four perennial grasses. I. Response to controlled temperature. *Agron. J.* 60: 155–158.

Balasko, J.A., Evers, G.W., and Duell, R.W. (1995). Bluegrasses, ryegrasses, and bentgrasses. In: *Forages, an Introduction to Grassland Agriculture* 5e (eds. R.F Barnes, D.A. Miller and C.J. Nelson), 357–371. Ames, IA: Iowa State University Press.

Bashaw, E.C. and Funk, C.R. (1987). Apomictic grasses. In: *Principles of Cultivar Development*, vol. 2. Crop sciences (ed. W.R. Fehr). New York: MacMillan.

Berg, C.C., McElroy, A.R., and Kunelius, H.T. (1996). Timothy. In: *Cool-Season Forage Grasses*. Agron. Monogr. 34 (ed. L.E. Moser), 643–664. Madison, WI: ASA, CSSA, SSSA.

Blake, C.T., Chamblee, D.S., and Woodhouse, W.W. Jr. (1966). Influence of some environmental and management factors on the persistence of ladino clover in association with orchardgrass. *Agron. J.* 58: 487–489.

Bootsma, A. (1984). Forage crop maturity zonage in the Atlantic region using growing degree-days. *Can. J. Plant. Sci.* 64: 329–338.

Bouton, J.H., Latch, G.C.M., Hill, N.S. et al. (2002). Reinfection of tall fescue cultivars with non-ergot alkaloid-producing endophytes. *Agron. J.* 94: 567–574.

Brink, G.E. and Casler, M.D. (2009). Meadow fescue, tall fescue, and orchardgrass response to nitrogen application rate. *Forage Grazinglands* https://doi.org/10.1094/FG-2009-0130-01-RS.

Brink, G.E., Casler, M.D., and Martin, N.P. (2010). Defoliation management effects on meadow fescue, tall fescue, and orchardgrass. *Agron. J.* 102: 667–674.

Buckner, R.C., Powell, J.B., and Frakes, R.V. (1979). Historical development. In: *Tall Fescue*. Agron. Monogr. 20 (eds. R.C. Buckner and L.P. Bush), 1–8. Madison, WI: ASA, CSSA, SSSA.

Burgess, P.L. and Grant, E.A. (1974). A grassland system for dairy cattle based on ensiled timothy (*Phleum pratense*) cultivars. In: *Proceedings of 12th International Grassland Congress, Vol. 3. Moscow* (eds. V.G. Iglovikov and A.P. Movsisyants), 78–86.

Cai, Q. and Bullen, M.R. (1991). Characterization of genomes of timothy (*Phleum pratense* L.). I. Karyotypes and C-banding patterns in cultivated timothy and two wild relatives. *Genome* 34: 52–58.

Carlson, I.T., Oram, R.N., and Surprenant, J. (1996). Reed canarygrass and other phalaris species. In: *Cool-Season Forage Grasses*. Agron. Monogr. 34 (ed. L.E. Moser), 569–604. Madison, WI: ASA, CSSA, SSSA.

Casler, M.D. (1988). Performance of orchardgrass, smooth bromegrass, and ryegrass in binary mixtures with alfalfa. *Agron. J.* 80: 509–514.

Casler, M.D. and Carlson, I.T. (1995). Smooth bromegrass. In: *Forages, an Introduction to Grassland Agriculture* 5e (eds. R.F Barnes, D.A. Miller and C.J. Nelson), 313–324. Ames, IA: Iowa State University Press.

Casler, M.D. and Drolsom, P.N. (1995). Registration of 'alpha' smooth bromegrass. *Crop Sci.* 35: 1508.

Casler, M.D., Undersander, D.J., Fredericks, C. et al. (1998). An on-farm test of perennial forage grass varieties under management intensive grazing. *J. Prod. Agric.* 11: 92–99.

Casler, M.D., Vogel, K.P., Balasko, J.A. et al. (2000). Genetic progress from 50 years of smooth bromegrass breeding. *Crop Sci.* 40: 13–22.

Casler, M.D., Vogel, K.P., Balasko, J.A. et al. (2001). Latitudinal and longitudinal adaptation of smooth bromegrass populations. *Crop Sci.* 41: 1456–1460.

Casler, M.D., Peterson, P.R., Hoffman, L.D. et al. (2002). Natural selection for survival improves freezing tolerance, forage yield, and persistence of festulolium. *Crop Sci.* 42: 1421–1426.

Casler, M.D., van Santen, E., Humphreys, M.W. et al. (2008). Remnant oak savanna acts as refugium for meadow fescue introduced during nineteenth century human migrations in the USA. In: *Molecular Breeding of Forage and Turf* (eds. T. Yamada and G. Spangenberg), 91–101. New York: Springer.

Casler, M.D., Brink, G.E., Cherney, J.H. et al. (2015). Registration of 'hidden valley' meadow fescue. *J. Plant Regist.* 9: 294–298.

Casler, M.D., Brink, G.E., and Cherney, J.H. (2017). Registration of 'Azov' meadow fescue. *J. Plant Regist.* 11: 9–14.

Cenci, C.A. (1980). Evaluation of differently adapted types of *Phleum pratense* L. *Z. Pflanzenzücht.* 85: 148–156.

Christie, B.R. and McElroy, A.R. (1995). Orchardgrass. In: *Forages, the Science of Grassland Agriculture*, 5e (eds. R.F Barnes, D.A. Miller and C.J. Nelson), 325–334. Ames, IA: Iowa State University Press.

Collins, M. (1991). Nitrogen effects on yield and forage quality of perennial ryegrass and tall fescue. *Agron. J.* 83: 588–595.

Collins, M. and Casler, M.D. (1990). Forage quality of five cool-season grasses. II. Species effects. *Anim. Feed Sci. Technol.* 27: 209–218.

Cuomo, G.J., Redfearn, D.D., Beatty, J.F. et al. (1999). Management of warm-season annual grass residue on annual ryegrass establishment and production. *Agron. J.* 91: 666–671.

Davies, A. (1988). The regrowth of grass swards. In: *The Grass Crop* (eds. M.B. Jones and A. Lazenby), 85–127. London: Chapman and Hall.

De Battista, J.P., Bouton, J.H., Bacon, C.W., and Siegel, M.R. (1990). Rhizome and herbage production of endophyte-removed tall fescue clones and populations. *Agron. J.* 82: 651–654.

Duell, R.W. (1985). The bluegrasses. In: *Forages – The Science of Grassland Agriculture*, 4e (eds. M.E. Heath, R.F Barnes and D.S. Metcalfe), 188–197. Ames, IA: Iowa State University Press.

Dunavin, L.S. (1975). Production of rye and ryegrass forage with sulfur-coated urea and ammonium nitrate. *Agron. J.* 67: 415–471.

Duncan, D.S., Krohn, A.L., Jackson, R.D., and Casler, M.D. (2013). Conservation implications of the introduction history of meadow fescue (*Festuca pratensis* Huds.) to the driftless area of the Upper Mississippi Valley, USA. *Plant Ecolog. Divers.* https://doi.org/10.1080/17550874.2013.851294.

Eastin, J.D., Teel, M.R., and Langston, R. (1964). Growth and development of six varieties of smooth bromegrass (*Bromus inermis* Leyss.) with observations on seasonal variation of fructosan and growth regulators. *Crop Sci.* 4: 555–559.

Edmond, D.B. (1964). Some effects of sheep treading on the growth of 10 pasture species. *N. Z. J. Agric. Res.* 7: 1–6.

Evers, G.W. (1995). Introduction to annual ryegrass. In: *Symposium on Annual Ryegrass. 31 Aug–1 Sep., 1995, Tyler, TX* (eds. G.W. Evers and L.R. Nelson), 1–6. College Station, TX: Texas Agricultural Experiment Station Bulletin. MP-1770.

Fribourg, H.A. and Bell, K.W. (1984). Yield and composition of tall fescue stockpiled for different periods. *Agron. J.* 76: 929–934.

Funk, C.R., White, R.H., and Breen, J.P. (1993). Importance of *Acremonium* endophytes in turf grass breeding and management. *Agric. Ecosyst. Environ.* 44: 215–232.

Gray, A. (1908). *Gray's Manual of Botany*, 7e (eds. B.L. Robinson and M.L. Fernald), 154–157. New York: Am. Book Co.

Harapiak, J.T., Malhi, S.S., Nyborg, M., and Flore, N.A. (1992). Soil chemical properties after long-term nitrogen fertilization of bromegrass: source and time of nitrogen application. *Commun. Soil Sci. Plant Anal.* 23: 85–100.

Hay, R.K.M. and Pedersen, K. (1986). Influence of long photoperiods on the growth of timothy (*Phleum pratense* L.) varieties from different latitudes in northern Europe. *Grass Forage Sci.* 41: 311–317.

Heide, O.M. (1982). Effects of photoperiod and temperature on growth and flowering in Norwegian and British timothy cultivars (*Phleum pratense* L.). *Acta Agric. Scand. Sect A* 32: 241–252.

Henning J. and Risner, N. (1990). Orchardgrass. University of Missouri. MU Extension Publication No. G6710.

Hitchcock, A.S. (1971). *Manual of the Grasses of the United States*, 2e. New York: Dover.

Hopkins, A.A. and Alison, M.W. (2006). Stand persistence and animal performance for tall fescue endophyte combinations in the South Central USA. *Agron. J.* 98: 1221–1226.

Hoveland, C.S. (1993). Importance and economic significance of Acremonium endophytes to performance of animals and grass plants. *Agric. Ecosyst. Environ.* 44: 3–12.

Hoveland, C.S., Hardin, D.R., Worley, P.C., and Worley, E.E. (1991). Steer performance on perennial vs. winter annual pastures in North Georgia. *J. Prod. Agric.* 4: 24–28.

Jung, G.A., Wilson, L.L., LeVan, P.J. et al. (1982). Herbage and beef production from ryegrass-alfalfa and orchardgrass-alfalfa pastures. *Agron. J.* 74: 937–942.

Jung, G.A., Van Wijk, A.J.P., Hunt, W.F., and Watson, C.E. (1996). Ryegrasses. In: *Cool-Season Forage Grasses*. Agron. Monogr. 34 (ed. L.E. Moser), 605–641. Madison, WI: ASA, CSSA, and SSSA.

Kallenbach, R.L., Bishop-Hurley, G.J., Massie, M.D. et al. (2003a). Stockpiled annual ryegrass for winter forage in the lower midwestern USA. *Crop Sci.* 43: 1414–1419.

Kallenbach, R.L., Bishop-Hurley, G.J., Massie, M.D. et al. (2003b). Herbage mass, nutritive value, and ergovaline concentration of stockpiled tall fescue. *Crop Sci.* 43: 1001–1005.

Kallenbach, R.L. (2015). Coping with tall fescue toxicosis: Solutions and realities. *J Anim Sci.* 93: 5487–5495.

Kimmons, C.A., Gwinn, K.D., and Bernard, E.C. (1990). Nematode reproduction on endophyte-infected and endophyte-free tall fescue. *Plant Dis.* 74: 757–761.

Krueger, C.R. and Scholl, J.M. (1970). Performance of bromegrass, orchardgrass, and reed canarygrass grown at five nitrogen levels and with alfalfa. Wisconsin Agricultural Experiment Station. Research Report 69.

Kunelius, H.T. and Halliday, L. (1989). Nutritive value and production of cool season grasses under two harvest regimes. In: *Proceedings of 16th International Grassland Congress 4–11 Oct. 1989, Nice, France* (ed. D. Desroches), 827–828. Montrogue: Dauer.

Latch, G.C.M. and Christenson, M.J. (1985). Artificial infection of grasses with endophytes. *Ann. Appl. Biol.* 107: 17–24.

Malhi, S.S., Nyborg, M., Harapiak, J.T., and Flore, N.A. (1991). Acidification of soil in Alberta by nitrogen fertilizers applied to bromegrass. In: *Plant-Soil Interactions at Low pH* (eds. R.J. Wright, V.C. Baligar and R.P. Murmann), 547–553. Amsterdam: Kluwer.

Marten, G.C. (1985). Reed canarygrass. In: *Forages – The Science of Grassland Agriculture*, 4e (eds. M.E. Heath, R.F Barnes and D.S. Metcalfe), 207–216. Ames, IA: Iowa State University Press.

Marten, G.C. (1989). Breeding forage grasses to maximize animal performance. In: *Contributions from Breeding Forage and Turf Grasses* (eds. D.A. Sleper, K.H. Asay and J.F. Pederson), 71–104. Madison, WI: ASA, CSSA, SSSA.

Marten, G.C., Clapp, C.E., and Larson, W.E. (1979). Effects of municipal wastewater effluent and cutting management on persistence and yield of eight perennial forages. *Agron. J.* 71: 650–658.

Marum, P., Hovin, A.W., and Marten, G.C. (1979). Inheritance of three indole alkaloids in reed canarygrass. *Crop Sci.* 19: 539–544.

Matches, A.G. (1979). Management. In: *Tall Fescue*. Agron. Monogr. 20 (eds. R.C. Buckner and L.P. Bush), 171–199. Madison, WI: ASA, CSSA, SSSA.

McCormick, M.E., Ward, J.D., Redfearn, D.D. et al. (2001). Supplemental dietary protein for grazing dairy cows: effect on pasture intake and lactation performance. *J. Dairy Sci.* 84: 896–907.

McElroy, A.R. and Kunelius, H.T. (1995). Timothy. In: *Forages – An Introduction to Grassland Agriculture* 5e (eds. R.F Barnes, D.A. Miller and C.J. Nelson), 305–311. Ames, IA: Iowa State University Press.

McKenzie, R.E. (1951). The ability of forage plants to survive early spring flooding. *Sci. Agric.* 31: 358–367.

Mooso, G.D., Feazel, J.I., and Morrison, D.G. (1990). Effect of sodseeding method on ryegrass-clover mixtures for grazing beef animals. *J. Prod. Agric.* 3: 470–474.

Morris, D.R., Weaver, R.W., Smith, G.R., and Rouquette, F.M. (1986). Competition for nitrogen-15-depleted ammonium nitrate between arrowleaf clover and annual ryegrass sown into bermudagrass sod. *Agron. J.* 78: 1023–1030.

Motazedian, I. and Sharrow, S.H. (1986). Defoliation effects on forage dry matter production of a perennial ryegrass-subclover pasture. *Agron. J.* 78: 581–584.

Narasimhalu, P., Black, W.N., McRae, K.B., and Winter, K.A. (1981). Effects of annual rate and timing of N fertilization on production of timothy, bromegrass, and reed canarygrass. *Can. J. Plant. Sci.* 61: 619–623.

Nihsen, M.E., Piper, E.L., West, C.P. et al. (2004). Growth rate and physiology of steers grazing tall fescue inoculated with novel endophytes. *J. Anim. Sci.* 82: 878–883.

Nuttall, W.F., Cooke, D.A., Waddington, J., and Robertson, J.A. (1980). Effect of nitrogen and phosphorus fertilizers on a bromegrass and alfalfa mixture grown under two systems of pasture management. I. Yield, percentage legume in sward, and soil tests. *Agron. J.* 72: 289–294.

Olson, G.L., Smith, S.R. and Phillips, T.D. (2015). 2015 Tall fescue and bromegrass report. University of Kentucky, Agricultural Experiment Station. Report PR-697. https://www.uky.edu/Ag/Forage/PR697.pdf (accessed 9 October 2019).

Oram, R. and Lodge, G. (2003). Trends in temperate Australian grass breeding and selection. *Aust. J. Agric. Res.* 54: 211–241.

Pysher, D. and Fales, S. (1992). Production and quality of selected cool-season grasses under intensive rotational grazing by dairy cattle. In: *Proceedings of the American Forage and Grassland Council, 1992, Grand Rapids, MI. 5–9 April, 1992*, 32–36. Georgetown, TX: AFGC.

Quin, B.F. (1979). A comparison of nutrient removal by harvested reed canarygrass and ryegrass *Lolium*-clover in plots irrigated with treated sewage effluent. *N. Z. J. Agric. Res.* 22: 291–302.

Read, J.C. and Anderson, S.J. (2003). Texas bluegrass. In: *Turfgrass Biology, Genetics, and Breeding* (eds. M.D. Casler and R.R. Duncan), 61–66. New York: Wiley.

Read, J.C. and Camp, B.J. (1986). The effect of the fungal endophyte *Acremonium coenophialum* in tall fescue on animal performance, toxicity, and stand maintenance. *Agron. J.* 78: 848–850.

Reynolds, J.H. and Smith, D. (1962). Trend of carbohydrate reserves in alfalfa, smooth bromegrass, and timothy grown under various cutting schedules. *Crop Sci.* 2: 333–336.

Riewe, M.E., Smith, J.C., Jones, J.H., and Holt, E.C. (1963). Grazing production curves I. Comparison of steer gains on gulf ryegrass and tall fescue. *Agron. J.* 55: 367–372.

Roberts, C.A. and Andrae, J.G. (2010). Fescue toxicosis and management. In: *Alliance of Crop, Soil, and Environmental Science*, 1–15. Madison, WI.

Roberts, C.A., Kallenbach, R.L., Hill, N.S. et al. (2009). Ergot alkaloid concentrations in tall fescue hay during production and storage. *Crop Sci.* 49: 1496–1502.

Rouquette, F.M. (1995). Grazing management and utilization of ryegrass. In: *Symposium on Annual Ryegrass. 31 Aug–1 Sep., 1995, Tyler, TX* (eds. G.W. Evers and L.R. Nelson), 62–82. College Station, TX: Texas Agricultural Experiment Station Bulletin. MP-1770.

Schmidt, S.P. and Osborn, T.G. (1993). Effects of endophyte-infected tall fescue on animal performance. *Agric. Ecosyst. Environ.* 44: 233–262.

Sheaffer, C.C. and Marten, G.C. (1995). Reed canarygrass. In: *Forages – An Introduction to Grassland Agriculture*, 5e (eds. R.F Barnes, D.A. Miller and C.J. Nelson), 335–343. Ames, IA: Iowa State University Press.

Sheaffer, C.C., Miller, D.W., and Marten, G.C. (1990). Grass dominance and mixture yield and quality in perennial grass-alfalfa mixtures. *J. Prod. Agric.* 3: 480–485.

Sheaffer, C.C., Peterson, P.E., Hall, M.H., and Stordahl, J.B. (1992). Drought effects on yield and quality of perennial grasses in the North Central United States. *J. Prod. Agric.* 5: 556–561.

Simons, A.B. and Marten, G.C. (1971). Relationship of indole alkaloids to palatability of *Phalaris arundinacea* L. *Agron. J.* 73: 915–919.

Sleper, D.A. and Buckner, R.C. (1995). The fescues. In: *Forages – The Science of Grassland Agriculture*, 5e (eds. R.F Barnes, D.A. Miller and C.J. Nelson), 345–356. Ames, IA: Iowa State University Press.

Sleper, D.A. and West, C.P. (1996). Tall fescue. In: *Cool-Season Forage Grasses*. Agron. Monogr. 34 (ed. L.E. Moser), 471–502. Madison, WI: ASA, CSSA, and SSSA.

Stuedemann, J.A. and Hoveland, C.S. (1988). Fescue endophyte: history and impact on animal agriculture. *J. Prod. Agric.* 1: 39–44.

Suzuki, M. (1989). Fructans in forage grasses with varying degrees of coldhardiness. *J. Plant Physiol.* 134: 224–231.

Terrell, E.E. (1968). *A Taxonomic Revision of the Genus Lolium*. USDA-ARS Tech. Bull. 1392. Washington, D.C: U.S. Goverment Printing Office.

Tor-Agbidye, J., Blythe, L.L., and Craig, A.M. (2001). Correlation of endophyte toxins (ergovaline and lolitrem B) with clinical disease: fescue foot and perennial ryegrass staggers. *Vet. Hum. Toxicol.* 43: 140–146.

Tremblay, G.G., Bélanger, G., and Drapeau, R. (2005). Nitrogen fertilizer application and developmental stage affect silage quality of timothy (*Phleum pratense* L.). *Grass Forage Sci.* 60: 337–355.

Van Santen, E. and Sleper, D.A. (1996). Orchardgrass. In: *Cool-Season Forage Grasses*. Agron. Monogr. 34 (ed. L.E. Moser), 503–534. Madison, WI: ASA, CSSA, and SSSA.

Vogel, K.P., Moore, K.J., and Moser, L.E. (1996). Bromegrasses. In: *Cool-Season Forage Grasses*. Agron. Monogr. 34 (ed. L.E. Moser), 535–567. Madison, WI: ASA, CSSA, SSSA.

Vose, P.B. (1959). The agronomic potentialities and problems of the canary grasses, *Phalaris arundinacea* L. and *Phalaris tuberosa* L. *Herb Abstr.* 29: 77–83.

Wedin, W.F. and Huff, D.R. (1996). Bluegrasses. In: *Cool-Season Forage Grasses*. Agron. Monogr. 34 (ed. L.E. Moser), 665–690. Madison, WI: ASA, CSSA, SSSA.

West, C.P. and Gwinn, K.D. (1993). Role of *Acremonium* in drought, pest, and disease tolerances of grasses. In: *Proceedings of 2nd International Symposium on Acremonium/Grass Interactions. 3–4 Feb* (eds. D.E. Hume, G.C. Latch and H.S. Easton), 131–140. Palmerston North, NZ: AgRes. Grassl. Res. Cent.

West, C.P., Izekor, E., Oosterhuis, D.M., and Robbins, R.T. (1988). The effect of *Acremonium coenophialum* on the growth and nematode infestation of tall fescue. *Plant Soil* 112: 3–6.

CHAPTER

17

Grasses for Arid and Semiarid Areas

Daren D. Redfearn, Associate Professor, *Agronomy, University of Nebraska, Lincoln, NE, USA*
Keith R. Harmoney, Range Scientist, *Kansas State University, Hays, KS, USA*
Alexander J. Smart, Professor and Rangeland Management Specialist, *South Dakota State University, Brookings, SD, USA*

Semiarid and arid grazing lands in North America are bordered on the east by approximately the 100th meridian and extend west through the Intermountain Region (Figure 17.1). These grazing lands are made up almost entirely of native rangeland, although seeded pastures have an important role in reclaiming depleted rangeland and restoring marginal cropland to pasture and forage production that is more economically and environmentally sustainable. Chapter 8 describes grassland zones and zones of adaptation for forage species. Refer to Hitchcock (1951) and Cronquist et al. (1977) for detailed botanic descriptions of grass species that are included in Chapter 17.

Cool-Season Grasses

Detailed information on cultivars and germplasms, release year and agencies, and unique adaptation characteristics for the wheatgrasses, wildryes, and other miscellaneous cool-season grasses can be found in Chapter 15 of the 6th edition of Vol. II (Berdahl and Redfearn 2007).

Wheatgrasses

Wheatgrasses are members of the Triticeae tribe and are among the most important grass species in temperate regions of the western US and Canada. Depending on the taxonomic authority, from 100 to 150 wheatgrass species have been described (Asay 1995). The taxonomy of wheatgrasses included in this chapter will correspond to the treatment described by Dewey (1982), Tzvelev (1983), and Barkworth and Dewey (1985). More than two-thirds of the wheatgrass species are native to Eurasia, with 22–30 species considered to be native to North America (Cronquist et al. 1977).

Most of the wheatgrasses are cross-pollinating **allopolyploids** that resulted from natural hybridization among genera and species of the Triticea tribe. This evolutionary process continues in modern times (Bowden 1965; Dewey 1984). The wheatgrasses include both bunchgrasses and rhizomatous types and encompass a wide range of morphologic diversity (Asay 1995).

Western Wheatgrass

Western wheatgrass is a native, long-lived, cross-pollinating perennial with strong creeping rhizomes and stems and leaves that are typically **glaucous** (Cronquist et al. 1977). It is distributed across the western two-thirds of the US and Canada, from Ontario to Alberta in the North (Hitchcock 1951) and into New Mexico and the Texas panhandle in the South (Alderson and Sharp 1994). Dewey (1975) concluded that thickspike wheatgrass and beardless wildrye, or closely related taxa, were progenitors of western wheatgrass, an octaploid made up of four different genomes.

Forages: The Science of Grassland Agriculture, Volume II, Seventh Edition.
Edited by Kenneth J. Moore, Michael Collins, C. Jerry Nelson and Daren D. Redfearn.

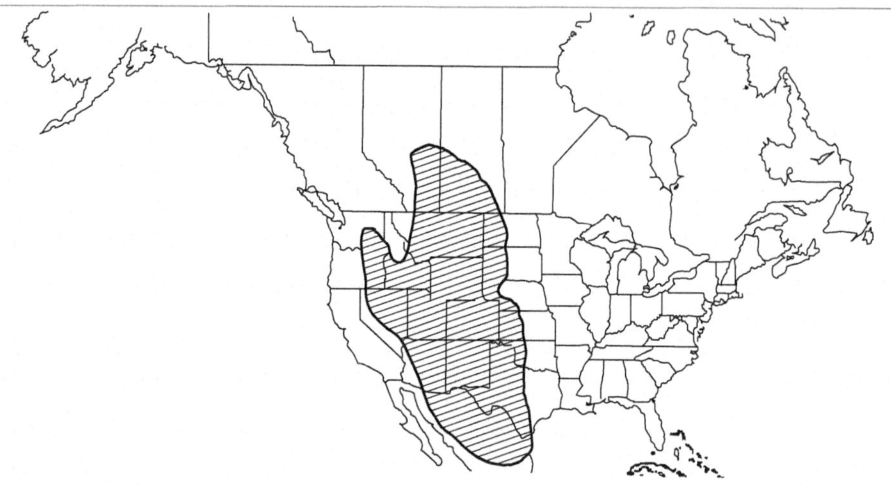

Fig. 17.1. Arid and semiarid grazing lands of North America.

Western wheatgrass withstands extended periods of drought and is more tolerant to grazing than most other native North American wheatgrass species (Cronquist et al. 1977). It is often the dominant grass in rangeland of the northern and central Great Plains, where it is commonly found in association with blue grama and the needlegrasses (Barker and Whitman 1988). In the Intermountain Region, it typically grows in communities with bluebunch wheatgrass, thickspike wheatgrass, and various shrubs (Asay 1995).

Western wheatgrass is adapted to a wide range of soil types and is often found on heavy alkaline soils characteristic of the bottoms of intermittent swales, shallow lakebeds, or areas subjected to periodic short-duration flooding (Rogler 1973). The species may be the most alkali tolerant of the North American wheatgrasses (Beetle 1955). Western wheatgrass was rated as one of the most promising of 174 grass, forb, and shrub species evaluated by the USDA-SCS in Wyoming and Montana for reclaiming saline seeps and other problem sites (Scheetz et al. 1981). Newell and Moline (1978) found digestibility of western wheatgrass was relatively high over the grazing season compared with other grasses. As with many other rhizomatous wheatgrasses, it is a relatively poor seed producer, and stands are slow to establish from seed (Asay 1995). Extensive rhizome development of established plants along with delayed **germination** of dormant seed in the soil often transform a poor stand into a relatively good stand after two to three years.

Thickspike Wheatgrass

This species is a native wheatgrass morphologically similar to western wheatgrass with aggressive rhizomes and stems and leaves that are typically glaucous. It is distributed from Michigan to British Columbia in the North and from Illinois west to Oregon, Nebraska, Colorado, and Nevada, in the South (Hitchcock 1951). The species is commonly known as northern wheatgrass in Canada. Thickspike wheatgrass is cross-pollinating and has an allotetraploid genome structure. The accepted taxonomic classification places thickspike wheatgrass along with what was traditionally known as streambank wheatgrass, Montana wheatgrass and Wyoming wheatgrass into *Elymus lanceolatus* (Dewey 1983). Snake River wheatgrass is a new taxon that is morphologically similar to bluebunch wheatgrass but has the same genome structure as thickspike wheatgrass (Asay 1995). Snake River wheatgrass has been tentatively treated taxonomically as *E. lanceolatus* ssp. *wawawai* (Carlson and Dewey 1987).

Thickspike wheatgrass is often found growing in sagebrush deserts and foothill woodlands in association with western wheatgrass and bluebunch wheatgrass. Compared with western wheatgrass, thickspike wheatgrass is more drought tolerant but tends to be less productive under more optimum soil water (Asay and Jensen 1996a). Thickspike wheatgrass does not persist as well as western wheatgrass under close and frequent grazing. It is most valuable for grazing in early- to mid-summer when nutritive value of early cool-season grasses such as crested wheatgrass is relatively low.

Bluebunch Wheatgrass

Bluebunch wheatgrass is native to the US typically with a bunchgrass growth habit under arid and semiarid conditions, but mildly rhizomatous types occur under more optimum moisture conditions (Asay 1995). It is one of the most valuable native grasses in the Intermountain Region and the Pacific Northwest (Asay 1995). Bluebunch wheatgrass is widely distributed from the western Dakotas across the western US, north into Canada, and south to the northern edge of the Sonoran Desert (Hitchcock 1951). The diploid form is most common, but an **autotetraploid** form also exists with distribution limited primarily to eastern Washington and adjacent northwest Idaho and Canada (Hartung 1946). The species is predominantly cross-pollinated, but varying degrees of self-fertility have been observed in the tetraploid form (Jensen et al. 1990). Beardless wheatgrass differs from bluebunch wheatgrass by absence of lemma awns and is grouped taxonomically with bluebunch wheatgrass (Dewey 1983).

Bluebunch wheatgrass is often found on dry plains and on rocky slopes and hills, but seldom on wet soils. It can be found in relatively pure stands and as a codominant with sagebrush species on arid sites. Bluebunch wheatgrass has excellent nutritive value and palatability, but stand density is often greatly reduced under moderate to high grazing pressure (Asay 1995). Close defoliation is damaging during early spring when energy reserves are low, and it is recommended that grazing be delayed until the late boot stage (Daer and Willard 1981; Miller et al. 1987). Grazing pressure should be adjusted to remove only 50% of the current forage production (Hafenrichter et al. 1968). Downy bromegrass, commonly known as cheatgrass, is highly competitive with bluebunch wheatgrass.

Slender Wheatgrass

Slender wheatgrass is described as a native, short-lived perennial bunchgrass, but plants with weak rhizomes are occasionally present (Cronquist et al. 1977). It is distributed from Newfoundland to Alaska in the North and from North Carolina to Arizona and California in the South. The species is self-fertile and has an **allotetraploid** genome structure (Asay and Jensen 1996a). Cronquist et al. (1977) treated bearded wheatgrass and several other variant taxa as a single species of slender wheatgrass.

Slender wheatgrass has good seedling vigor but is less drought tolerant than most other wheatgrasses (Asay 1995). It is found on dry to moderately moist roadsides, ditchbanks, streambanks, meadows and woodlands. Slender wheatgrass has potential for reclaiming saline seeps, though it is not as salt tolerant as tall wheatgrass (Scheetz et al. 1981). It is widely used as a companion crop during establishment of more permanent forages.

Crested Wheatgrass

Crested wheatgrass is an introduced, long-lived perennial bunchgrass. It is native to a wide area from central Europe and the Middle East through Central Asia to Siberia, China, and Mongolia (Dewey 1986). In the present taxonomic alignment, species of the crested wheatgrass complex are the sole members of the genus *Agropyron.* Dewey (1983) limited the crested wheatgrass complex to Fairway, Standard, and Siberian. The crested wheatgrass complex consists of an autoploid series of diploid, tetraploid, and hexaploid forms of the same genome. Diploids and tetraploids are distributed across the same general area in their native habitat, but diploids occur much more sporadically (Dewey 1986). The tetraploid Siberian form is more drought resistant and better adapted to sandy soils than the tetraploid Standard form or the diploid Fairway form (Asay et al. 1995). Hexaploid plants have larger leaves and are more robust than diploids and tetraploids. Hexaploids have been found in Turkey, Iran, and Kazakhstan, but are much less common than the diploid and tetraploid forms (Dewey and Asay 1975; Asay et al. 1990).

Crested wheatgrass was named in reference to its comb-like spike (Asay 1995). Floral characteristics are often used to differentiate among species in the complex, but **spike** morphology per se cannot be relied on for taxonomic classification. Variants with rhizomes are occasionally found, but crested wheatgrass is typically a bunchgrass with dense, erect culms from 15 to more than 100 cm tall.

Dillman (1946) and Rogler and Lorenz (1983) provide accounts of the introduction and early use of crested wheatgrass in the US, and Kilcher (1969) provides this information for Canada. In North America, the crested wheatgrasses are particularly well adapted to temperate regions that receive 30–45 cm of annual precipitation (Asay 1995). Crested wheatgrass is tolerant of grazing, and many of the areas seeded to crested wheatgrass in the late 1930's and early 1940's are still productive (Lorenz 1986). Mayland et al. (1992) estimated that over 6 million ha of crested wheatgrass are in production in North America. The tetraploid Standard form is most common in the US, and the diploid Fairway form is most prevalent in Canada.

Crested wheatgrass produces abundant, high-quality forage during the spring and early summer, but protein concentration and digestibility decline rapidly after heading (Newell and Moline 1978). In the northern Great Plains, crested wheatgrass traditionally provides good gains when used for spring and early summer grazing (Haferkamp et al. 2005), and is followed by fall grazing if regrowth is adequate. However, Karn et al. (1999) found season-long grazing of crested wheatgrass or native range in North Dakota produced comparable gains for yearling steers. The early practice of seeding crested wheatgrass in monoculture over large areas has raised environmental concerns. Recent seedings in North America have emphasized inclusion of other introduced and native species along with crested wheatgrass.

Intermediate Wheatgrass

This introduced forage species can be very productive. Intermediate wheatgrass is a high-yielding, moderately rhizomatous species native to an area from the steppes to the lower mountain areas of southern Europe, through the Middle East and central Asia, and to western Pakistan (Tzvelev 1983). Intermediate wheatgrass cross-pollinates as an autoallohexaploid with two partially homologous genomes and a distinct third genome (Dewey 1962). The current taxonomic treatment combines glabrous types and pubescent types into a single species with two subspecies (Dewey 1984).

In North America, intermediate wheatgrass is widely used for hay and pasture from Nebraska north to Manitoba and west to British Columbia and California (Asay 1995). It is adapted to temperate areas that receive at least 350 mm of annual precipitation and has a water requirement between that of smooth bromegrass and crested wheatgrass. Spike emergence of intermediate wheatgrass is approximately two weeks later than smooth bromegrass and crested wheatgrass. It is well suited for hay in mixtures with alfalfa under dryland conditions or limited irrigation where only one to two cuttings per year are made. Intermediate wheatgrass is still relatively immature when the first cutting of alfalfa is ready for harvest, and its erect stems help prevent alfalfa from lodging (Lawrence 1977).

Intermediate wheatgrass produces large seed and relatively high seed yields. Genotypes are currently being developed as a perennial grain crop (Kantar et al. 2016). Its relative ease of establishment and availability of seed have resulted in use of intermediate wheatgrass in conservation plantings such as the Conservation Reserve Program in the US. It is sensitive to continuous grazing at high stocking rates (Currie and Smith 1970), but has potential to produce abundant forage when grazed intermittently (Harmoney 2005). Lawrence and Ashford (1966) concluded that initial cutting in the boot stage of development should be avoided and cutting intervals should be at least six weeks apart for maximum herbage production and maintenance of root and crown mass associated with persistence. At dryland sites in the northern Great Plains, intermediate wheatgrass is often cut for hay followed by grazing of regrowth in the fall.

Much ecotypic variation is present among intermediate wheatgrass accessions that have been introduced to North America. Generally, **pubescent** types are better adapted than glabrous types to the southern limits of the species (Cornelius 1965) and to sites with more drought stress (Hafenrichter et al. 1968). In the Great Plains, pubescent types have good production potential into Kansas (Harmoney 2005), but may be limited by heat and drought further south (Gillen and Berg 2005; Malinowski et al. 2003). The pubescent types are characterized by pubescence on the outer **glumes**, lemmas, and occasionally on the leaves. Pubescence has relatively simple inheritance, and pubescent and glabrous types hybridize freely.

Tall Wheatgrass

Tall wheatgrass is a cross-pollinating bunchgrass, ranging in height from 1 to 1.5 m (Cronquist et al. 1977). It was introduced into the US from southern Europe and Asia Minor, where it is often found growing on saline and alkaline soils in meadows, salt marshes, and along seashores (Weintraub 1953). Tall wheatgrass, a decaploid with five genomes, shares common genomes with intermediate wheatgrass. Tall wheatgrass matures later than all other introduced wheatgrass species, heading about one week later than the native species, western wheatgrass.

Tall wheatgrass was introduced into North America from Turkey in 1909 but, was largely ignored until the 1930's (Weintraub 1953). It is used for hay and pasture throughout the Intermountain region, the northern and central Great Plains, and as far south as northern New Mexico and Arizona on sites that are subirrigated or receive a minimum of 350–400 mm of annual precipitation (Asay and Jensen 1996a). Although tall wheatgrass is slow to establish and not particularly drought tolerant, it produces forage and persists on sites that are too alkaline and saline for other productive crops.

In spite of its coarse leaves and stems, tall wheatgrass remains quite palatable during late summer. Protein and digestibility are sustained relatively well, and the grass has a potential grazing period of 11–12 weeks (Newell and Moline 1978). However, animal gains tend to be lower on tall wheatgrass compared to native western wheatgrass (Harmoney and Jaeger 2019). To maintain stands, it is recommended that 20 cm of stubble be left at the end of the season, and grazing be delayed in the following season until 25 cm of new growth has accumulated above the stubble (Asay and Knowles 1985).

The tall stature and bunchgrass growth habit of tall wheatgrass provide excellent nesting habitat and cover for upland game birds and waterfowl (Duebbert et al. 1981). The grass has been planted in single or double rows in the northern Great Plains to provide barriers against drifting snow (Hafenrichter et al. 1968). Tall wheatgrass has been used successfully in wide hybridization programs with wheat to transfer genes conditioning resistance to salinity, drought, and disease (Dewey 1984).

Quackgrass

Quackgrass is native to Europe and Asia, but it is widely adapted and distributed in temperate areas of every continent (Rogler 1973). An aggressive rhizomatous growth habit causes quackgrass to be considered a **noxious weed** in cultivated croplands, gardens, lawns, and orchards (Asay 1995). Quackgrass is a cross-pollinating hexaploid

(Dewey 1976) with high ecotypic variation that has been classified taxonomically into several subspecies.

Quackgrass is widely distributed in North America from Newfoundland to Alaska in the North and from North Carolina to Mexico in the South (Cronquist et al. 1977). It is less tolerant of drought than other wheatgrasses, with a water requirement similar to smooth bromegrass (Asay and Jensen 1996a). Quackgrass is tolerant of saline and alkaline soils and moderately tolerant of flooding. Its aggressive sod-forming characteristic makes quackgrass an excellent candidate for conservation plantings when soil stabilization is critical.

Digestibility of quackgrass forage compares favorably with other wheatgrasses (Casler and Goodwin 1998). Little plant breeding research has been conducted on quackgrass. Casler and Goodwin (1989) found that reduced rhizome development appeared to be conditioned by a single **homozygous** recessive gene. The USDA-ARS, Logan, UT has hybridized quackgrass with several other wheatgrasses (See section on bluebunch wheatgrass). Even so, it is illegal to purchase or intentionally plant quackgrass seed or rhizomes in much of North America, because many states consider it a **noxious weed**.

Wildryes

The wildrye grasses are members of the Triticea tribe and are currently grouped in the genera *Leymus, Elymus,* and *Psathyrostachys* as described by Dewey (1982), Tzvelev (1983), and Barkworth and Dewey (1985). The wildryes are distributed throughout the temperate regions of the world, primarily in Asia, Europe, and North America (Asay and Jensen 1996b). The wildryes, adapted to the same general areas as the wheatgrasses, are used extensively in rangeland improvement programs in North America for livestock production and wildlife habitat.

Great Basin Wildrye

Great Basin wildrye is a native, long-lived perennial bunchgrass that can form clumps up to 1 m in diameter. It is distributed in the North from Saskatchewan to British Columbia and in the South from Colorado to California (Cronquist et al. 1977). It is usually found growing in valley bottoms, along roadsides, streams, and in gullies. Great Basin wildrye is moderately tolerant of saline and alkaline soils and is adapted to areas receiving greater than 250 mm of annual precipitation.

Plants are coarse, erect, and can range in height from 0.7 to 2 m. Leaves predominate near the base of the plant and vary in color from green to blue-green. Plants are usually tetraploid in the western US and octaploid in western Canada (Asay and Jensen 1996b). Fertile hybrids occur between Great Basin wildrye and beardless wildrye where the two species grow in association.

Great Basin wildrye provides valuable forage for livestock during the fall and winter (Asay 1995) and is used as a component of mixtures to reseed depleted rangeland and disturbed areas (Asay and Jensen 1996b). On saline and alkaline soils, pastures of Great Basin wildrye seeded in monoculture can be grazed in spring and fall to complement tall wheatgrass pasture, which is grazed during the summer months. Standing forage provides excellent habitat for upland game birds and waterfowl and can serve as a **windbreak** and protection for young calves and lambs (Asay 1987).

Seed dormancy and poor seedling vigor have limited use of Great Basin wildrye in revegetation efforts (Evans and Young 1983). The species is sensitive to high-grazing pressure. To maintain productive stands, a 30-cm stubble height should be left after grazing and a 25-cm height after hay harvest (Asay 1995). A high incidence of ergot infection [caused by *Claviceps purpurea* (Fr.) Tul.] can occur in certain environments in which sclerotia in maturing spikes can cause abortions in grazing animals (Cronquist et al. 1977).

Beardless Wildrye

Beardless wildrye is a moderately long-lived perennial with strongly creeping rhizomes. It is native to the US and is distributed from Montana to Washington in the North and from Texas to Baja California in the South. It occurs in saline meadows and poorly drained alkaline sites. Beardless wildrye is a tetraploid, having the same two basic genomes as tetraploid Great Basin wildrye (Dewey 1984). Though the two species are morphologically distinct, they hybridize readily in nature. Beardless wildrye ranges in height from 30 to 70 cm and has stiff, rough-textured leaves (Cronquist et al. 1977).

Beardless wildrye is recommended for stabilizing saline seep areas. However, high levels of seed dormancy and slow-developing seedlings have restricted widespread use of the species.

Canada Wildrye

Canada wildrye is a native, relatively short-lived, self-pollinating perennial bunchgrass. It is widely distributed in North America from New Brunswick to Alaska in the North and from South Carolina to southern California in the South (Cronquist et al. 1977). It is commonly found along streambanks, dry to moist meadows, and roadsides. Much morphologic variation is found within the species. Plant height ranges from 80 to 150 cm, and leaves are rough-textured.

Canada wildrye has high seed yields, but tenacious awns make harvesting and seed processing difficult (Hafenrichter et al. 1949). Forage quality and palatability of canada wildrye decline rapidly after heading, and the species is not tolerant of heavy grazing pressure. Seedlings establish rapidly, and the grass has been used to stabilize disturbed areas with sandy soils. Seed shattering is minimal, and its nodding, awned **spikes** have created interest in ornamental uses. Extensive plantings of canada wildrye have not been made.

Russian Wildrye

Russian wildrye is an introduced, long-lived perennial bunchgrass that is tolerant of cold, drought, and grazing pressure. It is indigenous from Iran northward to the lower Volga River and lower Don River regions of Russia in the North, eastward into Siberia, and across Asia to Outer Mongolia and northern China (Dewey 1984). Russian wildrye has erect culms ranging in height from 60 to 120 cm, and leaves are mostly basal. The spike rachis becomes brittle at maturity, and seed shatters readily. Russian wildrye is a cross-pollinating diploid, although tetraploid forms have been found in nature (Asay and Jensen 1996b).

The first recorded introductions of russian wildrye into North America were in 1926 by the University of Saskatchewan and in 1927 by the US Department of Agriculture (Rogler and Schaaf 1963). Its value for grazing was recognized soon afterward (Rogler 1941). Smoliak and Johnston (1980) estimated in 1980 that approximately 400 000 ha of russian wildrye had been seeded in North America. It is primarily used as pasture to complement native rangeland in semiarid and arid portions of the northern Great Plains and Intermountain regions of the US and in the Prairie Provinces of Canada. Russian wildrye is best adapted to loam and clay soils and is moderately tolerant of saline and alkaline conditions (Rogler and Schaaf 1963). It is not well adapted to sandy soils.

Russian wildrye is best used for grazing, because its basal forage develops clumps that are difficult to harvest. It can be grazed season-long, but is most commonly stockpiled for fall grazing when its protein concentration and digestibility are higher than most other grasses (Knipfel and Heinrichs 1978; Haferkamp et al. 2005). Russian wildrye is often less palatable than other cool-season grasses early in the grazing season (Smoliak and Johnston 1980), and grass tetany can be a potential hazard with early-season growth (Karn et al. 1983; Asay and Mayland 1990). Relatively slow seedling development and associated risk of establishment failure is the major factor that limits more widespread use of the species.

Once established, russian wildrye is very persistent (Harmoney 2007), and is very competitive and may crowd out other forages in a mixture (Rogler and Schaaf 1963). Sowing legumes or other forages in alternate rows or cross-seeding is recommended to reduce early competition from russian wildrye (Kilcher 1982). Jefferson and Kielly (1998) concluded that a 30-cm row spacing for russian wildrye pasture at semiarid sites was more productive than the 60- to 90-cm spacing previously recommended (Lawrence and Heinrichs 1968).

Altai Wildrye

Altai wildrye, an introduced perennial with short creeping rhizomes, is long-lived and tolerant of cold and drought. It is native to western Siberia and the Altai mountain region between Siberia and Outer Mongolia (Asay and Jensen 1996b). In its native habitat, altai wildrye is found on semideserts, alkaline meadows, steppes, and river and lake valleys (Lawrence 1987).

Altai wildrye plants are coarse and erect, ranging in height from 60 to 120 cm (Asay and Jensen 1996b). Leaves are mostly basal and light-green to blue-green in color. Plants are considered alloautopolyploids that occur at a series of ploidy levels: $2n = 28$, 42, 56, and 84, with $2n = 84$ the most common (Dewey 1978).

Altai wildrye was introduced into Canada in the mid-1930's by the Canada Department of Agriculture (Lawrence 1976). It is well adapted to loam and clay soils of the Canadian prairies and is comparable to tall wheatgrass in tolerance to saline and alkaline soils (McElgunn and Lawrence 1973). Roots of altai wildrye can penetrate to 3–4 m and use water from a perched (high) water table (Lawrence 1987). Altai wildrye is compatible with alfalfa in mixed plantings at semiarid sites in southwest Saskatchewan (Lawrence and Ratzlaff 1985).

Altai wildrye is often used for late fall and winter grazing (Lawrence 1987). Its basal leaves maintain their nutritive quality at maturity better than most other cool-season grasses. Erect culms and leaves protrude above shallow snow and allow deep snow to bridge over the top of the leaves so that they are still accessible to grazing cattle. Seed dormancy and slow seedling development are the most serious limitations to more widespread use of altai wildrye (Asay 1995). The grass has limited use so far in the US.

Dahurian Wildrye

Dahurian wildrye is an introduced, short-lived perennial bunchgrass that is native to Siberia, Outer Mongolia, and northern China (Lawrence et al. 1990). Plant heights range from 1 to 1.5 m. Plants are self-pollinating and hexaploid with three distinct genomes (Lu and Bothmer 1992).

Dahurian wildrye establishes easily and is productive during its short stand life of three to four years in Saskatchewan and Alberta (Lawrence et al. 1990). It recovers rapidly after grazing or haying, has adequate digestibility, and is palatable to beef cattle. Though recommended as a short-rotation forage grass or for inclusion in mixed seedings to supplement early production of long-lived, slow-establishing forages, it has had limited use in North America.

Needlegrasses

Approximately 100 different species are classified as needlegrasses, many of which are native to North America (Stubbendieck and Jones 1996). Several species of the genus *Stipa* and *Nassella* are important perennial range grasses, but only green needlegrass has, so far, been successfully commercialized for seed production (Voigt and Sharp 1995). Impediments to use of the needlegrasses in commercial seed channels include seed dormancy and, in some cases, tenacious awns that make seed harvest and handling difficult (Vallentine 1989). Most needlegrasses provide good to fair forage for grazing. The needlegrasses are bunchgrasses with a panicle inflorescence. A sharp-pointed **callus** is present on one end of the floret of many species, and this pointed **callus** may injure grazing animals when seeds begin to develop.

Green Needlegrass

Green needlegrass is a native species and is widely distributed from Alberta and Saskatchewan southward to Illinois, Kansas, New Mexico, and Arizona (Sedivec and Barker 1997). It is adapted to loamy to heavy clay soils and also obtains good growth on sandy soils where water is not limiting or when the water table is shallow. Green needlegrass is palatable and is sought out by most all classes of livestock. The species decreases in frequency with grazing pressure.

Needle-and-Thread

Needle-and-thread is a native species that is widely distributed over the northern and central Great Plains and the Intermountain region of the US, as well as grasslands of the Prairie Provinces of Canada (Sedivec and Barker 1997). The species is often associated with sandy and rocky soils. Needle-and-thread is one of the most palatable of all native grasses if grazed before seeds develop a sharp-pointed **callus** ("seed needles"). Frequency of needle-and-thread plants will increase with moderate grazing but will decrease under long-term overuse.

Porcupine Grass

Porcupine grass occurs from southern Ontario to the southwest portion of the Northwest Territories in the North and from Pennsylvania, Missouri, and New Mexico in the South (Sedivec and Barker 1997). The species is commonly associated with sandy soils and is found on open hillsides and prairies if soil water is in moderate supply. Porcupine grass is only moderately palatable, even in the spring. Its sharp-pointed seed **callus** can be injurious to livestock.

Robust Needlegrass

This species is native to the central Rocky Mountains from Wyoming south to Colorado, Arizona, New Mexico, and extending into west Texas and adjacent Mexico (Hitchcock 1951; Young et al. 2003). The robust culms range in height from 1 to 1.5 m, but it is not a coarse grass. Leaves are drooping and narrow. Robust needlegrass is commonly known as "sleepy grass," because of its reputation for inducing narcosis and grazing avoidance, especially in horses, that is caused by a fungal endophyte (Jones et al. 2000). Robust needlegrass produces forage later in the growing season than most other cool-season grasses, and it is tolerant of salt and alkaline affected soils.

Texas Wintergrass

Texas wintergrass is an important native rangeland species, primarily in central and north Texas into southern Oklahoma, where it is one of two native, cool-season perennial grasses of major importance. Like some other cool-season grasses, it begins growth in the fall and continues through the winter into mid-spring, hence the name wintergrass. It is adapted to soil types ranging from sandy to clayey and grows to a height of 0.6–1.0 m under good growing conditions. Leaves are dark green and pubescent. Propagation is through self- or cross-pollinated seed under natural conditions. The balance between self- and cross-pollination depends on environmental conditions (Brown 1952). Under heavy grazing pressure, basal, axillary **florets** that have not exerted may be crucial to plant perenniation (Call and Spoonts 1989).

Other Cool-season Grasses

Indian Ricegrass

Indian ricegrass is a self-pollinated, highly variable, drought-resistant perennial bunchgrass native to North America (Jones and Nielson 1989). It is widely distributed from Manitoba to British Columbia in the North, across the western US, and south to northern Mexico (Hitchcock 1951). Indian ricegrass is found most frequently on droughty, coarse soils but is also adapted to clay soils (Jones 1990). Culms are erect, 35–80 cm tall, with a diffuse panicle that resembles rice inflorescence (Sedivec and Barker 1997). Indian ricegrass is closely related to the genus *Stipa*, and Johnson (1972) described sterile hybrids between indian ricegrass and 11 different *Stipa* species.

Indian ricegrass is one of the most palatable native grasses. When left uncut or not grazed, it cures well while maturing and provides valuable winter grazing. Indian ricegrass is a short-lived perennial and **overgrazing** greatly reduces its incidence on native sites (Rogler 1960). Stand maintenance requires frequent recruitment of new seedlings (Robertson 1976).

Rough Fescue

Rough fescue is an important native perennial bunchgrass found throughout the grasslands of Canada extending south to northern Oregon, Washington, Idaho, Montana, western Colorado, and northern North Dakota (Sedivec and Barker 1997). Rough fescue has been classified as two distinct species, with *Festuca campestris* found primarily in the foothills of Alberta and Montana and *Festuca hallii* found primarily in the prairie parklands of Alberta and Saskatchewan (Aiken and Darbyshire 1990). Rough fescue has few reproductive tillers and an abundance of basal leaves. Forage quality is high, relative to other cool-season grasses, and the grass cures well on the stem (Johnston 1962). Rough fescue is sensitive to defoliation, and its vigor declines with increased grazing pressure (Willms and Fraser 1992).

Fineleaf Fescues

Idaho fescue, arizona fescue, and greenleaf fescue are native, perennial bunchgrasses indigenous to the western US and western Canada, and several species provide valuable grazing for livestock and wildlife. Idaho fescue, sometimes called bluebunch fescue, is found from British Columbia to Alberta, south to central California and Colorado (Hitchcock 1951). Arizona fescue, commonly called pinegrass, resembles idaho fescue and is found from Nevada to Colorado, south to Texas and Arizona. Greenleaf fescue, commonly called mountain bunchgrass, is found in mountain meadows and on open slopes from British Columbia and Alberta, south to central California and Colorado. The fineleaf fescues become lignified and lose palatability in advanced stages of maturity.

Prairie Junegrass

Prairie junegrass is a native, cool-season perennial bunchgrass that is distributed throughout the northern and western halves of the US, southern Canada, and northern Mexico (Sedivec and Barker 1997). Nutritive value and palatability are high until **flowering**, but it loses palatability after flowering until the forage cures. Prairie junegrass is not a major component of most native pastures, and plant frequency declines with increased grazing pressure.

Bottlebrush Squirreltail

Bottlebrush squirreltail is a bristly-headed, native perennial bunchgrass that is found in many habitats from British Columbia to South Dakota, south throughout the mountain and plains states and into Mexico (Cronquist et al. 1977). Bottlebrush squirreltail is often the dominant grass in areas where **overgrazing** has eliminated more favorable range grasses. It is palatable early in the season, but bristly awns make bottlebrush squirreltail objectionable to grazing animals and may cause mouth and eye injury when seed heads begin to develop. Bottlebrush squirreltail has been used in reclamation of rangelands dominated by annual grassy weeds such as downy brome (cheatgrass) (Jones et al. 1998).

Texas Bluegrass

Texas bluegrass is a rhizomatous, perennial, cool-season grass that is native to southern Kansas, Oklahoma, Texas, and western Arkansas. Dense clusters of stems and leaves rise from long, slender rhizomes. Plants can grow up to 1.0 m on fertile soil, with numerous leaves 15–31 cm in length and 0.6 cm wide. The grass grows throughout the winter producing abundant forage, which is highly palatable. The species is **dioecious**, with separate male and female plants. It produces only limited quantities of seed, covered with pubescence that is difficult to remove. It is a valuable species in areas where it is native, but seeding is difficult. Consequently, establishment of cultivated stands for forage production use is limited.

Warm-Season Grasses

The major warm-season grasses discussed in this chapter are adapted to the arid and semiarid grazing lands of North America including western Canada, the central and western US and parts of Mexico. The east to west transition from tallgrass to shortgrass does not occur rapidly, though abrupt changes in vegetation type may occur due to differences in soil type and precipitation pattern. For example, sandy soils from western Nebraska to eastern New Mexico can support growth of tallgrass species, even though they are primarily classified as shortgrass or midgrass prairie areas. Still, the finer-textured soils in this region are not capable of supporting sustained production from tallgrasses, especially under moderate to heavy grazing pressure.

These species require **rest periods** of up to 40 days or more following a **grazing event**. However, rest periods can be reduced if the stubble height is retained at 25–40 cm after grazing. These grasses should not be grazed shorter than 20 cm after early September in the Great Plains (Anderson 2000).

The tallgrasses, including big bluestem, switchgrass, and indiangrass, are adapted to a wide range of climatic conditions in the east–west direction, but the genetic make-up changes in the north–south direction causing regional strains to differ dramatically. One noteworthy difference can be observed in growth and **flowering** response of sideoats grama (Figure 17.2).

Big bluestem has genetically and phenotypically different ecotypes along an east–west precipitation gradient as well. Many studies of traits of collections along the precipitation gradient at the same latitude showed morphologic differences in three distinct populations (Johnson et al. 2015). Similarly, Casler (2012) alluded to an east ecotype and a west ecotype sorted by drought tolerance in switchgrass.

Bluestems

Big bluestem is the most common and widely distributed bluestem of the 15 native *Andropogon* species in North America. Sand bluestem is an ecotype or subspecies adapted to sandy soils such as those that occur in the Nebraska sandhills and southward through Kansas and Oklahoma. Little bluestem, a native species that is better adapted to low-rainfall sites than the taller species, was formerly classified as an *Andropogon* but now is classified

Fig. 17.2. Adaptation of warm-season grass ecotypes found throughout the Great Plains is controlled by response to photoperiod. Sideoats grama strains grown at Manhattan, Kansas, were from left to right (A) North Dakota, (B) Nebraska, (C) Kansas, (D) Oklahoma, and (E) Texas. Measuring stick shows height in feet. *Source:* NRCS photo.

as a *Schizachyrium* species. Grasses that comprise the old-world bluestems (*Bothriochloa* spp., *Capillipedium* spp., and *Dichanthium* spp.) are all introduced.

Big Bluestem

This dominant species of the tallgrass prairie can make up to 80% of the vegetation at sites where it is well-adapted (Weaver 1968). It tolerates a wide range of soil types, except sandy soils. Plants grow in clumps, though most have short rhizomes (Stubbendieck et al. 2017). Big bluestem culms often reach 1–2 m tall whereas leaves tend to be basal with the leafy portion of the canopy seldom over 50–60 cm high. Big bluestem has an extensive root system that can be as deep as 2.5 m.

Recent improvements in forage yield and nutritive value of big bluestem (Mitchell et al. 2005), along with its wide range of adaptation, should increase its use for forage production, conservation practices, and reclamation projects in the central Great Plains. Big bluestem is much better suited to mid-summer grazing than switchgrass mainly due to the high vegetative-to-fertile stem ratio that results in increased leaf area. Another characteristic is numerous short basal buds capable of regrowth, though many of these do not break dormancy until the following year (Boe et al. 2004). It also does not elevate its growing point until later in development compared with switchgrass (Mitchell and Anderson 2008). Interestingly, much of the growth and development of both big bluestem (Mitchell et al. 1997) and sand bluestem (Hendrickson et al. 1998) are predictable based on accumulated heat units. Current cultivars are listed in Table 17.1.

Sand Bluestem

Sand bluestem grows well on sandy soils compared with big bluestem that is adapted to heavier soils. In contrast to big bluestem, sand bluestem has an extensive system of rhizomes that assist in the stabilization of sandy soils. Big bluestem and sand bluestem are interfertile (Newell and Peters 1961) and will cross-pollinate to produce plants that are intermediate between the two species.

Little Bluestem

Little bluestem is classified as a bunchgrass even though it can produce short rhizomes. It typically grows up to a height of about 1.0 m and can be found throughout the contiguous US, except for Nevada and states along the Pacific Coast. It is morphologically diverse and taxonomically complex (Voigt and MacLauchlan 1985). Little bluestem has more drought tolerance and can persist under grazing on heavier soils in drier climates than big bluestem, switchgrass, or indiangrass. However, this contrasts with observational data from the 1930s that tracked big bluestem and low persistence in plots during extended

Table 17.1 Switchgrass, big bluestem, and indiangrass cultivars and their regions of adaptation

Species/cultivar	Origin of collection	Type
Switchgrass		
Dacotah	North Dakota	Upland
Forestburg	South Dakota	Upland
Sunburst	South Dakota	Upland
Nebraska 28	Nebraska	Upland
Summer	Nebraska	Upland
Shelter	West Virginia	Upland
Pathfinder	Nebraska, Kansas	Upland
Trailblazer	Nebraska, Kansas	Upland
Blackwell	Oklahoma	Upland
Shawnee	Southern Illinois	Upland
Cave-in-Rock	Southern Illinois	Upland
Caddo	Oklahoma	Upland
Kanlow	Oklahoma	Lowland
Alamo	Texas	Lowland
Liberty	Nebraska, Oklahoma	Lowland
Tall bluestems		
Bison	North Dakota	Big bluestem
Bonilla	South Dakota	Big bluestem
Niagra	New York	Big bluestem
Champ	Nebraska, Iowa	Big bluestem
Pawnee	Nebraska	Big bluestem
Rountree	Iowa	Big bluestem
Kaw	Kansas	Big bluestem
Earl	Texas	Big bluestem
Sunnyview	South Dakota	Big bluestem
Bonanza	Nebraska	Big bluestem
Goldmine	Nebraska	Big bluestem
Goldstrike	Nebraska	Sand bluestem
Garden	Nebraska	Sand bluestem
Woodward	Oklahoma/New Mexico	Sand bluestem
Cherry	Nebraska	Sand bluestem
Elida	New Mexico	Sand bluestem
Indiangrass		
Tomahawk	North/South Dakota	
Holt	Nebraska	
Nebraska 54	Nebraska	
Oto	Nebraska, Kansas	
Rumsey	Illinois	
Osage	Kansas, Oklahoma	
Lometa	Texas	
Llano	New Mexico	
Chief	Nebraska	
Scout	Nebraska	
Warrior	Nebraska	

drought. Under this scenario, big bluestem had greater persistence than little bluestem, which nearly disappeared. Current cultivars are listed in Table 17.2.

Introduced Bluestems

This group of bluestems is comprised of *Bothriochloa*, *Capillipedium*, and *Dichanthium* species and are referred to collectively as old-world bluestems. These species are native to Africa, Australia, the Middle East, and southern and eastern Asia. They reproduce by **apomixis** (deWet and Harlan 1970) and are better adapted to fine-textured soils. In the US, yellow bluestem (*Bothriochloa ischaemum*) has received the most use for hay and grazing. Caucasian bluestem (*Bothriochloa bladhii*) also tolerates close and frequent grazing where it is used for summer pasture to complement tall fescue in the transition zone. It is somewhat more winter-hardy than most of the other yellow bluestems, but has slightly less drought tolerance than some (Coyne and Bradford 1985).

These species possess good forage potential for the southern Great Plains. These grasses respond well to fertilization, are drought and heat tolerant for the most part, withstand close grazing, and are palatable to cattle if old residual growth is not allowed to accumulate. Nutritive value of new growth of old-world bluestem can be similar to native bluestem species (Harmoney and Hickman 2012). However, old world bluestems are not closely related to the native big and little bluestem grasses found throughout the Great Plains. In the Southern Great Plains, several warm-season species are well-adapted to many of the same areas.

The old-world bluestems are extremely persistent plants that do well under dry and arid conditions. Often, they perform better than some native grasses under arid conditions (Eck and Sims 1984). These plants are also prolific seed producers. They accumulate seed banks that can result in new plants years after efforts to convert stands to other species. The old-world bluestems can reduce the growth and vigor of other grasses and legumes that are of higher quality and more palatable for livestock. Robertson et al. (2013) cited old world bluestem negative effects on plant biodiversity, insects, and wildlife. Some old-world bluestems also have a negative impact on growth of native grasses by altering soil microbiology (Wilson et al. 2012). As such, the old-world bluestems are often considered noxious, and at best invasive, when found in native rangelands.

Recently released cultivars of yellow bluestem (Bothriochloa spp.) (Table 17.2) have replaced 'King Ranch' bluestem because of its lack of **winterhardiness** and high susceptibility to leaf rust. No specific cultivars of caucasian bluestem are available.

Dichanthium species have limited interest in the US due primarily to lack of winterhardiness. 'Kleberg' bluestem is well-adapted to South Texas. However, it has poor seed quality (Mutz and Drawe 1983). 'Pretoria 90' has good quality (Conrad and Holt 1983) but produces little viable seed.

Panicgrasses

Switchgrass

Switchgrass is a native, perennial tallgrass that can grow on a variety of sites ranging from sand to clay-loam soils. Switchgrass also tolerates soil pH values in the range from 4.9 to 7.5 (Duke 1978). Switchgrass resembles a loose bunchgrass but has the capability to form a sod due to numerous short rhizomes. It has an erect growth habit and can range in height from 0.5 to 2.0 m tall. Nearly half of the tillers produce a fertile seedhead when adequate moisture is available.

Switchgrass has been separated into lowland and upland types. Lowland types are somewhat taller, coarser, and more resistant to rust (*Puccinia graminis*). Lowland types also have a more bunch-type growth habit with a more rapid growth rate than the upland types. Based on these type descriptions, lowland types can be found on floodplains and similar areas, whereas upland types are found in areas not subject to periodic flooding. Most switchgrass cultivars are either tetraploids or hexaploids (Riley and Vogel 1982). In the Great Plains, switchgrass has been planted in monocultures and mixtures since the 1950s. Recently, its use has increased in popularity in the central and eastern US as a source of summer pasture and potential as s biofuel (Vogel 2004). Numerous cultivars are commercially available (Table 17.1).

Kleingrass

Kleingrass is an introduced bunchgrass native to South and East Africa. The species has much genetic diversity (Lloyd and Thompson 1978), but the species adapted to the US represent only a small fraction of the diversity. The type adapted to Oklahoma and Texas is leafy (0.8–1.4 m tall), fine-stemmed, with variable coloration, pubescence, and growth habit. Kleingrass has an indeterminate growth habit and can produce tillers throughout the growing season. Thus, it can tolerate close, continuous grazing (Evers and Holt 1972) plus have excellent regrowth capacity. Current cultivars are listed in Table 17.2.

The Gramas

Blue Grama

Blue grama is a native bunchgrass with culm heights reaching 26–60 cm. It consists of primarily basal leaves with a large number of vegetative **shoots** of which only a small number become reproductive. It is adapted to the shortgrass region from Texas to Canada and into parts of the desert grassland. In northern areas, it is less productive than the associated cool-season grasses. However, in the

Table 17.2 Warm-season perennial grass cultivars

Species	Cultivars[a]
Bluestem, little	Aldous, Blaze, Camper, Cimarron, Pastura
Bluestem, old world	
Bothriochloa spp.	Caucasian or common, El Kan, Ganada, King Ranch, Plains, WW-B.Dahl, WW-Iron Master, WW-Spar
Dichanthium spp.	Angleton, Gordo, Kleburg, Medio, Pretoria 90, T-587
Buffalograss (seeded)	Bison, Comanche, Plains, Texoka, Topgun, Sharp's Improved
Dropseed, sand	N/A
Galleta	Viva
Grama, black	Nogal
Grama, blue	Alma, Hachita, Lovington
Grama, hairy	N/A
Grama, sideoats	Butte, El Reno, Haskell, Killdeer, Niner, Pierre, Premier, Trailway, Vaughn
Kleingrass	Selection 75, Verde
Lovegrass, boer	A-67 or common, A-84 or common, Catalina
Lovegrass, lehmann	A-68 or common, Kuivato, Puhuima
Lovegrass, sand	Bend, Mason, Neb. 27
Lovegrass, weeping	A-67 or common, Ermelo, Morpa, Renner
Sacaton, big	N/A
Sandreed, prairie	Goshen, Pronghorn

[a]Cultivars followed by "common" are usually available as common seed, i.e. cultivar not stated. All common seed may not be of that particular cultivar. Most listed cultivars are commercially available.

southern areas, it often occurs in nearly pure stands. It is well adapted to drought and close grazing.

Black Grama

Black grama is the most important native grassland species in the desert-grassland region. It is best adapted to coarse-textured soils and tends to grow in open stands. Leaves are short, thin, and inconspicuous with culms reaching 20–60 cm in height. Though being stoloniferous, most consider it as a bunchgrass because it produces few stolons. Individual clumps usually spread very slowly by **tillering**. 'Nogal' is the only cultivar available.

Stands of black grama can be easily damaged by prolonged drought or heavy, continuous stocking. Extensive stand damage is difficult to repair because stolons are extremely short (Valentine 1970). Average lifespan for black grama stands in south central New Mexico was only 2.2 year (Wright and Van Dyne 1976).

Sideoats Grama

Sideoats grama is a native species and the most widely distributed member of the genus. It is usually found in mixed stands. Sideoats grama has been divided into two distinct taxonomic groups based on growth form. *Bouteloua curtipendula* var. *curtipendula* is found throughout both the tallgrass and shortgrass regions (Gould 1959) and is primarily rhizomatous. *B. curtipendula* var. *caespitosa* has a bunch-type growth form and is not rhizomatous. The rhizomatous type has developed the ecotypic response to photoperiod that is commonly observed in many warm-season grass species that are native to the US (Hopkins 1941). Plants with the bunch-type growth form do not exhibit this photoperiod response and flower when moisture and temperature allow (McMillan 1961).

Cultivars of both types have been released, but bunch-type cultivars are no longer available. The great quantity of stems produced naturally to increase seed production results in lower palatability compared with low-stem production of the rhizomatous types. Cultivars of sideoats grama are available (Table 17.2).

Hairy Grama

Hairy grama typically grows in association with both blue grama and sideoats grama. Among the three, hairy grama is somewhat more tolerant of drought and low fertility. There are no cultivars available.

Lovegrasses

Sand Lovegrass

Sand lovegrass is the most important lovegrass of those native to the US. As its name implies, it is found only on sandy soils in the southern and central portions of the tallgrass prairie region. It has greater palatability than the introduced lovegrasses, but does not tolerate continuous, close grazing (Moser and Perry 1983). Where adapted, it is frequently included in grass blends for plantings on sandy sites. Cultivars are listed in Table 17.2.

Weeping Lovegrass

Weeping lovegrass is an introduced forage species native to Africa (Crider 1945) and is part of a large apomictic complex. This bunchgrass with long, narrow basal leaves has a mature plant height of 1–1.5 m, is winter hardy to northern Oklahoma and is best adapted to sandy soils. It has a long growing season for a warm-season grass and is extremely important in regions of Texas and Oklahoma where few cool-season perennial grasses are adapted.

Weeping lovegrass has relatively poor forage quality that can be overcome with intensive management (McIlvain and Shoop 1970; Dahl and Cotter 1984), such as fertilization with low rates of N and high stocking rates. Moreover, prescribed burning to remove accumulated litter can increase weeping lovegrass production under favorable growing conditions (McFarland and Mitchell 2000). Few improved cultivars are available (Table 17.2). However, 'Morpa' has a more favorable lignin: cellulose ratio that has produced greater animal gains than common lovegrass (Voigt et al. 1970).

Boer Lovegrass

This species was introduced from southern Africa. It has greater drought tolerance, but less cold tolerance than weeping lovegrass and grows to a height of 0.7–1.0 m. A few sexual strains of boer lovegrass will hybridize with weeping lovegrass (Voigt and Bashaw 1972), which demonstrates the close relationship between these two species. Boer lovegrass is the most palatable member of *Eragrostis curvula* and has been used extensively in revegetation of desert-shrubland regions (Cox et al. 1982).

Lehmann Lovegrass

Lehmann lovegrass is an **introduced species** but is distinctly different from boer and weeping lovegrass in that it has decumbent stems that root from the nodes. It also does not form the typical crown of the *E. curvula* complex. The weak, stoloniferous growth habit results in formation of an open sod. Lehmann lovegrass is easier to establish than boer lovegrass, but it also has less cold tolerance. Where adapted in the desert-grassland region, lehmann lovegrass persists and even expands from new seedings. It is also capable of invading stands of mesquite (*Prosopis glandulosa* Torr.) and many grasses (Cable 1971).

Other Perennial, Warm-season Grasses

Indiangrass

Indiangrass is a native, warm-season tallgrass with short rhizomes (Mitchell and Vogel, 2004). The rhizomes are generally shorter than 30 mm (McKendrick et al. 1975) resulting in a loose, bunch-type growth habit. Indiangrass ranges from 0.5 to 2.0 m tall with a brown- to black-colored compact panicle from 10- to 30-cm long.

Indiangrass is unique among most native warm-season grasses in that a considerable number of tillers are classified as biennial (McKendrick et al. 1975). Thus, vegetative tillers overwintering in one year will become reproductive the following year. However, the tillers do not appear to be obligate biennials since indiangrass will flower during the seeding year. Indiangrass is also distinctive among the native warm-season grasses in that it is **quiescent.** In dry conditions during growth, it can become dormant, with restored growth and development when moisture becomes available.

Indiangrass begins spring growth at about the same time as both big bluestem and switchgrass, but it does not mature as quickly. It normally flowers four to six weeks following floral emergence of switchgrass (McKendrick et al. 1975).

Recent genetic improvements in forage yield and nutritive value of indiangrass (Vogel et al. 2010), along with its wide range of adaptation, should increase its use for forage production, conservation practices, and reclamation projects in the central Great Plains. Currently, indiangrass is used primarily as a component in warm-season pasture and is almost always seeded in mixtures with big bluestem, switchgrass, and other C_4 grasses. Current cultivars are listed in Table 17.2.

Prairie Sandreed

Prairie sandreed is a native species limited to sandy soils of the central to northern Great Plains. Though it can be very productive, it is not tolerant of repeated defoliations (Mullahey et al. 1991). Nutritive value of prairie sandreed is lower than sand bluestem and more similar to little bluestem (Burzlaff 1971). Establishment of prairie sandreed can be difficult (Masters et al. 1990), but success can be improved by inoculating with vesicular-arbuscular mycorrihizae (Brejda et al. 1993).

Buffalograss

Buffalograss is a native species of the shortgrass region and is found from Texas to South Dakota. It is better adapted to soils having a high clay content. Thus, buffalograss is not well adapted to sandy soils. It is a low-growing species with drought resistance and spreads by numerous stolons that form a dense sod. Buffalograss can withstand heavy, close defoliation and can provide excellent forage for grazing. However, it is less productive than other warm-season grasses.

Buffalograss persists well with minimum management. This, coupled with its drought tolerance, has led to its increased use as a turfgrass. Though most new plantings are from seed, some turf cultivars are propagated vegetatively. Cultivars available are listed in Table 17.2.

Buffelgrass

This species is native to southern Africa and India and was introduced into Texas in the mid-1940s. It is best adapted to dry areas, such as semi-tropical arid regions, because it is deep-rooted which results in a high level of drought tolerance. Though it has importance worldwide, it has minimal use in the US because of limited cold tolerance. Like Old-World bluestems, it can become invasive.

Sand Dropseed

This native grass is the most widely distributed *Sporobolus* species in the US, but it is particularly important in the Great Plains and Southwest. This short-lived perennial bunchgrass propagates easily by producing many very small seeds. On abandoned, unseeded cropland, sand dropseed has produced favorable animal gains because it is readily accepted by grazing livestock.

Tobosagrass

Tobosagrass is a native, productive species that spreads by rhizomes and occurs predominantly in desert grassland and areas of the shortgrass region. It grows anytime during the frost-free period if moisture is available. Culms reach 30–75 cm tall, but produce few, if any, fertile seedheads.

Galleta

This native species is similar to tobosagrass but occurs across a region from Texas to Arizona and northward to Wyoming. It is adapted to the finer-textured soils that receive runoff from adjacent areas (Voigt and MacLauchlan 1985).

Curly Mesquite

Curly mesquite is a native, perennial, warm-season shortgrass that grows as a sod-forming grass that spreads by slender stolons. It is drought resistant and can withstand close, frequent defoliation (Voigt and MacLauchlan 1985). Plants grow to 10–25 cm tall. Though often confused with buffalograss, it produces male and female flowers on the same plant.

References

Aiken, S.G. and Darbyshire, S.J. (1990). *Fescue Grasses of Canada*. Publ. 1844/E, Agriculture Canada. Lethbridge, Alberta: Agricultural Research Station.

Alderson, J.A. and Sharp, W.C. (1994). *Grass Varieties in the United States*, USDA Agriculture Handbook 170. Washington, DC: United States Government Printing Office.

Anderson, B.E. (2000). Use of warm-season grasses by grazing animals. In: *Native Warm-Season Grasses: Research Trends and Issues*, CSSA Special Publications 30 (eds. K.J. Moore and B.E. Anderson), 147–158. Madison, WI: CSSA and ASA.

Asay, K.H. (1987). Revegetation in the sagebrush ecosystem. In: *Integrated Pest Management: State-of-the-Art in the Sagebrush Ecosystem*, USDA-ARS Publications 50 (ed. J.A. Onsager), 19–27. Washington, DC: United States Government Printing Office.

Asay, K.H. (1995). Wheatgrasses and wildryes: The perennial Triticeae. In: *Forages Vol. I: An Introduction to Grassland Agriculture*, 5e (ed. R.F Barnes, D.A. Miller, and C.J. Nelson), 373–394. Ames, IA: Iowa State University Press.

Asay, K.H. and Jensen, K.B. (1996a). Wheatgrasses. In: *Cool-Season Forage Grasses*, Agronomy Monograph 34 (ed. L.E. Moser, M.D. Casler and D.R. Buxton), 691–724. Madison, WI: ASA.

Asay, K.H. and Jensen, K.B. (1996b). Wildryes. In: *Cool-Season Forage Grasses*, Agronomy Monograph 34 (ed. L.E. Moser, M.D. Casler and D.R. Buxton), 725–748. Madison, WI: ASA.

Asay, K.H. and Knowles, R.P. (1985). The wheatgrasses. In: *Forages: The Science of Grassland Agriculture*, 4e (ed. M.E. Heath, R.F Barnes and D.S. Metcalfe), 166–176. Ames, IA: Iowa State University Press.

Asay, K.H. and Mayland, H.F. (1990). Genetic variability for elements associated with grass tetany in Russian wildrye. *J. Range Manage.* 43: 407–411.

Asay, K.H., Jensen, K.B., Dewey, D.R., and Hsiao, C.H. (1990). Genetic introgression among 6X, 4X, and 2X ploidy levels in crested wheatgrass. In: *Agronomy Abstracts*, 79. Madison, WI: ASA.

Asay, K.H., Johnson, D.A., Jensen, K.B. et al. (1995). Registration of 'Vavilov' Siberian crested wheatgrass. *Crop Sci.* 35: 1510.

Barker, W.T. and Whitman, W.C. (1988). Vegetation of the northern Great Plains. *Rangelands* 10: 266–272.

Barkworth, M.E. and Dewey, D.R. (1985). Genomically based genera in the perennial Triticeae of North America: Identification and membership. *Am. J. Bot.* 72: 767–776.

Beetle, A.A. (1955). Wheatgrasses of Wyoming. *Wyoming Agric. Exp. Stn. Bull.* 336.

Berdahl, J.D. and Redfearn, D.D. (2007). Grasses for arid and semiarid areas. In: *Forages Vol. II: The Science of Grassland Agriculture*, 6e (eds. R.F Barnes, C.J. Nelson and K.J. Moore et al.), 221–244. Ames, IA: Blackwell Publishing.

Boe, A., Keeler, K.H., Normann, G.A., and Hatch, S.L. (2004). Indigenous bluestems of the western hemisphere. In: *Warm-Season (C_4) Grasses*, Agronomy Monograph 45 (eds. L.E. Moser and L.E. Sollenberger), 873–908. Madison, WI: ASA, CSSA, SSSA.

Bowden, W.M. (1965). Cytotaxonomy of the species and interspecific hybrids of the genus *Agropyron* in Canada and neighboring areas. *Can. J. Bot.* 43: 1421–1428.

Brejda, J.J., Yocom, D.H., Moser, L.E., and Waller, S.S. (1993). Dependence of 3 Nebraska sandhills warm-season grasses on vesicular-arbuscular mycorrhizae. *J. Range Manage.* 46: 14–20.

Brown, W.V. (1952). The relationship of soil moisture to cleistogamy in *Stipa leucotricha*. *Bot. Gaz.* 112: 438–444.

Burzlaff, D.E. (1971). Seasonal variation of the in vitro dry matter digestibility of three sandhill grasses. *J. Range Manage.* 24: 60–63.

Cable, D.R. (1971). Lehmann lovegrass on the Santa Rita Experimental Range, 1937-1968. *J. Range Manage.* 24: 17–21.

Call, C.A. and Spoonts, B.O. (1989). Characterization and germination of chasmogamous and basal auxiliary cleistogamous florets in Texas wintergrass. *J. Range Manage.* 42: 111–119.

Carlson, J.R. and Dewey, D.R. (1987). Characterization of *Elymus lanceolatus* ssp. *wawawai* and its potential as a range forage grass. In: *Agronomy Abstracts*, 59. Madison, WI: ASA.

Casler, M.D. (2012). Switchgrass breeding, genetics, and genomics. In: *Switchgrass, Green Energy and Technology* (ed. A. Monti), 29–53. London: Springer-Verlag https://doi.org/10.1007/978-1-4471-2903-5_2.

Casler, M.D. and Goodwin, W.H. (1989). Genetic variation for rhizome growth traits in *Elytrigia repens* (L.) Nevski. In: *Proceedings of the 16th International Grassland Congress*, Nice, France. 4-11 Oct. 1989, 339–340. Lusignan, France: French Grassland Society.

Casler, M.D. and Goodwin, W.H. (1998). Agronomic performance of quackgrass and hybrid wheatgrass populations. *Crop Sci.* 38: 1369–1377.

Conrad, B.E. and Holt, E.C. 1983. Year-round grazing of warm-season perennial grasses. Texas Agricultural Experiment Station. MP-1540.

Cornelius, D.R. (1965). Latitude as a factor in wheatgrass variety response on California rangeland. In: *Proceedings of the 9th International Grassland Congress*, vol. 1, 471–473. San Paulo, Brazil. 8-19 Jan. 1965.

Cox, J.R., Morton, H.L., Johnsen, Jr., T.N. et al. 1984. Vegetation restoration in the Chihuahuan and Sonoran Deserts of North America. Rangelands 6 (3): 112–115.

Coyne, P.I. and Bradford, J.A. (1985). Some growth characteristics of four old world bluestems. *J. Range Manage.* 38: 27–33.

Crider, F.J. (1945). *Three Introduced Lovegrasses for Soil Conservation.* United States Department of Agriculture Circular No. 730.

Cronquist, A., Holmgren, A.H., Holmgren, N.H. et al. (1977). *Intermountain Flora, Vascular Plants of the Intermountain West. Vol. 6: The Monocotyledons.* New York: Columbia University Press.

Currie, P.O. and Smith, D.R. (1970). *Response of Seeded Ranges to Different Grazing Intensities*, USDA-Forest Service Production Research Report 112. Washington, DC: United States Government Printing Office.

Daer, T. and Willard, E.E. (1981). Total nonstructural carbohydrate trends in bluebunch wheatgrass related to growth and phenology. *J. Range Manage.* 34: 377–379.

Dahl, B.E., and Cotter, P.F. (1984). Management of Weeping Lovegrass in West Texas. Texas Technical Range and Wildlife Management. No. 5.

deWet, J.M.J. and Harlan, J.R. (1970). Apomixis, polyploidy, and speciation in *Dichanthium*. *Evolution* 24: 270–277.

Dewey, D.R. (1962). The genomic structure of intermediate wheatgrass. *J. Hered.* 53: 282–290.

Dewey, D.R. (1975). The origin of *Agropyron smithii*. *Am. J. Bot.* 62: 524–530.

Dewey, D.R. (1976). Derivation of a new forage grass from *Agropyron repens* X *Agropyron spicatum* hybrids. *Crop Sci.* 16: 175–180.

Dewey, D.R. (1978). Advanced generation hybrids between *Elymus giganteus* and *E. angustus*. *Bot. Gaz.* 139: 369–376.

Dewey, D.R. (1982). Genomic and phylogenetic relationships among North American perennial Triticeae. In: *Grasses and Grasslands: Systematics and Ecology* (eds. J.R. Estes et al.), 51–58. Norman, OK: University Oklahoma Press.

Dewey, D.R. (1983). New nomenclatural combinations in the North American perennial Triticeae (Gramineae). *Brittonia* 35: 30–33.

Dewey, D.R. (1984). The genomic system of classification as a guide to intergeneric hybridization with the perennial Triticeae. In: *Gene Manipulation in Plant Improvement. Proceedings of the 16th Stadler Genetics Symposium.* Columbia, MO. 19-21 March 1984 (ed. J.P. Gustafson), 209–279. New York: Plenum Publishing Corporation.

Dewey, D.R. (1986). Taxonomy of the crested wheatgrasses. In: *Crested Wheatgrass: Its Values, Problems and Myths. Utah State University Symposium.* Logan, UT. 3-7 Oct. 1983 (ed. K.L. Johnson), 31–44. Logan, UT: Utah State University.

Dewey, D.R. and Asay, K.H. (1975). The crested wheatgrasses of Iran. *Crop Sci.* 15: 844–849.

Dillman, A.C. (1946). The beginnings of crested wheatgrass in North America. *J. Am. Soc. Agron.* 38: 237–250.

Duebbert, H.F., Jacobson, E.T., Higgins, C.F. et al. 1981. Establishment of seeded grasslands for wildlife habitat in the prairie pothole region. USDI-Fish and Wildlife Service. Special Scientific Report – Wildlife No. 234.

Duke, J.A. (1978). The quest for tolerant germplasm. In: *Crop Tolerance to Suboptimal Land Conditions*, American Society of Agronomy Special Publications 32 (ed. G.A. Jung), 1–61. Madison, WI.

Eck, H.V. and Sims, P.L. (1984). Grass species adaptability in the southern high plains, Texas USA: a 36 year assessment. *J. Range Manage.* 37: 211–217.

Evans, R.A. and Young, J.A. (1983). 'Magnar' basin wildrye—germination in relation to temperature. *J. Range Manage.* 36: 395–398.

Evers, G.W. and Holt, E.C. (1972). Effects of defoliation treatments on morphological characteristics and carbohydrate reserves in kleingrass. *Agron. J.* 64: 17–20.

Gillen, R.L. and Berg, W.A. (2005). Response of perennial cool-season grasses to clipping in the southern plains. *Agron. J.* 97: 125–130.

Gould, F.W. (1959). Notes on apomixis in sideoats grama. *J. Range. Manage.* 12: 25–28.

Hafenrichter, A.L., Muellen, L.A., and Brown, R.L. (1949). *Grasses and Legumes for Soil Conservation in the Pacific Northwest*. USDA Misc. Publ. 678. Washington, DC: United States Government Printing Office.

Hafenrichter, A.L., Schwendiman, J.L., Harris, H.L. et al. (1968). *Grasses and Legumes for Soil Conservation in the Pacific Northwest and Great Basin States*, USDA Agriculture Handbook 339. Washington, DC: United States Government Printing Office.

Haferkamp, M.R., MacNeil, M.D., Grings, E.E., and Klement, K.D. (2005). Heifer production on rangeland and seeded forages in the northern Great Plains. *Rangeland Ecol. Manage.* 58: 495–604.

Harmoney, K.R. (2005). Growth responses of perennial cool-season grasses grazed intermittently. Online. *Forage and Grazinglands* https://doi.org/10.1094/FG-2005-0105-01-RS.

Harmoney, K.R. (2007). Persistence of heavily-grazed cool-season grasses in the central Great Plains. Online. *Forage and Grazinglands* https://doi.org/10.1094/FG-2007-0625-01-RS.

Harmoney, K.R. and Hickman, K.R. (2012). Comparing morphological development and nutritive value of Caucasian old world bluestem and native grasses. Online. *Forage and Grazinglands* https://doi.org/10.1094/FG-2012-0127-01-RS.

Harmoney, K. and Jaeger, J. (2019). Tall wheatgrass and western wheatgrass used for complementary cool-season forage systems. *Crop, Forage, and Turfgrass Management*. Online. doi:https://doi.org/10.2134/cftm2018.08.0065.

Hartung, M.E. (1946). Chromosome numbers in *Poa, Agropyron, and Elymus*. *Am. J. Bot.* 33: 516–531.

Hendrickson, J.R., Moser, L.E., Moore, K.J., and Waller, S.S. (1998). Morphological development of warm-season grasses in the Nebraska sandhills. *J. Range. Manage.* 51: 456–462.

Hitchcock, A.S. (1951). *Manual of the Grasses of the United States*, 2e. Rev. by A. Chase. USDA Misc. Publ. 200. Washington, DC: United States Government Printing Office.

Hopkins, H. (1941). Variations in the growth of side-oats grama grass at Hays, Kansas, from seed produced in various parts of the Great Plains region. *Trans. Kans. Acad. Sci.* 44: 86–95.

Jefferson, P.G. and Kielly, G.A. (1998). Reevaluation of row spacing/plant density of seeded pasture grasses for the semiarid prairie. *Can. J. Plant. Sci.* 78: 257–264.

Jensen, K.B., Zhang, Y.F., and Dewey, D.R. (1990). Mode of pollination of perennial species of the Triticeae in relation to genomically defined genera. *Can. J. Plant. Sci.* 70: 215–225.

Johnson, B.L. (1972). Polyploidy as a factor in the evolution and distribution of grasses. In: *The Biology and Utilization of Grasses* (eds. V.B. Youngner and C.M. McKell), 19–35. New York: Academic Press.

Johnson, L.C., Olsen, J.T., Tetreault, H. et al. (2015). Intraspecific variation of a dominant grass and local adaptation in reciprocal garden communities along a US Great Plains' precipitation gradient: implications for grassland restoration with climate change. *Evol. Appl.* 8: 705–723.

Johnston, A. (1962). Chemical composition of range forage plants of the *Festuca scabella* association. *Can. J. Plant. Sci.* 42: 105–115.

Jones, T.A. (1990). A viewpoint on Indian ricegrass: Its present status and future prospects. *J. Range Manag.* 43: 416–420.

Jones, T.A. and Nielson, D.C. (1989). Self–incompatibility in 'Paloma' Indian ricegrass. *J. Range Manage.* 42: 187–190.

Jones, T.A., Neilson, D.C., Ogle, D.G. et al. (1998). Registration of Sand Hollow squirreltail germplasm. *Crop Sci.* 38: 286.

Jones, T.A., Ralphs, M.H., Gardner, D.R., and Chatterton, N.J. (2000). Cattle prefer endophyte-free robust needlegrass. *J. Range Manage.* 53: 427–431.

Kantar, M.B., Tyl, C.E., Dorn, K.M. et al. (2016). Perennial grain and oilseed crops. *Annu. Rev. Plant Biol.* 67: 703–729.

Karn, J.F., Hofmann, L., and Lorenz, R.J. (1983). Russian wildrye nutritional adequacy and chemical composition. *Agron. J.* 75: 242–246.

Karn, J.F., Ries, R.E., and Hofmann, L. (1999). Season-long grazing of seeded cool-season pastures in the northern Great Plains. *J. Range Manage.* 52: 235–240.

Kilcher, M.R. (1969). Establishment and maintenance of seeded forage crops. In: *Canadian Forage Crops Symposium* (ed. K.F. Nielsen), 89–104. Saskatoon, SK: Modern Press.

Kilcher, M.R. (1982). Beef production from grass-alfalfa pastures grown in different stand patterns in a semiarid region of the Canadian prairies. *Can. J. Plant. Sci.* 62: 117–124.

Knipfel, J.E. and Heinrichs, D.H. (1978). Nutritional quality of crested wheatgrass, Russian wild ryegrass, and Altai wild ryegrass throughout the grazing season in southwestern Saskatchewan. *Can. J. Plant. Sci.* 58: 581–582.

Lawrence, T. (1976). Prairieland, Altai wild ryegrass. *Can. J. Plant. Sci.* 56: 991–992.

Lawrence, T. (1977). Productivity of intermediate wheatgrass, crested wheatgrass, and bromegrass as influenced by length of ley and pattern of seeding. *Forage Notes* 22 (1): 43–49.

Lawrence, T. 1987. Altai wild ryegrass. Canada Technical Bulletin No. 1983-3E. Agriculture Canada Research Station.

Lawrence, T. and Ashford, R. (1966). The productivity of intermediate wheatgrass as affected by initial harvest dates and recovery periods. *Can. J. Plant. Sci.* 46: 9–15.

Lawrence, T. and Heinrichs, D.H. (1968). Long-term effects of row spacing and fertilizer on the productivity of Russian wild ryegrass. *Can. J. Plant. Sci.* 48: 75–84.

Lawrence, T. and Ratzlaff, C.D. (1985). Evaluation of fourteen grass populations as forage crops for southwestern Saskatchewan. *Can. J. Plant. Sci.* 65: 951–957.

Lawrence, T., Jefferson, P.G., and Ratzlaff, C.D. (1990). James and Arthur, two cultivars of Dahurian wild ryegrass. *Can. J. Plant. Sci.* 70: 1187–1190.

Lloyd, D.L. and Thompson, J.P. (1978). Numerical analysis of taxonomic and parent-progeny relationships among Australian selections of *Panicum coloratum*. *Queensland J. Agric. Anim. Sci.* 35: 35–46.

Lorenz, R.J. (1986). Introduction and early use of crested wheatgrass in the Northern Great Plains. In: *Crested Wheatgrass: Its Values, Problems and Myths. Symposium Proceedings of the Utah State University*, Logan, UT. 3-7 Oct 1983 (ed. K.L. Johnson), 9–20. Logan, UT: Utah State University.

Lu, B.-R. and von Bothmer, R. (1992). Interspecific hybridization between *Elymus himalayanus* and *E. schrenkianus,* and other *Elymus* species (Triticeae: Poaceae). *Genome* 35: 230–237.

Malinowski, D.P., Hopkins, A.A., Pinchak, W.E. et al. (2003). Productivity and survival of defoliated wheatgrasses in the Rolling Plains of Texas. *Agron. J.* 95: 614–626.

Masters, R.A., Vogel, K.P., Reece, P.E., and Bauer, D. (1990). Sand bluestem and prairie sandreed establishment. *J. Range Manage.* 43: 540–544.

Mayland, H.F., Asay, K.H., and Clark, D.H. (1992). Seasonal trends in forage quality of crested wheatgrass. *J. Range Manage.* 45: 369–374.

McElgunn, J.D. and Lawrence, T. (1973). Salinity tolerance of Altai wild ryegrass and other forage grasses. *Can. J. Plant. Sci.* 53: 303–307.

McFarland, J.B. and Mitchell, R. (2000). Fire effects on weeping lovegrass tiller density and demographics. *Agron. J.* 92: 42–47.

McIlvain, E.H. and Shoop, M.C. 1970. Grazing weeping lovegrass for profit—8 keys. In: Proceedings of the 1st Weeping Lovegrass Symposium, Ardmore, OK, USA (28–29 April 1970).

McKendrick, J.D., Owensby, C.E., and Hyde, R.M. (1975). Big bluestem and indiangrass vegetative reproduction and annual reserve carbohydrate and nitrogen cycles. *Agro-Ecosyst.* 2: 75–93.

McMillan, C. (1961). Nature of the plant community. VI. Texas grassland communities under transplanted conditions. *Am. J. Bot.* 48: 778–785.

Miller, R.F., Seufert, J. M. and Haferkamp, M.R. 1987. The ecology and management of bluebunch wheatgrass (*Agropyron spicatum*): A review. Oregon State University. Agricultural Experiment Station Bulletin No. 669.

Mitchell, R.B. and Anderson, B.E. 2008. Switchgrass, big bluestem, and indiangrass for grazing and hay. University of Nebraska. Lincoln Extension No. 1908.

Mitchell, R. and Vogel, K.P. (2004). Indiangrass. In: *Warm-Season (C_4) Grasses*, Agronomy Monograph 45 (eds. L.E. Moser and L.E. Sollenberger), 937–953. Madison, WI: ASA, CSSA, SSSA.

Mitchell, R.B., Moore, K.J., Moser, L.E. et al. (1997). Predicting developmental morphology in switchgrass and big bluestem. *Agron. J.* 89: 827–832.

Mitchell, R.B., Vogel, K.P., Klopfenstein, T. et al. (2005). Grazing evaluation of big bluestems bred for improved forage yield and digestibility. *Crop Sci.* 45: 2288–2292.

Moser, L.E. and Perry, L.J. (1983). Yield, vigor, and persistence of sand lovegrass (*Eragrostis trichodes* [Nutt.] Wood) following clipping treatments on a range site in Nebraska. *J. Range Manage.* 36: 236–238.

Mullahey, J.J., Waller, S.S., and Moser, L.E. (1991). Defoliation effects on production and morphological development of little bluestem. *J. Range Manage.* 43: 497–500.

Mutz, J.L. and Drawe, D.L. (1983). Clipping frequency and fertilization influence herbage yields and crude protein of four grasses in south Texas. *J. Range Manage* 36: 582–585.

Newell, L.C. and Moline, W.J. 1978. Forage quality evaluations of twelve grasses in relation to season of grazing. University of Nebraska. Research Bulletin No. 283.

Newell, L.C. and Peters, L.V. (1961). Performance of hybrids between divergent types of big bluestem and sand bluestem in relation to improvement. *Crop Sci.* 1: 370–373.

Riley, R.D. and Vogel, K.P. (1982). Chromosome numbers of released cultivars of switchgrass, indiangrass, big bluestem, and sand bluestem. *Crop Sci.* 22: 1081–1083.

Robertson, J.H. (1976). The autecology of *Oryzopsis hymenoides*. *Mentzelia* 2 (18–21): 25–27.

Robertson, S., Hickman, K.R., Harmoney, K.R., and Leslie, D.M. Jr. (2013). Combining glyphosate with burning or mowing improves control of yellow bluestem (*Bothriochloa ischaemum*). *Rangeland Ecol. Manage.* 66: 376–381.

Rogler, G.A. (1941). Russian wild-rye, *Elymus junceus* Fisch. *J. Am. Soc. Agron.* 33: 266–267.

Rogler, G.A. (1960). Relation of seed dormancy of Indian ricegrass [*Oryzopsis hymenoides* (Roem. & Schult.) Ricker] to age and treatment. *Agron. J.* 52: 470–473.

Rogler, G.A. (1973). The wheatgrasses. In: *Forages: The Science of Grassland Agriculture* 3e (ed. M.E. Heath, D.S. Metcalfe and R.F Barnes), 221–230. Ames, IA: Iowa State University Press.

Rogler, G.A. and Lorenz, R.J. (1983). Crested wheatgrass—early history in the United States. *J. Range Manage.* 36: 91–93.

Rogler, G.A. and Schaaf, H.M. 1963. Growing Russian wildrye in the western states. United States Department of Agriculture Leaflet No. 313.Washington, DC.

Scheetz, J.G., Majerus, M.E., and Carlson, J.R. (1981). Improved plant materials and their establishment to reclaim saline seeps in Montana. In: *Agronomy Abstracts*, 96. Madison, WI: ASA.

Sedivec, K.K. and Barker, W.T. (1997). *Selected North Dakota and Minnesota Range Plants*. Fargo, ND: Extension Service, North Dakota State University.

Smoliak, S. and Johnston, A. (1980). Russian wildrye lengthens the grazing season. *Rangelands* 2: 249–250.

Stubbendieck, J. and Jones, T.A. (1996). Other cool-season grasses. In: *Cool-Season Forage Grasses*, Agronomy Monograph 34 (eds. L.E. Moser et al.), 765–780. Madison, WI: ASA.

Stubbendieck, J., Hatch, S.L., and Dunn, C.D. (2017). *Grasses of the Great Plains*. College Station, TX: Texas A & M University Press.

Tzvelev, N.N. (1983). Tribe 3. Triticeae Dum. In: *Grasses of the Soviet Union*. (Translated from Russian) (ed. A.A. Fedorov), 146–245. New Delhi, India: Amerind Publishing Co. Pvt. Ltd.

Valentine, K.A. 1970. Influence of Grazing Intensity on Improvement of Deteriorated Black Grama Range. New Mexico Agricultural Experiment Station Bulletin No. 553.

Vallentine, J.F. (1989). *Range Development and Improvements*, 3e. Provo, UT: Brigham Young University Press.

Vogel, K.P. (2004). Switchgrass. In: *Warm-Season (C_4) Grasses*, Agronomy Monograph 45 (eds. L.E. Moser and L.E. Sollenberger), 561–588. Madison, WI: ASA, CSSA, SSSA.

Vogel, K.P., Mitchell, R.B., Gorz, H.J. et al. (2010). Registration of 'Warrior', 'Scout', and 'Chief' indiangrass. *J. Plant Regist.* 4: 115–122.

Voigt, P.W. and Bashaw, E.C. (1972). Apomixis and sexuality in *Eragrostis curvula*. *Crop Sci.* 12: 843–847.

Voigt, P.W. and MacLauchlan, R.S. (1985). Native and other western grasses. In: *Forages: The Science of Grassland Agriculture*, 4e (eds. M.E. Heath, R.F Barnes and D.S. Metcalfe), 177–187. Ames, IA: Iowa State University Press.

Voigt, P.W. and Sharp, W.C. (1995). Grasses of the plains and southwest. In: *Forages Vol. I: An Introduction to Grassland Agriculture*, 5e (eds. R.F Barnes, D.A. Miller and C.J. Nelson), 395–408. Ames, IA: Iowa State University Press.

Voigt, P.W., Kneebone, W.R., McIlvain, E.H. et al. (1970). Palatability, chemical composition, and animal gains from selections of weeping lovegrass, *Eragrostis curvula* (Schrad.) Nees. *Agron. J.* 62: 673–676.

Weaver, J.E. (1968). *Prairie Plants and Their Environment*. Lincoln, NE: University of Nebraska Press.

Weintraub, F.C. (1953). *Grasses Introduced into the United States*, USDA-FS Agriculture Handbook 58. Washington, DC: United States Government Printing Office.

Willms, W.D. and Fraser, J. (1992). Growth characteristics of rough fescue (*Festuca scabrella* var. *campestris*) after three years of repeated harvest at scheduled frequencies and heights. *Can. J. Bot.* 70: 2125–2129.

Wilson, G.W.T., Hickman, K.R., and Williamson, M.M. (2012). Invasive warm-season grasses reduce mycorrhizal root colonization and biomass production of native prairie grasses. *Mycorrhiza* 22: 327–336.

Wright, G.R. and Van Dyne, G.M. (1976). Environmental factors influencing semidesert grassland perennial grass demography. *Southwest. Nat.* 21: 259–274.

Young, J.A., Clements, C.D., and Jones, T.A. (2003). Germination of seeds of robust needlegrass. *J. Range Manage.* 56: 247–250.

CHAPTER 18

Warm-Season Grasses for Humid Areas

Lynn E. Sollenberger, Distinguished Professor, *Agronomy, University of Florida, Gainesville, FL, USA*
João M.B. Vendramini, Associate Professor, *Agronomy, Range Cattle Research and Education Center, University of Florida, Ona, FL, USA*
Carlos G.S. Pedreira, Associate Professor, *Animal Science, University of São Paulo, Brazil*
Esteban F. Rios, *Agronomy, University of Florida, Gainesville, FL, USA*

Importance, Adaptation, and Use in Production Systems

Warm-season grasses predominate in subtropical and tropical climates and are the primary feed source for many livestock. They utilize the **C_4** carbon fixation pathway, associated with high-growth rates, high water and N-use efficiencies, and relatively low-nutritive value compared with **C_3** grasses. Conditions of high temperature, water stress, and saline soils generally favor C_4 over C_3 species, but C_4 grasses may even dominate temperate ecosystems to 50 °N latitude. In the US, C_4 grasses contribute mainly in the humid Southeast and as components of native grasslands across the Great Plains.

Characteristics

Anatomy and Morphology

The C_4 grasses have unique leaf anatomy closely associated with their photosynthetic metabolism (Dangler and Nelson 1999). This "Kranz," or wreath, anatomy includes thin-walled **mesophyll** cells that form a tight concentric ring around thick-walled **bundle sheath** cells that surround the **vascular bundles** (Sage 1999). This structure limits microbial access to cellular surface area, required for fiber digestion, thereby slowing and/or reducing degradability in the **rumen**. The C_4 grasses also exhibit greater integrity of the **epidermis** due to a "dove-tail," sinuous joint between cells, hindering the splitting of the leaf lengthwise during chewing and rumination (Wilson 1993). Thick-walled **sclerenchyma** cells often connect the epidermis to either the **vascular tissue** or the bundle sheath, forming a "girder" structure that confers rigidity to the leaf blade and hinders physical breakdown of tissue in the rumen (Wilson 1993). Dry matter **digestibility** of C_4 grasses averages 80–150 g kg^{-1} lower than C_3 grasses (Minson 1980). Concentration of **neutral detergent fiber** (NDF) in C_4 grasses is high and NDF digestibility is a major factor affecting overall forage digestibility of C_4 grasses (Oba and Allen 1999).

Tropical and **subtropical** grasses vary widely in morphology and growth habit, ranging from low-growing types such as bahiagrass and bermudagrass, which seldom grow taller than 60 cm (Gates et al. 2004), to the tall cultivars of elephantgrass, whose clumps may reach 7 m in height (Hanna et al. 2004b). Low-growing species are often stoloniferous and spread aggressively, with some producing rhizomes, which often results in greater cold tolerance. Root systems are normally fibrous

Forages: The Science of Grassland Agriculture, Volume II, Seventh Edition.
Edited by Kenneth J. Moore, Michael Collins, C. Jerry Nelson and Daren D. Redfearn.

and relatively shallow. Leaf:stem ratio of C_4 grasses can be significantly affected by daylength. Whereas many species are day-length neutral and produce stems and flowers throughout the growing season, others such as elephantgrass and guineagrass are **short-day plants** with prolonged vegetative tiller elongation, often resulting in low leaf percentage.

Physiology

In C_4 grasses, the primary carboxylation enzyme is phosphoenolpyruvate (**PEP**)-carboxylase, which has high affinity for CO_2 and discriminates against O_2 in the mesophyll cells. Four–carbon acids are CO_2 conveyors (Moore et al. 2004) to the bundle sheath cells where carbohydrate is synthesized in the Calvin cycle. **Photorespiration** and the associated loss of C is minimized due to the enriched concentration of CO_2 around Rubisco in the bundle **sheath**, resulting in photosynthetic rates of C_4 grasses that are nearly double those of C_3 grasses (Moore et al. 2004). Because of their highly efficient photosynthetic metabolism, C_4 water-use efficiency is twice that of C_3 grasses. This superiority is accentuated as temperatures increase from 20 (~2 x) to 35 °C (10 x) (Long 1999). Tropical grasses invest about half as much soluble protein in **Rubisco** as C_3 grasses (Brown 1978). Greater N-use efficiency is associated with better adaptation of C_4 grasses to low-N soils (Brown 1978) and reduced forage **nutritive value** (Kiniry et al. 1999). Across C_4 species, **crude protein** (CP) concentration averaged 40–60 g kg^{-1} less than for C_3 species, and occurrence of CP deficiency is much greater among livestock fed C_4 grasses (Minson 1980).

Reproductive Characteristics

Most tropical grass species have perfect flowers and, reproduce sexually through cross-pollination (allogamy) and/or asexually through **apomixis**. Tropical grasses have developed several mechanisms to foster allogamy over autogamy (self-pollination) (Sleper and Poehlman 2006). They include **self-incompatibility**, cytoplasmic-nuclear male sterility, genetic male sterility, and floral features such as **stigmas** that mature and exert before anthers (Sleper and Poehlman 2006). Mechanisms fostering allogamy are very useful tools for breeding tropical grasses that have small flowers and are difficult to emasculate for hand crossing.

Apomixis is an alternative route to sexual reproduction that creates individuals genetically identical to the female parent. This clonal mode of reproduction involves three alterations to the sexual female development (Asker & Jerling 1992): (i) unreduced **embryo** sac formation; (ii) the unreduced egg cell develops into an embryo without fertilization; and (iii) in most grass species, a male **germ** cell must fertilize the central cell for proper **endosperm** formation. Plants that reproduce exclusively by apomixis (obligate apomicts) are unusual in nature, and most apomictic plants maintain the ability to reproduce sexually (Asker and Jerling 1992). From an agronomic perspective, obligate apomixis provides a unique mechanism for developing superior hybrids and preserving them indefinitely without vigor loss. Ongoing research is studying relationships between structure, position, and function of recently-discovered apomixis-linked genes and the transfer of those genes to other crop species. The manipulation of the trait is currently having a direct impact on breeding of apomictic forage grasses (Jank et al. 2014).

A common feature of almost all tropical grasses is the link between mode of reproduction and cytogenetic characteristics (Comai 2005). Many tropical grasses are polyploids. In a group of related grass species of different ploidy levels, diploid plants are exclusively sexual, while polyploids could either be sexual and apomictic (facultative apomixis), or exclusively apomictic (obligate apomixis) (Naumova 1997).

Management and Utilization

Establishment

Factors affecting establishment of C_4 grasses are generally the same as other forage grasses. Many improved C_4 species (often hybrid cultivars), however, are propagated vegetatively because of infertile seed, poor seedling vigor, or because they do not breed true from seed. Propagation can be from tops, sprigs, stem cuttings, stolons, and rhizomes, provided that they are well-fertilized, vigorous, and mature, with viable growing points (Masters et al. 2004).

Fertilization

The combination of low soil nutrient levels, efficient nutrient extraction capability, and high total yield potential allows for significant yield responses when C_4 grasses are fertilized (Mathews et al. 2004). The C_4 grasses are usually more tolerant to acid soils than C_3 species. Reduction in herbage accumulation in soils with low pH is frequently associated with Al and Mn toxicity. The response of C_4 grasses to liming is variable, but there is a relationship between liming and response to fertilization. Adjei and Rechcigl (2004) observed 30% less herbage accumulation and less root-rhizome mass and reduced persistence when bahiagrass was fertilized without liming rather than with liming.

Nitrogen is the most limiting nutrient for growth of C_4 grasses and CP levels can be deficient for grazing livestock (Lima et al. 1999). Greater herbage accumulation and CP accompany increasing N fertilization (Wilkinson and Langdale 1974). The increase in production is associated with greater tiller number, tiller weight, leaves per tiller, and leaf appearance rate (Premazzi et al. 2003; Moreira et al. 2009).

Warm-season grasses have a minimal sufficiency requirement from 1.5 to 2.1 g kg^{-1} of P (Mathews et al. 2004; Silveira et al. 2011). Response to soil-P concentration varies with forage species, soil type, subsoil P, N fertilization, and soil texture (Mathews et al. 1998; Ibrikci et al. 1999). In tropical environments, P deficiencies are common for livestock consuming C_4 grasses.

The minimum sufficiency level for tissue-K concentration in warm-season grasses was reported to be 19 g kg^{-1} (Mathews et al. 2004); however, bermudagrass herbage accumulation did not increase when tissue-K concentration rose above 14 g kg^{-1} (Yarborough et al. 2017). Generally, K concentration exceeds the dietary needs of grazing ruminants, so deficiencies in animals are unlikely.

Micronutrient deficiencies can occur in warm-climate grasslands, but plant responses to micronutrient fertilization are inconsistent (Wright and Whitty 2011). In grazing systems, animal supplementation can be a feasible and cost-effective alternative to fertilization (Silveira et al. 2014). Soil and plant-tissue analyses are important tools to identify micronutrient deficiency.

Grazing Management

Flowering and growth characteristics and anatomic features of **C_4 plants** can make grazing management more challenging than for **C_3 plants**. Most C_3 perennial forage grasses are qualitative long-day plants (obligate long-day plants) that flower exclusively in spring and grow vegetatively throughout the remainder of the season. In contrast, most C_4 perennial grasses are daylength insensitive or short-day plants (Loch et al. 2004a). At low latitudes, where daylength varies little throughout the year, these plants elongate flowering stems throughout the growing season. Other C_4 grasses (e.g., bahiagrass and bermudagrass) are quantitative long-day plants, meaning that flowering is not synchronized and can occur whenever daylength is near or greater than the threshold (Loch et al. 2004a; Moore et al. 2004). Thus, grazing management is made more complicated for many C_4 grasses by flowering stem elongation throughout a significant part of the growing season. Contributing to the challenge of optimal grazing management is rapid growth rate associated with C_4 carbon fixation and high-growth temperatures that accelerate maturation and lignin deposition in cell walls. The anatomy of C_4 grass leaves reduces rate of forage particle size reduction and passage from the rumen (Coleman et al. 2004). These factors collectively increase the importance of timing of grazing and the complexity of managing grazing in C_4 grasslands.

Conservation

Stockpiling

Stockpiling is the most-used method of forage conservation in tropical and subtropical regions due to lesser cost and favorable climatic conditions. Stockpiling allows forage to accumulate in the absence of defoliation for subsequent utilization in a dry or cold season (Ruelke and Quesenberry 1983). Critical management factors include initiation date of stockpiling, duration of stockpiling period, and fertilization (Taliaferro et al. 1987; Evers et al. 2004; Wallau et al. 2015). Though some warm-season grasses like limpograss maintain nutritive value relatively well during stockpiling, supplementation is often needed to meet livestock nutrient requirements.

Hay

The chemical composition and morphology of warm-season grasses and weather-related constraints in humid environments reduce the likelihood of successful hay production when compared with cool-season forages. Drying rate of forage crops generally decreases with increased stem diameter (Clark et al. 1985). The large diameter and high proportion of stems for some C_4 species extends field drying time and increases probability of weather-related dry matter losses (Taliaferro et al. 2004). Frequent rain during the period of maximum forage growth limits the potential for drying forage for hay. Despite these limitations, a large number of warm-season grasses have been used successfully for hay (Sollenberger et al. 2004). Examples include bermudagrass, digitgrass, bahiagrass, stargrass, guineagrass, rhodegrass, and signalgrass.

Silage

Frequent rainfall events during the growing season make haylage or silage important alternative conservation methods to hay. In North America, silage is made primarily from corn and cool-season legumes, while a lesser amount is made from warm-season perennial grasses. High moisture concentration, low water-soluble carbohydrate concentration, and high buffering capacity relative to soluble carbohydrate supply may result in poor quality C_4 grass silage (Sollenberger et al. 2004). Six species of warm-season grass were ensiled after a six-week regrowth, and high pH, reduced lactate concentrations, and a relatively high incidence of mold were observed (Vendramini et al. 2010). Warm-season perennial grass should be wilted to dry matter (DM) concentrations near 400 g kg^{-1} before ensiling. **Wilted silage** stabilizes at a higher pH than wetter material and reduces the incidence of spoilage bacteria (Bates et al. 1989).

Additives such as sugarcane molasses and citrus pulp have been effective for enhancing silage fermentation. Bermudagrass wilted silage treated with sugarcane molasses had greater water-soluble carbohydrate, ***in vitro* true digestibility**, and lactate concentrations, and lower pH than untreated silage (Vendramini et al. 2016). Response to bacterial inoculants has been less consistent than to some other additives for warm-season perennial grasses (Vendramini et al. 2016).

Ecosystem Services

Benefits provided to society by an ecosystem are referred to as ecosystem services (Millennium Ecosystem Assessment

2005). In addition to providing livestock feed, grasslands enhance nutrient cycling, water quality, and soil-C storage (Sollenberger et al. 2019).

Nutrient Cycling

In grazed pastures, most nutrients are cycling in animal excreta or senescing plant material or litter (Dubeux et al. 2007). In warm, humid climates, animals are more likely to congregate around shade and watering points to minimize heat load. This results in a greater proportion of dung and urine being deposited in these areas, negatively affecting efficiency of nutrient cycling. Greater C: N ratio of C_4 than C_3 grass litter reduces nutrient mineralization and may result in nutrient **immobilization**. Half-life of C_4 grass litter decomposition is more than twice that of legumes (Dubeux et al. 2007). Longer litter half-life decreases nutrient turnover and nutrient supply.

Water Quality

Including bermudagrass as a component of year-round forage production systems reduced nitrate leaching to ground water relative to annual cropping systems (Woodard et al. 2002). Maintaining soil cover is critical to reducing rainwater runoff, and sediment and nutrient loss to surface water. Choosing a stocking rate that avoids overgrazing is the most important management factor facilitating accomplishment of these objectives (Sollenberger et al. 2012).

Soil Carbon

Perennial grasslands can be net C sinks (Soussana et al. 2010; Peichl et al. 2011). In warm-climate grasslands, conversion of native rangeland to bahiagrass pasture or bahiagrass-slash pine silvopasture increased soil organic C (Adewopo et al. 2014). Inclusion of legumes in bermudagrass (Wright et al. 2004) or other C_4 grass pastures (Fornara and Tilman 2008) increased soil-C and -N accrual. Grazing of bermudagrass led to greater levels of soil organic-C in the surface 15 cm of soil than ungrazed or hayed grassland, and soil organic-C concentration was greater for low than high stocking rate treatments (Wright et al. 2004; Franzluebbers and Stuedemann 2009). Generally, increased management intensity, association of legumes with grasses, and grazing vs haying of C_4 grasslands increases soil-C.

Perennial Grasses

Environmental adaptation and species characteristics of all forage grasses described in this chapter are shown in Tables 18.1 and 18.2.

Cynodon spp.

Cynodon originated in Africa, but it is cosmopolitan and one of the most widely used genera for forage, turf, and soil conservation. *Cynodon* species are sexual and most have some degree of **self-incompatibility**. Bermudagrass and stargrass are the two most important species, and both are sward-forming perennials (Sollenberger 2008). Both grasses require significant soil nutrient inputs to perform optimally and persist long term. Hybrid bermudagrasses and stargrasses can only be propagated vegetatively. It is recommended that *Cynodon* grasses be harvested at four-week intervals when growth is rapid; longer regrowth periods increase forage accumulation but result in lower nutritive value (Silva et al. 2015b). Attempts to grow legumes with *Cynodon* in the southern US have enjoyed few successes (Burton and DeVane 1992; Sollenberger, 2008).

Bermudagrass

In the US, bermudagrass is grown on approximately 12 million ha and is the most widely used perennial forage in southern states (Redfearn and Nelson 2003). It is best adapted south of a line connecting the southern boundaries of Virginia and Kansas (Burton and Hanna 1995). It is grown widely in tropical and subtropical regions of Africa, Asia, Australia, and the Americas. Bermudagrass is deeply rooted and drought tolerant. It produces rhizomes and is more persistent under grazing than stargrass (Hanna 1992). Bermudagrass grows best when mean daily temperatures are above 24 °C. Freezing temperatures will kill the leaves, but rhizomes survive in the dormant stage over winter, depending on the severity of cold, and begin growing the following spring. Bermudagrass grows best on well-drained soils with pH above 5.5, but the privately released bermudagrass ecotype Jiggs has shown greater tolerance to poorly drained soils (Aguiar et al. 2014). Vegetatively propagated cultivars are numerous and include 'Coastal', 'Grazer', 'Brazos', 'Tifton 85', 'Oklan', 'Midland 99', 'Alicia', and 'Russell'. Seed-propagated bermudagrass types are available commercially that are selections from common bermudagrass. Bermudagrasses have 18, 36, 45, and 54 chromosomes, but $4n = 36$ is most common (Hanna 1992; Taliaferro et al. 2004).

Stargrass

Because of lack of cold tolerance, stargrass in the US is limited to about 50 000 ha in South Florida. Stargrass contributes significantly in Central and South America, tropical Africa, and the Caribbean – areas where annual precipitation is >800 mm and temperatures do not fall below −4 °C. Stargrass lacks rhizomes but its robust stolons spread aggressively and form roots and shoots from the nodes. Stargrasses grow on a wide variety of soils but do best on moist, well-drained soils. Sustained productivity of stargrass requires greater attention to grazing management than bermudagrass because stargrass stem bases and stolons are important for reserve storage, but can can be removed under frequent, close defoliation

Table 18.1 General environmental adaptation characteristics of forage grasses adapted to subtropical and tropical climates

Common name	Climate	Frost tolerance	Winter tolerance	Soil drainage	Soil fertility
Bahiagrass	Humid subtropics	Moderate	High	Well to poor	Low
Bermudagrass	Humid subtropics	Low	High	Moderate to well	Moderate to high
Brachiariagrass hybrid (Mulato)	Humid tropics/subtropics	Low	Low	Moderate to well	Moderate to high
Buffelgrass	Semiarid/arid tropics/subtropics	Low	Low	Well	Moderate
Crabgrass	Warm temperate/subtropics/tropics	Low	NA[a]	Moderate to well	Moderate
Dallisgrass	Humid tropics/subtropics	Low	High	Moderate to poor	Moderate
Digitgrass	Humid tropics/subtropics	Low	Low	Moderate to poor	Moderate to high
Elephantgrass	Humid tropics/subtropics	Low	High	Moderate to well	Moderate to high
Guineagrass	Humid subtropics/tropics	Low	Low	Moderate to well	Moderate to high
Kikuyugrass	Humid subtropics/high elevation tropics	High	Moderate	Well	Moderate to high
Koroniviagrass	Humid tropics	Low	Low	Moderate to seasonally flooded	Moderate
Limpograss	Humid subtropics	Moderate	Moderate	Moderate to seasonally flooded	Moderate
Palisadegrass	Humid tropics	Low	Low	Moderate to well	Moderate
Pearlmillet	Warm temperature/subtropics	Low	NA[a]	Moderate to well	Moderate
Rhodesgrass	Humid tropics/subtropics	Moderate	Low	Moderate to poor	Moderate to high
Ruzigrass	Humid tropics	Low	Low	Moderate to well	Moderate to high
Setariagrass	Humid tropics/subtropics	Low	Low	Moderate	Moderate
Signalgrass	Humid tropics	Low	Low	Moderate to well	Moderate
Sorghum	Semiarid to humid	Low	NA[a]	Poor to well	Moderate to high
Stargrass	Humid tropics/subtropics	Low	Low	Moderate	Moderate to high

The basis for comparisons is relative to the broader population of C_4 species adapted to warm climates.
[a]NA = not applicable for annual species.

Table 18.2 General species characteristics of forage grasses adapted to subtropical and tropical climates

Common name	Perennial/ annual	Growth habit	Mode of propagation	Rate of establishment	Grazing tolerance	Nutritive value
Bahiagrass	Perennial	Decumbent	Seed	Slow	High	Low
Bermudagrass	Perennial	Intermediate	Vegetative/seed	Rapid	Moderate	Moderate
Brachiariagrass hybrid (Mulato)	Perennial	Upright	Seed	Moderate	Moderate	High
Buffelgrass	Perennial	Intermediate	Seed	Moderate	Moderate	Moderate
Crabgrass	Annual	Decumbent	Seed	Rapid	High	Moderate
Dallisgrass	Perennial	Decumbent	Seed	Slow	High	Moderate
Digitgrass	Perennial	Intermediate	Vegetative/seed	Moderate	Low	Moderate to high
Elephantgrass	Perennial	Upright	Vegetative	Moderate	Moderate	Moderate to high
Guineagrass	Perennial	Upright	Seed	Moderate	Moderate	Moderate to high
Kikuyugrass	Perennial	Decumbent	Vegetative/seed	Moderate	High	Moderate
Koroniviagrass	Perennial	Decumbent	Seed	Rapid	High	Moderate
Limpograss	Perennial	Upright	Vegetative	Moderate	Moderate	Moderate
Palisadegrass	Perennial	Upright	Seed	Rapid	Moderate	Moderate
Pearlmillet	Annual	Upright	Seed	Rapid	Moderate	Moderate to high
Rhodesgrass	Perennial	Intermediate	Seed	Moderate	Low	Moderate
Ruzigrass	Perennial	Intermediate	Seed	Rapid	Moderate	Moderate
Setariagrass	Perennial	Intermediate	Seed	Slow	Low	Moderate
Signalgrass	Perennial	Intermediate	Seed	Rapid	Moderate to High	Low to Moderate
Sorghum	Annual	Upright	Seed	Rapid	Moderate	Moderate to high
Stargrass	Perennial	Intermediate	Vegetative	Rapid	Moderate	Moderate

The basis for comparisons is relative to the broader population of C_4 species adapted to warm climates.

(Pitman 1991). Stargrass can produce high levels of **prussic acid**, but incidences of animal death are rare. Vegetatively propagated cultivars include 'Ona', 'Florico', 'Florona', 'Costa Rica', and 'Tifton 68'.

Digitaria spp.

Digitaria is a diverse genus native from Africa. *Digitaria* has a base chromosome number of x = 9 (Koyama 1987) and ploidy levels range from diploid to octoploid. Plant size varies considerably with genotype and environment, and inflorescences consist of digitate panicles of 3–10 racemes of uniform length. Annual *Digitaria* species are generally referred to as crabgrass and can be useful forage plants as well as potential weeds. The tropical and subtropical perennial digitgrasses are the most widely planted species for pastures (Pitman et al. 2004).

'Pangola' is the most-used cultivar of digitgrass, but 'Slenderstem', 'Transvala', 'Taiwan', and 'Survenola' were also used in tropical and subtropical regions (Chambliss 1999). These cultivars do not produce viable seed and are propagated vegetatively, but seed-propagated types are present in digitgrass. 'Pangola' is adapted to areas with rainfall exceeding 1000 mm and tolerates waterlogging and limited flooding. It is responsive to fertilization and adapted to soil pH from 4.5 to 8.0. Due to minimal cold tolerance, it is grown in warmer regions, such as the southern half of the Florida peninsula (Ocumpaugh and Sollenberger 1995). The initial widespread use of 'Pangola' was based on superior forage quality (Hodges and McCaleb 1959). 'Pangola' has been used for grazing, hay, and silage. A major constraint to use of 'Pangola' has been the lack of persistence under intense and frequent grazing. Rotational stocking is recommended to maintain competitive stands. The most important pests of 'Pangola' are stunt virus, armyworms (*Laphygma frugiperda* A&S and *Mocis repanda* F) and yellow sugarcane aphid (*Sipha flava* Forbes) (Pitman et al. 2004).

Paspalum spp

Bahiagrass

Bahiagrass is a broadly adapted, deep-rooted perennial that grows on a wide range of soils but prefers sandy soils. Growth is best at soil pH of 5.5–6.5. It can tolerate low pH, low fertility, and close grazing, and is resistant to many diseases (Burson and Watson 1995). Plants have thick, vigorous rhizomes and are very competitive. The stem, with a racemose panicle, is 10–60 cm tall.

Most bahiagrasses are apomictic and tetraploid but the most popular cultivar 'Pensacola' is a sexual diploid that is cross-pollinated. All are seed-propagated, but seed has various degrees of dormancy and germinates slowly. 'Tifton 9' bahiagrass was released as a sexual diploid with up to 47% greater herbage accumulation than 'Pensacola' (Burton 1989). Subsequently, 'Tifquick' was developed from 'Tifton 9' for faster and improved seed emergence due to reduced hard-seededness (Anderson et al. 2009). 'UF-Riata' is a selection from 'Pensacola' with greater cool-season productivity and less sensitivity to photoperiod (Interrante et al. 2009).

Bahiagrass is used primarily for grazing and occasionally for hay production. Persistence under low-fertility, drought, flooding, and continuous stocking makes bahiagrass a very reliable feed source for low-input beef cattle production (Gates et al. 2004). It is also used for low-maintenance turf and as ground cover on highway rights-of-way. Herbage accumulation is concentrated during summer; only 14% of forage production occurs from October through March (Mislevy and Everett 1981), primarily due to photoperiod sensitivity (Sinclair et al. 2003). Forage nutritive value is strongly affected by regrowth interval. The *in vitro* digestible DM concentration decreased from 599 to 432 $g\,kg^{-1}$ as regrowth interval increased from 6 to 15 weeks (Gates et al. 2004) and CP concentrations decreased from 120 to 70 $g\,kg^{-1}$ for 20- to 55-day regrowth intervals (Vendramini et al. 1999). The major pests of bahiagrass are mole cricket (*Scapteriscus* spp.) and fall armyworm (*Spodoptera frugiperda*).

Dallisgrass

Common dallisgrass is a deep-rooted perennial grass that grows in the southeastern US from Texas to North Carolina. It is best adapted to heavier-textured soils in areas with at least 900 mm annual rainfall, and it grows along streams and ditches in low rainfall areas. Plants grow in clumps with short rhizomes (Burson and Watson 1995). The inflorescence is an erect or nodding panicle with 2–11 spike-like racemes (Chase 1929) and ranges in height from 40 to 175 cm (Evers and Burson 2004). Common dallisgrass, the most widespread, is apomictic and has 2n = 5x = 50 chromosomes.

Dallisgrass is generally grazed and is usually found in mixed-species swards (Evers and Burson 2004). Robinson et al. (1988) reported yields from 3.4 to 13.9 $Mg\,ha^{-1}$ across soil types and N fertilizer levels. Nutritive value is considered to be greater than most C_4 perennial grass species. Average CP and *in vitro* **true digestibility** of 15 ecotypes in Louisiana were 110 and 610 $g\,kg^{-1}$, respectively (Venuto et al. 2003). Average daily gain of steers continuously stocked on dallisgrass pastures in Alabama was 0.32 $kg\,d^{-1}$. Ergot (*Claviceps paspali* F. Stevens and J.G. Hall) is the major disease problem in dallisgrass and infects the developing seed (Evers and Burson 2004). Cattle are most susceptible to ergot because of their preference for inflorescences.

Pennisetum and *Cenchrus* Species

Three perennial species from these genera are major contributors to forage production in the tropics and subtropics: elephantgrass or napiergrass, kikuyugrass, and buffelgrass. Taxonomic classification of these grasses has changed in recent years. Elephantgrass (2n = 4x = 28, allotetraploid and sexual) belongs to the secondary gene pool and both buffelgrass and kikuyugrass (both 2n = 4x = 36, autotetraploid and apomictic) belong to the tertiary gene pool (Jauhar 1981).

Buffelgrass

Buffelgrass (2n = 4x = 36, autotetraploid and apomictic), a drought-tolerant bunchgrass, is native from South Africa to India. It is important in south Texas, northern Mexico, and Australia, and can be found in many arid parts of the tropics and subtropics. It grows best in well-drained sandy loam soils and does not grow well in deep sands or poorly drained soils (Hanselka et al. 2004). 'Nueces' is a cultivar with improved productivity and better **cold resistance** due to rhizomes (Burson and Young 2001; Hanselka et al. 2004). Buffelgrass is relatively easy to establish from seed and produces good quality forage that is preferred by livestock. Though it is more tolerant of continuous stocking than most bunchgrasses, it does not tolerate continuous heavy stocking. Seed can spread voluntarily and through livestock to unintended areas.

Elephantgrass

This perennial was introduced from tropical Africa, has been distributed throughout Southeast Asia, and is widely used in South America and Africa. It is a robust, rhizomatous bunchgrass with greater herbage accumulation than most other grasses and legumes. Use of elephantgrass in the US is limited to southern Florida and Texas. It is sensitive to daylength and flowers under short days (usually <11 hour).

Plants can grow to 7 m in height and produce 30 $Mg\,ha^{-1}\,yr^{-1}$ DM in the US Gulf Coast, making elephantgrass an attractive bioenergy feedstock (Na et al. 2015). It uses water efficiently to produce biomass and minimizes NO_3-N leaching to groundwater

(Reyes-Cabrera et al. 2017). Dwarf genotypes such as 'Mott' grow to a maximum height of 1.6 m, are easier to manage under grazing, are very leafy, and produce high-quality forage (Sollenberger et al. 1988). Elephantgrass grows best in hot (30–35 °C), humid climates and well-drained soils. Propagation is generally by stems and rhizomes. Yield and persistence increase and forage nutritive value decreases with less frequent defoliation (Hanna et al. 2004b). Trailing legumes can be grown with tall elephantgrass and rhizoma peanut with dwarf elephantgrass.

Kikuyugrass

Originating from higher elevations in East and Central Africa, kikuyugrass is a low-growing, perennial sod-forming grass that produces both rhizomes and stolons. It grows well at 1000–3000 m elevation at latitudes of 0°–35° and at sea level at latitudes of 25°–35° on fertile soils. It does not grow well where temperatures exceed 30 °C. It usually requires at least 900 mm of rainfall, is adapted to well-drained soils, and is tolerant of low soil pH (4.5) and high salt (Hanna et al. 2004b). It can be propagated vegetatively or by seed. Progeny from seed are identical to the mother plant because of apomictic reproduction. Stems with inflorescences can be up to 9–15 cm tall. Seed are spread in the dung of animals grazing the mature seed heads. It is quite aggressive in its areas of adaptation and, therefore, legumes do not compete well in mixtures (Hanna et al. 2004b). Its growth habit and tolerance to heavy grazing make kikuyugrass ideal for pastures; however, it is also used for hay, green chop, silage, haylage, and turf. Nitrate toxicity can be a problem in forage receiving more than 50 kg ha^{-1} N per application, especially if CP concentration is above 230 g kg^{-1}. The *in vitro* **dry matter disappearance** (IVDMD) and CP of kikuyugrass forage are usually greater than 600 and 120 g kg^{-1}, respectively, when less than six weeks old. The major pest is "kikuyu yellows" caused by *Verrucalvus flavofaciens.*

Urochloa spp.

The common name "brachiariagrass" has been used for the species within the genus *Brachiaria* (Miles et al. 2004). This is now the genus *Urochloa*. Brachiariagrass species and cultivars within species can be difficult to distinguish vegetatively, and taxonomic distinction is better established by reproductive traits (Renvoize et al. 1996). Plants can be robust, tufted, and reach 2.5-m height such as 'Marandu' palisadegrass, or are low-growing stoloniferous types reaching 0.4- to 0.9-m height such as 'Llanero' koroniviagrass (do Valle et al. 2010).

Commercially important *Urochloa* species include palisadegrass, signalgrass, koroniviagrass, and ruzigrass (Miles et al. 2004). All except 'Kennedy' ruzigrass were direct selections from African germplasm. Brachiariagrass plant breeding programs quickly developed in the mid-1980s, mainly in Brazil, as the apomictic reproduction obstacle was overcome. Breeding goals have been to combine spittlebug (Homoptera: Cercopidae) resistance with adaptation to acid, infertile soils, (Miles et al. 2004). More recently, interspecific hybrids 'Mulato II' (Silva et al. 2016) and 'BRS Ipyporã' (do Valle et al. 2017) have been released as high-yielding cultivars with excellent quality (Mulato II) and increased resistance to spittlebug (BRS Ipyporã).

These four species originated from subhumid to humid environments, with dry seasons of about five months (Keller-Grein et al. 1996). Palisadegrass is better adapted to drier areas (seven-months dry season and annual rainfall as low as 600 mm). Little growth occurs below 15 °C (Pedreira et al. 2011). Ruzigrass is the least cold tolerant species (Miles et al. 2004). Spittlebug is the major biotic limitation to brachiariagrass productivity and persistence (Valério et al. 1996). Brachiariagrasses are well adapted to acid, infertile soils (Miles et al. 2004). They propagate easily and can produce in excess of 20 Mg DM ha^{-1} yr^{-1} (Pequeno et al. 2015). Their economic importance is greatest in tropical America, with estimated use on 99 million ha in Brazil alone (Jank et al. 2014). They occur mainly in low-input grazed pastures, but signalgrass is sometimes used for hay. Well-managed pastures can result in good daily animal gain (0.7 kg) and per unit land area gain (624 kg ha^{-1}) (Herling et al. 2011). Brachiariagrasses are used increasingly in crop rotations in Brazil, including silvopastoral systems, an option that has helped recover degraded grassland areas.

Other Perennial Grasses

Guineagrass

This upright perennial bunchgrass originated from Africa and has a free-growing plant height from 0.25 to 4.5 m (Muir and Jank 2004). Older cultivars 'Colonião', 'Sempre verde', and 'Tobiatã' are no longer recommended. More recent cultivars from EMBRAPA-Brazil include 'Mombaça', 'Tanzania-1', 'BRS-Zuri', 'Massai', and 'BRS-Quênia'. They have been bred or selected for specific edaphic adaptation or pest/disease resistance. Guineagrass is recommended for cultivation in areas with 800–1800 mm yr^{-1} of rainfall (Skerman and Riveros 1990). In the US, it is found in southern Texas and Florida. Guineagrasses can survive below-freezing temperatures for short periods (Skerman and Riveros 1990), but minimum temperatures for growth of most cultivars are 10–15 °C (Moreno et al. 2015). It is intolerant of poorly drained, acidic, or high-Al soils. It is highly responsive to N fertilizer, and well-fertilized guineagrass can produce in excess of 30 Mg DM ha^{-1} yr^{-1} (Jank et al. 1994).

Main uses of guineagrass are grazed pasture, **greenchop**, hay, and silage. Because plant morphology varies widely across cultivars, recommended post-grazing stubble (20–50 cm) is normally 50% of pre-grazing canopy height (Pedreira et al. 2015; Silva et al. 2015a). Shade tolerance allows it to fit well into agroforestry systems, including in association with eucalyptus (*Eucalyptus* sp.) in central Brazil (Andrade et al. 2003). Forage nutritive value is often greater than most C_4 grasses, making it an option for dairy and intensive beef production systems (Jank et al. 2010).

Johnsongrass

Johnsongrass, a tetraploid (2n = 4x = 40), is a rhizomatous perennial (Fribourg 1995). It is erect and resembles sudangrass, having thin stems and narrow leaves. It is native to the Mediterranean region of North Africa, was introduced to North America as a potential forage, and is now adapted naturally to areas south of a line from Iowa and Kansas and west to southern California (Hitchcock 1950). It also grows well throughout the tropics and subtropics, but it is a noxious weed and not planted. Johnsongrass spreads by seed and rhizomes. In grain sorghum fields, it can cross with cultivated sorghum and produce weedy outcrosses that are prone to seed shatter. Naturalized stands are grazed or used for hay (Hanna and Torres-Cardona 2001). Plants accumulate HCN-p and NO_3^-, especially when subjected to cold or drought stress (Ocumpaugh and Sollenberger 1995).

Limpograss

Limpograss is a perennial that was collected in southern Africa. It has stolons that give rise to erect-growing stems that may reach heights of 1.5 m. Leaves are small and narrow. The inflorescence consists of cylindrical racemes that are 6- to 10-cm long. The base chromosome number is x = 9, and diploid, tetraploid, and hexaploid types exist (Quesenberry et al. 1982). At least ten cultivars have been released from Florida and Brazil (Quesenberry et al. 2004); the most widely used is the tetraploid 'Floralta' and the most recent are 'Kenhy' and 'Gibtuck' (Quesenberry et al. 2018). Seed production is low, and vegetative establishment is widely practiced. Limpograss exhibits better growth during cool weather than do other C_4 grasses, is adapted to very poorly drained soils, and has relatively high digestibility that declines slowly with increasing maturity. Leaves and stems may become reddish or purple with maturity or cold stress. Reports are widespread of limpograss forage CP less than 70 g kg^{-1}, and pastures have been overseeded successfully with several legumes to address this limitation (Quesenberry et al. 2004). Most limpograss is grazed by beef cattle. It is well suited for use as stockpiled forage, but thick stems limit its use for hay. Nitrogen is required to maintain productive stands, but when limpograss is grazed, P and K requirements appear to be low (Adjei et al. 2001). Frequent, close grazing weakens stands, allowing invasion by common bermudagrass (Newman et al. 2003). Spittlebug and chinchbug may damage limpograss stands (Quesenberry et al. 2004).

Rhodesgrass

This stoloniferous perennial has erect or ascending stems up to 2 m tall and glabrous leaf blades (Bogdan 1977). The inflorescence is a digitate or subdigitate panicle with up to 20 spikes. The basic chromosome number is 10, and both diploid and tetraploid types exist. Diploids are generally subtropical in origin and tend to be more cold tolerant than tetraploids (Loch et al. 2004b). Among major C_4 grasses, rhodesgrass is one of the few cross-pollinating species. It is best suited to areas with 600–1200 mm of rainfall, although the tetraploids have extended into areas with 1500 mm rainfall. It is moderately drought tolerant, grows best at high temperatures, and is relatively tolerant of frost (Ivory and Whiteman 1978). It is killed by air temperatures of −10 °C, limiting its use in the US to the southern parts of Florida and Texas. Rhodesgrass is established from seed and requires moderate soil fertility to persist. It does not grow well on highly acid soils but, is more salt tolerant than many C_4 grasses (Loch et al. 2004b). It is used for soil conservation, reclamation of mined and degraded land, and forage. It is one of the best C_4 grasses for grazing by horses (*Equus caballus*) because it is well accepted and low in oxalate. It is less tolerant of heavy grazing than bermudagrass and stargrass, is often relatively short-lived, and is compatible with several legumes (Loch et al. 2004b).

Setaria

This perennial tropical grass has an upright growth habit, reaching a height of 3 m. The inflorescence is a cylindrical, spike-like panicle. Setaria is a polyploid complex ranging from diploid to decaploid with a base chromosome number of x = 9 (Jank and Hacker 2004). Important cultivars are primarily from East and Central Africa. Setaria is adapted to loamy and sandy soils in areas that are comparatively wet, with rainfall above 750 mm and no prolonged dry season. It survives mild frosts (Jank and Hacker 2004). Setaria is not used commercially in the US. It can be established from seed or vegetative cuttings, but establishment is slow. Setaria is used primarily as a pasture for dairy and beef cattle but can be harvested (Ghisi et al. 1994). Persistence is poor with high grazing pressure and low N fertility (Roe and Williams 1993). It can be grown with several legumes. Oxalate poisoning was reported when cattle that had not grazed setaria previously consumed it as their sole source of feed (Jank and Hacker 2004). Oxalate poisoning is more common in horses than in cattle.

Annual Grasses

Crabgrass

Digitaria ciliaris (Retz.) Koeler, referred to as southern crabgrass, and *Digitaria sanguinalis (L.) Scop.*, referred to as hairy crabgrass, overlap in temperate regions, while

D. ciliaris extends into the tropics (Webster 1987). The region of adaptation of crabgrass ranges from about 28 °N latitude and 87 °W longitude northward and westward to about 40 °N latitude and 100 °W longitude (Dalrymple et al. 1999). Forage production is limited by rainfall in areas with <600 mm annually, by flooding, and by soils below pH 5.0 or above pH 8.0 (Dalrymple et al. 1999). 'Red River' is the first known cultivar of crabgrass, originating from a naturally occurring ecotype in southern Oklahoma (Dalrymple 1999). Crabgrass has been used as pasture, hay, silage, and green chop and nutritive value is usually good (Ogden et al. 2005). Crabgrass can produce from 3300 to 10 000 kg DM ha^{-1} (Teutsch et al. 2005), but it generally is less productive than warm-season perennial grasses (Fike et al. 2005). Double cropping with crabgrass in the summer and small grains (Oklahoma) or annual ryegrass (Alabama) for cool-season forage, has provided quality pasture for young growing cattle.

Pearlmillet

This annual grass can grow to 5-m tall, but most improved cultivars are 2- to 3-m tall and produce abundant tillers. Believed to have originated in northern Africa, pearlmillet produces well on poor, droughty, and infertile soils, but also responds to high soil fertility and moisture. It grows best at soil pH of 5.5–7 and on well-drained loamy to sandy soils (Hanna et al. 2004a). It can be grown throughout most of the US, where it establishes quickly under warm (>22 °C) and moist soil conditions. The DM production increases linearly with N up to 200 kg ha^{-1}, and split applications are preferred. Production may increase up to 400 kg N ha^{-1} in tropical climates where the growing season is longer, depending on the response of the cultivar to daylength. Dry matter yields average 12–15 Mg ha^{-1} with 100 kg ha^{-1} N. Pearlmillet ($2n = 2x = 14$, diploid and sexual) is an important grain crop in India and Africa and is a popular annual forage crop for grazing and hay in the tropics and subtropics. It can be green chopped and used in cut-and-carry systems. Most improved cultivars provide about 120 days of grazing. Late plantings flower earlier due to shorter days, reducing forage production and quality. Pearlmillet does not produce hydrocyanic acid. Occasionally, NO_3^- poisoning occurs under high N fertilization and drought stress.

Sorghum

The sorghum species includes grain and forage sorghum types and now sudangrass. The *bicolor* sorghums, all diploids ($2n = 2x = 20$), are mainly grown as annuals, although technically they are perennials in regions where cold weather does not kill them. The origin of the sorghums is Africa (Pedersen and Rooney 2004). They are popular summer annuals in much of the US and throughout the tropics and subtropics and are used for grazing, hay, greenchop, and silage. The forage is usually highly digestible, especially in the vegetative stage. The forage sorghums are morphologically diverse, ranging from thick-stemmed with few tillers to thin-stemmed with many tillers (Hanna and Torres-Cardona 2001). Forage sorghum will grow in soils of moderate fertility and a pH of 6–7. It is less tolerant of acid soils, but it performs better on heavier and wetter soils than pearlmillet. Responses to fertilizer are similar. Stubble heights after harvesting should be 15 cm or taller (Fribourg 1995). When drought stressed or frosted, sorghums can accumulate HCN-p that can poison livestock. Consumption by livestock should be avoided during periods of plant stress. High N fertilization and drought stress can also cause NO_3^- accumulation and potential poisoning.

References

Adewopo, J.B., Silveira, M.L., Xu, S. et al. (2014). Management intensification impacts on soil and ecosystem carbon stocks in subtropical grasslands. *Soil Sci. Soc. Am. J.* 78: 977–986.

Adjei, M.B. and Rechcigl, J.E. (2004). Interactive effect of lime and nitrogen on bahiagrass pastures. *Soil Crop Sci. Soc. Fla. Proc.* 63: 52–56.

Adjei, M.B., Kalmbacher, R.S., and Rechcigl, J.E. (2001). Effect of P and K fertilizer on forage yield and nutritive value of Floralta limpograss. *Soil Crop Sci. Soc. Fla. Proc.* 60: 9–14.

Aguiar, A.D., Vendramini, J.M.B., Arthington, J.D. et al. (2014). Stocking rate effects on Jiggs bermudagrass pastures grazed by heifers receiving supplementation. *Crop Sci.* 54: 2872–2879.

Anderson, W.F., Gates, R.N., Hanna, W.W. et al. (2009). Recurrent restricted phenotypic selection for improving stand establishment of bahiagrass. *Crop Sci.* 49: 1322–1327.

Andrade, C.M.S., Garcia, R., Couto, L. et al. (2003). Desempenho de seis gramíneas solteiras ou consorciadas com o *Stylosanthes guianensis* cv. Mineirão e eucalipto em sistema silvipastoril. *R. Bras. Zootec.* 8: 1845–1850.

Asker, S.E. and Jerling, L. (1992). *Apomixis in Plants.* Boca Raton, FL, USA: CRC Press.

Bates, D.B., Kunkle, W.E., Chambliss, C.G., and Cromwell, R.P. (1989). Effect of dry matter and additives on bermudagrass and rhizoma peanut round bale silage. *J. Prod. Agric.* 2: 91–96.

Bogdan, A.V. (1977). *Tropical Pasture and Fodder Plants.* Longman, London: Tropical Agricultural Services.

Brown, R.H. (1978). A difference in N use efficiency in C_3 and C_4 plants and its implications in adaptation and evolution. *Crop Sci.* 18: 93–98.

Burson, B.L. and Watson, V.H. (1995). Bahiagrass, dallisgrass, and other *Paspalum* species. In: *Forages: An Introduction to Grassland Agriculture* (eds. R.F Barnes, D.A.

Miller and C.J. Nelson), 431–440. Ames, IA: Iowa State University Press.

Burson, B.L. and Young, B.A. (2001). Breeding and improvement of tropical grasses. In: *Tropical Forage Plants: Development and Use* (eds. A. Sotomayor-Rios and W.D. Pitman), 59–79. Boca Raton, FL: CRC Press.

Burton, G.W. (1989). Registration of 'Tifton 9' Pensacola bahiagrass. *Crop Sci.* 29: 1326.

Burton, G.W. and DeVane, E.H. (1992). Growing legumes with 'Coastal' bermudagrass in the lower coastal plain. *J. Prod. Agric.* 5: 278–281.

Burton, G.W. and Hanna, W.W. (1995). Bermudagrass. In: *Forages: An Introduction to Grassland Agriculture* (eds. R.F Barnes, D.A. Miller and C.J. Nelson), 421–429. Ames, IA: Iowa State University Press.

Chambliss, C.G. (1999). Digitgrass. In: *Florida Forage Handbook* (eds. C.G. Chambliss and M.B. Adjei), 36–38. Gainesville, FL: University of Florida, Cooperative Extension Service.

Chase, A. (1929). *The North American Species of Paspalum. Contributions from the United States National Herbarium*, vol. 28, Part 1. Washington, DC: United States Government Printing Office.

Clark, E.A., Crump, S.V., and Wijnheijmer, S.S. (1985). Morphological determinants of drying rate in forage legumes. In: *Proceedings of the American Forage and Grassland Conference.*, Hershey, PA. 3-6 Mar. 1985, 137–141. Georgetown, TX: American Forage and Grassland Council.

Coleman, S.W., Moore, J.E., and Wilson, J.R. (2004). Quality and utilization. In: *Warm-Season (C_4) Grasses*, Monograph 45 (eds. L.E. Moser, B.L. Burson and L.E. Sollenberger), 267–308. Madison, WI: ASA-CSSA-SSSA.

Comai, L. (2005). The advantages and disadvantages of being polyploid. *Nat. Rev. Genet.* 6: 836–846.

Dalrymple, R.L. (1999). Registration of 'Red River' crabgrass. *Crop Sci.* 41: 1998–1999.

Dalrymple, R.L., Mitchell, R., Flatt, B. et al. (1999). *Crabgrass for Forage: Management from the 1990s.* Publ. No. NF-FO-99-18. Ardmore, OK: Noble Found.

Dangler, N.G. and Nelson, T. (1999). Leaf structure and development in C_4 plants. In: *C_4 Plant Biology* (eds. R.F. Sage and R.K. Monson), 133–172. San Diego: Academic Press.

do Valle, C.B., Macedo, M.C.M., Euclides, V.P.B. et al. (2010). Gênero *Brachiaria*. In: *Plantas Forrageiras* (eds. D.M. da Fonseca and J.A. Martuscello), 30–77. Viçosa, Brazil: Editora UFV.

do Valle, C.B., Euclides, V.P.B., Montagner, D.B. et al. (2017). BRS Ipyporã ("belo começo" em guarani): híbrido de Brachiaria da Embrapa. Comunicado Técnico 12337. Embrapa. https://ainfo.cnptia.embrapa.br/digital/bitstream/item/159958/1/BRS-Ipypora-belo-comeco-em-guarani.pdf (accessed 16 October 2019).

Dubeux, J.C.B. Jr., Sollenberger, L.E., Mathews, B.W. et al. (2007). Nutrient cycling in warm-climate grasslands. *Crop Sci.* 47: 915–928.

Evers, G.W. and Burson, B.L. (2004). Dallisgrass and other *Paspalum* species. In: *Warm-Season (C_4) Grasses*, Monograph 45 (eds. L.E. Moser, B.L. Burson and L.E. Sollenberger), 681–714. Madison, WI: ASA-CSSA-SSSA.

Evers, G.W., Redmon, L.A., and Provin, T.L. (2004). Comparison of bermudagrass, bahiagrass, and kikuyugrass as a standing hay crop. *Crop Sci.* 44: 1370–1378.

Fike, J.H., Teutsch, C.D., and Ward, D.L. (2005). Warm-season grass production responses to site and defoliation frequency. *Forage Grazinglands* https://doi.org/10.1094/FG-2005-0824-01-RS.

Fornara, D.A. and Tilman, D. (2008). Plant functional composition influences rates of soil carbon and nitrogen accumulation. *J. Ecol.* 96: 314–322.

Franzluebbers, A.J. and Stuedemann, J.A. (2009). Soil-profile organic carbon and total nitrogen during 12 years of pasture management in the southern Piedmont USA. *Agric. Ecosyst. Environ.* 129: 28–36.

Fribourg, H.A. (1995). Summer annual grasses. In: *Forages: An Introduction to Grassland Agriculture* (eds. R.F Barnes, D.A. Miller and C.J. Nelson), 463–472. Ames, IA: Iowa State University Press.

Gates, R.N., Quarin, C.L., and Pedreira, C.G.S. (2004). Bahiagrass. In: *Warm-Season (C_4) Grasses*, Monograph 45 (eds. L.E. Moser, B.L. Burson and L.E. Sollenberger), 651–680. Madison, WI: ASA-CSSA-SSSA.

Ghisi, O., Alcantara, P.B., de Almeida, A.R.P., and Schammas, E.A. (1994). Agronomic and physiological evaluation of ten grasses under two fertilizer levels. *Bol. Ind. Anim.* 51: 35–42.

Hanna, W.W. (1992). *Cynodon dactylon* (L.) Pers. In: *Plant Resources of South-East Asia. No. 4* (eds. L.'t. Mannetje and R.M. Jones), 100–102. Wageningen, The Netherlands: Forages Pudoc Scientific Publishers.

Hanna, W.W. and Torres-Cardona, S. (2001). Pennisetums and sorghums in an integrated feeding system in the tropics. In: *Tropical Forage Plants: Development and Use* (eds. A. Sotomayor-Rios and W.D. Pitman), 193–200. Boca Raton, FL: CRC Press.

Hanna, W.W., Baltensperger, D.D., and Seetharam, A. (2004a). Pearl millet and other millets. In: *Warm-Season (C_4) Grasses* (eds. L.E. Moser, B.L. Burson and L.E. Sollenberger), 537–560. Madison, WI: ASA/CSSA/SSSA.

Hanna, W.W., Chaparro, C.J., Mathews, B.W. et al. (2004b). Perennial *Pennisetums*. In: *Warm-Season (C_4) Grasses* (eds. L.E. Moser, B.L. Burson and L.E. Sollenberger), 503–535. Madison, WI: ASA/CSSA/SSSA.

Hanselka, C.W., Hussey, M.A., and Ibarra, F. (2004). Buffelgrass. In: *Warm-Season (C_4) Grasses*, Agronomy Monograph no. 45 (eds. L.E. Moser, B.L. Burson and L.E. Sollenberger), 477–502. Madison, WI: ASA-CSSA-SSSA.

Herling, V.R., Pedreira, C.G.S., Luz, P.H.C. et al. (2011). Performance and productivity of Nellore steers on rotationally stocked palisadegrass (*Brachiaria brizantha*) pastures in response to herbage allowance. *J. Agric. Sci.* 149: 761–768.

Hitchcock, A.S. (1950). *Manual of Grasses of the United States*. Washington, DC: United States Government Printing Office.

Hodges, E.M. and McCaleb, J.E. (1959). Pasture development at the range cattle station. *Soil Crop Sci. Soc. Fla. Proc.* 19: 150–153.

Ibrikci, H., Hanlon, E.A., and Rechcigl, J.E. (1999). Inorganic phosphorus and manure effects on bahiagrass production on a spodosol. *Nutr. Cycling Agroecosyst.* 54: 259–266.

Interrante, S.M., Sollenberger, L.E., Blount, A.R. et al. (2009). Defoliation management of bahiagrass germplasm dry matter yield and herbage nutritive value. *Agron. J.* 101: 989–995.

Ivory, D.A. and Whiteman, P.C. (1978). Effects of environment and plant factors on foliar freezing resistance in tropical grasses. II. Comparison of frost resistance between cultivars of *Cenchrus ciliaris*, *Chloris gayana*, and *Setaria anceps*. *Aust. J. Agric. Res.* 29: 261–266.

Jank, L. and Hacker, J.B. (2004). Setaria. In: *Warm-Season (C_4) Grasses*, Monograph 45 (eds. L.E. Moser, B.L. Burson and L.E. Sollenberger), 785–807. Madison, WI: ASA-CSSA-SSSA.

Jank, L., Savidan, Y.H., Souza, M.T., and Costa, J.C.G. (1994). Avaliação do germoplasma de Panicum maximum introduzido da Africa: 1. Produção forrageira. *R. Bras. Zootec.* 23: 433–440.

Jank, L., Martuscello, J.A., Euclides, V.P.B. et al. (2010). *Panicum maximum*. In: *Plantas Forrageiras* (eds. D.M. da Fonseca and J.A. Martuscello), 166–196. Viçosa, Brazil: Editora UFV.

Jank, L., Barrios, S.C., do Valle, C.B. et al. (2014). The value of improved pastures to Brazilian beef production. *Crop Pasture Sci.* 65: 1132–1137.

Jauhar, P. (1981). *Cytogenetics and Breeding of Pearl Millet and Related Species*. New York: Alan R. Liss, Inc.

Keller-Grein, G., Maass, B.L., and Hanson, J. (1996). Natural variation in *Brachiaria* and existing germplasm collections. In: *Brachiaria: Biology, Agronomy, and Improvement* (eds. J.W. Miles, B.L. Maass, C.B. do Valle, et al.), 16–42. Cali, Colombia: CIAT and CNPGC/EMBRAPA, Campo Grande, MS, Brazil.

Kiniry, J.R., Tischler, C.R., and Van Esbroeck, G.A. (1999). Radiation use efficiency and leaf CO_2 exchange for diverse C_4 grasses. *Biomass Bioenergy* 17: 95–112.

Koyama, T. (1987). *Grasses of Japan and Its Neighboring Regions, and Identification Manual*. Tokyo: Kodansha.

da Lima, G.F., Sollenberger, L.E., Kunkle, W.E. et al. (1999). Nitrogen fertilization and supplementation effects on performance of beef heifers grazing limpograss. *Crop Sci.* 39: 1853–1858.

Loch, D.S., Adkins, S.W., Heslehurst, M.R. et al. (2004a). Seed formation, development, and germination. In: *Warm-Season (C_4) Grasses*, Monograph 45 (eds. L.E. Moser, B.L. Burson and L.E. Sollenberger), 95–143. Madison, WI: ASA-CSSA-SSSA.

Loch, D.S., Rethman, N.F.G., and van Niekerk, W.A. (2004b). Rhodesgrass. In: *Warm-Season (C_4) Grasses*, Monograph 45 (eds. L.E. Moser, B.L. Burson and L.E. Sollenberger), 833–872. Madison, WI: ASA-CSSA-SSSA.

Long, S.P. (1999). Environmental responses. In: *C_4 Plant Biology* (eds. R.F. Sage and R.K. Monson), 215–249. New York: Academic Press.

Masters, R.A., Mislevy, P., Moser, L.E., and Rivas-Pantoja, F. (2004). Stand establishment. In: *Warm-Season (C_4) Grasses*, Agronomy Monograph no. 45 (eds. L.E. Moser, B.L. Burson and L.E. Sollenberger), 145–177. Madison, WI: ASA-CSSA-SSSA.

Mathews, B.W., Tritschler, J.P. II, and Miyasaka, S.C. (1998). Phosphorus management and sustainability. In: *Grass for Dairy Cattle* (eds. J.H. Cherney and D.J.R. Cherney), 193–222. Wallingford, Oxon, UK: CAB International.

Mathews, B.W., Miyasaka, S.C., and Tritschler, J.P. (2004). Mineral nutrition of C_4 forage grasses. In: *Warm-season (C_4) grasses*. Agronomy Monograph 45 (eds. L.E. Moser et al.), 217–266. Madison, WI: ASA, CSSA, SSSA.

Miles, J.W., do Valle, C.B., Rao, I.M., and Euclides, V.P.B. (2004). Brachiariagrasses. In: *Warm-season (C_4) grasses*. Agronomy Monograph no. 45 (eds. L.E. Moser, B.L. Burson and L.E. Sollenberger), 745–783. Madison, WI: ASA-CSSA-SSSA.

Millennium Ecosystem Assessment (2005). *Ecosystems and Human Well-Being: Synthesis*. Washington, DC: Island Press.

Minson, D.J. (1980). Nutritional differences between tropical and temperate pastures. In: *Grazing Animals* (ed. F.H.W. Morley), 143–157. Amsterdam: Elsevier Science Publisher.

Mislevy, P. and Everett, P.H. (1981). Subtropical grass species response to different irrigation and harvest regimes. *Agron. J.* 73: 601–604.

Moore, K.J., Boote, K.J., and Sanderson, M.A. (2004). Physiology and developmental morphology. In: *Warm-season (C_4) grasses*. Agronomy Monograph 45 (eds. L.E. Moser, B.L. Burson and L.E. Sollenberger), 179–216. Madison, WI: ASA, CSSA, SSSA.

Moreira, L.M., Martuscello, J.A., Fonseca, D.M. et al. (2009). Perfilhamento, acúmulo de forragem e composição bromatológica do capim-braquiária adubado com nitrogênio. *R. Bras. Zootec.* 38: 1675–1684.

Moreno, L.S.B., Pedreira, C.G.S., Boote, K.J., and Alves, R.R. (2015). Base temperature determination of tropical *Panicum* spp. grasses and its effects on degree-day-based models. *Agric. For. Meteorol.* 186: 26–33.

Muir, J.P. and Jank, L. (2004). Guineagrass. In: *Warm-Season (C_4) Grasses*, Agronomy Monograph no. 45 (eds. L.E. Moser, B.L. Burson and L.E. Sollenberger), 589–621. Madison, WI: ASA-CSSA-SSSA.

Na, C.-I., Sollenberger, L.E., Erickson, J.E. et al. (2015). Management of perennial warm-season bioenergy grasses. I. Biomass harvested, nutrient removal, and persistence responses of elephantgrass and energycane to harvest frequency and timing. *Bioenergy Res.* 8: 581–589.

Naumova, T.N. (1997). Apomixis in tropical fodder crops: cytological and functional aspects. *Euphytica* 96: 93–99.

Newman, Y.C., Sollenberger, L.E., Fox, A.M., and Chambliss, C.G. (2003). Canopy height effects on vaseygrass and bermudagrass spread in limpograss pastures. *Agron. J.* 95: 390–394.

Oba, M. and Allen, M.S. (1999). Evaluation of the importance of the digestibility of neutral detergent fiber from forage: effects on dry matter intake and milk yield of dairy cows. *J. Dairy Sci.* 82: 589–596.

Ocumpaugh, W.R. and Sollenberger, L.E. (1995). Other grasses for the humid south. In: *Forages. Vol 1. An Introduction to Grassland Agriculture* (eds. R.F Barnes, D.A. Miller and C.J. Nelson), 441–449. Ames: Iowa State University Press.

Ogden, R.K., Coblentz, W.K., Coffey, K.P. et al. (2005). Ruminal in situ disappearance kinetics of dry matter and fiber in growing steers for common crabgrass forages sampled on seven dates in northern Arkansas. *J. Anim. Sci.* 83: 1142–1152.

Pedersen, J.F. and Rooney, W.L. (2004). Sorghums. In: *Warm-Season (C_4) Grasses*, Agronomy Monograph 45 (eds. L.E. Moser, B.L. Burson and L.E. Sollenberger), 1057–1079. Madison, WI: ASA, CSSA, and SSSA.

Pedreira, B.C., Pedreira, C.G.S., Boote, K.J. et al. (2011). Adapting the CROPGRO perennial forage model to predict growth of *Brachiaria brizantha*. *Field Crops Res.* 120: 370–379.

Pedreira, B.C., Pedreira, C.G.S., and Lara, M.A.S. (2015). Leaf age, leaf blade portion and light intensity as determinants of leaf photosynthesis in *Panicum maximum* (Jacq.). *Grassland Sci.* 61: 45–49.

Peichl, M., Leahy, P., and Kiely, G. (2011). Six-year stable annual uptake of carbon dioxide in intensively managed humid temperate grassland. *Ecosystems* 14: 112–126.

Pequeno, D.N.L., Pedreira, C.G.S., Sollenberger, L.E. et al. (2015). Forage accumulation and nutritive value of brachiariagrasses and Tifton 85 bermudagrass as affected by harvest frequency and irrigation. *Agron. J.* 107: 1741–1749.

Pitman, W.D. (1991). Management of stargrass pastures for growing cattle using visual pasture characteristics. Florida Agricultural Experiment Station Bulletin No. 884.

Pitman, W.D., Chambliss, C.G., and Hacker, J.B. (2004). Digitgrass and other species of *Digitaria*. In: *Warm-Season (C_4) Grasses*, Monograph 45 (eds. L.E. Moser, B.L. Burson and L.E. Sollenberger), 715–743. Madison, WI: ASA-CSSA-SSSA.

Premazzi, L.M., Monteiro, F.A., and Corrente, J.E. (2003). Tillering of Tifton 85 bermudagrass in response to nitrogen rates and time of application after cutting. *Sci. Agri.* 60: 565–571.

Quesenberry, K.H., Oakes, A.J., and Jessop, D.S. (1982). Cytological and geographical characterization of *Hemarthria*. *Euphytica* 31: 409–416.

Quesenberry, K.H., Sollenberger, L.E., and Newman, Y.C. (2004). Limpograss. In: *Warm-Season (C_4) Grasses*, Monograph 45 (eds. L.E. Moser, B.L. Burson and L.E. Sollenberger), 809–832. Madison, WI: ASA-CSSA-SSSA.

Quesenberry, K.H., Sollenberger, L.E., Vendramini, J.M.B. et al. (2018). Registration of Kenhy and Gibtuck limpograss hybrids. *J. Plant Regist.* 12: 19–24.

Redfearn, D.D. and Nelson, C.J. (2003). Grasses for southern areas. In: *Forages: An Introduction to Grassland Agriculture* (eds. R.F Barnes, D.A. Miller and C.J. Nelson), 149–169. Ames, IA: Iowa State Press.

Renvoize, S.A., Clayton, W.D., and Kabuye, C.H.S. (1996). Morphology, taxonomy, and natural distribution of *Brachiaria* (Trin.) Griseb. In: *Brachiaria: Biology, Agronomy, and Improvement* (eds. J.W. Miles, B.L. Maass and C.B. do Valle), 1–5. Cali, Colombia: CIAT and CNPGC/EMBRAPA, Campo Grande, MS, Brazil.

Reyes-Cabrera, J., Erickson, J.E., Leon, R.G. et al. (2017). Converting bahiagrass pastureland to elephantgrass bioenergy production enhances biomass yield and water quality. *Agric. Ecosyst. Environ.* 248: 20–28.

Robinson, D.L., Wheat, K.G., Hubbett, N.L. et al. (1988). Dallisgrass yield, quality, and nitrogen recovery responses to nitrogen and phosphorus fertilizers. *Commun. Soil Sci. Plant Anal.* 21: 1367–1379.

Roe, R. and Williams, R.W. (1993). Pasture studies on the podzolic soils on phyllites of the Gympie district, Queensland, CSIRO Div. Trop. Crops and Pastures. *Trop. Agron. Tech. Mem.* 82.

Ruelke, O.C. and Quesenberry, K.H. (1983). Effects of fertilization timing on the yields, seasonal distribution, and quality of limpograss forage. *Soil Crop Sci. Soc. Fla. Proc.* 42: 132–136.

Sage, R.F. (1999). Why C_4 photosynthesis? In: *C_4 Plant Biology* (eds. R.F. Sage and R.K. Monson), 3–16. San Diego: Academic Press.

Silva, S.C., Sbrissia, A.F., and Pereira, L.E.T. (2015a). Ecophysiology of C_4 forage grasses—understanding plant growth for optimizing their use and management. *Agriculture* 5: 598–625.

Silva, V.J., Pedreira, C.G.S., Sollenberger, L.E. et al. (2015b). Seasonal herbage accumulation and nutritive value of irrigated 'Tifton 85', Jiggs, and Vaquero bermudagrasses in response to harvest frequency. *Crop Sci.* 55: 2886–2894.

Silva, V.J., Pedreira, C.G.S., Sollenberger, L.E. et al. (2016). Canopy height and nitrogen affect herbage accumulation, nutritive value, and grazing efficiency of 'Mulato II' brachiariagrass. *Crop Sci.* 56: 2054–2061.

Silveira, M.L.A., Obour, A.K., Vendramini, J.M.B., and Sollenberger, L.E. (2011). Using tissue analysis as a tool to predict bahiagrass phosphorus fertilization requirement. *J. Plant Nutr.* 34: 2193–2205.

Silveira, M.L., Rouquette, F.M., Smith, G.R. et al. (2014). Soil fertility principles for warm-season perennial forages and sustainable pasture production. *Forage Grazinglands* 12: 1. https://doi.org/10.2134/FG-2013-0041-RV.

Sinclair, T.R., Ray, J.D., Mislevy, P., and Premazzi, L.M. (2003). Growth of subtropical forage grasses under extended photoperiod during short-daylength months. *Crop Sci.* 43: 618–623.

Skerman, P.J. and Riveros, F. (1990). *Tropical Grasses.* Rome: FAO.

Sleper, D.A. and Poehlman, J.M. (2006). *Breeding Field Crops*. Ames, Iowa: Blackwell Pub.

Sollenberger, L.E. (2008). Sustainable production systems for *Cynodon* species in the subtropics and tropics. In: *Proceedings of the Brazilian Society of Animal Science Meeting, 45th*, Lavras, Brazil. 22-25 July 2008, 85–100. Brasilia, Brazil: Sociedade Brasileira de Zootecnica.

Sollenberger, L. E., Prine, G.M., Ocumpaugh, W.R. et al. (1988). 'Mott' Dwarf Elephant grass: A high quality forage for the subtropics and tropics. Circular S-356, Agricultural Experiment Station, Institute of Food and Agricultural Sciences, University of Florida, Gainesville, FL.

Sollenberger, L.E., Reis, R.A., Nussio, L.G. et al. (2004). Conserved forages. In: *Warm-Season (C_4) Grasses*, Agronomy Monograph 45 (eds. L.E. Moser, B.L. Burson and L.E. Sollenberger), 355–388. Madison, WI: ASA, CSSA, and SSSA.

Sollenberger, L.E., Agouridis, C.T., Vanzant, E.S. et al. (2012). Prescribed grazing on pasturelands. In: *Conservation Outcomes from Pastureland and Hayland Practices: Assessment, Recommendations, and Knowledge Gaps* (ed. C.J. Nelson), 111–204. Lawrence, KS: Allen Press.

Sollenberger, L.E., Kohmann, M.M., Dubeux, J.C.B. Jr., and Silveira, M.L. (2019). Grassland management affects delivery of regulating and supporting ecosystem services. *Crop Sci.* 59 (2): 441–459.

Soussana, J.F., Tallec, T., and Blanfort, V. (2010). Mitigating the greenhouse gas balance of ruminant production systems through carbon sequestration in grasslands. *Animal* 4: 334–350.

Taliaferro, C.M., Coleman, S.W., and Claypool, P.L. (1987). Relative winter forage quality of selected bermudagrass cultivars. *Crop Sci.* 27: 1285–1290.

Taliaferro, C.M., Rouquette, F.M.J., and Mislevy, P. (2004). Bermudagrass and Stargrass. In: *Warm-Season (C_4) Grasses*, Agronomy Monograph 45 (eds. L.E. Moser, B.L. Burson and L.E. Sollenberger), 417–476. Madison, WI: ASA, CSSA, and SSSA.

Teutsch, C.D., Fike, J.H., and Tilson, W.M. (2005). Yield, digestibility, and nutritive value of crabgrass as impacted by nitrogen fertilization rate and source. *Agron. J.* 97: 1640–1646.

Valério, J.R., Lapointe, S.L., Kelemu, S. et al. (1996). Pests and diseases of *Brachiaria* species. In: *Brachiaria: Biology, Agronomy, and Improvement* (eds. J.W. Miles, B.L. Maass, C.B. do Valle, et al.), 87–105. Cali, Colombia: CIAT and CNPGC/EMBRAPA, Campo Grande, MS, Brazil.

Vendramini, J.M.B., Castro, F.G.F., Vieira, A.C. et al. (1999). Dry matter yield, chemical composition and nutritive value of *Paspalum notatum* cv. Tifton 9 at different ages of regrowth. *Scientia Agricola* 56: 225–234.

Vendramini, J.M.B., Adesogan, A.T., Silveira, M.L.A. et al. (2010). Nutritive value and fermentation parameters of warm-season grass silage. *Prof. Anim. Sci.* 26: 193–200.

Vendramini, J.M.B., Aguiar, A.D., Adesogan, A.T. et al. (2016). Effects of genotype, wilting, and additives on the nutritive value and fermentation of bermudagrass silage. *J. Anim. Sci.* 94: 3061–3071.

Venuto, B.C., Burson, B.L., Hussey, M.A. et al. (2003). Forage yield, nutritive value, and grazing tolerance of dallisgrass biotypes. *Crop Sci.* 43: 295–301.

Wallau, M.O., Sollenberger, L.E., Vendramini, J.M.B. et al. (2015). Herbage accumulation and nutritive value of limpograss breeding lines under stockpiling management. *Crop Sci.* 55: 2377–2383.

Webster, R.D. (1987). Taxonomy of *Digitaria* section *Digitaria* in North America (Poaceae:Paniceae). *Sida* 12: 209–222.

Wilkinson, S.R. and Langdale, G.W. (1974). Fertility needs of warm season grass. In: *Forage Fertilization* (ed.

D.A. Mays), 119–145. Madison WI: ASA, CSSA, and SSSA.

Wilson, J.R. (1993). Organization of forage plant tissues. In: *Forage Cell Wall Structure and Digestibility* (eds. H.G. Jung, D.R. Buxton, R.D. Hatfield, et al.), 1–32. Madison, WI: ASA-CSSA-SSSA.

Woodard, K.R., French, E.C., Sweat, L.A. et al. (2002). Nitrogen removal and nitrate-N leaching for forage systems receiving dairy effluent. *J. Environ. Qual.* 31: 1980–1992.

Wright, D.L. and Whitty, E.B. (2011). Fertilization of agronomic crops. United States Department of Agriculture. Florida Agricultural and Mechanical University Cooperative Extension Program.

Wright, A.L., Hons, F.M., and Rouquette, F.M. Jr. (2004). Long-term management impacts on soil carbon and nitrogen dynamics of grazed bermudagrass pastures. *Soil Biol. Biochem.* 36: 1809–1816.

Yarborough, J.K., Vendramini, J.M.B., Silveira, M.L. et al. (2017). Potassium and nitrogen fertilization effects on Jiggs bermudagrass herbage accumulation, root-rhizome mass, and tissue nutrient concentration. *Crop Forage Turfgrass Man.* https://doi.org/10.2134/cftm2017.04.0029.

CHAPTER

19

Forbs and Browse Species

David P. Belesky, Clinical Associate & Director, *Davis College Farm System, West Virginia University, Morgantown, WV, USA*
John W. Walker, Professor and Resident Director, *Texas A&M AgriLife Research & Extension Center, San Angelo, TX, USA*
Kimberly A. Cassida, Forage Extension Specialist, *Michigan State University, East Lansing, MI, USA*
James P. Muir, Professor, *Grassland Ecology, Texas A&M AgriLife Research & Extension Center, Stephenville, TX, USA*

Introduction

Focus on biodiversity, ecosystem function, and low-input approaches to pasture production has stimulated renewed interest in forbs, herbs, and nontraditional forage plants. Smith and Collins (2003) present a general and practical overview of forbs in agroecosystems, with a subsequent overview that includes browse species (Belesky et al. 2007). In our context, forbs are herbaceous dicotyledonous plants that are neither grass nor legume and are purported to have beneficial influences on overall pasture productivity and soil quality. This is based on lore that many forbs have (i) high-mineral concentrations, (ii) potentially beneficial health effects on livestock, (iii) the ability to create favorable **rhizosphere** conditions because of interactions of root architecture with the soil profile, and (iv) root exudates that interact with microbial populations to influence soil mineral dynamics (Foster 1988).

Forbs and browse (the leaf and young stem portion of woody plants and shrubs that can be consumed by livestock) typically help extend the temporal and spatial limits of forage availability, often as unmanaged parts of the plant community of a particular landscape. Whereas contributions to enhanced soil quality and livestock health remain to be explained, forbs and browse improve the seasonal distribution and availability of herbage by contributing to increased diversity of **plant functional groups**. Forbs and browse species acquire and use nutrients at different times and from different resource sites compared with co-occurring grasses and legumes in complex pasture communities.

Leaves and young shoots of trees and shrubs provide an estimated 75% of livestock feed in the tropics but are overlooked or ignored, for the most part, as a feed resource in temperate zones (Dupraz 1999). Production patterns, unlike those of companion grasses, often are not synchronized with weather patterns, especially conditions associated with soil water availability. The asynchrony of production and soil-water might be a product of the deep rooting characteristics of many forb and browse species.

Forbs

Many forbs are compatible with traditional range and pasture species and, depending on growth habit and

Forages: The Science of Grassland Agriculture, Volume II, Seventh Edition.
Edited by Kenneth J. Moore, Michael Collins, C. Jerry Nelson and Daren D. Redfearn.

tissue composition, could influence livestock selectivity and grazing behavior. Forbs can provide high-quality herbage at times when forage is inadequate in perennial grass-based pasture systems.

Brassicas

Improved cultivars and establishment methods, ability to extend the grazing season, and increased use as cover crops have led to renewed interest in brassicas as forage. The Brassicaceae (Cruciferae) family includes a number of species used as livestock feed since ancient times. *Brassica*, *Raphanus*, and *Sinapis* are the most widely used of the 51 genera (Tsunoda et al. 1980). The term brassica will be used to refer to species of this family with examples that include turnip, kale, rape, canola, swede, collards, mustard, and radish. These species are also known as crucifers.

Most forage brassicas are fast-growing, cold-tolerant, succulent (low dry matter, or DM, content), cool-season annuals or biennials. Turnip, swede, and radish have tuberous roots that can also be grazed. Brassicas can contain high concentrations of digestible energy and protein and readily retain these nutrients when stockpiled in the fall, which makes them useful for extending the grazing season. Low **cell wall content** coupled with high energy density produce rumen responses similar to those of concentrate feeds, and brassicas are often described as high-moisture concentrates.

Genetics and Species

Turnip

Turnip is a cool-season biennial that produces roots and foliage, both of which can be grazed by livestock. Up to 60% of total DM can be partitioned to roots. Root shape varies from globe-shaped to cylindrical, and cultivars differ in the proportion of fleshy root that is exposed above ground. Fleshy root biomass production is maximized by low seeding rates, earlier seeding date, stockpiling, and delayed grazing. Cultivars vary in the amount of DM that is allocated to roots (Kalmbacher et al. 1982; Rao and Horn 1986; Jung and Shaffer 1993). Multi-crowned cultivars (e.g. 'Appin') with improved regrowth potential are available and tolerate repeated grazing better than cultivars like 'Purple Top'. Turnips reach peak DM production 60–90 days after planting, and do not maintain leaf biomass as well after peak yield as do the other brassica species.

Rape

Forage rape is a biennial that produces a more fibrous and deeper root system than turnip, without the fleshy **tuber**. The leafy canopy varies among genotypes/cultivars from an erect, tall type with a significant proportion of stem tissue (e.g. 'Emerald') to dwarf cultivars that are leafier (e.g. 'Dwarf Essex'). Rape tolerates repeated grazing (Kalmbacher et al. 1982), and most regrowth comes from axillary buds on the stems. Most rape cultivars reach peak production in 90–120 days (Jung et al. 1986).

Swede or Rutabaga

Swedes are biennials that resemble turnips in growth form and management but require up to 120 days to reach peak production. Swede produces larger fleshy roots and shorter stems at low-plant density. At high plant density, stems can reach 75 cm in height. Swede does not regrow after grazing.

Kale

Forage kale includes marrow-stem and thousand-headed types. The former is taller and produces more stem, while the latter produces short, leafier stems. Kale is among the most cold tolerant of the brassicas and, therefore, the best suited for winter grazing. Thousand-headed kale (e.g. 'Premier') has excellent regrowth potential if grazing or cutting to less than 8 cm is avoided (Prestbye and Welty 1993), but marrow-stem kale (e.g. 'Gruner Ring') failed to regrow when cut to 10 cm (Kalmbacher et al. 1982). Kale planted in spring has greater yield potential than rape and turnip. In Canada, northern Europe, and the northern US kale requires the entire growing season (120 days) to reach maximum DM production (Fulkerson and Tossell 1972; Kunelius et al. 1987).

Collards

Forage collard is a biennial that is relatively new to the US. Cultivars such as 'Impact' are similar to thousand-headed kale in appearance and have cold tolerance similar to kale.

Mustard

Mustards are also relatively new to the forage brassica types available in the US, with interest driven by increasing availability of mustard seed. Species used as cover crops and forage include white or yellow mustard (*Sinapis alba* L.), brown or oriental mustard (*Brassica juncea* (L.) Czern.), black mustard (*Brassica nigra* (L.) W.D.J. Koch), and Ethiopian mustard (*Brassica carinata*). Some mustard cultivars are capable of producing high amounts of biologically active compounds such as glucosinolates. This class of compounds has attributes that, depending upon concentration, can be beneficial (e.g. biofumigant, antimicrobial, antioxidant capability) or detrimental (e.g. reduced-feed intake, goiter, **anemia**) (Bischoff 2016).

Forage Radish

Radish is an annual with peak forage production time varying from 60 to 80 days after planting. Oilseed radish cultivars ('Adagio', 'Colonel', 'Defender') have leafy tops with stubby, branched fleshy taproots. Radish cultivars developed specifically as cover crops ('Tillage Radish', 'Groundhog', 'Nitro') have leafy tops and extremely large fleshy taproots similar to the Daikon radish used as a vegetable. Cultivars developed for grazing include cabbage genetics ('Graza'), have fibrous roots with most growth devoted to leaves, and exhibit vigorous regrowth. All three of these types can be grazed. Because the fleshy-rooted types elevate the crown 10–15 cm. above the ground surface, regrowth after grazing usually does not occur, but Graza regrows readily and may be grazed multiple times per season if planted early. Oilseed and Daikon types bolt readily if planted early in the season and should not be planted until after mid-summer for forage use.

Brassica Hybrids

Chromosome number varies with species: *Brassica oleracea* = 18, *Brassica napus* = 38 and *Brassica campestris* var. *rapa* = 20. Many *Brassica*, *Sinapis* and *Raphanus* species will cross, and commercially available cultivars are often hybrids of one or more species. This makes categorization of cultivars challenging at best.

Uses

Depending on the species and days to maturity, brassicas can be planted in spring for summer-fall grazing or planted in midseason and stockpiled for fall–winter use. Planting by mid-July after harvest of an early crop such as winter wheat, can extend the grazing season into December or January, even in the northern US and Maritime Provinces of Canada. This late-season potential provides livestock producers with a rare opportunity to provide a high-quality forage resource to finishing or breeding livestock.

Brassicas are almost exclusively grazed. High-moisture content generally precludes practical use as a hay crop. There are a few reports of ensiled brassicas from Britain and New Zealand. Kale silage fed to lambs gave feed conversion and N-use-efficiency similar to alfalfa and red clover silages (Marley et al. 2007), but silage made from vegetative canola had unacceptably high pH that indicated poor fermentation (Neely et al. 2009). Mixing brassicas with other forage species, such as small grains, to reduce moisture content can improve fermentation (Moorby et al. 2003). There is no published research on brassicas as silage in North America, but producers do sometimes ensile them with mixed results.

The highly competitive brassica canopy can help control weeds and improve the seedbed for subsequent crop establishment in pasture renovation or cover crop programs. Radish cover crops suppress weeds (Lawley et al. 2011). Brassicas can also serve as a break crop for thin stands of alfalfa and other hay crops. Brassicas were interseeded successfully into sweet corn without affecting corn yield (Guldan et al. 1998). Cultivars of **fodder** radish that are resistant to sugarbeet nematodes and other brassicas can be used in rotation to control sugarbeet nematodes (*Hederodera schactii* Schmidt) in lieu of, or as a supplement to nematicides (Gardner and Caswell-Chen 1993). Most brassica cultivars are susceptible hosts and need to be plowed down prior to nematode reproduction. Recently developed radish cultivars that are resistant to sugar beet nematode effectively induce nematodes to hatch, but not reproduce. These resistant cultivars can be used in lieu of nematicides and can be grazed in the fall after soil temperatures reach 12 °C, without affecting nematode control (Yun et al. 1999; Gray and Koch 2002). Resistant radish cultivars (e.g. 'Adagio', 'Colonel', 'Arena', and 'Rimbo') should be used in place of susceptible radish cultivars and *Brassica* spp. on nematode-infested soils in rotations with sugar beet.

Adaptation

Brassicas and related species are adapted to a wide range of climate conditions, being grown mostly for summer-fall-winter grazing in cool, temperate climates and for winter-spring grazing in subtropical climates (Kalmbacher et al. 1982). Brassicas are often planted after cereals or other short-season crops in the northern US, Canada, northern Europe, and New Zealand.

Brassicas require fertile, well-drained soils and thrive over a wide range of soil textures. Soil pH should be 5.5–7.0. In general, salt tolerance is low, though rape has moderate salt tolerance. Turnip seed will germinate at soil temperatures as low as 4 °C. Brassicas grow well during cool weather and have good frost tolerance. When properly cold-hardened, the crop will grow with the tops surviving until air temperatures reach −6 to −9 °C. Roots can withstand temperatures as low as −9 to −12 °C (Dunavin 1987). Kale can survive −24 °C.

Nutritive Value

In contrast to perennial forages, brassicas maintain high nutrient value with maturity and stockpiling (Jung et al. 1986; Kunelius and Sanderson 1990; Wiedenhoeft and Barton 1994; Koch and Karakaya 1998). During autumn, *in vitro* dry matter digestibility (IVDMD) remained relatively unchanged (Jung et al. 1988) or increased (Sheldrick et al. 1981; Kunelius and Sanderson 1990) as harvest was delayed. The high-water content of brassica herbage raises concern for adequacy of DM intake, but its rapid digestion may compensate.

Foliage (leaves and stems) contains 140–250 g crude protein (CP) kg^{-1} DM, and fleshy roots of turnip and swede contain 80–120 g CP kg^{-1} DM. Brassicas are relatively low in fiber, readily digested, and provide high concentrations of energy for ruminant animals. **Metabolizable energy** ranges from 11 to 14 MJ kg^{-1} DM (Guillard and Allinson 1988). Nutritive value decreases with warm temperatures. Forage of brassicas planted in spring or early summer is higher in neutral detergent fiber (NDF) and acid detergent fiber (ADF) and lower in CP in July and August than is fall forage of similar age planted in midsummer (Rao and Horn 1986; Guillard and Allinson 1988; Wiedenhoeft and Barton 1994).

Mineral concentrations of brassicas meet recommendations for livestock dietary needs (Jung et al. 1986) though Cu deficiency occurred in cattle and sheep (Barry et al. 1981; Sharman et al. 1981). High foliar-Ca tends to produce a Ca: P ratio that is higher than recommended (Guillard and Allinson 1989b). Brassicas are relatively high in sulfur, which may induce polioencephalomalacia, especially if fed with other high-sulfur feeds or water sources (Arnold 2014). Grass tetany has been reported when grazing brassicas.

Brassicas contain a variety of **secondary metabolites** that can cause health problems that reduce animal DM intake and growth, including S-methyl cysteine sulfoxide (hemolytic **anemia**), glucosinolates (goiter) (Tookey et al. 1980; Barry et al. 1982; Gustine and Jung 1985; Lambert et al. 1987; Duncan 1990; Barry 2013), and nitrates (Cassida et al. 1995; Arnold 2014). Grazing immature rape (<45 days after planting) may cause **photosensitization** (Arnold 2014). The extent to which antiquality components in brassicas affect animal health in North American systems is unclear. Brassicas grown in mixtures with other forages to help dilute toxins, adapting animals to brassica pasture slowly, and offering mineral mixes containing iodine can all help minimize problems associated with grazing brassicas.

Cultural Practices

Brassica seed can be drilled or broadcast seeded. A no-till disk drill should be used when planting into stubble or sod. Planting is easier if loose straw and other coarse plant residue is removed from the soil surface. Burning is not recommended because the intact mulch or plant residue layer helps retain moisture and could increase success of summer plantings. Unless the soil surface can be kept moist for several days, for example, with irrigation, fields should be rolled and compacted after broadcast planting. Turnips and rape have been established by aerial seeding into small grain crops before final irrigation.

The **seedbed** should be firm and weed free, as recommended for perennial forages. After the seedling stage, brassicas are effective competitors and weeds are generally not troublesome. Compared with a tilled seedbed, seeding directly into residues of a previous crop or an herbicide-suppressed sod conserves soil, water, and energy, saves time, and reduces costs. In addition, residues are a source of high-fiber forage, needed when animals graze the succulent, low-fiber brassicas. Maintaining a sod reduces trampling losses, particularly when grazed during inclement weather (Jung et al. 1984; Koch et al. 1987).

Brassicas can be grown in mixture with other forage species. Mixtures, including italian ryegrass and crimson clover, had greater digestible organic matter and CP than did turnip or rape grown alone. Brassicas dominated early, with ryegrass dominating later cuttings (Dunavin 1987). Brassicas should not be planted in the same field for more than two consecutive years because of increased pest and disease problems.

Brassica seed are relatively small (200 000 seed kg^{-1}). Recommended seeding rates are 1.7–2.2 kg ha^{-1} for turnip, 4.0–4.5 kg ha^{-1} for rape and kale, and 15 kg ha^{-1} for fodder radish. A dense stand of fodder radish used for biological control of sugarbeet nematode is necessary for fibrous root proliferation and maximum nematode control. Otherwise, radish crop management is similar to that of turnip.

Planting date depends on when forage is needed in the production system, the species planted, and soil temperature. Brassica seed germinates between 10 and 35 °C, and turnip germinates better than rape or kale at 5 and 10 °C (Wilson et al. 1992). In northern areas, turnip, rape, or thousand-headed (stemless) kale can be planted in the spring after soil temperature reaches 10 °C to provide summer pasture (August–September). Swede or kale can be planted in spring, whereas rape, turnip, or a combination can be planted in midsummer providing fall-winter grazing (Jung et al. 1986; Kunelius et al. 1987).

Some brassicas, such as rape and kale, produce more forage than turnip and radish over the full season (Jung et al. 1986; Kunelius et al. 1987). Turnip, rape, and radish planted in August in the northern states, and in Canada, will produce less growth than the same crops planted in July (Kunelius et al. 1987; Koch et al. 2002).

Brassica seed respond to fertilizer N inputs (Jung et al. 1984; Guillard and Allinson 1989a). A minimum of 80 kg ha^{-1} of N is usually needed. More N may be necessary if residual soil-N is low, for example, if planted after a small grain crop or where straw was incorporated. Excess N accumulates as NO_3^- in herbage, but the high content of soluble carbohydrates may increase rate of NO_2^- production in the rumen and thus mitigate toxicity to livestock (Cassida et al. 1995). Young plants have a high P requirement (Jung et al. 1984). Brassicas are among the few forage species that do not form the root associations with mycorrhizae that help many other forage species obtain P (Brundrett 2009). Soil levels of P and K should be moderate or higher based on soil fertility analysis.

Brassicas require ample soil moisture; therefore, late-season irrigation may be necessary, particularly in semiarid and arid regions. About 30 cm of water over a 75-day period is required when planted as a second crop in central Washington (Heinemann et al. 1981).

Major Diseases and Pests

Disease impacts are more severe with spring plantings than with later plantings. Disease incidence was less with frequent harvest or grazing of brassicas (i.e. 60–90 days) than if forage was accumulated for longer periods (Jung et al. 1988). Most disease occurs near physiologic maturity. Most cultivars of turnip, rape, swede, and kale have low incidence of leaf spot (*Xanthomonas campestris*, *Alternaria* sp. and *Cercospora* sp.). Incidence of mildew (*Erysiphe cruciferarum*) was high on swede cultivars and on two rape cultivars, but low on turnip, kale, and other rape cultivars (Jung et al. 1986).

Flea beetles (*Phyllotreta cruciferae* Goeze and *Phyllotreta striolata* Fab.) are the most common insect problem of brassicas. Late-spring plantings are more likely to encounter insect attack than late-summer plantings. Flea beetle attack is less severe when turnip is planted into suppressed sod compared with a conventional seedbed (Jung et al. 1983). New Zealand turnip cultivars seem to be more resistant to flea beetle than European cultivars. Swede suffered less injury than turnip.

Aphids (*Hyadaphis pseudobrassicae* Davis) did not injure swede, rape or kale, but damaged some turnip cultivars, mainly between 60 and 90 days after planting (Jung et al. 1986). If needed, insecticides should be approved for use in grazed systems. If brassica use expands, disease and insect problems may develop, increasing the need for resistant cultivars.

Forage Production

Brassica can produce DM yields of 5–9 Mg ha^{-1} or more by 90 days after planting (Jung et al. 1986; Guillard and Allinson 1988; Koch et al. 2002). In some locations, brassica DM yields can be up to 12 Mg ha^{-1} (Jung et al. 1988). High yields depend on ample moisture and nutrient inputs. After grazing, at least 15–20 cm of rape or turnip stubble should remain to sustain regrowth.

Turnip and radish produce similar DM yields over 60–90 days; however, the fleshy roots of radish, constituting 25–40% of total radish DM, are not readily eaten by livestock, whereas turnip roots are highly palatable (Koch et al. 2002). Multiple cuttings increase top growth yield as a proportion of total production at the expense of roots. Rape produces higher top growth yields with multiple cuts than with stockpiling. Marrowstem kale, however, yields more when stockpiled than when cut multiple times (Kalmbacher et al. 1982).

Grazing Management

Low-DM content restricts use primarily to grazing, but grazing must be managed for optimal livestock performance. Some brassicas can be grazed as early as 45 days after planting (turnips, radishes), but most should not be grazed before 60 days of growth to reduce the risk of **photosensitization**. Animals should be allowed to adjust slowly, with the brassica component of the diet increasing gradually over 7–10 days. Limiting access with strip grazing or limiting daily duration on pasture will reduce herbage loss caused by trampling and fouling and improve utilization.

Animals grazing brassicas should have access to higher-fiber forages to increase DM intake, provide normal rumen function, and/or minimize the effects of antiquality factors of brassicas (Wikse and Gates 1987; Guillard and Allinson 1989b). Lambs should consume about 0.2 kg fiber d^{-1} and cows or steers 1–2 kg fiber d^{-1} from other forages when grazing brassicas. Growing lambs perform better when given access to residues from previous crops or hay supplement (Table 19.1). Lamb gains were generally more than 0.20 kg d^{-1} with higher-fiber supplements, whereas lambs without high-fiber supplements gained less than 0.15 kg d^{-1}. Supplementing lambs with hay while grazing brassicas improved daily gain during the first three weeks, but not later, indicating high-fiber forage may be most important during the adjustment stage (Reid et al. 1994).

Animals grazing brassicas had the highest intake of digestible DM and gains when they received 20–30% of the daily diet in a high-fiber form (Lambert et al. 1987; Pearce et al. 1991). Brassica-based diets supplemented with 25% and 50% hay had lower apparent digestibility, but higher gains of lambs compared with lambs fed only hybrid brassica (*Brassica rapa* L. × *B. rapa* subsp. *pekinensis* (Lour) Hanelt) (Cassida et al. 1994).

The soil surface tends to be devoid of cover after grazing monocultures of leafy brassicas. On erodible soils, a stemmy species can be interseeded in a mixture (e.g. rape with turnip), provided the pasture will not be overgrazed. A small grain such as oat or barley can be seeded along with brassicas and this is a common practice when brassicas are used in cover crop mixtures. Winter wheat, rye, or triticale can be used to extend production for spring grazing.

Livestock Performance

Brassicas are best for livestock with high-nutritional requirements. Grazing animal performance on brassicas is highly variable (Faix et al. 1980; Koch et al. 1987; Reid et al. 1994). A summary of New Zealand experiments with brassica crops showed growing lambs gained 0.095–0.147 kg d^{-1} (Nicol and Barry 1980), less than half that predicted from the nutritive composition (NRC 1985). Reduced voluntary intake of DM because of high-water content and plant toxins was implicated (Lambert et al. 1987; Duncan 1990); however, when brassica and hay were fed in various ratios, DM intake by lambs was not limited by brassica proportion (Cassida et al. 1994).

The change from high-fiber forage to low-fiber brassicas may create abnormal rumen fermentation, affecting performance (Lambert et al. 1987). Sod-seeded brassicas may have some residue from previous crops that is available for grazing (Table 19.1). Animals initially shun brassicas in favor of other forage types; therefore, even small amounts of residue may be adequate to help compensate for low

Table 19.1 Lamb performance grazing brassica pasture (summary of experiments)

Species	Planting method	Grazing dates	Supplement	ADG[a]	Source
Various	Conventional	Various	Unknown	0.10–0.15	Nicol and Barry (1980)
Turnip, rape	No-till; mid-July	23 Oct–14 Dec; 99 DAP[b]	None	0.05–0.09	Faix et al. (1980)
Turnip, radish	Conventional; late July	1 Nov–10 Jan; 90 DAP	None	0.11–0.12	Heinemann et al. (1981)
Rape, kale, fodder radish	Conventional; late July	Late Sept–mid-Nov; 60 DAP	None	0.11–0.15	Fitzgerald and Black (1984)
Turnip, kale	No-till; late July	Nov–early Dec; 90 DAP	None	0.02–0.05	Reid et al. (1994)
Turnip (Tyfon)	No-till; late July	Nov–early Dec; 90 DAP	Switchgrass 15–20% of DM	0.20–0.24	Reid et al. (1994)
Week 1–3	No-till; late July	Nov–early Dec; 90 DAP	No hay	0.36	Reid et al. (1994)
Week 1–3			Hay	0.42	
Week 4–7			No hay	0.30	
Week 4–7			Hay	0.24	
Turnip (Tyfon)	No-till; late July	Oct–early Dec; 70 DAP	Orchardgrass 11% of DM	0.21–0.25	Koch et al. (1987)
Turnip (Tyfon)	No-till; early Aug	Nov–mid-Dec; 92 DAP	Flatpea 12% DM	0.20–0.21	Rule et al. (1991)
Rape, turnip	Conventional; late May	Mid-Jul–late Sept; 58 DAP	Weeds 10–36% DM	0.20–0.23	Thomas et al. (1990)
Rape, turnip (Tyfon)	No-till; late Jul–mid-Aug	Mid-Oct–early Jan; 60–90 DAP	Straw/small grain regrowth 20–32% of DM	0.18	Koch et al. (2002)

[a]ADG, average daily live weight gain per head (kg $animal^{-1}$ d^{-1}).
[b]DAP, days after planting.

fiber in brassicas. Animals grazing brassicas supplemented with high-fiber dry forage gained better than animals grazing brassicas alone (Lambert et al. 1987).

Dry matter intake increased when rape content of a pen-fed diet was supplemented with up to 30% orchardgrass hay. Ideal DM intake of orchardgrass was 18% of the diet. Hay supplement increased average daily gain (ADG) of lambs during the first three weeks of grazing, but decreased ADG during weeks 4–7 (Reid et al. 1994). Carcass characteristics were similar for lambs grazing brassica crops or fattened on grass in drylot, but more time was required to reach target weight when grazing brassica (Fitzgerald and Black 1984; Koch et al. 2002). Lambs grazing radish and turnip with no grain supplement produced carcasses with acceptable yield and quality grade (Yun et al. 1999).

Chicory

Chicory, also known by many other names such as succory, blueweed, blue sailor, blue bunk, wild bachelor's button, hendibeh, and blue daisy, is a short-lived, perennial herb that is adapted throughout temperate regions. Chicory was mentioned in the early writings of Virgil, Horace, and Pliny. The term *chicory* is probably derived from "chicouryeh or chicourey," the name used by Arab physicians. Chicory (referred to as *agon*) was used as a green salad or for medicinal purposes by the ancient Egyptians, by the Greeks (*kichora* or *kichorea*), and by the Romans (*intybus agrestis*).

Chicory originated in the Mediterranean area (Vavilov 1992), primarily from Abyssinia (northeastern Africa). It is now indigenous to west Asia and southern Europe and became naturalized in North and South America (Simon et al. 1984). Chicory is a member of the Asterales Family; other important species include *Chichorium endiva*, *Chichorium spinosum*, *Chichorium glandulosum* Boiss; *Chichorium bottae* Defl., and *Chichorium calvum* Schulz-Bip. Early selection and improvement efforts were directed toward producing a table green or salad (witloof) endive.

Documented evidence of chicory growing in North America was made in 1758 by Governor Bowdoin of Massachusetts, but reference to chicory growing in herb gardens of tidewater Virginia was made in the early 1700s. The US imported large quantities of chicory root for use as a coffee additive (flavor) or substitute, especially in areas with a strong French influence. Chicory has been in use as a cattle feed in central Europe for the past 300 years (Brenchley 1920), but its use as a forage crop in North America is relatively new.

Morphology and Physiology

Chicory has a taproot with a large capacity for storage of inulin, a nonstructural carbohydrate polymer of fructose. Leaves arise from a basal **rosette** on vegetative plants and appear on stems that can be 1–2 m tall at flowering. The rosette type of the perennial plant, while similar in appearance to the wild biennial form, is cultivated as a green vegetable and is the source of the fodder-type chicory that is gaining popularity as a pasture forage resource.

The life cycle can be biennial (weed type) or short-lived perennial, both producing a compressed panicle consisting of 15–25 flowers that are generally self-incompatible and cross-pollinated. The fruit is an **achene**. The plant has nine chromosome pairs. Leaves are oblong but can vary from oblate to dentate, clasping, and **sessile**.

Chicory requires a long day (>14 hours) for flowering. The pale blue- to purple-colored flowers open in morning and close by noon. Plants are generally vegetative during the first growing season, but flowering can occur in the seeding year if plants experience intervals of drought or cold. An increasing number of flowering stems are expressed in subsequent years (Clapham et al. 2001). Some cultivars require a preceding cold period to flower (Ryder 1999). Flowers are present from June to September, and maturity varies such that vegetative and reproductive plants occur simultaneously in the population (Clapham et al. 2001).

'Grasslands Puna' chicory, the first commercial cultivar of the perennial forage type, was developed from multiple genotypes selected for dense, leafy, vigorous, and uniform shoot appearance (Rumball 1986). Chicory persists as a short-lived perennial and is compatible with traditional pasture grass and legume species (Jung et al. 1996; Belesky et al. 1999). Plants persist for four to six years in pastures, with stand loss influenced by sward management and weather. Canopy management is similar to that required for alfalfa. Chicory does not fix nitrogen.

General Adaptation

Chicory grows well in temperate and temperate-maritime climates and grows over a temperature range of 6–27 °C, with preference for cool weather. Basal rosettes begin growth early in spring and, in areas with no prolonged subfreezing temperatures, can remain green and productive throughout the year. Chicory persists under winter conditions in parts of northeastern US (Skinner and Gustine 2002) and maritime Canada (Kunelius and MacRae 1999). Leaf growth and stem development are most active when air temperatures are above 20 °C, with pollen **germination** active at 17–20 °C. Chicory requires deep, moderately to well-drained soil, but does tolerate wet periods. Chicory grows best at soil pH of 5.5–6.0 and can tolerate drought, salinity, and high nutrient conditions (Neel et al. 2002; Alloush et al. 2003). Minor pest problems occur with aphids and tomato fruit worms (*Heliothus armigera conferta*) (Hare et al. 1987). Chicory is generally free from foliar and root diseases.

Seeding and Management

Chicory can be sown into a prepared **seedbed** or sod-seeded with a planter that controls planting depth. Broadcast seedings are a less reliable means of stand establishment. Seed should not be sown deeper than 1 cm (Sanderson and Elwinger 2000a). Seeding rate should be about 2 kg ha^{-1} when seeded alone or 0.5–1 kg ha^{-1} when sown in mixtures. Soil should be cultipacked after seeding. Spring sowing is common, though chicory tolerates autumn sowing in the southern US (Pitman and Willis 1998). Soil temperature at seeding should be at least 10 °C.

Medium to high annual inputs of nutrients (75–150 kg N; 25–40 kg P; 20 kg K, and 20 kg S ha^{-1}) are required because chicory is very responsive to fertilizer, especially cations and P. Chicory responds to N fertilizer, but can accumulate high concentrations of **nitrate** (NO_3^-), especially in young, establishing plants in the seeding year.

Chicory is compatible with common forage grasses and legumes. Pure chicory stands are invaded by grasses and weeds in the Appalachian region of the eastern US. Establishing or maintaining pure stands of chicory is not practical from the standpoint of livestock nutrition. Chicory forage tends to have low concentrations of fiber (Turner et al. 1999; Belesky et al. 2000; Holden et al. 2000). Nutritive value depends on management (Table 19.2).

Chicory is usually managed as pasture, but it can also be ensiled (Kälber et al. 2012). Grazing should begin when the average canopy height reaches 15–20 cm to help control **bolting**. Livestock should be removed at 5-cm residual plant height and the sward rested for 25–30 days before grazing again. Plants in the seeding year may reach 15–20 cm in 80–100 days. If seedlings are grazed earlier, livestock will pull plants from the soil. Plants can winter heave if soil is wet. Frequent and intensive defoliation will accelerate chicory loss from the stand. Long intervals between grazing periods will help the stand persist (Volesky 1996).

Stems are often avoided by grazing livestock, but in grazing experiments in West Virginia, lambs preferred chicory flowers and consumed herbage with some reluctance. This raised questions about environmental and management influences on types and concentrations of secondary metabolites in chicory. Chicory contains lactupicrin and lactones, which may decrease palatability (Foster et al. 2001). Cultivars differ in concentrations and influence on livestock preference. Sheep and whitetail deer (*Odocoileus virginianus* Boddaert) avoided 'Forage Feast', which had greater concentrations of polyphenolics than 'Grasslands Puna' or 'Lacerta', (Foster et al. 2002), but goats displayed no preference among chicory cultivars that differed in sesquiterpene lactones (Cassida et al. 2010).

Cultivars grown in Pennsylvania (calcareous soil) contained less lactupicrin and polyphenolics than did plants grown in West Virginia (acidic soil matrix). Herbage yield (Table 19.2) ranged from 3.5 to 9.4 Mg ha^{-1} depending on location and management (Jung et al. 1996; Volesky 1996; Collins and McCoy 1997; Belesky et al. 1999, 2000). Cultivars available for forage applications include 'Grasslands Puna' (summer active), 'Forage Feast' (resists **bolting**), 'Good Hunt', 'Lacerta', 'Oasis', 'Puna II', 'La Niña' (cold tolerant), 'Ceres Grouse', and 'Chico'.

General advantages of forage chicory in production systems include high palatability, high production potential, persistence in mixtures, drought resistance, tolerance to acidic and saline soil conditions, rapid recovery from grazing, non-bloating, natural reseeding, good silage potential, and high mineral content. Disadvantages of chicory include the following: need for rotational stocking, need for N inputs, preference for better soil (but is productive on soils varying in chemical and physical properties), dormancy in moderate to severe winters, susceptibility to crown damage, and tainting of milk flavor if chicory makes up more than half the diet. There is no evidence of meat "off" flavor.

Plantain

Folklore attributes health-giving and restorative properties to plantain. There is a long record of use in pastures in Europe, with the intent to improve livestock health. Plantain probably originated in the Mediterranean region and is adapted globally in temperate regions and at high elevations of the tropics. The genus name refers to the anthropogenic form of dispersal ("of the foot"). Plantain is mentioned in Roman and Greek herbal pharmacopeia. Common name synonyms include broad-leaved plantain, ripple grass, waybread, slan-lus, snakeweed, Englishman's foot, white-man's foot, ribwort, and ribgrass.

Major species are *Plantago lanceolata*, *Plantago major*, and *Plantago psyllium*. Plantains are members of the Order Plantaginales and the Family Plantaginaceae. They are perennial herbs characterized by a **rosette** of oblate leaves and a large main root. Plants produce lateral crown buds but not very vigorously. Plantain is a prolific seed producer and does not reproduce vegetatively. Plants are gynodioeceous or androdioeceous (dioeceous plants), self-incompatible, and insect or wind pollinated. *P. lanceolata* has 6 or 12 chromosome pairs (Murin 1997).

General Adaptation

There is very little information on productivity and response of plantain in North America; most information was developed in New Zealand (Stewart 1996; Rumball et al. 1997). Plantain tolerates soils with high mineral

Table 19.2 Dry matter production and some nutritive value characteristics of forage chicory in North America

Management factor	DM (Mg ha^{-1})	IVDMD (g kg^{-1})	CP (g kg^{-1})	Source
Canopy removal				
Light vs. heavy	8.2–11.0	—	—	Jung et al. (1996)
	—	—	146–226	Holden et al. (2000)
Stocking				
Light vs. heavy	7.9–6.6	709	179–186	Volesky (1996)
Canopy strata (g 100 g^{-1})				
0–10 cm	37	746	162	McCoy et al. (1997)
10–20 cm	38	719	248	
>20 cm	25	741	293	
Nitrogen				
0–200 kg ha^{-1}	1.8–6.7	685–716	132–161	Collins and McCoy (1997)
0–400 kg ha^{-1}	3.5–4.8	—	—	Belesky et al. (1999)
		567–507	202–318	Turner et al. (1999)
Clipping frequency				
3-cut, Maritime Canada	6.4	—	—	Kunelius and MacRae (1999)
3-cut, Southern US	5.7	—	—	Pitman and Willis (1998)
3 or 6-wk, Southern US	6.5–8.3	631–630	184–166	Belesky et al. (1999)

nutrient content and pH ranging from 4.5 to 8, but not wet or saline soil conditions. *Plantago* species occur in the pioneer stage of site **succession** and often in pasture areas with a high degree of disturbance (e.g. lanes or trails). The species is generally a poor competitor for light (Kuiper and Bos 1992).

Seeding and Management

Established plants are productive for one to three years and up to four years if not grazed too closely. The species is responsive to N fertilizer but can be suppressed by competitors when grown in mixed pasture. Plantain generally contributes less than 20% of total DM to the sward, but pure-stand yields of up to 20 Mg ha^{-1} are possible. Plantain is not recommended for use in a pure stand because of weed encroachment. It is best to consider seeding it as part of a mixture with red or white clover, orchardgrass, or tall fescue. Perennial ryegrass is too competitive for plantain.

Plantain should be seeded at 2–4 kg ha^{-1} in mixture with grasses and legumes, or 2–3 kg ha^{-1} with brassicas. Plantain should not be planted more than 1-cm deep (Sanderson and Elwinger 2000b). Surface broadcast seeding is successful and can be done anytime, though Sanderson and Elwinger (2000a) found that plantain growth was slow and plants might not establish when seeded in autumn. Seed requires **cold stratification** (<5 °C for one to three months) to germinate and generally does not germinate in the dark. Grazing keeps plants vegetative and prevents seed stems from forming. Livestock performance is better on mixed swards rather than pure stands of plantain fertilized with N.

Cultivars include 'Lancelot' and 'Tonic'. Lancelot was selected from 'broad landraces in New Zealand' for a leafy, uniform and productive plant about 25–30 cm tall. Lancelot grows semierect with high-summer production and winter dormancy (Rumball et al. 1997). Lancelot can become prostrate with repeated grazing, yet, should be stocked rotationally at 20- to 25-day intervals (Stewart 1996). Tonic, also selected in New Zealand, is more suited to **set stocking**. It grows actively in winter, due in part to its parental origin (Portugal). In the northeastern US, Lancelot is more tolerant of winter conditions than is Tonic (Skinner and Gustine 2002).

Nutritive Value

Plantain provides the fiber needed for livestock grazing swards with leafy brassicas and chicory. However, **secondary metabolites** that influence rumen microorganisms may impair digestibility. Plantain contains up to 9% of DM of verbascoside, a phenylpropanoid glycoside that has anti-microbial, -fungal, and -helminthic properties (Fajer et al. 1992). Meat flavor is not affected by plantain. Concentrations of Ca, Mg, Na, P, Zn, Cu, and Co can be greater than or equal to those found in most grasses and legumes.

There is some evidence that Lancelot has the medicinal properties suggested in herbal literature. Jarzomski et al. (2000) showed that plant age, leaf age, and nutrient status influenced secondary metabolites in plantain, but herbivory did not. Botanic-medicinal constituents include iridoid glucoside (acubin) and derivatives with concentrations of up to 3% of DM. These are important biologically active compounds that can increase with drought stress. Plantain also contains polysaccharide hydrocolloids, which act as purgatives. Sorbitol is also present and could act as a palatability enhancer.

Other Forbs

Other forbs are used occasionally in seed mixtures for herbal leys and meadows, and these vary depending on location and reseeding objective. Foster (1988) cites "private farmer research" to point out the benefits of including herbs in British pasture-seed mixtures. Minimal experimental data for forbs exist for applications in US pastures.

Small burnet is noted for persistence in New Zealand hill-land pasture but has limited summer production. In the western US, an examination of burnet accessions showed yield potential similar to alfalfa in Utah (Peel et al. 2009). Burnet does not do well in higher-rainfall, leached soils, and is not tolerant of **overgrazing**.

Forage kochia has shown some potential as drought tolerant, perennial semi-shrub for browsing in the western US, but it is difficult to establish (Creech et al. 2013). Despite the common name, forage kochia is not the same species as the invasive weed.

Interest in buckwheat as a deep-rooted cover crop has also renewed interest in its use for grazing, hay, and silage. Buckwheat contains fagopyrin, an antiquality compound that can cause photosensitization in animals that consume it (Cheeke 1998). It has been ensiled and fed to dairy cattle successfully in Europe (Kälber et al. 2012). There is little published information from North American feeding systems, but numerous anecdotal reports of success from farmers and Cooperative Extension. To avoid **photosensitization**, Cornell University Extension (http://www.hort.cornell.edu/bjorkman/lab/buck/guide/forage.php) recommends limiting buckwheat to 30% or less of the ration if cattle are exposed to sunlight.

Sheep sorrel has wide adaptation and typically can be considered an indicator of acidic soil. Sheep sorrel has biologic properties that include antioxidant and anti-tumor functions (Leonard et al. 2006). Yarrow occurs in humid, temperate areas and is a weed in cropland, though considered an acceptable species for grazing lands in New Zealand. It is reported to accelerate blood clotting in sheep. The aforementioned forbs are purported to be rich in minerals, vitamins, and antioxidants. Seeding rates are 1 kg ha^{-1} or less in mixtures.

Browse

Worldwide, browse species occur on many thousands of hectares of extensively managed grassland, rangeland, and grazed forestland, including silvopasture. The areas vary in soil type, topographic features, and environmental characteristics, creating conditions that support a wide array of plant species differing in chemical composition and palatability. Browse is often less desirable as forage for domestic livestock than are co-occurring herbaceous plants; however, they do provide nutrients and cover at certain times of the year. **Browse** is defined as the leaves, tender twigs and stems of shrubs, woody vines, trees, cacti and other non-herbaceous vegetation that can be ingested by domestic and feral herbivores. Not only the plant parts mentioned but the fruit, pods and seeds of woody browse species provide valuable feed.

As many as 500–600 species of browse occur in western rangeland of the United States alone, making generalized statements about management and nutritive value of browse difficult. Of the 155 types of rangeland cover in the contiguous 48 states of the US, nearly half (43%) are characterized by the dominant shrub species on the site (Shiflet 1994).

Importance and Use in Range and Silvopasture Management Systems

Some browse species were introduced to improve seasonal distribution of herbage or nutritive value, but generally, browse encroachment reduces grazing animal productivity and enhances browser performance. Grazing is a practical way to manage browse and influence the floristic composition of grazing lands. Meeting some of the nutrient needs of the grazing animal may be a secondary benefit because the primary benefits often are wildlife habitat and environment stability.

Browse production and use requires understanding the ecologic function in a particular biome. For example, food, water, cover, and space are essential components of wildlife habitat. Browse provides food and cover (Robinette 1972) and influences water use and nutrient cycling in the landscape. About half of the indigenous mammal and bird species in North America are associated with woody plant cover and have specific habitat requirements (Yeager 1961). In fact, the common names of many shrub and wildlife species such as rabbitbrush and sage grouse (*Centrocercus urophasianus*) reflect this association.

Browse and herbaceous species compete for water and influence the amount and quality that flows from sites or reaches the groundwater (Thurow and Hester 1997). This is significant since water quality and conservation are important ecologic functions of rangelands. Table 19.3 shows examples of browse species by broad geographic regions of the US.

During the past 50–100 years, the floristic composition of natural grasslands and savannas in arid and semiarid

Table 19.3 List of important US browse species by region[a]

Common name	Scientific name
Northeast	
Autumn olive	*Elaeagnus umbellata* Thunb.
Quaking aspen	*Populus tremuloides* Michx.
Black locust	*Robinia pseudoacacia* L.
Multiflora rose	*Rosa multiflora* Thunb.
Southeast	
Yaupon holly	*Ilex vomitoria* Ait.
Blackjack oak	*Quercus marilandica* Muench.
Post oak	*Quercus stellata* Wangenh.
Mexican cliffrose	*Purshia mexicana* (D. Don) S.L. Welsh var. *mexicana*
Black locust	*Robinia pseudoacacia* L.
Multiflora rose	*Rosa multiflora* Thunb.
Littleleaf leadtree	*Lewucaena retusa* Benth.
Northwest	
Silver sagebrush	*Artemisia cana* Pursh
Fringed sagebrush	*Artemisia frigida* Willd.
Cudweed sagewort	*Artemisia ludoviciana* Nutt.
Black sagebrush	*Artemisia nova* A. Nels.
Budsage	*Artemisia spinescens* D.C. Eat.
Big sagebrush	*Artemisia tridentate* Nutt.
Snowberry	*Symphoricarpos albus* (L.) Blake
Fourwing saltbush	*Atriplex canescens* (Pursh) Nutt.
Shadscale saltbush	*Atriplex confertifolia* (Torr. & Frem.) S. Wats.
Saltbush	*Atriplex gardneri* Moq. D. Dietr.
Winterfat	*Ceratoides lanata* (Pursh) A. Meeuse & A. Smit
Spiny hopsage	*Grayia spinosa* (Hook.) Moq.
Black greasewood	*Sarcobatus vermiculatus* (Hook.) Torr.
Leadplant	*Amorpha canescens* Pursh
Serviceberry	*Amelanchier alnifolia* (Nutt.) Nutt. ex M. Roem.
True mountain mahogany	*Cercocarpus montanus* Raf.
Shrubby cinquefoil	*Potentilla fruticosa* L.
Antelope bitterbrush	*Purshia tridentata* (Pursh) DC.
Wild rose	*Rosa woodsii* Lindl.
Quaking aspen	*Populus tremuloides* Michx.
Wax currant	*Ribes cereum* Douglas
Southwest	
Skunkbrush sumac	*Rhus trilobata* Nutt. ex Torr. & A. Gray
Sand sagebrush	*Artemisia filifolia* Torr.
White bursage	*Ambrosia dumosa* (A. Gray) W.W. Payne
Fourwing saltbush	*Atriplex canescens* (Pursh) Nutt.
Black greasewood	*Sarcobatus vermiculatus* (Hook.) Torr.
Pointleaf manzanita	*Arctostaphylos pungens* Kunth
Guajillo	*Acacia berlandieri* Benth.
Gambel oak	*Quercus gambelii* Nutt.
Fendler ceanothus	*Ceanothus fendleri* A. Gray
Deerbrush	*Ceanothus integerrimus* Hook. & Arn.
True mountain mahogany	*Cercocarpus montanus* Raf.
Blackbrush	*Coleogyne ramosissima* Torr.
Wild rose	*Rosa woodsii* Lindl.

[a]This list is not all inclusive, but represents common browse plants that are considered fair or better forage for livestock.

rangelands has shifted to include more browse species because of the presence of domestic livestock (Archer 1989). In Texas alone, dense, woody plant cover on more than 88% of rangeland (35 million ha) restricts livestock production (Scifres 1980). Changes in climate, decreased fire frequency, overgrazing, and increased concentrations of atmospheric carbon dioxide (CO_2) are possible causes of increased woody plants in the plant community. Fire favors the occurrence of herbaceous over woody species, whereas grazing favors woody plants by reducing competition and increasing seed dispersal. Increased atmospheric CO_2 may provide a competitive advantage to shrubs because high CO_2 causes grasses to transpire less, thus slowing soil water depletion (Polley et al. 1997).

With the disappearance of natural predators on some winter range in the western US, mule deer (*Odocoileus hemionus*) and elk (*Cervus elaphus*) over-browsed and reduced willows in riparian areas (Kay 1997) and mountain big sagebrush (Wambolt and Sherwood 1999) and bitterbrush (Ganskopp et al. 1999) on uplands. These shrubs do not have chemical or mechanical characteristics such as thorns, pubescent leaves, or silica deposits that deter **overgrazing**. In some parts of the intermountain region of the western US, big sagebrush presence in the stand decreased when frequency of fire increased because downy cheatgrass invaded and spread among areas covered with species such as big sagebrush (Young and Allen 1997). Conversely, sagebrush increased in areas where natural fire intervals were disrupted (Bastian et al. 1995). Managed browse can help control invasive brushy species while, simultaneously, capturing nutritional benefits to support ruminant production. An example of this has been documented for the Edwards plateau of west-central Texas (USDA-NRCS 1994). Likewise, grazing livestock can be used to help manage early successional species, usually woody shrubs, on land areas disturbed by natural and anthropogenic events (Abaye et al. 2011).

Silvopasture is an integrated agro-ecosystem that involves the managed interaction of trees, forages, and livestock grown on the same site (Clason and Sharrow 2000). It is a common land management practice used throughout the world (Cubbage et al. 2012) that offers a means to diversify production and enhance income-generating possibilities on a limited land area. However, silvopasture is not widely used in North America. Grazing domestic livestock is an integral part of this system. In some areas, particularly the northeastern US, the practice was discouraged because of apparent negative impacts on soil and vegetation resulting from unmanaged livestock grazing in wooded areas. Damage included soil compaction, injury to trees, and the loss of tree regeneration attributed to livestock consuming new tree shoots and saplings. Conversely, with the emergence of new applied understanding of how to manage the soil-plant-livestock continuum, well-managed silvopasture can not only deliver plant and livestock products, but ecosystem benefits contributing to biologic diversity, soil and water resource conservation, and rural community and societal well-being.

New challenges associated with invasive plants, high land-ownership costs, and the need to sustain healthy and productive woodlands has led to renewed interest in silvopasture as a means to optimize agro-ecosystem production. Silvopasture differs from woodlot grazing in that the frequency and intensity of grazing is controlled to achieve desired objectives. New fencing systems, a better understanding of animal behavior, knowledge of forage plant responses to competition and the development of "**management intensive grazing**" practices facilitate sustainable management of complex silvopasture systems. The interaction is dynamic with changes occurring as the tree canopy matures and how that influences the micro-climate of the forest understory. Interactions vary during and among growing seasons with the seasonal patterns of plant canopy development and livestock grazing behavior.

Silvopasture involves the complex management of two distinctly different plant canopies, each aiming to optimize acquisition and use of sunlight, nutrients, water and space (Feldhake et al. 2005; Lindgren and Sullivan 2014). Trees modify the physical characteristics of a site and provide beneficial attributes such as shade and protection from wind and exposure that help minimize weather-related stresses on grazing livestock. Changes associated with canopy boundary-layer structure influence temperature, evapotranspiration and water availability to the forage canopy growing in the understory. Soil temperatures in a silvopasture site can be relatively cooler in summer and warmer in winter when compared to traditional open-pasture sites (Feldhake 2002). Lower temperatures during intense radiation that is likely to occur in mid-summer can benefit C_3 grasses and legumes (Lin et al. 1999), and higher temperatures during radiation-frost events can help extend the grazing season in early spring and late autumn (Feldhake 2002). Soil moisture in the root zone of a forage crop can be relatively greater and soil compaction less under tree canopies (Karki et al. 2009).

Woody and herbaceous plants, other than traditional forages, can be considered browse and managed as companion or cover crops in silvopasture. Efforts have been made across North America to understand silvopasture systems using regionally adapted tree species. Among those, with documented nutritional value for use in the eastern part of North America are honey locust, black locust, mimosa (Burner et al. 2008; Feldhake et al. 2008; Buergler et al. 2006) and bamboo (various cold-tolerant species), or a large grass of the Poaceae family (Halvorson et al. 2011).

Native North American bamboo (*Arundinarea* spp.), also known as cane, historically grew in vast near-monocultures (canebrakes) that provided a grazing resource until wiped out by **overgrazing**. Bamboo species are unique among grasses in retaining leaves over winter and upright growth habit that makes herbage accessible to livestock. Forage quality of bamboos grown in West Virginia was adequate for maintenance of adult goats. Slow stand establishment, and hence limited productivity, resulting from the need for vegetative propagation may limit usefulness (Halvorson et al. 2011). The leaves and young growth of some other woody species consumed by browse-consuming livestock in parts of the Appalachian region include honeysuckle, autumn olive and multiflora rose (Foster 2005). Sampson and Jespersen (1963) compiled a comprehensive summary of range and browse species growing in the Pacific west. Additional species suitable for browse in various regions of the US are listed in Table 19.3.

Tree seedpods are useful protein and energy resources for humans and livestock. In temperate North America, examples include millwood, honey locust and honey mesquite (Ansley et al. 2017) consumed by domesticated, wild and feral mammals. In warmer climates, cattle often consume green leguminous seed pods. The best-known examples are found among the *Leucaena* spp. (Austin et al. 1995).

Nutritive Value

Browse often appears to have high nutritive value based on standard methods of assessment. However, most woody plants have mechanical or chemical means to deter herbivory that compromise the value of standard nutrient analysis. Woody plants generally can tolerate removal of only a small amount of the annual browse production and, therefore, have evolved numerous mechanisms such as spines or secondary metabolic compounds that deter herbivory.

Based on the classification of rangeland forages by Huston and Pinchak (1991), browse species can provide the quality, level, or null component of the total forage resource. The quality component is made up of species that provide only a minor amount of forage but provide significantly higher amounts of nutrients than the bulk of **available forage**. For example, **mast** and current growth of many shrub species are classified as quality components. Representative deciduous shrubs are littleleaf leadtree, and mountain mahogany.

Species with fair- to good-quality forage throughout the seasons are the level component of the forage resource. Leaves of evergreen browse species such as fourwing saltbush, ashe juniper, or big sagebrush are examples. The amount of browse as an important component in wildlife diets is shown by low to moderate mortality of pronghorn antelope (*Antilocapra americana*) on areas with abundant sagebrush compared with high mortality in areas with very little sagebrush (Martinka 1967).

The null component is composed of species such as honey mesquite and tarbush that are not browsed unless the availability of other components is severely restricted. Species importance varies seasonally and with management. For example, at a proper stocking rate, most species listed as level components would probably be null components during the growing season.

Browse plants are a feed source for wild and domestic ruminants on range and certain intensively managed, forage-based livestock production enterprises such as meat-goats (*Capra hircus*) and silvopasture. Each ruminant species has characteristic anatomic features such as the structure of teeth, mouth, and tongue, rumen architecture, or grazing posture that facilitates acquisition and processing of browse plants. **Selective grazing** among browse species influences competition and persistence among plant species and has substantial effects on plant community structure.

Animal preference seems to depend on the number of secondary compounds (e.g. tannins) the plants contain. Fourwing saltbush and winterfat contain low amounts of harmful secondary compounds, have few physical barriers (such as spines), and are consumed preferentially by grazers (Holechek et al. 1990). Grazers avoid browse species such as creosotebush, mesquite, and one-seed juniper that have large concentrations of tannins, but there are exceptions. Mountain mahogany, though high in tannin, did not appear to affect forage intake or digestibility by goats when the species composed as much as 60% of the diet (Boutouba et al. 1990).

Generalization about nutritive value of browse is difficult because of variation for the same species among regions and among browse species within regions. Cook (1972) reported that browse species usually contain more P, CP, and carotene, and less digestible energy, than do grasses and forbs (Table 19.4). This changed as the season advanced and plants matured. Huston et al. (1981) reported similar results, except that browse had greater herbage digestibility compared with forbs and grasses (Table 19.4). The CP and IVDMD of browse in Appalachian hill lands (Table 19.4; Turner and Foster 2000) were comparable to values reported by Huston et al. (1981). Not all browse plants can supply this level of nutrition. Juniper, which can be an important component of goat diets during winter, has between 6 and 8 g CP kg^{-1} DM for the entire year (Huston et al. 1981). In general, the nutrient content of evergreen shrubs varies less than that of deciduous shrubs among seasons.

Antiquality Factors

Though nutrient composition of browse is an important consideration, of equal or greater importance are the

Table 19.4 Comparative seasonal nutrient content of forage from grass, forbs, and browse in different regions of the United States

		Spring	Summer	Fall	Winter
Edwards plateau region of Texas[a]					
			(% of dry weight)		
Crude protein	Grass	8	6	5	5
	Forbs	19	11	14	—
	Browse	16	11	09	—
Digestible organic matter	Grass	44	43	34	31
	Forbs	59	53	53	—
	Browse	70	64	58	—
Phosphorus	Grass	0.13	0.11	0.08	0.06
	Forbs	0.21	0.17	0.20	—
	Browse	0.22	0.10	0.09	—
Intermountain region of the west[b]					
Digestible protein	Grass	10.2	5.5	2.1	0.05
	Forbs	10.2	6.1	4.2	3.60
	Browse	10.4	7.9	5.7	4.80
			(kcal kg^{-1})		
Digestible energy	Grass	680	544	481	413
	Forbs	658	490	481	295
	Browse	653	472	318	249
			(%)		
Phosphorus	Grass	0.25	0.22	0.13	0.06
	Forbs	0.27	0.26	0.23	0.15
	Browse	0.30	0.28	0.24	0.18
Appalachian hill lands[c]					
Crude protein	Browse	22.7	15	20	—
IVDMD	Browse	70.3	61.7	65.5	—

[a]From Huston et al. (1981).
[b]From Cook (1972).
[c]From Turner and Foster (2000).

physical antiquality factors and components that reduce the utility and value of browse species. Browse species have greater amounts and a greater variety of plant **secondary metabolites** than do grasses. In some cases, these act as **antiquality factors** that cause selective browsing by mammals (Bryant et al. 1992). Structural antiquality components such as spines reduce bite mass and slow removal, thus reducing intake. Reduced intake has the greatest consequence in arid environments where primary production is low and where spines are more common (Laca et al. 2001).

Allelochemicals help plants avoid defoliation by reducing palatability (Launchbaugh 1996). The two most common are tannins (Haslam 1979) and terpenoids (Mabry and Gill 1979).

Tannins are soluble polymers able to combine and form precipitates with proteins. Tannin–protein complexes are generally resistant to **protease** catabolism (Van Soest 1982). There are two general groups of tannins: condensed and hydrolyzable. By binding with digestive **enzymes** and dietary proteins, the condensed tannins depress digestion. Tannins also depress intake either by reducing digestibility of the diet components or by the astringency of condensed tannins and short-term post-ingestive malaise (Landau et al. 2000).

Terpenoids consist of a collection of five-carbon units with a branched, isopentanoid skeleton that exhibits remarkable structural and functional diversity (Mabry and Gill 1979). Though terpenoids in browse species have a variety of functions, the most relevant are toxicity and

Table 19.5 Browse plants consumed by goats

Common name	Scientific name	Region	References
Cedar	*Juniperus ashei* Buchh. *Juniperus pinchotii* (Sudw.) van Melle	Edwards Plateau, Texas	Taylor et al. (1997)
Multiflora rose	*Rosa multiflora* Thunb.	Appalachian Region, North Carolina	Luginbuhl et al. (1999)
Gambel oak	*Quercus gambelii* Nutt.	Southern Rocky Mountains	Davis et al. (1975)
Blackberry	*Rubus* spp. L.	West Virginia hill land	Dabaan et al. (1997)
Greenbrier	*Smilax rotundifolia* L.	West Virginia hill land	Dabaan et al. (1997)
Live oak	*Quercus virginiana* P. Mill.	Central Texas	Magee (1957)
Post oak	*Quercus stellata* Wangenh.	Central Texas	Magee (1957)
Spanish oak	*Quercus falcata* Michx.	Central Texas	Magee (1957)
Blackjack oak	*Quercus marilandica* Muench.	Central Texas	Magee (1957)
Chaparral			
Scrub oak	*Quercus turbinella* Greene	Arizona	Severson and Debano (1991)
Buckbrush	*Ceanothus greggii* A. Gray		
Wright silktassel	*Garrya wrightii* Torr.		
Spiny redberry	*Rhamnus crocea* Nutt.		
Rough menodora	*Menodora scabra* A. Gray		
Mountain mahagony	*Cercocarpus betuloides* Nutt.		
Lespedeza			
Annual and perennial	*Kummerowia* spp. *Lespedeza cuneata* (Dum. Cours.) G. Don	Kansas, Missouri, Nebraska, Oklahoma	Hart (2000)

feeding deterrence. Terpenes and volatile oils in juniper reduce intake (Taylor et al. 1997); those in sagebrush decrease diet digestibility in sheep (Ngugi et al. 1995). High-protein supplements moderate intake suppression.

On average, cattle, sheep, and goats consume about 20%, 15%, and 60% browse in their respective diets each year (Van Dyne et al. 1980). Goats consume greater amounts of browse because of several adaptive mechanisms, including secretion of proline-rich saliva that binds tannins (Provenza and Malechek 1984), rumen microorganisms that degrade tannin-protein complexes (Brooker et al. 1994), and possibly greater activity of a mixed-function oxidase system because of a relatively large liver. In addition, goats have flexible and prehensile lips and a bipedal grazing stance that helps them overcome structural defenses of some browse species. Since goats utilize browse as a nutrient source, they can serve as biological weed control agents where browse species compete with other more desirable plants.

When desirable browse species are present, grazing events should occur during periods when preference for the target browse species is relatively high or when it is in the seedling or fruiting stage (Taylor et al. 1997). Goats can be used to control or slow encroachment by several undesirable shrubs (Table 19.5), leading to increased herbaceous plant production for livestock (Belesky and Wright 1994) or desirable browse production for wildlife (Hart 2000). Often, goats control undesirable plants most effectively when combined with other treatments such as fire, mechanical disturbance, or herbicides. Goats are generally more effective at controlling tannin-rich than terpenoid-rich browse species. Using goats as a biological weed control agent offers an environmentally benign method, coupled with production of marketable livestock products (Magee 1957; Hart 2001).

Summary

Forbs and browse can improve overall forage DM and seasonal distribution of yield, increase plant diversity, improve system function, provide natural remedies for livestock health, and improve nutritive value in terms of mineral composition. Using forbs and browse in pasture and silvopastoral systems increases flexibility and management options allowing grazing lands to be used for pasture, range, hay, silage conservation, mixed-species grazing, or medicinal and botanical production applications.

References

Abaye, A. O., Webb, D. M., Zipper, C. E. et al. (2011). Managing shrub-infested, post-mined pasturelands with goats and cattle: I. Effects on botanical composition and browse species. Virginia Cooperative Extension. Publication No. CSES-4.

Alloush, G.A., Belesky, D.P., and Clapham, W.M. (2003). Forage chicory: a plant resource for nutrient-rich sites. *J. Agron. Crop Sci.* 189: 96–104.

Ansley, R.J., Pinchak, W.E., and Owens, M.K. (2017). Mesquite pod removal by cattle, feral hogs and native herbivores. *Rangeland Ecol. Manage.* 70: 469–476.

Archer, S. (1989). Have southern Texas savannas been converted to woodlands in recent history? *Am. Nat.* 134: 545–561.

Arnold, M. (2014). *Brassicas: Be Aware of the Animal Health Risks*. University of Kentucky Cooperative Extension Service, ID-223.

Austin, M.T., Williams, M.J., Hammond, A.C. et al. (1995). Establishment, forage production and nutritive value of leucaena in central Florida. *Agron. J.* 87: 915–920.

Barry, T.N. (2013). The feeding value of forage brassica plants for grazing ruminant livestock. *Animal Feed Science and Technology* 181: 15–25.

Barry, T.N., Reid, T.C., Millar, K.R., and Sadler, W.A. (1981). Nutritional evaluation of kale (*Brassica oleracea*) diets: 2. Copper deficiency, thyroid function, and selenium status in young cattle and sheep fed kale for prolonged periods. *J. Agric. Sci.* 96: 269.

Barry, T.N., Manley, T.R., and Millar, K.R. (1982). Nutritional evaluation of kale (*Brassica oleracea*) diets: 4. Responses to supplementation with synthetic S-methyl-L-cysteine sulphoxide (SMCO). *J. Agric. Sci.* 99: 1–12.

Bastian, C.T., Jacobs, J.J., and Smith, M.A. (1995). How much sagebrush is too much: an economic threshold analysis. *J. Range Manage.* 48: 73–80.

Belesky, D.P. and Wright, R.J. (1994). Hill-pasture renovation using phosphate rock and stocking with sheep and goats. *J. Prod. Agric.* 7: 233–238.

Belesky, D.P., Fedders, J.M., Turner, K.E., and Ruckle, J.M. (1999). Productivity, botanical composition and nutritive values of swards including forage chicory. *Agron. J.* 91: 450–456.

Belesky, D.P., Turner, K.E., and Ruckle, J.M. (2000). Influence of nitrogen on productivity and nutritive values of forage chicory. *Agron. J.* 92: 472–478.

Belesky, D.P., Koch, D.W., and Walker, J. (2007). Forbs and browse species. In: *Forages: the Science of Grassland Agriculture*, 6e, vol. II (eds. R.F Barnes, C.J. Nelson, M. Collins, et al.), 257–276. Ames, IA: Blackwell Publishing.

Bischoff, K.L. (2016). Glucosinolates. In: *Nutraceuticals: Efficacy, Safety and Toxicity* (ed. R.C. Gupta), 551–554. Academic Press/Elsevier.

Boutouba, A., Holechek, J.L., Galyean, M.L. et al. (1990). Influence of two native shrubs on goat nitrogen status. *J. Range Manage.* 43: 530–534.

Brenchley, W.E. (1920). *Weeds of Farmland*. London: Longmans, Green and Co.

Brooker, J.D., O'Donovan, L.A., Skene, I. et al. (1994). *Streptococcus caprinus* sp. nov., a tannin-resistant ruminal bacterium from feral goats. *Lett. Appl. Microbiol.* 18: 313–318.

Brundrett, M.C. (2009). Mycorrhizal associations and other means of nutrition of vascular plants: understanding the global diversity of host plants by resolving conflicting information and developing reliable means of diagnosis. *Plant Soil* 320: 37–77.

Bryant, J.P., Reichardt, P.B., and Clausen, T.P. (1992). Chemically mediated interactions between woody plants and browsing mammals. *J. Range Manage.* 45: 18–24.

Buergler, A.L., Fike, J.H., Burger, J.A. et al. (2006). Forage nutritive value in an emulated silvopasture. *Agron. J.* 98: 1265–1273.

Burner, D.M., Carrier, D.J., Belesky, D.P. et al. (2008). Yield components and nutritive value of *Robinia pseudoacacia* and *Albizia julibrissin* in Arkansas, USA. *Agrofor. Syst.* 72: 51–62.

Cassida, K.A., Barton, B.A., Hough, R.L. et al. (1994). Feed intake and apparent digestibility of hay-supplemented Brassica diets for lambs. *J. Anim. Sci.* 72: 1623–1629.

Cassida, K.A., Barton, B.A., Hough, R.L. et al. (1995). Productivity and health of gestating ewes grazing tyfon pastures containing weeds. *J. Sustainable Agric.* 6: 81–95.

Cassida, K.A., Foster, J.G., and Turner, K.E. (2010). Forage characteristics affecting meat goat preferences for forage chicory cultivars. *Agron. J.* 102: 1109–1117.

Cheeke, P.R. (1998). *Natural Toxicants in Feeds, Forages, and Poisonous Plants*, 2e. Danville, IL: Interstate Publ., Inc.

Clapham, W.M., Fedders, J.M., Belesky, D.P., and Foster, J.G. (2001). Developmental dynamics of forage chicory (*Cichorium intybus* L.). *Agron. J.* 93: 443–440.

Clason, T.R. and Sharrow, S.H. (2000). Silvopastoral practices. In: *North American Agroforestry: An Integrated Science and Practice* (eds. H.E. Garrett, W.J. Rietveld and R.F. Fisher), 119–147. Madison, WI: ASA, CSSA, and SSSA.

Collins, M. and McCoy, J.E. (1997). Chicory productivity, forage quality, and response to nitrogen fertilization. *Agron. J.* 89: 232–238.

Cook, C.W. (1972). Comparative nutritive values of forbs, grasses and shrubs. In: *Wildland Shrubs, their Biology and Utilization*, General Tech. Rep. INT-1 (eds. C.M. McKell, J.P. Blaisdell and J.R. Goodin), 303–310. USDA Forest Service.

Creech, C.F., Waldron, B.L., Ransom, C.V., and Zobell, D.R. (2013). Factors influencing the field germination of forage kochia. *Crop Sci.* 53: 2202–2208.

Cubbage, F., Balmell, G., Bussoni, A. et al. (2012). Comparing silvopastoral systems and prospects in eight regions of the world. *Agrofor. Syst.* 86: 303–314.

Dabaan, M.E., Magadlela, A.M., Bryan, W.B. et al. (1997). Pasture development during brush clearing with sheep and goats. *J. Range Manage.* 50: 217–221.

Davis, G.G., Bartel, L.E., and Cook, C.W. (1975). Control of Gambel oak sprouts by goats. *J. Range Manage.* 28: 216–218.

Dunavin, L.S. Jr. (1987). Comparison of turnip-Chinese cabbage hybrid, rape, and rye, alone and in combination with annual ryegrass and crimson clover. *Agron. J.* 79: 591–594.

Duncan, A.J. (1990). Animal health implications of forage brassica use. *Brit. Grassl. Soc. Occ. Symp.*: 203–209. No. 24.

Dupraz, C. (1999). Fodder trees and shrubs in Mediterranean areas: browsing for the future? In: *Grassland and Woody Plants in Europe*. Proc. Intl. Occ. Sym. European Grassld. Fed. 27–29 May 1999, vol. 4 (eds. V.P. Papanastasis, J. Frame and A.S. Nastis), 145–157. Thessaloniki, Greece.

Faix, J.J., Stookey, J.M., Lewis, J.M. et al. (1980). Pastures of rye-ryegrass, rape, turnips, and tall fescue for lambs. Res. Rep., Ill. Agric. Exp. Stn. *Dixon Springs Agric. Ctr.* 8: 124–129.

Fajer, E.D., Bowers, M.D., and Bazzaz, F.A. (1992). The effect of nutrients and enriched CO_2 on production of carbon-based allelochemicals in *Plantago*: a test of the carbon/nutrient balance hypothesis. *Am. Nat* 140: 707–723.

Feldhake, C.M. (2002). Forage frost protection potential of conifer silvopastures. *Agric. For. Meteorol.* 112: 123–130.

Feldhake, C.M., Neel, J.P.S., Belesky, D.P., and Mathias, E.L. (2005). Light measurement methods related to forage yield in a grazed northern conifer silvopasture in the Appalachian region of eastern USA. *Agrofor. Syst.* 65: 231–239.

Feldhake, C.M., Belesky, D.P., and Mathias, E.L. (2008). Forage production under and adjacent to *Robinia pseudoacacia* in central Appalachia, West Virginia. In: *Toward Agroforestry Design: An Ecological Approach* (eds. S. Jose and A.M. Gordon), 55–66. Netherlands: Springer.

Fitzgerald, J.J. and Black, W.J.M. (1984). Finishing store lambs on green forage crops. I. A comparison of rape, kale and fodder radish as sources of feed for finishing store lambs in autumn. *Irish J. Agric. Res.* 23: 127–136.

Foster, L. (1988). Herbs in pastures. Development and research in Britain. 1850–1984. *Biol. Agric. Hort.* 5: 97–133.

Foster, J.G. (2005). Condensed tannins in leaves of woody plants in Appalachian pastures. *Proc. Am. Forage Grassl. Counc.* 14: 215–219.

Foster, J.G., Robertson, J.W., Bligh, D.P. et al. (2001). Variations in chemical composition among commercial cultivars of forage chicory. In: *Proceedings of the 2001 American Forage and Grassland Conference*, vol. 10, 326–330.

Foster, J.G., Fedders, J.M., Clapham, W.M. et al. (2002). Nutritive value and animal selection of forage chicory cultivars grown in central Appalachia. *Agron. J.* 94: 1034–1042.

Fulkerson, R.S. and Tossell, W.E. (1972). An evaluation of marrowstem kale. *Can. J. Plant Sci.* 52: 787–793.

Ganskopp, D., Svejcar, T., Taylor, F. et al. (1999). Seasonal cattle management in 3 to 5 year old bitterbrush stands. *J. Range Manage.* 52: 166–173.

Gardner, J. and Caswell-Chen, E.P. (1993). Penetration, development and reproduction of *Heterodera schachtii* on *Fagopyrum esculentum*, *Phacelia tanacetifolia*, *Raphanus sativum*, *Sinapis alba* and *Brassica oleracea*. *J. Nematol.* 25: 695–702.

Gray, F.A. and Koch, D.W. (2002). Trap crops. In: *Encyclopedia of Pest Management*, 852–854. NY: Marcel Dekker.

Guillard, K. and Allinson, D.W. (1988). Yield and nutrient content of summer- and fall-grown forage Brassica crops. *Can. J. Plant Sci.* 68: 721–731.

Guillard, K. and Allinson, D.W. (1989a). Seasonal variation in chemical composition of forage Brassicas. I. Mineral concentrations and uptake. *Agron. J.* 81: 876–881.

Guillard, K. and Allinson, D.W. (1989b). Seasonal variation in the chemical composition of forage Brassica. II. Mineral imbalances and anti-quality constituents. *Agron. J.* 81: 881–886.

Guldan, S.J., Martin, C.A., and Daniel, D.L. (1998). Interseeding forage brassicas into sweet corn: forage productivity and effect on sweet corn yield. *J. Sust. Agric.* 11: 51–58.

Gustine, D.L. and Jung, G.A. (1985). Influence of some management parameters on glucosinolate levels in Brassica forages. *Agron. J.* 77: 593–597.

Halvorson, J.J., Cassida, K.A., Turner, K.E., and Belesky, D.P. (2011). Nutritive value of bamboo as browse for livestock. *Renew. Agric Food Syst.* 26: 161–170.

Hare, M.D., Rolston, M.P., Crush, J.R., and Fraser, T.J. (1987). Puna chicory—a perennial herb for New Zealand pastures. *Proc. Agron. Soc. N. Z.* 17: 45–49.

Hart, S.P. (2000). Stocker goats for controlling sericea lespedeza. In: *Symposium Proceedings Sericea Lespedeza and the Future of Invasive Species*, 12–13. Manhattan: Kansas State University Department of Agronomy.

Hart, S.P. (2001). Recent perspective in using goats for vegetation management in the USA. *J. Dairy Sci.* 84 (E. Suppl): E170–E176.

Haslam, E. (1979). Vegetable tannins. In: *Biochemistry of Plant Phenolics*, Recent Advances in Phytochemistry (eds. T. Swain, J.B. Harborne and C.F. Van Sumere), 475–523. NY: Plenum Press.

Heinemann, W.W., Hinman, H.R. and Hanks, E.N. (1981). Turnips (Brassica rapa) and fodder radishes (Raphanus sativus) as forage crops for lambs. Washington State University Bulletin No. 0904.

Holden, L.A., Varga, G.A., Jung, G.A., and Shaffer, J.A. (2000). Comparison of 'Grassland Puna' chicory and orchardgrass for multiple harvests at different management levels. *Agron. J.* 92: 191–194.

Holechek, J.L., Munshikpu, A.V., Saiwana, L. et al. (1990). Influences of six shrub diets varying in phenol content on intake and nitrogen retention by goats. *Trop. Grassl.* 24: 93–98.

Huston, J.E. and Pinchak, W.E. (1991). Range animal nutrition. In: *Grazing Management an Ecological Perspective* (eds. D.D. Briske and R.K. Heitschmidt), 27–63. Portland, OR: Timber Press.

Huston, J.E., Rector, B.S., Merrill, L.B. et al. (1981). Nutritional value of range plants in the Edwards plateau region of Texas. Texas Agricultural Experiment Station. B-1357.

Jarzomski, C.M., Stamp, N.E., and Bowers, M.D. (2000). Effects of plant phenology, nutrients and herbivory on growth and defensive chemistry of plantain, *Plantago lanceolata*. *Oikos* 88: 371–379.

Jung, G.A. and Shaffer, J.A. (1993). Planting date and seeding rate effects on morphological development and yield of turnip. *Crop Sci.* 33: 1329–1334.

Jung, G.A., McClellan, W.L., Byers, R.A. et al. (1983). Conservation tillage for forage Brassicas. *J. Soil Water Conserv.* 38: 227–230.

Jung, G.A., Kocher, R.E., and Glica, A. (1984). Minimum-tillage forage turnip and rape production on hill land as influenced by sod suppression and fertilizer. *Agron. J.* 76: 404–408.

Jung, G.A., Byers, R.A., Panciera, M.T., and Shaffer, J.A. (1986). Forage dry matter accumulation and quality of turnip, swede, rape, Chinese cabbage hybrids, and kale in the eastern USA. *Agron. J.* 78: 245–253.

Jung, G.A., Shaffer, J.A., Stout, W.L., and Panciera, M.T. (1988). Harvest frequency effects on forage yield and a quality of rapes and rape hybrids. *Grass Forage Sci.* 43: 395–404.

Jung, G.A., Shaffer, J.A., Varga, G.A., and Everhart, J.R. (1996). Performance of 'Grassland Puna' chicory at different management levels. *Agron. J.* 88: 104–111.

Kälber, T., Kreuzer, M., and Leiber, F. (2012). Silages containing buckwheat and chicory: quality, digestibility and nitrogen utilization by lactating cows. *Archives of Animal Nutrition* 66: 50–65.

Kalmbacher, R.S., Everett, P.H., Martin, F.G., and Jung, G.A. (1982). The management of brassica for winter forage in the subtropics. *Grass Forage Sci.* 37: 219–225.

Karki, U., Goodman, M.S., and Sladden, S.E. (2009). Nitrogen source influences on forage and soil in young southern-pine silvopasture. *Agric. Ecosyst. Environ.* 131: 70–76.

Kay, C.E. (1997). Viewpoint: ungulate herbivory, willows, and political ecology in Yellowstone. *J. Range Manage.* 50: 139–145.

Koch, D.W. and Karakaya, A. 1998. Extending the grazing season with turnips and other brassicas. University of Wyoming. Cooperative Extension Service Bulletin No. B-1051.

Koch, D.W., Ernst, F.C. Jr., Leonard, N.R. et al. (1987). Lamb performance on extended-season grazing of tyfon. *J. Anim. Sci.* 64: 1275–1279.

Koch, D.W., Kercher, C., and Jones, R. (2002). Fall and winter grazing of brassicas—a value-added opportunity for lamb producers. *J. Sheep Goat Res.* 17: 1–13.

Kuiper, P.J.C. and Bos, M. (1992). *Plantago: A Multidisciplinary Study*. Ecological Studies No. 89. Berlin: Springer-Verlag.

Kunelius, H.T. and MacRae, K.B. (1999). Forage chicory persists in combination with cool-season grasses and legumes. *Can. J. Plant Sci.* 79: 197–200.

Kunelius, H.T. and Sanderson, J.B. (1990). Effect of harvest dates on yield and composition of forage rape, stubble turnip and forage radish. *Appl. Agric. Res.* 5: 159–163.

Kunelius, H.T., Sanderson, J.B., and Narasimhalu, P.R. (1987). Effect of seeding date on yields and quality of green forage crops. *Can. J. Plant Sci.* 67: 1045–1050.

Laca, E.A., Shipley, L.A., and Reid, E.D. (2001). Structural anti-quality characteristics of range and pasture plants. *J. Range Manage.* 54: 413–419.

Lambert, M.G., Abrams, S.M., Harpster, H.W., and Jung, G.A. (1987). Effect of hay substitution on intake and digestibility of forage rape (*Brassica napus*) fed to lambs. *J. Anim. Sci.* 65: 1639–1646.

Landau, S., Silanikove, N., Nitsan, Z. et al. (2000). Short-term changes in eating patterns explain the effects of condensed tannins on feed intake in heifers. *Appl. Anim. Behav. Sci.* 69: 199–213.

Launchbaugh, K.L. (1996). Biochemical aspects of grazing behavior. In: *The Ecology and Management of Grazing Systems* (eds. J. Hodgson and A.W. Illius), 159–184. New York: CAB International.

Lawley, Y., Well, R., and Teasdale, J. (2011). Forage radish cover crop suppresses winter annual weeds in fall and before corn planting. *Agron. J.* 103: 137–144.

Leonard, S.S., Keil, D., Mehlman, T. et al. (2006). Essiac tea: scavenging of reactive oxygen species and effects on DNA damage. *J. Ethnopharmacol.* 103: 288–296.

Lin, C.H., McGraw, R.L., George, M.F., and Garrett, H.E. (1999). Shade effects on forage crops with potential in temperate agroforestry practices. *Agrofor. Syst.* 44: 109–119.

Lindgren, P.M.F. and Sullivan, T.P. (2014). Response of forage yield and quality to thinning and fertilization of young forests: implications for silvopasture management. *Can. J. Forest Res.* 44: 281–289.

Luginbuhl, J.M., Harvey, T.E., Green, J.T. Jr. et al. (1999). Use of goats as biological agents for the renovation of pastures in the Appalachian region of the United States. *Agrofor. Syst.* 44: 241–252.

Mabry, T.J. and Gill, J.E. (1979). Sesquiterpene lactones and other terpenoids. In: *Herbivores their Interaction with Secondary Plant Metabolites* (eds. G.A. Rosenthal and D.H. Janzen), 501–537. NY: Academic Press.

Magee, A.C. (1957). Goats pay for clearing Grand Prairie rangelands. Texas Agricultural Experiment Station Miscellaneous Publication No. 206.

Marley, C.L., Fychan, R., Fraser, M.D. et al. (2007). Effects of feeding different ensiled forages on the productivity and nutrient-use efficiency of finishing lambs. *Grass Forage Sci.* 62: 1–12.

Martinka, C.J. (1967). Mortality of northern Montana pronghorn in a severe winter. *J. Wildlife Manage.* 31: 159–164.

McCoy, J.E., Collins, M., and Dougherty, C.T. (1997). Amount and quality of chicory herbage ingested by grazing cattle. *Crop Sci.* 37: 239–242.

Moorby, J.M., Evans, P.R., and Young, N.E. (2003). Nutritive value of barley/kale bi-crop silage for lactating dairy cows. *Grass Forage Sci.* 58: 184–191.

Murin, A. (1997). Karyotaxonomy of some medicinal and aromatic plants. *Thaiszia. J. Bot.* 7: 75–88.

National Research Council (NRC) (1985). *Nutrient Requirements of Sheep* (6th Rev. Ed.) Washington, D.C.: National Academy Press.

Neel, J.P.S., Alloush, G.A., Belesky, D.P., and Clapham, W.M. (2002). Influence of ionic strength on mineral composition, dry matter yield and nutritive value of forage chicory. *J. Agron. Crop Sci.* 188: 1–10.

Neely, C., Brown, J., Hunt, C. and Davis, J. (2009). Increasing the value of winter canola crops by developing ensiling systems (canolage) to produce cattle feed. Moscow, ID. University of Idaho Extension.

Ngugi, R.K., Hinds, F.C., and Powell, J. (1995). Mountain big sagebrush browse decreases dry matter intake, digestibility, and nutritive quality of sheep diets. *J. Range Manage.* 48: 487–492.

Nicol, A.M. and Barry, T.N. (1980). The feeding of forage crops. In: *Supplementary Feeding* (eds. K.R. Drew and P.F. Hennessy), 69–106. N. Z. Soc. Anim. Prod. Occ. Publ. No. 7.

Pearce, P.E., Hunt, C.W., Hall, M.H., and Loesche, J.A. (1991). Effects of harvest time and grass hay addition on composition and digestion of high- and low-glucosinolate rapeseed forage. *J. Prod. Agric.* 4: 411–416.

Peel, M.D., Waldron, B.L., and Mott, I.W. (2009). Ploidy determination and agronomic characterization of small burnet germplasm. *Crop Sci.* 49: 1359–1366.

Pitman, W.D. and Willis, C.C. (1998). Establishment and growth of chicory on west Louisiana coastal plain soils. Rosepine Research Station Report No. 10:47–49.

Polley, H.W., Mayeux, H.S., Johnson, H.B., and Tischler, C.R. (1997). Viewpoint: atmospheric CO_2, soil water, and shrub/grass ratios on rangelands. *J. Range Manage.* 50: 278–284.

Prestbye, L.S. and Welty, L.E. (1993). Evaluation of winter brassica varieties for forage production. *Montana Agric. Res.* 10(2): 11–14.

Provenza, F.D. and Malechek, J.C. (1984). Diet selection by domestic goats in relation to blackbrush twig chemistry. *J. Appl. Ecol.* 21: 831–841.

Rao, S.C. and Horn, F.P. (1986). Planting season and harvest date effects on dry matter production and nutritional value of *Brassica* spp. in the Southern Great Plains. *Agron. J.* 78: 327–333.

Reid, R.L., Puoli, J.R., Jung, G.A. et al. (1994). Evaluation of brassicas in grazing systems for sheep: I. Quality of forage and animal performance. *J. Anim. Sci.* 72: 1823–1831.

Robinette, W.L. (1972). Browse and cover for wildlife. In: *Wildland Shrubs, their Biology and Utilization* (eds. C.M. McKell, J.P. Blaisdell and J.R. Goodin), 69–76. USDA Forest Serv. Gen. Tech Rep INT-1.

Rule, D.C., Koch, D.W., Jones, R.R., and Kercher, C.J. (1991). *Brassica* and sugarbeet forages for lambs: growth performance of lambs and composition of forage and dock-fat fatty acids. *J. Prod. Agric.* 4: 29–33.

Rumball, W. (1986). 'Grasslands Puna' chicory (*Cichorium intybus* L.). *N. Z. J. Exp. Agric.* 14: 105–107.

Rumball, W., Keogh, R.G., Lane, G.E. et al. (1997). 'Grasslands Lancelot' plantain (*Plantago lanceolata* L.). *N.Z.J. Agric. Res.* 40: 373–377.

Ryder, E.J. (1999). *Lettuce, Endive and Chicory. Crop Production Science in Horticulture. No. 9*. Wallingford, UK: CABI Publishing.

Sampson, A.W. and Jespersen, B.S. (1963). California range brushlands and browse plants. University of California, Division of Agricultural Science. California Agricultural Experiment Station. Extension Service Manual 33.

Sanderson, M.A. and Elwinger, G.F. (2000a). Seedling development of chicory plantain. *Agron. J.* 92: 69–74.

Sanderson, M.A. and Elwinger, G.F. (2000b). Chicory and English plantain seedling emergence at different planting depths. *Agron. J.* 92: 1206–1210.

Scifres, C.J. (1980). *Brush Management Principles and Practices for Texas and the Southwest*. College Station, TX: Texas A&M University Press.

Severson, K.E. and Debano, L.F. (1991). Influence of Spanish goats on vegetation and soils in Arizona chaparral. *J. Range Manage.* 44: 111–117.

Sharman, G.A.M., Lawson, W.J., and Whitelaw, A. (1981). Potential growth-limiting factors in the Brassicae. *Anim. Prod.* 32: 383–384.

Sheldrick, R.D., Fenlon, J.S., and Lavender, R.H. (1981). Variation in forage yield and quality of three cruciferous catch crops grown in southern England. *Grass Forage Sci.* 36: 179–187.

Shiflet, T.N. (1994). *Rangeland Cover Types of the United States*. Denver, CO: Soc. Range Manage.

Simon, J.E., Chadwick, A.F., and Craker, L.E. (1984). *Herbs: An Indexed Bibliography 1971–1980. The Scientific Literature on Selected Herbs and Aromatic and Medicinal Plants of the Temperate Zone*. Hamden, CT: Archon Books.

Skinner, R.H. and Gustine, D.L. (2002). Freezing tolerance of chicory and narrow-leaf plantain. *Crop Sci.* 42: 2038–2043.

Smith, D.H. and Collins, M. (2003). Forbs. In: *Forages: An Introduction to Grassland Agriculture*, 6e, vol. I (eds. R.F Barnes, C.J. Nelson, K.J. Moore, et al.), 215–236. Ames, IA: Iowa State University Press.

Stewart, A.V. (1996). Plantain (*Plantago lanceolata*)—a potential pasture species. *Proc. N.Z. Grassld. Assoc.* 58: 77–86.

Taylor, C.A. Jr., Launchbaugh, K., Huston, E., and Straka, E. (1997). Improving the efficacy of goat grazing for biological juniper management. In: *1997 Juniper Symposium*. Texas Agric. Exp. Sta. Rpt. 97-1 (ed. C.A. Taylor Jr.), 5-17–5-22. San Angelo, TX.

Thomas, V.M., Kott, R.W., and Baldridge, D. (1990). Evaluation of rape and turnip forage for weaned lambs. *Sheep Res. J.* 6: 14–17.

Thurow, T.L. and Hester, J.W. (1997). Holistic perspective, rangeland hydrology and wildlife considerations in juniper management. How an increase or reduction in juniper cover alters rangeland hydrology. In: *1997 Juniper Symposium*, Texas Agric. Exp. Sta. Rpt. 97-1 (ed. C.A. Taylor Jr.), 9–22. San Angelo, TX.

Tookey, H.L., VanEtten, C.H., and Daxenbichler, M.E. (1980). Glucosinolates. In: *Toxic Constituents of Plant Foodstuffs* (ed. I.E. Liener), 103–142. New York: Academic Press.

Tsunoda, S., Hinata, H., and Gomez-Campo, C. (1980). *Brassica Crops and Wild Allies, Biology and Breeding*. Tokyo: Japan Science Society Press.

Turner, K.E. and Foster, J.G. (2000). Nutritive value of some common browse species. In: *Proceedings of the American Forage and Grassland Conference*, vol. 9, 241–245.

Turner, K.E., Belesky, D.P., and Fedders, J.M. (1999). Chicory effects on lamb weight gain and rate of in vitro organic matter and fiber disappearance. *Agron. J.* 91: 445–450.

USDA-NRCS (1994). *The Use and Management of Browse in the Edwards Plateau of Texas*, 7. Temple, TX.

Van Dyne, G.M., Brockington, N.R., Sxocs, Z. et al. (1980). Large herbivore subsystems. In: *Grasslands, Systems Analysis and Man* (eds. A.I. Breymeyer and G.M. Van Dyne), 269–253. Cambridge: Cambridge Press.

Van Soest, P.J. (1982). *Nutritional Ecology of the Ruminant*. Corvallis, OR: O & B Books.

Vavilov, N.I. (1992). *Origin and Geography of Cultivated Plants*. English Translated Edition. London: Cambridge Press.

Volesky, J.D. (1996). Forage production and grazing management of chicory. *J. Prod. Agric.* 9: 403–406.

Wambolt, C.L. and Sherwood, H.W. (1999). Sagebrush response to ungulate browsing in Yellowstone. *J. Range Manage.* 52: 363–369.

Wiedenhoeft, M.H. and Barton, B.A. (1994). Management and environment effects on *Brassica* forage quality. *Agron. J.* 86: 227–232.

Wikse, S. and Gates, N. (1987). Preventative health management of livestock that graze turnips. Washington State University. Cooperative Extension Service Bulletin No. EB 1453.

Wilson, R.E., Jensen, E.H., and Fernandez, G.C.J. (1992). Seed germination response for eleven forage cultivars of *Brassica* to temperature. *Agron. J.* 84: 200–202.

Yeager, L.E. (1961). Classification of North American mammals and birds according to forest habitat preference. *J. For.* 59: 671–674.

Young, J.A. and Allen, F.L. (1997). Cheatgrass and range science: 1930–1950. *J. Range Manage.* 50: 530–535.

Yun, L., Koch, D.W., Gray, F.A. et al. (1999). Potential of trap-crop radish for fall lamb grazing. *J. Prod. Agric.* 12: 559–563.

Part IV

FORAGE SYSTEMS

Pastureland and cropland are frequently integrated on the landscape with each field being used in the optimum way based on slope, soil type and other important factors. *Source:* Photo courtesy of Ken Moore.

There are many possibilities for designing productive, sustainable, and economic forage systems because different forage species are adapted to different areas. For beef production, the typical system is often based on grazing perennial pastures, since it is usually the most economic way to manage either a perennial cool-season or warm-season grass. In warmer areas, the most common systems are warm-season grass-based systems with winter annual forages meeting some of the needs for autumn, winter and early-spring forage. A major advantage for these areas is that there are substantial native grasslands available for grazing.

In the cooler regions, the forage systems shift to cool-season grass-based systems where summer annuals are often used to meet forage shortages. These areas also have a greater use of legumes in the forage systems. When selecting forages, major factors are forage yield and forage yield distribution, and forage nutritive value.

Forages: The Science of Grassland Agriculture, Volume II, Seventh Edition.
Edited by Kenneth J. Moore, Michael Collins, C. Jerry Nelson and Daren D. Redfearn.

Nearly always, forage shortages occur during times of the year when plant growth is slow or non-existent. Several possibilities for supplemental forage production are available. The most common choices are hay, silage, stockpiled forage for grazing, complementary forage systems, and grazing or feeding crop residues.

CHAPTER

20

Systems for Temperate Humid Areas

Jerome H. Cherney, Professor, *Soil & Crop Sciences, Cornell University, NY, USA*
Robert L. Kallenbach, Associate Dean, *Agriculture and Environment Extension, University of Missouri, MO, USA*
Valentín D. Picasso Risso, Assistant Professor, *Agronomy, University of Wisconsin, WI, USA*

Introduction

Large portions of the land area in the temperate humid zone are better suited to growing forage crops than grain crops. The primary livestock operations in the region involve dairy or beef cattle, and this chapter focuses on forage systems for these livestock. Forage systems are the integrated combination of animal, plant, soil, and other environmental components managed to achieve a productive **agroecosystem**. Forage crops provide one half or more of the total feed requirements of dairy cattle and most of the feed requirements for beef in the temperate humid zone. Forages provide a physical site to keep the animals part-time or full-time and are also one of the primary resources that allow effective nutrient management planning in the region.

The temperate humid zone is defined here as the region east of the 96th meridian and north of the Transition zone (Chapter 23). From a statistical standpoint, we have represented the temperate humid zone as whole states in our illustrations. Though not included in our statistical descriptions of the region, the eastern edge of Nebraska and Kansas, and the northern edge of Kentucky and Virginia are a part of the temperate humid zone. Conversely, the southern extremes of Missouri, Illinois, and West Virginia are considered part of the transition zone.

Although forage systems are complex, the basic principles involved are relatively simple: (i) match forage species with available soil resources, (ii) match quantity and quality of forage produced with the needs of specific animal classes, (iii) minimize nutrient management and water-quality problems, and (iv) match forage production with available storage options, and/or (v) match the forage production pattern with the grazing method. Integration is achieved by balancing the relative efficiencies of forage production, forage consumption and livestock production (Hodgson 1990).

Geography

Much of the landscape of the temperate humid region was shaped by glaciation, which also helped to create the wide range of soils in the region. The region has a spectrum of crop environments impacted by latitude and elevation. **Latitude** has a strong impact on forage species persistence due to variation in winterhardiness among perennial species. A 300 m increase in elevation can delay alfalfa maturation in the spring by as much as two weeks, with only 1 or 2 km of distance between sites. Accumulated growing degree days (base 10 °C) during a growing season in New York state range from less than 1400 to over 3000, due to a combination of latitude and

Forages: The Science of Grassland Agriculture, Volume II, Seventh Edition.
Edited by Kenneth J. Moore, Michael Collins, C. Jerry Nelson and Daren D. Redfearn.

elevation effects. With increased latitude and elevation, cold stress on perennial forages increases and a higher level of management is required for acceptable productivity as well as winter survival (Volenec and Nelson 2003).

Climate

The temperate humid zone ranges from hot summers in southern latitudes to cool summers in northern latitudes, and moderate to cold winters with freezing temperatures throughout the region. The average number of days at or exceeding 32 °C (90 °F) is 0 for some northeastern sites near the coast and exceeds 40 days for sites in Missouri, Illinois, and Indiana. Average annual minimum temperature ranges from about −15 °C (5 °F) for some east coast sites to −40 °C (−40 °F) for sites in northern Minnesota and Wisconsin. Average precipitation ranges from about 60 to 120 cm across the region. Not including high-elevation sites or locational anomalies such as lake effect snowfall, average snowfall in the region ranges from about 25 to 350 cm. Seasonal rainfall is generally adequate across the region, therefore temperature, as influenced by latitude and elevation, is the major climatic factor influencing species survival and production across the region.

Soils

Soils in the region range from deep, fertile, uniform loams in parts of the Midwest to much more variable, shallow soils in the Northeast, often with several different soil types located in the same field. Large tracts of uniform soil in the Midwest are ideal for precision agriculture applications, while the relatively small size and variability of fields in the Northeast make precision agriculture applications more challenging. Many of the soils in the Northeast have drainage limitations and are not suitable for most perennial forage species without artificial drainage improvements.

Land Use

Land use varies greatly across the region (Figure 20.1). Of the total 57 million ha of cropland in the temperate humid zone in 2012, less than 7 million ha were used for hay or haylage crops, less than 7 million ha of hay and haylage crops were in the rotation, about 12% of the total, ranging from 2% in Illinois to 76% in West Virginia. In general, a higher proportion of the cropland rotation in the Northeast is devoted to hay and haylage crops to support integrated livestock production, especially dairy. Most non-cropland areas consist of a mix of pasture and hay for cow-calf beef operations.

Hay cropland statistics do not address a major forage source in the region; corn silage. The temperate humid zone provided 62% of the total corn acreage in the US in 2012. While most corn in the grain belt is harvested as grain, a significant percentage of corn land in states outside the grain belt is harvested as silage for dairy production, as high as 93% in New Hampshire (Figure 20.2). Corn silage hectarage is negatively affected by high corn grain prices, which peaked in 2012. In New York, for example, corn silage as a percentage of corn land exceeded 50% before 2012 and again approached 50% in 2016, but dropped to a low of 36% in 2012, as farmers were attracted to high corn grain prices. The temperate humid zone typically accounts for over half of the total corn silage cropland in the US. This has a significant impact on feeding dairy cattle in the region (Allen et al. 2003).

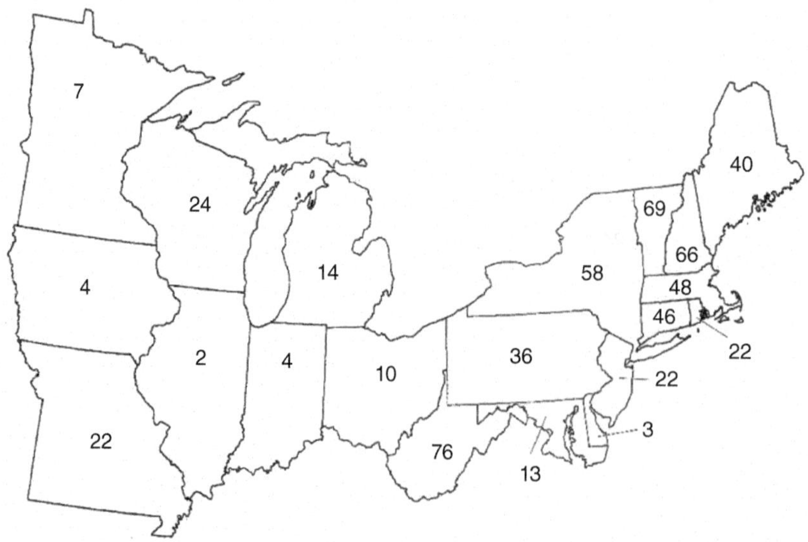

Fig. 20.1. Percentage of cropland that was hay/haylage in 2012 (NASS 2017).

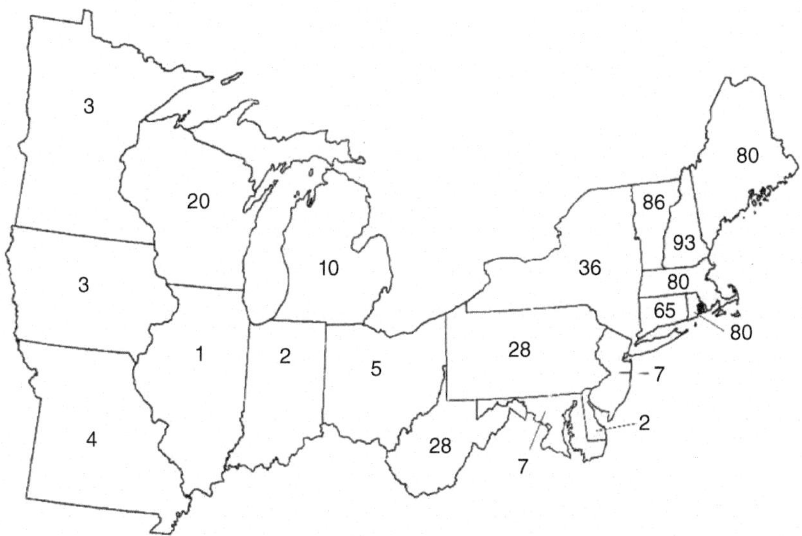

Fig. 20.2. Percentage of corn land that was harvested as silage in 2012 (NASS 2017).

Important Livestock Classes

Distribution of Dairy and Beef Animals in the Region

There are approximately 4.6 million dairy cows in the temperate humid zone, compared to 4.4 million beef cows (Figure 20.3). As a comparison, livestock totals for the entire US were 29 million beef cows and 9 million dairy cows as of Jan. 1, 2013 (NASS 2017). One half of all dairy cows in the US, but only 15% of all beef cows, are in this region which encompasses 18% of the total land area of the 50 states. In general, the northern and eastern portions of the temperate humid zone are primarily dairy (3.4 dairy:1 beef), while the southern portion is primarily beef (5.7 beef:1 dairy). The southern portion of the region is

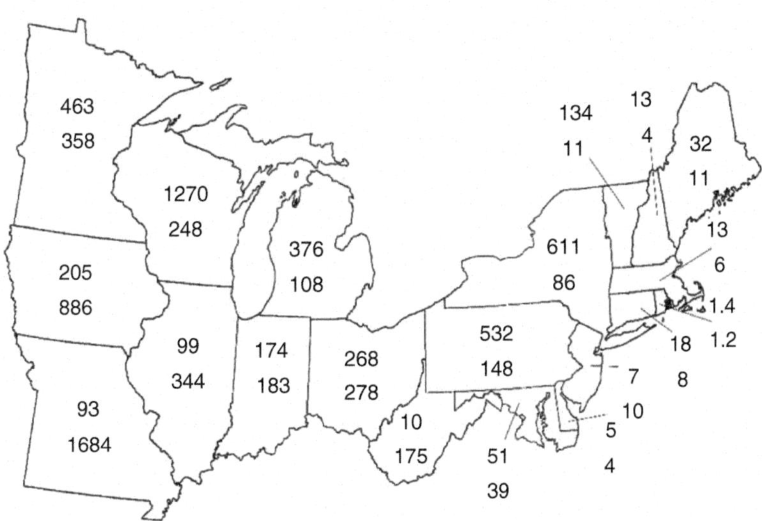

Fig. 20.3. Milk cows (upper value) and beef cows per state (1000 head) as of Jan. 1, 2013 (NASS 2017).

defined to include Illinois, Indiana, Iowa, Missouri, and West Virginia. Ohio is equally divided between beef and dairy and was not included in either the dairy or the beef regions for these livestock comparisons.

Livestock Density

Livestock density is becoming increasingly important from a manure management perspective. On a total land area basis, the highest density of beef cows occurs in Missouri and Iowa, while the highest density of dairy cows is in Wisconsin and Vermont (NASS 2017). A more critical statistic is the number of animals per ha of total cropland (Figure 20.4), because cropland is the major area potentially available for distribution of wastes. A different picture emerges from this view, with the Northeast (0.38 cows ha^{-1}) being considerably higher in animal density per unit of cropland than the rest of the region (0.14 cows ha^{-1}). The highest animal density per unit of cropland is in Vermont, followed closely by West Virginia and New York.

Total cropland figures do not address the issue of very large livestock operations and their ability to spread manure evenly over enough land area to not exceed proposed standards (Ribaudo et al. 2003). Considering the number of animals per unit of cropland on a particular farm, most temperate humid zone livestock farms own or rent enough cropland for manure application requirements, with the possible exception of counties in states surrounding the Chesapeake Bay (Ribaudo et al. 2003).

Important Forage Species

The principal forage species harvested for hay or silage in the region is alfalfa. On average, 35% of the hay cropland in the temperate humid zone in 2012 was in alfalfa and alfalfa mixtures (NASS 2017). As with other parameters, there is a wide range. In West Virginia, where hayland is almost exclusively used for beef, the hay cropland in alfalfa is 4%. Midwestern states average 50% of hayland in alfalfa, excluding Missouri which only has 7% alfalfa. Missouri, like West Virginia, is primarily focused on beef. In the Northeast on average, 22% of all hayland is alfalfa, as high as 28% for Pennsylvania and New York. Red clover, white clover and birdsfoot trefoil also are important legume species for the region (McGraw and Nelson 2003), but do not approach alfalfa in significance.

Though there is some use of C_4 warm-season grasses in the southern portion of the region, the majority of grasses grown in the region are C_3 cool-season grasses. Most of these grasses were originally introduced to the region from other parts of the world with the exception of reed canarygrass (Casler et al. 2009). Criteria for assessing perennial grass acceptability in the region include forage quality, overwintering potential, ability to withstand grazing, ability to withstand flooding or drought, and compatibility with alfalfa. Grasses are grown in the Midwest primarily as beef forage and are used sparingly in mixtures with alfalfa. In the Northeast, however, grasses are managed intensively for dairy cattle feed and, the vast majority of alfalfa is seeded in mixture with one or more perennial grasses. Approximately 85% of all alfalfa sown in New York is sown with a perennial grass (Karayilanli et al. 2016). Common grasses sown include orchardgrass, reed canarygrass, and smooth bromegrass in the Midwest, and timothy, orchardgrass, tall fescue, and reed canarygrass in the Northeast. In southern latitudes tall fescue is the primary cool-season perennial grass species.

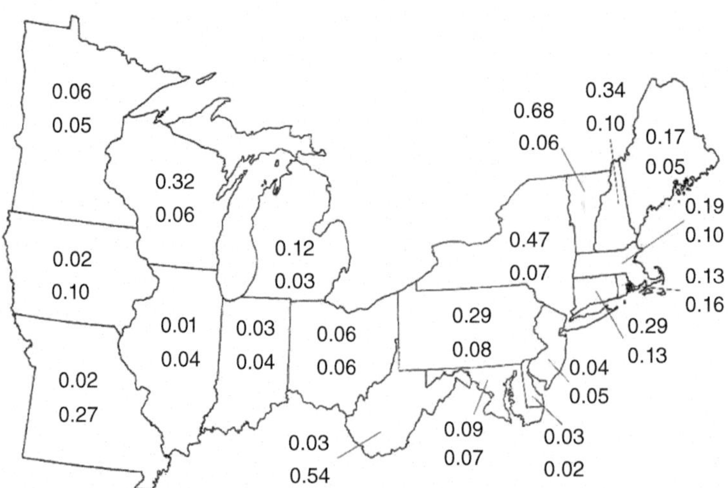

Fig. 20.4. Milk cows per ha of cropland (upper value) and beef cows per ha of cropland per state as of Jan. 1 2013 (NASS 2017).

Corn grown for silage is also a primary dairy forage in the temperate humid zone, and the region has helped to generate considerable interest in improving corn silage quality in the US (Schwab et al. 2003; Cherney et al. 2007; Ferraretto and Shaver 2012). Corn silage hybrid trials have been established in Michigan, Minnesota, New York, Ohio, Pennsylvania, and Wisconsin. Corn silage is a major component of dairy rations in the region, comprising up to 70% of the forage in the diet.

Forage Livestock Systems for Dairy Cattle

Forage Production

Knowledge of the basic components of forage and dairy management and an understanding of the main linkages between these components are necessary for integrating the system. Concept maps (read from top to bottom) are useful for this purpose (Figure 20.5). Due to great variability in soils and native drainage, and the need to overwinter forages in cold winter climates, species selection is key to the success of dairy forage systems in the temperate humid zone.

Perennial forage species and cultivar selection interact with site selection to provide a forage or combination of forages that are (i) persistent, and (ii) capable of producing acceptable yields of a desired quality (Figure 20.5). Selection of forage species that are best suited to a particular soil type, and forage use, will provide the most efficient land use. A program was developed to provide recommendations for perennial dairy-forage species in New York State based on soil type, drainage, and

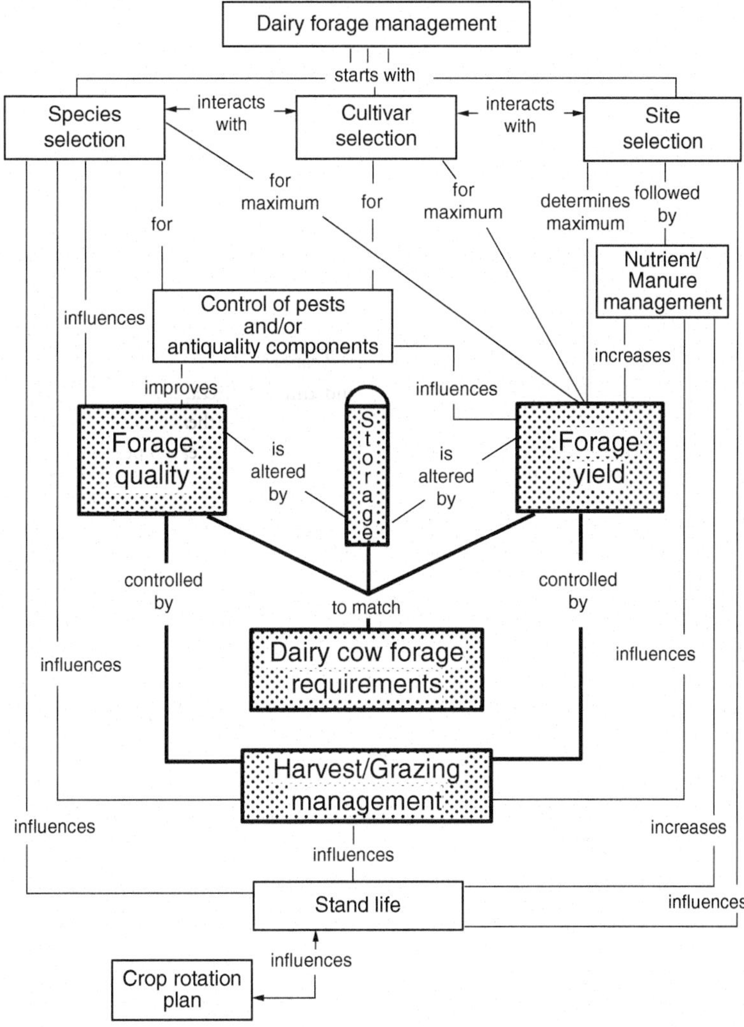

Fig. 20.5. Concept map of major linkages in a dairy-forage system (Cherney and Cherney 1993).

intended forage use through a world-wide-web interface (www.forages.org) (Cherney et al. 1998b). A version of this program is also available for Pennsylvania (http://www.forages.psu.edu/selection_tool/index.html). Recommendations provided are specific for pasture or stored feed and also specific for either lactating or dry cows. After the above requirements are met, the forage should be selected to match protein, fiber, and energy needs of the livestock.

Harvest/grazing management is the single most critical component of dairy forage management in the region, because it controls both forage yield and forage quality. The primary goal of harvest and nutrient management is not necessarily to maximize stand life, but to provide the necessary forage quality and yield while maintaining an acceptable stand life (Nelson et al. 2012). This goal can be achieved with multiple crops or combinations of crops. One combination of crops specific to this region is the planting of corn for silage after a spring harvest of alfalfa or grass (Roth et al. 1997; Smith et al. 1992).

General Description

Among classes of ruminant livestock in the region, dairy cattle have the highest nutrient requirements, and milk production is more sensitive to the balance of nutrients fed than is meat or wool production. Relatively low concentrate feed prices in the past allowed dairy managers in the temperate humid zone to increase profitability by using high-energy rations to increase milk production. This is not feasible in most other countries where concentrate feeds are expensive and milk marketing quotas may exist. As humans compete more directly with dairy cattle for edible grains, forages will become increasingly important in the region.

Important Management Considerations

Generally, the size of a farm operation is positively correlated with profitability. As farm size increases in the region, feasibility of grazing decreases and the tendency increases to adopt systems for year-round feeding of stored forages. Options for stored feed and for forage delivery systems are defined, in large part, by the size of the operation. As goals for production per animal increase, the amount of concentrate fed generally increases, thus increasing feed imports and resulting in more **nutrient balance** problems on dairy farms in the region (Wang et al. 1999). Large scale operations with maximum production efficiency are not a requisite for profitability. Smaller operations that control inputs and benefit from rotational stocking can be quite productive in the temperate humid zone. In a study of the profitability of grazing vs mechanical harvesting on New York State dairy farms, Gloy et al. (2002) concluded that farms with grazing systems were as profitable as farms relying totally on stored feed.

Harvest Methods and Strategies

High-quality forage is a necessity for both lactating cows and nutrient balance on temperate humid-zone farms. In a dairy feeding trial comparing alfalfa with headed orchardgrass in isonitrogenous rations, home grown forage accounted for 790 g kg^{-1} of N in alfalfa diets and only 230 g kg^{-1} of N in orchardgrass diets (Jonker et al. 2002). Harvest recommendations for legumes and grasses can be based on maturity stage or length of interval between harvests (Hall et al. 2007). For spring-harvested alfalfa, optimum neutral detergent fiber (NDF) concentration is approximately 400 g kg^{-1} harvested as hay or approximately 450 g kg^{-1} harvested as silage (Cherney and Sulc 1997). Optimum NDF of spring-harvested perennial grass is approximately 500 g kg^{-1} (Cherney and Cherney 1993). Alfalfa and grass with lower NDF concentrations than 350 or 450 g kg^{-1} respectively, may not have sufficient fiber to be used as a primary fiber source in a ration. While quality evaluation systems such as **Relative Feed Value** (RFV) and **Relative Forage Quality** (RFQ) (Undersander and Moore 2002) are very useful for evaluating harvested forage for feeding or sale, optimum NDF should be used for determining harvest dates (Cherney et al. 1994).

The Predictive Equations for Alfalfa Quality (**PEAQ**) system for estimating NDF concentration in alfalfa are effective throughout the temperate humid zone (Sulc et al. 1997). There is no simple, reliable method of estimating perennial grass quality. Grass matures in the spring before alfalfa and should be harvested prior to heading for optimum quality, giving rise to the rule of thumb "If you see the head, forage quality is dead." So, for optimum forage quality in the region, grass should be harvested first in the spring, followed by alfalfa-grass mixtures, followed by any pure alfalfa stands. A program is available to estimate NDF content of mixed alfalfa-grass stands based on alfalfa height and grass proportion in the mixture (www.forages.org) (Parsons et al. 2006). Often alfalfa is harvested first, resulting in poor-quality grass forage unacceptable for lactating cows. If corn is not planted when forage is at an optimum stage, which occurs frequently in the Northeast, it is recommended to stop planting corn and focus on forage harvesting.

Grazing Strategies

Management intensive rotational stocking is steadily expanding in use throughout the temperate humid zone (Paine et al. 1999). For dairy grazing systems in the region, it is advisable to supplement pasture (Muller and Fales 1998). Modeling results for grazing dairy herds in the Northeast US predict response to grain feeding is about 1.6 kg of milk per kg of grain dry matter supplement (McCall et al. 1999). Economic risk of grazing is reduced with grain supplementation, but it can have a negative impact on whole farm nutrient balance (Soder and Rotz 2001).

Integrating Forages into a Dairy Enterprise

A method to optimize use of forage resources and maximize profits on dairy farms in the region is based on the Pitt-Conway model (Figure 20.6) described by Conway (1993). The model is set up for a typical temperate humid zone dairy farm, with initial estimates developed for the upcoming year for each parameter using detailed, yet relatively simple, calculations. The typical dairy enterprise in the region is forage-based with minimal cash cropping, and grazing is a relatively minor component of the overall enterprise.

Estimation of Animal Forage Requirements

Average daily dry matter consumption per animal is estimated for each animal class as a basis for estimating total dry matter needs (Figure 20.6). Feed requirements differ by stage of lactation and by level of milk production, but dairy cattle will consume a relatively consistent amount of forage, depending on their body weight and forage quality, with the variable factor being concentrate supplement. Corn silage in the temperate humid zone is used as both

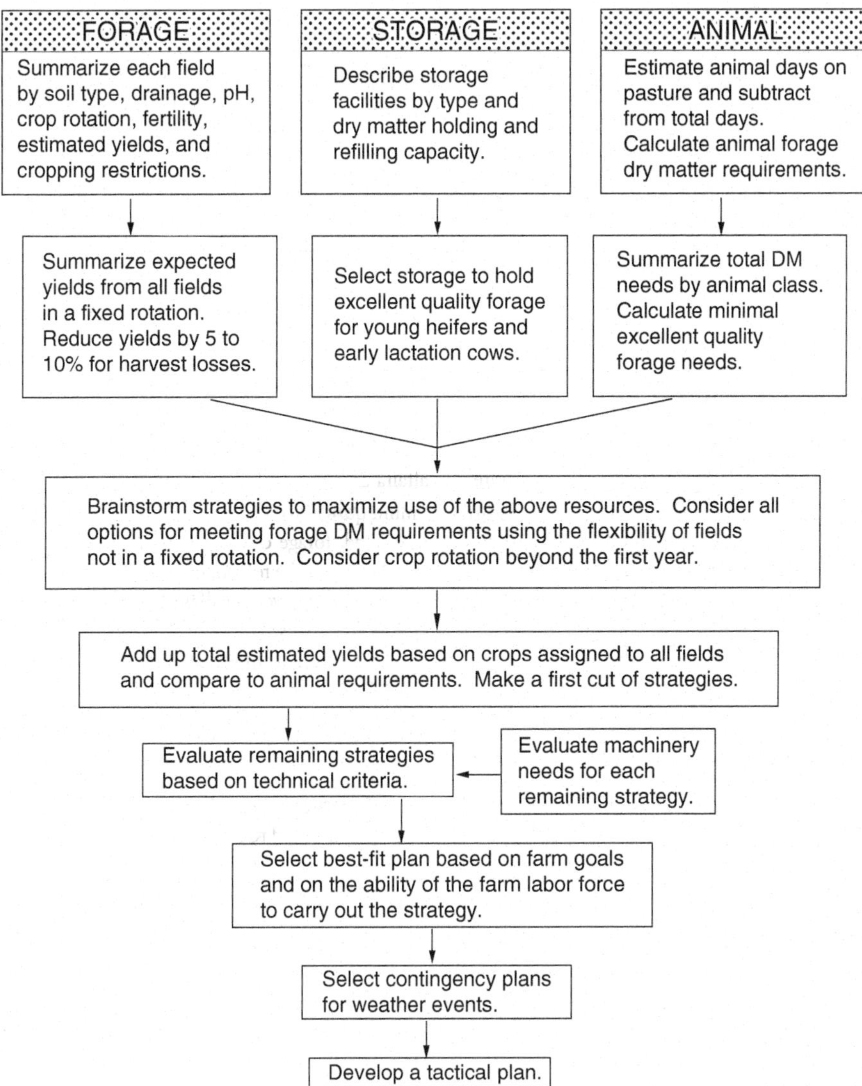

Fig. 20.6. Pitt-Conway dairy forage integration model for the Northeast. The objective is to match the quantity and quality of forage available with requirements of specific livestock classes and with available storage options (Conway 1993).

a fiber source and an energy source. The ratio of hay crop : corn silage in a balanced ration has a greater effect on energy and protein concentrate supplementation than on the total amount of forage (hay crop and corn silage) fed. Proper ration balancing is critical, and efficient utilization of forage is key to economic and environmental sustainability (Cherney and Mertens 1998). When developing actual tactical feeding plans, it is often advantageous to divide the cow class into four groups: early, mid, and late lactation, and dry, followed by ration balancing specific to each group.

Evaluation of Storage Capacity

Silos are described in terms of dimensions, filling strategy, calculated holding capacity, and practical holding capacity, including refills. While upright silos are still common in the western part of the region, bunker silos are more frequently used in the eastern part of the region. Flexible storage systems that fit in with available machinery and feeding systems currently on the farm are becoming very popular with smaller dairy operations. Limitations of each storage option are described in terms of forage moisture content, spoilage, unloading ease, and ability to separate forage based on quality (Figure 20.6). It is critical to have access to excellent quality forage throughout the year.

Estimation of Forage Production

Estimated yield of the previous crop is compared with yield estimates for other forage crops suited to the field, if the field will be rotated. Field use may be restricted by drainage, topography, distance from the home farm, previous pesticide applications, or government program regulations. All strategic plans for dairy-forage systems must be based on an adequate supply of high-quality forage. Some low-quality forage can be used in heifer or dry cow diets, as long as potassium content of such forage for dry cows is reasonably low (Cherney et al. 1998a).

Process of Integration

Dry matter produced on all fields in a fixed rotation is subtracted from the total animal requirement for forage dry matter (Figure 20.6). Several cropping strategies must be developed to meet the remaining dry matter needs from fields not in a fixed rotation. This involves asking a series of questions constrained only by farm size and specific storage and feeding facilities of the dairy enterprise under consideration. Example questions to consider can be found in Conway (1993). Strategies are evaluated based on overall farm goals, feasibility, and availability of necessary labor, machinery, storage structures, and feeding systems. After a "best fit" strategy is selected, a tactical plan for actually accomplishing the selected strategy is developed that should be coordinated with all other farm activities (Figure 20.6).

System Constraints

Forages grown on the farm can be an economical feed source for livestock in the region, but their true value is dependent on the cost and availability of on-farm and off-farm supplemental feeds. Efficient use of **feed supplements** is necessary to reduce the import of N and reduce losses to the environment (Rotz et al. 1999). High-pasture fertility can result in significant leaching of N (Stout et al. 2000). The dairy forage system should maximize the amount of forage in the diet in order to maintain a long-term P balance on the farm (Rotz et al. 2002).

Alfalfa and corn silage tend to have a high-quality image in the region compared to perennial grasses that leads to their preferred use, even on land not suited to these crops. Corn is frequently planted on land which cannot produce silage yields that are high enough to return costs of production. Alfalfa is often established on sites where it cannot persist because of soil constraints. These problems are due to a lack of integrated analysis of the forage management component of the dairy forage system. Another constraint to dairy forage systems in the region is the continuing focus of North American researchers on production per cow from alfalfa and corn silage. A major paradigm shift is necessary before grass-based systems enter the mainstream of dairying in the region (Fick and Clark 1998).

Forage Systems for Beef Cattle

Beef cattle operations can be classed into three main categories: cow-calf, stocker, and finishing. Cow-calf and stocker calf operations use forage crops to supply more than 90% of their nutrients, while cattle in the finishing stage obtain approximately 28% (Barnes and Nelson 2003). As a result, cow-calf and stocker calf production predominate in areas where row crop production is not feasible, while finishing operations tend to locate in semi-arid areas where feed grains are produced.

Cow-Calf Systems

Cow-calf operations tend to be small in the temperate humid zone, with an average herd size of about 26 mature cows in 2012 (NASS 2017). The main product produced from these operations is a weaned calf. Also, to be considered on these operations is the development of replacement heifers. The seasonal nutrient requirements for these types of herds varies depending on calving season, weaning date, breeding season, replacement animal needs, animal size, and production goals. In general, the cow herd needs the greatest nutrition from 30 days prior to calving through the first 60–90 days of lactation and again at rebreeding. Once pregnant, mature beef cows have fairly low nutritional requirements. Calves rely almost exclusively on their mother's milk for nutrients during the 60–90 days after birth (Blaser et al. 1986).

Once calves reach three to four months of age, they begin to consume forage and benefit from systems such as **creep grazing** or leader-follower systems that allow calves access to high-quality forage (Blaser et al. 1986).

Early Calves (Calves Born in January, February, or March)

Calves born early in the year typically have greater weaning weights in autumn and thus sell at a higher price than calves born later. Care of these calves conflicts less with other farm labor than do most systems, but the calves require more attention and housing at birth. Early calves are large enough to utilize peak milk flow from the cow that occurs on spring pasture; this contributes to good calf gains in spring.

Spring Calves (Calves Born in March, April, or May)

Calves born in spring typically wean at lighter weights in autumn and it often costs less to produce spring calves than early calves. Cost savings can be attributed to the fact that cows that calve in spring require less feed in winter. Calves born at this time of year are less likely to suffer from cold weather at birth and fewer facilities are needed to protect the herd from the elements at calving. In addition, cows are grazing high-quality grass during the breeding season in late spring which tends to improve conception rates. Conversely, peak milk flow from lush pasture often comes before calves are large enough to fully utilize it and cows that calve late often freshen after peak grazing.

Fall Calves (Calves Born from Late August to Early November)

Calves born in autumn are born at a time when the risk of severe weather is not a serious consideration. Cows are bred in early winter when heat stress is minimal, but feed quality must be maintained if high conception rates are to be expected. Fall calving systems require more and higher quality winter feed than do spring calving systems because cows are lactating during winter. However, the calves from a fall-calving herd can fully utilize lush spring growth and calves can be marketed in early summer when prices are typically high. In addition, calves are often sold before forage production from cool-season grasses declines in mid-summer which minimizes the need for high-quality forage at this time of year.

Stocker Calf Systems

Stocker calf producers usually purchase weaned calves that weigh 200–300 kg and sell them 60–180 days later to feedlots to be finished. Because livestock are frequently purchased or sold, livestock numbers can be managed to match seasonal fluctuations in forage supplies. In addition, because reproductive efficiency and replacement female animals are not required for these operations, the nutrient requirements and management of stocker calf operations is simpler to manage than cow-calf enterprises. However, because stocker animals are in a growing phase, their need for high-quality forage is generally higher than for cow-calf operations. Also, because stocker cattle are growing, forage consumption increases as animals become larger, so the forage system should provide for increasing forage supplies and/or decreased animal numbers over time.

Finishing Operations

While there is an abundance of feed grains in the western part of the temperate humid zone, most cattle finishing operations are located in the sub-humid and semi-arid regions of the US (NASS 2017). On an industry-wide basis, forages (and especially pasture) comprise a smaller amount of the nutrients used to feed finishing cattle than used to feed either cow-calf or stocker operations. The relatively low price of feed grains in the United States has made finishing animals on high-concentrate diets economically attractive. However, increasing concerns over dietary fat and other human health concerns may cause this to shift to back to greater level of forage being fed to finishing animals in the future (Robinson 2000). Maximum profitability in finishing systems typically corresponds with rapid animal gains so the forage required for these operations must be of high quality.

Forage Production

Pasture is often the cheapest source of nutrients for beef cattle and significant effort should be given to minimizing the amount of stored forage fed to beef animals (Bishop-Hurley and Kallenbach 2001). From a biologic perspective, there are three important concepts when planning a grazing system. They are:

1. Forage yield and yield distribution
2. Forage quality
3. Stand persistence or reliability

These concepts are discussed in the following section.

Forage Yield Distribution and Quality

Many beef producers consider yield the most important attribute for any forage. However, annual yield alone should not be used to select forages for pasture-based systems. For forage systems that rely on pasture, distribution of yield through the growing season and the number of annual grazing days provided by a forage species are far more important than total yield.

As an example, consider the two forages in Figure 20.7. Notice that forage "A" and forage "B" have the same annual yield. However, forage "A" produces 80% of its growth in May where forage "B" has a more even distribution of yield throughout the growing season. Forage "A" would work well in a stored-forage system, but forage "B" would be far superior for grazing.

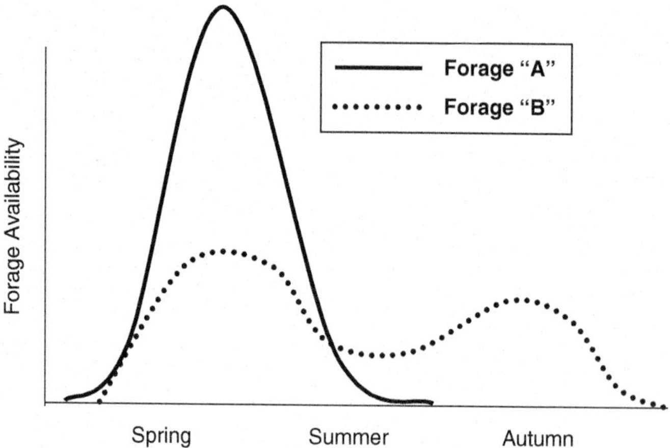

Fig. 20.7. Theoretical yield distribution of two different forage species.

Although forages vary in their seasonal yield distribution, no forage is productive during all seasons of the grazing year. An important principle for developing a productive forage program is utilizing the inherent differences in seasonal growth patterns to provide grazing for as much of the year as possible. Chapters 14, 15, 16, and 18 give more detailed information on the growth patterns of different forages for humid areas.

Many forages can be managed for beef cattle production. Legumes usually contain more energy and protein than grasses, but grasses typically persist longer than legumes. The most important management concept during the growing season is that forage must be kept in a vegetative stage to maximize forage quality and livestock production.

Keeping the grass in a vegetative stage of growth can be difficult to accomplish on a whole-farm basis, especially in late spring. During this time of year, grass growth often exceeds what the beef herd can consume. Grazing systems that include a number of divisions, or "paddocks", can help manage the forage supply. Paddocks that become over-mature can be "skipped" in the rotation and the herd "moved forward" to less mature paddocks. The skipped or mature paddocks should be harvested for hay or silage or grazed by dry cows or other livestock with low-nutrient requirements as soon as is feasible. The skipped paddocks can again be part of the rotation for the producing herd after the grass has been harvested and regrows later in the season.

Stand Persistence or Reliability

Many producers undervalue long-term stand persistence of many perennial forage species. Many tend to equate persistence with survival of individual plants, from a producer's perspective, there should be more interest in the "persistence of yield or productivity." In some cases, stand persistence may be the survival of individual plants, but in other instances, it may involve the natural reseeding capability of a species.

The soil environment is another factor that influences stand persistence. The most important aspects of the soil environment are soil depth, drainage, and fertility. As an example, alfalfa is one of the most productive and nutritious forages available on well-drained and fertile soils. However, it does not survive well on poorly drained soils and does not tolerate low-soil fertility. In this situation, a better choice might be to plant a different legume such as red clover or lespedeza, or a low-alkaloid cultivar of reed canarygrass and manage it to provide quality feed. Forages should also be selected for cold hardiness and drought tolerance.

Management also plays a vital role in stand persistence. Almost no forage can survive poor management and be productive. The major management factors that influence stand persistence are grazing/harvest frequency, residual leaf area after grazing/harvest, and planned **rest periods** for reseeding or fall growth.

Storage Methods

Though it is possible for producers to graze the beef herd the entire year, in a practical sense, most cow-calf operations need some stored feed during winter and during periods of drought. Stocker calf and finishing operations on pasture may need stored forage depending on when animals are received and marketed.

In many grazing systems, hay or silage production comes from excess spring growth from pastures. Perhaps the greatest difficulty with this approach is that the weather for making good-quality hay and silage is poor in spring due to frequent rains. Also, depending on how much the system relies on grazing to supply nutrients to the beef herd, excess pasture production may be insufficient to supply enough stored forage. For many small producers, the capital costs associated with stored forage are high and producers should consider purchasing stored forage (or other feeds) rather than to produce their own. Chapters 41 and 42 provide more detailed information on stored forage techniques.

Building a Forage System

Building a forage system for a beef herd is a complex process. However, using the three-step approach as outlined below can make the process easier.

Step 1: Begin with a "base forage." Typically, the "base forage" selected should be the legume-grass mixture, the one most likely to give the maximum number of grazing days annually. Cool-season grass and cool-season grass/legume pastures usually provide the greatest number of grazing days per unit of land area annually (Allen et al. 2000). Cool-season grasses break dormancy in early spring and produce about two-thirds of their annual production before mid-summer (Figure 20.8).

Step 2: Fill the production gaps with legumes or grasses. Although cool-season grasses and grass/legume mixture can supply much of the forage needed for a grazing system, they are often not productive in mid-summer. In the northern parts of the temperate humid zone this period is usually short, but in the southern parts of the zone it can last as long as eight to twelve weeks. These lulls in forage production are sometimes called "gaps."

The planned use of warm-season grasses and/or drought-tolerant legumes can fill this summer gap (Figure 20.8). Alfalfa, red clover, birdsfoot trefoil, and annual lespedeza provide the most forage in mid-summer. Annual warm-season grasses such as pearlmillet, sudangrass, sorghum x sudangrass hybrids, and immature corn can be used effectively for summer forage throughout the zone. In the southern part of the temperate humid zone, perennial warm-season grasses are used on a limited basis for summer grazing.

Most warm-season grasses provide grazing for only 30–120 days each summer in the temperate humid zone (Smart et al. 1995; Allen et al. 2000). The warm-season grass is effective for erosion control and wildlife benefits throughout the year. While warm-season forages can be useful in a planned grazing system, the goal should be on just enough for summer before going back to cool-season species with high quality for fall and winter grazing.

Step 3: Extend the grazing season. It is possible to extend the grazing season beyond the normal growing season. The advantage of extending the grazing season is lower feeding costs; it is about 50% cheaper to produce liveweight gain on pasture than on stored forage (Bishop-Hurley and Kallenbach 2001).

The most common way to extend the fall grazing period is by **stockpiling** forage (Figure 20.8). Stockpiling is a form of deferred grazing. Stockpiling usually involves fertilizing cool-season grasses, typically tall fescue, in early to mid-August with nitrogen and then allowing it to grow ungrazed until late autumn or winter. When the growing season ends, cattle can be moved to these reserve fields. Producers often choose tall fescue over other cool-season grasses for stockpiling because it: (i) produces more growth in autumn (Taylor and Templeton 1976),

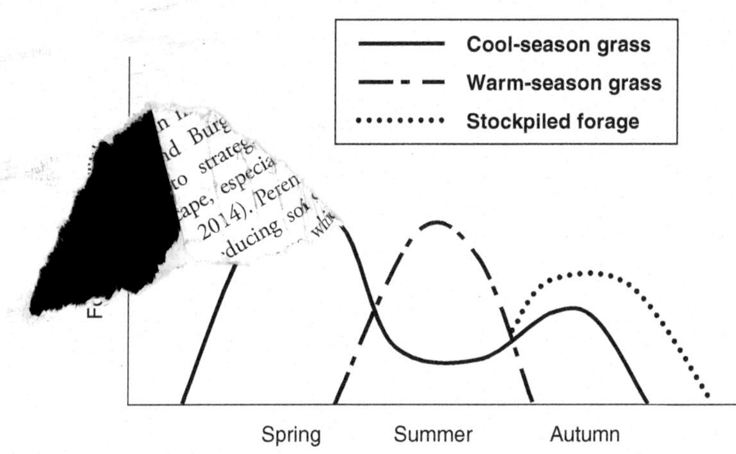

Fig. 20.8. Typical forage availability of cool-season grass, warm-season grass and stockpiled forage in the temperate humid zone.

and (ii) retains its nutritive value longer into winter (Taylor and Templeton 1976; Hitz and Russell 1998).

Annuals such as small grains can be used as cover crops to extend the grazing season. If planted in late summer, the small grains can provide forage for grazing in autumn and again in early spring. For growing livestock, winter annuals are often preferred over stockpiled perennials because energy values are higher. When double cropped with one of the annual warm-season grasses, this can be a productive system (Belesky et al. 1981).

Crop residues, primarily corn stalks, in the temperate humid zone, are yet another option for extending the grazing season. Corn stalks typically provide 40–60 days of inexpensive grazing in autumn (Klopfenstein et al. 1987). Livestock typically consume left-over grain, husks, and leaves but leave most of the stalks which are beneficial ground cover. The quality of corn stalks is usually low after harvest and declines incrementally thereafter. Corn residues are best suited to dry beef cows, but protein supplementation is usually needed to give acceptable performance.

Emerging Forage Livestock Systems

Small Ruminants

Small ruminants, primarily sheep and goats, currently supply about 2% of the meat consumed in the US (Horowitz 2006; Corum 2011). However, these two ruminant species supply 5% of the meat consumed worldwide (FAO 2017). In some cultures, lamb and/or goat products are desired for religious or ceremonial events; their production often fluctuates locally around these events.

Census of agriculture data indicate expansion in small herds of sheep and goats since 2002 (USDA 2012). In addition to meat/milk production, goat populations, have grown as a mechanism of natural brush control.

Forage Management for Small Ruminants

Sheep, goats, and cattle digest forages similarly as ruminants. However, the two former differ anatomically and behaviorally from the latter. Goats prefer woody, broadleaf plants (often called browse) and herbaceous forbs more than grass; frequently goats select the "weedy" plant species in pastures when given a choice (Hart 2001). Some producers take advantage of these differences by co-grazing cattle and small ruminants to maximize forage utilization. Land managers employ goats to control invasive brush before pasture renovation (Pitman 2007). Sheep production improves when legumes and forbs comprise a mixture with high-quality cool-season grasses during certain production phases (Ely 1995).

Internal parasites and hoof problems abound in small ruminant systems, especially when overgrazing occurs. Maintaining an adequate residual in pastures reduces parasite problems and foot-rot. Further, an adequate residual provides the leaf area often needed for rapid pasture regrowth. Stocking rate, grazing pressure, and pasture rest period affect overall animal and plant health in these small ruminant grazing systems that often include diverse plant species.

Ecosystem Services of Forages in Temperate Regions

As global population increases, agricultural systems will have to increase food productivity while reducing environmental impacts on soil, water, and climate (Foley et al. 2011). Environmental or **ecosystem services** are the benefits that humans obtain from ecosystems (both natural ecosystems and agroecosystems) including soil conservation, water quality, climate regulation, and biodiversity habitat (Daily 1997). Historically, forage systems have provided not only a source of feed for livestock and economic income for farmers, but also a wide range of environmental benefits to society as a whole.

These ecosystem services include reduced rates of soil erosion, reduced nitrate leaching to groundwater, increased soil organic matter content, improved nutrient supply through biologic N fixation, and reduced incidence of pests and diseases (Russelle et al. 2007). In this section, examples of those environmental benefits from temperate regions are presented. An in-depth review of Forages and the Environment is presented in Chapter 13.

Soil Conservation

Many landscapes in the temperate regions include topography where annual crop production places soils at high risk for erosion. Large areas of annual grain and oilseed crops leave bare soil exposed to rain and wind through several months of the year, which can contribute to soil erosion, nutrient leaching, and surface and groundwater eutrophication problems, seriously compromising water quality i[illegible]ost agricultural landscapes worldwide (Pimentel a[illegible]ess 2013). A solution to these problems is [illegible]ically place perennial forages into the lands[illegible]lly in more sensitive lands (Asbjornsen et al.[illegible]nial forages provide soil cover year-round, re[illegible]erosion, and they also have extensive root system[illegible]h absorb nutrients throughout most of the year, reducing nutrient leaching.

Long-term crop rotation studies which included perennial forages have shown the benefits of forages to soil conservation. For instance, the Wisconsin Integrated Cropping Systems Trial (https://wicst.wisc.edu) established in 1989 with annual and perennial cropping systems, has shown that soil organic carbon stocks decreased for annual row crop agriculture but increased for the rotationally grazed permanent pasture during the first 20 years of the study in the 0–60 cm depth (Sanford 2014). In another long-term crop rotation study in the temperate region of South America (Uruguay), soil erosion in crop-pasture rotations was reduced compared to continuous cropping both under tillage and no-tillage conditions, after 40 years (Garcia-Préchac et al. 2004).

Therefore, the different alternatives for feeding livestock (use of pastures, conserved forages, or grains) imply different cropping systems and land uses, which in turn, have consequences on the environmental impacts or benefits of the system. For instance, soil erosion per unit of land and per unit of beef was lower in beef production systems based on grazing grasslands and pastures than confined beef systems where grain accounted for a larger proportion of the diet (Picasso et al. 2014). Expanding the area in forages either through crop-pasture rotations or permanent pastures is a key to reduce soil erosion, therefore, conserving soil fertility and soil quality for future production cycles.

Nutrient Management and Water Quality

Nutrient Leaching

Nutrients like N and P leaching from agriculture landscapes (in addition to soil erosion) contribute to nutrient accumulation in water bodies like lakes, rivers, and the Gulf of Mexico, causing eutrophication and subsequent depletion of oxygen dissolved in water (**hypoxia**), killing fish and most aquatic life (a.k.a., "dead zones," Rabalais et al. 2002). Perennial forages consistently reduce this problem. For instance, the cool-season perennial forage intermediate wheatgrass reduced nitrate leaching in the second year of growth compared to spring or summer growth of winter or spring wheat in Michigan (Culman et al. 2013). A mixture of perennial cool-season grasses, smooth bromegrass and orchardgrass, had the lowest residual nitrate content in the whole soil profile (0–180 cm depth) compared to four other corn or soybean cropping systems with cover crops and cereals in Iowa (De Haan et al. 2017). Perennial forage legumes like kura clover intercropped as a living mulch with corn, reduced nitrate leaching as compared to the monoculture corn control in Wisconsin (Ochsner et al. 2010). All these examples illustrate the fact that perennial forages can be used in cropping systems in many different configurations to reduce nutrient leaching and, therefore, improve water quality.

Nutrient Balance

Intensive livestock management is often considered an economic necessity. With intensive management, an increased amount of crop imports often come onto the livestock farm. Simple nutrient balance calculations make it clear that when imports of a given nutrient into a farm ecosystem exceed exports in the form of meat, milk, or fiber, the nutrient will eventually accumulate and lead to water quality and other problems. Grazing livestock systems have less nutrient excesses per land area than confined intensive livestock systems (Picasso et al. 2014), and nutrient excesses are larger as the number of animals increase (Koelsch and Lesoing 1999). A manure management plan can reduce the severity of the problem but may not be able to solve it completely. Grasses have advantages over legumes when considering nutrient management because they can use excess N, support vehicle traffic, and tolerate saturated soils. Grass production for high-quality forage, however, requires a higher level of management compared to legumes.

Nutrient balance concerns must include the animal component. Number of animals per unit land and capacity of land to receive waste must be in harmony. Land-use regulations are being considered in the US, particularly in regions that supply water to large metropolitan centers. For example, water for New York City is transferred long distances from upstate New York through aqueducts. New York City has legal authority to restrict agricultural practices in the upstate watershed in order to control water quality.

Forages with high animal intake potential can reduce the amount of concentrate in the diet, such that forage management for optimum quality results in minimizing feed imports. Reduction in feed imports is a powerful tool to help balance nutrients in a livestock system (Koepf 1989). When a broader viewpoint than just the livestock enterprise is taken, however, minimizing feed imports to farms may not always be in society's best interest. Livestock can be viewed as environmentally-safe recyclers of many by-products from industries such as distillers, brewers, and bakers, which would otherwise have their own disposal problems (Pell 1992). The use of management intensive grazing practices in the region also has the potential to alleviate many nutrient management problems, especially those related to manure handling and storage.

Greenhouse Gas Balance and Climate Change

Concerns about climate change, caused by increased concentrations of **greenhouse gases** in the atmosphere (primarily carbon dioxide, **methane**, and nitrous oxide) have led to increased interest in estimations of greenhouse gas emissions of livestock production. The United Nations Food and Agriculture Organization (FAO) estimated that the global livestock sector contributes 14.5% of anthropogenic greenhouse gases, with a large range in the carbon footprint (i.e. emissions per unit of product) of livestock production depending on the production system and management. Therefore, livestock production presents challenges and opportunities for climate change mitigation (Gerber et al. 2013).

Livestock production contributes to CO_2 emissions through the burning of fossil fuels for farm machinery use, production of fertilizers and pesticides, and soil emissions from tillage for crop and forage production. Ruminants emit methane as a sub-product of enteric fermentation in the rumen, in addition to that arising from liquid manure management systems. Nitrous oxide is a very powerful greenhouse gas that is emitted after nitrogen fertilizers are applied in soils and also from manure systems (Gerber et al. 2013).

Improving efficiency of beef production, through improved feed quantity (i.e. intake) and quality (i.e. forage digestibility), reproductive efficiency (i.e. calving rate) and daily weight gain, can reduce greenhouse gas emissions from ruminants (Pelletier et al. 2010; Beauchemin et al. 2010). Furthermore, improving grazing efficiency can also significantly reduce emissions (Herrero et al. 2013; Picasso et al. 2014).

In dairy systems, improving production efficiency is also associated with reducing greenhouse emissions (Rotz et al. 2010). In grazing dairy systems, efficiency of production depends highly on the use of pasture forage, whereas increasing concentrate supply alone without efficient use of pasture may not reduce carbon footprint (Lizarralde et al. 2014).

Soils under forage production can also sequester and store carbon from the atmosphere, which is a key ecosystem service. When ruminants are managed in a way that restores and enhances grassland ecosystem function, the increased carbon stocks in the soil can increase soil microbial activity, carbon sequestration, and methane oxidation, which can create carbon negative budgets (Teague et al. 2016). However, on a global aggregate, this potential sequestration may offset only a fraction of emissions from grazing systems. Furthermore, since soils over time reach carbon equilibrium, this sequestration potential is time limited, and after a few decades no more carbon can be accumulated in soils (Garnett et al. 2017). Therefore, it is unlikely that grazing systems could sequester more total carbon than that emitted from livestock. Nevertheless, when accounting for the continuous conversion of land to row crops, livestock production based on well-managed forages and pastures is likely to have a lower carbon footprint than livestock production based mostly on grains (Teague et al. 2016).

Sustainable Intensification of Livestock Systems

Ecological intensification is the production of more agricultural product and ecosystem services relying on ecologic processes and biodiversity while reducing external inputs. Conventional intensification of livestock systems based on increased use of grain in diets, and therefore more fertilizers, pesticides, and fossil energy, has increased productivity substantially, while producing negative environmental impacts like increased soil erosion, decreased water quality, and biodiversity loss. Sustainable intensification of beef and dairy systems relies on efficient production and utilization of forages rather than just increased use of external inputs (Picasso et al. 2017; Llanos et al. 2018). Figure 20.9 illustrates

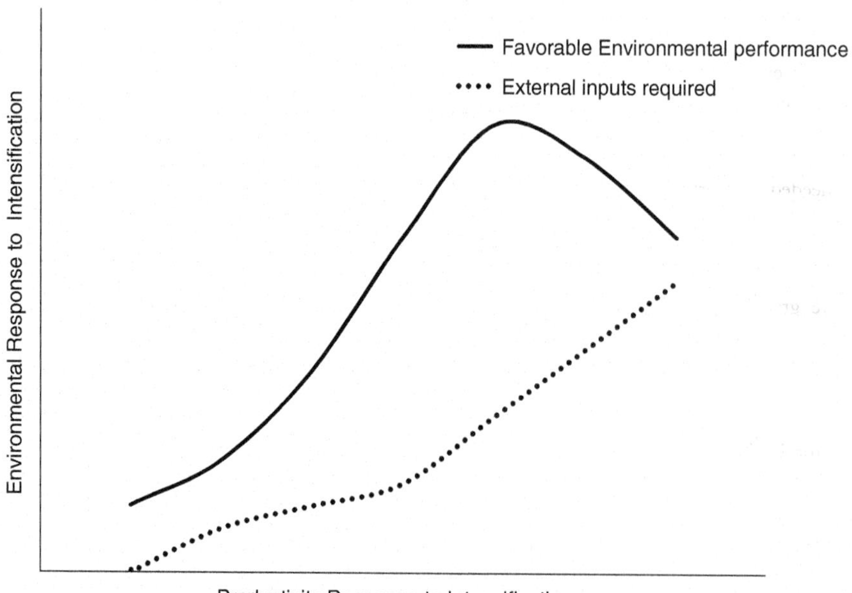

Fig. 20.9. Hypothetical intensification trajectories of forage-based livestock systems.

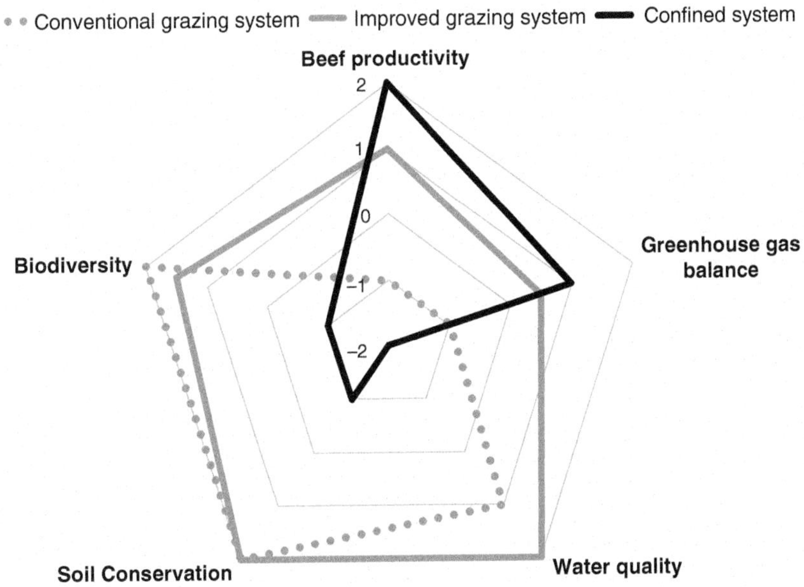

Fig. 20.10. Sustainability assessment of three hypothetical beef systems using multiple indicators. *Source:* Modified from Picasso et al. (2014).

hypothetical relationships in an intensification trajectory of livestock systems. As productivity increases, with low use of inputs and improved grazing and herd management, environmental performance also increases, up to an optimum. Further intensification based on external inputs can increase productivity, but with a greater cost in environmental performance.

Trade-offs may exist between different indicators of livestock system performance, even within the environmental dimension (Modernel et al. 2013). Therefore, a multidimensional assessment of sustainability of livestock production is needed to inform decision makers, accounting for multiple productivity, socio-economic, and environmental indicators (for example, see Figure 20.10). In temperate humid regions, there is great potential to improve grazing livestock systems productivity while reducing environmental impacts and conserving biodiversity (Picasso et al. 2014).

Summary

Livestock farming in the temperate humid zone is moving in two directions, large farms with more imported feeds, and smaller farms with more home-grown forages and grazing systems. Regardless of farm size, environmental issues as well as ecosystem services need to be addressed for long-term sustainability.

Forage-livestock management decisions now, and even more so in the future, will be affected, if not controlled, by environmental concerns. Forages can be considered the solution to many potential environmental problems in the region, with forage quality playing a key role in **nutrient balance** on the farm.

References

Allen, V.G., Fontenot, J.P., and Brock, R.A. (2000). Forage systems for production of stocker steers in the upper south. *J. Anim. Sci.* 78: 1973–1982.

Allen, M.S., Coors, J.G., and Roth, G.W. (2003). Corn silage. In: *Silage Science and Technology*, Agronomy Monographs 42 (eds. D.R. Buxton, R.E. Muck and J.H. Harrison), 547–608. Madison, WI: ASA, CSSA, and SSSA.

Asbjornsen, H., Hernandez-Santana, V., Liebman, M. et al. (2014). Targeting perennial vegetation in agricultural landscapes for enhancing ecosystem services. *Renewable Agric. Food Syst.* 29: 101–125.

Barnes, R.F and Nelson, C.J. (2003). Forages and grasslands in a changing world. In: *Forages: An Introduction to Grassland Agriculture*, 6e (eds. R.F Barnes, C.J. Nelson, M. Collins and K.J. Moore), 3–23. Ames, IA: Iowa State Press.

Beauchemin, K.A., Janzen, H.H., Little, S.M. et al. (2010). Life cycle assessment of greenhouse gas emissions from beef production in western Canada: a case study. *Agric. Syst.* 103: 371–379.

Belesky, D.P., Wilkinson, S.R., Dawson, R.N., and Elsner, J.E. (1981). Forage production of a tall fescue sod intercropped with sorghum x sudangrass and rye. *Agron. J.* 73: 657–660.

Bishop-Hurley, G.J. and Kallenbach, R.L. (2001). The economics of grazing beef cows during winter. In: *Proceedings of the American Forage and Grasslands Council, 2001*, Springdale, AR. 22-25 April, 2001, 274. Georgetown, TX: AFGC.

Blaser, R.E., Hammes, Jr., R.C., Fontenot, J.P. et al. (1986). Forage-animal management systems. Virginia Agricultural Experiment Station Bulletin No. 86–7.

Casler, M.D., Phillips, M.M., and Kohn, A.L. (2009). DNA polymorphisms reveal geographic races of reed canarygrass. *Crop Sci.* 49: 2139–2148.

Cherney, J.H. and Cherney, D.J.R. (1993). Annual and perennial grass production for silage. In: *Silage Production: From Seed to Animal. Proceedings from the National Silage Production Conference* NRAES-67. 23-25 Feb., Syracuse, NY, 9–17. Ithaca, NY: Northeast Regional Agricultural Engineering Service.

Cherney, D.J.R. and Mertens, D.R. (1998). Modelling grass utilization by dairy cattle. In: *Grass for Dairy Cattle* (eds. J.H. Cherney and D.J.R. Cherney), 351–371. Wallingford, UK: CAB International.

Cherney, J.H. and Sulc, R.M. (1997). Predicting first cutting alfalfa quality. In: *Silage: Field to Feedbunk. Proceedings from the North American Conference.* NRAES-99. 11-13 Feb., 1997, Hershey, PA, 53–65. Ithaca, NY: Northeast Regional Agricultural Engineering Service.

Cherney, J.H., Cherney, D.J.R., Chase, L.E. et al. (1994). Evaluating forages for dairy cattle. In: *Proceedings of the American Forage and Grasslands Council*, vol. 3, 74.

Cherney, J.H., Cherney, D.J.R., and Bruulesma, T.W. (1998a). Potassium management. In: *Grass for Dairy Cattle* (eds. J.H. Cherney and D.J.R. Cherney), 137–160. Wallingford, UK: CAB International.

Cherney, J.H., Reid, W.S., Elswit, D.M. et al. (1998b). World Wide Web Tool for Forage Species Recommendation Based on Soil Type. Agronomy Journal.

Cherney, D.J.R., Cherney, J.H., and Cox, W.J. (2007). Forage quality differences of corn hybrids as influenced by ensiling. *Forage Grazinglands* https://doi.org/10.1094/FG-2007-0918-01-RS.

Conway, J.F. (1993). Dynamic forage resource management. In: *Silage Production: From Seed to Animal. Proceedings from the National Silage Production Conference.* NRAES-67. 23-25 Feb., Syracuse, NY, 232–241. Ithaca, NY: Northeast Regional Agricultural Engineering Service.

Corum, J. (2011). A century of meat. *New York Times* (15 March).

Culman, S.W., Snapp, S., Ollenburger, M. et al. (2013). Soil and water quality rapidly responds to the perennial grain Kernza wheatgrass. *Agron. J.* 105: 735–744.

Daily, G.C. (1997). *Nature's Services: Societal Dependence on Natural Ecosystems*, 392. Washington: Island Press.

De Haan, R.L., Schuiteman, M.A., and Vos, R.J. (2017). Residual soil nitrate content and profitability of five cropping systems in Northwest Iowa. *PLoS One* 12 (3): e0171994.

Ely, D.G. (1995). Forages for sheep, goats, and rabbits. In: *Forages: Vol. II the Science of Grassland Agriculture*, 5e (eds. R.F Barnes, D.A. Miller and C.J. Nelson), 313–326. Ames, IA: Iowa State University Press.

Ferraretto, L.F. and Shaver, R.D. (2012). Meta-analysis: impact of corn silage harvest practice on intake, digestion and milk production by dairy cows. *Prof. Anim. Sci.* 28: 141–149.

Fick, G.W. and Clark, E.A. (1998). The future of grass for dairy cattle. In: *Grass for Dairy Cattle* (eds. J.H. Cherney and D.J.R. Cherney), 1–22. Wallingford, UK: CAB International.

Foley, J.A., Ramankutty, N., Brauman, K.A. et al. (2011). Solutions for a cultivated planet. *Nature* 478: 337–342.

Food and Agriculture Organization of the United Nations (2017). Food and agriculture data. http://www.fao.org/faostat/en/#data (accessed 10 October 2019).

Garcia-Préchac, F., Ernst, O., Siri-Prieto, G., and Terra, J.A. (2004). Integrating no-till into crop–pasture rotations in Uruguay. *Soil Tillage Res.* 77: 1–13.

Garnett, T., Godde, C., Muller, A. et al. (2017). Grazed and Confused? Ruminating on cattle, grazing systems, methane, nitrous oxide, the soil carbon sequestration question – and what it all means for greenhouse gas emissions. FCRN, University of Oxford.

Gerber, P.J., Steinfeld, H., Henderson, B. et al. (2013). *Tackling Climate Change Through Livestock — A Global Assessment of Emissions and Mitigation Opportunities*, 115. Rome: FAO.

Gloy, B.A., Tauer, L.W., and Knoblauch, W. (2002). Profitability of grazing versus mechanical forage harvesting on New York dairy farms. *J. Dairy Sci.* 85: 2215–2222.

Hall, M.W., Cherney, J.H., and Rotz, C.A. (2007). Saving forage as hay or silage. In: *Forage Utilization for Pasture-Based Livestock Production.* NRAES-173 (ed. E.B. Rayburn), 121–134. Ithaca, NY: Natural Resource, Agriculture, and Engineering Service.

Hart, S.P. (2001). Recent perspectives in using goats for vegetation management in the USA. *J. Dairy Sci.* 84: E170–E176.

Herrero, M., Havlík, P., Valin, H. et al. (2013). Biomass use, production, feed efficiencies, and greenhouse gas emissions from global livestock systems. *Proc. Natl. Acad. Sci. U.S.A.* 110 (52): 20888–20893.

Hitz, A.C. and Russell, J.R. (1998). Potential of stockpiled perennial forages in winter grazing systems for beef cows. *J. Anim. Sci.* 76: 404–415.

Hodgson, J.G. (1990). *Grazing Management: Science into Practice.* New York, NY: Wiley.

Horowitz, R. (2006). *Putting Meat on the American Table: Taste, Technology, Transformation*. Baltimore, MD: JHU Press.

Jonker, J.S., Cherney, D.J.R., Fox, D.G. et al. (2002). Orchardgrass versus alfalfa for lactating dairy cattle: production, digestibility and nitrogen balance. *J. Appl. Anim. Res.* 21: 81–92.

Karayilanli, E., Cherney, J.H., Sirois, P. et al. (2016). Botanical composition prediction of alfalfa-grass mixtures using NIRS: developing a robust calibration. *Crop Sci.* 56: 3361–3366.

Klopfenstein, T., Roth, L., Fernandez Rivera, S., and Lewis, M. (1987). Corn residues in beef production systems. *J. Anim. Sci.* 65: 1139–1148.

Koelsch, R. and Lesoing, G. (1999). Nutrient balance on Nebraska livestock confinement systems. *J. Anim. Sci.* 77 (Suppl. 215): 63–71.

Koepf, H.H. (1989). Nutrient recycling in balanced livestock-crop systems. In: *Sustaining the Smaller Dairy Farm in the Northeast* (ed. S.R.K.). New Milford, CT: Sunny Valley Foundation, Inc.

Lizarralde, C., Picasso, V., Rotz, A. et al. (2014). Practices to reduce milk carbon footprint on grazing dairy farms in southern Uruguay: case studies. *Sustainable Agric. Res.* 3 (2): 1–15. ISSN 1927-0518.

Llanos, E., Astigarraga, L., and Picasso, V. (2018). Energy and economic efficiency in grazing dairy systems under alternative intensification strategies. *Eur. J. Agron.* 92: 133–140.

McCall, D.G., Clark, D.A., Stachurski, L.J. et al. (1999). Optimized dairy grazing systems in the Northeast United States and New Zealand. I. Model description and evaluation. *J. Dairy Sci.* 82: 1795–1807.

McGraw, R.L. and Nelson, C.J. (2003). Legumes for northern areas. In: *Forages: An Introduction to Grassland Agriculture*, 6e (eds. R.F Barnes, C.J. Nelson, M. Collins and K.J. Moore), 171, 190. Ames, IA: Iowa State University Press.

Modernel, P., Astigarraga, L., and Picasso, V. (2013). Global versus local environmental impacts of grazing and confined beef production systems. *Environ. Res. Lett.* 8: 035052.

Muller, L.D. and Fales, S.L. (1998). Supplementation of cool-season grass pastures for dairy cattle. In: *Grass for Dairy Cattle* (eds. J.H. Cherney and D.J.R. Cherney), 335–350. Wallingford, UK: CAB International.

NASS (2017). USDA National Agricultural Statistics Service. https://quickstats.nass.usda.gov (accessed 10 October 2019).

Nelson, C.J., Redfearn, D.D., and Cherney, J.H. (2012. Chapter 4). Forage harvest management. In: *Conservation Outcomes from Pastureland and Hayland Practices: Assessment, Recommendations, and Knowledge Gaps* (ed. C.J. Nelson), 205–256. Lawrence, KS: Allen Press.

Ochsner, T.E., Albrecht, K.A., Schumacher, T.W. et al. (2010). Water balance and nitrate leaching under corn in Kura clover living mulch. *Agron. J.* 102: 1169–1178.

Paine, L.K., Undersander, D., and Casler, M.D. (1999). Pasture growth, production, and quality under rotational and continuous grazing management. *J. Prod. Agric.* 12: 569–577.

Parsons, D., Cherney, J.H., and Gauch, H.G. Jr. (2006). Estimation of preharvest fiber content of mixed alfalfa–grass stands in New York. *Agron. J.* 98: 1081–1089.

Pell, A.N. (1992). Does ration balancing affect nutrient management? In: *Cornell Nutrition Conference for Feed Manufacturers*. Ithaca, NY: Cornell University.

Pelletier, N., Pirog, R., and Rasmussen, R. (2010). Comparative life cycle environmental impacts of three beef production strategies in the upper Midwestern United States. *Agric. Syst.* 103: 380–389.

Picasso, V., Modernel, P., Becoña, G. et al. (2014). Sustainability of meat production beyond carbon footprint: a synthesis of case studies from grazing systems in Uruguay. *Meat Sci.* 98: 346–354.

Picasso, V., Schaefer, D., Modernel, P., and Astigarraga, L. (2017). Ecological intensification of beef grazing systems. *Grassland Sci. Eur.* 22: 218–220.

Pimentel, D. and Burgess, M. (2013). Soil erosion threatens food production. *Agriculture* 3: 443–463.

Pitman, W.D. (2007). Forage systems for warm humid region. In: *Forages: Vol. II the Science of Grassland Agriculture*, 6e (eds. R.F Barnes, C.J. Nelson, K.J. Moore et al.), 303–312. Blackwell Publishing.

Rabalais, N.N., Turner, R.E., and Wiseman W.J. (2002). Gulf of Mexico hypoxia, a.k.a. "the dead zone." *Annu. Rev. Ecol. Syst.* 33: 235–263.

Ribaudo, M., Gollehon, N., Aillery, M. et al. (2003). *Manure Management for Water Quality: Costs to Animal Feeding Operations of Applying Manure Nutrients to Land*, 90 p. USDA Agricultural Economic Report 824. Economic Research Service, Resource Economics Division. Washington, DC: USDA.

Robinson, J. (2000). *Why Grassfed Is Best*, 1e. Vashon, WA: Vashon Island Press.

Roth, G.W., Curran, W., Calvin, D. et al. (1997). *Considerations for Double-Cropping Corn Following Hay in Pennsylvania* Penn State Extension Service Publishing Agronomy Facts 56.

Rotz, C.A., Satter, L.D., Mertens, D.R., and Muck, R.E. (1999). Feeding strategy, nitrogen cycling, and profitability of dairy farms. *J. Dairy Sci.* 82: 2841–2855.

Rotz, C.A., Sharpley, A.N., Satter, L.D. et al. (2002). Production and feeding strategies for phosphorus management on dairy farms. *J. Dairy Sci.* 85: 3142–3153.

Rotz, C.A., Montes, F., and Chianese, D.S. (2010). The carbon footprint of dairy production systems through partial life cycle assessment. *J. Dairy Sci.* 93: 1266–1282.

Russelle, M.P., Entz, M.H., and Franzluebbers, A.J. (2007). Reconsidering integrated crop–livestock systems in North America. *Agron. J.* 99: 325–334.

Sanford, G.R. (2014). Perennial grasslands are essential for long term SOC storage in the Mollisols of the North Central USA. In: *Soil Carbon, Progress in Soil Science*, 281–288. (eds. A.E. Hartemink and K. McSweeney). Springer https://doi.org/10.1007/978-3-319-04084-4_29.

Schwab, E.C., Shaver, R.D., Lauer, J.G., and Coors, J.G. (2003). Estimating silage energy value and milk yield to rank corn hybrids. *Anim. Feed Sci. Technol.* 109: 1–18.

Smart, A.J., Undersander, D.J., and Keir, J.R. (1995). Forage growth and steer performance on Kentucky bluegrass vs. sequentially grazed Kentucky bluegrass-switchgrass. *J. Prod. Agric.* 8: 97–101.

Smith, M.A., Carter, P.R., and Imholte, A.A. (1992). No-till vs. conventional tillage for late-planted corn following hay harvest. *J. Prod. Agric.* 5: 261–264.

Soder, K.J. and Rotz, C.A. (2001). Economic and environmental impact of four levels of concentrate supplementation in grazing dairy herds. *J. Dairy Sci.* 84: 2560–2572.

Stout, W.L., Fales, S.L., Muller, L.D. et al. (2000). Water quality implications of nitrate leaching from intensively grazed pasture swards in the northeast US. *Agric. Ecosyst. Environ.* 77: 203–210.

Sulc, R.M., Albrecht, K.A., Cherney, J.H. et al. (1997). Field testing a rapid method for estimating alfalfa quality. *Agron. J.* 89: 952–957.

Taylor, T.H. and Templeton, W.C. (1976). Stockpiling Kentucky bluegrass and tall fescue forage for winter pasturage. *Agron. J.* 68: 235–239.

Teague, W.R., Apfelbaum, S., Lal, R. et al. (2016). The role of ruminants in reducing agriculture's carbon footprint in North America. *J. Soil Water Conserv.* 71: 156–164.

Undersander, D. and Moore, J.E. (2002). Relative forage quality. *Focus Forage Series* 4 (5): 1–2. http://www.uwex.edu/ces/crops/uwforage/FocusonForage.htm.

United States Department of Agriculture (2012). Census of agriculture. https://www.agcensus.usda.gov (accessed 10 October 2019).

Volenec, J.J. and Nelson, C.J. (2003). Environmental aspects of forage management. In: *Forages: An Introduction to Grassland Agriculture*, 6e (eds. R.F Barnes, C.J. Nelson, M. Collins and K.J. Moore), 99–124. Ames, IA: Iowa State Press.

Wang, S.J., Fox, D.G., Cherney, D.J.R. et al. (1999). Impact of dairy farming on well water nitrate level and soil content of phosphorus and potassium. *J. Dairy Sci.* 82: 2164–2169.

CHAPTER

21

Forage Systems for the Temperate Subhumid and Semiarid Areas

John R. Hendrickson, Research Rangeland Management Specialist,
USDA-Agricultural Research Service, Mandan, ND, USA
Corey Moffet, Research Rangeland Management Specialist,
USDA-Agricultural Research Service, Woodward, OK, USA

Introduction

Within North America, a majority of the temperate subhumid and semiarid zones are in the Great Plains. The defining factor separating the two zones is precipitation effectiveness (PE), which considers both precipitation and evaporation. Thornthwaite (1931) used this index to identify six different PE zones in North America. Historically, the separation between the subhumid and semiarid zones runs roughly along the 100th meridian (Thornthwaite 1941). However, more recent characterizations have moved both the semiarid and dry subhumid zones to the west (Figure 21.1). Using more recent data, the dry and moist subhumid zones lie roughly astride the 100th meridian with the moist subhumid zone to the east (Figure 21.1). The temperate subhumid zone comprises several million hectares (Table 21.1) and includes nearly-half of Texas, most of Oklahoma, Kansas, Nebraska, and the Dakotas and the western half of Minnesota (Figure 21.1). The temperate semiarid zone lies to the west and includes the eastern parts of Montana, Wyoming, Colorado, New Mexico, and west Texas (Figure 21.1). However, for this chapter, we will generally use the historical designations for subhumid and semiarid, which considers the 100th meridian as the delineator.

The Great Plains is a large expanse of land reaching from Mexico, across the interior of the United States and up into Canada (Trimble 1980). It stretches from the eastern slopes of the Rocky Mountains to the edge of the tallgrass prairie (Trimble 1980; Padbury et al. 2002). Elevation ranges from 1250 m along the Rockies to 300 m on their eastern border (Padbury et al. 2002). The Great Plains is generally characterized by gently rolling plains occasionally interrupted by wide river valleys, hills, or badlands (Padbury et al. 2002). The glaciated areas of the Northern Great Plains less variation in topography than the non-glaciated areas further south (Padbury et al. 2002). Although dominated by gently rolling landscapes, there are distinct geologic features contained within the region. These include the Black Hills of South Dakota, the Sandhills of Nebraska, the Flint Hills of Kansas, and the Edwards Plateau in Texas.

Within the Great Plains states, land use can vary widely. For example, Minnesota, has only 7.2% of its total land in farms being used as pasture or rangeland (Table 21.1). In contrast, over 90% of the total farmland in Wyoming and New Mexico is in pasture and rangeland. It is important to note that the western Great Plains states (Montana, Wyoming, Colorado, and New Mexico) are drier and the Great Plains geographic region does not cover the states

Forages: The Science of Grassland Agriculture, Volume II, Seventh Edition.
Edited by Kenneth J. Moore, Michael Collins, C. Jerry Nelson and Daren D. Redfearn.

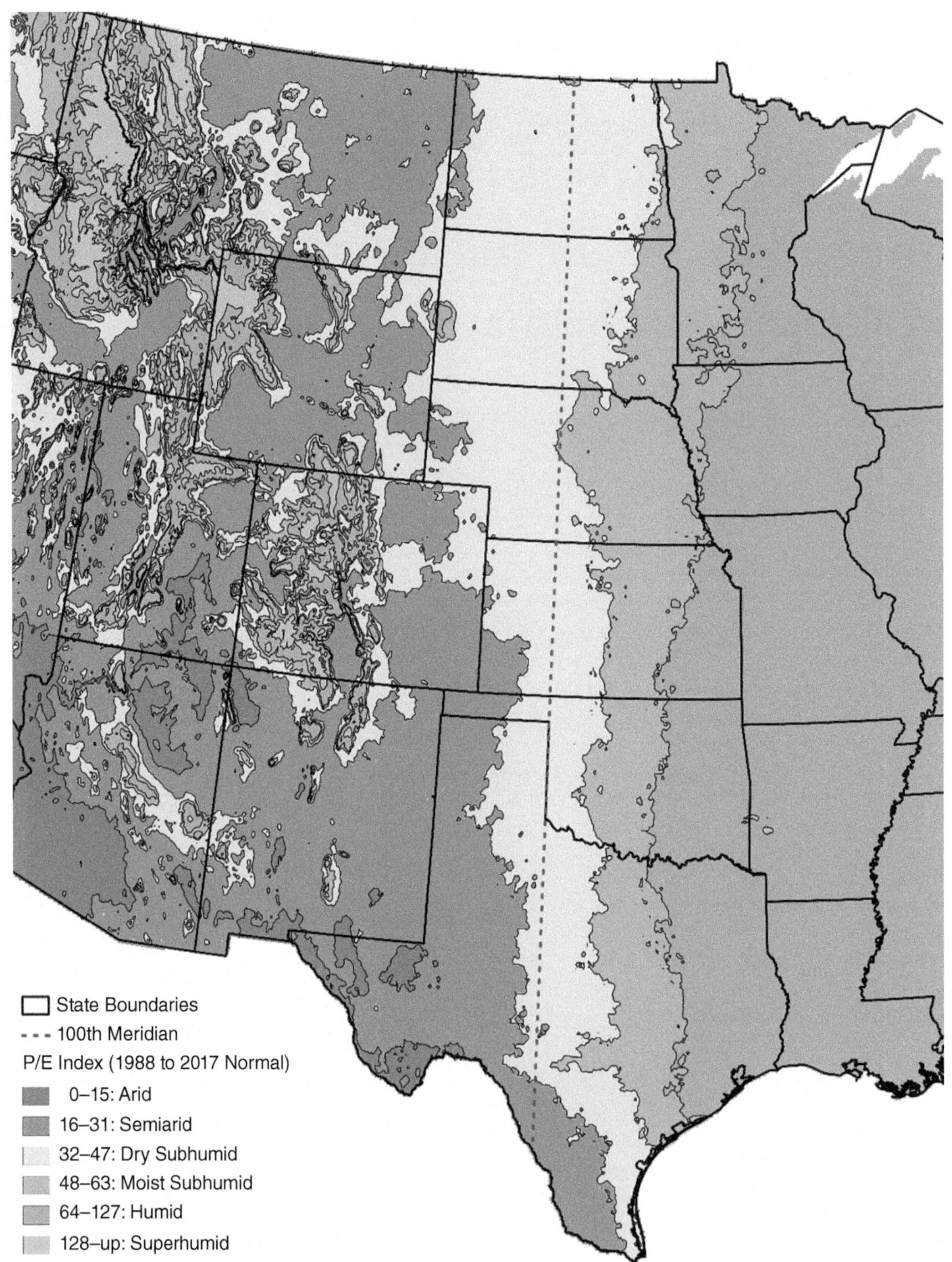

Fig. 21.1. Map of the central part of the United States including the Great Plains showing climatic classifications developed using the P/E index (Thornthwaite 1931) from climatic data from 1988 to 2017. *Source:* Figure developed by C. Moffet.

Table 21.1 Land use for states in the temperate subhumid and temperate semiarid regions (ha)

State	Pasture/grazingland that could be used for crops	Grassland/ rangeland	Pastured woodland	Total pastureland	Pastureland as a percent of total farmland
Colorado	427 615	19 223 470	826 838	20 477 923	64.2
Kansas	442 258	15 525 646	308 408	16 276 312	35.4
Minnesota	167 026	1 271 242	439 332	1 877 600	7.2
Montana	910 532	39 298 812	1 962 724	42 172 068	70.6
Nebraska	322 093	22 297 424	212 929	22 832 446	50.4
New Mexico	229 591	37 973 029	2 163 280	40 365 900	93.4
North Dakota	321 936	10 247 184	125 002	10 694 122	27.2
Oklahoma	1 001 543	19 451 870	1 731 414	22 184 827	64.5
South Dakota	518 702	22 545 069	180 751	23 244 522	53.7
Texas	2 845 326	90 287 767	5 759 183	98 892 276	76.0
Wyoming	282 997	27 203 663	360 419	27 847 079	91.7
Total	7 469 619	305 325 176	14 070 280	326 865 075	61.7
US total	12 802 847	415 309 280	27 999 006	456 111 133	49.9
% US total	48	61	32	55	58.0

Source: USDA-NASS (2014).

entirely. Doing so, would increase the percentage of total farmland that is devoted to pasture and rangeland. The percentage of land in pasture and rangeland can be influenced by the distinct geologic features mentioned earlier, but is primarily determined by climate.

Climate

Climate is the primary factor influencing forage systems in the temperate subhumid and semiarid zones. Climate determines the length of the growing season, biomass productivity and type of crops suitable for incorporating into forage systems. There are two climatic trends in the Great Plains that influence the management of forage systems. First, the length of the growing season decreases from south to north. For example, in southern areas of the temperate subhumid and semiarid zones, the frost-free period can range from 210 to 240 days (Orton 1974), but in the most northern areas, the frost-free period is usually only 90–120 days (Bavendick 1974). Because of the short growing season and the need to feed hay for significant parts of the year, hay quantity and nutritive value become especially important considerations in the Northern Great Plains (Lorenz 1976). Second, in the Plains states, precipitation often decreases from east to west, resulting in productivity per unit area often following the same trend. Increased precipitation in the eastern part of the Great Plains often results in a higher percentage of farmland being used for cropland, whereas in the Western Great Plains a greater percentage of the farmland is used for pastures (USDA 2014). In areas with mixed livestock and grain production, crop aftermath (corn or wheat stubble) can be used to meet forage needs, especially in the fall and winter.

Within both zones, the most significant attribute associated with climate is variability. Precipitation and temperature can show large variations not only spatially but also annually and even diurnally. Coping with this variation is a serious challenge for producers in the temperate semiarid region (Padbury et al. 2002).

Soils

Most soils in the temperate subhumid and semiarid zones are Mollisols, which are typical of soils that have developed under predominantly grassland vegetation (Soil Survey Staff 1999). Mollisols are thick, dark-colored soils with high-base saturation (>50%) that have organic-rich surface horizons and well-developed structure; they are not massive or hard when dry (Foth 1984). The high-natural fertility of Mollisols makes them desirable for crop production, though continuous cultivation has caused serious structural and erosion problems in some areas (Brady and Weil 1999). Mollisols are found from central Texas north into the Dakotas and Canada (Brady and Weil 1999). Other soil orders in the region include the less developed Entisols and Inceptisols, the more developed Alfisols, with their clay accumulation in the subsoil, the clayey shrinking and swelling Vertisols, and the dry Aridisols along the western edge of the region.

Livestock

Beef Cattle

Beef cattle far outnumber all other kinds of livestock combined in both the temperate subhumid and temperate semiarid regions (Table 21.2). In 2012, beef cattle made up over 75% of the ruminant livestock in the Great Plains,

Table 21.2 Beef cattle, dairy cattle, and sheep inventory of selected states for 1992 and 2012 (head)

	Beef		Dairy		Sheep	
State	1992	2012	1992	2012	1992	2012
Colorado	900 347	683 291	81 825	130 736	730 272	401 376
Kansas	1 434 017	1 270 538	85 132	131 688	206 566	62 541
Minnesota	381 869	357 826	609 034	463 312	221 777	126 506
Montana	1 506 445	1 439 653	22 409	13 947	634 361	236 646
Nebraska	1 857 347	1 730 112	83 295	54 628	151 777	71 771
New Mexico	631 738	461 595	110 422	318 878	460 700	89 745
North Dakota	837 716	881 662	74 885	17 876	217 240	64 607
Oklahoma	1 728 273	1 677 903	90 312	45 885	103 732	53 738
South Dakota	1 604 838	1 610 559	117 454	91 831	661 872	257 676
Texas	5 186 359	4 329 341	394 587	434 928	2 223 774	623 000
Wyoming	748 789	664 254	7 596	6 194	921 133	354 785
Total	16 817 738	15 106 734	1 676 951	1 709 903	6 533 204	2 342 391
US total	32 545 976	28 956 553	9 491 818	9 252 272	10 770 391	5 364 844
% US total	52	52	18	18	61	44

Source: USDA-NASS (2014).

which was greater than the 67% of ruminant livestock in 1994. The number of beef cattle has declined by about 11% over the last 20 years but data from 2018 (USDA-NASS 2019) shows national and regional beef numbers increasing to close to 1994 levels. The low beef cattle numbers in the 2012 Census of Agriculture were probably a result of the severe drought, which started in the Southern Great Plains in 2011 and spread north in 2012. Cow–calf production is common throughout both the temperate subhumid and temperate semiarid regions and cattle are grazed on both private and publicly held lands. Yearling or stocker cattle operations, however, are more prevalent in the temperate subhumid region.

Dairy Cattle

Dairy cattle numbers increased in Colorado, Kansas, New Mexico, and Texas but decreased in the remainder of the Great Plains during the 20-year period from 1992 to 2012 (Table 19.2). While the number of dairies is increasing in some nontraditional milk-producing areas, such as the western states, due to reduced land and labor costs, the overall number of dairy farms in the United States is decreasing, while the overall size of the farms has grown (Blayney 2002). The decline in dairy cow numbers was especially dramatic in North Dakota which saw a reduction of over 75% during the 20-year period between 1992 and 2012.

Sheep and Goats

The loss of the wool and mohair subsidy as part of the Farm Bill of 1996 resulted in a reduction in sheep and goat production. Sheep numbers have continued to decrease (Table 21.2) which has been the trend since 1942 when sheep numbers reached a peak of 56.2 million head. Up until 1977, USDA did not provide actual numbers for goats, but merely grouped goats with the sheep. The USDA annual reports indicate the number of angora goats clipped for mohair has decreased from almost 2 million head in 1992 to slightly less than 155 000 in 2012. Over 47% of the angora goats produced in 2012 were raised in Texas (USDA-NASS 2014). Meat goats have increased dramatically during the same period from 591 543 in 1992 peaking at 2 601 669 in 2007 and declining to 2 053 228 in 2012. Meat goats may be found in many states contained within the temperate subhumid and temperate semiarid regions. Texas, California, and Missouri reported the largest goat inventory in 2012 (USDA-NASS, Quick Stats 2019).

Horses

Horses are prevalent throughout the temperate subhumid and temperate semiarid regions, both for pleasure and work. There were over 3 million horses and ponies in the US in 2012 (USDA-NASS 2014). Between 45% and 60% of horses were owned for recreation (Raub 2018) suggesting that people own horses for pleasure. This impacts their thoughts about forage and grazing management. Horse numbers have been increasing for several years and this trend is likely to continue.

Important Forage Species

Additional information on the perennial cool-grasses including the wheatgrasses, wildryes, and other miscellaneous cool-season species can be found in Chapter 17.

Introduced Cool-Season Forages

Introduced cool-season forages have always been a critical part of temperate subhumid forage systems in the US. The introduced cool-season forages can be split into annual and perennial species. In the states of the central Great Plains (Texas, Oklahoma, Kansas, Nebraska, and the Dakotas) about 1.8 million acres of annual small grains were harvested for hay in 2012 (USDA-NASS 2014). Besides the small grains harvested for hay, grazing of winter wheat during winter and early spring is a common practice from Kansas to north Texas.

Other cool-season annual forages include annual ryegrass overseeded into warm-season perennial grass sods in the eastern part of the temperate subhumid region or for winter forage in Texas and Oklahoma. Triticale is an option for grazing and remains vegetative longer than wheat. In the northern portion of the temperate subhumid region, primarily North Dakota and Montana, spring barley is used as hay, silage, or occasionally grazed.

Interest in the forage potential of annual cover crops is increasing. The primary focus driving increased use of cover crops is the perceived conservation benefits (Myers and Watts 2015); however, they can also be an important source of livestock forage (Sanderson et al. 2018). Inclusion of legumes in cover-crop mixtures can enhance the forage quality (Hansen et al. 2015; Sanderson et al. 2018). An annual cool-season forage mixture that included a legume had consistently greater nutritional value than did a crested wheatgrass pasture in a study conducted near Sidney, Nebraska (Titlow et al. 2014).

By the early part of the twentieth century, many native grasslands in the Great Plains had either been converted to farmland or were not properly grazed. As a result, by the late 1930s, there were millions of acres of severely eroded or degraded pastures and croplands throughout the Great Plains (Vogel and Hendrickson 2019) but especially in the drier western portion of the region. Before the devastating drought of the 1930s, the only seeds widely available for pasture planting in the Great Plains were timothy and orchardgrass, both of which had poor drought tolerance (Vogel and Hendrickson 2019). Improving the degraded conditions found on croplands and grasslands required utilization of new or improved forage germplasm which was subsequently incorporated into grazing systems.

Both smooth bromegrass and crested wheatgrass were noted for their ability to survive during the 1930s drought (Vogel et al. 1996; Vogel and Hendrickson 2019). 'Lincoln' dairy smooth bromegrass was introduced in 1942 (Vogel and Hendrickson 2019) and has been a major cool-season forage in the Dakotas, Nebraska, Kansas, and northern Oklahoma. Crested wheatgrass, one of the earliest cool-season species to produce forage following winter, is credited for reducing hay use in its area of adaptation. Crested wheatgrass seed was distributed to producers in Wyoming, Montana and the Dakotas in the 1920s but 'Fairway', released in 1927 by the University of Saskatchewan, was one of the first cultivar releases (Vogel and Hendrickson 2019). Since the release of 'Lincoln' and 'Fairway' there have been 11 new cultivars of smooth bromegrass and 7 cultivars of crested wheatgrass released with the most recent smooth bromegrass release in 2014 and the most recent crested wheatgrass in 2003 (Vogel and Hendrickson 2019).

Other important introduced cool-season perennial forage grasses include intermediate wheatgrass, pubescent wheatgrass, tall wheatgrass, meadow bromegrass, orchardgrass, and creeping foxtail. The wheatgrasses demonstrate good cold tolerance, drought tolerance, and good dry matter production and nutritive value. However, they are relatively short-lived and susceptible to overgrazing (Hendrickson et al. 2005). Although primarily used in the Northern Great Plains, intermediate and pubescent wheatgrass have also been used in the Southern Great Plains to stabilize marginal land formerly in wheat production (Redmon et al. 1995) or to fill gaps between grazing of winter wheat and warm-season pastures (Malinowski et al. 2003). Tall wheatgrass is probably the most productive of all the wheatgrasses, and though coarse at maturity, can produce forage of good nutritive value when managed well (Asay and Jensen 1996). Both tall wheatgrass and intermediate wheatgrass monocultures outperformed switchgrass monocultures on non-irrigated plots in North Dakota (Monono et al. 2013). There have been 12 cultivar releases of intermediate wheatgrass and three releases of tall wheatgrass since 1945 (Vogel and Hendrickson 2019).

Meadow bromegrass is known for early forage production and rapid regrowth (Jensen et al. 2015). Meadow bromegrass produces 49% more biomass after defoliation than smooth bromegrass (Biligetu and Coulman 2010). While meadow bromegrass is currently best adapted to cooler and more moist areas (Alderson and Sharp 1994), cultivars better adapted to semiarid areas are being developed (Jensen et al. 2015).

Orchardgrass is used to a limited degree in the temperate subhumid region, but does not have the drought tolerance of tall or intermediate wheatgrass. 'Paiute' orchardgrass, which was released by the US Forest Service in 1981, may be the most drought tolerant of the orchardgrass cultivars but even this one generally requires at least 410 mm of annual precipitation (Bush et al. 2012).

Russian wildrye is a drought-tolerant, grazing-tolerant (Hendrickson and Berdahl 2003) cool-season grass seeded on approximately 400 000 ha in the Great Plains of the United States and Canada (Smoliak and Johnston 1980). Once established, russian wildrye provides excellent forage, especially for fall grazing (Lorenz 1977; Smoliak and Johnston 1980). There have been seven cultivars of russian wildrye released since 1960 (Vogel and Hendrickson 2019).

Both creeping and meadow foxtail are well-adapted in much of the temperate subhumid region. Creeping foxtail has become naturalized across the Great Plains and meadow foxtail across the northern half of the continental United States (Boe and Delaney 1996). Both are adapted to strongly acid and poorly drained soils and are used for pasture, hay, or silage. 'Garrison' creeping foxtail is a common cultivar in the Northern Great Plains.

Legumes can play an important role in forage systems in both the temperate subhumid and semiarid regions. Alfalfa is among the most important forage crops grown in the United States and about 46% of the United States alfalfa acres are planted in the temperate sub-humid and semi-arid regions. However, most alfalfa hectarage is planted within the northern part of the region (Figure 21.2). Alfalfa is generally used for hay or silage, though there are some grazing-tolerant cultivars available. In the Northern Great Plains, cultivars with a high degree of *Medicago sativa subsp. falcata* usually persist better under grazing (Bittman and McCartney 1994). In the colder climates of the Northern Great Plains, the ability to withstand the extreme low temperatures may be as important to persistence as the ability to withstand grazing (Hendrickson and Berdahl 2003). Incorporating alfalfa into grasslands increased total yield of the forage system by up to 185% (Hendrickson et al. 2008a, 2008b), but alfalfa mixtures with cool-season grasses appeared to be more sensitive to early-season water stress than did warm-season grass monocultures (Hendrickson et al. 2013). Alfalfa production in many parts of the Southern Great Plains is severely limited due to the presence of *Phymatotrichopsis omnivora* in the soil, an ascomycete fungus. The fungus is responsible for *Phymatotrichopsis* root rot disease (also referred to as cotton root rot or Texas root rot) in alfalfa (Mattupalli et al. 2018).

Other legumes of some importance include birdsfoot trefoil, sweetclover, red clover, and white clover. These legumes, except for sweetclover, do not have the drought

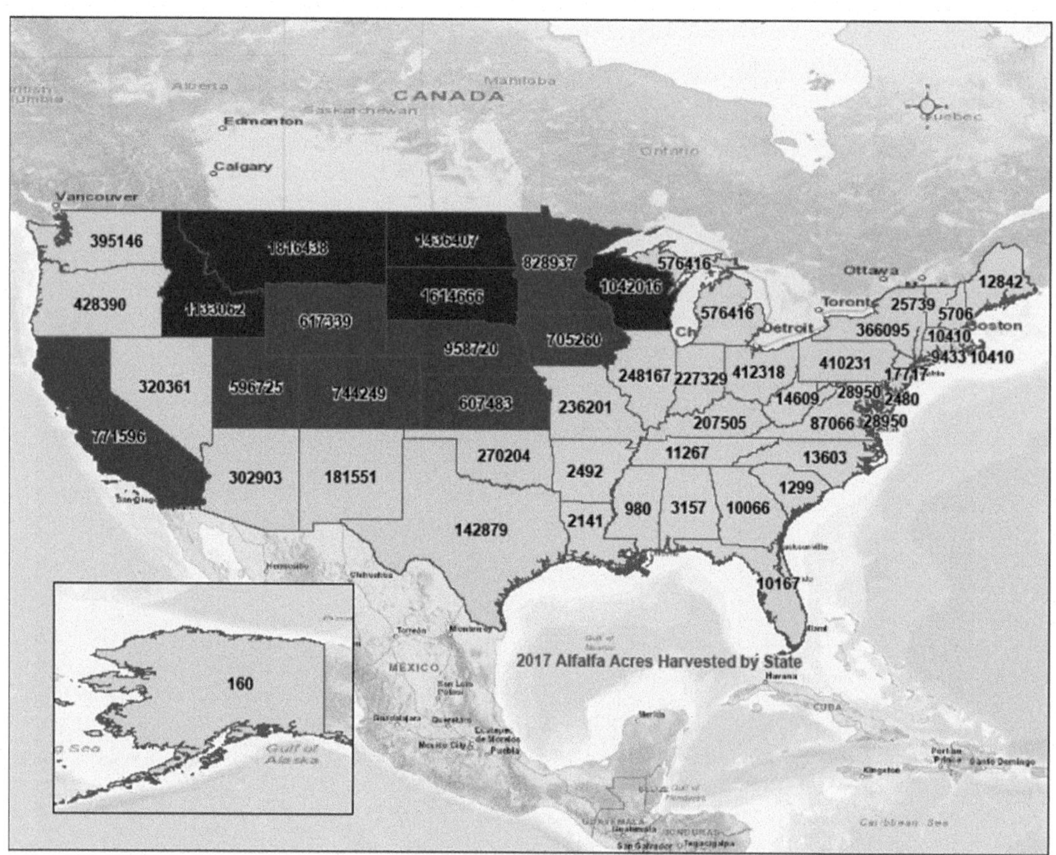

FIG. 21.2. Map of acres (acre = 0.40 hectares) of alfalfa harvested in each state using data from the 2017 Census of Agriculture (USDA-NASS 2019). The numbers in each state are the number of acres harvested. Darker colored states have the most harvested acres of alfalfa and the lighter colored states have the least. *Source:* Map developed by J.D. Carter.

tolerance necessary for production except in the extreme eastern portion of the temperate subhumid region. Sweetclover, which was introduced from Eurasia, has become naturalized in the region and is both drought tolerant and winter hardy.

With more focus on soil health, there has been increased interest in incorporating perennial forages into annual cropping systems. A five-year study in central North Dakota indicated that compared to a continuous wheat cropping system, including perennial forages reduced soil acidity and bulk density and increased particulate organic matter and water-stable aggregates (Liebig et al. 2018). Within the forage treatments, intermediate wheatgrass, alone or combined with alfalfa, reduced soil bulk density and increased particulate organic matter content compared to an alfalfa monoculture (Liebig et al. 2018). The same study indicated having an alfalfa or an alfalfa – intermediate wheatgrass mixture in place for four years reduced the amount of N needed on a subsequent wheat crop for at least four years (Franco et al. 2018). As soil health receives greater interest, the need for forages to address other aspects besides production and nutritional value will become more important.

Introduced Warm-Season Forages

Introduced perennial warm-season forages are important on lands dedicated for pasture or hay use. In general, these species are more difficult to establish and have lower nutritive value than annual forage crops, but with proper management and favorable weather, will persist as a dependable source of forage for many years. More information on warm-season forage management can be found in Chapter 17.

Bermudagrass is one of the more important introduced warm-season forage crops in the Southern Great Plains having been introduced to the US from Africa before the US was a country. Bermudagrass is well adapted to a wide range of soil types and soil pH values. Bermudagrass is an aggressively spreading plant that produces long stolons capable of rooting at nodes in contact with moist soil. The species responds well to fertilizer. There are many cultivars of bermudagrass, and selections of common bermudagrass can be propagated from seed. Hybrid cultivars, however, must be propagated vegetatively by transplanting small pieces of rooted stolons and rhizomes, known as sprigs, dug from an established field typically in the dormant season or soon after breaking dormancy in spring. Many bermudagrass cultivars have limited cold tolerance but 'Goodwell', 'Midland', 'Midland 99', 'Ozark', and 'Tifton 44', among the sprigged cultivars, and the seeded 'Guymon' and 'Wrangler' are cold-tolerant cultivars. Often two or more genetic lines or cultivars are included in blends of seeded bermudagrass. Seeded cultivars in south central Oklahoma, averaged over a three-year study, produced between 3400 and 5600 kg ha^{-1} without N fertilizer, and production increased between 9 and 15 kg ha^{-1} for every kg ha^{-1} of N applied up to 225 kg ha^{-1}. Crude protein (CP) and total digestible nutrients (TDNs) also increased with N applications in this range. The sprigged cultivars were more productive in general and more responsive to the applied N. Non-fertilized production for the sprigged cultivars varied between 5100 and 8500 kg ha^{-1} with a 15–19 kg ha^{-1} increase in production for every kg N ha^{-1} applied, but the effects of N fertilizer on nutritive value were similar to the seeded cultivars (Funderburg et al. 2011; Funderburg et al. 2012).

Old-world bluestems include a few warm-season bunch grass species introduced from Eurasia in the 1920s to 1960s. These grasses have good forage potential and have been planted on former cropland soils and along roadside rights-of-ways throughout Oklahoma and Texas. The old-world bluestems are more drought tolerant than bermudagrass and are more likely to be found in the western portions of the region. Several million hectares of old-world bluestem were planted beginning in the mid-1980s on highly erodible lands which had been previously used as cropland. In recent years, concern over the tendency of these grasses to escape the fields and roadsides where they were originally planted has greatly diminished their popularity in new plantings. Old-world bluestems include several selections from the species *Bothriochloa ischaemum* such as 'Plains' (a blend of 30 different accessions), 'WW-Spar', 'WW Ironmaster', 'King Ranch', and 'Ganada', the selection from *Bothriochloa bladhii*, 'WW-B. Dahl', and caucasian bluestem or *Bothriochloa caucasica*. 'WW-Ironmaster' was selected for high pH soils and is less susceptible to iron chlorosis.

Caucasian bluestem is not as drought tolerant as the other old-world bluestems, but when soil water is adequate, is among the most productive. WW-B. Dahl remains leafy and vegetative later into the growing season but, is less cold tolerant than other cultivars. King Ranch bluestem is also productive in the southern portion of the region, but it too lacks cold tolerance. In a study comparing the response of Plains and B. Dahl old-world bluestems, non-fertilized B. Dahl and Plains old-world bluestems produced more forage than many seeded bermudagrass cultivars and had similar forage production to many of the sprigged bermudagrass cultivars (Funderburg et al. 2011; Funderburg et al. 2012). However, Plains did not respond to added fertilizer N. WW-B. Dahl responded to small amounts of additional N, 3200 kg ha^{-1} for 76 kg ha^{-1} of applied N, but did not respond to greater amounts of N. The CP and TDN of both old-world bluestem cultivars increased in response to N application up to 225 kg ha^{-1}.

Weeping lovegrass was introduced into the US from South Africa during the late 1920s. It is well-adapted to a range of soils, but especially suited to sandy soils where it is important for soil conservation. Weeping lovegrass is a bunchgrass readily used by cattle during the early growing period but, after it matures in June, its palatability drops precipitously. If previous year's growth is not removed prior to spring grazing, utilization is reduced. Prescribed burning in spring is the preferred method to remove previous year's growth, increase nutritive value, production, and utilization by cattle (Klett et al. 1971). Fertilization will also increase nutritive value and production of weeping lovegrass (McMurphy et al. 1975; Rocateli 2017). Given these characteristics, it has been recommended that it be burned prior to initiation of spring growth, fertilized with 100 kg N ha^{-1}, and stocked heavily before palatability decreases (Rocateli 2017).

Johnsongrass was introduced into the US from Turkey in the early 1800s as a perennial forage crop. It is ubiquitous throughout much of the region, but has vigorous rhizomes and seed production such that it is listed as a **noxious weed** in several states. Other introduced warm-season perennial forage species occur within the region but are of limited local importance like kleingrass, wilman lovegrass, and dallisgrass.

Introduced annual warm-season grasses are best used on lands suited to cultivation and provide greater flexibility than perennial forages. These forages can be very productive and have high nutritive value. Sudangrass, sorghum × sudangrass hybrids, and forage sorghums, with adequate water and fertility, can produce large quantities of high-quality forage for grazing, hay, or silage. The yield potential under irrigation or in eastern Oklahoma is as high as 33 Mg ha^{-1}. However, 330 kg N ha^{-1} is needed to reach this level of production (Rocateli 2017). Pearlmillet is another common introduced warm-season forage used in the region.

Other warm-season annual forages include crabgrass and teff. Crabgrass is an aggressive annual that produces high-quality forage following wheat or other winter crops. This is not recommended in fields where crops that are not glyphosate tolerant may be grown as there are few postemergence herbicides available to control crabgrass (Warren et al. 2018). Crabgrass is used for grazing and hay and nutritive value remains high even as the crop reaches maturity (Ogden et al. 2005; Beck et al. 2007). Teff was introduced from east Africa and is a major cereal crop in Ethiopia. Teff seed is very small and the seed must not be placed too deeply in the soil. It germinates and establishes rather quickly and can be harvested 45–55 days after planting. The crop is most commonly used for hay but can also be grazed. Teff tends to be shallow rooted, especially early in the growing season and may be pulled out of the ground when grazed. The crop is drought tolerant and the nutritive value of the forage is similar to many other warm-season annual forages (Twidwell et al. 2002; Anderson and Volesky 2012).

Prussic acid poisoning can be a concern for grazing young or new growth of forage sorghums especially in fall, after drought, or following crop injury. Nitrate accumulation can also occur, especially under poor growing conditions or, when excess N is applied.

Native Grasses

Additional information on many of the native warm-season grasses including the bluestems, panicgrasses, and other miscellaneous warm-season species can be found in Chapter 17.

Native plant species used for livestock forage in the temperate subhumid and semiarid region are typically found growing within a community of adapted species rather than in monoculture. The native species include a tremendous variety from all the functional forms (e.g. warm-season tall grasses, cool-season annual grasses, warm-season short grasses, trees, and forbs). Importance of each species for provisioning forage in a community varies with climate and soils across the region. Within a given community, most species are sub-dominants with often only three to five key species responsible for providing 80–90% of the forage produced. The east-west mean annual precipitation gradient and, to a lesser extent, the north-south mean annual temperature gradient that occur across the Great Plains, in large part, determine which forage species occupy the native plant communities, their proportions and total productivity. The communities common to the temperate subhumid and semiarid region have been grouped into several broad rangeland types including tallgrass prairie, northern and southern mixed prairies, and shortgrass steppe (Lauenroth et al. 1999). Productivity of these native plant communities ranges across the region from less than 1500 kg ha^{-1} along the entire western edge to nearly 6000 kg ha^{-1} along the southeastern edge of the region where water and heat units are most plentiful. Along the north-eastern edge of the region production is typically between 3000 and 4000 kg ha^{-1} (Sala et al. 1988).

In the native tallgrass prairie, key forage species are warm-season tallgrasses such as big bluestem, switchgrass, indiangrass, and eastern gamagrass. Sand bluestem is an important tallgrass that is generally associated with sandy, loamy sand and sandy loam soils and with sandy regions such as the Nebraska Sand Hills. Of these species, eastern gamagrass begins growth earliest and big-bluestem is the latest. The tallgrasses are relatively tolerant of grazing, but eastern gamagrass, in part because it is extremely palatable, strongly preferred, and has high nutritive value, will decline and even disappear if continually grazed short. The forage value of the other tallgrass species is good, meeting the nutritional needs of most classes of livestock in spring and summer before stem elongation begins, but declines through the fall and by winter, forage quality is poor, lacking enough protein to meet the needs of grazing livestock.

In the shortgrass steppe, precipitation is insufficient in most years to support dense stands of tallgrasses and the dominant forages are warm-season short grasses including buffalograss and blue grama. These forages tend to be very tolerant of grazing and are palatable to all classes of livestock. The shortgrass steppe, while less productive than tallgrass prairie, produces forage composed mostly of leaf tissue with little stem tissue extending into the canopy. Therefore, the nutritive value of these forages is among the highest of the **native forages**, especially in the summer months. Blue grama and buffalograss will maintain good nutritive value with sufficient CP for many classes of livestock into the fall and winter months.

The northern and southern mixed-grass prairies are ecotones between the tallgrass prairie and the shortgrass steppe, and they contain species from both grassland types that contribute significantly to the forage of the mixed-grass prairie. The dominant forage species in the southern mixed-grass prairie, however, are warm-season mid-grasses such as little bluestem, sideoats grama, and curly mesquite, in the northern mixed-grass prairie, cool-season perennial grasses such as the wheatgrasses, needle-and-thread, and green needlegrass are dominant. In the southern mixed-grass prairie, cool-season perennial grasses such as texas wintergrass, canada wildrye, and texas bluegrass are scattered across the prairie, and may provide important high-quality forage as a supplement to the senesced warm-season forages during the late winter months before the warm-season grasses begin spring growth.

Forbs contribute the greatest floristic diversity to plant communities of the Great Plains and this functional group contributes greatly to forage quality as these species are often low in fiber and quite digestible. Another functional class of species that provide important forage for livestock, especially small ruminants, are shrubs and trees. Increases in the burn frequency since European settlement of these semiarid and subhumid temperate regions have resulted in increased abundance of many woody species; however, many are undesirable, such as the many *Juniperus* spp. becoming more prevalent across the southern plains.

Introduced, Invasive Species

One of the primary concerns with introducing non-native or even **native forages** into non-traditional areas on range and pasture lands is their invasion potential. Smooth bromegrass, for example, is reported to be a common invader of disturbed sites throughout the Great Plains (Howard 1996) though this may be less of a concern in well-managed pastures and rangelands (Vogel and Hendrickson 2019). Grant et al. (2009) evaluated US Fish and Wildlife Refuges in North and South Dakota and found most of the vegetation (45–49%) was comprised of smooth bromegrass, which the authors attributed to lack of disturbance. Also, in the northern portion of the region, crested wheatgrass has spread and is invading native rangelands (Henderson and Naeth 2005).

One of the major invasion threats in the northern part of the subhumid region is kentucky bluegrass. In the Northern Great Plains, kentucky bluegrass is present on more than 50% of the private acres sampled by the USDA-NRCS (Toledo et al. 2014). Although the original areas of concern with kentucky bluegrass invasion were in the subhumid portions of the temperate region, currently it is beginning to be a concern in the semiarid portion as well.

In the southern portions of the temperate region, old-world bluestems and sericea lespedeza are examples of invasive forage species of concern (Fulbright et al. 2013), while in the western or more semiarid part of the temperate region, annual bromegrasses, such as cheatgrass or Japanese brome, can be of major concern. Invasions of these annual bromegrasses often lead to an increase in wildfires further damaging native vegetation in the region (D'Antonio and Vitousek 1992). Though wildfires can be damaging, prescribed fire at the right time is an important control mechanism for woody invasive species such as juniper and mesquite (Wright 1988).

There are multiple concerns with invasive species. Invasive species can alter nitrogen (N) cycling and reduce carbon storage (Wedin and Tilman 1996), lower plant diversity (Henderson and Naeth 2005), and impact wildlife habitat and soil microbial diversity (Fulbright et al. 2013). Some invasive species, such as kentucky bluegrass, can alter forage distribution by shortening the grazing season rather than lengthening it. Because of these concerns, assessing invasion potential needs to be a high priority when designing forage systems for temperate regions.

Forage Systems

Harvest strategies for both grazed and conserved forages involve compromises and are dependent on climate, forage species, kind and class of grazing livestock, and the management skills of the producer. Forage of young plants is higher in nutritive value than that of more mature plants. This is because the cells of young plants are biochemically active, capturing and storing energy, synthesizing proteins and fats, and so on, while cells of older plants are low in biochemical activity (Huston and Pinchak 1991). Thus, for grazing and depending on the nutrient requirements of the target animal, it is often important to use some sort of grazing management strategy to provide forage of acceptable nutritive value for younger, growing animals or for lactating females. Hay harvest schedules are heavily dependent on weather conditions and, in many instances, hay is harvested at too mature a stage to meet the needs of animals with high-nutrient requirements. Prine and Burton (1956) and Knox et al. (1958) demonstrated the adverse effect on crude protein and the increasing levels of lignin in bermudagrass harvested at various stages of maturity.

Crude protein was reduced by nearly half and lignin concentration rose to 12% of the dry matter in bermudagrass cut at two-week intervals vs eight-weeks intervals. All forage species followed similar trends, although the effect was not as pronounced for cool-season grasses or legumes. Therefore, it is important that forages to be conserved as hay be harvested to optimize both quantity and nutritive value of the crop.

Many times, however, hay supplies are limited or nonexistent, hay is of low nutritive value, or forage may be in short supply due to drought and livestock must be supplemented. Two types of protein supplements are commonly used under these circumstances: (i) non-protein N sources (urea, biuret) and (ii) high-protein feeds such as alfalfa hay, cottonseed meal, or soybean meal (Holechek et al. 1998). Cottonseed meal is probably the most widely used supplement in the temperate subhumid and temperate semiarid regions. It typically contains 40–45% crude protein, thus only requiring approximately 0.5–0.9 kg $head^{-1}$ d^{-1} for cows and 0.1–0.2 kg $head^{-1}$ d^{-1} for ewes (Holechek et al. 1998).

Energy supplementation may also be required at times during the year due to drought or heavy snow. Barley and cracked corn are two of the more common energy supplements used under range conditions in the temperate subhumid and temperate semiarid regions. These feeds usually depress forage intake and serve as a substitute for range forage (Holechek et al. 1998). Alfalfa hay can also be used as an energy source and is particularly important when crude protein is also limiting. Holechek et al. (1998) suggested that due to the low cost per cow, mineral supplementation should be provided year-round by ranchers, rather than just during fall and winter when animals might be deficient in phosphorus.

Using corn residue is an important source of forage especially in the central and northern Corn Belt (Schmer et al. 2017). Redfearn et al. (2019) estimated the added economic value to the crop sector of grazing crop residues in Kansas, Nebraska and the Dakotas at $95 million. Grazing crop residues usually removes less than 40% of the total residue, which is important for soil conservation (Lorenz 1977), but grazing crop residues during spring may adversely affect water infiltration and soil bulk density (Rasby et al. 1998). A review by Rakkar and Blanco-Canqui (2018) suggested that grazing crop residues generally had a positive impact on soil nutrients and while it may increase some soil compaction parameters, residue grazing had little impact on crop yield. The composition of the crop residue (i.e. grain, leaf, or stalk), whether it was irrigated, and the days after harvest are all factors determining the nutritive value or crop residues (Rasby et al. 1998).

In addition to crop residues, many producers rely on winter range for forage during the winter months (Jordan et al. 2002). Grazed winter forages, however, may not meet the protein requirement of cows (Lardy et al. 1999; Jordan et al. 2002). Lardy et al. (1999) concluded that degradable intake protein was the first limiting nutrient before energy and undegradable intake protein for summer calving cows on autumn-winter range in the Nebraska Sandhills. Larson et al. (2009) found that cows grazing corn stalk residue produced calves with a greater birth and weaning weight than did cows grazing winter range. An inexpensive source of degradable intake protein may be the only requirement to maintain cows on low-quality winter forage (Jordan et al. 2002). Calf weaning weight was increased with using a 28% protein cube as a supplement (Larson et al. 2009). Switching degraded ranges from summer to winter use is a proven method of improving range condition in the Nebraska Sandhills, but the impact of dormant-season grazing on the vegetation has not been quantified.

Rangelands Only

Forage systems that almost exclusively utilize rangelands occur across much of the Great Plains are becoming increasingly important to the west, where other land uses are less common. These systems are also used in the southern Great Plains, where winters are less severe and standing dead forage is not snow-covered for extended periods of time. This system is also common in the Kansas Flint Hills and the Osage region of Oklahoma where soil conditions, rather than climate, limit other land use options. In general, rangelands are extensively managed with low inputs and use proper grazing management and prescribed fire as the two most important management practices. Other management practices such as fertilization, reseeding, or harvesting hay are generally impractical or uneconomical on rangelands. An exception to this is the annual harvesting of native grass hay from productive, yet fragile, wet meadows in the Nebraska Sandhills.

Rangelands provide grazing animals with a variety of plant species from which to select their diet; from low-nutritive value standing dead grass, moderate-nutritive value matured current year's growth, to the high-nutritive value young grass leaves and forbs. Thus, diet selection is dynamic resulting from changes in palatability among species available. Animals grazing these rangelands have different plant preferences and some animal species better match the vegetation in an environment than another species. Diet selection in these systems generally results in better diet quality than the **composite** quality of the forage on offer. Because all possible bites are not equal in value, increasing stocking rates can cause competition for high-quality bites reducing an animal's ability to select a quality diet and potentially reducing performance.

The objective of these systems is to harvest a diet that meets the nutritional needs of the grazing animals. These forage systems are composed of native vegetation, susceptible to degradation, and difficult to restore. If grazing is to be sustained, the forage use must be balanced with the need to maintain or improve the existing native plant community. This is accomplished by managing the vegetation to promote growth and persistence of desired plant species and suppressing the invasion and spread of undesirable plant species. Great Plains plant communities developed with both grazing and fire, burning every 2–20 years depending on the region (Guyette et al. 2012). These periodic fires kept woody species suppressed or relegated to fire protected sites, such as along steep escarpments, in canyons, and along rivers. Fire still plays an important role in keeping many undesirable species from spreading into rangelands although herbicides and mechanical practices have also been used. Hay fed on rangelands can be a vector for invasive weeds. The invasion risk depends on species being fed; for example, an introduced annual like sorghum-sudangrass is much less likely to escape into rangelands than old-world bluestem.

The most important grazing management decision in this system is managing the stocking rate and then providing a rotation for different timing of use for growing-season rest. The proper stocking rate is related to the quantity of forage produced each year. Rangeland productivity is related to site potential (defined by soil characteristics), the ecologic state or phase (which plant species occupy the site and in what proportion), previous year's weather (e.g., the bud bank), and current year's weather (e.g. the amount and timing of heat units and available water). Initially stocking rates should be set in relation to the proportion of different ecologic sites within a grazing unit, the state or phase in which the sites are found, objectives for livestock performance, patterns of land use (e.g. incomplete accessibility due to slope or distance from water), expected effects on the vegetation, and anticipated weather (e.g. expected annual precipitation). Median annual precipitation may be a more appropriate expectation for setting stocking rates than the mean in some areas. Median annual precipitation describes the amount of precipitation expected in half the years and does not give undo credit to exceptionally wet years that contribute little to the production in other years. The difference between mean and median precipitation is typically greater in dryer climates. Experience and changing management objectives are used in subsequent years to modify stocking rates. Stocking rates may also be adjusted, depending on when the grazing will occur and whether plants are susceptible to injury from grazing at that time. For example, a greater proportion of the annual production may be harvested if a good portion of the use will occur in the dormant season than if the grazing will all be done in the growing season.

In general, semiarid and subhumid temperate rangelands can be sustainably grazed using 40–50% of the key forage species produced each year (Holechek 1988). The rule of thumb of "take half, leave half" is generally adequate, but can be misapplied. When livestock graze, they "take" forage by consumption (i.e. direct intake), but also by trampling and fouling. Another source of "take" that should be considered is use by wildlife or damage by insects. The annual forage supply should be multiplied by an appropriate harvest efficiency factor to determine actual forage available for livestock consumption and 30% is a good initial value for this factor. At 30% harvest efficiency, 50% total use is achieved when 60% of all the forage used is consumed by livestock with the remaining 40% of use due to trampling, fouling, wildlife, and damage from pests. Grazing efficiency is used to describe the amount of forage that is consumed directly as a proportion of all the forage use (60% in the preceding example).

Grazing systems that utilize rangelands exclusively can be grouped into two categories; (i) breeding animal systems where mature animals graze the rangelands continually with their offspring from birth until weaning and (ii) growing animal systems where the growing animals graze the rangeland during periods when the nutritive value of the forage can support animal growth. In the breeding animal system, nutritional requirements change during the production cycle and are lowest after weaning and greatest during the first several months postpartum which can be aligned with changing nutritional value in the forage to minimize the need for nutritional supplements. On many rangelands, the winter months are protein deficient and a protein supplement is typically offered to meet the animal's maintenance requirement. This supplementation practice requires that the rangeland has enough low-quality forage available that energy will not become a limitation. In these systems, forage may become unavailable due to inclement weather and a supply of good-quality hay is typically used to supplement energy during these times. During the growing season, nutritive value is sufficient to support the greater nutritional needs of maintenance and lactation and no supplemental protein or energy is offered during this period. The most common of these systems is the cow-calf system, but in regions such as in the Edwards Plateau of Texas, where shrubs and forbs have a greater role in the plant communities, systems with sheep and goats are more common.

The growing animal systems are not year-round systems, rather they target the rangeland use to periods when the forage has the greatest quality. A fairly traditional system of this type would be the season-long stocking (SLS) system where light-weight stocker cattle graze the rangelands between late April and the first of October. In some regions, notably the Kansas Flint Hills and Osage region of Oklahoma, the SLS system has been replaced with other systems to better mitigate the declining forage quality as the season progresses. In regions where high-nutritive value is maintained in the forage throughout the growing season the SLS system is preferred. The first effort to replace the SLS system was an intensive early stocking (IES) system, where the period of use is shortened by half, and stock density is doubled, from the traditional SLS system. In the IES system, animal performance declines slightly but the small reduction in animal performance is more than compensated with the increased gains per hectare. Because the IES system kept the forage short and of high quality, researchers considered a next step in the evolution by adding a period a late-season grazing (LSG), where stock density was cut in half and the remaining cattle continued to graze until the first of October. To keep from over-grazing these ranges, the IES-LSG is followed in the next year by IES without LSG (the IES-LSG/IES rotation). In the tallgrass prairie where annual burning is acceptable, any of these stocker systems can be further improved by incorporating an annual fire before grazing begins. For tallgrasses, the best time to burn is just before the warm-season grasses green-up.

Introduced Forages Only

Depending on the north–south location in the temperate subhumid region, introduced-only forage systems may be based on either warm-season or cool-season perennial grass and often include both types. Introduced forages have been used in livestock production systems because of their grazing tolerance and response to fertilizer. Proper management begins with the selection of adapted species that are persistent and will produce acceptable dry matter yields of the desired nutritive value to meet the needs of the grazing livestock (Cherney and Allen 1995). Introduced forages in the temperate subhumid region generally respond well to fertilization if (i) pastures are irrigated or receive >625 mm average annual precipitation, (ii) hay meadows have equivalent soil moisture availability, or (iii) pastures or hay meadows are naturally sub-irrigated (Vallentine 1989).

In the temperate semiarid region, introduced forages must either have excellent adaptation to heat and drought (i.e. crested wheatgrass) or be irrigated. Vallentine (1989) noted that introduced cool-season perennial grasses responded better to (N) fertilization than did native cool-season grasses. While bermudagrass and old-world bluestem generally exhibit the best response to (N) fertilization, Vallentine (1989) also stated that the warm-season native tallgrasses and midgrasses such as the lovegrasses, switchgrass, bluestems, indiangrass, and sideoats grama responded more favorably to (N) than the shortgrasses such as blue grama and buffalograss. As in the more mesic areas of the United States, any supplemental fertilizer should be used with caution.

Bermudagrass and the old-world bluestems serve as the warm-season pasture base throughout Texas, Oklahoma, and Kansas. These warm-season perennial grass-based systems are generally stocked with cattle, sheep, or goats year-round with excess forage harvested as hay for winter feeding and supplemented with other purchased supplemental feedstuffs, as needed. Some producers, however, overseed the warm-season grass pastures with cool-season annuals to provide longer grazing seasons, improved pasture nutritive value, and reduced costs associated with winter feeding. Coffey and Moyer (1991) found that stocker cattle grazing cereal rye no-till drilled into bermudagrass sod, offered the potential for extending the grazing season and providing for more total cattle production than from bermudagrass alone. Volesky et al. (1996), likewise, evaluated interseeding rose clover and hairy vetch and noted that interseeding stands of old-world bluestem could reduce N fertilizer input, extend the grazing season, and enhance diet quality.

Further north in the temperate subhumid region, cool-season perennial grasses dominate introduced forage pasture systems. From Kansas north, smooth bromegrass pastures are used extensively. In a three-year study at Mandan, North Dakota, steer average daily gains from 'Lincoln' smooth bromegrass averaged 1.04 kg d^{-1}, which was greater than steers grazing native range (0.98 kg d^{-1}) or crested wheatgrass (0.97 kg d^{-1}) (Karn and Ries 2002). Other cool-season perennial species, such as the introduced wheatgrasses and orchardgrass, also provide pastures for grazing livestock. In Oklahoma (Reuter and Horn 1999), animal average daily gains were approximately 1 kg d^{-1} for 'Manska' pubescent wheatgrass, 'Paiute' orchardgrass, and 'Lincoln' smooth bromegrass for 56 days with beef gains of 116 kg ha^{-1}.

Though winter wheat throughout the United States is typically planted with grain harvest in mind, the crop is used extensively as a dual-purpose (grain + grazing) crop in much of the temperate subhumid region. Shelton (1888) was one of the first to report on the advantages of the dual-purpose use of wheat. Later, in their review of wheat grazing, Redmon et al. (1995) reported from 75% to 90% of the irrigated winter wheat in Texas was managed for cattle grazing and that 65% of the winter wheat planted in Kansas was used for fall and spring grazing. In Oklahoma, it is commonly reported that 50% or more of the winter wheat hectarage is grazed.

Complementary Livestock Systems

The word **complementary** means "serving to fill out or complete" or "mutually supplying each other's lack" (Merriam-Webster 1990). In the case of **complementary forage systems**, the term can have two similar but slightly different meanings. In more arid areas, complementary forage systems refer to the blending of both rangeland and introduced forages to provide a system that more fully meets the nutrient requirements of the grazing livestock, and thus more fully meets the manager's production goals. However, in more subhumid systems, complementary forage systems are systems that use forages with different seasons of growth, such as cool- and warm-season grasses sequentially to improve seasonal productivity (Moore et al. 2004).

In the Northern Great Plains, rangeland is the primary source of forage (Lorenz 1977) but the morphologic development of native grasses often requires using introduced forages in the spring and fall to complement the native range. In North Dakota, for example, weather often permits grazing in late April or early May. Delaying turnout by 30 days to late May or early June has been shown to increase production on native rangeland by 35% (Lorenz 1976). However, forage gaps are created in the early spring as well as the late summer or early fall when native grasses have senesced. Complementary forage systems combine different types of forages to fill these production and nutritional gaps. Including introduced grasses with native rangeland can "complement" or extend the grazing season on rangelands (Lorenz 1977). McIlvain (1976) noted the philosophy of complementing low-producing, rough grasslands with high-producing tame pastures opened the door to (i) fitting green forages into dry periods, (ii) opportunity grazing or resting of each forage resource for its proper development and use, (iii) avoidance of grazing during periods when poisonous plants were highly hazardous, (iv) use of **flushing** pastures, (v) use of breeding pastures, (vi) use of day–night rotation, (vii) use of dehydrated forages for concentrates, and (viii) use of high-quality feed as green creeps for calves or steers needing a rapid gain, and following them with cows or younger steers to clean up graze.

Derner and Hart (2010) found yearling Hereford heifers had two to four times the gain on crested wheatgrass or russian wildrye compared to shortgrass native range indicating these forages could fill forage gaps. The optimal ratio of crested wheatgrass pasture to native rangeland was determined to be 1 : 3.9 when estimated yields, costs and prices were considered (Hart et al. 1988). The use of complementary forages can double the carrying capacity for stockers and beef cows (Launchbaugh et al. 1978). Sims (1993) reported that a complementary grazing system using double-cropped winter wheat and annual forages reduced land requirements by 40% in Oklahoma. Therefore, a complementary forage system in either the temperate subhumid or temperate semiarid regions would offer multiple advantages to livestock producers.

Integrated Crop-Livestock Systems

Integrated crop-livestock systems have been proposed as a method to achieve agricultural sustainability while still maintaining productivity (Franzluebbers 2007; Martins et al. 2016). Research suggests that incorporating livestock into cropping systems has minimal negative impact and may actually increase subsequent crop production. However, there is less information on livestock performance in integrated systems or if there are differences among livestock breeds in their performance in integrated systems. Calves from cows grazing crop residues in winter had heavier birth and weaning weights than calves from cows grazing winter range only (Larson et al. 2009). As with most grazing systems, heavier stocking rates decreased individual cow body weight, but increased stocking rate did not seem to affect subsequent crop yield (Stalker et al. 2015). Even less is known about the performance of individual breeds within integrated crop-livestock systems. A study from Brazil found that ½ Angus by ½ Nellore steers performed better in integrated crop/livestock systems than did ½ Charolais by ¼ Angus and ¼ Nellore steers (Costa et al. 2017).

Incorporating forages into integrated crop-livestock systems provides new opportunities and challenges. Forages can be important aspects of these systems and provide a yield boost to subsequent crop production. In the semiarid temperate region, Franco et al. (2018) found that unfertilized spring wheat yields, following two years of alfalfa, were similar to fertilized continuous no-till spring wheat yields and that yield effects from three years of alfalfa in the crop rotation could last up to three years following stand termination. The same study indicated that a cool-season grass-alfalfa mixture could not only provide the yield benefits but enhance selected soil quality attributes more than alfalfa alone (Liebig et al. 2018).

Cover crops are often a critical part of integrated systems because producers are attracted to their soil health attributes but need livestock to increase the economic feasibility of using them. Since most cover crops are annuals, cover crop species can be adjusted yearly to address specific soil quality and animal performance objectives. In a semiarid portion of the Northern Great Plains, Sanderson et al. (2018) found that spring-planted cover crop mixtures yielded more, on average, than monocultures but produced less than the most productive monocultures. However, the same study found that late-season planted cover crops, produced little forage because of dry soil and erratic weather conditions. However, in South Dakota, legume cover crops planted in mid- to late-August following winter wheat harvest did show forage potential with crude protein concentrations ranging from 113 to 270 $g\,kg^{-1}$ on yields ranging from 933 to 4590 $kg\,ha^{-1}$ (Hansen et al. 2015).

Challenges for Future Forage Systems

Land Use

Southern Plains

During settlement of the southern plains, the typical family farm was far smaller than today and was generally diversified with both crop and livestock production. Farm mechanization and the resulting expansion in cultivated area occurred around the time much of the Great Plains entered a period of major drought. The drought resulted in poor crop establishment, vast areas of unprotected soil, and ultimately to significant soil losses by wind which earned parts of the region the moniker of "Dust Bowl" and this period is often referred to as the Dust Bowl Era. The period continues to shape land use in much of the southern Great Plains. Many of the lands broken out for cultivation were not well suited to the practice. These lands were productive during the initial years of cultivation because the soils still contained abundant organic matter and precipitation was above normal for the region. There was a belief, at this time, that "rain follows the plow".

Now, the land that remains in cultivation for crop production is primarily irrigated land and conservation tillage or no-till are the common tillage practices. Other lands have either naturally reverted back to mostly native species (commonly referred to in the region as "go-back land"), have been planted to a perennial forage crop such as bermudagrass or old-world bluestem, or the lands have been enrolled in the Conservation Reserve Program (CRP) and have been seeded to primarily perennial grasses. The CRP lands are not grazed or hayed unless allowed by emergency declaration. Over one quarter of all CRP acres enrolled in the US at the end of 2018 were in the southern plains states of Texas, Oklahoma, Kansas, and New Mexico.

Taylor et al. (2015), reported land used for hay and pasture was included in the agriculture class, but rangelands and CRP were included together in the grassland/shrubland class. Nonetheless, it is instructive that in recent years Great Plains land use changes predominately involved exchanges between the two largest land use classes, namely agriculture and grassland/shrubland. In the entire Great Plains, agriculture was the dominant land use at 46.4% and grassland/shrubland was a close second at 42.7% to 42.1% between 1973 and 1986. By 1992, the two classes were tied at 44.2% and by 2000, grassland/shrubland was the dominant land use at 44.4% while agriculture had dropped to 43.8%.

Some regions within the southern Great Plains experienced little to no change in land use such as the Edward Plateau in south-central Texas, the central Oklahoma/Texas Plains, and the Flint Hills of Kansas. Soil limitations in these areas hindered agricultural development during settlement and these lands are well suited to their current use. Other regions in the Southern Great Plains experienced great changes. For example, in the east central Texas plains about 3.8% of agricultural land was converted to the grassland/shrubland class between 1973 and 2000 and 1.7% of forest land was converted to agricultural land. The three greatest net changes between 1973 and 2000 in the east central Texas plains was the 3.7% increase in the grassland/shrubland class and the 2.7% and 1.8% losses in agricultural land and forestland, respectively. The western high plains experienced large net changes in almost exclusively the grassland/shrubland and agriculture land uses. Grassland/shrubland had a net increase of 5.7% and agriculture had a 5.8% net decrease. Most of these changes occurred between 1986 and 1992 when the CRP was implemented. The central Great Plains experienced agricultural land-use increases between 1973 and 1986 (primarily through conversion of grassland/shrubland), but this expansion was later reversed between 1986 and 1992 so that net land-use change over the 27-year study was relatively small with a net decrease in agricultural land and grassland/shrubland of 0.6% and 0.1%, respectively, and a 0.7% increase in developed land over this period with growth of cities like Oklahoma City, Wichita, KS, and Abilene, TX. The Texas Blackland Prairies ecoregion, which is home to parts of the Dallas-Fort Worth metroplex, Austin, and San Antonio has experienced rather large net land-use changes related to urban growth between 1973 and 2000. Agricultural lands and forests have had net losses of 5.6% and 0.2%, respectively, while developed land and grassland/shrubland have increased 3.8% and 0.9% respectively.

Northern Plains

One of the biggest challenges to forage and rangelands in the northern part of the subhumid and semiarid temperate region is the changes in land use that have occurred. Between 2006 and 2011, rates of conversion from grassland to corn/soybean cropland in the subhumid temperate region were 1–5.4% which is similar to deforestation rates in Brazil, Malaysia, and Indonesia (Wright and Wimberly 2013). Though there is some controversy about the Wright and Wimberly methodology (Laingen 2015) and interpretation of these changes (Kline et al. 2013), anecdotal accounts suggest that changes in land use are occurring, especially in more subhumid areas east of the Missouri river. Some estimates are about 203 000 ha of native prairie was converted to cropland in the Dakotas and Montana between 2002 and 2007 (Fargione et al. 2009).

Land conversion to and from cropland occurs continuously, but a major surge in land conversion was relatively recent. A rapid increase in crop prices between 2006 and 2009 resulted in a 64% gain in typical farm profitability (Swinton et al. 2011). The improved profitability increased pressure to find more land to farm. Between 2008 and 2012, most (77%) of the new cropland came from grassland (Lark et al. 2015), resulting in 2.3 million hectares of grassland being converted to cropland. While most land conversion has occurred in the Dakotas, the presence of biorefineries has resulted in cropland expansion in the temperate subhumid areas in Minnesota and Wisconsin (Wright et al. 2017).

Besides the price increase, increased precipitation, a longer growing season (DeKeyser et al. 2013), technological improvements such as irrigation, and precision agriculture have provided the means to increase cropland acreage (Stubbs 2015) or improve profits (Scharf et al. 2011). Technology has also reduced the management intensity needed in agricultural systems (Hendrickson et al. 2008a, 2008b) making it easier to operate larger acreages. Average farm size ranges from 209 acres in Wisconsin to 349 acres in Minnesota to 1268 acres in North Dakota (USDA-NASS 2014). The combination of improved profits, favorable climate, and technological advances provided an environment for changes in land use.

Land Fragmentation and Managing Small Units Sustainably

The 2012 census of agriculture shows significant increases in the number of farms in the counties surrounding major Texas urban centers including the Dallas-Fort Worth metroplex, Austin, San Antonio, and Houston. Around these large urban centers, agricultural land and forest is being lost to increases in developed land and grassland/shrubland. A significant number of people employed in these cities have purchased small acreage farms in nearby rural areas within commuting distance of their place of work. Some farms may be farther out from their work than is reasonable for a daily commute but can be easily visited on weekends.

The land is typically not cultivated and is used for keeping animals and forage production. This trend has resulted in land fragmentation and often the main objective of these small farms is not to make a profit but to provide the owners with a rural lifestyle and a place to keep livestock. The uncoupling of profits from the management of these small farms could promote grazing land improvement, but often, the forage demand of the animals kept will exceed the forage produced on these small farms and the deficit is supplemented with purchased feed and hay. Thus, many of these grazinglands are seldom rested, are over grazed, and a repository for imported nutrients.

Climate Change

Drought

Drought is a normal feature of a given climate that occurs when the precipitation received for a period is significantly less than normal. When the climate is stationary, these periods are necessarily offset by pluvial periods—periods with significantly more precipitation than normal. The Great Plains has had a very long history of alternating between drought and pluvial periods on an approximately decadal scale. Some of these periods stand out for the magnitude or duration of the deviation. Some historical examples of the decadal cycle oscillations include the droughts of the 1930s, 1950s, and 2010s with the more pluvial periods of the 1940s and the 1970s and 1980s. Proxy climate records, such as from tree-ring, lake sediment or dunal paleosol data, indicate these decadal drought cycles have persisted in the Great Plains for several hundred years and there is also evidence of century-scale mega-droughts such as those in the fourteenth, fifteenth, and sixteenth centuries. The decadal cycles seem to be modestly related to pan-Pacific sea surface temperature variation which explains about one-third of the variation in low-frequency precipitation (timescales longer than about six years) (Schubert et al. 2004).

Drought, especially widespread drought, has a significant effect on forage-based agriculture. The 2011 drought in the southern Great Plains was the worst single-year drought in the historical record for much of the southern Great Plains and the destocking that followed contributed to US beef cattle inventories reaching their lowest numbers since the 1950s. Drought not only impacts forage production and quality, but in regions that depend on earthen ponds for water may also impact accessibility to forage resources due to a lack of drinking water.

Future concern about drought has components of the natural variability in the historical and proxy records. Many climatologists are now predicting that we may be entering a period of climate change that may result in a shift in the central tendencies for precipitation around which decadal and longer-scale variation has persisted. While there appears to be consensus that much of the interior US is likely to see 1–2 °C warming over the next 40 years, predictions of changes in precipitation are more uncertain and growing season uncertainties are probably larger than dormant season uncertainty even with a 5% decrease in precipitation predicted for the south central US (Walthall et al. 2012).

Extreme Wetness

Rather than drought, some areas have experienced extreme wetness. Weather data from the Northern Great Plains Research Laboratory (USDA-ARS) located in Mandan, North Dakota showed that for a 75-year period (1916–1990), the average annual precipitation was 38.1 cm (High Plains Climate Center 2019). However, between 1991 and 2000, average annual precipitation increased 25% to 50.8 cm. The periods of 2001–2009 and 2011–2016 also had greater than average precipitation (44.4 and 47.6 cm yr^{-1}, respectively).

The increased precipitation did impact forages and grasslands. Several studies have documented an increase in kentucky bluegrass on rangelands between 1984 and 2004 (DeKeyser et al. 2015), a time that corresponds with the increased precipitation. Forage species grow now and are productive in areas previously considered too dry. The increase in precipitation may also mean that suggested stocking rates are too low for the current conditions. Producers need to be aware of, and react to, the increase in precipitation.

Summary

The goal of forage systems is for grazing animals to receive most, if not all, of their nutrition from forages that are standing in the pasture. The high cost of feeding hay and other supplements is the economic driver behind this goal. Because of potentially severe environmental constraints in the temperate subhumid and temperate semiarid regions associated with drought and limited growing seasons, the availability of emergency feedstuffs will also be required for contingency livestock feeding programs. Hay feeding and the use of supplements, however, should be considered tactical solutions to short-term problems such as drought or ice and/or snow-cover days. Supplementation should generally only be used for specific production goals such as heifer development, backgrounding stocker cattle, or when forage is in short supply.

The forage systems used in the sub-humid and semi-arid temperate regions are diverse and dynamic. Regional extremes of geography and annual weather extremes require producers to be adaptable. Some common examples of such adaptations include cover crop grazing, complementing rangeland with introduced grasses and alternative grazing systems. Producers should evaluate the potential for complementary use of warm- and cool-season forages, both native and introduced, to minimize production risks and improve potential net return of their livestock production systems. Development and utilization of a forage system should be a priority goal for all livestock producers.

References

Alderson, J. and Sharp, W.C. (1994). *Grass Varieties in the United States*. Agriculture Handbook No. 170. Washington, D.C.: Soil Conservation Service, United States Department of Agriculture.

Asay, K.H. and Jensen, K.B. (1996). Wheatgrasses. In: *Cool-Season Forage Grasses* (eds. L.E. Moser, D.R. Buxton and M.D. Casler), 691–724. Madison, WI: ASA, CSSA, and SSSA.

Bavendick, F.J. (1974). The climate of North Dakota. In: *Climates of the States: Volume II—Western States Including Alaska and Hawaii* (ed. Officials of the U.S. Department of Commerce), 811–825. Port Washington, NY: Water Information Center, Inc.

Beck, P.A., Hutchison, S., Stewart, C.B. et al. (2007). Effect of crabgrass (*Digitaria ciliaris*) hay harvest interval on forage quality and performance of growing calves fed mixed diets. *J. Anim. Sci.* 85: 527–535. https://doi.org/10.2527/jas.2006-358.

Biligetu, B. and Coulman, B. (2010). Quantifying the regrowth characteristics of three bromegrass (*Bromus*) species in response to defoliation at different development stages. *Grassland Sci.* 56: 168–176.

Bittman, S. and McCartney, D.H. (1994). Evaluating alfalfa cultivars and germplasms for pastures using the mob-grazing technique. *Can. J. Plant. Sci.* 74: 109–114.

Blayney, D.P. (2002). *The Changing Landscape of US Milk Production*. USDA SB978.

Boe, A. and Delaney, R.H. (1996). Creeping and meadow foxtail. In: *Cool-Season Forage Grasses* (eds. L.E. Moser, D.R. Buxton and M.D. Casler), 749–763. Madison, WI: ASA, CSSA, and SSSA.

Brady, N.C. and Weil, R.R. (1999). *The Nature and Properties of Soils*, 12e. Upper Saddle River, NJ: Prentice-Hall, Inc.

Bush, T., Ogle, D., St. John, L. et al. (eds.) (2012. Ed. (rev) St. John). *Plant Guide for Orchardgrass (Dactylis glomerata)*. Aberdeen, Idaho: USDA-Natural Resources Conservation Service, Aberdeen Plant Materials Center 83210.

Cherney, J.H. and Allen, V.G. (1995). Forages in a livestock system. In: *Forages, Volume I. An Introduction to Grassland Agriculture*, 5e (eds. R.F Barnes, D.A. Miller and C.J. Nelson), 175–188. Ames, IA: Iowa State University Press.

Coffey, K.P. and Moyer, J.L. (1991). Performance and forage intake by stocker cattle grazing rye in monoculture or no-till drilled into bermudagrass sod. Kansas State University. Agricultural Research, Report of Progress No. 628.

Costa, P.M., Barbosa, F.A., Alvarenga, R.C. et al. (2017). Performance of crossbred steers post-weaned in an integrated crop-livestock system and finished in a feedlot. *Pesq. Agropec. Bras.* 52: 355–365.

D'Antonio, C.M. and Vitousek, P.M. (1992). Biological invasions by exotic grasses, the grass/fire cycle, and global change. *Annu. Rev. Ecol. Syst.* 23: 63–87.

DeKeyser, E.S., Meehan, M., Clambey, G., and Krabbenhoft, K. (2013). Cool season invasive grasses in Northern Great Plains natural areas. *Nat. Areas J.* 33: 81–90.

DeKeyser, E.S., Dennhardt, L.A., and Hendrickson, J. (2015). Kentucky bluegrass (*Poa pratensis*) invasion in the Northern Great Plains: a story of rapid dominance in an endangered ecosystem. *Invasive Plant Sci. Manage.* 8: 255–261.

Derner, J.D. and Hart, R.H. (2010). Livestock responses to complementary forages in shortgrass steppe. *Great Plains Res.* 20: 223–228.

Fargione, J.E., Cooper, T.R., Flaspohler, D.J. et al. (2009). Bioenergy and wildlife: threats and opportunities for grassland conservation. *BioScience* 59 (9): 767–777.

Foth, H.D. (1984). *Fundamentals of Soil Science*, 7e. New York, NY: Wiley.

Franco, J.G., Duke, S.E., Hendrickson, J.R. et al. (2018). Spring wheat yields following perennial forages in a semiarid no-till cropping system. *Agron. J.* 110: 1–9.

Franzluebbers, A.J. (2007). Integrated crop–livestock systems in the southeastern USA. *Agron. J.* 99: 361.

Fulbright, T.E., Hickman, K.R., and Hewitt, D.G. (2013). Exotic grass invasion and wildlife abundance and diversity, South-Central United States. *Wildl. Soc. Bull.* 37: 503–509.

Funderburg, E.R., Biermacher, J.T., Moffet, C.A. et al. (2011). Effects of applying (N) on yield of introduced perennial summer grass cultivars in Oklahoma. *Forage and Grazinglands* 9 https://doi.org/10.1094/FG-2011-1223-01-RS.

Funderburg, E., Biermacher, J.T., Moffet, C.A. et al. (2012). Effects of applying five (N) rates on quality of nine varieties of introduced perennial forages. *Forage Grazinglands* 10 https://doi.org/10.1094/FG-2012-0517A-01-RS.

Grant, T.A., Flanders-Wanner, B., Shaffer, T.L. et al. (2009). An emerging crisis across northern prairie refuges: prevalence of invasive plants and a plan for adaptive management. *Ecol. Restor.* 27: 58–65.

Guyette, R.P., Stambaugh, M.C., Dey, D.C., and Muzika, R.M. (2012). Predicting fire frequency with chemistry and climate. *Ecosystems* 15: 322–335. https://doi.org/10.1007/s10021-011-9512-0.

Hansen, M.J., Owens, V.N., Beck, D., and Sexton, P. (2015). Suitability of legume cover crop mixtures in Central South Dakota for late-season forage. *Crop, Forage & Turfgrass Management* 1 (1): 1–7.

Hart, R.H., Waggoner, J.W. Jr., Dunn, T.G. et al. (1988). Optimal stocking rate for cow-calf enterprises on native range and complementary improved pastures. *J. Range Manage.* 41: 435–441.

Henderson, D.C. and Naeth, M.A. (2005). Multi-scale impacts of crested wheatgrass invasion in mixed-grass prairie. *Biol. Invasions* 7: 639–650.

Hendrickson, J.R. and Berdahl, J.D. (2003). Survival of 16 alfalfa populations space planted into a grassland. *J. Range Manage.* 56: 260–265.

Hendrickson, J.R., Berdahl, J.D., Leibig, M.A., and Karn, J.F. (2005). Tiller persistence of eight intermediate wheatgrass entries grazed at three morphological stages. *Agron. J.* 97: 1390–1395.

Hendrickson, J.R., Hanson, J.D., Tanaka, D.L., and Sassenrath, G. (2008a). Principles of integrated agricultural systems: introduction to processes and definition. *Renewable Agric. Food Syst.* 23: 265–271.

Hendrickson, J.R., Liebig, M.A., and Berdahl, J.D. (2008b). Responses of *Medicago sativa* and *M. falcata* type alfalfas to different defoliation times and grass competition. *Can. J. Plant. Sci.* 88: 61–69.

Hendrickson, J.R., Schmer, M.R., and Sanderson, M.A. (2013). Water use efficiency by switchgrass compared to a native grass or a native grass alfalfa mixture. *Bioenergy Res.* 6: 746–754.

High Plains Climate Center (2019). High Plains Regional Climate Center, CLIMOD page. http://climod.unl.edu (accessed 10 October 2019).

Holechek, J.L. (1988). An approach for setting the stocking rate. *Rangelands* 10: 10–14.

Holechek, J.L., Pieper, R.D., and Herbel, C.H. (1998). *Range Management Principles and Practices*, 3e. New Jersey: Prentice Hall.

Howard, J.L. (1996). Bromus inermus. In: *Fire Effects Information System [Online]*. U.S. Department of Agriculture, Forest Service, Rocky Mountain Research Station, Fire Sciences Laboratory (Producer) Available: https://www.fs.fed.us/database/feis/plants/graminoid/broine/all.html. Accessed Feb. 20, 2019.

Huston, J.E. and Pinchak, W.E. (1991). Range animal nutrition. In: *Grazing Management: An Ecological Perspective* (eds. R.K. Heitschmidt and J.W. Stuth), 27–63. Portland, OR: Timber Press.

Jensen, K.B., Singh, D., Bushman, B.S., and Robins, J.G. (2015). Registration of 'Arsenal' meadow bromegrass. *J. Plant Regist.* 9: 304–310.

Jordan, D.J., Klopfenstein, T.J., and Adams, D.C. (2002). Dried poultry waste for cows grazing low-quality winter forage. *J. Anim. Sci.* 80: 818–824.

Karn, J.F. and Ries, R.E. (2002). Free-choice grazing of native range and cool-season grasses. *J. Range Manage.* 55: 469–473.

Klett, W.E., Hollingsworth, D., and Schuster, J.L. (1971). Increasing utilization of weeping lovegrass by burning. *J. Range Manage.* 24: 22–24.

Kline, K.L., Singh, N., and Dale, V.H. (2013). Cultivated hay and fallow/idle cropland confound analysis of grassland conversion in the Western Corn Belt. *Proc. Natl. Acad. Sci. U.S.A.* 110: E2863–E2863.

Knox, F.W., Burton, G.W., and Baird, D.M. (1958). Effects of (N) rate and clipping frequency upon lignin content and digestibility of coastal bermudagrass. *Agric. Food Chem.* 6 (2): 217–218.

Laingen, C.R. (2015). Measuring cropland change: a cautionary tale. *Pap. Appl. Geogr.* 1: 65–72.

Lardy, G.P., Adams, D.C., Klopfenstein, T.J., and Clark, R.T. (1999). First limiting nutrient for summer calving cows grazing autumn-winter range. *J. Range Manage.* 52: 317–326.

Lark, T.J., Salmon, J.M., and Gibbs, H.K. (2015). Cropland expansion outpaces agricultural and biofuel policies in the United States. *Environ. Res. Lett.* 10 https://doi.org/10.1088/1748-9326/10/4/044003.

Larson, D.M., Martin, J.L., Adams, D.C., and Funston, R.N. (2009). Winter grazing system and supplementation during late gestation influence performance of beef cows and steer progeny. *J. Anim. Sci.* 87: 1147–1155.

Lauenroth, W.K., Burke, I.C., and Gutmann, M.P. (1999). The structure and function of ecosystems in the Central North American grassland region. *Great Plains Res.* 9: 223–259.

Launchbaugh, J.L., Owensby, C.E., Schwartz, F.L., and Corah, L.R. (1978). Grazing management to meet nutritional and functional needs of livestock. In: *Proceedings of the First International Rangeland Congress* (ed. D.N. Hyder), 541–546. Denver, CO.

Liebig, M.A., Hendrickson, J.R., Franco, J.G. et al. (2018). Near-surface soil property responses to forage production in a semiarid region. *Soil Sci. Soc. Am. J.* 82: 223–230.

Lorenz, R.J. (1976). Resource combination and technological innovations for beef cattle enterprises in the Northern Great Plains. In: *Integration of Resources for Beef Cattle production. Proc. Symp. Society of Range Management 29th Annual Meeting*, 37–52. Feb 16–20, 1976. Omaha, NE.

Lorenz, R.J. (1977). Complementary grazing systems for the Northern Great Plains. In: *Proc. the Range Beef Cow Symposium V*, 29–39. Dec. 12–14, 1977, Chadron, NE.

Malinowski, D.P., Hopkins, A.A., Pinchak, W.E. et al. (2003). Productivity and survival of defoliated wheatgrasses in the Rolling Plains of Texas. *Agron. J.* 95: 614–626.

Martins, A.P., Cecagno, D., Borin, J.B.M. et al. (2016). Long-, medium- and short-term dynamics of soil acidity in an integrated crop–livestock system under different grazing intensities. *Nutr. Cycling Agroecosyst.* 104: 67–77.

Mattupalli, C., Moffet, C.A., Shah, K., and Young, C. (2018). Supervised classification of RGB aerial imagery to evaluate the impact of a root rot disease. *Remote Sens.* 10: 917. https://doi.org/10.3390/rs10060917.

McIlvain, E.H. (1976). Seeded grasses and temporary pasture as a complement to native rangeland for beef cattle production. Integration of Resources for Beef Cattle Production Symposium. Proc. 29th Annual Meeting, Society for Range Management.

McMurphy, W.E., Denman, C.E., and Tucker, B.B. (1975). Fertilization of native grass and weeping lovegrass. *Agron. J.* 67: 233–236.

Monono, E.M., Nyren, P.E., Berti, M.T., and Pryor, S.W. (2013). Variability in biomass yield, chemical composition, and ethanol potential of individual and mixed herbaceous biomass species grown in North Dakota. *Ind. Crops Prod.* 41: 331–339.

Moore, K.J., Hintz, R.L., Wiedenhoeft, M.H. et al. (2004). Complementary grazing systems for beef cattle production. Leopold Center Completed Grant Reports No. 228.

Myers, R. and Watts, C. (2015). Progress and perspectives with cover crops: interpreting three years of farmer surveys on cover crops. *J. Soil Water Conserv.* 70: 125A–129A.

Ogden, R.K., Coblentz, W.K., Coffey, K.P. et al. (2005). Ruminal in situ disappearance kinetics of dry matter and fiber in growing steers for common crabgrass forages sampled on seven dates in northern Arkansas. *J. Anim. Sci.* 83: 1142–1152.

Orton, R.B. (1974). The climate of Texas. In: *Climates of the States: Volume II: Western States Including Alaska and Hawaii* (ed. Officials of the U.S. Department of Commerce). Port Washington, NY: Water Information Center, Inc.

Padbury, G., Waltman, S., Caprio, J. et al. (2002). Agroecosytems and land resources of the Northern Great Plains. *Agron. J.* 94: 251–261.

Prine, G.M. and Burton, G.W. (1956). The effect of (N) rate and clipping frequency upon yield, protein content and certain morphological characteristics of coastal bermudagrass. *Agron. J.* 48: 296–301.

Rakkar, M.K. and Blanco-Canqui, H. (2018). Grazing of crop residues: impacts on soils and crop production. *Agric. Ecosyst. Environ.* 258: 71–90.

Rasby, R., Selley, R. and Klopfenstein, T. (1998). Grazing crop residues. Nebraska Cooperative Extension Bulletin No. EC 98–278-B.

Raub, R.H. (2018). The back-yard horse owner and the equine industry at large. *J. Anim. Sci.* 96 (Suppl. 2): 33–34.

Redfearn, D.D., Parsons, J., Drewnoski, M. et al. (2019). Assessing the value of grazed corn residue for crop and cattle producers. *Agric. Envir. Lett.* 4: 180066. https://doi.org/10.2134/ael2018.12.0066.

Redmon, L.A., Horn, G.W., Krenzer, E.G. Jr., and Bernardo, D.J. (1995). A review of livestock grazing and wheat grain yield: boom or bust? *Agron. J.* 87: 137–147.

Reuter, R.R. and Horn, G.W. (1999). Cool-season perennial grasses as complementary forages to winter wheat pasture. *Prof. Anim. Sci.* 18: 44–51.

Rocateli, A. (2017). Fertilizing warm-season forages. In: *Oklahoma Forage and Pasture Fertility Guide, E-1021* (eds. B. Arnall and A. Rocateli), 32–43. Oklahoma Cooperative Extension Service Division of Agricultural Sciences and Natural Resources Oklahoma State University.

Sala, O.E., Parton, W.J., Joyce, L.A., and Lauenroth, W.K. (1988). Primary production of the central grassland region of the United States. *Ecology* 69: 40–45.

Sanderson, M., Johnson, H., and Hendrickson, J. (2018). Cover crop mixtures grown for annual forage in a semi-arid environment. *Agron. J.* 110: 525–534.

Scharf, P.C., Shannon, D.K., Palm, H.L. et al. (2011). Sensor-based nitrogen applications out-performed producer-chosen rates for corn in on-farm demonstrations. *Agron. J.* 103: 1683–1691. https://doi.org/10.2134/agronj2011.0164.

Schmer, M.R., Brown, R.M., Jin, V.L. et al. (2017). Corn residue use by livestock in the United States. *Agric. Envir. Lett.* 2: 160043. https://doi.org/10.2134/ael2016.10.0043.

Schubert, S.D., Suarez, M.J., Pegion, P.J. et al. (2004). Causes of long-term drought in the U.S. Great Plains. *J. Clim.* 17: 485–503.

Shelton, E.M. (1888). Experiments with wheat. Kansas Agricultural Experiment Station Research Report 4.

Sims, P.L. (1993). Cow weights and reproduction on native rangeland and native rangeland-complementary forage systems. *J. Anim. Sci.* 71: 1704–1711.

Smoliak, S. and Johnston, A. (1980). Russian wildrye lengthens the grazing season. *Rangelands* 2: 249–250.

Soil Survey Staff (1999). *Soil Taxonomy: A Basic System of Soil Classification for Making and Interpreting Soil Surveys*, 2e. USDA-NRCS Agricultural Handbook No. 436. Washington, D.C: U.S. Government Printing Office.

Stalker, L.A., Blanco-Canqui, H., Gigax, J.A. et al. (2015). Corn residue stocking rate affects cattle performance but not subsequent grain yield. *J. Anim. Sci.* 93: 4977–4983.

Stubbs, M. (2015). Irrigation in U.S. Agriculture : On - farm Technologies and Best Management Practices (No. 7–5700 R44158).

Swinton, S.M., Babcock, B.A., James, L.K., and Bandaru, V. (2011). Higher US crop prices trigger little area expansion so marginal land for biofuel crops is limited. *Energy. Policy* 39: 5254–5258.

Taylor, J.L., Acevedo, W., Auch, R.F. and Drummond, M.A. (2015). Status and trends of land change in the Great Plains of the United States-1973 to 2000. U.S. Geological Survey Professional Paper, 1794-B. http://dx.doi.org/10.3133/pp1794B.

Thornthwaite, C.W. (1931). The climates of North America: according to a new classification. *Geog. Rev.* 21: 633–655.

Thornthwaite, C.W. (1941). Climate and settlement in the Great Plains. In: *1941 Yearbook of Agriculture. Climate and Man* (ed. E. Raisz), 177–187. Washington, D.C.: US Department of Agriculture.

Titlow, A., Luebbe, M.K., Lyon, D.J. et al. (2014). Using dryland annual forage mixtures as a forage option for grazing beef cattle. *Forage and Grazinglands* 12: 1–6.

Toledo, D., Sanderson, M., Spaeth, K. et al. (2014). Extent of Kentucky bluegrass and its effect on native plant species diversity and ecosystem services in the Northern Great Plains of the United States. *Invasive Plant Sci. Manage.* 7: 543–552.

Trimble, D.E. (1980). *The Geologic Story of the Great Plains*. Geological Survey Bulletin 1493. Reston, VA: U.S. Geological Survey.

Twidwell, E.K., Boe, A. and Casper, D.P. (2002). Teff: A new annual forage grass for South Dakota? Extension Extra. Paper 278. http://openprairie.sdstate.edu/extension_extra/278 (accessed 22 November 2019).

USDA National Agricultural Statistics Service. (2014). 2012 Census of Agriculture. Report AC-12-A-51. Washington, DC. https://www.nass.usda.gov/AgCensus (accessed 10 October 2019).

USDA National Agricultural Statistics Service. (2019). Quick Stats. USDA–NASS. https://quickstats.nass.usda.gov/results/83473400-CED9-3E7D-BE9F-C372D6031701 (accessed 10 October 2019).

Vallentine, J.F. (1989). *Range Development and Improvements*, 3e. San Diego, CA: Academic Press, Inc.

Vogel, K.P. and Hendrickson, J.R. (2019). History of grass breeding for grazing lands in the Northern Great Plains of the USA and Canada. *Rangelands* 41: 1–16.

Vogel, K.P., Moore, K.J., and Moser, L.E. (1996). Bromegrasses. In: *Cool-Season Forage Grasses* (eds. L.E. Moser, D.R. Buxton and M.D. Casler), 535–567. Madison, WI: ASA, CSSA, SSSA.

Volesky, J.D., Mowrey, D.P., and Smith, G.R. (1996). Performance of rose clover and hairy vetch interseeded into Old World bluestem. *J. Range Manage.* 49: 448–451.

Walthall, C.L., Hatfield, J., Backlund, P. et al. (2012). *Climate Change and Agriculture in the United States: Effects and Adaptation*, 186. Washington, DC: USDA Technical Bulletin 1935.

Warren, J., Lalman, D., Mcgee, A. et al. (2018). *Managing Crabgrass in a Continuous Grazeout Wheat System.* PSS 2790. Stillwater, OK: Oklahoma Cooperative Extension Service. Division of Agricultural Sciences and Natural Resources, Oklahoma State University.

Merriam-Webster (1990). *Webster's Ninth New Collegiate Dictionary*. Springfield, MA: Merriam-Webster, Inc.

Wedin, D.A. and Tilman, D. (1996). Influence of nitrogen loading and species composition on the carbon balance of grasslands. *Science* 274: 1720–1723.

Wright, H.A. (1988). Role of fire in the management of southwestern ecosystems. In: *Effects of Fire Management of Southwestern Natural Resources*. USDA Forest Service General Technical Report RM-191 (ed. J.S. Krammes), 1–5.

Wright, C.K. and Wimberly, M.C. (2013). Recent land use change in the Western Corn Belt threatens grasslands and wetlands. *Proc. Natl. Acad. Sci. U.S.A.* 110: 4134–4139.

Wright, C.K., Larson, B., Lark, T.J., and Gibbs, H.K. (2017). Recent grassland losses are concentrated around US ethanol refineries. *Environ. Res. Lett* 12 044001.

CHAPTER

22

Systems for the Warm Humid Areas

William D. Pitman, Professor, *Louisiana State University Agricultural Center, Homer, LA, USA*
Montgomery W. Alison, Extension Forage Specialist, *Louisiana State University Agricultural Center, Winnsboro, LA, USA*

Introduction

Though warm-season grasses are adapted to a large portion of the US, they are extensively planted as the forage base of livestock production systems in this country primarily in the lower southeastern states and in Hawaii, Puerto Rico, and the US Virgin Islands. It is this forage base, rather than the specific geographic location, that characterizes the forage systems involved. The warm humid portion of the continental US (see Chapter 8 for comparisons of climate among regions) consists of the lower southeastern states of Alabama, Florida, Georgia, Louisiana, Mississippi, South Carolina, southern Arkansas, southeastern Oklahoma, eastern Texas, and coastal North Carolina largely within the Coastal Plain physiographic region. This area lies within USDA Plant Hardiness Zones 8 and 9 and east of a moisture transition area between longitudes 96 and 98 °W through Texas. Much of this region was naturally forested, and timber production remains a major land use and economic enterprise. Soils are diverse and often determine land use across this region. Pastures are often planted on the leached, infertile, acid soils characteristic of upland sites in such warm moist climates. Some rather unique pastureland areas within the region occur on alluvial bottomlands along rivers, some isolated prairie sites, flatwoods (Spodosol) soils, and a limited amount of organic soils.

In addition to forage production and forestry, crop production areas are interspersed within the landscape across the region, and these typically occur on the most productive soils. Despite the interspersed pattern of these distinct land uses, integrated production of livestock across the forage, crop, and forestland areas is rare within the region. A degree of integration is attained within areas of poultry production because pasture fertilization has been a primary use of litter from confined poultry production. Many production units, involving forage for either grazing livestock or for **stored forage** within the region, are rather small-scale operations often consisting of less than 40 ha. In contrast to the large number of small forage-based production units across the region, nine of the 25 largest cow-calf operations in the nation are in Florida and two are in Hawaii (NCBA 2017) supported by pastures of warm-season perennial grasses.

Environmental Factors Limiting Regional Forage Options

The climate characteristics of the region, which favor some forage species, provide distinct limitations for others resulting in opportunities for use of several different forage types (Table 22.1). The long, warm summer growing season of the lower southeastern US contributes to high-production potential of warm-season grasses. The

Forages: The Science of Grassland Agriculture, Volume II, Seventh Edition.
Edited by Kenneth J. Moore, Michael Collins, C. Jerry Nelson and Daren D. Redfearn.

Table 22.1 Use and opportunities for different classes of forage species in the warm-humid region

Forage class	Current use	Potential for increased use or productivity
Warm-season perennial grasses	Widely used	Productivity could generally be increased with more intensive management.
Warm-season annual grasses	Limited use	Extent of use could be effectively increased in production of specialty items such as forage-fed beef and grass-based dairy products.
Cool-season perennial grasses	Limited use	Novel-endophyte tall fescue cultivars appear to provide potential for increased use on selected, fertile sites.
Cool-season annual grasses	Widely used	Extent of use and management intensity could be increased to provide more cool-season pasture production from small grains and annual ryegrass.
Warm-season perennial legumes	Limited use	Localized benefits from both introduced and native species are possible on selected sites with appropriate management. Value of some species is higher for ecosystem contributions than for forage production.
Warm-season annual legumes	Limited use	Several species are adapted to various sites. Specialized uses such as wildlife food plantings and creep grazing provide some opportunity.
Cool-season perennial legumes	Limited use	Successes include white clover on some bottomland sites and localized alfalfa production. Increased use is limited by both plant adaptation and management requirements.
Cool-season annual legumes	Limited use	Adapted species provide tremendous potential, however risk of annual establishment failure limits use along with availability of economically priced nitrogen fertilizer.
Crucifers and other non-legume forbs	Essentially not used	Despite recent evaluations, potential contributions to perennial grass forage systems have not been identified, though some specialized uses have been proposed.

combination of warm summer temperatures and summer moisture can support dense grass growth on fertile sites. Competitiveness of the well-adapted warm-season grasses can limit opportunities for mixtures of pasture species, particularly when management for productive grasses includes nitrogen (N) fertilization and application of **selective herbicides** for broadleaf weed control.

The distinct winter season of the region, with much of the area susceptible to frost from approximately November until April, limits growth of distinctly warm-season forage species with the perennial pasture grasses typically dormant through the winter months. Although winter weather is typically mild, both temperature and rainfall are quite variable with substantial variability both inter-annually and inter-decadally (Konrad et al. 2013).

Winter temperatures restrict adaptation of some tropical grasses to southern Florida. Tropical climate conditions of Hawaii, Puerto Rico, and the US Virgin Islands can support a greater variety of tropical forage species with rainfall and soil type primarily determining species adaptation. While predictable dry seasons characterize many tropical environments, unpredictable periods of limited rainfall and short-term drought are characteristics of the lower southeastern states despite the humid conditions and typical annual rainfall of 1000–1500 mm (Konrad et al. 2013).

Temperature limitations to winter growth of warm-season forages in the lower southeastern states provide opportunities for use of a number of cool-season species for winter pasture. Cool-season annual species have been more successfully used in pastures than perennials which have often failed to survive the extended summer season, especially with competition from aggressive warm-season species. Dormancy of warm-season species, lack of adaptation of cool-season perennial species, and the necessity of establishing cool-season annual species each year make late fall and early winter the period when forage production is typically most limiting in the region.

Variable weather conditions, including lack of predictability of autumn rains for establishment of cool-season annuals, provide considerable risk to livestock enterprises that depend on cool-season forage production. More predictable winter and spring rainfall patterns across much of the region contribute to cool-season forage production in contrast to peninsular Florida where spring rainfall is often limited. Much of the cool-season forage growth typically occurs during March and April as temperatures and day length increase and spring rains occur. Though some hay is produced from this spring forage growth, low temperatures and frequent rains contribute to poor drying conditions that make production of high-quality hay unpredictable.

Important Forage Species

Grasses

Seed-propagated cultivars of bermudagrass and bahiagrass are widely naturalized across the lower southeastern states. Pastures dominated by these **naturalized grasses** along with planted stands of these grasses and improved cultivars of bermudagrass and bahiagrass provide the forage base for much of the region. Most of the improved cultivars of bermudagrass do not become naturalized because of their dependence on vegetative propagation. On moist, fertile sites dallisgrass is an important component of many pastures. Bahiagrass is particularly well adapted in subtropical, peninsular Florida, where several more-tropical grasses including stargrass, limpograss, digitgrass, and rhodesgrass have also been effectively used (Chambliss 1999). On organic soils in southern Florida, St Augustinegrass has been the preferred pasture grass. Guineagrass is a genetically variable species that is widely distributed in tropical pastures because of both widespread planting and naturalization. Grasses in the genera *Panicum*, *Cynodon*, *Digitaria*, and *Brachiaria* have been used as pasture grasses in the tropical areas in recent years, while elephantgrass has been successfully used in cut-and-carry systems. Kikuyugrass is used for pastures at high elevations in Hawaii.

Hybrid bermudagrass cultivars provide considerable production capacity and flexibility for forage systems across the lower southeastern states. Cultivars range from dense-growing types which limit competition to more open-growing types with full canopies over relatively sparse cover at soil level. This variation in growth form can affect management options including weed control and interseeding of other species. Relative production potential among cultivars depends on the environment. In general, the high-production potential allows effective responses to a range of N fertilizer rates and frequencies. Forage quality and yield can be manipulated by fertilizer input levels and by length of growing period to produce high-quality forage under high-fertilization levels and frequent harvest or higher yields of lower quality forage with less frequent harvest. Bermudagrass can be effectively harvested as stored forage or grazed pasture during the growing season. In some environments, even stockpiled bermudagrass forage for grazing early in the dormant season can be effectively utilized.

Though the seeded forage bermudagrass cultivars and bahiagrass do not generally have production potentials as high as those of the superior hybrid bermudagrasses at high-N fertilizer rates, they respond well to moderate-N fertilizer rates and provide harvest flexibility with potential for either stored forage or grazed pasture. The other widely distributed, perennial, warm-season grass, dallisgrass, is restricted primarily to moist, fertile sites. Dallisgrass can be managed to provide grazing earlier in spring, and retain nutritive value of accumulated forage late in the growing season, to make unique contributions to forage systems on appropriate sites within the region. Limpograss is particularly well suited for such accumulation of growth for late-season grazing or even as dormant-season stockpiled forage in peninsular Florida when supplemented to overcome low-protein levels.

Some warm-season annual grasses are used for specific purposes in specialized production systems. Pearl millet, sorghum-sudan hybrids, and corn are important harvested forages. Crabgrasses are widely naturalized weedy annuals that can be effectively used as high-quality forages for a short grazing season.

Cool-season annual grasses are important winter pasture plants in the lower southeastern states. Annual ryegrass and the small grains are extensively used for forage in winter and spring. Annual ryegrass provides a widely useful option because of wide adaptation and relative ease of establishment from sod-seeding with minimal or no disturbance of dormant perennial grass sod (Figure 22.1) or from surface broadcasting on prepared seedbeds. The cool-season perennial grasses typically fail to survive the summer season across most of the region. In the northern extent, especially on moist sites, some tall fescue cultivars are useful as cool-season pastures.

Legumes

The most widely used legumes in the lower southeastern states are various cool-season species of the genus *Trifolium* (the true clovers). These are generally planted into the sod of perennial grasses in autumn, sometimes in mixture with cool-season annual grasses, to provide grazing during the dormant season of the warm-season perennial grasses. Both the species planted, and the extent of such plantings, differ across the region and from year to year. Success of these plantings is also variable among locations and years as illustrated by results of Han et al. (2012). On moist sites (Figure 22.2), the season of production and stand life can be greater for some intermediate white clover cultivars than for the other adapted clovers that are predominately annuals (Brink et al. 1999). Some of these cool-season legumes, particularly white clover, are also adapted at higher elevations of Hawaii with appropriate management (Smith 1989).

A few warm-season legume species are adapted in the lower southeastern states, but their use is somewhat localized. Sericea lespedeza can provide a sustainable, perennial legume stand for use as pasture or hay and has shown anthelmintic properties (Lange et al. 2006) and other ruminal effects (Muir 2011) when used as a pasture or hay crop for small ruminants. The annual lespedezas typically produce variable and decreasing stands when populations of the legume are based on natural reseeding. Several additional warm-season annual legumes including aeschynomene, alyceclover, cowpea, and lablab are

Fig. 22.1. Sod seeded annual ryegrass emerging in bermudagrass pasture in fall prior to frost.

Fig. 22.2. White clover persisting in moist bottom area and not on hillside.

adapted to various sites across the region (Thro et al. 1991). These species provide potential for specialized use in forage systems, as illustrated by Pitman et al. (2015), by providing biologic N fixation and forage of high nutritive value. Successful use of these legumes often requires more intensive management than for the cool-season legumes. Rhizoma peanut is a valuable hay plant on well-drained soils in Florida and southern portions of Alabama and Georgia. Alternatives with this legume have increased as new cultivars have become available in recent years (Muir et al. 2010; Dubeux et al. 2017). On the flatwoods of peninsular Florida, aeschynomene and carpon desmodium can add N and improve the forage quality of warm-season grass pastures during the summer (Aiken et al. 1991). A large number of tropical legumes can be potential forage species for various soils and management situations in the tropical locations. These are not widely used, and most are not readily available within the region despite recognized potential. Legumes have been planted in pastures in Hawaii for several decades. These range from herbaceous *Desmodium* species to shrubby *Desmanthus* species and small trees in the genus *Leucaena*.

Important Livestock Classes

In the lower southeastern states, cow–calf production is the major livestock enterprise, although several livestock

Table 22.2 Forage-based enterprises in the warm-humid region: forage base, extent of use, production requirements, and limitations

Enterprise	Primary forage base	Extent of use	Key production requirements	Major limitations
Beef cow-calf	Warm-season perennial grass pasture with winter hay or over-seeded cool-season annuals	Widely used	Fertilizer and weed control	Input-based and seasonal forage production
Beef stocker and heifer development	Cool-season annuals	Secondary	Annual planting and fertilizer	Risk with both pastures and cattle market
Hay production targeting beef cattle	Warm-season perennial grasses and annual ryegrass	Widely used	Inputs higher than for grazed pasture plus harvest equipment and expertise required	Cost and unpredictable harvest-period weather
Hay production for high-quality hay markets	Early-growth bermudagrass, alfalfa, rhizoma peanut	Limited and localized	Management required and site requirements for the legumes	Unpredictable harvest weather and limited availability of appropriate sites for the legumes
Sheep and goats	Warm-season grasses with hay seasonally	Limited	Grazing management	Livestock management and marketing requirements
Dairy	Annual grasses	Limited	Maintaining high nutrient intake	Economics of production
Forage-fed beef	Annual grasses	Limited	Maintaining high nutrient intake	Maintaining continuous availability of adequate quality forage, marketing of the high-value product, and limited number of processing facilities

production options and hay enterprises provide market avenues for forage crops (Table 22.2). The warm-season perennial grasses that provide the base resource for the grazing livestock industries throughout the warm-humid region are particularly well suited for cow–calf production. Stocker steers and replacement heifers, both beef and dairy, are associated grazing enterprises that are typically provided the best **available forage** quality from recent regrowth of perennial grasses or plantings of annual forages. Milk production is of localized importance in the region. Concentrate feeds and considerable amounts of harvested forage, some imported from outside the region, are utilized with lactating dairy cows in addition to grazed forages. Small-scale production of grass-fed, finished beef is supported by niche markets in isolated locations within the region. Similarly, niche markets support sheep and goat production in some locations within the region. Forages are also important sources of nutrients for an expanding horse industry in the region. Forages provide a greater proportion of nutrients in the production of meat and milk in some tropical situations than in typical livestock production systems in other parts of the US. While commercial livestock enterprises in these tropical locations benefit from extended periods of forage availability, livestock maintained for household production of meat and milk are sometimes provided forages as the sole source of nutrients. Beef and dairy cattle are important grazing livestock in the included tropical locations, although horses, goats, and sheep are the primary grazing livestock in some areas.

Forage-Livestock Systems

Warm-Season Perennial Grasses for Cow-Calf Production

Most of the warm-season perennial-grass pastures in the region are used in cow–calf production (Hoveland 1992). This industry is characterized by a large number of small herds and a very small number of large herds of up to several thousand cows. Cow–calf production systems can use the high potential productivity of most warm-season perennial forage grasses, and these systems can accommodate the variable quality of these plant species. Milk of high nutritional value can be provided to nursing calves by cows grazing forage of lower nutritional value. Body condition can fluctuate considerably through the year for the mature beef cow without greatly affecting the value of the marketable product, the weaned calf. Even this substantial flexibility in quality of forage that can be used by mature cows is often not enough to avoid limitations in productivity in many cow–calf systems in the warm humid region. Management decisions can result in high weaning weights with high-input levels, while contrasting approaches produce low weaning weights and low calf-crop percentages, often under excessive stocking rates. Optimal production levels depend on several factors and are affected by fluctuating weather conditions along with interactions of available forage resources and their management.

The typical relationship between calf weight and price per unit of weight at the time of sale is inverse, so maximizing weaning weight does not necessarily maximize profit. Managing the forage and cow herd to produce a high percentage calf crop is often a key aspect in developing a profitable cow–calf production system. Long-term patterns of marketing and prices have developed in the region with peak numbers of weaned calves and lowest prices typically coinciding during autumn. In response, in recent years, some cow–calf producers in the region have retained ownership of calves following weaning. Such changes in marketing strategy, as illustrated in Figure 22.3, can alter the economics of production and forage management since additional weight can be efficiently gained by lighter calves when grazing high-quality pasture after weaning. Acceptance of lighter weaning weights may allow earlier weaning, which in turn may be managed to increase body condition of thin cows before winter and reduce supplementation needs.

Benefits from additional approaches to cow–calf production using different systems based on warm-season perennial grasses can be substantial and are due to several factors. Decisions regarding extent of intensification, desired levels of forage availability and resulting cow condition, breeding season, and forage species can differ substantially among successful enterprises. Diversity of associated enterprises within the farm or simply within the area can alter cow–calf production systems across the region. Some rather simple intensive systems have been based on heavily fertilized hybrid-bermudagrass pastures for warm-season grazing with excess summer forage production harvested as high-quality hay for winter feed. These systems are highly vulnerable to fluctuations in growing-season rainfall, with short-term moisture deficits reducing forage availability and untimely rains providing hazards for hay harvest. However, such systems have economic advantages in areas where confined animal feeding, including the rather widespread southern broiler industry, can contribute economical sources of nutrients for pasture fertilization.

Use of low-quality hay caused by rain damage, excessive plant maturity, or deterioration during storage, can still provide acceptable results for mature beef cows where by-products of cotton and soybean processing provide locally available, economical sources of protein supplements. Such supplementation has allowed widespread use of low-quality hay from several different warm-season grass species. Sugarcane processing provides high-moisture molasses by-products useful for energy supplementation and as a carrier for other nutrients, such as non-protein N, for economical nutrition programs in portions of the region. Other locally available supplements include rice bran and hulls, citrus pulp, and sugarcane bagasse. Crop stubble and volunteer annual grasses following harvest of crops also contribute to cow herd nutrition in the region, although this practice is not used extensively. Even in this humid region, where most pastures consist of introduced species, native grasslands in some locations are important grazing resources for cow–calf production. These include some marsh and flatwoods rangelands in peninsular Florida and remnant coastal marshland in Louisiana.

Where introduced, warm-season perennial grasses and other economically priced, locally available sources of nutrients do not meet the nutrient needs of the cow herd during particular seasons, adapted annual forage species are available. These are often either expensive or have high management requirements.

The approach of fall seeding annual ryegrass into perennial warm-season grass sod is used extensively in the lower southeastern states, but this forage production is limited primarily to late winter and spring (Cuomo and Blouin 1997). Seeding ryegrass on a prepared seedbed adds cost but provides more days of grazing than sowing directly into a sod (Table 22.3) primarily because the pastures can be stocked earlier, typically by late fall or early winter (Utley et al. 1976). The forage quality of the annuals often supports greater profitability from enterprises such as stocker cattle or milk production than from cow–calf production. Fall calving can be used with cool-season annual forages to produce particularly high weaning weights in some portions of the region. Increased

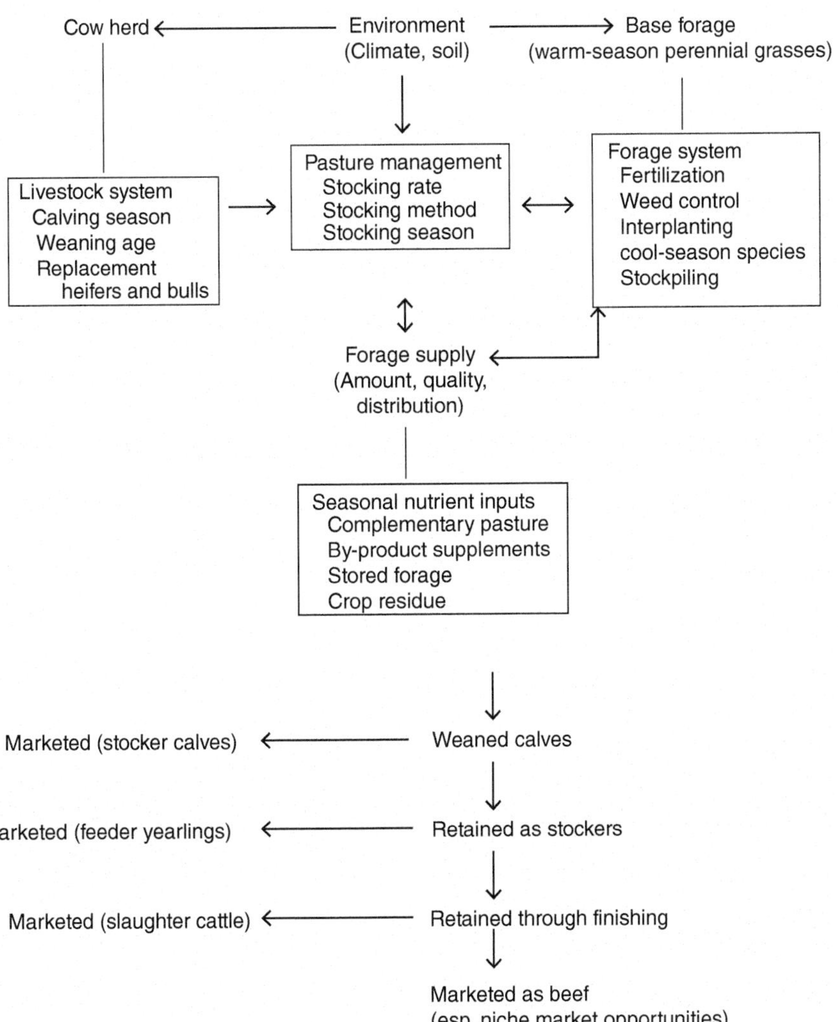

Fig. 22.3. Conceptual model of pasture-based cow–calf production in the warm humid region.

Table 22.3 Impact of seedbed management on annual ryegrass performance

Land preparation	DM yield at 1st harvest[a]	Steer grazing[b]
	kg ha^{-1}	d ha^{-1}
Prepared seedbed	1960	489
Sod-seeded	1000	222

[a]Adapted from Cuomo and Blouin (1997).
[b]Adapted from Utley et al. (1976).

dependence on cool-season annuals generally results in a system more vulnerable to variations in weather during fall and winter. Nitrogen contributions from cool-season legumes can help offset the limitations of additional management and risk from the annual grass forages.

Stockpiling forage of some warm-season perennial grasses is a less expensive approach than hay production or cool-season annuals for extending the grazing season into the early portion of the forage deficit period. This period coincides with the typical gap between the end of the growing season for the warm-season grasses and availability of forage for grazing from cool-season annual forages. Some tropical grasses such as limpograss are particularly useful for stockpiling because acceptability to grazing livestock is maintained even as forage nutritive value declines. Bermudagrass can be effectively stockpiled for only short periods of time under humid conditions.

Considerable planning and management are required to manage warm-season grass pastures for three distinct objectives in autumn. These are: (i) to have some pastures heavily grazed by early October for over-seeding, (ii) some pastures with forage available for grazing from early October through November, and (iii) late-summer fertilized pastures with grazing deferred (stockpiled) until late November or December. Except in peninsular Florida, bermudagrass appears to be superior to bahiagrass for such stockpiled forage (Gates et al. 2001; Evers et al. 2004).

Depending on location, rainfall, and management approach, warm-season grasses may be expected to provide from only about six months of grazing to essentially year-round grazing in some humid tropical locations. Distribution of available forage, both during the growing season and into the early dormant season, depends on several management and environmental factors. Grazing pressure, N fertilization, and moisture availability largely determine the supply of warm-season perennial pasture grasses, except during periods of winter dormancy. Thus, theoretical growth curves of individual warm-season pasture grasses can be greatly altered by extent of forage utilization, date and amount of fertilizer application, and the amount and distribution of rainfall.

As suggested by Hoveland (1986) three decades ago, the potential for increased productivity and profitability from these warm-season perennial grass–based cow–calf systems is substantial. Market objectives, livestock production cycles, and forage resources can be coordinated more closely for increased efficiency. Increased management for use of cool-season legumes on appropriate sites may be particularly successful when combined with a potentially responsive stage in the livestock production cycle and appropriate market opportunity. Similarly, the timing and amount of N fertilizer applied to warm-season grasses can be much more effectively manipulated than typically occurs. Though rather frequent applications of N fertilizer throughout the growing season are often effective for intensively managed bermudagrass, bahiagrass can respond to a single application of a high rate of N fertilizer with efficient uptake and storage in plant-base tissue for growth through periods of even longer than a complete growing season (Pitman et al. 1992).

Financial options, including budgeting and cash flow, differ considerably between the high investment and high maintenance needs of hybrid bermudagrasses and the more resilient, yet responsive bahiagrass. Pastures of these distinctly different grasses are often managed similarly. As demonstrated with the tropical *Cynodon* species (Pitman 1991), animal responses from bermudagrass and stargrass pastures can be greatly influenced by grazing management. Stocking for a light grazing pressure can allow high levels of leaf consumption and improved individual animal performance. Sequential stocking with livestock such as first-calf heifers grazed first, followed by the mature cow herd, could substantially enhance results from some intensively managed bermudagrass pastures without decreased forage utilization efficiency. Additional integration of cow–calf enterprises with field crops, other livestock enterprises, forestry, and even wildlife offers substantial economic opportunities in various portions of the region. These examples illustrate opportunities to enhance pasture-based, cow–calf systems in the warm humid region.

The ecologic processes naturally favoring **succession** of grassland to woodland and forest on most sites in the warm humid region result in the need for knowledgable pasture management. Inputs of soil amendments and suppression of undesired plants are required to maintain the desired grassland vegetation. Economic viability of such input-dependent systems requires increasing efficiency with the increasing costs of fuel and other inputs.

Intensive Forage Systems for Young Growing Cattle

The large number of weaned calves in the region and the demand for replacement beef and dairy cows provide a variety of opportunities for converting forage into weight gain for young cattle. Systems range from confined feeding of harvested forage as greenchop and silage, primarily to dairy replacement heifers, to grazing of warm-season perennial grasses at high stocking rates. Direct competition for suitable land between cash crops and most high-quality harvested-forage options limits this approach in most of the region. Highly stocked bermudagrass pastures typically produce high rates of gain per unit area but only modest individual animal gains.

The most widespread systems use cool-season annual forages to provide high-quality pasture for grazing primarily from late winter through spring. Some of the most productive enterprises are based on annual ryegrass, which typically provides a one- to two-month longer period of grazing in the spring than do the small grains and some annual clovers. The cool-season annual grasses can generally be provided enough N fertilizer to support higher livestock carrying capacities than do the legumes. Several different clover species can be used in various parts of the region, with some species limited to specific soil types. These legumes can be used in mixtures with cool-season annual grasses to provide very high-quality pasture with limited additional N fertilizer required. Intensive management of all system components, including pastures, livestock, and marketing is required for optimal results.

Productive high-quality warm-season annual grasses are available for most of the region. As with the cool-season annuals, stands of these grasses can produce higher maximum individual animal gains than are typically obtained from warm-season perennial grasses. Even so, use of such forages for summer grazing is quite limited. Cost, management requirements, short duration of the grazing period, and adverse effects of the typical heat and humidity on grazing livestock combine to limit use of the warm-season annual grasses.

Forage quality of both bermudagrass and bahiagrass can be adequate to support acceptable gains of young growing cattle when the plants are maintained in a vegetative stage with rapid growth. Maintaining pastures at such a growth stage while also providing adequate forage availability for optimal levels of intake is very difficult with bahiagrass. Bermudagrass can be managed for the desired forage quality and animal performance; however, stocking rates and pasture utilization often must be lower than economically optimal levels. Integration of stocker and cow–calf enterprises in leader–follower stocking systems could provide the initial grazing period with forage of high quality and high availability to the stocker cattle and a subsequent grazing period for mature cows.

Interest in retained ownership of calves following weaning has resulted in a variety of approaches (Figure 22.3). Contract grazing of southeastern calves on rangelands and winter-grazed wheat fields in the Southern Great Plains has been one option. Established prices for gain in such contracts provide a basis for comparisons of potential profitability of various forage production options within the warm humid region. Stocker grazing on cool-season annual forages planted on cropland between warm-season cash crops is a very under-utilized approach providing considerable opportunity with some cropping systems.

Other Forage-Livestock Systems

Livestock in several other types of enterprises in the region utilize forage to various extents. In contrast to the cow–calf industry and the stocker cattle industry, these other livestock enterprises are not distinctly forage based. Dairy cattle and horses consume large amounts of forage in the region; however, concentrate feeds typically supply large proportions of the nutrients for these livestock enterprises. Production of sheep and goats in the region is limited but could expand to meet food demands from recently changing population patterns. Widespread planting of wildlife food plots has also created a market for forage seed and an opportunity for increased benefits with improved forage management.

Milk production in the warm humid region uses large amounts of forage. Forage quality, particularly nutrient density, can be a limitation of forages for high-producing dairy cows (Bull 1995). The typical approach has been to meet nutrient requirements using concentrate feeds and, perhaps, some high-quality forage such as imported alfalfa hay. Lower-quality local roughage is often used primarily for fill. Locally produced harvested forages, including green chop and silage from warm-season annual grasses or ryegrass, are used to some extent. Seasonal grazing of ryegrass and crabgrass forage, as available, is used by some small dairies. Management of dry cows is often a more forage-based portion of the dairy operation. Due to limited land availability, some large confinement operations do not graze even dry cows.

An actual forage-based system of seasonal milk production from pastures of annual ryegrass with cows in production only during periods of pasture growth can serve a niche market for products from grass-fed livestock. The dairy industry in the warm humid region has been in transition in recent years with economics apparently favoring increased capacity, additional mechanization, and confined feeding. The number of dairy farms has decreased, whereas the average number of cows per farm has increased substantially. On-farm forage production for cows in these larger herds is often limited.

Horse population has increased in the lower southeastern states in recent years with increased amounts of forage needed. Commercial enterprises are included with racehorses and performance horses of substantial economic importance in localized areas across the region. Pleasure horses are widely dispersed throughout the Southeast. A large proportion of these horses is typically maintained on pastures. The pastures, however, generally provide only a portion of the nutrients, and grain and purchased hay typically are provided. High-quality alfalfa hay from outside the region and some of the best-quality bermudagrass hay produced within the region receive premium prices as feed for horses.

In Florida and the southern portions of neighboring states, rhizoma peanut has become a preferred hay for many horse producers. The highly discriminating market with premium prices for acceptable hay in the horse industry has stimulated local hay producers to improve production practices and product quality. Management of horse pastures is typically limited except for routine clipping. The primary interest often becomes maintaining high levels of ground cover because close grazing by horses and traffic effects from excessive stocking rates contribute to erosion or seasonally muddy conditions. Sod-forming grasses such as common bermudagrass and bahiagrass are often selected for horse pastures primarily because of grazing tolerance and ground-cover characteristics.

Environmental Impacts of Forage Management

The diversity of **plant functional groups**, including cool- and warm-season species, grasses and legumes, and annual and perennial species in various combinations, provide opportunities for a great range of forage-based enterprises. Among these forage plants are species planted to provide a variety of **ecosystem services** such as erosion control, wildlife habitat, and N fixation. Expansion of forage-livestock systems to include additional objectives is readily within the capabilities of the plant species available.

Regional pasture fertilization has provided only a minor contribution to the "dead zone" extending from the mouth of the Mississippi River across more than 20 000 km^2 of the Gulf of Mexico (Hendy 2017). Yet these N losses from fields and pastures provide a regional illustration of the potential drastic, unintended, cumulative effects of individually minor environmental impacts. Of even more direct consequence, the well adapted, productive, sod-forming, perennial grasses directly supporting the regional pasture-based cow-calf industry have been implicated in region-wide decline of some grassland birds including the iconic northern bobwhite (Palmer et al. 2011).

Rather recent emphasis has nurtured the perspective that intensive agriculture based on inorganic inputs, including fertilizer nutrients and pesticides, has reduced soil health, ecosystem sustainability, and production efficiency (Teague 2015). Potential contributions of pastures to unintended environmental impacts and input inefficiency include the most critical inputs for maintenance of productivity of intensively managed perennial warm-season grass pastures, N fertilizer and selective pasture herbicides. Reduced availability of some N sources has complicated pasture fertilization decisions resulting in lower nutrient-use efficiency as illustrated by Connell et al. (2011).

The number and changing availability and relative price of pasture herbicides increases the complexity of weed-control decisions and consequences as shown by recent evaluations (Butler and Muir 2006; Han and Twidwell 2017). At the pasture scale, complexity of pasture fertilization and pesticide decisions with changing product availabilities underscores the value of timely information for appropriate decision-making and the increasing need for effective extension input. At the larger scale of ecosystems or landscapes, diversity of grasslands and their spatial arrangement within a farm or landscape can affect various ecosystem functions including conservation of biodiversity, physical and chemical fluxes in ecosystems, and pollution mitigation (Gibon 2005).

The numerous and diverse plant species useful as forage plants across the region provide opportunity to develop pasture systems contributing a diversity of ecosystem functions in relation to landscape position, seasonal cover requirements, environmental concerns, economic opportunities, and other community needs in addition to livestock production requirements. Mechanisms to introduce the value of such approaches and programs to provide appropriate incentives for adoption of landscape-scale management and resulting benefits appear to be additional incremental steps yet to be taken to verify efficacy of such broad landscape management. Available forage resources and previously evaluated native grassland species of the region not currently used can be combined to provide a tremendous diversity of grassland and appropriate plants for deployment within diverse crop, forest, and grassland landscapes across the southeastern states.

References

Aiken, G.E., Pitman, W.D., Chambliss, C.G., and Portier, K.M. (1991). Responses of yearling steers to different stocking rates on a subtropical grass-legume pasture. *J. Anim. Sci.* 69: 3348–3356.

Brink, G.E., Pederson, G.A., Alison, M.W. et al. (1999). Growth of white clover ecotypes, cultivars, and germplasms in the Southeastern USA. *Crop Sci.* 39: 1809–1814.

Bull, L.S. (1995). Forages for dairy cattle. In: *Forages: Vol. II. The Science of Grassland Agriculture*, 5e (eds. R.F Barnes, D.A. Miller and C.J. Nelson), 295–301. Ames, IA: Iowa State University Press.

Butler, T.J. and Muir, J.P. (2006). Coastal bermudagrass (*Cynodon dactylon*) yield response to various herbicides. *Weed Technol.* 20: 95–100.

Chambliss, C.G. (ed.) (1999). *Florida Forage Handbook*. Gainesville, FL: University of Florida.

Connell, J.A., Hancock, D.W., Durham, R.G. et al. (2011). Comparison of enhanced-efficiency nitrogen fertilizers for reducing ammonia loss and improving bermudagrass forage production. *Crop Sci.* 51: 2237–2248.

Cuomo, G.J. and Blouin, D.C. (1997). Annual ryegrass forage mass distribution as affected by sod-suppression and tillage. *J. Prod. Agric.* 10: 256–260.

Dubeux, J.C.B. Jr., Blount, A.R.S., Mackowiak, C. et al. (2017). Biological N_2 fixation, belowground responses, and forage potential of rhizoma peanut cultivars. *Crop Sci.* 57: 1027–1038.

Evers, G.W., Redmon, L.A., and Provin, T.L. (2004). Comparison of bermudagrass, bahiagrass, and kikuyugrass as a standing hay crop. *Crop Sci.* 44: 1370–1378.

Gates, R.N., Mislevy, P., and Martin, F.G. (2001). Herbage accumulation of three bahiagrass populations during the cool season. *Agron. J.* 93: 112–117.

Gibon, A. (2005). Managing grassland for production, the environment and the landscape. Challenges at the farm and the landscape level. *Livest. Prod. Sci.* 96: 11–31.

Han, K.J., Alison, M.W., Pitman, W.D., and McCormick, M.E. (2012). Contributions of overseeded clovers to bermudagrass pastures in several environments. *Crop Sci.* 52: 431–441.

Han, K.J. and Twidwell, E.K. (2017). Herbage mass and nutritive value of bermudagrass influenced by non-growing-season herbicide application. *Agron. J.* 109: 1024–1030.

Hendy, I. (2017). Gulf of Mexico 'dead zone' is already a disaster – but it could get worse. Phys Org. https://phys.org/news/2017-08-gulf-mexico-dead-zone-disaster.html (accessed 10 October 2019).

Hoveland, C.S. (1986). Beef-forage systems for the southeastern United States. *J. Anim. Sci.* 63: 978–985.

Hoveland, C.S. (1992). Grazing systems for humid regions. *J. Prod. Agric.* 5: 23–27.

Konrad, C.E., Fuhrmann, C.M., Billiot, A. et al. (2013). Climate of the southeast USA: past, present, and future. In: *Climate of the Southeast United States: Variability, Change, Impacts, and Vulnerability* (eds. K.T. Ingram, K. Dow, L. Carter and J. Anderson), 8–42. Washington, DC: Island Press.

Lange, K.C., Olcott, D.D., Miller, J.E. et al. (2006). Effect of sericea lespedeza (*Lespedeza cuneata*) fed as hay, on natural and experimental *Haemonchus contortus* infections in lambs. *Vet. Parasitol.* 141: 273–278.

Muir, J.P. (2011). The multi-faceted role of condensed tannins in the goat ecosystem. *Small Ruminant Res.* 98: 115–120.

Muir, J.P., Butler, T.J., Ocumpaugh, W.R., and Simpson, C.E. (2010). 'Latitude 34', a perennial peanut for cool, dry climates. *J. Plant Regist.* 4: 106–108.

NCBA (2017). *Directions*, 21e. National Cattlemen's Beef Association. www.beefusa.org (accessed 31 July 2017).

Palmer, W., Terhume, T., Dailey, T. et al. (2011). *National Bobwhite Conservation Initiative… the Unified Strategy to Restore Wild Quail*. Tall Timbers Research Station, Tallahassee, FL: National Bobwhite Technical Committee.

Pitman, W.D. (1991). *Management of Stargrass Pastures for Growing Cattle Using Visual Pasture Characteristics*. Bulletin No. 884. Gainesville, FL: Florida Agricultural Experiment Station.

Pitman, W.D., Portier, K.M., Chambliss, C.G., and Krestchmer, A.E. Jr. (1992). Performance of yearling steers grazing bahia grass pastures with summer annual legumes or nitrogen fertilizer in subtropical Florida. *Trop. Grassl.* 26: 206–211.

Pitman, W.D., Walker, R.S., Scaglia, G. et al. (2015). Summer legumes for creep grazing in cow-calf production on bermudagrass pastures. *J. Agric. Sci.* 7: 8–17.

Smith, B. (1989). A case study of white clover/ryegrass introductions into kikuyugrass on a commercial cattle ranch in Hawaii. In: *Persistence of Forage Legumes* (eds. G.C. Marten, A.G. Matches, R.F Barnes, et al.), 387–392. Madison, WI: ASA, CSSA, and SSSA.

Teague, W.R. (2015). Toward restoration of ecosystem function and livelihoods on grazed agroecosystems. *Crop Sci.* 55: 2550–2556.

Thro, A.M., Mooso, G.D., Friesner, G.D. et al. (1991). *Adaptation and Yield of Summer Pasture Legumes in Louisiana*. Bulletin No. 825. Baton Rouge, LA: Louisiana Agricultural Experiment Station.

Utley, P.R., Marchant, W.H., and McCormick, W.C. (1976). Evaluation of annual ryegrass forages in prepared seedbeds and overseeded into perennial sods. *J. Anim. Sci.* 42: 16–20.

CHAPTER

23

Systems for Humid Transition Areas

Renata N. Oakes, Assistant Professor, *Department of Forage Management and Systems, University of Tennessee, Knoxville, TN, USA*
Dennis W. Hancock, Center Director, *USDA-Agricultural Research Service, US Dairy Forage Research Center, Madison, WI, USA*

Introduction

Much of the east central region of the United States is referred to as the humid transition zone. The region includes USDA Hardiness Zones 6b and 7a (US Department of Agriculture 2012a) in an area spanning from 34 to 38 °N **latitude** and 78 to 96 °W longitude (Figure 23.1). This region is described as a humid transition zone because of its warm-temperate climate (humid mesothermal). The region has an overall mean temperature of 13–18 °C and a frost-free period of 180–210 d yr^{-1}, of which 70–100 d yr^{-1} will have maximum temperatures above 30 °C. Annual precipitation in this region averages 1000–1400 mm, which exceeds average annual potential evapotranspiration but this precipitation is not evenly distributed and can be highly variable from year to year, often resulting in moderate to severe droughts.

The forage systems employed in this region are largely determined by soil characteristics and topography. The soils of the humid transition area are typically non-glaciated, moderately to strongly weathered, and acidic Ultisols (U.S. Department of Agriculture 1998). The landscape is dominated by hilly, highly erodible slopes and more than one-half of the land is in **Land Capability Class** III or higher. Though significant areas of Class I and II cropland and high-nutrient Alfisol soils are common, particularly in alluvial river bottoms in this region, these areas are typically used for row crops rather than forage production.

Before European settlement, the humid transition zone was largely dominated by deciduous forest, with some **savanna** and intermittent **prairie** areas on drier sites. Forests and savannas were cleared for timber and cropland. Intensive tillage often resulted in severe soil erosion, with some areas losing more than 25 cm of topsoil. Slope, fertility, drainage, and other soil characteristics can vary widely even within a given farm. Consequently, a diversity of forage species is needed to best adapt to the assorted microenvironments.

There is also significant diversity among the ca. 350 000 forage-based livestock operations in the humid transition zone (US Department of Agriculture 2017). One of the most common forage-based livestock enterprises is the cow-calf production system, with about 5 million beef cows within the region. Other forage-based livestock operations in this area include goats (900 000), dairy cattle (300 000), sheep (200 000), and about 2 million horses. In addition to the 16 million ha of pastureland in this region, there are nearly 4 million ha of hay harvested each year. Most of these forage-based livestock farms are small with two-thirds of the beef cattle operations reporting gross farm sales of less $25 000 (US Department of Agriculture 2012b). The low income from these enterprises results in relatively limited ability to invest

Forages: The Science of Grassland Agriculture, Volume II, Seventh Edition.
Edited by Kenneth J. Moore, Michael Collins, C. Jerry Nelson and Daren D. Redfearn.

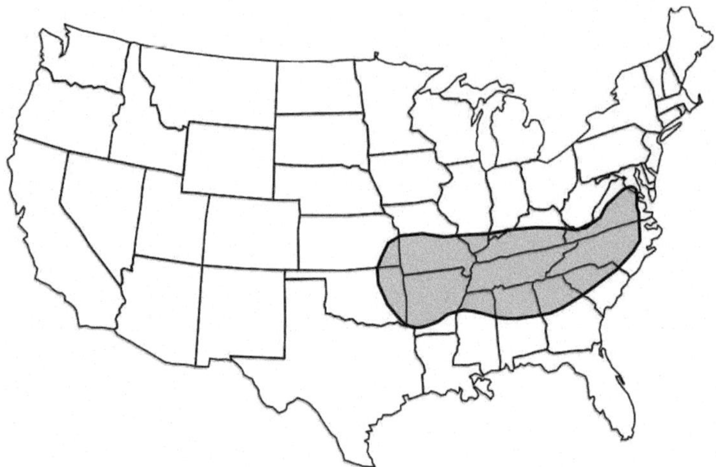

Fig. 23.1. Humid transition zone in the United States.

in forage renovations or highly efficient management practices. Primary income on forage/livestock farms in the region is, in many cases, from off-farm employment which competes for management time on the farm.

Important Forage Species

The humid transition zone is a region of significant livestock production, especially cow–calf operations. Mild fall and winter temperatures and warm humid summer temperatures characterize weather in this region. **Cool-season plant** productivity can be limited by periods of drought and high temperatures during warmer months but, in general, these species dominate pastures during spring and fall. **Warm-season plants**, primarily grasses, are frequently used during hot and humid weather conditions to complement cool-season grass forage availability. Good growing conditions for many cool-season species allow extension of the grazing season and reduce supplemental feed costs in this region.

Most forage species cultivated in this region are introduced, but extensive research in the area has shown that native warm-season species can be advantageous when adequate management is provided. Though establishment of these species is often slow, they can be beneficial due to their high productivity, adequate nutritive value, high-drought tolerance, and ability to increase soil and water quality.

Cool-Season Grasses

Tall fescue is the predominant cool-season perennial grass in this region, especially due to its tolerance to heat and soil moisture stress. Tall fescue has high-nutritive value in early spring and early fall, especially when compared with other cool-season grasses.

Tall fescue is often infected with an endophytic fungus, *Epichloë coenophiala,* that produces alkaloids that are toxic to livestock (Ball et al. 1993). Toxicity to ruminant animals, results from accumulation of these alkaloids in leaves, stems, and seed heads. The syndrome resulting from the subclinical effects is referred to as **fescue toxicosis** and, is more evident during warmer months. Cattle diagnosed with the symptoms exhibit reduced feed intake, weight gain, reproductive efficiency, and milk production as well as lower heat tolerance, hyperthermia, and increased respiration rate (Yates 1983). Many years ago, it was discovered that elimination of the fungal endophyte eliminated the negative effects of tall fescue toxicosis.

The presence of the endophyte, however, also enhances tall fescue drought tolerance, pest resistance (Sleper and West 1996), and plant **persistence** in this region. Therefore, endophyte-free cultivars may not be as productive or persistent as infected cultivars. More recently, tall fescue cultivars containing a novel endophyte have been developed that produce livestock weight gains similar to endophyte-free cultivars, while maintaining the environmental stress tolerance characteristics of earlier endophyte infected tall fescue. These cultivars are proving to be a viable alternative for pasture renovation in this area.

Orchardgrass is also a very common cool-season perennial grass in the humid transition zone. It grows as a bunchgrass with low drought tolerance. Stand persistence can also be an issue, especially on soils lacking adequate fertility and proper grazing management. A good alternative to overcome these issues is to plant orchardgrass mixed with cool-season legumes in order to achieve high **forage quality** while extending stand life (Smith et al. 1973).

Timothy is a cool-season perennial grass that is more commonly cultivated in the northern portion of the humid transition zone. It grows best under cooler temperatures and is primarily used as a hay crop but, can also be productive in pasture mixtures. Kentucky bluegrass, like timothy, is not as commonly grown in the southern portion of the transition zone due to a limitation in production when average monthly temperatures are 24 °C or higher (Hartley 1961).

Annual ryegrass is very popular in pastures across the humid transition zone due to ease of establishment, high-forage quality and adaptability to various soil types (Evers et al. 1997). Most forage-type cultivars in this species are true annuals, also known as 'Westerwold' types, which are usually planted in the fall for winter grazing, often overseeded into existing bermudagrass pastures. Perennial ryegrass is occasionally planted in the northern areas of this region. Despite offering high forage yield, tolerance to close grazing and high-nutritive value, perennial ryegrass does not tolerate severe winter temperatures (0 °C or lower) and summer drought (Jung et al. 1996), so persistence has been a major limitation to its use in the humid transition area.

Warm-Season Grasses

Bermudagrass is the predominant warm-season perennial grass in the humid transition zone and is characterized by a **rhizomatous** and **stoloniferous** growth habit. Its high tolerance to drought, high temperature, and high-grazing pressure are major contributors to its competitiveness and persistence in the region. It was first introduced to the US in the early 1800s, and has been recognized as an important forage crop since the 1940s due to the development of high yielding cultivars and the increase in the livestock industry (Harlan et al. 1970).

In the southern portion of the transition zone, bermudagrass is often found volunteering in tall fescue pastures. During summer, bermudagrass predominates while tall fescue dominates pastures during spring and fall. Most of the time, pastures with mixtures of bermudagrass and tall fescue have a reduced summer slump due to the complementary growth patterns of the two species. With good management, these mixed systems can provide near year-round grazing in this region.

Many native warm-season grasses such as switchgrass, big bluestem and indiangrass, can be used by grazing livestock. Switchgrass can produce twice as much forage as tall fescue and provides good production during summer in the transition zone, especially early in the season when its nutritive value is highest. Big bluestem has higher drought tolerance than other native warm-season grasses and is well adapted to excessively drained soils with low water-holding capacity (Anderson 2000). Indiangrass is known to mature later than other native warm-season grasses and it is commonly seeded in mixtures with switchgrass or big bluestem. One of the main drawbacks of native warm-season grasses is their generally slow establishment. Successful establishment of these species requires close attention to management, including adequate weed control.

Warm-season annual grasses are frequently used in this region to provide summer feed, especially when higher quality forage is needed. Sorghum-sudangrass hybrids are most commonly used for silage and hay, and are characterized by rapid growth, heat and drought tolerance and high productivity (Pedersen and Toy 1997). Pearlmillet is well adapted to poor, droughty, and infertile soils, but also responds well to high fertility and adequate moisture (Hanna et al. 2004). Vegetative pearlmillet has a high leaf: stem ratio and is generally highly digestible and readily consumed by livestock, with high concentrations of protein and low fiber and lignin (Hanna et al. 2004). Teff has a short growing cycle, reaching maturity three months after planting (Stallknecht et al. 1993). It is tolerant of both drought and excessive moisture conditions, has lower fertilization requirements than some other species, and has high regrowth potential after harvesting (Girma et al. 2012). Crabgrass has been volunteering in pastures and for many years was considered a weed. Most recently, crabgrass has become a common warm-season annual grass for grazing. It is a vigorous grass with high leaf : stem ratio, high animal intake and excellent nutritive value (Bosworth et al. 1980). Its common occurrence throughout the transition zone, combined with its high-forage quality, has the potential to increase forage production during the summer slump caused by cool-season species, therefore, many crabgrass cultivars have been improved and selected for high-quality and high-forage productivity.

Legumes

Alfalfa production in the humid transition zone has been limited due to climatic limitations and low fertility and pH soils. Enthusiasm for the crop is growing, especially for use in grazing systems, and there is potential for further expansion as these problems are overcome (Lacefield et al. 2009). Recently, many alfalfa cultivars have been developed to allow more flexible grazing schedules because they are more tolerant of continuous stocking and abusive grazing than hay types (Brummer 2006). In addition, alfalfa growing in mixtures with cool- and warm-season grasses can increase the forage quality of pastures plus provide a more stable forage yield throughout the season compared with producing these species in monocultures. When considering all these factors, expansion of alfalfa production in this region is possible and should be explored.

White clover is one the most adapted forage legumes in this region. It provides high-quality forage throughout the entire growing season with high protein concentrations and high digestibility, and is usually grown in a mixture with grasses for grazing (Gibson and Cope 1985). Like other legumes, it is able to fix atmospheric nitrogen, providing enough nitrogen to itself and any companion grass, when grown in a mixture. White clover can cause **bloat**, a disorder characterized by the accumulation of gas in the rumen. Seeding pastures to grass-legume mixtures is the most effective and least costly method to avoid this condition.

Red clover is commonly grown as a hay crop and tolerates acid or poorly drained soils better than alfalfa (Pederson and Quesenberry 1998). It also is a very good choice to grow with winter cereal grains due to its tolerance to shading (Blaser et al. 2006). Red clover persistence can be reduced under continuous stocking. Rotational stocking provides greater yields and better distribution of production than continuous stocking in mixed pastures (Hay and Hunt 1989). Recent studies have shown that red clover could be used to complement established white clover and grass mixtures. The complementary growth of red and white clover can maximize yield and increase the persistence of clovers under grazing (Eriksen et al. 2012).

Warm-season legumes are not widely utilized in the transition zone due to slow establishment, limited selection of seeded varieties and lack of persistence under grazing (Hoveland 2000). However, warm-season legumes have the potential to provide fixed nitrogen during times of limited forage availability, increasing grazing sustainability during summer. Cowpea has historically been used as a cover crop for soil conservation and soil fertility improvement and is generally planted for wildlife feed (Muir 2002), but recent studies have shown its potential to improve crude protein (CP) levels for cattle in the mid- to late-summer period when the quality of perennial grasses typically declines (Foster et al. 2009).

Important Livestock Classes for the Region

Among all livestock classes, beef cattle are considered the most important for the transition zone. Table 23.1 shows the numbers of cattle in the humid transition zone from 1980 to 2017. Beef cattle are the primary class produced in the region, especially beef cow-calf operations (Figure 23.2). The number of dairy cattle in the region has declined significantly during the last 30 years, as have the number of dairy operations in the US (Figure 23.2).

Beef Cow–Calf

Most beef cattle in this region are produced by relatively small cow-calf operations. As a general description, small-scale farms have an annual gross income of less than $250 000, though most researchers define small-scale cow-calf operations as those with fewer than 100 beef cows (US Department of Agriculture 2011). Using this criterion, more than 90% of all US beef cattle operations can be classified as small-scale, and the majority of these farms in the humid transition zone have fewer than 50 beef cows (US Department of Agriculture 2011). Also, most of these small-scale farms must rely on off-farm income (Hoppe et al. 2010).

Cow nutrition plays an important role in maintaining adequate reproduction, health, and growth of both cows and calves. For increased profitability, the primary source of nutrition for these animals should be grazed or harvested forages paired with seasonal supplementation to ensure adequate productivity (Hersom 2007).

Most of these cow-calf operations choose to manage their herds for a controlled-breeding season so that calving occurs during a two- to three-month calving season. The actual dates of the two- to three-month calving vary from the northern to southern extent of the humid transition area but, fall-calving cows generally calve from September through early December and spring-calving cows will calve from January through early April. Approximately 80% of cow-calf producers, using a defined calving season, follow spring calving. Spring calving is often utilized because calving coincides with high availability of cool-season forages for lactation. Spring calving also allows producers to avoid higher costs of winter feeding since weaning and marketing of calves occur prior to winter (Griffith et al. 2015).

Table 23.1 Number of cattle in the humid transition zone and in the US

	All cattle	Beef cows	Milk cows
Year		Million	
1980	21.07	7.12	1.00
1990	17.44	6.56	0.77
2000	17.27	6.67	0.52
2017	12.20	8.96	0.34
US, 2017	94.40	31.72	9.40
% US[a]	13	28	3.6

Source: National Agricultural Statistics Service (2018).
[a]Inventory of region expressed as a percentage of US total in 2017.

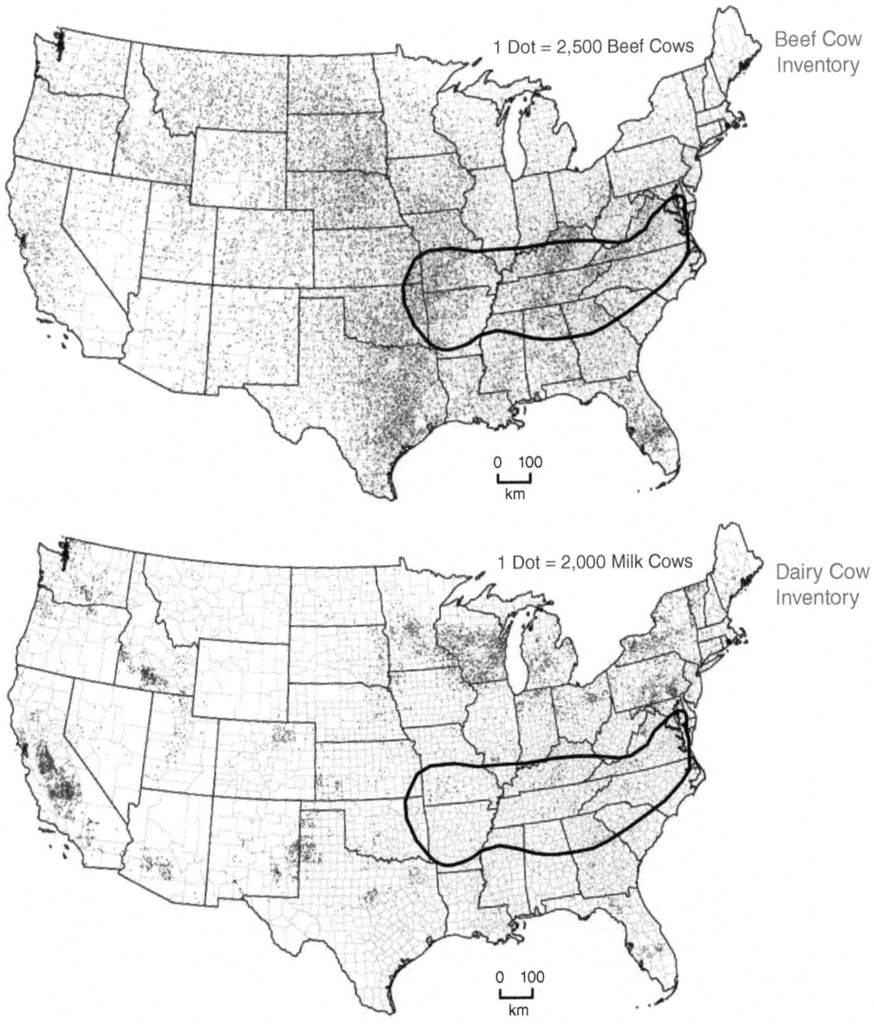

Fig. 23.2. Concentration of beef and dairy cows in the continental United States and within the humid transition area.

Though spring calving is the most utilized system in the region, many producers choose to calve in the fall. Under fall calving, weaning occurs during the warm and drier months of the year, which is also when calf prices are usually at their seasonal highest. Nutritional requirements for cows are higher during winter with fall calving. Tall fescue, as the predominant forage in the region, is most productive during spring and fall, with decreased productivity under high temperatures and limited rainfall during the summer. That can be problematic for spring calving operations since forage availability can be reduced. One solution is to designate pastures of warm-season grass to meet the needs of these cows and calves.

In the humid transition zone, weaning and marketing with spring calving occurs in September, as compared to April for fall-born calves. Thus, fall born calves are marketed earlier in the season, when prices per pound are generally higher. However, the cost of switching calving seasons might exceed the increased revenue due to marketing season (Griffith et al. 2015).

The fluctuation of cool- and warm-season grass productivity can make it difficult to determine the optimum farm carrying capacity. If stocking rates are based on forage production during summer, the higher spring growth can be undergrazed, resulting in mature and low-quality pastures. However, if stocking rates are based on forage production during spring, plant stands and future productivity, may be reduced due to overgrazing. A common solution to this problem is to buy or sell animals throughout the year, depending on forage productivity (Bates 1995).

Management strategies such as **stockpiling**, which is the accumulation of forage for grazing at a later time (Fribourg and Bell 1984), can also be beneficial to extending the grazing season, reducing the amount of hay fed to cattle, and improving profitability of cow-calf operations in this region (Nave et al. 2016). Stockpiled tall fescue is lower in cost compared with other feed sources and can be used to maintain livestock for less than the cost of hay (Bishop-Hurley and Kallenbach 2001). The success of stockpiling depends on the accumulation period, choice of species, and nutrient management. Nitrogen fertilization prior to stockpiling forage can result in greater yield and nutritive value during the winter (Rayburn et al. 1979). These authors found that higher yields were obtained with earlier stockpiling periods, but it compromised nutritive value. However, Cuomo et al. (2005) found that stockpiling later in the fall reduced yields and that application of N increased CP concentration.

Stocker Beef Operations

Stocker cattle production represents an important enterprise in the humid transition zone. Stocker cattle are weaned calves that are generally grown from less than 275 kg to approximately 375–425 kg. Stocker operations usually manage these cattle for periods of 45–100 days, which offers producers flexibility in size and scope depending on available forage systems, market conditions, geography, and weather variation (Banta et al. 2016).

Most forage systems for stocker cattle in this region are based on toxic endophyte-infected tall fescue interseeded with white or red clover. The toxic endophyte-infected tall fescue contains ergot alkaloids, which reduces stocker cattle body weight gains, especially during summer (Roberts and Andrae 2004). In an effort to minimize these risks, a study looking at integrating bermudagrass into tall-fescue based pastures for stocker cattle concluded that the lower digestibility of the bermudagrass component limited steer body weight gain as much as did grazing the endophyte-infected tall fescue/clover pastures (Kallenbach et al. 2012). These authors concluded that the benefits of moving animals to nontoxic pastures to reduce tall fescue toxicosis are limited unless the forage quality of the alternative pasture is similar to or higher than the endophyte-infected tall fescue.

To alleviate some of the negative effects of tall fescue toxicosis, forage breeders developed endophyte-free, non-toxic cultivars. However, plant persistence is lower, and these cultivars require a higher level of management to maintain productivity (Parish et al. 2003). Beginning in the late 1990s, researchers infected endophyte-free seedlings of tall fescue with non-ergot alkaloid-producing strains of *E.pichloë coenophiala* and found that these novel endophyte-infected tall fescue cultivars did not exhibit tall fescue toxicosis while retaining the persistence and stress tolerance traits (Bouton et al. 2002). Additionally, studies have shown that stocker cattle had similar performance on novel endophyte tall fescue cultivars as those grazing endophyte-free tall fescue (Parish et al. 2003).

Dairy

In the humid transition zone, dairy production has been continuously declining. From 1995 until 2018, total milk production in the nine southeastern- most US states in the humid transition area decreased by an average of 46% (Bernard 2019). Dairy farms milking 500 or fewer cows declined sharply in these states during this period, while only one these nine states (Georgia) had an increase in milk production. This reflected a shift of dairy production further south into the warm humid areas and a substantial increase in milk production per cow (Bernard 2019).

The warm climate in the humid transition area is favorable to pasture-based dairies and this, combined with the local milk deficit, has led many to consider the viability of expanding dairy systems (Hill et al. 2008). However, heat stress negatively affects profitability of the US dairy industry, with a negative effect on milk yield (Tao et al. 2012). The economic loss caused by heat stress across the US is estimated to be $2 billion (Key et al. 2014; Ferreira et al. 2016). This effect is most acute in Southeast, where heat, humidity, and prolonged summer conditions result in heat stress for six to eight months of the year (Tao et al. 2012).

Integrated Crop-Livestock Systems

Grazing lands can be very diverse ecosystems, and one of the main benefits from increased plant diversity in these systems is higher overall productivity (Soder et al. 2006) and increased soil quality and health. Most of the time, higher yields can be observed with higher plant diversity due to the ability of mixed systems to use resources more efficiently than a monoculture cropland (Hector et al. 2005).

Grazing animals play a key role in modifying plant diversity in grazing lands. Livestock can be very selective when grazing, preferring either plant parts or plant species (Soder et al. 2006). Also, changes in plant morphology may occur due to over-grazing, such as decreased leaf/stem ratio; or due to under-grazing where the canopy starts producing reproductive stems decreasing the overall nutritive value. Therefore, it is extremely important to understand the interrelationship between plant and animal systems in a mixed and diverse grazing land system.

Winter cover crops have been identified as important components of diversified crop and forage rotations

(Snapp et al. 2005), that can reduce soil erosion and nutrient losses through runoff from these systems. Also, these cover crops can make forage available during the cooler months, providing an additional economic benefit while increasing diversity in cropping systems.

Extending the grazing season beyond August and September requires adequate forage resources. Integrating grazing into cropping systems requires crops that complement these beef production systems (Senturklu et al. 2018). Within an integrated crop-livestock system, crop residues can be grazed by livestock, increasing sustainability and efficiency of these operations. The economic benefits generated from cover crops can be enhanced through increased animal production and reduced supplemental feeding costs, while enhancing soil quality and increasing long-term environmental benefits, therefore adding both short- and long-term economic value within their operations (Fae et al. 2009).

Challenges and Opportunities

Forage producers in the humid transition zone face many economic, social, and environmental factors that challenge the long-term sustainability of their operations. Producers on these generally small farms derive a relatively low level of profitability from the forage-based enterprises that limits their ability to expand, adopt more efficient practices, or implement strategies to reduce environmental impacts. Additionally, many of these farms are complicated by family ownership arrangements or short-term lease agreements. Thus, it is often impractical for producers to renovate or make significant changes to the farm if a substantial upfront investment is required and/or returns occur over the long-term. Investment in the land and management system is further complicated by expansion of urban areas. Profitability of the farms on the rural-urban interface is challenged by increased land values and property taxes, as well as regulations and land use restrictions.

Yet, the growing population base in the humid transition zone has also created new market options for producers in this region. Consumers willing to pay a premium for locally-grown or specific production practices are often within a short drive of the producers in this region. Several producers have diversified their farm operations to include higher-value products, such as meat, dairy, and clothing products that are directly marketed to consumers. This has often resulted in greater profitability and a cashflow that allows renovations and the adoption of more efficient technologies.

Use of Poultry Litter and Other Animal Manures

In addition to the forage-based livestock industries, the humid transition zone is the United States' largest poultry-producing region, and also has a sizeable swine industry. In 2016, this area produced over 18 billion eggs, 5 billion broilers, 75 million turkeys, and 10 million hogs (US Department of Agriculture 2017). Much of the feed for these animals is imported from off-farm grain production systems. Consequently, the large scale of these animal production systems results in a sizable import of nutrients. In the humid transition zone, these nutrients are most commonly applied to pastures and hayfields in the form of manure.

These animal manures are rich in N, P_2O_5, K_2O, Ca, Mg, S, and other nutrients essential for crop growth and are excellent sources of fertility for the soils in this region. However, the composition of these animal manures is highly variable. Figure 23.3 provides a summary of the N, P_2O_5, K_2O content of various types of litter from chicken production systems in Georgia. Even among manures from similar animal classes, the nutrient content can vary by more than 30% from published values (Ritz et al. 2014). Factors affecting nutrient content in manure include clean-out frequency, storage and handling practices, feed composition, type of bedding material, use of amendments and **ammonia** volatilization control measures, and other factors. Each lot of animal manure differs enough that every lot should be tested for nutrient content.

When animal manures are applied to provide comparable rates of plant available N, P_2O_5, and K_2O, forage crop yields and quality generally equal those where only commercial fertilizer is used (Franzluebbers et al. 2004). However, animal manures contain a ratio of N : P_2O_5 K_2O that typically ranges from around 3 : 3 : 3 to 2 : 3 : 2 (Council for Agricultural Science and Technology 2006), while pastures and hayfields normally remove these nutrients from the soil at a ratio from 3 : 1 : 3 to 3 : 1 : 4 (International Plant Nutrition Institute 2013). Moreover, only 50–60% of the N applied as animal manure will be plant available in the season of application, since 10–25% of the N is commonly lost to volatilization and 25–30% of the N is held by organisms or in insoluble forms in the soil (Cabrera et al. 1993; Havlin et al. 1999). Consequently, producers often apply rates of manure required to meet short-term plant-available N needs, resulting in over-application of P_2O_5. In addition, producers may apply higher rates of manures to pastures and hayfields near the poultry or hog houses to minimize transportation costs. Consequently, high levels of soluble phosphorus (P) can be present near the soil surface of many pastures and hayfields, resulting in a rapid increase in soil-test P (Figure 23.4). Though these levels pose little or no risk to the forage plants or to livestock consuming the forage, the high soluble-P levels in and on the soil can pose environmental risks. Intense rainfall on these sites can result in P loading in runoff water, which can ultimately impair freshwater quality. Soluble-P in runoff water can remain elevated for over 18 months after manure application (Pierson et al. 2001).

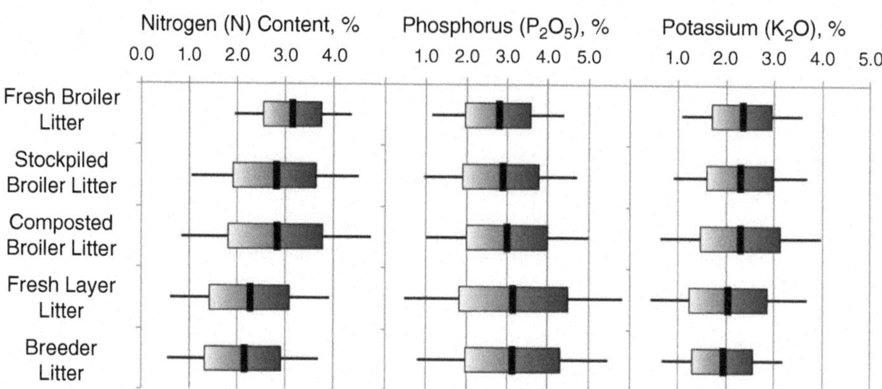

Fig. 23.3. The N, P_2O_5, and K_2O content in samples of different types of poultry litter submitted to the University of Georgia Agricultural and Environmental Services Laboratory over the course of 24 months (Ritz et al. 2014). Broilers are chickens raised for meat, while layers are hens producing eggs, and breeders are hens producing chicks for either broiler or layer operations. The average (black vertical lines), typical expected range (shaded bars), and the extent of what is considered low or high for a species (extent of horizontal black lines) for each nutrient in the five types of poultry litter.

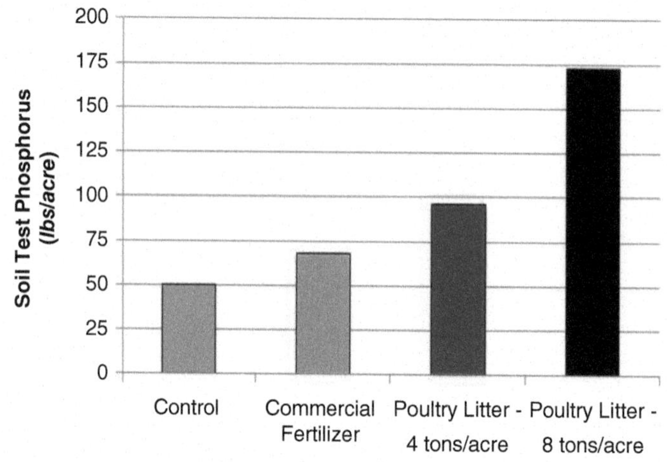

Fig. 23.4. Soil test phosphorus under bermudagrass test plots in northwest Georgia after four years of poultry litter application (Gaskin and Risse 2001).

State and federal agencies have instituted manure application regulations and mandated nutrient management plans to lower the risk of contamination of surface and groundwater by nutrients and other pollutants. If P levels and other risk factors are high, producers are required to reduce manure applications to a level equivalent to the P_2O_5 recommended by the soil test.

Portions of the other products fed or applied in the animal operation will also end up in the manure. Aluminum sulfate, or alum, is often added to bedding in poultry houses at a rate of 1–2 tons/20 000 birds in the flock to reduce **ammonia** levels in the facility (Moore et al. 2000). Treating litter with alum reduces ammonia volatilization by up to 70% (Moore et al. 2000), but additional benefits of alum application include greater forage yields (Moore and Edwards 2005) and a 75% reduction in P runoff compared to untreated litter (Moore and Edwards 2007). Poultry producers have also historically fed compounds such as roxarsone (3-nitro-4-hydroxyphenylarsonic acid) and nitarsone (4-nitrophenylarsonic acid) to prevent disease and improve weight gains (Jones 2007). Consequently, arsenic (As) has been commonly found in poultry litter at elevated levels. Long-term applications of poultry litter from facilities where these arsenic-containing compounds were fed have resulted in higher concentrations of As in the soil relative to where poultry litter was not applied (3–4 mg kg^{-1} vs. 1.5 mg kg^{-1} As). However, these levels are below levels of environmental concern (Ashjaei et al. 2011) and are well below the Environmental Protection Agency's (EPA) loading limits established for other land applications, such as municipal wastes (20 mg kg^{-1} As; US Environmental Protection Agency 1993). The companies that sold roxarsone (Alapharma, a subsidiary of Pfizer, Inc.) and nitarsone (Zoetis, Inc.) voluntarily suspended sales in 2011 and 2015, respectively (US Food and Drug Administration 2015).

Poultry litter and other manures also contain low concentrations of ionophore antibiotics, such as monensin (which is also fed to cattle under the trade name Rumensin®). Monensin sorbs to the soil, but long-term application of monensin has shown the amount of this sorption to be finite (Doydora et al. 2017). To date, there are no data that demonstrate an environmental risk due to acute concentrations of monensin, but examinations of chronic exposure to sub-acute levels of monensin are not available (Hansen et al. 2009).

Significant concentrations of natural hormones, such as testosterone and estrogen-like compounds, occur in animal manures, and their fate in the environment is a concern. These hormones can end up in surface and groundwater systems at concentrations that can have negative effects on aquatic species, as well as the wildlife and humans who consume the water (Ying et al. 2002). The concentration of 17ß-estradiol and testosterone is approximately 55 and 30 μg kg^{-1}, respectively, in broiler litter and 70 and 25 μg kg^{-1}, respectively, in layer litter (Lorenzen et al. 2004). Measurements of runoff from pastures in the Southern Piedmont that were treated with poultry litter have been observed to have concentrations of estradiol at 120–820 μg kg^{-1} and concentrations of testosterone at 50–920 μg kg^{-1} (Finlay-Moore et al. 2000). These concentrations are diluted substantially in most streams and rivers, but some research (Routledge et al. 1998) suggests that concentrations of these hormones in freshwater ecosystems or drinking water should be kept under 10 μg kg^{-1} to minimize risks.

Weed Pressure Challenges

An increase in weed pressure following the application of poultry litter or other animal manures is a common observation by producers in the humid transition zone. Numerous studies have conclusively established that poultry litter and other animal manures generally contain no significant quantities of viable weed seed (e.g. Harmon and Keim 1934; Mitchell et al. 1993; Rasnake 1995). Nonetheless, Franzluebbers et al. (2004) confirmed on-farm observations that poultry litter can increase broadleaf weed pressure when applied to hayfields and, to a lesser degree, overgrazed pastures compared to the use of commercial fertilizer (Figure 23.5). However, when poultry litter is applied to rotationally stocked pastures, stand thickness and weed pressure were similar to where commercial fertilizer had been used. It is thought that larger particle sizes, in some types of poultry litter, may shade forage plants underneath the particle and provide an ideal mix of organic matter, nutrients, and moisture for weed seeds to germinate. A relatively low percentage of time of canopy closure over the soil surface in hayfields and overgrazed pastures relative to rotationally stocked pastures is also likely a potential contributing factor.

Legume Persistence

Historically, legumes have often not been used as extensively in pastures and hayfields in the humid transition zone as in other areas (Hoveland 1989). Legume productivity in this area has been unstable and uneconomical compared to N-fertilized grass monocultures (Burns and Standaert 1985), particularly when abundant N could be sourced as poultry litter or other animal manures. Much of the inconsistent performance of annual and perennial legumes in the humid transition zone can be attributed to the poor distribution of rainfall, limited water holding capacity of the soils in the region, poor soil fertility and water infiltration, and the legacy of severe soil erosion in past management history (Hoveland 1989). Persistence of legume stands under these conditions is further reduced by **selective grazing** by grazing livestock, a lack of herbicide options to selectively kill broadleaf weeds in the presence of legume species, and disease and insect pests that reduce carbohydrate reserve development (Hoveland 1989; Beuselinck et al. 1994).

Plant-breeding efforts within the humid transition zone have greatly increased the persistence of perennial legumes in this area. Cultivars of white clover (e.g. Bouton et al. 2005; Bouton et al. 2017), red clover (e.g. Taylor 2008a), alfalfa (e.g. Bouton et al. 1991), and sericea lespedeza (e.g. Mosjidis 2001) selected under the challenges of the humid transition zone resulted in many new, hardier cultivars being available to producers in the area (Taylor 2008b).

Summary

The humid transition area's climate allows for a longer growing season and the use of diverse forage species, which offer a competitive advantage for forage-based livestock production systems. Though it is home to many forage-based livestock farms, most of these are small farms with limited profitability and cashflow. Additionally, this region is challenged by the dominance of toxic endophyte-infected tall fescue pastures. Though there is

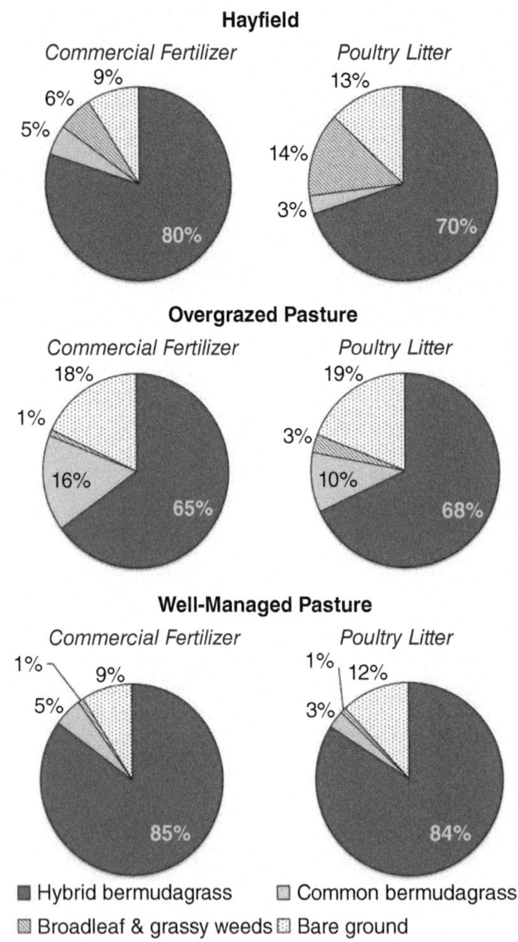

Fig. 23.5. Forage composition in hayfields and pastures in northeast Georgia at the end of three years of fertilizing with commercial fertilizer or poultry litter. *Source:* Adapted from Franzluebbers et al. (2004).

great opportunity to renovate these pastures and replant with novel-endophyte tall fescue or other non-toxic species, the tight economic margins of these livestock enterprises and challenging landholding arrangements often limit these changes.

Still, there are many opportunities to improve the economic efficiency of cow-calf production systems through economies of scale and expand stocker cattle production systems in this region. Integrating cropping and livestock production systems may prove a successful strategy to facilitate this transition in some areas. Additionally, there is also significant opportunity for expansion of low-cost, pasture-based dairy production in this region. The abundance of nutrients from the high concentration of large-scale poultry operations in this region is supplying and could further supply nutrient needs for expanding these forage-based livestock systems at a very low cost. However, steps to limit the environmental impact of the use of these nutrients will need to be taken.

This area also has an advantage in being relatively close to a large portion of the US population. With opportunities to directly market products to consumers, producers in this area have the potential to add value and profitability to their production systems. Balancing the expansion of these direct-to-consumer opportunities and the steps to protect the environment will likely influence the changes in these areas over the next several years.

References

Anderson, B.E. (2000). Use of warm-season grasses by grazing livestock. In: *Native Warm-Season Grasses: Research Trends and Issues*, CSSA Spec. Publ, vol. 30 (eds. K.J. Moore and B.E. Anderson), 147–157. Madison, WI: CSSA and ASA https://doi.org/10.2135/cssaspecpub30.c10.

Ashjaei, S., Miller, W.P., Cabrera, M.L., and Hassan, S.M. (2011). Arsenic in soils and forages from poultry litter-amended pastures. *Int. J. Environ. Res. Public Health* 8: 1534–1546. https://doi.org/10.3390/ijerph8051534.

Ball, D.M., Pedersen, J.F., and Lacefield, G.D. (1993). The tall-fescue endophyte. *Am. Sci.* 81: 370–379.

Banta, J.P., Hersom, M.J., Lehmkuhler, J.W. et al. (2016). An overview of cow-calf production in the Southeast: forage systems, cow numbers, and calf marketing strategies. *J. Anim. Sci.* 94 (1): 60. https://doi.org/10.2527/ssasas2015-124.

Bates, G.E. (1995). An overview of controlled grazing. http://utbfc.utk.edu/Forage-Grazing%20Management-Controlled%20Grazing.html (accessed 10 October 2019).

Bernard, J.K. (2019). Dairy extension programs in the southern region: finding novel ways to meet the needs of our producers. *Appl. Anim. Sci.* 35: 1–7. https://doi.org/10.15232/aas.2018-01781.

Beuselinck, P.R., Bouton, J.H., Lamp, W.O. et al. (1994). Improving legume persistence in forage crop systems. *J. Prod. Agric.* 7: 287–322. https://doi.org/10.2134/jpa1994.0311.

Bishop-Hurley, G.J. and Kallenbach, R.L. (2001). The economics of grazing beef cows during winter. In: *Proceedings of American Forage Grassland Council* (ed. T. Terrill), 274. Springdale, AR.

Blaser, B.C., Gibson, L.R., Singer, J.W., and Jannink, J.L. (2006). Optimizing seeding rates for winter cereal grains and frost-seeded red clover intercrops. *Agron. J.* 98: 1041–1049. https://doi.org/10.2134/agronj2005.0340.

Bosworth, S.C., Hoveland, C.S., Buchanan, G.A., and Anthony, W.B. (1980). Forage quality of selected warm-season weed species. *Agron. J.* 72: 1050–1054. https://doi.org/10.2134/agronj1980.00021962007200060044x.

Bouton, J.H., Smith, S.R., Wood, D.T. et al. (1991). Registration of 'Alfagraze' alfalfa. *Crop Sci.* 31: 479–479. https://doi.org/10.2135/cropsci1991.0011183X003100020052x.

Bouton, J.H., Latch, G.C.M., Hill, N.S. et al. (2002). Reinfection of tall fescue cultivars with non-ergot alkaloid–producing endophytes. *Agron. J.* 94: 567–574. https://doi.org/10.2134/agronj2002.0567.

Bouton, J.H., Woodfield, D.R., Caradus, J.R., and Wood, D.T. (2005). Registration of 'Durana' white clover. *Crop Sci.* 45: 797–797. https://doi.org/10.2135/cropsci2005.0797.

Bouton, J.H., Motes, B., Wood, D.T. et al. (2017). Registration of 'renovation' white clover. *J. Plant Regist.* 11: 218–221. https://doi.org/10.3198/jpr2016.11.0063crc.

Brummer, E.C. (2006). Grazing-tolerant alfalfa cultivars have superior persistence under continuous and rotational stocking. *Forage Grazinglands* https://doi.org/10.1094/FG-2006-0825-01-RS.

Burns, J.C. and Standaert, J.E. (1985). Productivity and economics of legume-based vs. nitrogen-fertilized grass-based pastures in the United States. In: *Proceedings of Trilateral Workshop*, Palmerston North, New Zealand. 30 Apr.–4 May 1984 (ed. R.F Barnes), 56–71. Washington, DC: USDA-ARS.

Cabrera, M.L., Chiang, S.C., Merka, W.C. et al. (1993). Nitrogen transformations in surface-applied poultry litter: effect of litter physical characteristics. *Soil Sci. Soc. Am. J.* 57: 1519–1525. https://doi.org/10.2136/sssaj1993.03615995005700060021x.

Council for Agricultural Science and Technology (2006). *Biotechnological Approaches to Manure Nutrient Management*. Issue Paper No. 33. Ames, IA: Council for Agricultural Science and Technology.

Cuomo, G.J., Rudstrom, M.V., Peterson, P.R. et al. (2005). Initiation date and nitrogen rate for stockpiling smooth bromegrass in the North-Central USA. *Agron. J.* 97: 1194–1201. https://doi.org/10.5513/JCEA01/19.3.2181.

Doydora, S.A., Sun, P., Cabrera, M. et al. (2017). Long-term broiler litter amendments can alter the soil's capacity to sorb monensin. *Environ. Sci. Pollut. Res.* 24: 13466–13473. https://doi.org/10.1007/s11356-017-8727-9.

Eriksen, J., Askegaard, M., and Soegaard, K. (2012). Complementary effects of red clover inclusion in ryegrass-white clover swards for grazing and cutting. *Grass Forage Sci.* 69: 241–250. https://doi.org/10.1111/gfs.12025.

Evers, G.W., Smith, G.R., and Hoveland, C.S. (1997). Ecology and production of annual ryegrass. In: *Ecology, Production, and Management of Lolium for Forage in the USA*, CSSA Spec. Publ. 24 (eds. F.M. Rouquette and L.R. Nelson), 29–43. Madison, WI: CSSA https://doi.org/10.2135/cssaspecpub24.c3.

Fae, G.S., Sulc, R.M., Barker, D.J. et al. (2009). Integrating winter annual forages into a no-till corn silage system. *Agron. J.* 101: 1286–1296. https://doi.org/10.2134/agronj2009.0144.

Ferreira, F.C., Gennan, R.S., Dahl, G.E., and De Vries, A. (2016). Economic feasibility of cooling dry cows across the United States. *J. Dairy Sci.* 99: 9931–9941. https://doi.org/10.3168/jds.2016-11566.

Finlay-Moore, O., Hartel, P.G., and Cabrera, M.L. (2000). 17-β estradiol and testosterone in soil and runoff from grasslands amended with broiler litter. *J. Environ. Qual.* 29: 1604–1611. https://doi.org/10.2134/jeq2000.00472425002900050030x.

Foster, J.L., Adesogan, A.T., Carter, J.N. et al. (2009). Annual legumes for forage systems in the United States Gulf Coast region. *Agron. J.* 101: 415–421. https://doi.org/10.2134/agronj2008.0083x.

Franzluebbers, A.J., Wilkinson, S.R., and Stuedemann, J.A. (2004). Bermudagrass management in the Southern Piedmont USA: X. Coastal productivity and persistence in response to fertilization and defoliation regimes. *Agron. J.* 96: 1400–1411.

Fribourg, H.A. and Bell, K.W. (1984). Yield and composition of tall fescue stockpiled for different periods. *Agron. J.* 76: 929–934. https://doi.org/10.2134/agronj1984.00021962007600060016x.

Gaskin, J.W. and Risse, L.M. (2001). Evaluating agronomic value and environmental safety of poultry litter on a bermudagrass hayfield. In: *Proceedings of the International Symposium: Addressing Animal Production and Environmental Issues* (ed. G.B. Havenstein). October 3–5, 2001. Research Triangle Park, NC.

Gibson, P.B. and Cope, W.A. (1985). White clover. In: *Clover Science and Technology*, Agronomy Monograph 25 (ed. N.L. Taylor), 471–490. Madison, WI: ASA, CSSA, SSSA https://doi.org/10.2134/agronmonogr25.c20.

Girma, K., Reinert, M., Ali, M.S. et al. (2012). Nitrogen and phosphorus requirements of teff grown under dryland production system. *Crop Manage.* 11 https://doi.org/10.1094/CM-2012-0319-02-RS.

Griffith, A.P., Boyer, C.N., Henry, G.W. et al. (2015). *Fall Versus Spring Calving: Considerations and Profitability Comparison*. University of Tennessee Extension W 419.

Hanna, W.W., Baltensperger, D.D., and Seetharam, A. (2004). Pearl millet and other millets. In: *Warm-Season (C4) Grasses*, Agronomy Monograph 45 (eds. L.E. Moser, B.L. Burson and L.E. Sollenberger), 537–560.

Madison, WI: ASA, CSSA, SSSA https://doi.org/10.2134/agronmonogr45.c15.

Hansen, M., Krogh, K.A., Bjorklund, E., and Brandt, A. (2009). Environmental risk assessment of ionophores. *Trends Anal. Chem.* 28: 534–542.

Harlan, J.R., de Wet, J.M.J., and Rawal, K.M. (1970). Geographic distribution of the species of *Cynodon* L. C. Rich (*Gramineae*). *East Afr. Agric. For. J.* 36: 220–226. https://doi.org/10.1080/00128325.1970.11662465.

Harmon, G.W. and Keim, F.D. (1934). The percentage and viability of weed seeds recovered in the feces of farm animals and their longevity when buried in manure. *J. Am. Soc. Agron.* 26: 762–767.

Hartley, W. (1961). Studies on the origin, evolution, and distribution of the Gramineae. IV. The genus *Poa* L. *Aust. J. Bot.* 9: 152–161. https://doi.org/10.1071/BT9730201.

Havlin, J.L., Beaton, J.D., Tisdale, S.L., and Nelson, W.L. (1999). *Soil Fertility and Fertilizers: An Introduction to Nutrient Management*, 6e. Upper Saddle River, NJ: Prentice Hall, Inc.

Hay, R.J.M. and Hunt, W.F. (1989). Competition from associated species on white and red clover in grazed swards. In: *Persistence of Forage Legumes* (eds. G.C. Marten, A.G. Matches, R.F Barnes, et al.), 311–326. Madison, WI: ASA, CSSA, SSSA https://doi.org/10.2134/1989.persistenceofforagelegumes.c22.

Hector, A., Schmid, B., Beierkuhnlein, C. et al. (2005). Plant diversity and productivity experiments in European grasslands. *Science* 286: 1123–1127. https://doi.org/10.1126/science.286.5442.1123.

Hersom, M. (2007). Basic nutrient requirements of beef cows. http://edis.ifas.ufl.edu/an190 (accessed 10 October 2019).

Hill, N., Hancock, D., Cabrera, M. and Blount, A. (2008). Improved efficiency of pasture-based dairies using complementary pasture species and irrigation scheduling. Online SARE Reporting System: LS07-196. https://projects.sare.org/sare_project/ls07-196/ (accessed 22 November 2019).

Hoppe, R.A., MacDonald, J.M., and Korb, P. (2010). *Small Farms in the United States: Persistence Under Pressure*. Darby, PA: Diane Pub. Co.

Hoveland, C.S. (1989). Legume persistence under grazing in stressful environments of the United States. In: *Persistence of Forage Legumes* (eds. G.C. Marten et al.), 375–385. Madison, WI: ASA, CSSA, and SSSA https://doi.org/10.2134/1989.persistenceofforagelegumes.c22.

Hoveland, C.S. (2000). Achievements in management and utilization of southern grasslands. *J. Range Manage.* 53: 17–22. https://doi.org/10.2307/4003387.

International Plant Nutrition Institute (2013). IPNI Crop nutrient removal calculator. http://www.ipni.net/article/IPNI-3296 (22 November 2019).

Jones, F.T. (2007). A broad view of arsenic. *Poult. Sci.* 86: 2–14. https://doi.org/10.1093/ps/86.1.2.

Jung, G.A., Van Wijk, A.J.P., Hunt, W.F., and Watson, C.E. (1996). Ryegrasses. In: *Cool-Season Forage Grasses*, Agronomy Monograph 34 (eds. L.E. Moser, D.R. Buxton and M.D. Casler), 605–641. Madison, WI: ASA, CSSA, SSSA https://doi.org/10.2134/agronmonogr34.c19.

Kallenbach, R.L., Crawford, R.J. Jr., Massie, M.D. et al. (2012). Integrating bermudagrass into tall fescue-based pasture systems for stocker cattle. *J. Anim. Sci.* 90: 387–394. https://doi.org/10.2527/jas.2011-4070.

Key, N., Sneeringer, S. and Marquardt, D. (2014). Climate change, heat stress, and U.S. dairy production. United States Department of Agriculture, Economic Research Report No. 175.

Lacefield, G.D., Ball, D.M., Hancock, D. et al. (2009). *Growing Alfalfa in the South*. National Alfalfa and Forage Alliance.

Lorenzen, A., Hendel, J.G., Conn, K.L. et al. (2004). Survey of hormone activities in municipal biosolids and animal manures. *Environ. Toxicol.* 19: 216–225. https://doi.org/10.1002/tox.20014.

Mitchell, C.C., Walker, R.H., and Shaw, P.P. (1993). Are there weeds in broiler litter? *Highlights Agric. Res.* 40 (4): 4. Alabama Agricultural Experiment Station, Auburn University, Alabama.

Moore, P.A. Jr., Daniel, T.C., and Edwards, D.R. (2000). Reducing phosphorus runoff and inhibiting ammonia loss from poultry manure with aluminum sulfate. *J. Environ. Qual.* 29: 37–49. https://doi.org/10.2134/jeq2000.00472425002900010006x.

Moore, P.A. Jr. and Edwards, D.R. (2005). Long-term effects of poultry litter, alum-treated litter, and ammonium nitrate on aluminum availability in soils. *J. Environ. Qual.* 34: 2104–2111. https://doi.org/10.2134/jeq2004.0472.

Moore, P.A. Jr. and Edwards, D.R. (2007). Long-term effects of poultry litter, alum-treated litter, and ammonium nitrate on phosphorus availability in soils. *J. Environ. Qual.* 36: 163–174. https://doi.org/10.2134/jeq2004.0472.

Mosjidis, J.A. (2001). Registration of 'AU Grazer' sericea lespedeza. *Crop Sci.* 41: 262–262. https://doi.org/10.2135/cropsci2001.411262x.

Muir, J.P. (2002). Hand-plucked forage yield and quality and seed production from annual and short-lived perennial warm-season legumes fertilized with composted manure. *Crop Sci.* 42: 897–904. https://doi.org/10.2135/cropsci2002.8970.

National Agricultural Statistics Service (2018). Livestock national and county data. https://www.nass.usda.gov/ (22 November 2019).

Nave, R.L.G., Barbero, R.P., Boyer, C.N. et al. (2016). Nitrogen rate and initiation date effects on stockpiled tall fescue during fall grazing in Tennessee. *Crop Forage Turfgrass Manage.* 2: 1–8. https://doi.org/10.2134/cftm2015.0174.

Parish, J.A., McCann, M.A., Watson, R.H., and Paiva, N.N. (2003). Use of nonergot alkaloid-producing endophytes for alleviating tall fescue toxicosis in stocker cattle. *J.Anim. Sci.* 81: 2856–2868. https://doi.org/10.2527/2003.81112856x.

Pedersen, J.F. and Toy, J.J. (1997). Forage yield, quality, and fertility of sorghum × sudan grass hybrids in A1 and A3 cytoplasm. *Crop Sci.* 37: 1973–1975. https://doi.org/10.2135/cropsci1997.0011183X003700060049x.

Pederson, G.A. and Quesenberry, K.H. (1998). Clovers and other forage legumes. In: *Plant and Nematode Interactions*, Agronomy Monograph 36 (eds. K.R. Barker, G.A. Pederson and G.L. Windham), 399–425. Madison, WI: ASA, CSSA, SSSA https://doi.org/10.2134/agronmonogr36.c19.

Pierson, S.T., Cabrera, M.L., Evanylo, G.K. et al. (2001). Phosphorus and ammonium concentrations in surface runoff from grassland fertilized with broiler litter. *J. Environ. Qual.* 30: 1784–1789. https://doi.org/10.2134/jeq2001.3051784x.

Rasnake, M. (1995). *Weed Seed in Poultry Litter: Should Farmers Be Concerned? Soil Science News and Views.* Lexington, KY: University of Kentucky Plant and Soil Sciences Department. Accessed from http://uknowledge.uky.edu/pss_views/100.

Rayburn, E.B., Blaser, R.E., and Wolf, D.D. (1979). Winter tall fescue yield and quality with different accumulation periods and N rates. *Agron. J.* 71: 959–963. https://doi.org/10.2134/agronj1979.00021962007100060017x.

Ritz, C.W., Vendrell, P.F. and Tasistro, A. (2014). Poultry litter sampling. University of Georgia Extension Bulletin 1270. http://extension.uga.edu/publications/detail.html?number=B1270 (accessed 10 October 2019).

Roberts, C. and Andrae, J. (2004). Tall fescue toxicosis and management. *Crop Manage.* https://doi.org/10.1094/CM-2004-0427-01-MG.

Routledge, E.J., Sheahan, D., Desbrow, C. et al. (1998). Identification of estrogenic chemicals in STW effluent: 2. In vivo responses in trout and roach. *Environ. Sci. Technol.* 32: 1559–1565. https://doi.org/10.1021/es970796a.

Senturklu, S., Landblom, D.G., Maddock, R. et al. (2018). Effect of yearling steer sequence grazing of perennial and annual forages in an integrated crop and livestock system on grazing performance, delayed feedlot entry, finishing performance, carcass measurements, and systems economics. *J. Anim. Sci.* 96: 2204–2218. https://doi.org/10.1093/jas/sky150.

Sleper, D.A. and West, C.P. (1996). Tall fescue. In: *Cool Season Grasses* (eds. L.E. Moser et al.), 471–502. Madison, WI: American Society of Agronomy https://doi.org/10.2134/agronmonogr34.c15.

Smith, D., Jacques, A.V.A., and Balasko, J.A. (1973). Persistence of several temperate grasses grown with alfalfa and harvested two, three, or four times annually at two stubble heights. *Crop Sci.* 13: 553–556. https://doi.org/10.2135/cropsci1973.0011183X001300050017x.

Snapp, S.S., Swinton, S.M., Labarta, R. et al. (2005). Evaluating cover crops for benefits, costs, and performance within cropping system niches. *Agron. J.* 97: 322–332. https://doi.org/10.2134/agronj2005.0322.

Soder, K.J., Rook, A.J., Sanderson, M.A., and Goslee, S.C. (2006). Interaction of plant species diversity on grazing behavior and performance of livestock grazing temperate region pastures. *Crop Sci.* 47: 416–425. https://doi.org/10.2135/cropsci2006.01.0061.

Stallknecht, G.F., Gilbertson, K.M., and Eckhoff, J.L. (1993). *Teff: Food Crop for Humans and Animals* (eds. J. Janick and J.E. Simon), 231–234. New York: Wiley.

Tao, S., Thompson, I.M., Monteiro, A.P.A. et al. (2012). Effect of cooling heat-stressed dairy cows during the dry period on insulin response. *J. Dairy Sci.* 95: 5035–5046. https://doi.org/10.3168/jds.2012-5405.

Taylor, N.L. (2008a). Registration of 'FreedomMR' red clover. *J. Plant Regist.* 2: 205–207. https://doi.org/10.3198/jpr2007.12.0688crc.

Taylor, N.L. (2008b). A century of clover breeding developments in the United States. *Crop Sci.* 48: 1–13. https://doi.org/10.2135/cropsci2007.08.0446.

U.S. Department of Agriculture (1998). Dominant soil orders. U.S. Department of Agriculture, Natural Resources Conservation Service. https://www.nrcs.usda.gov/wps/portal/nrcs/main/soils/survey/class (accessed 10 October 2019).

U.S. Department of Agriculture (2011). Small-scale U.S. cow-calf operations. Animal health and plant health inspection service, April 2011.

U.S. Department of Agriculture (2012a). USDA plant hardiness zone map. Agricultural Research Service, U.S. Department of Agriculture. http://planthardiness.ars.usda.gov (accessed 10 October 2019).

U.S. Department of Agriculture (2012b). The 2017 Census of Agriculture. National Agricultural Statistics Service, U.S. Department of Agriculture. https://www.agcensus.usda.gov (accessed 10 October 2019).

U.S. Department of Agriculture (2017). Quick stats database. National Agricultural Statistics Service, U.S. Department of Agriculture. https://www.nass.usda.gov/Quick_Stats (accessed 10 October 2019).

U.S. Environmental Protection Agency (1993). Standards for the use or disposal of sewage sludge, final rules. 40 CFR Part 503. *Fed. Regist.* 58: 9387–9404.

U.S. Food and Drug Administration (2015). Arsenic-based animal drugs and poultry. https://www.fda.gov/animalveterinary/safetyhealth/productsafetyinformation/ucm257540.htm (accessed 10 October 2019).

Yates, S.G. (1983). Tall fescue toxins. In: *Handbook of Naturally Occurring Food Toxicants* (ed. M. Recheigel), 249–273. Boca Raton, FL: CRC Press.

Ying, G.G., Kookana, R.S., and Ru, Y.J. (2002). Occurrence and fate of hormone steroid in the environment. *Environ. Int.* 28: 545–551. https://doi.org/10.1016/S0160-4120(02)00075-2.

CHAPTER

24

Forage Systems for Arid Areas

Daniel H. Putnam, Forage Extension Specialist, *Department of Plant Sciences, University of California, Davis, CA, USA*
Tim DelCurto, Professor and Nancy Cameron Chair, *Range Beef Cattle Nutrition and Management, Department of Animal and Range Sciences, Montana State University, Bozeman, MT, USA*

Introduction

Although many western arid regions appear superficially bereft of greenery, forages are an important, if not dominant, component of the agricultural landscape. At least one third of the land area of the continental United States may be categorized as arid, semiarid or Mediterranean, including Arizona, California, Colorado, Idaho, Montana, Nevada, New Mexico, Oregon, Utah, Washington and Wyoming, and portions of transition states such as Oklahoma, Texas, Kansas and Nebraska, up to the Dakotas. While irrigation is critical to this region, extensive rain-fed forage systems are also vital to the forage-livestock system.

The term "arid" belies a diverse set of agro-ecologic zones. This region encompasses extensive grasslands, rugged mountains, fertile irrigated valleys with either temperate or Mediterranean climates, searing deserts, cold, dry, high-elevation grassy plains, and productive forests. Though most of this region is generally characterized by its low rainfall, there are some coastally-influenced transitional regions in the West which can be described as rain forests.

History

Cattle, grazing systems and harvested forage crops have played a critical role in the early history of the western US. The movement of cattle, horses, and people in the high plains, rangelands and deserts of the West in search of forages or markets is the stuff of Hollywood legend (Figure 24.1). During the Mexican period in the early 1800s Southwest, cattle from extensive grazing operations were the major product of the sleepy rancheros and cattle barons of the western territories of the expanding United States. Hides were routinely shipped from western ports of San Francisco and Portland around South America to shoe manufacturers in Boston and New York. Cattle were moved from seasonal grazing areas from Texas to Montana.

During the California Gold Rush in the 1850s, alfalfa was introduced and proved instantaneously successful on the fertile irrigated fields of California and other western states (Figure 24.2). It quickly moved eastward to the rapidly-settling territories from Utah, Colorado, Kansas, and the Pacific Northwest. Within a few years, alfalfa became the predominant forage crop in the West, unlike most other crops that moved west from eastern settlements. According to the USDA-NASS (2019) agricultural census, at the turn of the twentieth century, 98% of the nation's alfalfa was grown west of the Mississippi River, and today, Western States still account for about 50% of the Nation's alfalfa production with California, Idaho, and Montana as the leading producers.

Forages: The Science of Grassland Agriculture, Volume II, Seventh Edition.
Edited by Kenneth J. Moore, Michael Collins, C. Jerry Nelson and Daren D. Redfearn.

Fig. 24.1. Grazing beef cows in Montana. Movement of cattle and other livestock in search of available forages has been a historical feature of western forage systems and remains important today. *Source:* Photo, T. DelCurto.

Fig. 24.2. Stacking alfalfa hay, Fresno County, CA, circa late 1800s. Hay was historically stacked outside in large loafs in many western regions using horsepower. Alfalfa moved quickly from its introduction in California circa 1850 to other western states, where it became an important crop in the expanding West in the nineteenth and early twentieth centuries (Post card image, early twentieth century).

Hay and Cattle as Commodities

One of the key attributes of western forage systems was the early development of animals and hay as commodities, and the movement of both forages and animals to take advantage of seasonal availability of forages and markets.

In nineteenth century America, eastern farmsteads were characterized by small diversified operations, where the forages were primarily consumed by animals on the farm. However, the size, diversity, and climate of many western states necessitated larger farmscapes, and the long-distance movement of both cattle and forages. As dairying developed around the population centers of Seattle, San Francisco and Los Angeles, Denver and Salt Lake City, hay was grown in other areas and moved to dairies and feedlots near cities via horse, boats on rivers, and later rail and trucks. Similarly, beef cattle and sheep have historically moved long distances from remote rural grazing areas to city consumers.

This dominance of extensive systems, and the commercialization of hay and beef production was an important distinguishing feature of western forage production from the earliest period and remains the case today. For example, more than 90% of the alfalfa hay currently is sold off-farm in most western states today, as contrasted with eastern rain-fed regions, where a small minority (usually less than 20%) of forages enter commerce. Additionally, cattle are frequently moved between grazing areas, from cow-calf operations to rangeland and to centralized feedlots in western forage systems.

Forage Systems in Arid Zones

Forage systems in Arid Zones are characterized primarily by the limitation of water, thus, forage systems can be divided into two major categories: (i) rainfed, mostly seasonally-available forages harvested primarily by grazing in extensive rangelands, forests, pastures, chaparral, and coastal or foothill seasonal rain-fed grasses, and (ii) irrigated forages, often grown intensively in rotations with other irrigated crops, usually for cash (Figure 24.3).

Western forage agroecosystems originate with the primary producers (forage crops) which are dependent upon a natural-resource matrix consisting of soil, air, and rainfall, circumscribed by limitations such as temperature, elevation, labor and availability of irrigation water (Figure 24.3). Natural resources are supplemented by additions of irrigation water, fertilizers, pesticides, and inputs of machinery, fencing, labor and energy (fossil fuel and electricity) to produce forage crops or range

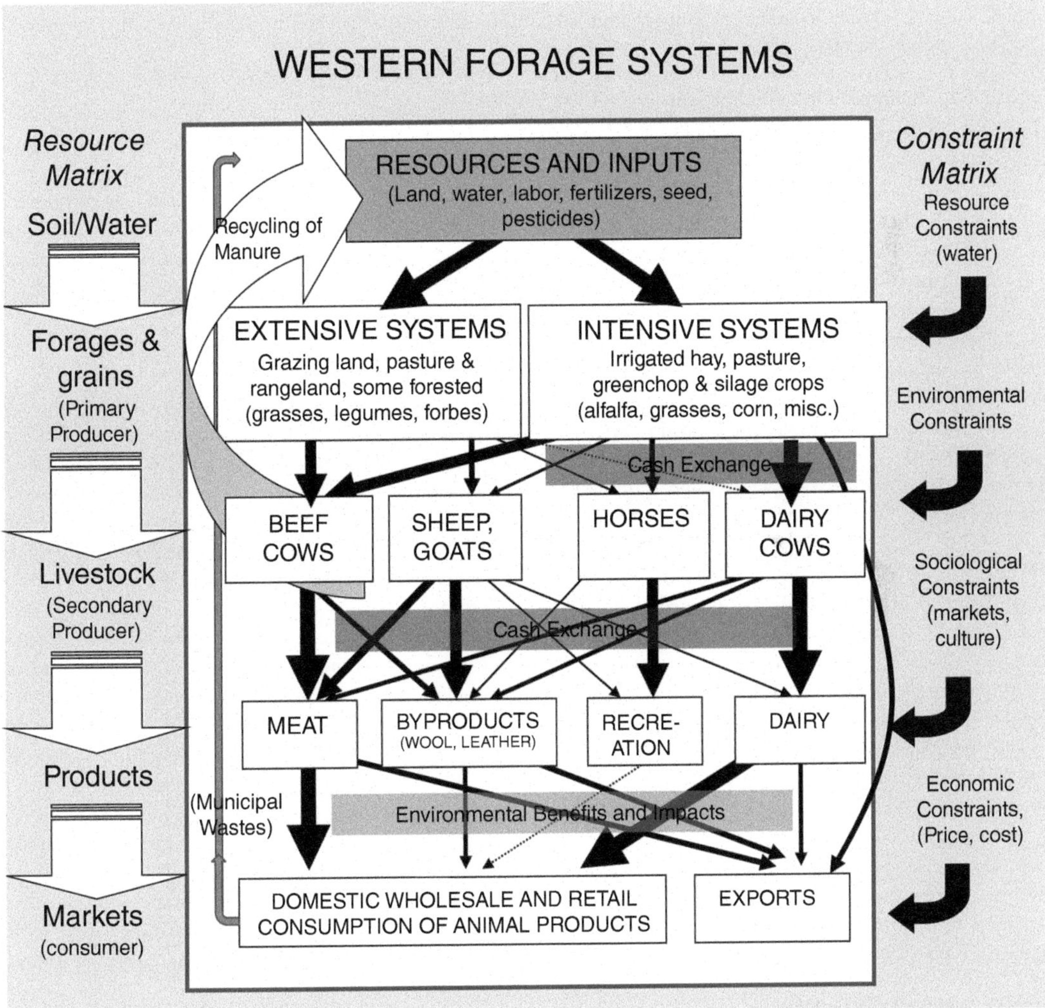

Fig. 24.3. A model of forage-livestock systems common in western US arid regions, divided into extensive rainfed grazing systems (left) and intensive harvested irrigated systems (right). Thickness of arrow indicates importance of flow. The major constraint in western systems is water both for irrigated and non-irrigated systems.

pasture and eventually the food products of milk, meat, and axillary benefits such as wool, leather, and recreation (Figure 24.3).

Irrigated and non-irrigated forages are distinct but interrelated systems. For example, range-fed beef cows are also frequently supplemented with irrigated hay and dairy replacement heifers are frequently raised on rainfed grazing land. However, there are clear physical demarcations between non-irrigated forages and irrigated forage crops. Irrigated forages often compete with other potential cash crops (corn, potatoes, rice, cotton, tomatoes, wheat, and even orchards, vineyards, and specialty crops) for arable land, and more importantly, water resources. Rainfed-forage crops are usually grown on land with limitations to crop production and must contend with periodic droughts, environmental constraints and regulations and government policy.

Forages Support Major Agricultural Sectors

These two distinct forage systems are the basis of several important economic sectors, which produce a range of products for different markets, both export and domestic (Figure 24.3). The most important of these are beef and dairy, but horses have increased in importance in recent years due to urbanization of the West. Sheep and goat products are a minor component of Western systems but, are still important in some areas; sheep are commonly moved throughout western states to take advantage of seasonal pasture or winter-grazing of grasses or alfalfa. Recently increased demand for high-quality wool has boosted interest in range sheep production systems because of the dominance of Merino and Rambouillet fine wool breeds.

Constraints

A significant challenge of both range and feedlot-style animal husbandry is completion of the nutrient cycle from manures back to plant production without harming the environment. Though these systems are theoretically sustainable when water resources are renewable and manures cycled through forages, both the beef and dairy forage systems face significant challenges due to manure management issues and water limitations for forage production.

The dependence upon irrigation water, frequent movement of cattle and hay, extensive rangeland grazing areas, intensive animal units, and the predominance of a cash hay system are distinguishing features of western forage systems compared with other regions.

Rainfed Range-Beef Systems

Seasonal rainfall produces large quantities of harvestable forages temporarily available for grazing. This is the basis for one of the major classes of forage systems in the West: rangeland. The classes of animals involved are predominately beef, sheep and horses. Some dairies depend upon grazing for some of their feed, but these are a small minority, and are primarily located on the more humid coastal regions of Washington, Oregon, and California and on irrigated pastures inland. All beef forage systems, however, depend more heavily upon grazing under both irrigated and rain-fed conditions.

Beef cattle production is extremely important to the rural regions of the western US. However, western beef production only accounts for approximately 20% of the US beef cow/calf production (DelCurto et al. 2017). In addition, feedlot production is limited and primarily located in regions with cheap grain sources (e.g. Nebraska or eastern Colorado) or near areas with cheap by-products that can reduce the cost of feedlot rations such as the Columbia Basin region of NE Oregon and SE Washington, or the Central and Imperial Valleys of California.

Importance of Public Lands

The other distinguishing feature of the western beef industry is the reliance on arid regions that are predominantly under federal jurisdiction. Many of the states in the western US have 50% or more of their land area managed by the Bureau of Land Management and the USDA Forest Service or other state/federal agencies. The utilization of these lands involves significant regulation and constraints dictated by policy.

Rangeland Forage Resources

Unlike other meat animal industries, such as swine and poultry, the beef industry in the western US is highly dependent upon rapidly changing arid environments and the resulting changes in forage supply and quality. The seasonal forage quality and supply are often not well synchronized with beef cattle nutritional requirements (Figure 24.4). Thus, the western beef industry is very extensive in its land use, with optimal management being a function of the resources on each ranching unit, and the success in matching the type of cow and/or production expectations to the available resources. Successful beef producers are not necessarily the ones who wean the heaviest calves, or who obtain 95% conception or provide the most optimal winter nutrition. Instead, the successful producers are the ones who demonstrate economic viability despite the multiple economic and public pressures on the industry.

A vast majority of the land area in the Western United States fits the general classification of "rangeland," meaning that it is not suitable for tillage due to arid conditions, shallow/rocky soils, high elevations or short growing seasons. From the arid rangelands in the Northern Great Basin (cold high deserts) to arid low-elevation rangelands in Southern New Mexico (Low Deserts), ranchers are faced with limited forage resources and challenging nutritional calendars. Arid, high-elevation rangelands are also characterized by dynamic, highly variable climates that

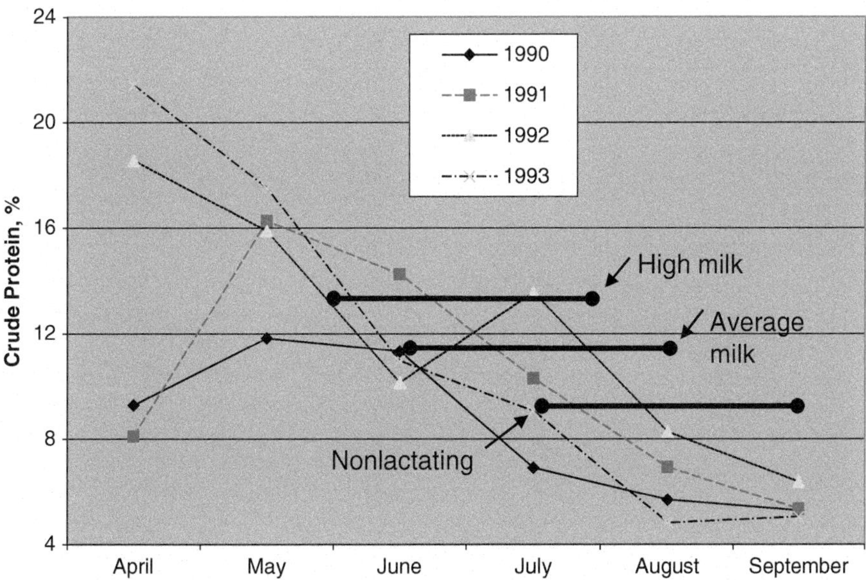

FIG. 24.4. Seasonal patterns of crude protein availability in diets selected by beef cattle grazing Northern Great Basin native rangelands, illustrating changes in forage quality over the season. Crude protein requirements (NASEM 2016) are indicated for high milking (9.1 kg d^{-1}), average milking (4.5 kg d^{-1}) and nonlactating, gestating beef cows (499 kg body weight). Forage supply and quality are sometimes in excess, but frequently insufficient for grazing animals.

change drastically from season to season and year to year. For example, the CP content of diets selected by cattle in the Northern Great Basin differs dramatically across seasons and years from 1990 to 1993 (DelCurto et al. 2000; Figure 24.4). The extremes in CP content were, in turn, related to wide ranges of crop-year precipitation averaging 158, 246, 231, and 524 mm for 1990, 1991, 1992, and 1993, respectively (40-year average = 277 mm). The extreme fluctuations of precipitation also have significant effects on forage yield, with 1990–1992 yields averaging 240 kg ha^{-1}, whereas 1993 forage production was 580 kg ha^{-1}. Thus, beef managers must adapt to wide ranges of both forage quality and quantity.

Seasonal Needs

Because of the dynamic nature of arid rangelands in terms of forage quality, forage availability and environmental extremes (e.g. snow cover, precipitation, temperature), cattle body weight and condition commonly change during winter grazing. DelCurto et al. (1991) found similar patterns of cow weight and body condition change when supplemented with alfalfa were fed to beef cattle winter grazing sagebrush steppe rangelands. However, the magnitude of response was dramatically different between consecutive years due to observed changes in forage quality, forage availability and environmental stress imposed on the grazing cattle. Likewise, other researchers in the western US have indicated variable results with supplementing free-range beef cattle consuming stockpiled forage due to dramatic changes in forage resources and(or) environmental conditions (Bowman and Sowell 1997; DelCurto et al. 2000 Rittenhouse et al. 2000). While these examples do not adequately describe all the considerations needed for supplementing grazing livestock (Clanton 1982), they do point out some of the complexities in achieving optimal response to supplementation for western beef systems.

Winter Feed Needs

Seasonal deficiencies in nutrient (protein/energy) concentrations frequently occur in arid and high-elevation rangelands (Figure 24.4). Producers dependent on rangeland forage resources must develop strategies to maximize the use of forage resources and supplemental inputs while maintaining acceptable levels of beef cattle production. Likewise, high-elevation and high-latitude beef cattle operations are likely to have significant periods of snow accumulation, which necessitate feeding of harvested forages (Brandyberry et al. 1994). In the Pacific Northwest and Intermountain West, many producers feed 1500–3000 kg of hay per mature cow during the winter-feeding period. The success of producers in these regions may depend on their ability to find economic alternatives to winter feeding of purchased hays, such as use of stockpiled forages and crop residues. Stockpiled forages are pasture and/or rangeland forages that are deferred for use until late fall or winter months. However, like dormant range forages, stockpiled forage and crop residues are low-quality roughages that require nutritional inputs for optimal use (DelCurto et al. 2000; Kunkle et al. 2000).

Strategies for Supplemental Feeding

One of the most important goals of economic sustainable livestock production in the western United States is to avoid nutritional inputs such as harvested winter feeds and supplements, unless, absolutely necessary. Therefore, the first goal of a manager should be to match the biologic cycle of the cow herd, and, associated nutritional demands, to the forage resources available (DelCurto et al. 2000).

Calving Date

Calving date (or breeding season) sets the biologic cycle which, in turn, determines the nutritional cycle of the cow herd in relationship to forage resources (Kartchner et al. 1979). The Western beef cattle industry is generally dominated by spring-calving practices, following the "55 days to grass" philosophy (Figure 24.5). This concept would target calving to occur about 55 days before grasses green up in the spring. The **gestation** length in beef cattle is approximately 284 days. Therefore, cattle will be exposed to green, highly nutritious, forage for approximately 25 days before they need to conceive in early summer to stay on a 365-day calving interval and give birth the following spring. The goal is to match the cows' nutritional requirements to range forage quality, so a producer might plan for calving to coincide with the onset of green forage (McInnis and Vavra 1997).

Additionally, calves may also benefit from the "55 day before grass" calving approach. A typical beef calf does not become a fully functioning ruminant until 90–120 days of age. However, cows pass their peak lactation period at day 70–90. As a result, calf performance after 90 days will depend to a greater degree on the forage quality available to the calf. Thus, a calf born March 1 will be effectively utilizing the better-quality forage available in June. In contrast, a calf born May 1 cannot effectively utilize forage resources until August, a time at which forage quality and yield are lower than in spring. Because of the vast differences in calf nutritional needs from day 90 to weaning, the earlier-born calf will have weaning weight advantages that greatly outweigh the 60-day difference in age. If higher weaning weights are a measure of economic importance, then the "55 days before grass" philosophy may be the best approach (Figure 24.5).

Weaning Weights

Weaning weights have changed drastically in the US, increasing from approximately 180 kg in 1967 to greater than 260 kg in 2017. This is related to increased use of continental European breeds, greater selection on growth traits and general improvements in management efficiency. If the goal is to market spring calves in the fall,

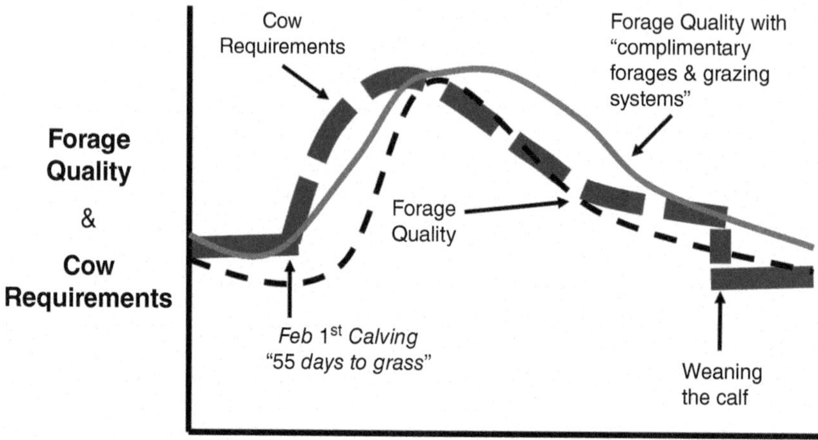

Fig. 24.5. In arid and semi-arid regions of the western US, cow requirements (thick gray dashed line) are often only approximately matched to the forage quality (black dashed line). Successful beef producers enhance nutrition by using complementary forages (e.g. multiple species that yield high quality at different times) and grazing systems that take advantage of topographically-induced changes in forage species composition and nutritional quality (solid line).

then this change in production efficiency has improved the economic potential.

High weaning weights can cause difficulties. The opportunities to put on post-weaning weight have become more limited with the heavier weaning weight calf. For example, if a spring calving beef cow/calf producer weans his calves in late October at 272 kg (600 lb), he/she may choose to sell in the fall market or retain calves over the winter-feeding period. With only marginal gains of 0.45–0.90 kg per head per day, this producer will come out of the winter-feeding period (120–150 days) with 350–400 kg yearlings. This restricts opportunities to place these animals on spring grass to lower feed costs. To fit market standards the yearlings need to be placed in the feedlot (avg. 90 days) with an expected gain of 200–250 kg and a target end weight of 600–650 kg. Therefore, spring calving cow/calf production with high weaning weights may limit opportunities for forage-based stocker cattle systems.

Producers wishing to retain ownership of calves after weaning should consider calving dates more strongly. Weaning weight takes on less significance. If a producer wishes to decrease costs per cow, moving the calving date to better match the range/pasture forage quality with cow nutrient demands may effectively reduce costs associated with supplementing cows during forage nutrient deficiencies. Any advantage of heavier weaning weight is reduced, but the producer has more opportunities to capture gains in the stocker, backgrounding and finishing phases.

Time of Weaning

Traditionally, beef producers in the western region have weaned calves at approximately seven months of age, which usually coincides to late October or November for spring-calving herds. However, there is some evidence that early weaning benefits both calves and dams. Early-weaned calves outgained late-weaned calves by 10 kg from September 12 to October 12 in one study in Eastern Oregon, despite going through the stress of weaning and adjusting to new feed (Figure 24.6, Turner and DelCurto 1991). Early-weaned calves were removed from their dams on September 12 and put on hay meadow regrowth and supplemented with 0.9 kg of barley and 0.5 kg of cottonseed. Similar calves remained on range forage with their dams until October 12 and then were managed with the early-weaned calves. On November 12, all calves were fed meadow hay and received 1 kg of barley and 0.5 kg of cottonseed meal throughout the winter. The early-weaned calves out-gained late-weaned calves by an additional 14 kg and were 23 kg heavier. Late-weaned calves compensated somewhat over the remainder of the winter but were still 11 kg lighter on April 12 (Figure 24.6).

A number of factors need to be considered when deciding if early weaning is appropriate. First, forage quality must be limiting to the point that calf gain will be reduced, and cows will likely lose body condition if they continue to nurse from late-August to the October or November weaning date. If forage quality and quantity are not limiting, there is really no advantage to early weaning. The real advantage of early-weaning is to improve the weight and body condition of the cows from late-summer to the beginning of the winter-feeding period by reducing their nutrient requirements (NASEM 2016). In addition, the producer must provide adequate forage/nutrition to the early-weaned calf. However, for producers who frequently have limited nutritional options during the late-summer and fall period, early-weaning may provide an alternative that allows for more efficient management relative to a dynamic arid-rangeland environment.

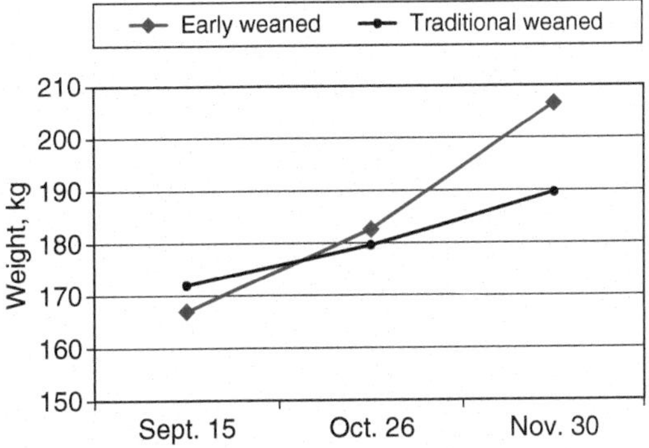

FIG. 24.6. The influence of early weaning vs traditional weaning on calf weights with beef cattle grazing rangeland forages (Turner and DelCurto 1991).

Winter Feeding Strategies

Beef cattle producers in the western US and especially the intermountain and Pacific Northwest are limited by high winter-feed costs, and production frequently becomes unprofitable due to the need to purchase supplemental hay. Many producers currently feed 1.4–2.8 Mg of harvested forage or supplements to mature cows during the winter-feeding period (Rittenhouse 1970, Horney et al. 1996). This represents costs of \$200–\$500 per cow per year and may be greater than 50% of the total input costs per cow for the year. Feeding costs can often reach \$2 per day during winter periods. Therefore, producer profitability is closely related to the ability to reduce winter feed costs while maintaining acceptable levels of beef cattle production.

Rake Bunch Hay

"Rake bunch hay" is one alternative to traditional winter management. With this system, hay is cut, cured, then raked into small piles, 30–60 kg with a bunch rake, and left in the field. The forage is then strip grazed, using New Zealand-type electric fences, throughout the winter. In a 10-year study, cows wintered on rake bunch hay came out of the winter period in better condition than traditionally fed cows and did not require supplements or additional hay (Turner and DelCurto 1991). Conception rates, calving interval, weaning weights, and attrition rates were equal between control and treatment groups, but the cost of winter-feeding "rake bunch" hay was $30–$40 less per head than the traditional feeding of harvested hay (Turner and DelCurto 1991).

Winter Grazing

Winter grazing of "stockpiled" forage requires deferring grazing of irrigated pasture forage or native range forage to the fall or winter months. Like rake-bunch hay, winter grazing may decrease winter feed cost by $20–$30 dollars per cow during mild to average years. To effectively utilize winter grazing in a management program, the producer must have access to the animals to accommodate supplementation programs. The range forage base will be dormant and, as a result, will likely need some level of supplementation (Kartchner 1981). Water must be available throughout the fall or winter grazing period, though the cow can effectively utilize snow when available.

In addition, the grazing area must be relatively free of significant snow accumulation during most years. Indirect benefits of winter grazing relate to the increased management opportunities of traditional hay meadows for spring and early summer grazing. Fall and winter grazing is an alternative use of native rangelands which has minimal impact on the plants as compared to traditional spring and summer grazing. This is particularly true with high-elevation, higher latitude rangelands. Nonlactating, gestating cows are generally better distributed over the grazing area and travel greater distances from water, making better use of slopes and more uniform use of the grassland area. A more thorough discussion of winter grazing (DelCurto et al. 2000; DelCurto and Olson 2010) is available.

Winter grazing of alfalfa or grasses in warmer Mediterranean or desert regions is common, since forages are highly productive during these periods, but grazing can cause compaction damage on wet soils.

Grass and Cereal Straws

Another alternative to traditional winter management would be the use of grass-seed residues or cereal straws as a winter-feed resource (Turner et al. 1995). Currently, Oregon's Grass Seed Industry produces over 1 million Mg of crop residues. In most cases, grass-seed residues should not be considered a complete feed for wintering mature beef cows. Instead, grass-seed straws should be tested, and supplements formulated to meet the cows' nutritional requirements while maximizing the use of this low-quality roughage.

Irrigated Grazing-hay systems utilizing grain forages

Though the use of winter wheat for grazing is widely practiced throughout the Central Southern Great Plains of the US (Texas, Oklahoma), it is not widely practiced in the intermountain West. However, the use of small grains and other alternative forages provides some viable alternatives to winter-feeding of hay. Drake and Orloff (2005) found that fall and early-spring grazing yields of triticale, ryegrass, barley, and wheat forages made a significant contribution to irrigated pasture systems, and enabled producers to save purchased resources. Additionally, fall grazing of triticale did not significantly reduce subsequent hay yields, depending upon planting, grazing, and harvest dates (Drake and Orloff 2005). Small-grain irrigated forage systems fit with rotations of alfalfa, specialty crops and pasture for beef producers in the Intermountain West.

Western Dairy Systems

Unlike earlier times when beef and sheep dominated, dairying has become a dominant component of western forage systems in the past four decades, creating large demand for high-quality hays, corn silage and other forages. The US dairy production industry has traditionally centered around the Midwestern states of Wisconsin and Minnesota, and eastern states of New York and Pennsylvania. However, in 1970, western states produced about 17% of the US milk supply, by 2016, nearly 45% of US milk was produced in western states, led by California, Idaho, New Mexico, and Texas (Figure 24.7). New Mexico alone increased milk production 24-fold over that same period and the western states collectively by 365% over the 46 years.

This trend of expansion of western dairying was driven largely by low cost of production, expansion of population centers in the West, exports of milk products, and the availability of high-quality alfalfa hay and feed by-products for dairy rations. However, in the 2010–2020 decade, western dairy production has begun to level off (Figure 24.7) due to volatile prices and unprofitability, high costs, environmental and labor regulations and water limitations.

Western dairy production is characterized by intensive large-scale feedlot-style dairying where harvested forages, and purchased grains and concentrates are brought to animals and rations balanced using analytic information. Predominant forage crops are alfalfa (hay, haylage, and **greenchop**), corn or sorghum silage, with some small grain and grass silages. Most (>96%) of the forage crops are irrigated.

Average herd size during the second decade of the twenty-first century in California was over 1200 cows,

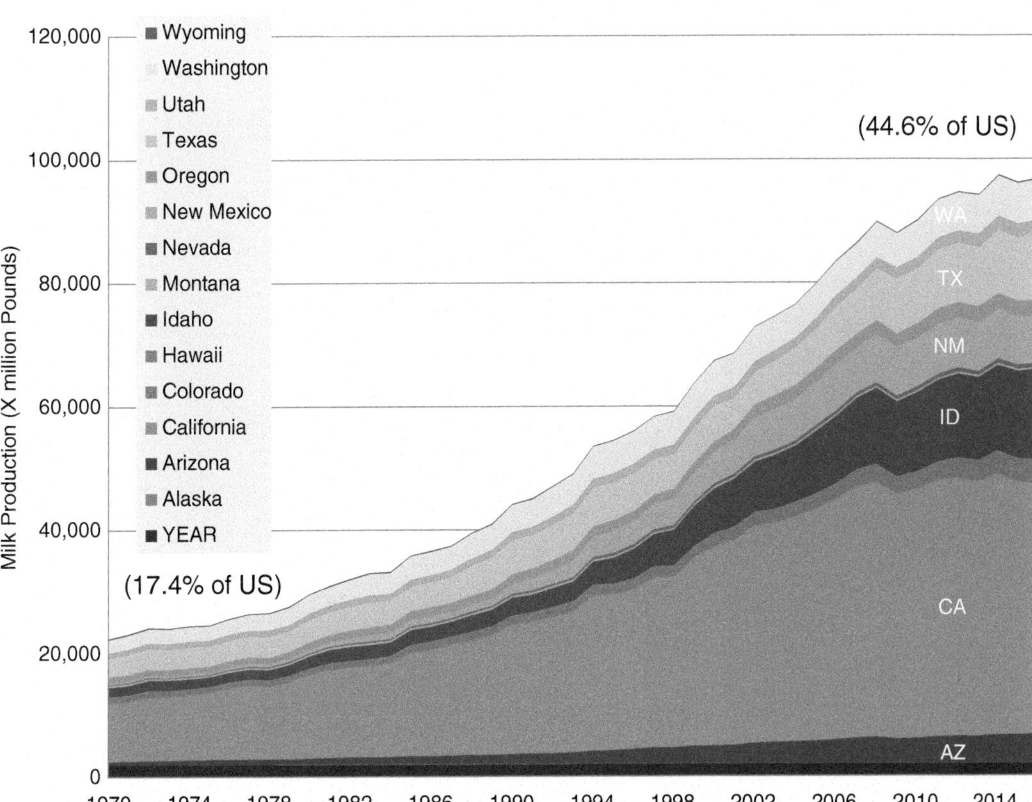

FIG. 24.7. Growth in western state's milk production, 1970–2016 (USDA-NASS).

and much larger units (1500–4000 cows) are common throughout the West. Average production per cow in western states has been slightly above the national average of 10 300 kg yr^{-1} in 2016. Though smaller integrated grazing and organic dairies can be found in some regions, these make up less than 5–10% of the production units in the region.

Irrigated Forage Crops

A wide range of crops can be grown when irrigation is available in many areas of western states. While states such as Idaho are known for specialty crops like potatoes, Washington for apples, and California for raisins, lettuce, wine, almonds and tomatoes, forage crops are a dominant component of the irrigated systems in each of these regions. In California, for example, alfalfa occupied the second largest acreage in 2017, and forages as a whole account for over 20% of agricultural water use of the state. Alfalfa, miscellaneous grass hay, irrigated pasture, and corn silage are the key forage crops grown under irrigation in western states.

Irrigated Alfalfa

Alfalfa is, unquestionably, the most important irrigated forage grown throughout the arid zones (Figure 24.8). It is primarily grown under irrigation, though some rainfed alfalfa systems are found in the eastern transition zones of Montana, Utah and Wyoming. Alfalfa is often the most important of all field crops in all western states (see Table 24.1).

In western regions, alfalfa is primarily grown as a monoculture, whereas alfalfa-grass mixtures are more common in eastern and midwestern US states (see Putnam and Summers 2008 for a review of western production methods). In most western regions, over 90% of the crop is sold as a cash crop and moved from point of production to animal facilities. Arizona produces the highest alfalfa yield of any US state, approximately 19 Mg DM ha^{-1} $year^{-1}$. California produces more total alfalfa hay and forage than any US state, producing approximately 5–7 million Mg yr^{-1}. In 2012, western states accounted for about 50% of the nation's alfalfa production (Figure 24.9). Alfalfa

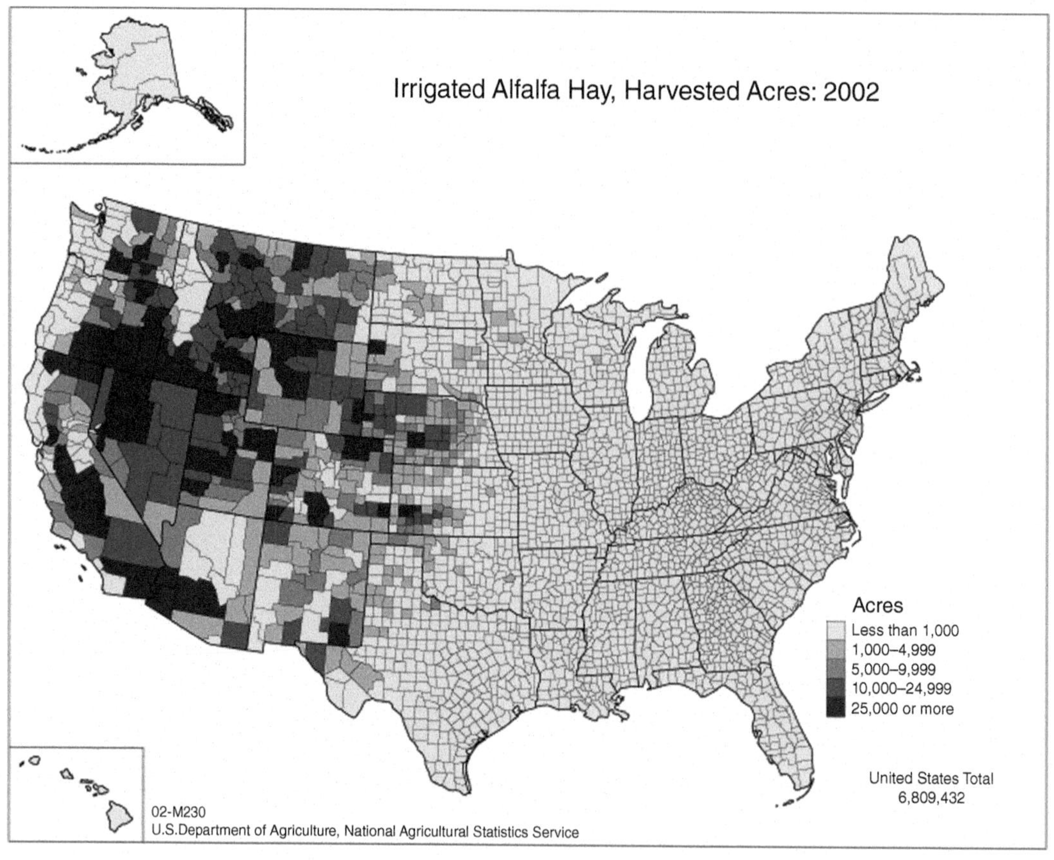

Fig. 24.8. Irrigated alfalfa production in the western United States (USDA-NASS 2012).

Table 24.1 Average alfalfa yield, percentage irrigated, approximate number of harvests, fall dormancy (FD) classes of the cultivars grown, and persistence estimates for 12 US western states (USDA-NASS yield data average 2015–2017, and information from forage specialists from western states)

State	Ave. yield (t/a)	Acreage under irrigation (%)	Cuts/Year	FD classes grown in state	Stands replaced every (yr)
Arizona	8.5	98	8–10	7–11	3
California	6.9	99	3–10	3–11	3–5
Colorado	3.8	89	1–4	2–4	3–8
Idaho	4.3	79	2–5	2–6	5
Montana	2.0	50	1–3	1–4	7–15
Nevada	4.5	100	3–4	3–5	8
N. Mexico	4.8	90	3–8	3–9	3–5
Oregon	4.6	80	3–5	2–5	6
Texas	4.7	90	3–5	5–9	6
Utah	4.2	67	1–7	3–6	6–20
Washington	5.1	95	3–5	3–6	3–5
Wyoming	2.7	68	1–4	2–4	4

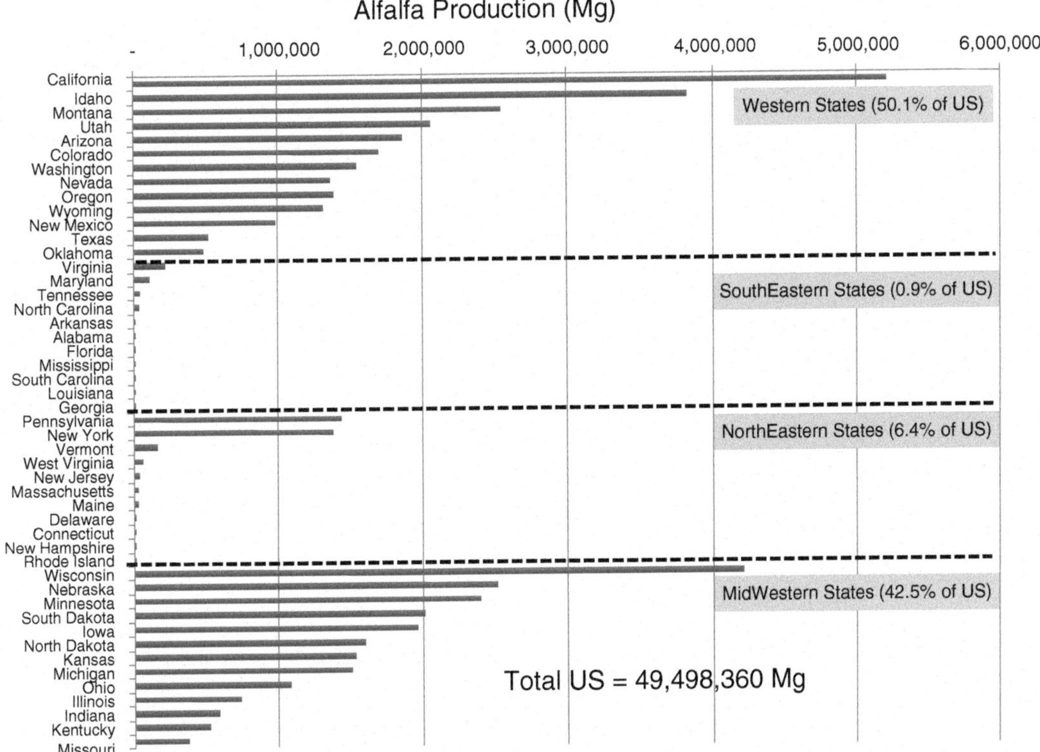

Fig. 24.9. US alfalfa forage production (hay, greenchop, and haylage) by region and state (USDA-NASS 2012).

is currently the third or fourth most important crop economically in the United States, behind corn, soybeans and sometimes wheat, depending upon year (note: "hay" which includes alfalfa and grass hay has been the #3 economic crop for many years).

Though production methods for alfalfa are similar in many respects to other regions, there are several important distinguishing features in the West. The key differences are the predominance of irrigation, the arid climatic conditions conducive to good curing conditions and high forage quality, and the generally long production season in most regions. The wide range of production methods (Table 24.2) throughout the West is determined by differences in elevation, latitude, climate, and length of season.

Irrigation Methods

The predominance of irrigation dictates alfalfa production practices in western states. Two types of irrigation predominate: (i) sprinkler (pressurized) systems, and (ii) surface (gravity-fed) irrigation systems. Subsurface drip irrigation (SDI) has been implemented commercially but, is currently practiced on less than 3% of production fields (Figure 24.10).

Sprinkler Systems

Sprinkler systems include continuous move center pivots and linear overhead sprinklers (Figure 24.10), wheel lines (which irrigate during a set period of time and then are moved), moveable hand lines, solid set (buried) pipe, and traveling gun-type sprinklers. Sprinklers are common on lighter soils, higher elevation sites, sites with only well water (no surface water), and areas where land leveling is not feasible. Water availability and electricity or fuel costs for pumping are major constraints as well as system maintenance. Wind losses are a major disadvantage of sprinklers but, can be mitigated by low-pressure overhead irrigation systems with better low-profile nozzle systems. Overhead sprinklers (pivots and linears) and SDI have the advantage of the ability to apply very small amounts of water frequently, which makes them particularly appropriate for sandy sites (where smaller irrigation amounts are recommended), and areas with limited water supply. Sprinklers are most common in high-elevation sites where land leveling is impractical or no surface water is available.

Table 24.2 Irrigated pasture area in Western States and US

State	Irrigated Pasture 2002	2007	2012	Percent of Irrigated land 2002	2007	2012
	(ha)			(%)		
Arizona	43 769	52 680	26 098	4.9	6.4	3.1
California	760 302	741 911	490 553	9.6	10.2	6.7
Colorado	411 906	571 192	406 654	18.9	24.9	19.3
Idaho	458 432	432 671	320 782	16.2	15.1	10.5
Montana	419 455	455 045	420 660	26.9	29.2	28.4
Nevada	212 001	188 052	126 589	39.7	37.4	22.6
New Mexico	190 627	181 776	90 214	29.1	28.0	15.3
Oregon	491 801	511 453	363 479	34.7	38.3	28.7
Utah	310 776	346 939	250 382	39.8	44.1	29.3
Washington	153 227	146 399	83 433	9.2	9.2	5.4
Wyoming	581 258	525 541	418 965	60.5	51.3	41.2
Western States:	4 033 554	4 153 659	2 997 809	18.8	20.1	14.5
US	4 977 214	5 062 201	3 729 847	9.9	9.8	7.2

Source: 2012 USDA Census of Agriculture.

Fig. 24.10. Overhead (linear) irrigation system on alfalfa in northern Arizona. Center pivots, linears, wheel-lines or hand-move sprinklers, and surface (check flood) irrigation systems dominate western irrigated forages (see Chapter 27).

Surface Irrigation Systems

Surface irrigation systems include "check flood" irrigation systems, dead level basins, and bedded alfalfa. All surface irrigation practices require a high degree of attention to land preparation, land leveling, soil texture and drainage issues.

Check Flood Systems

Check flood systems consist of very gently sloping fields (<0.5% slope), with small (e.g. 18–28 cm) levies spaced 6–30 m apart, with runs from 160 to 800 m long. Water moves down these "checks," covering the whole surface with between 76 and 200 mm of water at each irrigation, typically in an 6–16 hour period. Land must be leveled prior to crop establishment (usually using laser or GPS systems). These systems predominate in lower elevation valleys (such as California's Great Central Valley, Central Idaho, Arizona and Imperial Valley, California), heavy soil types, and in regions with plentiful surface water from canals or rivers. Key constraints are a greater land preparation requirement and water volume requirements, and less ability to easily control runoff or deep percolation. Applications of small amounts of water (e.g. 4–5 cm) per irrigation are impossible.

Dead Level Basins

Dead level basins are systems which utilize smaller fields designed with high border levies and leveled to 0% slope. Large volumes of water are pushed onto the field within a few hours completely flooding the field and allowing it to infiltrate into the root zone. This method is not as common as check flood irrigation, requires good infiltration characteristics, and high flow rates.

"Bedded Alfalfa"

Bedded alfalfa is practiced on very heavy soil types, and more closely resembles furrow irrigation (Figure 24.11). Alfalfa is planted on flat beds, from 15-152 cm in width, with furrows from 12-30 cm deep. In the purest form of bedded alfalfa, water moves down the furrows, and infiltrates laterally into the bed through capillary action. Modifications include "corrugation" with shallower furrows, the water over-tops the beds, more closely resembling check flood irrigation. The advantage of bedded alfalfa is that water drains from the alfalfa crowns, reducing disease incidence, improving oxygen movement into heavy soils, and preventing "scald" (death of plants due to high temperatures under water). There is an added advantage in that wheel traffic on alfalfa crowns is avoided, since equipment can only travel in the furrows (Figure 24.11).

Subsurface Drip Irrigation (SDI)

Drip irrigation has been widely adapted in tree, vine, vegetable crops and some row crops like corn and cotton, and alfalfa growers have been working to develop SDI systems for alfalfa. In this system, high-quality drip tape with emitters spaced approximately 30–60 cm apart are buried 20–60 cm below ground, 90–120 cm apart. This system has the advantage of delivering water daily to the crop, avoiding temporary droughts. Growers and researchers in Arizona and California have reported increases in yield of 20–30% but high cost, maintenance, and rodent damage have been major constraints (Putnam et al. 2017).

Stand Establishment

Alfalfa is planted on soils ranging from beach sands to heavy clay and clay-loam soils in western regions. The control of water applications enables production of alfalfa on heavy soils in the West – a practice not feasible in heavy rainfall areas, where frequent rains cause root diseases.

Deep ripping (60–140 cm) of heavy soils to break up compacted layers is common on some sites, utilizing heavy machinery with deep ripper-plows or shanks. Plowing, disking, and harrowing to create a fine but firm seedbed is important, to allow enough loose soil to cover the seed, but not allow the seed to be planted too deep in "fluffy" soils. Seeding rates of 15–25 kg ha^{-1} are generally recommended, with seeding depth 0.6–1.2 cm depending upon soil type. Seeding time is critical for alfalfa, with fall seeding (September-October) recommended in southern regions, and summer seeding or spring seeding in higher elevation regions. Irrigating up new stands is recommended, even on flood irrigated fields, but some growers seed and allow winter rains to bring up stands where feasible. Herbicides are commonly used during stand establishment.

Cultivar Selection

A wide range of alfalfa cultivars are grown in the West. These range from dormant cultivars (Fall Dormancy classes 2-4) in the northern and higher elevation regions, through non-dormant lines (FD 8-11) in the southern and low elevation deserts (Figure 24.12). Semi-dormant lines (FD 5-7) are grown in intermediate zones. In many western states, such a diversity of elevations and climates is found that a single cultivar recommendation is not possible – cultivars ranging from very dormant to non-dormant are grown. Cultivars are chosen for yield, persistence, pest resistance, and potential for high quality. Generally non-dormant (FD ratings 8-11) cultivars are higher yielding but have lower persistence and so are not grown in colder regions. Due to the market demand for high quality, growers in some regions choose a lower dormancy cultivar (and lower yielding variety) simply to improve the quality of the forage (Putnam and Orloff 2003) (Figure 24.12).

Pest Management

Many alfalfa fields in western states are treated with herbicides during stand establishment, usually post-emergence. The development of "roundup ready" cultivars (fully deregulated in 2011) have had an impact on weed control methods, though conventional systems are also used. Since pure alfalfa is demanded by the dairy markets, alfalfa is usually sown alone without a companion crop, though some growers allow weeds to develop and harvest them with the first harvest. Organically-grown alfalfa is grown on about 2–4% of the crop area, and weeds are a major challenge.

Insecticides are commonly used in the cash hay regions of western states. The key times of application are during the early growth periods (from February through May) for control of alfalfa weevil and Egyptian alfalfa weevil,

Fig. 24.11. Bedded alfalfa, Imperial Valley, California. This method resembles furrow irrigation systems for vegetables, moves water down the field effectively, and prevents excess water from damaging alfalfa seedlings and crowns on heavy soil types.

Fig. 24.12. Fall dormancy of western-grown alfalfa ranges from very dormant (class 2-4) to taller, very non-dormant cultivars (class 9-11) which are winter-active and produce year-round forage supplies (10 cuts/year). *Source:* D. Putnam, photo.

and a complex of aphids, including the blue alfalfa aphid, spotted alfalfa aphid, and pea aphid. Additionally, insecticides are frequently used to control a summer complex of Lepidoptera insects including alfalfa caterpillar, beet armyworm, and fall armyworm during late summer and fall periods.

Harvesting Methods and Harvest Schedules

Hay is harvested from 2 to 12 times per year in western states. Two to three-cut systems dominate in colder regions of Montana and Wyoming and high-elevation valleys, whereas on the low deserts of Arizona, New Mexico and California, alfalfa is harvested beginning in January, through December, totaling 10–12 cuts per year in some areas.

Although cutting schedules are determined to a considerable degree by weather, especially in the spring, harvest intervals between 21 and 40 days are common (typically 28 days). Logistics of irrigation water, contracted harvesting, and labor are major factors in determining harvest schedules. Forage quality has a large influence on cutting schedules and growers must continually balance the economic value of quality of shorter harvest intervals with the higher yield of longer intervals in different regions (Putnam and Orloff 2003; Orloff and Putnam 2004).

Over the past 40 years, western states have developed highly-mechanized methods for handling alfalfa and other forage crops as hay crops. This is driven by the cash-hay nature of the business, large field and ranch size (hundreds to a thousand hectares of alfalfa on a single farm is common), and the necessity to transport hay long distances. This is contrasted with eastern regions where hay and haylage is used mostly on the farm where it is produced. Almost all western states incorporate methods in which hay is handled or stacked with no hand labor. Swathers (3.7–5.5 m width) cut and windrow the hay using disc- or sickle-type cutter. Hay is raked once or twice. Hay packages are either small rectangular (22–68 kg) bales, large rectangular bales (340–900 kg) or, less commonly in this region, round bales. Round balers are primarily used when the hay is used on-farm whereas rectangular bales predominate in regions where hay is transported. Hay is commonly baled at night to take advantage of the dew, which softens the leaves in arid climates, reducing leaf loss. Mechanical bale gathering machines (bale wagons, accumulators, or harrow beds) are used to pick up bales in the field for roadsiding and subsequent pickup. A modified fork-lift called a *squeeze* is commonly used to pick up blocks of bales from the roadside and load and unload trucks. Storage can be in hay barns (often without sides), or outside using tarps to preserve quality. Haylage (alfalfa silage) and greenchop harvesting methods are utilized near dairy regions but, comprise less than 2% of total production in this region.

Markets and Quality

Alfalfa hay production in western states is characterized by high quality. The combination of pure alfalfa stands, good weed and pest management, with generally favorable harvesting weather and high demand for quality by dairy producers and horse owners creates the conditions for high-quality production. Price incentives for high-quality range from $25/MT to $100/MT.

The dairy industry utilizes 50–80% of the alfalfa in most western states, followed by horse, export, beef and sheep industries. Dairy producers demand pure alfalfa of low-fiber content, high protein and digestibility. Much of the hay is tested for the dairy industry, often more than once. Though the value of quality changes over the years and by market region, we have calculated differences of approximately $0.6–$0.9 per 1 g kg^{-1} in acid detergent fiber (ADF) across years and regions (Putnam 2004).

Horse owners and feed stores account for a large segment of the commercial hay market. Alfalfa mixtures with grasses, or medium-quality alfalfa (weed free, but cut in a later stage of development), or pure grass hays (oat hay, bermudagrass, timothy, or cool-season grass mixtures) are preferred. Beef producers utilize medium-grade alfalfa hays due to price constraints, as well as a range of other hay products.

Corn, Sorghum, and Small Grain Silage

Corn silage is an important forage in western regions where dairies predominate, and in some beef-production areas. Over the past 20–30 years, corn silage has grown from a minor forage crop in western states to a major contributor. In the largest western dairy states, California and Idaho, corn silage production has increased more than fourfold over a 46-year period (Figure 24.13). California and Idaho are the second and third largest producers of corn silage in the US, following Wisconsin, which is number one. Corn silage is highly valued for its high productivity under irrigation, high water-use efficiency, and high feed energy content compared with other coarse forages. Corn silage plays a critical role for large dairies in crop rotations and the recycling of animal manures.

Sorghum is grown as a silage crop on minor acreages in some areas affected by drought, and in western Texas, California, and New Mexico, and has advantages in cost of production, relative tolerance of periodic drought, and lower N fertilizer requirement. Small grain forages, particularly wheat, triticale, barley and oats are commonly grown as silage crops in rotation with corn in dairy regions to improve silage supply and to recycle animal wastes.

Other Harvested Forages

Though alfalfa hay and corn silage are the predominate forages in most western regions, a wide range of other forages are produced. These include sudangrass, oat hay, bermudagrass, kleingrass, and cool-season grasses

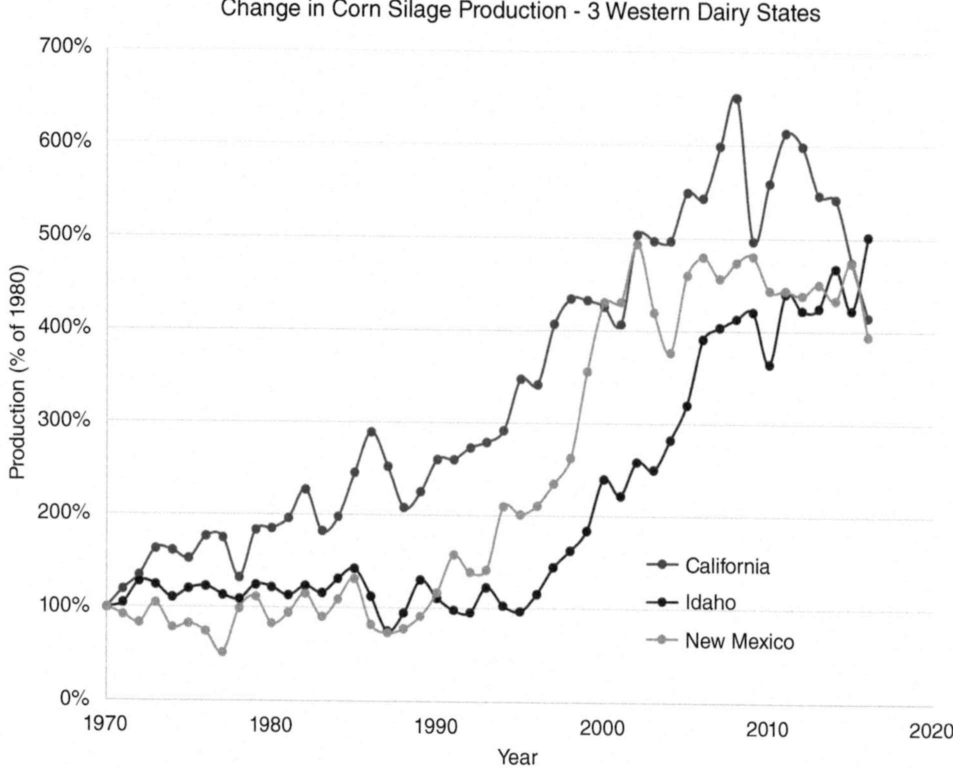

Fig. 24.13. Changes in corn silage production in three western dairy states, 1980–2016. In 2016, production was 7.3, 6.5, 1.6 million Mg (70% moisture silage) for CA, ID, and NM, respectively (USDA-NASS 2019). California was until recently second only to Wisconsin in corn silage production.

(timothy, ryegrass, orchardgrass) and mixed pasture grasses, grown under irrigation. Cool-season grasses are highly preferred by domestic horse markets, and timothy for export. The higher rainfall areas of eastern Washington, Oregon and California utilize grasses for grass-fed and organic dairies and for grazing beef operations.

Irrigated Pasture

Irrigated pasture forms a significant component of western forage systems. Though production area has declined 25% from 2002 to 2012, pasture still consists of a high percentage of the irrigated land in western states, totaling an average of 14.5% of the irrigated land in 11 western states in 2012 (Table 24.2). This decline can be traced to economic factors and competition from other crops.

Exports of Hay

Exporting of hay to other countries is now a common feature of western hay production and, has become much more important in recent years. While hay exports are a minor component of forage production in the US as a whole (less than 5% of production), the equivalent of approximately 17% of alfalfa hay and more than 45% of the grass hays produced in the seven western-most states were exported in 2016. This trend has seen dramatic increases in the past ten years (Figure 24.14). Coarse hays and straw products (timothy, sudangrass, bermudagrass, kleingrass, grass-seed straw from the Oregon grass seed industry, and rice straw) are also exported. The dominant markets for hay have been Japan, China, and other countries in the Far East, but recent markets have developed in the Middle East (UAE and Saudi Arabia). Total hay exports were about 4.8 million MT in 2016, consisting of 56% alfalfa and 44% grass hays (Figure 24.14).

Challenges and Opportunities in Western Forage Systems

Forage systems have a long history in the West and consist of two interconnected sectors, irrigated forages and rangelands. Forages dominate land areas in many parts of the West and are a critical economic force in the farming systems in this region. It is a sector in transition, beset by

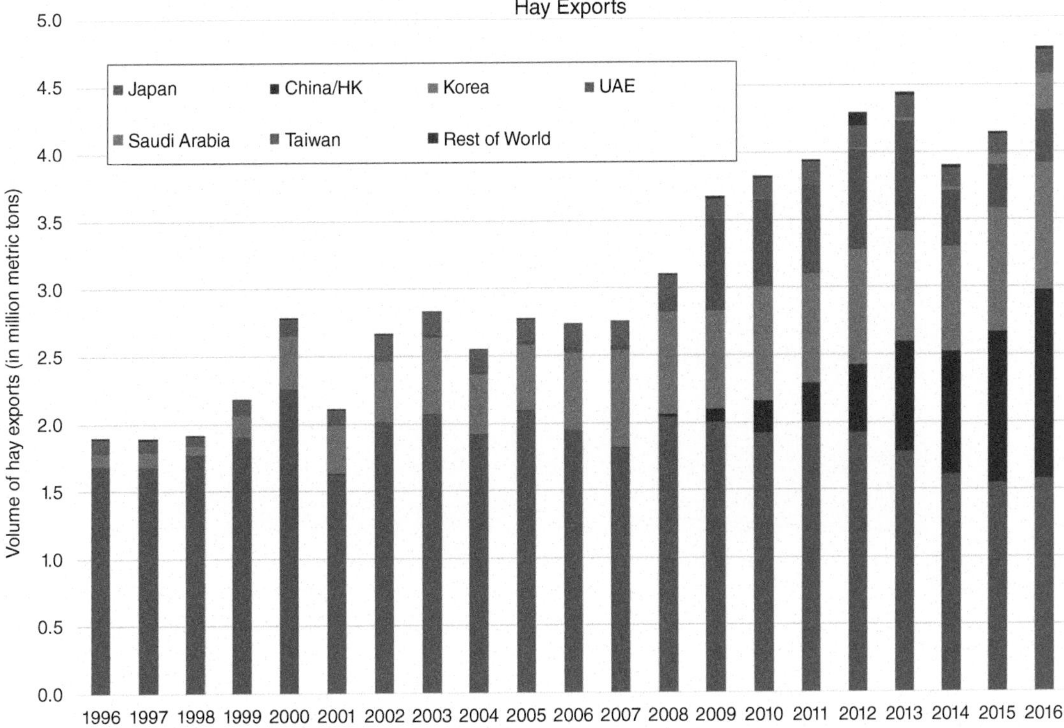

Fig. 24.14. Alfalfa and grass hay exports from western US ports, 1996–2016 (US Dept. of commerce Data).

many challenges: economic, policy, and environmental. Water limitations, regulations and restrictions on rangeland grazing will continue to challenge range producers. While beef production systems have predominated historically, the rise of western dairying in large units, accompanied by water restrictions, rapid urbanization, and environmental pressures have been important trends that will likely continue. Water and economic limitations are likely to slow the growth of dairying in the West in the third decade of the twenty-first century.

The Importance of Water

Drought and limitations of water are the most important limiting factors for western forage production systems. There are numerous examples of curtailment of irrigation water for forages due to wildlife habitat decisions, protection of endangered species, competition with "higher value" crops, transfer to urban uses, groundwater limitations, or water-conflict litigation.

However, irrigated forage crops, particularly alfalfa, offer some significant advantages with regards to water. This is due to its deep-rooted characteristics, high yields, and ability to be "deficit irrigated," that is, to produce some yield with less than full water allocations. This is not true with many other fruiting and seed-producing crops which must obtain close to full water supplies to be productive. Given the importance of water, continued investigations regarding optimizing forage production with limited water resources are essential for the future of forages in this region.

Importance of Forages to the Environment

Though little recognized, forage crops offer a wide range of environmental services. Alfalfa has the potential to provide significant positive benefits to the environment due to N_2 fixation, wildlife habitat, prevention of nitrate and particulate contamination, and soil conservation (Putnam et al. 2001). Alfalfa and pasture are major sources of wildlife habitat (Figure 24.15); 28% of California's wild species use alfalfa for nesting, cover, or feeding (Putnam et al. 2001). Swainson's Hawk utilizes alfalfa 10 times more than surrounding landscapes, and small grain forage is important nesting habitat for tri-colored blackbirds in dairy regions.

Likewise, rangeland grazing systems can provide environmental benefits by coexisting with wildlife, suppressing weeds and fire suppression, but management is key to these benefits. Development of sustainable

Fig. 24.15. Stag grazing in an alfalfa field in northern California, with wheel-line irrigation system in the background. Alfalfa, rangeland and pasture are important wildlife habitat throughout western regions. *Source:* Steve Orloff, photo.

grazing strategies that maintain or enhance vegetation and biologic value are needed throughout the western US (DelCurto et al. 2005; Vavra 2005). Likewise, development of forage production strategies that complement wildlife and riparian habitat are needed. In addition, the use of domestic livestock to encourage positive vegetation change shows promise and warrants further investigation (DelCurto and Olson 2010).

Non-irrigated forage production, particularly on public rangelands, also presents a unique set of challenges. Western rangelands have historically been managed to accommodate livestock production. However, Congress has altered the framework that governs federal land management with the passage of the Multiple Use Act (1968), National Environmental Policy Act (1969), Clean Water Act (1972) and the Threatened and Endangered Species Act (1973). Water quality of rangeland watersheds has been challenged from many quarters (Tate et al. 1999), and rangeland production is highly dependent upon evolving government policy and regulation.

Though forages are an important part of farming systems in western states, their continued production will depend upon the ability to develop systems that maintain or enhance biologic diversity, ecosystem services, conserving soil, water, and wildlife habitat while providing profits for producers. Compared with many other agricultural enterprises, forage crops have unique attributes to offer the general public to enhance sustainability and environmental quality.

References

Bowman, J.G.P. and Sowell, B.F. (1997). Delivery method and supplement consumption by grazing ruminants: a review. *J. Anim. Sci.* 75: 543–550.

Brandyberry, S.D., DelCurto, T. and Angell, R.A. (1994). Winter grazing as a grazing management program for northern Great Basin Rangelands. Oregon Agricultural Experiment Station, Special Report No. 935.

Clanton, D.C. (1982). Crude protein in range supplements. In: *Protein Requirements for Cattle. Symposium* (ed. F.N. Owens), 228–234. Oklahoma State Univ. MP-109, Stillwater.

DelCurto, T. and Olson, K.C. (2010). Issues in grazing livestock nutrition. In: *Proc. 4th Grazing Livestock Nutrition Conference* (eds. B.W. Hess, T. DelCurto, J.G.P. Bowman and R.C. Waterman), 1–10. Champaign, Il: West. Sect. Am. Soc. Anim. Sci.

DelCurto, T., Angell, R.A, Barton, R.K. et al. (1991). Alfalfa supplementation of beef cattle grazing winter sagebrush-steppe range forage. Oregon Agricultural Experiment Station, Special Report No. 880.

DelCurto, T., Olson, K.C., Hess, B., and Huston, E. (2000). Optimal supplementation strategies for beef cattle consuming low-quality forages in the Western United States. *J. Anim. Sci. Symposium Proc.* 77: 1–16.

DelCurto, T., Porath, M., Parsons, C.T., and Morrison, J.A. (2005). Management strategies for sustainable beef cattle grazing on forested rangelands in the Pacific northwest. Invited synthesis paper. *J. Range Ecol. Manage.* 58: 119–127.

DelCurto, T., Murphy, T. and Moreaux, S. (2017). Demographics and Long-Term Outlook for Western US Beef, Sheep and Horse Industries and Their Importance for the Forage Industry. Proceedings of the Western Alfalfa and Forage Symposium, Reno, NV, USA (28–30 November 2017).

Drake, D.J. and Orloff, S.B. (2005). Simulated grazing effects on triticale forage yield. *Forage and Grazinglands* https://doi.org/10.1094/FG-2005-0314-01-RS.

Horney, M.R., DelCurto, T., Stamm, M.M. et al. (1996). Early-vegetative meadow hay versus alfalfa hay as a supplement for cattle consuming low-quality roughages. *J. Anim. Sci.* 74: 1959.

Kartchner, R.J. (1981). Effects of protein and energy supplementation of cows grazing native winter range forage on intake and digestibility. *J. Anim. Sci.* 51: 432.

Kartchner, R.J., Rittenhouse, L.R., and Raleigh, R.J. (1979). Forage and animal management implications of spring and fall calving. *J. Anim. Sci.* 48: 425.

Kunkle, W.E., Johns, J.T., Poore, M.H., and Herd, D.B. (2000). Designing supplementation programs for beef cattle fed forage-based diets. *J. Animal Sci. Symposium Proc.* 77: 1–11.

McInnis, M.L. and Vavra, M. (1997). Dining out: Principles of range cattle nutrition. Oregon State University Agricultural Experiment Station, Special Report No. 979:7.

National Academies of Sciences, Engineering and Medicine (2016. 8th rev. ed.). *Nutrient Requirements of Beef Cattle.* Washington, DC: National Academy Press.

Orloff, S.B. and Putnam, D.H. (2004). Balancing yield, quality and persistence. Proceedings of the National Alfalfa Symposium, San Diego, CA, USA (13–15 December 2004).

Putnam, D.H. (2004). Forage quality testing and markets. In: National Alfalfa Symposium Proceedings, 13–15 December, 2004. San Diego, CA. http://alfalfa.ucdavis.edu (accessed 31 October 2019).

Putnam, D.H. and Orloff, S.B. (2003). Using varieties or cutting schedule to achieve quality hay – what are the tradeoffs? Proceedings of the 33rd California Alfalfa and Forage Symposium, Monterey, CA, USA (17–18 December 2003).

Putnam, D.H. and Summers, C.G. (2008). *Irrigated Alfalfa Management for Mediterranean and Desert Zones.* Publication 3512. University of California Agriculture and Natural Resources.

Putnam, D.H., Russelle, M., Orloff, S. et al. (2001). Alfalfa, wildlife and the environment, the importance and benefits of alfalfa in the 21st century. California Alfalfa and Forage Association. http://agric.ucdavis.edu/files/242006.pdf (accessed 10 October 2019).

Putnam, D.H., Montazar, A., Zaccaria, D. et al. (2017). Agronomic practices for subsurface drip irrigation in alfalfa. Proceedings of the Western Alfalfa and Forage Symposium, Reno, NV, USA (28–30 November 2017).

Rittenhouse, L.R., Clanton, D.C., and Streeter, C.L. (1970). Intake digestibility of winter-range forage by cattle with and without supplements. *J. Anim. Sci.* 31: 1215.

Tate, K.W., Dahlgren, R.A., Singer, M.J. et al. (1999). Monitoring water quality on California rangeland watersheds: timing is everything. *California Agric.* 53: 44–48.

Turner, H.A. and DelCurto, T. (1991). Nutritional and managerial considerations for range beef cattle production. *Vet. Clin. N. Am. Food Anim. Pract.* 7 (1): 95–125.

Turner, B., Conklin, F., Cross, T. et al. (1995). Feeding Oregon's grass straw to livestock: Economic and nutritional considerations. Oregon Agricultural Experiment Station Special Report No. 952.

USDA-NASS (2019). United States Department of Agriculture, National Agriculture and Statistics Service 2018 Summary. Washington, D.C. https://www.nass.usda.gov/Publications/Todays_Reports/reports/cropan19.pdf.

Vavra, M. (2005). Livestock grazing and wildlife: developing compatibilities. *Rangeland Ecol. Manage.* 58: 128–134.

PART V

FORAGE PRODUCTION AND MANAGEMENT

Potassium deficiency symptoms on alfalfa. Legumes such as alfalfa need high levels of major nutrients to produce optimum yields. *Source*: Photo courtesy of Mike Collins.

The most widely used forage species are not native to their primary areas of use, so it is commonly necessary to establish or reestablish vigorous stands in order to achieve optimum productivity and forage quality. Forage establishment, fertilization and management are especially complex because of the large number of species available, each with its own unique characteristics and management needs. As discussed in Parts I and II, for example, legumes have the capacity to fix atmospheric N, which impacts their own management needs as well as those of mixed grass-legume stands.

Conditions of inadequate or excessive water supply occur frequently in forage systems because pasture and harvested forages are often found on soils with limitations

Forages: The Science of Grassland Agriculture, Volume II, Seventh Edition.
Edited by Kenneth J. Moore, Michael Collins, C. Jerry Nelson and Daren D. Redfearn.

to high-value row crop production. Irrigation is an economic option in some cases, especially in highly productive situations where high-value forages such as alfalfa hay or corn silage are being produced.

Weeds and insects can detract from forage productivity, shorten stand life, and reduce forage quality and/or palatability. Herbicides and insecticides can be effective tools in many cases, but this part will also show how proper management can help minimize the negative effects of these pests in forage systems. Much of the effort of plant breeders has been directed at improving forage production and limiting the negative effects of stresses such as drought, diseases, and insects. This part provides the most current information on these important aspects of profitable forage systems.

Chapter 25

Forage Establishment and Renovation

Marvin H. Hall, Professor, *Crop and Soil Sciences, Pennsylvania State University, University Park, PA, USA*
Yoana C. Newman, Associate Professor, *Plant and Earth Science, University of Wisconsin River Falls, River Falls, WI, USA*
Jessica A. Williamson, Assistant Professor, *Crop and Soil Sciences, Pennsylvania State University, University Park, PA, USA*

Thick, vigorous stands are essential for high forage yield. Obtaining such stands via a new establishment or adding a new species to an existing forage stand (i.e. renovation) depends on a proper seeding practice, a favorable seedbed and expected environmental conditions. The more productive the soil and the better the **seedbed** conditions, the greater is the potential for successful establishment of highly productive stands.

Preparing for Successful Establishment

Key factors for obtaining thick, vigorous stands include proper soil pH and fertility, seedbed preparation, avoiding herbicide residues from previous crops, selection of high-quality seed, seeding at the right time, use of good seeding techniques with equipment precisely adjusted for seeding rate and depth, and adequate control of weeds and insects. Preparation for forage establishment ideally begins as much as two years prior to the actual seeding. This is especially the case for **no-till seedings** where lime and fertilizer are top-dressed and it takes time for the lime to react.

Lime and Fertility Adjustments

Persistent, high-yielding stands are always associated with a favorable pH and highly fertile soil. Failure to achieve an adequate pH on acid soils is the greatest limitation to high forage production since it affects the availability of other nutrients both during and after establishment. **Lime** corrects soil acidity and, depending on the liming material used, supplies calcium (Ca) and magnesium (Mg). Proper soil pH also enhances growth of desirable microorganisms and reduces toxic effects of aluminum (Al) and manganese (Mn).

The second key is to provide adequate nutrients for the desired forage mixture based on soil tests (see Chapter 26). Soil samples should be taken to the depth of the plow layer, usually about 15 cm, for seeding into a tilled seedbed. For no-till seedings, two sets of soil samples should be taken; one at the 0- to 5-cm depth to determine surface pH and a second to about 15 cm. The 0- to 5-cm sample is critical for fields that have been in no-till for three or more years. Repeated surface applications of N fertilizers to no-till crops frequently result in acidification of the upper layer of soil.

Meeting Lime Requirements

Lime reacts slowly to increase the pH, so it should be applied and thoroughly mixed into the plow layer 6–12 months prior to seeding. In preparation for no-till seedings, surface applications of lime should be made one

Forages: The Science of Grassland Agriculture, Volume II, Seventh Edition.
Edited by Kenneth J. Moore, Michael Collins, C. Jerry Nelson and Daren D. Redfearn.

to two years ahead of seeding to allow for movement of the lime into the soil profile, which is especially slow on heavy textured soils. An annual crop can be used while the lime is working. Compared with incorporated lime and conventional seeding, surface liming and no-till seeding resulted in less vigorous forage seedlings and slower establishment, but seedling densities after 30 days were similar (Koch and Estes 1986). Annual yields following establishment for both seeding methods were similar, indicating that initial differences did not affect later performance.

Meeting Plant Nutrient Requirements

Phosphorus (P) is especially critical during establishment. Banding 15–30 kg ha^{-1} P at a depth of 2.5–5.0 cm directly below the seed promotes rapid root development and seedling establishment, especially on soils testing low to medium in P (Sheard 1980; Teutsch et al. 2000). Fertilizer should not be in direct contact with the seed because it has salt-like characters and can inhibit germination. Banded application of P fertilizer below or near the seed is more effective than broadcast because it is strategically placed to be available to newly forming plant roots where it is in higher concentration. It takes two to four times more broadcast P to give the same growth response as banded P (McKenzie 2013). Requirements of young seedlings for potassium (K) are relatively low, but K is very important for yield and persistence of established stands.

Application of fertilizer N at establishment increases yield of both grasses and legumes when soils are low in N (less than 15 mg soil NO_3^- per kg soil) or soil organic matter is less than 1.5% (Hannaway and Shuler 1993; Seguin et al. 2001). However, when soil NO_3^- levels are greater than 15 mg kg^{-1} and conditions are favorable for effective nodulation (soil pH 6.2–7.5) and high populations of appropriate *Rhizobium* bacteria are present, preplant-N fertilizer did not increase legume yield and reduced both rhizobial infection and N fixation (Vough et al. 1982; Eardly et al. 1985). Usually N application hinders legume seedling development in renovation seedings because it stimulates growth of existing grasses and weeds, leading to excessive shading of young seedlings and undue competition for water, light, and soil nutrients.

Weed Control and Herbicide Carryover

A strategy for weed management to kill existing plants and reduce the seedbank during establishment needs to be developed and generally initiated long before the crop is seeded. Residual herbicides from previous crops must also be considered. Triazine carryover may be encountered with forage seedings following corn, and its residual activity can be exacerbated if lime is applied immediately prior to seeding the forage crop (Lowder and Weber 1982).

Aminopyralid, a relatively new herbicide registered for use in forage systems, may interfere with spring establishment after fall applications. Stand failures have been observed for cool-season forage legumes, and their planting is not recommended following applications of soil residual herbicides such as aminopyralid, imazapic, and metsulfuron unless some reduction in stand density is acceptable (Renz 2010). Some tillage prior to seeding helps mix residual herbicide, **lime** and fertilizer into the soil and reduces potential herbicide injury.

Allelopathy and Autotoxicity

Allelopathy has been implicated in poor establishment of new forage stands (VanToai and Linscott 1979; Peters and Zam 1981; Young and Bartholomew 1981; Cope 1982; Hall and Henderlong 1989). Such allelopathic toxins from previous plants affect certain seedlings and may explain some difficulty in establishing alfalfa after alfalfa (Jennings and Nelson 2002a, 2002b) or interseeding birdsfoot trefoil into tall fescue sod (Luu et al. 1982). It is best to grow one or more annual crops that are not affected by the toxin on the field before replanting alfalfa or before no-till planting legumes into a sod crop. Alfalfa autotoxicity may be a greater problem and take longer to dissipate on heavier soils and in locations with low rainfall (Jennings and Nelson 2002b; Seguin et al. 2002). Toxins associated with tall fescue swards were largely eliminated by burning the sod prior to seeding birdsfoot trefoil (Luu et al. 1982).

Species and Cultivar Selection

The species or mixture of species to plant and the cultivar(s) selected will affect establishment and production for the life of the stand. Compared to mixtures, stands of one grass or one legume are generally easier to establish and manage for hay and silage production. For example, several herbicides are available for weed control in pure stands, but few are available for mixed legume–grass stands. Proper harvest timing can be difficult with mixtures when recommended stages of maturity for cutting each component do not coincide. And, if not properly managed, the grass portion of the mixture can become dominant over time and subsequently lower the feeding value of the hay or silage.

Legume–grass mixtures require a high level of management to maintain a proper balance in a mixture, but they offer many advantages over pure stands because:

1. Legumes fix atmospheric nitrogen (N_2) into a usable form, reducing the need for N fertilizer.
2. Legume–grass mixtures generally have higher crude protein (CP) concentrations and digestibility than a pure grass.
3. Legumes extend the grazing of pastures further into the warm, dry midsummer period than do cool-season grasses alone.
4. Once established, perennial legume–grass mixtures are more competitive with weeds than are pure stands.

5. Mixtures provide better protection against **plant heaving**, winter injury, and soil erosion than do pure stands of legumes.
6. Mixtures are easier to field cure as hay than are pure legume stands.
7. Mixtures tolerate a wider range of soil conditions.
8. Mixtures containing 40% or more grass tend to reduce legume bloat in grazing cattle.
9. Mixtures containing 40% or more legumes tend to reduce the possibility of NO_3^- poisoning and grass tetany.
10. Grasses reduce **lodging** of legumes.

Principles for composing mixtures include the following: (i) Keep the mixture simple. One grass and one legume are often sufficient; seldom are more than four species justified. (ii) Plants should have similar maturity dates, have compatible growth characteristics, and be adapted to the intended use. Species that germinate quickly and have vigorous seedling development improve early production, but more vigorous species can crowd out others. (iii) Pasture mixtures should include species that mature together and have similar **palatability**, since unpalatable species will not be grazed and soon will dominate.

Seed should have high cultivar purity and be of high quality (Chapter 32). Most recent cultivars have superior agronomic characteristics that economically justify their selection over older, lower-cost cultivars that have less disease and insect resistance and lower yield potential.

Establishing from Seed

Most forage plantings are from seed since it is convenient, cost effective and has a high probability of success if done correctly. In contrast, some species have low seed production or other limitations leading to vegetative propagation as the preferred method. This leads to differences in time of planting and early management to assure a good stand of vigorous forage plants.

Seeding at Proper Time

Due to climate factors that affect both soil preparation and seedling vigor, there are given windows in the season when seeding is most successful. Obtaining rapid germination and vigorous seedling growth are goals for achieving a stand that is dense, weed-free and productive.

Cool-Season Forages

The two primary seeding periods for cool-season species are late winter–early spring (late February to early May) and late summer–early fall (August to mid-October). Spring seedings are most common, particularly in the northern half of the US and into Canada. Soil moisture is generally adequate, evaporation is low, and soil moisture is retained longer during the establishment period than with late-summer seedings. However, early seeding into cold, wet soils can result in slow or poor germination, high seedling loss due to fungal diseases, and weak stands. Seeding late in the spring can cause young seedings to fail due to stresses from high temperatures, lack of moisture during summer and high weed competition. The surface 2.5–5.0 cm of soil can dry quickly, and young seedlings can desiccate.

Advantages of late summer–early fall seedings include: (i) Less competition from weeds reduces need for herbicides; (ii) Seedings can be made after early-harvested crops such as small grains or vegetables and have full production the following year; (iii) Workload does not compete during the busy spring season; (iv) Liming, fertilization, and tillage are done during drier weather, thus reducing risk of soil compaction; and (v) **Damping-off** diseases (*Pythium* spp. and *Rhizoctonia* spp.) are not usually a problem.

In addition to adequate soil moisture, late summer–early fall seedings need sufficient time and accumulation of heat units before a **killing frost** to help ensure survival over winter. Consequently, seedings should be made early enough to allow at least six weeks for growth (seedlings 7–10 cm tall or to 4-leaf stage) before a killing frost (−4 °C night temperature). This means seeding July 20–August 1 in areas such as the northern portions of Minnesota, Wisconsin, Michigan, and New York to as late as October 15 in parts of the southern and southwestern US. In central Pennsylvania, yields during the following year were reduced by 177, 118, and 85 kg ha^{-1} per day of delay in planting after August 1 for alfalfa, red clover and birdsfoot trefoil, respectively (Hall 1995a). At the same location, when planted before August 30, orchardgrass and perennial ryegrass did not obtain maximum yield.

Legumes planted in late summer–early fall have few problems with damping-off diseases but are more susceptible to sclerotinia crown and stem rot (*Sclerotinia trifoliorum* Eriks) (Rhodes and Myers 1987). Early fall seedings are less susceptible than late fall seedings.

Warm-Season Forages

Much like cool-season forage species, warm-season species are established during spring and/or early summer to avoid summer drought and improve establishment before cool temperatures slow or stop their growth in the fall (Table 25.1). In general, warm-season species have lower seedling vigor from seedings than cool-season species and several are established using stem, stolon, or rhizome fragments Table 25.1.

Seeding Depth for Tilled Seedbeds

Optimum seeding depth varies with forage species, soil type (sandy, clay, or loam), soil moisture, time of seeding, and firmness of seedbed (Table 25.2). Seed of forage grasses and legumes are relatively small and have a very small supply of stored food to support the developing

Table 25.1 Establishment characteristics of selected warm-season forages

Species	Propagation method	Timing	Ease of establishment
		Legumes	
Aeschynomene	Seed	Spring to early summer	Moderate
Alyceclover	Seed	Spring to early summer	Moderate
Leucaena	Seed	Early summer	Difficult
Rhizoma peanut	Vegetative	January–March	Difficult
		Perennial grasses	
Bermudagrass	Vegetative	February–July	Easy
Bahiagrass	Seed	Early to late spring	Difficult
Big bluestem	Seed	Early to late spring	Difficult
Dallisgrass	Seed	Early to late spring	Difficult
Switchgrass	Seed	Early to late spring	Moderate
		Annual grasses	
Sorghum	Seed	Late spring to early summer	Easy
Sudangrass	Seed	Late spring to early summer	Easy
Millet	Seed	Late spring to early summer	Easy

Table 25.2 Percentage of established plants per 100 seeds planted in spring in Wisconsin at various depths in soils differing in texture

	Sand			Clay			Loam		
Depth (cm)	1.2	2.5	5.0	1.2	2.5	5.0	1.2	2.5	5.0
Species				*Established plants (%)*					
Alfalfa	71	73	40	52	48	13	59	55	16
Red clover	67	66	27	40	35	7	47	45	13
Bromegrass	71	64	29	56	37	6	68	50	19
Orchardgrass	61	56	13	60	26	1	56	39	16

Source: Sund et al. (1966).

seedling. Seed planted deeper in the soil depend on seed reserves longer before becoming self-sufficient. Seed placed too deep are less likely to emerge and be vigorous, whereas seed placed on the surface, at very shallow depths, or in a loose or cloddy seedbed, often do not have adequate seed-to-soil contact to absorb water and germinate.

A firm seedbed is essential for proper seed placement, good seed-to-soil contact, and successful establishment. Seed should be covered with enough soil to provide moist conditions for germination but not so deep that the shoot cannot reach the surface before reserves have been exhausted. In general, best results in humid areas are obtained when seed is placed 0.5–1.5 cm deep. More arid areas may require a slightly greater seeding depth, but seed placed deeper than 2.5 cm may not emerge or may be so weakened that survival is reduced (Table 25.2). More specifically, optimum seeding depths are 0.6–1.2 cm on clay and loam soils and 1.2–2.5 cm on sandy soils (Table 25.2). Shallower depths are better for early spring seedings because moisture is usually abundant, and the surface of the soil is warmer than at lower levels. Deeper depths are recommended for late spring and summer seedings when soils are warmer, but soil moisture conditions may be less favorable.

Seed-to-Soil Contact

Forage seed need to absorb at least their own weight in water before germination is initiated (Hall 1995b). The absorbed water generally moves into the seed from surrounding soil. Adequate seed-to-soil contact achieves maximum water movement into the seed in the shortest time. Field situations that do not promote good seed-to-soil contact (cloddy or loose soil) generally result in extended germination periods and variable emergence. Firming the soil with a cultipacker or press wheels on a grain drill after seeding increases seed-to-soil contact and improves seedling emergence compared to not firming, especially with light-textured soils (Tesar and Marble 1988).

Seeding Rates

Recommended seeding rates for a given species or mixture vary from state to state, and in some cases even within states, due to differences in soils, climate, and typical establishment methods. It is common for no more than one third of the sown seed to produce seedlings, and for only about half of those to survive the first year (Vough et al. 1982; Hall 1995c; Nelson et al. 1998). In addition, increasing seeding rates of alfalfa did not increase seeding-year yield or stand longevity (Hall et al. 2004). Consequently, recommendations on rates are usually given in ranges. For example, the recommended rate for pure seedings of alfalfa in Maryland is 17–20 $kg\,ha^{-1}$, Iowa 13–22 $kg\,ha^{-1}$, central California 22–34 $kg\,ha^{-1}$, and Southern California 28–39 $kg\,ha^{-1}$. Selecting a specific seeding rate within the recommended ranges depends on several factors.

Seed Size

Forage species vary greatly in the seed size. For example, tall fescue seed is approximately five times larger than timothy seed. Improved forage establishment has been linked to larger seed size (McKersie et al. 1981; Sanderson et al. 2002). However, the larger and more practical effect of seed size is achieving a proper seeding rate. Seed is usually sold on a weight basis, whereas forage seeding devices typically dispense seed based on volume and not number of seeds. Consequently, it is important to calibrate seeders for each species. Using the same seeder setting for timothy and tall fescue would result in either too much timothy or too little tall fescue.

Soil Type and Fertility

Lower rates can be used when seeding on light, sandy soils if there is adequate soil moisture because seedling emergence has less physical resistance to emergence than in heavy soils. Higher rates should be used on low-fertility soils because seedlings will tend to be less vigorous, more seedling losses are likely to occur, and plants will be smaller with less spread of the crowns.

Amount and Distribution of Rainfall

In areas of limited rainfall and on light, sandy soils, lower seeding rates are used since soil moisture may not be sufficient to support denser populations. Also, seedling losses due to damping-off diseases are generally lower than in more moist conditions, and competition from weeds may be lower.

Condition of the Seedbed and Seeding Method

Lower rates can be used when seeding into well-prepared seedbeds where uniform seeding depth can be controlled, and soil coverage and good seed-to-soil contact can be achieved. Higher rates are recommended when seedbed preparation and seeding techniques are not optimal.

Seed Quality

Higher rates are necessary if the seed lot has low germination, a high percentage of hard seed, or a low percentage of pure seed. The term **pure live seed** (PLS) is used to express the quality of the seed, even though it is usually not listed on the seed analysis label. The percentage PLS is determined by multiplying the percentage of seed purity by the percentage of germination and dividing by 100. The certified seed label shows the tested percentage for both that is based on analysis.

The PLS provides one way to compare the quality of different seed lots. For example, if one lot of seed is 98% pure and has 90% germination, the PLS is 88.2%. Another lot with 90% pure seed and 98% germination would also have 88.2% PLS. However, the first lot contains only 2% of something other than pure seed; the second lot has 10%. The label will list the other components such as other crop seeds, weed seeds, or inert matter (see Chapter 32).

Inert matter refers to dirt, sand, stones, flower parts, stems, and other non-seed items that are in the seed bag. Some forage grass species have inherently high amounts of inert matter because seeds are enclosed by dried flower structures called palea and lemma. In some species, for example, smooth bromegrass or big bluestem, a relatively large palea and lemma surround the seed which makes the seed 'fluffy' and can create problems with accurate seed metering during planting.

Inoculation of Legume Seed

All legume seed should be inoculated with the proper strain of N-fixing bacteria (*Rhizobium* spp.) before seeding. Much of the alfalfa and clover seed being marketed today is **preinoculated**, i.e. has bacteria already applied. The seed should be planted before the date when inoculant becomes ineffective which is listed on the tag. All legume seed should be inoculated with fresh inoculant prior to seeding including seed in which the preinoculation has expired. It is critical to use inoculant with the proper strain of **rhizobia** because a strong specific relationship exists between the bacteria strain and the species of the host plant.

Heat, direct sunlight, and drying are detrimental to survival of rhizobia inoculant. Inoculant should be stored in a refrigerator and preinoculated seed should be stored in a cool, dry place from time of purchase until time of use. Also, the seed dealer should have stored the inoculant and pre-inoculated seed in similar conditions. The expiration date should be checked before purchasing any packaged inoculant or preinoculated seed.

Other Seed Treatments

Lime-coated seed improves the pH environment immediately surrounding the seed and is beneficial to bacterial growth and infection in areas of acid soils (White

1970). On soils near neutral in pH, the lime coat has not been helpful in improving stands or yields (Barnes 1978; Tesar and Huset 1979; Sheaffer et al. 1988). Seed treatment with a systemic fungicide such as metalaxyl [N-(2,6-dimethylphenyl)-N-(methoxy-acetyl)-alanine methyl ester] provides seedling protection against damping-off, seed rot, and root rot organisms (*Phytophthora* spp.) (Sheaffer et al. 1988).

Mycorrhizae Addition to New Forage Seedings

Mycorrhizae are fungi, different from rhizobia bacteria, that have a symbiotic relation with the roots of some specific forage species. The mycorrhizae attach to the roots and function as root extensions to assist the plant in nutrient (primarily phosphorus) uptake while the plant provides the fungi with sugars from photosynthesis (Allen et al. 2011). It was previously assumed that mycorrhizae played an insignificant role in the life of plants. However, more recent research has shown that some plant-mycorrhizae relationships are common with many range species and provide large benefits to the plant, especially in P deficient or saline soils (Ashrafi et al. 2014).

Mycorrhizae inoculant products are now commercially available that can be applied to the soil prior to planting or mixed with the seed at planting. However, to date, positive efficacy of these products has not been demonstrated with many cool-season forage species commonly grown in humid areas of North America.

Seeding with or Without a Companion Crop

A **companion crop** of small grain or a small grain–pea mixture can be used with spring seedings of legumes and grasses in northern latitudes to provide early ground cover while the forage seedlings become established, help reduce wind and water erosion, and deter weed invasion (Wollenhaupt et al. 1995; Sheaffer et al. 2014). It also provides a usable crop for grain, bedding, silage, or pasture. Any spring-seeded small grain can be used as a companion crop, but spring oat is least competitive and the most satisfactory (Chapko et al. 1991).

Companion crops compete with the forage seedlings for light, moisture, and nutrients (Hall et al. 1995; Hoy et al. 2002). But weeds, especially broadleaf types, can be even more competitive and compete over a longer time period than a grain companion crop (Laskey and Wakefield 1978). Spring-seeded small grains should be sown at rates less than 75% of the recommended rate for grain production to reduce competition for light, water during dry periods, and to decrease the likelihood of lodging of the companion crop in wet seasons.

To reduce competition, the companion crop should not be fertilized with high rates of N. Establishment is favored by early killing of the spring companion crop with an herbicide or by removal as silage or pasture (Curran et al. 1993). Also, insects that feed on forage seedlings below the companion crop canopy are not easily controlled with an insecticide (Stout et al. 1992).

Planting forages in the fall with winter grains is often not preferred because the long fall-development period allows the companion crop to become very competitive with the forage. In contrast, if forages are seeded into companion crops such as winter wheat in early spring, the small grain can be harvested or grazed during winter to reduce competition. Winter grains are excellent quality feed, but grazing must be discontinued before stem jointing if they are to be harvested for grain. Following harvest of the grain companion crop, the straw and tall stubble should be removed as soon as possible to avoid shading and smothering of young forage seedlings.

Spring-seeded forage legumes can be harvested later in summer if they reach the early bloom stage. However, red clover should be harvested to reduce flowering to maximize winter survival. Most cool- and warm-season grasses will remain vegetative in growth habit and can be grazed lightly in late summer. Higher seeding rates of the forage are recommended when seeding in spring without a companion crop since more plants are needed to obtain good ground cover and high yields the seeding year and to reduce weed invasion (Brown and Stafford 1970; Lanini et al. 1991).

Seeding perennial forages, especially cool-season grasses, in late summer after a small-grain harvest is often more practical than seeding in the spring with a companion crop. The seedlings do not have to compete with a companion crop or weeds, and the forage crop will not have added competition in the spring. Success of a late-summer seeding depends on adequate rain between the time of grain harvest and forage seeding. Ideally, weeds will germinate soon after the grain harvest and are destroyed with tillage operations or herbicides prior to forage seeding. Winter annual weeds, such as common chickweed and henbit, can be a problem with late-summer seedings, especially during mild winters. When this occurs, herbicides may be needed and should be applied at an early stage of weed development (see Chapter 28).

Herbicides are replacing companion crops for weed management during spring establishment of forage monocultures (Brothers et al. 1994). Preemergence herbicides, in combination with postemergence herbicides, if necessary, can give weed control and higher forage yields in the seeding year than normally obtained with companion crops (Hall et al. 1995). Also, when herbicides are used to effectively control weeds, forage seeding rates can often be reduced (Hoveland 1970; Moline and Robinson 1971).

Presently available herbicides are not registered for use with grass–legume mixtures during establishment. Decisions on whether to use a companion crop or an herbicide during forage establishment should be based on site-specific conditions such as erosion potential, weed populations, and forage needs during the establishment year (Hoy et al. 2002).

Seeding Method Dictated by Soil Preparation

Common terminology used to describe forage seeding methods includes *conventional tilled, frost, no-till, broadcast, fluid* or *suspension, drilled, band,* and *cultipacker.* These terms are not mutually exclusive. For example, conventional seedings might be broadcast-seeded or drill-seeded. Seeding methods in this chapter are categorized by the degree of soil tillage (tilled or no-till) conducted prior to seeding.

Tilled Seedbed

The purposes of tilling a **seedbed** are to loosen the soil, reduce existing vegetation, bury surface weed seeds, incorporate lime and fertilizer into the soil, and provide a smooth surface for seeding and harvesting operations. Disadvantages of a tilled seedbed are greater loss of soil moisture during tillage and increased potential for soil erosion until the crop is established (Wollenhaupt et al. 1995).

Inadequate tillage that leaves too much surface residue or trash may result in shallow seed placement due to seeding machinery not being able to cut through the residue. Cloddy or trashy seedbeds are usually too rough or uneven for uniform seed placement and soil clods are too coarse for good seed-to-soil contact. Excessive tillage results in poor seedling emergence due to fluffy, powdery seedbeds that dry out quickly or crust following rainfall. Crusting is particularly a problem with small-seeded legumes. Leaving small clods or soil granules can be beneficial to prevent soil crusting.

Broadcast Seeding

Broadcast seeding includes many seeding techniques, all of which spread the seed evenly on the soil surface. This contrasts with techniques that place the seed in distinct rows. Examples of broadcast methods are:

- *Cultipacker seeding.* Cultipacker seeders consist of fore and aft corrugated rollers with seed-metering boxes mounted between them (Figure 25.1). The first roller firms the soil into shallow corrugations; seed is dropped ahead of the second roller that covers and firms the soil around the seed. These seeders are widely used on tilled **seedbeds** because of the shallow depth of seed placement and good seed-to-soil contact.

  On medium- and heavier-textured soils, some seed remain on the top and sides of the ridges as well as at the bottom of the corrugations. Thus, seed are distributed across a range of depths, from on the soil surface to about 2.5 cm deep. On sandy soils, most seed falls to the bottom of the corrugations, giving deeper coverage. Using cultipacker seeders on heavy soils that have been finely tilled, increases the potential for crusting. In addition, soil coverage and seed-to-soil contact may be limited with cultipacker seeders when

Fig. 25.1. Using a cultipacker seeder on a tilled seedbed gives good depth control. The first corrugated roller (barely visible in photo) firms the seedbed while leaving grooves and depressions. The seed is dropped between the rollers, after which the second roller (in picture), which is offset, covers the seed and firms the soil over the seed to give good seed-to-soil contact.

heavy crop residues are present or when planting lightweight grass seed such as smooth bromegrass or wheatgrass.

- *Other broadcast seedings*. The advantages of cultipacker seeding can be accomplished with a single corrugated roller to firm the tilled soil into shallow channels. Seed can then be broadcast on the soil and covered by another trip across the field with the roller. Seed can be broadcast by using a spinner seeder, grain drill with small-seeded legume and grass seed attachments, large-capacity sprayers, or seeding from the air. Grain drills with boxes for small seeds often drop the seed from a height of about 60 cm. If using a grain drill, seeding should be done with the opener discs raised out of the soil to reduce placing seed too deep.

Distributing seed through sprayers is a relatively new technique often referred to as fluid or suspension seeding. It is very effective for broadcasting seed uniformly over large areas in a short time, usually by custom applicators. However, placement and efficacy of legume inoculant after fluid seeding has been questioned because the fluid either decreases the viability of the **rhizobia** or washes it off the seed.

Seeding in Rows

Seeding in rows that are 15–20 cm apart is effective since the grass seedlings will tiller and legumes will have crown development to display leaf area between the rows. This also allows seeding methods that are designed for seed placement and ensuring good seed-soil contact.

- *Grain drilling*. Grain drills with special boxes for small seeds and seed tubes extending to ground level can accurately meter the seed, but depth of seeding is hard to control allowing a considerable amount of seed to be covered too deeply, especially if the furrow openers (disks, shovels, or hoes) are set too deeply and the seed is dropped before or near the openers. Seed that falls beneath soil thrown up by the openers are usually covered with too much soil for the seedlings to emerge. If the seed furrow is too deep, seed falling to the bottom of the furrow may fail to emerge, especially if additional soil is washed into the furrow by rain. Higher seeding rates should be used to compensate for these losses.

Drills with presswheels to firm the soil around the seed generally assure good seed-soil contact and provide excellent results if a uniform shallow depth can be maintained (Figure 25.2). They also work better than cultipacker seeders when some crop residue remains on the soil surface. Drills without presswheels should be followed with a cultipacker to ensure adequate seed-to-soil contact. In some areas, producers lightly harrow to cover the seed with soil rather than use a cultipacker, but a cultipacker also firms the soil and provides better seed-to-soil contact.

- *Band seeding*. Some grain drills are equipped with attachments that place fertilizer in a band directly below but not in contact with the seed. A band of fertilizer is placed 3–6 cm deep in the opened furrow, after which, some soil can flow back into the furrow to cover the fertilizer. Seed tubes extend to within

Fig. 25.2. Press wheels on a drill seeder close the slit into which the seed has been dropped and firm the soil to give good seed-to-soil contact. A large box in front of the smaller seed box can be used for seeding a companion crop or dispensing fertilizer.

5–10 cm of the soil surface about 30–40 cm behind the openers to place the seed about 1–2 cm deep in a band a few cm above the band of fertilizer. Seed is dropped well behind the openers, thus separating it from the fertilizer band below. Best results are obtained with drills having presswheels that roll on the band to firm the soil around the seed and provide good seed-to-soil contact. This is especially important on light textured soils to reduce drying of the surface soil and seed. Band seeding is superior to other seeding methods, especially on soils with low-fertility levels and during seasons when environmental conditions are less favorable (Brown et al. 1960; Cramer and Jackobs 1963).

No-Till Seedbed

Technology and equipment are now available to establish excellent forage stands consistently without tillage (Figure 25.3). No-till seeding leaves vegetative or plant residue on the soil that reduces soil erosion, thus leaving the field in better physical condition and conserves soil moisture for germination and growth of the new seedling. No-till seeding also reduces fuel, labor, and time requirements for tillage, avoids problems of soil crusting that frequently occur with tilled seedbeds, allows seeding on an already firm seedbed, and keeps stones on or below the soil surface. However, no-till seedings do not uniformly smooth the soil surface to facilitate seeding and harvesting, and the residue can exacerbate sclerotinia crown and stem rot in legumes (Rhodes and Myers 1987).

Killing Vegetation Prior to No-Till Seeding

No-till seedings can be made into a variety of ground covers. Late-summer seeding into small-grain stubble is common because it allows great flexibility in terms of matching seeding dates, soils, and weather conditions. Many weed seeds will germinate within two to three weeks following small-grain harvest and can be killed with a nonselective herbicide such as paraquat (1,1′-dimethyl-4,4′-bipyridinium ion) or glyphosate [isopropylamine salt of N-(phosphonomethyl) glycine]. A second application should be made at seeding if a second flush of weeds has germinated.

Seeding can be made anytime within a three- to six-week period after herbicide application, depending on when soil moisture and other conditions are right. Since soil tillage is not required, seedings can be made much sooner following a rain, and the soil will retain more moisture for the new seeding than with a tilled seedbed. Stubble of the small grain and dead weeds conserve moisture, reduce soil erosion, and provide protection for the seedlings from intense sunlight and damage from blowing sand or soil particles.

Seedings can be made into the killed stubble of summer annual grasses, particularly sudangrass or sorghum-sudangrass hybrids and foxtail millet. These grasses grow vigorously during the summer and compete effectively with weeds. Allelopathic toxins, especially from sorghum species, may also inhibit germination and development of weeds. Thus, summer annuals are good **smother crops** in preparation for forage seedings. However, since they do grow vigorously, they can deplete

Fig. 25.3. Seeding forages into soybean stubble using a no-till drill. The disc openers in front till a narrow area, allowing the seed to be placed ahead of the presswheels that regulate seeding depth and firm the seed and soil.

soil moisture if not harvested or killed three to six weeks prior to the forage seeding.

When perennial broadleaf weeds are present, selective translocated herbicides such as 2,4-D (2,4-dichlorophenoxyacetic acid) and dicamba (2-methoxy-3,6-dichlorobenzoic acid) will provide control, but treatment must be at least 30 days prior to seeding, and six months is better (Campbell 1976). Where both broadleaf and grass perennials are present, a translocated herbicide such as glyphosate should be used; with these herbicides a seeding delay of one to three weeks will give increased weed control during emergence and improved forage stands (Mueller-Warrant and Koch 1980; Welty et al. 1981).

Killing an old sod with a herbicide in preparation for no-till seeding does not kill insects directly, but drastically changes the food supply and habitat of the insect population, making young no-till seedlings especially vulnerable. Stands are often improved when insecticides are applied at seeding (Kalmbacker et al. 1979; Vough et al. 1982). Insect populations vary considerably, and the potential problem is difficult to predict, so some producers consider insecticides to be justified insurance.

Frost Seeding

Forage seed can be broadcast on the soil surface of fall-sown cereals or existing forage stands without tillage in late winter (i.e. frost seeding). Freezing and thawing action (honeycombing of the soil surface with ice crystals) leaves small pockets for broadcast seeding. Later rains will cause soil movement to cover the seed with soil and create good seed-to-soil contact. This works best in thin grain stands or where the grass sward has been grazed closely during fall and early winter to expose some soil surface which reduces competition and improves seed-soil contact.

Red clover is an excellent species to frost seed because it has excellent seedling vigor, adds nitrogen to the system, establishes better via frost seeding than most other legumes and the seed is relatively inexpensive, but success is possible with most species (Undersander et al. 2001). Frost seeding is generally less successful than seeding in rows with a no-till drill (Moorhead et al. 1994), but it is low cost. Frost seeding is most successful during periods in late winter when the ground is freezing and thawing daily, and the soil surface is moist.

Seeding in Rows

There are many different types of no-till seeders that place the seed in rows. While these machines differ widely in price and design, all can produce satisfactory stands when properly adjusted and operated. Some features that improve reliability of no-till seeders include:

1. Heavy enough to ensure proper penetration into the soil.
2. Rolling or power-driven coulters to cut through layers of thatch or mulch.
3. Double-disc or other types of seed placement units that line up precisely behind the slit created by the cutting coulter (Figure 25.4).
4. Depth bands, wheels, or other mechanisms to control seed depth on each seeding unit.
5. Units that operate independently to follow the soil terrain.
6. Presswheels to ensure good seed-to-soil contact. The importance of presswheels increases as sand content of the soil increases, especially when soil moisture is low. In clay soils with high moisture content, presswheels can even be detrimental. In these cases, it may be desirable to remove the presswheel, leaving the furrow open for the seed to germinate and develop at the bottom of the open slit. Closing the slit with a presswheel may cover the seed too deeply and restrict emergence.

Establishing from Vegetation (Sprigging)

Vegetative propagation, termed **sprigging**, is the main method of planting forages that do not have viable sexual seed. This requires planting aboveground stems, stolons, or rhizomes that have one or more nodes that are capable of producing **axillary buds** and adventitious roots.

Some warm-season perennials such as hybrid bermudagrass, limpograss, and perennial peanut rely on this type of vegetative propagation (Prine et al. 1986; Baseggio et al. 2015a). This method also allows for the staging of plant growth for harvesting or grazing (Hanna and Anderson 2008). While vegetative propagation permits the reproduction and commercialization of superior hybrid genotypes, it comes with added cost due to the special equipment required, cost of maintaining the nursery that will supply the vegetative parts, bulkiness of the materials and cost of added labor.

Vegetative Propagation by Rhizomes

This process requires digging of dormant rhizomes (Figure 25.5) and immediate planting. Factors to consider are rhizome planting rate and depth. The quicker a fully productive stand is desired, the higher the planting rate that should be used. A high rate of planted rhizomes (volume of $10\,m^3\,ha^{-1}$) in the case of 'Tifton 85' bermudagrass resulted in higher shoot density than medium ($5\,m^3\,ha^{-1}$) or low ($2.5\,m^3\,ha^{-1}$) rates, but medium rates provided the greatest cost-benefits (Baseggio et al. 2015b).

Planting depth of the rhizomes is a compromise between shallow planting favoring sprouting and emergence and deep planting aiding in winter survival and moisture retention. If planted too deep, the new growth may use all its reserves and die before emerging (Stichler and Bade 2003). Under dry-land conditions, planting rhizomes up to 12 cm deep is adequate, whereas a shallower

Fig. 25.4. Close-up photo showing double-disc openers and presswheel on the no-till drill in Figure 25.3. Note the large tube above the disc openers for delivery of fertilizer, followed by the smaller tube for delivering forage seed behind the openers after soil has flowed back into the disturbed area. The presswheels help control depth and firm the disturbed area to ensure good soil-to-seed contact.

(a)

(b)

Fig. 25.5. Rhizome digging apparatus (a) and transferring dug rhizomes into semi-trailer (b).

depth of 6 cm can be used under irrigated conditions (Baseggio et al. 2015b).

Vegetative Propagation by Stolons

With adequate moisture conditions, stolons of bermudagrass produce more shoots than do rhizome fragments, which allows for quicker ground cover (Fernandez 2003). Harvesting stolons for planting does not require digging like for rhizomes. Use of stolons over rhizomes also has the advantage of lower cost, greater shoot production, and more planted area per area of nursery. Nevertheless, planting with stolons is associated with higher frequency of establishment failures due to stolon desiccation in challenging environmental conditions.

Each sprig of bermudagrass stolon should have six or more nodes for planting. This is usually achieved when donor plants have 8–12 weeks of regrowth and measure 60 cm in circumference. Larger stolons contain more carbohydrate reserves which allows greater success. Larger stolons that grow larger before cutting produce a higher number of nodes and accumulate more reserves for better rooting. Planting rates can range from 1500 to 2500 kg ha^{-1}, with planting rates of 2000 kg ha^{-1} or greater being recommended to achieve rapid and successful establishment (Baseggio et al. 2015b).

Planting depth is usually from 6 to 12 cm. The harvested stolons have few roots and experience rapid desiccation of the planted material. Thus, the planted stolons require frequent and abundant rainfall over a relatively long period of time for increased establishment success. Strategies for harvesting and planting stolons require that planting equipment be set to ensure that part of the stolon containing the **axillary bud** remains uncovered (Baseggio et al. 2014).

Renovating and Planting into Existing Forage Stands

Renovation is the improvement of a forage stand by the partial or complete destruction of the sod, plus liming, fertilizing, weed control, and seeding to establish or re-establish desired forage plants without an intervening crop. Productivity of most permanent and **naturalized pastures** in the humid US and Canada with desired species can be greatly improved, without renovating, by using proper fertilization and grazing management. However, a change in the forage species present may be needed in some situations. Renovation of these stands could improve production more than twofold, depending on soil characteristics and the condition of the existing sod (Evers 1985; Carlassare and Karsten 2002).

No-till technology enables renovation without plowing or disking through management and use of herbicides that suppress existing vegetation. If properly adjusted, no-till drills can place forage seeds at the proper depth and ensure good seed-to-soil contact in the old sod. While most pasture improvements using sod-seeding have taken place in humid regions, the technique has also been used effectively on some western rangelands (Willard and Schuster 1971).

Frequently, legumes are added to existing grass forages during renovation. Red clover is more easily established than most legumes due to seedling vigor and shade tolerance and should usually be a part of most mixtures (Decker and Dudley 1976). It produces high yields the year of seeding but seldom persists more than three years under grazing. Species like birdsfoot trefoil and kura clover are slower to establish but stands persist longer than red clover.

A combination of legume species should provide good production for the first year or two from one species, e.g. red clover, with longer-lasting results from another. Since red clover is so competitive, no more than 2.5 kg seed ha^{-1} should be used in mixtures with other legumes. Species competition can be reduced by seeding species in alternate rows, but this requires a seeder with two legume hoppers or dividers in a single box (Decker et al. 1976). Seeding rates for sod-seeding are normally the same as for conventional seedings. Increased rates may be justified when competition is high and available moisture is low (Groya and Sheaffer 1981; Sheaffer and Swanson 1982).

Suppressing Existing Vegetation

Fields being prepared for renovation benefit from intensive, close grazing or close clipping during the fall prior to a late-winter seeding or, four to six weeks prior to a late-summer seeding. The weakened sward should be allowed to regrow sufficiently to allow for effective herbicide retention. Vegetation suppression prior to seeding conserves soil moisture and markedly reduces competition with the young seedlings during establishment. Herbicides for suppression can be applied broadcast or in bands if an adequate stand of desirable grasses is present and if the renovation is being made primarily for introducing legumes. Broadcast applications of paraquat or glyphosate are recommended if productive species are not present in the existing sod and both grass and legume species are being seeded in late summer.

Selection of paraquat or glyphosate and the application rate depend on the type of vegetation present. If the pasture has been closely grazed or mowed, plants should produce 2.5–5.0 cm of regrowth before spraying to obtain more effective control. This should be followed by an application of paraquat at the time of seeding for control of weeds that germinated or regrew after the herbicide application four to six weeks earlier.

Band application of herbicide entails applying a band of spray over the seed rows so that about half of the field surface area is sprayed; for example, spray bands 10-cm wide with the later seeding to be in 20-cm rows allows unsprayed bands of 10 cm between each row. Advantages of band application are: (i) Fewer weeds, especially broadleaf weeds, and summer-annual grasses are near and compete with the new seedlings. (ii) The unsprayed area continues to grow and serve as a companion crop. (iii) The unsprayed area provides pasture for grazing much sooner than if all the sod is suppressed or killed. However, caution must be exercised to avoid grazing damage or pugging, in which the hoof action of animals under wet conditions compacts soil and damages the new seedlings. (iv) Pasture production later in the seeding year is of higher quality because legumes can support animal gains that are near double those of unimproved grass swards (Decker et al. 1976; Taylor and Jones 1982; Badger 1983). (v) Band spraying allows desirable species to remain in the pasture while adding other desirable species.

Improving Quality of Tall Fescue Pastures

The sulfonylurea broad leaf herbicide Metsulfuron-methyl {Methyl 2-[[[[(4-methoxyl-6-methyl-1,3,5-triazin-2-yl)-amino]carbonyl]amino]sulfonyl] benzoate} suppresses seedhead emergence in tall fescue (Aiken et al. 2012; Goff et al. 2012) resulting in a more favorable leaf-to-stem ratio and higher quality. In addition, ergovaline, a substance toxic to grazing livestock, is produced by a fungal endophyte (Chapters 16 and 35) in older cultivars of tall fescue (primarily Kentucky 31 E+). As tall fescue becomes mature and produces seedheads, the concentration of ergovaline is increased within the seedhead of the plant. Suppressing the emergence of tall fescue seedheads provides several benefits to the pasture and grazing livestock.

Suppressing the emergence of the seedhead in tall fescue can reduce its population within pastures. For example, Goff et al. (2014) showed a reduction in tall fescue population when metsulfuron was applied early in the spring (mid-March in central Kentucky) compared to a fall and late-April application. They attributed this to the likelihood that the spring application was made when tall fescue plants were most susceptible to stresses during the season.

Goff et al. (2014) also found that increasing the rate of application of Chaparral™ herbicide (potassium salts (2-pyridine carboxylic acid, 4-amino-3,6-dichloro) and metsulfuron) from 35 g to 140 g ha^{-1} in early spring doubled the loss of tall fescue crowns. Across all application timings, density of tall fescue crowns declined linearly ($r^2 = 0.89$) as concentration of herbicide increased from 35 g to 140 g ha^{-1}. They also reported that populations of other desirable forage species increased when the percentage of tall fescue crowns declined.

Treating tall fescue with metsulfuron to reduce seedhead emergence improved overall forage quality including higher crude protein and relative forage quality. Treatment reduced acid detergent fiber, neutral detergent fiber and ergot alkaloid levels of the spring harvest. This was likely a result of the tall fescue remaining in a vegetative state (Israel et al. 2016). There was no difference in forage quality of treated and non-treated tall fescue at the summer harvest (Israel et al. 2016), indicating that forage quality improvements from metsulfuron application result primarily from seedhead suppression.

Williamson et al. (2016) reported improvements in pasture carrying capacity, stocking rate and average daily gain when mixed cool-season perennial pastures were treated with metsulfuron to suppress extension of seedheads. Preferential selective grazing of tall fescue in treated pastures was reported over other cool-season perennial forages (Aiken et al. 2012; Williamson et al. 2016).

No-Tilling Annuals into Pastures

Both winter- and summer-annual species can be seeded to increase yield and extend the grazing season (Hart et al. 1971; Hoveland and Carden 1971; Templeton and Taylor 1975; Hoveland et al. 1978; Belesky et al. 1981; Rao et al. 2000). The practice is very successful in southern latitudes with fall seedings of cool-season annuals such as annual ryegrass into warm-season perennial grasses such as bermudagrass. Seeding should be delayed until cool night temperatures slow the growth of the warm-season grass and reduce its competitiveness.

In areas of the US, between Interstate Highways 40 and 80, the greatest advantage from winter annuals, such as wheat, triticale, or annual covers, is for earlier spring grazing, while further south, fall and winter grazing of sorghum or sorghum × sudangrass are feasible. Seeding annuals into heavy grass swards allows continued grazing during extended periods of precipitation without the soil damage that often occurs when annuals are seeded into tilled fields (Decker 1970).

Winter-annual forages (annual ryegrass, clovers, and small grains) sod-seeded into bermudagrass pastures work especially well since there is minimum competition between the two species (Fribourg and Overton 1973; Decker et al. 1974). Management of such combinations to lengthen the grazing season with a higher quality forage in late fall and winter is relatively simple, and yield potentials are higher than from bermudagrass pastures alone. A disadvantage is that seedings must be made each fall when soil moisture may be limiting.

It is also possible to sod-seed cool-season perennial forages such as white clover mixed with tall fescue or orchardgrass into dormant warm-season grass swards (Fribourg and Overton 1979; Fribourg et al. 1979). The two types of perennials, with contrasting seasons of growth, can extend the overall grazing season without annual sod-seeding. But management of these mixtures of perennials is more complicated than merely adding a winter-annual species. Grazing management of each species and rates and timing of N applications are critical because the cool-season perennial species must survive during the summer when warm-season grasses are most competitive.

Spring seeding of summer-annual forages such as sudangrass or pearlmillet into perennial cool-season swards can improve summer production which provides improved forage distribution throughout the grazing season, but also adds more challenges. Seedings are made in late spring when soil moisture is generally low. In addition, to achieve the potential of the summer annual, close grazing and/or an herbicide application may be required to weaken the existing cool-season sward before seeding the annual (Hart et al. 1971; Belesky et al. 1981).

No-Tilling Cool-Season Species into Warm-Season Species

Warm-season perennial species in the southern US can be **overseeded** with winter annual grasses (Alava 2013; Bartholomew 2013) or legumes (Evers 2005; Biermacher et al. 2012) for winter and spring grazing. This is a common practice to extend the forage supply by 6–12 weeks (Freeman et al. 2014). While the cool-season forage brings added benefits of yield and quality, there is a cost to warm-season production due to competition in spring by the cool-season species (Alava 2013).

The first step in **overseeding** legumes is to determine the pH to learn if and how much lime is needed. Surface applied lime needs time (several months to a year) to lower the pH. The warm-season grass should be grazed very short and the area disked lightly before planting the cool-season species. The light disking action will mix any lime remaining on the surface into the upper layer of soil and open the sod without killing it, which will provide for better cool-season seed-soil contact. The planting of cool-season forages with a small-grain drill should take place after the first frost that reduces growth of the warm-season species, and a good rain.

The large box on the no-till drill can be used to plant small grains or the smaller box for clovers and ryegrass. Rates for overseeding small grains are 95–100 kg ha^{-1} to a seeding depth of 2.5–3 cm. For ryegrass, the recommended rate is 25–30 kg ha^{-1} at a seeding depth of 0.5–1 cm. Studies with annual ryegrass and bermudagrass show that warm spring temperatures will require early removal of cool-season grass to reduce competition and allow vigorous bermudagrass regrowth in spring (Alava 2013).

Management of New Seedings

Optimizing growth of new forage seedlings by minimizing weed and insect competition, maintaining optimum soil fertility, and employing optimum harvest management is critical to maintain plant vigor. The most productive stands result when weed competition is minimized (Brothers et al. 1994). Mowing weeds for control in new seedings should be delayed until weed growth is above the cutting height. This avoids branching and regrowth that can result in even greater competition. Most forages regrow from crown buds near or below soil level (sweet clover and annual lespedeza are exceptions) and are not usually seriously damaged by low and infrequent clipping (Sheaffer 1983).

New seedings are also subject to winter damage. Pasturing or clipping new seedings should end four to six weeks before a **killing frost**, to allow plants to harden for winter. Less winter damage will occur with moderate grazing just before or after frost, after plants are hardy, than by continuous grazing until cold temperatures terminate growth. Top growth serves as protective mulch for young plants to help insulate the soil during the winter and minimize frost heaving.

References

Aiken, G.E., Goff, B.M., Witt, W.W. et al. (2012). Steer and plant responses to chemical suppression of seedhead emergence in toxic endophyte-infected tall fescue. *Crop Sci.* 52: 960–969.

Alava, E. (2013). Annual ryegrass removal effects on regrowth of overseeded bermudagrass. 'Tifton 85' bermudagrass grazing management effects on animal performance and pasture characteristics. PhD dissertation. University of Florida. http://ufdcimages.uflib.ufl.edu/UF/E0/04/53/56/00001/ALAVA_E.pdf (accessed 10 October 2019).

Allen, M., Swenson, W., Querejeta, J.I. et al. (2011). Ecology of mycorrhizae: a conceptual framework for complex interactions among plants and fungi. *Annu. Rev. Phytopathol.* 41: 271–300.

Ashrafi, E. and Zahedi, M. (2014). Co-inoculations of arbuscular mycorrhizal fungi and rhizobia under salinity in alfalfa. *Soil Sci. Plant Biol.* 60: 619–629. https://doi.org/10.1080/00380768.2014.936037.

Badger, T.H. (1983). Evaluation of sod-seeded pasture renovation. Master Thesis. University of Maryland.

Barnes, D.K. (1978). Effects of treated seed on inoculation, establishment and yield of alfalfa. In: *Proceedings 8th Annual Alfalfa Symposium*, 52–55. Davis, CA: Certified Alfalfa Seed Council.

Bartholomew, P.W. (2013). Productivity of annual and perennial cool-season grasses established in warm-season pasture by no-till overseeding or by conventional tillage and sowing. *Forage Grazinglands* https://doi.org/10.1094/FG-2013-0621-01-RS.

Baseggio, M., Newman, Y.C., Sollenberger, L.E. et al. (2014). Stolon type and soil burial effects on 'Tifton 85' bermudagrass establishment. *Crop Sci.* 54: 2386–2393.

Baseggio, M., Newman, Y.C., Sollenberger, L.E. et al. (2015a). Stolon planting rate effects on Tifton 85 bermudagrass establishment. *Agron. J.* 107: 1287–1294.

Baseggio, M., Newman, Y.C., Sollenberger, L.E. et al. (2015b). Planting rate and depth effects on 'Tifton 85' bermudagrass establishment using rhizomes. *Crop Sci.* 55: 1338–1345.

Belesky, D.P., Wilkinson, S.R., Dawson, R.N., and Elsner, J.E. (1981). Forage production of a tall fescue sod intercropped with sorghum x sudangrass and rye. *Agron. J.* 73: 657–660.

Biermacher, J.T., Reuter, R., Kering, M. et al. (2012). Expected economic potential of substituting legumes for nitrogen in bermudagrass pastures. *Crop Sci.* 52: 1923–1930. https://doi.org/10.2135/cropsci2011.08.0455.

Brothers, B.A., Schmidt, J.R., Kells, J.J., and Hesterman, O.B. (1994). Alfalfa establishment with and without spring-applied herbicides. *J. Prod. Agric.* 7: 494–501.

Brown, C.S. and Stafford, R.F. (1970). Get top yields from alfalfa seedings. *Better Crops Plant Food* 54 (2): 16–18.

Brown, B.A., Decker, A.M., Sprague, M.A. et al. (1960). Band and broadcast seeding of alfalfa-bromegrass in the Northeast. Maryland Agricultural Experiment Station, Bulletin No. A-108 and Northeast Regional Publication No. 41.

Campbell, M.H. (1976). Effect of timing of glyphosate and 2,2-DPA application on establishment of surface-sown pasture species. *Aust. J. Exp. Agric. Anim. Husb.* 16: 491–499.

Carlassare, M. and Karsten, H.D. (2002). Species contribution to seasonal productivity of a mixed pasture under two sward grazing height regimes. *Agron. J.* 94: 840–850.

Chapko, L.B., Brinkman, M.A., and Albrecht, K.A. (1991). Oat, oat-pea, barley, and barley-pea for forage yield, forage quality and alfalfa establishment. *J. Prod. Agric.* 4: 360–365.

Cope, W.A. (1982). Inhibition of germination and seedling growth of eight forage species by leachates from seeds. *Crop Sci.* 22: 1109–1111.

Cramer, S.G. and Jackobs, J.A. (1963). Establishment and yield of late-summer alfalfa seedings as influenced by placement of seed and phosphate fertilizer, seeding rate, and row spacing. *Agron. J.* 55: 28–30.

Curran, B.S., Kephart, K.D., and Twidwell, E.K. (1993). Oat companion crop management in alfalfa establishment. *Agron. J.* 85: 998–1003.

Decker, A.M. (1970). Crop combinations produce feed more efficiently. *Better Crops Plant Food* 54 (1): 6–9.

Decker, A.M. and Dudley, R.F. (1976). Minimum tillage establishment of five forage species using five sod-seeding units and two herbicides. In: *Proceedings of International Symposium on Hill Lands*, 3–9 Oct. (ed. J. Luchok, J.D. Cawthorn and M.J. Breslin), 140–146. Morgantown, WV: West Virginia University Books.

Decker, A.M., Retzer, H.J., and Dudley, R.F. (1974). Cool-season perennials vs. cool-season annuals sod seeded into a bermudagrass sward. *Agron. J.* 66: 381–383.

Decker, A.M., Vandersall, J.H., and Clark, N.A. (1976). Pasture renovation with alternate row sod-seeding of different legume species. In: *Proceedings of International Symposium on Hill Lands*, 3–9 Oct. (ed. J. Luchok, J.D. Cawthorn and M.J. Breslin), 146–149. Morgantown, WV: West Virginia University Books.

Eardly, B.D., Hannaway, D.B., and Bottomley, P.J. (1985). Nitrogen nutrition and yield of seedling alfalfa as affected by ammonium nitrate fertilization. *Agron. J.* 77: 57–62.

Evers, G.W. (1985). Forage and nitrogen contributions of arrowleaf and subtropical clovers overseeded on bermudagrass and bahiagrass. *Agron. J.* 77: 960–963.

Evers, G.W. (2005). A guide to overseeding warm-season perennial grasses with cool-season annuals. *Forage Grazinglands* 3: 1. https://doi.org/10.1094/FG-2005-0614-01-MG.

Fernandez, O.N. (2003). Establishment of *Cynodon dactylon* from stolon and rhizome fragments. *Weed Res.* 43: 130–138. https://doi.org/10.1046/j.1365-3180.2003.00324.x.

Freeman, S., Poore, M.H., Glennon, H.M., and Shaeffer, A.D. (2014). Winter annual legumes overseeded into seeded bermudagrass (*Cynodon dactylon*): productivity, forage composition, and reseeding capability. *Forage Grazinglands* 12: 1. https://doi.org/10.2134/FG-2013-0060-RS.

Fribourg, H.A. and Overton, J.R. (1973). Forage production on bermudagrass sods overseeded with tall fescue and winter annual grasses. *Agron. J.* 65: 295–298.

Fribourg, H.A. and Overton, J.R. (1979). Persistence and productivity of tall fescue in bermudagrass sods subjected to different clipping managements. *Agron. J.* 71: 620–624.

Fribourg, H.A., McLaren, J.B., Barth, K.M. et al. (1979). Productivity and quality of bermudagrass and orchardgrass-ladino clover pastures for beef steers. *Agron. J.* 71: 315–320.

Goff, B.M., Aiken, G.E., Witt, W.W. et al. (2012). Ergovaline recovery from digested residues of grazed tall fescue seedheads. *Crop Sci.* 52: 1437–1440.

Goff, B.M., Aiken, G.E., Witt, W.W. et al. (2014). Timing and rate of Chaparral treatment affects tall fescue seedhead development and pasture grass plant densities. Online. *Forage Grazinglands* https://doi.org/10.2134/FG-2013-0001-RS.

Groya, F.L. and Sheaffer, C.C. (1981). Establishment of sod-seeded alfalfa at various levels of soil moisture and grass competition. *Agron. J.* 73: 560–565.

Hall, M.H. (1995a). Plant vigor and yield of perennial cool-season forage crops when summer planting is delayed. *J. Prod. Agric.* 8: 233–238.

Hall, M.H. (1995b). *How an Alfalfa Plant Develops*. Woodland, CA: Certified Alfalfa Seed Council.

Hall, M.H. (1995c). Seeding rate effects on alfalfa plant density, yield, and quality. *Am. Forage Grassl. Counc. News* 4 (4): 6–8.

Hall, M.H. and Henderlong, P.R. (1989). Alfalfa autotoxicity fraction characterization and initial separation. *Crop Sci.* 29: 425–428.

Hall, M.H., Curran, W.S., Werner, E.L., and Marshall, L.E. (1995). Evaluation of weed control practices during spring and summer alfalfa establishment. *J. Prod. Agric.* 8: 360–365.

Hall, M.H., Nelson, C.J., Coutts, J.H., and Stout, R.C. (2004). Effect of seeding rate on alfalfa stand longevity. *Agron. J.* 96: 717–722.

Hanna, W.W. and Anderson, W.F. (2008). Development and impact of vegetative propagation in forage and turf bermudagrasses. *Agron. J.* 100: 103–107. https://doi.org/10.2134/agronj2006.0302c.

Hannaway, D.B. and Shuler, P.E. (1993). Nitrogen fertilization in alfalfa production. *J. Prod. Agric.* 6: 80–85.

Hart, R.H., Retzer, H.J., Dudley, R.F., and Carlson, G.E. (1971). Seeding sorghum and sudangrass hybrids into tall fescue sod. *Agron. J.* 63: 478–480.

Hoveland, C.S. (1970). Establishing sericea lespedeza at low seeding rate with a herbicide. *Ala. Agric. Exp. Stn. Circ.* 174: 1–12.

Hoveland, C.S. and Carden, E.L. (1971). Overseeding winter annual grasses in sericea lespedeza. *Agron. J.* 63: 333–334.

Hoveland, C.S., Anthony, W.B., McGuire, J.A., and Starling, J.G. (1978). Beef cow-calf performance on coastal bermudagrass overseeded with winter annual clovers and grasses. *Agron. J.* 70: 418–420.

Hoy, M.D., Moore, K.J., George, J.R., and Brummer, E.C. (2002). Alfalfa yield and quality as influenced by establishment method. *Agron. J.* 94: 65–71.

Israel, T.D., Bates, G.E., Mueller, T.C. et al. (2016). Effects of aminocyclopyrachlor plus metsulfuron on tall fescue yield, forage quality, and ergot alkaloid concentration. *Weed Technol.* 30: 171–180.

Jennings, J.A. and Nelson, C.J. (2002a). Rotational interval and pesticide effects on establishment of alfalfa after alfalfa. *Agron. J.* 94: 786–791.

Jennings, J.A. and Nelson, C.J. (2002b). Zone of auto-toxic influence around established alfalfa plants. *Agron. J.* 94: 1104–1111.

Kalmbacker, R.S., Minnick, D.R., and Martin, F.G. (1979). Destruction of sod-seeded legume seedlings by the snail *(Polygyra cereola)*. *Agron. J.* 71: 365–368.

Koch, D.W. and Estes, G.O. (1986). Liming rate and method in relation to forage establishment: crop and soil chemical responses. *Agron. J.* 78: 567–571.

Lanini, W.T., Orlaff, S.B., Vargas, R.N. et al. (1991). Oat companion crop seeding rate effect on alfalfa establishment, yield and weed control. *Agron. J.* 83: 330–333.

Laskey, B.C. and Wakefield, R.C. (1978). Competitive effects of several grass species and weeds on the establishment of birdsfoot trefoil. *Agron. J.* 70: 146–148.

Lowder, S.W. and Weber, J.B. (1982). Atrazine efficacy and longevity as affected by tillage, liming, and fertilizer type. *Weed Sci.* 30: 273–280.

Luu, K.T., Matches, A.G., and Peters, E.J. (1982). Allelopathic effects of tall fescue on birdsfoot trefoil as influenced by N fertilization and seasonal changes. *Agron. J.* 74: 805–808.

McKenzie, R.H. (2013). Phosphorus fertilizer application in crop production. *Alberta Agric. For. Agric. Facts*: 542–543.

McKersie, B.D., Tomes, D.T., and Yamamoto, S. (1981). Effect of seed size on germination, seedling vigor, electrolyte leakage and establishment of bird's-foot trefoil (*Lotus corniculatus* L.). *Can. J. Plant. Sci.* 61: 337–343.

Moline, W.J. and Robinson, L.R. (1971). Effects of herbicides and seeding rates on the production of alfalfa. *Agron. J.* 63: 614–617.

Moorhead, A.J.E., White, J.G.H., Jarvis, P. et al. (1994). Effect of sowing method and fertilizer application on establishment and first season growth of Caucasian clover. *Proc. N. Z. Grassl. Assoc.* 56: 91–95.

Mueller-Warrant, G.W. and Koch, D.W. (1980). Establishment of alfalfa by conventional and minimum-tillage seeding techniques in a quackgrass-dominant sward. *Agron. J.* 72: 884–889.

Nelson, C.J., Hall, M.H., and Coutts, J.H. (1998). Seeding rate effects on self-thinning of alfalfa. *Proc. Am. Forage Grassl. Conf.* 7: 6–10.

Peters, E.J. and Zam, A.H.B.M. (1981). Allelopathic effects of tall fescue genotypes. *Agron. J.* 73: 56–58.

Prine, G.M., Dunavin, L.S., Moore, J.E., and Roush, R.D. (1986). Registration of 'Florigraze' rhizoma peanut. *Crop Sci.* 26 (5): 1084–1085.

Rao, S.C., Coleman, S.W., and Volesky, J.D. (2000). Yield and quality of wheat, triticale, and elytricum forage in the southern plains. *Crop Sci.* 40: 1308–1312.

Renz, M. (2010). Establishment of forage grasses and legumes after fall herbicide applications. *Forage Grazinglands* 8 (1) https://doi.org/10.1094/FG-2010-0806-01-RS.

Rhodes, L.H. and Myers, D.K. (1987). Sclerotinia crown and stem rot of alfalfa. The Ohio State University. Plant Diseases AC 16.

Sanderson, M.A., Skinner, R.H., and Elwinger, G.F. (2002). Seedling development and field performance of prairiegrass, grazing bromegrass, and orchardgrass. *Crop Sci.* 42: 224–230.

Seguin, P., Sheaffer, C.C., Ehlke, N.J. et al. (2001). Nitrogen fertilization and rhizobial inoculation effects on kura clover growth. *Agron. J.* 93: 1262–1268.

Seguin, P., Sheaffer, C.C., Schmitt, M.A. et al. (2002). Alfalfa autotoxicity: effects of reseeding delay, original stand age and cultivar. *Agron. J.* 94: 775–781.

Sheaffer, C.C. (1983). Seeding year harvest management of alfalfa. *Agron. J.* 75: 115–119.

Sheaffer, C.C. and Swanson, D.R. (1982). Seeding rates and grass suppression for sod-seeded red clover and alfalfa. *Agron. J.* 74: 355–358.

Sheaffer, C.C., Hall, M.H., Martin, N.P. et al. (1988). Effects of seed coating on forage legume establishment in Minnesota. Minnesota Agricultural Experiment Station, Bulletin No. 584–1988.

Sheaffer, C.C., Martinson, K.M., Wyse, D.L., and Moncada, K.M. (2014). Companion crops for alfalfa establishment. *Agron. J.* 106: 309–314. https://doi.org/10.2134/agronj2013.0250.

Sheard, R.W. (1980). Nitrogen in the P band for forage establishment. *Agron. J.* 72: 89–97.

Stichler, C. and Bade, D. (2003). *Forage Bermudagrass: Selection, Establishment and Management*. E-179 4-03. College Station: Texas A&M University.

Stout, W.L., Byers, R.A., Leath, K.T. et al. (1992). Effects of weed and invertebrate control on alfalfa establishment in oat stubble. *J. Prod. Agric.* 5: 349–352.

Sund, J.M., Barrington, G.P., and Scholl, J.M. (1966). Methods and depths of sowing forage grasses and legumes. In: *Proceedings 10th International Grassland Congress, Helsinki, Finland*, 319–323.

Taylor, T.H. and Jones, L.T. Jr. (1982). Persistence and productivity of sod-seeded legumes compared with nitrogen fertilized grass sod. In: Agron. Abstr., 129. Madison, WI: ASA.

Templeton, W.C. Jr. and Taylor, T.H. (1975). Performance of big flower vetch seeded into bermudagrass and tall fescue swards. *Agron. J.* 67: 709–712.

Tesar, M.B. and Huset, D. (1979). Lime coating and inoculation of alfalfa. In: *Proceedings 16th Central Alfalfa Improvement Conference*, 1.

Tesar, M.B. and Marble, V.I. (1988). Alfalfa establishment. In: *Alfalfa and Alfalfa Improvement*. Agronomy Monograph 29 (eds. A.A. Hanson et al.), 303–322. Madison, WI: ASA.

Teutsch, C.D., Sulc, R.M., and Barta, A.L. (2000). Banded phosphorus effects on alfalfa seedling growth and productivity after temporary waterlogging. *Agron. J.* 92: 48–54.

Undersander, D.J., West, D.C., and Casler, M.D. (2001). Frost seeding into aging alfalfa stands: sward dynamics and pasture productivity. *Agron. J.* 93: 609–619.

VanToai, T.V. and Linscott, D.L. (1979). Phytotoxic effect of decaying quackgrass *(Agropyron repens)* residues. *Weed Sci.* 27: 595–598.

Vough, L.R., Decker, A.M., and Dudley, R.F. (1982). Influence of pesticide, fertilizers, row spacings, and seeding rates on no-tillage establishment of alfalfa. In: *Proceedings 14th International Grassland Congress, 15 June 1981, Lexington, KY* (eds. J.A. Smith and V.W. Hays), 547–550. New Zealand Grassland Association, Wellington, New Zealand. Westview Press.

Welty, L.E., Anderson, R.L., Delaney, R.H., and Hensleigh, P.F. (1981). Glyphosate timing effects on establishment of sod-seeded legumes and grasses. *Agron. J.* 73: 813–817.

White, J.G.H. (1970). Establishment of lucerne *(Medicago sativa* L.) in uncultivated country by sod seeding and oversowing. In: *Proceedings 11th International Grassland Congress, 13–23 April* (ed. M.J.T. Norman), 134–138. Surfers Paradise, Queensland, Australia.

Willard, E.W. and Schuster, J.L. (1971). An evaluation of an interseeded sideoats grama stand four years after establishment. *J. Range Manage.* 24: 223–226.

Williamson, J.A., Aiken, G.E., Flynn, E.S., and Barrett, M. (2016). Animal and pasture responses to grazing management of chemically suppressed tall fescue in mixed pastures. *Crop Sci.* 56: 2861–2869.

Wollenhaupt, N.C., Bosworth, A.A., Doll, J.D., and Undersander, D.J. (1995). Erosion from alfalfa established with oat under conservation tillage. *Soil Sci. Soc. Am. J.* 59: 538–543.

Young, C.C. and Bartholomew, D.P. (1981). Allelopathy in a grass–legume association: I. Effects of *Hemarthria altissima* (Poir.) Stapf. and Hubb. Root residues on the growth of *Desmodium intortum* (Mill.) Urb. and *Hemarthria altissima* in a tropical soil. *Crop Sci.* 21: 770–774.

CHAPTER

26

Fertilization and Nutrient Management

David J. Barker, Professor, *Horticulture and Crop Science, The Ohio State University, Columbus, OH, USA*
Steven W. Culman, Professor, *School of Environment and Natural Resources, The Ohio State University, Columbus, OH, USA*

Forage fertilization is among the management options having the greatest influence on forage yield and quality. The key objective in forage fertilization programs is to produce abundant, high-quality forage that minimizes the need for supplemental feed and optimizes forage utilization by livestock. Fertilization can also have significant off-farm impacts resulting from losses to air and water. Producers now face increasing regulation, as they optimize benefits on-farm, and minimize impacts off-farm.

Soil Fertility and Soil Health

Increasingly, we are taking a holistic approach to soil management, and that encompasses the concepts of both soil fertility and **soil health**. Soil microorganisms, pH, **soil structure**, **soil texture**, organic matter, water, and many other soil characteristics influence the provision of nutrients for forage growth. For example, soil compaction can limit forage root growth, **nutrient** uptake, legume nodulation, oxygen diffusion, and nitrogen efficiency. Any of these factors can limit total forage growth. Simply adding more fertilizer to compensate for an impaired soil characteristic, such as compaction, cannot be considered acceptable soil management. Good soil management is necessary for productive forage growth.

Soil fertility can be defined as the ability of the soil to provide the nutrients required for plant growth. While the primary component of soil fertility is the abundance of 18 key nutrients in desired amounts, biological (organic matter, soil microbes, etc.), chemical (pH, oxygen, etc.), and physical soil factors (porosity, compaction, drainage, etc.) also affect the supply of nutrients to plants. It can be as important to manage these factors as it is to manage the concentrations of soil nutrients by fertilization.

Soil health encompasses the ability of soil to provide ecosystem services that humans require. In most cases for forage-producing soils, the predominant service is the supply of forage for livestock (meat, dairy, fiber, hides, etc.). There are numerous other services provided by grasslands, and although most landowners are not remunerated for those services, this provision is increasingly being valued by society. These services include: (i) serving as a reservoir of nutrients (including a trap for nutrients prevented from reaching waterways), (ii) serving as a host for pollinators, (iii) erosion control, (iv) providing a water reservoir (infiltration), (v) water harvesting (surface-collected municipal water and quality water for streams and rivers), (vi) sustaining biodiversity (host for rare and native plant and animal species), (vii) providing a medium for soil biodiversity (including macro-fauna, such as earthworms), (viii) carbon sequestration (including absorption of CH_4), and (ix) esthetic purposes (including agri-tourism, landscape appearance, etc.).

Forages: The Science of Grassland Agriculture, Volume II, Seventh Edition.
Edited by Kenneth J. Moore, Michael Collins, C. Jerry Nelson and Daren D. Redfearn.

Soil Sampling

Accurate fertilizer recommendations depend on accurate information on the nutrient status of the production area. Many samples are needed from a given land area to accurately assess normal soil variation. Four variables should be considered when taking samples: (i) the spatial distribution of samples across the landscape, (ii) the depth of sampling, (iii) the time of year when samples are taken, and (iv) the frequency with which an area is sampled.

Soil samples used for nutrient recommendations should be taken from the same depth that was used in the research that generated the recommendations, normally the upper 15–20 cm (Vitosh et al. 1995). When sampling sites subjected to little or no inversion tillage, including those in established forages or for no-till crops, additional samples should be taken at 0–10 cm to assess acidification of the soil surface to make appropriate lime recommendations. Many southern states base fertilizer and lime recommendations for forages on a 0–15 cm depth sample, whereas other states recommend sampling at 0–7.5 or 0–10 cm (Hodges and Kirkland 1994). Prior to seeding or **sprigging**, most public laboratories recommend sampling the plow depth (0–15 cm). In fairly uniform areas smaller than 10 ha, a composite sample of 15–20 cores taken in a random or zigzag pattern is usually sufficient. Areas larger than 10 ha are normally subdivided.

Grid Sampling

Considerable variation in soil fertility may occur within hay and pasture fields due to soil type and topography that can be mapped using grid sampling (Leep et al. 2000). Data collected from a grid pattern can be used to identify variability in nutrient concentrations (Figure 26.1) that can be used for site-specific (within field) adjustments of fertilizer application. Grid points can be from 9 m up to 100 m; 8–10 cores are usually combined per point. Grid points, each representing 0.4 ha, were found to be economic on Michigan commercial alfalfa fields (Leep et al. 2002). Grid sampling is increasingly being used in pasture and hay fields even though nutrient levels in pastures are highly variable due to uneven distribution of dung and urine.

Areas with a uniform soil type, soil texture, and only moderate slope can be sampled together. Soil surveys may help identify distinct boundaries in soil types or management zones that may not be readily noticeable, yet soil fertility can vary significantly even where no manure or fertilizer has been applied. Grid sampling may be used to help define management zones, especially for pastures.

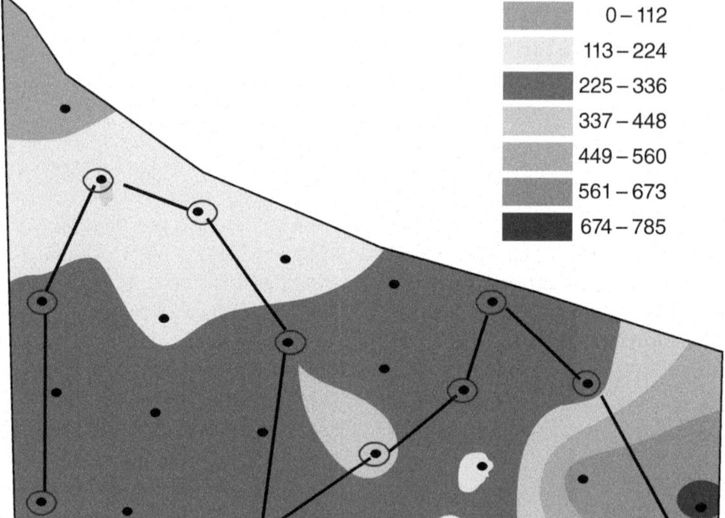

Fig. 26.1. Soil phosphorus map for a 2.5 ha pasture developed from 26 separate samples taken at grid points. The black line indicates the zigzag pattern and the points circled in black indicate where samples were collected and composited into one sample to represent the field.

Soil fertility levels in pastures and hay meadows with a history of manure or fertilizer application are usually highly variable. For example, after applying poultry litter for several years at rates of 1.1–7.3 Mg ha^{-1}, it took nearly twice as many samples (28 vs. 18) to obtain test results within 10% of the true value at high soil test phosphorus (P) levels (>150 mg kg^{-1}) than at low levels (<150 mg kg^{-1}) (Daniels et al. 2001).

Sampling Pastures

Uneven patterns of livestock defecation and grazing create spatial variability in the nutrient status of pastures. Nutrients ingested during grazing of a pasture tend to become concentrated in animal resting and drinking areas. Tall fescue pastures in Georgia accumulated 18% (low potassium, or K, rate) to 59% (high K rate) of the fertilizer K near the water source, shade, and mineral feeder (Wilkinson et al. 1989). The higher rate of redistribution and accumulation occurred with the higher fertility rate. Greater soil K was found near resting areas on pastures with high endophyte infection levels because animals spent less time grazing and congregated more near shade in the pastures than did those on low-endophyte pasture.

Mathews et al. (1994) found soil K levels in bermudagrass pastures in Florida ranged from 14 to 181 mg kg^{-1} Levels were greatest near shade and water, leading them to recommend zonal soil sampling of grazed pastures to more accurately assess the actual soil nutrient status. In grazed bermudagrass pastures in Oklahoma, Raun et al. (1998) found differences in surface soil test results for both mobile and immobile nutrients in soil samples taken less than 1 m apart. For example, among field locations, P recommendations ranged from 0 to 31 kg ha^{-1}, whereas K recommendations ranged from 0 to 108 kg ha^{-1}, but no P or K would have been recommended based on a field composite sample.

Spatial variability must be considered in nutrient-management planning, in developing sequential soil-sampling programs, and in development of sound regulations and public policy regarding nutrient application. Optical sensors are now available for scanning plants ahead of variable-rate applicators for N fertilization. Failure to consider variability in soil nutrient status can result in lost profitability and unrealistic environmental constraints to sound nutrient use.

Sampling over Time

To establish trends, samples should be collected at the same depth and time of year. Temporal variability can be minimized by sampling at similar moisture and temperature conditions each year. Fall sampling is preferred if lime applications are anticipated. The sampling pattern (zigzag, grid, random, or serpentine) should also be consistent to reduce errors associated with spatial variability.

In general, soil samples should be collected every three years from pastures, and fields used for silage or hay production should be sampled every two years to monitor changes in soil nutrients. Frequent sampling is also recommended in fields with low (<7 $cmol^{+}$ kg^{-1}) soil **cation exchange capacity** that are receiving manure application.

Interpretation of Test Results

Most soil testing and plant analysis laboratories use standardized procedures to determine fertilizer recommendations that are consistent with the regional climate and soil characteristics. Soil test values to a defined depth and yield responses to a range of levels of each nutrient are determined in field tests to develop a calibration curve. Most states use a rating scale of the soil test to determine fertilizer recommendations to meet a specified yield goal (Table 26.1).

Categories vary among labs and states because different extractants are used, depending on the parent material, to determine the amount of each nutrient available to the plant. Also, agronomic responses to nutrient applications in field tests used for calibration vary among soils and geographic areas (Hanlon 2001).

Plant Tissue Sampling

Plant tissue nutrient analyses identify deficiencies during the growing season and can serve as a nutrient management tool. Plant tissue analysis is more reliable than soil

Table 26.1 Example soil fertility rating scale used in Alabama and Florida public laboratories

Soil test rating	Expected crop yield potential (%)	Expected benefit from fertilization
Very low (VL)	<50	Very high
Low (L)	50–75	High
Medium (M)	75–100	Medium
High (H)	100	Low
Very high (VH)	100	None
Extremely high (EH)	100	None

Source: Hanlon (2001).

analyses for identifying sulfur (S) and **micronutrient** deficiencies. Nitrate (NO_3^-) analyses can identify potential animal toxicity, which can occur in some forage species under persistent drought, high manure-application rates, or with very high nitrogen fertilizer rates. With high nitrate, livestock producers can defer grazing or harvest and store high-nitrate hay separately for mixing with lower-nitrate feeds. Colorimetric 'quick tests' are available for nitrate, but these are qualitative. Suspect forages should be tested by quantitative laboratory tests.

Plant Parts to Sample

For alfalfa and other legumes, the top 15 cm of the plant should be sampled. For grasses, the entire aboveground portion can be collected, avoiding soil. Stubble and residues from previous growth should also be avoided. For silage crops such as corn or sorghum-sudangrass, the uppermost mature leaves should be sampled. A reliable sample should include at least 25–30 leaves.

Samples should be submitted in paper bags after air drying to prevent molding during shipment. If possible, air-dried forage samples should be submitted within 24–48 hours after collection.

When to Sample

Samples for assessment of nutrient deficiencies or imbalances should be taken just prior to **flowering** or whenever a nutrient imbalance is suspected. Separate samples should be collected from healthy or productive areas for comparison. Companion soil samples that match plant tissue samples can aid diagnosis of deficiencies or imbalances.

Interpretation of Results

Table 26.2 illustrates some nutrient sufficiency ranges for several forages grown in the US. When nutrient levels are above the critical range, response to the addition of a given nutrient is considered unlikely. Factors such as drought, strong soil acidity, soil compaction, shallow soil depth, a perched water table, diseases, herbicide injury, and insect damage also affect tissue composition and can make interpretation difficult.

Environmental Impacts of Nitrogen and Phosphorus

Nitrogen (N)

Fertilization with N is invariably associated with losses of N to the atmosphere and waterways. Losses can result from **volatilization**, denitrification, and leaching (Figure 26.2). Though many factors affect these losses, such as the fertilizer form, soil type, temperature, and rainfall, the predominant factor is the rate of N-fertilizer applied. Applications below 100 kg N ha^{-1} typically have tolerable environmental impacts, while applications exceeding 200 kg N ha^{-1} can result in excessive losses from pasture due to leaching, volatilization, and denitrification. Environmental impacts can be minimized by a number of management options including use of less labile N products, split applications of low rates, avoiding areas near waterways, and by avoiding unfavorable climatic conditions. To prevent off-site movement (leaching and/or runoff) of applied N, applications should be made just prior to, or during, active forage growth.

Phosphorus (P)

Although P is relatively insoluble, it can become an environmental issue in cases where erosion results in soil-absorbed P reaching waterways. Where **soil erosion** risk is minimal, much of the P lost from the site as runoff will likely be in the water-soluble form. On tilled soils, or where erosion is a concern, more of the P-loss will likely be as sediment-bound P. Runoff losses of P from pastures and hay meadows are normally below 0.5–1 kg of P ha^{-1} yr^{-1}, or less than 1% of P applied, assuming P rates of 50–100 kg ha^{-1} yr^{-1} (115–230 kg P_2O_5 ha^{-1} yr^{-1}). The natural loss of P to surface waters before human settlement has been estimated at about 0.4 kg P ha^{-1} yr^{-1} (National Research Council 2000; Snyder and Bruulsema 2002). With excessive applications of P, such as where excessive manure applications are made or where poorly timed or unincorporated (by gentle rains, subsurface placement, or tillage) applications are made, an increased risk of runoff loss of P to surface water exists.

Losses of N and P to surface waters can lead to over-enrichment and eutrophication of aquatic ecosystems (Sharpley et al. 1999), accelerate algae blooms, reduce dissolved oxygen in water bodies, and contribute to fish kills. Eutrophic conditions may also increase the cost of water purification by treatment plants.

State and Federal Regulation

In watersheds where N and P have impaired or threatened water quality, local, state, and national authorities are increasingly resorting to a regulatory approach for control of nutrient problems. This can include mandatory certification of fertilizer applicators, and the requirement of a comprehensive nutrient management plan. To date, most of the affected watersheds are impacted by a high concentration of **confined animal-feeding operations (CAFO)**, and large areas of annual cropping. Comprehensive nutrient management plans require strict attention to the rates, sources, placement, and timing of all nutrients.

Forage producers in many states have access to university-developed tools such as the P-index (Snyder et al. 1999) and N-index, which can help determine the risk of off-site movement of N and P. Comprehensive plans also involve calculation of the import and export of nutrients from fields and farms as feed, hay, live animals, milk, manure, and fertilizer. Careful consideration of nutrient needs and the potential for nutrient surplus development, as well as evaluation of the risk for offsite N and P losses, are helping forage producers prioritize use

Table 26.2 Nutrient sufficiency ranges or critical values for selected forage and hay crops

Forage	N	P	K	Ca	Mg	S	Fe	Mn	Zn	Cu	B	Reference
	(g 100 g^{-1} dry weight)						(ppm)					
Alfalfa	3.00–5.00	0.25–0.70	2.00–3.50	0.80–3.00	0.25–1.00	0.25–0.50	30–250	25–100	20–70	4–30	20–80	Campbell (2000)
Coastal bermudagrass	2.00–2.60	0.20–0.40	1.5–2.30	0.25–0.50	0.10–0.25	0.15–0.25	50–200	20–300	15–70	4–20	5–15	Campbell (2000)
Hybrid bermuda-grasses	2.00–4.00	0.25–0.60	2.00–3.00	0.25–0.40	0.18–0.30	0.18–0.50	50–350	25–300	20–50	5–25	6–30	Snyder (1998)
Corn	2.76–3.75	0.25–0.50	1.75–2.75	0.30–0.60	0.16–0.40	0.16–0.50	50–250	19–75	19–75	3–15	5.1–40	Kelling et al. (2000)
Kentucky bluegrass	2.60–3.20	0.28–0.36	2.00–2.40		0.40–0.48	0.16–0.24						Kelling and Matocha (1990)
Clovers		0.20–0.40	1.20–2.40			0.12–0.19						Kelling and Matocha (1990)
Red clover	3.01–4.5	0.29–0.60	1.8–3.0	2.01–2.60	0.22–0.60	0.27–0.30	31–250	31–120	18–80	8–15	31–80	Little and McCutcheon (1999)
Cool–season grasses	3.21–4.20	0.24–0.35	2.61–3.50	0.51–0.90	0.11–0.30	0.21–0.25	51–200	51–150	20–50	3–5	8–12	Little and McCutcheon (1999)
Dallisgrass	2.3–3.0	0.28–0.30	2.10–2.40									Kelling and Matocha (1990)
Tall fescue	3.20–3.60	0.34–0.40	2.80–3.20									Kelling and Matocha (1990)
Johnsongrass	1.60–1.80	0.20–0.25	1.6–1.80									Kelling and Matocha (1990)
Millet	2.20–3.00	0.22–0.30	2.30–3.80			0.14–0.18						Kelling and Matocha (1990)
Annual and perennial ryegrasses	3.50–4.00	0.36–0.44	2.8–3.20									Kelling and Matocha (1990)
Pangolagrass	1.70–2.00	0.16–0.24	1.60–2.00			0.20–0.25						Kelling and Matocha (1990)
Sorghum–sudangrass	1.08–2.20	0.20–0.30	1.90–2.80			0.14–0.18						Kelling and Matocha (1990)

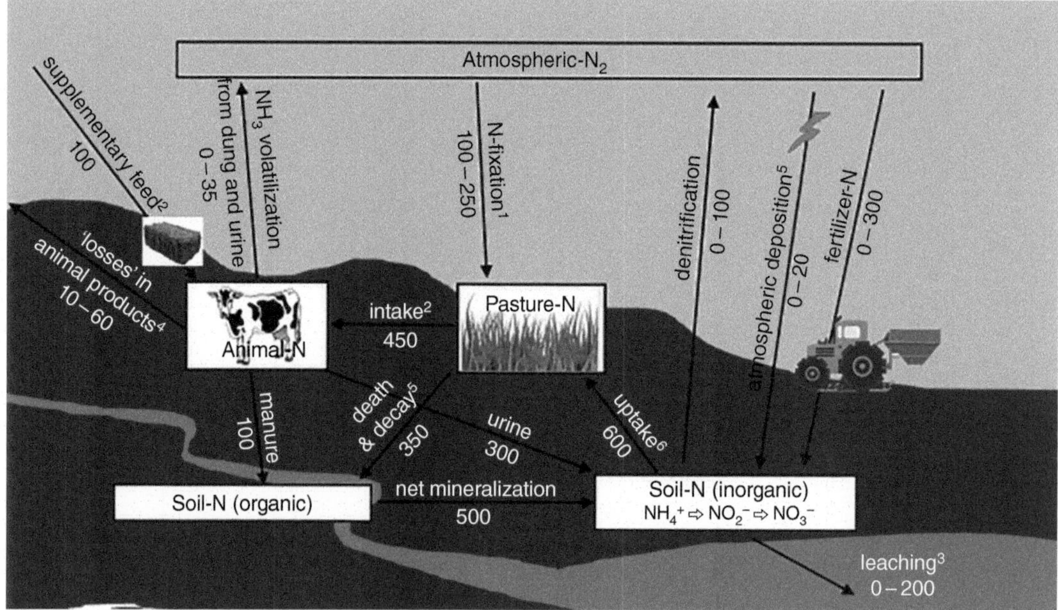

Fig. 26.2. Nitrogen cycle, showing the approximate rate (kg N ha^{-1} yr^{-1}) for a high producing dairy pasture. *Source*: Adapted from Steele (1982).

of nutrient resources where they can be safely and most efficiently used.

Nutrient Removal with Pasture vs Hay Production

Loss of nutrients (nutrient removal) from a pasture or hayfield is an inevitable consequence of finished hay/silage or livestock being removed as products from the production area. Grazing, on the other hand, recycles most of the nutrients as manure. Nutrient removal quantities from grazing and haying systems are given in Tables 26.3 and 26.4. Redmon (1996) reported that 10–15% of P and 0.2–0.3% of K are removed by grazing livestock compared with bermudagrass harvested as hay at a yield of 8.96 Mg ha^{-1} and orchardgrass at a yield of 6.72 Mg ha^{-1}. Continued removal of nutrients without replacement is not sustainable, and one of the most important strategies in fertilization is to replace nutrients lost by removal.

Nutrient Cycles

Nutrients such as N, P, and K are present in many different forms in soil, only a few of which are available to plants. Nutrient availability is dependent on biologic, geologic, and chemical processes. Potassium availability is largely under geologic control through **weathering**, which can provide substantial K for many plants, though removal of harvested crops such as alfalfa, ryegrass, and bermudagrass may deplete soil K. Nitrogen is an example of a nutrient under strong biologic control (Figure 26.2).

Table 26.3 Nutrient removal per unit yield of major forages (dry matter basis)

Forage	N	P_2O_5	K_2O
	kg (1000 kg)$^{-1}$		
Alfalfa	28	7	30
Annual ryegrass	39	8	31
Bahiagrass	22	6	17
Bermudagrass	23	6	25
Bromegrass	18	7	30
Clover-grass	25	7	30
Fescue	19	9	26
Orchardgrass	25	9	31
Sorghum-sudangrass	20	8	29
Timothy	19	7	31
Vetch	28	8	23
Corn silage (67% water)	4	2	4

Source: PPI/PPIC (2002).

Some elements such as P are under largely chemical control. Most soils strongly bind P, releasing only a small amount to the **soil solution** in a form that is available for plant uptake.

Used or lost nutrients must be replaced to maintain long-term productivity. Nutrients can be replaced by using manure or fertilizers. The various sources of

Table 26.4 Approximate quantities of nutrients removed under grazing and haying systems

	Nutrient removal (kg ha^{-1})				
System	N	P	K	Ca	Mg
Grazing[a]	14	3.8	0.84	7.28	0.84
Bermudagrass[b]	224	24.6	161.00	71.00	35.00
Orchardgrass[c]	195	35.8	241.00	39.00	21.00

Source: Redmon (1996).

[a]Assumes one 227-kg calf weaned per ha on a year-round basis. If more than 1 ha yr^{-1} is required to produce the calf, the values would be reduced.

[b]Assumes 8.96 Mg ha^{-1} hay yield with all forage removed as hay crop.

[c]Assumes 6.72 Mg ha^{-1} hay yield with all forage removed as hay crop.

nutrients behave differently in soil, and it is important to understand how to manage nutrients effectively.

Nitrogen

Forage Plant Requirements for Nitrogen

Adequate N nutrition is essential for the establishment and good early growth of grass and legume seedlings, for high forage yield and for stand maintenance. Nitrogen is usually the most abundant nutrient in plants (Tables 26.2 and 26.3) and is the most frequently deficient plant nutrient.

Role of Nitrogen in Forage Plants

Nitrogen is essential for amino acid and protein synthesis and in the formation of nucleic acids. Nitrogen is also an integral part of the chlorophyll molecule involved in photosynthesis. Adequate N is associated with vigorous plant growth and dark green color. Nitrogen is readily mobilized from older tissues to newer tissues. A deficiency of N results in a stunted, chlorotic (yellow) appearance, causes carbohydrates to accumulate, and reduces protein content. Under severe N deficiency, older leaves will turn yellow and then brown, beginning at the leaf tip and progressing along the leaf midrib toward the base of the plant.

Forms of Nitrogen in Soil

Most soils contain between 0.3 g kg^{-1} and 4 g kg^{-1} total N. Much of this N is bound in soil organic matter. The remainder (2–5%) is present as ammonium, nitrite, and nitrate (Tisdale et al. 1993). Nitrogen mineralization from organic matter, inorganic N immobilization by soil microorganisms as they decompose organic matter with a high C : N ratio (>20 : 1), volatilization of **ammonia**, and **nitrification**, leaching, and denitrification are all part of the complex N cycle (Figure 26.2). Rates of these processes affect N uptake and utilization efficiency and the amount of N in the soil.

Soil tests for plant-available N are not very predictive, so N fertilization should be based on forage yield goals, N removal, experience, and science-based recommendations. Nitrogen needs of nonlegume forages is provided by fertilizer and/or manure. Ammonium held by negatively charged clay and organic matter is converted to nitrate by special soil bacteria. Nitrate-N moves downward with the leaching front or upward with the capillary rise of moisture in drying soils.

Nitrogen Sources for Forage Plants

Nitrogen sources used on forages in the US include urea (45-0-0), ammonium nitrate (34-0-0), urea-ammonium nitrate solutions (32-0-0 or 28-0-0), and to a lesser extent, ammonium sulfate (21-0-0-24) and anhydrous ammonia (82-0-0). Diammonium phosphate (18-46-0) also provides N. Both ammonium and nitrate are readily used by forages after they enter the soil. Manure contains N in both organic and inorganic forms. Most of the inorganic N in manure is present as ammonium.

Organic matter is a more important source of N in some northern environments than in the southern US, where soils with less than 2% organic matter are common.

Ammonia losses from urea-containing N sources can be large when used in the late spring through summer under humid conditions (Chapman 1984; Robinson and Dabney 1986; Eichhorn 1989; Robinson 1996; Haby 2002). Urea and urea-ammonium nitrate solutions are about 80–90% as effective as ammonium nitrate or ammonium sulfate. Calcium nitrate and sodium nitrate are as effective as ammonium nitrate, but they are less commonly used because of their higher cost. Volatilization losses from urea-containing N sources vary with weather conditions, thus yields can be as much as one-third less than those from the same amount of N as ammonium nitrate (Robinson and Dabney 1986).

Nitrogen Fixation

The atmosphere contains 78% N_2 gas, but this N is not available to plants. Rhizobia (Rhizobium) bacteria in nodules on legume roots convert N_2 to a form plants can use. Clovers and vetches can fix between 20 and 200 kg of N ha^{-1} yr^{-1} under ideal conditions, and alfalfa can fix 50–400 kg of N ha^{-1}. Inoculation of seed before planting provides the proper Rhizobium species in sufficient numbers to efficiently nodulate the developing plant (Snyder and Hankins 1987).

Nitrogen Fertilization

Nitrogen is the nutrient that can be most readily managed to influence forage yields. Timing, rate, and source of N can be varied to match grazing, hay, or silage production goals and the soil and rainfall environment for the specific forage species (Robinson 1996; Haby 2002). Factors affecting the response to fertilizer N include botanic composition, age of sward, seasonal distribution of fertilizer N, method of harvesting, frequency and height of defoliation, and whether removal is by grazing or by cutting.

Yield response of perennial cool-season grasses to fertilizer N generally ranges from 11 to 56 kg dry matter (DM) per kg N, with the greatest incremental response at lower N rates. For example, a study in Iowa on smooth bromegrass showed that the first 67 kg ha^{-1} N application resulted in 47 kg DM per kg N, but the next 67-kg increment resulted in 24 kg DM per kg N (Schaller et al. 1979). Zemenchik and Albrecht (2002) found mean N use efficiency ranged from 12 to 18 kg forage DM per kg N for kentucky bluegrass, 9 to 16 for smooth bromegrass, and from 11 to 28 for orchardgrass. Across several studies, the DM yield increased almost 20 kg ha^{-1} for each kg of N ha^{-1} applied to 'Coastal' bermudagrass (Figure 26.3; Table 26.5). Average yield across studies had not reached a plateau even at an N rate of 1200 kg ha^{-1}.

Based on more than 20 years of research and experience, Eichhorn (1995) offered lime and fertilizer recommendations (illustrated in Table 26.6), for annual production of 13–22 Mg of bermudagrass hay per hectare on Coastal Plain soils. Cool-season grasses such as tall fescue and annual ryegrass have lower yield potentials than warm-season grasses such as bahiagrass and dallisgrass, and they generally require less N: 200–300 kg N ha^{-1}.

Fertilizer N recovery by southern forage grasses typically ranges from 50% to 75% at recommended N rates. Nitrogen accumulation and loss by leaching or denitrification are minimal (Robinson 1990).

Excess Accumulation of Nitrogen in Forage Plants

High plant tissue levels of nitrate may accumulate under conditions of high soil nitrate, drought, low-light intensity, herbicide stress, or deficiencies of other plant nutrients such as molybdenum (Mo). At high feed nitrate-N levels, excess nitrite produced in the rumen is absorbed into the blood, where it forms methemoglobin and limits oxygen supply to the animal (See Chapter 47). The critical nitrate-N concentration varies depending on forage-intake level, general animal condition, and whether the change in diet to high-nitrate forage is gradual or abrupt.

Robinson (1995) reported that nitrate accumulation in ryegrass, millet, bermudagrass, dallisgrass, and bahiagrass is unlikely when N rates do not exceed 100 kg ha^{-1} in a single application.

Effect of Nitrogen Fertilizers on Soil pH

Nitrification of ammonium-N releases hydrogen ions and increases soil acidity. As nitrate moves through the soil profile, leaching of cations such as Ca, Mg, and K occurs. Unless lime is added to counter this effect, acidification of surface and subsoil can occur (Adams 1984).

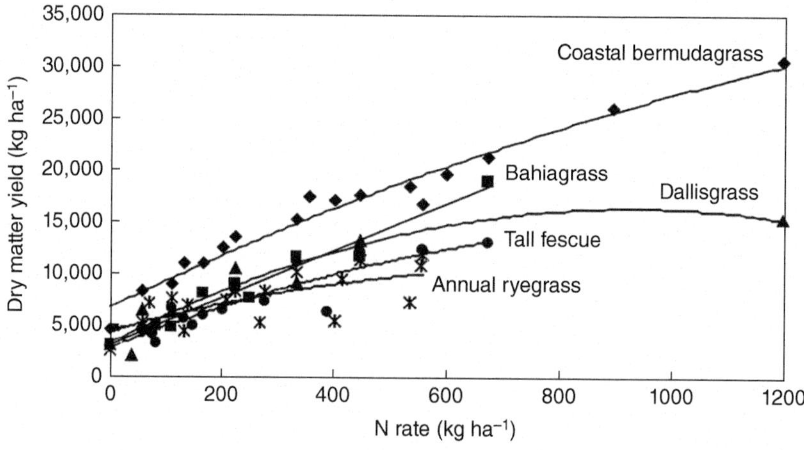

FIG. 26.3. Forage grass average response to N rate in the South. Data from 15 'Coastal' bermudagrass, 5 bahiagrass, 2 dallisgrass, 6 fescue, and 3 annual ryegrass published reports. *Source:* See Table 26.5 for data and sources for Coastal bermudagrass.

Table 26.5 Nitrogen response by 'Coastal' bermudagrass at different levels of nitrogen fertilization (annual DM production in kg ha^{-1} yr^{-1})

	Location – cuttings/yr														
	AL-5	AL-5	AL–	AL-4	AR-4	AR-4	GA-4	GA-	LA-3	LA-5	LA-5	LA-4 or 5	TX-	TX-	TX-4
N rate (kg ha^{-1} yr^{-1})	Greenville fine sandy loam	Greenville fine sandy loam-irrigated	Fine sandy loam to silt loam	Norfolk sandy loam	Captina silt loam	Captina silt loam-irrigated	Tifton loamy sand	—	Severn v. fine sandy loam	Gilead fine sandy loam	Olivier silt loam	Sandy loams	—	Lufkin fine sandy loam-irrigated	Gallime fine sandy loam
0	3 360	3 696	2 912	2 397	3 270	4 480		5 510	8 781		5 152	2 330	6 003	5 981	6 003
56											6 048				
112			8 736	6 496	8 758	10 483	11 984			6 496	7 616	6 496		9 811	
134								9 744							13 597
168	9 610	10 931													
202												10 349		13 283	
224			12 768	10 304	11 939	14 426	16 778	13 597	14 000	9654	10 976				
336	14 426	17 718			15 142	18 502				12 947		10 326			
358															17 046
403									20 608			14 381			
448			16 464	16 106	17 875	20 160	22 781	15 053	20 675	14 627	15 904	15 187		13 283	
538					18 256	21 616									18 010
560										17 181		17 427			
599												17 046	15 277		
672	18 816	21 997			19 085	22 669				19 062		18 682		23 856	
896											20 384			27 104	
1 198													30 464		
	Evans et al. (1961)	Evans et al. (1961)	Cope (1970)	Scarsbrook (1970)	Huneycutt et al. (1988)	Huneycutt et al. (1988)	Burton and Jackson (1962)	Cited by Overman and Wilkinson (1992)	Robinson and Dabney (1986)	Eichhorn (1989)	Robinson (1990)	Eichhorn and Huffman (1991)	Cited by Wilkinson and Langdale (1974)	Pratt and Darst (1984)	Haby (2002)
Study duration	4-yr avg.	4-yr avg.	5 sites, 3 - to 8-yr avg.	2-yr avg.	5-yr avg.	5-yr avg.	5-yr avg.		6-yr avg.	5-yr avg.	7-yr avg.	Multiple sites, over 23 yr			3-yr avg.

Table 26.6 Aglime and fertilizer recommendations for annual production of bermudagrass hay of 13 000–22 000 kg ha^{-1} : 4500 kg ha^{-1} harvest^{-1}

Nutrient	Source	Rate	Timeliness of application
Aglime			
Calcium (Ca) & magnesium (Mg)	High Mg calcitic or dolomitic	2 000–4 000 kg ha^{-1}, depending on quality and soil pH	Late fall or winter, every 5 yr, depending on soil pH
Fertilizer			
N	Ammonium nitrate	224 kg ha^{-1} (34-0-0)	Initially apply blended fertilizer (630 kg ha^{-1}) March 15 to April 15. After hay is harvested in May, reapply the blended fertilizer for the next harvest in June. Reapply fertilizer at similar rate after each harvest for each additional harvest desired.
N & P	Diammonium phosphate	70 kg ha^{-1} (18-46-0)	
N & sulfur (S)	Ammonium sulfate	112 kg ha^{-1} (21-0-0-24)	
K	Muriate of potash (KCl)	224 kg ha^{-1} (0-0-60)	
N, P, K, & S	Fertilizer blend	630 kg ha^{-1} (18-5-21-4)	

Source: Eichhorn (1995).

Grass–Legume Mixture Interactions with Applied Nitrogen

Grass-clover swards have a variable response to applied N because clover growth varies more widely from site to site and year to year. When clover is sparse or conditions are unfavorable, herbage yield and response to fertilizer N will be similar to an all-grass sward. When clover dominates the sward, herbage yield is relatively high without fertilizer N (sometimes greater than 10 Mg DM ha^{-1}), and the response to fertilizer N is small (Whitehead 1995). Increasing N fertilization of grass-legume mixtures favors grasses over legumes because grasses lack the ability to fix their own N. Timely clipping or defoliation can reduce competition between grasses and legumes for light. Legumes use inorganic soil N preferentially rather than fix N (Follett and Wilkinson 1995). Dobson and Beaty (1977) reported that inclusion of white clover with cool- or warm-season grasses in Georgia increased forage yields when N was applied at an amount up to 112 kg N ha^{-1}. In a 10-year study, Tesar (1982) reported similar yields of smooth bromegrass, orchardgrass, reed canarygrass, and tall fescue fertilized annually with ammonium nitrate applied at 150 kg N ha^{-1} compared with binary mixtures with 'Vernal' alfalfa. Overseeding both cool- and warm-season grass pastures with legumes often improves the seasonal forage distribution of DM over grass alone. In Texas, the amount of fertilizer N required by grasses to replace the clover DM contribution to the total grass-plus-clover yield was 127, 211, 160, and 254 kg N ha^{-1} for arrowleaf clover-bermudagrass, arrowleaf clover-bahiagrass, subterranean clover-bermudagrass, and subterranean clover-bahiagrass mixtures, respectively (Evers 1985).

Phosphorus

Forage Plant Requirements for Phosphorus

Phosphorus is essential for plant growth. Phosphorus is required for energy transformation processes in all plant and animal life. Most forage plants require between 2.5 and 6.0 g P kg^{-1} DM for optimum yield and quality (Table 26.2). Symptoms of P deficiency include stunted growth, purple stems and leaves, reduced yield, and others.

Role of Phosphorus in Forage Plants

Once taken up by the plant root, P may be stored or transported to other plant parts. It occurs and is transported as both inorganic and organic forms. Phosphorus influences synthesis and degradation of starch and transport of nutrients from soil through roots and to aboveground plant parts. Organic P is found in proteins, nucleic acids, adenosine triphosphate (ATP), lipids, esters, and **enzymes**. During photosynthesis, light energy is absorbed by chlorophyll, reducing nicotinamide adenine dinucleotide phosphate (NADP) to ATP. Numerous plant biosynthetic reactions are driven by NADP and ATP, which serve as energy donors.

Inadequate P hinders carbohydrate utilization. When P is most limiting, there is a reduction in leaf expansion rate, surface area, and number. Shoot and root growth are both reduced by P deficiency, but shoots are affected more. Movement of nutrients within plants depends largely on active transport across cell membranes, which requires energy. Phosphorus deficiency limits movement of P and other mobile nutrients from older tissues.

Large amounts of P are required for seed formation and development. Inadequate P reduces seed size, seed number, and seed viability.

Forms of Phosphorus in Soil

Phosphorus is taken up mainly as primary orthophosphate ($H_2PO_4^-$), but some is also absorbed as secondary orthophosphate (HPO_4^{2-}) at higher soil pH levels. Other P sources must be transformed by microbial and chemical processes to these forms for plant uptake.

Only about 4 kg ha^{-1} of P is in the soil solution and plant available in the upper 15–20 cm of soil at any given time. Soluble P is readily converted to insoluble compounds containing iron (Fe), aluminum, and Ca, which have low plant availability. These same reactions occur whether the P comes from apatite mineral, P fertilizer, manure, or organic matter.

Replenishment of **soil solution** P must take place several times a day for most crops to achieve optimum yields. Maintenance of medium to high levels of extractable P helps ensure that roots will have continual access to soluble orthophosphate.

In North America, 43% of soils tested medium or low in P in the fall of 2000 to the spring of 2001 (PPI/PPIC 2001). In southern states, 29 to–72% of soils tested medium or low in P.

Phosphorus Sources

The most common P fertilizers currently used are diammonium phosphate (18-46-0), monoammonium phosphate (11-52-0), and ammonium polyphosphate (10-34-0 or 11-37-0). Triple superphosphate (0-46-0), and regular superphosphate (0-20-0) no longer have widespread use because more concentrated P sources are more economical to transport and spread. All these commercial fertilizer sources are at least 70% water soluble and are agronomically equal in providing P for plants.

Naturally occurring phosphate rock (about 15% P) is the basic material used in all P fertilizer production. Finely ground phosphate rock is used in some countries to provide P for plants, but it becomes available very slowly, and large quantities (~2 000 kg ha^{-1}) must be applied before a plant response can be expected.

Manures contain P and other major nutrients (Table 26.7) but contain P primarily in the organic form. Many types of manure have 0.4–5% of their total P in water-extractable or water-soluble forms. Some liquid swine manures may contain as much as 60% of their P as readily available, dissolved phosphate (Snyder and Bruulsema 2002).

Forage Plant Response to Phosphorus

Cool-season annual grasses and legumes are considered highly responsive to P fertilization (Westerman et al. 1984; Robinson 1996). Warm-season perennial grasses such as Coastal bermudagrass are also responsive to high levels of soil test P. On a low-P Lilbert loamy fine sand soil near Tyler, Texas, the DM yield was increased by almost 6.8 Mg ha^{-1} yr^{-1} with 90 kg of P_2O_5 ha^{-1} yr^{-1} (Hillard et al. 1992b) (Figure 26.4). In addition to DM yield increases, P fertilization can increase crude protein yields, digestible DM yields, and total digestible nutrient yields per harvest (Figure 26.5).

Coastal bermudagrass and other warm-season grass yields typically increase as soil-test P levels increase up to an optimum of 20–25 mg kg^{-1} by Mehlich 1, Mehlich 3, or pH 4.2 ammonium acetate + EDTA (ethylenediamine-tetraacetic acid) (former Texas A&M method) extraction methods. This contrasts with the findings of Eichhorn (1996) that forage yield was maximized when Bray 2 soil-test P levels were 112 mg kg^{-1} at the 0- to 3-cm depth and 70 mg kg^{-1} at the 0- to 15-cm depth (Figure 26.6) (Eichhorn 1996).

Annual ryegrass yields responded to P fertilization at P_2O_5 rates up to 360 kg ha^{-1} yr^{-1} on a low-P soil (17 mg kg^{-1} Bray 2 P) (Figure 26.7) (Robinson and Eilers 1996). Soil-test P levels did not increase unless P_2O_5 rates exceeded 90 kg ha^{-1} yr^{-1}. Phosphorus fertilization also increased annual ryegrass yield 20–80% in Texas (Hillard et al. 1992a) and 80–103% in Florida (Rechcigl 1992) on low-P soils. Haby (1995) reported that at very low soil test P levels, rates of 134 kg of P_2O_5 ha^{-1} were needed for optimum yields and recommended that P rates be higher for annual ryegrass when the soil pH is below 5.0 and toxic levels of aluminum are present.

Bermudagrass hay yields may exceed 15 Mg ha^{-1} yr^{-1} of DM. With a tissue P concentration of 2.5 g kg^{-1} DM, such yields would remove 38 kg P ha^{-1} yr^{-1} (86 kg P_2O_5 ha^{-1} yr^{-1}). Annual ryegrass yields are frequently above 6.5 Mg ha^{-1} yr^{-1}. With a tissue P concentration of 3.6 g kg^{-1} DM, 23 kg of P ha^{-1} yr^{-1} (54 kg of P_2O_5 ha^{-1} yr^{-1}) would be removed by annual hay or silage harvests.

Annual fertilization of arrowleaf clover with 57 kg of P ha^{-1} (131 kg P_2O_5 ha^{-1}) on a low-P soil (Bray 1 P < 20 mg kg^{-1}) in Oklahoma raised annual forage yields to 6.9 Mg ha^{-1}, 122% (3.8 Mg ha^{-1}) above the no-P control yield (Westerman et al. 1984). A portion of this response was due to increased N fixation by the clover. Plant P uptake was also stimulated by fertilization

Table 26.7 Nutrient content of manures (in kg unit^{-1} wet basis)

Type	TKN	P_2O_5	K_2O	Ca	Mg	S
Dairy						
Fresh (kg Mg^{-1})	5	3	4	2	1	1
Paved surface scraped (kg Mg^{-1})	5	3	5	3	1	1
Liquid manure (kg $(1000\ kg)^{-1}$)[a]	23	14	21	10	5	3
Lagoon liquid (kg ha-cm^{-1})[b]	61	34	86	30	15	11
Anaerobic lagoon sludge (kg ha-cm^{-1})[b]	7	10	4	5	2	2
Beef						
Fresh (kg Mg^{-1})	6	4	5	3	1	1
Paved surface scraped (kg Mg^{-1})	7	5	7	3	2	1
Unpaved feedlot (kg Mg^{-1})	13	8	10	7	3	3
Lagoon liquid (kg ha-cm^{-1})[b]	37	34	57	11	8	—
Lagoon sludge (kg $(1000\ kg)^{-1}$)[a]	38	51	15	36	5	—
Broiler						
Fresh (kg Mg^{-1})	13	9	6	5	2	1
House litter (kg Mg^{-1})	36	39	23	21	4	8
Stockpiled litter (kg Mg^{-1})	18	40	17	27	4	6
Duck						
Fresh (kg Mg^{-1})	14	12	9	—	—	—
House litter (kg Mg^{-1})	10	9	7	11	2	2
Stockpiled litter (kg Mg^{-1})	12	21	11	14	2	3
Goat						
Fresh (kg Mg^{-1})	11	6	9	—	—	—
Horse						
Fresh (kg Mg^{-1})	6	3	6	6	1	1
Layers						
Fresh (kg Mg^{-1})	13	11	6	21	2	2
Undercage paved (kg Mg^{-1})	14	16	10	22	3	4
Deep pit (kg Mg^{-1})	19	28	15	43	3	5
Liquid (kg $(1000\ kg)^{-1}$)[a]	62	59	37	35	7	8
Lagoon liquid (kg ha-cm^{-1})[b]	79	20	118	11	3	23
Lagoon sludge (kg $(1000\ kg)^{-1}$)[a]	26	92	13	71	7	12
Rabbit						
Fresh (kg Mg^{-1})	12	12	7	10	2	1
Sheep						
Fresh (kg Mg^{-1})	11	5	10	7	2	2
Unpaved (kg Mg^{-1})	7	6	10	12	4	3
Swine						
Fresh (kg Mg^{-1})	6	5	5	4	1	1
Surface scraped (kg Mg^{-1})	7	6	5	6	1	1
Liquid manure (kg $(1000\ kg)^{-1}$)[a]	31	22	17	9	3	5
Lagoon liquid (kg ha-cm^{-1})[b]	60	23	59	11	4	4
Lagoon sludge (kg $(1000\ kg)^{-1}$)[a]	22	49	7	16	4	8
Turkey						
Fresh (kg Mg^{-1})	14	13	6	14	1	—
House litter (kg Mg^{-1})	26	32	19	18	3	5
Stockpiled litter (kg Mg^{-1})	18	36	17	21	4	5

Source: PPI/PPIC (2002).

Notes: TKN, total Kjeldahl nitrogen. Approximate nutrient contents are given. Have materials analyzed for nutrient content before using.

North Carolina mean waste analysis 1981–1990 supplied by J.C. Barker, NCSU Department of Biological and Agricultural Engineering.

[a]Kilograms per thousand kilograms of manure liquid (slurry).

[b]Kilograms per hectare-centimeter. Estimated total lagoon liquid includes total liquid manure plus average annual lagoon surface rainfall surplus; does not account for seepage.

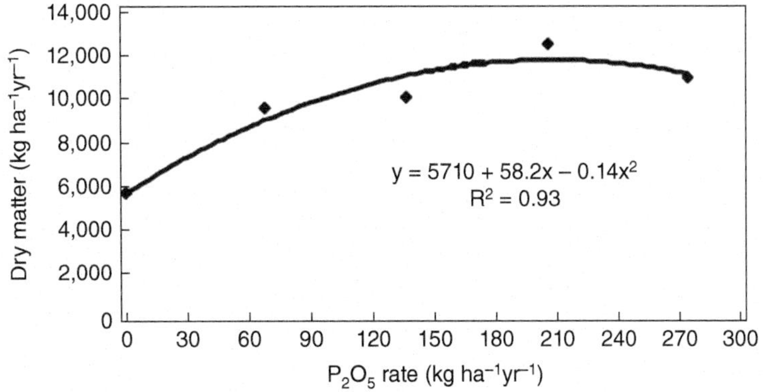

Fig. 26.4. 'Coastal' bermudagrass DM yield response to P fertilization on a low-P soil in Texas (3-year average). *Source:* Hillard et al. (1992b).

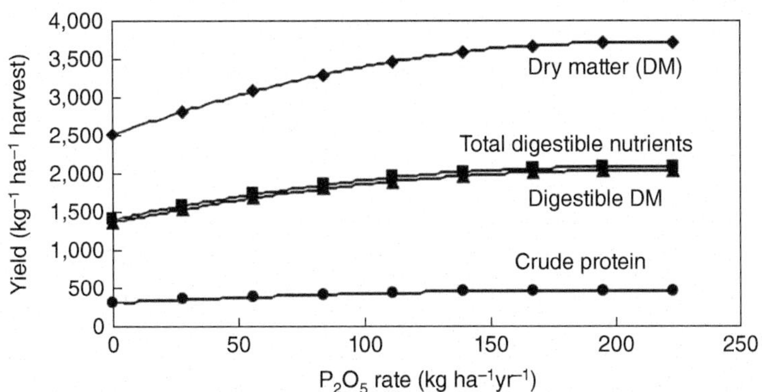

Fig. 26.5. 'Coastal' bermudagrass per harvest DM and nutritive yield response to P fertilization on a low-P soil in Louisiana (five-year average). *Source:* Eichhorn (1996).

with 57 kg of P ha^{-1} and resulted in an annual P removal of 13.5 kg of P ha^{-1} (31 kg P_2O_5 ha^{-1}), a 211% increase over the no-P control.

Mixtures of seven different legumes with coastal bermudagrass on a Lynchburg sand in Georgia receiving nine different fertilization treatments produced 6.7 Mg of DM ha^{-1}, which was equivalent to that of Coastal bermudagrass monoculture fertilized with 112 kg of N ha^{-1} (Burton and DeVane 1992). When P and K rates were doubled to 47 and 84 kg ha^{-1}, respectively, mixture yields increased by 1 030 kg DM ha^{-1} and increased crude protein yield. Haby (2002) found that during the seeding year, yields of alfalfa were increased by P fertilization up to 60 kg ha^{-1} (137 kg of P_2O_5 ha^{-1}) at low soil P levels. In succeeding years, the study showed that P rates of 30–40 kg ha^{-1} (69–92 kg of P_2O_5 ha^{-1}) annually would be adequate.

To compensate for harvest removal of P, or to build soil test P levels, 6–14 kg of P ha^{-1} above the crop removal rates are required to raise soil-test P levels by 1 mg kg^{-1}.

Phosphorus Application

It is desirable to incorporate applied P in the upper 0–15-cm zone during establishment, but there is no evidence that surface-applied P on established forages limits fertilization response (Mays et al. 1980; Gelderman et al. 2002). The frequency of P application for warm-season perennial grasses is not critical. Eichhorn (1996) found that an initial application of 840 kg of P_2O_5 ha^{-1} resulted in yields equivalent to annual or split applications of 168 kg ha^{-1} in a seven-year study with bermudagrass. Single annual applications of P are sufficient as long as adequate soil-test P levels are maintained.

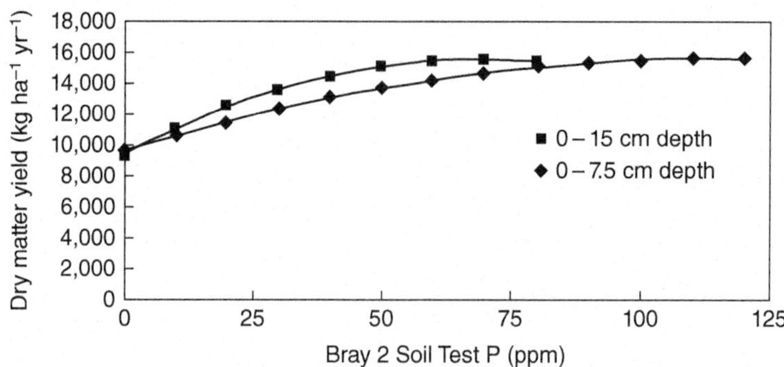

Fig. 26.6. 'Coastal' bermudagrass response to soil test P in Louisiana. *Source:* Eichhorn (1996).

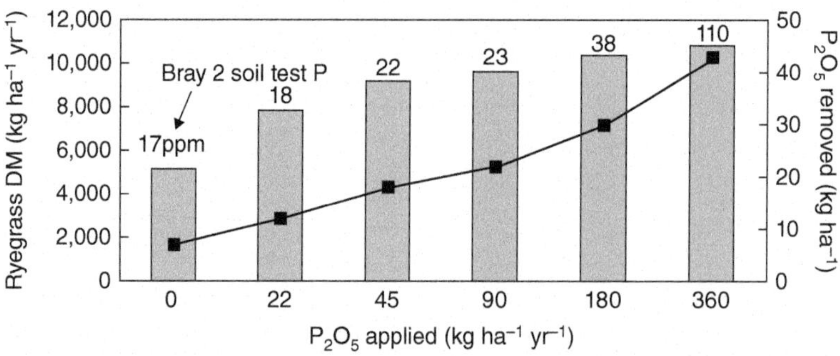

Fig. 26.7. Effects of P fertilization on annual ryegrass forage yield (bars), P removal in harvest (line), and soil test P levels in Louisiana. *Source:* Robinson and Eilers (1996).

Environmental Implications of Phosphorus Fertilization

Continued applications in excess of crop use heighten the risk of P loss to surface waters with negative environmental impacts. In watersheds with CAFO, it is not unusual to find soils in the very high to excessive P range.

Although scientists have considered P immobile in soils when it is applied at agronomic rates, leaching can occur when the soil P sorptive capacity (P saturation) has been exceeded. Erosion and runoff losses of P bound to soil particles, and as manure or fertilizer P, can be limited by good nutrient-management planning, appropriate best-management practice adoption, and use of other conservation measures.

Potassium

Forage Plant Requirements for Potassium

Potassium is required in higher concentrations than any other mineral element besides N (Robinson 1996) (Table 26.2). Potassium removal by hay crops in 13 southern states accounts for an average of about 44% of the total K removed by all crops (PPI/PPIC 2002; Snyder 2003). For sustainable production of high-yielding Coastal bermudagrass, at least 200–400 kg of K ha^{-1} are required annually (Robinson 1996). A hay yield of 18 Mg ha^{-1} alfalfa removes 448 kg of K ha^{-1}, while a 10.6 Mg ha^{-1} orchardgrass yield removes 291 kg of K ha^{-1} (Vough 2000).

Role of Potassium in Forage Plants

Potassium is involved in activating enzymes and catalyzes several physiologic reactions in plants. Potassium regulates opening and closing of stomata, enabling leaves to exchange carbon dioxide, water vapor, and oxygen with the atmosphere. Its role in photosynthesis is complex and includes activation of enzymes involved in ATP production. Potassium plays a major role in the transport of water and nutrients through the xylem. It is required for every major step in protein synthesis, and it also activates starch synthetase which is responsible for starch synthesis. Deficiency reduces **translocation** of nitrate, phosphate, Ca, Mg, amino acids, and starch.

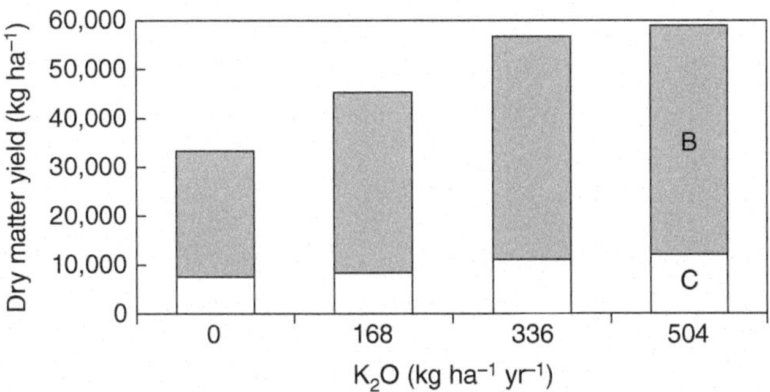

Fig. 26.8. Potassium fertilization increases 'Coastal' bermudagrass (B) and clover (C) yields in Texas. Initial soil test K was very low. *Source:* Adapted from Cripps et al. (1988).

Adequate levels of K improve the yield and quality of many crops, helps plants use water more efficiently, and decreases the potential for losses in production associated with drought. Soil K levels below 150 mg kg^{-1} may contribute to loss of alfalfa stands, particularly on loam and sandy loam soils (Vitosh et al. 1995; Bagg 2000). Adequate K is especially important in increasing the stand longevity of Coastal bermudagrass (Keisling et al. 1979; Eichhorn et al. 1987). Inadequate K predisposes bermudagrass to leaf spot disease organisms (*Helminthosporium cynodontis* Marg.) and can lead to winterkill of crowns, reduced number and vigor of rhizomes, and thinned stands (Robinson 1985; Eichhorn et al. 1987). When clover is overseeded on grass pastures, the yield of both grass and clover can be increased with adequate K nutrition (Cripps et al. 1988; Leep 1989) (Figure 26.8).

Forms of Potassium in Soil

Most soils contain as much as 20 Mg of total K ha^{-1}, but only small amounts (<2%) are available to plants over the growing season. In simple terms, soil K exists in three forms: unavailable, slowly available, and available. The unavailable K is found in soil minerals and is only slowly released as soil minerals weather. The slowly available form is 'fixed' or trapped in the lattice of certain clay minerals. Many of the highly weathered soils in the South have low amounts of the clays that fix K. Based on work reported by Robinson (1985), loess soils may require higher K rates than other soils to overcome K fixation and to increase plant K uptake.

Available K is held on cation exchange sites and in the soil solution. Potassium is taken up from the soil by plants in the ionic form (K^+). Uptake of K by plant roots is limited by the supply of K^+ to root surfaces, and most of the K^+ comes in contact with roots through diffusion (Robinson 1985).

Most soils contain less than 10 kg of K ha^{-1} in the soil solution at any given time. As solution K is leached or taken up by plants, the K on the cation exchange sites reacts to establish a new equilibrium with the soil-solution K. Potassium fertilization should be managed to raise extractable soil-test levels to the high range, followed by maintenance applications to replace harvest removal and leaching losses. From 7 to 13 kg of K ha^{-1} (8–16 kg of K_2O ha^{-1}) above the crop harvest removal of K is required to raise soil test K levels by 1 mg kg^{-1}.

Potassium Sources for Forage Plants

The principal fertilizer K source for forages is KCl (0-0-60 or 0-0-62). Other sources include potassium sulfate (0-0-50-18), sulfate of potash-magnesia (0-0-22-22; also contains 11% Mg), and potassium nitrate (13-0-44). Manures can also provide significant quantities of K (Table 26.7).

Forage Plant Response to Potassium

Large amounts of K are removed by harvest of many warm-season and cool-season forages (Table 26.3). Fertilization with K can provide significant yield increases, as illustrated with Coastal bermudagrass in Figure 26.9. Adequate K nutrition also improves water- and N-use efficiency (Figure 26.10).

The effects of 467 kg of K ha^{-1} yr^{-1} (560 kg of K_2O ha^{-1} yr^{-1}) on yields of 15 different bermudagrasses and on soil test levels were evaluated by Hallmark et al. (1986). The soil was an Iberia silty clay loam and initially had 170 ppm of ammonium acetate-extractable K. One-half of the K (233 kg of K ha^{-1}) was applied in early March and the remainder after the second harvest. With four harvests totaling more than 27 Mg ha^{-1}, 467 kg of K ha^{-1} yr^{-1} annually did not maintain soil test K levels on this relatively fine-textured soil.

Splitting K applications to better match K demand may improve forage yields and may help maintain soil test

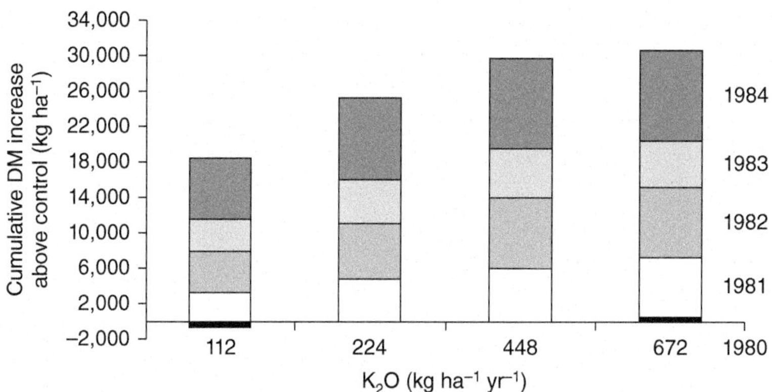

Fig. 26.9. 'Coastal' bermudagrass cumulative forage yield increase with K on a low-K soil in Louisiana. Initial soil test K was low: 47 mg kg^{-1} ammonium acetate extraction. *Source:* Eichhorn et al. (1987).

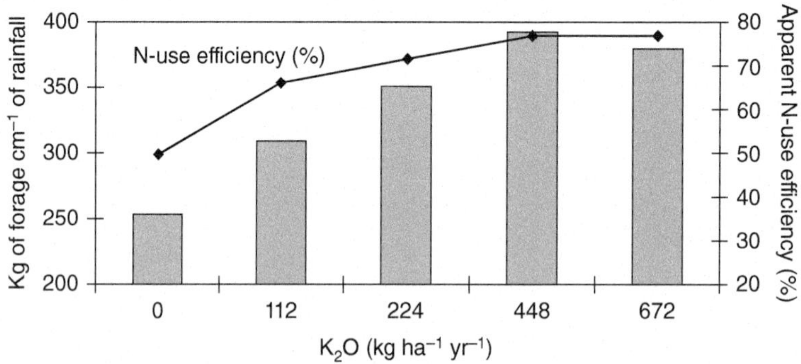

Fig. 26.10. Potassium improves water use efficiency and nitrogen use efficiency by 'Coastal' bermudagrass in Louisiana. Initial soil test K was low: 47 mg kg^{-1} ammonium acetate extraction. *Source:* Eichhorn (1996).

K levels. Soil K on two sandy soils in Texas declined with increased harvest removal of K, indicating that 278 kg of K ha^{-1} in four equal increments was insufficient to maintain soil K fertility with bermudagrass yields above 11 Mg ha^{-1} (Nelson et al. 1983). Day and Parker (1985) speculated that some applied K was leached on a sandy loam soil, since K applications exceeded K removal by 73 kg ha^{-1} and soil test K declined at 0–20 cm and 20–40-cm depths. Eichhorn et al. (1987) found that soil K fertility (0–15 cm and 15–30-cm depths) on a sandy Coastal Plain soil in Louisiana was improved only when 448 kg of K ha^{-1} were applied in four equal increments or where 672 kg of K ha^{-1} were applied in any number of increments. Maximum forage yields (~15 Mg ha^{-1} yr^{-1}), with a K removal rate of 367 kg of K ha^{-1} yr^{-1}, were achieved with either rate, irrespective of application interval.

Maximum annual ryegrass yields occurred with 600 kg of K ha^{-1}, although 90% of the maximum occurred with 112 kg of K ha^{-1} on a soil with 49 ppm K (ammonium acetate extractable) (Robinson and Eilers 1996). At the highest K application rate, the crop removed 405 kg of K ha^{-1} yr^{-1}. Soil test K levels declined over time at or below 300 kg K ha^{-1} yr^{-1}. Robinson (1996) stated that 'annual ryegrass is capable of luxury consumption of K, and that maintaining even medium soil test levels may not be practical where high-yielding ryegrass crops are harvested.'

Potassium nutrition of grass-legume swards is more complicated than with pure species, especially on acid soils. This is due, in part, to the competition for K in these associations. With liberal K application (372 kg K ha^{-1}), alfalfa had only 8% less K than the grass in a mixture (Blaser and Kimbrough 1968). On this low to medium-K soil, as the fertilizer K was reduced, the alfalfa absorbed much less K than the grass. Without an application of K, the alfalfa had 74% less K in herbage tissue than the associated grass.

When clover is overseeded on grass pastures, the yield of both can be increased with adequate K nutrition (Cripps et al. 1988) (Figure 26.8). Legume persistence in mixtures with grasses is increased by maintaining high soil test K, pH, and other essential nutrients. Forage legumes are generally believed to be more effective at using subsoil K than grasses (Lanyon and Smith 1985).

Implications of Excess Potassium in Forage Plants

Very high rates of K fertilization increase K uptake and concentration in plants and can decrease uptake of Ca and Mg. Forage yield, yield response to other nutrients, and animal performance can be depressed if K is not kept in balance with other nutrients (Robinson 1985). Adequate P fertilization can help offset negative effects of high K because it stimulates Mg uptake and helps decrease the potential for grass tetany (Lock et al. 2002). Excessive K in forage may induce metabolic alkalosis when fed to prepartum dairy cows, which reduces their ability to maintain Ca homeostasis, increasing the incidence of milk fever (Goff and Horst 1997).

Soil pH, Acidity, Alkalinity, and Salinity

Forage crops vary in their requirements for soil pH levels for optimum performance. When pH falls outside these levels, plant growth may be adversely affected. The topsoil in pasture and hay fields with acid subsoils should be maintained at higher soil pH than those with neutral or alkaline subsoils to minimize chances for nutrient deficiencies associated with acid soil conditions. When drainage of arid region soils is impeded, and the surface evaporation becomes excessive, soluble salts tend to accumulate in the surface horizon. Such intrazonal soils are designated as halomorphic and have been classified as saline, saline-alkali, or nonsaline-alkali (Brady and Weil 2001).

Lime Requirement of Soils

Lime requirements depend on soil pH, soil-buffering capacity, and target pH (Adams 1984). Soil buffering capacity rises with increased cation exchange capacity, clay content, and soil organic matter level. Many labs use both a buffer pH and a soil pH to arrive at a lime recommendation (Kidder et al. 1988).

Soil pH and Nutrient Availability

Soil pH affects mineralization of N, P, and S from organic matter. It also influences the forms of P in the soil solution and P availability to plants. The effective cation exchange capacity of highly weathered soils increases as the soil pH rises, which in turn, affects the capacity to retain K and other cations. Except for Mo and Cl, the availability of most other micronutrients decreases as soil pH increases. The availability of chloride is not affected by soil pH. Molybdenum availability increases as soil pH increases.

Soil pH and Legume Growth

Legume species differ greatly in tolerance of low soil pH and in response to lime. Sanfoin, sulla, alfalfa, and sweetclover are generally considered to be the most acid-sensitive forage legumes and respond well to lime (Adams and Pearson 1967). Red clover, white clover, and alsike clovers are reported to be tolerant to acid soils (Helyar and Anderson 1970; Weeks and Lathwell 1967). Birdsfoot trefoil can tolerate acid soils and, in general, does not require lime when soil pH is above 5.8 (Leep and Tesar 1981). Legumes requiring a higher soil pH for optimum growth are usually associated with nodule bacteria (*Rhizobium*), which require a high supply of Mo.

Soil pH and Soil Microorganisms

Soil pH above 6.5 is optimum for most legumes to encourage N fixation by the symbiotic bacteria and to enhance availability of soil Mo required for symbiotic N fixation and nitrate reductase. Maintaining soil pH above 5.5–5.8 for most grasses and above 6.5 for most legumes encourages bacterial decomposition of organic matter and nutrient release.

Liming

Robinson (1996) suggested that forage grasses in the southern US can produce optimum DM yields when soil pH is between 5.0 and 5.5. Soil pH above 5.5 helps prevent Al and Mn toxicity, which could damage forage legumes on certain soils testing high in extractable Mn. Most cool-season grasses are tolerant to relatively low soil pH, except for smooth bromegrass and reed canarygrass. Regional differences in plant response to liming appear to exist, in part, due to soluble Al and Mn variations in soils at the same pH (Pearson and Hoveland 1974).

Effect of Lime Particle Size on Neutralizing Value

Lime particle size determines reactivity. Purity, relative to pure calcium carbonate, describes the capacity of liming materials to neutralize acidity.

Lime particles coarser than 10-mesh do little to alter pH and are considered to be ineffective. Particles smaller than 50 to 60 mesh (0.25–0.3 mm) are more reactive, and particles finer than 100 mesh (0.15 mm) are considered quite reactive and will begin to neutralize soil acidity upon contact (Barber 1984). Most methods of lime classification are based on purity and fineness of grind. Many terms – such as *relative neutralizing value, effective neutralizing value, effective liming material,* and *effective calcium carbonate equivalence* – have been used, and formulas have been developed to classify lime quality. Haby and Leonard

(2002) found finer-sized particles to be more efficient in neutralizing soil acidity, in raising soil pH to higher levels, and in maintaining a pH against reacidification longer than the same weight of coarser-ground limestone from the same source. Better agricultural limestone (aglime) had 94% of the particles passing a 0.25-mm screen (60 mesh), whereas the lower-quality aglime had 28% of the particles within 2.36–0.85 mm (8–20 mesh), 26% within 0.85–0.25 mm (20–60 mesh), and 40% passing 0.25 mm (60 mesh).

Liming Materials

Calcitic and dolomitic (>6% Mg) aglime are the two most common liming materials. Other materials may be used if they have adequate purity (high calcium carbonate equivalence) and the particles are fine enough to provide reaction within a few weeks to several months. Hydrated lime (slaked lime or calcium hydroxide) and burnt lime (calcium oxide) are excellent liming materials, but they are difficult to handle and spread, can burn **foliage**, and are rarely used in liming pastures and hay meadows. Gypsum is a good Ca and S source but has no acid-neutralizing value.

Effect of Lime on Forage Plants

The threshold pH (highest pH at which lime is recommended) and the target soil pH for forages vary with location. Recommended target pH values range from 5.5 to 6.5 for most forage grasses, from 6.0 to 7.0 for legume-grass mixtures, and from 6.5 to 7.0 for alfalfa (Kidder et al. 1988).

Forage grasses may respond to lime when the pH is below 5.5–5.8, and clovers and other legumes may not respond when the pH is above 6.5–6.8. A soil pH of 6.5–6.8, depending upon the subsoil pH, is recommended for alfalfa in the midwestern US (Vitosh et al. 1995). Liming for other forage legumes from 6.0 to 6.8 is also recommended in the midwestern US. However, some of the yield response may be due to improved nodulation (Munns 1965). Liming an acid Coastal Plain soil in Texas to raise the soil pH from 4.7 to 5.7 increased annual ryegrass DM yield by nearly 4.5 Mg ha^{-1} (Haby et al. 1996). Phillips and Snyder (1988) found that liming an established bahiagrass pasture on a soil with a pH of 5.2 (0–15 cm) increased DM yield from 0.7 to 1.2 Mg ha^{-1} yr^{-1}. The greater responses occurred when Mg was also provided as dolomitic lime or as potassium-magnesium-sulfate. Phillips et al. (1995) found a slight depression in Coastal bermudagrass yield with liming on a sandy loam with a pH of 5.9 (0–15 cm). Cripps et al. (1988) reported significant yield increases by Coastal bermudagrass whenever soil pH fell below about 5.5. Eichhorn (1989), on the other hand, observed maximum Coastal bermudagrass yield when soil pH ranged between 4.7 and 5.3.

Forage response to liming is site and species specific. Therefore, liming decisions should be based on local university research whenever possible.

Effect of Lime on Soil

Liming acid soils decreases soil acidity and soluble aluminum and manganese (Mn) levels, increases Mo and P availability, provides Ca and Mg (dolomitic sources), helps promote microbial activity, and contributes to improved soil aggregation, structure and porosity.

Surface lime applications not incorporated by tillage primarily affect the upper few centimeters of soil (Figure 26.11). A long time is needed for surface-applied

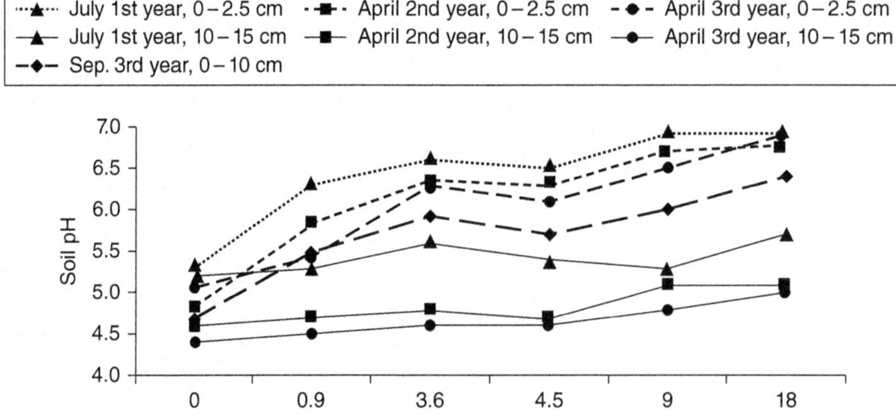

FIG. 26.11. Soil pH response at different depths to surface lime applications on a 'Coastal' bermudagrass pasture. *Source:* Adapted from Young et al. (1984).

lime to affect soil deeper than 10–15 cm because lime is only slightly soluble in water. It is best to establish a desirable pH in the upper 15–20 cm by incorporating lime before forage establishment.

Role of Calcium and Magnesium in Forage Plants

Calcium is taken up by plant roots as the Ca^{2+} cation. Calcium stimulates root and leaf development, forms compounds that are part of cell walls, helps strengthen plant structure, helps reduce nitrate in the plant, helps activate several plant enzymes, and helps maintain cell pH balance by neutralizing organic acids.

When soils are properly limed, Ca deficiency is unlikely. Common Ca sources include aglime, gypsum (22% Ca), marl (24% Ca), hydrated lime (46% Ca), and burnt lime (60% Ca). Normal superphosphate, which is 50% gypsum, and to a lesser extent, triple superphosphate, can also add Ca to the soil.

Magnesium (Mg) is taken up by plants as the Mg^{2+} cation. It is the central atom in the chlorophyll molecule, and it aids in phosphate metabolism, plant respiration, and in the activation of many enzyme systems.

An imbalance between Ca and Mg on low cation-exchange capacity soils may accentuate a Mg deficiency. When the ratio of Ca to Mg becomes too high, plants may take up less Mg. Magnesium deficiency can also be accentuated by high K rates or by high availability of ammonium-N when the soil is marginally sufficient in Mg.

The ratio of K to Mg in soil is important. Failure to maintain an adequate supply of Mg can lead to reduced absorption of Mg by grazing animals, resulting in low blood serum Mg levels and a heightened risk of grass tetany (Chapter 47). Low soil temperatures and high K levels, with certain forages, can contribute to a K:Mg imbalance. High soil P levels can help increase Mg uptake by forages, help raise blood serum Mg levels, and lower the risk of grass tetany (Lock et al. 2002).

Application of Mg, either as dolomitic lime, when lime is needed, or as potassium magnesium sulfate (11% Mg) can increase forage yields (Phillips and Snyder 1988) on low-Mg soils and help improve fertilizer N-use efficiency (Phillips et al. 1995).

Sulfur

Role in Forage Plants

Sulfur is taken up by plants primarily as the sulfate anion (SO_4^{-2}) but can also enter leaves as sulfur dioxide (SO_2) from the air. Sulfur is part of every living cell and a constituent of two amino acids that form proteins. It is involved in biosynthesis of enzymes and vitamins, promotes nodulation (by symbiotic bacteria) for N fixation by legumes, is important in seed production, and is necessary for chlorophyll formation, although it is not part of the chlorophyll molecule. The critical plant tissue level of S in alfalfa is estimated to be 2 g S kg^{-1} DM in plant parts from 35 cm in height to the bud stage of growth (Leep 1999).

Sulfur in Soil

Because of its negative charge, sulfate is not held on cation exchange sites in the soil. It moves readily with soil water and can leach. Certain soils accumulate sulfate in the subsoil, which can be available to deep-rooted forages. Much of the S in soils is associated with soil organic matter. Through biologic transformations, similar to those affecting N, sulfate and sulfate compounds are released from soil organic matter. Warm, moist conditions promote these S transformations. Leaching of sulfate-S is more likely on sandy than on loamy or clayey soils.

Extractable sulfate-S soil tests are available but, are not considered as reliable as other nutrient tests. Sulfate levels change with soil temperature and moisture levels. A lack of sufficient extractable sulfate in the topsoil may lead to response to S application, particularly early in the growing season. Later, roots usually extend into the subsoil where sulfate may have accumulated.

Sources

Soil organic matter is the primary soil S source, and more than 95% of the S in soil is found tied up in soil organic matter. Animal manures range from less than 0.2 to 3.0 g S kg^{-1} DM. In some parts of the country, sulfur dioxide and other gases dissolved in rain or snow can contribute up to 5–10 kg of S ha^{-1} yr^{-1}.

Commercial sources of S include: elemental S (>85% S), ammonium sulfate (24% S), ammonium thiosulfate (26% S), ammonium polysulfide (40–50% S), potassium sulfate (18% S), potassium magnesium sulfate (22% S), gypsum (12–18% S), magnesium sulfate (14% S), and potassium thiosulfate (17% S). Elemental S must be converted to sulfate by soil bacteria before being used by plants. Relatively high rates (50–200 kg ha^{-1}) of elemental S may be needed to provide adequate sulfate, and three to six weeks may be needed under ideal conditions for conversion to occur.

Plant Response to Applied Sulfur

Historically, most soils in North America have been marginal to adequate in S, however, implementation of the Clean Air Acts during 1960–1990, has reduced atmospheric deposition of S in industrialized states to the extent that some soils are now considered to be deficient in S. Soils below 2% organic matter can be S deficient. Many southern forages, particularly high-yielding forage grasses receiving high N rates, can remove 25–50 kg of S ha^{-1} yr^{-1}. Response to S fertilization is likely when the forage tissue analyses show levels below the sufficiency range in Table 26.2 or when the N : S ratio in the forage exceeds about 17 : 1.

Most soils in the US supply adequate S for plant growth; however, sandy soils that are low in organic matter and are subject to excessive leaching may not. Alfalfa growing rapidly on such soils, at cool temperatures when mineralization of S is slow, is most likely to be deficient (Vitosh et al. 1995). Kelling and Speth (2001) reported a four-year yield response from a one-time S application at preplanting of 84 kg S ha^{-1} for alfalfa grown on a sandy loam soil in Wisconsin. Phillips and Curtis (1989) found that 100 kg of S ha^{-1} yr^{-1} increased coastal bermudagrass yield and N recovery at N rates from 224 to 448 kg ha^{-1}, whether applied as gypsum or elemental S. Additional S rate and timing research with Coastal bermudagrass on a Coastal Plain soil in Arkansas showed that DM yield was increased by over 980 kg ha^{-1} yr^{-1}, and apparent fertilizer N recovery was greatest when at least 27 kg of sulfate-S ha^{-1} yr^{-1}, or 27 kg of sulfate-S ha^{-1} yr^{-1} plus 12 kg of Mg ha^{-1} yr^{-1}, were applied in the spring (Phillips et al. 1995).

Excess Sulfur

Excess S could trigger nutritional imbalances in livestock, according to Haby (2002). Excessively high levels of S can depress copper absorption in horses (Little 1999).

Micronutrients

Boron

Clover, alfalfa, and other legumes are considered to have a higher boron (B) requirement than most forage grasses. Haby (2002) reported an alfalfa yield increase of over 3.9 Mg ha^{-1} with application of 3.36 kg B ha^{-1}, regardless of soil pH. Hot water-extractable B test levels were not significantly related to alfalfa yield response.

Some midwestern states recommend applying up to 3 kg of elemental B ha^{-1} on alfalfa grown on sandy soils or highly weathered soils low in organic matter (Vitosh et al. 1995). Boron sufficiency ranges are 31–80 mg kg^{-1} DM for alfalfa sampled from the top 15 cm of growth. Many southern states recommend about 1 kg of B ha^{-1} for clovers and other legumes, and up to 3 kg of B ha^{-1} annually for alfalfa. Common fertilizer sources for B include sodium tetraborate (14–20% B), Solubor® (20% B), and liquid boron (10% B).

Cobalt

Cobalt (Co) is essential for rhizobia to carry out nitrogen fixation and is found in vitamin B_{12}. The range of Co found in the upper 15 cm of soils ranges from 2 to 90 kg ha^{-1}. Areas of the country where Co deficiencies may occur are on sandy Spodosols in New England and in the southeastern US. In these areas, as little as 2.2 kg Co ha^{-1} used as fertilizer can increase the Co concentration of forage plants above requirements for grazing animals (Reid and Jung 1974).

Molybdenum

Forage legume response to Mo is closely related to acidic soil pH. The potential response to Mo addition increases with increasing soil acidity. Mo deficiencies occur on acid prairie soils in the Midwest. Sufficiency ranges for alfalfa tissue sampled for Mo from the top 15 cm of forage legume growth are between 1 and 5 ppm Production of most forage legumes may be limited on soils with a pH < 5.5 because of low Mo availability in the southern US (Mortvedt and Anderson 1982). Molybdenum deficiency in legumes results in N deficiency because of impaired symbiotic N fixation.

Treating legume seed with Mo at planting can prevent deficiency on Mo-deficient or low-pH soils. Rates from 14 to 50 g of Mo ha^{-1} may be adequate.

Although specific forage legume toxicity due to Mo is not known, excess Mo in forage legumes (>5–10 ppm) can cause problems with copper metabolism in sheep. When Mo is applied at appropriate rates on low-Mo soils, induced copper deficiency in animals is not considered to be very likely.

Manganese

Manganese deficiency is not considered a significant problem in the production of forage legumes. Manganese toxicity in the US is probably of greater concern, especially on soils below pH 5.2. Normal sufficiency ranges for Mn sampled from the top 15 cm of alfalfa growth is between 31 and 100 mg kg^{-1} DM. Although soil levels for Mn toxicity are not well established, more than 100 mg kg^{-1} DM of extractable Mn could pose potential toxicity concerns for forage legumes.

Iron

Iron is present in several peroxidase, catalase, and cytochrome oxidase enzymes and in ferredoxin, which is involved in oxidation-reduction reactions and is important in chlorophyll formation in plants. Plants grown on calcareous soils in arid areas are most likely to exhibit Fe deficiencies. Iron deficiency is not common in forage plants; however, excess Fe in plants may induce zinc (Zn) deficiencies in corn grown on organic soils. High Fe accumulation was found in the nodes of corn showing a Zn deficiency grown on organic soil (Lucas and Knezek 1972).

Zinc

Zinc is present in several dehydrogenase, proteinase, and peptidase enzymes. Zinc promotes growth hormones, starch formation, seed maturation, and production. Deficiencies are most likely on soils such as peat, muck, or mineral soils with pH greater than 6.5. The range of Zn found in the upper 15 cm of soils is from 25 to 700 kg ha^{-1}. Corn and sorghum may respond to Zn addition. Sufficiency ranges of Zn for corn sampled from the ear leaf at initial silking are from 20 to 70 $mg\,kg^{-1}$ DM Sources of Zn commonly used for correcting deficiencies in plants include zinc sulfate (23–36% Zn), zinc-ammonia complex (10% Zn), zinc oxysulfates (variable percentages of Zn), zinc oxide (50–80% Zn), and zinc chelate (9–14% Zn). Zinc fertilizer recommendations in the Midwest states are based upon both Zn soil test levels and soil pH.

References

Adams, F. (1984). Crop response to lime in the southern United States. In: *Soil Acidity and Liming*, 2e (ed. F. Adams), 211–266. Madison, WI: ASA, CSSA, and SSS.

Adams, F. and Pearson, R.W. (1967). Crop response to lime in the southern United States and Puerto Rico. In: *Soil Acidity and Liming*. Agronomy Monograph 12 (eds. R.W. Pearson and F. Adams), 161–206. Madison, WI: American society of Agronomy.

Bagg, J. (2000). *Alfalfa Winter Kill Risk Factors*. Lindsay, Ontario, CA: Ontario Ministry of Agriculture and Food Newsletter.

Barber, S.A. (1984). Liming materials and practices. In: *Soil Acidity and Liming*, 2e (ed. F. Adams), 171–210. Madison, WI: American society of Agronomy, Crop Science Society of Agronomy, and Soil Science Society of Agronomy.

Blaser, R.E. and Kimbrough, E.L. (1968). Potassium nutrition of forage crops with perennials. In: *The Role of Potassium in Agriculture* (eds. V.J. Kilmer, S.E. Younts and N.C. Brady), 423–445. Madison, WI: American Society of Agronomy.

Brady, N.C. and Weil, R.R. (2001). *The Nature and Properties of Soils*, 13e. Upper Saddle River, NJ: Prentice Hall.

Burton, G.W. and DeVane, E.H. (1992). Growing legumes with Coastal bermudagrass in the lower Coastal Plains. *J. Prod. Agric.* 5: 278–281.

Burton, G.W. and Jackson, J.E. (1962). Effect of rate and frequency of applying six nitrogen sources on Coastal bermudagrass. *Agron. J.* 54: 40–43.

Campbell, C.R. (2000). Reference sufficiency ranges for plant analysis in the southern region of the United States. Southern Cooperative Service Bulletin No. 394.

Chapman, S.L. (1984). Commercial nitrogen sources for forages. In: *Proceedings of 1984 Forage and Grassland Conference*. January 23–26, 1984. Houston, TX, 333–343. American Forage and Grassland Council.

Cope, J.T., Jr. (1970). Response of cotton, corn, and bermudagrass to rates of N, P, and K. Alabama Agricultural Experiment Station, Circular No. 181.

Cripps, R.W., Young, J.L., Bell, T.L., and Leonard, A.T. (1988). Effects of lime and potassium application on Arrowleaf clover, Crimson clover, and Coastal bermudagrass yields. *J. Prod. Agric.* 1: 309–313.

Daniels, M.B., DeLaune, P., Moore, P.A. Jr. et al. (2001). Soil phosphorus variability in pastures: implications for sampling and environmental management strategies. *J. Environ. Qual.* 30: 2157–2165.

Day, J.L. and Parker, M.B. (1985). Fertilizer effects on crop removal of P and K in 'Coastal' bermudagrass forage. *Agron. J.* 77: 110–114.

Dobson, J.W. and Beaty, E.R. (1977). Forage yields of five perennial grasses with and without white clover at four nitrogen rates. *J. Range Manage.* 30: 461–465.

Eichhorn, M.M., Jr. (1996). Phosphorus fertilization of Coastal bermudagrass grown on Darley gravelly fine sandy loam soil. In: *1996 Agronomy Research Report* (ed. M.M. Eichhorn), 125–142. Hill Farm Research Station/Louisiana Agricultural Experiment Station.

Eichhorn, M.M., Jr. (1989). Effects of fertilizer nitrogen rates and sources on Coastal bermudagrass grown on Coastal Plain soil. Louisiana Agricultural Experiment Station Bulletin No. 797.

Eichhorn, M.M., Jr. (1995). Bermudagrasses. *Louisiana Agric.* 38 (3): 12–13.

Eichhorn, M.M. Jr. and Huffman, D.C. (1991). Fertilizer nitrogen requirements for maximum economic yield of Coastal bermudagrass hay. *Louisiana Agric.* 35 (3–4): 19.

Eichhorn, M.M., Jr., Nelson, B.D., Amacher, M.C. et al. (1987). Effects of fertilizer potassium on Coastal bermudagrass grown on Coastal Plain soil. Louisiana Agricultural Experiment Station Bulletin No. 782.

Evans, E.M., Ensminger, L.E., Doss, B.D., and Bennett, O.L. (1961). *Nitrogen and Moisture Requirements of Coastal Bermudagrass and Pensacola Bahia*, 19. Alabama Agricultural Experiment Station Bull. 337.

Evers, G.W. (1985). Forage and nitrogen contributions of arrowleaf and subterranean clovers overseeded on bermudagrass and bahiagrass. *Agron. J.* 77: 960–963.

Follett, R.F. and Wilkinson, S.R. (1995). Nutrient management of forages. In: *Forages Volume II: The Science of Grassland Agriculture*, 5e (eds. R.F Barnes, D.A. Miller and C.J. Nelson), 55–82. Ames, IA: Iowa State University Press.

Gelderman, R.H., Gerwing, J.R., and Twidwell, E. (2002). Point-injected phosphorus effects on established cool-season grass yield and phosphorus content. *Agron. J.* 94: 48–51.

Goff, J.P. and Horst, R.L. (1997). Effects of the addition of potassium or sodium, but not calcium to prepartum ratios on milk fever on dairy cows. *Dairy Sci.* 80 (1): 176–186.

Haby, V.A. (1995). Soil management and fertility practices for annual ryegrass. In: *Symposium on Annual Ryegrass*. August 31–September 1, 1995. Tyler, TX, 24–42. Texas Agricultural Experiment Station MP-1770.

Haby, V.A. (2002). Soil fertility and management of acid Coastal Plain soils for crop production. *Commun. Soil Sci. Plant Anal.* 33 (15–18): 2497–2520.

Haby, V.A. and Leonard, A.T. (2002). Limestone quality and effectiveness for neutralizing soil acidity. *Commun. Soil Sci. Plant Anal.* 33 (15–18): 2935–2948.

Haby, V.A., Hillard, J.B., and Clary, G. (1996). Liming acid soils for ryegrass production. *Better Crops* 80 (1): 14–15.

Hallmark, W.B., Amacher, M.C., and Brown, L.P. (1986). How much soil potassium do bermudagrasses require? *Louisiana Agric.* 29 (3): 22–23.

Hanlon, E.A. (2001). Procedures used by state soil testing laboratories in the southern region of the United States. Southern Cooperative Series Bulletin No.190-C.

Helyar, K.R. and Anderson, A.J. (1970). Response of five pasture species to phosphorus, lime, and nitrogen on an infertile acid soil with a high phosphate sorption capacity. *Aust. J. Agric. Res.* 21: 677–692.

Hillard, J.B., Haby, V.A., and Hons, F.M. (1992a). Annual ryegrass response to limestone and phosphorus on an ultisol. *J. Plant Nutr.* 15 (8): 1253–1268.

Hillard, J.B., Haby, V.A., and Hons, F.M. (1992b). Effects of limestone and phosphorus on nutrient availability and Coastal bermudagrass yield on an ultisol. *Commun. Soil Sci. Plant Anal.* 23 (1&2): 175–188.

Hodges, S.C. and Kirkland, D. (1994). Pastures and forages. In: *Soil Sampling Procedures for the Southern Region of the United States*. Southern Cooperative Series Bulletin No. 337 (eds. W.O. Thom and W. Sabbe), 25–30. Lexington, KY: University of Kentucky Agricultural Experiment Station.

Huneycutt, H.J., West, C.P. and Phillips, J.M. (1988). Responses of bermudagrass, tall fescue, and fescue-clover to broiler litter and commercial fertilizer. Arkansas Agricultural Experiment Station Bulletin No. 913.

Keisling, T.C., Roquette, F.M. Jr., and Matocha, J.E. (1979). Potassium fertilization influences on Coastal bermudagrass rhizomes, roots, and stand. *Agron. J.* 71: 892–894.

Kelling, K.A. and Matocha, J.E. (1990). Plant analysis as an aid in fertilizing forage crops. In: *Soil Testing and Plant Analysis*, 3e (ed. R.L. Westerman), 603–644. Soil Science Society of America Book Series, No. 3.

Kelling, K.A. and Speth, P.E. (2001). Sulfur Status and Responses on Wisconsin Alfalfa. Wisconsin Fertilizer Conference Proceedings. http://www.uwex.edu/ces/forage/wfc/proceedings2001/sulfur.htm (accessed 10 October 2019).

Kelling, K.A., Combs, S.M., and Peters, J.B. (2000). *Using Plant Analysis as a Diagnostic Tool*. Publication No. 6. University of Wisconsin Extension.

Kidder, G., Sabbe, W.E., and Parks, C.L. (1988). *Procedures and Practices Followed by Southern State Soil Testing Laboratories for Making Lime Recommendations*. Southern Cooperative Series Bulletin No. 332. Institute of Food and Agricultural Sciences, University of Florida.

Lanyon, L.E. and Smith, F.W. (1985). Potassium nutrition of alfalfa and other forage legumes: temperate and tropical. In: *Potassium in Agriculture* (ed. R.D. Munson), 861–893. Madison, WI: American society of Agronomy, Crop Science Society of America, and Soil Science Society of America.

Leep, R.H. (1989). *Improving Pastures in Michigan by Frost Seeding*. Extension Bulletin E-2185. Cooperative Extension Service, Michigan State University.

Leep, R.H. (1999). Sulfur, an essential nutrient for forage crops. *Michigan Dairy Rev.* 4 (1): 4–9. Michigan State University.

Leep, R.H. and Tesar, M.B. (1981). *Growing Birdsfoot Trefoil in Michigan*. Extension Bulletin E-1536. Cooperative Extension Service, Michigan State University.

Leep, R.H., McNabnay, M., Warncke, D. et al. (2000). Variability in soil factors in Michigan commercial alfalfa fields. In: *Proceedings of 5th International Conference on Precision Agriculture*, 123. Bloomington, MN. July 16–19, 2000.

Leep, R.H., Min, D.H., and DeYoung, J.R. (2002). Site-specific versus whole field management of soil fertility in alfalfa. In: *Proceedings of 2002 American Forage and Grasslands Conference*. Minneapolis, MN. July 14–17, 2002.

Little, C. (1999). *Feeding Horses*. Fact Sheet, ANR-6-99, Agriculture and Natural Resources. The Ohio State University.

Little, C. and McCutcheon, J. (1999). *Fertility Management of Meadows*. Fact Sheet, ANR-5-99, Agriculture and Natural Resources. The Ohio State University.

Lock, T.R., Kallenbach, R.L., Blevins, D.G. et al. (2002). Adequate soil phosphorus decreases the grass tetany potential of tall fescue pasture. *Crop Manage.*: 8. Plant Management Network On-line. https://doi.org/10.1094/CM-2002-0809-01-RS.

Lucas, R.E. and Knezek, B.D. (1972). Climatic and soil conditions promoting micronutrient deficiencies in plants. In: *Micronutrients in Agriculture* (ed. J.J. Mortvedt), 265–288. Madison, WI: Soil Science Society of America.

Mathews, B.W., Sollenberger, L.E., Nkedi-Kizza, P. et al. (1994). Soil sampling for monitoring potassium distribution in grazed pastures. *Agron. J.* 86: 121–126.

Mays, D.A., Wilkinson, S.R., and Cole, C.V. (1980). Phosphorus nutrition of forages. In: *The Role of Phosphorus in Agriculture* (eds. F.E. Khasawneh, E.C. Sample and E.J. Kamprath), 805–846. Madison, WI: American society of Agronomy, Crop Science Society of America, and Soil Science Society of America.

Mortvedt, J.J. and Anderson, O.E. (1982). *Forage Legumes: Diagnosis and Correction of Molybdenum and Manganese Problems*. Southern Cooperative Series Bulletin 278. University of Georgia Agricultural Experiment Station.

Munns, D.N. (1965). Soil acidity and growth of a legume. I. Interactions of lime with nitrogen and phosphate on growth of *Medicago sativa* L. and *Trifolium subterraneum* L. *Aust. J. Agric. Res.* 16: 733–741.

National Research Council (2000). *Clean Coastal Waters: Understanding and Reducing the Effects of Nutrient Pollution*. Washington, DC: National Academy Press.

Nelson, L.R., Keisling, T.C., and Rouquette, F.M. Jr. (1983). Potassium rates and sources for Coastal bermudagrass. *Soil Sci. Soc. Am. J.* 47 (5): 963–966.

Overman, A.R. and Wilkinson, S.R. (1992). Model evaluation for perennial grasses in the Southern United States. *Agron. J.* 84 (3): 523–529.

Pearson, R.W. and Hoveland, C.S. (1974). Lime needs of forage crops. In: *Forage Fertilization* (ed. D.A. Mays), 301–322. Madison, WI: American Society of Agronomy.

Phillips, J.M. and Curtis, C.C. (1989). Nitrogen and sulfur recovery of Coastal bermudagrass. In: *Proceedings of 1989 Forage and Grassland Conference*. May 22–25, 1989. Guelph, Ontario, Canada, 250–254. American Forage and Grassland Council.

Phillips, J.M. and Snyder, C.S. (1988). Effect of limestone and magnesium on bahiagrass yield. Quality, nutrient concentration and uptake and soil test levels. Arkansas Agricultural Experiment Station Bulletin No. 914.

Phillips, J.M., Snyder, C.S., Gbur, E.E. et al. (1995). Yield quality and nitrogen recovery of Coastal bermudagrass as affected by limestone rates and sulfur-magnesium additions. Arkansas Agricultural Experiment Station Report Series No. 329.

PPI/PPIC (2001). *Soil Test Levels in North America: Summary Update*, 17. PPI/PPIC/FAR Technical Bulletin 2001-1. Potash & Phosphate Institute.

PPI/PPIC (2002). *Plant Nutrient Use in North American Agriculture: Producing Food and Fiber, Preserving the Environment, and Integrating Organic and Inorganic Sources*, 117. PPI/PPIC/FAR Technical Bulletin 2002-1. Potash & Phosphate Institute.

Pratt, J.N. and Darst, B.C. (1984). Effects of selected plant nutrients on yield, chemical composition, and drought tolerance of coastal and other hybrid bermudagrasses (*Cynodon dactylon* L.). In: *Proceedings of 1984 American Forage and Grassland Conference*, 290–294. Houston, TX. January 23–26, 1984.

Raun, W.R., Solie, J.B., Johnson, G.V. et al. (1998). Microvariability in soil test, plant nutrient, and yield parameters of bermudagrass. *Soil Sci. Soc. Am. J.* 62: 683–690.

Rechcigl, J.E. (1992). Response of ryegrass to limestone and phosphorus. *J. Prod. Agric.* 5: 602–607.

Redmon, L.A. (1996). *Selecting Forages for Nutrient Recycling*. Production Technology Bulletin 96-36. Oklahoma State University.

Reid, R.L. and Jung, G.A. (1974). Effects of elements other than nitrogen on nutritive value of forage. In: *Forage Fertilization* (ed. D.A. Mays), 395–435. Madison, WI: American Society of Agronomy.

Robinson, D.L. (1985). Potassium nutrition of forage grasses. In: *Potassium in Agriculture* (ed. R.D. Munson), 895–914. Madison, WI: American society of Agronomy, Crop Science Society of America, and Soil Science Society of America.

Robinson, D.L. (1990). Nitrogen utilization efficiency by major forage grasses in Louisiana. *Louisiana Agric.* 33 (3): 21–24.

Robinson, D.L. (1995). The role of soil fertility in forage-livestock production. *Louisiana Agric.* 38 (3): 18–19.

Robinson, D.L. (1996). Fertilization and nutrient utilization in harvested forage systems—southern forage crops. In: *Nutrient Cycling in Forage Systems*. March 7–8, 1996. Columbia, MO (eds. R.E. Joost and C.A. Roberts), 65–92. Potash & Phosphate Institute and the Foundation for Agronomic Research.

Robinson, J.L.R. and Dabney, S.M. (1986). Comparing ammonium nitrate and urea for Louisiana crop production. *Louisiana Agric.* 30 (1): 18–19. 23.

Robinson, D.L. and Eilers, T.L. (1996). Phosphorus and potassium influences on annual ryegrass production. *Louisiana Agric.* 39 (2): 10–11.

Scarsbrook, C.E. (1970). Regression of nitrogen uptake on nitrogen added from four sources applied to grass. *Agron. J.* 62: 618–620.

Schaller, F.W., Voss, R.D., and George, J.R. (1979). *Fertilizing Pasture*, 5. Iowa State University Cooperative Extension Service. Pm-869.

Sharpley, A.N., Daniel, T., Sims, T. et al. (1999). *Agricultural Phosphorus and Eutrophication*, 42. U.S. Department of Agriculture, Agricultural Research Service, ARS-149.

Snyder, C.S. (1998). *Plant Tissue Analysis—A Valuable Nutrient Management Tool*, 4. News & Views, June 1998. Potash & Phosphate Institute.

Snyder, C.S. (2003). Removal of potassium in hay harvests—a huge factor in nutrient budgets. *Better Crops* 87 (4): 3–5. Potash & Phosphate Institute.

Snyder, C.S. and Bruulsema, T.W. (2002). Nutrients and environmental quality. In: *Plant Nutrient Use in North American Agriculture: Producing Food and Fiber, Preserving the Environment, and Integrating Organic and Inorganic Sources*. 117 pp. PPI/PPIC/FAR Technical Bulletin 2002-1, 45–68. Potash & Phosphate Institute.

Snyder, C.S. and Hankins, B.J. (1987). *Forage Legume Inoculation*, 4. University of Arkansas Cooperative Extension Service. FSA 2035.

Snyder, C.S., Bruulsema, T.W., Sharpley, A.N., and Beegle, D.B. (1999). *Site-Specific Use of the Environmental Phosphorus Index Concept. Site-Specific Management Guideline-1*, 4. Potash & Phosphate Institute.

Steele, K.W. (1982). Nitrogen in grassland soils. Chapter 3 In: *Nitrogen Fertilizers in New Zealand Agriculture* (ed. P.B. Lynch), 29–45. Wellington: Ray Richards Publisher for New Zealand Institute of Agriculture Science.

Tesar, M.B. (1982). *Forage Management Notebook*, 3e. East Lansing, MI: Michigan State University Press.

Tisdale, S.L., Nelson, W.L., Beaton, J.D., and Havlin, J.L. (1993). *Soil Fertility and Fertilizers*, 5e. New York: Macmillan Publishing.

Vitosh, M.L., Johnson, J.W., and Mengel, D.B. (1995). *Tri-State Fertilizer Recommendations for Corn, Soybeans, Wheat and Alfalfa*. Extension Bulletin E-2567. Michigan State University Extension.

Vough, L.M. (2000). *Nutrient Manager: Focus on Potassium*, vol. 7, 1. University of Maryland.

Weeks, M.E. and Lathwell, D.J. (1967). Crop response to lime in the northeastern United States. In: *Soil Acidity and Liming*. Agronomy 12 (eds. R.W. Pearson and F. Adams), 233–263. Madison, WI: American Society of Agronomy.

Westerman, R.L., Silvertooth, J.C., Barreto, H.J., and Minter, D.L. (1984). Phosphorus and potassium effects on yield and nutrient uptake in arrowleaf clover. *Soil Sci. Soc. Am. J.* 48: 1292–1296.

Whitehead, D.C. (1995). *Grassland Nitrogen*. Wallingford, UK: CAB International.

Wilkinson, S.R. and Langdale, G.W. (1974). Fertility needs of the warm-season grasses. In: *Forage Fertilization* (ed. D.A. Mays), 119–145. Madison, WI: American Society of Agronomy.

Wilkinson, S.R., Stuedemann, J.A., and Belesky, D.P. (1989). Soil potassium distribution in grazed KY-31 tall fescue pastures as affected by fertilization and endophytic fungus infection level. *Agron. J.*: 508–512.

Young, J.L., Bell, T.L., Leonard, A.T., and Cripps, R.W. (1984). Effects of lime and K on Arrowleaf clover and Coastal bermudagrass. In: *Proceedings of 1984 Forage and Grassland Conference*. January 23–26, 1984. Houston, TX, 307–311. American Forage and Grassland Council.

Zemenchik, R.A. and Albrecht, K.A. (2002). Nitrogen use efficiency and apparent nitrogen recovery of kentucky bluegrass, smooth bromegrass, and orchardgrass. *Agron. J.* 94 ((4): 421–428.

CHAPTER

27

Irrigation and Water Management

L. Niel Allen, Associate Professor and Irrigation Specialist, *Utah State University, Logan, UT, USA*
Jennifer W. MacAdam, Professor of Plants, Soils and Climate, *Utah State University, Logan, UT, USA*

Irrigation is the application of water to the soil to sustain or improve crop production. Irrigation management considers the method of water application, the timing of water application, and how much water to apply.

Need and Extent of Forage Irrigation

In arid and semi-arid regions of the world, irrigation makes it possible to grow forage crops and/or improve forage yields. The western US is a good region for irrigated forage crop production from an agronomic and economic perspective. There are domestic and international markets for machine-harvested irrigated forage crops and grazed irrigated land is key to many livestock operations. The total irrigated area of alfalfa in the US in 2012 was about 2 340 000 ha, primarily in the west (Figure 27.1). Highest alfalfa yields in the US also occur in the arid western states under irrigation where good drying conditions provide an opportunity to harvest high-quality forages without rain damage. Some irrigation of forage crops occurs in humid regions to increase yields, particularly during dry periods.

Area of irrigated pasture and some rangeland in the US is 1 509 000 ha (Figure 27.2). Total irrigated acreage in 2012 was about 22 590 000 ha (Figure 27.3), thus irrigated alfalfa and pastures comprise about 17% of the irrigated land in the US (USDA-NASS 2014).

In addition to alfalfa and pastures, irrigated forage land includes small grains and silage corn. In some countries, forage crops account for an even higher percentage of total irrigation; e.g. 38% of the irrigated land in Australia is used for forage crops (Australian Bureau of Statistics 2016). Worldwide, irrigated pastures and forage fields account for 7% of the total irrigated land (FAO 2014).

Irrigated pastures are prevalent in the higher elevations of the western US which have short growing seasons that limit production of alfalfa. For example, in the Upper Colorado River Basin (Utah, Wyoming, New Mexico, and Colorado) there are approximately 631 700 ha irrigated land, with 450 400 ha (72%) used as irrigated pasture and 108 900 ha (17%) used for irrigated alfalfa (Wyoming Water Development Commission 2010; Utah DNR 2013; Colorado DWR 2013).

Both cool-season and warm-season grasses are common in irrigated areas. Cool-season grasses produce well during spring and early summer, during the vegetative and reproductive stages of growth. In locations with temperate climates (winter, spring, summer, and fall), regrowth after the first cutting or grazing has a lower water-use efficiency (yield per unit of water use) than during early spring growth as the pasture comes out of winter dormancy (Volesky and Berger 2010). In contrast, irrigated warm-season grasses such as bermudagrass and sudan grasses do well in hot, arid locations such as the Imperial Valley of California that can produce both cool- and warm-season grasses. In general, perennial forages

Forages: The Science of Grassland Agriculture, Volume II, Seventh Edition.
Edited by Kenneth J. Moore, Michael Collins, C. Jerry Nelson and Daren D. Redfearn.

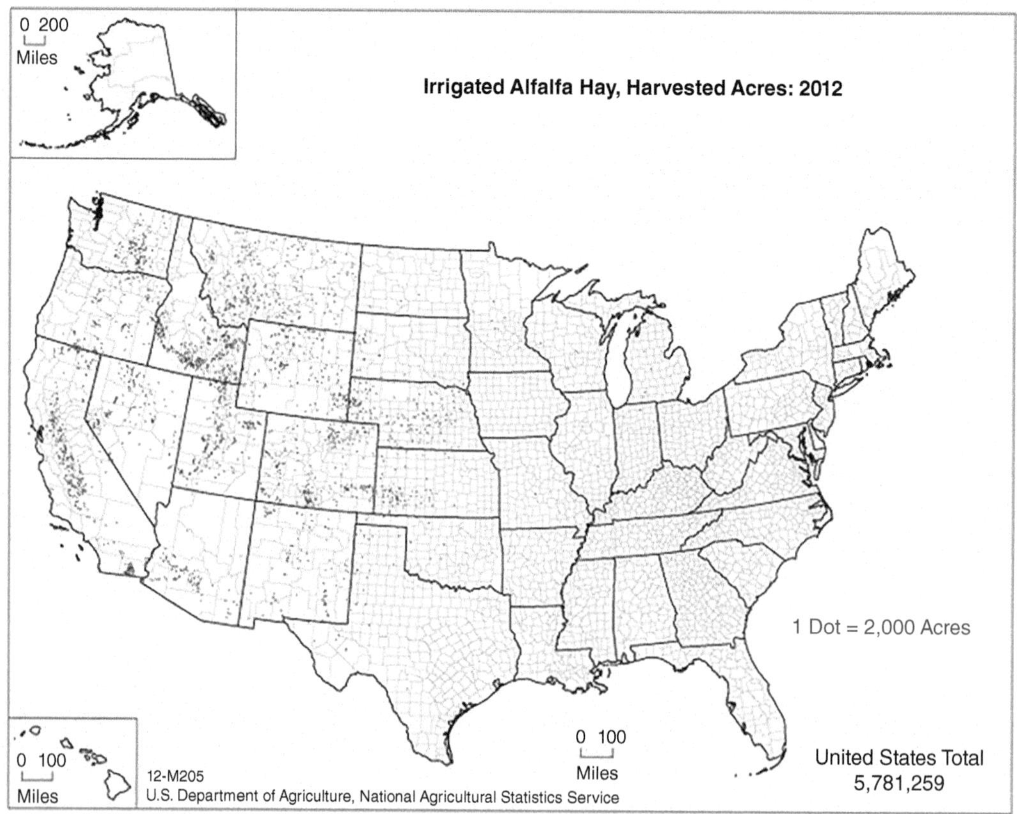

Fig. 27.1. Irrigated alfalfa distribution in the United States, total of 2 339 592 ha (1 dot = 810 ha).

are drought tolerant, and though yields decrease with seasonal water shortages, most forages survive the stress and maintain an adequate stand for future production (Orloff et al. 2014).

In addition to grasses and alfalfa, irrigated forage legumes including birdsfoot trefoil, cicer milkvetch, sainfoin, and clovers produce well in many climates. These legumes are often grown with grasses, in which case, they can improve yield and feed quality, and benefit soil fertility (Sleugh et al. 2000; MacAdam and Griggs 2006).

Sources of Water

Irrigation, the largest water user in the western US, uses both surface and groundwater. In the western US, policies and laws on water rights vary from state to state, but generally, surface and groundwater rights are based on the prior appropriation doctrine; i.e. first in time (beneficial water use), first in water right priority. A notable exception is the groundwater law in California where landowners can drill wells on their property. The water user generally has a right to a reasonable share of the groundwater for use on the landowner's land that overlies the basin. However, this privilege has resulted in significant water supply problems. California is addressing the problems associated with excessive groundwater pumping and now requires entities to develop groundwater management plans (California State Legislature 2014).

Some areas rely on surface water, and some areas use both groundwater and surface water. Overall, in the US, 57% of agricultural irrigation water is from surface sources. The western states use a higher percentage of surface water, including reservoirs, than other parts of the US (Maupin et al. 2014). For example, 90% of Wyoming's irrigation water is from surface sources, while 75% of Texas's irrigation water is from groundwater. In many areas, groundwater pumping has exceeded recharge rates, lowering the water table, which can result in increased pumping costs, soil subsidence, and saltwater intrusion.

Forage Water Use

Forage water use and evapotranspiration (ET) have been studied for decades. Measuring or estimating forage water use is important for proper irrigation management. Water use or ET is usually expressed as depth per unit time, such as mm d^{-1}.

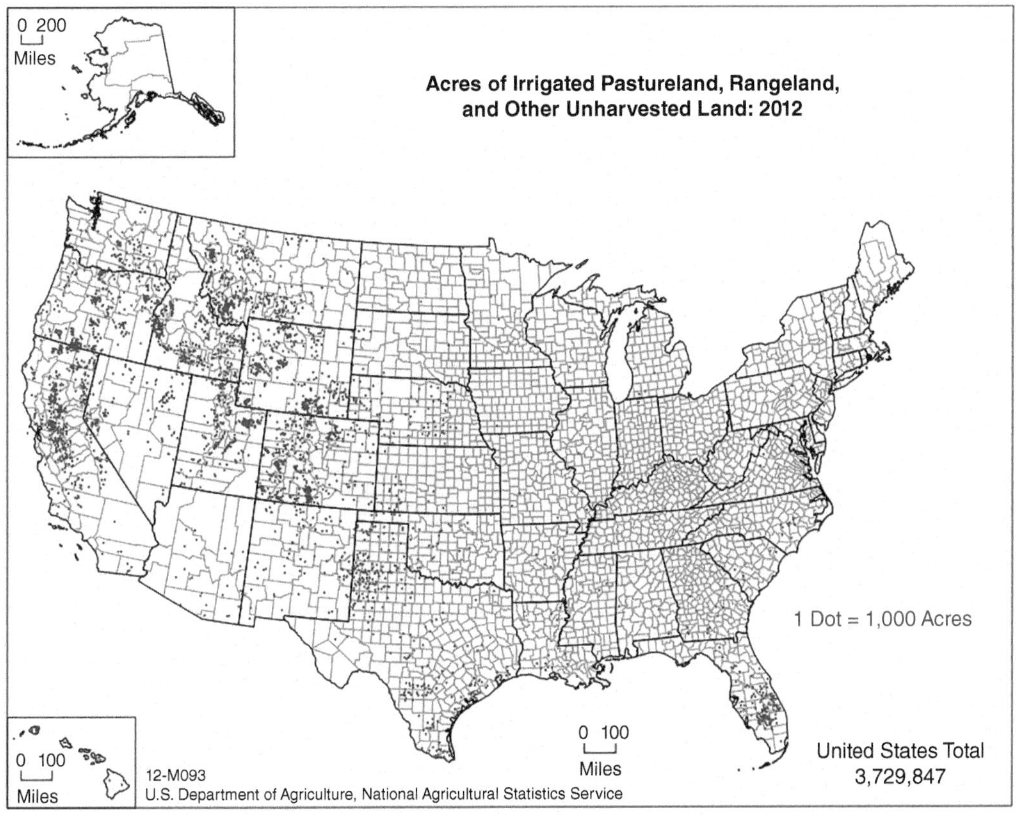

Fig. 27.2. Irrigated pasture distribution in the US, total of 1 509 415 ha (1 dot = 405 ha).

Measurement Methods

Water use or ET is measured using the following methods:

- Soil water budgets. Calculations of ET = precipitation + irrigation – drainage + change in soil moisture for a given time period.
- Direct measurements using weighing lysimeters. Lysimeters are containers growing vegetation, which are weighed on an hourly or daily basis, and changes in weight are converted to a depth of water use. This, along with water added to or drained from the lysimeter, allows calculation of ET.
- Vapor flux measurement using instrumentation. One measure of vapor flux leaving a vegetative area is based on eddy covariance. This method relies on instrumentation that rapidly measures direction and velocity of air movement (eddies) along with water vapor in the air.
- Remotely sensed thermal and spectral data. Information from sensors and cameras (obtained from satellites, drones, or aircraft) is used to calculate the energy used to evaporate water in a vegetated area.

Estimation Methods

Water use by plants is driven by the gradient in water vapor pressure from inside the leaf where the relative humidity is near 100% to the outside of a leaf where the humidity is lower and variable. Water use is limited to the available soil moisture and controlled by plant characteristics such as leaf area (Chapter 6). The vapor pressure deficit (evaporation potential) is greater under climatic conditions such as high solar radiation, high temperature, low humidity, and high wind speed. There are interactions that need to be considered since the conditions may partially offset one another.

Many methodologies to predict crop ET have been developed since the 1930s (Jensen and Allen 2016). A crop coefficient (Kc) is the ratio of expected maximum ET at a particular stage of crop growth relative to a reference ET (ETr), such as evaporation from a theoretical crop or a free-water surface (Hanks 1985). Currently, the most commonly used method to calculate ET is to multiply the ETr by a crop coefficient (Kc) (Allen et al. 1998).

There are two reference ET definitions in use, ETo short crop (grass) and ETr tall crop (alfalfa). The ETo

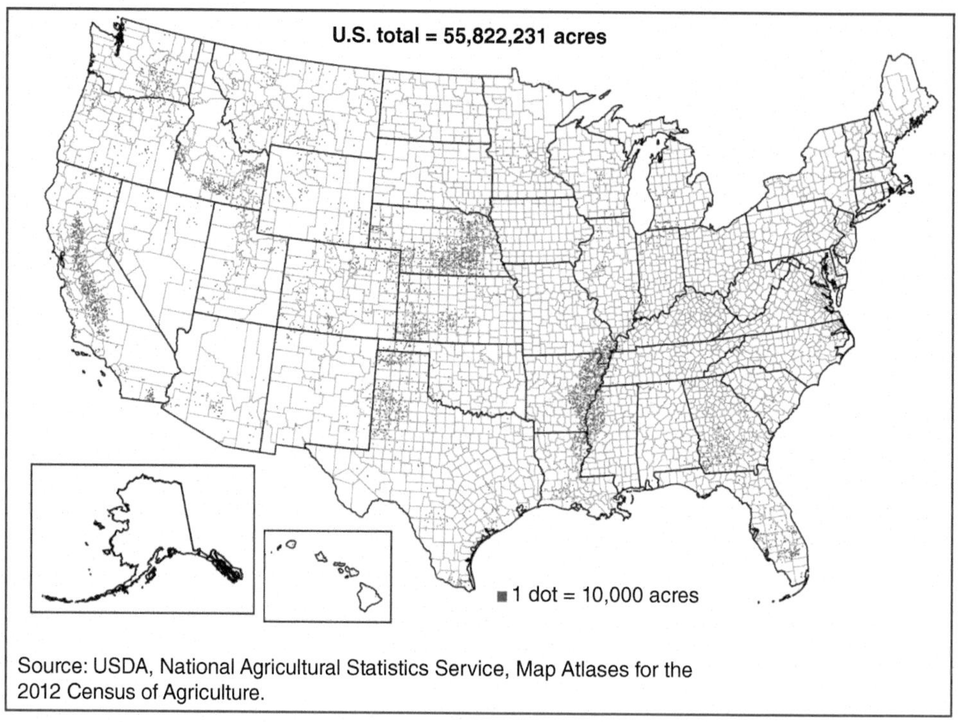

Fig. 27.3. Irrigated crop distribution in the US, total of 22 590 455 ha (1 dot = 4050 ha).

rates are lower than the ETr rates because a short crop uses less water due to less leaf area and less air mixing within the crop canopy. A standardized energy-based Penman-Monteith equation has been developed that can be used to calculate both ETr and ETo (Jensen and Allen 2016). Potential ET values for specific crops at specific times (ETc) are calculated by the following formula:

$$\text{ETc} = \text{Kc} \times \text{ETo}$$

Crop coefficients (Kc), based on species and growth stage of the crop, are generally calculated on a daily basis for the entire growing season (Figures 27.4 and 27.5). Because ETo and ETr are different, the crop coefficients are also different. For example, the Kc for a mid-season pasture that has been stocked continuously is 0.78 based on ETr (USBR 2017), and the Kc for a mid-season pasture that has been stocked rotationally is 1.05 because it is based on ETo (Allen et al. 1998). Some agriculture weather networks report both ETo and ETr, so it is important to know which is reported and/or applicable. The selection of ETo or ETr is based on calibration of Kc values to crops of regional interest. Worldwide, ETo is commonly used; ETr is used in regions of the US where decades of crop water-use research have been based on alfalfa (a tall, deep-rooted common forage crop) as a reference crop (ETr).

Many data sources provide reference ET on a spatial and near-real-time basis. Examples of weather networks that provide reference ET for agricultural use and irrigation scheduling include:

- USBR Agricultural Meteorological Network (AgriMet) – The AgriMet network consists of over 70 agricultural weather stations located throughout the Pacific Northwest (data available for states of Washington, Oregon, Idaho, Nevada, Montana, and Utah). AgriMet uses ETr as a reference (USBR 2017).
- The California Irrigation Management Information System (CIMIS) helps growers develop efficient water budgets and irrigation strategies. CIMIS is a program of the California Department of Water Resources (DWR) that includes a network of over 145 automated weather stations in California. CIMIS uses ETo as a reference, and Kc values from FAO 56 are applicable (California DWR 2017).
- Arizona Meteorological Network (AZMET) has 30 agricultural weather stations in Arizona and reports ETo (Univ. Arizona 2017).

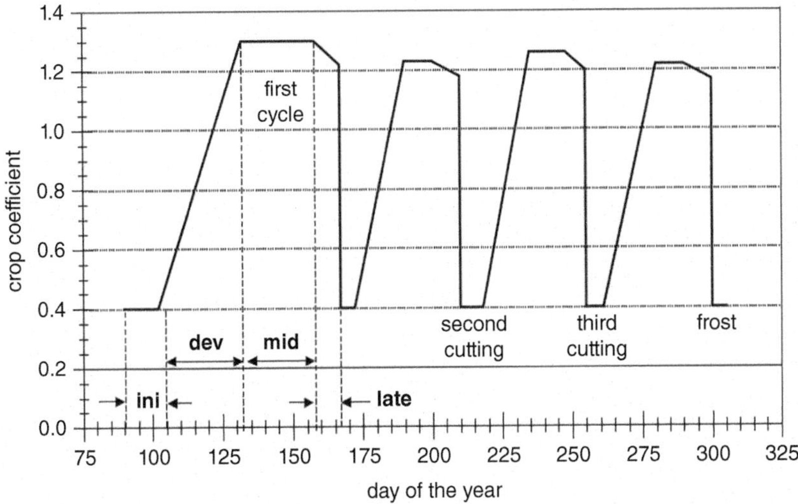

Fig. 27.4. Example crop coefficient curve for alfalfa in southern Idaho (from FAO 56), day of year is Julian day.

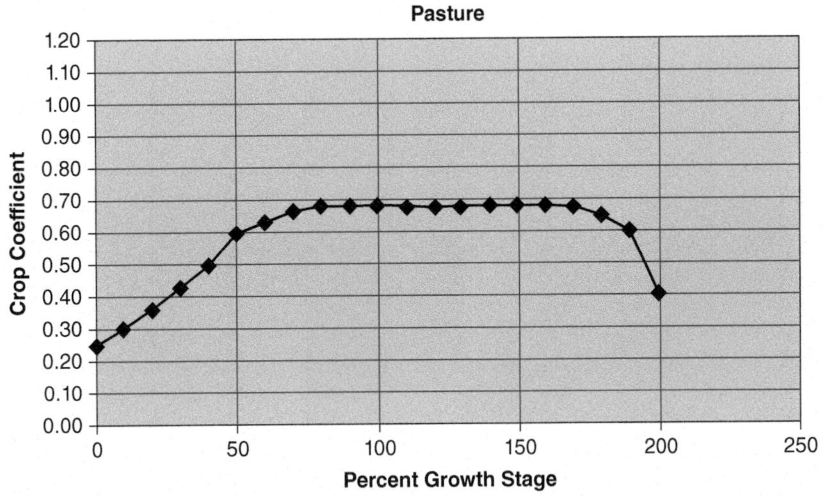

Fig. 27.5. Pasture crop coefficients based on ETr (tall crop reference) where 60% is 10–15 cm growth and 100% growth stage is heading. *Source:* From USBR AgriMet. From USBR (2017) AgriMet (Developed by Agricultural Research Service, Kimberly, Idaho, Feb. 16, 1994).

- Colorado Agricultural Meteorological network (CoAgMET) has 94 agricultural weather stations in Colorado and reports ETr (Colorado State Univ. 2017).
- Nebraska Agricultural Water Management Network (NAWMN) has dozens of weather stations and offers help determining crop coefficients and irrigation scheduling. The network reports ETo and appropriate crop coefficients (Univ. Nebraska 2017).

Irrigation Methods and Efficiencies

Sprinkler, surface, and drip (trickle or micro-irrigation) are irrigation systems used for forages. Sprinkler irrigation provides about 57% of agricultural irrigation in the US and includes center pivots, hand lines, wheel lines, big guns, and fixed sprinkler systems (Figure 27.6). Surface irrigation provides about 35% of the irrigation in the US and includes flood, furrow, basin, corrugations, contour,

Fig. 27.6. Photographs of wheel line (upper left), pivot with long drop-nozzles to limit evaporative loss (upper right), and two types of flood irrigation systems. *Source:* Photos courtesy of L. Niel Allen and Robert W. Hill, Utah State University, Logan, Utah.

border irrigation, and sub-surface (controlling the water table to irrigate from the bottom of the crop root zone). Drip irrigation provides about 8% of the irrigation in the US (USDA-NASS 2014). Though not common in forage crops, there are buried-drip systems irrigating alfalfa, pastures, and silage corn along with a few surface-drip systems on row crops grown for silage.

The most practical irrigation method is determined by considering topography, soils (profile, texture, drainage, chemistry, and salinity), system costs, water availability, energy availability, crops, crop rotation, harvest method(s), and labor. Irrigation system costs depend on the irrigation method, pumping and piping costs, operation and management costs, energy costs, labor costs, water source, and irrigated land area. An economic analysis is recommended to determine the feasibility of irrigation and the selection of an irrigation system. Some systems may have lower capital costs, but higher labor and operation costs.

In 2018, capital costs for an irrigation system ranged from a few hundred dollars per ha to over \$2500 ha^{-1}. In general, costs are lower for surface, intermediate for sprinkler and highest for drip systems. However, each system has a large variation in costs, so irrigation system costs are site-specific. In the US, the trend is toward fewer surface and more sprinkler irrigation systems (USDA-NASS 2014). However, there are large irrigated areas and irrigation projects that are well suited for surface irrigation and have high efficiencies due to irrigation water conveyance and distribution, soils, well-leveled or graded fields, and good water supplies.

Efficiency of Systems

Not all irrigation water applied is available to the crop. Major losses from field irrigation include deep percolation (water draining below the root zone), surface runoff, evaporation and drift from sprinklers, irrigation systems leaks, seepage and evaporation from ditches. Water from non-consumptive irrigation losses can remain in the system as groundwater to become available downstream for irrigation or other purposes via groundwater pumping.

Field irrigation efficiency (fraction of delivered water that is beneficially consumed at field level) is the percentage of the irrigation water applied that is stored in the root zone and available for crop use. Field irrigation efficiencies depend on irrigation system design and water management. In general, drip irrigation has the highest efficiency, followed by sprinkler and then surface systems.

Some irrigation systems are used to apply fertilizer or other chemicals (referred to as chemigation) to enhance forage production. These systems inject the chemicals into the irrigation water. Only irrigation systems with high uniformity of application should be used for chemigation.

Surface (also called gravity) irrigation is defined as water distributed to the field by gravity flow. It is the oldest and still most common irrigation method: 39% of irrigation in the US (USDA-NASS 2014) and 77% worldwide (ICID 2017). Surface irrigation is common in **riparian** areas along mountain streams and rivers where irrigation has been practiced for many generations.

Cost for surface irrigation is generally lower (dependent on land leveling costs), but irrigation uniformities and efficiencies can be lower than other methods. Efficiency of surface irrigation can be improved with proper system construction and operation. Management includes irrigation scheduling, field leveling, flowrates, and set times. Some of the most productive irrigated areas such as Yuma County, Arizona and Imperial County, California are surface irrigated. Surface irrigation efficiencies typically range from 60% to 85%. Most field irrigation losses from flood irrigation are due to deep percolation and runoff.

Types of Surface Irrigation

Border irrigation consists of rectangular strips of land in a field, generally 30–60 m in width and 200–800 m in length, separated by small dikes. There is no slope across the width of the strip and a small slope on the length. High flows (140–280 $l\,s^{-1}$) are introduced at the top of the strip and shut off when water has advanced to a predetermined location. The field strips are generally open on the lower side to drain, but can be closed or blocked, if managed to prevent excessive ponding and crop injury.

Basin irrigation is a leveled, closed basin often about 4 ha in area that will be planted. A high flow of water (300–600 $l\,s^{-1}$) is discharged into the basin until a predetermined depth is reached. These systems can be very efficient because no runoff occurs and water percolates into the root zone in a few hours.

Furrows or corrugations are narrow and shallow channels in the field used to convey and direct the water across the field. In forage crops, the corrugations or furrows are generally spaced about 0.7 m apart and are quite shallow.

Sub-irrigation is practical in some level areas with a high-water table. Some forages are irrigated by controlling the level of the water table. The water table is artificially raised into the crop root zone to replace plant water use and then drained after the soil is wetted. This is done by filling and draining ditches that are spaced 20–40 m apart, depending on soil hydrology.

Types of Sprinkler Irrigation

Sprinkler irrigation is suitable for forages and is adaptable to most soils and topography. The systems are flexible in design and operation. Sprinkler irrigation efficiencies typically range from 70% to 85% of the water reaching the soil, with center pivots and linear-move systems being the most efficient. Primary field irrigation losses are from deep percolation, evaporation, and drift.

Periodic-move sprinkler systems include hand-move, wheel lines, and drag-move (K-line or pod sprinklers). Systems are generally moved twice daily with intervals between irrigations from a few days to more than a week. These systems can include small sprinklers (10- to 26-m spacing) and big gun sprinklers (30 m or greater spacing).

Stationary sprinkler systems include buried pipe with upright sprinklers or above-ground systems with sprinklers (often-called solid set sprinklers) that are in place for a single or multiple irrigation seasons.

Continuous-move systems include center pivots with varying lengths, linear-move, and traveling big gun sprinklers. These systems usually provide good uniformity due to the continuously moving sprinklers. Sprinkler spacing can be less than 1 m for closely spaced sprays on drop tubes, to more than 10 m on impact sprinklers. Center pivots are common for forage crops due to adaptability for many soils and topography, low labor requirements, and uniformity of application. They are not well suited for some fields due to unique field dimensions, variable soils, and topography. Pivots irrigate a circular area or arc of a circle. Most of the corner areas in a rectangular or square field can be irrigated by corner pivot systems, but the system costs per ha are higher than for the circle portion of the pivot.

Selecting the Appropriate System

Selection of the proper forage irrigation system includes consideration of capital and operation costs (energy and labor), compatibility with harvesting and/or grazing management, and suitability for soils, topography, crops, water supply, and climate. Sprinkler systems are used on about 63% of the irrigated land in the US, of which about 80% are center pivots (USDA-NASS 2014).

Trickle (drip) irrigation provides water to the soil by light, frequent irrigations, and generally only wets a small proportion of the soil surface. Drip irrigation components include water source, pumps, filtration, mainlines, manifolds, pressure regulators, and drip laterals. Drip laterals (16–35 mm dia., 0.10–0.45 mm wall thickness, and lay-flat or round) have equally spaced emitters (0.15–0.60 m) that control discharge (0.26–3.80 $l\,h^{-1}$).

Subsurface drip irrigation can be used for forages to allow normal harvesting or grazing. Subsurface drip irrigation (placed 20–30 cm deep at 0.75- to 1.25-m line spacing) of alfalfa provides high-irrigation efficiency (85–90%), reduces soil surface evaporation, and allows for field operations during or soon after irrigation. Subsurface drip irrigation can increase yields and decrease water use (Kandelous et al. 2012; Lamm et al. 2012). The use of subsurface irrigation is primarily an economic decision and is not suitable for all applications. In some areas, drip irrigation can result in high-soil salinity due to the high irrigation efficiency of drip irrigation and limited leaching (Lamm et al. 2011).

Forage Responses to Irrigation

Decades of research have demonstrated a strong relationship between soil water availability and forage yield response (Smeal et al. 1991; Montazar and Sadeghi 2008; Neal et al. 2009). This relationship is dependent on good management that includes adequate fertility and pest and disease control. Most perennial forage crops are drought-tolerant and will produce or at least survive under drought. The maximum yield of some forage crops adapted to dryland or limited irrigation is less than species adapted to full irrigation.

Pasture irrigation management during drought or times of water shortage can include the following irrigation practices and methods:

- Early-season water is the most important for both deep-rooted legumes like alfalfa and for cool-season grasses.
- Limit N fertilization of grasses to the level of expected yield.
- Maintain health of the stand-by not grazing to a short stubble height, especially when forages are drought-stressed.
- Consider the species of grass if deficit irrigation conditions are expected. Tall wheatgrass and intermediate wheatgrass yield less and lose stand cover faster over time under full compared with partial-season irrigation. In contrast, tall fescue and orchardgrass both yield well under full irrigation, but orchardgrass stands thin over time under partial-season irrigation. Summer-active tall fescue yields and survives well under both full and partial-season irrigation, as does alfalfa. Tall fescue would be a good choice for pasture plantings where irrigation availability varies from year to year (Orloff et al. 2016).
- Manage irrigation scheduling to maximize use of available water by giving priority to irrigating the most vigorous and productive forage stands.

Similarly, there are irrigation practices for alfalfa that can be used during water-short years.

- Alfalfa uses water most efficiently during spring growth (Orloff et al. 2004; Putnam 2012; Putnam et al. 2017).
- Irrigation system efficiency influences the amount of water applied per unit of yield. Slight under-irrigation results in higher irrigation system efficiencies and provides higher irrigation water-use efficiency (more yield per volume of water applied) (Hanson et al. 2007).
- It is better to stop irrigating alfalfa than to provide a small amount of water that produces little or no growth.
- Alfalfa is generally considered drought-tolerant. When soil moisture is very low, most alfalfa cultivars go into dormancy in response to drought with no long-term effect. Healthy, well-established alfalfa can withstand several months with no irrigation without plant mortality or reduced yield in subsequent years (Orloff et al. 2014)
- Prioritize irrigation on the most productive alfalfa fields (e.g. best soils, best alfalfa stem density, fewest weeds, youngest alfalfa stand, best irrigation systems, best cultivar).
- Prioritize on land with the most efficient irrigation application.

Irrigation, Soil Fertility, and Yield Relationships

For forages, yield is related to crop water use and fertility (Koenig et al. 1999). Forage production results from photosynthesis, which requires a sufficient supply of soil water. Forage grass responses can be as high as 100 kg of forage produced per kg of N applied (Koenig et al. 2002). These responses require adequate water for the plant. While grass pastures depend on N replacement to maintain good yields, alfalfa and other forage legumes obtain N through symbiosis with soil microorganisms fixing dinitrogen from the air, and they add N to the soil when roots and leaves die and decay.

Adequate soil water from precipitation or irrigation can increase the yield benefits of fertilization. A leaching fraction of irrigation may periodically be needed to reduce the concentration of soluble salts in irrigated soils, but over-irrigation will also leach fertilizer below the root zone, particularly nitrate-N (NO_3^-) which does not adsorb to mineral particles and organic matter. Waterlogged soils also reduce availability of plant nutrients in soils.

Irrigation Scheduling

Irrigation scheduling describes when to irrigate and how much water to apply. Irrigation scheduling strategically replaces soil water that is depleted by crop water use and soil water evaporation. It improves production, can conserve water, and increase profits. Understanding some basic terms, definitions and concepts will improve irrigation management choices.

Soil Water

Figure 27.7 shows soil moisture ranges, and general soil moisture definitions are stated below:

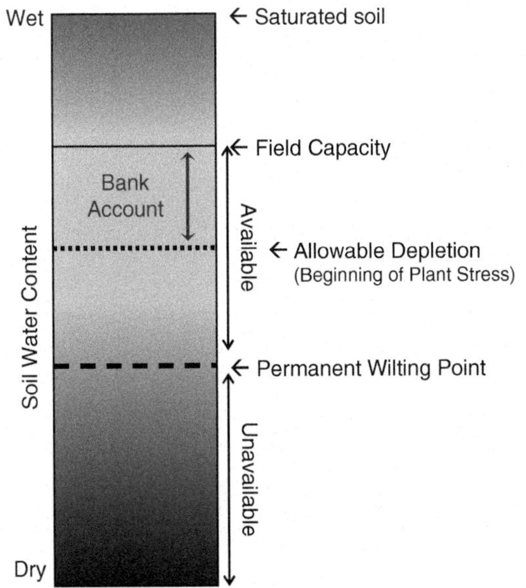

Fig. 27.7. Classification system for soil water held in soil voids. Volumes of water between saturated soil, field capacity and permanent wilting point differ according to soil characteristics and root distribution of the crop.

- *Saturation*: At saturation, all pore space in the soil is filled with water and no air is present. Most agricultural soils have 40–50% voids (pore space) per volume of soil.
- *Field capacity*: Soil water content after water has drained by gravity. Field capacity of most agriculture soils ranges between 20% (sand) and 40% (clay) of the volume of the soil.
- *Permanent wilting point*: Soil water content when plants or crops cannot obtain water from the soil. Permanent wilting point ranges between 7% (sand) and 24% (clay) water by volume for most agriculture soils.
- *Available (usable) water*: Soil water contents between field capacity and permanent wilting point. Though plants can access the water, the water potential (energy required to extract water) of soil water increases as soil water content approaches permanent wilting point. Plant stress occurs when plants cannot maintain adequate water uptake for optimal growth. Water stress reduces plant cell water turgor, reduces photosynthesis, growth, and reproduction.
- *Allowable depletion*: Soil water content available to crops without causing enough stress to reduce yield or crop quality. Allowable depletion depends on the crop type, growth stage, and climate. Allowable depletion can range between 25% of available water for crops (75% remaining in soil) that are very sensitive to soil moisture stress to over 50% of available water for crops that are less sensitive to water stress. A guideline for allowable depletion fraction is 50–60% for forages.

Generally, forage crops growing in sandy soils need to be irrigated more frequently with less irrigation depth than finer-textured soils. Silt soils have a medium drainage rate and infiltration rate, and clay soils drain slowly, have a low infiltration rate, and higher field capacity because of smaller pores, but greater overall total pore space. Clay and silt soils have similar available water-holding capacity. Good soil structure (clusters of soil particles) and high organic matter help improve infiltration rate, drainage, and available water holding capacity.

Soil Moisture Budget

The maximum days between irrigations is based on the available water holding capacity of the soil, the crop rooting depth, and the crop ET. For example, the maximum irrigation interval for a pasture with a rooting depth of 0.75 m, recommended allowable depletion of 60% total available water holding capacity, and a daily ET of 5 mm d^{-1} is illustrated in Table 27.1 for different soil types. Sandy soils need to be irrigated more frequently.

Figure 27.8 shows daily calculated ET and measured precipitation of a pasture near Randolph, Utah, US and Figure 27.9 shows the same data as cumulative potential pasture ET and precipitation; the difference between ET and precipitation is net irrigation need. Figure 27.10 shows the soil moisture in the root zone daily along with properly scheduled irrigations to replace the depletion (vertical bars). Irrigation intervals in this example range

Table 27.1 Example of maximum irrigation interval for a pasture with a rooting depth of 0.75 m, recommended allowable depletion of 60% of the total capacity to hold available water, and a daily ET of 5 mm d^{-1}

Soil Type	Available water (mm mm^{-1})	Root zone available soil water (mm)	Depletion maximum (mm)	Irrigation interval (d)
Sand	0.05	37.5	22.5	4
Fine sandy loam	0.08	60	36	7
Loam	0.16	120	72	14

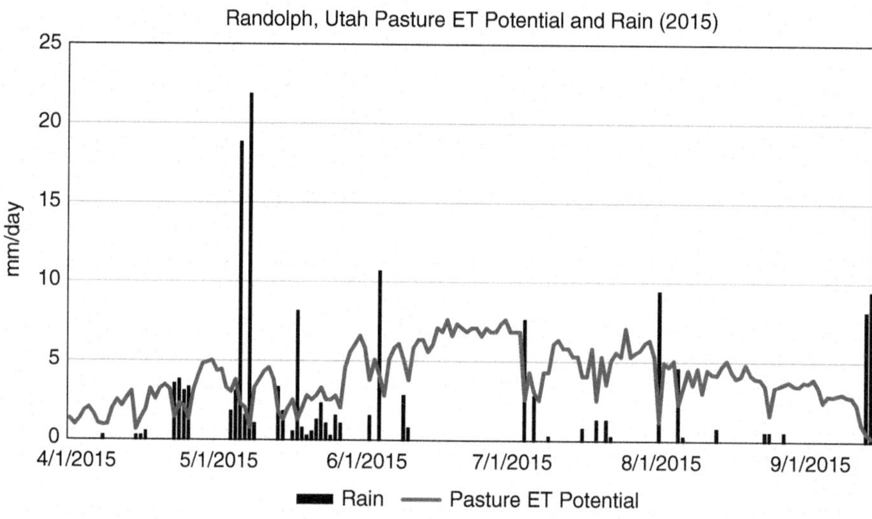

Fig. 27.8. Example of daily ET potential of a grass pasture with vertical bars to indicate amounts of rainfall events. Note the variation in ET among seasons and following rainfall events. Figures 27.8 through 27.10 are from Randolph, Utah 2015 weather data, courtesy of Utah Climate Center.

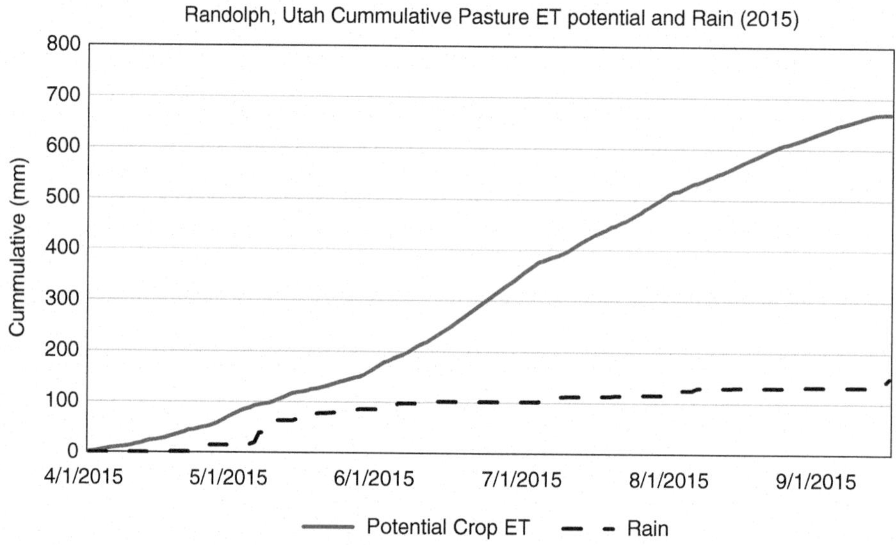

Fig. 27.9. Example of cumulative potential crop ET and rain. The difference is the irrigation need.

between 10 days during summer to over a month in spring when ET is lower, and rainfall occurs.

The objective for scheduling is to maintain soil moisture between field capacity and minimum allowable depletion so pasture plants experience minimal water stress. This example works well for a periodic-move sprinkler system to replace the readily available soil moisture, which in this case, is 72 mm of water in the 0.75-m-deep root zone (750-mm root zone × 0.16 mm mm^{-1} available water × 0.6 fraction readily available).

The actual irrigation applied would be more than 72 mm to account for irrigation inefficiencies. If the efficiency is 80%, the required irrigation would be 90 mm (72 mm divided by 0.8). For sprinkler-periodic-move systems it may take about a week to irrigate a field; therefore, the first irrigation can occur prior to depletion

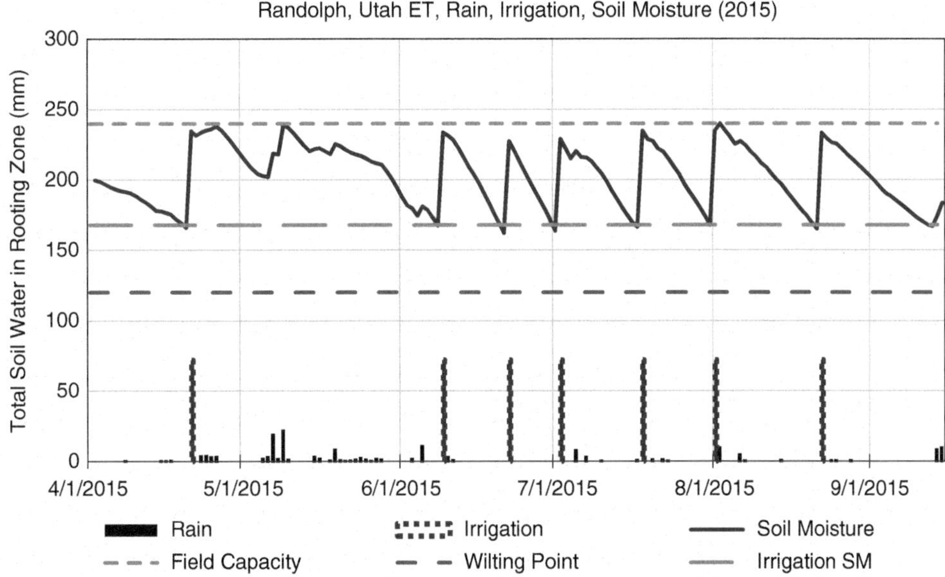

Fig. 27.10. Example of irrigation scheduling based on a soil moisture budget.

of the available soil moisture, and some of the field may be slightly stressed before the first round of irrigation is completed. After the first irrigation, irrigation intervals for each set are the same.

Crop rooting depth is a function of the plant rooting nature, soil properties, and depth to the water table. Rooting depth of alfalfa can range from 1.2 to 1.8 m, whereas most grasses have a rooting depth less than 1 m. Hard soil layers and high-water tables can restrict rooting depth. Though some roots are deeper, these two general rooting depths are sufficient for most irrigation management purposes. For management purposes, it is best to irrigate based on fluctuations in the top meter or less of soil. That region generally has the most roots, the best soil conditions and the greatest use of water.

Soil Moisture Monitoring

Irrigation scheduling can be based on monitored soil moisture levels. This method can be used independently or as a check on the scheduling method based on soil moisture budget. Though it is common for growers to estimate soil moisture by feel, appearance, or time between irrigation events, soil moisture can be most accurately and effectively monitored using a soil moisture monitoring system (Robinson et al. 2008). Due to a limited number of soil measurement devices used, sensors should be located to represent the overall field.

The soil moisture readings need to be converted to volume or depths of water so that the target irrigation amount can be applied. Sensors can be connected to data loggers to record the measurements and/or data can be sent via radio or cell phone technology to irrigation controllers, computers, or other communication devices.

Types of Soil Moisture Detectors

Porous Blocks

This common method uses porous blocks of gypsum, fiberglass or ceramic that are buried in the soil at rooting depth where water moves in or out until equilibrium is reached with water in the soil. Electrodes in the block are used to measure the electrical conductivity (EC) of water in the block, which is assumed to be equal to that of the soil water. These readings are not a direct reading of soil water and they are sensitive to soil salinity, which alters water conductivity.

Neutron Probes

A neutron probe involves lowering a radioactive source and receiver into long, narrow access tubes installed in fields. The effect of soil water on neutrons emitted from the radiated soil is measured (neutrons lose energy when they collide with hydrogen). These systems are expensive, require operator certification training, and need calibration; however, they can be practical for some applications.

Dielectric Methods

These probes sense dielectric properties of soil and water to monitor soil moisture. They consist of two or more electrodes inserted into the soil or circular electrodes inside a polyvinyl chloride (PVC) access tube. Most electrodes are not affected by salts or temperature and are easy to install. There are several measurement approaches relying upon the dielectric properties of the soil, including time domain reflectometry (TDR) and frequency domain (FD). The TDR system is accurate, and the cost has decreased significantly. Output is usually in percentage soil water by volume.

Plant Stress Monitoring

Monitoring plant canopy temperature with remote temperature sensors can help determine water stress in crops (Blonquist Jr. et al. 2009; Wang et al. 2010). Water loss by ET cools the plant and so non-water-stressed plants during the day have a lower leaf temperature than water-stressed plants. These indicators can also be used to assess distribution problems with an irrigation system.

Irrigation Amounts

Irrigation depths and set times can be calculated using the mass-volume balance equation:

$q \times t = A \times d$, *where* q = flowrate (e.g. $m^3\ s^{-1}$; $l\ min^{-1}$), t = irrigation time (e.g. h, min), A = irrigated area (e.g. ha, m^2), and d = irrigation depth (e.g. mm, cm). Note: × indicates multiplication. This equation can be rearranged and solved for other unknowns:

$$d = q \times t/A$$

$$t = d \times A/q$$

$$A = t \times q/d$$

As an example, knowing the time(t) required to apply a specific amount of water is important and determined from irrigation depth (d), irrigation flow rate (q), and area irrigated (A). Thus, the irrigation set time required to apply 0.1 m of irrigation depth to 2 ha (20 000 m^2), with a flowrate of 0.1 $m^3\ s^{-1}$ is 5.6 hours. This method can be applied to surface, sprinkler, or drip irrigation. It is important to use proper unit conversions and dimensions.

Soil Salinity

In arid regions where irrigation is prevalent, both irrigation and soil water contain a higher level of dissolved salts (total dissolved solids or TDS) than in more humid regions. Salts from erosion, dissolution, and mineralization have accumulated over thousands of years as evaporation in excess of precipitation has left salts in soils, groundwater, and water bodies. Increasing concentrations of salts in the soil water decreases the ability of crops to extract water from the soil, resulting in decreased yield when compared to potential yield. Crop ET removes water from soil but almost no dissolved salts, which results in increased salinity in the soil water.

Salinity control usually requires a two-step process; reclamation leaching removes accumulated salts (applicable to new ground or neglected soils), while maintenance leaching removes salts as they accumulate in the soil. A comprehensive manual on agriculture salinity assessment and management is available (Wallender and Tanji 2011).

A common indicator of dissolved salts is increased electrical conductivity (EC) of the soil water (dissolved salts increase the EC of a solution). Threshold values of EC have been determined for specific crops (Figure 27.11). The thresholds correspond to a measurable yield decrease. As a general guideline, a yield decrease of 10% or less due to salt accumulation is recommended. TDS can be approximated from the EC of water by TDS ($mg\ L^{-1}$) = 640 × EC (milli-mhos cm^{-1}). The value of 640 is dependent on the type of dissolved solids but, is appropriate for most irrigation and drainage water.

In general, irrigators have little choice concerning the quality of their irrigation water. For example, the waters of the Colorado River at Yuma, Arizona can have salinity levels of about 800 $mg\ l^{-1}$. This is equivalent to adding 8000 kg of dissolved salts in the 10 000 m^3 of irrigation water applied. To maintain the balance of salt in the soil water, an equivalent amount of soluble salt needs to be leached or precipitated from the root zone.

Leaching is accomplished by applying more water than crop ET to move salts deeper within the soil profile. A smaller amount may be precipitated naturally to a less soluble form due to the chemistry of the soil and salinity concentration in the soil water. Another method to manage salinity is to maintain a high soil-moisture level via irrigation. The salinity of the soil water increases as soil water is drawn out by ET while the salt content of the soil remains the same. Reducing soil water content increases the proportion of soil water that is unavailable due to adhesion to soil particles and high-osmotic potential.

The quantity of water required for leaching depends on many factors, including irrigation uniformity, irrigation frequency, soil texture, soil and irrigation water chemistry, salt precipitation or dissolution, and crop sensitivity to salinity. A traditional model assumes that the waterfront from each irrigation pushes soluble salts down, resulting in increasing soil-water salinity with depth (Rhoades 1974). The leaching ratio (LR) is estimated based on the EC of the irrigation water (ECiw) and the allowable EC of the soil-water extract (ECe) (dependent on crop sensitivity, Figure 27.11). The leaching ratio is the amount of irrigation applied in excess of crop water need. For example, a LR of 0.1 would require 1.1 times the crop irrigation requirement to include water for leaching. A common equation is as follows:

LR = ECiw ÷ (5 × [ECe-ECiw]), where five assumes the EC of the drainage water is five times higher than the EC of the soil-water extract.

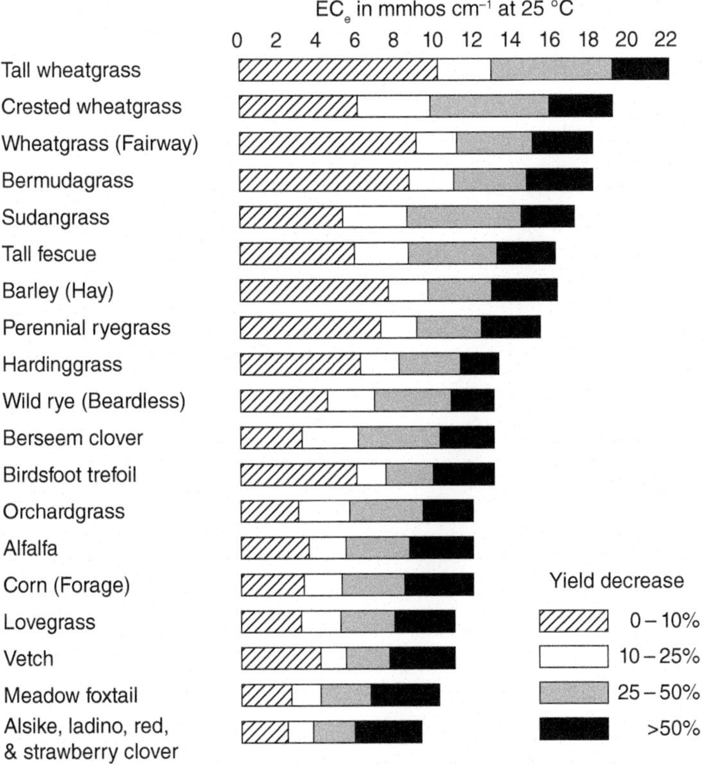

Fig. 27.11. Salinity tolerance, based on electrical conductivity of a saturated soil paste (EC_e) (James et al. 1982), shows that the range to the 10% yield loss is very dependent on the forage crop species. Further yield reductions tend to be smaller, more linear and not as dependent on the forage species (Maas and Hoffman 1977).

Leaching can occur during the irrigation season by applying more water than is required to replace crop ET, or leaching can occur from precipitation outside the growing season, during the irrigation season, or during periods of the year when crop water requirements are low. Rain and snow should be considered in determining the amount of irrigation needed for salinity leaching. For example, measurement of soil salinity in autumn, and again in spring, can provide data to guide salinity management. There are salinity sensors and soil moisture sensors that provide in-situ measurements.

Drainage of Soils

Waterlogging of soils occurs from poor subsurface drainage resulting in a high-water table, poor surface drainage (low spots in fields), impervious layers such as hard clay layers and rock formations, and/or precipitation or irrigation rates that exceed soil water infiltration capacity. Drainage can be an important part of soil water management for both irrigated and non-irrigated forage production. Poorly drained soils can contribute to soil salinity and can be the result of soils with a high concentration of sodium, which impedes water infiltration.

Poorly drained and waterlogged soils have low O_2 levels or **hypoxia**, which reduce root respiration and inhibit cellular metabolism and root growth (Vaughan et al. 2002; Sairam et al. 2009; Gurovich and Oyarce 2015). These conditions can restrict root development, reduce yield, and even kill crops. As discussed, drainage is also needed to maintain the proper salinity level in the soil when irrigating with water that has a high salt content.

Poorly-drained alfalfa stands can suffer chlorosis, root **hypoxia**, suppression of N_2 fixation, stunted root development, root and crown rot, scalding from ponded water, and increased weeds. Waterlogging may increase insect pressure, decrease nutrient uptake, increase disease organisms such as phytophthora and nematodes, and result in plant death and stand reduction. The best remedy is to avoid planting alfalfa on land that experiences flooding or has drainage issues. In some cases, it is possible to improve drainage by subsurface drains and land leveling. There are alfalfa cultivars that have some resistance to phytophthora and nematodes, or are more adaptable to wet soils because they produce more fibrous roots (Vaughan et al. 2002; An et al. 2004).

Many grasses are better adapted to waterlogged soils and soils with high-water tables than alfalfa, partially due to their shallower rooting system. It is impractical to provide artificial drainage to some irrigated pastures, and in these cases, choosing well-adapted native vegetation can provide forage. As an example, **sedges** and rushes can provide livestock forage and wildlife habitat in poorly-drained riparian areas.

Good-quality introduced forage species adapted to periodically wet soil conditions are reed canarygrass, perennial ryegrass, and creeping meadow foxtail (Wheaton 1993). Birdsfoot trefoil and white clover can be grown as monocultures or mixtures in poorly drained soils (Teakle et al. 2006; Rogers et al. 2008). Table 27.2 lists the relative tolerance of selected forage species on waterlogged and saline sodic soils (Ogle and St. John 2010; Ogle et al. 2011; Wallender and Tanji 2011).

When natural soil drainage is not adequate, it may be economic to use buried drainage pipes or open drains to convey excess water away from the root zone soil, along with land leveling or drainage contouring to prevent ponding or excessive accumulation of water on the soil surface. In 2012, about 25% of US cropped land was drained by manmade improvements; this includes 20 million ha of subsurface (tile) drains and about 17 million ha of improved surface drainage (USDA-NASS 2012). In the US, USDA soil surveys provide information about water table, drainage, salinity, soil texture, and soil permeability (USDA-NRCS 2018). Worldwide, large drainage projects for improved land use and agriculture production have been constructed on about 150 million ha in the last 50 years (Gurovich and Oyarce 2015).

Other Irrigation Considerations

Drought and Limited Water

In recent decades, due to droughts and climate change, irrigation research has focused on improving drought tolerance, reducing water use, and limited or deficit irrigation on perennial forage crops (Lindenmayer et al. 2008; Volesky and Berger 2010; Orloff et al. 2014). Deficit irrigation can conserve water during drought to provide water for municipal and other critical uses.

While some crops need nearly a full irrigation supply for economic production, alfalfa and many grasses can produce harvestable crops with normal precipitation or partial season irrigation. These forage crops can go dormant and survive on very limited water and resume growth when water is resupplied. Deficit irrigation of forage crops also provides a water management tool that can benefit agricultural producers through water sales or improved utilization of limited irrigation water.

Environmental Considerations

Irrigation management of pastures and forages helps maintain and improve water quality in streams and rivers. Where irrigated pastures are adjacent to streams, controlling runoff and preventing excessive grazing along riverbanks are important. These practices help maintain healthy riparian areas for wildlife and good quality water for aquatic life. Many cooperative projects, including fish screens on irrigation diversions, streambank stabilization, off-stream livestock watering facilities, and **riparian** buffer fencing have been incorporated by ranchers and farmers to enhance the environment.

Irrigation and Soil Water Quality

Plants obtain required nutrients and fertilizers through water. Plants require trace amounts of B, Cu, Fe, Mn, Mo, Ni, Cl, and Zn; but at higher levels, these micronutrients can be toxic. Other elements Cr and Se accumulate in plants and are essential for livestock at trace amounts; but higher levels can be toxic. Other nonessential elements such as As, Cd, Hg, and Pb can also be harmful to plants and livestock at high concentrations. High concentrations of these elements are only a problem in localized areas and when water sources become polluted.

Species Selection for Low Soil and Water Quality

With proper species selection, adapted forages can be grown on marginal lands with high water tables, poor drainage, alkaline, or saline soils, and/or poor irrigation water quality. In salt-affected soils or when irrigating with high salinity water, forage species should be selected based on tolerance to salts (Figure 27.11). Yield in a riparian area can be enhanced by irrigation and proper fertilization. While native grasses exist in these environments, seeded introduced species can sometimes enhance forage quality and yield.

Irrigation System Maintenance

In addition to irrigation scheduling, irrigation system maintenance is important for forage production. For sprinkler irrigation, maintenance includes replacing worn sprinklers or nozzles, replacing pressures regulators, and fixing leaks by replacing gaskets and drains. For surface irrigation, maintenance can include periodic leveling of fields prior to planting or replanting, controlling weeds along ditch banks, cleaning ditches, and fixing leaks in water control structures. For drip irrigation, maintenance includes fixing leaks in drip lines, controlling rodents, maintaining filters and pressure regulator systems, and replacing drip tubing and emitters when needed.

Table 27.2 Relative waterlogging tolerance of selected forage species

Excellent–Good	Good–Fair	Fair–Poor
Legumes		
Alsike clover	Arrowleaf clover	Alfalfa
Berseem clover	Ball clover	Cicer milkvetch
Birdsfoot trefoil	Kura clover	Crimson clover
Persian clover	Red clover	Crownvetch
Strawberry clover	Subterranean clover	Rose clover
White clover		Sainfoin
Grasses		
Creeping foxtail	Intermediate wheatgrass	Big bluestem
Creeping red fescue	Kentucky bluegrass	Bluebunch wheatgrass
Eastern gamagrass	Orchardgrass	Crested wheatgrass
Limpograss	Perennial ryegrass	Green needlegrass
Meadow fescue	Smooth bromegrass	Indiangrass
Meadow foxtail	Switchgrass	Meadow bromegrass
Redtop		Siberian wheatgrass
Reed canarygrass		St. Augustine grass
Tall fescue		
Tall wheatgrass		
Timothy		
Western wheatgrass		

Source: Adapted from MacAdam and Barta (2007).

Notes: Excellent–good: tolerates chronically wet, poorly drained soils for extended periods. Good–fair: some tolerance to waterlogging and variable/imperfect soil drainage. Fair–poor: tolerates wet soil and waterlogging for only short periods of time.

Summary

Irrigation is important for forage production in arid and semi-arid regions and in humid regions during drought. Production of irrigated alfalfa hay for export has been criticized as "exporting the water" used by the crop. However, the reality is that producers will grow profitable crops. It takes only about 3% of cropped land in the US to produce the marketed fruits and vegetables. Additionally, some agricultural land should not be tilled annually because of erodibility or other limitations, but perennial forages on the same land can be used for meat and milk production. The total area of permanent pasture is more than 100 times the area used for production of fruits and vegetables. Good management of irrigated forages can reduce water use and, provide benefits to many other water users.

References

Allen, R.G., Pereira, L.S., Raes, D., and Smith, M. (1998). *Crop Evapotranspiration: Guidelines for Computing Crop Water Requirements*. Rome: FAO Irrigation and Drainage Paper 56. FAO.

An, Y., Cheng, F.Y., Wang, J. et al. (2004). Studies on waterlogging tolerance of semi-fall and non-fall dormant alfalfa cultivars. *Grassl. China* 26: 31–36.

Australian Bureau of Statistics (2016). Water Use on Australian Farms, 2015–16. www.abs.gov.au/AUSSTATS/abs@.nsf/Lookup/4618.0Main+Features12015-16?OpenDocument (accessed 14 October 2019).

Blonquist, J.M. Jr., Norman, J.M., and Bugbee, B. (2009). Automated measurement of canopy stomatal conductance based on infrared temperature. *Agric. For. Meteorol.* 149: 2183–2197.

California DWR (2017). California irrigation management and information system (CIMIS), California Department of Water Resources. http://www.cimis.water.ca.gov (accessed 14 October 2019).

California State Legislature (2014). Sustainable Groundwater Management Act, Collectively AB 1739, SB 1319, and SB 1168. Sections 10750–10756 of the California Water Code.

Colorado DWR (2013). Geographic information system data of irrigated lands and water rights: Colorado's decision support system. Division of Water Resources. http://cdss.state.co.us/basins/Pages/BasinsHome.aspx (accessed 14 October 2019).

Colorado State University (2017). Colorado agricultural metrological weather network. https://coagmet.colostate.edu (accessed 14 October 2019).

FAO (2014). Irrigated crops. Aquastat, FAO's global water information system, FAO, Rome. http://www.fao.org/nr/water/aquastat/didyouknow/index3.stm (accessed 14 October 2019).

Gurovich, L. and Oyarce, P. (2015). New approaches to agricultural land drainage: a review. *Irrig. Drain. Sys. Eng.* 4: 135. https://doi.org/10.4172/2168-9768.1000135.

Hanks, R.J. (1985). Crop coefficients for transpiration. In: *Advances in Evapotranspiration*, 431–438. ASAE Pub. 14–85. St. Joseph, MI: Amer. Soc. Agr. Eng.

Hanson, B., Putnam, D., and Snyder, R. (2007). Deficit irrigation of alfalfa as a strategy for providing water for water-short areas. *Agr. Water Manage.* 93: 73–80.

ICID (2017). Agricultural water management for sustainable rural development. International Commission on Irrigation and Drainage Annual Report 2016–2017. http://www.icid.org/ar_2016.pdf (accessed 10 October 2019).

James, D.W., Hanks, R.J., and Jurinak, J.J. (1982). *Modern Irrigated Soils*. New York, NY: Wiley.

Jensen, M.E. and Allen, R.G. (2016). *Evaporation, Evapotranspiration, and Irrigation Water Requirements*, 2e. Amer. Soc. Civ. Eng. Manual of Practice 70.

Kandelous, M.M., Kamai, T., Vrugt, J.A. et al. (2012). Evaluation of subsurface drip irrigation design and management parameters for alfalfa. *Agr. Water Manage.* 109: 81–93.

Koenig, R., Hurst, C., Barnhill, J. et al. (1999). Fertilizer management for alfalfa. Utah State University Cooperative Extension No. AG-FG-01.

Koenig, R.T., Nelson, M., Barnhill, J. et al. (2002). Fertilizer management for grass and grass-legume mixtures. Utah State University Cooperative Extension No. AG-FG-03.

Lamm, F.R., Harmoney, K.R., Aboukheira, A.A. et al. (2011). Subsurface drip irrigation of alfalfa. University of California. http://ucanr.edu/sites/adi/Publications (accessed 14 October 2019).

Lamm, F.R., Harmoney, K.R., Aboukheira, A.A., and Johnson, S.K. (2012). Alfalfa production with subsurface drip irrigation in the central Great Plains. *Trans. Amer. Soc. Agr. Biol. Eng.* 55: 1203–1212.

Lindenmayer, B., Hansen, N., Crookston, M. et al. (2008). Strategies for reducing alfalfa consumptive water use. Colorado State University. Hydrology Days, 52–61.

Maas, E.V. and Hoffman, G.J. (1977). Crop salt tolerance–current assessment. *J. Irrig. Drain. Amer. Soc. of Civil Eng.* 103: 115–134.

MacAdam, J.W. and Barta (2007). Adapted from forages. In: *Forages, Vol. II: The Science of Grassland Agriculture*, 6e (eds. R.F Barnes, C.J. Nelson, K.J. Moore, et al.). Hoboken, NJ: Wiley.

MacAdam, J.W. and Griggs, T.C. (2006). Performance of birdsfoot trefoil, white clover, and other legume-grass mixtures under irrigation in the intermountain West USA. *Proc. N.Z. Grassl. Assoc.* 68: 355–359.

Maupin, M.A., Kenny, J.F., Hutson, S.S. et al. (2014). Estimated use of water in the United States in 2010. United States Geological Survey Circular 1405. http://pubs.usgs.gov/circ/1405 (accessed 14 October 2019).

Montazar, A. and Sadeghi, M. (2008). Effects of applied water and sprinkler irrigation uniformity on alfalfa growth and hay yield. *Agr. Water Manage.* 95: 1279–1287.

Neal, J.S., Fulkerson, W.J., Lawrie, R., and Barchia, I.M. (2009). Difference in yield and persistence among perennial forages used by the dairy industry under optimum and deficit irrigation. *Crop Pasture Sci.* 60: 1071–1087.

University of Nebraska (2017). Nebraska Agricultural Water Management Network (NAWMN), University of Nebraska, Lincoln. https://water.unl.edu/category/nawmn (accessed 14 October 2019).

Ogle, D. and St. John, L. (2010). Plants for saline to sodic soil conditions. NRCS Plant Materials Technical Note No. 9a.

Ogle, D., St. John, L., Stannard, M. et al. (2011). Pasture and range seedings: Planning-installation-evaluation-management. NRCS Technical Note No. 10.

Orloff, S.B., Putnam, D.H., Hanson, B. et al. (2004). Controlled deficit irrigation of alfalfa (*Medicago sativa*): A strategy for addressing water scarcity in California. Proceedings of the 4th International Crop Science Congress, Brisbane, Australia (26 September 2004).

Orloff, S., Bali, K. and Putnam, D.H. (2014). Deficit irrigation of alfalfa and grasses: What are the impacts/options? Proceedings of the California Alfalfa & Grains Symposium, Long Beach, California, USA (10–12 December 2014). http://alfalfa.ucdavis.edu (accessed 14 October 2019).

Orloff, S.B., Brummer, E.C., Shrestha, A., and Putnam, D.H. (2016). Cool-season perennial grasses differ in tolerance to partial-season irrigation deficits. *Agron. J.* 108: 692–700.

Putnam, D.H. (2012). Strategies for the improvement of water-use efficient irrigated alfalfa systems. Proceedings of the California Alfalfa & Grains Symposium, Sacramento, California, USA (10–12 December 2012).

Putnam, D.H., Radawich, J., Zaccaria, D. et al. (2017). Partial-season alfalfa irrigation: An effective strategy for a future of variable water supplies. University of California Alfalfa and Forage News. http://ucanr.edu/blogs/blogcore/postdetail.cfm?postnum=22908 (accessed 14 October 2019).

Rhoades, J.D. (1974). Drainage for salinity control. In: *Drainage for Agriculture*. Agron. Monogr. 17 (ed. J. Van

Schilfgaarde), 433–461. Madison, WI: Amer. Soc. Agron.

Robinson, D.A., Campbell, C.S., Hopmans, J.W. et al. (2008). Soil moisture measurement for ecological and hydrological watershed-scale observatories: a review. *Vadose Zone J.* 7: 358–389.

Rogers, M.E., Colmer, T.D., Frost, K. et al. (2008). Diversity in the genus *Melilotus* for tolerance to salinity and waterlogging. *Plant Soil* 304: 89–101.

Sairam, R.K., Dharmar, K., Chinnusamy, V., and Meena, R.C. (2009). Waterlogging-induced increase in sugar mobilization, fermentation, and related gene expression in the roots of mung bean (*Vigna radiata*). *J. Plant Physiol.* 166: 602–616.

Sleugh, B., Moore, K.J., George, J.R., and Brummer, E.C. (2000). Binary legume–grass mixtures improve forage yield, quality, and seasonal distribution. *Agron. J.* 92: 24–29.

Smeal, D., Kallsen, C.E., and Sammis, T.W. (1991). Alfalfa yield as related to transpiration, growth stage and environment. *Irrig. Sci.* 12: 79–86.

Teakle, N.L., Real, D., and Colmer, T.D. (2006). Growth and ion relations in response to combined salinity and waterlogging in the perennial forage legumes *Lotus corniculatus* and *Lotus tenuis*. *Plant Soil* 289 (1–2): 369–383.

Univ. Arizona (2017). Arizona Meteorological Network. https://cals.arizona.edu/azmet (accessed 14 October 2019).

USBR (2017). Cooperative Agricultural Weather Network, AgriMet. U.S. Department of Interior, Bureau of Reclamation. https://www.usbr.gov/pn/agrimet (accessed 14 October 2019).

USDA-NASS (2012). 2012 Census of Agriculture. https://www.agcensus.usda.gov/Publications (accessed 14 October 2019).

USDA-NASS (2014). 2013 Farm and ranch irrigation survey, Vol. 3, Special Studies, Part 1, AC-12-SS-1, Table 26. https://www.agcensus.usda.gov/Publications/2012/Online_Resources/Farm_and_Ranch_Irrigation_Survey (accessed 14 October 2019).

USDA-NRCS (2018). Web soil survey. https://websoilsurvey.sc.egov.usda.gov/App/HomePage.htm (accessed 14 October 2019).

Utah DNR (2013). Water related land use. Utah Department. of National Resources, Division of Water Resources. http://gis.utah.gov/data/water-data-services (accessed 14 October 2019).

Vaughan, L.V., MacAdam, J.W., Smith, S.E., and Dudley, L.M. (2002). Root growth and yield of differing alfalfa rooting populations under increasing salinity and zero leaching. *Crop Sci.* 42: 2064–2071.

Volesky, J.D. and Berger, A.L. (2010). Forage production with limited irrigation. University of Nebraska, Lincoln. http://extensionpublications.unl.edu/assets/html/g2012/build/g2012.htm#target (accessed 14 October 2019).

Wallender, W.W. and Tanji, K.K. (2011). *Agricultural Salinity Assessment and Management*. Amer. Soc. Civil Eng. Manual and Reports on Engineering Practice, 2nd Rev.

Wang, X., Yang, W., Wheaton, A. et al. (2010). Automated canopy temperature estimation via infrared thermography: a first step towards automated plant water stress monitoring. *Comput. Electron. Agr.* 73: 74–83.

Wheaton, H.N. (1993). Reed canarygrass, ryegrass and garrison creeping foxtail. University of Missouri Cooperative Extension No. G4649.

Wyoming Water Development Commission, Basin Planning Program (2010). Green River Basin Plan. Prepared by WWC Engineering, AECOM, ERO Resources Corp.

Chapter 28

Weed Management

Robert A. Masters, Rangeland Scientist (Retired), *Corteva Agriscience, Indianapolis, IN, USA*
Byron B. Sleugh, Forage Agronomist, *Corteva Agriscience, Indianapolis, IN, USA*
E. Scott Flynn, Forage Agronomist, *Corteva Agriscience, Lee's Summit, MO, USA*

Introduction

Forage management systems involve applying practices that manipulate plant interactions to enable land management goals to be achieved. Application of incorrect practices can accelerate forage stand invasion by unwanted plants, or weeds, that degrade the forage resource and hinder attainment of management goals. Weeds influence the structure and function of forage-based ecosystems whether forages are grown in cropland, pasture, rangeland, or grassland communities.

Weeds interfere with forage establishment, yield, and quality by competing for resources (i.e. light, space, nutrients, or water) and/or by producing and releasing allelochemicals (Smith 1991) that inhibit forage growth. Weeds often reduce the feed value of forages and can be unpalatable or toxic to livestock (Marten et al. 1987). Weeds often possess antiquality or poisonous characteristics that can impact livestock performance (Cheeke 1998). Some weed species may be more tolerant or resistant to biotic or abiotic stresses than forage species, which gives them a competitive advantage and can accelerate the invasion process (Ball et al. 2007).

Weeds usually have a negative effect on forages, but they can provide some benefits. Weeds provide organic matter and nutrients that can improve soil quality, provide cover and food for wildlife and insects, and can serve as forage when more desirable forage species are not available. Weeds can be attractive to grazing livestock depending on the availability of other forages, stage of development of the forage and weed species, and grazing pressure. The nutritive value of some weeds can be comparable to commonly used forage crops (Marten and Andersen 1975; Ball et al. 2007). Nutritive value of many weeds in the **vegetative growth** stage can be similar to desirable forages. Dutt et al. (1982) reported that dandelion and white cockle were palatable and did not reduce the nutritive value of an alfalfa-grass hay mixture. Yield and quality of quackgrass forage was similar to that of reed canarygrass, smooth bromegrass, and timothy (Sheaffer et al. 1990). Dandelion, white campion, canada thistle, jerusalem artichoke, and perennial sowthistle had a nutritive value equal or superior to that of alfalfa, though palatability of some of these species was low (Marten et al. 1987). Carolina geranium, virginia pepperweed, wild oats, downy brome, and little barley had an ***in vitro* dry matter disappearance** (IVDMD) equal or superior to rye, tall fescue and ladino clover (Bosworth et al. 1985). Some members of the *Amaranthus* genus have forage nutritive value equal to, or better than, commonly used forages. The potential for nitrate poisoning exists when Amaranthus species are harvested or grazed at early growth stages (Sleugh et al. 2001), while perirenal edema may be an issue when grazed in the summer and fall (Casteel et al. 1994; Last et al. 2007). Typically, the

Forages: The Science of Grassland Agriculture, Volume II, Seventh Edition.
Edited by Kenneth J. Moore, Michael Collins, C. Jerry Nelson and Daren D. Redfearn.

duration for which weeds provide acceptable quality forage is relatively short and, with maturation, they reduce the quality and quantity of the forage resource.

Even though weeds may be consumed by livestock, managers need to consider the potential negative impacts on grazing animals. Weeds can also affect grazing distribution and pasture utilization. "As heterogeneity of vegetation and topography increase, so does the variation in the use of the area by grazing animals" (Vallentine 2000). Three months after herbicide application, cattle residence time in herbicide-treated pastures was 1.3–5 times greater than in areas in the same pastures not treated with herbicide (Sather et al. 2013). Scifres et al. (1983) observed increased grazing in sites treated with 2,4-D, picloram, or tebuthiuron compared to non-treated areas. The implication of preferential grazing of herbicide-treated areas is that when only a portion of a pasture is treated action should be taken to prevent over-utilization and degradation of the treated area.

Weeds are often indicators of broader deficiencies in a forage management program. Inappropriate grazing management practices, nutrient deficiencies, or use of poorly adapted forages are among the factors that can facilitate weed invasion. Weeds often invade a plant community because prevailing management practices put desirable plants at a competitive disadvantage. The best defense against weed invasion is maintaining a vigorous and diverse plant community comprised of adapted forages that are managed to optimize persistence and productivity and maximize utilization of available resources by the forage species. Tracy and Sanderson (2004) reported "consistent negative relationships between forage species diversity and weed abundance." Maintaining a diverse community of desirable species is recommended because increasing plant diversity leaves fewer niches available for occupation by invading weeds.

A process to develop programs to manage weeds in forage management programs will be presented in this chapter. The aim is to provide ecologic principles that can be applied to a wide variety of forage production systems. Use of these principles will increase efficacy of the weed management program adopted. Steps in this process include: (i) setting management goals and objectives; (ii) selecting desired plant community (DPC) attributes; (iii) assessing site characteristics and management history; (iv) determining biology and **ecology** of forage and weed species; (v) identifying appropriate sequences and combinations of forage resource management practices that expedite development and maintenance of the desired forage community; and (vi) periodic monitoring of the plant community to enable rapid adjustment in practices to further improve the forage resource.

Setting Goals and Objectives

An important part of natural resource planning is setting realistic goals and appropriate management objectives to achieve those goals. Establishing objectives provides the land manager with a framework within which to prioritize management actions, make decisions, and plan and organize management activities to attain goals.

Desired Plant Community

Achievement and maintenance of a DPC can serve as a goal for forage resource management programs. The DPC concept originated within the United States Department of Interior-Bureau of Land Management. The Society for Range Management, Task Group on Unity in Concepts and Terminology (1995) defined DPC as "of the several plant communities that may occupy a site, the one that has been identified through a management plan to best meet the plan's objectives for the site. It (the DPC) must protect the site at a minimum." This concept recognizes that changes in plant community composition on a site can progress along multiple successional trajectories and result in different outcomes. Factors that influence these outcomes include past management, plant and animal dispersal from adjacent areas, climate, disturbance regimes (White and Jentsch 2001), and species added to the community. The DPC concept is consistent with prevailing state and transition models of vegetation change (Westoby et al. 1989 and Briske et al. 2005). The DPC is an appealing concept because it empowers land managers to define the DPC and establish objectives directed at assembling that plant community.

Succession moves along a trajectory that is driven by naturally occurring and human-induced processes (Connell and Slatyer 1977; Huston and Smith 1987). Attaining the DPC involves managing plant community succession. This requires knowledge of the three components of succession: site availability; differential species availability; and species performance (Table 28.1) (Pickett et al. 1987, 2009). Succession can be managed by using designed disturbance, controlled colonization, and controlled species performance. Designed disturbance alters successional trajectories by creating or restricting sites available for plant occupation. The timing and magnitude of disturbances on a site can affect weed species recruitment, diversity and reproductive output (Hobbs and Huenneke 1992; Renne and Tracy 2013). Controlled colonization is the intentional alteration of sites to facilitate germination and establishment of desirable species. Controlled species performance involves manipulating growth and reproduction of plant species to promote species of interest and inhibit undesirable species. Biological and chemical weed control, grazing, mowing, fertilization, and planting competitive species are examples of practices than can influence differential species performance. A generalized model describes processes by which succession can be managed using various tools in appropriate sequences and combinations to achieve a DPC (Figure 28.1) (Masters et al. 1996).

Table 28.1 General causes of ecological succession, contributing processes, and modifying factors (Pickett et al. 1987)

General causes	Contributing process	Modifying factors
Site availability	Disturbance	Size, severity, time, dispersion
Species availability	Dispersal	Landscape configuration, dispersal agents
	Propagules	Land use, time since last disturbance
	Resources	Soil, topography, site history
Species performance	Ecophysiology	Germination response, assimilation rates, growth rates, genetic differentiation
	Life history	Allocation, reproductive timing, mode of reproduction
	Stress	Climate, site history, prior occupants
	Competition	Competition, herbivory, resource availability
	Allelopathy	Soil chemistry, microbes, neighboring species
	Herbivory	Climate, predators, plant defenses, patchiness

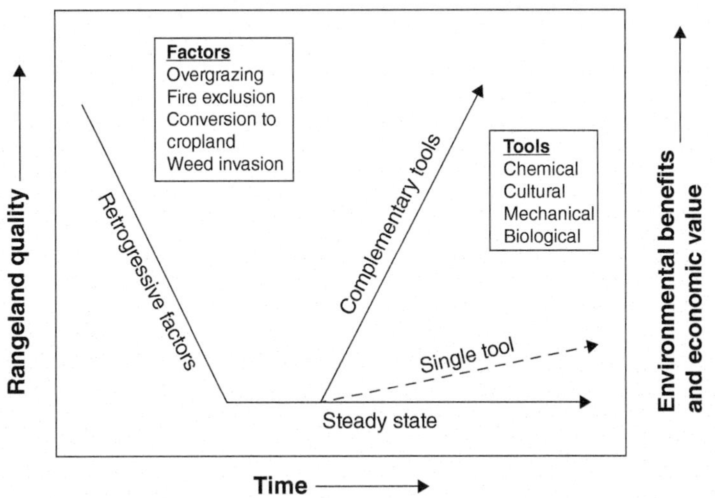

Fig. 28.1. Generalized community succession model for grasslands. Retrogression leads to a steady-state condition of low productivity. Reliance on a single technology results in slow grassland recovery rate. Sequential application of complementary and possibly synergistic management practices accelerates progress toward higher quality grasslands (Masters et al. 1996).

Site Assessment

Site assessment includes determining floristic composition, soil physical and chemical characteristics, topographic attributes, prevailing climatic regime, and past management history. An inventory of species comprising the plant community and their distribution and abundance is critical to forage resource improvement. Accurate identification of weed species is essential to developing strategies to reduce their adverse impacts. Likewise, determining the extent and distribution of desired forage species enables a manager to decide whether existing populations of forage species are sufficient for recovery with proper management or if reintroduction of forage species is required.

Assessment of the physical and chemical characteristics of the soils and of soil quality (Herrick 2000; Franzluebbers 2002) on the site is important because soils are the foundation upon which the plant community is built. Knowledge of soil pH, organic matter content, fertility, textural classes, structure, and depth to parent material will enable the development of a plan to alleviate soil constraints that could limit attainment of the DPC.

Slope and aspect (Gillingham 1973; Holland and Steyn 1975; Amezaga 2004; Bennie et al. 2006) and landscape position (Harmoney et al. 2001) influence forage resource productivity. Level to gently sloping terrain is often more amenable and responsive to management inputs than steeply sloping terrain. Aspect of the slope can affect plant community composition. In the Northern Hemisphere, north-facing slopes tend to be wetter and cooler than south-facing slopes. Important climatic variables that require consideration include average amount and seasonal distribution of precipitation, seasonal temperature fluctuations, and length of the growing season.

The influence of management history on site conditions should be understood so past mistakes and their adverse impacts on the plant community can be avoided as a new management plan is designed and implemented. It is critical that the causes of weed invasion and population expansion, including awareness of past disturbance regimes, be understood so management tactics can be deployed to reverse the invasion process (Hobbs and Norton 1996). Increases in weed invasion and diversity of weeds is more noticeable in pastures with a history of disturbance (Renne and Tracy 2013). The presence and spread of weeds are often symptomatic of underlying management deficiencies that must be corrected before sustainable improvement of the forage resource can be fully realized.

Biology and Ecology of Key Species

Developing effective forage management programs requires an understanding of the biology and ecology of both undesirable and desirable plant species in the community. Knowledge of plant demography, seedling recruitment, plant growth and development, and methods of reproduction is important to developing strategies to effectively manage weeds and improve forage species performance.

Plants are classified as annuals, biennials, and perennials based on life span, season of growth, and method of reproduction (Monaco et al. 2002). Annual plants complete their lifecycle within one year or a single growing season. Summer annuals, such as common ragweed, spiny amaranth, marshelder, camphorweed, prickly lettuce, bitter sneezeweed, common sowthistle, common lambsquarters, annual broomweed, russian thistle, buffalobur, barnyard grass, crabgrasses, and foxtails, germinate in the spring or summer and complete their life cycle by autumn. Winter annuals, such as hoary cress, henbit, chickweed, downy brome, japanese brome, foxtail barley, rescue grass, and six weeks fescue, germinate in the autumn or winter, and complete their life cycle the following spring or early summer. Annuals are usually the first plants to colonize a site following severe disturbance, such as tillage or repeated overgrazing. **Seed** production perpetuates annual plant species so reducing the amount of seeds produced is a usually a critical component of weed management programs.

Biennial plants (musk thistle, bull thistle, plumeless thistle, wild carrot, wild parsnip, common mullein, and common burdock) complete their life cycle in two consecutive growing seasons. Seeds of biennials usually germinate in the late summer or early fall, remain vegetative, usually as rosettes, and accumulate energy reserves through the winter and early spring. Low temperatures during the winter vernalize plants and prepare them for flowering. During the second year, plants bolt (elevate reproductive shoots), flower, produce seed, and die. Like annuals, biennials rely on seed production to persist. The difference between a biennial and a winter annual can be confusing. Generally, biennials live longer, are larger and produce more seed than annuals (Monaco et al. 2002).

Perennials persist for more than two years and are further classified as simple, creeping, or woody (Monaco et al. 2002). Simple perennials (dandelion, buckhorn plantain, and broadleaf dock) are herbaceous plants that reproduce by seed and do not spread vegetatively. Creeping perennials are herbaceous plants that reproduce by seed and by vegetative means including stolons (bermudagrass), rhizomes (quackgrass, johnsongrass, field bindweed), spreading root system containing adventitious **shoot** buds (leafy spurge and canada thistle), or **tubers** (nutsedge and jerusalem artichoke). Herbaceous perennials usually dieback to the soil surface at the end of the growing season and top growth resumes the following spring from perennating tissues located below ground.

Woody plants are trees and shrubs that have secondary growth and are perennials (Schweingruber et al. 2007). Scifres (1980) defined a shrub as "a plant that has persistent, woody stems and a relatively low growth habit and generally produces several basal shoots instead of a single bole." Most woody species can be categorized as one of four growth forms: upright, single-stemmed trees; shrubs or trees with a creeping growth habit; multi-stemmed shrubs; and those plants that grow as vines or canes (Scifres 1980). Original-growth oaks, ashes, elms, honey locust, and osage orange are examples of single-stemmed upright growth forms. Wild plum, buckbrush, and dogwoods that possess multiple shoots arising from a spreading root system are examples of creeping growth habit. Multi-stemmed brush often results from a single-stemmed tree that was incompletely controlled in some manner. Blackberry and multiflora rose are examples of plants that exhibit the vine growth habit. Upon removal of the apical meristem of these plants, new shoots may arise from the base of the plant located belowground or new shoots can arise from cut stems, otherwise known as canes, which take root where they make soil contact. Succulents, such as *Opuntia* spp. occur in many forms, but are generally fleshy with thick, water-retaining stems.

Weed Designations

Weeds can be designated as native, invasive, and/or noxious (Sutherland 2004). Native weeds are those that were endemic and part of North American plant communities before Europeans colonized the continent. Invasive and noxious weeds represent species of special concern. Invasive weeds are exotic species that pose a threat because they often lack natural enemies to limit population growth and have high growth and reproductive rates that promote rapid population expansion. Though all invasive plants are exotic, not all exotic plants are invasive. Noxious weeds refer to those plant species for which control is mandated by law or regulation.

Weed Management Strategies

Prevention, control, and eradication are three basic weed management strategies. Prevention is probably the most economic and practical way to manage weeds. One means of prevention is the removal of weed seed and vegetative material from farming implements before preparing a **seedbed** and planting seed that is not contaminated with weed seed. Control is the process of minimizing weed interference with desirable plants. Eradication involves complete elimination of a weed and requires removal of living plants and destruction of seed in the soil. In practice, eradication is difficult to achieve except on a small scale where a weed outbreak is quickly recognized, and intense management efforts are feasible (Rejmanek and Pitcairn 2002).

Preventing weed introduction by restricting movement of propagules from infested areas can minimize invader dispersal into new habitats. Early detection followed by swift, intensive, and aggressive implementation of control measures during the early invasion lag phase (Figure 28.2) are essential to eliminate an invader. Once the invasion process enters the exponential phase, eradication of the invader is usually not a realistic goal. Instead, emphasis is often directed to reducing impact of the invader and keeping it from dominating the plant community and altering ecosystem processes. After the weed has reached its maximum abundance, containment (keeping the weed population from expanding into new habitats) or restoration efforts are management options to consider.

Integrated Weed Management

Repeated use of a single measure to control weeds will not provide sustained control and will likely open niches for other undesirable species to occupy unless desirable plant species are present to fill the vacant niches. Where desirable species are either not present or in low abundance, plant community recovery will be slow or may not occur without revegetation (Masters et al. 1996).

Instead of relying on one single control measure, integrated pest management (IPM) emphasizes the sequential application of complementary or synergistic control measures in an economically and ecologically effective manner (Pimentel et al. 2005; Ehler 2006). Entomologists developed IPM during the late 1950s in response to problems created by excessive use of insecticides (Thill et al. 1991). Two common definitions of IPM are: (Aiken et al. 2012) a combination of biological, chemical, and cultural methods for maintaining pests below economic crop injury thresholds (Burn et al. 1987) or (Allen and Collins 2002) non-chemical pest control measures to reduce reliance on chemical pesticides (Goldstein 1978). An IPM program should be developed from interdisciplinary efforts that gather information about: the ecologic basis of the pest problem; how to make the crop environment unfavorable for pests; when pesticide treatments are needed based on population dynamics of the pest and natural enemies; and benefits and risks of the IPM strategy for agriculture and society (Pimentel 1982).

Integrated weed management (IWM) evolved from the concept of IPM in agricultural crops. IWM is the application of technologies in a mutually supportive manner, and selected, integrated, and implemented with consideration of economic, ecologic, and sociologic consequences (Walker and Buchanan 1982; Swanton et al. 2008). Thill et al. (1991) defined IWM as the integration of effective, environmentally safe, and sociologically acceptable

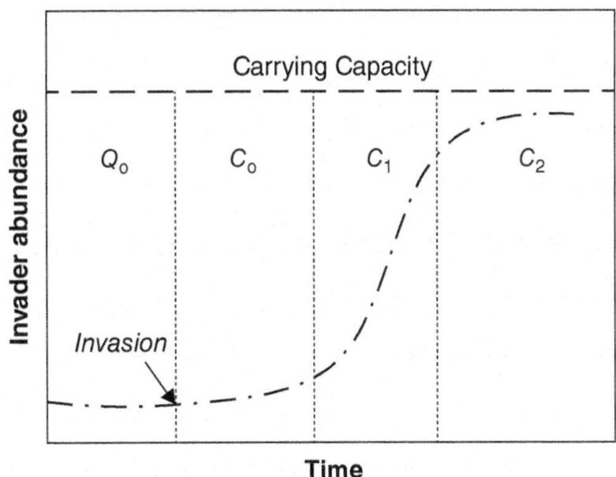

Fig. 28.2. Weeds invade and increase in abundance overtime. Phases of weed invasion and priorities for action at each phase: Q_0 – quarantine priority phase; C_0 – eradication priority phase; C_1 – control priority phase (exponential growth phase); C_2 – maximum population level, effective control unlikely without massive resource inputs. Ease of treatment declines and difficulty and cost increase moving from left to right (Hobbs and Humphries 1995).

control tactics that reduce weed interference below the economic injury level. Liebman and Gallandt (1997), considered IWM as the "use of many little hammers" that alone do not provide the level of weed control desired, but when used in concert in a systematic manner achieve the desired level of weed control. IWM emphasizes management of ecosystem function (energy flow and nutrient cycling) and structure (species composition) rather than a specific weed or control method (Scifres 1980). With this in mind, the goal of weed management should be to renovate or restore degraded weed-infested communities so that they resist weed invasion and can better meet land use objectives (Masters et al. 1996; Funk et al. 2008).

IWM provides a process for managing weeds that is ecosystem-centric and not specific to a species or weed control technology. Frequently, the stated or implied goal of IWM is pesticide-use reduction. But this is not consistent with the basic concept of IWM, which is a sustainable approach to managing weeds by combining biological, cultural, mechanical and chemical methods that minimize economic, health and environmental risks (US Congress, Office of Technology Assessment 1993). All available control methods should be considered during development of IWM programs and those selected should enable attainment of weed management objectives.

Biological Control

Biological control of weeds is the planned use of living organisms to reduce the reproductive capacity, density, and effect of weeds (Quimby et al. 1991). Biological control can involve any of three strategies: conservation; augmentation; and importation of natural enemies. Conservation is manipulating the environment to enhance the effect of existing natural enemies and is usually used to manage native weeds. Augmentation employs periodic release of natural enemies and is restricted to managing weeds in high-value food crops because it requires costly repeated intervention. Importation, also known as classical biological control, is the planned relocation of natural enemies of exotic weeds from their native habitats onto weeds in their naturalized habitats. This strategy seeks to reestablish weed and natural enemy interactions that can reduce the weed population to an acceptable level in the new environment. Synchrony in the life cycles of host plant and agent, adaptation of the agent to a new climate and habitats, ability of the agent to find the host at varying densities, capacity of the agent to reproduce rapidly, and the nature, extent, and timing of the damage caused by the biocontrol agent are among the factors that determine biological control agent efficacy.

Genetic variation in populations of the natural enemy and the invasive plant can influence the success of biological control programs (Roush 1990). High levels of genetic variability in insect traits could improve the insect's ability to better adapt to the new environment. Identification of important genetic variation and its maintenance in importation, mass-rearing, and release should enhance chances of success for biological control programs. Biologic diversity is usually highest in the center of origin of a taxon (Vavilov 1992) and the greatest genetic variation in natural enemies may be found in the areas of weed origin (Zwolfer et al. 1976).

Molecular biology offers tools to quantify invasive plant genetic diversity and to better match natural enemies with the target invasive plant (Nissen et al. 1995; Paterson et al. 2009; Gaskin et al. 2011). Taxonomists, evolutionary biologists and breeders use molecular techniques to measure plant genetic diversity and determine how plants are related. DNA-based molecular marker techniques offer an approach to quantify invasive plant genetic diversity in native and introduced habitats and provide a better understanding of the complex relationships between invasive plants and potential biological control agents. This information could provide insights into the geographic origins of invasive plants and provide a means to direct the search for compatible biological control agents.

Success of biological weed control during the past 200 years has been variable. Winston et al. (2014) documented 2042 biologic control projects that span 130 countries and 551 biocontrol agents targeting 224 weeds. There have been some phenomenally successful biological control projects including control of *Opuntia* spp. in Australia by the moth *Cactoblastis cactorum* and control of St. Johnswort in the Pacific Northwest of the United States by *Chrysolina quadrigemina* and *Chloantha hyperici*. Sheppard (1992) reported 72 examples worldwide where weed biological control programs had been underway for a sufficient period to assess control. Of these programs, 28% resulted in control that could be rated as sometimes complete. In contrast, no control was achieved in 35% of these programs even though biological control agents were successfully established. Important factors that contribute to the limited success of biological weed control programs include a high level of genetic diversity in the target species, limited compatibility of agents with the invasive plant genotype, and opportunistic predation and parasitism of biological control agents in introduced environments (Sheppard 1992). About one third of biological control agents were estimated by Winston et al. (2014) to be successfully established on host plants.

The release of imported biological control agents on invasive plants is not without risk (Louda et al. 1997; Strong and Pemberton 2000; Follett and Duan 2000; Van Wilgen et al. 2013). By its very nature, classical biological control involves release of exotic organisms to control other exotic organisms. Use of native relatives of the exotic weed species by the introduced natural enemy is a potential detrimental unintended effect of biological control. The seed head weevil, *Rhinocyllus conicus* Froel., introduced from Europe into North America to control musk thistle, has been found to be successfully reproducing on flower heads of several native *Cirsium* species in California (Turner et al. 1987). Additionally, the weevil reduced seed production of native *Cirsium* species at several locations in the central Great Plains (Louda et al. 1997). *C. cactorum,* released to control exotic *Opuntia* spp., threatens native *Opuntia* spp. in Mexico and the United States (Strong and Pemberton 2000). Once released into a new environment, little can be done to restrict biological control agent distribution or host affinity. Monitoring potential biological control agents for expanded host range, host shifts, and effects on related non-target plants is critical. Pemberton (2000) determined that the greatest risk is where the weed targeted for biological control has close relatives that are native in the community where the agents will be released (e.g., thistles in *Carduus* and *Cirsium* genera). Harm to native plants can be reduced by targeting weeds with few or no close relatives in the country or broad region where the exotic weed occurs.

Chemical Control

Herbicides are assigned to groups according to their chemistry and mechanism of action (Ross and Lembi 1999) (Table 28.2). Mechanism of action refers to the system, process, or tissue affected by the herbicide. Herbicides are usually selective within certain rates, methods of application, and environmental conditions. Foliar-active herbicides are absorbed by leaves or stems and many may be translocated in the plant. These herbicides may remain active once moved into the soil. Soil-active herbicides are absorbed from the soil water solution by roots. Herbicides can also be categorized as to whether they are applied before planting and before or after weed emergence.

Herbicides are important tools to control weeds in forage production systems and can be a catalyst to expedite desired vegetation change and attainment of forage management objectives. Potential adverse effects on desirable plants and cost are concerns associated with herbicide use. Rate, timing, and frequency of application, and mechanism of action influence herbicide selectivity and can be manipulated to alleviate adverse impacts of herbicides on desirable plants.

A variety of herbicides are currently available to provide options to control weeds, renovate pasture and rangeland communities, and minimize potential negative effects of herbicides on desired plants (Table 28.2). Synthetic auxin herbicides, such as phenoxy-carboxylates, benzoates, pyridine-carboxylates, or pyridyloxy-carboxylates, are commonly used herbicides in pastures and on rangeland. These herbicides include, 2,4-D, dicamba, picloram, aminopyralid, and triclopyr. Aminopyralid is a synthetic auxin herbicide that provides preemergence and postemergence control of many broadleaf noxious and invasive plants. Aminopyralid is effective at rates between 53 and 120 g acid equivalent/ha, which is about 1/4 to 1/20 the use rates of several other rangeland and pasture herbicides. Undesirable plants in the *Ascription, Ambrosia, Carduus, Centaurea, Cirsium, Croton, Solanum,* and *Vernonia* genera are among those controlled by aminopyralid.

Herbicides that inhibit aromatic amino acid synthesis (glyphosate) and branched chain amino acid synthesis (imidazolinone and sulfonylurea herbicides) are important rangeland and pasture herbicides. Glyphosate is a nonselective, foliar-applied systemic herbicide that controls a wide variety of annual, biennial and perennial grasses, **sedges**, broadleaf weeds and woody plants. It is not active in the soil due to strong adsorption to soil particles. Selectivity of this herbicide is achieved by timing application when the target weeds have emerged and are growing, and desirable plants are dormant. Glyphosate enters the plant through the foliage. Environmental conditions (e.g. high light intensity, high temperature, low humidity, and low soil moisture) that decrease plant cuticle hydration and increase plant water stress will reduce uptake and slow herbicide transport to sites of action. Imazapic is an imidazolinone herbicide that is phytotoxic at low rates. Imazapic provides preemergence and postemergence weed control. A unique attribute of imazapic is the ability to control various annual grass and broadleaf weed species during establishment of desirable native warm-season grasses, forbs, and legumes (Masters et al. 1996; Beran et al. 1999a,b; Washburn and Barnes 2000a). Imazapic was found to control tall fescue and promote the return of remnant native warm-season grasses in Kentucky grasslands (Washburn et al. 1999; Washburn and Barnes 2000b).

Metsulfuron-methyl, a sulfonylurea herbicide, can be used to reduce tall fescue seed head formation while having little adverse effect on other perennial cool-season grasses (Aiken et al. 2012; Williamson et al. 2016). When applied to endophyte-infected tall fescue, metsulfuron-methyl can mitigate effects of fescue toxicosis in cattle. Seed head suppression with metsulfuron-methyl improved cattle average daily gain (Aiken et al. 2012; Goff et al. 2012), increased cattle conception rates by 8–25%, and increased calf weaning weight by 10–25 kg (Boyer 2015). Animal performance was improved because of reduced consumption of highly toxic seed heads and higher crude protein, *in vitro* dry matter digestibility, and **water-soluble carbohydrates** of tall fescue that was maintained in a vegetative stage of growth because of seed head suppression following treatment with metsulfuron (Aiken et al. 2012; Goff et al. 2012, 2015). Metsulfuron-methyl can maintain endophyte-free and novel endophyte tall fescue in a vegetative stage and preserve forage quality in the summer.

Indaziflam is a cellulose-biosynthesis-inhibiting herbicide that provides excellent control of invasive winter

Table 28.2 Selected herbicides registered in the US as of 2019 for use on rangeland, pastures, and alfalfa[a]

Chemical family	Common name	Chemical name	Mechanism of action name and classification (HRAC/WSSA[b])	Plants controlled[c]	Activity[d]	Application timing[e]
Arylpicolinate	Forpyrauxifen-benzyl (Rinskor)	2-pyridinecarboxylic acid, 4-amino-3-chloro-6- (4-chloro-2-fluoro-3-methoxy-phenyl)-5-fluoro-, phenyl methyl ester	Synthetic Auxin (O/4)	B	F, S	POST
Benzoic acid	Dicamba	3,6-dichloro-2-methoxybenzoic acid	Synthetic Auxin (O/4)	B	F, S	PRE, POST
Benzonitrile	Bromoxynil	3,5-dibromo-4-hydroxybenzonitrile	Photosystem II inhibitors (C3/6)	B	F	POST
Bipyridilium	Paraquat	1,1′-dimethyl-4,4′-bipyridinium ion	Photosystem-1-eletron diversion (D/22)	B, G	F	POST
Carbamothioate	EPTC	S-ethyl dipropyl carbamothioate	Fatty Acid and Lipid Biosynthesis Inhibitors (N/8)	B, G	S	PPI
Cyclohexanediones	Clethodim	(E,E)-(±)-2-[1-[[(3-chloro-2-propenyl) oxy]imino]propyl]-5-[2-(ethylthio) propyl]-3-hydroxy-2-cyclohexen-1-one	ACCase inhibitor (A/1)	G	F	POST
	Sethoxydim	2-[1-(ethoxyimino)butyl]-5-[2-(ethylthio) propyl]-3-hydroxy-2-cyclohexen-1-one	ACCase inhibitor (A/1)	G	F	POST
Dinitroaniline	Trifluralin	2,6-dinitro-N,N-dipropyl-4-(trifluoromethyl)benzenamine	Microtubule inhibitors (K1/3)	B, G	S	PPI, PRE
Imidazolinone	Imazethapyr	2-[4,5-dihydro-4-methyl-4-(1-methylethyl)-5-oxo-1H-imidazol-2-yl]-5-ethyl-3 -pyridinecar-boxylic acid	Acetolactate Synthase (ALS) Inhibitors (B/2)	B, G	F, S	PRE, POST

	Imazamox	2-[4,5-dihydro-4-methyl-4-(1-methylethyl)-5-oxo-1H-imidazol-2-yl]-5-(methoxymethyl)-3-pyridinecarboxylic acid	Acetolactate Synthase (ALS) Inhibitors (B/2)	B, G	F, S	PRE, POST
	Imazapic	2-[4,5-dihydro-4-methyl-4-(1-methylethyl)-5-oxo-1H-imidazol-2-yl]-5-methyl-3-pyridine-carboxylic acid	Acetolactate Synthase (ALS) Inhibitors (B/2)	B, G	F, S	PRE, POST
	Imazapyr	2-[4,5-dihydro-4-methyl-4-(1-methylethyl)-5-oxo-1H-imidazol-2-yl]-3-pyridinecarboxylic acid	Acetolactate Synthase (ALS) Inhibitors (B/2)	B, G	F, S	PRE, POST
Phenoxy acid	2,4-D	(2,4-dichlorophenoxy)acetic acid	Synthetic Auxin (O/4)	B	F	POST
	2,4-DB	4-(2,4-dichlorophenoxy)butanoic acid	Synthetic Auxin (O/4)	B	F	POST
	MCPA	(4-chloro-2-methylphenoxy)acetic acid	Synthetic Auxin (O/4)	B	F	POST
Phenylurea	Diuron	N′-(3,4-dichlorophenyl)-N,N-dimethylurea	Photosystem II inhibitors (C2/7)	B, G	F, S	PRE, POST
	Tebuthiuron	N-[5-(1,1-dimethylethyl)-1,3,4-thiadiazol-2-yl]-N,N′-dimethylurea	Photosystem II inhibitors (C2/7)	B, G	F, S	PRE, POST
Pyridine carboxylic acid	Aminopyralid	3,6-dichloro-pyridinecarboxylic acid	Synthetic Auxin (O/4)	B	F, S	PRE, POST
	Clopyralid	3,6-dichloro-2-pyridinecarboxylic acid	Synthetic Auxin (O/4)	B	F, S	PRE, POST
	Fluroxypyr	4-amino-3,5-dichloro-6-fluoro-2-pyridyloxyacetic acid methylheptyl ester	Synthetic Auxin (O/4)	B	F, S	POST
	Picloram	4-amino-3,5,6-trichloro-2-pyridinecarboxylic acid	Synthetic Auxin (O/4)	B	F, S	PRE, POST
	Triclopyr	[(3,5,6-trichloro-2-pyridinyl)oxy]acetic acid	Synthetic Auxin (O/4)	B	F, S	POST
s-Triazine	Hexazinone	3-cyclohexyl-6-(dimethylamino)-1-methyl-1,3,5-triazine-2,4(1H,3H)-dione	Photosystem II inhibitors (C1/5)	B, G	F, S	PRE, POST
	Metribuzin	4-amino-6-(1,1-dimethylethyl)-3-(methylthio)-1,2,4-triazin-5(4H)-one	Photosystem II inhibitors (C1/5)	B	F, S	PRE, POST
Sulfonylurea	Chlorsulfuron	2-chloro-N-[[(4-methoxy-6-methyl-1,3,5-triazin-2yl)amino]carbonyl] benzenesulfonamide	Acetolactate Synthase (ALS) Inhibitors (B/2)	B, G	F, S	PRE, POST
	Metsulfuron-methyl	2-[[[[(4-methoxy-6-methyl-1,3,5-triazin-2yl)amino]carbonyl]amino]sulfonyl]benzoic acid	Acetolactate Synthase (ALS) Inhibitors (B/2)	B, G	F, S	PRE, POST

Table 28.2 (*Continued.*)

Chemical family	Common name	Chemical name	Mechanism of action name and classification (HRAC/WSSA[b])	Plants controlled[c]	Activity[d]	Application timing[e]
	Nicosulfuron	2-[(4,6-dimethoxypyrimidin-2-yl) carbamoylsulfamoyl]-N,N-bis(trideuteriomethyl) pyridine-3-carboxamide	Acetolactate Synthase (ALS) Inhibitors (B/2)	B, G	F, S	PRE, POST
Sulfonyl amino carbonyl-triazolinone	Propoxycarbazone sodium	Sodium (2-methoxycarbonylphenyl)sulfonyl-(4-methyl-5-oxo-3-propoxy-1,2,4-triazole-1-carbonyl)azanide	Acetolactate Synthase (ALS) Inhibitors (B/2)	B, G	F	POST
Not assigned	Glyphosate	N-(phosphonomethyl)glycine	Enolpyruvyl Shikimate-3-Phosphate (EPSP) Synthase Inhibitors (G and 9)	B, G	F, S	POST

[a]Herbicide Action Committee (http://hracglobal.com/tools/classification-lookup)
[b]HRAC/Weed Science Society of America herbicide classification designation.
[c]B = broadleaf species and G = grass species
[d]F = taken up by plant foliage and S = has activity in the soil
[e]PRE = applied before plant emergence, POST = applied after plant emergence, PPI = pre-plant incorporated

annual grasses (Sebastian et al. 2016a,b, 2017). Indaziflam provided control (84–99%) of a broad spectrum of invasive cool-season annual grasses two years after treatment that was superior to control with imazapic (36%). In addition, biomass and richness of desirable plant species two years after indaziflam application increased on western US rangeland sites.

Tebuthiuron is a photosynthetic inhibitor that is used for controlling annual grasses, perennial cool-season grasses and woody species. This herbicide can be used to selectively remove or suppress cool-season grasses and rejuvenate stands of warm-season perennial grasses. Tebuthiuron applied at 0.9 kg ha^{-1} reduced cool-season grass yields by over 60% and increased warm-season grass yields by 50–300% the growing season after an autumn application (Hillhouse et al. 2015).

Herbicides can either be broadcast-applied or applied to individual plants. Broadcast treatments can be applied using ground equipment or aerially by fixed-wing aircraft or helicopter. Individual plant treatments can be efficient, cost-effective alternatives to broadcast applications to control trees, shrubs, vines, and patches of herbaceous plants. The high degree of selectivity achieved with individual plant treatments provides a means to reduce injury to desirable plants caused by the herbicide. Individual plant treatments include foliar sprays, basal sprays, cut-surface, injection, cut-stump applications, and soil treatment (Table 28.3) (Bovey 2001). Foliar sprays involve application of diluted herbicide solution directly to the plant foliage. The spray should be applied after full leaf expansion when the plants are actively growing. Thorough coverage of the foliage with the spray solution is critical to optimize herbicide efficacy. Low-volume or streamline basal sprays can be used on larger woody plants. Basal sprays involve applying mixtures of 20–30% herbicide in 70–80% oil carrier (diesel fuel, kerosene, or crop oils). Low-volume basal sprays are applied to the lower 35–50 cm of the trunk including the root collar area. These treatments are effective in controlling trees up to 15 cm in diameter. In streamline applications, a band 7–10 cm wide around the entire stem is applied near the ground line. This application technique is most effective to control trees with smooth bark that are less than 10 cm in diameter. Basal sprays can be applied any time during the year, except when snow or water prevents spraying to the ground line.

When trees are larger than 15 cm in diameter, control can be achieved by applying herbicide solution to notches cut in the tree bark, directly injecting the herbicide solution into the tree, or to the cut surface after the tree is cut down. A cut-stump spray (20–30% herbicide and 70–80% oil) is applied to the cut stump surface of a felled tree. The stump should be treated as soon after cutting as possible and enough spray solution should be applied to thoroughly wet the sides of the stump and cut surface, including the cambium ring along the inner bark. Like basal sprays, the cut-stump sprays can be applied any time of the year providing snow or water does not hinder spray contacting the cut surface.

Soil-applied liquid or pellet formulations of some herbicides provide effective control of individual plants. Picloram liquid concentrate controls *Juniperus* species when applied to the soil inside the drip line of trees less than 5 m tall at a rate of 1 1/3 ml per 30 cm of tree height. Tebuthiuron or hexazinone pellets applied to the soil can control *Quercus* and *Juniperus* species.

Cultural Control

Cultural practices include fire, grazing, haying, revegetation, plant competition, liming, and fertilization. These methods are generally aimed at enhancing desirable vegetation to minimize weed invasion.

Fire, climate, and herbivory were the primary forces responsible for the formation and maintenance of grassland ecosystems in North America (Pyne 1984). North American grassland fire regimes were shaped by sources of ignition (lightning and humans) and climate (Pyne 1984). As with any disturbance, fire effects on ecosystems are influenced by frequency, intensity, season of occurrence, and interactions with other disturbances. In general, mesic grasslands burned more frequently than **xeric** grasslands. Frequent fires (every three to five years) in grasslands with precipitation: evaporation ratios of 1 or greater are necessary to prevent woody vegetation establishment (Bragg and Hulbert 1976).

Fire is an essential practice to meet land management objectives for many plant communities in North America (Pyne 1984). Prescribed fire can be applied to control woody plants (Bragg and Hulbert 1976), increase the nutritive value of grasses (Mitchell et al. 1994), reduce exotic cool-season grasses such as kentucky bluegrass and smooth bromegrass (Mitchell et al. 1996), increase tiller density (McFarland and Mitchell 2000), and increase grass seed production (Masters et al. 1993). Fire interactions with other disturbance factors such as grazing, promote rangeland community complexity and, increase biologic diversity (Fuhlendorf and Engle 2001).

Most animals have preferences for grazing certain plant species. Selectivity by herbivores alters competitive interactions within plant communities (Crawley 1983). Excessive cattle grazing without periodic rest can selectively reduce forage grass vigor and give weeds the competitive advantage. Grazing should be closely monitored and managed since having too high a grazing intensity can increase the number of weed species, the density of emerged weeds seedlings, and the density of the weed seed bank (Schuster et al. 2016). Using livestock that prefer to graze or browse weeds can shift plant community composition toward more desired species

Table 28.3 Methods of applying selected herbicides for brush control

	Application methods					
Herbicide	Foliar spray	Basal spray	Cut-surface	Injection	Cut-stump	Soil treatment
2,4-D	X	X	X	X	X	
Clopyralid	X					
Dicamba	X		X	X	X	X
Fosamine	X					
Glyphosate	X		X	X	X	
Hexazinone						X
Metsulfuron	X					
Picloram	X		X	X	X	X
Tebuthiuron						X
Triclopyr	X	X			X	

Source: Modified after Bovey (2001) and Monaco et al. (2002).

(Walker 1994). In some situations, sheep or goat grazing (Walker 1994; Lym et al. 1997) can control leafy spurge. Milk thistle and syrian thistle were not toxic when eaten by goats and preconditioned goats spent 50% more time consuming the thistles than non-conditioned goats (Arviv et al. 2016). Goats will preferentially consume musk thistle (Holst et al. 2004). Popay and Field (1996) reviewed the effectiveness of using different livestock types with distinctly different dietary preferences to control weeds and woody plants.

Reseeding is an important step in building weed resistant plant communities on sites depleted of desirable species. Masters et al. (2004) provides a process to follow to increase probability of plant establishment success. Planting should occur when conditions are conducive for seed germination and seedling growth. Establishing desirable grasses, forbs, and legumes may suppress invasive plants, enhance plant community resistance to further invasion, and improve forage production and quality (Masters et al. 1996; Lym and Tober 1997; Bottoms and Whitson 1998; Whitson and Koch 1998).

Selecting plant species for site revegetation is a critical consideration when designing a DPC if the desired species are not present in sufficient abundance to enable regeneration within an acceptable timeframe. Jones and Johnson (1998) described an integrated approach for making decisions about how to select plant materials for rangeland revegetation. Site potential, desired landscape, seeding objectives, appropriate plant species, invasive plants, and economic limitations are among the key components of the decision-making process.

A challenge often faced by land managers considering revegetation is whether to use native or introduced plant materials. Should local native ecotypes (Linhart and Grant 1996), native or exotic plant cultivars with improved agronomic traits developed by breeding programs (Casler et al. 1996; Jones 2003), and mixed populations of hybrid genotypes be used in revegetation programs? One perspective is that rather than emphasizing individual species, the focus of revegetation programs should be on establishing functional groups (Walker 1992) that maintain ecosystem processes (Noss 1991). Johnson and Mayeux (1992) argue, "no special quality should be attributed to a species labeled as a native, rather the focus should be on ecosystems as self-sustaining in terms of physiognomic structure and functional processes in which various species .... are interchangeable." To increase plant community resistance to weed invasion, it is important to select multiple species to be planted that are compatible and productive (Tracy and Sanderson 2004; Sanderson et al. 2007). Planting species that resist pests and persist under grazing and environmental stress enables development of plant communities that resist weed interference and are resilient to weed invasion.

Fertilizers can be applied to grasslands to alleviate nutrient deficiencies in plants and increase forage yield and quality. For optimum forage grass response, N application should be timed to coincide with rapid growth periods. C_3 grasses should be fertilized earlier in the spring than C_4 grasses. If C_4 grasses are fertilized too early, C_3 grasses and forbs will be encouraged, which will put C_4 grasses at a competitive disadvantage. Timing of fertilizer application is critical in swards containing both C_3 and C_4 grasses, such as tallgrass prairie, where enhancing the C_3 component may not be desired. Combining management practices that reduce the C_3 component and enhance the C_4 component can improve tallgrass prairie. Burning and fertilizing tallgrass prairie in mid-May with 67 kg N and 23 kg P ha^{-1} increased big bluestem dry matter production and crude protein concentration (Mitchell et al. 1994).

Fertilizing rangeland west of the 100th Meridian may be less effective because of reduced precipitation. Grass yield response to N is greatest when soil water is adequate.

Amount of N applied to a grassland system should equal the N removal capabilities of the desired species in grassland ecosystems. Though grass dry matter production increases at high N rates, N recovery (kg plant N kg^{-1} N applied) generally declines as N rate increases (Singer and Moore 2003). In Wisconsin, orchardgrass had greater N-use-efficiency and apparent N recovery than smooth bromegrass or kentucky bluegrass when multiple fertilizer applications were made with total annual rates of up to 336 kg N ha^{-1} (Zemenchik and Albrecht 2002). In New Jersey, orchardgrass removed more N than smooth bromegrass in a three-cut system, primarily by capturing more N during the autumn growth period (Singer and Moore 2003). If high levels of residual N remain in the system after the dominant grasses have concluded growth, undesirable plants may utilize the excess N and reduce the quality of the forage resource. Fertilizer should be applied at the rate and date that optimizes forage stand production and resistance to weed invasion.

Mechanical Control

Mechanical treatments involve either removal of the aerial portions of the weed or removal of enough of the root and crown to weaken or kill the plant (Vallentine 1989). Annuals and some biennials and perennials can be suppressed or controlled by mowing before floral structures mature and viable seeds form. Mowing perennial herbaceous or woody plants that reproduce vegetatively can exacerbate weed interference by stimulating production of new stems from vegetative buds below the cut surface. However, perennial plants that reproduce vegetatively can be severely damaged or killed by tillage, bulldozing, root-plowing, or grubbing (Vallentine 1989). The high cost of these more energy intensive and highly disruptive mechanical treatments limits their widespread use. If mowing is to be used, the timing and frequency should be carefully planned to fit the management objectives. A single mowing of canada thistle failed to produce long-term control and led to increased shoot density and biomass the following year (Grekul and Bork 2007).

Integrating Multiple Weed Control Tactics

Adapting the basic concepts of IWM to forage resource management depends on the setting. Forage systems are comprised of cropland, annual pastures, improved perennial pastures, and native rangeland. The management of these different forage systems is either intensive or extensive. Annual pastures are often intensively managed, requiring tillage, seeding, fertilization, weed management, and regular livestock movement. Under these conditions, it is cost effective to invest in improved perennial forage pastures that can benefit from over-seeding, fertilization, weed management, and temporary fencing. The goal of intensive management is to increase livestock production per animal or unit area or to improve forage production and utilization by increasing inputs of labor, materials, and/or capital (Allen and Collins 2002). The increased cost of production should be offset by greater economic return. In contrast, rangeland is often extensively managed because land values and forage production potential per unit area does not justify the cost of management inputs.

There are several examples of integrated strategies used to manage weeds to improve pasture and rangeland communities (Table 28.4). Competition with forages can intensify the impact of biocontrol organisms on the target weed. Musk thistle reduction was greatest where the weevils, *Trichosirocalus horridus* Panzer and *R. conicus* Froelich, infested thistle growing in association with tall fescue (Kok et al. 1986). The competitive interaction between the musk thistle and tall fescue coupled with the stress imposed by the weevils reduced musk thistle seed production. They suggested that tall fescue and other competitive grasses or broadleaf species should be used in concert with biologic agents to further improve musk thistle control.

Efforts to assess the compatibility of insect biocontrol agents and herbicides during development of integrated management systems are increasing (Nelson and Lym 2003). Revegetation has been a common component of integrated approaches because it is essential that desirable plant species, rather than another weed species, fill the niches vacated by the controlled invader. Herbicides and tillage were used to suppress dalmatian toadflax and St. Johnswort (Gates and Robocker 1960), cheatgrass (Eckert and Evans 1967), and medusahead (Young et al. 1969) in early attempts to prepare degraded rangeland sites for revegetation with cool-season grasses. Fertilization can improve weed control with herbicides. Grass yield and quality increased when herbicide application for control of canada thistle was combined with fertilization even though there was a decline in forb yield and associated quality (Grekul and Bork 2007). The adverse effect of herbicides on forbs and resulting impacts on forage yield and quality is mitigated when pastures contain an abundance of forbs (Bork et al. 2007). Annual spring fertilization extended canada thistle control with herbicides and appeared to be linked to enhanced competiveness of the forage (Grekul and Bork 2007).

Approaches that include herbicide application and establishing monoculture stands of exotic and native perennial grasses have been successfully used to suppress leafy spurge and improve forage production on rangeland. Exotic cool-season grasses were planted in a tilled **seedbed** following broadcast applications of glyphosate and 2,4-D in North Dakota (Lym and Tober 1997). The planted grasses most effective in suppressing leafy spurge were 'Bozoisky' russian wildrye and 'Luna' pubescent wheatgrass in Wyoming, and 'Rebound' smooth bromegrass and 'Reliant' intermediate wheatgrass in North Dakota.

Table 28.4 Examples of integrated strategies for control of invasive plants on rangeland

Invasive plant	Strategy components	Citation
Russian knapweed	Tillage, herbicide, and revegetation	Bottoms and Whitson (1998), Benz et al. (1999)
Downy brome (cheatgrass)	Tillage, herbicide, and revegetation	Eckert and Evans (1967)
	Herbicide and grazing	Whitson and Koch (1998)
Yellow starthistle	Herbicide, revegetation, and biocontrol	Enloe and DiTomaso (1999)
	Herbicide and fire	DiTomaso et al. (2006)
Musk thistle canada thistle	Biocontrol and grass competition	Kok et al. (1986)
	Herbicide and revegetation	Wilson and Kachman (1999)
Leafy spurge	Herbicide and biocontrol	Nelson and Lym (2003)
	Tillage, herbicide, and revegetation	Lym and Tober (1997)
	Grazing and herbicide	Lym et al. (1997)
	Herbicide, fire, and revegetation	Masters and Nissen (1998), Masters et al. (2001)
St. Johnswort	Tillage and revegetation	Gates and Robocker (1960)
Perennial pepperweed	Mowing and herbicide	Renz and DiTomaso (1999)
Dalmatian toadflax	Tillage and revegetation	Gates and Robocker (1960)
Medusahead mimosa	Tillage, herbicide, and revegetation	Young et al. (1969)
	Herbicide, bulldozing, biocontrol, and fire	Paynter and Flanagan (2004)

In Nebraska, an IWM strategy that suppressed leafy spurge and associated vegetation to facilitate planting and establishment of mixed swards of native warm-season grasses and legumes was developed (Masters et al. 2001). Conceptually, these assemblages of plant species should more fully use resources on degraded rangeland and preempt resource use by less desirable species, including leafy spurge. The strategy consisted of herbicide application in the fall followed by burning the herbaceous standing crop and planting mixtures of native species without tillage in the spring. Glyphosate and imazapic were used to suppress existing resident vegetation and expedite establishment of the planted species mixtures. Glyphosate controlled cool-season grasses that were growing at the time of application, but provided no residual weed control. Imazapic provided residual control of leafy spurge, annual grasses, and annual broadleaf plants and was tolerated by the planted warm-season grasses (big bluestem, little bluestem, sideoats grama, and indiangrass) and legumes (illinois bundleflower and purple prairie clover).

Converting pasture dominated by toxic endophyte-infected (E+) tall fescue to novel or endophyte-free tall fescue while preventing re-infestation of E+ plants can be expedited by integrating chemical and cultural practices. The "spray-smother-spray" (S-S-S) strategy requires herbicide application to control E+ tall fescue and weeds, followed by planting an annual grass forage to "smother" or restrict emergence of undesirable vegetation, and then herbicide application before planting (Bagegni et al. 1994). Hill et al. (2010) implemented the S-S-S strategy and then examined the in-field survival of toxic tall fescue seeds to determine the potential for re-infestation during the year of renovation. It was determined that split applications of herbicide as a part of the S-S-S strategy prevented E+ escapes, but spring seed production during the establishment year did contribute to re-infestation as E+ tall fescue seeds germinated. An early spring mowing to prevent seed head formation and split applications of glyphosate (1.68 kg active ingredient ha^{-1}) applied six weeks before planting and again one day prior to planting 'Jesup MaxQ' tall fescue resulted in less than 2.5% E+ tall fescue escapes.

Adaptive Management

Adaptive management is implementing a land resource management plan and then learning by monitoring impacts of that management approach and then adjusting management tactics based on what is learned (Williams 2011). Adaptive management can complement integrated programs to manage weeds in forage production systems. This approach requires establishing management goals, developing and implementing management programs based on objectives designed to achieve the goals, monitoring and assessing impacts of management efforts, and modifying plant management tactics in light of new information (Randall 1997; Jacobson et al. 2006). Adaptive management is an integrated, multidisciplinary approach to deal with the uncertainty associated with natural resource management (Gunderson 1999). To be successful, weed management programs must be compatible with and integrated into forage resource management programs. Effective weed management programs must consider other management components that impinge upon the forage resource. Integrating management tactics in the proper sequence and combination is essential to the economic and ecologic sustainability of forage resource management programs.

References

Aiken, G.E., Goff, B.M., Witt, W.W. et al. (2012). Steer and plant responses to chemical suppression of seed-head emergence in toxic endophyte-infected tall fescue. *Crop Sci.* 52: 960–969.

Allen, V.G. and Collins, M. (2002). Grazing management systems. In: *Forages; an Introduction to Grassland Agriculture*, 6e (eds. R.F Barnes, C.J. Nelson, M. Collins, et al.), 473–501. Ames: Iowa State University Press.

Amezaga, I., Mendarte, S., Albizu, I. et al. (2004). Grazing intensity, aspect, and slope effects on limestone grassland structure. *J. Range Manage.* 57: 606–612.

Arviv, A., Muklada, H., Kigel, J. et al. (2016). Targeted grazing of milk thistle (*Silybum marianum*) and Syrian thistle (*Notobasis syriaca*) by goats: preference following preconditioning, generational transfer, and toxicity. *Appl. Anim. Behav. Sci.* 179: 53–59.

Bagegni, A.M., Kerr, H.D., and Sleper, D.A. (1994). Herbicides with crop competition replace endophytic tall fescue (*Festuca arundinacea*). *Weed Technol.* 8: 689–695.

Ball, D.M., Hoveland, C.S., and Lacefield, G.D. (2007). *Southern Forages, Modern Concept for Forage Crop Management*. IPIN: Norcross.

Bennie, J., Hill, M.O., Baxter, R., and Huntley, B. (2006). Influence of slope and aspect on long-term vegetation change in British chalk grasslands. *J. Ecol.* 94: 355–368.

Benz, L.J., Beck, G., Whitson, T.D., and Koch, D.W. (1999). Reclaiming Russian knapweed infested rangeland. *J. Range Manage.* 52: 351–356.

Beran, D.D., Masters, R.A., and Gaussoin, R.E. (1999a). Grassland legume establishment with imazethapyr and imazapic. *Agron. J.* 91: 592–596.

Beran, D.D., Gaussoin, R.E., and Masters, R.A. (1999b). Native wildflower establishment with imidazolinone herbicides. *Hort. Sci.* 34: 283–286.

Bork, E.W., Grekul, C.W., and DeBruijn, S.L. (2007). Extended pasture forage sward response to Canada thistle (*Cirsium arvense*) control using herbicides and fertilization. *Crop Prot.* 26: 1546–1555.

Bosworth, S.C., Hoveland, C.S., and Buchanan, G.A. (1985). Forage quality of selected cool-season weed species. *Weed Sci.* 34: 150–154.

Bottoms, R.M. and Whitson, T.D. (1998). A systems approach for the management of Russian knapweed (*Centaurea repens*). *Weed Technol.* 12: 363–366.

Bovey, R.W. (2001). *Woody Plants and Woody Plant Management*. New York, NY: Marcel Dekker.

Boyer, W.F. (2015). Cow-calf response to seed head suppressed tall fescue pastures in southern Missouri. Ph.D. Dissertation. Missouri State University.

Bragg, T.B. and Hulbert, L.C. (1976). Woody plant invasion of unburned Kansas bluestem prairie. *J. Range Manage.* 29: 19–24.

Briske, D.D., Fuhlendorf, S.D., and Smeins, F.E. (2005). State-and-transition models, thresholds, and rangeland health: a synthesis of ecological concepts and perspectives. *Rangeland Ecol. Manage.* 58: 1–10.

Burn, A.J., Coaker, T.H., and Jepson, P.C. (1987). *Integrated Pest Management*. San Diego, CA: Academic Press.

Casler, M.D., Pedersen, J.F., Eizenga, G.C., and Stratton, S.D. (1996). Germplasm and cultivar development. In: *Cool-Season Forage Grasses* (eds. L.E. Moser, D.R. Buxton and M.D. Casler), 413–469. American Society of Agronomy Monographs 34.

Casteel, S.W., Johnson, G.C., Miller, M.A. et al. (1994). *Amaranthus retroflexus* (redroot pigweed) poisoning in cattle. *J. Am. Vet. Med. Assoc.* 204: 1068–1070.

Cheeke, P.R. (1998). *Natural Toxicants in Feeds, Forages, and Poisonous Plants*, 2e. Interstate Publishers, Inc.

Connell, J.H. and Slatyer, R.O. (1977). Mechanisms of succession in natural communities and their role in community stability and organization. *Am. Nat.* 111: 1119–1144.

Crawley, M.J. (1983). *Herbivory: The Dynamics of Animal-Plant Interactions*. Oxford: Blackwell Scientific.

DiTomaso, J.M., Kyser, G.B., Miller, J.R. et al. (2006). Integrating prescribed burning and clopyralid for the management of yellow starthistle (*Centaurea solstitialis*). *Weed Sci.* 54: 757–767.

Dutt, T.E., Harvey, R.G., and Fawcett, R.S. (1982). Feed quality of hay containing perennial broadleaf weeds. *Agron. J.* 74: 673–676.

Eckert, R.E. and Evans, R.A. (1967). A chemical-fallow technique for control of downy brome and establishment of perennial grasses on rangeland. *J. Range Manage.* 20: 35–41.

Ehler, L.E. (2006). Integrated pest management (IPM): definition, historical development and implementation, and the other IPM. *Pest Manage. Sci.* 62: 787–789.

Enloe, S. and DiTomaso, J. (1999). Integrated management of yellow starthistle on California rangeland. *Proc. California Weed Sci. Soc.* 51: 24–27.

Follett, P.A. and Duan, J.J. (eds.) (2000). *Nontarget Effects of Biological Control*. Dordrecht, The Netherlands: Kluwer.

Franzluebbers, A.J. (2002). Soil organic matter stratification ratio as an indicator of soil quality. *Soil Tillage Res.* 66: 95–106.

Fuhlendorf, S.D. and Engle, D.M. (2001). Restoring heterogeneity on rangelands: ecosystem management based on evolutionary grazing patterns. *BioScience* 51: 625–632.

Funk, J.L., Cleland, E.E., Suding, K.N., and Zavaleta, E.S. (2008). Restoration through reassembly: plant traits and invasion resistance. *Trends Ecol. Evol.* 23: 695–703.

Gaskin, J.F., Bon, M.C., Cock, M.J. et al. (2011). Applying molecular-based approaches to classical biological control of weeds. *Biol. Control* 58: 1–21.

Gates, D.H. and Robocker, C. (1960). Revegetation with adapted grasses in competition with dalmatian toadflax and St. Johnswort. *J. Range Manage.* 13: 322–326.

Gillingham, A.G. (1973). Influence of physical factors on pasture growth on hill country. *Proc. N. Z. Grassland Assoc.* 35: 77–85.

Goff, B.M., Aiken, G.E., Witt, W.W. et al. (2012). Steer consumption and ergovaline recovery from in vitro digested residues of tall fescue seedheads. *Crop Sci.* 52: 1437–1440.

Goff, B.M., Aiken, G.E., Witt, W.W. et al. (2015). Forage nutritive 474 value and steer responses to grazing intensity and seedhead suppression of endophtyte-free tall fescue in mixed pictures. *Prof. Anim. Sci.* 31: 120–129.

Goldstein, J. (1978). *The Least Is Best Pesticide Strategy*. Emmaus, PA: The JG Press.

Grekul, C.W. and Bork, E.W. (2007). Fertilization augments Canada thistle (*Cirsium arvense)* control in temperate pastures with herbicides. *Crop Prot.* 26: 668–676.

Gunderson, L. (1999). Resilience, flexibility and adaptive management -antidotes for spurious certitude? *Conserv. Ecol.* 3: 7.

Harmoney, K.R., Moore, K.J., Brummer, E.C. et al. (2001). Spatial legume composition and diversity across seeded landscapes. *Agron. J.* 93: 992–1000.

Herrick, J.E. (2000). Soil quality: an indicator of sustainable land management? *Appl. Soil Ecol.* 15: 75–83.

Hill, N.S., Andrae, J.G., Durham, R.G., and Hancock, D.W. (2010). Herbicide treatments to renovate toxic endophyte infected tall fescue pastures with 'Jesup'Maxq. *Crop Sci.* 50: 1086–1093.

Hillhouse, H.L., Schacht, W.H., Masters, R.A. et al. (2015). Tebuthiuron use in restoring degraded tallgrass prairies and warm-season grass pastures. *Am. Midl. Nat.* 173: 99–109.

Hobbs, R.J. and Huenneke, L.F. (1992). Disturbance, diversity, and invasion: implications for conservation. *Conserv. Biol.* 6 (3): 324–337.

Hobbs, R.J. and Humphries, S.E. (1995). An integrated approach to the **ecology** and management of plant invasions. *Conserv. Biol.* 9: 761–770.

Hobbs, R.J. and Norton, D.A. (1996). Towards a conceptual framework for restoration ecology. *Restor. Ecol.* 4: 93–110.

Holland, P.G. and Steyn, D.G. (1975). Vegetational responses to latitudinal variations in slope angle and aspect. *J. Biogeogr.* 2: 179–183.

Holst, P.J., Allan, C.J., Campbell, M.H., and Gilmour, A.R. (2004). Grazing of pasture weeds by goats and sheep: nodding thistle (*Carduus nutans*). *Aust. J. Exp. Agric.* 44: 547–551.

Huston, M. and Smith, T. (1987). Plant succession: life history and competition. *Am. Nat.* 130: 168–198.

Jacobson, S.K., Morris, J.K., Sanders, J.S. et al. (2006). Understanding barriers to implementation of an adaptive land management program. *Conserv. Biol.* 20: 1516–1527.

Johnson, H.B. and Mayeux, H.S. (1992). Viewpoint: a view on species additions and deletions and the balance of nature. *J. Range Manage.* 45: 322–333.

Jones, T.A. (2003). The restoration gene pool concept: beyond the native versus non-native debate. *Restor. Ecol.* 11: 281–290.

Jones, T.A. and Johnson, D.A. (1998). Integrating genetic concepts into planning rangeland seedings. *J. Range Manage.* 51: 594–606.

Kok, L.T., McAvoy, T.J., and Mays, W.T. (1986). Impact of tall fescue grass and Carduus thistle weevils on the growth and development of musk thistle (*Carduus nutans*). *Weed Sci.* 34: 966–971.

Last, R.D., Hill, J.H., and Theron, G. (2007). An outbreak of perirenal oedema syndrome in cattle associated with ingestion of pigweed (*Amaranthus hybridus* L.): clinical communication. *J. S. Afr. Vet. Assoc.* 78: 171–174.

Liebman, M. and Gallandt, E.R. (1997). Many little hammers: ecological management of crop-weed interactions. *Ecol. Agri.* 14: 291–343.

Linhart, Y.B. and Grant, M.C. (1996). Evolutionary significance of local genetic differentiation in plants. *Annu. Rev. Ecol. Syst.* 27: 237–277.

Louda, S., Kendall, D., Connor, J., and Simberloff, D. (1997). Ecological effects of an insect introduced for the biological control of weeds. *Science* 277: 1088–1090.

Lym, R.G. and Tober, D.A. (1997). Competitive grasses for leafy spurge (*Euphorbia esula*) reduction. *Weed Technol.* 11: 787–792.

Lym, R.G., Sedivec, K.K., and Kirby, D.R. (1997). Leafy spurge control with angora goats and herbicides. *J. Range. Manage.* 50: 123–128.

Marten, G.C. and Andersen, R.N. (1975). Forage nutritive value and palatability of 12 common weeds. *Crop Sci.* 15: 821–827.

Marten, G.C., Sheaffer, C.C., and Wyse, D.L. (1987). Forage nutritive value and palatability of perennial weeds. *Agron. J.* 79: 980–986.

Masters, R.A. and Nissen, S.J. (1998). Revegetating leafy spurge (*Euphorbia esula* L.)-infested grasslands with native tallgrasses. *Weed Technol.* 12: 381–390.

Masters, R.A., Mitchell, R.B., Vogel, K.P., and Waller, S.S. (1993). Influence of improvement practices on big bluestem and indiangrass seed production in tallgrass prairies. *J. Range Manage.* 46: 183–188.

Masters, R.A., Nissen, S.J., Gaussoin, R.E. et al. (1996). Imidazolinone herbicides improve restoration of Great Plains grasslands. *Weed Technol.* 10: 92–403.

Masters, R.A., Beran, D.D., and Gaussoin, R.E. (2001). Restoring tallgrass prairie species mixtures on leafy spurge-infested rangelands. *J. Range Manage.* 54: 362–369.

Masters, R.A., Mislevy, P., Moser, L.E., and Rivas-Pantoja, F. (2004). Stand establishment. In: *Warm-Season (C_4) Forage Grasses* (eds. L.E. Moser, L. Sollenberger and B. Burson), 145–177. American Society of Agronomy Monographs No. 45.

McFarland, J.B. and Mitchell, R.B. (2000). Fire effects on weeping lovegrass tiller density and demographics. *Agron. J.* 92: 42–47.

Mitchell, R.B., Masters, R.A., Waller, S.S. et al. (1994). Big bluestem production and forage quality responses to burning date and fertilizer in tallgrass prairies. *J. Prod. Agric.* 7: 355–359.

Mitchell, R.B., Masters, R.A., Waller, S.S. et al. (1996). Tallgrass prairie vegetation response to spring burning dates, fertilizer, and atrazine. *J. Range Manage.* 49: 131–136.

Monaco, T.J., Weller, S.C., and Ashton, F.M. (2002). *Weed Science – Principles and Practices*. New York, NY: Wiley.

Nelson, J.A. and Lym, R.G. (2003). Interactive effects of *Aphthona nigriscutis* picloram plus 2,4-D in leafy spurge (*Euphorbia esula*). *Weed Sci.* 51: 118–124.

Nissen, S.J., Masters, R.A., Lee, D.J., and Rowe, M.L. (1995). DNA-based marker systems to determine genetic diversity of weedy species and their application to biocontrol. *Weed Sci.* 43: 504–513.

Noss, R.F. (1991). From endangered species to biodiversity. In: *Balancing on the Brink of Extinction: The Endangered Species Act and Lessons from the Future* (ed. K. Kolm), 227–246. Washington, DC: Island Press.

Paterson, I.D., Downie, D.A., and Hill, M.P. (2009). Using molecular methods to determine the origin of weed populations of *Pereskia aculeata* in South Africa and its relevance to biological control. *Biol. Control* 48: 84–91.

Paynter, Q. and Flanagan, G.J. (2004). Integrating herbicide and mechanical control treatments with fire and biological control to manage an invasive wetland shrub, *Mimosa pigra*. *J. Appl. Ecol.* 41: 615–629.

Pemberton, R.W. (2000). Predictable risk to native plants in weed biological control. *Oecologia* 125: 489–494.

Pickett, S.T.A., Collins, S.L., and Armesto, J.J. (1987). Models, mechanisms and pathways of succession. *Bot. Rev.* 53: 335–371.

Pickett, S., Cadenasso, M.L., and Meiners, S.J. (2009). Ever since Clements: from succession to vegetation dynamics and understanding to intervention. *Appl. Veg. Sci.* 12: 9–21.

Pimentel, D. (1982). Perspectives of integrated pest management. *Crop Prot.* 1: 5–26.

Pimentel, D., Zuniga, R., and Morrison, D. (2005). Update on the environmental and economic costs associated with alien-invasive species in the United States. *Ecol. Econ.* 52: 273–288.

Popay, I. and Field, R. (1996). Grazing animals as weed control agents. *Weed Technol.* 10 (1): 217–231.

Pyne, S.J. (1984). *Introduction to Wildland Fire*. New York, NY: Wiley.

Quimby, P.C., Bruckart, W.L., DeLoach, C.J. et al. (1991). Biological control of rangeland weeds. In: *Noxious Range Weeds* (eds. L.F. James, J.O. Evans, M.H. Ralphs and R.D. Child), 84–102. Boulder, CO.: Westview Press.

Randall, J.M. (1997). Defining weeds of natural areas. In: *Assessment and Management of Plant Invasions* (eds. J.O. Luken and J.W. Thieret), 18–25. New York, NY: Springer.

Rejmanek, M. and Pitcairn, M.J. (2002). When is eradication of exotic pest plants a realistic goal? In: *Turning the Tide: The Eradication of Invasive Species* (eds. C.R. Veitch and M.N. Clout), 249–253. IUCN SSC Invasive Species Specialist.

Renne, I.J. and Tracy, B.F. (2013). Disturbance intensity, timing and history interact to affect pasture weed invasion. *Basic Appl. Ecol.* 14: 44–53.

Renz, M. and DiTomaso, J. (1999). Biology and control of perennial pepperweed. *Proc. California Weed Sci. Soc.* 51: 13–16.

Ross, M.A. and Lembi, C.A. (1999). *Applied Weed Science*. Upper Saddle River, NJ: Prentice Hall.

Roush, R.T. (1990). Genetic variation in natural enemies: Critical issues for colonization in biological control. In: *Critical Issues in Biological Control* (eds. M. Mackauer, L.E. Ehler and J. Roland), 263–287. Andover, England: Intercept Ltd.

Sanderson, M.A., Goslee, S.C., Soder, K.J. et al. (2007). Plant species diversity, ecosystem function, and pasture management – a perspective. *Can. J. Plant. Sci.* 87: 479–487.

Sather, B.C., Kallenbach, R.L., Sexten, W.J., and Bradley, K.W. (2013). Evaluation of cattle grazing distribution in response to weed and legume removal in mixed tall fescue (*Schedonorus phoenix*) and legume pastures. *Weed Technol.* 27: 101–107.

Schuster, M.Z., Pelissari, A., de Moraes, A. et al. (2016). Grazing intensities affect weed seedling emergence and the seed bank in an integrated crop-livestock system. *Agri. Ecosys. Environ.* 232: 232–239.

Schweingruber, F.H., Börner, A., and Schulze, E.D. (2007). *Atlas of Woody Plant Stems: Evolution, Structure, and Environmental Modifications*. Springer Science & Business Media.

Scifres, C.J. (1980). *Brush Management Principles and Practices for Texas and the Southwest*. College Station, TX: Texas A&M Univ. Press.

Scifres, C.J., Scifres, J.R., and Kothmann, M.M. (1983). Differential grazing use of herbicide-treated areas by cattle. *J. Range Manage.* 36: 65–69.

Sebastian, D.J., Nissen, S.J., and De Souza Rodrigues, J. (2016a). Pre-emergence control of six invasive winter annual grasses with imazapic and indaziflam. *Invasive Plant Sci. Manage.* 9: 308–316.

Sebastian, D.J., Sebastian, J.R., Nissen, S.J., and Beck, K.G. (2016b). A potential new herbicide for invasive annual grass control on rangeland. *Rangeland Ecol. Manage.* 69: 195–198.

Sebastian, D.J., Fleming, M.B., Patterson, E.L. et al. (2017). Indaziflam: a new cellulose biosynthesis inhibiting herbicide provides long-term control of invasive winter annual grasses. *Pest Manage. Sci.* 73: 2149–2162.

Sheaffer, C.C., Wyse, D.L., Marten, G.C., and Westra, P.H. (1990). The potential of quackgrass for forage production. *J. Prod. Agric.* 3: 256–259.

Sheppard, A.W. (1992). Predicting biological weed control. *Trends Ecol. Evol.* 7: 290–296.

Singer, J.W. and Moore, K.J. (2003). Nitrogen removal by orchardgrass and smooth bromegrass and residual soil nitrate. *Crop Sci.* 43: 1420–1426.

Sleugh, B.B., Moore, K.J., Brummer, E.C. et al. (2001). Forage nutritive value of various amaranth species at different harvest dates. *Crop Sci.* 41: 466–472.

Smith, A.E. (1991). The potential importance of allelopathy in the pasture ecosystem, a review. *Adv. Agron.* 1: 27–37.

Strong, D.R. and Pemberton, R.W. (2000). Biological control of invading species: risk and reform. *Science* 288: 1969–1970.

Sutherland, S. (2004). What makes a weed a weed: life history traits of native and exotic plants in the USA. *Oecologia* 141: 24–39.

Swanton, C.J., Mahoney, K.J., Chandler, K., and Gulden, R.H. (2008). Integrated weed management: knowledge-based weed management systems. *Weed Sci.* 56: 168–172.

Task Group on Unity in Concepts and Terminology (1995). New concepts for assessment of rangeland condition. *J. Range Manage.* 48: 271–282.

Thill, D.C., Lish, J.M., Callihan, R.H., and Bechinski, E.J. (1991). Integrated weed management - a component of integrated pest management: a critical review. *Weed Technol.* 5: 648–656.

Tracy, B.F. and Sanderson, M.A. (2004). Forage productivity, species evenness and weed invasion in pasture communities. *Agric. Ecosyst. Environ.* 102: 175–183.

Turner, C.E., Pemberton, R.W., and Rosenthal, S.S. (1987). Host utilization of native *Cirsium* thistles (Asteraceae) by the introduced weevil *Rhinocyllus conicus* (Coleoptera: Curculionidae) in California. *Environ. Entomol.* 16: 111–115.

U.S. Congress, Office of Technology Assessment (1993). *Harmful Non-Indigenous Species in the United States*. Washington, D.C.: U.S. Government Printing Office.

Vallentine, J.F. (1989). *Range Development and Improvements*, 3e. San Diego, CA: Academic Press.

Vallentine, J.F. (2000). *Grazing Management*. Elsevier.

Van Wilgen, B.W., Moran, V.C., and Hoffmann, J.H. (2013). Some perspectives on the risks and benefits of biological control of invasive alien plants in the management of natural ecosystems. *Environ. Manage.* 52: 531–540.

Vavilov, N.I. (1992). *Origin and Geography of Cultivated Plants*. New York, NY: Cambridge Press.

Walker, B.H. (1992). Biological and ecological redundancy. *Conserv. Biol.* 6: 18–23.

Walker, J.W. (1994). Multi-species grazing: the ecological advantage. *Sheep Res. J.* Special Issue: 52–64.

Walker, R.H. and Buchanan, G.A. (1982). Crop manipulation in integrated weed management systems. *Weed Sci.* 30: 17–24.

Washburn, B.E. and Barnes, R.G. (2000a). Native warm-season grass and forb establishment using imazapic and 2,4-D. *Native Plants J.* 1: 61–68.

Washburn, B.E. and Barnes, T.G. (2000b). Postemergence tall fescue (*Festuca arundinacea*) control at different growth stages with glyphosate and AC 263,222. *Weed Technol.* 14: 223–230.

Washburn, B.E., Barnes, T.G., and Sole, J.D. (1999). No-till establishment of native warm-season grasses in tall fescue fields. *Ecol. Restor.* 17: 144–147.

Westoby, M., Walker, B., and Noy-Meir, I. (1989). Opportunistic management for rangelands not at equilibrium. *J. Range Manage.* 42: 266–274.

White, P.S. and Jentsch, A. (2001). The search for generality in studies of disturbance and ecosystem dynamics. In: *Progress in Botany* (ed. J.W.K.), 399–450. Springer Berlin Heidelberg.

Whitson, T.S. and Koch, D.W. (1998). Control of downy brome (*Bromus tectorum*) with herbicides and perennial grass competition. *Weed Technol.* 12: 391–396.

Williams, B.K. (2011). Adaptive management of natural resources - framework and issues. *J. Environ. Manage.* 92: 1346–1353.

Williamson, J.A., Aiken, G.E., Flynn, E.S., and Barrett, M. (2016). Animal and pasture responses to grazing management of chemically suppressed tall fescue in mixed pastures. *Crop Sci.* 56: 2861–2869.

Wilson, R.G. and Kachman, S.D. (1999). Effect of perennial grasses on Canada thistle (*Cirsium arvense*) control. *Weed Technol.* 13: 83–87.

Winston, R.L., Schwarzländer, M., Hinz, H.L. et al. (2014). *Biological Control of Weeds: A World Catalogue of Agents and their Target Weeds*, 5e, 838. Morgantown, WV, USA: USDA Forest Service, Forest Health Technology Enterprise Team, FHTET-2014-04.

Young, J.A., Evans, R.A., and Eckert, R.E. (1969). Wheatgrass establishment with tillage and herbicides in a Mesic medusahead community. *J. Range Manage.* 22: 151–155.

Zemenchik, R.A. and Albrecht, K.A. (2002). Nitrogen use efficiency and apparent nitrogen recovery of Kentucky bluegrass, smooth bromegrass, and orchardgrass. *Agron. J.* 94: 421–428.

Zwolfer, H., Ghani, M.A., and Rao, V.P. (1976). Foreign exploration and importation of natural enemies. In: *The Theory and Practice of Biological Control* (eds. C.B. Huffaker and P.S. Messenger), 189–207. New York, N.Y: Academic Press.

CHAPTER

29

Insect Management

R. Mark Sulc, Professor, *Horticulture and Crop Science, The Ohio State University, Columbus, OH, USA*
William O. Lamp, Professor, *Entomology, University of Maryland, College Park, MD, USA*
G. David Buntin, Professor, *Entomology, University of Georgia, Griffin, GA, USA*

Introduction

Biodiversity of insects is increasingly recognized as a vital component of natural functions on farms, many of which contribute to stability and resilience for crop production (Scherr and McNeely 2008). Biodiversity is not just the sum of all species, but also the variety of life at all its levels, from genes to ecosystems, as well as the ecologic and evolutionary processes that sustain it. The biodiversity concept places emphasis on important ecologic processes, such as plant productivity, nutrient cycling, and decomposition of debris, all of which have value to humans. Such processes of value are called ecosystem services and, are provided by the complete variety of life on farms. Insects and related invertebrates play major roles in providing ecosystem services on farms, especially as natural enemies of pests (biologic control), pollinators of economic plants, and decomposition of plant and animal debris for soil-building. Of course, they also provide negative services as pests of crops and animals, which is the focus of this chapter.

Forages provide favorable habitat and food sources for a wide array of insect species. Estimates of loss in forage dollar value due to insect pests range from 5–10% yr^{-1}, although outbreak infestations can cause losses of 40–75% and occasionally total loss of yield. Insect pests are so varied in their life cycles and habits that feeding injury caused by one or more species to a particular forage species may occur any time throughout the year. In this chapter, we discuss the concept of ecologic intensification using insects in forage systems, insect pest damage in forages, and the principles of insect pest management in forages, focusing on concepts that can be applied across a wide range of pests and forage crops, and including the potential implications of climate change to insect pest management.

Ecologic Intensification Using Insects in Forage Systems

In recent years, ecologic intensification has been suggested as a means to better meet our growing agricultural needs while reducing inputs and enhancing ecosystem services of value to crop and pest management (Gaba et al. 2014). Intensification is accomplished by supporting natural ecosystem processes to the benefit of agriculture. For example, producers can aid the development of species that suppress pests (Bommarco et al. 2013). The planting and management of forage crops on farms may be used to support ecologic intensification not only in the forage crop, but also across the farm as a whole.

Alfalfa provides an appropriate example for the on-farm benefits of biodiversity. Entomologists have long

Forages: The Science of Grassland Agriculture, Volume II, Seventh Edition.
Edited by Kenneth J. Moore, Michael Collins, C. Jerry Nelson and Daren D. Redfearn.

noted the large number of arthropod species (insects, spiders, and related taxa) in alfalfa. For example, a survey conducted by Pimentel and Wheeler (1973) of the arthropods found in a New York State alfalfa community documented 591 species. Of these species, only 21 (3.6%) are considered pests, and of those, only 4 (<1%) are key pests requiring monitoring. Herbivores comprised 53% of the community while natural enemies comprised 47%. Most of the herbivores that were found were incidental (i.e. not feeding on alfalfa) or secondary (i.e. likely suppressed by natural enemies). Other herbivores included 26 species of bees, as well as many other valuable pollinators such as butterflies, true flies, and beetles. Natural enemies included predators, such as species like dragonflies that readily move between crops foraging for flying insects, and parasitoids, such as species that kill aphids and caterpillars both in and out of the alfalfa field. In addition to providing food resources for natural enemies, alfalfa can serve as a reproductive habitat for beneficial species like predatory lady beetles, lacewings, and ground beetles. Consequently, management of pest insects requires care to minimize negative impact on the insect community as a whole, thus requiring the approach of **Integrated Pest Management** (IPM).

Management Concept for Insect Pests

Insects migrate into forages from adjacent areas, or are introduced accidentally, and may or may not interfere with forage production goals. Producers and advisors tend to consider insect pests only at the specific times when they interfere with production goals. This has led to reliance on insecticides, which usually provide quick and effective short-term suppression of insect infestations. Reliance on any one method of suppression can lead to secondary pest outbreaks, pest resurgence, evolution of pest resistance, and exposure to health and environmental risks associated with pesticide use.

IPM involves all components of decision making and planning of the overall forage crop production system. Smith and van den Bosch (1967) initially used the term IPM, and their description of the concept remains one of the best. Recognizing that the losses associated with insect pests are the result of the interaction between the pest, the crop, and the environment (the pest triangle; Fig. 13.1 in Fick et al. 2003), IPM can be illustrated as the processes associated with a series of three pest triangles (Figure 29.1). At the top, before a cropping season begins, the producer expects losses from major insect pests (key pests, as described below). Through pre-season planning, specific measures may be taken (e.g. selection of a resistant variety) to prevent the outbreak of key pests, thus modifying the expectation of loss. As the crop develops, specific environmental or crop conditions or unusual patterns of pest populations may result in the unexpected outbreak of a pest. The pest density at which the expected losses from the pest equal the costs associated with a specific management action, is called the economic injury level (EIL). To prevent pest populations from exceeding the EIL, action thresholds (AT; also called economic thresholds, ET) have been developed to define, in the development of a pest population, the point at which the management action should be taken. However, in IPM, the use of pesticides is limited to emergency situations, and the emphasis is on prevention of significant losses from pests. The realized losses at the end of the season are an important source of information during post-season evaluation, thus affecting the planning and expected losses for the next season.

Crop damage depends on complex interactions involving the insect pest and the crop, which in turn are influenced by biotic agents, abiotic factors, and crop management practices. An obvious example of a management decision affecting the potential for insect damage is planting an insect-resistant/tolerant variety vs a susceptible variety. Fertilization, harvest, and grazing management, crop rotation, and other management practices influence the potential impact of pests in forage stands. In turn, insect management, or lack thereof, influences crop production decisions, often in subtle ways. For example, failing to manage insect pests or providing a late rescue treatment of outbreak infestations can result in delayed crop development, reduced yield, and/or reduced stand persistence thus influencing harvest timing and possibly encroachment of weeds due to reduced crop vigor. Thus, insect pest management requires the balancing of multiple objectives (e.g. yield, environmental quality, and economics).

Arthropod Pests of Forages and Plant Injury

Pests in Forages

Table 29.1 lists major species of insects and their scientific names that are pests of forage crops in North America. Out of the many thousands of insect species that inhabit forages, only 77 species are considered pests, that is, species having the potential to cause economic losses. Of those 77 species, only five are considered key pests. On alfalfa, the potato leafhopper, alfalfa weevil, spotted alfalfa aphid, and alfalfa snout beetle are pest problems nearly every year over large areas and are the main focus of preventive management practices (Lamp et al. 1991; Metcalf and Metcalf 1993; Shields et al. 1999; Undersander et al. 2004). Recently the bermudagrass stem maggot has become a key pest of bermudagrass in the southern US. The remaining 72 species occasionally or rarely present a significant pest problem, thus managers must scout fields for outbreaks and respond as needed.

Host Specificity of Forage Pests

Though a few insect pest species inhabit only certain forage crops, most pest species are generalists and are found within at least several crop species in either the legume or grass groups. Nearly half (48%) of the insect species in

Table 29.1 List of insect species known as pests of forage crops in North America

Group	Common name	Order: Family	Species	Crops attacked	State or region[a]	Type of injury	Pest Status[b]
Ants	Red harvester ant	Hymenoptera: Formicidae	*Pogonomyrmex barbatus*	All	SC, SW	Defoliation, seed-feeding	O
	Red imported fire ant	Hymenoptera: Formicidae	*Solenopsis invicta*	All	SE, SC, CA	Human and animal sting	O
Aphids	Spotted alfalfa aphid	Homoptera: Aphididae	*Therioaphis maculata*	Legumes	US	Sap feeding of phloem	K
	Pea aphid	Homoptera: Aphididae	*Acyrthosiphon pisum*	Legumes	US	Sap feeding of phloem	O
	Blue alfalfa aphid	Homoptera: Aphididae	*Acyrthosiphon kondoi*	Legumes	SE, SC, SW, CA	Sap feeding of phloem	O
	Cowpea aphid	Homoptera: Aphididae	*Aphis craccivora*	Legumes	NE, SE, NC, SC, SW, CA	Sap feeding of phloem	R
	Greenbug	Homoptera: Aphididae	*Schizaphis graminum*	Grasses	US	Sap feeding of phloem	O
Blister beetles	Striped blister beetle	Coleoptera: Meloidae	*Epicauta vittata*	All	US	Toxic to livestock	R
	Black blister beetle	Coleoptera: Meloidae	*Epicauta pennsylvanica*	All	US	Toxic to livestock	R
Caterpillars	Alfalfa webworm	Lepidoptera: Crambidae	*Loxostege cereralis*	Legumes	US	Defoliation	R
	Garden webworm	Lepidoptera: Crambidae	*Achyra rantalis*	Legumes	US	Defoliation	O
	Yellowstriped armyworm	Lepidoptera: Noctuidae	*Spodoptera ornithogalli*	Legumes	US	Defoliation – new seedings	O
	Western yellowstriped armyworm	Lepidoptera: Noctuidae	*Spodoptera praefica*	Legumes	NW, SW, CA	Defoliation – new seedings	O
	Army cutworm	Lepidoptera: Noctuidae	*Euxoa auxiliaris*	All	US	Defoliation	R
	Variegated cutworm	Lepidoptera: Noctuidae	*Peridroma saucia*	All	US	Defoliation	O
	Claybacked cutworm	Lepidoptera: Noctuidae	*Agrotis gladiaria*	All	NE, SE, NC, SC	Defoliation	R
	Dingy cutworm	Lepidoptera: Noctuidae	*Feltia ducens*	All	US	Defoliation	R
	Bronzed cutworm	Lepidoptera: Noctuidae	*Nephelodes minians*	Grasses	NE, NC, NW	Defoliation	R
	Green cloverworm	Lepidoptera: Erebidae	*Hypena scabra*	Legumes	NE, SE, NC, SC	Defoliation	R
	True armyworm	Lepidoptera: Noctuidae	*Mythimna unipunctata*	All	NE, SE, NC, SC	Defoliation	R
	Alfalfa looper	Lepidoptera: Noctuidae	*Autographa californica*	Legumes	US	Defoliation	R
	Fall armyworm	Lepidoptera: Noctuidae	*Spodoptera frugiperda*	All	NE, SE, NC, SC, SW, CA	Defoliation – new seedings	O, K
	Beet armyworm	Lepidoptera: Noctuidae	*Spodoptera exigua*	All	SE, SC, SW, CA, NW	Defoliation – new seedings	O
	Alfalfa caterpillar	Lepidoptera: Pieridae	*Colias eurytheme*	Legumes	US	Defoliation	O
	Clover hayworm	Lepidoptera: Pyralidae	*Hypsopygia costalis*	Legumes	US	Defoliation	R
	Range caterpillar	Lepidoptera: Saturniidae	*Hemileuca oliviae*	Grasses	SW	Defoliation	O

Table 29.1 (*continued*)

Group	Common name	Order: Family	Species	Crops attacked	State or region[a]	Type of injury	Pest Status[b]
	Clover head caterpillar	Lepidoptera: Tortricidae	*Grapholita interstinctana*	Clovers	NE, SE, NC, SC	Defoliation	R
	Striped grass looper	Lepidoptera Erebidae	*Mocis latipes*	Grasses	SE, SC, NE, NC	Defoliation	R
Chinch bugs	Common chinch bug	Heteroptera: Lygaeidae	*Blissus leucopterus leucopterus*	Grasses	US	Sap feeding of crowns and stems	O
	Hairy chinch bug	Heteroptera: Lygaeidae	*Blissus leucopterus hirtus*	Grasses	NE	Sap feeding of crowns and stems	O
	Southern chinch bug	Heteroptera: Lygaeidae	*Blissus insularis*	Grasses	SE, SC	Sap feeding of crowns and stems	O
Crickets	Striped ground cricket	Orthoptera: Gryllidae	*Allonemobius fasciatus*	All	NE, SE, NC	Defoliation	R
	Mormon cricket	Orthoptera: Tettigonidae	*Anabrus simplex*	Legumes	NW, SW	Defoliation	O
	Tawny mole cricket / Southern mole cricket	Orthoptera: Gryllotalpidae	*Scapteriscus abbreviatus*	Grasses	SE	Defoliation, root feeding	O
Grasshoppers	Differential grasshopper	Orthoptera: Acrididae	*Melanoplus differentialis*	All	NE, NC, SC, SW	Defoliation	O
	Redlegged grasshopper	Orthoptera: Acrididae	*Melanoplus femurrubrum*	All	US	Defoliation	R
	Migratory grasshopper	Orthoptera: Acrididae	*Melanoplus sanguinipes*	All	US	Defoliation	O
	Twostriped grasshopper	Orthoptera: Acrididae	*Melanoplus bivittatus*	All	US	Defoliation	O
Leafhoppers	Potato leafhopper	Homoptera: Cicadellidae	*Empoasca fabae*	Legumes	NE, SE, NC	Sap feeding of stems	K
Leafminers	Alfalfa blotch leafminer	Diptera: Agromyzidae	*Agromyza frontella*	Alfalfa	NE, NC	Leaf mining	R
Mites	Cereal rust mite / Timothy mite	Acari: Eriohyoidea	*Abacarus hystrix*	Grasses	NE, NC	Sap feeding of mesophyll	O
	Clover mite	Acari: Tetranychidae	*Bryobia praetiosa*	Legumes	US	Sap feeding of mesophyll	O
	Winter grain mite	Acarina: Penthaleidae	*Penthaleus major*	Grasses	SE, SC, NE	Sap feeding of mesophyll	O
Plant bugs	Tarnished plant bug	Heteroptera: Miridae	*Lygus lineolaris*	Legumes	US	Sap feeding, especially of flowers	O
	Alfalfa plant bug	Heteroptera: Miridae	*Adelphocoris lineolatus*	Legumes	NE, NC	Sap feeding, especially of flowers	O
	Western tarnished plant bug	Heteroptera: Miridae	*Lygus hesperus*	Legumes	SW, CA, NW	Sap feeding, especially of flowers	O
	Black grass bug	Heteroptera: Miridae	*Labops hesperius*	Grasses	NC, SW, CA, NW	Sap feeding of mesophyll	O

	Meadow plant bug	Heteroptera: Miridae	*Leptopterna dolabrata*	Grasses	NE, SE, NC, NW	Sap feeding of mesophyll	R
	Black grass bug	Heteroptera: Miridae	*Irbisia pacifica*	Grasses	CA, NW	Sap feeding of mesophyll	O
	Garden fleahopper	Heteroptera: Miridae	*Halticus bractatus*	Legumes	NE, SE, NC, SC, SW, CA	Sap feeding of mesophyll	R
Seed chalcids	Clover seed chalcid	Hymenoptera: Eurytomidae	*Bruchophagus gibbus*	Clovers	US	Seed feeding	O
	Alfalfa seed chalcid	Hymenoptera: Eurytomidae	*Bruchophagus roddi*	Alfalfa	US	Seed feeding	O
	Trefoil seed chalcid	Hymenoptera: Eurytomidae	*Bruchophagus kolobovae*	Trefoil	US	Seed feeding	O
Spittlebugs	Meadow spittlebug	Homoptera: Cercopidae	*Philaenus spumarius*	Legumes	NE, NC, NW, CA	Sap feeding of xylem	O
	Two-lined spittlebug	Homoptera: Cercopidae	*Prosapia bicincta*	All	SE, SC	Sap feeding of xylem	O
Stem borers	Clover stem borer	Coleoptera: Languriidae	*Languria mozardi*	Legumes	US	Stem boring	R
	Southwestern corn borer	Lepidoptera: Crambidae	*Diatraea grandiosella*	Grasses	SE, SC, SW	Stem boring	R
	Sugarcane borer	Lepidoptera: Crambidae	*Diatraea saccarhalis*	Grasses	SE, SC, FL	Stem boring	R
	European corn borer	Lepidoptera: Crambidae	*Ostrinia nubilalis*	Grasses	NE, SE, NC, SC	Stem boring	R
	Lesser cornstalk borer	Lepidoptera: Pyralidae	*Elasmopalpus lignosellus*	Grasses	SE, SC, SW, CA	Stem boring	R
	Frit fly	Diptera: Chloropidae	*Oscinella* spp.	Grasses	SE, SC, NE, NC	Stem boring	R
	Bermudagrass stem maggot	Diptera: Muscidae	*Atherigona reversura*	Grasses	SE, SC	Stem borer	K
	Bahiagrass borer	Coleoptera: Cerambycidae	*Derobrachus brevicollis*	Grasses	SE, SC	Stem borer	R
Stink bugs	Conchuela	Heteroptera: Pentatomidae	*Chlorochroa ligata*	All	SW, CA, NW	Sap feeding of stems and flowers	R
	Brown stink bug	Heteroptera: Pentatomidae	*Euschistus servus*	All	US	Sap feeding of stems and flowers	R
	Say stink bug	Heteroptera: Pentatomidae	*Chlorochroa sayi*	All	SW, CA, NW	Sap feeding of stems and flowers	R
	Redshouldered stink bug	Heteroptera: Pentatomidae	*Thyanta accerra*	All	NE, SE, NC, SC, SW	Sap feeding of stems and flowers	R
Treehoppers	Three-cornered alfalfa hopper	Homoptera: Membracidae	*Spissistilus festinus*	Legumes	SE, SC, SW, CA	Sap feeding of phloem, girdling	O

Table 29.1 (*continued*)

Group	Common name	Order: Family	Species	Crops attacked	State or region[a]	Type of injury	Pest Status[b]
Weevils	Lesser clover leaf weevil	Coleoptera: Curculionidae	*Hypera nigrirostris*	Clovers	US	Defoliation	R
	Clover leaf weevil	Coleoptera: Curculionidae	*Hypera punctata*	Legumes	US	Defoliation	O
	Alfalfa weevil	Coleoptera: Curculionidae	*Hypera postica*	Alfalfa	US	Defoliation	K
	Clover root curculio	Coleoptera: Curculionidae	*Sitona hispidulus*	Legumes	US	Root feeding	O
	Alfalfa snout beetle	Coleoptera: Curculionidae	*Otiorhynchus ligustici*	Alfalfa	NE	Root feeding	K
	Clover seed weevil	Coleoptera: Curculionidae	*Tychius picirostris*	Clovers	SE, SC, NW	Defoliation/seed feeding	O
	Clover root borer	Coleoptera: Curculionidae	*Hylastinus obscurus*	Legumes	NE, NC, NW	Root feeding	O
White grubs	May or June beetles	Coleoptera: Scarabaeidae	*Phyllophaga spp*	Grasses	US	Root feeding	O
	Chafer beetles	Coleoptera: Scarabaeidae	*Cyclocephala spp*	Grasses	US	Root feeding	O
	Green June beetle	Coleoptera: Scarabaeidae	*Cotinis nitida*	Grasses	NE, SE, NC, SC	Root feeding	O

[a]Region names include US, continental United States, southern Canada, and northern Mexico; NE, northeastern states and eastern provinces; SE, southeastern states; NC, north central states and central provinces; SC, south central states; SW, southwestern states and northern Mexico; CA, California; NW, northwestern states and western provinces; FL, Florida.

[b]Pest status: K, key; O, occasional; R, rare.

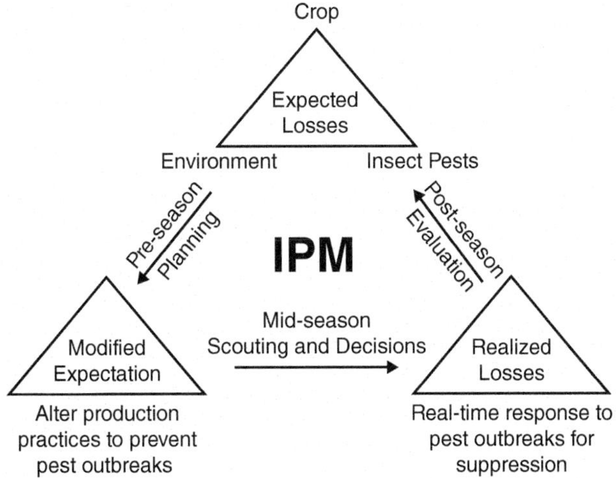

Fig. 29.1. A series of pest triangles to illustrate the concept of IPM.

Table 29.1 are found only in legume crops, 20% only in grass crops, and 32% in both. By far, the most serious pest problems in forage crops occur in alfalfa, though many of the species can occur at densities below the economic threshold in other forages.

Pest Distributions

Many (~40%) of the pest species are distributed over most, if not all, of North America (Table 29.1). Some are limited to specific but sometimes broad regions while a few have very localized distributions. Thus, the potential insect pest problems for any given area are more limited than suggested by the list of 77 species. Not given in Table 29.1 is whether the species is native or exotic. Exotic insect species that have been accidentally introduced into North America include the alfalfa weevil, all the legume-feeding aphids, and many of the caterpillars. In the absence of their natural enemies, many exotic species were initially key pests, and some remain major pests in some regions. In some cases, however, introductions of natural enemies have resulted in suppression of pest densities through biologic control of species such as the alfalfa weevil that has generally lower densities now than in the 1960s and 1970s (Ratcliffe and Flanders 1998). Likewise, introduction of entopathogenic nematodes has reduced the pest status of southern mole cricket in bermudagrass hay fields, though the insect remains a damaging pest of turf where the threshold for damage is much lower. Some important pests are native, such as potato leafhopper, but alfalfa was introduced, so the combination of the forage crop and the pest is new. Consequently, the crop is not adapted to be naturally resistant like native plants that have evolved with the pest. This suggests that breeding programs may enhance **resistance** of alfalfa to such pests, as the case with use of glandular-haired alfalfa for potato leafhopper resistance (Sulc et al. 2014, 2015).

Types of Pests and Plant Injury

Several insect species are pests, not because of their feeding on the plant but rather, because of their direct effect on livestock (Williams 2010). For example, fire ants sting livestock and blister beetles may be toxic to horses when incorporated within the harvested forage. Yet, most of the insect pests cause plant injury directly through their feeding behavior. Pest species injure forage crops through defoliation, stem boring, sap feeding, root feeding, or seed feeding (Table 29.1).

Defoliators

These pests, including all caterpillars and grasshoppers, preferentially consume leaves, and thus directly reduce both forage yield and quality. Some species, such as armyworms and alfalfa weevil especially, feed on new growth, causing delayed growth and morphologic development or death of young plants. Because defoliation reduces leaf area and the ability of plants to fix carbon by photosynthesis, the amount of root reserve storage is reduced and plant regrowth potential in subsequent growth cycles may be reduced (Hutchins et al. 1990).

Stem Borers

Stem boring pests are mostly the larval stage of flies or moths. Stem bores may kill the apical meristem causing reduced growth and excessive branching of infested stems. The bermudagrass stem maggot kills the apical whorl thereby preventing stem elongation and reducing yield. A study by Bentley and Clements (1989) found that suppression of the stem-boring frit fly (*Oscinella* spp.) increased the number of tillers in italian ryegrass during stand establishment.

Sap Feeders

These pests can feed on the contents in xylem (e.g. spittlebugs), phloem (e.g. aphids), or mesophyll cells within leaves (e.g. black grass bugs). Each has its own impact on the plant physiology. In addition, most sap-feeding pests found in forage crops inject substances in their saliva that cause additional physiologic disruptions to the plant (Manglitz and Ratcliffe 1988). For example, potato leafhopper saliva, along with the mechanical probing by their mouthparts, causes changes in the structure of the phloem and xylem within stems, leading to malfunction of the xylem and reduced flow of water from the roots to the foliage (Ecale and Backus 1995). Thus, alfalfa plants injured by potato leafhoppers appear shorter and have reduced rates of photosynthesis (Lamp et al. 2004). In addition, the blocked phloem reduces the transport of sugars and other important chemicals to the roots, thus affecting regrowth and plant survival (Lamp et al. 2001).

Distortion of leaves and stems is a typical symptom of certain sap feeders, such as blue alfalfa aphid or alfalfa plant bug, and may be used to diagnose the cause of injury. Mesophyll feeders, such as winter grain mite and timothy mite, remove chlorophyll from individual cells within leaves, causing a stippled or speckled appearance on leaves.

Root Feeders

Root feeders are easily overlooked because of their belowground activity. Yet, they can cause significant injury to the root tissue and function, as well as provide points of entry for pathogenic microorganisms to enter root tissue, as occurs with clover root curculio larvae feeding on alfalfa roots (Leath and Hower 1993; Hower et al. 1995). Scarab grubs feed on roots of most forage crops and can cause chronic injury but sometimes cause severe injury resulting in stand loss. Because of the difficulty in sampling these pests, and because management options are limited to the insect life stages that occur above the ground, few (or "no" if this is more accurate) practical options currently exist to manage these pests. Host plant resistance is currently the best potential option for suppressing root-feeding insects.

Flower and Seed Feeders

Some pests are especially important when growing forage crops for seed production (Metcalf and Metcalf 1993). Insect pest control is especially difficult when seed production depends on insect pollinators. Some species feed directly on seeds, such as seed chalcids, and are managed by specific cultural control practices. For example, seed chalcid abundance can be reduced by burying or composting chaff and light seed at harvest. Others, such as the plant bugs and stink bugs, feed on flowers and thus impact seed production indirectly. None of this feeding activity is of practical importance in production of forage seed.

Livestock Pests

There are numerous arthropod pests of livestock. Livestock pests do not usually directly injury pasture plants. Most, such as lice, mites, ticks, and parasitic flies, spend their entire life cycle in close association with the host. But some are pests of livestock during the adult stage after developing in pastures as immature stages. Several biting flies, such as the horn fly, stable fly, and horse flies, feed on livestock blood as adults. Larvae of horn flies develop in manure and stable fly and horse fly larvae develop in decaying plant matter. In this way, they aid in nutrient cycling within a pasture. Many dung-feeding scarab beetles also occur in pastoral systems where they feed on manure and aid in nutrient and carbon cycling (Kaufman and Woods 2011). Insecticide use for plant and livestock pests can be very detrimental to these beneficial insects. Imported fire ants are also a pest because they can sting young and newborn animals and people during hay harvest. Their mounds prevent forage growth and can also damage hay harvesting equipment. Furthermore, the imported fire ant is a quarantine pest in the US, which may interfere with the movement of hay to non-quarantine areas. (USDA-ARS 2017).

Components of Insect Damage

Insect pests cause plant injury, as described above, that in turn leads to "damage," defined as the economic loss caused by insect activity (Pedigo et al. 1986). Damage to forage crops reduces forage yield and/or quality.

Impact on Forage Yield

Insect impact on forage yield can be described using a yield component analysis (Volenec et al. 1987), where

$$\begin{aligned}\text{Yield per unit area} &= \text{plants per unit area}\\ &\quad \times \text{shoots per plant}\\ &\quad \times \text{mass per shoot}\end{aligned}$$

These yield components help describe the immediate, short-term effects of insect damage within a growth cycle vs long-term effects on future productivity. Reduced mass per shoot is usually the most immediate result of insect injury (e.g. alfalfa weevil, potato leafhopper, caterpillars, grasshopper feeding). In some cases, shoots per plant can be reduced relatively quickly but, more often, this component is affected over a longer period as the physiologic status of the injured plant deteriorates. Such is the case with leafhoppers, aphids, and root feeders.

Plant density per unit area can be affected in a short time under severe infestations of certain insect species, such as alfalfa snout beetle in alfalfa and scarab grubs in grass pastures. However, this component usually reflects cumulative insect injury over a long period, especially when combined with other biotic and abiotic stressors (see Complex Interactions below).

The impact on future productivity is important because most forage crops are perennials which are harvested multiple times and over several growing seasons. The long-term impact of insect feeding is very difficult to assess and is influenced by many factors. Thus, most assessments of insect-induced damage in forage crops have focused on yield loss over a relatively short period of time, providing the sole basis for most EIL and action thresholds.

Stand Establishment Phase

Yield reductions through reduced shoots per plant and shoot mass may result from insect feeding during the establishment phase, but reduction in plant density is potentially more detrimental, especially for forage species lacking the ability to spread vegetatively (e.g. alfalfa, red clover). Several insect species can injure or even kill legume seedlings, for example, aphids, armyworms, grasshoppers, crickets, and even potato leafhoppers (Flinn and Hower 1984; Rogers et al. 1985; Manglitz and Ratcliffe 1988). Stand-establishment failures of legumes are often attributed to other causes when, in fact, insects and slugs are partly or solely responsible, especially in conservation tillage plantings (Byers et al. 1983). Little information is available on insect injury to forage grasses during establishment, although infestation by stem-boring frit fly (*Oscinella* spp.) may reduce the number of tillers in italian ryegrass (Bentley and Clements 1989).

Established Stands

Defoliators and sap feeders (see Table 29.1 for examples) cause reductions in mass per shoot (Hutchins et al. 1990; Buntin 1991), a primary factor leading to immediate forage yield loss. Foliar feeding can also lead to lower yield in subsequent growth cycles (Hutchins et al. 1990). Stubble feeding by insects during early regrowth reduces stem length, delays plant maturity, and suppresses rate of stem initiation, thus reducing **shoot** density and shoot mass (Buntin and Pedigo 1985; Hutchins et al. 1990). Little is known about alfalfa root-feeding pests on shoots per plant and mass per shoot, though forage yield reductions have been reported (Berberet and Dowdy 1989). However, large infestations of mole crickets, bahiagrass borers and green June beetle grubs can cause fairly rapid stand loss of bermudagrass, bahiagrass and tall fescue pastures, respectively, and stand loss is more pronounced under drought than wet conditions (Flanders and Cobb 2000). Defoliators and sap feeders often reduce or delay morphologic development (Hutchins et al. 1990), which potentially reduces yield in subsequent cycles and perhaps a lost harvest interval, resulting in lower total seasonal yield (Hutchins and Pedigo 1990).

Plant mortality directly affects stand persistence and yield of crown-forming forage species such as legumes (Chapters 14 and 15). Unless there is a severe infestation, relatively few insect pests reduce plant density directly (Beuselinck et al. 1994); however, insects do act in concert with other biotic or abiotic stressors to reduce persistence. For example, weed and alfalfa weevil infestations accelerated stand decline of alfalfa in Oklahoma (Berberet et al. 1987; Dowdy et al. 1993). Insects can also reduce stolon or rhizome production of clone-forming plant species, which might reduce stand persistence. For species dependent on reseeding for persistence in the sward, insect injury to seed production might reduce stand persistence; however, the impact of reduced seed production on stand persistence and plant demography is unclear (Beuselinck et al. 1994).

Impact on Forage Quality

The impact of insect pests on forage quality has received far less attention than the impact on yield. As a general rule, insect pests have a relatively small effect on forage quality in terms of nutrient concentration, (e.g. grams of crude protein per kilogram of forage dry matter). For example, potato leafhopper feeding reduces carotene, crude protein, ash, calcium, and phosphate concentrations in alfalfa (Smith and Medler 1959; Kindler et al. 1973). Though insects have a small effect on nutrient concentration in the plant, they can cause large reductions in nutrient yield per unit of land area because of reduced biomass production (Hutchins et al. 1989).

Though sap feeders like the potato leafhopper cause small reductions in crude protein or mineral concentration in forages, the stem stunting caused by those pests increases leaf-to-stem ratio, which may have a favorable effect on other quality traits. For example, fiber concentration may be lowered and leaf and stem digestibility can be enhanced slightly by sap feeders (Hutchins et al. 1990; Mackun and Baker 1990).

Defoliators cause a moderate reduction in crude protein concentration and forage digestibility largely because the loss of leaf area (high-quality material) is associated with less stem production (lower-quality material) (Hutchins et al. 1990). As mentioned earlier, foliar feeding can result in delayed morphologic development, that is, the insect-damaged forage is less mature on any given date than undamaged forage. So, when harvested on the same date, forage damaged by foliar feeding may actually have higher quality than insecticide-treated forage that is more mature. On the other hand, reduced plant vigor due to insect feeding often allows greater weed encroachment into forage stands (Berberet et al. 1987; Buntin 1989), which probably has a greater impact than the direct effects of insect injury on the quality of forage harvested. Stem boring damage by bermudagrass stem maggot causes a 7% reduction in relative feed quality of late-season bermudagrass hay due to a decrease in **total digestible nutrients** (TDN) and slightly lower dry matter intake (DMI). A small increase in crude protein actually occurs in the damaged bermudagrass hay, as a function of dilution of desirable carbohydrates (Baxter et al. 2017).

Predicting the economic impact of changes in forage quality due to insect feeding involves many factors, but two stand out. There is no common market for forage that includes its on-farm value as a nutrient source for livestock, and classes of animals differ in their ability to physiologically use forages. Least-cost rationing algorithms, based on linear programming, can be used to estimate feed-value EILs. Hutchins and Pedigo (1998) compared alfalfa forage damaged by potato leafhopper vs healthy alfalfa forage in feed rations and found that the difference in insect population levels causing economic injury differed by 129% among animal types. They concluded that calculation of economic-injury levels should include the effect of the insect on final feed value for the specific livestock consuming the forage.

Complex Interactions

Multiple-pest interactions have been identified that affect forage yield, persistence, and quality. Insect damage may have its largest detrimental effect by accentuating damage caused by other biotic and abiotic stresses such as weed competition, diseases, nematodes, drought, and waterlogging stress (Chamblee et al. 1983; Berberet et al. 1987; Schroeder et al. 1988; Pederson et al. 1991; Summers and Gilchrist 1991; Moellenbeck et al. 1992; Dowdy et al. 1993; Barta et al. 2002). Some combined stresses have simple additive effects on plant damage, whereas others have synergistic effects where the damage from combined stresses is greater than the sum of their individual damages. Insect management needs to be viewed as an integral component within the overall framework of crop management decision making, and there is a need to develop integrated programs that consider all types of pests.

Steps to Managing Insect Pests in Forages

The goal of IPM is to stabilize pest populations below an acceptable damage level using techniques that minimize health and environmental hazards while optimizing economic benefits for producers and consumers. This implies management and planning, using preventative practices wherever possible, rather than simply responding as pest outbreaks occur. The following sections outline important principles and practices of insect management in forage crops in achieving these objectives.

Diversity of Organisms in the Forage Crop

As discussed above, there are few specific combinations of insect species, host crops, and environments that lead to economic losses on a consistent basis in forage ecosystems. Most noncrop organisms are beneficial in some way and help maintain a balance in the agroecosystem. Environmentally sound pest management is an attempt to maintain this biodiversity.

Identifying Pests and Understanding their Biology

Correctly identifying insect species and understanding their biology are critical in accurately assessing the potential for economic damage. For example, five species of leafhoppers are common in alfalfa during the summer in the North Central and northeastern United States, but only one, the potato leafhopper, consistently causes economic damage to the crop. Correct identification is very important when a new exotic invasive insect pest is detected. It took several months in 2010 to determine that damage in bermudagrass was something new, that a maggot species was causing the damage and to identify the species among the hoard of small flies normally found in permeant pasture systems (Baxter et al. 2014). Knowledge of a pest's overwintering capability is also important in identifying pest management tactics. If a pest can overwinter locally, natural enemies will be maintained, and pest management may be accomplished by cultural practices such as stubble destruction or rotation. Pests that cannot overwinter and are migratory, such as fall armyworm and potato leafhopper, are more difficult to predict in their timing, location, and especially severity of infestation. Management of migratory pests usually involves adult trapping and in-field sampling around the expected time of arrival and applying control measures based on established action thresholds. Regional trapping networks and use of mobile apps can assist in monitoring a pest's arrival and spread, which helps to time sampling activities.

Pests are easier to control at certain life stages than others. Environmental conditions and crop stage may foster or retard development of pest populations and damage. For example, on **rangelands** where cool-season grasses dominate, damage by bigheaded and migratory grasshoppers may be severe and permanent if not controlled early (Hewitt and Onsager 1983). In contrast, on southwestern rangelands, high densities of grasshoppers early in the season often will not influence total herbage production because they occur before summer rainfall stimulates growth of the later-maturing warm-season grasses that dominate these areas (Thompson and Gardner 1996).

Managing Pest Suppression

All pest populations are suppressed directly by natural enemies and indirectly by competitors. The conditions within a crop can be managed to take advantage of this natural suppression. Practices encouraging natural enemies include avoiding the use of insecticide or more complex activities such as ensuring favorable habitat for natural enemies, especially following harvest. In addition, each crop management decision, such as the selection of species and variety, the fertility regime, planting date, and the harvest schedule can influence pest populations and their damage. Plant damage can be prevented or reduced by practices contributing to a vigorous forage stand (Kitchen et al. 1990). Three approaches are often cited to prevent the buildup of pest populations: host resistance, biologic control, and cultural control.

Host Resistance

Host plant **resistance** is the innate genetic ability of crops to avoid or minimize the impact of pests. General types of host resistance to insects are (i) antibiosis, in which the plant causes the death of the insect pest; (ii) antixenosis (also called nonpreference), in which insects avoid feeding on the plant; and (iii) tolerance, in which the plant can withstand the insect injury without significant damage. Because of its low cost and potential for suppressing plant-feeding herbivores, host resistance is a particularly powerful tool, especially in low-value crops such as forages and when resistance to multiple pests is combined in the same cultivar or species. Good examples are the resistance to pathogens and insects now available in alfalfa cultivars. Partial resistance in bermudagrass to bermudagrass stem maggot has been identified and is related to stem diameter, with thicker stemmed varieties being less susceptible to infestation than thinner stemmed varieties (Baxter et al. 2015). A phenomenon related to plant **resistance** in certain forage grasses like tall fescue and perennial ryegrass is the presence of an endophytic fungus that may provide protection from insect pests. Removal of the endophyte makes the plant more susceptible to insect injury (Young et al. 2015)

Cultural Control

Crop production practices can be modified to suppress the insect pest population by making the habitat less favorable for pest survival or reproduction. Practices that maximize plant health such as good soil fertility, balanced plants nutrients, and proper pH levels allow forage plants to better tolerate pest stresses. Harvest management is a useful practice for avoiding or minimizing insect damage. For example, in the north, where alfalfa weevils overwinter only as adults and lay eggs in the spring, early harvest of forages is often recommended to avoid injury caused by late-developing alfalfa weevil larvae. In southern regions, alfalfa weevil eggs overwinter, hatch in early spring, and cause damage to plants at stages too early for harvest to be an effective approach. Decreasing the fall stubble in alfalfa fields through fall harvesting or grazing when weevil eggs are laid sometimes decreases or delays larval damage in early spring in southern regions. Early cutting of infested hay is recommended to minimize plant damage by the bermudagrass stem maggot (Baxter et al. 2017).

Cutting alfalfa closely, along with forage removal from entire fields at one time, reduces the residual food source for potato leafhopper survival between cuttings. Including grasses in mixtures with alfalfa can also help reduce insect damage (Lamp 1991; Lamp et al. 1994; Roda et al. 1997; DeGooyer et al. 1999, Baxter et al. 2017). Crop rotation and fall or early spring planting dates can reduce the presence of insect species at the time of planting or during seedling development, therefore improving the success of establishment. For example, late summer seeding of alfalfa avoids potato leafhopper injury during the seedling stage since that pest usually disappears from alfalfa fields by late August. Crop rotation is effective in delaying migration and buildup of root-feeding weevil populations in susceptible hosts. The root weevils only colonize the susceptible host in the fall, so the first year of a spring seeding is weevil free. At the landscape level, planning of crop proximities can reduce the migration of potential pests between fields.

Biologic Control

Using predators, parasites, and pathogens to reduce insect pest populations is the basis of biologic control. Classical biologic control involves the importation and release of exotic natural enemies, usually host-specific parasites. For example, many potential natural enemies of alfalfa weevil have been introduced from Europe, and the five parasites and one pathogen listed in Table 29.2 have become important in suppressing populations below economic levels. Biologic control has also been very successful in reducing damage by the alfalfa leafminer to non-economic levels in most areas where it occurs (Manglitz and Ratcliffe 1988). Starting in the 1980s, an entomopathogenic nematode, *Steinernema scapterisci,* a parasitic wasp, *Larra bicolor*, and parasitic fly, *Ormia depleta*, were released in Florida in an attempt to control the invasive tawny and southern mole crickets. The nematode and wasp have spread naturally and have reduced mole cricket populations enough to prevent damaging infestation in grass pastures in the southeastern US. (Held and Cobb 2016).

Another type of biologic control in forages is habitat manipulation, in which care is taken by the producer to allow survival and spread of naturally occurring predators, such as providing forest habitat near fields (Landis et al. 2000). Most importantly, producers should recognize the value of these natural enemies for suppressing pest populations because they often naturally prevent outbreaks. For example, pea aphid populations can develop very rapidly in the spring, but typically a number of natural enemies, including lady beetles, damsel bugs, parasitic wasps, and fungal pathogens (Table 29.2), limit their density to sub-economic levels.

Monitoring Pest Populations

Even with the best preventive measures, potentially damaging infestations of insect pests do occur. Thus, for those forages with potential for significant damage, producers need to monitor fields to determine the current density of the pest relative to the density at which control steps are economically justified. To monitor a pest population, one needs to sample the species in an appropriate way and

Table 29.2 Common natural enemies of pests associated with alfalfa in eastern North America

Species of natural enemy	Order: Family	Source[a]	Examples of pests attacked	Mode of action
Nabis alternatus and related species	Hemiptera: Nabidae	N	Alfalfa weevil, potato leafhopper, pea aphid	Predator: Immature and adult damsel bugs attack and kill many insects, especially caterpillars and aphids
Coleomegilla maculata and related species	Coleoptera: Coccinellidae	N	Alfalfa weevil, potato leafhopper, pea aphid, blue alfalfa aphid	Predator: Larvae and adult lady beetles feed and kill a wide range of herbivorous insects, especially aphids
Bathyplectes curculionis, Bathyplectes anurus	Hymenoptera: Ichneumonidae	I	Alfalfa weevil	Parasite: Adult females lay eggs into alfalfa weevil larvae; developing parasite kills host
Oomyzus incertus	Hymenoptera: Eulophidae	I	Alfalfa weevil	Parasite: Adult females lay eggs into large alfalfa weevil larvae just before pupation; developing parasite kills host
Microctonus colesi, Microctonus aethiopoides	Hymenoptera: Braconidae	I	Alfalfa weevil	Parasite: Adult females lay eggs into large alfalfa weevil larvae or adults; developing parasite kills adult
Anagrus nigriventris	Hymenoptera: Mymaridae	N	Potato leafhopper	Parasite: Adult wasps lay eggs inside the eggs of leafhoppers; developing larvae kill host
Aphidius ervi	Hymenoptera: Aphidiidae	I	Pea aphid, blue alfalfa aphid	Parasite: Adult females lay eggs into pea aphids; developing parasite kills host
Zoophthora phytonomi	Entomophthora: Entomophthoraceae	I	Alfalfa weevil	Pathogen: Disease-producing organisms infect and kill alfalfa weevil larvae
Erynia radicans	Entomophthora: Entomophthoraceae	N	Potato leafhopper	Pathogen: Disease-producing organisms infect and kill leafhoppers
Erynia neoaphidis	Entomophthora: Entomophthoraceae	?	Pea aphid, blue alfalfa aphid	Disease-producing organisms infect and kill aphids

[a]N, native; I, introduced.

at an appropriate time. Guidelines vary by insect species, crop, and region and are available through local and state extension services.

Knowledge of the developmental stage of the insect and crop is also important for making decisions to respond to an insect outbreak. For example, alfalfa weevil damage is usually greatest during the larval stage. A predominance of late-instar alfalfa weevil larvae and pupae indicate that little additional damage will be sustained by the crop and that insecticide application is not likely to be cost-effective, especially if an early harvest or grazing is possible.

Monitoring should also involve assessment of the health of the pest and any biologic suppression occurring in the field, which if significant, may preclude the need for insecticide treatment. For example, insecticide applications may be recommended on the sole basis of the number of alfalfa weevil larvae present, when in fact the larvae are being weakened and killed by biologic agents and thus do not pose an economic threat to the crop.

Strategies to Respond to Pest Outbreaks

When monitoring reveals the potential for economic damage, then there are usually two choices one should consider for suppressing pest populations in forage crops: an insecticide application or early cutting or grazing. Many factors are important in the decision to apply insecticide, such as the chemical, formulation, rate, label restrictions, weather conditions, time to harvest, potential impact on adjacent areas, and cost of treatment versus the benefit. A number of newer more selective, but more expensive, insecticides have been registered for use in forage/pasture systems in North America. But one challenge to IPM in forage crops and generally in all crops is that the very low cost of pyrethroid insecticides make preventive application of these insecticides to avoid the risk of pest damage financially attractive regardless of the status of pest populations. Unnecessary preventive use of inexpensive insecticides increases the risk of the development of pest resistance to these insecticides.

Several factors affect the use of early harvesting as a response to insect presence, including the impact on the physiologic state of the crop, handling of the forage relative to other farming activities, and expected weather conditions. Insecticide use and early harvesting are designed as last-minute efforts to prevent pest-induced losses, but both have several potential negative impacts that must be considered.

Evaluation and Refinement

Effective pest management programs adjust to match changing conditions and should be continually evaluated within the overall management system. Good record keeping can pinpoint successes and failures in space and time, and smaller "check plots" can help demonstrate what might have happened with a different management decision than what was carried out on the rest of the field. Such information provides a sound basis for making adjustments to correct past deficiencies and for incorporating new developments and new technologies.

Looking to the Future

Several important trends will affect insect pest management in forage crops in the future. Though some deal particularly with insects and other types of pests, others are present in agriculture as a whole. Yet, all will likely affect the way insect pests are managed in the future, as described below.

Integration of Crop Protection as Part of Crop Production

Insect pests are affected by each management practice, whether directed at pests or at the crop. Thus, as information on these interactions is gained, more comprehensive management systems that include both pest and crop management will be developed. In addition, the recognition of the long-term effects of pest injury on stand persistence will help to integrate pest and crop management.

Economic, Sociologic, and Technologic Changes

Numerous changes continue to occur within our society and in scientific knowledge that will affect pest management. Some examples include the following: the development of genetically modified crops, the development of new pesticides that are less toxic to humans and more compatible with the environment, the expansion in the organic foods market, and growing interest in ecologically compatible farming methods vs the continuation of specialization that tends to reduce diversity in production systems. New understanding of pest behavior and plant defense responses may provide more ecologically based pest-suppression strategies and technologies.

Development of Farm and Site-Specific IPM

Each farm and each field have conditions that specifically affect the optimum application of IPM. As this information is developed, and as geographic information systems provide better data, IPM practices can be customized to more specific conditions. A better understanding of how to construct farm habitats that enhance natural suppression of pests would contribute to this.

Availability of Real-Time Information

Weather and certain pest-population information are currently available to growers on a real-time basis to enhance decision making. As new and more robust technologies are developed, the usefulness of this information will be expanded and data on additional pest species and natural enemies will become available.

Invasion of New Pests

The history of insect pest management has many examples of the movement of pests into new areas. Indeed, most of the insect pests listed in Table 29.1 are exotic in North America. Despite strong efforts to prevent such movement, this trend will continue as international travel and trade are expanded. Careful attention will be needed to locate and respond to new invasions.

Adaptation of Pest Species to Management

Pests will continue to evolve in response to pest management practices, including pesticide resistance, breakdown of host plant resistance, and cultural practices that disrupt habitat or food sources. Thus, new management techniques will be needed to maintain current levels of insect pest management.

Insect pest management is an integral component of a well-managed forage crop system. Some pest problems are predictable, and efforts should be focused on preventing their development. Other pest problems are unpredictable and thus require responsive efforts to suppress pest densities. The basic principles of insect pest management will not change, yet the specific decision-making processes for dealing with pest problems will change in response to new technology, new pests, changes in current pests, and new tools for integrating information into crop management.

Climate Change

The scientific community has concluded, from available evidence, that the Earth's climate is changing in response to anthropogenic production of greenhouse gasses in our atmosphere (McCarthy 2009). The rising level of CO_2 and other gasses in our atmosphere is resulting in the capture of thermal radiation, resulting in global warming. Rising temperature is also increasing the rate of evaporation in the hydrologic cycle, resulting in the intensification of storms and drought cycles. Though the combined effects of temperature and precipitation change on insects are unclear, evidence is growing of the following: (i) pest distributions are changing, (ii) seasonal phenology of pests is changing, (iii) developmental rates and fitness of pests are changing, (iv) coupled interactions between pests and their plant hosts, as well as between pests and their natural enemies, may be disconnected, and (v) phenologic disruptions within insect communities may occur. As an example, from alfalfa, first detection of spring migration of potato leafhopper has moved up an average of 10 days between 1951 and 2012, related to changes in air temperatures (Baker et al. 2015).

Another effect of increasing atmospheric CO_2 is to modify rates and products of photosynthesis, resulting in changes in plant growth and including the suggestion of increased crop yields in some areas. These complex physiologic changes are likely to lead to indirect effects on pests, such as affecting host plant defenses, host plant nutrition for herbivores, and cues used by pests for life history events.

As a result of these changes to pest biology and ecology, our adaptation to maintain insect-derived ecosystems services requires us to understand the short-term ecologic effects and mechanisms, recognize the long-term evolutionary consequences (including extinctions), protect endangered species, conserve natural habitats, educate students and the public, monitor for the loss of key services, and provide an appropriate response to ecologic loss. Our adaptation to changing pest populations and their damage is best accomplished by recognizing the concept of IPM and continue to apply its principles, including (i) to prepare for the expected (local) range expansion and extirpations of pests and their natural enemies, (ii) to use degree-day and other models to predict phenologic changes such as first occurrence in the spring, number of generations per year, and timing for monitoring and responsive measures, (iii) to be alert for subtle changes within crops, such as disconnected interactions leading to pest outbreaks, and (iv) to design agricultural landscapes to be resilient to expected changes in the climate. Climate change is complicated, but IPM is robust enough to suggest appropriate response by using science to evaluate the current and projected changes.

References

Baker, M.B., Venugopal, P.D., and Lamp, W.O. (2015). Climate change and phenology: *Empoasca fabae* (Hemiptera: Cicadellidae) migration and severity of impact. *PLoS One* 10 (5): e0124915.

Barta, A.L., Sulc, R.M., Ogle, M.J., and Hammond, R.B. (2002). Interaction between flooding or drought stress and potato leafhopper injury in alfalfa. *Plant Health Prog.* https://doi.org/10.1094/PHP-2002-0502-01-RS.

Baxter, L.L., Hancock, D.W., and Hudson, W.G. (2014). The bermudagrass stem maggot (*Atherigona reversura* Villeneuve): a review of current knowledge. *Forage Grazinglands* 12 (1) https://doi.org/10.2134/FG-2013-0049-RV.

Baxter, L.L., Hancock, D.W., Hudson, W.G. et al. (2015). Response of selected bermudagrass cultivars to bermudagrass stem maggot damage. *Crop Sci.* 55 (6): 2682–2689. https://doi.org/10.2135/cropsci2015.12.0828.

Baxter, L.L., Hancock, D.W., Hudson, W.G. et al. (2017). Managing bermudagrass stem maggots. University of Georgia. Cooperative Extension Service Bulletin No. 1484.

Bentley, B.R. and Clements, R.O. (1989). Impact of time of sowing on pest damage to direct-drilled grass and the mode of attack by dipterous stem borers. *Crop Prot.* 8: 55–62.

Berberet, R.C. and Dowdy, A.K. (1989). Insects that reduce persistence and productivity of forage legumes in the USA. In: *Persistence of Forage Legumes* (eds. G.C. Marten, A.G. Matches, R.F Barnes, et al.), 481–500. Madison, WI: American Society of Agronomy.

Berberet, R.C., Stritzke, J.F., and Dowdy, A.K. (1987). Interactions of alfalfa weevil (Coleoptera: Curculionidae) and weeds in reducing yield and stand of alfalfa. *J. Econ. Entomol.* 80: 1306–1313.

Beuselinck, P.R., Bouton, J.H., Lamp, W.O. et al. (1994). Improving persistence in forage crop systems. *J. Prod. Agric.* 7: 311–322.

Bommarco, R., Kleijn, D., and Potts, S.G. (2013). Ecological intensification: harnessing ecosystem services for food security. *Trends Ecol. Evol.* 28: 230–238.

Buntin, G.D. (1989). Competitive interactions of alfalfa and annual weeds as affected by alfalfa weevil (Coleoptera: Curculionidae) stubble defoliation. *J. Entomol. Sci.* 24: 78–83.

Buntin, G.D. (1991). Effect of insect damage on the growth, yield, and quality of Sericea lespedeza forage. *J. Econ. Entomol.* 84: 277–284.

Buntin, G.D. and Pedigo, L.P. (1985). Development of economic injury levels for last stage variegated cutworm (Lepidoptera: Noctuidae) larvae in alfalfa stubble. *J. Econ. Entomol.* 78: 1341–1346.

Byers, R.A., Mangan, R.L., and Templeton, W.C. Jr. (1983). Insect and slug pests in forage legume seedings. *J. Soil Water Conserv.* 38: 224–226.

Chamblee, D.S., Lucas, L.T., and Campbell, W.V. (1983). Ladino clover persistence as affected by physical management and use of pesticides. In: *Proc. Int. Grassl. Congr., 14th, Lexington, KY*. 15–24 June 1981 (eds. J.A. Smith and V.W. Hays), 584–587. Boulder, CO: Westview Press.

DeGooyer, T.A., Pedigo, L.P., and Rice, M.E. (1999). Effect of alfalfa-grass intercrops on insect populations. *Environ. Entomol.* 28: 703–710.

Dowdy, A.K., Berberet, R.C., Stritzke, J.F. et al. (1993). Interaction of alfalfa weevil (Coleoptera: Curculionidae), weeds, and fall harvest options as determinants of alfalfa productivity. *J. Econ. Entomol.* 86: 1241–1249.

Ecale, C.L. and Backus, E.A. (1995). Mechanical and salivary aspects of potato leafhopper probing in alfalfa stems. *Entomol. Exp. Appl.* 77: 121–132.

Fick, G.W., Lamp, W.O., and Collins, M. (2003). Integrated pest management in forages. In: *Forages: An Introduction to Grassland Agriculture*, 6e (eds. R.F Barnes, C.J. Nelson, M. Collins and K.J. Moore), 295–313. Ames: Iowa State Press.

Flanders, K.L. and Cobb, P.P. (2000). Biology and control of the green June beetle. Alabama Cooperative Extension Service Bulletin No. ANR-991.

Flinn, P.W. and Hower, A.A. (1984). Effects of density, stage and sex of the potato leafhopper, *Empoasca fabae* (Homoptera: Cicadellidae) on seedling alfalfa growth. *Can. Entomol.* 116: 1543–1548.

Gaba, S., Bretagnolle, F., Rigaud, T., and Philippot, L. (2014). Managing biotic interactions for ecological intensification of agroecosystems. *Front. Ecol. Evol.* 2: 2.

Held, D. and Cobb, P.P. (2016). Biology and control of mole crickets. Alabama A&M and Auburn University. Cooperative Extension Service Bulletin No. ANR-0176.

Hewitt, G.B. and Onsager, J.A. (1983). Control of grasshoppers on rangeland in the United States: a perspective. *J. Range Manage.* 36: 202–207.

Hower, A.A., Quinn, M.A., Alexander, D., and Leath, K.T. (1995). Productivity and persistence of alfalfa in response to clover root curculio (Coleoptera: Curculionidae) injury in Pennsylvania. *J. Econ. Entomol.* 88: 1433–1440.

Hutchins, S.H. and Pedigo, L.P. (1990). Phenological disruption and economic consequence of injury to alfalfa induced by potato leafhopper (Homoptera: Cicadellidae). *J. Econ. Entomol.* 83: 1587–1594.

Hutchins, S.H. and Pedigo, L.P. (1998). Feed-value approach for establishing economic-injury levels. *J. Econ. Entomol.* 91: 347–351.

Hutchins, S.H., Buxton, D.R., and Pedigo, L.P. (1989). Forage quality of alfalfa as affected by potato leafhopper feeding. *Crop Sci.* 29: 1541–1545.

Hutchins, S.H., Buntin, G.D., and Pedigo, L.P. (1990). Impact of insect feeding on alfalfa regrowth: a review of physiological responses and economic consequences. *Agron. J.* 82: 1035–1044.

Kaufman, P.E. and Woods, L.A. (2011). Indigenous and exotic dung beetles (Coleoptera: Scarabaeidae and Geotrupidae) collected in Florida cattle pastures. *Annals Entomol. Soc. Amer.* 105 (2): 225–231. https://doi.org/10.1603/AN11121.

Kindler, S.D., Kehr, W.R., Ogden, R.L., and Schalk, J.M. (1973). Effect of potato leafhopper injury on yield and quality of resistant and susceptible alfalfa clones. *J. Econ. Entomol.* 66: 1298–1302.

Kitchen, N.R., Buckholz, D.D., and Nelson, C.J. (1990). Potassium fertilizer and potato leafhopper effects on alfalfa growth. *Agron. J.* 82: 1069–1074.

Lamp, W.O. (1991). Reduced *Empoasca fabae* (Homoptera: Cicadellidae) density in oat-alfalfa intercrop systems. *Environ. Entomol.* 20: 118–126.

Lamp, W.O., Nielsen, G.R., and Dively, G.P. (1991). Insect pest-induced losses in alfalfa: patterns in Maryland and implications for management. *J. Econ. Entomol.* 84: 610–618.

Lamp, W.O., Nielsen, G.R., and Danielson, S.D. (1994). Patterns among host plants of potato leafhopper, *Empoasca fabae* (Homoptera: Cicadellidae). *J. Kansas Entomol. Soc.* 67: 354–368.

Lamp, W.O., Nielsen, G.R., Quebedeaux, B., and Wang, Z. (2001). Potato leafhopper (Homoptera: Cicadellidae) injury disrupts basal transport of ^{14}C-labelled photoassimilates in alfalfa. *J. Econ. Entomol.* 94: 93–97.

Lamp, W.O., Nielsen, G.R., Fuentes, C.B., and Quebedeaux, B. (2004). Feeding site preference of potato leafhopper (Homoptera: Cicadellidae) on alfalfa and its effect on photosynthesis. *J. Agric. Urban Entomol.* 21: 25–38.

Landis, D., Wratten, S.D., and Gurr, G. (2000). Habitat manipulation to conserve natural enemies of arthropod pests in agriculture. *Annu. Rev. Entomol.* 45: 472–483.

Leath, K.T. and Hower, A.A. (1993). Interaction of *Fusarium oxyporum* f. sp. *medicaginis* with feeding activity of clover root curculio larvae in alfalfa. *Plant Dis.* 77: 799–802.

Mackun, I.R. and Baker, B.S. (1990). Insect populations and feeding damage among birdsfoot trefoil-grass mixtures under different cutting schedules. *J. Econ. Entomol.* 83: 260–267.

Manglitz, G.R. and Ratcliffe, R.H. (1988). Insects and mites. In: *Alfalfa and Alfalfa Improvement* (eds. A.A. Hansonet, D.W. Barnes and R.R. Hill), 671–704. Madison, WI: American Society of Agronomy.

McCarthy, J.J. (2009). Reflection on: our planet and its life, origins, and futures. *Science* 326: 1646–1655.

Metcalf, R.L. and Metcalf, R.A. (1993). *Destructive and Useful Insects*, 5e. New York: McGraw-Hill.

Moellenbeck, D.J., Quisenberry, S.S., and Colyer, P.D. (1992). *Fusarium* crown-rot development in alfalfa stressed by three-cornered alfalfa hopper (Homoptera: Membracidae) feeding. *J. Econ. Entomol.* 85: 1442–1449.

Pederson, G.A., Windham, G.L., Ellsbury, M.M. et al. (1991). White clover yield and persistence as influenced by cypermethrin, benomyl, and root-knot nematode. *Crop Sci.* 31: 1297–1302.

Pedigo, L.P., Hutchins, S.H., and Higley, L.G. (1986). Economic injury levels in theory and practice. *Annu. Rev. Entomol.* 31: 341–368.

Pimentel, D. and Wheeler, A.G. Jr. (1973). Species and diversity of arthropods in the alfalfa community. *Environ. Entomol.* 2: 659–668.

Ratcliffe, E.B. and Flanders, K.L. (1998). Biological control of alfalfa weevil in North America. *Integr. Pest Manage. Rev.* 3: 225–242.

Roda, A.L., Landis, D.A., and Coggins, M.L. (1997). Forage grasses elicit emigration of adult potato leafhopper (Homoptera: Cicadellidae) from alfalfa–grass mixtures. *Environ. Entomol.* 26: 745–753.

Rogers, D.D., Chamblee, D.S., Mueller, J.P., and Campbell, W.V. (1985). Fall no-till seeding of alfalfa into tall fescue as influenced by time of seeding and grass and insect suppression. *Agron. J.* 77: 150–157.

Scherr, S.J. and McNeely, J.A. (2008). Biodiversity conservation and agricultural sustainability: towards a new paradigm of 'ecoagriculture' landscapes. *Phil. Trans. R. Soc. B* 363: 477–494.

Schroeder, P.C., Brandenburg, R.L., and Nelson, C.J. (1988). Interaction between moisture stress and potato leafhopper (Homoptera: Cicadellidae) damage in alfalfa. *J. Econ. Entomol.* 81: 927–933.

Shields, E.J., Testa, A., Miller, J.M., and Flanders, K.L. (1999). Field efficacy and persistence of the ento-mopathogenic nematodes *Heterorhabditis bacteriophora* 'Oswego' and *H. bacteriophora* 'NC' on alfalfa snout beetle larvae (Coleoptera: Curculionidae). *Environ. Entomol.* 28: 128–136.

Smith, D. and Medler, J.T. (1959). Influence of leafhoppers on the yield and chemical composition of alfalfa hay. *Agron. J.* 51: 118–119.

Smith, R.F. and van den Bosch, R. (1967). Integrated control. Pp. 295–340. In: *Pest Control* (eds. W.W. Kilgore and R.L. Doutt). New York: Academic Press.

Sulc, R.M., McCormick, J.S., Hammond, R.B., and Miller, D.J. (2014). Population responses to potato leafhopper (Hemiptera:Cicadellidae) to insecticide in glandular-haired and non-glandular-haired alfalfa cultivars. *J. Econ. Entomol.* 107: 2077–2087.

Sulc, R.M., McCormick, J.S., Hammond, R.B., and Miller, D.J. (2015). Forage yield and nutritive value responses to insecticide and host resistance in alfalfa. *Crop Sci.* 55: 1346–1355.

Summers, C.G. and Gilchrist, D.G. (1991). Temporal changes in forage alfalfa associated with insect and disease stress. *J. Econ. Entomol.* 84: 1353–1363.

Thompson, D.C. and Gardner, K.T. (1996). Importance of grasshopper defoliation period on southwestern blue grama-dominated rangeland. *J. Range Manage.* 49: 494–498.

Undersander, D., Becker, R., Cosgrove, D. et al. (2004). *Alfalfa Management Guide*. NCR547. Madison, WI: ASA, CSSA, SSSA.

USDA-ARS (2017). Imported Fire Ants. https://www.aphis.usda.gov/aphis/ourfocus/planthealth/plant-pest-and-disease-programs/pests-and-diseases/imported-fire-ants (accessed 10 October 2019).

Volenec, J.J., Cherney, J.H., and Johnson, K.D. (1987). Yield components, plant morphology, and forage quality of alfalfa as influenced by plant population. *Crop Sci.* 27: 321–326.

Williams, R.E. (2010). *Veterinary Entomology: Livestock and Companion Animals*, xxvii. Boca Raton, FL, USA: CRC Press, Taylor and Francis Group. ISBN 978-1-4200-6849-8.

Young, C.A., Hume, D.E., and McCulley, R.L. (2015). Fungal endophytes of tall fescue and perennial ryegrass: pasture friend or foe? *J. Animal Sci.* 91: 2379–2394.

PART VI

FORAGE IMPROVEMENT

The goal of plant breeding programs is the improvement of forage species for yield, disease resistance, forage quality and other important characteristics. *Source:* Photo courtesy of Mike Collins.

As we have seen, plant species grown for forage vary remarkably in life cycle, growth habit, morphology, composition and other traits. Each trait is controlled at a genetic level and plant breeders have developed cultivars of many of these species with improved forage characteristics. Since before modern history, humans have been identifying and selecting forage plants with superior traits and using them to feed livestock. Based on the discoveries of Mendel, plant breeders have been developing genetic tools over the last century for accelerating plant improvement.

Forage plants vary in their reproductive physiology and mating strategies, and different species require different approaches to improvement. The majority of forage species are self-incompatible and plants must be intermated to produce viable seed. The selection strategies and cultivar development procedures used for these species varies from those that are self-pollinated. Mechanisms for

Forages: The Science of Grassland Agriculture, Volume II, Seventh Edition.
Edited by Kenneth J. Moore, Michael Collins, C. Jerry Nelson and Daren D. Redfearn.

pollination also vary with some species requiring insect pollinators and others relying mostly on wind-borne pollen. Polyploidy occurs in many forage species and can be a further complication in cultivar development.

There have been a number of significant advances in plant improvement in the last decade. New biotechnologies and molecular genetic approaches are enabling more rapid development of new forage plant cultivars. The use of molecular markers and genomic selection are two relatively new approaches being used in forage improvement. Once a new cultivar is developed, its seed needs to be increased to make it available for commercial production. Producing seed of any plant cultivar requires special production practices to ensure genetic purity and seed quality.

This part covers the fundamentals of forage plant genetics and how they are used to develop new and improved cultivars. It also covers some of the new and exciting developments in plant breeding and genetics related to forage improvement. Finally, the processes involved in producing seed of new cultivars and ensuring their purity and quality are covered.

Forage Breeding

Michael D. Casler, *Research Geneticist, USDA-ARS, U.S. Dairy Forage Research Center, Madison, WI, USA*
Kenneth P. Vogel, *Research Geneticist (retired), USDA-ARS, Lincoln, NE, USA*

Introduction

Plant breeding is human-directed evolution. This process has been used to develop all major crops and their respective races, strains, or cultivars. Although humans have successfully manipulated the genetic resources of plants for several thousand years, the science of genetics and breeding was not developed until the twentieth century. Breeding work on a few forage crops began in the early part of the twentieth century (Wilkins and Humphreys 2003). Initial work was focused on developing strains that had improved establishment, persistence, high forage yields, and good insect and disease resistance. These remain essential attributes of cultivated forages (Burton 1986). Since the 1960s, when laboratory procedures became more highly developed and amenable to high-throughput approaches, breeding objectives have expanded to include improving forage digestibility and removing or reducing the concentration of antiquality constituents.

A pasture or hay field consists of a population of plants. The characteristics of individual plants of the same forage species vary widely for cross-pollinated species (e.g. alfalfa, ryegrass, and switchgrass), but generally less so for self-pollinated (e.g. common vetch and lespedeza), vegetatively propagated (e.g. bermudagrass), and apomictic species (e.g. kentucky bluegrass and dallisgrass). The **phenotype** of individual plants growing in a field or breeding nursery is expressed in a specific environment. Each phenotype (P) results from genetic expression of a genotype (G) as affected by its environment (E) and can be described by the equation:

$$P = G + E + GxE \text{ (Interaction Effect)}$$

Plants that are genetically identical such as those of vegetatively propagated cultivars of bermudagrass may differ in size and other characteristics when grown in different environments such as different field sites. Plant breeders use genetic manipulation or breeding to change the genetic characteristics of plant populations so that, when planted or transplanted in fields or pastures, the phenotypic mean of the bred plants represents an improvement over the original populations or older cultivars (Figure 30.1). In its simplest form, this involves selecting the "best" individuals and advancing them to the next generation by isolated intercrossing or selfing. If heritability is moderate to high, then most of the difference between the population mean and the mean of the selected individuals will be realized as the "gain" in the next generation.

Changing plant populations by breeding is a multi-step process that includes assembling and evaluating germplasm sources, selecting plants with the desired phenotypes, mating the selected plants, and evaluating the progeny in small plots, hay fields, pastures, and seed production fields (Table 30.1). Each phase takes five or more years with perennial species. Often the process

Forages: The Science of Grassland Agriculture, Volume II, Seventh Edition.
Edited by Kenneth J. Moore, Michael Collins, C. Jerry Nelson and Daren D. Redfearn.

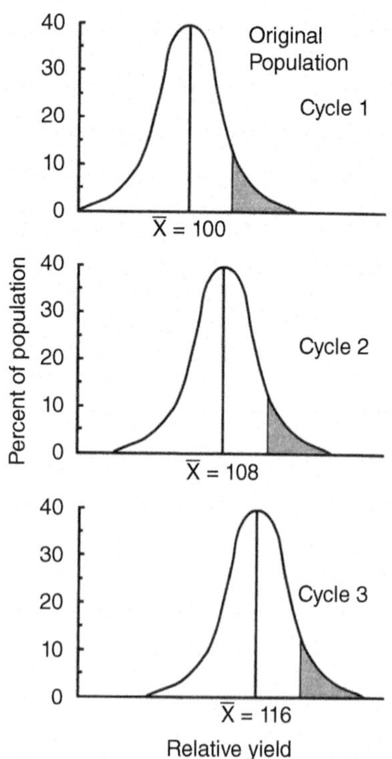

Fig. 30.1. The theoretical effects on forage yield from three cycles of restricted, recurrent selection (RRPS). Response to selection for other traits such as seed yield, forage quality, and disease resistance would be similar in a carefully planned and implemented breeding program.

of selection and mating needs to be repeated generation after generation (Figure 30.1; Table 30.1) because gains per generation are often small for complex traits and heritability is not always moderate to high. New cultivars achieved by breeding are released as accumulative, step-wise genetic gains in economic value.

Breeding Objectives

Forage breeding can improve the value of forage to livestock producers by solving specific production system problems. Production system problems can include inadequate forage quantity or low-quality forage during specific periods of the year, lack of persistence, and losses in yield and quality due to insects and diseases. It is important to identify and characterize specific production system problems before initiating a forage breeding program. In some cases, it may be easier to solve these problems by incorporating additional or new species into a production system than by improving multiple deficiencies in an existing species through breeding. For example, if forage quality appears to be low, it is necessary to first determine if the problem is due to antiquality factors such as alkaloids, low cell wall digestibility, or some other factor (Vogel and Sleper 1994). Breeding is most successful when the goal is clearly defined and the selection methods are available to differentiate among phenotypes for the specific traits under selection.

Forage breeders attempt to modify plants for traits that have economic value. These include forage yield, forage quality, resistance or tolerance to abiotic and biotic stresses, and improved establishment and persistence. Breeders have improved establishment capability by breeding for increased seedling vigor, a complex trait affected by seed size, seed quality, germination rate, emergence rate, relative growth rate and other physiological processes (McKell 1972). Substantial genetic gains in establishment have been made in some species, such as kura clover and reed canarygrass. In other species, improvements in establishment have occurred through improved seed quality and agronomic establishment practices including the use of pesticides for weed and insect control (Vogel et al. 1989).

Breeding objectives must be developed with the agricultural context and the environments well defined (Figure 30.2). Management and environmental factors that impose stress on a plant must be considered in developing breeding objectives and specific screening procedures to ensure that breeding goals will be achieved. In some cases, this may involve dissecting complex traits into simpler traits: for example, using a growth chamber or modified freezer to screen plants for persistence in harsh field environments. In other cases, management factors can dictate breeding objectives: for example, the concentration and type of toxic alkaloids in reed canarygrass are a critical breeding objective for grazing systems, but have no relevance for hay management, because alkaloids are metabolized during hay drying.

Persistence

Persistence is an economically important trait for perennial forages because the cost of establishment (including the associated loss of production) is amortized over the number of years the stand persists. Breeders have selected and bred for persistence using germplasm adapted to the climatic conditions of the target region and by breeding for resistance or tolerance to biotic and abiotic stresses (Hanson and Carnahan 1956; Vogel et al. 1989). Adapted germplasm can be obtained by using germplasm accessions that are native to the intended region of use or introduced from an area of the world with similar climate and soils. Improving adaptation to abiotic stresses such as drought, heat, wet soils, and other stresses are most effectively solved via breeding by incorporating germplasm adapted to those environmental conditions.

Table 30.1 Research phases and timetable for a perennial forage breeding program

Phase	Year 1	Year 2	Year 3	Year 4	Year 5
Phase 1: Germplasm collection and evaluation	Establish germplasm evaluation nurseries	Evaluate forage yields, quality, and other traits	2nd year of evaluation	Identify superior plants and move to crossing blocks, initial seed harvest	Harvest seed. Use seed in Phase 2 Synthetic populations may need to be random mated for several generations
Phase 2: Recurrent selection breeding program	Establish selection nurseries using seed from selected germplasm sources	Evaluate forage yields, quality, and other traits	2nd year of evaluation	Identify superior plants and move to crossing blocks, initial seed harvest	Harvest seed, repeat cycle in breeding program; use seed to plant regional trials
Phase 3: Regional small plot trials	Plant trials	Harvest trials	Harvest trials	Summarize data; begin seed increase of best strains for field trials	Seed harvested from increase nurseries
Phase 4: Grazing trials or field scale trials of advanced lines	Plant pastures or field trials	Grazing trial or field scale harvests	Grazing trial or field scale harvests	Increase best strain for release	Release seed to seed growers

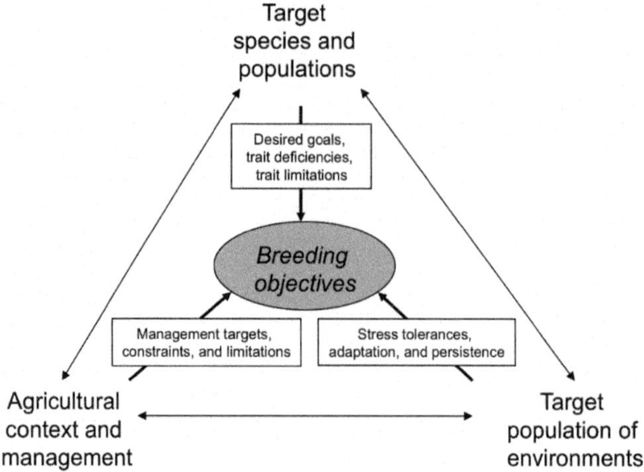

Fig. 30.2. Illustration of the trilateral relationship of target species, agricultural context, and target population of environments in determining forage breeding objectives.

Insect and Disease Resistance

Diseases and insects can affect forage yield, quality, and utilization by livestock in addition to persistence. Breeding for insect and disease resistance requires team efforts of entomologists and/or pathologists and breeders. Screening for resistance or tolerance under controlled conditions identifies genetically superior individuals. Resistant or tolerant plants are intercrossed or selfed, their progeny are screened, selections are made, and the process is repeated until populations with adequate levels of resistance are obtained (Figure 30.1). This process has been used to improve resistance or tolerance to diseases and insects in many grass and legume forages (Barnes et al. 1988; Casler et al. 1996). Almost all current alfalfa cultivars have resistance to several insects and diseases (National Alfalfa Alliance 2004).

Forage Yield

Forage yield has been and continues to be a main objective of forage breeders, with significant improvements made in some species (Barnes et al. 1988; Vogel et al. 1989; Casler et al. 1996; Wilkins and Humphreys 2003). In general, genetic gains from breeding for yield in forages have been less than those achieved for grain yields in cereals. A significant portion of the genetic gains for grain yields has been achieved by increasing the percentage of the total biomass that is grain, i.e. the harvest index. With forages, the physiological processes that result in increased above-ground biomass must be genetically improved. Further, this genetic increase in forage yield must be achieved while maintaining forage quality and its acceptability by livestock (Casler et al. 1996; Casler and Vogel 1999; Vogel and Jung 2001).

Seed Production

Although seed is not the principal use of forage plants, cultivars must have adequate seed production to be commercially viable. Significant improvements have been made in seed production in many species, particularly where a specific problem such as shattering can be overcome (Vogel et al. 1989). Increased seed yield should be a breeding objective if low seed yields adversely affect the economic availability of seed.

Forage Utilization and Quality

Forages are utilized mainly by ruminants, and quality of forages can significantly affect both milk and meat production. Forage quality is improved by reducing or eliminating antiquality compounds or by increasing forage digestibility, depending on the specific needs and limitations of a forage species (Vogel et al. 1989; Casler et al. 1996; Casler and Vogel 1999; Vogel and Jung 2001). For example, breeding to reduce the levels of undesirable alkaloids in reed canarygrass has led to broader acceptance and demand for this species (Casler 2010). Breeding for improved forage digestibility has significantly improved livestock productivity for improved cultivars compared to controls (Vogel and Sleper 1994; Casler and Vogel 1999; Vogel and Jung 2001; Wilkins and Humphreys 2003). Increased quality can increase net return to a livestock producer by increasing body weight gain or milk production without requiring additional investment for more livestock (Vogel and Sleper 1994; Casler and

Vogel 1999). Higher yield can increase net return, but the producer may need more livestock to use the additional forage.

Mode of Reproduction

The breeding system used to improve a species is determined by its mode of reproduction (Fehr 1987; Allard 1999). The mode of reproduction also limits the types of cultivars that can be produced. Depending on the species, forages can be propagated vegetatively or by seed, and seed can be produced via sexual or asexual (apomixis) mechanisms. Sexual species can be completely self- or cross-pollinated, or a combination of the two. Pollen of cross-pollinated species can be transferred either by wind, especially in grasses, or insects. Fortunately, the reproductive biology is already known for many important species (Tables 30.2 and 30.3) (Hanson and Carnahan 1956; Fehr and Hadley 1980).

Pollination Systems

Inflorescence structure and physiology can determine if a species is self- or cross-pollinated (Allard 1999). Dioecious species such as buffalograss have **staminate** and **pistillate** flowers on different plants and, of necessity, are cross-pollinated. Monoecious species such as eastern gamagrass are also cross-pollinated because they have staminate and pistillate flowers borne in separate locations on the same plant. Differences in time of pollen and pistil maturity also can result in cross-pollination or outcrossing. Restrictions on outcrossing, which enhance inbreeding, usually involve cleistogamy, i.e. fertilization before the bud opens. In grasses, cleistogamy occurs while the inflorescence is still enclosed in the upper leaf sheath, i.e. at boot stage.

Self-incompatibility or self-sterility mechanisms enforce cross-pollination in plants with perfect flowers. Incompatibility in plants can be defined as the inability of functional male and female gametes to produce normal seed following **pollination** (Brewbaker 1957; de Nettancourt 1977). The genotype of a pollen grain or its gametes is recognized as compatible or incompatible by the female flower (Dodds et al. 1997). If the genetic relationship between the pollen grain, or male gamete, and the **stigma** or **style** of the female flower is incompatible, the pollen grain will be rejected and fail to effect fertilization. Self-incompatibility systems in plants are analogous to recognition systems like antibody-antigen systems in animals. Self-incompatibility occurs in both legumes and grasses.

When no information is available, some basic tests can be conducted to determine the mode of reproduction (Allard 1999). The species is probably cross-pollinated and self-incompatible if covering the inflorescences with a bag prior to pollination, or physically isolating plants reduces or eliminates seed set. If seed are produced and the progeny are phenotypically very similar, the plants are either self-pollinated or apomictic. If some seed are produced and progeny are phenotypically variable, the parents likely are heterozygous plants of a primarily cross-pollinated species with some self-fertility. Self-pollinated plants can be crossed or hybridized with other unrelated plants of the same species by

Table 30.2 Modes of pollination, life cycle, ploidy level, and pollinators of some forage legumes

Forage legume	Life cycle	Ploidy[a] Level	Pollination system	Primary pollinator
Alfalfa	Perennial	4x, 2x	Cross	Leafcutter bees[b], Honey bees[b]
Alsike clover	Perennial – short lived	2x	Cross	Honey bees
Arrowleaf clover	Winter annual	2x	Cross	Honey bees
Berseem clover	Annual	2x	Cross	Honey bees
Birdsfoot trefoil	Perennial	2x, 4x	Cross	Honey bees, Bumble bees[b]
Cicer milkvetch	Perennial	8x	Cross	Bumble bees
Common vetch	Winter annual	2x	Self	—
Crimson clover	Winter annual	2x	Cross	Honey bees
Kura clover	Perennial	2x, 4x, 6x	Cross	Honey bees
Lespedeza	Annual	2x	Self	—
Medics	Annual	2x	Self	—
Red clover	Perennial – short lived	2x	Cross	Bumble bees
Subterranean clover	Winter Annual	2x	Self	—
Sweetclover	Biennial	2x	Cross	Honey Bees
White clover	Perennial – short lived	4x	Cross	Honey bees, Bumble bees

[a]Ploidy level, 2x = diploid, 4x = tetraploid, 6x = hexaploid, 8x = octaploid.
[b]Leafcutter bees (*Megachile rotundata*), Honey bees (*Apis mellifera*), Bumble bees (*Bombus* spp.).

Table 30.3 Modes of pollination, life cycle, and chromosome number of various forage grasses

Forage grass	Life cycle	Chromosome number	Pollination system[a]
Crested wheatgrass	perennial	28	cross – wind
Smooth bromegrass	perennial	28, 56	cross – wind
Tall fescue	perennial	42	cross – wind
Perennial ryegrass	perennial	14	cross – wind
Reed canarygrass	perennial	14, 28	cross – wind
Orchardgrass	perennial	28	cross – wind
Bermudagrass	perennial	30, 36	cross – wind
Switchgrass	perennial	36, 72	cross – wind
Big bluestem	perennial	60 (6x)	cross – wind
Buffalograss	perennial	20, 40, 50, 60	cross – wind
Weeping lovegrass	perennial	40	self (< 5% cross)
Bahiagrass	perennial	20, 40	cross – wind, or apomictic
Dallisgrass	perennial	40, 50, 60	apomictic
Buffelgrass	perennial	26, 32, 40, 54	apomictic
Pearl millet	annual	14	cross – wind, CMS
Sorghum	annual	20	self; cross with CMS
Maize	annual	20	cross – wind

[a]CMS = cytoplasmic male sterility.

using emasculation methods, which are either chemical (e.g. ethanol) or physical (e.g. using forceps to remove anthers).

Impacts on Breeding

The two main components of the breeding process are selection and mating/hybridization (Fehr 1987; Allard 1999). Forage species that reproduce via self-pollination are primarily cool-season grasses or annual legumes (Hanson and Carnahan 1956). To make a controlled mating, flowers of self-pollinated species need to be emasculated (anthers removed) prior to pollen shed. Each emasculated plant, inflorescence, or flower needs to be bagged or isolated to prevent unintentional crossing. Pollen from the selected male plant is transferred to the **stigma** of the emasculated flower when it is receptive, usually when the flower is fully open. Monoecious or dioecious cross-pollinated plants can be physically isolated or their flowers must be bagged for controlled matings. Pollen must be transferred by hand.

Cross-pollinated species with perfect flowers, i.e. containing both anthers and pistils, often have varying degrees of self-incompatibility (Knox et al. 1986; Vogel and Burson 2004). Plants of completely self-incompatible species can be intermated without contamination by mutual bagging of the parents or placing the plants together. The same process is usually used with cross-pollinated plants that have some self-compatibility because few seed are produced as a result of self-pollination. Detailed mating procedures for the major forage grasses and legumes have been developed (Fehr and Hadley 1980; Cope and Taylor 1985; Viands et al. 1988).

Seed of advanced populations of legumes can be produced using insect pollinators in isolated plots in the field or in cages in the field or greenhouse (see Chapter 32). Controlled cross-pollination by hand is used to produce seed of selected alfalfa parents for production of synthetic populations (Viands et al. 1988). Seed of wind-pollinated species can be produced in isolated nurseries or fields. Isolation distances and procedures to restrict foreign pollen differ among species but are known for most forages (Fehr and Hadley 1980).

Cytology and Ploidy

Many forage species are cytologically complex due to a wide range of chromosome numbers and ploidy levels among and within species (Hanson and Carnahan 1956; Cleveland 1985; McCoy and Bingham 1988). The chromosome number and meiotic chromosome behavior of a species must be known before a breeding program is initiated (Vogel and Pedersen 1993). A trait that may be simply inherited in a diploid like perennial ryegrass may have more complex inheritance in a hexaploid such as tall fescue due to the larger number of segregating genes and due to gene duplication between genomes. In polyploids such as alfalfa, there is the potential to have quadrivalent (four-at-a-time) or higher levels of chromosome pairing at meiosis that can affect the inheritance of traits. Plants of the same species with different ploidy levels are often not cross-compatible. Sometimes, the ploidy **pollination** barrier can be crossed by brute force, simply making hundreds of crosses and by using *in vitro* embryo rescue methods to save any embryos that are produced in these matings.

Germplasm

Genes available for plant breeders to use via conventional breeding methods are those accumulated by a species during its evolutionary history. A germplasm accession is a distinct genetic entity, which often is based on seed or plants collected at a specific site and stored in a gene bank or in a private collection. Genetic variation (or variation among plants of a species for specific **alleles** and their frequency) exists among germplasm accessions collected from different regions (ecotype variation), among accessions of an ecotype (population variation), and among plants of a population collected from a specific site (within-population variation) (Vogel and Pedersen 1993).

Plant breeders select plants from these natural sources to use as parents in breeding programs. There is sufficient genetic variation in most forage species to allow genetic improvements in desired traits (Vogel et al. 1989; Vogel 2000; Wilkins and Humphreys 2003). Breeders must continually collect, assemble, and evaluate germplasm for the environments of interest (Asay 1991; Rumbaugh 1991). Germplasm resources can be from *ex-situ* or *in-situ* sources. *Ex situ* sources are seed banks such as those in the USDA National Plant Germplasm system (USDA 2004) and from other breeding programs. *In situ* sources are from regions or sites where the species is growing and reproducing naturally in either private or public ownership. *Ex situ* germplasm sources are easily accessed whereas *in situ* sites require collection trips or expeditions at the proper times for seed or rhizome collection. The germplasm base must be adapted to latitude and climatic conditions where the cultivar products of the breeding program will be used. Latitude determines natural daylength of a site during the growing season which, in turn, regulates flowering time and dormancy.

For forage species with little or no previous breeding effort, direct selection of a superior accession or "ecotype selection" can lead to the rapid development and release of excellent cultivars (Vogel and Pedersen 1993). Ecotype selection is initiated by collecting an array of accessions for the specified region. For native species, this method is most effective if the germplasm is collected from the intended region of use. For introduced species, germplasm is collected and assembled from areas of the world that are climatic analogs of the target area. Both native and introduced accessions can be obtained from *in situ* collections or *ex situ* collections stored in germplasm banks.

Collected or acquired germplasm is first evaluated in replicated trials. Seed supplies of germplasms can be limited, and seed collected from native stands is often of low quality due to environmental conditions during seed development. It is preferred to maximize seed germination and seedling survival by starting seedlings in a greenhouse and then transplanting seedlings into space-planted plots in evaluation nurseries.

Multiple locations are preferred for germplasm evaluation, and the traits measured will vary with species and objectives (Figure 30.2). Data from evaluation nurseries are used to select the best local ecotypes or accessions and, in some instances, the superior plants within the best accessions. Selected plants of many perennial grasses can be moved to **polycross** or multiple-plant crossing nurseries simply by transplanting clonal pieces or ramets. An outstanding accession can be increased for testing and release as a cultivar without additional breeding work. Examples of cultivars developed by direct increase of germplasm accessions are 'Kentucky 31' tall fescue and 'Lincoln' smooth bromegrass (Alderson and Sharp 1994). When accessions are increased for release without additional selection, only the genetic variation among accessions is utilized. Both within and among accession genetic variation is used if selection is made within accessions and the best plants are used to produce a new population by intercrossing them in a polycross nursery. Strains produced by polycrossing require several years of testing before release.

The above system also is used to develop elite populations for use in breeding systems. Plants intercrossed in a polycross nursery produce Syn 1 (synthesis generation 1) seed. The Syn 1 should be advanced by one or more generations of random mating in a polycross or seed-increase nursery before beginning the selection process. This insures the population is approximately at random mating equilibrium (Falconer 1981) so that observed phenotypic differences among plants are due to **additive genetic effects** rather than **heterosis** (Vogel and Pedersen 1993).

Breeding Systems and Methods

A major objective of a breeding system is to reduce or identify the environmental effects on the phenotype so that true genetic differences among plants and families can be determined or estimated, i.e. to maximize heritability of a trait. Another objective is to intercross selected parents to achieve maximum genetic gains per year. The theoretical and practical efficiencies of an array of breeding systems available to forage breeders have been extensively reviewed (Sleper 1987; Vogel and Pedersen 1993; Casler and Brummer 2008).

Self-Pollinated Species

Breeding systems for self-pollinated forages are adapted from self-pollinated crops such as wheat. Depending on the degree of self-pollination, germplasms of these species usually consist of a mixture of highly inbred genotypes. Initially, there is extensive testing of parent lines from which superior parents are selected and mated to produce

Table 30.4 Comparison of pedigree and bulk breeding system for self-pollinated crops

Generation	Pedigree system	Bulk system
0	A cross is made between two homozygous plants; F_1 seed produced.	A cross is made between two homozygous plants; F_1 seed produced
1	F_1 plants produce F_2 seed.	F_1 plants produce F_2 seed.
2	Individual F_2 plants grown. Best plants selected and seed harvested on individual plants basis.	F_2 plants grown and seed harvested as a bulk.
3	F_2 family rows of F_3 plants grown. Best plants in best rows selected and harvested on individual plant basis.	F_3 plants grown and seed harvested as a bulk.
4	F_3 family rows of F_4 plants grown. Best plants in best rows selected and harvested on individual plant basis.	F_4 plants grown as a bulk. Individual inflorescences (or heads) harvested, threshed, and packaged on a single head basis.
5	F_5 family rows grown. Seed harvested on a family row basis. Selected lines given a number and advanced for testing.	F_5 rows planted from single head seed packets. Selections made on a single head row basis. Selected rows given a line number and advanced for testing.
6	Advanced testing and increase on a numbered line basis.	Advanced testing and increase on a numbered line basis.

Source: Derived from Fehr (1987).

F_1 seed. F_1 (filial generation 1) seed is used to plant the next generation which is naturally selfed to produce F_2 seed and so on. In a cross between two homozygous parents, the F_2 is the first segregating generation. Two principal breeding systems, bulk or pedigree, are used for self-pollinated species (Fehr 1987; Allard 1999). They differ in how the segregating generations are handled in the F_2 and subsequent generations.

In the bulk breeding method, individuals of the F_2 are harvested and seed is bulked to produce the F_3 and so on (Fehr 1987; Allard 1999) (Table 30.4). No selection is made during these segregating generations. In the F_4 or later generations when the plants in the bulk population are more than 80% homozygous, seed from selected plants are harvested individually and given line designations. After testing, the superior lines are advanced in generation for additional testing and subsequent release as a cultivar.

In the pedigree method, individual F_2 plants are coded and their progeny are subsequently tracked separately during the segregating generations (Fehr 1987; Allard 1999; Table 30.4). Selection occurs at each generation until the lines are almost completely homozygous (F_5 or F_6 generation). At this time, testing and release proceeds as in the bulk breeding method. The pedigree method enables the breeder to test the segregating lines each generation and discard undesirable lines, but it takes significantly more labor and land area than the bulk breeding method for the same number of crosses.

Cross-Pollinated Species

The most effective breeding systems for cross-pollinated forage grasses are based on fundamental principles of population and quantitative genetics and utilize **recurrent selection** or repeated generations of breeding to achieve desired goals (Figure 30.1). Objectives are to change the population mean for specific traits by increasing the frequency of desirable genes for those traits that are limiting or require improvement (Figure 30.2). Improved populations are generally released as synthetic cultivars.

Restricted, recurrent **phenotypic selection** (RRPS) and among-and-within-family selection (AWFS) are two popular recurrent selection systems for forage species (Vogel and Pedersen 1993; Casler and Brummer 2008). The RRPS system is an efficient form of mass selection for perennial forages (Burton 1974, 1982). In RRPS, a space-planted evaluation nursery with 1000 or more plants is established and subdivided into blocks or grids of 20–50 plants each to reduce the effect of within-field environmental variation (Casler 1992). Figure 30.3 illustrates a small portion of an RRPS selection nursery of reed canarygrass, in which ovine judges are used to help the breeder select plants that are palatable and desirable for advancement to the next generation.

Once the breeder is confident in the selection decisions, selected plants are dug up and split into multiple vegetative propagules for establishment of the polycross block, usually using a randomized block design. Polycrossing the selected plants doubles the expected genetic gain from selection as compared to traditional mass selection where only the female parents are selected (Vogel and Pedersen 1993; Casler and Brummer 2008). An equal amount of seed from each plant (genotype) in the polycross is bulked and is used to start the next cycle of selection. The polycross nursery also is used to produce

Fig. 30.3. A spaced-plant selection nursery of reed canarygrass (distance between plants is 0.9 m), showing differences in consumption (acceptability) by sheep, which is due to differences in alkaloid concentration and type. To control environmental and animal variation, the nursery was divided into grids by fences, one of which is shown in the background. *Source:* Photo by A.W. Hovin, University of Minnesota.

seed for yield tests and can serve as the source of **breeder seed**. The advantages of RRPS are that it is an easy breeding system to use, requires minimum time intervals per cycle, utilizes all the **additive genetic variation**, and because of the large number of plants that are intermated, it minimizes the potential for inbreeding depression.

The AWFS breeding system utilizes both among- and within-family genetic variation and allows the breeders to change the plot-testing methods (Vogel and Pedersen 1993; Casler and Brummer 2008). The system is usually initiated with a single cycle of RRPS. Seed are harvested from each plant in the cycle-1 polycross nursery and bulked by female genotype to create half-sib or polycross families. All seed from a single plant has the same maternal parent, but its male parents are the other plants in the polycross nursery, hence it is half-sib seed. The seed lots are used to establish a replicated evaluation nursery of spaced plants, row plots, or drilled sward plots, depending on the breeding objectives. Complex traits, such as forage yield, generally require more complex plots, such as drilled sward plots, to generate accurate yield estimates (Casler et al. 1996; Casler and Brummer 2008).

After two or more years of evaluation, the best families are identified and the following year individual plants within the best families are evaluated. The best plants from the best families are then selected for polycrossing. About 5 to 10% of the total plants in the nursery are polycrossed and the process is repeated the next generation. This breeding method has advantages over RRPS for traits such as forage yield that are highly influenced by environment effects. Because family records are maintained, the rate of inbreeding can be monitored and is under some control by the breeder.

Forage Hybrids

The previous breeding methods utilize additive genetic variation or simply the accumulated effects of desirable genes. In general, perennial forage breeders have not capitalized on the non-additive genetic variation, i.e. heterosis, even though substantial heterosis for traits such as forage yield exists in many species. Hybrids for commercial use have not been developed for most perennial forages because of the inability to effectively emasculate large numbers of plants in seed production fields. A notable exception is bermudagrass, for which a very effective sprigging system for propagating F_1 hybrids allows bermudagrass breeders to utilize all forms of genetic variation within the species and for interspecies hybrids.

Methods to produce hybrids of forages propagated by seed include first-generation chance hybrids, self-incompatibility hybrids, cytoplasmic male-sterile hybrids, apomictic hybrids, and hybrids produced by the use of male-gametocides (Burton 1986; Vogel et al. 1989). First generation chance hybrids, self-incompatible hybrids, and apomictic hybrids have been produced for a limited number of grasses. Hybrid cultivars of forage sorghum and alfalfa are currently being produced using cytoplasmic male sterility (Velde et al. 2002; Sun et al. 2003). Breeding procedures similar to those used to produce hybrid maize could be used to produce hybrid forage cultivars if methods to more effectively control pollination on a field scale could be developed. Breeders of perennial forage species have an advantage over maize breeders in that they may maintain their parent plants indefinitely through **vegetative propagation**.

Apomixis

Apomixis is an asexual form of reproduction where a seed develops without the union of a female and male gamete (Hanna and Bashaw 1987; Bashaw and Hanna 1990). Apomixis mimics sexual reproduction in that a female "gametophyte," i.e. an embryo sac, is usually formed in an **ovule**. However, the apomictic embryo sac develops from a vegetative or somatic cell in the ovule, so the nuclei in the sac have an unreduced chromosome number and all the chromosomes are from the maternal plant. The "egg cell" in an apomictic embryo sac is capable of initiating mitosis and developing into an embryo without being fertilized. Consequently, the seed and progeny developed from this embryo are exact replicas of the female parent, unless a **mutation** occurs or an unreduced egg is fertilized (both very rare events).

Apomixis is nature's way of cloning plants by seed, in effect similar to propagating plants with buds, stolons, or rhizomes. Except for kentucky bluegrass, many grasses that reproduce apomictically originated in tropical or subtropical regions. Breeding systems for improving apomictic forages are unique and, in general, are not useful for improving sexual species (Hanson and Carnahan 1956; Bashaw 1980; Bashaw and Funk 1987). In nature, most apomictic species also produce some sexual offspring so they are known as facultative apomicts. Apomixis can be either an impediment or a valuable tool to genetic improvement depending on whether a large number of different polymorphic genotypes occur naturally within the apomictic species and if sexual plants exist within the species to which crosses can be made to produce genetic variants.

In breeding programs, superior, naturally occurring apomictic ecotypes are identified using the ecotype evaluation and selection procedure (Bashaw and Funk 1987; Hanna and Bashaw 1987). The most vigorous and productive ecotypes are selected, increased, tested, and released as new cultivars. The success of ecotype selection for apomictic species is improved if there is a large amount of genetic variation between and within ecotypes. Additional gains are made via sexual recombination in facultative apomictic species such as kentucky bluegrass and buffelgrass, in which some genotypes are capable of producing sexual seed, creating new genetic combinations.

Selection, Testing, and Cultivar Development

Heritability is the proportion of the total phenotypic variation among plants that is due to genetic differences. For many important forage traits such as yield or digestibility that are controlled by many genes, heritability is usually 0.30 or lower. This means that only 30% of the total phenotypic variation is due to genetic differences among individuals; i.e. 70% of the observed variability is environmental variation (Vogel and Sleper 1994; Vogel 2000). The efficiency of plant breeding could be greatly enhanced if breeders could directly measure true genetic differences or identify genotypes, increasing the heritability of the trait. Genetic gain per year is directly proportional to heritability.

Selection and Testing Procedures

Breeders are concerned with genetic gain per year, because they are under pressure to solve existing problems that limit livestock production or profitability and because funding is always limited. There are three general approaches that breeders use to increase gain per year, each of which carries a cost. First is to increase heritability, which can sometimes be accomplished by improving plot methods, increasing replication, and controlling spatial or environmental variation. Second is to increase selection intensity by using larger populations of plants. Third is to reduce the cycle or generation time to the minimum required time for adequate trait measurements.

Breeders need to have efficient evaluation methods to detect differences among phenotypes. A lot of forage breeding evaluation is done using visual scores, but potential progress is limited because it is not possible to visually score forage quality and other complex traits. Biological assays and technologies are used routinely to quantify complex traits. For example, the use of Near Infrared Reflectance Spectroscopy (NIRS) enables breeders to rapidly measure many quality traits on thousands of forage samples per year (Vogel and Jung 2001).

Each forage breeding product, i.e. strain or experimental cultivar, needs to be thoroughly tested in the target environments under management conditions in which it will be used. This requires field plot research and pasture trials replicated at multiple locations to best simulate the environments and management practices that will be encountered on working farms (Table 30.1; Figure 30.2). Grass breeders have relied extensively on evaluations using small plot trials that have been managed for hay production even though most forage grasses are used in pastures. More grazing trials need to be conducted in the future to ensure that improved cultivars are adapted to the grazing environment (Casler and Vogel 1999).

All forage species are subject to environmental limitations, with a fairly well-defined range of adaptation. Some species are adapted to such a broad range of environments that individual cultivars cannot be used across the entire range, due to limitations in stress tolerances. Switchgrass and alfalfa are two prime examples. Both are adapted to climates that range from subtropical or Mediterranean to temperate environments with long and harsh winters. These characteristics create genotype x environment (GE) interactions, the phenomenon by which some cultivars are superior in some environments, but other cultivars are superior in other environments. Breeders are highly challenged by GE interactions, chiefly by the need to understand the underlying environmental factors that regulate the interaction (e.g. temperature, moisture, mineral deficiencies, photoperiod) and to carefully match germplasm resources, breeding locations, and testing locations with the appropriate environmental conditions to maximize the probability of cultivar success.

Genomic Prediction and Selection

With the development of DNA marker systems that are capable of generating large numbers of markers across the entire genome on thousands of individual plants, genomic prediction and selection has become a viable tool for many forage breeders (Hayes et al. 2013; Lin et al. 2014). Briefly, this methodology involves the creation of a training population from the breeding program, consisting of several hundred genotypes or families. Each genotype or family is evaluated for both the trait of interest, e.g. forage yield, and DNA markers that ideally represent the entire genome. Predictive equations developed from the DNA markers to predict forage yield can be sufficiently accurate to allow predictions from these equations to serve as a surrogate for forage yield (Ramstein et al. 2016). This allows the breeder to conduct two or three cycles of selection for complex traits, such as forage yield, using greenhouse-grown seedlings, drastically reducing generation time and increasing genetic gain per year by up to three times that possible using traditional field-based selection (Resende et al. 2014).

Cultivar Development and Release

Forage cultivars released for use in production agriculture include clonal cultivars, line cultivars, open-pollinated cultivars of cross-pollinated species, synthetic cultivars, hybrid cultivars, composite cultivars, and apomictic cultivars. The types vary because of differences in the reproductive systems of forage species and the different breeding methods used to develop improved cultivars (Fehr 1987).

Clonal cultivars consist of a single clone or very similar clones propagated by vegetative propagules. 'Coastal' bermudagrass is an example of an asexual grass cultivar (Alderson and Sharp 1994). Line cultivars are groups of plants that are very closely related and have a coefficient of parentage greater than 0.87. These cultivars are usually self-pollinated and trace to a single plant selected at the F_3 or later generation. 'Revenue' slender wheatgrass traces to seed from a single selected plant and is an example of a line cultivar of a species that is largely self-pollinated (Alderson and Sharp 1994). Open-pollinated cultivars consist of diverse populations of plants of normally cross-pollinated species, generally selected for uniformity for traits such as flowering time, but have some variation for other traits. They are produced by cross-pollination in isolation. 'Lincoln' smooth bromegrass is an example.

Synthetic cultivars are developed for cross-pollinated species by intercrossing several selected genotypes or parent clones grown in isolation. The parent lines are designated the Syn 0 generation (Allard 1999). The Syn 1 generation is grown from seed produced by intercrossing Syn 0 plants in isolation. Progeny of the Syn 1 are the Syn 2 generation, etc. In practice, Syn 1 seed is usually produced by the breeder and labeled as "breeder's seed," Syn 2 seed is often labeled as "foundation seed," while Syn 3 or later generations are the commercial certified seed.

Single-cross hybrid cultivars are the first generation (F_1) progenies from a cross of two inbred lines. Mating two single crosses produces a double-cross hybrid. Maize silage hybrids are hybrid cultivars. Intercrossing several populations, lines, or accessions to produce a highly heterogeneous population is a method used to create composite cultivars. 'Cimarron' little bluestem is an example of a composite cultivar (Alderson and Sharp 1994). Composite cultivars are becoming increasingly popular for use in revegetation of native rangelands, prairies, and savannas, offering the advantage of high levels of genetic variation to allow these populations the opportunity to evolve with a changing and variable environment (Jones 2003; Jones and Monaco 2007). Seed production, seed certification, and cultivar protection issues are addressed in Chapter 32.

Summary, Needs, and Future Outlook

Forage breeders have created thousands of new cultivars that represent significant improvements in forage traits, translated into improved livestock production, farm profitability, and broader adaptation of individual forage species. Despite these successes, many challenges remain. Long-term funding for forage breeding is probably the single greatest challenge, due to the length of time required to produce a new cultivar and the need for long-term continuity in the breeding program. As breeders have made progress and developed more elite and narrower gene pools over time, they are additionally challenged by the need for more refined and more relevant breeding systems and methods. Many of the early gains were achieved by relatively easy and simple elimination of unadapted germplasm and plants with low vigor or highly susceptible to disease, insects, and stresses. As germplasm has improved, it has become more elite and more homogeneous, challenging breeders to use more relevant plot methods, such as sward plots instead of spaced plants, and family-based selection methods that allow for increased replication or selection intensity, as well as possible incorporation of genomic DNA marker technologies.

One of the greatest challenges for the next generation of forage breeders will be the development of cultivars that can either resist the negative impacts of climate change or adapt to changing environments. One of the benefits of composite cultivars is the diversity that can be built into the cultivar, providing potential levels of genetic diversity that exist within multiple-origin polycrosses, which are an integral part of the restoration gene pool concept (Jones 2003; Jones and Monaco 2007). For many forage crops, in which livestock producers depend on stands being viable and sustainable for 10 years or more, composite cultivars that contain more diversity than simply polycrosses of a small number of genotypes, may provide sufficient genetic diversity to allow stands to evolve in response to local soil types or environmental fluctuations. We have known since the 1960s that the diversity within forage-plant populations can allow those populations to undergo natural selection and adapt to changing environmental conditions (Casler et al. 1996). Diverse populations of many forage species are capable of responding with natural selective shifts in response to soil acidity, heavy metals, air pollutants, salinity, extreme temperatures, drought, and several other stress factors. Breeders should be cognizant of the potential and opportunities for forage cultivars to adapt to changing environments, building in genetic diversity to allow this phenomenon to occur as long as it can be done without sacrificing overall performance. For species such as alfalfa, cultivar turnover is rapid and new cultivars are available every few years. In this scenario, breeders do not require this level of diversity within any given cultivar, but it should be incorporated into the breeding program in such a way to allow breeders to keep up with the stress tolerances required to meet the challenges of changing environments.

References

Alderson, J. and Sharp, W.C. (1994). *Grass Varieties in the United States. Agricultural Handbook No. 170*. Washington, DC: Soil Conservation Service, USDA.

Allard, R.W. (1999). *Principles of Plant Breeding*, 2e. New York, NY: Wiley.

Asay, K.H. (1991). Contributions of introduced germplasm in the development of grass cultivars. In: *Use of Plant Introductions in Cultivar Development. Part 1. CSSA Special Publication 17* (eds. H.L. Shands and L.E. Wisner), 115–125. Madison, WI: CSSA.

Barnes, D.K., Goplen, B.P., and Baylor, J.E. (1988). Highlights in the USA and Canada. In: *Alfalfa and Alfalfa Improvement*. Agron. Monogr. 29 (eds. A.A. Hanson, D.K. Barnes and R.R. Hill), 1–24. Madison WI: ASA.

Bashaw, E.C. (1980). Apomixis and its application in crop improvement. In: *Hybridization of Crop Plants* (eds. W.R. Fehr and H.H. Hadley), 45–63. Madison, WI: ASA.

Bashaw, E.C. and Funk, C.R. (1987). Apomictic grasses. In: *Principles of Cultivar Development: Vol. 2, Crop Species* (ed. W.R. Fehr), 40–82. New York, NY: MacMillan.

Bashaw, E.C. and Hanna, W.W. (1990). Apomictic reproduction. In: *Reproduction Versatility in the Grasses* (ed. G.P. Chapman), 100–130. Cambridge: Cambridge University Press.

Brewbaker, J.L. (1957). Pollen cytology and self-incompatibility systems in plants. *J. Heredity* 48: 271–277.

Burton, G.W. (1974). Recurrent restricted phenotypic selection increases forage yields of Pensacola bahiagrass. *Crop Sci.* 14: 831–835.

Burton, G.W. (1982). Improved recurrent restricted phenotypic selection improves Bahia forage yields. *Crop Sci.* 22: 1058–1061.

Burton, G.W. (1986). Developing better forages for the south. *J. Anim. Sci.* 63: 63–65.

Casler, M.D. (1992). Usefulness of the grid system in phenotypic selection for smooth bromegrass fiber concentration. *Euphytica* 63: 239–243.

Casler, M.D. (2010). Genetics, breeding and ecology of reed canarygrass. *Int. J. Plant Breeding.* 4(1): 30–36.

Casler, M.D. and Brummer, E.C. (2008). Theoretical expected genetic gains for among-and-within-family selection methods in perennial forage crops. *Crop Sci.* 48: 890–902.

Casler, M.D. and Vogel, K.P. (1999). Accomplishments and impact from breeding for increased forage nutritional value. *Crop Sci.* 39: 12–20.

Casler, M.D., Pedersen, J.F., Eizenga, G.C., and Stratton, S.D. (1996). Germplasm and cultivar development. In: *Cool-Season Forage Grasses*. Agron. Mono. 34 (eds. L.E. Moser, D.R. Buxton and M.D. Casler), 413–469. Madison, WI: ASA.

Cleveland, R.W. (1985). Reproductive cycle and cytogenetics. In: *Clover Science and Technology*. Agron. Mono. 25 (ed. N.L. Taylor), 71–105. Madison, WI: ASA.

Cope, W.A. and Taylor, N.L. (1985). Highlights in the USA and Canada. In: *Clover Science and Technology*. Agron. Mono. 25 (ed. N.L. Taylor), 323–405. Madison, WI: ASA.

Dodds, P.N., Clark, A.E., and Nebigin, E. (1997). Molecules involved in self-incompatibility in flowering plants. *Plant Breed. Rev.* 15: 19–42.

Falconer, D.S. (1981). *Introduction to Quantitative Genetics*, 2e. New York, NY: Longman.

Fehr, W.R. (1987). *Principles of Cultivar Development*, vol. 1. New York, NY: MacMillan.

Fehr, W.R. and Hadley, H.H. (eds.) (1980). *Hybridization of Crop Plants*. Madison, WI: ASA.

Hanna, W.W. and Bashaw, E.C. (1987). Apomixis: its identification and use in plant breeding. *Crop Sci.* 27: 1136–1139.

Hanson, A.A. and Carnahan, H.L. (1956). Breeding perennial forage grasses. United States Department of Agriculture. Technical Bulletin No. 1145.

Hayes, B.J., Cogan, N.O.I., Pemberton, L.W. et al. (2013). Prospects for genomic selection in forage plant species. *Plant Breed.* 132: 133–143.

Jones, T.A. (2003). The restoration gene pool concept: beyond the native versus non-native debate. *Restor. Ecol.* 11: 281–290.

Jones, T.A. and Monaco, T.A. (2007). A restoration practitioner's guide to the restoration gene pool concept. *Ecol. Rest.* 25: 12–19.

Knox, R.B., Williams, E.G., and Dumas, C. (1986). Pollen, pistil and reproductive function in plants. *Plant Breed. Rev.* 4: 9–80.

Lin, Z., Hayes, B.J., and Daetwyler, H.D. (2014). Genomic selection in crops, trees and forages: a review. *Crop Pasture Sci.* 65: 1177–1191.

McCoy, T.J. and Bingham, E.T. (1988). Cytology and cytogenetics of alfalfa. In: *Alfalfa and Alfalfa Improvement*. Agron. Mono. 29 (eds. A.A. Hanson, D.K. Barnes and R.R. Hill), 737–776. Madison, WI: ASA.

McKell, C.M. (1972). Seedling vigor and seedling establishment. In: *The Biology and Utilization of Grasses* (eds. V.B. Younger and C.M. McKell), 76–89. New York, NY: Academic Press.

National Alfalfa Alliance (2004). Fall dormancy & pest resistance ratings for alfalfa varieties 2004/2005 Edition. http://www.alfalfa.org/pdf/Alfalfa%20variety%20leaflet.pdf (accessed 10 October 2019).

de Nettancourt, D. (1977). *Incompatibility in Angiosperms*. New York, NY: Spring-Verlag.

Ramstein, G.P., Evans, J., Kaeppler, S.M. et al. (2016). Accuracy of genomic prediction in switchgrass (*Panicum virgatum* L.) improved by accounting for linkage disequilibrium. Genes, genomes. *Genetics* 6: 1049–1062.

Resende, R.M.S., Casler, M.D., and de Resende, M.V. (2014). Genomic selection in forage breeding: accuracy and methods. *Crop Sci.* 54: 143–156.

Rumbaugh, M.D. (1991). Plant introductions: the foundation of north American forage legume cultivar development. In: *Use of Plant Introductions in Cultivar Development. Part 1. CSSA Specical Publication 17* (eds. H.L. Shands and L.E. Wisner), 69–102. Madison, WI: CSSA.

Sleper, D.A. (1987). Forage grasses. In: *Principles of Cultivar Improvement* (ed. W.R. Fehr), 161–208. New York, NY: Macmillan.

Sun, P., Velde, M. and Gardner, D.B. (2003). Alfalfa hybrids having at least 75% hybridity. US Patent Application No. 20030172410.

USDA-ARS (2004). National plant germplasm system. GRIN. http://www.ars-grin.gov/npgs/index.html (accessed 10 October 2019).

Velde, M., Undersander, D., Sun, P. et al. (2002). Forage yield response of alfalfa to percent hybridism. In: *North American Alfalfa Improvement Conference, Sacramento, CA, July 28–31, 2002*, 28.

Viands, D.R., Sun, P., and Barnes, D.K. (1988). Pollination control: mechanical and sterility. In: *Alfalfa and Alfalfa Improvement*. Agron. Mono. 29 (eds. A.A. Hanson, D.K. Barnes and R.R. Hill), 931–960. Madison, WI: ASA.

Vogel, K.P. (2000). Improving warm-season grasses using selection, breeding, and biotechnology. In: *Native Warm-Season Grasses: Research Trends and Issues. CSSA Special Publication 30* (eds. K.J. Moore and B. Anderson), 83–106. Madison, WI: CSSA.

Vogel, K.P. and Burson, B. (2004). Breeding and genetics. In: *Warm-Season Grasses* (eds. L.E. Moser, L. Sollenberger and B. Burson), 51–96. Madison, WI: ASA-CSSA-SSSA Monograph.

Vogel, K.P. and Jung, H.G. (2001). Genetic modification of herbaceous plants for feed and fuel. *Crit. Rev. Plant Sci.* 20: 15–49.

Vogel, K.P. and Pedersen, J.F. (1993). Breeding systems for cross-pollinated perennial grasses. *Plant Breed. Rev.* 11: 251–274.

Vogel, K.P. and Sleper, D.A. (1994). Alteration of plants via genetics and plant breeding. In: *Forage Quality, Evaluation, and Utilization* (ed. G.C. Fahey Jr. et al.), 891–921. Madison WI: ASA.

Vogel, K.P., Gorz, H.J., and Haskins, F.A. (1989). Breeding grasses for the future. In: *Contributions from Breeding Forage and Turf Grasses. CSSA Special Publication 15* (eds. D.A. Sleper, K.H. Asay and J.F. Pedersen), 105–122. Madison, WI: CSSA.

Wilkins, P.W. and Humphreys, M.O. (2003). Progress in breeding perennial forage grasses for temperate agriculture. *J. Agric. Sci.* 140: 129–150.

CHAPTER

31

Biotechnology and Molecular Approaches to Forage Improvement

E. Charles Brummer, Professor, *University of California, Davis, CA, USA*
Zeng-Yu Wang, Professor, *Qingdao Agricultural University, Qingdao, China*

Introduction

Biotechnology may be defined as the targeted genetic manipulation of plants to produce novel traits and improved germplasm. Biotechnologic methods represent a broad collection of laboratory techniques that have been developed to investigate, understand, and manipulate the genetic constitution of plants and animals. Traditional plant breeding (Chapter 30) has been highly effective at improving forage crops for many traits. Biotechnologies such as genetic transformation, gene editing, and marker-assisted selection provide additional tools to breeders that can increase the efficiency of cultivar development and lead to the incorporation of novel traits currently unavailable in a particular crop species.

A plant of a given genotype will produce a range of phenotypes depending on the environment in which it is grown and on its developmental stage. Determining how genes interact with each other and with the environment to control various phenotypes is the ultimate goal of modern genetics. The journey from gene to phenotype (Figure 31.1) is complex for most traits. For agronomically important traits like biomass yield or **winterhardiness**, many genes interact to form the ultimate phenotype. Understanding the relationship between genotype and phenotype enables biotechnologies to be used to improve crop production.

The science supporting the suite of tools called biotechnology derives from genomics, genetics, molecular biology, and plant breeding. Broadly speaking, **genomics** is the study of the genome, which is the entire DNA sequence, including all of the genes, of a plant. Genomics technologies are making the identification of genes routine and providing unparalleled opportunities to manipulate the genomes of crop plants for both basic scientific understanding and for improved traits.

Biotechnologies are used in two major ways for plant improvement. First, they are used to identify genes, or chromosomal regions likely to contain the genes, that control traits of interest. This enables the use of genetic markers to select for desirable alternate forms of genes, called **alleles**, more effectively within breeding populations. Second, once identified, genes can be inserted into any plant's genome to produce novel products or to enhance the ability of a plant to survive stresses. If specific changes in a gene are known to produce a desired effect, then genome editing methods can be used to make that targeted alteration.

Basic Genetics

The genetic material of a cell is deoxyribonucleic acid (DNA), which is packaged into chromosomes residing in the cell's nucleus. Plants also have a small amount of

Forages: The Science of Grassland Agriculture, Volume II, Seventh Edition.
Edited by Kenneth J. Moore, Michael Collins, C. Jerry Nelson and Daren D. Redfearn.

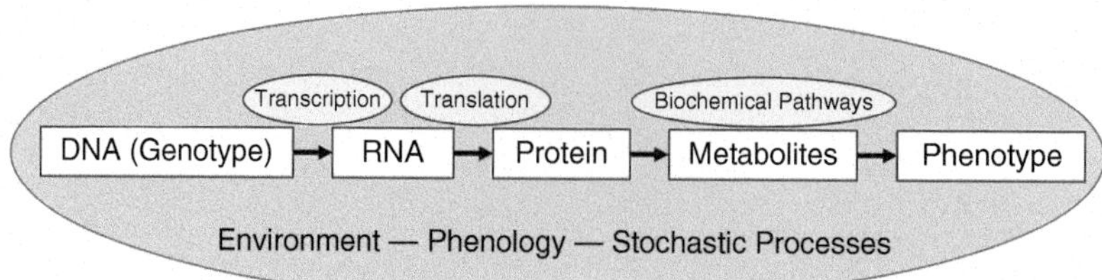

Fig. 31.1. The path from genotype (DNA) to phenotype (what we observe) includes transcription of DNA to RNA, translation of RNA to proteins, and the interaction of proteins in various biochemical pathways to produce metabolites that collectively add up to the ultimate, visible phenotype. Though this canonical pathway is drawn as linear and one-way, feedback loops exist to repress transcription or translation or to alter the flux through pathways. This all occurs within the environment across developmental stages of a plant both of which further influence how this pathway operates. Further variation is due to stochastic processes.

DNA in the mitochondrion and the chloroplast. DNA consists of four nucleotides: adenosine, cytosine, guanine, and thymine (or A, C, G, and T), arranged in long linear chains. Two parallel chains are bound together by chemical bonds between A and T or between C and G to form a helical structure. The DNA on a chromosome contains both genic and non-genic sequences. All cells of a plant contain the same DNA sequence.

The linear order of nucleotides provides the information needed to produce proteins. Each set of three nucleotides specifies a particular amino acid. In order to produce proteins, a cell copies one of the DNA strands of a gene to make a single-stranded RNA molecule. Transcription of DNA to RNA does not happen continuously but only occurs when the particular protein coded by a given segment of DNA is needed by the cell. The RNA molecule is read by cellular machinery to specify a chain of amino acids that are linked together to form a protein molecule, which is then folded into the correct conformation in order to become functional. Many proteins function within biochemical networks as enzymes to produce compounds (metabolites) necessary for cellular growth, maintenance, and defense. The ultimate result of the activity and interaction of various genes and their downstream products is the observable phenotype. Analogously to the genome representing all DNA within a plant, the **transcriptome** refers to all the transcribed RNA sequences, the **proteome** to all the proteins produced, and the **metabolome** to the complete set of metabolites produced by a plant. Experimentally, plants are often grown under a particular set of conditions – for example, drought and no drought – and the transcriptome, proteome, and/or genome analyzed under each condition and compared. Differences between these conditions could indicate genes, proteins, or metabolites that are related to the condition analyzed, in this case, drought. Further information on modern genetics may be found in elementary genetics texts (e.g. Griffiths et al. 2015).

Genomes and Ploidy

Ploidy describes the number of chromosomes and chromosome sets a plant contains in its nucleus. One complete set of chromosomes is called a **genome**. Typically, plants have an even number of chromosomal sets, leading to regular pairing during meiosis. A diploid species has two copies of its genome. Many forage species are **polyploid**, having more than two sets of chromosomes. **Autopolyploid** species have three or more copies of a single genome. Cultivated alfalfa and orchardgrass are autotetraploids, containing four sets of chromosomes, though diploids of both species are present in the wild. In contrast, **allopolyploid** species evolve when two partially related or unrelated species hybridize, each of which contributes its own genome. The two parental genomes that now constitute the allopolyploid do not pair with one another during meiosis. Thus, allopolyploids consist of two or more separate sub-genomes that each contain at least two copies. White clover is an allotetraploid and tall fescue is an allohexaploid.

Various methods have been developed to analyze chromosome number and ploidy. A rapid and inexpensive method to determine the ploidy of a large number of accessions is **flow cytometry**. With this method, nuclei are isolated from young plant leaves, fluorescently stained, and evaluated on the flow cytometer, which assesses the amount of DNA in a cell based on the intensity of staining. Knowing ploidy is essential to effectively using germplasm in a breeding program because hybridization between plants of different ploidy levels cannot be done easily. Part of the USDA-National Plant Germplasm System collection of alfalfa was assessed by flow cytometry to discriminate between diploid and tetraploid accessions.

The complete DNA sequence of a species' genome serves as a useful framework for all other genomic activities. Unfortunately, for most forage crops, no genome sequences exist and those that do are rather rudimentary. A draft genome has been assembled for italian ryegrass (Knorst et al. 2019b). For perennial ryegrass, the transcriptome was sequenced, enabling the construction of a partial genome sequence based on alignment of the gene sequences to barley, which is closely related to ryegrass (Byrne et al. 2015). Among forage legumes, the model legume and minor forage crop, barrel medic has been sequenced and assembled to very high standards (Young et al. 2011). This genome has been used to align genetic maps of alfalfa (e.g. Li et al. 2014b) or draft sequences of red clover (de Vega et al. 2015). The rapidly decreasing cost of sequencing will, undoubtedly, lead to most forage crops having sequenced and assembled genomes in the near future. Even then, comparative genomics – drawing information from related species to support or augment what is known for any given species – is a very powerful tool for breeders and geneticists to exploit (Figure 31.1).

Genome sequences represent just one of many sources of genomic data that can be used to support breeding activities. Making sense of large datasets and using one resource to assist with others – such as comparing genetic maps with DNA sequences – requires a significant amount of bioinformatics support. Consequently, databases compiling whole genome sequences, transcriptomes, genetic maps, quantitative trait locus (QTL) information, and even phenotypic information have been developed to make these data useful for research and breeding. Databases such as the Legume Information System (http://www.comparative-legumes.org), Gramene (http://www.gramene.org), and the Alfalfa Breeders' Toolbox (https://www.alfalfatoolbox.org/) are some of these resources that help researchers apply genomics and mapping information to forage crops.

Breeding with Genetic Markers

A goal of plant breeding is to select desirable plants as quickly and as accurately as possible. Some traits may not be measurable for several years – multi-year persistence, for example – and others may be expensive or time consuming to assay on large numbers of plants, such as biomass yield or forage nutritive value across multiple harvests over several years. Traditionally, plant breeders have used correlated traits to indirectly select for difficult traits; for example, if plant height were correlated with biomass yield, then selection could be made on height, which is easier to measure, indirectly improving forage yield. However, not all desired traits can be selected using a highly correlated trait. DNA-based genetic markers serve the role of a correlated trait and once identified, the genetic marker that is linked to a gene controlling a trait of interest can be used for selection. Since the genetic marker represents variation in a plant's genome, it can be assayed at any developmental stage and from any tissue, including when plants are seedlings. At least theoretically, this can shorten the selection cycle dramatically, rapidly accelerating genetic gain.

Genetic Marker Systems

Genetic markers first became available for crop improvement in the mid-1980s (Beckmann and Soller 1986). Though they are widely applied to forage crop research today, their direct use in developing improved cultivars is just beginning. Because DNA sequences differ at many points in the genome among plants within a species, DNA-based markers are plentiful and increasingly inexpensive to assay. The most common differences observed between DNA sequences are variant nucleotides at specific genomic locations. For example, on a given strand of DNA, one plant may carry the nucleotide base sequence "AG**A**TA" while another may carry "AG**G**TA." The difference between the two plants for that particular nucleotide is termed a "single nucleotide polymorphism" or "SNP." This nucleotide locus has two alleles, either "A" or "G." In addition to SNPs, insertions or deletions (indels) of short to long sequences of DNA may also be observed among plants of a given species. Based on differences in the DNA nucleotide sequences, genetic marker systems have been developed to assay the polymorphisms. Because of their frequency throughout the entire genome, SNPs are the most widely used DNA markers at the current time. Older markers systems, such as simple sequence repeat (SSR) markers, which are based on different numbers of short repeated sequences among plants (such as "AATAAT" in one plant and "AATAATAATAATAAT" in another), are still in use, particularly for variety identification (Annicchiarico et al. 2016).

Identifying the polymorphisms is the first step in developing useful genetic markers. Next, breeders need to assay a plant's DNA for the marker allele(s) it carries. Many genetic marker types and marker assays have been developed over the past 40 years. Platforms have been developed to assay thousands (or millions) of loci throughout the genome simultaneously (for a recent review, see Torkamaneh et al. 2018); others can inexpensively evaluate one or a few markers on thousands of plants. The use of particular platforms depends on project goals, budgets, and timeliness of data generation.

Sequence-Based Markers/Technologies

Perhaps the greatest recent advance in marker technologies has been the development of whole-genome sequence-based markers. Grouped collectively under the descriptive names of Randomly Amplified DNA (RAD) (Miller et al. 2007) or Genotyping-by-Sequencing (GBS) (Elshire et al. 2011), these technologies are used to partially sequence the genomes of every plant being evaluated in order to directly identify SNP and possibly indel markers directly from the sequence. Sequence-based markers can also be developed by sequencing targeted genome regions (Choi et al. 2009; Ali et al. 2016). For instance, exome capture (Choi et al. 2009) is a method to preferentially separate gene sequences from the entire genome, and then sequence just this selected part. These methods require a significant amount of bioinformatics to analyze the millions of short DNA sequences to determine the marker genotypes for all individuals being evaluated. Though pipelines to handle these data are being developed and improved, the bioinformatic needs of GBS-type marker platforms are a bottleneck for many breeding and genetics programs (Torkamaneh et al. 2018).

Array-Based Markers

Array-based marker systems assay particular genomic loci that have been previously identified. An array can contain tens to millions of loci simultaneously. Generally, these systems work by means of fluorescence, with a differently colored dye representing the sequence difference at a locus (e.g. "A" vs "G"). Thus, plants with one nucleotide ("A") will show a red fluorescence at a specific locus while those with another nucleotide ("G") will show green. A heterozygote with both alleles will be intermediate. As one example, an Infinium array was used in alfalfa to assess genetic diversity (Li et al. 2014a). Arrays produce high-quality data but can be expensive to develop and use.

Single Marker Analyses

Commonly, mapping projects begin by using one of the genome-wide technologies to identify one or a few markers associated with traits of interest. Those markers then need to be converted into a new marker assay that can be used singly on many genotypes in a marker-assisted breeding program. A commonly used method to assess single genetic markers is the Kompetitive Allele Specific PCR (KASP™) marker assay (Semagn et al. 2014). Single marker assays are useful to screen large numbers of plants for a specific polymorphism very cheaply. As with all marker systems, markers that can unambiguously distinguish between alleles at a locus ("A" vs. "G," for example) and differentiate all possible genotypes (for example, in a diploid, among AA, AG, GG; or in an autotetraploid, among AAAA, AAAG, AAGG, AGGG, and GGGG) are the most useful.

Using Markers for Germplasm Diversity and Varietal Identification

Besides their use in genetic mapping (see below), molecular genetic markers can be used to assess relationships among cultivars, germplasm accessions, populations, and/or species. Understanding the partitioning of genetic variation between and within populations and clarifying how different germplasms are related to one another can improve both the management and use of germplasm resources; as examples, an analysis of diploid alfalfa germplasm (Şakiroğlu et al. 2010) and the population genetic structure of North American switchgrass germplasm (Evans et al. 2015, 2018) (Figure 31.2). An important aspect of cultivar registration and plant breeder's rights is the necessity to confirm the distinctiveness of new cultivars. DNA profiling using SSR or SNP markers has been applied to many species, recently in alfalfa (Annicchiarico et al. 2016; Julier et al. 2018) and in ryegrasses (Pembleton et al. 2016).

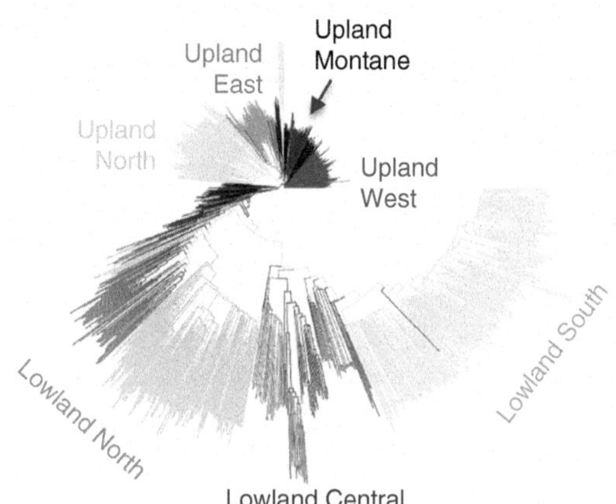

FIG. 31.2. Genetic distances among 1169 genotypes from throughout the native range of switchgrass in North America. This figure demonstrates the value of exome-capture SNP markers for classifying germplasm.

Using Markers for Paternity Analysis

Many forage crops are outcrossing, pollinated by wind or insects. Consequently, only records of maternal parents are kept in most breeding programs. Paternity analysis, based on DNA markers, would theoretically improve genetic gain because both parents of selected plants could be ascertained (Riday 2011). Paternity analysis improved selection for seed yield in red clover (Vleugels et al. 2014) and was beneficial for a timothy polycross breeding program (Tanaka et al. 2018). Extending paternity analysis to autotetraploids demonstrated more self-fertilization than expected in progeny derived from an alfalfa polycross (Riday et al. 2013).

Using Markers for Breeding and Genetics

Linking Markers to Genes

In order to use genetic markers as a proxy for a phenotypic trait, the marker needs to be associated with the trait, most commonly been done using linkage analysis. When particular marker alleles are associated with a particular phenotype, the marker and trait are said to be genetically linked. In some cases, the genetic marker variant may actually be the cause of the phenotypic variation, in which case, this is a perfect marker (Andersen and Lübberstedt 2003) but generally, the marker is only linked to the trait.

Since the mid-1990s, genetic maps have been developed for many forage crops. As marker systems have improved, genetic maps with thousands of markers have been created not only in major temperate species, e.g. alfalfa (Li et al. 2014b) and perennial ryegrass (Velmurugan et al. 2016), but also in tropical forage species, e.g. *Brachiaria decumbens* (Worthington et al. 2016) and napiergrass (Paudel et al. 2018). A genetic map can be used to locate loci controlling traits. Conceptually, if a subset of plants in a population has a particular marker **allele** (e.g. a longer SSR allele or a "G" SNP allele) and also a particular phenotype (e.g. tall), then the marker and the trait are linked (Figure 31.3). Many agronomically important traits exhibit quantitative inheritance; they are controlled by many genes whose expression is affected by the environment. Locating genes controlling **quantitative traits**, especially if no gene has a large effect, requires extensive evaluation of the individuals within a segregating population in several locations and/or years. Each individual locus controlling a quantitative trait is called a *quantitative trait locus* or QTL.

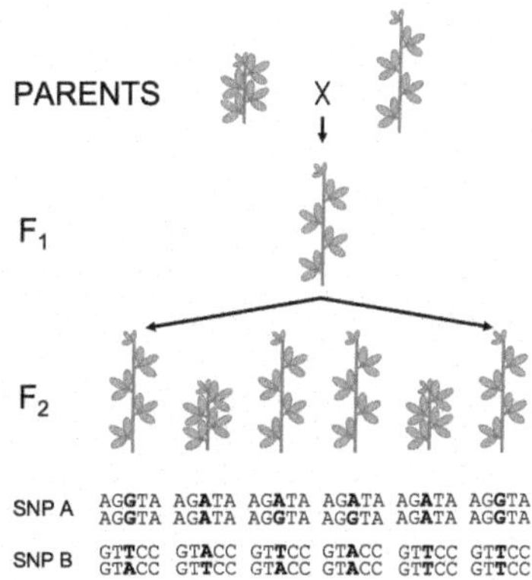

Fig. 31.3. Illustration of linkage mapping. Two parental plants are hybridized to create an F_1 individual which is self-fertilized to produce a segregating F_2 population. The essence of linkage mapping is that a number of markers are screened on the population, and their genotypes are tested for association with each other and with traits. In this case, SNP A is linked to stem length: short plants are all "AA"; tall plants are either "AG" or "GG." Obviously, "G" is linked to "tall" and "A" to short, and the allele governing "tall" is dominant because both "AG" and "GG" genotypes fall into the tall group. SNP B is not linked to SNP A or to the trait; either SNP allele can be present in either size class. A property of many genetic markers is that they are co-dominant, enabling the identification of heterozygotes even if their phenotype is no different than one homozygous class.

Mapping single genes or major QTL for traits has been done for seed yield in white clover (Barrett et al. 2005), apospory in *B. decumbens* (Worthington et al. 2016) (Figure 31.4), jumbo pollen in alfalfa (Tavoletti et al. 2000), and many others. In alfalfa, QTL have been mapped for forage yield (Robins et al. 2007), winterhardiness and fall dormancy (Brouwer et al. 2000; Li et al. 2015a), and numerous other traits (reviewed in Li and Brummer 2012 and Hawkins and Yu 2018). Identifying the exact gene that the marker is associated with requires additional research, but if identified, it enables more directed improvement of a trait using transformation (see below). The major limitation of QTL mapping is that it only identifies some of the genes involved in trait expression, and in many cases, is unable to identify the genes controlling the majority of phenotypic variation for the trait.

Other methods may also be used to identify marker-trait associations. **Genome-wide association studies (GWAS)** can be used to find desirable markers in germplasm collections or breeding populations. Exome capture was used to genotype SNP markers for a GWAS of flowering time in switchgrass (Grabowski et al. 2017), and GBS was used to identify **resistance** to verticillium wilt in alfalfa (Yu et al. 2017). A **bulked segregant analysis** (Michelmore et al. 1991) can be used to compare marker alleles in contrasting bulked samples of several individuals each, for example, plants with and without resistance to a disease. This approach was recently used to identify a bacterial wilt QTL in italian ryegrass by using pooled DNA sequencing (Knorst et al. 2019a). Marker alleles that differed between pools were used to create KASP markers for marker-assisted selection of resistance. Translating marker-trait associations found in experimental settings into markers useful in a breeding context is not always as straightforward (Xu and Crouch 2008).

Selection mapping (Wisser et al. 2008) takes advantage of shifts in allele frequencies that occur during natural or artificial selection. Marker loci genetically linked to the genes controlling the trait under selection shift in their allele frequency over generations of selection. Therefore, by comparing a selected population with the population from which it was selected for allele frequency shifts at loci throughout the genome, regions that have been affected by selection can be identified. Alfalfa populations selected for more and less fall dormancy were used to

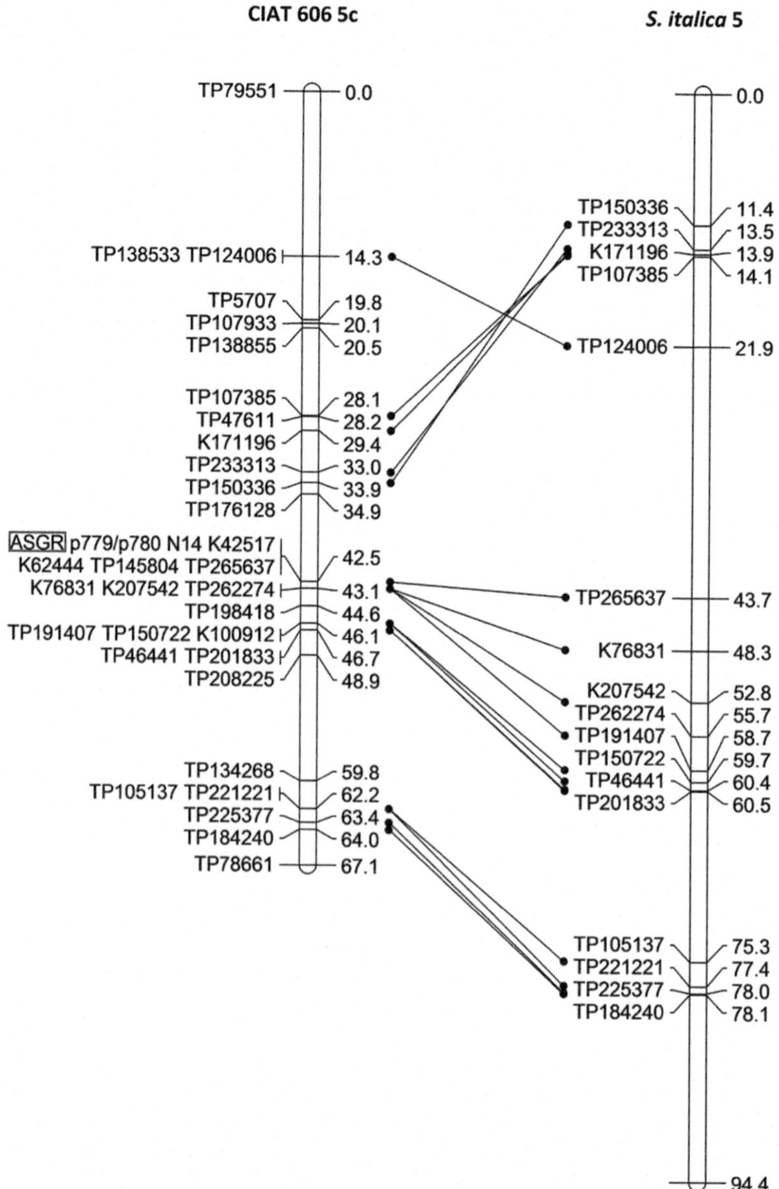

Fig. 31.4. Genetic map composed of GBS markers of Linkage Group 5c in *Brachiaria decumbens* compared with Linkage Group 5 in *Setaria italica*, a model grass species. The *Brachiaria* map includes the location of an Apospory-Specific Genomic Region (ASGR). *Source:* From Worthington et al. (2016). Reprinted with permission of the Genetics Society of America.

identify markers associated with dormancy (Munjal et al. 2018). This methodology has potential applicability for many traits that have been selected in breeding programs, though it is particularly useful if only one single trait has been targeted, since selection on multiple traits will obscure which loci are involved with which traits.

Because of the difficulty of identifying loci that have a small effect on a particular trait, an alternate selection method for using markers to improve traits is **genomic selection** (Hayes et al. 2013). The basic premise of genomic selection is that every marker throughout the genome has a potentially small effect on the phenotype (or is linked to a gene that has the effect). Using one of a variety of statistical methods, breeders can generate genome-wide markers on all individuals in a population, measure phenotypes on those plants, and then construct a mathematical model to relate genotype to phenotype. Subsequent selection can be based on genotypes in the absence of phenotypes. Genome-wide markers generated using arrays, GBS, or exome capture have been used for genomic selection in alfalfa (Li et al. 2015b; Annicchiarico et al. 2015), switchgrass (Ramstein et al. 2016), and perennial ryegrass (Fè et al. 2016; Grinberg et al. 2016; Pembleton et al. 2018; Faville et al. 2018).

Population Based Genotyping

Many forage crops are outcrossing and consequently, cultivars are developed and sold as populations (i.e. synthetics or open pollinated populations). Because inbred lines are rarely available and may not be possible to develop, at least easily, plants being evaluated are highly **heterozygous** and every plant is different. Therefore, genotyping individual plants is often of less importance than genotyping populations. Though a group of individuals from a population may carry different genotypes at some locus (e.g. for a SNP locus, AA, AG, or GG), when comparing two or more populations, each may contain both alleles. Thus, what differentiates populations, and certainly this is true of most advanced breeding populations, is not that one contains one allele and the other a different allele, but that they differ in allele frequencies. Methods to estimate allele frequencies of populations have been developed (e.g. Byrne et al. 2013; Munjal et al. 2018), which have been useful to identify markers linked to fall dormancy related genes in alfalfa (Munjal et al. 2018) and for genomic selection for heading date in ryegrass (Fè et al. 2015).

Genetic Transformation and Gene Editing

Identifying Genes for Use in Transformation

Applying markers to breeding programs is one main way that genomics can be applied to forage improvement. The second main way is to directly insert and/or modify specific genes to create a desirable phenotype. For these technologies to be successfully used, the gene or genes necessary to produce the desired phenotype must be known. A number of methods can be used to identify a gene. Genetic mapping discussed above is one way to start to figure out which genes are of interest. Using standard molecular biology techniques, genes have been cloned in several forage crops, including those involved in nitrogen fixation (Endre et al. 2002), cold tolerance (Monroy et al. 1993), **drought tolerance** (Zhang et al. 2005), cell wall composition (Tu et al. 2010), condensed tannin production (Hancock et al. 2012), seed physical dormancy (Chai et al. 2016), and axillary bud formation and plant architecture (Gou et al. 2017).

In addition to mapping and standard molecular biology experiments, genes can be identified in a given forage crop by looking at what is known from another species, whether forages or not. For example, dissection of flowering pathways in *Arabidopsis* identified a suite of genes, some of which, were then isolated and characterized in switchgrass (Niu et al. 2016). Genes can also be identified from gene expression analyses, experiments that determine which genes are actively transcribed between genotypes differing for some trait or for a single genotype grown under contrasting conditions. By profiling thousands of genes simultaneously, these gene expression profiles can help hone in on a small number of likely candidate genes, as was done in comparing alfalfa germplasms with and without tolerance to salinity (Sandhu et al. 2017).

Genetic Transformation

Genetic transformation refers to the directed insertion of a gene into the genome to create a transgenic plant. Genetic transformation is used when a desired trait does not exist in either the plant species of interest or a related species with which that species can be hybridized. The inserted gene is termed a transgene. The gene can come from any other organism, though today, most often genes are derived from other plants, or in some cases, from the same species (cisgenics). The terms **genetic engineering** and genetically modified organism (GMO) usually refer specifically to plant transformation technologies and not to other biotechnologies.

Two processes are needed to produce transgenic plants. First, the gene needs to be introduced into the plant and stably integrated into its nuclear genome, and second, transformed cells need to be regenerated into whole plants. Transformation can be effected by *Agrobacterium tumefaciens* mediated gene transfer, by particle bombardment, by electroporation of protoplasts, and by other less common methods. Though plant transformation has been investigated for 30 years or more, substantial impediments to developing transformants still exist in many plant species (Altpeter et al. 2016).

Agrobacterium tumefaciens, a naturally occurring soil bacterium that causes crown gall disease in many ornamental and fruit plants, has been modified so that it inserts a transgene into the plant rather than tumor forming genes it inserts naturally. Incubating leaf, stem, or petiole sections with *Agrobacterium* containing the transgene enables the bacteria to infect the cut edges of the plant tissue and insert the transgene into the plant's DNA. Though *Agrobacterium* is not able to naturally infect grasses, techniques have been refined so that *Agrobacterium* mediated gene transfer has been used successfully in several grasses including switchgrass (Xi et al. 2009) and ryegrass (Bettany et al. 2003). Figure 31.5 illustrates the process of *Agrobacterium*-mediated transformation of alfalfa.

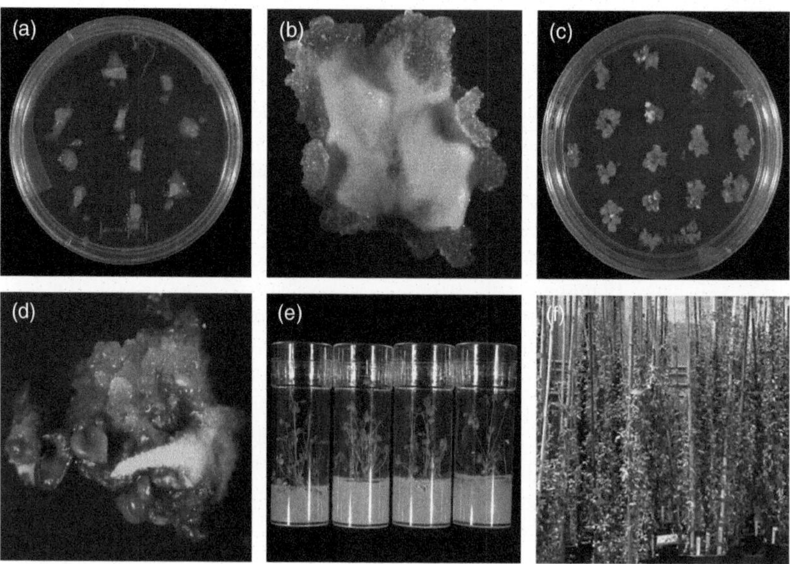

Fig. 31.5. Transgenic alfalfa plants obtained after *Agrobacterium*-mediated transformation. (a) Leaf pieces infected by *Agrobacteria*. (b) Callus formation after *Agrobacterium* transformation and antibiotic selection. (c) Transformed calli from (b). (d) Regenerating shoots from transformed calli. (e) *In vitro* transgenic alfalfa plantlets. (f) Transgenic alfalfa plants growing in the greenhouse.

Particle bombardment involves the use of tiny gold or tungsten particles coated with DNA containing the transgene (Altpeter et al. 2005). The particles are propelled at high speed into plant tissue using an instrument called a "gene gun". DNA on particles that penetrate the nucleus can become integrated into the nuclear genome. Other technologies successfully used in forage transformation include direct gene transfer to protoplasts and silicon carbide whisker mediated transformation (Wang and Ge 2006). Numerous genes have been inserted into a diverse range of forage crops, including alfalfa, white clover, subterranean clover, perennial ryegrass, tall fescue, switchgrass, bahiagrass, and others (see Wang and Brummer 2012).

Tissue Culture and Regeneration

After plant cells are transformed, they need to be manipulated to divide and eventually regenerate into whole plants. Following transformation by any of the methods described above, the plant tissue is placed on media to kill the *Agrobacterium* and/or non-transformed plant cells and to induce the transformed plant cells to multiply into a uniform mass called *callus*. Callus cultures will be grown on media containing plant hormones to promote adventitious shoot and root formation, leading to a regenerated plant. Some callus can form somatic embryos, which contain both root and shoot axes similar to normal seeds. Somatic embryos are also called artificial seeds because they can be removed from callus cultures, dried and stored, and later germinated on culture media (McKersie and Brown 1996).

The site at which a transgene integrates into the nuclear genome can affect its expression, but targeting the transgene to a precise genomic region is not yet possible in plants. However, this may soon be possible using the gene editing technologies described in the next section. Therefore, multiple transgenic plants are regenerated followed by traditional plant breeding evaluation and selection procedures to identify the best performing transgenic plants. Selection is based on the stability of transgene expression and the realization of the desired phenotype under different environmental conditions and in different genetic backgrounds.

Gene Editing

Gene editing tools have been developed to precisely alter genes in specific ways. Gene editing works by creating a double strand break in a plant's DNA at a targeted site in the genome, and then taking advantage of the plant's inherent DNA repair machinery to introduce mutations and/or short segments of DNA at the breakpoint (Baltes and Voytas 2015; Adli 2018). Several gene editing methods have been developed, including zinc finger nucleases (ZFN), transcription activator-like effector nucleases (TALEN), and clustered regularly interspaced short palindromic repeat DNA sequences (CRISPR). CRISPR is the most versatile of the technologies at the current time (Adli 2018). Currently, the genetic machinery to create the desired change at a specific location – that is, a nuclease and a guide template for repair – is incorporated into the plant's genome using traditional transformation technology described above. After the desired change is made, the inserted genes enabling editing can be removed by screening segregating progeny that only have the desired edited gene.

Unlike common genetic transformation, gene editing involves specific changes at a discrete point in the genome, focused even down to the individual nucleotide level. Making changes in the genome that reduce or "knock out" the function of a particular gene is the most easily realized application of gene editing methods. As with other aspects of biotechnology and genomics, gene editing in forage crops has lagged behind other crops, but it is beginning to be applied successfully. An obvious target for gene editing is the lignin biosynthetic pathway because the disruption of genes could lead to altered lignin composition and/or quantity in plant cell walls. CRISPR/Cas9 was used to decrease lignin concentration in switchgrass, which had the positive result of increasing **sugar** release from cell wall carbohydrates, useful for biofuel or for animal feed (Park et al. 2017). Because switchgrass is allotetraploid, the 4-coumarate : coenzyme A ligase (Pv4Cl1) gene is present in four copies, two in each subgenome. Successful editing required changes to all four alleles and this was achieved. Gene editing in alfalfa, an autotetraploid crop, faces similar problems due to four alleles at a locus. Nevertheless, successful gene editing has been reported (Gao et al. 2018).

Generating Value-Added Traits

The goal of plant breeders is to incorporate unique traits into adapted plant materials to produce a better cultivar for the farmer and the consumer. Addition of value-added traits has traditionally been accomplished using **phenotypic selection**, which is confined to genes available in the existing germplasm. Biotechnological methods enable the breeder to efficiently incorporate and manipulate genes from any organism, greatly increasing the scope of value-added traits available for breeding. Forage crops have lagged behind the major grain and oilseed crops for incorporation of value-added traits via transgenesis due to the smaller market for forage seed and the large number of forage grass and legume species. The high cost of biotechnology prevents its use on species without broad-based and sophisticated sales markets. Therefore, major species such as alfalfa, white clover, and perennial ryegrass are the most likely to be improved via biotechnology. The first commercialized transgenic forage crop was glyphosate tolerant alfalfa, with Roundup Ready™ cultivars (Beazley et al. 2009), commercialized initially in 2005, pulled from the market pending additional environmental study, and finally on the market in 2011 (Wang and Brummer 2012). The second commercialized transgenic trait in forages was low lignin alfalfa, created by downregulating the caffeoyl CoA 3-*O*-methyltransferase (CCOMT) gene (Barros et al. 2019). Many other genes have been transformed into forage crops for experimental and/or proof-of-concept testing. The likelihood of commercialization is low given current regulatory hurdles.

Potentials and Limitations of Biotechnology

A major limitation of biotechnology is the ability of breeders and seed companies to commercialize new cultivars carrying transgenes (Wang and Brummer 2012). Transgenic plants are initially regulated in the United States, meaning that they need to be evaluated for safety prior to being "de-regulated." Field testing of transgenic plants must be conducted in accordance with current governmental regulations. In the United States, three agencies could be involved in regulation. The Animal and Plant Health Inspection Service (APHIS) of the USDA is responsible for regulating field testing of plant pests. The Environmental Protection Agency (EPA) is involved if the transgenic plant could pose problems for natural ecosystems, such as becoming a weed. The Food and Drug Administration (FDA) regulates food products, and thus, for most forage plants, would not be involved with regulation.

Formerly, many transgenic plants were developed using small fragments of bacterial or viral DNA in the construction of the transgene vector, and therefore, APHIS became the key regulatory agency for transgenic crops. However, today biotechnologists are developing purely plant-based constructs, and in many cases, constructs derived solely from the plant species being transformed. The transgenic plants resulting from these so-called "cisgenic" methods do not have any plant pest components; consequently, APHIS does not have regulatory authority over these products. Whether they will be able to be commercialized in the US is still not clear; the European Union considers them analogous to transgenics, preventing their commercialization (van Hove and Gillund 2017).

Debate on Transgenic Technology

The debate on transgenic technology has raged ever since the first transgenic crops were postulated, and it shows no signs of abating now. Far better platforms for discussing the topic are available outside of this volume. Here, we will simply discuss several ramifications of regulation. Perhaps the issue most difficult to simply ignore in forages is the issue of gene flow. Many major forage species are cross-pollinated perennials, and some are relatives of major grain or oilseed crops. The problem of escape from cultivation and the potential hazards that could occur in natural ecosystems via gene flow if significant weeds developed needs to be considered carefully in contexts where it could occur. Other issues, such as those related to human health, are of less importance in the case of a forage crop that humans do not consume; health of livestock or horses consuming the crop does need to be considered, however.

The regulatory status of transgenics does not appear likely to change in the US or elsewhere, meaning that few products are likely to come to market through that channel unless the seed market is substantial and the product can penetrate a large proportion of that market. Whether cisgenics, or possibly more likely, gene edited plants can be commercialized is less clear (Parrott 2018). The impediments to commercialization in the US, at least for some products in these classes, do not appear problematic. However, even if products are approved in the US for sale, they may still be prevented from being grown in other markets in the world. And, since seed markets are increasingly international, products that are only able to be marketed in some countries are less desirable than ones that can be marketed everywhere. Realizing the benefits that could come from these technologies still appears to be an uphill climb.

Discussion of Costs and Benefits of Biotechnology and Molecular Markers

Regardless of regulatory hurdles, whether marker-assisted selection, transgenics, or gene editing will result in sufficient genetic improvement to cover the costs of the technology is a question that needs to be considered carefully. The forage seed industry in the US is mainly a commodity business, with most seed sold as a low-margin product. Marketing expensive proprietary forage varieties as value-added cultivars is frequently difficult because inexpensive commodity seed is often easily available. To sell farmers high-cost technology products will require a new model for seed production, marketing, and sales. If the products of these technologies fail to recoup their development, regulatory, and marketing costs, then they have little future in forage crop improvement. Biotechnology expands the range of traits that breeders can manipulate and may improve the efficiency of trait manipulation. However, these techniques do not replace plant breeding but only provide additional tools for plant breeders to use to develop and test improved germplasm in likely target environments. Thus, simply shifting resources from breeding to biotechnology will not lead to improved cultivars, particularly in those forage species on which little breeding work is being conducted in the first place.

Summary

Modern genetics form the science which underpins genomics and biotechnology. By understanding how genes interact with each other and with the environment to produce phenotypes, plant breeders can develop and select superior genotypes more effectively, leading to better cultivars. Until now, breeders have essentially been working with a black box, not knowing what genes their selected plants contained. Biotechnology will identify at least some of the genes contributing to agronomically important traits and will enable insertion of novel genes for new traits, both of which will assist the breeding process. Significant progress in developing improved cultivars can be realized by coupling the most recent advances in biology with modern breeding programs to deliver new traits to the producer and consumer.

References

Adli, M. (2018). The CRISPR tool kit for genome editing and beyond. *Nat. Commun.* 9 (1): 1911.

Ali, O.A., O'Rourke, S.M., Amish, S.J. et al. (2016). RAD Capture (Rapture): Flexible and efficient sequence-based genotyping. *Genetics* 202 (2): 389–400.

Altpeter, F., Baisakh, N., Beachy, R. et al. (2005). Particle bombardment and the genetic enhancement of crops: myths and realities. *Mol. Breed.* 15 (3): 305–327.

Altpeter, F., Springer, N.M., Bartley, L.E. et al. (2016). Advancing crop transformation in the era of genome editing. *The Plant Cell* 28 (7): 1510–1520.

Andersen, J.R. and Lübberstedt, T. (2003). Functional markers in plants. *Trends Plant Sci.* 8 (11): 554–560.

Annicchiarico, P., Nazzicari, N., Li, X. et al. (2015). Accuracy of genomic selection for alfalfa biomass yield in different reference populations. *BMC Genomics* 16 (1): 1020.

Annicchiarico, P., Nazzicari, N., Ananta, A. et al. (2016). Assessment of cultivar distinctness in alfalfa: a comparison of genotyping-by-sequencing, simple-sequence repeat marker and morphophysiological observations. *Plant Genome* 9 https://doi.org/10.3835/plantgenome2015.10.0105.

Baltes, N.J. and Voytas, D.F. (2015). Enabling plant synthetic biology through genome engineering. *Trends Biotechnol.* 33: 120–131.

Barrett, B., Baird, I., and Woodfield, D. (2005). A QTL analysis of white clover seed production. *Crop Sci.* 45: 1844–1850.

Barros, J., Temple, S., and Dixon, R.A. (2019). Development and commercialization of reduced lignin alfalfa. *Curr. Opin. Biotechnol.* 56: 48–54.

Beazley, K.A., Ferreira, K.L., Fitzpatrick, S.N. et al. (2009). Glyphosate tolerant alfalfa events and methods for detection. United States Patent No. 7,566,817.

Beckmann, J.S. and Soller, M. (1986). Restriction fragment length polymorphisms and genetic improvement of agricultural species. *Euphytica* 35 (1): 111–124.

Bettany, A., Dalton, S., Timms, E. et al. (2003). Agrobacterium tumefaciens-mediated transformation of *Festuca arundinacea* (Schreb.) and *Lolium multiflorum* (Lam.). *Plant Cell Rep.* 21 (5): 437–444.

Brouwer, D.J., Duke, S.H., and Osborn, T.C. (2000). Mapping genetic factors associated with winter hardiness, fall growth, and freezing injury in autotetraploid alfalfa. *Crop Sci.* 40 (5): 1387–1396.

Byrne, S., Czaban, A., Studer, B. et al. (2013). Genome Wide Allele Frequency Fingerprints (GWAFFs) of populations via genotyping by sequencing. *PloSOne* 8 (3): e57438. https://doi.org/10.1371/journal.pone.0057438.

Byrne, S.L., Nagy, I., Pfeifer, M. et al. (2015). A synteny-based draft genome sequence of the forage grass *Lolium perenne*. *Plant J.* 84 (4): 816–826.

Chai, M., Zhou, C., Molina, I. et al. (2016). A Class II KNOX gene, KNOX4 controls seed physical dormancy. *Proc. Natl. Acad. Sci. U.S.A.* 113 (25): 6997–7002.

Choi, M., Scholl, U.I., Ji, W. et al. (2009). Genetic diagnosis by whole exome capture and massively parallel DNA sequencing. *Proc. Natl. Acad. Sci. U. S. A.* 106 (45): 19096–19101.

De Vega, J.J., Ayling, S., Hegarty, M. et al. (2015). Red clover (*Trifolium pratense* L.) draft genome provides a platform for trait improvement. *Sci. Rep.* 5: 17394.

Elshire, R.J., Glaubitz, J.C., Sun, Q. et al. (2011). A robust, simple genotyping-by-sequencing (GBS) approach for high diversity species. *PLoS One* 6 (5): e19379.

Endre, G., Kereszt, A., Kevei, Z. et al. (2002). A receptor kinase gene regulating symbiotic nodule development. *Nature* 417 (6892): 962.

Evans, J., Crisovan, E., Barry, K. et al. (2015). Diversity and population structure of northern switchgrass as revealed through exome capture sequencing. *Plant J.* 84 (4): 800–815.

Evans, J., Sanciangco, M.D., Lau, K.H. et al. (2018). Extensive genetic diversity is present within North American switchgrass germplasm. *The Plant Genome* 11: 170055.

Faville, M.J., Ganesh, S., Cao, M. et al. (2018). Predictive ability of genomic selection models in a multi-population perennial ryegrass training set using genotyping-by-sequencing. *Theor. Appl. Genet.* 131 (3): 703–720.

Fè, D., Cericola, F., Byrne, S. et al. (2015). Genomic dissection and prediction of heading date in perennial ryegrass. *BMC Genomics* 16 (1): 921.

Fè, D., Ashraf, B.H., Pedersen, M.G. et al. (2016). Accuracy of genomic prediction in a commercial perennial ryegrass breeding program. *The Plant Genome* 9 (3).

Gao, R., Feyissa, B.A., Croft, M., and Hannoufa, A. (2018). Gene editing by CRISPR/Cas9 in the obligatory outcrossing Medicago sativa. *Planta* 247 (4): 1043–1050.

Gou, J., Fu, C., Liu, S. et al. (2017). The miR156-SPL4 module regulates axillary bud formation and shoot architecture. *New Phytol.* 216: 829–840.

Grabowski, P.P., Evans, J., Daum, C. et al. (2017). Genome-wide associations with flowering time in switchgrass using exome-capture sequencing data. *New Phytol.* 213 (1): 154–169.

Griffiths, A.J.F., Wessler, S.R., Carroll, S.B., and Doebley, J. (2015). *An Introduction to Genetic Analysis*, 11e. New York, NY: Freeman/Worth.

Grinberg, N.F., Lovatt, A., Hegarty, M. et al. (2016). Implementation of genomic prediction in *Lolium perenne* (L.) breeding populations. *Front. Plant Sci.* 7: 133.

Hancock, K.R., Collette, V., Fraser, K. et al. (2012). Expression of the R2R3-MYB transcription factor TaMYB14 from *Trifolium arvense* activates proanthocyanidin biosynthesis in the legumes *Trifolium repens* and *Medicago sativa*. *Plant Physiol.* 159 (3): 1204–1220.

Hawkins, C. and Yu, L.X. (2018). Recent progress in alfalfa (*Medicago sativa L.*) genomics and genomic selection. *Crop J.* 6 (6): 565–575.

Hayes, B.J., Cogan, N.O., Pembleton, L.W. et al. (2013). Prospects for genomic selection in forage plant species. *Plant Breed.* 132 (2): 133–143.

van Hove, L. and Gillund, F. (2017). Is it only the regulatory status? Broadening the debate on cisgenic plants. *Environ. Sci. Eur.* 29 (1): 22.

Julier, B., Barre, P., Lambroni, P. et al. (2018). Use of GBS markers to distinguish among lucerne varieties, with comparison to morphological traits. *Mol. Breed.* 38 (11): 133.

Knorst, V., Byrne, S., Yates, S. et al. (2019a). Pooled DNA sequencing to identify SNPs associated with a major QTL for bacterial wilt resistance in Italian ryegrass (*Lolium multiflorum* Lam.). *Theor. Appl. Genet.* 132 (4): 947–958.

Knorst, V., Yates, S., Byrne, S. et al. (2019b). First assembly of the gene-space of *Lolium multiflorum* and comparison to other Poaceae genomes. *Grassland Sci.* 65: 125–134. https://doi.org/10.1111/grs.12225.

Li, X. and Brummer, E.C. (2012). Applied genetics and genomics in alfalfa breeding. *Agronomy* 2 (1): 40–61.

Li, X., Han, Y., Wei, Y. et al. (2014a). Development of an alfalfa SNP array and its use to evaluate patterns of population structure and linkage disequilibrium. *PLoS One* 9 (1): e84329.

Li, X., Wei, Y., Acharya, A. et al. (2014b). A saturated genetic linkage map of autotetraploid alfalfa (*Medicago sativa L.*) developed using genotyping-by-sequencing is highly syntenous with the *Medicago truncatula* genome. *G3: Genes, Genomes, Genet.* 4 (10): 1971–1979.

Li, X., Alarcón-Zúñiga, B., Kang, J. et al. (2015a). Mapping fall dormancy and winter injury in tetraploid alfalfa. *Crop Sci.* 55 (5): 1995–2011.

Li, X., Wei, Y., Acharya, A. et al. (2015b). Genomic prediction of biomass yield in two selection cycles of a tetraploid alfalfa breeding population. *The Plant Genome* 8 (2).

McKersie, B.D. and Brown, D.C. (1996). Somatic embryogenesis and artificial seeds in forage legumes. *Seed Sci. Res.* 6 (3): 109–126.

Michelmore, R.W., Paran, I., and Kesseli, R.V. (1991). Identification of markers linked to disease-resistance genes by bulked segregant analysis: a rapid method to detect markers in specific genomic regions by using segregating populations. *Proc. Natl. Acad. Sci. U.S.A.* 88 (21): 9828–9832.

Miller, M.R., Dunham, J.P., Amores, A. et al. (2007). Rapid and cost-effective polymorphism identification and genotyping using restriction site associated DNA (RAD) markers. *Genome Res.* 17 (2): 240–248.

Monroy, A.F., Castonguay, Y., Laberge, S. et al. (1993). A new cold-induced alfalfa gene is associated with enhanced hardening at subzero temperature. *Plant Physiol.* 102 (3): 873–879.

Munjal, G., Hao, J., Teuber, L.R., and Brummer, E.C. (2018). Selection mapping identifies loci underpinning autumn dormancy in alfalfa (*Medicago sativa*). *G3: Genes, Genomes, Genet.* 8 (2): 461–468.

Niu, L., Fu, C., Lin, H. et al. (2016). Control of floral transition in the bioenergy crop switchgrass. *Plant Cell Environ.* 39: 2158–2171.

Park, J.-J., Yoo, C.G., Flanagan, A. et al. (2017). Defined tetra-allelic gene disruption of the 4-coumarate: coenzyme A ligase 1 (Pv4CL1) gene by CRISPR/Cas9 in switchgrass results in lignin reduction and improved sugar release. *Biotechnol. Biofuels* 10: 284.

Parrott, W. (2018). Outlaws, old laws and no laws: the prospects of gene editing for agriculture in United States. *Physiol. Plant.* 164 (4): 406–411.

Paudel, D., Kannan, B., Yang, X. et al. (2018). Surveying the genome and constructing a high-density genetic map of napiergrass (*Cenchrus purpureus* Schumach). *Sci. Rep.* 8 (1): 14419.

Pembleton, L.W., Drayton, M.C., Bain, M. et al. (2016). Targeted genotyping-by-sequencing permits cost-effective identification and discrimination of pasture grass species and cultivars. *Theor. Appl. Genet.* 129 (5): 991–1005.

Pembleton, L.W., Inch, C., Baillie, R.C. et al. (2018). Exploitation of data from breeding programs supports rapid implementation of genomic selection for key agronomic traits in perennial ryegrass. *Theor. Appl. Genet.* 131 (9): 1891–1902.

Ramstein, G.P., Evans, J., Kaeppler, S.M. et al. (2016). Accuracy of genomic prediction in switchgrass (*Panicum virgatum L.*) improved by accounting for linkage disequilibrium. *G3: Genes, Genomes, Genet.* 6 (4): 1049–1062.

Riday, H. (2011). Paternity testing: A non-linkage based marker-assisted selection scheme for outbred forage species. *Crop Sci.* 51 (2): 631–641.

Riday, H., Johnson, D.W., Heyduk, K. et al. (2013). Paternity testing in an autotetraploid alfalfa breeding polycross. *Euphytica* 194 (3): 335–349.

Robins, J.G., Bauchan, G.R., and Brummer, C. (2007). Genetic mapping forage yield, plant height, and regrowth at multiple harvests in tetraploid alfalfa. *Crop Sci.* 47 (1) https://doi.org/10.2135/cropsci2006.07.0447.

Şakiroğlu, M., Doyle, J.J., and Charles Brummer, E. (2010). Inferring population structure and genetic diversity of broad range of wild diploid alfalfa (*Medicago sativa* L.) accessions using SSR markers. *Theor. Appl. Genet.* 121: 403. https://doi.org/10.1007/s00122-010-1319-4.

Sandhu, D., Cornacchione, M.V., Ferreira, J.F., and Suarez, D.L. (2017). Variable salinity responses of 12 alfalfa genotypes and comparative expression analyses of salt-response genes. *Sci. Rep.* 7: 42958.

Semagn, K., Babu, R., Hearne, S., and Olsen, M. (2014). Single nucleotide polymorphism genotyping using Kompetitive Allele Specific PCR (KASP): overview of the technology and its application in crop improvement. *Mol. Breed.* 33 (1): 1–14.

Tanaka, T., Tamura, K.I., Ashikaga, K. et al. (2018). Marker-based paternity test in polycross breeding of timothy. *Crop Sci.* 58 (1): 273–284.

Tavoletti, S., Pesaresi, P., Barcaccia, G. et al. (2000). Mapping the jp (jumbo pollen) gene and QTLs involved in multinucleate microspore formation in diploid alfalfa. *Theor. Appl. Genet.* 101 (3): 372–378.

Torkamaneh, D., Boyle, B., and Belzile, F. (2018). Efficient genome-wide genotyping strategies and data integration in crop plants. *Theor. Appl. Genet.* 131 (3): 499–511.

Tu, Y., Rochfort, S., Liu, Z. et al. (2010). Functional analyses of caffeic acid O-methyltransferase and cinnamoyl-CoA-reductase genes from perennial ryegrass (*Lolium perenne*). *The Plant Cell* 22 (10): 3357–3373.

Velmurugan, J., Mollison, E., Barth, S. et al. (2016). An ultra-high density genetic linkage map of perennial ryegrass (Lolium perenne) using genotyping by sequencing (GBS) based on a reference shotgun genome assembly. *Ann. Bot.* 118 (1): 71–87.

Vleugels, T., Cnops, G., and Roldán-Ruiz, I. (2014). Improving seed yield in red clover through marker assisted parentage analysis. *Euphytica* 200 (2): 305–320.

Wang, Z.Y. and Brummer, E.C. (2012). Is genetic engineering ever going to take off in forage, turf and bioenergy crop breeding? *Ann. Bot.* 110 (6): 1317–1325.

Wang, Z.-Y. and Ge, Y. (2006). Recent advances in genetic transformation of forage and turf grasses. *In: Vitro Cell. Dev. Biol. Plant* 42 (1): 1–18.

Wisser, R.J., Murray, S.C., Kolkman, J.M. et al. (2008). Selection mapping of loci for quantitative disease resistance in a diverse maize population. *Genetics* 180 (1): 583–599.

Worthington, M., Heffelfinger, C., Bernal, D. et al. (2016). A parthenogenesis gene candidate and evidence for segmental allopolyploidy in apomictic *Brachiaria decumbens. Genetics* 203 (3): 1117–1132.

Xi, Y., Fu, C., Ge, Y. et al. (2009). Agrobacterium-mediated transformation of switchgrass and inheritance of the transgenes. *Bioenergy Res.* 2 (4): 275–283.

Xu, Y. and Crouch, J.H. (2008). Marker-assisted selection in plant breeding: from publications to practice. *Crop Sci.* 48: 391–407. https://doi.org/10.2135/cropsci2007.04.0191.

Young, N.D., Debellé, F., Oldroyd, G.E. et al. (2011). The Medicago genome provides insight into the evolution of rhizobial symbioses. *Nature* 480 (7378): 520.

Yu, L.X., Zheng, P., Zhang, T. et al. (2017). Genotyping-by-sequencing-based genome-wide association studies on Verticillium wilt resistance in autotetraploid alfalfa (*Medicago sativa L.*). *Mol. Plant Pathol* 18 (2): 187–194.

Zhang, J., Broeckling, C.D., Blancaflor, E.B. et al. (2005). Overexpression of WXP1, a putative *Medicago truncatula* AP2 domain-containing transcription factor gene, increases cuticular wax accumulation and enhances drought tolerance in transgenic alfalfa (*Medicago sativa*). *Plant J.* 42: 689–707.

Chapter

32

Seed Production

Jeffrey J. Steiner, Associate Director, *Global Hemp Innovation Center, Oregon State University, Corvalis, OR, USA*
Tim L. Springer, Research Agronomist, *USDA-Agricultural Research Service, Woodward, OK, USA*

Introduction

The primary objectives of most forage breeding programs are to increase forage yield and quality. Though these traits are certainly important, forage cultivars are not necessarily preferred by producers because of them alone. A successful forage cultivar must be adapted to its environment, of value to the end-user, produce an abundance of good quality seeds, and once planted have quick emergence and establishment (Harlan 1960).

In the past, when farmers needed seeds for planting, they allowed hay fields or pastures to flower and produce seeds. Seed production was a co-product of forage production. As adapted varieties became available, grazing and harvested forage systems became more specialized, and the dedicated seed trade expanded, the seeds of many forage species were produced as specialty crops in regions geographically removed from the areas where the seeds were utilized for forage production; regions where environmental conditions are more conducive to dependable yields of high-quality seeds.

Today, seed producers and the seed sales industry provide consumers with dependable quantities of high-performing seeds that add value to the production systems where used. Special rules are followed by seed producers to help ensure that the genetic quality of cultivars remains true to the genetic composition as originally developed by plant breeders. Also, specialized production and **seed conditioning** practices are needed to assure that the seeds produced are as pure as possible from physical contaminants such as weed and other crop seeds.

Most forage plants are grazed or cut frequently for hay or silage and rarely allowed to complete their life cycles to produce seeds. When forage plants are allowed to produce seeds, they are exposed to additional physiologic pressures in the field that may not normally affect performance when grown for forage. Many of the flowers produced may fail to set viable seeds as a result of disease or insect infestation or due to a lack of pollination if cross-pollination is required. Once seeds begin to mature after fertilization, environmental stresses related to weather may lower seed quality and the ability of seeds to produce vigorous plants after seeding. Numerous problems may exist during seed harvest and conditioning, the cleaning processes used to separate the forage seeds grown from other crops seeds, weed seeds, and inert materials, that reduce the actual amount of crop seed that will be eventually sold and planted.

Many of these problems can be reduced by growing forage grass and legumes species for seeds in environments that are conducive to reproduction. But even for the same species, production practices may vary greatly across the regions where they are grown for seeds (Figure 32.1).

Forages: The Science of Grassland Agriculture, Volume II, Seventh Edition.
Edited by Kenneth J. Moore, Michael Collins, C. Jerry Nelson and Daren D. Redfearn.

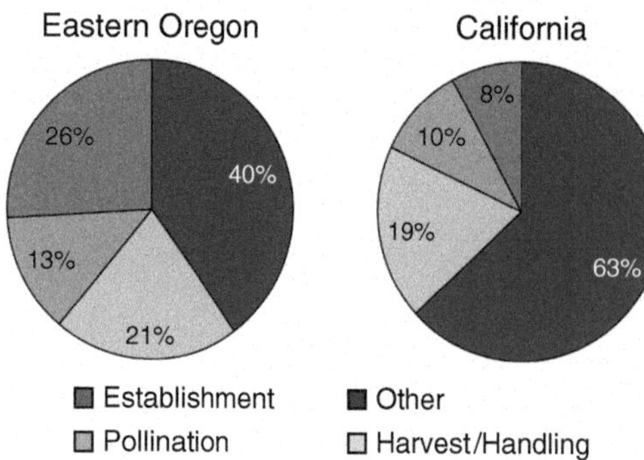

Fig. 32.1. A comparison of the relative costs of different production operations for alfalfa seed grown in eastern Oregon and central California. Notice the cost of pollination is much greater in eastern Oregon than California, but the cost of operations, including irrigation and insect management, are much greater in California than eastern Oregon. Growers also need to consider where the greatest gains in production efficiency for their operation can be achieved and consider alternative management practices to increase net income.

Regional Seed Production in North America

Most temperate grass seed production occurs in the northwestern United States with the majority of perennial ryegrass, italian ryegrass, tall fescue, red fescue, and bentgrass grown in western Oregon. Significant acreage of kentucky bluegrass seed is produced in central and eastern Oregon, eastern Washington, western Idaho, and western Canada. Some forage legume species are still grown for seed in areas other than the western United States. Seed production of most of the improved red clover cultivars occurs in the west, but uncertified red clover seed is also produced in Illinois, Indiana, Ohio, Missouri, and central Canada. Similarly, most of the production area of alfalfa seed shifted in the late 1940s from the Great Plains states to California and regions of the Pacific Northwest east of the Cascade Mountains, and the western Canadian provinces, but some acreage still remains in Nebraska, Kansas, and Oklahoma.

Most of the white clover seed is produced in California's Sacramento Valley with more limited acreage in Western Oregon and Idaho. However, much of the white clover seed used in the United States is imported from New Zealand. Most crimson clover seed is produced in the Willamette Valley of western Oregon, but some seed production of this species as well as that for korean and japanese bushclover and timothy are produced in the midwestern or southeastern United States (Wheeler and Hill 1957; Youngberg and Buker 1985). Most birdsfoot trefoil seed is produced in Minnesota, Wisconsin, Michigan, and adjacent regions of Canada. Specialty winter-annual forage legumes such as common vetch, rose clover, subterranean clover, arrowleaf clover, and egyptian clover are produced in limited amounts in either California or western Oregon.

Most native warm-season grass seed production occurs in the central and southern Great Plains region of the United States. Native grass seed is harvested from rangeland where the "wild" harvest will consist of a mixture of important rangeland species such as big bluestem, little bluestem, indiangrass, and switchgrass. Seed production of improved cultivars of these four native species, as-well-as blue and sideoats grama, buffalograss, sand bluestem, sand lovegrass, and eastern gamagrass also occurs in this region. Seed production of most introduced warm-season grasses is typically done in the areas of major use for the species. Bahiagrass and dallisgrass seed production can be found in the southeastern United States as well as Australia, weeping lovegrass and yellow bluestem seed production mostly occurs in the southern Great Plains region, and bermudagrass seed production typically occurs in Arizona, southern California, and Oklahoma.

Native grass seeds from the Intermountain Great Basin region and those from west of the Sierra Nevada and Cascade mountains are typically produced as specialized seed crops in the California Great Central Valley including beardless wildrye, purple needlegrass, and sandberg bluegrass. Produced in western Oregon are tuffed hairgrass, blue wildrye, and roemer's fescue, and in central Oregon, Idaho, and in eastern Washington are desert wheatgrass, mountain brome, and western wheatgrass.

Seeds of annual warm-season grasses, such as forage sorghum, sudangrass, sorghum-sudangrass hybrids, and pearlmillet are primarily developed and grown by commercial seed companies in the southern High Plains and Southwest states where the growing conditions are characterized by fertile soils, hot and dry days, and adequate available irrigation water (Karper and Quinby 1947).

Environmental Variability and Seed Quality

Consistent production of high-quality forage seeds depends upon the ways cultural practices are applied to optimize the reproductive expression of each species. Since most forage seed crops are grown in regions different from where the seeds will be utilized, numerous aspects of the seed production environment must be considered in relation to plant growth and reproduction. It is critical to maintain the genetic integrity of cultivars as the volumes of seeds are increased in each successive generation of seed production. Each crop cultivar starts from a small amount of seeds developed and tested by public or private plant breeders. The original **breeder seed** is increased until enough quantity is available for the market place to meet user demand. Factors such as photoperiod and seasonal temperature ranges along with rainfall patterns

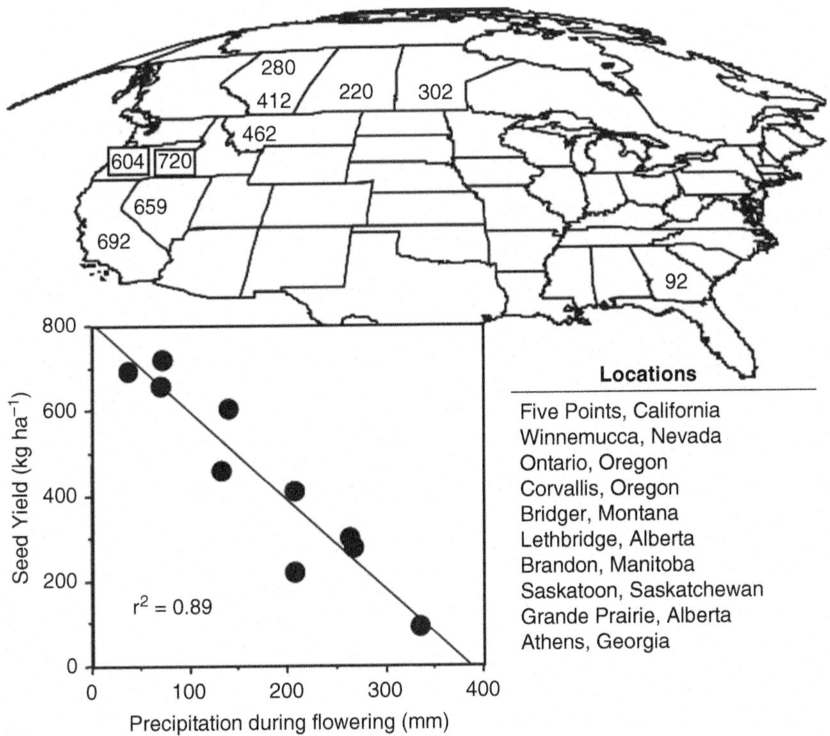

Fig. 32.2. The effect of precipitation during the reproductive period on alfalfa seed yield. Seed yields shown are based on reported average annual yields for the general production area. Precipitation amounts are the annual average amounts for the time period for alfalfa reproduction in the area.

have traditionally determined the regions where cultivars are successfully produced. However, economic and social concerns may also influence the kinds of farming practices that can be used and the capacity of the seed industry to produce consistent amounts of high-quality seeds in a given seed production environment.

The geographic shift of forage grass and legume seed production from the humid mid-western to arid western United States was due to more dependable weather conditions during seed maturation and harvest (Rincker et al. 1988). As greater amounts of seeds of improved cultivars from specialized seed production were available in the marketplace, farmers used less on-farm produced seeds. The climatic conditions of the west are best characterized by relatively warm summer temperatures and low humidity, as well as dry periods during harvest time in late-summer (Youngberg and Buker 1985). Frequent or untimely rainfall in the Midwestern states contributed to reduced seed yields and frequent crop failures. For example, alfalfa seed yield is reduced as the amount of precipitation received during the seed production period increases, so summer arid environments are best for seed production (Figure 32.2). However, even in the western United States, some specific regional climates may be more favorable to seed production than others. High ambient air temperatures during reproduction may reduce seed yields by decreasing the number of seeds successfully formed in each flower. Furthermore, the number of alfalfa florets setting pods decreases when the air temperature exceeds 38 °C, with 27 °C being the optimal temperature for seed production. Seed production usually increases as the relative humidity decreases, but low humidity associated with high temperatures can make easy-to-shatter seed crops more difficult to harvest because pods can dehisce with shattered seeds falling to the ground at late crop maturation time or during harvest. High-light intensity is also essential for good seed production as well as clear skies and warm temperatures that can favor insect pollinator activity that is needed for cross-pollinated species such as the clovers and alfalfa.

Care must be taken when choosing a region for seed production to fit the reproductive physiology of the cultivar, as well as maintenance of genetic quality. The differential effects of environmental factors such as temperature and photoperiod length may shift the genetic composition of a cultivar during seed production. For example, the temperatures in the southern California and Arizona seed production regions are characteristically mild in winter and hot in summer, compared to the cold winter weather characteristic of the mid-western or upper Great Lakes regions. When cold-resistant alfalfa cultivars are grown for seed in mild climates, the plants produced from those seeds of those populations are shifted to taller ones than the original parent material, with greater autumn plant growth and greater susceptibility to winter injury than plants grown from seeds produced in climates more like those where the cultivar was developed (Garrison and Bula 1961). As the number of seed production generations increases in regions different from the regions of adaptation, the resulting **genetic shift** is toward plants more similar to cultivars adapted to the seed production region. A reverse genetic shift may occur if warm climate southern cultivars are increased for several generations in much colder regions.

Another type of genetic shift may occur from the different flowering responses to photoperiod length of individual plants in a population (Garrison and Bula 1961; Bula et al. 1964; Taylor et al. 1990). Seed production of red clover cultivars adapted to northern regions at southern latitudes may result in a disproportionate production of non-winter-hardy genotypes that are not adapted to cold continental winter climates. This phenomenon can be useful for producing cultivars that are strongly vegetative at southern latitudes. Some cultivars of white clover and birdsfoot trefoil may produce more seeds when grown at northern latitudes where longer daylengths induce more flowering than when grown for seed at southern latitudes.

There has been increasing demand for native grass species to meet conservation planting needs, including vegetation restoration following wildfires. Depending upon the location of the restoration site, planting specifications may require the use of locally grown ecotypes or plant materials from similar ecoregions. The reasoning behind this is to ensure the reestablished plants will perform similarly to those that were established before the fire, and the community will have a similar natural ecologic function.

Specialized Management for Reproduction and Quality

The cultural practices employed for seed production are different from those used for conventional forage production and serve the purpose of optimizing the yield of pure, high-quality seeds. These may include special planting methods and plant population configurations, soil fertility requirements, and other management operations that benefit seed production. Forage seed crops may have irrigation requirements, weed control methods, and insect and disease pest management practices very different than those when grown for forage. In addition, special practices not used for forage production, such as pre-bloom herbage removal and pollinator management may be required to enhance reproductive development and seed yield. As the seed plant changes throughout the season, so various growth factors need to be considered and the timing of different practices be applied to optimize seed yield (Figure 32.3).

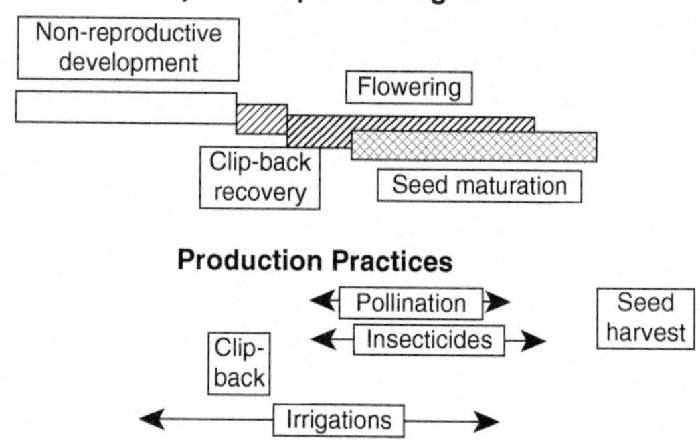

Fig. 32.3. The relationship of alfalfa seed crop-development stages with the timing of seed production practices. Notice how multiple management practices must be considered at the same time. Improper irrigation timing or amount influences the balance of vegetative and reproductive development, and improperly managed insect pests can affect the effectiveness of pollinators that significantly impact final seed yield.

Stand Establishment

Most forage crops grown for seed are planted in rows 15–122 cm apart and at very low seeding rates. Conventional grain drills are typically used to place seed in rows of less than 30 cm, whereas, single-row, box-type, planters are used to plant seed in rows of greater than 30 cm. Spaced plantings similar to those used for vegetable crops have been utilized in alfalfa to increase seed production by minimizing intra-row plant competition, while aerial broadcast seedings are used in large white clover seed fields in California.

Soil-Nutrient Management

Grass seed crops generally need supplemental nitrogen (N) fertilization (Youngberg and Buker 1985). The optimal time of N application depends upon the growth habit of the crop and local soil conditions where grown (Young et al. 1995a,b). Some grasses primarily produce their flower buds during the summer and thus respond to one N application in spring. Other species, such as some cool-season grasses, initiate their flower primordia during the fall or winter and respond when N applications are split between the autumn and spring. However, the optimal amounts and application times of N for cool-season grasses vary for different species. The total amount of applied N depends upon residual amounts of soil N from previous fertilizer applications or from previous crops in the rotation cycle, such as following legume crops.

Legume seed crops should have specific **rhizobia** inoculant applied to seeds at the time of planting to ensure that adequate N fixation occurs for plant growth. Legumes need adequate levels of phosphorus, potassium, calcium, and other nutrients to ensure proper growth. The amount of each nutrient required for seed production may be less than that needed for maximal forage production. High phosphorus levels in white clover seed fields may reduce seed yield because **vegetative growth** is favored over reproductive development (Clifford 1987). Low levels of the micronutrient boron in the soil may reduce seed yields in legumes such as white clover (Johnson and Wear 1967).

Local soil tests used in conjunction with local extension service or crop consultant recommendations can help determine the amounts of each nutrient needed for adequate plant growth. Soil pH may greatly influence legume seed crop success. Many legume seed crops are very sensitive to acid soil conditions (Steiner and Alderman 2003). For example, crimson clover is very sensitive to low pH conditions and will not adequately fix atmospheric N unless excessive soil acidity is modified with lime. Strawberry clover is better adapted to alkaline than to acid soils. These factors should be considered when choosing a suitable region or soil series for forage legume seed crops such as these.

Spring Growth Herbage Management

Removal of early spring plant growth by grazing or haying prior to or during early flowering is a common practice for some annual and perennial forage legumes species grown for seed (Rincker and Rampton 1985; Rincker et al. 1988). Herbage removal helps synchronize uniform flowering in early summer when insect pollinators are most active, disrupts the life cycles of some insect pests that attack the flowers, and provides the seed grower additional income from the sale of the herbage as hay or from grazing fees. White clover and subterranean clover are commonly grazed by sheep (*Ovis aies* L.), while alfalfa and red clover herbage is removed mechanically. Crimson clover and kura clover do not typically have herbage removed, or seed yields could be greatly reduced.

The optimal time for herbage clip-back in California alfalfa seed fields is between late-March and mid-May when new crown buds are about one-half centimeter long (Jones and Pomeroy 1962). Red clover seed fields in western Oregon have herbage removed around late-May, but the amount of available soil water, number of accumulated heat units during spring, and root health of stands greatly affects optimal timing for maximal seed yield (Steiner et al. 1995).

Some temperate grass seed crops such as annual and perennial ryegrass are grazed with sheep during the autumn and winter before **reproductive primordium** have elongated. The timing of herbage removal is critical, and must be performed before the inflorescences begin rapid development or seed yield will be reduced (Young et al. 1995a,b). Plant growth regulator use on forage grass seed crops has been investigated but, in general, has not produced beneficial results.

Soil-Water Management

In general, it is important that plant growth be balanced with active floral development to achieve good seed production. Specific water requirements vary greatly by species and depend upon soil water holding capacity and soil depth, amount and pattern of natural precipitation, air temperature, and wind speed that can affect evapotranspiration and growing season length.

Excessive non-reproductive vegetative growth of some forage legume species can reduce seed yields, so properly managed water application timing and amount applied is important to favor reproduction (Figure 32.4). If alfalfa, white clover, and birdsfoot trefoil are over-watered, **vegetative growth** will be unrestricted and seed yields will either be greatly reduced or unimproved compared to plants grown without irrigation (Taylor et al. 1959; Steiner et al. 1992; Oliva et al. 1994; Garcia-Diaz and Steiner 2000). Shallow-rooted white clover plants need more frequent applications of water than deep-rooted

Fig. 32.4. Surface irrigation for alfalfa seed production. Siphon pipes direct water into furrows between the planted rows of alfalfa plants that are grown on low raised beds. Forage legume seed crops must be managed to favor reproduction over herbage growth. Most forage legume seed crops have reduced seed yields if they are overwatered.

alfalfa plants when grown in California. White clover grown with supplemental irrigation is best adapted to soils with low water holding capacities in order to reduce excessive plant vegetative growth. Alfalfa with a long taproot may utilize water from deep within the soil profile and survive much longer dry periods than either white or red clover. Red clover responds well to large water application amounts at early bloom in western Oregon, while birdsfoot trefoil does not require irrigation.

Pollination and Pollinator Management

Most forage crops require some form of pollination and fertilization for seed set to occur. After a pollen grain reaches a receptive **stigma**, it germinates. The germinating pollen grain produces a pollen tube that grows down the **style** to the ovule entering the embryo sac. Within the embryo sac, one of the two sperm nuclei fuses with the egg nucleus forming the first cell of the new plant, the second sperm nucleus fuses with one or more polar nuclei forming the endosperm, thus completing double fertilization (Dumas and Gaude 1993).

Most grass species such as tall fescue, perennial ryegrass, and crested wheatgrass are cross-pollinated by wind. Vast clouds of pollen can be seen in spring blowing from seed fields making physical isolation distances between different cultivars within a species necessary to reduce genetic contamination between cultivars (Figure 32.5). Research has shown that pollen from domestic creeping bentgrass can be transferred to plants of the same species 21 km away and plants of wild *Agrostis* species as far as 14 km (Watrud et al. 2004). As transgenic forage varieties are developed, concerns have been raised about transgenes being spread to non-genetically modified plants, such as with alfalfa (Greene et al. 2015).

Some important grasses such as kentucky bluegrass, dallisgrass, buffelgrass, yellow bluestem, and eastern gamagrass reproduce by apomixis from which seeds are produced without fertilization of the egg nucleus. There is no pollen-mediated transfer of genetic material between different cultivars within these species, so genetic contamination during seed production in this manner is not generally a concern.

Cross-pollinated forage legumes require an insect pollinator to manipulate the flower so fertilization can be accomplished (McGregor 1976). In alfalfa, the pistil of the flower is held under pressure within the keel and fertilization cannot occur until the keel is opened, releasing the reproductive column and allowing it to strike the standard flower in an action called tripping. Other common forage legume flowers do not have this mechanism and therefore are not tripped. Instead, bees force the reproductive column out of the keel in a piston-like, reversible action that brings the pollen in contact with the insect. When a pollinator visits a flower, pollen from that flower either adheres to small hairs or sticks to its body and is transferred between flowers of different plants as the insect forages. The amount of forage legume seed produced is directly related to pollinator activity (McGregor 1976).

The western honeybee (*Apis mellifera* L.) is an important domesticated pollinator of many forage legume seed crops that require cross-pollination. A high percentage of the pollinating insects used in California alfalfa and white clover seed fields are honeybees. Honeybees are also used for most clover, vetch, and birdsfoot trefoil seed production. A general recommendation is to use an average of 3.7–5 two-story high honeybee hives per hectare for adequate pollination. Some situations may require the use of more than five hives per hectare, but the cost of additional hives may prevent an economic return. When using honeybees in large fields, it is best to have the hives spread no more than 160 m apart with groups of 12–18 colonies

Fig. 32.5. A pollen cloud released from crested wheatgrass. *Source:* Image provided by Ken Moore.

at each apiary site. Each hive should have about one-half square meters of brood covered with adult bees and an actively laying queen for adequate pollination.

In eastern Oregon and Washington, Idaho, Nevada, and Canada, the leafcutter bee (*Megachile rotundata* L.) is used for alfalfa seed production. Leafcutter bees have also been used in combination with honeybees in California. Leafcutter bees nest in grooved, laminated boards composed of wood, particle board, or polystyrene plastic or other solid materials with drilled holes that serve as nests. The blocks or nesting boards are placed in portable shelters that can be moved from field to field to facilitate pollination. A unique characteristic of the leafcutter bee is that it collects pollen or nectar to provision the brood cell. The egg is laid on the mixture of nectar and pollen in each individual cell. Because the leafcutter bee forages for pollen, it is a very effective pollinator of alfalfa (Rincker et al. 1988). Approximately one female leafcutter bee is required per four square meters of alfalfa flowers (Bohart 1967). Approximately 20 000 healthy prepupae cells of leafcutter bees are adequate for pollinating alfalfa seed fields.

The alkali bee (*Nomia melanderi* L.) is a native pollinator that has been used in alfalfa seed production in California and south-central Washington. These bees nest in the soil and have been cultured in gravel and soil beds lined with plastic so that bed moisture can be controlled (Frick et al. 1960). Salt is often added to the surface of the soil to maintain proper soil moisture by reducing evaporation and to control weeds (Rincker et al. 1988). Alkali bee beds are difficult to move and are expensive to establish. Approximately 10 m^2 of well-populated nesting sites is required per hectare of alfalfa seed field (McGregor 1976).

Bumblebees (*Bombus* spp.) are also very effective pollinators in alfalfa and all other forage legume seed crops. Bumblebees are very difficult to culture but are abundant in seed fields that are adjacent to wild areas along roadsides, ditches, and wooded areas. They nest primarily in undisturbed soil, but also in abandoned rodent holes. Only impregnated female bumblebees over-winter and they are very particular about nesting conditions. The solitary nature of bumblebees makes them difficult to domesticate, but nesting can be encouraged (Heinrich 1979). All bees, and especially the wild bees, are very sensitive to insecticides so pest management plans must consider how to avoid seriously damaging pollinators.

Weed Management

Weed control methods vary for different seed crops and the kinds of weeds found in specific fields. Weeds may be a serious problem when stands are thin because the competitive nature of weeds lowers crop seed yields and can delay harvest. Also, depending on weed seed shape and size, weed seeds that cannot be easily removed by **seed conditioning** (the cleaning of seeds after harvest) may render the crop unsaleable.

Depending on planting method, appropriate strategies must be employed to control weeds. Effective crop rotations, properly prepared seed beds, and planting at a time that allows rapid and uniform establishment of the seed crop stand can increase the effectiveness of herbicides. When seed crops are planted in wide rows such as with alfalfa, mechanical cultivation methods can be employed to physically control weeds or help the crop plants to gain a competitive advantage over the weeds. Many fall-planted temperate grass seed crops are seeded under a narrow band of activated charcoal sprayed over the planting row at planting time (Lee 1973). Herbicides that adsorb to charcoal are then sprayed over the field and activated by precipitation or irrigation. Weeds and volunteer crop seeds between the treated rows are killed by the herbicide, but the crop emerges unharmed from under the charcoal-treated row because the herbicide is bound to the charcoal and rendered inactive. Early-maturing weeds in several perennial legume seed crops can be controlled by removal of winter and early-spring herbage by grazing, mowing, or hay harvest. Mowing is also an effective method to reduce weed competition. Different combinations of management practices including rotation crops and establishment methods affect the kinds of weeds that grow in grass seeds fields, and can influence the reservoir of weed seeds that are found in the soil weed seed bank (Medeiros and Steiner 2002).

Preemergence herbicides are often important management tools for achieving good initial crop stands after planting. Postemergence herbicides are effective when specific sensitive weeds are selectively controlled by applications to establish seed crop stands. Herbicide effectiveness for controlling weeds can be greatly enhanced when used in combination with other control methods to significantly reduce weed establishment (Lee 1965).

Dodder, a parasitic plant, is one of the most serious weeds to control in legume seed production fields. Since this weed is classified as a primary noxious weed, there is no tolerance for dodder seeds found in seed lots that are to be sold. This weed must be aggressively eradicated from seed fields. Seed fields are monitored from the ground and air to identify infested areas that are then sprayed locally and burned before the dodder plants produce seeds. There are specific herbicides that can reduce dodder infestations if properly applied. Most of the few remaining dodder seeds harvested with the seed crop can be removed during **seed conditioning** using special separation equipment. However, each additional conditioning procedure reduces the amount of clean seed produced and adds to the final cost of production (Purdy et al. 1961; Youngberg and Buker 1985).

There are problems associated with relying on herbicides as the only means of controlling weeds, particularly in perennial seed crops. These include the danger of induced herbicide resistance in weed populations and shifts in weed populations toward those species that are not susceptible to commonly used herbicides. Resistant populations of annual ryegrass have developed in wheat fields where selective grass herbicides have been continually used to control grassy weeds in wheat. Due to shifts in the predominate kinds of weeds found in grass seed fields, cheatgrass and smooth bromegrass have become the predominate competitors in some seed fields.

The judicious use of those herbicides that continue to be registered in combination with cultural practices that reduce weed establishment, such as crop rotation and alternating the kinds of herbicides used, offer the best possibilities for effective control in seed fields. There are some cases where effective herbicides are no longer available because of human and wildlife health risks. In addition, due to the relatively small acreage of seed crops grown, the expense to chemical companies to register new chemicals is often not warranted compared to major crops such as corn, soybean, and cotton. The USDA Minor Crop Pest Management IR-4 Program provides assistance to help ensure new and more effective crop protection products are developed and made available to minor and specialty crop producers, including forage crops grown for seed.

Insect Pest Management

For both grass and legume seed crops, cultural practices can often be employed to control certain insect pests or reduce the need for insecticides. Shorter perennial grass seed rotations have reduced the incidence of bluegrass billbug (*Sphenophorus parvulus* Gyllenhal) damage as was also found in older orchardgrass seed stands. The subterranean sod webworm (*Chrysoteuchia topiaria* Zeller) can attack kentucky bluegrass and reduce stand longevity, but the use of deep-rooted cultivars reduces crop damage compared to those cultivars with shallow root systems.

In the Pacific Northwest, burning of grass seed field crop residues (Figure 32.6) after harvest often controlled some insect pests in long rotation stands, but the accumulation of charcoal on the soil surface also bound insecticides making them ineffective when applied at labeled application rates (Kamm and Montgomery 1990). The market change from public cultivars grown in long rotations to a far greater number of privately developed cultivars grown in shorter rotations has mitigated the negative impact of perennial grass seed crop pests, regardless of the reduction in grass seed crops burned after harvest in the Pacific Northwest. With the reduction in field burning of post-harvest straw, the increase in soil organic matter and earth worm populations has reduced the effectiveness of chemically treated baits used to control invertebrate slugs (*Derocerus reticulatum* Mueller). Earthworms are immune to the pesticide and carry the bate carrier into their burrows at night where it is out of reach of the slugs (Gavin et al. 2012).

Fig. 32.6. Open field burning was once a common practice in the Pacific Northwest for disposing of grass straw after seed harvest. Burning effectively controlled some disease, insect, and weed pests when perennial grass seed stands were kept in production for 10 years or more. Growers have changed their postharvest straw management and weed control practices since legislation to restrict field burning was enacted in the 1990s. Shorter rotations have greatly reduced the threat of some disease and insect pests.

In the past, organophosphate insecticides were frequently used in legume seed production. Depending on the seed production region and crop, one-to-as-many as four applications were required to adequately control pests such as Western tarnished plant bug (lygus) (*Lygus hesperus* (Hemiptera: Miridae)), aphids (*Aphis* spp.), and leaf-hoppers (*Erythroneura* spp.). More recently, pyretheroid-type insecticides are widely used and are replacing the organophosphates. Special care must be taken to use pyretheroid-type insecticides properly because some insect pests can quickly develop resistance to these types of chemicals. When utilizing herbage from seed fields for livestock feed, careful attention must be given to product registrations to be sure such use is allowed.

Certain insect pests are not controlled by available insecticides, so other methods need to be used. There are no chemicals available to control the clover seed midge (*Dasineura leguminicola* Lintner, Cecidomyiidae) in red clover, but its life cycle can be disrupted by herbage removal during the early bloom period in spring. Alfalfa seed chalcid (*Bruchophagus roddi*, Eurytomidae) damage in alfalfa can be reduced by incorporating residues into the soil after seed harvest using cultivation and irrigation to rot infested seeds; performing spring clip-back of herbage in early spring delays initial crop flowering time that, in turn, removes available host sites to females trying to oviposit. Also, not extending the pollination period into late-summer reduces the build-up of chalcid populations in later flower cycles. This pest has become a more significant pest in California since pyrethroid insecticides have replaced organophosphates.

Control of alternate nesting sites in weeds can also be used to control some insect pests. A dogfennel weed control program in winter and early-spring, both alongside and within red clover seed fields, removes feeding and oviposition sites for lygus and thus reduces pest pressure later in the season (Kamm 1987). This combined weed and insect control practice can probably be used for other legume seed crops where lygus is a problem.

Insect control must be carefully planned and done in a manner that optimizes pest control, minimizes harm to insect pollinators and beneficial insects, and avoids chemical trespass from the field where the product is applied. Similar as the situation with herbicides, forage seed growers have a diminishing number of insecticides available for insect pest management. Decline in honeybee populations have reached a critical juncture, so care must be exerted to help reduce any stress that could be due to insecticide use, particularly since these insects are critical to legume seed production.

Additionally, significant changes in climate may influence the distribution and abundance of insects (Cannon 1998). For example, it has been suggested that some aphids may cause more severe damage to plants due to increased settling times and reproduction under elevated CO_2 concentrations (Awmack et al. 1996; Smith 1996), and other aphid species may not be affected at all by elevated CO_2 concentrations (Salt et al. 1996). Regardless of these facts, seed producers need to be aware of changing environmental conditions and adjust their seed production practices to meet new challenges.

Diseases and Their Management

The best way to control diseases in forages is by using disease-resistant cultivars. However, most cultivars are genetically selected for their performance in the region where they will be grown for forage, and not necessarily where grown for seed. Such cultivars are often exposed to different kinds of disease organisms or amounts of severity in the seed production regions, compared to where they are grown for forage.

Plant pathogens that affect grass seed production can attack both foliar and reproductive plant parts. Important inflorescence diseases include ergot (*Claviceps purpurea* (Fr. Tul.)), blind seed (*Phialea temulenta* Prill. & Delacr.), and the nematode seed gall (*Anguina* spp.) (Hardison 1963). The epidemiology of these diseases has become better understood, but there are no completely effective chemical controls (Alderman 1991). Blind seed disease is controlled through cultural practices including timely harvesting to avoid seed shatter, removal of lightweight (including infected) seed from the field, planting seed at least 1 cm deep, plowing, rotation with a non-susceptible crop, and the use of late-maturing cultivars (Alderman 2001). Seed conditioning is important because it can mechanically remove ergot and seed galls. Choke disease (*Epichloë typhina* (Pers.) Tul. & C. Tul) is more problematic since plants are systemically infected (Pfender and Alderman 2003). However, most inflorescence diseases can be managed though crop rotation with a non-susceptible host.

Stem rust (*Puccinia graminis subsp. graminicola* Urban) is a devastating disease in perennial ryegrass and tall fescue seed production (Pfender 2001). Typically, several fungicide applications are required annually to control this disease. Epidemic development is strongly dependent on weather conditions (Pfender 2003), so optimal timing for fungicide applications differs among years and locations. Seed fields must be carefully monitored to assess when the disease is present. Fungicides must be applied before disease severity reaches damaging levels because after symptoms are obvious, infections become less responsive to fungicides. Weather-based plant growth models have been developed to determine the time of optimal fungicide applications to improve disease control with minimized fungicide use (Pfender and Upper 2015).

The kinds of diseases that impact forage legume seed crops are primarily the same as those that attack forage crops. Most important are the root rotting diseases caused by (*Fusarium* ssp.) and (*Phytophthora* spp.) that affect stand persistence. The resistance of improved red clover cultivars to *Fusarium oxysporum* Schlecht. Emend. Snyder and Hansen grown in mid-western state forage fields does not reduce the impact of *Fusarium solani* (Mart.) Sacc. that is prominent in Oregon seed production fields (Steiner et al. 1997). Wilt caused by *Verticillium albo-atrum* Reinke and Berthold has become an important problem in some alfalfa seed producing regions because it can be seed-borne. When red clover and other legume seed fields are not properly rotated, *Sclerotinia* spp. can become a serious problem with long-term implications. Northern anthracnose (*Kabatiella caulivora* (Kirchn) Karak.), a seed-borne disease of crimson clover and red clover, can be controlled by seed treatment and proper planting time in the autumn (Leach 1962).

Seed Harvest

Many forage species have a tendency to shatter before all seeds on a plant are mature or produce flowers in an indeterminate fashion so that seeds of different stages of maturity are present on the same plant at the same time. Legume seed produced in naked pods, such as birdsfoot trefoil and vetches, tend to dehisce when ripe, propelling seeds in all directions. Most all forage legumes and grasses are susceptible to **seed shattering** once the seeds have dried.

Forage seed crops are harvested with a combine, either by direct combining of the standing mature crop or by first windrowing the crop before shattering begins and then combining the windrowed material at a later time (Rincker et al. 1988). The windrow method is chiefly used for temperate grass, clover, and other forage legumes because it allows cutting the crop when the foliage is slightly green and before shattering has begun. Once the mature inflorescences with seed have dried adequately for proper threshing, a special continuous belt attachment mounted to a combine gently picks up the cut windrow with minimal shaking that greatly reduces shatter loss. When direct combining is involved, a chemical desiccant may be used to aid plant drying without windrowing. Monitoring of seed moisture content during maturation can help determine the proper time to harvest forage seed crops by maximizing seed maturity and minimizing seed shattering (Klein and Harmond 1971).

There are also specialty harvest machines for forage seed crops with unique growing features. Subclover produces flowers near the ground and its pods often develop below the soil surface, similar to peanut. Once the seeds have matured, special harvesters developed in Australia vacuum up the pods and loose soil, separate the pods from soil, and then thresh the seeds.

Unlike temperate grass and legume species, the inflorescences of many native warm-season grass species not only have indeterminate ripening, but the seeds are also chaffy which makes harvest and cleaning difficult. Specialized harvesting equipment like the Woodward Flail-Vac seed harvester (Figure 32.7) was designed to harvest native and introduced chaffy-seeded grasses (Dewald and Beisel 1983; Dewald et al. 1985, 1993), and the crop stripping combine header that was designed to harvest a variety of crop seeds has been used successfully to harvest native grass seeds (Shelbourne and McCredie 1995).

Fig. 32.7. Stripper harvesters are designed to make multiple harvests from the same native grass stand in the same year. These machines come in widths from 1 to 8 m to fit different kinds of terrain. They can be powered by hydraulic motors, combustion engines, or tractor power take-off systems and can be mounted on a combine, tractors, jeeps, and all-terrain vehicles, as well as hand-held units. This technology greatly increases the ease of harvest of chaffy native seeds.

References

Alderman, S.C. (1991). Assessment of ergot and blind seed diseases of grasses in the Willamette Valley of Oregon. *Plant Dis.* 75: 1038–1041.

Alderman, S.C. (2001). Blind seed disease. United States Department of Agriculture, Miscellaneous Publication No. 1567.

Awmack, C.S., Harrington, R., Leather, S.R., and Lawton, J.H. (1996). The impacts of elevated CO_2 on aphid–plant interactions. *Aspects Appl. Biol.* 45: 317–322.

Bohart, G.E. (1967). Management of wild bees. In: *Beekeeping in the United States*, U.S. Department Agriculture, Agriculture Handbook 335, 109–118. Washington, D.C: U.S. Government Printing Office.

Bula, R.J., May, R.G., Garrison, C.S. et al. (1964). Growth responses of white clover progenies from five diverse geographic location. *Crop Sci.* 4: 295–297.

Cannon, R.J.C. (1998). The implications of predicted climate change for insect pests in the UK, with emphasis on non-indigenous species. *Global Chang. Biol.* 4: 785–796.

Clifford, P.T.P. (1987). Producing high seed yields from high forage producing white clover cultivars. *J. Appl. Seed Prod.* 5: 1–9.

Dewald, C.L. and Beisel, V.A. (1983). The woodward flail-vac seed stripper. *Trans. Am. Soc. Eng.* 26: 1027–1029.

Dewald, C.L., Berg, W.A., and Sims, P.L. (1985). New seed technology for old farmlands. *J. Soil Water Conserv.* 40: 277–279.

Dewald, C.L., Sims, P.L. and White, L.M. (1993). The Woodward flail-vac seed stripper – recent progress in harvesting chaffy seed. Proceedings of the International Grassland Congress, Palmerston North, New Zealand (8–21 February 1993).

Dumas, C. and Gaude, T. (1993). Progress towards understanding fertilization in angiosperms. In: *The Molecular Biology of Flowering* (ed. B.R. Jordan), 185–217. Wallingford, UK: C.A.B. International.

Frick, K.E., Porter, H., and Weaver, H. (1960). Development and maintenance of alkali bee nesting sites. *Washington Agric. Exp. Station Circ.* 366.

Garcia-Diaz, C.A. and Steiner, J.J. (2000). Birdsfoot trefoil seed production: II. Plant-water status on reproduction and seed yield. *Crop Sci.* 40: 449–456.

Garrison, C.S. and Bula, R.J. (1961). Growing seeds of forages outside their regions of use. In: *Seed*, USDA Yearbook Agr, 401–406.

Gavin, W.E., Mueller-Warrant, G.W., Griffith, S.M., and Banowetz, G.M. (2012). Removal of molluscididal bait pellets by earthworms and its impact on control of the gray field slug (*Derocerus reticulatum* Mueller) in western Oregon grass seed fields. *Crop Prot.* 42: 94–101.

Greene, S.L., Kesoju, S.R., Martin, R.C., and Kramer, M. (2015). Occurrence of transgenic feral alfalfa (*Medicago sativa* subsp. sativa L.) in alfalfa seed production areas in the United States. *PLoS One* 10 (12): e0143296. https://dot.org/10.1371/journal.pone.0143296.

Hardison, J.R. (1963). Commercial control of *Puccinia striiformis* and other rusts in seed crops of *Poa pratensis* by nickel fungicides. *Phytopathology* 53: 209–216.

Harlan, J.R. (1960). Breeding superior forage plants for the Great Plains. *J. Range Manage.* 13: 86–89.

Heinrich, B. (1979). *Bumblebee Economics*. Cambridge, MA: Harvard University Press.

Johnson, W.C. and Wear, J.I. (1967). Effect of boron on white cloverr (*Trifolium repens*, L.) seed production. *Agron. J.* 59: 205–206.

Jones, L.G. and Pomeroy, C.R. (1962). Effect of fertilizer, row spacing and clipping on alfalfa seed production. *California Agric.* 16: 312–217.

Kamm, J.A. (1987). Impact of feeding by *Lygus hesperus* (Heteroptera: Miridae) on red clover grown for seed. *J. Econ. Entomol.* 80: 1018–1021.

Kamm, J.A. and Montgomery, M.L. (1990). Reduction of insecticide activity by carbon residue produced by burning grass seed fields after harvest. *J. Econ. Entmol.* 83: 55–58.

Karper, R.E. and Quinby, J.R. (1947). Sorghum–its production, utilization and breeding. *Econ. Bot.* 1: 355–371.

Klein, L.M. and Harmond, J.E. (1971). Seed moisture - a harvest timing index for maximum yields. *Trans. Am. Soc. Agric. Eng.* 14: 124–126.

Leach, C.M. (1962). *Kabatiella caulivora*, a seed-borne pathogen of *Trifolium incarnatum* in Oregon. *Phytopathology* 52: 1184–1190.

Lee, W.O. (1965). Herbicides in seed bed preparation for establishment of grass seed fields. *Weeds* 13: 293–297.

Lee, W.O. (1973). Clean grass seed crops established with activated carbon bands and herbicides. *Weed Sci.* 21: 537–541.

McGregor, S.E. (1976). *Insect Pollination of Cultivated Crop Plants*, USDA-ARS Agric. Handb. 496. Washington, DC: U.S. Department of Agriculture U.S. Government Printing Office.

Medeiros, R.B. and Steiner, J.J. (2002). Influence of temperate grass seed rotation systems on weed seed soil bank composition. *Rev. Bras. Sementes* 24: 118–128.

Oliva, R.N., Steiner, J.J., and Young, W.C. III (1994). White clover seed production. II. Soil and plant water status on seed yield and yield components. *Crop Sci.* 34: 768–774.

Pfender, W.F. (2001). Host range differences between populations of *Puccinia graminis* subsp. *graminicola* obtained from perennial ryegrass and tall fescue. *Plant Dis.* 85: 993–998.

Pfender, W.F. (2003). Prediction of stem rust infection favorability, by means of degree-hour wetness duration, for perennial ryegrass seed crops. *Phytopathology* 93: 467–477.

Pfender, W.F. and Alderman, S.C. (2003). Evaluation of post-harvest burning and fungicides to reduce the polyetic rate of increase of choke disease in orchardgrass seed production. *Plant Dis.* 87: 375–379.

Pfender, W.F. and Upper, D. (2015). A simulation model for epidemics of stem rust in ryegrass seed crops. *Phytopathology* 105: 45–56.

Purdy, L.H., Harmond, J.E., and Welch, G.B. (1961). Special processing and treatment of seeds. In: *Seeds*, The Yearbook of Agriculture, 322–330. Washington, DC: U.S. Government Printing Office.

Rincker, C.M. and Rampton, H.H. (1985). Seed Production. In: *Clover Science and Technology*, Agronomy Monograph, vol. 25 (ed. N. Taylor), 417–443. Madison, WI, ASA-CSSA-SSSA.

Rincker, C.M., Marble, V.L., Brown, D.E., and Johansen, C.A. (1988). Seed production practices. In: *Alfalfa and Alfalfa Improvement Monograph* (eds. A.A. Hanson, D.K. Barnes and R.R. Hill Jr.), 985–1021. Madison, WI: American Society of Agronomy.

Salt, D.T., Fenwick, P., and Whittaker, J.B. (1996). Interspecific herbivore interactions in a high CO_2 environment: root and shoot feeding on *Cardamine*. *Oikos* 77: 326–330.

Shelbourne, K.H. and McCredie, P.J. (1995). Crop stripping apparatus. United States Patent No. 5,419,107.

Smith, H. (1996). The effects of elevated CO_2 on aphids. *Antenna* 20: 109–111.

Steiner, J.J. and Alderman, S.C. (2003). Red clover seed production: VI. Effect of soil pH adjusted by lime application. *Crop Sci.* 43: 624–630.

Steiner, J.J., Hutmacher, R.B., Gamble, S.D. et al. (1992). Alfalfa seed water management: I. Crop reproductive development and seed yield. *Crop Sci.* 32: 476–481.

Steiner, J.J., Leffel, J.A., Gingrich, G., and Aldrich-Markham, S. (1995). Red clover seed production III. Effect of hay harvest time under varying environments. *Crop Sci.* 35: 1667–1675.

Steiner, J.J., Smith, R.R., and Alderman, S.C. (1997). Red clover seed production: IV. Root rot resistance under forage and seed production systems. *Crop Sci.* 37: 1278–1282.

Taylor, S.A., Haddock, J.L., and Pedersen, M.W. (1959). Alfalfa irrigation for maximum seed production. *Agron. J.* 51: 357–360.

Taylor, N.L., Rincker, C.M., Garrison, C.S. et al. (1990). Effect of seed multiplication regimes on genetic stability of Kenstar red clover. *J. Appl. Seed Prod.* 8: 21–27.

Watrud, L.S., Lee, E.H., Fairbrother, A. et al. (2004). Evidence for landscape-level, pollen-mediated gene flow from genetically modified creeping bentgrass with *CP4 EPSPS* as a marker. *Proc. Natl. Acad. Sci.* 101: 14533–14538.

Wheeler, W.A. and Hill, D.D. (1957). *Seed Production of Grassland Seeds*. Princeton, NJ: Van Nostrand.

Young, W.C. III, Chilcote, D.O., and Youngberg, H.W. (1995a). Annual ryegrass seed yield response to grazing during early stem elongation. *Agron. J.* 88: 211–215.

Young, W.C. III, Youngberg, H.W., and Chilcote, D.O. (1995b). Spring nitrogen rate and timing influence on seed yield components of perennial ryegrass. *Agron. J.* 88: 947–951.

Youngberg, H.W. and Buker, R.J. (1985). Grass and legume seed production. In: *Forages*, 4e (eds. R.F Barnes, D.A. Miller and C.J. Nelson), 72–79. Ames, IA: Iowa State University Press.

PART VII

FORAGE QUALITY

Orchardgrass seedhead in spring. Forage quality usually declines as plant shoots mature. Boot to early head stages are frequently recommended for forage harvest as a good compromise between dry matter yield and forage quality.
Source: Photo courtesy of Mike Collins.

Forage quality describes how well forages meet the nutritional requirements of the animals consuming them. Forages generally provide the lowest cost feeding systems for livestock, so forage quality is an extremely important topic. The fibrous nature of forages compared with concentrate feeds, and the generally slower digestion rates for forages, mean that even small differences can have major effects on nutrient intake and, thus, on animal performance. Understanding forage quality requires knowledge about how forage composition and structure differences impact animal response.

Key factors that can be altered through management to improve forage quality include maturity stage at harvest, forage species selection and nutrient management.

Forages: The Science of Grassland Agriculture, Volume II, Seventh Edition.
Edited by Kenneth J. Moore, Michael Collins, C. Jerry Nelson and Daren D. Redfearn.

Managers must be cognizant of these factors to provide the needed forage quality for their particular animal production system. Forage characteristics impact forage digestibility and voluntary intake, both of which have very large effects on animal response. Energy supply is often the most limiting factor in animal performance, so this aspect of forage assessment has received a great deal of research attention. Even if nutrient concentrations would otherwise be adequate, animal performance can be negatively affected by antiquality components that limit forage digestion or intake, or in more extreme cases, lead to chronic illnesses or death.

CHAPTER

33

Carbohydrate and Protein Nutritional Chemistry of Forages

Ronald D. Hatfield, Research Plant Physiologist, *USDA-Agricultural Resource Service, US Dairy Forage Research Center, Madison, WI, USA*
Kenneth F. Kalscheur, Research Dairy Scientist, *USDA-ARS US Dairy Forage Research Center, Madison, WI, USA*

Introduction

Forages provide important nutritional components for the healthy development of a wide range of ruminants and pseudo-ruminants (e.g. horses, rhinoceroses, rabbits). In addition, perennial forages can play an important role in providing biomass for bioenergy production while stabilizing and improving landscapes that have been degraded from poor management. Being able to effectively utilize a wide range of plants, with varying amounts of fiber content, allows for the better utilization of agricultural lands with sustainable practices to protect soils on various landscapes. Nutritional value of forages is dependent upon the chemical makeup of individual components; their composition and structure as it relates to utilization by the animal. Each of the nutritional components is dependent on biochemical synthesis by the plant. Understanding this biochemical process is critical as we seek ways to alter the plant through genetic manipulation (traditional trait selection or molecular biology) or by the application of post-harvest treatments.

The most nutritionally relevant molecules in forages are carbohydrates, proteins, and lipids. Nucleotide-based molecules (e.g. DNA, RNA, tRNA) are critical for the functioning of the plant, but are not of sufficient quantity to make major contributions to animal nutrition. Non-nutritional components, lignin, phenolics, and polyphenolics (**tannins**) are not degraded by the animal to provide energy, but their interactions can limit availability of nutritional components. Additional components that are important to the functioning of the plant and can have nutritional impact are vitamins and minerals. A recent review by Weiss details the importance of minerals and vitamins for dairy cows (Weiss 2014). Earlier reviews by Spears (1994) and Fleming (1973) provide good coverage of this topic for a wide range of forages. Our attention in this chapter will be on proteins and carbohydrates present in forage cells and their significance to ruminant nutrition.

Forage Protein

Types and Amino Acid Composition

Of all the nutritional components in forages, **enzyme** proteins are the most important – at least to the health and growth of the plant. All other components (carbohydrates, lipids, and proteins themselves) are dependent upon the activity of enzymes for their biosynthesis. It is a well-coordinated metabolic process that allows the plant to grow and accumulate components of nutritional

Forages: The Science of Grassland Agriculture, Volume II, Seventh Edition.
Edited by Kenneth J. Moore, Michael Collins, C. Jerry Nelson and Daren D. Redfearn.

Table 33.1 Amino acid composition of Fraction 1 protein (it has been shown that Fraction 1 protein is essentially RuBisCo) and total protein from different plant species

	Amino Acid (g/kg protein)						
	Fraction 1 protein				Total forage protein		
Amino acid	Alfalfa	Soybean	Sugar beet	Spinach	Alfalfa	Annual ryegrass	Barley leaf
Cys	13	37	32	14	14	30	17
Met	21	21	23	14	18	26	18
His	24	31	22	61	26	29	25
Ile	52	43	44	34	38	26	49
Leu	90	91	81	65	94	49	91
Lys	62	59	55	82	66	63	66
Phe	56	56	38	34	61	35	59
Tyr	40	53	38	31	44	20	39
Thr	56	52	56	49	51	67	52
Try	17	23	15	21	—	—	—
Val	66	48	56	62	68	72	64
Ala	66	64	81	62	61	73	66
Arg	68	73	59	140	59	79	57
Asp	93	88	98	61	94	78	90
Glu	111	110	122	72	101	101	110
Gly	51	52	77	68	57	48	57
Pro	45	40	42	40	53	22	55
Ser	42	33	47	37	45	66	43

Source: Adapted from Lyttleton (1973) and Sheen (1991).

value for all ruminants and pseudo-ruminants. Chief among this metabolic machinery is ribulose bisphosphate carboxylase/oxygenase (**RuBisCO**), the primary enzyme involved in CO_2 fixation, producing glucose phosphate (Glc-1-P), a starting point for all other basic materials whether energy or building blocks. RuBisCO is the most abundant protein accounting for approximately 40–60% of the total plant leaf protein. The remaining proteins (40–60%) are divided into a complex mixture of some 20 000 different proteins over the development of the plant. This, in turn, makes it difficult to change the protein concentration and/or the composition of proteins to improve their nutritional benefits to animals. Proteins in forage have no unique structural or compositional properties that set them apart from other herbaceous plants (Table 33.1) (Lyttleton 1973).

Seed proteins can have distinct properties and amino acid compositions, compared with leaf proteins, that vary among plant species. Forages are typically harvested at a **vegetative** stage, therefore having only vegetative proteins, the bulk of which is RuBisCO. Corn and small grains are exceptions that are ensiled after full grain development, but before the grain matures completely (Wildman 2002; Ferraretto and Shaver 2015). Because RuBisCO makes up a major portion of the total protein and is a highly conserved protein among higher plants, it is not surprising that amino acid compositions are similar among most herbaceous plants (Table 33.1). This also makes it difficult to significantly alter protein composition, either by simple genetic selection or molecular biology techniques, in order to improve nutritional benefits to the animal. The total protein content and the amino acid composition available to the animal at the time of feeding are key nutritional considerations.

Genetic engineering to insert genes into forages for proteins resistant to degradation would seem to be a natural solution to improve amino acid composition and protein utilization. Low levels of gene expression for high sulfur proteins have been achieved in alfalfa (e.g. Schroeder et al. 1991; Tabe et al. 1995). The extensive disulfide cross-linking found in these proteins limits their degradation (Wallace 1983). Even though this direct approach seems reasonable, expression of foreign proteins in plants is usually quite low. To add 1% of a resistant protein or one with unique amino acid composition (e.g. methionine or lysine) to a dairy cow diet containing 50% alfalfa forage dry matter (DM) would require protein expression of 2% of the DM or 10% of the protein in alfalfa. This level of expression seems unlikely even if targeted to a specific organelle such as the vacuole (Tabe et al. 1995).

However, recent advances in gene expression techniques may hold promise for reaching this level in the future (Dirks-Hofmeister et al. 2013; Chen and Lai 2015).

Losses During Forage Harvest and Storage

Physical Losses

The level of protein in forage available to the animal is always highest standing in the field and begins to fall once the forage is cut. Preserving this protein during harvest and storage for future use is a critical challenge. With the highest concentrations of herbage protein in the leaves, harvest losses of leaf material can significantly reduce forage **crude protein** (CP) level and lead to limitations in livestock production. Mechanical losses of the leaf fraction of legumes are always higher than grasses, reducing the quantity and concentrations of preserved nutrients. Leaf losses are directly related to DM content of the foliage (Nelson et al. 1989; Nelson and Satter 1992) and are much greater when preserved as hay vs **silage**. Rotz (1995) modeled factors influencing forage losses during harvest and storage and reported losses during harvest of dry hay of 23–27% of the DM while losses were only about half as great for silage harvest. Alfalfa hay consistently has lower CP content than silage, reflecting this greater leaf loss (Nelson and Satter 1992). There is evidence that, despite these losses, protein preserved in hay is better utilized than that in silage (Broderick 1995; Vagnoni and Broderick 1997).

Harvest mechanization has not changed appreciatively over recent decades except that equipment has become bigger and faster. Koegel et al. (1992) developed a new mechanized system for forage harvesting to improve hay harvest and preservation. Extensive shredding (maceration) of herbage to the extent of partial juice extraction resulted in more rapid drying rates and increased recovery of leaves (Koegel et al. 1992). Macerated alfalfa was more digestible, producing about 15% more energy from macerated than conventional alfalfa forage (Hong et al. 1988). Macerated hay also had slightly higher **ruminal-undegraded protein** (RUP) as estimated from *in vitro* incubations (Yang et al. 1993). The severe disruption of the physical structure of alfalfa plants allowed more extensive access to cell contents and cell wall (CW) carbohydrates. Unfortunately, the process required slower harvesting speeds and limited swath size to achieve maximum maceration and minimum drying times. No commercial version of the macerator has yet been developed and marketed.

Recently, prototype alfalfa harvesters have been developed to strip the leaves from standing plants. Since the majority of proteins in the alfalfa are in the leaves (25–30% of leaf DM) and does not change appreciatively until seed set, separating leaves from stems results in a high-protein fraction (leaves) and a high-fiber fraction (stems). This allows harvest intervals to be longer and less frequent saving on harvest inputs (Shinners et al. 2007; Muck et al. 2010). The challenge, however, is the handling and packaging of the leaves so they can be preserved to maintain high protein levels. Harvesting a protein-rich fraction from alfalfa may provide new avenues for the increased use of alfalfa on agricultural landscapes. Utilizing alfalfa, a perennial forage, in crop rotations with annual crops like corn and soybean could improve the sustainability and productivity of agricultural systems.

Formation of NPN

One of the challenges of forage preservation is limiting the formation of **nonprotein nitrogen** (NPN). With true protein values highest in the living plant, conversion to NPN is the result of **proteolysis** that occurs once the forage is cut and during wilting. Rapid proteolysis begins with the breakdown of plant membranes and the release of proteases from the vacuole (McDonald et al. 1991). These proteases represent a complex mixture of enzymes that, as a group, have pH and temperature optima of approximately 5.5 and 45–55 °C, respectively (McKersie 1981). During wilting forage to 35–45% DM, as is often the case for silage production, NPN can increase from 8% to 18% of total N even under good field drying conditions (Carpintero et al. 1979). Proteolysis will continue in the swath as long as DM content of the forage remains less than approximately 70% (Muck 1991). This is an additional concern for hay production, especially if drying conditions are less than ideal, as often found in the cool more humid regions of agricultural production.

Silage production methods often result in less weather damage and dry matter losses compared with hay. However, limiting protein degradation is dependent upon silage fermentation producing sufficient acid to reach suitable low pH values to prevent or at least reduce proteolytic activity (McDonald Henderson 1962; Playne and McDonald 1966). Though proteolysis decreases markedly at pH 3.8–4.0, it does not stop. Some advantage can be gained by more rapid pH drop, such as occurs when silage is inoculated with lactic acid bacteria or when fermentable sugars are added, but these effects are small in terms of preserving silage protein (Borreani et al. 2018). Elevated temperature and slow pH decline during the ensiling process exacerbates **proteolysis**. Slower pH decline can also be due to the buffering capacity of the forage ensiled, being slower in highly buffered forages (Playne and McDonald 1966).

Since low pH is beneficial to slowing proteolysis, acid treatments of silage suppressed, but did not stop, NPN formation (Vagnoni and Broderick 1997). Nagel and Broderick (1992) found that applying formic acid to alfalfa wilted to 40% DM reduced silage NPN by one-third and increased milk yield 3.4 kg day^{-1} and protein yield 110 g day^{-1} when fed to early lactation

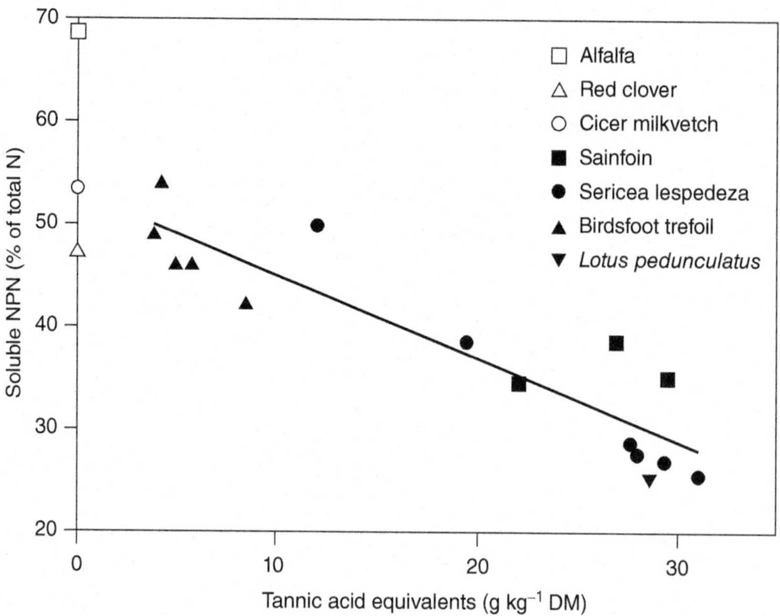

Fig. 33.1. Protein linear regression for tannin-containing forages of soluble nonprotein N (NPN), as a proportion of total N (Y), on condensed tannin concentration (X) 45 days after ensiling samples of seven legume forage species (mean data from 1988 only; Albrecht and Muck 1991): $Y = 54.8–0.875\ X$; $r^2 = 0.799$, $P < 0.01$.

cows. Non-fermentative acidification through the addition of acid (e.g. formic acid) may be a practical way to reduce silage NPN. Specific protein degradation during ensiling may actually improve the nutrient value of the silage in some situations. Corn silage, a prominent forage in many regions of the US, especially in dairy regions, can benefit from protein degradation. A factor that can limit nutrient value is the presence of the protein Zein in corn grain. Zeins coat starch grains in corn and can limit access of amylases to starch molecules resulting in poor utilization as an energy source. Plant proteases degrade the zeins during silage storage but it is a slow process (Hoffman et al. 2011) and silages utilized before 30 days of fermentation may not have optimal starch utilization (Hoffman et al. 2011).

Plant Compounds Altering Protein Breakdown

Condensed tannins, found in certain legumes resulted in lower NPN than alfalfa and other non-tannin legumes (Albrecht and Muck 1991). Forage tannins reduce ruminal protein degradation resulting in increased ruminal **bypass protein** (Broderick and Albrecht 1997). The reduction in degradation rate and increase in ruminal escape were proportional to tannin concentration (Figure 33.1). Birdsfoot trefoil selected for variable amounts of condensed tannins improved protein utilization and decreased milk urea when replacing alfalfa in dairy rations (Hymes-Fecht et al. 2013).

Red clover is a unique forage producing silage with 30–50% less NPN than alfalfa (Papadopoulos and McKersie 1983; Albrecht and Muck 1991). Though it is a non-tannin forage like alfalfa, free amino-acid release from alfalfa extracts was five-times more rapid than from red clover extracts (Jones et al. 1995a) and mixing red clover extract with alfalfa extract suppressed free amino-acid release (Jones et al. 1995b). It has been shown that red clover contains a polyphenol oxidase (PPO) that produces *o*-quinones from *o*-diphenols inhibiting proteolysis during ensiling (Sullivan and Hatfield 2006). Ruminal *in vitro* degradability of red clover protein was lower than that in alfalfa (Broderick and Albrecht 1997). In feeding studies with lactating cows, milk yield per cow was similar on diets containing either red clover or alfalfa silage; however, N efficiency was higher, and urinary N, fecal N, and fecal DM excretion were substantially reduced on red clover silage (Broderick et al. 2000). Protein compositions are similar between red clover and alfalfa (Eppendorfer 1977) suggesting that perhaps the PPO/*o*-diphenol system in red clover not only inhibits proteolysis but may also modify non-protease proteins to decrease their rapid degradation in the rumen. Although red clover forage has better N-use efficiency, it tends to have lower DM yields, poorer persistence and generally has reduced feed intake compared to alfalfa. Improving the agronomic characteristics of red clover is an important area of ongoing research (De Vega et al. 2015).

Alternatively, alfalfa can be genetically modified to produce PPO enzymes (Sullivan et al. 2003; Schmitz et al. 2007). Incorporation of genes for *o*-diphenol production in alfalfa is more challenging than insertion of PPO genes but it is possible to produce appropriate *o*-diphenol substrates (Sullivan 2014, 2017). Biosynthesis of *o*-diphenols along with PPO in alfalfa could result in a greater reduction in the formation of NPN in alfalfa silages and increased effective protein utilization.

Carbohydrates

The major source of **digestible energy** in forages comes from carbohydrates. Forage carbohydrates can be broken into structural and **nonstructural carbohydrate** (Figure 33.2) groups, with roles that are independent of physical and chemical properties of the individual carbohydrates. **Structural carbohydrates** (often referred to as fiber) are polysaccharides that make up the cell wall and function in providing structural integrity to the plant. There is limited turnover of structural carbohydrates once they are incorporated into the cell wall. **Nonstructural** types include all other carbohydrates including

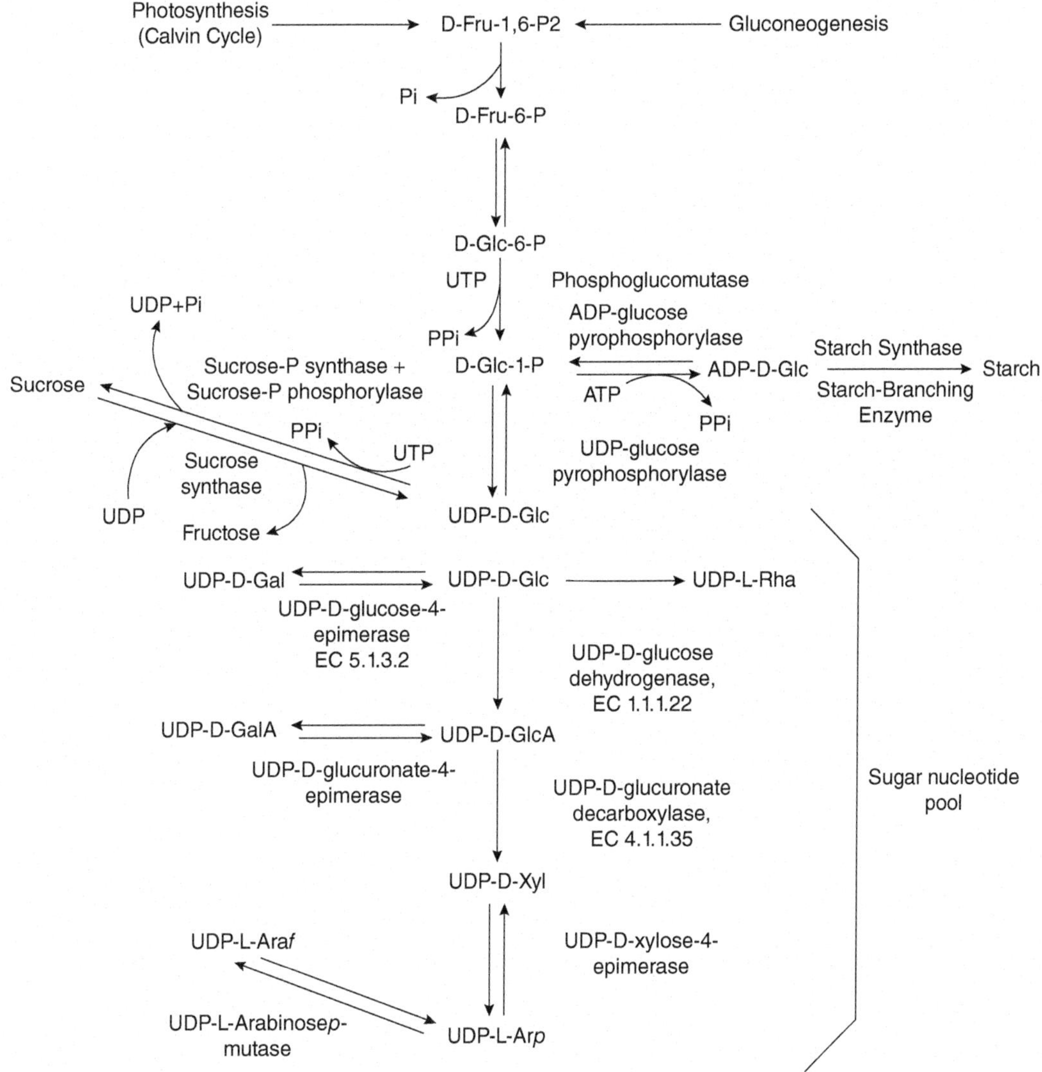

Fig. 33.2. Sugar synthesis pathway for the formation of activated glucose and its conversion to sucrose, starch, and sugar nucleotides for the formation of structural polysaccharides.

the building blocks (activated monosaccharides) for structural carbohydrates and storage molecules such as starch, fructans, and sucrose.

NonStructural Carbohydrates

Biosynthesis of individual carbohydrates whether monosaccharides, oligosaccharides, or polysaccharides begins with CO_2 fixation during **photosynthesis**. Glucose is one of the primary products of photosynthesis and leads to the formation of all other **sugars** and polysaccharides predominantly through the molecular conversion of activated glucose (UDP-Glc) (Figure 33.2).

Of particular importance is the critical role sucrose plays in the metabolic processes ranging from transport to storage. Sucrose is a non-reducing disaccharide formed by a glycosidic linkage between the anomeric carbons (C1 on glucose and C2 on fructose). This creates an oligosaccharide without an open anomeric carbon rendering the molecule metabolically inactive and making sucrose a perfect carbohydrate for transport throughout the plant and for storage. Sucrose also functions as the primary donor for the synthesis of starch, fructans, and cellulose synthesis through the action of a membrane-bound sucrose synthase (Delmer 1999; Li et al. 2014).

Due to their unique structural characteristics, fructans may have roles in addition to energy reserves for the plant. Recent work indicates potential roles in dealing with abiotic stresses such as drought and cold tolerance (Valluru and Van den Ende 2008; Livingston et al. 2009) through their unique ability to interact with plant membranes. Fructans are localized in the vacuole and are generally found in both leaf and stem tissues (Pontis 1990). In temperate grasses that accumulate fructans, there is a trend for higher levels of accumulation in the lower parts of the plant, especially the stem (Table 33.2).

Utilization of NonStructural Carbohydrates in Forages

For most forages, harvest and utilization is limited to the vegetative tissues. Corn silage, silage sorghum, and cereals used for silage are exceptions for which harvesting takes place at grain fill to optimize starch storage and energy value to the animal when fed. For these forage materials, the accumulated starch reaches a constant level and little change occurs except for limited loss during the ensiling process. For other forages whether harvested as hay, silage, or grazed, the level of **nonstructural carbohydrates** is variable due to the metabolic flux through the different pools (Figure 33.3). This flux is influenced by the time of day, physiologic stage of development, and the state of the plant in its environment (level of abiotic and biotic stresses).

Table 33.2 Distribution of fructans among plant tissues in temperate grasses timothy and smooth bromegrass

	Fructan g/kg dry weight			
Plant	Leaf	Sheath	Stem	Stem base
Timothy	12	14	55	305
Smooth bromegrass	87	100	132	203

Source: Adapted from Smith (1967).

Plants were at early anthesis. Stem bases represent partially elongated internodes (8–10 cm).

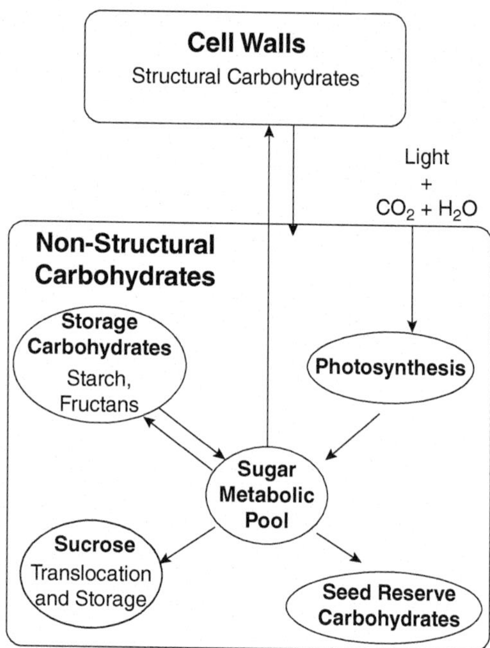

Fig. 33.3. Carbohydrate pools with flux between pools within developing forages.

Forage Cell Walls

Structure of Forage Cell Walls

Cell walls of forages represent large reserves of potential energy. The challenge is to recover this energy whether it is by ruminants or enzymatic conversion in biomass to bioenergy. Not all components of the cell wall are available as energy especially for ruminants. **Cellulose** is the most abundant polysaccharide in forage cell walls (Delmer 1999; Somerville 2006). It is a polymer of glucose units linked by β-1,4 linkages. Hydrogen bonding between the cellulose molecules holds the **microfibrils** and fibrils together and creates a hydrophobic surface that allows hydrogen bonding to other molecules such as xylans, proteins, and **lignin**.

Other structural non-cellulosic polysaccharides make up the remaining CW matrix. By far the most abundant polysaccharides in this group are the xylans, polysaccharides composed of a β-1,4-linked xylose backbone with branch substitutions of arabinose, glucuronic acid, and xylose. Some of the arabinose residues of grass xylans are substituted with ferulic acid (FA) and, to a lesser extent, *p*-coumaric acid (*p*-CA) ester linked to the C5 hydroxyl (Hatfield et al. 2017). The ferulates can form linkages to one another, creating diferulate structures that serve to cross-link arabinoxylan chains in grass cell walls (Ralph et al. 1995; Hatfield et al. 2017) (Figure 33.4). Similar ferulate structures are not known to exist in legumes. Xyloglucans consist of β-1,4 glucose chains with xylose and fucose substitutions and are important in legume cell-wall growth and development (Hatfield 1994). Pectins are the most complex group with a potentially highly substituted galacturonosyl backbone. They are found in both legumes and grasses though much more prominent in legume cell walls (Aman 1994; Hatfield 1994).

Lignin and hydroxycinnamic acids are non-carbohydrate components of forage cell walls (Hatfield 1994) that strongly reduce degradability of the cell wall. In grasses, the majority of *p*-coumarates are ester-linked to monolignol building blocks of lignin (Ralph et al. 1994) though their role is not clearly understood. Grasses appear to esterify coniferyl and sinapyl alcohols through the activity of a *p*-coumaryl transferase before export to the wall and incorporation into lignin (Hatfield et al. 2008; Hatfield and Marita 2010). Due to the electro-chemical properties of *p*-CA, it does not undergo free-radical coupling reactions to become incorporated into lignin. However, the monolignol portion of the *p*-CA-monolignol conjugate does undergo free-radical coupling reactions to add to lignin. The esterified ferulates and diferulates of arabinoxylans in grasses can act as nucleation sites for lignin polymerization (Ralph et al. 1995; Hatfield et al. 2017).

Protein is present in all plant cell walls, though at low concentrations (Aman 1994). Legume cell walls contain larger amounts of protein than are found in grasses. There is a wide array of proteins with both structural and enzymatic functions; e.g. peroxidases and oxidases (Verdonk et al. 2012). Lipid content of forage cell walls is limited, but epidermal tissue does develop a waxy **cutin** layer on the exterior surface of the plant. Forage species that evolved in hot, dry environments have thicker cutin layers than common cool-season forages.

Cell Wall Development

Plant CWs represent from $350\,g\,kg^{-1}$ to over $700\,g\,kg^{-1}$ of the DM in forages depending on the species and developmental stage. The concentration of CW material is greater in stem material of forages than leaves, with the difference between these plant parts being much greater in legumes than grasses (Hatfield 1992; Wilson 1994). As forages mature, they accumulate increasing amounts of CW material and this is mostly a function of the shift in leaf to stem ratio of forages as they mature and to the accumulation of lignin.

For legumes, CW concentration of leaves changes little with age, whereas stems undergo cambial growth, which adds thick-walled **xylem** and **sclerenchyma** tissue during development (Wilson 1993). This xylem tissue accounts for the majority of the CW material in legumes at typical harvest maturities (Jung and Engels 2002). In contrast to the legumes, both leaves and stems of grasses deposit additional CW material as they mature. Rather than adding new tissues as occurs in legume stems, the increased concentration of CW material in grasses is due to deposition of thick secondary walls in all grass tissues except for mesophyll and phloem (Wilson 1993). While some non-xylem tissues in legumes develop thickened CWs, this pattern of development accounts for less accumulation of CW material in legumes than occurs in grasses. Sclerenchyma fiber that surrounds **vascular bundles** is the predominate contributor of CW material in grasses (Engels and Schuurmans 1992).

Altering Cell Wall Composition

Altering CW composition can be achieved through genetic selection, molecular modification (gene manipulation) or by taking advantage of natural **mutations**. Most selection work has focused on a general decrease in total CW amounts as a proportion of the total DM in a given forage to increase overall digestion (Casler 1999). However, this approach leads to decreased total biomass. At similar lignin concentration, smooth bromegrass plants with higher levels of ferulate cross-linking exhibited reduced neutral detergent fiber (NDF) digestibility (Casler and Jung 1999) indicating the possibility of selection for reduced ferulate cross-linking. Ferulate cross-linking was decreased in corn using a transposon system to select plants with decreased total CW ferulates (Jung and Phillips 2010; Hatfield et al. 2018). Backcrossed materials were generated and used in a feeding trial with dairy cows. Though the selected materials had relatively small decreases in ferulate cross-linking, the ensiled forages were more digestible than conventional corn silage. This would indicate that even small decreases in the extent of CW cross-linking could positively influence digestibility (Hatfield et al. 2018). A direct gene approach was attempted using anti-sense (RNAi) of the genes suspected to regulate the ferulate transferases which are responsible for formation of ferulated arabinosyl branches on arabionxylans (Piston et al. 2010). Though this approach resulted in reduced CW ferulates, the plant materials were not tested to determine if there was an increase in digestibility.

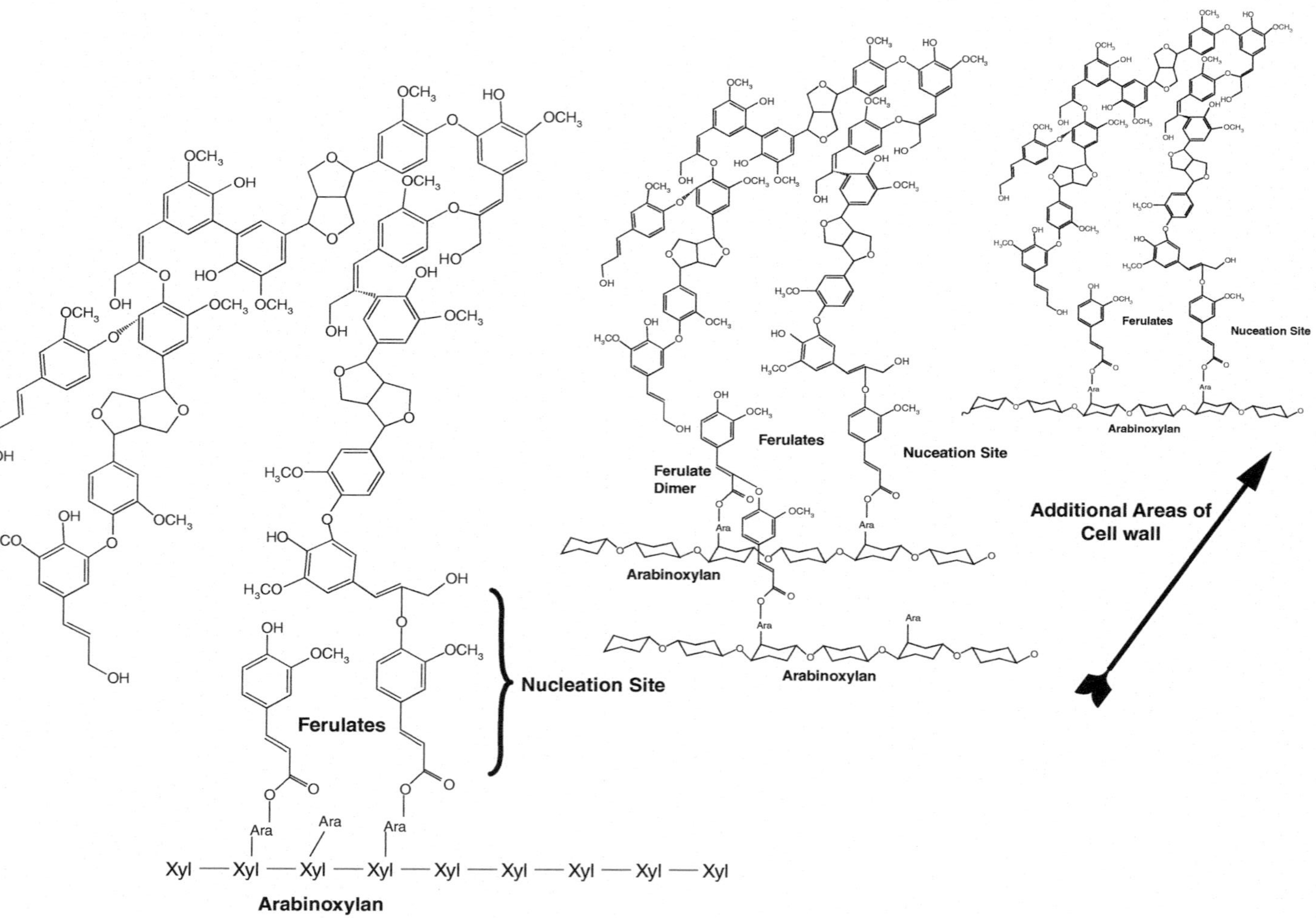

Fig. 33.4. Diagram of ferulates attached to arabinoxylans, formation of ferulate dimmers and the role in cross-linking.

Decreasing lignification of CWs would be expected to have a positive impact upon digestibility. The **brown midrib (BMR)** trait of corn (Jorgenson 1931) is a natural lignin mutation that impacts a caffeic acid *O*-methyl transferase (COMT) that results in a reduction in CW lignin accumulation. There are also BMR mutants of sorghum (BMR-6) and millet that behave in a similar fashion with reduced lignin (Sattler et al. 2010). The genes impacted in sorghum also appear to reduce lignin content and have the same general phenotype though the gene impacted appears to be coniferyl alcohol dehydrogenase (CAD), the final step in the formation of coniferyl alcohol, one of the monlignol building blocks. In all these cases, the change in lignin has been modest so as not to cause severe stunting of the forages. A similar strategy was undertaken to modify alfalfa by suppression of the caffeoyl CoA O-methyltransferase (CCoAOMT) or COMT gene to decrease the synthesis of lignin without causing a suppression of biomass accumulation (Guo et al. 2001a,b; Marita et al. 2002). Lignin was decreased by 8–10% and digestibility increased by 15–20%. An advantage of having lower lignin at a given stage of development is that harvest can be delayed for a short period of time allowing an increase in biomass production without a significant decrease in quality.

Composition of structural carbohydrates is harder to change through molecular processes, even though most of the genes responsible for the different polysaccharides have been identified. The complexity of polysaccharides (multiple sugars and multiple linkages) requires an array of enzymes to synthesize even some of the simplest polysaccharides. Cellulose, though it is a homopolymer of glucose with a single type of linkage, requires a complicated arrangement of cellulose synthases to produce cellulose microfibrils at the plasmalemma (Li et al. 2014). In addition, sucrose synthase that is required to produce UDP-Glc from sucrose once it is transported to the cellulose synthesis complex must be coordinated with the other parts of the complex and may have a regulatory role in cellulose synthesis (Fujii et al. 2010). The genes for all the enzymes involved in cellulose synthesis have been identified but coordinated regulation of these components over the development of the plant is not fully understood.

Biosynthesis of the non-cellulosic polysaccharides involves a complex array of genes controlling the production of enzymes generally known as glycosyl- and acyl- transferases. For example, the production of glucuronoarbinoxylans (GAX) substituted with FA/*p*CA, typically found in grasses, requires enzymes that include five glycosyltransferases (GTs) and two acyltransferases (ATs) (Faik 2010; Rennie and Scheller 2014). More highly branched and complex polysaccharides that make up the pectic fraction of the walls would require a much greater number of glycosyltransferases. Most of the genes involved in biosynthesis of all the structural polysaccharides have been identified. However, an understanding of their control and coordination remains to be determined.

Another approach is to alter the supply of specific **sugar** nucleotides needed for structural polysaccharide synthesis (Figure 33.2). By limiting the formation of specific uridine disphosphate glucose (UDP)-sugars, it is thought that final polysaccharide composition could be skewed in specific directions. Such an approach was used to alter the formation of arabinofuranosyl from arabinopyranosyl intermediates. The enzyme responsible for this is a mutase and down regulation of the gene resulted in decreased arabinofuranose (Araf) incorporation into the wall. Because Araf is needed for ferulate attachment to GAX molecules, these genetically modified plants had reduced total ferulates (Rancour et al. 2015). Unfortunately, degradation by xylanase of the modified CW was lower than the control. This may have been due to tighter bonding of the modified xylans to cellulose due to structural changes. It indicates that there is potential to alter CWs in a specific manner but will require a thorough understanding of the implications and outcomes upon utilization.

There have been efforts to select for specific types of polysaccharides such as pectin that are neutral detergent soluble fiber (NDSF) of alfalfa (Tecle et al. 2006). Selection for increased pectin in alfalfa resulted in increased digestibility. Additional work would have to be done to determine how far alfalfa plants could be pushed in a direction of increased digestibility since there seemed to be a correlated decrease in other CW components. The advantage would be replacing slowly digested CW polysaccharides with more rapidly digested polysaccharides.

One area that remains a mystery, is how the CW is assembled, creating the well-organized structure needed to support the plant. As discussed above, much is known about the genes and enzymes needed to produce the individual components that make up the CW but our understanding of what processes and controls occur outside the plasma membrane remains unclear. It is the organization of the CW that has a large impact upon how it can be degraded during rumination. The Klason or acetyl bromide lignin content of legume and grass CWs is similar when compared at similar stages of physiologic development (Hatfield et al. 1994). However, digestion patterns of these walls are quite different. Legumes tend to have a rapid initial degradation followed by a sharp decline to a plateau. Grasses, on the other hand, have a slow initial rate of degradation that tends to persist for long periods of time. It has been proposed that the organization of grass CWs is dependent upon the formation of ferulated xylans that are hydrogen bound to cellulose and the ferulates act as nucleation sites for lignification (Hatfield et al. 2017). Due to the placement of ferulates in clustered groups as opposed to randomly distributed, lignin may form in patches within the CW. This would leave the wall partially open to enzyme attack from rumen microbes but it would be a slow process since areas would only be partially exposed. There would need to be coordinated enzyme activity in the CW to remove some of the arabinose residues on GAX to allow hydrogen bonding to cellulose and to other xylans. The released arabinose could act as a feedback indicator to signal the release of monolignols building blocks for lignin to occur at this time. This process involving GAX, FA, FA dimers and lignin formation would be going throughout the CW.

Summary

Strategies to improve the utilization of forages rely on understanding the chemical components that make up forages and, most importantly, the interactions among these components. For example, it is the interactions of all the wall components that control the organization of the CW as it is being formed. This is one area that remains virtually unknown in terms of how the final wall comes to be organized as it is to support the plant. The biosynthesis of components dictates their final levels and how these components are structured together to form functioning tissues. Altering one component, through genetic selection or molecular manipulation, may impact several other components. Therefore, strategies to improve forages must be aware of these interrelationships and strive to move all in a direction that has a positive impact. The improvement of forages for ruminant nutrition cannot rely upon the selection of one single parameter, but should strive to produce the best plant, meeting overall nutritional needs as well as agronomic concerns for production.

References

Albrecht, K.A. and Muck, R.E. (1991). Proteolysis in ensiled forage legumes that vary in tannin concentration. *Crop Sci.* 31: 464–469.

Aman, P. (1994). Composition and structure of cell wall polysaccharides in forages. In: *Forage Cell Wall Structure and Digestibility* (eds. H.G. Jung, D.R. Buxton, R.D. Hatfield, et al.), 183–199. Madison: ASA-CSSA-SSSA.

Borreani, G., Tabacco, E., Schmidt, R.J. et al. (2018). Silage review: factors affecting dry matter and quality losses in silages. *J. Dairy Sci.* 101: 3952–3979. https://doi.org/10.3168/jds.2017-13837.

Broderick, G.A. (1995). Performance of lactating dairy cows fed either alfalfa silage or alfalfa hay as the sole forage. *J. Dairy Sci.* 78: 320–329. https://doi.org/10.3168/jds.S0022-0302(95)76640-1.

Broderick, G.A. and Albrecht, K.A. (1997). Ruminal in vitro degradation of protein in tannin-free and tannin–containing forage legume species. *Crop Sci.* 37: 1884–1891. https://doi.org/10.2135/cropsci1997.0011183X003700060037x.

Broderick, G.A., Walgenbach, R.P., and Sterrenburg, E. (2000). Performance of lactating dairy cows fed alfalfa or red clover silage as the sole forage. *J. Dairy Sci.* 83: 1543–1551. https://doi.org/10.3168/jds.S0022-0302(00)75026-0.

Carpintero, C.M., Henderson, A.R., and McDonald, P. (1979). The effect of some pretreatments on proteolysis during the ensiling of herbage. *Grass Forage Sci.* 34: 311–316. https://doi.org/10.1111/j.1365-2494.1979.tb01483.x.

Casler, M.D. (1999). Correlated responses in forage yield and nutritional value from phenotypic recurrent selection for reduced fiber concentration in smooth bromegrass. *Theor. Appl. Genet.* 99: 1245–1254. https://doi.org/10.1007/s001220051330.

Casler, M.D. and Jung, H.G. (1999). Selection and evaluation of smooth bromegrass clones with divergent lignin and etherified ferulic acid concentration. *Crop Sci.* 39: 1866–1873. https://doi.org/10.2135/cropsci199.3961866x.

Chen, Q. and Lai, H. (2015). Gene delivery into plant cells for recombinant protein production. *Biomed. Res. Int.* 2015: 932161. https://doi.org/10.1155/2015/932161.

De Vega, J.J., Ayling, S., Hegarty, M. et al. (2015). Red clover (*Trifolium pratense* L.) draft genome provides a platform for trait improvement. *Sci. Rep.* 5: 17394. https://doi.org/10.1038/srep17394.

Delmer, D.P. (1999). Cellulose biosynthesis: exciting times for a difficult field of study. *Annu. Rev. Plant Physiol. Plant Mol. Biol.* 50: 245–276.

Dirks-Hofmeister, M.E., Kolkenbrock, S., and Moerschbacher, B.M. (2013). Parameters that enhance the bacterial expression of active plant polyphenol oxidases. *PLoS One* 8: e77291. https://doi.org/10.1371/journal.pone.0077291.

Engels, F.M. and Schuurmans, J.L.L. (1992). Relationship between structural development of cell walls and degradation of tissues in maize stems. *J. Sci. Food Agric.* 59: 45–51.

Eppendorfer, W.H. (1977). Amino acid composition and nutritional value of Italian ryegrass, red clover and lucerne as influenced by application and content of nitrogen. *Sci. Food Agric.* 28: 607–614. https://doi.org/10.1002/jsfa.2740280705.

Faik, A. (2010). Xylan biosynthesis: news from the grass. *Plant Physiol.* 153: 396–402. https://doi.org/10.1104/pp.110.154237.

Ferraretto, L.F. and Shaver, R.D. (2015). Effects of whole-plant corn silage hybrid type on intake, digestion, ruminal fermentation, and lactation performance by dairy cows through a meta-analysis. *J. Dairy Sci.* 98: 2662–2675. https://doi.org/10.3168/jds.2014-9045.

Fleming, G.A. (1973). Mineral composition of herbage. In: *Chemistry and Biochemistry of Herbage* (eds. G.W. Butler and R.W. Bailey), 529–566. London/New York: Academic Press.

Fujii, S., Hayashi, T., and Mizuno, K. (2010). Sucrose synthase is an integral component of the cellulose synthesis machinery. *Plant Cell Physiol.* 51: 294–301. https://doi.org/10.1093/pcp/pcp190.

Guo, D., Chen, F., Inoue, K. et al. (2001a). Downregulation of caffeic acid 3-O-methyltransferase and caffeoyl CoA 3-O-methyltransferase in transgenic alfalfa. Impacts on lignin structure and implications for the biosynthesis of G and S lignin. *Plant Cell* 13: 73–88.

Guo, D., Chen, F., Wheeler, J. et al. (2001b). Improvement of in-rumen digestibility of alfalfa forage by genetic manipulation of lignin O-methyltransferases. *Transgenic Res.* 10: 457–464.

Hatfield, R.D. (1992). Carbohydrate composition of alfalfa cell walls isolated from stem sections differing maturity. *J. Agric. Food. Chem.* 40: 424–430. doi.org/10.1021/jf00015a012.

Hatfield, R.D. (1994). Cell wall polysacchride interactions and degradability. In: *Forage Cell Wall Structure and Digestibility* (eds. H.J. Jung, D.R. Buxton, R.D. Hatfield, et al.), 285–313. Madison: ASA-CSSA-SSSA.

Hatfield, R.D. and Marita, J.M. (2010). Enzymatic processes involved in the incorporation of hydroxycinnamates into grass cell walls. *Phytochem. Rev.* 9: 35–45. https://doi.org/10.1007/s11101-010-9165-1.

Hatfield, R.D., Jung, H.G., Ralph, J. et al. (1994). A comparison of the insoluble residues produced by the Klason lignin and acid detergent lignin procedures. *J. Sci. Food Agric.* 65: 51–58. https://doi.org/10.1002/jsfa.2740650109.

Hatfield, R.D., Marita, J.M., and Frost, K. (2008). Characterization of p-coumarate accumulation, p-coumaroyl transferase, and cell wall changes during the development of corn stems. *J. Sci. Food Agric.* 88: 2529–2537. https://doi.org/10.1002/jsfa.3376.

Hatfield, R.D., Rancour, D.M., and Marita, J.M. (2017). Grass cell walls: a story of cross-linking. *Front. Plant Sci.* 7 https://doi.org/10.3389/fpls.2016.02056.

Hatfield, R.D., Jung, H., Marita, J.M., and Kim, H. (2018). Cell wall characteristics of a maize mutant selected for decreased ferulates. *Am. J. Plant Sci.* 9: 446–466. https://doi.org/10.4236/ajps.2018.93034 F.

Hoffman, P.C., Esser, N.M., Shaver, R.D. et al. (2011). Influence of ensiling time and inoculation on alteration of the starch-protein matrix in high-moisture corn. *J. Dairy Sci.* 94 (10) https://doi.org/10.3168/jds.2010-3562.

Hong, B.J., Broderick, G.A., Koegel, R.G. et al. (1988). Effect of shredding alfalfa on cellulolytic activity, digestibility, rate of passage, and milk production. *J. Dairy Sci.* 71: 1546–1555. https://doi.org/10.3168/jds.S0022-0302(88)79718-0.

Hymes-Fecht, U.C., Broderick, G.A., Muck, R.E., and Grabber, J.H. (2013). Replacing alfalfa or red clover silage with birdsfoot trefoil silage in total mixed rations increases production of lactating dairy cows. *J. Dairy Sci.* 96: 460–469. https://doi.org/10.3168/jds.2012-5724.

Jones, B.A., Hatfield, R.D., and Muck, R.E. (1995a). Characterization of proteolysis in alfalfa and red clover. *Crop Sci.* 35: 537–541. https://doi.org/10.2135/cropsci1995.0011183X003500020043x.

Jones, B.A., Muck, R.E., and Hatfield, R.D. (1995b). Red clover extracts inhibit legume proteolysis. *J. Sci. Food Agric.* 67: 329–333. https://doi.org/10.1002/jsfa.2740670309.

Jorgenson, R.L. (1931). Brown midrib in maize and its lignage relations. *J. Am. Soc. Agron.* 23: 549–557.

Jung, H.G. and Engels, F.M. (2002). Alfalfa stem tissues: cell wall deposition, composition, and degradablity. *Crop Sci.* 42 (2) https://doi.org/10.2135/cropssci2002.0524.

Jung, H.G. and Phillips, R.L. (2010). Putative seedling ferulate ester (*sfe*) maize mutant: morphology, biomass yield, and stover cell wall composiiton and rumen degradability. *Crop Sci.* 50: 403–418. https://doi.org/10.2135/cropsci2009.04.0191.

Koegel, R.G., Straub, R.J., Shinners, K.J. et al. (1992). An overview of physical treatments of lucerne performed at Madison, Wisconsin, for improving properties. *J. Agric. Eng. Res.* 52: 183–191.

Li, S., Bashline, L., Lei, L., and Gu, Y. (2014). Cellulose synthesis and its regulation. *Arabidopsis Book* 12: e0169. https://doi.org/10.1199/tab. 0169.

Livingston III, D.P., Hincha, D.K., and Heyer, A.G. (2009). Fructan and its relationship to abiotic stress tolerance in plants. *Cell. Mol. Life Sci.* 66: 2007–2023. https://doi.org/10.1007/s00018-009-0002-x.

Lyttleton, J.W. (1973). Proteins and nucleic acids. In: *Chemistry and Biochemistry of Herbage* (eds. G.W. Butler and R.W. Bailey), 63–105. London/New York: Academic Press.

Marita, J.M., Ralph, J., Hatfield, R.D. et al. (2002). Modifications in lignin of transgenic alfalfa downregulated in caffeic acid 3-O-methyltransferase and caffeoyl coenzyme A 3-O-methyltransferase. In: *Plant Biology 2002 Conference, Denver, CO*, 300.

McDonald Henderson, P.A.R. (1962). Buffering capacity of herbage samples as a factor in ensilage. *J. Sci. Food Agric.* 13: 395–400. doi.org/10.1002/jsfa.2740130709.

McDonald, P., Henderson, A.R., and Heron, S.J.E. (1991). *The Biochemistry of Silage*. New York: Wiley.

McKersie, B.D. (1981). Proteinases and peptidases in alfalfa herbage. *Can. J. Plant Sci.* 61: 53–60. https://doi.org/10.4141/cjps81-008.

Muck, R.E. (1991). Silage fermentation. In: *Mixed Cultures in Biotechnology* (eds. J.G. Zeikus and E.A. Johnson), 171–204. New York, NY: McGraw-Hill.

Muck, R.E., Shinners, K. and Duncan, J.A. 2010. Ensiling characteristics of alfalfa leaves and stems. ASABE Annual International Meeting, Pittsburg, PA. Paper No. 10086613.

Nagel, S.H. and Broderick, G.A. (1992). Effect of formic acid or formaldehyde treatment of alfalfa silage on nutrient utilization by dairy cows. *J. Dairy Sci.* 75: 140–145.

Nelson, W.F. and Satter, L.D. (1992). Impact of alfalfa maturity and preservation method on milk production by cows in early lactation. *J. Dairy Sci.* 75: 1562–1570. https://doi.org/10.3168/jds.S0022-0302(92)77914-4.

Nelson, M.L., Headley, D.M., and Loesche, J.A. (1989). Control of fermentation in high-moisture baled alfalfa by inoculation with lactic acid-producing bacteria: II. Small rectangular bales. *J. Anim. Sci.* 67: 1586–1592.

Papadopoulos, Y.A. and McKersie, B.D. (1983). A comparison of protein degradation during wilting and ensiling of six forage species. *Can. J. Plant Sci.* 63: 903–912. https://doi.org/10.4141/cjps83-114.

Piston, F., Uauy, C., Fu, L.H. et al. (2010). Down-regulation of four putative arabinoxylan feruloyl transferase genes from family PF02458 reduces ester-linked ferulate content in rice cell walls. *Planta* 231: 677–691. https://doi.org/10.1007/s00425-009-1077-1.

Playne, M.J. and McDonald, P. (1966). The buffering constituents of herbage and of silage. *J. Sci. Food Agric.* 17: 264–268. https://doi.org/10.1002/jsfa.2740170609.

Pontis, H.G. (1990). Fructans. In: *Methods in Plant Biochemistry* (ed. P.M. Dey), 353–370. New York: Academic Press.

Ralph, J., Hatfield, R.D., Quideau, S. et al. (1994). Pathway of *p*-coumaric acid incorporation into maize lignin as revealed by NMR. *J. Am. Chem. Soc.* 116: 9448–9456.

Ralph, J., Grabber, J.H., and Hatfield, R.D. (1995). Lignin-ferulate crosslinks in grasses: active incorporation of ferulate polysaccharide esters into ryegrass lignins. *Carbohydr. Res.* 275: 167–178. https://doi.org/10.1016/0008-6215(95)00237-N.

Rancour, D.M., Hatfield, R.D., Marita, J.M. et al. (2015). Cell wall composition and digestibility alterations in *Brachypodium distachyon* achieved through reduced expression of the UDP-arabinopyranose mutase. *Front. Plant Sci.* 6 https://doi.org/10.3389/fpls.2015.00446.

Rennie, E.A. and Scheller, H.V. (2014). Xylan biosynthesis. *Curr. Opin. Biotechnol.* 26: 100–107. https://doi.org/10.1016/j.copbio.2013.11.013.

Rotz, C.A. (1995). Loss models for forage harvest. *Trans. Am. Soc. Agric. Eng.* 38: 1621–1631.

Sattler, S.E., Funnell-Harris, D.L., and Pedersen, J.F. (2010). Brown midrib mutations and their importance to the utilization of maize, sorghum, and pearl millet lignocellulosic tissues. *Plant Sci.* 178: 229–238. https://doi.org/10.1016/j.plantsci.2010.01.001.

Schmitz, G.E., Sullivan, M.L., and Hatfield, R.D. (2007). Three polyphenoloxidases from red clover (*Trifolium pratense*) differ in enzymatic activities and activation properties. *J. Agric. Food. Chem.* 56: 272–280.

Schroeder, H.E., Khan, M.R.I., Knibb, W.R. et al. (1991). Expression of a chicken ovalbumin gene in three lucerne cultivars. *Aust. J. Plant Physiol.* 18: 495–505.

Sheen, S.J. (1991). Comparison of chemical and functional properties of soluble leaf proteins from four plant species. *J. Agri. Food Chem.* 39 (4): 681–685.

Shinners, K.J., Herzmann, M.E., Binversie, B.N., and Digman, M.F. (2007). Harvest fractionation of alfalfa. *Trans. ASABE* 50: 713–718.

Smith, D. (1967). Carbohydrates in grasses. II: Sugar and fructosan composition of stem bases of bromegrass and timothy at several growth stages and in different plant parts at anthesis. *Crop Sci.* 7: 62–67.

Somerville, C. (2006). Cellulose synthesis in higher plants. *Annu. Rev. Cell Dev. Biol.* 22: 53–78.

Spears, J.W. (1994). Minerals in forages. In: *Forage Quality, Evaluation, and Utilization* (ed. G.C. Fahey Jr.), 281–317. Madison, WI: American Society of Agronomy, Inc Crop Science Society of America, Inc, Soil Science Society of America, Inc.

Sullivan, M.L. (2014). Perennial peanut (*Arachis glabrata* Benth.) leaves contain hydroxycinnamoyl-CoA:tartaric acid hydroxycinnamoyl transferase activity and accumulate hydroxycinnamoyl-tartaric acid esters. *Planta* 239: 1091–1100. https://doi.org/10.1007/s00425-014-2038-x.

Sullivan, M.L. (2017). Identification of bean hydroxycinnamoyl-CoA:tetrahydroxyhexanedioate hydroxycinnamoyl transferase (HHHT): use of transgenic alfalfa to determine acceptor substrate specificity. *Planta* 245: 397–408. https://doi.org/10.1007/s00425-016-2613-4.

Sullivan, M.L. and Hatfield, R.D. (2006). Polyphenol oxidase and *o*-diphenols inhibit postharvest proteolysis in red clover and alfalfa. *Crop Sci.* 46: 662–670.

Sullivan, M., Thoma, S., Samac, D., and Hatfield, R. (2003). Cloning of red clover and alfalfa polyphenol oxidase genes and expression of active enzymes in transgenic alfalfa. In: *Molecular Breeding of Forage and Turf* (eds. A. Hopkins, Z.-Y. Wang, R. Mian et al.),

189–195. Dordrecht, Netherlands: Kluwer Academic Publishers.

Tabe, L., Khan, M.R.I., Wardley-Richardson, T. et al. (1995). Biotechnological approaches to improving the nutritive value and performance of pasture legumes. *J. Anim. Sci.* 73: 2752–2759. https://doi.org/10.2527/1995.7392752x.

Tecle, I.Y., Viands, D.R., Hansen, J.L., and Pell, A.N. (2006). Response from selection for pectin concentration and indirect response in digestibility of alfalfa. *Crop Sci.* 46: 1081–1087. https://doi.org/10.2135/cropsci2005.05-0087.

Vagnoni, D.B. and Broderick, G.A. (1997). Effects of supplementation of energy or ruminally undegraded protein to lactating cows fed alfalfa hay or silage. *J. Dairy Sci.* 80: 1703–1712. https://doi.org/10.3168/jds.S0022-0302(97)76102-2.

Valluru, R. and Van den Ende, W. (2008). Plant fructans in stress environments: emerging concepts and future prospects. *J. Exp. Bot.* 59: 2905–2916. https://doi.org/10.1093/jxb/em164.

Verdonk, J.C., Hatfield, R.D., and Sullivan, M.L. (2012). Proteomic analysis of cell walls of two developmental stages of alfalfa stems. *Front. Plant Sci.* 3 https://doi.org/10.3389/Fpls.2012.00279.

Wallace, R.J. (1983). Hydrolysis of 14C-labelled proteins by rumen microorganisms and by proteolytic enzymes prepared from rumen bacteria. *Br. J. Nutr.* 50: 345–355.

Weiss, W.P. (2014). Minerals and vitamins for diary cows: Magic bullets or just bullets?. https://ecommons.cornell.edu/bitstream/handle/1813/36547/3.Weiss.Manuscript.pdf?sequence=1.

Wildman, S.G. (2002). Along the trail from fraction I protein to rubisco (ribulose bisphosphate carboxylase-oxygenase). *Photosynth. Res.* 73: 243–250. https://doi.org/10.1023/A:1020467601966.

Wilson, J.R. (1993). Organization of forage plant tissues. In: *Forage Cell Wall Structure and Digestibility* (eds. H.G. Jung, D.R. Buxton, Hatfield R.D., et al.), 1–27. Madison: ASA-CSSA-SSSA.

Wilson, J.R. (1994). Cell wall characteristics in relation to forage digestion by ruminants. *J. Agric. Sci.* 122: 173–182. https://doi.org/10.1017/S0021859600087347.

Yang, J.H., Broderick, G.A., and Koegel, R.G. (1993). Effect of heat treating alfalfa hay on chemical composition and ruminal in vitro protein degradation. *J. Dairy Sci.* 76: 154–164. https://doi.org/10.3168/jds.S0022-0302(93)77334-8.

CHAPTER

34

Digestibility and Intake

David R. Mertens, President, *Mertens Innovation & Research LLC, Belleville, WI, USA;* Rsearch Dairy Scientist (Retired), *US Dairy Forage Research Center, Agricultural Research Service, USDA, Madison, WI, USA*

Richard J. Grant, President and Research Scientist, *The William H. Miner Agricultural Research Institute, Chazy, NY, USA*

Introduction

Digestibility and intake are major determinants of forage quality, which can be defined as the relative performance of animals when fed forages for **ad libitum intake**. Animal performance is the product of nutrient concentration, intake, digestibility, and metabolic efficiency of absorbed nutrients. Though nutrient concentration, particularly dietary fiber or plant cell wall concentration, influences intake potential and digestibility, it is the animal's response to these forage characteristics that ultimately determines the nutritional quality of forages. **Voluntary intake** typically accounts for most of the variation in animal productivity among forages, but digestibility is important because feces are the greatest loss of ingested nutrients.

Forages present a unique challenge to an animal's capacity to ingest and digest nutrients. Forages have a large volume in relation to dry matter (DM) weight; that is, they are bulky and are difficult to digest in comparison with grains and concentrate feeds. Both limitations of forages are related to the higher concentration of dietary fiber that they contain compared with concentrate feeds (Van Soest 1967). For **ruminants**, fiber is a nutritional term that is defined as the indigestible and slowly digesting fraction of feeds that occupy space in the gastrointestinal (GI) tract. Theoretically, insoluble dietary fiber is defined by its nutritional properties, but practically it is defined by the chemical method used to measure it. Of the methods currently available, amylase-treated neutral detergent fiber is the only official method (Mertens 2002) used routinely to estimate the total fiber in plant cell walls. Though they are not synonymous terms, *fiber, nutritional fiber,* or *neutral detergent fiber* (NDF) will be used to represent plant cell walls throughout this chapter because cell walls are not routinely measured.

Compared with grains, the higher fiber content of forages results in greater bulkiness, not because the specific density of fiber is higher than non-fiber components, but because the fiber in cell walls encloses space that increases forage volume in relation to weight. This is analogous to the volume-to-weight relationship of a room enclosed by walls, the so-called "hotel concept" of Van Soest (1994). The volume of the building (plant tissue) is much greater than that indicated by the weight of the demolished rubble (thoroughly masticated plant tissue) that comprises it. Because fibrous feeds are bulkier, the mass of forage that can be consumed or stored in the GI tract of the animal is lower than that for concentrates.

In addition to intake, the fiber in plant cell walls also affects the total DM digestibility (DMD) of forages. Fiber consists primarily of polysaccharides linked by beta bonds and associated with **lignin**. Mammalian digestive enzymes for carbohydrates react exclusively with alpha

Forages: The Science of Grassland Agriculture, Volume II, Seventh Edition.
Edited by Kenneth J. Moore, Michael Collins, C. Jerry Nelson and Daren D. Redfearn.

bonds; therefore, they are ineffective in hydrolyzing the beta-linked polysaccharides in plants that are predominantly in cell walls. Bacteria and other microorganisms synthesize enzymes that can hydrolyze the beta bonds, and bacterial fermentation is the mechanism used by animals to digest these polysaccharides to products that can be absorbed and metabolized. However, bacterial fermentation of beta-linked polysaccharides is a slow process compared with the digestion of alpha-linked polysaccharides or other nutrients such as proteins and fats. Efficient digestion of beta-linked polysaccharides requires both bacterial fermentation and long retention times (>30 hours) in the GI tract.

These chemical bonds and the linkages between cell wall polysaccharides and lignin are primary determinants of fiber digestibility. The proportion of various plant tissue types also affects digestibility and varies with plant species, plant anatomical part, and stage of growth. Ruminal **digestion** is often greatest for **mesophyll** and **phloem**, followed by epidermis and **parenchyma sheath**, **sclerenchyma**, and finally lignified vascular tissue (Minson 1990).

Digestive Adaptations of Herbivores

Ruminants

Ruminants, such as cattle, sheep, goats, deer, and buffalo, are cud-chewing, foregut fermenters that have a compartmentalized stomach in which symbiotic microbial fermentation occurs prior to mammalian digestion in the intestines. The stomach is compartmentalized into a rumen, reticulum, **omasum**, and **abomasum**. Gastric secretions of acid and enzymes occur in the abomasum, which connects to the small intestine. The omasum joins the reticulum to the abomasum and functions to absorb water and strain reticulo-rumen effluent so that only small particles pass out. Microbial fermentation by bacteria, protozoa, and fungi occurs in the first two compartments that are indistinctly separated in most ruminants and are sometimes called the reticulo-rumen. The reticulo-rumen is large, typically comprising 15% of a ruminant's body weight. The larger volume of the reticulo-rumen in bigger ruminants allows a slower turnover of particles and longer retention times of fiber. Thus, larger ruminants such as water buffalo are more efficient digesters of fiber and can survive and be productive on lower-quality forages than smaller ruminants such as sheep.

The environment of the rumen is warm (ca 39 °C), **anaerobic**, and liquid (< 150 g DM kg^{-1}). Buoyant larger particles float to the top of ruminal contents, and in larger ruminants, ruminal contents are distinctly biphasic, consisting of a mat of large particles floating on a pool of liquid and small particles. Ruminants regurgitate particles from the upper layer of the rumen and rechew them, referred to as chewing the cud or rumination, to obtain a particle size that can pass out of the reticulum (Poppi et al. 1981). Strong contractions of the rumen wall move contents horizontally along the midline of the rumen to the posterior, during which digested dense particles settle to the bottom and undigested large particles float to the top before circling back to the anterior for passage or rumination. Entanglement of small forage particles within the large particle mat in the rumen may be a crucial mechanism influencing how quickly particles, particularly of grasses, pass out of the rumen (Kammes and Allen 2012).

The anaerobic environment of the rumen limits fermentation to microorganisms that are facultatively or strictly anaerobic (many fiber fermenters are strict anaerobes that are poisoned by even small amounts of oxygen). During aerobic respiration, carbon chains in organic nutrients can be oxidized completely to carbon dioxide and water: for example, $1\ C_6H_{12}O_6 + 6\ O_2 \rightarrow 6\ CO_2 + 6\ H_2O$. However, during anaerobic fermentation the oxidation is incomplete resulting in the production of carbon dioxide, methane, and volatile fatty acids (VFAs), which are primarily acetic, propionic, and butyric acids. For example, the net reaction for anaerobic hydrolysis to acetic acid is $1\ C_6H_{12}O_6 \rightarrow 1\ CO_2 + 1\ CH_4 + 2\ C_2H_4O_2$. In addition, a portion of the carbonaceous nutrient is incorporated into microbial cells as they reproduce and grow. Ruminants can digest the microbial cells and absorb the VFA and use them as synthetic precursors or energy sources. However, methane is lost energy to the animal, reflecting the inefficiency of fermentative digestion. Both carbon dioxide and methane must be absorbed and respired from the lungs or belched from the rumen to prevent bloating of the animal.

The acids produced during microbial fermentation must be buffered or neutralized to prevent ruminal contents from becoming acidic and detrimentally affecting digestion. Ruminants secrete large amounts of buffers in their saliva when resting and especially when chewing during eating and rumination. With adequate diets, chewing and rumination stimulate secretion of salivary buffers that maintain ruminal pH above 6.2. However, high intakes of diets containing a large proportion of soluble carbohydrates and minimal fiber result in excessive acid production and reduced salivary buffer secretion. This is often associated with shifts in the microbial population, decreases in fiber digestion, and changes in the pattern of VFA production, which affect the performance of the animal.

Cecal Fermenters

Nonruminant herbivores, such as horses, rabbits, and some rodents, are hindgut fermenters with an enlarged **cecum** or large intestine in which microbial fermentation occurs on **digesta** after mammalian enzymatic digestion. Though some fiber is digested, the retention time of particles is much shorter than in ruminants, and fiber digestion is less efficient. In addition, the fermentation environment is less well regulated by the animal, and many nutrients that might stimulate microbial fermentation have been digested in the small intestines. For example, most amino acids are digested and absorbed in the small intestine, and urea is cycled into the large intestine to provide the nitrogen needed for microbial production and fermentation. Though the VFA produced by hindgut fermentation can be absorbed and used by the animal, the utilization of protein in microbial cells is lower than in ruminants because they are excreted in the feces without digestion.

Ingestive Adaptations of Herbivores

Herbivores vary in ingestive behavior from browsers, which consume primarily fruits, seeds, leaves, buds, and young **shoots**, to grazers, which consume coarse roughages such as mature grasses in bulk. Though both ruminants and nonruminants fall on the continuum between browsers and grazers, the classification system is not linear. Browsers tend to be more selective in what they eat than grazers, but there can be nonselective browsers, such as elephants, and very discriminating grazers, such as duikers. The degree of selectivity in ingestive behavior seems to be inversely related to body size.

Smaller herbivores, whether ruminant or nonruminant, eat more selectively. This may be related to the relatively higher energy requirements and faster rates of passage of herbivores with smaller body weight (BW). The (GI) contents of herbivores are roughly proportional to $BW^{1.0}$ (Parra 1978; Van Soest 1994). Animal energy requirements are proportion to $BW^{0.75}$ over a wide range of animal species and body weights (Brody 1945; Kleiber 1961), and this function is used to define **metabolic body weight** ($MBW = BW^{0.75}$). An animal's MBW is less than its BW because MBW represents the weight of those tissues in the body that metabolize most of the nutrients that are absorbed. GI contents, hair, skin, and bone are examples of BW components with little or no metabolic activity. Given that nutrient requirements are higher per unit of BW than is gut capacity, small herbivores meet their nutrient needs by eating more DM per unit of BW than large herbivores, which results in higher rates of passage. Thus, they must be more selective and consume higher-quality forages or plant parts to meet their energy requirements.

In general, ruminants and nonruminants also differ in initial **mastication**. Ruminants, especially larger ruminants, tend to swallow large particles that can be selectively retained in the rumen. These larger particles are regurgitated in a cud and rechewed during rumination. Swallowing large particles has the effect of increasing the time the forage is retained in the rumen and allows more complete digestion of slowly digesting fiber. Most hindgut fermenters, except those that are very large, meet their nutrient needs by ingesting relatively large quantities of forage to obtain the more readily available nutrients in plant cell contents and pass the fiber through the GI tract relatively rapidly without extensive digestion. Thus, nonruminants chew feeds very thoroughly during eating to release cell contents and generate small particles that maximize rate of passage.

Differences in ingestive behavior affect both digestibility and intake. When grazing, herbivores will vary in the type and quantity of forage consumed. Even when fed in controlled environments, smaller ruminants may select different forage parts under *ad libitum* feeding conditions. Because they have faster rates of passage and swallow smaller particles, smaller ruminants may also digest forages differently from larger ones. This does not mean digestibility measurements using sheep cannot be reliable estimates of digestion in cattle, but it does suggest the forage digestibilities by the two species are not equivalent or interchangeable, especially at *ad libitum* levels of forage intake. However, the ranking of forage intake and digestibility in cattle and sheep are usually similar. This is especially true when digestibility is measured under the traditional protocol in which intake is limited to maintenance levels (intake that allows animals to maintain BW). Measuring digestibility at maintenance levels of intake allows more accurate comparisons among forages because it minimizes the variation associated with differences in feed intake among animals.

Digestive Processes

Because domestic and wild ruminants are the major consumers of forages the remainder of this chapter will focus on their digestion and intake of forages. However, most of the discussion is also applicable to nonruminant herbivores because the processes of digestion are similar, except for the order and location of fermentative digestion, and most of the factors affecting intake and digestibility of forages apply to all herbivores.

Mastication

Physical breakdown of particles by mastication, during eating or rumination, is an important first step in the digestion process. Particle size reduction is necessary for passage of undigested residues from the reticulo-rumen, but it is also crucial for disruption of **cuticle** and cell wall membranes to release cell contents and allow access for microbes and enzymes. Though reducing particle size can increase the surface area for digestion, the effect is probably not as great in forage tissues as in more dense concentrates. For spherical solids the surface area is proportional to $mass^{0.67}$. Thus, the surface area per unit of mass is much greater for small spheres compared with large ones. However, plant tissues are not solid spheres but are more like hollow cubes. Bacteria appear to enter plant cells and digest them from the inside outward rather than outside inward (probably because lignification is greatest on the outside surface of most plant cells). Thus, mastication of plant tissues may be more important for cell wall disruption and access than for increasing surface area. Chemical structure limits digestibility, but physical factors also affect the digestion of forages.

Microbial Fermentation

In ruminants, typically 60% of total DM digestion occurs in the reticulo-rumen. The primary nutrients fermented in the rumen are proteins and carbohydrates. Lipids are not digested to any appreciable extent, but the excess hydrogen produced during anaerobic fermentation can hydrogenate unsaturated fatty acids in forages. Normally, fatty acids are completely hydrogenated, which explains why the fat in ruminant meat is saturated. Under conditions of low ruminal pH, polyunsaturated fats are only partially hydrogenated, which produces bioactive fatty acids such as C18:2 trans-10, cis-12 that have dramatic effects on lipid metabolism in the animal, resulting in milk fat depression.

Fermentation of protein is unique because it involves both degradation and synthesis. Typically, 60–90% of the crude protein (CP) in forages will be degraded in the rumen. Dried forages generally have lower protein degradabilities than fresh pasture or fermented silages. Much of the protein fermented in the rumen is converted to ammonia, though a portion is absorbed as amino acids and used for microbial reproduction and growth. Some ammonia is used by bacteria to synthesize amino acids, but a significant portion is absorbed by the animal and detoxified by converting it to urea, which can either be excreted in the urine or recycled into the rumen. When ruminal ammonia is excreted in the urine as urea, it represents a loss of potential forage protein, which is a negative result of foregut fermentative digestion of protein. These losses are greatest for forages high in crude protein, especially those with a large fraction of soluble protein or with rapid rates of degradation. However, recycling of urea into the rumen or large intestine provides a means of recapturing some of the degraded protein into microbial protein. Recycling of urea is especially important when forages low in protein content are consumed because microorganisms require nitrogen. When nitrogen in the ruminal contents cannot meet microbial requirements, fermentation of fiber is reduced, and the animal responds with lower DM intake and digestibility.

Most soluble carbohydrates in forages are almost completely fermented in the reticulo-rumen. In addition, 90% or more of the total digestion of fiber by ruminants also occurs in the rumen. The fermentation of starch in the rumen is variable depending on source and particle size. Starch in corn and sorghum seed is fermented very slowly (typically with rates very similar to fiber), but the starches of barley, oat, and wheat seeds are fermented much more rapidly. Processing of starches by fine grinding or heating to gelatinize the starch can greatly increase their fermentation rates. The end products of carbohydrate fermentation are carbon compounds in microbial cells, VFA, carbon dioxide, methane, and heat. Microbial cells and VFA can be digested and used by the animal, but the carbohydrate fermented to carbon dioxide, methane, and heat is lost. Though fermentative digestion is inefficient, it results in a net gain in nutrients to the animal because fiber, which is not hydrolyzed by mammalian enzymes, is converted to microbial cells and VFA that are used by the animal. In addition, the microbial protein that is synthesized from ammonia and feed amino acids is high in nutritional value and often complements the **protein quality** of forages.

The reticulo-rumen acts as a continuously stirred fermentation chamber into which feed is added and digesta is removed by passage. Thus, fermentative digestion is the result of competition between digestion and passage. Material is fermented only while it is retained in the rumen. As intake increases, the rate of passage of digesta increases, and the retention time in the rumen for fermentative digestion decreases (Thornton and Minson 1973). Thus, ruminants with high levels of intake digest their feed less completely because some material that could be digested passes out of the rumen before digestion is complete. Because more undigested material passes to the lower GI tract, the site of digestion is shifted.

Mammalian Enzymatic Hydrolysis

Digestion by mammalian enzymes secreted by the animal begins in the abomasum. Proteases and acid are secreted that initiate digestion of protein in microbial cells or in feeds that escaped the rumen unfermented. The acid conditions of the abomasum may also degrade some polysaccharides making them more digestible in the small intestine or more fermentable in the large intestine. The acidity of digesta leaving the abomasum is neutralized in the small intestine where digesta becomes alkaline. Amylase secreted by the pancreas and a complement of proteases secreted by the small intestine function to complete the digestion of starch and protein.

Determining Digestibility

The most direct way to determine DMD is to measure DM consumed and excreted in the feces and calculate the proportion of DM that disappeared between intake and excretion:

$$\text{DMD (kg DM kg DM}^{-1}) = (\text{kg DMI} - \text{kg FDM}) / \text{kg DMI} \quad (34.1)$$

where DMI is DM consumed and FDM is fecal DM excreted.

The DMD determined by Eq. (34.1) is an apparent digestibility because, in addition to undigested feed, feces

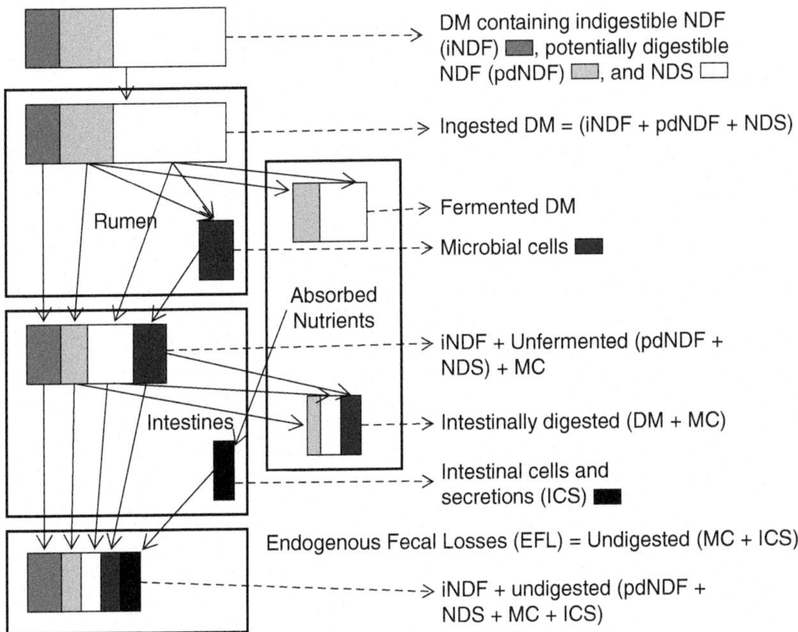

Fig. 34.1. Illustration of the conceptual transformation of forage DM, which consists of indigestible and potentially digestible NDF and neutral detergent solubles (NDS), as it is digested in ruminants and excreted in the feces.

can contain endogenous fecal losses (EFLs), which arise from undigested microbial cells generated by GI fermentations and GI secretions and cell sloughing (Figure 34.1). To determine DM **true digestibility** requires that EFL be measured and subtracted from fecal DM excretion:

$$\text{DMTD (kg DM kg DM}^{-1}) = [\text{kg DMI} - (\text{kg FDM} - \text{kg EFL})]/\text{kg DMI} \quad (34.2)$$

where DMTD is DM true digestibility, DMI is DM consumed, FDM is fecal DM excreted, and EFL is endogenous fecal loss of DM.

For most constituents, apparent digestibility will always be smaller than true digestibility because feces contain EFL. To determine true digestibility directly, the endogenous losses of a feed component must be zero or be measured by some method that can distinguish between endogenous losses and undigested feed in the feces. True digestibility equals apparent digestibility for forage constituents such as fiber because they have no EFL; however, many important feed components such as DM, organic matter (OM), protein, fat, neutral detergent solubles, and possibly starch and soluble carbohydrates have associated endogenous secretions.

The difference between true and apparent digestibility is important when deriving equations or models to estimate digestibility of any nutrient. Also important is the distinction between digested nutrient and nutrient digestibility. These two concepts are not interchangeable, and it would be less confusing if standard nomenclature and abbreviations based on historical precedent were used. Nutrient digestibility is the fraction of the *nutrient* that is digested; that is, it is the digestion coefficient of the nutrient. Historically, nutrient digestibility has most often been used as a suffix with a capitalized letter *D*, for example, DMD or CP digestibility (CPD). Thus, nutrient digestibility would have the abbreviation NutrD with the SI units of kg Nutr kg Nutr^{-1} of nutrient:

$$\text{NutrD (kg Nutr kg Nutr}^{-1}) = (\text{kg Intake\_Nutr} - \text{kg Fecal\_Nutr})/\text{kg Intake\_Nutr} \quad (34.3)$$

where Intake_Nutr is the intake of a specified nutrient and Fecal_Nutr is the excretion of the specified nutrient.

Traditionally, the term *digestible* was used to indicate the amount of a nutrient in feed DM that was digested during an animal trial or *in vitro* fermentation. However, this was a misnomer because *digestible* indicates what *can be* digested, but these methods actually measure what *was digested.* Thus, "digested" is the correct term and the National Research Council (NRC 2001) used a prefix lowercase letter *d* (more correctly *td* for truly digested) for this concept; e.g. digested CP (dCP). Thus, digested nutrient would have the abbreviation dNutr with the SI units of g kg^{-1} of DM:

$$\text{dNutr (g kg}^{-1}\text{ DM)} = (\text{g Intake\_Nutr} - \text{g Fecal\_Nutr})/\text{kg Intake\_DM} \quad (34.4)$$

where Intake_Nutr is the intake of a specified nutrient, Fecal_Nutr is the excretion of the specified nutrient, and Intake_DM is the intake of DM.

Equations (34.3) and (34.4) can be used to calculate *in vivo, in situ,* or *in vitro* apparent or **true digestibility** of any nutrient.

Though NutrD is not the same as dNutr (with the exception of DM where digested DM equals DM digestibility), there is a relationship between them:

$$\text{dNutr (g kg}^{-1}\text{ DM)} = \text{Nutr (g kg}^{-1}\text{ DM)} \times \text{NutrD (kg Nutr kg Nutr}^{-1}\text{)} \quad (34.5)$$

where Nutr is the concentration of a specified nutrient in DM of feeds and NutrD is the fractional digestion coefficient of the specified nutrient. (Note: × in equations indicates multiplied by.)

The distinction between dNutr and NutrD becomes crucial when trying to describe digestibility mathematically or when trying to develop prediction equations. Only dNutr values can be summed to determine DMD, and the relationships of nutrient concentration to dNutr or NutrD are distinctly different.

Lucas Test of Nutritional Uniformity

Equation (34.5) also demonstrates that dNutr is a function of nutrient concentration in DM. For example, the Lucas test described by Van Soest (1994) postulates that dCP should be related linearly to CP concentration across all types of feeds, if it is a nutrient with uniform availability. Van Niekerk et al. (1967) reported one of many published equations relating dCP to CP:

$$\text{dCP (g kg}^{-1}\text{ DM)} = -32.6 + 0.94 \times \text{CP (g kg}^{-1}\text{ DM)} \quad (34.6)$$

with r = 0.994 and SE_{yx} = 0.58. This equation provides an indirect approach for estimating the true digestibility and EFL of CP. The slope of this equation is the average true digestibility of CP, which indicates that 0.94 of CP is truly digested. The intercept is an estimate of the EFL of CP because it indicates the excretion of 32.6 g dCP kg^{-1} DM when no CP is consumed (CP = 0). The uniform availability of CP, as evidenced by the high correlation and small standard error of regression (SE_{yx}) between dCP and CP, leads to three biologic conclusions (Figure 34.2a). First, in the vast majority of forages and feeds (excluding those in which CP is bound by tannins or contained in Maillard products in heat-damaged feeds), CP is almost completely digested in the total GI tract. Second, it is possible to have negative digestibilities when intake of a nutrient is less than its endogenous excretion. Third, because CPD is related to CP concentration, differences in digestion coefficients are relevant only when the CP concentrations in diets are similar.

In addition, the relationship between dCP and CP also demonstrates that CPD cannot be a linear function of CP. Knowing that CPD = dCP (g kg^{-1} DM)/CP (g kg^{-1} DM), we can rearrange the previous equation into the correct formulation for predicting CPD by multiplying each side of the equation by CP^{-1}:

$$\text{CPD (kg CP kg CP}^{-1}\text{)} = \text{dCP/CP} = -32.6 \times \text{CP}^{-1} + 0.94 \quad (34.7)$$

The regression coefficients in Figure 34.2b obtained by Van Niekerk et al. (1967) differ slightly from Eq. (34.7) because the variances of CP and CP^{-1} are different, which affects least squares regression solutions. These equations demonstrate that both dCP and CPD are related to CP concentration, but the functional form of the relationship is different; that is, the first is linear and the second is reciprocal (Figure 34.2). This may explain why attempts to estimate NutrD from linear functions of chemical composition are often unsatisfactory; that is, the functional form is incorrect.

Figure 34.1 also illustrates the difference between *in vitro* DMD (IVDMD) and *in vitro* DM true digestibility (IVDMTD). The determination of IVDMD and IVDMTD involve two different *in vitro* procedures and measure two different nutritional entities. The traditional method for determining IVDMD is the two-stage Tilley and Terry (T&T) technique that consists of a 48-hour incubation in buffered ruminal fluid followed by a 48-hour incubation in an acid pepsin solution (Tilley and Terry 1963). This method has been highly correlated in several experiments with *in vivo* apparent DMD measured at maintenance levels of feed intake. The T&T measurement of IVDMD is apparent because most of the microbial residues generated during microbial fermentation remain in the undigested DM residue.

Van Soest et al. (1966) replaced the second stage of the T&T technique with ND extraction, which

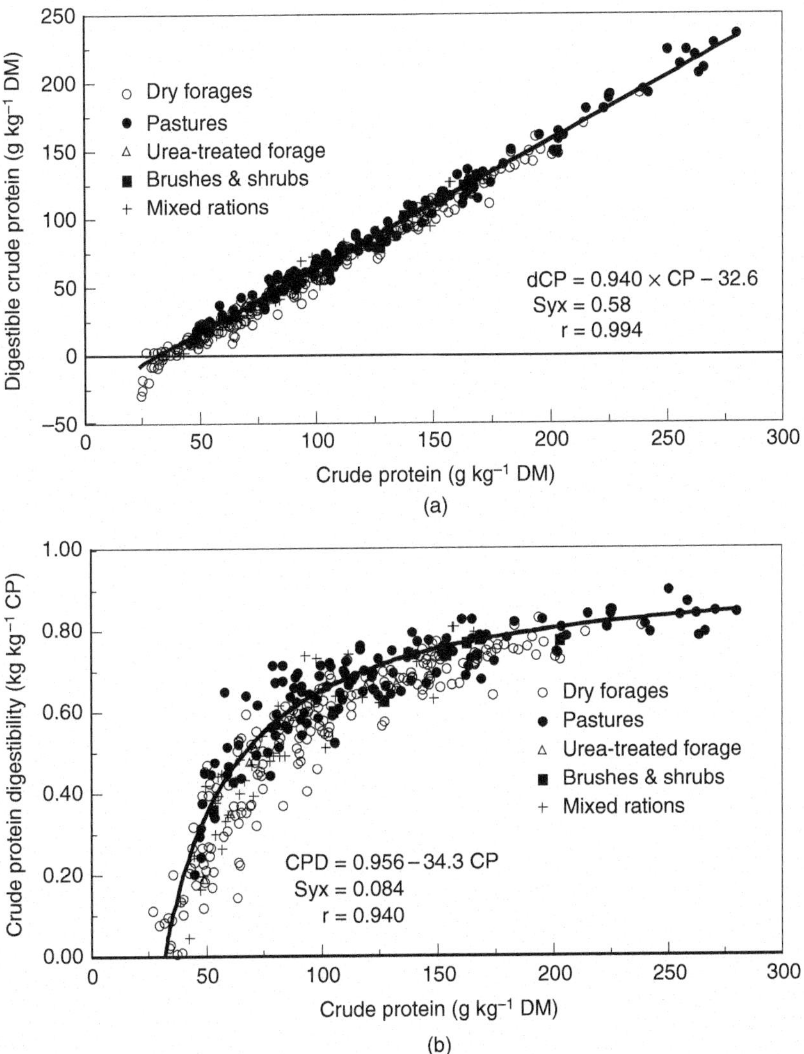

Fig. 34.2. Relationship of digestible (digested) crude protein *(a)* and crude protein digestibility *(b)* to crude protein concentration. *Source:* Adapted from Van Niekerk et al. (1967).

removes intestinal cells and secretions and microbial debris, and leaves a residue that is undigested NDF after 48 hours of fermentation. Both IVDMTD and *in vitro* NDF digestibility (NDFD) can be determined in this procedure by assuming that truly undigested *in vitro* DM residue equals undigested *in vitro* NDF residue:

$$\text{IVDMTD} = (\text{g Initial\_DM} - \text{g truly\_undig\_IV\_DM\_Res}) / \text{g Initial\_DM}$$

therefore,

$$\text{IVDMTD} = (\text{g Initial\_DM} - \text{g undig\_IV\_NDF\_Res}) / \text{g Initial\_DM} \quad (34.8)$$

where Initial_DM is the amount of dry sample fermented, truly_undig_IV_DM_Res is the *in vitro* DM residue remaining after fermentation, and undig_IV_NDF_Res is the undigested *in vitro* NDF residue after fermentation and neutral detergent extraction.

To calculate *in vitro* NDFD:

$$IVNDFD = (g\ Initial\_NDF - g\ undig\_IV\_NDF\_Res)/g\ Initial\_NDF \quad (34.9)$$

where IVNDFD is *in vitro* NDF digestibility, g Initial_ NDF is g Initial_DM × NDF ($g\,kg^{-1}$ DM)/1000, and undig_IV_NDF_Res is the undigested *in vitro* NDF residue after fermentation and neutral detergent extraction.

Measuring digested NDF (dNDF) or NDFD using *in vitro* or *in situ* methods is an attractive alternative to prediction equations based on chemical composition. However, *in vitro* or *in situ* techniques also have drawbacks. The fermentation vessel (bags, tubes, flasks, etc.); ruminal inoculum (single vs multiple donors, donor diet, fluid vs blended solids, etc.); time (24, 30, vs 48 hours); material particle size (2-, 1-, 0.5-mm screen, etc.); carbon dioxide equilibration system (closed, released, continuous gassing, etc.); and buffer, as well as their interactions, can all influence results. Finally, there is the issue of how well *in situ* or *in vitro* determination mimics total tract digestion in animals. *In vitro* and *in situ* methods measure primarily ruminal digestion and high correlations between 48 hours T&T IVDMD and *in vivo* DMD measured at maintenance level of intake do not necessarily mean a 1: 1 equality of values. Furthermore, *in vitro* techniques other than T&T may yield different estimates of IVDMD.

Conceptual Description of Digestibility

Van Soest and Wine (1967) developed the NDF method to isolate the total insoluble fiber in feeds that was responsible for the variable DMD in forages. Van Soest (1967) observed that neutral detergent solubles (NDS), the complement of NDF [NDS ($g\,kg^{-1}$ DM) = 1000 − NDF ($g\,kg^{-1}$ DM)], had uniform nutritional availability across all forages because the standard error of regression was small and the r^2 was greater than 0.90 when concentration of apparently digested NDS (dNDS) in DM (measured at maintenance levels of intake) is regressed against the concentration of NDS in DM:

$$dNDS\ (g\,kg^{-1}\ DM) = -129 + 0.98 \times NDS\ (g\,kg^{-1} DM) \quad (34.10)$$

where dNDS is apparently digested NDS. However, Van Soest (1967) observed that NDF did not have uniform availability and the dNDF of a feed had to be determined or predicted for each individual material:

$$dNDF\ (g\,kg^{-1}\ DM) = NDFD \times NDF\ (g\,kg^{-1}\ DM) \quad (34.11)$$

He concluded that the effects of lignification were confined to NDF, and Goering and Van Soest (1970) reported that NDFD (kg NDF kg NDF^{-1}) could be estimated from the ratio of lignin to acid detergent fiber. From these concepts, Van Soest (1967) developed a simple summative equation for explaining and predicting the digestibility when forages are thoroughly chewed and soluble matter is almost completely digested:

$$dDM\ (g\,kg^{-1}\ DM) = dNDF\ (g\,kg^{-1}\ DM) + dNDS\ (g\,kg^{-1}\ DM) \quad (34.12)$$

where dNDS is apparently digested NDS, therefore

$$dDM\ (g\,kg^{-1}\ DM) = NDFD \times NDF + 0.98 \times NDS - 129 \quad (34.13)$$

Because NDS = (1000 − NDF) and dDM = DMD × 1000 (this relationship is only true for DM, but not for any other constituent), the simple summative equation can be solved to show that DMD is a function of only NDF and its digestibility (NDFD):

$$DMD\ (kg\ DM\ kg\ DM^{-1}) = .851 + (0.98 - NDFD) \times NDF\ (g\,kg^{-1}\ DM) \quad (34.14)$$

Replace the above with:

$$DMD\ (g\,kg\ DM^{-1}) = 851 - (0.98 - NDFD) \times NDF\ (g\,kg^{-1}\ DM)$$

This equation has the same form as regression Eqs. (DMD = a + b × NDF) that are often used to estimate DMD from fiber concentrations; however, (0.98 − NDFD) is variable, whereas the regression coefficient (b) is a constant parameter that reflects the average NDFD of the regression population. However, this derivation suggests that expressing DMD as a linear function of fiber concentration has a biologic basis.

Estimation of Digestibility Using Summative Equations

The simple summative equation can be used to illustrate the unique characteristics of forages that affect DMD (Table 34.1). At comparable maturities, it demonstrates that the high DMD of legumes results substantially from their low NDF concentration, whereas grasses achieve similar DMD by having higher NDFD. Corn silage provides the highest DMD by having both low NDF concentration and high NDFD. Most of the improvement in the quality of conserved forages has been accomplished by decreasing the maturity at harvest.

Table 34.1 Using the simple summative equation to estimate digestibility of forages

Component	Legume hay	Grass hay	Corn silage
NDF[a]	400	550	400
NDFD[b]	0.450	0.625	0.600
Digested NDF (dNDF = NDFD × NDF)[a]	180	344	240
NDS[a]	600	450	600
dNDS = 0.98 × NDS[a]	588	441	588
True DM digestibility[a]	768	785	828
Endogenous fecal DM excretion[a]	−129	−129	−129
Apparent DM digestibility$_{1\times}$[a]	639	656	699

[a]g kg^{-1} DM.
[b]kg NDF kg NDF^{-1}.

The major factor that improves digestibility when more immature forages are harvested is a reduction in fiber concentration, though the increased fiber digestibility of immature forages is also beneficial. However, there is a limit to the value of using maturity differences to improve DMD because harvesting forages at very immature stages reduces yields and longevity. The simple summative equation indicates that DMD can also be improved by changing NDFD. A 10% increase in NDFD for legumes (from 0.450 to 0.495 kg NDF kg NDF^{-1}) has the same effect on dDM as a 9.3% reduction in NDF concentration (from 400 to 366 g kg^{-1} DM). This explains why most of the effort in improving forage quality is focused on fiber and factors that limit its digestibility.

The National Research Council (NRC 2001) adopted a more complex summative equation to estimate total digestible nutrients (TDNs) in all types of feeds. This approach was developed by Conrad et al. (1984) and Weiss (1992), who modified the simple summative equation by (i) fractionating NDS into CP, ether extract, and nonfiber carbohydrates; (ii) relating the NDFD to lignin using a complex surface law function; and (iii) adjusting the EFL to correspond to TDN. The equation also includes a processing adjustment factor for the nonfibrous carbohydrate fraction to account for the effect of physical and heat processing on starch digestibility (NRC 2001):

$$\mathrm{TDN}_{1\times} = \mathrm{tdCP} + \mathrm{tdFA} \times 2.25 + \mathrm{tdNFC} + \mathrm{tdNDF} - 70 \qquad \text{(NRC Eq. 2-5)}$$

$$\mathrm{tdCP\ (forages)} = (e^{-.012 \times \mathrm{ADICP/CP}}) \times \mathrm{CP} \qquad \text{(NRC Eq. 2-4b)}$$

$$\mathrm{tdFA} = 1.00 \times (\mathrm{EE} - 1) \qquad \text{(NRC Eq. 2-4d)}$$

$$\mathrm{tdNFC} = \mathrm{PAF} \times 0.98 \times \mathrm{NFC} \qquad \text{(NRC Eq. 2-4a)}$$

$$\mathrm{tdNDF} = 0.75 \times [(\mathrm{NDF} - \mathrm{NDICP}) - \mathrm{L}] \times (1 - [\mathrm{L}/(\mathrm{NDF} - \mathrm{NDICP})]^{2/3}) \qquad \text{(NRC Eq. 2-4e)}$$

where $\mathrm{TDN}_{1\times}$ is TDN at maintenance level of intake, td is truly digested, FA is fatty acid, EE is ether extract, NFC is nonfibrous carbohydrates = 100 − ash − CP − EE − (NDF − NDICP), ADICP is acid detergent insoluble CP expressed as a percentage of DM, NDICP is neutral detergent insoluble CP expressed as a percentage of DM, PAF is processing adjustment factor, L is lignin, and 70 is the EFL of TDN (g kg^{-1} TDN).

An alternative way of calculating tdNDF using an *in vitro* assay was also provided (NRC 2001): tdNDF = IVNDFD × NDF, where IVNDFD is *in vitro* NDF digestibility measured after 48 hours of fermentation.

The primary benefits of separating NDS used in the simple summative equation for DMD into protein, fat, and nonfibrous carbohydrates in the more complex summative equation for TDN are to account for the added energy density of fatty acids (by multiplying them by 2.25) and to exclude **ash**, which provides no energy value. However, separating NDS into its components has little impact on evaluating forage nutritive value because (i) the average true digestibility of proteins, fats, and nonfibrous carbohydrates are only slightly less than the 0.98 for NDS; (ii) forages contain little fat; and (iii) the effect of ash is partially accounted for by the difference in endogenous losses between DM and TDN (129 g kg^{-1} DM and 70 g kg^{-1} TDN, respectively).

Digestion Kinetics

Summative equations of Van Soest (1967) or NRC (2001) are static estimates of digestibility that assume that rates of passage are constant, which is a reasonable assumption for maintenance levels of intake. However, rates of passage can vary tremendously among animals at productive levels of intake, especially for lactating dairy cows and small growing animals that may eat more than 30 g DM kg^{-1} of BW daily. Tyrrell and Moe (1975) observed that each multiple of intake above maintenance resulted in a decrease of 40 g kg^{-1} TDN because the retention time for digestion in the GI tract decreases as intake and passage rate increases.

Static estimates of digestion are inadequate when rates of digestion are low, and rates of passage are variable. Rates of fiber digestion are similar in magnitude to rates of passage and may vary considerably, which results in low and variable extents of digestion for fiber when intake increases. Though protein digestion is completed very efficiently in the intestines, the rates of ruminal fermentation of insoluble protein in forages reported by the NRC (2001) averaged $0.091\ h^{-1} \pm 0.041$. Grass forages had rates of insoluble protein digestion that were 40–50% of legumes, and silages were 60–70% of those for hays. The magnitude of these differences can have a significant effect on the site of protein digestion and may alter extent of digestion when rate of passage is rapid. Clearly, dynamic differences among feeds and animals must be used to determine digestibility at productive levels of intake (Mertens and Ely 1982). However, to predict digestibility, the kinetics of digestion and passage of forages must be determined.

Waldo et al. (1972) discovered that the key to describing digestion kinetics of fibrous carbohydrates was identifying potentially digestible fractions that have homogeneous digestion attributes. They realized that the asymptotic digestion of cellulose indicated that some of the cellulose was not potentially digestible (Waldo 1969) and that this fraction must be subtracted to obtain a potentially digestible fraction that was kinetically homogenous and follows first-order digestion kinetics. The research of Smith et al. (1972), Mertens (1973), and Traxler et al. (1998) established that a part of the NDF in feeds is indigestible (iNDF) and will not be digested even if left in the anaerobic fermentative environment of the GI tract indefinitely. Indigestible NDF is different from undigested NDF at any fermentation time, which contains iNDF and some potentially digestible NDF (pdNDF) that is not degraded because retention time is not long enough to achieve complete NDF digestion (Figure 34.1).

To determine pdNDF, iNDF must be estimated and subtracted from total NDF. The iNDF can be estimated by the asymptote of fermentation curves or by long-term *in situ* or *in vitro* fermentations. The fermentation time needed to estimate iNDF when digestion is 99% complete can be approximated by dividing the fractional digestion rate into 4.6; for example, for a rate of $0.10\ h^{-1}$ this will take 46 hours and for a rate of $0.05\ h^{-1}$ it will take 92 hours (Mertens 1993). Some argue that measurement of the asymptote of digestion is irrelevant because feeds do not remain in the GI tract for these long periods of time. However, the determination of the asymptote of digestion has nothing to do with the average retention time of feeds in the rumen. It is necessary to define the potential digestible fraction so that rate of digestion can be determined. By definition, a fractional digestion rate applies only to the portion of forage that is potentially digestible; therefore, measuring a rate of digestion on total NDF (iNDF + pdNDF) creates the illogical situation that a digestion rate is determined on a fraction that cannot be digested, that is, iNDF.

Waldo's (1969) hypothesis that some of the cellulose (or fiber in general) may not be digestible because it remained after six days of fermentation changed our concept of NDF digestion completely. We changed from a one-pool model of NDF with variable NDFD to a new two-pool model that contained iNDF and pdNDF with a fractional first-order rate of digestion (Figure 34.3). Initial experiments used $uNDF_{72}$ to estimate $iNDF_2$ for the two-pool model, but Mertens (1977) reported that when longer fermentations (144 hours) were used to estimate iNDF, there appeared to be two digestible pools (three pools in total). Raffrenato and Van Amburgh (2010) suggest that if $uNDF_{240}$ is used to estimate $iNDF_3$ then a proposed 3-pool model of NDF digestion is appropriate (Figure 34.3). Mertens and Ely (1982) proposed a model of NDF digestion, particle size reduction, and passage that included 3 pools in NDF and 3 particle sizes of NDF in the rumen. Though it appears that most accurate description of NDF digestion kinetics may require fast and slow digesting pools, the size of the slowly digesting pool is small or even zero in some feeds. The practical utility of the 3-pool model for NDF digestion remains to be established and it seems questionable to use a complex model of fiber digestion without as similarly complex model of NDF particle size reduction and passage.

The concept of iNDF has interesting consequences for the interpretation of dNDF or tdNDF, which are equivalent, in summative equations (Eq. (34.11) and NRC Eq. 2-4e). Digestion kinetics suggest that dNDF is more accurately a function of pdNDF because digested NDF can only arise from that which is potentially digestible:

$$\text{dNDF} = \text{pdNDFD} \times \text{pdNDF} \tag{34.15}$$

where pdNDFD is pdNDF digestibility (kg pdNDF kg pdNDF^{-1}).

Smith et al. (1972), Mertens (1973), and Traxler et al. (1998) observed that iNDF is related to acid detergent lignin (ADL) concentration: iNDF = FCP × ADL, where ADL is determined using 72% sulfuric acid and FCP is ADL plus the proportion of fibrous carbohydrates protected by ADL and made resistant to anaerobic fermentation. Because pdNDF = NDF − FCP × ADL and dNDF = tdNDF = pdNDFD ×(NDF − FCP × ADL), then

$$\text{dNDF} = \text{tdNDF} = \text{pdNDFD} \times \text{NDF} - \text{pdNDF} \times \text{FCP} \times \text{ADL} \tag{34.16}$$

Note that pdNDFD is larger than NDFD because it is the digestibility of the NDF that is potentially digestible.

Because NDFD = tdNDF/NDF, Eq. (34.16) can be solved for NDFD by dividing each side of the equation

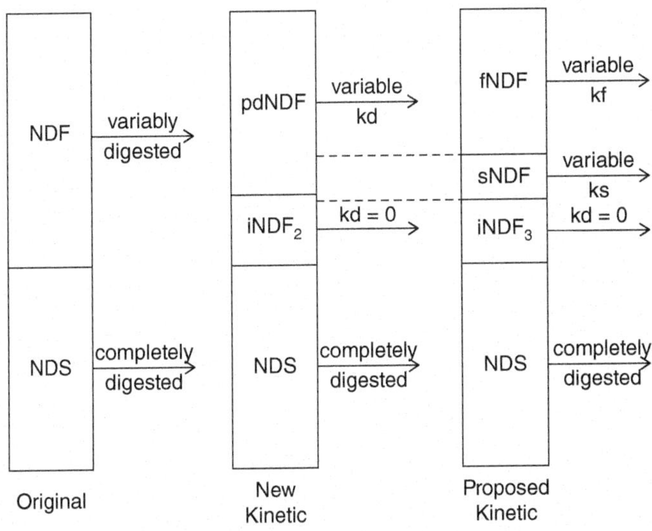

Fig. 34.3. Changes in the concepts and models of NDF and feed digestibility; where NDS = neutral detergent solubles, pdNDF = potentially digestible NDF, iNDF = indigestible NDF, fNDF = fast-digestion NDF, sNDF = slow-digesting NDF and k = fractional rate for each pool (Mertens 2016).

by NDF^{-1}:

$$NDFD = tdNDF/NDF = pdNDFD - pdNDFD \times FCP \times ADL/NDF \quad (34.17)$$

Equation (34.17) indicates that dNDF is a linear function of ADL; a reciprocal function of NDF (NDF^{-1}). This explains why DMD has been observed empirically to be a linear function of lignin; however Eq. (34.17) suggests that NDFD should be a function of the ADL/NDF ratio, which may explain the poor linear relationship often observed between NDFD and ADL. It is interesting that both the empiric equation of Goering and Van Soest (1970) and the theoretic equation of Conrad et al. (1984) used ratios of lignin to fiber to predict NDFD. Equation (34.17) provides a testable alternative.

Previous equations have presented NDFD as a static variable based on steady-state conditions of digestion and passage. The dynamic (changes with time) mathematical model for the first-order digestion of NDF *in vitro* or *in situ* is

$$NDF\_Res(t) = pdNDF \times e^{-kd \times (t-lag)} + iNDF \quad (34.18)$$

where NDF_Res(t) is the NDF residue remaining at any time = t, kd is the fractional digestion rate, t is time in hours, and lag is discrete lag time, also in hours, before digestion begins.

Kinetic data are collected by determining the NDF residue remaining after various times of *in vitro* or *in situ* fermentation (typically 0, 3, 6, 12, 18, 24, 36, 48, 72, and 96 hours). Model parameters can be estimated by either logarithmic transformation and linear regression or non-linear regression (Mertens and Loften 1980). Typical kinetic parameters are given in Table 34.2. In general, legumes have faster fractional rates of digestion and a larger fraction of the total NDF as iNDF. Immature forages have faster rates than mature ones with a smaller proportion of the NDF as iNDF.

Digestion kinetics are determined by measuring residues that can disappear only by digestion; that is, escape of residues by passage is not allowed because measurements are made within closed vessels (tubes, flasks, or indigestible bags). Where digestion and passage are occurring simultaneously (as it happens in animals), digestibility can be described mathematically as the proportion of total disappearance due to digestion and passage that can be attributed to digestion only. If both digestion and passage are assumed to be simple first-order processes, then true digestibility (TD) of any potentially digestible component is its fractional rate of digestion divided by the sum of the rates of digestion and passage (Waldo et al. 1972; Mertens 1993):

$$TD = kd/(kd + kp) \quad (34.19)$$

Table 34.2 Kinetic digestion parameters for neutral detergent fiber measured *in vitro*

Plant characteristic	Rate[a] (h^{-1})	pdNDF[b] ($g\,kg^{-1}$)	iNDF[c] ($g\,kg^{-1}$)	Lignin[d] ($g\,kg^{-1}$)	iNDF/lignin ($g\,g^{-1}$)
Legume average[e]	0.116	195	200	96	2.37
Grass average[e]	0.096	351	190	62	2.98
Immature forage average[e]	0.149	303	97	42	2.70
Mature forage average[e]	0.059	283	290	108	2.80
Spring, stage 3.5[f]	0.094	236	344	102	3.37
Spring, stage 4.3[f]	0.104	243	370	111	3.33
Spring, stage 5.7[f]	0.084	224	443	121	3.66
Summer, stage 3.5[f]	0.127	247	320	95	3.37
Summer, stage 4.3[f]	0.115	253	366	113	3.24
Summer, stage 5.7[f]	0.098	237	434	124	3.50

[a] First-order fractional rate of digestion.
[b] Potentially digestible neutral detergent fiber.
[c] Indigestible neutral detergent fiber.
[d] Acid detergent lignin (72% sulfuric acid method).
[e] Smith et al. (1972).
[f] Sanderson et al. (1989).

where kd is fractional rate of digestion (h^{-1}) and kp is fractional rate of passage (h^{-1}).

This simple relationship illustrates several interesting nutritional consequences. The effect of increasing passage rate is small for feed components with large rates of digestion ($>0.20\,h^{-1}$) but is large for feed components with digestion rates similar in magnitude to passage rates. This suggests that depression in DMD associated with high levels of intake is related primarily to decreased fiber digestion because fiber digests slowly in relation to its rate of passage.

Factors Affecting Forage Digestibility

Plant species, maturity, growth environment of the plant, chemical composition, genetic differences, and plant anatomy and morphology may all play a role in determining the digestibility of forages. The impact of these factors on digestibility is the focus of forage evaluation. However, other factors affect digestibility that are important not only in assessing forage evaluation, but also in determining forage utilization when it is fed in productive situations.

Animal Factors

Though forage characteristics determine potential digestibility, animal factors and other dietary components can dramatically affect forage digestion. In some situations, animal and dietary factors can overwhelm the intrinsic digestibility of the forage. Differences in the level of intake and corresponding rates of passage can affect digestibility, as has been discussed. In addition, selection among forage components by the animal can affect digestibility. When fed at restricted levels of intake, animals typically eat all the material provided, and digestibility reflects the composition of the forage that was fed. However, when forage is offered in excess, animals often select the more desirable portions (typically those lower in fiber), and the composition of the forage that actually consumed and digested differs from that offered to the animal.

Dietary Factors

Dietary factors that are extrinsic to the forage can also affect digestibility. Reducing particle size by fine chopping, grinding, or pelleting forages can alter rumination, the ruminal fermentative environment, and increase rate of passage, resulting in decreased digestibility. In addition, deficiencies of essential nutrients for microbial fermentation, such as nitrogen, phosphorus, and sulfur can also limit digestion. If forage is fed alone, without supplementation, the impact of limiting nutrients on digestion should be measured as a part of forage evaluation. However, if forage is fed with other feeds, measuring digestibility without supplementation would underestimate the nutritive value of forages that were deficient in nutrients when they are fed in mixed diets.

Digestion of forage requires an active microbial population with fiber-digesting capability. In mixed forage and concentrate diets, associative effects between diet ingredients can result in situations in which digestion is limited by microbial or enzymatic capacity and not by intrinsic properties of the forage. For example, when starch or rapidly fermentable carbohydrates are fed, ruminal pH can drop below 6.2 and depress fiber digestion. In addition, microorganisms may preferentially utilize the more easily digestible carbohydrates and delay digestion of fiber until the more digestible substrates are fermented. When evaluating forages, it is important to minimize associative effects. However, in the practical feeding of forages, both positive and negative associative effects must be considered (Huhtanen 1991).

Importance of Intake in Assessing Forage Quality

The product of intake and digestibility (digested DM intake) is the primary determinant of animal productivity, which defines forage quality. Crampton et al. (1960) were the first to propose that the feeding value of forages could be described by a nutritive value index that was the product of intake and digestibility. Relative forage value was developed as an analogous measure of forage quality that was based on the prediction of intake and digestibility from fiber concentrations in forages. Typically, 60–90% of the variation in digested DM or energy intake is related to differences in intake (Crampton et al. 1960; Reid 1961). Thus, the intake potential of a forage is the most important element in determining forage quality. Unfortunately, it is the most difficult forage attribute to determine because actual intake is a function of the forage, the animal, and the feeding circumstance.

Concepts of Intake Regulation

Numerous factors affect intake (Figure 34.4), and many books and reviews have been written on the subject (see Balch and Campling 1962; Conrad 1966; Baumgardt 1970; Bines 1971; Weston 1985; NRC 1987; Mertens 1994; Forbes 1995). In this chapter, discussion focuses on ways in which intake regulation affects forage quality evaluation. Much of the confusion in the literature about the relationships between forage characteristics and intake is related to lack of recognition that forage, animal, and feeding characteristics can each affect *ad libitum* intake, making it difficult to assign a **forage intake potential** (Figure 34.4). The relationships between intrinsic forage characteristics and intake can be clarified by understanding quantitative concepts of intake regulation in ruminants that are based on biologic principles and defined mechanisms. These concepts can serve as a basis for identifying feeding situations in which the potential intake of forages can be measured.

Physiologic Intake Regulation

When animals are fed high-energy forages or rations that are palatable, low in bulk, and readily digested, intake is regulated typically to meet the energy demands of the animal (Jones 1972; Journet and Remond 1976). This is especially true for mature animals with low energy requirements. Physiologic regulation of intake suggests that animals regulate intake such that the product of intake and energy concentration in the diet will equal the animal's energy demand. This relationship can be described by a simple algebraic equation: $I_e \times E = R$ and

$$I_e = R/E \tag{34.20}$$

where I_e is intake (kg DM d^{-1}) expected when energy demand is regulating intake, E is the digested energy concentration of the diet (Mj kg^{-1} DM), and R is the animal's energy requirement or demand (Mj d^{-1}).

It is important to recognize that intake is not the only response in Eq. (34.19) that can be varied by the animal. If dietary energy concentration is too low and I_e cannot be adjusted sufficiently to meet animal intake requirements, the animal may reduce its energy requirement by reducing productivity or using body reserves. Thus, the animal effectively changes its output to match allowable energy input when given a stressful situation in which forage quality is low, relative to the animal's energy requirement.

Physical Intake Limitation

The fill limitation concept of intake regulation indicates that when animals are fed palatable forages or rations that are low in digested energy and high in bulk (filling effect), intake is limited by the fill-processing capacity of the animal (Campling 1970). Physical limitations to intake indicate that the product of intake and the diet's filling effect equals the animal's fill-processing capacity. This mechanism of intake limitation can be described by a simple algebraic equation, $I_f \times F = C$, that can be rearranged to solve for intake:

$$I_f = C/F \tag{34.21}$$

where I_f is intake (kg DM d^{-1}) expected when fill is limiting intake, F is the volume of the filling effect (l kg^{-1} DM) of the diet, and C is the animal's daily fill-processing capacity (l d^{-1}).

It should be noted that fill is probably related to volume and not weight because stretch receptors in the rumen, which react to changes in volume, are signals for intake regulation. Most importantly, it should be noted that the daily fill-processing capacity is a flux or flow (l d^{-1}) and not a volume of ruminal contents. Assuming that digestion and passage are first-order processes, the daily fill-processing capacity is the product of reticuloruminal volume and the combined effects of rates of consumption, rumination, passage, and digestion (flux = pool × fractional rate). Thus, intake limitation associated with bulky forages could be related to the time available for eating or ruminating, factors affecting rate of passage and changes in rate of digestion. As with energy requirement, fill-processing capacity is not fixed because animals will attempt to increase capacity by increasing the pool or rates of disappearance to meet nutrient needs for

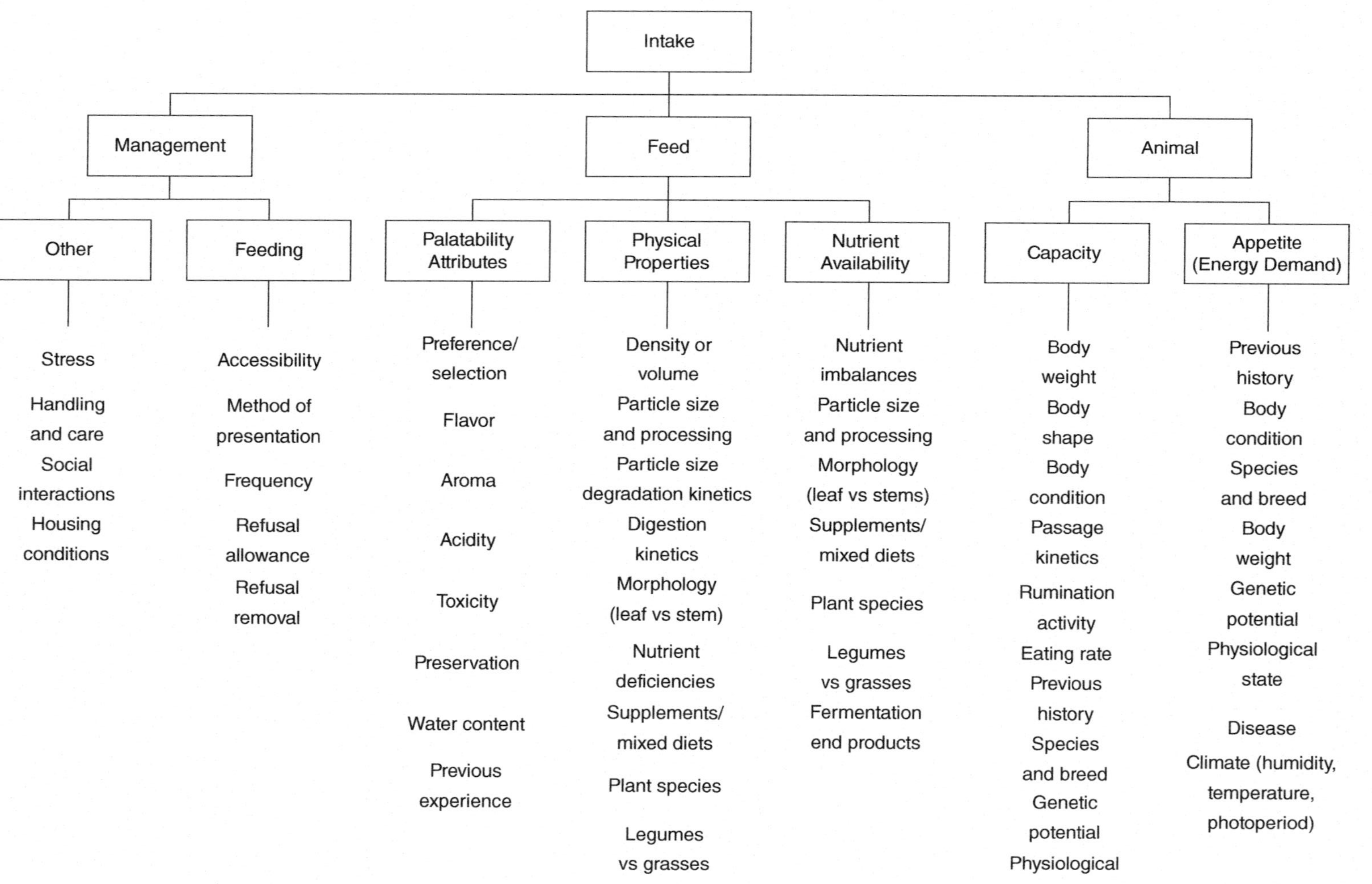

Fig. 34.4. Factors affecting voluntary feed intake. (Mertens 1994).

production and survival when extremely bulky diets are fed, and energy demand is high. Though the animal can increase its fill-processing capacity, maximizing it stresses the animal in its attempt to meet its energy demands when a low-quality diet is fed.

Combining Intake Regulation Mechanisms

Equations (34.20) and (34.21) indicate that intake is a linear function of animal characteristics (energy demand or fill-processing capacity). However, both equations also indicate that intake is a reciprocal function of feed characteristics (energy or filling effect), which suggests a curvilinear relationship between intake and forage or dietary characteristics. Given that energy availability and filling effects are inversely related, Eqs. (34.20) and (34.21) result in two curvilinear lines that intersect (Figure 34.5). Assuming that NDF is a feed characteristic that is directly related to the filling effect of the diet and inversely related to energy availability, it can be used to describe the intake mechanisms on a common scale (Figure 34.6).

The intersection of energy demand and filling effect curves describes a unique occurrence in which the animal both maximizes fill processing capacity and meets its target energy demand (point *g* in Figure 34.6). Because the two lines intersect curving upward, the intercept will always result in the maximum DM intake. Mertens (1987, 1994) suggested that this intersection defines the fiber characteristics of optimal rations that both maximize forage intake and meet target levels of production. However, the upper limit for fill-processing capacity must allow the animal to meet its maximum production potential without stress. Mertens (1987) observed that NDF intake of 12 g kg^{-1} BW d^{-1} is the optimal value for fill-processing capacity in most feeding situations. Animals will eat more NDF than this when fed high-fiber diets, but this will also reduce performance below their maximum potential.

The physical and physiologic mechanisms of intake control provide limits at the opposite extremes of forage quality. When high-energy, low-fill diets or forages are fed, physiologic energy demand regulates intake, whereas when high-fill, low-energy diets or forages are offered, physical fill limits intake. With the exception of the intersection point, there are two solutions for intake (I_e and I_f) for each NDF concentration (e.g. points *i* and *j* in Figure 34.6 for forages containing 550 g NDF kg^{-1} DM). When fed a diet or forage with 550 g NDF kg^{-1} DM, a cow with low-energy demand (15 kg of milk d^{-1}) would need intake at point *i* to meet energy demands, but would be limited by fill at intake point *j* (Figure 34.6). Though the animal may make compromises between C and R to arrive at an actual intake between points *i* and *j*, the simplest mathematical description of these mechanisms is that the lesser of the two intake limits will define the predicted intake for a specified animal-feed combination. Predicted intake will be defined by the intake mechanism that is most limiting, that is, the minimum of the two mechanisms (e.g. point *j* in Figure 34.6):

$$I_p = \min(I_e, I_f) \qquad (34.22)$$

where I_p is predicted intake and I_e and I_f are defined in Eqs. (34.19) and (34.20), respectively.

This simple mathematical description of intake can be completed by assuming that all other factors (such as

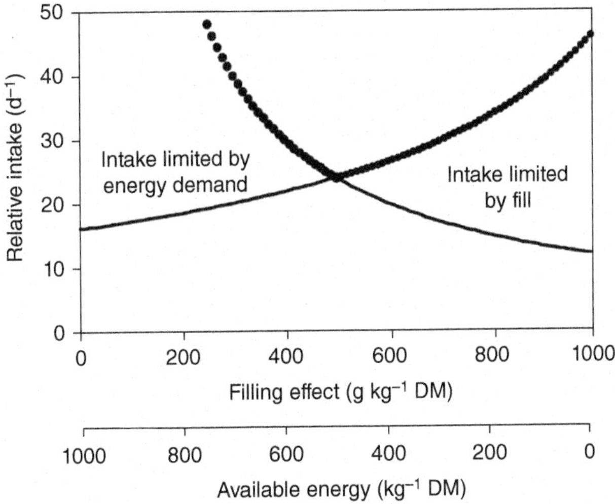

FIG. 34.5. Graphic depiction of simple algebraic descriptions of the physical and physiologic mechanisms of intake regulation.

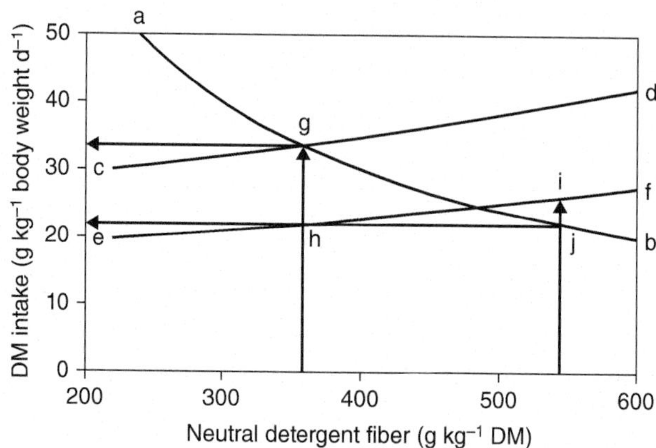

Fig. 34.6. Illustration of the consequences of the physical and physiologic mechanisms of intake regulation based on NDF. Line *a–b* represents fill-limited intake assuming a fill-processing constraint of 12 g kg^{-1} body weight d^{-1}. Line *c–d* represents the energy demand of a dairy cow producing 30 kg d^{-1} of milk. Line *e–f* represents a cow producing 15 kg d^{-1} of milk. Points *g* and *h* indicate the expected intakes of cows producing 30 or 15 kg of milk daily, respectively, when fed a diet containing 360 g of NDF kg^{-1} DM. Intake *h* does not represent the intake potential of the diet because it is limited by the cow's energy demand. Points *i* and *j* indicate that intake of the diet containing 550 g of NDF kg^{-1} DM is limited by fill. Points *h* and *j* illustrate how a low producing cow can have the same intake for a low and high fiber forage because the mechanisms of intake regulation differ.

feeding situation, animal interactions, feed **palatability**, etc.) have a multiplicative effect on predicted intake. Then actual intake can be defined as

$$I_a = I_p \times M \tag{34.23}$$

where M is the multiplier associated with factors affecting intake that are not related to physical or physiologic regulation.

Equations (34.20)–(34.23) quantify the most commonly accepted theories of intake regulation in ruminants. This simple framework of intake regulation (Mertens 1994) suggests that

1. an animal's basic drive to eat (appetite) is determined by its genetic potential and physiologic state, which defines its energy demand.
2. when the diet contains adequate concentrations of available energy, protein, vitamins, and minerals, the animal consumes feed at a level that matches its appetite (energy demand), and animal potential is the limit for intake.
3. when diets with inadequate energy value are offered, the animal consumes feed at a level that matches its gut capacity, so that intake is restricted by the filling effect of the diet and the fill-processing constraint of the animal.
4. psychogenic stimuli associated with palatability, social interactions, disease, and feeding management modify the dominant roles of physical limitation and physiologic regulation on intake.

This simple conceptual framework of intake regulation fits many observed intake responses and is useful in identifying factors affecting forage intake responses and describing the conditions under which forage intake potential can be measured accurately.

Not only is the simple framework valuable in developing ration-formulation systems (Mertens 1987), it can also serve as the starting point for more complex models of intake regulation. Fisher (1987) developed a model that mathematically integrates the physiologic and physical mechanisms in a continuous function, and other scientists have proposed more complex qualitative and quantitative models for predicting intake (Baldwin et al. 1977; Pienaar et al. 1980; Black et al. 1981; Bywater 1984; Illius and Gordon 1991, Ketelaars and Tolkamp 1991; Pittroff and Kothmann 1999). The objective in this chapter is not to discuss the details of intake regulation, but to use a general framework of intake regulation to discuss the difficulties in measuring and interpreting forage intake potential.

Measuring the Intake Potential of Forages

Mathematical description of the physical and physiologic mechanisms of intake regulation demonstrates how forages containing 360 or 550 g NDF kg^{-1} DM, which should have different intake potentials, could obtain similar intakes if fed to animals producing 15 kg of milk daily because different mechanisms regulate intake in each instance. Intake *h* for the hay with 360 g NDF kg^{-1} DM is limited by the energy demand of the animal (Figure 34.6), whereas intake *j* for the hay with 550 g NDF kg^{-1} DM is limited by the filling effect of the forage (Figure 34.6). Conversely, it is possible to measure different intakes when feeding the same forage due to differences in the animal's energy demand. When a forage containing 360 g NDF kg^{-1} DM is fed to animals with energy demands for daily production of 15 or 30 kg of milk, two *ad libitum* intakes can be expected (e.g. points *h* and *g*, respectively in Figure 34.6). Intake *h* is limited by the energy demand of the animal whereas intake *g* is limited by the filling effect of the forage. When energy demand of the animal limits the intake *h* of a high-quality forage, the measurement does not represent the intake potential of the forage (which is intake *g* in Figure 34.6). Some scientists suggest that voluntary intake, as it is commonly measured, never determines forage intake potential because it always depends on both the animal and the forage.

Variability in the animal, forage, and feeding circumstance makes measuring intake potential of a forage difficult. However, simple physical and physiologic mechanisms of intake regulation can help to describe the conditions during which intake potential can be measured. Most nutritionists would define **forage intake potential** as the maximum possible intake when forage characteristics, and not those of the animal or feeding circumstance, limit intake. When the filling effect of a forage is high (i.e. NDF concentration is >500 g kg^{-1} DM), its maximum intake is measured because fill (Eq. 34.21) usually is limiting (Figure 34.6). Because the simple mechanisms result in upward curving lines that intersect, the maximum intake for any level of animal energy demand occurs when demand matches the fill-processing constraint. Thus, the best measure of forage intake potential would occur when the forage is fed to an animal that maximizes its filling effect and meets its energy demand (intake *g* for a forage containing 360 g NDF kg^{-1} DM in Figure 34.6). Determining this potential would require an experimental design in which each forage is fed to several groups of animals varying in target energy demand. Animals with the highest intake would define the intake potential. This design is impractical due to the time, feed, and expense involved.

Because the intersection of the two intake mechanisms increases as the energy demand increases, a more practical approach to measuring the intake potential of forages would be to use animals with high-energy demands, such as young growing animals or lactating females. Using this approach, the energy demand of the animals would not limit the intake so the true intake potential of the forage would be measured. When animals with high-energy demand are fed forages with a high filling effect, they compromise by decreasing productivity and increasing fill-processing capacity. Thus, the intake potential of these feeds may be overestimated, but this error is typically much less than the error of underestimating the intake of high-quality forages when animals with low-energy demands are used. In effect, the intake potential of a forage can be measured only when the filling effect of the forage and not the energy demand of the animal limits intake. It is also evident that all extraneous sources of variation should be minimized so only the effects of the forage on intake are measured.

Factors Affecting Intake

Though the true intake potential of a forage may not be measured, *ad libitum* intake is often measured as an indicator of forage quality. Many factors can influence the measurement of intake and must be taken into account when interpreting results, especially when comparing forage intakes among experiments.

Animal Characteristics Affecting Intake

Differences in intake among individual animals fed the same forage introduces significant variation to the measurement of intake both within and among experiments. The differences in intakes among forages can be determined more accurately when variation due to animals is removed. For example, intake generally increases with the size of the animal, and intake is typically expressed as a proportion of the animal's body weight or metabolic body weight to adjust for differences among animals. Mertens (1994, Fig. 7) demonstrated the effects of expressing forage intake as a function of BW or BW$^{0.75}$ and suggested that the intake potential of forages, which is constrained by daily fill-processing capacity, should be expressed in terms of BW to adjust for differences among animals. However, differences in size do not account for all the differences in intake among animals. Osbourn et al. (1974) demonstrated that significant variation in intake among forages was removed by feeding animals a reference forage and using the intake response to it as a covariate to account for individual animal variation. Intake measurements were made using different animals in each of the two years. When forage intakes were adjusted for animal differences measured by the intake of the reference forage, the variations in intake at all concentrations of NDF were greatly diminished, but the effect was largest for low-fiber forages in which animal energy demand probably limited intake. In addition, the differences between years were removed (Figure 34.7). Additional animal variation may have been removed by expressing intake as a function of BW instead of metabolic body weight. Abrams et al. (1987) also observed intake differences among forages

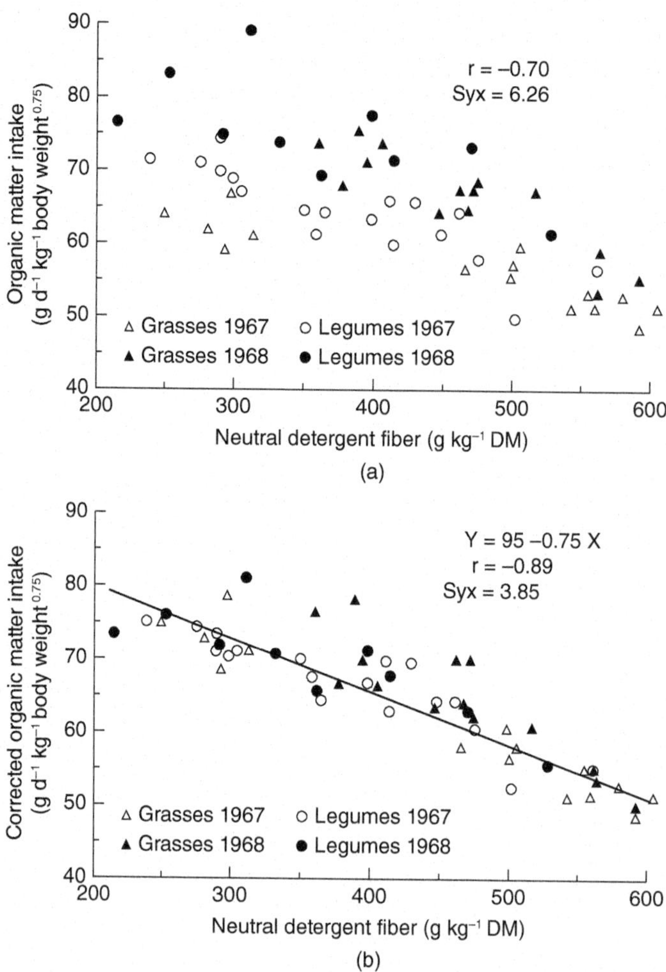

Fig. 34.7. Correcting intake for differences among animals using a reference hay *(b)* improves the relationship between intake and NDF *(a)*. *Source:* Adapted from Osbourn et al. (1974).

were measured more accurately when animal effects on intake were removed.

Alternatively, variation in intake due to animals can be reduced by using a reference animal to measure intake. This approach was used to measure ingestibility (i.e. intake response) in the French system of feed evaluation (INRA 1989). Ingestibility was determined by feeding all test forages to similar reference animals, thereby minimizing variation associated with animal differences. French fill units (FU) are expressed as reciprocals of measured intake, though the term is a misnomer because FU include factors, such as palatability and energy concentration, other than fill that influence intake. The linear relationship between FU and NDF for forages observed by Mertens (1994) suggests that, in general, the FU system is related to the filling effect of the forage.

Though using reference animals or covariate intakes using a reference forage can reduce animal variation, they do not guarantee that the intake potential of forages is measured unless the reference animals are selected to have high-energy demands that would not limit forage intake. Similarly, the reference feed used to measure a covariate intake should be high in energy to measure differences in energy demand.

Dietary or Forage Factors

Palatability is defined broadly as any characteristic of a feed affecting its acceptability, usually associated with the gustatory, olfactory, or visual senses. Palatability affects the preference for a feed when several are available (Marten 1969; Black et al. 1989) and also the rate of eating and intake when a single feed is offered (Baumont et al. 2000). Preference, which is defined as the relative acceptability of a feed when animals are given the choice among two or more feeds that are available in a cafeteria-style feeding situation, is a more specific intake attribute than palatability that may not indicate a change in intake when a feed is fed alone.

Equation (34.23) attempts to define palatability as a multiplicative factor that is independent of the forage's available energy or filling effect. Mertens (1994) postulated that the differences in intake among feeds with the same NDF concentration could be attributed to differences in palatability. Palatability was calculated as the multiplier needed for various types of forage to adjust for deviations in intake from the linear relationship between NDF and FU reported by INRA (1989). Compared with fresh forage, the multiplicative factors were 0.92 for barn-dried hay and direct-cut, finely chopped silage with additives; 0.88 for field-dried hay and direct-cut, finely chopped silages without additive; 0.86 for wilted, finely chopped silages; 0.84 for slightly rain-damaged hays; 0.82 for extensively rain-damaged hays; 0.79 for direct-cut, medium-chopped silages with additives; 0.57 for direct-cut, flail-chopped silages with additives; and 0.55 for direct-cut, flail-chopped silages without additives. These coefficients provide a quantitative estimate of palatability that can be used to evaluate the intake potential of forages for sheep with low energy demands. These palatability estimates may be closer to 1.00 for lactating cows that have high-energy demands and express less selectivity.

Feeding Factors

To measure *ad libitum* intake, animals must be offered more feed than they will consume. The degree of excess feeding is often described as the proportion of the feed offered that the animal refuses to consume. Unfortunately, allowing some forage to be refused also allows the animal to sort and selectively consume what is offered. Some researchers limit the refusal to a constant amount per day, but this may allow greater selection of low-quality forages because they have lower intakes and the proportion of refusal is higher. It is more typical to limit refusals to 50–100 g kg^{-1} of feed offered. In general, animals allowed higher levels of refusal tend to eat more because they can selectively consume the feed. Zemmelink (1980) concluded that variation in refusal levels within and among trials results in measurements of intake that are not comparable. He observed that the ranking of intakes of tropical forages varied with the level of refusal. The impact of refusal level on intake suggests that it should be evaluated and reported for each forage or treatment with the results of the intake experiment. The heterogeneity of the forage (leaves versus stems or small vs large fragments) increases the animal's ability to select and the relative importance of refusal level. If heterogeneous materials are fed in excess, the animal typically selects the less-fibrous portions. Thus, the composition of the forage eaten does not match the forage offered, and any relationship of intake or digestibility to the composition of the forage offered may be biased.

Predicting Intake Potential Using Forage Characteristics

Equation (34.22) indicates that only one of the mechanisms of intake regulation functions for any concentration of NDF and energy demand of the animal (except at the intersection). Thus, the overall relationship between NDF concentration and intake is a combined discontinuous function of physiologic and physical regulation (e.g. Figure 34.6, line connecting *c*, *g*, and *b*). Therefore, the relationship is not linear and linear regression is not a valid approach for predicting actual forage intake over a range in forage quality in which both energy demand and fill-processing capacity limit intake. Depending on how intake was measured, Figure 34.6 suggests that the relationship of NDF and intake can be positive (when NDF ranges from 200 to 360 g kg^{-1} DM) to zero (over the full range of NDF) to negative (when NDF ranges from 360 to 600 g kg^{-1} DM).

Given the reciprocal nature of the relationship, intake potential (measured by animals with high-energy demands) would be most accurately related to the reciprocal of feed characteristics, for example, 1/NDF. Though intake potential is the best description of the forage characteristic, predicting actual intake in a given forage-animal-feeding situation would require the use of comprehensive models of intake regulation. More sophisticated chemical analyses or new technologies such as near infrared reflectance spectroscopy, which measure only forage properties, cannot address the basic problem in predicting intake – it depends as much on the animal and feeding situation, which are not measured, as it does on feed characteristics, which are measured. The complex nature of intake regulation and the fact that it results from the interaction of the animal with the forage and feeding circumstance precludes the prediction of actual intake based solely on forage characteristics.

Animal Management Environment and Chewing Response

As dietary NDF content increases, animals will typically spend more time eating, have longer meal length, and practice greater sorting behavior (Beauchemin 1991). In contrast, as NDF digestibility increases, chewing time per unit of NDF often decreases. The chewing index, expressed as minutes of chewing elicited per kilogram of DM, ordinarily decreases as forage NDF digestibility increases, the forage particle length is shortened, or NDF content decreases (Jensen et al. 2016). Chewing response is governed by physical as well as chemical attributes of the forages. Important physical properties include particle size and dimensions, and fragility and rate of particle breakdown when chewed. Chemical properties include moisture content, which aids in swallowing, lower NDF content, and lower lignin concentration and crosslinking which are associated with more effective mastication.

Italian researchers have focused on the chewing and eating process in dairy cows and observed that cattle tend to chew forages while eating just enough to swallow the **bolus** (Schadt et al. 2012). Generally, larger feed particles were chewed to a threshold size that was suitable for bolus formation and deglutition. When measured, using a combination of wet sieving and image analysis, these researchers found that the swallowed bolus particle size was approximately 10–11 mm. Though the offered forages varied from 9.7 to 43.5 mm in size, the bolus mean size was quite similar. The feeds offered included ryegrass hay of various lengths, grass silage, corn silage, and a total mixed ration.

Feeding long-cut silage or dry hay particles to dairy cows does not necessarily boost particle size in the rumen beyond the size of the swallowed bolus of feed. Rather, the particle size of the swallowed bolus is directly related to forage factors such as lignin, NDF, and moisture content (Rinne et al. 2002; Schadt et al. 2012). Importantly, forages that are higher in NDF concentration and(or) have longer particle size effectively lengthen the time required to consume feed. In the study by Schadt et al. (2012), the chews per gram of NDF varied from 0.4 to 3.5. Longer particle size may lengthen the time needed to consume a meal. Depending on feed bunk management and the resulting level of competition for feed, having too great a particle size of the forage may be disadvantageous for high-producing cows.

Summary

With few exceptions, intake and digestibility are the prime determinants of forage quality. Intake determines the input of nutrients for the animal's use, and digestibility defines how much of the input nutrient is absorbed. The fiber portion of forages is most difficult to digest and ultimately determines the extent of DMD. Fiber is predominantly plant cell wall material that consists of complex polysaccharides and lignin that cannot be digested by mammalian enzymes. Thus, herbivores have developed a symbiotic relationship with anaerobic microorganisms in the GI tract to ferment fiber and produce volatile fatty acids and microbial cells that they can digest.

Because fiber limits digestibility, summative equations that partition feeds into fiber and nonfiber components can be used to predict digestibility of most forages from chemical composition. At maintenance levels of intake, the digestibility of most nonfiber constituents of forages is constant and nearly complete. However, the digestibility of fiber is variable, and its digestibility must be measured or predicted from lignin concentration or *in vitro* or *in situ* digestibility to predict DM digestibility. Though summative predictions are useful, they are static estimates of digestibility, which do not account for variability due to interactions among rates of digestion and passage that occur at productive levels of animal performance when mixed forage and concentrate rations are fed. For this reason, digestion kinetics of fiber will become the basis for forage evaluation in the future. Digestion kinetics, used in dynamic models of animal intake and digestion, will become the basis for evaluating forage and formulating rations that optimize forage utilization.

Intake has the greatest impact on forage quality because it is responsible for 60–90% of the variation in digested DM intake. However, voluntary intake, as it is commonly measured, has limited value in describing forage quality because it is affected not only by the forage, but also by the animal and feeding circumstance. Simple algebraic equations can be used to describe the physical and physiologic mechanisms of intake regulation that provide insights into the factors affecting intake that are important in forage evaluation. For forage evaluation, the estimate of importance is the intake potential of the forage, which is the maximum intake not limited by the energy demand of the animal. Though it can never be certain that an observed voluntary intake is an estimate of the intake potential of a forage, the use of reference animals with high-energy demands and covariate intakes of a reference forage can improve our ability to estimate intake potential.

List of Abbreviations

ADL—Acid detergent lignin (72% sulfuric acid method), g kg^{-1} DM
BW—Body weight
CP—Crude protein
dCP—Digested crude protein, g kg^{-1} DM
DM—Dry matter
DMD—Dry matter digestibility, kg DM kg DM^{-1}
dNDF—Digested neutral detergent fiber, g kg^{-1} DM
dNDS—Digested neutral detergent solubles, g kg^{-1} DM
dNutr—Generic digested nutrient, g kg^{-1} DM
EFL—Endogenous fecal losses
iNDF—Indigestible neutral detergent fiber, g kg^{-1} DM
IVDMD—*In vitro* dry matter digestibility measured by the two-stage Tilley and Terry technique
IVDMTD—*In vitro* dry matter true digestibility measured by the two-stage Van Soest technique
MBW—Metabolic body weight
NDF—Generic neutral detergent fiber determined by a variety of methods
NDFD—Neutral detergent fiber digestibility, kg kg^{-1} NDF
NDS—Neutral detergent solubles = (1000 − NDF)

NutrD—Generic nutrient digestibility, kg Nutr kg $Nutr^{-1}$
pdNDF—Potentially digestible neutral detergent fiber, g kg^{-1} DM
T&T—Tilley and Terry *in vitro* technique
TDN—Total digestible nutrients, g kg^{-1} DM
VFA—Volatile fatty acids, primarily acetic, propionic, and butyric

References

Abrams, S.M., Harpster, H.W., Wangness, P.J. et al. (1987). Use of a standard forage to reduce effects of animal variation on estimates of mean voluntary intake. *J. Dairy Sci.* 70: 1235–1240.

Balch, C.C. and Campling, R.C. (1962). Regulation of voluntary intake in ruminants. *Nutr. Abstr. Rev.* 32: 669–686.

Baldwin, R.L., Koong, L.J., and Ulyatt, M.J. (1977). A dynamic model of ruminant digestion for evaluation of factors affecting nutritive value. *Agric. Syst.* 2: 255–288.

Baumgardt, B.R. (1970). Control of feed intake in the regulation of energy balance. In: *Physiology of Digestion and Metabolism in the Ruminant* (ed. A.T. Philipson), 235–253. England: Oriel Press Ltd. Newcastle upon Tyne.

Baumont, R., Prache, S., Meuret, M., and Morand-Fehr, P. (2000). How forage characteristics influence behaviour and intake in small ruminants: a review. *Livestock Prod. Sci.* 64: 15–28.

Beauchemin, K.A. (1991). Ingestion and mastication of feed by dairy cattle. *Vet. Clin. North Am. Food Anim. Pract.* 7: 439–463.

Bines, J.A. (1971). Metabolic and physical control of food intake in ruminants. *Proc. Nutr. Soc.* 30: 116–122.

Black, J.L., Beever, D.E., Faichney, G.J. et al. (1981). Simulation of the effect of rumen function on the flow of nutrients from the stomach of sheep: part 1—description of a computer program. *Agric. Syst.* 6: 195–219.

Black, J.L., Colebrook, W.F., Gherardi, S.G. et al. (1989). Diet selection and the effect of palatability on voluntary feed intake by sheep. pp. 139–151. In: Proceedings of the 50th Minnesota Nutrition Conference, Minnesota Agriculture Extension Services, St. Paul, Minnesota, USA (19–20 September 1989).

Brody, S. (1945). *Bioenergetics and Growth*. New York: Hafner Publishing Co., Inc.

Bywater, A.C. (1984). A generalized model of feed intake and digestion in lactating cows. *Agric. Syst.* 13: 167–186.

Campling, R.C. (1970). Physical regulation of voluntary intake. In: *Physiology of Digestion and Metabolism in the Ruminant* (ed. A.T. Philipson), 226–234. Newcastle upon Tyne, England: Oriel Press Ltd.

Conrad, H.R. (1966). Symposium on factors influencing the voluntary intake of herbage by ruminants: physiological and physical factors limiting intake. *J. Anim. Sci.* 25: 227–235.

Conrad, H.R., Weiss, W.P., Odwongon, W.O., and Shockey, W.L. (1984). Estimating net energy of lactation from components of cell solubles and cell walls. *J. Dairy Sci.* 67: 427–436.

Crampton, E.W., Donefer, E., and Lloyd, L.E. (1960). A nutritive value index for forages. *J. Anim. Sci.* 19: 538–544.

Fisher, D.S., Burns, J.C., and Pond, K.R. (1987). Modelling ad libitum dry matter intake by ruminants as regulated by distension and chemostatic feedbacks. *J. Theor. Biol.* 126: 407–418.

Forbes, J.M. (1995). *Voluntary Food Intake and Diet Selection in Farm Animals*. Wallingford, UK: CAB International.

Goering, H.K. and Van Soest, P.J. (1970). *Forage Fiber Analyses*. USDA Agric. Handbook No. 379. Washington, DC: US Government Printing Office.

Huhtanen, P. (1991). Associative effects of feeds in ruminants. *Norw. J. Agric. Sci.* 5: 37–57.

Illius, A.W. and Gordon, I.J. (1991). Prediction of intake and digestion in ruminants by a model of rumen kinetics integrating animal size and plant characteristics. *J. Agric. Sci. Camb.* 116: 145–157.

INRA (Institut National de la Recherche Agronomique) (1989). *Ruminant Nutrition: Recommended Allowances and Feed Tables* (ed. R. Jarrige). Paris: John Libbey Eurotext.

Jensen, L.M., Markussen, B., Nielsen, N.I. et al. (2016). Description and evaluation of a net energy intake model as a function of dietary chewing index. *J. Dairy Sci.* 99: 8699–8715.

Jones, G.M. (1972). Chemical factors and their relation to feed intake regulation in ruminants: a review. *Can. J. Anim. Sci.* 52: 207–239.

Journet, M. and Remond, B. (1976). Physiological factors affecting the voluntary intake of feed by cows: a review. *Livestock Prod. Sci.* 3: 129–146.

Kammes, K.L. and Allen, M.S. (2012). Rates of particle size reduction and passage are faster for legume compared with cool-season grass, resulting in lower rumen fill and less effective fiber. *J. Dairy Sci.* 95: 2011–2012.

Ketelaars, J.J.M.H. and Tolkamp, B.J. (1991). Toward a new theory of feed intake regulation in ruminants. Ph.D. dissertation. Wageningen University & Research.

Kleiber, M. (1961). *The Fire of Life. An Introduction to Animal Energetics*. New York: Wiley.

Marten, G.C. (1969). Measurement and significance of forage palatability. In: *Proc. National Conference on Forage Quality Evaluation and Utilization* (eds. R.F

Barnes, D.C. Clanton, C.H. Gordon, et al.), D1–D55. Lincoln: Nebraska Center for Continuing Education.

Mertens, D.R. (1973). Application of theoretical mathematical models to cell wall digestion and forage intake in ruminants. Ph.D. dissertation. Cornell University.

Mertens, D.R. (1977). Dietary fiber components: relationship to the rate and extent of ruminal digestion. *Federation Proceedings* 36 (2): 187–192.

Mertens, D.R. (1987). Predicting intake and digestibility using mathematical models of ruminal function. *J. Anim. Sci.* 64: 1548–1558.

Mertens, D.R. (1993). Rate and extent of digestion. In: *Quantitative Aspects of Ruminant Digestion and Metabolism* (eds. J.M. Forbes and J. France), 13–51. Wallingford, UK: CAB International.

Mertens, D.R. (1994). Regulation of forage intake. In: *Forage Quality, Evaluation, and Utilization* (ed. G.C. Fahey Jr.), 450–493. Madison, WI: American Society of Agronomy.

Mertens, D.R. (2002). Gravimetric determination of amylase-treated neutral detergent fiber in feeds with refluxing in beakers or crucibles: collaborative study. *J. AOAC Int.* 85: 1217–1240.

Mertens, D.R. (2016). Using uNDF to predict dairy cow performance and design rations. pp. 12–19 Proc. Four-State Dairy Nutrition and Management Conf. June 15 & 16, 2016. Dubuque, IA.

Mertens, D.R. and Ely, L.O. (1982). Relationship of rate and extent of digestion to forage utilization. *J. Anim. Sci.* 54: 895–905.

Mertens, D.R. and Loften, J.R. (1980). The effect of starch on forage fiber digestion and kinetics in vitro. *J. Dairy Sci.* 63: 1437–1446.

Minson, D.J. (1990). *Forage in Ruminant Nutrition*. San Diego, CA: Academic Press, Inc.

NRC (National Research Council) (1987). *Predicting Feed Intake of Food-Producing Animals*. Washington, DC: National Academy Press.

NRC (National Research Council) (2001). *Nutrient Requirements of Dairy Cattle*. 7th rev. ed. Washington, DC: National Academy Press.

Osbourn, D.F., Terry, R.A., Outen, G.E., and Cammell, S.B. (1974). The significance of a determination of cell walls as the rational basis for the nutritive evaluation of forages. In: *Proc. XII Int. Grassl. Congr.*, 11–20 June, 1974, Moscow, vol. 3, 374–380. Moscow: Izda-telbstvo.

Parra, R. (1978). Comparison of foregut and hindgut fermentation in herbivores. In: *The Ecology of Arboreal Folivores* (ed. G.G. Montgomery), 205–230. Washington, DC: Smithsonian Institution.

Pienaar, J.P., Roux, C.Z., Morgan, P.J.K., and Grattarola, L. (1980). Predicting voluntary intake of medium quality roughages. *S. Afr. J. Anim. Sci.* 10: 215–225.

Pittroff, W. and Kothmann, M.M. (1999). Nutritional ecology of herbivores. Proceedings of the 5th International Symposium on the Nutrition of Herbivores. American Society of Animal Science. Savoy, Illinois, USA.

Poppi, D.P., Minson, D.J., and Ternouth, J.H. (1981). Studies of cattle and sheep eating leaf and stem fractions of grasses. III. The retention time in the rumen of large feed particles. *Aust. J. Agric. Res.* 32: 123–137.

Raffrenato, E. and Van Amburgh, M.E. (2010). Development of a mathematical model to predict sizes and rates of digestion of a fast and slow degrading pool and the indigestible NDF fraction. In: *Proc. Cornell Nutr. Conf. For Feed Manu*, 52–65. East Syracuse, NY.

Reid, J.T. (1961). Problems of feed evaluation related to feeding dairy cows. *J. Dairy Sci.* 11: 2122–2133.

Rinne, M., Huhtanen, P., and Jaakkola, S. (2002). Digestive processes of dairy cows fed silages harvested at four stages of grass maturity. *J. Anim. Sci.* 80: 1986–1998.

Sanderson, M.A., Hornstein, J.S., and Wedin, W.F. (1989). Alfalfa morphological stage and its relations to in situ digestibility of detergent fiber fractions of stems. *Crop Sci.* 29: 1315–1319.

Schadt, I., Ferguson, J.D., Azzaro, G. et al. (2012). How do dairy cows chew? –particle size analysis of selected feeds with different particle length distributions and of respective ingested bolus particles. *J. Dairy Sci.* 95: 4707–5118.

Smith, L.W., Goering, H.K., and Gordon, C.H. (1972). Relationships of forage compositions with rates of cell wall digestion and indigestibility of cell walls. *J. Dairy Sci.* 55: 1140–1147.

Thornton, R.F. and Minson, D.J. (1973). The relationship between apparent retention time in the rumen, voluntary intake, and apparent digestibility of legume and grass diets in sheep. *Aust. J. Agric. Res.* 24: 889–898.

Tilley, J.M.A. and Terry, R.A. (1963). A two-stage technique for the in vitro digestion of forage crops. *J. Brit. Grassl. Soc.* 18: 104–111.

Traxler, M.J., Fox, D.G., Van Soest, P.J. et al. (1998). Predicting forage indigestible NDF from lignin concentration. *J. Anim. Sci.* 76: 1469–1480.

Tyrrell, H.F. and Moe, P.W. (1975). Symposium on production efficiency in the high producing cow. Effect of intake on digestive efficiency. *J. Dairy Sci.* 58: 1151–1163.

Van Niekerk, B.D.H, Smith, D.W.W.Q. and Oost-huysen, D. (1967). The relationship between crude protein content of South African feeds and its apparent digestion by ruminants. Proceedings of The South African Society of Animal Production.

Van Soest, P.J. (1967). Development of a comprehensive system of feed analyses and its application to forages. *J. Anim. Sci.* 26: 119–128.

Van Soest, P.J. (1994). *Nutritional Ecology of the Ruminant*, 2e. Cornell University Press.

Van Soest, P.J. and Wine, R.H. (1967). The use of detergents in analysis of fibrous feeds: IV. Determination of plant cell-wall constituents. *J. AOAC* 50: 50–55.

Van Soest, P.J., Wine, R.H., and Moore, L.A. (1966). Estimation of the true digestibility of forages by the in vitro digestion of cell walls. In: Proc. 10th Int. Grassl. Congr. Section 2, 438–441.

Waldo, D.R. (1969). Factors influencing the voluntary intake of forages. In: *Proc. National Conference on Forage Quality Evaluation and Utilization* (eds. R.F Barnes, D.C. Clanton, C.H. Gordon, et al.), E1–E22. Lincoln: Nebraska Center for Continuing Education.

Waldo, D.R., Smith, L.W., and Cox, E.L. (1972). Model of cellulose disappearance from the rumen. *J. Dairy Sci.* 55: 125–129.

Weiss, W.P., Conrad, H.R., and Pierre, N.R.S. (1992). A theoretically based model for predicting total digestible nutrient values of forages and concentrates. *Anim. Feed Sci. Technol.* 39: 95–110.

Weston, R.H. (1985). The regulation of feed intake in herbage-fed ruminants. *Proc. Nutr. Soc. Aust.* 10: 55–62.

Zemmelink, G. (1980). Effect of selective consumption on voluntary intake and digestibility of tropical forages. Agricultural Research Report. The Netherlands.

CHAPTER

35

Plant Chemistry and Antiquality Components in Forage

Nicholas S. Hill, Professor, *Crop and Soil Sciences, The University of Georgia, Athens, GA, USA*
Craig A. Roberts, Professor, *Plant Sciences, University of Missouri, Columbia, MO, USA*

Introduction

Why should this book include a chapter on the chemistry of plant compounds that can lower forage intake, digestion, and animal performance? It is the "**antiquality**" **constituents** that frequently define management and utilization of forages. This chapter discusses significant organic components of forages worldwide that impact selection of the plants for forage management systems and in animal utilization and animal production. Included in the discussion are the tannins and phytoestrogens in many legumes; cyanogenic compounds in white clover, sorghum, and related grasses; alkaloids in a host of forages that affect intake, digestion, and animal health; and two nonphysiologic amino acids, **mimosine** and *S*-methyl-L-cysteine sulfoxide (SMCSO). Physiologic amino acids and **protein quality** are discussed in Chapter 47, especially as they relate to **bloat**. The discussion addresses these antiquality components from the standpoint of their chemistry, genetics, and impact on the environment, forage management decisions, and animal responses.

Polyphenols

Chemistry and Forumulae

By modern definition, **tannins** are polyphenolic compounds capable of precipitating proteins. The word *tannin* originated in the late eighteenth century to describe plant compounds that are used in tanning animal hides. Tannins are secondary products of the shikimic acid pathway. They are commonly grouped into two categories. The first, hydrolyzable tannins, consists of a phenolic acid, such as gallic acid, and a hexose, such as glucose (Figure 35.1). **Condensed tannins** are much larger than hydrolyzable tannins and have molecular weights ranging from 1 to 20 kDa (Min and Hart 2003). Condensed tannins are the most relevant to forage quality and livestock performance.

Tannins and polyphenols occur in leaves, stems, roots, and flowers of many forages. They are especially prevalent in forage legumes. As shown in Table 35.1, condensed tannins are present in many tissues of some legumes, yet in only select tissues of other legumes, such as alfalfa. Their presence in the alfalfa seed coat, though absent in the other tissues, indicates a mechanism to produce tannin that may be down-regulated as a function of plant development.

Role and Impact in Plants

The role of tannins and polyphenols relates to interaction with the biotic environment. Anthocyanidins add pigments to flower petals, attracting insects to assist in pollination.

Forages: The Science of Grassland Agriculture, Volume II, Seventh Edition.
Edited by Kenneth J. Moore, Michael Collins, C. Jerry Nelson and Daren D. Redfearn.

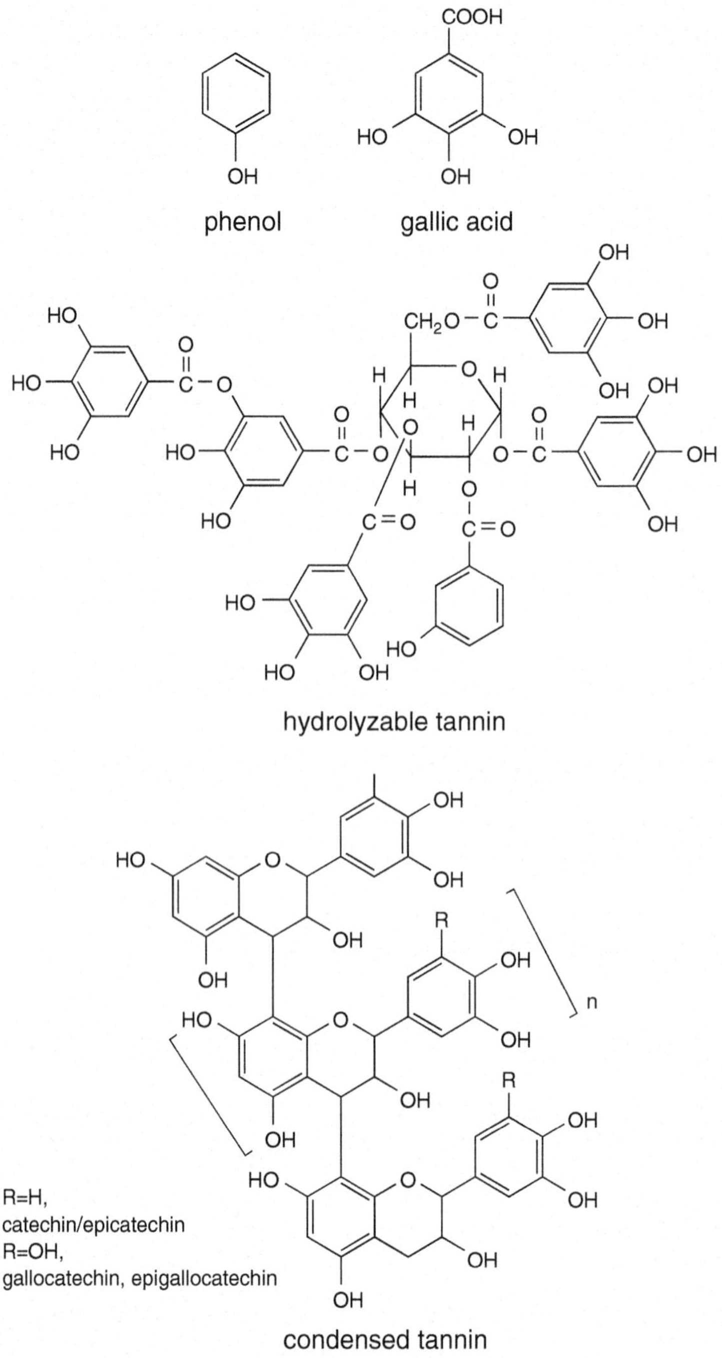

phenol

gallic acid

hydrolyzable tannin

R=H,
catechin/epicatechin
R=OH,
gallocatechin, epigallocatechin

condensed tannin

Fig. 35.1. Hydrolyzable and condensed tannins of plants.

Table 35.1 Chemical characteristics of condensed tannins found in various forage legumes

Legume	Concentration ($g\,kg^{-1}$ DM)[a]	MW (kDa)[a]	Procyanin: Prodelphinidin[a]	Protein binding[b]	
				Plant	Enzymes
Birdsfoot trefoil	48	1.8–2.1	67 : 30	+	−
Big trefoil	77	2.1–3.1	19 : 64	+	+
Sainfoin	29–38	1.6–3.3	81 : 19	+++	−
Alfalfa (seed only)	0.5	—	90 : 10	−	−
Sericea lespedeza	46	14–20	20 : 80	+	+++

[a]Adapted from Min and Hart (2003).
[b]Adapted from Petersen and Hill (1991), McMahon et al. (2000), Frutos et al. (2004), McAllister et al. (2005).
Note: Tannin affinity to plant proteins and ruminal enzymes indicated by + and −.

Condensed tannins and some other polyphenols play a more defensive role by preventing browsing and increasing bird resistance. They also demonstrate antimicrobial activity and are associated with disease resistance.

Genetics

Condensed tannin production is controlled by a single, dominant gene (Dalrymple et al. 1984). In an extensive survey of forage legumes (Goplen et al. 1980), tannins were not present in populations of diverse genotypes of alfalfa, including 28 perennial *Medicago* spp., both diploid and autotetraploid, that had been mutated chemically. Nor were tannins expressed in the leaves of 33 species of annual *Medicago* or in 30 species of *Trigonella*. Crownvetch, sericea lespedeza, birdsfoot trefoil, rabbit foot clover, large hop clover, small hop clover, and many species of sainfoin (*Onobrychis* spp.) have tannins.

In sericea lespedeza, high-tannin cultivars can contain several times the condensed tannin concentration as low-tannin types. Experimental germplasm of birdsfoot trefoil has an extreme range of condensed tannin varying from 0% to 13% DM (Roberts et al. 1993). When crossed with rhizomatous genotypes from Morocco, birdsfoot trefoil not only expresses rhizomes, it may also express twice as much condensed tannin (Wen et al. 2003).

Exotic germplasm varies in condensed tannin concentration by region of origin. Lowest concentrations were found in accessions from Egypt, Spain, Iran, Turkey, and Uzbekistan and highest concentrations in accessions from Ethiopia (Roberts et al. 1993). Only moderate concentrations have been reported in accessions from South America, which is not surprising because South America is not regarded as a center of origin for birdsfoot trefoil.

Concentration as Affected by Environment and Management

In addition to the effect of genetics, concentrations of tannins and phenols are affected by environment and management. Many researchers report seasonal fluctuation of condensed tannins. Tannin concentrations in sericea lespedeza increased greatly under warm temperatures, with greatest increases in high-tannin cultivars (Fales 1984). In birdsfoot trefoil, condensed tannin concentration fluctuates during the spring (Wen et al. 2003) and decreases from summer to autumn (Roberts et al. 1993).

Condensed tannins may increase under low soil fertility. For acid soils in New Zealand, tannins in big trefoil were 2.0–3.2% dry matter (DM) when S and P were adequate, but 5.1–7.8% when S and P were limiting (Barry and Forss 1983). Phenols in pearlmillet respond similarly (Sinha and Chatterjee 1994). Boron deficiency increased total phenol in pearlmillet grain three-fold over plants with moderate boron nutrition, and phenols of pearlmillet grain may be high when soil boron reaches toxic levels. Tannins in birdsfoot trefoil increase when this species is grown in combination with a companion grass (Wen et al. 2003).

Effect on Animals and Ecosystem

Some plant species contain condensed tannins capable of preventing bloat but have little effect on other forage quality parameters (McMahon et al. 2000). Condensed tannins in other species have adverse effects on microbial populations and digestive enzymes which catalyze the necessary reactions for normal ruminal function. The ability of condensed tannins to disrupt the digestive process is dependent upon their ability to attach to proteins. Condensed tannins bind to proteins via hydrogen bonds (Hagerman and Butler 1981). Their affinity to proteins depends upon the degree of polymerization of the monomeric units (i.e. molecular weight), the number of terminal hydroxyl groups on the B ring of the monomeric units in the tannin polymer, the molecular weight and amino acid profile of proteins, the relative abundance of each, and the chemical environment in which they are present (Figure 35.1, Table 35.1). Our ability to provide predictive measures as to how condensed tannins affect forage quality is complicated by the fact that affinity of tannins to proteins is affected by interactions among these variables. Therefore, it is necessary to understand some basic principles of tannin/protein chemistry to develop predictive measures of their chemical and biologic significance.

Tannin polymers contain numerous hydroxyl groups that result in hydrogen bonding with proteins, but the affinity of condensed tannins to proteins is dependent upon the size of the tannin polymer and the ratio of prodelphinidin to procyanin subunits (Hagermann and Butler 1981; Foo et al. 1996, 1997) (Figure 35.1). The high-molecular weight tannins found in sericea lespedeza non-competitively bind to ruminal digestive enzymes (Bell et al. 1965; Petersen and Hill 1991) suggesting that feed additives with greater affinity to tannins could improve forage digestibility. Tannins have particular affinity to proteins rich in proline (Hagerman and Butler 1981; Ropiak et al. 2017). Proline rich proteins and polyalcohols inhibit the effects of tannins on digestive enzymes (Hagerman and Butler 1981) and addition of calfskin gelatin, hemoglobin, or ovalbumin increased total ruminal digestion and rate of digestion of sericea lespedeza *in vitro* (Petersen and Hill 1991). Low-molecular weight tannins found in other forage species tend to be rich in catechin and/or epicatechin and do not react with digestive enzymes (Min and Hart 2003). The low-molecular weight tannins tend to form hydrogen bonds with plant proteins resulting in increased ruminal protein bypass, a measure of ruminal nitrogen efficiency (see Chapter 45). Tannins in birdsfoot trefoil have low-molecular weights and vary considerably in their procyanin:prodelphinidin ratios (Hedqvist et al. 2000). Forage protein digestibility in birdsfoot trefoil decreases as the prodelphinidin subunits increase.

Generally speaking, high concentrations (above 5% DM) of condensed tannins reduce voluntary feed intake and depress digestion efficiency of ruminants (Barry and Duncan 1984), likely due to low palatability caused by their astringent flavor. Condensed tannin concentrations are correlated with concentrations of other quality and antiquality components in forages. In birdsfoot trefoil, tannin concentrations are negatively correlated with hydrogen cyanide (HCN) (Ross and Jones 1983), **digestibility**, and **crude protein** (Miller and Ehlke 1996) and are positively correlated with lignin. This explains, at least in part, why high-tannin forages have reduced digestibility.

Suggested optimum concentrations for improved animal performance differ because composition of proanthocyanidins differs among forages, as do their affinities for protein, due to variation among analytic methods. Concentrations between 20 and 40 g condensed tannins kg^{-1} DM are ideal for forage crops in general (Aaerts et al. 1999). A smaller range has been suggested for big trefoil (Barry et al. 1986), and a much larger range has been proposed for birdsfoot trefoil (Miller and Ehlke 1994).

Effect on Environment

Condensed tannins, but not some other polyphenols, prevent decomposition of plant litter by preventing nitrification (Baldwin et al. 1983). Additionally, polyphenols and tannins have been shown to produce allelopathic effects.

Solution

In grazing systems, tannin intake can be limited by increasing low-tannin species or cultivars in the pasture. Plant breeders can reduce concentrations of tannins in high-tannin legumes such as big trefoil and sericea lespedeza. Additional efforts to limit the effect of tannin include removal of tannins from feeds and altering microbial activity in the livestock.

Phytoestrogens

Phytoestrogens are another important group of phenolic compounds in forages. Usually the relative estrogenic activity is low, about 10^4 less than diethylstilbestrol. Animal reproductive response is not often observed even though the number of phytoestrogens ingested may be high. Partial disruption of the reproduction process frequently occurs but goes unnoticed at subclinical levels. The most studied of several compounds include the coumestans, isoflavones, and isoflavans (Figure 35.2). These compounds are products of phenylpropanoid-acetate metabolism.

Management

Coumestans are primarily reported in alfalfa and white clover with **coumestrol**, 4′-methoxycoumestrol, and sativol occurring in alfalfa, and coumestrol, repensol, and trifoliol in white clover (Figure 35.2). Concentrations of these isoflavonoid phytoestrogens increased in the presence of foliar diseases. Phytoestrogens were very low or not detected in healthy plants but were detected in plants infected with pathogenic fungi (Wong and Latch 1971). Lesion size and amount of common leaf spot, *Pseudopeziza medicaginis* (Lib.) Sacc., and rust, *Uromyces striatus* Schroet., were positively correlated with coumestrol concentration in alfalfa (Loper et al. 1967). *Phoma medicaginis* Malbr. and Roum. and *Leptosphaeruline trifolii* (Rostr.) Petr. also had increased coumestrol content in green forage, end-of-season dry stems and fruits, and deceased forage yield. Apparently, much of the phytoestrogen coumestans associated with the forage are accumulated in the fungal mycelium and spores associated with the forage and are not directly related to the growth and development of the forage per se.

Biochanin A, daidzein, formononetin, and genistein are the most important isoflavonoid phytoestrogens found in red clover, subterranean clover, and white clover.

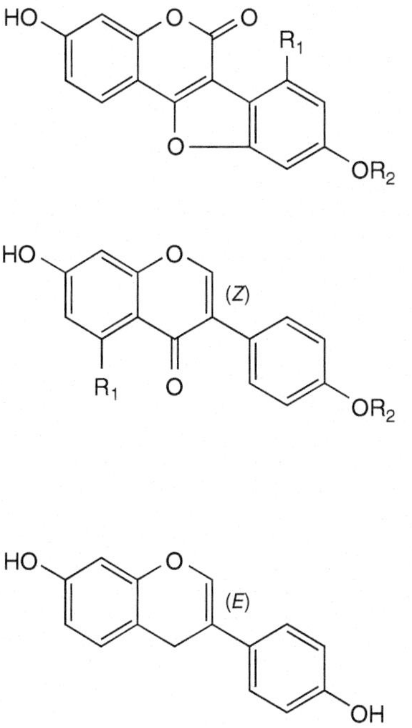

Coumestans	R_1	R_2
coumestrol	H	H
4′-methoxycoumestrol	H	CH_3
repensol	OH	H
trifoliol	OH	CH_3

Isoflavones	R_1	R_2
diaidzein	H	H
formononetin	H	CH_3
genistein	OH	H
biochanin A	OH	CH_3

Isoflavan
Equol

Fig. 35.2. Examples of phytoestrogens.

Owl-headed clover, red clover, and subterranean clover had greatest concentrations of these phytoestrogens of the species evaluated (Vetter 1995). Genistein was very low in all species but subterranean clover. Biochanin A was highest across species tested, with daidzein and formononetin intermediate in concentration. In general, the stem fraction was lowest in phytoestrogen, with the leaf and flower tissues about two-fold higher. McMurray et al. (1986) had found similar differences between leaf and stem tissue of red clover because leaf tissue had higher concentrations of formononetin than petiole or stem tissue. Expanding lamina had the highest concentration of formononetin followed in descending order by expanded lamina, petioles, and stems. Maximum concentration of isoflavones was attained at completion of cell expansion regardless of plant age (Gildersleeve et al. 1991). Estrogenic substances in tuberous roots of kwao krua increased more than three-fold with increased age from 6 to 12 months and varied with geographic location (Tubcharoen et al. 2003), indicating the importance of understanding the forage species and tissue in animal diets when implicating phytoestrogens in animal reproduction maladies.

With increased time to first harvest of red clover, formononetin concentration decreased by up to 40% (Sarelli et al. 2003). Hay making reduced formononetin at least 50% (Kelly et al. 1979), whereas ensiling red clover increased the phytoestrogens initially but reduced them after 180 days of storage.

Physiology

Gildersleeve et al. (1991) ranked subterranean clover seedlings for potential phytoestrogen. Seedlings sampled 42 days after seeding or shoots from 21-days regrowth tissue on plants harvested at 42 days had isoflavone content similar to field-grown plants. These assays have allowed for selection that led to development of cultivars with lowered estrogenic potential. Rumball et al. (1997) selected for seven generations, reducing formononetin level to less than half of the initial level, and increasing ewe ovulation, conception, and lambing rates by a significant amount.

Dear et al. (2003) developed transgenic subterranean clover tolerant to the herbicide bromoxynil. Some constructs were herbicide tolerant with respect to herbage yield, but genistein and biochanin A increased 68% and 106%, respectively. Also, one construct had reduced seed yield and reduced hard seed production, both negative agronomic qualities. These results emphasize the importance of assessing many agronomic qualities prior to release of a new cultivar.

The phytoestrogens discussed here occur mainly as glucosides, and normally the sugar has malonate or methylmalonate hemiesters (Rijke et al. 2001). A glycosidase enzyme present in the leaf is released upon maceration and causes the release of the phytoestrogen from the glycoside. The contribution of genistein and biochanin A to ruminant reproduction problems is not known but is probably very small. Metabolism of biochanin A and genistein produce nonestrogenic breakdown products in the rumen whereas metabolism of formononetin yields mainly equol and lesser amounts of daidzein. Equol is found in the blood and has greater estrogenic activity than formononetin.

Animal Response

Information on negative impacts of phytoestrogens on animal performance has been reviewed extensively by Adams (1995). More current research has focused on soybean products and their potential role in both animal and human diets. Subterranean clover, red clover, and soybean are economic species most studied for their estrogenic activity. Cattle and sheep grazing these species exhibit impaired fertility accompanied by signs of cervical mucus discharge, an enlarged uterus, swollen vulva, cystic ovaries, irregular estrus, and anestrus (Adams 1995). Clover disease, a syndrome found in ewes grazing subterranean clover, results in very low lambing rates, uterine prolapse, and dystocia. Rams are typically unaffected reproductively by phytoestrogens; however, death of wethers grazing subterranean clover has resulted.

Some research has focused on the positive impacts phytoestrogens may have on growth, carcass traits, and meat quality in finishing pigs (Payne et al. 2001a) and commercial broilers (Payne et al. 2001b). Corn plus soy protein concentrate diets with added isoflavones fed to growing finishing pigs increased dressing percentage, lean-to-fat ratios, and ham weight when compared with corn-soy protein concentrate diets, which are low in isoflavones. Similar results were found in pigs consuming a corn-soybean meal diet (C-SBM). However, when isoflavones were increased two and three times the levels found in the base C-SBM diet, no effects on growth performance or carcass traits were observed. The authors of the study suggested that isoflavones decrease fat and increase lean, but not at levels typically found in C-SBM diets, and thus, would not be of practical benefit to producers. Similar trends were observed when broilers consumed similar diets.

Phytoestrogens as Human Nutraceuticals

Hormone replacement therapy (HRT) uses synthetic estrogenic compounds to alleviate estrogenic deficiencies in post-menopausal women. However, there are concerns over the side effects of conventional HRT and the medical community is searching for natural alternatives to synthetic estrogens (Beck et al. 2005). Phytoestrogens have chemical structures and molecular weights similar to 17β-estradiol and behave as selective estrogen receptor modulators – compounds that provide benefits to health but do not have the adverse effects common to synthetic estrogens (Virk-Baker et al. 2010). The molecular mechanisms through which some phytoestrogens have selective estrogenic activity have been elucidated (An et al. 2001), but the term "phytoestrogen" may be a pharmacologic misnomer since they also have androgen and progesterone effects (Beck et al. 2005).

In vitro studies demonstrate that phytoestrogens are potent inhibitors of human bladder, cervical, prostate, breast, and renal cancer cells (Singh et al. 2006; Yashar et al. 2005; Raffoul et al. 2006; Sasamura et al. 2002). Genistein treatments in animal experiments have found mixed results as anti-carcinogen therapies, but there are two human clinical studies in which isoflavone extracts reduced breast cancer recurrence (Shu et al. 2009) and mortality rates (Fink et al. 2007; Shu et al. 2009), and one reported case of reduced progression of prostate cancer (Jarred et al. 2002). A patent was awarded in 2016 for red clover products containing isoflavones for use in treatment of post-menopausal women (US patent 20160000747 A1). Thus, farming legumes for phytoestrogen medicinal purposes may provide opportunities for alternative uses of forages.

Cyanogenic Glucosides

Cyanogenic compounds (cyanogenic glucosides) are a group of nitrogenous compounds found in selected forages, notably white clover, sorghum, and indiangrass. When the plant tissue is disrupted, enzymatic hydrolysis releases HCN, a sugar, and a keto compound (Figure 35.3). Cyanogenesis protects against herbivory, a character of some interest for researchers with the intent of expressing cyanogenic compounds to improve pest and disease resistance. A lethal dose of HCN ranges between 0.5 and 3.5 $mg\,kg^{-1}$ body weight (BW) (Solomonson 1981). Dhurrin levels in sorghum and linamarin and lotaustralin in white clover have been modified through plant breeding. Linamarin is also the principal cyanogenic compound in cassava.

Tissue Accumulation

Cyanogen concentration in white clover and sorghum has been related to many environmental factors, and also responds to plant growth stage. Stochmal and Oleszek (1997) found the mean air temperature four days prior to sampling was negatively correlated with cyanogen content. Plants contained the highest concentration of cyanogen at temperatures below 15 °C in the spring and fall but decreased very significantly as temperatures increased during the summer growing season. Results such as these, raise the question of function of cyanogenic glucosides in plants. Possibly with slow plant growth

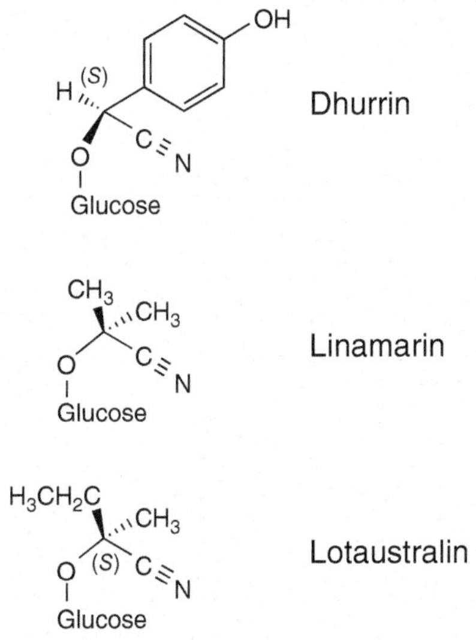

Fig. 35.3. Cyanogenic glucosides dhurrin, linamarin, and lotaustralin.

under low temperatures, there may be more differentiation, cyanogenic glucoside production, and accumulation in order to "defend" against herbivores. As temperatures rise and growth rate increases, herbivory would be less as a percentage of the total shoot. This growth-differentiation balance hypothesis has the premise of a physiologic balance between growth and differentiation (linamarin production). If herbivory can be minimized, this hypothesis would support the idea of reducing or uncoupling linamarin production with low temperatures and low forage growth rates and maximize the growth rate throughout the growing season to improve forage yields.

In sorghum, dhurrin content is greatest in young plants and in regrowth tissue (Busk and Moller 2002). Generally, less cyanogenic glucoside was found with increased DM accumulation. Indiangrass contains cyanogenic glucoside, and the young shoots, less than 20 cm in height, may be dangerous to grazing cattle. In addition to poor persistence under short grazing, the presence of higher levels of the toxin is reason to avoid grazing prior to shoot heights of 40–100 cm. In rubber tree leaves, linamarin is localized exclusively in the vacuole of mesophyll and epidermal cells, but linamarase was located only in the apoplast (Gruhnert et al. 1994). Linamarase was also found in the apoplast of white clover.

Genetics

Biosynthesis of dhurrin is catalyzed by two cytochrome-P450 enzymes, CYP79A1 and CYP71E1, and a soluble uridine diphosphoglucose (UDPG)-glucosyltransferase (Busk and Moller 2002). This results in a complex multi-genic inheritance of dhurrin in sorghum (Gorz et al. 1987) with the presence of genes modifying dhurrin content on at least half of the chromosome pairs in a plant. In white clover cyanogen, production is based only on two pairs of genes. Cyanogenesis may not occur in individual plants because (i) they lack the cyanogenic glucoside, (ii) they lack the cyanogenic ß-glucosidase needed to release the HCN, or (iii) they contain neither the cyanogenic glucosides nor the ß-glucosidase. One gene, *Ac*, regulates presence or absence of the cyanogenic glucosides. The *Ac* gene is dominant, and *acac* plants have at least two steps in the biosynthesis blocked. The *Li* is the structural gene for the presence of linamarase, the ß-glucosidase, to cleave the HCN from linamarin. *Li* is also dominant.

A logical breeding objective may be to inhibit cyanogenic glycoside accumulation in leaf tissues without affecting their production in other plant parts. **Anti-sense genes** to cyanogenic CYP97D1 and CYP97D2 gene fragments were engineered and differentially expressed in leaf and **tuber** tissues of genetically transformed cassava using cab1 or patatin promoters, respectively (Siritunga and Sayre 2004). The result was reduced cyanogenic glycosides in the respective tissues of transformed plants. Similar strategies may prove useful in reducing cyanogenic compounds in leaves and stems of forages while maintaining their presence in seeds and roots for their anti-herbivory benefits.

Biosynthesis

Tyrosine and valine are the amino acid substrates for biosynthesis of dhurrin and linamarin, respectively. A generalized biosynthetic scheme proceeds from the amino acid → *N*-hydroxyamino acid → α-nitro-carboxylic acid → aci-nitro compound → *E*-aldoxime → *Z*-aldoxime → nitrile → *α*-hydroxynitrile, and then glucosylation to the cyanogenic glucoside. Biosynthesis of dhurrin in sorghum is highly channeled (Moller and Conn 1980). The enzyme system has high level of substrate specificity for the amino acid, and the amino acid hydroxylation step is the rate-limiting reaction. Tattersall et al. (2001) have transferred the entire biosynthetic pathway for dhurrin from sorghum to mouseear cress. The genetically engineered mouseear cress plants were able to synthesize and store amounts of dhurrin similar to those found in sorghum. Importantly, the incorporation of tyrosine into dhurrin did not cause any apparent physiologic problems for the plant but did confer resistance to an insect pest.

Animal Response

Cyanide poisoning is most common in animals grazing forage sorghums and white clover or cassava, which is commonly used in the tropical regions of Africa, Latin America, and Asia (Soto-Blanco et al. 2001). Enzymatic breakdown of the glycoside by the glycosidases is initiated when the fresh plant tissue is ingested and macerated by the animal. Mortality of animals under more intensive management systems is rare because cyanophoric plants are typically offered in processed form, as opposed to a fresh and uncooked form (Cheeke 1995). Acute cyanide toxicosis is caused by the inhibition of cytochrome oxidase, a respiratory enzyme in cells. When this enzyme is blocked, cells suffer from rapid ATP deprivation, causing labored breathing, excitement, gasping, convulsions, paralysis, staggering, and death (Cheeke 1995). Prunasin, found primarily in serviceberry and chokecherry, can be degraded in the rumen through enzymatic hydrolysis of the glycosidic bond followed by non-enzymatic dissociation of HCN from the aglycone (Majak 1992). Hydrocyanic acid can then be further detoxified by rumen microorganisms; however, it is more likely to be absorbed across the rumen wall into the circulatory system where it rapidly binds to macro-molecules and is toxic to the animal (Majak 1992). Miserotoxin, detected in varieties of timber milkvetch, is metabolized by a two-step process. First, in the rumen, microorganisms hydrolyze the glycosidic bond and release the aglycone. During this step, the glycoside itself is not considered to be harmful. The second step is enzymatic oxidation of the aglycone to 3-nitropropionic acid, a potent inhibitor of the Krebs cycle, by hepatic alcohol dehydrogenase (Majak 1992). Cyanide concentrations above 500 $\mu g\, g^{-1}$ are considered dangerous for grazing animals; however, concentrations as low as 200 $\mu g\, g^{-1}$ can also be hazardous for hungry animals grazing under drought conditions (Bertram et al. 2003). Although cyanide is exceptionally toxic, acute toxicosis occurs only when the detoxification processes are overloaded (Cheeke 1995). Direct toxicosis due to HCN is not the only negative impact linamarin (from white clover) may have on grazing animals because cyanide also predisposes the animal to selenium deficiency (Ayres et al. 2001).

Alkaloids

Alkaloids are basic nitrogenous compounds, produced primarily by plants, that have significant pharmacologic activity. Compounds considered alkaloids are so grouped because of the basic nitrogen atom in their structure and not their structure per se. They are generally considered to be secondary metabolites. This section of the chapter will discuss alkaloids in four different forage situations: The indole alkaloids of *Phalaris* spp., unsaturated pyrrolizidine and ergot alkaloids in fungal endophyte-infected tall fescue and ryegrass, quinolizidine alkaloids in lupines, and the indolizidine alkaloid of black patch on red clover.

Indols

Physiology

The *Phalaris* species reed canarygrass and hardinggrasss grow well in poorly drained wet soils but are also drought tolerant. However, they have not been widely used in areas of adaptation due to poor animal performance. Several alkaloids in the forage are responsible for a wide range of adverse animal responses from low voluntary intake to diarrhea, neurologic disorders, and cardiac failure. The alkaloids are mainly derived from tryptophan and have the indole ring intact. The principal groups of alkaloids in *Phalaris* are gramine, tryptamine derivatives, and ß-carboline derivatives. Hordenine, a phenol derivative, is also present, and more recently the oxindoles, coerulescine and horsfiline (Anderton et al. 1998), and the furanobisindole, phalarine (Anderton et al. 1999), have been isolated from sumolgrass. Toxicity of the latter two alkaloids has not been determined, but with attempts to introduce this species and circumstantial evidence suggesting association with equine toxicity, such evaluations must be completed (Colegate et al. 1999).

The biosynthesis of the indole alkaloids is by decarboxylation of tryptophan to tryptamine through tryptophan decarboxylase followed by mono- or di-methylation of the amino-N or hydroxylation and methylation at C_5. A *N*-methyltransferase catalyzes the methylation using *S*-adenosylmethionine as the methyl donor. The rate-limiting enzyme in the scheme seems to be the pyridoxal phosphate-dependent tryptophan decarboxylase (Mack et al. 1988). Tryptophan is also the precursor for the simple ß-carboline alkaloids which are formed by decarboxylation to tryptamine followed by N-alkylation, ring closure, and appropriate methylation and hydroxylation to methyl and methoxy derivatives. The predominate tryptamine alkaloids in reed canarygrass and hardinggrass are gramine, *N,N*-dimethyltryptamine (DMT), and 5-methoxy-*N,N*-dimethyltryptamine (Figure 35.4). The alkaloids present vary among plant populations but are stable within a population. Total alkaloid concentrations range from near 0 to over 27 $g\, kg^{-1}$ DM (Simons and Marten 1971). The simple indol alkaloid gramine has been reported to accumulate to 10 $g\, kg^{-1}$ DM (Coulman et al. 1977) and 5-methoxy-*N*, *N*-dimethyltryptamine to be as high as 4 $g\, kg^{-1}$ DM (Majak et al. 1979).

Concentrations of alkaloids in different plant parts vary and may change rapidly. In dry seed of hardinggrass, no free indole compounds were detected, but tryptamine and *N*-methyltryptamine were detected at day three of germination and reached a maximal concentration at day five (Mack et al. 1988). Concentrations of tryptophan were greatest on day six after initiation of germination, and the DMT concentration was greatest on day eight. Alkaloids in reed canarygrass are more concentrated in the leaf blades than other plant parts (Marten 1973). The upper portion of the leaf blade had higher alkaloid concentrations than the basal portion. Total alkaloid concentration decreased in tissues in the following order: upper portion of blade > lower portion of leaf blade > leaf

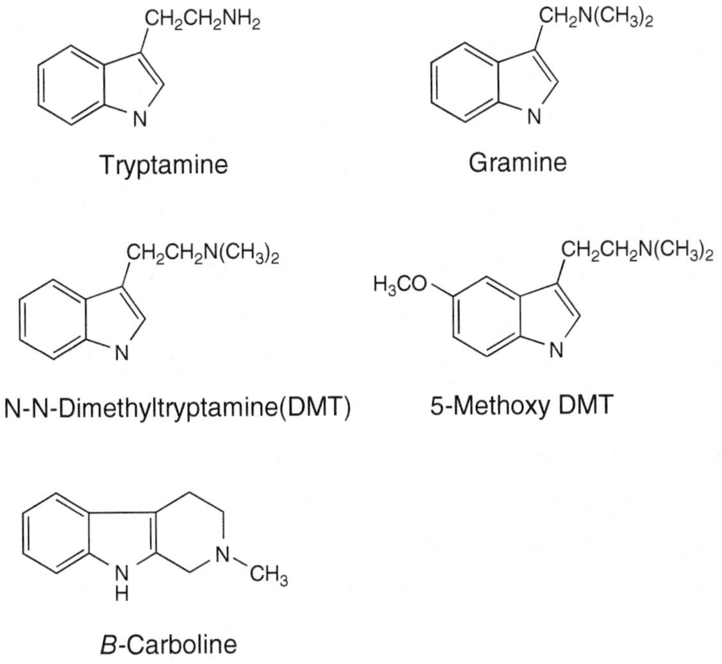

Tryptamine

Gramine

N-N-Dimethyltryptamine(DMT)

5-Methoxy DMT

B-Carboline

Fig. 35.4. Significant alkaloids of *Phalaris* spp.

sheath > stem > rhizomes ≅ roots > seed. Alkaloid content decreased 40–50% as maturity advanced in both first-growth and regrowth tissue over a 20- to 30-day period (Marten 1973). Regrowth tissue had greater alkaloid concentration than first-growth herbage (Majak et al. 1979), presumably because of the greater proportion of leaf tissue in the regrowth herbage.

Tissue alkaloid concentrations increased as soil nitrogen increased (Marten 1973; Majak et al. 1979), and ammonium-N sources increased alkaloids more than other N sources. Higher growth temperature and shading increased alkaloid concentration by 25–50% (Marten 1973). However, moisture stress had the greatest effect on alkaloid concentration, increasing it an average of 121% (Marten 1973).

Genetics

Plant genotypes have been identified that have quantitatively and qualitatively altered alkaloid content (Marten 1973; Marum et al. 1979; Oram et al. 1985). Genetic analyses fit the enzymatic data for the biosynthetic model for these alkaloids (Mack et al. 1988). The conclusions are that a two-gene model best fit the inheritance data. One locus controls biosynthesis for the tryptamine and carboline alkaloids, and the second locus controls biosynthesis of the methoxylated derivatives (Marum et al. 1979). Alleles at both loci must be homozygous recessive for a gramine-accumulating genotype. Much progress has been made in reducing the tryptamine-type alkaloids, and selection is ongoing to reduce gramine in *Phalaris* species to below toxic levels. Bellevue reed canarygrass released from Canada in 1995 had undetectable amounts of tryptamine and ß-carboline alkaloids and less than $2\,g\,kg^{-1}$ of gramine (Coulman 1995). The gramine level of approximately $1.5\,g\,kg^{-1}$ is below the $2\,g\,kg^{-1}$ level considered to be the threshold at which lamb performance is reduced (Marten et al. 1981).

Animal Response

Animal responses include acute sudden death syndrome, a neurologic/staggers syndrome, and reduced voluntary intake response. The tyramine alkaloids, tyramine and *N*-methyltyramine, may be responsible for sudden death. Sheep grazing rapidly growing pastures may exhibit signs of cardiac failure followed by death within one to two days following ingestion. Death results from cardiac arrest and ventricular fibrillation. Anderton et al. (1994) showed that $300–400\,mg\,kg^{-1}$ BW *N*-methyltyramine isolated from hardinggrass and given orally to sheep caused cardiac distress. Hordenine appears to be less problematic in causing cardiotoxicosis. A second form of sudden death syndrome has been reported in sheep and cattle that are acutely exposed to *Phalaris aquatica* pastures during lush **vegetative growth** (Bourke et al. 2005). However, this sudden death syndrome is a form of ammonia toxicity rather than indole alkaloid related.

Tryptamine alkaloids, DMT and 5-methoxy-DMT, are the most likely causes of the staggers syndrome, but the ß-carbolines and 5-hydroxy-DMT cannot be ignored as causative agents in this syndrome. Bourke et al. (1988) showed that 5-methoxy-DMT and gramine induced the clinical signs of the nervous disorder. Animal responses to the two alkaloids were similar, but 5-methoxy-DMT was 10–100 times more potent than gramine. Tryptamine alkaloids are serotonin receptor agonists, and the signs produced in sheep are consistent with the alkaloid effect on serotoninergic receptors in the brain and spinal cord (Bourke et al. 1990). Marten et al. (1981) associated tryptamines and ß-carbolines with diarrhea in sheep, and this may also be a sign of the chronic phase of the disease. As a protective measure, animals not exhibiting signs of *Phalaris* staggers may be treated with a cobalt drench, and in Australia cobalt is given as intraruminal pellets.

Roe and Mottershead (1962) reported low voluntary intake of reed canarygrass associated with the presence of tryptamine alkaloids. Marten and colleagues at the University of Minnesota demonstrated that palatability of reed canarygrass was negatively correlated with alkaloid concentration in the herbage (Marten 1973; Marten et al. 1981). They concluded that alkaloid concentrations above 2 g kg^{-1} of DM reduced gain of grazing lambs. This was the stimulus for much of the genetic work to lower the tryptamine alkaloids in *Phalaris*. Casler and Vogel (1999) have called this one of the more remarkable success stories in breeding cool-season grasses and cultivars with low gramine and no tryptamine or ß-carboline alkaloids have dominated the marketplace for reed canarygrass.

Alkaloids and Endophytes of Cool-Season Grasses

Many cool-season grasses have developed mutualistic relationships with clavicepitaceous fungal endophytes of the *Epichloe* genus (formerly *Acremonium* or *Neotyphodium*). The endophytes colonize aboveground plant tissues and are transmitted among generations via the seed. Endophytes can produce a range of bioactive alkaloids that may benefit the plant host. The bioactive alkaloids include the aminopyrrolizines (n-formyl and n-acetyl lolines), pyrrolizine (peramine), ergot (lysergic amides and ergopeptine), and indolediterpines (lolitrem) alkaloids (Figure 35.5). These bioactive alkaloids are found within economically important *Lolium* grass species that are infected with *Epichloe* endophytes. The properties of the alkaloids vary, with the loline and peramine alkaloids providing resistance to insect herbivory (Popay et al. 1995; Popay and Lane 2000; Rowan et al. 1986; Schardl et al. 2007) and the ergot and lolitrem alkaloids providing resistance to mammalian herbivory (Gallagher et al. 1984; Lyons et al. 1986; Hill et al. 1994).

Tall fescue, perennial, and annual ryegrass cultivars adapted to the US, New Zealand, and Australia are derived from European progenitors with narrow genetic bases. Consequently, their endophytes likewise express little genetic diversity (Easton 2007). However, genetic diversities of endophytes are broader than those found in common cultivars used in these countries. Exploration of tall fescue and ryegrass germplasms in Europe, Asia, and Africa have found distinct endophyte strains that vary in their ability to produce alkaloids (Takach and Young 2015). The chemotypes of endophytes are associated with morphotypes of their tall fescue hosts. Rhizomatous tall fescues have longer rhizomes than continental types and originate from the Iberian Peninsula, Continental morphotypes are typical of those bred for the US, NZ, and Australia, and the summer dormant Mediterranean morphotypes are genomically distinct from the rhizomatous and Continental types. Continental and rhizomatous tall fescues are hexaploids while Mediterranean morphotypes may contain hexaploid, octaploid, or decaploid genomic structures. Endophytes within Continental tall fescues are capable of producing peramine, loline, and ergot alkaloids, and endophytes from rhizomatous tall fescue produce peramine and loline but some produce very low levels of ergot alkaloids. Endophytes from the Mediterranean morphotypes appear to have the greatest diversity of chemotypes and offer opportunities for discovery of non-ergot alkaloid producing types. However, Mediterranean endophytes may or may not produce lolitrem alkaloids (Table 35.2; Figure 35.6).

Physiology

There are different physiologic effects of different host/endophyte associations (Bush et al. 1993; Malinowski and Belesky 2000). In competitive or water-stressed environments endophyte-infected plants had lower dry weight and tiller number but a higher net growth rate than endophyte-free plants (Belesky et al. 1989; Hill et al. 1991; Assuero et al. 2000). Malinowski et al. (1997) demonstrated that meadow fescue infected with *Eciton uncinatum* Gams Petrini & Schmidt had a greater potential to adapt to drought by adjusting to soil-water depletion earlier than endophyte-free plants. A similar osmotic adjustment was measured for *Epichloë coenophialum*–infected tall fescue plants when compared with controls with or without inoculation by root-knot nematode *Meloidogyne marylandi* (Elmi et al. 2000). Nematode survival was nearly zero in soil with endophyte-infected plants growing, suggesting that the presence of the endophyte enhances the persistence of tall fescue in *M. marylandi*– infested soils subject to water stress by minimizing the numbers of nematodes as well as enhancing drought tolerance of

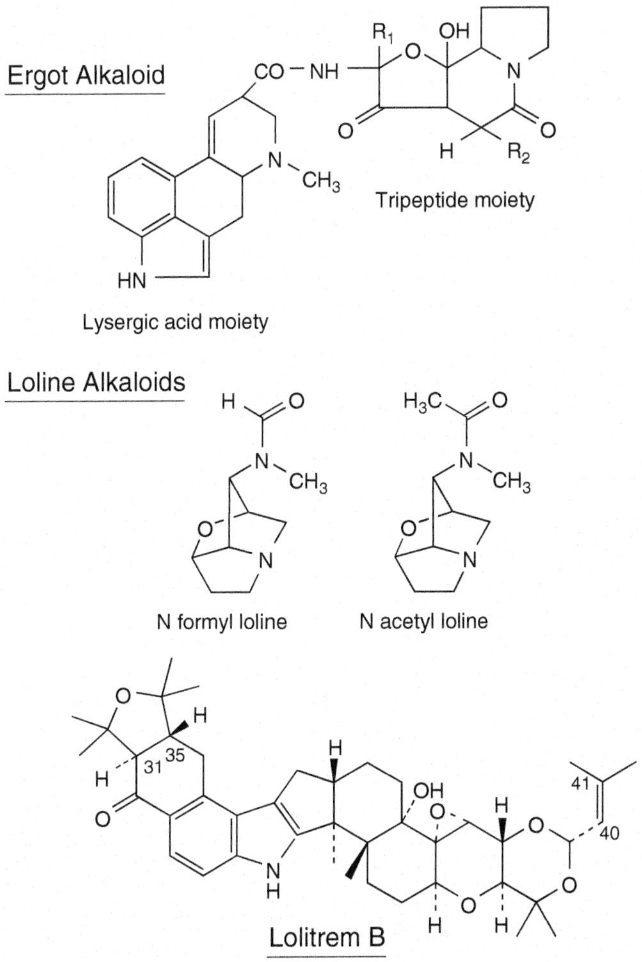

Fig. 35.5. Fungal endophyte alkaloids in found in *Epichloe* spp. infected grasses.

Table 35.2 Alkaloid chemotypes of endophytes isolated from Continental and Mediterranean tall fescue

Endophyte	Germplasm type	Alkaloid type			
		Peramine	Loline	Ergot	Lolitrem
E. coenophialum e19	Continental	+	+	+	−
E. coenophialum Fe45119	Continental	+	+	+	−
E. coenophialum e4163	Continental	+	+	+	−
E. coenophialum AR542	Mediterranean	+	+	−	−
E. coenophialum AR584	Mediterranean	+	+	−	−
Epichloe spp FaTG-2	Mediterranean	+	−	+	+
Epichloe spp FaTG-2	Mediterranean	+	−	+	−
Epichloe spp FaTG-3	Mediterranean	+	+	−	−
Epichloe spp FaTG-3	Mediterranean	+	+	−	−
Epichloe spp FaTG-4	Mediterranean	+	−	+	−
Epichloe spp FaTG-4	Mediterranean	+	−	+	−

Source: Adapted from Takach and Young (2015).

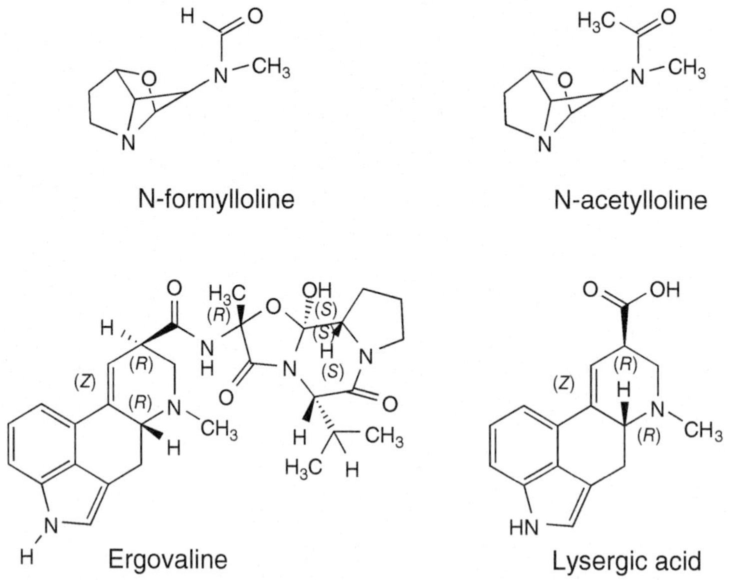

N-formylloline

N-acetylloline

Ergovaline

Lysergic acid

Fig. 35.6. Ergot alkaloids accumulate in endophyte-infected grasses.

the grass. Endophyte-infected tall fescue plants had smaller-diameter roots than control plants (Malinowski et al. 1999), and with P deficiency root diameter decreased and root hair length increased more than in control plants. Plants also accumulated more P from P-deficient soils, suggesting an adaptation mechanism for endophyte-infected plants under nutrient and water stresses. Also, endophyte-infected plants had greater shoot and root mass than endophyte-free plants and an increased root-to-shoot ratio for high-yielding genotypes (Belesky and Fedders 1995).

Mycelia of *Epichloe* are usually present in the intercellular spaces of the leaf sheath and stem, are sometimes found in the leaf blade, and appear to be absent in the root (Christensen et al. 1998) (Figure 35.7). Location of the endophyte has implications for the translocation of the alkaloids because all are fungal products and are toxic in different tissues consumed by herbivores. Alkaloid accumulation in a symbiota is dependent upon the host and specific endophyte involved (Siegel et al. 1990).

There are regulatory interactions by plant hosts controlling expression of endophyte metabolites. This was demonstrated by examining maternal and paternal effects on endophytes and alkaloid synthesis. Pollen parent influences the extent by which endophytes can produce ergot alkaloids in the maternal offspring containing a common endophyte (Adcock et al. 1997; Roylance et al. 1994; Easton et al. 2002). Thus, nuclear plant genes control the level to which endophytes can express their alkaloid production capacity. Furthermore, alkaloid production was either not-related to, or marginally related to, the mass of endophyte within the grass hosts (Hiatt and Hill 1997; Easton et al. 2002). The mechanisms of plant-regulated alkaloid expression are not fully understood but, could be one of many components of the symbiosis including antagonism between plant and endophyte, capability of an endophyte to produce alkaloids, deficiencies of plant metabolites needed for alkaloid production, and/or plant morphology.

Morphology

Ergovaline accumulated in perennial ryegrass to the greatest level in pseudostems and to a lesser extent in leaves of vegetative plants (Davies et al. 1993). The greatest amount of the endophyte is also found in the pseudostems. Accumulation of ergot alkaloids in vegetative tall fescue is greatest in the crown and base of the plant, with decreased concentrations in the stem and leaves and highest concentrations usually found in the seed. Translocation of ergot alkaloids must occur from the crown and **pseudostem** to the leaf blade since little endophyte is found in the blade.

Environment

Environmental factors (temperature, moisture, and nutrient stress) can impact alkaloid concentrations in herbage (Easton 2007). Herbage alkaloid concentration increases in late spring but diminishes in late autumn and winter

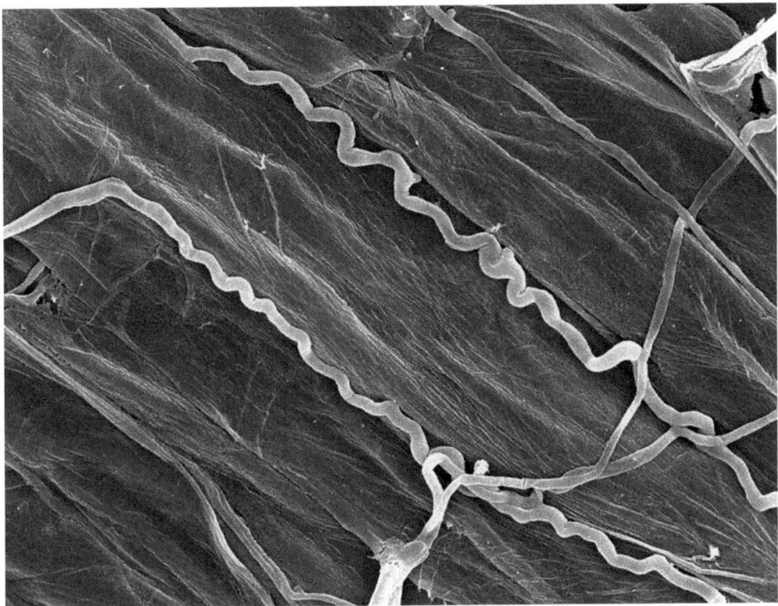

Fig. 35.7. A freeze-fracture scanning electron micrograph illustrating *Epichloe coenophialum* growing in between tall fescue plant cells. *Source:* Photo taken by H.J. Ju, courtesy of N.S. Hill.

months (Justus et al. 1997; Kallenbach et al. 2003; Rogers et al. 2011).

Plant/Endophyte Manipulation

Non-toxic endophytes can be inserted into improved forages that maintain the agronomic benefits of improved cultivars without the anti-mammalian toxins. These strategies have been successfully employed in both perennial ryegrass and tall fescue (Fletcher 1999; Bouton et al. 2002). However, compatibility between the endophyte and its grass host cannot be taken for granted as successful transmission from one generation to the next may be compromised in specific circumstances (Bouton et al. 2002). Agronomic practices must be carefully prescribed as some fungicides partially kill or compromise endophyte health in developing seed (Chynoweth et al. 2012; Hill and Brown 2000), harvesting immature seed reduces endophyte viability (Hill et al. 2005), and environmental conditions affect how quickly endophytes precede seed embryo mortality during seed storage (Welty et al. 1987). Interactions of the variables affecting endophyte viability necessitates vigilance during seed production and storage and, requires continual monitoring to make sure cultivar/endophyte purity is maintained.

Seed companies utilizing non-toxic endophyte technologies recognize that errors in product development and/or seed production could threaten the reputation of non-toxic endophyte technology. Endophyte technology requires additional testing parameters for seed quality beyond that required by law. The companies participate in an umbrella organization called the "Alliance for Grassland Renewal" (http://grasslandrenewal.org/) in which seed quality parameters for endophyte technology are standardized. Current requirements are that seed be infected with 70% or more viable endophyte with less than 5% toxic types. In return, the "Alliance" provides educational events for producers who are interested in purchasing non-toxic cultivars for pasture renovation purposes.

Animal Response

Two endophyte-related livestock disorders are of significant economic importance: ryegrass staggers and fescue toxicosis. Ryegrass staggers occurs in sheep grazing endophyte-infected perennial ryegrass (*Lolium perenne/Epichloe lolii*) containing lolitrem B (Fletcher 1982; Prestidge 1993). The disorder primarily affects sheep during summer and autumn when forage is limited, and livestock are forced to graze close to the ground. Toxicity is acute and caused by excitation of livestock. Affected animals develop a staggered gait, lose balance, and become crushed by cohorts or drown in ponds. In some cases, livestock starve because they are unable to walk to forage or water.

Fescue toxicosis, on the other hand, tends to be a sub-acute poisoning due to the ingestion of ergot alkaloids in endophyte-infected tall fescue (Hill et al. 1994; Ayers et al. 2009). Poor reproduction and low-weight gains are the main effects of fescue toxicosis, though sloughing of hooves, ear tips and tails occur in cattle. Hard necrotic fat accumulates in mesenteric tissues that can alter organ function (Stuedemann et al. 1975). The vasoconstrictive properties of the toxins prevent normal dissipation of body heat. Cattle spending an excessive amount of time in shade and/or water in summer, expressing rough or bronzed hair coats, and having muddy hindquarters from laying in wallows are clear signs they are exposed to toxic tall fescue. Most of these problems are seen with cattle and, to a lesser extent, sheep. Pregnant mares generally exhibit reproductive complications during gestation and parturition (prolonged gestation, weak foals, stillbirths, and agalactia), rather than a decrease in feed intake or weight gain. However, many of the complications in mare reproduction can be alleviated by removing pregnant mares from endophyte-infected fescue pastures at around 300 days of gestation (Cross et al. 1995).

Ergovaline, is the most prevalent ergopeptine alkaloid found in tall fescue and perennial ryegrass tissues (Lyons et al. 1986) and compared to other ergot alkaloids has a higher affinity to dopamine and serotonin receptors, known physiologic agonists for the fescue toxicosis syndrome (Dyer 1993; Foote et al. 2011). Hence, it became the most studied of the ergot alkaloids and was considered the primary toxin causing the syndrome. However, it was later found that ergovaline is metabolized by ruminal microbes to lysergic amide end products (Ayers et al. 2009). The lysergic amides are actively transported across rumen epithelia (Hill et al. 2001; Ayers et al. 2009) and are postulated to have a significant role in the fescue toxicosis syndrome. This presents a conundrum and complicates methods of assessing plant materials for toxin potential, and whether one or both alkaloid classes are the causative agent(s). High-resolution metabolomics was utilized to analyze the plasma and urine metabolomes of beef cattle while grazing toxic or non-toxic endophyte-infected tall fescue to test for metabolic differences associated with ergot alkaloid exposure. The implementation of high-resolution metabolomics allows for sensitive detection of metabolic aberrations and molecular changes associated with pathologies specific to cattle. Lysergic acid and ergonovine were the predominant plasma alkaloids but no ergovaline was either circulating or excreted from grazing steers (Mote et al. 2017). They found that grazing toxic tall fescue pastures perturbed tryptophan, tyrosine, and glycerophospholipid metabolism. Additional high-resolution metabolomics studies are needed to develop putative metabolite-based explanation(s) for complications seen in fescue toxicosis, from which targeted livestock therapies might be developed.

Management

Obviously, from the above discussion, the level of endophyte infestation is important for toxin production and animal response. Also, management of soil fertility will alter alkaloid accumulation in symbiota. High nitrogen increases ergot alkaloid levels 2–20 times (Azevedo et al. 1993). However, increased nitrogen fertility does not increase loline alkaloids (Bush et al. 1993). There has been much discussion on change of infestation levels in pastures or hay fields over time. At the end of a three-year grazing period, the tall fescue stand density remained acceptable regardless of grazing pressure (Gwinn et al. 1998). Endophyte infestation levels in the low-grazing pressure treatment were not changed over the experimental period. Also, at high-infestation levels (>80%), grazing pressure treatments did not change infestation level. However, in the high- and medium-grazing pressure treatments on pastures with intermediate infestation levels (25% and 60% infestation), infestation increased by 20–30%. The authors concluded that the influence of moderate- or high-grazing pressure on a pasture influenced endophyte infestation level, and the resulting potential toxicosis must be considered when making management decisions.

Endophyte-mediated alkaloid toxicity is best managed by reducing toxin ingestion. Livestock often prefer to consume energy-rich diets. When pasture plants are in the vegetative stage of growth livestock prefer to graze leaf blades, but when seedheads are present they prefer seeds over other plant components (Bedell 1968). Seeds happen to have the greatest concentration of toxins so mowing in late spring to prevent seedhead production is a sound management strategy to minimize alkaloid ingestion. Overgrazing pastures results in grazing of pseudostems, another alkaloid-rich component of the plant. Thus, removing livestock from pastures during periods of low forage availability will prevent toxin consumption. Adding other pasture species, especially legumes that complement forage production, enables livestock to acquire part of their dietary needs from a non-toxic source (Coffey et al. 1990) and feeding energy or protein supplements will improve livestock gains when grazing endophyte-infected tall fescue pastures (Elizalde et al. 1998). Removing livestock from pastures in summer will minimize heat exposure and overcome some of the vasoconstrictive effect of the toxins. However, it must be kept in mind these management variables do not eliminate toxicosis and imposing the same management treatments on non-toxic pastures will provide similar livestock responses.

The most effective option for eliminating endophyte-mediated toxic syndromes is to replace existing pastures with cultivars infected with non-toxic endophytes (Fletcher 1999; Bouton et al. 2002). Superficially, this seems like a simple task, but toxic endophytes might persist in fields if care is not taken to prevent the potential for incidental contamination. Depending on growing conditions, seed produced in a toxic pasture may fall to the soil surface and, both seed and endophyte, be viable the following year (Hill et al. 2010). Preventing seedhead production is of paramount importance to permit successful pasture renovation. Currently, there are two options used to renovate toxic tall fescue pastures with non-toxic endophyte cultivars. The first, referred to as the spray-smother-spray technique, begins by using herbicides to kill existing tall fescue in the spring of the year, followed by a smother crop in the summer. Smother crops may be summer annual forages to help offset lack of forage production during the renovation year, or a cash crop. Herbicides are applied a second time after harvest/utilization of the smother crop to kill plants which escaped the spring applied herbicide treatment. Tall fescue can be established after the smother crop for fall and winter establishment. The second option is referred to the spray-spray-plant method. In this case, seedheads are mowed in early summer to prevent seed production. Then, six weeks prior to planting, the pasture is sprayed with herbicide to kill the existing tall fescue and a second application made six weeks later to kill any remnant plants missed by the first application. The six-week waiting period is important to permit adequate regrowth for the second herbicide application to be effective. Tall fescue can be no-till drilled into the dead sod one week after the final herbicide application.

It is absolutely vital not to over utilize the renovated pasture in the summer following establishment regardless of the method of renovation. The anti-herbivory effects of toxicosis are absent in the renovated tall fescue pastures, so animals are more likely to overgraze non-toxic tall fescue. Producing hay, which allows a **rest period** between harvests, during the first summer after establishment is a good practice to follow. Care must also be taken to prevent overgrazing of pastures even after the first year of renovation.

Quinolizidine Alkaloids

Physiology

Lupine poisoning is caused by ingestion of quinolizidine alkaloids (QA) produced by plants. Of more than 200 QA-producing *Lupinis* species described, only four are used extensively for forage production: Narrow leaf lupine, white lupine, yellow lupine, and pearl lupine. There are another 10–20 that are used for small-scale forage production or seed protein production. Lupine seeds may contain up to 40% protein and 20% lipids, thus, being of high-feed quality, but they may also contain over 3% QA (Ruiz and Sotelo 2001). Their predominant use is as fodder crops grown in grass mixtures that are cut for hay or silage. Over 100 different bicyclic, tricyclic, and tetracyclic QAs have been identified in lupines (Wink et al. 1995). Lupinine and epilupinine are the most abundant bicyclic alkaloid, and cytisine and angustifoline are examples of tricyclic alkaloids (Figure 35.8). Some of the most abundant tetracyclic alkaloids are anagyrine, baptifoline, 13α-hydroxylupanine, lupanine, multiflorine, and sparteine (Wink et al. 1995; Ainouche et al. 1996). Each cultivated lupine species has a unique QA profile, but lupanine and lupinine are the most common (Table 35.3) (Frick et al. 2017).

The QAs are obviously very diverse, and their biosynthesis is not well understood. They are metabolic products of lysine, and the diamine, cadaverine, is likely involved. Lysine is decarboxylated to cadaverine, and the C_2 of lysine becomes randomized as C_1 and C_5 of cadaverine. Cadaverine undergoes oxidative deamination by copper amine oxidase to yield 5-aminopentanal which spontaneously cyclizes to Δ^1-piperideine (Frick et al. 2017). Wink et al. (1980) proposed that three molecules of cadaverine are converted in the chloroplast to 17-oxosparteine. 17-Oxosparteine is the immediate precursor for lupanine. In this pathway, the tetracyclic lupanine serves as a precursor for the tricyclic alkaloids as well as the skeleton for the wide range of tetracyclic substituted alkaloids. Grafting high QA shoots onto low QA rootstock and vice versa confirmed that shoot tissue controls the expression of QA synthesis (Lee et al. 2007). The bicyclic alkaloids may be formed by a separate pathway or by early release of a two-cadaverine intermediate from the tetracylic pathway. Lysine decarboxylase and the enzyme involved in cyclization of the quinolizidine skeleton are in the chloroplasts of leaves (Wink and Hartmann 1982). The acyltransferases for tiglic and *p*-coumaric acid in quinolizidine alkaloid ester biosynthesis are most likely in the mitochondria and the **cytosol**, respectively (Suzuki et al. 1996). After synthesis, the alkaloids are translocated via the phloem throughout the plant where they accumulate principally in epidermal tissue and especially the seed and seedpod (Wink and Mende 1987). Ripe seed contain most of the alkaloid present in the plant, but the composition of the alkaloid fraction differs between leaf and seed tissue (Van Wyk et al. 1995). Alkaloid levels decrease during germination and seedling development and increase in vegetative tissue prior to flowering, with a rapid accumulation in the pod and seed at maturity. In high-alkaloid plants of blue lupine, yellow lupine, and white lupine, the biogenic polyamines were present in greater amounts than in the low-alkaloid plants, whereas the basic amino acids were higher in the low-alkaloid plants than in the high-alkaloid plants (Aniszewski et al. 2001).

Plant health and the environment affect alkaloid accumulation in plant tissues. Nitrogen-fixing plants and plants grown with high available soil nitrogen had greater amounts of alkaloids (Barlog 2002). Alkaloid accumulation was greater in the seed with ammonium nitrate fertilizer than ammonium sulfate or calcium nitrate. In leaf tissue, alkaloid content was negatively correlated with magnesium concentration, and if both nitrogen and magnesium fertilizer were applied, the seed alkaloid content was reduced slightly. The effect of magnesium and nitrogen fertilizers on vegetative accumulation of alkaloids

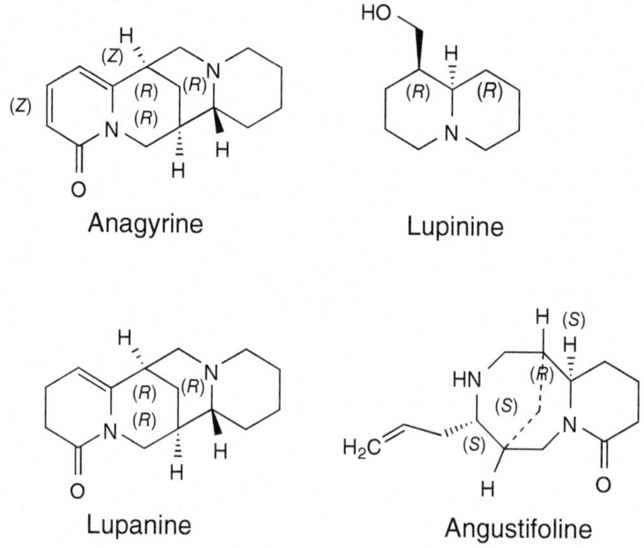

Fig. 35.8. Examples of quinolizidine alkaloids from lupines.

Table 35.3 Major quinolizidine alkaloids identified in seeds of lupine, species

Lupine species	Major quinolizidine alkaloid (% of total alkaloids)
L. angustifolius	Lupanine (70%), 13α-hydroxylupanine (12%), angustifoline (10%)
L. albus	Lupanine (70%), albine (15%), 13α-hydroxylupanine (12%), multiflorine (3%)
L. luteus	Lupinine (60%), sparteine (30%)
L mutabilis	Lupanine (46%), sparteine (16%), 3β-dydroxylupanine (12%), 13α-hydroxylupanine (7%)

Source: Adapted from Frick et al. (2017).

was dependent upon the stage of growth and weather conditions. Severe potassium deficiency increased seed alkaloid concentrations greatly in low-alkaloid lines of blue lupine but not in high-alkaloid lines (Gremigni et al. 2001). The predominant alkaloid in the low lines was lupanine and in the high lines 13-hydroxylupanine. Foliar application of Fe and Mn to hartweg's lupine increased growth and seed alkaloid content (Tawfik 1997), again emphasizing the interaction of environment on growth and alkaloid accumulation. Water stress during **vegetative growth** increased alkaloid content but decreased alkaloid content when stress occurred at flowering (Christiansen et al. 1999).

Total alkaloid concentration was lower in regrowth tissue than in the original leaf tissue (Ralphs and Williams 1988); however, breakage of leaf cells stimulated alkaloid accumulation (Johnson et al. 1989). Within a whole plant with some damaged leaves and undamaged leaves, the undamaged leaves had greater accumulation of alkaloids than control plants. Chloroplasts are involved in at least the first steps in quinolizidine alkaloid biosynthesis, and the amount of alkaloid accumulation is altered by light, with maximum concentrations obtained five to eight hours into the photoperiod and minimum levels measured at night (Wink and Hartmann 1982).

Genetics

Lupines are an extremely diversified group of plants with many natural hybrids among the taxa collected, and most species have not been domesticated. Genetic improvement of forage and seed lupines has been occurring for the last 100 years, and it was the discovery of low-alkaloid or alkaloid-free mutants that has contributed most significantly to lupines as a cultivated crop. The quinolizidine alkaloids of lupines have a bitter taste; the low-alkaloid or alkaloid-free lines have become known as sweet lupines and the high-alkaloid lines as bitter lupines. Compared to other antiquality traits in forages, little is known about the biochemical synthesis and genetic regulation of QA's. Three mutant genes, PauperM1 (Lin et al. 2008), iucundus (iuc) (Bunsupa et al. 2011), and O-tigloyltransferase (HMT/HTLase) (Chen et al. 2007) control expression for the low alkaloid trait. All genes for the low-alkaloid condition are recessive. Many collections of lupines are being made around the world and being evaluated for adaptation to specific environments, for quality and quantity of herbage for animal feed, and for the low-alkaloid trait. Germplasm evaluation is complicated by the fact that there are no simple analytic techniques available to quantify low-alkaloid phenotypes (Lin et al. 2008). Thus, generation of molecular markers and use of marker assisted selection has become the preferred method for selecting for low alkaloid types, which are then phenotyped using laboratory methods. One study examined elite breeding lines of *Lupinus angustifolius* over 12 environments and three years (Beyer et al. 2015). The lines ranged from 78 to 1312 $mg\,kg^{-1}$ in QA alkaloid concentration but the low- and high-alkaloid traits were stable over locations and years as there were no genotype × location, genotype × year, or genotype × location × year interactions. Heritability for the QA trait was >95%. Thus, the low-alkaloid trait is very stable once identified.

An example of the kind of improvement that is being made is the release of a low-alkaloid, early-flowering variety adapted to very acid soils in low-rainfall areas (Cowling et al. 2000). As the low-alkaloid lines become more readily available in all commercial species, more emphasis is placed on disease resistance to improve agronomic performance. Different species are adapted to many locally different environmental niches around the world, and because low-alkaloid lines are locally accepted with good yield, lupines will contribute greatly to local agriculture economies (Christiansen et al. 1999).

Animal Response

Animal responses to quinolizidine alkaloid toxicosis are variable due to the many different alkaloids produced by the large number of varieties within the *Lupinus* genus. Sheep are most susceptible to lupine toxicity, a neurologic disease; however, horses and cattle are also at risk. This variable response among livestock classes may be attributed to the differences in eating patterns. Horses and cattle will readily consume velvet lupine and silky lupine when the plants are immature but not when mature. On the contrary, if legumes are present, sheep will consume the mature plant.

The most common symptoms observed with lupine toxicity are abortions, stillbirths, or congenital defects, such as crooked calf disease, cleft palates, and distorted/malformed spines. The critical time period for toxin consumption is 40–70 days of gestation with malformations related to a reduction in fetal movement (Shupe et al. 1968; Panter and Keeler 1993). Silky lupine, tailcup lupine, lunara lupine, sulfur lupine, and velvet lupine are among the most common species associated with crooked-calf disease. Panter et al. (2001) observed losses when steers consumed an estimated 486–648 g of silvery lupine within 24 hours. Alkaloid concentrations ranged from 0.7 to 2.5 mg 100 mg^{-1} plant DM. Although death occurred at these levels, concentrations that produce a toxic state will vary with species and stage of growth of the plant.

Indolizidine Alkaloids

Slaframine and Slobbers

There are three indolizidine alkaloids from plants that are of special significance. Castanospermine is found in Blackbean, also called Moreton Bay Chestnut, an Australian rain forest and riverine tree. Castanospermine is an inhibitor of protein glycosylation intensively studied as a medicinal against viral disease development. Swainsonine and slaframine are associated with plants consumed by livestock. Swainsonine is the primary **toxicant** in locoweed (Taylor et al. 2003). Slaframine is a product of fungal growth primarily on red clover and may cause significant reduction in animal performance.

Slobbers is an animal disease caused mainly by ingestion of red clover infected with the fungus *Rhizoctonia leguminicola* Gough & E.S. Elliott and has been observed in cattle, sheep, and horses. *R. leguminicola* causes a disease known as blackpatch on the leaves of red clover. Disease development occurs mainly in the warm humid weather of the second or later forage cuttings that are allowed to go beyond full bloom. The fungus on diseased forage, hay or pasture, will produce slaframine levels as high as 50–100 $\mu g\,g^{-1}$ DM with the result being slobbers disease.

Lysine is the precursor for slaframine biosynthesis, and a key intermediate is 1-ketooctahydroindolizine. This compound is a common intermediate for both slaframine and swainsonine biosynthesis. 1-Ketooctahydroindolizine is reduced, hydroxyalated, transaminated, and esterified to yield slaframine (Figure 35.9).

Animal Response

Slaframine causes excessive salivation, or slobbers, in livestock consuming herbage infected with *R. leguminicola*. After absorption in the digestive tract, slaframine is metabolized by a hepatic microsomal flavoprotein oxidase to a ketoimine (Hagler and Croom 1989), and this may be the active compound acting as a cholinergic agonist to stimulate exocrine glands. Other signs of slobbers that are more readily identified with the parasympathetic nervous system include, but are not limited to, **lacrimation**, feed refusal, decreased milk production, weight loss, stiff joints, hypothermia, bloat, **polyuria**, and diarrhea. Administration of slaframine does not produce all the signs of slobbers in cattle. Broquist et al. (1984) suggested that some of the signs of slobbers may be attributed to swainsonine found in cultures of *R. leguminicola*. Croom et al. (1995) suggested that only the excessive salivation of slobbers is caused by slaframine, and the other signs are from swainsonine or other metabolites.

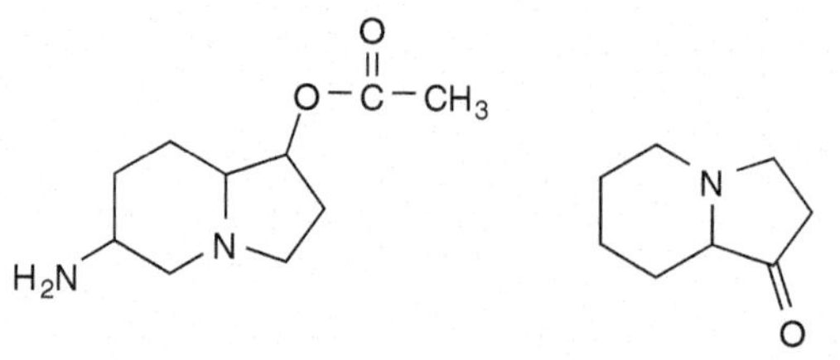

slaframine

1-ketooctahydroindolizine

Fig. 35.9. The indolizidine alkaloids slaframine and 1-ketooctahydroindolizine are precursors for slaframine and swainsonine, respectively.

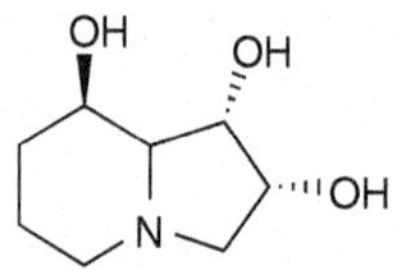

Swainsonine

Fig. 35.10. Chemical structure of swainsonine.

Swainsonine in Locoweed

Locoweed poisoning, or "locoism," is caused by *Astragalus* and *Oxytropis* plant species. "Loco" is a Spanish word for crazy, and locoism was first recognized in 1542 when horses in an expedition led by Hernando De Soto developed strange behavior when grazing native vegetation in his Westernmost venture (Cook et al. 2009). There are over 350 *Astragalus* and 22 *Oxytropis* species, of which 24 species produce swainsonine, the cause of locoism (Figure 35.10).

Recently, a fungal endophyte, *Embellisia* spp. *Pleosporaceae*, was isolated from *Astragalus* and *Oxytropis* spp. and shown to produce swainsonine. The fungus is present in stems, leaves, flowers, and seed of locoweed plants (Braun et al. 2003; Yu et al. 2010) and was shown to be solely responsible for the synthesis of swainsonine and thus, the toxicity of the plant (McLain-Romero et al. 2004a; Ralphs et al. 2008). Fungal isolates from *Oxytropis* species are genetically and morphologically similar, but those from Astragalus species are diverse (Belfon and Creamer 2003). The fungus is passed to the next generation via the seed coat and seed coat removal results in fungus- and swainsonine-free plants (McLain-Romero et al. 2004b).

Animal Response

The clinical signs of locoweed poisoning are a slow staggering gait, rough hair coat, staring gaze, emaciation, lack of muscular coordination, and nervousness (Cook et al. 2009) Generally symptoms become more severe as livestock spend time on locoweed pastures and can progress to clinical signs associated with poisoning, including a slow staggering gait, a rough hair coat, a staring gaze, emaciation, lack of muscular coordination, and extreme nervousness. Chronic exposure results in decreased libido, infertility, abortion, water belly, fetal skeletal deformities, cardiovascular disease and death.

The mechanism of toxicity is swainsonine mimicry of the sugar mannose and inhibition of mannosidase enzymes. Mannose is important for metabolism of glycoproteins, a class of proteins which serve as cell receptors that regulate cell function. Inhibition of mannosidases eventually causes cell death. Horses are the most sensitive livestock class to locoism but sheep and cattle are affected as well. Locoism toxicity is reversible by removing affected animals from areas infested with *Astragalus* and *Oxytropis* to allow animals to break down the toxic compounds in approximately 28 days. However, neurologic damage may be permanent when livestock are exposed for long durations.

Amino Acids

Leucaena and Mimosine

Mimosine (Figure 35.11), a nonprotein amino acid [ß{-3-hydroxy-4-(1H) pyridone}α-amino propionic acid], that accumulates in some species of the forage legume leucaena has curtailed its use as a forage in tropical areas. Leucaena would be an excellent tropical forage because of its high crude protein content, high yields, and resistance to drought, if the mimosine toxicity could be reduced.

Garcia et al. (1996) summarized several publications for the nutritive value and forage productivity of leucaena and reported median values of 22% crude protein, 39.5% neutral detergent fiber, 35.1% acid detergent fiber, 1.45% K, and 2.14% mimosine on a forage DM basis (stem and leaf petiole and blade). However, digestible energy is very low for this forage, averaging about $12\ MJ\ kg^{-1}$ DM, but yield may be as high as $25\ Mg\ ha^{-1}$ (Hammond 1995).

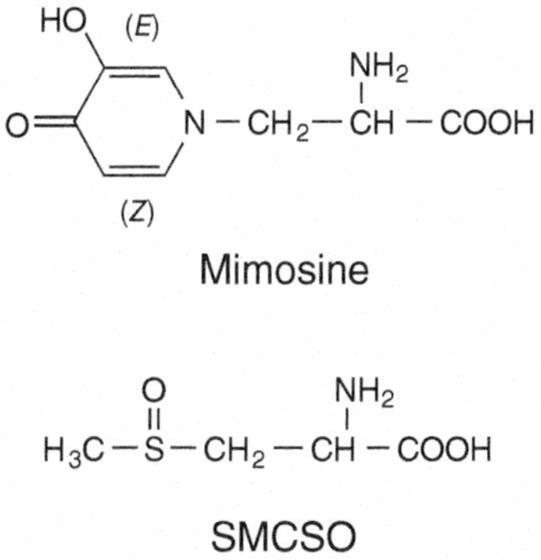

HO (E)

$$O{=}C_5H_3N{-}CH_2{-}CH(NH_2){-}COOH$$

(Z)

Mimosine

$$H_3C{-}S({=}O){-}CH_2{-}CH(NH_2){-}COOH$$

SMCSO

FIG. 35.11. Mimosine from *Leucaena* spp. and SMCSO from *Brassica* spp.

Mimosine is biosynthesized from 3,4 dihydroxypridine and *O*-acetylserine through a cysteine synthase. This reaction may be a competitive branch point for cysteine formation, but leucaena forage contains sufficient cysteine for dietary requirements (Garcia et al. 1996). Mimosine is found also in the nodules and root exudates of leucaena, and some of the *Rhizobium* strains nodulating leucaena used mimosine as a carbon and nitrogen source (Soedarjo and Borthakur 1998).

Animal Response

Many of the early studies including leucaena in the diet, restricted leucaena intake to reduce potential mimosine toxicity. Animals that have mimosine detoxifying bacteria in the rumen can safely consume greater amounts of leucaena. In pastures with leucaena planted in rows 3 m apart on 0%, 25%, and 100% of the paddocks, cattle gained 90, 127, and 205 kg steer^{-1} yr^{-1}, respectively (Quirk et al. 1990). This dramatic increase in animal performance demonstrated the great potential of using this tropical legume as a forage.

The reader is referred to the review of Hammond (1995) for details on leucaena toxicosis. Important to the use of leucaena as a forage is the presence of ruminal bacteria capable of detoxifying 3,4-dihydroxypyridine (3,4-DHP) and 2,3-DHP. These bacteria, *Synergistes jonesii,* may be inoculated into cattle by simple means (see Hammond 1995). However, the ability of an animal to maintain a ruminal population capable of detoxifying the DHP compounds appears dependent upon maintaining leucaena in the diet. Not only are growth rates of grazing animals increased when 3,4-DHP–degrading bacteria are present, but reproductive performance is improved. Akingbade et al. (2001) demonstrated greater daily gain of pregnant goats during gestation and high kid birthweight of indigenous goats inoculated with DHP-degrading rumen bacteria.

Brassica and SMCSO

Kale poisoning or hemolytic anemia from consumption of kale fodder has been known for 70 years, but only for about half this time has it been associated with SMCSO (Figure 35.11). Though SMCSO is found in many *Brassica* and *Allium* species, few of these are used for forage. In plant tissues, SMCSO may account for most of the soluble S and a significant amount of the nonprotein N. Concentrations of SMCSO in kale and rape generally are about 10 g kg^{-1} DM and increase during winter in the UK (Bradshaw and Borzucki 1981). With high soil-sulfate levels, increased soil nitrogen increased SMCSO but, at low soil sulfate levels, SMCSO levels in plant tissues decreased with increased available nitrogen (McDonald et al. 1981).

Animal consumption of SMCSO causes decreased feed intake and subsequent decreased growth. Amounts of SMCSO that will cause toxicity are dependent upon the basal diet, but with the concentrations generally found in kale, toxic amounts are frequently ingested (Benevenga et al. 1989). McCaffery (2002) reported the primary cause of death in sheep grazing canola during drought was SMCSO. Details of the toxic entity have not been determined, but with appropriate genetic manipulations, forages with reduced SMCSO levels should be developed and greatly minimize or eliminate the toxicity in ruminant animals.

Summary and Conclusions

The adverse effects of plant-related toxic secondary chemical compounds on livestock performance are long recognized. Their presence in plant species, that are otherwise agronomically acceptable, has limited their use as forage resources. Progress towards resolving livestock-related disorders has come via traditional gene technology and plant breeding methods, but likely will be enhanced with modern molecular genetic manipulation. Some plant species developed mutualistic associations with microorganisms that, together, interact to produce anti-hebivory compounds. Mutualism between the trophic species may include enhanced agronomic adaptations beyond the anti-herbivory chemicals produced by the association. Understanding the ecochemical and evolutionary significance of these associations is providing new avenues towards forage improvement by manipulating the mutualistic association. Modern metabolomic methods are providing insight into gut-related chemical transformations, hepatic transport, and modes of action of antiquality chemical derivatives. These technologies may provide new insights as to pharmacologically related strategies for toxin remediation. Finally, sufficient evidence exists indicating antiquality components of forages have biomedical implications. Use of all the aforementioned technologies may lead to novel biomedical uses of secondary antiquality phytochemicals found in forages, thus, transforming plant species from low-value antiquality forages to value-added **nutraceuticals**.

References

Aaerts, A.U., Barry, T.N., and McNabb, W.C. (1999). Polyphenols and agriculture: beneficial effects of proanthocyanidins in forages. *Agric. Ecosyst. Environ.* 75: 1–12.

Adams, N.R. (1995). Detection of the effects of phytoestrogens on sheep and cattle. *J. Anim. Sci.* 73: 1509–1515.

Adcock, R.A., Hill, N.S., Bouton, J.H. et al. (1997). Symbiont regulation and reducing ergot alkaloid concentration by breeding endophyte-infected tall fescue. *J. Chem. Ecol.* 23: 691–704.

Ainouche, A., Greinwald, R., Witte, L., and Huon, A. (1996). Seed alkaloid composition of *Lupinus tassilicus* Maire (Fabaceae: Genisteae) and comparison with its related rough seeded lupin species. *Biochem. System. Ecol.* 24: 405–414.

Akingbade, A.A., Nsahlai, I.V., Bonsi, M.L.K. et al. (2001). Reproductive performance of South African indigenous goats inoculated with DHP-degrading rumen bacteria and maintained on *Leucaena leucocephala*/grass mixture and natural pasture. *Small Ruminant Res.* 39: 73–85.

An, J., Tzagarakis-Foster, C., Scharschmidt, T.C. et al. (2001). Estrogen receptor beta-selective transcriptional activity and recruitment of coregulators by phytoestrogens. *J. Biol. Chem.* 276: 17808–17814.

Anderton, N., Cockrum, P.A., Walker, D.W., and Edgar, J.A. (1994). Identification of a toxin suspected of causing sudden death in livestock grazing phalaris pastures. In: *Plant-Associated Toxins: Agricultural, Phyto-Chemical And Ecological Aspects* (eds. S.M. Colegate and P.R. Dorling), 269–274. Wallingford, UK: CAB International.

Anderton, N., Cockrum, P.A., Colegate, S.M. et al. (1998). Oxindoles from *Phalaris coerulescens*. *Phytochemistry* 48: 437–439.

Anderton, N., Cockrum, P.A., Colegate, S.M. et al. (1999). Phalarine, a furano*bis*indole alkaloid from *Phalaris coerulescens*. *Phytochemistry* 51: 153–157.

Aniszewski, T., Ciesiolka, D., and Gulewicz, K. (2001). Equilibrium between basic nitrogen compounds in lupin seeds with differentiated alkaloid content. *Phytochemistry* 57: 43–50.

Assuero, S.G., Matthew, C., Kemp, P.D. et al. (2000). Morphological and physiological effects of water deficit and endophyte infection on contrasting tall fescue cultivars. *N. Z. J. Agric. Res.* 43: 49–61.

Ayers, A.W., Hill, N.S., Rottinghaus, G.E. et al. (2009). Ruminal metabolism and transport of tall fescue ergot alkaloids. *Crop Sci.* 49: 2309–2316.

Ayres, J.F., Murison, R.D., Turner, A.D., and Harden, S. (2001). A rapid semi-quantitative procedure for screening hydrocyanic acid in white clover (*Trifolium repens* L.). *Aust. J. Exp. Agric.* 41: 515–521.

Azevedo, M.D., Welty, R.E., Craig, A.M., and Bartlett, J. (1993). Ergovaline distribution, total nitrogen and phosphorous content of two endophyte-infected tall fescue clones. In: *Proc. 2nd Int. Symp. Acremonium/Grass Interactions* (eds. D.E. Hume, G.C.M. Latch and H.S. Easton), 59–62, Feb. 2–5. AgResearch, Grassland Research Centre, Palmerston North N.Z.

Baldwin, I.T., Olson, R.K., and Reiners, W.A. (1983). Protein binding phenolics and the inhibition of nitrification in sub-alpine balsam fir soils. *Soil Biol. Biochem.* 15: 419–423.

Barlog, P. (2002). Effect of magnesium and nitrogenous fertilizers on the growth and alkaloid content in *Lupinus angustifolius* L. *Aust. J. Agric. Res.* 53: 671–676.

Barry, T.N. and Duncan, S.J. (1984). The role of condensed tannins in the nutritional value of *Lotus pedunculatus* for sheep. 1. Voluntary intake. *Brit. J. Nutr.* 51: 485–491.

Barry, T.N. and Forss, D.A. (1983). The condensed tannin content of vegetative *Lotus pedunculatus,* its regulation by fertilizer application, and effect upon protein solubility. *J. Sci. Food Agric.* 34: 1047–1056.

Barry, T.N., Manley, T.R., and Duncan, S.J. (1986). The role of condensed tannins in the nutritional value of *Lotus pedunculatus* for sheep. 4. Site of carbohydrate and protein digestion as influenced by dietary reactive tannin concentration. *Brit. J. Nutr.* 55: 123–137.

Beck, V., Rohr, U., and Jungbauer, A. (2005). Phytoestrogens derived from red clover: an alternative to estrogen replacement therapy? Steroid Biochem. *Mol. Biol.* 94: 499–518.

Bedell, T.E. (1968). Seasonal forage preferences of grazing cattle and sheep in western Oregon. *J. Range Man.* 21: 291–297.

Belesky, D.P. and Fedders, J.M. (1995). Tall fescue development in response to *Acremonium coenophialum* and soil acidity. *Crop Sci.* 35: 529–553.

Belesky, D.P., Stringer, W.C., and Hill, N.S. (1989). Influence of endophyte and water regime upon tall fescue accessions. I. Growth characteristics. *Ann. Bot.* 63: 495–503.

Belfon, R. and Creamer, R. (2003). RAPD-PCR analysis of fungal endophyes of locoweed. *Phytopathology* 93: S7.

Bell, T.A., Etchells, J.L. and Smart, W.W.G. (1965). Pectinase and cellulose enzyme inhibitor from sericea lespedeza and certain other plants. *Botanical Gazette* 126 (1): 40–45.

Benevenga, N.J., Case, G.L., and Steele, R.D. (1989). Occurrence and metabolism of *S*-methyl-L-cysteine and *S*-methyl sulfoxide in plants and their toxicity and metabolism in animals. In: *Toxicants of Plant Origin*, vol. III (ed. P.R. Cheeke), 203–228. Boca Raton, FL: CRC Press.

Bertram, J.D., Sneath, R.J., Taylor, K.M. et al. (2003). Cyanide (prussic acid) and nitrates in sorghum crops: Risk management. Queensland Government, Department of Primary Industries, Animal and Plant Health. https://www.business.qld.gov.au/industries/farms-fishing-forestry/agriculture/land-management/health-pests-weeds-diseases/livestock/cyanide-nitrate-sorghum (accessed 23 November 2019).

Beyer, H., Schmalenberg, A.K., Jansen, G. et al. (2015). Evaluation of variability, heritability and environmental stability of seed quality and yield parameters of *L. angustifolius*. *Field Crops Res.* 174: 40–47.

Bourke, C.A., Carrigan, M.J., and Dixon, R.J. (1988). Experimental evidence that tryptamine alkaloids do not cause *Phalaris aquatica* sudden death syndrome in sheep. *Aust. Vet. J.* 65: 218–220.

Bourke, C.A., Carrigan, M.J., and Dixon, R.J. (1990). The pathogenesis of the nervous syndrome of *Phalaris aquatica* toxicity in sheep. *Aust. Vet. J.* 67: 356–358.

Bourke, C.A., Colegate, S.M., Rendell, D. et al. (2005). Peracute ammonia toxicity: a consideration in the pathogenesis of *Phalaris aqualtica* 'Polioencephalomalacia-like sudden death' poisoning of sheep. *Aust. Vet. J.* 83: 168–171.

Bouton, J.H., Latch, G.C.M., Hill, N.S. et al. (2002). Reinfection of tall fescue cultivars with non-ergot alkaloid-producing endophytes. *Agron. J.* 94: 567–574.

Bradshaw, J.E. and Borzucki, R. (1981). The effect of cultivar and harvest date on the chemical composition and digestibility of fodder kale. *J. Sci. Food Agric.* 32: 965–972.

Braun, K., Romero, J., Liddell, C. et al. (2003). Production of swainsonine by fungal endophytes of locoweed. *Mycol. Res.* 107: 980–988.

Broquist, H.P., Mason, P.S., Hagler, W.M. Jr., and Harris, T.M. (1984). Identification of swainsonine as a probable contributory mycotoxin in moldy forage toxicosis. *Appl. Environ. Microbio.* 48: 386–388.

Bunsupa, S., Okada, T., Saito, K., and Yamazaki, M. (2011). An acyltransferase-like gene obtained by differential gene expression profiles of quinolizidine alkaloid-producing and nonproducing cultivars of *Lupinus angustifolius*. *Plant Biotech.* 28: 89–94.

Busk, P.K. and Moller, B.L. (2002). Dhurrin synthesis in sorghum is regulated at the transcriptional level and induced by nitrogen fertilization in older plants. *Plant Physiol.* 129: 1222–1231.

Bush, L.P., Fannin, F.F., Siegel, M.R. et al. (1993). Chemistry, occurrence and biological effects of saturated pyrrolizidine alkaloids associated with endophyte-grass interactions. *Agric. Ecosyst. Environ.* 44: 81–102.

Casler, M.D. and Vogel, K.P. (1999). Accomplishments and impact from breeding for increased forage nutritional value. *Crop Sci.* 39: 12–20.

Cheeke, P.R. (1995). Endogenous toxins and mycotoxins in forage grasses and their effects on livestock. *J. Anim. Sci.* 73: 909–918.

Chen, Y., Lee, L.S., Luckett, D.J. et al. (2007). A quinolizidine alkaloid O-tigloyltraqnsferase gene in wild and domesticated white lupin (*Lupinus albus*). *Ann. Appl. Biol.* 151: 357–362.

Chynoweth, R.J., Rolston, M.P., Kelly, M., and Grbavac, N. (2012). Control of blind seed disease (*Gloeotinia temulenta*) in perennial ryegrass (*Lolium perenne*) seed crops and implications for endophyte transmission. *Agron. N. Z.* 42: 141–148.

Christensen, M.J., Easton, H.S., Simpson, W.R., and Tapper, B.A. (1998). Occurrence of the fungal endophyte *Neotyphodium coenophialum* in leaf blades of tall fescue and implications for stock health. *N. Z. J. Agric. Res.* 41: 595–602.

Christiansen, J.L., Raza, S., and Ortiz, R. (1999). I: White lupin (*Lupinus albus* L.) germplasm collection and preliminary in situ diversity assessment in Egypt. *Genet. Resour. Crop Evol.* 46: 169–174.

Colegate, S.M., Anderton, N., Edgar, J. et al. (1999). Suspected blue canary grass (*Phalaris coerulescens*) poisoning of horses. *Aust. Vet. J.* 77: 537–538.

Coffey, K.P., Lomas, L.W., and Moyer, J.L. (1990). Grazing and subsequent feedlot performance by steers that grazed different types of fescue pasture. *J. Prod. Ag.* 3: 415–420.

Cook, D., Ralphs, M., Welch, K., and Stegelmeier, B. (2009). Locoweed poisoning in livestock. *Rangelands* 31: 16–21.

Coulman, B.E. (1995). Bellevue reed canarygrass (*Phalaris arundinacea* L.). *Can. J. Plant Sci.* 74: 473–474.

Coulman, B.E., Woods, D.L., and Clark, K.W. (1977). Distribution within the plant, variation with maturity, and heritability of gramine and hordenine in reed canarygrass. *Can. J. Plant Sci.* 57: 771–777.

Cowling, W.A., Gladstones, J.S., and Knight, R. (2000). Lupin breeding in Australia. In: *Proc. 3rd Int. Food Legumes Res. Conf*, 541–547. Adelaide, Aust. 22–26 Sept. 1997. Kluwer Acad. Pub., Dordrecht, The Netherlands.

Croom, W.J. Jr., Hagler, W.M. Jr., Froetschel, M.A., and Johnson, A.D. (1995). The involvement of slaframine and swainsonine in slobbers syndrome: a review. *J. Anim. Sci.* 73: 1499–1508.

Cross, D.L., Redmond, L.M., and Strickland, J.R. (1995). Equine fescue toxicosis: signs and solutions. *J. Anim. Sci.* 73: 899–908.

Dalrymple, E.J., Goplen, B.P., and Howarth, R.E. (1984). Inheritance of tannins in birdsfoot trefoil. *Crop Sci.* 24: 921–923.

Davies, E., Lane, G.A., Latch, G.C.M. et al. (1993). Alkaloid concentrations in field-grown synthetic perennial ryegrass endophyte associations. In: *Proc. 2nd Int. Symp. Acremonium/Grass Interactions* (eds. D.E. Hume, G.C.M. Latch and H.S. Easton), 72–76, Feb. 2–5. AgResearch, Grassland Research Centre, Palmerston North N.Z.

Dear, B.S., Sandral, G.A., Spencer, D. et al. (2003). The tolerance of three transgenic subterranean clover (*Trifolium subterraneum* L.) lines with the *bxn* gene to herbicides containing bromoxynil. *Aust. J. Agric. Res.* 54: 203–210.

Dyer, D.C. (1993). Evidence that ergovaline acts on serotonin receptors. *Life Sci.* 53: PL223–PL228.

Easton, H.S. (2007). Grasses and Neotyphodium endophytes: co-adaptation and adaptive breeding. *Euphytica* 154: 295–306.

Easton, H.S., Latch, G.C.M., Tapper, B.A., and Ball, O.J. (2002). Ryegrass host genetic control of concentrations of endophyte-derived alkaloids. *Crop Sci.* 42: 51–57.

Elmi, A.A., West, C.P., Robbins, R.T., and Kirkpatrick, T.L. (2000). Endophyte effects on reproduction of a root-knot nematode (*Meloidogyne marylandi*) and osmotic adjustment in tall fescue. *Grass Forage Sci.* 55: 166–172.

Elizalde, J.C., Cremin, J.D., Faulkner, D.B., and Merchen, N.R. (1998). Performance and digestion by steers grazing tall fescue and supplemented with energy and protein. *J. Anim. Sci.* 76: 1691–1701.

Fales, S.L. (1984). Influence of temperature on chemical composition and IVDMD of normal and low-tannin sericea lespedeza. *Can. J. Plant Sci.* 65: 637–642.

Fink, B.N., Steck, S.E., Wolff, M.S. et al. (2007). Dietary flavonoid intake and breast cancer survival among women on Long Island. *Cancer Epidem. Biomarkers Prev.* 16: 2285–2292.

Fletcher, L.R. (1982). Observations of ryegrass staggers in weaned lambs grazing different ryegrass pastures. *N. Z. J. Exp. Agric.* 10: 203–207.

Fletcher, L.R. (1999). "Non-toxic" endophytes in ryegrass and their effect on livestock health and production. *Grassland Res. Pract.* 7: 133–139.

Foo, L.Y., Newman, R., Waghorn, G.C. et al. (1996). Proanthocyanidins from *Lotus corniculatus. Phyochemistry* 41: 617–621.

Foo, L.Y., Newman, R., Waghorn, G.C., and Ulyatt, M.J. (1997). Proanthocyanidins from *Lotus pedunculatus. Phytochemistry* 45: 1689–1696.

Foote, A.P., Harmon, D.L., Strickland, J.R. et al. (2014). Effect of ergot alkaloids on contractility of bovine right ruminal artery and vein. *J. An. Sci.* 89: 2944–2949.

Frick, K.M., Kamphuis, L.G., Diddique, K.H.M. et al. (2017). Quinolizidine alkaloid biosynthesis in lupins and prospects for grain quality improvement. *Front. Plant Sci.* 8: 87. https://doi.org/10.3389/fpls.2017.00087.

Frutos, P., Hervas, G., Giraldez, F.J., and Mantecon, A.R. (2004). Review. Tannins and ruminant nutrition. Spanish. *J. Agr. Res.* 2: 191–202.

Gallagher, R.T., Hawkes, A.D., Steyn, P.S., and Vleggaar, R. (1984). Tremogenic neurotoxins from perennial ryegrass causing ryegrass staggers disorder of livestock: structure and elucidation of lolitrem B. *J. Chem. Soc. Commun.* 9: 614–616.

Garcia, G.W., Ferguson, T.U., Neckles, F.A., and Archibald, K.A.E. (1996). The nutritive value and forage productivity of *Leucaena leucocephala. Anim. Feed Sci. Tech.* 60: 29–41.

Gildersleeve, R.R., Smith, G.R., Pemberton, I.J., and Gilbert, C.L. (1991). Detection of isoflavones in seedling subterranean clover. *Crop Sci.* 31: 889–892.

Goplen, B.P., Howarth, R.E., Sarkar, S.K., and Lesins, K. (1980). A search for condensed tannins in annual and perennial species of *Medicago, Trigonella,* and *Onobrychis. Crop Sci.* 20: 801–804.

Gorz, H.J., Haskins, F.A., Morris, R., and Johnson, B.E. (1987). Identification of chromosomes that condition dhurrin content in sorghum seedlings. *Crop Sci.* 27: 201–203.

Gremigni, P., Wong, M.T.F., Edwards, N.K. et al. (2001). Potassium nutrition effects on seed alkaloid concentrations, yield and mineral content of lupins (*Lupinus angustifolius*). *Plant Soil* 234: 131–142.

Gruhnert, C., Biehl, B., and Selmar, D. (1994). Compartmentation of cyanogenic glucosides and their degrading enzymes. *Planta* 195: 36–42.

Gwinn, K.D., Fribourg, H.A., Waller, J.C. et al. (1998). Changes in *Neotyphodium coenophialum* infestation levels in tall fescue pastures due to different grazing pressures. *Crop Sci.* 38: 201–204.

Hagerman, A.E. and Butler, L.G. (1981). The specificity of proanthocyanidin-protein interactions. *J. Biol. Chem.* 256: 4494–4497.

Hagler, W.M. Jr. and Croom, W.J. Jr. (1989). Slaframine: occurrence, chemistry and physiological activity. In: *Toxicants of Plant Origin*, vol. I (ed. P.R. Cheek), 257–279. Alkaloids: CRC Press, Boca Raton, FL.

Hammond, A.C. (1995). Leucaena toxicosis and its control in ruminants. *J. Anim. Sci.* 73: 1487–1492.

Hedqvist, H., Mueller-Harvey, I., Reed, J.D. et al. (2000). Characterisation of tannins and in vitro protein digestibility of several *Lotus corniculatus* varieties. *Anim. Feed Sci. Tech.* 87: 41–56.

Hiatt, E.E. and Hill, N.S. (1997). *Neotyphodium coenophialum* mycelial protein and herbage mass effects on ergot alkaloid concentration in tall fescue. *J. Chem. Ecol.* 12: 2721–2736.

Hill, N.S., Andrae, J.G., Durham, R.G., and Hancock, D.W. (2010). Herbicide treatments to renovate toxic endophyte infected tall fescue pastures with 'Jesup' Maxq. *Crop Sci.* 50: 1086–1093.

Hill, N.S., Belesky, D.P., and Stringer, W.C. (1991). Competitiveness of tall fescue as influenced by *Acremonium coenophialum*. *Crop Sci.* 31: 185–190.

Hill, N.S. and Brown, E. (2000). Endophyte viability in seedling tall fescue treated with fungicides. *Crop Sci.* 40: 1490–1491.

Hill, N.S., Bouton, J.H., Hiatt, E.E., and Kittle, B. (2005). Seed maturity, germination, and endophyte relationships in tall fescue. *Crop Sci.* 45: 859–863.

Hill, N.S., Thompson, F.N., Dawe, D.L., and Stuedemann, J.A. (1994). Antibody binding of circulating ergot alkaloids in cattle grazing tall fescue. *Am. J. Vet. Res* 55: 419–424.

Hill, N.S., Thompson, F.N., Stuedemann, J.A. et al. (2001). Ergot alkaloid transport across ruminant gastric tissues. *J. Anim. Sci.* 79: 542–549.

Jarred, R.A., Keikha, M., Dowling, C. et al. (2002). Induction of apoptosis in low to moderate grade human prostate carcinoma by red clover-derived dietary isoflavones. *Am. Assoc. Cancer Res.* 11: 1689–1696.

Johnson, N.D., Rigney, L.P., and Bentley, B.L. (1989). Short-term induction of alkaloid production in lupins. Differences between N_2-fixing and nitrogen-limited plants. *J. Chem. Ecol.* 15: 2425–2444.

Justus, M., Witte, L., and Hartmann, T. (1997). Levels and tissue distribution of loline alkaloids in endophyte-infected *Festuca pratensis*. *Phytochemistry* 44: 51–57.

Kallenbach, R.L., Bishop-Hurley, G.J., Massie, M.D. et al. (2003). Herbage mass, nutritive value, and ergovaline concentration of stockpiled tall fescue. *Crop Sci.* 43: 1001–1005.

Kelly, R.W., Hay, R.J.M., and Shackell, G.H. (1979). Formononetin content of "Grasslands Pawera" red clover and its oestrogenic activity to sheep. *N. Z. J. Exp. Agric.* 7: 131–134.

Lee, M.J., Pate, J.S., Harris, D.J., and Atkins, C.A. (2007). Synthesis, transport and accumulation of quinolizidine alkaloids in *Lupinus albus* L. and *L. angustifolius* L. *J. Exp. Bot.* 58: 935–946. https://doi.org/10.1093/jxb/erl254.

Lin, R., Renshaw, D., Luckett, D. et al. (2008). Development of a sequence-specific PCR marker linked to the gene "*pauper*" conferring low alkaloids in white lupin (*Lupinus albus*) for marker assisted selection. *Mol. Breed.* 23: 153–161.

Loper, G.M., Hanson, C.H., and Graham, J.H. (1967). Coumestrol content of alfalfa as affected by selection for resistance to foliar diseases. *Crop Sci.* 7: 189–192.

Lyons, P.C., Plattner, R.D., and Bacon, C.W. (1986). Occurrence of peptide and clavine ergot alkaloids in tall fescue grass. *Science* 232: 487–489.

Mack, J.P.G., Mulvena, D.P., and Slayton, M. (1988). N,N-dimethyltryptamine production in *Phalaris aquatica* seedlings. *Plant Physiol.* 88: 315–320.

Majak, W. (1992). Metabolism and absorption of toxic glycosides by ruminants. *J. Range Manage.* 45: 67–71.

Majak, W., McDiarmid, R.E., van Ryswyk, A.L. et al. (1979). Alkaloid level in reed canarygrass grown on wet meadows in British Columbia. *J. Range Manage.* 32: 322–326.

Malinowski, D.P. and Belesky, D.P. (2000). Adaptations of endophyte-infected cool-season grasses to environmental stresses: mechanisms of drought and mineral stress tolerance. *Crop Sci.* 40: 923–940.

Malinowski, D., Leuchtmann, A., Schmidt, D., and Nosberger, J. (1997). Growth and water status in meadow fescue is affected by *Neotyphodium* and *Phialophora* species endophytes. *Agron. J.* 89: 673–678.

Malinowski, D.P., Brauer, D.K., and Belesky, D.P. (1999). The endophyte *Neotyphodium coenophialum* affects root morphology of tall fescue grown under phosphorus deficiency. *J. Agron. Crop Sci.* 183: 53–60.

Marten, G.C. (1973). Alkaloids in reed canarygrass. In: *Anti-Quality Components of Forages* (ed. A.G. Matches), 15–31. CSSA Spec. Publ. 4. CSSA, Madison, WI.

Marten, G.C., Jordan, R.M., and Hovin, A.W. (1981). Improved lamb performance associated with breeding for alkaloid reduction in reed canarygrass. *Crop Sci.* 21: 295–298.

Marum, P., Hovin, A.W., and Marten, G.C. (1979). Inheritance of three groups of indole alkaloids in reed canarygrass. *Crop Sci.* 19: 539–544.

McAllister, T.A., Martinez, T., Bae, H.D. et al. (2005). Characterization of condensed tannins purified from legume forages: chromophore production, protein precipitation, and inhibitory effects on cellulose digestion. *J. Chem. Ecol.* 31: 2049–2067.

McCaffery, D. (2002). Risks in grazing or feeding canola. New South Wales Agriculture Agnote DPI-418.

McDonald, R.C., Manley, T.R., Barry, T.N. et al. (1981). Nutritional evaluation of kale (*Brassica oleracea*) diets. III. Changes in plant composition induced by soil fertility practices, with special reference to SMCSO and glucosinolate concentrations. *J. Agric. Sci. Camb.* 97: 13–23.

McLain-Romero, J., Creamer, R., Zepeda, H. et al. (2004a). The toxicosis of *Embellisia* fungi from locoweed (*Oxytropis lambertii*) is similar to locoweed toxicosis in rats. *J. Anim. Sci.* 82: 2169–2174.

McLain-Romero, J., Padilla, J., and Creamer, R. (2004b). Microscopic localization of *Embellisia* in seed and seedlings of southern speckle pod locoweed. *Phytopathology* 94: S69.

McMahon, L.R., McAllister, T.A., Berg, B.P. et al. (2000). A review of the effects of forage condensed tannins on ruminal fermentation and bloat in grazing cattle. *Can. J. Plant Sci.* 80: 469–485.

McMurray, C.H., Laidlaw, A.S., and McElroy, M. (1986). The effect of plant development and environment on formononetin concentration in red clover (*Trifolium pretense* L.). *J. Sci. Food Agric.* 37: 333–340.

Miller, P.R. and Ehlke, N.J. (1994). Condensed tannin relationships with in vitro forage quality analyses for birdsfoot trefoil. *Crop Sci.* 34: 1074–1079.

Miller, P.R. and Ehlke, N.J. (1996). Condensed tannins in birdsfoot trefoil: genetic relationships with forage yield and quality in NC-83 germplasm. *Euphytica* 92: 383–391.

Min, B.R. and Hart, S.P. (2003). Tannins for suppression of internal parasites. *J. Anim. Sci.* 81E (Suppl. 2): E102–E109.

Moller, B.L. and Conn, E.E. (1980). The biosynthesis of cyanogenic glucosides in higher plants. Channeling of intermediates in dhurrin biosynthesis by a microsomal system from *Sorghum bicolor* (Linn) Moench. *J. Biol. Chem* 255: 3049–3056.

Mote, R.S., Hill, N.S., Uppal, K. et al. (2017). Metabolomics of fescue toxicosis in grazing beef steers. *Food Chem. Toxicol.* 105: 285–299.

Oram, R.N., Schroeder, H.E. and Culvenor, R.A. (1985). Domestication of Phalaris aquatica as a pasture grass. *Proceedings of the 15th International Grassland Congress*, Kyoto, Japan (24–31 August 1985).

Panter, K.E. and Keeler, R.F. (1993). Quinolizidine and piperidine alkaloid tratogens from poisonous plants and their mechanism of action in animals. *Vet. Clin. North. Am. Food Anim.* 9: 33–40.

Panter, K.E., Mayland, H.F., Gardner, D.R., and Shewmaker, G. (2001). Beef cattle losses after grazing *Lupinus argenteus* (silvery lupine). *Vet. Hum. Toxicol.* 43: 279–282.

Payne, R.L., Bidner, T.D., Southern, L.L., and Geaghan, J.P. (2001a). Effects of dietary soy isoflavones on growth, carcass traits, and meat quality in growing-finishing pigs. *J. Anim. Sci.* 79: 1230–1239.

Payne, R.L., Bidner, T.D., Southern, L.L., and McMillin, K.W. (2001b). Dietary effects of soy isoflavones on growth and carcass traits of commercial broilers. *Poultry Sci.* 80: 1201–1207.

Petersen, J.C. and Hill, N.S. (1991). Enzyme inhibition by sericea lespedeza tannins and the use of supplements to restore activity. *Crop Sci.* 31: 827–832.

Popay, A.J. and Lane, G.A. (2000). The effect of crude extracts containing loline alkaloids on two New Zealand insect pests. In: *The Grassland Conference 2000 – 4th International Neotyphodium/Grass Interactions Symposium* (eds. V.H. Paul and P.D. Dapprich), 471–475. Abteilung Soest, Soest, Germany: Universitat-Gesamthochschule Paderborn.

Popay, A.J., Hume, D.E., Mainland, R.A., and Saunders, C.J. (1995). Field resistance to Argentine stem weevil (*Listronotus bonariensis*) in different ryegrass cultivars infected with an endophye deficient in lolitrem B. *N. Z. J. Agric. Res.* 38: 519–528.

Prestidge, R.A. (1993). Causes and control of perennial ryegrass staggers in New Zealand. *Agric. Ecosys. Env.* 44: 283–300.

Quirk, M.F., Paton, C.J., and Bushell, J.J. (1990). Increasing the amount of leucaena on offer gives faster growth rates of grazing cattle in South East Queensland. *Aust. J. Exp. Agric.* 30: 51–54.

Ralphs, M.H. and Williams, C. (1988). Alkaloid response to defoliation of velvet lupine (*Lupinus leucophyllus*). *Weed Tech.* 2: 429–432.

Ralphs, M.H., Creamer, R., Baucom, D. et al. (2008). Relationship between the endophyte Embellisia spp. and the toxic alkaloid swainsonine in major locoweed species (Astragalus and Oxytropis). *J. Chem. Ecol.* 34: 32–38.

Rijke, E., Zafra, G.A., Ariese, F. et al. (2001). Determination of isoflavone glucoside malonates in *Trifolium pretense* L. (red clover) extracts: quantification and stability studies. *J. Chromat.* 932: 55–64.

Roberts, C.A., Beuselinck, P.R., Ellersieck, M.R. et al. (1993). Quantification of tannins in birdsfoot trefoil germplasm. *Crop Sci.* 33: 675–679.

Roe, R. and Mottershead, B. (1962). Palatability of *Phalaris arundinacea* L. *Nature* 193: 255–256.

Rogers, W.M., Roberts, C.A., Andrae, J.G. et al. (2011). Seasonal fluctuation of ergovaline and total ergot alkaloid concentrations in tall fescue regrowth. *Crop Sci.* 51: 1291–1296.

Ropiak, H.M., Lachmann, P., Ramsay, A. et al. (2017). Identification of structural features of condensed tannins that affect protein aggregation. *PLoS One* 12 (1):e0171768. Doi:https://doi.org/10.1371/journal.pone.0170768.

Ross, M.D. and Jones, W.T. (1983). A genetic polymorphism for tannin production in *Lotus corniculatus* and its relationship to cyanide polymorphism. *Theor. Appl. Genet.* 64: 263–268.

Raffoul, J.J., Wang, Y., Kucuk, O. et al. (2006). Genistein inhibits radiation-induced activation of NF-κB in prostate cancer cells promoting apoptosis and G 2/M cell cycle arrest. *BMC Cancer* 6: 107. https://doi.org/10.1186/1471-2407-6-107.

Rowan, D.D., Hunt, M.B., and Gaynor, D.L. (1986). Peramine, a novel insect feeding deterrent from ryegrass infected with the endophyte *Acremonium loliae*. *J. Chem. Soc. Chem. Commun.* 12: 935–936.

Roylance, J.T., Hill, N.S., and Agee, C.S. (1994). Ergovaline and peramine production in endophyte-infected tall fescue: independent regulation and effects of plant and endophyte genotype. *J. Chem. Ecol.* 20: 2171–2183.

Ruiz, M.A. and Sotelo, A. (2001). Chemical composition, nutritive value, and toxicology evaluation of Mexican wild lupins. *J. Agric. Food Chem.* 49: 5336–5339.

Rumball, W., Keogh, R.G., Miller, J.E., and Claydon, R.B. (1997). 'Grasslands G27' red clover (*Trifolium pratense* L.). *N. Z. J. Agric. Res.* 40: 369–372.

Sarelli, L., Tuori, M., Saastamoinen, I. et al. (2003). Phytoestrogen content of birdsfoot trefoil and red clover: effect of growth stage and ensiling method. *Acta Agric. Scand. Section A.* 53: 58–63.

Sasamura, H., Takahashi, A., Miyao, N. et al. (2002). Inhibitory effect on expression of angiogenic factors by antiangiogenic agents in renal cell carcinoma. *Br. J. Cancer* 86: 768–773.

Schardl, C.L., Grossman, R.B., Nagabhyru, P. et al. (2007). Loline alkaloids: currencies of mutualism. *Phytochemistry* 68: 980–996.

Shu, X.O., Zheng, Y., Cai, H. et al. (2009). Soy food intake and breast cancer survival. *JAMA* 302: 2437–2443.

Shupe, J.L., Binns, W., James, L.F., and Keller, R.F. (1968). A congenital deformity in calves induced by the maternal consumption of lupin. *Aust. J. Agric. Res.* 19: 335–340.

Siegel, M.R., Latch, G.C.M., Bush, L.P. et al. (1990). Fungal endophyte-infected grasses: alkaloid accumulation and aphid response. *J. Chem. Ecol.* 16: 3301–3315.

Simons, A.B. and Marten, G.C. (1971). Relationship of indole alkaloids to palatability of *Phalaris arundinacea* L. *Agron. J.* 63: 915–919.

Singh, A.V., Franke, A.A., Blackburn, G.L., and Zhou, J.-R. (2006). Soy phytochemicals prevent orthotopic growth and metastasis of bladder cancer in mice by alterations of cancer cell proliferation and apoptosis and tumor angiogenesis. *Cancer Res.* 66: 1851–1858.

Sinha, P. and Chatterjee, C. (1994). Influence of boron on yield and grain quality of pearl millet (*Pennisetum glaucum*). *Indian J. Agric. Sci.* 64: 836–840.

Siritunga, D. and Sayre, R. (2004). Engineering cyanogen synthesis and turnover in cassava (*Manihot esculenta*). *Plant Mol. Biol.* 56: 661–669.

Soedarjo, M. and Borthakur, D. (1998). Mimosine, a toxin produced by the tree-legume *Leucaena* provides a nodulation competition advantage to mimosine-degrading *Rhizobium* strains. *Soil Biol. Biochem.* 30: 1605–1613.

Solomonson, L.P. (1981). Cyanide as metabolic inhibitor. In: *Cyanide in Biology* (eds. B. Vennesland, E.E. Conn, C.J. Knowles, et al.), 11–28. London: Academic Press.

Soto-Blanco, B., Gorniak, S.L., and Kimura, E.T. (2001). Physiopathological effects of the administration of chronic cyanide to growing goats: a model for ingestion of cyanogenic plants. *Vet. Res. Commun.* 25: 379–389.

Stochmal, A. and Oleszek, W. (1997). Changes of cyanogenic glucosides in white clover (*Trifolium repens* L.) during the growing season. *J. Agric. Food Chem.* 45: 4333–4336.

Stuedemann, J.A., Wilkinson, S.R., Williams, D.J. et al. (1975). Long-term broiler litter fertilization of tall fescue pastures and health and performance of beef cows. In: *Proc. 3rd Int. Symp*, 264. Livestock Wastes.

Suzuki, H., Koike, Y., Murakoshi, I., and Saito, K. (1996). Subcellular localization of acyltransferases for quinolizidine alkaloid biosynthesis in *Lupinus*. *Phytochemistry* 42: 1557–1562.

Takach, J.E. and Young, C.A. (2015). Alkaloid genotype diversity of tall fescue endophytes. *Crop Sci.* 54: 667–678.

Tattersall, D.B., Bak, S., Jones, P.R. et al. (2001). Resistance to an herbivore through engineered cyanogenic glucoside synthesis. *Science* 293: 1826–1828.

Tawfik, A.A. (1997). Response of *Lupinus hartwegii* Lindl. and its alkaloid content to foliar application of some chelated micronutrients. *Assiut J. Agric. Sci.* 28: 57–65.

Taylor, J.B., Strickland, J.R., Krehbiel, C.R. et al. (2003). Disposition of swainsonine in sheep following acute oral exposure. In: *Poisonous Plants and Related Toxins* (eds. T. Acamovic, C.S. Stewart and T.W. Pennycott), 102–107. Wallingford, Oxfordshire, UK: CABI Publishing.

Tubcharoen, S., Chaunchom, S., Saardrak, K. et al. 2003. The evaluation of estrogenic substance on 3 cultivars *Pueraria mirifica* tuberous roots at 6, 9 and 12 months growth for utilization in livestock production. *Proceedings of the 41st Kasetsart University Annual Conference*, Bangkok, Thailand (3–7 February 2003).

Van Wyk, B., Greinwald, R., and Witte, L. (1995). Alkaloid variation in the *Lupinus pusillus* group (Fabaceae: tribe Genisteae). *Biochem. System. Ecol.* 23: 533–537.

Vetter, J. (1995). Isoflavones in different parts of common *Trifolium* species. *J. Agric. Food Chem.* 43: 106–108.

Virk-Baker, M.K., Nagy, T.R., and Barnes, S. (2010). Role of phytoestrogens in cancer therapy. *Planta Med* 76: 1132–1142.

Welty, R.E., Azevedo, M.D., and Cooper, T.M. (1987). Influence of moisture content, temperature, and length of storage on seed germination and survival of endophytic fungi in seeds of tall fescue and perennial ryegrass. *Phytopathology* 77: 893–900.

Wen, L., Roberts, C.A., Williams, J.E. et al. (2003). Condensed tannin concentration of rhizomatous and non-rhizomatous birdsfoot trefoil in grazed mixures and monocultures. *Crop Sci.* 43: 302–306.

Wink, M. and Hartmann, T. (1982). Diurnal fluctuation of quinolizidine alkaloid accumulation in legume plants and photomixotrophic cell suspension cultures. *Z. Naturforsch.* 37: 369–375.

Wink, M. and Mende, P. (1987). Uptake of lupanine by alkaloid-storing epidermal cells of *Lupinus polyphyllus*. *Planta Med.* 53: 465–469.

Wink, M., Hartmann, T., and Witte, L. (1980). Enzymatic synthesis of quinolizidine alkaloids in lupin chloroplasts. *Z. Naturforsch.* 35: 93–97.

Wink, M., Meibner, C., and Witte, L. (1995). Patterns of quinolizidine alkaloids in 56 species of the genus *Lupinus*. *Phytochemistry* 38: 139–153.

Wong, E. and Latch, G.C.M. (1971). Effect of fungal diseases on phenolic contents of white clover. *N. Z. J. Agric. Res.* 14: 633–638.

Yashar, C.M., Spanos, W.J., Taylor, D.D., and Gercel-Taylor, C. (2005). Potentiation of the radiation effect with genistein in cervical cancer cells. *Gynecol. Oncol.* 99: 199–205.

Yu, Y., Zhao, Q., Wang, J. et al. (2010). Swainsonine-producing fungal endophytes from major locoweed species in China. *Toxicon* 56: 330–338.

CHAPTER

36

Laboratory Methods for Evaluating Forage Quality

William P. Weiss, Professor, *Animal Sciences, Ohio Agricultural Research and Development Center, The Ohio State University, Wooster, OH, USA*
Mary Beth Hall, Research Animal Scientist, *US Dairy Forage Research Center, USDA – Agricultural Research Service, Madison, WI, USA*

The nutritional value of forage is a function of its chemical composition and **digestibility**, and its effects on dry matter (DM) intake. Laboratory methods are available that can either measure or be used to estimate the concentrations of nutrients, digestibility, and intake potential of forages. Figure 36.1 contains a flow chart with many common quality measures. Proper sampling and analytic techniques are essential to obtain accurate, precise, and repeatable values. Accuracy, however, can be difficult to define if the analytic method measures a fraction that lacks true standards (e.g. fiber fractions). The utility and applicability of any method should be assessed by comparing performance of animals fed the test forage with the results estimated from laboratory data.

Sampling Methods

The goal of sampling is to obtain the same analytic values when a small portion of a larger pool of forage (i.e. population) is analyzed as would be obtained if the entire population were analyzed. However, with forages it is often difficult to sample the entire population or the population is so small, it has limited value. For example, only a small portion of stored **silage** is available to be sampled on one given day (e.g. the exposed face of a bunker silo). If sampled correctly, the sample should represent what was fed that day but may not represent the silage available for feeding the following day or the following week because that silage was not included in the sample. Furthermore, because of the heterogeneous nature of many forages, obtaining a representative sample can be extremely difficult and sampling error or sampling variation can be substantial. When sampling, a person collects particles, not nutrients and, for many forages, the nutrient composition, size, shape, and density of different particles vary greatly. For example, large pieces of alfalfa are usually stem which is high in fiber and low in protein and smaller particles are often leaves with high protein and low fiber. If the sample contained fewer small particles than the population, the fiber concentration of the sample would be greater than the true value for the silage. For corn silage sampled on multiple farms over a 12-month period, sampling variation was approximately the same magnitude as true variation for neutral detergent fiber (NDF) (St-Pierre and Weiss 2015). For alfalfa silage, over a 12-month period, true variation was about two-thirds of total variation and sampling variation comprised about one-third of the variation in NDF concentration. Proper sampling techniques are needed to reduce sampling variation and obtain a sample that truly represents the forage, and protocols for sampling are available (Weiss et al. 2014) and are discussed briefly below.

Forages: The Science of Grassland Agriculture, Volume II, Seventh Edition.
Edited by Kenneth J. Moore, Michael Collins, C. Jerry Nelson and Daren D. Redfearn.

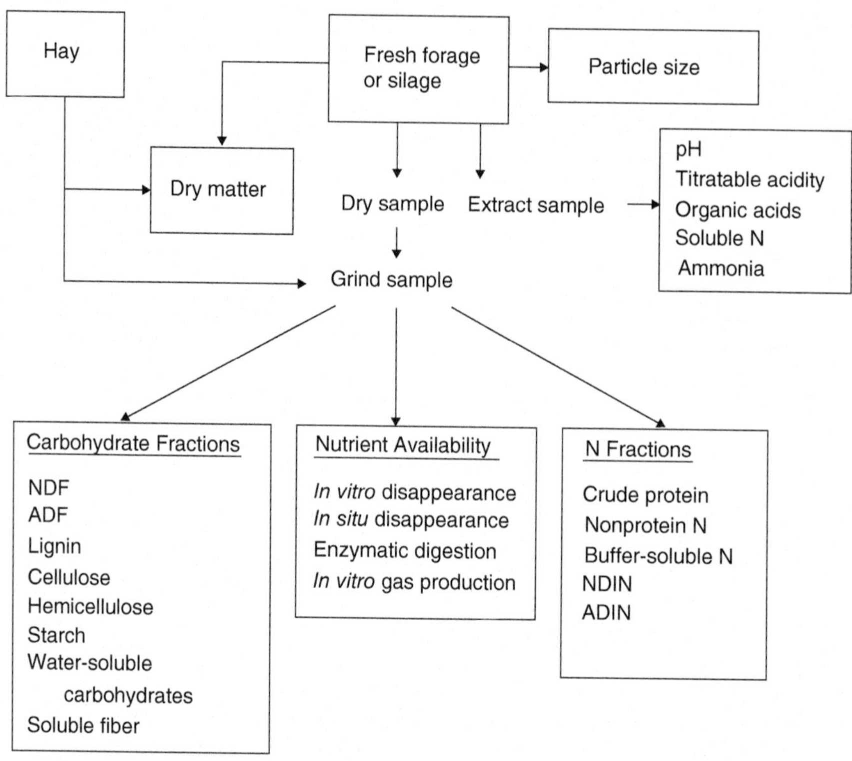

Fig. 36.1. Useful analytic measures for evaluating forage quality. NDIN = neutral detergent insoluble N; ADIN = acid detergent insoluble N.

Standing Crops and Pasture

For agronomic research, a plot is usually the experimental unit, and statistically appropriate sampling protocols have been developed. The entire plot is available for random sampling, and samples can be taken over time to examine temporal patterns. Grazing animals, however, do not randomly select pasture material (Coffey et al. 1991; Coblentz et al. 2002). Animals with esophageal or rumen **cannulas** are needed to obtain samples of pasture material actually consumed by grazing animals. Samples taken from surgically modified animals will better reflect the quality of the forage consumed, but their use is limited to research situations. To account for the high-ash concentration caused by saliva contamination, analytic data should be expressed on an organic matter basis. Hand sampling the paddock is often the only viable option. The goal of such sampling is to estimate the value of the forage consumed by the animal. Samples that mimic the forage selected by grazing animals should be collected. After sample collection, stubble height should resemble that left by grazing animals, and plant species not consumed by animals should not be sampled.

Chopped Forages and Silage

Forage samples can be collected as either chopped forage when it is placed in the silo or as silage when it is removed from the silo. The only advantage of sampling chopped forage is that the population from which samples are collected is broad (material from an entire field or cutting or all the material put into a silo is available for sampling). However, fresh or wilted forage samples are subject to considerable enzymatic and microbial activity, and artifacts can develop during storage and processing. Also, pre-ensiled samples may not accurately reflect the composition of the final silage because fermentation causes biologically important changes in certain quality measures.

A sample taken from a silo usually represents a very small portion of the silage because only a small proportion of the total silage mass is accessible to sampling at a specific point in time. Continuous sampling during the feed-out phase and compositing the samples should result in a representative sample of the silage fed. However, that process will cause a lag in obtaining population data. Samples should be taken on a regular schedule and the data composited (i.e. a rolling mean) over a period of time. The optimum number of data points used in the rolling mean is not known at this time, but a rolling mean of two or three samples is probably appropriate (e.g. if silage was sampled monthly; data from June to July would be averaged and when data from August is available, data from June is removed, and July and August data are averaged). When there is reason to believe that the population has changed (e.g. silage is from a different cutting) a new rolling mean should be started.

During the sampling process, care should be exercised to avoid losing smaller particles. For example, taking a grab sample often results in large particles being captured while smaller particles drop through the fingers. The use of a scoop or similar device is recommended.

Hay

Properly made **hay** that is protected from precipitation is stable, and quality does not change greatly during storage. Different hays (e.g. different cuttings, fields, etc., are generally aggregated into *lots*) can be segregated, and usually the entire population is available for sampling. Baled hay should be sampled with a sampling probe designed specifically for hay. The cutting surface of the probe should be sharp to avoid preferential sampling of leaf material. Cores should be taken from the small end of rectangular bales and from the lateral side of round bales. Based on variation in quality measurements among bales of alfalfa hay, 12 large rectangular bales (ca. 400 kg) or 20 small bales (ca. 30 kg) of hay within a defined population represented by a lot need to be sampled to obtain a representative sample (Sheaffer et al. 2000). The sampling requirements for round bales are probably similar to that for large rectangular bales. One core sample from each bale should be taken and composited into a single sample. To avoid sampling errors, a sample splitter should be used to obtain subsamples.

Sample Processing

Every sample-processing procedure has the potential to alter quality measurements, but it is essential that sample processing maintain important differences in quality measures among forages. Ideally, samples would be analyzed immediately after collection and without drying, but this is not usually practical. Samples of dry hay (<180 g kg^{-1} of moisture) do not require drying and are stable, hence analytic results from hay samples closely resemble the true composition of the hay. Silage and fresh forage samples are not stable. Several assays must be conducted on ground samples, and usually, most of the moisture must be removed to facilitate grinding. Because analytic artifacts can be produced during storage and drying, analytic data from samples of silage and fresh forage may not reflect the true composition of the original material.

Sample Storage

Silage samples deteriorate over time when stored at room temperature; hence they must be dried or frozen shortly after sampling. Concentrations of carbohydrate and N fractions in alfalfa and corn silage are not altered substantially by freezing (−25 °C). However, with fresh forage, sample freezing followed by oven-drying (55–70 °C) altered concentrations of carbohydrate and N fractions in fresh plant material (Kohn and Allen 1992; Deinum and Maassen 1994). Most of the changes caused by freezing were the result of increased N contamination of the NDF fraction.

Sample Drying

Wet samples need not be dried before analysis, but grinding and storing wet samples is difficult, especially for commercial labs that need high sample throughput and rapid analysis time. Samples can be air-dried (<30 °C), oven-dried (40°–100 °C), microwaved, or lyophilized. All these procedures produce artifacts.

Lyophilization is considered the sample-drying method that affects sample composition least, but it requires expensive equipment, and several days may be needed to dry a sample. Most studies comparing drying methods use lyophilization as the reference method. Compared with lyophilization, other drying methods usually increase fiber concentrations (Deinum and Maassen 1994; Doane et al. 1997) and decrease crude protein (CP) concentrations (Alomar et al. 1999). A portion of the decrease in CP is likely loss of ammonia and other volatiles that occur during oven-drying but the magnitude of the loss is greater than would be expected if only volatiles were lost. Increased fiber and decreased water-soluble carbohydrate (WSC) concentrations in air-dried samples are caused by continued plant **respiration** (fresh samples) or spoilage (silage samples) during the slow drying process. The increased fiber concentrations in oven-dried samples can also be a result of reduced solubility of certain N compounds in detergent solutions causing increased N contamination of the fiber fractions. Expressing concentrations of fiber on an N-free basis removes much of the differences between lyophilized and oven-dried samples.

Air-drying or low-temperature oven-drying of fresh forage increases the proportion of nonprotein N (NPN) and buffer soluble N compared with lyophilized samples because of proteolytic activity of endogenous proteases during the slow drying process. Fresh forage samples dried at moderate temperatures (ca. 50 °C) have similar NPN concentrations but lower concentrations of buffer soluble N than lyophilized samples. Drying silage samples at moderate temperatures has little effect on NPN concentrations and causes only small decreases in N solubility compared with lyophilized samples, probably because most of the N in silage is NPN. High-temperature drying (>70 °C) causes substantial changes in N fractions in fresh and ensiled samples and should be avoided. Microwave-drying usually changes N fractions more than drying in a conventional oven at moderate temperatures.

In vitro disappearance of dry matter and fiber does not differ greatly between samples that have been lyophilized, air-dried, oven-dried at moderate temperature, or microwaved (Deinum and Maassen 1994; Doane et al. 1997). Air-drying and oven-drying at moderate temperatures can reduce the rate of protein breakdown measured by various techniques. Microwave-drying has been shown to alter the ruminal *in situ* or *in vitro* fermentation degradability of protein (e.g. in canola meal; Sadeghi and Shawrang 2006a) and ruminal *in situ* degradation of **starch** (e.g. in corn grain; Sadeghi and Shawrang 2006b). Effects of microwave drying can vary substantially by device, its wattage, and how evenly and the degree to which the sample is heated.

Sample Grinding

Fineness of grind can affect both accuracy and precision of analytic data. Large particles have a lower surface area-to-mass ratio than small particles, which reduces interaction with reagents, and small particles can be lost during filtration steps in gravimetric assays. Concentrations of fiber in forages are negatively correlated with particle size of the sample (Ehle 1984); therefore, official methods (i.e. AOAC International) for gravimetric assays include recommended grinding procedures that should be followed. In contrast, use of an abrasion mill with a 1-mm screen or Wiley mill with a 0.5-mm screen are recommended for assays such as for starch or WSCs in which the measured analyte is extracted. Samples for detergent fiber methods should be ground through a Wiley mill (or similar device) with a 1-mm screen. Analytic precision for *in vitro* ruminal digestion and enzymatic digestion assays increases as particle size decreases, and generally grinding through a 1-mm screen is recommended (Weiss 1994). Fine grinding increases the rate of *in vitro* and enzymatic disappearance and can reduce differences among samples that may be biologically meaningful (Robles et al. 1980). Conversely, coarse grinding may exaggerate differences among samples.

Chemical Analyses

Dry Matter

On-Farm Methods

Because dry matter concentrations of silages vary more day to day (St-Pierre and Weiss 2015) than nutrients such as fiber or protein, dry matter should be assayed more frequently. On-farm methods must be quick, reasonably accurate, simple, and not require expensive equipment. Three common devices used for on-farm dry matter measurements are microwave ovens, forced heated air dryers (e.g. Koster Tester), and electronic moisture testers. If used properly, all give precise and reasonably accurate values for alfalfa and corn silages (Oetzel et al. 1993). Microwave and forced-air drying overestimated dry matter slightly, and the electronic tester overestimated dry matter in alfalfa but underestimated it with corn silage.

Laboratory Methods

Dry matter determination in laboratories has a high requirement for accuracy, but because essentially every sample obtained by a lab must be analyzed for dry matter, the method must be inexpensive and rapid. Dry matter can be measured as the mass remaining after oven-drying at temperatures between 60 and 104 °C for 3–48 hours (time required increases as oven temperature decreases) or after lyophilization over a period of at least two days. Water can be measured using **toluene distillation**, gas chromatography, or Karl Fischer titration (Petit et al. 1997). **Toluene distillation** (after correcting for volatile fatty acids and ethanol) is generally considered the reference method (Table 36.1). Though this method is accurate, it is not conducive to commercial application because of time requirements and the generation of toxic waste. Karl Fischer titration is relatively simple, but water concentrations tend to be underestimated in silages. Gas chromatography is accurate but is time consuming and requires expensive equipment.

Oven-drying at 60–104 °C is accurate for hay and fresh forage samples but causes losses of volatile organic compounds from silage. Direct oven-drying (100 °C) of moist samples usually overestimates dry matter of silages by 10–50 $g\,kg^{-1}$ compared with toluene distillation (Petit et al. 1997; Porter and Murray 2001). Free glucose and starch values can be reduced and Maillard products produced when moist corn silage is dried at 105 °C (Hall and Mertens 2008). When silage samples are oven-dried at about 60 °C, the difference between oven- and toluene-determined dry matter concentrations is decreased because of reduced loss of organic matter or incomplete removal of water. Reasonably accurate dry matter concentrations for silages have been obtained by oven-drying at 104 °C for six hours (Thiex and Richardson 2003).

Nitrogen Fractions

Official methods are available for determination of total N concentration in forages (AOAC International 2016 www.aoac.org). Kjeldahl analysis is accurate, but is hazardous and generates toxic waste. Combustion methods are accurate and do not generate waste, but the equipment is expensive. The concentration of N is multiplied by a conversion factor (usually 6.25) to calculate CP concentration. Near infrared reflectance spectroscopy (NIRS) can be used to determine N and CP, but must be calibrated accurately to a combustion or chemical method.

Table 36.1 Dry matter concentrations of silages determined using different methods

	Dry matter (g kg^{-1})
Grass, grass–legume mixtures[a]	
Toluene distillation	313
Oven (65 °C)	310
Karl Fischer titration	336
Gas chromatography	318
Grass[b]	
Toluene distillation	259
Oven (100 °C)	244
Oven (85 °C)	248
Oven (60 °C)	252
Gas chromatography	262
Corn silage[c]	
Toluene distillation	242
Oven (100 °C)	220
Alfalfa[d]	
Toluene distillation	294
Oven (90 °C)	277

[a]From Petit et al. (1997).
[b]From Porter and Murray (2001).
[c]From Haigh (1979).
[d]From Clancy et al. (1977).

Nitrogen in plants can be divided into protein and NPN fractions. Plant proteins include enzymes (mostly in leaves and stems) and storage proteins (mostly in seeds). Peptides, amino acids, and inorganic N compounds are the major components of NPN. High-quality forages often provide a substantial amount of the total CP in a diet and accurate estimation of its ability to provide metabolizable protein is essential to accurate diet formulation. Metabolizable protein is the protein (actually amino acids) absorbed in the small intestine of ruminants. Amino acids flowing to the intestine include the amino acids in feeds that were not degraded in the rumen (rumen undegradable protein) and microbial protein that was synthesized by ruminal microbes from rumen degradable protein in the feeds. To obtain estimated metabolizable protein concentrations, most ruminant nutrition models divide total N into an A fraction, which readily goes into solution and is assumed to be rapidly digested; a B fraction composed of true protein; and an unavailable C fraction. Measurement of the A, B, and C fractions has not been standardized and *in situ*, *in vitro*, and solubility methods have been used (Sniffen et al. 1992; NRC 2001, 2016). Mathematical models then estimate rumen degradable, rumen undegradable, and metabolizable protein using the A, B, and C fractions, plus some additional inputs such as dry matter intake. Additional research is needed to improve current methods and develop new methods to estimate nutritionally relevant protein fractions in forages and other feeds.

Forages ensiled with inadequate moisture or hays baled with excessive moisture are subject to spoilage by aerobic microorganisms. Metabolism of these microbes can generate substantial heat which damages the protein in the forage and reduces the digestibility of both protein and energy. Concentrations of acid detergent insoluble nitrogen (ADIN) are a good indicator of the extent of heat damage and the reduction in apparent digestibility of N in forages. Accurate equations are available that use ADIN to estimate digestibility of protein in heat-damaged forages (NRC 2001).

Fiber Fractions

The concentration of fiber in forage is arguably the most important measure of quality. Fiber concentrations are correlated with both forage digestibility and intake. The detergent system (Van Soest et al. 1991) is now the standard for forage fiber analyses. The detergent system partitions feeds into fractions that can be used to predict the nutritive value of a feed mechanistically, but does not divide feeds into chemically pure entities (Van Soest 1994). Van Soest relied heavily on the **Lucas method** to test the uniformity of feed fractions. The Lucas test is a statistical method developed by H.L. Lucas to determine feed fractions that have similar digestibility regardless

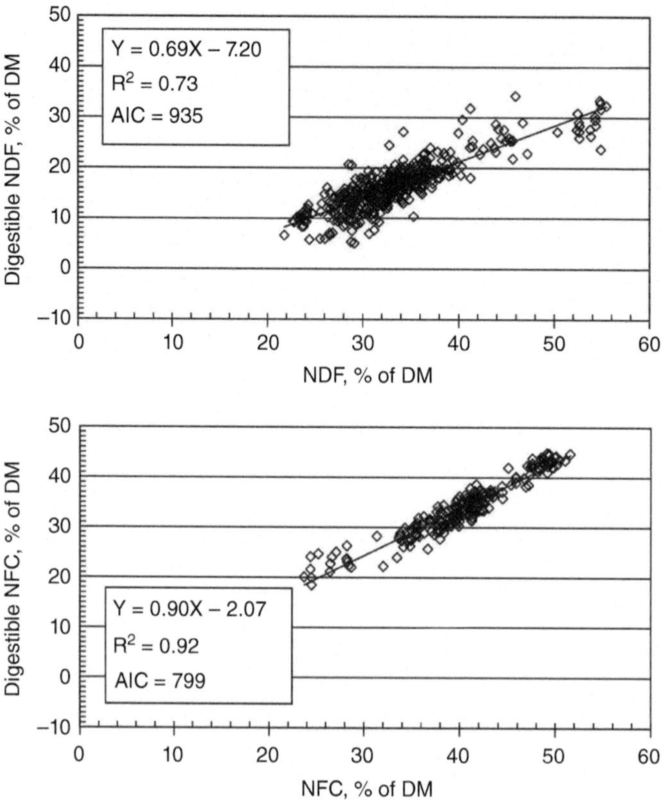

Fig. 36.2. Lucas test results showing that the non-fiber carbohydrate fraction is a uniform digestible fraction while the NDF fraction is not. *Source:* Data from numerous experiments conducted by W. Weiss lab.

of the feedstuff. The concentrations of a nutrient are regressed on the concentrations of digestible nutrient and if the data fit a linear function well (high R^2 and low Akaike's information criterion [AIC]) with a slope greater than 0 but less than or equal to 1 and the intercept is equal or less than 0, the fraction is considered uniform. As shown in Figure 36.2, non-fiber carbohydrate fits a linear function (similar results were observed for protein and starch) indicating it is a uniform fraction; whereas the fit for NDF was substantially poorer suggesting it is not uniform. Indeed, neither NDF nor Acid Detergent Fiber (ADF) are uniform fractions, but most components of the neutral detergent-soluble material approximate a uniform digestible fraction whereas acid detergent-soluble material does not. Because of this, NDF is more meaningful nutritionally than ADF. After a satisfactory ring test (Mertens 2002) with 12 laboratories (the repeatability standard deviation was 0.37–2.24 among laboratories), the AOAC approved the neutral detergent method with amylase and sodium sulfite. The Ankom filter-bag NDF, ADF, and lignin methods are growing in popularity despite some concerns about their values as compared to filter crucible-based methods. NDF is a chemically diverse fraction containing mostly carbohydrates but can have substantial concentrations of **ash** and nitrogen-containing compounds. In many samples, ash and N compounds may be intrinsic components of the cell wall but, in other situations, some of the ash and N should be considered contaminants. The addition of sulfite removes much, but not all, of the contaminating protein in NDF and is included in the reference method for NDF (Mertens 2002). The NDF residue can be assayed for N which is usually then converted to CP by multiplying by 6.25. That value can be subtracted from NDF to yield CP-free NDF. Likewise, NDF residue can be ashed and the ash subtracted to yield NDF_{OM} (NDF on an organic matter basis), which may be important with samples that are heavily soil contaminated. The NDF in most forages contains 10–30 g kg^{-1} ash and 20–100 g kg^{-1} CP; however, significant soil contamination and heat damage can

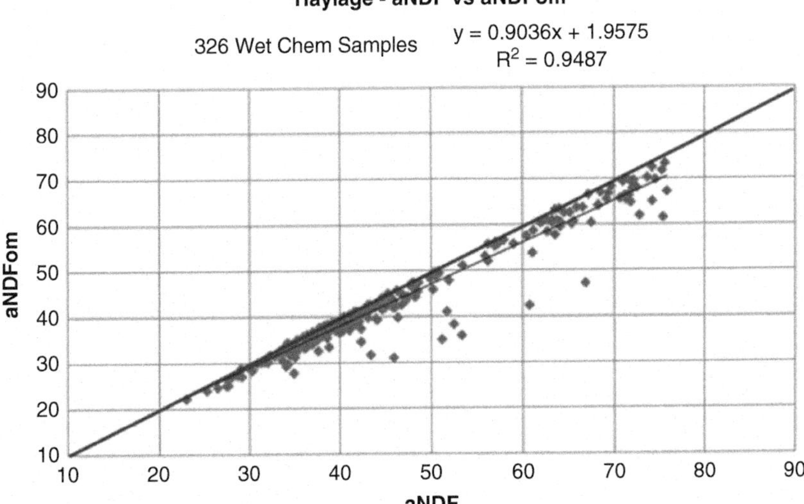

Fig. 36.3. Relationship between NDF and NDF corrected for ash (NDFom) for mixed legume-grass silage samples. For samples of hay crop silages (legumes, grasses, and mixtures). On average, NDFom is approximately 3 percentage units lower than NDF but for certain samples, the difference is large. The difference is correlated with the ash concentration of the sample. *Source:* Data provided by Dairyland Laboratories, Inc., Arcadia, WI.

increase concentrations of ash and CP in NDF, respectively. For most forages, adjusting values to an ash-free, CP-free, or ash and CP-free basis is not necessary because recommendations and feeding standards are based on uncorrected NDF values and for most forages, ash and CP are minor and relatively consistent contaminants (Figure 36.3). However, under certain conditions, contamination can be substantial. From a data set of 800 samples of hay crop silage provided by a commercial lab, in 9% of the samples NDF was more than 50 g kg^{-1} units higher than NDF_{OM}. The ash concentration of hay crop forages is positively correlated with the difference between NDF and NDF_{OM} (Figure 36.3). Analyzing NDF_{OM} in samples with higher than average **ash** concentrations may be useful but, even in those cases, the nutritional implications of high neutral detergent insoluble ash is not known. In other words, we do not know whether formulating diets for NDF_{OM} is an improvement over formulating for NDF. Laboratories are now often reporting NDF_{OM} in addition to or in place of NDF. ***In vivo*** digestibility of NDF_{OM} and CP-free NDF is greater than NDF but the size of the fraction is smaller; therefore, differences between concentrations of digestible NDF (e.g. digestible NDF vs digestible NDF_{OM}) is smaller than the differences in concentrations of NDF and NDF_{OM} would suggest (Tebbe et al. 2017). Whether *in vitro* **NDF digestibility** (IVNDFD) differs between NDF and NDF_{OM} is not known. The effect on digestibility likely mitigates the importance of the difference in concentrations among the different NDF terms.

The concentration of lignin is related to fiber indigestibility, however, developing accurate lignin methods that can be used for large numbers of samples has proven difficult. Differences in concentrations of lignin monomers (*p*-coumaric and ferulic acids in grasses), types of bonds (ether or C—C), and linkages to hemicellulose and other forage components all contribute to these analytic problems (Van Soest 1994; Brinkmann et al. 2002). A comparison of three lignin methods – acid detergent lignin (ADL), thioglycolic acid (TGA) and acetyl bromide (AB) – indicated that ADL had greater lignin values than the other two methods. Amino acid analysis revealed that the ADL residue contained significant amounts of amino acids (20–90 g kg^{-1} of dry matter (DM) in beech wood and 140–180 g kg^{-1} of DM in beech leaves). This would not invalidate use of ADL according to Van Soest's nutritional classification, assuming that these amino acids are indigestible.

Lignin × 2.4 and the ratio of lignin to NDF raised to the 2/3 power have both been used to estimate the indigestible and, inversely, the digestible fraction of NDF (NRC 2001, 2016). Estimates of digestible NDF are useful to identify any portion that may potentially be fermented and supply nutrients to the animal. Alternative methods for measuring indigestible neutral detergent fiber (INDF) have been explored. Extended ruminal *in situ* incubations (288 hours) to determine INDF showed that it gave values different for individual forage species than predicted by permanganate lignin × 2.4, though the average slope was 2.4 (Huhtanen et al. 2006). *In vivo* evaluation of the 288 hours *in situ* INDF showed that it was highly correlated with total tract organic matter digestibility. Extended *in vitro* fermentations (240 hours) have been suggested to estimate INDF (Van Amburgh et al. 2015), however, work relating these values to responses in the animal remains to be examined.

Measures of Silage Fermentation Quality

The type and extent of silage fermentation affects nutrient losses during ensiling as well as animal performance. Common measures of fermentation quality include pH and concentrations of organic acids and ammonia. Analyses must be conducted on wet samples, and samples should be frozen or analyzed as soon as possible after sampling to minimize changes that occur when silage is exposed to oxygen.

The compounds of interest are first extracted with water, and the pH is measured using a pH meter. Organic acids can be measured using **high performance liquid chromatography (HPLC)**, gas chromatography, or specifically configured autoanalyzers (Ward 2000). The HPLC methods are accurate for all organic acids, but instrument costs and sample throughput, limits their use. Volatile organic acids (acetic, propionic, butyric) are accurately measured using gas chromatography and, with appropriate derivatization, other organic acids can also be assayed with gas chromatography. Colorimetric assays based on the enzyme lactic dehydrogenase are accurate for lactic acid. Lactic dehydrogenase is isomer specific. About equal amounts of D- and L-lactic acid are usually found in silage, so if only L-lactic acid is measured that value must be multiplied by two to obtain an estimate of total lactic acid. Colorimetric assays or Kjeldahl-type distillation can be used to measure ammonia. Because ammonia and volatile fatty acids are lost upon oven or microwave drying, silage samples should not be dried prior to analysis of these compounds.

Mycotoxins

Mycotoxins are fungal compounds that cause toxic effects when consumed by animals (Sinha and Bhatnager 1998). The chemistry, structure, and toxicologic effects of these compounds are immensely diverse. More than 20 specific toxins have been identified in forages but many others may be present. Reference methods usually involve elaborate extraction procedures followed by HPLC or **gas-liquid chromatography** and can be extremely expensive. Enzyme-linked immunosorbent assays (ELISA) have been developed for some mycotoxins and their accuracy has improved (Urusov et al. 2015) but these methods have not been evaluated extensively with forages. Mycotoxins are not usually distributed randomly within a mass of forage so sampling error can be high. Repeated sampling decreases sampling error but increases analytic costs. Another major problem with mycotoxin analysis is interpretation of the data because the clinical effects of many mycotoxins on ruminants are not known.

Other Nutrients

Non-fiber carbohydrates tend to make up a smaller proportion of hay crop forage than does fiber, yet can provide a substantial portion of the digestible energy. Non-fiber carbohydrates include starch, WSC, and neutral detergent-soluble fiber. Starch is most commonly measured in assays in which the starch is first gelatinized with heat and moisture (most common), or with alkali or dimethyl sulfoxide which can hydrolyze enzyme-resistant starch. Then the α-(1,4), α-(1,6) linked polysaccharide is specifically hydrolyzed to glucose with the enzymes α-amylase and amyloglucosidase. The released glucose is measured with a glucose-specific assay such as the colorimetric glucose oxidase-peroxidase assay (Karkalas 1985). Starch is calculated as 0.9 multiplied by the difference in glucose values between the enzymatically-treated sample and the free glucose present in a sample with no enzymatic treatment. The 0.9 corrects for the water added when the starch was hydrolyzed. Failure to correct for free-glucose yields inflated starch values. Use of a starch assay that combines use of acidic buffer and specific measurement of glucose is recommended (Hall 2015).

WSCs in forages include monosaccharides, sucrose, oligosaccharides, and fructans, the last of which is present in cool-season grasses. The WSC represent the most rapidly available carbohydrates. Samples are extracted with water and the clarified extract is analyzed for its carbohydrate content. A comparison of the phenol-sulfuric acid assay (Dubois et al. 1956) and a reducing sugar assay (Lever 1973) as detection methods showed that the phenol-sulfuric acid assay gave WSC values closer to that of high performance ion chromatography (Hall 2014). Sucrose is commonly used as a standard; however, because different carbohydrates give different responses in the assay, the carbohydrate that predominates in samples should be used as the standard. The diversity of carbohydrates in WSC typically precludes finding an ideal standard.

Neutral detergent-soluble fiber includes pectins and mixed linkage beta-glucans. Generally, this fraction accounts for cell wall carbohydrates not found in NDF. Similar to WSC, the diversity in composition of the carbohydrates in this fraction make it challenging to analyze. A by-difference assay has been suggested to estimate this fraction as the 80% ethanol-insoluble fraction minus NDF, both corrected for ash and CP, and minus starch

(Hall et al. 1999). As with any by-difference estimate, variability or errors in measurement of each component will contribute to the soluble fiber value.

The concentrations of minerals in forages are usually not a primary measure of forage quality. Ash concentration is related inversely with energy concentrations and can be used as an index of soil contamination which reduces availability of some trace minerals. Overall, total ash is a valuable measurement to evaluate quality of forages. Measurement of ether extract has limited value for forages because fatty acids are essentially the only component of ether extract that provides digestible energy. In forages, typically less than 50% of the ether extract consists of fatty acids. Direct measurement of fatty acids is preferred, but because most forages have low concentrations of fatty acids ($<15\,g\,kg^{-1}$) fatty acid analysis is seldom warranted for routine quality evaluation.

Physical Measurements

Physical characteristics include particle size and fragility. Forage particle size in animal diets can affect both forage utilization and ruminal health. Terms such as physically effective fiber or simply effective fiber are used to describe the ability of a forage or diet to maintain a healthy rumen through encouraging **rumination** and selective retention of particles in the rumen. Finely chopped or ground forages may not stimulate adequate chewing to maintain proper rumen pH. Particle size of processed forages (e.g. silages and chopped hay) is measured by sieving through different screens. The standard method uses dried samples, five different-sized screens, and a mechanical shaker (ANSI 1998). Probably the most common method of evaluating forage particle size is a hand-operated shaker with either 2 (1.9 and 0.8 cm) or 3 (1.9, 0.8 and 0.4 cm) sieves (Penn State Particle Separator) that is available for on-farm use with wet samples (Kononoff et al. 2003; Heinrichs 2013). Though widespread consensus has not yet been reached regarding the best method to measure effective fiber, different indices have utility and research continues in this area (Zebeli et al. 2012).

Forage fragility can be measured in various ways including: the amount of energy needed to grind a specific mass of forage, particle size reduction of dried forages before and after ball-milling, and **shear force** using an instrument that measures the force (i.e. pressure) required to cut or tear a forage particle (Mir et al. 1995; Iwaasa et al. 1996; Cotanch et al. 2007). Forage fragility may improve estimation of the ability of a forage to stimulate chewing (i.e. effective fiber) and it may have value in improving the accuracy of estimated fiber digestibility. Water-holding capacity is measured with a pycnometer or by measuring the difference in mass between a dry sample and a sample that has soaked in buffer for a period of time.

Quality Measurements Unique to Corn Silage

In contrast to most other forages, corn silage provides substantial quantities of both forage fiber and starch. Consequently, analyses for fiber, starch, particle size of kernels and forage, and starch degradability can be useful for describing its nutritional value. **Kernel** processing scores are measurements unique to corn silage. The corn silage processing score requires dried corn silage to be sieved in a vertical mechanical shaker (Ferreira and Mertens 2005). The starch passing through the 4.75 mm sieve is estimated to be that located in kernels that were adequately processed. A goal is to have more than 50% of the starch in particles <4.75 mm, with preferred values being 70% or more <4.75 mm. An alternative field procedure that must be used on freshly chopped corn silage is to place a quantity of corn silage in a basin of water and agitate it by hand. The floating forage portion of the material is scooped out by hand, the water carefully poured off, and the grain that remains in the bottom of the basin evaluated. Adequately processed corn silage should have at least 95% of the kernels cracked, and 70% equal to or smaller than 1/3 to 1/4 kernel size (personal communication, Dr L. Kung, University of Delaware).

Near Infrared Reflectance Spectroscopy

Though instrumentation costs are high, per sample operating costs are low and analysis time is short for NIRS. A sample is introduced into a near infrared reflectance (NIR) spectrometer, reflectance of radiation in the infrared region (700–2500 nm) is measured, and a spectrum is generated. The spectra from feeds with known chemical composition (i.e. standards) are used to generate equations that are used to estimate the composition of unknown samples. Complex mathematical and statistical methods are used to generate and evaluate the equations (Shenk and Westerhaus 1994). NIRS has been used to estimate all measures of forage quality and new uses are being developed frequently (see review; Stuth et al. 2003). Accuracy of NIRS analysis is usually high for most macronutrients in forages (differences between NIR and the reference method is equal to or less than sampling error). NIRS can be used to accurately estimate *in vitro* fiber digestibility (e.g. Hoffman et al. 1999; DeBoever et al. 2002). Minor components including lignin and tannins have also been accurately estimated using NIRS when good calibration equations have been developed. Because NIRS detects the organic portion of the forage, it is generally not accurate enough for estimating mineral concentrations in forages, though concentrations of certain minerals may be correlated with certain organic molecules (Clark et al. 1987). Accuracy of NIRS is influenced by sample preparation (e.g. grinding and drying), thus test samples should be prepared using the same method used for calibration standards. Appropriate calibration equations are essential to obtain high accuracy with NIRS (the test forage must be similar to the calibration forages). Lower analytic costs and faster turnaround time for NIRS analysis may increase the number of samples that can be analyzed, thereby decreasing sampling error. New uses of NIRS to evaluate forage quality will undoubtedly continue to be developed.

In Vitro Methods

Attempts to predict *in vivo* digestibility using *in vitro* fermentations were first attempted in the early twentieth century, but poor anaerobic technique and inadequate buffers resulted in DM disappearance values significantly lower than those observed *in vivo*. By the early 1960s, several *in vitro* methods had been developed, including the two-stage Tilley and Terry (1963) method that is still widely used today with some modifications. A small amount of substrate is incubated anaerobically with ruminal fluid and buffer for 48 hours, microbial activity is halted, the residue is digested with acid-pepsin, and then the DM remaining is weighed. Other buffers have been developed (Goering and Van Soest 1970; Marten and Barnes 1980). Goering and Van Soest (1970) replaced the acid-pepsin step with neutral detergent extraction. This assay measures IVNDFD and *in vitro* true dry matter digestibility (IVTDMD). An equation is used to estimate *in vivo* digestibility from IVTDMD. Standardization is essential to ensure reproducible results. The factors that affect the accuracy and precision of *in vitro* results, including anaerobic technique, source, timing of collection, and handling of ruminal fluid inoculum, particle size, and buffer, have been reviewed (Weiss 1994; Kitessa et al. 1999).

The development of the Ankom system (Daisy II, Ankom, Macedon, NY) has increased the commercial applicability of *in vitro* assays. In the Tilley-Terry methods, each sample is fermented in a separate flask, but in the Ankom system feeds are placed into filter bags, and the bags share a common environment. Associative effects can be readily studied in the Ankom system, a potential advantage, but this can also be a disadvantage because inhibitory compounds found in some samples can affect the results of all samples in the batch. In the Ankom system, samples (0.25–0.5 g) in individual filter bags are placed in a 4-l vessel with buffer and rumen fluid and rotated gently during the fermentation. At the end of the fermentation, samples are rinsed with minimal agitation until the rinse water is clear. Inconsistent rinsing methods increase the variation in Ankom results.

Several studies have compared the Ankom method with a Tilley-Terry-like method or with an *in situ* technique (Robinson et al. 1999; Mabjeesh et al. 2000; Wilman and Adesogan 2000). Generally, the Ankom results did not differ from those obtained with the reference method. However, Wilman and Adesogan (2000) found that the Ankom method resulted in higher estimates of IVDMD than the Tilley-Terry method. Robinson et al. (1999) reported higher NDF digestibilities with an Ankom system (48 hours) than with the *in situ* method. Mabjeesh et al. (2000) reported that the Ankom system yielded higher digestibility estimates for some concentrate feeds but values for forages were similar to the reference method.

Evaluation of NDF digestibility is important for assessing the nutritional value of the forage to the animal. Samples are typically ground through the 1 mm screen of a Wiley mill or 2 mm screen of an abrasion mill. Forage IVNDFD is commonly measured at 24, 30, or 48 hours of *in vitro* fermentation, with earlier hours being more sensitive to differences in rate, and later hours more sensitive to extent of digestibility (Figure 36.4). The length of the lag time, the time in which no disappearance of NDF has occurred, can influence IVNDFD at later time points simply by delaying the start of disappearance. The variability inherent in the IVNDFD analysis averaged $\pm 50\,g\,kg^{-1}$ units from the mean on 30-hour values measured within a lab (Hall and Mertens 2012). The 48-hour values appeared to have slightly less variation. This would indicate that using IVNDFD with greater precision than $\pm 50\,g\,kg^{-1}$ is not warranted with current methods. Sample IVNDFD did rank consistently within fermentation run.

In vitro starch digestibility is estimated by the same fermentation approach as IVNDFD, but with a coarser grind to the samples and a seven hours incubation. A coarser grind such as using a 4 mm screen is important to partially maintain the structure of any grains present, as the physical and protein matrices around the **starch granules** affect starch digestibility. Starch digestibility by rumen microbes *in vitro* is affected by particle size of the material (Richards et al. 1995). Data relating *in vitro* starch digestibility to *in vivo* ruminal starch disappearance are lacking and at this time, this assay should be used to rank feeds.

Gas Production

Measurement of gas produced from *in vitro* fermentations is an indirect method of monitoring the progress of fermentation. *In vitro*, methane, carbon dioxide, and hydrogen gases are produced directly by microbes during fermentation, and carbon dioxide gas is produced indirectly by the neutralization of fermentation acids by the bicarbonate buffer in the fermentation medium (Figure 36.5). Accordingly, as fermentation gases or acids increase in the fermentation vessel, total **gas production** increases. Commonly, gas is typically measured using pressure transducers in sealed vessels in a computerized system (Pell and Schofield 1993) or by the use of solenoid valves that discharge gas at pre-set pressure levels, and the number of valve openings and the interval between openings are used to calculate cumulative gas production (Williams 2000). Blanks and standards are necessary to ensure that changes in barometric pressure can be used to

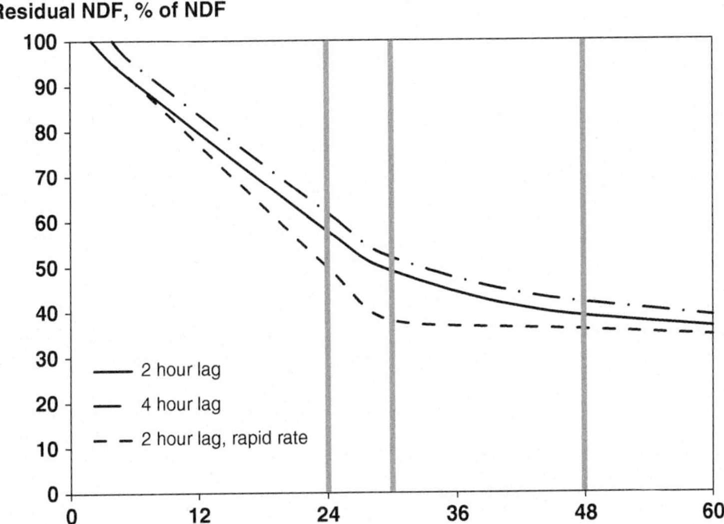

Fig. 36.4. Examples of *in vitro* NDF fermentation curves. The samples with two and four hours lag differ only in lag time. With different lag times and different rates or extents of NDF fermentation, the timing of sampling can change how samples compare. Gray vertical bars represent 24, 30, and 48 hour sampling times; lag is the time at which the NDF begins to disappear.

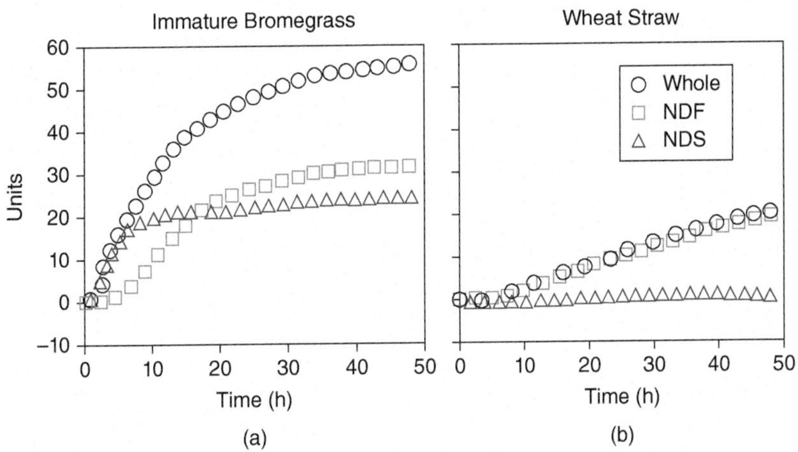

Fig. 36.5. Gas production from NDF, neutral detergent soluble material (NDS), and the whole forage (Schofield and Pell 1995).

correct gas readings (Schofield 2000). In sealed systems, provision of adequate head space relative to sample and fermentation medium amounts is important to limit the absolute increase in pressure. An advantage of gas production methods is the ability to obtain numerous time points on a given sample that can then be used to describe the rate kinetics of the fermentation. Extensive reviews of gas measurement have been published (Schofield 2000; Williams 2000)

The results of gas production measurements have been handled as a single pool or mathematically parsed into "fast" and "slow" pools. Because the gas produced comes from all fermented substrate, it has not been possible to relate the values to specific compositional fractions in feeds. Fermentation of protein or pentoses, and the variety of organic acids produced by rumen microbes produce variable amounts of gas that can influence results. Thus, the amount and profile of the volatile fatty acids produced must be taken into account.

As with conventional *in vitro* measurements, sample size, particle size, inoculum source, and pH control affect results obtained from **gas production** measurements (Schofield 2000; Williams 2000). Gas production methods are sensitive to changes in atmospheric pressure, temperature, stirring, and end-product formation. Gas production has been used to calculate fermentation of soluble fractions by incubating the whole forage and its extracted NDF fraction (Figure 36.5) (Schofield and Pell 1995; Kennedy et al. 1999). The validity of this approach is predicated on three assumptions: (i) extraction does not affect the digestion kinetics of the extracted fraction, (ii) associative effects between soluble and insoluble fractions are small, and (iii) compounds that inhibit digestion are not removed by the extraction procedure. Measuring NDF digestibility at the end of the fermentation can be used to assess the validity of these assumptions. The assumptions were true for temperate grasses and maize (Doane et al. 1997) but not for tropical grasses and legumes (Kennedy et al. 1999). As with many assays, extrapolating results obtained from small glass containers to whole animals has many potential pitfalls.

References

Alomar, D., Fuchslocher, R., and Stockebrand, S. (1999). Effects of oven- or freeze-drying on chemical composition and NIR spectra of pasture silage. *Anim. Feed Sci. Technol.* 80: 309–319.

American National Standards Institute (ANSI) (1998). *Method of Determining and Expressing Particle Size of Chopped Forage*. ASAE S424.1, p. 578. St. Joseph, MI: ASAE.

Brinkmann, K., Blaschke, L., and Polle, A. (2002). Comparison of different methods for lignin determination as a basis for calibration of near-infrared reflectance spectroscopy and implications of lignoproteins. *J. Chem. Ecol.* 28: 2483–2501.

Clancy, M., Wangsness, P.J., and Baumgardt, B.R. (1977). Effects of moisture determination method on estimates of digestibility and intakes of conserved alfalfa. *J. Dairy Sci.* 60: 216–223.

Clark, D.H., Mayland, H.F., and Lamb, R.C. (1987). Mineral analysis of forages with near infrared reflectance spectroscopy. *Agron. J.* 79: 485–490.

Coblentz, W.K., Coffey, K.P., Turner, J.E. et al. (2002). Comparisons of in situ dry matter disappearance kinetics of wheat forages harvested by various techniques and evaluated in confined and grazing steers. *J. Dairy Sci.* 85: 854–865.

Coffey, K.P., Moyer, J.L., Lomas, L.W., and Turner, K.E. (1991). Technical note: sampling technique and drying method effects on chemical composition of tall fescue or fescue-ladino clover pasture samples. *J. Anim. Sci.* 69: 423–428.

Cotanch, K.W., Grant, R.J., Darrah, J. et al. (2007). Development of a method for measuring forage fragility. *J. Dairy Sci.* 90 (suppl. 1): 563. (abstract).

DeBoever, J.L., Vanacker, J.M., and DeBrabander, D.L. (2002). Rumen degradation characteristics of nutrients in maize silages and evaluation of laboratory measurements and NIRS as predictors. *Anim. Feed Sci. Technol.* 101: 73–86.

Deinum, B. and Maassen, A. (1994). Effects of drying temperature on chemical composition and in vitro digestibility of forages. *Anim. Feed Sci. Technol.* 46: 75–86.

Doane, P.H., Pell, A.N., and Schofield, P. (1997). The effect of preservation method on the neutral detergent soluble fraction of forages. *J. Anim. Sci.* 75: 1140–1148.

Dubois, M., Gilles, K.A., Hamilton, J.K. et al. (1956). Colorimetric method for determination of sugars and related substances. *Anal. Chem.* 28: 350–356.

Ehle, F.R. (1984). Influence of particle size determination on fibrous feed components. *J. Dairy Sci.* 67: 1482–1488.

Ferreira, G. and Mertens, D.R. (2005). Chemical and physical characteristics of corn silages and their effects on in vitro disappearance. *J. Dairy Sci.* 88: 4414–4425.

Goering, H.K. and Van Soest, P.J. (1970). *Forage Fiber Analysis (Apparatus, Reagents, Procedures, and some Applications)*, Agric. Handbook No. 379. Washington, DC: ARS-USDA.

Haigh, P.M. (1979). A short note on the relationship between oven and toluene dry matter in maize silage. *J. Sci. Food Agric.* 30: 543–545.

Hall, M.B. (2014). Selection of an empirical detection method for determination of water-soluble carbohydrates in feedstuffs for application in ruminant nutrition. *Anim. Feed Sci. Technol.* 198: 28–37.

Hall, M.B. (2015). Determination of dietary starch in animal feeds and pet food by an enzymatic-colorimetric method: collaborative study. *J. AOAC Int.* 98: 397–409.

Hall, M.B. and Mertens, D.R. (2008). Technical note: effect of sample processing procedures on measurement of starch in corn silage and corn grain. *J. Dairy Sci.* 91: 4830–4833.

Hall, M.B. and Mertens, D.R. (2012). A ring test of in vitro neutral detergent fiber digestibility: analytical variability and sample ranking. *J. Dairy Sci.* 95: 1992–2003.

Hall, M.B., Hoover, W.H., Jennings, J.P., and Miller Webster, T.K. (1999). A method for partitioning neutral detergent soluble carbohydrates. *J. Sci. Food Agric.* 79: 2079–2086.

Heinrichs, J. (2013). The Penn State Particle Separator. Penn State University. Extension Bulletin No. DSE 2013-186.

Hoffman, P.C., Brehm, N.M., Bauman, L.M. et al. (1999). Prediction of laboratory and in situ protein fractions in legume and grass silages using near-infrared reflectance spectroscopy. *J. Dairy Sci.* 82: 764–770.

Huhtanen, P., Nousiainen, J., and Rinne, M. (2006). Recent developments in forage evaluation with special reference to practical applications. *Agric. Food Sci.* 15: 293–323.

Iwaasa, A.D., Beauchemin, K.A., Buchanan-Smith, J.G., and Acharya, S.N. (1996). A shearing technique measuring resistance properties of plant stems. *Anim. Feed Sci. Technol.* 57: 225–237.

Karkalas, J.J. (1985). An improved enzymatic method for the determination of native and modified starch. *J. Sci. Food Agric.* 36: 1019–1027.

Kennedy, P.M., Lowry, J.B., and Conlan, L.L. (1999). Isolation of grass cell walls as neutral detergent fibre increases their fermentability for rumen microorganisms. *J. Sci. Food Agric.* 79: 544–548.

Kitessa, S., Flinn, P.C., and Irish, G.G. (1999). Comparison of methods used to predict the in vivo digestibility of feeds in ruminants. *Aust. J. Agric. Res.* 50: 825–841.

Kohn, R.A. and Allen, M.S. (1992). Storage of fresh and ensiled forages by freezing affects fibre and crude protein fractions. *J. Sci. Food Agric.* 58: 215–220.

Kononoff, P.J., Heinrichs, A.J., and Buckmaster, D.R. (2003). Modification of the Penn State forage and total mixed ration particle separator and the effects of moisture content on its measurements. *J. Dairy Sci.* 86: 1858–1863.

Latimer, G.W. Jr. (2016). *Official Methods of Analysis of AOAC International*, 20e. Rockville, MD: AOAC International.

Lever, M. (1973). Colorimetric and fluorometric carbohydrate determination with p-hydroxybenzoic acid hydrazide. *Biochem. Med.* 7: 274–281.

Mabjeesh, S.J., Cohen, M., and Arieli, A. (2000). In vitro methods for measuring the dry matter digestibility of ruminant feedstuffs: comparison of methods and inoculum source. *J. Dairy Sci.* 83: 2289–2294.

Marten, G.C. and Barnes, R.F (1980). Prediction of energy digestibility of forages with in vitro rumen fermentation and fungal enzyme systems. In: *Standardization of Analytical Methodology for Feeds* (eds. W.C. Pigden, C.C. Balch and M. Graham), 61–71. Ottawa: International Development Research Center.

Mertens, D.R. (2002). Gravimetric determination of amylase-treated neutral detergent fiber in feeds with refluxing in beakers or crucibles: collaborative study. *J. AOAC Int.* 85: 1217–1240.

Mir, P.S., Mir, Z., Broersma, K. et al. (1995). Prediction of nutrient composition and in vitro dry matter digestibility from physical characteristics of forages. *Anim. Feed Sci. Technol.* 55: 275–285.

NRC (2001). *Nutrient Requirements of Dairy Cattle. 7th Revised edn.* Washington, DC: National Academy Press.

NRC (2016). *Nutrient Requirements for Beef Cattle. 8th Revised edn.* Washington, DC: National Academy Press.

Oetzel, G.R., Villalba, F.P., Goodger, W.J., and Nordlund, K.V. (1993). A comparison of on-farm methods for estimating the dry matter content of feed ingredients. *J. Dairy Sci.* 76: 293–299.

Pell, A.N. and Schofield, P. (1993). Computerized monitoring of gas production to measure forage digestion in vitro. *J. Dairy Sci.* 76: 1063–1073.

Petit, H.V., LaFreniere, C., and Veira, D.M. (1997). A comparison of methods to determine dry matter in silages. *J. Dairy Sci.* 80: 558–562.

Porter, M.G. and Murray, R.S. (2001). The volatility of components of grass silage on oven drying and the inter-relationship between dry-matter content estimated by different analytical methods. *Grass Forage Sci.* 56: 405–411.

Richards, C.J., Pedersen, J.F., Britton, R.A. et al. (1995). In vitro starch disappearance procedure modifications. *Anim. Feed Sci. Technol.* 55: 35–45.

Robinson, P.H., Mathews, M.C., and Fadel, J.G. (1999). Influence of storage time and temperature on in vitro digestion of neutral detergent fibre at 48 h, and comparison to in sacco neutral detergent fibre digestion. *Anim. Feed Sci. Technol.* 80: 257–266.

Robles, A.Y., Belyea, R.L., Martz, F.A., and Weiss, M.F. (1980). Effect of particle size upon digestible cell wall and rate of in vitro digestion of alfalfa and orchardgrass forages. *J. Anim. Sci.* 51: 783–790.

Sadeghi, A.A. and Shawrang, P. (2006a). Effects of microwave irradiation on ruminal degradability and in vitro digestibility of canola meal. *Anim. Feed Sci. Technol.* 127: 45–54.

Sadeghi, A.A. and Shawrang, P. (2006b). Effects of microwave irradiation on ruminal protein and starch degradation of corn grain. *Anim. Feed Sci. Technol.* 127: 113–123.

Schofield, P. (2000). Gas production methods. In: *Farm Animal Metabolism and Nutrition: Critical Reviews* (ed. J.P.F. D'Mello), 209–232. Wallingford, UK: CAB International.

Schofield, P. and Pell, A.N. (1995). Measurement and kinetic analysis of the neutral detergent-soluble carbohydrate fraction of legumes and grasses. *J. Anim. Sci.* 73: 3455–3463.

Sheaffer, C.C., Martin, N.P., Jewett, J.G. et al. (2000). Sampling requirements for forage quality characterization of rectangular hay bales. *Agron. J.* 92: 64–68.

Shenk, J.S. and Westerhaus, M.O. (1994). The application of near infrared reflectance spectroscopy (NIRS) to forage analysis. In: *Forage Quality, Evaluation and Utilization* (eds. G.C. Fahey, M. Collins, D.R. Mertens and L.E. Moser), 406–449. Madison, WI: ASA, CSSA, and SSSA.

Sinha, K.K. and Bhatnager, D. (1998). *Mycotoxins in Animal and Food Safety*. New York: Marcel Dekker.

Sniffen, C.J., O'Connor, J.D., Van Soest, P.J. et al. (1992). A net carbohydrate and protein system for evaluating cattle diets: II. Carbohydrate and protein availability. *J. Anim. Sci.* 70: 3562–3577.

St-Pierre, N.R. and Weiss, W.P. (2015). Partitioning variation in nutrient composition data of common feeds and mixed diets on commercial dairy farms. *J. Dairy Sci.* 98: 5004–5015.

Stuth, J., Jama, A., and Tolleson, D. (2003). Direct and indirect means of predicting forage quality through near infrared reflectance spectroscopy. *Field Crop Res.* 84: 45–56.

Tebbe, A.W., Faulkner, M.J., and Weiss, W.P. (2017). Effect of partitioning the nonfiber carbohydrate fraction and neutral detergent fiber method on digestibility of carbohydrates by dairy cows. *J. Dairy Sci.* 100: 6218–6228.

Thiex, N. and Richardson, C.R. (2003). Challenges in measuring moisture content of feeds. *J. Anim. Sci.* 81: 3255–3266.

Tilley, J.M.A. and Terry, R.A. (1963). A two-stage technique for the in vitro digestion of forage crops. *J. Br. Grassl. Soc.* 18: 104–111.

Urusov, A., Zherdev, A., Petrakova, A. et al. (2015). Rapid multiple immunoenzyme assay of mycotoxins. *Toxins* 7: 238.

Van Amburgh, M.E., Collao-Saenz, E.A., Higgs, R.J. et al. (2015). The Cornell net carbohydrate and protein system: updates to the model and evaluation of version 6.5. *J. Dairy Sci.* 98: 6361–6380.

Van Soest, P.J. (1994). *Nutritional Ecology of the Ruminant*. Ithaca, NY: Cornell University Press.

Van Soest, P.J., Robertson, J.B., and Lewis, B.A. (1991). Methods for dietary fiber, neutral detergent fiber, and nonstarch polysaccharides in relation to animal nutrition. *J. Dairy Sci.* 74: 3583–3597.

Ward, R.T. (2000). Fermentation analysis: use and interpretation. In: *Tri-State Dairy Nutrition Conference*, 117–135. Ft. Wayne, IN.

Weiss, W.P. (1994). Estimation of digestibility of forages by laboratory methods. In: *Forage Quality, Evaluation and Utilization* (eds. G.C. Fahey, M. Collins, D.R. Mertens and L.E. Moser), 644–681. Madison, WI: ASA, CSSA, and SSSA.

Weiss, W.P., Hill, C.T., and St-Pierre, N. (2014). Proper sampling and sampling scheduling can prevent reduced milk yields. In: *Proceedings of Tri-State Dairy Nutrition Conference*, 149–162.

Williams, B.A. (2000). Cumulative gas-production techniques for forage evaluation. In: *Forage Evaluation in Ruminant Nutrition* (eds. D.I. Givens, E. Owens, R.F.E. Axford and H.M. Omed), 189–211. Wallingford, UK: CAB International.

Wilman, D. and Adesogan, A.T. (2000). A comparison of filter bag methods with conventional tube methods of determining the in vitro dry matter digestibility of forages. *Anim. Feed Sci. Technol.* 84: 33–47.

Zebeli, Q., Aschenbach, J.R., Tafaj, M. et al. (2012). Invited review: role of physically effective fiber and estimation of dietary fiber adequacy in high-producing dairy cattle. *J. Dairy Sci.* 95: 1041–1056.

CHAPTER

37

Animal Methods for Evaluating Forage Quality

Eric S. Vanzant, Associate Professor, *University of Kentucky, Lexington, KY, USA*
Robert C. Cochran, Professor, *Kansas State University, Manhattan, KS, USA*
Wayne K. Coblentz, Research Dairy Scientist/Agronomist, *USDA-Agricultural Research Service, Marshfield, WI, USA*

The words ***forage quality*** can evoke a variety of images. Examples include color and aroma, physiologic maturity, leaf-to-stem ratio, or concentrations of particular **nutrients**. In most situations, there is a legitimate association between the identified characteristic(s) and forage quality; however, none of these represent the essence of forage quality. Fundamentally, forage quality is the ability of forages to support maintenance and production functions in the animals to which they are fed. Though forages contain constituents of all nutrient classes, their primary contribution to animal sustenance is through provision of nitrogenous components and energy. Thus, our primary focus when describing forage quality is on these components. The most accurate representation of the ability to support energetic needs (maintenance and production) in ruminant feedstuffs is the **net energy** (NE) value (Blaxter 1989). The world's major feeding systems for ruminants are founded upon some adaptation of the net energy concept. Protein systems are largely predicated on knowledge of the ruminal and postruminal fates of nitrogenous constituents. However, direct characterization of net energy values and nitrogen utilization is expensive, time consuming, and technically challenging. Thus, both laboratory (see Chapters 34 and 36) and animal methods have been developed to predict nutritive value (e.g. net energy) or provide insight into the potential of a forage to support maintenance and (or) productive functions.

Direct Measurement of Livestock Performance

Though not well suited to the development of standardized feeding systems, simple gravimetric (e.g. body weight gain) or subjective (e.g. body condition change) measurements are easier to complete than formal energy balance measurements and can provide useful insights into forage quality.

Live Weight Gain

Measurement of live weight change is conceptually simple (i.e. final weight minus starting weight); however, there are several experimental factors that investigators must manage properly to minimize variability and to produce reliable inferences. Animal-to-animal variation is generally quite large and is potentially larger than pasture-to-pasture variation (Mott and Lucas 1953). Therefore, it is important to consider animal numbers closely when individuals serve as experimental units.

Forages: The Science of Grassland Agriculture, Volume II, Seventh Edition.
Edited by Kenneth J. Moore, Michael Collins, C. Jerry Nelson and Daren D. Redfearn.

Working with **empty body weight** changes in cattle, Garrett and Hinman (1969) found that six to eight or more animals per treatment was adequate for gravimetric determination of empty body composition differences. Because reticulo-ruminal fill ranges from 10% to 23% of live weight (Lawrence 2002) and varies substantially relative to time of feeding, the use of increased numbers of animals is justified when individual animals serve as experimental units and data are collected on a live weight basis. Moreover, this illustrates the value of standardizing the time of weighing relative to feeding (Lawrence 2002). Cook and Stubbendieck (1986) concluded that five pasture replications per treatment will do an acceptable job of detecting gain differences of 20% or larger in grazing trials using growing cattle. However, it is important to recognize that such recommendations are heavily dependent on variation among experimental units, which itself is highly variable. For example, standard deviations for average daily gain (ADG) of growing cattle from recent *Journal of Animal Science* reports where animal group served as the experimental unit ranged from about 0.10 to 0.25 $kg\,d^{-1}$ (Hersom et al. 2015; Vendramini et al. 2015; Wilson et al. 2015; Wilson et al. 2016). Power analysis indicates that, to detect a 0.2 $kg\,d^{-1}$ difference in ADG with $\alpha = 0.05$ and $\beta = 0.80$, 6 replicates per treatment are sufficient at the lower level of variation (0.10 $kg\,d^{-1}$), whereas 26 replicates per treatment would be required for studies at the higher end (0.25 $kg\,d^{-1}$) of the variation range. This highlights the importance of using realistic local estimates of variation for power analysis and of using appropriate methodologic and statistic procedures to control variation among experimental units.

Despite these realities, logistical constraints often result in the use of fewer (two to four) pasture replicates. Precision can be increased when replications are limited by conducting trials over multiple years to increase the number of observations per treatment. However, this approach frequently uncovers year by treatment interactions, which require interpretation of treatment effects within years and, thereby, fails to improve the replication issue.

Length of period used for measuring weight change often corresponds to the production situation in which the data will ultimately be applied. In general, it is difficult to draw meaningful statistical inferences from live weight change data collected over periods shorter than one month unless large numbers of replicates are used. Live weight change is preferably determined over periods of 45–60 days or longer. Precision of live weight change data will be improved by longer periods of measurement. Because weight differences among treatments are often small, they can easily be masked by errors in weighing methodology, such as confounding specific treatments with weighing time. Often, cattle are confined overnight without food or water prior to weighing in an effort to reduce animal to animal variability associated with differences in gut fill, but Brown et al. (1993) reported that coefficients of variation for shrunk weights and associated gains were generally even larger than observed for full weights and associated gains. As a result, they were unable to conclude that overnight fasting offered any improvement over weighing without fasting. If fasting is used, the length of fast should be standardized among treatments. Similarly, weights collected without fasting should be standardized relative to major feeding events and should be obtained without bias toward any specific treatment. These standards should be maintained throughout the course of a given experiment.

Body Condition Score Change

Body condition scoring systems have been officially adopted by both the NRC Dairy (NRC 2001) and Beef (NRC 2016) committees under the assumption that condition scores effectively estimate tissue energy reserves. Forage quality effects on body condition score change are measured by subjective evaluation of cattle (visual and/or palpated) at the beginning and end of an experimental period and assignment of a rank score at each point. Typically, dairy cattle are scored on a 1–5 scale (NRC 2001) and beef cattle on a 1–9 scale (NRC 2016). For both, lower numbers are associated with lower tissue energy reserves. Often, multiple observers independently assign condition scores. Condition score data are typically collected coincident with live weight measurements using similar replication and trial lengths as those described for measuring live weight gain. Coefficients of variation calculated from data published on over 61 000 measurements of weight and condition score from different breeds, ages, and parities of cattle (Arango et al. 2002) suggest that the similarity in replication is likely justified (Berndtson 1991).

Milk Production

Milk yield is a more sensitive indicator of the nutritional value of the diet than is live weight gain over relatively short time periods (Stuedemann and Matches 1989). Milking cows daily is routine when working with lactating dairy cows, even under grazing conditions. Thus, animal behavior is often affected only minimally, which improves precision relative to measurements with beef cattle, and allows for daily measurement of milk production and regular sampling for milk composition. As few as four dairy cows per dietary treatment can be used to detect a lactation response if the cows are fed in confinement. However, this type of evaluation is often conducted in conjunction with digestion trials or other evaluations requiring intense sampling. Animal numbers are usually increased when the primary objective is to demonstrate a lactation response to experimental diets in an applied production context. Dairy cattle are not usually bred to calve seasonally, and considerations must be made to normalize the stage of lactation and parity across treatments. Measurements of milk production in grazing beef cows are usually made only periodically, rather than daily. Because beef cattle calve seasonally, it is less critical to normalize stage of lactation across treatments; however, normalization of parity may be necessary. A variety of methods have been used to estimate milk production in beef cows because they are typically unaccustomed to manual milking and, therefore, do not readily exhibit milk "let down" under experimental conditions. Stuedemann and Matches (1989) suggested the following rank order (from least desirable to most desirable) for some common methods used with beef cattle: hand milking, weigh-suckle-weigh, and machine milking after injection with oxytocin (to stimulate milk "let down"). The criteria used to rank these procedures were their accuracy and their relationship to calf weaning weight.

Diet Selection

Ruminants, particularly grazing ruminants, have been shown to exhibit significant diet selectivity. As a result, considerable effort has been devoted to characterizing both the nutritional characteristics (Holechek et al. 1982b) and botanic composition (Holechek et al. 1982a) of grazed forage. Similarly, composition of forage consumed in confinement must be determined in order to accurately calculate nutrient digestion or balance.

To monitor diet selection in grazing ruminants (Holechek et al. 1982a), collection of samples from esophageally or ruminally **fistulated** animals is considered to be the most accurate procedure (Cochran and Galyean 1994). Fistulation at either site yields acceptable results, though each procedure has its limitations. More labor is required during collection events when using ruminally fistulated animals, whereas esophageally fistulated animals typically require more day-to-day attention and care.

Before sampling, fistulated animals should be adapted to dietary conditions representative of those in the pasture(s) being evaluated. Ideally, they should graze with the animals that their samples are intended to mimic, or at least receive the same dietary inputs (forage plus supplements, if fed). Though recommendations in the literature vary, Holechek et al. (1982b) suggested the use of four animals per treatment and a collection period of four days (one sample collected per animal per day) as reasonable for determining nutrient composition of grazed forage. They also suggested altering the time of day at which samples are collected over the course of the collection period. Because extrusa collected from fistulated livestock often has high moisture content (particularly when grazing lush, growing forage), it is desirable to **lyophilize** rather than oven dry the samples to help minimize formation of artifacts during drying (Cochran and Galyean 1994).

Permitting diet selection in confinement studies requires offering forage in excess of ***ad libitum*** consumption (commonly 5–10% above *ad libitum*), though amounts as high as 35% above *ad libitum* have been suggested (Burns et al. 1994). Following adaptation to the forage in question (typically 10–14 days), daily weights of feed offered and refused are recorded (for 7 ± 3 days) and both the material offered and refused are sampled (Cochran and Galyean 1994). Nutrient composition of the selected diet is calculated as follows:

$$\begin{aligned}&\text{Nutrient concentration (g kg}^{-1}\text{ DM) in forage consumed}\\&= [(\text{feed offered, kg DM} \times \text{nutrient in feed offered,}\\&\text{g kg}^{-1}\text{ DM}) - (\text{feed refused, kg DM} \times \text{nutrient in}\\&\text{feed refused, g kg}^{-1}\text{ DM})] \div (\text{kg DM consumed})\end{aligned} \tag{37.1}$$

where DM is dry matter.

Forage Intake

Considering forage quality as the ability of forage to provide for animals' needs, voluntary forage intake is paramount because of both its direct relationship with nutrient supply and its indirect relationship with nutrient concentrations. For example, voluntary intake among forages is typically positively correlated with both forage digestibility and crude protein concentration. In addition, from a purely functional standpoint, nutrient balance studies intended to elucidate forage quality involve the acquisition of accurate intake measurements.

Measurements of *ad libitum* forage intake in confinement employ essentially the same procedures as those used in the determination of diet selection. Furthermore, because opportunity for diet selection will influence diet quality, and therefore voluntary intake, the choice of feeding level (% of *ad libitum* consumption for confinement studies, forage availability levels for grazing studies) requires careful consideration (Cochran and Galyean 1994). Reliable determination of voluntary intake in confinement is assumed to require a minimum of four animals per treatment (Burns et al. 1994). However, it is preferable to determine the number of replications needed for meaningful inference by using statistical data (means and variances or standard deviations) reported in similar experiments (Berndtson 1991). Power analysis can be readily conducted using procedures available in common statistical packages (Casler 2017).

Estimation of voluntary forage intake by grazing animals is considerably more complex. In the most straightforward approach, estimating intake in grazing animals is essentially the reverse of the approach used to determine diet digestion. That is, instead of offering a known amount of feed, determining the corresponding fecal output, and subsequently calculating the amount that disappeared during gastrointestinal transit, forage intake estimation uses knowledge of fecal output and diet digestibility to back-calculate forage consumption. Forage intake for a nonsupplemented grazing animal would be calculated as:

$$\text{Forage intake (kg DM d}^{-1}) = (\text{fecal DM output, kg d}^{-1}) \div [1 - (\text{g DM digested/kg diet DM})] \quad (37.2)$$

Several procedures can be used to determine the inputs to Eq. (37.2) (fecal output and diet digestion) and have been detailed in various reviews (Cordova et al. 1978; Burns et al. 1994). The most common procedures used for determining fecal DM output are total fecal collection (using a fecal collection bag and harness) and estimation through the use of external markers. In the latter case, if the daily dose of marker and the concentration of marker in the feces are known, the output can be calculated as:

$$\text{Fecal output (g DM d}^{-1}) = \text{marker consumed (units d}^{-1}) \div \text{fecal marker concentration (units g}^{-1}\text{ fecal DM)} \quad (37.3)$$

Marker concentration is measured in dried, ground fecal material usually collected as "grab" samples over a period of time (typically several days, with one or more samples collected per day) beginning after the marker reaches equilibrium in the gastrointestinal tract. It typically requires more than four days to reach equilibrium (Cochran and Galyean 1994).

Digestion coefficients used in Eq. (37.2) are usually determined by use of internal marker ratios (see below) or the use of an *in vitro* fermentation assay (see Chapter 34). For accurate predictions of *in vivo* digestion, a reliable regression relationship should be established for the specific *in vitro* assay to be used and *in vivo* measurements of digestion for the forage(s) of interest. *In vitro* assays that have not been so calibrated are used effectively by many laboratories to provide relative rankings of digestibility values; however, this is a distinctly different application than attempting to specifically predict the *in vivo* digestion of a diet under a unique set of conditions. Relationships between *in vitro* and *in vivo* digestibility are particularly problematic when attempting to predict *in vivo* digestion of forages fed with supplements that may elicit associative effects.

Rate and Extent of Digestion

In Vivo – Total Tract Digestion

Of the major routes by which energy is lost during the utilization of forage-based diets in ruminants, fecal energy loss is typically the largest. The difference between total energy consumed and fecal energy excreted, as a proportion of energy consumed, is digestible energy (DE), the first major partition in classical energy schemes (NRC 1981). The term *apparent* is often appended to DE because some components of fecal matter are of endogenous origin (e.g. sloughed epithelial cells, microbial debris, etc.), thus yielding a digestion estimate that is numerically less than true disappearance. Nonetheless, apparent DE is a useful estimate of the proportion of a feedstuff broken down and absorbed during transit through the gastrointestinal tract (i.e. the extent of digestion). Similar measurements, though not as precise as apparent DE from an energy-partitioning standpoint, are apparent DM or organic matter (OM) digestion. The latter two are determined in a manner similar to that for DE, except that feed and fecal materials are not combusted in a **bomb calorimeter** for determination of their energy content. These measurements are very closely correlated and numerically very similar; Moir (1961) reported that dry matter digestibility (DMD) and DE seldom differed by more than two percentage units for a wide range of diets.

In its most fundamental form, calculation of digestion entails direct determination of the amount of a specific nutrient or diet component consumed (e.g. DM, OM, N, etc.; calculated as diet DM intake × concentration of the nutrient in the diet DM) and the amount of the corresponding nutrient or diet component excreted in the feces (i.e. fecal DM output × concentration of the nutrient in the fecal DM). Digestion of a given nutrient (or diet component) is calculated as:

$$\text{Nutrient digestion coefficient} = (\text{kg nutrient consumed} - \text{kg nutrient excreted}) \div \text{kg nutrient consumed} \quad (37.4)$$

Measurement of the quantity and composition of the forage consumed typically follows an adaptation period of 10–14 days. Feed offered and refused is weighed and subsampled over a period of 7 ± 3 days; shorter collection periods (4–5 days) and adaptation periods (<10 days) are more likely to compromise accuracy. Because of the delay associated with passage of dietary residues through the gastrointestinal tract, the collection of feed offered, feed refused (i.e. orts), and fecal output are typically staggered. A standard protocol suggested by Cochran and Galyean (1994) recommends initiation of data collection for feed, **orts**, and feces on three successive days relative to the beginning of the experiment.

Fecal output and subsampling are typically accomplished by confining experimental animals in a collection stall or by fitting animals with bags that hold a day's fecal output (Cochran and Galyean 1994). Because of variation in time of defecation and resulting "endperiod" errors (Blaxter et al. 1956), total collections must be conducted over several days. A potential exception is when fecal output is being determined by using an external marker and grab sampling. In this case, a marker is delivered for sufficient time before the collection period begins to permit marker equilibration in the gastrointestinal tract. Then, samples of feces are collected per rectum (i.e. fecal grab samples; commonly several hundred grams of wet weight per sample) at least once per day throughout the collection period. In some cases, researchers will attempt to address diurnal variation in fecal marker concentration by collecting samples at different times. Commonly, these samples will be collected at evenly spaced intervals (e.g. every six or eight hours), with the times of collection advanced each day of the collection period such that, by the end of the period, a comprehensive set of samples from around the clock (for example, every other hour of a hypothetical 24-hour period) is available for determining average fecal marker concentrations. If diurnal variation is adequately addressed by more frequent sampling within a day, a shorter collection period than that used in total collection experiments may be acceptable.

Another way to determine diet digestibility is to use indigestible internal markers. Internal markers are inherent components of feedstuffs that are quantitatively recoverable in the feces. Numerous internal markers have been examined for the determination of digestibility (Cochran et al. 1987). If diet consumption is quantitatively determined, internal markers can be used in the same manner as external markers for the determination of fecal output (see Eq. 37.3). However, the more common application is when forage intake is unknown (for example, under grazing conditions) and an estimate of diet digestion is desired. In this situation, the relative concentrations of the internal marker in the feed and feces and of some dietary nutrient or component in the feces and feed can be used to calculate digestibility as follows:

$$\text{Nutrient digestion coefficient} = 1 - [(\text{g marker kg}^{-1}\ \text{feed DM} \div \text{g marker kg}^{-1}\ \text{fecal DM}) \times (\text{g nutrient kg}^{-1}\ \text{fecal DM} \div \text{g nutrient kg}^{-1}\ \text{feed DM})] \quad (37.5)$$

The number of animals recommended in the literature for digestibility determinations frequently falls between three and six animals per treatment (Cochran and Galyean 1994). However, the minimum of four per treatment suggested by Burns et al. (1994) is probably a reasonable benchmark. As noted for intake, the number of replicates needed to achieve a given power of statistical inference can be determined by the procedure described by Berndtson (1991).

In Vivo – Site and Extent of Digestion

The processes and efficiencies of digestion vary in the different segments of the ruminant gastrointestinal tract. For example, fiber fermentation in the ruminoreticulum is more efficient than hindgut fermentation given that the resultant microbial matter (especially the protein fraction) is available for digestion and absorption in the small intestine. Consequently, quantifying site and extent of nutrient digestion can provide insight into issues of great relevance to the evaluation of forage quality. A detailed accounting of the contributions of site of digestion studies to ruminant nutrition has been published by Merchen et al. (1997).

Conceptually, determining site and extent of digestion is similar to determining total tract digestion – we measure the quantity of specific components that flow into and out of the gastrointestinal segments of interest. Typically, these segments include the pregastric and gastric segments (ruminoreticulum, omasum, abomasum), the small intestine, and the hindgut (**cecum**, large intestine). To accomplish this, sampling ports (**cannulas**) can be surgically placed at the beginning (proximal **duodenum**) and end (terminal ileum) of the small intestine to allow quantification of the flow of constituents of interest past these sites. The concentrations of dietary constituents of interest are typically used in conjunction with measured concentrations of some **digesta** flow marker to calculate digestion across the various segments of interest. This calculation is similar to Eq. (37.5) except that the concentrations of marker and nutrient/diet component are expressed relative to the material entering and exiting the particular segment(s) of interest. Alternatively, *in situ* approaches can be used to assess disappearance of certain constituents within certain segments of the tract. This strategy, which involves placing samples of feeds into porous bags, is typically used for estimating the extent of ruminal disappearance (discussed below). However, this approach has been modified to allow for estimating disappearance in the postgastric portions of the gastrointestinal tract, using a procedure known as the mobile **nylon bag technique** (Stern et al. 1997). In this approach, feed samples, isolated in sealed, porous bags, are introduced at the beginning of the gastrointestinal segment of interest and removed at the distal end of the segment. Challenges associated with the mobile bag technique (Ross et al. 2013) have stimulated interest in other methods, particularly for assessing small intestinal amino acid digestibility. Because of similarities in small intestinal digestion between ruminants and nonruminants, one promising approach involves the use of a bioassay in which amino acid disappearance is determined in feedstuffs or *in situ* residues using roosters which have been cecectomized to eliminate fermentative degradation (Titgemeyer et al. 1990; Boucher et al. 2009). Nonetheless, even with justified interest in alternative approaches, *in vivo* measurement of nutrient disappearance remains the standard for assessing site and extent of digestion.

Though determination of the site and extent of digestion is conceptually similar to that of total tract digestion (see preceding section), there are additional important considerations. Issues related to gastrointestinal cannulations in ruminants have been reviewed by Harmon and Richards (1997). Duodenal cannulas are generally preferred to omasal or abomasal cannulas because of difficulties with cannula patency, representative sampling of digesta, and other complications. **Cannulas** vary widely in construction materials and design, and many are available commercially. Major considerations include the flexibility of the construction materials (flexible designs can minimize tissue irritation, but are often subject to increased difficulties in maintaining placement), tissue compatibility, durability, and cannula type (re-entrant vs T-type). Most commonly, researchers use a simple T-type cannula, from which, only a portion of the duodenal digesta is collected. One concern with simple T-type cannulas is that samples may not accurately represent the digesta flowing through the intestine. The use of gated T-type cannulas offers a good compromise, allowing representative sampling with minimal complications. Because motility of the ileum and cecum present greater possibility for postsurgical complications, simple T-type cannulas are generally recommended for ileal cannulations.

A variety of digesta flow markers (internal and external) have been evaluated for use in determining the site and extent of digestion. Marker choice can be affected by specific dietary and experimental conditions (Titgemeyer 1997). Primary difficulties with flow markers include incomplete recovery, variation in outflow from the rumen, and dealing with unrepresentative samples. Marker recovery should be quantitative and can be validated through the use of re-entrant cannulas or through quantitative fecal collection. However, with fecal collection alone, it is impossible to determine the site of marker disappearance if recoveries are less than 100%. Diurnal variation in marker outflow is common and is typically accounted for by increasing the frequency of dosing and ensuring that samples are obtained at a variety of times relative to feeding. A common strategy when collecting samples for this application is to collect samples at 6-hour intervals across a 72-hour period, advancing collection time by 2 hours each day so that samples represent every other hour of a hypothetical 24-hour period. Dual-phase marker systems have been recommended on occasion to account for some of the problems associated with nonrepresentative samples. Limitations of this approach have been discussed by Titgemeyer (1997).

To identify microbial contribution of nutrients at the small intestine, microbial markers are typically used. Besides selection of an appropriate marker (reviewed by Broderick and Merchen 1992), researchers must take care to obtain a representative sample of ruminal microorganisms to determine marker concentrations. Concentrations of markers will vary among animals, diets, different microbial populations, and with time after feeding. Thus, microbial samples should be obtained from the individuals from which intestinal samples are obtained and at a variety of times after feeding (three to four samples between feedings). Additionally, variation exists in the composition of fluid vs particle associated bacteria. Thus, to obtain representative samples of microorganisms, whole ruminal contents are typically processed through a blender to remove bacteria adhered to particles and then subjected to differential centrifugation for microbial isolation (Cecava et al. 1990).

Endogenous contributions (e.g. sloughed epithelial cells, luminal secretions) are typically accounted for using literature values. These corrections are of particular importance for calculating undegraded feed protein flow with low-protein diets (Vanzant et al. 1996).

Based on a review of the variation in typical response variables from flow studies, Titgemeyer (1997) indicated that flow studies should typically have six replications per treatment.

In Situ – Rate and Extent of Ruminal Digestion

Description of Technique and Overview

Current ruminant feed evaluation systems account for interactions between protein and energy when determining nutritive value (NRC 2001; CSIRO 2007; Van Duinkerken et al. 2011; Volden 2011; NRC 2016). Such models typically require extensive inputs regarding pool sizes and degradation rates of various forage components. Depending on the degree of complexity, such kinetic data can be estimated using *in situ* procedures, including partitioning specific forage components into subfractions based on relative susceptibility to microbial degradation in the rumen (Figure 37.1) and estimation of a discrete lag time (L), ruminal disappearance rate (K_d), and potential extent of ruminal disappearance. Effective degradability, an estimate of ruminal disappearance at a specific passage rate, can also be calculated from equations (Ørskov and McDonald 1979) using these kinetic data. Though modeling efforts typically require pool sizes and kinetic information specifically for protein and carbohydrate fractions, considerable research has been conducted in which degradation rates of DM or fiber have been used as part of the portrayal of forage quality.

Though the fundamental *in situ* procedure used to collect such information is considered well established, methodological variation abounds in the execution of this procedure (Michalet-Doreau and Ould-Bah 1993; Broderick and Cochran 2000). Such variation contributes substantially to differences among laboratories in measurements conducted on standard forages and has served as the basis for the call for standardization in technique (AFRC 1992; Michalet-Doreau and Ould-Bah 1993; Broderick and Cochran 2000). Key issues for which recommendations are available (Broderick and Cochran 2000) include basal diet (type and level of feeding),

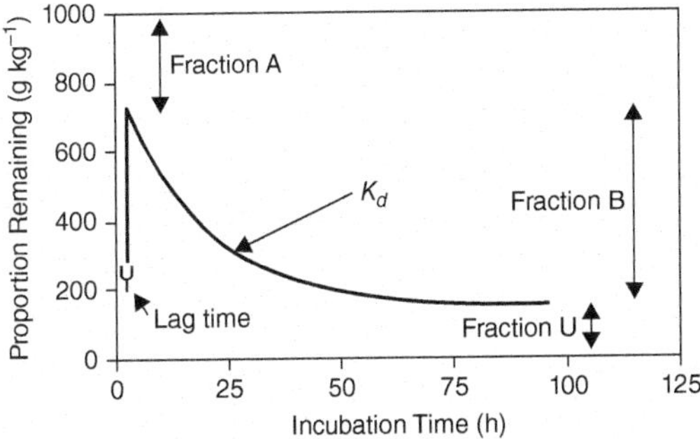

Fig. 37.1. Forage disappearance kinetics based on component susceptibility to ruminal attack and decay. Fraction A is the immediately soluble portion, fraction B is the portion that disappears from the rumen at a measurable disappearance rate (K_d), and fraction U is the portion considered unavailable or indigestible in the rumen.

bag characteristics (material, pore size, and sample size-to-surface area ratio), grind size, animal species, replication (numbers of animals, days, and bags), incubation conditions (preincubation, position in the rumen, entry or removal order, and incubation times), rinsing procedure, microbial correction, and use of standard feeds. Though laboratories frequently differ on the manner in which these items are executed, the basic elements of the *in situ* procedure are often similar among laboratories.

In general, this procedure entails weighing dried, ground forage into synthetic bags with a pore size small enough to prevent migration of solid particles in or out of the bag when it is suspended in the rumen. Sealed bags containing the experimental forage are suspended in the rumen and then withdrawn (or additional bags are added if working in reverse order) after specified incubation times. The percentage of the component remaining or disappearing from the bag at each incubation time can then be used to establish a decay or disappearance curve from which kinetic parameters can be calculated using one of several statistical approaches.

Calculations

The model for the kinetics of disappearance from *in situ* bags is normally assumed to follow first-order kinetics. Modifications are usually made to accommodate a lag time (L) and to exclude that portion of the forage component that is indigestible (Mertens and Loften 1980; Moore and Cherney 1986). The most commonly used model is

$$R = D_0\, e^{-K_d(t-L)} + U \qquad (37.6)$$

where

R = the fraction of the forage component remaining after t hours of ruminal incubation when $t > L$ (if $t < L$, then $R = D_0 + U$),
D_0 = potentially digestible fraction ($D_0 = R - U$ at $t < L$),
L = discrete lag time, and U = indigestible fraction.

Methods of Calculation

Generally, two methods are used to calculate kinetic parameters. One method is the ln-linear procedure (LTP) in which the remaining portion of the potentially digestible residue (R − U) is transformed using the natural logarithm function and then regressed on t. The working linear equation for this method is

$$\ln(R - U) = \ln(D_{int}) - K_d t \qquad (37.7)$$

where $\ln(D_{int})$ = intercept at t = 0 hour, and L is estimated as $[\ln(R-U) - \ln(D_{int})]/-K_d$ after time intervals are excluded where (R − U) does not differ statistically from (R − U) at t = 0 hour (Moore and Cherney 1986).

An alternative approach is a direct nonlinear least squares procedure (NLP), where R is regressed against t and parameters are estimated by iterative adjustment until a least-squares convergence criterion is met. Regardless of the method of calculation (LTP or NLP) used, parameter estimates, such as fractions A, B, and U; the potential extent of disappearance, K_d; and effective ruminal degradability can then be evaluated statistically using analysis of variance. An alternative, but less common approach, is to simply evaluate the relative disappearance at each time point by analysis of variance. Though this approach permits relative comparisons among treatments, it ignores the kinetics of the underlying processes and therefore fails to yield information necessary for quantitative analysis of forage quality.

Important Considerations

Among the methodologic considerations described above, correction for microbial attachment to *in situ* residues is of particular importance. The importance of microbial contamination of residues is influenced by a number of factors, including feedstuff composition, incubation time, and the nutrient of interest. Failure to correct for microbial attachment can lead to substantial errors in estimates of ruminal (and postruminal) disappearance, particularly for fibrous feeds, and for nitrogenous components (Vanzant et al. 1998; Arroyo and González 2013).

Definition of Kinetic Terms

Forage components can be partitioned into three fractions based on relative susceptibility to ruminal disappearance.

Fraction A (Figure 37.1) is usually defined as the immediately soluble portion or that portion of the specific forage component that disappears from *in situ* bags at a rate too fast to measure. When NLP is used to estimate kinetic parameters, fraction A is frequently estimated from the *y*-intercept of the disappearance curve [calculated as 1000 − (B + U)], but can also be determined directly as the proportion that washes out of filled bags subjected to standardized water rinsing in a washing machine (Broderick and Cochran 2000) without ruminal incubation ($t = 0$ hour). If fraction A is estimated by LTP, it must be determined directly because the difference between ln (R − U) and ln (D_{int}) is the basis for calculating L.

The B fraction comprises that portion of any forage component disappearing slowly (at a measurable rate), and the disappearance rate constant, K_d, is associated with the ruminal disappearance of this fraction. When NLP is used to estimate kinetic parameters, fraction B is usually equated with D_0 in Eq. (37.6), leaving a working model of

$$R = B\,e^{-K_d(t-L)} + U \tag{37.8}$$

Fraction U is considered unavailable or indigestible in the rumen and is usually determined by one of two methods. In one method, the indigestible portion of a specific forage component is measured after an extended t. In practice, indigestibility is usually estimated for DM or fiber components after 72–96 hours of ruminal incubation. Considerably shorter incubation times may be suitable for estimating indigestibility of N. A second approach is mathematical; fraction U can be calculated directly by NLP from the asymptotic disappearance curve. In either case, fraction U can be used to calculate the potential extent of disappearance (1000 − U). For determination of N-disappearance kinetics, acid detergent insoluble N (ADIN) has occasionally been used as an estimate of fraction U (Broderick 1994).

Effective Degradability

Though *in situ* procedures may be used to generate kinetic observations as inputs into dynamic models, the goal of an *in situ* analysis in many situations is to estimate the effective rumen degradability of some specific forage component (DM, N, neutral detergent fiber [NDF], etc.). Effective degradability is an estimate of ruminal digestibility at a specific passage rate and, as a result, incorporates passage rate in its calculation:

$$\text{Effective degradability} = A + B\,[K_d/(K_d + K_p)] \tag{37.9}$$

where K_p = ruminal passage rate (Ørskov and McDonald 1979). Ideally, actual experimental observations are used for both passage rate and rate of degradation; however, it is often the case that effective degradability is estimated using measured rates of digestion and assumed passage rates. As a result, it is important to recognize the potential impact of such decisions on estimates of effective degradability. In general, at any given value of K_d, the estimate of effective degradability will decrease as passage rate increases. However, the effects of passage rate increase when K_d is slow, and approaches K_p in magnitude. This has been illustrated and described by Mertens (1987) and Broderick (1994), and this approach has been adapted in Figure 37.2 to cover the normal ranges of passage rate and K_d encountered in most *in situ* evaluations for a forage component with fraction A = 300 g kg^{-1} and fraction B = 500 g kg^{-1}.

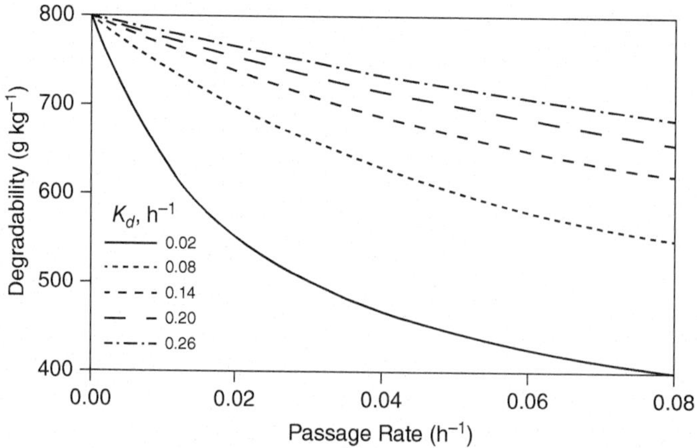

Fig. 37.2. Effects of passage rate on estimates of effective ruminal degradability at five rates of disappearance (K_d) for a forage component when fraction A = 300 g kg^{-1} and fraction B = 500 g kg^{-1}. *Source:* Adapted from illustrations and discussions in Mertens (1987) and Broderick (1994).

Balance Trials

Energy Balance

Advances in our understanding of the applicability of the laws of thermodynamics to biological systems and of ruminant digestive physiology and metabolism, combined with methodological and technological advances, prompted the convention of expressing nutritive value and requirements in units of energy (Blaxter 1989). Taken together, these advances led to the concept of net energy, whereby energetic value of a feedstuff is expressed in terms of its net ability to meet maintenance energy demands for a given class of livestock and/or to increase energy recovered in the form of some product (e.g. tissue energy gain, lactation energy output, etc.). In recent years, dynamic models have gained widespread acceptance (NRC 2001; NRC 2016). These models accommodate advancing knowledge of the interplay between major dietary constituents in determining nutritive value yet continue to adhere to the net nutrient concept. The fundamental procedure employed for determining the net energy value of all feedstuffs, including forages, is referred to as an energy balance trial.

The term *energy balance* is derived from the fact that, in conducting such a measurement, the total energy input and sources of energy loss are quantified, with the remainder (i.e. balance) of energy input recovered in some measurable product. The starting point for an energy balance measurement is the determination of total intake energy (IE), which is the product of daily diet intake and a diet's heat of combustion (i.e. **gross energy** concentration). In the classical partitioning of IE (NRC 1981), major sources of energy loss include the following energies: fecal (FE), urinary (UE), gaseous (GE), and heat (HE). As described above (see *In Vivo* Digestion section), quantification of diet consumption, fecal output, and the **gross energy** concentration in the feed and feces permits the calculation of DE (by Eq. 37.4). The DE value provides an estimate of the proportion of IE that is made available (if not already present in an absorbable form) and absorbed during gastrointestinal transit. To estimate the proportion of IE actually available to contribute to maintenance or productive functions (AFRC 1993), the FE, UE, and GE are subtracted from IE. The resulting quantity is referred to as metabolizable energy (ME). Urinary energy loss is commonly determined by total collection of urine output (typically concomitant with the collection of feces for determination of DE) and combustion of the urine in a bomb calorimeter. Most metabolism stalls are specifically constructed with adaptations that facilitate total collection of urine; however, in situations where such stalls are not available, specially designed urinals may be required that are fitted to the animal (Cochran and Galyean 1994). In the latter case, application of suction may significantly facilitate urine collection with some devices. Because methane is the predominant combustible gas produced during ruminal fermentation (AFRC 1993), determination of GE focuses upon the measurement of methane production. A common method for quantifying methane production is by monitoring gaseous flux when animals are subjected to respiration calorimetry (Blaxter 1989). Alternatively, indirect techniques are available (e.g. micrometeorologic technique, Harper et al. 1999; sulfur hexafluoride tracer gas technique, Boadi et al. 2002) that are suitable for use with grazing livestock. Rough estimates of methane production can also be calculated from volatile fatty acid production data (Ørskov and Ryle 1990).

Subtracting all major sources of energy loss (FE, UE, GE, and HE) from IE yields recovered energy (RE). Heat energy loss (attributable to energy expended in maintaining vital body functions as well as to the inefficiency of ME use for different functions) can be determined with direct calorimetry or indirect respiration calorimetry or can be estimated by difference (i.e. ME − RE = HE) when independent techniques are used to estimate RE. Classically, the net energy (NE) value of a feedstuff is defined as the ΔRE/ΔIE (NRC 1981). The forms in which RE is evident (e.g. tissue, milk, etc.) are several, and characterization of the ΔRE/ΔIE relationship requires careful accounting of all such forms. Over the range of intake from fasting to *ad libitum* consumption, this relationship (ΔRE/ΔIE) is curvilinear (NRC 1981; Blaxter 1989). However, this relationship (Figure 37.3) has historically been approximated as two straight lines intersecting at the point at which RE = 0 (i.e. the point of energy equilibrium) (NRC 1981). The line below energy equilibrium is used to define the NE_m (NE for maintenance), whereas the line above maintenance is used to define the NE_r (NE for promoting RE). In situations in which the form of RE is strictly tissue energy gain, the NE_r is identified as NE_g

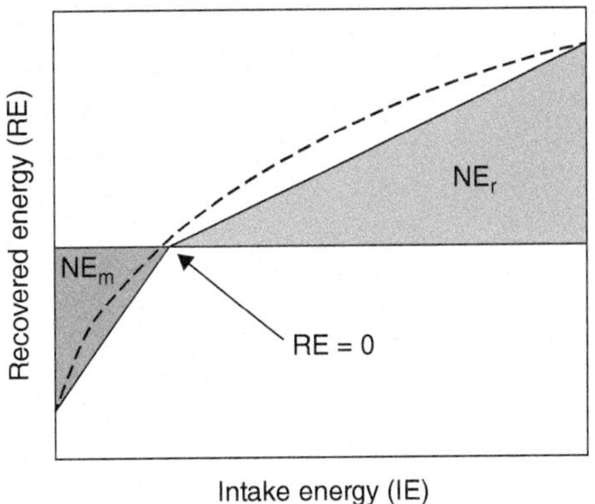

Fig. 37.3. Relationship of recovered energy (RE) to intake energy (IE). *Source:* Adapted from NRC (1981).

(NE for gain). Because ME is used with varying efficiency in promoting RE of different forms, a single NE value is generally not tenable (NRC 1981), though assigning a single NE value to a feed has proven to be feasible for mature, lactating dairy cows because efficiency of ME use for maintenance, lactation, and increasing body condition are similar when cows are lactating (NRC 1981). The procedure used to establish NE values is known as a *difference trial* (NRC 1981). The name derives from the fact that complete energy balance determinations are conducted using at least two levels of IE (e.g. *ad libitum* consumption and the point where RE = 0), thus allowing one to divide the difference in RE by the difference in IE. To facilitate application of these values in production settings (i.e. to yield NE concentration per unit of feed DM), the change in feed DM consumption is commonly substituted for the ΔIE. Conceptually, the AFRC (1993) adheres to the same net energy concept as that employed by the NRC (1981, 1996, 2001). However, their fundamental unit of measurement is ME determined at maintenance, which is subsequently adjusted for intake level and efficiency of ME use (which varies according to function) to calculate a nutritive value functionally equivalent to NE (AFRC 1993).

Nitrogen Balance

The efficiency of utilization of dietary N is an important consideration for optimizing the efficiency of ruminant production. Additionally, we are concerned with minimizing N excretion into the environment. Ultimately, our goal is to capture as much N as possible in the animals' tissues or products. This can be determined by direct measurement of N captured in the product(s) of interest. Conversely, in the classical N-balance experiment, N retention is determined indirectly by subtracting known sources of N loss from N inputs. In this case, the major contributors to N loss include fecal, urinary, **scurf**, and eructated NH_3; however, in the conventional N balance experiment, fecal and urinary N are typically subtracted from N input (Owens and Zinn 1988). If all inputs and outputs could be measured with complete accuracy, the difference approach would give accurate estimates of N retention. Unfortunately, input is often overestimated (through losses rather than actual consumption) and output is often underestimated (due to incomplete accounting or losses that occur between excretion and measurement), and as a result, N-balance data frequently overestimate N retention (Asplund 1979). An additional potential source of error is changes in concentrations of short-term storage proteins (e.g. plasma proteins) in the body (Owens et al. 2014). Despite the limitations, N-balance studies occupy a significant, useful niche in the evaluation of forage quality.

The typical N-balance experiment requires quantitative measurement of feed intake and fecal and urinary output, along with N determinations (typically by Kjeldahl or Dumas methods) of each of these constituents. Generally, the same guidelines that apply to energy balance studies, with respect to length of feeding periods, animal numbers, and sample collection, also apply to N-balance studies. Additionally, care must be exercised in handling urine samples because of the potential for **volatilization** of N as NH_3. Though NH_3-N is also present in fecal samples, it is generally disregarded because available reports (Martin 1966; Hinnant and Kothmann 1988; Karn 1991) suggest that N losses from feces are minimal and are relatively unaffected by ambient conditions or drying method (freeze-drying, oven drying, or microwave-drying). However, additional research is required to definitively assess N losses that occur with typical sample-processing methods. When comparing N losses from feed and fecal samples that were rapidly frozen and freeze-ground, McLeod (2003, unpublished data) found that significant N losses can occur with oven and freeze-drying and that these losses are dependent on feed characteristics. The primary factor affecting N loss from urine is pH. Data from Martin (1966) indicate that urine samples should be acidified to a pH of about 2.0 to minimize NH_3 **volatilization**. In addition to protonating NH_3, acidification of urine will help minimize microbial growth and sample deterioration. Generally, a concentrated acid (e.g. H_3PO_4) is desirable to minimize dilution of the sample, thereby minimizing potential problems with sensitivity of the N analysis. Typically, urine is collected into containers to which acid has been added, weighed, and sampled daily. Samples are usually stored refrigerated or frozen prior to analysis. Fecal samples are also collected daily and, typically, are refrigerated, frozen, or oven-dried prior to N determination.

Metabolic Indicators

Nutrients presented to, and absorbed from, the gastrointestinal tract are suitable targets for assay in body fluids or tissues. Concentrations of nutrients in body fluids (e.g. milk, blood, urine) commonly reflect differences in dietary nutrient concentrations. Of particular interest, is the concentration of urea N in body fluids. Diets with excessive protein are commonly characterized by high concentrations of urea in the blood, milk, and urine. Hammond et al. (1994) concluded that blood urea N was an effective tool for monitoring excesses or deficiencies of protein in forage-based diets consumed by beef cattle. Similarly, Jonker et al. (1998) indicated that milk urea nitrogen values from 10 or more dairy cattle would yield meaningful information relative to the appropriateness of protein levels in dairy diets.

Conclusion

Numerous methods are available that employ animals in the assessment of forage quality. Some of these procedures provide information needed to address very specific goals (e.g. monitoring protein adequacy), some serve as useful contributors to the efforts to accurately predict nutritive value, whereas some are simply viewed as relative indicators of nutritive value. In all cases, it is helpful to be mindful of the fact that the fundamental goal in conducting any forage quality assay is to help define the ability of forage to support maintenance and production functions in ruminants. Though this is most accurately represented at present by NE values, the variety of other assays available can also serve a useful role in helping point us toward our fundamental endpoint.

References

AFRC (1992). Nutritive requirements of ruminant animals: protein. *Nutr. Abstr. Rev.* 62: 787–835.

AFRC (1993). *Energy and Protein Requirements of Ruminants. An Advisory Manual Prepared by the AFRC Technical Committee on Responses to Nutrients*. Wallingford, UK: CAB International.

Arango, J.A., Cundiff, L.V., and Van Vleck, L.D. (2002). Genetic parameters for weight, weight adjusted for body condition score, height, and body condition score in beef cows. *J. Anim. Sci.* 80: 3112–3122.

Arroyo, J. and González, J. (2013). Effects of the ruminal comminution rate and microbial contamination of particles on accuracy of in situ estimates of ruminal degradability and intestinal digestibility of feedstuffs. *J. Anim. Physiol. Anim. Nutr.* 97: 109–118.

Asplund, J.M. (1979). Interpretation and significance of nutrient balance studies. *J. Anim. Sci.* 49: 826–831.

Berndtson, W.E. (1991). A simple, rapid and reliable method for selecting or assessing the number of replicates for animal experiments. *J. Anim. Sci.* 69: 67–76.

Blaxter, K.L. (1989). *Energy Metabolism in Animals and Man*, 1e. Cambridge: Cambridge University Press.

Blaxter, K.L., Graham, N.M.C., and Wainman, F.W. (1956). Some observations on the digestibility of food by sheep, and on related problems. *Br. J. Nutr.* 10: 69–91.

Boadi, D.A., Wittenberg, K.M., and Kennedy, A.D. (2002). Validation of the sulphur hexafluoride (SF6) tracer gas technique for measurement of methane and carbon dioxide production by cattle. *Can. J. Anim. Sci.* 82: 125–131.

Boucher, S., Calsamiglia, S., Parsons, C. et al. (2009). Intestinal digestibility of amino acids in rumen undegradable protein estimated using a precision-fed cecectomized rooster bioassay: I. Soybean meal and SoyPlus. *J. Dairy Sci.* 92: 4489–4498.

Broderick, G.A. (1994). Quantifying protein quality. In: *Forage Quality, Evaluation, and Utilization* (eds. G.C. Fahey, M. Collins, D.R. Mertens and L.E. Moser), 200–228. Madison, WI: University of Nebraska, Lincoln/American Society of Agronomy, Crop Science Society of America, and Soil Science Society of America.

Broderick, G.A. and Cochran, R.C. (2000). In vitro and in situ methods for estimating digestibility with reference to protein degradability. In: *Feeding Systems and Feed Evaluation Models* (eds. M.K. Theodorou and J. France). Wallingford, UK: CAB International.

Broderick, G.A. and Merchen, N.R. (1992). Markers for quantifying microbial protein synthesis in the rumen. *J. Dairy Sci.* 75: 2618–2632.

Brown, M.A., Aiken, G.E., Brown, A.H. Jr. et al. (1993). Evaluation of overnight shrinkage in reduction of variability of weights and gains of beef cattle. *Prof. Anim. Sci.* 9: 120–126.

Burns, J.C., Pond, K.R., and Fisher, D.S. (1994). Measurement of forage intake. In: *Forage Quality, Evaluation, and Utilization* (eds. G.C. Fahey, M. Collins, D.R. Mertens and L.E. Moser), 494–532. Madison, WI: University of Nebraska, Lincoln/American Society of Agronomy, Crop Science Society of America, and Soil Science Society of America.

Casler, M.D. (2017). Power and replication - designing powerful experiments. In: *Applied Statistics in Agricultural, Biological, and Environmental Sciences* (eds. B. Glaz and K.M. Yeater), 51–61. Madison, WI, USA: American Society of Agronomy, Crop Science Society of America, and Soil Science Society of America.

Cecava, M.J., Merchen, N.R., Gay, L.C., and Berger, L.L. (1990). Composition of ruminal bacteria harvested from steers as influenced by dietary energy level, feeding frequency, and isolation techniques. *J. Dairy Sci.* 73: 2480–2488.

Cochran, R.C. and Galyean, M.L. (1994). Measurement of in vivo forage digestion by ruminants. In: *Forage Quality, Evaluation, and Utilization* (eds. G.C. Fahey, M. Collins, D.R. Mertens and L.E. Moser), 613–643. Madison, WI: University of Nebraska, Lincoln/American Society of Agronomy, Crop Science Society of America, and Soil Science Society of America.

Cochran, R.C., Vanzant, E.S., Jacques, K.A. et al. (1987). Internal markers. In: *Proceedings of the Grazing Livestock Nutrition Conference* (eds. M.B. Judkins, D.C. Clanton, M.K. Petersen and J.D. Wallace), 39–48. Laramie: University of Wyoming.

Cook, C.W. and Stubbendieck, J. (eds.) (1986). Livestock selection and management in range research. In: *Range Research: Basic Problems and Techniques*, 133–154. Denver, CO: Society for Range Management.

Cordova, F.J., Wallace, J.D., and Pieper, R.D. (1978). Forage intake by grazing livestock: a review. *J. Range Manage.* 31: 430–438.

CSIRO (2007). *Nutrient Requirements of Domesticated Ruminants*. Collingwood, VIC, Australia: CSIRO Publishing.

Garrett, W.N. and Hinman, N. (1969). Re-evaluation of the relationship between carcass density and body composition of beef steers. *J. Anim. Sci.* 28: 1–5.

Hammond, A.C., Bowers, E.J., Kunkle, W.E. et al. (1994). Use of blood urea nitrogen concentration to determine time and level of protein supplementation in wintering cows. *Prof. Anim. Sci.* 10: 24–31.

Harmon, D.L. and Richards, C.J. (1997). Considerations for gastrointestinal cannulations in ruminants. *J. Anim. Sci.* 75: 2248–2255.

Harper, L.A., Denmead, O.T., Freney, J.R., and Byers, F.M. (1999). Direct measurements of methane emissions from grazing and feedlot cattle. *J. Anim. Sci.* 77: 1392–1401.

Hersom, M., Imler, A., Thrift, T. et al. (2015). Comparison of feed additive technologies for preconditioning of weaned beef calves. *J. Anim. Sci.* 93: 3169–3178.

Hinnant, R.T. and Kothmann, M.M. (1988). Collecting, drying, and preserving feces for chemical and microhistological analysis. *J. Range Manage.* 41: 168–171.

Holechek, J.L., Vavra, M., and Pieper, R.D. (1982a). Botanical composition determination of range herbivore diets: a review. *J. Range Manage.* 35: 309–315.

Holechek, J.L., Vavra, M., and Pieper, R.D. (1982b). Methods for determining the nutritive quality of range ruminant diets: a review. *J. Anim. Sci.* 54: 363–376.

Jonker, J.S., Kohn, R.A., and Erdman, R.A. (1998). Using milk urea nitrogen to predict nitrogen excretion and utilization efficiency in lactating dairy cows. *J. Dairy Sci.* 81: 2681–2692.

Karn, J.F. (1991). Chemical composition of forage and feces as affected by microwave oven drying. *J. Range Manage.* 44: 512–515.

Lawrence, T.L.J. (2002). *Growth of Farm Animals*. Wallingford, UK: CABI Publishing.

Martin, A.K. (1966). Some errors in the determination of nitrogen retention of sheep by nitrogen balance studies. *Br. J. Nutr.* 20: 325–337.

McLeod, K. (2003). *Personal communication*. University of Kentucky, KY.: Department of Animal and Food Sciences, Lexington.

Merchen, N.R., Elizalde, J.C., and Drackley, J.K. (1997). Current perspective on assessing site of digestion in ruminants. *J. Anim. Sci.* 75: 2223–2234.

Mertens, D.R. (1987). Predicting intake and digestibility using mathematical models of ruminal function. *J. Anim. Sci.* 64: 1548–1558.

Mertens, D.R. and Loften, J.R. (1980). The effect of starch on forage fiber digestion kinetics in vitro. *J. Dairy Sci.* 63: 1437–1446.

Michalet-Doreau, B. and Ould-Bah, M.Y. (1993). In vitro and in sacco methods for the estimation of dietary nitrogen degradation in the rumen: a review. *Anim. Feed Sci. Technol.* 40: 57–86.

Moir, R.J. (1961). A note on the relationship between the digestible dry matter and the digestible energy content of ruminant diets. *Aust. J. Exp. Agric. Anim. Husb.* 1: 24–26.

Moore, K.J. and Cherney, J.H. (1986). Digestion kinetics of sequentially extracted cell wall components of forages. *Crop Sci.* 26: 1230–1235.

Mott, G.O. and Lucas, H.L. (1953). The design, conduct, and interpretation of grazing trials on cultivated and improved forages. In: *Proceedings of the 6th International Grassland Congress*, 1380–1385.

National Research Council (1996). *Nutrient Requirements of Beef Cattle*, 7e. Washington, DC: National Academy Press.

National Research Council (NRC) (1981). *Nutritional Energetics of Domestic Animals*. Washington, DC: National Academy Press.

National Research Council (NRC) (2001). *Nutrient Requirements of Dairy Cattle*, 7th rev. ed. Washington, DC: National Academy Press.

National Research Council (NRC) (2016). *Nutrient Requirements of Beef Cattle*, 8th rev. ed. Washington, DC: National Academy Press.

Ørskov, E.R. and McDonald, I. (1979). The estimation of protein degradability in the rumen from incubation measurements weighted according to rate of passage. *J. Agric. Sci.* 92: 499–503.

Ørskov, E.R. and Ryle, M. (1990). *Energy Nutrition in Ruminants*. Essex, England: Elsevier Science Publishers LTD.

Owens, F.N. and Zinn, R. (1988). Protein metabolism of ruminant animals. In: *The Ruminant Animal Digestive Physiology and Nutrition* (ed. D.C. Church), 227–249. Englewood Cliffs, NJ: Prentice-Hall.

Owens, F., Qi, S., and Sapienza, D. (2014). Invited review: applied protein nutrition of ruminants—current status and future directions. *Prof. Anim. Sci.* 30: 150–179.

Ross, D., Gutierrez-Botero, M., and Van Amburgh, M. (2013). Development of an in vitro intestinal digestibility assay for ruminant feeds. In: *Proceedings of the Cornell Nutrition Conference*, 190–202. Syracuse, NY.

Stern, M.D., Bach, A., and Calsamiglia, S. (1997). Alternative techniques for measuring nutrient digestion in ruminants. *J. Anim. Sci.* 75: 2256–2276.

Stuedemann, J.A. and Matches, A.G. (1989). Measurement of animal response in grazing research. In: *Grazing Research: Design, Methodology, and Analysis* (ed. G.C. Marten), 21–35. Madison, WI: Crop Science Society of America.

Titgemeyer, E.C. (1997). Design and interpretation of nutrient digestion studies. *J. Anim. Sci.* 75: 2235–2247.

Titgemeyer, E., Merchen, N., Han, Y. et al. (1990). Assessment of intestinal amino acid availability in cattle by use of the precision-fed cecectomized rooster assay. *J. Dairy Sci.* 73: 690–693.

Van Duinkerken, G., Blok, M., Bannink, A. et al. (2011). Update of the Dutch protein evaluation system for ruminants: the DVE/OEB2010 system. *J. Agric. Sci.* 149: 351–367.

Vanzant, E.S., Cochran, R.C., Titgemeyer, E.C. et al. (1996). In vivo and in situ measurements of forage protein degradation in beef cattle. *J. Anim. Sci.* 74: 2773–2784.

Vanzant, E.S., Cochran, R.C., and Titgemeyer, E.C. (1998). Standardization of in situ techniques for ruminant feedstuff evaluation. *J. Anim. Sci.* 76: 2717–2729.

Vendramini, J.M.B., Sanchez, J.M.D., Cooke, R.F. et al. (2015). Stocking rate and monensin supplemental level effects on growth performance of beef cattle consuming warm-season grasses. *J. Anim. Sci.* 93: 3682–3689.

Volden, H. (ed.) (2011). *NorFor – The Nordic Feed Evaluation System*. EAAP Pub. No. 130. Wageningen, The Netherlands: Wageningen Academic Publishers.

Wilson, T.B., Faulkner, D.B., and Shike, D.W. (2015). Influence of late gestation drylot rations differing in protein degradability and fat content on beef cow and subsequent calf performance. *J. Anim. Sci.* 93: 5819–5828.

Wilson, T.B., Long, N.M., Faulkner, D.B., and Shike, D.W. (2016). Influence of excessive dietary protein intake during late gestation on drylot beef cow performance and progeny growth, carcass characteristics, and plasma glucose and insulin concentrations. *J. Anim. Sci.* 94: 2035–2046.

CHAPTER

38

Predicting Forage Quality

Debbie J. Cherney, Professor, *Animal Science, Cornell University, Ithaca, NY, USA*
David Parsons, Professor, *Crop Science, Swedish University of Agricultural Sciences (SLU), Umeå, Sweden*

Introduction

Forage use in ruminant diets has always been important, but increasing concerns about environmental issues, and economic sustainability has made prediction of **forage quality** an increasingly important part of animal management. Animal performance is the best indicator of forage quality (Mott 1973). Unfortunately, animal performance is not easy to measure, and we certainly cannot measure a forage's true quality before that forage is fed to the animal. There are many instances where a prediction of forage quality is needed, however, before that forage can be fed to the animal. Predicting forage quality is important when marketing hay, predicting when to harvest a first cutting, moving animals onto pasture or to another pasture, predicting the quality of a silage, predicting how well a forage will perform in a given animal's ration, and increasingly, for use in soil nutrient management. The required indices may vary with the intended use. We have gained much knowledge since the first edition of Forages (Hughes et al. 1951), and without question our knowledge will continue to grow. Objectives of this chapter are to give the reader a historic analysis of forage quality analyses and why prediction of forage quality is more important than ever.

Evolution of Forage Quality and Definitions

Forage quality has been defined as production per animal, only if animal potential is not limiting performance, no energy or protein (including amino acids) is fed, and forage is offered *ad libitum* (Mott and Moore 1970). Forage quality is affected by factors including nutritive value and intake. Nutritive value is a contribution of concentration of a particular nutrient times the digestibility of that nutrient, along with the nature of digested products (Moore et al. 2007). Forage **intake** is the amount consumed under the set of conditions outlined above. If production per animal is not available, **voluntary intake** of dry matter (using the above limitations), digestible DM, organic matter (OM) and digestible OM are used as measures of forage quality (Sollenberger and Cherney 1995).

Intake is positively related to accessibility and **acceptability** of the forage, and negatively related to the amount of time that the forage is retained in the **reticulo-rumen** (Sollenberger and Cherney 1995).

As hay markets became more than local and hay buyers and sellers were interacting across country, and even across the world, it became desirable to have a single index that was recognizable across a wide distribution of producers, marketers, and buyers. An index that reflects relative differences in forage quality across a wide spectrum of forages would provide a more objective basis for hay pricing (Moore 1994). An American Forage and Grassland Council (AFGC) Hay Marketing Task Force was assigned with the task of developing such an index. An ideal index of forage qualify would be a single number that represented the combination of a forage's potential for voluntary intake

Forages: The Science of Grassland Agriculture, Volume II, Seventh Edition.
Edited by Kenneth J. Moore, Michael Collins, C. Jerry Nelson and Daren D. Redfearn.

and nutritive value and allowed for relative comparisons among forages differing in **genotype**, season of growth, and maturity. The AFGC Hay Marketing Task Force developed the relative feed value (RFV) that was intended to achieve this objective (Rohweder et al. 1978).

Van Soest et al. (1994) stated, "and **acid detergent fiber (ADF)** are more consistently associated with digestibility while other components particularly hemicellulose and **neutral-detergent fiber (NDF)** are more closely related to voluntary intake." Based on this statement, most forage laboratories adopted ADF and NDF as routine analyses and use them to predict DM **digestibility** and DM intake, respectively. Observed RFV is determined by two animal responses, dry matter intake (DMI % of body weight [BW]) and **digestible** dry matter (DDM) concentration (% of DM). Therefore, RFV is calculated from predicted values for both DMI and DDM based on laboratory analyses for NDF and ADF, respectively. The system was designed such that a full bloom alfalfa hay having 53% NDF and 41% ADF on a dry matter (DM) basis would have a value of 100. Prediction equations for calculating RFV (Linn et al. 1987) are as follows:

$$\text{DMI, \% of BW} = 120/\text{NDF, g}/100\text{ g DM})$$

$$\text{DDM, g}/100\text{ g DM} = 88.9 - .779 \times (\text{ADF, g}/100\text{ g DM})$$

$$\text{RFV} = \text{DMI} \times \text{DDM}/1.29$$

Where NDF = neutral detergent fiber.

These equations are also the current equations used by the National Forage Testing Association (NFTA; Undersander et al. 1993).

The divisor, 1.29, was chosen so that the RFV of full bloom alfalfa has a value of 100. Larger RFV values indicate greater overall quality relative to mature alfalfa.

The RFV system has been used successfully in the alfalfa hay market for over 30 years. Currently the USDA lists quality standards for hay; categories of Prime, 1, 2, 3, 4, and 5 are given and values for CP (crude protein), ADF, NDF, and RFV are listed, using RFV as the base (Table 38.1). The RFV system, however, lacks an ability to accurately evaluate **grasses** in comparison to **legumes**. Forage fiber constituents in grasses do not have the same characteristics as those in legumes. The RFV index is based on predictions of intake and digestibility from laboratory assays that are often not highly correlated (Moore and Coleman 2001).

Moore et al. (2007) developed a similar system to RFV, calling it quality index (QI). The difference between the two is well described by Moore and Undersander (2002b). In QI, the reference base is in terms of energy required for maintenance of a defined animal, rather than the arbitrary forage used in RFV. Neither RFV or QI works well for predicting DMI and cannot be used in nutritional models.

No matter how sound RFV and QI are conceptually, the accuracy of predicted RFV or QI values is dependent on the equations used to predict DMI, DDM, and organic matter digestibility (OMD) from NDF, ADF, and *in vitro* OM digestion, respectively. Predictive relationships between NDF and DMI, and ADF and DDM reported in the literature have often been poor (Moore and Coleman 2001).

Van Soest et al. (1994) reported that NDF and ADF accounted for only 58% and 56% of the variability in DMI and DDM, respectively, in a diverse data set (n = 187). Further, they noted that neither ADF nor NDF were related to DDM in **aftermath** cuttings (r = −0.20). Abrams (1988) found that more than half of the error in predicting forage digestibility from ADF was associated with selection of an unacceptable equation. Current NFTA equations used to predict RFV often underestimate RFV of higher-quality grasses, and give unacceptable estimates in many cases (Moore et al. 1996). The RFV of high-quality grasses is underestimated because DMI is underestimated (Moore et al. 2007). The NFTA intake prediction equation is based on the assumption that NDF intake is a constant 1.2% of BW. Intake of NDF is not, however, a constant 1.2% of BW (Moore et al. 2007). The constant (1.2 ± 0.1% of BW) seems to have originated from work by Mertens (1987) that determined daily NDF intake in diets of dairy cows producing maximum 4% fat corrected milk (FCM) (very high producing animals). Moore and Undersander (2002b) suggested that extrapolation of data of animals fed high concentrate diets to forages fed alone should not be done, again suggesting that NDF is not correlated closely with DMI.

Others have reported NDF intake to be variable for grasses and legumes fed alone and for mixed diets for lactating cows (Moore et al. 2007). With alfalfa, only a very small correlation between observed DMI and DMI predicted from NDF using the NFTA equation was observed (Moore et al. 2007).

Mertens (1994), however, had a reasonable explanation for why NDF was not always correlated with DMI. Mertens (1992) developed the aNDF-Energy Intake system, refining the concept that NDF and net energy of lactation (NEL) could serve as "proxies" for the filling effect and available energy in intake regulation (Figure 38.1). When intake is limited by metabolic needs (high-energy diets), intake is positively correlated with NDF concentration; when fill limits intake (high NDF diets) NDF is negatively correlated with intake. Over a wide range of forages, the correlation would approach zero. This is not because NDF is not related to intake, but rather that the relation is not linear. (Mertens 1994). Feeding an animal a high-quality forage will not always result in NDF being the factor controlling intake Mertens (1994).

Table 38.1 Quality standards for hay

	Chemical component			
	CP[a]	ADF	NDF	
Quality standard		% of DM		RFV
Prime	>19	<31	<40	>151
1	17–19	31–35	40–46	151–125
2	14–16	36–40	47–53	124–103
3	11–13	41–42	54–60	102–87
4	8–10	43–45	61–65	86–75
5	<8	>45	>65	<75

Source: Adapted from Hay Market Task Force of American Forage and Grassland Council (Moore 1994).
[a]CP = Crude protein; ADF = Acid detergent fiber; NDF = Neutral detergent fiber; RFV = Relative feed value.

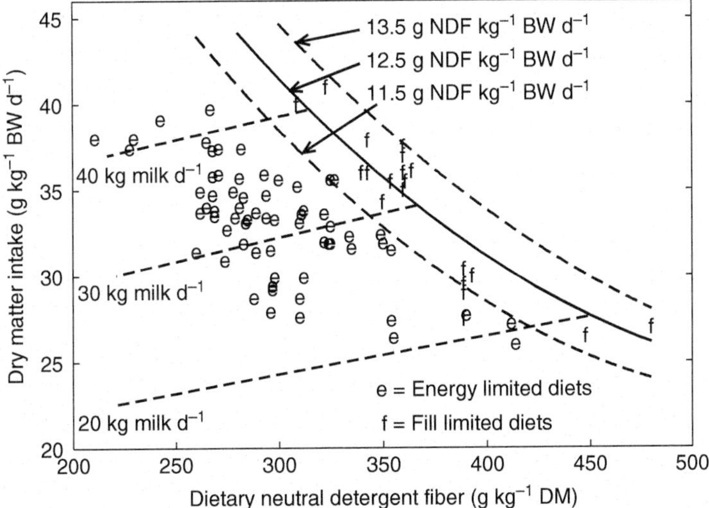

Fig. 38.1. Comparisons of the Energy-Intake to the observed intakes of dairy cows fed different diets at different production levels. *Source:* Adapted from Mertens (1992).

Moore and Undersander (2002a) proposed a new index, relative forage quality (RFQ), with new prediction equations that could be used in place of RFV. Initially, two sets of prediction equations were recommended for calculating RFQ, one for legumes and another for grasses (Moore and Undersander 2002a). An important feature of this newer system was the greater number of variables included and especially the key role of NDF digestibility in estimating both **nutritive value** and DMI.

Calculation of RFQ uses the appropriate set of equations (legume or grass) as follows:

$$RFQ = DMI \times TDN/1.23$$

Where **TDN = total digestible nutrients**

Digestibility is calculated from ADF as TDN, which assumes that ADF has a constant relationship to digestibility. There is, however, considerable variation in the digestibility of the dry matter relative to the ADF content as discussed previously. Nutrient Requirements of Dairy Cattle (NRC 2001) recognizes this and recommends the use of digestible fiber instead of ADF. Relative Forage Quality was developed as a result of this new information and ability to quickly get the data needed (Moore et al. 2007).

The RFQ system has the flexibility to update prediction equations as new data become available. In addition, local equations can and should be used (Moore et al. 2007).

Total digestible nutrients for alfalfa, clovers, and legume/grass mixtures are calculated from the NRC recommendations (NRC 2001) using *in vitro* estimates of digestible NDF (not those calculated from lignin) as follows:

$$TDN_{legume} = (NFC \times 0.98) + (CP \times 0.93) + (FA \times 0.97 \times 2.25) + [NDFn \times (NDFD/1000)] - 7$$

Where:

CP = crude protein (% of DM)
EE = ether extract (% of DM)
FA = fatty acids (% of DM) = ether extract − 1 (unless FA < 1, in which case FA = 0).
NDF = neutral detergent fiber (% of DM)
NDFCP = neutral detergent fiber crude protein
NDFn = nitrogen free NDF = NDF − NDFCP, otherwise estimated as NDFn = NDF × 0.93
NDFD = 48-hour *in vitro* NDF digestibility (% of NDF)
NFC = non fibrous carbohydrate (% of DM) = 100 − (NDFn + CP + EE + ash)

Dry matter intake calculations for alfalfa, clover, and legume/grass mixtures are as follows:

$$DMI_{legume} = 120/NDF + (NDFD - 45) \times 0.374/1350 \times 100$$

(Moore et al. 2007). Forty-five is an average value for fiber digestibility of alfalfa and alfalfa/grass mixtures. Where DMI is expressed as % of BW, NDF as % of DM and NDFD as % of NDF.

Relative forage quality is calculated as follows:

$$RFQ = (DMI_{leg}, \% \text{ of BW}) \times (TDN_{leg}, \% \text{ of DM}) \div 1.23$$

Total digestible nutrients for warm and cool season grasses are calculated as:

$$TDN_{grass} = (NFC \times 0.98) + (CP \times 0.87) + (FA \times 0.97 \times 2.25) + (NDFn \times NDFDp/100) - 10$$

Where terms are as defined previously and

$$NDFDp = 22.7 + 0.664 \times NDFD$$

(Moore and Undersander 2002b)

Dry matter intake calculations for warm- and cool-season grasses are as follows:

$$DMI_{grass} = -2.318 + 0.442 \times CP - 0.01 \times CP^2 - 0.0638 \times TDN + 0.000922 \times TDN^2 + 0.18 \times ADF - 0.00196 \times ADF^2 \times CP \times ADF$$

(Moore et al. 2007)

Where DMI is expressed as a percentage of BW, and CP, ADF, and TDN are expressed as a percentage of DM.

Relative forage quality can be found using:

$$RFQ = (DMI_{grass}, \% \text{ of BW}) \times (TDN_{grass}, \% \text{ of DM}) \div 1.23$$

(Moore et al. 2007).

The RFV continues to be an important tool in the marketing of forage, and in forage quality education (Moore et al. 2007).

Forage-testing reports may or may not include an estimate of voluntary intake. If RFV or RFQ are predicted, however, an estimate of intake is required. Between RFV and RFQ, RFV remains the more popular of the two, largely because the primary hay markets are dominated by alfalfa hay and buyers/sellers have been reluctant to change a value they already know that works for them. Numerous analysis laboratories will calculate and print both. The RFQ has other advantages over RFV that should be mentioned.

Advantages of RFQ over RFV:

1. RFQ may be translated into energy requirements for maintenance and production.
2. Development of a new index provides the opportunity for flexibility in choice of equations for predicting DMI and TDN; these equations should be specific for different types of forage.
3. Associative effects between forages and concentrates that influence forage intake and digestibility can be predicted from estimates of forage TDN intake when fed alone (Moore et al. 2007).

Crude Protein

In our current definition of forage quality, **crude protein** (CP) is not included. Protein, however, is an essential component of herbivore nutrition. CP reflects a forage's potential to provide protein that herbivores require. The term "crude protein" does not refer to or measure how much actual protein is present in forages. Forage CP is based on a chemical analysis of the amount of nitrogen (N), which is the building block for amino acids that make up proteins. This is then used to estimate the amount of true protein and non-protein nitrogen. To calculate CP, total N is determined, then multiplied by 6.25 (forage proteins contain about 16% N; 100% DM/6.25 = 16% protein). When the percentage of CP is low, adequate levels of rumen bacteria for forage digestion cannot be sustained, and the animal's intake and digestibility are reduced. If, however, a forage is very high in CP, much of the CP may be lost in the urine as urea, and also be unavailable to the ruminant.

Though CP is important and a commonly used measure of feed quality, as our understanding of forage quality grows, it has become apparent that while analysis of CP is fine for some situations, it is not refined enough to deal with the more sophisticated programs to feed ruminants in high-production situations. Ruminants have specific requirements for amino acids rather than for proteins. However, evaluating protein sources and formulating rations still requires characterization of proteins in terms of their degradability in the rumen. The ultimate goal of feed models is to formulate rations that meet the precise nitrogen needs of rumen microbes as well as the ruminant animal. Licetra et al. (1996) described a methodology for separating N and protein fractions (Table 38.2) that meshed well with the commonly accepted fiber method of Van Soest (1994). This method of nitrogen fractionation has become popular among modelers and many quality analysis laboratories run these protein analyses. Many laboratories use **near infrared reflectance spectroscopy** (NIRS) to determine these fractions.

Predicted Values

Increasingly, predicting total forage value requires that forage testing reports include estimates of nutritive value, including CP and one or more expressions of available energy (e.g. DDM, TDN, digestible energy (DE), and net energy (NE)). Currently many laboratories use the calculations reported in the Nutrient Requirements for Dairy Cattle (2001) for estimating energy available to the animals or TDN. The TDN denotes the sum of the **digestible protein**, digestible **non-structural carbohydrates (NSC)**, digestible NDF and 2.25 × the digestible fat in the form of a summative equation. Chemical constituents can be determined in the laboratory. The summative equation of Weiss et al. (1992) is being used successfully to estimate TDN concentrations in feeds, forages, and mixed diets. It includes estimates of truly digestible NFC, truly digestible CP, truly digestible fatty acids, truly digestible NDF, and metabolic fecal output excretion. As with the Van Soest summative equation (Van Soest 1994), the Weiss equation requires an estimate of NDF digestibility. The Weiss equation uses lignin concentration to predict digestibility of NDF. This equation was adopted by the NRC Dairy committee (NRC 2001).

$$TDN_{1x}\,(\%) = tdNFC + tdCP + (tdFA \times 2.25) - 7$$

Where TDN_{lx} is truly **digestible nutrients** at maintenance intake,

tdNFC is truly digestible non fibrous carbohydrates

tdCP is truly digestible CP

tdFA is truly digestible fatty acids.

Individual components are estimated as follows (NRC 2001):

$$tdNFC = -0.98\,(100\text{–}[[NDF\text{–}NDICP] + CP + EE + Ash]) \times PAF$$

where PAF = processing adjustment factors (Table 2.1, NRC 2001)

$$tdCPf = CP \times \exp[-1.2 \times (ADICP/CP)]$$

Table 38.2 Suggested partitioning of nitrogen and protein fractions in forages

Fraction	Definition or composition	Enzymatic degradation	Classification
Non-protein nitrogen (NPN)	Peptides, nitrate, non-essential amino acids, etc.	Not applicable	A
True protein	Forage plant proteins (typically contain about 16% N)		
True soluble protein (Buffer soluble protein [BSP])	Buffer soluble and precipitable	Fast	B1
Insoluble protein	IP-NDIP	Variable	B2
Neutral detergent soluble protein			
ND insoluble protein (NDIP)	ND insoluble protein but soluble in AD	Variable to slow	B3
AD insoluble protein (ADIP)	ADIP or ADIN Lignified N, heat damaged protein	Indigestible	C

Source: Adapted from Licitra et al. (1996).

where f = forage

$$tdFA = FA$$

$$tdNDF = 0.75 \times (NFD_n - ADL) \times (1 - ADL/NDF_n)^{0.0667}$$

where NDICP = ND insoluble N × 6.25, ADICP = AD insoluble N × 6.25, FA = fatty acids and NDFn = NDF – NDICP (NRC 2001).

The current TDN prediction equations are adapted from NRC (2001).

Moore et al. (2007) modified these equations specifically using forage legumes and grasses.

Legume:

$$DMI_{legume} = [(0.012 \times 1350)/(NDF/100) + (NDFD - 45) \times .374]/1350 \times 100$$

$$TDN_{legume} = NFC \times .98 + CP \times .93 + FA \times .97 \times 2.25 + NDFn \times NDFD/100 - 7$$

Grass:

$$DMI_{grass} = -2.318 + .442 \times CP - .0100 \times CP^2 - .0638 \times TDN + .000922 \times TDN^2 + .180 \times ADF - .00196 \times ADF^2 - .00529 \times CP \times ADF$$

$$TDN_{grass} = NFC \times .98 + CP \times .87 + FA \times .97 \times 2.25 + NDFn \times NDFDp/100 - 10$$

where

DMI = DM intake, percentage of BW
TDN = total digestible nutrients, g 100 g^{-1} DM
OM = organic matter = 100 – ash, g 100 g^{-1} DM
CP = crude protein, g 100 g^{-1} DM; maximum for CP is 16% for grass DMI.
EE = ether extract, g 100 g^{-1} DM
FA = fatty acids = EE – 1, g 100 g^{-1} DM
NFC = nonfiber carbohydrates = OM – CP – EE – NDFn, g 100 g^{-1} DM
ADF = acid detergent fiber, g 100 g^{-1} DM
NDF = neutral detergent fiber, g 100 g^{-1} DM
NDFn = NDF free of CP = .93 × NDF, g 100 g^{-1} DM
NDFD = 48 hour *in vitro* NDF digestion, g 100 g^{-1} NDF
NDFDp = 22.7 + .642 × NDFD, g 100 g^{-1} (for grasses only)
(Moore et al. 2007).

Available Energy and Digestibility

Available energy and digestibility cannot be measured in the laboratory and are estimated from chemical composition, as noted above. Most energy values are predicted from fiber analyses because fiber is negatively related to the animal's ability to digest and use nutrients in the feed. Various groups have developed equations for predicting energy (Moore et al. 2007). It should be recognized that the source and accuracy of NRC (2001) values are also unknown.

Forage Quality for Use in Predicting Quality through Models

Increasingly, any one index of forage quality gives insufficient information on a forage to be useful. Dynamic models such as the Cornell Net Carbohydrate and Protein System (Van Amburgh et al. 2017) are increasingly being used to formulate rations for specific groups of animals. These models rely on accurate descriptions of animal, environment, feed, and forage composition, and management conditions on a particular farm. Accurate and rapid analyses of representative samples of forages and feeds on the farm are essential for using models in management decisions. NIRS has made it possible to obtain nutritive value data quickly. If appropriate equations are available to predict forage quality from nutritive value, potential exists for successful application of models.

Numerous models for predicting DMI have become increasingly complex and advanced. Models such as the simple model proposed by NRC (2001), which only includes animal factors, are being increasingly replaced by more complex models such as the Nordic Feed Evaluation System (Norfor; Volden et al. 2011), total dry matter index (TDMI; Huhtanen et al. 2011), and another by Zom et al. (2012a,b). Jensen et al. (2015) does a good job of evaluating these European Models. In the US, models such as the Cornell Net Carbohydrate and Protein System (Version 7.0; Van Amburgh et al. 2017) and the total tract DM digestibility (TTDMD) (Lopes et al. 2015; Lopes and Combs 2015) are being used to better estimate the DMI of animals.

Tools and the Digital Age

Laboratory and quick estimates of forage nutritive value would have been impossible without the advent of near NIRS. Norris et al. (1976) showed that the technology could be used to evaluate quality components in forages. By 1983, commercial companies were selling NIRS along with prediction and calibration techniques (Shenk and Westerhaus 1994). NIRS has been used to evaluate forage quality since that time and NIRS use in forage nutritive value and quality prediction has become ubiquitous. This method provides rapid turn-around times on forage analysis that are not possible using other currently available methods.

Properly timed spring forage harvest sets the stage for harvest management for the rest of the season for pasture, hay or silage fields. Measures such as digestibility, fiber digestibility, RFQ, milk per ton, and milk per acre are possibilities for harvest timing. Fiber digestibility is very highly correlated with *in vitro* true digestibility (Cherney et al. 2006), so they are equivalent in usefulness. This relationship would likely only be significantly impacted by varieties with radically different forage nutritive value, such as brown-midrib mutants which may have 15% units higher fiber digestibility than normal varieties. Also, dry matter yield per acre is highly correlated with milk yield per acre. None of the above parameters replace use of the compromise between forage yield and quality that using forage maturity/harvest stage provides (Cherney et al. 2006).

One definition of forage is a crop that can meet the effective fiber needs of an animal when fed as the primary forage source in the diet. Forage grasses and legumes need to be harvested at optimal fiber content for the class of livestock being fed; therefore, NDF is the most useful for predicting harvest date target in the US. There is a relatively small range in optimal NDF for some classes of animals, such as lactating dairy cows, making correct harvest management decisions critical relative to nutritive value. A reliable method to estimate the fiber content of grass and alfalfa-grass mixtures would help producers in timing harvesting operations to optimize the nutritive value of the harvested forage (Cherney et al. 2006).

Methods for estimating grass growth and alfalfa growth exist (Parsons et al. 2006, Forages.org). The target harvest date window for pure alfalfa is relatively narrow, regardless of the target NDF level (Cherney and Sulc 1997). Methods to predict harvest date must be quick and accurate. Numerous attempts have been made to improve upon the alfalfa quality prediction standard set by Kalu and Fick (1981). Current approaches to forage quality predictions for alfalfa include age and weather-based equations, mean stage, and leafiness (Fick et al. 1994). Mean stage by count (MSC) or mean stage by weight have worked very well to estimate forage quality (Kalu and Fick 1983; Moore and Moser 1995), but are laborious and time consuming from the producer standpoint.

A more practical approach to predicting alfalfa forage quality is based on the development of the most mature stem maximum stage of development (MAXSTAGE) and the height of the most mature stem (MAXHT) (Moore et al. 1990). This is a rapid method for estimating alfalfa quality based on Wisconsin predictive equations for alfalfa quality (PEAQ; Hintz and Albrecht 1991). It has been successfully field tested over a wide range of environments (Sulc et al. 1997). Parsons et al. (2006) found that MAXHT on its own provides a good estimate of pre-harvest NDF and, including MAXSTAGE provides little or no extra accuracy. Models are also available that take into account different stubble heights (Parsons et al. 2009). Plant height and maturity stage along with growing degree days (GDD) for estimating alfalfa fiber content has been suggested (Allen and Beck 1996). Cherney and Sulc (1997) suggested a method for predicting first cutting alfalfa fiber concentration based on early sampling and analysis, followed by use of predictive equations. Accurately predicting legume forage quality is hindered by variability in environmental conditions which greatly affect quality (Cherney and Cherney 2002). Parsons et al. (2006) developed an on-line tool (Forages .org) that allows for estimation of current NDF and target harvest height for pure alfalfa. The user provides the current maximum height (tallest stem), planned stubble height, and target NDF at harvest (Parsons et al. 2006). The tool then predicts the day on which the alfalfa crop will reach the target NDF.

It is particularly difficult to assess standing quality of pure grass without actually analyzing samples. Grass morphology changes are often not obvious until heading, and once grass has headed it is often high in NDF, limiting its use as a forage for high-producing animals. Often a calendar date works as well as any morphologic indicators for spring harvest. Latitudinal and temperature differences make this method ineffective for prediction over diverse locations or regrowths (Van Soest 1994). Some research has concentrated on equations to predict NDF using GDD (Cherney et al. 2001). The benefit of GDD32 over day of the year is that it better explains differences in growth between years with different weather patterns. Prediction equations developed from several locations and years did not accurately predict NDF or CP. However, regression equations developed to predict NDF at individual locations, years, and fertility levels had low mean square error (MSE) and correlations (R^2) above 0.90. These equations could be used to assist producers in developing harvest strategies to optimize forage nutritive value. Proximity to weather stations was critical in accurate equation development, however, so producers may not be able to rely on weather stations unless they are within a few kilometers. The Parsons et al. (2012) tool allows estimation of current NDF and target harvest height for several grass species. Similar to the alfalfa tool developed earlier (Parsons et al. 2006), the user provides the current grass canopy height, planned stubble height, estimated yield and target NDF at harvest. The equations to predict DM loss were promising and warrant further field testing. Because the equations were based on grass height rather than yield, the equations will give producers an indication of the fraction of yield lost even if the estimate of potential yield is not accurate (Parsons et al. 2012). Proficient models can be developed for predicting grass NDF based on explanatory variables. These may not have, however, good predictive ability for years and situations different to those for which the models were developed. Models that are accurate for estimating grass NDF concentration may continue to be difficult to develop over a range of locations, years, and management practices (Parsons et al. 2012).

Being able to estimate harvest time for mixed legume/grass stands is more difficult than either grass or legume (alfalfa) alone. Variability in forage composition is the primary concern when feeding alfalfa–grass mixtures, which is strongly impacted by the ratio of grass to legume (Parsons et al. 2006). Increasingly, alfalfa is sown with a perennial grass, providing greater yield constancy in variable environments (Tracy et al. 2016). Forage quality can be optimized by harvesting at the prime time for the mixture in a particular field (Parsons et al. 2006, Table 38.3). Monitoring botanic composition of alfalfa–grass fields preharvest, as well as estimating composition of alfalfa–grass forage post-harvest, aids forage producers in managing alfalfa–grass use in the field and in animal diets. Knowledge of the alfalfa content of mixtures in fields also permits estimation of biologic N fixation for nutrient management purposes (Cela et al. 2015).

Botanic composition in mixed stands of alfalfa and grass is a critical parameter in equations estimating harvest fiber concentration for dairy rations. Composition is difficult to estimate by visual observation.

Manual and semi-automated approaches have been attempted to estimate botanic composition in legume–grass stands. Visual estimation methods have been used historically as the principal method. Rayburn and Green (2014) developed a visual reference guide for mixed stands of clover and grass to help calibrate the eye for human field estimation, but estimates can vary from person to person. Himstedt et al. (2010) developed equations using logit-transformed legume coverage in a statistical model, and selected a multivariate model predicting legume dry matter contribution with effects including logit-transformed legume coverage, total biomass in the sample, and their interaction.

Equation testing on field samples yielded a strong relationship with legume dry matter contribution for clover–grass mixes ($R^2 = 0.98$). Practicality of such an equation for field use is questionable, however, without further investment in technology such as field spectroscopy, given the need for the total dry matter biomass variable.

Predicting proportions of legume and grass in forage mixtures using NIRS has been attempted numerous times. Past studies evaluating forage botanic composition via NIRS analysis has used the terms real, true, or natural mixtures to describe samples generated from species grown in mixtures in the same stand (Karayilanli et al. 2016). Almost all attempts to calibrate NIRS for estimating proportions of species in mixtures have been successful. Validation was successful when validation samples were a subset of the calibration samples (closed population) (Coleman et al. 1990; Moore et al. 1990; Wachendorf et al. 1999).

Cougnon et al. (2013) recommended a calibration strategy based on diverse hand-sorted samples rather than many artificial samples that have similar spectral information. These authors reported that calibration equations for prediction of grass–legume composition based only on artificial samples produced unsatisfactory results.

Karayilanli et al. (2016) developed an NIRS method to estimate percentage composition of binary alfalfa–grass mixtures in the Northeastern United States. Alfalfa–grass samples were collected across New York state over four growing seasons and hand separated into grass and alfalfa samples. One sample was oven dried and another subset was separately ensiled. Samples were mixed to range from 0% to 100% alfalfa for NIRS calibration, with a total

Table 38.3 Estimated stand NDF of a mixed alfalfa-grass stand based on alfalfa height and the percent grass in the stand

	% Grass in the stand (dry matter basis)								
Max. alfalfa height(cm)	10	20	30	40	50	60	70	80	90
36	23.5	26.7	29.9	33.1	36.3	39.5	42.7	45.9	49.1
41	25.1	28.3	31.5	34.7	37.9	41.1	44.3	47.5	50.7
43	25.9	29.1	32.3	35.5	38.7	41.9	45.1	48.3	51.5
48	27.6	30.8	34.0	37.2	40.4	43.6	46.8	50.0	53.2
53	29.2	32.4	35.6	38.8	42.0	45.2	48.4	51.6	54.8
58	30.9	34.1	37.3	40.5	43.7	46.9	50.1	53.3	56.5
79	37.5	40.7	43.9	47.1	50.3	53.5	56.7	59.9	63.1
84	39.1	42.3	45.5	48.7	51.9	55.1	58.3	61.5	64.7
90	40.8	44.0	47.2	50.4	53.6	56.8	60.0	63.2	66.4

Source: Adapted from Parsons et al. (2006).
Target NDF for each mixture is highlighted.

about 750 individual samples from three years used for calibration of three NIRS instruments and samples from a fourth year used for validation. Grass composition was predicted with good precision and accuracy showing biases of 2.49 and standard errors of prediction (SEP) of 5.06, with R^2 of 0.972, using the equation developed across multiple instruments. With selection of a robust set of calibration samples over many environments, NIRS can be used to determine the botanical composition of fresh-dried or ensiled-dried alfalfa–grass samples, and replicate scans from multiple instruments can be combined to develop a single calibration that will perform with equal efficiency across different instruments (Figure 38.2). Producers will be able to accurately estimate alfalfa and grass percentages in harvested and stored forage mixtures using NIRS. Commercial forage testing laboratories can provide grass percentage as an analysis option to the general public using the calibrations. Single cameras were used in all tests. The technique would require further testing under variable field conditions to evaluate potential field use (Karayilanli et al. 2016).

In a field study, Post et al. (2007) related plant canopy spectral reflectance (wavelengths 680 and 705 nm in the second derivative spectra) with alfalfa fraction in a mixed stand ($R^2 = 0.6 - 0.7$, $n = 95$). The approach is promising for further investigation, and potentially for post-calibration field use (Post et al. 2007). Others have implemented variations on canopy spectral reflectance to predict stand composition with promising results (Kawamura et al. 2011). However, spectral technology is only now becoming accessible for end users (McRoberts et al. 2016). There is a commercial product, Yara N sensor, which is a spectroradiometer ranging from 400 to 1000 nM. It is mostly used for estimation of N content, however, some work on using it to estimate forage CP and NDF has been done with success (Zhou and Parsons 2018).

More sophisticated image intelligence processing methods such as artificial and texture classification (Sabeenian and Palanisamy 2010) have been tested to discriminate between vegetation types (e.g. crops and weeds). Methods that permit crop–weed discrimination in real time, combined with robotic herbicide application and cultivation systems, currently play an important role in precision agriculture. Local binary patterns (LBP), commonly known for their use in facial recognition (Ahonen et al. 2006), provide a powerful, robust, computationally efficient method for texture classification in image analysis (Ojala et al. 2002). Digital image analysis in mixed stands could reduce botanic composition uncertainty and improve spring harvest management decisions (McRoberts et al. 2016).

Under field conditions and with different image acquisition devices, illumination variability and color variability are high. McRoberts et al. (2016) attempted to develop a practical, farmer-accessible method that can be applied to estimate stand composition (i.e. grass and alfalfa dry matter fractions in binary mixes) under variable field conditions. A method was developed that combined digital image analysis using LBP with statistical modeling

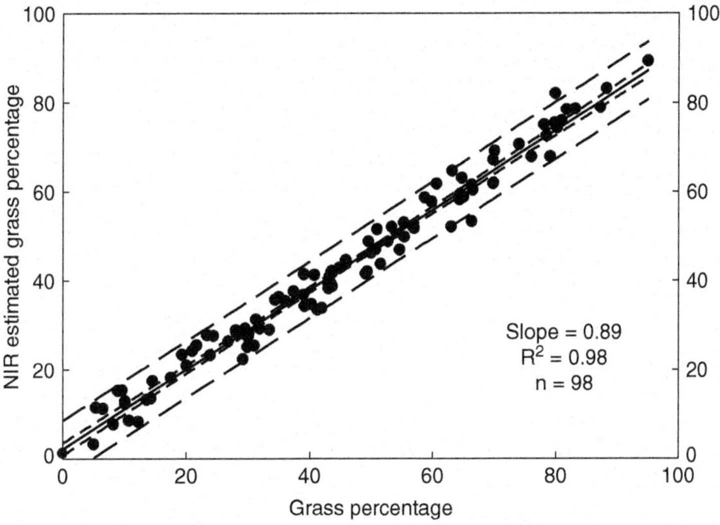

Fig. 38.2. The NIRS estimated vs actual grass percentage for the validation set across three instruments. Inner short-dash lines are the 95% confidence bands, and the outer inner long-dash lines are the 95% prediction bands. Prediction bands are approximately ±6.3%. *Source:* Adapted from Karayilanli et al. (2016).

to estimate alfalfa–grass stand composition. The sampling process and its implementation with several processing approaches to estimate stand composition were used to develop this method.

Mixed stands were sampled (n = 168) in farmers' fields in New York. A digital image was captured of standing samples at 5-Megapixel resolution, and alfalfa and grass height relationships were recorded. After clipping representative samples at 10-cm above ground level, samples were manually separated into alfalfa and timothy, and dried to calculate fractions on a dry matter basis. Uniform rotation invariant LBP were extracted from whole images and 64 × 64-pixel tiles and, were used to develop regression equations estimating grass fraction. An iterative process selected most accurate LBP operator settings.

Predictive accuracy in whole image models using tile LBP histogram averages was highest for models generated from LBP tile histogram bin means, with pairwise correlations between tile model-generated grass coverage estimates and sample grass fraction up to 0.895. LBP are effective in differentiating alfalfa and grass under field conditions, because the method is robust to changes in color and illumination. Furthermore, key LBP histogram bins (e.g. symmetric edges) strongly differentiate alfalfa and grass in tiles. The LBP method is promising based on this study, but further evaluation under diverse field conditions, including different cameras and grass species, is still necessary to assess usefulness.

Bias and Accuracy

Variation due to sampling is most often the largest contributor to error in the analysis of forage for quality (Cherney and Cherney 2003). Appropriate sampling of the forage, whether it be pasture, hay, greenchop or silage, is critical because analytic results are useless without it. The goal for a good sample is accuracy and precision. Sampling accuracy refers to the ability to get a sample that represents the whole field, silo, or bunker that actually represents what is there. The more variability that there is, the larger the number of samples that must be taken. Laboratory accuracy is a measure of how well that laboratory's samples compare with some accepted standard reference. If using models to predict nutrient values, the models should be as unbiased as possible.

Conclusion

Many chemical characterization assays are designed to estimate *in vivo* digestibilities. Numerous researchers have concluded, however, that chemical assays are not as closely related to *in vivo* assays as are *in vitro* assays (Moore and Coleman 2001). Relationships between a single chemical assay and an *in vivo* assay can be difficult because of the immense complexity of biologic systems, both feed and animal. It is clear, however, that there are situations, such as hay marketing and pre-harvest decisions, where a single chemical assay is as good or better than a more complex analysis. Understanding of biologic systems through the use of use of models and new technologies, such as satellite and drone imagery, is and will be a major path forward in forage research and improvements in forage research. Predictive equations continue to require evaluation against animal performance. *In vitro* digestibilities using rumen fluid continues to be important. Welfare concerns will make it increasingly difficult to use fistulated animals and research using modern techniques needs to evaluate alternative methods to using these animals in the development of predictive equations (Cherney and Cherney 2003).

References

Abrams, S.M. (1988). Sources of error in predicting digestible dry matter from the acid-detergent fiber content of forages. *Anim. Feed Sci. Technol.* 2: 205–208.

Ahonen, T., Hadid, A., and Pietikainen, M. (2006). Face description with local binary patterns: application to face recognition. *IEEE Trans. Pattern Anal. Mach. Intell.* 28: 2037–2041.

Allen, M. and Beck, J. (1996). Relationship between spring harvest alfalfa quality and growing degree days. In: *Proceedings of the National Alfalfa Symposium, 26th. Certified Alfalfa Seed Council*, Davis, CA., East Lansing, MI. 4–5 Mar. 1996, 16–25.

Cela, S., Ketterings, Q.M., Czymmek, K.J. et al. (2015). Long-term trends of nitrogen and phosphorus mass balances on New York dairy farms. *J. Dairy Sci.* 98: 7052–7070. https://doi.org/10.3168/jds.2015-9776.

Cherney, D.J.R., Dewing, D.R., and Cherney, J.H. (2001). Prediction of reed canarygrass quality as influenced by N fertilization and maturity. *J. Dairy Sci.* 84 (Suppl. 1): 345.

Cherney, D.J.R., Cherney, J.H., and Parsons, D. (2006). Forage quality assessment of spring forages. In: *Proceedings of the Cornell Nutrition Conference*, Oct., Syracuse, NY, 125–133.

Cherney, J.H. and Cherney, D.J. (2002). Forage legume quality: animal productivity and sustainability issues. *Journal of Crop Prod.* 5: 261–284.

Cherney, J.H. and Cherney, D.J.R. (2003). Assessing silage quality. In: *Silage Science and Technology*, Agronomy Monograph no. 42 (eds. D.R. Buxton, R.E. Muck and J.H. Harrison), 141–198. Madison, WI: American Society of Agronomy.

Cherney, J.H. and Sulc, R.M. (1997). Predicting first cutting alfalfa quality. In: *Silage: Field to Feedbunk. Proceedings from the North American Conference*, NRAES-99, 53–65. Ithaca, NY. Hershey, PA. 11–13 Feb. 1997: Northeast Regional Agricultural Engineering Service.

Coleman, S.W., Christiansen, S., and Shenk, J.S. (1990). Prediction of botanical composition using NIRS calibrations developed from botanically pure samples. *Crop Sci.* 30: 202–207.

Cougnon, M., Van Waes, C., Dardenne, P. et al. (2013). Comparison of near infrared reflectance spectroscopy calibration strategies for the botanical composition of grass-clover mixtures. *Grass Forage Sci.* 69: 167–175.

Fick, G.W., Wilkens, P.W., and Cherney, J.H. (1994). Modeling forage quality changes in the growing crop. In: *Forage Quality, Evaluation, and Utilization* (ed. G.C. Fahey Jr.), 757–795. Madison, WI: ASA, CSSA, and SSSA.

Himstedt, M., Fricke, T., and Wachendorf, M. (2010). The relationship between coverage and dry matter contribution of forage legumes in binary legume–grass mixtures. *Crop Sci.* 50: 2186–2193.

Hintz, R.W. and Albrecht, K.A. (1991). Prediction of alfalfa chemical composition from maturity and plant morphology. *Crop Sci.* 31: 1561–1565.

Hughes, H.D., Heath, M.E., and Metcalfe, D.S. (eds.) (1951). *Forages, the Science of Grassland Agriculture.* Ames: Iowa State College Press.

Huhtanen, P., Rinne, M., Mantysaari, P., and Nousiainen, J. (2011). Integration of the effects of animal and dietary factors on total dry matter intake of dairy cows fed silage-based diets. *Animal* 5: 691–702.

Jensen, L.M., Nielsen, N.I., Nadeau, E. et al. (2015). Evaluation of five models predicting feed intake by dairy cows fed total mixed rations. *Livestock Sci.* 176: 91–103. https://doi.org/10.1016/j.livsci.2015.03.026.

Kalu, B.A. and Fick, G.W. (1981). Quantifying morphological development of alfalfa for studies of herbage quality. *Crop Sci.* 21: 267–271.

Kalu, B.A. and Fick, G.W. (1983). Morphological stage of development as a predictor of alfalfa herbage quality. *Crop Sci.* 23: 1167–1172.

Karayilanli, E., Cherney, J.H., Sirois, P. et al. (2016). Botanical composition prediction of alfalfa-grass mixtures using NIRS: developing a robust calibration. *Crop Sci.* 56: 3361–3366. https://doi.org/10.2135/cropsci2016.04.0232.

Kawamura, K., Watanabe, N., Sakanoue, S. et al. (2011). Waveband selection using a phased regression with a bootstrap procedure for estimating legume content in a mixed sown pasture. *Grassland Sci.* 57: 81–93.

Licetra, G., Hernandez, T.M., and Van Soest, P.J. (1996). Standardization of procedures for nitrogen fractionation of ruminant feeds. *Anim. Fd. Sci. Technol.* 57: 347–358.

Linn, J.G., Martin, N.P., Howard, W.T., and Rohweder, D.A. (1987). Relative feed value as a measure of forage quality. *Minnesota Forage Update* 12 (4): 2–4.

Lopes, F.K.R. and Combs, D.K. (2015). Validation of an approach to predict total-tract fiber digestibility using a standardized in vitro technique for different diets fed to high-producing dairy cows. *J. Dairy Sci.* 98: 2596–2602. https://doi.org/10.3168/jds.2014-8665.

Lopes, F., Cook, D.E., and Combs, D.K. (2015). Validation of an in vitro model for predicting rumen and total-tract fiber digestibility in dairy cows fed corn silages with different in vitro neutral detergent fiber digestibilities at 2 levels of dry matter intake. *J. Dairy Sci.* 98: 574–585. https://doi.org/10.3168/jds.2014-8661.

McRoberts, K.C., Benson, B.M., Mudrak, E.L. et al. (2016). Application of local binary patterns in digital images to estimate botanical composition in mixed alfalfa–grass fields. *Comput. Electron. Agric.* 123 (April): 95–103. ISSN 0168-1699, https://doi.org/10.1016/j.compag.2016.02.015.

Mertens, D.R. (1987). Predicting intake and digestibility using mathematical models of ruminal function. *J. Ani. Sci.* 64: 1548–1558.

Mertens, D.R. (1992). Nonstructural and structural carbohydrates. In: *Large Dairy Herd Management*, 219–235. Champaign, IL: Am. Dairy Sci. Assoc.

Mertens, D.R. (1994). Regulation of forage intake. In: *Forage Quality, Evaluation, and Utilization* (ed. G.C. Fahey Jr. et al.), 450–493. Madison, WI: ASA, CSSA, and SSSA.

Moore, J.E. (1994). Forage quality indices. In: *Forage Quality, Evaluation, and Utilization* (ed. G.C. Fahey Jr. et al.), 967–998. Madison, WI: ASA, CSSA, and SSSA.

Moore, J.E. and Coleman, S.W. (2001). Forage intake, digestibility, NDF, and ADF: how well are they related? In: *Proceedings American Forage and Grassland Council*, Springdale, Arkansas, vol. 10, 238–242.

Moore, J.E. and Undersander, D.J. (2002a). A proposal for replacing relative feed value with an alternative: relative forage quality. In: *Proceedings of the American Forage and Grassland Council*, Bloomington, MN. 14–17 July 2002 (ed. K. Cassida), 171–175. Georgetown, TX: American Forage and Grassland Council.

Moore, J.E. and Undersander, D. (2002b). Relative Forage Quality: An Alternative to Relative Feed Value and Quality Index. In: *Proceedings 13th Annual Florida Ruminant Nutrition Symposium*, 16–32.

Moore, J.E., Burns, J.C., and Fisher, D.S. (1996). Multiple regression equations for predicting relative feed value of grass hays. In: *Proceedings of the American Forage and Grassland Council*, Vancouver, BC (ed. M.J. Williams), 35–139.

Moore, J.E., Adesogan, A.T., Coleman, S.W., and Undersander, D.J. (2007). Predicting forage quality. In: *Forages: The Science of Grassland Agriculture* (eds. R.F Barnes, C.J. Nelson, K.J. Moore and M. Collins), 553–568. Ames: Blackwell.

Moore, K.J. and Moser, L.E. (1995). Quantifying developmental morphology of perennial grasses. *Crop Sci.* 35: 37–43.

Moore, K.J., Roberts, C.A., and Fritz, J.O. (1990). Indirect estimation of botanical composition of alfalfa–smooth bromegrass mixtures. *Agron. J.* 82: 287–290. https://doi.org/10.2134/agronj1990.00021962008200020022x.

Mott, G.O. (1973). Evaluating forage production. In: *Forages: The Science of Grassland Agriculture*, 3e (eds. M.E. Heath, D.S. Metcalfe, R.F Barnes, et al.), 126–135. Iowa State University.

Mott, G.O. and Moore, J.E. (1970). Forage evaluation techniques in perspective. In: *Proceedings of the National Conference Forage Quality, Evaluation, and Utilization*, L1–L10. Lincoln, NE: Center for Continuing Education.

National Research Council (NRC) (2001). *Nutritional Requirements of Dairy Cattle*. Washington, DC: National Academy Press.

Norris, K.H., Barnes, R.F, Moore, J.E., and Shenk, J.S. (1976). Predicting forage quality by infrared reflectance spectroscopy. *J. Anim. Sci.* 43: 1354–1354.

Ojala, T., Pietikainen, M., and Maenpaa, T. (2002). Multiresolution gray-scale and rotation invariant texture classification with local binary patterns. *IEEE Trans. Pattern Anal. Mach. Intell.* 24: 971–987.

Parsons, D., Cherney, J.H., and Gauch, H.G.J. (2006). Estimation of spring forage quality for alfalfa in New York state. *Forage and Grazinglands*. https://doi.org/10.1094/FG-2006-0323-01-RS.

Parsons, D., Cherney, J.H., and Peterson, P.R. (2009). Preharvest neutral detergent fiber concentration of alfalfa as northern New York influenced by stubble height via either bring along some supplies food lion king live. *Agron J.* 101 (4): a769–a774. https://doi.org/10.2134/agronj2008.0174x.

Parsons, D., McRoberts, K., Cherney, J.H. et al. (2012). Preharvest neutral detergent fiber concentration of temperate perennial grasses as influenced by stubble height. *Crop Sci.* 52: 923–931. https://doi.org/10.2135/cropsci2011.09.0478.

Post, C., DeGloria, S., Cherney, J., and Mikhailova, E. (2007). Spectral measurements of alfalfa/grass fields related to forage properties and species composition. *J. Plant Nutr.* 30: 1779–1789. https://doi.org/10.2134/agronj2005.0326.

Rayburn, E.B. and Green, J.T. (2014). Visual reference guide for estimating legume content in pastures. *Forage and Grazinglands* https://doi.org/10.2134/FG-2011-0176-DG.

Rohweder, D.A., Barnes, R.F, and Jorgenson, N.J. (1978). Proposed hay grading standards based on laboratory analysed for evaluating quality. *J. Anim Sci.* 47: 747–759.

Sabeenian, R. and Palanisamy, V. (2010). Crop and weed discrimination in agricultural field using MRCSF. *Int. J. Sign. Imaging Syst. Eng.* 3: 61–69.

Shenk, J.S. and Westerhause, M.O. (1994). The application of near infrared reflectance spectroscopy (NIRS) analysis. In: *Forage Quality, Evaluation, and Utilization* (ed. G.C. Fahey Jr. et al.), 406–449. Madison, WI: ASA, CSSA, and SSSA.

Sollenberger, L.E. and Cherney, D.J.R. (1995. Chapt. 7). Evaluating forage production and quality. In: *Forages, Vol. II: The Science of Grassland Agriculture*, 5e (eds. R.F Barnes, C.J. Nelson and D. Miller), 97–110. Ames: Iowa State University Press.

Sulc, R.M., Albrecht, K.A., Cherney, J.H. et al. (1997). Field testing a rapid method for estimating alfalfa quality. *Agron. J.* 89: 952–957.

Tracy, B.F., Albrecht, K., Flores, J. et al. (2016). Evaluation of alfalfa-tall fescue mixtures across multiple environments. *Crop Sci.* 56: 2026–2034.

Undersander, D., Mertens, D.R., and Thiex, N. (1993). *Forage Analyses Procedures*. Omaha, NE: National Forage Testing Association.

Van Amburgh, M.E., Ortega, A.F., Fessenden, S.W. et al. (2017). The amino acid content of rumen microbes, fee, milk, and tissue after multiple hydrolysis times and implications for the CNCPS. In: *Proceedings of the Cornell Nutrition Conference* , Oct. 17-19, Syracuse, NY, 16.

Van Soest, P.J. (1994). *Nutritional Ecology of the Ruminant*, 2e. Ithaca, NY: Cornell University Press.

Volden, H., Nielsen, N.I., Åkerlind, M. et al. (2011). Prediction of voluntary feed intake. In: *The Nordic Feed Evaluation System* (ed. H. Volden), 113–126. Wageningen, The Netherlands: Wageningen Academic Publishers.

Wachendorf, M., Ingwersen, B., and Taube, F. (1999). Prediction of the clover content of red clover and white clover-grass mixtures by near-infrared reflectance spectroscopy. *Grass and Forage Sci.* 54: 87–90.

Weiss, W.P., Conrad, H.R., and Pierre, N.R.S. (1992). A theoretically based model for predicting total digestible nutrient values of forages and concentrates. *Anim. Feed Sci. Technol.* 39: 95–110.

Zhou, Z. and Parsons, D. (2018). Estimation of yield and height of legume-grass swards with remote sensing in northern Sweden. In: *Proceedings of the Conference: 27th European Grassland Federation General Meeting, June,*

2018 (eds. D. Hennessy, M. O'Donovan, E. Kennedy, et al.), Cork, Ireland.

Zom, R.L.G., André, G., and van Vuuren, A.M. (2012a). Development of a model for the prediction of feed intake by dairy cows: 1. Prediction of feed intake. *Livest. Sci.* 143: 43–57.

Zom, R.L.G., André, G., and van Vuuren, A.M. (2012b). Development of a model for the prediction of feed intake by dairy cows 2. Evaluation of prediction accuracy. *Livest. Sci.* 143: 58–69.

CHAPTER 39

Factors Affecting Forage Quality

Kenneth J. Moore, Charles F. Curtiss Distinguished Professor, *Agriculture and Life Sciences, Iowa State University, Ames, IA, USA*
Andrew W. Lenssen, Professor, *Agronomy, Iowa State University, Ames, IA, USA*
Steven L. Fales, Emeritus Professor, *Agronomy, Iowa State University, Ames, IA, USA*

Quality of forage is defined by its capacity to supply nutrients to support the nutritional needs of livestock (Chapter 38). It is determined by the chemical, biologic, and physical properties of the forage as they relate to the ability of an animal to derive nutritional value from it (Buxton et al. 1995; Cochran et al. 2007). **Forage quality** can be separated conceptually into forage nutritive value and forage **intake** (Fales and Fritz 2007; Moore et al. 2007). Forage **nutritive value** refers to the chemical composition of the forage and its ability to provide energy, protein, minerals, vitamins, and other nutrients to support maintenance and production. It is determined largely by the chemical composition and biologic properties of the forage. Forage intake is the amount of forage that an animal will consume, usually on a daily basis, and is limited by both chemical and physical factors (Mertens 2007; Riaz et al. 2014). Together, forage nutritive value and intake determine the ability of forage to meet the daily nutritional requirements of an animal and thus its quality.

★This chapter is dedicated to our late friend and colleague Steven L. Fales who was a leading researcher in assessing and understanding factors that influence forage quality. He was the senior author on the chapter of the same title in the 6th edition of Forages and some of his writing from that chapter has been incorporated into this one.

Forage Consumed

Unlike feed grains, forages vary widely in their composition and feeding value. Animals consuming forages ingest much of the aboveground portion of the plant whether grazing or consuming harvested or conserved herbage. The forage that is available for consumption depends on several factors that are constantly changing and consequently impacting its nutritional value.

There is a hierarchy of factors that interact to influence forage quality (Figure 39.1) (Moore and Jung 2001). These include the environment in which it is growing, the genetics of the plants comprising the sward or pasture, expression of these factors at the cellular, anatomic and morphologic levels, and a number of post-harvest factors including harvest and storage variables (Chapters 40–42). In this chapter, we describe how pre-harvest factors interact to influence the quality of forage.

Morphology

All plants are comprised of specialized organs such as stems, leaves, roots, and inflorescences. Each organ has a different tissue structure and cellular composition that influence its forage quality. Forage consumed by livestock is comprised mostly of leaves and stems, both vegetative structures, but may include reproductive structures when

Forages: The Science of Grassland Agriculture, Volume II, Seventh Edition.
Edited by Kenneth J. Moore, Michael Collins, C. Jerry Nelson and Daren D. Redfearn.

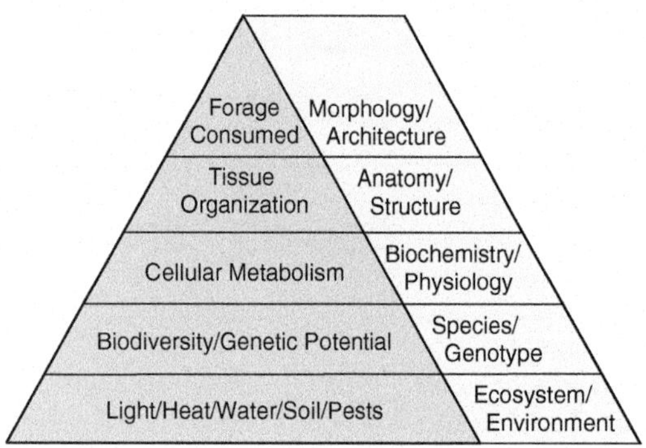

Fig. 39.1. Hierarchy of factors influencing forage quality. The nutritive value of forage is subject to the environment in which it is growing and the genetics of the plant; the so-called Genotype x Environment (GxE) interaction. The plant responds to the changing growth environment at the cellular, anatomic and ultimately morphologic levels to influence the composition of herbage that is available for harvest. The GxE interaction and its influence on forage quality can be manipulated through management practices designed to alter the growth response of the plant to the changing environment.

harvested or grazed at a mature stage of development. It is this differential assembly of plant organs and their proportions within the forage consumed by animals that provide the most visible indicator of forage quality.

Leaves are the most active site of metabolism within a plant and consequently have the highest concentrations of readily degradable metabolites. Van Soest (1994) referred to these metabolites as cell solubles and determined their digestion in the rumen was near complete (98%). Typically, leaves also contain a lower volume of the types of tissues that are particularly recalcitrant to digestion such as conducing vessels and other vascular tissues. The primary tissues comprising leaves include thin-walled mesophyll, the site of carbon fixation and photosynthesis in C_3 plants, the vascular tissue associated with veins, and the leaf epidermis (Moore et al. 2004). Vascular tissues within the leaves are the least degradable and become even less so as the leaf ages (Grabber et al. 1992; Twidwell et al. 1990).

Stems of most forage plants are relatively rigid and structurally support the display of leaves and inflorescences. They generally contain more lignified tissues than does leaf tissue and thus have a lower digestibility (Moore and Jung 2001). These vascular tissues include the **xylem** and **phloem** and are associated with solute movement within the plant.

Leaf: Stem Ratio

The contrasting forage quality of grasses and legumes is largely explained by the relative composition and digestibility of leaves and stems. Legume leaves are generally highly digestible due to their high concentration of cell solubles, particularly protein and other nitrogenous compounds (Andrzejewska et al. 2017). Grass leaves generally have a much higher fiber concentration and are more slowly and less completely digested than legume leaves (Twidwell et al. 1988).

However, the leaves of both grasses and legumes are much more digestible and have a higher nutritional quality than their stems. Therefore, the ratio of leaves to stems in forage is a good predictor of quality of the forage consumed.

Grasses

Grass leaves are divided into blades and **sheaths** that are joined at the collar region (Chapter 2). Once elongated, the sheath envelops the stem with the blade subtending from the collar with the sheath. The composition and digestibility of the sheath is more similar to the leaf blade than it is to the stem, but because it is closely associated with the stem, it is sometimes included as part of that fraction (Cherney et al. 1983). This is reasonable since prehension causes breakage of the blade above the more fibrous sheath causing leaf sheaths to be consumed in conjunction with the stem sections they surround.

The relationship between the quantity of leaf and stem tissues in grasses is strongly influenced by the morphologic development of the plant as described in more detail below. Therefore, the relative quantities of leaves and stems are highly influenced by stage of maturity (Smart et al. 2001). In early vegetative stages of grass development, the tillers consist primarily of leaves. Once stem elongation begins, leaf growth slows and the ratio of leaves to stem tissues declines relatively quickly.

An effective strategy for managing the quality of grasses for forage is limiting the quantity of stems in the herbage by harvesting or grazing when the plants have a higher proportion of leaves. Defoliating or grazing grasses prior to boot stage, before stems are fully developed, will ensure a greater proportion of leaf tissue and thus higher forage quality, but also results in lower dry matter yield. Suppressing stem development of grasses with the application of plant growth regulators such as amidochlor or Chaparral™ herbicide maintains protein and digestibility at higher levels (Roberts and Moore 1990; Aiken et al. 2012).

Legumes

Legume leaves have a relatively high nutritive value compared to grasses. They are composed of leaflets and petioles (Chapter 3). Petioles are short stalks that attach leaflets to the stem and are generally of lower quality than leaflets (Wilman and Atimimi 1984). Compared to grass leaf blades, legume leaflets maintain quality better as the plant matures (Minson 1990). Albrecht et al. (1987) reported *in vitro* dry matter (DM) digestibility of alfalfa leaves remained at about $800\,g\,kg^{-1}$ over a wide range of growth stages. However, the digestibility of stems declined from approximately $750\,g\,kg^{-1}$ one week before early bud stage to 500 to $600\,g\,kg^{-1}$ at advanced maturity stages depending on the year measured. In that same study, the leaf-stem ratio declined from about 1.4 one week prior to early bud stage to between 0.50 and 0.75 five weeks later depending on the year. The nutritive value of legumes is therefore determined primarily by degree of leafiness.

Fractional harvest of leaves from legumes produces herbage with high nutritional value. Using a modified flail harvester to collect mainly leaves, Andrzejewska et al. (2017) increased the proportion of leaves in harvested alfalfa from 45.9% to 65.8% at early bloom stage. Crude protein in the harvested herbage increased from 195 to $221\,g\,kg^{-1}$ DM and *in vitro* true digestibility from 79.7% to 84.2%. These gains in nutritional value occurred at the expense of dry matter yield, mainly stem tissue, which decreased from 567 to $277\,g\,m^{-2}$.

Alfalfa genotypes with an increased number of leaflets produced per leaf have been developed through selection and breeding (Buxton 1996). While some multifoliolate genotypes have a higher leaf: stem ratio than typical trifoliolate genotypes, improvements in nutritive value have been inconsistent (Volenec and Cherney 1990). In some studies, the multifoliolate trait has been associated with higher concentrations of protein and amino acids in the complete forage (Linqing et al. 1998).

Stratification

The proportion of leaves and stems varies vertically within a forage canopy as does the relative age of tissues within the canopy. Harvest or grazing height thus has an important impact on the quality of forage harvested. In legumes, Buxton et al. (1985) found quality of grass stem bases was lower than for younger segments. In alfalfa, *in vitro* DM digestibility of forage increased by $20\,g\,kg^{-1}$ from the lowest (oldest) to the highest node. This occurred concomitant with a higher leaf: stem ratio at upper nodes.

In grasses, Nave et al. (2015) reported higher neutral detergent fiber (NDF) in lower strata (5–15 cm) of mixed cool-season grass canopies that were predominantly vegetative tall fescue compared to that from higher strata (15–25 cm). However, this was not consistent and occurred in only 32% of the treatments sampled. Differences in digestibility of NDF between the strata were even less consistent. Brink et al. (2007) also reported lower concentrations of NDF in upper segments of temperate grass canopies with a mean difference of $50\,g\,kg^{-1}$ between the upper and lower canopy.

Development

The **morphology** of a forage plant changes as the plant matures (Buxton 1996). Plants progress through a predictable series of vegetative and reproductive stages that involve the successive accumulation of leaves, stems, and reproductive organs (Chapter 7). Once established the grasses and legumes pass through an annual cycle of growth. Growth in spring is comprised of vegetative tissues. As the plants mature, the ratio of leaves to stems decreases resulting generally in a decrease in forage quality.

Grasses

Grasses can be conceptualized as a meta-population of tillers (Briske 2007). Each tiller is composed of a series of stem internodes divided by nodes from which tillers develop from axillary buds. Early in the development of a grass tiller the stem remains differentiated, but internodes remain unelongated at the base of the plant. Leaves continue to arise from the nodes while the stem remains compact (Moore and Moser 1995). Thus, spring growth in grasses consists almost entirely of leaves until the proper daylength is exceeded and the **shoot apex** differentiates to initiate development of the inflorescence. This is followed by internode elongation and transition to reproductive development. Once the stem begins to elongate in response to environmental stimuli, the leaf: stem ratio is gradually reduced with negative implications for forage quality. Additionally, unlike legumes, the nutritive value of grass leaves declines significantly as they age so not only is the proportion of leaves decreasing, the quality of the leaves is also declining during the transition to reproductive development.

Another aspect of grass development affecting the quality of forage is that not all tillers at any point in time are in the same stage of maturity (Chapter 7) with many remaining vegetative. Depending on the species, the distribution of growth stages in the composite forage can be quite broad (Moore and Moser 1995; Mitchell et al. 1998). The overall quality of forage then reflects this heterogeneity since tillers in early stages of development higher in forage quality. One way to characterize the condition is by determining mean stage of development by count (MSC). Each tiller in a representative sample is scored for maturity. The average stage of tillers collected is used to quantify development. Several studies have reported a strong relationship between various measures of quality and MSC (Mitchell et al. 2001; Waramit et al. 2012). In general, as a grass matures its quality declines nonlinearly with the greatest rate of decrease occurring during stem elongation and early reproductive stages (Figure 39.2).

Legumes

Growth and development of legumes occur differently than for grasses with both leaf and stem development occurring throughout the **vegetative growth** stage

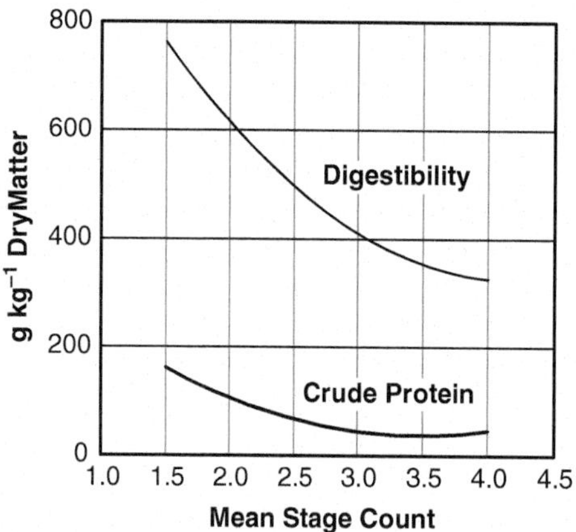

Fig. 39.2. *In vitro* digestibility and crude protein concentration of switchgrass in relation to mean stage count. Growth stages were determined according to Moore et al. (1991). *Source:* Calculated from equations published by Mitchell et al. (2001).

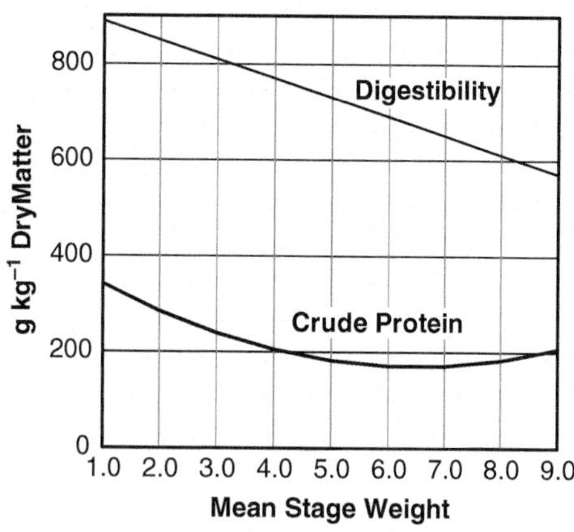

Fig. 39.3. *In vitro* true digestibility and crude protein concentration of alfalfa in relation to mean stage weight. Growth stages were determined according to Kalu and Fick (1981). *Source:* Calculated from equations published by Kalu and Fick (1981).

(Chapter 7). Kalu and Fick (1981) developed a widely used system for quantifying the growth and development of alfalfa. Overall, there are nine stages ranging from early vegetative (0) to ripe seed pod (9). Vegetative growth stages, i.e. no buds or flowers, are based on stem length of a representative sample. Reproductive stages are based on the development of buds, flowers, and seed. Their research demonstrated that regression equations, calculated on a weight of each stem (including leaves) rather than count basis, provided excellent predictions of digestibility and crude protein from mean stage (Figure 39.3).

The system of Kalu and Fick (1981) for quantifying morphologic development of alfalfa has been adapted for use in staging other legumes that have upright growth. Ohlsson and Wedin (1989) developed an analogous system for staging red clover development and evaluated its use for predicting forage quality. Acceptable regression equations were developed for predicting forage quality parameters including digestibility and crude protein based on mean stage calculated by either count or weight. Prostrate growing legumes like white clover tend to remain vegetative due to prostrate stems (stolons) which produce leaves on long peduncles, so the stem is rarely removed, and forage quality does not decrease as rapidly as for other legumes.

The nutritional quality of legumes declines with maturity in much the same way as grasses, though at a relatively lower rate. This decrease reflects changes in morphology and associated differences in composition of the leaf and stem tissues associated with maturation. Underlying these changes in morphology are shifts in anatomy and these account for much of the observed loss in forage quality.

Anatomy

The structure and composition of plant cells comprising the various tissues of forage plants differ markedly. Forage plants are an assembly of tissues organized into organs that make up the plant. Plant cells are differentiated and aggregated into multiple tissues that perform different and multiple functions within the plant. All plant cells have cell walls but, in some like mesophyll cells, they are thin and easily digested. In others, such as the vascular xylem and phloem cells, the walls undergo secondary thickening. The epidermal cells have some secondary thickening but also form waxy layers on the outside that restrict rapid digestion in the rumen. These processes partition more of the cellular dry matter into relatively less degradable **cell wall constituents** resulting in an increase in fiber concentration and a decrease in digestibility (Moore and Hatfield 1994).

Though there is considerable variation among species, all land plants have three principal tissue systems – dermal (protective) tissues, conducting tissues, and the remaining non-protective, non-conducting tissues that make up the fundamental or ground tissue (Von Sachs 1890). These tissues occur in different proportions in different parts of the plant and, in turn, make up the organs of the plant leaf, stem, root, flower, and fruit. Due to both physical and chemical characteristics, the degradability of these tissues by rumen microbes varies from essentially 0% for thick-walled cells of the xylem to 100% of the mesophyll cells. The primary inhibiting factor is degree of lignification (Valente et al. 2016), although cell density can also inhibit tissue degradability (Akin 1989).

Lignin deposition occurs primarily in tissues that have ceased growth and function in either conduction (xylem, phloem), mechanical support (**sclerenchyma**), or protection (**epidermis**). Sclerenchyma and xylem cells are generally considered undegradable, whereas mesophyll cells are almost completely degradable. Epidermal cells and **parenchyma** bundle **sheath** cells in C_4 grasses vary in degradability (Akin 1989). Amounts and proportions of these tissues vary among forage species and organs (Table 39.1). Though proportions of xylem and sclerenchyma are similar in warm- and cool-season grasses, parenchyma bundle **sheath** and epidermal cells are higher and mesophyll lower in warm-season grasses (Twidwell et al. 1990). Stem anatomy is similar for warm- and cool-season grasses, and both contain high proportions (25–38%) of sclerenchyma and xylem that are highly lignified and not degradable. Degradability of stem parenchyma, which can represent 50% or more of stem tissue, varies among species and decreases with tissue age (Akin 1989; Grabber et al. 1992).

Differences between warm- and cool-season grasses are exemplified in Figure 39.4. Scanning electron micrographs of leaf blades of a C_4 grass (bermudagrass) and a C_3 grass (orchardgrass) after incubation in rumen fluid clearly show the difference in degradability between the

Table 39.1 Tissue proportions in organs of tropical and temperate grasses and legumes

	Proportion of tissue in cross-sectional area (%)								
	Thin-walled				Thick-walled				
Forage type	EPI	MES	PAR	PHL	PF	COL	PBS	VT	SCL
Tropical species									
C_4 grass[a]									
Blade	22	31	14	<1			24	6	2
Midrib	6	12	57	1			14	7	3
Sheath	4	7	66	1			7	9	6
Stem	2	2	75	1			0	12	8
C_3 grass[b]									
Blade	30	51	0	<1			14	3	1
Midrib	24	44	0	1			20	7	3
Stem	3	12	68	1			0	9	7
Legume[c]									
Blade	9	88	0	<1	2	0		1	0
Midrib	9	83	0	3	2	0		3	0
Petiole	4	17	63	2	8	2		4	0
Stem	4	6	42	9	5	4		30	0
Temperate species									
C_3 grass[d]									
Blade	23	66	2	<1			5	3	1
Midrib	13	66	13	<1			4	3	1
Sheath		86		<1			0	4	10
Stem	2	2	75	<1			0	9	12
Legume[e]									
Blade	19	77	0	1	1	0		2	0
Midrib	17	73	0	3	3	0		4	0
Petiole	11	36	33	5	4	1		10	0
Stem	5	14	40	4	3	2		32	0

Source: Adapted from Wilson (1993).

EPI, epidermis; MES, mesophyll; PAR, parenchyma; PHL, phloem; PF, phloem fibers; COL, collenchyma; PBS, parenchyma bundle sheath; VT, vascular tissue; SCL, sclerenchyma.

[a]Switchgrass.

[b]Lax panicgrass.

[c]Pencilflower.

[d]Annual ryegrass.

[e]Alfalfa.

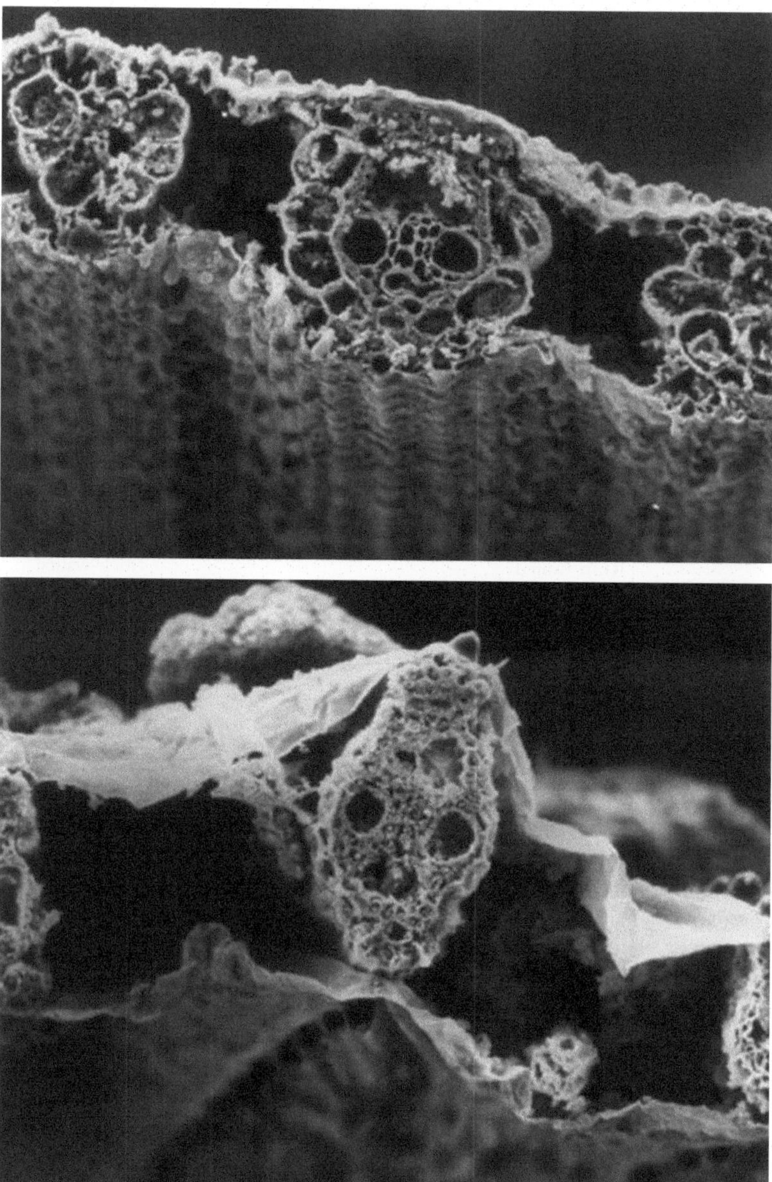

Fig. 39.4. Scanning electron micrographs of cross sections of bermudagrass leaf blades (top) and orchardgrass leaf blades (bottom) after incubation in rumen fluid for 48 hours. *Source:* Image used with permission of D.E. Akin.

two species. Bermudagrass is generally less degraded, with remnants of the epidermal and parenchyma bundle sheath cell walls remaining, as well as recalcitrant vascular bundle tissues and cuticle. In contrast, orchardgrass shows a total loss of all tissues except vascular bundles and cuticle. In addition to having tissues that are more digestible, orchardgrass leaves have a cool-season anatomy, with a lower density of vascular bundles compared with the bermudagrass.

Compared with grasses, there is relatively little information available on legume anatomy as related to forage quality. Cross-sections of leaves of temperate and tropical legumes show roughly similar anatomies, with large areas (up to 88%) of highly digestible **mesophyll** tissue

(Table 39.1). Young legume stems consist of a ring of xylem and phloem and can be highly digestible, but as they mature a thickened lignified ring of vascular and surrounding cells develops, providing a barrier to degradation. Unlike grasses, it appears legume stem parenchyma does not become lignified with maturity and is almost completely degradable, perhaps partially accounting for the greater voluntary intake of legumes compared with grasses (Demarquilly and Jarrige 1974).

Cellular Composition and Metabolism

Plant biochemical processes occur within the cytoplasm and membranes of cells. These processes tend to be localized within specific cells and tissues that have specific functions within the plant. Photosynthesis primarily occurs within specialized parenchyma cells localized within leaf tissues. These mesophyll cells contain chloroplasts that capture energy from sunlight and the complement of enzymes necessary to use it to reduce carbon dioxide to form sugars. Photosynthetic enzymes account for about 50% of the proteins in plants causing leaves to be more valued than other plant organs. In addition to proteins and amino acids, the cytoplasm contains sugars, storage polysaccharides such as starch and fructan, organic acids, lipids, pectins, water-soluble vitamins, and minerals. The biochemical composition of the cytoplasm varies with cell type and function, and nearly all cytoplasmic constituents are completely digestible by ruminants (Van Soest 1994).

The plant cell wall is a rigid composite material, composed of **cellulose** fibrils embedded in a matrix of **hemicellulose**, pectic materials, and lignin. It also contains water, various organic acids, and phenolic compounds, all of which enable the wall to play an important role in ion exchange, water balance, protection from **biotic** stresses, and cell recognition. Cell wall composition varies with plant species and maturity, as well as with organs and tissues within the plant. The middle lamella is the first layer formed during cell division and makes up the thin outer wall of the cell that is shared with adjacent cells. The highly digestible primary cell wall, formed next to the middle lamella during cell elongation, is composed almost entirely of polysaccharides, including cellulose, hemicellulose, pectic compounds, and glycoproteins.

Secondary wall deposition begins when cell enlargement is completed. Strands of cellulose, consisting of polymers of beta-D-glucose molecules, are laid down in mesh-like layers, forming microfibers through H-bonding. These microfibers are then gradually intertwined with lignin resulting in the formation of a rigid wall that not only is extremely strong but, is also very resistant to attack by rumen microbes.

The digestibility of plant cell walls varies considerably among different cell types and tissues (Moore and Hatfield 1994). From a nutritional perspective, plant cell walls comprise most of the fiber of forages, which is defined as the complex of dietary nutrients which are relatively resistant to digestion. Composed mostly of cellulose, hemicelluloses, and lignin, fiber in forages is slowly and only partially digested by ruminants. Thus, concentration of indigestible fiber limits the energy value of most forages (Buxton and Redfearn 1997; Moore et al. 2017). Managing to reduce total fiber concentration or improving its degradability are the most effective means of increasing digestible energy of a forage (Moore et al. 1993).

Biodiversity and Genetics

Plant Community

Diversity among species allow a mosaic of different forages to be distributed widely over the earth, and the quality of forage varies with the species that are adapted to the climate and soils of a given region. Species adaptation also greatly influences approaches to forage management. In North America, forage systems in the humid East are characterized by monocultures or binary mixtures of introduced species that are managed intensively. Maintaining forage productivity and quality involves selecting, establishing, and managing the most desirable plants. Traditionally, it is expected that over time the stand will deteriorate and need to be renovated or re-established.

In drier western regions, forage systems include multiple plant species, usually native, and are managed to sustain plant diversity as a part of an ecosystem. Whereas forage management in the East has been more allied with an industrial approach, western forage managers have taken a more ecologic approach. However, increasing public interest in environmental and economic sustainability has led many eastern agriculturists to follow an ecosystem approach to forage management that emphasizes persistence and resilience of system productivity (Nelson and Moser 1994).

Forage plants often grow in a community comprising two or more plant species. The relative proportion of each species within a plant community varies with season and weather (Chapter 8; Sleugh et al. 2000), plant interactions (Chapter 9), soil spatial variation within the area being used to produce the forage (Harmoney et al. 2001; Guretzky et al. 2004), and a host of other factors that affect interspecific competition (Nelson and Moser 1994).

Forage plant communities vary in species diversity from simple monocultures grown for hay and silage to complex polycultures that occur in more extensive pasture and rangeland systems. Each species within a plant community provides nutrients in varying quantities to livestock with varying bioavailability. Some species within a forage plant community may possess antiquality factors that are detrimental to animals consuming them (Chapter 35). Since these latter species are generally not desired, they are, by definition, weeds and their reduction or elimination from the community is often a priority (Chapter 28).

In more extensive forage systems with complex plant communities, managing the species composition is often a key for improving forage quality (DiTomaso 2000) usually by encouraging growth of more desirable species at the expense of more detrimental ones. Grazing and other defoliation management including prescribed burning are common cultural practices used to this end in rangeland environments (Mitchell et al. 1994). However, in more intensively managed systems, tillage, herbicides, and interseeding are often used to achieve a more favorable composition of species (Chapters 25–28).

Breeding for Forage Quality Traits

Thus far, all traits affecting forage quality discussed are under genetic control. Traditionally, intraspecific variation in expression of forage quality traits has been used for improving forage quality through selection and breeding (Chapter 30). Indeed, plant breeders and geneticists have exploited this variation to develop cultivars of many species with improved forage quality characteristics (Casler and Vogel 1999). Now, interspecific variation, including transgenic materials and molecular approaches, have been used to reduce lignin in alfalfa (Chapter 31).

Digestibility

The development of reliable and economic laboratory methods for predicting forage digestibility enabled plant breeders to develop forage genotypes with improved energy availability (Casler 2001) and cultivars of several forage species with improved digestibility (Casler and Vogel 1999). Casler and Vogel (1999) reported breeding improvements in *in vitro* digestibility across a range of forage crops. They reported improvements in *in vitro* dry matter digestibility (IVDMD) of 8–45 $g\,kg^{-1}$ per cycle of selection. The mechanisms by which these improvements occurred varied among species and cultivars.

In surveying the literature, Casler (2001) found improvements in digestibility resulted from changes in chemical composition, anatomy, and morphology and were not often related to changes in maturity. Using a model for true digestibility, Moore et al. (1993) examined the composition and digestibility of several cycles of switchgrass selected for IVDMD. They found selection for improved IVDMD did not alter the concentration of fiber, but rather increased fiber digestibility without affecting the rate at which it is digested. They concluded that a reasonable approach for improving the digestibility of warm season grasses is to select for reduced concentrations of indigestible fiber.

The brown midrib (bmr) mutation either found or induced in certain annual forage grasses is associated with improved forage digestibility (Cherney et al. 1991). Brown midrib mutations have been found in maize, chemically induced in sorghum and pearl millet, and produced through genetic manipulation (backcrossing) in sudangrass. Mutant plants are characterized by the presence of a reddish-brown pigment, which is associated with lignified tissue and is most visible in the leaf midrib.

Brown-midrib **mutations** cause changes in the chemical composition of mutant genotypes, most notably in the phenolic components of the cell walls (Fritz et al. 1990; Cherney et al. 1991). Typically, bmr genotypes have 5–50% lower lignin concentration than their normal counterparts. Compared with their normal counterparts, bmr mutations reduced permanganate lignin concentrations by 10 $g\,kg^{-1}$ DM for pearl millet, 12 $g\,kg^{-1}$ for sorghum, and 20 $g\,kg^{-1}$ for maize. Reduction in lignin concentration of bmr genotypes has been attributed to altered activity of certain enzymes in the lignin biosynthetic pathway (Halpin et al. 1998; Bout and Vermerris 2003).

Improvements in digestibility of bmr forages have been associated with greater extent of fiber digestibility (Wedig et al. 1988; Gerhardt et al.1994) that were related to reduced lignin concentration and qualitative changes in fiber composition. Fritz et al. (1990) reported that lignin concentration was lower in leaves and stems of bmr sorghum than a normal near-isogenic genotype and that the bmr lignin contained less trans-p-coumaric acid. The extent of fiber digestion was higher in the bmr genotype, while the rate of digestion was similar, suggesting that chemical changes in the composition of lignin led to greater accessibility of the fiber to digestive enzymes.

In a study evaluating the use of an ensiled bmr sorghum cultivar for use in rations of lactating dairy cattle, Grant et al. (1995) reported that bmr sorghum silage supported similar milk production as corn and alfalfa silages, and was higher than for a sorghum cultivar without the bmr trait. *In situ* fiber digestibility of the bmr sorghum was 653 $g\,kg^{-1}$ compared to 608 $g\,kg^{-1}$ for the normal sorghum. Commercial bmr varieties of both corn and sorghum are being marketed on the basis of improved fiber digestibility.

Environment

The growing environment within a region determines what forage species are adapted and generally imposes constraints to the system used to produce and utilize them (Chapter 8). Forage production systems fall along a continuum from extensive to intensive and the growing environment generally determines which species, systems, and management approaches are used. Extensive systems are most often used in areas such as rangelands with severe limitations to plant growth. These systems are typically limited by low precipitation and access to affordable water for irrigation. Rangelands managed for livestock production in the arid west are representative of an extensive management approach. These extensive systems are typically managed to encourage favorable plant communities (Chapter 24). They are largely located on public lands in the western states and generally have few inputs and multiple management goals in addition to livestock production, including water harvesting, conservation, and recreation.

Intensive management relies on a more agronomic approach with use of purchased inputs to adapt the environment to needs of a single or few forage species. These systems require intensive management, high investment in inputs and usually focus on productivity as the primary goal. Irrigated alfalfa hay production is an example of a high-value forage crop that is managed intensively with adequate fertilizer and pesticide applications. It is one of the few forage species where elite genetics are available through relatively frequent cultivar releases.

Corn and sorghum grown for silage are additional examples of forage crops grown under intensive management. Improved pastures, and hay or silage grown for on-farm or local use are more intensively managed than range, but receive fewer inputs than forages produced as commodities. The quality of forage produced across the spectrum of management intensity varies and is tightly coupled with the intended use of the forage.

The genotype of a plant determines how it will respond to stimuli encountered within its growing environment. Weather has a very large impact on plant growth and development because many plant processes are directly affected by temperature, light, and moisture. Weather is largely unpredictable and highly variable within a growing area so plants must have genetic plasticity to respond to these changes in order to survive. Other **abiotic** factors also influence plant growth and development. These include the physical environment in which it grows such as edaphic and hydrologic properties of the soil and related topographic features of the landscape (Wilson et al. 2016).

Growing plants must also respond to other organisms that exist within their environment. These include microbes, other plants, and animals. There are many microorganisms present and essential to the normal functioning of forage ecosystems including symbionts that contribute directly to the health of forage plants. Others, however, interfere with and are detrimental to plant growth processes to which the plant must respond to survive. Genetic resistance to plant diseases is fundamental to plant survival and natural variation in mechanisms for resisting or tolerating infection is present in many forage species. Larger organisms such as insects and larger herbivores can also influence forage quality through direct and indirect effects associated with their feeding. Plant responses to herbivory are also coded at the genetic level.

Ability of a plant to respond to variation in environmental conditions determines its range of adaptation. The many ways that plants respond to environmental variation have consequences for forage quality. Any factor that affects the growth and development of a forage plant influences its nutritive value. Thus, the environment in which a forage plant grows has a strong influence on forage quality. These effects are mediated through plant responses that occur at various levels within the hierarchy of factors that affect forage quality (Figure 39.1).

Climatic Effects

Species Adaptation

Over the long term, climate, defined as the meteorologic conditions – including wind, temperature, and precipitation – that prevail in a particular region, is a major factor determining quality of forage produced, mainly because it determines which species are present (Woodward 1987). Regional and seasonal temperatures in particular, determine the geographic distribution of C_3 and C_4 species, with C_4 types being more prevalent in warm climates (Nelson and Moser 1994).

The fact that species' adaptation to climate is such a major determinant of potential forage quality within a given region, suggests potential impacts of global climate change on forage species distribution, quality of the forage and, ultimately, the kind of animal agriculture possible in a given region (Chapter 8).

Temperature

Plant temperature is influenced by complex interactions between plants and their environment. High irradiation bleaches leaves lowering their protein content and raises leaf and canopy temperature, increases **transpiration** leading to water stress, and lowers soil water content. Water stressed leaves close their stomata to reduce transpiration and grass leaves roll up to reduce exposed leaf area while leaf angles become more vertical. Thus, actual tissue temperature can vary widely within a single plant at any given time. Effects of air temperature on forage growth, development, and chemical composition was first described by Alberda (1965) for perennial ryegrass. Temperatures below the optimum for growth led to accumulation of soluble sugars because photosynthesis was reduced less by suboptimum temperatures than was growth. At supra-optimal temperatures, soluble sugar concentrations tended to decrease, but not as consistently as they responded to sub-optimal temperature. Optimal temperatures for growth are near 20 °C for cool-season species and between 30 and 35 °C for warm-season species (Cooper and Tainton 1968). When harvested at a particular stage of growth, yields are normally highest when forages are grown at temperatures in the lower part of their optimal range (Fick et al. 1988).

Temperature has a greater effect on forage quality than any other environmental variable. Temperature stress reduces growth and development of the plant such that nutrients are concentrated in tissues present when the stress occurred (Nahar et al. 2015). Differences in forage digestibility due to high temperature stress appear to be due mainly to decreases in stem digestibility, though smaller decreases in leaf digestibility also occur (Deinum and Dirven 1975). Higher ambient temperatures during growth are normally associated with decreased dry matter digestibility, and this is most commonly attributed to higher concentrations of cell wall constituents (Ford et al. 1979).

Temperature can alter the carbohydrate status of metabolic sinks by increasing respiration at high temperatures to maintain tissue or slowing individual metabolic reactions by reducing active transport across membranes, by changing the concentration of different enzymes (through modification of gene expression) and by changing enzyme activity. Cell wall synthesis is a metabolic **sink** for photosynthate and there is considerable evidence of a greater conversion of photosynthate to structural components at higher temperatures (Ford et al. 1979). Supra-optimal temperatures also tend to decrease leaf size and therefore the leaf : stem ratio, decrease stem diameter, and increase lignification (Fick et al. 1988).

The negative effects of high temperature on forage quality are difficult to study in the field because of inevitable confounding of temperature with plant ontogeny and maturation during most growing seasons. Controlled-environment studies are helpful to isolate temperature from other variables. Reductions in forage digestibility induced by high temperature appear to be closely related to an increase in indigestible cell wall (Fales 1986). Akin et al. (1987) grew tall fescue plants under an increasing, then decreasing, temperature regimen in a controlled environment. Anatomy of leaf blades was evaluated after incubation in rumen fluid for 48 hours (Figure 39.5). When leaf tissue was grown under a diurnal temperature regimen of 20 °C/18 °C, the epidermis, mesophyll, parenchyma bundle sheath, and phloem were completely removed during *in vitro* digestion. Leaf tissue grown at 30 °C/27 °C showed only a slight digestion of mesophyll, while the remainder was relatively undigested. Higher amounts of lignified vascular tissue and higher concentrations of both p-coumaric and ferulic acid were present in leaves grown at 30 °C/27 °C.

Light

Energy from the sun drives almost every known physical and biologic cycle on the earth, and the potential amount of light reaching the earth's surface varies with time of year, latitude, and atmospheric conditions. Plants respond to the total amount of energy received, the spectral quality of the light, and the duration of light, or photoperiod. The primary effect of photoperiod on forage quality is its role in the induction of reproductive growth. Most cool-season forage species are "long-day" plants that flower in response to increasing photoperiod during the late spring. Many also require vernalization by exposure to low temperatures during winter to induce the photoperiod response and morphologic change (Chapter 3).

Both the quantity and spectral quality of radiation received by plants are influenced by canopy structure (Ballare et al. 1991). Radiation in the blue and red wavelengths is absorbed in photosynthesis and decreases with depth of canopy, while the relative proportion of far-red wavelengths increases (Holmes 1981). These differential changes in radiation quality within the canopy can influence morphogenetic responses such as promotion of leaf and stem elongation (Ballare et al. 1991) and inhibition of tillering in grasses (Casal et al. 1987).

Forage plants frequently are exposed to shade, which has a greater effect on morphologic development and yield than on forage quality. Leaves developing in shade tend to be longer, thinner, and narrower than leaves developing in full sunlight (Kephart and Buxton 1993). Shading causes a decline in photosynthetic rates, with a greater decrease occurring in C_4 species than in C_3 species. Shading tends to reduce cell wall concentration in most forages, presumably through a reduction in photosynthate available for secondary wall development (Wilson and Wong 1982; Kephart and Buxton 1993). Effects of shading itself on digestibility, however, are not consistent across species.

Radiation quality can affect digestibility of forage plants. Jung and Russell (1991) grew orchardgrass and birdsfoot trefoil under fluorescent-incandescent lamps (control) and under low-pressure sodium lamps (no **ultraviolet** or blue wavelengths). Fiber digestibility of the orchardgrass decreased under the sodium lamps. Fiber digestibility of the birdsfoot trefoil, however, was decreased by the fluorescent-incandescent lamps. In both cases, differences in fiber digestibility were unrelated to changes in lignification.

Moisture

Stress can be induced in forage plants by both excesses and deficits of soil moisture. Waterlogged soils can be rapidly depleted of O_2 by root respiration resulting in an anoxic condition for root systems. Anoxia can have a profound effect on growth and therefore forage yield, but there is little evidence that it impacts forage quality.

Moisture stress alters the physiology of a forage plant in ways that have a profound influence on forage quality. In general, drought results in an improvement in forage quality because new growth is inhibited, nutrients are concentrated and the development of the plant slows or ceases (Grant et al. 2014).

Drought is more common than excess moisture in most forage-producing areas, and water deficit more frequently limits forage yield than other factors. Water deficit causes stomatal closure, reduces **transpiration**

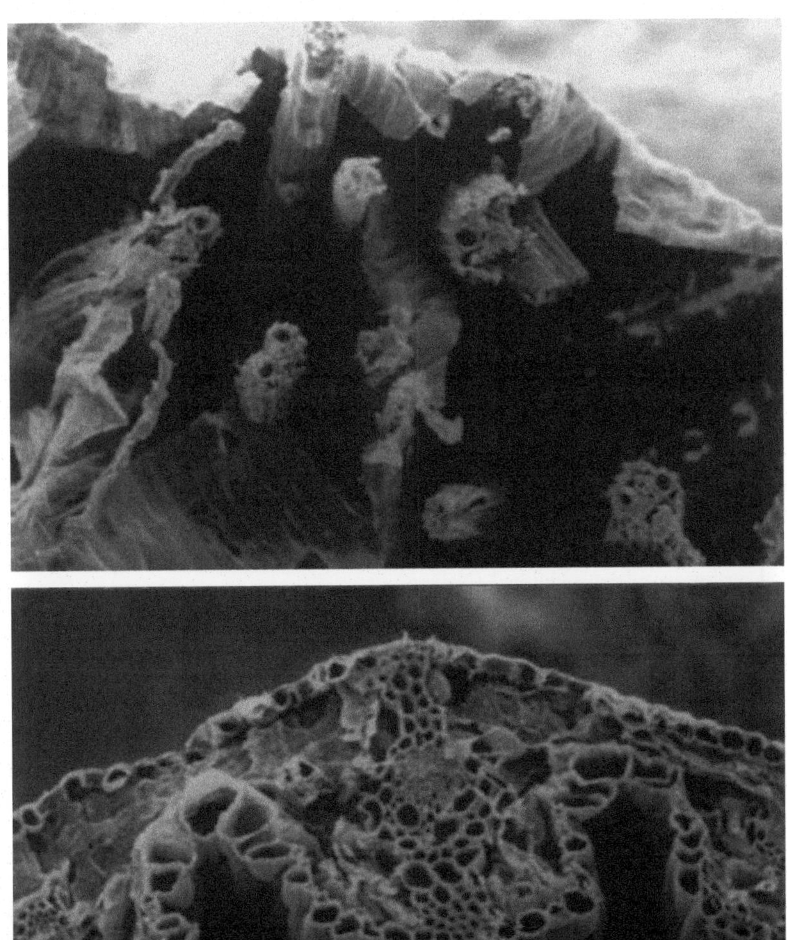

Fig. 39.5. Scanning electron micrographs of tall fescue leaves grown at 20 °C/18 °C (top) and 30 °C/27 °C (bottom) after being incubated with rumen microorganisms for 48 hours. Note the lower digestion of tissue grown at the higher temperatures. *Source:* Akin et al. (1987).

rates, and results in elevated leaf temperature. Because of reduced **turgor** pressure, cell enlargement is reduced under drought stress. However, cell division appears to be less sensitive to water deficit than cell enlargement (Levitt 1980). Concentrations of N in herbage show no consistent relation to water stress (Buxton and Fales 1994). Other physiologic responses to moisture deficit include acceleration of leaf senescence, a reduction in tiller production, and death of established tillers (Turner 1979). Both N and soluble carbohydrates are exported from leaves as they senesce, potentially contributing to reduced forage quality.

In general, water-deficit stress has less effect on forage quality than on growth and development, primarily because it tends to delay maturity. The forage quality trait most affected by water deficit appears to be decreased concentration of NDF due to a reduction in the amount of photosynthate incorporated into cell walls (Peterson et al. 1992; Sheaffer et al. 1992). The limited photosynthate available is used to provide higher concentrations of sugars and ions for osmotic adjustment (Buxton and Fales 1994). Jensen et al. (2003) studying different ryegrass and orchardgrass cultivars, found the primary effect of water stress was to increase nutritional value by increasing concentrations of crude protein and digestible NDF.

In addition to these effects on nutrient concentrations and availability, moisture stress can also lead to the accumulation of compounds that are toxic to ruminants. Nitrate toxicity is of particular concern in drought-stressed sorghum (Mayland et al. 2007). The reduction of nitrate to ammonia is inhibited causing a buildup within plant tissues, particularly in stem tissues (Gleadow et al. 2016).

Edaphic Effects

The amount and availability of different soil nutrients affect both the production and the quality of forages. Extensive reviews of the influence of soil nutrients on forage growth and quality have been published (Noller and Rhykerd 1974; Buxton and Fales 1994; Spears 1994). Of all the nutrients, N has the most dramatic impact on plant growth whether it is supplied through mineralization of soil organic matter, symbiotic bacteria, or applications of N fertilizer. In intensively managed situations involving monoculture grasses, large applications of N can significantly enhance yields, but can also result in environmental degradation due to the mobility of nitrate in water. Much of the N absorbed by plants is incorporated into amino acids and proteins. Nitrogen is absorbed and transported primarily as nitrate (NO_3^-) in grasses, before being reduced to ammonium (NH_4^+) in leaves and incorporation into amino acids and protein. When growth of grasses is restricted by drought, cold stress, limited radiation or mineral deficiency, NO_3^- can accumulate to toxic levels for livestock. Applications of N fertilizer to grasses tend to reduce soluble carbohydrate concentration and increase water content. At very high rates, it can cause an imbalance of energy and protein that results in poor animal performance, reproductive problems, and other metabolic disorders (Messman et al. 1991). Applications of N fertilizer generally have little effect on forage digestibility, though there is some evidence that rate of fiber digestion can be increased by N fertilization, especially in situations where N concentration in the forage tissue was low prior to fertilization (Messman et al. 1991).

Most of the P in animals is contained in the skeleton, where it functions as a storage pool to maintain the balance of P in blood (Minson 1990). Requirements for P are similar for forages and the animals that consume them, and concentrations of P in forages generally are adequate for animal performance. The exception is for soils that are very low in available P, where deficiencies can cause a variety of animal symptoms, including low intake, poor performance, breeding disorders, bone-chewing, and rickets (Minson 1990). Grasses are generally lower in P concentration than legumes, and cool-season grasses tend to have higher P concentrations than warm-season grasses. Because of the positive effect of P fertilization on the proportion of legumes in mixed grass–legume swards, an indirect effect of P can be an improvement in forage quality simply by changing botanic composition. Otherwise, there is no evidence that concentrations of soil P have any effect on intake or digestibility.

Potassium is required in higher amounts by plants than animals, and therefore concentrations of K in the soil that are suboptimal for plant growth generally do not limit animal performance. Conversely, high soil K is related to excess K uptake in cool-season grasses, wherein K interferes with uptake of Ca, Mg, and Na, resulting in an imbalance and grass tetany in animals. This potentially fatal metabolic disorder is associated with a rapid decrease in serum Mg and occurs primarily in spring pasture situations (Grunes et al. 1970). A more detailed discussion of pasture nutritional quality for grazing livestock is presented in Chapter 45.

Biotic Effects

Depending on **edaphic** and environmental conditions, forage quality will be determined by the mix of plant species growing within a given field, and modified by various weeds, disease organisms, and insects, which either directly or indirectly can alter the nutritional value of the harvested forage. Forage quality is therefore highly dynamic, changing over the course of the growing season as different plants pass through stages of morphologic development and are impacted by diseases and insects.

Insect Pests

Insects can significantly alter both the yield and quality of forages (Buxton and Fales 1994). Insect feeding generally lowers the quantity of forage but, has variable impacts on forage quality (Buxton and Fales 1994). Insects that preferentially feed on leaves such as the alfalfa weevil reduce leaf biomass and consequently, the leaf: stem ratio. However, because overall plant development is stunted, including stem growth, nutritive value does not change appreciably unless plants are severely defoliated. Insects that feed on assimilates such as the potato leafhopper cause a reduction in stem growth thereby increasing the leaf : stem ratio and nutritive value (Hutchins et al. 1990). There are conflicting reports, however, relative to the impacts of insect feeding on forage quality and these appear to be related to the severity of the infestation, intensity of feeding, and their impact on the physiology and development of the plant.

Insects usually feed on leaves and young stems of legumes that have direct effects on growth and loss of leaf tissues of high quality. Until recently, potato leafhopper had been a serious insect pest of alfalfa during the summer and needed chemical treatment. Yield was reduced and quality was reduced due to loss of protein from damaged leaves and also from leaf senescence and dropping. New cultivars with glandular hairs on the leaves that deter the females from feeding have been an effective control that allows growth to continue and quality of leaves to be retained.

Alfalfa weevil removes the highly digestible mesophyll cells prior to the first harvest which decrease forage quality and yield due to lower leaf digestibility and reduced photosynthesis effects on yield. Removing the fall growth reduces egg laying by the weevil and kills many eggs, which overall delays the infection from spring-laid eggs such that an early harvest kills the young larvae and retains forage quality. In addition, biocontrol with parasitic insects has lessened damage.

In general, insects present fewer challenges to grasses than to legumes, but are subject to losses in yield and quality from grasshoppers, armyworm and some other insects. These are often seasonal, usually associated with a dry summer when growth is reduced, and cause reductions in both yield and quality.

Plant Pathogens

At high levels of infection, diseases will cause a marked reduction in both yield and quality. Yield reductions can be direct like with rust on orchardgrass leaves that also reduces digestibility of the infected tissue. Indirectly, orchardgrass plants weakened by rust often die in late summer of fail to overwinter causing future losses.

In contrast, many diseases on legumes occur on roots such as phytophthora in alfalfa or a complex mix of pathogens with red clover and birdsfoot trefoil that reduce stress resistance and ability to winter harden sufficiently to persist. Birdsfoot trefoil overcomes this problem of short lifespan by heavy seed production and white clover produces new roots from stolons that perennate. When plants of red clover are lost, the stand can be reseeded. But with loss of alfalfa, the area needs to be out of alfalfa for a year to minimize problems with autotoxicity before reseeding.

Weeds

Weeds are very competitive for soil resources and radiation, resulting in decreased yield of forages; but weeds also provide biomass that will compensate for some of the forage yield loss. Many weeds are palatable and nutritious when immature, and their net effect on forage quality is often minimal. However, if weeds are toxic or unpalatable and are present in sufficient quantity, they will have a negative impact on animal performance (Cords 1973).

Fungal Endophyte

The fungal endophyte *Neotyphodium coenophialum* (Morgan-Jones and Gams.) Glenn, Bacon, and Hanlin is associated with tall fescue, one of the most widely grown cool-season forage grasses in the US. Cattle grazing forage from most endophyte-infected tall fescue cultivars suffer from fescue toxicity, due mainly to ergot alkaloids present in the forage (Bush et al. 2007). A more detailed discussion of this forage-related animal disorder is presented in Chapter 47.

An effective management to minimize effects of the endophyte is having 30% or more of legume in the pasture. This higher quality forage helps offset the effects in the associated grass. New cultivars of tall fescue are available that contain a derivative of the fungus that does not produce the toxic compounds, yet gives the desirable vigor for the plants to persist during drought periods and colder winters.

Mycorrhiza

Arbuscular mycorrhiza infection of roots can improve mineral availability and uptake in plants which can also enhance the quality of forage. Inoculation of barley with arbuscular mycorrhiza improved the *in vivo* digestibility of barley by mature goats (Sabia et al. 2015). Herbage from inoculated barley had higher protein content and lower concentrations of acid detergent fiber (ADF) and acid detergent lignin (ADL) than that which was not inoculated. In the same study, however, inoculation of berseem clover had minor effects on composition and digestibility. In another study, inoculated alfalfa exposed to high atmospheric CO_2 had improved forage quality due largely to changes in chemical composition of leaves (Baslam et al. 2014).

Management Implications

Forage quality is a dynamic characteristic influenced by ecologic, edaphic, and environmental variables, yet management can play a major role. Three major approaches to managing forages for optimal forage quality include managing the plant community, managing the environment, and managing plant development.

The species makeup of a plant community can be managed through species and cultivar selection and by weed control. Options for the producer are limited primarily by the ecologic adaptation of the species to the local site (see Part III). It is important for producers to regularly assess the condition of forage stands with regard to the proportion of desirable species present and the need for weed management. Renovation of a poor pasture or hayfield can significantly improve both forage production and quality; but eventually, it may be necessary to re-establish the sward (Chapter 25).

Environmental modification can be accomplished through fertilization or irrigation. When based on soil tests, fertilization can correct mineral imbalances that affect not only plant growth, but animal health as well (Buxton and Fales 1994). Maintaining adequate fertility encourages the vigor and persistence of desired species and thus discourages the encroachment of undesirable species. Irrigation of forages is a less-common management practice, mainly due to cost issues, but it can be used to modify the environment so that high-quality forages can be grown where they otherwise are not adapted (e.g. irrigated alfalfa in arid regions). In situations where lack of moisture is limiting growth, irrigation will also encourage the production of new, high-quality herbage. Forage fertilization and irrigation are covered in more detail in Chapters 26 and 27, respectively.

Manipulating plant development is the most commonly used management tool for influencing forage quality. Because the natural process of plant maturation involves a decline in leaf : stem ratio as well as a decline in quality of the stem component, timely defoliation to minimize the proportion of stem tissue will positively influence quality. In using harvest management to optimize quality, it is important to take into account stage of morphologic development, the physiologic status of the plant with regard to carbohydrate reserves, and, particularly for grasses, the location of growing points. Ill-timed harvests can have a significant negative impact on regrowth, depending on species and time of year. This topic is discussed in more detail in Chapters 2, 3, and 7 in Part I and the different systems for mechanical harvesting and for grazing are described in Parts VIII and IX, respectively.

As an alternative to defoliation, exogenous growth-regulating compounds, used initially to suppress grass growth along roadsides, have been found to be effective in reducing stem development and enhancing forage quality (Fales et al. 1990; Roberts and Moore 1990; Aiken et al. 2012). These materials are particularly useful for inhibiting reproductive spring growth in cool-season grasses, which frequently becomes stemmy and too mature. When applied at proper rates and times, the materials can completely inhibit development of stems without affecting leaves. Certain growth regulators are specific enough to manipulate the botanic composition of binary mixtures of grasses and legumes (Fritz et al. 1987; Fales and Hoover 1990). Growth regulators are potentially a very good management tool, particularly in situations for which high-forage quality is more important than high yield. However, they have not been employed widely by producers because of contemporary concerns over the use of chemicals in the food system and availability due to high costs for companies to obtain federal approval to label chemicals for such use.

Summary and Conclusions

Though the nutritional requirements of a given class of livestock are relatively constant, forage quality, defined as the ability of a forage to meet those requirements, is highly variable. Plant species used for forage have been selected by humans because of their nutritional characteristics, their growth characteristics, and because they are adapted to local climate and soils. Within an ecologic region, adapted forage species show a wide range in inherent quality; within a species or cultivar, quality varies considerably over time, declining most rapidly with advancing maturity. Quality loss with maturity is exacerbated by the elevated ambient temperatures that normally occur simultaneously.

Most forage producers seek to produce forage that meets the nutritional needs of livestock as uniformly as possible through the growing season. Achieving this requires the selection and establishment of adapted forage species and managing the forage through timely harvests that foster the capture of a high proportion of nutrients while maintaining the vigor of the plants. This requires a good understanding of the characteristics and growth patterns of the forages used and how they are impacted by various stresses, including environment and defoliation itself. For this reason, forage management to achieve consistently high quality is one of the most challenging of all agricultural endeavors.

References

Aiken, G.E., Aiken, B.M., Goff, W.W. et al. (2012). Steer and plant responses to chemical suppression of seedhead emergence in toxic endophyte-infected tall fescue. *Crop Sci.* 52: 960–969.

Akin, D.E. (1989). Histological and physical factors affecting digestibility of forages. *Agron. J.* 81: 17–25.

Akin, D.E., Fales, S.L., Rigsby, L.L., and Snook, M.E. (1987). Temperature effects on leaf anatomy, phenolic acids, and tissue digestibility in tall fescue. *Agron. J.* 79: 271–275.

Alberda, T. (1965). The influence of temperature, light intensity, and nitrate concentration on dry matter production and chemical composition of *Lolium perenne* L. *Neth. J. Agric. Sci.* 13: 335–360.

Albrecht, K.A., Wedin, W.F., and Buxton, D.R. (1987). Cell-wall composition and digestibility of alfalfa stems and leaves. *Crop Sci.* 27: 735–741.

Andrzejewska, J., Ignaczak, S., Albrecht, K.A., and Surucu, M. (2017). Fractional harvest of perennial legumes can improve forage quality and their exploitation. In: *Grassland Resources for Extensive Farming Systems in Marginal Lands: Major Drivers and Future Scenarios. Proceedings of the 19th Symposium of the European Grassland Federation, Alghero, Italy, 7–10 May 2017* (eds. C. Porqueddu, A. Franca, G. Lombardi, et al.).

Ballare, C.L., Scopel, A.L., and Sanchez, R.A. (1991). Photocontrol of stem elongation in plant neighbourhoods: effects of photon fluence rate under natural conditions of radiation. *Plant Cell Environ.* 14: 57–65.

Baslam, M., Antolin, M.C., Gogorcena, Y. et al. (2014). Changes in alfalfa forage quality and stem carbohydrates induced by arbuscular mycorrhizal fungi and elevated atmospheric CO_2. *Ann. Appl. Biol.* 164: 190–199.

Bout, S. and Vermerris, W. (2003). A candidate-gene approach to clone the sorghum brown midrib gene encoding caffeic acid O-methyltransferase. *Mol. Gen. Genomics* 269: 205–214.

Brink, G.E., Casler, M.D., and Hall, M.B. (2007). Canopy structure and neutral detergent fiber differences among temperate perennial grasses. *Crop Sci.* 47: 2182–2189.

Briske, D.D. (2007). Plant interactions. In: *Forages: The Science of Grassland Agriculture*, 6e (eds. R.F Barnes, C.J. Nelson, K.J. Moore and M. Collins), 105–122. Ames, IA: Blackwell Publishing Professional.

Bush, L., Roberts, C.A., and Schultz, C. (2007). Plant chemistry and antiquality components in forage. In: *Forages: The Science of Grassland Agriculture*, 6e (eds. R.F Barnes, C.J. Nelson, K.J. Moore and M. Collins), 509–528. Ames, IA: Blackwell Publishing Professional.

Buxton, D.R. (1996). Quality-related characteristics of forages as influenced by plant environment and agronomic factors. *Anim. Feed Sci. Technol.* 59: 37–49.

Buxton, D.R. and Fales, S.L. (1994). Plant environment and quality. In: *Forage Quality, Evaluation and Utilization* (eds. G.C. Fahey Jr., M. Collins, D.R. Mertens and L.E. Moser), 155–199. Madison, WI: American Society of Agronomy Monograph Series.

Buxton, D.R. and Redfearn, D.D. (1997). Plant limitations to fiber digestion and utilization. *J. Nutr.* 127: 814S–818S.

Buxton, D.R., Hornstein, J.S., Wedin, W.F., and Marten, G.C. (1985). Forage quality in stratified canopies of alfalfa, birdsfoot trefoil, and red clover. *Crop Sci.* 25: 273–279.

Buxton, D.R., Mertens, D.R., Moore, K.J. et al. (1995). Forage quality for ruminants: plant and animal considerations. *Prof. Anim. Sci.* 11: 121–131.

Casal, J.J., Sanchez, R.A., and Deregibus, V.A. (1987). Tillering responses of *Lolium multiflorum* plants to changes of red/far-red ratios typical of sparse canopies. *J. Exp. Bot.* 38: 1432–1439.

Casler, M.D. (2001). Breeding forage crops for increased nutritional value. *Adv. Agron.* 71: 51–107.

Casler, M.D. and Vogel, K.P. (1999). Accomplishments and impact from breeding for increased forage nutritional value. *Crop Sci.* 39: 12–20.

Cherney, J.H., Marten, G.C., and Goodrich, R.D. (1983). Rate and extent of cell wall digestion of total forage and morphological components of oats and barley. *Crop Sci.* 23: 213–216.

Cherney, J.H., Cherney, D.J.R., Akin, D.E., and Axtell, J.D. (1991). Potential of brown-midrib, low lignin mutants for improving forage quality. *Adv. Agron.* 46: 157–198.

Cochran, R.C., Coblentz, W.K., and Vanzant, E.S. (2007). Animal methods for evaluating forage quality. In: *Forages: The Science of Grassland Agriculture*, 6e (eds. R.F Barnes, C.J. Nelson, K.J. Moore and M. Collins), 541–552. Ames, IA: Blackwell Publishing Professional.

Cooper, J.P. and Tainton, N.M. (1968). Light and temperature requirements for the growth of tropical and temperate grasses. *Herbage Abstr.* 38: 167–176.

Cords, H.P. (1973). Weeds and alfalfa hay quality. *Weed Sci.* 21: 400–401.

Deinum, B. and Dirven, J.G.P. (1975). Climate, nitrogen, and grass. 6. Comparison of yield and chemical composition of some temperate and tropical grass species grown at different temperatures. *Neth. J. Agric. Sci.* 23: 69–82.

Demarquilly, C. and Jarrige, R. (1974). The comparative nutritive value of grasses and legumes. *Vaxtodling* 28: 38–41.

DiTomaso, J.M. (2000). Invasive weeds in rangelands: species, impacts, and management. *Weed Sci.* 48: 255–265.

Fales, S.L. (1986). Effects of temperature on fiber concentration, composition and in vitro digestion kinetics of tall fescue. *Agron. J.* 78: 963–966.

Fales, S.L. and Fritz, J.O. (2007). Factors affecting forage quality. In: *Forages: The Science of Grassland Agriculture*, 6e (eds. R.F Barnes, C.J. Nelson, K.J. Moore and M. Collins), 569–580. Ames, IA: Blackwell Publishing Professional.

Fales, S.L. and Hoover, R.J. (1990). Chemical regulation of alfalfa/grass mixtures with imazethapyr. *Agron. J.* 82: 5–9.

Fales, S.L., Hill, R.R., and Hoover, R.J. (1990). Chemical regulation of growth and forage quality of cool-season grasses with imazethapyr. *Agron. J.* 82: 9–17.

Fick, G.W., Holt, D.A., and Lugg, D.G. (1988). Environmental physiology and crop growth. In: *Alfalfa and Alfalfa Improvement* (eds. A.A. Hanson, D.K. Barnes and R.R. Hill Jr.), 163–194. Madison, WI: American Society of Agronomy Monograph Series.

Ford, C.W., Morrison, I.M., and Wilson, J.R. (1979). Temperature effects on lignin, hemicellulose and cellulose in tropical and temperate grasses. *Aust. J. Agric. Res.* 30: 621–634.

Fritz, J.O., Moore, K.J., and Roberts, C.A. (1987). Chemical regulation of quality and botanical composition of alfalfa-grass mixtures. In: *Proceedingsof the American Forage and Grassland Conference, 1987*, 166–170.

Fritz, J.O., Moore, K.J., and Jaster, E.H. (1990). Digestion kinetics and cell wall composition of brown midrib sorghum × sudangrass morphological components. *Crop Sci.* 30: 213–219.

Gerhardt, R.L., Fritz, J.O., Moore, K.J., and Jaster, E.H. (1994). Digestion kinetics and composition of normal and brown midrib sorghum morphological components. *Crop Sci.* 34: 1353–1361.

Gleadow, M., Ottman, M.J., Kimball, B.A. et al. (2016). Drought-induced changes in nitrogen partitioning between cyanide and nitrate in leaves and stems of sorghum grown at elevated CO_2 are age dependent. *Field Crops Res.* 185: 97–102.

Grabber, J.H., Jung, G.A., Abrams, S.M., and Howard, D.B. (1992). Digestion kinetics of parenchyma and sclerenchyma cell walls isolated from orchardgrass and switchgrass. *Crop Sci.* 32: 806–810.

Grant, R.J., Haddad, S.G., Moore, K.J., and Pedersen, J.F. (1995). Brown midrib sorghum silage for midlactation dairy cows. *J. Dairy Sci.* 78: 1970–1980.

Grant, K., Kreyling, J., Dienstbach, L.F.H. et al. (2014). Water stress due to increased intra-annual precipitation variability reduced forage yield but raised forage quality of a temperate grassland. *Agric. Ecosyst. Environ.* 186: 11–22.

Grunes, D.L., Stout, P.R., and Brownell, J.R. (1970). Grass tetany of ruminants. *Adv. Agron.* 22: 231–374.

Guretzky, J.A., Moore, K.J., Burras, C.L., and Brummer, E.C. (2004). Distribution of legumes along gradients of slope and soil electrical conductivity in pastures. *Agron. J.* 96: 547–555.

Halpin, C., Holt, K., Chojecki, J. et al. (1998). Brown-midrib maize (bm1) – a mutation affecting the cinnamyl alcohol dehydrogenase gene. *Plant J.* 14: 545–553.

Harmoney, K.R., Moore, K.J., Brummer, E.C. et al. (2001). Spatial legume composition and diversity across seeded landscapes. *Agron. J.* 93: 992–1000.

Holmes, M.G. (1981). Spectral distribution of radiation within plant canopies. In: *Plants and the Daylight Spectrum* (ed. H. Smith), 147–158. London, UK: Academic Press.

Hutchins, S.H., Buntin, G.D., and Pedigo, L.P. (1990). Impact of insect feeding on alfalfa regrowth: a review of physiological responses and economic consequences. *Agron. J.* 82: 1035–1044.

Jensen, K.B., Waldron, B.L., Asay, K.H. et al. (2003). Forage nutritional characteristics of orchardgrass and perennial ryegrass at five irrigation levels. *Agron. J.* 95: 668–675.

Jung, H.G. and Russell, M.P. (1991). Light source and nutrient regime effects on fiber composition and digestibility of forages. *Crop Sci.* 31: 1065–1070.

Kalu, B.A. and Fick, G.W. (1981). Quantifying morphological development of alfalfa for studies of herbage quality. *Crop Sci.* 21: 267–271.

Kephart, K.D. and Buxton, D.R. (1993). Forage quality responses of C3 and C4 perennial grasses in reduced irradiance. *Crop Sci.* 32: 1033–1038.

Levitt, J. (1980). *Response of Plants to Environmental Stresses*, 2e, vol. II. New York, NY: Academic Press.

Linqing, Y., Maotai, H., Zhaolan, W., and Wei, Z. (1998). Study on stabilization and agricultural traits of multifoliate alfalfa. *Grassld. China* 3: 6–8.

Mayland, H.F., Cheeke, P.R., Majak, W., and Goff, J.P. (2007). Forage-induced animal disorders. In: *Forages: The Science of Grassland Agriculture*, 6e (eds. R.F Barnes, C.J. Nelson, K.J. Moore and M. Collins), 687–707. Ames, IA: Blackwell Publishing Professional.

Mertens, D.R. (2007). Digestibility and intake. In: *Forages: The Science of Grassland Agriculture*, 6e (eds. R.F Barnes, C.J. Nelson, K.J. Moore and M. Collins), 487–507. Ames, IA: Blackwell Publishing Professional.

Messman, M.A., Weiss, W.P., and Erickson, D.O. (1991). Effects of nitrogen fertilization and maturity of bromegrass on in situ ruminal digestion kinetics of fiber. *J. Anim. Sci.* 69: 1151–1161.

Minson, D.J. (1990). *Forage in Ruminant Nutrition*. San Diego, CA: Academic Press Inc.

Mitchell, R.B., Masters, R.A., Waller, S.S. et al. (1994). Big bluestem production and forage quality responses to burning date and fertilizer in tallgrass prairies. *J. Prod. Agric.* 7: 301–359.

Mitchell, R.B., Redfearn, D.D., Moser, L.E. et al. (1998). Tiller demographics and leaf area index of four perennial grasses. *Agron. J.* 90: 47–53.

Mitchell, R., Fritz, J., Moore, K. et al. (2001). Predicting forage quality in switchgrass and big bluestem. *Agron. J.* 93: 118–124.

Moore, K.J. and Hatfield, R.D. (1994). Carbohydrates and forage quality. In: *Forage Quality, Evaluation, and Utilization* (ed. G.C. Fahey Jr.), 229–280. Madison, WI: American Society of Agronomy, Inc.

Moore, K.J. and Jung, H.G. (2001). Lignin and fiber digestion. *J. Range Manage.* 54: 420–430.

Moore, K.J. and Moser, L.E. (1995). Quantifying developmental morphology of perennial grasses. *Crop Sci.* 35: 37–43.

Moore, K.J., Moser, L.E., Vogel, K.P. et al. (1991). Describing and quantifying growth stages of perennial forage grasses. *Agron. J.* 83: 1073–1077.

Moore, K.J., Vogel, K.P., Hopkins, A.A. et al. (1993). Improving the digestibility of warm season perennial grasses. In: *Proceedings of XVI International Grassland Congress*, 447–448.

Moore, K.J., Boote, K.J., and Sanderson, M.A. (2004). Physiology and developmental morphology. In: *Warm Season Forage Grasses*. Agron. 45 (eds. L.E. Moser, L.E. Sollenberger and B. Burson), 179–216.

Moore, J.E., Adesogan, A.T., Coleman, S.W., and Undersander, D.J. (2007). Predicting forage quality. In: *Forages: The Science of Grassland Agriculture*, 6e (eds. R.F Barnes, C.J. Nelson, K.J. Moore and M. Collins), 553–568. Ames, IA: Blackwell Publishing Professional.

Moore, K.J., Archontoulis, S.V., and Lenssen, A.W. (2017). Simple models for describing ruminant herbivory. Chapter 4. In: *Herbivore*, ISBN 978-953-51-4901-9 (ed. V.D.C. Shields). Rijeka, Croatia: InTech.

Nahar, K., Hasanuzzaman, M., Ahamed, K.U. et al. (2015). Plant responses and tolerance to high temperature stress: role of exogenous phytoprotectants. In: *Crop Production and Global Environmental Issues* (ed. K. Hakeem), 385–435. Cham: Springer.

Nave, R.L.G., Sulc, R.M., Barker, D.J., and St-Pierre, N. (2015). Changes in forage nutritive value among vertical strata of a cool-season grass canopy. *Crop Sci.* 54: 2837–2845.

Nelson, C.J. and Moser, L.E. (1994). Plant factors affecting forage quality. In: *Forage Quality, Evaluation, and Utilization* (ed. G.C. Fahey Jr.), 115–154. Madison, WI: ASA, CSSA, and SSSA.

Noller, C.H. and Rhykerd, C.L. (1974). Relationship of nitrogen fertilization and chemical composition of forage to animal health and performance. In: *Forage Fertilization* (ed. D.A. Mays), 363–394. Madison, WI: American Society of Agronomy Monograph Series.

Ohlsson, C. and Wedin, W.F. (1989). Phenological staging schemes for predicting red clover quality. *Crop Sci.* 29: 416–420.

Peterson, P.R., Sheaffer, C.C., and Hall, M.H. (1992). Drought effects on perennial forage legume yield and quality. *Agron. J.* 84: 774–779.

Riaz, M.Q., Südekum, K.H., Clauss, M., and Jayanegaraet, A. (2014). Voluntary feed intake and digestibility of four domestic ruminant species as influenced by dietary constituents: a meta-analysis. *Livestock Sci.* 162: 76–85.

Roberts, C.A. and Moore, K.J. (1990). Chemical regulation of tall fescue reproductive development and quality with amidochlor. *Agron. J.* 82: 523–526.

Sabia, E., Claps, S., Napolitano, F. et al. (2015). In vivo digestibility of two different forage species inoculated with arbuscular mycorrhiza in Mediterranean red goats. *Small Ruminant Res.* 123: 83–87.

Sheaffer, C.C., Peterson, P.R., Hall, M.H., and Stordahl, J.B. (1992). Drought effects on yield and quality of perennial grasses in the North Central United States. *J. Prod. Agric.* 5: 556–561.

Sleugh, B., Moore, K.J., George, J.R., and Brummer, E.C. (2000). Binary legume–grass mixtures improve forage yield, quality, and seasonal distribution. *Agron. J.* 92: 24–29.

Smart, A.J., Schacht, W.H., and Moser, L.E. (2001). Predicting leaf/stem ratio and nutritive value in grazed and nongrazed big bluestem. *Agron. J.* 93: 1243–1249.

Spears, J.W. (1994). Minerals in forages. In: *Forage Quality, Evaluation, and Utilization* (ed. G.C. Fahey Jr.), 281–317. Madison, WI: ASA, CSSA, and SSSA.

Turner, N.C. (1979). Drought resistance and adaptation to water deficits in crop plants. In: *Stress Physiology in Crop Plants* (eds. H. Mussell and R.C. Staples), 343–372. New York, NY: Wiley.

Twidwell, E.K., Johnson, K.D., Cherney, J.H., and Volenec, J.J. (1988). Forage quality and digestion kinetics of switchgrass herbage and morphological components. *Crop Sci.* 28: 778–782.

Twidwell, E.K., Johnson, K.D., Patterson, J.A. et al. (1990). Degradation of switchgrass anatomical tissue by rumen microorganisms. *Crop Sci.* 30: 1321–1328.

Valente, T.N.P., da Silva Lima, E., Gomes, D.I. et al. (2016). Anatomical differences among forage with respect to nutrient availability for ruminants in the tropics: a review. *Afr. J. Agric. Res.* 11: 1585–1592.

Van Soest, P.J. (1994). *Nutritional Ecology of the Ruminant*, 2e. Ithaca, NY: Cornell University Press.

Volenec, J.J. and Cherney, J.H. (1990). Yield components, morphology, and forage quality of multifoliolate alfalfa phenotypes. *Crop Sci.* 30: 1234–1238.

Von Sachs, J. (1890). *The History of Botany, 1530 to 1860*. Oxford: Clarendon Press.

Waramit, N., Moore, K.J., and Fales, S.L. (2012). Forage quality of native warm-season grasses in response to nitrogen fertilization and harvest date. *Anim. Feed Sci. Technol.* 174: 46–59.

Wedig, C.L., Jaster, E.H., and Moore, K.J. (1988). Effect of brown midrib and normal genotypes of sorghum x sudangrass on ruminal fluid and particulate rate of passage from the rumen and extent of digestion at various sites along the gastrointestinal tract in sheep. *J. Anim. Sci.* 66: 559–565.

Wilman, E. and Atimimi, M.A.K. (1984). The in vitro digestibility and chemical composition of plant parts in white clover, red clover, and lucerne during primary growth. *J. Sci. Food Agric.* 35: 133–138.

Wilson, J.R. (1993). Organization of forage plant tissues. In: *Forage Cell Wall Structure and Digestibility* (eds. H.G. Jung, D.R. Buxton, R.D. Hatfield and J. Ralph), 1–32. Madison, WI: ASA, CSSA, and SSSA.

Wilson, J.R. and Wong, C.C. (1982). Effects of shade on some factors influencing nutritive quality of green panic and siratro pastures. *Aust. J. Agric. Res.* 33: 937–950.

Wilson, D.M., Gunther, T.P., Schulte, L.A. et al. (2016). Variety interacts with space and time to influence switchgrass quality. *Crop Sci.* 56: 773–785.

Woodward, F.I. (1987). *Climate and Plant Distribution*. Cambridge: Cambridge University Press.

PART VIII

FORAGE HARVESTING AND UTILIZATION

Harvesting alfalfa silage for an Ohio dairy. *Source*: Photo courtesy of Mike Collins.

Part VIII reviews harvest, packaging and storage practices for forage as hay or silage, or when used as a biofuel. When a cow prehends, or a sheep or horse takes a bite of pasture, the quality of the forage consumed is usually higher than the same forage if harvested mechanically. Quality of cut forage decreases due to continued metabolism during early drying and to physical losses. Thus, a major goal is to expedite moisture loss to a safe level to package or ensile. Retention of leaves, the highest quality part, is critical during harvesting, raking, packaging, transport and storage. New technologies include rotary mowers for faster cutting, crushers or crimpers to speed stem drying, and improved windrow shape for rapid moisture loss to lower the probability for rain damage.

Forages: The Science of Grassland Agriculture, Volume II, Seventh Edition.
Edited by Kenneth J. Moore, Michael Collins, C. Jerry Nelson and Daren D. Redfearn.

About 50 years ago the large round baler, especially for grass-based forages, revolutionized haymaking and labor needs, but the relatively low-density bales require storage inside or wrapping in plastic to retain quality. More recently, machinery for large high-density square bales has decreased transport cost of marketed hay, especially internationally. New silage harvesters increase the proportion of cut or smashed corn kernels to improve silage fermentation and nutritional value. Biomass of forages and other grasses can be processed to energy compounds like ethanol that recycle CO_2 and reduce use of fossil fuel resources.

CHAPTER

40

Post-Harvest Physiology

Wayne K. Coblentz, Research Dairy Scientist/Agronomist, *USDA-Agricultural Research Service, US Dairy Forage Research Center, Marshfield, WI, USA*

Introduction

Weather and management constraints, as well as the intended use of the harvested forage, all influence the forage-harvest system selected by the producer. Maximum retention of dry matter (DM) from harvested forage crops is usually achieved at moistures intermediate between the standing fresh crop and dry hay (Figure 40.1). Conserving forage at intermediate moisture avoids the potentially large storage losses that can occur with high-moisture silages, as well as the large field losses associated with very dry hay. Management strategies for conserving forage crops should be based on a thorough understanding of post-harvest physiologic processes.

Pre-Harvest Physiologic Status of Forage Plants

Water-Soluble Carbohydrates

Three chemical constituents or characteristics profoundly affect forage preservation; these include: (i) moisture concentration, as illustrated in Figure 40.1, (ii) non-structural carbohydrates, a large subset of which are frequently measured on the basis of solubility in water (WSC); and (iii) buffering capacity. For water-soluble carbohydrates (WSC), there is a strong species effect, with warm-season annuals, such as corn, exhibiting greater concentrations of WSC than annual cool-season grasses (Table 40.1; Adesogan and Newman 2010). Among cool-season perennial grasses, which included ryegrass, orchardgrass, meadow fescue, and timothy, Jones (1970) and Humphreys (1989) found the greatest concentrations of WSC in ryegrass, and the least in orchardgrass. Initial concentrations of WSC in alfalfa often range from 60 to 70 $g\,kg^{-1}$ (Coblentz and Muck 2012), which is somewhat greater than normally reported for perennial warm-season grasses, such as bermudagrass (Table 40.1).

Numerous other weather and management factors affect WSC in forage plants, including stage of growth, time of day, frost events, N fertilization, growth temperature conditions, and rain damage and/or poor or extended wilting conditions. It has been long understood that forage composition changes diurnally (Lechtenberg et al. 1971; Smith 1974) with concentrations of non-structural carbohydrates increasing during daylight hours as a result of photosynthetic activity. For example, red clover cut during early morning had greater moisture and less total non-structural carbohydrates (TNC) than the same forage cut in late afternoon (Figure 40.2) (Owens et al. 2002). There have been numerous attempts to leverage this typical diurnal pattern of nonstructural carbohydrate accumulation/oxidation, particularly with respect to improving animal preference, silage fermentation, and utilization of forage N by beef and dairy cattle. For example, sheep, goats, and beef steers preferred alfalfa hay if it had been cut at sundown (TNC = 56 $g\,kg^{-1}$) rather than at sunup (TNC = 46 $g\,kg^{-1}$) (Fisher et al. 2002). Similarly, cattle preferred tall fescue cut during the afternoon (TNC = 92 $g\,kg^{-1}$) over that cut during the morning (TNC = 74 $g\,kg^{-1}$) (Fisher et al. 1999). Greater animal preference for one forage over another does not

Forages: The Science of Grassland Agriculture, Volume II, Seventh Edition.
Edited by Kenneth J. Moore, Michael Collins, C. Jerry Nelson and Daren D. Redfearn.

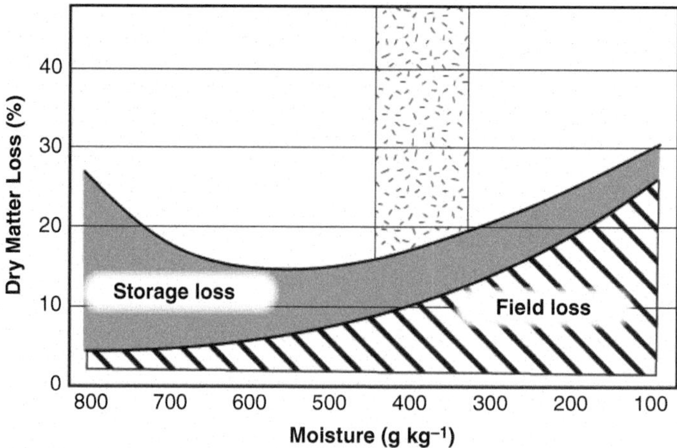

Fig. 40.1. The general relationship between forage moisture concentration and the amount of dry matter loss during the harvest (field) and storage phases. *Source:* Adapted from Collins and Owens (2003).

Table 40.1 Species effects on concentrations of water-soluble carbohydrates in forages

Crop	WSC ($g\,kg^{-1}$)
Warm-season annuals	
Corn	100–200
Forage sorghum	100–200
Sudan, Sorghum-sudan, Millet	100–150
Soybean	20–40
Cowpea	50–80
Cool-season annuals	
Rye, Oats, Wheat, Triticale	80–120
Ryegrass	80–120
Cool-season perennials	
Alfalfa	40–70
Warm-season perennials	
Bermudagrass, Stargrass	20–40
Bahiagrass	<50
Limpograss	<50
Perennial peanut	10–40

Source: Compiled by Adesogan and Newman (2010).

necessarily lead to improved intake or performance in the absence of a choice.

A similar concept leveraging predictable diurnal changes in plant sugars has been evaluated for baled silages, specifically to improve DM intake and/or efficiency of N use by steers or dairy cattle. Because the fermentation of baled silages is often restricted relative to chopped silages (Nicholson et al. 1991), elevated concentrations of WSC associated with an evening mowing also have the potential to improve silage fermentation by increasing the substrate pool. In studies conducted with perennial warm-season grasses, an evening harvest was effective in increasing TNC pools compared to those observed for morning-harvested baled silages, but production of significant concentrations of ethanol confounded detection of improvements

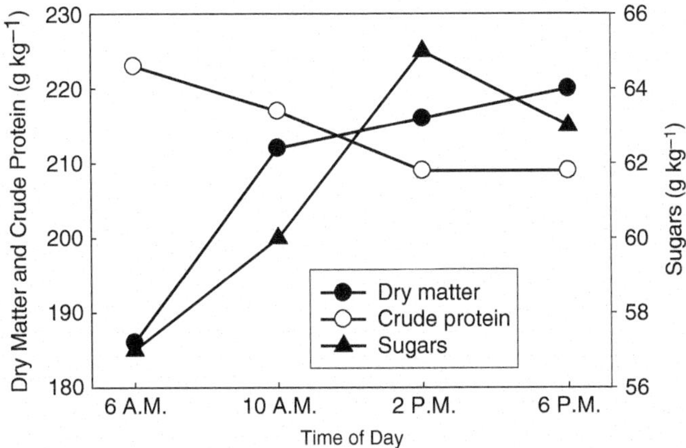

Fig. 40.2. Daily fluctuations in dry matter, crude protein and total non-structural carbohydrate concentrations of red clover herbage in Wisconsin. *Source:* Adapted from Owens et al. (2002).

Table 40.2 Effects of N fertilization with urea (46-0-0) on concentrations of water-soluble carbohydrates (WSC) in fall-grown oat

	WSC	
N fertilization rate	Coblentz et al. (2014a)	Coblentz et al. (2017)
kg N ha^{-1}	g kg^{-1}	
Urea (46-0-0)		
0	159	212
20	149	196
40	145	171
60	133	160
80	132	135
100	—	144
Bedded-pack manure[a,b]		
102	160	—
195	153	—
Dairy slurry[b]		
105	—	182
209	—	161

Source: Adapted from Coblentz et al. (2014a), Coblentz et al. (2017).
[a]Primary bedding source was wooden shavings.
[b]Mean application rate of N over two years, irrespective of availability to the plant.

in silage fermentation (Huntington and Burns 2007; Sauvé et al. 2010). Despite these responses, Huntington and Burns (2007) reported increased DM intakes and improved efficiency of N use by beef steers consuming evening-harvested switchgrass and eastern gamagrass baled silages. Conceptually, a similar trial was conducted with evening- and morning-harvested alfalfa (Brito et al. 2008) that yielded clearer and more consistent responses. Evening-harvested baled silages (469 g kg^{-1} moisture) exhibited improved fermentation characteristics, reduced concentrations of most fiber components, as well as greater DM intakes, milk production, and improved N-use efficiency when offered to late-lactation cows without supplemental concentrate feeds. Similarly,

Tremblay et al. (2014) reported greater concentrations of lactic acid, lower NH_3-N, and a more acidic final pH whenever evening mowing and/or wide swathing yielded a non-structural carbohydrate advantage of >10 g kg^{-1} for alfalfa forages ensiled in mini-silos.

Nitrogen fertilization can also have strong effects on accumulation of WSC in forages, though clear elucidation of these effects in experiments are often confounded by other management practices, such as multiple forage harvests, split fertilizer applications, etc. Two recent studies with fall-grown oat have been largely void of these confounding issues, and have demonstrated a strong negative relationship between N fertilization rate and WSC (Table 40.2; Coblentz et al. 2014a; Coblentz et al. 2017). Fall-grown oat has consistently demonstrated very high WSC that is strongly associated with stem tissue (Contreras-Govea and Albrecht 2006), and with specific coordination of frost events with the late-stem elongation or boot stages of growth in a unique form of the winter-hardening response similar to that exhibited by winter-annual cereals (Coblentz et al. 2013b). Under these conditions, similar concentrations of neutral detergent fiber (NDF) can be observed across wide time intervals (mid-September to early November), as the normal effects of advancing plant maturity are offset by dilution with WSC.

Buffering Capacity

In simplest terms, buffering capacity can be defined as the natural, inherent resistance of any forage to a pH change. Within that context, buffering capacity is species specific (Table 40.3), and is lower in corn and cool-season grasses compared to legumes, thereby making the former easier to ensile. McDonald et al. (1991) has identified the greater organic acid (malic, citric, quinic, malonic, and glyceric) and protein concentrations within legumes as the principle reasons for elevated buffering capacity. For fresh italian ryegrass, malic and citric acids were identified as the most important buffering organic acids, while very high concentrations of glyceric acid were found in red clover (Playne and McDonald 1966); after silage fermentation, these organic acids were largely metabolized, yielding acetic and lactic acids that greatly increased buffering capacity of silage relative to fresh forage. Muck and Walgenbach (1985) determined that the buffering capacity of alfalfa leaf and stem tissue both declined with plant maturity, but buffering capacity was much greater for the leaf, which declined at a slower rate with plant maturity than did stem tissue. As such, the relative proportions of leaf to stem heavily influence whole-plant estimates of buffering capacity. Other factors that affect leaf/stem proportions, such as rain damage that induces leaf shatter, may also lower buffering capacity (Coblentz and Muck 2012). Conversely, factors increasing buffering capacity include fertilization with P or K (Muck and Walgenbach 1985) and wrapping delays prior to sealing silage bales in polyethylene film that induce **spontaneous heating** (Coblentz et al. 2016).

Field Drying of Forages

Physical Changes

Stomata, Cuticle, and Plant Hydrology

Water is a major constituent of the cytoplasm and is involved in the maintenance of **turgor** and stomatal movement (Fiscus and Kaufmann 1990). Symplastic water inside the plasmalemma makes up 50–95% of total plant water, while the remainder, apoplastic water, includes that located in the xylem, cell walls and intercellular spaces. Cell walls contain 50–400 g kg^{-1} water, and standing forage contains ~800 g kg^{-1} water at the early reproductive stages of growth, but often becomes drier at more advanced maturities. For example, timothy declined from >800 to <700 g kg^{-1} between the early boot and anthesis maturity stages (Savoie 1988). Similarly, Edmisten et al. (1998) evaluated four cereal-grain forages (oat, rye, barley, and wheat) across three years, and reported a mean moisture concentration of 801 g kg^{-1} at the vegetative stage of growth that decreased across six growth stages, reaching a minimum of 463 g kg^{-1} at hard dough. This has particular relevance for silage production, suggesting that these cereal-grain forages could be direct-cut without wilting beginning at the milk stage of grain development.

Field curing is essential for hay production, and wilting is used widely in silage production, largely to avoid problematic fermentations commonly associated with overly wet silages. The primary plant factors affecting drying of mowed forage include both stomatal and cuticular resistance. Open stomata may occupy about 1% of the surface area of leaves, which have greater stomatal densities than stems. Estimates of the numbers of stomata on grass leaves are 16–335 mm^{-2} compared with 211–700 mm^{-2} on alfalfa (Harris and Tullberg 1980). Stomata close when crop moisture reaches about 700 g kg^{-1} (Savoie et al. 1984). Approximately 20–30% of water loss occurs prior to stomatal closure, when stomatal resistance is low (Harris and Tullberg 1980; Jones and Harris 1980).

The cuticle is comprised of **cutin**, an extracellular, insoluble polymer, as well as waxes. Water loss through intact plant cuticles is slow (Martin and Juniper 1970). Cuticular resistance is high (2500–10 000 s m^{-1}) in crop plants (Macdonald and Clark 1987), and it is generally so much greater than stomatal resistance that the former is often disregarded as a route of water loss. The majority of cuticular water loss from cut forage is through openings caused by cutting or mechanical conditioning, as well as the use of desiccant materials, such as potassium carbonate.

Compared with a typical forage stem, leaves are relatively thin and have low internal resistance to drying.

Table 40.3 Buffering capacity of some forage species

Crop	Range	Mean or Median
	mEq kg^{-1} DM	
Corn[a]	149–225	185
Timothy[a]	188–342	265
Fall oat (boot)[b]	296–345	318
Orchardgrass[a]	247–424	335
Red clover[a]	—	350
Italian ryegrass[a]	265–589	366
Alfalfa (full bloom)[c]	357–383	369
Fall oat (headed)[b]	365–378	372
Perennial ryegrass[a]	257–558	380
Alfalfa (1/3 bloom)[c]	428–473	438
Alfalfa[a]	390–570	472
White clover[a]		512
Alfalfa plant parts		
Alfalfa stem (early flower)[d]	—	365
Alfalfa stem (late vegetative)[d]	—	582
Alfalfa leaf (early flower)[d]	—	723
Alfalfa leaf (late vegetative)[d]	—	858

[a]Compiled by McDonald et al. (1991).
[b]Adapted from Coblentz et al. (2015).
[c]Adapted from Coblentz et al. (2014b).
[d]Adapted from Muck and Walgenbach (1985).

Legume leaflets and grass leaf laminae generally dry faster than their respective stem components. In one study, detached leaf blades of annual ryegrass dried three times faster than stems during the early part of the drying cycle and seven times faster during the last phase (Jones 1979). Mechanical conditioning treatments, such as crushing rollers or rotating tines, are included on many mowers, primarily to increase stem drying rates. Oftentimes, this approach is also coupled with wide swathing, at least in part for the same purpose. Furthermore, mowed grass forage usually dries faster than legume forage, and inclusion of grass in binary mixtures can increase the drying rates of the legume component, presumably by altering swath structure (Collins 1985). Species differences in moisture loss rates among cool-season grasses have not been observed consistently. Tetlow and Fenlon (1978) reported respective drying rates for annual ryegrass, perennial ryegrass, and tall fescue of 0.111, 0.079, and 0.131 kg water kg^{-1} plant water h^{-1}. Similarly, both leaf and stem components of tall fescue dried faster than the same components of perennial ryegrass (Jones and Prickett 1981). In contrast, results of a two-year study conducted in Wisconsin suggested that drying rates for meadow fescue, orchardgrass, and reed canarygrass harvested at the same relative maturity are affected only minimally by species (Brink et al. 2014).

Transpiration

Moisture loss from mowed forages typically exhibits a rapid early phase, followed by a slower phase or phases that comprise most of the total curing time (Jones 1979). Slower drying that begins with stomata closure is partially alleviated by conditioning, and often is subdivided to include a final phase of tightly held water, largely confined within the stem (Undersander and Saxe 2013). The initial, rapid phase begins with mowing, and continues while the stomata remain open (Macdonald and Clark 1987). A large portion of the early moisture lost from stems and **pseudostems** appears to exit the plant via the leaf after moving through the xylem. Increased transpirational flux increases the tension on xylem water and provides a gradient for movement of water from tissues surrounding the xylem. Jones (1979) and Harris and Tullberg (1980) illustrated the importance of this route of water loss when they showed that drying of grass pseudostems slowed when leaf laminae were removed. Recently, extension recommendations (Undersander and Saxe 2013; Penn State University 2017) have encouraged producers interested only in preservation of their forage crop as silage to leverage this initial, rapid, natural **transpiration** by mowing (often without conditioning) in wide swaths, thereby reducing the time required for forage moisture to decline from ~800 g kg^{-1} to approximately

$650\,g\,kg^{-1}$ in a concept often referred to as "haylage in a day." Shinners and Herzmann (2006) demonstrated additional improvement in drying rates by combining conditioning with wide swaths, but this option is not consistent with many wide, tractor-mounted mowers driven by front-located, power-take-off configurations. Within this specific context, a wide swath may be the single most important factor maximizing initial drying rates, thereby preserving sugars and starches, and it is more important than conditioning for good haylage management (Undersander and Saxe 2013). However, conditioning to break stems and disrupt the waxy cuticle layer is necessary for harvest as dry hay.

Biochemistry

Respiration

Under aerobic storage conditions, the complete oxidation (respiration) of hexose sugars to CO_2 and water, with concomitant production of heat, can be summarized as (Rees 1982):

$$C_6H_{12}O_6 + 6O_2 \rightarrow 6CO_2 + 6H_2O$$
$$+ 2.83\text{ MJ heat energy.}$$

Forage quality usually declines under these conditions because readily digestible carbohydrates are metabolized, with both plant enzymes and microbial growth contributing to respiration within the cut forage. Many species of epiphytic microorganisms, including bacteria, molds, and yeasts reside on cut plants, and contribute to microbial respiration losses. These losses of sugars can be significant in scope; for example, wheat forage harvested at the early heading stage of growth and wilted in swaths for 20 hours had $196\,g\,kg^{-1}$ of WSC compared to $242\,g\,kg^{-1}$ when wilting began (Williams et al. 1995). Respiration rates are usually greatest in freshly cut crops, decline as the forage dries, and become nil at $<400\,g\,kg^{-1}$ moisture (Rucker and Knabe 1980). At a given moisture level, respiration rates are greater at higher ambient temperatures. At $700\,g\,kg^{-1}$ moisture, the rate of organic matter loss from wilting grass increased from 1.9% day^{-1} at 10 °C to 4.7% day^{-1} at 25 °C (Rucker and Knabe 1980). In addition, respiratory losses have been described on the basis of temperature and forage moisture by the following equation:

$$\text{DM loss (mg g}^{-1}\text{ crop DM h}^{-1})$$
$$= t\,(0.02231 - 0.000124\,m + 0.00000027\,m^2),$$

where t = temperature (°C) and m = moisture concentration ($g\,kg^{-1}$) (Rees 1982).

Changes in Nitrogenous Compounds

Losses in nitrogen (N) from forage crops during wilting or hay curing are normally small compared with losses of carbohydrates, but some autolysis of protein into non-protein nitrogen (NPN) forms occurs during wilting (Rotz and Muck 1994). Proteolysis occurs both during field wilting and during silage preservation. As such, protein is converted to NPN forms, including amino-N, volatile N, and amide-N, that are lost or used less efficiently by animals compared to true protein. These forms of protein breakdown have long been understood, and the extent of autolysis is exacerbated by the length of wilting time, as well as by moist or poor wilting conditions (Macpherson 1952; Kemble and Macpherson 1954). More recent work continues to support these concepts. After alfalfa and smooth bromegrass were wilted to moisture concentrations suitable for ensiling, 30–49% of total crude protein (CP) was in the NPN fraction (Kohn and Allen 1995), which represented a threefold increase over forage that was freeze-dried immediately after cutting. Similarly, Carpintero et al. (1979) observed decreased concentrations of protein-N within a ryegrass-clover mixture that were exacerbated by extended wilting times and/or moist (100% relative humidity) wilting conditions. After a 144-hour wilt under the moist conditions, concentrations of NH_3-N were 22-fold greater than in fresh cut-forage. Formation rates of NPN are low when crop moisture falls below $400\,g\,kg^{-1}$ (Papadopoulos and McKersie 1983), but are species specific, with greater (more extensive) proteolysis occurring for alfalfa compared to red clover or birdsfoot trefoil. Concentrations of NPN have been related to wilting time by the following expression:

$$\text{NPN (g kg}^{-1}\text{ total N)}$$
$$= 116.1 + 3.2\,h - 0.01168\,h^2, R^2 = 0.88,$$

where h = wilting time in hours (Harrison et al. 2003).

Rain Damage

Leaching

The high probability of rain damage favors selection of chopped or baled silage over hay harvesting systems for preserving yield and quality of stored forages in humid regions. Over a four-year period in Wisconsin during which alfalfa hay was harvested and cured weekly during the growing season, measurable rainfall occurred during about 50% of the hay harvests (Collins 1989). On average, DM losses during alfalfa hay harvests were increased by 50% via exposure to rain damage. Respiration, leaching, and physical shattering contribute to losses in forage yield and quality due to rain damage during hay curing. Physical shattering can occur by direct impact of rain droplets, or from tedding/raking operations after rewetting.

Leaching of soluble forage components during rainfall events will reduce forage nutritive value by preferentially removing soluble constituents, particularly sugars (Collins 1982; Coblentz and Muck 2012), though final concentrations of WSC may be buoyed by hydrolysis of starch (Coblentz and Muck 2012). Regardless, fiber components, which are largely inert during rainfall events, can increase sharply by severe leaching of sugars and other soluble constituents during hay curing (Table 40.4) (Collins 1982; Collins 1985). Furthermore, any loss of sugars is problematic within the context of maintaining an adequate pool of substrate to support silage fermentation. This indirect response to wetting is so consistent that fiber components, such as NDF, have been used as internal markers in controlled studies to measure DM losses more effectively than traditional gravimetric techniques, provided all shattered leaves are recovered prior to analysis (Scarbrough et al. 2004; Coblentz and Jokela 2008). These dynamics often have clear, negative effects on *in vitro* or *in situ* measures of DM digestibility that are not confounded by reduced voluntary intakes and/or slower passage rates through the digestive system (Collins 1982; Collins 1985; Turner et al. 2003; Scarbrough et al. 2005). Based on the described increases in NDF and reduced non-fiber carbohydrates, similar reductions in estimates of energy density calculated by the commonly used summative approach also are a mathematical certainty (NRC 2001).

Though some CP is lost to leaching, disproportionately larger losses of sugars, and other more soluble, non-N constituents often results in a mildly positive effect on concentrations of CP in rain-damaged forages, but this response has not been consistent across all forages and conditions, and is dependent on complete recovery of shattered leaf tissue prior to analysis (Collins 1982; Turner et al. 2003; Scarbrough et al. 2006; Coblentz and Muck 2012). While overall effects of rain damage on CP concentrations may be mildly positive, the internal partitioning of CP may be affected, and is dependent on forage species, as well as the timing of rainfall. Scarbrough et al. (2006) reported strong linear increases in proportions of CP associated with the cell wall (neutral-detergent-insoluble CP) when 0–76 mm of simulated rainfall was applied to orchardgrass forage wilted to 153 or 41 g kg^{-1} moisture at the time rain was applied. As a result, estimates of rumen degradable protein also declined linearly with rainfall amount. However, this response was not observed when rainfall was applied to wet (674 g kg^{-1}) orchardgrass, or when it was applied to wilting bermudagrass possessing the C4 photosynthetic pathway.

Table 40.4 Effects of leaching on the composition of alfalfa, red clover, and birdsfoot trefoil hays

	NDF	Total sugars	TNC	CP	IVDMD
Treatment[a,b]			g kg^{-1}		
Alfalfa					
Control	323	99	122	155	715
Wet 48 h	341	84	107	187	710
Wet 24 and 48 h	384	59	80	182	692
Red clover					
Control	291	126	157	146	758
Wet 48 h	327	111	127	169	726
Wet 24 and 48 h	399	79	52	175	670
Birdsfoot trefoil					
Control	310	130	152	137	713
Wet 48 h	360	106	134	139	702
Wet 24 and 48 h	408	35	96	152	664

Source: Adapted from Collins (1982).

Abbreviations: NDF, neutral-detergent fiber; TNC, total nonstructural carbohydrates; CP, crude protein; and IVDMD, *in-vitro* DM disappearance.

All shattered leaves were recovered prior to analysis; therefore, results reflect only respiratory and leaching losses.

[a]Forages all were collected for analysis 96 hours post-cutting.

[b]Control forages were wilted for 96 hours without wetting; wetted forages received 2.5 cm of deionized water per application from a sprinkler can.

Factors Affecting Leaching and DM Losses

The magnitude of the rainfall-induced leaching losses increases as hay dries (Rucker and Knabe 1980; Scarbrough et al. 2005), and can be affected strongly by forage species (Collins 1982; Scarbrough et al. 2005). Mechanical conditioning may also increase the extent of DM loss and forage quality reduction (Kormos and Chestnutt 1968). Physical openings caused by conditioning may ease rewetting, and the additional handling required to complete drying may increase shatter losses. Smith and Brown (1994) also have suggested that field drying may cause cracks in the leaf cuticle, allowing deeper penetration by rain, thereby facilitating more aggressive leaching.

The interactive effects of forage species and moisture concentration within the wilting forage at the time rainfall occurs has been illustrated by Scarbrough et al. (2005) (Figure 40.3). When simulated rainfall was applied to freshly mowed orchardgrass (674 $g\,kg^{-1}$ moisture) in 12.5-mm increments ranging up to 76 mm, DM losses were maximized at only 19 $g\,kg^{-1}$; however, when the forage was dry enough to bale (153 $g\,kg^{-1}$ moisture), DM losses increased with rainfall amount to 86 $g\,kg^{-1}$, exhibiting strong linear and quadratic effects. Under the same experimental conditions, bermudagrass forage wilted to 130 $g\,kg^{-1}$ moisture was far more stable under simulated rainfall, with a maximum DM loss of only 21 $g\,kg^{-1}$ after 64 mm of simulated rainfall was applied. Minimal losses of DM from bermudagrass following as much as 76 mm of rainfall were attributed to the lower concentrations of soluble, non-structural carbohydrates typically found within perennial warm-season grasses (Van Soest 1982; Moore and Hatfield 1994).

Animal Responses to Rain-Damaged Forages

While several studies have evaluated the effects of natural or simulated rainfall on DM loss or concentrations of various chemical constituents within the forage, there have been relatively few that have examined animal responses to rain-damaged hays. Early work by Guilbert and Mead (1931) showed that mowed bur clover receiving 0, 8, or 20 mm of rainfall yielded DM digestibility coefficients from wether sheep of 64.2%, 59.2%, and 55.7%, respectively, when intake was constrained at maintenance. Similar responses to rain damage were observed for digestibility coefficients of CP, N-free extract, and ether extract, but crude fiber digestibility was not affected. In contrast, Montgomery et al. (1974) reported a 10.9-percentage unit decrease in DM digestibility for 'Coastal' bermudagrass that received 71 mm of natural rainfall compared to no rain, but this depression was attributed primarily to a 30.0-percentage unit decrease in digestibility of hemicellulose (65.3% vs. 35.3%). Follow-up work with lactating cows showed a 12% reduction in *ad-libitum* intake (2.17% vs. 1.91% of bodyweight), as well as a concomitant 1.0 $kg\,d^{-1}$ reduction in production of 4% fat-corrected milk. Similarly, Turner et al. (2004) reported a 9% reduction in voluntary intake by growing steers offered tall fescue hays receiving 23 or 72 mm of rainfall during wilting compared to no rain; however, digestibility coefficients for DM, organic matter, and NDF were greater for rain-damaged hays, which was assumed to be related to increased total tract retention time; however, this could not be verified analytically. Though data are limited, a soaking rainfall event will likely suppress voluntary intake of hay by roughly 10%, but this animal response may also slow passage rate through the total tract, potentially erasing differences in DM digestibility observed typically with *in-vitro* or *in-situ* procedures, and conceivably yielding digestible DM intakes that are similar.

Hay Storage

Equilibrium Moisture Concepts

Baling hay that is too moist can lead to spontaneous heating, mold development, DM losses, and deleterious changes in nutritive value. Traditional recommendations for safe storage are about 200 $g\,kg^{-1}$ moisture for conventional rectangular bales (Collins et al. 1987; Rotz and Muck 1994; Collins 1995) and between 160 and 180 $g\,kg^{-1}$ for large-round bales, though these thresholds have trended lower in recent years (Undersander and Saxe 2013). During storage, bales continue to lose moisture until they reach 80–150 $g\,kg^{-1}$; the final moisture concentration is dependent on environmental conditions, especially relative humidity, and storage structure (Pitt 1990). The final moisture concentration will remain relatively constant unless water is absorbed from the ground, rain, or humid air (Table 40.5; Hill et al. 1976; Pitt 1990).

Plant Enzymatic Activity

Respiration during hay storage includes both microbial and plant cell components. Plant respiration is very limited between 200 and 273 $g\,kg^{-1}$ moisture (Wood and Parker 1971). Since hay is normally baled at moisture concentrations in this range (<300 $g\,kg^{-1}$), plant enzymatic activity should be minimal. Respiratory rate (mg CO_2 g^{-1} DM h^{-1}) is a positive, linear function of forage moisture concentration (Wood and Parker 1971), and within typical ambient ranges, enzymatic activity increases in a positive relationship with temperature (Honig 1979). Plant respiration and heat release are maximized at about 30–35 °C, which is easily attainable in moist hays, but further temperature increases to 40–45 °C likely will cause cell death. Stage of maturity at harvest has little effect on the respiration rate of moist hay during storage (Wood and Parker 1971).

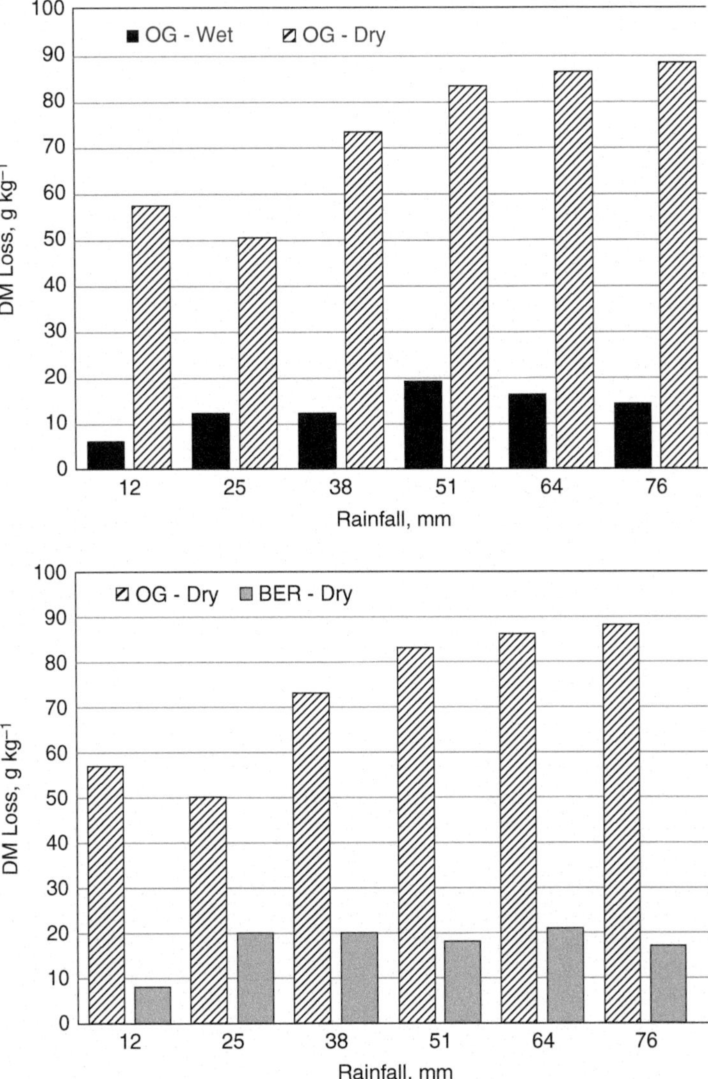

Fig. 40.3. Contrast of DM losses for freshly mowed orchardgrass forages (674 g kg^{-1}; OG–Wet) compared to orchardgrass wilted to a suitable moisture concentration for baling (153 g kg^{-1}; OG-Dry) and then damaged by simulated rainfall (top). Similarly (bottom), greater DM losses from OG-Dry are contrasted with those from dry bermudagrass (130 g kg^{-1}; BER-Dry). *Source:* Adapted from Scarbrough et al. (2005).

Microbial Activity

Microbial activity generally decreases as hay dries, being low in dry hays. Microbial respiration is responsible for most of the spontaneous heating and DM losses observed during hay storage (Wood and Parker 1971; Hlödversson and Kaspersson 1986; Rotz and Muck 1994). Moisture and temperature are major factors affecting stored-forage microbial populations, but their effects are difficult to separate in a practical context (Roberts 1995). Overall bacterial populations are generally stable during the first two or three weeks of storage (Hlödversson and Kaspersson 1986), but specific populations may change within that time period, as evidenced by declining ratios of mesophilic to thermophilic, and gram-negative to gram-positive bacteria (Kaspersson et al. 1984). Significant fungal growth in moist hay requires the relative

Table 40.5 Final moisture concentrations of baled hay as affected by air temperature and relative humidity

	Relative Humidity (%)			
Temperature	30	50	70	90
°C	$g\,kg^{-1}$			
21	100	130	210	390
27	80	120	200	380
29	70	100	180	370
35	50	80	160	360

Source: Adapted from Pitt (1990) and Hill et al. (1976).

humidity to be at least 70% and temperature to be at or above 20 °C (Rees 1982). Unlike bacteria, populations of fungi increase sharply during the first two to three weeks of hay storage. During this time period, shifts occur in dominant fungi populations from field genera, such as *Fusarium* and *Cladosporium,* to storage types, such as *Aspergillus, Penicillium,* and *Scopulariopsis brevicaulis.* The latter of these has been specifically associated with the high N content of clover hay (Hlödversson and Kaspersson 1986).

Spontaneous Heating

Controlling Factors

Spontaneous heating in forage is the most obvious practical consequence of plant and microbial respiration in hay; because of this phenomenon, internal bale temperatures increase, and the nutritive value, energy density, and digestibility of affected hays decline. Factors contributing to the relative amount of spontaneous heating include: (i) bale moisture; (ii) bale type, because larger packages include more DM within each bale, and have less surface area per unit of DM, thereby restricting dissipation of moisture and heat; (iii) bale density, in part, because denser bales condense more DM per unit volume and lose heat less rapidly than looser bales; (iv) environmental factors, such as relative humidity, ambient temperature, and air movement; (v) storage site, because well-ventilated bales heat less; and (vi) use of preservatives to control microbial growth.

Heating Characteristics

The temporal heating characteristics of baled hays often exhibit predictable patterns. For conventional rectangular (~45-kg) bales of alfalfa hay packaged at 297 and 202 $g\,kg^{-1}$ of moisture (Figure 40.4) (Coblentz et al. 1996), internal bale temperature spiked initially, presumably as a result of plant cell respiration (Kaspersson et al. 1984; Roberts 1995). After four to five days, bale temperatures typically subside, but this lull is often followed by a prolonged period (two to five weeks) of undesirable heating that is associated with respiration by storage microorganisms. The initial temperature spike associated with plant respiration also may initiate proliferation of storage microorganisms (Sussman 1976). Similar temporal-temperature trends have been reported for alfalfa hay by Nelson (1966), and for bermudagrass hay packaged in small rectangular bales (Figure 40.4) (Coblentz et al. 2000). Larger hay packages exhibit greater internal bale temperatures that persist longer than similar hays baled in conventional 45-kg bales (Collins et al. 1987; Montgomery et al. 1986).

Moisture concentration at the time of baling is the most important factor affecting **spontaneous heating**; bales packaged at greater moisture concentrations reach higher temperatures that persist longer than drier hays (Figure 40.4). The heating degree day (HDD) concept, using a base temperature of 30 or 35 °C, is often used to integrate the magnitude and duration of heating, thereby creating a single response variable that better describes heat accumulation during storage relative to a single-point-in-time measurement of temperature (Nelson 1966; Nelson 1968; Buckmaster et al. 1989). A positive, linear relationship between cumulative HDD and moisture concentration at baling has been observed for alfalfa (Nelson 1966; Nelson 1968) and bermudagrass (Coblentz et al. 2000) in small bales.

During the last quarter-century, large-round or large-square bales have replaced many of the small (~20–45-kg) hay packages that were primarily moved and stacked by hand. Due to their greater size, weight, and often density, large hay packages are far more susceptible to spontaneous heating than similar hays packaged in small bales (Montgomery et al. 1986; Collins et al. 1987; Collins 1995). The effects of bale size on measures of heating were demonstrated by Coblentz and Hoffman (2009a) in a study where alfalfa-orchardgrass hay was baled in 0.9, 1.2, or 1.5-m diameter round bales at moistures ranging from 93 to 466 $g\,kg^{-1}$. Maximum internal bale temperatures during storage ranged as high as 77 °C, where temperatures above 70 °C are normally generated by oxidative reactions, rather than by heat from microbial and plant respiration (Festenstein 1971). Regardless

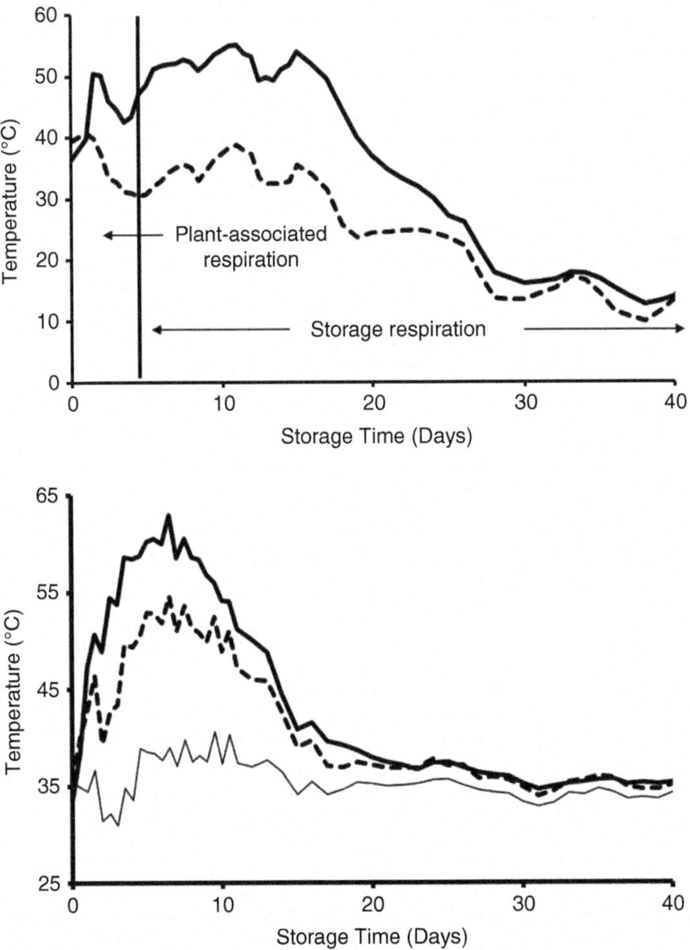

Fig. 40.4. Typical temperature vs storage time curves for alfalfa hay (top) and bermudagrass (bottom) packaged in conventional rectangular bales and stored under cover. Alfalfa was baled at 297 (solid line) and 202 (dashed line) g kg^{-1} moisture and stored for 40 days. Bermudagrass hay was packaged at 325 (solid line), 248 (dashed line, long spaces), and 178 (thin, solid line) g kg^{-1} moisture and stored under roof for 40 days. *Source:* Adapted from Coblentz et al. (1996); Coblentz et al. (2000).

of bale diameter, maximum internal bale temperatures increased linearly with bale moisture exhibiting very high coefficients of determination ($r^2 > 0.94$) (Figure 40.5). Slopes did not differ across bale diameters, indicating maximum internal bale temperature in large-round bales increased by about 0.9 °C for every 10 g kg^{-1} of initial bale moisture. However, the biggest concern with very large hay packages is reflected in assessments of HDD accumulation (Figure 40.5), where 0.9 and 1.2-m diameter bales retained the linear relationship with bale moisture described previously for conventional rectangular bales, but 1.5-m diameter bales were less able to readily dissipate heat, and the relationship lost linearity, increasing at an escalated rate with bale moisture (HDD = $[0.99 \times \text{moisture}^2] - 82$; $R^2 = 0.957$).

Under experimental conditions, temperature measurements are often made on individual (isolated) bales, not on large stacks that are difficult to replicate. In production situations, where stacks of large-round or square bales are created for efficient storage, another layer of complexity is added to heating dynamics, further restricting dissipation of heat and water from the hay. Normally, spontaneous combustion occurs near the outside of such haystacks because concentrations of oxygen are higher, thereby

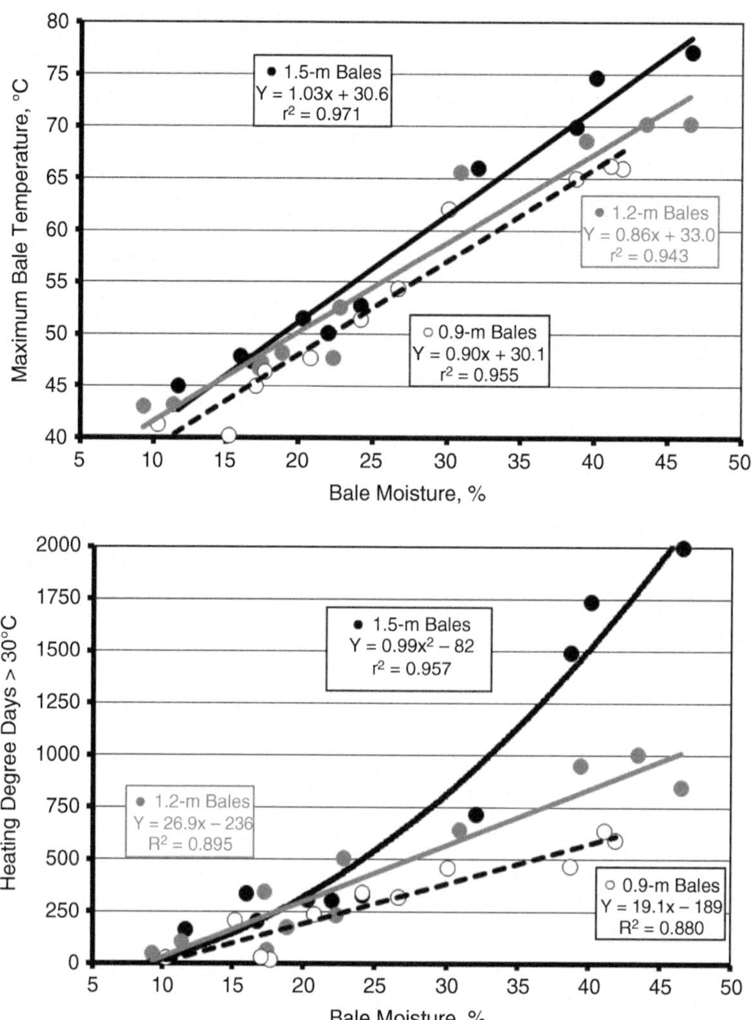

Fig. 40.5. Regressions of maximum internal bale temperature (top) and **heating degree days** >30 °C accumulated during storage (bottom) for alfalfa-orchardgrass hays packaged in large-round bales at 0.9, 1.2, and 1.5-m diameters (Coblentz and Hoffman 2009a).

increasing the probability of combustion. Because storage conditions vary widely, it is impossible to know what threshold of initial heating is likely to lead to spontaneous combustion; however, many recommendations suggest 65 °C as a marker of serious risk.

Other management factors that affect heating characteristics in bales include bale density (Nelson 1966; Buckmaster et al. 1989) and the use of preservatives, such as ammoniation (Lines et al. 1996) or formulations based on propionic acid (Sheaffer and Clark 1975; Rotz et al. 1991; Coblentz et al. 2013a). Theoretically, bale density and spontaneous heating are positively related; however, this effect is primarily a function of the larger quantity of DM in each bale, as well as reduced capacity to dissipate heat, rather than any substantial change in the heat produced per unit of forage DM (Nelson 1966; Rotz and Muck 1994). From a practical standpoint, this effect is not always observed. A mean density difference of 22 kg m^{-3} for bales packaged at five moisture concentrations ranging from 178 to 325 g kg^{-1} did not result in detectable differences in heating characteristics in conventional rectangular bales of bermudagrass hay (Coblentz et al. 2000).

The use of propionic acid to control spontaneous heating has a long history of success (Sheaffer and

Clark 1975; Rotz et al. 1991; Coblentz et al. 2013a), though its application will not completely eliminate heating, and effectiveness varies considerably with formulation and specific conditions (Rotz et al. 1991). As hay production has moved away from small hay packages, use of propionic-acid-based preservatives has been very successful in applications to large-square bales (~285 kg) of alfalfa-orchardgrass hay (Coblentz et al. 2013a) (Figure 40.6), but recent experiments with large-round bales (~570 kg) have been largely unsuccessful in suppressing internal bale temperatures (Coblentz and Bertram 2012). Reasons for these inconsistencies remain unclear, but may be related to some combination of: (i) bale size differences; (ii) poor control of preservative application rate; or (iii) the hygroscopic nature of propionic acid (Rotz et al. 1991), which tends to retard dissipation of water and maintain low levels of heating for extended periods of time, particularly in large-round bales (Coblentz and Bertram 2012). Generally, propionic-acid-based preservatives are most effective in hays only marginally too moist (<250 g kg^{-1}) to undergo satisfactory storage, and thresholds for effectiveness likely decline in an inverse relationship with bale size.

Dry Matter Losses

Losses of DM during spontaneous heating occur primarily through respiration of non-structural carbohydrates to carbon dioxide, water, and heat by storage fungi (Hlödversson and Kaspersson 1986). Losses of DM are small, and sometimes not detectable in hays baled near 150 g kg^{-1} moisture and stored for one to two months, but can reach 150 g kg^{-1} or more in poorly managed large-round bales (Coblentz and Hoffman 2009a). Numerous studies (Buckmaster et al. 1989; Rotz and Abrams 1988; Rotz et al. 1991) have shown that DM losses increase with moisture concentration at baling, and considerable evidence indicates that this relationship is linear for alfalfa (Nelson 1966; Nelson 1968), bermudagrass (Coblentz et al. 2000), and mixed hays containing primarily cool-season grasses (Hlödversson and Kaspersson 1986). Similarly, recoveries of DM from large-round bales of alfalfa-orchardgrass hay following storage declined linearly with maximum internal bale temperature or HDD in relationships consistently exhibiting very high coefficients of determination ($r^2 > 0.83$; Coblentz and Hoffman 2009a).

Non-structural Carbohydrates

During storage, non-structural carbohydrates are respired or oxidized; for example, TNC in alfalfa hay baled at 297 g kg^{-1} moisture declined in a negative, linear relationship with HDD, where TNC (g kg^{-1}) = [−0.086 × HDD] + 59.5; r^2 = 0.769 (Coblentz et al. 1997a). However, individual pools of various components comprising TNC are fluid; reducing sugars increased by 46% and 42% after four days of storage in alfalfa hays baled at 297 and 202 g kg^{-1} of moisture, respectively, while concentrations of TNC and non-reducing sugars declined over the same time period (Table 40.6).

Fiber Components

In general, the effects of modest **spontaneous heating** on the nutritive value of the cured forage are very similar to those created by leaching during rainfall events. Forage fiber components, such as NDF, acid detergent fiber (ADF), crude fiber, and lignin remain relatively stable during bale storage (Rotz and Muck 1994); however, their concentrations increase indirectly in heated hays, primarily due to oxidation of non-structural carbohydrates as described previously. Quality-retention ratios developed by Buckmaster et al. (1989) suggest that the recovery of ADF and ash in heated hays is nearly complete. Conceptually, these relationships are very close as evidenced by successful use of changes in ADF or WSC concentrations to predict DM losses in moist hays (Hlödversson and Kaspersson 1986). Furthermore, within any given lot or field of hay whose species composition is similar, there are positive, linear relationships between concentrations of most fiber components (NDF, ADF, hemicellulose, and lignin) and measures of spontaneous heating, such as maximum temperature or HDD. Coefficients of determination for these linear regressions are usually quite high (generally >0.7), as demonstrated for both bermudagrass (Coblentz et al. 2000; Turner et al. 2002) and alfalfa (Coblentz et al. 1996). With excessive heating, such as that attainable only in poorly-managed, large hay packages, relationships of fiber components with HDD become nonlinear, and concentrations of NDF may actually decline as hemicellulose becomes reactive and loses its normal analytic properties (Goering et al. 1973; Weiss et al. 1986; Coblentz and Hoffman 2009b; Coblentz and Bertram 2012).

Crude Protein

Concentrations of CP in heated hays are highly dependent on time since baling (Rotz and Muck 1994). During the earlier phases of storage (<60 days), concentrations of CP often increase (Rotz and Abrams 1988; Coblentz et al. 2000; Turner et al. 2002) in a similar manner to that exhibited by fiber components, largely due to preferential oxidation of non-structural carbohydrates. For conventional rectangular bales of bermudagrass, these increases are linear functions of both moisture concentration at baling and measures of spontaneous heating (Coblentz et al. 2000; Turner et al. 2002). For large hay packages (Coblentz et al. 2010), concentrations of CP in alfalfa-orchardgrass hays increased by

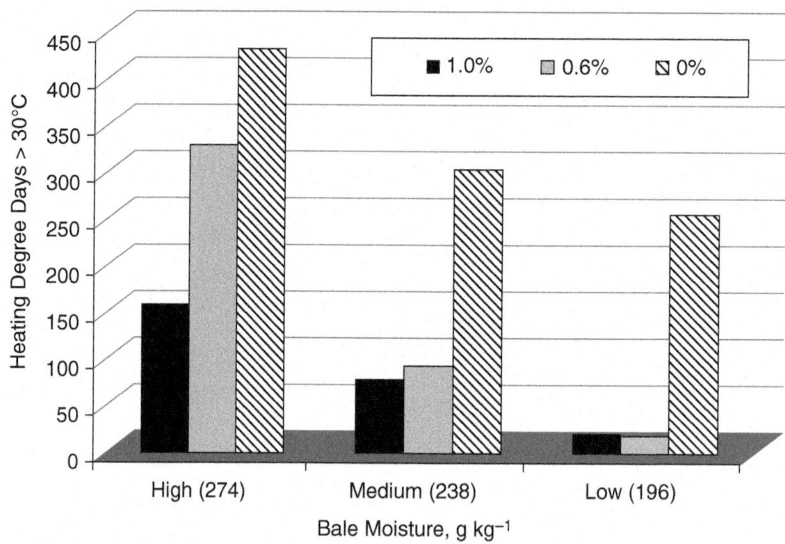

Fig. 40.6. Effects of applying a propionic-acid-based preservative at three rates (0%, 0.6%, or 1.0%) on heating degree day >30 °C accumulation within 285-kg large-rectangular bales of alfalfa-orchardgrass hay made at initial bale moistures of 196 (Low), 238 (Medium), and 274 (High) g kg^{-1}. *Source:* Adapted from Coblentz et al. (2013a).

Table 40.6 Concentrations of total non-structural carbohydrates, non-reducing sugars, and reducing sugars in alfalfa hays baled at 297 (high) and 202 g kg^{-1} (low) of moisture and sampled over time in storage

	Total non-structural carbohydrate		Non-reducing sugars		Reducing sugars	
Storage time	High	Low	High	Low	High	Low
d			g kg^{-1}			
h	59.6	72.9	27.5	41.1	23.0	17.2
4	56.3	51.7	9.0	16.0	33.5	24.5
11	36.7	39.5	7.8	12.0	17.0	18.0
22	27.1	35.9	2.9	8.3	15.2	18.6
60	20.7	42.1	4.4	9.9	6.9	21.8

Source: Adapted from Coblentz et al. (1997a).

10 ± 6.7 g kg^{-1} across 96 large-round bales incurring a wide range of heating. However, the overall relationship with HDD was poor and described with a negative slope ($\Delta CP = [-0.0053 \times HDD] + 13$; $r^2 = 0.149$), where ΔCP = change in CP concentration during storage (post-storage CP − pre-storage CP). Furthermore, the greatest increases in CP occurred with only modest heating (HDD < 500). Over long-term storage, CP may decrease by 2.5 g CP kg^{-1} DM per month as a result of slow **volatilization** of ammonia contained in the hay or produced through microbial respiration (Rotz and Muck 1994).

Heat-Damaged Protein

Maillard or nonenzymatic browning reactions occur actively in hays and silages in association with spontaneous heating. Carbohydrates are degraded in the presence of water and amines or amino acids to form indigestible polymers containing approximately 110 g N kg^{-1} DM that also possess many of the physical properties of lignin (Van Soest 1982). Water is a catalyst within Maillard reactions; the maximum catalytic effect occurs at about 300 g kg^{-1} moisture (Van Soest 1982), but the reaction is active between 200 and 700 g kg^{-1} moisture (Goering et al. 1973). Normally, Maillard reaction damage is assessed by measuring the N that is insoluble in acid-detergent solution and, can be expressed on N or CP basis (ADIN or ADICP, respectively). Some "native" ADICP (usually $<100\,\mathrm{g\,kg^{-1}}$ CP) is present in all forages, but on a percentage of CP basis, concentrations of ADICP are often greater in mature grasses, particularly perennial warm-season grasses (Vanzant et al. 1996; Coblentz et al. 1998; Scarbrough et al. 2002). With spontaneous heating, ADICP increases linearly above this native baseline, and the linear nature of this response has been demonstrated for both alfalfa (Coblentz et al. 1996) and bermudagrass (Coblentz et al. 2000; Turner et al. 2002) hays packaged in conventional (~45-kg) rectangular bales. Bermudagrass appears to be more susceptible to heat damage than alfalfa, based on greater responses per unit of heating (Figure 40.7). Unlike fiber components, increases in ADICP are largely the direct result of artifact N formation, though some increases will occur indirectly as a result of slow **volatilization** of soluble N components (Rotz and Muck 1994), and from preferential oxidation of non-structural carbohydrates.

Foundational work on heat damage in forages conducted by Goering et al. (1973) suggests that hemicellulose is a reactive structural carbohydrate, exhibiting an inverse relationship with ADICP whenever heat damage occurs (Figure 40.8). In practice, this relationship hasn't been observed in field studies with small conventional hay bales (Coblentz et al. 2000); instead, hemicellulose and ADICP often increase in parallel in response to spontaneous heating. However, recent experiments with large round bales (Figure 40.9) suggest that temperatures and/or heat increments attained in small hay packages are not sufficient to reproduce the foundational effects described by Goering et al. (1973), but similar responses do occur in large hay packages that can attain much greater HDD or internal bale temperatures during storage (Coblentz et al. 2010).

It has long been understood that natural spontaneous heating (Bechtel et al. 1945; Miller et al. 1967; Montgomery et al. 1986; McBeth et al. 2001) or externally applied (steam) heating (Broderick et al. 1993) reduces the apparent total-tract digestibility of CP in ruminants. In addition, ruminal disappearance rates are slowed and estimates of rumen degradable CP are decreased (Broderick et al. 1993; Coblentz et al. 1997b; McBeth et al. 2003). Within limits of modest spontaneous heating, these relationships appear to be inverse, linear functions of heating (Coblentz et al. 1997b; Coblentz et al. 2010), but are explained by higher-ordered polynomials when extreme heating, such as that attainable in large hay packages, is included in the regression (Coblentz et al. 2010). The effects of heating on digestion coefficients for ADICP are somewhat counterintuitive. Generally, with little or no heating, digestion coefficients for "native" ADICP oscillate closely around nil (Broderick et al. 1993; McBeth et al. 2001; Coblentz et al. 2013a), but they can increase linearly to $>400\,\mathrm{g\,kg^{-1}}$ ADICP in heated hays (McBeth et al. 2001; Coblentz et al. 2013a), thereby suggesting that the ADICP accumulated via heat damage retains some bioavailability.

Digestibility/Energy Estimates

Digestibility of DM or OM decreases with spontaneous heating, largely due to the oxidation of highly digestible non-structural carbohydrates, a greater proportion of which are lost during storage compared to unheated hays. While McBeth et al. (2001) reported that digestibility coefficients for NDF and hemicellulose measured in lambs consuming heated bermudagrass hays declined linearly with HDD, other studies (Coblentz and Hoffman 2009b) found little relationship between fiber digestibility and spontaneous heating, unless visibly charred samples were included in the analysis. Generally, estimates of energy density also are depressed, and can be severe ($>110\,\mathrm{g\,kg^{-1}}$ total digestible nutrients [TDN]) with excessive spontaneous heating (Figure 40.10; Coblentz and Hoffman 2010). This response also occurs mostly through losses of non-structural carbohydrates oxidized during storage; however, contributions of CP digestibility to the total TDN estimate also are reduced via increased ADICP as explained for forages by: truly digestible $\mathrm{CP} = \mathrm{CP} \times e^{(-1.2 \times \mathrm{ADICP/CP})}$, where ADICP is expressed on a $\mathrm{g\,kg^{-1}}$ DM basis (NRC 2001). The close, linear relationship between the change in concentrations of TDN (ΔTDN) and maximum internal bale temperature ($\Delta\mathrm{TDN} = -3.80x + 163$; $r^2 = 0.954$) suggests initial peak temperature may have a more direct effect on energy losses than sustained elevated temperatures within the bale, which would be reflected in cumulative HDD. Another factor contributing to the reduced TDN in heated hays is changing concentrations of ash, which are inert during heating, make no contribution to energy estimates, and normally increase as an indirect consequence of respiration. A survey of 3612 hay or haylage samples submitted to private laboratories revealed that losses of energy due to heating exceeded $40\,\mathrm{g\,kg^{-1}}$ TDN in 50% of the samples evaluated, and losses exceeded $80\,\mathrm{g\,kg^{-1}}$ TDN in 16% of the samples, suggesting significant economic loss to many forage producers, in part as a result of problematic moisture management in dry hays and haylages (Yan et al. 2011).

Silage Storage

Plant Enzymatic Activity

The immediate goal of silage fermentation is to reduce oxygen concentrations so that natural anaerobic processes can preserve silage. Oxygen trapped in any silage mass is consumed by respiration of functioning plant cells and microbes which are adhered to forage plants. Entrapped oxygen can be consumed by respiratory processes of

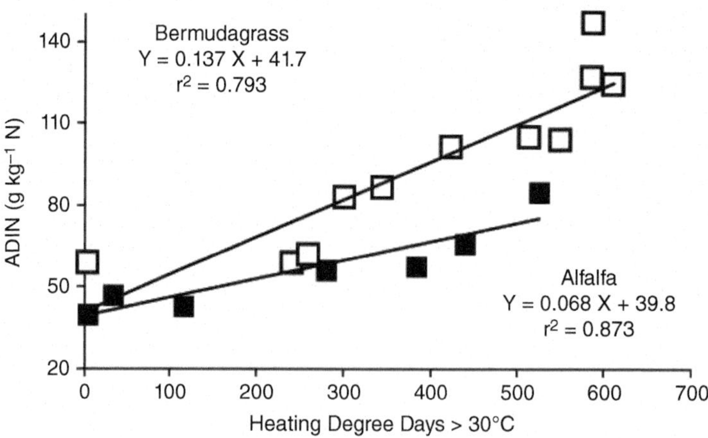

Fig. 40.7. Linear regressions of acid-detergent insoluble N (ADIN) on **heating degree days** >30 °C for alfalfa and bermudagrass. *Source:* Adapted from Coblentz et al. (1994); Coblentz et al. (2000).

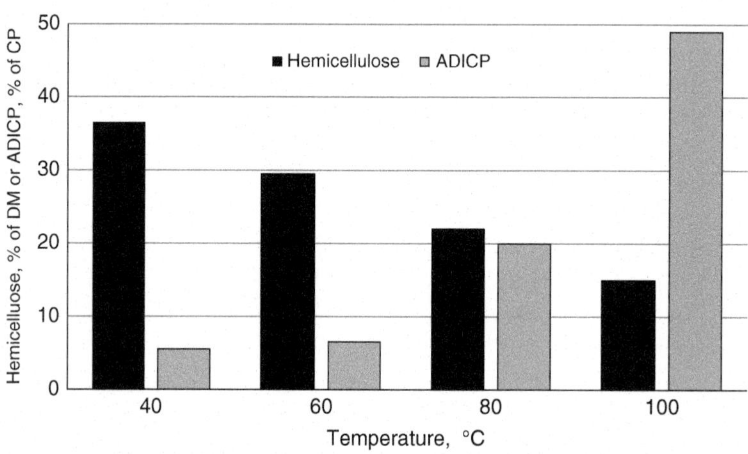

Fig. 40.8. Summary of foundational work by Goering et al. (1973), and further discussed by Van Soest (1982) illustrating the inverse relationship between hemicellulose and ADICP concentrations in heated forages. Forages were heated in 10-ml flasks for 24 hours at 530 g kg^{-1} of moisture.

plant cells in a matter of hours following sealing (Pitt et al. 1985), and the (unavoidable) DM losses associated with depletion of this oxygen are usually minimal (1–2%) (McDonald et al. 1991). Excessive amounts of oxygen in poorly consolidated (packed) silage causes prolonged respiration that wastes plant sugars required as fermentable substrate by lactic-acid producing bacteria (LAB) and other desirable silage microorganisms. Other consequences of prolonged respiration include depressed DM digestibility and energy density, and formation of Maillard products facilitated in part by slow dissipation of heat from silage coupled with the catalytic function of water in browning reactions (Goering et al. 1973; Coblentz et al. 2016). Silage management techniques such as fine chopping, rapid filling, and proper packing are largely designed to increase the rate of oxygen depletion. Suggested target densities have been established, partly with the goal of reducing air trapped in the silo. Targets for precision-chopped silages are 243 kg DM m^{-3} (15 lb DM ft^{-3}), which has been further refined to 712 kg (as is)/m^3 (44 lb ft^{-3}) on the basis of varying porosity of the silage to air (Holmes and Muck 2008).

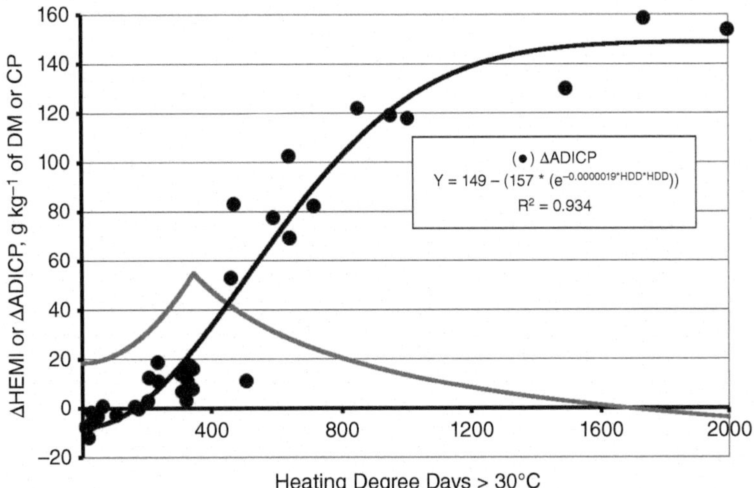

Fig. 40.9. Regressions of the change in acid-detergent insoluble CP (g kg^{-1} N; ΔADICP) and hemicellulose (g kg^{-1} DM; ΔHEMI) on heating degree days >30 °C (HDD) accumulated during storage of large-round bales of alfalfa-orchardgrass hays. Solid black line represents regression of ΔADICP on HDD, while solid gray line represents ΔHEMI. *Source:* Adapted from Coblentz et al. (2010).

Targets for round-bale silages are less dense, such as the >162 kg DM m^{-3} (10 lb DM ft^{-3}) threshold suggested by Jennings (2011), which can be exceeded routinely by experienced operators as reported by Rhein et al. (2005) for large-round bales of orchardgrass packaged at 456 g kg^{-1} moisture (218 kg DM m^{-3}; 13.5 lb DM ft^{-3}). For silages made in large-round bales, practices that maximize the number of forage layers per bale, such as slowing the ground speed of the baler and reducing the swath density, will generally increase bale density, reduce trapped air, and limit the associated respiratory heating.

Plant cells are often damaged during mechanical harvesting, thereby releasing proteolytic plant enzymes that break down proteins to peptides and free amino acids. After ensiling, concentrations of NPN can vary over a wide range, depending on forage species, pH, time from ensiling, temperature, and the moisture concentration of the forage. Alfalfa proteins, particularly ribulose 1,5-bisphosphate carboxylase oxygenase (Messman et al. 1994), are sensitive to proteolysis, whereas proteins in red clover and cicer milkvetch are more resistant (Albrecht and Muck 1991; McKersie 1985). Red clover derives its resistance to proteolysis from the action of o-quinones formed by polyphenol oxidase (Albrecht and Muck 1991; Jones et al. 1995a; Jones et al. 1995b). Alternatively, forages containing tannins undergo less proteolysis than those lacking tannins (Albrecht and Muck 1991), and ruminal degradation rates of CP, as well as estimates of effective rumen degradability decline linearly in hays and silages as a function of tannin concentration (Coblentz and Grabber 2013). Moisture concentration is positively related to initial proteolysis rates, which decline from 0.9 mg N g^{-1} DM h^{-1} at 800 g kg^{-1} to almost nil when the forage moisture is less than 240 g kg^{-1} (Muck 1987). Tracey (1948) described the optimum pH range for leaf proteases as 5.0 to 6.0; similarly, the optimum pH for proteolytic activity in legumes is about 6.0 (McKersie 1985), and proteolysis decreases as pH declines during fermentation. Proteolytic activity can continue to occur at pH levels as low as 4.0, and some acid hydrolysis of proteins also is possible (McDonald 1982). Storage temperature also can affect proteolysis positively (Muck and Dickerson 1988). Other enzymes released at harvest can convert hemicellulose (Dewar et al. 1963) and non-structural carbohydrates to simple sugars (Rotz and Muck 1994), thereby increasing the substrate supply for fermentation (Bolsen 1995).

Microbial Activity

Overview

In the preservation of silage, LAB, enterobacteriaceae, yeasts, molds, and clostridia all play major roles in determining the final quality of the silage. The competitive effects of these microorganisms can result in different types of silage fermentations, previously classified for grass silages in Europe into five broad categories: (i) lactate; (ii) wilted; (iii) acetate; (iv) butyrate; and (v) chemically restricted or sterilized (McDonald and Edwards 1976) (Figure 40.11). In this example, lactate, acetate, and butyrate silages were very wet (810–830 g kg^{-1}), while

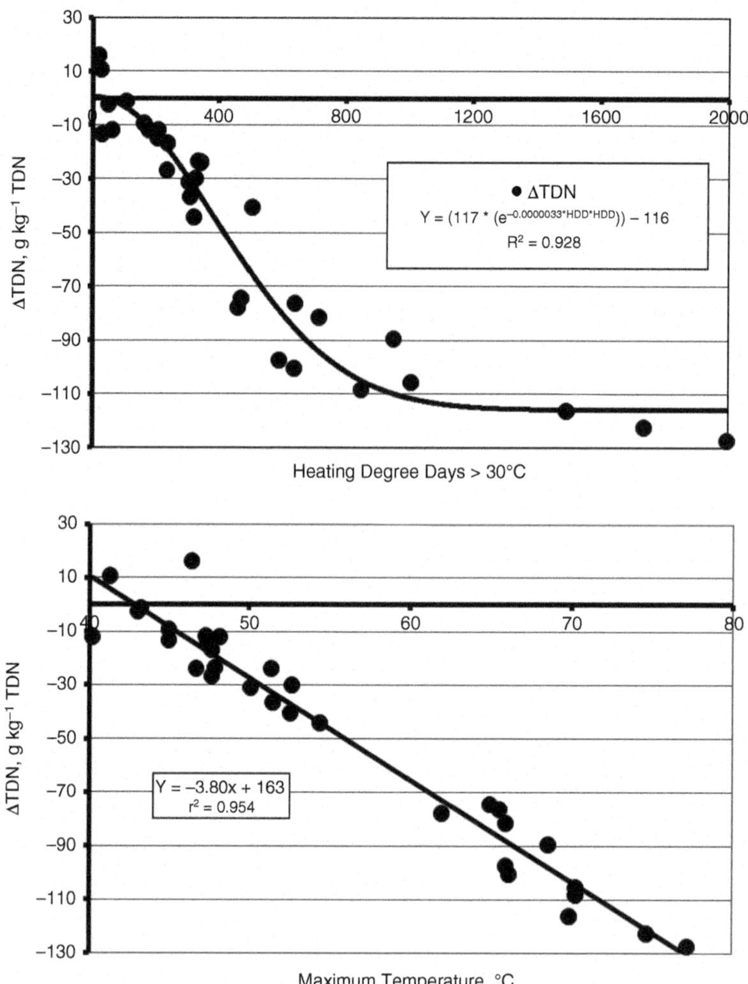

Fig. 40.10. Change in energy density (ΔTDN) within large-round bales of alfalfa hay as affected by **heating degree days** >30 °C (top) and maximum internal bale temperature (bottom). *Source:* Adapted from Coblentz and Hoffman (2010).

the wilted silage represented only limited desiccation (692 g kg^{-1} moisture). Distinguishing characteristics of the lactate silage were high concentrations of lactic acid present in a disproportionately large ratio to acetic acid (2.83: 1), as well as a very acidic final pH (3.9), essentially no butyric acid, and only minimal residual WSC. Wilted silage can also be considered a lactate fermentation, but the total production of fermentation acids was restricted by desiccation, though the lactic/acetic-acid ratio remained similar (2.46: 1). Other notable traits included increased residual (unfermented) WSC, and improved preservation of protein-N compared to the direct-cut lactate silage. Acetate silage exhibited large concentrations of acetic acid (97 g kg^{-1} DM), and a very low lactic/acetic-acid ratio (0.35: 1). Butyrate silages are the product of secondary clostridial activity, and exhibited low concentrations of lactic acid and residual WSC, a less acidic final pH than other silage types (pH = 5.2), and high concentrations of butyric acid and ammonia-N. Generally, chemically restricted silages are not produced in the United States, but have been produced by adding formic acid and formalin to the forage at ensiling, essentially sterilizing the silage, such that final concentrations of protein-N and WSC largely reflect the initial composition of the forage (740 N and 133 g kg^{-1} DM, respectively). Of these silage types, lactate fermentations are most desirable, largely due to the relative strength of lactic acid, and its associated capacity for attaining the lowest possible pH in the silage. It should be noted that it is uncommon to ensile direct-cut or unwilted grasses or legumes in the United States unless the moisture concentration of the standing crop is <700 g kg^{-1}; however, ensiling these wet forages has been historically commonplace throughout most of northern Europe. As such, most lactate-dominated fermentations throughout the United States would be considered "wilted" within the previously defined classification system.

Moisture Effects

As suggested previously (Figure 40.11), the moisture concentration of the silage is positively related to the production of silage fermentation acids (Muck 1990; Nicholson et al. 1991). An excellent illustration of this concept for round-bale silages of alfalfa ensiled at 520 (ideal) and 620 (wet) g kg^{-1} moisture was reported by Nicholson et al. (1991) (Figure 40.12). Bales produced at the greater moisture concentration exhibited more rapid production of fermentation acids, as well as a more rapid and extensive pH drop than did bales produced at an ideal moisture for baled silages (450–550 g kg^{-1}; Shinners 2003). From a management perspective, this relationship has cautionary limits; moisture recommendations for ensiling precision-chopped, highly buffered legumes, such as alfalfa, are commonly ≤700 g kg^{-1}, largely to avoid problematic clostridial fermentations (Muck et al. 2003). Due in part to the heterogeneous nature of baled silages, this threshold should be reduced further to 600 g kg^{-1} for this specific preservation environment (Coblentz et al. 2016). Conversely, for very dry silages, production of fermentation acids can be minimal, and silage preservation is primarily a function of continued oxygen exclusion; Coblentz et al. (2014b) reported excellent preservation, but little or no production of lactic acid when the moisture concentration of baled alfalfa silages was <450 g kg^{-1}. Muck (1988) compiled the problematic aspects of ensiling dry chopped forages, which included: (i) greater weather risks; (ii) larger harvest losses; (iii) lower specific heat; (iv) greater porosity; (v) greater difficulty in packing; and (vi) greater susceptibility to spontaneous heating and aerobic deterioration.

Lactic Acid Bacteria

Initially, populations of LAB are low, particularly relative to those of enterobacteriaceae (Lin et al. 1992); however, their numbers increase markedly on alfalfa during the first 48 hours of wilting (Muck 1989), as well as in response to chopping (McDonald et al. 1991). High temperature and high forage moisture concentrations both positively affect LAB populations on wilting alfalfa (Muck 1989). Without chopping action, fermentation is restricted. A distinctive characteristic of round-bale silages, which are usually ensiled without particle size reduction, is that the plant sugars serving as substrate for fermentation must diffuse from inside the plant to reach LAB adhered to the outside surface of forage plants. This process typically limits both the rate and extent of fermentation (Nicholson et al. 1991; Muck et al. 2003; Savoie and Jofriet 2003). **Homofermentative** and **heterofermentative** types of LAB exist, but the homofermentative types are generally preferred because they ferment glucose and other simple plant sugars to lactic acid only. Heterofermentative types yield lactic acid, various alcohols, acetic acid, and CO_2, and are less efficient at conserving both DM and energy (McDonald et al. 1991). These two types of LAB compete under natural conditions and inoculation with homofermentative LAB can help to bias the fermentation toward the often desired homofermentative pathway (Bolsen 1995). In recent years, production of acetic acid through the heterofermentative pathway, usually following inoculation with *Lactobacillus buchneri*, has been used to improve aerobic stability or bunk life of many silages, particularly corn, sorghum, and small-grain silages that often have greater residual sugar concentrations than other forages. A meta-analysis (Kleinschmit and Kung Jr. 2006) indicated that treatment of these forages with *L. buchneri* consistently improved aerobic stability, but also increased final pH and acetic acid, while reducing lactic acid, yeasts, and residual WSC. For corn silages, numbers of yeasts were related negatively to concentrations of acetic acid in a linear relationship, where yeasts ($\log_{10}$ cfu g^{-1}) = (− 0.761 × acetic acid (%) + 5.43; $r^2 = 0.66$). Recently, there have been increased efforts to combine the benefits of homofermentative and heterofermentative bacteria in cocktail-type inoculants designed to improve both fermentation and aerobic stability.

Enterobacteria

Enterobacteria are primarily anaerobic bacteria whose populations on alfalfa and corn are greater than observed for LAB (Lin et al. 1992). Their most distinguishing characteristic is their primary fermentation product, acetic acid, which is problematic. Acetic acid is less desirable because it is a weaker acid than lactic acid, and the resulting silage pH may not be acidic enough to prevent undesirable secondary fermentations by clostridia. Use of *L. buchneri* is often avoided as an inoculant on alfalfa for this reason. Furthermore, silages dominated by acetic acid have poorer recovery of DM and may be less acceptable to livestock than lactic-acid dominated silages (Bolsen 1995). Generally, enterobacteria do not affect silage quality greatly. They are active mainly during the early stages of fermentation when the silage pH is still relatively high; their optimum pH is about 7.0, and populations decline rapidly when the pH falls to 5.0 (McDonald et al. 1991).

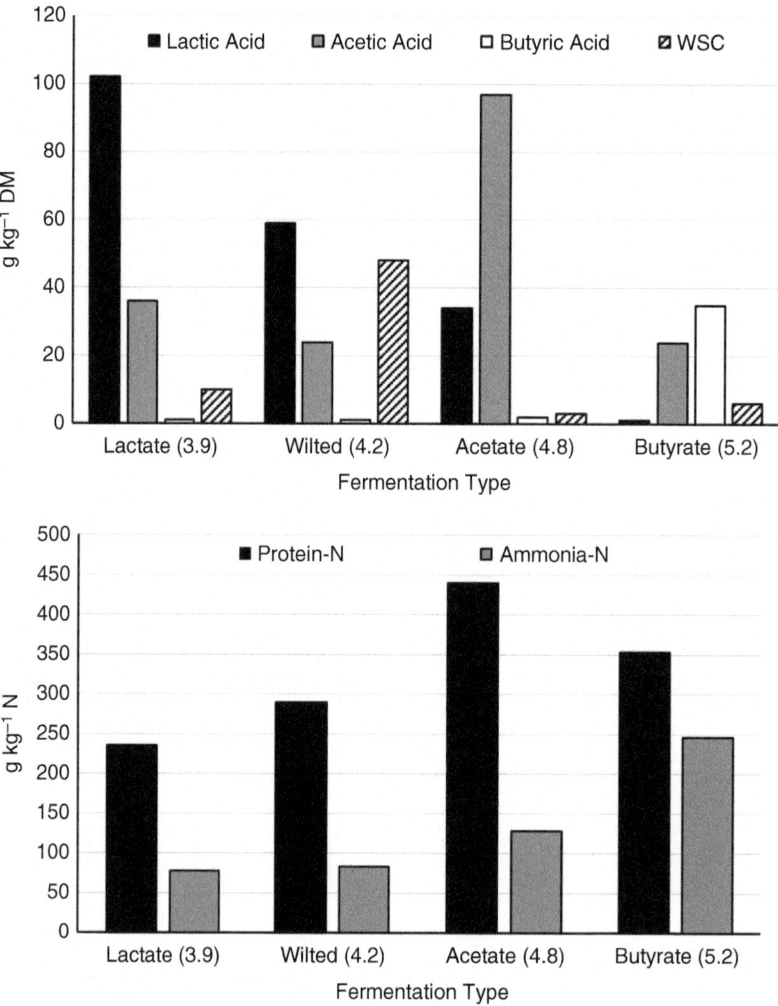

Fig. 40.11. Concentrations of fermentation acids and residual water-soluble carbohydrates (WSC) (top) and protein- and ammonia-N (bottom) for four grass-silage fermentation types in Europe. Silages ranged from 810 to 830 g kg^{-1} moisture for lactate, acetate, and butyrate silages, and 692 g kg^{-1} moisture for wilted silage. Final pH of each silage appears parenthetically on the x-axis. *Source:* Adapted from McDonald and Edwards (1976).

Clostridia

The most common undesirable silage fermentation is facilitated by clostridial spores and, is often associated with wet forages ensiled at >700 g kg^{-1} moisture. Other conditions creating risk for clostridial fermentations include: (i) direct-cutting; (ii) immaturity and/or rapid plant growth; (iii) high buffering capacity; (iv) limited sugar to support fermentation; (v) contamination with dirt or manure; (vi) high-protein concentrations; (vii) legumes; and (viii) non-homogenous forages. Clostridia ferment sugars and organic acids to butyric acid, and can have proteolytic activity, resulting in the production of ammonia and amines, and the decarboxylation of amino acids (McDonald et al. 1991). Frequently, these are referred to as saccharolytic and proteolytic types, respectively. Silages affected by clostridia have poor DM recovery (McDonald et al. 1991), and are not readily consumed by livestock (Bolsen 1995). Wilting to <700 g kg^{-1} moisture greatly reduces the incidence of clostridial fermentations in chopped silages (Muck et al. 2003), as does achieving a rapid pH decline. The optimum pH for clostridial growth is 7.0–7.4, and they are

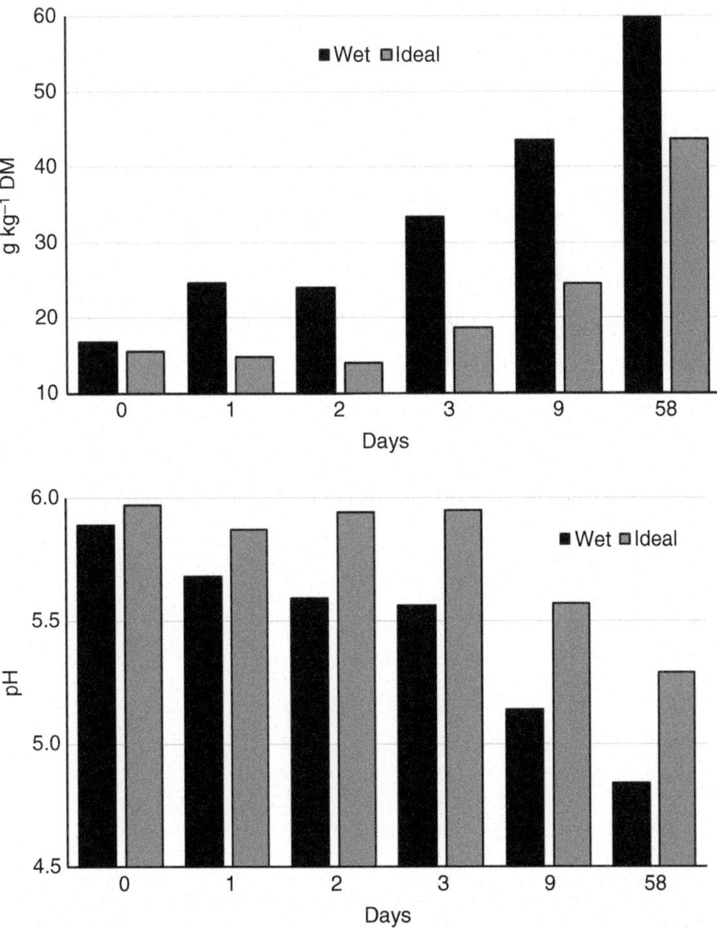

Fig. 40.12. Effects of bale moisture and fermentation time on concentrations of total fermentation acids (top), and pH depression (bottom) for alfalfa/grass baled silages made at 520 (ideal) and 620 (wet) g kg^{-1}. *Source:* Adapted from Nicholson et al. (1991).

largely inhibited by acidic conditions (Leibensperger and Pitt 1987; McDonald et al. 1991). Anaerobic conditions are often considered a requirement for clostridial growth in silages; however, some of these bacteria can grow in the presence of oxygen, provided the oxidation-reduction potential is low (McDonald et al. 1991). This heavy dependence on anaerobiosis should not be leveraged for control of clostridia; poor integrity of any silo with respect to oxygen exclusion will initiate aerobic deterioration, which is not consistent with good silage management.

Other Considerations

Other undesirable microbial activity within silages include the action of yeasts and molds, which generally require the presence of oxygen, though anaerobic activity of yeasts is frequently associated with ensiling high-sugar crops and the (undesirable) production of ethanol (McDonald et al. 1991). The integrity of silo walls (plastic, bunker, or upright) is critical to limiting oxygen entry to the silage and preventing aerobic deterioration, but high silage bulk density also will limit deterioration by reducing the permeability of the silage mass to air (Pitt 1990; Holmes and Muck 2008), which is particularly important with exposure to air during feedout.

Silage Quality

The general trends described previously for heated hays also apply to heated silages. Spontaneous heating is frequently associated with low-forage moisture concentrations (<500 g kg^{-1} moisture) at ensiling. This is especially problematic when subsequent exclusion of oxygen is not effective (Muck 1988), such as observed in many tower silos where the only packing action is derived from the weight of the silage. Generally, spontaneous heating occurs at the expense of the most digestible plant constituents, resulting in greater concentrations of fiber components, lower energy density, reduced digestibility, and increased ADICP. Merchen and Satter (1983) reported that alfalfa silage ensiled at $307\,g\,kg^{-1}$ moisture reached similar maximum temperatures (~41–42 °C) as did alfalfa silages wilted to only 597 and $704\,g\,kg^{-1}$ moisture prior to ensiling. Temperatures of the wetter silages declined steadily after four to five weeks, whereas the temperature of the dry silage did not decline for eight to nine weeks. The concentration of ADICP in the dry silage was $129\,g\,kg^{-1}$ CP, which was 63–72% greater than the wetter silages. A survey study conducted in Pennsylvania indicated that about 40% of the hay-crop silages had significant elevation in heat-damaged CP; however, there was little or no evidence of heat damage in the corn silages evaluated in the same survey (Goering and Adams 1973). In addition, long-term, low levels of oxygen access to silages, such as that facilitated by hand-tying (sealing) preformed bags for silage bales, have been correlated to increased incidence of listeria in sheep in Europe (McDonald et al. 1991). In contrast to dry silages, forages ensiled at $>700\,g\,kg^{-1}$ moisture are problematic for several different reasons, including excessive effluent production, which is a potential environmental hazard and particularly problematic with tower-type silos. From a nutritional perspective, clostridial fermentations are undesirable because of extensive DM losses, dissipation of energy, low concentrations of residual WSC and desirable fermentation acids, creation of "sour silage," and extensive conversion of protein to NPN forms, particularly ammonia and amines (Mahanna and Chase 2003). Though these silages may exhibit good aerobic stability, they also are associated with reduced palatability and voluntary intake. Erdman (1993) has summarized the effects of various silage factors on voluntary forage intake, concluding that several fermentation products of wet silages, including acetic acid, soluble amines, ammonia, and butyric acid affect intake negatively.

Summary

Numerous plant microbial and physiologic processes occur during forage harvest and storage, and many are deleterious. One notable exception is the anaerobic fermentation of forages into silages dominated by production of lactic acid. These processes are influenced by pre-harvest factors, as well as harvest and storage variables. Avoidance of rain damage and facilitation of rapid drying to minimize other losses are key considerations in hay production to avoid excessive plant enzymatic and microbial activity. Using silage to preserve forage greatly reduces negative effects of rain during field and storage phases compared with hay. Rapid elimination of oxygen and adequate maintenance of anaerobic conditions in silage minimizes aerobic respiration by plant enzymes and microbes.

References

Adesogan, A.T. and Newman, Y.C. (2010). Silage harvesting, storing, and feeding. University of Florida. IFAS Extension Publication No. SS-AGR-177.

Albrecht, K.A. and Muck, R.E. (1991). Proteolysis in ensiled forage legumes that vary in tannin concentration. *Crop Sci.* 31: 464–469.

Bechtel, H.E., Shaw, A.O., and Atkeson, F.W. (1945). Brown alfalfa hay – its chemical composition, and nutritive value in dairy rations. *J. Dairy Sci.* 28: 35–48.

Bolsen, K.K. (1995). Silage: basic principles. In: *Forages Vol. II, the Science of Grassland Agriculture*, 5e (eds. R.F Barnes, D.A. Miller and C.J. Nelson), 163. Ames, IA: Iowa State University Press.

Brink, G.E., Digman, M.F., and Muck, R.E. (2014). Field drying-rate differences among three cool-season grasses. *Forage Grazinglands* 12 (1) https://doi.org/10.2134/FG-2013-2014-RS.

Brito, A.F., Tremblay, G.F., Bertrand, A. et al. (2008). Alfalfa cut at sundown and harvested as baleage improves milk yield of late-lactation dairy cows. *J. Dairy Sci.* 91: 3968–3982.

Broderick, G.A., Yang, J.H., and Koegel, R.G. (1993). Effect of steam heating alfalfa hay on utilization by lactating dairy cows. *J. Dairy Sci.* 76: 165–174.

Buckmaster, D.R., Rotz, C.A., and Mertens, D.R. (1989). A model of alfalfa hay storage. *Trans. Amer. Soc. Agric. Eng.* 32: 30–36.

Carpintero, C.M., Henderson, A.R., and McDonald, P. (1979). The effect of some pretreatments on proteolysis during the ensiling of herbage. *Grass Forage Sci.* 34: 311–315.

Coblentz, W.K. and Bertram, M.G. (2012). Effects of a propionic acid-based preservative on storage characteristics, nutritive value, and energy content for alfalfa hays packaged in large round bales. *J. Dairy Sci.* 96: 2521–2535.

Coblentz, W.K. and Grabber, J.H. (2013). In situ protein degradation from alfalfa and birdsfoot trefoil hays and silages as influenced by condensed tannin concentration. *J. Dairy Sci.* 96: 3120–3137.

Coblentz, W.K. and Hoffman, P.C. (2009a). Effects of bale moisture and bale diameter on spontaneous heating, dry matter recovery, in-vitro true digestibility, and in-situ disappearance kinetics of alfalfa-orchardgrass hays. *J. Dairy Sci.* 92: 2853–2874.

Coblentz, W.K. and Hoffman, P.C. (2009b). Effects of spontaneous heating on fiber composition, fiber digestibility, and in situ disappearance kinetics of neutral detergent fiber for alfalfa-orchardgrass hays. *J. Dairy Sci.* 92: 2875–2895.

Coblentz, W.K. and Hoffman, P.C. (2010). Effects of spontaneous heating on estimates of TDN for alfalfa-orchardgrass hays packaged in large-round bales. *J. Dairy Sci.* 93: 3377–3389.

Coblentz, W.K. and Jokela, W.E. (2008). Estimating losses of dry matter from wetted alfalfa-orchardgrass mixtures using cell-wall components as internal markers. *Crop Sci.* 48: 2481–2489.

Coblentz, W.K. and Muck, R.E. (2012). Effects of natural and simulated rainfall on indicators of ensilability and nutritive value for wilting alfalfa forages sampled before preservation as silage. *J. Dairy Sci.* 95: 6635–6653.

Coblentz, W.K., Fritz, J.O., and Bolsen, K.K. (1994). Performance comparisons of conventional and laboratory-scale alfalfa hay bales in small haystacks. *Agron. J.* 86: 46–54.

Coblentz, W.K., Fritz, J.O., Bolsen, K.K., and Cochran, R.C. (1996). Quality changes in alfalfa hay during storage in bales. *J. Dairy Sci.* 79: 873–885.

Coblentz, W.K., Fritz, J.O., Bolsen, K.K. et al. (1997a). Relating sugar fluxes during bale storage to quality changes in alfalfa hay. *Agron. J.* 89: 800–807.

Coblentz, W.K., Fritz, J.O., Cochran, R.C. et al. (1997b). Protein degradation in response to spontaneous heating in alfalfa hay by in situ and ficin methods. *J. Dairy Sci.* 89: 700–713.

Coblentz, W.K., Fritz, J.O., Fick, W.H. et al. (1998). In situ dry matter, nitrogen, and fiber degradation of alfalfa, red clover, and eastern gamagrass at four maturities. *J. Dairy Sci.* 81: 150–161.

Coblentz, W.K., Turner, J.E., Scarbrough, D.A. et al. (2000). Storage characteristics and nutritive value changes in bermudagrass hay as affected by moisture content and density of rectangular bales. *Crop Sci.* 40: 1375–1383.

Coblentz, W.K., Hoffman, P.C., and Martin, N.P. (2010). Effects of spontaneous heating on forage protein fractions and in situ disappearance kinetics of crude protein for alfalfa-orchardgrass hays packaged in large-round bales. *J. Dairy Sci.* 93: 1148–1169.

Coblentz, W.K., Coffey, K.P., Young, A.N., and Bertram, M.G. (2013a). Storage characteristics, nutritive value, energy content, and in vivo digestibility of moist, large rectangular bales of alfalfa-orchardgrass hay treated with a propionic acid-based preservative. *J. Dairy Sci.* 96: 2521–2535.

Coblentz, W.K., Nellis, S.E., Hoffman, P.C. et al. (2013b). Unique interrelationships between fiber composition, water-soluble carbohydrates, and in-vitro gas production for fall-grown oat forages. *J. Dairy Sci.* 96: 7195–7209.

Coblentz, W.K., Jokela, W.E., and Bertram, M.G. (2014a). Cultivar, harvest date, and nitrogen fertilization affect production and quality of fall oat. *Agron. J.* 106: 2075–2086.

Coblentz, W.K., Muck, R.E., Borchardt, M.A. et al. (2014b). Effects of dairy slurry on silage fermentation characteristics and nutritive value of alfalfa. *J. Dairy Sci.* 97: 7197–7211.

Coblentz, W.K., Muck, R.E., and Cavadini, J.S. (2015). Fermentation of fall-oat balage over winter in northern climates. *Crop Forage Turfgrass Manage.* https://doi.org/10.2134/cftm2014.0110.

Coblentz, W.K., Coffey, K.P., and Chow, E.A. (2016). Storage characteristics, nutritive value, and fermentation characteristics of alfalfa packaged in large-round bales and wrapped in stretch plastic film after extended time delays. *J. Dairy Sci.* 99: 3497–3511.

Coblentz, W.K., Akins, M.S., Cavadini, J.S., and Jokela, W.E. (2017). Net effects of nitrogen fertilization on the nutritive value and digestibility of oat forages. *J. Dairy Sci.* 100: 1739–1750.

Collins, M. (1982). The influence of wetting on the composition of alfalfa, red clover, and birdsfoot trefoil hay. *Agron. J.* 74: 1041–1044.

Collins, M. (1985). Wetting effects on the yield and quality of legume and legume-grass hays. *Agron. J.* 77: 936–941.

Collins, M. (1989). Conditioning effects and field variation in dry matter concentration of alfalfa hay. In: *Proc. Intern. Grassl. Congr., 16th. Nice, France. 4–11 October, 1989*, vol. 2, 1005–1006. French Grassland Society.

Collins, M. (1995). Hay preservation effects on yield and quality. In: *Post-Harvest Physiology and Preservation of Forages*. CSSA Special Publication No. 22 (eds. K.J. Moore and M.A. Peterson), 67. Madison WI: American Society of Agronomy, Crop Science Society of America, Soil Science Society of America.

Collins, M. and Owens, V.N. (2003). Preservation of forage as hay and silage. In: *Forages: An Introduction to Grassland Agriculture*, 6e, vol. I (eds. R.F Barnes, C.J. Nelson, M. Collins and K.J. Moore), 443–471. Ames, IA: Iowa State University Press.

Collins, M., Paulson, W.H., Finner, M.F. et al. (1987). Moisture and storage effects on dry matter and quality losses of alfalfa in round bales. *Trans. Amer. Soc. Agric. Eng.* 30: 913–917.

Contreras-Govea, F.E. and Albrecht, K.A. (2006). Forage production and nutritive value of oat in autumn and early summer. *Crop Sci.* 46: 2382–2386.

Dewar, W.A., McDonald, P., and Whittenbury, R. (1963). The hydrolysis of grass hemicelluloses during ensilage. *J. Sci. Food Agric.* 14: 411–417.

Edmisten, K.L., Green, J.T. Jr., Mueller, J.P., and Burns, J.C. (1998). Winter annual small grain forage potential. I. Dry matter yield in relation to morphological characteristics of four small grain species at six growth stages. *Commun. Soil Sci. Plant Anal.* 29: 867–879.

Erdman, R. (1993). Silage fermentation characteristics affecting feed intake. In: *Proceedings of NRAES*

National Silage Production Conference. Syracuse, NY. 23-28 February1993. NRAES, Ithaca, NY, 210–219.

Festenstein, G.N. (1971). Carbohydrates in hay on self-heating to ignition. *J. Sci. Food Agric.* 22: 231–234.

Fiscus, E.L. and Kaufmann, M.R. (1990). The nature and movement of water in plants. In: *Irrigation of Agricultural Crops* (eds. B.A. Stewart and D.R. Nielsen), 191–241. Madison, WI: American Society of Agronomy.

Fisher, D.S., Mayland, H.F., and Burns, J.C. (1999). Variation in ruminants' preference for tall fescue hays cut either at sundown or at sunup. *J. Anim. Sci.* 77: 762–768.

Fisher, D.S., Mayland, H.F., and Burns, J.C. (2002). Variation in ruminant preference for alfalfa hays cut at sunup and sundown. *Crop Sci.* 42: 231–237.

Goering, H.K. and Adams, R.S. (1973). Frequency of heat-damaged protein in hay, hay-crop silage, and corn silage. *J. Anim. Sci.* 37: 295.

Goering, H.K., Van Soest, P.J., and Hemken, R.W. (1973). Relative susceptibility of forages to heat damage as affected by moisture, temperature, and pH. *J. Dairy Sci.* 56: 137–143.

Guilbert, H.R. and Mead, S.W. (1931). The digestibility of bur clover as affected by exposure to sunlight and rain. *Hilgardia* 6 (1): 1–12.

Harris, C.E. and Tullberg, J.N. (1980). Pathways of water loss from legumes and grasses cut for conservation. *Grass Forage Sci.* 35: 1–11.

Harrison, J.H., Huhtanen, P., and Collins, M. (2003). Grass silage. In: *Silage Science and Technology* (eds. D.R. Buxton, J.H. Harrison and R.E. Muck). (665–748). Madison, WI: ASA, CSSA, and SSSA.

Hill, J.D., Ross, I.J. and Barfield, B.J. (1976). The use of vapor pressure deficit to predict drying time for alfalfa hay. Annual Meeting of American Society of Agricultural Engineers. Paper No. 76-3040.

Hlödversson, R. and Kaspersson, A. (1986). Nutrient losses during deterioration of hay in relation to changes in biochemical composition and microbial growth. *Anim. Feed Sci. Tech.* 15: 149–165.

Holmes, B.J., and Muck, R.E. (2008). Packing bunker and pile silos to minimize porosity. Focus on Forage. Vol. 10. No. 1. University of Wisconsin Extension, Madison, WI.

Honig, H. (1979). Mechanical and respiration losses during prewilting of grass. In: *Forage Conservation in the 80's. Occasional Symp. No. 11* (ed. C. Thomas), 201. London: British Grassland Society Janssen Services.

Humphreys, M.O. (1989). Water-soluble carbohydrates in perennial ryegrass breeding. I. Genetic differences among cultivars and hybrid progeny grown as spaced plants. *Grass Forage Sci.* 44: 231–236.

Huntington, G.B. and Burns, J.C. (2007). Afternoon harvest increases readily fermentable carbohydrate concentration and voluntary intake of gamagrass and switchgrass baleage by beef steers. *J. Anim. Sci.* 85: 276–284.

Jennings, J.A. (2011). Baled silage for livestock. University of Arkansas Cooperative Extension Service No. #FSA3051-PD-4-11RV.

Jones, D.I.H. (1970). The ensiling characteristics of different herbage species and varieties. *J. Agric. Sci.* 75: 293–300.

Jones, L. (1979). The effect of stage of growth on the rate of drying of cut grass at 20 °C. *Grass Forage Sci.* 34: 139–144.

Jones, L. and Harris, C.E. (1980). *Plant and Swath Limits to Drying. Occasional Symposium No. 11*, 53–60. British Grassland Society.

Jones, L. and Prickett, J. (1981). The rate of water loss from cut grass of different species dried at 20 °C. *Grass Forage Sci.* 36: 17–23.

Jones, B.A., Muck, R.E., and Hatfield, R.D. (1995a). Screening legume forages for soluble phenols, polyphenol oxidase, and extract browning. *J. Sci. Food Agric.* 67: 109–112.

Jones, B.A., Muck, R.E., and Hatfield, R.D. (1995b). Red clover extracts inhibit legume proteolysis. *J. Sci. Food Agric.* 67: 329–333.

Kaspersson, A., Hlödversson, R., Palmgren, U., and Lindgren, S. (1984). Microbial and biochemical changes occurring during deterioration of hay and preservative effect of urea. *Swed. J. Agric. Res.* 14: 127–133.

Kemble, A.R. and Macpherson, H.T. (1954). Liberation of amino acids in perennial rye grass during wilting. *Biochem. J* 58: 46–49.

Kleinschmit, D.H. and Kung, L. Jr. (2006). A meta-analysis of effects of *Lactobacillus buchneri* on the fermentation and aerobic stability of corn and grass and small-grain silages. *J. Dairy Sci.* 89: 4005–4013.

Kohn, R.A. and Allen, M.S. (1995). Effect of plant maturity and preservation method on in vitro protein degradation of forages. *J. Dairy Sci.* 78: 1544–1551.

Kormos, J. and Chestnutt, D.M.B. (1968). (1) Measurement of dry matter losses in grass during the wilting period. (2) effects of rain, mechanical treatment, maturity of grass and some other factors. *Rec. Agric. Res. (Northern Ireland)* 27: 59–65.

Lechtenberg, V.L., Holt, D.A., and Youngberg, H.W. (1971). Diurnal variation in nonstructural carbohydrates, in vitro digestibility, and leaf to stem ratio of alfalfa. *Agron. J.* 63: 719–723.

Leibensperger, R.Y. and Pitt, R.E. (1987). A model of clostridial dominance in ensilage. *Grass Forage Sci.* 42: 297–317.

Lin, C., Bolsen, K.K., Brent, B.E. et al. (1992). Epiphytic microflora on alfalfa and whole-plant corn. *J. Dairy Sci.* 75: 2484–2493.

Lines, L.W., Koch, M.E., and Weiss, W.P. (1996). Effect of ammoniation on the chemical composition of alfalfa hay baled with varying concentrations of moisture. *J. Dairy Sci.* 79: 2000–2004.

Macdonald, A.M. and Clark, E.A. (1987). Water and quality loss during field drying of hay. In: *Advances in Agronomy*, 407–437. Academic Press.

Macpherson, H.T. (1952). Changes in nitrogen distribution in crop conservation. *J. Sci. Food Agric.* 3: 365–367.

Mahanna, W. and Chase, L.E. (2003). Practical applications and solutions to silage problems. In: *Silage Science and Technology* (eds. D.R. Buxton, J.H. Harrison and R.E. Muck), 855–895. Madison, WI: ASA, CSSA, and SSSA.

Martin, J.T. and Juniper, B.E. (1970). *The Cuticles of Plants*. New York, NY: St. Martin's Press.

McBeth, L.J., Coffey, K.P., Coblentz, W.K. et al. (2001). Impact of heating-degree-day accumulation during bermudagrass hay storage on nutrient utilization by lambs. *J. Anim. Sci.* 79: 2698–2703.

McBeth, L.J., Coffey, K.P., Coblentz, W.K. et al. (2003). Impact of heating-degree-day accumulation during storage of bermudagrass hay on in situ degradation kinetics from steers. *Anim. Feed Sci. Technol.* 108: 147–158.

McDonald, P. (1982). The effect of conservation processes on the nitrogenous components of forages. In: *Forage Protein in Ruminant Animal Production. Occasional Publication No. 6* (eds. D.J. Thompson, D.E. Beever and R.G. Gunn), 41. Haddington: British Society of Animal Production, D. & J. Croal, Ltd.

McDonald, P. and Edwards, R.A. (1976). The influence of conservation methods on digestion and utilization of forages by ruminants. *Proc. Nutr. Soc.* 35: 201–211.

McDonald, P., Henderson, A.R., and Heron, S.J.E. (1991). *The Biochemistry of Silage*, 2e. Bucks, England: Chalcombe Publication.

McKersie, B.D. (1985). Effect of pH on proteolysis in ensiled legume forage. *Agron. J.* 77: 81–86.

Merchen, N.R. and Satter, L.D. (1983). Changes in nitrogenous compounds and sites of digestion of alfalfa harvested at different moisture contents. *J. Dairy Sci.* 66: 789–801.

Messman, M.A., Weiss, W.P., and Koch, M.E. (1994). Changes in total and individual proteins during drying, ensiling, and ruminal fermentation of forages. *J. Dairy Sci.* 77: 492–500.

Miller, L.G., Clanton, D.C., Nelson, L.F., and Hoehne, O.E. (1967). Nutritive value of hay baled at various moisture contents. *J. Anim. Sci.* 64: 343–347.

Montgomery, C.R., Nelson, B.D., Morgan, E.B., and Johns, D.M. (1974). Nutritional losses in hay due to rain during curing. *La. Agric.* 17: 8–9.

Montgomery, M.J., Tineo, A., Bledsoe, B.L., and Baxter, H.D. (1986). Effect of moisture content at baling on nutritive value of alfalfa orchardgrass hay in conventional and large round bales. *J. Dairy Sci.* 69: 1847–1853.

Moore, K.J. and Hatfield, R.D. (1994). Carbohydrates and forage quality. In: *Forage Quality, Evaluation, and Utilization. 13-15 April 1994, Nat. Conf. on Forage Quality, Evaluation, and Utilization* (eds. G.C. Fahey, M. Collins, D.R. Mertens and L.E. Moser), 229–280. Madison, WI: University of Nebraska, Lincoln. ASA-CSSA-SSSA.

Muck, R.E. (1987). Dry matter level effects on alfalfa silage quality. I. Nitrogen transformations. *Trans. Amer. Soc. Agric. Eng.* 30: 7–14.

Muck, R.E. (1988). Factors influencing silage quality and their implications for management. *J. Dairy Sci.* 71: 2992–3002.

Muck, R.E. (1989). Initial bacterial numbers on lucerne prior to ensiling. *Grass Forage Sci.* 44: 19–25.

Muck, R.E. (1990). Dry matter level effects on alfalfa silage quality. II. Fermentation products and starch hydrolysis. *Trans. Amer. Soc. Agric. Eng.* 33: 373–380.

Muck, R.E. and Dickerson, J.T. (1988). Storage temperature effects on proteolysis in alfalfa silage. *Trans. Amer. Soc. Agric. Eng.* 31: 1005–1009.

Muck, R.E. and Walgenbach, R.P. (1985). Variations in alfalfa buffering capacity. American Society of Agricultural Engineering. ASAE Paper No. 85-1535.

Muck, R.E., Moser, L.E., and Pitt, R.E. (2003). Post-harvest factors affecting ensiling. In: *Silage Science and Technology* (eds. D.R. Buxton et al.), 251–304. Madison, WI: ASA, CSSA, and SSSA.

Nelson, L.F. (1966). Spontaneous heating and nutrient retention of baled alfalfa hay during storage. *Trans. Amer. Soc. Agric. Eng.* 9: 509–512.

Nelson, L.F. (1968). Spontaneous heating, gross energy retention, and nutrient retention of high-density alfalfa hay bales. *Trans. Amer. Soc. Agric. Eng.* 11: 595–600, 607.

Nicholson, J.W.G., McQueen, R.E., Charmley, E., and Bush, R.S. (1991). Forage conservation in round bales or silage: effect on ensiling characteristics and animal performance. *Can. J. Anim. Sci.* 71: 1167–1180.

NRC (2001). *Nutrient Requirements of Dairy Cattle*, 7e. Washington, DC: National Academy Press.

Owens, V.N., Albrecht, K.A., and Muck, R.E. (2002). Protein degradation and fermentation characteristics of unwilted red clover and alfalfa silage harvested at various times during the day. *Grass and Forage Sci.* 57: 329–341.

Papadopoulos, Y.A. and McKersie, B.D. (1983). A comparison of protein degradation during wilting and ensiling of six forage species. *Can. J. Plant. Sci.* 63: 903–912.

Penn State University (2017). Haylage in a day. Penn State Extension, University Park, PA.

Pitt, R.E. (1990). *Silage and Hay Preservation.* Cornell Univ. Coop. Ext. Bull. NRAES-5. Ithaca, N.Y: Northeast Regional Agricultural Engineering Service.

Pitt, R.E., Muck, R.E., and Leibensperger, R.Y. (1985). A quantitative model of the ensilage process in lactate silages. *Grass Forage Sci.* 40: 279–303.

Playne, M.J. and McDonald, P. (1966). The buffering constituents of herbage and silage. *J. Food Sci. Agric.* 17: 264–268.

Rees, D.V.H. (1982). A discussion of sources of dry matter loss during the process of haymaking. *J. Agric. Eng. Res.* 27: 469–479.

Rhein, R.T., Coblentz, W.K., Turner, J.E. et al. (2005). Aerobic stability of wheat and orchardgrass round-bale silages during winter. *J. Dairy Sci.* 88: 1815–1826.

Roberts, C.A. (1995). Microbiology of stored forages. In: *Post-Harvest Physiology and Preservation of Forages. CSSA Special Publication No. 22* (eds. K.J. Moore and M.A. Peterson), 21. Madison, WI: American Society of Agronomy, Crop Science Society of America, Soil Science Society of America.

Rotz, C.A. and Abrams, S.M. (1988). Losses and quality changes during alfalfa hay harvest and storage. *Trans. Amer. Soc. Agric. Eng.* 31: 350–355.

Rotz, C.A. and Muck, R.E. (1994). Changes in forage quality during harvest and storage. In: *Forage Quality, Evaluation, and Utilization. 13-15 April 1994, Nat. Conf. on Forage Quality, Evaluation, and Utilization* (eds. G.C. Fahey, M. Collins, D.R. Mertens and L.E. Moser), 828–868. Madison, WI: University of Nebraska, Lincoln. ASA-CSSA-SSSA.

Rotz, C.A., Davis, R.J., Buckmaster, D.R., and Allen, M.S. (1991). Preservation of alfalfa hay with propionic acid. *Appl. Eng. Agric.* 7: 33–40.

Rucker, G. and Knabe, O. (1980). Non-mechanical field losses in wilting grasses as influenced by different factors. In: *Proceedings of 13th International Grassland Congress, Leipzig. 1977*, 1379–1381.

Sauvé, A.K., Huntington, G.B., Whisnant, C.S., and Burns, J.C. (2010). Intake, digestibility, and nitrogen balance of steers fed gamagrass baleage topdressed at two rates of nitrogen and harvested at sunset and sunrise. *Crop Sci.* 50: 427–437.

Savoie, P. (1988). Hay tedding losses. *Can. Agric. Eng.* 30: 39–42.

Savoie, P. and Jofriet, J.C. (2003). Silage storage. In: *Silage Science and Technology* (eds. D.R. Buxton, R.E. Muck and J.H. Harrison), 405–468. Madison, WI: American Society of Agronomy, Crop Science Society of America, and Soil Science Society of America.

Savoie, P., Patley, E., and Dupuis, G. (1984). Interactions between grass maturity and swath width during hay drying. *Trans. Amer. Soc. Agric. Eng.* 27: 1679–1683.

Scarbrough, D.A., Coblentz, W.K., Coffey, K.P. et al. (2002). Effects of summer management and fall harvest date on ruminal in situ degradation of crude protein in stockpiled bermudagrass. *Anim. Feed Sci. Technol.* 96: 119–133.

Scarbrough, D.A., Coblentz, W.K., Humphry, J.B. et al. (2004). Estimating losses of dry matter in response to simulated rainfall for bermudagrass and orchardgrass forages using plant cell wall components as markers. *Agron. J.* 96: 1680–1687.

Scarbrough, D.A., Coblentz, W.K., Humphry, J.B. et al. (2005). Evaluation of dry matter loss, nutritive value, and in situ dry matter disappearance for wilting orchardgrass and bermudagrass forages damaged by simulated rainfall. *Agron. J.* 97: 604–614.

Scarbrough, D.A., Coblentz, W.K., Ogden, R.K. et al. (2006). Nitrogen partitioning and estimates of degradable intake protein in wilting orchardgrass and bermudagrass hays damaged by simulated rainfall. *Agron. J.* 98: 85–93.

Sheaffer, C.C. and Clark, N.A. (1975). Effects of organic preservatives on the quality of aerobically stored high moisture baled hay. *Agron. J.* 67: 660–662.

Shinners, K.J. (2003). Engineering principles of silage harvesting management. In: *Silage Science and Technology* (eds. D.R. Buxton, R.E. Muck and J.H. Harrison), 361–404. Madison, WI: American Society of Agronomy, Crop Science Society of America, and Soil Science Society of America.

Shinners, K.J. and Herzmann, M.E. (2006). Wide-swath drying and post cutting processes to hasten alfalfa drying. ASABE Annual International Meeting, Paper No. 061049.

Smith, D. (1974). Diurnal variations of nonstructural carbohydrates in the individual parts of switchgrass shoots at anthesis. *J. Range Management.* 27: 466–469.

Smith, D.M. and Brown, D.M. (1994). Rainfall-induced leaching and leaf losses from drying alfalfa forage. *Agron. J.* 86: 503–510.

Sussman, A.S. (1976). Activators of fungal spore germination. In: *The Fungal Spore. Form and Function* (eds. D.J. Weber and W.M. Hess), 101–139. New York, NY: Wiley.

Tetlow, R.M. and Fenlon, J.S. (1978). Pre-harvest desiccation of crops for conservation. 1. Effect of steam and formic acid on the moisture concentration of lucerne, ryegrass, and tall fescue before and after cutting. *J. Br. Grassl. Soc.* 33: 213–222.

Tracey, M.V. (1948). Leaf protease of tobacco and other plants. *Biochem. J* 42: 281–287.

Tremblay, G.F., Morin, C., Béllanger, G. et al. (2014). Silage fermentation of PM- and AM-cut alfalfa wilted in wide and narrow rows. *Crop Sci.* 54: 439–452.

Turner, J.E., Coblentz, W.K., Scarbrough, D.A. et al. (2002). Changes in nutritive value of bermudagrass hay during storage. *Agron. J.* 94: 109–117.

Turner, J.E., Coblentz, W.K., Scarbrough, D.A. et al. (2003). Changes in nutritive value of tall fescue hay as affected by natural rainfall and moisture concentration at baling. *Anim. Feed Sci. and Technol.* 109: 47–63.

Turner, J.E., Coblentz, W.K., Coffey, K.P. et al. (2004). Effects of natural rainfall and spontaneous heating on voluntary intake, digestibility, in situ disappearance kinetics, passage kinetics, and ruminal fermentation characteristics of tall fescue hay. *Anim. Feed Sci. Technol.* 116: 15–33.

Undersander, D. and Saxe, C. (2013). Field drying forage for hay and haylage. Focus on Forage. Vol. 12. No. 5. University of Wisconsin Extension, Madison, WI.

Van Soest, P.J. (1982). *Nutritional Ecology of the Ruminant*. Ithaca, NY: Cornell University Press.

Vanzant, E.S., Cochran, R.C., Titgemeyer, E.C. et al. (1996). In vivo and in situ measurements of forage protein degradation in beef cattle. *J. Anim. Sci.* 74: 2773–2784.

Weiss, W.P., Conrad, H.R., and Shockey, W.L. (1986). Amino acid profiles of heat-damaged grasses. *J. Dairy Sci.* 69: 1824–1836.

Williams, C.C., Froetschel, M.A., Ely, L.O., and Amos, H.E. (1995). Effects of inoculation and wilting on the preservation and utilization of wheat forage. *J. Dairy Sci.* 78: 1755–1765.

Wood, J.G.M. and Parker, J. (1971). Respiration during the drying of hay. *J. Agric. Eng. Res.* 16: 179–191.

Yan, R., Undersander, D.J., and Coblentz, W.K. (2011). Prediction of total digestible nutrient losses from heat damaged hays and haylages with NIR. *Forage and Grazinglands.* https://doi.org/10.1094/FG-2011-0208-01-RS.

Chapter 41

Hay Harvest and Storage

C. Alan Rotz, Agricultural Engineer, *USDA-Agricultural Research Service, US Dairy Forage Research Center, Madison, WI, USA*
Kevin J. Shinners, Professor, *Agricultural Engineering, University of Wisconsin, Madison, WI, USA*
Matthew Digman, Assistant Professor, *Agricultural Engineering, University of Wisconsin, River Falls, WI, USA*

Introduction

Forage is harvested and preserved as either **hay** or **silage**. Though the use of silage is growing, haymaking remains the major process for conserving forage in the US (NASS 2016). Many techniques are used to make hay. Hay is normally baled at a moisture content below 20% so that the forage is stable for long-term storage. Hay can also be successfully stored between 20% and 35% moisture when treatments are used to preserve the hay. Finally, forage can be baled at 45–55% moisture and sealed in a plastic wrap for fermentation and preservation as silage (see Chapter 42 for information on **silage preservation**).

A primary objective in haymaking is to preserve the dry matter (DM) yield and nutrient content of standing forage for later use. Hay may be stored up to one year, and sometimes longer, for animal feed or other use. Physical, biological, and chemical processes during harvest and storage (see Chapter 40) cause DM and nutrient losses. Good management and proper equipment use can help reduce losses and preserve forage quality.

Other important considerations in hay production include resource use and production cost. Resource inputs include machinery, labor, and energy. Due to the high cost of each, these resources must be used efficiently to minimize the production cost. Maintaining a sustainable animal industry requires efficient and low-cost production of forages.

Hay production processes include cutting, conditioning, **swath** manipulation, baling, handling, storage, and preservation. Each of these will be discussed, including their effects on forage quality and resource requirements.

Cutting

Forage crop cutting machines use sickle cutterbar or rotary disk cutting devices. For many years, the sickle cutterbar mower has been a reliable and relatively inexpensive cutting device (Figure 41.1). Though it is still used, this cutting device has largely been replaced by the rotary disk, which offers greater harvest capacity (Figure 41.1). With faster speeds, productivity is enhanced, and less labor and tractor time are required per ton of hay produced. Greater productivity allows forage producers to harvest quickly when weather and crop quality are suitable. However, the purchase cost per unit width is greater and repair costs may be greater as the machine ages (ASABE 2015). Disk mowers also require more power (ASABE 2015). Thus, a larger tractor may be required, and more fuel will be consumed per hour of use (Rotz and Sprott 1984). Yield

Forages: The Science of Grassland Agriculture, Volume II, Seventh Edition.
Edited by Kenneth J. Moore, Michael Collins, C. Jerry Nelson and Daren D. Redfearn.

losses due to cutting (without a conditioning device) are normally small (Table 41.1). Disk mowers may, at times, cause slightly more loss than do sickle cutterbar mowers, but under most conditions, losses are similar (Rotz and Sprott 1984; Koegel et al. 1985; Shinners et al. 1991; Rotz and Muck 1994;). All things considered, disk mowers provide efficient and cost-effective cutting of forage crops.

Conditioning

Treatments used to reduce the forage plant's resistance to moisture loss are commonly known as conditioning. Numerous mechanical, chemical, and other techniques have been evaluated, but mechanical conditioning has proven most practical for wide use. Nearly all hay mowers used today include a mechanical conditioning device. Conditioning may not be used sometimes when harvesting high-moisture forage for silage (see Chapter 42) but, is always a beneficial practice when harvesting forages for hay.

Interactions occur between the effectiveness of a conditioning treatment and the plant, swath, and environmental conditions during field curing (Iwan et al. 1993; Rotz 1995). Conditioning is most effective when the plant is the primary constraint to drying. For this to occur, a thin or fluffy swath is needed, and the weather must be favorable (see Chapter 40). The field-drying rate of hay is most highly correlated to solar radiation level with effects from ambient air temperature, humidity, and soil moisture level (Rotz 1995). Conditioning of the plant has little or no effect when other drying conditions are poor. The effectiveness of conditioning is also influenced by crop species and maturity (Rotz 1995).

Mechanical

Mechanical conditioning uses either rolls or impellers to crush, crimp or abrade forage stems to speed drying (Figure 41.1). Mechanical conditioning has consistently been shown to dry forage crops more quickly than unconditioned crops (Shinners and Herzmann 2006; Shinners et al. 2006; Shinners and Friede 2017a). Many combinations of intermeshing and non-intermeshing rubber and steel rolls can be used. A crimping device passes the crop between intermeshing, non-contacting rolls, which bend and break the stem at about 5 cm intervals. A crushing device uses intermeshing rolls that crush the stems in several locations along the stems' length. Plant moisture evaporates more easily from these splits and breaks in the epidermis. For alfalfa, roll conditioning tends to be more effective in the first harvest of spring growth than in later harvests due to a finer stem structure in late summer regrowth (Table 41.2; Rotz et al. 1987). Common roll designs provide similar improvements in drying when properly adjusted (Rotz and Sprott 1984; Shinners et al. 1991; Shinners et al. 2006). The most important adjustment is roll clearance, which should be set to slightly less than the average stem diameter (Shinners et al. 2006).

Impeller conditioners use rotating tines or brushes to scratch the plant **cuticle.** Because stems are not as broken or crushed, they remain stiffer, creating a well-aerated swath or **windrow**. This conditioner type is best suited for harvesting grass with greater throughput in high-yielding or entangled crops. Comparisons between impeller and roll designs show varied effects with a trend toward faster drying with impeller conditioning (Rotz and Sprott 1984; Greenlees et al. 2000). However, when impeller conditioners are adjusted to provide leaf loss in alfalfa similar to that obtained with properly adjusted roll conditioners, drying rates favor roll conditioning (Shinners et al. 2006).

More intensive conditioning can be obtained by passing the forage between rolls with small clearances and/or rotating at different speeds. Differential roll speed causes abrasion and tearing of the stem, which increases the rate of moisture loss, but may increase leaf loss on leafy crops like alfalfa. Intensive conditioning by passing the crop through a set of rolls with a 57% speed differential has improved drying rate of alfalfa (Morissette and Savoie 2012) and switchgrass (Shinners and Friede 2017a). In both of these studies, drying in a wide swath by **tedding** was as effective as intensive conditioning at improving drying rate. Furthermore, combining intensive conditioning and wide-swath drying produced the fastest drying rate (Shinners and Friede 2017a).

Cutting and conditioning are normally combined using a mower-conditioner (Figure 41.2). In a properly adjusted mower-conditioner, losses vary from 1% to 6% of DM yield with this loss being only 1–2% over cutting alone (Rotz and Sprott 1984; Koegel et al. 1985; Shinners et al. 1991; Greenlees et al. 2000). Impeller and roll devices adjusted for aggressive conditioning may double this loss in alfalfa (Greenlees et al. 2000). Material lost during cutting and conditioning is mostly leaves. Because leaves are higher in **crude protein** (CP) and lower in fiber than stem material, a typical loss causes a small decrease in CP and increase in fiber concentration in the remaining forage (Table 41.1; Rotz and Muck 1994).

Other Techniques

Other techniques have been evaluated to speed forage drying, but practical and economic constraints have prevented wide commercial use. The most severe form of mechanical conditioning is referred to as maceration. Plant stems are shredded removing the drying restraints imposed by the internal cell structure, epidermis, and cuticle. Macerated alfalfa can dry to a moisture suitable for baling in four to six hours with favorable drying conditions (Krutz et al. 1979; Shinners et al. 1987). Macerated forage contains many fine particles that are very susceptible to loss during field curing and harvest. To reduce potential loss, shredded material has been

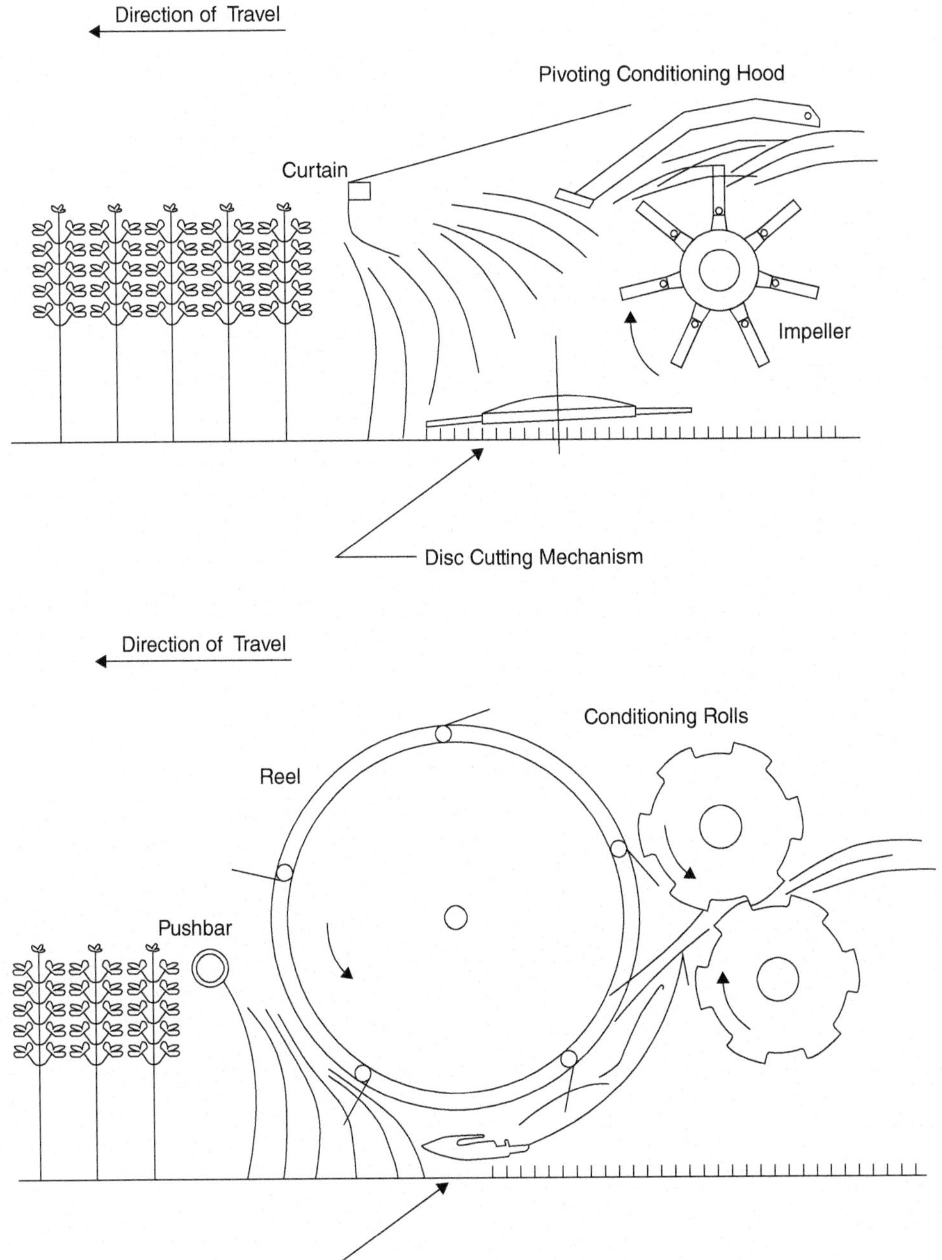

Fig. 41.1. Schematic of sickle cutterbar with intermeshing roll conditioner (bottom) and rotary disk cutting device with impeller conditioner (top) as used in a mower-conditioner (Shinners 2003).

pressed into a mat that is laid on the field surface for drying (Koegel et al. 1988; Savoie et al. 1993). Shredding also improves forage digestibility and animal intake and performance (Savoie 2001). Despite these benefits, equipment complexity has prevented commercial use.

Chemical treatments can improve drying by opening stomata, desiccating the plant prior to cutting, or by modifying the epicuticlar waxes. A number of chemical treatments have been evaluated with varying effectiveness on the drying rate of grass and legume forages (Rotz

Table 41.1 Typical dry matter losses and nutrient changes during hay harvest and storage operations

	Dry matter loss (% DM)		Change in nutrient concentration ($g\,kg^{-1}$)		
	Average	Range	CP	NDF	TDN
Legume crops					
Mowing	1	1–3	−4.0	6.0	−7.0
Mowing and conditioning	2	1–4	−7.0	12.0	−14.0
Tedding	3	2–8	−5.0	9.0	−12.0
Swath inversion	1	1–3	0.0	0.0	0.0
Raking	5	1–20	−5.0	10.0	−12.0
Baling, small square	4	2–6	−9.0	15.0	−20.0
large square	3	1–4	−7.0	10.0	−15.0
large round	6	3–9	−17.0	30.0	−40.0
Hay storage, inside	5	3–9	−7.0	21.0	−21.0
Outside	15	6–30	0.0	50.0	−70.0
Grass crops					
Mowing	1	1–2	0.0	0.0	0.0
Mowing and conditioning	1	1–2	0.0	0.0	0.0
Tedding	1	1–3	−2.0	4.0	−4.0
Swath inversion	1	1–3	0.0	0.0	0.0
Raking	5	1–20	−3.0	5.0	−6.0
Baling, small square	4	2–6	−5.0	9.0	−10.0
large square	3	1–4	−4.0	7.0	−8.0
large round	6	3–9	−10.0	18.0	−20.0
Hay storage, inside	5	3–9	−13.0	32.0	−18.0
Outside	12	5–22	0.0	80.0	−48.0

Source: Adapted from Rotz and Muck (1994).

Table 41.2 Typical field curing times for alfalfa harvested in a temperate climate of the northern US using various conditioning and swath manipulation practices

Conditioning treatment	Swath treatment	Field curing time[a] (days)			
		Cut 1	Cut 2	Cut 3	Average
None	Wide, raked	8.7	6.7	6.0	7.0
Mechanical	Narrow, none	7.1	6.9	5.7	6.6
Mechanical	Wide, raked	5.3	5.2	5.5	5.3
Mechanical	Tedded and raked	4.7	4.7	5.5	4.8

[a]25-year average drying times to baling moisture content for simulated alfalfa harvest in central Pennsylvania (Rotz 2004).

1995). The most effective treatment has been the application of aqueous solutions of potassium and sodium carbonates (Rotz and Davis 1986; Rotz et al. 1987). When sprayed on the crop as it is cut, the drying rate of a first cutting of alfalfa was increased by about 40% with up to a 120% increase for mid-summer harvests (Rotz et al. 1987; Iwan et al. 1993). The requirement for large volumes of solution, increased costs, plant discoloration, and concern for chemical residues have prevented commercial use.

Swath Manipulation

As forage dries in the field, the top of the swath, exposed to sunlight and air, dries more rapidly than the shaded bottom lying close to the moist soil. Manipulation of the swath can speed the drying process by moving the wetter material to the upper surface. Swath manipulation can also spread the hay over more field surface for greater exposure to radiant solar energy and drying air. Operations used in haymaking to manipulate the swath include tedding, raking, swath inversion, and merging.

Fig. 41.2. Rotary disk mower-conditioner cutting a grass crop. *Source:* Courtesy of John Deere Ottumwa Works.

Tedding

A common tedder design uses rotating tines to stir, spread, and fluff the swath. Tedding can be done anytime during field curing, but it is best to do so when the crop is above 40% in moisture content to reduce leaf loss. The stirring or fluffing aspect of tedding typically reduces field-curing time up to half a day (Table 41.2; Pattey et al. 1988; Savoie and Beauregard 1990a). Tedders are sometimes used to spread narrow swaths formed by the mower-conditioner over the entire field surface. When done soon after cutting, the average curing time may be reduced up to two days compared to drying in a narrow swath (Shinners and Herzmann 2006) (Table 41.2). Tedding may also allow more uniform drying throughout the swath.

Tedding fresh grass or alfalfa causes a small loss (about 1% of crop yield; Savoie 1988). Greater loss occurs in wilted crops, particularly in alfalfa with as much as 10% loss at 300 g kg^{-1} moisture. Furthermore, with 70% of the loss being higher quality leaf material, tedding can affect the quality of the remaining forage (Table 41.1; Rotz and Muck 1994). Tedding may also increase raking loss. When a light crop (less than 2.5 t DM ha^{-1}) is spread over the field surface, raking loss can be more than double that when narrower swaths are raked (Rotz and Abrams 1988).

Tedding operations increase fuel, labor, and machinery operation costs. The decision to use tedding must be made by comparing the probable loss from more time lying in the field to the known loss and cost of tedding (Rotz and Savoie 1991). If tedding is used, the best time to ted for low loss and to obtain the best drying benefit is early in the drying process, following rainfall, or early in the morning when the crop is wet with dew.

Raking

Raking is primarily used to reduce swath width for pickup, but proper timing may also improve drying. Rake designs include parallel-bar, wheel, and rotary rakes. Parallel-bar and wheel rakes are similar in that they roll swaths into narrower windrows (Figure 41.3). As the forage is rolled, the wetter material from the bottom of the swath is wrapped on the outside of the windrow. A rotary rake uses horizontal rotating tines to sweep the swath into a windrow. The sweeping action tends to create a well-aerated windrow, but substantial and consistent differences in drying have not been measured among rake designs (Savoie et al. 1982).

Raking provides a 10–20% increase in drying rate on the day of raking (Rotz 1995). Following this initial improvement, the heavy swath formed by raking can impede drying. For minimal loss and optimum drying, hay should be raked when the moisture is between 300 g kg^{-1} and 400 g kg^{-1}. Raking in the morning of the day of baling can reduce the field curing time by one or two hours. Alternatively, a narrow swath can be formed by the mower-conditioner with no raking or other manipulation during field curing. However, this slows drying, requiring up to two more days in the field. In humid climates, more time in the field increases the probability of rain damage and rewetting (Rotz and Savoie 1991; Shinners et al. 1991).

Raking causes a loss of 1–20% of crop yield (Rotz and Muck 1994). Substantial differences in loss among rake designs are not reported, but losses caused by a wheel rake were slightly greater than that of a parallel-bar rake (Buckmaster 1993). With the sweeping action of a rotary rake, more forage particles may be lodged in the stubble contributing to greater loss (Rotz and Muck 1994).

Fig. 41.3. Wheel rake (top), rotary rake (middle) and windrow merger (bottom) used to combine and narrow hay swaths for baler pickup. *Source:* Courtesy of Kuhn North America, Inc.

Raking loss increases as the crop dries, particularly below a moisture content of 300 g kg^{-1}. When the crop is spread over much of the field surface, loss increases because it is more difficult to gather. The loss consists of similar amounts of leaf and stem material (Rotz and Abrams 1988; Buckmaster 1993), so the quality of the remaining forage is conserved (Table 41.1; Rotz and Muck 1994). Leaf loss and the resulting effect on quality may be greater though when the raked crop is drier than 300 g kg^{-1} moisture (Buckmaster 1993).

Swath Inversion and Merging

Swath inverting machines provide a more-gentle method for manipulating field-curing forage. Though many machine designs are used, a pickup device normally lifts the swath onto a platform or moving belts. The swath is turned and dropped back to the soil surface inverted from its original position (Savoie and Beauregard 1990b). Exposing the wetter bottom layer of the swath increases the drying rate of alfalfa by about 15% on the day the treatment is applied, and this may reduce field-curing time by a few hours. This small improvement normally does not justify the added costs of the operation.

Windrow mergers use a similar design, but they are used to convey one or more swaths into a narrow windrow for pickup (Figure 41.3). With this pickup and conveying of the crop, rocks are less likely to be carried into the windrow. Though mergers are primarily intended for use before a forage harvester, some producers prefer them to rakes in hay production to reduce rock and dirt contamination.

Raking or merging is often a required component of haymaking. Increases in equipment, fuel, and labor resource inputs and the resulting increased costs are normally justified to reduce harvest time. By combining swaths, balers can be operated at their full capacity for a faster and more efficient harvest. With appropriate timing, these operations may also improve drying.

Baling and Handling

After the crop has dried to a suitable moisture content, the hay must be collected, compressed, and packaged for handling and storage. Most hay is baled, and many baler designs are available that produce either square or round bales. Other packages such as low-density stacks and cubes are sometimes used, but they constitute a small portion of the hay produced.

Square Bales

Square balers are often categorized by the bale size produced, defined as either small or large bales. Small bale use has been in decline, but it remains a viable package for small beef or dairy operations, equestrian markets, or uses in landscaping. Small square bales have a cross-section of 36 × 46 or 40 × 55 cm with bale density typically between 130 and 200 kg m^{-3}. Large bales are typically configured with three cross-sections: 80 × 70; 120 × 90, and 120 × 120 cm. Large bale density is greater at 180–220 kg m^{-3}. Dimensions and density of large bales are intended to better utilize the volume and weight limits of commercial trailers for long-distance transport. A baler classification system has been developed that segments balers into seven different classes based on bale cross-section, density, and maximum plunger load (Anonymous 2017).

A relatively small tractor of 25–60 kW is sufficient to power small square balers. A baling operation using small bales can require up to five people for operating equipment and handling bales, but this labor can be reduced with laborsaving devices. A bale thrower mounted at the exit of the baler can deposit bales into a trailing wagon, reducing the need for handling and stacking on the wagon. Automatic bale wagons can collect and aggregate bales creating storage stacks without hand labor.

Larger bales now dominate the dairy and commercial hay markets, especially in the western US. Large square balers offer greater harvest capacity, more dense bales, and more efficient use of labor. They can harvest four to seven times the amount of hay per hour as small square balers. To ensure a uniform, dense bale, the large balers use a pre-compression chamber to form a uniform flake before it is sent to the bale chamber for compression (Figure 41.4). A double-tie knotter system is used to restrain the dense bales. High-density large balers are now available that increase bale density by more than 30%. More power is required with a minimum tractor size of up to 200 kW recommended for the largest balers. Large tractors and special equipment are needed for lifting, transporting, and feeding these large, heavy bales.

Loss during hay baling varies between 2% and 5% of available DM; typically equally divided between pickup and chamber losses (Table 41.1; Koegel et al. 1985; Rotz and Abrams 1988; Shinners et al. 1992; Shinners et al. 1996). The smaller diameter pickup used in bottom feed balers provides gentler lifting of the hay, which can reduce pickup loss by 15% (Shinners et al. 1992). Chamber loss increases with drier hay (Shinners et al. 1996), but the bottom feed design can reduce this loss by about 15% (Shinners et al. 1992). With large square balers, loss is reduced by over 50% compared to small square balers, and the loss is less sensitive to over-dried hay (Shinners et al. 1996).

Though dry hay baled with a larger baler has low leaf loss (Shinners et al. 1996), hay customers do not accept shattered leaves even if they are captured in the bale. In arid climates, hay may be baled at night to allow dew to soften brittle leaves. Another option is to apply steam to the hay as it is baled. Steam generated by a diesel-fired boiler located between the baler and tractor is applied at the baler pick-up (Staheli 1997). The steam quickly softens the plant tissue, resulting in reduced

Fig. 41.4. Cutaway schematic of a large square hay baler illustrating the windrow pickup, cross conveyor, feeder tines, power drive, plunger, and bale chamber. *Source:* Courtesy of Deere and Company, Ottumwa, IA, US.

losses and greater bale density (Shinners and Schlesser 2014). Compared to baling at night with dew, steam rehydration has reduced alfalfa DM loss from 1.2% to 0.5% and increased bale density by more than 20%. Baler loss is mostly high-quality leaf material, so excessive loss has more effect on the quality of the remaining forage compared to other machine losses (Table 41.1; Rotz and Muck 1994).

Large Round Bales

More hay, straw, and biomass are packaged in large round bales than any other bale form. Bales produced are 120–180 cm in diameter and 120–160 cm in width with a density of 150–200 kg m^{-3}. The hay windrow is lifted by a pickup and rolled into a bale in the compression chamber (Figure 41.5). Variable and fixed chamber designs are used (Freeland and Bledsoe 1988). The dominant design is the variable chamber where belts apply constant pressure on the rolling hay to form a bale of relatively uniform density. Fixed chambers apply pressure only near bale completion, which forms a low density or soft core in the center of the bale. Round balers must stop to wrap each completed bale with twine or woven mesh wrap. Wrapping with the mesh wrap is faster than twine, increasing productivity by about 24% (Shinners et al. 2009a).

Completed bales are normally transported to the storage site with a wagon or truck, but individual bale transport can be used on small operations. When bales are not transported simultaneously with harvest, one person can perform all operations. Large round balers require more power than small square balers. Recommended minimum tractor sizes vary with baler size from 30 to 60 kW. Power requirement is related to baler design with fixed chamber balers requiring about 50% more energy (Freeland and Bledsoe 1988; ASABE 2015). Fuel and labor requirements are largely dependent on the method of transport. When bales are hauled to a storage site on wagons, fuel use is comparable to that of small bale systems and labor use is cut in half.

Losses from round balers vary among baler designs. Pickup losses would be expected to be similar to that of other baler designs, but greater loss has been reported (Shinners et al. 1996). For a variable chamber baler, chamber loss is about 40% greater than that of a small square baler. For a fixed chamber baler, the loss can be three times as much (Koegel et al. 1985). Chamber loss increases with decreasing moisture content at the time of baling (Shinners et al. 1996). Wrapping with a net wrap reduces chamber loss by 1–2% DM over the use of twine (Shinners et al. 2009a).

Baler Precutter

Both large square and round balers can be equipped with a precutter, which consists of a set of stationary knives between the pick-up and bale chamber. A helical feed rotor sweeps the crop past these knives, producing size-reduction. Typical spacing between the knives, ranges between 40 and 80 mm, though actual cut-length is typically much longer. Precutting reduces stem length to improve feed mixing and animal intake. Mixing time and mixer fuel use were reduced by up to 50% with precut alfalfa and corn **stover** round bales (Jones et al. 2010). Bale density is increased by 1–5% with the use of a precutter (Borreani and Tabacco 2006; Lötjönen and Paappanen 2013), and baler power requirement is increased by 4–7% (Trembley et al. 1997; Shinners and Friede 2017b).

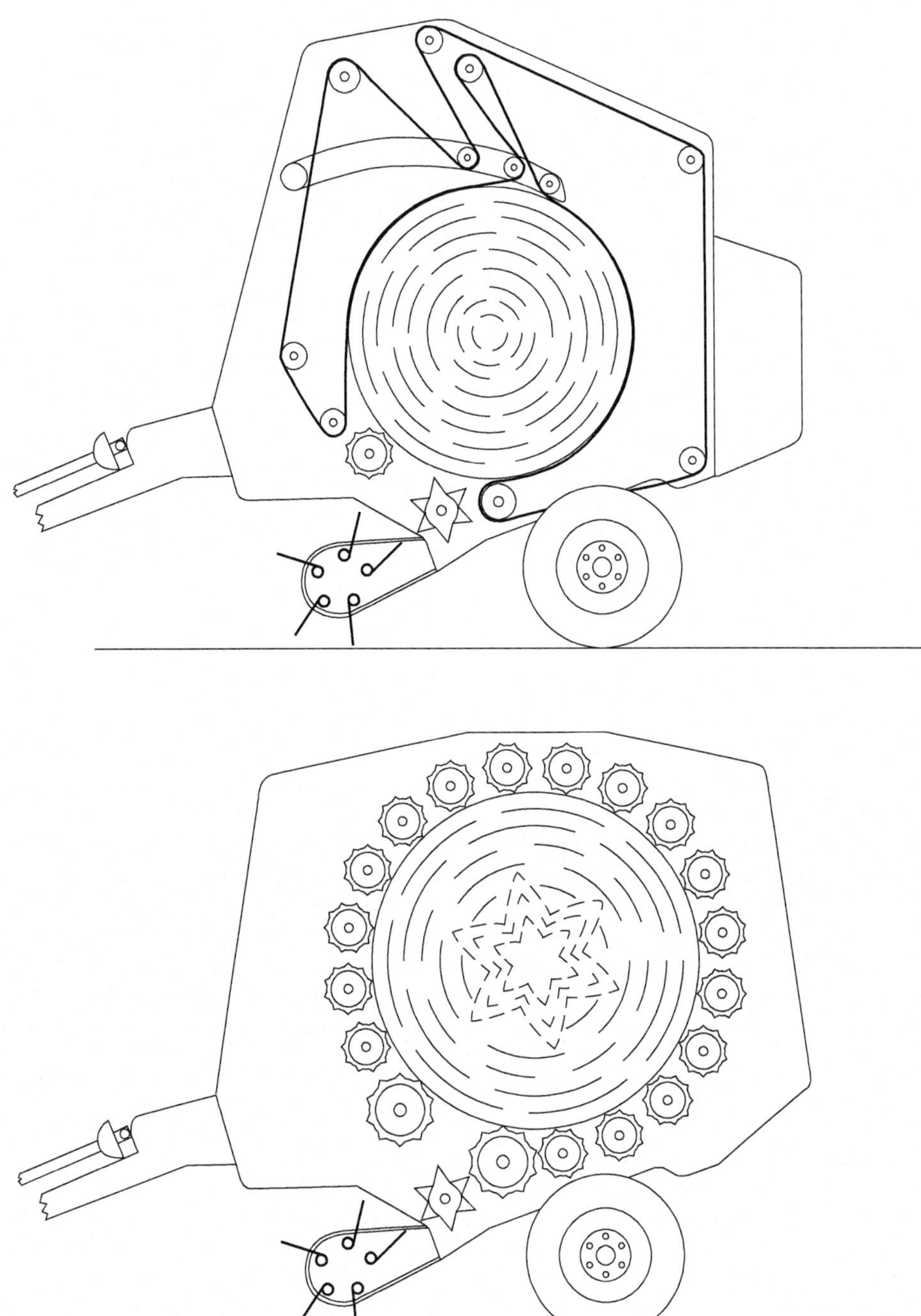

Fig. 41.5. Schematic cross-section of variable chamber (top) and fixed chamber (bottom) large round hay balers (Shinners 2003).

Yield Mapping

Precision agriculture requires continuous monitoring of crop yields during harvest. Monitoring of hay yield is challenging, but it is being done by correlating continuous moisture, weight, and position measurements during baling. Crop moisture is a crucial component. Three types of moisture sensors are used including conductance, capacitance, and microwave transmission. Each technology uses the effect of moisture content on the electromagnetic properties of hay. Variations in these properties also result from factors such as temperature and bale density, so in-field calibration is necessary.

Though bale weight is a more direct measurement than moisture, the dynamics of weighing bales on-the-go limit accurate measurement of static weight. Static bale weight estimates are improved by proprietary algorithms that employ moving averages and weighing bales during certain events in the baler (Wild and Auernhammer 1999; Shinners et al. 2000). Most of these systems require calibration and careful maintenance to ensure accurate predictions of static weight and moisture and, consequently, guide the user through the process via a calibration "wizard" or step-by-step procedure.

To build a yield map, the electronic calculations combine bale weight and moisture with time (throughput), harvest width, and spatial location. Harvested width, time, and location are determined using information from a Global Navigation Satellite System receiver. These data are formatted in an open or proprietary way that can be saved on a data card or transmitted through onboard telematics systems. Data are post-processed by computer or web-based mapping software, and they can be correlated to individual bales via radio-frequency identification (RFID) tags (Figure 41.6).

Storage

Hay is stored inside a shelter or outside with varying levels of protection from the weather. During storage, microorganisms on the hay respire, transforming DM to heat and gases, which leave the hay, causing DM and nutrient losses (see Chapter 40). In dry hay, stored under cover, respiration is low with a 4–5% loss of the stored DM. Hay stored outside experiences this loss plus additional **weathering** loss on exposed hay.

Inside Storage

Square bales are most often stored inside a shed. Round bales can be but are often not. A common shed design is a pole barn enclosed on at least three sides. A roof without sidewalls is also used, but this design provides less protection. The size of shed required depends on bale size, shape, and density and the height of the stack. A typical shed requirement for small square bales is about 1.5 m^2 of floor area for each MT DM of hay stored. With the increased density of large square bales, the requirement may be similar or greater since bales may not be stacked as high. Round bales require more space (1.8–2.5 $m^2\ t^{-1}$ DM) since they cannot be stacked as tightly or perhaps as high.

For protected hay, the major factor influencing preservation is hay moisture content entering storage (see Chapter 40). The amount of heating and associated loss during the first month increases from about 1% DM loss in hay of 150 $g\,kg^{-1}$ moisture to 8% DM loss in 300 $g\,kg^{-1}$ moisture hay (Rotz et al. 1991). As density is increased, moisture content must be reduced to maintain the same heating and loss (Buckmaster et al. 1989). Thus, high-density bales from large square balers must be baled at a lower moisture content compared to small bales (Table 41.3; Shinners et al. 1996; Shinners 2000). For small bales containing more than 250 $g\,kg^{-1}$ moisture or large dense bales with more than 20% moisture, excessive heating and spontaneous combustion can occur (see Chapter 40). After the initial heating and drying phase during the first month of storage, hay is relatively stable, but a small loss of about 0.3% DM per month continues throughout storage.

Nutritive value changes are relatively small in dry hay. Typically, changes during six months of storage do not exceed a 1–2% decrease in digestible DM with little change in protein content and a small increase in fiber content (Table 41.1). Only in high-moisture hay do substantial changes occur that affect the diet and performance of animals consuming the forage (Buckmaster et al. 1989).

Outside Storage

Round bales are often stored outside to eliminate the investment in a storage structure. Square bales are sometimes stacked outdoors, but they are normally covered with a tarp for protection. Reported DM losses in large round bales stored outside vary from 3% to 40% (Rotz and Muck 1994). Microorganisms on the hay primarily cause this loss, and biologic activity is greater when the hay is moist and warm. Therefore, loss is less in hay stored over winter periods in northern climates or in hay stored in arid climates where the hay remains relatively dry. Losses range from 0.5% to 1.5% DM/month of storage in drier climates and 1.0 to 3.0% DM/month in wetter climates. In general, the DM lost is the most digestible portion of hay, which increases fiber concentration (Table 41.1). Crude protein concentration may increase or decrease, and available protein decreases with the loss of more soluble protein (Rotz and Muck 1994).

Storage conditions affect losses by controlling exposure of the hay to moisture (Harrigan and Rotz 1994). Rain on exposed bales increases the moisture content in the outer 10–20 cm to between 25% and 40% with a small increase at the core (Harrigan et al. 1994; Shinners et al. 2009a). Increased moisture leads to greater microbial activity and loss. The center of bales stored outside is preserved similarly to hay stored in a shed, but much greater loss occurs in the outer layers exposed to rain and damp soil. Setting

Fig. 41.6. Yield mapping can be done by monitoring bale weight, moisture content, harvest width, and spatial location using a Global Navigation Satellite System receiver and bales can be labeled using radio frequency identification (RFID) tags. *Source:* Courtesy of Harvest Tec, Inc., Hudson, WI, US.

bales on rock pads or another material to break soil or sod contact reduces storage loss by about 3% DM (Rotz and Muck 1994; Shinners et al. 2009a). Normally, a tarp cover provides the best protection for a stack of bales stored outside (Table 41.3). When placed on a well-drained surface, this method produces DM and nutrient losses similar to inside storage (Harrigan et al. 1994).

A plastic cover or wrap around the circumference of the bale greatly reduces moisture accumulation, particularly when bales are elevated. This protection reduces **weathering** loss by up to 35% (Rotz and Muck 1994; Shinners et al. 2009a). Covers are more beneficial in legume hay than in predominantly grass hay (Rodriguez et al. 1994). Reported storage benefits of a woven plastic net wrap are variable but generally show benefit in humid climates (Shinners et al. 2009a). A breathable film wrap may reduce losses in round bales stored outdoors by shedding precipitation yet allowing internal moisture to exit

Table 41.3 Typical dry matter losses (%) during six months of storage for various storage methods and bale sizes

	Bale size[a]		
	Small	Medium	Large
Square bales			
Shed, 160 g kg^{-1} moisture	3.5	5.0	5.0
Shed, 200 g kg^{-1} moisture	5.0	7.0	7.0
Round bales			
Shed, 180 g kg^{-1} moisture	4.0	4.0	4.0
Covered stack	5.5	5.0	4.5
Plastic wrap, elevated	7.0	6.0	5.0
Net wrap, elevated	7.0	6.0	5.0
Twine wrap, elevated	13.0	9.0	7.0
Twine wrap, on soil	16.0	12.0	10.0

[a]Square bales are approximately 36 × 46 × 130 cm for small, 80 × 90 × 180 cm for medium, and 120 × 120 × 240 cm for large. Round bale sizes are 1.2, 1.5, and 1.8 m in diameter for small, medium, and large, respectively.

through microscopic pores. Compared to net-wrapped bales stored outdoors, the breathable film has significantly reduced losses and cattle strongly preferred this hay (Shinners et al. 2013). Covering the top of bales with an edible material such as feed grade tallow has successfully reduced DM loss (White 1997; Hand et al. 1994), but practical and economic constraints have prevented adoption.

Storage and Feeding Loss

Storage method can also affect loss during feeding. If hay is protected inside a shed, feeding loss will typically be 5% DM or less. For outside storage with little or no protection, weathered hay on the outer circumference of the bale is high in fiber content, low in digestible nutrients, and thus more likely to be rejected by animals (Harrigan and Rotz 1994). For *ad libitum* fed hay, feeding losses of up to 17% are reported with feeding loss similar to storage loss that occurred prior to feeding (Baxter et al. 1986). When hay is chopped and blended in a total mixed ration, animals have less opportunity for rejection, so feeding loss should be relatively low regardless of storage method (Harrigan et al. 1994).

Storage facilities and bale covers increase production costs. The economic return from investing in hay protection depends upon the climate during storage and how the hay is fed (Harrigan et al. 1994). If large amounts of hay are fed to high-producing dairy cows, hay quality is very important and the economic benefit for protecting the hay is high. If small amounts of hay are fed in a total mixed ration, the quality is less critical and the economic benefit of protecting hay is less. However, for long-term planning of dairy and other high-producing animal operations, shed storage is normally among the most economic options regardless of how the hay is fed.

High-Moisture Hay Preservation

Harvest losses can be reduced by baling hay at moisture contents greater than that required for stable preservation of untreated hay. Baling moist hay reduces baler chamber losses, providing a small improvement in harvested yield and quality. Raking and pickup losses may also be reduced a small amount. Field curing time on the average is reduced, which reduces the potential for rain damage. With all these factors combined, harvested yield is increased an average of 7% (Rotz et al. 1992). However, moist hay deteriorates rapidly in storage offsetting the benefit of reduced field losses unless it is treated to enhance preservation. Potential treatments include baled silage, chemical or biologic additives, and drying of hay.

Baled Silage

Baled silage (aka baleage) is an alternative method of producing forage that aims to overcome the challenges of making high-quality dry hay during difficult weather conditions. Bales are typically baled at 450–550 g kg^{-1} moisture and then wrapped with stretch plastic film to create the anaerobic conditions needed for silage fermentation (see Chapter 42). Compared to dry hay, baleage has shorter field drying time, less risk of weather damage, and fewer field losses. Baleage can be made from either large round or square bales, though bales are often made smaller to account for the greater bale weight associated with elevated moisture. Silage balers have options to reduce crop build-up on machine components and often include a precutter.

Silage bales can be wrapped individually or in a tube. Wrapping bales individually takes more time and requires more plastic film (Shinners et al. 2009b), but these bales can be moved and stored similarly to dry bales

Fig. 41.7. Large round balers can be equipped with a device that provides a sealed plastic wrap for producing bale silage with hay harvested at 45–55% moisture content. *Source:* Courtesy of Kuhn North America, Inc.

provided film integrity is maintained. Round balers can be equipped with an integrated wrapper that eliminates post-baling wrapping and this helps start the fermentation process sooner. However, bales must be handled carefully so as not to damage the film (Figure 41.7).

Additives

Materials used for the preservation of high-moisture hay have included organic acids, buffered acid mixtures, anhydrous ammonia, and microbial inoculants. Propionic and similar organic acids normally reduce mold growth and heating of high-moisture hay when applied at rates of 1–2% of hay weight (Rotz et al. 1991). Acid treatments reduce loss in moist hay during the first month of storage, but the loss remains greater than in dry hay. Due to lower heating, acid-treated hay does not dry as much during storage as untreated moist hay. The more moist environment may maintain a higher level of microbial activity even with the acid treatment (Rotz et al. 1991; Shinners 2000). Over a six-month storage period, the loss in acid-treated hay catches up, providing little difference in DM loss and nutrient changes between treated and untreated high-moisture hays.

Organic acids cause corrosion in balers and bale handling equipment. To reduce corrosion, buffered acid products are commonly used where the acid is blended with ammonia or other compatible chemical to increase the pH of the treatment. Buffered mixtures may be as effective as propionic acid when equivalent amounts of propionate are applied, but may be less effective when applied at rates below 1% of hay weight (Rotz et al. 1991).

Anhydrous ammonia is a very effective hay preservative (Rotz et al. 1986; Mir et al. 1991). Storage DM loss is reduced to the level in dry hay when hay with up to 350 $g\,kg^{-1}$ moisture is wrapped in plastic and ammonia is applied at 1% or more of its weight. The ammonia prevents heating, and it may eliminate mold development while the hay is covered. The ammonia adds non-protein nitrogen to the forage and often increases fiber-bound nitrogen. With less loss of non-structural carbohydrates, the increase in fiber content that normally occurs during storage is reduced (Rotz et al. 1986), but some hemicellulose loss may occur (Lines et al. 1996). Though anhydrous ammonia provides very effective hay preservation, animal, and human safety concerns deter its use (Lines et al. 1996; Mir et al. 1991). Urea provides a similar and safer treatment. Urea application has effectively lowered bale temperature, nonenzymatic browning, and visible mold with no adverse effects on DM intake, milk production, and digestion by dairy cows (Alhadhrami et al. 1993; Rotz et al. 1990).

Microbial inoculants are sometimes applied to moist hay. Inoculation with a few forms of lactobacillus, in combination with other bacteria and enzymes, has shown no effect on mold, color, heating, DM loss, or quality change in high-moisture hay (Rotz et al. 1988; Shinners 2000). Applications of *Pediococcus pentosaceus* and bacillus-based products have improved the visual appraisal of moist hay following storage with minor and inconsistent effects on heating of hay, forage nutritive content, and animal performance relative to untreated hay at similar moisture contents (Wittenberg 1995; Emanuele et al. 1992; Tomes et al. 1990). To obtain a response from this treatment, the added bacterial population apparently must be much greater than the natural occurring bacterial populations (Wittenberg 1995). This implies that a poor response can be expected when the crop is field-cured under poor drying conditions.

Drying

Both ambient and heated air-drying systems can stabilize moist hay for long-term storage (Plue and Bilanski 1990; Arinze et al. 1996). Energy and labor must be carefully managed to maintain acceptable production costs. Square bales are normally tightly stacked around or over air-carrying ducts (Parker et al. 1992; Rotz and Muhtar 1992; Arinze et al. 1996). Ambient unheated air forced through the stack can dry hay of up to 300 g kg^{-1} moisture over a two to three week period (Plue and Bilanski 1990; Rotz and Muhtar 1992; Misener 1993). This approach is most practical when hay is stacked and dried during long-term storage without additional handling of the hay. Drying with ambient air can greatly reduce or eliminate heating of hay and provide DM losses and nutritive changes similar to that found in hay stored with less than 200 g kg^{-1} moisture (Rotz and Muhtar 1992).

Adding heat to the air greatly reduces drying time. Hay stacks of 250–400 g kg^{-1} initial moisture content can be dried to 120–150 g kg^{-1} moisture in 17–37 hours using air temperatures around 45 °C with an energy requirement of 800–1700 MJ t^{-1} of dry hay (Arinze et al. 1996). Rapid drying can produce premium hay with little DM loss and nutritive change, but care is required to prevent brittle hay through over drying (Arinze et al. 1996).

Production Systems

On a given farm, hay production is a series of processes from cutting through storage and feeding. This production system must be developed with the goal of obtaining optimum hay quality considering end use of the forage and production cost. Many factors must be considered when developing a system for a given farm (Rotz 2001). Factors like the investment in equipment and structures, losses and nutritive changes, and the labor and fuel requirements are important considerations. Other factors like timeliness of harvest, reducing the number of machine operations, and labor availability may also be considered. Best options vary with climatic regions, crops grown, and farm management styles.

Hay for Animal Feed

Hay is primarily produced for animal feed with most used to feed beef and dairy cattle and horses. The nutritive value is important, particularly when feeding dairy cattle to maintain feed intake and production.

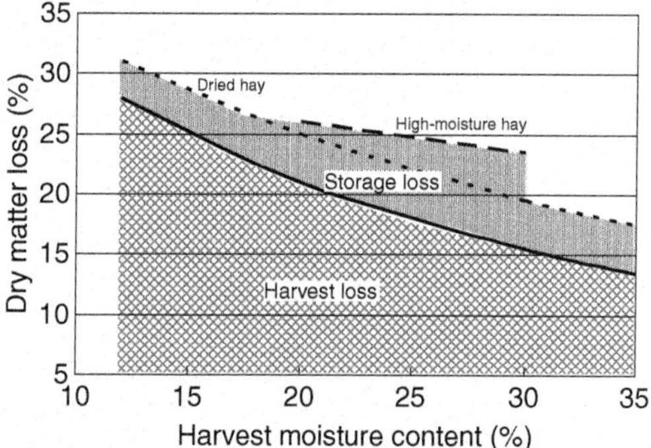

Fig. 41.8. Typical harvest and storage losses for hay harvested in small square bales or large round bales and stored in a shed. Dried hay represents that naturally dried in the field or mechanically dried in storage, and high moisture hay is not dried but treated for preservation.

Total loss in hay production is typically between 20% and 30% of initial crop DM (Figure 41.8). Harvest losses are relatively high, but storage losses are low if inside storage is used. This loss is primarily leaves and soluble carbohydrates extracted from the plant tissue. The loss typically causes a substantial increase in fiber concentration with the potential for a small loss or gain in CP concentration (Rotz and Muck 1994). These changes often increase the need for supplemental feeds and may limit animal intake and performance. The loss of feed value to animals consuming the forage is, therefore, greater than the DM loss, but this is highly dependent upon the nutrient needs of the animals fed (Buckmaster et al. 1990).

Cellulosic Biomass Feedstocks

A growing use for forage is cellulosic biomass feedstocks for use in energy production and other industrial uses (see Chapter 43). Biomass feedstock systems encompasses harvest, baling, storage, transport, and processing, typically on a very large-scale. For instance, a cellulosic **biorefinery** annually producing 114 million liters of ethanol requires over 450 000 large-square bales (Wang et al. 2017). Large-square bales are often preferred because they provide the greatest package density for transport, which has been identified as having the greatest impact on feedstock costs (Kenney et al. 2014).

Much of the equipment used to harvest hay and forage for animal feed is also used to harvest biomass crops like perennial grasses (i.e. switchgrass), annual grasses (i.e. sorghum) or crop residues (i.e. corn stover). To minimize costs, and because nutritional composition is not critical, biomass crops are typically harvested only once per year, so yields will be much greater compared to typical animal forage. High-yields can challenge cutting machine productivity (Hartley et al. 2011; Mathanker and Hansen 2015) and crop drying rate (Shinners et al. 2010; Pari et al. 2015; Shinners and Friede 2017a). Most biomass will be shipped from production sites to large, centralized biorefineries. Modifications have been made to large-square balers to increase the density of biomass bales so that weight-limited transport is achievable with most biomass crops (Shinners and Friede 2017b). The economics of harvesting crop residues is challenged by low relative yields and the many field operations needed. To reduce operations and costs, modifications have been made to grain combine harvesters to allow single-pass harvest of the crop residue with a towed baler (Webster et al. 2012; Keene et al. 2013).

References

Alhadhrami, G., Huber, J.T., Harper, J.M., and Al-Dehneh, A. (1993). Effect of addition of varying amounts of urea on preservation of high moisture alfalfa hay. *J. Dairy Sci.* 76: 1375–1386.

Anonymous (2017). Classifications for square balers. https://www.farm-equipment.com/articles/13770-agco-introduces-classification-system-for-square-balers (accessed 23 November 2019).

Arinze, E.A., Sokhansanj, S., Schoennau, G.J., and Trauttmansdorff, F.G. (1996). Experimental evaluation, simulation and optimization of a commercial heated-air batch drier: part 1, drier functional performance, product quality, and economic analysis of drying. *J. Agric. Eng. Res.* 63: 301–314.

ASABE (2015). *D497, Agricultural Machinery Management Data. ASABE Standards*, 62e. St. Joseph, MI: American Society of Agricultural and Biological Engineers.

Baxter, H.D., Bledsoe, B.L., Montgomery, M.J., and Owen, J.R. (1986). Comparison of alfalfa-orchardgrass hay stored in large round bales and conventional rectangular bales for lactating cows. *J. Dairy Sci.* 69: 1854–1864.

Borreani, G. and Tabacco, E. (2006). The effect of a baler chopping system on fermentation and losses of wrapped big bales of alfalfa. *Agron. J.* 98 (1): 1–7.

Buckmaster, D.R. (1993). Alfalfa raking losses as measured on artificial stubble. *Trans. ASAE* 36: 645–651.

Buckmaster, D.R., Rotz, C.A., and Mertens, D.R. (1989). A model of alfalfa hay storage. *Trans. ASAE* 32: 30–36.

Buckmaster, D.R., Rotz, C.A., and Black, J.R. (1990). Value of alfalfa losses on dairy farms. *Trans. ASAE* 33: 351–360.

Emanuele, S.M., Horton, G.M.J., Baldwin, J. et al. (1992). Effect of microbial inoculant on quality of alfalfa hay baled at high moisture and lamb performance. *J. Dairy Sci.* 75: 3084–3090.

Freeland, R.S. and Bledsoe, B.L. (1988). Energy required to form large round hay bales-effect of operational procedure and baler chamber type. *Trans. ASAE* 31: 63–67.

Greenlees, W.J., Hanna, H.M., Shinners, K.J. et al. (2000). A comparison of four mower conditioners on drying rate and leaf loss in alfalfa and grass. *Appl. Eng. Agric.* 16: 15–21.

Hand, R.K., Goonewardene, L.A., Winchell, W. et al. (1994). Effects of top covering and ground base on round bale hay quality over winter. *Can. J. Anim. Sci.* 74: 371–374.

Harrigan, T.M. and Rotz, C.A. (1994). Net, plastic, and twine-wrapped large round bale storage loss. *Appl. Eng. Agric.* 10: 189–194.

Harrigan, T.M., Rotz, C.A., and Black, J.R. (1994). A comparison of large round bale storage and feeding systems on dairy farms. *Appl. Eng. Agric.* 10: 479–491.

Hartley, B.E., Gibson, J.M, Thomasson, J.A. et al. (2011). Machine performance of forage harvesting equipment on high-tonnage sorghum. American Society of Agricultural and Biological Engineering. Paper No. 1111549.

Iwan, J.M., Shanahan, J.F., and Smith, D.H. (1993). Impact of environmental and harvest management variables on alfalfa forage drying and quality. *Agron. J.* 85: 216–220.

Jones, S.Q., DeRouchey, J.M., Waggoner, J.W. et al. (2010). Precutting round alfalfa and cornstalk bales decreases time and fuel required for bale breakup in a vertical mixer. *Kansas Agric. Exp. Stn. Res. Rep.* https://doi.org/10.4148/2378-5977.2882.

Keene, J.R., Shinners, K.J., Hill, L.J. et al. (2013). Single-pass baling of corn Stover. *Trans. ASABE* 56 (1): 33–40.

Kenney, K.L., Hess, J.R., Stevens, N.A. et al. (2014). Biomass logistics. In: *Bioprocessing of Renewable Resources to Commodity Bioproducts* (eds. V.S. Bisaria and A. Kondo), 29–42. Wiley.

Koegel, R.G., Straub, R.J., and Walgenbach, R.P. (1985). Quantification of mechanical losses in forage harvesting. *Trans. ASAE* 28: 1047–1051.

Koegel, R.G., Shinners, K.J., Fronczak, F.J., and Straub, R.J. (1988). Prototype for production of fast-drying forage mats. *Appl. Eng. Agric.* 4: 126–129.

Krutz, G.W., Holt, D.A., and Miller, D. (1979). For fast drying of forage crops. *Agric. Eng.* 60: 16–17.

Lines, L.W., Koch, M.E., and Weiss, W.P. (1996). Effect of ammoniation on the chemical composition of alfalfa hay baled with varying concentrations of moisture. *J. Dairy Sci.* 79: 2000–2004.

Lötjönen, T. and Paappanen, T. (2013). Bale density of reed canary grass spring harvest. *Biomass Bioenergy* 51: 53–59.

Mathanker, S.K. and Hansen, A.C. (2015). Impact of *miscanthus* yield on harvesting cost and fuel consumption. *Biomass Bioenergy* 81: 162–166.

Mir, Z., Jan, E.Z., Robertson, J.A. et al. (1991). Effects of ammoniation of brome-alfalfa and alfalfa hay, stored as large round bales on preservation and feed quality. Can. *J. Anim. Sci.* 71: 755–765.

Misener, G.C. (1993). Comparison of strategies for ambient-air drying of large round bales. *Drying Technol.* 11: 1107–1114.

Morissette, R. and Savoie, P. (2012). Effect of maceration, tedding and windrow handling sequence on hay drying. Canadian Society for Bioengineering. Paper No. NABEC/CSBE 12-035.

NASS (2016). Quick Stats, Hay and haylage production (2016). National Agricultural Statistics Service. https://quickstats.nass.usda.gov (accessed 14 October 2019).

Pari, L., Assirelli, A., Acampora, A. et al. (2015). A mower-conditioner to improve fiber sorghum plant drying. *Appl. Eng. Agric.* 31 (5): 733–740.

Parker, B.F., White, G.M., Lindley, M.R. et al. (1992). Forced-air drying of baled alfalfa hay. *Trans. ASAE* 35: 607–615.

Pattey, E., Savoie, P., and Dube, P.A. (1988). The effect of a hay tedder on the field drying rate. *Can. Agric. Eng.* 30: 43–50.

Plue, P.S. and Bilanski, W.K. (1990). On-farm drying of large round bales. *Appl. Eng. Agric.* 6: 418–421.

Rodriguez, A.A., Kaercher, M.J., and Rust, S.R. (1994). Storage method and nutritive value of large round bales harvested with different proportions of legumes. *J. Prod. Agric.* 7: 501–504.

Rotz, C.A. (1995). Field curing of forages. In: *Post-Harvest Physiology and Preservation of Forages*. Special Publication 22 (eds. K.J. Moore and M.A. Peterson), 39–66. Madison, WI: Crop Science Society of America.

Rotz, C.A. (2001). Mechanization: planning and selection of equipment. In: *Proceedings of the XIX International Grassland Congress* (ed. S.C. da Silva), 763–768, February 11-21, Sao Pedro, Sao Paulo, Brazil.

Rotz, C.A. (2004). The Integrated Farm System Model: A Tool for Developing more Economically and Environmentally Sustainable Farming Systems for the Northeast. Proceedings of the 2004 Northeast Agricultural and Biological Engineering Conference. The Pennsylvania State University (27–30 June 2004).

Rotz, C.A. and Abrams, S.M. (1988). Losses and quality changes during alfalfa hay harvest and storage. *Trans. ASAE* 31: 350–355.

Rotz, C.A. and Davis, R.J. (1986). Sprayer design for chemical conditioning of alfalfa. *Trans. ASAE* 29: 26–30.

Rotz, C.A. and Muck, R.E. (1994). Changes in forage quality during harvest and storage. In: *Forage Quality, Evaluation, and Utilization* (ed. G.C. Fahey Jr.), 828–868. Madison, WI: American Society of Agronomy.

Rotz, C.A. and Muhtar, H.A. (1992). Ambient air drying of baled hay. In: *Proceedings of the 1992 Forage and Grassland Conference American Forage and Grassland Council*, 103–107, P.O. Box 94, Georgetown, TX.

Rotz, C.A. and Savoie, P. (1991). Economics of swath manipulation during field curing of alfalfa. *Appl. Eng. Agric.* 7: 316–323.

Rotz, C.A. and Sprott, D.J. (1984). Drying rates, losses and fuel requirements for mowing and conditioning alfalfa. *Trans. ASAE* 27: 715–720.

Rotz, C.A., Sprott, D.J., Davis, R.J., and Thomas, J.W. (1986). Anhydrous ammonia injection into baled forage. *Appl. Eng. Agric.* 2: 64–69.

Rotz, C.A., Abrams, S.M., and Davis, R.J. (1987). Alfalfa drying, loss and quality as influenced by mechanical and chemical conditioning. *Trans. ASAE* 30: 630–635.

Rotz, C.A., Davis, R.J., Buckmaster, D.R., and Thomas, J.W. (1988). Bacterial inoculants for preservation of alfalfa hay. *J. Prod. Agric.* 1: 362–367.

Rotz, C.A., Thomas, J.W., Davis, R.J. et al. (1990). Preservation of alfalfa hay with urea. *Appl. Eng. Agric.* 6: 679–686.

Rotz, C.A., Davis, R.J., Buckmaster, D.R., and Allen, M.S. (1991). Preservation of alfalfa hay with propionic acid. *Appl. Eng. Agric.* 7: 33–40.

Rotz, C.A., Buckmaster, D.R., and Borton, L.R. (1992). Economic potential of preserving high moisture hay. *Appl. Eng. Agric.* 8: 315–323.

Savoie, P. (1988). Hay tedding losses. *Can. Agric. Eng.* 30: 151–154.

Savoie, P. (2001). Intensive mechanical conditioning of forages: a review. *Can. Biosyst. Eng.* 43: 2.1–2.12.

Savoie, P. and Beauregard, S. (1990a). Predicting the effects of hay swath manipulation on field drying. *Trans. ASAE* 33: 1790–1794.

Savoie, P. and Beauregard, S. (1990b). Hay windrow inversion. *Appl. Eng. Agric.* 6: 138–142.

Savoie, P., Rotz, C.A., Bucholtz, H.F., and Brook, R.C. (1982). Hay harvesting system losses and drying rates. *Trans. ASAE* 25: 581–585,589.

Savoie, P., Binet, M., Choiniere, G. et al. (1993). Development and evaluation of a large-scale forage mat maker. *Trans. ASAE* 36: 285–291.

Shinners, K.J. (2000). Evaluation of methods to improve storage characteristics of large square bales in a humid climate. *Appl. Eng. Agric.* 16: 341–350.

Shinners, K.J. (2003). Engineering principles of silage harvesting equipment. In: *Silage Science and Technology*, ASA Monograph, vol. 42 (eds. D.R. Buxton, R.E. Muck and J.H. Harrison). Madison, WI: ASA.

Shinners, K.J. and Friede, J.C. (2017a). Enhancing the drying rate of switchgrass. *Bioenergy Res.* 10 (3): 603–612.

Shinners, K.J. and Friede, J.C. (2017b). Energy requirements to produce high-density large-square biomass bales. Submitted to Energies.

Shinners, K.J. and Herzmann, M.E. (2006). Wide-swath drying and post cutting processes to hasten alfalfa drying. American Society of Agricultural and Biological Engineering. Paper No. 061049.

Shinners, K.J. and Schlesser, W.M. (2014). Reducing baler losses in arid climates by steam re-hydration. *Appl. Eng. Agric.* 30 (1): 11–16.

Shinners, K.J., Koegel, R.G., and Straub, R.J. (1987). Drying rates of macerated alfalfa mats. *Trans. ASAE* 30: 909–912.

Shinners, K.J., Koegel, R.G., and Straub, R.J. (1991). Leaf loss and drying rate of alfalfa as affected by conditioning roll type. *Appl. Eng. Agric.* 7: 46–49.

Shinners, K.J., Straub, R.J., and Koegel, R.G. (1992). Performance of two small rectangular baler configurations. *Appl. Eng. Agric.* 8: 309–313.

Shinners, K.J., Straub, R.J., Huhnke, R.L., and Undersander, D.J. (1996). Harvest and storage losses associated with mid-size rectangular bales. *Appl. Eng. Agric.* 12: 167–173.

Shinners, K.J., Barnett, N.G. and Schlesser, W.M. (2000). Measuring mass-flow-rate on forage cutting equipment. American Society of Agricultural Engineers. Paper No. 001036.

Shinners, K.J., Wuest, J.M., Cudoc, J.E. et al. (2006). Intensive conditioning of alfalfa: drying and leaf loss. American Society of Agricultural and Biological Engineering. Paper No. 061051.

Shinners, K.J., Huenink, B.M., Muck, R.E., and Albrecht, K.A. (2009a). Storage characteristics of large round alfalfa bales: dry hay bales. *Trans. ASABE* 52 (2): 409–418.

Shinners, K.J., Huenink, B.M., Muck, R.E., and Albrecht, K.A. (2009b). Storage characteristics of large round and square alfalfa bales: low-moisture silage bales. *Trans. ASABE* 52 (2): 401–407.

Shinners, K.J., Boettcher, G.C., Muck, R.E. et al. (2010). Harvest and storage of two perennial grasses as biomass feedstocks. *Trans. ASABE* 53 (2): 359–370.

Shinners, K.J., Boettcher, G.C., Schaefer, D.M., and Troutman, A.M. (2013). Cattle preference for hay from round bales with different wrap-types. *Prof. Anim. Sci.* 29: 665–670.

Staheli, D.H. (1997). Dew simulation for hay baling operations, purpose, process and control. In: *Proceedings of the 27th California Alfalfa Symposium*, 50–56. Davis, CA: Cooperative Extension, University of California Davis.

Tomes, N.J., Soderlund, S., Lamptey, J. et al. (1990). Preservation of alfalfa hay by microbial inoculation at baling. *J. Prod. Agric.* 3: 491–497.

Trembley, D., Savoie, P., and LePhal, Q. (1997). Power requirements and bale characteristics for a fixed and a variable chamber baler. *Can. Agric. Eng.* 39 (1): 73–76.

Wang, Y., Ebadian, M., Sokhansanj, S. et al. (2017). Impact of the biorefinery size on the logistics of corn stover supply–a scenario analysis. *Appl. Energy* 198: 360–376.

Webster, K., Darr, M.J. and Peyton, K.S. (2012). Single-pass baling productivity and grain logistics analysis. American Society of Agricultural and Biological Engineering. Paper No. 12-1337653.

White, J.S. (1997). Nutrient conservation of baled hay by sprayer application of feed grade fat with and without barn storage. *Anim. Feed Sci. Technol.* 65: 1–4.

Wild, K. and Auernhammer, H. (1999). A weighing system for local yield monitoring of forage crops in round balers. *Comput. Electron. Agric.* 23 (2): 119–132.

Wittenberg, K.M. (1995). Efficacy of *Pediococcus pentosaceus* for alfalfa forage exposed to precipitation during field wilting. *Can. J. Anim. Sci.* 75: 303–308.

CHAPTER

42

Silage Production

Richard E. Muck, Agricultural Engineer, *USDA-Agricultural Research Service, US Dairy Forage Research Center, Madison, WI, USA*
Limin Kung Jr., Professor, *Animal and Food Sciences, University of Delaware, Newark, DE, USA*
Michael Collins, Professor and Director Emeritus, *Division of Plant Sciences, University of Missouri, Columbia, MO, USA*

Ensiling

Silage is an important method for preservation of forages for livestock feeding in the US (Klopfenstein et al. 2013) and around the world (Wilkinson and Toivonen 2003). Preservation of forage by ensiling has been practiced for at least 3000 year (McDonald et al. 1991; Woolford 1984). Murals in Egypt, dating from 1000 to 1500 BCE, suggest that whole-plant cereal crops were preserved using ensiling. **Silos** dating from 1200 BCE have been found in the ruins of Carthage. In addition, various early manuscripts in the Mediterranean area note the importance of sealing the crop for good preservation.

Ensiling appears to have been a relatively localized phenomenon until the 1800s. In 1842, Grieswald published the first recommendations for making fresh grass silage (McDonald et al. 1991). His recommendations are still recognized as important: filling the silo rapidly, packing the crop well, and effectively sealing out air. In 1877, Goffart, a French farmer, published the first book on ensiling, detailing his experiences in making whole-plant corn silage. About one year later, an English translation was published in the US, stirring great interest in ensiling corn in North America. Ensiling was also furthered by the invention of the tower silo by FH King in Wisconsin in 1889 (Woolford 1984). By the 1900s ensiling was a common, though not dominant, means of preserving crops in both Europe and North America.

Preservation Mechanisms

Ensiling uses two primary mechanisms to preserve a moist crop: an anaerobic environment and a **fermentation** of plant sugars to lactic and other organic acids producing a low pH. An anaerobic environment is essential to prevent the growth of aerobic spoilage microorganisms (including molds, yeasts, and aerobic bacteria) because many of these microorganisms can grow at low pH (<4.0) but require oxygen. Thus, the sealing of a silo is critical to achieving and maintaining an anaerobic environment. Any oxygen remaining in the silo after sealing is usually used up by plant and microbial respiration within a few hours.

A low pH reduces the activity of plant enzymes and inhibits growth of undesirable anaerobic bacteria. Inhibition of clostridial bacteria is most critical to successful silage preservation. These bacteria produce butyric acid and amines from fermentation of sugars or lactic acid and amino acids, respectively. Such fermentations cause losses of dry matter (DM), reduce nutritive value, and reduce silage intake by ruminants.

Forages: The Science of Grassland Agriculture, Volume II, Seventh Edition.
Edited by Kenneth J. Moore, Michael Collins, C. Jerry Nelson and Daren D. Redfearn.

Generally, **lactic acid bacteria** (LAB) already present on the crop lower pH by fermenting plant sugars, primarily producing lactic acid, as well as acetic acid, ethanol, and other products. Beyond lowering pH, the lactic and acetic acids at sufficient levels are themselves inhibitory to undesirable aerobic bacteria and fungi, respectively. Natural fermentation can be assisted by inoculating the crop with selected LAB or by adding an acid to immediately reduce pH (Oliveira et al. 2017; Muck et al. 2018).

Importance of Ensiling

Farmers have two options for storing forages: ensiling or haymaking. The predominant means in a region varies by climate and, to some extent, by tradition, technology available, and use. In haymaking, the largest losses occur during harvesting, with little loss during storage if the crop is sufficiently dry. In ensiling, harvest losses are reduced, but storage losses increase. Hay is more marketable than silage, whereas the handling of silage is more easily mechanized. Countries with predominantly dry climates, such as the US and Australia, preserve most of their forage as **hay** (Table 42.1). In contrast, most northern European countries store forages as silage due to their wet climates.

In many countries, it appears that silage production is increasing relative to hay (Wilkinson and Toivonen 2003). In western Europe, silage was approximately 40% of harvested forage production in 1975, whereas today, it accounts for 67%. In the US, silage- (largely corn silage) to-hay ratios have remained constant, but legume silage production increased approximately 50% between 1984 and 2000 (Wilkinson and Toivonen 2003).

Success in ensiling crops depends on five general areas: the crop, harvest management, silo type, silo management, and **silage additives**.

Table 42.1 Estimated production of silage and hay in selected countries in 2000

Country	Hay	Silage		
		Grass	Corn	Other
		(million Mg DM)		
Australia	4.5	0.9	0.3	0.04
Austria	1.9	1.6	1.2	0.2
Belgium	0.9	1.1	2.4	na
Bulgaria	1.8	na[a]	0.3	0.1
Canada	45.0	na	2.8	4.8
Chile	0.6	1.3	na	na
Czech Republic	1.7	7.3	2.6	0.5
Denmark	0.07	0.8	0.6	0.8
Finland	0.6	1.8	na	0.02
France	22.5	6.1	16.8	5.3
Germany	2.0	8.6	14.6	3.2
Ireland	1.0	5.1	0.04	na
Italy	15.1	0.2	6.9	0.4
Japan	1.5	2.2	1.1	0.07
New Zealand	0.4	0.6	0.3	0.02
Norway	0.08	2.3	na	0.1
Poland	8.6	2.1	2.2	2.6
Slovakia	0.8	1.8	0.6	1.0
South Africa	1.5	0.3	1.9	0.8
Spain	3.1	1.7	0.7	0.2
Sweden	0.4	3.6	0.03	0.02
Switzerland	1.7	0.3	0.4	na
The Netherlands	0.3	4.3	2.9	0.07
Turkey	1.5	na	0.8	0.1
United Kingdom	2.5	9.4	1.1	0.4
United States	138.0	1.7	32.4	9.0

Source: Adapted from Wilkinson and Toivonen (2003).
[a]na = Estimate not available.

Crop Factors Influencing Ensiling

Chemical Composition

Certain crops, such as whole-crop corn, have a reputation as being easy to ensile. In contrast, alfalfa is generally viewed as being difficult to ensile. Three characteristics of a crop may explain such perceptions: **non structural carbohydrates**, **buffering capacity**, and moisture concentration.

Non structural Carbohydrates

The crop provides the substrates, primarily sugars that the LAB ferment to produce lactic acid and other products. Glucose is the most universally fermented sugar by the various species of LAB found on plants. However, all common plant monosaccharides and disaccharides can be fermented by at least some strains of LAB. Also, some plant organic acids such as citric and malic may be fermented.

Generally, the LAB found on forage plants are not able to ferment larger oligosaccharides or polysaccharides such as cellulose and hemicellulose and the storage carbohydrate starch. One exception is fructan, the storage carbohydrate in cool-season grasses. Recent research suggests that bacteria play a role in fructan hydrolysis during ensiling, and three effective *Lactobacillus* strains have been isolated (Merry et al. 1995; Winters et al. 1998). However, most LAB isolated lack this ability.

Even if LAB cannot directly use plant polysaccharides, some polysaccharides are hydrolyzed by plant enzymes during ensiling, producing monosaccharides and disaccharides that can be fermented. The most common sources are the non structural polysaccharides, starch and fructan, that are hydrolyzed by plant amylases and fructan hydrolases, respectively.

Buffering Capacity

Forage crops also contain compounds that resist pH decline. This resistance is called buffering capacity. The most common definition of buffering capacity is the meq

H^+ kg^{-1} crop DM needed to decrease pH from 6.0, the typical crop pH at ensiling, to 4.0, the final pH required for an anaerobically stable, unwilted cool-season grass silage. Unfortunately, this is not a universal definition. Typical variants are the milliequivalents of lactic acid required to reach pH 4.0 and/or the milliequivalents of acid starting from the actual crop pH at ensiling, not pH 6.0.

The buffering capacity is attributed primarily to salts of various anions such as organic acids (e.g. citric, malic, malonic), phosphates, sulfates, nitrates, and chlorides. Some amino acids also cause buffering in this pH range (Playne and McDonald 1966). In general, forages with higher mineral concentrations have higher buffering capacities.

Moisture Concentration

The moisture concentration of the crop at ensiling affects the rate and extent of fermentation. A drier crop has a higher concentration of solutes dissolved in the residual plant moisture, raising osmotic pressure. Higher osmotic pressure reduces overall microbial growth rate, raises the critical pH that is inhibitory to microbial growth, and thus reduces the quantity of sugar needed to be fermented for an anaerobically stable silage. LAB are more tolerant of high osmotic pressure than other bacteria on forage crops so that ensiling a drier crop helps inhibit **clostridia** and other undesirable bacteria while promoting the dominance of LAB. Ensiling overly wet crops increases opportunities for clostridia and enterobacteria, which may lead to excessive acid concentrations and losses. Beyond fermentation effects, crops ensiled too wet may produce effluent. Crops ensiled too dry are more prone to heating and spoilage.

Species Differences

The type of crop, crop maturity, and environmental factors affect the buffering capacity and amount of sugar at harvest. Silage crops are divided into five general groups: annual and **perennial** C_3 (cool-season) grasses, annual and perennial C_4 (warm-season) grasses, and legumes. Perennial grasses and legumes are generally ensiled at **vegetative** to boot or early bloom stages. Silages from annual cool-season grasses such as wheat, oat, and barley can be harvested at those maturities but are more often ensiled later, between the boot and soft dough stages. Silages from annual warm-season grasses such as corn and sorghum are harvested normally in the milk stage.

Buffering capacities vary considerably within and across species (Table 42.2). Legumes tend to be highest. Cool-season and warm-season grasses often have similar ranges.

Buffering capacity declines with maturity in silage crops (Muck and Walgenbach 1985; Muck et al. 1991). The sources of buffering (salts and amino acids) are diluted by the increasing levels of insoluble components (e.g. cell wall, insoluble seed carbohydrates) as the plant matures. This suggests that annual grasses, which are harvested at later reproductive stages, should require less sugar for successful fermentation than perennial grasses and legumes, which are harvested at more vegetative stages.

Buffering capacity is also affected by soil fertility and moisture stress. High fertility increases mineral uptake and buffering capacity whereas moisture stress decreases buffering capacity (Melvin 1965; Playne and McDonald 1966; Muck and Walgenbach 1985).

Cool-season grasses are highest in soluble sugars, and warm-season grasses tend to be lowest (Table 42.3). However, reported values vary widely, depending on species and environmental conditions around the time of harvest. In addition, fructan in cool-season grasses is often

Table 42.2 Buffering capacities of various forage species

		Buffering capacity (*meq kg*$^{-1}$ *DM*)	
Species	Number of samples	Range	Mean
Cool-season grass[a]			
Timothy	2	188–342	265
Orchardgrass	5	247–424	335
Italian ryegrass	11	265–589	366
Perennial ryegrass	13	257–558	380
Warm-season grass[b]			
Corn	39	148–351	260
Sweet sorghum	18	176–430	297
Sorghum × sudangrass	16	333–571	458
Pearl millet	6	315–520	393
Kikuyugrass	11	269–496	388
Legume[a]			
Alfalfa	9	390–570	472

[a]McDonald et al. (1991).
[b]Kaiser and Piltz (2002).

Table 42.3 Typical concentrations of nonstructural carbohydrates in perennial forages

	Temperate legumes	Cool-season grasses	Warm-season grasses
Category		(*g kg*$^{-1}$ DM)	
Soluble sugars	20–50	30–60	10–50
Starch	10–110	0–20	10–50
Fructan	—[a]	30–100	—

Source: Adapted from Moore and Hatfield (1994).
[a]Not present in significant quantities.

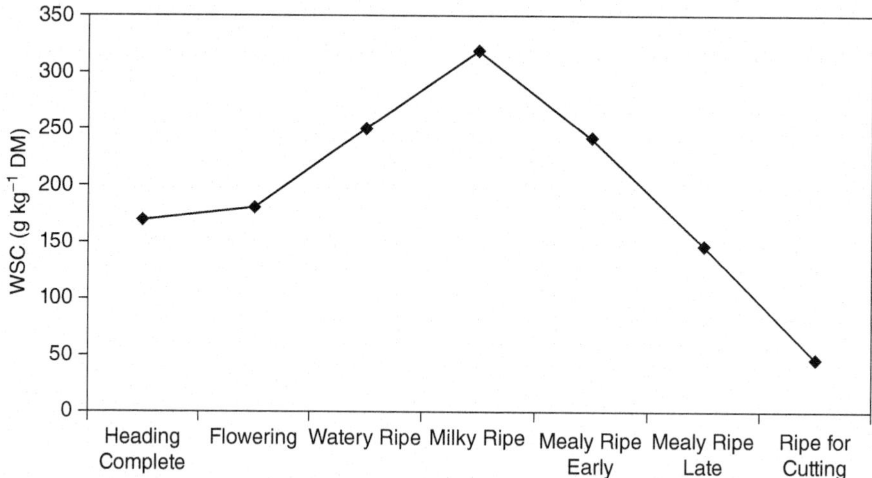

Fig. 42.1. Average WSC concentration in four cultivars of whole barley plants at different stages of maturity. *Source:* Adapted from Edwards et al. (1968).

hydrolyzed to a greater extent than starch in legumes and warm-season grasses during harvesting and ensiling. In part, this may be due to LAB fermenting fructan.

Maturity effects on sugar levels of forage crops are inconsistent. Leading up to the full-bloom stage for legume crops, water-soluble carbohydrate (WSC) concentrations tend to decline with time (Smith 1973). In contrast, WSC concentrations in cool-season grasses increase with advancing maturity (McDonald et al. 1991). In whole-crop cereals (both cool- and warm-season), WSC concentrations increase until the milk stage of seed development and then decrease as the seed develop (Figure 42.1).

Environmental factors also affect crop ensilability. High soil fertility and shade reduce WSC concentrations, whereas drought has the opposite effect (Buxton and Fales 1994). Concentrations of WSC vary diurnally, being highest in late afternoon and lowest in early morning. Higher temperatures in tropical climates negatively affect silage production by impacting growth of silage bacteria (Gulfam et al. 2017) and because WSC concentrations are generally lower in herbaceous forages grown at higher temperatures.

Whole-crop cereals are ideal crops for ensiling due to high WSC and low buffering capacities and generally attain silages with pH levels at or below 4.0. Among perennial forages, cool-season grasses are easiest to ensile, and their ease of preservation as silage increases as they mature due to increasing WSC levels and decreasing buffering capacity. Compared with cool-season grasses, generally lower WSC concentrations make warm-season perennial grasses more difficult to ensile. High buffering capacities and low WSC concentrations make legume crops the most difficult to ensile. Because sugar concentrations and buffering capacity both decline with maturity, ease of preservation of legume silage changes little with crop maturity.

A final difference among species is the moisture concentration at harvest. Most annual grasses when harvested at optimum maturity have a moisture concentration less than 700 g kg^{-1} and may be directly ensiled successfully in a wide range of silo types. Perennial forages are usually cut when the moisture concentrations in the standing crop are high (800–900 g kg^{-1}). At such moisture concentrations, most perennial forages are susceptible to clostridial fermentation unless sugar concentrations are high. Approaches to avoid clostridial fermentation in perennial forage silages depend on climate. Where 1- or 2-d windows occur frequently for field drying, perennial forages are typically mown, wilted in the field to 750 g moisture kg^{-1} or less, and then chopped and ensiled. In tropical areas, where daily showers occur and in cool, damp climates as found in northern Europe, farmers often ensile perennial forages with little or no wilting. Under these conditions, monitoring of sugar concentration to ensure it is high enough and using silage additives may be necessary to minimize clostridial activity.

Harvesting Issues

Moisture

Moisture concentration profoundly affects fermentation and subsequent nutritive value of silage because of its effects on microbial growth. In standing forage, lack of metabolizable nutrients (e.g. WSC) and dry plant surfaces

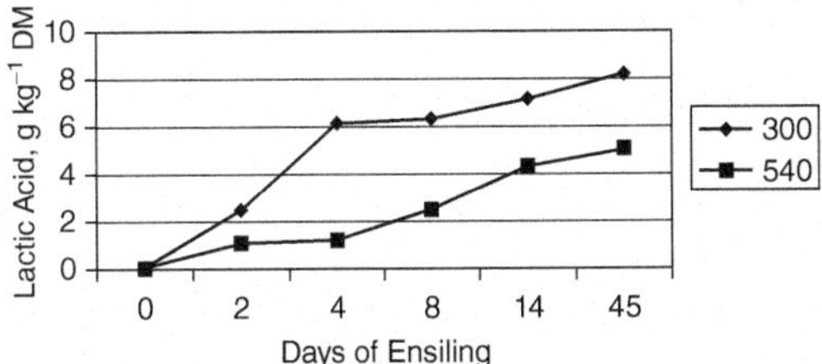

Fig. 42.2. Lactic acid production in alfalfa ensiled at two dry matter concentrations (g kg^{-1}). *Source:* Adapted from Whiter and Kung (2001).

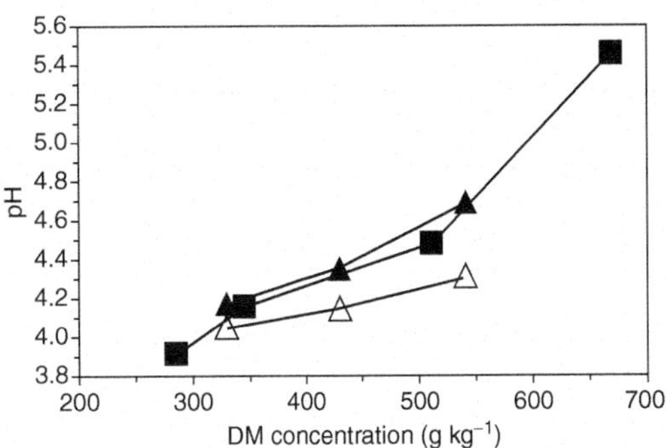

Fig. 42.3. Final pH of alfalfa ensiled over a range of dry matter concentrations with additional glucose to promote a fermentation ended by low pH. *Source:* Data from Muck (1987) *(squares)*, and Jones et al. (1992) *(triangles)*.

prevent the growth of microorganisms that are naturally found on the crop. However, chopping releases nutrients and moisture that encourage microorganisms to multiply. Assuming forage has been packed tightly and quickly to remove air, LAB on these surfaces use the moisture and WSC for growth and produce lactic acid and other products.

In forages dried to below 650 g moisture kg^{-1} (or > 350 g DM kg^{-1}) before chopping, LAB become stressed, and the rate and amount of lactic acid production are decreased (Figure 42.2). When moisture concentration is below 200–300 g kg^{-1}, bacterial-growth is completely inhibited. The net effect is that drier silages ferment less of the available sugar, form fewer fermentation products, and have higher pH values (Figure 42.3).

Conversely, excess moisture (>700 g kg^{-1}), while stimulating growth of LAB, also encourages the growth of some undesirable bacteria. Clostridia, in particular, thrive in wet conditions and can dominate the ensiling process in grasses and alfalfa, resulting in a low-quality product.

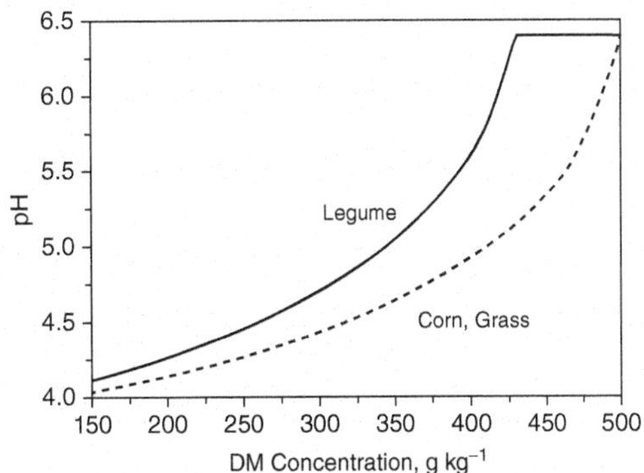

Fig. 42.4. The pH below which the growth of *Clostridium tyrobutyricum* ceases as a function of the dry matter concentration of the ensiled crop. *Source:* Based on Leibensperger and Pitt (1987).

The pH that must be achieved to avoid significant clostridial growth is lower for wetter silage (Figure 42.4). Thus, because corn silage readily ferments to pH levels below 4.0, clostridial activity is rare. In contrast, the high-buffering capacity and low sugar concentration in alfalfa typically require wilting the crop to less than 700 g moisture kg^{-1} (or >300 DM kg^{-1}) to inhibit clostridial growth.

High moisture concentrations are also undesirable because compaction in the silo may produce seepage (effluent) losses, which contain high levels of soluble nutrients. Silage effluents are as environmentally damaging to surface waters as manure slurries. The moisture concentration needed to avoid effluent varies by silo structure, ranging from less than 700 g moisture kg^{-1} for bunker and pressed bag silos to less than 550 g kg^{-1} for large tower silos.

Silage porosity, that is, the fraction of gas voids, is inversely related to its moisture concentration. More porous silages are more susceptible to spoilage by aerobic microorganisms when oxygen is present at filling and emptying or if the silo is not sealed well. The growth of aerobic microbes is accompanied by significant heat production. More heat is required to raise the temperature of water in plant matter than of gasses or plant DM, so the temperature of a wet silage is increased less than that of a dry silage for a given amount of heat production. High temperatures (>35 °C) reduce silage quality by stimulating the **Maillard browning reaction** in which amino acids are bound to carbohydrates. Excessive heating binds these amino acids irreversibly and thus decreases the availability of protein to the animal. Recommended minimum moisture concentrations for different silo types take potential spoilage and heating problems into account.

Length of Cut

The optimal particle or chop length for ensiled forage is a compromise between longer particles to meet the requirement for "physically effective **fiber**" by the animal and the need for short particles that pack well and exclude air from the silo. Lactating dairy cows need effective fiber that stimulates cud chewing and production of saliva. Saliva contains large quantities of buffers that help maintain optimum rumen pH for fiber-digesting bacteria. Chop-length settings of 10–13 mm (3/8–1/2 in.) for unprocessed corn and legume silages and 19 mm (3/4 in.) for kernel-processed corn silage have been suggested (Shaver 1993, 2003). Since those recommendations, a new **kernel** processing technology, shredlage, has been marketed. For shredlage, the chop length is set to 26–30 mm (1.0–1.2 in.). Theoretically, even longer particles would further increase saliva production, but these particles would pack poorly, especially when moisture concentration is below 600 g kg^{-1}. For very dry silages, a shorter chop length may be warranted to ensure good packing and adequate silage density.

Kernel Processing

The protective **pericarp** of intact corn kernels must be broken to provide access for rumen microbes and digestive enzymes that digest **kernel** starch. The addition of a kernel processor to a forage harvester accomplishes this task. In this process, the chopped particles are passed between two rollers set 1–3 mm apart that break the kernels. Shredlage rolls are set a similar distance apart, but the shredlage rolls are cross-grooved and run at a greater speed differential than traditional kernel processing rolls. With either technology, well-processed corn silage should have at least 95% of its kernels broken, and the cob should be broken into six or more small pieces.

Predicted improvement in starch digestion in dairy cows due to kernel processing corn silage increases with decreasing whole-plant moisture concentration as the kernel matures (Schwab et al. 2003). If whole-plant moisture exceeds 700 g kg^{-1}, processing will probably have minimal benefits and could increase seepage. Processing effects on fiber digestion have been inconsistent. Improvements in milk production from processing have been observed in some studies (Bal et al. 2000) but not in others (Weiss and Wyatt 2000), probably due to variation in the stage of corn maturity, amount of starch in the diet, forage particle size, and stage of lactation of the cows. Processing corn silage has also improved pack density and aerobic stability (Johnson et al. 2002).

With shredlage, there are expected animal effects from both the longer forage particles and the broken kernels. Ferraretto and Shaver (2012) compared shredlage silage with conventional kernel-processed silage (3 mm roll gap and a 19 mm cut length) and found that cows fed shredlage tended to consume slightly more DM and produce slightly more milk than those fed conventional corn silage. A later study (Vanderwerff et al. 2015), using **brown midrib (BMR)** shredlage corn silage, produced higher milk yields but had the same DM intake levels compared with conventional silage. Rumination time was unaffected, and there was no apparent effect on physically effective NDF for the shredlage treatment.

Temperature

The temperature of the crop at ensiling will affect the dominant microbial species, the speed of fermentation, and the products of fermentation. Laboratory silage research studies have typically been run at room temperature (20–25 °C). However, the breadth of crop temperatures at ensiling in commercial settings extend, for example, from whole-plant corn or high-moisture corn grain being ensiled frozen in northern climates to tropical grasses being ensiled at temperatures greater than 40 °C. If a crop is frozen at ensiling, fermentation will be delayed until the crop thaws in the silo. Most LAB species have an optimum temperature for growth between 27 and 38 °C (Yamamoto et al. 2011). Ensiling at cooler temperatures decreases the rate of fermentation as indicated in the study of Zhou et al. (2016) where whole-plant corn was ensiled at 5, 10, 15, 20 and 25 °C. After 7 d ensiling, the pH of the 25 °C silage was similar to its final pH (60 d) whereas the pH in the silage stored at 5 °C did not begin to decline until after d 7. At d 60, the 5 °C silage was at pH 4.3 whereas the other silages were at 4.0 or lower. The highest ratio of lactic-to-acetic acids occurred in the 5 °C silage while the 20 °C silage had the highest levels of lactic and acetic acids, and the 25 °C silage had the lowest lactic-to-acetic acid ratio. *Lactobacillus buchneri*, a heterolactic LAB, was dominant in the 20 and 25 °C silages after d 7, explaining the low lactic-to-acetic acid ratios in those silages, but was not observed in the cooler silages. Yeast levels at 60 d were below detectable levels at 20 and 25 °C whereas yeast counts were 10^4 cfu/g at the lower temperatures. Lower populations of yeasts in warmer silages (35–40 °C vs. 20 °C) have been observed in several other studies (Borreani et al. 2018). Napiergrass silage ensiled at 50 °C had much lower lactic and acetic acid levels than silage ensiled at 30 °C and had a pH of 5.3 compared with pH levels of 4.6 and 4.2 in silages ensiled at 30 and 40 °C, respectively (Gulfam et al. 2017). The authors were able to isolate heat-tolerant bacterial strains that improved fermentation at 50 °C, though only one strain performed similarly at both 40 and 50 °C.

These studies provide some observations of storage temperature effects on ensiling with different crops. At this time, few generalizations can be made except that the overall speed of fermentation increases as temperature is raised to approximately 35–40 °C. Further research is needed to better understand the effects of temperature on microbial dynamics in the silo, particularly as climate change alters ensiling temperatures.

Silo Types

Drive-over Piles and Bunker Silos

Common methods for silage production range from low-cost covered piles to permanent concrete or steel structures. The most common silos worldwide are piles placed on the ground, concrete pad, or asphalt and covered with plastic. A variant of this is the bunker silo with walls on two or three sides (Figure 42.5). Crops are commonly ensiled at 600–700 g moisture kg^{-1} in these silos. Ensiling at higher moistures than these is common in northern Europe and in tropical areas but requires facilities to collect and dispose of effluent.

The capital cost of pile silos is low, needing only plastic to seal out air. However, the large surface area increases the risk of significant spoilage losses compared with those in more permanent structures where a concrete or steel wall reduces air contact.

Losses are minimized by decreasing crop porosity, maintaining the integrity of the plastic seal, and removing silage from the face during feedout at high rates. Porosity is inversely correlated to the density and moisture concentration of the crop. Density in these types of silos is highly variable (approximately 100–400 kg DM m^{-3}) (Muck and Holmes 2000) and is determined by how the crop is packed during filling. Porosity is decreased by spreading each load thinly over the silo surface, using a heavy tractor for packing, increasing the depth of silage, and increasing packing time per unit wet weight.

The plastic must adhere tightly to the crop to minimize storage losses. In North America, used tires are commonly used to weight the plastic and create a tight seal, but sand, soil, and a wide variety of other materials can also be used. When the surface is left uncovered, spoilage losses of 40% or more occur in the top 50-cm layer of silage (Bolsen et al. 1993). Mesh tarpulins of various types are a recent development for securing plastic. These tarps are normally secured with gravel-filled bags butted against each other along the walls, seams, and across the width of silos. This not only keeps the plastic secured to the crop but also provides some additional protection against animal and hail damage.

Pressed Bag Silos

Use of pressed bags, another type of horizontal silo, is increasing in North America because of the low cost, variable capacity, and the ability to segregate silages by quality (Figure 42.6). Bags may be placed on bare ground but, unloading of bags in wet regions is easier if the bags are located on concrete or asphalt.

There are a wide variety of bagging machines and bag sizes. Nominal bag diameters are 1.8–3.6 m, and standard lengths are 30, 60, and 90 m. Bags are filled through a slot in the bagging machine by a set of rotating fingers (Figure 42.7). Both tractor-powered and self-propelled bagging machines are available. The bagging machine is pushed forward as the bag is filled. Silage density in the bag is regulated by varying the force needed to push the machine forward using external cable tension between the front and back of the bag, tractor brakes, and/or internal chains or cables.

Fig. 42.5. Typical bunker silo covered with polyethylene and used tires.

Fig. 42.6. Pressed bag silos.

Fig. 42.7. Bagging machine showing the inlet to the bags.

The goal in filling the bag is to obtain a dense but smooth bag surface surrounding the finished product. Excessive density can lead to an irregular surface on the filled bag, creating passageways for air to move back rapidly from the open face. This exposes more of the silage to oxygen soon after opening, increasing the opportunity for spoilage and heating. Silo bags commonly have parallel vertical lines (stretch marks) along the side. These help in obtaining an optimum density. If the density is too low, the distance between lines will be less than recommended. The converse is true if the distance is greater than recommended.

The sealing of a bag after filling generally provides an excellent seal. This poses a problem if there is no mechanism for carbon dioxide and other fermentation gases to escape. Commercial valves may be installed on a bag or a slit cut in a bag to allow gas escape. These need to be closed after approximately one week of ensiling.

Bag silos can produce an excellent fermentation because the crop becomes anaerobic rapidly, is protected from rainfall during filling, and should maintain the seal from oxygen exposure during storage. However, the polyethylene is the only seal and is susceptible to puncturing by birds, animals, and hail. Also, the surface-to-volume ratio is higher than for bunker or pile silos. Consequently, monitoring and patching plastic is more critical than for other horizontal silos.

Tower Silos

Tower silos are of three common types: concrete stave (Figure 42.8), poured concrete, and oxygen-limiting steel (Figure 42.9). Though these are more costly to build than other silo types, they are more permanent and are present on more than half of all dairy farms in the US (Anonymous 2002).

Filling is accomplished by blowing the crop into the top of the silo. In concrete stave silos, the unloader, located at the top of the silo, blows silage through doors located in the side of the silo and down a chute (Figure 42.8). In oxygen-limiting steel silos, the unloader is at the bottom of the silo. Poured concrete silos may be set up for either type of unloading mechanism.

Fig. 42.8. Concrete stave tower silos showing blower pipes for filling on the left side of each silo and the unloading chute for the left silo.

Fig. 42.9. Oxygen-limiting steel tower silos.

The weight of the crop being ensiled compacts material beneath it in the silo and produces the final silage density. Smaller-diameter silos have lower densities because of the greater relative contribution of wall friction. Taller silos achieve higher densities than shorter silos. Densities at the bottom of tower silos are such that the crop needs to be less than 600 to 650 g moisture kg^{-1} to avoid effluent production.

The upper surface of upright concrete silos is usually left open to the air. Spoilage may affect a 1-m depth of this loose material, and it is commonly discarded when emptying begins. The walls of older silos may need to be relined or the seals on doors fixed if substantial spoilage is evident. In oxygen-limiting silos, a breather bag at the top of the silo prevents oxygen from entering the silage, under normal storage conditions, while permitting gases in the silo to expand and contract due to diurnal heating and cooling. As this type of silo is emptied from the bottom, the silage slides down and some air enters the silo, equal to the volume of silage removed.

Wrapped Bales

The wrapping of large, round or rectangular bales with multiple layers of stretch polyethylene film is becoming more popular as an ensiling practice. It is most prevalent in Europe (Anonymous 2002; Wilkinson and Toivonen 2003). Bales may be wrapped individually or wrapped in lines end-to-end (Figure 42.10). Also, in a process similar to bag silage, round bales may be placed end-to-end in a bag.

These systems have many of the same advantages and disadvantages of pressed bag silage. Additional benefits include (i) allowing a farmer to make hay under good conditions and silage when rainy conditions prevail, and (ii) allowing silage to be bought, sold, and transported as individually wrapped bales.

Management of the plastic is essential for good preservation. A minimum of four layers of 25-μm stretch polyethylene film is needed. More is desirable for long storage periods or in warm climates to maintain plastic integrity and minimize losses in these conditions. Like pressed bag silage, monitoring for and patching holes is critical to minimize spoilage.

The long forage particles in wrapped bales do not ferment as well as chopped forage in other systems. Some balers have stationary knives to cut forage in 40- to 100-mm lengths, depending on the model, which should

Fig. 42.10. Wrapping large round bales with stretch polyethylene film using an inline wrapper to make silage.

Table 42.4 Recommended DM concentrations for ensiling and typical DM losses for different silo types

Silo type	Recommended DM range ($g\,kg^{-1}$)	Typical range of DM losses (%)	Expected DM loss, good management (%)
Drive-over pile	300–400	10–35	15
Bunker silo	300–400	8–30	12
Pressed bag	300–400	3–40	10
Concrete stave tower	350–450	5–15	10
Oxygen-limiting tower	450–550	3–12	6
Wrapped bale, individual	400–700	3–40	8
Wrapped bale, line	400–700	3–40	10

improve fermentation. Even so, wilting legumes such as alfalfa to 600 g moisture kg^{-1} or less is recommended to avoid clostridial fermentation.

Losses

Tower silos, particularly oxygen-limiting silos, are the most consistent at preserving the crop with low DM losses (Table 42.4). Wrapped bales and bag silos can produce similar results when plastic is maintained without holes. Pile and bunker silos are usually sealed less effectively than wrapped bales and bag silos so losses are typically higher, but the reduced surface-to-volume ratio of these bigger silos prevents the catastrophic losses that can occur in bags and wrapped bales.

Storage/Feeding Management Issues

Losses during storage consist of fermentation losses and microbial respiration of oxygen entering the silo. Fermentation losses (typically 1–4%) are considered unavoidable and are primarily the result of CO_2 production during fermentation of hexoses to acetic acid or ethanol. However, such losses can be reduced by bacterial inoculants as discussed later. The most significant losses during storage and emptying are losses from aerobic microbial respiration. Minimizing silage exposure to oxygen minimizes respiration losses. Prior to opening the silo, plastic quality, seal integrity and silage porosity affect respiration losses. During the emptying process, silage porosity, feedout rate, and feedout surface influence respiration losses.

Plastic Quality

Storage losses from all silos types except towers depend on plastic quality and management. For most of the twentieth century, plastic meant low-density polyethylene (LDPE). While LDPE is permeable to oxygen (178 000 cm^3 $\mu m\,m^{-2}\,d^{-1}$; Pitt 1986), oxygen permeability can be reduced to acceptable rates by increasing film thickness. The most common thicknesses range from 25 to 200 μm, with 25 μm being typical of stretch films used to wrap bales and 100–200 μm for covering bunker or pile silos. As LDPE thickness increases, Savoie (1988) calculated that losses under films would decrease: 24.4–3.2 $g\,kg^{-1}$ DM per 30 d storage for 25–200 μm, respectively. In wrapped bales, it is not possible to use the thicker films so six to eight layers of 25 μm LDPE are generally recommended (Coblentz and Akins 2018).

At the end of the twentieth century, technology advancements allowed the co-extrusion of polyethylene with other resins to form films having less than 10% of the oxygen permeability of LDPE. These films are generically called oxygen barrier films. The first commercial oxygen barrier films for silos were based on a polyamide (PA) layer surrounded by polyethylene. While the product formulations varied, these new films reduced losses. A review of 41 trials by Wilkinson and Fenlon (2014) found losses under the film were on average 195 $g\,kg^{-1}$ for LDPE films and 114 $g\,kg^{-1}$ for the oxygen barrier films. Inedible silage in the top surface was reduced by 77 $g\,kg^{-1}$ and aerobic stability increased 60 h with the oxygen barrier films.

When oxygen barrier films based on polyamide were commercialized, one issue surfaced: problems of fragility in handling some of the films (Borreani et al. 2018). This has led to the development of high-oxygen barrier films based on ethylene-vinyl alcohol copolymer (EVOH). These EVOH films are less permeable to oxygen and have better mechanical properties than the earlier oxygen barrier films. Early results have found better DM recovery, less spoiled silage and improved aerobic stability compared to PA-based films (Borreani et al. 2018).

Seal Integrity

Silos are not hermetically sealed, so some movement of oxygen into silos during storage is unavoidable. Diurnal heating and cooling cause pressure differences that expel gases from a silo during the day and draw air in at night. Wind passing over a silo creates a pressure differential between the windward and leeward sides of a silo that draws air into the silo. Also, if a plastic cover is not held tightly to the silage, the wind may cause it to act like a bellows pumping air into a silo. Polyethylene and concrete allow a slow diffusion of oxygen. After active fermentation in the silo, the gas atmosphere in silage may be 900 $ml\,l^{-1}$ or more CO_2. Because CO_2 is heavier in air, it moves downward to the bottom, where it may exit if openings allow, thus pulling outside air into the top. One or more of these factors will cause a slow continuation of respiration losses in even the best-sealed silos.

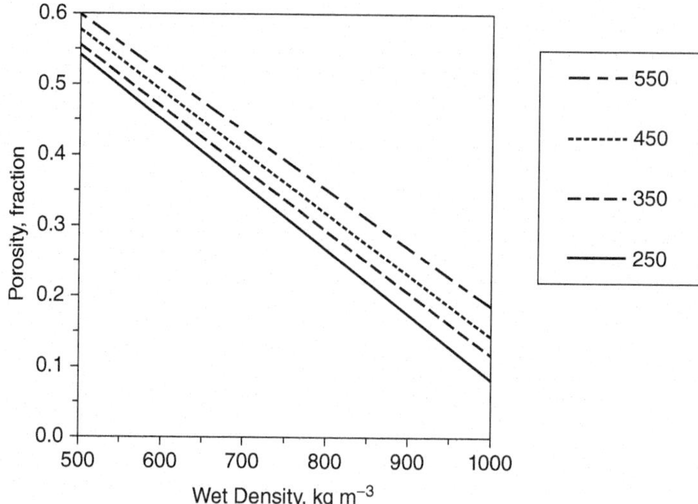

Fig. 42.11. Porosity in silage as a function of density and DM concentration (250, 350, 450, and 550 g kg^{-1}) of the silage.

Holes in plastic sheeting or cracks in silo walls allow oxygen to penetrate at a rate that is proportional to the area of the hole, the porosity of silage near the hole, and duration of the exposure. Porosity is a function of density and DM concentration of the silage (Figure 42.11). In all silo types, ensiling forage that is too dry leads to increased porosity and thus susceptibility to spoilage losses. In pile, bunker, and bag silos, packing management also determines density and subsequent effects on respiration losses when holes occur.

Feedout Rate

When the silo is opened, oxygen is present at the open face and diffuses into the silage from the face. In bunker silos with above average densities, measurable oxygen concentrations have been observed 1 m back from the face in several studies (Honig 1991; Weinberg and Ashbell 1994). Typical feedout recommendations in the northern US for bunker silos are 15 cm d^{-1}. At that rate, silage would be exposed to oxygen for almost one week before removal. As gas measurements have not been made in other silo types, recommended feedout rates are inversely proportional to the average density among silo types suggesting that seven or more days of oxygen exposure are typical prior to feeding in all silo types except for individually wrapped bales that are used when opened.

The effect of the feedout rate on respiration losses in silage near the face has not been measured directly. Modeling of microbial respiration at the silo face indicates a nonlinear relationship between losses and feedout rate (Figure 42.12). This suggests that substantial losses can occur during silo emptying when feedout rates are low.

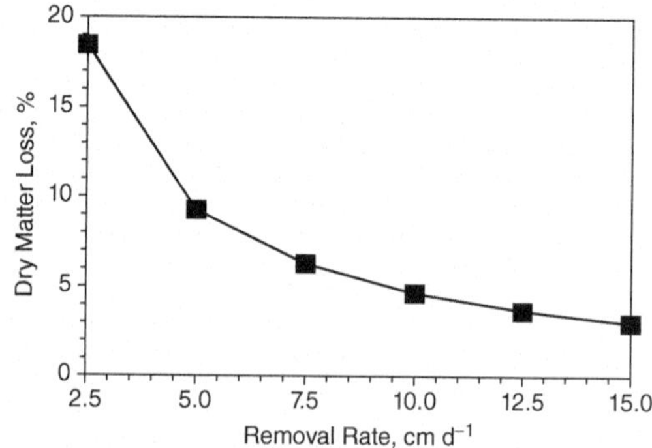

Fig. 42.12. Simulated DM loss during emptying as affected by removal rate from a bunker silo. A 350 g DM kg^{-1} corn silage at a density of 640 kg silage m^{-3} was assumed. *Source:* Adapted from Pitt and Muck (1993).

Much circumstantial evidence indicates that low feedout rates lead to heating and excessive spoilage of silages.

Feedout Surface

Tower silos are emptied with specialized unloaders that leave a smooth feedout face, but this is not necessarily the case with pile, bunker, or bag silos. In North America, front-mounted buckets on tractors, skid-steers, or industrial loaders are used most frequently to unload these silos often creating a ragged face and may open seams for more rapid oxygen ingress from the open face.

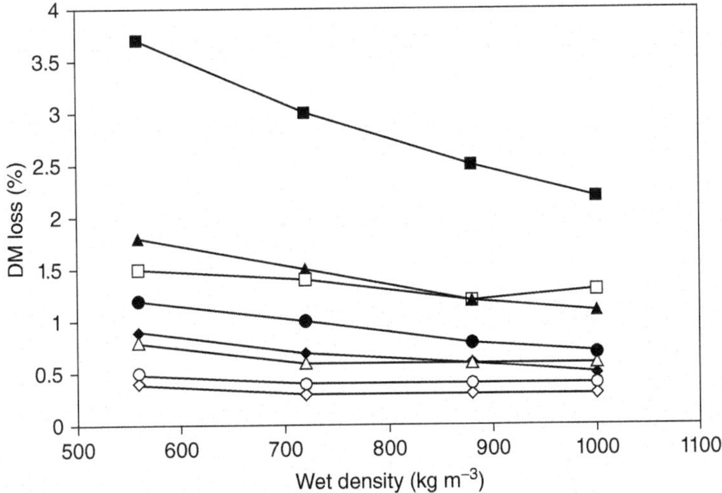

Fig. 42.13. Predicted 25-yr average difference in DM loss between silo unloaders (bucket method loss minus milling type loss) as affected by silage density, unloading rate (cm d^{-1} = 5, *square*; 10, *triangle*; 15, *circle*; 20, *diamond*), and crop (alfalfa = *filled symbol*; corn = *open symbol*). *Source:* Adapted from Muck and Rotz (1996).

Various specialized bunker silo unloaders are commercially available: block cutters, milling devices, grab buckets, etc. A milling device was found to reduce the surface area on the face of a bunker silo by 9% in corn silage and 26% in alfalfa silage compared with that on a well-managed skid-steer face (Muck and Huhnke 1995). This device also reduced oxygen concentration in the silage behind the face (up to 1 m) by 12–22 ml l^{-1} compared with a conventional bucket unloader. The effect in alfalfa was greater than in corn silage because the greater proportion of long particles in alfalfa silage made it very difficult to make a smooth face with a bucket unloader. Extending those results, Muck and Rotz (1996) predicted that a milling device would provide modest but significant reductions in DM loss with a greater response for low-density silages or slow feedout rates (Figure 42.13).

Additives

Fermentation in the silo is often a relatively uncontrolled process leading to less than optimal preservation of nutrients. Silage additives may improve silage fermentation and/or aerobic stability during feedout. Some common reasons for using additives during the ensiling process are to:

- Inhibit growth of aerobic microorganisms (especially those associated with aerobic instability, such as lactate-assimilating yeasts, and poor hygiene, such as *Listeria monocytogenes*)
- Inhibit growth of undesirable anaerobic organisms (e.g. enterobacteria and clostridia)
- Inhibit activity of plant and microbial proteases and deaminases
- Improve the supply of fermentable substrates for LAB.
- Add beneficial microorganisms to dominate fermentation
- Supply or release nutrients to stimulate growth of beneficial microorganisms
- Alter ensiling conditions to optimize fermentation (e.g. absorbents)
- Form beneficial end products that stimulate animal intake and productivity
- Improve nutrient and DM recovery

Inoculants

Many bacteria, generally LAB, have been used as microbial inoculants to improve silage fermentation. The effects on silage characteristics will vary by the strain(s) of bacteria used in an inoculant. In the twentieth century, most species in silage inoculants (e.g. *Lactobacillus plantarum*, *Pediococcus* spp.) were homolactic LAB. Homolactic bacteria produce only lactic acid from glucose fermentation. This fermentation via the Embden-Meyerhof-Parnas pathway is desirable because it yields high recoveries of energy (99.3%) and DM (100%) and converts all of the glucose into lactic acid, a relatively strong organic acid (McDonald et al. 1991). In contrast, heterolactic LAB produce multiple end products including lactic acid, ethanol, acetic acid, and CO_2, because these organisms lack the enzyme fructose-diphosphate aldolase. Energy recoveries are still high (≥98%), but DM recoveries are reduced (≥76%). Today, however, many homolactic inoculant species have been reclassified as facultative heterolactics. The facultative heterolactic LAB ferment glucose solely to lactic acid like homolactics, but in contrast to obligate homolactics they possess phosphoketolase, which allows them to ferment pentoses, producing lactic and acetic acids.

Today, there are three classes of inoculants: homolactic and/or facultative heterolactic LAB, obligate heterolactic LAB, and combination products containing obligate heterolactic LAB plus facultative heterolactic and/or homolactic LAB. Each class addresses different issues that may be of concern to the producer.

The primary goal of the homolactic inoculant (whether homolactic or fermentative heterolactic species are used) is to guarantee a fast, efficient fermentation in the silo. Some of these inoculants contain multiple species of LAB to take advantage of potential synergistic actions. In general, populations of enterococci and pediococci grow faster than the lactobacilli when pH is high (>5.0), and oxygen is present. However, below pH 5.0, populations of *Enterococcus* species decrease sharply relative to species such as *L. plantarum* and *Pediococcus pentosaceus* (Bolsen et al. 1992b; Lin et al. 1992). Thus, *Enterococcus* species alone are generally unable to improve silage quality (Cai et al. 1999). Pediococci are also common inoculant species because of their tolerance of low moisture conditions.

When effective, inoculation with homolactic LAB results in a faster rate of fermentation, less proteolysis, more lactic acid, less acetic, propionic, and butyric acids, less ethanol, a lower pH, and greater recovery of energy and DM. These benefits primarily come from the inoculant bacteria overwhelming the natural LAB, making for a more efficient conversion of sugars to lactic acid. Less proteolysis results because clostridia, enterobacteria, and plant proteases are inhibited by rapid acidification. Inhibition of clostridia also reduces butyric acid production. A meta-analysis of 130 articles indicated that the effectiveness of these inoculants on silage quality varied by crop (Oliveira et al. 2017). Inoculation reduced silage pH in temperate and tropical grasses and legumes but not in corn, sorghum, and sugarcane. Acetic acid was reduced in all silage crops except legumes. DM recovery was 2.8 percentage units higher in inoculated grass silages compared to untreated whereas no benefit was found in corn and sorghum silages. In sugarcane silages, these inoculants reduced DM recovery by 2.4 percentage units.

Beyond improving silage fermentation, homolactic LAB inoculants have also improved animal performance. Kung and Muck (1997) summarized reports indicating positive effects of inoculants on intake, gain, and milk production. Where milk production benefited, the average increase was 1.4 kg d^{-1} cow^{-1}. Summarizing their research results Bolsen et al. (1992a) reported that inoculants improved feed efficiency by 1.8%, and steers gained an additional 1.6 kg body weight Mg^{-1} crop ensiled. A recent meta-analysis of 31 lactating dairy cattle studies found inoculation increased raw milk production by 0.37 kg d^{-1} cow^{-1} with trends for increased DM intake, milk fat and milk protein but no effect on feed efficiency (Oliveira et al. 2017). Forage type did not affect the cattle response to inoculated silage.

The reasons for improved animal performance from these inoculants are not clear. Hypotheses include the inhibition of detrimental microorganisms such as yeasts, molds and *Listeria* as well as potential probiotic effects. Ruminal *in vitro* studies comparing treated and untreated silages have found the treated silage reduced methane production in one study (Jalc et al. 2009) and increased rumen microbial biomass in another (Contreras-Govea et al. 2011). More recently, Muck et al. (2012) found that *L. plantarum* inoculation of alfalfa silage increased milk production and decreased milk urea, suggesting that the treatment increased rumen microbial biomass. These studies provide evidence of altered rumen fermentation in a direction supporting higher milk production but still do not explain what factor(s) in the inoculated silage is causing the animal/rumen response.

Homolactic LAB have not generally been successful in inhibiting yeasts that cause aerobic spoilage because lactic acid itself has poor antifungal characteristics. This lack of success has led to other species appearing in inoculants. For example, *Propionibacteria* are able to convert lactic acid and glucose to acetic and propionic acids that are more inhibitory to yeasts and molds than lactic acid. However, few published studies have shown improved aerobic stability from addition of these bacteria (Flores-Galaraza et al. 1992; Dawson et al. 1998), probably because *Propionibacteria* are strict anaerobes, grow slowly, and are relatively acid intolerant.

Since the beginning of the twenty-first century, *L. buchneri*, an obligate heterolactic LAB, has been marketed widely as an inoculant for improving the aerobic stability of silages. This organism converts lactic acid to acetic acid, 1,2-propanediol, and ethanol under anaerobic conditions when the pH is low (Elferink et al. 2001). Increased aerobic stability has been reported in a variety of silages treated with *L. buchneri*, both at laboratory and fieldscale (e.g. Kung and Ranjit 2001; Kung et al. 2003; Muck 2004; Kristensen et al. 2010; Tabacco et al. 2011; Queiroz et al. 2013), and is thought to be due to the inhibitory effect of increased acetic acid levels on yeasts. Increases in aerobic stability are strain specific (Muck 2004) and dose dependent (Ranjit and Kung 2000). A meta-analysis of laboratory studies (Kleinschmit and Kung 2006) found aerobic stability in corn silage was

25 hours for untreated, 35 hours for silage treated with *L. buchneri* at 100 000 cfu g^{-1} fresh forage or less, and 503 hours when treated at >100 000 cfu g^{-1}.

While acetic acid production is beneficial for inhibiting yeasts and molds, production of acetic acid from lactic acid results in CO_2 production and thus a loss of DM. Also, high acetic acid concentrations have raised concerns about potential negative effects on intake and animal performance. The meta-analysis of Kleinschmit and Kung (2006) did find reduced DM recovery of approximately 1 percentage unit in laboratory silos, reflecting additional fermentation loss. However, at field-scale, one would expect the increased fermentation loss to be more than offset by reduced respiration losses from less aerobic microbial activity. A summary of animal trials in the review of Muck et al. (2018) reported no effect of *L. buchneri* treatment on DM intake compared to untreated silage. Kristensen et al. (2010) summarized results from 39 farms, finding no detrimental effects of *L. buchneri* inoculation on intake, milk production, health or reproduction. Overall, *L. buchneri* inoculant strains appear to have no effect on animal intake and performance and likely a positive effect on DM recovery at farm-scale.

Because the homolactic inoculant species and *L. buchneri* have different roles in improving silages, combination inoculants containing both types of strains have been studied and are now being marketed. The goal is to gain the rapid domination of silage fermentation and improvements in animal performance of the homolactic strains while adding the aerobic stability improvements of *L. buchneri*.

The most consistent effect observed in these combination inoculants has been an improvement in aerobic stability. For example, Schmidt and Kung (2010) found that corn silage treated with *L. buchneri* with or without *P. pentosaceus* had increased acetic acid concentrations, reduced yeast numbers, and improved aerobic stability (44 hours for control silage vs 136 hours for *L. buchneri* silage). However, results were not consistent across the five locations used in this study. In a recent review (Muck et al. 2018), most studies on combination inoculants reported improved aerobic stability compared to untreated. However, there were some exceptions such as Arriola et al. (2011), who found similar populations of *L. buchneri* in both treated and untreated corn silages and suggested that this may have contributed to the absence of an inoculant effect in this study.

Evidence of the effects of the homolactic strain(s) in these combination inoculants is less well documented. In the earliest study of a combination inoculant (*L. buchneri* plus, *L. plantarum*, and *P. pentosaceus*), Driehuis et al. (2001) reported that the combination inoculant had a rapid pH decline over the first 14 d of ensiling similar to the treatment with *L. plantarum* and *P. pentosaceus*. Only later in ensiling was the presence of *L. buchneri* in the combination inoculant observed by increases in acetic acid. Thus, the combination inoculant did work as expected with the homolactic LAB dominating early fermentation and *L. buchneri* converting lactic acid to acetic after the end of normal silage fermentation. Reich and Kung (2010) investigated three homolactic strains for pairing with *L. buchneri* (*Pediococcus acidilactici*, *P. pentosaceus*, and *L. plantarum*). All three pairs improved aerobic stability compared to that of the untreated silage. The pairs with the *Pediococcus* strains had the lowest ethanol concentrations whereas the pairs with *P. acidilactici* and *L. plantarum* had the highest DM recoveries. These results suggest the homolactic strains were influencing early fermentation though not uniformly across the three inoculant treatments.

Effects of combination inoculants on animal production are less certain because few studies have been published. *In situ* and *in vitro* ruminal assays have not provided consistent results but indicate that improvements in fiber digestibility are possible (Reich and Kung 2010; Muck et al. 2018). Currently, it is not certain whether this is a matter of selecting the right strains or some other issue.

The development of new inoculants is expected to continue. There are likely strains yet to be discovered that will better inhibit detrimental microorganisms, increase aerobic stability and improve silage utilization in livestock.

With inoculants, the producer needs to remember that these microorganisms are only effective if delivered alive to the crop. Storing the inoculant appropriately and proper application are important. Windle and Kung (2016) stressed the importance of water temperature, time in the application tank, and water pH in affecting the numbers of LAB present on treated silage. The authors recommended that application tank water should be kept below 35 °C. It has also been suggested that non-chlorinated water be used when applying inoculants. Inoculants must also be mixed thoroughly with the water carrier so there is even dispersal throughout the forage mass.

Enzymes

A variety of enzymes, particularly those breaking down plant fiber and starch, have been used as silage additives. Plant fiber–digesting enzymes (cellulases and hemicellulases) are the most widely used enzyme additives. Pectinases, cellobiase, amylases, and glucose oxidase are others that have been included in additives. Today, enzymes are generally found in combination with inoculants rather than as standalone additives.

Fiber-digesting enzymes could provide additional substrate for fermentation by partially hydrolyzing plant cell walls (cellulose and hemicellulose) to produce soluble sugars. This would be particularly advantageous for perennial forages where pH might not otherwise be low enough to prevent clostridial activity. However, the rate of cellulose hydrolysis must be sufficiently fast to provide sugars while the LAB are still actively growing.

Partial digestion of the plant cell wall may also improve rate and/or extent of DM digestibility in the ruminant. For an improvement in digestibility, a change in the association of various cell wall components must occur.

Cell wall–degrading enzymes have been shown to hydrolyze cellulose and hemicellulose in trials (Muck and Kung 1997; Muck et al. 2018). This has helped to lower pH where substrate limited fermentation. These enzymes have been less successful in improving digestibility and animal performance than might be expected (Kung and Muck 1997; Muck et al. 2018).

A new approach has been to select an LAB strain that produces cell-wall degrading enzyme activity. A strain of *L. buchneri* was found that produces ferulic acid esterase (Nsereko et al. 2008). This enzyme breaks ferulic acid-sugar ester linkages in plant cell walls, bonds that limit plant cell wall digestion in the rumen. Inoculants with this strain are on the market. While a promising approach, published research to date has shown mixed results, and the reasons for the variability are not clear as yet (Muck et al. 2018).

Nonprotein Nitrogen

Both ammonia and urea have been used as silage additives, particularly to improve corn silage quality. Ammonia has been applied as anhydrous ammonia or in mixtures with water or molasses. Ammonia additions have resulted in (i) addition of an economic source of crude protein (Huber et al. 1979); (ii) reduced heating and spoilage during storage and feeding (Britt and Huber 1975); and (iii) decreased protein degradation in the silo (Johnson et al. 1982). Urea has also been added to corn silage (5–6 kg Mg^{-1}) as an economical source of crude protein. However, beneficial effects of urea on aerobic stability and proteolysis have not been well substantiated. Whenever ammonia or urea is added to the diet, special attention should be made to ensure that degradable and undegradable protein requirements are balanced for the target ruminant animal.

Application of anhydrous ammonia should be at 8–10 kg N Mg^{-1} forage DM. This will increase the crude protein concentration in corn silage by 50–60 g kg^{-1} DM. Excess ammonia (14–18 kg N Mg^{-1} DM) may result in poor fermentation (because of a prolonged buffering effect), and both the poor fermentation and high-ammonia concentrations can reduce animal performance. The Cold-flo method is the simplest way to apply ammonia. Gaseous ammonia is supercooled in a converter box and about 80–85% becomes liquid.

Anhydrous ammonia should not be added to corn forage below 580–600 g moisture kg^{-1} because fermentation is restricted in drier material and binding of ammonia to the forage is poorer. If forage moisture is below this level, water–ammonia or molasses–ammonia mixes should be used. Rates and application methodology for molasses–ammonia mixes should be as recommended by the manufacturer.

Acids and Their Salts

Many acids have been added to forages at ensiling to alter silage fermentation. Much research has been conducted in Europe using formic acid as a silage additive, and it has been a popular means to avoid clostridial activity in unwilted cool-season grass silages. Formic acid immediately reduces pH to 4.7–4.8 and allows natural fermentation to decrease pH further.

However, in the US, the use of acids other than propionic acid is uncommon. Propionic acid inhibits growth of yeasts and molds, improving aerobic stability. Undissociated propionic acid has good antifungal properties, and the fraction of propionic acid left undissociated depends on pH (Lambert and Stratford 1999). At the pH of standing crops, 6.5, only about 1% of the acid is in the undissociated form whereas at a pH of 4.8 about 50% of the acid is undissociated. The undissociated acid functions both by staying active on the surface of microorganisms, competing with amino acids for space on active sites of enzymes, and by altering the cell permeability of microorganisms.

Like other acids, propionic acid is corrosive. Thus, the acid salts (e.g. calcium, sodium, and ammonium propionate) have been used in some commercial products to form a "buffered" acid. The antifungal properties of propionic acid and its salts parallel their solubility in water. Among these salts, ammonium propionate is most soluble in water (90%), followed by sodium propionate (25%) and calcium propionate (5%).

Currently, in the US, there are many buffered propionic acid products with relatively low suggested application rates (0.5–2.0 g kg^{-1} fresh weight). Often other antimycotic agents (e.g. sorbic, benzoic, citric, and acetic acids) are added. In several experiments with such additives, application rates of 2–3 g kg^{-1} were needed to consistently improve aerobic stability of corn silage (Kung et al. 2000; Kung et al. 1998).

In Europe, sorbates and benzoates are becoming more common in silage additives for improving aerobic stability. Also, nitrites and hexamine are more commonly included in silage additives than in the US. These latter chemicals have been added to inhibit clostridial growth.

Troubleshooting

Effluent

In many areas, unfavorable conditions make wilting of forage crops difficult or impossible. Crops with high moisture (>700 g kg^{-1}) can have large nutrient losses from poor fermentation and excessive production of effluent. This effluent is also a potential contaminant to waterways because of its high nutrient concentration.

Two primary approaches are used to control this problem: (i) collection and land spreading and (ii) mixing absorbents with forages to decrease moisture concentration and reduce effluent. Cereal straw (Offer and Alrwidah 1989), alfalfa cubes (Fransen and Strubi 1998), cereal grains (Jones et al. 1990), and beet pulp (Ferris and Mayne 1994) have been used for this purpose. Jones and Jones (1996) concluded that the use of high-fiber material (e.g. straw and paper) to reduce silage effluent had little practical value because it reduced the nutritive value of silage. Inclusion of cereal grains was not always successful, and practical difficulties such as the need to pre-roll or grind discouraged this practice. Inclusion of sugar beet pulp was chosen as a good alternative. Overall, successful addition of absorbents is difficult, requiring increased labor at ensiling and uniform distribution throughout the silage mass.

Silo Gas

Various forms of nitrogen oxide are formed during fermentation, primarily by enterobacteria using nitrate as an electron acceptor in place of oxygen. These nitrogen oxides are collectively referred to as silo gas. Inhalation of even small quantities of nitrogen dioxide (NO_2) and nitrogen tetraoxide (N_2O_4) can lead to chronic pulmonary problems and be fatal. Formation of silo gas occurs within four to six hours of silo filling and may continue for a two- to three-week period. During this time, special care should be taken around fermenting feeds to avoid inhalation by humans, livestock, and pets. Along with CO_2, the nitrogen oxide gases are heavy and tend to settle in low areas. Some gases smell like bleach, but others are odorless. Some gases may also be yellow or brownish, whereas others are colorless. Yellow or reddish-brown staining of equipment or silage may sometimes be observed.

To avoid silo gas, stay away from silos for at least three weeks or more after filling. Ventilate upright silos before entering and use a chemical detector to ensure safety. Never enter an enclosed silo without having another person nearby.

In addition to these safety issues with silage gases, recent work has focused on the production of volatile organic compounds (VOCs) that can negatively affect air quality. Hafner et al. (2013) reviewed the literature on this topic and found great variability in the types and amounts of VOCs produced. For corn silage, alcohols, especially ethanol, were the dominant VOC. In other cases, acids such as acetic acid were important.

Animal Performance

Numerous studies have investigated potential correlations between end products of silage fermentation and ruminant productivity. Conflicting evidence suggests that diets high in moisture from fermented feeds may decrease DM intake. The 1989 National Research Council (NRC) requirements for dairy cattle (National Research Council 1989) reported that DM intake declines by 0.02% of body weight for each $10\,g\,kg^{-1}$ increase in ration moisture above $500\,g\,kg^{-1}$. However, in a review of 392 lactating cow diets, Holter and Urban (1992) found no relationship between DM intake and ration moisture when moisture was greater than $500\,g\,kg^{-1}$. Though Rook and Gill (1990) reported moderately strong negative correlations between intake and acetic, butyric and total volatile fatty acids, Steen et al. (1998) reported only very weak correlations between these variables. Some silages also contain biogenic amines, and these compounds have sometimes been implicated in poor animal production.

The end products of clostridial fermentations may also have negative effects on animal performance and health. Because clostridial silages are often high in free amino acids and ammonia, excessive consumption of these end products can lead to asynchrony of optimal ruminal fermentation because of excessive amounts of rapidly available ammonia N. High levels of butyric acid in silage may also contribute to problems of cows in early lactation that are in negative energy balance as butyric acid is converted in the rumen wall to beta hydroxy butyrate, a ketone body. High levels of ketones in blood can lead to the metabolic disease state known as ketosis. Garrett Oetzel (Univ. of Wisconsin, personal communication, 2003) suggested limiting the intake of butyric acid by dairy cows to less than $50\,g\,d^{-1}$ to avoid metabolic problems. Transition cows should receive no butyric acid in their rations.

Silages that are aerobically unstable heat and spoil primarily because yeasts assimilate lactic acid. Incorporating spoiled silage from the top layer of a bunker silo into steer diets markedly reduced DM intake, nutrient digestion, and adjusted daily gain (Whitlock 1999). Feeding hot, spoiling feeds has been implicated as the reason for poor intake and milk production on many dairy farms. Surprisingly, there is no "rule of thumb" for describing the degree of spoilage required to cause decreases in animal performance.

Silages sometimes contain mycotoxins that can be extremely toxic to animals and humans (Whitlow and Hagler 2002; Ma et al. 2017). Mycotoxins have been suggested as causes of abortions, reduced intake, poor reproduction, and low milk production. Mycotoxins may be on the crop at ensiling, but their production and control in silage are not well understood. General recommendations for limiting their occurrence include minimizing plant disease (e.g. damage to the corn ear or stalk), rapid filling and tight packing of silos, and using silage preservatives designed to inhibit the growth of molds. Obtaining representative samples of forage from large silos for analyses of mycotoxins presents a challenge because they are not usually uniformly distributed throughout the silo.

Overall, silages that contain undesirable levels of fungal metabolites should be completely removed from the diet of lactating cows or minimized at the very least. Specifically, in the case of silages with mycotoxins, use of binders may be useful. However, to date, no products have been approved by the FDA for treatment of mycotoxicosis.

References

Anonymous (2002). Hoard's dairyman continuing market study, 2002. Hoard's Dairyman, Fort Atkinson, WI.

Arriola, K.G., Kim, S.C., and Adesogan, A.T. (2011). Effect of applying inoculants with heterolactic or homolactic bacteria on the fermentation and quality of corn silage. *J. Dairy Sci.* 94: 1511–1516.

Bal, M.A., Shaver, R.D., Jirovec, A.G. et al. (2000). Crop processing and chop length of corn silage: Effects on intake, digestion, and milk production by dairy cows. *J. Dairy Sci.* 83: 1264–1273.

Bolsen, K.K., Sonon, R.N., Dalke, B. et al. (1992a). Evaluation of inoculant and NPN silage additives: A summary of 26 trials and 65 farm-scale silages. In: *Kansas Agricultural Experiment Station Research Reports of Programs*, vol. 651, 101–102. Kansas State University, Manhattan.

Bolsen, K.K., Lin, C., Brent, B.E. et al. (1992b). Effect of silage additives on the microbial succession and fermentation process of alfalfa and corn silages. *J. Dairy Sci.* 75: 3066–3083.

Bolsen, K.K., Dickerson, J.T., Brent, B.E. et al. (1993). Rate and extent of top spoilage losses in horizontal silos. *J. Dairy Sci.* 76: 2940–2962.

Borreani, G., Tabacco, E., Schmidt, R.J. et al. (2018). Silage review: Factors affecting dry matter and quality losses in silages. *J. Dairy Sci.* 101: 3952–3979.

Britt, D.G. and Huber, J.T. (1975). Fungal growth during fermentation and refermentation of non-protein nitrogen treated corn silage. *J. Dairy Sci.* 58: 1666–1674.

Buxton, D.R. and Fales, S.L. (1994). Plant environment and quality. In: *Forage Quality, Evaluation, and Utilization* (ed. G.C. Fahey Jr. et al.), 155–199. Madison, WI: American Society of Agronomy.

Cai, Y., Kumai, S., Zhang, J. et al. (1999). Comparative studies of lactobacilli and enterococci associated with forage crops as silage inoculants. *Anim. Sci. J.* (4): 188–194.

Coblentz, W.K. and Akins, M.S. (2018). Silage review: recent advances and future technologies for baled silages. *J. Dairy Sci.* 101: 4075–4092.

Contreras-Govea, F.E., Muck, R.E., Mertens, D.R., and Weimer, P.J. (2011). Microbial inoculant effects on silage and in vitro ruminal fermentation, and microbial biomass estimation for alfalfa, BMR corn, and corn silages. *Anim. Feed Sci. Technol.* 163: 2–10.

Dawson, T.E., Rust, S.R., and Yokoyama, M.T. (1998). Improved fermentation and aerobic stability of ensiled, high moisture corn with the use of *Propionibacterium acidipropionici*. *J. Dairy Sci.* 81: 1015–1021.

Driehuis, F., Oude Elferink, S.J.W.H., and Van Wikselaar, P.G. (2001). Fermentation characteristics and aerobic stability of grass silage inoculated with *Lactobacillus buchneri*, with or without homofermentative lactic acid bacteria. *Grass Forage Sci.* 56: 330–343.

Edwards, R.A., Donaldson, E., and MacGregor, A.W. (1968). Ensilage of whole-crop barley. I. Effects of variety and stage of growth. *J. Sci. Food Agric.* 19: 656–660.

Elferink, S.J.W.H.O., Krooneman, J., Gottschal, J.C. et al. (2001). Anaerobic conversion of lactic acid to acetic acid and 1,2-propanediol by *Lactobacillus buchneri*. *Appl. Envir. Microbiol.* 67: 125–132.

Ferraretto, L.F. and Shaver, R.D. (2012). Effect of corn shredlage on lactation performance and total tract starch digestibility by dairy cows. *Prof. Anim. Sci.* 28: 639–647.

Ferris, C.P. and Mayne, C.S. (1994). The effects of incorporating sugar-beet pulp with herbage at ensiling on silage fermentation, effluent output and in-silo losses. *Grass Forage Sci.* 49: 216–228.

Flores-Galaraza, R.O., Glatz, B.A., Bern, C.J., and Van Fossen, L.D. (1992). Preservation of high-moisture corn by microbial fermentation. *J. Food Prot.* 48: 407–411.

Fransen, S.C. and Strubi, F.J. (1998). Relationships among absorbents on the reduction of grass silage effluent and silage quality. *J. Dairy Sci.* 81: 2633–2644.

Gulfam, A., Guo, G., Tajebe, S. et al. (2017). Characteristics of lactic acid bacteria isolates and their effect on the fermentation quality of Napier grass silage at three high temperatures. *J. Sci. Food Agric.* 97: 1931–1938.

Hafner, S.D., Howard, C., Muck, R.E. et al. (2013). Emission of volatile organic compounds from silage: compounds, sources, and implications. *Atmos. Environ.* 77: 827–839.

Holter, J.B. and Urban, W.E. (1992). Water partitioning and intake prediction in dry and lactating dairy cows. *J. Dairy Sci.* 75: 1472–1479.

Honig, H. (1991). Reducing losses during storage and unloading of silage. In: *Forage Conservation Towards 2000* (eds. G. Pahlow and H. Honig), 116–128. Braunschweig, Germany: Landbauforschung Völkenrode.

Huber, J.T., Foldager, J., and Smith, N.E. (1979). Nitrogen distribution in corn silage treated with varying levels of ammonia. *J. Anim. Sci.* 48: 1509–1515.

Jalc, D., Laukova, A., Varadyova, Z. et al. (2009). Effect of inoculated grass silage on rumen fermentation and lipid metabolism in an artificial rumen (RUSITEC). *Anim. Feed Sci. Technol.* 151: 55–64.

Johnson, C.O.L.E., Huber, J.T., and Bergen, W.G. (1982). Influence of ammonia treatment and time of ensiling on proteolysis in corn-silage. *J. Dairy Sci.* 65: 1740–1747.

Johnson, L.M., Harrison, J.H., Davidson, D. et al. (2002). Corn silage management: Effects of maturity, inoculation, and mechanical processing on pack density and aerobic stability. *J. Dairy Sci.* 85: 434–444.

Jones, R. and Jones, D.I.H. (1996). The effect of in-silo effluent absorbents on effluent production and silage quality. *J. Agric. Eng. Res.* 64: 173–186.

Jones, D.I.H., Jones, R., and Moseley, G. (1990). Effect of incorporating rolled barley in autumn-cut ryegrass silage on effluent production, silage fermentation and cattle performance. *J. Agric. Sci.* 115: 399–408.

Jones, B.A., Satter, L.D., and Muck, R.E. (1992). Influence of bacterial inoculant and substrate addition to alfalfa ensiled at different dry matter contents. *Grass Forage Sci.* 47: 19–27.

Kaiser, A.G. and Piltz, J.W. (2002). Silage production from tropical forages in Australia. In: *The XIIIth International Silage Conference* (eds. L.M. Gechie and C. Thomas), 48–61. Auchincruive, Scotland, UK: Scottish Agricultural College.

Klopfenstein, T.J., Erickson, G.E., and Berger, L.L. (2013). Maize is a critically important source of food, feed, energy, and forage in the USA. *Field Crops Res.* 153: 5–11.

Kleinschmidt, D. and Kung, L. (2006). A meta-analysis of the effects of *Lactobacillus buchneri* on the fermentation and aerobic stability of corn and grass and small-grain silages. *J. Dairy Sci.* 89: 4005–4013.

Kristensen, N.B., Sloth, K.H., Højberg, O. et al. (2010). Effects of microbial inoculants on corn silage fermentation, microbial contents, aerobic stability, and milk production under field conditions. *J. Dairy Sci.* 93: 3764–3774.

Kung, L. Jr. and Muck, R.E. (1997). Animal response to silage additives. In: *Silage: Field to Feedbunk*, vol. 99, 200–210. Hershey, PA: Northeast Regional Agricultural Engineering Service.

Kung, L. and Ranjit, N.K. (2001). The effect of *Lactobacillus buchneri* and other additives on the fermentation and aerobic stability of barley silage. *J. Dairy Sci.* 84: 1149–1155.

Kung, L., Sheperd, A.C., Smagala, A.M. et al. (1998). The effect of preservatives based on propionic acid on the fermentation and aerobic stability of corn silage and a total mixed ration. *J. Dairy Sci.* 81: 1322–1330.

Kung, L., Robinson, J.R., Ranjit, N.K. et al. (2000). Microbial populations, fermentation end-products, and aerobic stability of corn silage treated with ammonia or a propionic acid-based preservative. *J. Dairy Sci.* 83: 1479–1486.

Kung, L., Taylor, C.C., Lynch, M.P., and Neylon, J.M. (2003). The effect of treating alfalfa with *Lactobacillus buchneri* 40788 on silage fermentation, aerobic stability, and nutritive value for lactating dairy cows. *J. Dairy Sci.* 86: 336–343.

Lambert, R.J. and Stratford, M. (1999). Weak-acid preservatives: modelling microbial inhibition and response. *J. Appl. Microbiol.* 86: 157–164.

Leibensperger, R.Y. and Pitt, R.E. (1987). A model of clostridial dominance in ensilage. *Grass Forage Sci.* 42: 297–317.

Lin, C., Bolsen, K.K., Brent, B.E. et al. (1992). Epiphytic microflora on alfalfa and whole-plant corn. *J. Dairy Sci.* 75: 2484–2493.

Ma, Z.X., Amaro, F.X., Romero, J.J. et al. (2017). The capacity of silage inoculant bacteria to bind aflatoxin B1 in vitro and in artificially contaminated corn silage. *J. Dairy Sci.* 100: 7198–7210.

McDonald, P., Henderson, A.R., and Heron, S.J.E. (1991). *The biochemistry of silage*, 2e. Marlow, Bucks, UK: Chalcombe Publications.

Melvin, J.F. (1965). Variations in the carbohydrate content of lucerne and the effect on ensilage. *Aust. J. Agric. Res.* 16: 951–959.

Merry, R.J., Winters, A.L., Thomas, P.I. et al. (1995). Degradation of fructans by epiphytic and inoculated lactic-acid bacteria and by plant enzymes during ensilage of normal and sterile hybrid ryegrass. *J. Appl. Bacteriol.* 79: 583–591.

Moore, K.J. and Hatfield, R.D. (1994). Carbohydrates and forage quality. In: *Forage Quality, Evaluation, and Utilization* (eds. G.C. Fahey Jr. et al.), 229–280. Madison, WI: American Society of Agronomy.

Muck, R.E. (1987). Dry matter level effects on alfalfa silage quality: I. Nitrogen transformations. *Trans. ASAE* 30: 7–14.

Muck, R.E. (2004). Effects of corn silage inoculants on aerobic stability. *Trans. ASAE* 47: 1011–1016.

Muck, R.E. and Holmes, B.J. (2000). Factors affecting bunker silo densities. *Appl. Eng. Agric.* 15: 613–619.

Muck, R.E. and Huhnke, R.L. (1995). Oxygen infiltration from horizontal silo unloading practices. *Trans. ASAE* 38: 23–31.

Muck, R.E. and Kung, L. Jr. (1997). Effects of silage additives on ensiling. In: *Silage: Field to feedbunk*, vol. 99, 187–199. Hershey, PA: Northeast Regional Agricultural Engineering Service.

Muck, R.E. and Rotz, C.A. (1996). Bunker silo unloaders: An economic comparison. *Appl. Eng. Agric.* 12: 273–280.

Muck, R.E. and Walgenbach, R.P. (1985). Variation in alfalfa buffering capacity. American Society of Agricultural Engineering. Paper No. 85-1535.

Muck, R.E., O'Kiely, P., and Wilson, R.K. (1991). Buffering capacities in permanent pasture grasses. *Irish J. Agric. Res.* 30: 129–141.

Muck, R.E., Broderick, G.A., Faciola, A.P. et al. (2012). Milk production response to feeding alfalfa silage inoculated with *Lactobacillus plantarum*. XVI International Silage Conference, Hämeenlinna, Finland (2–4 July 2012).

Muck, R.E., Nadeau, E.M.G., McAllister, T.A. et al. (2018). Silage review: recent advances and future uses of silage additives. *J. Dairy Sci.* 101: 3980–4000.

National Research Council (1989). *Nutrient requirements of dairy cattle*, 6e. Washington, DC: National Academy Press.

Nsereko, V.L., Smiley, B.K., Rutherford, W.M. et al. (2008). Influence of inoculating forage with lactic acid bacterial strains that produce ferulate esterase on ensilage and ruminal degradation of fiber. *Anim. Feed Sci. Technol.* 145: 122–135.

Offer, N.W. and Alrwidah, M.N. (1989). The use of absorbent materials to control effluent loss from grass-silage: Experiments with pit silos. *Res. Develop. Agric.* 6: 77–82.

Oliveira, A.S., Weinberg, Z.G., Ogunade, I.M. et al. (2017). Meta-analysis of effects of inoculation with homofermentative and facultative heterofermentative lactic acid bacteria on silage fermentation, aerobic stability, and the performance of dairy cows. *J. Dairy Sci.* 100: 4587–4603.

Pitt, R.E. (1986). Dry matter losses due to oxygen infiltration into silos. *J. Agric. Eng. Res.* 35: 193–205.

Pitt, R.E. and Muck, R.E. (1993). A diffusion model of aerobic deterioration at the exposed face of bunker silos. *J. Agric. Eng. Res.* 55: 11–26.

Playne, M.J. and McDonald, P. (1966). The buffering constituents of herbage and of silage. *J. Sci. Food Agric.* 17: 264–268.

Queiroz, O.C.M., Arriola, K.G., Daniel, J.L.P., and Adesogan, A.T. (2013). Effects of 8 chemical and bacterial additives on the quality of corn silage. *J. Dairy Sci.* 96: 5836–5843.

Ranjit, N.K. and Kung, L. Jr. (2000). The effect of *Lactobacillus buchneri*, *Lactobacillus plantarum*, or a chemical preservative on the fermentation and aerobic stability of corn silage. *J. Dairy Sci.* 83: 526–535.

Reich, L.J. and Kung, L. Jr. (2010). Effects of combining *Lactobacillus buchneri* 40788 with various lactic acid bacteria on the fermentation and aerobic stability of corn silage. *Anim. Feed Sci. Technol.* 159: 105–109.

Rook, A.J. and Gill, M. (1990). Prediction of the voluntary intake of grass silages by beef cattle. 1. Linear-regression analyses. *Anim. Prod.* 50: 425–438.

Savoie, P. (1988). Optimization of plastic covers for stack silos. *J. Agric. Eng. Res.* 41: 65–73.

Schmidt, R.J. and Kung, L. Jr. (2010). The effects of *Lactobacillus buchneri* with or without a homolactic bacterium on the fermentation and aerobic stability of corn silages made at different locations. *J. Dairy Sci.* 93: 1616–1624.

Schwab, E.C., Shaver, R.D., Lauer, J.G., and Coors, J.G. (2003). Estimating silage energy value and milk yield to rank corn hybrids. *Anim. Feed Sci. Tech.* 109: 1–18.

Scudamore, K.A. and Livesey, C.T. (1998). Occurrence and significance of mycotoxins in forage crops and silage: A review. *J. Sci. Food Agric.* 77: 1–17.

Shaver, R. (1993). Troubleshooting problems with carbohydrates in dairy rations. *Vet. Med.* 88: 1001–1008.

Shaver, R.D. (2003). Impact of vitreousness, processing, and chop length on the utilization of corn silage by dairy cows. In: *Proc. Wis. Forage Counc. Ann. Mtg*, 14–22, Wisconsin Dells, WI.

Smith, D. (1973). The nonstructural carbohydrates. In: *Chemistry and Biochemistry of Herbage*, vol. 1 (eds. G.W. Butler and R.W. Bailey), 105–155. London: Academic Press Inc.

Steen, R.W.J., Gordon, F.J., Dawson, L.E.R. et al. (1998). Factors affecting the intake of grass silage by cattle and prediction of silage intake. *Anim. Sci.* 66: 115–127.

Tabacco, E., Piano, S., Revello-Chion, A., and Borreani, G. (2011). Effect of *Lactobacillus buchneri* LN4637 and *Lactobacillus buchneri* LN40177 on the aerobic stability, fermentation products, and microbial populations of corn silage under farm conditions. *J. Dairy Sci.* 94: 5589–5598.

Vanderwerff, L.M., Ferraretto, L.F., and Shaver, R.D. (2015). Brown midrib corn shredlage in diets for high-producing dairy cows. *J. Dairy Sci.* 98: 5642–5652.

Weinberg, Z.G. and Ashbell, G. (1994). Changes in gas composition in corn silages in bunker silos during storage and feedout. *Canadian Agric. Eng.* 36: 155–158.

Weiss, W.P. and Wyatt, D.J. (2000). Effect of oil content and kernel processing of corn silage on digestibility and milk production by dairy cows. *J. Dairy Sci.* 83: 351–358.

Whiter, A.G. and Kung, L. (2001). The effect of a dry or liquid application of *Lactobacillus plantarum* MTD1 on the fermentation of alfalfa silage. *J. Dairy Sci.* 84: 2195–2202.

Whitlock, L.A. (1999). Effect of level of surface spoilage in the diet on feed intake, nutrient digestibilities, and ruminal metabolism in growing steers fed a whole-plant corn silage-based diet. Master thesis. Kansas State University.

Whitlow, L.W. and Hagler, W.M. (2002). Mycotoxins in feeds. *Feedstuffs* 74: 68–78.

Wilkinson, J.M. and Fenlon, J.S. (2014). A meta-analysis comparing standard polyethylene and oxygen barrier film in terms of losses during storage and aerobic stability of silage. *Grass Forage Sci.* 69: 385–392.

Wilkinson, J.M. and Toivonen, M.I. (2003). *World Silage.* Painshall, Lincoln, UK: Chalcombe Publications.

Windle, M.C. and Kung, L. Jr. (2016). Factors affecting the numbers of expected viable lactic acid bacteria in inoculant applicator tanks. *J. Dairy Sci.* 99: 9334–9338.

Winters, A.L., Merry, R.J., Muller, M. et al. (1998). Degradation of fructans by epiphytic and inoculant lactic acid bacteria during ensilage of grass. *J. Appl. Microbiol.* 84: 304–312.

Woolford, M.K. (1984). *The silage fermentation.* New York, NY: Marcel Dekker, Inc.

Yamamoto, Y., Gaudu, P., and Gruss, A. (2011). Oxidative stress and oxygen metabolism in lactic acid bacteria. In: *Lactic Acid Bacteria and Bifidobacteria: Current Progress in Advanced Research* (eds. K. Sonomoto and A. Yokota), 91–102. Norfolk, UK: Caister Scientific Press.

Zhou, Y., Drouin, P., and Lafrenière, C. (2016). Effect of temperature (5–25°C) on epiphytic lactic acid bacteria populations and fermentation of whole-plant corn silage. *J. Appl. Microbiol.* 121: 657–671.

CHAPTER

43

Biomass, Energy, and Industrial Uses of Forages

Matt A. Sanderson, Research Agronomist and Research Leader (Retired), *USDA-Agricultural Research Service, State College, PA, USA*
Paul Adler, Research Agronomist, *Pasture Systems and Watershed Management Research Unit, USDA-Agricultural Research Service, University Park, PA, USA*
Neal P. Martin, Director (Retired), *US Dairy Forage Research Center, USDA-Agricultural Research Service, Madison, WI, USA*

Introduction

Society relies heavily on nonrenewable energy sources such as coal, oil, and natural gas. Shifting from fossil carbon (C) sources to contemporary C sources for energy and industrial products has been termed a shift to a **bio-based economy**. The discovery of new uses for forages has enhanced the value of these perennial crops beyond their traditional uses for animal feed and conservation. Converting plants and other biologic materials into biofuels, industrial products, and human-use products has been termed the **biorefinery** concept (Figure 43.1). Relying on contemporarily fixed-C rather than fossil sources as the feedstock for these new products is a renewable approach.

Biomass generally refers to the organic matter from plants and, in terms of energy production, includes herbaceous and woody crops along with their residues (McKendry 2002a; Brown and Brown 2014). Biofuels derived from this organic matter include alcohols, ethers, esters, and other chemicals. The term **biofuels** is often used interchangeably when referring to fuels for electricity or liquid fuels for transportation. When derived from biomass, these fuels are referred to as **second-generation biofuels** as opposed to the **first-generation biofuels** derived from sugars and oils of arable crops.

Before World War II, forages fueled agriculture even in industrialized countries. In 1920, the 27 million horses and draft animals in the US, fed mainly hay and pasture (i.e. herbaceous biomass), pulled plows and transported goods and people (Vogel 1996). By the 1950s, agriculture was largely mechanized, and fossil fuels provided nearly all of the energy inputs. The **lignocellulose** in forage crops represents a vast and renewable source of biomass feedstock for conversion into liquid fuels, thermochemical products, and other energy-related end products (Figure 43.1; US DOE 2011). With the appropriate technologies and processes for biomass production and conversion implemented economically, forages could once again fuel agriculture.

In this chapter, we address the use of perennial forage crops for the production of these alternative products and how management practices may differ from traditional forage uses.

Forages: The Science of Grassland Agriculture, Volume II, Seventh Edition.
Edited by Kenneth J. Moore, Michael Collins, C. Jerry Nelson and Daren D. Redfearn.

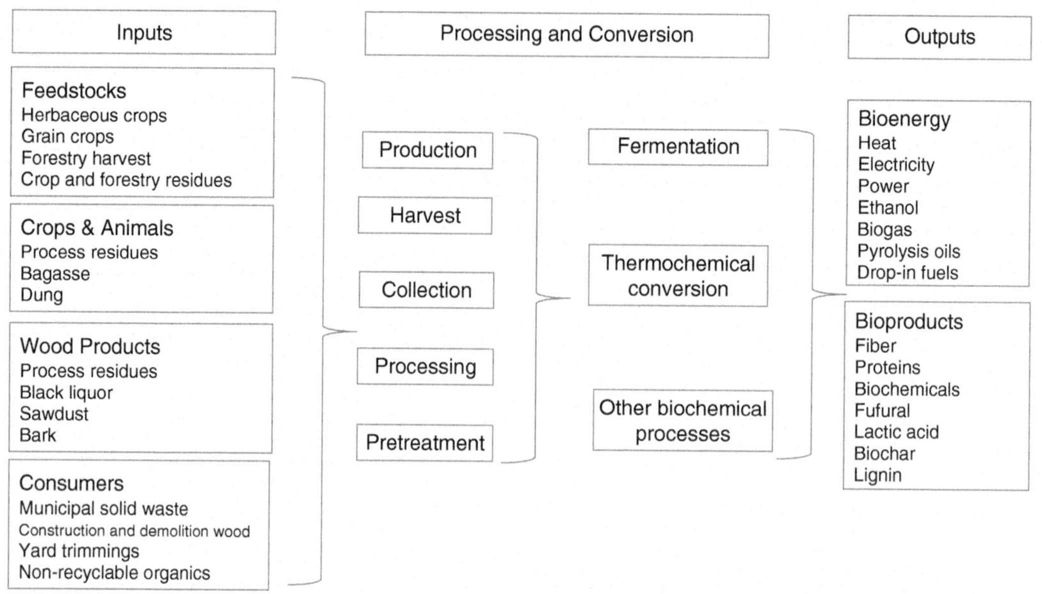

Fig. 43.1. Illustration of the biorefinery concept, that is, the use, processing, and conversion of biomass feedstock to energy and other products.

Forage Species for Biofuels

Some of the most extensively studied perennial species for biomass feedstock production include switchgrass, *Miscanthus* spp., energy cane, napiergrass, reed canarygrass, and alfalfa (Table 43.1).

Switchgrass

Switchgrass is perhaps the most studied herbaceous energy crop because of its adaptability across many environments, suitability for marginal and erosive land, relatively low water and nutrient requirements (Hendrickson et al. 2013), and potential environmental benefits (McLaughlin et al. 2002; Parrish and Fike 2005). Traditional varieties adapted for biomass production include 'Alamo' (adapted to the southern United States) and 'Cave-in-Rock' (along with 'Shawnee'; adapted to the mid-Atlantic, Northeast, and Midwest regions; Sanderson et al. 2012) (Figure 43.2). Varieties developed specifically for biomass energy use include 'BoMaster' (Burns et al. 2008), 'Liberty' (Vogel et al. 2014), and 'Cimmaron' (Wu and Taliaferro 2009). Switchgrass also performed well as a biomass crop in Europe (Elbersen et al. 2004). Research continues on genetic improvement of switchgrass for agronomic and biofuels traits and environmental benefits such as increases in soil organic-C (Casler 2012).

Miscanthus

Beginning in the 1980s, European bioenergy research focused on *Miscanthus* species as biomass feedstock for combustion steam plants (Lewandowski et al. 2003). The genus *Miscanthus* includes C_4 rhizomatous grasses native to Asia, northern India, and Africa, which are winter hardy in temperate areas of Europe (Heaton et al. 2010). *Miscanthus* research in the US accelerated in the early 2000s (Heaton et al. 2010). Illinois research demonstrated yields twice that of switchgrass for 8–10 years (Table 43.2; Arundale et al. 2014). Giant *Miscanthus* has relatively low N fertility needs (Heaton et al. 2010); however, it is a sterile triploid and must be planted and established vegetatively (Scordia et al. 2015).

Field-plot trials of *Miscanthus* across Europe demonstrated yields greater than 40 Mg ha^{-1} (Table 43.2). *Miscanthus giganteus* hybrids performed better in mid- and southern Europe (Germany southward), whereas *Miscanthus sinensis* hybrids performed better in northern Europe. In general, *M. giganteus* and *Miscanthus sacchariflorus* will not perform well where winter soil temperatures fall below −3 °C at a 5-cm soil depth (Clifton-Brown et al. 2002) and do not tolerate drought.

Reed Canarygrass

Reed canarygrass is a perennial C_3 grass that is well adapted to northern temperate climates and does well on wet soils (Wrobel et al. 2009; Heinsoo et al. 2011; Tahir et al. 2011; Table 43.1). Similar to switchgrass, reed canarygrass can be slow to establish, and yields are low in the seeding year. Reed canarygrass may also

Table 43.1 Examples of biomass yields of selected perennial forages in the United States

Species	Location	Yield range			Site-years	References
		Low	High	Average		
			($Mg\ ha^{-1}$)		(no.)	
Alfalfa	Minnesota	8.4	12.7	11.2	9	Sheaffer et al. (2000)
Bermudagrass[a]	Florida	6.9	23.2		3	Silveira et al. (2013)
Energy cane[b]	Florida	23.9	39.2		2	Silveira et al. (2013)
Napiergrass[c]	Florida	29.2	42.7		3	Silveira et al. (2013)
Reed canarygrass[d]	Iowa/Wisconsin	4.2	12.6	7.5	8	Tahir et al. (2011)

[a]One location for three years; average of four cultivars. 270–450 $kg\ N\ ha^{-1}\ yr^{-1}$; three to seven harvests per year.
[b]One location for two years. 200 $kg\ N\ ha^{-1}\ yr^{-1}$; one harvest per year.
[c]One location for three years. 270–450 $kg\ N\ ha^{-1}\ yr^{-1}$; two to four harvests per year.
[d]Three locations for two or three years. 112 $kg\ N\ ha^{-1}\ yr^{-1}$; one harvest in autumn or two harvests in spring and autumn.

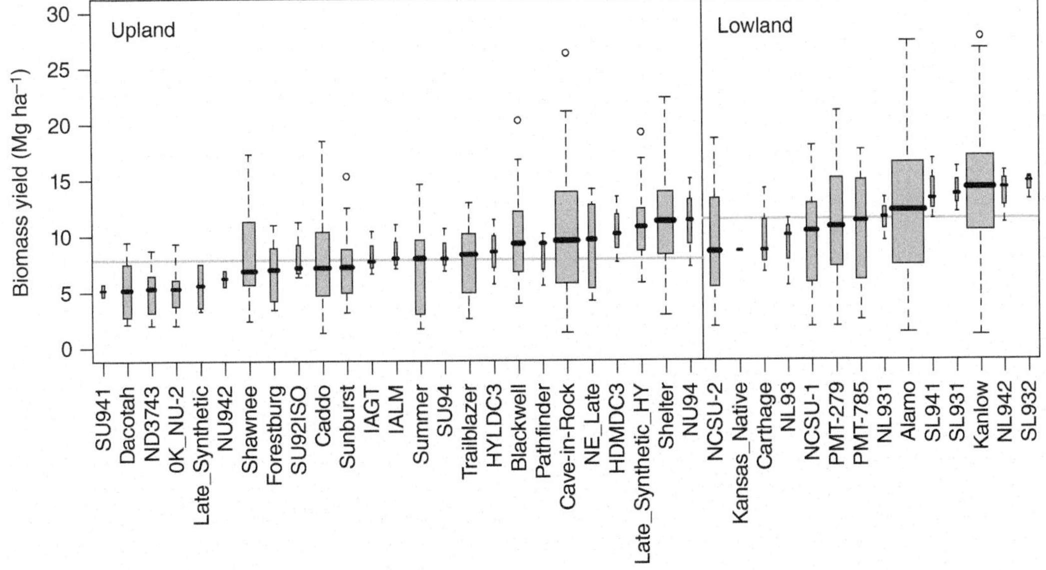

Fig. 43.2. Biomass yield of upland and lowland switchgrasses in several environments in the US. *Source:* From Wullschleger et al. (2010).

contain relatively high concentrations (>100 $g\ kg^{-1}$ of dry matter) of **ash**, which can be reduced by postponing harvest over winter until the next spring (Landström et al. 1996). In North America, however, some consider reed canarygrass an invasive species, especially in native wetlands (Galatowitsch et al. 1999). The invasiveness of reed canarygrass appears unrelated to development of improved varieties via plant breeding (Jakubowski et al. 2011).

Napiergrass and Energy Cane

The sub-tropical climate of the lower southern US, Puerto Rico, and Hawaii favors the tall perennial tropical grasses such as napiergrass and energy cane. Because of their photoperiod sensitivity, these grasses continue vegetative growth late into the season until stopped by frost (Anderson et al. 2016). The long, warm growing season and high rainfall in these areas provide conditions for high yields ranging from 20 to 50 $Mg\ ha^{-1}$ dry matter (Fedenko et al. 2013; Silveira et al. 2013) (Table 43.1).

Alfalfa

An innovative system of using alfalfa for both biomass feedstock and a high-quality animal feed has been proposed. The undeveloped system separates the leaves for high-value products, high-protein feed and uses the

Table 43.2 Example biomass yields of *Miscanthus* at several US and European locations

Location	Low	High	Average	Site years	References
		(Mg ha^{-1})			
Illinois	14.7	31.1	23.4	[a]	Arundale et al. (2014)
Texas	2.8	5.7	4.6	5	Kiniry et al. (2013)
Arkansas	4.5	11.8	8.5	3	Kiniry et al. (2013)
Missouri	10.7	49.7	23.0	6	Kiniry et al. (2013)
North Carolina	1.6	21.3	20.8	4	Palmer et al. (2014)
Denmark	1.4	18.2	09.1	3	Clifton-Brown et al. (2001)
England	0.9	18.7	08.5	3	Clifton-Brown et al. (2001)
Germany	3.0	29.1	13.4	3	Clifton-Brown et al. (2001)
Sweden	0.4	24.7	11.5	3	Clifton-Brown et al. (2001)
Portugal	7.5	40.9	25.2	3	Clifton-Brown et al. (2001)

[a]Data are statistically modeled averages of 8–10 years at seven locations in Illinois. Peak biomass yields at individual locations in specific years ranged from 21 to more than 40 Mg ha^{-1}.

lower-quality stems, which are high in cell-wall polysaccharides, which can be broken down and converted to ethanol (Samac et al. 2006). The proposed system included two-cut harvest management to optimize economics, yield of stem and leaf, and wildlife habitat. Genetic selection efforts concentrated on lines developed for stiff stems with increased internode length to be grown under infrequent harvest (Lamb et al. 2003, 2007).

Energy balance and potential farmer profits of four energy production systems within the Midwest are shown in Table 43.3 (Vadas et al. 2008). Continuous corn showed most profit, but the alfalfa-corn rotation had significant advantages in efficiency of energy production, decreased soil erosion and less nitrogen leaching than corn. In the US, corn production historically has been subsidized with attendant effects on commodity and (first generation) biofuel prices. Currently, alfalfa production is not subsidized; however, as a second-generation biofuel crop it may benefit from policies such as the renewable fuel standard. A new system based on using a growth regulator, prohexadione-calcium, on alfalfa interseeded under corn reduced soil erosion potential of corn and increased alfalfa yield by eliminating low establishment year yield of alfalfa (Grabber 2016).

Other Species

Several other perennial grasses, some not necessarily forage grasses, have been suggested as potential energy plants, including big bluestem, indiangrass, timothy, common reed and giant reed (both common and giant reed are considered invasive species in the US), cordgrass, bermudagrass, and coastal panicgrass (El-Bassam 2010). Each has unique growth and adaptation characteristics that give them potential as **bioenergy crops**.

Polycultures

Polycultures or mixtures of many species from several functional groups may be a less intensive alternative to monocultural bioenergy cropping if multiple ecosystem services, such as reduced pest pressure and more efficient use of resources, can be generated (Tilman et al. 2006; DeHaan et al. 2010). Field experiments with polycultures in different environments have shown biomass yields nearly equal or sometimes much less than monocultures of grasses (Griffith et al. 2011; Johnson et al. 2013; Zilverberg et al. 2014). A survey of farm fields planted to a switchgrass monoculture or a polyculture of prairie plant species along with a field plot experiment comparing mixtures of 1–30 prairie plant species indicated no relationship between the number of species in the plantings and biomass yields (Dickson and Gross 2015). Multilocation trials in Minnesota and North Dakota, however, indicated that some mixtures of prairie plants can be as productive as switchgrass monocultures (Jungers, Clark et al. 2015). Very little research with polycultures has addressed multiple ecosystem services beyond simply biomass productivity.

The farmer's ultimate decision to select and grow a specific bioenergy crop or combination of crops will depend on site-specific variables such as soils, climate, and management in addition to broader societal variables such as market forces and national energy and agricultural policies (Dale et al. 2011). At the national scale, a widely dispersed renewable energy industry would benefit from having several types of feedstocks distributed regionally, nationally, and temporally. A cellulosic biorefinery that can process multiple feedstocks (e.g. wheat straw, *Miscanthus* bales, and mixed species biomass) could enable

Table 43.3 Example energy balance and potential farmer profits of four bioenergy crop production systems in the Midwest

Item	Corn no stover[a]	Corn with stover	Alfalfa-corn	Switchgrass
Total energy inputs (MJ ha^{-1})	79 800	81 400	48 000	9 800
Total energy outputs (including co-products[b]; MJ ha^{-1})	111 800	181 800	139 600	111 000
Net energy yield (outputs minus inputs; MJ ha^{-1})	32 000	100 400	91 000	101 200
Efficiency (outputs divided by inputs; MJ ha^{-1})	1.40	2.23	2.91	11.33
Profits – medium commodity price scenario ($ ha^{-1})	0.35	66.82	28.44	54.14
Profits – high commodity price scenario ($ ha^{-1})	333.93	457.48	349.42	242.38

Source: Adapted from Vadas et al. (2008)

Data are estimates from enterprise budget analyses.

[a]Cropping systems were: corn grown for grain ethanol only; corn grown for grain ethanol and stover collected to produce cellulosic ethanol; alfalfa grown in two-year rotation with corn to produce grain, stover, and alfalfa stem ethanol; switchgrass grown for cellulosic ethanol only.

[b]Co-products include distillers grains from corn ethanol, animal feed from alfalfa leaves, and excess electricity from cellulosic ethanol fermentation waste products.

an expanded fuelshed based on a diverse agricultural landscape (Robertson et al. 2011).

Management for Bioenergy Cropping

An advantage of using existing forages as bioenergy crops is that farmers are familiar with their agronomic management and already have the machinery, technology, and infrastructure needed to establish, manage, harvest, store, and transport them. Forage crops offer additional flexibility in management, because they can be used for biomass or forage and the land can be returned to other uses or put into crop rotation.

Key management practices for bioenergy crop production (common also to forage crop production) broadly include: (i) rapid establishment to realize economically harvestable biomass in the seeding year, (ii) very efficient use of fertilizers (especially N) and native soil fertility to reduce energy inputs, and (iii) harvest management to either maximize biomass yield or optimize yield, stand life, and biomass feedstock quality.

Planting and Establishment

Planting and establishment requirements for most forage crops have been covered in other chapters. Generally, planting and establishment recommendations for switchgrass, alfalfa, and reed canarygrass grown as forage crops apply to these same crops under **bioenergy crop** management. For switchgrass biomass production, a stated goal is to produce 50% of full production potential at the end of the planting year and 75–100% of full production in the year after planting. Currently, *Miscanthus* must be established vegetatively because it is a sterile triploid plant.

Fertility Management

Because manufacture of N fertilizer requires large amounts of energy (~35 MJ of natural gas per kg of NH_3), its efficient use is paramount in bioenergy crop production. Thus, the use of a N-fixing crop, such as alfalfa, offers advantages in terms of N fertilizer use.

Traditional fertilizer recommendations for forage production may not apply to production of bioenergy feedstock. For example, research on N fertilization of switchgrass for biomass feedstock indicates variable responses depending on management and soils (Hong et al. 2014). Recommendations of soil fertility for switchgrass management in the mid-Atlantic region of the US include: maintain pH above 5.0; apply 50 kg P ha^{-1} when soil test P is low; apply 100 kg K ha^{-1} when soil test K is low to medium; apply 50 kg N ha^{-1} in spring for a one-cut harvest system or 50 kg in spring and 50 kg after the first harvest for a two-cut system (Parrish and Fike 2005). In the Midwest US, recommendations for switchgrass include supplying 10–12 kg ha^{-1} of N for each Mg of yield and up to 20 kg P ha^{-1} at very low soil test P levels.

Harvest Management

Because traditional forage quality attributes differ from those of biomass feedstock quality, harvest management for biomass feedstock emphasizes yield and persistence. For example, proposed management of alfalfa as a biomass energy crop includes only two harvests, whereas traditional forage management involves harvests at bud to early flower stages to optimize yield, nutritive value, and persistence (Lamb et al. 2003; Samac et al. 2006). An alternative harvest method, called field fractionation, has been developed to produce a high-protein feedstuff from alfalfa leaves separated in the field and high-fiber products (including biomass feedstock) from the stems (Shinners et al. 2007). The new field harvest and fractionation technology could enable the on-farm production of several high-value products with relatively low capital equipment costs.

Harvest recommendations for switchgrass for maximum biomass yields include a single fall harvest in the south central US (Sanderson et al. 1999), north central US (Casler and Boe 2003), and Quebec, Canada (Madakadze et al. 1999). Though two harvests per year may be needed for maximum production of upland switchgrass cultivars in the mid-Atlantic and southeastern US, this may not be economic due to the relatively low yield on the second harvest (Parrish and Fike 2005). Developmental stages of full panicle emergence to post-anthesis were recommended as the optimal time to harvest switchgrass in the midwestern US (Vogel et al. 2002). Compared with an August harvest, delaying harvest until after frost resulted in yield losses of 1–2 Mg ha^{-1} in the Midwest. Delaying a single harvest until late winter or early spring reduces the concentrations of N, alkali elements, and moisture in grasses grown for biomass and may be appropriate for biomass crops in some instances. Delayed harvest, however, reduced biomass yields of *Miscanthus* by 35% (Lewandowski et al. 2003). In Pennsylvania, delaying switchgrass harvest over winter until the following spring reduced mineral element concentrations but also reduced biomass yields almost 40% (Adler et al. 2006).

Enabling flexibility in harvest management of bioenergy crops would allow diversified farmers to satisfy multiple goals, such as providing quality forage for livestock as well as biomass and other byproducts. Harvest alternatives would provide farmers flexibility to respond to potential fluctuations in future feedstock markets.

Use of Conservation Lands for Feedstock Production

Land in the Conservation Reserve Program (CRP) has been suggested as a potential, readily available resource for biomass feedstock production in the US (U.S. DOE 2011). The CRP pays owners and operators to set aside environmentally sensitive lands. Assessing the quality of the feedstock and developing management systems consistent with maintaining the environmental benefits of the CRP are key considerations in its potential use for bioenergy.

In 2017, there were 9.5 million ha of CRP land concentrated mainly in the central plains and midwestern US (USDA Farm Service Agency 2017a). Of the total CRP area, 1.3 million ha were planted to CP-1 mixtures (introduced grasses), 2.3 million ha were planted to CP-2 mixtures (native grasses), and 1.5 million ha were classified as CP-10 (established grass), which could potentially be available for biomass feedstock production. The remaining 4.4 million ha were in trees, wildlife habitat, or other conservation practices.

Little is known about the plant composition or amount of biomass produced on CRP grasslands. In the northeastern US, a two-year survey identified 285 herbaceous plant species on CRP and other lands used for conservation purposes. The total plant species richness ranged from 12 to 60 species with a mean of 34 per 0.1 ha while aboveground biomass ranged from 0.5 to 12.9 with a mean of 6.6 Mg ha^{-1}, increasing with tall C_4 prairie grass cover (Adler et al. 2009). In Minnesota, biomass yield ranged from 0.5 to 5.7 Mg ha^{-1} on CRP lands and did not decline over three years with a late fall harvest (Jungers et al. 2013) nor did plant community composition change (Jungers et al. 2015).

Managed harvesting of CRP is permitted with the condition that environmental benefits be maintained or enhanced (USDA Farm Service Agency 2017b). Several CRP practices are eligible for managed harvest including CP-1 and CP-2 once cover is fully established, but no more frequently than one out of every three years. To protect ground nesting wildlife, managed harvesting is not allowed during the primary nesting or brood-rearing season. In addition, a payment reduction of 25% is assessed for the acreage harvested.

Life Cycle Assessment

Life cycle assessment was developed as a tool to quantify the material flows of a product cycle, and has been used to quantify the environmental impacts of bioenergy production, with the most common analysis focusing on the energy balance (Schmer et al. 2008) and GHG (greenhouse gas) emissions (Adler et al. 2007). The energy balance for switchgrass production considers the energy content of the biomass minus the fossil energy used in production (i.e. the net energy production from the system). Biomass can be directly combusted, or the cellulose fraction can be converted to ethanol and the lignin fraction combusted or land applied as an amendment (Adler et al. 2015). Producing ethanol from crops results in an energy ratio (ratio of energy output vs energy input; values greater than one imply energy output greater than input) of >5 for switchgrass (Schmer et al. 2008) and *Miscanthus* (Wang et al. 2012), and ~4.8 for corn stover, and ~4.3 for sugarcane (Wang et al. 2012), compared with 1.6 from corn grain (Wang et al. 2012). While transportation fuels have been a priority target for biomass, displacing fuel oil may be the best use of the limited biomass resource in the Northeastern US (Wilson et al. 2012), where there can be greater displacement of liquid fuels and significant savings to consumers.

Feedstock production contributes a large portion of GHG emissions associated with biofuel production, more than 50% for switchgrass (Adler et al. 2012). Nitrogen (N_2O and GHG emissions associated with production of N fertilizers) and soil carbon are the largest source and sink of GHG emissions, respectively, associated with feedstock production (Adler et al. 2007). For switchgrass production, N_2O emissions and GHG emissions associated with N fertilizer production account for more than 80% of the GHG emissions (Adler et al. 2012). With a lower requirement of N fertilizer (Maughan et al. 2012), *Miscanthus* could greatly reduce emissions associated with feedstock production (Adler 2017).

There are many types of marginal lands (Richards et al. 2014) which can have different effects on the life cycle assessment (LCA) of feedstock production. While drought prone sites have lower switchgrass yields, poorly drained sites may have similar yields to prime lands (Casler et al. 2017), resulting in lower inputs per unit of production; however N_2O emissions may be higher (Saha et al. 2017). On marginal lands with perennial grass vegetation, such as those in CRP, there is less potential for further sequestration of soil C (Gelfland 2011). Prior vegetation, soil texture, and climate can all affect GHG emissions leading to many options for the landscape design to optimize GHG emissions (Field et al. 2017).

Bioenergy Conversion

In contrast with petroleum refineries, which use oil as the feedstock, a biorefinery converts biomass feedstock into a number of high-value chemicals and energy. Optimally, the biorefinery also finds uses for by-products to provide additional income sources and to minimize wastes and emissions.

Conversion Methods

The three broad categories of converting lignocellulosic biomass to different energy or chemical end products include biochemical processes, thermochemical methods, or **direct combustion** (McKendry 2002b). To produce ethanol, plant cell walls are chemically or biochemically digested to simple fermentable sugars such as glucose. Typically, the biomass is pretreated to reduce feedstock size, to facilitate the breakdown of hemicellulose to simple sugars, and to expose the cellulose to allow greater access by enzymes. The feedstock is then hydrolyzed and fermented before the fermented product is distilled to obtain ethanol (Brown and Brown 2014). The lignin remaining after separation from sugars can be used to fuel the process (Sun and Cheng 2002). Approximately 200–250 l of ethanol can be produced from 1 Mg of a dry biomass, such as switchgrass, depending on process efficiencies.

Thermochemical conversion processes include **pyrolysis**, gasification, and **liquefaction** (McKendry 2002b), which can be used to convert biomass to methanol, **synthesis gas**, and pyrolysis oils. **Gasification** converts all carbon to a synthetic gas, mainly hydrogen (H_2) and carbon monoxide (CO), which is then burned or converted chemically to other products.

Larger-scale direct combustion includes processes where herbaceous biomass is burned in industrial-sized boilers to produce steam and generate electricity. Mixing biomass and coal together for combustion, known as co-firing, can help reduce sulfur emissions, allow for flexibility in using different fuels, and can alleviate some problems of biomass combustion associated with ash and minerals fouling the combustor (Brown and Brown 2014).

Chemical Composition and Fuel Quality of Feedstock

The efficiency and end products of the various conversion processes depend on the chemical composition of the biomass. Biomass contains higher concentrations of inorganic elements such as K and Ca compared with fossil fuels such as coal (Table 43.4). High concentrations of alkali metals enhance the formation of fusible ash, which causes slagging and fouling of boilers used for direct combustion (Miles et al. 1996) and disrupts fluidized bed combustion systems. Feedstocks high in N and ash reduce hydrocarbon yields during thermochemical conversion. **Pyrolysis** oils obtained from feedstock having high-ash concentrations are higher in Cl and K. Burning these pyrolysis oils corrodes turbines used to generate electricity (McKendry 2002a,b). Total ash concentrations in forages usually decrease as forages mature. Thus, harvesting forages at late maturity stages or overwintering biomass *in situ* would minimize the concentrations of inorganic elements in the feedstock.

Combustion of lignin from forage crops used for biomass contributes energy to the thermochemical conversion process (Sun and Cheng 2002). Lignin, however, reduces the availability of cellulose and other structural polysaccharides in forage plants and reduces ethanol yields during the biochemical process of fermentation (Sun and Cheng 2002). Pretreatment of **lignocellulose**, for example, with anhydrous ammonia under pressure or with a steam explosion, may increase the conversion efficiency by physically disrupting the fiber.

Table 43.4 Comparison of the chemical composition of selected biomass feedstocks

Constituent	Reed canarygrass	Switchgrass	*Miscanthus*	Hybrid poplar	Coal
Energy (GJ Mg^{-1})	17.9	17–19	17–19	19	27–30
Ash (g kg^{-1})	64.0	57	45–58	5–15	82.0
K (g kg^{-1})	12.3	05	05–10	0.3	00.2
S (g kg^{-1})	01.7	1.2	1.2	0.3	05.5
Cl (g kg^{-1})	05.6	1.4	—	—	00.2

Source: Data from Jenkins et al. (1998).

Modern plant-breeding and molecular biology techniques can be used to improve the chemical composition of forages for use as biomass and co-products (Casler 2012). This technology will enable plant breeders, conversion chemists, and engineers to tailor bioenergy crops for specific conversion processes, higher energy yields, and development of new co-products.

Other Industrial Products from Forages

Wet fractionation of forage crops, such as alfalfa, adds value to biomass through the spin-off of co-products. The fractionation process consists of expressing high-value juice from fresh herbage, leaving a reduced-moisture fraction high in **lignocellulose** that can be used as biomass feedstock (Koegel and Straub 1996). The juice fraction can be used in animal feed (Jorgensen and Koegel 1988) or further processed to isolate protein suitable for both food-grade and feed-grade concentrates.

Other high-value juice products include xanthophyll concentrates for poultry feeds, plant and animal growth stimulants, cosmetic substances, and pharmaceuticals. Additional industrial products from alfalfa include lactic acid, used as a food ingredient and preservative, enzymes such as phytase, **cellulase**, and alpha-amylase, and biodegradable plastics (Saruul et al. 2000). To date, transgenic alfalfa cultivars have been produced that contain Mn-dependent lignin peroxidase for biopulping, alpha-amylase for converting starch to sugar, phytase for releasing P from phytic acid, and cellulase for the conversion of cellulose to sugars. Alfalfa cellulose fermented with a preparation of *Ruminococus albus* resulted in an adhesive that could replace phenol-formadehyde resin used in plywood and other wood products (Weimer 2003).

Summary

Forages are sustainable feedstocks for energy and industrial products. Several forage species can be grown as bioenergy crops in a wide range of environments as long-term stands, in rotations with cash crops, or on marginal and environmentally sensitive lands. Traditional forage management practices can be applied to biomass feedstock production; however, emerging innovative technologies (e.g. harvest methods, conversion techniques, new plant cultivars) will likely require equivalent management innovations.

Currently, cellulosic biofuels cost more to produce than starch-derived biofuels and fossil fuels. The need for pretreatment and current high costs of enzymes contributes to the higher costs. Cellulosic biofuel production costs are very sensitive to feedstock costs emphasizing the need for highly efficient crop production and harvest methods. Biomass yield, harvest and transport costs, conversion efficiency, and cost of fossil fuel used to produce the biofuel determine the economics of bioenergy production and vary across the US. Developing new co-products and valuing environmental benefits of bioenergy crops could open new avenues in the bioeconomy and parity among biofuels and fossil fuels.

References

Adler, P.R. (2017). Life cycle greenhouse gas emissions from *Miscanthus* production at farm scale. *ASA 2017 Meeting Abstr.* Poster Number 1238.

Adler, P.R., Sanderson, M.A., Boateng, A.A. et al. (2006). Biomass yield and biofuel quality of switchgrass harvested in fall or spring. *Agron. J.* 98: 1518–1525.

Adler, P.R., Del Grosso, S.J., and Parton, W.J. (2007). Life-cycle assessment of net greenhouse-gas flux for bioenergy cropping systems. *Ecol. Appl.* 17: 675–691.

Adler, P.R., Sanderson, M.A., Weimer, P.J., and Vogel, K.P. (2009). Plant species composition and biofuel yields of conservation grasslands. *Ecol. Appl.* 19: 2202–2209.

Adler, P.R., Del Grosso, S.J., Inman, D. et al. (2012). Mitigation opportunities for life cycle greenhouse gas emissions during feedstock production across heterogeneous landscapes. In: *Managing Agricultural Greenhouse Gasses* (ed. M. Liebig), 203–219. New York: Elsevier.

Adler, P.R., Mitchell, J.G., Pourhashem, G. et al. (2015). Integrating biorefinery and farm biogeochemical cycles offsets fossil energy and mitigates soil carbon losses. *Ecol. Appl.* 25: 1142–1156.

Anderson, W.F., Sarath, G., Edme, S. et al. (2016). Dedicated herbaceous biomass feedstock genetics and development. *Bioenergy Res.* 9: 399–411.

Arundale, R.A., Dohleman, F.G., Heaton, E.A. et al. (2014). Yields of *Miscanthus* × *giganteus* and *Panicum virgatum* decline with stand age in the Midwestern USA. *GCB Bioenergy* 6: 1–13.

Brown, R.C. and Brown, T.R. (2014). *Biorenewable Resources: Engineering New Products from Agriculture*, 2e. Wiley Blackwell.

Burns, J.C., Godschalk, E.B., and Timothy, D.H. (2008). Registration of 'BoMaster' switchgrass. *J. Plant Registrations* 2: 31–32.

Casler, M.D. (2012). Switchgrass breeding, genetics, and genomics. In: *Switchgrass, Green Energy and Technology* (ed. A. Monti), 29–53. London: Springer-Verlag.

Casler, M.D. and Boe, A.R. (2003). Cultivar × environment interactions in switchgrass. *Crop Sci.* 43: 2226–2233.

Casler, M.D., Sosa, S., Hoffman, L. et al. (2017). Biomass yield of switchgrass cultivars under high- versus low-input conditions. *Crop Sci.* 57: 821–832.

Clifton-Brown, J.C., Lewandowski, I., Andersson, B. et al. (2001). Performance of 15 *Miscanthus* genotypes at five sites in Europe. *Agron. J.* 93: 1013–1019.

Clifton-Brown, J.C., Lewandowski, I., Bangerth, F., and Jones, M.B. (2002). Comparative responses to water stress in stay-green, rapid- and slow senescing genotypes of the biomass crop, *Miscanthus*. *New Phytol.* 154: 335–345.

Dale, V.H., Kline, K.L., Wright, L. et al. (2011). Interactions among bioenergy feedstock choices, landscape dynamics and land use. *Ecol. Appl.* 21: 1039–1054.

DeHaan, L.R., Weisberg, S., Tilman, D., and Fornara, D. (2010). Agricultural and biofuel implications of a species diversity experiment with native perennial grassland plants. *Agric. Ecosyst. Environ.* 137: 33–38.

Dickson, T.L. and Gross, K.L. (2015). Can the results of biodiversity-ecosystem productivity studies be translated to bioenergy production? *PLoS One* 10 e0135253. Doi:10.137/journal.pone.0135253.

El-Bassam, N. (2010). *Handbook of Bioenergy Crops: A Complete Reference to Species, Development and Applications*. London: Earthscan Publications Ltd.

Elbersen, H.W., Christian, D.G., El-Bassam, N. et al. (2004). A management guide for planting and production of switchgrass as a biomass crop in Europe. In: *Proceedings of the 2nd World Conference on Biomass for Energy, Industry, and Climate Protection*, 10–14 May, 2004, 140–142. Rome, Italy.

Fedenko, J.R., Erickson, J.E., Woodard, K.R. et al. (2013). Biomass production and composition of perennial grasses grown for bioenergy in a subtropical climate across Florida, USA. *Bioenergy Res.* 6: 1082–1093.

Field, J.L., Evans, S.G., Marx, E. et al. (2017). High resolution techno-ecological modelling of a bioenergy landscape to identify greenhouse gas mitigation opportunities in bioethanol production. *Nat. Energy* 3: 211.

Galatowitsch, S.M., Anderson, N.O., and Ascher, P.D. (1999). Invasiveness in wetland plants in temperate North America. *Wetlands* 19: 733–755.

Gelfand, I. (2011). Carbon debt of Conservation Reserve Program (CRP) grasslands converted to bioenergy production. *Proc. Natl. Acad. Sci. U.S.A.* 108: 13864–13869.

Grabber, J.H. (2016). Prohexadione-calcium improves stand density and yield of alfalfa interseeded into silage corn. *Agron. J.* 108: 726–735.

Griffith, A.P., Epplin, F.M., Fuhlendorf, S.D., and Gillen, R. (2011). A comparison of perennial polycultures and monocultures for producing biomass for biorefinery feedstock. *Agron. J.* 103: 617–627.

Heaton, E.A., Dohleman, F.G., Miguez, A.F. et al. (2010). *Miscanthus*: a promising biomass crop. *Adv. Bot. Res.* 56: 76–156.

Heinsoo, K., Hein, K., Melts, I. et al. (2011). Reed canary grass yield and fuel quality in Estonian farmers' fields. *Biomass Bioenergy* 35: 616–625.

Hendrickson, J.R., Schmer, M.R., and Sanderson, M.A. (2013). Water use efficiency by switchgrass compared to a native grass or native grass alfalfa mixture. *Bioenergy Res.* 6: 746–754.

Hong, C.O., Owens, V.N., Bransby, D.I. et al. (2014). Switchgrass response to nitrogen fertilizer across diverse environments in the USA: a regional feedstock partnership report. *Bioenergy Res.* 7: 777–788.

Jakubowski, A.R., Casler, M.D., and Jackson, R.D. (2011). Has selection for improved agronomic traits made reed canarygrass invasive? *PLoS One* 6 e25757 doi:https://doi.org/10.1371/journal.pone.0025757.

Jenkins, B.M., Baxter, L.L., Miles, T.R. Jr., and Miles, T.R. (1998). Combustion properties of biomass. *Fuel Process. Technol.* 54: 17–46.

Johnson, G.A., Wyse, D.L., and Sheaffer, C.C. (2013). Yield of perennial herbaceous and woody biomass crops over time across three locations. *Biomass Bioenergy* 58: 267–274.

Jorgensen, N.A. and Koegel, R.G. (1988). Wet fractionation processes and products. In: *Alfalfa and Alfalfa Improvement*, Agronomy Monograph, vol. 29 (eds. A.A. Hanson et al.), 553–566. Madison, WI: ASA, CSSA, and SSSA.

Jungers, J.M., Fargione, J.G., Sheaffer, C.C. et al. (2013). Energy potential of biomass from conservation grasslands in Minnesota, USA. *PLoS One* 8 e61209.

Jungers, J.M., Clark, A.T., Betts, K. et al. (2015). Long-term biomass yield and species composition in native perennial bioenergy cropping systems. *Agron. J.* 107: 1627–1640.

Jungers, J.M., Sheaffer, C.C., Fargione, J., and Lehman, C. (2015). Short-term harvesting of biomass from conservation grasslands maintains plant diversity. *GCB Bioenergy* 7: 1050–1061.

Kiniry, J.R., Anderson, L., Johnson, M.-V. et al. (2013). Perennial biomass grasses and the Mason-Dixon line: comparative productivity across latitudes in the southern Great Plains. *Bioenergy Res.* 6: 276–291.

Koegel, R.G. and Straub, R.J. (1996). Fractionation of alfalfa for food, feed, biomass, and enzymes. *Trans. Am. Soc. Agric. Biol. Eng.* 39: 769–774.

Lamb, J.F.S., Sheaffer, C.C., and Samac, D.A. (2003). Population density and harvest maturity effects on leaf and stem yield in alfalfa. *Agron. J.* 95: 635–641.

Lamb, J.F.S., Jung, H.J.G., Sheaffer, C.C., and Samac, D.A. (2007). Alfalfa leaf protein and stem cell wall polysaccharide yields under hay and biomass management systems. *Crop Sci.* 47: 1407–1415.

Landström, S., Lomakka, L., and Andersson, S. (1996). Harvest in spring improves yield and quality of reed canary grass as a bioenergy crop. *Biomass Bioenergy* 11: 333–341.

Lewandowski, I., Scurlock, J.M.O., Lindvall, E., and Christou, M. (2003). The development and current status of perennial rhizomatous grasses as energy crops in the US and Europe. *Biomass Bioenergy* 25: 335–361.

Madakadze, I.C., Stewart, K., Peterson, P.R. et al. (1999). Switchgrass biomass and chemical composition for biofuel in eastern Canada. *Agron. J.* 91: 696–701.

Maughan, M., Bollero, G., Lee, D.K. et al. (2012). *Miscanthus × giganteus* productivity: the effects of management in different environments. *GCB Bioenergy* 4: 253–265.

McKendry, P. (2002a). Energy production from biomass (part 1): overview of biomass. *Bioresour. Technol.* 83: 37–46.

McKendry, P. (2002b). Energy production from biomass (part 2): conversion technologies. *Bioresour. Technol.* 83: 47–54.

McLaughlin, S.B., De La Torre Ugarte, D.G., Garten, C.T. Jr. et al. (2002). High-value renewable energy from prairie grasses. *Environ. Sci. Technol.* 36: 2122–2129.

Miles, T.R., Miles, T.R. Jr., Baxter, L.L. et al. (1996). Boiler deposits from firing biomass fuels. *Biomass Bioenergy* 10: 125–138.

Palmer, I.E., Gehl, R.J., Ranney, T.G. et al. (2014). Biomass yield, nitrogen response, and nutrient uptake of perennial bioenergy grasses in North Carolina. *Biomass Bioenergy* 63: 218–228.

Parrish, D. and Fike, J. (2005). The Biology and Agronomy of switchgrass for Biofuels. *Crit. Rev. Plant Sci.* 24: 423–459.

Richards, B.K., Stoof, C.R., Cary, I., and Woodbury, P.B. (2014). Reporting on marginal lands for bioenergy feedstock production—a modest proposal. *Bioenergy Res.* 7: 1060–1062.

Robertson, G.P., Hamilton, S.K., Del Grosso, S.J., and Parton, W.J. (2011). The biogeochemistry of bioenergy landscapes: carbon, nitrogen, and water considerations. *Ecol. Appl.* 21: 1055–1067.

Saha, D., Rau, B.M., Kaye, J.P. et al. (2017). Landscape control of nitrous oxide emissions during the transition from conservation reserve program to perennial grasses for bioenergy. *GCB Bioenergy* 9: 783–795.

Samac, D.A., Jung, H.J.G., and Lamb, J.F.S. (2006). Development of alfalfa (*Medicago sativa* L.) as a feedstock for production of ethanol and other bioproducts. In: *Alcoholic Fuels* (ed. S. Minteer), 79–98. Boca Raton, FL: CRC Press.

Sanderson, M.A., Read, J.C., and Reed, R.L. (1999). Harvest management of switchgrass for biomass feedstock and forage production. *Agron. J.* 91: 5–10.

Sanderson, M.A., Schmer, M.R., Owens, V. et al. (2012). Crop management of switchgrass. In: *Green Energy and Technology* (ed. A. Monti), 87–112. London: Springer-Verlag.

Saruul, P., Somers, D.A., and Samac, D.A. (2000). Synthesis of biodegradable plastics in alfalfa plants. In: *Proceedings of 7th North American Alfalfa Improvement Conference*. 16–19 July 2000, 296. Madison, WI.

Schmer, M.R., Vogel, K.P., Mitchell, R.B., and Perrin, R.K. (2008). Net energy of cellulosic ethanol from switchgrass. *Proc. Natl. Acad. Sci. U.S.A.* 105: 464–469.

Scordia, D., Zanetti, F., Varga, S.S. et al. (2015). New insights into the propagation methods of switchgrass, *Miscanthus* and giant reed. *Bioenergy Res.* 8: 1480–1491.

Sheaffer, C.C., Martin, N.P., Lamb, J.F.S. et al. (2000). Leaf and stem properties of alfalfa entries. *Agron. J.* 92: 733–739.

Shinners, K.J., Herzmann, M.E., Binversie, B.N., and Digman, M.F. (2007). Harvest fractionation of alfalfa. *Trans. Am. Soc. Agric. Biol. Eng.* 50: 713–718.

Silveira, M.L., Vendramini, J.M.B., Sui, X. et al. (2013). Screening warm-season bioenergy crops as an alternative for phytoremediation of excess soil P. *Bioenergy Res.* 6: 469–475.

Sun, Y. and Cheng, J. (2002). Hydrolysis of lignocellulosic materials for ethanol production: a review. *Bioresour. Technol.* 83: 1–11.

Tahir, M.H.N., Casler, M.D., Moore, K.J., and Brummer, E.C. (2011). Biomass yield and quality of reed canarygrass under five harvest management systems for bioenergy production. *Bioenergy Res.* 4: 111–119.

Tilman, D., Hill, J., and Lehman, C. (2006). Carbon-negative biofuels from low-input high-diversity grassland biomass. *Science* 314: 1598–1600.

U.S. DOE (Department of Energy) (2011). U.S. billion-ton update: biomass supply for a bioenergy and bioproducts industry. https://www1.eere.energy.gov/bioenergy/pdfs/billion_ton_update.pdf (accessed 14 October 2019).

USDA Farm Service Agency (2017a). Conservation Reserve Program Statistics. https://www.fsa.usda.gov/programs-and-services/conservation-programs/reports-and-statistics/conservation-reserve-program-statistics/index. (accessed 14 October 2019).

USDA Farm Service Agency (2017b). FSA Handbook, Agricultural Resource Conservation Program 2-CRP (Revision 5). https://www.fsa.usda.gov/FSA/webapp?area=home&subject=empl&topic=hbk (accessed 14 October 2019).

Vadas, P.A., Barnett, K.H., and Undersander, D.J. (2008). Economics and energy of ethanol production from alfalfa, corn and switchgrass in the upper midwest, USA. *Bioenergy Res.* 1: 44–55.

Vogel, K.P. (1996). Energy production from forages (or American agriculture—Back to the future). *J. Soil Water Conserv.* 51: 137–139.

Vogel, K.P., Brejda, J.J., Walters, D.T., and Buxton, D.R. (2002). Switchgrass biomass production in the Midwest: harvest and nitrogen management. *Agron. J.* 94: 413–420.

Vogel, K.P., Mitchell, R.B., Casler, M.D., and Sarath, G. (2014). Registration of 'Liberty' switchgrass. *J. Plant Reg.*: 242–247.

Wang, M., Han, J., Dunn, J.B. et al. (2012). Well-to-wheels energy use and greenhouse gas emissions of ethanol from corn, sugarcane and cellulosic biomass for US use. *Environ. Res. Lett.* 7 045905.

Weimer, P.J. (2003). Wood adhesives containing solid residues of biomass fermentations. Patent Application Number 0108.03.

Wilson, T.O., McNeal, F.M., Spatari, S. et al. (2012). Densified biomass can cost-effectively mitigate greenhouse gas emissions and address energy security in thermal applications. *Environ. Sci. Technol.* 46: 1270–1277.

Wrobel, C., Coulman, B.E., and Smith, D.L. (2009). The potential use of reed canarygrass (*Phalaris arundinacea* L.) as a biofuel crop. *Acta Agric. Scand. Sect. B* 59: 1–18.

Wu, Y. and Taliaferro, C.M. (2009). Switchgrass cultivar. US Patent 20090300977. United States Patent Office, Washington, D.C.

Wullschleger, S.D., Davis, E.B., Borsuk, M.E. et al. (2010). Biomass production in switchgrass across the United States: database description and determinants of yield. *Agron. J.* 102: 1158–1168.

Zilverberg, C., Johnson, J.W.C., Owens, V. et al. (2014). Biomass yield from planted mixtures and monocultures of native prairie vegetation across a heterogeneous farm landscape. *Agric. Ecosyst. Environ.* 186: 148–159.

PART IX

PASTURE MANAGEMENT

Beef cows and calves on a permanent pasture in central Kentucky. *Source*: Photo courtesy of Mike Collins.

When separated into individual disciplines, both forage management and animal management are relatively straightforward processes. Pasture management refers to the interactions that occur when forages and animals are managed collectively. This typically results in numerous trade-offs. The basis for most common pasture management strategies within a forage-livestock system is the reality that as forage maturity (yield) increases, forage nutritive value (quality) decreases. Nearly every pasture management strategy imagined, researched, or implemented addresses this reality.

There is a wealth of science behind pasture design and stocking methods to balance forage growth with livestock nutritional needs. Often, forage quality is either too high

Forages: The Science of Grassland Agriculture, Volume II, Seventh Edition.
Edited by Kenneth J. Moore, Michael Collins, C. Jerry Nelson and Daren D. Redfearn.

or too low to match animal requirements. Providing supplemental forage to offset forage supply or nutritional deficiencies is not fool-proof because some forages can have toxic effects on animals that either lower animal production or, in extreme cases, result in death. Animal nutritional requirements over time for a production class of livestock are usually more consistent than either forage yield or forage quality. This somewhat simplifies the nutritional aspects for meeting energy, protein, or mineral deficiencies of grazing livestock.

CHAPTER

44

Pasture Design and Grazing Management

Lynn E. Sollenberger, Distinguished Professor, *Agronomy, University of Florida, Gainesville, FL, USA*
Yoana C. Newman, Associate Professor, *Plant and Earth Science, University of Wisconsin River Falls, Madison, WI, USA*
Bisoondat Macoon, Research Professor, *Mississippi State University, Raymond, MS, USA*

Importance of Grazing Management and Pasture Design

Grazing management is defined as the manipulation of grazing in pursuit of a specific objective or set of objectives (Allen et al. 2011). There are often multiple objectives in addition to forage production including forage-use efficiency, plant persistence, production per animal and per unit of land area, economic return, and delivery of ecosystem services (Sollenberger et al. 2012). Pasture design relates to pasture and/or paddock size and shape, slope and aspect of grazing units, and location of feeding, watering, shade, and handling facilities. The goal of pasture design is to achieve a livestock distribution that positively affects pasture utilization, plant diversity, watershed function, and control of animal wastes and nutrient flows. This chapter will (i) describe plant responses to defoliation and the mechanisms underpinning them, (ii) define the key grazing management choices and their potential impact on a grazing system, and (iii) review the elements of effective pasture design.

Defoliation and Plant Response

Grazing Vs Cutting

Grazing and mechanical harvesting affect forage swards differently. Grazing livestock exert a pulling force, possibly disturbing the root system or even uprooting the plant. They also selectively consume plant species, tillers within a plant, and plant parts within a tiller. In contrast, defoliation by clipping is instantaneous for all plants and tillers to the selected stubble height. Defoliation by grazing is also accompanied by treading and excreta deposition that present impacts unlike those associated with mechanical harvesting (Mikola et al. 2009). Thus, clipping and grazing are very different and plant responses to clipping may not be indicative of the response to grazing (Gastal and Lemaire 2015).

Immediate Responses to Defoliation

Gastal and Lemaire (2015) state two guiding principles for understanding plant responses to defoliation. First, defoliation disturbs the carbohydrate supply for plant growth by removing photosynthetic tissues, and second, plant growth processes operate to maintain plants in a dynamic equilibrium with their environment such that resource use is optimal for growth and reproduction. When defoliation removes leaf tissue, reduced photosynthesis limits carbohydrate available to support growth, and a series of physiologic responses ensues to restore homeostatic growth (Richards 1993). These responses are short lived, and if defoliation is infrequent or lenient,

Forages: The Science of Grassland Agriculture, Volume II, Seventh Edition.
Edited by Kenneth J. Moore, Michael Collins, C. Jerry Nelson and Daren D. Redfearn.

leaving a significant portion of the leaf area, then restoration of carbohydrate supply and growth patterns will occur before another defoliation event (Chapman and Lemaire 1993).

Photosynthesis

When plants are grazed, instantaneous reduction of photosynthesis occurs and translocation of previously fixed C is temporarily stopped (Richards 1993). The proportional reduction in photosynthesis exceeds the proportion of leaf area removed because residual leaf is often older than the average of the pre-grazing canopy and had previously been shaded (Gold and Caldwell 1989).

Root Processes

Root elongation ceases within 24 hours after removal of 40–50% or more of the forage shoot mass, and some fine roots may also die and begin to decompose soon after defoliation (Jarvis and Macduff 1989). Biologic N fixation by legumes and nutrient absorption by most plants decline rapidly after defoliation (Richards 1993). The rate of nitrate absorption by perennial ryegrass roots declined within 30 minutes after removal of 70% of forage mass and reached levels less than 40% of pre-defoliation rates within 2 hours (Clement et al. 1978).

Resource Allocation

Carbon supply is diminished due to the reduction in photosynthesis, but plants compensate for the reduced supply. The amount of photosynthate allocated to roots is reduced, and the proportion that is exported from photosynthetically active leaves to actively growing shoot meristematic regions increases (Richards 1993). These compensatory processes begin within hours after defoliation and contribute to more rapid replacement of photosynthetic leaf area. Nitrogen allocation patterns are similar to those described for C. After defoliation of perennial ryegrass, previously absorbed N was preferentially allocated to regrowing leaves, and 80% of the N originated from remaining stubble (Ourry et al. 1988). The remainder came from the root system. These processes appear to be **sink** driven and provide for rapid recovery from grazing by defoliation-tolerant plants (Richards 1993).

Short-Term Responses to Defoliation

After the immediate responses to defoliation, subsequent processes are set in place that lead to restoration of a positive whole-plant C balance. This phase of recovery requires up to several weeks. Richards (1993) suggested that two main processes contribute to increased "carbon gain capacity" after defoliation. These processes include reestablishment of the photosynthetic canopy and increases in the photosynthetic capacity of remaining foliage.

Reestablishment of Positive Whole-Plant Carbon Balance

The most important factor affecting rapid restoration of the leaf canopy is the presence of active shoot meristems. It is not until the plant has enough leaf area to provide the plant with adequate photosynthetic capacity for maintenance and growth that the plant begins reserve replenishment and initiation of root growth. In the case of ryegrass, this occurs when about 75% of a new leaf has regrown (Fulkerson and Donaghy 2001). Leaf expansion results from expansion of already formed cells, and their presence serves as a strong sink for remobilized C and N soon after defoliation (Briske 1986). During this intermediate period, remobilized and current photosynthate continue to be preferentially allocated to the regrowing shoots until their demand is met. Dependence of regrowth on stored reserves is thought by some to persist longer for N than C because of the delay in resumption of N uptake until the plant achieves a positive C balance (Culvenor et al. 1989).

Compensatory Photosynthesis

Another factor affecting canopy recovery, though less important than rapid reestablishment of photosynthetic canopy area, is the potential for increased photosynthetic rates of leaves remaining after defoliation. Compensatory photosynthesis may reflect the ability of mature leaves to rejuvenate their photosynthetic capacity to the higher levels of younger leaves or of younger leaves to slow the normal decline in photosynthetic capacity with aging (Richards 1993).

Long-Term Responses to Defoliation

Plants exhibit physiologic and morphologic responses to defoliation. Physiologic responses generally occur over short-time scales, whereas morphologic responses are generally longer term and are associated with sustained, more severe defoliation (Chapman and Lemaire 1993). Like morphologic responses, plant reserves become more important with extended periods of relatively severe defoliation.

Morphologic Responses

Plants and swards have the capacity to adapt their structure to defoliation, i.e. they exhibit plasticity of sward structure (Gastal and Lemaire 2015) or **phenotypic plasticity** (Nelson 2000). Plasticity of sward structure is reversible and includes changes in size, structure, and spatial positioning of organs (Huber et al. 1999). For example, optimization of canopy leaf area at lower defoliation height may be achieved through a decrease in mean tiller mass and an increase in tiller population density (Matthew et al. 2000). There are limits to phenotypic sward plasticity, however, and if grazing becomes too severe, leaf area, substrate supply, and tiller production are decreased and tiller survival diminished (Matthew et al. 2000), weakening the stand.

Morphologic responses to defoliation are an important part of an "avoidance" mechanism (Briske 1986) that reduces the probability of defoliation for individual plants. There is considerable variation among species or even among cultivars within a species in the extent of phenotypic plasticity for particular traits (Gibson et al. 1992; Shepard et al. 2018), and this can be related to grazing tolerance. Less erect tiller angle, shorter stems, and greater herbage bulk density have been associated with phenotypic plasticity. These changes can result in greater post-grazing residual leaf mass and leaf area index, increasing rate of refoliation and decreasing dependence on stored reserves during regrowth (Hodgkinson et al. 1989; Mullenix et al. 2016; Shepard et al. 2018).

Plant Reserve Status

The mobilization of C and N reserves and their supply to growing leaves is a direct effect of defoliation (Alderman et al. 2011a; Gastal and Lemaire 2015). The importance of reserves in the regrowth and persistence of perennial grasses under defoliation has long been recognized (Richards 1993) but is not without controversy. Some consider plant reserves to play a limited role in regrowth (Humphreys 2001). Others suggest that reserves are used for regrowth only for a few days after defoliation (Richards 1993), yet reduction in storage tissues has occurred from one to several weeks (Skinner et al. 1999; Alderman et al. 2011b). Fulkerson and Slack (1994) observed a positive correlation between water-soluble carbohydrate content (g per plant or $g\,m^{-2}$) in perennial ryegrass stubble and leaf growth in the six days following defoliation; regrowth was more closely correlated with stubble carbohydrate content than concentration ($g\,kg^{-1}$). For several cool-season grasses, including perennial ryegrass, it was only during the first 2–6 days after defoliation that remobilization of reserves was the primary source of C and N for regrowth (Thornton et al. 2000). Thereafter, the plant became progressively more dependent on current assimilate for growth and replenishment of reserves. If current assimilation rates recover quickly to support plant needs, as described earlier for plants demonstrating phenotypic plasticity, the role of reserves is relatively small. However, in some environments and, with particular combinations of forage species and management, reserves are very important.

A general rule is the more stressful the conditions (e.g. heavier grazing, colder winter temperatures, prolonged drought, lower light environments), the more likely reserves will play a significant role in response to defoliation. Under stressful conditions, reserve *content* in storage organs, a function of both storage organ mass and reserve concentration, is more responsive to defoliation than is reserve *concentration* (Ortega-S et al. 1992; Chaparro et al. 1996). These results support the conclusion that some forage species depend significantly on stored reserves to sustain growth during extended periods of severe defoliation.

Grazing Management Choices

A goal of grazing management is to achieve canopy conditions and forage productivity that result in optimal levels of animal performance (Hodgson 1990). Manipulation of grazing intensity, stocking method, and timing of grazing are the primary means of achieving the desired canopy characteristics.

Grazing Intensity

Grazing intensity relates to the severity of grazing. Measures of grazing intensity are animal or pasture based or both. **Stocking rate** (animal units ha^{-1}) is the most common animal-based measure of grazing intensity. Pasture- or sward-based measures include forage mass, canopy height, and canopy light interception. **Forage allowance** and **grazing pressure** incorporate both pasture and animal measures (Allen et al. 2011).

Importance

The selection of grazing intensity is more important than any other grazing management decision (Sollenberger et al. 2012) due to its prominent role in determining forage plant productivity and persistence (Newman et al. 2003b; Hernández Garay et al. 2004), animal performance (Sollenberger and Vanzant 2011), and profitability of the grazing operation (Table 44.1). Understanding the relationship of grazing intensity to pasture and animal performance is crucial for the long-term success of the forage–livestock enterprise.

Effects on Pasture Attributes and Animal Performance

Increasing grazing intensity consistently (>90% of experiments reporting these responses) decreases forage mass and forage allowance, but the effect on forage accumulation rate depends on forage species, grazing frequency, and the environment (Figure 44.1; Sollenberger et al. 2012). In 66% of studies reporting forage **nutritive value** responses to grazing intensity, nutritive value increased with greater stocking rates and when swards were grazed to shorter rather than taller stubble heights (Table 44.1; Hernández Garay et al. 2004; Jones and LeFeuvre 2006; Sollenberger et al. 2012). In more intensively grazed swards, leaf proportion of the forage mass is greater and average age of regrowth is younger because of shorter intervals between grazing bouts or slower regrowth following heavy defoliation (Roth et al. 1990; Pedreira et al. 1999; Dubeux et al. 2006).

Continuously stocked limpograss pastures grazed to a 20-cm stubble had greater leaf, stem, and total bulk density, and crude protein concentration than

Table 44.1 Stocking rate effects on pre-grazing forage mass and nutritive value, forage allowance, and animal performance of weanling bulls rotationally stocked on stargrass (*Cynodon nlemfuensis* Vanderyst) pastures

Stocking rate (head ha^{-1})	Forage mass (Mg ha^{-1})	Forage allowance (kg kg^{-1})	Crude protein (g kg^{-1})	*In vitro* digestion (g kg^{-1})	Neutral detergent fiber (g kg^{-1})	Average daily gain (kg)	Gain ha^{-1} (kg)
2.5	6.6	7.6	134	586	774	0.68	500
5.0	4.5	2.7	140	593	762	0.54	760
7.5	2.7	1.2	151	599	749	0.31	550
Polynomial contrast	Linear	Linear, quadratic	Linear	Linear	Linear	Linear, quadratic	Quadratic

Source: Adapted from Hernandez-Garay et al. (2004).

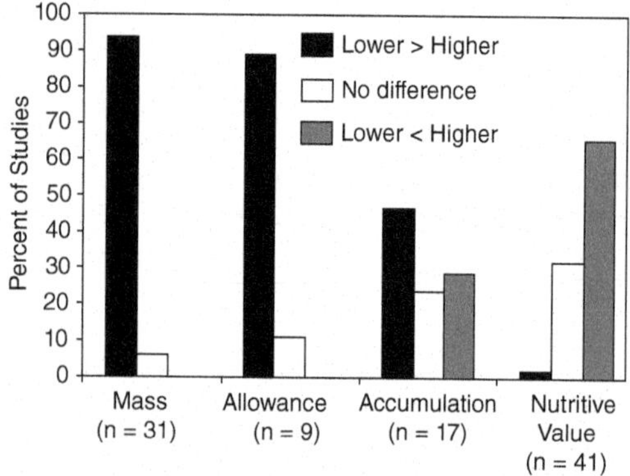

Fig. 44.1. Percentage of studies showing responses to higher and lower grazing intensity for experiments reviewed that reported data based on measures of forage mass, forage allowance, forage accumulation, and forage nutritive value. Number of experiments for each data set is indicated in parentheses. "Higher" and "lower" refer to grazing intensity (i.e. higher or lower stocking rate). *Source:* Adapted from Sollenberger et al. (2012).

pastures grazed to 40 or 60 cm (Newman et al. 2003a). Though greater leaf proportion and bulk density are often positively associated with animal daily gain (Burns and Sollenberger 2002), *accessibility* of leaf to grazing herbivores may be more important than *abundance* of leaf (Sollenberger and Burns 2001). Limpograss canopies grazed to a 40-cm height had lower bulk density and greater livestock daily gain than those grazed to 20 cm, attributable to greater opportunity for leaf selection in the less-dense 40-cm sward (Newman et al. 2002).

The main objective of numerous grazing-intensity studies during the past five decades has been to describe the individual animal performance response as a function of stocking rate or grazing pressure (Sollenberger and Vanzant 2011). All authors agree that performance per animal declines as stocking rate increases across a wide range of stocking rates, but there are different perspectives among authors on the shape of the curve (Jones and Jones 1997).

The differential effects of grazing intensity on forage nutritive value and forage mass underlie this relationship. Above some forage mass threshold, perhaps 2 Mg ha^{-1} for temperate and 4 Mg ha^{-1} for tropical swards, animals are able to select a diet of their choice in a sustainable daily grazing time (6–9 hours) (Burns et al. 1989; Hernández Garay et al. 2004), and forage mass has little causative influence on animal response. With extreme understocking, however, daily animal production may be reduced due to accumulation of mature and senescent forage. Newman et al. (2002) showed that lightly grazed canopies had more trampled and lodged forage, and gains were lower than with moderately grazed swards. In contrast, as stocking rate is increased, at some point forage intake decreases sufficiently to cause a shift in use of consumed energy away from maximum daily animal growth and toward meeting the animals' maintenance requirement (Burns et al. 2004). The consequence is reduced gain per animal (Figure 44.2). Thus, the influence of forage quantity on animal performance is greater at low levels than at high levels of forage mass or allowance. When forage mass or allowance were not limiting, forage nutritive value explained 56–77% of variation in performance per animal (Duble et al. 1971; McCartor and Rouquette 1977). Based on a meta-analysis, forage nutritive value (i) sets the upper limit for average daily gain, (ii) determines the slope of the regression of daily gain on stocking rate, and (iii) establishes the forage mass at which daily gain plateaus (Sollenberger and Vanzant 2011). In contrast, forage quantity determines the proportion of potential daily gain that is achieved and is the primary determinant of the pattern of the daily gain response (negative) to increasing stocking rate.

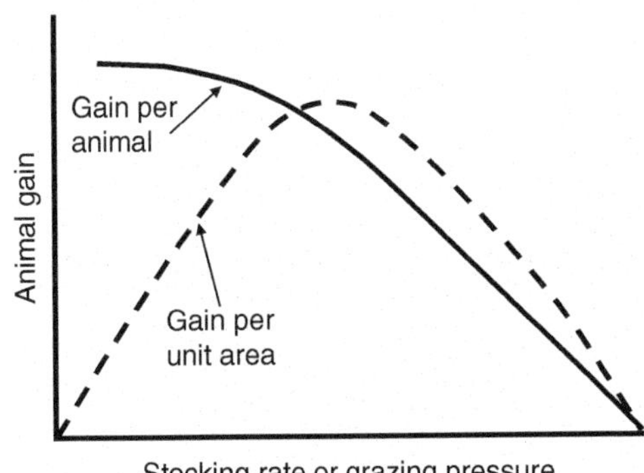

Fig. 44.2. The relationship of gain per animal and gain per hectare with stocking rate or grazing pressure. *Source:* Adapted from Mott and Moore (1985).

In contrast to its effect on individual animal gain, increasing stocking rate on a previously underutilized pasture causes animal gain per hectare to increase up to some maximum (Figure 44.2; Hernandez Garay et al. 2004). Increasing stocking rates above this level causes production per hectare to decline because animals can consume only enough forage to support lower levels of daily gain (Figure 44.3).

Stocking Method

Stocking method is the manner that animals are stocked or have access to a number of pastures (grazing management units) and paddocks (pasture subdivisions, if present) during the grazing season. Choice of stocking method is distinctly separate from that of grazing intensity; thus, a particular stocking method may be used across a wide range of stocking rates or grazing pressures.

Many stocking methods have been described (Allen et al. 2011), but they conform to or derive from one of two types: continuous or some form of rotational (also called intermittent) stocking. Under **continuous stocking**, animals have unlimited and uninterrupted access to the grazing area throughout the period when grazing occurs. If the number of animals used is fixed, it is referred to as **set stocking. Rotational stocking** uses alternating periods of stocking and rest among two or more paddocks in a pasture (Figure 44.4). The objective is to balance rest and stocking periods to achieve an efficient and uniform defoliation of the pasture. **Stocking period** length is set to leave a target stubble height or residual leaf area, and the optimum height is dictated by the grazing tolerance of the forage species and the nutrient requirements of the grazing animal. The **rest period** is, likewise, species dependent and is set to allow the maximum forage accumulation rate without compromising persistence or unduly compromising nutritive value (Pedreira et al. 1999).

A literature synthesis found that 71% of studies comparing rotational and continuous stocking reported no difference in forage nutritive value, but 81% reported an advantage in pasture carrying capacity for rotational stocking (Sollenberger et al. 2012). This advantage averaged approximately 30% and was due, in part, to greater forage accumulation rate for rotational stocking, attributable to greater average leaf area index and younger average leaf age than on continuously stocked pastures (Parsons et al. 1988). Rotational stocking also resulted in greater homogeneity of forage utilization than continuous stocking which reduced spot overgrazing and increased the proportion of pasture area experiencing a longer linear growth phase (Figure 44.5; Saul and Chapman 2002; Hunt et al. 2007; Barnes et al. 2008). Sixty-six percent of studies comparing animal performance on rotationally and continuously stocked pastures showed no difference in daily animal performance, and 69% showed no difference in production per unit land area (Sollenberger et al. 2012). The latter response was dependent on research methodology, because when stocking methods were compared using a **variable stocking** rate technique, rotationally stocked pastures achieved greater animal production per unit land area in more than 40% of experiments (Sollenberger et al. 2012).

Though greater uniformity of excreta deposition is often attributed to rotational stocking, this response is environment specific. Continuous stocking was compared with rotational stocking with 1-, 3-, 7-, and 21-days grazing periods and a 21-days rest period (Dubeux et al. 2014). Soil nutrients accumulated for all stocking methods in the surface 8 cm of soil in zones near shade and water. Air temperature, wind speed, and temperature–humidity index explained 49% of the variation in time cattle spent under shade, confirming the importance of the environment in animal behavior. Results of this, and other studies (Mathews et al. 1994), support a conclusion that the greatest benefit of rotational stocking in terms of uniformity of nutrient return in excreta is likely to occur in temperate environments or during cool seasons (Dubeux et al. 2007).

Rotational stocking can occur in several forms. The least-complex is **alternate stocking** where two paddocks are used for the rotation. At the other extreme is **strip stocking**, more commonly used in pasture-based dairies with forages such as alfalfa or hybrid bermudagrass. Paddocks in strip stocking are smaller than for other rotational systems, and the grazing period is usually a fraction of a day to 2 days. Strip stocking minimizes daily variability in diet nutritive value because the residence period in the paddock is short and selectivity is reduced. The **first–last grazer** approach to rotational stocking is used when animals with different nutritional requirements are grazed sequentially on a given paddock. Animals with higher nutritional requirements are allowed

Fig. 44.3. The weanling bulls in the foreground were stocked at 7.5 head ha^{-1} on stargrass pastures for a 300-days grazing season while the bull in the background was on a pasture stocked at 2.5 head ha^{-1}. Average daily gain was 0.31 and 0.68 kg for animals from high- and low- stocking rate treatments, respectively (Hernández Garay et al. 2004). *Source:* Photo by Lynn Sollenberger, University of Florida.

Fig. 44.4. Rotational stocking applied to a 'Florakirk' bermudagrass pasture. *Source:* Photo by Lynn Sollenberger, University of Florida.

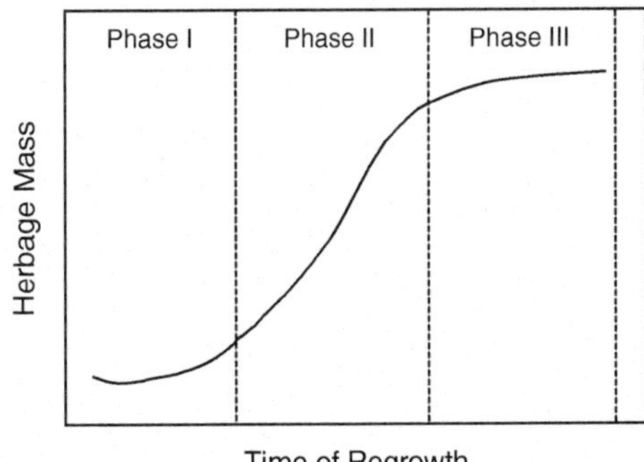

Fig. 44.5. Accumulation of forage mass during a regrowth period follows a sigmoid curve as the canopy develops from low mass (Phase 1: low accumulation rate) to intermediate mass (Phase 2: high accumulation rate) to high mass (Phase 3: little or no net accumulation due to balance between new growth and senescence). *Source:* Adapted from Saul and Chapman (2002).

first access, making it possible to achieve targeted rates of daily gain for two classes of livestock grazing the same pasture. Other methods in use, but less widely adopted, are **frontal stocking** and **creep stocking**. Frontal stocking is a type of strip stocking in which a sliding fence is pushed by cattle and gradually exposes new forage. A back fence prevents the animals from accessing the previously grazed areas. This approach has resulted in uniform grazing and defoliation of close to 100% of tillers (Volesky 1994). Creep stocking uses lactating–nursing animal pairs. A higher-quality forage is available adjacent to the base rotation module, and this forage is available to the nursing animals through the use of creep gates.

The use of continuous stocking is widespread throughout the US. Reasons include fewer management decisions (Bertelsen et al. 1993), rotational stocking is not required for persistence of some species (e.g. tall fescue and bahiagrass), and no consistent advantage in animal performance has been documented for rotational stocking (Sollenberger et al. 2012). Continuous stocking is a common practice in extensive grazing systems, including shortgrass rangeland (Hart and Ashby 1998) and mixed-grass prairies (Guillen et al. 2000). In these locations, dividing pastures and moving cattle may not be practical or economic. If stocking rates are moderate, animals have opportunities for selection when continuously stocked (Vallentine 2001).

In some cases, stocking method may be used strategically to control weeds in planted pastures. Vaseygrass, a bunchgrass weed, becomes stemmy and unpalatable during the rest period in rotationally stocked limpograss pastures, resulting in avoidance by cattle, seed set, and an increase in weed density and cover. Under continuous stocking, as new vaseygrass leaves emerged they were readily consumed by cattle, and vaseygrass plant density and cover decreased (Newman et al. 2003b). Yet, in other cases, control of a weed was achieved by rotational stocking. In tall fescue (Hoveland et al. 1997) and 'Callie' bermudagrass (Mathews et al. 1994) pastures, rotational stocking allowed the preferred species to shade common bermudagrass during the **rest period** and favored persistence of tall fescue and Callie.

Still for other species, such as alfalfa, production and persistence may not be sustained under continuous stocking (Schlegel et al. 2000), though there are large differences among cultivars (Brummer and Moore 2000). In considering persistence, it should be noted that continuous stocking does not imply a high stocking rate. In some cases, where continuous stocking has been implicated in stand loss, **overstocking** may have been more directly responsible.

Timing of Grazing

Use of a particular management practice may be effective at some times or under certain conditions but not others. The choice of timing for defoliation may be influenced by stage of plant regrowth following defoliation. An example is the degree to which reserves have been restored prior to onset of winter or a dry season. In other situations, timing of defoliation is influenced by reproductive tiller formation or elevation of apical meristems (Matches and Burns 1995). Stand losses of smooth bromegrass and timothy growing with alfalfa have been associated with defoliation during the critical period between stem elongation and heading growth stages (Casler and Carlson 1995). Early-maturing timothy cultivars persisted better with alfalfa than did late-maturing cultivars (Casler and Walgenbach 1990). Similarly, defoliation that removes shoot apices of switchgrass often reduces tiller density and, if not followed by a long regrowth period, may compromise stand persistence (Anderson et al. 1989).

Closure date of late-season grazing can be critical for annual or short-lived perennial species that rely on natural reseeding for stand regeneration. In northeastern Texas, most cultivars of annual ryegrass grazed until late April produced satisfactory volunteer stands the following autumn (Evers and Nelson 2000). Later grazing greatly reduced inflorescence density and seed weight per spike and decreased volunteer seedling density. Similarly, seed yield of the summer-annual legume aeschynomene was greatly reduced if autumn grazing continued after first flower (Chaparro et al. 1991).

There is diurnal variation in forage nutritive value that may influence recommended timing of defoliation. These changes are associated with accumulation of photosynthate during the day (Fisher et al. 2002). They showed that nutritive value and animal preference were greater for hays cut in the afternoon compared with those harvested in the morning. In strip-stocking systems where animals are moved daily (e.g. lactating dairy cows), it may be advantageous to move animals to new paddocks in the afternoon/early evening so that the larger meal that usually follows transition to a new grazing area is composed of forage of the greatest possible nutritive value.

Pasture Design in Grazing Systems

Design of pastures depends on a number of factors, including landscape characteristics and intensity/complexity of grazing management. Pasture design is particularly important when considering responses that are affected by distribution of livestock across the landscape. These include efficient utilization of forage, sustaining plant diversity, maintaining riparian control and watershed function, avoiding animal waste and excess nutrient flow into water bodies, supporting stream bank stability, and provision of ecosystem services.

Fencing to define the boundaries of pastures and **paddocks** can be permanent or temporary. Permanent fences often require less management once installed but initial costs are greater than for temporary fences, and management flexibility is reduced.

Paddock Number, Size, and Shape

Choice of stocking method is a major determinant of pasture design. In rotational stocking, the optimum number of paddocks depends on grazing management objectives and type of animal production system. The required number of paddocks in a rotational stocking scheme can be calculated as: rest period ÷ grazing period + 1. The literature is not clear on the benefits of greater vs fewer number of paddocks in rotational stocking. Greater number of paddocks had a positive effect on forage accumulation rate or pasture carrying capacity in approximately 50% of research comparisons but had no effect on forage nutritive value in 75% of comparisons of more vs fewer paddocks (Sollenberger et al. 2012). In a study with grazing periods of 1, 3, 7, or 21 days (all with a 21-days rest period), there was no effect of number of paddocks on forage accumulation rate, crude protein, or *in vitro* digestibility of bahiagrass (Stewart et al. 2005). More vs fewer paddocks in rotationally stocked swards definitely reduces day-to-day variation in diet nutritive value, and in some cases, it increases uniformity of excreta deposition (most likely in cooler environments) and homogeneity of forage utilization across the pasture. The latter contributes to greater pasture carrying capacity.

Increasing popularity of rotational stocking methods with many paddocks in the rotation is facilitated by the easy availability, improved technology, and economic benefits of temporary fencing. A form of high-density rotational stocking with long rest intervals (60 days or more) between grazing events is called **mob grazing** in the popular press. While the International Forage and Grazing Terminology Committee does not include mob grazing as official terminology, they define **mob stocking** as "a method of stocking at a high grazing pressure for a short time to remove forage rapidly as a management strategy" (Allen et al. 2011). It is useful to note that the definition of mob stocking does not reference length of rest interval between grazing events, thus it should not be confused with the informal term mob grazing. While mob grazing is practiced in various forms by growers, this method of grazing is poorly defined and its source unclear. Practitioners of mob grazing claim numerous pasture and animal benefits (Gompert 2010). Some recommend that achieving 60% trampling of the standing forage mass is the optimum level for increasing soil organic matter and nutrient concentration (Peterson and Gerrish 1995), but data are currently lacking to substantiate these claims.

The optimum size of paddocks depends on many factors, including management objectives and number of paddocks desired, land availability and terrain, and herd size relative to stocking density desired. Distribution of animals in the landscape is affected by stocking density and this can be manipulated to make more effective use of pasture resources (Hunt et al. 2007) as well as reduce potential for grassland degradation that may result from patch grazing (Barnes et al. 2008). Nutrient distribution in pastures benefits from smaller paddock sizes in some environments (Dubeux et al. 2009). Temporary fencing can be used to allow flexibility of subdivisions within permanent boundaries of paddocks or pastures in cases where management based on forage allowance is desired. It may be desirable to have larger pastures with low-productivity forage systems while highly productive pastures may be better managed by division into smaller paddocks.

Shape of pastures is determined by management objectives, land availability, and landscape features. It has been suggested that, within practical limits, square pastures allow more grazing efficiency than other shapes by allowing greater forage intake in less grazing time, less energy expenditure incurred during grazing, and reduced loss of forage due to trampling. It is recommended that long, narrow pastures be avoided, mostly because they tend to increase the potential for patch grazing. Irregularly shaped pastures are sometimes the only option when dictated by terrain and landscape constraints and are often practical for commercial pasture-based livestock production. In research settings, however, consistency in shape and size of paddocks is important, especially to minimize variation in implementing non-treatment management practices like applying fertilizer.

Slope and Aspect

Topography and aspect can affect type of vegetation and timing of forage readiness, for example, in the northern hemisphere north-facing slopes are likely to start growing later in spring than south-facing slopes. Steepness of slope can affect herbage accumulation rate with lower slopes having greater herbage accumulation; this can affect grazing behavior of animals and should be considered in pasture design (López et al. 2003). Topography affects distribution of livestock on pasture and plays a key role in deposition of dung and urine (Rowarth et al. 1992), which is closely associated with time spent in a portion of the landscape (Dubeux et al. 2009). Most livestock species prefer easy access to forage and to minimize expenditure of energy during grazing. As a result, they spend more time on flat areas than on slopes, and time spent on slopes decreases with increasing slope (Rowarth et al. 1992). A key consideration regarding slope and aspect in designing and laying out pastures is erosion control and nutrient runoff. Fences should be erected across slopes rather than up and down. Animals, especially cattle and horses, patrol fence lines and the resulting paths developed by their hoof action become natural channels for water flow. Additionally, management practices like spraying herbicide in fence lines increase likelihood of channels forming.

Shade and Water Placement

In addition to topography, shade sources, water sources, and feed and mineral salt sources placement affect distribution of livestock across pastures because they tend to congregate at these locations (Sollenberger et al. 2012). Shade, both natural and artificial, is useful to livestock management because it allows animals to escape from heat. Dubeux et al. (2009) reported that cattle spend a disproportionate amount of time in shade during warm weather. Management concerns center around unequal distribution of nutrients, bacteria, and other contaminants of pasture due to the concentration of urine and feces in areas frequented by cattle. Shade, water, and feed sources can be used as a management tool to, for example, lure animals away from natural water sources such as streams where congregation by livestock may damage banks and cause other undesirable environmental effects (Belsky et al. 1999; Agouridis et al. 2005).

When locating water sources, how far animals have to travel should also be considered. Many practitioners suggest a rule of thumb that livestock not walk more than 400 m to water. Where the distance is greater, animals may spend more time near the water source leading to overgrazing and excessive nutrient build up.

Factors Affecting Choice of Grazing Management

In practice, choice of a grazing management is much more complex than identifying the combination of intensity, method, and timing that maximizes forage accumulation, daily animal performance, or production per unit of land area. Other key considerations are risk and economic return to the producer, long-term pasture persistence, environmental impact, and whether or not the level of decision making associated with a given management practice suits the interests of the practitioner.

Looking forward, production potential may play a lesser role in management decisions, and environmental impact and delivery of ecosystem services may assume a greater importance. Thus, the effects of grazing management on soil nutrient redistribution and accumulation (Dubeux et al. 2007), nutrient runoff and leaching, soil compaction and erosion (da Silva et al. 2003), surface water and groundwater quality, and C sequestration may well be critical factors that affect future recommendations of grazing management in forage–livestock systems (Sollenberger et al. 2012).

References

Agouridis, C.T., Workman, S.R., Warner, R.C., and Jennings, G.D. (2005). Livestock grazing management impacts on stream water quality: a review. *J. Am. Water Resour. Assoc.* 41: 591–606.

Alderman, P.D., Boote, K.J., and Sollenberger, L.E. (2011a). Regrowth dynamics of 'Tifton 85' bermudagrass as affected by nitrogen fertilization. *Crop Sci.* 51: 1716–1726.

Alderman, P.D., Boote, K.J., Sollenberger, L.E., and Coleman, S.W. (2011b). Carbohydrate and nitrogen reserves relative to regrowth dynamics of 'Tifton 85' bermudagrass as affected by nitrogen fertilization. *Crop Sci.* 51: 1727–1738.

Allen, V.G., Batello, C., Berretta, E.J. et al. (2011). An international terminology for grazing lands and grazing animals. *Grass Forage Sci.* 66: 2–28.

Anderson, B., Matches, A.G., and Nelson, C.J. (1989). Carbohydrate reserves and tillering of switchgrass following clipping. *Agron. J.* 81: 13–16.

Barnes, M.K., Norton, B.E., Maeno, M., and Malechek, J.C. (2008). Paddock size and stocking density affect spatial heterogeneity of grazing. *Rangeland Ecol. Manage.* 61: 380–388.

Belsky, A.J., Matzke, A., and Uselman, S. (1999). Survey of livestock influences on stream and riparian ecosystems in the western United States. *J. Soil Water Conserv.* 54: 419–431.

Bertelsen, B.S., Faulkner, D.B., Buskirk, D.D., and Castree, J.W. (1993). Beef cattle performance and forage characteristics of continuous, 6-paddock, and 11-paddock grazing systems. *J. Anim. Sci.* 71: 1381–1389.

Briske, D.D. (1986). Plant responses to defoliation: morphological considerations and allocation priorities. In: *Rangelands: A Resource Under Siege* (eds. J.W. Joss et al.), 425–427. UK: Cambridge University Press.

Brummer, E.C. and Moore, K.J. (2000). Persistence of perennial cool-season grass and legume cultivars under continuous grazing by beef cattle. *Agron. J.* 92: 466–471.

Burns, J.C., Lippke, H., and Fisher, D.S. (1989). The relationship of herbage mass and characteristics to animal responses in grazing experiments. In: *Grazing Research: Design, Methodology, and Analysis* (ed. G.C. Marten), 7–19. Madison, WI: CSSA, Special Publication No. 16.

Burns, J.C., McIvor, J.G., Villalobos, L. et al. (2004). Grazing systems for C4 grasslands: a global perspective. In: *Warm-Season (C4) Grasses* (eds. L.E. Moser et al.). Madison, WI: ASA, CSSA.

Casler, M.D. and Carlson, I.T. (1995). Smooth bromegrass. In: *Forages: An Introduction to Grassland Agriculture* (eds. R.F Barnes et al.), 313–324. Ames: Iowa State University Press.

Casler, M.D. and Walgenbach, R.P. (1990). Ground cover potential of forage grass cultivars mixed with alfalfa at divergent locations. *Crop Sci.* 30: 825–831.

Chaparro, C.J., Sollenberger, L.E., and Linda, S.B. (1991). Grazing management effects on aeschynomene seed production. *Crop Sci.* 31: 197–201.

Chaparro, C.J., Sollenberger, L.E., and Quesenberry, K.H. (1996). Light interception, reserve status, and persistence of clipped Mott elephantgrass swards. *Crop Sci.* 39: 649–655.

Chapman, D.F. and Lemaire, G. (1993). Morphogenetic and structural determinants of plant growth after defoliation. In: *Grasslands for Our World* (ed. M.J. Baker), 55–64. Wellington, New Zealand: SIR Publishing.

Clement, C.R., Hopper, M.J., Jones, L.H.P., and Leafe, E.L. (1978). The uptake of nitrate by *Lolium perenne* from flowing nutrient solution. II. Effect of light, defoliation, and relationship to CO_2 flux. *J. Exp. Bot.* 29: 1173–1183.

Culvenor, R.A., Davidson, I.A., and Simpson, R.J. (1989). Regrowth by swards of subterranean clover after defoliation. 1. Growth, non-structural carbohydrate and nitrogen content. *Ann. Bot.* 64: 545–556.

Dubeux, J.C.B. Jr., Stewart, R.L. Jr., Sollenberger, L.E. et al. (2006). Spatial heterogeneity of herbage response to management intensity in continuously stocked Pensacola bahiagrass pastures. *Agron. J.* 98: 1453–1459.

Dubeux, J.C.B. Jr., Sollenberger, L.E., Mathews, B.W. et al. (2007). Nutrient cycling in warm-climate grasslands. *Crop Sci.* 47: 915–928.

Dubeux, J.C.B. Jr., Sollenberger, L.E., Gaston, L.A. et al. (2009). Animal behavior and soil nutrient distribution in continuously stocked Pensacola bahiagrass pastures managed at different intensities. *Crop Sci.* 49: 1453–1459.

Dubeux, J.C.B. Jr., Sollenberger, L.E., Vendramini, J.M.B. et al. (2014). Stocking method, animal behavior, and soil nutrient redistribution: how are they linked? *Crop Sci.* 54: 2341–2350.

Duble, R.L., Lancaster, J.A., and Holt, E.C. (1971). Forage characteristics limiting animal performance on warm-season perennial grasses. *Agron. J.* 63: 795–798.

Evers, G.W. and Nelson, L.R. (2000). Grazing termination date influence on annual ryegrass seed production and reseeding in the southeastern USA. *Crop Sci.* 40: 1724–1728.

Fisher, D.S., Mayland, H.F., and Burns, J.C. (2002). Variation in ruminant preference for alfalfa hays cut at sunup and sundown. *Crop Sci.* 42: 231–237.

Fulkerson, W.J. and Donaghy, D.J. (2001). Plant-soluble carbohydrate reserves and senescence – key criteria for developing an effective grazing management system for ryegrass-based pastures: a review. *Aust. J. Exp. Agric.* 41 (2): 261–275.

Fulkerson, W.J. and Slack, K. (1994). Leaf number as a criterion for determining defoliation time for *Lolium perenne*. 1. Effect of water-soluble carbohydrates and senescence. *Grass Forage Sci.* 49: 373–377.

Gastal, F. and Lemaire, G. (2015). Defoliation, shoot plasticity, sward structure and herbage utilization in pasture: review of underlying ecophysiological processes. *Agriculture* 5: 1146–1171.

Gibson, D., Casal, J.J., and Deregibus, V.A. (1992). The effect of plant density on shoot and leaf lamina angles in *Lolium multiflorum* and *Paspalum dilatatum*. *Ann. Bot.* 70: 69–73.

Gold, W.G. and Caldwell, M.M. (1989). The effects of the spatial pattern of defoliation on regrowth of a tussock grass. II. Canopy gas exchange. *Oecologia* 81: 437–442.

Gompert, T. (2010). The power of stock density. Proceedings of the Nebraska Grazing Conference, Kearney, NE, USA (20 July 2010).

Guillen, R.L., Eckroat, J.A., and McCollum, F.T. III (2000). Vegetation response to stocking rate in southern mixed-grass prairie. *J. Range Manage.* 53: 471–478.

Hart, R.H. and Ashby, M.M. (1998). Grazing intensities, vegetation, and heifer gains: 55 years on shortgrass. *J. Range Manage.* 51: 392–398.

Hernández Garay, A., Sollenberger, L.E., McDonald, D.C. et al. (2004). Nitrogen fertilization and stocking rate affect stargrass pasture and cattle performance. *Crop Sci.* 44: 1348–1354.

Hodgkinson, K.C., Ludlow, M.M., Mott, J.J., and Baruch, Z. (1989). Comparative responses of the savanna grasses *Cenchrus ciliaris* and *Themeda triandra* to defoliation. *Oecologia* 79: 45–52.

Hodgson, J. (1990). *Grazing Management: Science into Practice*. New York: Wiley.

Hoveland, C.S., McCann, M.A., and Hill, N.S. (1997). Rotational vs. continuous stocking of beef cows and calves on mixed endophyte-free tall fescue-bermudagrass pasture. *J. Prod. Agric.* 10: 245–250.

Huber, H., Lukács, S., and Watson, M.S. (1999). Spatial structure of stoloniferous herbs: an interplay between structural blueprint, ontogeny and phenotypic plasticity. *Plant Ecol.* 141: 107–115.

Humphreys, L.R. (2001). International grassland congress outlook: an historical review and future expectations. In: *Proceedings of the International Grasslands Congress, 19th, São Pedro, Brazil, 10–21 February 2001* (eds. J.A. Gomide et al.), 1085–1087. Piracicaba, Brazil: Brazilian Society of Animal Husbandry.

Hunt, L.P., Petty, S., Cowley, R. et al. (2007). Factors affecting the management of cattle grazing distribution in northern Australia: preliminary observations on the effect of paddock size and water points. *Rangeland J.* 29: 169–179.

Jarvis, S.C. and Macduff, J.H. (1989). Nitrate nutrition of grasses from steady-state supplies in flowing solution culture following nitrate deprivation and/or defoliation. I. Recovery of uptake and growth and their interactions. *J. Exp. Bot.* 40: 965–975.

Jones, R.J. and Jones, R.M. (1997). Grazing management in the tropics. In: *Proceeding of the International Grasslands Congress, 18th, Winnipeg and Saskatoon, Canada. 8–17 June 1997. Grasslands 2000, Toronto*, 535–542.

Jones, R.J. and LeFeuvre, R.P. (2006). Pasture production, pasture quality and their relationships with steer gains on irrigated, N-fertilised pangola grass at a range of stocking rates in the Ord Valley, Western Australia. *Trop. Grasslands* 40: 1–13.

López, I.F., Hodgson, J., Hedderly, D.I. et al. (2003). Selective defoliation by sheep according to slope and plant species in the hill country of New Zealand. *Grass Forage Sci.* 58: 339–349.

Matches, A.G. and Burns, J.C. (1995). Systems of grazing management. In: *Forages: The Science of Grassland Agriculture* (eds. R.F Barnes et al.), 179–192. Ames: Iowa State University Press.

Mathews, B.W., Sollenberger, L.E., Kunkle, W.E. et al. (1994). Dairy heifer and bermudagrass pasture responses to rotational and continuous stocking. *J. Dairy Sci.* 77: 244–252.

Matthew, C., Assuero, S.G., Black, C.K., and Sackville Hamilton, N.R. (2000). Tiller dynamics of grazed swards. In: *Grassland Ecophysiology and Grazing Ecology* (eds. G. Lemaire et al.), 127–150. New York: CABI Publisher.

McCartor, M.M. and Rouquette, F.M. Jr. (1977). Grazing pressures and animal performance from pearl millet. *Agron. J.* 69: 983–987.

Mikola, J., Setälä, H., Virkajärvi, P. et al. (2009). Defoliation and patchy nutrient return drive grazing effects on plant and soil properties in a dairy cow pasture. *Ecol. Monogr.* 79: 221–244.

Mott, G.O. and Moore, J.E. (1985). Evaluating forage production. In: *Forages: The Science of Grassland Agriculture* (eds. R.F Barnes et al.), 97–110. Ames, IA: Iowa State University Press.

Mullenix, M.K., Sollenberger, L.E., Wallau, M.O. et al. (2016). Sward structure, light interception, and rhizome-root responses of rhizoma peanut cultivars and germplasm to grazing management. *Crop Sci.* 56: 899–906.

Nelson, C.J. (2000). Shoot morphological plasticity of grasses: leaf growth vs. tillering. In: *Grassland Ecophysiology and Grazing Ecology* (eds. G. Lemaire et al.), 101–126. New York: CABI Publisher.

Newman, Y.C., Sollenberger, L.E., Kunkle, W.E., and Chambliss, C.G. (2002). Canopy height and nitrogen supplementation effects on performance of heifers grazing limpograss. *Agron. J.* 94: 1375–1380.

Newman, Y.C., Sollenberger, L.E., and Chambliss, C.G. (2003a). Canopy characteristics of continuously stocked limpograss swards grazed to different heights. *Agron. J.* 95: 1246–1252.

Newman, Y.C., Sollenberger, L.E., Fox, A.M., and Chambliss, C.G. (2003b). Canopy height effects on vaseygrass and bermudagrass spread in limpograss pastures. *Agron. J.* 95: 390–394.

Ortega-S., J.A., Sollenberger, L.E., Bennett, J.M., and Cornell, J.A. (1992). Rhizome characteristics and canopy light interception of grazed rhizoma peanut pastures. *Agron. J.* 84: 804–809.

Ourry, A., Boucaud, J., and Salette, J. (1988). Nitrogen mobilization from stubble and roots during re-growth of defoliated perennial ryegrass. *J. Exp. Bot.* 39: 803–809.

Parsons, A.J., Johnson, I.R., and Harvey, A. (1988). Use of a model to optimize the interaction between frequency and severity of intermittent defoliation and to provide a fundamental comparison of the continuous and intermittent defoliation of grass. *Grass Forage Sci.* 43: 49–59.

Pedreira, C.G.S., Sollenberger, L.E., and Mislevy, P. (1999). Productivity and nutritive value of 'Florakirk' bermudagrass as affected by grazing management. *Agron. J.* 91: 796–801.

Peterson, P.R. and Gerrish, J.R. (1995). *Grazing Management Affects Manure Distribution by Beef Cattle*, 170–174. Lexington: Proceedings of American Forage Grassland Council.

Richards, J.H. (1993). Physiology of plants recovering from defoliation. In: *Grasslands for Our World* (ed. M.J. Baker), 46–54. Wellington, New Zealand: SIR Publishing.

Roth, L.D., Rouquette, F.M. Jr., and Ellis, W.C. (1990). Effects of herbage allowance on herbage and dietary attributes of Coastal bermudagrass. *J. Anim. Sci.* 68: 193–205.

Rowarth, J.S., Tillman, R.W., Gillingham, A.G., and Gregg, P.E.H. (1992). Phosphorus balances in grazed hill-country pastures: the effect of slope and fertilizer input. N.Z. *J. Agric. Res.* 35: 337–342.

Saul, G.R. and Chapman, D.F. (2002). Grazing methods, productivity and sustainability for sheep and beef pastures in temperate Australia. *Wool Technol. Sheep Breed.* 50: 449–464.

Schlegel, M.L., Wachenheim, C.J., Benson, M.E. et al. (2000). Grazing methods and stocking rates for direct-seeded alfalfa pastures: I. Plant productivity and animal performance. *J. Anim. Sci.* 78: 2192–2201.

Shepard, E.M., Sollenberger, L.E., Kohmann, M.M. et al. (2018). Grazing management affects Ecoturf rhizoma peanut forage performance and canopy structure. *Crop Sci.* https://doi.org/10.2135/cropsci2015.02.0090.

da Silva, A.P., Imhoff, S., and Corsi, M. (2003). Evaluation of soil compaction in an irrigated short-duration grazing system. *Soil Tillage Res.* 70: 83–90.

Skinner, R.H., Morgan, J.A., and Hanson, J.D. (1999). Carbon and nitrogen reserve remobilization following defoliation: nitrogen and elevated CO_2 effects. *Crop Sci.* 39: 1749–1756.

Sollenberger, L.E. and Burns, J.C. (2001). Canopy characteristics, ingestive behaviour and herbage intake in cultivated tropical grasslands. In: *Proceedings of the International Grassland Congress, 19th, São Pedro, Brazil. 10–21 February 2001* (eds. J.A. Gomide et al.), 321–327. Piracicaba, Brazil: Brazilian Society of Animal Husbandry.

Sollenberger, L.E. and Vanzant, E.S. (2011). Interrelationships among forage nutritive value and quantity and individual animal performance. *Crop Sci.* 51: 420–432.

Sollenberger, L.E., Agouridis, C.T., Vanzant, E.S. et al. (2012). Prescribed grazing on pasturelands. In: *Conservation Outcomes from Pastureland and Hayland Practices: Assessment, Recommendations, and Knowledge Gaps* (ed. C.J. Nelson), 111–204. Lawrence, KS: Allen Press.

Stewart, R.L. Jr., Dubeux, J.C.B. Jr., Sollenberger, L.E. et al. (2005). Stocking method affects plant responses of Pensacola bahiagrass pastures.Online. *Forage Grazinglands* https://doi.org/10.1094/FG-2005-1028-01-RS.

Thornton, B., Millard, P., and Bausenwein, U. (2000). Reserve formation and recycling of carbon and nitrogen during regrowth of defoliated plants. In: *Grassland Ecophysiology and Grazing Ecology* (eds. G. Lemaire, J. Hodgson and A. de Moraes), 85–99. New York: CABI Publisher.

Vallentine, J.F. (2001). *Grazing Management*, 2e. San Diego, California: Academic Press.

Volesky, J.D. (1994). Tiller defoliation patterns under frontal, continuous, and rotation grazing. *J. Range Manage.* 47: 215–219.

CHAPTER

45

Grazing Animal Nutrition

Gregory Lardy, Department Head, *Animal Sciences, North Dakota State University, Fargo, ND, USA*
Richard Waterman, Research Animal Scientist, *USDA-ARS, Fort Keogh Livestock and Range Research Laboratory, MT, USA*

What Makes Grazing Animals Unique?

Herbivorous animals are anatomically designed to make use of a variety of structural polysaccharides (cellulose and hemicellulose) found in forages throughout the world. These animals have the unique ability to make use of a carbohydrate source that is of limited nutritional value to humans and convert it into a nutrient-dense, highly digestible sources of protein (meat and milk products). Given that much of the land mass on Earth is not suitable for cultivation (mountainous, arid, etc.), having a means of utilizing the wide variety of forages produced on such land is important for human welfare.

Herbivores rely on gastrointestinal microflora to utilize forages through fermentation processes. These microflorae have the necessary enzymes, which allow these species to break down the cellulose and hemicellulose found in forages and, through fermentation, to produce products such as short-chain fatty acids which are an important source of energy for the host species. In ruminants, the microflora also provide the host with other nutrients, including lipids, amino acids, vitamins, and minerals, in addition to the short-chain fatty acids.

Herbivores can be separated into two major categories based on where these fermentation processes occur. The first category is pre-gastric fermenters, or ruminants, (e.g. cattle, sheep, goats, deer) in which most of the fermentation processes occur in the rumen-**reticulum** complex, prior to gastric digestion which occurs in the stomach and small intestine. Ruminants can utilize **rumen** microorganisms as a source of amino acids, lipids, vitamins, and minerals as the microbial cells pass out of the rumen and into the stomach and small intestine. Post-gastric fermenters such as horses have their fiber-digesting microbial population in the lower part of the gastrointestinal tract. In post-gastric fermenters, the animal has the first opportunity to digest the various fibrous feedstuffs it has ingested prior to exposing the feed to fermentation.

Symbiosis of the Grazing Animal and Enteric Microflora

In herbivorous animals, it is especially important to recognize the **symbiotic** relationship between the host species and the hundreds of species of bacteria, protozoa, and fungi which inhabit the gastrointestinal tract. Managers and livestock owners must be cognizant of this relationship as they evaluate nutritional programs for grazing animals.

Particularly for ruminants, a manager must understand the nutrient requirements of the rumen microflora in order to optimize ruminal fermentation first. Then, if additional nutrients are needed, managers can provide them for the animal.

Microbial populations which inhabit the rumen are dynamic and responsive to dietary changes. When

Forages: The Science of Grassland Agriculture, Volume II, Seventh Edition.
Edited by Kenneth J. Moore, Michael Collins, C. Jerry Nelson and Daren D. Redfearn.

the diet is made up largely of forages, fibrolytic (fiber fermenting) microbial species will be predominant. If the diet changes to include large proportions of starch (cereal grains) then amylolytic (starch fermenting) species will be predominant.

Fibrolytic bacteria can utilize forages because they synthesize both **cellulase** and hemicellulase enzyme complexes. These enzymes are capable of hydrolyzing ß-glycosidic bonds found in cellulose and hemicellulose. The enzyme complex is affixed to the extracellular membrane of the microbe and as a result, the disaccharide and monosaccharide hydrolysis products are close to the cell membrane for absorption. Other nutrients that are required by fibrolytic species in addition to structural polysaccharides include ammonia, minerals, vitamins, branched chain fatty acids, and other micronutrients. The rumen complex also serves to create the optimum environment for the microbes by maintaining anaerobic conditions, an optimal range of pH, osmolality, and temperature.

What Nutrients Do Grazing Livestock Require?

Tabular presentations of nutrient requirements of the various classes of livestock are available from a number of references (NASEM 2016; NRC 2001, 2007a,b). This chapter provides the most up-to-date description and information related to nutrient requirements for beef cattle. Similar information is available for sheep and goats, horses, and dairy cattle (NRC 2001, 2007a,b).

Energy

Requirements for energy vary depending on a variety of animal factors including body weight, milk production level, growth rate, as well as environmental factors such as temperature and wind speed, among others. In the case of grazing animals, it is important to note that much of the energy that becomes available to the animal is in the form of volatile fatty acids produced either in the rumen (in the case of ruminants) or from the large intestine (in the case of hind gut fermenters). The volatile fatty acids produced during the fermentation process are then utilized by the animal to make glucose in the liver. In many cases, a substantial majority of herbivore energy requirements may be met by these volatile fatty acids. The amount of energy required by the animal increases dramatically during lactation. Peak lactation in cattle generally occurs 8–12 weeks post calving and this coincides with peak nutrient requirements. Higher levels of milk production also increase nutrient requirements. As stage of pregnancy advances, nutrient demand by the developing fetus also increases. The fetus grows exponentially during the last trimester of pregnancy resulting in rapid increases in the nutrients required as the dam nears parturition. Figure 45.1 shows the changes in NE_m requirements across a production calendar for a 550 kg beef cow.

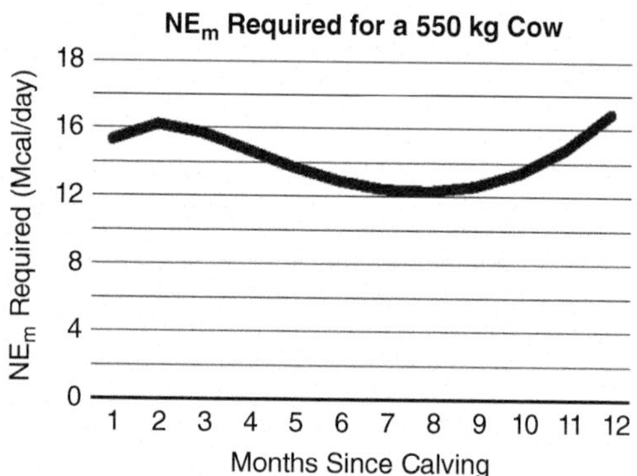

Fig. 45.1. Changes in the net energy for maintenance requirements of beef cows across the production calendar.

Energy requirements are also affected by air temperature and wind speed, as well as hair coat conditions. Animals can experience cold or heat stress both of which increase their energy requirement. However, the temperatures at which animals experience either cold or heat stress can vary. Therefore, it is not possible to give absolute temperatures at which these stresses occur. It should be noted that herbivores have evolved grazing under extreme weather conditions ranging from bitter cold to extreme heat and humidity. To reduce the potential negative effects of cold or heat stress, managers must be prepared to assist livestock with acclimation, provide increased levels of nutrition, and to provide shelter such as **windbreaks**, shade, or bedding as needed.

Protein

Protein requirements for **ruminant** animals are complex. This is because the resident microbial population in the forestomach has a protein requirement which is distinctly separate from the protein requirements of the animal (NASEM 2016). The rumen microbes' requirement for protein must be met for them to effectively ferment and break down the cellulose and hemicellulose found in forages. Some rumen microbial species require specific amino acids while others can utilize ammonia as a source of nitrogen (hence the ability of ruminants to utilize urea as a portion of their dietary protein).

Protein requirements for beef cattle are expressed as metabolizable protein. These requirements vary with factors such as body size, as well as stage of pregnancy, lactation status and level, and growth rate, in a manner similar to how NE_m requirements vary across the production calendar (NASEM 2016). Figure 45.2 shows the changes in metabolizable protein requirements for a 550 kg beef cow.

In addition to the microbial protein, which is produced during fermentation, the ruminant animal receives

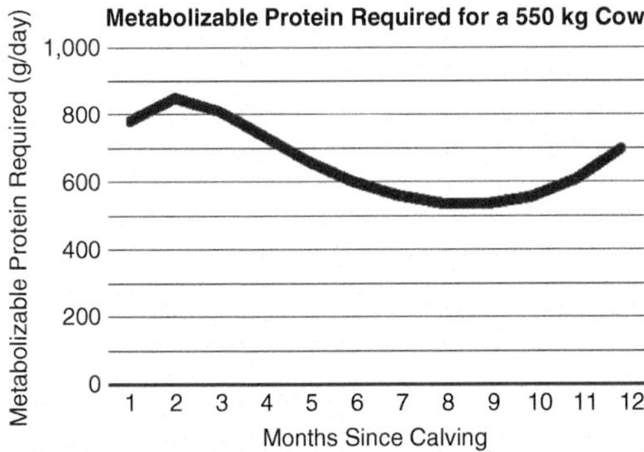

Fig. 45.2. Changes in metabolizable protein requirements of beef cows across the production calendar. Assumes a 550 kg mature Angus cow 60 months of age, a calf with 40-kg expected birth weight, peak milk production of 8 kg daily, milk composition (4% fat, 3.4% protein, 8.3% solids not fat), 8.5 weeks at peak milk. *Source:* NASEM (2016).

protein from undegraded feed protein that passes through the rumen without being fermented. Microbial protein is high-quality protein meaning it contains relatively high proportions of essential amino acids such as lysine and methionine in ratios similar to the animal's nutrient requirement. In addition, microbial protein is highly digestible. Depending on the nutrient requirements of the animal, the combination of the undegraded feed protein and the microbial protein may be sufficient for the animal and supplementation may not be required.

In cases where supplementation is required, it is important to ensure the needs of the rumen microorganisms are met first. In many cases, this strategy will result in an increase in fermentation and subsequently, an increase in the supply of microbial protein to the animal. If additional protein is needed to meet animal needs once supplementation is provided to the rumen microorganisms and rumen fermentation is optimized, it can be provided through a variety of supplements.

Water

Water is often overlooked as a nutrient for livestock. Water quality and quantity can affect performance as well as the health and wellbeing of grazing livestock. Water plays an important role in regulation of body temperature, growth, reproduction, lactation, digestion, metabolism, excretion, maintenance of mineral homeostasis, hearing, and eyesight. Water requirements can be influenced by several factors, including pregnancy, lactation, rate and composition of gain, activity, diet type, dry matter intake, and environmental temperature (NASEM 2016).

Grazing livestock receive some water from the pastures and other forages that they consume. Consequently, the entire water requirement of the animal does not need to be met with free water. Green, lush forages, such as rapidly growing pasture or green-chopped forages, have higher moisture contents than dry hays or dormant forages.

Level of milk production directly impacts water requirements for grazing animals. In most production systems, lactation usually occurs at the same time as environmental temperatures are increasing (summer), necessitating the need for increased water supplies as ambient temperatures rise. Environment also dramatically influences water requirements. Higher environmental temperatures and/or arid climates greatly increase the need for water for the grazing animal. Greater levels of physical activity will increase water consumption. Water consumption is approximately three-fold greater than dry matter intake and can range as high as seven-fold.

Water quality is also an important consideration for grazing livestock (Ayers and Westcot 1985; NASEM 2016). Grazing livestock consume water from a variety of sources including ponds, streams, lakes, other surface water sources, as well as tanks fed by shallow wells. Water quality varies from source to source, season to season, and year to year. Five criteria are often used to assess water quality for both humans and livestock. These include **organoleptic** properties (odor and taste); physiochemical properties (pH, total dissolved solids, total dissolved oxygen, and hardness); presence of toxic compounds (**heavy metals**, toxic minerals, organophosphates, and hydrocarbons); presence of excess minerals or compounds (nitrates, sodium, sulfates, and iron); and the presence of bacteria (NASEM 2016). Drought conditions can have a negative impact on water quality by concentrating minerals and salts found in surface water sources as evaporation occurs. In addition, drought conditions generally increase the development of algae blooms which can be harmful to livestock and wildlife.

Water contamination can also occur as a result of poor livestock management. Allowing cattle access to riparian areas, allowing cattle to defecate in water sources, or mismanagement of manure applications to cropland can all negatively impact water quality resulting in microbial pathogens, nitrates, and other pollutants entering water sources.

Grazing animals have access to a variety of water sources which vary in quality. Therefore, blanket recommendations on whether one source is better than another are generally not possible. In some areas of the country, water from shallow wells is inferior for livestock production due to excessive levels of sulfates, nitrates, and/or other contaminants. Other demonstration projects have indicated that water from streams or dugouts may also be problematic at certain times (Surber et al. 2003). This study showed that, when given the choice, livestock will generally prefer water from stock tanks, rather than dugouts. Livestock producers should evaluate water quality on a periodic basis to assess if quality and quantity are adequate for the class of livestock in question. Surber et al. (2003) also noted that there may be an economic advantage to fencing stock ponds and dugouts and installing siphon tubes from the water source to a nearby stock tank. This operation is fairly simple and will restrict cattle from wading into the pond or stock dam to drink. In addition, it also keeps cattle from defecating and urinating in the water source, which will be beneficial for water quality and ultimately beneficial to cattle health.

Minerals

Table 45.1 gives the mineral requirements for beef cattle along with the maximum tolerable concentrations. Mineral nutrition of the grazing animal is important, and supplements are usually required to maximize animal productivity. The macro minerals sodium, phosphorus, and magnesium can be deficient in many types of forage (NASEM 2016). Sodium and chloride are almost always deficient in forage. Forages are typically rich in calcium and potassium. Phosphorus concentration in forages can be variable and is influenced primarily by plant species and soil fertility. If soil fertility is adequate and forage quality is high, forage phosphorus can be sufficient to meet the animal's phosphorus requirement without supplementation. Magnesium often has low in availablility in new-growth forage. **Hypomagnesemic tetany** (commonly referred to as grass tetany) can occur when livestock graze lush forages and pastures fertilized with high rates of N and K, which has been associated with increased incidence of grass tetany (NASEM 2016). Magnesium is typically supplemented to cattle grazing lush growth forages, usually during spring growth, to prevent grass tetany. The micro minerals zinc, copper, and/or selenium have been shown to be deficient in forages across many large geographic regions. In some areas, selenium toxicity has been reported as well. Cobalt, manganese, and iodine can also be deficient and are usually supplemented to grazing animals.

Legumes tend to have higher concentrations of minerals than grasses, thus pastures with legumes will provide greater mineral intake levels than non-leguminous forages (NASEM 2016). The mineral profiles of cool- and warm-season grasses are similar at comparable physiologic stages of growth. The concentration of minerals in forages does not decline greatly as the plant becomes reproductive and lower in nutritional quality. Rather, the greatest factor influencing mineral nutrition of grazing animals is forage intake. As forage matures, its consumption by most grazing animals, particularly ruminants, will decrease due to increased fiber levels. For example, forage containing 6 mg kg^{-1} copper and consumed at 2.2% of the animal's body weight when the forage is vegetative, compared with a consumption of 1.6% of body weight when the forage is mature, would result in substantially lower copper intakes for the more mature forage at the same forage copper concentration. Reduced intake as the forage matures is the most influential factor affecting forage mineral intake by grazing animals.

Vitamins

Fat-soluble vitamins are often provided either through supplemental feeding or injection to ensure that deficiencies do not occur. Of the fat-soluble vitamins, vitamin A is the most likely to be limiting. This is particularly true in low-quality, low-digestibility forages. Green vegetative forages are generally adequate in vitamin A. Water-soluble vitamins need to be supplemented to nonruminant grazing animals to meet their vitamin requirements. However, ruminants do not require supplemental water-soluble vitamins because rumen microflora typically synthesize them in adequate quantities to meet their requirement.

Nutritional Requirements During the Production Cycle

The first task in developing nutrition programs for grazing animals is understanding their nutritional requirements. Nutrient requirements of grazing animals are cyclical, meaning they vary depending on the physiologic status of the animal (pregnancy, stage of lactation; Figures 45.1 and 45.2). Physiologic state, size, genetic potential for production (growth, lactation, etc.), and environment all play a role in determining nutrient requirements. Nutrient requirements are lowest during mid gestation, following weaning. Requirements for protein and energy gradually increase throughout gestation, as the fetus grows. Approximately 70% of fetal growth occurs during the last third of pregnancy (NASEM 2016). This results in increased nutrient requirements to support the rapidly growing fetus (energy and protein) during this time period. As the cow approaches parturition, nutrient requirements (energy and protein) increase. Nutrient requirements during pregnancy are also affected by the number of fetuses being carried by the gestating dam (NRC 2007a). For instance, ewes carrying twin lambs have higher nutrient requirements than ewes carrying singletons.

Lactation demands the greatest nutrient requirements that herbivorous animals will undergo. Protein and energy requirements are directly related to level of milk production. In beef cows, for example, milk production typically peaks six to eight weeks following parturition and then declines gradually for the remainder of the lactation period (NASEM 2016). Genetic potential for milk production varies with breed type (Jenkins and Ferrell 1992). Some breeds were developed for the sole purpose of producing milk (e.g. Holstein, Jersey, Brown Swiss), others developed as dual-purpose animals (meat and milk; e.g. Simmental), and others with primary emphasis on meat production (e.g. Angus, Hereford). Therefore, milk production potential also varies. Large quantities of

Table 45.1 Mineral requirements and maximum tolerable concentrations (dry matter basis) for beef cattle

Mineral	Unit	Growing and finishing cattle	Cows: Gestating	Cows: Lactating	Maximum tolerable concentrations
Cobalt	mg/kg	0.15	0.15	0.15	25.0
Copper	mg/kg	10	10	10	40
Iodine	mg/kg	0.50	0.50	0.50	50
Iron	mg/kg	50	50	50	500
Magnesium	%	0.10	0.12	0.20	0.40
Manganese	mg/kg	20	40	40	1000
Potassium	%	0.60	0.60	0.70	2.0
Selenium	mg/kg	0.10	0.10	0.10	5.0
Sodium	%	0.06–0.08	0.06–0.08	0.10	–
Sulfur	%	0.15	0.15	0.15	0.3–0.5
Zinc	mg/kg	30	30	30	500

Source: Adapted from NAS (2016).

high-quality forage are required to maintain high levels of milk production. In situations where milk production is relatively high but forage nutrients are relatively low, the animal will typically lose weight and have a lower probability of achieving pregnancy due to negative energy balance.

Nutrient requirements of growing animals, whether grazing or in confinement, are defined by physiologic potential of the animal for growth and forage nutrient levels. Of particular interest from a grazing perspective, is forage digestibility and protein concentration of the forage, since nutrient supply generally limits productivity. Greater digestibility and protein concentration will support greater levels of growth. For growing steers, energy, and protein requirements increase with increased average daily weight gain. Mineral and vitamin requirements do not change drastically with changes in animal growth.

Nutrient requirements of grazing horses are also affected by the level of work expected. At maintenance, nutrient requirements of mature horses are relatively low but as level of activity or work increases, energy requirements also increase. NRC (2007b) defines various levels of activity for horses, including light exercise, moderate exercise, heavy exercise, and very heavy exercise. Light exercise includes recreational riding and showing; moderate exercise examples would be general ranch work, frequent showing, and training or breaking; heavy exercise would include race training, frequent ranch work, and polo; while very heavy exercise would include racing and multi-day endurance events. Care should be taken to be sure that the level of nutrition provided in the forage, along with the supplement program, match the level of work activity expected from the horse. Failure to do so will result in loss in body weight and condition, as well as poor animal health and wellbeing.

Larger animals require greater amounts of nutrients on a daily, weekly, or annual basis compared with smaller animals. Animal size affects decisions that a grazing manager makes affecting stocking rate. Larger animals will consume more forage. Unfortunately, many livestock producers have largely ignored this fact when planning grazing programs. This has resulted in a tendency to stock pastures using only animal numbers rather than taking into account both animal size and animal numbers in stocking rate planning processes.

Matching Animal Requirements to the Forage Resource

One of the most fundamental concepts in sound grazing nutritional management is the ability to match animal requirements to the forage resource. This concept is of greater importance in resource-limiting situations and semi-arid and arid rangeland environments, but given the increasing consumer interest and emphasis on sustainability (White et al. 2015), it is likely to receive more attention in all production scenarios as we move forward.

In the past, some managers have taken the approach of selection for greater and greater productivity (measured by various growth traits, weaning weight, or milk production level) with little emphasis on ensuring that the selected animals are adapted to the local environment and forage conditions. This may result in increased reliance on supplemental feed inputs of various sorts, increased production costs, and lowered profitability.

Because grazed forages generally provide the least expensive source of nutrients, emphasis should be placed on following practices which reduce reliance on purchased feed inputs and focusing selection pressure on animals best suited for the environment which they will be placed. In some cases, following this principle, will mean smaller animals with lower levels of milk production as nutrient demand increases with increasing body size and milk production level. This type of selection pressure will reduce the overall nutrient requirements of the animal and allow them to better adapt to their forage environment. This approach will also better prepare for drought conditions, which can necessitate the purchase of a variety of feedstuffs to use as supplements or forage replacements.

Matching animals to the forage resource requires a good understanding of the cyclical nature of forage quality (e.g. when are the forages at their peak in terms of nutritive value) and attempts to match this with peak animal nutrient requirements (which tends to occur approximately six to eight weeks after parturition in beef cattle; Adams et al. 1996). Other considerations also come into play as a manager determines what is best for their particular operation but considering the timing of nutrient availability, at a minimum, provides a place to start.

Devising Supplements for Grazing Animals

The best approach to nutritional management of grazing animals is to maximize forage utilization, promoting the growth and activity of fibrolytic microbes in the gastrointestinal tract: in turn providing essential nutrients to the host animal. Supplements should be designed to augment limitations in nutrient supply from the available forage with the goal of reducing feed inputs by managing forages to more closely match the nutritional requirements of the animal. Management can influence the nutritional requirements of the animal by changing production parameters such as the length of lactation, timing of parturition, and/or timing of weaning as well as animal size, productivity level, and growth rate. In this system, forage species selection and grazing management options would be employed to match nutrient requirements of animals throughout the year. However, various climatic, plant physiologic, and economic factors often require some form of nutrient supplementation to achieve production goals. When devising supplementation programs, emphasis should be given to accentuating forage use by the gut microflora and then the animal's use of the forage nutrients, fermentation end products, and microbial biomass.

Targeted or strategic supplementation regimes in forage-based production systems rely on historic knowledge of forage and animal parameters as well as current knowledge of environmental, animal, and forage conditions which the animals are now encountering. Supplementation strategies should be adaptable and flexible in order to match the limitations in grazed forage supply at any point during the production cycle. The two main factors that drive forage production in extensive systems are sunlight and precipitation. Obviously, precipitation is less pertinent in irrigated forage production systems. Fluctuating environmental conditions greatly influence the quantity and quality of grazable biomass. Grazing livestock managers should seek a greater understanding of the impact of environmental conditions in order to better implement supplementation regimes which are better matched to different stages of the production cycle to optimize both forage nutrient utilization and animal production.

Supplementation Strategies to Enhance Grazing Animal Performance

To accomplish the physiologic functions of maintenance, growth, reproduction, and lactation, grazing animal diets must provide nutrients in the form of energy, amino acids, fatty acids, minerals, and vitamins. The amount of nutrients consumed is a function of forage intake and maturity of the plant being consumed. Therefore, when first evaluating a grazing system, managers should consider how well the available forage matches the grazing animal's nutrient requirements. Supplementation is the complementary addition of essential nutrients for a given production setting to optimize forage utilization, meet animal nutrient requirements, and achieve the desired animal performance for a specific production goal.

Mineral and vitamin supplements are generally required for most grazing animals. This is most often accomplished by offering a free-choice mineral supplement. Typically, physical form (i.e. molasses-based tub) or chemical composition (sodium chloride concentration) is used to control intake.

Essential fatty acids rarely need to be supplemented because the grazed forage provides the required amount of unsaturated fatty acids. However, recent research has shown that performance may be enhanced by increasing polyunsaturated fatty acid intake. There is interest in defining the optimum ratio of omega-6 and omega-3 fatty acids and production benefits induced by essential fatty acid supplementation. Providing supplemental vegetable oil (high in linoleic acid) resulted in a greater percentage of heifers pubertal at the beginning of the breeding season (Lammoglia et al. 2000) and improved calf survival (Lammoglia 1999a,b). First calf heifers supplemented with fish meal (0.4% added fat) tended to have improved first-service conception rates compared to diets containing no fish meal (Burns et al. 2002). Forages have a high proportion of linoleic acid, an essential fatty acid. Though reproductive improvements have occurred from supplementing 0.25 kg or less of supplemental fat, similar in fatty acid composition to the oils in forages, achieving increased fat consumption by the animal through forage species selection or management will likely prove difficult as overall changes in fatty acid content and composition are relatively small (Boufaïed et al. 2003). Achieving increased fat consumption through the grazed forage would require a near doubling of forage lipid content and/or the ability of the plant to synthesize omega-3 fatty acids. As a result, it seems more likely that oil or fat supplementation will be used to enhance reproduction rather than efforts to boost forage fatty acid concentration. Generally, prepartum fat supplementation has been more effective than postpartum supplementation (Funston 2004).

The greatest potential to enhance animal productivity, and concomitantly the greatest nutritional need of grazing animals, is through management of grazed forage to maximize digestible energy and protein content. Because acid detergent fiber (ADF) and NE_m are inversely related, forage management practices should strive to minimize ADF content. Forage protein is composed predominantly of photosynthetic proteins that are rapidly and extensively fermented by gut microflora and extensively degraded by mammalian proteases. Therefore, forage protein has a high bioavailability of amino acids. Forage management practices should strive to maximize forage crude protein. In most instances, management practices that minimize ADF concomitantly maximize protein content.

Energy and protein requirements can be calculated for the various stages of animal production. The maintenance (NE_m) requirement is 0.077 Mcal kg^{-1} of metabolic body weight. A 635-kg nonlactating, early-gestation cow would require 11 Mcal NE_m d^{-1} (NASEM 2016). Using the estimates of NE_m for timothy at the early vegetative growth stage (1.38 Mcal kg^{-1}) and seed (0.86 Mcal kg^{-1}) stage of growth, this cow would have to consume 1.3% and 2.0% of her body weight, respectively, to meet her maintenance energy requirements. During the last 60–90 days of gestation this cow would require 15.5 Mcal NE_m d^{-1} due to the increased energy demand related to accelerated fetal growth rate. Contrasting the two qualities of timothy above, this cow would have to consume 1.8% and 2.8% of her body weight, respectively, to meet her energy needs. During the first three to four months of lactation, the cow would have to consume 18.3 Mcal NE_m d^{-1} to meet the increased energy demand of lactation. This would require the cow to consume 2.1% and 3.4% of her body weight, respectively. The cow would likely not be able to consume this amount of forage, since gut fill limits forage intake as fiber content increases. As evidenced in these examples, the reproducing cow would be able to meet her energy requirement throughout the production cycle with a combination of high-quality and low-quality forage that coordinated with her energy requirement. Figure 45.3 is the estimated Mcal NE_m kg^{-1} and ADF concentration of forage required to meet the annual production cycle needs of a 635-kg cow. When the cow is not lactating and in early gestation, her energy requirement is lowest and maximum forage ADF is highest. During late gestation and early lactation, her energy demands are greatest and maximum forage ADF is lowest. The demand for energy should be matched to the forage quality (ADF) available to the animal.

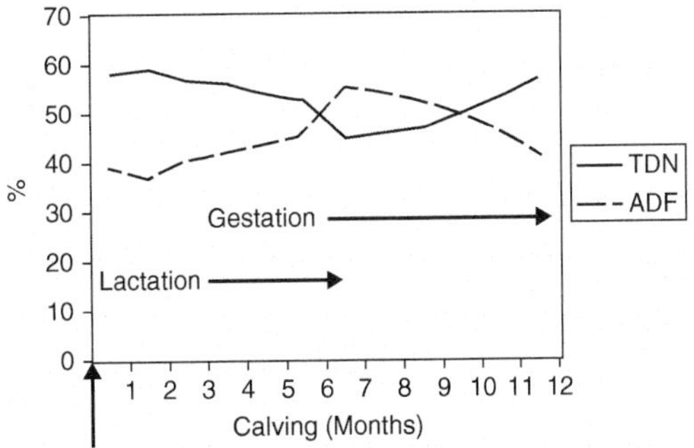

Fig. 45.3. Total digestible nutrient (TDN) requirement and maximum forage acid detergent fiber (ADF) concentration allowable for a 635-kg cow after calving (Kerley and Lardy 2007).

Protein requirements for nonruminant and ruminant herbivores differ. Nonruminants, such as the horse, must meet their amino acid requirement from the forage protein consumed because fiber digestion, and the associated microbial growth, occurs past the site of amino acid absorption in the small intestine. Amino acids are required for maintenance, growth, reproduction, and lactation. The quantity of amino acids consumed is a function of forage protein concentration and amino acid profile. Forage protein concentration is largely influenced by management, growth stage of the forage, and forage species. Comparison of the amino acid profile in forage protein to the profile of tissue provides an assessment of amino acids most likely limiting maintenance, growth, or productive functions. Using the horse example, the amino acids most limiting are lysine and threonine, while methionine and tryptophan may also be limiting (NRC 2007b). When low-forage protein concentration and/or heightened animal production level warrants feeding supplemental protein to nonruminant herbivores, protein sources should be selected that are rich in the limiting amino acids.

As discussed previously, the forage protein consumed by ruminants is extensively fermented by rumen microflora. Consequently, the amino acid profile of the forage has little resemblance to the amino acids available for absorption by the animal in the small intestine. Rather, forage protein supports growth of the rumen microflora. The dominant species of bacteria present in the rumen of grazing ruminants are fibrolytic species. These species require ammonia as their sole source of nitrogen. The ammonia is supplied primarily through protein fermentation by proteolytic species.

The microflora in the rumen synthesize most of amino acids digested and absorbed by the animal. The amount of microbial protein synthesized is dependent upon various rumen characteristics and functions, but an average microbial crude protein production of 81 g N/day with a mean efficiency of 19 g N/kg of organic matter truly fermented in the rumen has been reported (NASEM 2016). The amino acid profile of bacterial protein is very similar to the animal tissue amino acid profile, making the bacterial protein an ideal source of protein for the animal. Because photosynthetic proteins are similar across most forage species and are extensively fermented by the rumen microflora, the limiting amino acids are similar across forage diets. Methionine is regarded as the most limiting amino acid in forage diets, followed by arginine, lysine, histidine, and threonine (Titgemeyer and Löest 2001). When productive functions require greater quantities of amino acids than that supplied by microbial protein, protein sources should be selected that are high in rumen undegradable methionine, arginine, lysine, histidine, and threonine. There is potential to increase the rumen undegradable amino acid concentration in forages. Warm-season grasses contain greater quantities of rumen undegradable protein than cool-season grasses. However, the protein level of these forages is typically too low for the quantity of rumen-undegradable protein to have substantial nutritional implications. Tannins, present in some legumes and grasses, can protect high enough proportions of protein from fermentation in the rumen to have positive nutritional effects. It is important that **tannin** or other methods used to protect protein from microbial proteolysis dissociate at an acidic pH, allowing the animal to digest the protein once past the rumen, and that the tannins do not result in inadequate ammonia concentrations in the rumen due to a limited quantity of degradable protein.

Two-year-old heifers have been shown to respond to supplemental rumen-undegradable protein. When bloodmeal was fed (0.2 kg head^{-1} d^{-1}) to two-year-old lactating beef heifers grazing endophyte-infected tall fescue during the spring, their weight gain was improved (Forcherio et al. 1995). Backgrounding calves grazing alfalfa and fed bloodmeal (0.1 kg head^{-1} d^{-1}) had improved daily gains. Similar results were obtained with calves grazing warm-season grass pastures and supplemented with a methionine hydroxy analog. Growing calves and growing/lactating cows would potentially have greater amino acid requirements than could be provided by microbial protein flowing to the small intestine. However, it is only when amino acid requirements are elevated due to a protein-demanding physiologic function that rumen-undegradable protein supplementation would be beneficial. The first approach in protein nutrition of the grazing ruminant is to supply adequate ammonia for the rumen microflora, and the second approach is to supply limiting amino acids post-ruminally to the animal.

Supplementing protein to non-ruminant herbivores or rumen-undegradable protein to ruminants requires consideration of amino acid composition of the protein. The amino acid profile of the protein should reflect the amino acids limiting in the diet.

Supplementation strategies should complement the nutrient composition of the grazed forage and improve overall nutrient utilization of the forages being consumed. Supplements are used to deliver limiting nutrients but can also be used as a management tool to improve utilization of forages in various production settings ranging from high-forage production settings (i.e. improved pastures) to larger extensive rangeland pastures by attracting animals into underutilized areas. Supplements should enhance not discourage consumption of the basal forages. Supplements should also enhance or not interfere with natural grazing patterns.

Optimizing the Supplement Composition

As discussed previously, grazing animal nutrition should be based upon the foundational strategy that the forage will supply all necessary nutrients. This strategy is based upon the assumption that standing forage is the most economical feed resource available and that the forage quality is adequate to meet nutritional requirements. When one or both assumptions are not true, provision of supplemental nutrients would be beneficial. Designing the nutritional supplement for grazing animals requires an assessment of their nutritional needs, deficiencies in the forage diet, and an understanding of how the supplemental nutrient form influences forage use by the animals as discussed above.

Base forages for grazing systems can include cool-season or warm-season grasses or legumes or some combination of these forage types. In situations where native range is the primary forage, it will be predominated by either cool- or warm-season species, depending on location and environment. Generally, cool-season grasses dominate northern Great Plains ecosystems whereas warm-season grasses dominate southern Great Plains ecosystems. Both types of grasses have growth cycles which include vegetative growth followed by flowering and seeding (Chapter 7). Grass digestibility decreases as plants reach the reproductive stage and dormancy. The primary differences between cool- and warm-season grasses are the date that they begin growth (cool-season grasses being earlier than warm-season grasses) and their protein concentration (cool-season grasses typically contain more protein per unit of dry matter). The fiber (neutral detergent fiber, or NDF, and ADF) concentration of the plant increases and fiber digestibility decreases due to secondary cell wall formation as the plant matures (Chapter 39). As fiber fermentability declines, digestibility, and energy value of the forage declines. Protein concentration of the plant also decreases as the plant reaches the reproductive stage of growth. Therefore, animals grazing forage will consume less energy and protein as the plant moves from vegetative to reproductive growth stage. Depending on nutrient requirements, animals grazing mature (reproductive) forage are more likely to require supplemental energy and/or protein than animals grazing vegetative forage.

Supplemental Energy

Energy drives cellular processes, and the need for supplemental energy should be determined prior to other nutrients. As discussed previously in this chapter, forage digestibility is negatively correlated to ADF content. Using cattle as a model, digestible dry matter, or total digestible nutrients (TDN), which is highly correlated with digestible dry matter, can be calculated from ADF using the equation DDM (%) $= 88.9 - (0.779 \times$ ADF, %). The net energy for maintenance and lactation (NE_m or NE_l) of grasses can be predicted from ADF using the equation $NE_m = 0.9996 - (0.0112 \times ADF)$ (National Forage Testing Association n.d.). Predicting the energy density of the forage and the forage intake of the animal will allow the need for supplemental energy, if any, to be determined. The greatest energy demand of the cow (545 kg mature weight) is in the second month after calving. This cow would require 16.1 Mcal NE_m d^{-1}. Using the example of timothy at two different forage qualities contrasted previously, and the expected intake of 12.6 kg, the cow would consume 17.5 Mcal NE_m from the vegetative timothy and 10.8 Mcal NE_m from the seed-stage timothy. The poorer-quality timothy forage would require an additional 5.3 Mcal NE_m d^{-1} for the cow to maintain body weight and milk production.

The primary adjustment the cow would make in the above example, if not provided a nutritional supplement, would be to lose weight while attempting to maintain sufficient milk production to ensure the survival of her offspring. Assuming tissue loss has 7 Mcal NE_m kg^{-1}, the 5.3 Mcal NE_m deficit would result in the cow losing approximately 0.75 kg d^{-1}. This level of weight loss would continue until the energy demand, the combination of body tissue maintenance and lactation energy expenditures, was equal to energy consumed. If the weight loss continued from parturition until breeding (three months), the cow would lose approximately 67.5 kg which is equivalent to a 1.5 unit decrease in body condition. To avert the potential negative effects of body weight loss, supplemental energy would be required.

The form of supplemental nutrient affects the efficiency of forage use by the animal as discussed previously. There are three primary forms of supplemental energy that could be offered: starch, digestible fiber, or lipid. Using corn (2.2 Mcal NE_m kg^{-1}), soybean hulls (2.0 Mcal NE_m kg^{-1}), and soybean oil (4.8 Mcal NE_m kg^{-1}) as examples, 2.4 kg of corn, 2.7 kg of soybean hulls, or 1.1 kg of soybean oil must be fed to meet the cow's energy deficiency. The most appropriate choice is one that minimizes cost per unit of energy and enhances forage use by the cow while taking into consideration various aspects of supplement storage, delivery, and management.

The cow could consume 1.1 kg of corn (0.3% of body weight) without negative associative effects on ruminal fermentation. At levels of high-starch supplements provided at greater than 0.3% of body weight, the rumen ecosystem experiences a reduction in ruminal pH which negatively impacts fibrolytic bacteria and ultimately decreases fiber digestion. Assuming the timothy forage contained 2% lipid, up to 0.5 kg of soybean oil could be fed. This level of supplemental oil would result in the total diet containing 6% fat. Diets greater than 6% fat can result in decreased fiber digestibility. Functionally, supplement formulation would most likely require that fat be substantially less than 0.5 kg due to the mechanical problems of supplement blending, storage, and pelleting. Typically, price limits the amount of fat supplementation to grazing animals. Soybean hulls are an example of a by-product from oilseed processing that is high in fiber (66% neutral detergent fiber). Soybean hulls are unique in that the fiber is highly fermentable but yet does not contribute to the negative associative effects that can be encountered with supplements which are high in starch. Consequently, soybean hulls would not cause negative associative effects similar to supplementing corn above 0.3% of the animal's body weight. To meet the energy deficit, the cow would need to consume 2.7 kg of soybean hulls, or the combination of 1.1 kg corn and 1.4 kg of soybean hulls. The decision to use only soybean hulls or the combination of corn plus soybean hulls would be dependent upon costs of the two ingredients and expenses and complications associated with feeding the blend vs a single ingredient. Other sources of energy supplements exist. One of the more common sources are coproducts of the ethanol industry (Lardy and Anderson 2014). These products include both wet and dry distillers' grains plus solubles. The choice to use wet or dry products will depend on a variety of factors including storage facilities, handling equipment, and transportation costs to name a few. Care should be taken to reduce or limit losses that occur when these supplements are fed on the ground. Bunks and other feeding equipment can reduce product loss. Ethanol coproducts are an excellent source of energy and protein and the coproducts have good availability in many regions of the country. Livestock producers should request a nutrient analysis or conduct regular sampling for nutrient composition to adequately balance nutrient intake.

Supplemental Protein

The greatest protein demand for this cow parallels energy demand trends and is 1.3 kg of crude protein daily. Contrasting the two qualities of timothy, the vegetative growth would contain 14% crude protein and the seed stage of growth would contain only 6%. Using the forage intake of 12.9 kg, the intake of crude protein would be

1.8 and 0.8 kg for the high- and low-quality timothy, respectively. The cow would be able to meet her crude protein requirement with the high-quality timothy. Poor-quality warm-season grasses can contain even lower crude protein levels than poor-quality cool-season grasses, such as the timothy example used in this illustration. Dormant winter range can contain 4–5% crude protein which would result in even lower levels of crude protein intake.

Protein nutrition should be separated into the nitrogen requirements of the rumen microflora and the host animal. In the examples of the poor-quality timothy and winter range, both forages would likely be limiting in adequate nitrogen supply for the microflora. As discussed previously, microbes responsible for fermenting fiber require ammonia as their sole source of nitrogen. To meet their minimum requirement for nitrogen, forages should contain approximately 6–7% crude protein. This estimate is for a non-lactating, non-gestating cow. Those developing rations or supplements should keep in mind that protein requirements increase with lactation and gestation. Any supplemental protein form can be used in raising the diet crude protein content to 7% as long as the protein is degradable by the protein-fermenting bacteria, which will produce ammonia. Urea or natural protein could be used. In most cases, results have been better when natural protein has been used rather than urea or another nonprotein nitrogen source. Urea can be used to supply a portion of the supplemental protein for grazing ruminants but generally should not be used as the sole source of supplemental protein. In general, supplemental urea should not exceed 25% of the rumen degradable protein content of supplements being delivered on a daily basis and should not exceed 15% when fed on two or three times a week basis in extensive production systems due to concerns related to supplement palatability associated with high levels of urea.

Mature grazing ruminants can often meet their amino acid requirements through the microbial protein synthesized in the rumen when forage quality is adequate in energy. Consequently, ammonia provision to maximize microbial growth is important. Mature nonruminants will also be able to meet their amino acid requirements from the forage protein. On occasion, forage digestibility can be sufficiently low or the demand for growth and/or lactation demands can be great enough to require greater amounts of amino acids than are supplied by the rumen microflora. The rumen microflora produce protein to facilitate mitotic division, or growth. The growth rate of the microflora is a function of fermentable substrate supply and dilution rate from the rumen. An average value of 81 g of bacterial crude protein is produced per kilogram of organic matter fermented. Using the timothy example, the high-quality timothy forage would yield 0.8 kg of metabolizable protein: [(dry matter intake × digestibility of the forage × assumed 70% digested in the rumen × 81 g of bacterial protein per kilogram fermented) + (the forage protein × 21.5% of the forage protein being undegradable)], and the cow's requirement would be 0.4 kg. Using the winter range forage example, the predicted yield of metabolizable protein would be 0.4 kg, if 0.4 kg of supplemental protein was fed to provide a minimal 7% crude protein in the diet. Because the microbial protein production would be adequate to meet the cow's amino acid requirement, the supplemental concern would be to feed a degradable nitrogen source to facilitate ammonia production. As an example, approximately 0.9 kg of soybean meal would be needed to supply the required supplemental crude protein.

The first goal in protein nutrition of grazing ruminant animals is to maximize microbial protein yield. In mature animals, the combination of microbial protein and forage undegradable protein will usually suffice without the need to supplement amino acids to the animal.

Growing animals and grazing dairy cows will often respond to protein supplementation due to their greater demand for amino acids to support protein synthesis for growth and milk protein synthesis. The literature is not conclusive, but typically for ruminants the first-limiting amino acids in grazing situations are methionine, followed by arginine, lysine, histidine, and, potentially, threonine with no order of limitation intended (Titgemeyer and Löest 2001). In some cases, cattle growth has been improved with supplemental rumen-undegradable protein (Klopfenstein 1996). Likewise, reproductive tract scores of developing heifers and growth rates of lactating two-year-old cows have been improved by provision of rumen-undegradable protein.

Generally, the most effective approach is to manage forage to contain sufficient protein for microbial and animal requirements. When this is not feasible, the nutritional approaches should be to maximize the microbial protein output and supply the animal with the limiting amino acids through rumen-undegradable protein or rumen-stable amino acids.

Supplemental Minerals and Vitamins

Mineral and vitamin supplementation should be based upon expected, if not measured, deficiencies in the forage. Caution should also be exercised to ensure that the supplemental minerals being fed are bioavailable and that they are delivered in a consumable format.

Other Limiting Nutrients

Other limiting nutrients that influence animal performance are derived by the lack of essential microbial fermentation end products specifically in ruminants grazing mature forages. Forages generally tend toward a high-ruminal fermentation of acetate compared to propionate (Cronje et al. 1991). This imbalance can lead to metabolic disorders in the host ruminant that ultimately leads to insufficient cellular uptake of glucose and other nutrients and lowers animal performance. Supplementing sources of additional propionate to the rumen has been shown to improve cellular uptake of glucose and other nutrients and improve reproductive efficiency (Mulliniks et al. 2011).

Species Differences in Relation to Environmental Conditions and Forage Utilization

Cattle are primary consumers of forages across the world and there are two species classifications, *Bos taurus* and *Bos indicus*. The way these two species can tolerate different environmental conditions greatly influences how they graze forages. Therefore, it is important to not only match forage resources to how well they meet animal requirements but also to consider the type of animal being produced in a given environment. For example, *B. indicus* cattle are more heat tolerant and will generally utilize forages better in tropical/warmer environments than *B. taurus* species would. *B. taurus* cattle are better adapted to temperate climates and are more winter hardy than *B. indicus*.

Summary

Grazing animals require a wide array of nutrients including energy, protein, water, vitamins, and minerals. Requirements for energy and protein vary with stage of production with peak lactation representing the greatest nutrient requirements. Ruminant animals have an additional layer of complexity due to the resident microbial population present in the rumen. Optimizing ruminal fermentation will ensure the animal has a solid nutritional foundation and will ultimately decrease supplementation costs. Managers should consider both animal requirements and the supply of nutrients from the forage when designing supplemental nutrition programs. Taking time to consider ways to match animal requirements with the nutrients available from the forage will generally result in lower supplementation costs in the long term.

References

Adams, D.C., Clark, R.T., Klopfenstein, T.J., and Volesky, J.D. (1996). Matching the cow with forage resources. *Rangelands* 18: 57–62.

Ayers, R. S. and Westcot, D. W. (1985). Food and Agriculture Organization of the United Nations. Irrigation and Drainage Paper. http://www.fao.org/docrep/003/t0234e/T0234E00.htm#TOC (accessed 14 October 2019).

Boufaïed, H., Chouinard, P.Y., Tremblay, G.F. et al. (2003). Fatty acids in forages. I. Factors affecting concentrations. *Can. J. Anim. Sci.* 83: 501–511.

Burns, P.D., Bonnette, T.R., Engle, T.E., and Whittier, J.C. (2002). Effects of fishmeal supplementation on fertility and plasma omega-3 fatty acid profiles in primiparous, lactating beef cows. *Prof. Anim. Sci.* 18: 373–379.

Cronje, P.B., Nolan, J.V., and Leng, R.A. (1991). Acetate clearance rate as a potential index of the availability of glucogenic precursors in ruminants fed on roughage-based diets. *Br. J. Nutr.* 66: 301–312.

Forcherio, J.C., Catlett, G.E., Patterson, J.A. et al. (1995). Supplemental protein and energy for beef cows consuming endophyte-infected tall fescue. *J. Anim. Sci.* 73: 3427–3436.

Funston, R.N. (2004). Fat supplementation and reproduction in beef females. *J. Anim. Sci.* 82 (E-Suppl): E154–E161.

Jenkins, T.G. and Ferrell, C.L. (1992). Lactation characteristics of nine breeds of cattle fed various quantities of dietary energy. *J. Anim. Sci.* 70: 1652–1660.

Kerley, M.S. and Lardy, G.P. (2007). *Grazing Animal Nutrition in Forages: The Science of Grassland Agriculture*. 6th rev. ed. Ames, IA: Blackwell Publishing.

Klopfenstein, T. (1996). Need for escape protein by grazing cattle. *Anim. Feed Sci. Technol.* 60: 191–199.

Lammoglia, M.A., Bellows, R.A., Grings, E.E., and Bergman, J.W. (1999a). Effects of prepartum supplementary fat and muscle hypertrophy genotype on cold tolerance in newborn calves. *J. Anim. Sci.* 77: 2227–2233.

Lammoglia, M.A., Bellows, R.A., Grings, E.E. et al. (1999b). Effects of feeding beef females supplemental fat during gestation on cold tolerance in newborn calves. *J. Anim. Sci.* 77: 824–834.

Lammoglia, M.A., Bellows, R.A., Grings, E.E. et al. (2000). Effects of dietary fat and sire breed on puberty, weight, and reproductive traits of F1 beef heifers. *J. Anim. Sci.* 78: 2244–2252.

Lardy, G. and Anderson, V. (2014). Feeding Coproducts of the Ethanol Industry to Beef Cattle. NDSU Extension. AS-1242. https://www.ag.ndsu.edu/publications/livestock/feeding-coproducts-of-the-ethanol-industry-to-beef-cattle (accessed 14 October 2019).

Mulliniks, J.T., Cox, S.H., Kemp, M.E. et al. (2011). Protein and glucogenic precursor supplementation: A nutritional strategy to increase reproductive and economic output. *J. Anim. Sci.* 89: 3334–3343.

National Academies of Sciences, Engineering and Medicine (2016). *Nutrient Requirements of Beef Cattle* 8th rev. ed. Washington, DC: National Academy Press.

National Forage Testing Association n.d.. Estimates of energy availability. https://81f07d1b-f7c6-4303-85b9-9822425f1146.filesusr.com/ugd/24f64f_4502c4143ec34a6ba1c4940c6147fd07.pdf

National Research Council (2001). *Nutrient Requirements of Dairy Cattle*: 7th rev. ed. Washington, DC: The National Academies Press.

National Research Council (2007a). *Nutrient Requirements of Small Ruminants: Sheep, Goats, Cervids, and New World Camelids*. Washington, DC: The National Academies Press.

National Research Council (2007b). *Nutrient Requirements of Horses*: 6th rev. ed. Washington, DC: The National Academies Press.

Surber, G., Williams, K., and Manoukian, M. (2003). Drinking water quality for beef cattle: An environment friendly and production management enhancement technique. Montana State University Extension Service. https://extension.usu.edu/rangelands/ou-files/Drinking_Water_Quality.pdf (accessed 14 October 2019).

Titgemeyer, E.C. and Löest, C.A. (2001). Amino acid nutrition: demand and supply in forage-fed ruminants. *J. Anim. Sci.* 79 (E_Suppl): E180–E189.

White, R.R., Brady, M., Capper, J.L. et al. (2015). Cow–calf reproductive, genetic, and nutritional management to improve the sustainability of whole beef production systems. *J. Anim. Sci.* 93: 3197–3211.

CHAPTER 46

Grazing Animal Behavior

Karen L. Launchbaugh, Heady Professor, *Rangeland Ecology, University of Idaho, Moscow, ID, USA*

Overview

Every **herbivore** is born with behavioral predispositions and physical abilities that influence their foraging decisions. As herbivores grow, they gain experience and knowledge about habitat quality and refine foraging skills. Animals learn about their foraging environment through their own experiences and from other members of their herd or flock serving as social models for appropriate behavior. Foraging behaviors of an animal therefore result from complex and ongoing interactions between genetic and environmental factors. The intertwined actions of inheritance and experience lead to adaptive foraging behavior.

Herbivores make thousands of foraging decisions each day. The cumulative result of these decisions is how herbivores acquire enough nutrients to grow and reproduce while evading lethal consumption of toxic plants. To walk this biologic tightrope, herbivores must appropriately decide what to eat (**diet selection**), where to eat (feeding site selection), and how much to eat (**intake**). These foraging decisions are integrated with other activities, such as drinking, ruminating, resting, and avoiding predators.

Livestock grazing behavior is an immensely important process because it simultaneously influences the animal's nutritional well-being, the composition and productivity of the forage resource, and many aspects of management. Grasslands are a plentiful source of chemical energy in the form of polymeric matrices of cellulose and hemicellulose, lignin, and other polymers. Grazing herbivores evolved variants of modified digestive systems capable of extracting energy from plant cell walls through a symbiotic relationship with gut microflora and microfauna in anaerobic fermentation processes. Fore and hindgut fiber fermenters exhibit quite different grazing behaviors because of structural and functional properties of their gastrointestinal tracts.

Important to larger mammalian herbivores was also the evolution of complex herd social behaviors that impact all aspects of grazing and animal management. What, where, and how much an herbivore consumes therefore results from internal **morphologic** and **physiologic** attributes interacting with external social and environmental conditions. A clear understanding of the animal and environmental characteristics that drive foraging behaviors allows ranchers and **grassland** farmers the opportunity to change animal behavior and improve forage management systems to enhance animal nutrient intake and manage the ecologic impacts of grazing on the foraging environment.

This chapter provides a review of plant attributes and animal characteristics that affect foraging decisions and characterize the behavioral processes that guide these decisions. This review begins by examining the basis for foraging decisions at the bite, plant, and **feeding station** level (Figure 46.1). Important behavioral responses at this hierarchic level include diet selection (i.e. selective choices among plants and plant parts) and subsequent ingestive behaviors (i.e. prehending, biting, and chewing). The selection of appropriate patches and feeding sites that result in landscape-use patterns is also discussed (Figure 46.1). The deliberate and careful modification

Forages: The Science of Grassland Agriculture, Volume II, Seventh Edition.
Edited by Kenneth J. Moore, Michael Collins, C. Jerry Nelson and Daren D. Redfearn.

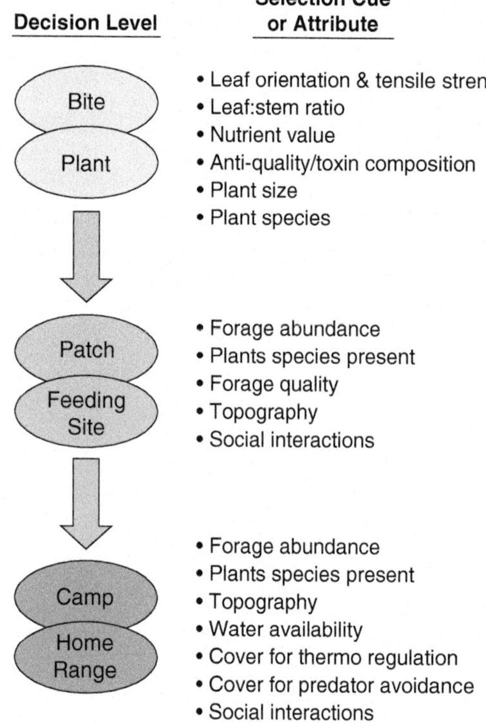

Fig. 46.1. Foraging decisions occur along a hierarchical continuum from bite to home range. At each of these decision levels, animals respond to varying selection cues related to plant and habitat attributes. Understanding the relevant selection cue for specific foraging decisions is necessary to develop management plans that are sensitive to animal behavior processes.

of animal attributes and forage characteristics is based on an understanding of foraging behavior at these hierarchic scales and could yield new and efficient options for adaptive forage management. Revealed through this chapter are practices that could be applied to alter diet selection or habitat use patterns in grazing herbivores to meet management objectives.

Diet Selection Based on Digestive Consequences

"Good" foraging environments, as described throughout this text, are those replete with green, leafy plants that are easy for the animal to harvest and full of readily digestible forms of energy and nutrients. In these foraging environments, it is not difficult for herbivores to make appropriate foraging decisions. Unfortunately, most foraging environments are tremendously complex and often inhospitable places for mammalian herbivores. These environments may contain nutritious plants, but there is immense variation in the nutritional value and toxic properties of these plants. To further complicate matters, levels of nutrients and toxins in plants vary both spatially and temporally (Provenza and Balph 1990). Fortunately, grazing animals possess adaptive behaviors that allow them to choose and eat forages that are more nutritious and less toxic than the average of the available forage resource (Arnold 1981; Provenza 1995; Cruz and Ganskopp 1998; Lyman et al. 2011; Villalba et al. 2011).

Grazing behavior, though complex and difficult to explain, stems from the basic tenet that the consequences of foraging (i.e. nutritional enhancement or toxicity) direct an animal's foraging decisions. Consequence-based learning is uniquely applied to diet selection through the formation of conditioned flavor aversions and preferences. The principles of learned food aversions and preferences were first outlined by John Garcia and colleagues (Garcia et al. 1974; Garcia et al. 1985) and subsequently applied to grazing animals by Frederick Provenza (Provenza et al. 1992; Provenza 1995). In essence, animals acquire preferences for foods if positive digestive consequences follow ingestion, such as (i) energy or protein enrichment (Villalba and Provenza 1996; Villalba and Provenza 1999); (ii) recovery from nutritional deficiency (Garcia et al. 1967); or, (iii) recovery from illness (Phy and Provenza 1998). Alternatively, herbivores form aversions to foods when their consumption is followed by gastrointestinal distress, particularly if this post-ingestive feedback stimulates an emetic system causing "nausea" (Provenza et al. 1992; Provenza 1995).

The key result of these flavor-consequence relationships is that the hedonic value of flavor is modified by post-ingestive consequences. In other words, a plant tastes "good" because its consumption made the herbivore feel "better" when eaten in the past. Likewise, plants taste "bad" when the herbivore feels ill or unsatisfied after eating them (Provenza 1995). Plants do not have inherently good or bad flavors; the value of flavor is determined by post-ingestive consequences, and this value changes as forage value waxes and wanes.

Once the hedonic value of a plant is established, the animal uses its senses of smell and sight to differentiate among plants with high hedonic value (i.e. "good" flavor) or aversive foods. Searching and selective grazing are cognitive processes that can be further reinforced by interactions with other animals, including humans. The resulting behavior patterns lead to increased consumption of forages that are likely to yield nutritional benefits and limited consumption of toxic or low-quality plants (Distel and Villalba 2018). Forage preference is, therefore, a moving target that depends on internal and external factors that set the expression of digestive feedback and create the basis for decisions (Figure 46.2). The internal conditions that direct foraging decisions include animal attributes that affect ingestion, digestion, and metabolism. Environmental conditions, social interactions with peers,

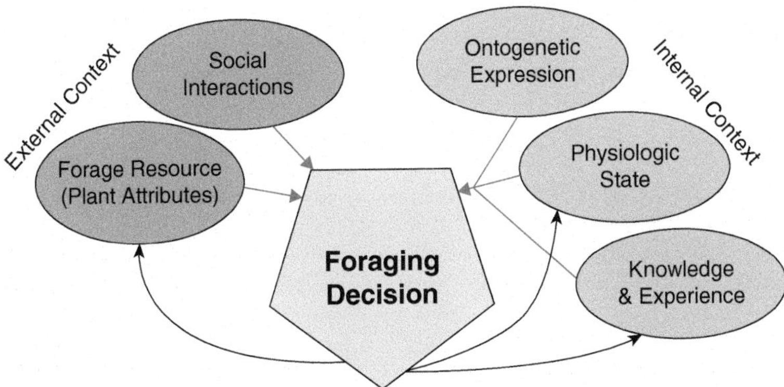

Fig. 46.2. Foraging decisions are directed by many factors and conditions that are external to the grazing animal, such as social interactions with other herbivores or plant quality and antiquality factors. The environment inside the animal also directs foraging decisions through physical and physiologic capabilities, current physiologic state, and acquired knowledge and experience. Collectively these factors set the "external" and "internal" context for foraging.

and plant attributes set the external context for foraging (Distel and Villalba 2018).

Ontogenetic Expression of Diet Selection

Forage preference and intake vary by herbivore species, breed, and individual. Much of this observed variation in diet preferences can be traced to inherited morphologic and physiologic characteristics. These animal attributes determine the digestive consequences of foraging and, therefore, shape foraging decisions. Morphologic characteristics are unquestionably inherited and clearly influence diet selection (Hofmann 1989). Inherited digestive characteristics contributed to observed differences among breeds of livestock in their ability to digest dry matter and energy from similar diets (Phillips 1961; Beaver et al. 1989). The inheritance of enzyme systems involved in digestion is well documented (Velázquez and Bourges 1984). This may explain why absorption of minerals (Green et al. 1989) and nutrients (Beaver et al. 1989) during digestion differs among animal breeds.

Enzyme systems necessary for detoxification of some plant **allelochemicals** are also strongly inherited (e.g. fluoroacetates in range plants; Mead et al. 1985). Research on cattle (Herbel and Nelson 1966; Winder et al. 1996), sheep (Warren et al. 1984), and goats (Warren et al. 1984; Pritz et al. 1997) has revealed that animal breeds also differ in diet preferences. Furthermore, there is some evidence that selection pressures placed on domestic animals can alter their ability to harvest nutrients in specific environments. For example, compared with their US counterparts, New Zealand Holsteins selected for productivity under grazing have anatomic and behavioral adaptations that facilitate grazing: larger mouths and longer grazing times per day (Bryant 1990).

Foraging Decisions Based on Physiologic State

To forage successfully, grazing animals must possess a foraging system that is responsive to their changing energy and nutritive requirements (Provenza 1996; Provenza et al. 1998). These nutrient demands vary depending on what was eaten earlier in the day, ambient external conditions, and animal physiologic state. Grazing animals are also able to modify their diet selection depending on their current nutritional or physiologic condition (Provenza 1995; Provenza et al. 1998). When the animal's need for a specific nutrient is high (e.g. protein), preferences for foods containing the nutrient are high. For example, lambs fed diets deficient in sodium, energy, or protein show a strong preference for foods high in sodium, energy, or protein, respectively (Villalba and Provenza 1996; Villalba and Provenza 1999). When needs have been met, preference declines.

Animals on a high plane of nutrition will often be more selective and choose diets different from those of animals in a deficient state (Murden and Risenhoover 1993). This could result from varying digestive or metabolic capacities. Animals in poor body condition may have a decreased ability to detoxify consumed allelochemicals (Foley et al. 1995). Thus, an animal's nutritional state and body condition influence its incentive to seek out and eat particular plants (Distel and Villalba 2018).

Knowledge and Experience in Making Foraging Decisions

The memory of flavor-consequences association is critical to the development of adaptive foraging habits. To accomplish appropriate foraging decisions, animals must learn and remember which foods are nutritious and which are toxic. Grazing animals express a remarkable ability to remember which foods were satisfying, were not particularly reinforcing, or caused illness. Lambs exposed to barley with their mothers for as little as 30 minutes at 6 months of age, readily consumed this "familiar preferred" food two years later when offered barley again (Green et al. 1984).

As animals mature and gain foraging experience, they also become more efficient at harvesting forage. For example, lambs with experience grazing potted crested wheatgrass plants ingested grass more efficiently (i.e. more grams per minute of foraging) than their counterparts raised on potted shrub plants (Flores et al. 1989). Similarly, goats raised in a shrubland environment with blackbrush consumed browse more efficiently than goats raised in drylot conditions (Distel and Provenza 1991). Foraging experience may also adapt anatomic characteristics, such as rumen size and papillae development (Ortega-Reyes et al. 1992).

Social Interactions Affecting Diet Selection

Grazing livestock are gregarious creatures that live in social groups where dietary information can be easily passed from experienced to inexperienced animals. Young livestock, therefore, do not require perfect and complete dietary information at birth. Learning from their mothers may begin even before young herbivores take their first bites because flavors in uterine fluid can influence food aversions (Smotherman 1982). Mother's milk is also a source of information for young livestock. For example, Nolte and Provenza (1992) found that orphan lambs raised on onion-flavored milk preferred onion-flavored feed later in life.

As animals begin to forage, their mother is an important model (Thorhallsdottir et al. 1990a). Lambs quickly learn to avoid the same harmful novel foods their mother had previously been trained to avoid and to consume novel alternatives readily consumed by their mother (Mirza and Provenza 1990; Mirza and Provenza 1994). Nursing calves began to eat substantial quantities of locoweed (Pfister, unpublished observations) and low larkspur (Pfister and Gardner 1999) on the same day as their grazing mothers, suggesting that calves mimicked their mother's diet. Research to tease out maternal influences through genetics (nature) vs social modeling (nurture) indicates that the mother's role in learning is more influential than genetic inheritance (Glasser et al. 2009).

Young livestock can also learn appropriate food choices from other adult animals and peers (Thorhallsdottir et al. 1990b). For example, inexperienced heifers began grazing sooner (0.5 hour vs 2.1 hour) compared with similar heifers grazing along with mature experienced cows (Costa et al. 2016). This effect dissipated quickly, and no difference was apparent over a 3-day period.

Social interactions may also facilitate the extinction of aversions. Ewes and lambs averse to a pelleted ration ingested more of the ration when feeding with non-averse peers than when feeding alone (Provenza and Burritt 1991). Cattle also consumed more of a toxic plant they had been conditioned to avoid when feeding with non-averted peers (Lane et al. 1990).

Animals are, however, more influenced by their own dietary experiences than by social models. Lambs consistently avoided a food after experiencing toxicosis even if their mothers readily consumed the food (Provenza et al. 1993; Pfister et al. 1993; Glasser et al. 2009). Calves that initially ate low larkspur with their mothers sharply curtailed consumption a few days later (Pfister and Gardner 1999), perhaps because of adverse feedback.

Plant Attributes that Influence Digestive Consequences

Forage plants vary continuously and radically in nutritive quality or toxicity depending on time of day or year and location in the landscape. Grazing animals must use foraging tactics that allow them to track the nutritive quality of plants that vary spatially (Cook and Harris 1950; Launchbaugh et al. 1990), seasonally (Cook and Harris 1950; Sims et al. 1971), and even daily (Fisher et al. 1999).

Animals sense plant quality through conditioned preferences during which stronger preferences develop for plants with more positive digestive feedback (Mehiel and Bolles 1984). Preferences are also higher for plants with relatively rapid digestive feedback, that is, those that digest quickly (Kyriazakis and Oldham 1997). This could explain why grazing preference generally correlates to dry matter digestibility and why forbs with high ratios of soluble and fermentable carbohydrates are preferred to those with low ratios.

Excessive or deficit amounts of nutrients (i.e. energy, protein, minerals) cause **palatability** to decrease (Provenza 1995; Distel and Villalba 2018). Protein and energy are important resources, but excessive protein can cause dramatic decreases in preference and intake (Provenza 1995; Villalba and Provenza 1997). The balance of protein to energy can also have a strong influence on palatability. Food preference generally declines if there is too much protein relative to energy or vice versa or digestion rates of protein and energy are not similar (Kyriazakis and Oldham 1997).

Forage plants also possess a wide variety of chemical and physical properties that reduce forage value and serve as grazing deterrents. From the animal's perspective, the effects of antiquality factors can be expressed along a continuum from those that reduce the forage nutrient or energy yield to those producing toxic or ill effects. How strongly a plant attribute affects diet selection or intake, therefore, depends on the magnitude, timing, or nature of digestive feedback (Launchbaugh et al. 2001). Villalba et al. (2011) found that lambs previously fed birdsfoot trefoil, alfalfa, or tall fescue containing condensed tannins, saponins, or ergotamine, respectively, reduced their consumption of the forage containing each plant secondary compound. The result is that animals form strong aversions to plants when their consumption is quickly followed by intense illness causing nausea.

Feeding deterrents may be important to promote survival of plant species under grazing by herbivores ranging in mass from nematodes to large mammals. Deterrent compounds may cause digestive distress resulting in reduced palatability though conditioned aversions, or these organic compounds may reduce the eating drive by moderating hunger and satiety, thereby slowing grazing and shortening grazing time. Behavior-modifying substances are also found in associations between plants and microorganisms. Ergo-alkaloids in tall fescue/perennial ryegrass affect neuroreceptors, including those for gamma amino butyric acid (GABA), dopamine, serotonin, and melatonin (Porter and Thompson 1992), and affect many aspects of grazing. Angus calves from Missouri, where tall fescue is abundant, showed less reduction in intake on endophyte-infected tall fescue than similar cattle from Oklahoma, both compared with nontoxic fescue (Johnson et al. 2014). Recent work has shown that genetic differences exist in negative cattle response to toxic tall fescue related to two genes, DRD2 and XKR4 (Bastin et al. 2014; Campbell et al. 2014). Cows homozygous for DRD2, a dopamine receptor, and not for XKR4 had higher serum prolactin levels than cows without this profile.

There is also some evidence that neurotransmitters present in healthy herbage, such as GABA in alfalfa, alter grazing and other behavior. Legume phytoestrogens, which increase in response to infection by plant pathogens, impact reproductive behavior, thus altering energy demand and grazing behavior (McDonald 1995). Plant compounds can also alter grazing behavior by affecting populations of intestinal parasites that lower eating drive by initiating the synthesis of gut wall neuropeptides involved in the regulation of hunger and satiety (Sykes and Coop 2001). Villalba et al. (2013) found that lambs exposed to *Haemonchus contortus* preferentially grazed the tanniferous legume sainfoin over a nontanniferous alternative, cicer milkvetch (82% vs 60%), following exposure to the parasite. Lyman et al. (2011) speculated that animals grazing separate birdsfoot trefoil (condensed tannins), endophyte-infected tall fescue (alkaloids), and alfalfa (saponins) in different sequences learned the postingestive benefits of the sequence of ingestion.

Selection and Rejection of Patches of Feeding Stations

Varying preference for individual plants and selective behavior aimed at preferred species can be revealed as patch, feeding station, and feeding site preferences at the spatial scales of pastures or landscapes. Individual animals and herds preferentially seek sites with abundant desirable vegetation and avoid sites with inadequate or less-desirable forage choices (reviewed by Senft et al. 1987; Illius and Gordon 1993).

Just as animals form preferences or aversions for foods based on consequences of consumption, they form likes and dislikes for foraging sites based on the consequences of foraging in these places. Places that provide high-quality food, water, or appropriate thermal regimes are preferred over those that offer no positive reinforcement (see reviews in Schechter and Calcagnetti 1998, and Tzschentke 1998). Conditioned habitat preferences are also likely formed to places that provide escape from fear (such as that caused by predators), pain (such as that induced by electric shock or insect pests), stress, hunger, and excessive heat and cold, or nausea (see Schechter and Calcagnetti 1998; Tzschentke 1998; Scaglia and Boland 2014b).

A grazing preference would be formed for habitats where satiety, relief of thirst, thermal neutrality, freedom from pain, comfort, a sense of security, or rest were experienced. Aversions would conversely be formed for habitats where animals experienced hunger, excessive heat or cold, pain, stress, illness, weariness, or fear. For example, it is widely recognized that animals can become averted to handling facilities if the movement through these facilities is associated with pain and fear (Grandin and Dessing 1998). Alternatively, animals can form place preferences and easily move through handling facilities associated with a food reward (Hutson 1981).

Where animals graze across a landscape is influenced by social models including mother and peers (Howery et al. 1998) and by an animal's previous grazing experiences. For example, Bailey et al. (2010a) demonstrated that Brangus cattle raised on Chihuahuan desert rangeland grazed further from water, used larger areas, and selected higher quality diets than Brangus cattle raised in south Texas when both groups grazed Chihuahuan desert pastures unfamiliar to them.

Landscape use patterns also have a genetic basis. It is well known that some breeds use steeper slopes (Bailey et al. 2010b) or distances farther from water (Russell et al. 2012) than other breeds and that these differences in landscape use patterns yield differences in diet quality. Variation in landscape use among individuals of the same breed can also be traced to genetic differences. Bailey et al. (2015) found that 20–24% of phenotypic variation in use of steep slopes and high elevations by cattle was explained by just a few specific chromosomal regions.

Foraging animals may bypass patches or feeding stations for a number of reasons, most unknown to humans. Patches of herbage in or around dung deposits may be avoided by grazing animals, possibly as an evolutionary consequence of endoparasitism (Cooper et al. 2000).

Aversion to fresh dung has been attributed to emanating vapors of indole and skatole (Mulla and Ridsdill-Smith 1986). In contrast, aversion to grazing in and around recent urine deposits is transitory. Aversion to herbage contaminated with fresh urine deposits may be related to the presence of alarm substances (Boissy et al. 1998) and behavioral effects modulated by pheromones (Marinier et al. 1988). It is more difficult to explain the aversion of cattle to patches of herbage surrounding dung and urine deposits long after volatiles have dissipated. These patches may be taller, greener, leafier, darker, and richer in preferred species than surrounding areas, characteristics that are normally associated with preferred patches or feeding stations. Animals may extend the zone of repugnance surrounding recent dung and urine spots when offered generous herbage allowances or are nearing satiety, or they may shrink the rejected area when hungry. In some grazing systems, animals may reject herbage on as much as half of the pasture area, drastically changing grazing behavior (Marsh and Campling 1970).

Foraging Strategies and Ingestive Behavior

Foraging mechanisms that maximize net energy gain per unit foraging time lead to enhanced fitness of the herbivores to their grazing range. Energy maximization resides at the center of nearly all optimal foraging theorems with a few notable exceptions, including nutrient maximization and time minimization (Stephen and Krebs 1986). Optimal foraging models have proven to be useful in describing the general patterns and constraints of forage selection by large generalist herbivores (Westoby 1974; Owen-Smith and Novellie 1982; Belovsky 1984). However, these models have not clearly defined the processes that herbivores use to achieve optimal diets (Stephen and Krebs 1986). In addition, these models have not very successfully predicted between- or within-animal variability (Illius and Gordon 1993; Cruz and Ganskopp 1998). Only a small proportion of animals apparently graze at the optimum range allowed by the sward at any feeding station (Rook et al. 2002). With the broad goal of "energy maximization," animals are apparently afforded decision opportunities to adapt to specific foraging environments and accomplish additional foraging goals such as sampling and maintaining a diverse diet (Westoby 1974).

Cattle at feeding stations graze in horizons down from the sward surface. In managed pastures, and when depth of grazing is not restricted, these horizons may be 10–15 cm in alfalfa (Dougherty et al. 1988) or generally about 0.7 of sward height (Griffiths et al. 2003a). Grazing the upper horizons of swards permits deeper bites, larger bite volumes, and heavier bites, with the latter also being dependent on herbage density. Grazing upper horizons ensures bites are of high-densities of metabolizable energy and nutrients because herbage quality typically declines down through the canopy (Buxton et al. 1985). When herbage is depleted at feeding stations, animals seek unexploited feeding stations (Laca et al. 1994). Time spent at feeding stations is an index of sward structure, grazing intensity, and herbage allowance. Such foraging strategy may be interpreted as an intuitive attempt to maximize metabolizable energy intake with minimum effort (Wallis de Vries and Daleboudt 1994), in accordance with foraging theory (Stephen and Krebs 1986).

Foraging fiber digesters are primarily energy limited, and grazing behavior is driven by the need to maximize rates of energy intake with the minimum energy expenditure. Instantaneous rate of dry matter intake is a useful index of ingestion, including eating drive, physical and chemical attributes of pastures, and environmental influences on both plant and animal components. Therefore, instantaneous intake rates afforded by specific patches of vegetation may constitute foraging consequences that direct preferences at the patch or feeding station level.

Exposure to fresh patches stimulates grazing drive as indicated by rates of biting and intake over the first minute or so after initiation of grazing (Forbes and Hodgson 1985; Dougherty et al. 1989a). Griffiths et al. (2003b) indicated that cattle exposed to ungrazed but familiar patches graze cautiously for 20–30 bites before attaining full bite depth. Exploration and exploitation of already selected patches, therefore, involve some cognitive processes that modulate hunger and satiety.

Griffiths et al. (2003b), investigating ryegrass patch selection by grazing dairy cows based on bites per patch and residence time per patch, found that cows preferred the taller of two vegetative swards, chose swards of equal regrowth with shorter stem stubble over swards with taller stem stubble, and harvested more bites from patches with deeper regrowth horizons than shorter ones. Griffiths et al. (2003a) ranked their three patch-appraisal cues in order of importance: depth of regrowth, sward maturity, and sward height.

In high-biomass swards, the distance to the next feeding station may be as little as one step (Ruyle and Dwyer 1985). In depleted or sparse pastures, the distance and, therefore, interval between feeding stations increases. Longer travel times between feeding stations slow the rate of intake and may lead to longer grazing times or limited daily nutrient intake. Scaglia and Boland (2014a) found that heifers grazing at low-stocking densities spent more time walking on annual ryegrass and mixed ryegrass-clover pastures compared with separate pastures of the grass and clover. The same study found that grazing time was greater for heifers grazing ryegrass alone than for those grazing the clover mix.

Consequently, higher herbage allowances are needed on sparse pastures to achieve the same daily intake (Dougherty et al. 1992). Grazing animals decide when and where to move to a new feeding station or patch in search of greater reward, but the stimulus is not apparent. Interestingly, Griffiths et al. (2003a) concluded that patch-grazing activity was not influenced by apparently visible grazing opportunities about 1 m distant, at least in the short term. Departure from a grazed patch to another patch may be an intuitive response related to the one that determines when it is more energy efficient for grazing animals to cease foraging and conserve energy.

Herbivores also prefer forages that can be consumed quickly (Kenney and Black 1984; Distel et al. 1995; Illius et al. 1999). Boland and Scaglia (2011) reported that steers in adjacent monocultures of tall fescue and alfalfa spent less time grazing than those on a mixed sward of the same species, presumably, because animals were able to select their preferred forages more easily when offered separately.

Illius et al. (1999) formalized these concepts of foraging efficiency into an "intake-rate maximization" model and clearly demonstrated that intake rate was positively correlated with forage preference. Furthermore, research has confirmed that some grazing animals (e.g. sheep) prefer plants that yield high amounts of green leaf mass per bite (Prache et al. 1998). Leaf mass, however, was not an important cue in patch selection by dairy cows on perennial ryegrass (Griffiths et al. 2003a). Plants that cannot be harvested quickly and those from which herbivores cannot effectively separate high- and low-quality tissue (i.e. select leaves from stem) are less preferred (Griffiths et al. 2003a). These ingestion patterns suggest that forage preference is affected by patch architecture and the spatial distribution among tissues of various nutrients (Buxton et al. 1985).

It may be assumed that the coordination of canopy structure, canopy nutrient profiles, and processes of grazing is a consequence of coevolution of grasslands and large mammalian herbivores (Janis et al. 2002). Intervention and selection by humans may have altered the importance of diet selectivity and other grazing behaviors over time as indicated by the behavior of feral cattle (Hernandez et al. 1999). Conscious decision-making behavior may dominate grazing in rangeland settings while intuitive behaviors may be involved in grazing of highly managed monocultures of uniform sward structure. Grazing behaviors, such as those involved in bite and patch selection, may not be evident in **rotational grazing** systems operating at high-stock densities (say 200 lactating dairy cows per day per hectare) and with high-nutrient demand and high degree of pasture utilization.

Social Interactions Affecting Selection of Feeding Areas

Herd socialization influences grazing behavior of free-ranging cattle. When management permits development of hierarchies within herds, herd behavior may override certain individual grazing behaviors (Rook and Huckle 1995). The "herd boss" may initiate many activities, including the commencement and cessation of grazing (Phillips and Rind 2002). Consequently, the amount and quality of herbage ingested by individuals within herds may be different from that of solo grazers.

Livestock in grazing systems operating at low-stocking rates with minimal human interference might be expected to exhibit strong herd tendencies, though herds may be smaller in number (Hernandez et al. 1999). Suppression of herd behaviors may be a consequence of intensive grazing systems. Grazing systems operating at high stocking rates and densities and where livestock are frequently shifted may suppress herd behavior and recognition of spatial cues, possibly raising overall productivity.

Once herds have settled in a grazing area, individuals usually select feeding stations near their herd mates. Peripheral vision may be limited in head-down grazing, which reduces perception of movement on the horizon. This response may be a survival behavior related to the "many eyes" counter to predation (Roberts 1996). When first exposed to new pastures, individuals may graze shoulder to shoulder, somewhat like their behavior in "frontal grazing" systems (Volesky et al. 1994). This behavior may be an expression of an "aggregation" response seen in cattle and equines in the presence of biting flies (Schmidtmann 1985). Conversely, livestock secure and unfamiliar with predators and used to high stocking rates under **rotational grazing** management (i.e. at low allowances and/or high utilization rates) may disperse more widely when first introduced to fresh swards.

Behavior in the Presence of Other Species

Grazing behavior of individuals and herds of one livestock species may be affected by the presence of another species of livestock or wildlife. Indirect interactions between different grazing species that modify grazing behavior of herds and individuals are important when species compete for forage resources. Bonding within and between grazing species is apparent in co-grazing, multi-species livestock systems (Hulet et al. 1989). Bonds may form between herd species and other species and the bonding of sheep with donkeys, goats, and sheep dogs is used for predator control (Hulet et al. 1989). In the case of co-grazing cattle and horses, cattle graze in rougher areas that are ignored by horses because the structure of the pasture suits grazing mechanics of cattle and because they are not deterred by the presence of horse dung (Arnold 1980).

Visual Cues and Spatial Memory Affecting Grazing Behavior

In familiar pastures, grazers may employ landscape cues to stimulate spatial memories and to select or reject patches or feeding stations (Edwards et al. 1996). Grazing starts when adjacent or visible herd mates are in a grazing attitude; i.e. head down and moving slowly or not moving at all. The feeding stations that best suit their eating apparatus, body mass, muzzle, incisor arcade, tongue, etc., apparently invoke memory. Visual sward cues may be used to select patches or feeding stations upon entering unfamiliar pastures; cattle, for example, may seek the tallest visible patch as their first feeding station. These patches afford large bite masses and high rates of dry matter intake (Distel et al. 1995).

Foraging choices (i.e. cognitive processes) based on seeking or avoiding individual plants are revealed in spatial patterns where animals seek and remember locations or sites with good forage resources (Bailey et al. 1989; Bailey and Sims 1998). Visual cues greatly enhanced the ability of free-ranging livestock to more efficiently locate and consume nutritious food patches in pastures and landscapes (Edwards et al. 1997; Howery et al. 2000). Grazing animals also use visual cues associated with the locations of forage, cover, and water resources, as well as predators and other environmental hazards (Senft et al. 1987; Bailey et al. 1996). Manipulating natural and artificial visual cues could be used to address animal distribution problems under field conditions. For example, training cattle to recognize specific images and associate them with food to "lure" them to underutilized rangeland (Bailey and Welling 1999).

Temporal Aspects of Grazing Behavior

Grazing time each day is the single most important parameter of grazing behavior because it is the primary variable that free-ranging animals use to adjust for high energy demand (e.g. lactation) or compensate for pasture constraints that lower intake rate (e.g. low density and allowances of energy). Hunger and satiety are accepted mechanisms for the regulation of start, slowing, and cessation of grazing. Grazing time is quite difficult to measure and quantify because grazing activities are sporadic and are influenced by herd activity. Grazing time of ruminants is limited by the time needed for rumination and other social activities. Rumination times typically increase with grazing time and with declining quality of ingesta. Cattle grazing pasture and range typically have two major grazing meals each day, one commencing at first light and one starting in the late afternoon; sometimes a few individuals may have a short grazing bout around solar noon. Ingestion of herbage with a slow rate of passage usually means that ruminants have fewer but longer grazing bouts each day, whereas animals grazing herbage with a fast rate of passage have several shorter grazing bouts per day (Coleman and Phillips 1991).

Grazing ecologists, nutritionists, and managers operate on 24-hour time periods, but the slow rate of passage of digesta through the gastrointestinal tract of large herbivorous mammals means that the amount and quality of ingesta may impact grazing for at least four days, possibly more, if **toxicants** are involved. Ruminants grazing forage with a high rate of passage, such as alfalfa or wheat pastures, have little residual effect on grazing the day following ingestion, whereas herbage with slow rates of passage, such as tall fescue, have a negative and diminishing effect on intake over subsequent days (Dougherty et al. 1989a, 1989b). One might expect equines to have less carryover effect on their grazing because they do not have a restriction to flow of digesta, such as the reticulo-omasal orifice of ruminants, and they accelerate the rate of passage of digesta of low-quality diets.

Grazing at night may occur only in animals with high demand for nutrients or in animals when stress (from toxicants, ambient heat, biting flies, etc.) has limited grazing time during the day (Stricklin et al. 1976). Feeding hay, silage, and other supplements affects grazing and other behaviors, with the primary effects being on reduction of the grazing time per day and on spatial and temporal patterns (Phillips and Rind 2002). Sheahan et al. (2013) found that supplementation in the afternoon did not affect grazing by dairy cows whereas supplementation in the morning reduced grazing during that period.

Conclusions and Management Applications

Livestock managers on pasture and rangeland should be aware of how their management actions affect animal behavior if they are to optimize productivity, profit, and sustainability of grazed ecosystems and enterprises. Behavioral changes of individuals and herds are the first observable responses of grazing livestock to changing situations and are cues managers could use to fine-tune their grazing practices. Noting behavioral adjustments to environmental, nutritional, disease, and other stresses and making appropriate management decisions is critical in risk management of grazing systems. Finally, diet selection and landscape use are malleable behaviors that can be shaped to achieve land management goals through an understanding of the consequences that sustain grazing behavior patterns.

References

Arnold, G.W. (1980). Behavioural aspects of mixed grazing. In: *Proceedings of a Workshop on Mixed Grazing (September 1980)* (eds. T. Nolan and J. Connolly), 140–143. Ireland: Galway.

Arnold, G.W. (1981). Grazing behavior. In: *Grazing Animals. World Animal Science B1* (ed. F.H.W. Morley), 79–104. New York, NY: Elsevier Science Publishing Co.

Bailey, D.W. and Sims, P.L. (1998). Association of food quality and locations by cattle. *J. Range Manage.* 51: 2–8.

Bailey, D.W. and Welling, G.R. (1999). Modification of cattle grazing distribution with dehydrated molasses supplement. *J. Range Manage.* 52: 575–582.

Bailey, D.W., Rittenhouse, L.R., Hart, R.H. et al. (1989). Association of relative food availabilities and locations by cattle. *J. Range Manage.* 42: 480–482.

Bailey, D.W., Gross, J.E., Laca, E.A. et al. (1996). Mechanisms that result in large herbivore grazing distribution patterns. *J. Range Manage.* 49: 386–400.

Bailey, D.W., Thomas, M.G., Walker, J.W. et al. (2010a). Effect of previous experience on grazing patterns and diet selection of Brangus cows in the Chihuahuan Desert. *Rangeland Ecol. Manage.* 63: 223–232.

Bailey, D.W., Marta, S., Jensen, D. et al. (2010b). Genetic and environmental influences on distribution patterns of beef cattle grazing foothill rangeland. In: *Proceedings, Western Section*, vol. 61, 64–66. American Society of Animal Science.

Bailey, D.W., Lunt, S., Lipka, A. et al. (2015). Genetic influences on cattle grazing distribution: association of genetic markers with terrain use in cattle. *Rangeland Ecol. Manage.* 68 (2): 142–149.

Bastin, B.C., Houser, A., Bagley, C.P. et al. (2014). A polymorphism in XKR4 is significantly associated with serum prolactin concentrations in beef cows grazing tall fescue. *Anim. Genet.* 45: 439–441.

Beaver, E.D., Williams, J.E., Miller, S.J. et al. (1989). Influence of breed and diet on growth, nutrient digestibility, body composition and plasma hormones of Brangus and Angus steers. *J. Anim. Sci.* 67: 2415–2425.

Belovsky, G.E. (1984). Herbivore optimal foraging: a comparative test of three models. *Am. Nat.* 124: 97–115.

Boissy, A., Terlouw, C., and Le Neindre, P. (1998). Presence of cues from stressed conspecifics increases reactivity to aversive events in cattle: evidence for the existence of alarm substances in urine. *Physiol. Behav.* 63: 489–495.

Boland, H.T. and Scaglia, G. (2011). Giving beef calves a choice of pasture-type influences behavior and performance. *Prof. Anim. Sci.* 27: 160–166.

Bryant, A.M. (1990). Present and future grazing systems. *Proc. N.Z. Anim. Prod. Soc.* 50: 35–42.

Buxton, D.R., Hornstein, J.S., Wedin, W.F., and Marten, G.C. (1985). Forage quality in stratified canopies of alfalfa, birdsfoot trefoil and red clover. *Crop Sci.* 26: 180–184.

Campbell, B.T., Kojima, C.J., Cooper, T.A. et al. (2014). A single nucleotide polymorphism in the dopamine receptor D2 gene may be informative for resistance to fescue toxicosis in Angus-based cattle. *Anim. Biotechnol.* 25: 1–12.

Coleman, S.W. and Phillips, W.A. (1991). Behavior of cattle grazing either dormant native grass or winter wheat. *J. Anim. Sci.* 69 (suppl. 1): 284.

Cook, C.W. and Harris, L.E. (1950). The nutritive value of range forage as affected by site and stage of maturity. Utah Agricultural Experiment Station Bulletin No. 344.

Cooper, J., Gordon, I.J., and Pike, A.W. (2000). Strategies for the avoidance of faeces by grazing sheep. *Appl. Anim. Behav. Sci.* 69: 15–33.

Costa, J.H.C., Costa, W.G., Weary, D.M. et al. (2016). Dairy heifers benefit from the presence of an experienced companion when learning how to graze. *J. Dairy Sci.* 99: 562–568.

Cruz, R. and Ganskopp, D. (1998). Seasonal preferences of steers for prominent northern Great Basin grasses. *J. Range Manage.* 51: 557–565.

Distel, R.A. and Provenza, F.D. (1991). Experience early in life affects voluntary intake of blackbrush by goats. *J. Chem. Ecol.* 17: 431–450.

Distel, R.A. and Villalba, J.J. (2018). Use of unpalatable forages by ruminants: the influence of experience with the biophysical and social environment. *Animals* 8 (56): E56.

Distel, R.A., Laca, E.A., Griggs, T.C., and Demment, M.W. (1995). Patch selection by cattle: maximization of intake rate in horizontally heterogeneous pastures. *Appl. Anim. Behav. Sci.* 45: 11–21.

Dougherty, C.T., Smith, E.M., Bradley, N.W. et al. (1988). Ingestive behaviour of beef cattle grazing alfalfa (*Medicago sativa* L.). *Grass Forage Sci.* 43: 121–130.

Dougherty, C.T., Cornelius, P.L., Bradley, N.W., and Lauriault, L.M. (1989a). Ingestive behavior of beef heifers within grazing sessions. *Appl. Anim. Behav. Sci.* 23: 341–351.

Dougherty, C.T., Bradley, N.W., Cornelius, P.L., and Lauriault, L.M. (1989b). Short-term fasts and the ingestive behaviour of grazing cattle. *Grass Forage Sci.* 44: 295–302.

Dougherty, C.T., Bradley, N.W., Lauriault, L.M. et al. (1992). Allowance-intake relations of cattle grazing vegetative tall fescue. *Grass Forage Sci.* 47: 211–219.

Edwards, G.R., Newman, J.A., Parsons, A.J., and Krebs, J.R. (1996). The use of spatial memory by grazing animals to locate food patches in spatially heterogenous environments: an example with sheep. *Appl. Anim. Behav. Sci.* 50: 147–160.

Edwards, G.R., Newman, J.A., Parsons, A.J., and Krebs, J.R. (1997). Use of cues by grazing animals to locate food patches: an example in sheep. *Appl. Anim. Behav. Sci.* 51: 59–68.

Fisher, D.W., Mayland, H.L., and Burns, J.C. (1999). Variation in ruminants' preference for tall fescue hays cut either at sundown or at sunup. *J. Anim. Sci.* 77: 762–768.

Flores, E.R., Provenza, F.D., and Balph, D.F. (1989). Role of experience in the development of foraging skills of lambs browsing the shrub serviceberry. *Appl. Anim. Behav. Sci.* 23: 271–278.

Foley, W.J., McLean, S., and Cork, S.J. (1995). Consequences of biotransformation of plant secondary metabolites on acid-base metabolism in mammals: a final common pathway? *J. Chem. Ecol.* 21: 721–743.

Forbes, T.D.A. and Hodgson, J. (1985). Comparative studies of the influence of sward conditions on the ingestive behavior of cows and sheep. *Grass Forage Sci.* 40: 69.

Garcia, J., Ervin, F.R., Yorke, C.H., and Koelling, R.A. (1967). Conditioning with delayed vitamin injections. *Science* 155: 716–718.

Garcia, J., Hankins, W.G., and Rusiniak, K.W. (1974). Behavioral regulation of the milieu interne in man and rats. *Science* 1985: 824–831.

Garcia, J., Lasiter, P.A., Bermudez-Rattoni, F., and Deems, D.A. (1985). A general theory of aversion learning. In: *Experimental Assessments and Clinical Applications of Conditioned Food Aversions* (eds. N.S. Braveman and P. Bronstein), 8–21. New York, NY: The New York Academy of Sciences.

Glasser, T.A., Ungar, E.D., Landau, S.Y. et al. (2009). Breed and maternal effects on the intake of tannin-rich browse by juvenile domestic goats (*Capra hircus*). *Appl. Anim. Behav. Sci.* 119 (1–2): 71–77.

Grandin, T. and Dessing, M.J. (1998). Genetics and behavior during handling, restraint, and herding. In: *Genetics and the Behavior of Domestic Animals* (ed. T. Grandin), 113–144. New York, NY: Academic Press.

Green, G.C., Elwin, R.L., Mottershead, B.E., and Lynch, J.J. (1984). Long-term effects of early experience to supplementary feeding in sheep. *Proc. Aust. Soc. Anim. Prod.* 15: 373–375.

Green, L.W., Baker, J.F., and Hardt, P.F. (1989). Use of animal breeds and breeding to overcome the incidence of grass tetany: a review. *J. Anim. Sci.* 67: 3463–3469.

Griffiths, W.M., Hodgson, J., and Arnold, G.C. (2003a). The influence of sward canopy structure on foraging decisions by grazing cattle. I. Patch selection. *Grass Forage Sci.* 58: 12–124.

Griffiths, W.M., Hodgson, J., and Arnold, G.C. (2003b). The influence of sward canopy structure on foraging decisions by grazing cattle. II. Regulation of bite depth. *Grass Forage Sci.* 58: 125–137.

Herbel, C.H. and Nelson, A.B. (1966). Species preference of Hereford and Santa Gertrudis cattle on a southern New Mexico range. *J. Range Manage.* 19: 177–181.

Hernandez, L., Barral, H., Halffter, G., and Colon, S.S. (1999). A note on the behavior of feral cattle in the Chihuahuan Desert of Mexico. *Appl. Anim. Behav. Sci.* 63: 259–267.

Hofmann, R.R. (1989). Evolutionary steps of ecophysio-logical adaptation and diversification of ruminants: a comparative view of their digestive system. *Oecologia* 78: 443–457.

Howery, L.D., Provenza, F.D., Banner, R.E., and Scott, C.B. (1998). Social and environmental factors influence cattle distribution on rangeland. *Appl. Anim. Behav. Sci.* 55: 231–244.

Howery, L.D., Bailey, D.W., Ruyle, G.B., and Renken, W.J. (2000). Cattle use visual cues to track food locations. *Appl. Anim. Behav. Sci.* 67: 1–14.

Hulet, C.V., Anderson, D.M., Smith, J.N. et al. (1989). Bonding of goats to sheep and cattle for protection from predators. *Appl. Anim. Behav. Sci.* 22: 261–267.

Hutson, G.D. (1981). Food preferences of sheep. *Aust. J. Exp. Agric. Anim. Husb.* 21: 575–582.

Illius, A.W. and Gordon, I.J. (1993). Diet selection in mammalian herbivores: constraints and tactics. In: *An Interdisciplinary Approach to Foraging Behavior* (ed. R.N. Hughes), 157–181. Boston, MA: Blackwell Science Publishers.

Illius, A.W., Gordon, I.J., Elston, D.A., and Milne, J.D. (1999). Diet selection in goats: a test of intake-rate maximization. *Ecology* 80: 1008–1018.

Janis, C.M., Damath, J., and Theodor, J.M. (2002). The origins and evolution of north American grassland biomes: the story from the hoofed mammals. *Palaeogeogr. Palaeoclimatol. Palaeoecol.* 177: 183–198.

Johnson, J.S., Bryant, J.K., Scharf, B. et al. (2014). Regional differences in the fescue toxicosis response of *Bos taurus* cattle. *Int. J. Biometeorol.* 59: 385–396.

Kenney, P.A. and Black, J.L. (1984). Factors affecting diet selection by sheep. I: potential intake rate and acceptability of feed. *Aust. J. Agric. Res.* 35: 551–563.

Kyriazakis, I. and Oldham, J.D. (1997). Food intake and diet selection in sheep: the effect of manipulating the rates of digestion and carbohydrates and protein in the food offered as a choice. *Br. J. Nutr.* 77: 243–254.

Laca, E.A., Distel, R.A., Griggs, T.C., and Demment, M.W. (1994). Effects of canopy structure on patch depression by grazers. *Ecology* 75: 706–716.

Lane, M.A., Ralphs, M.H., Olsen, J.D. et al. (1990). Conditioned taste aversion: potential for reducing cattle loss to larkspur. *J. Range Manage.* 43: 127–131.

Launchbaugh, K.L., Stuth, J.W., and Holloway, J.W. (1990). Influence of range site on diet selection and nutrient intake of cattle. *J. Range Manage.* 43: 109–116.

Launchbaugh, K.L., Provenza, F.D., and Pfister, J.A. (2001). Herbivore response to anti-quality factors in forages. *J. Range Manage.* 54: 431–440.

Lyman, T.D., Provenza, F.D., Villalba, J.J., and Wiedmeier, R.D. (2011). Cattle preferences differ when endophyte-infected tall fescue, birdsfoot trefoil, and alfalfa are grazed in different sequences. *J. Anim. Sci.* 89: 1131–1137.

Marinier, S.L., Alexander, A.J., and Waring, G.H. (1988). Flehman behaviour in the domestic horse: discrimination of conspecific odours. *Appl. Anim. Behav. Sci.* 19: 227–237.

Marsh, R. and Campling, R.C. (1970). Fouling of pastures by dung. *Herb. Abstr.* 40: 123–130.

McDonald, M.F. (1995). Effects of plant oestrogens in ruminants. *Proc. Nutr. Soc. N.Z.* 20: 43–51.

Mead, R.J., Oliver, A.J., King, D.R., and Hubach, P.H. (1985). The co-evolutionary role of fluroacetate in plant-animal interactions in Australia. *Oikos* 44: 55–60.

Mehiel, R. and Bolles, R.C. (1984). Learned flavor preferences based on caloric outcome. *Anim. Learn. Behav.* 12: 421–427.

Mirza, S.N. and Provenza, F.D. (1990). Preference of the mother affects selection and avoidance of foods by lambs differing in age. *Appl. Anim. Behav. Sci.* 28: 255–263.

Mirza, S.N. and Provenza, F.D. (1994). Socially induced food avoidance in lambs: direct or indirect material influence. *J. Anim. Sci.* 72: 899–902.

Mulla, M.S. and Ridsdill-Smith, J.T. (1986). Chemical attractants tested against the Australian bush fly *Musca vetustissima* (Diptera: Muscidae). *J. Chem. Ecol.* 12: 261–270.

Murden, S.B. and Risenhoover, K.L. (1993). Effects of habitat enrichment on patterns of diet selection. *Ecol. Appl.* 3: 497–505.

Nolte, D.L. and Provenza, F.D. (1992). Food preferences in lambs after exposure to flavors in milk. *Appl. Anim. Behav. Sci.* 32: 381–389.

Ortega-Reyes, L., Provenza, F.D., Parker, C.F., and Hatfield, P.G. (1992). Drylot performance and ruminal papillae development of lambs exposed to a high concentrate diet while nursing. *Small Rumin. Res.* 7: 101–112.

Owen-Smith, N. and Novellie, P. (1982). What should a clever ungulate eat? *Am. Nat.* 119: 151–178.

Pfister, J.A. and Gardner, D.R. (1999). Consumption of low larkspur (*Delphinium nuttallianum*) by cattle. *J. Range Manage.* 52: 378–384.

Pfister, J.A., Astorga, J.B., Panter, K.E., and Molyneux, R.J. (1993). Maternal locoweed exposure in utero and as a neonate does not disrupt taste aversion learning in lambs. *Appl. Anim. Behav. Sci.* 36: 159–167.

Phillips, G.D. (1961). Physiological comparisons of European and Zebu steers. I. Digestibility and retention times of food and rate of fermentation of rumen contents. *Res. Vet. Sci.* 2: 202.

Phillips, C.J.C. and Rind, M.I. (2002). The effects of social dominance on the production and behavior of grazing dairy cows offered forage supplements. *J. Dairy Sci.* 85: 51–59.

Phy, T.S. and Provenza, F.D. (1998). Sheep fed grain prefer foods and solutions that attenuate acidosis. *J. Anim. Sci.* 76: 954–960.

Porter, J.K. and Thompson, F.N. (1992). Effects of fescue toxicosis on reproduction in livestock. *J. Anim. Sci.* 70: 1594–1603.

Prache, S., Roguet, C., and Petit, M. (1998). How degree of selectivity modifies foraging behaviour of dry ewes on reproductive compared to vegetative sward structure. *Appl. Anim. Behav. Sci.* 57: 91–108.

Pritz, R.K., Launchbaugh, K.L., and Taylor, C.A. Jr. (1997). Effects of breed and dietary experience on juniper consumption by goats. *J. Range Manag.* 50: 600–606.

Provenza, F.D. (1995). Postingestive feedback as an elementary determinant of food preference and intake in ruminants. *J. Range Manage.* 48: 2–17.

Provenza, F.D. (1996). Acquired aversions as the basis for varied diets of ruminants foraging on rangelands. *J. Anim. Sci.* 74: 2010–2020.

Provenza, F.D. and Balph, D.F. (1990). Applicability of five diet selection models to various foraging challenges ruminants encounter. In: *Behavioural Mechanisms of Food Selection. Vol. 20: NATO ASI Series G: Ecological Sciences* (ed. R.N. Hughes), 423–459. Heidelberg, Germany: Springer-Verlag.

Provenza, F.D. and Burritt, E.A. (1991). Socially induced diet preference ameliorates condition food aversion in lambs. *Appl. Anim. Behav. Sci.* 31: 229–236.

Provenza, F.D., Pfister, J.A., and Cheney, C.D. (1992). Mechanisms of learning in diet selection with reference to phytotoxicosis in herbivores. *J. Range Manag.* 45: 36–45.

Provenza, F.D., Lynch, J.J., and Nolan, J.V. (1993). The relative importance of mother and toxicosis in the selection of foods by lambs. *J. Chem. Ecol.* 19: 313–323.

Provenza, F.D., Villalba, J.J., Cheney, C.D., and Werner, S.J. (1998). Self-organization of foraging behavior: from simplicity to complexity without goals. *Nutr. Res. Rev.* 11: 199–222.

Roberts, G. (1996). Why individual vigilance declines as group size increases. *Anim. Behav.* 51: 1077.

Rook, A.J. and Huckle, C.A. (1995). Synchronization of ingestive behaviour by grazing dairy cows. *Anim. Sci.* 60: 25–30.

Rook, A.J., Harvey, A., Parsons, A.J. et al. (2002). Effect of long-term changes in relative resource availability on dietary preference of grazing sheep for perennial ryegrass and white clover. *Grass Forage Sci.* 57: 54–60.

Russell, M.L., Bailey, D.W., Thomas, M.G., and Witmore, B.K. (2012). Grazing distribution and diet quality of Angus, Brangus, and Brahman cows in the Chihuahuan Desert. *Rangeland Ecol. Manage.* 65: 371–381.

Ruyle, G.B. and Dwyer, D.D. (1985). Feeding stations of sheep as an indicator of diminished forage supply. *J. Anim. Sci.* 61: 349–353.

Scaglia, G. and Boland, H.T. (2014a). Spatial arrangement of forages affects grazing behavior of beef heifers continuously stocked at low stocking rate. *Crop Sci.* 54: 1227–1237.

Scaglia, G. and Boland, H.T. (2014b). The effect of bermudagrass hybrid on forage characteristics, animal performance, and grazing behavior of beef steers. *J. Anim. Sci.* 92: 1228–1238.

Schechter, M.D. and Calcagnetti, D.J. (1998). Continued trends in conditioned place preference literature from 1992 to 1996, inclusive with a cross-indexed bibliography. *Neurosci. Biobehav. Rev.* 22: 827–846.

Schmidtmann, E.T. (1985). The face fly, *Musca autumnalis* De Geer, and aggregation behavior in Holstein cows. *Vet. Parasitol.* 18: 203–208.

Senft, R.L., Coughenour, M.B., Bailey, D.W. et al. (1987). Large herbivore foraging and ecological hierarchies. *Bioscience* 37: 789–799.

Sheahan, A.J., Gibbs, S.J., and Roche, J.R. (2013). Timing of supplementation alters grazing behavior and milk production response in dairy cows. *J. Dairy Sci.* 96: 477–483.

Sims, P.L., Loveii, G.R., and Hervey, D.F. (1971). Seasonal trends in herbage nutrient production of important sandhill grasses. *J. Range Manage.* 24: 35–37.

Smotherman, W.P. (1982). Odor aversion learning by the rat fetus. *Physiol. Behav.* 29: 769–771.

Stephen, D.W. and Krebs, J.R. (1986). *Foraging Theory*. Princeton, NJ: Princeton University Press.

Stricklin, R., Wilson, L.L., and Graves, H.B. (1976). Feeding behavior of Angus and Charolais-Angus cows during summer and winter. *J. Anim. Sci.* 43: 721–732.

Sykes, A.R. and Coop, R.L. (2001). Interaction between nutrition and gastrointestinal parasitism in sheep. *N.Z. Vet. J.* 49: 222–226.

Thorhallsdottir, A.G., Provenza, F.D., and Balph, D.F. (1990a). The role of mother in the intake of harmful foods by lambs. *Appl. Anim. Behav. Sci.* 25: 35–44.

Thorhallsdottir, A.G., Provenza, F.D., and Balph, D.F. (1990b). Ability of lambs to learn about novel foods while observing or participating with social models. *Appl. Anim. Behav. Sci.* 25: 25–33.

Tzschentke, T.M. (1998). Measuring reward with the conditioned place preference paradigm: a comprehensive review of drug effects, recent progress, and new issues. *Prog. Neurobiol.* 56: 613–672.

Velázquez, A. and Bourges, H. (eds.) (1984). *Genetic Factors in Nutrition*. New York, NY: Academic Press.

Villalba, J.J. and Provenza, F.D. (1996). Preference for flavored wheat straw by lambs conditioned with intraruminal administrations of sodium propionate. *J. Anim. Sci.* 74: 2362–2368.

Villalba, J.J. and Provenza, F.D. (1997). Preference for flavoured foods by lambs conditioned with intraruminal administration of nitrogen. *Br. J. Nutr.* 78: 545–561.

Villalba, J.J. and Provenza, F.D. (1999). Nutrient-specific preferences by lambs conditioned with intraruminal infusions of starch, casein, and water. *J. Anim. Sci.* 77: 378–387.

Villalba, J.J., Provenza, F.D., Clemensen, A.K. et al. (2011). Preference for diverse pastures by sheep in response to intraruminal administrations of tannins, saponins and alkaloids. *Grass Forage Sci.* 66: 224–236.

Villalba, J.J., Miller, J., Hall, J.O. et al. (2013). Preference for tanniferous (*Onobrychis viciifolia*) and non-tanniferous (*Astragalus cicer*) forage plants by sheep in response to challenge infection with *Haemonchus contortus*. *Small Ruminant Res.* 112: 199–207.

Volesky, J.D., De Archaval, F., O'Ferrell, W.C. et al. (1994). A comparison of frontal, continuous and rotation grazing systems. *J. Range Manage.* 47: 210–214.

Wallis de Vries, M.F. and Daleboudt, C. (1994). Foraging strategy of cattle in patchy grassland. *Oecologia* 100: 98–106.

Warren, L.E., Ueckert, D.N., and Shelton, J.M. (1984). Comparative diets of Rambouillet, barbado, and karakul sheep and Spanish and angora goats. *J. Range Manage.* 37: 172–179.

Westoby, M. (1974). The analysis of diet selection by large generalist herbivores. *Am. Nat.* 108: 290–304.

Winder, J.A., Walker, D.A., and Bailey, C.C. (1996). Effect of breed on botanical composition of cattle diets on Chihuahuan desert range. *J. Range Manage.* 49: 209–214.

CHAPTER

47

Forage-Induced Animal Disorders

Tim A. McAllister, Principal Research Scientist, *Agricultural and Agri-Food Canada, Lethbridge, AB, Canada*
Gabriel Ribeiro, Assistant Professor and Saskatchewan Beef Industry Chair, *Veterinary Medicine, Faculty of Veterinary Medicine, University of Calgary, Calgary, AB, Canada*
Kim Stanford, Research Scientist, *Alberta Agriculture and Forestry, Lethbridge, AB, Canada*
Yuxi Wang, Research Scientist, *Agriculture and Agri-Food Canada, Lethbridge, AB, Canada*

Pasture Bloat

Forages are a major source of nutrients for herbivores around the world. In the US and Canada about 110 million cattle, 7.4 million sheep, 1.4 million goats, and 7.4 million horses depend on forages for all or part of their nutritional needs (Table 47.1). Even in the Canadian grain-based intensive beef production system, 80% of the feed provided to the cattle herd is forage (Legesse et al. 2015). Sometimes the balance of nutrients or presence of secondary compounds in the forage can have negative effects on health. This chapter presents some of these forage-induced health problems, including **bloat**, milk fever, **grass tetany**, laminitis, **nitrate poisoning**, mineral imbalances, and effects of toxic secondary compounds.

Pasture Bloat

Description

Pasture bloat occurs when the production of gas, from **fermentation** in the **rumen** exceeds the ability of ruminants to expel the gas produced. With some **legume** forages such as alfalfa and clover, fermentation in the rumen is very rapid, producing large quantities of gas in a short period. For example, steers fed fresh alfalfa can produce gas at a rate of $2\,l\,min^{-1}$. Under normal conditions, this gas collects in the free space at the top of the rumen and is expelled by **eructation**. With pasture bloat, the gas coalesces in small bubbles such that the eructation mechanism is inhibited by frothy rumen contents (Figure 47.1). Receptors in the rumen wall sense that the area is exposed to liquid rather than free gas, so the esophagus remains closed, preventing eructation. As a result, the gas and bubbles remain trapped in the rumen fluid and the rumen swells (Figure 47.2). This expansion restricts contractions of the diaphragm that inflate the lungs and death ensues by suffocation. If the **cannula** plug is opened in rumen-fistulated cattle, a large proportion of the rumen contents are expelled (Figures 47.3 and 47.4).

Risk among Forages

With fresh alfalfa, the concentration of chloroplast fragments in the rumen of cattle is associated with the

Forages: The Science of Grassland Agriculture, Volume II, Seventh Edition.
Edited by Kenneth J. Moore, Michael Collins, C. Jerry Nelson and Daren D. Redfearn.

Table 47.1 Approximate numbers of domestic beef and dairy cattle, sheep, goats, and horses in the United States and Canada in 2015

	Number of animals (in millions)	
Animal	United States[a]	Canada[b]
Dairy	9.3	0.95
Beef	89.0	12.0
Sheep	5.30	0.87
Goats	2.62	0.30
Horses	6.9	0.96

[a]US data are from USDA Agricultural Census and horse data are from http://www.horsecouncil.org/ahcstats.html.

[b]Canadian Agricultural Census and horse data are from www.equestrian.ca.

Fig. 47.1. An example of normal rumen fluid *(left)*, where gases have been expelled, and frothy rumen fluid *(right)* showing trapped fermentation gases.

occurrence of bloat (Majak et al. 1983). This led to a general theory that frothy bloat occurs as result of an excess of small feed particles in the rumen that contribute to the formation of the stable froth. At the same time, these small particles have a large surface area that is rapidly colonized and digested by rumen microorganisms, accelerating the production of gas. Recent work has suggested that methanogens are more abundant in bloated cattle and that carbohydrate metabolism of the **microbiome** is compromised (Pitta et al. 2016).

Forages can be classified as bloat causing, low risk, or bloat safe (Table 47.2). However, even so-called bloat-safe forages can cause bloat under certain conditions. Grasses are usually bloat-safe if not overly lush or immature. Bloat is most commonly associated with alfalfa and clovers with the risk being higher for alfalfa. Clover flowers can contain **condensed tannins** which may lower the risk of bloat. Bloat potential of forages is related to the ease with which they are digested by rumen microbes. Bloat-causing forages are digested rapidly, whereas bloat-safe forages are digested more slowly (Fay et al. 1980) and this difference has been used successfully to select for bloat-resistant alfalfa (Berg et al. 2000).

Forage Management

Stage of growth or crop maturity is the most important factor in preventing pasture bloat on bloat causing or low-risk species. The risk of bloat is highest at the vegetative (or prebud) stage and decreases progressively as the plant matures to full bloom. Alfalfa harvested at the vegetative stage causes the highest incidence of bloat, which declines at the bud stage and is virtually absent when alfalfa is in full flower (Thompson et al. 2000). The decrease in the leaf-to-stem ratio that occurs with advancing maturity also decreases chloroplast particles in the rumen. A leaf-to-stem ratio of <1 : 2 (on a dry weight basis) could be used as an indicator of a low potential for bloat in alfalfa. The importance of forage maturity should also be considered when managing grass–legume mixtures because cattle may preferentially select legumes over grass within the mixture. Also, grazing cattle consume the upper layers of the sward first and this horizon has a higher proportion of leaf. In alfalfa-rich stands the switch in preference from grass to legumes can occur two to three days after cattle are introduced to the pasture, so cattle should continue to be monitored for bloat during this time. Generally, the incidence of bloat is markedly reduced if the bloat-causing legume comprises less than 50% of the forage mixture (Majak et al. 2003). This explains why most pasture mixes produce stands that contain no more than 20–30% alfalfa.

Pasture Mixes

Seeding rates, fertilization, and grazing management can help maintain a 50 : 50 mixture of grass and alfalfa. Nitrogen fertilization promotes grass growth at the expense of alfalfa. Though alfalfa–grass mixtures may be seeded to promote the desired proportion, selective grazing may still allow excessive intake of alfalfa, resulting in bloat. The ideal companion grass should have the same seasonal growth pattern and regrowth characteristics as alfalfa. The principles of competition in mixtures of grass and alfalfa have been thoroughly reviewed by Chamblee and Collins (1988).

Birdsfoot trefoil, annual lespedeza, sericea lespedeza, and sainfoin are bloat safe. These plants contain tannins, which at low concentrations complex with the cytoplasmic proteins and prevent the formation of stable foam. Lespedeza and trefoils often contain sufficiently high

Fig. 47.2. An example of severe distension of the rumen resulting from the inability to expel microbial fermentation gases. The animal is shown before *(left)* and after *(right)* the onset of frothy bloat.

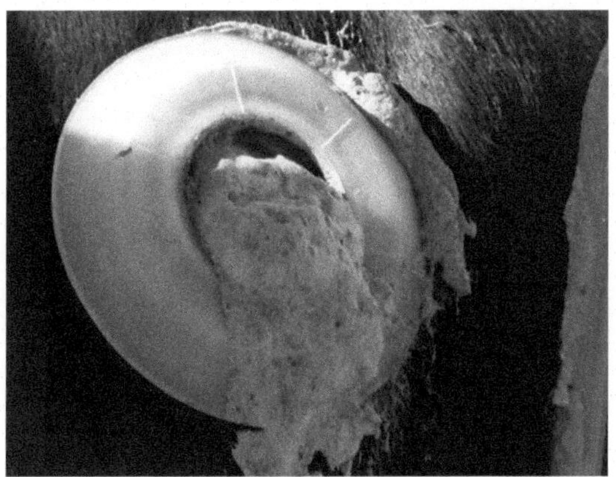

Fig. 47.3. The release of frothy rumen contents of alfalfa when the cannula is opened in a bloated cow. The degree of bloat was 2 on a scale of 1–5. *Source:* Sottie (2014), with permission.

tannin levels to inhibit animal performance by reducing feed intake and protein digestibility. Though **saponins** are foaming agents that occur in most bloat-producing legumes, earlier studies suggested they did not have a significant role in bloat (Majak et al. 1980), but triterpenoid saponins may still play a role through effects on **rumen motility** (Mathison et al. 1999).

Alfalfa bloat is reduced, but not eliminated if sainfoin comprises 10–20% of the grazing diet (McMahon et al. 1999). This reduction has been attributed to the condensed tannins binding proteins and preventing the formation of the stable foam in the rumen, a relationship that has also been observed to reduce the bloat potential of wheat forage (Malinowski et al. 2015). Original sainfoin varieties did not compete well with alfalfa, but new sainfoin cultivars have been developed that compete with and persist in alfalfa pastures (Acharya et al. 2013). Growing these new cultivars in mixed pastures with alfalfa, virtually eliminated the occurrence of pasture bloat (Sottie et al. 2014). However, bloat incidence with alfalfa was not reduced by a commercially available source of condensed tannin (extract of *Schinopsis quebracho-colorado*) administered either in drinking water (5 g l^{-1}) or when sprinkled on top of the **hay** (75 g head^{-1} d^{-1}) (Majak et al. 2004a). These findings suggest that both the type and concentration of condensed tannin influences the ability of these phenolic compounds to prevent bloat.

Dry Matter Intake

Ensuring uniform and regular DM intake can play a key role in avoiding bloat in cattle on legume-dominated pastures. Bloat is less likely to occur if animals are turned out to legume-dominated pasture in the afternoon as opposed to the morning (Hall and Majak 1995). Animals should be fed coarse hay before being introduced to legume-dominated pastures so as to reduce hunger and prevent them from gorging and over-eating the lush legume forage. Bloat risk increases dramatically if grazing is interrupted by overnight removal of animals from legume pastures or if grazing is impeded by inclement weather. Factors that alter normal grazing habits generally result in more intensive, shorter feeding periods that

Fig. 47.4. Alfalfa bloat with a degree of 4, which could rapidly escalate to a degree of 5. Through diligent animal care, extreme cases of bloat can be avoided to prevent animal distress.

Table 47.2 Bloat-causing, low-risk, and bloat-safe forages used as pasture

Bloat-causing	Low-risk	Bloat-safe
Alfalfa	Arrowleaf clover	Sainfoin
Sweetclover	Spring wheat	Birdsfoot trefoil
Red clover	Oat	Cicer milkvetch
White clover	Rape (canola)	Crownvetch
Alsike clover	Perennial ryegrass	Lespedeza
Winter wheat	Berseem clover	Fall rye
	Persian clover	Most perennial grasses

Note: Classification is based on published reports. Insufficient data for species not listed.

can increase the incidence of bloat (Majak et al. 1995). Grazing alfalfa that has been swathed and wilted reduces risk of bloat. Compared with feeding a fresh swath, wilting a swath for 24–48 hours can significantly reduce bloat and may even eliminate it if moisture content is sufficiently reduced (Majak et al. 2001). This response likely reflects wilting induced alterations in the soluble proteins that are responsible for the formation of stable foam in the rumen.

Weather

Weather conditions affect the incidence of bloat in cattle fed fresh alfalfa (Hall et al. 1984). For example, on non-irrigated pasture, bloat occurred more frequently on a day preceded by lower maximum and minimum temperatures of 1–2 °C. This daily change in bloat frequency was not seen under irrigated pasture conditions, since the foliage was always quite high in water content.

In general, bloat can occur at any time during the year, even immediately after a frost (Hall and Majak 1991). The fall peak may be caused by frequent heavy dew or frost. Immediately after a killing frost, which for alfalfa is −9 °C (McKenzie and McLean 1982), soluble proteins may remain high and contribute to bloat. However, if the alfalfa is allowed to stand for a few days after the frost, moisture content and soluble protein contents can decline, reducing the risk of bloat.

Animal Variability

Individual cattle may differ in their susceptibility to pasture bloat, which is related to the rate of passage of particles in the liquid phase of the rumen contents as well as differences in grazing behavior. Frequent bloaters have a slower clearance rate than less-susceptible animals (Majak et al. 1986). Highly susceptible and less-susceptible lines of cattle were selected in New Zealand, but a simple genetic marker could not be identified to distinguish the two lines to allow efficient selection (Cockrem et al. 1983). Nevertheless, it is a good practice to cull bloat-susceptible animals from a breeding herd.

Prevention

Oils and detergents are effective for the prevention and treatment of pasture bloat because they act as surfactants to break down the frothy condition in the rumen contents. Effective detergents include poloxalene, the active ingredient in products such as Bloat Guard® (Hall et al. 1994). However, prevention of bloat is not absolute as variability in the intake of the product can occur if it is fed-free choice. Water-soluble products containing pluronic detergents are available in New Zealand and Australia. For example, when Blocare® 4511 (Ancare©, NZ) was administered at a concentration of 0.06% in the drinking water, bloat was completely prevented under conditions of extreme risk, provided the treated water was the only source of drinking water (Stanford et al. 2001). A new water-soluble product called Alfasure™ is available and has been shown to be completely effective at preventing bloat (Majak et al. 2004a). Water consumption can be reduced during periods of high rainfall due to increased moisture levels in consumed forage, so dosages should be adjusted accordingly. Alfasure can also be sprayed on alfalfa or injected intraruminally to prevent bloat (Majak et al. 2005). Ionophores such as monensin can also reduce the occurrence of pasture bloat (Hall et al. 2001) but are not as effective as pluronic detergents.

Milk Fever

The genetic selection and breeding of dairy cows over the last century has dramatically increased milk production per cow, and consequently, the requirements for Ca in the diets of lactating cows. The sudden increase in the demand for Ca as dairy cows' transition from late gestation to early lactation presents a severe metabolic challenge. The onset of lactation places such a large demand on the Ca homeostatic mechanisms of the body that most dairy cows develop some degree of hypocalcemia at calving. Hypocalcemia is the decrease in plasma Ca concentrations, as a result of the delayed or inadequate adaptation to increased Ca demand. In some cases, plasma Ca concentrations become too low (total blood Ca < $1.4\,\text{mmol}\,\text{l}^{-1}$) to support nerve and muscle function, resulting in parturient paresis or milk fever (clinical hypocalcemia; DeGaris and Lean 2009). Subclinical hypocalcemia (i.e. total blood Ca 1.4–$2.0\,\text{mmol}\,\text{l}^{-1}$ and without clinical signs) is estimated to affect approximately 40–60% of older cows (Reinhardt et al. 2011). Clinical and subclinical hypocalcemia affect herd profitability due to their direct and/or indirect association to increased risk of metabolic and infectious disease, reproductive disorders, early culling, and losses in milk production. Several factors are consistently associated with increased incidence of milk fever, including advancing age, breed, and diet. Milk fever is rare in first lactation cows, with incidence increasing dramatically in third and subsequent lactations. Jersey cows are also more susceptible to milk fever than Holsteins (Horst et al. 1997; DeGaris and Lean 2009; Reinhardt et al. 2011).

Mechanism

Ordinarily, the cow replaces Ca lost to milk by parathyroid hormone (PTH)-mediated homeostatic mechanisms such as: renal re-absorption of urinary calcium, bone Ca resorption, and increased renal production of vitamin D, 1,25-dihydroxyvitamin D [1,25(OH)2D] to enhance intestinal absorption of dietary Ca (Ramberg et al. 1984). In cows with milk fever, bone and kidney tissues have lost their sensitivity to PTH. Recent research (Goff et al. 2014) suggests PTH resistance is the result of diet-induced metabolic alkalosis (see below). Alkaline blood pH causes a conformational change in the PTH receptor located on the surface of bone and kidney cells that interferes with binding of the hormone. Calcium homeostatic processes are disrupted, and milk fever develops.

Hypomagnesemia can also interfere with the ability of PTH to act on target tissues. When PTH binds its receptor, it initiates activation of adenylate cyclase, resulting in production of adenosine 3′,5′-cyclic monophosphate (cyclic AMP). Adenylate cyclase requires Mg as a cofactor thus, hypomagnesemia can cause tissues to become nonreceptive to PTH and hypocalcemia ensues. High K and N concentrations in forage can interfere with ruminal Mg absorption. In general, the best approach to avoid hypomagnesemia in dairy cows is to increase dietary Mg content to 3–$4\,\text{g}\,\text{kg}^{-1}$ DM in dry cow and early lactation diets through the addition of inorganic Mg sources.

Excessive dietary P is another risk factor for milk fever. Dietary P and Ca concentrations affect blood PO_4 concentration which is regulated directly by 1,25(OH)2D and indirectly by the PTH/Ca negative feedback loop. Pre-calving diets high in P seem to negatively affect Ca homeostasis (DeGaris and Lean 2009).

Blood pH and Ionic Balance

Blood pH depends on the number of cations and anions entering the blood from the diet (Stewart 1983). The major dietary cations and the charge they carry are Na (+1), K (+1), Ca (+2), and Mg (+2). The major anions and their charges found in feeds are Cl (−1), SO_4 (−2), and phosphate (−3). The relative numbers of absorbable cations and anions in the diet, known as the dietary cation–anion difference (DCAD) can alter the pH of the blood. High-DCAD diets alkalinize the blood (metabolic alkalosis) and predispose cows to milk fever and subclinical hypocalcemia. Metabolic alkalosis is largely the result of a diet that supplies more cations than anions to the blood. Based on that, many DCAD equations were developed to predict the risk of a pre-calving diet causing milk fever. A meta-analysis study by Lean et al. (2006) concluded that the DCAD equation using milliequivalents of $(Na^+ + K^+) - (Cl^- + S_2^-)$ was most effective at predicting milk fever risk. The major dietary cation of concern in dairy pre-calving diets is excess K^+. High dietary K concentrations increase DCAD, predisposing cows to milk fever.

Manipulating Dietary Cation–Anion Balance

The K content of forages can often range from 14 to 41 $g\,kg^{-1}$ DM. Plant tissue K concentrations of 21–23 $g\,kg^{-1}$ DM achieve maximum yields of most grasses and legumes but, can reach higher concentrations with heavy fertilization. In heavily fertilized fields, cool-season grasses can often contain 30–35 $g\,K\,kg^{-1}$ DM. Small-grain forages, such as oat, rye, and to a lesser extent, wheat, can be particularly high in K. Fresh sugarcane, used in many tropical regions, is also high in K. High-K forages create a challenge to balance prepartum low DCAD diets. In a survey by Lanyon (1980), the average alfalfa sample contained 31 $g\,K\,kg^{-1}$ DM, a level likely to contribute to hypocalcemia in dairy cows. In the past, alfalfa and other legumes were often left out of dry cow diets in favor of grasses because it was falsely believed the high-Ca content of alfalfa was the cause of milk fever. Using conservative agronomic practices, with the application of manure and K fertilizers based on soil tests (Arnold and Lehmkuhler 2014), it is possible to grow forages with lower K content, thereby reducing the risk of milk fever. Nutritionists can offset the alkalinizing effects of K^+ to a certain extent by adding sources of Cl^- or SO_4^- ("anionic salts") to the prepartum diet of dairy cows. However, if the DCAD of the diet anion sources is greater than 250 $mEq\,kg^{-1}$, then anion supplementation will not be effective (Oetzel 2000).

The anionic salts used to lower DCAD can also be unpalatable. Therefore, the best strategy to reduce DCAD is to incorporate low-K forages in the diet during the prepartum period. Even better is to use a low-K, high-Cl forage that can further reduce dietary DCAD. Depending on agronomic practices, chloride concentration in forages can range from 2.5 to 12 $g\,kg^{-1}$ DM. The K content of forage also decreases with maturity, though the increase in neutral detergent fiber (NDF) and lignin can lower forage intake and digestibility. To estimate optimal DCAD, mineral analysis through wet ashing of the forage is the gold standard. The feasibility of using NIRS to predict total ash, Ca, P, K, Mg, and Cl in forage samples was demonstrated, but the accurate prediction of Na and S was not achieved (Halgerson et al. 2004; Tremblay et al. 2009). Even though Na and S were not accurately predicted, Tremblay et al. (2009) showed that the NIRS could be successfully used to predict DCAD, likely due to associations between minerals and organic compounds.

Corn is a warm-season grass, and corn silage tends to contain 10–14 $g\,K\,kg^{-1}$ DM. Corn silage is the basis for most prepartum (close-up, dry) cow rations in the US. Some other warm-season grasses, such as switchgrass, eastern gamagrass, and indiangrass, tend to be low in K, but their use is limited by their high NDF and low digestibility.

Balancing prepartum diets to achieve intakes of Ca in the range of 50–70 $g\,d^{-1}$, Mg intakes of 40–50 $g\,d^{-1}$, P intakes of <35 $g\,d^{-1}$, and a DCAD between +150 and − 150 $mEq\,kg^{-1}$ DM has been recommended to reduce the risk of milk fever (DeGaris and Lean 2009). As metabolic acidosis promoted by low-DCAD diets reduces urine pH, urinary pH can be used as a biologic marker to identify optimal dietary DCAD levels. Diet DCAD should be adjusted to achieve an average urinary pH of between 6.0 and 7.0 or 5.5 and 6.5 for Jersey cows (Oetzel 2000). The management of DCAD in the dry cow involves many other aspects and for more details the reader is referred to Oetzel (2000).

Grass Tetany – Hypomagnesemia

Symptoms and Economic Losses

Grass tetany is a Mg deficiency of ruminants usually associated with grazing of fast-growing cool-season grasses during spring. Terms also used for this disorder include spring tetany, grass staggers, wheat-pasture poisoning and milk tetany (Arnold and Lehmkuhler 2014). Clinical signs include nervousness, rapid respiration, aggressiveness, stiff- and high-stepping gait, bellowing, muscle tremors and convulsions. Animals suffering from tetany may be hypersensitive to stimuli such as loud noises (Soni and Shukula 2012). In milder cases, blood or urine tests can confirm hypomagnesmia. Cases may also be asymptomatic and recognized only after sudden death occurs. Annual death losses in the US are estimated at \$50–\$150 million (Mayland and Sleper 1993). Grass tetany occurs in all classes of cattle and sheep but, is most prevalent among older females in early lactation and may be triggered by a stressor such as cold weather. Magnesium must be supplied daily because it is excreted in urine and milk and cation balances of fast-growing spring grass may contribute to decreased absorption of magnesium (Arnold and Lehmkuhler 2014).

Blood Cation Concentrations

Plasma Mg levels are normally 18–32 $mg\,Mg\,l^{-1}$. Blood plasma or serum less than 18 $mg\,Mg\,l^{-1}$ and urine values less than 20 $mg\,Mg\,l^{-1}$ (Puls 1994) are indicators of hypomagnesemia, and concern should be expressed when any of these values fall below 15 $mg\,Mg\,l^{-1}$ (Mayland 1988; Vogel et al. 1993). The dramatic signs of tetany are clearly evidenced at levels below 10 $mg\,l^{-1}$. Measurement of Mg levels in blood may not be conclusive as rupture of muscle fibers in the later stages of the condition may release Mg into the blood.

Grass tetany is a complex disorder. First, Mg requirements are greater for lactating than for nonlactating cattle and are greater for older than for younger animals. Second, bovine breeds differ in susceptibility to grass

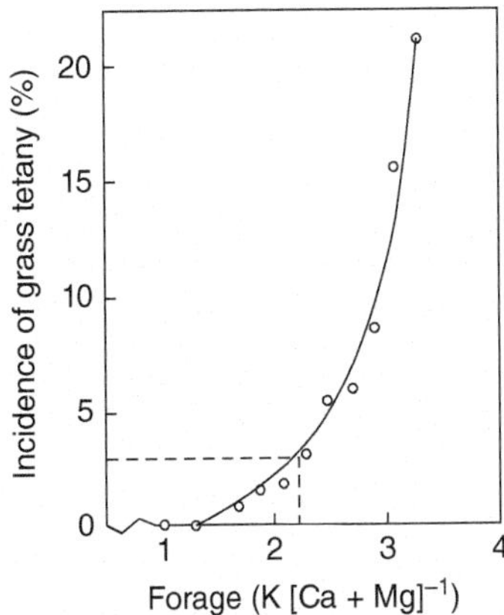

Fig. 47.5. Relationship between the forage K/(Ca + Mg), calculated on an equivalent basis and the relative incidence of grass tetany in The Netherlands. *Source:* Adapted from Kemp and 'tHart (1957).

tetany with Brahman and Brahman crossbreeds being the most tolerant and European breeds being more susceptible (Greene et al. 1989). Third, many factors influence Mg concentration and availability in the herbage with high-soil K levels that negatively affect soil Mg uptake by plants being a principal factor (Cherney et al. 2002). Also, low levels of readily available energy, Ca and P or high levels of organic acids, fatty acids, and N in the ingested herbage can reduce the absorption and retention of Mg by the animal. Concentrations greater than 2.0 g Mg kg^{-1} DM or a milliequivalent ratio of less than 2.2 for K (Ca + Mg)$^{-1}$ are considered safe (Grunes and Welch 1989). This ratio (Figure 47.5) is the best single indicator of the risk of forage causing grass tetany.

Prevention

Agronomic practices to reduce the risk of tetany include adding legumes to pasture mixes and the conservative application of N and K fertilizers that are aligned with needs based on soil tests (Arnold and Lehmkuhler 2014). Beneficial husbandry practices include assigning less-susceptible breeds (Braman types) and classes (non-lactating) of livestock to tetany-prone pastures (Greene et al. 1989). A general recommendation for prevention is to provide a high-magnesium mineral supplement to pregnant grazing ruminants for at least 30 days prior to parturition (Arnold and Lehmkuhler 2014).

Laminitis – Lameness

Foot problems in cattle and horses are attributed to several causes, one of which is nutritional. An excess of readily fermentable carbohydrates ingested during the grazing of lush pasture grasses or consuming grains can produce acidosis and, ultimately, lead to laminitis or sore feet (Nocek 2001). The etiology of laminitis is not fully understood, but it seems to accompany advanced stages of acidosis and is a complex and costly disorder (Longland and Cairns 2000).

Laminitis in horses is described as a noninfectious inflammation or failure of the attachment between the distal phalanx and the inner hoof wall (Pollitt 2005). It is often associated with excessive ingestion of high-starch grain, but is also commonly associated with grass founder, which is caused by ingestion of rapidly digestible plants by horses and ponies (Longland and Cairns 2000). The problem has been attributed to the ingestion of fructans in the form of the inulin-like sugar, raffinose. This **sugar** can cause laminitis by interfering with insulin function.

Laminitis in cows is associated with a shortage of roughage in the ration. It often starts with a disturbance of digestion and accumulation of excess amounts of readily fermentable carbohydrates, which predisposes cows to laminitis. This is followed by an excessive accumulation of lactic acid, imbalances of microflora, and decreases in ruminal pH (Westwood et al. 2003; Whitaker et al. 2004). Laminitis also occurs in cows on pasture and is the most expensive disease condition affecting dairy herds in New Zealand and Australia. Forages associated with this condition are characterized as low in **fiber**, high in water content, and having a rapid rate of rumen fermentation. However, in pasture-based dairy systems other potential causes of laminitis also need to be considered. For example, walking long distances to the milking parlor on poorly constructed or badly maintained roads, or poorly drained soils can also affect the prevalence and severity of laminitis (Olmos et al. 2009).

Nitrate Poisoning

Description

Nitrate (NO_3^-) accumulates in plant tissue because of luxury uptake of soil N when plant metabolism of N is slow or even stopped. This condition is promoted by cool temperature, cloudy weather, drought, frost or other physiologic stress that slows growth. Plant NO_3^- itself is not normally toxic, but it is reduced to nitrite (NO_2^-) in the rumen. If the NO_2^- is not reduced further to NH_4^+, the accumulated NO_2^- can cause several conditions detrimental to animal health (Bush et al. 1979). These include the formation of methemoglobin which impairs blood O_2 transport, abortion, and vasodilation leading to cardiovascular collapse (Singer 1972). This condition is referred to as methemoglobinemia (see below).

Intensive applications of N fertilizer to silage crops often results in appreciable amounts of NO_3^- in the herbage. During ensiling, NO_3^- is almost completely transformed to or partially degraded to ammonia and nitrous oxide, with nitrite and nitric oxide occurring as intermediates (Spoelstra 1985). NO_3^- and NO_2^- should be measured in high-risk crops before feeding.

Toxicity

Silofiller disease is an illness of farm workers caused by inhalation of nitric and nitrous oxides from fermenting forages containing high N concentrations (Wright and Davison 1964). These gases are heavier than air and accumulate just above the silage in upright silos. Good air ventilation will reduce the health hazard of these noxious gases.

The most common effect of NO_3^- poisoning is the formation of methemoglobin that occurs when NO_2^- oxidizes the ferrous iron of blood hemoglobin to ferric iron (Deeb and Sloan 1975). This produces a chocolate-brown colored blood that cannot release O_2 to body tissue. As the toxicity intensifies, the brownish-colored blood casts a brownish discoloration to the nonpigmented areas of skin and the mucous membranes of the nose, mouth, and vulva. Clinical signs progress with staggering, rapid pulse, frequent urination, and labored breathing, followed by collapse, coma, and death. Sublethal toxicity may be evidenced by abortion in pregnant females. Animals suffering from excess NO_3^- intake may be given drenches of methylene blue, which can turn methemoglobin back into normal hemoglobin. Affected animals should be given an alternative source of forage and the inclusion of concentrates in the diet may accelerate the conversion of NO_3^- through to NH_4^+ in the rumen.

Causes

The rate and degree of NO_3^- reduction in the rumen depend on the microflora present and the amount of energy available for continued reduction of NO_3^- to NO_2^- to NH_4^+ and its use for the synthesis of amino acids by ruminal microbes. This interaction of available energy and NO_3^- reduction results in differential sensitivity to forage NO_3^- levels (Table 47.3). Sheep receiving a low-energy diet have been poisoned by as little as 500 mg NO_3^--N kg^{-1} DM, whereas with high-energy diets, 8000 mg NO_3^--N kg^{-1} DM had no effect (Singer 1972). Other data show that 2000 mg NO_3^--N kg^{-1} DM in forage was the threshold limit for cattle (Bush et al. 1979), whereas no problem occurred with feeding forage containing up to 4600 mg NO_3^--N kg^{-1} DM (Davison et al. 1964). A level of 4600 mg NO_3^--N kg^{-1} DM was recommended as a safe level.

Suspect forages should be tested for NO_3^-. Be aware that these test levels may be reported as nitrate (NO_3^-) or as nitrate nitrogen (N) on an elemental N basis. Thus, a value of 4400 NO_3^- ppm (nitrate) is equivalent to a value of 1000 ppm NO_3^--N (nitrate nitrogen).

Although risky, feeding additional forage low in NO_3^- will dilute NO_3^- intake. Energy supplements also can accelerate the reduction of NO_3^- to NH_4^+, possibly due to an increase in the use of NH_4^+ to form amino acids for microbial protein synthesis. Several rumen bacteria, protozoa and even possibly anaerobic fungi may have the ability to reduce NO_3^- to NH_4^+. In addition, there is evidence that microbial populations can adapt to more readily reduce NO_3^-, especially if concentrations are increased gradually over time (Nolan et al. 2016). Such an approach may enable ruminants to **graze** higher NO_3^- forages, but this would still be risky due to variable intake

Table 47.3 Livestock responses to nitrate-N in forage and drinking water

Nitrates	Comments
Forage[a] (mg N kg^{-1})	
<1000	Safe to feed under most conditions.
1000–1500	Safe to feed to nonpregnant animals. Limit use for pregnant animals to 50% of total DM intake.
1500–2000	Safely fed if limited to 50% of total DM intake.
2000–3500	Forage should be limited to 35–40% of the total DM intake. Do not feed to pregnant animals.
3500–4000	Forage should be limited to 25% of the total DM intake. Do not feed to pregnant animals.
>4000	Forage is potentially toxic. DO NOT FEED.
Water[b] (mg N l^{-1})	
<20	Generally considered safe.
20–40	Caution: Consider additive effect of NO_3-N in feed.
40–200	Greater caution especially if feed contains >1000 mg NO_3-N.
>200	Acute toxicity and some death losses in swine and ruminants.

[a]Forage information from Holland and Kezar (1990).
[b]Water information from Cash et al. (2002).

among grazing animals and the possibility that methemoglobin may accumulate in the circulatory system over time (Godwin et al. 2015).

Levels of NO_3^- in growing plants may be higher in the morning than the afternoon (Fisher et al. 2002). This likely reflects the use of NO_3^- by the plant to synthesize amino acids for growth as photosynthesis becomes more active. Consequently, afternoon cutting will reduce nitrate concentrations in the hay and retain more sugar than if cut in the morning. However, such a practice will not offer protection if the plant is not metabolically active as would be the case during drought or as a result of frost damage.

Forage Management

Sudangrass, sorghums, corn, and small grains are often implicated in NO_3^- poisoning. These crops are often heavily fertilized with N and when subject to drought, frost, or other stress, may accumulate large concentrations of NO_3^-. Perennial grasses generally are less of a problem because they usually receive lower levels of N fertilizer and have greater tolerance of drought and frost. Legumes obtain reduced forms of N through symbiosis with *Rhizobium* and generally do not contribute to NO_3^- poisoning. However, alfalfa grown under low or high temperature or water stress may contain in excess of 1000 mg N kg^{-1} DM as nitrate.

Diurnal cycling of NO_3^- has been noted from anecdotal information in which a forage rape (brassica hybrid) pasture receiving large amounts of manure was toxic to grazing swine in the morning, but not in the afternoon (Herb Simon, Glenn Allen, Alaska, personal communication, November 5, 2003). Similar trends have been noted by Fisher et al. (2002). The NO_3^- supply contributes to the formation of N products in the presence of soluble sugars produced by photosynthesis. Concentrations of these sugars are minimal in the morning, allowing the accumulation of NO_3^- in the forage. The reverse is true in the afternoon.

Mineral Elements: Deficiency, Toxicity, and Interactions

Phosphorous Deficiency

Phosphorous (P) deficiency in diets of grazing animals has been reported in many regions of the world and is usually associated with cattle grazing pastures on P deficient soils (common in tropical areas). A clinical sign of P deficiency is depraved appetite (pica), characterized by animals chewing on bones, wood, and rocks. The osteophagia promoted by P deficiency can often lead to death from botulism when the bones are contaminated with *Clostridium botulinum*. Other symptoms often observed are peg leg (cracking of the joints during movement, stiffness, and lameness), reduced feed intake, low average daily gains and milk production, and impaired fertility. Dietary supplementation of P deficient forage with P supplements is usually the quickest way of overcoming the negative impacts of P deficiency. Applying P fertilizer to P-deficient soils can also increase forage yield and improve its P content. When P intake from forage fails to meet requirements, P supplements should be included in the diet to optimize animal productivity. More details on P in forages and in **ruminant** nutrition are reported by Minson (1990) and Karn (2001).

Copper, Molybdenum, Sulfur and Interactions

Copper (Cu) deficiency significantly affects ruminant livestock production in large areas of North America (Gooneratne et al. 1989) and elsewhere (Grace 1983), often in areas where soils are naturally high in molybdenum (Mo) and sulfur (S) (Kubota and Allaway 1972). In ruminant nutrition, the two-way and three-way interactions between these three elements are unique in their complex effects on animal health.

Sheep consuming a complete diet, low in S and Mo and with modest Cu (12–20 mg kg^{-1} DM), can succumb to Cu toxicity. Sheep grazing another pasture of similar Cu concentration, but high in Mo and S will produce Cu-deficient lambs showing clinical signs of sway-back disease (Suttle 1991). Copper-deficient cattle and sheep appear **unthrifty** and exhibit poor growth and reproduction. Copper deficiency reduces the level of melanin pigments in hair and wool so that normally dark-colored fibers will be white or gray (Grace 1983; Grace and Clark 1991). Low dietary Cu levels may limit effectiveness of the immune system in animals. Normal copper levels are necessary to maintain the structural integrity of DNA during oxidative stress (Pan and Loo 2000; Picco et al. 2004).

Forages growing on soil or peat high in Mo can produce a scouring disease known as molybdenosis. In the presence of S, high intake of Mo can induce a Cu deficiency due to formation of insoluble Cu-Mo-S complexes in the digestive tract that reduce the absorption of Cu. Several pathways exist by which two-way and three-way interactions of Cu, Mo, and S mediate their effects on ruminants (Suttle 1991). Most clinical signs attributed to the three-way interaction are the same as those produced by simple Cu deficiency and probably arise from impaired Cu metabolism. An appropriate Cu : Mo ratio is desirable to minimize the effects of Mo on Cu metabolism and ADG (Dias et al. 2013). The tolerable risk threshold of Cu : Mo depends on Mo concentration – declining from 5 : 1 to 2 : 1 as pasture Mo concentrations increase from 2 to 10 mg Mo kg^{-1} DM. The appropriate dietary Cu : Mo ratios must be limited to values close to 6 : 1 for sheep, and ratios >10 : 1 can cause severe clinical Cu toxicosis (NRC 2005); whereas this ratio for cattle should be >3 : 1. Using this risk avoidance approach has been only partially successful, as the true nature of the interactions among these minerals is still not fully understood (Suttle 1991).

Clinical signs in cattle grazing high Mo forage have been summarized by Majak et al. (2004b).

Cattle are more sensitive than sheep to molybdenosis; however, sheep are more susceptible to Cu toxicity (Suttle 1991). Sheep should not be allowed to graze pastures that recently received poultry or swine manure, especially if Cu salts are fed to poultry or swine to control worms or used in foot baths or for foot problems. The practice of using $CuSO_4$ or $ZnSO_4$ in foot baths in dairies is also leading to increased Cu and Zn in waste products and recipient pasture and forage lands.

Polioencephalomalacia (PEM), which is caused by necrosis of the cerebrocortical region of the brain of cattle, sheep and goats can occur if diets contain high levels of S (greater than 0.40% of diet DM). It is suggested that S is reduced to H_2S by ruminal bacteria, which is toxic and can exert toxic effects if absorbed into the blood stream. Large quantities of distiller's grains (DGS) with potentially high-soluble S from ethanol production are available for ruminant feed. In general, DGS contain high but highly variable content of S 0.31–1.93% (average 0.69%) of DM. Feeding a large quantity of such feed will contribute to increased dietary intake of S, which will have negative effects on animal performance due to the production of undesirable H_2S (Uwituze et al. 2011). This can have subsequent negative effects on Cu metabolism (Spears et al. 2011). It is suggested that wet DGS may be more prone to conversion of S to H_2S in the rumen than dried DGS (Sarturi et al. 2013) and supplemental Cu may improve feed efficiency in cattle consuming diets containing 60% dried DGS (Felix et al. 2012). Supplemental Mo has been speculated as a potential means of alleviating the effects of excess dietary S as well, but Kessler et al. (2012) showed that added dietary Mo failed to bind excess S in the rumen and resulted in aggravated toxic effects as a result of both high dietary S and Mo. Water sources should also be monitored to ensure that they do not contain high levels of S.

Copper-depleted animals appear to respond equally well to the administration of Cu in supplements, oral **boluses** or pellets, and injections. Copper fertilization of Cu deficient pastures should be done carefully because the range between plant sufficiency and plant toxicity is quite small. It is generally accepted that organic trace minerals are more bioavailable, resulting in better animal performance, health, production immune response and stress alleviation than their inorganic salts. Tribasic Cu chloride is also more bioavailable than $CuSO_4$ when added to diets high in the Cu antagonists Mo and S (Spears et al. 2004).

Selenium Deficiency and Toxicity

Herbage–Se concentrations are marginal to severely deficient for herbivores in many areas of the world. These areas include the Pacific Northwest and the eastern one-third of the US (Kubota and Allaway 1972; Mayland et al. 1989). Herbage Se concentrations of 0.03 mg Se kg^{-1} are generally considered adequate. However, 0.1 mg Se kg^{-1} may be necessary when high S in the herbage reduces Se availability to the animal. Climate conditions and management practices that favor high-forage yield may dilute Se concentrations to potentially deficient levels in herbage.

Selenium deficiency most commonly occurs in young calves and calving cows but, is also seen in adult cattle. Selenium deficiency causes white muscle disease in lambs, calves, and colts. The young may be born dead or suddenly die within a few days after birth as Se levels in the milk of the dam were too low to rectify the deficiency. A delayed form of white muscle disease occurs in young animals, whereas a third form is identified as impaired health in animals of all ages. Injectable Se, often with vitamin E, or oral supplementation (selenized salt or Se boluses) can be used to meet animal requirements. Addition of Se to deficient soils has been shown to increase the Se content of alfalfa in a dose-dependent manner and is useful as a management tool to improve the Se status of weaned beef calves (Hall et al. 2013) and lactating dairy cows (Séboussi et al. 2016).

In many semiarid areas of the world, grasses and forbs contain adequate (0.03–0.1 mg Se kg^{-1} DM) to toxic (>5 mg Se kg^{-1} DM) levels of Se for grazing animals. These areas include desert, prairie, and plains regions (cretaceous geology) of North America where Se toxicity is often observed in grazing animals. Some plants in these areas can accumulate 100–1000 mg Se kg^{-1} DM. Animals eating these generally unpalatable plants will likely succumb to Se toxicosis. Grasses, small grains, and some legumes growing on the Se-rich cretaceous geologic materials may contain 5–20 mg Se kg^{-1} DM. Some animals eating this herbage may die, but most are likely to develop chronic sclerosis called *alkali disease*, in which there is hair loss and hoof tissues become brittle. In these instances, some animals may develop a tolerance for as much as 25 mg Se kg^{-1} DM.

A second chronic disorder in ruminants called *blind staggers* also occurs in these areas. This disorder, while historically attributed to Se, is more likely caused by excess S. High sulfate levels in the drinking water and ingested herbage have led to the occurrence of blind staggers (Beke and Hironaka 1991). Changing to high-quality, low-sulfate water and forage reduces the risk.

Under conditions of marginally available soil Se, increased S reduces the uptake of Se by plants and the bioavailability of dietary Se to animals. However, when high concentrations of Se are present in soils, the addition of S has little effect in reducing Se uptake by plants and subsequent toxicity in animals. Replacing high-Se forage with low-Se forage is the most effective way of countering Se toxicity.

Cobalt, Iodine, Zinc

Cobalt (Co) deficiencies in grazing herbivores have been identified in many areas of the world (McDowell and Arthington 2005). Cobalt is a cofactor in vitamin B_{12}, which is required in processes of energy metabolism in ruminants. Concentration of cobalt in forages is affected by soil properties, plant species, stage of maturity, yield, pasture management and climate. Forages grown on poorly drained soils tend to have higher concentrations of cobalt (MacNaeidhe 2001). Over-liming soils to increase the pH above 6.0 will reduce the availability of cobalt and may lead to deficiency. Most forages and feedstuffs fed to dairy and beef do not contain adequate quantities of cobalt to support the rumen and animal requirements. In addition, the body's capacity to store vitamin B_{12} is limited. Consequently, cobalt must be continuously supplemented in beef and dairy diets. The signs of Co deficiency include a transient unthriftiness and anemia leading to severely reduced feed intake and eventually death. Compared to plasma, levels of vitamin B_{12} in liver may be a more useful indicator of cobalt status, where concentrations below 300 nmol kg^{-1} fresh weight are considered marginal (Suttle 2004). Two other conditions sometimes attributed to Co deficiency are ovine (sheep) white liver disease and phalaris staggers (Graham 1991), which have been attributed to alkaloids in the forage (see below).

Pasture herbage levels of at least 0.11 and 0.08 mg Co kg^{-1} DM will provide adequate Co for sheep and cattle, respectively, but the mechanism by which oral Co supplementation prevents staggers is unknown. Cobalt injections or oral supplements can also be given to animals. Acceptable cobalt sources include: cobalt carbonate; cobalt sulfate; cobalt chloride, cobalt glucoheptonate and cobalt propionate. Pastures may also be fertilized with cobalt sulfate. Manganese and Fe in feed are antagonists to Co absorption (Grace 1983).

Plants do not require iodine (I), and herbage in the northern half of the US is generally deficient for animal requirements. Some signs of iodine deficiency include reduced fertility, enlarged thyroid (goiter), and stillborn, weak, and/or hairless calves. Ruminants fed large amounts of brassicas, especially turnips, containing glucosinolates that prevent uptake of iodine by the thyroid gland can cause hypothyroidism and goiter. The use of iodized salt can easily meet the iodine needs of animals on pasture.

Zinc (Zn) concentration in pasture plants ranges from 10 to 70 mg kg^{-1} DM but is most often in the 10–30 mg kg^{-1} DM range. Legumes are generally higher in zinc than grasses. Cows require 30–40 mg zinc kg^{-1} DM with diets containing 2–10 mg zinc kg^{-1} DM considered deficient. Cattle grazing forage having 15–20 mg Zn kg^{-1} DM gained weight faster when supplemented with additional Zn (Mayland et al. 1980). Blood Zn levels were higher in supplemented than in control animals, but the difference was too small to be useful as a diagnostic tool. High Cu levels will exacerbate Zn deficiency because they are absorbed through common pathways. Similarly, excess Zn can cause Cu deficiency. Mineral supplements should be formulated with a Cu : Zn ratio of 1 : 2 or 1 : 3 to avoid these interactions.

Fluorosis and Silicosis

Plants do not require fluorine (F), but herbage generally contains 1–2 mg kg^{-1} DM, which is adequate for bone and tooth development in animals. At higher levels, the development of fluorosis is influenced by the age, species, dietary form, and duration of exposure. Fetuses and young animals are most susceptible to excess F. Dietary concentrations in excess of 5 mg F kg^{-1} DM result in mottling of tooth enamel or even structural weaknesses; otherwise, long-term intakes of 30 mg F kg^{-1} d^{-1} may be tolerated by ruminants before bone abnormalities appear (Underwood 1977). In areas of endemic fluorosis, plants may be contaminated by naturally fluoridated dust from rock phosphate or other smelters. During manufacture of superphosphate and di-calcium phosphate, 25–50% of the original F is lost. Excess F may also be absorbed by plants that are sprinkler irrigated with thermal groundwater. Rock phosphate supplement and naturally fluoridated drinking water are the primary dietary sources of excess F.

Grasses contain more silicon (Si) than legumes and account for the large amount of Si ingested by grazing animals. Silicon may be needed in trace amounts by animals. While not required by herbage plants, it is known to increase disease and insect resistance in many horticultural plants. Silicon adversely affects **forage quality** and may affect animal performance and selectivity of plants. Silicon serves as a varnish on the cell walls, complexes microelements that reduce their availability to rumen flora, and inhibits the activity of cellulases and other digestive enzymes (Shewmaker et al. 1989). The net effect of forage Si is to reduce DM digestibility by three percentage units for each 10 g kg^{-1} DM Si present (Van Soest and Jones 1968).

Silicon is also responsible for urolithiasis (urinary calculi or range water belly) in steers. Incidences of water belly are associated with reduced water intake and urine volume and only weakly related to herbage Si. Steers are more sensitive because castration often reduces internal diameter of the ureter. Management strategies in high-Si areas include stocking only heifers and providing adequate drinking water. If feasible, the Ca : P ratio in the diet should be reduced and urine acidified by supplementing animals with ammonium chloride NH_4Cl (Stewart et al. 1991).

Natural Toxicants in Forages

Plants are protected against herbivores by such physical defenses as leaf hairs, spines, thorns, highly lignified tissue, and growth habits (e.g. prostrate form) and by chemical defenses that include a wide array of chemicals that are toxic or poisonous. These chemicals may be synthesized by the plant itself or produced by symbiotic or mutualistic fungi growing with the plant. These are usually secondary compounds (e.g. **alkaloids**) that do not function directly in cellular metabolism but, are apparently synthesized to contribute to the plant's defensive arsenal.

Chemicals synthesized by fungi, known as *mycotoxins*, may be produced by fungi living on or in forage plants. Mycotoxins are responsible for many disorders of grazing animals. For example, an endophytic fungi, *Neotyphodium coenophialum* (formerly *Acremonium coenophialum*), of tall fescue produces ergot alkaloids that cause **fescue foot**, summer **fescue toxicosis**, and reproductive disorders, while an **endophyte** in perennial ryegrass produces lolitrems that cause ryegrass staggers (see Chapter 35). In the US, total livestock-related losses attributed to this tall fescue endophyte are estimated between $500 million and $1 billion a year (Ball et al. 1993). The economic impact of an array of poisonous plants on livestock production in the western US is estimated in the hundreds of millions of dollars annually (James et al. 1988).

Plant toxins can be classified into several major categories, including alkaloids, glycosides, proteins and amino acids, and phenolics (tannins). Alkaloids are bitter substances containing N in a heterocyclic ring structure. There are hundreds of different alkaloids, which are classified according to the chemical structure of the N-containing ring(s). For example, the pyrrolizidine alkaloids in *Senecio* species have a pyrrolizidine nucleus of two five-membered rings, whereas ergot alkaloids have an indole ring structure. Lupines (*Lupinus* L.) contain quinolizidine alkaloids, which are based on two six-membered rings. Glycosides are composed of a carbohydrate (**sugar**) portion linked to a noncarbohydrate group (aglycone) by an ether bond. Examples are **cyanogenic** glycosides, glucosinolates, saponins, and **coumarin** glycosides. Their toxicity is associated with the aglycone, such as cyanide in cyanogenic glycosides. Glycosides are hydrolyzed by enzymatic action, releasing the aglycone, often when plant tissues are damaged by wilting, freezing, mastication, or trampling. A good example of this is the production of toxic cyanide from sudangrass after a frost. The lysis of plant cells releases the glycoside from storage **vacuoles**, allowing it to be hydrolyzed by enzymes in the **cytosol**, releasing free cyanide.

Many toxic amino acids also occur in plants. One of the best known is **mimosine**, a toxic amino acid in the tropical forage legume, leucaena (Mimosaceae). Others include lathyrogenic amino acids in *Lathyrus* L., indospicine in hairy indigo, and the brassica anemia factor, which is caused by S-methylcysteine sulfoxide, a metabolic product of forage brassicas.

Phenolic compounds, including condensed and hydrolyzable tannins, are substances containing aromatic rings with one or more hydroxyl groups. The hydroxyl groups are chemically reactive and can react with functional groups of proteins to form indigestible complexes. The tannin–protein complexes are astringent and reduce feed intake (Min et al. 2003). All plants contain phenolic compounds. In some cases, their type or concentration may cause negative animal responses. These include reduced feed intake and protein digestibility of birdsfoot trefoil and sericea lespedeza. Oak (*Quercus* spp.) poisoning is caused by tannins in oak browse. Many tree legumes used in tropical agroforestry can contain sufficient levels of tannins to impair animal performance. At moderate concentrations, some tannins can also be beneficial to the animal preventing pasture bloat (see above), protecting proteins from ruminal degradation, or lowering the survival of intestinal parasites. Some herbivores also have evolved to produce proline rich proteins in their saliva which bind to tannins during ingestion and reduce their biologic activity (see below).

Toxins and Animal Disorders Associated with Forage Legumes

Phytoestrogens occur in grass and forage legumes such as *Phalaris* spp., alfalfa, red clover, and subterranean clover. Phytoestrogens reduce sheep fertility and cause various abnormalities of genitalia. Plant breeders have developed low-estrogen cultivars of subterranean clover, greatly reducing animal losses.

Toxins associated with specific forage legume species will be briefly described. Further detail is provided by Cheeke (1988).

Red Clover

When infected with the black patch fungus (*Rhizoctonia leguminicola* Gough and ES Elliot), red clover hay may contain the indolizidine alkaloid, slaframine. Slaframine is a cholinergic agent that causes excessive salivation (clover slobbers), eye discharge, bloat, frequent urination, and watery diarrhea. These effects are due to stimulation of the autonomic nervous system. The fungal infection and potential to cause toxicity develops most rapidly in periods of high humidity. Prompt removal of the toxic forage from livestock generally alleviates all signs of intoxication.

White Clover

This legume may contain cyanogenic glycosides that may confer some resistance to slugs and other pests. Cyanogens in white clover are below toxic levels for livestock, but they may reduce DM intake during midsummer.

Alsike Clover

Poisoning from alsike clover has been reported in Canada and the northern US, especially with horses. Though not proven, circumstantial evidence strongly suggests the poisoning is caused by alsike clover (Nation 1991). Toxicity signs include photosensitization, neurologic effects such as depression and stupor, and liver damage. In some cases, the liver is extremely enlarged, whereas in others it is shrunken and fibrotic.

Sweetclover

Sweetclover causes significant animal health problems in North America. It contains coumarin glycosides, which are converted to dicoumarol by mold growth during hay storage. Dicoumarol is an inhibitor of vitamin K metabolism in animals, thus causing an induced vitamin K deficiency. Sweetclover poisoning causes a pronounced susceptibility to prolonged bleeding and hemorrhaging, due to the essential role of vitamin K in blood clotting. Wet, humid weather that favors mold growth during curing of sweetclover hay increases the likelihood of poisoning. Cattle are the main livestock affected.

Moldy sweetclover hay should not be fed to animals or should be used with caution. Ammoniation of stacked hay with anhydrous ammonia reduces dicoumarol levels. Animals with signs of sweetclover poisoning are treated with injections of vitamin K. Low-coumarin cultivars of sweetclover have been developed and should be used in areas where sweetclover poisoning is a problem. Coumarin has a sweet to vanillin-like odor and, is responsible for the characteristic smell of sweetclover.

Other Forages

Additional forage legumes contain various toxins. As mentioned above, birdsfoot trefoil and lespedeza contain tannins. Crownvetch contains glycosides of 3-nitropropionic acid, which are metabolized in ruminants to yield NO_2^-. Concentrations are rarely enough to cause poisoning, but the glycosides contribute to reduced intake of crownvetch. Cicer milkvetch, a minor forage legume in the northern US, has caused photosensitization in cattle and sheep (Marten et al. 1987, 1990).

Common Vetch

The seeds of common and hairy vetch contain toxic lathyrogenic amino acids, which cause damage to the nervous system, with signs such as convulsions and paralysis. This occurs primarily in nonruminants that consume seeds as a contaminant of grain. Poisoning of ruminants consuming hairy vetch forage has been reported in the US (Kerr and Edwards 1982) and South Africa (Kellerman et al. 1988). Signs include severe dermatitis, skin edema, conjunctivitis, corneal ulcers, and diarrhea. About 50% of affected animals die. The toxic agent in hairy vetch has not been conclusively identified but, is likely related to cyanamide toxicity (Kamo et al. 2015).

Lathyrus spp.

Many plants in this genus contain toxic amino acids that cause neurologic problems and skeletal defects known as lathyrism, which is caused by a non-protein amino acid (β-ODAP). Flatpea, a forage crop for degraded soils such as reclaimed strip-mined areas, is nearly free of toxicity (Foster 1990), but Foster noted that "the question of flatpea toxicity must be answered conclusively before this plant can be recommended for use by livestock producers." For example, sheep fed flatpea hay showed typical signs of neurolathyrism (Rasmussen et al. 1993).

Lupinus spp.

Plants in the genus *Lupinus* contain a variety of alkaloids of the quinolizidine class. The sweet lupines such as *Lupinus albus* and *Lupinus angustifolius* contain low levels of various alkaloids (e.g. cytisine, sparteine, lupinine, lupanine). These alkaloids cause feed refusal and neurologic effects. Sheep are frequently poisoned by wild lupines on rangelands, because they avidly consume the seedpods. On rangelands in western North America, there are many species of wild lupines that are toxic to livestock. Some species (e.g. silky lupine, tail cup lupine, spurred lupine) contain anagyrine, an alkaloid that is teratogenic in cattle. It causes crooked calf disease if consumed by pregnant cows during days 40–70 of gestation. Severe skeletal deformations in the fetuses may occur. This alkaloid does not occur in genetically improved *Lupinus* spp. In Australia, sweet lupines (low alkaloid) are extensively grown as a grain crop and sheep are grazed on the lupine **stubble** after harvest. Often the stems of lupines are infected with *Phomopsis leptostromiformis* (Köhn) Bubak, a fungus that produces toxic phomopsins. These mycotoxins cause liver damage, including fatty liver and necrosis, eventually leading to liver failure and death. This condition is referred to as lupinosis.

Leucaena

Leucaena contains a toxic amino acid, mimosine. In the rumen, mimosine is converted to various metabolites, including 3,4-dihydroxypyridine (DHP). Both mimosine and DHP are toxic to ruminants, causing dermatitis, hair loss, and poor growth (mimosine), and goitrogenic (thyroid-inhibitory) effects. Australian researchers (Jones and Megarrity 1986) learned that Hawaiian ruminants, adapted to a leucaena diet, had mimosine-degrading rumen bacteria that eliminated the toxicity. These bacteria have now been introduced into cattle in Australia, allowing leucaena to be used as a productive source of high-protein forage (Quirk et al. 1988). Hammond et al. (1989) in Florida also reported detoxification of **mimosine** by use of natural or introduced rumen microbes. Recent evidence suggests that there may be several rumen bacteria capable of degrading these toxins (Derakhshani et al. 2016).

Other Tropical Legumes

Many legume plants in this group contain toxic factors, probably to act as a grazing deterrent or for pest resistance. Many *Indigofera* spp. contain the toxic amino acid indospicine (Aylward et al. 1987). The jackbean contains canavanine, an amino acid analog of arginine (Cheeke 1998). Generally, the effects of these toxins are sufficiently diluted by the consumption of other forages during grazing that toxicities do not occur.

Toxins and Animal Disorders Associated with Grasses

In contrast with other herbaceous plants, grasses are generally not well-defended chemically. Most grasses have coevolved with grazing animals and by growth habit survive frequent defoliation. Hence, there are a few intrinsic toxins in common grasses. More frequent are mycotoxins produced by fungi living in or on grasses. Fungi living within plant tissues or tissue spaces and showing no external signs are called *endophytes*. Many livestock syndromes are attributed to endophyte toxins. Examples of toxins intrinsically present in grasses without an associated fungus are the alkaloids of *Phalaris* spp., cyanogens in forage sorghums (e.g. sudangrass), and oxalates in many tropical grasses. These have been reviewed by Cheeke (2005).

Phalaris Poisoning

A neural disorder (phalaris staggers) and a sudden death syndrome can occur in cattle and sheep grazing or consuming hay of *Phalaris* spp. (Alden et al. 2014). Phalaris staggers is characterized by convulsions and other neurologic signs due to brain damage, culminating in death (Bourke et al. 1988). The syndrome is caused by tryptamine alkaloids believed to inhibit serotonin receptors in specific brain and spinal cord nuclei (Bourke et al. 1990). These tryptamine alkaloids are also responsible for the low **palatability** of the grass and poor performance of animals on reed canarygrass pastures (Marten et al. 1976). As many as four different chemicals, including a cardio-respiratory toxin, thiaminase, and amine cosubstrate, cyanogenic compounds, and NO_3^- compounds, have been implicated (Bourke and Carrigan 1992). Cultivars of reed canarygrass, bred for low alkaloid concentrations, give improved animal productivity (Marten et al. 1981; Wittenberg et al. 1992). Cobalt supplementation has also been proposed as a preventative, but the effectiveness of this strategy is still inconclusive.

Hydrocyanic Acid Poisoning

Forage sorghums such as sudangrass contain cyanogenic glycosides from which free cyanide can be released by enzymatic action. Damage to the plant from wilting, trampling, frost, drought stress, and so on, results in the breakdown of the cellular structure, exposing the glycosides to the hydrolyzing enzymes and formation of free cyanide. Cyanide inhibits the enzyme cytochrome oxidase that is needed for oxidative respiration in the animal. The risks of high concentrations of glycosides or cyanide in the plant are increased with N fertilization, immaturity, and frost damage (Wheeler et al. 1990).

Signs of poisoning include labored breathing, excitement, gasping, convulsions, paralysis, and death. The likelihood of acute cyanide poisoning may be greater when feeding sorghum hay than when grazing fresh plants because of more rapid DM intake. Ground and pelleted sorghum hay may be especially toxic because of the rapid rate of cyanide intake and release (Wheeler and Mulcahy 1989). Ensiling markedly reduces the cyanide risk.

Oxalate Poisoning

Many tropical grasses contain high levels of oxalate, which when ingested by ruminants, complexes dietary Ca and forms insoluble Ca oxalate. This leads to disturbances in Ca and P metabolism involving excessive mobilization of bone mineral. The demineralized bones become fibrotic and misshapen, causing lameness and "bighead" in horses. Ruminants are less affected, but prolonged grazing by cattle and sheep of some tropical grass species can result in severe hypocalcemia, resulting in Ca deposits in the kidneys and kidney failure. Tropical grasses that have high oxalate levels include *Setaria* spp., *Brachiaria* spp., buffelgrass, 'Pangola' digitgrass, and kikuyugrass. Providing mineral supplements high in Ca to grazing animals overcomes the adverse effect of oxalates in grasses.

Facial Eczema

Facial eczema of grazing ruminants is a classic example of secondary or hepatogenous photosensitization due to liver damage. Facial eczema is a major problem of sheep and cattle on perennial ryegrass pastures in New Zealand and has been reported sporadically in other countries. The fungus *Pithomyces chartarum* (Berk. and M.A. Curtis) M.B. Ellis grows on the dead litter in ryegrass pastures and produces large numbers of spores. The spores contain a hepatotoxin, sporidesmin, which is only slowly broken down in the liver. Spores consumed during grazing lead to sporidesmin-induced liver damage. The damaged liver is unable to metabolize phylloerythrin, a metabolite of chlorophyll breakdown, which then accumulates in the blood. Phylloerythrin is a photodynamic agent that reacts with sunlight, causing severe dermatitis of the face, udder, and other exposed areas. There are species differences in susceptibility to sporidesmin. For example, goats are much more resistant to facial eczema than sheep, probably because of a faster rate of sporidesmin detoxification in the liver (Smith and Embling 1991). Facial eczema can be prevented by avoiding pastures infected with *P. chartarum*, reducing *P. chartarum* pasture populations through the application of substituted thiabendazole fungicides, and/or by feeding cattle high daily oral doses of zinc, an approach that poses its own risks as near toxic levels must be administered (Di Menna et al. 2009).

Mycotoxicosis

Seed heads of many grasses are susceptible to infection with *Claviceps purpurea* and other *Claviceps* species that can form ergot alkaloids. In the US, dallisgrass poisoning is the major cause of *Claviceps* ergotism. Ergot alkaloids cause vasoconstriction and reduced blood supply to the extremities, resulting in sloughing of ear tips, tail, and hooves. There are also neurologic effects, including hyperexcitability, incoordination, and convulsions. Ergotism can be avoided by preventing seed set in grasses.

Fescue Toxicosis

Tall fescue cultivars (older cultivars and most turf types) are likely infected with the endophytic fungus *N. coenophialum*. The primary ergot alkaloid produced is ergovaline which confers the plant with tolerance to grazing and other stresses but is responsible for reduced animal performance (Klotz 2015) and three types of livestock disorders when forage or seed is consumed. These disorders include fescue foot, summer fescue toxicosis, and fat necrosis. These occur because of the inhibition of prolactin secretion by the pituitary gland and vasoconstriction of blood vessels in the extremities. Newer tall fescue cultivars have been developed that contain an endophytic fungus that provides acceptable tolerance to grazing and high temperature, but is not toxic to livestock (Nihsen et al. 2004). The vasoconstriction properties of ergot alkaloids may be counteracted if clovers are included in the pasture as isoflavones in these forages can promote vasodilation (Flythe and Aiken 2017).

Ryegrass Staggers

Besides facial eczema (described above), two other major syndromes are perennial ryegrass staggers and annual ryegrass toxicity. Perennial ryegrass staggers is caused by compounds called tremorgens. Affected animals exhibit various degrees of incoordination and other neurologic signs (head shaking, stumbling and collapse, severe muscle spasms), particularly when disturbed or forced to run. Even in severe cases, there are no pathologic signs in nervous tissue, and upon a change of feed, affected animals usually spontaneously recover. The growth rate of the animal is also reduced (Fletcher and Barrell 1984). In Australia and New Zealand, ryegrass staggers occurs in sheep, cattle, horses, and deer. It has been reported in sheep and cattle in California (Galey et al. 1991). It also occurs in sheep grazing winter forage and **stubble** residue of endophyte-enhanced turf-type ryegrasses in Oregon.

The main causative agents of ryegrass staggers are a group of potent tremorgens called *lolitrems*, the most important of which is lolitrem B (Gallagher et al. 1984). Lolitrem B is a potent inhibitor of neurotransmitters in the brain. The lolitrems are produced by an endophytic fungus, *Neotyphodium* (formerly *Acremonium*) *lolii*, which is often present in perennial ryegrass. Turf cultivars of both tall fescue and perennial ryegrass are often deliberately infected with endophytes, because the endophyte increases plant vigor and stress tolerance, in part by producing ergot alkaloids (*N. coenophialum*) and tremorgens (*N. lolii*) that are deleterious to livestock.

In Australia and South Africa, annual ryegrass toxicity is a significant disorder of livestock. It has an interesting etiology, involving annual ryegrass, a nematode, and bacteria. Though the neurologic signs are superficially similar, annual ryegrass toxicity and ryegrass staggers are totally different disorders. In contrast to the temporary incoordination seen with ryegrass staggers, permanent brain damage occurs with annual ryegrass toxicity. The neurologic damage is evidenced by convulsions of increasing severity that terminate in death.

Annual ryegrass toxicity is caused by corynetoxins, which are chemically similar in structure to the tunicamycin antibiotics. Corynetoxins are produced by a *Clavibacter* spp. (formerly designated *Corynebacterium* spp.). This bacterium parasitizes a nematode (*Anguina agrostis*) that infects annual ryegrass. Ryegrass is toxic only when infected with the bacteria-containing nematode *A. agrostis*.

The parasitized nematodes infect the seedling shortly after germination, and the larvae are passively carried up the plant as the plant stem elongates. They invade the florets, producing a nematode gall instead of seed. When consumed by animals, corynetoxins from the bacteria inhibit an enzyme involved in glycoprotein synthesis, leading to defective formation of various blood components of the reticulo-endothelial system. This impairs cardiovascular function and vascular integrity, causing inadequate blood supply to the brain.

Corynetoxins have been identified in other grasses besides annual ryegrass, including *Polypogon* and *Agrostis* spp. (Finnie 1991; Bourke et al. 1992). Annual ryegrass toxicity can be avoided by not allowing animals to graze mature ryegrass containing seed heads or by clipping pastures to prevent seed head development. In Australia, these measures are often impractical because of the extensive land areas involved.

Other Grass Toxins

Kikuyugrass is a widely used tropical forage that is occasionally toxic to livestock (Peet et al. 1990). Clinical signs include depression, drooling, muscle twitching, convulsions, and sham drinking which can appear as an inability to swallow (Newsholme et al. 1983). **Rumen motility** is lost, and severe damage occurs to the mucosa of the rumen and **omasum**. In many but not all cases, kikuyugrass poisoning occurs when the pasture is invaded by armyworm (*Spodoptera exempta*). The causative agent has not been identified, and it is not conclusively known if the armyworm has a role in the toxicity.

Photosensitization of grazing animals often occurs with *Panicum* and *Brachiaria* spp. (Bridges et al. 1987; Cornick et al. 1988; Graydon et al. 1991). The condition is usually accompanied by accumulation of salt crystals in and around the bile ducts in the liver. Miles et al. (1991) have shown that the crystals are formed from metabolites of **saponins**, which are common constituents of *Panicum* spp. and *Brachiaria* spp. The crystals impair biliary excretion, leading to elevated phylloerythrin levels in the blood, causing secondary (hepatic) photosensitization. In Brazil, most cases of hepatogenous photosensitization are caused by *Brachiaria decumbens*, however *Brachiaria brizantha*, *Brachiaria humidicola* and *Brachiaria ruziziensis* can also cause poisoning (Riet-Correa et al. 2011). Cattle with symptoms of hepatogenous photosensitization should be removed from toxic pastures, and kept in the shade, with provision of feed and water. However, removing cattle from *Brachiaria* spp. can be a challenge as they are often the only pastures available on many farms in the tropics. For example, of the 60 million ha of cultivated pastures in the Brazilian Cerrado, 51 million ha (85%) consist of *Brachiaria* spp.

Toxins in Other Forages

Other forages including buckwheat, spineless cactus, saltbush, and forage brassicas (e.g. kale, rape, cabbage, and turnip) used in ruminant production can contain toxins. *Brassica* spp. contain glucosinolates (goitrogens) and can result in a condition known as brassica anemia. Glucosinolates are primarily of concern in brassicas grown for seed, such as mustard. Forage brassicas contain a toxin, S-methylcysteine sulfoxide (SMCSO) a compound that has been termed the "brassica anemia factor." Ruminants often develop severe hemolytic anemia on kale or rape pastures, and growth is reduced. The SMCSO is metabolized in the rumen to dimethyl disulfide, an oxidant that destroys the red blood cell membrane leading to anemia, hemoglobinuria (red urine), liver and kidney damage, and often mortality. Because the SMCSO content of brassicas increases with plant maturity, it is not advisable to graze mature brassica during late winter in temperate areas, though grazing immature rape may induce photosensization (rape scald), predominantly in sheep (Vermunt et al. 1993). Avoiding the use of S and high N in fertilizer reduces SMCSO levels and toxicity. Brassica anemia is reviewed by Cheeke (1998) and Smith (1980). The high-sulfur content of brassicas may also lead to polioencephalomalacia (PEM) which is symptomized by blindness, lack of coordination, circling convulsions and death. Injections of thiamine are used to treat PEM.

Other Health Problems

Acute bovine pulmonary emphysema or atypical interstitial pneumonia may occur when cattle are moved from sparse dry pasture to lush grass, legume, or brassica pasture. The abrupt change in pasture type results in a disturbance in the rumen microbes, resulting in excessive conversion of the amino acid tryptophan to a metabolite, 3-methyl indole (3-MI). The 3-MI is absorbed and is toxic to the lung tissue, causing pulmonary edema and emphysema (Carlson and Breeze 1984). The condition, also called summer pneumonia or fog fever is often fatal. Provision of supplementary feed before moving cattle on to lush meadows is helpful in preventing the disorder.

Blister beetles (*Epicauta* spp.) contain cantharidin, a toxin that causes irritation of the lining in the digestive tract, and at high enough doses, is lethal to horses (Helman and Edwards 1997). Alfalfa (especially that grown east of 100th meridian in the US and Canada) is often associated with blister beetle infestation. Crop scouting should be more intensive near field edges, where blister beetles tend to congregate. The toxicity to horses depends on insect species and sex, the horse's size and condition, and the number of insects ingested with the hay. Intake of 1 mg cantharidin kg^{-1} body weight is considered the lethal threshold. Concentrations of cantharidin are highest in male beetles in the "vittata" group, or the "stripped" beetle group. Ingestion of 75 stripped male beetles could be fatal for a 375-kg horse.

Animal Metabolism of Plant Toxins

Plants and animals have coevolved. As plants developed the enzymatic means to synthesize defensive chemicals, animals evolved detoxification mechanisms to overcome the plant defenses. The most fundamental of these are the drug-metabolizing enzyme systems of the liver, such as the cytochrome P450 system. This enzyme system (also called the mixed function oxidase system) oxidizes hydrophobic, nonpolar substances such as plant toxins by introducing a hydroxyl (–OH) group to change the chemical. The hydroxyl group increases the water solubility of the compound, mainly by providing a site to react (conjugate) with other water-soluble compounds such as amino acids (e.g. glycine), peptides (glutathione), and sugars (e.g. glucuronic acid). These conjugated compounds are much less toxic and can be excreted in the urine. Differences in susceptibility among livestock species to plant toxins are due mostly to differences in liver metabolism. In contrast, some toxins are bioactivated, or made more toxic, because of liver metabolism (e.g. **aflatoxin**, slaframine). The relative rates at which the active metabolites are formed and detoxified determine the extent of cellular damage.

Browsing animals such as sheep and goats are generally more resistant to plant toxins because they have been exposed to greater concentrations of plant toxins during their evolution than grazing animals such as horses and cattle. Sheep and goats find plants containing toxins more palatable than do cattle and horses (Cheeke 2005). Sometimes, as with pyrrolizidine alkaloids in *Senecio* spp., the resistance of sheep and goats is due to a lower rate of conversion of the compounds to the toxic metabolites in the liver (Cheeke 1988).

Browsing animals are also better able than grazers to resist adverse effects of dietary tannins and phenolic compounds, which are common constituents of **shrubs**, trees, and other browse plants. For example, deer, which are browsers, have salivary tannin-binding proteins, absent in sheep and cattle, which counteract the astringent effects of tannins (Austin et al. 1989). Mehansho et al. (1987) reviewed the roles of salivary tannin-binding proteins as animal defenses against plant toxins. Resistance to tannin astringency results in tannin-containing plants being more palatable to browsers than to grazers, which lack the tannin-binding proteins.

In ruminants, metabolism of toxins by rumen microbes is an important factor in altering sensitivity to plant toxins. In some cases, for example, cyanogenic glycosides and the brassica anemia factor, the toxicity is increased by rumen microbial fermentation. Sometimes, for example, mimosine or oxalate toxicity, the compounds are detoxified by microbial metabolism. The toxic amino acid mimosine has been of particular interest in this regard. Similar mechanisms may also reduce the sensitivity of ruminants to some mycotoxins such as deoxynivalenol (DON) produced by *Fusarium graminearum*. As discussed earlier, the successful use of leucaena as a high-protein forage was not possible in Australia and many other areas until ruminants were dosed with mimosine-degrading bacteria. The effective bacteria are transferred orally as uninoculated animals graze plants covered with slobber from animals having the effective organism.

References

Acharya, S., Sottie, E., Coulman, B. et al. (2013). New sainfoin populations for bloat-free alfalfa pasture mixtures in Western Canada. *Crop Sci.* 53: 2283–2293. https://doi.org/10.2135/cropsci2012.10.0591.

Alden, R., Hackney, B., Weston, L.A., and Quinn, J.C. (2014). Phalaris toxicoses in Australian livestock production systems: prevalence, aetiology and toxicology. *J. Toxins* 1: 7.

Arnold, M. and Lehmkuhler, J. (2014). Hypomagnesmic tetany or "Grass Tetany" Agriculture and Natural Resources. Paper No. 173. http://uknowledge.uky.edu/anr_reports/173 (accessed 14 October 2019).

Austin, P.J., Suchar, L.A., Robbins, C.T., and Hagerman, A.E. (1989). Tannin-binding proteins in saliva of sheep and cattle. *J. Chem. Ecol.* 15: 1335–1347.

Aylward, J.H., Court, R.D., Haydock, K.P. et al. (1987). *Indigofera* species with agronomic potential in the tropics: rat toxicity studies. *Aust. J. Agric. Res.* 38: 177–186.

Ball, D.M., Pedersen, J.F., and Lacefield, G.D. (1993). The tall-fescue endophyte. *Am. Sci.* 81: 370–379.

Beke, G.J. and Hironaka, R. (1991). Toxicity to beef cattle of sulfur in saline well water: a case study. *Sci. Total Environ.* 101: 281–290.

Berg, B.P., Majak, W., McAllister, T.A. et al. (2000). Bloat in cattle grazing alfalfa cultivars selected for a low initial rate of digestion: a review. *Can. J. Plant. Sci.* 80: 493–502.

Bourke, C.A. and Carrigan, M.J. (1992). Mechanisms underlying *Phalaris aquatica* "sudden death" syndrome in sheep. *Aust. Vet. J.* 69: 165–167.

Bourke, C.A., Carrigan, M.J., and Dixon, R.J. (1988). Experimental evidence that tryptamine alkaloids do not cause *Phalaris aquatica* sudden death syndrome in sheep. *Aust. Vet. J.* 65: 218–220.

Bourke, C.A., Carrigan, M.J., and Dixon, R.J. (1990). The pathogenesis of the nervous syndrome of *Phalaris aquatica* toxicity in sheep. *Aust. Vet. J.* 67: 356–358.

Bourke, C.A., Carrigan, M.J., and Love, S.C.J. (1992). Flood plain staggers, a tunicaminyluracil toxicosis of cattle in northern New South Wales. *Aust. Vet. J.* 69: 228–229.

Bridges, C.H., Camp, B.J., Livingston, C.W., and Bailey, E.M. (1987). Kleingrass *(Panicum coloratum)* poisoning in sheep. *Vet. Pathol.* 24: 525–531.

Bush, L., Boling, J., and Yates, S. (1979). Animal disorders. In: *Tall Fescue* (eds. R.C. Buckner and L.P. Bush), 247–292. Madison, WI: American Society of Agronomy.

Carlson, J.R. and Breeze, R.G. (1984). Ruminal metabolism of plant toxins with emphasis on indolic compounds. *J. Anim. Sci.* 58: 1040–1049.

Cash, D., Funston, D.R., King, M. et al. (2002). Nitrate toxicity of Montana forages. Montana State University Extension Service, MontGuide, No. 200205 AG.

Chamblee, D.S. and Collins, M. (1988). Relationship with other species in a mixture. In: *Alfalfa and Alfalfa Improvement* (ed. A.A. Hanson), 439–466. Madison, WI: American Society of Agronomy.

Cheeke, P.R. (1988). Toxicity and metabolism of pyrrolizidine alkaloids. *J. Anim. Sci.* 66: 2343–2350.

Cheeke, P.R. (1998). *Natural Toxicants in Feeds, Forages, and Poisonous Plants*. Upper Saddle River, NJ: Prentice Hall.

Cheeke, P.R. (2005). *Applied Animal Nutrition: Feeds and Feeding*. Upper Saddle River, NJ: Prentice Hall.

Cherney, J.H., Mikhailova, E.A., and Cherney, D.J.R. (2002). Tetany potential of orchardgrass and tall fescue as influenced by fertilization with dairy manure or commercial fertilizer. *J. Plant Nutr.* 25: 1501–1525.

Cockrem, F.R.M., McIntosh, J.T., and McLaren, R. (1983). Selection for and against susceptibility to bloat in dairy cows: a review. *Proc. N. Z. Soc. Anim. Prod.* 43: 101–106.

Cornick, J.L., Carter, G.K., and Bridges, C.H. (1988). Kleingrass: associated hepatotoxicosis in horses. *J. Am. Vet. Med. Assoc.* 193: 932–935.

Davison, K.L., Hansel, W., Krook, L. et al. (1964). Nitrate toxicity in dairy heifers. I. Effects on reproduction, growth, lactation and vitamin a nutrition. *J. Dairy Sci.* 47: 1065–1073.

Deeb, B.S. and Sloan, K.W. (1975). Nitrates, nitrites, and health. Illinois Agricultural Experiment Station Bulletin No. 750.

DeGaris, P.J. and Lean, I.J. (2009). Milk fever in dairy cows: a review of pathophysiology and control principles. *Vet. J.* 176: 58–69.

Derakhshani, H., Corely, S.W., and Jassim, R.A. (2016). Isolation and characterization of mimosine, 3, 4 DHP and 2,3 DHP degrading bacteria from commercial rumen inoculum. *J. Basic Micribiol.* 56: 580–585.

Di Menna, M.E., Smith, B.L., and Miles, C.O. (2009). A history of facial eczema (pithomycotoxicosis) research. *N. Z. J. Agric. Res.* 52 (4): 345–376.

Dias, R.S., López, S., Montanholi, Y.R. et al. (2013). A meta-analysis of the effects of dietary copper, molybdenum, and sulfur on plasma and liver copper, weight gain, and feed conversion in growing-finishing cattle. *J. Anim. Sci.* 91: 5714–5723.

Fay, J.P., Cheng, K.-J., Hanna, M.R. et al. (1980). In vitro digestion of bloat-safe and bloat-causing legumes by rumen microorganisms: gas and foam production. *J. Dairy Sci.* 63: 1273–1281.

Felix, T.L., Weiss, W.P., Fluharty, F.L., and Loerch, S.C. (2012). Effects of copper supplementation on feedlot performance, carcass characteristics, and rumen sulfur metabolism of growing cattle fed diets containing 60% dried distillers grains. *J. Anim. Sci.* 90: 2710–2716.

Finnie, J.W. (1991). Corynetoxin poisoning in sheep in the southeast of South Australia associated with annual beard grass (*Polypogon monspeliensis*). *Aust. Vet. J.* 68: 370.

Fisher, D.W., Mayland, H.F., and Burns, J.C. (2002). Variation in ruminant preference for alfalfa hays cut at sunup and sundown. *Crop Sci.* 42: 231–237.

Fletcher, L.R. and Barrell, G.K. (1984). Reduced liveweight gains and serum prolactin levels in hoggets grazing ryegrasses containing *Lolium* endophyte. *N. Z. Vet. J.* 32: 139–140.

Flythe, M. and Aiken, G. (2017). Promote growth and animal health with isoflavones in red clover and other legumes. http://uknowledge.uky.edu/cgi/viewcontent.cgi?article=1070&context=forage_kca (accessed 14 October 2019).

Foster, J.G. (1990). Flatpea (*Lathyrus sylvestris*): a new forage species? A comprehensive review. *Adv. Agron.* 43: 241–313.

Galey, F.D., Tracy, M.L., Craigmill, A.L. et al. (1991). Staggers induced by consumption of perennial ryegrass in cattle and sheep from northern California. *J. Am. Vet. Med. Assoc.* 199: 466–470.

Gallagher, R.T., Hawkes, A.D., Steyn, P.S., and Vleggaar, R. (1984). Tremorgenic neurotoxins from perennial ryegrass causing ryegrass staggers disorder of livestock: structure elucidation of lolitrem B. *J. Chem. Soc. Chem. Commun.*: 614–616.

Godwin, I.R., Li, L., Luijben, K. et al. (2015). The effects of chronic nitrate supplementation on erythrocytic methaemoglobin reduction in cattle. *Anim. Prod. Sci.* 55: 611–616. https://doi.org/10.1071/AN13366.

Goff, J.P., Liesegang, A., and Horst, R.L. (2014). Diet-induced pseudohypoparathyroidism: a hypocalcemia and milk fever risk factor. *J. Dairy Sci.* 97: 1520–1528.

Gooneratne, S.R., Buckley, W.T., and Christensen, D.A. (1989). Review of copper deficiency and metabolism in ruminants. *Can. J. Anim. Sci.* 69: 819–845.

Grace, N.D. (1983). *The Mineral Requirements of Grazing Ruminants*. Palmerston North, NZ: New Zealand Society of Animal Production, Occasional Publ. No. 9.

Grace, N.D. and Clark, R.G. (1991). Trace element requirements, diagnosis and prevention of deficiencies in sheep and cattle. In: Physiological Aspects of Digestion and Metabolism in Ruminants: Proc. 7th Int. Symp. Ruminant Physiol., 321–346. Academic Press, Inc.

Graham, T.W. (1991). Trace element deficiencies in cattle. *Vet. Clin. North Am. Food Anim. Pract.* 7: 153–215.

Graydon, R.J., Hamid, H., Zahari, P., and Gardiner, C. (1991). Photosensitization and crystal-associated cholangiohepatopathy in sheep grazing *Brachiaria decumbens*. *Aust. Vet. J.* 68: 234–236.

Greene, L.W., Baker, J.F., and Hardt, P.F. (1989). Use of animal breeds and breeding to overcome the incidence of grass tetany: a review. *J. Anim. Sci.* 67: 3463–3469.

Grunes, D.L. and Welch, R.M. (1989). Plant contents of magnesium, calcium, and potassium in relation to ruminant nutrition. *J. Anim. Sci.* 67: 3486–3494.

Halgerson, J.L., Sheaffer, C.C., Martin, N.P. et al. (2004). Near-infrared reflectance spectroscopy prediction of leaf and mineral concentrations in alfalfa. *Agron. J.* 96: 344–351.

Hall, J.W. and Majak, W. (1991). Relationship of weather and plant factors to alfalfa bloat in autumn. *Can. J. Anim. Sci.* 71: 861–866.

Hall, J.W. and Majak, W. (1995). Effect of time of grazing or cutting and feeding on the incidence of alfalfa bloat in cattle. *Can. J. Anim. Sci.* 75: 271–273.

Hall, J.W., Majak, W., Williams, R.J., and Howarth, R.E. (1984). Effect of daily weather conditions on bloat in cattle fed fresh alfalfa. *Can. J. Anim. Sci.* 64: 943–950.

Hall, J.W., Walker, I., and Majak, W. (1994). Evaluation of two supplements for the prevention of alfalfa bloat. *Can. Vet. J.* 35: 702–704.

Hall, J.W., Majak, W., McAllister, T.A., and Merrill, J.K. (2001). Efficacy of Rumensin controlled release capsule (CRC) for the control of alfalfa bloat in cattle. *Can. J. Anim. Sci.* 81: 281–283.

Hall, J.A., Bobe, G., Hunter, J.K. et al. (2013). Effect of feeding selenium-fertilized alfalfa hay on performance of weaned beef calves. *PLoS One* 8 (3): e58188. https://doi.org/10.1371/journal.pone.0058188.

Hammond, A.C., Allison, M.J., Williams, M.J. et al. (1989). Prevention of leucaena toxicosis of cattle in Florida by ruminal inoculation with 3-hydroxy-4-(1H)-pyridone-degrading bacteria. *Am. J. Vet. Res.* 50: 2176–2180.

Helman, R.G. and Edwards, W.C. (1997). Clinical features of blister beetle poisoning in equids: 70 cases (1983–1996). *J. Am. Vet. Med. Assoc.* 211: 1018–1021.

Holland, C. and Kezar, W. (1990). *Pioneer Forage Manual: A Nutritional Guide*. Des Moines, IA: Pioneer Hi-Bred International Inc.

Horst, R.L., Goff, J.P., Reinhardt, T.A., and Buxton, D.R. (1997). Strategies for preventing milk fever in dairy cattle. *J. Dairy Sci.* 80: 1269–1280.

James, L.F., Ralphs, M.H., and Nielsen, D.B. (1988). *The Ecology and Economic Impact of Poisonous Plants on Livestock Production*. Boulder, CO: Westview Press.

Jones, R.J. and Megarrity, R.G. (1986). Successful transfer of DHP-degrading bacteria from Hawaiian goats to Australian ruminants to overcome the toxicity of leucaena. *Aust. Vet. J.* 63: 259–262.

Kamo, T., Sakurai, S., Yamanashi, T., and Todoroki, Y. (2015). Cynamide is biosynthesized from L-canavanine in plants. *Sci. Rep.* 5: 10527. https://doi.org/10.1038/srep10527.

Karn, J.F. (2001). Phosphorus nutrition of grazing cattle: a review. *Anim. Feed Sci. Technol.* 89: 133–153.

Kellerman, T.S., Coetzer, J.A.W., and Naude, T.W. (1988). *Plant Poisonings and Mycotoxicoses of Livestock in Southern Africa*. Cape Town: Oxford University Press.

Kemp, A. and 'tHart, M.L. (1957). Grass tetany in grazing milking cows. *Neth. J. Agric. Sci.* 5: 4–17.

Kerr, L.A. and Edwards, W.C. (1982). Hairy vetch poisoning in cattle. *Vet. Med. Small Anim. Clin.* 77: 257–261.

Kessler, K.L., Olson, K.C., Wright, C.L. et al. (2012). Effects of supplemental molybdenum on animal performance, liver copper concentrations, ruminal hydrogen sulfide concentrations, and the appearance of sulfur and molybdenum toxicity in steers receiving fiber-based diets. *J. Anim. Sci.* 90: 5005–5012.

Klotz, J.L. (2015). Activities and effects of ergot alkaloids on livestock physiology and production. *Toxins* 7: 2801–2821.

Kubota, J. and Allaway, W.H. (1972). Geographic distribution of trace element problems. In: *Micronutrients in Agriculture* (eds. J.J. Mortvedt, P.M. Giordano and W.L. Lindsay), 525–554. Madison, WI: American Society of Agronomy.

Lanyon, L.E. (1980). Pennsylvania alfalfa growers program alfalfa mineral relationships. Proceedings of the Annual Conference of Pennsylvania Forage and Grassland Council, Pennsylvania, USA (24–25 November 1980.

Lean, I.J., DeGaris, P.J., McNeil, D.M., and Block, E. (2006). Hypocalcemia in dairy cows: meta-analysis and dietary cation anion difference theory revisited. *J. Dairy Sci.* 89: 669–684.

Legesse, G., Beauchemin, K.A., Ominski, K.H. et al. (2015). Greenhouse gas emissions of Canadian beef production in 1981 as compared with 2011. *Anim. Prod. Sci.* 56: 153–168. https://doi.org/10.1071/AN15386.

Longland, A.G. and Cairns, A.J. (2000). Laminitis: Fructans and their implications in the etiology of laminitis. http://members.aol.com/wdds1/horsetalk/dh-sci1lam.htm (accessed 14 October 2019).

MacNaeidhe, F.S. (2001). Pasture management and composition as a means of minimising mineral disorders in organic livestock. In: Proceedings of the 5th NAHWOA Workshop, November 2001, 102–112. Rødding, Denmark.

Majak, W., Howarth, R.E., Fesser, A.C. et al. (1980). Relationships between ruminant bloat and the composition of alfalfa herbage. II. Saponins. *Can. J. Anim. Sci.* 60: 699–708.

Majak, W., Howarth, R.E., Cheng, K.-J., and Hall, J.W. (1983). Rumen conditions that predispose cattle to pasture bloat. *J. Dairy Sci.* 66: 1683–1688.

Majak, W., Hall, J.W., Rode, L.M., and Kalnin, C.M. (1986). Rumen clearance rates in relation to the occurrence of alfalfa bloat in cattle. 1. Passage of water-soluble markers. *J. Dairy Sci.* 69: 1560–1567.

Majak, W., Hall, J.W., and McCaughey, W.P. (1995). Pasture management strategies for reducing the risk of legume bloat in cattle. *J. Anim. Sci.* 73: 1493–1498.

Majak, W., Hall, J.W., and McAllister, T.A. (2001). Practical measures for reducing risk of alfalfa bloat in cattle. *J. Range Manage.* 54: 490–493.

Majak, W., Garland, G.J., and Lysyk, T.J. (2003). The effect of herbage mixtures of alfalfa and orchardgrass on the incidence of bloat in cattle. *Can. J. Anim. Sci.* 83: 827–829.

Majak, W., Garland, G.J., Lysyk, T.J., and Olson, M.E. (2004a). Efficacy of water-soluble feed supplements for the prevention of bloat in cattle. *Can. J. Anim. Sci.* 84: 155–157.

Majak, W., Steinke, D., McGillivray, J., and Lysyk, T. (2004b). Clinical signs in cattle grazing high molybdenum forage. *J. Range Manage.* 57: 269–274.

Majak, W., Lysyk, T.J., Garland, G.J., and Olson, M.E. (2005). Efficacy of Alfasure™ for the prevention and treatment of alfalfa bloat in cattle. *Can. J. Anim. Sci.* 85: 111–113.

Malinowski, D.P., Pinchak, W.E., Min, B.R. et al. (2015). Phenolic compounds affect bloat potential of wheat forage. *Crop, Forage Turfgrass Manage.* 1 https://doi.org/10.2134/cftm2015.0146.

Marten, G.C., Jordan, R.M., and Hovin, A.W. (1976). Biological significance of reed canarygrass alkaloids and associated palatability to grazing sheep and cattle. *Agron. J.* 68: 909–914.

Marten, G.C., Jordan, R.M., and Hovin, A.W. (1981). Improved lamb performance associated with breeding for alkaloid reduction in reed canarygrass. *Crop Sci.* 21: 295–298.

Marten, G.C., Ehle, F.R., and Ristau, E.A. (1987). Performance and photosensitization of cattle related to forage quality of four legumes. *Crop Sci.* 27: 138–145.

Marten, G.C., Jordan, R.M., and Ristau, E.A. (1990). Performance and adverse response of sheep during grazing of four legumes. *Crop Sci.* 30: 860–866.

Mathison, G.W., Soofi-Siawash, R., Klita, P.T. et al. (1999). Degradability of alfalfa saponins in the digestive tract of sheep and their rate of accumulation in rumen fluid. *Can. J. Anim. Sci.* 79: 315–319.

Mayland, H.F. (1988). Grass tetany. In: *The Ruminant Animal: Its Physiology and Nutrition*, 511–523, 530–531 (ed. D.C. Church). Englewood Cliffs, NJ: Prentice-Hall.

Mayland, H.F. and Sleper, D.A. (1993). Developing a tall fescue for reduced grass tetany risk. Proceedings of the International Grassland Congress, Palmerston, New Zealand (8–12 February 1993).

Mayland, H.F., Rosenau, R.C., and Florence, A.R. (1980). Grazing cow and calf responses to zinc supplementation. *J. Anim. Sci.* 51: 966–974.

Mayland, H.F., James, L.F., Panter, K.E., and Sonderegger, J.L. (1989). Selenium in seleniferous environments. In: *Selenium in Agriculture and the Environment*. Special Publ no. 23 (ed. L.W. Jacobs), 15–50. Madison, WI: Soil Science Society of America.

McDowell, L.R. and Arthington, J.D. (2005). *Minerals for Grazing Ruminants in Tropical Regions*, 4e, 33–34. Animal Science Department, Centre for Tropical Agriculture, University of Florida.

McKenzie, J.S. and McLean, G.E. (1982). The importance of leaf frost resistance to the winter survival of seedling stands of alfalfa. *Can. J. Plant. Sci.* 62: 399–405.

McMahon, L.R., Majak, W., McAllister, T.A. et al. (1999). Effect of sainfoin on *in vitro* digestion of fresh alfalfa and bloat in steers. *Can. J. Anim. Sci.* 79: 203–212.

Mehansho, H., Butler, L.G., and Carlson, D.M. (1987). Dietary tannins and salivary proline-rich proteins: interactions, induction, and defense mechanisms. *Annu. Rev. Nutr.* 7: 423–440.

Miles, C.O., Munday, S.C., Holland, P.T. et al. (1991). Identification of a sapogenin glucuronide in the bile of sheep affected by *Panicum dichotomiflorum* toxicosis. *N. Z. Vet. J.* 39: 150–152.

Min, B.R., Barry, T.N., Attwood, G.T., and McNabb, W.C. (2003). The effect of condensed tannins on the nutrition and health of ruminants fed fresh temperate forages: a review. *Anim. Feed Sci. Technol.* 106: 3–19.

Minson, D.J. (1990). *Forage in Ruminant Nutrition*, 230–264. Academic Press/Harcourt Brace Jovanovich.

Nation, P.N. (1991). Hepatic disease in Alberta horses: a retrospective study of "alsike clover poisoning" (1973–1988). *Can. Vet. J.* 32: 602–607.

National Research Council (2005). *Mineral Tolerance of Animals*. 2nd rev. ed. Washington, DC: National Academies Press.

Newsholme, S.J., Kellerman, T.S., Van Der Westhuizen, G.C.A., and Soley, J.T. (1983). Intoxication of cattle on kikuyu grass following army worm (*Spodoptera exempta*) invasion. *Onderstepoort J. Vet. Res.* 50: 157–167.

Nihsen, M.E., Piper, E.L., West, C.P. et al. (2004). Growth rate and physiology of steers grazing tall fescue inoculated with novel endophytes. *J. Anim. Sci.* 82: 878–883.

Nocek, J.E. (2001). The link between nutrition, acidosis, laminitis and environment. http://members.aol.com/wdds1/horsetalk/lamin-9.htm (accessed 14 October 2019).

Nolan, J.V., Godwin, I.R., de Raphélis-Soissan, V., and Hegarty, R.S. (2016). Managing the rumen to limit the incidence and severity of nitrite poisoning in nitrate-supplemented ruminants. *Anim. Prod. Sci.* 56: 1317–1329. https://doi.org/10.1071/AN15324.

Oetzel, G.R. (2000). Management of dry cows for the prevention of milk fever and other mineral disorders. *Vet. Clin. North Am. Food Anim. Pract.* 16: 369–386.

Olmos, G., Boyle, L., Horan, B. et al. (2009). Effect of genetic group and feed system on locomotion score, clinical lameness and hoof disorders of pasture-based Holstein–Friesian cows. *Animal* 3 (1): 96–107.

Pan, Y.J. and Loo, G. (2000). Effect of copper deficiency on oxidative DNA damage in Jurkat T-lymphocytes. *Free Radical Biol. Med.* 28: 824–830.

Peet, R.L., Dickson, J., and Hare, M. (1990). Kikuyu poisoning in goats and sheep. *Aust. Vet. J.* 67: 229–230.

Picco, S.J., Abba, M.C., Mattioli, G.A. et al. (2004). Association between copper deficiency and DNA damage in cattle. *Mutagenesis* 19 (6): 453–456.

Pitta, D.W., Pinchak, W.E., Indugu, N. et al. (2016). Metagenomic analysis of the rumen microbiome of steers with wheat-induced frothy bloat. *Front. Microbiol.* 7 https://doi.org/10.3389/fmicb.2016.00689.

Pollitt, C.C. (2005). Laminitis research at the Australian Equine Laminitis Research Unit. Parts 1 and 2. https://www.researchgate.net/publication/251595507_Advances_in_Laminitis_research_at_the_Australian_Equine_Laminitis_Research_Unit (accessed 2 December 2019)

Puls, R. (1994). *Mineral Levels in Animal Health: Diagnostic Data*, 2e. Clearbrook, BC: Sherpa International Publisher.

Quirk, M.F., Bushell, J.J., Jones, R.J. et al. (1988). Live-weight gains on leucaena and native grass pastures after dosing cattle with rumen bacteria capable of degrading DHP, a ruminal metabolite from leucaena. *J. Agric. Sci.* 111: 165–170.

Ramberg, C.F.J., Johnson, E.K., Fargo, R.D., and Kronfeld, D.S. (1984). Calcium homeostasis in cows, with special reference to parturient hypocalcemia. *Am. J. Physiol.* 246: R698–R704.

Rasmussen, M.A., Allison, M.J., and Foster, J.G. (1993). Flatpea intoxication in sheep and indications of ruminal adaptation. *Vet. Hum. Toxicol.* 35: 123–127.

Reinhardt, T.A., Lippolis, J.D., McCluskey, B.J. et al. (2011). Prevalence of subclinical hypocalcemia in dairy herds. *Vet. J.* 188: 122–124.

Riet-Correa, B., Castro, M.B., Lemos, R.A.A. et al. (2011). *Brachiaria* spp. poisoning of ruminants in Brazil. *Pesq. Vet. Bras.* 31 (3): 183–192.

Sarturi, J.O., Erickson, G.E., Klopfenstein, T.J. et al. (2013). Effect of sulfur content in wet or dry distillers grains fed at several inclusions on cattle growth performance, ruminal parameters and hydrogen sulfide. *J. Anim. Sci.* 91: 4849–4860.

Séboussi, R., Tremblay, G.F., Ouellet, V. et al. (2016). Selenium-fertilized forage as a way to supplement lactating dairy cows. *J. Dairy Sci.* 99 (7): 5358–5369.

Shewmaker, G.E., Mayland, H.F., Rosenau, R.C., and Asay, K.H. (1989). Silicon in C-3 grasses: effects on forage quality and sheep preference. *J. Range Manage.* 42: 122–127.

Singer, R.H. (1972). The nitrate poisoning complex. Proceedings of the United States Animal Health Association, Miami Beach, FL, USA (5–10 November 1972).

Smith, R.H. (1980). Kale poisoning: the brassica anemia factor. *Vet. Rec.* 107: 12–15.

Smith, B.L. and Embling, P.P. (1991). Facial eczema in goats: the toxicity of sporidesmin in goats and its pathology. *N. Z. Vet. J.* 39: 18–22.

Soni, A.K. and Shukula, P.C. (2012). Hypomagnesmic tetany in cow calves: a case study. *Environ. Ecol.* 30 (4A): L 1601–L 1602.

Sottie, E.T. (2014). Characterization of new sainfoin populations for mixed alfalfa pastures in western Canada. PhD thesis. University of Lethbridge.

Sottie, E.T., Acharya, S.N., McAllister, T. et al. (2014). Alfalfa pasture bloat can be eliminated by intermixing with newly-developed sainfoin population. *Agron. J.* 106: 1470–1478. https://doi.org/10.2134/agronj13.0378.

Spears, J.W., Kegley, E.B., and Mullis, L.A. (2004). Bioavailability of copper from tribasic copper chloride and copper sulfate in growing cattle. *Anim. Feed Sci. Technol.* 116: 1–13.

Spears, J.W., Lloyd, K.E., and Fry, R.S. (2011). Tolerance of cattle to increased dietary sulfur and effect of dietary cation-anion balance. *J. Anim. Sci.* 89: 2502–2509.

Spoelstra, S.F. (1985). Nitrate in silage. *Grass Forage Sci.* 40: 1–11.

Stanford, K., Wang, Y., Berg, B.P. et al. (2001). Effects of alcohol ethoxylate and pluronic detergents on the development of pasture bloat in cattle and sheep. *J. Dairy Sci.* 74: 167–176.

Stewart, P.A. (1983). Modern quantitative acid-base chemistry. *Can. J. Physiol. Pharmacol.* 61: 1444–1461.

Stewart, S.R., Emerick, R.J., and Pritchard, R.H. (1991). Effects of dietary ammonium chloride and variations in Ca to phosphorus ratio on silica urolithiasis in sheep. *J. Anim. Sci.* 69: 2225–2229.

Suttle, N.F. (1991). The interactions between copper, molybdenum, and sulfphur in ruminant nutrition. *Annu. Rev. Nutr.* 11: 121–140.

Suttle, N. (2004). Assessing the needs of cattle for trace elements. *In Practice* 26: 553–561.

Thompson, D.J., Brooke, B.M., Garland, G.J. et al. (2000). Effect of stage of growth of alfalfa on the incidence of bloat in cattle. *Can. J. Anim. Sci.* 80: 725–727.

Tremblay, G.F., Nie, Z., Bélanger, G. et al. (2009). Predicting timothy mineral concentrations, dietary cation-anion difference, and grass tetany index by near-infrared reflectance spectroscopy. *J. Dairy Sci.* 92: 4499–4506.

Underwood, E.J. (1977). *Trace Elements in Human and Animal Nutrition*, 4e. New York, NY: Academic Press.

Uwituze, S., Parsons, G.L., Schneider, C.J. et al. (2011). Evaluation of sulfur content of dried distillers grains with solubles in finishing diets based on steam-flaked corn or dry-rolled corn. *J. Anim. Sci.* 89: 2582–2591.

Van Soest, P.J. and Jones, L.H.P. (1968). Effect of silica in forages upon digestibility. *J. Dairy Sci.* 51: 1–5.

Vermunt, J.J., West, D.M., and Cooke, M.M. (1993). Rape poisoning in sheep. *N. Z. Vet. J.* 41: 151–152.

Vogel, K.P., Gabrielsen, B.C., Ward, J.K. et al. (1993). Forage quality, mineral constituents, and performance of beef yearling grazing two crested wheatgrasses. *Agron. J.* 85: 584–590.

Westwood, C.T., Bramlayand, E., and Lean, I.J. (2003). Review of the relationship between nutrition and lameness in pasture-fed dairy cattle. *N. Z. Vet. J.* 51: 208–218.

Wheeler, J.L. and Mulcahy, C. (1989). Consequences for animal production of cyanogenesis in sorghum forage and hay: a review. *Trop. Grasslands* 23: 193–202.

Wheeler, J.L., Mulcahy, C., Walcott, J.J., and Rapp, G.G. (1990). Factors affecting the hydrogen cyanide potential of forage sorghum. *Aust. J. Agric. Res.* 41: 1093–1100.

Whitaker, D.A., Macrae, A.I., and Burrough, E. (2004). Disposal and disease rates in British dairy herds between April 1998 and March 2002. *Vet. Record* 155: 43–44, 45–47.

Wittenberg, K.M., Duynisveld, G.W., and Tosi, H.R. (1992). Comparison of alkaloid content and nutritive value for tryptamine- and carboline-free cultivars of reed canarygrass (*Phalaris arundinacea* L.). *Can. J. Anim. Sci.* 72: 903–909.

Wright, M.J. and Davison, K.L. (1964). Nitrate accumulation in crops and nitrate poisoning in animals. *Adv. Agron.* 16: 197–247.

CHAPTER

48

Grazing Systems and Strategies

Michael Collins, Professor and Director Emeritus, *Plant Sciences, University of Missouri, Columbia, MO, USA*
Kenneth J. Moore, Distinguished Professor, *Agronomy, Iowa State University, Ames, IA, USA*
C. Jerry Nelson, Professor Emeritus, *Plant Sciences, University of Missouri, Columbia, MO, USA*
Daren D. Redfearn, Associate Professor, *Agronomy, University of Nebraska, Lincoln, NE, USA*

Grazing Systems, Methods, and Tactics

Over 40 years ago, agricultural economists began calling for a "systems" approach to better understand interrelationships of all the production and marketing aspects of the beef industry (Purcell 1977). In the context of this chapter, an *ecologic system* is defined as an assemblage of living organisms in association with their physical and chemical environment (Odum 1971). Thus, a **grazing system** in 2011 was considered as "a defined, integrated combination of soil, plant, animal, social and economic features, stocking (grazing) method(s) and management objectives designed to achieve specific results or goals" (Allen et al. 2011). Today, there are expanded and genuine public concerns relative to interactions with global change, broadened environmental and wildlife management, understanding animal rights and markedly increased interest in food quality and safety. Today, the design of a grazing system must consider providing a much wider range of inputs and outputs in a way that is sustainable. These factors need research to develop quantitative guidelines to measure effectiveness of a grazing system.

Grazing management is "the manipulation of grazing in pursuit of a specific objective or set of objectives," whereas a **stocking method** is "a defined procedure or technique to manipulate animals in space and time to achieve a specific objective(s)" (Syn. Grazing method). In each case, the specific objectives will consider the inputs of resources and the output of desired goals which will differ among animal **species** (beef, dairy, lamb, goat, etc.) and types of forage in the system.

Each grazing system is unique, but the principles that function within systems can be applied to other locations and situations. Desired outcomes include economic objectives, production goals for plants and animals, enhancement of environmental quality and may include objectives such as recreation, preservation, and esthetics of open spaces (Kallenbach and Collins 2018).

System components are highly interactive such that when examined in isolation they rarely, if ever, function as they would within the context of the system. Systems are influenced by microbes, arthropods, earthworms, target and non-target plant and animal species, and humans. They must respond to economic factors, regulatory issues,

Forages: The Science of Grassland Agriculture, Volume II, Seventh Edition.
Edited by Kenneth J. Moore, Michael Collins, C. Jerry Nelson and Daren D. Redfearn.

and political policies. Systems include dynamics of energy flows, nutrient cycles, water relationships, and carbon flux. Interactions among the dynamic and changing **biotic** and **abiotic** components can be classed as competitive requiring compromise, complementary requiring optimization, or neutral.

System design should accommodate positive environmental effects, acceptable well-being of animals and the health attributes of their products that will play an increasing role in the future. If these are not satisfied, the system must be relocated to more appropriate environments or have management improvement. As human populations grow and social concerns continue to increase, the integration of forage–livestock systems with production of crop and forest products will likely increase.

Understanding and managing these integrated systems will likely depend on mathematic modeling based on global positioning systems (GPS) and use of drones or other ways to have rapid and frequent measurements of mass and quality of the grassland resource. Telemetry and GPS can provide real-time information on animal health, grazing behavior and stress levels of the animals. These data and outcomes will provide information needed for optimal output of the sustainable system.

Grazing Resources

Types of Grazed Ecologic Systems

Grazed ecologic systems, or grazing lands, can be organized into natural or imposed ecosystems and given designations that provide the basis for land-use mapping units (Allen et al. 2011). These ecosystems include **pastureland**, **cropland**, **forestland**, and **rangeland**. Within these designations, ecologic land types can be further described, including **desert**, **prairie**, **savanna**, **steppe**, **marshland**, **tundra**, **grassland**, **shrubland**, and **meadow** (see Chapter 8).

Both pastureland and cropland are imposed ecosystems. Pastureland is intended primarily for grazing by exotic and **indigenous** species, but requires management to prevent successional processes that would allow it to develop into other ecosystems (Figure 48.1). Cropland frequently offers grazing opportunities with crop residues, crops such as wheat during specific growth periods, crops that can include grazing in the harvest management strategy or for weed control.

Beef production systems in the Midwest have historically been comprised of traditional corn, soybean, and wheat cropping systems with cattle integrated into these systems based on the availability of nearby grasslands. Common strategies include grazing perennial cool-season grasses during the spring and fall, warm-season perennial grasses during the summer, along with corn residues during fall and winter or small grains during the winter and early spring. Annual forages have been used to fill forage needs during these grazing periods.

Rangeland, even when **exotic plant** species are present, is managed within the framework of a natural system (Figures 48.1). Forestland includes both native forests and specific **species** in plantations, and each may offer opportunities for grazing. Interest in **agroforestry** and **grazable forestland** is increasing as the need to intensify productivity and increase CO_2 sequestration, especially when trees are small or widely spaced allowing light in the understory to support grassland species. Landscapes frequently contain components of two or more of these **grazingland** types, and appropriate grazing systems should make use of these opportunities.

In the early 1900s, geographic distribution of livestock production was part of the mixed farm operations of crops and livestock that existed in most humid areas of the US and Canada. This allowed animal feed production and waste management via manure applications on cropland in a somewhat-closed system on the same farm. Markets for animals were small and dispersed that facilitated marketing and met personal preferences for livestock production and consumption of animal products.

In the 1960s–1980s, livestock industries became more specialized and began to shift away from population centers. Gradually, they became concentrated with large numbers away from urban areas and the associated challenges of increased regulatory and nutrient management issues in addition to environmental issues associated with dust, odors, availability of and impact on water quality. This allowed the high-value land to be concentrated on crop production (Figures 48.2 and 48.3).

The Temperate Steppe and Tropical/Subtropical Steppe ecoregions (Chapter 8) are dominated by vast areas of productive, native grasslands (Figure 48.1). These regions support approximately one-half of the beef cows and heifers and one-half of the breeding sheep in the US (Figures 48.2a and 48.3b). Conversely, only a small percentage of US dairy cows are located in these regions (Figure 48.3a). This is largely because the US-dairy industry has focused primarily on total confinement or intensively managed **pasture**-based systems. As such, about two-thirds of US dairy cows are found in the Hot Continental (regions 220 and M220), Warm Continental (regions 210 and M210), and Mediterranean (regions 260 and M260) ecoregions (see Chapter 8 for descriptions), where planted forages, primarily irrigated alfalfa are predominant.

Concentrated feedlot operations with several thousand fattening beef cattle developed during the 1970s to 2000 (Figure 48.4). In 2018, 74% of beef cattle in finishing

Fig. 48.1. Examples of grazing resources across North America. Forage crop grazing or harvest is often the optimum land use where restrictions of slope, shallow soils, poor drainage, frequent droughts, stoniness etc. limit row crop production. (a) carpon desmodium, a warm-season perennial legume, in Florida, (b) permanent grass pasture on hill land in Kentucky, (c) prairie rangeland in eastern Kansas, (d) prairie rangeland in Manitoba, Canada, (e) rangeland in northern California, and (f) rangeland in northwestern New Mexico. Rainfall is the primary limiting factor to production in many of these areas. Photo credit: Photo a courtesy of Al Kretschmer, Jr., University of Florida, photos b–f courtesy of Michael Collins, University of Missouri.

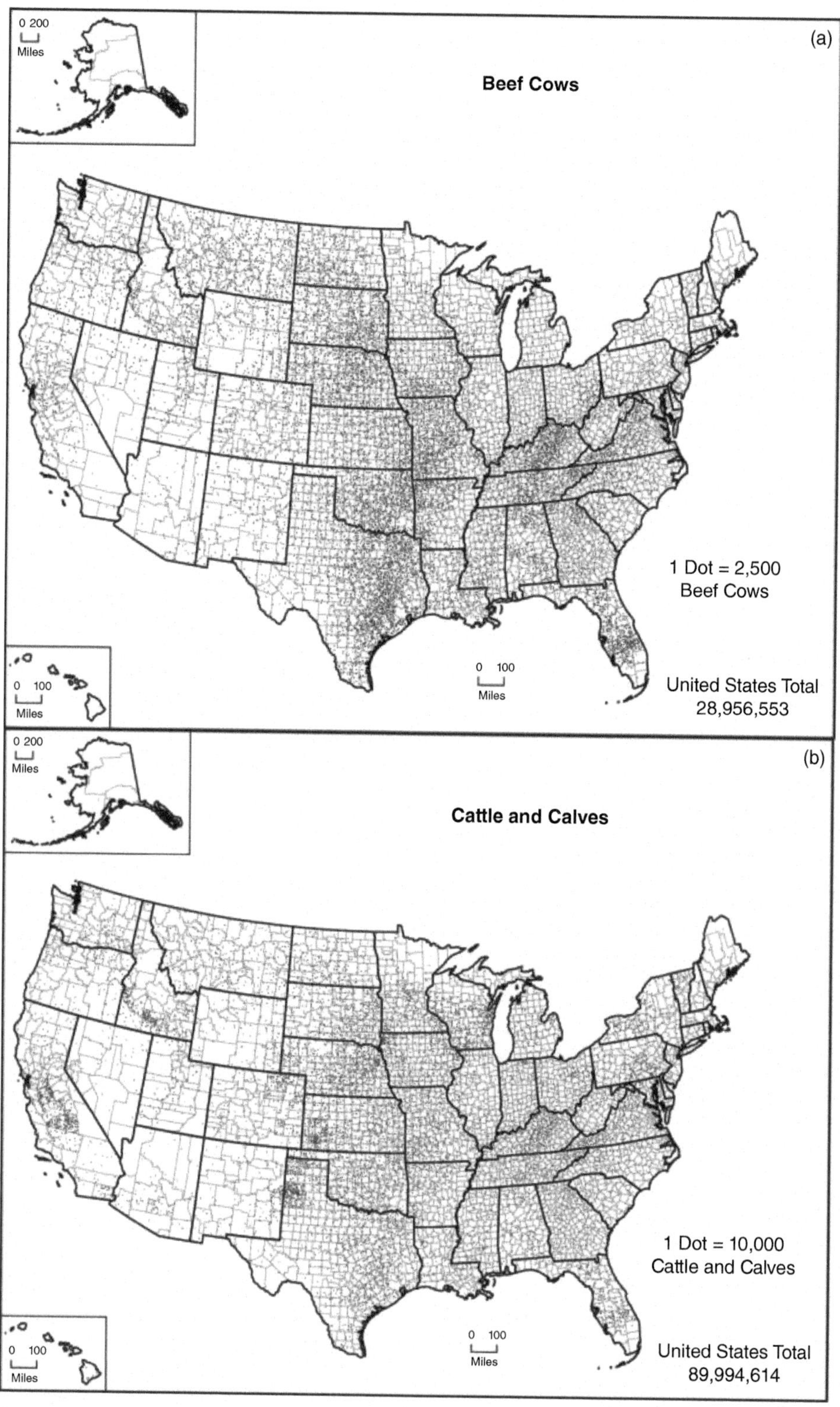

Fig. 48.2. Geographic distribution of beef cows and all cattle and calves in the United States in 2012 (USDA-NASS 2014). One dot = (a) 2500 beef cows; (b) 5000 cattle and calves. Inventory (2012 US total) = (a) 28 957 000 beef cows; (b) 89 995 000 cattle and calves.

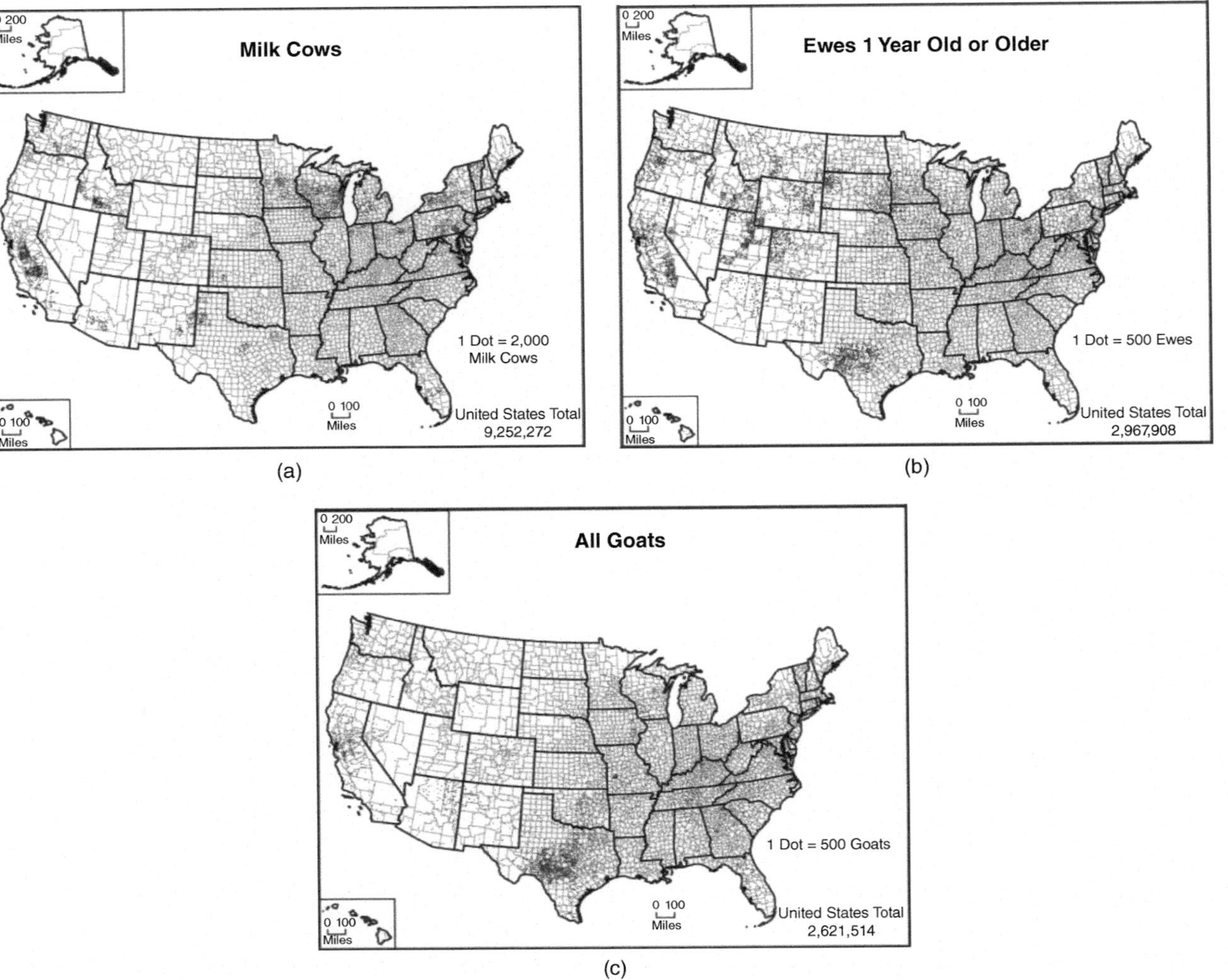

Fig. 48.3. Geographic distribution of milk cows, ewes, and goats in the US in 2012 (USDA NASS 2014). One dot = (a) 2000 milk cows; (b) 500 ewes and 500 goats. Inventory (2012 US total) = (a) 9 252 000 milk cows; (b) 2 968 000 ewes; (c) 2 600 000 goats.

Fig. 48.4. Beef and dairy cattle production systems require year-round feeding for reproductive females and seasonal feeding for weaned calves, backgrounding cattle, and finishing cattle. (a) beef cows and calves on **permanent pasture** in Kentucky, (b) grazing dairy cows arrayed behind an electrified tape while grazing high-quality pasture in Ireland, (c) yearling beef steers grazing dwarf elephantgrass, also called napiergrass, in Florida, (d) beef cattle on rangeland in Queensland, Australia, (e) finishing cattle in a beef feedlot in western Kansas, (f) a creep gate to allow calves access to high quality wheat pasture in the background while dams graze perennial grass-legume pastures in Arkansas, (g) beef cattle grazing an annual sorghum-sudangrass **hybrid** pasture in Mississippi, and (h) beef cows and calves grazing alfalfa pasture in Kentucky. Photo credit: Photo c courtesy of Lynn Sollenberger, University of Florida, photo f courtesy of Chuck West, Texas Tech University, photo h courtesy of Jimmy Henning, University of Kentucky, photos a, b, d, e, and g Michael Collins, University of Missouri.

feedlots of the US were located in Texas, Nebraska, Kansas, and Colorado (USDA NASS 1982–2018). These feedlot locations benefit from a favorable climate for maximizing **animal performance**, availability of hay and silage (**roughage**), and proximity to irrigated crop regions where feed grains are grown for finishing rations. Feedlots also benefit because nutrient and waste management are facilitated in regions of low precipitation.

Forage–Livestock Systems

Cow-Calf production

During the past 100 years the population of beef cows in the US has declined from 44.7 million in 1902 (USDA 1903) to 32.5 million in 2018 (USDA 1982–2018). Similar reductions have occurred in Canada. Over the last 25 years, populations have cycled somewhat but have been relatively stable (Figure 48.5). In the early 1900s, beef cow production in the US was concentrated in Texas and the central states of Iowa, Kansas, Nebraska, Missouri, and Illinois. Today, Texas, Nebraska and Missouri lead in numbers of beef cows, whereas the Corn Belt region of Iowa, Illinois, Indiana, and Ohio has declined dramatically primarily due to increased land values from specialization on crop production because soils and climate are very suitable for corn and soybean production.

Today, due to the land capabilities for grassland production, there are distinct areas of calf production, **backgrounding** of weaned calves and finally locations of feedlots for finishing. Currently about 45% of the US beef cow herd is located in the prairie (250) and steppe (310 and 330) regions of Texas, Oklahoma, Kansas, Nebraska, North and South Dakota, and Montana (USDA NASS 1982–2018) (Figure 48.2a). In 2018, about one-quarter of the cow herd was located in the Southern Region, which is important in calf production. After weaning, many calves are transported to other owners for use on wheat pastures and other resources to increase body size and reach weights of 360–430 kg in preparation to be sold to the feedlot operator for finishing at about 500 kg. The number of cattle on feed, mainly in feedlots in Nebraska, Kansas, and Colorado with a capacity of 1000 or more, totaled 16.1 million in 2007, 14.4 million in 2012, and 11.4 million in Oct. 2018. Specialization in finishing in large lots is made possible by transport of grain for feeding and a favorable climate for animal health and management of animal waste.

Cow–Calf Systems

Cow–calf production depends heavily on grazed forage which uses lower productivity land sites. Forage systems are generally based on perennials, but can be augmented by grazing crop residues, aftermath from harvested forages, and some **annuals** such as crabgrass and sudangrass for summer or winter wheat for winter grazing. The goal is to maintain the cow to provide and wean a healthy and vigorous calf each year to sell. Longevity of the cow is important and with good management a cow can produce 8–11 calves. Some heifer calves are raised for replacement cows.

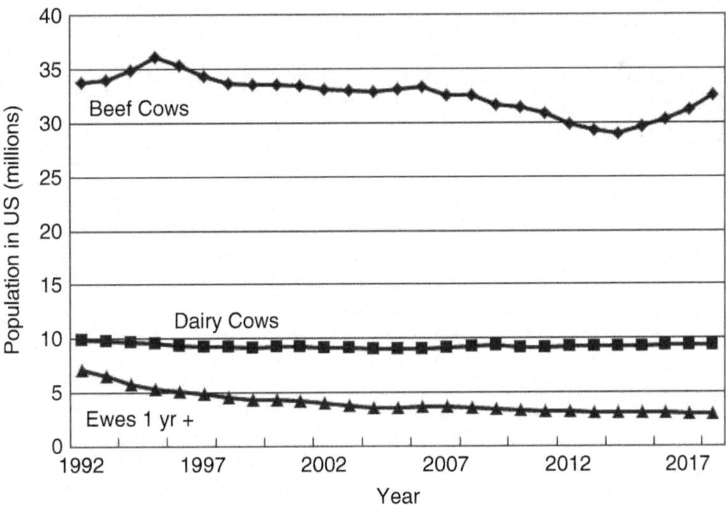

Fig. 48.5. Population trends of beef cows, dairy cows, and ewes in the US, including Hawaii and Alaska, from 1992 to 2018 (USDA NASS 1982–2018).

Corn residue grazing is an important component in many integrated production systems within the Corn Belt, especially the western Corn Belt (Schmer et al. 2017). The amount of residue is directly related to corn grain yield. For each bushel of corn yield, about 8 kg of dry leaf and husk are produced. Cattle grazing corn stalks select the grain first, then the husk, leaf, cob, and stalk. Beef cows can be wintered on corn residue with free-choice mineral supplementation. Corn residue can be used to background growing calves when supplemented with distiller's grains. It is common for cattle to successfully **graze** corn residue with 2–3 cm of snow cover.

Nutritional requirements of the brood cow vary seasonally (see Kallenbach and Collins 2018: Chapter 20 in Forages, Vol. 1). High-quality forage diets allow heifers to breed at 18 months of age and support the developing calf. Subsequent milk production and high-quality forage are critical for the growing calves to reach maximum weaning weights (Figure 48.4). In addition to forage, co-products from industries such as distiller's grain from ethanol production can **supplement** nutritional needs when forage quantity or quality is low.

Southeastern US

In southeastern states an extended grazing season is practical and near year-round grazing systems are possible. Systems are often based on warm-season **perennial** grasses such as bermudagrass and bahiagrass (Chapters 15, 18, and 22) that are generally productive from April to October. The grazing season is lengthened by sequential use of several cool-season annuals, either seeded directly into perennial sods (often bermudagrass) or planted separately to better match cow and calf nutritional requirements. In southeast Alabama, interseeding cereal rye, arrowleaf clover and crimson clovers into 'Coastal' bermudagrass in late September provided a 43% longer pasture season than when no **interseeding** was done and increased gain per ha of calves by 91% (Hoveland et al. 1978). Late winter- and early spring-born calves in Georgia were heavier at weaning when either bermudagrass or bahiagrass was interseeded with annual ryegrass in late October (Hill et al. 1985), but calves from bermudagrass systems weighed more than those from bahiagrass systems in the control and interseeding treatment. Cows grazing bahiagrass alone tended to have a lower pregnancy rate than cows on the other systems.

The eastern transition area between North and South provides opportunities to use both cool- and warm-season perennial grasses. Systems are generally based on the cool-season component, primarily tall fescue, but can include orchardgrass, bromegrasses, kentucky bluegrass, and other grasses (See Chapters 14, 16, and 20). Legumes such lespedeza, birdsfoot trefoil, white clover and especially red clover are interseeded into the grass pasture to provide N, improve forage quality and provide better seasonal distribution of production. Warm-season perennial grasses, e.g. switchgrass or big bluestem, can provide complementary forages to increase grazing opportunities in midsummer when growth of cool-season **species** declines. Because of the prevalence of tall fescue, presence (E+) or absence (E–) of the **endophyte** *Epichloë coenophiala* becomes a dominant factor due to its negative influence on calving percentages and gain.

Transition Zone

In Missouri, cow-calf pairs sequence-grazed tall fescue in spring and 'Tifleaf-1' pearlmillet during summer. Cows and calves that grazed 'Mozark' (E–) tall fescue for the 132-day season gained more weight than those grazing KY-31 (E+) (Rhodes et al. 1991). Switching from tall fescue to Tifleaf-1 pearlmillet during the summer increased calf-weaning weights by about 14 kg. Aiken et al. (2012), in Kentucky, treated endophyte-infected tall fescue-kentucky bluegrass pastures in late March or early April with 88 g ha^{-1} of amino-pyralid and metsulfuron as Chaparral™ herbicide (Dow AgroSciences; Indianapolis, IN). Treatment reduced seedhead **density** to less than 7 m^{-2} from 69 to 113 m^{-2} and improved steer average daily gain (ADG) during an April through June **grazing period** to an average of 0.93 kg $head^{-1}$ d^{-1} from a range of 0.55 to 0.79 kg for untreated pastures. Serum prolactin levels were consistently lower for steers grazing treated pastures, but overall levels were much higher in 2010, when air temperatures were higher, than in 2009 and the authors did not conclude that pasture treatment alleviated fescue toxicosis.

Great Plains

The Great Plains historically comprised much of US native grassland, including the tallgrass prairies of the Flint Hills of Kansas, the Osage Hills of eastern Oklahoma, the Sandhills of Nebraska, and the midgrass and shortgrass prairies ranging from Texas into Canada (Figure 48.1). Historically, this vast region was dominated by native grasses; however, current systems for cows and calves are based on both native and introduced forages. Cows and calves in Nebraska that sequentially grazed smooth bromegrass, switchgrass, and big bluestem had higher total seasonal gains than similar cattle grazing bromegrass for the entire season (Anderson 1988).

In the Nebraska Sandhills, about half of the ranches contain some subirrigated meadow that offers opportunities for grazing in spring (Adams et al. 1994) prior to moving to warm-season pastures in late May. Profitability was increased for cows that grazed these subirrigated meadows during winter and again during the pre-breeding season in May compared with wintering on range, meadow hay, or subirrigated meadow followed by hay during May. Weaning weights of calves were increased about 5 kg by grazing subirrigated meadow in May compared with feeding hay.

The Great Plains region also includes much of the wheat-growing area in the US, and wheat pastures have long been important in livestock grazing systems. Wheat pasture can be either "grazed out" by leaving cattle on throughout spring or "dual-purpose" by removing cattle earlier in spring to allow normal or nearly normal grain production. In central Oklahoma, grazing wheat pasture for six hours twice weekly was an effective **protein** supplement for cows and calves winter-grazing native tallgrass prairie or grazing non-native 'Plains' bluestem (Coleman et al. 2001). Plains bluestem provided 2.5 times the **carrying capacity** of the native prairie, but increased productivity was offset by increased production costs.

Replacement Heifers

Calves developed for use as replacement beef heifers should gain at a rate that achieves about 65% of their expected mature body weight (BW) by the time they are bred. For optimum performance, heifers should reach about 85% of their mature BW by the time of calving. When **replacement heifers** in Louisiana grazed cool-season annuals from November to late May, followed by common bermudagrass, and then **hybrid** millet (Humes 1973), more than 90% calved during the first 90 days of the calving season. However, only about one-half of the heifers calved if wintered on common bermudagrass or ryegrass hay plus 0.9–1.36 kg d^{-1} of cottonseed meal-salt, followed by grazing common bermudagrass–white clover pasture and then common bermudagrass during summer and autumn. In the humid region of east Texas, short periods of slowed or no growth in yearling heifers after the breeding season but with adequate forage during the period prior to calving did not affect rebreeding of first-calf heifers (Rouquette et al. 1990).

Stocker and Backgrounding Systems

Following weaning (generally at 180–270 kg), calves are typically either backgrounded as stocker cattle to reach 360–430 kg prior to entering the finishing phase or quality heifers are retained for the breeding herd (Figure 48.4). The stocker phase is generally regarded as the time after weaning to increase body size and transition to the final finishing phase. As such, the length of this period can vary dramatically. Information is not readily available on numbers and distribution of stocker cattle in the US because steers from the dairy industry and cattle imported from Mexico and Canada are also included.

Eastern US

In the eastern US, most calves are sold at weaning and leave the region (Figure 48.2a, b), often to **graze** southern Great Plains wheat pastures prior to finishing. Calves can also be backgrounded either on-farm or at collection points to capture the inexpensive gains possible when grazing high-quality forages and to improve health of yearlings prior to shipping to feedlots.

In southern Virginia, performance of stocker steers that sequence-grazed tall fescue and caucasian old world bluestem from November, after weaning, to October was greater (554 kg ha^{-1}; 0.58 kg $steer^{-1}$ d^{-1}) than in systems that sequence-grazed tall fescue with kentucky bluegrass–white clover (454 kg ha^{-1}; 0.47 kg $steer^{-1}$ d^{-1}), orchardgrass–alfalfa with kentucky bluegrass–white clover (472 kg ha^{-1}; 0.49 kg $steer^{-1}$ d^{-1}); or kentucky bluegrass–white clover with double-cropped winter rye and soybean–foxtail millet for silage (487 kg ha^{-1}; 0.51 kg $steer^{-1}$ d^{-1}) (Allen et al. 2000).

In Wisconsin, at least in years of favorable growth, there was no difference in either gain per ha or per animal in steers continuously stocked on kentucky bluegrass from 1 May to 11 September compared with steers that sequence-grazed kentucky bluegrass (1 May–15 June) then switchgrass (15 June–25 July) and back to kentucky bluegrass (15 July–11 September) (Smart and Undersander 1991).

Great Plains

The Flint Hills of eastern Kansas which includes the largest remaining remnants of the native tallgrass prairie, is historically where steers from Texas and other regions were fattened on the lush **native forage** (Figure 48.1). Steers that grazed native tall grasses from May to mid-July at a stocking rate of 0.7 ha $steer^{-1}$ gained less total weight per steer than did steers that were stocked at 1.4 ha $steer^{-1}$ from May to October because of the longer grazing season, but daily gains and gain per hectare were greater for intensive early stocking (Smith and Owensby 1978). Condition of the range was also improved by removal of steers in July. In Nebraska, steers gained 40% more over the April to November grazing season if they sequentially grazed cool- and then warm-season forages compared with season-long grazing on cool-season forages (Anderson 1988). In Oklahoma, during a 103-day grazing season, crossbred yearling steers had similar total seasonal daily gains from native forages sequence-grazed with 'Iron Master' old world bluestem compared with eastern gamagrass sequence-grazed with the same old world bluestem (Gillen et al. 1999). The eastern gamagrass system supported higher stocking rates and produced more than twice the beef per ha.

Forage System Effects on Finishing Performance and Carcass Characteristics

Dietary experience and performance of cattle at various phases of production can influence performance during later stages of production as well as carcass merit and composition (Allen et al. 1996). For example, steers finished on annual ryegrass pasture or ryegrass pasture and Coastal bermudagrass hay had higher beta-carotene concentration of ribeye steaks and ground beef than steers finished on a feedlot diet in Alabama (Simonne et al. 1996). Consumer panels could not distinguish between steaks from feedlot vs pasture-finished animals, but ground beef was preferred from steers finished on the feedlot.

In Virginia, performance and carcass characteristics of Angus steers and heifers were affected as much or more by forage consumed during the previous winter stocker phase as by the forage fed during finishing (Allen et al. 1996). Wintering cattle on stockpiled tall fescue–alfalfa or alfalfa–orchardgrass hay generally resulted in higher body weights at harvest and more desirable carcass characteristics than winter systems using tall fescue alone or with red clover.

Feeding stocker cattle in West Virginia to achieve gains of 0.23, 0.46, or 0.69 kg d^{-1} during winter had little effect on subsequent gains during finishing on forages or on meat quality (Neel et al. 2003a). Final weights were higher for steers that gained fastest over winter, but all carcasses were graded Select.

In Missouri, steers were backgrounded during spring and summer on pastures of low-endophyte tall fescue or smooth bromegrass, both interseeded with birdsfoot trefoil and alfalfa, using **rotational stocking** with 3, 12, or 24 paddocks (Terrill et al. 1994). Systems and stocking methods had no effect on daily gains during the pasture phase or on carcass characteristics or daily gain during the following feedlot phase.

Grazing continuously stocked smooth bromegrass all season or smooth bromegrass from May–June followed by sudangrass in July–August had similar cattle performance and economic returns in Nebraska (Sindt et al. 1991), but continuously stocking bromegrass from 3 May to 20 November resulted in the lowest production costs. Feeding escape protein-ionophore **supplement** increased gain on pasture and lowered production costs.

Forage Finishing Systems

The availability of ample supplies of grain has allowed development of a profitable grain-fattened beef-finishing industry in the US. This industry, along with the slaughter and meat-packing industry, is concentrated in the Great Plains (Regions 310 and 330; Chapter 8, Figure 8.2). In 2017, 32.2 million cattle were slaughtered commercially in the US (USDA NASS 1982–2018).

Though most stockers are finished in feedlots with grain as a major part of the **ration**, there is renewed interest in forage finishing systems based on consumer concerns for diet and health issues, chemical residues, antibiotic resistance, and demands for lower-fat diets (Figure 48.4). Other issues include environmental concerns, nutrient, and manure management challenges due to concentrating large numbers of livestock on feed and perceptions regarding animal care and well-being. Consumer preferences and perceptions also influence demand for different meat products in the market.

Consumers are willing to pay more for grass-fed beef, but it is more difficult to produce. Challenges for systems that depend exclusively on forage for finishing cattle include uneven seasonal distribution of high-quality forages, maintaining a consistent year-round supply of animals to processors, inconsistencies in the beef product, potentials for off flavors and undesirable color of both lean and fat, longer time required to reach an acceptable weight and grade, and access to specialized slaughter and processing facilities.

The types of forage used in finishing can influence meat quality, but the mechanisms are not well understood. For instance, beef from animals fed perennial ryegrass silage had better overall quality in terms of color, **lipid** oxidation, and alpha tocopherol levels than did beef from animals fed corn silage (O'Sullivan et al. 2002).

Intake of digestible energy (component of **digestible dry matter [DDM]**) is generally the limiting factor when finishing cattle on forage diets that tend to be bulky. Feeding supplemental energy as a **concentrate** to livestock on grazed forages can increase total voluntary intake over forage alone, but the increase is less than expected since less forage is consumed (Minson 1990). Supplementing energy has little effect on voluntary intake of high-quality forage, but it can lead to large increases in total energy intake and animal performance in cattle grazing mature or tropical forages.

Autumn-weaned angus steers in Virginia were wintered on forages and were finished on **grass** pastures without or with legumes (Allen et al. 1996). One-half of steers from each pasture system received grain supplements at 1% of body weight from July until harvest in October. The other half received no grain on pasture and were fed a finishing diet of corn silage plus a protein supplement of soybean meal (SBM) from October to harvest in January. Feeding grain on pasture increased total gain by 42 kg $steer^{-1}$ and gave a conversion rate of 6.7 kg of grain kg $gain^{-1}$. Final body weight and carcass characteristics were lowest with full-season grazing without grain followed by feeding corn silage with the best with cattle finished with grain on pasture.

High-energy corn and sorghum silages, generally supplemented with **crude protein**, have been used successfully to finish cattle. Grass silages are generally lower in energy concentration than corn or sorghum silages (Hammes et al. 1964). Daily gains and carcass grades of steers fed corn silage plus cottonseed meal were similar to those fed a high-grain diet, and both exceeded gains of steers fed grass–legume silages (Hammes et al. 1964). Reducing the moisture concentration of the grass–alfalfa silage led to increased gains and quality grades of the carcass compared with high-moisture silage.

Dairy Cattle

The number of milking cows in the US has declined from 17.1 million in 1902 (USDA 1903) to 9.3 million in 2012 (USDA-NASS 2014), but has been relatively stable for the past 25 years (Figure 48.3a). In the early 1900s, about one-third of the total US dairy herd was located in Illinois, Iowa, New York, Pennsylvania, and Wisconsin (USDA 1903). Today, dairying is concentrated in California, Wisconsin, New York, Pennsylvania, and Minnesota (Figure 48.3a), primarily due to the adaptation and use of alfalfa. Though historically important, today fewer than 10% of the US milk cows are in the humid Southeast.

The long-term trend in which milk production in California had increased nearly every year since 1970 reached 18.7 B kg in 2008 and, with minor fluctuations, leveled off at 18.0 B kg in 2017 (USDA 1982–2018). Other states, including Idaho, New York, Texas, and Wisconsin, increased total production during this period from a combined level of 25.0–32.6 B kg during the same time. Other states, including Colorado, Kansas, Indiana, Iowa, Michigan, Minnesota, South Dakota, have also increased milk production substantially during this period.

Milk production is very sensitive to heat load on the cow and depends on high-quality forage. California currently hosts over 18% of the total US herd, but it is now decreasing due to the high cost of cooling the cows and production of alfalfa is not competitive for water and land costs with nuts and fruit crops. That location is continuing to shift away, particularly to New Mexico and Idaho.

Lactating dairy cows have traditionally been fed high-concentrate diets, and as herd size of the dairy operations has increased, cows are maintained year-round in drylot, particularly in western states. These types of operations still require large quantities of silages produced locally and/or hay from local production or transported from some distance. Interest in grazing-based dairies has increased in the last 15 years, particularly for smaller herds with less investment in infrastructure and in regions with comparative advantages for producing high-quality forages and pastures.

Table 48.1 Percentage of dairy operations by type and herd size

	Herd Size (number of cows)			
Operation type	Very small (fewer than 30)	Small (30–99)	Medium (100–4)	Large (500 or more)
			%	
Conventional	17.7	55.3	77.3	93.8
Grazing	22.6	5.1	2.8	0.7
Combination of conventional and grazing	47.8	31.1	13.1	3.2
Organic	8.5	8.5	6.5	2.3

Source: USDA (2016). The "Other category" was very small and has been omitted from the table. Estimates are for the 2013 calendar year. States Surveyed: West, CA, CO, ID, TX, WA; East, IN, IA, KY, MI, MN, MO, NY, OH, PA, VT, VA, WI.

In 1941, it was estimated that grazed forage provided 75% of the feed units consumed by lactating dairy cows in New York during the 159-day average summer (Warren and Williamson 1941). However, during the last 25 years, the industry in the Northeast has relied primarily on mechanically harvested forages. In 1991, about 69% of the dairy cow ration was forage, but only about 15% was grazed (Seaney 1996).

USDA-APHIS surveyed the dairy industry in 17 northeastern and western US states representing 77% of dairy operations and 80% of the dairy cows in the country (USDA 2016) (Table 48.1). Grazing dairies, defined as those for which grazing provided most forage consumed by lactating cows, made up nearly one-quarter of the very small dairies but less than 1% of large dairies, which were 94% conventional drylot operations.

On average, of producers surveyed from the western states (Table 48.1), 66.8% were conventional, 5.1% were grazing dairies, 19.3% used a combination of these two systems, and 7.0% were organic (USDA 2016). For the 12 eastern states in the survey, percentages were 58.0, 6.9, 27.2, and 27.2, respectively.

Lactating Herd

Soil and weather conditions in the humid southeast often are not conducive for production of high-quality forages, and acceptance of grazing systems for lactating cows has been limited. The greatest opportunity for exploiting pasture may be during the cool season, when temperate species can provide high-quality herbage and heat stress is minimal. In Florida, where most dairies have large herds in conventional systems, two year-round pasture systems were compared with confined housing for lactating Holstein cows (Fontaneli et al. 2005). Pasture systems were (i) bermudagrass in summer followed by winter rye and annual ryegrass in winter or (ii) pearlmillet in summer followed by rye–ryegrass–red clover–crimson clover in winter. Cows on pasture received concentrate supplement at an average rate of 1 kg per 2.25 kg of milk produced. During a 280-day postpartum period, cows in free-stall housing produced 20% more milk (29 vs. 24 kg d^{-1}) and lost less weight and body condition than grazing cows. Pasture systems had lower costs of milk production associated with lower feed costs.

Lactating cows on cereal rye and annual ryegrass with fertilizer N or on crimson clover and red clover during winter in Florida produced similar milk yields (Macoon et al. 2011). Cows produced more milk per day when stocked at 2.5 rather than 5 cows ha^{-1} but production ha^{-1} was greater at the higher stocking rate. Forage DM intake was greater, and cows gained weight during the three-month study at the lower stocking rate, but lost weight at the high stocking rate. Daily supplementation with a mixed concentrate consisting mainly of hominy feed, soybean hulls, whole cottonseed, and citrus pulp at 0.29 or 0.4 kg kg^{-1} milk had no effect. These authors concluded that successful winter grazing systems for moderate-producing lactating cows could be achieved in Florida using N-fertilized annual grasses.

In parts of this region, pastures using rhizoma peanut, a perennial legume, are an alternative summer forage to bermudagrass. Cows grazing rhizoma peanut produced more milk, but bermudagrass pastures supported more cows per hectare and, thus, greater milk yield per ha (Fike et al. 2003). Increasing supplementation of cows from 0.33 to 0.5 kg d^{-1} per kg of milk had a greater positive effect on milk production for cows grazing bermudagrass than for those grazing rhizoma peanut because bermudagrass had lower substitution of grain for forage.

A seasonal, pasture-based dairy project in Missouri with 120 crossbred cows milking from mid-February to mid-December had a five-year average milk production of 5253 kg cow^{-1} (Horner et al. 2012). Forages included mainly endophyte-free tall fescue and perennial ryegrass with some summer annuals. Feed and labor were the major costs. Data suggested that a 150-cow unit, based primarily on cool-season, perennial pastures, could be operated economically by a typical family. However, larger units were projected to be more profitable.

Fifteen dairy farms with 37–135 cows were compared in New York before and after adoption of intensive pasturing systems (Emmick and Toomer 1991). Mean length of **grazing season** was 178 days, with an average of 0.42 ha allotted per cow. Average savings in production costs due to grazing were $153 per cow per year and $3.44 per 100 kg milk produced, indicating intensive pasture use could reduce input costs and increase overall profitability. However, the grazing season in the Northeast is typically only six to seven months, and the most desirable pasture species varies according to environment.

In Minnesota, grazing dairies tend to be small (<100 cows) with a maximum cow inventory of about 300 cows, primarily due to the need for large areas of pasture. In the primary dairy area of Minnesota, smooth bromegrass, orchardgrass, and timothy are the dominant pasture species. Legumes important for grazing include alfalfa, red and white clover, and kura clover. Most pastures are stocked rotationally with a beginning grazing height of 20–25 cm and an ending height of 6–12 cm (Loeffler et al. 1996). **First-last stocking** methods are sometimes used in which milking cows graze the upper canopy to a height of 12–20 cm after which dry cows and heifers consume the remaining forage.

Among dairy farmers in Wisconsin who have adopted management-intensive rotational stocking, farms grazed an average of 100 ha with an average of 52 cows, somewhat fewer than for the average confinement operation (Jackson-Smith and Barham 2000). Forage mixtures are commonly used, and over 80% of dairies report that legume percentage is less than 50%. Grasses, including kentucky bluegrass, smooth bromegrass, timothy, and quackgrass, provide most of the forage but legumes such as alfalfa, clovers, and birdsfoot trefoil are also used. Most managers move cows to a new paddock daily, and about 25% combine dry cows and heifers to graze paddocks following the milking herd. Most also produce forage for hay, haylage, or silage.

The trend toward increasing cow numbers for US dairies continues. A USDA-APHIS survey of 17 western and eastern states (USDA 2016) indicated that very small operations (less than 30 head) had a 24% reduction in cow numbers during the previous five years compared with 24% increase for large operations (more than 500 head) during the same period. The increase in dairies with a large inventory of cows affects the pattern of forage use. Small dairies tend to use pasture and forages produced on the farm, whereas large dairies in the West rely mainly on green chop, silage, and high-quality dry hay purchased off the farm.

Assessment of 292 small dairy farms in New York, Oregon, and Wisconsin indicated that organic dairies and conventional dairies using rotational grazing had lower incidences of cow health problems with pneumonia and mastitis (Richert et al. 2013). Of the farms included,

70.9% had 20–99 cows, 13.4% had between 100 and 199 cows, and 9.9% had more than 200 cows.

Non-lactating Cows

Dry cows should be fed a relatively low-energy **ration** with adequate but not excessive levels of protein, vitamins, and minerals. Overly fat cows are predisposed to calving problems, including displaced abomasums, dystocia, and **ketosis** (Heinrichs et al. 1996). For health and metabolic reasons, dry cows should eat 1.6–1.8 kg d^{-1} of forage dry matter per 100 kg of BW in a **ration** containing 70–90% forage. Thus, pasture systems can play a major role in feeding the dry cowherd. Grass or grass-dominated mixtures are recommended for dry cows because legumes may exceed quality needs of dry cows. On a pasture-based dairy, dry cows can effectively serve as last grazers following the lactating herd as first grazers.

Replacement Heifers

Forage intake is important in early stages of heifer growth to hasten development and function of the rumen. Grazing can be used effectively for dairy heifers between 10 and 22 months of age (Van Horn et al. 1980), thereby reducing total costs per heifer calved, including feed, labor, and vaccines. A replacement heifer must gain about 0.7 kg d^{-1} to calve at 24 months, placing greater demands on grazed forage supply and high quality.

Holstein heifers in Michigan were rotationally stocked using four paddocks (10- to 12-days grazing periods) or 12 paddocks (3- to 5-days grazing periods) on smooth bromegrass–alfalfa, smooth bromegrass–birdsfoot trefoil, and smooth bromegrass grown alone with 224 kg N ha^{-1} applied per season. The stocking rate averaged 3.7, 180–230 kg heifers ha^{-1} (Barclay et al. 1996). Grazing began in mid-May and lasted 134–155 days. Daily gains on the bromegrass-N treatment were lower than the grass–legume mixtures using any number of paddocks. Averaged across number of paddocks, daily gains for bromegrass with birdsfoot trefoil or alfalfa were 1.07 and 1.02 kg d^{-1}, respectively.

In Florida, 'Callie' bermudagrass pastures were grazed by 250-kg holstein heifers during summer (Mathews et al. 1994). Heifers received 1.5 kg d^{-1} of corn–SBM, pastures were stocked continuously or rotationally (3 or 15 paddocks). **Grazing method** had little effect on calf daily gain, which averaged 0.5 kg d^{-1}, which was below the target gain for developing replacement heifers. In a subsequent Florida study, 225-kg holstein calves received no supplement and gained 0.63 and 0.58 kg d^{-1} on 'Florigraze' and 'Arbrook' rhizoma peanut, respectively (Hernández Garay et al. 2004). Gains approaching 1 kg d^{-1} were achieved with 350-kg yearling beef heifers grazing rhizoma peanut. **Overseeding** warm-season perennials in the Southeast with cool-season forages such as oat, rye, ryegrass, and clovers provides opportunity for excellent animal performance during the cool season.

Sheep Systems

The US sheep industry has declined dramatically during the past century from about 64 million head in 1902 (USDA 1903) to 5.4 million sheep and lambs in 2012 (USDA NASS 2014). Much of the decrease occurred since World War II as numbers declined from over 55 million head of ewes and lambs to 3.0 million breeding females today (Figure 48.2b). Currently, the largest populations are centered in Texas, California, Colorado, Wyoming, Utah, South Dakota, Idaho, and Montana.

Primary reasons for decreases are increased predatory death losses and a loss of market demand for sheep meat products and wool. However, with developing ethnic markets due to immigration the demand for meat products is expected to increase.

Ewes and Lambs

Approximately 15% of US breeding sheep are located in northern and east central states (Figure 48.2b). Production systems are largely dependent on a mix of planted pastures, **crop residues**, harvested forages, and grain concentrates (Figure 48.6).

In southwest Virginia, Notter et al. (1991) compared systems for ewes and lambs by (i) grazing a kentucky bluegrass–white clover mix from early spring until autumn weaning; (ii) grazing the same kentucky bluegrass–white clover mix from early spring to May, when lambs were weaned and then allowed to rotationally graze either alfalfa or ladino clover until autumn; and (iii) feeding a high-concentrate diet in drylot from early spring until autumn. Lambs from system i had the lowest ADG, were leanest, and weights among lambs were most variable. In contrast, lambs in system iii had the highest ADG and were fattest. Lambs from system ii were intermediate, with no distinguishable effect between grazing alfalfa or ladino clover.

In Minnesota, performance of ewes and lambs rotationally grazing a five-paddock system (7-day graze, 28-day **rest**) of either alfalfa or smooth bromegrass was equal during the first 15 grazing days (12–27 June), but thereafter (28 June–28 July) weight gain of lambs grazing alfalfa was superior to those grazing bromegrass. In Virginia, cows and ewes with their offspring grazed together or separately as cow-calf pairs (6 per pasture) or ewe-lamb units (6 ewes and 11 lambs) on kentucky bluegrass–white clover pastures, from April to October (Abaye et al. 1994). Lambs from **mixed stocking** reached target weights for weaning before those that grazed alone, and ewes were in better condition at breeding on mixed-stocked pastures (Figure 48.6). Calf weights at weaning, 212 kg average, were not affected by the presence of sheep. White clover

Fig. 48.6. Sheep, goats, horses, and other herbivores are important in many areas for meat, milk, fiber, weed management, and recreation. Sheep grazing permanent pasture in Kentucky (a) Mixed stocking, photos (b and e), takes advantage of differences in grazing behavior of different animal species. Goats, photo (c), readily graze forage grasses and legumes but also browse some weed species avoided by sheep and cattle and can help maintain a more desirable botanical composition in the pasture. Photo (d) illustrates goat browsing of sumac and pine that helps to control these species in pastures. Photo (e) illustrates that "lawns and roughs" often created by horses grazing closely in some areas while avoiding other areas of the pasture. Mixed stocking with cattle, which graze more uniformly, can make more complete use of the available herbage and reduce the frequency of mowing needed to maintain horse pastures. Millions of bison, (photo F), grazed North American prairies and their range extended into the Appalachian region. Photo credit: Photo b courtesy of Chuck West, Texas Tech University, photos a, c, d, e, and f courtesy of Michael Collins, University of Missouri.

was more abundant where cattle grazed alone than where sheep grazed alone, but sheep reduced weeds in pastures. **Mixed stocking** did not increase total animal production per ha.

Approximately one-third of US breeding sheep flocks are located in the northern Great Plains and Intermountain Regions, with concentrations in South Dakota, Wyoming, Montana, Idaho, and Colorado (Figure 48.2b). The dominant forage base is native rangeland. Native rangeland and planted forages together provide about two-thirds of the nutrients annually with the remainder provided by hay, concentrates, and small amounts of **crop residues** that are fed when grazing is not feasible (Shapouri 1991). **Continuous stocking** is commonly used with some variation of rotational stocking that may or may not include herding (Glimp and Swanson 1994).

Planted forages play less of a role in this region than in the eastern US. For example, in Montana, Thomas et al. (1995) evaluated the merits of continuous (one herd, one pasture) and rotational (one herd, eight pastures) stocking of a mixed stand of alfalfa and orchardgrass from mid-May to mid-August. Grazing method had only minor effects, whereas gains per animal and per ha were well above those expected from grazing adjoining rangeland.

About 15% of the US breeding flocks are located in the Southern Plains and Mountain Regions, with the greatest proportion in Texas (Figure 48.2b). The dominant forage base is native rangeland, with approximately 80% of annual nutrients derived from grazing and the remaining 20% derived from hay and concentrates (Shapouri 1991). The longer grazing season in this region allows year-round grazing if sufficient forage is available.

The Great Basin and Intermountain Regions contribute about 12% of the US breeding ewe flocks (Figure 48.2b) where federal rangelands are an important forage base (Shapouri 1991). A typical grazing system consists of wintering and lambing on the lower elevation, shrub-dominated, salt desert floors. In spring, the sheep are slowly herded through mid-elevation, sagebrush grasslands to higher elevation summer pastures before returning in autumn (American Sheep Industry Association 1997).

Finishing Lambs

The primary states for production of market lambs are Colorado, California, and Texas, whereas the primary slaughter facilities are in Colorado, Iowa, California, and Texas. Post-weaning finishing systems vary from a mix of grazed forages and supplemental grain concentrates to full feeding of harvested forages and grain concentrates in drylots.

About 10% of all US market lambs reside in the North and East Central Regions, where grazed forages have historically played a major role in systems for lamb production from weaning to slaughter. Murphy et al. (1994) examined finishing systems in Ohio, designed to add about 20 kg to lambs initially weighing 29 kg. Grazing rotationally stocked ryegrass for 42 days followed by concentrate feeding (113 days) gave the highest daily gain whereas grazing alfalfa produced leaner animals. Productive capacity of *Brassica* forages is high (Reid et al. 1994) but lamb performance is highly variable, possibly because *Brassicas* have inherent variation in concentration of metabolic inhibitors (Cox-Ganser et al. 1994).

Most lambs in the western US are finished using forage–grain concentrate diets but, in some cases, forages play a key role in finishing systems. For example, in Wyoming, animal gains and key carcass traits from autumn grazing of several irrigated annual forages, including turnips, tyfon, rape, sugar beets and radishes, were all equal to those of animals fed a forage–grain concentrate diet in drylot (Koch et al. 2002).

Goat Systems

In 2012, 92% of all sheep and goat operations in the US were family or individually owned. As with sheep, Texas and California are also among the leading states in goat production (Figure 48.2c). Southeastern states, including Tennessee, Missouri, Georgia, North Carolina, and Kentucky are also important in goat production, but do not have large sheep herds (USDA NASS 2014). Total US goat inventory declined from 3.14 million in 2007 to 2.62 million in 2012. Of the latter, 2.05 million were meat goats.

Small scale goat operations, with less than 500 goats, made up 99.7% of all goat farms in 2007 (USDA-NASS 1982–2018). Two-thirds of small-scale producers of meat goats had been in operation 10 years or less (USDA-APHIS 2011). Conversely, about two-thirds of producers of fiber goats had 10 years or more of experience. Many small producers were unaware that several diseases of goats and sheep can be **transmitted** to humans. Even so, about 16% of the goats are milking goats which is likely to increase since there is a growing market for milk for infants and others who are intolerant to lactose in cow milk.

Dry matter (DM) intake by goats is less affected by acid detergent fiber (ADF) level than that of sheep or cattle (Riaz et al. 2014). Goats are actually browsers and often eat brush and twigs of trees (Figure 48.6). Webb et al. (2011) found that **mixed stocking** of goats with cattle on reclaimed mine land in Virginia reduced branch length of autumn olive and botanic contribution of sericea lespedeza compared with cattle grazing alone. These authors suggested stocking one to three goats per **animal unit** of cattle in this system. Krueger et al. (2014) found that grazing goats alone or with cattle reduced blackberry leaf and stem **density** on rhizoma peanut-grass pastures in Florida. When female goats grazed tall fescue-clover pastures alone or with dairy heifers in Indiana, goats maintained their weight while reducing weed presence (Dennis et al. 2015). Heifer seasonal gains were not affected but were lower than typically seen for confinement-fed heifers regardless of treatment.

Integrated Production Systems

Crop and Livestock Systems

Historically, integration of livestock and cropping farms characterized US and Canadian agriculture (Figure 48.7). Regions of specialization within ideal climates led to development of monocultures for crop production to maximize output and profitability. Livestock production shifted regionally to areas not as suitable for crops, except concentrated animal feeding facilities developed in or near regions that emphasize mainly feed grains. Benefits included economies of scale and regional comparative advantages for specialized industries.

Today, however, the challenges of public concerns about climate change, animal welfare and sustainability are increasing. Issues such as environmental quality, pest management, use of nonrenewable natural resources including water and energy, and issues of biosecurity have reawakened interest in more diversified farms and landscapes (Sulc and Tracy 2007). Benefits of integrated cropping and livestock forage systems include improved nutrient management, increased soil organic matter, improved soil chemical, physical, and biologic properties, reduced soil erosion, suppression of weeds, insects, and disease cycles, and diversified income sources (Figure 48.7a–f).

Two crop-livestock systems as simulated at the farm level consisting of corn, alfalfa hay, and pastures for stocker steers were compared in Virginia (Luna et al. 1994). In one system, corn for silage and alfalfa for hay were each grown for 5 years in a 10-year, two-paddock rotation. The control system of tall fescue with N fertilizer provided stockpiled fall and winter grazing for steers and the same grass with red clover provided hay and additional grazing. Steer grazing and corn grain production were integrated into the alfalfa-corn rotation followed by wheat and foxtail millet and two years of alfalfa.

The integrated system improved daily gains of steers, required less N fertilizer within the overall system, and used fewer pesticides and herbicides, but more hay feeding was required. Steers from the more integrated system were heavier following finishing but both systems had similar profitability.

In the semiarid High Plains of west Texas where cotton is widely grown in monoculture, an integrated cotton–forage system used 25% less irrigation water, required 36% less N fertilizer, lowered inputs of pesticides and herbicides, and improved profitability compared with the cotton monoculture system (Allen et al. 2012). One-half of the integrated system was 'WW-B. Dahl' old world bluestem, which was grazed and then harvested for **seed** late in the season, with the other half divided equally into two paddocks where cotton was grown in alternate rotation with wheat and rye.

Over 10 years, cotton lint yields did not differ between the monoculture and the no-till planted cotton in the integrated system. However, year-to-year variation was greater for cotton in the integrated system, possibly because of late season rainfall on the no-till cotton. Steers grazed 185 days with average gains of 0.79 kg d^{-1}. The integrated system reduced soil erosion potential (Collins 2003) and enhanced soil microbial activity and microbial C (Acosta-Martínez et al. 2004, 2010).

No-till planting cotton into the grazed rye provided protection to young cotton plants from hail and blowing sand. Reduced N fertilization saved energy for both the manufacture of N fertilizer and its transport and application (Whitehead 1995). Profitability of the two systems was similar initially, but was greater for the monoculture after higher yielding cotton cultivars were introduced into the study. However, as irrigation water became more limiting, the integrated system became increasingly more profitable (Johnson et al. 2013).

Sulc and Tracy (2007) identified constraints to adoption of integrated systems including the tradition of single enterprise farming, ease of management of simple systems, additional managerial and labor input for diversified systems, a lack of appreciation among producers of system-level performance of integrated systems, and limited incentives for the diversity and environmental conservation benefits of integrated systems.

Silvopastoral Systems

Integrating livestock grazing, forage production, and management of woodland resources is becoming more important to maximize resource use and improve profitability. Issues of income diversification, greater **carbon sequestration**, improved wildlife habitat, and enhanced recreational opportunities add to interest in agroforestry and **silvopasture** systems (Figure 48.7f).

In Arkansas, a hay crop of mixed-forage species yielded 5500 kg ha^{-1} in a cropping system with clustered loblolly pine trees grown for saw timber that were surrounded by slower growing "nurse" trees of shortleaf pine (Francis et al. 2003). The hay crop was a mixture of tall fescue, dallisgrass, common bermudagrass, and bahiagrass, and was interseeded with white clover. When trees were tall enough to withstand grazing, cattle were introduced. Prior to harvest of the saw timber, the sale of cattle, hay, and pulpwood from the nurse trees provided a diverse source of income.

(a) (b) (c) (d) (e) (f)

Fig. 48.7. Forages and livestock are integrated into the agricultural landscape even in areas dominated by row crop production systems. Photo (a) illustrates strip intercropping using corn and alfalfa planted on the contour. Corn, in rotation with alfalfa, utilizes residual fixed-N to produce grain and silage that are important in dairy rations. The perennial component, alfalfa, produces high quality forage and reduces soil erosion from tilled areas of the field. Photo (b) illustrates the use of sloping areas on the farm as permanent pasture while suitable areas are used for row crop production. Photo (c) illustrates this pattern on a larger scale with woodlots, forages, and row crops all visible across the landscape. **Crop residues** following corn harvest, photo (d), provide low-cost nutrients for beef brood cows during fall and winter. Photo (e) illustrates a crop rotation system involving corn, soybean, and red clover forage. Photo (f) illustrates a silvopasture system in which corn has been planted between tree rows in a pine plantation for grazing by cattle. Photo credit: Photo b, Kenneth Moore, Iowa State University, photos a, c, d, e, and f, Michael Collins, University of Missouri.

The Appalachian Region contains numerous small farms composed of open pastures and woodlots, but forage quality under trees may not be as high as that grown in full sunlight. However, due to less temperature and drought stress, some forage grown under hardwoods may be of higher quality than that grown in open pastures. In West Virginia, farm income could be increased by using forages grown in lightly shaded wooded areas. Lambs grazing within a hardwood silvopastoral system performed similarly to those grazing in open pasture (Neel et al. 2003b).

Most dairy farmers in New York who had woodland reported at least some management of the resource that included sugarbush or tree planting for establishing **shelterbelts** and hedgerows, planting fencelines and roadsides, and establishing a variety of valuable timber species (Teel and Lassoie 1991).

Summary

Livestock grazing systems are defined by multiple factors, including climate, soil, and forage resources, type of livestock enterprise(s), proximity to and timing of markets, and environmental concerns. Grazing systems are also shaped by the lifestyle, timing of goals and both plant and animal experience of the manager. Within systems, an array of different stocking methods (Chapter 44) can be employed to optimize animal performance, to meet their varying nutritional needs, and to manage the forage resource for yield, quality, and persistence.

Though the "system" is an integration of soil, plant, animal, environment, economic, and other factors, most research currently focuses on animal and plant performance with major attention to profitability. Future research must focus on understanding the dynamics of the interrelationships of a greater number of components and outcomes from grazing systems. This will require long-term studies by multidisciplinary teams and likely modeling approaches to better understand the direct and indirect effects of management options.

Few such opportunities exist under present funding sources and current abilities to commit resources and researchers over long time periods. It is only through such research, however, that needed information can be generated to understand the principles and dynamics of grazing systems and to apply this knowledge to address issues of growing national and global concern. These issues include protection of soil, water, air and other natural resources. Healthy food is needed to feed a growing global population in a sustainable way that provides a level of profitability that adequately supports an agricultural population and a healthy national economy.

Acknowledgment

The authors gratefully acknowledge the contributions of Drs. Vivien Gore Allen, Rodney K. Heitschmidt and Lynn E Sollenberger to the 6th Edition of Forages, The Science of Grassland Agriculture.

References

Abaye, A.O., Allen, V.G., and Fontenot, J.P. (1994). Influence of grazing cattle and sheep together and separately on animal performance and forage quality. *J. Anim. Sci.* 72: 1013–1022.

Acosta-Martínez, V., Zobeck, T.M., and Allen, V. (2004). Soil microbial, chemical, and physical properties in continuous cotton and integrated crop-livestock systems. *Soil Sci. Soc. Am. J.* 68: 1875–1884.

Acosta-Martínez, V., Bell, C.W., Morris, B.E.L. et al. (2010). Long-term soil microbial community and enzyme activity responses to an integrated cropping–livestock system in a semi-arid region. *Agric. Ecosyst. Environ.* 137: 231–240.

Adams, D.C., Clark, R.T., Coady, S.A. et al. (1994). Extended grazing systems for improving economic returns from Nebraska sandhills cow/calf operations. *J. Range Manage.* 47: 258–263.

Aiken, G.E., Goff, B.M., Witt, W.W. et al. (2012). Steer and plant responses to chemical suppression of seedhead emergence in toxic endophyte-infected tall fescue. *Crop Sci.* 52: 960–969.

Allen, V.G., Fontenot, J.P., Kelly, R.F., and Notter, D.R. (1996). Forage systems for beef production from conception to slaughter: III. Finishing systems. *J. Anim. Sci.* 74: 625–638.

Allen, V.G., Fontenot, J.P., and Brock, R.A. (2000). Forage systems for production of stocker steers in the upper south. *J. Anim. Sci.* 78: 1973–1982.

Allen, V.G., Batello, C., Berretta, E.J. et al. (2011). An international terminology for grazing lands and grazing animals. *Grass Forage Sci.* 66: 2–28.

Allen, V.G., Brown, C.P., Kellison, R. et al. (2012). Integrating cotton and beef production in the Texas southern High Plains: I. water use and measures of productivity. *Agron. J.* 104: 1625–1642.

American Sheep Industry Association (1997). SPH forages chapter. In: *In Sheep Production Handbook*, 101–132. Denver, CO: C & M Press.

Anderson, B. (1988). Sequential grazing of cool-warm-cool season perennial grasses: the concept and practice. In: *Proceedings of Innovative Grazing Systems: A Joint Symposium*. AFGC, April 11–14, 1988. Baton Rouge, LA, 274–281. Belleville, PA: American Forage and Grassland Council.

Barclay, D.J., Moline, W.J., and Richie, H.D. (1996). Two grazing systems and three forage species for rotational grazing in Northern Michigan. *Proc. Am. Forage Grassl. Conf.* 5: 100.

Coleman, S.W., Phillips, W.A., Volesky, J.D., and Buchanan, D. (2001). A comparison of native tallgrass prairie and plains bluestem forage systems for cow-calf production in the Southern Great Plains. *J. Anim. Sci.* 79: 1697–1705.

Collins, J. (2003). Agricultural phosphorus in an integrated crop/livestock system in the Texas High Plains. Master thesis. Texas Tech University.

Cox-Ganser, J.M., Jung, G.A., Pushkin, R.T., and Reid, R.L. (1994). Evaluation of brassicas in grazing systems for sheep: II. Blood composition and nutrient status. *J. Anim. Sci.* 72: 1832–1841.

Emmick, D.L. and Toomer, L.F. (1991). The economic impact of intensive grazing management of fifteen dairy farms in New York state. In: *Proceedings of American Forage and Grassland Conference*, Columbia, MO, April 1–4, 1991, 10–22. Georgetown, TX: AFGC.

Fike, J.H., Staples, C.R., Sollenberger, L.E. et al. (2003). Pasture forages, supplementation rate, and stocking rate effects on dairy cow performance. *J. Dairy Sci.* 86: 1268–1281.

Fontaneli, R.S., Sollenberger, L.E., Littell, R.C., and Staples, C.R. (2005). Performance of lactating dairy cows managed in pasture-based or in freestall barn feeding systems. *J. Dairy Sci.* 88: 1264–1276.

Francis, P.B., Colvin, R.J., and Zeide, B. (2003). Forage production in agroforestry systems with structured tree clusters. *Proc. Am. Forage Grassl. Conf.* 12: 293–297.

Gillen, R.L., Berg, W.A., DeWald, D.L., and Sims, P.L. (1999). Sequence grazing systems on the southern plains. *J. Range Manage.* 52: 583–589.

Glimp, H.A. and Swanson, S.R. (1994). Sheep grazing and riparian and watershed management. In: *The Role of Sheep Grazing in Natural Resource Management*. Sheep Res. J. Spec. Issue: 1994 (ed. M. Shelton), 65–71. Englewood, CO: American Sheep Industry Association.

Hammes, R.C. Jr., Fontenot, J.P., Bryant, H.T. et al. (1964). Value of high-silage rations for fattening beef cattle. *J. Anim. Sci.* 23: 795–801.

Heinrichs, A.J., Ishler, V.A. and R.S Adams. 1996. Feeding and managing dry cows. Pennsylvania State Extension Circular No. 372.

Hernández Garay, A., Sollenberger, L.E., Staples, C.R., and Pedreira, C.G.S. (2004). Florigraze and Arbrook rhizoma peanut as pasture for growing Holstein heifers. *Crop Sci.* 44: 1355–1360.

Hill, G.M., Utley, P.R., and McCormick, W.C. (1985). Evaluation of cow-calf systems using ryegrass sod-seeded in perennial pastures. *J. Anim. Sci.* 61: 1088–1094.

Horner, J., Milhollin, R. and Prewitt, W. (2012). Economics of pasture-based dairies. University of Missouri Extension Bulletin No. M168.

Hoveland, C.S., Anthony, W.B., McGuire, J.A., and Starling, J.G. (1978). Beef cow-calf performance on Coastal bermudagrass overseeded with winter annual clovers and grasses. *Agron. J.* 70: 418–420.

Humes, P.E. (1973). Pasture management programs for replacement beef heifers. *Louisiana Agric.* 16 (4): 3.

Jackson-Smith, D. and Barham, B. (2000). The changing face of Wisconsin dairy farms. University of Wisconsin. Program on Agricultural Technology Studies Research Report No. 7.

Johnson, P., Zilverberg, C.J., Allen, V.G. et al. (2013). Integrating cotton and beef production in the Texas southern High Plains: III. An economic evaluation. *Agron. J.* 105: 929–937.

Kallenbach, R.L. and Collins, M. (2018). Grazing management systems. In: *Forages – An Introduction to Grassland Agriculture*, 7e, vol. I (eds. M. Collins, C.J. Nelson, K.J. Moore and R.F Barnes), 341–356. Wiley.

Koch, D.W., Kercher, C., and Jones, R. (2002). Fall and winter grazing of Brassicas: A value-added opportunity for lamb producers. *Sheep Goat Res. J.* 17: 1–13.

Krueger, N.C., Sollenberger, L.E., Blount, A.R. et al. (2014). Mixed stocking by cattle and goats for blackberry control in rhizome peanut–grass pastures. *Crop Sci.* 54: 2864–2871.

Loeffler, B., Murray, H., Johnson, D.G. et al. (1996). Knee deep in grass: A survey of twenty-nine grazing operations in Minnesota. University of Minnesota Extension Service Bulletin No. BU-06693.

Luna, J., Allen, V., Fontenot, J. et al. (1994). Whole farm systems research: An integrated crop and livestock systems comparison study. *Am. J. Alter. Agric.* 9: 57–63.

Macoon, B., Sollenberger, L.E., Staples, C.R. et al. (2011). Grazing management and supplementation effects on forage and dairy cow performance on cool-season pastures in the southeastern United States. *J. Dairy Sci.* 94: 3949–3959.

Mathews, B.W., Sollenberger, L.E., and Staples, C.R. (1994). Dairy heifer and bermudagrass pasture responses to rotational and continuous stocking. *J. Dairy Sci.* 77: 244–252.

Minson, D.J. (1990). *Forage in Ruminant Nutrition*. San Diego: Academic Press.

Murphy, T.A., Loerch, S.C., McClure, K.E., and Solomon, M.B. (1994). Effects of grain or pasture finishing systems on carcass composition and tissue accretion rates of lambs. *J. Anim. Sci.* 72: 3138–3144.

Neel, J.P.S., Clapham, W.M., Fontenot, J.P., and Lewis, P.E. (2003a). Animal performance and carcass characteristics of pasture-finished steers as influenced by winter gain. *Proc. Am. Forage Grassl. Conf.* 12: 313–317.

Neel, J.P.S., Clapham, W.M., Fontenot, J.P., and Lewis, P.E. (2003b). Forage nutritive value and performance of lambs in a silvo-pastoral system. *Proc. Am. Forage Grassl. Conf.* 12: 303–307.

Notter, D.R., Kelly, R.F, and McClaugherty, F.S. (1991). Effects of ewe breed and management system on efficiency of lamb production: II. Lamb growth, survival and carcass characteristics. *J. Anim. Sci.* 69: 22–33.

Odum, E.P. (1971). *Fundamentals of Ecology*, 3e. Philadelphia: W.B. Saunders.

O'Sullivan, A., O'Sullivan, K., Galvin, K. et al. (2002). Grass silage versus maize silage effects on retail packaged beef quality. *J. Anim. Sci.* 80: 1556–1563.

Purcell, W. (1977). A systems view of the beef industry. In: *Forage-Fed Beef: Production and Marketing Alternatives in the South*. Bull. 220 (eds. J. Stuedemann, D. Huffman, J. Purcell and O. Walker), 1. Athens, GA: Southern Cooperative Series.

Reid, R.L., Puoli, J.R., Jung, G.A. et al. (1994). Evaluation of Brassicas in grazing systems for sheep: I. Quality of forage and animal performance. *J. Anim. Sci.* 72: 1825–1830.

Rhodes, M.T., Grigsby, K.N., Larson, B.T. et al. (1991). Influence of rotation from low- or high-endophyte tall fescue to pearl millet on cow-calf performance. In: *Proceedings of American Forage and Grassland Conference* Columbia, MO, April 1–4, 1991, 141–144. Georgetown, TX: AFGC.

Riaz, M.Q., Südekum, K.-H., Clauss, M., and Jayanegara, A. (2014). Voluntary feed intake and digestibility of four domestic ruminant species as influenced by dietary constituents: A meta-analysis. *Livestock Sci.* 162: 76–85.

Richert, R.M., Cicconi, K.M., Gamroth, M.J. et al. (2013). Risk factors for clinical mastitis, ketosis, and pneumonia in dairy cattle on organic and small conventional farms in the United States. *J. Dairy Sci.* 96: 4269–4285.

Rouquette, F.M. Jr., Florence, M.J., Long, C.R. et al. (1990). Growth and development of F-1 (Brahman × Hereford) heifers under various short-term grazing pressures. In: *Forage and Livestock Research— 1990*. Research Center Technical Report 90-1, 166. Overton: Texas Agriculture Experiment Station.

Schmer, M.R., Brown, R.M., Jin, V.L. et al. (2017). Corn residue use by livestock in the United States. *Agric. Environ. Lett.* 2: 160043. https://doi.org/10.2134/ael2016.10.0043.

Seaney, R.R. (1996). Pasture production and management in New York. Cornell University Beef Production Reference Manual. Fact Sheet 2200.

Shapouri, H. (1991). Sheep production in 11 western states. Economic Research Service, USDA. Staff Report No. AGES 9150.

Simonne, A.H., Green, N.R., and Bransby, D.I. (1996). Consumer acceptability and beta-carotene content of beef as related to cattle finishing diets. *J. Food Sci.* 61: 1254–1256.

Sindt, M., Klopfenstein, T., and Stock, R. (1991). Production systems to increase summer gain. In: *Proceedings of American Forage and Grassland Conference*. Columbia, MO, April 1–4, 1991, 265–268. Georgetown, TX: AFGC.

Smart, A. and Undersander, D. (1991). Cattle performance on rotated pastures of switchgrass and Kentucky bluegrass. In: *Proceedings of American Forage and Grassland Conference* Columbia, MO, April 1–4, 1991, 248–251. Georgetown, TX: AFGC.

Smith, E.F. and Owensby, C.E. (1978). Intensive-early stocking and season-long stocking of Kansas Flint Hills range. *J. Range Manage.* 31: 14–17.

Sulc, R.M. and Tracy, B.F. (2007). Integrated Crop–Livestock Systems in the U.S. Corn Belt. *Agron. J.* 99: 335–345.

Teel, W.S. and Lassoie, J.P. (1991). Woodland management and agroforestry potential among dairy farmers in Lewis County, New York. *For. Chron.* 67: 236–242.

Terrill, J.S., Martz, F.A., Morrow, R.E., and Gerrish, J.R. (1994). Finishing beef cattle backgrounded on different forage systems. *Proc. Am. Forage Grassl. Conf.* 3: 223.

Thomas, V.M., Kott, R.W., and Ditterline, R.W. (1995). Sheep production response to continuous and rotational stocking on dryland alfalfa/grass pasture. *Sheep Goat Res. J.* 11: 122–126.

USDA (2016). Dairy cattle management practices in the United States, 2014. United States Department of Agriculture, Animal and Plant Health Inspection Service.

USDA (United States Department of Agriculture) (1903). *Yearbook of the United States Department of Agriculture-1902*. Washington, DC: U.S. Government Publishing Office.

USDA (United States Department of Agriculture)-APHIS (2011). *Small-Scale U.S. Goat Operations*. Fort Collins, CO: USDA-Animal and Plant Health Inspection Service.

USDA NASS (1982–2018). *Agricultural Statistics*. Washington, DC: U.S. Government Publishing Office.

USDA-NASS (United States Department of Agriculture) (2014). Census of Agriculture 2012. https://www.nass.usda.gov/AgCensus (accessed 14 October 2019).

Van Horn, J.J., Marshall, S.P., Floyd, G.T. et al. (1980). Complete rations for growing dairy replacements utilizing by-product feedstuffs. *J. Dairy Sci.* 63: 1465–1474.

Warren, S.W. and Williamson, P.S. (1941). The cost of pasture. Cornell University. Agricultural and Economic Publication No. 356.

Webb, D.M., Abaye, A.O., Teutsch, C.D. et al. (2011). Mixed grazing goats with cattle on reclaimed coal mined lands in the Appalachian Region: effects on forage standing biomass, forage botanical composition and autumn olive (*Elaeagnus umbellata* Thunb.). *Sheep Goat Res. J.* 26: 26–31.

Whitehead, D.C. (1995). *Grassland Nitrogen*. Guildford, UK: CAB International, Biddles Ltd.

Common and Scientific Names of Forages

The common and scientific names of grasses, legumes, and other plants referred to in this book are given below. Common names are listed alphabetically followed by the scientific name. Many forage plants are known by more than one common name, but scientific names are unique and therefore unambiguous. The first part of the scientific name, the genus, is always capitalized; the second part, the species epithet, is written entirely in lowercase. The scientific name is completed by including the abbreviation of the name of the person or authority who named the species. Multiple authorities may be listed for a species. If two authorities are cited, the first to have classified the species is enclosed in parentheses and the most recent to have revised the classification is not. When two authorities are separated by "ex," the first name corresponds to the person who originally described the species and the second is the one who formally published the name.

The following abbreviations and symbols are used: syn. = synonym, in which case there are two accepted scientific names; X before the genus name indicates a hybrid between two genera; x before the species epithet or subspecies name refers to a hybrid between two species or two subspecies.

Scientific names continue to change as new technologies for classification are developed. The wheatgrasses and wildryes provide an excellent example of the ever-evolving nature of plant taxonomy. There have been several historic efforts that separated *Thinopyrum, Pascopyrum*, and *Psuedoroegneria* from the *Elymus* genera. Stubbendieck et al. (2017)[1] concluded that many of the characteristics used to separate species of these genera were based on cytogenetic and plant hybridization studies and not solely on morphologic characteristics. The scientific names listed here were verified against the USDA-ARS Germplasm Resources Information Network (GRIN) database,[2] which is regularly updated. A few species not listed in the GRIN database were verified using the USDA-NRCS Plants Database.[3]

aeschynomene—*Aeschynomene americana* L.
alemangrass—*Echinochloa polystachya* (Kunth) Hitchc.
alfalfa—*Medicago sativa* L.
algarrobo mesquite—*Prosopis pallida* (Humb. & Bonpl. ex Willd.) Kunth
alkali sacaton—*Sporobolus airoides* (Torr.) Torr.
alpine timothy—*Phleum alpinum* L.
alsike clover—*Trifolium hybridum* L.
altai wildrye—*Leymus angustus* (Trin.) Pilger
alyceclover—*Alysicarpus vaginalis* (L.) DC.
american jointvetch—*Aeschynomene americana* L.
american sloughgrass—*Beckmannia syzigachne* (Steud.) Fernald
annual bluegrass—*Poa annua* L.
annual broomweed—*Gutierrezia sarothrae* (Pursh) Britton & Rusby
annual canarygrass—*Phalaris canariensis* L.
annual lespedeza—*Kummerowia striata* (Thunb.) Schindl.
annual medics—*Medicago* spp.

Forages: The Science of Grassland Agriculture, Volume II, Seventh Edition.
Edited by Kenneth J. Moore, Michael Collins, C. Jerry Nelson and Daren D. Redfearn.

annual ryegrass—*Lolium multiflorum* Lam.
annual yellow sweetclover—*Melilotus officinalis* (L.) Lam.
antelope bitterbrush—*Purshia tridentata* (Pursh) DC.
arctic wheatgrass—*Elymus macrourus* (Turcz.) Tzvelev
arizona fescue—*Festuca arizonica* Vasey
arrowleaf clover—*Trifolium vesiculosum* Savi
artichoke—*Cynara cardunculus* L.
ashe juniper—*Juniperus ashei* J. Buchholz
asian pigeonwings—*Clitoria ternatea* L
australian jointvetch—*Aeschynomene falcata* (Poir.) DC.
autumn olive—*Elaeagnus umbellata* Thunb.
axillaris—*Macrotyloma axillare* (E. Mey.) Verdc.
bahiagrass—*Paspalum notatum* Flüggé
ball clover—*Trifolium nigrescens* Viv.
bamboo—*Bambusa* spp.
barley—*Hordeum vulgare* L.
barnyard grass—*Echinochloa crus-galli* (L.) P. Beauv.
barrel medic—*Medicago truncatula* Gaertn.
bearded wheatgrass—*Elymus caninus* (L.) *L.* (syn. *Agropryron subsecundum* var. *andinum*)
bearded wheatgrass—*Pseudoroegneria spicata* (Pursh) Á. Löve (syn. *Agropyron inerme* [Scribn. & J. G. Sm.] Rydb.)
beardless wildrye—*Leymus triticoides* (Buckley) Pilg.
bermudagrass—*Cynodon dactylon* (L.) Pers.
berseem clover—*Trifolium alexandrinum* L.
big bluestem—*Andropogon gerardii* Vitman
big deervetch—*Lotus crassifolius* (Benth.) Greene (syn. *Hosackia crassifolia* Benth.)
big quakinggrass—*Briza maxima* L.
big sacaton—*Sporobolus wrightii* Munro ex Scribn.
big sagebrush—*Artemisia tridentata* Nutt.
big sandreed—*Calamovilfa gigantea* (Nutt.) Scribn. & Merr.
big trefoil—*Lotus uliginosus* Schkuhr (syn. *L. pedunculatus* Cav.)
bigflower clover—*Trifolium michelianum* Savi
bigflower vetch—*Vicia grandiflora* Scop.
bird vetch—*Vicia cracca* L.
birdsfoot trefoil—*Lotus corniculatus* L.
bitter sneezeweed—*Helenium amarum* (Raf.) H. Rock
bitter vetch—*Vicia ervilia* (L.) Willd.
bitterbrush—*Purshia tridentata* (Pursh) DC.
black grama—*Bouteloua eriopoda* (Torr.) Torr.
black greasewood—*Sarcobatus vermiculatus* (Hook.) Torr.
black locust—*Robinia pseudoacacia* L.
black medic—*Medicago lupulina* L.
black oat—*Avena strigosa* Schreb.
black sagebrush—*Artemisia nova* A. Nelson
black walnut—*Juglans nigra* L.
blackberry—*Rubus* spp.
blackbrush—*Coleogyne ramosissima* Torr.
blackjack oak—*Quercus marilandica* Münchh.
blacksamson—*Echinacea angustifolia* DC.
bladygrass—*Imperata cylindrica* (L.) Raeusch.
blue grama—*Bouteloua gracilis* (Kunth) Lag. ex Griffiths
blue lupine—*Lupinus angustifolius* L.
blue panicgrass—*Panicum antidotale* Retz.
blue wildrye—*Elymus glaucus* Buckley
bluebonnet—*Lupinus subcarnosus* Hook.
bluebunch wheatgrass—*Pseudoroegneria spicata* (Pursh) Á. Löve
bluejoint reedgrass—*Calamagrostis canadensis* (Michx.) P. Beauv.
bluetop—*Calamagrostis canadensis* (Michx.) P. Beauv.
boer lovegrass—*Eragrostis curvula* (Schrad.) Nees
bottlebrush squirreltail—*Elymus elymoides* (Raf.) Swezey
brassica—*Brassica* spp.
brassica hybrid—*Brassica rapa* L. × *B. rapa* subsp. *Brassica campestris* var. *rapa*
broadleaf plantain—*Plantago major* L.
broom millet—*Panicum miliaceum* L. subsp. *miliaceum*
broomcorn millet—*Panicum miliaceum* L. subsp. *miliaceum*
broomrape—*Orobanche* spp.
broomsedge—*Andropogon virginicus* L.
brown beetle grass—*Diplachne fusca* (L.) P. Beauv. ex Roem. & Schult. subsp. *fusca* (syn. *Leptochloa fusca* (L.) Kunth)
brownseed paspalum—*Paspalum plicatulum* Michx.
browntop millet—*Urochloa ramosa* (L.) T. Q. Nguyen (syn. *Panicum ramosum* L.)
buckbrush—*Symphoricarpos orbiculatus* Moench
buckhorn plantain—*Plantago lanceolata* L.
buckthorn—*Ceanothus greggii* A. Gray
buckwheat—*Fagopyrum esculentum* Moench
budsage—*Picrothamnus desertorum* Nutt. (syn. *Artemisia spinescens* D. C. Eaton)
buffalo clover—*Alysicarpus vaginalis* (L.) DC.
buffalobur—*Solanum rostratum* Dunal
buffalograss—*Bouteloua dactyloides* (Nutt.) Columbus (syn. *Buchloe dactyloides* [Nutt.] Engelm.)
buffelgrass—*Cenchrus ciliaris* L.
bull thistle—*Cirsium vulgare* (Savi) Ten. (syn. *Cirsium lanceolatum* [L.] Scop.)
bundleflower—*Desmanthus bicornutus* S. Watson
burclover—*Medicago polymorpha* L.
burnet—*Sanguisorba minor* Scop.
bur medic—*Medicago polymorpha* L.
bush muhly—*Muhlenbergia porteri* Scribn. ex Beal
butterfly pea—*Centrosema virginianum* (L.) Benth. (Chapter 13)
butterfly pea—*Clitoria ternatea* L. (Chapter 2)
buttonclover—*Medicago orbicularis* (L.) Bartal.
button medic—*Medicago orbicularis* (L.) Bartal.
caatinga stylo—*Stylosanthes seabrana* B.L. Maass & 't Mannetje

cabbage—*Brassica oleracea* L.
caleypea—*Lathyrus hirsutus* L.
california cordgrass—*Spartina foliosa* Trin.
california needlegrass—*Achnatherum occidentale* (Thurb.) Barkworth subsp. *californicum* (Merr. & Burtt Davy) Barkworth
calopo—*Calopogonium mucunoides* Desv.
camphorweed—*Heterotheca subaxillaris* (Lam.) Britton & Rusby
canada bluegrass—*Poa compressa* L.
canada thistle—*Cirsium arvense* (L.) Scop.
canada wildrye—*Elymus canadensis* L.
canola—*Brassica* spp.
capitate—*Stylosanthes capitata* Vogel
caribbean stylo—*Stylosanthes hamata* (L.) Taub.
carolina geranium—*Geranium carolinianum* L.
carpon desmodium—*Desmodium heterocarpon* (L.) DC.
cassava—*Manihot esculenta* Crantz
catclaw acacia—*Senegalia greggii* (A. Gray) Britton & Rose (syn. *Acacia greggii* A. Gray)
caucasian bluestem—*Bothriochloa bladhii* (Retz.) S.T. Blake
cedar—*Juniperus* spp.
centipedegrass—*Eremochloa ophiuroides* (Munro) Hack.
centro—*Centrosema pubescens* Benth.
centurion—*Centrosema pascuorum* Mart. ex Benth.
cereal rye—*Secale cereale* L.
cheatgrass—*Bromus tectorum* L.
chess—*Bromus secalinus* L.
chewings fescue—*Festuca rubra* L.
chickpea—*Cicer arietinum* L.
chicory—*Cichorium intybus* L.
chinese cabbage—*Brassica rapa* L. subsp. *pekinensis* (Lour) Hanelt
chokecherry—*Prunus virginiana* L.
cicer milkvetch—*Astragalus cicer* L.
clover—*Trifolium* spp.
cluster clover—*Trifolium glomeratum* L.
coastal panicgrass—*Panicum amarum* Elliott
cocksfoot—*Dactylis glomerata* L.
cogongrass—*Imperata cylindrica* (L.) Raeusch.
collards—*Brassica oleracea* L. var. *viridis* L.
common barley—*Hordeum vulgare* L.
common burdock—*Arctium minus* (Hill) Bernh.
common carpetgrass—*Axonopus fissifolius* (Raddi) Kuhlm.
common dallisgrass—*Paspalum dilatatum* Poir.
common chickweed—*Stellaria media* (L.) Vill.
common lambsquarters—*Chenopodium album* L.
common lespedeza—*Kummerowia striata* (Thunb.) Schindl.
common mullein—*Verbascum thapsus* L.
common ragweed—*Ambrosia artemisiifolia* L.
common reed—*Phragmites australis* (Cav.) Trin. ex Steud.
common rivergrass—*Scolochloa festucacea* (Willd.) Link
common sainfoin—*Onobrychis viciifolia* Scop.
common sow thistle—*Sonchus oleraceus* L.
common velvetgrass—*Holcus lanatus* L.
common vetch—*Vicia sativa* L.
coralberry—*Symphoricarpos orbiculatus* Moench
cordateleaf—*Vicia sativa* subsp. *cordata* (Wulfen ex Hoppe) Batt.
cordgrass—*Spartina* spp.
corn—*Zea mays* L.
cotton—*Gossypium hirsutum* L.
cottontop—*Digitaria californica* (Benth.) Henrard.
couchgrass—*Elytrigia repens* (L.) Desv. ex Nevski (syn. *Elymus repens* (L.) Gould subsp. *repens*)
cow vetch—*Vicia cracca* L.
cowpea—*Vigna unguiculata* (L.) Walp.
crabgrass—*Digitaria ciliaris* (Retz.) Koeler
creeping beggarweed—*Desmodium incanum* DC.
creeping bentgrass—*Agrostis stolonifera* L. var. *palustris* (Huds.) Farw.
creeping foxtail—*Alopecurus arundinaceus* Poir.
creeping meadow foxtail—*Alopecurus arundinaceus* Poir.
creeping paspalum—*Paspalum scrobiculatum* L.
creeping signalgrass—*Urochloa humidicola* (Rendle) Morrone & Zuloaga
creeping vigna—*Vigna parkeri* Baker
creosote bush—*Larrea tridentata* (Sessé & Mocino ex DC.) Coville
crested wheatgrass—*Agropyron cristatum* (L.) Gaertn.
crimson clover—*Trifolium incarnatum* L.
crownvetch—*Securigera varia* (L.) Lassen
cudweed sagewort—*Artemisia ludoviciana* Nutt.
curly dock—*Rumex crispus* L.
curly mesquite—*Hilaria belangeri* (Steud.) Nash
dahurian wildrye—*Elymus dahuricus* Turcz. ex Grieseb.
dallisgrass—*Paspalum dilatatum* Poir.
dalmatian toadflax—*Linaria dalmatica* Mill.
dandelion—*Taraxacum officinale* F. H. Wigg.
darnel—*Lolium temulentum* L.
deerbrush—*Ceanothus integerrimus* Hook. & Arn.
deertongue—*Panicum clandestinum* L. (syn. *Dichanthelium clandestinum* [L.] Gould)
desmanthus—*Desmanthus virgatus* (L.) Willd.
digitgrass—*Digiteria eriantha* Steud.
dodder—*Cuscuta* spp.
dogfennel—*Eupatonium capillifolium* (Lam.) Small
dotted gayfeather—*Liatris punctata* Hook.
downy bromegrass—*Bromus tectorum* L.
downy chess—*Bromus tectorum* L.
dune wildrye—*Leymus mollis* (Trin.) Pilg.
durum wheat—*Triticum turgidum* L. subsp. *durum* (Desf.) van Slageren
dutch clover—*Trifolium repens* L.
eastern gamagrass—*Tripsacum dactyloides* (L.) L.

eastern red cedar—*Juniperus virginiana* L.
egyptian clover—*Trifolium alexandrinum* L.
elephantgrass—*Cenchrus purpureus* (Schumach.) Morrone (syn. *Pennisetum purpureum* Schumach.)
emmer wheat (wild)—*Triticum turgidum* L. subsp. *dicoccon* (Schrank) Thell.
endive—*Cichorium endivia* L.
energy cane—*Saccharum* spp.
evenia aeschynomene—*Aeschynomene evenia* C. Wright
evenia jointvetch—*Aeschynomene evenia* C. Wright
falcon pea—*Lathyrus cicera* L.
fendler ceanothus—*Ceanothus fendleri* A. Gray
fenugreek—*Trigonella foenum-graecum* L.
field bindweed—*Convolvulus arvensis* L.
field bromegrass—*Bromus arvensis* L.
field paspalum—*Paspalum laeve* Michx.
field pea—*Pisum sativum* L. var. *arvense* (L.) Poir.
fine stem stylo—*Stylosanthes guianensis* (Aubl.) Sw.
fine stem stylo—*Stylosanthes guianensis* (Aubl.) Sw. var. *intermedia* (Vogel) Hassl.
fingergrass—*Digiteria eriantha* Steud.
flaccidgrass—*Cenchrus flaccidus* (Griseb.) Morrone (syn. *Pennisetum flaccidum* Griseb.)
flatpea—*Lathyrus sylvestris* L.
flatpod peavine—*Lathyrus cicera* L.
florida beggarweed—*Desmodium tortuosum* (Sw.) DC.
florida paspalum—*Paspalum floridanum* Michx.
fodder radish—*Raphanus sativus* L.
forage kochia—*Bassia prostrata* (L.) A. J. Scott
forage radish—*Raphanus sativus* L.
fourwing saltbush—*Atriplex canescens* (Pursh) Nutt.
foxtail barley—*Hordeum jubatum* L.
foxtail millet—*Setaria italica* (L.) P. Beauv.
fringed sagebrush—*Artemisia frigida* Willd.
galleta—*Pleuraphis jamesii* Torr. (syn. *Hilaria jamesii* [Torr.] Benth.)
gambel oak—*Quercus gambelii* Nutt.
garbanzo—*Cicer arietinum* L.
german velvetgrass—*Holcus mollis* L.
giant dropseed—*Sporobolus giganteus* Nash
giant foxtail—*Setaria faberi* R.A.W. Herrm.
giant miscanthus—*Miscanthus* x *giganteus* Greef and Deuter ex Hodk. & Renvoize
giant reed—*Arundo donax* L.
giant stargrass—*Cynodon aethiopicus* Clayton & J.R. Harlan
giant wildrye—*Leymus condensatus* (J. Presl) Á. Löve
gliricidia—*Gliricidia sepium* (Jacq.) Kunth
glycine—*Neonotonia wightii* (Wight & Arn.) J.A. Lackey
goatgrass—*Aegilops tauschii* Coss.
golden timothy—*Setaria sphacelata* (Schumach.) Stapf & C. E. Hubb.
goldenrod—*Solidago altissima* L.
grandiflora vetch—*Vicia grandiflora* Scop.
grasspea—*Lathyrus sativus* L.
grain sorghum—*Sorghum bicolor* (L.) Moench
great basin wildrye—*Leymus cinereus* (Scribn. & Merr.) Á. Löve
green needlegrass—*Nassella viridula* (Trin.) Barkworth
green panicgrass—*Panicum maximum* Jacq. (syn. *Megathyrsus maximus* [Jacq.] B. K. Simon & S. W. L. Jacobs)
green sprangletop—*Disakisperma dubium* (Kunth) P. M. Peterson & N. Snow (syn. *Leptochloa dubia* [Kunth] Nees)
greenbrier—*Smilax rotundifolia* L.
greenleaf desmodium—*Desmodium intortum* (Mill.) Urb.
greenleaf fescue—*Festuca viridula* Vasey
guajillo—*Senegalia berlandieri* (Benth.) Britton & Rose (syn. *Acacia berlandieri* Benth.)
guineagrass—*Panicum maximum* Jacq. (syn. *Megathyrsus maximus* [Jacq.] B. K. Simon & S. W. L. Jacobs)
hairy chess—*Bromus commutatus* Schrad.
hairy crabgrass—*Digitaria sanguinalis* (L.) Scop.
hairy grama—*Bouteloua hirsuta* Lag.
hairy indigo—*Indigofera hirsuta* L.
hairy vetch—*Vicia villosa* Roth
halogeton—*Halogeton glomeratus* (Bieb.) Ledeb.
hardinggrass—*Phalaris aquatica* L.
heath aster—*Symphyotrichum ericoides* (L.) G. L. Nesom (syn. *Aster ericoides* L.)
hemp sesbania—*Sesbania exaltata* (Raf.) Rybd.
henbit—*Lamium amplexicaule* L.
herdgrass—*Phleum pratense* L.
hetero desmodium—*Desmodium heterophyllum* (Willd.) DC.
honey locust—*Gleditsia triacanthos* L.
honey mesquite—*Prosopis glandulosa* Torr.
honeysuckle—*Lonicera periclymenum* L.
hop clover—*Trifolium aurem* Pollich
horsenettle—*Solanum carolinense* L.
hungarian clover—*Trifolium pannonicum* Jacq.
hungarian vetch—*Vicia pannonica* Crantz
hyacinth bean—*Lablab purpureus* (L.) Sweet
hybrid poplar—*Populus* spp.
idaho fescue—*Festuca idahoensis* Elmer
illinois bundleflower—*Desmanthus illinoensis* (Michx.) MacMill. ex B. L. Rob. & Fernald
indian mustard—*Brassica juncea* (L.) Czern.
indian ricegrass—*Achnatherun hymenoides* (Roem. & Schult.) Barkworth
indiangrass—*Sorghastrum nutans* (L.) Nash
inland saltgrass—*Distichlis spicata* (L.) Greene var. *stricta* (Torr.) Scribn.
intermediate wheatgrass—*Thinopyrum intermedium* (Host) Barkworth & D. R. Dewey
italian millet—*Setaria italica* (L.) P. Beauv.
italian ryegrass—*Lolium multiflorum* Lam.
jackbean—*Canavalia ensiformis* (L.) DC.

japanese bromegrass—*Bromus japonicus* Houtt.
japanese lawngrass—*Zoysia japonica* Steud.
japanese millet—*Echinochloa frumentacea* Link
jerusalem artichoke—*Helianthus tuberosus* L.
jimsonweed—*Datura stramonium* L.
johnsongrass—*Sorghum halepense* (L.) Pers.
jointvetch—*Aeschynomene elaphroxylon* (Guill. & Perr.) Taub.
juniper—*Juniperus* spp.
kaimi clover—*Desmodium incanum* (Sw.) DC.
kale—*Brassica oleracea* L.
kentucky bluegrass—*Poa pratensis* L.
kenya white clover—*Trifolium semipilosum* Fresen.
kikuyugrass—*Cenchrus clandestinus* (Hochst. ex Chiov.) Morrone (syn. *Pennisetum clandestinum* Hochst. ex Chiov.)
kleingrass—*Panicum coloratum* L.
knotgrass—*Paspalum distichum* L.
kochia—*Kochia scoparia* (L.) Schrad.
kohlrabi—*Brassica oleracea* L. var. *gongylodes* L.
korean lespedeza—*Kummerowia stipulacea* (Maxim.) Makino
koroniviagrass—*Urochloa humidicola* (Rendle) Morrone & Zuloaga
kretschmer's peanut—*Arachis kretschmeri* Krapov. & W.C. Greg.
kudzu—*Pueraria montana* (Lour.) Merr.
kura clover—*Trifolium ambiguum* M. Bieb.
kwao krua—*Pueraria candollei* Graham ex Benth. var. *mirifica* (Airy Shaw & Suvat.) Niyomdham (syn. *Pueraria mirifica* Airy Shaw & Suvat.)
lablab bean—*Lablab purpureus* (L.) Sweet
ladino clover—*Trifolium repens* L.
lanceleaf ragweed—*Ambrosia bidentata* Michx..
lappa clover—*Trifolium lappaceum* L.
large hop clover—*Trifolium campestre* Schreb.
lax panicgrass—*Steinchisma laxum* (Sw.) Zuloaga (syn. *Panicum laxum* Sw.)
leadplant—*Amorpha canescens* Pursh
leafy spurge—*Euphorbia esula* L.
lehmann lovegrass—*Eragrostis lehmanniana* Nees
leucaena—*Leucaena leucocephala* (Lam.) de Wit
limpograss—*Hemarthria altissima* (Poir.) Stapf & C.E. Hubb.
little barley—*Hordeum pusillum* Nutt.
little bluestem—*Schizachyrium scoparium* (Michx.) Nash
little burclover—*Medicago minima* (L.) Bartal.
little bur medic—Medicago polymorpha L.
little quakinggrass—*Briza minor* L.
littleleaf leadtree—*Leucaena retusa* Benth.
live oak—*Quercus virginiana* Mill.
loblolly pine—*Pinus taeda* L.
locoweed—*Oxytropis* spp.
longleaf rushgrass—*Sporobolus compositus* (Poir.) Merr.
lotononis—*Lotononis bainesii* Baker
low larkspur—*Delphinium nuttallianum* Pritz.
lucerne—*Medicago sativa* L.
lupine—*Lupinus angustifolius* L.
maize—*Zea mays* L.
mandan ricegrass—X *Achnella caduca* (Beal) Barkworth
marshelder—*Cyclachaena xanthiifolia* (Nutt.) Fresen.
matua bromegrass—*Bromus catharticus* Vahl var. *catharticus* (syn. *Bromus unioloides* Kunth)
meadow barley—*Hordeum brachyantherum* Nevski
meadow bromegrass—*Bromus riparius* Rehmann
meadow dropseed—*Sporobolus compositus* (Poir.) Merr.
meadow fescue—*Festuca pratensis* Huds.
meadow foxtail—*Alopecurus pratensis* L.
medusahead—*Taeniatherum caput-medusae* (L.) Nevski
mesquite—*Prosopis* spp.
mexican cliffrose—*Purshia mexicana* (D. Don) S.L. Welsh
mikes clover—*Trifolium michelianum* Savi
millwood honey locust—*Gleditsia triacanthos* L.
mimosa—*Albizzia julibrissin* Scop.
miscanthus—*Miscanthus sinensis* Andersson
missouri goldenrod—*Solidago missouriensis* Nutt.
mouseear cress—*Arabidopsis thaliana* (L.) Heynh.
mountain big sagebrush—*Artemisia tridentata* Nutt. subsp. *vaseyana* (Rydb.) Beetle
mountain bromegrass—*Bromus carinatus* Hook. & Arn. var. *marginatus* (Steud.) Barkworth & Anderton (syn. *Bromus marginatus* Nees ex Steud.)
mountain mahogany—*Cercocarpus betuloides* Nutt.
mountain muhly—*Muhlenbergia montana* (Nutt.) Hitchc.
multiflora rose—*Rosa multiflora* Thunb.
musk thistle—*Carduus nutans* L.
mustard—*Sinapis and Brassica* spp.
naivasha stargrass—*Cynodon plectostachyus* (K. Schum.) Pilg.
napiergrass—*Pennisetum purpureum* (Schumach.) Morrone
narrowleaf birdsfoot—*Lotus tenuis* Waldst. & Kit. ex Willd. (syn. *Lotus glaber* Mill.)
narrowleaf trefoil—*Lotus tenuis* Waldst. & Kit. ex Willd. (syn. *Lotus glaber* Mill.)
narrowleaf vetch—*Vicia sativa* L. subsp. *nigra* (L.) Ehrh. (syn. *Vicia angustifolia* L.)
needle-and-thread—*Hesperostipa comata* (Trin.& Rupr.) Barkworth
needlegrass—*Hesperostipa comata* (Trin.& Rupr.) Barkworth
needlegrasses—*Stipa* spp.
nimblewill—*Muhlenbergia schreberi* J.F. Gmel.
oak—*Quercus* spp.
oat—*Avena sativa* L.
old world bluestems—*Bothriochloa* spp.

one-seed juniper—*Juniperus monosperma* (Engelm.) Sarg.
onion—*Allium cepa* L.
orchardgrass—*Dactylis glomerata* L.
ovalifolium—*Desmodium heterocarpon* (L.) DC. subsp. *ovalifolium* [Prain] H. Ohashi
owl-headed clover—*Trifolium alpestre* L.
palisadegrass—*Urochloa brizantha* (Hochst. ex A. Rich.) R.D. Webster
pampasgrass—*Cortaderia selloana* (Schult. & Schult. f.) Asch. & Graebn.
pangolagrass—*Digitaria eriantha* Steud.
paragrass—*Urochloa mutica* (Forssk.) T. Q. Nguyen
partridge pea—*Chamaecrista fasciculata* (Michx.) Greene
pea—*Pisum sativum* L.
peanut—*Arachis hypogaea* L.
pearlmillet—*Cenchrus americanus* (L.) Morrone (syn. *Pennisetum americanum* [L.] Leeke)
pencil flower, sidebeak—*Stylosanthes biflora* (L.) Britton et al.
pencilflower—*Stylosanthes scabra* Vogel
pennsylvania smartweed—*Persicaria pensylvanica* (L.) M. Gómez (syn. *Polygonum pensylvanicum* L.)
perennial horsegram—*Macrotyloma axillare* (E. Mey.) Verdc.
perennial peanut—*Arachis glabrata* Benth. var. *glabrata*
perennial pepperweed—*Lepidium latifolium* L.
perennial ryegrass—*Lolium perenne* L.
perennial sow thistle—*Sonchus arvensis* L.
perennial soybean—*Neonotonia wightii* (Wight & Arn.) J. A. Lackey
persian clover—*Trifolium resupinatum* L.
persimmon—*Diospyros virginiana* L.
phasey bean—*Macroptilium lathyroides* (L.) Urb.
pigeonpea—*Cajanus cajan* (L.) Huth
pinegrass—*Calamagrostis rubescens* Buckley
pineland three-awn—*Aristida stricta* Michx.
pinto peanut—*Arachis pintoi* Krapov. & W.C. Greg.
plains bluestem—*Bothriochloa ischaemum* (L.) Keng
plains bristlegrass—*Setaria leucopila* (Scribn. & Merr.) K. Schum.
plains sunflower—*Helianthus petiolaris* Nutt.
plantain—*Plantago lanceolata* L.
plantain—*Plantago* spp.
plumeless thistle—*Carduus acanthoides* L.
pointleaf manzanita—*Arctostaphylos pungens* Kunth
porcupinegrass—*Hesperostipa spartea* (Trin.) Barkworth
post oak—*Quercus stellata* Wangenh.
povertygrass—*Danthonia spicata* (L.) Beauv. ex Rowm. & Schult.
prairie cordgrass—*Spartina pectinata* Link
prairie dropseed—*Sporobolus heterolepis* (A. Gray) A. Gray
prairie junegrass—*Koeleria macrantha* (Ledeb.) Schult.
prairie sandreed—*Calamovilfa longifolia* (Hook.) Scribn.
prickly lettuce—*Lactuca serriola* L.
prickly pear—*Opuntia ficus-indica* (L.) Mill.
proso millet—*Panicum miliaceum* L.
prostrate pigweed—*Amaranthus blitoides* S. Watson
pubescent wheatgrass—*Thinopyrum intermedium* (Host) Barkworth & D.R. Dewey subsp. *barbulatum* (Schur) Barkworth & D.R. Dewey
purple three-awn—*Aristida purpurea* Nutt.
purple prairie clover—*Dalea purpurea* Vent.
purple vetch—*Vicia benghalensis* L.
purpletop—*Tridens flavus* (L.) Hitchc.
quackgrass—*Elytrigia repens* (L.) Gould subsp. *repens*
quaking aspen—*Populus tremuloides* Michx.
quakinggrass—*Briza media* L.
rabbit foot clover—*Trifolium arvense* L.
rabbitbrush—*Chrysothamnus* spp.
radish—*Raphanus sativus* L.
rape—*Brassica napus* L.
rattailgrass—*Sporobolus indicus* (L.) R. Br. var. *capensis* Engl.
red clover—*Trifolium pratense* L.
red fescue—*Festuca rubra* L.
red oatgrass—*Themeda triandra* Forssk.
red three-awn—*Aristida purpurea* Nutt. var. *longiseta* (Steud.) Vasey (syn. *Aristida longiseta* Steud.)
redberry juniper—*Juniperus pinchotii* Sudw.
redroot pigweed—*Amaranthus retroflexus* L.
redtop—*Agrostis gigantea* Roth
reed canarygrass—*Phalaris arundinacea* L.
rescuegrass—*Bromus catharticus* Vahl var. *catharticus* (syn. *Bromus unioloides* Kunth)
rhizoma peanut—*Arachis glabrata* Benth. var. *glabrata*
rhodesgrass—*Chloris gayana* Kunth
ribbed paspalum—*Paspalum malacophyllum* Trin.
ribbongrass—*Phalaris arundinacea* L.
rice—*Oryza sativa* L.
ripgutgrass—*Bromus diandrus* Roth var. *rigidus* (Roth) Sales (syn. *Bromus rigidus* Roth)
robust needlegrass—*Achnatherum robustum* (Vasey) Barkworth
rose clover—*Trifolium hirtum* All.
rose-of-sharon—*Hypericum calycinum* L.
rough fescue—*Festuca campestris* Rydb. (syn. *F. hallii* [Vasey] Piper)
rough menodora—*Menodora scabra* A. Gray
roughpea—*Lathyrus hirsutus* L.
roundleaf cassia—*Chamaecrista rotundifolia* (Pers.) Greene
rubber tree—*Hevea brasiliensis* (Willd. ex A. Juss.) Müll. Arg.
ruby milkvetch—*Oxytropis riparia* Litv.
rushes—*Juncaceae* spp.
russian knapweed—*Rhaponticum repens* (L.) Hidalgo

russian sainfoin—*Onobrychis transcaucasica* Grossh.
russian thistle—*Salsola tragus* L.
russian wildrye—*Psathyrostachys juncea* (Fisch.) Nevski
ruzigrass—*Urochloa ruziziensis* (R. Germ. & C.M. Evrard) Crins
rye—*Secale cereale* L.
sagebrush—*Artemisia tridentata* Nutt.
sainfoin—*Onobrychis viciifolia* Scop.
saltbush, gardiner's—*Atriplex gardneri* (Moq.) D. Dietr.
saltbush—*Atriplex* spp.
sand blackberry—*Rubus cuneifolius* Pursh
sand bluestem—*Andropogon hallii* Hack
sand dropseed—*Sporobolus cryptandrus* (Torr.) A. Gray
sand lovegrass—*Eragrostis trichodes* (Nutt.) Alph. Wood
sand paspalum—*Paspalum setaceum* Michx.
sand sagebrush—*Artemisia filifolia* Torr.
sandhill muhly—*Muhlenbergia pungens* Thurb.
sauvages—*Calopogonium caeruleum* (Benth.) C. Wright
scrobicgrass—*Paspalum scrobiculatum* L.
scrub oak—*Quercus turbinella* Greene
sedges—*Cyperaceae* spp.
sericea lespedeza—*Lespedeza cuneata* (Dum. Cours.) G. Don
serviceberry—*Amelanchier alnifolia* (Nutt.) Nutt. ex M. Roem.
sesban—*Sesbania sesban* (L.) Merr.
sesbania—*Sesbania sesban* (L.) Merr.
setaria—*Setaria sphacelata* (Schum.) Stapf & C.E. Hubb.
shadscale saltbush—*Atriplex confertifolia* (Torr. & Frém.) S. Watson
sheep fescue—*Festuca ovina* L.
sheep sorrel—*Rumex acetosella* L.
shell-leaf penstemon—*Penstemon grandiflorus* Nutt.
shepherd's purse—*Capsella bursa-pastoris* (L.) Medik.
shrub lespedeza—*Lespedeza bicolor* Turcz.
shortleaf pine—*Pinus echinata* Mill.
shrubby cinquefoil—*Dasiphora fruticosa* (L.) Rydb. subsp. *fruticosa* (syn. *Potentilla fruticosa* L.)
shrubby stylo—*Stylosanthes scabra* Vogel
siberian sainfoin—*Onobrychis arenaria* (Kit.) DC.
siberian wheatgrass—*Agropyron fragile* (Roth) P. Candargy
siberian wildrye—*Elymus sibiricus* L.
sicklepod milkvetch—*Astragalus falcatus* Lam.
sideoats grama—*Bouteloua curtipendula* (Michx.) Torr.
signalgrass—*Urochloa decumbens* (Stapf) R.D. Webster
silky lupine—*Lupinus sericeus* Pursh
silver bluestem—*Bothriochloa saccharoides* (Sw.) Rydb.
silver sagebrush—*Artemisia cana* Pursh
silverleaf desmodium—*Desmodium uncinatum* (Jacq.) DC.
single-flowered vetch—*Vicia articulata* Hornem.
singletary pea—*Lathyrus hirsutus* L.
siratro—*Macroptilium atropurpureum* (DC.) Urb.
sixweeks fescue—*Vulpia octoflora* (Walter) Rydb.
skunkbush sumac—*Rhus aromatica* Aiton var. *trilobata* (Nutt.) A. Gray (syn. *Rhus trilobata* Nutt. ex Torr. & A. Gray)
slash pine—*Pinus elliottii* Engelm.
slender grama—*Bouteloua repens* (Kunth) Scribn.
slender oat—*Avena barbata* Pott ex Link
slender wheatgrass—*Elymus trachycaulus* (Link) Gould ex Shinners
small burnet—*Sanguisorba minor* Scop.
small hop clover—*Trifolium dubium* Sibth.
smooth bromegrass—*Bromus inermis* Leyss.
smutgrass—*Sporobolus indicus* (L.) R. Br. var. *indicus*
snail medic—*Medicago scutellata* (L.) Mill.
snowberry—*Symphoricarpos albus* (L.) S. F. Blake
snow-on-the-mountain—*Euphorbia marginata* Pursh
soft chess—*Bromus hordeaceus* L.
sorghum—*Sorghum bicolor* (L.) Moench
sourclover—*Melilotus indicus* (L.) All.
southern crabgrass—*Digitaria ciliaris* (Retz.) Koeler
soybean—*Glycine max* (L.) Merr.
spanish oak—*Quercus falcata* Michx.
speargrass—*Heteropogon contortus* (L.) Beauv. ex Roem. & Schult.
spike cordgrass—*Spartina foliosa* Trin.
spike dropseed—*Sporobolus contractus* Hitchc.
spike muhly—*Muhlenbergia wrightii* Vasey ex J. M. Coult.
spiny amaranth—*Amaranthus spinosus* L.
spiny hopsage—*Grayia spinosa* (Hook.) Moq.
spiny redberry—*Rhamnus crocea* Nutt.
spotted bur medic—*Medicago arabica* (L.) Huds.
spotted burclover—*Medicago arabica* (L.) Huds.
spotted knapweed—*Centaurea stoebe* L.
spur lupine—*Lupinus arbustus* Douglas ex Lindl.
spurred butterfly pea—*Centrosema virginianum* (L.) Benth.
squirreltailgrass—*Elymus elymoides* (Raf.) Swezey
st. augustine grass—*Stenotaphrum secundatum* (Walter) Kuntze
st. john'swort—*Hypericum perforatum* L.
standard crested wheatgrass—*Agropyron desertorum* (Fisch. ex Link) Schult.
stargrass—*Cynodon nlemfuensis* Vanderyst
stargrass, giant—*Cynodon aethiopicus* Clayton & J. R. Harlan
stiff sunflower—*Helianthus pauciflorus* Nutt.
strawberry clover—*Trifolium fragiferum* L.
streambank wheatgrass—*Elymus lanceolatus* (Scribn. & J. G. Sm.) Gould subsp. *riparius* (Scribn. & J. G. Sm.) Barkworth
striate clover—*Trifolium striatum* L.
striate lespedeza—*Kummerowia striata* (Thunb.) Schindl.
stylo—*Stylosanthes guianensis* (Aubl.) Sw.

subterranean clover—*Trifolium subterraneum* L.
succory—*Cichorium intybus* L.
sudangrass—*Sorghum bicolor* (L.) Moench
sugarbeet—*Beta vulgaris* L.
sugarbush—*Acer saccharum* Marshall
sugarcane—*Saccharum officinarum* L.
sulla (sulla sweetvetch)—*Hedysarum coronarium* L.
sumac—*Rhus* spp.
sunolgrass—*Phalaris coerulescens* Desf.
sunflower—*Helianthus annuus* L.
swede—*Brassica napus* L. subsp. *rapifera* Metzg.
sweet sorghum—*Sorghum bicolor* (L.) Moench
switchgrass—*Panicum virgatum* L.
tailcup lupine—*Lupinus argenteus* Pursh var. *heteranthus* (S. Watson) Barneby (syn. *Lupinus caudatus* Kellogg)
tall dropseed—*Sporobolus compositus* (Poir.) Merr.
tall fescue—*Festuca arundinacea* Schreb. (syn. *Schedonorus arundinaceus* [Schreb.] Dumort.)
tall oatgrass—*Arrhenatherum elatius* (L.) P. Beauv ex J. & C. Presl subsp. *elatius*
tall tickclover—*Desmodium tortuosum* (Sw.) DC.
tall wheatgrass—*Thinopyrum ponticum* (Podp.) Barkworth & D. R. Dewey
tangier pea—*Lathyrus tingitanus* L.
tanglehead—*Heterogon contortus* (L.) Beauv. ex Roem. & Schult.
tansymustard—*Descurainia pinnata* (Walter) Britton
tarbush— *Flourensia cernua* DC.
tardio stylo—*Stylosanthes guianensis* (Aubl.) Sw. var. *pauciflora* M.B. Ferreira & Sousa Costa
teff—*Eragrostis tef* (Zuccagni) Trotter
teosinte—*Zea mays* subsp. *mexicana* (Schrad.) H.H. Iltis
teramnus—*Teramnus labialis* (L. f.) Spreng.
texas bluegrass—*Poa arachnifera* Torr.
texas lupine—*Lupinus subcarnosus* Hook.
texas wintergrass—*Nassella leucotricha* (Trin. & Rupr.) R. W. Pohl
thickspike wheatgrass—*Elymus lanceolatus* (Scribn. & J. G. Sm.) Gould
threadleaf sedge—*Carex filifolia* Nutt.
threetip sagebrush—*Artemisia tripartita* Rydb.
ticklegrass—*Agrostis hyemalis* (Walter) Britton et al.
timber milkvetch—*Astragalus miser* Douglas
timothy—*Phleum pratense* L.
tobacco—*Nicotiana tabacum* L.
tobosagrass—*Pleuraphis mutica* Buckley
torpedograss—*Panicum repens* L.
townsville stylo—*Stylosanthes humilis* Kunth
triticale—X *Triticosecale* Wittmack
tropical carpetgrass—*Axonopus compressus* (Sw.) Beauv.
tropical kudzu—*Neustanthus phaseoloides* (Roxb.) Benth. (syn. *Pueraria phaseoloides* [Roxb.] Benth.)
true mountain mahogany—*Cercocarpus montanus* Raf.
tunisgrass—*Sorghum bicolor* (L.) Moench subsp. *verticilliflorum* (Steud.) de Wet ex Wiersema & J. Dahlb. (syn. *Sorghum arundinaceum* [Desv.] Stapf
turf timothy—*Phleum nodosum* L. (syn. *Phleum bertolonii* DC.)
turkey foot—*Andropogon gerardii* Vitman
turnip—*Brassica rapa* L. subsp. *rapa*
tyfon— *Brassica rapa* L.
variegated alfalfa—*Medicago sativa* subsp. *varia* (Martyn) Arcang. (syn. *M. media* Pers.)
vaseygrass—*Paspalum urvillei* Steud.
veldtgrass—*Ehrharta calycina* Sm.
velvet mesquite—*Prosopis velutina* Wooton
velvetbean, florida—*Mucuna pruriens* (L.) DC. var. *utilis* (Wall. ex Wight) Baker ex Burck
velvetleaf—*Abutilon theophrasti* Medik.
vernalgrass, sweet—*Anthoxanthum odoratum* L.
vervain, hoary—*Verbena stricta* Vent.
vetch—*Vicia* spp.
vetiver grass—*Chrysopogon zizanioides* (L.) Roberty
vigna—*Vigna adenantha* (G. Mey.) Maréchal et al.
villosa jointvetch—*Aeschynomene villosa* Poir.
vine mesquitegrass—*Hopia obtusa* (Kunth) Zuloaga & Morrone (syn. *Panicum obtusum* Kunth)
virginia pepperweed—*Lepidium virginicum* L.
wagner pea—*Lathyrus sylvestris* L.
water hemlock—*Cicuta maculata* L.
wax currant—*Ribes cereum* Douglas
weeping lovegrass—*Eragrostis curvula* (Schrad.) Nees
western ironweed—*Vernonia baldwinii* Torr.
western ragweed—*Ambrosia psilostachya* DC.
western wheatgrass—*Pascopyrum smithii* (Rydb.) Barkworth & D. R. Dewey
wheat—*Triticum aestivum* L.
white bursage—*Ambrosia dumosa* (A. Gray) W. W. Payne
white campion—*Silene latifolia* Poir.
white clover—*Trifolium repens* L.
white cockle—*Silene latifolia* Poir.
white locoweed—*Astragalus allochrous* A. Gray
white lupine—*Lupinus albus* L.
white sweetclover—*Melilotus albus* Medik.
whorled milkweed—*Asclepias verticillata* L.
wild carrot—*Daucus carota* L. subsp. *carota*
wild oat—*Avena fatua* L.
wild parsnip—*Pastinaca sativa* L.
wild rose—*Rosa woodsii* Lindl.
wildbean, trailing—*Strophostyles helvola* (L.) Elliott
wildrice, cultivated—*Zizania palustris* L.
willow—*Salix* spp.
wilman lovegrass—*Eragrostis superba* Peyr.
wimmera ryegrass—*Lolium rigidum* Gaudin
winter bentgrass—*Agrostis hyemalis* (Walter) Britton et al.
winter pea—*Pisum sativum* L.

winter vetch—*Vicia villosa* Roth subsp. *villosa*
winterfat—*Krascheninnikovia lanata* (Pursh) A. Meeuse & A. Smit
witchweed—*Striga* spp.
woolly finger grass—*Digitaria milanjiana* (Rendle) Stapf
wright silktassel—*Garrya wrightii* Torr.
yarrow—*Achillea millefolium* L.
yaupon holly—*Ilex vomitoria* Sol. ex Aiton
yellow alfalfa—*Medicago sativa* subsp. *falcata* (L.) Arcang. (syn. *M. falcata* L.)
yellow bluestem—*Bothriochloa ischaemum* (L.) Keng var. *ischaemum*
yellow lupine—*Lupinus luteus* L.
yellow nutsedge—*Cyperus esculentus* L.
yellow rocket—*Barbarea vulgaris* W. T. Aiton
yellow star thistle—*Centaurea solstitialis* L.
yellow sweetclover—*Melilotus officinalis* (L.) Lam.
yorkshire fog—*Holcus lanatus* L.
zigzag clover—*Trifolium medium* L.
zornia—*Zornia latifolia* Sm.
zoysia—*Zoysia japonica* Steud.

Notes

1. Stubbendieck, J., Hatch, S.L., and Dunn, C.D. (2017). *Grasses of the Great Plains*. Texas A&M AgriLife Research and Extension Service Series. Texas A&M University Press. College Station, TX. p. 260.
2. The Germplasm Resources Information Network (GRIN) Taxonomy (http://www.ars-grin.gov/). National Plant Germplasm System, USDA-ARS, Beltsville, Maryland. URL Verified May 30, 2019.
3. The PLANTS Database, Version 3.1 (http://plants.usda.gov). National Plant Data Center, USDA-NRCS, Baton Rouge, LA. URL Verified May 30, 2019.

Glossary

abiotic—Nonliving components of the environment, such as water, solar radiation, O_2, organic compounds, and soil nutrients. *See* biotic.

abomasum—The fourth compartment of the ruminant stomach, comprising the true stomach, in which digestive processes occur similar to those found in the nonruminant stomach.

arbuscular mycorrhizae—A fungus that penetrates cortical cells of roots forming a branched appendage called an arbuscule. The mycorrhizae extend from the root to help plants capture nutrients such as phosphorus, sulfur, nitrogen and micronutrients from the soil.

acceptability—Readiness with which animals select and ingest a forage; sometimes used interchangeably to mean either palatability or voluntary intake.

achene—A dry, one-seeded indehiscent fruit.

acid detergent fiber (ADF)—Insoluble residue following extraction of herbage with acid detergent (Van Soest); cell wall constituents minus hemicellulose.

acid detergent insoluble nitrogen (ADIN)—Nitrogen in the residue after a feed is extracted with acid detergent solution. Concentration is negatively correlated with feed nitrogen digestibility and ADIN >15% of nitrogen is an indicator of heat damage.

acid detergent lignin (ADL)—Lignin remaining in the residue following extraction with acid detergent and 24 N sulfuric acid.

acid pepsin—Used in second stage of *in vitro* forage digestion, 2 g of 0.1 $g\,kg^{-1}$ pepsin in 1 l of 0.1 M HCl.

additive genetic effects—The accumulative effects of multiple-gene actions on a complex trait that depend on the number of alleles affecting the trait.

additive genetic variation—The breeding value of an individual is the sum of the additive effects of its genes; variation in breeding value among plants in a population.

***ad libitum* intake**—Consumption of a feed or forage by an animal when offered in excess of what the animal can consume. Generally, when fed to have a daily excess of 15% of feed remaining.

adventitious roots—Roots that emerge from nodes at the base of vertical tillers and the nodes of rhizomes and stolons. They become the dominant root system for established grasses.

aerenchyma—Plant tissues with large intercellular spaces or channels that facilitate gas exchange, often transporting O_2 from the surface to roots under saturated or anaerobic conditions.

aerobic—Pertaining to life or processes occurring in free O_2 or in O_2 concentrations normal in air (21% O_2). Opposite of anaerobic.

aerobic respiration—Respiration in the presence of O_2 that is more efficient in terms of ATP production than anaerobic respiration.

aflatoxin ($C_{17}H_{12}O_6$)—A polynuclear substance derived from molds; a known carcinogen. Produced by a fungus occurring on peanuts, corn, and other plants, especially seeds.

aftermath—Residue and/or regrowth of plants (forage) used for grazing after harvesting of a crop.

agro-ecosystem—An ecosystem managed for food and/or fiber production.

agroforestry—Land-use system in which woody perennials are grown for wood or nut production in association with agricultural crops, with or without animal production.

agro-silvo-pastoral—Land-use system in which woody perennials are grown in association with pasture or forage crops used for livestock production.

agrostology—Study of grasses; their classification, management, and utilization.

Forages: The Science of Grassland Agriculture, Volume II, Seventh Edition.
Edited by Kenneth J. Moore, Michael Collins, C. Jerry Nelson and Daren D. Redfearn.

alkaloid—A class of basic organic compounds with nitrogen in their structure; a secondary product of plant metabolism such as perloline, produced by tall fescue.
alkaloids—*See* alkaloid.
allele—Any of a group of possible mutational forms of a gene.
allelochemical—A plant compound that yields deleterious effects on other plants or when consumed by animals (when the term is used to describe plant–animal interactions).
allelomimicry—Mimicking of another, for example, when a foal mimics its dam's grazing behavior.
allelopathy—The positive or negative influence of one living plant upon another due to secretion of chemical substances. *See* autotoxicity.
allopolyploid—Species having two or more partially related or unrelated genomes that typically do not pair during meiosis.
allotetraploid—Hybrid species with a chromosome set four times that of a haploid resulting from both chromosome sets of each parent being present.
alternate stocking—The repeated grazing and resting of forage using two paddocks in succession.
ammonia (NH_3)—Simple compound containing nitrogen that can be produced from the breakdown of amino acids during fermentation in the rumen or silo.
amylopectin—A type of starch used as a storage compound that consists of glucose molecules connected in a branching structure. Amylopectin has a high molecular weight and is insoluble in water at room temperature. Branching occurs through α-1,6 linkages from an amylose backbone. Amylopectins are more easily dissolved and digested than amylose.
amylose—A type of starch that consists of a linear chain of glucose molecules in which linkages are exclusively α-1,4. It is less soluble in water, and has a relatively low molecular weight relative to amylopectin.
anaerobic—Living in the absence of free O_2; the opposite of aerobic.
anaerobic respiration—Respiration in the absence of O_2 that is less efficient in terms of ATP production than aerobic respiration.
anemia—A condition of animals characterized by a lack of hemoglobin or a deficiency of red blood cells, which limits O_2 supply to body tissues.
animal day—One day's tenure upon pasture by one animal. Not synonymous with animal unit day.
animal days per hectare—Unit to express total tenure of animals upon a unit of pasture. Usage: Typically expressed in terms of a longer time period: for example, animal d ha^{-1} yr^{-1}.
animal month—One month's tenure upon pasture by one animal. Usage: Not synonymous with animal unit month.
animal performance—Production per animal (weight change or animal products) per unit of time.
animal unit—One mature nonlactating bovine weighing 500 kg and fed at a maintenance level, or the equivalent, expressed as $(\text{weight})^{0.75}$ in other kinds or classes of animals (cf. standard livestock unit).
animal unit day—The amount of dry forage consumed by one animal unit per 24-hours period. Animal unit day is used to express the quantity of forage intake for a period of time and may be extrapolated to other time periods, such as week, month, or year (cf. animal unit, forage intake unit).
animal unit month (AUM)—The amount of feed or forage required by an animal unit for one month; tenure of one animal unit for a period of one month. Not synonymous with animal month.
annual—Plants completing their life cycle in less than one year. Summer annuals germinate in the spring, produce seed in summer or fall, and then die. Winter annuals germinate in the fall, overwinter, and grow and produce seed the following spring or summer.
anoxia—Oxygen deficiency.
anthesis—The period when a flower (in grass, the lemma and palea) is open, the anthers and stigma are mature, and pollen is shed. In self-pollinated plants, this occurs before flower opening.
anthropogenic—Caused by or associated with humankind.
antiquality constituents—Chemical compounds that have negative effects on forage intake or produce negative responses in animals consuming the forage.
antiquality factors—See antiquality constituents.
anti-sense gene—A gene that transcribes an RNA segment complementary to a protein-coding mRNA with which it hybridizes and, thereby, blocks its translation into protein.
apical dominance—Inhibiting effect of a terminal bud upon the development of lateral buds.
apomixis—Formation of viable embryos without union of male and female gametes.
aquaporins—Integral membrane proteins that form channels in the cell plasma membrane specifically for the rapid movement of water into or out of the cell.
asexual—Reproduction by cell division or spore formation without the union of individuals or gametes.
ash—The residue remaining after complete burning of combustible matter; consists mainly of minerals in oxidized form.
atomic absorption spectroscopy—Observation by means of an optical device (spectroscope) of the wavelength and intensity of electromagnetic radiation (light) absorbed by various materials. Particular elements absorb well-defined wavelengths on an atomic level.

ATP (adenosine triphosphate)—The molecule containing chemical energy synthesized during respiration that can be used within the cell.

auricle—Earlike projections at the base of the grass leaf blade.

autotetraploid—Individual with an additional set of chromosomes identical to the parent that results in four copies of a single genome from a doubling of the parental chromosomes.

autotoxicity—A specific type of allelopathy in which adult plants interfere with the germination and development of seedlings of the same species.

autotroph—A plant that is able to synthesize its own organic food supply, especially by photosynthesis.

available forage—That portion of the forage, expressed as weight of forage per unit land area, that is accessible for consumption by a specified kind, class, sex, size, age, and physiological status of grazing animal (cf. forage allowance, forage mass).

awn—Bristle-like structure originating from the lemma or glume of a grass flower.

axillary bud—Meristematic apex located in the junction of the leaf and stem; gives rise to tillers in grasses and to branches and flowers in dicots.

backgrounding—Intensive management of young cattle, post-weaning, using forages to facilitate maximum performance before animals are moved to a feedlot.

bacteroid—Nitrogen-fixing organelle derived from rhizobia bacteria residing in the root nodules of host legume plants.

biennial—A plant that completes its life cycle in two years. A true biennial grows vegetatively during the first growing season, then produces seed and dies during the second.

bio-based economy—System based on the sustainable production of energy and industrial products from renewable resources rather than from fossil carbon sources (e.g. petroleum).

biodiversity—The variability among living organisms on the earth, including the variability within and between species and within and between ecosystems. It describes the natural biological wealth that undergirds human life and well-being, and reflects the interrelatedness of genes, species, and ecosystems.

bioenergy crop—Crop grown for industrial energy production through either direct combustion or by conversion to another fuel (i.e. bioethanol, biodiesel).

biofuel—Fuels, such as alcohols, ethers, esters, and other chemicals, derived from contemporary biological materials. Used interchangeably when referring to fuels for electricity or liquid fuels for transportation.

biogeochemical cycles—The pathways by which a chemical element moves through biotic and abiotic compartments of an ecosystem.

biomass—(i) The weight of living organisms (plants and animals) in an ecosystem at a given point in time. (ii) Refers to the organic matter from plants. (iii) In terms of energy production, generally refers to the organic matter from plants and includes herbaceous and woody crops along with their residues.

biome—An ecologic region that is often defined according to its predominant vegetation, such as grassland, temperate deciduous forest, or desert.

biopore—Soil pore created by plant roots, insects, or soil fauna.

biorefinery—Facility or operation where plant biomass and other biological materials are processed and converted into multiple end products including biofuels, industrial products, and chemicals.

biotic—Living components of the environment, such as higher plants, algae, microorganisms, nematodes, worms, insects, birds, and mammals.

blade—The flat, expanded part of a leaf above the sheath or petiole; the major photosynthetic organ.

bloat—Excessive accumulation of gases in the rumen of animals causing distension because normal escape through the esophagus is impaired.

body weight, empty—Conceptually, weight of an animal when the alimentary tract is empty; equal to live weight minus gut contents.

body weight, shrunk—Body weight after a period of fasting (no feed and/or water, usually overnight or for 24 hours) to reduce variation in gut-fill contribution to body weight. *See also* metabolic body weight.

bolting—A sudden onset of reproductive growth; results in rapid stem elongation and production of a flowering structure at the top.

bolus—A wad of herbage accumulated in the mouth from a number of bites in preparation for swallowing. (Plural: boli.)

bomb calorimetry—Process whereby a substance is completely oxidized in 25–30 atm of O_2 to determine gross energy (GE) content based on heat released.

boot stage—Growth stage when a grass inflorescence is enclosed by the sheath of the uppermost leaf.

bound water—Water that is incapable of forming into ice crystals because it is held tightly to cellular constituents.

brace root—Root originating at nodes above ground but penetrating the soil.

bract—A modified or reduced leaf subtending a flower or inflorescence.

bran—Outer wall (pericarp) of cereal grain; co-product from converting the grain to flour.

breeder seed—Seed or vegetative propagating material that is the source for initial and reoccurring increases of foundation seed. Controlled by the organization that developed it.

brown midrib—In maize *(br)* and sorghum *(bmr),* a single recessive gene character resulting in the dark brown coloration of the back side of the leaf midrib and under the leaf sheaths; associated with reduced lignin content of the plant.

browse—*n.* Leaf and twig growth of shrubs, woody vines, trees, cacti, and other non-herbaceous vegetation available for animal consumption. *v.* To browse. The consumption of browse *in situ* by animals (cf. forage, graze).

buffer stocking—The practice of using temporary fencing to adjust pasture area available to animals.

buffering capacity—The ability of a solution to resist changes in pH.

bulked segregant analysis—A method to analyze pooled samples of plants with and without a particular trait in order to identify associated genetic markers.

bulliform cells—Rows of cells in the upper epidermis of grass leaves that are large and somewhat thin-walled. They decrease in diameter when drought stressed causing the leaf blade to roll inward to reduce transpiration and radiation absorbance.

bunch-type growth habit—Plant development, especially grasses, in which new tillers emerge vertically along the stem while remaining enclosed in the sheath; tillering at or near the soil surface without production of rhizomes or stolons.

bundle sheath—A sheath of one or more layers of parenchymatous or of sclerenchymatous cells surrounding a vascular bundle in the leaf.

bypass protein—Dietary protein that passes from the rumen to the abomasum without being degraded by rumen microorganisms.

C_3 plant—A plant employing ribulosebisphosphate carboxylase as the primary CO_2-capturing enzyme, with the first product being a 3-carbon acid. These plants display photorespiration.

C_4 plant—A plant employing phosphoenolpyruvate carboxylase as the primary CO_2-capturing enzyme, with the first product being a 4-carbon acid. These plants do not display photorespiration.

callus—Soft tissue consisting of undifferentiated cells on a cut surface of a plant or from cell division in tissue culture.

cannula—A tubular device inserted into a body cavity, duct, or vessel (e.g. esophagus or rumen); mainly used to divert digesta or to allow digesta sampling.

canopy architecture (structure)—The spatial (three-dimensional) physical arrangement of leaves and stems of different species making up a pasture sward.

capsule—A dry, dehiscent fruit containing two or more seed.

carbohydrates, nonstructural—Sugars, starches, fructan, and other soluble carbohydrates found in the cell contents, as contrasted with structural carbohydrates in the cell walls. Considered available to support life processes.

carbohydrates, structural—Carbohydrates found in the cell walls (e.g. hemicellulose, cellulose); considered not available to support life processes.

carbon sequestration—The net removal of CO_2 from the atmosphere into long-lived pools of carbon in terrestrial ecosystems. The pools can be living, aboveground biomass (e.g. trees); products with a long, useful life created from biomass (e.g. lumber); living biomass in soils (e.g. roots and microorganisms); or recalcitrant organic and inorganic carbon in soils and deeper subsurface environments.

carrying capacity—The maximum stocking rate that will achieve a target level of animal performance, with a specified stocking method, that can be applied over a defined time period without deterioration of the ecosystem. Carrying capacity is not static from season-to-season or year-to-year and may be defined over fractional parts of years. The "average" carrying capacity refers to the long-term carrying capacity averaged over years, whereas the "annual" carrying capacity refers to a specific year.

caryopsis—Small, one-seeded, dry fruit with a thin pericarp surrounding and adhering to the seed; the seed (grain) or fruit of grasses.

cataphyll—The reduced, often scaly leaf structure located at each node on a rhizome.

cation exchange capacity—The weak electrostatic charge of soil particles, resulting from loss of H^+ ions, which attracts soil cations, holding them in a plant-available form.

caudex—An underground stem base of an herbaceous plant that is usually woody and from which new branches can arise.

cecum—Intestinal pouch located at the junction of large and small intestines. Site of post-gastric fermentation in nonruminant herbivores.

cellulase—Enzyme that hydrolyzes cellulose to hexose units.

cellulose—A carbohydrate formed from glucose that is linked by ß-1,4 bonds. It is a major constituent of plant cell walls.

cell wall constituents—Compounds that make up or constitute the cell wall, including cellulose, hemicellulose, lignin, and minerals (ash).

cell wall content—The proportion of plant material made up of cell walls as opposed to cell contents, usually determined by solubility differential.

chapparal—An area of grassland in a semiarid region characterized by a mixture of woody shrubs, scrub trees, and short-stature herbaceous species, mainly grasses.

chasmogamy—Opening of a mature flower in the normal way to ensure pollination and fertilization, either self- or cross-pollinated.

chemostatic—A theory describing regulation of feed intake based on blood levels of components that signal the hypothalamus gland.

chilling injury—Temporary reduction in photosynthesis and plant growth of sensitive plants following exposure to temperatures just above freezing.

chloroplast—Cellular organelle where photosynthesis occurs.

chlorosis—Yellowing or blanching of leaves and other parts of chlorophyll-bearing plants; usually caused by a mineral deficiency, temperature stress, or a pathogen.

chromic oxide—A completely indigestible chemical (Cr_2O_3) used as a marker to estimate forage intake.

cladode—A leaf-like flattened branch or stem, usually associated with cacti. Photosynthesis occurs in the cladode using CAM metabolism.

cleistogamy—The condition of having flowers, often small and inconspicuous, that are self-pollinated before the flower opens or the flower may never open.

clone—Progeny produced asexually from a single original individual by vegetative propagation, usually by cuttings or natural propagation of axillary buds, bulbs, tubers, or rhizomes.

clostridia—Gram-positive, spore-forming, anaerobic bacteria of the genus *Clostridium* that typically cause butyric acid formation in silage, resulting in low quality.

cofiring—A process in which biomass is mixed with coal and burned together in a direct combustion system to produce steam and generate electricity.

cold resistance—Ability of plants to resist cold ($<0\,°C$) temperature stress. Often manifested as cold avoidance or cold tolerance.

cold stratification—A seed treatment that simulates winter by keeping imbibed seed in a cool, moist environment for a period of time that reduces seed dormancy and increases germination when planted in normal conditions.

cold tolerance—Ability of plants to tolerate stresses when exposed to temperatures below 0 °C.

coleoptilar node—The first node above the seed on a developing grass seedling where the coleoptile is attached.

coleoptile—Specialized leaf consisting of a modified sheath that is attached to the coleoptilar node. It elongates through the seed coat of grasses and protects the shoot and its tender leaves as they are pushed through the soil to emerge above ground.

coleorhiza—Protective sheath around the radicle within the seed. The radicle grows through the coleorhiza during germination, after which the coleorhiza seals around the root, reducing pathogen entry to the seed.

collar—Area at the junction of the grass leaf blade and sheath.

companion crop—A rapidly establishing annual crop such as oat or spring wheat sown at a reduced rate along with a forage legume or grass that emerges and develops slowly. The companion crop helps control early weed growth and soil erosion (preferred term over nurse crop).

competition—Mutually adverse effects on plants utilizing a shared resource in short supply.

composite—A family of dicotyledonous plants characterized by having a head-type inflorescence with many flowers borne on a large receptacle. Examples include dandelion, sunflower, and asters.

concentrate—Feeds such as grains that are low in fiber and high in total digestible nutrients that supply primary nutrients (protein, carbohydrate, and fat).

condensed tannins—Complex polymers of flavanoid subunits that occur primarily in cell vacuoles and bind with proteins, metal ions, and polysaccharides to influence herbage nutritive value and physiological processes in grazers.

confined animal feeding operation (CAFO)—Defined by US-EPA for regulatory purposes as a facility where a large number of animals (minimum number depending on animal species and other factors) are stabled or confined, fed for at least $45\,d\,yr^{-1}$ and where no vegetation is sustained.

Conservation Reserve Program (CRP)—A voluntary federal conservation program to remove highly erodible land from crop production. Landowners can receive annual rental payments and cost-share assistance to establish long-term, resource-conserving covers on eligible farmland.

continuous grazing—Not a recommended term because animals do not graze continuously. If used, it is synonymous with continuous stocking.

continuous stocking—A method of grazing livestock on a specific unit of land where animals have unrestricted and uninterrupted access throughout the time period when grazing is allowed (cf. rotational stocking, set stocking). The duration of the grazing period should be defined.

controlled grazing—Not an acceptable term; sometimes erroneously used for increased grazing management.

cool-season plant—Plant species that grow best during cool, moist periods of the year. They commonly have temperature optimums of 15–25 °C and exhibit C_3 photosynthesis.

coprophagy—Eating of dung. Certain non-ruminant herbivores such as rabbits practice autocoprophagy, consuming their own feces to obtain vitamins and other nutrients arising from fermentation that occurs in the gastrointestinal tract past the site of active absorption.

corm—A solid, swollen stem base of a grass plant that functions as a storage tissue and usually occurs underground.

cortex—Ground tissue between the vascular system and epidermis of roots and stems that consists mainly of highly vacuolated, thin walled parenchyma cells

cotyledon—Specialized seed leaf in the embryo; two in dicots and only one in monocots (*see* scutellum). Serves as a storage organ in dicots and in nutrient absorption in monocots such as grasses.

coumarin—A white, crystalline compound with a vanilla-like odor that gives sweetclover its characteristic odor. An antiquality component of sweetclover that is converted by mold to dicoumarol which is an anticoagulant that can cause bleeding disease in cattle and horses.

cover crop—An annual planted to stabilize the topsoil and reduce water runoff and erosion between successive annual crops, usually grown over winter and grazed or harvested.

coumestrol—Estrogenic compound occurring naturally in forage crops, especially in ladino clover, strawberry clover, and alfalfa.

creep stocking—The practice of allowing juvenile animals to graze areas of high-quality forage that their dams cannot access at the same time.

crimped—Rolled between corrugated rollers, especially fresh forage, to crack stems and facilitate drying.

critical leaf area index—The leaf area index at which 95% of light is intercepted by the canopy. *See* leaf area index.

cropland—Land devoted to the production of cultivated crops. May be used to produce forage crops (cf. forage crop). Provides basis for land-use mapping unit.

cropland pasture—Cropland on which grazing occurs but is generally of limited duration.

crop residue—Portion of plants remaining after seed harvest; mainly of grain crops, such as corn stover, or of small-grain straw and stubble.

crown—The persistent base of an herbaceous perennial plant located between the soil surface and mowing/grazing height. It is comprised of stem bases, including tiller and rhizome buds.

crude fiber—Coarse, fibrous portions of plants, such as cellulose, that are partially digestible and relatively low in nutritional value. In chemical analysis, it is the residue obtained after boiling plant material with dilute acid and then with dilute alkali.

crude protein (CP)—An estimate of protein concentration in a feed or forage that is based on total nitrogen (N) concentration. Generally calculated as total N × 6.25.

cultivar—A variety, strain, or race that has originated and persisted under cultivation or was specifically developed for the purpose of cultivation. For cultivated plants, it is the equivalent of botanical variety.

cuticle—A waxy layer secreted by epidermal cells on the outer surface on plants.

cutin—A waxy, somewhat waterproof material that provides an outer covering on plants.

cyanogenesis—The release of hydrocyanic acid (HCN) from plant tissues in the process of chemical change. *Cyanogenetic* is the adjective form.

cyathium—An inflorescence in which a single female flower and several male flowers are enclosed within a cluster of modified leaves.

cyme—A branched inflorescence that is relatively flat-topped and in which the central flower opens first.

cytogenetics—Study of chromosomal structure and behavior within either a somatic or reproductive cell.

cytosol—The liquid portion (matrix) of the cell protoplasm in which organelles, proteins, salts, and sugars are dissolved or suspended.

damping-off—Seedling disease characterized by necrosis of stem or hypocotyl tissue near the soil surface. Caused by several different fungal organisms, often in poorly drained soils.

day-neutral plant—A plant that flowers independently of photoperiod. Often flowers at a particular developmental stage.

days per hectare, animal—*See* animal days per hectare.

deferment—The postponement or delay of grazing to achieve a specific management objective (cf. deferred stocking, rotational deferred). A strategy aimed at providing time for plant reproduction, establishment of new plants, restoration of plant vigor, a return to environmental conditions appropriate for grazing, or the accumulation of forage for later use.

deferred stocking—The deferment of grazing in a non-systematic rotation with other land units (cf. deferment).

defoliation—Application of a chemical or cultural practice to make leaves fall from a plant prematurely or removal of the leaves (tops) from a plant by cutting or grazing.

dehiscence, seed—Natural opening or release of a seed as the result of splitting of a seedpod or other containing structure, generally as a result of drying.

denitrification—The reduction of NO_3^- to N_2O, NO_2, and N_2, which are lost to the atmosphere.

density—The number of individuals per unit area or the relative closeness of individuals to one another.

dental pad—Cornified gum tissues of upper jaw that replace the upper incisors in ruminants.

desert—Land on which the vegetation is absent or sparse, usually shrubby, and is characterized by an arid, hot to cool climate.

desertification—The ecologic process by which an ecosystem becomes desert-like and degrades to a more xeric, less productive condition.

determinate—Growth habit characterized by termination of vegetative growth by differentiation of the shoot apex into an inflorescence.

diaspore—A seed unit of grasses consisting of one or more spikelets that tend to remain attached to one another during harvest and processing resulting in a seed unit that consists of more than one floret.

diet learning—Learning processes in animals that lead to the selection of familiar feedstuffs and refusal of toxic ones for ingestion.

diet selection—Expression of diet learning during feeding.

digesta—Contents of the digestive tract of animals. Consists mainly of undigested and partially digested food.

digestibility—Proportion of dry matter or a defined constituent digested within the digestive tract of the animal.

digestible—Subject to digestion in the gastrointestinal system of an animal.

digestible dry matter (DDM)—Dry weight of feed consumed minus dry weight of feces, expressed as percentage of feed dry matter consumed.

digestible energy (DE)—Feed-intake gross energy minus fecal energy, expressed as calories per unit feed dry matter consumed.

digestible nutrients—Portion of nutrients consumed that is digested and taken into the animal body. This may be either apparent or true digestibility; generally applied to energy and protein.

digestible protein—Feed protein minus feces protein (N content times 6.25), expressed as a percentage of amount in feed.

digestion—The conversion of complex, generally insoluble foods to simple substances that are soluble in water.

dioecious—Male and female flowers are located on different individuals of a species, that is, male and female plants.

direct combustion—Burning. The transformation of biomass fuel into heat, chemicals, and gases through chemical combination of hydrogen and carbon in the fuel with oxygen in the air.

distal esophageal sphincter—A constriction in the esophagus that separates the esophagus from the reticulum and regulates reflux.

diurnal—The repeated change in a process during a 24-hours period, mainly due to the differences between day and night.

dormancy—A period of arrested growth and development due to physical or physiological factors.

dough stage—Seed development stage at which endosperm is pliable, like dough (e.g. soft, medium, hard); usually used when 50% of seed on an inflorescence are in this stage of development.

drought avoidance—Ability of a plant to avoid water deficit stress. Often manifested by increased rooting or reduced transpiration so the plant is not stressed.

drought resistance—Ability of plants to resist injury caused by drought stress. Often manifested by drought avoidance or drought tolerance.

drought tolerance—Ability of plants to survive and produce yield under conditions of water deficit stress.

dry matter (DM)—The mass of a plant remaining after oven drying to constant weight at a temperature slightly above the boiling point of water.

dry matter disappearance (DMD)—(i) *Grazing:* forage present at the beginning of a stocking period plus growth during the period, minus forage present at the end of the period. (ii) *Digestibility:* loss in dry weight of forage exposed to *in vitro* digestion.

dry matter intake, daily—Amount of dry matter ingested by an animal on a daily basis.

duodenum—The first part of the small intestine where digesta is further broken down chemically in preparation for absorption of nutrients by the small intestine.

ecodormancy—Dormancy of a tissue or plant caused by environmental constraints such as water or temperature stress. Plant or tissue growth resumes when the environmental constraint is alleviated.

ecological intensification—Optimal management of natural ecological functions to increase ecosystem performance.

ecologically based pest management—An integrated pest management system that eliminates the use of broad-spectrum pesticides.

ecology—The study of communities of living things and the relationships between organisms and their environment.

ecosystem—A living community and all the factors in its nonliving environment. Can be natural or managed for a purpose such as forest, grassland, or agricultural crops.

ecosystem services—Benefits provided by ecosystems including provisioning services of food and water, regulating services of flood and pest control, cultural services of recreational opportunities, and supporting services of nutrient cycling.

ecotype—A variety or strain within a given species that maintains its distinct identity by adaptation to a specific environment.

edaphic—Relating to or influenced by the soil.

effluent—The liquid, which contains some nutrients and other solids, that is lost from silage.

embolism—The formation of air pockets in xylem vessels and/or tracheids that interrupts the flow of water in plants, mainly trees.

embryo—A young plant that exists in an arrested state of development within a seed.

embryo axis—The growing parts of the seed, including the radicle, coleorhiza, coleoptilar node, and epicotyl.

endodormancy—Also known as innate dormancy, is associated with physiological and biochemical inhibitors of growth that develop and exist within the dormant tissue. This tissue will not grow even if placed in growth-conducive conditions.

endophyte—An organism that lives at least part of its life cycle within a host plant, as a parasite or symbiont.

endosperm—Nutritive tissue in seed plants formed within the embryo sac by the union of a male nucleus with two polar nuclei of the female.

ensilage—*See* silage.

enzyme—A protein that catalyzes one or more biochemical reactions within a living cell.

epicotyl—The shoot, or upper portion of the plant, that is above the cotyledonary node.

epidermis—The outermost cells of the primary plant body, usually consisting of a single layer but sometimes several layers thick.

epinasty—Increased growth on upper surface of a plant organ or part (especially the leaf), causing it to bend downward.

equilibrium moisture—A hay-curing term that describes the moisture concentration at which no further exchange of water occurs between the drying crop and the atmosphere.

eructation—Belching of fermentation gases in the rumen through the esophagus and mouth.

escape protein—*See* bypass protein.

ether extract—Fats, waxes, oils, and similar plant components that are extracted with warm ether in chemical analysis.

evapotranspiration—The sum of water loss due to evaporation from soil and transpiration from plants.

exotic plant—An introduced plant that is not fully naturalized or acclimated.

extensive grazing management—Grazing management that utilizes relatively large land areas per animal and a relatively low level of labor, resources, or capital (cf. management intensive grazing).

extravaginal tiller—A lateral tiller that penetrates the leaf sheaths of the parent tiller.

facilitated recurrent selection—A type of recurrent selection in which genetic male sterility is maintained in the population to maintain heterozygosity and genetic diversity and to permit the recombination and shifting of gene frequencies.

fatty acids—Organic compounds with the general formula R-COOH. In common forages, R usually is between 1 and 17 carbons with one, two, or three hydrogens. Fatty acids are the primary source of digestible energy provided by plant lipids.

fecal index—Indirect method of estimating digestibility of dry matter by determining the differences in concentration of an indigestible indicator in the feed and feces.

feed supplements—Feeds added to a ruminant's diet to provide the daily requirement for one or more nutrients.

feeding deterrents—Chemical and physical barriers of ingestion.

feeding station—An area that can be grazed by an animal without taking another step.

feeding value—Characteristics that make feed valuable to animals as a source of nutrients; the combination of chemical, biochemical, physical, and organoleptic characteristics of forage that determine its potentials to produce meat, milk, wool, or work. Considered by some as synonymous with nutritive value. *See* forage quality.

feedstock—Raw materials, such as bioenergy crops and agricultural residues and wastes, used in the production of biofuels or other industrial products.

fermentation—Anaerobic chemical transformation induced by activity of enzyme systems of microorganisms, such as yeast that produce CO_2 and alcohol from sugar or, in silage, that produce organic acids such as lactic and propionic acids from sugars.

fescue foot—Syndrome characterized by red and swollen skin at junction of the hoof in cattle grazing tall fescue, along with loss of appetite and emaciation. Sloughing off of hoofs, tail tips, and ear tips may occur in advanced stages.

fescue toxicosis—The collective animal syndromes associated with intake of endophyte-infected tall fescue. These include fescue foot, fat necrosis, what is called summer syndrome, and other related disorders.

festucoid—A type of grass seedling development where there is a long coleoptile and generally little or no subcoleoptilar internode elongation. Characteristic of grasses belonging to the Festucoideae subfamily.

fiber—A nutritional entity that is relatively resistant to digestion and is slowly and only partially degraded by herbivores. Forage fiber is composed of structural polysaccharides, cell wall proteins, and lignin.

first-generation biofuel—Refers to biofuels manufactured from sugars and oils of arable food crops via conventional processes (e.g. ethanol from corn grain).

first-last stocking—A method of utilizing two or more groups of animals, usually with different nutritional requirements, to graze sequentially on the same land area. (*Syn.* leader-follower, preference follower, top and bottom grazing. First-last stocking is the preferred term.) If more than two groups of animals are grazed sequentially, this would be described as "first, second, and last stocking."

fistula—A surgical opening, duct, or passage from a cavity or hollow organ of the body.

fixed stocking—Not a recommended term. If used, it is synonymous with set stocking.

flag leaf—The uppermost leaf on a fruiting (fertile) grass culm; the leaf immediately below the inflorescence or seedhead.

flash stocking—*See* mob stocking.

flooding tolerance—Ability of plants to tolerate injury associated with anoxia and other stresses caused by lack of soil air and water inundation.

floret—The grass flower. The lemma and palea and enclosed stamens and pistil (later with included caryopsis).

florigen—A compound or signal that is stimulated by changes in day length that regulates flowering of plants.

flow cytometry—A method to determine the ploidy or DNA content of a plant by counting cells with stained nuclei.

flowering—The physiologic stage of a grass plant in which anthesis (blooming) occurs, or in which flowers are visible in nongrass plants.

fluorescence—The emission of electromagnetic radiation from a body, resulting from the absorption of incident radiation and persisting only as long as the stimulating radiation is continued.

flushing—Improving the nutrition of female breeding animals, such as by providing high-quality forage or energy concentrates prior to and at the beginning of the breeding season, as a means of stimulating ovulation.

fodder—Coarse grasses such as corn and sorghum harvested when the seed and leaves are green or live; cured and fed in its entirety as forage.

foliage—The green or live leaves of growing plants; plant leaves collectively. Often used in reference to aboveground development of forage plants.

follicle—A dry, dehiscent fruit that splits along one side.

forage—*n.* Edible parts of plants, other than separated grain, that can provide feed for grazing animals or that can be harvested for feeding. Includes browse, herbage, and mast. *v.* To search for or to consume forage (cf. browse, graze).

forage allowance—The relationship between the weight of grazable forage dry matter per unit land area and the number of animal units or forage intake units at any one point in time; a forage-to-animal relationship. The inverse of grazing pressure. May be expressed as forage mass per animal unit or forage intake unit (forage mass per animal unit at a specific time). This definition can be specific to *herbage* or *browse* by substituting these terms in place of *forage*.

forage crop—A crop of cultivated plants or plant parts, other than separated grain, produced to be grazed or harvested for use as feed for animals.

forage intake potential—Maximum intake possible when forage characteristics rather than animal characteristics limit forage consumption.

forage intake unit—An animal with a rate of forage consumption equal to 8.8 kg dry matter per day (kg DM d^{-1}).

forage mass—The total dry weight of forage per unit area of land, usually that above ground or at a defined reference level (cf. available forage, herbage). This definition can be specific to *herbage* or *browse* mass by substituting these terms for *forage*.

forage quality—Characteristics that make forage valuable to animals as a source of nutrients; the combination of chemical and biological characteristics of forage that determines its potential to produce meat, milk, wool, or work. Considered by some as synonymous with feeding value and nutritive value.

forb—Any herbaceous broadleaf plant, including legumes, that is not a grass and is not grasslike (cf. legume, grass, grasslike plant).

forest grazing—The combined use of forestland or woodland for both wood production and animal production by grazing of the coexisting indigenous forage, or vegetation that is managed like indigenous forage.

forestland—Land on which the vegetation is dominated by forest or, if trees are lacking, the land bears evidence of former forest and has not been converted to other vegetation. Used as basis for land-use mapping unit.

forest range—A forest ecosystem that produces, at least periodically, an understory of natural herbaceous or shrubby plants that can be grazed or browsed.

forward creep—A method of creep stocking in which dams and offspring rotate through a series of paddocks with offspring as first grazers and dams as last grazers. A specific form of first-last stocking (cf. first-last stocking).

fragipan—A compact, dense layer in a soil that restricts vertical water movement and root penetration.

free water—Tissue water that is not associated with cellular constituents, which can freeze when temperatures decline below 0 °C.

frontal stocking—A stocking method that allocates forage within a land area by means of a sliding fence that livestock can advance to gain access to ungrazed forage.

fructan—A polymer of fructose that includes one glucose molecule. Synthesized from sucrose, the polymer is water soluble, stored in vacuoles, and largely functions as an osmoticum or storage carbohydrate.

gasification—A chemical or heat process to convert a solid fuel to a gaseous form.

gas-liquid chromatography (GLC)—An analytical method in which a sample (usually after extraction) is injected onto a heated column containing specific materials that allow volatile compounds in the sample to migrate through the column at unique rates. As the compounds elute from the column, they are detected and quantified by various methods.

gastric lysozymes—Enzymes secreted by the stomach that degrade bacterial cell walls.

gene—The physical and functional unit of heredity; a segment of chromosome, plasmid, or DNA molecule that encodes a functional protein or RNA molecule.

gene splicing—The addition of a new gene on a chromosome.

genetic engineering—Transfer of selected genes between species and genera using a vector, without using gametes.

genetic shift—A change in the gene frequency, sometimes impacting the phenotype, from the original seed lot.

genetic transformation—The directed insertion of a gene into the genome to create a transgenic plant.

genome—The hereditary material of a cell comprising an entire chromosomal set as found in each nucleus or organelle of a given species.

genome-wide association studies (GWAS)—A method of identifying genetic markers associated with a trait of interest by evaluating large, diverse germplasm collections or breeding populations.

genomic selection—A method to select plants based only on genome-wide genetic markers using a statistical model previously developed that assigns phenotypic effects to each marker allele.

genomics—The study of the genome.

genotype—Genetic makeup of an individual or group.

genotypic plasticity—The ability of a plant population to change, especially for heterogeneous species, due to differential survival based on morphological and physiological attributes (nonreversible) or to differences in seed genetics due to cross-pollination (reversible).

genus—A taxonomic category that designates a closely related and definable group of plants, including one or more species.

germ—*Biology:* a small organic structure or cell from which a new organism may develop. *Seed:* refers to the embryo.

germination—Resumption of active growth of a seed that results in rupture of the seed coat and emergence of the radicle.

germplasm—The living substance of the cell nucleus that determines the hereditary properties of organisms and that transmits these properties to the next generation.

gestation—The time interval in mammals between conception and birth during which the offspring develop.

glandular stomach—A stomach that secretes enzymes and acids from its glands. The abomasum of ruminants, for example.

glaucous—Pale green or blue-gray color often the result of a powdery, waxy layer.

glumes—The pair of bracts on the grass spikelet that contain the floret or florets.

gluten—A mixture of plant proteins occurring in cereal grains, chiefly corn and wheat; substance in wheat flour that gives cohesiveness to dough.

grain grade—Market standard established to describe the amount of contamination, grain damage, immaturity, test weight, and marketable traits.

grain maturity—Stage of growth after which no further dry matter is accumulated in the grain.

grain-to-stalk ratio—Threshed grain weight per stalk weight.

grana—Stacks of thylakoid membranes in the stroma of chloroplasts that are the site of light reactions in photosynthesis.

grass—Members of the plant family Poaceae.

grassed waterway—A natural or constructed watercourse consisting of vegetation and designed to accommodate concentrated flows without erosion. Grassed waterways are capable of sustaining higher in-channel velocities than bare areas because the vegetation protects the soil by covering it and reducing water velocity.

grassland—Land on which the vegetation is dominated by grasses (cf. pastureland, rangeland).

grassland agriculture—A farming system that emphasizes the importance of grasses, legumes, and other forages in livestock and land management.

grasslike plant—Vegetation that appears similar to grass and is usually a member of the families Cyperaceae (sedges) or Juncaceae (rushes).

grass tetany—Livestock disorder caused by low levels of blood magnesium. Also known as hypomagnesemic tetany.

gravitropism—Orientation of plant growth relative to the force of gravity; e.g. the downward growth of plant roots in response to gravity.

grazable forestland—Forestland that produces, at least periodically, sufficient understory vegetation that can be grazed. Forage is indigenous or, if introduced, it is managed as though it were indigenous. *Syn.:* grazable woodland, woodland range, forest range.

graze—*v.* The consumption of forage *in situ* by animals (cf. browse). This verb should be used in the active form, with the animal as the subject, and not in the passive voice: that is, cattle graze; people do not graze cattle.

grazed horizon—The layer of a sward that is removed in a grazing event.

grazer—Animal that grazes standing herbage.

grazier—A person who manages a grazing system.

grazing, high-intensity/low-frequency—A grazing system in which the forage on individual pastures is removed by grazing in a relatively short period (high-intensity, days) and the pasture is not grazed again for a relatively long period (low-frequency, weeks).

grazing capacity—Synonym for carrying capacity.

grazing cycle—See stocking cycle.

grazing event—The length of time that an animal grazes without stopping (cf. grazing period).

grazing intensity—A general term that refers to the severity of grazing. Specific measures of grazing intensity include forage mass, canopy height (under continuous stocking) or post-grazing stubble height (under rotational stocking) of the pasture, forage allowance, grazing pressure, or stocking rate.

grazing land—Any vegetated land that is grazed or has the potential to be grazed.

grazing land management—The manipulation of the soil–plant–animal complex of the grazing land in pursuit of a desired result.

grazing management—The manipulation of animal grazing in pursuit of a defined objective.

grazing management unit—The grazing land area used to support a group of grazing animals for a grazing season. It may be a single area or it may have a number of subdivisions (cf. paddock, pasture).

grazing method—See stocking method.

grazing period—See stocking period.

grazing pressure—The relationship between the number of animal units or forage intake units and the weight of forage dry matter per unit area at any one point in time; an animal-to-forage relationship. May be expressed as animal units or forage intake units to forage mass.

grazing season—See stocking season.

grazing system—A defined, integrated combination of animal, plant, soil, and other environmental components and the stocking method(s) by which the system is managed to achieve specific results or goals.

greenchop—Forage cut and removed from the field to be fed fresh to livestock without a specific plant moisture content goal for removal.

green manure crop—A crop, usually an annual, planted with the primary purpose of incorporation into the soil to increase content of organic matter and/or nitrogen if a legume. A portion of these crops is sometimes grazed or harvested prior to incorporation of the remaining biomass.

greenhouse gases—Chemical compounds or gases that allow sunlight (which is mainly visible) and ultraviolet radiation to pass through the atmosphere unimpeded. Some sunlight absorbed by the earth's surface is reradiated as infrared radiation (heat) and absorbed by the greenhouse gases, trapping the heat in the atmosphere. Greenhouse gases include CO_2, ammonia, methane, nitrous oxide, and ozone.

grinding resistance—A laboratory method that measures the amount of energy (i.e. electrical work) required to grind a specific quantity of feed through a specific mesh size. Grinding resistance is negatively correlated with digestibility.

groat—The caryopsis (kernel) of oats after the husk (lemma and palea) has been removed.

gross energy (GE)—The amount of heat released when a substance is completely oxidized in a bomb calorimeter containing 25–30 atm of O_2.

growth respiration—The portion of aerobic respiration used for growth processes such as the synthesis of cell walls.

hard seed—Seed that does not germinate due to seed coat characters that limit water absorption until the coat ages or is made permeable by abrasion or microbial activity in the soil.

hay—Forage preserved by field drying to moisture levels low enough to prevent microbial activity that leads to spoilage.

haylage—Product resulting from ensiling forage with about 45% moisture in the absence of O_2.

head—A type of inflorescence in which several individual flowers are attached directly to an enlarged receptacle.

heading—The stage of development of a grass plant between initial emergence of the inflorescence from the last leaf (boot) and the time the inflorescence is fully exerted.

heating degree days—Daily increment by which the internal bale temperature exceeds a designated threshold temperature (usually 30 °C) summed over the entire interval the bale is in storage.

heavy metal—Those metals that have densities in soil $>5.0\ Mg\ m^{-3}$. These include the elements Cd, Co, Cr, Cu, Fe, Hg, Mn, Mo, Ni, Pb, and Zn, which can be taken up by plants.

hemicellulose—Polysaccharides associated with cellulose and lignin in the cell walls of green plants. It differs from cellulose in that it is soluble in alkali and, with acid hydrolysis, gives rise to uronic acid, xylose, galactose, and other carbohydrates, as well as glucose.

herbaceous—A plant that dies back to the ground each year as opposed to a woody plant.

herbage—The biomass of herbaceous plants, other than separated grain, generally above ground but including edible roots and tubers.

herbivore—An animal, insect, or other higher organism that subsists primarily on plants or plant products.

herbivory—Consumption of plant material by animals.

heterofermentative—Microorganisms that produce only 50% lactic acid and considerable amounts of ethanol, acetic acid, and CO_2 while fermenting glucose.

heterogenous population—A population of plants with different genotypes. Individual genotypes can be homozygous or heterozygous.

heterosis—Improved yield or vigor obtained from crossbreeding genetically different plants. *Syn.:* hybrid vigor.

heterotroph—An organism that is not capable of synthesizing its own organic food supply, so it is dependent on another organism or its products as an energy source.

heterozygous—Situation when unlike alleles exist at one or more corresponding loci of an individual. (Opposite of homozygous.)

high-intensity grazing—This is a relative concept and not an acceptable term (cf. grazing management, management intensive grazing).

high moisture silage—Silage prepared from plant material without wilting or otherwise drying before ensiling; often containing 70% or more moisture.

high performance liquid chromatography (HPLC)—An analytical procedure in which a sample (usually after extraction) is injected into a liquid solvent that is forced through a column with a specific packing material under high pressure. Compounds migrate through the column at different rates and are detected as they elute from the column so that they can be quantified.

histochemistry—The chemistry of cells and tissues.

homofermentative—Microorganisms that produce nearly all lactic acid while fermenting glucose.

homozygous—Identical alleles are at corresponding loci on chromosomes. A plant can be homozygous at one, many, or all loci.

hybrid—First-generation progeny resulting from the controlled cross-fertilization between individuals that differ in one or more genes.

hydraulic lift—A mechanism where plants with tap roots in very dry soils redistribute water from moist, deep soil layers to dry soil near the soil surface. Hydraulic lift can occur at night, when the stomata of the deep-rooted plant are closed.

hydrolyzable tannin—A large molecule, composed of glucose units esterified with phenolic groups, that occurs mainly in fruit pods and plant galls. With metabolic degradation, the products are easily absorbed and can be toxic to ruminants.

hyper-accumulator—A plant that takes up and retains much higher quantities of an element than is normal for that species. Usually used in reference to remediation of disturbed soils and with reference to heavy metals or xenobiotic materials.

hypocotyl—The region of the embryonic axis located between the cotyledonary node and the radicle.

hypomagnesemic tetany—*See* grass tetany.

hyponastic—Increased growth on lower surface of a plant organ or part (especially the leaf), causing it to bend upward.

hypoxia—Condition of low O_2 concentration.

hypsodont teeth—Teeth with relatively large crowns and short roots that are characteristic of herbivores.

ice sheet—A relatively thin ice layer covering the soil surface and the plants present at that location. Restricts atmospheric gas exchange, leading to hypoxia.

immobilization—Binding of soil-N by carbon, typically when the C:N ratio exceeds 10–12.

incompatibility (self-incompatibility)—Genetically controlled physiological processes that inhibit or prevent self-fertilization.

indeterminate—Growth habit characterized by continuation of vegetative growth of the apical meristem while lateral apices differentiate into inflorescences.

indigenous—Originating or produced naturally in a particular region or environment; native.

induction—The change in status of a shoot apex that gives it potential to flower. Response is stimulated by exposure to a prolonged cold period. (*See* vernalization.)

inflorescence—The reproductive structure that contains the flowers or spikelets of a plant.

infrared—Electromagnetic radiation with wavelengths longer than 700 nm and less than 1 mm.

inoculation—Introducing or placing a microorganism or bacteria on a plant part, especially rhizobia bacteria being placed in or on a legume seed.

in situ—In the natural or original position.

intake—Quantity of forage consumed by animal during a specified period. Usually expressed in units of kilograms per day (kg d^{-1}).

integrated pest management—An approach to pest management based on biological knowledge of the pest and host, observations of conditions in the field, and economic assessment of alternative controls. The goal is to select the best control procedure including biological, cultural, genetic, and chemical methods.

intercalary meristem—A zone of cell division and cell elongation in grass shoots that is not part of the shoot apex. Functions mainly for extension growth of leaf blades, leaf sheaths, and culm internodes.

intermittent stocking—A method that imposes grazing for indefinite periods at irregular intervals.

interseeding—*See* sod-seeding.

intravaginal tiller—An upright tiller that emerges at the collar of the subtending leaf and does not penetrate the leaf sheaths of a parent tiller.

introduced species—A species not part of the original fauna or flora of the area in question, that is, brought from another geographical region by human activity.

invasive species—An alien species whose introduction does or is likely to cause economic or environmental harm or harm to human health.

in vitro—In glass, outside the living body.

***in vitro* dry matter disappearance (IVDMD)**—A gravimetric measurement of the amount of dry matter lost upon filtration following the incubation of forage in test tubes with rumen microflora. Usually expressed as a percentage of the dry sample weight.

***in vitro* dry matter digestibility (IVDMD)**—*See* *in vitro* dry matter disappearance.

***in vitro* gas production**—Method used to estimate rumen fermentation of feeds. The amount of gas produced when feeds are incubated in vessels containing rumen fluid is proportional to the amount of mass fermented.

in vivo—In a living organism, such as in an animal or plant.

***in vivo* nylon bag technique**—System of determining dry matter disappearance of forage contained in a fine mesh nylon bag, after placing the bag in the rumen of a fistulated animal for a specified period of time, usually 48 hours.

keel—Two fused petals within a legume flower that enclose the stamens and pistil.

kernel—Agronomic term for mature ovule of a grass plant that has the ovary wall fused to it. Botanic term is caryopsis.

ketosis—A pathological accumulation of ketone bodies in an organism.

killing frost—A temperature that affects the shoot apex enough to stop growth but not kill all the leaves; generally considered to be about −4.5 °C for upright legumes that have the apex near the top of the canopy.

lacrimation—Secretion or shedding of tears, usually associated with excessive, profuse tearing.

lactic acid bacteria (LAB)—A group of related bacteria with complex nutritional requirements, and lacking many biosynthetic capabilities, that produce lactic acid during carbohydrate fermentation. Lactic acid bacteria are important in silage preservation.

land capability class—A classification of soils or landscapes by the USDA-NRCS based on suitability for cultivation and necessity for conservation practices.

latitude—The angular distance north or south from the earth's equator measured in degrees.

leaf area index—The ratio of leaf surface area of plants to the land area on which the plants are growing. A measure of the relative density of leaves within a canopy.

leghemoglobin—An O_2 carrier in legume root nodules used to capture and supply O_2 for respiration while maintaining a low O_2 concentration within nodule cells.

legume—Members of the plant family Fabaceae.

lemma—Outer or lower covering of the grass floret that is usually larger and heavier than the palea.

ley—The forage component of a crop rotation that includes cultivated grain crops.

life cycle analysis—A comprehensive evaluation of the environmental and economic impacts of products, materials, or processes through quantifying their energy and material flows at all stages across their full life cycle from materials acquisition to manufacturing, use, and disposal.

life cycle assessment—*See* life cycle analysis.

lignin—An organic chemical, of very low digestibility, that strengthens and hardens the walls of plant cells, especially those of vascular tissues and the epidermis.

lignocellulose—Plant materials made up primarily of lignin, cellulose, and hemicellulose.

ligule—The membrane-like projection on the inner side of the leaf sheath arising at the collar.

lime—Pulverized limestone, which provides $CaCO_3$ when applied to the soil to reduce acidity.

lipid—An organic compound that contains long-chain aliphatic hydrocarbons and their derivatives, such as fatty acids, alcohols, amines, amino alcohols, and aldehydes; includes waxes, fats, and derived compounds.

liquefaction—Production of liquid fuels from the reaction of biomass with certain gases at high temperatures and pressures with a catalyst in the absence of air.

lodging—The falling down of a crop due to either stalk breakage or uprooting.

lodicules—Small sacs in the base of the grass flower that expand to help force open the lemma and palea at anthesis to facilitate cross pollination.

long-day plant—A plant that flowers under long photoperiods (short nights). Cf. short-day plant.

low-moisture silage—Silage prepared from relatively dry plant material, usually below 50% moisture.

Lucas method—A statistical method used to determine the true digestibility of a nutrient. The concentration of a digestible nutrient (dependent variable) is regressed on the concentration of a nutrient (independent variable). If the fraction is nutritionally uniform, the data fit a straight line and the slope is an estimate of true digestibility of the nutrient.

lumen—Inner space of a tubular structure (e.g. esophagus).

lyophilize—A procedure that removes water directly from ice in a frozen sample by evaporation in a vacuum (also known as freeze-drying).

macroclimate—Climate occurring over a large geographic scale that is independent of local topography and vegetation.

maillard browning reaction—Refers to the reaction between reducing sugars and exposed amino groups in proteins to form a complex that undergoes a series of reactions to produce brown polymers. Higher temperatures and basic pH favor the reaction. The process reduces the digestibility of the reactants.

maintenance respiration—The portion of aerobic respiration used to support ongoing functions of nongrowing tissues; that is, it does not contribute directly to growth.

major land resource area—A region defined by the USDA-NRCS as a major soil group having distinctive physical features (e.g. topography or hydrology) that determine the dominant land use.

management intensive grazing—Management of grazing designed to increase animal production or forage utilization per unit area or production per animal through knowledge-based use of stocking rates, forage utilization, labor, resources, or capital. (Preferred term to intensive grazing management; cf. extensive grazing management.)

marshland—Flat, wet, treeless land usually covered by water and dominated by marsh grasses, indigenous rushes, sedges, or other grasslike plants.

mast—Fruits and seed of shrubs, woody vines, trees, cacti, and other nonherbaceous vegetation available for animal consumption.

mastication—Initial chewing prior to swallowing; in ruminants, chewing the cud after regurgitation of a bolus.

meadow—A tract of grassland where productivity of indigenous or introduced forage species is modified due to characteristics of the landscape position or hydrology (cf. grassland, pasture, pastureland, rangeland). May be characterized as hay meadow, native meadow, mountain meadow, wet meadow, or other designations.

meristem—A localized group of dividing cells from which tissue systems (e.g. root, shoot, leaf, inflorescence) are derived.

meristematic—Small, undifferentiated, rapidly dividing cells from which other cells and tissues arise.

mesocotyl—An alternative term for subcoleoptilar internode of grasses, the one between the scutellar node and the coleoptilar node.

mesophyll—Thin-walled leaf cells that contain chloroplasts and are located between the upper and lower epidermis.

metabolic body weight—Basal metabolic rate (energy expenditure per unit body weight per unit time; i.e. kcal heat/weight/day) varies as a function of a fractional power of body weight, usually determined to be body weight raised to the 0.75 power.

metabolizable energy (ME)—Digestible energy (DE) less the energy lost in urine and as methane from the rumen.

metabolome—The complete set of metabolites produced by a plant.

methane—A gas (CH_4) produced naturally by respiration under anaerobic conditions such as in the rumen or a wetland.

microbiome—The community of microorganisms that inhabit a particular environment.

microclimate—The local, rather uniform climate of a specific place or habitat, such as within and near to a plant canopy, compared with the climate of the entire area of which it is a part.

microfibril—An aggregation of cellulose molecules as found in cell walls.

micronutrient—Plant nutrient found in relatively small amounts ($<100\ mg\ kg^{-1}$) in plants. Examples are B, Cl, Cu, Fe, Mn, Mo, Ni, Co, and Zn.

microsward—A population of herbaceous plants resulting from collecting and arranging species or sowing species in small containers in order that they may be presented to grazing livestock for short-term measures of diet selection or grazing behavior.

middlings—A co-product of flour milling that contains varying proportions of bran, germ, and residual endosperm.

milk stage—In grain (seed), the stage of development following pollination in which the endosperm appears as a whitish liquid that is somewhat like milk.

mimosine—A nonprotein amino acid in leucaena leaves that is metabolized in the rumen to 3-hydroxy-4(1H)-pyridone (DHP), which is a potent goitrogen, a depilatory agent, and reduces feed intake.

mineralization—Conversion of nutrients (especially N) by soil microorganisms from either organic or inorganic sources into plant-available forms.

mitochondria—Organelles in all living cells where aerobic respiration produces ATP from carbohydrates and lipids.

mixed stocking—Grazing by two or more species of grazing animals on the same land unit, not necessarily at the same time but within the same grazing season.

mob stocking—Grazing by a relatively large number of animals at a high stocking density for a short time.

moisture equilibrium—The condition reached by a sample when it no longer takes up moisture from, or gives up moisture to, the surrounding atmosphere.

monecious—Female (pistil) and male (stamens) organs are in separate flowers but on the same plant, as in corn.

monogastric—Simple-stomached animals that lack significant capability to digest structural carbohydrate (fiber).

morphology—The features comprising the form and structure of an organism or any of its parts.

multifunctional agriculture—Agriculture that has roles or functions in addition to producing food and fiber. These might include agriculture's contribution to long-term food security, the viability of rural areas, cultural heritage, land conservation, maintenance of agricultural landscapes, recreation and biological diversity.

mutation—A heritable change in the genetic material of a cell.

mycorrhiza—A fungus, usually filamentous, that grows in association with the roots of a plant in a symbiotic or mildly pathogenic relationship.

NADH (nicotinamide adenine dinucleotide)—An electron acceptor that functions in respiration to reduce other compounds.

NADPH (nicotinamide adenine dinucleotide phosphate)—An electron acceptor that functions in photosynthesis to reduce other compounds.

native forages—Forage species indigenous to an area, not introduced from another environment or area.

naturalized grasses—Non-indigenous grasses that have become established in and more or less adapted to a given region.

naturalized pasture—Plants introduced from other countries that have become established in and more or less adapted to a given region by long continued growth there. The name is appropriate for pastures made up of plants such as white clover, kentucky bluegrass, and bermudagrass.

near infrared reflectance spectroscopy (NIRS)—A method of forage quality analysis based on spectrophotometry measurements of reflected radiation at wavelengths in the near infrared region (780–2500 nm).

nematodes—Elongated, cylindrical worms that feed on plants in soil.

net energy (NE)—Metabolizable energy minus the energy lost in the heat increment.

net primary production—The net assimilation (corrected for respiration) of atmospheric CO_2 into plant mass per area per unit time.

neurohormones—Chemicals associated with functions of neural systems.

neutral detergent fiber—Insoluble residue following extraction with a neutral detergent solution, consisting mainly of cell wall constituents (cellulose, hemicellulose, and lignin) of low biological availability.

neutral detergent fiber digestibility (NDF)—The digestibility of neutral detergent fiber determined as the difference in NDF in a forage before and after *in vivo* or *in vitro* digestion.

neutral detergent insoluble nitrogen (NDIN)—Nitrogen in the residue after a feed is extracted with neutral detergent solution.

niche complementarity—Growth of organisms or species in environments that supply resources necessary for existence but in manners that complement environments of others, thereby providing a mechanism for species coexistence.

nitrate (NO_3^-)—A common form of nitrogen in the soil that is readily taken up by plants.

nitrate poisoning—*See* nitrate toxicity.

nitrate toxicity—Condition in animals resulting from ingestion of feed high in nitrate (NO_3^-). Actual toxicity is from nitrite (NO_2), which results from NO_3^- reduction to NO_2 in the rumen. Nitrite reduces the O_2 carrying capacity of the blood.

nitrification—Oxidation of NH_4^+ to NO_3^- by the free-living soil bacteria *Nitrosomonas* and *Nitrobacter*.

nitrogen fixation—Process by which atmospheric nitrogen (N_2) is made available to plants by rhizobia that reduce it to two molecules of NH_3.

nitrogen-free extract (NFE)—The highly digestible portion of a plant, consisting mostly of carbohydrates, that remains after the protein, ash, crude fiber, ether extract, and moisture content have been determined.

nonfibrous carbohydrate (NFC)—Made up of starch, simple sugars, and soluble fiber. NSC is calculated by difference [100 − ((% NDF − (% NDF − % CP) + % CP + % Fat +% Ash))].

nonnutritive fiber—That portion of fiber in a feed that is not digestible and hence is of no nutritive value.

nonprotein nitrogen—Soluble fraction of nitrogen in the plant. Includes inorganic forms of nitrogen and that contained in low molecular weight compounds such as ammonia, free amino acids, and nucleic acids.

nonselective grazing—Utilization of forage by grazing animals so that all forage species and/or all plants within a species are grazed (cf. mob grazing). Nonselective grazing is generally encouraged by high stocking rates or high stocking densities for short periods. In practice, nonselective grazing is rarely achieved.

nonstructural carbohydrates—*See* carbohydrates, nonstructural.

no-till seeding—Establishing a crop directly into soil that has not been tilled.

noxious weed—An undesirable plant for which control measures are required by law.

nuclear magnetic resonance (NMR)—A type of radio frequency or microwave spectroscopy used for quantitative analysis and measuring rates of reactions of chemicals in solution.

nutraceutical—A food, or parts of a food, that provide medical or health benefits, including the prevention and treatment of disease.

nutrient—In a strict sense, an element of the periodic table that is essential for plant or animal life. More broadly, a food constituent of a diet that sustains life.

nutrient, animal—Food constituent or group of food constituents of the same general chemical composition required for support of animal life.

nutrient balance—Net difference between the cumulative additions of nutrients to a system (field, farm, or watershed) and losses from that system.

nutritive value—Chemical composition, digestibility, and nature of digested products of a forage.

nutritive value index (NVI)—Daily amount of digestible forage per unit of metabolic body size relative to a standard forage.

occlusal—Related to grinding surfaces of teeth.

omasum—The third chamber of the ruminant stomach, where the partially digested contents are mixed to a more or less homogeneous state.

omnivore—An organism that eats both plants and animals.

ontogenetic—Resulting from the combined and indistinguishable genetic and experiential effects that give rise to the origin and development of an individual.

opaque-2—An endosperm mutant of maize associated with suppressed prolamine production in the endosperm, resulting in increased lysine content of the protein fraction.

organic reserve—Sugars, polysaccharides, amino acids, and storage proteins accumulated in vegetative tissues that can be translocated to and used by other plant organs.

organoleptic—Sensible characteristics such as color, texture, and odor used in describing hay.

orts—Rejected feedstuffs left under conditions of *ad libitum* stall feeding.

osmolarity—Molar concentration of dissolved substances.

osmotica—Solutes that lower the osmotic potential of plant cells to increase water uptake in growing tissues, to reduce water loss in plants exposed to drought or freezing stress, and to lower the freezing point during cold stress.

osmotic adjustment—Lowering of osmotic potential due to net accumulation of solutes in the cells, often synthesized in response to water stress. It is considered a beneficial drought tolerance mechanism in some plants.

osmotic potential—The chemical activity of water attributable to solutes (units are megapascals, MPa).

ovary—The organ that produces the female gametes and contains one or more ovules (one in grasses). Develops into the single-seeded grass fruit (caryopsis) or legume (pod and contents).

overseed, overseeding—Adding seed to a pasture or forage field for the purpose of improving production without complete reestablishment.

overstocking—The placing of a number of animals on a given area that will result in overuse if continued to the end of the planned stocking period. Not to be confused with overgrazing, because an area may be overstocked for a short period (mob grazed), but the animals are removed before the area is overused. However, continued overstocking will lead to overgrazing.

ovule—The structure in the ovary that contains the egg cell. Develops into a seed when mature.

paddock—A grazing area that is a subdivision of a grazing management unit and is enclosed and separated from other areas by a fence or barrier (cf. grazing management unit, pasture).

palatability—Preference based on plant characteristics eliciting a choice between two or more forages or parts of the same forage, conditioned by the animal and environmental factors that stimulate a selective intake response.

palatable—Acceptable for consumption by an animal.

palea—The inner or upper leaf-like covering of the grass floret, generally more membranous and smaller than the lemma.

panicle—A grass inflorescence in which the spikelets are attached to pedicels on a subdivided or branched axis. A branched inflorescence.

panicoid—A type of grass seedling development where there is a short coleoptile and considerable elongation of the subcoleoptile internode, which pushes the coleoptilar node, coleoptile, and enclosed shoot through the soil. Characteristic of grasses belonging to the Panicoideae subfamily. (Cf. festucoid.)

paradormancy—Dormancy caused by biochemical factors from outside the dormant tissue. An example is apical dominance of the shoot apex over an axillary meristem on the same shoot.

parenchyma—A tissue of higher plants consisting of living cells with thin walls that are active in photosynthesis and/or storage. Generally, very high in digestibility.

pasturage—Not a recommended term (cf. forage, pasture). The recommended definition of pasture refers to a specific kind of grazing management unit, not that which is consumed, which is forage.

pasture—A type of grazing management unit enclosed and separated from other areas by fencing or other barriers and devoted to the production of forage for harvest primarily by grazing. Cf. grazing management unit, paddock, pastureland.

pasture, rotation—A fenced pasture area used for a few seasons and then plowed for other crops.

pasture, supplemental—A crop used to provide grazing for supplemental use, usually during periods of low production of permanent or rotational pastures.

pasture, tame—Grazing lands planted with introduced or domesticated forage species that may receive periodic cultural treatments such as renovation, fertilization, and weed control.

pasture, temporary—A field of crop or forage plants grazed for only a short period, usually not more than one crop season.

pastureland—Land devoted to the production of indigenous or introduced forage for harvest primarily by grazing. Pastureland generally must be managed to arrest successional processes (cf. pasture). Provides basis for land-use mapping unit. Pastureland can include some grassland.

Predictive Equations for Alfalfa Quality (PEAQ)—A system for estimating the quality of alfalfa by measuring the length of the tallest stem and determining the growth stage of the most mature stem and using these values in predictive equations to estimate forage quality (e.g. RFV, NDF).

pectin—A group of water-soluble carbohydrate substances found in the cell walls and intercellular tissues of many plants. It functions to help keep the walls of adjacent cells joined together.

pedicel—The stalk that attaches an individual flower to the peduncle. Some flowers are sessile and do not have a pedicel; they are directly attached to the peduncle.

peduncle—The main stalk that connects the pedicel of a flower or inflorescence to the stem.

PEP (phosphoenolpyruvate carboxylase)—The enzyme that captures CO_2 for photosynthesis in C_4 plants.

perennation—The self-perpetuation of a plant by reseeding or vegetative reproduction.

perennial—Plant that lives for more than one year through persisting organs, such as rhizomes, stolons, or crowns that contain tiller buds.

pericarp—The ripened and variously modified walls of a plant ovary, especially those contributing the outer layer of a cereal caryopsis.

pericycle—A layer of parenchyma or sclerenchyma cells in the root that separates the vascular tissue on the inside from the cortex on the outside and initiates root branches.

period of occupation—The length of time that a specific land area is occupied, whether by one animal group or by two or more animal groups in succession. It differs from stocking period in that grazing may or may not be involved, such as feeding hay on pasture in winter when ice prevents grazing. (Cf. first-last stocking, forward creep stocking, period of stay).

period of stay—The length of time a particular animal group occupies a specific land area (cf. first-last stocking, forward creep stocking, period of occupation). When two or more animal groups are involved the "period of stay" defines the fractional part of the "period of occupation" for each.

perloline—A naturally occurring plant alkaloid that interferes with cellulose digestion by rumen microorganisms; commonly associated with tall fescue.

permanent pasture—Pasture composed of perennial or self-seeding annual plants kept indefinitely for the purpose of grazing.

persistence, plant—The ability of perennial plants to remain alive and productive over long periods of time.

persistence, stand—The ability of a forage stand to be productive for several years. This can occur due to plant persistence (e.g. alfalfa or red clover) or natural reseeding (e.g. annual lespedeza or birdsfoot trefoil).

petiole—The stalk of a leaf that attaches the blade to the stem.

petiolule—The stalk that attaches a leaflet to the petiole of a compound leaf.

phenology—Plant development phenomena that are correlated with climatic conditions.

phenotype—The observable plant or plant part that results from the interaction of the genotype and the environment. *See* genotype.

phenotypic plasticity—Ability of a plant to adapt to environmental cues such as stress or management practices by changing its morphology and physiology. This change is generally reversible.

phenotypic selection—Selection of desired plant traits based on visible or measurable plant characteristics, such as plant morphology, persistence, yield, color, and so on.

phloem—A conducting tissue present in vascular plants, chiefly used for transport of sugars and other organic food materials. Fully developed phloem consists of sieve tubes and parenchyma, generally with companion cells.

photon—A particle of light.

photoperiod—Duration of daily exposure to light that regulates certain growth processes.

photoperiodism—Regulation of plant response by the duration of daylight.

photorespiration—Special respiratory activity in cool-season plants during the light period due to O_2 reaction instead of CO_2 in the photosynthetic pathway; no useful form of energy is derived.

photosensitization—A noncontagious disease resulting from the abnormal reaction of light-colored skin to sunlight after a photodynamic agent has been absorbed through the animal's system. Grazing certain kinds of vegetation or ingesting certain molds under specific conditions causes photosensitization.

photosynthesis—Process by which carbohydrates are produced from CO_2 and water in chloroplasts or chlorophyll-bearing cell granules using the energy of sunlight.

photosynthetically active radiation (PAR)—Radiation within the broad spectral range of solar radiation with wavelengths from 400 to 700 nm that is absorbed by chloroplasts for use in photosynthesis.

phototropism—Directional growth response of plants to light.

phyllochron—The interval of time between appearances of successive leaves sometimes used as an index for describing vegetative growth.

physiology—Processes in the plant that involve metabolism, dynamics of growth and development, and the interactions with the environment.

phytic acid—A unique organic chemical in seeds that contains bound phosphorus that can interfere with the absorption of iron, zinc and calcium and may promote mineral deficiencies.

phytochrome—A protein pigment that changes its conformation in response to changes in red and far-red light it absorbs. Major pigment involved with monitoring length of the photoperiod or amount of light penetration through a canopy.

phytomer—The unit of structure of a grass tiller. It consists of a leaf blade and sheath, the internode, node, and associated axillary bud below the point of sheath attachment.

pistil—The female reproductive structure of a flower comprised of the stigma, style, and ovary.

pistillate—Term for plants, inflorescences, or flowers having pistils only (female).

pith—Usually a continuous central strand of spongy tissue of parenchyma cells in the stems of most vascular plants.

plant functional group—A set of species or collection of organisms that have characteristics in common with lifeform, rather than taxonomic relationships, contributing to similar ecosystem effects.

plant heaving—Upward movement of overwintering plants caused by the alternate freezing and thawing of wet soils.

plasmodesmata—Tube-like connections through cell walls that allow substances such as sugars, proteins, or minerals to move between living cells.

plasticity—Capacity of plants to change shape or metabolism in response to changing environmental conditions.

ploidy—Refers to the number of complete sets of chromosomes (genomes) in a diploid (2x) or haploid (1x) cell. Ploidy levels for tetraploids and hexaploids are 4x and 6x, respectively.

plumule—The embryonic shoot of a germinating seed that develops into the epicotyl (shoot).

pollination—The process of moving pollen to the stigma of the same plant (self-pollination) or to a different plant (cross-pollination).

polycross—Designed system for genetic improvement using open, random pollination and intermating of selected plants via wind or insects. The polycross is isolated from other plants of the same species to prevent pollination from undesired genotypes.

polymorphism—Presence of genetic variation within a population to provide different forms among the members upon which natural selection can operate.

polyploid—Having more than two sets of chromosomes.

polyurea—Abnormally frequent or excessive production of urine.

prairie—Nearly level or rolling grassland, originally treeless, and usually characterized by fertile soil.

prebloom—The plant stage or period immediately preceding the stage of maximum flowering.

preinoculated—Legume seed that has been coated with the appropriate rhizobia bacteria prior to reaching the farmer.

primordia—Embryonic tissue of organs such as a leaf or axillary bud at early development stages. Plural of primordium.

primordium—A group of cells formed on the shoot apex that is the initial structure that will develop into a leaf or axillary bud.

progeny testing—Growing and evaluating offspring to determine the desirability of the selected parent plant.

protease—Enzyme that causes the degradation or hydrolysis of proteins to amino acids or other simpler substances.

protected cultivar—A cultivar that is released and granted a certificate of plant variety protection under the legal statutes of the United States or some other country. The owner of a protected cultivar can exclude others from selling it, reproducing it, importing it, exporting it, or using it in producing a hybrid or different cultivar.

protein—A class of compounds composed of amino acids that occur in all living organisms. Some are important biocatalysts (enzymes), some are structural components of body tissues such as muscle, hair, and collagen, and others are nitrogen storage compounds.

protein, crude—Protein content estimated by multiplying total nitrogen (N) content by 6.25 because proteins average about 16% N.

protein quality—Refers to the balance of essential amino acids in the protein, as well as the biological availability of the protein. In general, many cereals are low in lysine relative to animal needs and thus have low quality for monogastric animals.

proteoid roots—A root with dense clusters of rootlets of limited growth that can symbiotically fix atmospheric N_2 and mobilize mineral P from organic layers in soil, obtain Fe and Mn from alkaline soils, and take up organic forms of N.

proteolysis—The hydrolysis of proteins into peptides and amino acids by cleavage of peptide bonds.

proteolytic enzyme assay—A laboratory procedure used to estimate rumen degradable protein. Feeds are incubated with specific proteases and loss of true protein is measured.

proteome—All the proteins produced by a plant.

proximate analysis—Analytical system for feedstuffs that includes the determination of ash, crude fiber, crude protein, ether extract, moisture (dry matter), and nitrogen-free extract.

prussic acid—A poison produced as a glucoside by several plant species, especially sorghums. Also called hydrocyanic acid because cyanide is released during metabolism.

pseudostem—The rolled or folded sheaths of successive leaves of a vegetative grass tiller that form the lower portion of the canopy.

pubescent—Covered with fine, soft, short hairs, or trichomes.

pulses—The edible seed of various leguminous crops (such as peas, beans, lentils); collectively, legume plants that produce edible seeds.

pure live seed—Percentage of seeds within a unit of seeds that are of the desired species/cultivar and will germinate.

put-and-take animal—*See* grazer.

put-and-take stocking—The use of variable animal numbers during a grazing period or a season, with a periodic adjustment in animal numbers in an attempt to maintain desired sward management criteria, that is, a desired quantity of forage, degree of defoliation, or grazing pressure.

pyrolysis—The thermal decomposition of biomass at high temperatures (>200 °C) in the absence of air.

quantitative trait—A trait controlled by many genes whose expression is affected by the environment; e.g. crop yield.

quiescent—State of suspended development of an organism in response to unfavorable environmental conditions. Quiescent seed, for example, will begin growth once environmental conditions (temperature, moisture) are conducive whereas dormant seed will not.

raceme—An unbranched inflorescence where the spikelets are attached directly to the rachis by pedicels.

races, pathogen—A group within a species of pathogens that infect a given set of plant cultivars.

rachis—The central axis of an inflorescence.

radicle—The embryonic root of seed plants that emerges first through the seed coat during germination (primary root). It develops as the main taproot of legumes, forbs, and other species but is short-lived in grasses and is replaced with an adventitious root system.

range—Land supporting indigenous vegetation that is grazed, or that has the potential to be grazed, and is managed as a natural ecosystem. Range includes grazable forestland and rangeland.

rangeland—Land on which the indigenous vegetation (climax or natural potential) is predominantly grasses, grasslike plants, forbs, or shrubs and is managed as a natural ecosystem. If plants are introduced, they are managed as indigenous species. Provides basis for land-use mapping unit. Rangelands include natural grasslands, savannas, shrublands, many deserts, tundras, alpine communities, marshes, and meadows.

range management—The science of maintaining maximum forage production, generally with natural vegetation, without jeopardy to other resources or uses of the land.

ration—The total amount of feed (diet) allotted to one animal for a 24-hour period.

ration stocking—Confining animals to an area of grazing land to provide the daily allowance of forage per animal (cf. strip stocking).

recurrent selection—Breeding system used to improve the frequency of desired alleles for one or more traits by crossing among the best plants generation after generation.

redox potential—The oxidation potential of a soil expressed in millivolts. Soils with reduced O_2 levels have less electrical potential or ability to transfer electrons from organic or inorganic compounds to oxidants (such as O_2) with the subsequent production of energy.

reducing sugars—Sugars that have the ability to donate electrons to copper cations to produce copper metal (a reducing process). Glucose, fructose, and maltose are reducing sugars, whereas sucrose, raffinose, melibiose, and stachyose are nonreducing sugars.

relative feed value (RFV)—An index of forage quality based on its predicted digestible dry mater intake relative to that of a standard forage (full bloom alfalfa).

relative forage quality (RFQ)—An adaptation and refinement of RFV that uses an alternative method for predicting digestible dry matter intake. It better predicts performance of animals consuming forage with higher and more digestible fiber, such as grass hay.

relative growth rate (RGR)—Dry weight increase in a time interval in relation to the initial weight.

replacement heifers—Immature female cattle being raised to replace cows in the herd.

reproductive primordium—The early, recognizable cells that will differentiate into a reproductive organ that will produce flowers.

reseeding annual—A forage that completes its life cycle in one growing season and produces seed from which it may reestablish the following growing season and allowing it to be managed as a perennial.

residue biomass—The forage that remains following removal or utilization of part of the biomass by grazing, harvesting, burning, or other means.

resins—Sticky to brittle plant products from essential oils that sometimes possess marked odors; more common with woody vegetation than with herbaceous vegetation. Used in medicines, varnishes, and so on.

resistance—The ability of a plant or crop to grow and produce even though heavily inoculated or actually infected or infested with a biotic pest, or to survive a period of abiotic stress such as drought, cold, or heat.

respiration—Energy-producing biochemical reactions in cells that utilize O_2 to oxidize carbohydrates and lipids to produce ATP, CO_2, and water.

rest—To leave an area of grazing land ungrazed or unharvested for a specific time, such as a year, a growing season, or a specified period required within a particular management practice (cf. rest period). *Syn.:* spell.

rest period—The length of time that a specific land area is allowed to rest (cf. rest). *Syn.:* spelling period.

reticulo-rumen—The forestomach compartment of the ruminant digestive system that is comprised of the rumen and reticulum. Site of microbial digestion of nonstructural carbohydrates.

reticulum—The second stomach in ruminants.

reverse peristalsis—Waves of contractions of the esophagus wall that carry boli of ingesta from the rumen back to the mouth.

rhizobia—Bacteria of the genus *Rhizobium* that form nodules on legume roots and symbiotically fix N_2 from the air into forms useful to the plant host.

rhizodeposits—Sloughed cells, mucilage, and exudates originating from roots to the surrounding soil.

rhizomatous—Having modified stems (rhizomes) with definite nodes and internodes located below ground.

rhizome—A belowground horizontal stem, with scalelike leaves (cataphylls) and axillary buds at the nodes, that can develop new tillers or rhizomes.

rhizosphere—Zone of soil occupied and influenced by plant roots.

rind—The epidermis and sclerenchyma tissue on the outer surface of stems of corn, sorghum, and other grass plants.

riparian—Land area adjacent to a natural waterway.

riparian buffers—Strips of grass, shrubs, and/or trees along the banks of rivers and streams. They filter polluted runoff and provide a transition zone between water and human land use.

rosette—A form of plant resulting in a radiating cluster of leaves, usually close to the ground at the base of a plant.

rotational deferred—Systematic rotation of deferment among land areas within a grazing management unit.

rotational deferred stocking—A stocking management system that uses a systematic rotation of deferment among land areas within a grazing management unit.

rotational grazing—Not a recommended term. If used, it is synonymous with the preferred term, rotational stocking.

rotational stocking—A grazing method that utilizes recurring periods of grazing and rest among two or more paddocks in a grazing management unit throughout the period when grazing is allowed (cf. continuous stocking). The lengths of the grazing and of the rest periods should be defined.

roughage—Animal feeds that are relatively high in fiber and low in digestible nutrients and protein.

rubisco (ribulose-1,5-bisphosphate carboxylase/oxygenase)—The enzyme that captures CO_2 for photosynthesis in C_3 plants, but it also may react with O_2 when CO_2 concentration is low. It is very abundant in plants, constituting about 40% of the soluble protein in C_3 plant leaves.

rumen—First and largest compartment of the stomach of a ruminant or cud-chewing animal. Site of microbial fermentation.

rumen degradable protein (RDP)—Crude protein in a feed that is broken down in the rumen.

rumen motility—Movements of digesta promoted by contractions of the rumen wall.

rumen undegraded protein (RUP)—Crude protein in a feed that escapes degradation in the rumen and passes to the intestine. It may or may not be digested in the intestines.

ruminant—A suborder of mammals having a complex multichambered stomach; uses forages primarily as feedstuffs.

rumination—Regurgitation and remastication of food in preparation for true digestion in ruminants.

saponin—Any of various plant glucosides that form soapy colloidal solutions when mixed and agitated with water.

savanna—Grassland with scattered trees or shrubs; often a transitional type between true grassland and forestland, and accompanied by a climate with alternating wet and dry seasons.

scarification—Process of scratching or abrading of the seed coat of certain species to allow uptake of water and gases as an aid to seed germination.

sclerenchyma—Strengthening tissue made up of cells with heavily lignified cell walls; supports and protects the softer tissues of the plant.

scurf—Small flakes of dry tissue shed from the epidermal covering of an animal.

scutellar node—The node of the embryo axis in developing grass seedlings where the scutellum (cotyledon) is attached. Designated as the first node.

scutellum—The single cotyledon in a monocot.

seasonal stocking—Grazing restricted to one or more specific seasons of the year.

second-generation biofuel—Refers to biofuels manufactured from biomass feedstocks of nonfood crops (e.g. ethanol from switchgrass) or crop residues. Also known as advanced biofuels.

secondary metabolite—Chemical substance produced by an organism and often stored in the vacuole that is not involved in the fundamental metabolic pathways that sustain life. In plants, often serves as a defense mechanism against other organisms.

sedge—A grasslike plant, generally with a three-sided stem, that is a member of the Cyperaceae family.

seed—*n.* Ripened (mature) ovule consisting of an embryo, a seed coat, and a supply of food that, in some species, is stored in the endosperm. *v.* To sow, as to broadcast or drill small-seeded grasses and legumes or other crops.

seedbed—Upper portion of the soil into which seeds are placed for germination and growth.

seed conditioning—Mechanical processes used to remove undesirable materials including debris and other crop and weed seeds from harvested raw seed, so as to create pure planting seeds of a crop species.

seed shatter—The dispersal of seed from the reproductive structure upon becoming ripe.

selective herbicides—Chemicals applied to vegetation for control of plant growth with effects targeted to specific plant types. In grass pastures, selective herbicides for control of broadleaf weeds are commonly used.

selection mapping—A method to identify genes controlling a trait by evaluating marker allele frequency changes during selection.

selective grazing—Expression of diet learning by grazing herbivores.

seminal roots—Roots of a grass seedling that emerge from the cotyledonary node shortly after germination but live for only a few weeks. There are generally 3–4 seminal roots per seedling.

senescence—The natural process of aging during which plant tissues alter physiological activity to redistribute nonstructural proteins, carbohydrates, nucleic acids, and mineral nutrients from plant organs preceding death.

sequence stocking—The grazing of two or more land units in succession that differ in forage species composition. Sequence stocking takes advantage of differences among forage species and species combinations, grown in separate areas for management purposes, to extend grazing seasons, enhance forage quality and/or quantity, or achieve some other management objective.

sessile—Directly attached to a central axis without a stalk.

set stocking—The practice of allowing a fixed number of animals on a fixed area of land during the time when grazing is allowed (cf. variable stocking).

shattering, seed—The dispersal of mature seed either before harvest due to dehiscence or during harvest due to mechanical treatment.

shear force—The amount of force (pressure) required to tear a forage particle. Shear force is correlated with the amount of work required by a ruminant to masticate forage, and it may be correlated with digestibility.

sheath—The tubular basal portion of the grass leaf that encloses the stem on reproductive tillers.

shelterbelt—*See* windbreak.

shoot—Collectively, the aboveground organs of a plant. A stem and connected leaves (may also include flowers and reproductive structures) that arises from the seed or an axillary bud. Often used for dicots; tiller is used for grasses.

shoot apex—Meristematic area at the end of a stem that initiates leaf primordia, nodes, internode initials, and axillary buds; differentiates into an inflorescence in grasses and other determinate plants.

short-day plant—A plant that flowers under short photoperiods (long nights).

shrub—Any low-growing, woody plant that produces multiple stems.

shrubland—Land on which the vegetation is dominated by shrubs.

silage—Forage preserved at low pH in a succulent condition due to production of organic acids by partial anaerobic fermentation of sugars in the forage.

silage additive—Material added to forage at the time of ensiling to enhance favorable fermentation processes.

silage preservative—*See* silage additive.

silo—A structure or container designed to contain forage and exclude air during silage fermentation.

silvo-pastoral—Preferred term is forest grazing.

silvopasture—A combination of trees, improved pasture plants, and grazing livestock in a carefully defined agroforestry practice that is an integration of intensive animal husbandry, silviculture, and forage management.

sink—Area of metabolic activity or storage; place where organic materials and nutrients are translocated.

smother crop—Strongly competitive crop that is grown in monoculture to control weeds until it is harvested, grazed or used as green manure.

sodbound—The condition when the upper soil profile is filled with live and dead roots, making it impermeable to water and low in productivity due to lack of available nitrogen.

sod seeding—Mechanically placing seed, usually legumes or small grains, directly into a grass sod.

soil—The layer(s) of generally loose mineral and/or organic material that are affected by physical, chemical, and/or biological processes at or near the planetary surface and usually holds liquids, gases and biota and support plants.

soil aggregates—A group of primary soil particles that cohere to each other more strongly than to other surrounding particles.

soil erosion—The wearing away of the land surface by rain or irrigation water, wind, ice, or other natural or anthropogenic agents.

soil health—The continued capacity of soil to function as a vital living ecosystem that sustains plants, animals, and humans.

soil matric potential—The potential energy of water that is tightly adsorbed to the charged surface of soil mineral particles.

soil organic carbon—Carbon in a mineral soil derived from decomposed plant and animal residues, root exudates, soil microorganisms, and soil biota.

soil quality—The capacity of a specific kind of soil to function, within natural or managed ecosystem boundaries, to sustain plant and animal productivity, maintain or enhance water and air quality, and support human health and habitation.

soil solution—Layer of water covering soil particles in which nutrients are dissolved prior to uptake by plants.

soil structure—The combination or arrangement of primary soil particles into secondary units, or peds.

The secondary units are characterized on the basis of size, shape, and grade (degree of distinctness).

soil texture—The relative proportions of the various soil separates in a soil with the classes being clay, clay loam, loam, loamy sand, sand, sandy clay, sandy clay loam, sandy loam, silt, silty clay, silty clay loam, and silt loam.

solute or osmotic potential—The potential energy of water molecules to move from a dilute solution to a more concentrated solution across a semipermeable membrane, such as the plasma membrane of a cell.

species—A taxonomic category that ranks immediately below a genus and includes closely related and morphologically similar individuals that can interbreed.

species epithet—The second word of the binomial name used to indicate a plant species.

spike—An inflorescence in which the spikelets are attached directly (sessile) to the rachis.

spikelet—The basic unit in the grass inflorescence consisting of two subtending glumes (generally) enclosing one or more florets.

spontaneous heating—Natural process whereby moist forages undergo respiration in the presence of oxygen, yielding carbon dioxide, water, and heat.

sprigging—Vegetative propagation by planting stolons or rhizomes (sprigs) in furrows or holes in the soil.

stamen—The male portion of the flower that produces pollen. Consists of an anther borne on a filament.

staminate—Plants, inflorescences, or flowers having only stamens (male).

standard livestock unit (SLU)—The equivalent of a nonlactating bovine weighing 500 kg used to measure stocking rate in grazing studies. Body weight to the 0.75 power is used to convert bovine to other weights and the 0.90 power is used between sheep or goats and cattle.

starch—Insoluble but readily digested storage carbohydrate, such as amylose and amylopectin, formed from hundreds of linked glucose units.

starch granules—The fundamental unit in which starch is deposited in amyloplasts in cells of storage tissue of many higher plants. Granules are insoluble in cold water and have a characteristic size and shape depending on the plant species that produced them.

steppe—Semiarid grassland characterized by short grasses occurring in scattered bunches with other herbaceous vegetation and occasional woody species.

stigma—The tip part of the pistil where the pollen germinates after it is deposited or captured.

stocker—Young cattle, post-weaning, generally being grown on forage diets to increase size before going to concentrate feed rations in feedlots.

stocking cycle—The time elapsed between the beginning of one stocking period and the beginning of the next stocking period in the same paddock where the forage is regularly grazed and rested. One stocking cycle includes one stocking period plus one rest period.

stocking density—The relationship between the number of animals and the specific unit of land being grazed at any one time (cf. stocking rate). May be expressed as animal units or forage intake units per unit of land area (animal units at a specific time per area of land).

stocking density index—The reciprocal of the fraction: land available to the animals at any one time per land available to the animals for the entire grazable period.

stocking method—A defined procedure or technique of stocking management designed to achieve a specific objective(s). One or more stocking methods can be utilized within a grazing system.

stocking period—The length of time that grazing livestock or wildlife occupy a specific land area (cf. grazing event).

stocking plan—The number and kind of livestock assigned to one or more given management areas or units for a specific period.

stocking rate—The relationship between the number of animals and the grazing management unit utilized over a specified time period (cf. stocking density). May be expressed as animal units or forage intake units per unit of land area (animal units over a described time period per area of land).

stockpiling—To allow forage to accumulate for grazing at a later period. Forage is often stockpiled for autumn and winter grazing after or during dormancy or semi-dormancy, but stockpiling may occur at any time during the year as a part of a management plan. Stockpiling can be described in terms of deferment and forage accumulation.

stolon—Aboveground lateral stems having nodes at which buds can form with the potential of developing into new plants. A structure used for vegetative reproduction.

stoloniferous—A plant species that reproduces itself by growing prostrate stems (stolons) at or just above the soil surface that subsequently produce new plants from buds at its tips or nodes.

stored forage—Commonly refers to forage that has been harvested and processed for retention of nutritive value such as through drying of hay or fermentation of silage.

stocking season—(i) The time period during which grazing can normally be practiced each year or portion of each year. (ii) On US public lands, an established period for which grazing permits are issued. It may be the whole year or a very short time span, and it is normally a function of forage mass and climate. In this context, the vegetative growing season may be only a part of the stocking season.

stover—The matured, cured stalks of such crops as corn or sorghum from which the grain has been removed. A type of roughage.

stratification—Process of exposing imbibed seeds to cool temperature conditions to break seed dormancy.

strip stocking—Confining animals to an area of grazing land to be grazed in a relatively short period of time, where the paddock size is varied to allow access to a specific land area (cf. ration grazing). Strip stocking may or may not be a form of rotational stocking, depending on whether or not specific paddocks are utilized for recurring periods of grazing and rest (cf. rotational stocking).

stroma—The aqueous inner matrix of chloroplasts where reduction of CO_2 to fixed carbon structures occurs; also the location of thylakoid membranes.

structural carbohydrate—*See* carbohydrates, structural.

stubble—The basal portion of the stems of herbaceous plants left standing after grazing or harvest.

style—The stalk of the pistil. Connects the stigma and ovary.

subcoleoptilar internode—The internode between the scutellar node and the coleoptilar node on a developing grass seedling. Sometimes referred to as the mesocotyl.

suberin—A lipophilic macromolecule that forms a protective barrier in specialized plant cell walls.

subtropical—Region or area of transition between temperate and tropical climates on earth.

succession—Plant succession is the directional, non-seasonal, cumulative change in type of plant species that occupy a given area through time involving the processes of colonization, establishment, and extinction of plant species.

sugar—Low molecular weight carbohydrate that includes mono- and disaccharides active in cellular metabolism and transport within a plant.

supplement—A nutritional additive (salt, protein, phosphorus, and so on) intended to improve nutritional balance and remedy deficiencies of the diet.

sward—A population of herbaceous plants characterized by relatively short habit of growth and relatively continuous ground cover, including both aboveground and belowground parts.

swath—A layer of forage material left by mowing machines or self-propelled windrowers. Swaths are wider than windrows and have not been subjected to raking.

symbiotic—A mutually beneficial relationship between two organisms, such as the relationship between legumes and rhizobia bacteria.

synthesis gas—A mixture of carbon dioxide, carbon monoxide, and hydrogen.

T—Soil loss tolerance or allowable soil loss, defined as the maximum level of annual soil erosion that will permit a high level of crop productivity to be maintained economically and indefinitely. The T value is operationally defined by the USDA-Natural Resources Conservation Service (NRCS) in terms of long-term average annual soil losses as estimated by the universal soil loss equation (USLE).

tannin—Broad class of soluble polyphenols that occur naturally in many forage plants. They commonly condense with protein to form a leather-like substance that is insoluble and of low digestibility.

tedding—A mechanical fluffing of a cut forage in the field to aid in drying.

terminal—Of or relating to an end or extremity; growing at the end of a branch or stem.

terminal meristem—Meristematic area at the end of a stem or root. Sometimes called the growing point, but the preferred term is shoot or root apex. The shoot apex initiates leaf primordia, nodes, internode initials, and axillary buds. It later differentiates into an inflorescence in grasses and other determinate plants.

tester animals—Animals of like kind and similar physiological condition used in grazing experiments to measure animal performance or pasture quality; usually assigned to a treatment for the duration of the grazing season, versus "grazer" animals, which may be assigned temporarily to graze excess forage.

thermoneutral zone—Temperature range within which an animal is able to maintain core temperature without expenditure of energy (in cattle: 15–25 °C).

thylakoid membrane—Location of the photosynthetic chlorophyll and carotenoid pigments in the chloroplast. Site of capture and conversion of solar energy to the chemical energy of ATP and NADPH.

tiller—A series of phytomers consisting of a single growing point, a stem, leaves, roots, nodes, dormant buds, and if reproductive, the inflorescence

tillering—Adding new vegetative growth from axillary buds, especially those of grasses, from the base of the plant.

toluene distillation—A laboratory method used to measure the water concentration of a sample by mixing the sample with toluene and collecting the water by distillation.

total digestible nutrients (TDN)—Sum total of the digestibility of the organic components of plant material and/or seed; for example, crude protein + NFE + crude fiber + fat.

total nonstructural carbohydrates (TNC)—*See* carbohydrates, nonstructural.

toxicants—The preferred term for describing toxins, substances that are poisonous to living organisms.

toxoid—Toxin that has been treated to be rendered nontoxic but that will still induce the formation of antibodies.

trace element—*See* micronutrient.

transcriptome—All the transcribed RNA sequences of a plant.

translocation—Movement of organic nutrients within a plant from regions of synthesis (leaves) or deposition (storage organs) to sites of utilization (meristems or seed).

transmit or transmitted—Spread abroad through infection.

transpiration—Water that passes from the soil through the plant xylem and ultimately escapes to the atmosphere, primarily via the stomata.

trichome—A filamentous outgrowth of a leaf epidermal cell; an epidermal hair structure or pubescence.

trophic level—A category of individual organisms that is defined by their position in the food chain.

tropical—Related to or having characteristics of the tropics; having a frost-free climate with temperatures high enough to support year-round plant growth.

true digestibility—The proportion of a forage that is actually digested in the alimentary tract of an animal. It differs from apparent digestibility by excluding fecal matter arising from microbial and animal cells in the calculation of dry matter disappearance.

tuber—An underground stem that is usually short and fleshy with scale-like leaves that bear axillary buds, for example, a potato.

tundra—Land areas in arctic and alpine regions devoid of large trees, varying from bare ground to various types of vegetation consisting of grasses, sedges, forbs, dwarf shrubs and trees, mosses, and lichens.

tunica—Layer of cells that cover the tip of the shoot apex that divide to expand the apex or to initiate cells that develop the leaf primordium, the progenitor of a leaf blade.

turgor—The force (typically positive) that cellular water exerts on the cell wall to drive cell growth. Assists in keeping mature tissue and organs expanded so they do not show wilting.

ultraviolet—Radiation wavelengths shorter than 400 nm.

umbel—A type of inflorescence in which individual flowers are attached to the tip of a peduncle by pedicels of equal length.

undergrazing—Utilizing pasture forage with grazing animals at a rate less than that required for optimum animal production and/or forage production.

utilized metabolizable energy (UME)—The amount of metabolizable energy (ME) grown as forage that is eaten by grazing animals. The UME may be expressed either per animal on a daily basis (megajoules [MJ] per head) or per unit area over a specified time period (gigajoules [GJ] per hectare). UME is a measure of output from a forage system based on the estimated energy requirements of animals and the energy value of forage.

utricle—A small, one-seeded indehiscent fruit with a thin membranous wall.

vacuole—A large organelle, up to 90% of cell volume, surrounded by a single membrane and containing water and dissolved salts, pigments, and other organic compounds; water uptake into the vacuole drives cell expansion by generating turgor pressure.

variable stocking—The practice of allowing a variable number of animals on a fixed area of land during the time when grazing is allowed (cf. set stocking).

variety—*See* cultivar.

vascular bundle—An elongated strand containing phloem and xylem, the conducting tissues of plants that transport food and water, respectively.

vascular tissue—Conducting tissue with vessels or ducts.

vegetative—Nonreproductive plant parts (leaf and stem) in contrast to reproductive plant parts (flower and seed) in developmental stages of plant growth. The nonreproductive stage in plant development.

vegetative cover—A soil cover of plants, irrespective of species.

vegetative filter strip—Area of close-growing plants next to cropland designed to remove sediment, organic material, nutrients, and chemicals carried in runoff or irrigation waste water. Strips are planted in riparian areas along streams, ponds, and lakes and are important management tools around sinkholes and agricultural drainage wells.

vegetative growth—Growth of nonreproductive plant parts prior to the onset of reproductive development. Growth during the nonreproductive stage in plant development.

vegetative propagation (reproduction)—(i) In seed plants, reproduction by means other than seeds. (ii) In lower forms, reproduction by vegetative spores, fragmentation, or division of the plant body.

vegetative storage proteins (VSPs)—Proteins that display preferential synthesis and accumulation within vegetative storage organs that are utilized as a reserve during reactivation of shoot meristems.

veld—Grasslands of eastern and southern Africa that are usually level and mixed with trees and shrubs; or grasslands similar to the African veld.

vernalization—A cold treatment required by shoot apices of certain plant species in order for them to initiate flowering.

volatilization—Process where some applied fertilizers become converted to free ammonia (NH_3) gas, which is lost to the atmosphere.

voluntary intake—*See ad libitum* intake.

warm-season plant—Plant species that grow best during warm periods of the year. They commonly have temperature optimums of 30–35 °C and exhibit C_4 photosynthesis.

water-soluble carbohydrates—Nonstructural carbohydrates (mostly simple sugars) that are soluble in water. Quantification of water-soluble carbohydrates is often used as a measure of the substrate available for silage fermentation.

water potential—The energy or force of water that causes it to flow from a compartment where it is relatively pure to compartments with a lower water potential, often in response to accumulated salts. A relative measure of plant water status. Ranges from values just less than zero for well-watered plants and soils to more negative values as water-deficit stress increases.

water-use efficiency (WUE)—A way to compare the use of water per unit of dry matter production. It is usually calculated as the mass of water (kg) transpired through the plant and evaporated by the soil that is needed to produce a kg of plant dry matter during a defined time period.

weathering—The loss of quality in a crop due to the effects of weather on the product or process.

wilted silage—Silage prepared from plant material at intermediate moisture levels, usually between 50% and 70%.

windbreak—A planting of trees, shrubs, or other vegetation, usually perpendicular or nearly so to the principal wind direction, to protect soil, crops, homesteads, roads, etc., against the effects of winds, such as wind erosion and the drifting of soil and snow.

windrow—The narrow band of forage material remaining after raking a swath or field of forage in preparation for baling or chopping.

winterhardiness—Ability of a plant to survive winter.

xeric—Extremely dry.

xylem—The portion of vascular tissue that has thick, lignified cell walls and is specialized for the movement of water and minerals.

Index

Note: Page numbers in *italic* refer to figures, those in **bold** refer to tables. Glossary terms are in **bold**.

Forages: The Science of Grassland Agriculture, Volume II, Seventh Edition.
Edited by Kenneth J. Moore, Michael Collins, C. Jerry Nelson and Daren D. Redfearn.